Organic Chemistry

Eighth Edition

William H. Brown
Beloit College, Emeritus

Brent L. Iverson
University of Texas, Austin

Eric V. Anslyn
University of Texas, Austin

Christopher S. Foote
University of California, Los Angeles

Chapter 29 was originally contributed by
Bruce M. Novak
University of Texas at Dallas

CENGAGE Learning

Australia • Brazil • Mexico • Singapore • United Kingdom • United States

CENGAGE Learning

Organic Chemistry, **Eighth Edition**
William H. Brown, Brent L. Iverson, Eric V. Anslyn, Christopher S. Foote

Product Director: Dawn Giovanniello

Product Manager: Courtney Heilman

Content Developer: Peter McGahey

Product Assistant: Anthony Bostler

Marketing Manager: Ana Albinson

Content Project Manager: Teresa L. Trego

Art Director: Sarah B. Cole

Manufacturing Planner: Judy Inouye

Production Service: MPS Limited

Photo Researcher: Lumina Datamatics

Text Researcher: Lumina Datamatics

Copy Editor: MPS Limited

Text Designer: Pier Design Company

Cover Designer: Pier Design Company

Cover Image: © Philippe Reichert/Getty Images

Compositor: MPS Limited

Library of Congress Control Number: 2016952760

Student Edition:

ISBN: 978-1-305-58035-0

Loose-leaf Edition:

ISBN: 978-1-305-86554-9

Cengage Learning
20 Channel Center Street
Boston, MA 02210
USA

Cengage Learning is a leading provider of customized learning solutions with employees residing in nearly 40 different countries and sales in more than 125 countries around the world. Find your local representative at **www.cengage.com**.

Cengage Learning products are represented in Canada by Nelson Education, Ltd.

To learn more about Cengage Learning Solutions, visit **www.cengage.com**.

Purchase any of our products at your local college store or at our preferred online store **www.cengagebrain.com**.

Printed in the United States of America
Print Number: 01 Print Year: 2016

Dedication

This Eighth Edition is dedicated to the memory of our dear friend and colleague, Christopher Foote. Chris's insights, encouragement, and dedication to this project can never be replaced. His kind and nurturing spirit lives on in all who are lucky enough to have known him.

About the Authors

William H. Brown is an Emeritus Professor of Chemistry at Beloit College, where he has twice been named Teacher of the Year. His teaching responsibilities included organic chemistry, advanced organic chemistry, and special topics in phar-macology and drug synthesis. He received his Ph.D. from Columbia University under the direction of Gilbert Stork and did postdoctoral work at the California Institute of Technology and the University of Arizona.

Brent L. Iverson received his B.S. from Stanford University and his Ph.D. from the California Institute of Technology. He is a University Distinguished Teaching Professor and the Dean of the School of Undergraduate Studies at the University of Texas at Austin as well as a respected researcher. Brent's research spans the interface of organic chemistry and molecular biology. His group has developed several patented technologies, including an FDA-approved treatment for late-stage anthrax.

Eric V. Anslyn received his B.S. from California State University, Northridge, and his Ph.D. from the California Institute of Technology. He is the Norman Hackerman Professor and a University Distinguished Teaching Professor at the University of Texas at Austin. Eric's research focuses on the physical and bioorganic chemistry of synthetic and natural receptors and catalysts.

Christopher S. Foote received his B.S. from Yale University and his Ph.D. from Harvard University. His scholarly credits include Sloan Fellow; Guggenheim Fellow; ACS Baekland Award; ACS Cope Scholar; Southern California Section ACS Tolman Medal; President, American Society for Photobiology; and Senior Editor, Accounts of Chemical Research. He was a Professor of Chemistry at UCLA.

Contents in Brief

Preface / xxiii

Acknowledgements / xxix

1. Covalent Bonding and Shapes of Molecules / 1

Things You Should Know I: General Conclusions from Quantum Mechanics / 69

2. Alkanes and Cycloalkanes / 72
3. Stereoisomerism and Chirality / 127
4. Acids and Bases / 170
5. Alkenes: Bonding, Nomenclature, and Properties / 206

Things You Should Know II: Nucleophiles and Electrophiles / 228
Things You Should Know III: Reaction Mechanisms / 232

6. Reactions of Alkenes / 240
7. Alkynes / 297
8. Haloalkanes, Halogenation, and Radical Reactions / 330

Things You Should Know IV: Common Mistakes in Arrow Pushing / 369

9. Nucleophilic Substitution and β-Elimination / 374
10. Alcohols / 437
11. Ethers, Epoxides, and Sulfides / 492
12. Infrared Spectroscopy / 535
13. Nuclear Magnetic Resonance Spectroscopy / 558
14. Mass Spectrometry / 606
15. An Introduction to Organometallic Compounds / 630
16. Aldehydes and Ketones / 652

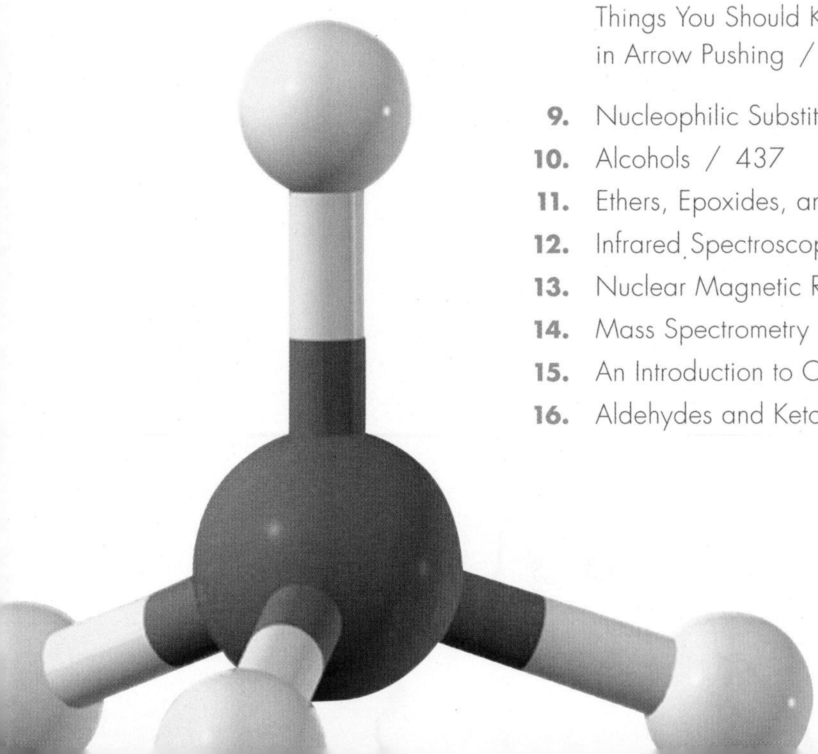

17. Carboxylic Acids / 725

Things You Should Know V: Carboxylic Acid Derivative Reaction Mechanisms / 759

18. Functional Derivatives of Carboxylic Acids / 762
19. Enolate Anions and Enamines / 823
20. Dienes, Conjugated Systems, and Pericyclic Reactions / 894
21. Benzene and the Concept of Aromaticity / 938
22. Reactions of Benzene and Its Derivatives / 995
23. Amines / 1037
24. Catalytic Carbon-Carbon Bond Formation / 1095
25. Carbohydrates / 1134
26. Lipids / 1170
27. Amino Acids and Proteins / 1199
28. Nucleic Acids / 1238
29. Organic Polymer Chemistry / 1264

Appendices:
1. Thermodynamics and the Equilibrium Constant / A-1
2. Major Classes of Organic Acids / A-2
3. Bond Dissociation Enthalpies / A-3
4. Characteristic ^{1}H-NMR Chemical Shifts / A-4
5. Characteristic ^{13}C-NMR Chemical Shifts / A-5
6. Characteristic Infrared Absorption Frequencies / A-6
7. Electrostatic Potential Maps / A-7
8. Summary of Stereochemical Terms / A-8
9. Summary of the Rules of Nomenclature / A-11
10. Organic Chemistry Reaction Roadmaps / Insert

Glossary / G-1

Index / I-1

Contents

Preface xxiii

Acknowledgements xxix

1 Covalent Bonding and Shapes of Molecules 1

1.1 Electronic Structure of Atoms 2

1.2 Lewis Model of Bonding 7

HOW TO: Quickly Figure Out Formal Charge 15

HOW TO: Draw Lewis Structures from Condensed Structural Formulas 17

1.3 Functional Groups 19

1.4 Bond Angles and Shapes of Molecules 24

1.5 Polar and Nonpolar Molecules 27

MCAT PRACTICE: PASSAGE AND QUESTIONS Fullerenes 27

1.6 Quantum or Wave Mechanics 29

1.7 A Combined Valence Bond and Molecular Orbital Theory Approach to Covalent Bonding 32

CONNECTIONS TO BIOLOGICAL CHEMISTRY Phosphoesters 40
HOW TO: Quickly Recognize the Hybridization and Geometry of Atoms 45
1.8 Resonance 45
HOW TO: Draw Curved Arrows and Push Electrons in Creating Contributing Structures 46
1.9 Molecular Orbitals for Delocalized Systems 51
MCAT PRACTICE: PASSAGE AND QUESTIONS VSEPR and Resonance 55
1.10 Bond Lengths and Bond Strengths in Alkanes, Alkenes, and Alkynes 56
STUDY GUIDE 57 • **PROBLEMS** 61

Things You Should Know I: General Conclusions from Quantum Mechanics 69

2 Alkanes and Cycloalkanes 72

2.1 The Structure of Alkanes 73
2.2 Constitutional Isomerism in Alkanes 74
2.3 Nomenclature of Alkanes and the IUPAC System 77
2.4 Cycloalkanes 82
2.5 Conformations of Alkanes and Cycloalkanes 85
HOW TO: Draw Alternative Chair Conformations of Cyclohexane 96
2.6 *Cis,Trans* Isomerism in Cycloalkanes and Bicycloalkanes 99
HOW TO: Convert Planar Cyclohexanes to Chair Cyclohexanes 100
MCAT PRACTICE: PASSAGE AND QUESTIONS Tetrodotoxin 105
2.7 Physical Properties of Alkanes and Cycloalkanes 107
2.8 Reactions of Alkanes 110
2.9 Sources and Importance of Alkanes 112
CHEMICAL CONNECTIONS Octane Rating: What Those Numbers at the Pump Mean 114
STUDY GUIDE 115 • **PROBLEMS** 119

3 Stereoisomerism and Chirality 127

3.1 Chirality—The Handedness of Molecules 128
3.2 Stereoisomerism 129
HOW TO: Draw Chiral Molecules 130
3.3 Naming Chiral Centers—The *R,S* System 134
HOW TO: Assign R or S Configuration to a Chiral Center 136
3.4 Acyclic Molecules with Two or More Stereocenters 137
HOW TO: Quickly Draw and Recognize Enantiomers and Diastereomers 143
3.5 Cyclic Molecules with Two or More Chiral Centers 144
3.6 Tying All the Terminology Together 147
3.7 Optical Activity—How Chirality Is Detected in the Laboratory 149

3.8 The Significance of Chirality in the Biological World 153

CONNECTIONS TO BIOLOGICAL CHEMISTRY Chiral Drugs 155

MCAT PRACTICE: PASSAGE AND QUESTIONS Amino Acid Stereochemistry 155

3.9 Separation of Enantiomers—Resolution 157

STUDY GUIDE 160 • **PROBLEMS** 164

4 Acids and Bases 170

4.1 Arrhenius Acids and Bases 170

4.2 Brønsted-Lowry Acids and Bases 171

4.3 Acid Dissociation Constants, pK_a, and the Relative Strengths of Acids and Bases 178

4.4 The Position of Equilibrium in Acid-Base Reactions 180

HOW TO: Calculate the Equilibrium Constants for Acid-Base Reactions 181

CONNECTIONS TO BIOLOGICAL CHEMISTRY The Ionization of Functional Groups at Physiological pH 182

4.5 Thermochemistry and Mechanisms of Acid-Base Reactions 184

4.6 Molecular Structure and Acidity 188

MCAT PRACTICE: PASSAGE AND QUESTIONS Acid-Base Equilibria 193

4.7 Lewis Acids and Bases 193

STUDY GUIDE 196 • **PROBLEMS** 199

5 Alkenes: Bonding, Nomenclature, and Properties 206

5.1 Structure of Alkenes 208

HOW TO: Calculate the Index of Hydrogen Deficiency 208

5.2 Nomenclature of Alkenes 211

5.3 Physical Properties of Alkenes 217

CHEMICAL CONNECTIONS The Case of the Iowa and New York Strains of the European Corn Borer 217

5.4 Naturally Occurring Alkenes—Terpene Hydrocarbons 218

CONNECTIONS TO BIOLOGICAL CHEMISTRY The Importance of Cis Double Bonds in Fats Versus Oils 220

STUDY GUIDE 222 • **PROBLEMS** 223

Things You Should Know II: Nucleophiles and Electrophiles 228

Things You Should Know III: Reaction Mechanisms 232

6 Reactions of Alkenes 240

6.1 Reactions of Alkenes—An Overview 241

6.2 Organic Reactions Involving Reactive Intermediates 242

6.3 Electrophilic Additions 244

6.4 Hydroboration-Oxidation 264

6.5 Oxidation 268

HOW TO: Write a Balanced Half-Reaction 270

6.6 Reduction 273

CONNECTIONS TO BIOLOGICAL CHEMISTRY Trans Fatty Acids: What They Are and How to Avoid Them 276

6.7 Molecules Containing Chiral Centers as Reactants or Products 277

STUDY GUIDE 282 • **PROBLEMS** 287

7 Alkynes 297

7.1 Structure of Alkynes 298

7.2 Nomenclature of Alkynes 298

7.3 Physical Properties of Alkynes 300

7.4 Acidity of 1-Alkynes 301

7.5 Preparation of Alkynes 301

7.6 Electrophilic Addition to Alkynes 305

7.7 Hydration of Alkynes to Aldehydes and Ketones 307

7.8 Reduction of Alkynes 312

7.9 Organic Synthesis 314

STUDY GUIDE 318 • **PROBLEMS** 322

8 Haloalkanes, Halogenation, and Radical Reactions 330

8.1 Structure 331

8.2 Nomenclature 331

8.3 Physical Properties of Haloalkanes 332

8.4 Preparation of Haloalkanes by Halogenation of Alkanes 336

8.5 Mechanism of Halogenation of Alkanes 340

CHEMICAL CONNECTIONS Freons 344

8.6 Allylic Halogenation 348

8.7 Radical Autoxidation 353

MCAT PRACTICE: PASSAGE AND QUESTIONS Antioxidants 354

8.8 Radical Addition of HBr to Alkenes 356

STUDY GUIDE 359 • **PROBLEMS** 363

Things You Should Know IV: Common Mistakes in Arrow Pushing 369

9 Nucleophilic Substitution and β-Elimination 374

9.1 Nucleophilic Substitution in Haloalkanes 376

9.2 Mechanisms of Nucleophilic Aliphatic Substitution 377

9.3 Experimental Evidence for S_N1 and S_N2 Mechanisms 381
9.4 Analysis of Several Nucleophilic Substitution Reactions 397
9.5 β-Elimination 400
9.6 Mechanisms of β-Elimination 402
9.7 Experimental Evidence for E1 and E2 Mechanisms 404
9.8 Substitution Versus Elimination 410
9.9 Analysis of Several Competitions Between Substitutions and Eliminations 415
MCAT PRACTICE: PASSAGE AND QUESTIONS Solvents and Solvation 417
9.10 Neighboring Group Participation 418
CONNECTIONS TO BIOLOGICAL CHEMISTRY Mustard Gases and the Treatment of Neoplastic Diseases 420
STUDY GUIDE 422 • **PROBLEMS** 427

10 Alcohols 437

10.1 Structure and Nomenclature of Alcohols 438
10.2 Physical Properties of Alcohols 441
CONNECTIONS TO BIOLOGICAL CHEMISTRY The Importance of Hydrogen Bonding in Drug-Receptor Interactions 442
10.3 Acidity and Basicity of Alcohols 445
10.4 Reaction of Alcohols with Active Metals 446
10.5 Conversion of Alcohols to Haloalkanes and Sulfonates 447
10.6 Acid-Catalyzed Dehydration of Alcohols 454
10.7 The Pinacol Rearrangement 459
MCAT PRACTICE: PASSAGE AND QUESTIONS Pinacol Rearrangement 462
10.8 Oxidation of Alcohols 463
CHEMICAL CONNECTIONS Blood Alcohol Screening 466
CONNECTIONS TO BIOLOGICAL CHEMISTRY The Oxidation of Alcohols by NAD^+ 470
MCAT PRACTICE: PASSAGE AND QUESTIONS Alcohol Oxidations 472
10.9 Thiols 473
STUDY GUIDE 477 • **PROBLEMS** 483

11 Ethers, Epoxides, and Sulfides 492

11.1 Structure of Ethers 493
11.2 Nomenclature of Ethers 493
11.3 Physical Properties of Ethers 494
11.4 Preparation of Ethers 496
11.5 Reactions of Ethers 500
11.6 Silyl Ethers as Protecting Groups 503

11.7 Epoxides: Structure and Nomenclature 505

11.8 Synthesis of Epoxides 505

11.9 Reactions of Epoxides 510

MCAT PRACTICE: PASSAGE AND QUESTIONS Benzo[a]pyrene 513

11.10 Ethylene Oxide and Epichlorohydrin: Building Blocks in Organic Synthesis 515

11.11 Crown Ethers 517

11.12 Sulfides 518

STUDY GUIDE 520 • **PROBLEMS** 526

12 Infrared Spectroscopy 535

12.1 Electromagnetic Radiation 535

12.2 Molecular Spectroscopy 536

12.3 Infrared Spectroscopy 537

12.4 Interpreting Infrared Spectra 543

12.5 Solving Infrared Spectral Problems 552

STUDY GUIDE 552 • **PROBLEMS** 554

13 Nuclear Magnetic Resonance Spectroscopy 558

13.1 Nuclear Spin States 559

13.2 Orientation of Nuclear Spins in an Applied Magnetic Field 559

13.3 Nuclear Magnetic "Resonance" 561

13.4 An NMR Spectrometer 563

13.5 Equivalent Hydrogens 565

13.6 Signal Areas 567

13.7 Chemical Shift 568

13.8 Signal Splitting and the $(n + 1)$ Rule 572

13.9 The Origins of Signal Splitting 574

13.10 Stereochemistry and Topicity 582

CHEMICAL CONNECTIONS Magnetic Resonance Imaging 584

13.11 ^{13}C-NMR 585

13.12 Interpretation of NMR Spectra 587

HOW TO: Solve NMR Spectral Problems 591

STUDY GUIDE 593 • **PROBLEMS** 597

14 Mass Spectrometry 606

14.1 A Mass Spectrometer 607

14.2 Features of a Mass Spectrum 610

14.3 Interpreting Mass Spectra 613

CONNECTIONS TO BIOLOGICAL CHEMISTRY Mass Spectrometry of Biological Macromolecules 621

14.4 Mass Spectrometry in the Organic Synthesis Laboratory and Other Applications 623

STUDY GUIDE 623 • **PROBLEMS** 625

15 An Introduction to Organometallic Compounds 630

15.1 Organomagnesium and Organolithium Compounds 631

15.2 Lithium Diorganocopper (Gilman) Reagents 636

15.3 Carbenes and Carbenoids 639

MCAT PRACTICE: PASSAGE AND QUESTIONS Inorganic Coordination Compounds 643

STUDY GUIDE 644 • **PROBLEMS** 647

16 Aldehydes and Ketones 652

16.1 Structure and Bonding 653

16.2 Nomenclature 653

16.3 Physical Properties 657

16.4 Reactions 658

16.5 Addition of Carbon Nucleophiles 659

16.6 The Wittig Reaction 666

16.7 Addition of Oxygen Nucleophiles 670

16.8 Addition of Nitrogen Nucleophiles 678

MCAT PRACTICE: PASSAGE AND QUESTIONS Pyridoxine (Vitamin B_6): A Carrier of Amino Groups 682

16.9 Keto-Enol Tautomerism 684

16.10 Oxidation 688

16.11 Reduction 691

CONNECTIONS TO BIOLOGICAL CHEMISTRY NADH: The Biological Equivalent of a Hydride Reducing Agent 694

HOW TO: Retrosynthetically Dissect an Amine into the Proper Starting Materials for a Reductive Amination 695

16.12 Reactions at an α-Carbon 699

STUDY GUIDE 701 • **PROBLEMS** 710

17 Carboxylic Acids 725

17.1 Structure 725

17.2 Nomenclature 726

17.3 Physical Properties 729

CHEMICAL CONNECTIONS From Willow Bark to Aspirin and Beyond 730

17.4 Acidity 732

17.5 Preparation of Carboxylic Acids 736

17.6 Reduction 736

CHEMICAL CONNECTIONS *Industrial Synthesis of Acetic Acid—Transition Metal Catalysis* 737

17.7 Esterification 738

Mechanism 17.1 Formation of a Methyl Ester Using Diazomethane 740

17.8 Conversion to Acid Chlorides 740

CHEMICAL CONNECTIONS *Esters as Flavoring Agents* 741

MCAT PRACTICE: PASSAGE AND QUESTIONS *Permethrin and Bifenthrin* 742

17.9 Decarboxylation 743

Connections to Biological Chemistry Ketone Bodies and Diabetes Mellitus 745

STUDY GUIDE 746 • **PROBLEMS** 750

Things You Should Know V: Carboxylic Acid Derivative Reaction Mechanisms 759

18 Functional Derivatives of Carboxylic Acids 762

18.1 Structure and Nomenclature 763

CHEMICAL CONNECTIONS *From Cocaine to Procaine and Beyond* 765

CHEMICAL CONNECTIONS *From Moldy Clover to a Blood Thinner* 766

18.2 Acidity of Amides, Imides, and Sulfonamides 768

Connections to Biological Chemistry The Unique Structure of Amide Bonds 769

18.3 Characteristic Reactions 770

18.4 Reaction with Water: Hydrolysis 774

CHEMICAL CONNECTIONS *Mechanistic Alternatives for Ester Hydrolysis: S_N2 and S_N1 Possibilities* 779

18.5 Reaction with Alcohols 786

18.6 Reactions with Ammonia and Amines 788

18.7 Reaction of Acid Chlorides with Salts of Carboxylic Acids 790

18.8 Interconversion of Functional Derivatives 790

MCAT PRACTICE: PASSAGE AND QUESTIONS *β-Lactam Antibiotics* 791

18.9 Reactions with Organometallic Compounds 793

18.10 Reduction 796

STUDY GUIDE 801 • **PROBLEMS** 808

19 Enolate Anions and Enamines 823

19.1 Formation and Reactions of Enolate Anions: An Overview 823

19.2 Aldol Reaction 825

19.3 Claisen and Dieckmann Condensations 832

19.4 Claisen and Aldol Condensations in the Biological World 838

CHEMICAL CONNECTIONS *Drugs That Lower Plasma Levels of Cholesterol* 839

19.5 Enamines 841
19.6 Acetoacetic Ester Synthesis 845
19.7 Malonic Ester Synthesis 849
19.8 Conjugate Addition to α,β-Unsaturated Carbonyl Compounds 851
19.9 Crossed Enolate Reactions Using LDA 860
MCAT PRACTICE: PASSAGE AND QUESTIONS Ibuprofen: The Evolution of an Industrial Synthesis 864
STUDY GUIDE 866 • **PROBLEMS** 874

20 Dienes, Conjugated Systems, and Pericyclic Reactions 894

20.1 Stability of Conjugated Dienes 894
20.2 Electrophilic Addition to Conjugated Dienes 898
20.3 UV-Visible Spectroscopy 904
20.4 Pericyclic Reaction Theory 909
CHEMICAL CONNECTIONS Curry and Cancer 909
20.5 The Diels-Alder Reaction 911
20.6 Sigmatropic Shifts 920
STUDY GUIDE 925 • **PROBLEMS** 930

21 Benzene and the Concept of Aromaticity 938

21.1 The Structure of Benzene 939
21.2 The Concept of Aromaticity 943
HOW TO: Recognize Aromatic Compounds: Criteria and Caveats 953
21.3 Nomenclature 954
21.4 Phenols 957
*MCAT PRACTICE: PASSAGE AND QUESTIONS Capsaicin, "Some Like It Hot"** 961
21.5 Reactions at a Benzylic Position 966
STUDY GUIDE 970 • **PROBLEMS** 976

22 Reactions of Benzene and Its Derivatives 995

22.1 Electrophilic Aromatic Substitution 996
22.2 Disubstitution and Polysubstitution 1006
22.3 Nucleophilic Aromatic Substitution 1013
STUDY GUIDE 1017 • **PROBLEMS** 1022

23 Amines 1037

23.1 Structure and Classification 1037
23.2 Nomenclature 1039
23.3 Chirality of Amines and Quaternary Ammonium Ions 1042

23.4 Physical Properties 1043

CHEMICAL CONNECTIONS The Poison Dart Frogs of South America 1043

23.5 Basicity 1045

MCAT PRACTICE: PASSAGE AND QUESTIONS The Planarity of —NH_2 Groups on Heterocyclic Rings 1049

23.6 Reactions with Acids 1053

23.7 Preparation 1056

23.8 Reaction with Nitrous Acid 1058

23.9 Hofmann Elimination 1067

23.10 Cope Elimination 1069

STUDY GUIDE 1071 • **PROBLEMS** 1078

24 Catalytic Carbon-Carbon Bond Formation 1095

24.1 Carbon-Carbon Bond-Forming Reactions from Earlier Chapters 1096

24.2 Organometallic Compounds and Catalysis 1097

24.3 The Heck Reaction 1098

24.4 Catalytic Allylic Alkylation 1104

24.5 Palladium-Catalyzed Cross-Coupling Reactions 1107

24.6 Alkene Metathesis 1112

STUDY GUIDE 1115 • **PROBLEMS** 1119

25 Carbohydrates 1134

25.1 Monosaccharides 1135

25.2 The Cyclic Structure of Monosaccharides 1139

CHEMICAL CONNECTIONS l-Ascorbic Acid (Vitamin C) 1141

25.3 Reactions of Monosaccharides 1143

CHEMICAL CONNECTIONS Testing for Glucose 1148

MCAT PRACTICE: PASSAGE AND QUESTIONS Fucose 1150

25.4 Disaccharides and Oligosaccharides 1151

CHEMICAL CONNECTIONS A, B, AB, and O Blood Group Substances 1153

25.5 Polysaccharides 1154

CHEMICAL CONNECTIONS High-Fructose Corn Syrup 1155

25.6 Glucosaminoglycans 1156

STUDY GUIDE 1158 • **PROBLEMS** 1163

26 Lipids 1170

26.1 Triglycerides 1171

26.2 Soaps and Detergents 1174

Connections to Biological Chemistry FAD/FADH$_2$: Agents for Electron Transfer in Biological Oxidation-Reductions: Fatty Acid Oxidation 1176

26.3 Prostaglandins 1178

26.4 Steroids 1180

26.5 Phospholipids 1185

CHEMICAL CONNECTIONS Snake Venom Phospholipases 1186

26.6 Fat-Soluble Vitamins 1187

MCAT PRACTICE: PASSAGE AND QUESTIONS Vitamin K, Blood Clotting, and Basicity 1190

STUDY GUIDE 1191 • **PROBLEMS** 1194

27 Amino Acids and Proteins 1199

27.1 Amino Acids 1199

27.2 Acid-Base Properties of Amino Acids 1203

27.3 Polypeptides and Proteins 1208

27.4 Primary Structure of Polypeptides and Proteins 1209

27.5 Synthesis of Polypeptides 1216

27.6 Three-Dimensional Shapes of Polypeptides and Proteins 1221

CHEMICAL CONNECTIONS Spider Silk 1225

STUDY GUIDE 1227 • **PROBLEMS** 1232

28 Nucleic Acids 1238

28.1 Nucleosides and Nucleotides 1239

28.2 The Structure of DNA 1241

CHEMICAL CONNECTIONS The Search for Antiviral Drugs 1244

28.3 Ribonucleic Acids 1248

CHEMICAL CONNECTIONS The Fountain of Youth 1249

28.4 The Genetic Code 1251

28.5 Sequencing Nucleic Acids 1253

CHEMICAL CONNECTIONS DNA Fingerprinting 1257

STUDY GUIDE 1259 • **PROBLEMS** 1261

29 Organic Polymer Chemistry 1264

29.1 The Architecture of Polymers 1265

29.2 Polymer Notation and Nomenclature 1265

29.3 Molecular Weights of Polymers 1267

29.4 Polymer Morphology—Crystalline Versus Amorphous Materials 1267

29.5 Step-Growth Polymerizations 1268

CHEMICAL CONNECTIONS Stitches That Dissolve 1274

29.6 Chain-Growth Polymerizations 1275

CHEMICAL CONNECTIONS Organic Polymers That Conduct Electricity 1279

MCAT PRACTICE: PASSAGE AND QUESTIONS The Chemistry of Superglue 1286

CHEMICAL CONNECTIONS Recycling of Plastics 1291

STUDY GUIDE 1293 • **PROBLEMS** 1296

Appendices:

1. Thermodynamics and the Equilibrium Constant A-1
2. Major Classes of Organic Acids A-2
3. Bond Dissociation Enthalpies A-3
4. Characteristic ^{1}H-NMR Chemical Shifts A-4
5. Characteristic ^{13}C-NMR Chemical Shifts A-5
6. Characteristic Infrared Absorption Frequencies A-6
7. Electrostatic Potential Maps A-7
8. Summary of Stereochemical Terms A-8
9. Summary of the Rules of Nomenclature A-11
10. Organic Chemistry Reaction Roadmaps Insert

Glossary G-1

Index I-1

List of Mechanisms

Chapter 6 Reactions of Alkenes

6.1 Electrophilic Addition of HBr to 2-Butene **(Section 6.3A)**
6.2 Acid-Catalyzed Hydration of Propene **(Section 6.3B)**
6.3 Carbocation Rearrangement in the Addition of HCl to an Alkene **(Section 6.3C)**
6.4 Addition of Bromine with Anti Stereoselectivity **(Section 6.3D)**
6.5 Halohydrin Formation and Its Anti Stereoselectivity **(Section 6.3E)**
6.6 Oxymercuration-Reduction of an Alkene **(Section 6.3F)**
6.7 Hydroboration **(Section 6.4)**
6.8 Oxidation of a Trialkylborane by Alkaline Hydrogen Peroxide **(Section 6.4)**
6.9 Formation of an Ozonide **(Section 6.5B)**

Chapter 7 Alkynes

7.1 Addition of HBr to an Alkyne **(Section 7.6B)**
7.2 $HgSO_4/H_2SO_4$ Catalyzed Hydration of an Alkyne **(Section 7.7B)**
7.3 Reduction of an Alkyne by Sodium in Liquid Ammonia **(Section 7.8C)**

Chapter 8 Haloalkanes, Halogenation, and Radical Reactions

8.1 Radical Chlorination of Ethane **(Section 8.5B)**
8.2 Allylic Bromination of Propene Using NBS **(Section 8.6A)**
8.3 Radical Initiated Non-Markovnikov Addition of HBr to Alkenes **(Section 8.8)**

Chapter 9 Nucleophilic Substitution and β-Elimination

9.1 An S_N2 Reaction **(Section 9.2A)**
9.2 An S_N1 Reaction **(Section 9.2B)**
9.3 Rearrangement During Solvolysis of 2-Chloro-3-phenylbutane **(Section 9.3F)**
9.4 E1 Reaction of 2-Bromo-2-methylpropane **(Section 9.6A)**
9.5 E2 Reaction of 2-Bromobutane **(Section 9.6B)**

9.6 E2 Reaction of meso-1,2-Dibromo-1,2-diphenylethane **(Section 9.7C)**
9.7 E2 Reaction of the Enantiomers of 1,2-Dibromo-1,2-diphenylethane **(Section 9.7C)**
9.8 E2 Reaction of *cis*-1-Chloro-2-isopropylcyclohexane **(Section 9.7C)**
9.9 Hydrolysis of a Sulfur Mustard—Participation by a Neighboring Group **(Section 9.10)**

Chapter 10 Alcohols

10.1 Reaction of a 3° Alcohol with HBr—An S_N1 Reaction **(Section 10.5A)**
10.2 Reaction of a 1° Alcohol with HBr—An S_N2 Reaction **(Section 10.5A)**
10.3 Rearrangement upon Treatment of Neopentyl Alcohol with HCl **(Section 10.5A)**
10.4 Reaction of a Primary Alcohol with PBr_3 **(Section 10.5B)**
10.5 Acid-Catalyzed Dehydration of 2-Butanol—An E1 Reaction **(Section 10.6)**
10.6 Acid-Catalyzed Dehydration of an Unbranched Primary Alcohol **(Section 10.6)**
10.7 The Pinacol Rearrangement of 2,3-Dimethyl-2,3-butanediol (Pinacol) **(Section 10.7)**
10.8 Chromic Acid Oxidation of an Alcohol **(Section 10.8A)**
10.9 Swern Oxidation, Starting at the Point of the Chlorosulfonium Ion **(Section 10.8C)**
10.10 Dess-Martin Oxidation **(Section 10.8D)**
10.11 Oxidation of a Glycol by Periodic Acid **(Section 10.8E)**
10.12 Oxidation of an Alcohol by NAD^+ **(Section 10.8E)**

Chapter 11 Ethers, Epoxides, and Sulfides

11.1 Acid-Catalyzed Intermolecular Dehydration of a Primary Alcohol **(Section 11.4B)**
11.2 Acid-Catalyzed Addition of an Alcohol to an Alkene **(Section 11.4C)**
11.3 Acid-Catalyzed Cleavage of a Dialkyl Ether **(Section 11.5A)**
11.4 Epoxidation of an Alkene by RCO_3H **(Section 11.8C)**
11.5 Acid-Catalyzed Hydrolysis of an Epoxide **(Section 11.9A)**
11.6 Nucleophilic Opening of an Epoxide Ring **(Section 11.9B)**

Chapter 14 Mass Spectrometry

14.1 McLafferty Rearrangement of a Ketone **(Section 14.3E)**
14.2 McLafferty Rearrangement of a Carboxylic Acid **(Section 14.3F)**

Chapter 15 An Introduction to Organometallic Compounds

15.1 Formation of Dichlorocarbene and Its Reaction with Cyclohexene **(Section 15.3B)**
15.2 The Simmons-Smith Reaction with an Alkene **(Section 15.3C)**

Chapter 16 Aldehydes and Ketones

16.1 Grignard Reagent Reacting with Formaldehyde **(Section 16.5A)**
16.2 Organolithium Reagent Reacting with a Ketone **(Section 16.5B)**
16.3 Alkyne Anion Reacting with a Ketone **(Section 16.5C)**
16.4 Formation of a Cyanohydrin **(Section 16.5D)**
16.5 The Wittig Reaction **(Section 16.6)**

16.6 Base-Catalyzed Formation of a Hemiacetal **(Section 16.7B)**
16.7 Acid-Catalyzed Formation of a Hemiacetal **(Section 16.7B)**
16.8 Acid-Catalyzed Formation of an Acetal **(Section 16.7B)**
16.9 Formation of an Imine from an Aldehyde or a Ketone **(Section 16.8A)**
16.10 Base-Catalyzed Equilibration of Keto and Enol Tautomers **(Section 16.9A)**
16.11 Acid-Catalyzed Equilibration of Keto and Enol Tautomers **(Section 16.9A)**
16.12 Pinnick Oxidation **(Section 16.10A)**
16.13 Sodium Borohydride Reduction of an Aldehyde or a Ketone **(Section 16.11A)**
16.14 Wolff-Kishner Reduction **(Section 16.11E)**
16.15 Acid-Catalyzed α-Halogenation of a Ketone **(Section 16.12C)**
16.16 Base-Promoted α-Halogenation of a Ketone **(Section 16.12C)**

Chapter 17 Carboxylic Acids

17.1 Formation of a Methyl Ester Using Diazomethane **(Section 17.7B)**
17.2 Decarboxylation of a β-Ketocarboxylic Acid **(Section 17.9A)**
17.3 Decarboxylation of a β-Dicarboxylic Acid **(Section 17.9B)**

Chapter 18 Functional Derivatives of Carboxylic Acids

18.1 Hydrolysis of an Acid Chloride **(Section 18.4A)**
18.2 Acid-Catalyzed Ester Hydrolysis **(Section 18.4C)**
18.3 Hydrolysis of an Ester in Aqueous Base (Saponification) **(Section 18.4C)**
18.4 Hydrolysis of an Amide in Aqueous Acid **(Section 18.4D)**
18.5 Hydrolysis of an Amide in Aqueous Base **(Section 18.4D)**
18.6 Hydrolysis of a Cyano Group to an Amide in Aqueous Base **(Section 18.4E)**
18.7 Reaction of an Acid Chloride and Ammonia **(Section 18.6A)**
18.8 Reaction of an Ester with a Grignard Reagent **(Section 18.9A)**
18.9 Reduction of an Ester by Lithium Aluminum Hydride **(Section 18.10A)**
18.10 Reduction of an Amide by Lithium Aluminum Hydride **(Section 18.10B)**

Chapter 19 Enolate Anions and Enamines

19.1 Base-Catalyzed Aldol Reaction **(Section 19.2A)**
19.2 Acid-Catalyzed Aldol Reaction **(Section 19.2A)**
19.3 Acid-Catalyzed Dehydration of an Aldol Product **(Section 19.2A)**
19.4 Claisen Condensation **(Section 19.3A)**
19.5 Alkylation of an Enamine **(Section 19.5A)**
19.6 Michael Reaction—Conjugate Addition of Enolate Anions **(Section 19.8A)**

Chapter 20 Dienes, Conjugated Systems, and Pericyclic Reactions

20.1 1,2- and 1,4-Addition to a Conjugated Diene **(Section 20.2A)**
20.2 The Claisen Rearrangement **(Section 20.6A)**
20.3 The Cope Rearrangement **(Section 20.6B)**

Chapter 21 Benzene and the Concept of Aromaticity

21.1 Kolbe Carboxylation of Phenol **(Section 21.4E)**

Chapter 22 Reactions of Benzene and Its Derivatives

22.1 Electrophilic Aromatic Substitution—Chlorination **(Section 22.1A)**
22.2 Formation of the Nitronium Ion **(Section 22.1B)**
22.3 Friedel-Crafts Alkylation **(Section 22.1C)**
22.4 Friedel-Crafts Acylation—Generation of an Acylium Ion **(Section 22.1C)**
22.5 Nucleophilic Aromatic Substitution via a Benzyne Intermediate **(Section 22.3A)**
22.6 Nucleophilic Aromatic Substitution by Addition-Elimination **(Section 22.3B)**

Chapter 23 Amines

23.1 Formation of the Nitrosyl Cation **(Section 23.8)**
23.2 Reaction of a 2° Amine with the Nitrosyl Cation to Give an *N*-Nitrosamine **(Section 23.8C)**
23.3 Reaction of a 1° Amine with Nitrous Acid **(Section 23.8D)**
23.4 The Tiffeneau-Demjanov Reaction **(Section 23.8D)**
23.5 The Hofmann Elimination **(Section 23.9)**
23.6 The Cope Elimination **(Section 23.10)**

Chapter 24 Catalytic Carbon-Carbon Bond Formation

24.1 The Heck Reaction **(Section 24.3B)**
24.2 The Catalytic Cycle for Allylic Alkylation **(Section 24.4A)**
24.3 The Catalytic Cycle of Cross-Coupling **(Section 24.5A)**

Chapter 26 Lipids

26.1 Oxidation of a Fatty Acid $—CH_2—CH_2—$ to $—CH{=}CH—$ by FAD **(Section 26.2C)**

Chapter 27 Amino Acids and Proteins

27.1 Cleavage of a Peptide Bond at Methionine by Cyanogen Bromide **(Section 27.4B)**
27.2 Edman Degradation—Cleavage of an *N*-Terminal Amino Acid **(Section 27.4B)**

Chapter 29 Organic Polymer Chemistry

29.1 Radical Polymerization of a Substituted Ethylene **(Section 29.6A)**
29.2 Ziegler-Natta Catalysis of Ethylene Polymerization **(Section 29.6B)**
29.3 Homogeneous Catalysis for Ziegler-Natta Coordination Polymerization **(Section 29.6B)**
29.4 Initiation of Anionic Polymerization of Alkenes **(Section 29.6D)**
29.5 Initiation of Anionic Polymerization of Butadiene **(Section 29.6D)**
29.6 Initiation of Cationic Polymerization of an Alkene by $HF \cdot BF_3$ **(Section 29.6D)**
29.7 Initiation of Cationic Polymerization of an Alkene by a Lewis Acid **(Section 29.6D)**

Preface

Learning Through Understanding

The best way to master organic chemistry is to develop an intuitive understanding of basic principles, then understand how to apply those principles to new and increasingly complex situations.

The learning of organic chemistry is analogous to putting together a complex jigsaw puzzle and at least two distinct approaches can be imagined. The first approach, reflecting the way many students unfortunately approach their first organic chemistry class, is to memorize each new piece of the puzzle individually, with no context, and then try to blindly guess where it is to be placed by trial-and-error. We can all agree that a much better approach to finishing a jigsaw puzzle (as well as learning organic chemistry) is to keep the overall picture of the completed puzzle in mind and then evaluate each new piece being considered in the context of where it might fit into the emerging image. This book has been written in alignment with the far more effective second approach to learning by providing a strong foundation of basic concepts at the beginning, then constantly referring to a "big picture" understanding as each new concept and functional group is presented in context. To accomplish this, the eighth edition builds upon two key innovations from previous editions that teach students *how* to learn two of the most important elements of organic chemistry, namely mechanisms and synthesis.

- **Mechanisms** We present a revolutionary paradigm for learning organic chemistry mechanisms. Students are introduced to a small set of individual mechanism elements, and importantly, *when* each of these mechanism elements is to be used. The four most important of these elements are *"Make a bond between a nucleophile and an electrophile," "Break a bond to create stable molecules or ions," "Add a proton,"* or *"Take a proton away."* Reaction mechanisms throughout the book are written in stepwise fashion and described as logical combinations of the individual mechanism elements. This new approach not only simplifies the learning of mechanisms for students but also makes it easier for them to recognize similarities and differences between related reactions. Most important, this approach makes the *prediction* of reaction mechanisms a straightforward, multiple-choice situation in which the correct mechanism element for a given step of each new reaction mechanism is systematically chosen from a small menu of choices.

Reaction Roadmap

- **Synthesis** We present another important innovation in organic chemistry learning that we refer to as the Organic Chemistry Reaction Roadmap. It is a graphical representation of the different organic reactions taught in the context of the important functional groups. The functional groups of an organic chemistry roadmap are analogous to cities on a real roadmap, and the reactions are like the roads between those cities. Arrows are used to represent known routes between functional groups, and the reagents required to bring about each reaction are written next to the corresponding arrow. Multistep synthesis questions are often very challenging for organic chemistry students even though synthesis is at the core of organic chemistry as a discipline. The power of the organic chemistry reaction roadmap is that it helps students visualize the reactions that are appropriate to interconvert key functional groups in multistep synthesis problems. The construction and use of organic chemistry reaction roadmaps are introduced in the end-of-chapter problems beginning in Chapter 6 and presented in complete form inside the back cover of this book, which students can tear out and use.

What's New for the Eighth Edition?

There are four new features in the eighth edition, each intended to enhance dramatically student engagement and learning. We are excited to introduce almost 100 videos, which are lessons from the authors in an audio/visual format that take students through the working of example problems and important graphics. In other words, most of the key lessons from the book. We also introduce three other new features designed to enhance student learning and test preparation: (1) "**Things You Should Know**," (2) **Margin bullets** emphasizing and reinforcing key foundational ideas that recur throughout the book, and (3) the most comprehensive **end-of-chapter "Study Guides"** of any text that recap the key ideas, lessons, homework problems, and reactions of each chapter.

- **Videos** The authors are particularly excited by the addition of almost 100 short videos in MindTap to accompany the eighth edition. These new audio/visual features are specifically intended to bring the material in the text to life. The videos provide students with an on-demand resource that covers much of the most difficult and important material. Icons in the margin of the text indicate to students when a video is available to better explain an example problem or key figure.
  - In the videos, Eric Anslyn and Brent Iverson, the two lead authors, discuss the material in an interactive fashion, reminiscent of the NPR program "Car Talk" with Click and Clack, the Tappet Brothers. While undoubtedly not quite as amusing as that classic program, the students should get a feeling for the love of organic chemistry shared by both authors, their strong bond of friendship, and of course most important—key ways to think about organic chemistry.
  - This feature is forward looking, because the authors presume that textbooks will become increasing interactive, having aspects to their structure that bring the authors and students closer and more connected, while also making the topics increasingly more interesting, live, and thereby easier to grasp. After all, experiencing a video explanation that includes dynamic graphics as well as verbal discussions and banter is substantially more engaging and informative than reading and examining graphics on the written page. We believe that this extensive collection of groundbreaking new videos will help propel the college textbook publishing industry into a future in which student learning is supported by a seamless combination of text, videos, and assigned problems.
- **Things You Should Know** There are important transitions between distinct topics in the sequence of the book that are good places to stop, reflect, and take

stock of what has been learned, while looking forward to the next set of chapters. At such transitions, we have now included five primers that we call "Things You Should Know." By pulling these key topics out from the chapters, we keep the flow of the chapters moving, while emphasizing topics that permeate several of the future chapters. If students read and fully digest these primers, they will be well prepared to approach the discipline of organic chemistry including all the material in subsequent chapters.

  - For example, after Chapter 1 we have "Things You Should Know I, General Conclusions from Quantum Mechanics," which primarily recaps lessons and principles from quantum mechanics or other essential pieces of information that the students must use to understand molecular structure, properties, reactions and mechanisms. Such lessons include "Delocalization of charge over a larger area is stabilizing," and "Delocalization of π electron density over a larger area is stabilizing." Note that several of these lessons and principles cannot be fully explained unless students move on to advanced classes where the basis of these lessons are taught, yet we still need students to use them in introductory organic chemistry courses.
  - As another example, after Chapter 5 we have a Things You Should Know section that contrasts the terminology "Brønsted/Lowry Acids/Bases," "Lewis Acids/Bases," and "Nucleophiles/Electrophiles." All of these terms can be used to describe similar kinds of chemical reactions, and understanding which is appropriate in a given scenario is often unclear to students. We give straightforward guidelines for when to use the different terms.
  - As a third example, prior to Chapter 17 we summarize the common reaction types that are about to be explored in the next three chapters in a primer called "Things You Should Know V, Carboxylic Acid Derivative Reaction Mechanisms."

- **Margin Bullets** To emphasize that many of the fundamental lessons of organic chemistry continue to guide the discussions of topics in future chapters, we now place in the margins a series of bullet points derived from the first primer "Things You Should Know I, General Conclusions from Quantum Mechanics." The idea is to emphasize for students that the same fundamental principles are used to explain patterns of molecular structure, properties and reactivity throughout the text.
- **End-of-Chapter "Study Guides"** One of the largest structural changes to the book is a reimagining of the synopsis sections of the chapters into concise end-of-chapter "Study Guides." Instead of simply a recap of the material, it is now laid out in a fashion that takes the students through topics with simple bulleted lessons, explanations, and with graphic summaries all of the key reactions related to that specific topic. Representative end-of-chapter problems are listed for each section, directing students to questions that test the knowledge being summarized. These chapter summaries can be seen as "crib notes" that the students carry with them, continually refer to, and guide their study when preparing for exams. As an added bonus, there are two related appendices that students find particularly helpful: "Summary of Stereochemical Terms" and "Summary of the Rules of Nomenclature," both of which compile all the information students need to know about these important topics in a convenient, single location.

Other Important Features

Several important additional features of the book have been retained because they foster student learning.

- **Survival Skills** Mastering organic chemistry requires the development of certain intellectual skills. To this end, fifteen How To boxes highlight "survival skills" for organic chemistry students, including How To Quickly Figure Out Formal Charge, How To Quickly Recognize the Hybridization and Geometry of Atoms, How To Quickly Draw and Recognize Enantiomers and Diastereomers, How To Retrosynthetically Dissect an Amine into the Proper Starting Materials for a Reductive Amination, and How To Recognize Aromatic Compounds: Criteria and Caveats.
- **MCAT Preparation** A significant number of students taking organic chemistry will be taking standardized tests such as the MCAT, DAT, or PCAT. Often, organic chemistry content on the MCAT or similar exam is in the form of passages followed by a series of multiple-choice questions. Learning to answer questions based on passages requires students to develop increased reading comprehension and analytical skills. Ours is the first text to aid students in developing these skills by introducing an extensive series of passages followed by several thought-provoking multiple-choice questions in almost every chapter (MCAT Practice: Passage and Questions). The passages cover interesting applications of organic chemistry principles as well as biological and chemical topics. Thus, far from being just test preparation, these passages add considerable enrichment to the material being presented.
- **Unique Chapter Sequence** Carbonyl chemistry (Chapters 16–19) is placed earlier than in most texts so that professors have time to teach this material to the majority of students in an organic chemistry class who are geared toward a life science degree and/or career in the health professions. Carbonyl chemistry is fundamental to the chemistry of living systems, and connections between carbonyl chemistry and the chemistry of carbohydrates is highlighted earlier in the book. This latter change mirrors the increasing importance of carbohydrate chemistry as well as other biological molecules on the MCAT.
- **Applications to Synthesis** Organic chemistry enables the synthesis of thousands of useful molecules. Synthetic applications of the reactions covered in this text are emphasized throughout, partly through the many challenging end-of-chapter synthesis problems, the goal of which is to demonstrate to students how synthetic organic chemistry is used in pharmaceutical research and in the production of useful pharmaceuticals. The text provides applications of the reactions to the synthesis of important molecules, such as Valium, fluoxetine (Prozac), meperidine (Demerol), albuterol (Proventil), tamoxifen, and sildefanil (Viagra). Multistep synthesis problems challenge students to develop their own multistep synthetic plan for converting a relatively simple starting material into a more complex target molecule. Multistep synthesis is supported by the Organic Chemistry Reaction Roadmaps as well as an expanded description of retrosynthetic analysis in multiple chapters. We include tips on recognizing when to use certain reactions, such as those involving enolates in the construction of complex structures.
- **Applications to Biology** The application of organic chemistry principles to important biological molecules is integrated where appropriate to establish a bridge with biochemistry courses. In particular, Connections to Biological Chemistry gives special attention to those aspects of organic chemistry that are essential to understanding the chemistry of living systems. For example, the organic chemistry of amino acids is highlighted beginning in Section 3.8, along with the importance of alkene geometry to both membrane fluidity and nutrition. How hydrogen bonding is involved with drug-receptor interactions (Section 10.2) is discussed. Importantly, these Connections to Biological Chemistry features have been added throughout the book, not just at the end, because not all

instructors make it through the biological chemistry chapters at the end of the text. Relevance to practical application is also emphasized in an expanded array of essays titled Chemical Connections. Topics include medicines such as penicillins and cephalosporins (MCAT Practice: Section 18.8), food supplements such as antioxidants (Section 8.7), and materials science concepts such as spider silk (Chemical Connections: Section 27.6). These sections provide a bridge between the theory of organic chemistry and well-known, current, practical applications.

Supporting Materials

Please visit **http://www.cengage.com/chemistry/BIAF/organicchem8e** for information about student and instructor resources for this text.

Acknowledgements

This book is the product of collaboration of many individuals—some obvious, others not so obvious. It is with gratitude that we herein acknowledge the contributions of the many.

Dawn Giovanniello as product director has masterfully guided the revision of the text. Peter McGahey has been a rock of support as content developer. We so appreciate his ability to set challenging but manageable schedules for us and then his constant encouragement as we worked to meet those deadlines. Others at the Cengage Learning organization have helped to shape our words into this text, including Teresa Trego, content project manager and Sarah Cole, art director. Lynn Lustberg of MPS Limited served as our production manager. Also, many thanks to Jordan Fantini of Denison University, who read all of the page proofs for this book with an eye for accuracy.

We are also indebted to the many reviewers of our manuscript who helped shape its contents. With their guidance, we have revised this text to better meet the needs of our and their students.

Eighth Edition

Mark Erickson	Hartwick College
Eric Helms	State University of New York, Geneseo
Nicole Horenstein	University of Florida
Jane Kalkikanda	State University of New York, Oneonta
Steven Neal	The University of Tennessee, Knoxville
Louise Stracener	Edgewood College

Seventh Edition

Philip Albiniak	Ball State University
Andrew Frazer	University of Central Florida
Katie Hailer	Montana Tech
Eric Helms	State University of New York College at Geneseo
Patrick Jokiel	The College of New Jersey
Steven Kass	University of Minnesota
Susan Klein	Manchester College
Thuy Le	The College of New Jersey
Deborah Lieberman	University of Cincinnati
Barbara Mayer	California State University, Fresno

Donna Nelson	University of Oklahoma
Hasan Palandoken	California Polytechnic State University
Lucas Tucker	Siena College

Sixth Edition

Thomas Albright	University of Houston
Zachary D. Aron	Indiana University
Valerie Ashby	University of North Carolina
B. Mikael Bergdahl	San Diego State University
Robert Boikess	Rutgers University
Jean Chmielewski	Purdue University
Elizabeth Harbron	The College of William and Mary
Arif Karim	University of California, Los Angeles
Susan King	University of California, Irvine
Mark Lipton	Purdue University
Allan Pinhas	University of Cincinnati
Owen Priest	Northwestern University
Jonathan Stoddard	California State University, Fullerton

Fifth Edition

Jon Antilla	University of Southern Florida
Christopher Bielawski	University of Texas
Alan Campion	University of Texas
David Cartrette	South Dakota State University
H. J. Peter de Lijser	California State University, Fullerton
Malcolm Forbes	University of North Carolina
John Grutzner	Purdue University
Robert C. Kerber	SUNY, Stony Brook
Spencer Knapp	Rutgers University
Paul Kropp	University of North Carolina
Deborah Lieberman	University of Cincinnati
James Mack	University of Cincinnati
Felix Ngassa	Grand Valley State University
Milton Orchin	University of Cincinnati
Allan Pinhas	University of Cincinnati
Suzanne Ruder	Virginia Commonwealth University
Laurie Starkey	California State Polytechnic University, Pomona
Qian Wang	University of South Carolina
Alexander Wei	Purdue University
Laurie Witucki	Grand Valley State University
Lei Zhu	Florida State University

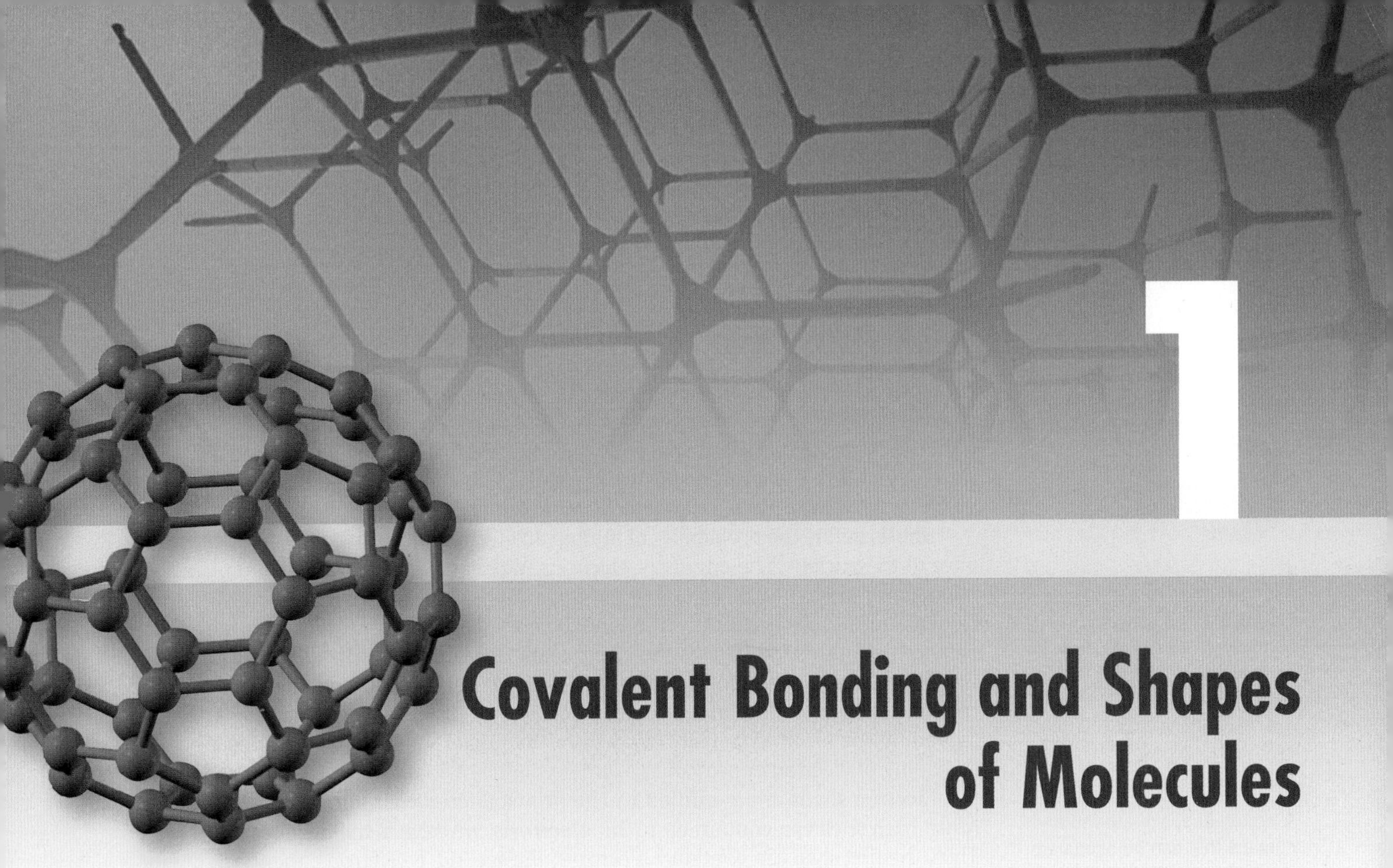

1

Covalent Bonding and Shapes of Molecules

A model of the structure of diamond, one form of pure carbon. Each carbon is bonded to four other carbons at the corners of a tetrahedron. **Inset:** a model of fullerene (C_{60}). See "MCAT Practice: Fullerenes."

Outline

1.1 Electronic Structure of Atoms
1.2 Lewis Model of Bonding
HOW TO: Quickly Figure Out Formal Charge
HOW TO: Draw Lewis Structures from Condensed Structural Formulas
1.3 Functional Groups
1.4 Bond Angles and Shapes of Molecules
1.5 Polar and Nonpolar Molecules
1.6 Quantum or Wave Mechanics
1.7 A Combined Valence Bond and Molecular Orbital Theory Approach to Covalent Bonding
HOW TO: Quickly Recognize the Hybridization and Geometry of Atoms
1.8 Resonance
HOW TO: Draw Curved Arrows and Push Electrons in Creating Contributing Structures
1.9 Molecular Orbitals for Delocalized Systems
1.10 Bond Lengths and Bond Strengths in Alkanes, Alkenes, and Alkynes

According to the simplest definition, **organic chemistry** is the study of the compounds of carbon. Perhaps its most remarkable feature is that most organic compounds consist of carbon and only a few other elements—chiefly, hydrogen, oxygen, and nitrogen. Chemists have discovered or made well over 10 million compounds composed of carbon and these three other elements. Organic compounds are all around us—in our foods, flavors, and fragrances; in our medicines, toiletries, and cosmetics; in our plastics, films, fibers, and resins; in our paints and varnishes; in our glues and adhesives; in our fuels and lubricants; and, of course, in our bodies and the bodies of all living things.

Let us review how the elements of C, H, O, and N combine by sharing electron pairs to form bonds, and ultimately molecules. No doubt, you have encountered much of this initial material in previous chemistry courses; however, the chapters that follow require your ability to use this knowledge fluently.

1.1 Electronic Structure of Atoms

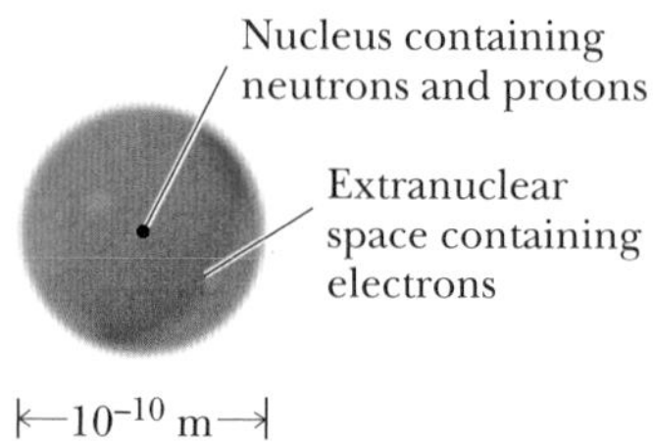
Nucleus containing neutrons and protons
Extranuclear space containing electrons
10^{-10} m

FIGURE 1.1 A schematic view of an atom. Most of the mass of an atom is concentrated in its small, dense nucleus.

An atom contains a small, dense nucleus made of neutrons and positively charged protons. Most of the mass of an atom is contained in its nucleus. The nucleus is surrounded by an extranuclear space containing negatively charged electrons. The nucleus of an atom has a diameter of 10^{-14} to 10^{-15} meters (m). The electrons occupy a much larger volume with a diameter of approximately 10^{-10} m (Figure 1.1).

Shells define the probability of finding an electron in various regions of space relative to the nucleus. The energy of electrons in the shells is quantized. **Quantization** means that only specific values of energy are possible, rather than a continuum of values. These shells occur only at quantized energies in which three important effects balance each other. The first is the electrostatic attraction that draws the electrons toward the nucleus; the second is the electrostatic repulsion between the electrons; and the third is the wavelike nature of an electron that prefers to be delocalized, thereby spreading the electron density away from the nuclei. **Delocalization** describes the spreading of electron density over a larger volume of space.

Electron shells are identified by the principal quantum numbers 1, 2, 3, and so forth. Each shell can contain up to **$2n^2$ electrons**, where n is the number of the shell. Thus, the first shell can contain 2 electrons; the second, 8 electrons; the third, 18 electrons; the fourth, 32 electrons; and so on (Table 1.1). Electrons in the first shell are nearest to the positively charged nucleus and are held most strongly by it; these electrons are lowest in energy. Electrons in higher-numbered shells are farther from the positively charged nucleus and are held less strongly.

Shells are divided into subshells designated by the letters *s, p, d,* and *f,* and within these subshells, electrons are grouped in orbitals (Table 1.2). An **orbital** is a region of space that can hold two electrons and has a specific quantized energy. The first shell contains a single orbital called a 1*s* orbital. The second shell contains one *s* orbital and three *p* orbitals. The three 2*p* orbitals reflect orthogonal angular momentum states in three-dimensional space. **Orthogonal** in this context results in 90° angles between the orbitals, but in all cases, orthogonal also means that the orbitals have no net overlap. As a point of reference, to discuss the 2*p* orthogonal orbitals, we consider them to be directed along the *x*-, *y*-, and *z*-axes and give them designations, $2p_x$, $2p_y$, and $2p_z$. The third shell contains one 3*s* orbital, three 3*p* orbitals, and five 3*d* orbitals. The shapes of *s* and *p* orbitals are shown in Figures 1.8 and 1.9 and are described in more detail in Section 1.6B.

Shell

A region of space around a nucleus that can be occupied by electrons, corresponding to a principal quantum number.

Quantized

Having discrete values for energy and momentum.

Delocalization

The spreading of electron density over a larger volume of space.

Orbital

A region of space that can hold two electrons.

Orthogonal

Having no net overlap.

Table 1.1 Distribution of Electrons in Shells

Shell	Number of Electrons Shell Can Hold	Relative Energies of Electrons in These Shells
4	32	higher
3	18	↑
2	8	
1	2	lower

Table 1.2 Distribution of Orbitals in Shells

Shell	Orbitals Contained in That Shell
3	$3s$, $3p_x$, $3p_y$, $3p_z$, plus five $3d$ orbitals
2	$2s$, $2p_x$, $2p_y$, $2p_z$
1	$1s$

A. Electron Configuration of Atoms

The electron configuration of an atom is a description of the orbitals its electrons occupy. Every atom has an infinite number of possible electron configurations. At this stage, we are concerned primarily with the **ground-state electron configuration**—the electron configuration of lowest energy. We determine the ground-state electron configuration of an atom by using the following three rules.

Ground-state electron configuration
The lowest-energy electron configuration for an atom or a molecule.

Rule 1: The Aufbau ("Build-Up") Principle. According to the **Aufbau principle**, orbitals fill in order of increasing energy, from lowest to highest. In this course, we are concerned primarily with the elements of the first, second, and third periods of the Periodic Table. Orbitals fill in the order $1s$, $2s$, $2p$, $3s$, $3p$, and so on.

Aufbau principle
Orbitals fill in order of increasing energy, from lowest to highest.

Rule 2: The Pauli Exclusion Principle. The **Pauli exclusion principle** requires that only two electrons can occupy an orbital and that their spins must be paired. To understand what it means to have paired spins, recall from general chemistry that just as the Earth has a spin, electrons have a quantum mechanical property referred to as spin. And just as the Earth has magnetic north (N) and south (S) poles, so do electrons. As described by quantum mechanics, a given electron can exist in only two different spin states. Two electrons with opposite spins are said to have **paired spins**.

Pauli exclusion principle
No more than two electrons may be present in an orbital. If two electrons are present, their spins must be paired.

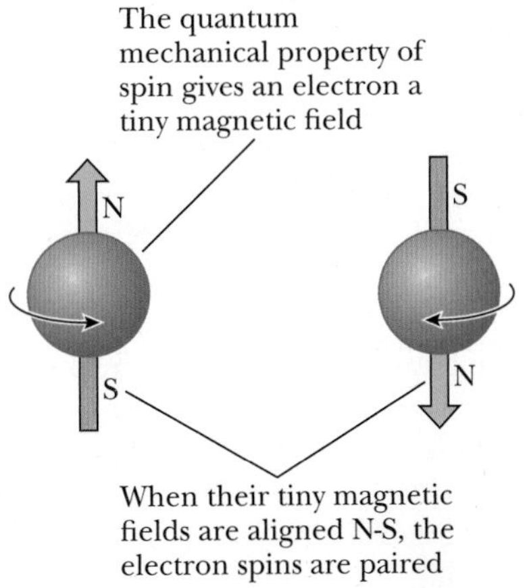
The quantum mechanical property of spin gives an electron a tiny magnetic field
N
S
S
N
When their tiny magnetic fields are aligned N-S, the electron spins are paired

When filling orbitals with electrons, place no more than two in an orbital. For example, with four electrons, the $1s$ and $2s$ orbitals are filled and are written $1s^2\,2s^2$. With an additional six electrons, the set of three $2p$ orbitals is filled and is written $2p_x^2\,2p_y^2\,2p_z^2$. Alternatively, a filled set of three $2p$ orbitals may be written $2p^6$.

Rule 3: Hund's Rule. **Hund's rule** has two parts. The first part states that when orbitals of equal energy (called degenerate) are available but there are not enough electrons to fill all of them completely, then one electron is added to each orbital before a second electron is added to any one of them. The second part of Hund's

Hund's rule
When orbitals of equal energy are available but there are not enough electrons to fill all of them completely, one electron is put in each before a second electron is added to any.

Table 1.3 Ground-State Electron Configurations for Elements 1–18

First Period*			Second Period			Third Period		
H	1	$1s^1$	Li	3	[He] $2s^1$	Na	11	[Ne] $3s^1$
He	2	$1s^2$	Be	4	[He] $2s^2$	Mg	12	[Ne] $3s^2$
			B	5	[He] $2s^2\ 2p^1$	Al	13	[Ne] $3s^2\ 3p^1$
			C	6	[He] $2s^2\ 2p^2$	Si	14	[Ne] $3s^2\ 3p^2$
			N	7	[He] $2s^2\ 2p^3$	P	15	[Ne] $3s^2\ 3p^3$
			O	8	[He] $2s^2\ 2p^4$	S	16	[Ne] $3s^2\ 3p^4$
			F	9	[He] $2s^2\ 2p^5$	Cl	17	[Ne] $3s^2\ 3p^5$
			Ne	10	[He] $2s^2\ 2p^6$	Ar	18	[Ne] $3s^2\ 3p^6$

*Elements are listed by symbol, atomic number, and simplified ground-state electron configuration.

rule states that the spins of the single electrons in the degenerate orbitals should be aligned. Recall that electrons have a negative charge; partially filling orbitals as much as possible minimizes electrostatic repulsion between electrons. After the $1s$ and $2s$ orbitals are filled with four electrons, a fifth electron is added to the $2p_x$ orbital, a sixth to the $2p_y$ orbital, and a seventh to the $2p_z$ orbital. Only after each $2p$ orbital contains one electron is a second electron added to the $2p_x$ orbital. Carbon, for example, has six electrons, and its ground-state electron configuration is $1s^2\ 2s^2\ 2p_x^{\,1}\ 2p_y^{\,1}\ 2p_z^{\,0}$. Alternatively, it may be simplified to $1s^2\ 2s^2\ 2p^2$. Table 1.3 shows ground-state electron configurations of the first 18 elements of the Periodic Table.

Chemists routinely write **energy-level diagrams** that pictorially designate where electrons are placed in an electron configuration. For example, the energy-level diagram for the electron configuration of carbon, $1s^2$, $2s^2$, $2p^2$, shows three energy levels, one each for the $1s$, $2s$, and $2p$ orbitals. Moving up in the diagram means higher energy. Electrons in these diagrams are drawn as arrows. The Aufbau principle tells us to place the first four electrons in the $1s$ and $2s$ orbitals, and the Pauli exclusion principle tells us to pair the two electrons in each orbital (shown as arrows with opposing directions). The remaining two electrons are left to go into the $2p$ level, and because there are three such orbitals, the second part of Hund's rule tells us to place these electrons in different orbitals with their spins aligned (shown as arrows pointing in the same direction). We will use energy-level diagrams later in this chapter to explain bonding and throughout the book when discussing relative energies of orbitals.

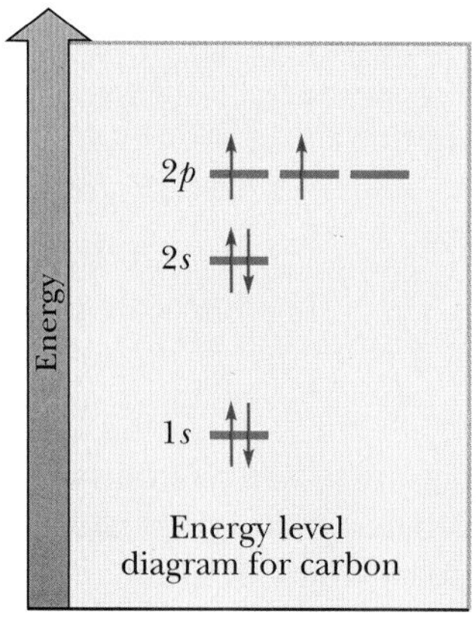

Example 1.1 Electron Configurations

Write the ground-state electron configuration for each element showing the occupancy of each p orbital. For (c), write the energy-level diagram.

(a) Lithium **(b)** Oxygen **(c)** Chlorine

Solution

(a) Lithium (atomic number 3): $1s^2\ 2s^1$

(b) Oxygen (atomic number 8): $1s^2\ 2s^2\ 2p_x{}^2\ 2p_y{}^1\ 2p_z{}^1$

(c) Chlorine (atomic number 17): $1s^2\ 2s^2 2p_x{}^2\ 2p_y{}^2\ 2p_z{}^2 3s^2\ 3p_x{}^2\ 3p_y{}^2\ 3p_z{}^1$

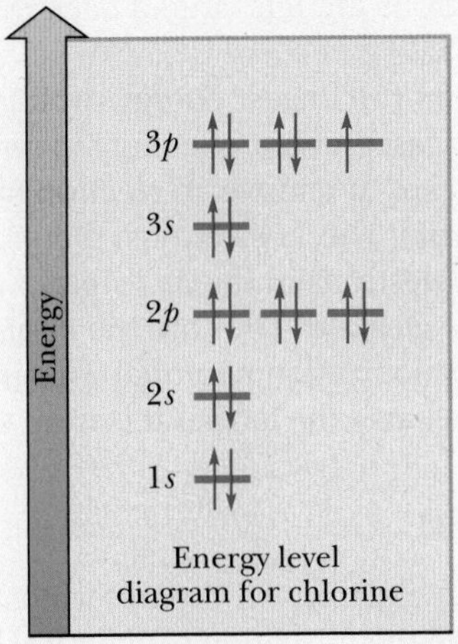

Energy level diagram for chlorine

Problem 1.1

Write and compare the ground-state electron configurations for each pair of elements.

(a) Carbon and silicon **(b)** Oxygen and sulfur **(c)** Nitrogen and phosphorus

Watch a video explanation

B. The Concept of Energy

In the discussion of energy-level diagrams, the lines were drawn on the diagram to depict relative energy. In the energy-level diagram for carbon, the $1s$ level is the reference and the $2s$ and $2p$ levels are placed higher on the diagram relative to it. But you may be asking, "How is energy defined?"

Energy is the ability to do work. The higher in energy an entity is, the more work it can perform. If you hold an object above the ground, it is unstable relative to when it is lying on the ground. You expend energy lifting the object, and this energy is stored in the object as potential energy. The **potential energy** can be released when the object is released. The higher you hold the object, the more energy the object stores and the greater the impact the object will have when it hits the ground.

The force that restores the object to its resting state on the ground is the gravitational attraction of the object to the Earth. Interestingly, the farther the object is from the Earth, the easier it is to take the object even farther from the Earth. As an extreme example, thousands of miles above the Earth the object has incredibly large potential energy and could wreak serious damage to a building if dropped. But at that distance, it is relatively easy to remove the object farther from the Earth because the gravitational attraction is weak.

Energy
The ability to do work.

Potential energy
The energy that can be released if given an opportunity.

We can generalize this example to chemical structures. Unstable structures possess energy waiting to be released. When a structure is higher in energy, the more energy it has stored. When that energy is released, work can be done. In chemistry, released energy is very often harnessed to do work, such as the burning of gasoline to drive the pistons in an internal combustion engine. In chemical reactions carried out in the laboratory, the release of energy commonly just heats up the reaction vessel.

Ground state

The lowest energy state of a system.

Excited state

A state of a system at higher energy than the ground state.

Ionization potential

The energy needed to remove an electron from an atom or a molecule.

Valence electrons

Electrons in the valence (outermost) shell of an atom.

Valence shell

The outermost occupied electron shell of an atom.

Lewis dot structure

The symbol of an element surrounded by the number of dots equal to the number of electrons in the valence shell of the atom.

Let's return to the energy-level diagram of carbon. In the **ground state** of carbon, the electrons are placed in accordance with the quantum chemistry principles (e.g., Aufbau principle, Pauli exclusion principle, and Hund's rule) that dictate the lowest energy form of carbon. If we place the electrons in a different manner (as an example, only one electron in 2*s* and three electrons in 2*p*), we would have a higher energy state of carbon, referred to as an **excited state**. All of nature seeks its lowest energy state; when the electrons are rearranged back to this ground state, energy is released.

Note that the electrons in the lowest energy orbital, 1*s*, are held tightest to the nucleus. It would take the largest amount of energy to remove these electrons relative to the others. The energy it takes to remove an electron from an atom or a molecule is called the **ionization potential**. The 1*s* electrons, therefore, have the highest ionization potential; however, the electrons in the 2*p* levels of carbon are the farthest from the nucleus and are held the weakest. They are the easiest to remove from the atom and therefore have the lowest ionization potential. This is analogous to it being easier to remove an object from the Earth the farther it is from the surface.

C. Lewis Dot Structures

Chemists often focus on the electrons in the outermost shell of the atom because these electrons are involved in the formation of chemical bonds and in chemical reactions. Carbon, for example, with the ground-state electron configuration $1s^2\,2s^2\,2p^2$, has four outer-shell electrons. Outer-shell electrons are called **valence electrons**, and the energy level in which they are found is called the **valence shell**. To illustrate the outermost electrons of an atom, chemists commonly use a representation called a **Lewis dot structure**, named after the American chemist Gilbert N. Lewis (1875–1946), who devised it. A Lewis dot structure shows the symbol of the element surrounded by the number of dots equal to the number of electrons in the outer shell of an atom of that element. In Lewis dot structures, the atomic symbol represents the core (i.e., the nucleus and all inner shell electrons). Table 1.4 shows Lewis dot structures for the first 18 elements of the Periodic Table.

The noble gases helium and neon have filled valence shells. The valence shell of helium is filled with two electrons; that of neon is filled with eight electrons. Neon and argon have in common an electron configuration in which the *s* and *p* orbitals of their valence shells are filled with eight electrons. The valence shells of all other elements shown in Table 1.4 contain fewer than eight electrons.

Bettmann/Corbis

Gilbert N. Lewis (1875–1946) introduced the theory of the electron pair that extended our understanding of covalent bonding and of the concept of acids and bases. It is in his honor that we often refer to an "electron dot" structure as a Lewis structure.

Table 1.4 Lewis Dot Structures for Elements 1–18*

1A	2A	3A	4A	5A	6A	7A	8A
H·							He:
Li·	Be:	B:	·C:	·N:	:O:	:F:	:Ne:
Na·	Mg:	Al:	·Si:	·P:	:S:	:Cl:	:Ar:

*These dots represent electrons from the valence shell. They are arranged as pairs or single electrons in accordance with Hund's rule.

For C, N, O, and F in period 2 of the Periodic Table, the valence electrons belong to the second shell. With eight electrons, this shell is completely filled. For Si, P, S, and Cl in period 3 of the Periodic Table, the valence electrons belong to the third shell. This shell is only partially filled with eight electrons; the 3*s* and 3*p* orbitals are fully occupied, but the five 3*d* orbitals can accommodate an additional ten electrons.

1.2 Lewis Model of Bonding

In 1916, Lewis devised a beautifully simple model that unified many of the observations about chemical bonding and reactions of the elements. He pointed out that the chemical inertness of the noble gases indicates a high degree of stability of the electron configurations of these elements: helium with a valence shell of two electrons ($1s^2$), neon with a valence shell of eight electrons ($2s^2\ 2p^6$), and argon with a valence shell of eight electrons ($3s^2\ 3p^6$). The tendency of atoms to react in ways that achieve an outer shell of eight valence electrons is particularly common among second-row elements of Groups 1A–7A (the main-group elements) and is given the special name **octet rule**.

Octet rule

Group 1A–7A elements react to achieve an outer shell of eight valence electrons.

Example 1.2 The Octet Rule

Show how the loss of an electron from a sodium atom leads to a stable octet.

Solution

The ground-state electron configurations for Na and Na^+ are:

$$\text{Na (11 electrons): } 1s^2\ 2s^2\ 2p^6\ 3s^1$$

$$\text{Na}^+ \text{ (10 electrons): } 1s^2\ 2s^2\ 2p^6$$

Thus, Na^+ has a complete octet of electrons in its outermost (valence) shell and has the same electron configuration as neon, the noble gas nearest it in atomic number.

Problem 1.2

Show how each chemical change leads to a stable octet.

(a) Sulfur forms S^{2-}. **(b)** Magnesium forms Mg^{2+}.

A. Formation of Chemical Bonds

According to Lewis's model, atoms interact in such a way that each participating atom acquires a completed outer-shell electron configuration resembling that of the noble gas nearest to it in atomic number. Atoms acquire completed valence shells in two ways.

1. An atom may become ionic (i.e., lose or gain enough electrons to acquire a completely filled valence shell). An atom that gains electrons becomes an **anion** (a negatively charged ion), and an atom that loses electrons becomes a **cation** (a positively charged ion). A positively charged ion and a negatively charged ion attract each other. This attraction can lead to the formation of ionic crystals such as sodium chloride, in which each positive ion is surrounded by negative ions and vice versa, in a definite geometric arrangement that depends on the crystal. When atoms are held together primarily by attraction of oppositely charged ions, we say

Anion

An atom or a group of atoms bearing a negative charge.

Cation

An atom or a group of atoms bearing a positive charge.

Ionic interaction

Attraction between oppositely charged ions.

Covalent bond

A chemical bond formed between two atoms by sharing one or more pairs of electrons.

that an **ionic interaction** exists between them. (This ionic interaction is often referred to as an ionic bond.)

2. An atom may share electrons with one or more other atoms to complete its valence shell. A chemical bond formed by sharing electrons is called a **covalent bond**.
3. Bonds may be partially ionic and partially covalent; these bonds are called **polar covalent bonds**. Polar covalent bonds are defined more precisely in the next section.

B. Electronegativity and Chemical Bonds

Electronegativity

A measure of the force of an atom's attraction for electrons.

How do we estimate the degree of ionic or covalent character in a chemical bond? One way is to compare the electronegativities of the atoms involved. **Electronegativity** is a measure of an atom's attraction for electrons that it shares in a chemical bond with another atom. The most widely used scale of electronegativities (Table 1.5) was devised by Linus Pauling in the 1930s.

Table 1.5 Electronegativity Values for Some Atoms (Pauling Scale)

1A	2A	3B	4B	5B	6B	7B	8B	8B	8B	1B	2B	3A	4A	5A	6A	7A
								H 2.1								
Li 1.0	Be 1.5											B 2.0	C 2.5	N 3.0	O 3.5	F 4.0
Na 0.9	Mg 1.2											Al 1.5	Si 1.8	P 2.1	S 2.5	Cl 3.0
K 0.8	Ca 1.0	Sc 1.3	Ti 1.5	V 1.6	Cr 1.6	Mn 1.5	Fe 1.8	Co 1.8	Ni 1.8	Cu 1.9	Zn 1.6	Ga 1.6	Ge 1.8	As 2.0	Se 2.4	Br 2.8
Rb 0.8	Sr 1.0	Y 1.2	Zr 1.4	Nb 1.6	Mo 1.8	Tc 1.9	Ru 2.2	Rh 2.2	Pd 2.2	Ag 1.9	Cd 1.7	In 1.7	Sn 1.8	Sb 1.9	Te 2.1	I 2.5
Cs 0.7	Ba 0.9	La 1.1	Hf 1.3	Ta 1.5	W 1.7	Re 1.9	Os 2.2	Ir 2.2	Pt 2.2	Au 2.4	Hg 1.9	Tl 1.8	Pb 1.8	Bi 1.9	Po 2.0	At 2.2

<1.0 | 1.5 – 1.9 | 2.5 – 2.9 | 1.0 – 1.4 | 2.0 – 2.4 | 3.0 – 4.0

Bettmann/Corbis

Linus Pauling (1901–1994) was the first person to receive two unshared Nobel prizes. He received the 1954 Nobel Prize in Chemistry for his contributions to our understanding of chemical bonding. He received the 1962 Nobel Peace Prize for his efforts on behalf of international control of nuclear weapons testing.

On the Pauling scale, fluorine, the most electronegative element, is assigned an electronegativity of 4.0 and all other elements are assigned values in relation to fluorine. As you study the electronegativity values in this table, note that they generally increase from left to right within a period of the Periodic Table and generally decrease from top to bottom within a group. Values increase from left to right because the increasing positive charge on the nucleus results in a greater force of attraction for the atom's valence electrons. Electronegativity decreases from top to bottom because the increasing distance of the valence electrons from the nucleus results in a lower attraction between the nucleus and these electrons.

Let's further analyze the trends in the Periodic Table we just discussed. As you proceed from left to right in a row of the Periodic Table, the atoms get smaller. This contraction occurs because as you go across a row, the electrons are placed in the same shell, but the charge on the nuclei is increasing, thereby pulling the electrons in closer. This means that the orbitals get lower in energy as you move from left to right in the table and that the atoms hold their electrons tighter. It therefore takes more energy to remove the electrons from atoms as you move toward the right in the Periodic Table (with some exceptions), meaning that these atoms have a higher first ionization potential.

In contrast, consider adding—rather than removing—an electron to the atoms. For example, when an electron is added to the halogens (Group 7A), energy is released because these atoms achieve a noble gas configuration. The energy released upon addition of an electron is called the **electron affinity**, which becomes more favorable as you move from left to right in a row of the Periodic Table. In contrast, as you proceed down a column in the Periodic Table, the principal quantum levels increase and the outermost electrons are farther from the nuclei, are held less tightly, and have lower ionization potentials. The atoms also have decreasing electron affinities. Because the electronegativity of an atom reflects its tendency to hold on to and to acquire electrons, the phenomenon arises from a combination of ionization potentials and electron affinities.

Electron affinity
Energy added or released when an electron is added to an atom or a molecule.

When combining the trends of moving from left to right and up to down in the Periodic Table, you can conclude that fluorine must have the most tightly held electrons of any atom that can make bonds. Further, fluorine most tightly holds any electron that it gains during ion formation or covalent bond formation. Hence, fluorine has the highest electronegativity of any atom.

Example 1.3 Electronegativity

Judging from their relative positions in the Periodic Table, which element in each set is more electronegative?

(a) Lithium or carbon **(b)** Nitrogen or oxygen **(c)** Carbon or oxygen

Solution

All of the elements in these sets are in the second period of the Periodic Table. Electronegativity in this period increases from left to right.

(a) C > Li **(b)** O > N **(c)** O > C

Problem 1.3

Judging from their relative positions in the Periodic Table, which element in each set is more electronegative?

(a) Lithium or potassium **(b)** Nitrogen or phosphorus **(c)** Carbon or silicon

Formation of Ions

Ions are formed by the transfer of electrons from the valence shell of an atom of lower electronegativity to the valence shell of an atom of higher electronegativity. As a rough guideline, we say that ions will form if the difference in electronegativity between interacting atoms is 1.9 or greater. As an example, ions are formed from sodium (electronegativity 0.9) and fluorine (electronegativity 4.0). In the following equation, we use a single-headed (barbed) curved arrow to show the transfer of one electron from sodium to fluorine.

$$\mathrm{Na}\cdot + \cdot\ddot{\underset{\cdot\cdot}{\mathrm{F}}}: \longrightarrow \mathrm{Na}^{+}:\ddot{\underset{\cdot\cdot}{\mathrm{F}}}:^{-}$$

In forming Na^+F^-, the single 3*s* valence electron of sodium is transferred to the partially filled valence shell of fluorine:

$$\mathrm{Na}(1s^2\,2s^2\,2p^6\,3s^1) + \mathrm{F}(1s^2\,2s^2\,2p^5) \longrightarrow \mathrm{Na}^+(1s^2\,2s^2\,2p^6) + \mathrm{F}^-(1s^2\,2s^2\,2p^6)$$

As a result of this transfer of one electron, both sodium and fluorine form ions that have the same electron configuration as neon, the noble gas nearest each in atomic number. The attraction between ions is what permits ionic salts such as sodium fluoride to form a strong crystal lattice and gives them a high melting point.

Covalent Bonds

A covalent bond is formed between atoms that share one or more pairs of electrons to give a noble gas configuration to each atom. The simplest example occurs in the hydrogen molecule. When two hydrogen atoms bond, the single electrons from each combine to form an electron pair. This shared pair completes the valence shell of each hydrogen. According to the Lewis model, a pair of electrons in a covalent bond functions in two ways simultaneously: it is shared by two atoms and at the same time fills the outer (valence) shell of each. We use a line between the two hydrogens to symbolize the covalent bond formed by the sharing of a pair of electrons.

$$\text{H}\cdot + \cdot\text{H} \longrightarrow \text{H:H} \quad \text{Symbolized H—H}$$

$$\Delta H^0 = -435 \text{ kJ } (-104 \text{ kcal})/\text{mol}$$

In this pairing, a large amount of energy is released, meaning that two hydrogen atoms are unstable relative to H_2. The same amount of energy, called the **bond dissociation enthalpy** (BDE, also known as the bond dissociation energy) would have to be absorbed to break the bond.

Later in the chapter, we see that electrons have both wave and particle character (Section 1.6). When bonds are formed by the sharing of two electrons between adjacent atoms, the system becomes more stable because the wave character of the electrons is stabilized relative to two separate atoms. The wave that represents the electrons in a bond is partially concentrated in the space between the two nuclei, leading to repulsion between these electrons. In contrast, the electrons are attracted to each nucleus and shield the repulsion between the two positively charged nuclei. The lowering of the energy of the wave character of the electrons along with their added attraction to each nucleus is balanced with the repulsion between the nuclei and between the electrons. This balance results in an optimum internuclear distance called the bond length.

Bond length

The distance between nuclei in a covalent bond in picometers (pm; 1 pm = 10^{-12} m) or Å (1 Å = 10^{-10} m).

The distance between nuclei participating in a chemical bond is called the **bond length**. Every covalent bond has a characteristic bond length. In H—H, it is 74 pm (picometer; 1 pm = 10^{-12} m). We use SI units of picometers in this book; many chemists still use Å (Ångstroms); 1 Å = 100 pm.

Because each bond requires two electrons, a maximum of four bonds can form with second-row atoms. For each unshared pair of electrons on an atom (called a **lone pair**), one fewer bond is possible.

In many situations, filled valence shells can be satisfied only when bonded atoms share more than two electrons. In these cases, multiple covalent bonds form between the same two atoms. For example, four electrons shared between two atoms form a double bond. Six shared electrons form a triple bond.

A good way to distinguish covalent bonds from ionic attraction is the fact that covalent bonds have defined geometries and connectivities resulting from the sharing of electrons. In other words, the number and positions of atoms taking part in covalent bonds are defined. In crystals, the positions of the atoms associated through ionic attraction depend on the particular crystal lattice.

Polar Covalent Bonds

Although all covalent bonds involve the sharing of electrons, they differ widely in the degree of sharing. Homonuclear diatomics such as H_2, N_2, O_2, and F_2 share the electrons equally between the two atoms and are said to have nonpolar covalent bonds. Many compounds such as HCl and H_2O share the electrons in the bond unequally and are said to contain polar covalent bonds. The polarity in the bond increases with increasing difference in electronegativity between the bonded atoms (Table 1.6).

Nonpolar covalent bond

A covalent bond between atoms whose difference in electronegativity is less than approximately 0.5.

A covalent bond between carbon and hydrogen, for example, is classified as **nonpolar covalent** because the difference in electronegativity between these two atoms is 2.5 − 2.1 = 0.4.

Table 1.6 Classification of Chemical Bonds

Difference in Electronegativity Between Bonded Atoms	Type of Bond
Less than 0.5	Nonpolar covalent
0.5 to 1.9	Polar covalent
Greater than 1.9	Ions formed

An example of a **polar covalent bond** is that of H—Cl. The difference in electronegativity between chlorine and hydrogen is $3.0 - 2.1 = 0.9$. An important consequence of the unequal sharing of electrons in a polar covalent bond is that the more electronegative atom gains a greater fraction of the shared electrons and acquires a partial negative charge, indicated by the symbol $\delta-$. The less electronegative atom has a smaller fraction of the shared electrons and acquires a partial positive charge, indicated by the symbol $\delta+$. Alternatively, we show the direction of bond polarity using an arrow with the arrowhead pointing toward the negative end and a plus sign on the tail of the arrow at the positive end.

$$\overset{\delta+}{\mathrm{H}}\!-\!\overset{\delta-}{\mathrm{Cl}} \qquad \overset{+\!\!\longrightarrow}{\mathrm{H{-}Cl}}$$

Polar covalent bond

A covalent bond between atoms whose difference in electronegativity is between approximately 0.5 and 1.9.

Example 1.4 Types of Bonds

Classify each bond as nonpolar covalent or polar covalent or state that ions are formed.

(a) O—H **(b)** N—H **(c)** K—Br **(d)** C—Mg

Solution

Based on differences in electronegativity between the bonded atoms, three of these bonds are polar covalent and one involves the formation of ions.

Bond	Difference in Electronegativity	Type of Interaction
(a) O—H	$3.5 - 2.1 = 1.4$	Polar covalent
(b) N—H	$3.0 - 2.1 = 0.9$	Polar covalent
(c) K—Br	$2.8 - 0.8 = 2.0$	Ions formed
(d) C—Mg	$2.5 - 1.2 = 1.3$	Polar covalent

Problem 1.4

Classify each bond as nonpolar covalent or polar covalent or state that ions are formed.

(a) S—H **(b)** P—H **(c)** C—F **(d)** C—Cl

Electronegativity varies somewhat depending on the chemical environment and oxidation state of an atom; therefore, these rules are only guidelines. Lithium iodide, for example, has a high melting point (449°C) and boiling point (1180°C) characteristic of ionic compounds. Yet, based on the difference in electronegativity between these two elements of 2.5 − 1.0 = 1.5 units, we would classify LiI as a polar covalent compound. In particular, carbon in compounds is not atomic carbon, and it often behaves as if its electronegativity is greater than that shown in Table 1.5.

Example 1.5 Bond Polarity

Using the symbols δ− and δ+, indicate the direction of polarity in each polar covalent bond.

(a) C—O **(b)** N—H **(c)** C—Mg

Solution

Electronegativity is given beneath each atom. The atom with the greater electronegativity has the partial negative charge; the atom with the lesser electronegativity has the partial positive charge.

(a) $\overset{\delta+}{C}—\overset{\delta-}{O}$ (2.5 3.5) **(b)** $\overset{\delta-}{N}—\overset{\delta+}{H}$ (3.0 2.1) **(c)** $\overset{\delta-}{C}—\overset{\delta+}{Mg}$ (2.5 1.2)

Problem 1.5

Using the symbols δ− and δ+, indicate the direction of polarity in each polar covalent bond.

(a) C—N **(b)** N—O **(c)** C—Cl

Bond dipole moment (μ)

A measure of the polarity of a covalent bond. It is the product of the charge on either atom of a polar covalent bond times the distance between the nuclei.

The polarity of a covalent bond is measured by a vector quantity called a **bond dipole moment** and is given the symbol μ (Greek mu). Bond dipole moment is defined as the product of the charge, e (either the δ+ or δ− because each is the same in absolute magnitude), on one of its atoms times the distance, d, separating the two atoms. Note that the dipole moment (μ) is not simply the difference between the electronegativities of the atoms, as students sometime mistakenly believe. The SI unit for a dipole moment is the coulomb · meter, but they are commonly reported instead in a derived unit called the debye (D: 1 D = 3.34×10^{-30} C · m). Table 1.7 lists bond dipole moments for the types of covalent bonds we deal with most frequently in this course.

Table 1.7 Average Bond Dipole Moments of Selected Covalent Bonds

Bond	Bond Dipole (D)	Bond	Bond Dipole (D)	Bond	Bond Dipole (D)
↔		↔		↔	
H—C	0.3	C—F	1.4	C—O	0.7
H—N	1.3	C—Cl	1.5	C=O	2.3
H—O	1.5	C—Br	1.4	C—N	0.2
H—S	0.7	C—I	1.2	C≡N	3.5

C. Lewis Structures for Molecules and Polyatomic Ions

The ability to write Lewis structures for molecules and polyatomic ions is a fundamental skill for the study of organic chemistry. The following guidelines will help you do this. As you study these guidelines, look at the examples in Table 1.8.

1. Determine the number of valence electrons in the molecule or ion. To do this, add the number of valence electrons contributed by each atom. For ions, add one electron for each negative charge on the ion and subtract one electron for each positive charge on the ion. For example, the Lewis structure for a water molecule, H_2O, must show eight valence electrons: one from each hydrogen and six from oxygen. The Lewis structure for the hydroxide ion, HO^-, must also show eight valence electrons: one from hydrogen, six from oxygen, plus one for the negative charge on the ion.
2. Determine the connectivity (arrangement) of atoms in the molecule or ion. Except for the simplest molecules and ions, this connectivity must be determined experimentally because alternative possibilities may lead to **isomers**. Isomers are different compounds with the same molecular formula. For example, there are two different compounds with the molecular formula C_4H_{10}: one with four carbons connected in a row and one in which there are three in a row and the fourth branches off. We discuss isomers extensively in Section 2.2 and Chapter 3.

Isomers
Different compounds with the same molecular formula.

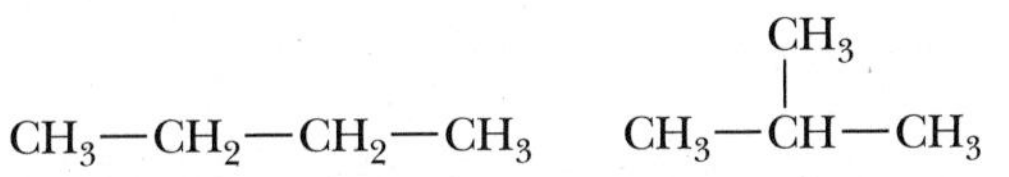

$CH_3—CH_2—CH_2—CH_3$
$CH_3—CH(CH_3)—CH_3$

The two isomers of C_4H_{10}

For some molecules and ions given as examples in the text, you are asked to propose an arrangement of atoms. For most, however, you are given the experimentally determined arrangement.

3. Connect the atoms with single bonds. Then arrange the remaining electrons in pair so that each atom in the molecule or ion has a complete outer shell. Each hydrogen atom must be surrounded by two electrons. Each atom of carbon, oxygen, nitrogen, and halogen must be surrounded by eight electrons (per the octet rule).

Table 1.8 Lewis Structures for Several Compounds*

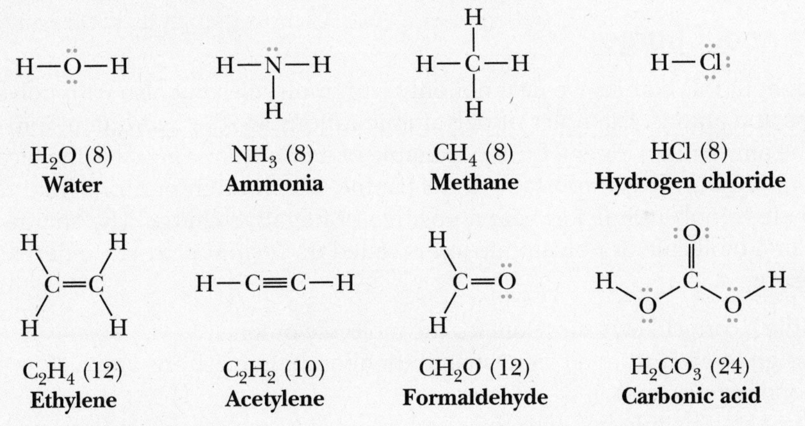

H_2O (8)
Water
NH_3 (8)
Ammonia
CH_4 (8)
Methane
HCl (8)
Hydrogen chloride
C_2H_4 (12)
Ethylene
C_2H_2 (10)
Acetylene
CH_2O (12)
Formaldehyde
H_2CO_3 (24)
Carbonic acid

*The number of valence electrons is shown in parentheses.

Bonding electrons

Valence electrons involved in forming a covalent bond (i.e., shared electrons).

Nonbonding electrons

Valence electrons not involved in forming covalent bonds. Also called unshared pairs or lone pairs.

4. Each pair of electrons (**bonding electrons**) shared between two atoms is shown as a single line between the atoms. Each unshared pair of electrons (often called a **lone pair** or **nonbonding electrons**) is shown as a pair of dots.
5. If two atoms share only a single pair of electrons, they form a single bond and a single line is drawn between them. If two pairs of electrons are shared between two atoms, they form a double bond (two lines). If three pairs of electrons are shared between two atoms, they form a triple bond (three lines).

Table 1.8 shows Lewis structures, molecular formulas, and names for several compounds. The number of valence electrons each molecule contains is shown in parentheses. Notice in these molecules that each hydrogen is surrounded by two valence electrons and that each carbon, nitrogen, oxygen, and chlorine is surrounded by eight valence electrons. Furthermore, each carbon has four bonds, nitrogen has three bonds and one unshared pair of electrons, oxygen has two bonds and two unshared pairs of electrons (lone pairs), and chlorine (and other halogens) has one bond and three unshared pairs of electrons.

Example 1.6 Lewis Structures

Draw Lewis structures showing all valence electrons for these molecules.

(a) CO_2 **(b)** CH_3OH **(c)** CH_3Cl

Solution

The number of valence electrons each molecule contains appears under the Lewis structure.

(a)

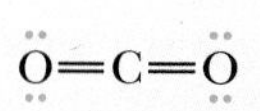

Carbon dioxide
(16 valence electrons)

(b)

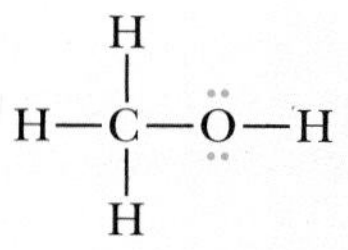

Methanol
(14 valence electrons)

(c)

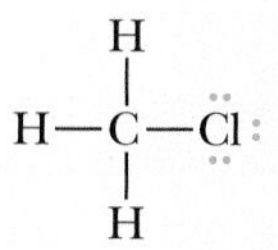

Chloromethane
(14 valence electrons)

Watch a video explanation

Problem 1.6

Draw Lewis structures showing all valence electrons for these molecules.

(a) C_2H_6 **(b)** CS_2 **(c)** HCN

D. Formal Charge

Formal charge

The charge on an atom in a polyatomic ion or molecule.

Throughout this course, we deal not only with molecules but also with polyatomic cations and anions. Examples of polyatomic cations are the hydronium ion, H_3O^+, and the ammonium ion, NH_4^+. An example of a polyatomic anion is the bicarbonate ion, HCO_3^-. It is important to determine which atom or atoms in a neutral molecule or polyatomic ion bear a positive or negative charge. The charge on an atom in a molecule or polyatomic ion is called its **formal charge**. To derive a formal charge:

1. Write a correct Lewis structure for the molecule or ion.
2. Assign to each atom all its unshared (nonbonding) electrons and half its shared (bonding) electrons.
3. Compare this number with the number of valence electrons in the neutral, unbonded atom. Remember that for a neutral unbonded atom, the positive charge

on the nucleus equals the number of its surrounding electrons. So when there are more positive charges on the nucleus than counterbalancing negative charges from the surrounding electrons, that atom has a positive formal charge. Conversely, if the number of electrons assigned to a bonded atom is greater than that assigned to the unbonded atom, the atom has a negative formal charge.

$$\text{Formal charge} = \begin{array}{l}\text{Number of valence}\\ \text{electrons in the neutral,}\\ \text{unbonded atom}\end{array} - \left(\begin{array}{l}\text{All unshared}\\ \text{electrons}\end{array} + \begin{array}{l}\text{One half of all}\\ \text{shared electrons}\end{array}\right)$$

4. The sum of all the formal charges is equal to the total charge on the molecule or ion.

HOW TO

Quickly Figure Out Formal Charge

The preceding discussion provides a step-by-step process that always works for calculating formal charge. However, there are bonding-related patterns to formal charge that, when recognized, allow us to figure out the charge very quickly. When an atom possessing an octet of electrons has one bond more than its neutral state, the atom is positively charged. Conversely, when that atom has one bond less than its neutral form, it is negatively charged. For example, it is well known that water (H_2O) is neutral, while hydronium (H_3O^+) is positive and hydroxide (OH^-) is negative. Hence, we find that oxygen with an octet of electrons and two bonds is neutral. But with three bonds, it is positive, and with only one bond, it is negative. This trend holds true with other molecules as well. An oxygen with two bonds, as in acetone, is neutral, but when a ketone oxygen has a third bond, the oxygen is positive. An oxygen with a single bond is negative (such as an alkoxide).

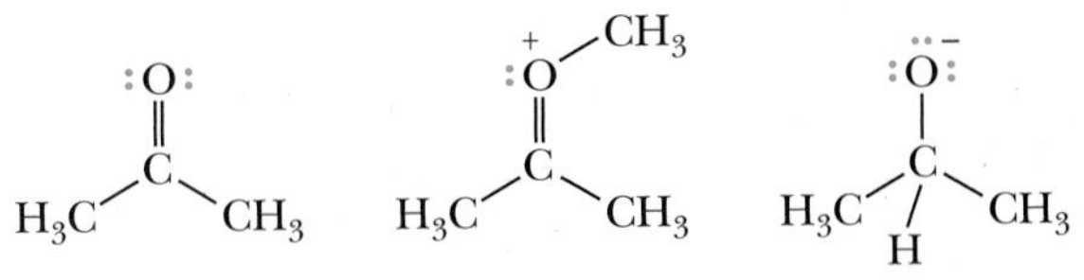

acetone

an alkoxide

Let's look at nitrogen to see if the same trend holds true. For example, nitrogen with an octet and three bonds is neutral (such as ammonia, NH_3). When nitrogen has four bonds, it is positive, and when it has two bonds, it is negative.

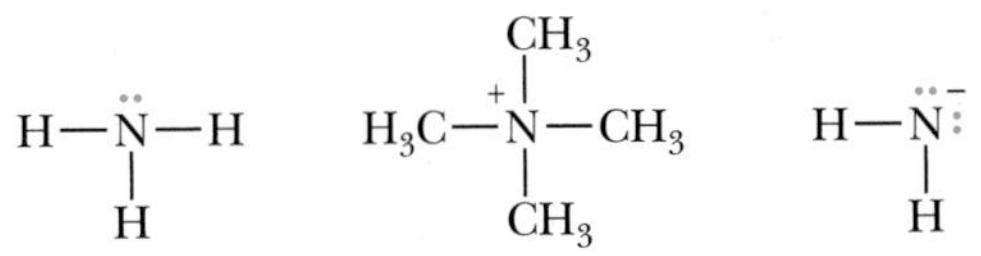

The last trend to note is that atoms in the same column of the Periodic Table have the same charge when they possess the same number of bonds. For example, sulfur atoms have charges as shown below, which are analogous to oxygen.

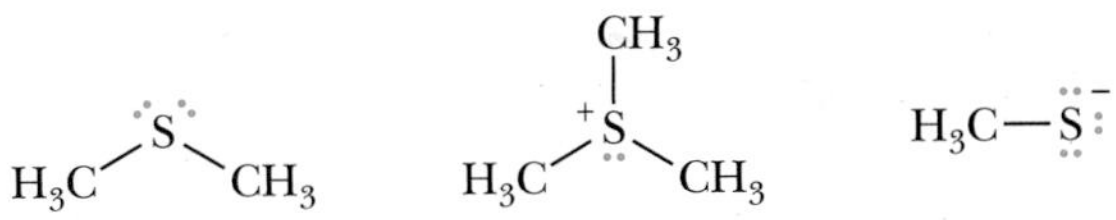

In summary, all you need to remember is the neutral form of each atom and then recall that if an octet is retained, one bond more than the neutral form will impart a positive charge and one bond less will impart a negative charge.

Example 1.7 Formal Charge

Draw Lewis structures for these ions and show which atom in each bears the formal charge.

(a) H_3O^+ **(b)** HCO_3^-

Solution

(a) The Lewis structure for the hydronium ion must show 8 valence electrons: 3 from the three hydrogens, 6 from oxygen, minus 1 for the single positive charge. An oxygen atom has 6 valence electrons. The oxygen atom in

(*Continued*)

H_3O^+ is assigned 2 unshared electrons and 1 from each shared pair of electrons, giving it a formal charge of $6 - (2 + 3) = +1$.

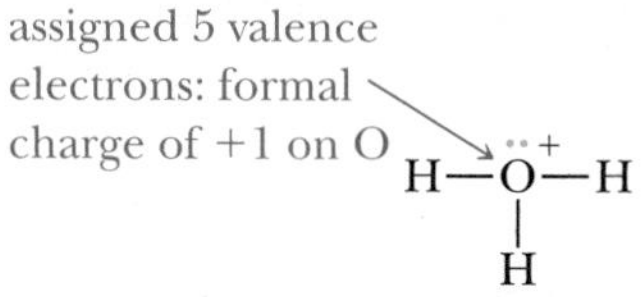

(b) The Lewis structure for the bicarbonate ion must show 24 valence electrons: 4 from carbon, 18 from the three oxygens, 1 from hydrogen, plus 1 for the single negative charge. Loss of a hydrogen ion from carbonic acid (Table 1.8) gives the bicarbonate ion. Carbon is assigned 1 electron from each shared pair and has no formal charge ($4 - 4 = 0$). Two oxygens are assigned 6 valence electrons each and have no formal charges ($6 - 6 = 0$). The third oxygen is assigned 7 valence electrons and has a formal charge of $6 - (6 + 1) = -1$.

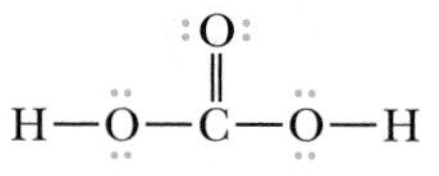

Carbonic acid, H_2CO_3

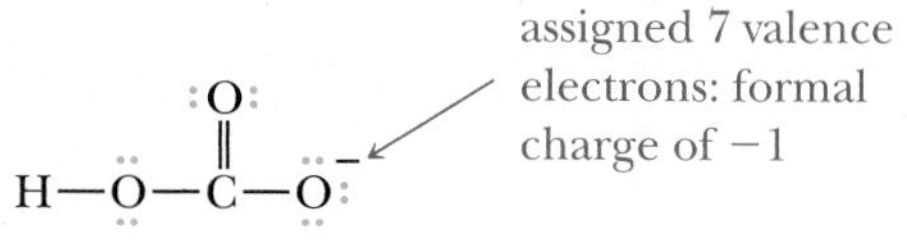

Bicarbonate ion, HCO_3^-

Problem 1.7

Draw Lewis structures for these ions and show which atom (or atoms) in each bears the formal charge.

(a) $CH_3NH_3^+$ **(b)** CO_3^{2-} **(c)** OH^-

When writing Lewis structures for molecules and ions, you must remember that elements of the second period, including carbon, nitrogen, oxygen, and fluorine, can accommodate no more than 8 electrons in the four orbitals ($2s$, $2p_x$, $2p_y$, and $2p_z$) of their valence shells. Following are two Lewis structures for nitric acid, HNO_3, each with the correct number of valence electrons (24):

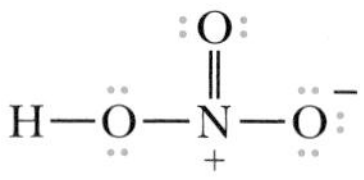

An acceptable Lewis structure

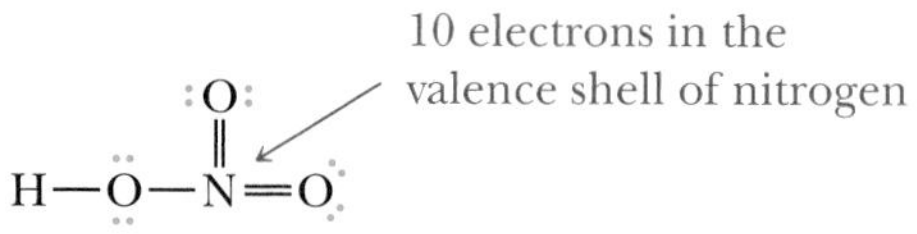

Not an acceptable Lewis structure

The structure on the left is an acceptable Lewis structure. It shows the required 24 valence electrons, and each oxygen and nitrogen has a completed valence shell of eight electrons. Further, it shows a positive formal charge on nitrogen and a negative formal charge on one of the oxygens. Note that the sum of the formal charges on the acceptable Lewis structure for HNO_3 is zero. The structure on the right is not an acceptable Lewis structure. Although it shows the correct number of valence electrons, it places 10 electrons in the valence shell of nitrogen.

After enough practice counting valence electrons, you will begin to recognize functional groups based on the numbers of bonds and lone pairs on the atoms. For example, hydrogen has one bond and no lone pairs, neutral carbon with an

HOW TO

Draw Lewis Structures from Condensed Structural Formulas

Drawing Lewis structures from condensed structural formulas is a survival skill for organic chemistry students. You should follow three steps to draw a correct structure.

1. From a structural formula, obtain information about which atoms are bonded to each other in a molecule. Connect all of the appropriate atoms with single bonds (single lines) first.

Example: $CH_3CH_2CH_2COOCH_3$

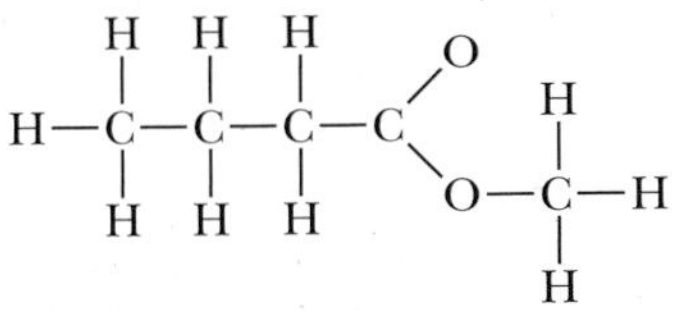

Comment: The difficult part of this structure is deciding how to arrange the two oxygen atoms. Using the arrangement shown will produce a stable structure with filled valences for all of the atoms after you have completed Step 3 below. With practice, you will begin to recognize common functional groups (Section 1.3) such as the carboxylic ester group ($—COOCH_3$ in this example). If you are unsure, you must draw the different possibilities you are considering and, upon completing the structure, determine which one produces the stable structure with the maximum number of filled valence shells around the atoms.

2. Determine how many electrons have been used for the bonds and how many remain. Add all of the additional valence electrons for each atom that does not already have a filled valence shell due to the single bonds. Remember to assign one electron to each atom taking part in a single bond for the purpose of counting valence electrons around atoms. Keep track of any formal charges that may be present in the condensed structural formula (the present example has none).

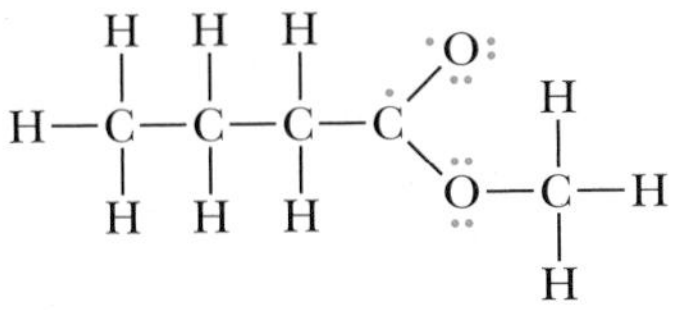

Comment: Recall that each neutral carbon atom has four valence electrons and each neutral oxygen atom has six valence electrons. After all of the single bonds in the molecule have been taken into account, the carbon atom bonded to both oxygen atoms has a single electron left over (4 total electrons − 3 single bonds = 1 electron left over), the oxygen atom bonded only to one carbon atom has five electrons left over (6 total electrons − 1 single bond = 5 electrons left over), and the other oxygen atom has four electrons left over (6 total electrons − 2 single bonds = 4 electrons left over).

3. Add multiple bonds to eliminate unpaired electrons. Draw the remaining nonbonding electrons as lone pairs.

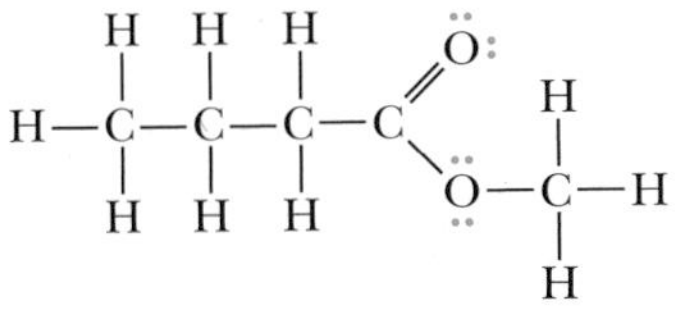

Comment: The only unpaired electrons were on carbon and oxygen, leading to one new bond being formed.

The Lewis structure is now complete. The good news is that drawing Lewis structures gets easier with practice.

octet has four bonds and no lone pairs, neutral nitrogen with an octet has three bonds and one lone pair, neutral oxygen has two bonds and two lone pairs, and neutral halogens have one bond and three lone pairs. When counting bonds for this analysis, double bonds count as two bonds and triple bonds count as three bonds. For atoms with a formal charge, the number of bonds and lone pairs is altered. For example, positively charged nitrogen has four bonds and no lone pairs, positively charged oxygen has three bonds and one lone pair, and positively charged carbon has three bonds and no lone pairs (carbon has an unfilled valence shell). Negatively charged carbon has three bonds and one lone pair, negatively charged nitrogen has two bonds and two lone pairs, and negatively charged oxygen has one bond and three lone pairs.

The guidelines for formal charge can be generalized. Any atom with an octet plus one bond more than its neutral form has a positive formal charge. Any atom with an octet and one bond less than its neutral form has a negative charge. For example, oxygen is neutral with an octet and two bonds, such as in water. Therefore, any time that oxygen has an octet and three bonds, it will be positive, but with an octet and only one bond, it will be negative. We will see in the next section some atoms that have an octet and two bonds more than their neutral form and hence have a plus two formal charge.

E. Exceptions to the Octet Rule

The Lewis model of covalent bonding focuses on valence electrons and the necessity for each atom other than H participating in a covalent bond to have a completed valence shell of eight electrons. Although most molecules formed by main-group elements (Groups 1A–7A) have structures that satisfy the octet rule, there are important exceptions to this rule.

One group of exceptions consists of molecules containing atoms of Group 3A elements. The following graphic is a Lewis structure for BF_3. In this uncharged covalent compound, boron is surrounded by only six valence electrons. Aluminum trichloride is an example of a compound in which aluminum, the element immediately below boron in Group 3A, has an incomplete valence shell. Because their valence shells are only partially filled, trivalent compounds of boron and aluminum exhibit a high reactivity with compounds that have extra electrons, enabling them to fill their octets (Section 4.7).

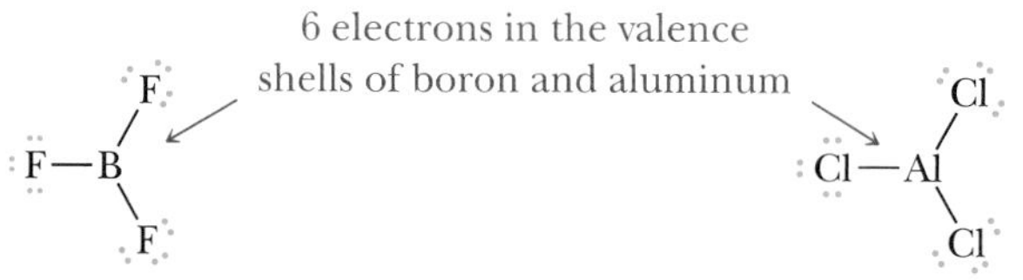

Boron trifluoride **Aluminum trichloride**

F. Dative Bonds

In the Lewis structures we've focused on thus far, the dot structures were built from neutral groups or atoms with the goal of achieving an octet for C, N, O, or the halogens. This led to bonds between adjacent atoms that were created by the sharing of two electrons, one coming from each atom participating in that bond. The sharing of electrons occurs for covalent and polar covalent bonds. However, there are some cases where the construction of a Lewis structure from neutral groups or atoms to achieve an octet gives rise to bonds where the two electrons come from only a single group or atom while the other group or atom involved in that bond does not contribute any electrons. Such bonds are referred to as dative or coordinate covalent.

In many cases, the compound that does not contribute electrons has only six to start (e.g., the **B** and **Al** examples), and the compound contributing the two electrons to the bond has a lone pair of electrons. Further, it is common that this gives rise to a bond that has formal charges. The following **B-N** bond arises from the neutral molecules boron trifluoride (BF_3) and ammonia (NH_3). In the compound shown, the **B** and **N** both have an octet. The bond generated is dative, and one common way to show this is to use an arrow that indicates which atom is donating the two electrons that are shared between the atoms (here, the **N** is donating to the **B**).

$$H_3\overset{\oplus}{N} - \overset{\ominus}{B}F_3 \quad \text{alternatively drawn as} \quad H_3N \rightarrow BF_3$$

1.3 Functional Groups

Carbon combines with other atoms (e.g., H, N, O, S, halogens) to form structural units called **functional groups**. Functional groups are important for three reasons. First, they allow us to divide organic compounds into classes. Second, they exhibit characteristic chemical reactions. A particular functional group, in all compounds that contain it, undergoes the same types of chemical reactions. Third, functional groups serve as a basis for naming organic compounds.

We introduce here several of the functional groups we encounter early in this course. At this point, our concern is only with pattern recognition. We shall have more to say about the structure and properties of these functional groups in following chapters. A complete list of the major functional groups we study in this text is presented on the inside front cover.

Functional group
An atom or a group of atoms within a molecule that shows a characteristic set of physical and chemical properties.

A. Alcohols

The functional group of an **alcohol** is an —OH (**hydroxyl**) group bonded to a tetrahedral carbon atom (a carbon having single bonds to four other atoms). Here is the Lewis structure of ethanol. Note that an —OH group is called a hydroxyl, but when attached to a carbon, that combination is referred to as an alcohol group, and the compound is similarly called an alcohol.

Alcohol
A compound containing an —OH (hydroxyl) group bonded to a tetrahedral carbon atom.

Hydroxyl group
An —OH group.

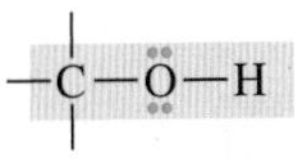

Functional group

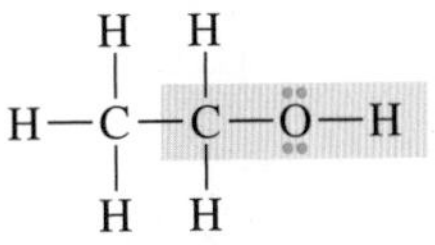

Ethanol
(an alcohol)

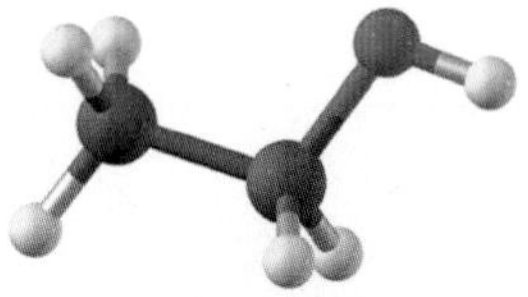

We can also represent this alcohol in a more abbreviated form called a condensed structural formula. In a **condensed structural formula**, CH_3 indicates a carbon bonded to three hydrogens, CH_2 indicates a carbon bonded to two hydrogens, and CH indicates a carbon bonded to one hydrogen. Unshared pairs of electrons are generally not shown in a condensed structural formula. Thus, the condensed structural formula for the alcohol with molecular formula C_2H_6O is $CH_3—CH_2—OH$. It is also common to write these formulas in an even more condensed manner by omitting all single bonds: CH_3CH_2OH.

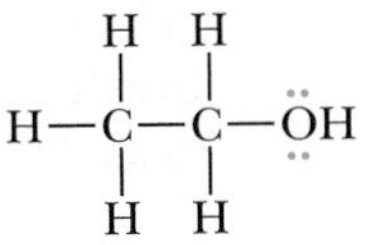

Lewis structure

$CH_3—CH_2—OH$

Condensed structural formula

CH_3CH_2OH

Fully condensed structural formula

Alcohols are classified as **primary (1°)**, **secondary (2°)**, or **tertiary (3°)** depending on the number of carbon atoms bonded to the carbon bearing the —OH group.

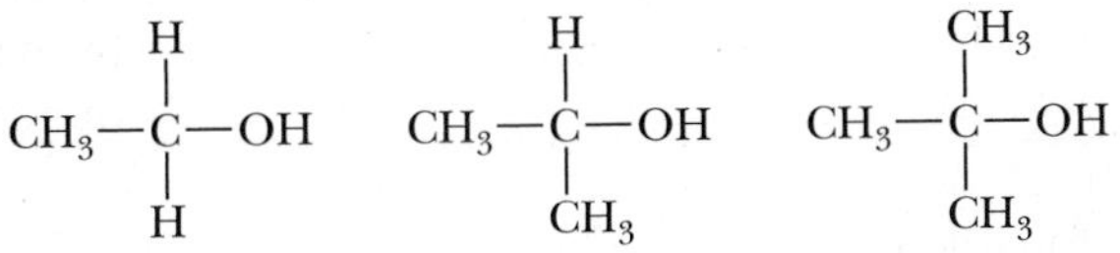

A 1° alcohol A 2° alcohol A 3° alcohol

Primary (1°)
A compound containing a functional group bonded to a carbon atom bonded to only one other carbon atom and two hydrogens.

Secondary (2°)
A compound containing a functional group bonded to a carbon atom bonded to two other carbon atoms and one hydrogen atom.

Tertiary (3°)
A compound containing a functional group bonded to a carbon atom bonded to three other carbon atoms.

Example 1.8 Condensed Structural Formulas I

Draw Lewis structures and condensed structural formulas for the two alcohols with the molecular formula C_3H_8O. Classify each as primary, secondary, or tertiary.

Solution

Begin by drawing the three carbon atoms in a chain. The oxygen atom of the hydroxyl group may be bonded to the carbon chain in two ways: either to an end carbon or to the middle carbon.

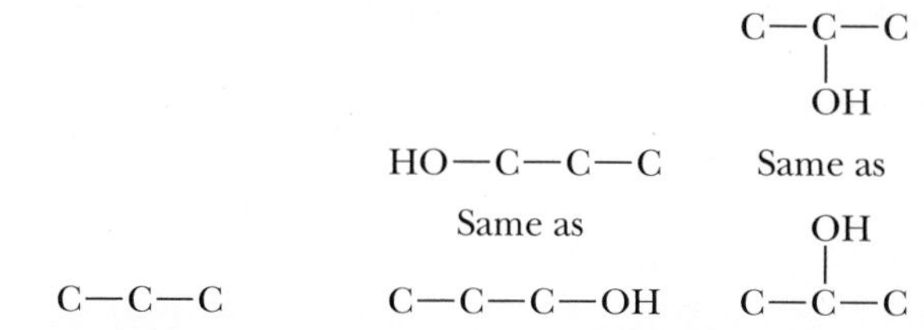
Same as
Same as

The carbon chain   Only two locations for the OH group

Finally, add seven more hydrogens for a total of eight shown in the molecular formula. Show unshared electron pairs on the Lewis structures but not on the condensed structural formulas.

Lewis structures	Condensed structural formulas	
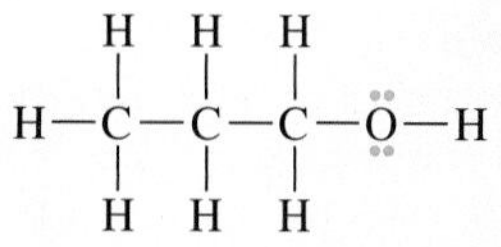	$CH_3CH_2CH_2OH$ A primary alcohol	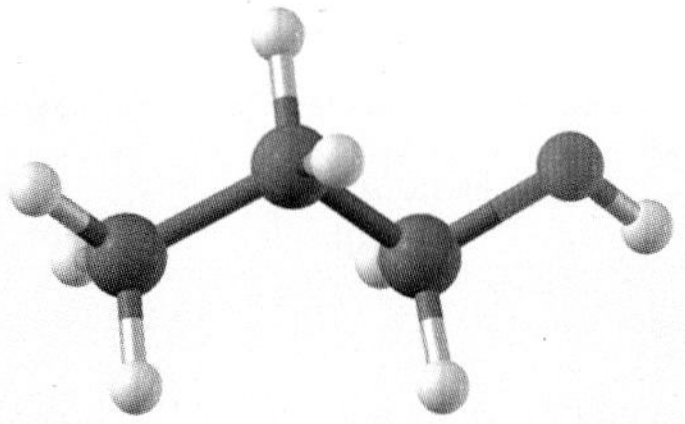
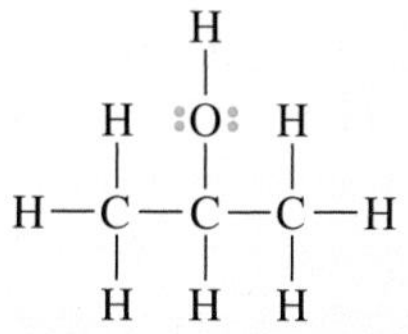	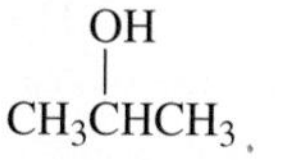 A secondary alcohol	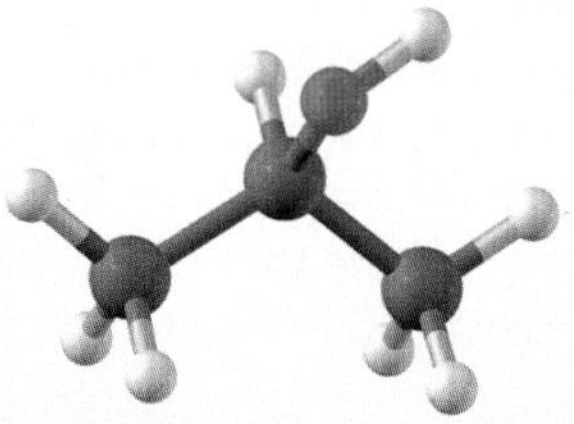

The secondary alcohol, whose common name is isopropyl alcohol, is the major component in rubbing alcohol. We also describe other functional groups such as halides as primary, secondary, and tertiary. For example,

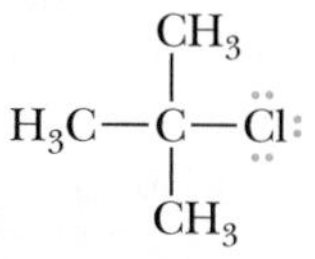 is a tertiary halide.

Problem 1.8

Draw Lewis structures and condensed structural formulas for the four alcohols with the molecular formula $C_4H_{10}O$. Classify each alcohol as primary, secondary, or tertiary.

B. Amines

The functional group of an amine is an **amino group**, a nitrogen atom bonded to one, two, or three carbon atom(s) by single bonds. In a **primary (1°) amine**, nitrogen is bonded to one carbon atom. In a **secondary (2°) amine**, it is bonded to two carbon atoms, and in a **tertiary (3°) amine**, it is bonded to three carbon atoms. Notice that this classification scheme is different from that used with alcohols and halides.

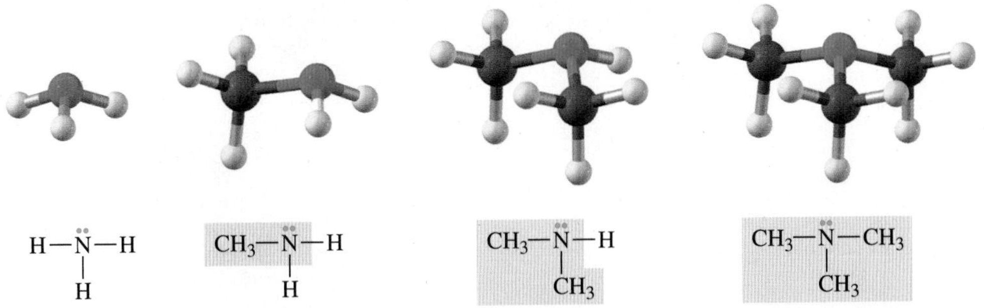

Ammonia — Methylamine (a 1° amine) — Dimethylamine (a 2° amine) — Trimethylamine (a 3° amine)

Amino group
A functional group containing a nitrogen atom bonded to one, two, or three carbon atom(s) by single bonds.

Primary (1°) amine
An amine in which nitrogen is bonded to one carbon and two hydrogens.

Secondary (2°) amine
An amine in which nitrogen is bonded to two carbons and one hydrogen.

Tertiary (3°) amine
An amine in which nitrogen is bonded to three carbons.

Example 1.9 Condensed Structural Formulas II

Draw condensed structural formulas for the two primary amines with the molecular formula C_3H_9N.

Solution

For a primary amine, draw a nitrogen atom bonded to two hydrogens and one carbon.

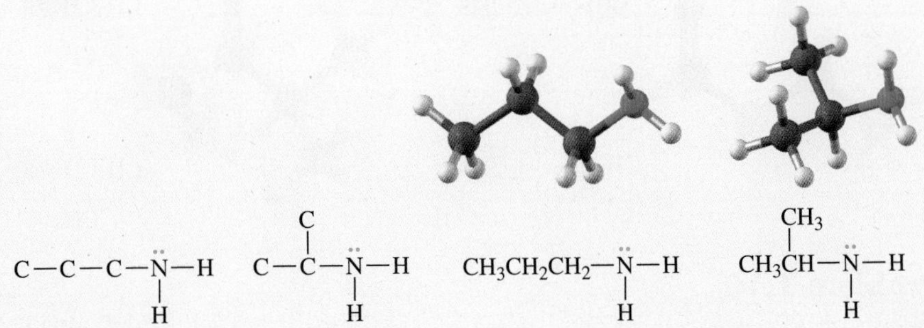

The three carbons may be bonded to nitrogen in two ways.

Add seven hydrogens to give each carbon four bonds and give the correct molecular formula.

Problem 1.9

Draw structural formulas for the three secondary amines with the molecular formula $C_4H_{11}N$.

C. Aldehydes and Ketones

The functional group of both aldehydes and ketones is the **C=O (carbonyl) group**. In formaldehyde, CH_2O, the simplest **aldehyde**, the carbonyl carbon is bonded to two hydrogens. In all other aldehydes, it is bonded to one hydrogen and

Carbonyl group
A C=O group.

Aldehyde
A compound containing a —CHO group.

Ketone

A compound containing a carbonyl group bonded to two carbons.

one carbon. In a condensed structural formula, the aldehyde group may be written showing the carbon-oxygen double bond as —CH=O; alternatively, it may be written —CHO. In a **ketone**, the carbonyl carbon is bonded to two carbon atoms.

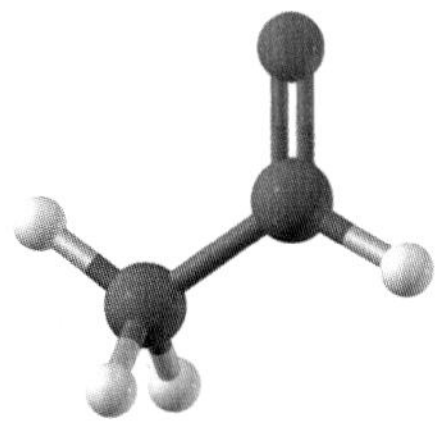

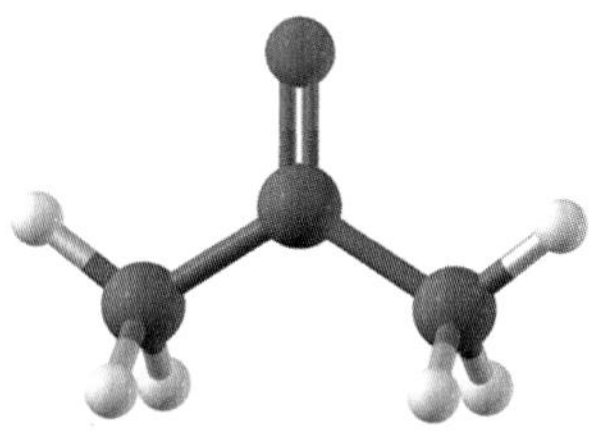

Functional group | **Acetaldehyde** (an aldehyde) | Functional group | **Acetone** (a ketone)

Example 1.10 Condensed Structural Formulas III

Draw condensed structural formulas for the two aldehydes with the molecular formula C_4H_8O.

Solution

First, draw the functional group of an aldehyde and then add the remaining carbons, which may be bonded to the carbonyl group in two ways. Finally, add seven hydrogens to complete the four bonds to each carbon.

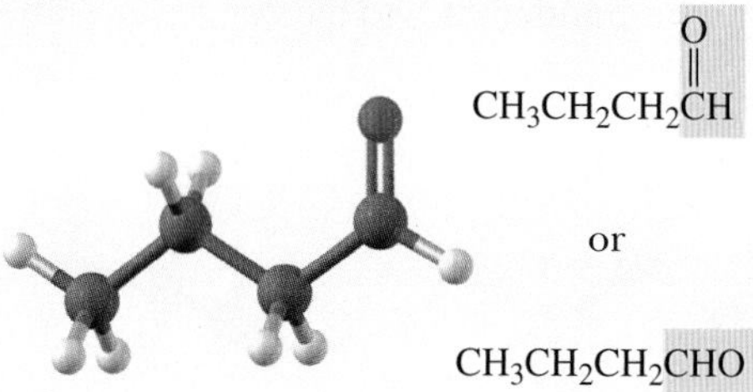

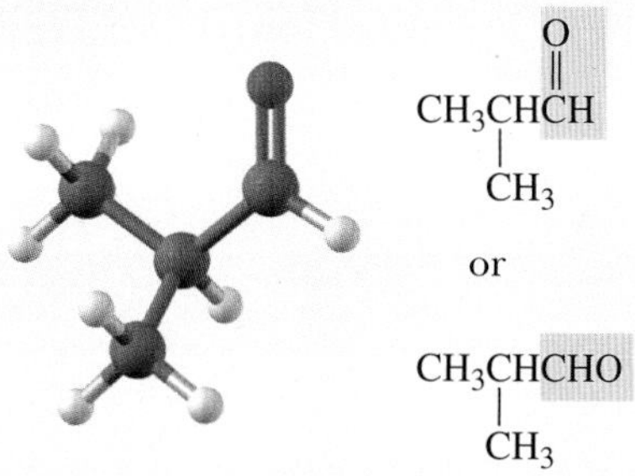

Problem 1.10

Draw condensed structural formulas for the three ketones with the molecular formula $C_5H_{10}O$.

D. Carboxylic Acids

Carboxylic acid

A compound containing a carboxyl, —COOH, group.

Carboxyl group

A —COOH group.

The functional group of a **carboxylic acid** is a **—COOH** (**carboxyl**: *carb*onyl + hydr*oxyl*) **group**. In a condensed structural formula, a carboxyl group may also be written $—CO_2H$.

Functional group | **Acetic acid** (a carboxylic acid)

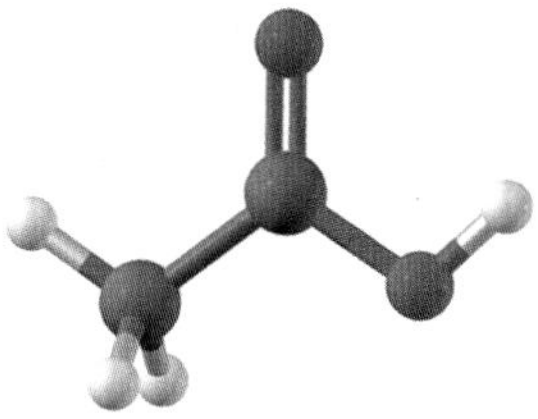

Example 1.11 Condensed Structural Formula IV

Draw a condensed structural formula for the single carboxylic acid with molecular formula $C_3H_6O_2$.

Solution

The only way the carbon atoms can be bonded is three in a chain, and the —COOH group must be on an end carbon of the chain.

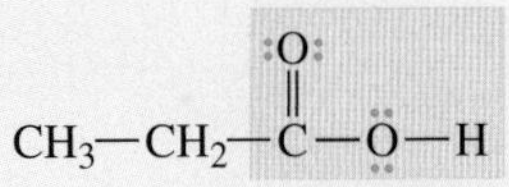

or CH_3CH_2COOH

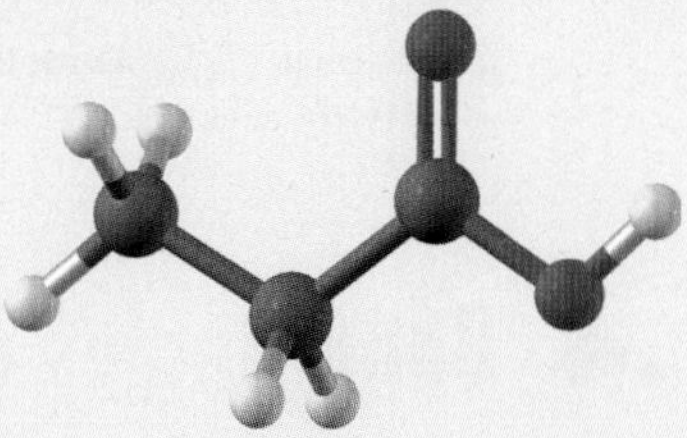

Problem 1.11

Draw condensed structural formulas for the two carboxylic acids with the molecular formula $C_4H_8O_2$.

E. Carboxylic Esters

A **carboxylic ester**, commonly referred to as an **ester**, is a derivative of a carboxylic acid in which the hydrogen of the carboxyl group is replaced by a carbon-containing group.

Carboxylic ester
A derivative of a carboxylic acid in which H of the carboxyl group is replaced by a carbon.

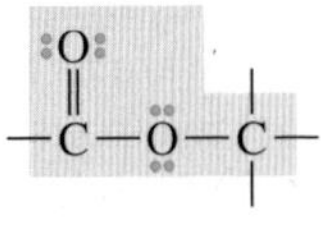

Functional group

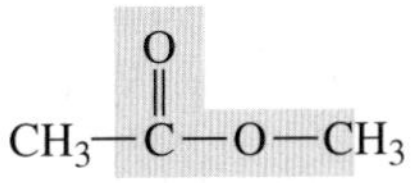

Methyl acetate (an ester)

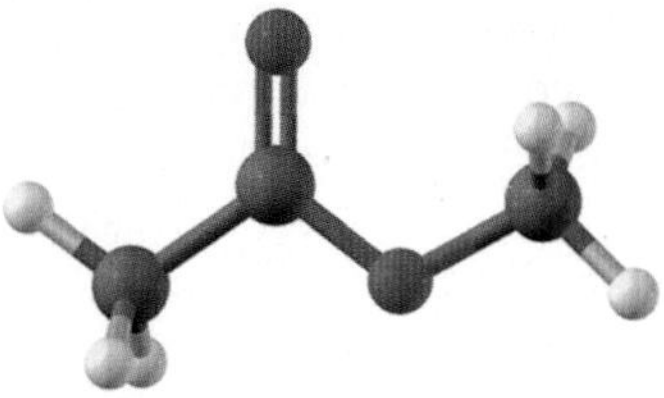

Example 1.12 Condensed Structural Formula V

The molecular formula of methyl acetate is $C_3H_6O_2$. Draw the structural formula of another ester with this same molecular formula.

Solution

There is only one other ester with this molecular formula. Its structural formula is

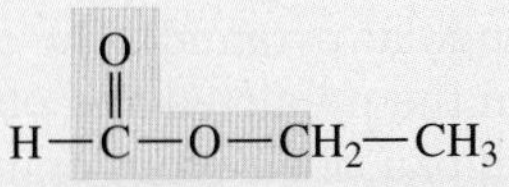

(We will usually write this $HCOOC_2H_5$.)

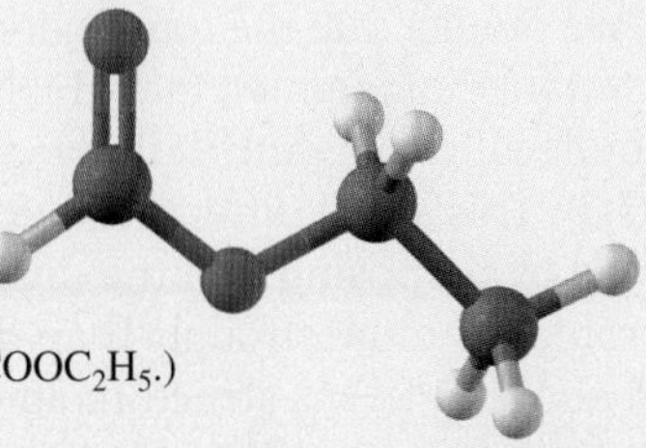

Problem 1.12

Draw structural formulas for the four esters with the molecular formula $C_4H_8O_2$.

F. Carboxylic Amides

Carboxylic amide
A derivative of a carboxylic acid in which the —OH is replaced by an amine.

A **carboxylic amide**, commonly referred to as an **amide**, is a derivative of a carboxylic acid in which the —OH of the carboxyl group is replaced by an amine. As the model shows, the group is planar, something we will explain later.

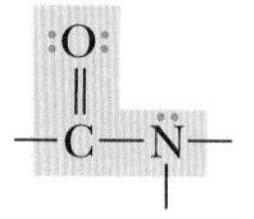
Functional group

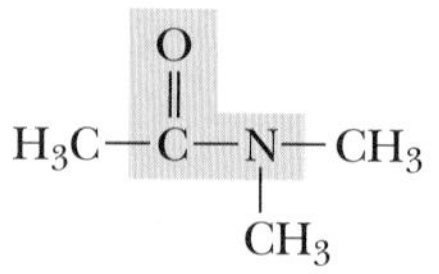
Dimethylacetamide (an amide)

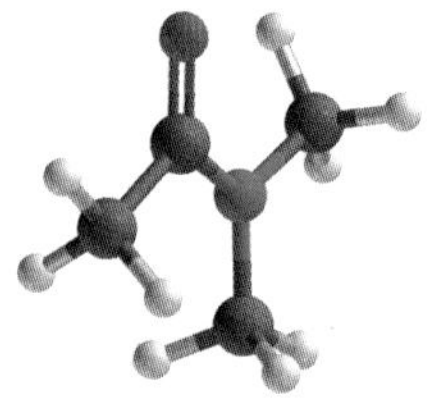

1.4 Bond Angles and Shapes of Molecules

VSEPR
A method for predicting bond angles based on the idea that electron pairs repel each other and stay as far apart as possible.

Tetrahedral
A bonding arrangement in which an atom is bonded to four atoms located at the corners of a tetrahedron.

Pyramidal
A bonding arrangement in which an atom is bonded to three atoms in a triangular pyramid.

In Section 1.2, we used a shared pair of electrons as the fundamental unit of a covalent bond and drew Lewis structures for several molecules and ions containing various combinations of single, double, and triple bonds. We can predict bond angles in these and other molecules and ions in a very straightforward way using a concept referred to as **valence-shell electron-pair repulsion (VSEPR)**. VSEPR is based on the electrons in an atom's valence shell. These valence electrons may be involved in the formation of single, double, or triple bonds, or they may be unshared (lone pair). Each combination creates a negatively charged region of space, and because "like" charges repel each other, the various regions of electron density around an atom will spread out so that each is as far away from the others as possible.

We use VSEPR in the following way to predict the shape of a methane molecule, CH_4. The Lewis structure for CH_4 shows a carbon atom surrounded by four regions of electron density, each of which contains a pair of electrons forming a bond to a hydrogen atom. According to VSEPR, the four regions radiate from carbon so that they are as far away from each other as possible. This occurs when the angle between any two pairs of electrons is 109.5°. Therefore, we predict all H—C—H bond angles to be 109.5° and the shape of the molecule to be **tetrahedral** (Figure 1.2). The H—C—H bond angles in methane have been measured experimentally and found to be 109.5°, identical to those predicted.

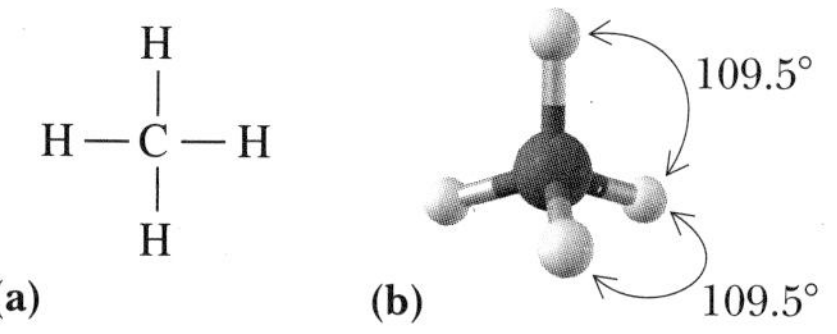

FIGURE 1.2 A methane molecule, CH_4. (a) Lewis structure and (b) shape.

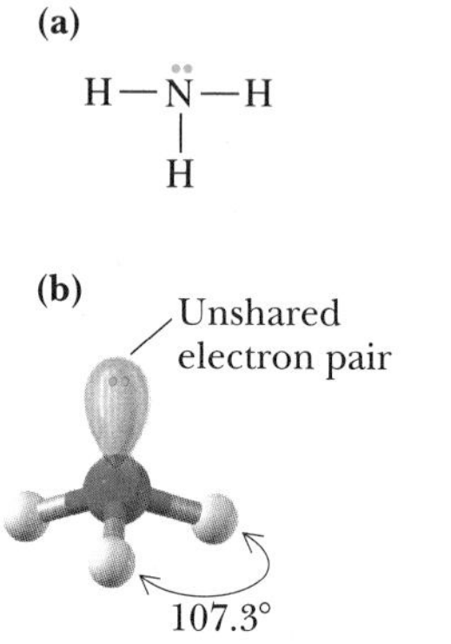

FIGURE 1.3 An ammonia molecule, NH_3. (a) Lewis structure and (b) shape.

We predict the shape of an ammonia molecule, NH_3, in the same manner. The Lewis structure of NH_3 shows nitrogen surrounded by four regions of electron density. Three regions contain single pairs of electrons forming covalent bonds with hydrogen atoms. The fourth region contains an unshared pair of electrons (Figure 1.3). Using VSEPR, we predict that the four regions of electron density around nitrogen are arranged in a tetrahedral manner, that H—N—H bond angles are 109.5° and that the shape of the molecule is **pyramidal** (like a triangular pyramid). The observed bond angles are 107.3°. This small difference between the predicted and observed angles can be explained by proposing that the unshared pair of electrons on nitrogen repels adjacent electron pairs more strongly than do bonding pairs.

Figure 1.4 shows a Lewis structure and a ball-and-stick model of a water molecule. In H_2O, oxygen is surrounded by four regions of electron density. Two of these regions contain pairs of electrons used to form single covalent bonds to the two hydrogens; the remaining two contain unshared electron pairs. Using VSEPR, we predict that the four regions of electron density around oxygen repel each other and are arranged in a tetrahedral manner. The predicted H—O—H bond angle is 109.5°. Experimental measurements show that the actual bond angle is 104.5°, a value smaller than that predicted. This difference between the predicted and observed bond angles can be explained by proposing, as we did for NH_3, that unshared pairs of electrons repel adjacent pairs more strongly than do bonding pairs. Note that the distortion from 109.5° is greater in H_2O, which has two unshared pairs of electrons, than it is in NH_3, which has only one unshared pair.

A general prediction emerges from this discussion of the shapes of CH_4, NH_3, and H_2O molecules. If a Lewis structure shows four regions of electron density around a central atom, VSEPR predicts a tetrahedral distribution of electron density and bond angles of approximately 109.5°.

In many of the molecules we shall encounter, an atom is surrounded by three regions of electron density. Figure 1.5 shows Lewis structures and ball-and-stick models for formaldehyde, CH_2O, and ethylene, C_2H_4.

According to VSEPR, a double bond is treated as a single region of electron density. In formaldehyde, carbon is surrounded by three regions of electron density: two regions contain single pairs of electrons forming single bonds to hydrogen atoms, while the third region contains two pairs of electrons forming a double bond to oxygen. In ethylene, each carbon atom is also surrounded by three regions of electron density: two contain single pairs of electrons, and the third contains two pairs of electrons.

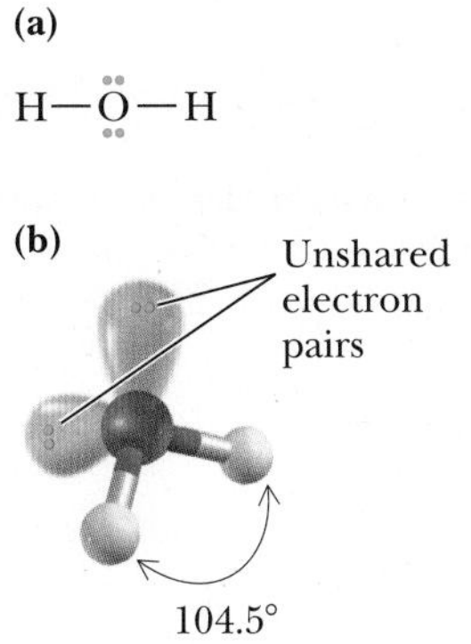
(a)
H—O—H
(b)
Unshared electron pairs
104.5°

FIGURE 1.4 A water molecule, H_2O. (a) Lewis structure and (b) shape.

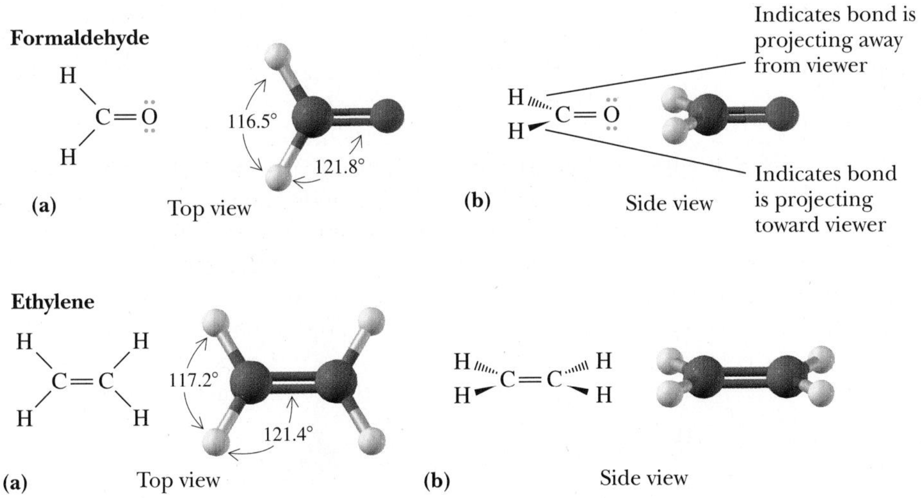
Formaldehyde
116.5°
121.8°
(a)
Top view
Indicates bond is projecting away from viewer
(b)
Side view
Indicates bond is projecting toward viewer
Ethylene
117.2°
121.4°
(a)
Top view
(b)
Side view

FIGURE 1.5 Shapes of formaldehyde, CH_2O, and ethylene, C_2H_4. Molecules shown from (a) top view and (b) side view. Note that chemists commonly use solid wedges to represent bonds projecting toward the viewer and broken wedges for bonds projecting away from the viewer.

Three regions of electron density around an atom are farthest apart when they are coplanar (in the same plane) and make angles of 120° with each other. Thus, the predicted H--C—H and H—C—O bond angles in formaldehyde and the predicted H—C—H and H—C—C bond angles in ethylene are all 120° and the atoms are coplanar. The experimentally measured angles are quite close to this prediction, as shown in Figure 1.5.

In still other types of molecules, a central atom is surrounded by only two regions of electron density. Figure 1.6 shows Lewis structures and ball-and-stick models of carbon dioxide, CO_2, and acetylene, C_2H_2.

In carbon dioxide, carbon is surrounded by two regions of electron density: each contains two pairs of electrons and forms a double bond to an oxygen atom. In acetylene, each carbon is also surrounded by two regions of electron density. One contains a single pair of electrons and forms a single bond to a hydrogen atom, and the other contains three pairs of electrons and forms a triple bond to a carbon atom. In each case, the two regions of electron density are farthest apart if they form a straight line through the central atom and create an angle of 180°. Both carbon dioxide and acetylene are **linear** molecules. Predictions of VSEPR are summarized in Table 1.9.

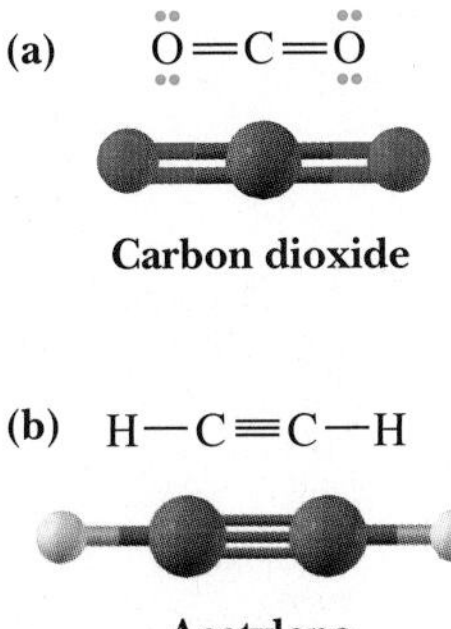
(a) O=C=O
Carbon dioxide
(b) H—C≡C—H
Acetylene

FIGURE 1.6 Shapes of (a) carbon dioxide, CO_2, and (b) acetylene, C_2H_2, molecules.

Table 1.9 Predicted Molecular Shapes (VSEPR)

Regions of Electron Density Around Central Atom	Predicted Distribution of Electron Density	Predicted Bond Angles	Examples
4	Tetrahedral	109.5°	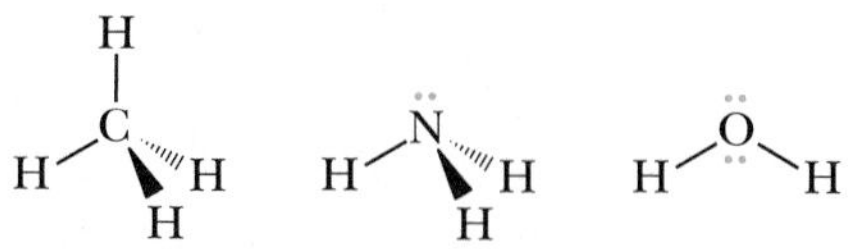
3	Trigonal planar	120°	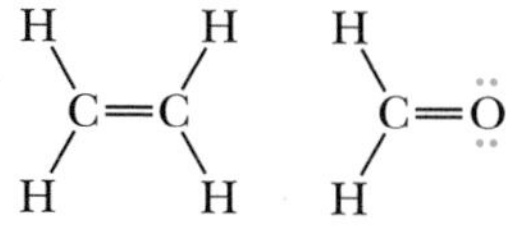
2	Linear	180°	$O=C=O$ $H-C\equiv C-H$

Example 1.13 VSEPR

Predict all bond angles in these molecules.

(a) CH_3Cl **(b)** $CH_2{=}CHCl$

Solution

(a) The Lewis structure for CH_3Cl shows carbon surrounded by four regions of electron density. Therefore, we predict the distribution of electron pairs about carbon to be tetrahedral, all bond angles to be 109.5°, and the shape of CH_3Cl to be tetrahedral. The actual H—C—Cl bond angle is 108°.

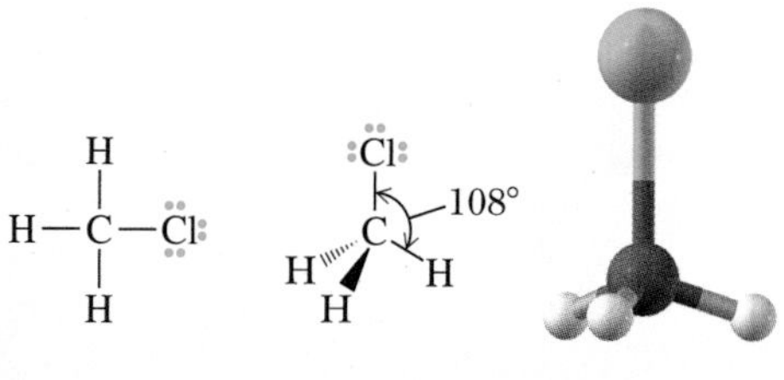

(b) The Lewis structure for $CH_2{=}CHCl$ shows each carbon surrounded by three regions of electron density. Therefore, we predict all bond angles to be 120°. The actual C—C—Cl bond angle is 122.4°.

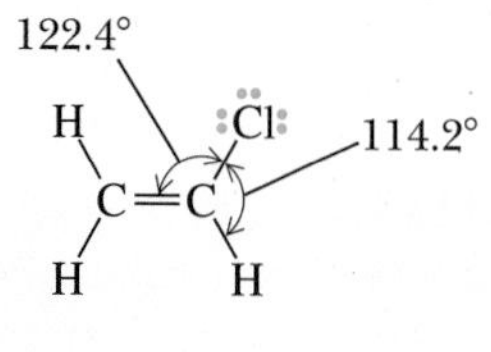

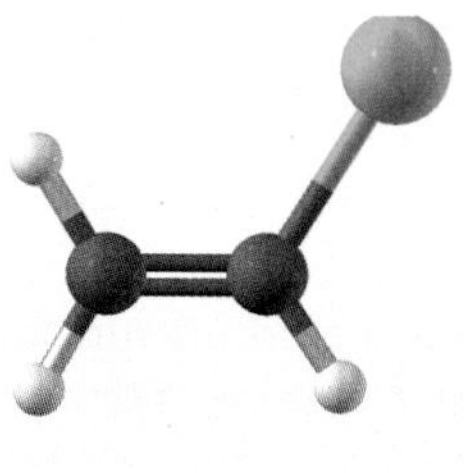

Problem 1.13

Predict all bond angles for these molecules.

(a) CH_3OH **(b)** PF_3 **(c)** H_2CO_3

1.5 Polar and Nonpolar Molecules

We can now combine our understanding of bond polarity (Section 1.2B) and molecular geometry (Section 1.4) to predict the polarity of polyatomic molecules. As we shall see, to be polar, a molecule must have one or more polar bonds. But as we shall also see, not every molecule with polar bonds is polar.

To predict whether a molecule is polar, we need to determine (1) whether the molecule has polar bonds and (2) what the arrangement of its atoms is in space (using VSEPR, for example). The **molecular dipole moment (μ)** of a molecule is the vector sum of its individual bond dipoles. Sometimes the bond dipoles exactly cancel each other due to a molecule's geometry. In carbon dioxide, for example, each C—O bond is polar with oxygen, the more electronegative atom, bearing a partial negative charge and with carbon bearing a partial positive charge. Because carbon dioxide is a linear molecule, the vector sum of its two bond dipoles is zero; therefore, the dipole moment of a CO_2 molecule is zero. Boron trifluoride is planar with bond angles of 120°. Although each B—F bond is polar, the vector sum of its bond dipoles is zero and BF_3 has no dipole moment. Carbon tetrachloride is tetrahedral with bond angles of 109.5°. Although it has four polar C—Cl bonds, the vector sum of its bond dipoles is zero and CCl_4 also has no dipole moment.

Molecular dipole moment (μ)
The vector sum of individual bond dipoles.

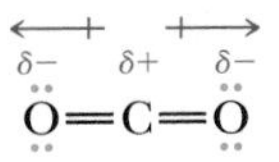

Carbon dioxide
$\mu = 0$ D

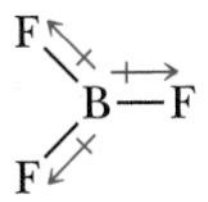

Boron trifluoride
$\mu = 0$ D

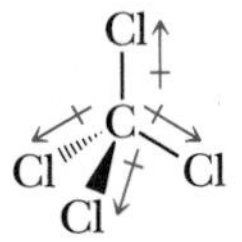

Carbon tetrachloride
$\mu = 0$ D

Other molecules, such as water and ammonia, have polar bonds and dipole moments greater than zero; they are polar molecules. Each O—H bond in a water molecule and each N—H bond in ammonia are polar, with oxygen and nitrogen, the more electronegative atoms, bearing a partial negative charge and each hydrogen bearing a partial positive charge.

MCAT Practice: *Passage and Questions*

Fullerenes

Many of the questions on the Medical College Admission Test (MCAT) start with a passage followed by a series of multiple-choice questions. After studying the passage, select the best answer to each question in the group. If you are not certain of an answer, eliminate the alternatives that you know to be incorrect and select an answer from the remaining ones.

For centuries, elemental carbon was thought to have only two forms: graphite and diamond. The scientific world was startled in 1985 when Richard Smalley of Rice University and Harry W. Kroto of the University of Sussex, UK, and their coworkers announced that they had detected a new form of carbon with the molecular formula C_{60}. The molecule C_{60} resembles a soccer ball (see the figure to the right); it has 12 five-membered rings and 20 six-membered rings arranged such that each five-membered ring is surrounded by 5 six-membered rings. The structure reminded its discoverers of a geodesic dome, a structure invented by the innovative American engineer and philosopher R. Buckminster Fuller. Therefore, the official name of this new allotrope of carbon became known as *fullerene*. Some chemists also call C_{60} "buckyball." Kroto, Smalley, and Robert F. Curl were awarded the 1996 Nobel Prize in Chemistry for this work.

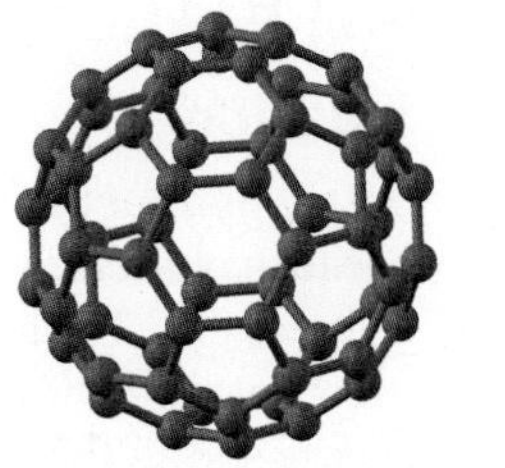

(Continued)

Questions

A. The geometry of carbon in diamond is tetrahedral, while carbon's geometry in graphite is trigonal planar. What is the geometry of the carbons in C_{60}?

1. They are all tetrahedral.
2. They are all trigonal planar.
3. They are all pyramidal with bond angles near 109.5°.
4. They are not perfectly trigonal planar but have an extent of pyramidalization.

B. Because of their spherical shape, C_{60} molecules are used as nanoscale ball bearings in grease and lubricants. We can estimate the size of these ball bearings by examining C—C bond distances. Carbon-carbon bond distances vary between approximately 120 pm (pm = picometers) and 155 pm. What is the approximate diameter of C_{60}?

1. 10 pm
2. 100 pm
3. 1000 pm
4. 10,000 pm

C. What best describes the C—C—C bond angles in C_{60}?

1. They are exactly 120°.
2. They are a bit larger than 120°.
3. They are a bit smaller than 120°.
4. They are near 109.5°.

The charge densities are easily computed by modern desktop computer programs such as Spartan. Here are electrostatic potential maps that display the computed electronic charge density in water and ammonia. In these models, red represents negative charge and blue represents positive charge. In agreement with the dipole moment diagram and our expectations, the more electronegative atom has substantial negative charge in both molecules. For more information on how to interpret these plots, see Appendix 7.

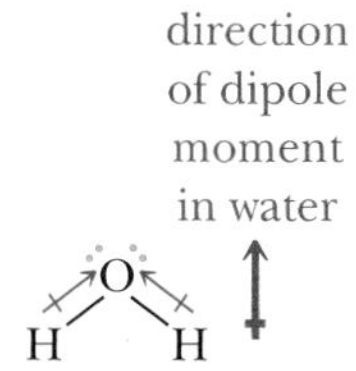
direction
of dipole
moment
in water
O
H
H

Water
μ = 1.85 D

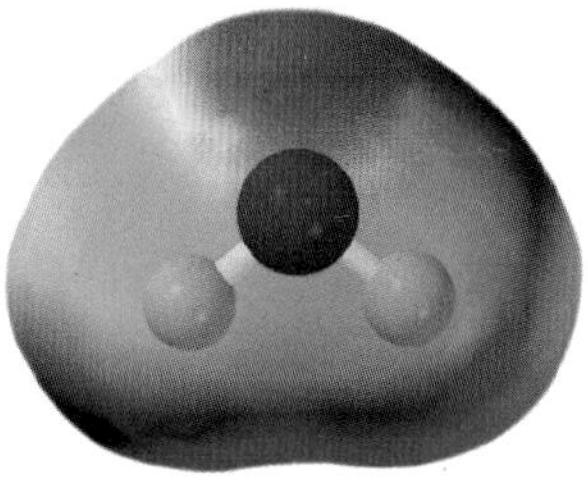

An electrostatic potential map (elpot) of a water molecule.

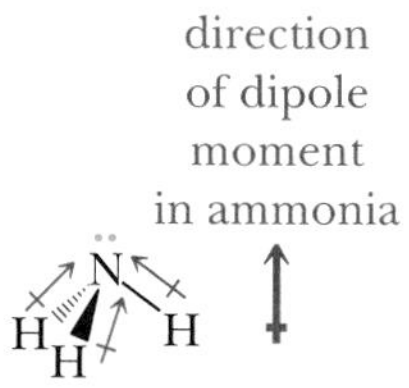
direction
of dipole
moment
in ammonia
N
H
H
H

Ammonia
μ = 1.47 D

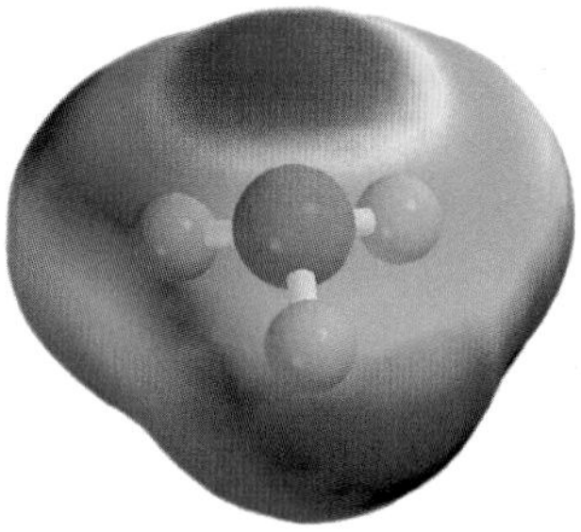

An electrostatic potential map (elpot) of an ammonia molecule.

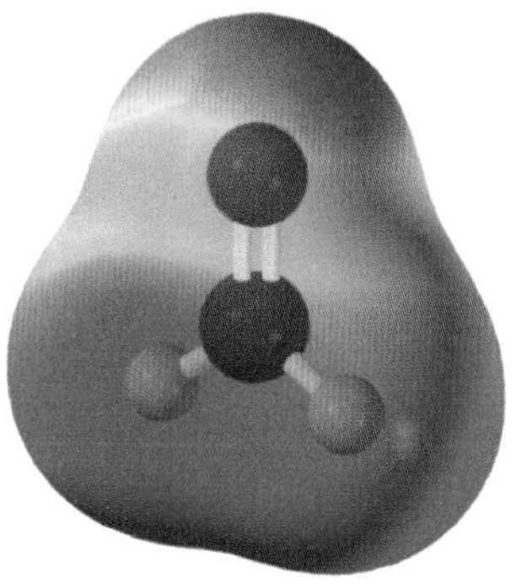

An electrostatic potential map (elpot) of a formaldehyde molecule.

Example 1.14 Molecular Dipoles

Which of these molecules are polar? For each that is polar, specify the direction of its dipole moment.

(a) CH_3Cl **(b)** CH_2O **(c)** C_2H_2

Solution

Both chloromethane, CH_3Cl, and formaldehyde, CH_2O, have polar bonds and, because of their geometries, are polar molecules. Because of its linear geometry, acetylene, C_2H_2, has no dipole moment. The experimentally measured dipole moments are shown. The electrostatic potential map (elpot) of formaldehyde clearly shows this charge distribution.

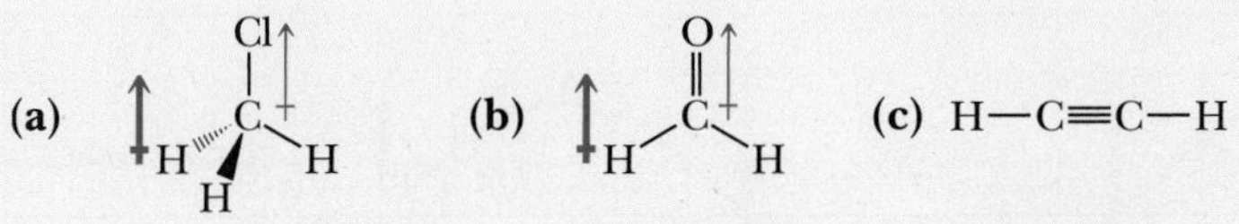

Chloromethane $\mu = 1.87$ D

Formaldehyde $\mu = 2.33$ D

Acetylene $\mu = 0$ D

Problem 1.14

Watch a video explanation

Which molecules are polar? For each that is polar, specify the direction of its dipole moment.

(a) CH_2Cl_2 (b) HCN (c) H_2O_2

1.6 Quantum or Wave Mechanics

Thus far in this chapter, we have concentrated on the Lewis model of bonding and on VSEPR. The Lewis model deals primarily with the coordination numbers of atoms (the number of bonds a given atom can form), and VSEPR deals primarily with bond angles and molecular geometries. Although each is useful in its own way, neither gives us any means of accounting in a quantitative or even semiquantitative way for the reasons atoms combine in the first place to form covalent bonds with the liberation of energy. At this point, we need to study an entirely new approach to the theory of covalent bonding, one that provides a means of understanding not only the coordination numbers of atoms and molecular geometries but also the energetics of chemical bonding.

A. Moving Particles Exhibit the Properties of a Wave

The beginning of this new approach to the theory of covalent bonding was provided by Albert Einstein (1879–1955), a German-born American physicist. In 1905, Einstein postulated that light consists of photons of electromagnetic radiation. The energy, E, of a photon is proportional to the frequency, ν (Greek nu), of the light. The proportionality constant in this equation is Planck's constant, h.

$$E = h\nu$$

In 1923, the French physicist Louis de Broglie followed Einstein's lead and advanced the revolutionary idea that if light exhibits properties of particles in motion, then a particle in motion should exhibit the properties of a wave. He proposed that a particle of mass m and speed v has an associated wavelength λ (Greek lambda), given by the equation

$$\lambda = \frac{h}{mv} \quad \text{(the de Broglie relationship)}$$

Illustrated in Figure 1.7 is a wave that might result from plucking a guitar string. The mathematical equation that describes this wave is called a **wave equation**. The numerical value(s) of the solution(s) of a wave equation may be positive (corresponding to a wave crest), negative (corresponding to a wave trough), or zero. The sign of the numerical value of the wave equation is called the phase, and changes in sign are referred to as phase changes. Phase changes occur at nodes. A **node** is any point where the value of a solution of a wave equation is zero. A **nodal plane** is any plane perpendicular to the direction of propagation that runs through a node. Shown in Figure 1.7 are three nodal planes.

Node
A point in space where the value of a solution of a wave equation is zero.

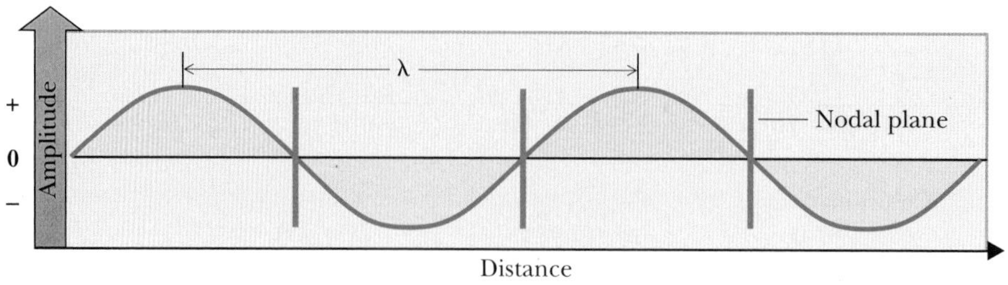

FIGURE 1.7 Characteristics of a wave associated with a moving particle. Wavelength is designated by the symbol λ.

Quantum mechanics

The branch of science that studies particles and their wavelike properties.

Wave function

A set of solutions to the Schrödinger equation that define the energy of an electron in an atom and the region of space it may occupy.

Phasing

Sign of the wave function at particular coordinates in space, either plus or minus. Phasing is often represented as different colors, such as red and blue used in this text.

Erwin Schrödinger built on the idea of de Broglie and in 1926 proposed an equation that could be used to describe the wave properties associated with an electron in an atom or a molecule. **Quantum mechanics (wave mechanics)** is the branch of science that studies particles and their associated waves. Solving the Schrödinger equation gives a set of solutions called **wave functions**. Each wave function ψ (Greek psi) is associated with a unique set of quantum numbers and with a particular atomic or molecular orbital. A wave function occupies three-dimensional space and is called an **orbital**. Each orbital can contain no more than two electrons. The value of ψ^2 is proportional to the probability of finding an electron at a given point in space. Looked at in another way, the value of ψ^2 at any point in space is proportional to the electron density at that point. A plot of electron density (ψ^2) in a given orbital theoretically reaches to infinity but becomes vanishingly small at long distances from the nucleus. Notice that although the value of ψ at any point can be positive or negative, the value of ψ^2 will always be positive in an orbital. In other words, the electron density in two regions of an orbital will be equal if those regions have the same absolute value of ψ, regardless of whether that value is negative or positive. Of course, ψ^2 will be zero at a node. We often represent orbitals as a solid or mesh with the surface representing the volume within which some amount (such as 95%) of the electron density is contained.

For most aspects of organic chemistry, it is best to consider the wavelike properties of electrons. In this text, we concentrate on wave functions and shapes associated with *s* and *p* atomic orbitals because they are the orbitals most often involved in covalent bonding in organic compounds. The wave nature is reflected by the orbital in which the electron resides. We should not think about finding an electron as a particle at a particular location in space; rather, we should consider the electron density in various regions of space as the square of the amplitude of the orbital at that position in space (see previous paragraph).

When we describe orbital interactions, we are referring to interactions of waves. Waves interact constructively or destructively (adding or subtracting, respectively). When two waves overlap, positive **phasing** adds constructively with positive phasing, as does negative phasing with negative phasing. However, positive and negative phasing also add destructively, meaning they cancel. For waves on the ocean or on a plucked guitar string, this characteristic is sometimes referred to as waves adding "in phase" and "out of phase."

There is one reason to consider the particle nature of electrons, and that is charge. Negatively charged particles, such as electrons, are attracted to positively charged particles, such as nuclei, while being repulsed by other negative particles, such as other electrons.

It is convenient to add up the electron densities in all of the orbitals in a molecule and then determine which areas of a molecule have larger and smaller amounts of electron density. In general, the greater electron density is on the more electronegative atoms, especially those with lone pairs. Relative electron density distribution in molecules is important because it allows us to identify sites of chemical reactivity. Many reactions involve an area of relatively high electron density on one molecule reacting with an area of relatively low electron density on another molecule. It is convenient to keep track of overall molecular electron density distributions using computer graphics. This text presents electrostatic potential maps (elpots) in which areas of relatively high calculated electron density are shown in red and areas of relatively low calculated electron density are shown in blue, with intermediate electron densities represented by intermediate colors. The water, ammonia, and formaldehyde molecules in Section 1.5 are examples of electrostatic potential maps (elpots).

B. Shapes of Atomic *s* and *p* Orbitals

All s orbitals have the shape of a sphere, with the center of the sphere at the nucleus. Shown in Figure 1.8 are three dimensional shapes (plots of ψ) for 1*s* and 2*s* orbitals. These orbitals are completely symmetrical along all axes. We present orbitals that were calculated using the Schrödinger equation, as well as the common cartoons that organic chemists use to represent them. Mesh diagrams are used throughout this book because they allow you to see the inner features of orbitals, such as the change in phase in the 2*s* orbital. You should not consider the lines of the mesh as trajectories of electrons, but rather that the mesh shows a surface within which about 95% of the electron density exists.

Shown in Figure 1.9 are the three-dimensional shapes (plots of ψ) of the three 2*p* orbitals, combined in one diagram to illustrate their relative orientations in space. Each 2*p* orbital consists of two lobes arranged in a straight line with the nucleus in the middle. The three 2*p* orbitals are mutually perpendicular and are designated $2p_x$, $2p_y$, and $2p_z$. The sign of the wave function of a 2*p* orbital is positive in one lobe, zero at the nucleus, and negative in the other lobe. The plus or minus is simply the sign of the mathematical function ψ_{2p} and has no relationship to energy or electron distribution. These signs are shown by blue or red colors; however, these colors should not be confused with the colors used to represent charge density. Recall that the value of ψ^2 is always positive, so the probability of finding electron density in the (+) lobe of a 2*p* orbital (i.e., value of ψ is positive) is the same as that of finding it in the (−) lobe (i.e., value of ψ is negative). Again, electron density is zero at the node.

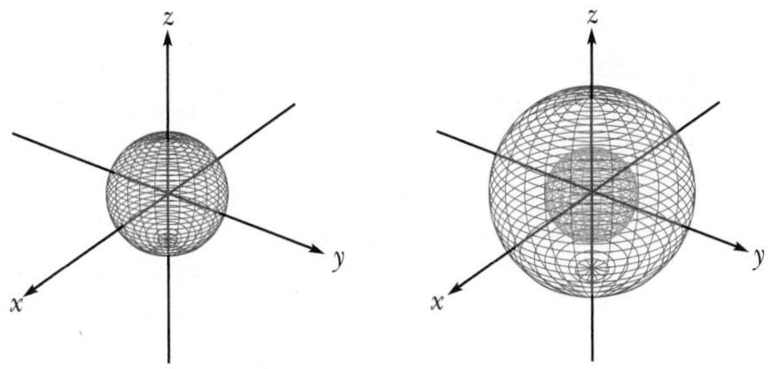

(a) 1*s* and 2*s* orbitals computed using the Schrödinger equation

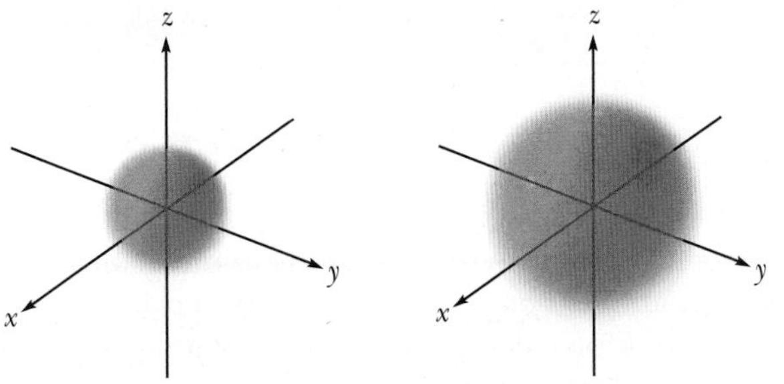

(b) Cartoon representations of 1*s* and 2*s* orbitals

FIGURE 1.8 (a) Calculated and (b) cartoon representations of the 1*s* and 2*s* atomic orbitals showing an arbitrary boundary containing about 95% of the electron density. Note that the 2*s* orbital has a phase change within the spherical electron density.

FIGURE 1.9 Three-dimensional representations of the $2p_x$, $2p_y$, and $2p_z$ atomic orbitals and their orientation in space relative to one another. Nodal planes are shaded. Note that the lobes of the computed orbitals (a) are more spherical in appearance than the bowling-pin-like cartoon drawings (b) chemists routinely use to represent them.

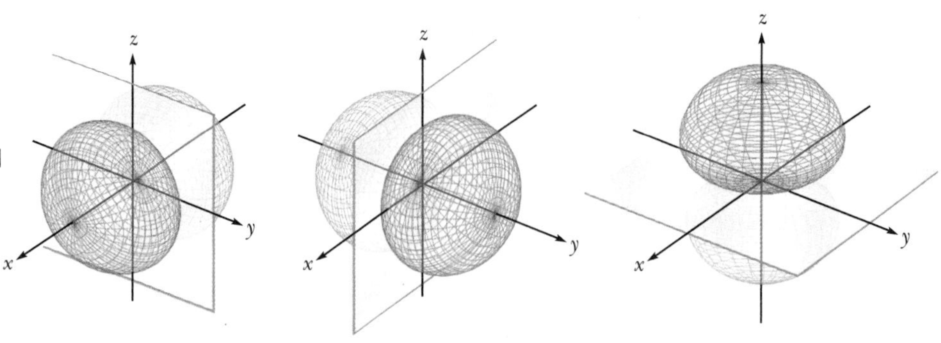

(a) $2p$ orbitals computed using the Schrödinger equation

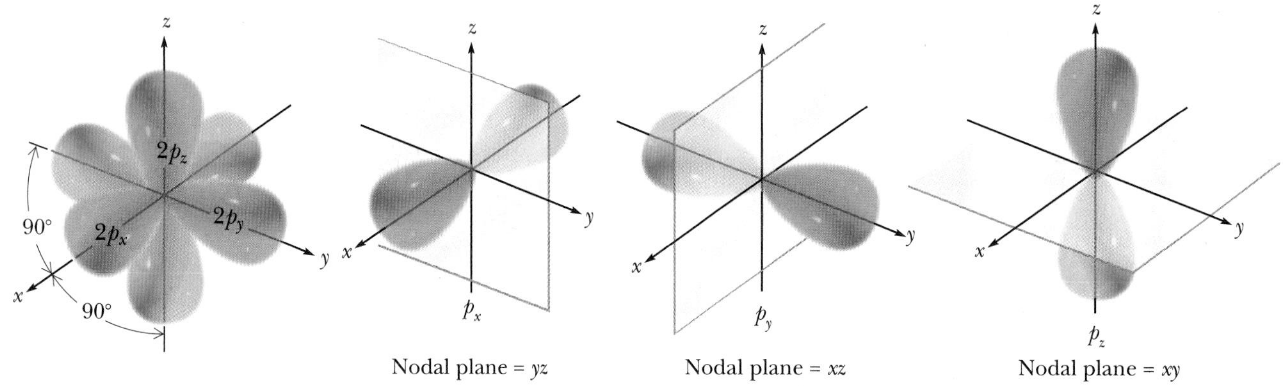

(b) Cartoon representations of $2p$ orbitals

Besides providing a way to determine the shapes of atomic orbitals, the Schrödinger equation also provides a way to approximate the energetics of covalent bond formation. These approximations have taken two forms: (1) valence bond (VB) theory and (2) molecular orbital (MO) theory. Both theories of chemical bonding use the methods of quantum mechanics, but each makes slightly different simplifying assumptions. At sufficiently high levels of theory, both models converge. The VB approach provides the most easily visualized description of single bonds, while the MO method is most convenient for describing multiple bonds and for carrying out detailed calculations on computers. In practice, the model for bonding that most organic chemists use is a combined VB/MO theory, and this is the approach used in this book.

1.7 A Combined Valence Bond and Molecular Orbital Theory Approach to Covalent Bonding

A. Molecular Orbital Theory; Formation of Molecular Orbitals

Molecular orbital (MO) theory

A theory of chemical bonding in which electrons in molecules occupy molecular orbitals that extend over the entire molecule and are formed by the combination of the atomic orbitals that make up the molecule.

Molecular orbital (MO) theory begins with the hypothesis that electrons in atoms exist in atomic orbitals and assumes that electrons in molecules exist in molecular orbitals. Just as the Schrödinger equation can be used to calculate the energies and shapes of atomic orbitals, molecular orbital theory assumes that the Schrödinger equation can also be used to calculate the energies and shapes of molecular orbitals. Following is a summary of the rules used in applying molecular orbital theory to the formation of covalent bonds.

1. Combination of n atomic orbitals (mathematically adding and subtracting wave functions) forms a set of n molecular orbitals (new wave functions); that is, the number of molecular orbitals formed is equal to the number of atomic orbitals combined. When only MO theory is used to model bonding in organic compounds the molecular orbitals are spread over all atoms in a molecule or ion whose atomic orbitals are properly aligned to overlap with one another.
2. Just like atomic orbitals, molecular orbitals are arranged in order of increasing energy. It is possible to calculate reasonably accurate relative energies of a set of molecular orbitals. Experimental measurements such as those derived from molecular spectroscopy can also be used to provide very detailed information about the relative energies of molecular orbitals.
3. Filling of molecular orbitals with electrons is governed by the same principles as the filling of atomic orbitals. Molecular orbitals are filled beginning with the lowest energy unoccupied molecular orbital (the Aufbau principle). A molecular orbital can accommodate no more than two electrons, and their spins must be paired (the Pauli exclusion principle). When two or more molecular orbitals of equal energy are available, one electron is added to each before any equivalent orbital is filled with two electrons.

To illustrate the formation of molecular orbitals, consider the shapes and relative energies of the molecular orbitals arising from combination of two 1*s* atomic orbitals. With only two atoms, the resulting molecular orbitals are simple to visualize, whereas with more atoms, the orbitals become increasingly complex (see Section 1.9). Combination by addition of their wave functions (a process referred to as in-phase addition) gives the molecular orbital shown in Figure 1.10(a). As with atomic orbitals, a molecular orbital is visualized as a plot of its wave function (ψ) in three-dimensional space. When electrons occupy this bonding molecular orbital, electron density is concentrated in the region between the two positively charged nuclei and serves to offset the repulsive interaction between them. The molecular orbital we just described is called a sigma

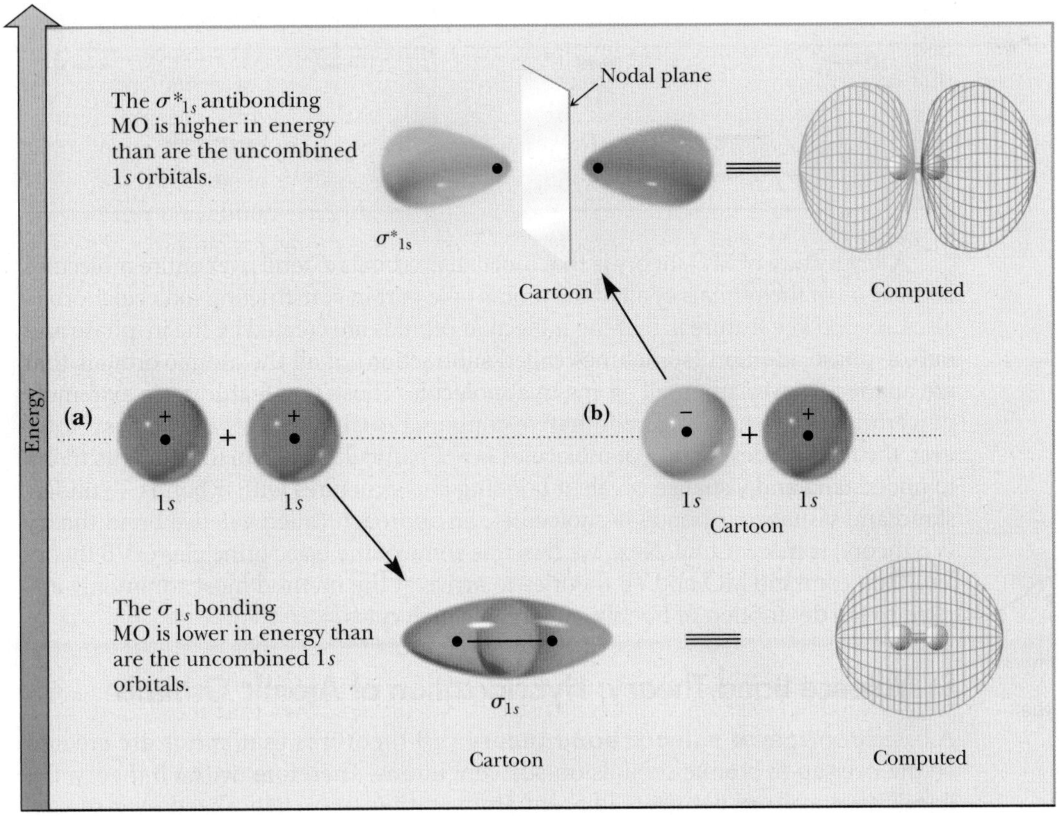

FIGURE 1.10 Molecular orbitals (plots of ψ) derived from combination of two 1*s* atomic orbitals: (a) combination by addition and (b) combination by subtraction. Electrons in the bonding MO spend most of their time in the region between the two nuclei and bond the atoms together. Electrons in the antibonding MO lead to repulsion between nuclei and decrease bonding.

bonding molecular orbital and is given the symbol σ_{1s} (pronounced sigma one ess). A **bonding molecular orbital** is an orbital in which electrons have a lower energy than they would in the isolated atomic orbitals. A **sigma (σ) bonding molecular orbital** is an orbital in which electron density lies between the two nuclei, along the axis joining them, and is *cylindrically symmetric* about the axis.

Bonding molecular orbital
A molecular orbital in which electrons have a lower energy than they would in isolated atomic orbitals.

Sigma (σ) bonding molecular orbital
A molecular orbital in which electron density is concentrated between two nuclei, along the axis joining them, and is cylindrically symmetric.

Combination of two 1s atomic orbitals by subtraction of their wave functions (a process referred to as out-of-phase addition) gives the molecular orbital shown in Figure 1.10(b). If electrons occupy this orbital, electron density is concentrated outside the region between the two nuclei. There is a nodal plane, or plane of zero electron density, between the atoms. This molecular orbital is called a sigma antibonding molecular orbital and is given the symbol σ^*_{1s} (pronounced sigma star one ess). An **antibonding molecular orbital** is an orbital in which the electrons in it have a higher energy (are more easily removed) than they would have in the isolated atomic orbitals. Population of this orbital with electrons actually causes repulsion of the nuclei involved. An asterisk (*) is used to indicate that a molecular orbital is antibonding.

Antibonding molecular orbital
A molecular orbital in which electrons have a higher energy than they would in isolated atomic orbitals.

The **ground state** of an atom or a molecule is its state of lowest energy. In the ground state of a hydrogen molecule, the two electrons occupy the σ_{1s} MO with paired spins. An **excited state** is any electronic state other than the ground state. In the lowest excited state of the hydrogen molecule, one electron occupies the σ_{1s} MO, and the other occupies the σ^*_{1s} MO. There is no net bonding in this excited state, and dissociation will result. Energy-level diagrams of the ground state and the lowest excited state of the hydrogen molecule are shown in Figure 1.11. Under normal circumstances, we do not have to consider antibonding orbitals because they are unoccupied.

Ground state
The lowest energy state of an atom or a molecule.

Excited state
Any electronic state other than the ground state. Will be higher in energy than the ground state.

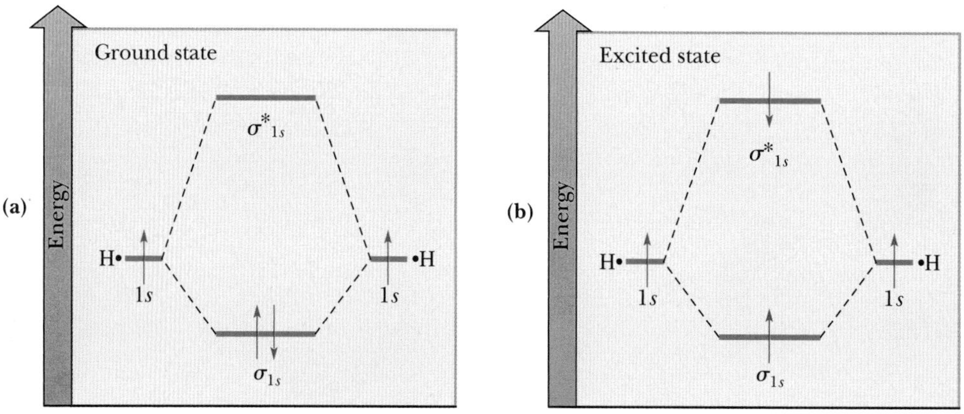

FIGURE 1.11 A molecular orbital energy diagram for the hydrogen molecule, H_2. (a) Ground state and (b) lowest excited state.

A key feature of MO theory is that molecular orbitals extend over entire molecules because all of the orbitals of all of the atoms take part in constructing molecular orbitals. A second key feature is that the molecular orbitals are created by the in-phase and out-of-phase addition (sometimes called subtraction) of all the atomic orbitals that are aligned to overlap on all atoms in a molecule. These two features are extremely powerful when generating quantitative computational models of molecules. However, the full MO description of molecules is not particularly useful for students trying to understand and visualize covalent bonding and structures with σ bonds. To understand and visualize σ bonds in molecules, an approach called valence bond theory (VB theory) is more useful. Next, we describe some of the basic principles of VB theory and then combine MO and VB theories to arrive at the method most commonly applied to the description of bonding in organic molecules.

Valence bond theory
A model of bonding that places electron pairs between adjacent atoms to create bonds.

B. Valence Bond Theory; Hybridization of Atomic Orbitals

A basic principle of **valence bond theory** (VB theory) is that bonds are created by the overlap of atomic orbitals on adjacent atoms. Therefore, with VB theory, the bonds are localized between adjacent atoms rather than delocalized over several atoms as in MO theory. This model correlates with Lewis pictures where two electrons are visualized between atoms as a bond. To represent the bonds, lines are drawn between the atoms. However, the localization of bonds between atoms presents a problem for second-period elements. In forming covalent bonds, atoms of carbon, nitrogen, and oxygen (all second-period elements) use 2*s* and 2*p* atomic orbitals. The three 2*p* atomic orbitals are at angles of 90° to each other (Figure 1.9), and if atoms of second-period elements used these orbitals to form covalent bonds, we would expect bond angles around each to be approximately 90°. However, we rarely observe bond angles of 90° in organic molecules. What we find instead are bond angles of approximately 109.5° in molecules with only single bonds, 120° in molecules with double bonds, and 180° in molecules with triple bonds, as shown in Table 1.9.

To account for the observed bond angles in a way that is intuitive for chemists, Linus Pauling proposed **that atomic orbitals for each atom should be thought of as combining to form new atomic orbitals, called hybrid orbitals, which then interact to form bonds by overlapping with orbitals from other atoms**. The hybrid orbitals have the bond angles we observe around each atom, so molecular structure and bonding based on the overlap of hybrid orbitals provides an intuitive understanding. Being able to construct organic molecules from the overlap of hybrid orbitals is an essential organic chemistry survival skill.

Hybrid orbital
An atomic orbital formed by the combination of two or more atomic orbitals.

Hybrid orbitals are formed by combinations of atomic orbitals, a process called **hybridization**. Mathematically, this is accomplished by combining the wave functions of the 2*s* (ψ_{2s}) and three 2*p* (ψ_{2p_x}, ψ_{2p_y}, ψ_{2p_z}) orbital wave functions. The number of hybrid orbitals formed is equal to the number of atomic orbitals combined. Elements of the second period form three types of hybrid orbitals, designated sp^3, sp^2, and *sp*, each of which can contain up to two electrons.

Hybridization
The combination of atomic orbitals of different types.

sp^3 Hybrid Orbitals—Bond Angles of Approximately 109.5°

The mathematical combination of the 2*s* atomic orbital and three 2*p* atomic orbitals forms four equivalent **sp^3 hybrid orbitals** described by four new wave functions. Plotting ψ for the four new wave functions gives a three-dimensional visualization of the four sp^3 hybrid orbitals. Each sp^3 hybrid orbital consists of a larger lobe pointing in one direction and a smaller lobe of opposite sign pointing in the opposite direction. The axes of the four sp^3 hybrid orbitals are directed toward the corners of a regular tetrahedron, and sp^3 hybridization results in bond angles of approximately 109.5° (Figure 1.12). Note that each sp^3 orbital has 25% *s*-character and 75% *p*-character because those are the percentages of the orbitals combined when constructing them (one 2*s* orbital, three 2*p* orbitals). We will refer back to these percentages several times in the text.

sp^3 Hybrid orbital
A hybrid atomic orbital formed by the combination of one *s* atomic orbital and three *p* atomic orbitals.

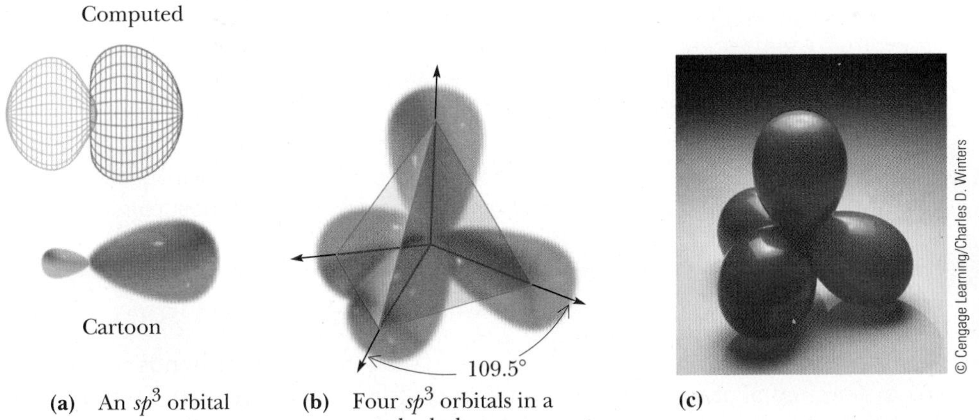
Computed
Cartoon
109.5°
© Cengage Learning/Charles D. Winters

(a) An sp^3 orbital (b) Four sp^3 orbitals in a tetrahedral arrangement (c)

FIGURE 1.12 sp^3 Hybrid orbitals. (a) A single sp^3 hybrid orbital in computed and cartoon form. (b) Three-dimensional cartoon representation of four sp^3 hybrid orbitals centered on the same atom and directed toward the corners of a regular tetrahedron. (c) If four balloons of similar size and shape are tied together, they will naturally assume a tetrahedral geometry.

Atoms with four sp^3 hybrid atomic orbitals are referred to as **sp^3 hybridized**, or as having a hybridization state of sp^3. A diagram depicting this hybridization shows the creation of four orbitals of equal energy.

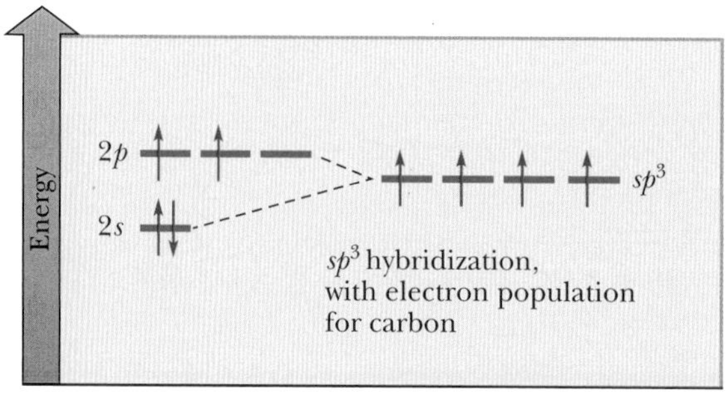

Population of these orbitals with electrons follows the standard rules for populating atomic orbitals. For example, for C with four valence electrons, each sp^3 orbital contains a single electron. Carbon can therefore make four bonds, one with each sp^3 hybrid. Nitrogen, with five valence electrons, would have a lone pair and three singly occupied sp^3 orbitals. The orbital populations of O and F follow in an analogous fashion.

You must remember that superscripts in the designation of hybrid orbitals tell you how many atomic orbitals have been combined to form the hybrid orbitals. You know that the designation sp^3 represents a hybrid orbital because it shows a combination of *s* and *p* orbitals. The superscripts in this case tell you that *one s* atomic orbital and *three p* atomic orbitals are combined in forming the hybrid orbital. Do not confuse this use of superscripts with that used in writing a ground-state electron configuration, as for example $1s^22s^22p^5$ for fluorine. In the case of a ground-state electron configuration, superscripts tell you the number of electrons in each orbital or set of orbitals.

In Section 1.2, we described the covalent bonding in CH_4, NH_3, and H_2O in terms of the Lewis model, and in Section 1.4, we used VSEPR to predict bond angles of approximately 109.5° in each molecule. Now let us consider the bonding in these molecules in terms of the overlap of hybrid atomic orbitals. To bond with four other atoms with bond angles of 109.5°, carbon uses sp^3 hybrid orbitals. Carbon has four valence electrons, and one electron is placed in each sp^3 hybrid orbital. Each partially filled sp^3 hybrid orbital then overlaps with a partially filled 1*s* atomic orbital of hydrogen to form the four sigma (σ) bonds of methane, and hydrogen atoms occupy the corners of a regular tetrahedron (Figure 1.13). We address how to create and model these σ bonds in Section 1.7C.

In bonding with three other atoms, the five valence electrons of nitrogen are distributed so that one sp^3 hybrid orbital is filled with a pair of electrons (the lone pair) and the other three sp^3 hybrid orbitals have one electron each. Overlapping of these partially filled sp^3 hybrid orbitals with 1*s* atomic orbitals of three hydrogen atoms produces an NH_3 molecule (Figure 1.13).

In bonding with two other atoms, the six valence electrons of oxygen are distributed so that two sp^3 hybrid orbitals are filled and the remaining two have

FIGURE 1.13 Orbital overlap pictures of methane, ammonia, and water.

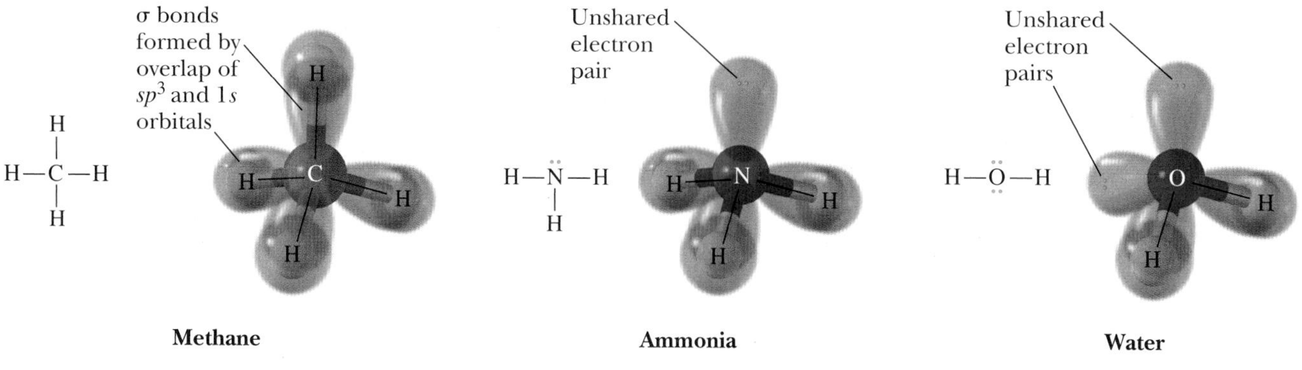

one electron each. Each partially filled sp^3 hybrid orbital overlaps with a 1*s* atomic orbital of hydrogen, and hydrogen atoms occupy two corners of a regular tetrahedron. The remaining two sp^3 hybrid orbitals, each occupied by an unshared pair of electrons, are directed toward the other two corners of the regular tetrahedron (Figure 1.13).

sp^2 Hybrid Orbitals—Bond Angles of Approximately 120°

The mathematical combination of one 2*s* atomic orbital wave function and two 2*p* atomic orbital wave functions forms three equivalent **sp^2 hybrid orbital** wave functions. Because they are derived from three atomic orbitals, sp^2 hybrid orbitals always occur in sets of three. As with sp^3 orbitals, each sp^2 hybrid orbital (three-dimensional plot of ψ) consists of two lobes, one larger than the other. The axes of the three sp^2 hybrid orbitals lie in a plane and are directed toward the corners of an equilateral triangle; the angle between sp^2 hybrid orbitals is 120°. The third 2*p* atomic orbital (remember $2p_x$, $2p_y$, $2p_z$) is not involved in hybridization (its wave function is not mathematically combined with the other three) and remains as two lobes lying perpendicular to the plane of the sp^2 hybrid orbitals. Figure 1.14 shows three equivalent sp^2 orbitals along with the remaining unhybridized 2*p* atomic orbital. Each sp^2 orbital has 33% *s*-character and 67% *p*-character (one 2*s* orbital, two 2*p* orbitals).

sp^2 Hybrid orbital

A hybrid atomic orbital formed by the combination of one *s* atomic orbital and two *p* atomic orbitals.

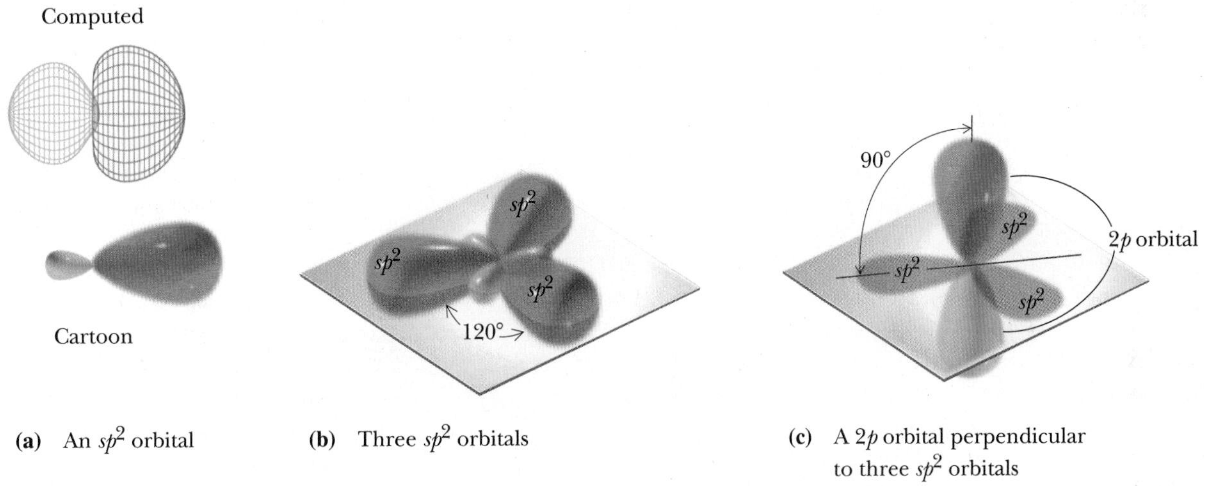

Computed
Cartoon
sp^2
sp^2
sp^2
120°
90°
sp^2
sp^2
sp^2
2*p* orbital
(a) An sp^2 orbital
(b) Three sp^2 orbitals
(c) A 2*p* orbital perpendicular to three sp^2 orbitals

FIGURE 1.14 sp^2 Hybrid orbitals and a single *p* orbital on an sp^2 hybridized atom.
(a) A single sp^2 hybrid orbital in computed and cartoon form.
(b) Three sp^2 hybrized orbitals in a trigonal planar arrangement.
(c) The lone *p* orbital.

An atom possessing three sp^2 hybrid orbitals and a single *p* atomic orbital is referred to as **sp^2 hybridized**, or as having a hybridization state of sp^2. The energy-level diagram shows the creation of the three hybrid orbitals and a remaining 2*p* orbital. With C, all four orbitals contain a single electron. Therefore, a carbon atom makes three bonds with the sp^2 hybrids and one bond with a *p* orbital. With N, a lone pair is residing in an sp^2 hybrid orbital, and with O, two lone pairs are residing in sp^2 orbitals. Verify the orbital population of O for yourself by placing six valence electrons into the hybridized orbital diagram shown.

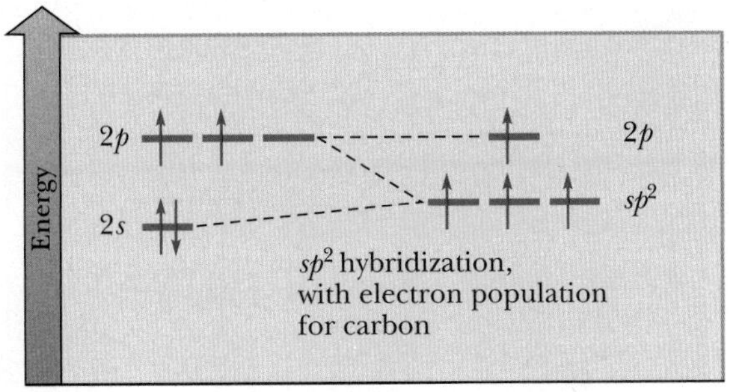

Energy
2*p*
2*p*
2*s*
sp^2
sp^2 hybridization, with electron population for carbon

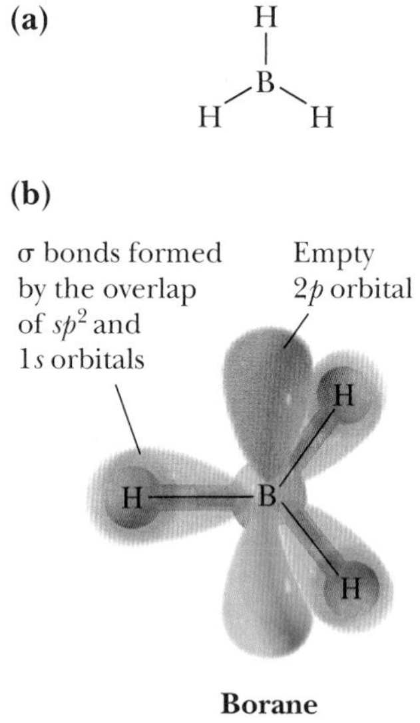
(a)
H
B
H
H
(b)
σ bonds formed by the overlap of sp^2 and 1s orbitals
Empty 2p orbital
H
H
B
H
Borane

FIGURE 1.15 Covalent bond formation in borane, BH_3. (a) Lewis structure for borane. (b) Orbital overlap picture of borane.

In Section 1.2, we covered a few apparent exceptions to the octet rule, where boron compounds such as BF_3 are common examples. Analogously, VSEPR tells us that BH_3 is trigonal planar, with 120° H—B—H bond angles. Therefore, sp^2 hybridization is the appropriate descriptor for B in such structures. Boron has three valence electrons, and one electron is placed in each sp^2 hybrid orbital. Each partially filled sp^2 hybrid overlaps with a 1*s* hydrogen orbital, containing one electron, to form three B—H σ bonds (Figure 1.15). The unhybridized 2*p* atomic orbital is empty. Note that BH_3 is a highly reactive molecule and generally dimerizes in solution (see Section 6.4).

sp Hybrid Orbitals—Bond Angles of Approximately 180°

The mathematical combination of one 2*s* atomic orbital and one 2*p* atomic orbital produces two equivalent **sp hybrid orbital** wave functions. Because they are derived from two atomic orbitals, *sp* hybrid orbitals always occur in sets of two. The three-dimensional plot of ψ shows that the two *sp* hybrid orbitals lie at an angle of 180°. The axes of the unhybridized 2*p* atomic orbitals are perpendicular to each other and to the axis of the two *sp* hybrid orbitals. In Figure 1.16, *sp* hybrid orbitals are shown on the *x*-axis and unhybridized 2*p* orbitals are on the *y*- and *z*-axes. Each *sp* orbital has 50% *s*-character and 50% *p*-character because those are the percentages of the orbitals combined when constructing them (one 2*s* orbital, one 2*p* orbital).

You may have noticed by now that the cartoon representations of sp^3, sp^2, and *sp* orbitals all look the same. Although the computed *sp* orbital is the most spherical and the sp^3 orbital is the most *p*-like with sp^2 between, chemists often do not attempt to render these differences in drawings.

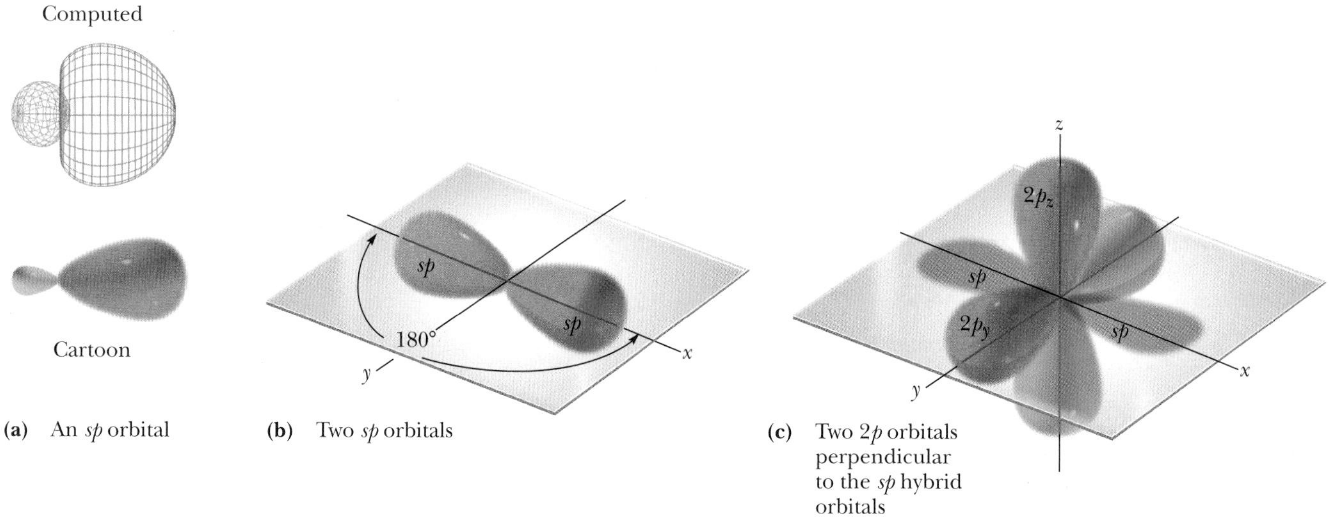
Computed
Cartoon
(a) An sp orbital
sp
sp
180°
x
y
(b) Two sp orbitals
z
$2p_z$
sp
$2p_y$
sp
x
y
(c) Two 2p orbitals perpendicular to the sp hybrid orbitals

FIGURE 1.16 *sp* Hybrid orbitals and two 2*p* orbitals on an *sp* hybridized atom. (a) A single *sp* hybrid orbital in computed and cartoon form. (b) Two *sp* hybrid orbitals in a linear arrangement. (c) The two 2*p* orbitals in perpendicular orientations to the *sp* hybrid orbitals.

An atom possessing two *sp* hybrid orbitals and two 2*p* orbitals is called ***sp* hybridized**, and the energy-level diagram for a carbon atom is shown below.

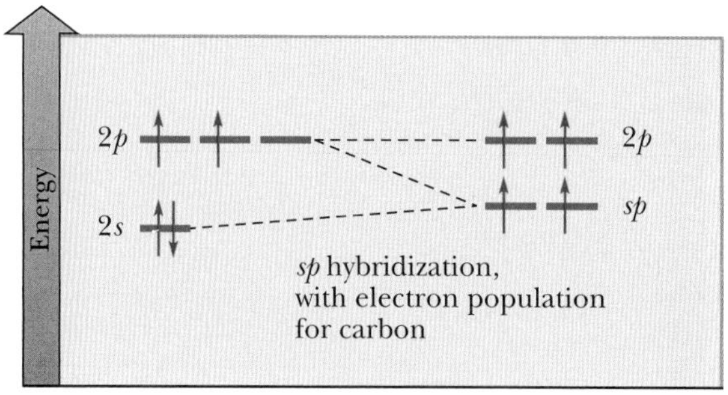
Energy
2p
2s
2p
sp
sp hybridization, with electron population for carbon

sp Hybrid orbital

A hybrid atomic orbital formed by the combination of one *s* atomic orbital and one *p* atomic orbital.

C. An Analysis of S and P

Compounds containing S and P are often drawn with more bonds than are allowed by the octet rule; therefore, the hybridization states of these atoms are difficult to assign. For example, compare trimethylphosphine to phosphoric acid. The P in trimethylphosphine has three bonds and a lone pair of electrons and is sp^3 hybridized. However, a common depiction of phosphoric acid has five bonds to P and is a well-accepted representation. The creation of five bonds has been explained by invoking the use of 3*d* orbitals on phosphorus to accommodate the additional bond. However, the use of 3*d* orbitals for bonding in such structures is in debate, and recent calculations indicate that 3d orbitals are too high in energy to take part in hybridization and therefore bonding for most compounds of phosphorus. The currently accepted depiction assumes sp^3 hybridization and gives the P an octet and a positive formal charge (the symbol ≡ means an equivalent structure drawn two different ways). The oxygen involved in the double bond of the common depiction has one bond and a negative formal charge in the alternative depiction.

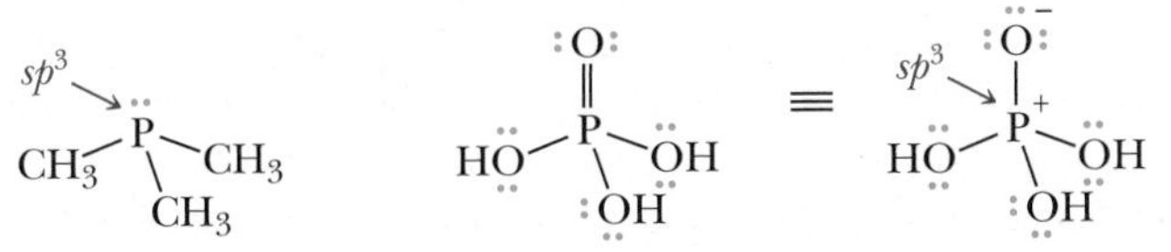

Trimethylphosphine (obeys octet rule) **Phosphoric acid**

Sulfur, another third-period element, likewise is commonly depicted with varying numbers of bonds. Because S is below O in the Periodic Table, we can predict it will be neutral with two bonds and an octet and is sp^3 hybridized (see hydrogen sulfide). Dimethylsulfoxide is commonly drawn with 10 valence electrons and a double bond to oxygen, but is better considered as the alternative charge-separated structure with a positive formal charge on an sp^3 hybridized S and a negative formal charge on O. Finally, sulfuric acid has a similar dichotomy, either being drawn as having 12 valence electrons on S or the increasingly accepted picture with an sp^3 S having an octet with a plus two formal charge. Although inconsistent with the recent calculations that indicate sp^3 hybridization in molecules containing S and P atoms, these will be drawn in the rest of this book with five or six bonds, in accord with historical and traditional depictions.

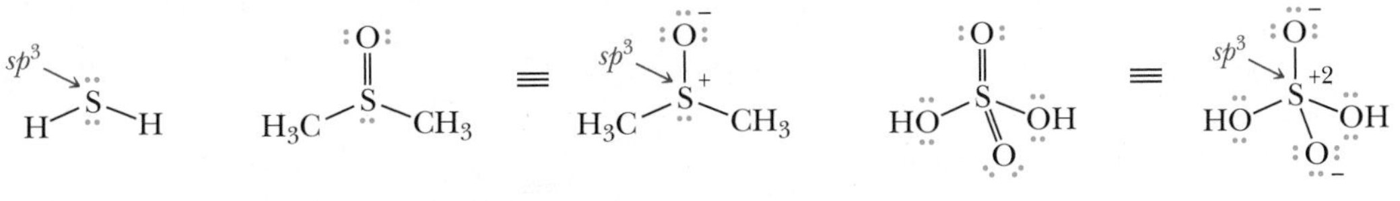

Hydrogen sulfide **Dimethyl sulfoxide** **Sulfuric acid**

D. Combining Valence Bond (VB) and Molecular Orbital (MO) Theories: The Creation of σ and π Bonding and Antibonding Orbitals

At the beginning of this section (1.7A), we noted that the molecular orbitals of H_2 are created by adding and subtracting wave functions for the atomic 1*s* orbitals on the individual hydrogen atoms. We also noted that the exercise of adding and subtracting atomic orbitals such as 1*s*, 2*s*, and 2*p* to create molecular orbitals is one principle of MO theory (MOT) where the resulting molecular orbitals are spread across atoms in the entire molecule. Yet, orbitals spread across many atoms in a molecule are often difficult to visualize. Therefore, it is convenient to consider orbitals to be localized between adjacent atoms as in valence bond theory or slightly delocalized over three or more atoms (as is done in Section 1.9). The most common model for bonding in organic compounds uses the MOT notions of addition and subtraction to create the molecular orbitals, and also includes some principles from valence bond theory.

Valence bond theory views bonding as arising from electron pairs localized between adjacent atoms. These pairs of electrons create bonds. Further, organic chemists commonly use the atomic orbitals involved in the three hybridization states of atoms (sp^3, sp^2, and sp) to create the orbitals that hold these electrons because doing so allows the resulting orbitals to match the experimentally determined geometries around the atoms. Therefore, hybridization is also a VB theory concept. But how do we make the orbitals that contain the electrons and that reside between adjacent atoms? This is where we return to MO theory.

To create orbitals that are localized between adjacent atoms we add and subtract (also called in- and out-of-phase addition) the atomic orbitals on the adjacent atoms, which are aligned to overlap each other. For example, let's consider methane, CH_4 (Figure 1.17). The sp^3 hybrid orbitals point at the 1s hydrogen orbitals; therefore, we add and subtract these atomic orbitals to create the molecular orbitals. As with H_2, one resulting molecular orbital is lower in energy than the two separate atomic orbitals and is called the bonding σ orbital. The other resulting molecular orbital is higher in energy than the two atomic orbitals and is antibonding. Only the lower-energy orbital is populated with electrons in methane. Population of the σ bonding orbital results in what we call a σ bond between the C and the H. Each of the four C—H bonds in methane is created in the manner discussed here. Also, although we created this picture for the C—H bonds in methane, we will view all C—H bonds in other organic structures in the same way. In other words, even with sp^2 and sp hybridized carbons, we think of their C—H bonds as looking similar to those in Figure 1.17.

Connections to Biological Chemistry

Phosphoesters

Many biochemical structures, including DNA, are in part made up of derivatives of phosphoric acid. These derivatives are referred to as phosphoesters. The P atoms in these structures are nearly always depicted with five bonds. However, as just described, the modern view of such structures involves charge separation and sp^3 hybridization at P. Because the five-bond representation is historically the most widely spread depiction, this depiction is how we render such structures throughout this book. Yet, you should keep in mind that the alternative is now considered more correct.

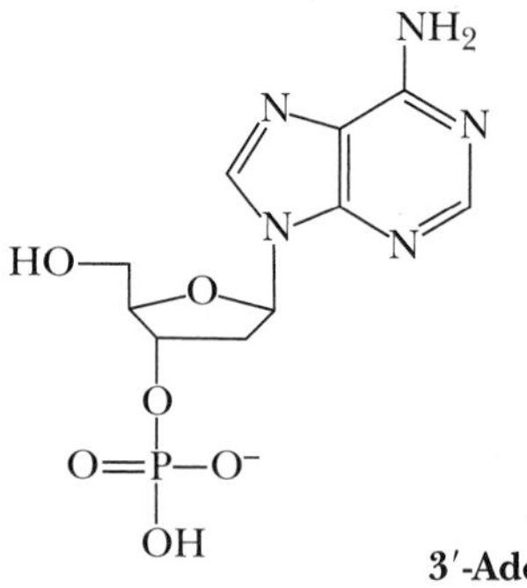

Common depiction

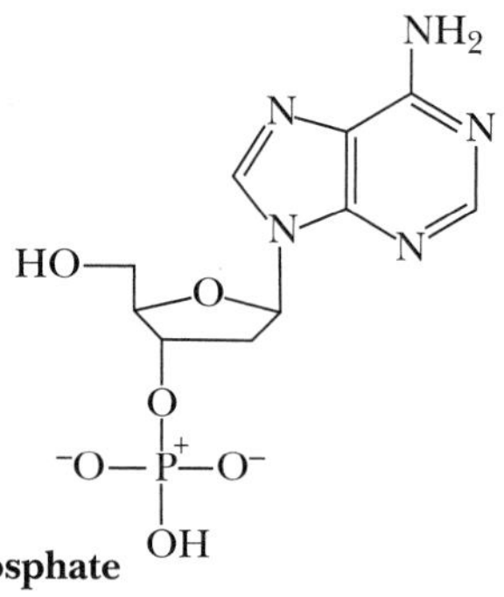

More correct

3′-Adenine monophosphate

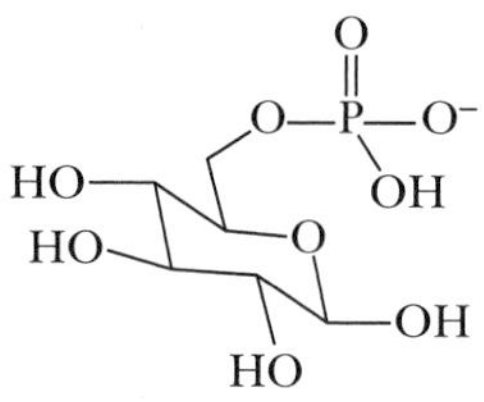

Common depiction

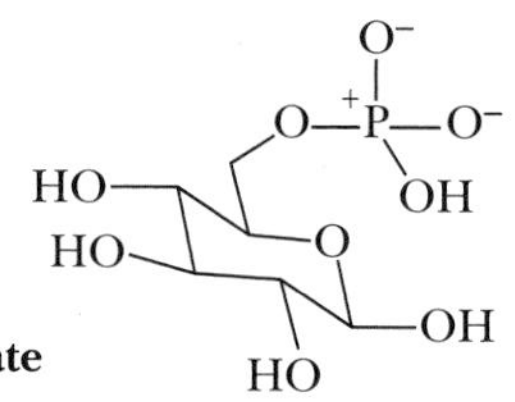

More correct

Glucose 6-phosphate

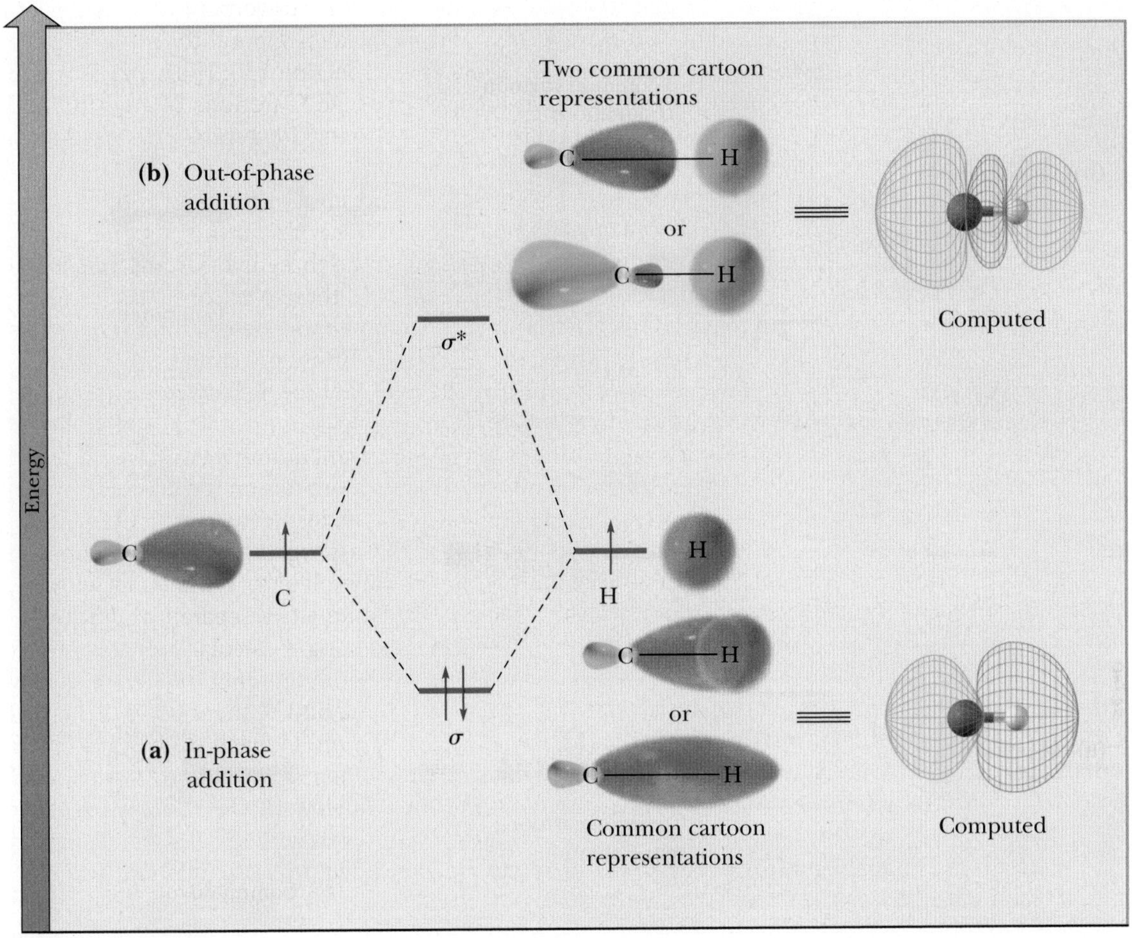

FIGURE 1.17 Molecular orbital mixing diagram for the creation of any C—H σ bond. (a) In-phase addition of a C hybrid orbital (either sp^3, sp^2, or sp) with a H 1s orbital forms a σ orbital that is lower in energy than the two starting orbitals. When the resulting orbital is populated with two electrons, a σ bond results. (b) Addition of the orbitals in an out-of-phase manner (meaning reversing the phasing of one of the starting orbitals) leads to an antibonding σ^* orbital.

It is important to note in Figures 1.17 and 1.10, as well as in the orbital mixing diagrams (Figures 1.18, 1.21, and 1.25), that the energy of the antibonding orbital goes up further than the drop experienced by the bonding orbital. This is the reason population of the antibonding orbital with electrons leads to cleavage of the bond. Early, less accurate theoretical methods for modeling bonding found the increase and decrease in energy of these respective orbitals to be identical, but the bonding approach described here correctly predicts the relative energies.

An identical approach used to create C—H σ bonds is used to create C—C σ bonds. For example, whenever a C—C bond exists in an organic structure, we consider the overlap of hybrid orbitals on the two carbons. As shown in Figure 1.18, the overlap of two sp^3 hybrid orbitals on the individual carbons creates σ bonding and antibonding molecular orbitals. Only the bonding orbital is populated with electrons, thereby creating a carbon-carbon σ bond. We consider all C—C σ bonds to consist of orbitals similar to those in Figure 1.18.

For example, the simplest two-carbon compound ethane (CH_3CH_3) contains one C—C σ bond and six C—H σ bonds. As shown in Figure 1.19, we consider these bonds to arise from overlap of H 1s and C sp^3 orbitals, while the actual bonding orbitals appear as shown in Figures 1.17 and 1.18.

Let's now examine compounds with a double bond. Wherever there is a double bond, sp^2 hybridization should be considered for the atoms involved. For example, second-period elements use a combination of an sp^2 hybrid orbital and the

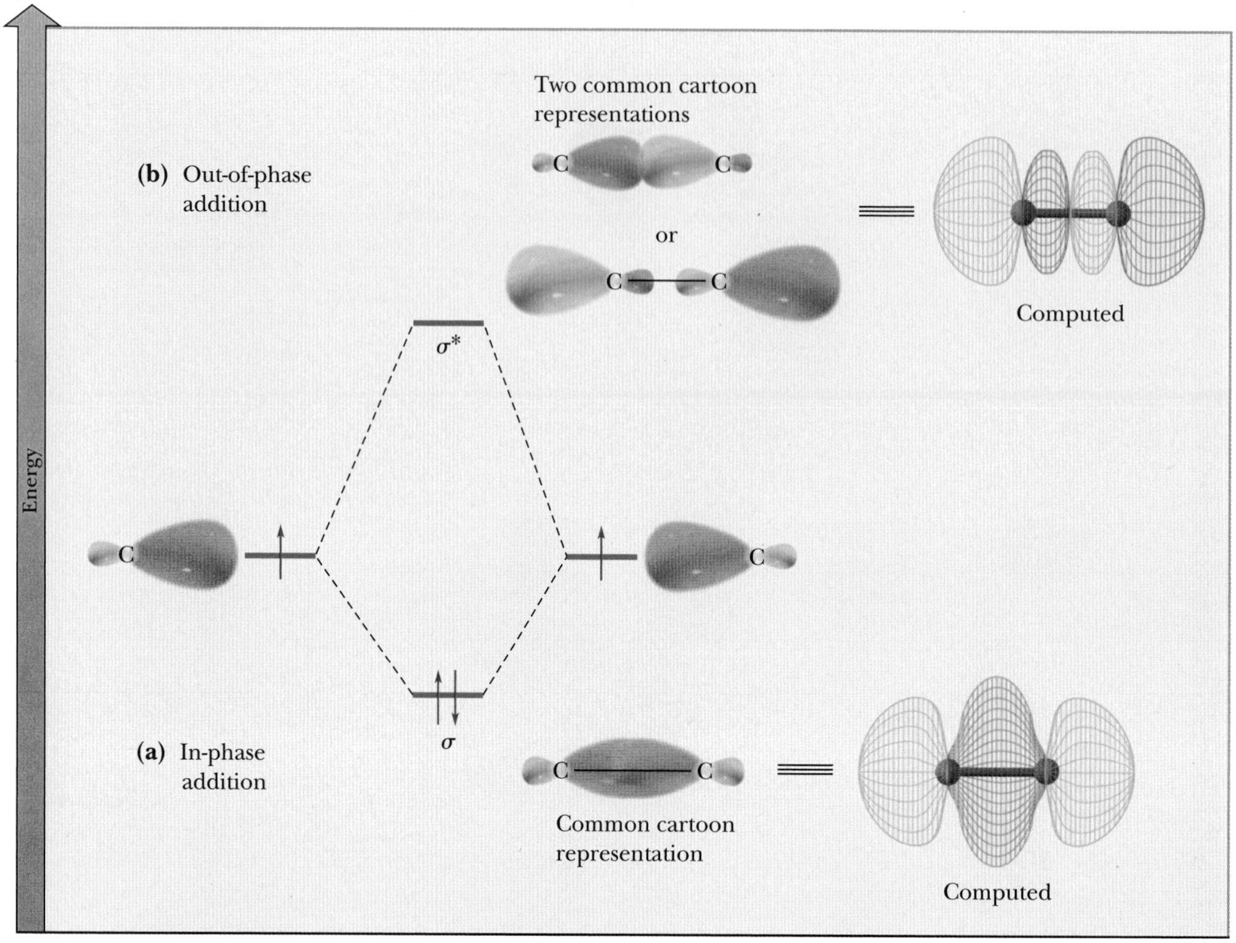

FIGURE 1.18 Molecular orbital mixing diagram for the creation of any C—C σ bond. (a) In-phase addition of two C hybrid orbitals (either sp^3, sp^2, or sp orbital) forms a σ orbital that is lower in energy than the two starting orbitals. When the resulting orbital is populated with two electrons, a bond results. (b) Addition of the orbitals in an out-of-phase manner (meaning reversing the phasing of one of the starting orbitals) leads to an antibonding σ^* orbital.

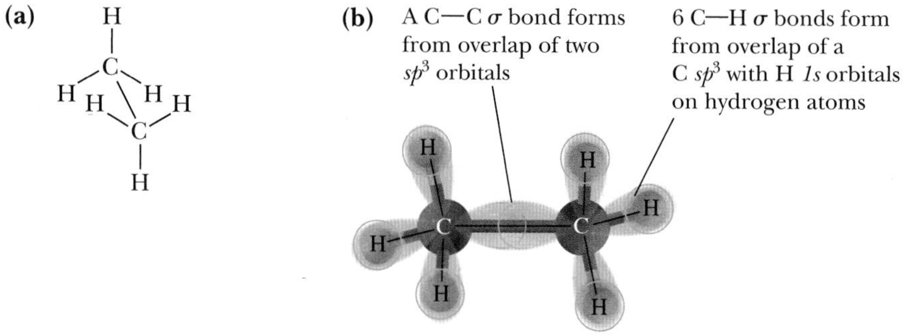

FIGURE 1.19 (a) Lewis structure for ethane (CH_3CH_3). (b) Overlap of sp^3 hybrid orbitals on adjacent carbons forms a C—C σ bond (see Figure 1.18), and overlap of carbon sp^3 hybrid orbitals with hydrogen $1s$ orbitals gives C—H σ bonds (see Figure 1.17).

Pi (π) bonding molecular orbital

A molecular orbital with a nodal plane that cuts through both atomic nuclei, with electron density concentrated above and below the nodal plane.

unhybridized $2p$ atomic orbital to form double bonds. Consider ethylene, C_2H_4, whose Lewis structure is shown in Figure 1.20(a). A σ bond between the carbons in ethylene is formed by overlapping sp^2 hybrid orbitals along a common axis as shown in Figure 1.18(b). Each carbon also forms σ bonds with two hydrogens (as in Figure 1.17).

The combination of parallel $2p$ atomic orbitals by in-phase and out-of-phase addition of their wave functions to give a **pi (π) bonding molecular orbital** and a pi antibonding molecular orbital (π^*) is shown in Figure 1.21. A π bonding

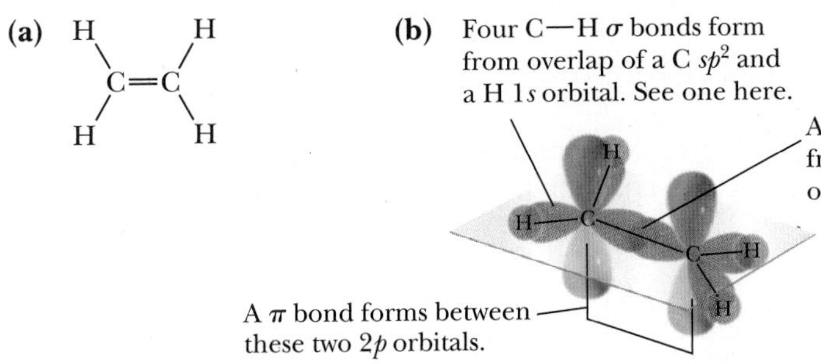
(a)
H
H
C=C
H
H
(b) Four C—H σ bonds form from overlap of a C sp^2 and a H 1s orbital. See one here.
A C—C σ bond forms from overlap of two sp^2 orbitals.
A π bond forms between these two 2p orbitals.

FIGURE 1.20 Covalent bond formation in ethylene (CH_2CH_2). (a) Lewis structure. (b) Overlap of sp^2 hybrid orbitals on adjacent carbons forms a C—C σ bond (see Figure 1.18), and overlap of carbon sp^2 hybrid orbitals on carbons with 1s orbitals on hydrogens gives C—H σ bonds (see Figure 1.17). Further, overlap of parallel 2p orbitals on the adjacent carbons gives a π bond (see Figure 1.21).

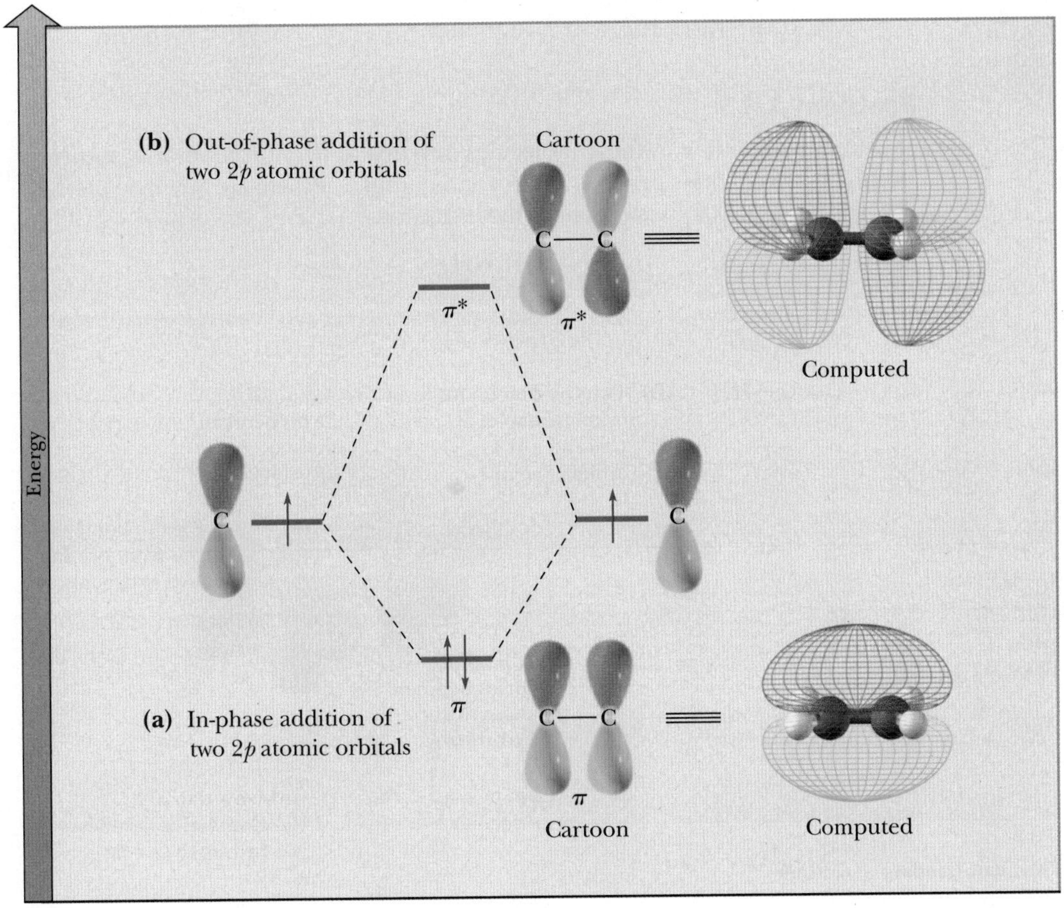
Energy
(b) Out-of-phase addition of two 2p atomic orbitals
Cartoon
π*
π*
Computed
C
C
π
(a) In-phase addition of two 2p atomic orbitals
π
Cartoon
Computed

Watch a video explanation

FIGURE 1.21 Molecular orbital mixing diagram for the creation of any C—C π bond. (a) Addition of two p atomic orbitals in phase leads to a π orbital that is lower in energy than the two separate starting orbitals. When populated with two electrons, the π orbital gives a π bond. (b) Addition of the p orbitals in an out-of-phase manner (meaning a reversal of phasing in one of the starting orbitals) leads to a π* orbital. Population of this orbital with one or two electrons leads to weakening or cleavage of the π bond, respectively.

molecular orbital has a nodal plane that cuts through both atomic nuclei, with electron density above and below the nodal plane concentrated between the nuclei. We picture all isolated π bonds between carbons to have orbitals such as those of Figure 1.21.

Finally, let's examine triple bonds. Wherever there is a triple bond, *sp* hybridization is appropriate for the atoms involved. Figure 1.22 shows an orbital overlap diagram for acetylene, C_2H_2. A carbon-carbon triple bond consists of one σ bond formed by overlapping *sp* hybrid orbitals and two π bonds. Overlapping a pair of parallel 2*p* atomic orbitals gives one π bond. Overlapping the other pair of parallel 2*p* atomic orbitals (perpendicular to the first pair) gives the second π bond.

The relationship among the number of atoms bonded to carbon, orbital hybridization, and types of bonds involved is summarized in Table 1.10.

Example 1.15 Hybridization and Bonding

Describe the bonding in 1-methoxypropanone ($CH_3OCH_2COCH_3$) in terms of (a) hybridization of C and O, (b) type of bonds between C and O, and (c) type of orbitals that hold the lone electron pairs on O.

Solution

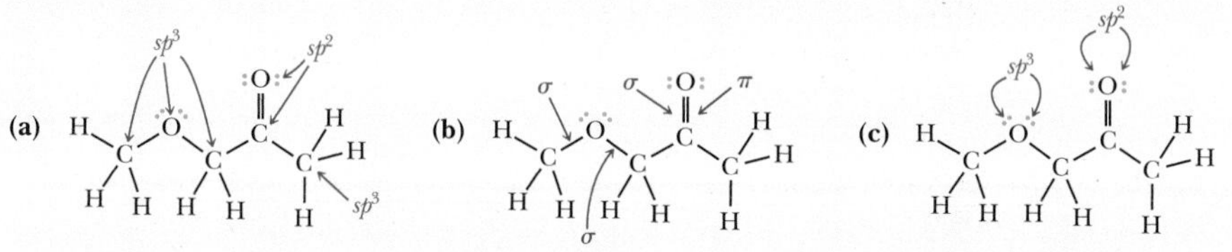
(a)
sp3
sp2
sp3
(b)
σ
σ
π
σ
(c)
sp2
sp3

Problem 1.15

Describe the bonding in these molecules in terms of hybridization of C and N and the types of bonds between carbon and nitrogen. If there are any lone pairs, describe what type of orbital contains these electrons.

(a) $CH_3CH{=}CH_2$ **(b)** CH_3NH_2

FIGURE 1.22 Covalent bond formation in acetylene. (a) Lewis structure. (b) Overlap of *sp* hybrid orbitals on adjacent carbons forms a C—C σ bond (see Figure 1.18), and overlap of carbon *sp* hybrid orbitals with hydrogen 1*s* orbitals gives C—H σ bonds (see Figure 1.17). Further, overlap of parallel 2*p* orbitals on the adjacent carbons gives a C—C π bond (see Figure 1.21). Two such π bonds exist in acetylene.

(a) H—C≡C—H

(b)
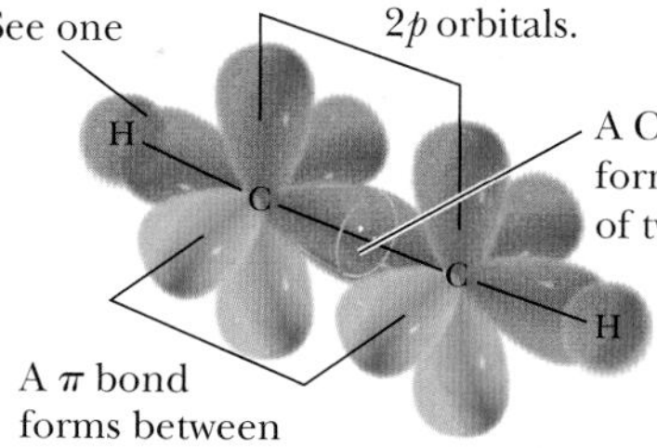
Two C—H σ bonds form from overlap of a C sp and a H 1s orbital. See one here.
A π bond forms between these two 2p orbitals.
A C—C σ bond forms from overlap of two sp orbitals.
A π bond forms between these two 2p orbitals.
H
C
C
H

Table 1.10 Covalent Bonding of Carbon

Groups Bonded to Carbon	Orbital Hybridization	Predicted Bond Angles	Types of Bonds to Carbon	Example	Name
4	sp^3	109.5°	Four σ bonds	$H_3C{-}CH_3$	**Ethane**
3	sp^2	120°	Three σ bonds and one π bond	$H_2C{=}CH_2$	**Ethylene**
2	sp	180°	Two σ bonds and two π bonds	H—C≡C—H	**Acetylene**

HOW TO

Quickly Recognize the Hybridization and Geometry of Atoms

All of the insights into geometric shapes and orbitals concerning the valence-shell electron-pair repulsion rule, hybridization, and a Valence Bond/Molecular Orbital Theory (VB/MOT) picture of bonding can be tied together with some very simple rules. When C, N, and O have an octet of electrons and are involved in making only single bonds to other atoms (irrespective of the number of single bonds), these atoms are sp^3 hybridized. When C, N, and O have an octet of electrons and are involved in double bonds, these atoms are sp^2 hybridized (unless the atom makes two double bonds; in which case they are *sp* hybridized). Finally, when C, N, and O have an octet of electrons and are involved in triple bonds, they are *sp* hybridized.

After you recognize hybridization, assigning geometry is straightforward. Atoms bonded to only one other atom (called a terminal atom) are not assigned a geometry, irrespective of their hybridization. Atoms with sp^3 hybridization adopt tetrahedral, pyramidal, or bent geometries depending on whether they are bonding to four, three, or two other atoms, respectively. Atoms with sp^2 hybridization adopt trigonal planar geometries or bent geometries depending on whether they are bonded to three or two other atoms, respectively. Atoms with *sp* hybridization adopt only linear geometries. The series of examples compiled below illustrate the points presented here. In these examples, the CH_3 groups are sp^3 and tetrahedral.

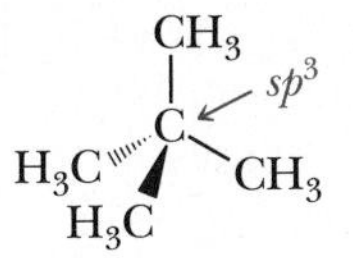

tetrahedral central carbon

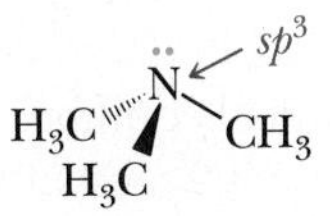

pyramidal nitrogen

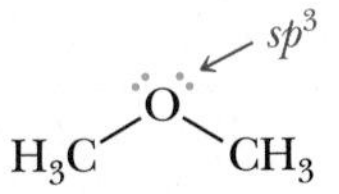

bent oxygen

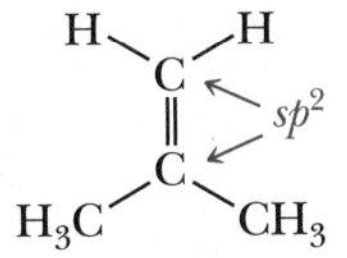

both sp^2 carbons are trigonal planar

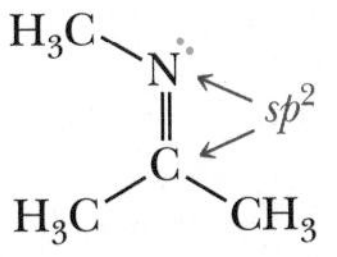

the carbon is trigonal planar, while the nitrogen is bent

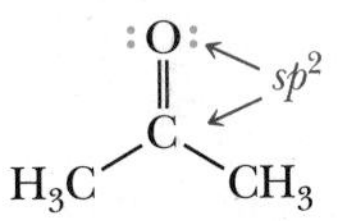

the carbon is trigonal planar, and no geometry is assigned to the oxygen

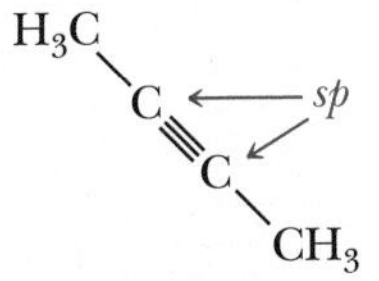

both *sp* carbons are linear

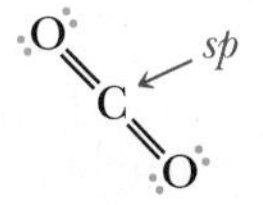

linear central carbon

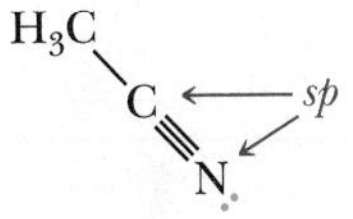

the carbon is linear, and no geometry is assigned to the nitrogen

1.8 Resonance

As chemists developed a deeper understanding of covalent bonding in organic compounds, it became obvious that, for a great many molecules and ions, no single Lewis structure provides a truly accurate representation. For example, Figure 1.23 shows three Lewis structures for the carbonate ion, CO_3^{2-}, each of which shows carbon bonded to three oxygen atoms by a combination of one double bond and two single bonds. Each Lewis structure implies that one carbon-oxygen bond is different from the other two. However, this is not the case. Experiments showed that all three carbon-oxygen bonds are identical.

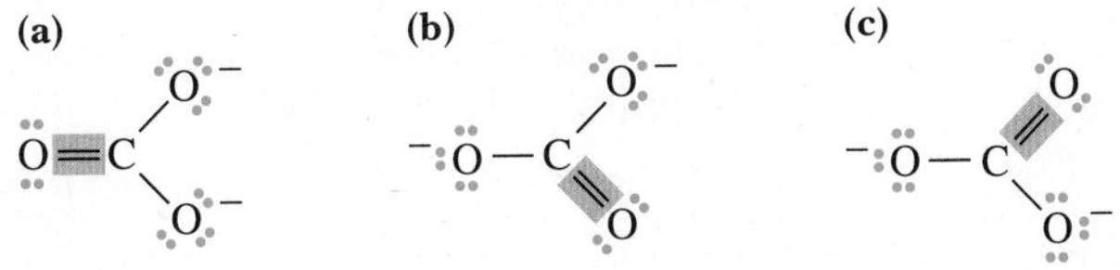

FIGURE 1.23 (a–c) Three Lewis structures for the carbonate ion.

Resonance

A theory that many molecules are best described as a hybrid of several Lewis structures.

Contributing structures

Representations of a molecule that differs only in the distribution of valence electrons.

Resonance hybrid

A molecule, an ion, or a radical described as a composite of a number of contributing structures.

Double-headed arrow

A symbol used to show that structures on either side are resonance contributing structures.

The problem for chemists, then, was how to describe the structure of molecules and ions for which no single Lewis structure was adequate and yet still retain Lewis structures. As an answer to this problem, Linus Pauling proposed the theory of resonance.

A. Theory of Resonance

The theory of **resonance** was developed primarily by Pauling in the 1930s. According to this theory, many molecules and ions are best described by writing two or more Lewis structures and considering the real molecule or ion to be a composite of these structures. Individual Lewis structures are called **contributing structures**. They are also sometimes referred to as **resonance structures** or **resonance contributors**. We show that the real molecule or ion is a **resonance hybrid** of the various contributing structures by interconnecting them with **double-headed arrows**. Do not confuse the double-headed arrow with the double arrow used to show chemical equilibrium. As we explain shortly, resonance structures are not in equilibrium with each other.

Three contributing structures for the carbonate ion are shown in Figure 1.24. These three contributing structures are said to be equivalent; they have identical patterns of covalent bonding.

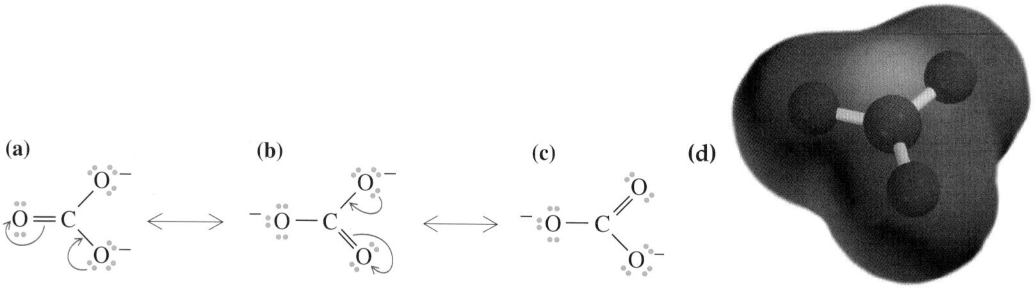

FIGURE 1.24 (a–c) The carbonate ion represented as a resonance hybrid of three equivalent contributing structures. Curved arrows show the redistribution of valence electrons between one contributing structure and the next. (d) An electrostatic potential map of a carbonate ion shows that the negative charge is indeed distributed equally among the three oxygens.

The use of the term *resonance* for this theory of covalent bonding might suggest that bonds and electron pairs are constantly changing back and forth from one position to another over time. This notion is not correct. The carbonate ion, for example, has one and only one real structure. The problem is ours—how do we draw that one real structure? The resonance method is a way to describe the real structure and at the

HOW TO

Draw Curved Arrows and Push Electrons in Creating Contributing Structures

Notice in Figure 1.24 that the only difference among contributing structures (a), (b), and (c) is the position of valence electrons. To generate one resonance structure from another, chemists use a symbol called a **curved arrow**. *The arrow indicates where a pair of electrons originates (the tail of the arrow) and where it is positioned in the next structure (the head of the arrow).*

A curved arrow is nothing more than a bookkeeping symbol used to keep track of electron pairs, or, as some call it, **electron pushing**. Do not be misled by its simplicity. Electron pushing will help you see the relationship among contributing structures. Later in the course, it will help you follow bond-breaking and bond-forming steps in organic reactions. Stated directly, electron pushing is a survival skill in organic chemistry.

Following are contributing structures for the nitrite and acetate ions. Curved arrows show how the contributing structures are interconverted. For each ion, the contributing structures are equivalent.

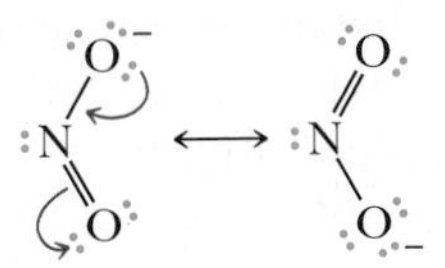

Nitrite ion
(equivalent contributing structures)

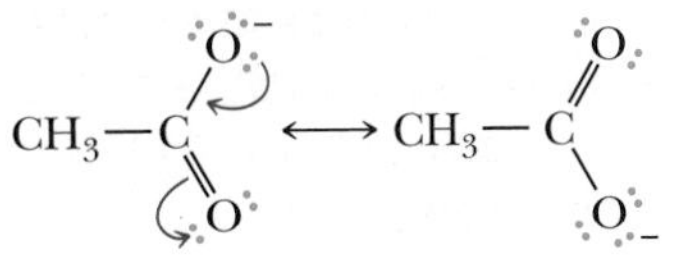

Acetate ion
(equivalent contributing structures)

A common mistake is to use curved arrows to indicate the movement of atoms or positive charges. This is not correct. Curved arrows must be used only to show the repositioning of electron pairs.

When you draw curved arrows to indicate the creation of a new contributing structure, the arrows always start on either a double (or triple) bond or a lone pair of electrons, as shown in the examples above. Further, the arrows should end at an atom that can accept a bond or should create a lone pair of electrons on an atom. Often when a new bond to an atom is created, one of the existing bonds to that atom must break so as not to exceed the octet rule. In the two examples given above, the central N of nitrite and C of acetate both acquire one bond and break one bond when you redistribute valence electrons between the contributing structures.

same time retain Lewis structures with electron-pair bonds. Thus, although we realize that the carbonate ion is not accurately represented by any one contributing structure shown in Figure 1.24, we continue to represent it by one of these for convenience. We understand, of course, that what is intended is the resonance hybrid.

Curved arrow
A symbol used to show the redistribution of valence electrons in resonance contributing structures or reaction mechanisms, symbolizing movement of two electrons.

Example 1.16 Contributing Structures I

Draw the contributing structure indicated by the curved arrow(s). Show all valence electrons and all formal charges.

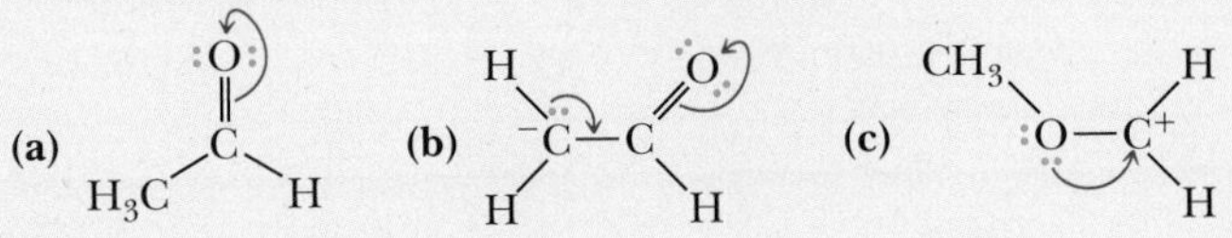

Solution

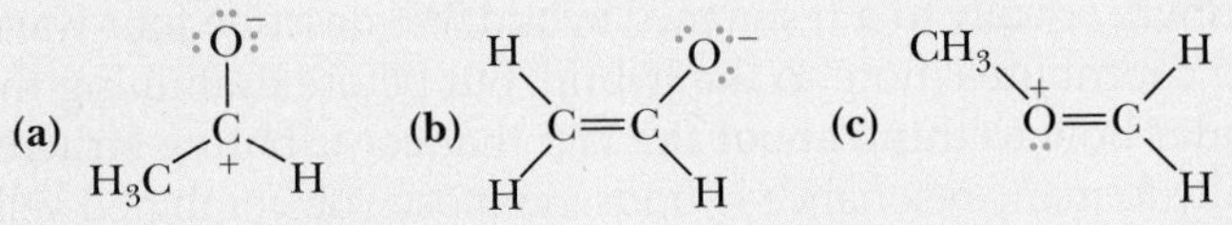

Problem 1.16

Draw the contributing structure indicated by the curved arrows. Show all valence electrons and all formal charges.

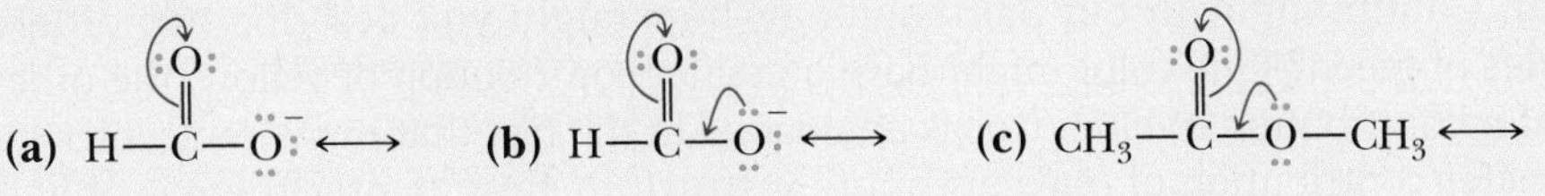

Watch a video explanation

B. Rules for Writing Acceptable Contributing Structures

Certain rules must be followed when writing acceptable contributing structures:

1. All contributing structures must have the same number of valence electrons.
2. All contributing structures must obey the rules of covalent bonding; no contributing structure may have more than two electrons in the valence shell of hydrogen or

more than eight electrons in the valence shell of a second-period element. Third-period elements, such as phosphorus and sulfur, may be drawn indicating up to 12 electrons in their valence shells (however, see Section 1.7C).

3. The positions of all nuclei must be the same in all contributing structures; that is, contributing structures differ only in the distribution of valence electrons.
4. All contributing structures must have the same number of paired and unpaired electrons.

Example 1.17 Contributing Structures II

Which sets are valid pairs of contributing structures?

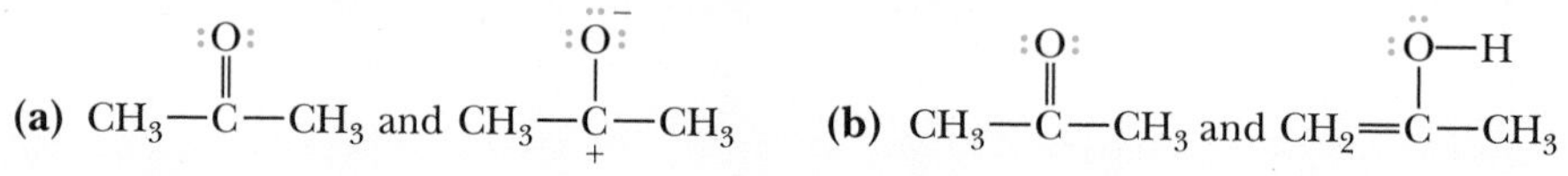

Solution

(a) These are valid contributing structures. They differ only in the distribution of valence electrons.

(b) These are not valid contributing structures. They differ in the connectivity of their atoms.

Problem 1.17

Which sets are valid pairs of contributing structures?

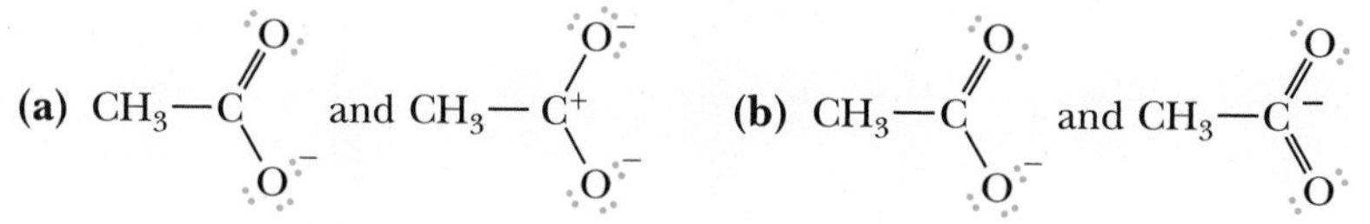

C. Estimating the Relative Importance of Contributing Structures

Not all structures contribute equally to a resonance hybrid. We describe four ways to predict which structure contributes more to the hybrid. But before examining these preferences, let's consider how to think about the fact that contributing structures may contribute unequally to many resonance hybrids. Suppose you combined yellow and blue paint to make green paint. This is analogous to writing two structures that contribute equally to the final resonance hybrid. The contributing structures do not interconvert back and forth; rather, they are used to represent only one structure—just as the green paint you mixed is not yellow one moment and blue another moment. Continuing with our paint analogy, how might you describe two different shades of green? One color might have a greater contribution of yellow; the other, a greater contribution of blue. This is analogous to the fact that many molecules have a greater contribution of one contributing structure. Importantly, there is only one actual way that the electrons are distributed in the molecule, but the limitations of how we draw molecules with dot pairs and lines require us to rely on examining two or more depictions—just as many colors of paint mixed in different proportions give only one final color.

The following preferences will help you estimate the relative importance of the various contributing structures. In fact, we can rank structures by the number of these preferences they follow. Those that follow the most preferences contribute the most to the hybrid, and any structure that violates all four of these preferences can be ignored.

Preference 1: Filled Valence Shells

Structures in which all atoms have filled valence shells (completed octets) contribute more than those in which one or more valence shells are unfilled. For example, the following are the contributing structures for Example 1.16(c) and its solution.

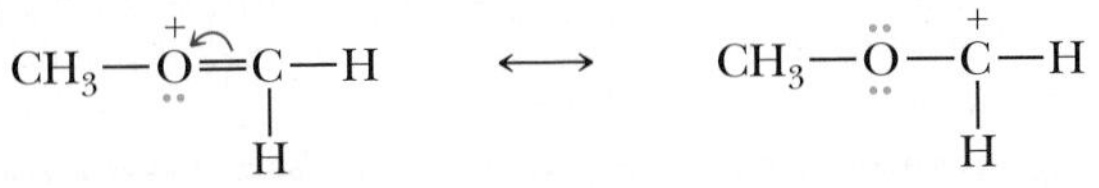

Greater contribution: both carbon and oxygen have complete valence shells

Lesser contribution: carbon has only six electrons in its valence shell

Preference 2: Maximum Number of Covalent Bonds

Structures with a greater number of covalent bonds contribute more than those with fewer covalent bonds. In the illustration for preference 1, the structure on the left has eight covalent bonds and makes the greater contribution to the hybrid. The structure on the right has only seven covalent bonds. Yet, this rule should never lead you to write structures that exceed eight valence electrons (some exceptions exist, such as S and P discussed in Section 1.7C).

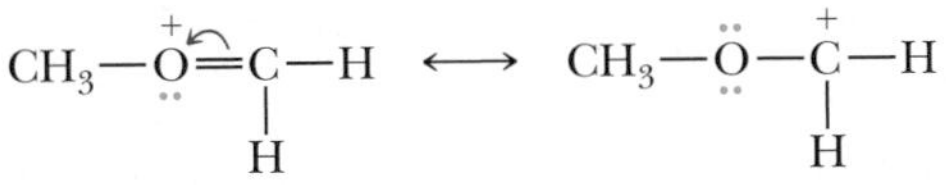

Greater contribution: eight covalent bonds

Lesser contribution: seven covalent bonds

Preference 3: Least Separation of Unlike Charges

Structures that involve separation of unlike charges contribute less than those that do not involve charge separation because separation of charges costs energy.

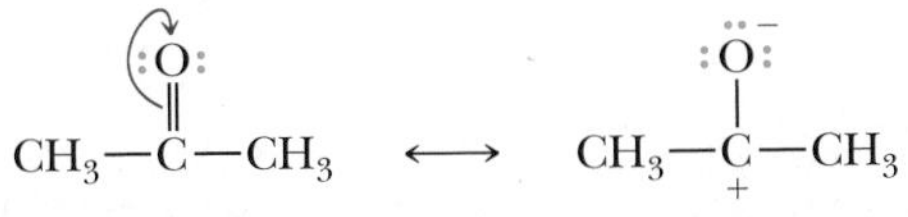

Greater contribution: no separation of unlike charges

Lesser contribution: separation of unlike charges

Preference 4: Negative Charge on a More Electronegative Atom

Structures that carry a negative charge on a more electronegative atom contribute more than those with the negative charge on a less electronegative atom. Conversely, structures that carry a positive charge on a less electronegative atom contribute more than those that carry the positive charge on a more electronegative atom. Following are three contributing structures for acetone:

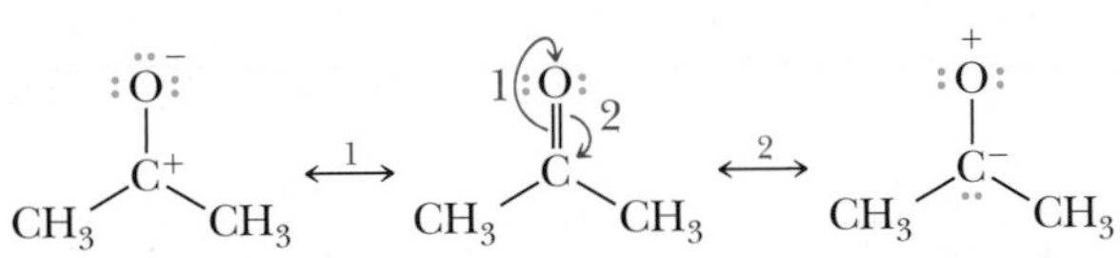

(a) Lesser contribution

(b) Greater contribution

(c) Should not be drawn

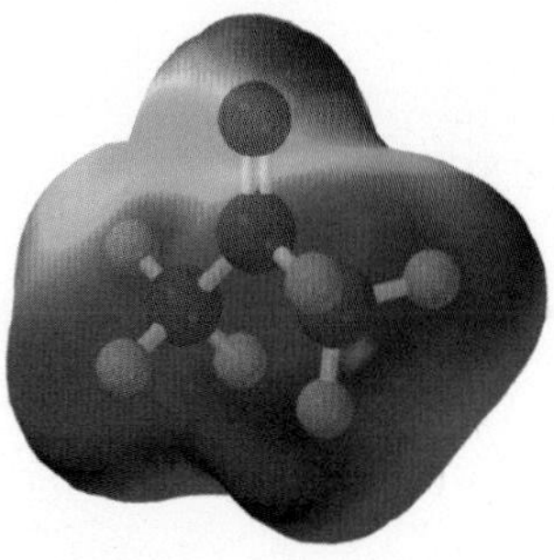

An electrostatic potential map of an acetone molecule.

Structure (b) makes the largest contribution to the hybrid. Structure (a) contributes less because it involves separation of unlike charge and because carbon has an incomplete octet. Nevertheless, on structure (a), the more electronegative O atom has the negative charge and the less electronegative C atom has the positive charge. Structure (c) violates all four preference rules and should not be drawn, and arrow 2 on structure (b) can be ignored.

It is important to realize that if resonance structures contribute unequally, the actual structure of the hybrid most resembles the structure that contributes most. The electrostatic potential map of acetone shows the negative charge (red) on oxygen and the positive charge (blue) on carbon in agreement with the results we derive from the resonance treatment.

Example 1.18 Relative Contributions to Resonance

Estimate the relative contribution of the members in each set of contributing structures.

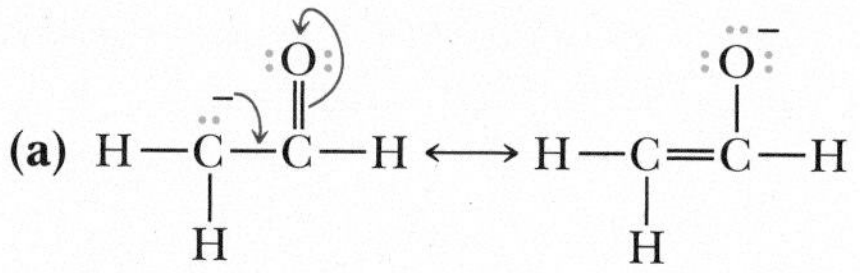

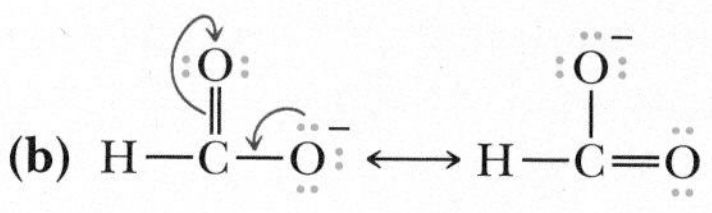

Solution

(a) The structure on the right makes a greater contribution to the hybrid because it places the negative charge on oxygen, the more electronegative atom.

(b) The structures are equivalent and make equal contributions to the hybrid.

Problem 1.18

Estimate the relative contribution of the members in each set of contributing structures.

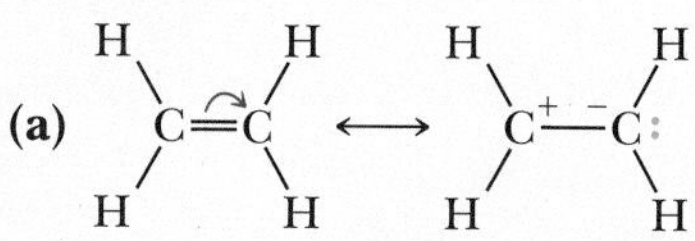

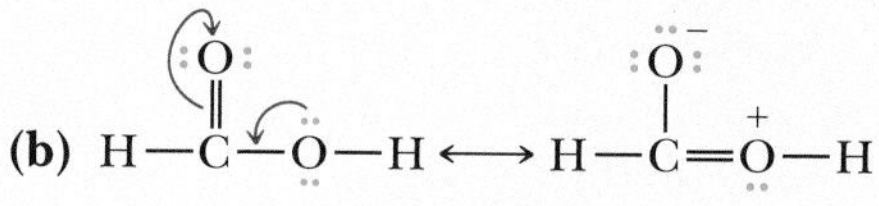

To emphasize, do not confuse resonance contributing structures with equilibration among different species. A molecule described as a resonance hybrid is not equilibrating among individual electron configurations of the contributing structures. Rather, the molecule has only one structure, which is best described as a hybrid of its various contributing structures.

The contributing structures for acetone drawn in the preceding section give us insight into the charge distribution of the molecule. In the lesser contributing structure, there is a positive charge on the carbon and a negative charge on the oxygen, as emphasized in the electrostatic potential map shown for acetone. We could come to the same conclusion by recognizing that the C=O bond is polar, with partial positive and negative charges on C and O, respectively (as we saw in Section 1.5). Given the insights into bonding orbitals developed in the previous section, we should now be able to use our hybrid VB/MO theory to visualize this polarization. In other words, the σ and π orbitals discussed for ethylene in Section 1.7C should be slightly perturbed for acetone to reflect the polarization in the C=O

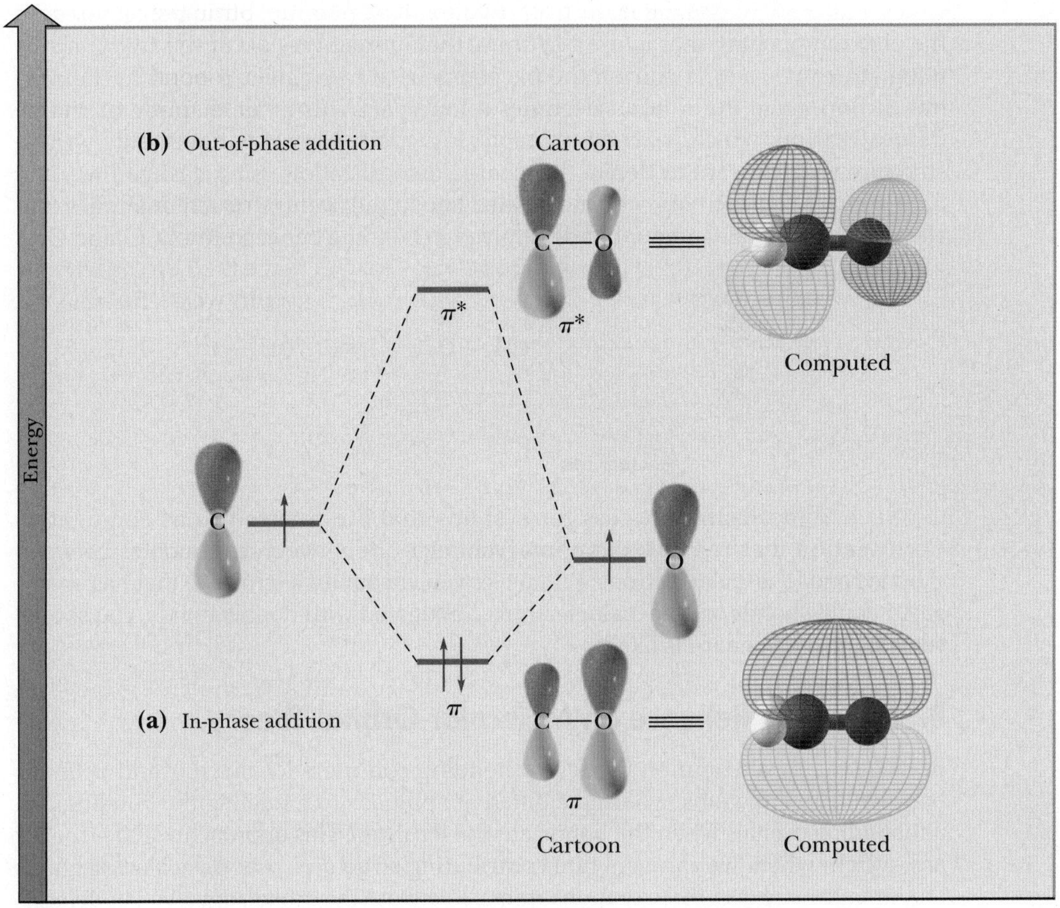

FIGURE 1.25 Molecular orbital mixing diagram for the creation of any C—O π bond. (a) Addition of two $2p$ orbitals in-phase leads to a π orbital that is lower in energy than the two separate starting orbitals. When populated with two electrons, the π orbital gives a bond that has greater electron density on O compared to C. This is not visually evident in the computed orbital because an O $2p$ orbital is smaller than a C $2p$ orbital. (b) Out-of-phase addition leads to a π* orbital that has greater orbital character on C (easily seen in the computed orbital).

that is absent in a C═C. Here, only the π bond of C═O will be analyzed because the lesson for the σ bond is identical.

Figure 1.25 shows the MO theory mixing diagram for the creation of the π and π* orbitals of C═O. The only difference between this diagram and that of Figure 1.21 is that the energy level of the O $2p$ orbital is placed lower than that of the C $2p$ orbital. The $2p$ orbitals for O are lower in energy because O is more electronegative than C. In MO theory, when the starting atomic orbitals are not equal in energy, the resulting molecular orbitals most resemble (have a greater contribution from) the atomic orbital that is closest in energy. Hence, the bonding π orbital has a larger contribution from the O, while the antibonding π orbital has a larger contribution from the C. Because only the bonding orbital is occupied, the electron density in the bond is concentrated on O. In the antibonding orbital, there is considerable orbital character on C, thereby placing a partial positive charge on this atom (a fact that will become important when we start analyzing reactions on C═O groups). Hence, the simple polarization concepts (Section 1.5), the notion of resonance (Section 1.8), and the VB/MO model of bonding (Section 1.7) all lead to the same conclusions.

1.9 Molecular Orbitals for Delocalized Systems

A. Resonance Revisited

As discussed in Section 1.8, resonance gives us an approach to depicting bonding when more than one Lewis structure is possible. The real bonding and electron distribution in the molecule is a weighted average of the various contributing structures. For example, let's again examine the acetate anion, which has a 50/50 contribution from two equivalent contributing structures. In the electron pushing that is used to show the interconversion of the two contributing structures, a lone pair on the oxygen is used to create a π bond to the central carbon while the π bond becomes a lone pair. This is an example of charge **delocalization**, which is always a stabilizing effect. Chemists commonly write a structure that is meant to depict this charge delocalization using dashed lines and a $-1/2$ charge on each oxygen. The dashed line in the average structure implies four electrons (a lone pair and a π bond) involved in bonding between the O, C, and O.

Delocalization

The spreading of charge and/or an electron density over a larger volume of space.

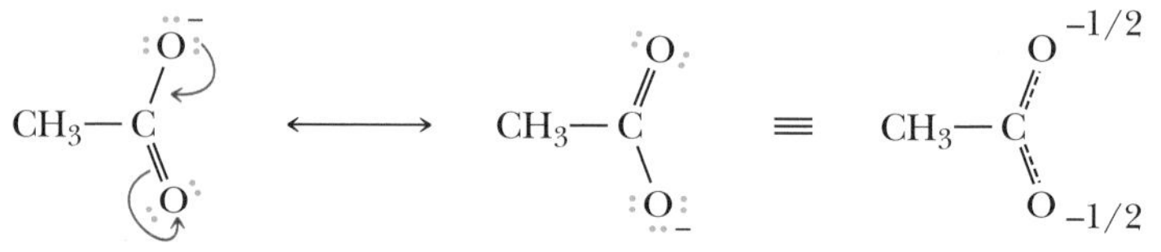
CH3—C
CH3—C
CH3—C
−1/2
−1/2

Acetate ion

Such delocalization occurs for systems that are referred to as conjugated. **Conjugation** means a lack of an intervening atom between π bonds or between π bonds and lone-pair electrons. Whenever you encounter a structure that has conjugated double bonds or lone-pair electrons conjugated with double bonds, you should consider that resonance is likely.

Conjugation

Lack of atoms between π bonds or between π bonds and lone-pair electrons.

B. A Greater Reliance on Molecular Orbital Theory

Because the most common model for bonding considers localized σ and π bonds between adjacent atoms, we need a different approach to modeling the bonding and antibonding orbitals when the bonds are not localized. This approach will be true for any system where we can generate contributing structures that delocalize π bonds and/or lone pairs over three or more atoms. To picture the orbitals involved in delocalized systems, we use only MO theory, dropping the VB theory localization principle.

MO theory tells us to add and subtract *all* the atomic orbitals in a molecule to create the bonding and antibonding molecular orbitals. Hence, the resulting bonding and antibonding orbitals are necessarily delocalized over all atoms in the molecule when the atomic orbitals on those atoms are properly aligned to overlap. Because the resulting molecular orbitals are necessarily delocalized, this model for bonding is particularly useful for delocalized systems. In addition, because we consider delocalization only for molecules containing π bonds adjacent to π bonds and/or π bonds adjacent to lone pairs, we will use only the pure MO theory model of bonding for such systems.

The manner in which the in-phase and out-of-phase addition occurs to create the bonding and antibonding orbitals of delocalized systems is beyond the scope of this book; therefore, only the results are given here. In Figure 1.26, we show the molecular orbitals for an acetate ion. Some features of these orbitals are worth discussing. The first is to remember that although the orbitals are spread over more than two atoms, each orbital can hold only two electrons. Second, note that the central orbital for acetate anion has the two electrons in p orbitals that are only on the oxygens. This electron distribution is in agreement with the contributing structures of acetate, which place the lone pairs on only the oxygens and −1/2 of a charge on each oxygen.

C. Hybridization Considerations in Light of Resonance and MO Theory

Given the discussions of delocalization and MO theory in the last two sections, we need to reconsider the assignment of hybridization to various atoms. For example, what is the hybridization of each oxygen in acetate? If you consider only an individual

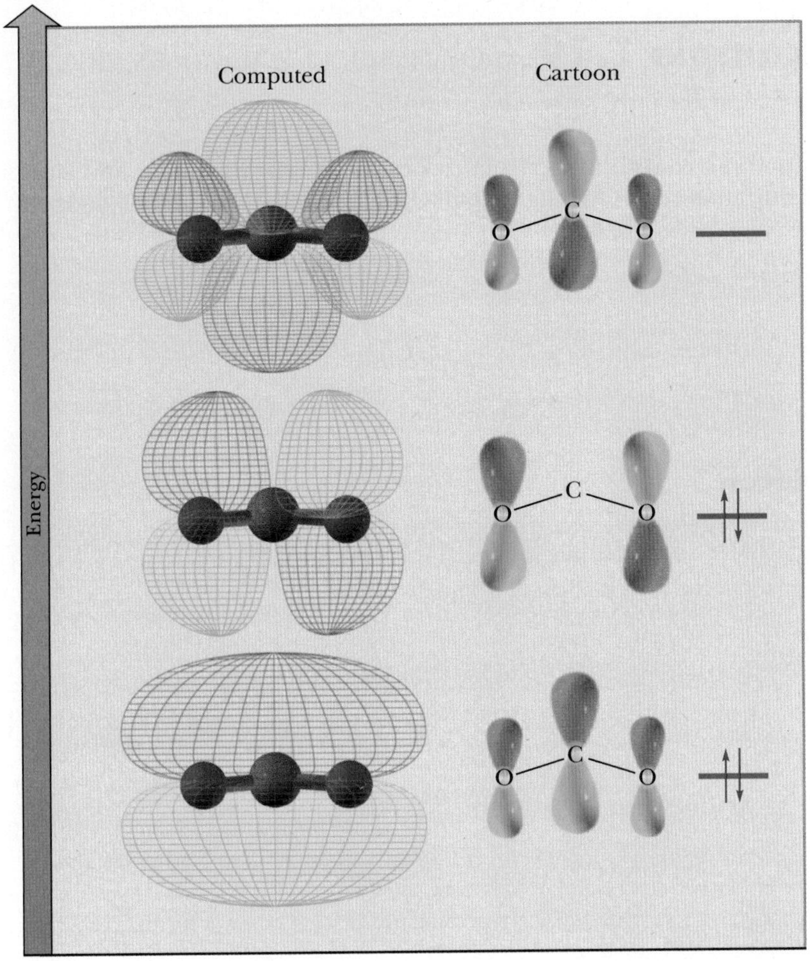

Watch a video explanation

FIGURE 1.26 An example of molecular orbitals for delocalized systems: the three π orbitals of acetate anion. Only the lowest two are populated with electrons.

contributing structure as indicated below, the conclusion is that one oxygen is sp^3 and one is sp^2. However, the other contributing structure would flip-flop the hybridization conclusion.

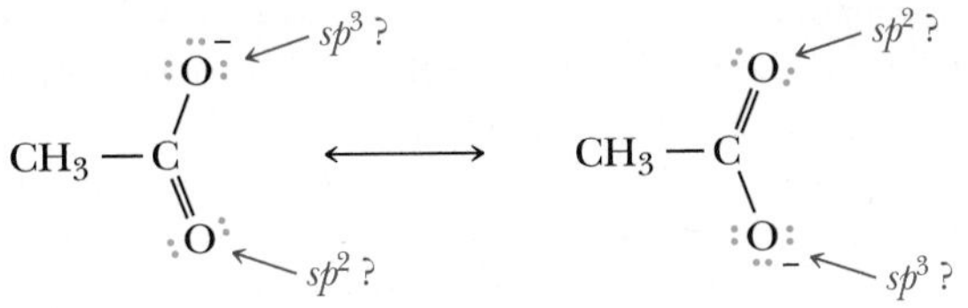

Examining the π bonds in the contributing structures allows the correct hybridization assignment to be made. For an atom to be involved in π bonding, it must have a $2p$ orbital. Of the two hybridization choices, only sp^2 atoms have $2p$ orbitals. Therefore, in acetate, both oxygens are sp^2 hybridized. This means that two of the lone pairs on the oxygens are residing in sp^2 hybrid orbitals. The third lone pair on each oxygen is in a $2p$ orbital that is mixing with $2p$ orbitals on the carbon and the alternate oxygen to create the three delocalized molecular orbitals of Figure 1.24(a). In fact, the electron pushing used to interconvert the two contributing structures of acetate shows lone pairs being used to create π bonds (see Section 1.9A), and it should not be surprising that these lone pairs are in $2p$ orbitals. In general, whenever atoms are involved in resonance that has one or more contributing structures involving a π bond to that atom, the atoms must be sp^2 hybridized (or on rare occasions, sp hybridized, but never sp^3 hybridized).

Example 1.19 Contributing Structures and Hybridization

Draw three contributing structures of the following amide and state the hybridization of the highlighted O, C, and N. In which orbitals do the three lone pairs drawn reside?

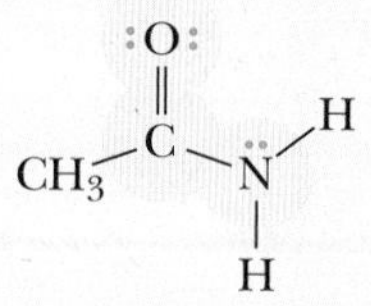

Solution

There is a reasonable contributing structure (C), which places a double bond between the carbon and the nitrogen. Therefore, all three atoms are sp^2. This means that the lone pair on N is in a $2p$ orbital.

A. B. C.

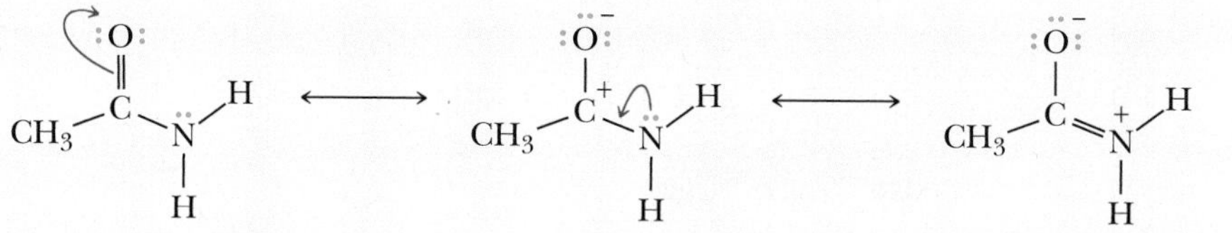

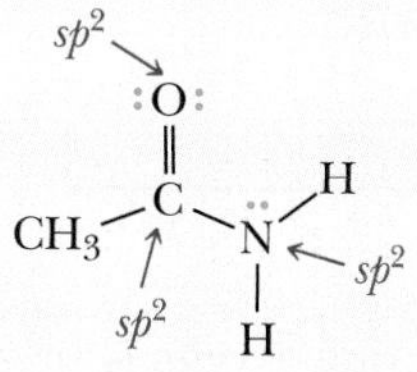

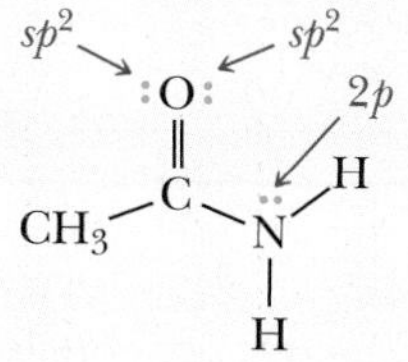

Watch a video explanation

Problem 1.19

Draw three contributing structures of the following compound (called guanidine) and state the hybridization of the four highlighted atoms. In which orbitals do the three lone pairs drawn reside?

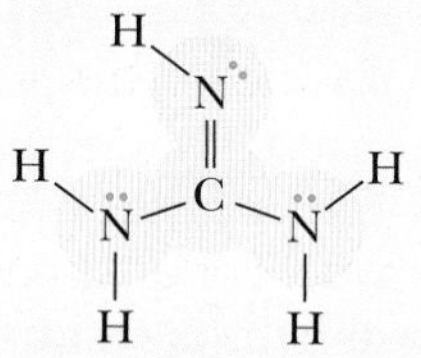

guanidine

MCAT Practice: *Passage and Questions*

VSEPR and Resonance

The concept of valence-shell electron-pair repulsion (VSEPR) is presented in introductory organic chemistry as a way to predict molecular geometries. The idea behind VSEPR is that areas of electron density repel each other so that the geometry of bonds and/or lone pairs of electrons around any one atom places these areas as far apart as possible. For four areas of electron density, a tetrahedral geometry is predicted. For three areas of electron density, a trigonal planar geometry is predicted. Two areas of electron density lead to a linear geometry.

VSEPR is simply a predictive tool, but in some cases, it gives an incorrect prediction. In these instances, additional insights into bonding are necessary, such as resonance. Interestingly, several important situations are critical to biochemistry where VSEPR breaks down. Two examples are shown here.

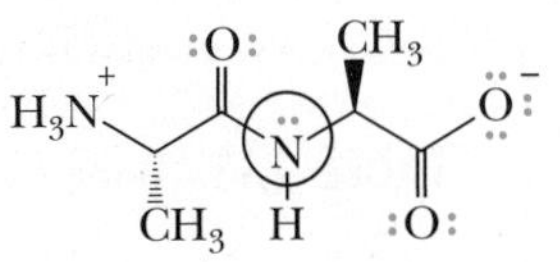

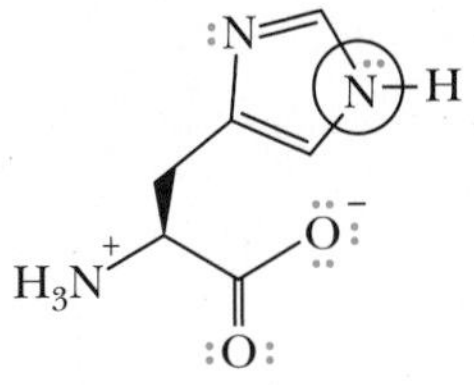

Histidine

The circled nitrogen atoms are predicted by VSEPR to be tetrahedral in geometry because each appears to have four areas of electron density: three bonds and a single lone pair of electrons. However, in both cases, structural analysis has revealed that the atoms actually have a trigonal planar geometry.

Questions

A. What is the hybridization state of the circled nitrogens? What kind of orbital contains the lone pairs identified in these circles?

1. *sp*, 2*p*
2. sp^2, sp^2
3. sp^3, 2*p*
4. sp^2, 2*p*

B. The molecule shown on the right in the example in the right column is the amino acid histidine, and the five-membered ring is known as aromatic. An aromatic ring has 2, 6, 10, 14, etc., electrons placed in 2*p* orbitals around a ring. Indicate which of the following statements must therefore be true.

1. There are a total of six electrons in the pi system (defined as electrons in 2*p* orbitals), including the lone pair on the ring N that is not circled.
2. There are a total of six electrons in the pi system, including the lone pair on the ring N atom that is circled.
3. The lone pair on the ring N atom that is not circled resides in an sp^2 orbital on an sp^2 hybridized nitrogen atom.
4. Statements 2 and 3 are both true.

C. Which of the following are reasonable contributing structures for the amide bond of the molecule shown on the left in the example above?

FIGURE 1

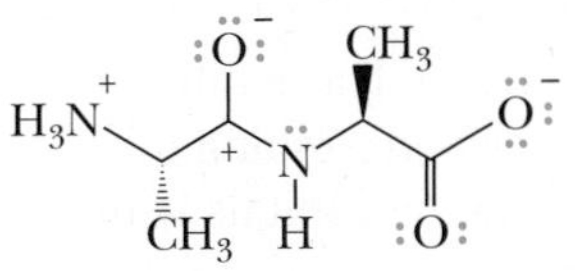

FIGURE 2

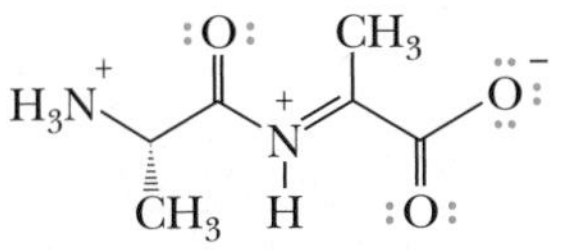

FIGURE 3

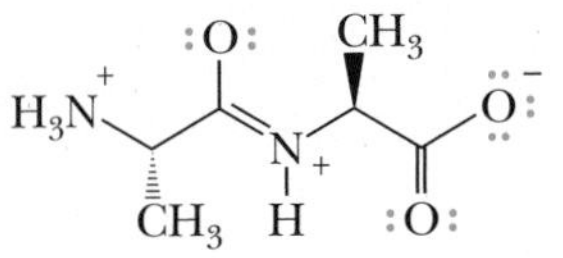

1. Figure 1
2. Figure 2
3. Figure 3
4. Both Figures 1 and 3.

D. The following structure is called imidazolium. Which of the following statements about imidazolium are true?

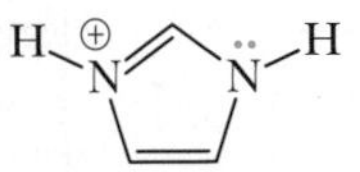

Imidazolium

a. Both nitrogens are sp^2 hybridized, and the lone pair of electrons is in a 2*p* orbital.
b. The nitrogen on the right is sp^3 hybridized, while the nitrogen on the left is sp^2 hybridized. The lone pair of electrons is in an sp^3 hybrid orbital.
c. The molecule has an equivalent contributing structure not shown.
d. The molecule has no reasonable contributing structures.

1. Statements a and c are true.
2. Statements a and d are true.
3. Statements b and c are true.
4. Statements b and d are true.

1.10 Bond Lengths and Bond Strengths in Alkanes, Alkenes, and Alkynes

Values for bond lengths and bond strengths (bond dissociation enthalpies) for ethane, ethylene, and acetylene are given in Table 1.11.

Table 1.11 Bond Lengths and Bond Strengths for Ethane, Ethylene, and Acetylene

Name	Formula	Bond	Bond Orbital Overlap	Bond Length (pm)	Bond Strength [kJ (kcal)/mol]
Ethane	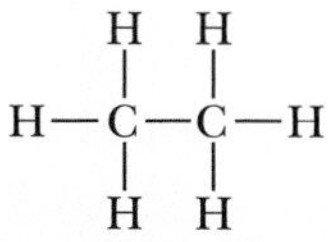	C—C	sp^3–sp^3	153.2	376 (90)
		C—H	sp^3–1s	111.4	422 (101)
Ethylene	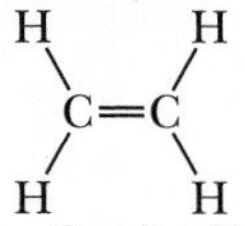	C—C	sp^2–sp^2, 2p–2p	133.9	727 (174)
		C—H	sp^2–1s	110.0	464 (111)
Acetylene	H—C≡C—H	C—C	sp–sp, two 2p–2p	121.2	966 (231)
		C—H	sp–1s	109.0	566 (133)

As you study Table 1.11, note the following points:

1. Carbon-carbon triple bonds are shorter than carbon-carbon double bonds, which in turn are shorter than carbon-carbon single bonds. This order of bond lengths exists because there are three versus two versus one bond holding the carbon atoms together.
2. The C—H bond in acetylene is shorter than that in ethylene, which in turn is shorter than that in ethane. The relative lengths of these C—H bonds are determined by the percent *s*-character in the hybrid orbital of carbon forming the σ bond with hydrogen. The greater the percent *s*-character of a hybrid orbital, the closer electrons in it are held to the nucleus and the shorter the bond because *s* electrons are on average closer to the nucleus than are *p* electrons. The relative lengths of C—H single bonds correlate with the fact that the percent *s*-character in an *sp* orbital is 50%, in an sp^2 orbital is 33.3%, and in an sp^3 orbital is 25%. Also, because electrons in *s* orbitals are bound more tightly than those in *p* orbitals, the more *s*-character in a bond, the stronger it is. This is the first of several times we will refer to the percent *s*-character in hybrid orbitals. It has a large influence on a variety of molecular properties.
3. There is a correlation between bond length and bond strength: The shorter the bond, the stronger it is. A carbon-carbon triple bond is the shortest C—C bond and is also the strongest. The carbon-hydrogen bond in acetylene is the shortest and also the strongest.
4. Although a C—C double bond is stronger than a C—C single bond, it is not twice as strong. By the same token, a C—C triple bond is stronger than a C—C single bond, but it is not three times as strong. These differences arise because the overlap of orbitals lying on the same axis and forming σ bonds is more efficient (gives a greater bond strength) than overlap of orbitals lying parallel to each other and forming π bonds.

Study Guide

1.1 Electronic Structure of Atoms

- Atoms consist of a small, dense nucleus and electrons distributed about the nucleus in regions of space called **shells**.
  - *Each shell can contain as many as $2n^2$ electrons, where n is the number of the shell.*
  - *Each principal energy level is subdivided into regions of space called **orbitals**. The **valence shell** is the outermost occupied shell, and it contains the **valence electrons**. Valence electrons are important because they take part in chemical bonding.*

P **1.1, 1.21, 1.22, 1.23**

- The **Lewis dot structure** of an atom shows the symbol of the atom surrounded by a number of dots equal to the number of electrons in the valence shell of the atom.

Problem 1.1

1.2 Lewis Model of Bonding

- According to the Lewis model of covalent bonding, atoms bond together in such a way that each atom participating in a chemical bond acquires a completed valence-shell electron configuration resembling that of the noble gas nearest it in atomic number.
  - *Anions and cations attract each other but do not form bonds with defined directionality.*
  - *A **covalent bond** is a chemical bond formed by the sharing of electron pairs between adjacent atoms.*
  - *The tendency of main-group elements (Groups 1A–7A) to achieve an outer shell of eight valence electrons is called the **octet rule**.*

P **1.2**

- **Electronegativity** is a measure of the force of attraction by an atom for electrons it shares in a chemical bond with another atom. In the context of the Periodic Table, the electronegativity of elements increases from left to right with a period (increasing positive charge of the nucleus) and decreases from the top to bottom within a group (increasing distance of electrons from the nucleus).
  - *A **nonpolar covalent bond** is a covalent bond in which the difference in electronegativity of the bonded atoms is less than 0.5.*
  - *A **polar covalent bond** is a covalent bond in which the difference in electronegativity of the bonded atoms is between 0.5 and 1.9.*
    - *In a polar covalent bond, the more electronegative atom bears a partial negative charge ($\delta-$) and the less electronegative atom bears a partial positive charge ($\delta+$).*
    - *A polar bond has a bond dipole moment equal to the product of the absolute value of the partial charge times the distance between the dipolar charges (the bond length).*

P **1.3–1.5, 1.24–1.26, 1.33–1.37, 1.66**

- An acceptable **Lewis structure** for a molecule or an ion must show (1) the correct connectivity of atoms, (2) the correct number of valence electrons, (3) no more than two electrons in the outer shell of hydrogen and no more than eight electrons in the outer shell of any second-period element, and (4) all formal charges.
  - *There are some apparent exceptions to the octet rule: neutral compounds of boron and aluminum can have only six valence electrons.*

P **1.6, 1.7, 1.27, 1.32, 1.64, 1.65**

▶ **Problem 1.6**

1.3 Functional Groups

- **Functional groups** are characteristic structural units by which we divide organic compounds into classes and that serve as a basis for nomenclature.
  - *Functional groups are also sites of chemical reactivity. A particular functional group generally undergoes the same types of chemical reactions in whatever compound it occurs.*

P **1.8–1.12, 1.41–1.47**

1.4 Bond Angles and Shapes of Molecules

- Bond angles of molecules and polyatomic ions can be predicted using Lewis structures and **valence-shell electron-pair repulsion (VSEPR)**.
  - *For atoms surrounded by four regions of electron density, VSEPR predicts bond angles of* $109.5°$*; for atoms surrounded by three regions of electron density, it predicts bond angles of* $120°$*; and for two regions of electron density, it predicts bond angles of* $180°$.

P **1.13, 1.38–1.40, 1.67, 1.68**

1.5 Polar and Nonpolar Molecules

The **dipole moment** (μ) of a molecule is the vector sum of the individual bond dipoles.

- *Predicting a molecular dipole moment requires a detailed understanding of molecular geometry as well as of bond polarity.*
- *Depending on the three-dimensional shape of the molecule, the individual bond dipoles might reinforce each other, cancel each other, or something in between.*

P **1.14, 1.49, 1.50**

▶ **Problem 1.14**

1.6 Quantum or Wave Mechanics

- **Quantum mechanics** is the branch of science that studies particles and their associated waves. It provides a way to determine the shapes of atomic orbitals and to quantify the energetics of covalent bond formation.

1.7 A Combined Valence Bond and Molecular Orbital Theory Approach to Covalent Bonding

- According to **molecular orbital (MO)** theory, combination of n atomic orbitals gives n molecular orbitals.
  - *Molecular orbitals are divided into sigma (σ) and pi (π) bonding and antibonding molecular orbitals. These orbitals are arranged in order of increasing energy, and their order of filling with electrons is governed by the same rules as for filling atomic orbitals.*
  - *Although useful for quantitative calculations on computers, molecular orbital theory alone does not provide an intuitive understanding of σ bonds in complex molecules.*

P **1.61**

- For an intuitive understanding of bonding in molecules, we use molecular orbital theory concepts (in-phase and out-of-phase addition of overlapping orbitals to give bonding and antibonding orbitals) in combination with **valence bond theory**.
  - *Valence bond theory involves the combination of atomic orbitals on each atom before considering bonding. This combination process is called* ***hybridization****, and the resulting atomic orbitals are called* ***hybrid orbitals****.*
    - *The combination of one $2s$ atomic orbital and three $2p$ atomic orbitals produces four equivalent* ***sp^3 hybrid orbitals****, each directed toward a corner of a regular tetrahedron at angles of 109.5°.*
    - *The combination of one $2s$ atomic orbital and two $2p$ atomic orbitals produces three equivalent* ***sp^3 hybrid orbitals****, the axes of which lie in a plane at angles of 120°.*
    - *The combination of one $2s$ atomic orbital and one $2p$ atomic orbital produces two equivalent* ***sp hybrid orbitals****, the axes of which lie at an angle of 180°.*

P **1.15, 1.55, 1.62**

- S and P atoms are commonly depicted with more than eight valence electrons, invoking participation of $3d$ electrons.
  - *Recent calculations reveal that in many of these cases, the S and P atoms are best thought of as sp^3 hybridized with a formal charge, rather than involving $3d$ orbitals.*
- In the combined molecular orbital theory/valence bond theory approach, bonding in organic molecules is thought of as the in-phase addition (overlapping to create bonding orbitals) and out-of-phase addition, also referred to as subtraction (to create antibonding orbitals) of the hybridized (and possibly any unhybridized $2p$) atomic orbitals on adjacent atoms.
  - *All C—C, C—O, and C—N single bonds are sigma (σ) bonds formed by the overlapping of hybrid orbitals.*
  - *All C—H, O—H, N—H single bonds are sigma (σ) bonds formed by overlapping hybrid orbitals on C, O, or N with the $1s$ orbital of H.*
  - *All C=C, C=O, C=N, N=N, and N=O double bonds are a combination of one sigma (σ) bond formed by overlapping hybrid orbitals and one pi (π) bond formed by overlapping parallel, unhybridized $2p$ orbitals.*
  - *All C≡C and C≡N triple bonds are a combination of one sigma (σ) bond formed by the overlap of sp hybrid orbitals and two pi (π) bonds formed by the overlap of two sets of parallel, unhybridized $2p$ orbitals.*

P **1.15, 1.57–1.61, 1.63, 1.71, 1.72, 1.76**

▶ **Figure 1.21**

1.8 Resonance

- According to the **theory of resonance**, molecules and ions for which no single Lewis structure is adequate are best described by writing two or more **contributing structures**. The real molecule or ion is a **resonance hybrid** of the various contributing structures.
  - *Double-headed arrows are drawn between contributing structures to describe the hybrid. Do not confuse the double-headed arrow with the two single-headed arrows used to show chemical equilibrium. Contributing structures used to indicate resonance are not equilibrating with each other.*
  - *The most important contributing structures have (1) filled valence shells, (2) a maximum number of covalent bonds, (3) the least separation of unlike charges, and (4) any negative charge on the more electronegative atom and/or any positive charge on the less electronegative atom.*
  - *Curved arrows show the manner in which valence electrons are redistributed from one contributing structure to the next. Use of curved arrows in this way is commonly referred to as **electron pushing**. Curved arrows always show movement of electron pairs, never atoms.*

P **1.16–1.18, 1.51–1.55, 1.69, 1.70, 1.74**

▶ **Problem 1.16**

1.9 Molecular Orbitals for Delocalized Systems

- Most examples of molecules described by more than one resonance contributing structure have charge and/or electron **delocalization**, a stabilizing effect in which charge and/or electron density are spread over more than two atoms.
  - *Delocalization occurs in molecules that have **conjugation**. A π bond and a lone pair of electrons or multiple π bonds without an intervening atom are conjugated.*
  - *The delocalized electrons are in molecular orbitals formed from overlapping $2p$ orbitals on three or more adjacent atoms.*
    - *If atoms are taking part in delocalization as described by resonance contributing structures, they must have $2p$ orbitals; so they must be sp^2 hybridized or, in rare cases, sp hybridized.*

P **1.19, 1.73, 1.75, 1.77**

▶ **Figure 1.26 and Example 1.19**

1.10 Bond Lengths and Bond Strengths in Alkanes, Alkenes, and Alkynes

- The greater the number of bonds between two atoms, the shorter the bond length and the greater the bond strength.
  - *Carbon-carbon triple bonds are shorter and stronger than carbon-carbon double bonds, which are shorter and stronger than carbon-carbon single bonds.*
- The more *s*-character in the hybridized orbital taking part in a bond, the shorter and stronger is the bond.

Problems

Red numbers indicate applied problems.

Electronic Structure of Atoms

1.20 Write the ground-state electron configuration for each atom. After each atom is its atomic number in parentheses.

(a) Sodium (11) **(b)** Magnesium (12) **(c)** Oxygen (8) **(d)** Nitrogen (7)

1.21 Identify the atom that has each ground-state electron configuration.

(a) $1s^2\,2s^2\,2p^6\,3s^2\,3p^4$ **(b)** $1s^2\,2s^2\,2p^4$

1.22 Define valence shell and valence electron.

1.23 How many electrons are in the valence shell of each atom?

(a) Carbon **(b)** Nitrogen **(c)** Chlorine **(d)** Aluminum

Lewis Structures and Formal Charge

1.24 Judging from their relative positions in the Periodic Table, which atom in each set is more electronegative?

(a) Carbon or nitrogen **(b)** Chlorine or bromine **(c)** Oxygen or sulfur

1.25 Which compounds have nonpolar covalent bonds, which have polar covalent bonds, and which have ions?

(a) LiF **(b)** CH_3F **(c)** $MgCl_2$ **(d)** HCl

1.26 Using the symbols $\delta-$ and $\delta+$, indicate the direction of polarity, if any, in each covalent bond.

(a) C—Cl **(b)** S—H **(c)** C—S **(d)** P—H

1.27 Write Lewis structures for these compounds. Show all valence electrons. None of them contains a ring of atoms.

(a) Hydrogen peroxide, H_2O_2 **(b)** Hydrazine, N_2H_4 **(c)** Methanol, CH_3OH

1.28 Write Lewis structures for these ions. Show all valence electrons and all formal charges.

(a) Amide ion, NH_2^- **(b)** Bicarbonate ion, HCO_3^- **(c)** Carbonate ion, CO_3^{2-}
(d) Nitrate ion, NO_3^- **(e)** Formate ion, $HCOO^-$ **(f)** Acetate ion, CH_3COO^-

1.29 Complete these structural formulas by adding enough hydrogens to complete the tetravalence of each carbon. Then write the molecular formula of each compound.

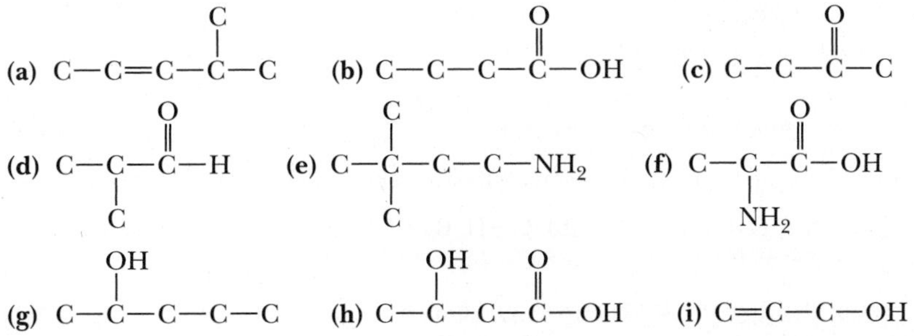

1.30 Some of these structural formulas are incorrect (i.e., they do not represent a real compound) because they have atoms with an incorrect number of bonds. Which structural formulas are incorrect? Which atoms in them have an incorrect number of bonds?

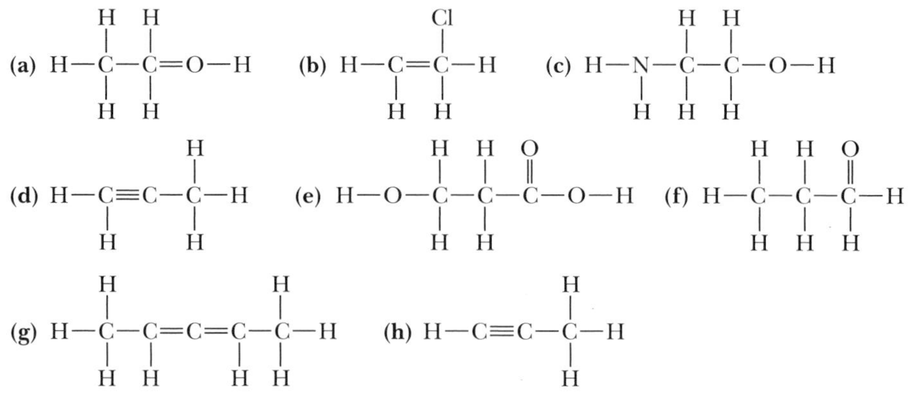

1.31 Following the rule that each atom of carbon, oxygen, and nitrogen reacts to achieve a complete outer shell of eight valence electrons, add unshared pairs of electrons as necessary to complete the valence shell of each atom in these ions. Then assign formal charges as appropriate.

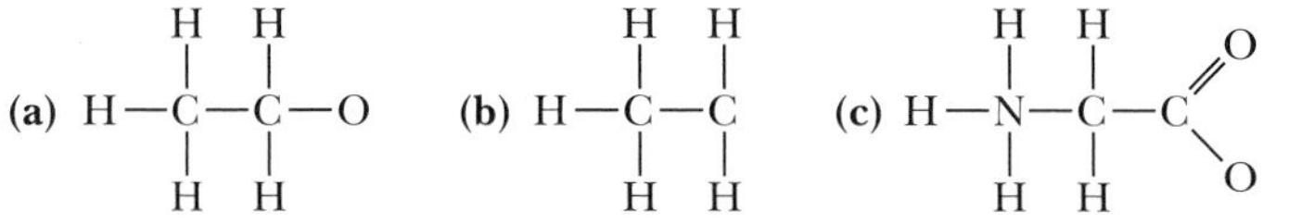

1.32 Following are several Lewis structures showing all valence electrons. Assign formal charges in each structure as appropriate.

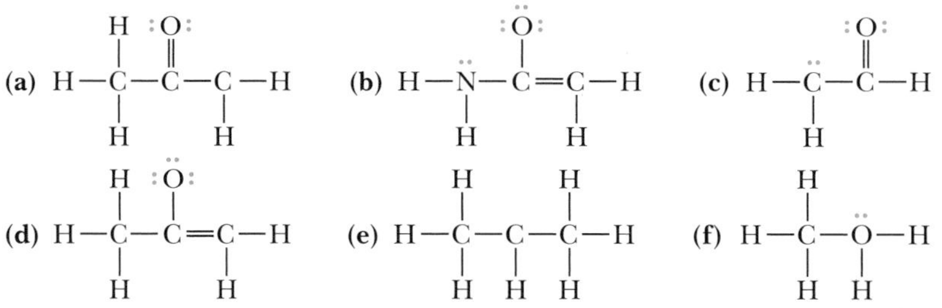

Polarity of Covalent Bonds

1.33 Which statements are true about electronegativity?

(a) Electronegativity increases from left to right in a period of the Periodic Table.
(b) Electronegativity increases from top to bottom in a column of the Periodic Table.
(c) Hydrogen, the element with the lowest atomic number, has the smallest electronegativity.
(d) The higher the atomic number of an element, the greater its electronegativity.

1.34 Why does fluorine, the element in the upper right corner of the Periodic Table, have the largest electronegativity of any element?

1.35 Arrange the single covalent bonds within each set in order of increasing polarity.

(a) C—H, O—H, N—H **(b)** C—H, B—H, O—H **(c)** C—H, C—Cl, C—I
(d) C—S, C—O, C—N **(e)** C—Li, C—B, C—Mg

1.36 Using the values of electronegativity given in Table 1.5, predict which indicated bond in each set is more polar, and using the symbols $\delta+$ and $\delta-$, show the direction of its polarity.

(a) CH_3—OH or CH_3O—H **(b)** CH_3—NH_2 or CH_3—PH_2
(c) CH_3—SH or CH_3S—H **(d)** CH_3—F or H—F

1.37 Identify the most polar bond in each molecule.

(a) $HSCH_2CH_2OH$ (b) $CHCl_2F$ (c) $HOCH_2CH_2NH_2$

Bond Angles and Shapes of Molecules

1.38 Use VSEPR to predict bond angles about each highlighted atom.

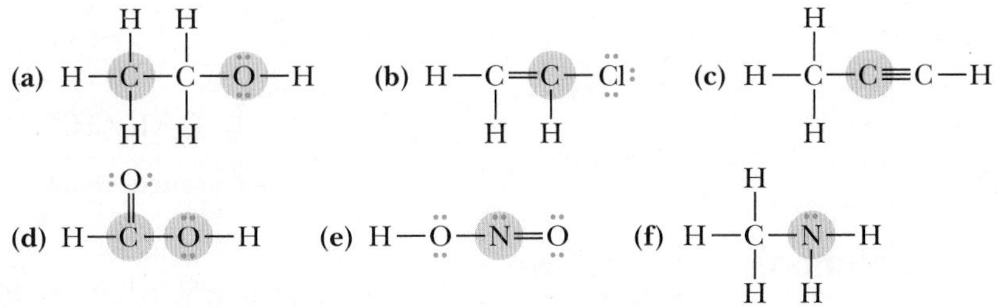

1.39 Use VSEPR to predict bond angles about each atom of carbon, nitrogen, and oxygen in these molecules.

(a) $CH_3-CH{=}CH_2$ (b) 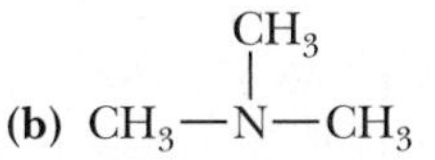 (c) 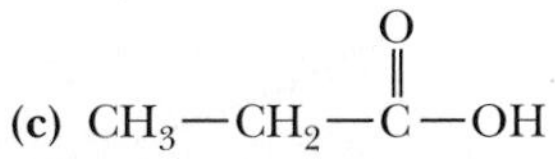

(d) $H_2C{=}C{=}CH_2$ (e) $H_2C{=}C{=}O$ (f) $CH_3-CH{=}N-OH$

1.40 Use VSEPR to predict the geometry of these ions.

(a) NH_2^- (b) NO_2^- (c) NO_2^+ (d) NO_3^-

Functional Groups

1.41 Draw Lewis structures for these functional groups. Show all valence electrons on each.

(a) Carbonyl group (b) Carboxyl group (c) Hydroxyl group
(d) Ester group (e) Amide group

1.42 Draw condensed structural formulas for all compounds with the molecular formula C_4H_8O that contain

(a) A carbonyl group (there are two aldehydes and one ketone).
(b) A carbon-carbon double bond and a hydroxyl group (there are eight).

1.43 What is the meaning of the term *tertiary* (3°) when it is used to classify alcohols? Draw a structural formula for the one tertiary (3°) alcohol with the molecular formula $C_4H_{10}O$.

1.44 What is the meaning of the term tertiary (3°) when it is used to classify amines? Draw a structural formula for the one tertiary (3°) amine known as Hünig's base (*N,N*-diisopropylethylamine).

1.45 Draw structural formulas for

(a) The four primary (1°) amines with the molecular formula $C_4H_{11}N$.
(b) The three secondary (2°) amines with the molecular formula $C_4H_{11}N$.
(c) The one tertiary (3°) amine with the molecular formula $C_4H_{11}N$.

1.46 Draw structural formulas for the three tertiary (3°) amines with the molecular formula $C_5H_{13}N$.

1.47 Draw structural formulas for

(a) The eight alcohols with the molecular formula $C_5H_{12}O$.
(b) The eight aldehydes with the molecular formula $C_6H_{12}O$.
(c) The six ketones with the molecular formula $C_6H_{12}O$.
(d) The eight carboxylic acids with the molecular formula $C_6H_{12}O_2$.
(e) The nine carboxylic esters with the molecular formula $C_5H_{10}O_2$.

1.48 Identify the functional groups in each compound.

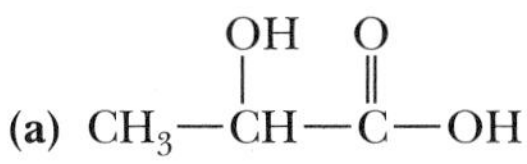

Lactic acid

(b) $HO—CH_2—CH_2—OH$

Ethylene glycol

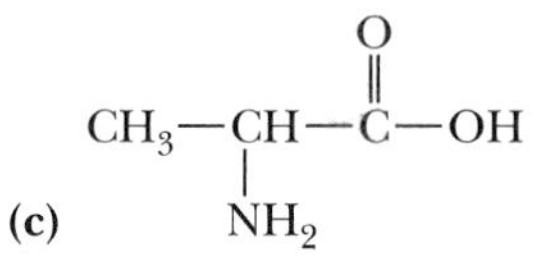

Alanine

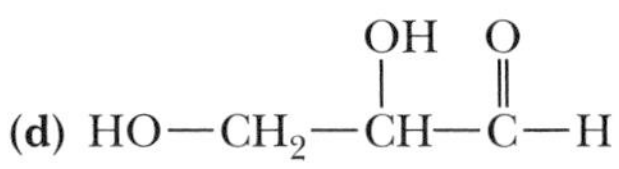

Glyceraldehyde

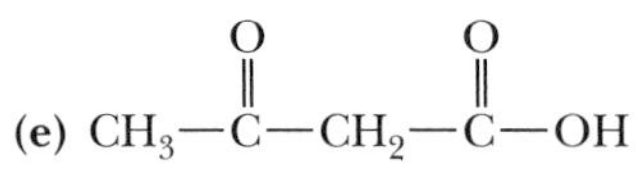

Acetoacetic acid

(f) $H_2NCH_2CH_2CH_2CH_2CH_2CH_2NH_2$

1,6-Hexanediamine

Polar and Nonpolar Molecules

1.49 Draw a three-dimensional representation for each molecule. Indicate which ones have a dipole moment and in what direction it is pointing.

(a) CH_3F **(b)** CH_2Cl_2 **(c)** CH_2ClBr **(d)** $CFCl_3$ **(e)** CCl_4
(f) $CH_2{=}CCl_2$ **(g)** $CH_2{=}CHCl$ **(h)** $HC{\equiv}C{-}C{\equiv}CH$ **(i)** $CH_3C{\equiv}N$
(j) $(CH_3)_2C{=}O$ **(k)** $BrCH{=}CHBr$ (two answers)

1.50 Tetrafluoroethylene, C_2F_4, is the starting material for the synthesis of the polymer polytetrafluoroethylene (PTFE), one form of which is known as Teflon. Tetrafluoroethylene has a dipole moment of zero. Propose a structural formula for this molecule.

Resonance and Contributing Structures

1.51 Which statements are true about resonance contributing structures?

(a) All contributing structures must have the same number of valence electrons.
(b) All contributing structures must have the same arrangement of atoms.
(c) All atoms in a contributing structure must have complete valence shells.
(d) All bond angles in sets of contributing structures must be the same.

1.52 Draw the contributing structure indicated by the curved arrow(s). Assign formal charges as appropriate.

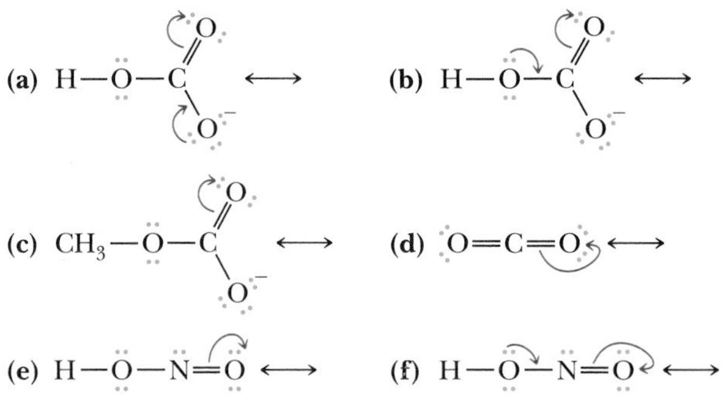

1.53 Using VSEPR, predict the bond angles about the carbon and nitrogen atoms in each pair of contributing structures in Problem 1.52. In what way do these bond angles change from one contributing structure to the other?

1.54 In Problem 1.52, you were given one contributing structure and were asked to draw another. Label pairs of contributing structures that are equivalent. For those sets in which the contributing structures are not equivalent, label the more important contributing structure.

1.55 Are the structures in each set valid contributing structures?

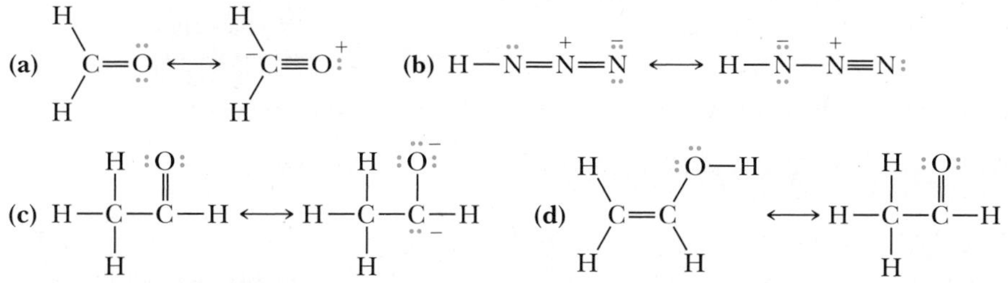

Valence Bond Theory

1.56 State the orbital hybridization of each highlighted atom.

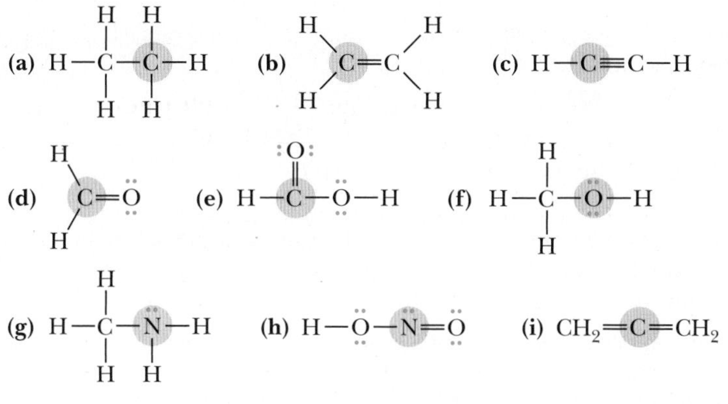

1.57 Describe each highlighted bond in terms of the overlap of atomic orbitals.

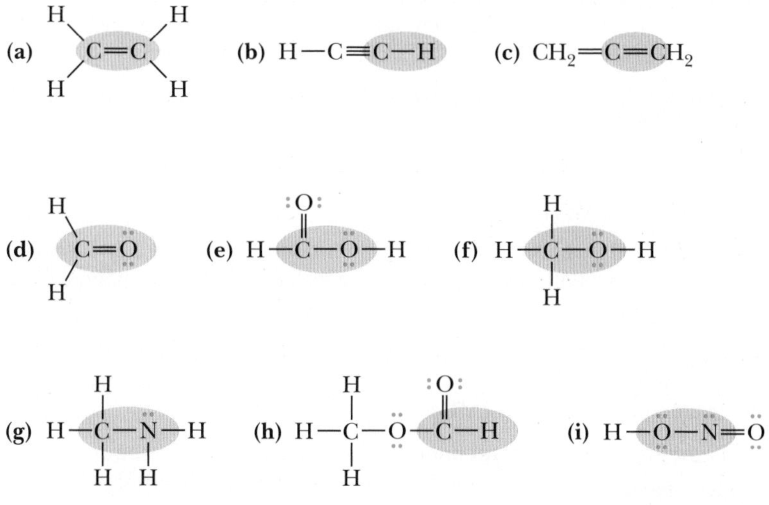

1.58 Following is a structural formula of the prescription drug famotidine, marketed by McNeil Consumer Pharmaceuticals Co. under the name Pepcid. The primary clinical use of Pepcid is for the treatment of active duodenal ulcers and benign gastric ulcers. Pepcid is a competitive inhibitor of histamine H_2 receptors that reduces both gastric acid concentration and the volume of gastric secretions.

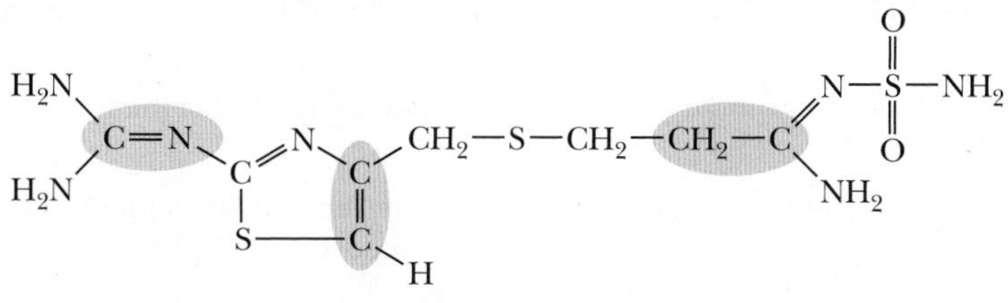

(a) Complete the Lewis structure of famotidine showing all valence electrons and any formal positive or negative charges.
(b) Describe each circled bond in terms of the overlap of atomic orbitals.

1.59 Draw a Lewis structure for methyl isocyanate, CH_3NCO, showing all valence electrons. Predict all bond angles in this molecule and the hybridization of each C, N, and O.

Combined MO/VB Theory

1.60 What is the hybridization of the highlighted atoms in the following structures? What are your estimates for the bond angles around these highlighted atoms? In each case, in what kind of orbital does the lone pair of electrons on the nitrogen reside?

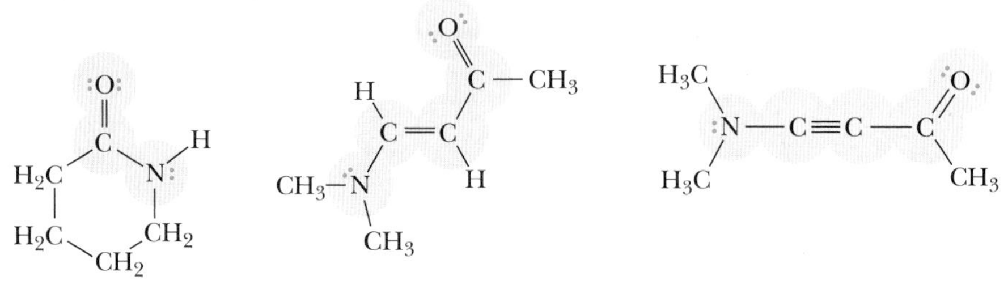

1.61 Using cartoon representations, draw a molecular orbital mixing diagram for a C—O σ bond. In your picture, consider the relative energies of C and O and how this changes the resulting bonding and antibonding molecular orbitals relative to a C—C σ bond.

1.62 In what kind of orbitals do the lone-pair electrons on the oxygen of acetone reside? Are they in the same plane as the methyl —CH_3 groups, or are they perpendicular to the methyl —CH_3 groups?

1.63 Draw the delocalized molecular orbitals for the following molecule. Are both π bonds of the triple bond involved in the delocalized orbitals?

$$CH_3—C{\equiv}C—CH{=}CH_2$$

Additional Problems

1.64 Why are the following molecular formulas impossible?

(a) CH_5 **(b)** C_2H_7

1.65 Each compound contains both ions and covalent bonds. Draw the Lewis structure for each compound. Show with dashes which are covalent bonds and show with charges which are ions.

(a) Sodium methoxide, CH_3ONa
(b) Ammonium chloride, NH_4Cl
(c) Sodium bicarbonate, $NaHCO_3$
(d) Sodium borohydride, $NaBH_4$
(e) Lithium aluminum hydride, $LiAlH_4$

1.66 Predict whether the carbon-metal bond in these organometallic compounds is nonpolar covalent, polar covalent, or ionic. For each polar covalent bond, show the direction of its polarity using the symbols $\delta+$ and $\delta-$.

(a) 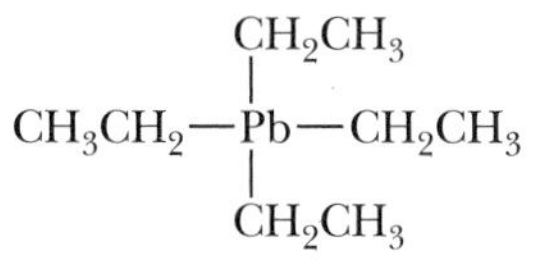
Tetraethyllead

(b) $CH_3—Mg—Cl$
Methylmagnesium chloride

(c) $CH_3—Hg—CH_3$
Dimethylmercury

1.67 Silicon is immediately under carbon in the Periodic Table. Predict the geometry of silane, SiH_4.

1.68 Phosphorus is immediately under nitrogen in the Periodic Table. Predict the molecular formula for phosphine, the compound formed by phosphorus and hydrogen. Predict the H—P—H bond angle in phosphine.

1.69 Draw a Lewis structure for the azide ion, N_3^-. (The order of atom attachment is N—N—N, and they do not form a ring.) How does the resonance model account for the fact that the lengths of the N—N bonds in this ion are identical?

1.70 Cyanic acid, HOCN, and isocyanic acid, HNCO, dissolve in water to yield the same anion on loss of H^+.

(a) Write a Lewis structure for cyanic acid.
(b) Write a Lewis structure for isocyanic acid.
(c) Account for the fact that each acid gives the same anion on loss of H^+.

Looking Ahead

1.71 In Chapter 6, we study a group of organic cations called carbocations. Following is the structure of one such carbocation, the *tert*-butyl cation.

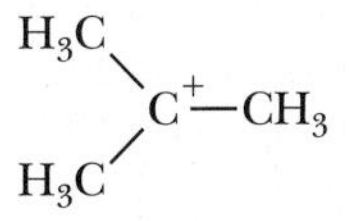

***tert*-Butyl cation**

(a) How many electrons are in the valence shell of the carbon bearing the positive charge?
(b) Using VSEPR, predict the bond angles about this carbon.
(c) Given the bond angle you predicted in (b), what hybridization do you predict for this carbon?

1.72 Many reactions involve a change in hybridization of one or more atoms in the starting material. In each reaction, identify the atoms in the organic starting material that change hybridization and indicate the change. We examine these reactions in more detail later in the course.

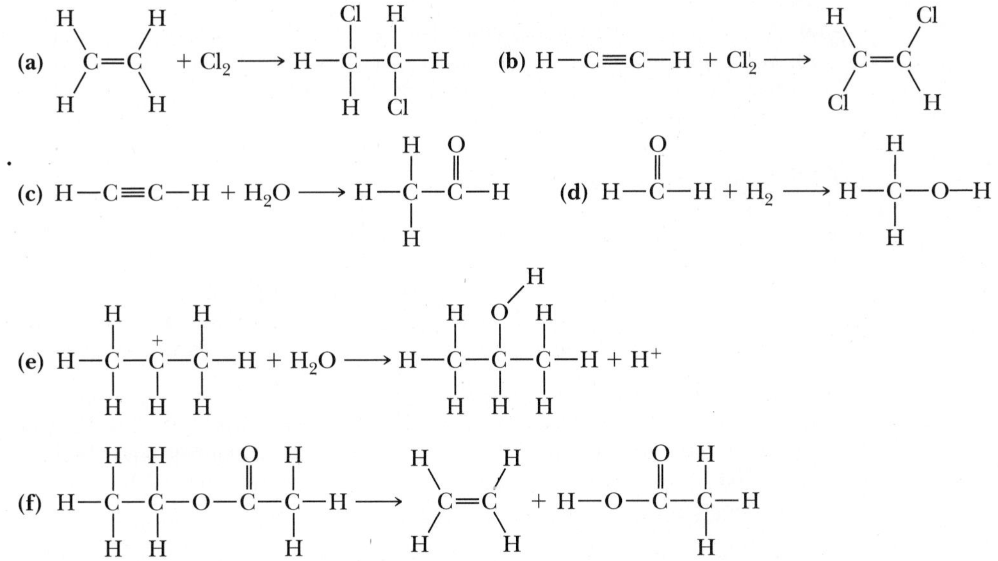

1.73 Following is a structural formula of benzene, C_6H_6, which we study in Chapter 21.

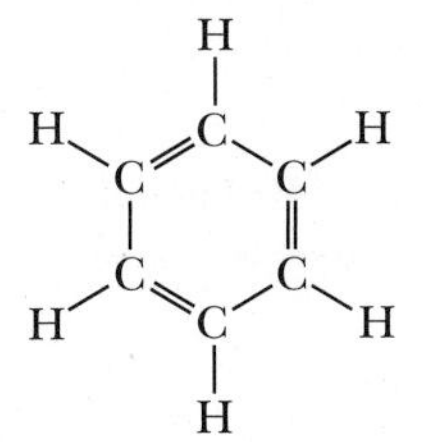

(a) Using VSEPR, predict each H—C—C and C—C—C bond angle in benzene.
(b) State the hybridization of each carbon in benzene.
(c) Predict the shape of a benzene molecule.
(d) Draw important resonance contributing structures.

1.74 Following are three contributing structures for diazomethane, CH_2N_2. This molecule is used to make methyl esters from carboxylic acids (Section 17.7C).

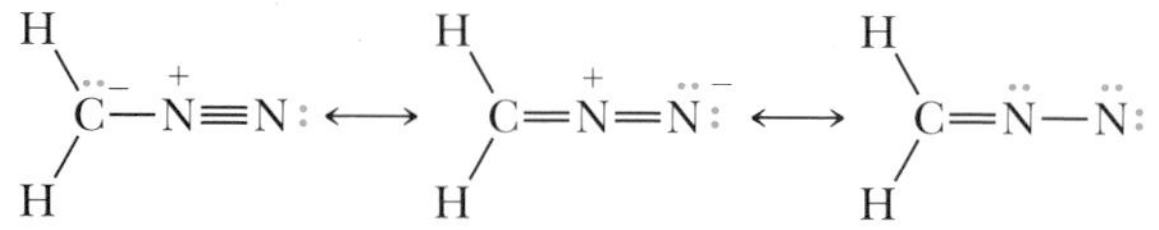

(a) Using curved arrows, show how each contributing structure is converted to the one on its right.

(b) Which contributing structure makes the largest contribution to the hybrid?

1.75 **(a)** Draw a Lewis structure for the ozone molecule, O_3. (The order of atom attachment is O—O—O, and they do not form a ring.) Chemists use ozone to cleave carbon-carbon double bonds (Section 6.5C).

(b) Draw four contributing resonance structures; include formal charges.

(c) How does the resonance model account for the fact that the length of each O—O bond in ozone (128 pm) is shorter than the O—O single bond in hydrogen peroxide (HOOH, 147 pm) but longer than the O—O double bond in the oxygen molecule (123 pm)?

Molecular Orbitals

1.76 The following two compounds are isomers; that is, they are different compounds with the same molecular formula. We discuss this type of isomerism in Chapter 5.

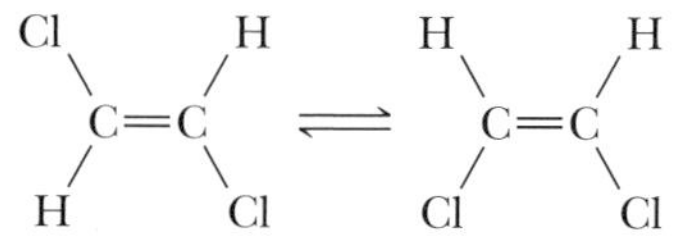

(a) Why are these different molecules that do not interconvert?

(b) Absorption of light by a double bond in a molecule excites one electron from a π molecular orbital to a π^* molecular orbital. Explain how this absorption can lead to interconversion of the two isomers.

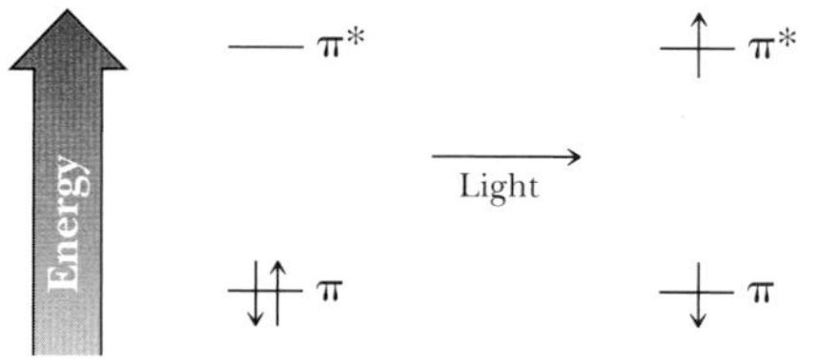

1.77 In future chapters, we will encounter carbanions—ions in which a carbon atom has three bonds and a lone pair of electrons and bears a negative charge. Draw another contributing structure for the allyl anion. Now using cartoon representations, draw the three orbitals that represent the delocalized π system (look at Figure 1.26 for a hint). Which of the three orbitals are populated with electrons?

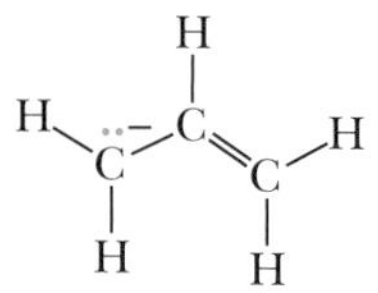

Allyl anion

1.78 Describe the bonding in PCl_5 without using *d* orbitals. As a hint, the geometry of PCl_5 is as shown.

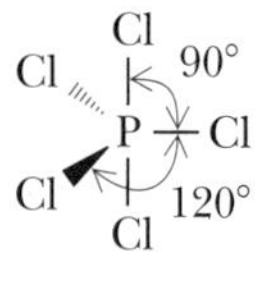

Things You Should Know I

General Conclusions from Quantum Mechanics

Between various chapters in this book we have placed small sections called "Things You Should Know." This is the first of six in total. This first one reviews a few basic ideas from Chapter 1 that will be used throughout this book, but it also presages several issues coming in the next few chapters. Very important: Your goal should be to understand, not memorize, the material presented in your organic chemistry course. Thoroughly understanding the following key principles presented in these sections will allow you to develop an intuitive feel for organic chemistry that avoids the necessity of resorting to the far less effective use of extensive memorization.

The following principles have a basis in quantum mechanics, and as such, at this level of your education should be accepted and then used as a foundation for building your understanding throughout the course. If you want to learn more about their underpinnings, then you are encouraged to take courses in quantum mechanics. Importantly for now, these simple ideas explain a great deal about the structures and properties of organic molecules, as well as the characteristic ways in which they react. As appropriate, these principles will be mentioned in the margins of various future chapters as a reminder of just how extensively we fall back on these notions to explain organic chemistry.

Predicting Structure and Bonding

1. **In most stable molecules, all the atoms will have filled valence shells.** This means that C, N, O, and the halogens will have eight electrons in their valence shells, and H atoms will have two electrons in their valence shells. This simple yet powerful principle predicts the type of bonds created (single, double, or triple) and how many lone pairs are found around the different atoms of a molecule. In general, an atom surrounded by four atoms/lone pairs will have a tetrahedral geometry, an atom surrounded by three atoms/lone pairs will have a trigonal planar geometry, and an atom surrounded by two atoms/lone pairs will have a linear geometry. You will encounter a small number of molecules containing an atom such as a C atom with only six or seven electrons in its valence shell. Atoms such as this

with only a partially filled valence shell are noteworthy and highly reactive. Note, however, that you can *never* overfill the valence shell of any atom in a molecule such as placing *more* than eight electrons in the valence shells of C, N, or O.

2. **Five- and six-membered rings are the most stable**. Molecules often contain rings of connected atoms, and by far the most common are five- and six-membered rings because the required bond angles for these rings require the least distortion (have the least strain).
3. **There are two possible arrangements of four different groups around a tetrahedral atom**. The two different arrangements are mirror images of each other, a property referred to as chirality and often compared to handedness.

Predicting Stability and Properties

1. **The most important question in organic chemistry is, "Where are the electrons?"** The answer is that electrons are generally in higher amounts around the more electronegative atoms (e.g., F, Cl, O, N). The electronegative atoms pull electron density away from the less electronegative atoms (e.g., C, H) to which they are bonded. Thus, understanding electronegativity provides a simple method of deciding which portions of a molecule have relatively high electron density and which portions have relatively low electron density. Understanding electron density distributions in molecules—in other words, where the electrons are—allows the prediction of most molecular properties and reactions because one of the key principles of organic chemistry is that molecules with relatively electron-rich and/or electron-poor areas have important and characteristic properties and patterns of reactivity. Looking forward, we will return to this idea repeatedly, and apply various terms to the electron rich and poor species: Lewis bases or nucleophiles, and Lewis acids or electrophiles, respectively.
2. **Delocalization of charge over a larger area is stabilizing**. The majority of molecules you will encounter will be neutral, but many carry negative or positive charges because they contain an imbalance in their total number of electrons and protons. In general, charges can impart reactivity to the molecules that possess them. Localized charges are the most destabilizing (highest Gibbs free energy). Delocalizing the charge over a larger area through interactions such as resonance, inductive effects, and hyperconjugation (you will learn these terms soon), is stabilizing (lowering the Gibbs free energy). In addition, it is more stabilizing to have more negative charge on a more electronegative atom (e.g., O) and more positive charge on a less electronegative atom (e.g., C).
3. **Delocalization of unpaired electron density over a larger area is stabilizing**. Most molecules you will encounter will only have atoms with filled valence shells and therefore an even number of electrons. These electrons will have paired spins. However, especially in Chapter 8, you will encounter some molecules with an unpaired electron (radicals). In general, unpaired electron density is destabilizing (higher Gibbs free energy), dramatically increasing the reactivity of the molecules that possess it. Highly localized unpaired electron density is the most destabilizing. Delocalizing the unpaired electron density over a larger area through interactions such as resonance and hyperconjugation is stabilizing (lowering the Gibbs free energy).
4. **Delocalization of π electron density over a larger area is stabilizing**. Pi electron density delocalization occurs through overlapping $2p$ orbitals, so to take part in π electron density delocalization, atoms must be sp^2 or sp hybridized and reside in the same plane. π electron delocalization can involve large numbers of such atoms. Pi electron density cannot delocalize onto or through sp^3 hybridized atoms since an sp^3 atom has no $2p$ orbital. Aromaticity is a special type of π electron density delocalization involving rings and a specific number of π electrons, and is the most stable form of π electron density delocalization.

While some of the terms discussed here are not yet familiar to you (e.g., radicals, aromaticity, and nucleophiles), the overarching issue presented here is that delocalization (spreading, so to speak) of charges and electrons over larger numbers of atoms is found to be stabilizing by a quantum mechanical analysis. This unifying principle will guide your understanding of many structures and reactions involved in organic chemistry.

Predicting Reactions

1. **Reactions will occur if the products are more stable than the reactants and the energy barrier is low enough**. Reactions will be favorable if the products are of lower Gibbs free energy than the starting materials—for example, if stronger bonds are made than are broken, if a weaker acid or base is formed in the product, or if more molecules are created than consumed. A favorable reaction will occur if the energy barrier (Gibbs free energy of activation) has no step in the mechanism containing a transition state of such high energy that it cannot be formed at the temperature being used.
2. **Functional groups react the same in different molecules**. Chemists classify groups of atoms that take part in characteristic reactions as functional groups, such as —OH groups, carbonyls, haloalkanes, or enolates. Although functional groups can influence each other in some complex molecules, functional groups serve as the most generally useful organizing principle in organic chemistry. This is because functional groups mostly react the same, albeit in different molecules. Recognizing functional groups and understanding their characteristic reactions are key to being able to predict reactions in molecules you have never seen before. This is why the majority of chapters in this book focus on specific functional groups.
3. **A reaction mechanism describes the sequence of steps occurring during a reaction**. Gaining an intuitive understanding of reaction mechanisms allows the accurate prediction of products for new reactions, a key goal of any organic chemistry course. Most mechanisms involve combinations of the four elementary steps: (1) make a bond; (2) break a bond; (3) add a proton; or (4) take a proton away. Learning how to predict which step (or perhaps some other elementary step) is appropriate at a given stage in a mechanism requires recognition of the properties of the participating molecules. When writing mechanisms, arrows are used to indicate the redistribution of electrons during each step. We have a separate "Things You Should Know" section on just this topic coming up in a few chapters, along with practice problems.
4. **Most bond-making steps in reaction mechanisms involve nucleophiles reacting with electrophiles**. Nucleophiles are molecules that have a lone pair or bond that can donate electrons to make a new bond, usually corresponding to an area of relatively high electron density. Electrophiles contain atoms that can accept the new bond, usually corresponding to areas of relatively low electron density or even an unfilled valence shell. Note that often a bond is broken in the electrophile to make room for the new bond being made. We have another "Things You Should Know" section on just this topic coming up in a few chapters.

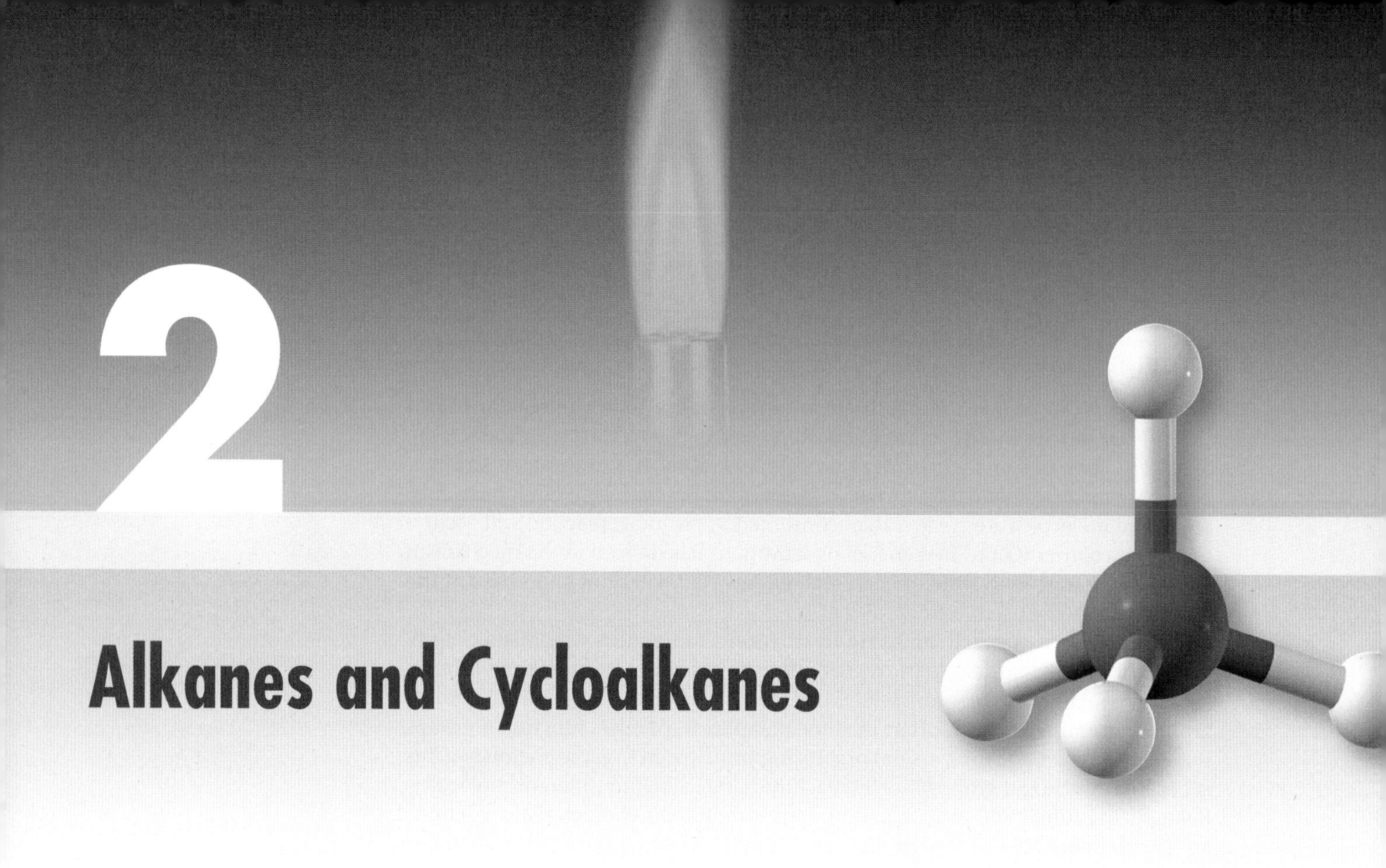

Bunsen burners burn natural gas, which is primarily methane with small amounts of ethane, propane, and butane (Section 2.9A). **Inset:** a model of methane, the major component of natural gas.

2

Alkanes and Cycloalkanes

Outline

2.1 The Structure of Alkanes
2.2 Constitutional Isomerism in Alkanes
2.3 Nomenclature of Alkanes and the IUPAC System
2.4 Cycloalkanes
2.5 Conformations of Alkanes and Cycloalkanes
HOW TO: Draw Alternative Chair Conformations of Cyclohexane
2.6 *Cis,Trans* Isomerism in Cycloalkanes and Bicycloalkanes
HOW TO: Convert Planar Cyclohexanes to Chair Cyclohexanes
2.7 Physical Properties of Alkanes and Cycloalkanes
2.8 Reactions of Alkanes
2.9 Sources and Importance of Alkanes

In this chapter, we begin our study of organic compounds with the physical and chemical properties of alkanes, the simplest types of organic compounds. Actually, alkanes are members of a larger group of organic compounds called hydrocarbons. A **hydrocarbon** is a compound composed of only carbon and hydrogen. Figure 2.1 shows the four classes of hydrocarbons, along with the characteristic pattern of bonding between the carbon atoms in each.

Alkanes are **saturated hydrocarbons**; that is, they contain only carbon-carbon single bonds. In this context, *saturated* means that each carbon has the maximum number of hydrogens bonded to it. We often refer to alkanes as **aliphatic hydrocarbons** because the physical properties of the higher members of this class resemble those of the long carbon-chain molecules we find in animal fats and plant oils (Greek: *aleiphar,* fat or oil).

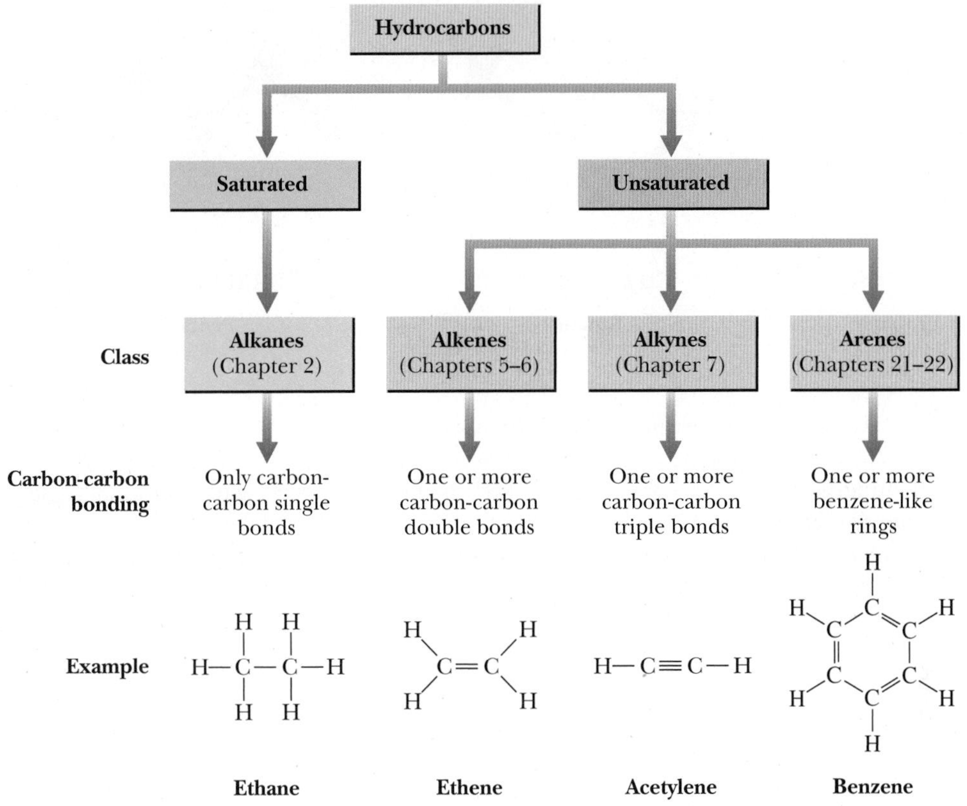

FIGURE 2.1 The four classes of hydrocarbons.

A hydrocarbon that contains one or more carbon-carbon double bonds, triple bonds, or benzene rings, is classified as an **unsaturated hydrocarbon**. We study alkanes (saturated hydrocarbons) in this chapter. We study alkenes and alkynes (both unsaturated hydrocarbons) in Chapters 5, 6, and 7, and we study arenes (also unsaturated hydrocarbons) in Chapters 21 and 22.

2.1 The Structure of Alkanes

Methane (CH_4) and ethane (C_2H_6) are the first two members of the alkane family. Figure 2.2 shows Lewis structures and molecular models for these molecules. The shape of methane is tetrahedral, and all H—C—H bond angles are 109.5°. Each carbon atom in ethane is also tetrahedral, and all bond angles are approximately 109.5°.

Although the three-dimensional shapes of larger alkanes are more complex than those of methane and ethane, the four bonds about each carbon are still arranged in a tetrahedral manner and all bond angles are approximately 109.5°.

The next three alkanes are propane, butane, and pentane. In the following representations, these hydrocarbons are drawn first as condensed structural formulas that show all carbons and hydrogens. They are also drawn in an even more abbreviated form called a **line-angle formula**. In a line-angle formula, each vertex and line ending represents a carbon atom. Although we do not show hydrogen atoms in line-angle formulas, we assume they are there in sufficient numbers to give each carbon four bonds.

Line-angle formula
An abbreviated way to draw structural formulas in which vertices and line endings represent carbons.

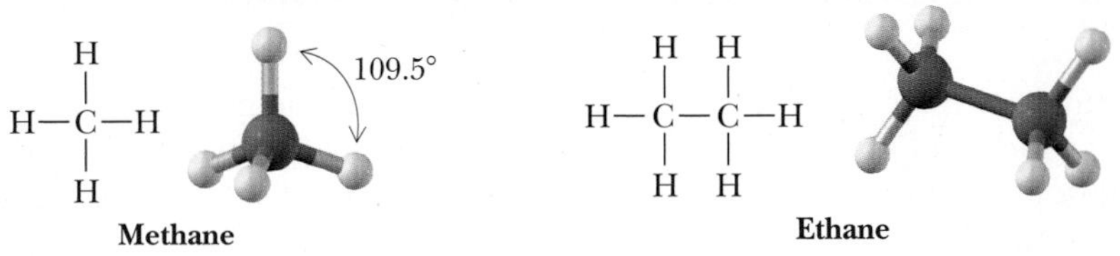

FIGURE 2.2 Methane and ethane. Lewis structures and ball-and-stick models.

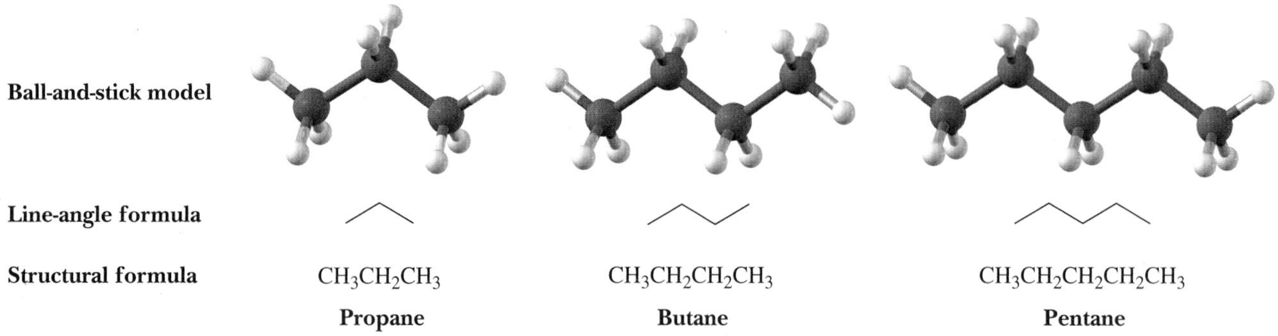
Ball-and-stick model
Line-angle formula
Structural formula
CH3CH2CH3
CH3CH2CH2CH3
CH3CH2CH2CH2CH3
Propane
Butane
Pentane

We can write structural formulas for alkanes in still another abbreviated form. The structural formula of pentane, for example, contains three CH_2 (methylene) groups in the middle of the chain. We can collect them and write the structural formula of pentane as $CH_3(CH_2)_3CH_3$. Table 2.1 gives the names and molecular formulas of the first 20 alkanes. Note that the names of all these alkanes end in *-ane*. We will have more to say about naming alkanes in Section 2.3.

Table 2.1 Names, Molecular Formulas, and Condensed Structural Formulas for the First 20 Alkanes with Unbranched Chains

Name	Molecular Formula	Condensed Structural Formula	Name	Molecular Formula	Condensed Structural Formula
Methane	CH_4	CH_4	Undecane	$C_{11}H_{24}$	$CH_3(CH_2)_9CH_3$
Ethane	C_2H_6	CH_3CH_3	Dodecane	$C_{12}H_{26}$	$CH_3(CH_2)_{10}CH_3$
Propane	C_3H_8	$CH_3CH_2CH_3$	Tridecane	$C_{13}H_{28}$	$CH_3(CH_2)_{11}CH_3$
Butane	C_4H_{10}	$CH_3(CH_2)_2CH_3$	Tetradecane	$C_{14}H_{30}$	$CH_3(CH_2)_{12}CH_3$
Pentane	C_5H_{12}	$CH_3(CH_2)_3CH_3$	Pentadecane	$C_{15}H_{32}$	$CH_3(CH_2)_{13}CH_3$
Hexane	C_6H_{14}	$CH_3(CH_2)_4CH_3$	Hexadecane	$C_{16}H_{34}$	$CH_3(CH_2)_{14}CH_3$
Heptane	C_7H_{16}	$CH_3(CH_2)_5CH_3$	Heptadecane	$C_{17}H_{36}$	$CH_3(CH_2)_{15}CH_3$
Octane	C_8H_{18}	$CH_3(CH_2)_6CH_3$	Octadecane	$C_{18}H_{38}$	$CH_3(CH_2)_{16}CH_3$
Nonane	C_9H_{20}	$CH_3(CH_2)_7CH_3$	Nonadecane	$C_{19}H_{40}$	$CH_3(CH_2)_{17}CH_3$
Decane	$C_{10}H_{22}$	$CH_3(CH_2)_8CH_3$	Eicosane	$C_{20}H_{42}$	$CH_3(CH_2)_{18}CH_3$

Alkanes have the general molecular formula C_nH_{2n+2}. Thus, given the number of carbon atoms in an alkane, we can determine the number of hydrogens in the molecule and its molecular formula. For example, decane, with ten carbon atoms, must have $(2 \times 10) + 2 = 22$ hydrogen atoms and a molecular formula of $C_{10}H_{22}$.

2.2 Constitutional Isomerism in Alkanes

Constitutional isomers

Compounds with the same molecular formula but a different connectivity of their atoms.

Constitutional isomers are compounds that have the same molecular formula but different structural formulas. By "different structural formulas," we mean that constitutional isomers differ in the kinds of bonds they have (single, double, or triple) and/or in the connectivity of their atoms.

For the molecular formulas CH_4, C_2H_6, and C_3H_8, only one connectivity is possible. For the molecular formula C_4H_{10}, two connectivities are possible. In one of these, named butane, the four carbons are bonded in a chain; in the other, named 2-methylpropane, three carbons are bonded in a chain with the fourth carbon as a branch on the chain.

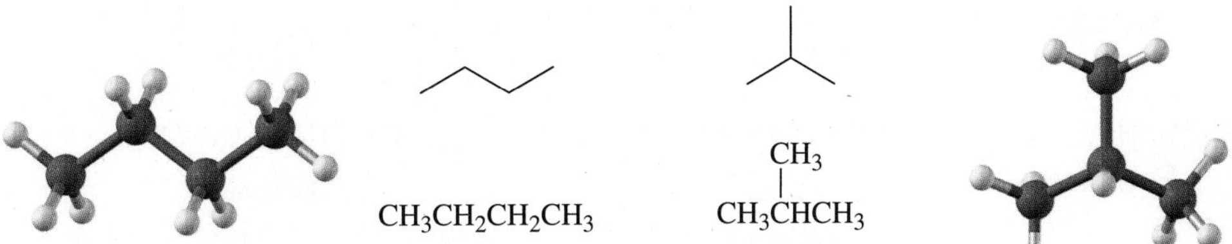
CH3CH2CH2CH3
CH3
CH3CHCH3

Butane (bp −0.5°C) **2-Methylpropane** (bp −11.6°C)

Butane and 2-methylpropane are constitutional isomers; they are different compounds and have different physical and chemical properties. Their boiling points, for example, differ by approximately 11°C.

In Section 1.3, we encountered several examples of constitutional isomers. We saw, for example, that there are two alcohols with the molecular formula C_3H_8O, two aldehydes with the molecular formula C_4H_8O, and two carboxylic acids with the molecular formula $C_4H_8O_2$.

To determine whether two or more structural formulas represent constitutional isomers (i.e., different compounds with the same molecular formula), write the molecular formula of each and then compare them. All compounds that have the same molecular formula but different structural formulas (different connectivities of their atoms) are constitutional isomers.

Example 2.1 Constitutional Isomers

Do the condensed formulas in each pair represent the same compound or constitutional isomers?

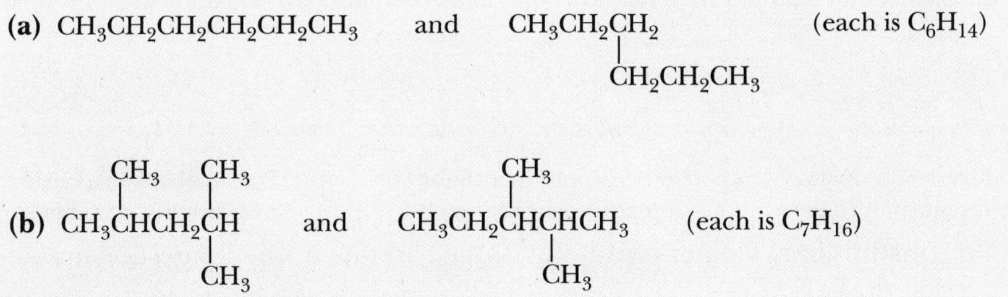
(a) CH3CH2CH2CH2CH2CH3 and CH3CH2CH2
CH2CH2CH3
(each is C6H14)
(b) CH3 CH3
CH3CHCH2CH
CH3
and
CH3
CH3CH2CHCHCH3
CH3
(each is C7H16)

Solution

(a) The molecules are drawn here as both condensed structural formulas and line-angle formulas. Each formula has an unbranched chain of six carbons; the two are identical and represent the same compound.

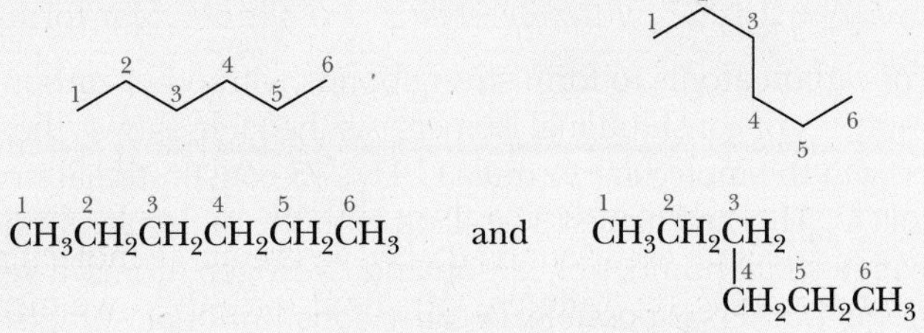
1 2 3 4 5 6
CH3CH2CH2CH2CH2CH3
and
1 2 3
CH3CH2CH2
4 5 6
CH2CH2CH3

(Continued)

(b) Each formula has a chain of five carbons with two $—CH_3$ branches. Although the branches are identical, they are at different locations on the chains; these formulas represent constitutional isomers.

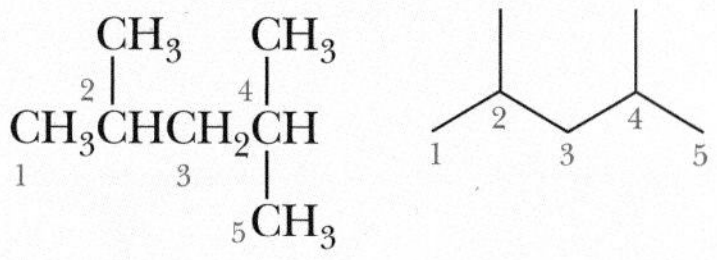

and

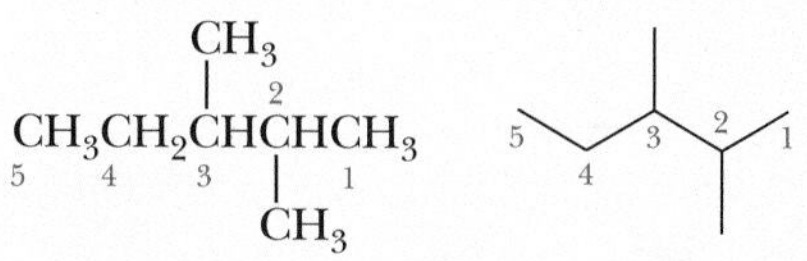

Problem 2.1

Do the line-angle formulas in each pair represent the same compound or constitutional isomers?

(a) 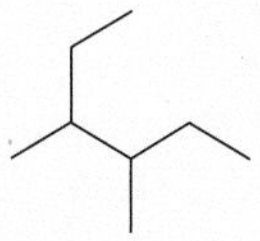 and 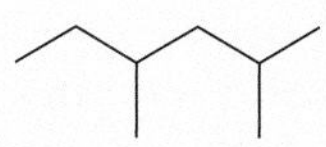

(b) 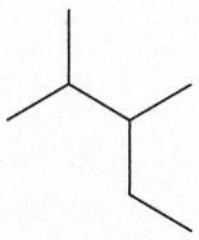 and 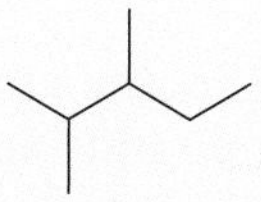

Example 2.2 Line-Angle Formulas

Write line-angle formulas for the five constitutional isomers with the molecular formula C_6H_{14}.

Solution

In solving problems of this type, you should devise a strategy and then follow it. Here is one such strategy. First, draw a line-angle formula for the constitutional isomer with all six carbons in an unbranched chain. Then draw line-angle formulas for all constitutional isomers with five carbons in a chain and one carbon as a branch on the chain. Finally, draw line-angle formulas for all constitutional isomers with four carbons in a chain and two carbons as branches.

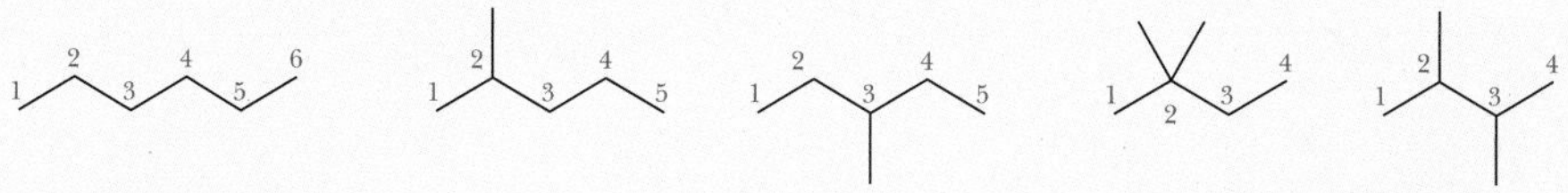

Six carbons in an unbranched chain

Five carbons in a chain; one carbon as a branch

Four carbons in a chain; two carbons as branches

No constitutional isomers with only three carbons in the longest chain are possible for C_6H_{14}.

Problem 2.2

Draw line-angle formulas for the three constitutional isomers with the molecular formula C_5H_{12}.

Carbon Atoms	Constitutional Isomers
1	0
5	3
10	75
15	4,347
25	36,797,588

The ability of carbon atoms to form strong bonds with other carbon atoms results in a staggering number of constitutional isomers. As the table shows, there are 3 constitutional isomers with the molecular formula C_5H_{12}, 75 constitutional isomers with the molecular formula $C_{10}H_{22}$, and almost 37 million with the molecular formula $C_{25}H_{52}$.

Thus, for even a small number of carbon and hydrogen atoms, a very large number of constitutional isomers is possible. Because constitutional isomers have different chemical properties, a rich diversity of chemistry is possible within these sets.

2.3 Nomenclature of Alkanes and the IUPAC System

A. The IUPAC System

Ideally, every organic compound should have a name from which its structural formula can be drawn. For this purpose, chemists have adopted a set of rules established by the International Union of Pure and Applied Chemistry (IUPAC).

The IUPAC name of an alkane with an unbranched chain of carbon atoms consists of two parts: (1) a prefix that indicates the number of carbon atoms in the chain and (2) the suffix *-ane* to show that the compound is a saturated hydrocarbon. Table 2.2 gives the prefixes used to show the presence of 1 to 20 carbon atoms.

The first four prefixes listed in Table 2.2 were chosen by the IUPAC because they were well established in the language of organic chemistry. In fact, they were well established even before there were hints of the structural theory underlying the discipline. For example, the prefix *but-* appears in the name *butyric acid,* a compound of four carbon atoms formed by air oxidation of butter (Latin: *butyrum,* butter). Prefixes to show five or more carbons are derived from Greek or Latin numbers. See Table 2.1 for the names, molecular formulas, and condensed structural formulas for the first 20 unbranched alkanes.

Table 2.2 **Prefixes Used in the IUPAC System to Show the Presence of 1 to 20 Carbon Atoms in an Unbranched Chain**

Prefix	Number of Carbon Atoms	Prefix	Number of Carbon Atoms
meth-	1	undec-	11
eth-	2	dodec-	12
prop-	3	tridec-	13
but-	4	tetradec-	14
pent-	5	pentadec-	15
hex-	6	hexadec-	16
hept-	7	heptadec-	17
oct-	8	octadec-	18
non-	9	nonadec-	19
dec-	10	eicos-	20

The IUPAC name of an alkane with a branched chain consists of a parent name that indicates the longest chain of carbon atoms in the compound and substituent names that indicate the groups bonded to the parent chain.

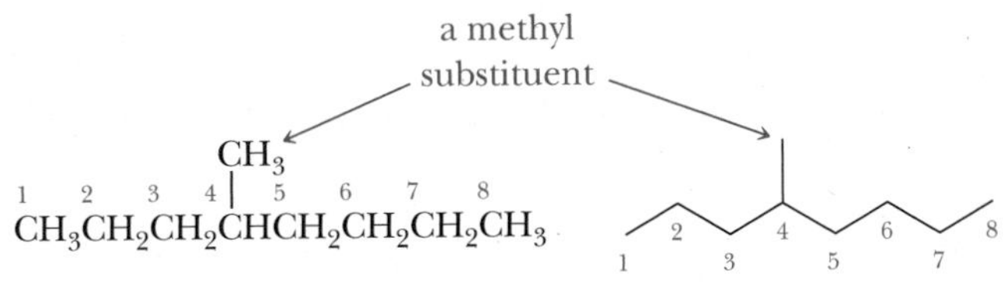

4-Methyloctane

Alkyl group

A group derived by removing a hydrogen from an alkane; given the symbol R—.

A substituent group derived from an alkane by the removal of a hydrogen atom is called an **alkyl group**; it is commonly represented by the symbol R—. We name alkyl groups by dropping the *-ane* from the name of the parent alkane and adding the suffix *-yl*. The substituent derived from methane, for example, is methyl, CH_3—, and that derived from ethane is ethyl, CH_3CH_2—.

The rules of the IUPAC system for naming alkanes follow:

1. The name for an alkane with an unbranched chain of carbon atoms consists of a prefix showing the number of carbon atoms in the chain and the ending *-ane*.
2. For branched-chain alkanes, select the longest chain of carbon atoms as the parent chain; its name becomes the root name.
3. Give each substituent on the parent chain a name and a number. The number shows the carbon atom of the parent chain to which the substituent is bonded. Use a hyphen to connect the number to the name.

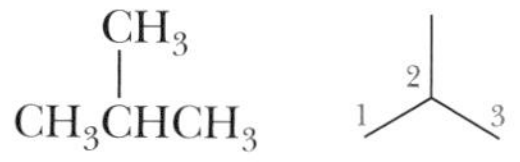

2-Methylpropane

4. If there is one substituent, number the parent chain from the end that gives the substituent the lower number.

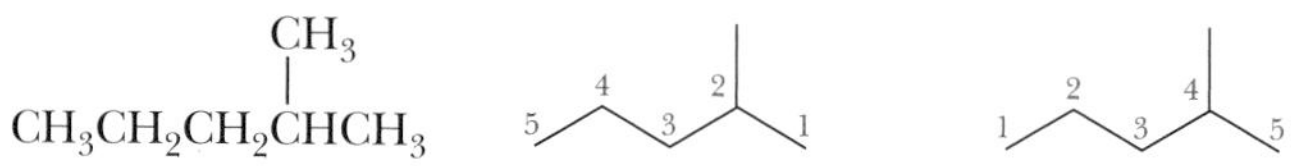

2-Methylpentane (not 4-Methylpentane)

5. If there are two or more identical substituents, number the parent chain from the end that gives the lower number to the substituent encountered first. The number of times the substituent occurs is indicated by the prefix *di-, tri-, tetra-, penta-, hexa-,* and so on. A comma is used to separate position numbers.

2,4-Dimethylhexane (not 3,5-Dimethylhexane)

6. If there are two or more different substituents, list them in alphabetical order and number the chain from the end that gives the lower number to the substituent encountered first. If there are different substituents in equivalent positions on opposite ends of the parent chain, give the substituent of lower alphabetical order the lower number.

3-Ethyl-5-methylheptane (not 3-Methyl-5-ethylheptane)

7. The prefixes *di-, tri-, tetra-,* and so on, are not included in alphabetizing. Alphabetize the names of the substituents first and then insert these prefixes. In the following example, the alphabetizing parts are *ethyl* and *methyl,* not ethyl and dimethyl.

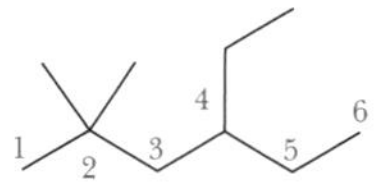

4-Ethyl-2,2-dimethylhexane
(not 2,2-Dimethyl-4-ethylhexane)

8. Where there are two or more parent chains of identical length, choose the parent chain with the greater number of substituents.

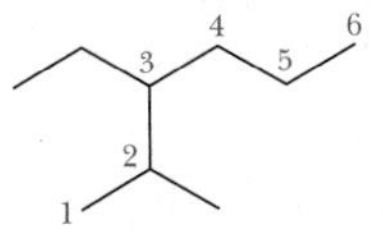

3-Ethyl-2-methylhexane

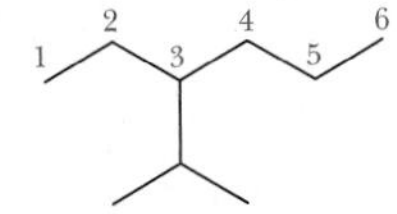

(not 3-Isopropylhexane)

Substituents are named following this same set of rules. Those with unbranched chains are named by dropping *-ane* from the name of the parent alkane and replacing it with *-yl*. Thus, unbranched alkyl substituents are named *methyl, ethyl, propyl, butyl, pentyl,* and so forth. Substituents with branched chains are named according to rules 2 and 3. The IUPAC names and structural formulas for unbranched and branched alkyl groups containing one to five carbon atoms are given in Table 2.3. Also given in parentheses are common names for the alkyl substituents. Their common names are so deeply entrenched in organic chemistry that in the official IUPAC nomenclature system, it is acceptable to use either the formal IUPAC name (such as 1-methylethyl) or the common name (in this case, isopropyl) for the alkyl substituents given in Table 2.3.

Table 2.3 Names for Alkyl Groups with One to Five Carbons. Common Names and Their Abbreviations Are Given in Parentheses

Name	Condensed Structural Formula	Name	Condensed Structural Formula
Methyl (Me)	$-CH_3$	1,1-Dimethylethyl (*tert*-butyl, *t*-Bu)	$-\overset{CH_3}{\underset{CH_3}{C}}CH_3$
Ethyl (Et)	$-CH_2CH_3$		
Propyl (Pr)	$-CH_2CH_2CH_3$	Pentyl	$-CH_2CH_2CH_2CH_2CH_3$
1-Methylethyl (isopropyl, iPr)	$-\underset{CH_3}{CH}CH_3$	3-Methylbutyl (isopentyl)	$-CH_2CH_2\underset{CH_3}{CH}CH_3$
Butyl (Bu)	$-CH_2CH_2CH_2CH_3$	2-Methylbutyl	$-CH_2\underset{CH_3}{CH}CH_2CH_3$
2-Methylpropyl (isobutyl, iBu)	$-CH_2\underset{CH_3}{CH}CH_3$	2,2-Dimethylpropyl (neopentyl)	$-CH_2\overset{CH_3}{\underset{CH_3}{C}}CH_3$
1-Methylpropyl (*sec*-butyl, *s*-Bu)	$-\underset{CH_3}{CH}CH_2CH_3$		

Example 2.3 IUPAC Nomenclature I

Write the IUPAC and common names for these alkanes.

(a) 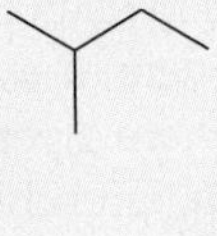

(b) 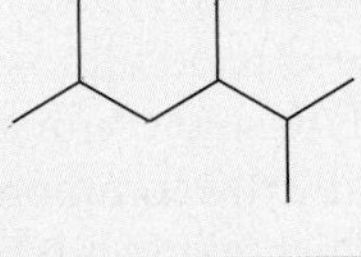

(Continued)

Solution

Number the longest chain in each compound from the end of the chain toward the substituent that is encountered first. For (a), the longest chain is four carbons (a butane) with a methyl group on carbon 2. For (b), the longest chain is seven carbons (a heptane), with substituents on carbons 2 and 4.

(a)

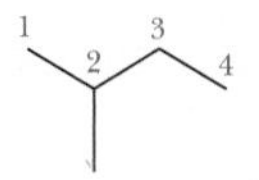

2-Methylbutane

(b)

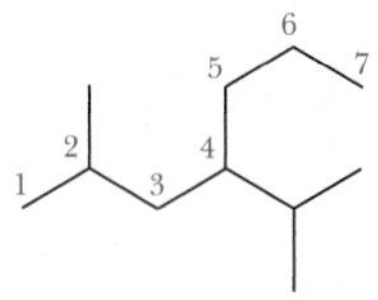

2-Methyl-4-(1-methylethyl)heptane or **4-Isopropyl-2-methylheptane**

Problem 2.3

Write IUPAC names for these alkanes.

(a)

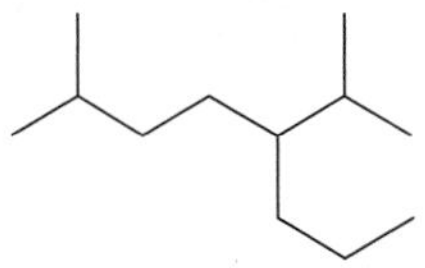

(b)

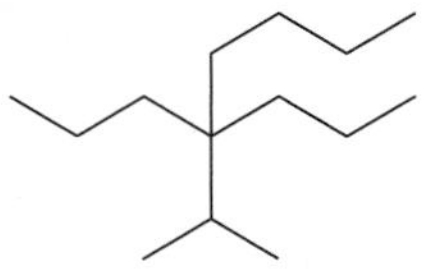

B. Common Names

In an alternative system known as common nomenclature, the total number of carbon atoms in an alkane, regardless of their arrangement, determines the name. The first three alkanes are methane, ethane, and propane. All alkanes with the molecular formula C_4H_{10} are called butanes, all those with the molecular formula C_5H_{12} are called pentanes, all those with the molecular formula C_6H_{14} are called hexanes, and so forth. The fact that an alkane chain is unbranched is sometimes indicated by the prefix *n*- (normal); an example is *n*-pentane for $CH_3CH_2CH_2CH_2CH_3$. For branched-chain alkanes beyond propane, *iso*- indicates that one end of an otherwise unbranched chain terminates in a $(CH_3)_2CH$— group and *neo*- indicates that it terminates in —$C(CH_3)_3$. Following are examples of common names.

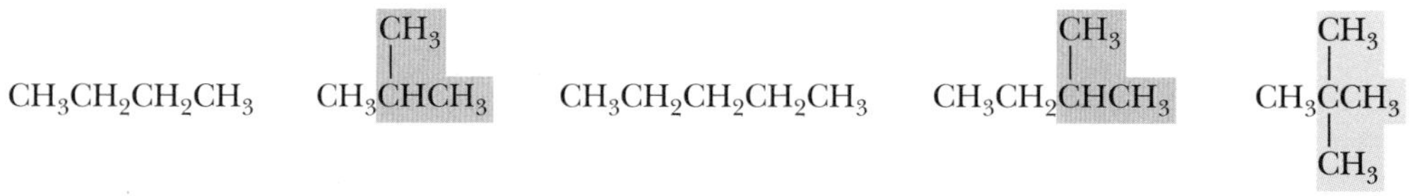

Butane | **Isobutane** | **Pentane** | **Isopentane** | **Neopentane**

This system of common names has no good way of naming other branching patterns; for more complex alkanes, it is necessary to use the more flexible IUPAC system of nomenclature.

In this text, we concentrate on IUPAC names. However, we also use common names, especially when the common name is used almost exclusively in everyday discussions among chemists. When both IUPAC and common names are given in the text, we give the IUPAC name first, followed by the common name in parentheses. In this way, you should have no doubt about which name is which.

C. The IUPAC System—A General System of Nomenclature

The naming of alkanes and cycloalkanes in Section 2.3A illustrates the application of the IUPAC system of nomenclature to a specific class of organic compounds. Now let us describe the general approach of the IUPAC system. The name we give to any compound with a chain of carbon atoms consists of three parts: a **prefix**, an **infix** (a modifying element inserted into a word), and a **suffix**. Each part provides specific information about the structure of the compound.

1. The prefix indicates the number of carbon atoms in the parent chain. Prefixes that show the presence of 1 to 20 carbon atoms in an unbranched chain are given in Table 2.2.
2. The infix indicates the nature of the carbon-carbon bonds in the parent chain.

Infix	Nature of Carbon-Carbon Bonds in the Parent Chain
-an-	all single bonds
-en-	one or more double bonds
-yn-	one or more triple bonds

3. The suffix indicates the class of compound to which the substance belongs.

Suffix	Class of Compound
-e	hydrocarbon
-ol	alcohol
-al	aldehyde
-amine	amine
-one	ketone
-oic acid	carboxylic acid

Example 2.4 IUPAC Nomenclature II

Following are IUPAC names and structural formulas for four compounds:

(a) $CH_2{=}CHCH_3$ **Propene**

(b) CH_3CH_2OH **Ethanol**

(c) $CH_3CH_2CH_2CH_2\overset{\overset{\displaystyle O}{\|}}{C}OH$ **Pentanoic acid**

(d) $HC{\equiv}CH$ **Ethyne**

Divide each name into a prefix, an infix, and a suffix and specify the information about the structural formula that is contained in each part of the name.

Solution

(a) prop-en-e
- prop: three carbon atoms
- en: a carbon-carbon double bond
- e: a hydrocarbon

(b) eth-an-ol
- eth: two carbon atoms
- an: only carbon-carbon single bonds
- ol: an alcohol

(c) pent-an-oic acid
- pent: five carbon atoms
- an: only carbon-carbon single bonds
- oic acid: a carboxylic acid

(d) eth-yn-e
- eth: two carbon atoms
- yn: a carbon-carbon triple bond
- e: a hydrocarbon

D. Classification of Carbon and Hydrogen Atoms

We classify a carbon atom as primary (1°), secondary (2°), tertiary (3°), or quaternary (4°) depending on the number of carbon atoms bonded to it. A carbon bonded to one carbon atom is a primary carbon; a carbon bonded to two carbon atoms is a secondary carbon, and so forth. For example, propane contains two primary carbons and one secondary carbon; 2-methylpropane contains three primary carbons and one tertiary carbon; and 2,2,4-trimethylpentane contains five primary carbons, one secondary carbon, one tertiary carbon, and one quaternary carbon.

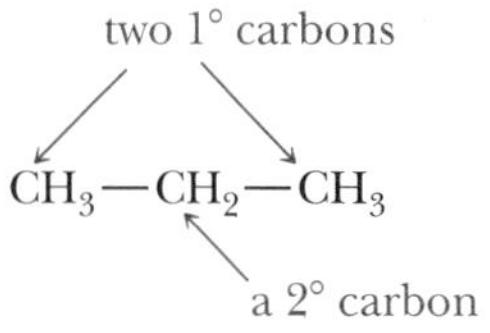
two 1° carbons
$CH_3-CH_2-CH_3$
a 2° carbon

Propane

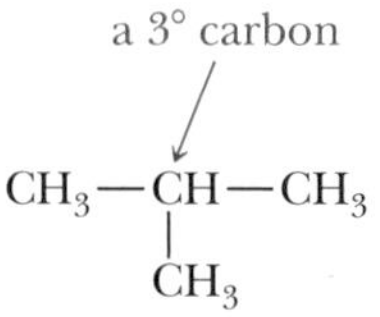
a 3° carbon
$CH_3-CH-CH_3$
CH_3

2-Methylpropane

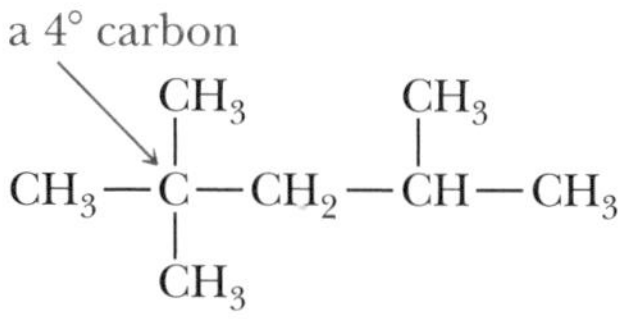
a 4° carbon
CH_3
CH_3
$CH_3-C-CH_2-CH-CH_3$
CH_3

2,2,4-Trimethylpentane

Hydrogens are also classified as primary, secondary, or tertiary depending on the type of carbon to which each is bonded. Those bonded to a primary carbon are classified as primary hydrogens, those bonded to a secondary carbon are secondary hydrogens, and those bonded to a tertiary carbon are tertiary hydrogens.

2.4 Cycloalkanes

A hydrocarbon that contains carbon atoms joined to form a ring is called a **cyclic hydrocarbon**. When all carbons of the ring are saturated, the hydrocarbon is called a **cycloalkane**.

Cycloalkane

A saturated hydrocarbon that contains carbons joined to form a ring.

A. Structure and Nomenclature

Cycloalkanes of ring sizes from 3 to over 30 are found in nature, and in principle, there is no limit to ring size. Five-membered rings (cyclopentanes) and six-membered rings (cyclohexanes) are especially common and will receive special attention. Figure 2.3 shows structural formulas of cyclopropane, cyclobutane, cyclopentane, and cyclohexane. When writing structural formulas for cycloalkanes, chemists rarely show all carbons and hydrogens. Rather, they use line-angle formulas to represent cycloalkane rings. Each ring is represented by a regular polygon that has the same number of sides as there are carbon atoms in the ring. For example, chemists represent cyclobutane by a square, cyclopentane by a pentagon, and cyclohexane by a hexagon.

FIGURE 2.3 Examples of cycloalkanes.

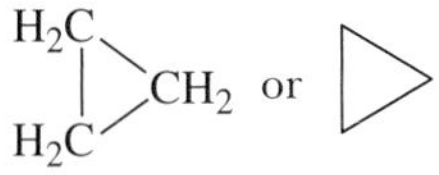

Cyclopropane
C_3H_6

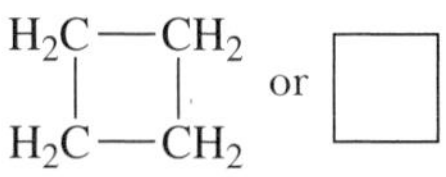

Cyclobutane
C_4H_8

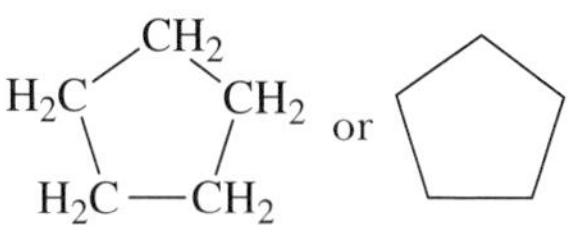

Cyclopentane
C_5H_{10}

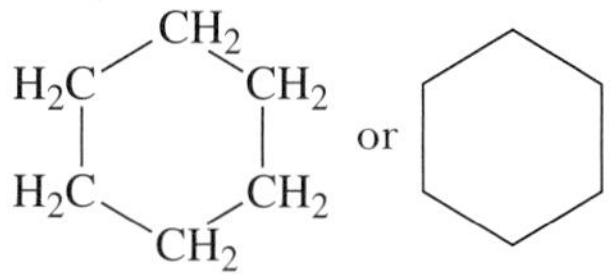

Cyclohexane
C_6H_{12}

Cycloalkanes contain two fewer hydrogen atoms than an alkane with the same number of carbon atoms and have the general formula C_nH_{2n}. For example, compare the molecular formulas of cyclohexane, C_6H_{12}, and hexane, C_6H_{14}.

To name a cycloalkane, prefix the name of the corresponding open-chain alkane with *cyclo-* and name each substituent on the ring. If there is only one substituent on the cycloalkane ring, there is no need to give it a number. If there are two substituents, number the ring by beginning with the substituent of lower alphabetical order. If there are three or more substituents, number the ring to give the substituents the lowest set of numbers and list them in alphabetical order.

Example 2.5 IUPAC Nomenclature III

Write the molecular formula and the IUPAC name for each cycloalkane.

(a) 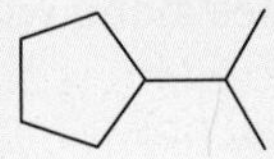 **(b)** 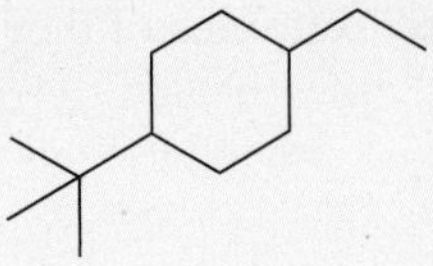 **(c)** 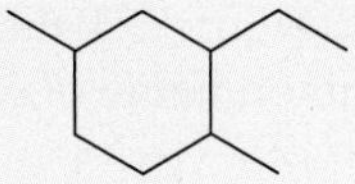

Solution

(a) First, replace each vertex and line terminus with a carbon and then add hydrogens as necessary to give each carbon four bonds. The molecular formula of this compound is C_8H_{16}. Because there is only one substituent on the ring, there is no need to number the atoms of the ring. This compound's name is (1-methylethyl)cyclopentane. The substituent also could be named isopropyl, giving the alternative IUPAC name isopropylcyclopentane.

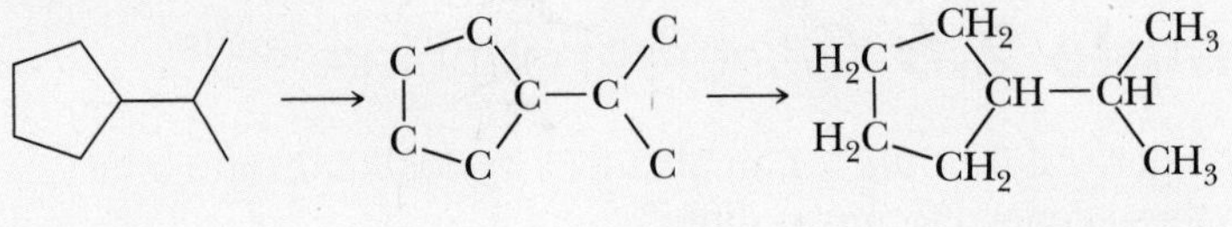

C_8H_{16}

(b) The two substituents are ethyl and 1,1-dimethylethyl, and the IUPAC name of the cycloalkane is 1-ethyl-4-(1,1-dimethylethyl)cyclohexane. The substituents also could be named ethyl and *tert*-butyl, giving the cycloalkane the alternative IUPAC name 1-*tert*-butyl-4-ethylcyclohexane. Its molecular formula is $C_{12}H_{24}$.

(c) Number the ring to give the three substituents the lowest set of numbers and then list them in alphabetical order. The name of this compound is 2-ethyl-1,4-dimethylcyclohexane, and its molecular formula is $C_{10}H_{20}$.

Problem 2.4

Watch a video explanation

Combine the proper prefix, infix, and suffix and write the IUPAC name for each compound.

(a) $CH_3\overset{\overset{O}{\|}}{C}CH_3$ **(b)** $CH_3(CH_2)_3\overset{\overset{O}{\|}}{C}H$ **(c)** 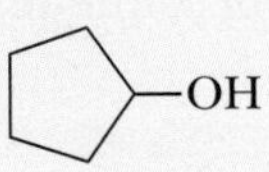 **(d)** 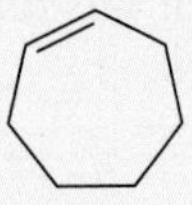

Problem 2.5

Write the molecular formula, IUPAC name, and common name for each cycloalkane.

(a) 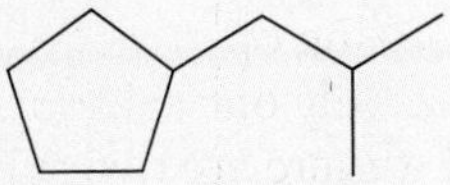 **(b)** 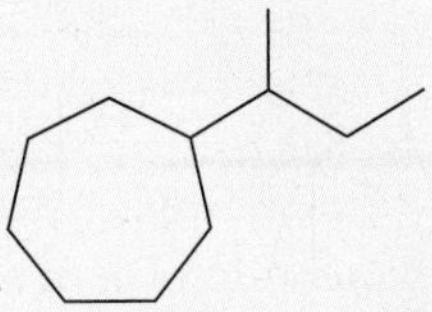 **(c)** 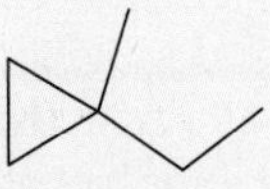

B. Bicycloalkanes

Bicycloalkane
An alkane containing two rings that share two carbons.

An alkane that contains two rings that share two carbon atoms is classified as a **bicycloalkane**. The shared carbon atoms are called **bridgehead carbons**, and the carbon chain connecting them is called a **bridge**. The general formula of a bicycloalkane is C_nH_{2n-2}. Figure 2.4 shows three examples of bicycloalkanes along with the IUPAC and common name of each.

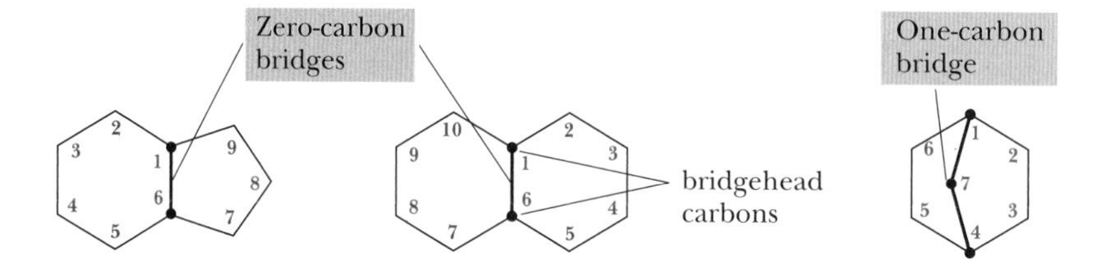

Zero-carbon bridges
One-carbon bridge
bridgehead carbons

Bicyclo[4.3.0]nonane (Hydrindane) **Bicyclo[4.4.0]decane (Decalin)** **Bicyclo[2.2.1]heptane (Norbornane)**

FIGURE 2.4 Examples of bicycloalkanes.

Example 2.6 General Formulas

Write the general formula for an alkane, a cycloalkane, and a bicycloalkane. How do these general formulas differ?

Solution

General formulas are C_nH_{2n+2} for an alkane, C_nH_{2n} for a cycloalkane, and C_nH_{2n-2} for a bicycloalkane. Each general formula in this series has two fewer hydrogens than the previous member of the series.

Problem 2.6

Write molecular formulas for each bicycloalkane, given its number of carbon atoms.

(a) Hydrindane (9 carbons) **(b)** Decalin (10 carbons)

(c) Norbornane (7 carbons)

The IUPAC has a set of rules for naming bicycloalkanes, but they are beyond the scope of this text. We will continue to refer to most bicycloalkanes by their common names.

Example 2.7 Molecular Formulas

Following are line-angle formulas and common names for three bicyclic compounds. Write the molecular formula of each compound.

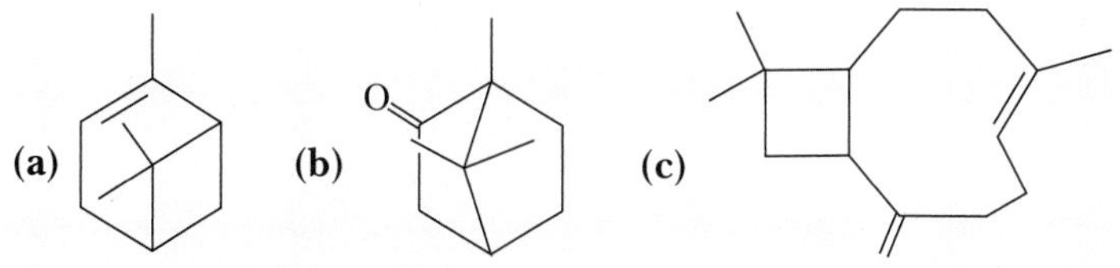

(a)
(b)
O
(c)

α-Pinene **Camphor** **Caryophyllene**

Solution

(a) The molecular formula of α-pinene is $C_{10}H_{16}$. α-Pinene is a major component, often as high as 65% by volume, of pine oil and turpentine.

(b) The molecular formula of camphor is $C_{10}H_{16}O$. Camphor, obtained from the camphor tree, *Cinnamonium camphora*, is used in the manufacture of certain plastics, lacquers, and varnishes.

(c) The molecular formula of caryophyllene is $C_{15}H_{24}$. Caryophyllene is one of the fragrant components of oil of cloves.

Problem 2.7

Following are the line-angle formulas and names of four bicycloalkanes. Write the molecular formula of each compound. Which of these compounds are constitutional isomers?

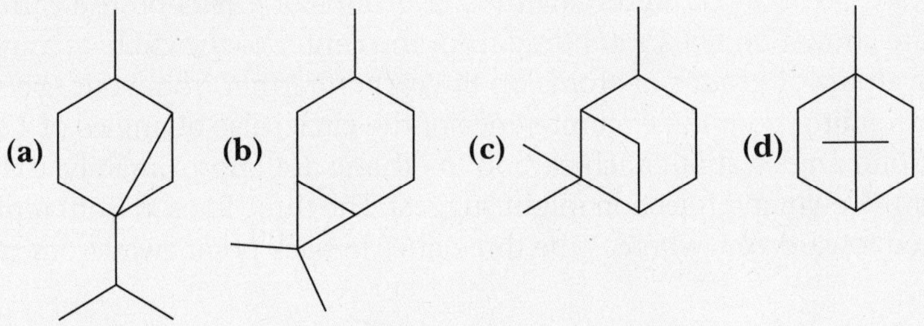

Thujane Carane Pinane Bornane

2.5 Conformations of Alkanes and Cycloalkanes

Structural formulas are useful for showing the connectivity of atoms in a molecule. However, they usually do not show three-dimensional shapes. As chemists try to understand more about the relationships between structure and the chemical and physical properties of compounds, it becomes increasingly important to know more about the three-dimensional shapes of molecules.

In this section, we ask you to look at molecules as three-dimensional objects and to visualize not only bond angles but also distances between various atoms and groups of atoms within the molecules. We also describe intramolecular strain, which we divide into three types: torsional strain, steric strain, and angle strain. We urge you to build models (either physically or with desktop modeling programs such as Spartan) of the molecules discussed in this section so that you become comfortable in dealing with them as three-dimensional objects and understand fully the origins of the various types of intramolecular strain.

Strain

An instability within a structure associated with higher internal energy.

Strain, a key concept in organic chemistry, is a measure of the energy stored in a compound due to a structural distortion. Chemicals are physical entities, each possessing an optimal structure, as does any macroscopic object. If you perturb the optimal structure, you put a strain on the system. A tree branch bending in the wind is a strained form of the branch. A bent branch is at higher energy than an undisturbed branch, and it releases that energy when relaxing between wind gusts. In this section, we describe strained forms of alkanes that interconvert with more optimal structures. The interconversion between the strained and relaxed structures results from collisions and thermal motions. Some structures are permanently strained and do not readily have a pathway to release that strain. Analogously, a compressed spring is strained, and unless the compression is removed, the spring is stuck in the strained form. Energy is stored in the compressed spring waiting to be released if given an opportunity. The opportunity to release permanently stored strain in a chemical structure comes from chemical reactions, and near the end of this chapter, we analyze combustion (burning) as a means of measuring this kind of strain.

Conformation
Any three-dimensional arrangement of atoms in a molecule that results from rotation about a single bond.

Staggered conformation
A conformation about a carbon-carbon single bond in which the atoms or groups on one carbon are as far apart as possible from atoms or groups on an adjacent carbon.

Newman projection
A way to view a molecule by looking along a carbon-carbon single bond to help evaluate the relative orientations of attached groups.

A. Alkanes

Alkanes of two or more carbons can be twisted into a number of different three-dimensional arrangements of their atoms by rotating about one or more carbon-carbon bonds. Any three-dimensional arrangement of atoms that results from rotation about single bonds is called a **conformation**. Figure 2.5(a) shows a ball-and-stick model of a **staggered conformation** of ethane. In this conformation, the three C—H bonds on one carbon are as far apart as possible from those bonds on the adjacent carbon. Figure 2.5(b), called a **Newman projection** (named for Melvin Newman, of Ohio State University, who developed these projections), is a shorthand way to represent this conformation of ethane. In a Newman projection, a molecule is viewed down the axis of a C—C bond. The three atoms or groups of atoms on the carbon nearer your eye are shown on lines extending from the center of the circle at angles of 120°. The three atoms or groups of atoms on the carbon farther from your eye are shown on lines extending from the circumference of the circle, also at angles of 120°. Remember that bond angles about each carbon in ethane are approximately 109.5° and not 120°, as this Newman projection might suggest. The three lines in front represent bonds directed toward you, whereas the three lines in back point away from you.

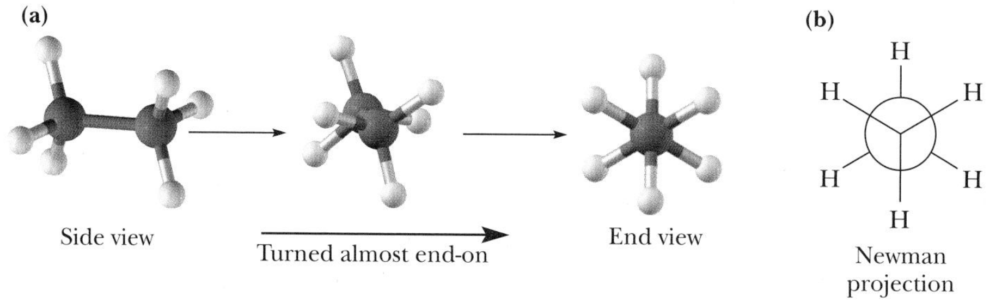

FIGURE 2.5 A staggered conformation of ethane. (a) Ball-and-stick models and (b) Newman projection.

Eclipsed conformation
A conformation about a carbon-carbon single bond in which the atoms or groups on one carbon are as close as possible to the atoms or groups on an adjacent carbon.

Figure 2.6 shows a ball-and-stick model and a Newman projection for an **eclipsed conformation** of ethane. In this conformation, the three C—H bonds on one carbon are as close as possible to the three C—H bonds on the adjacent carbon. In other words, hydrogen atoms on the back carbon are eclipsed by the hydrogen atoms on the front carbon (just as the sun is eclipsed when the moon passes in front of it). Different conformations are often called **conformational isomers** or **conformers**.

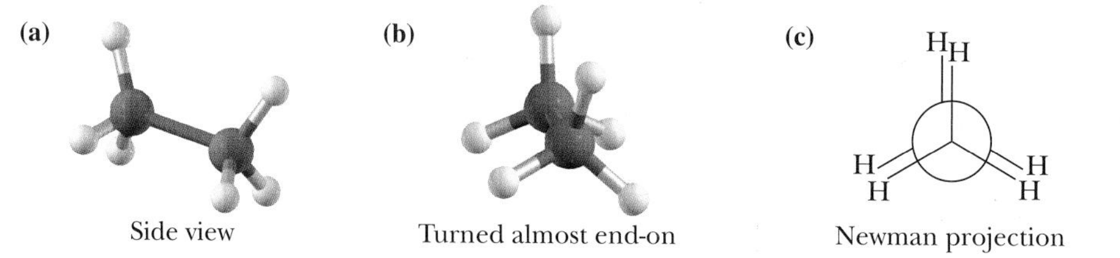

FIGURE 2.6 An eclipsed conformation of ethane. (a, b) Ball-and-stick models and (c) Newman projection.

Dihedral angle
The angle created by two intersecting planes.

If we are to discuss energy relationships among conformations, it is convenient to define the term *dihedral angle*. A **dihedral angle**, θ (Greek theta), is the angle created by two intersecting planes, each plane defined by three atoms. In the Newman projection of the eclipsed conformation of ethane in Figure 2.7(a), two H—C—C planes are shown. The angle at which these planes intersect (the dihedral angle) is 0°.

A staggered conformation in which the dihedral angle of the two H—C—C planes is 60° is illustrated in Figure 2.7(b).

In principle, there are an infinite number of conformations of ethane that differ only in the degree of rotation about the carbon-carbon single bond. Because there is a small energy barrier between conformations, rotation is not completely free. As we shall see, the lowest energy (the most stable) conformation of ethane is a staggered conformation. The highest energy (the least stable) conformation is an eclipsed conformation. At room temperature, ethane molecules undergo collisions with sufficient energy so that the energy barrier between extreme conformations can be crossed and rotation about the carbon-carbon single bond from one conformation to another occurs rapidly.

The difference in energy between an eclipsed conformation and a staggered conformation of ethane is approximately 12.6 kJ (3.0 kcal)/mol (approximately 4.2 kJ (1.0 kcal)/mol for each eclipsed H/H) and is referred to as torsional strain. **Torsional strain** arises when nonbonded atoms achieve conformations via rotations around bonds where they are closer to one another than in a more stable alternate conformation. For example, when the nonbonded atoms are eclipsed rather than staggered, the torsional strain would be considered an eclipsed-interaction strain. In ethane, for example, torsional strain occurs when pairs of hydrogens H(4)-H(6), H(5)-H(8), and H(3)-H(7) on adjacent carbons are forced into eclipsed positions. The models shown here represent only one of the three different but equivalent eclipsed conformations of ethane.

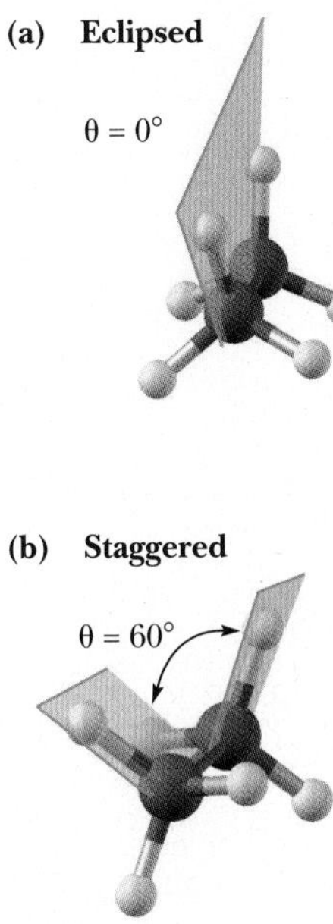

FIGURE 2.7 Dihedral angles in ethane. (a) An eclipsed conformation and (b) a staggered conformation.

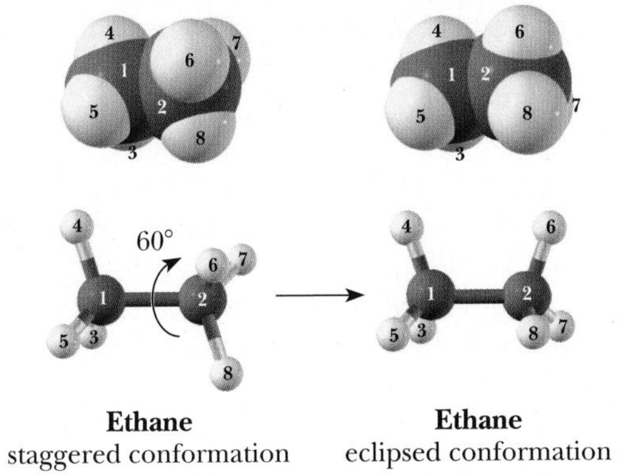

Torsional strain
Strain that arises when nonbonded atoms separated by three bonds are forced from a staggered conformation to an eclipsed conformation. Torsional strain is also called eclipsed-interaction strain.

Figure 2.8 shows the relationship between energy and dihedral angle for the conformations of ethane. All energy diagrams in this book use *energy* as a vertical axis. Several types of energy—potential energy, Gibbs free energy, and enthalpy—are

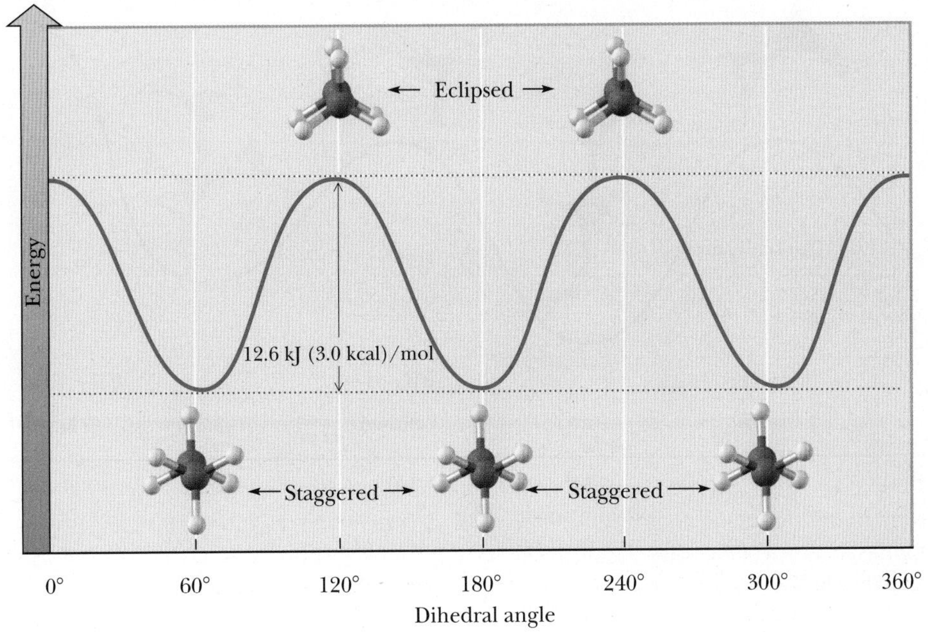

FIGURE 2.8 The energy of ethane as a function of dihedral angle. The eclipsed conformations are approximately 12.6 kJ (3.0 kcal)/mol higher in energy than the staggered conformations.

important in various contexts. The diagrams for all of these types of energy are often nearly indistinguishable, and in most cases, "energy" will do for the concepts we are introducing. When it is necessary to be more precise about the type of energy, we will do so and explain why.

There has been disagreement over the years as to the origin of the torsional strain in the eclipsed conformations of ethane. It was originally thought that this strain was the result of the repulsion between eclipsed hydrogen nuclei; they are separated by 255 pm in a staggered conformation but by only 235 pm in an eclipsed conformation. Alternatively, it has been held that the torsional strain was the result of repulsion between the electron clouds of the adjacent C—H bonds. Theoretical molecular orbital calculations, however, suggest that the energy difference between the conformational extremes arises not from destabilization of the eclipsed conformation, but rather from stabilization of the staggered conformation. This stabilization of the staggered conformation arises because of a small donor-acceptor interaction (donation of electron density from a filled orbital into an empty acceptor orbital) between a filled C—H bonding MO of one carbon and the empty or unfilled C—H antibonding MO on the adjacent carbon with which it is aligned. This donor-acceptor stabilization is lost when a staggered conformation is converted to an eclipsed conformation.

Anti conformation

A conformation about a single bond in which two groups on adjacent carbons lie at a dihedral angle of 180°.

Gauche conformation

A conformation about a single bond of an alkane in which two groups on adjacent carbons lie at a dihedral angle of 60°.

Watch a video explanation

Next, let us look at the conformations of butane viewed along the bond between carbons 2 and 3. For butane, there are two types of staggered conformations and two types of eclipsed conformations. The staggered conformation in which the methyl groups are the maximum distance apart ($\theta = 180°$) is called the **anti conformation**; the staggered conformation in which they are closer together ($\theta = 60°$) is called the **gauche conformation**. In one eclipsed conformation ($\theta = 0°$), methyl is eclipsed by methyl. In the other ($\theta = 120°$), methyl is eclipsed by hydrogen. Figure 2.9 shows the energy relationships for rotation from $-180°$ to 180°. Note that both the gauche and anti conformations of butane are staggered conformations, yet the gauche conformations are approximately 3.8 kJ (0.9 kcal)/mol higher in energy than the anti conformation.

In dealing with the relative stabilities of the various conformations of butane, we again encounter torsional strain. In addition, we encounter two other types of strain:

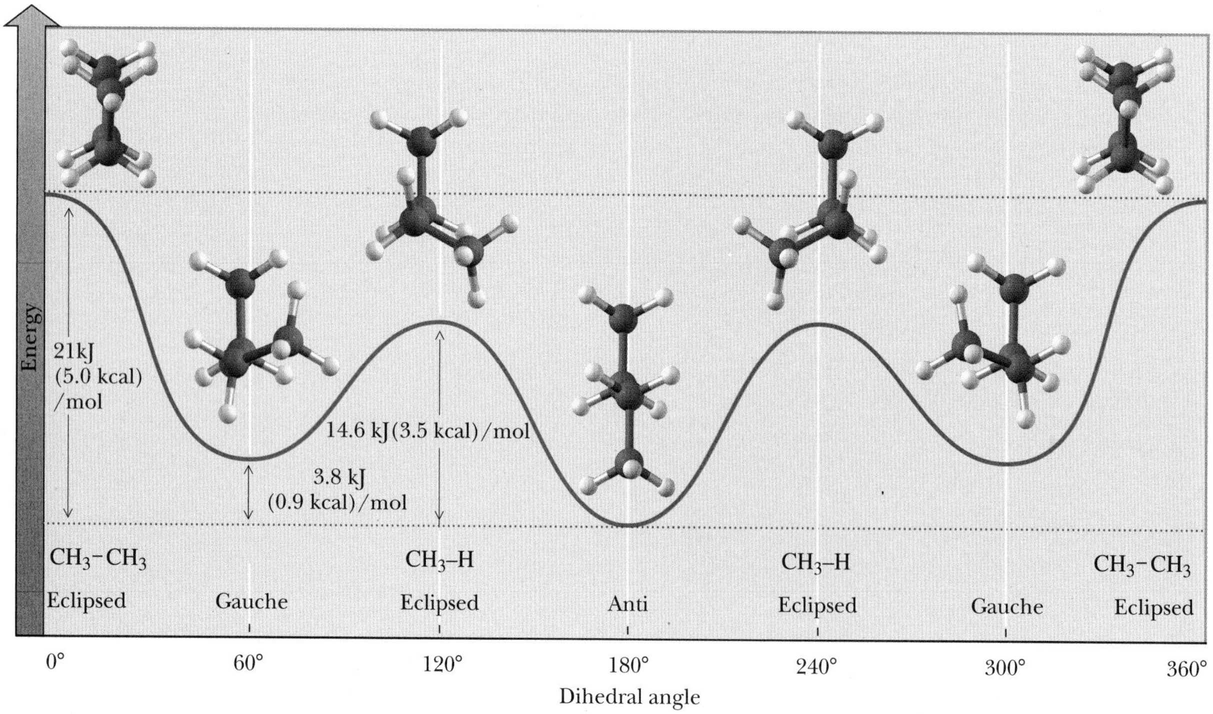

FIGURE 2.9 The energy of butane as a function of the dihedral angle about the bond between carbons 2 and 3. The lowest energy conformation occurs when the two methyl groups are the maximum distance apart ($\theta = 180°$). The highest energy conformation occurs when the two methyl groups are eclipsed ($\theta = 0°$).

angle strain and steric strain. **Angle strain** results when a bond angle in a molecule is either expanded or compressed compared to its optimal values. Strain can also occur when bond lengths are forced to become shorter or longer than normal. In general, bond stretching is not as easy as bond angle bending. **Steric strain** (also called nonbonded interaction or van der Waals strain) results when nonbonded atoms separated by four or more bonds are forced closer to each other than their atomic (contact) radii allow—that is, when they are forced to smash into each other. The total of all types of strain can be calculated by molecular mechanics programs; angles and bond lengths are chosen in an iterative procedure so that an optimum geometry with the lowest total energy is calculated. Such routines are included with popular desktop programs such as Spartan. Molecular mechanics calculations can determine the lowest energy arrangement of atoms in a given conformation of a molecule, a process referred to as energy minimization.

Angle strain
The strain that arises when a bond angle is either compressed or expanded compared to its optimal value.

Steric strain
The strain that arises when nonbonded atoms separated by four or more bonds are forced closer to each other than their atomic (contact) radii would allow. Steric strain is also called nonbonded interaction strain or van der Waals strain.

Let us illustrate the origins of both angle strain and steric strain by comparing the anti ($\theta = 180°$) and eclipsed ($\theta = 0°$) conformations of butane. In an energy-minimized conformation of anti butane, the C—C—C bond angle is 111.9° and all H—C—H bond angles are between 107.4° and 107.9°.

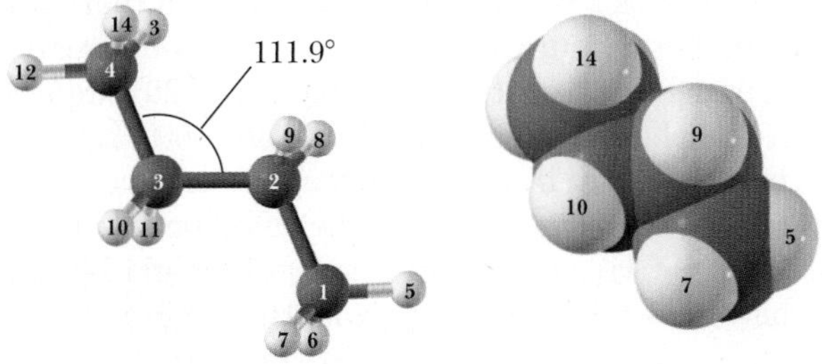
111.9°

Now consider the eclipsed conformation. Figure 2.10(a) shows models of an eclipsed butane conformation in which C—C—C bond angles are set at 111.9°, the same value as in the energy-minimized anti conformation. Notice that in this eclipsed conformation, hydrogens H(5)-H(13) and H(5)-H(14) are forced to within 178 pm of each other, a distance closer than their contact radii would allow. When the energy of eclipsed butane is minimized [Figure 2.10(b)], the distance between these same pairs of hydrogens increases to 231 pm (which reduces steric strain). At the same time, the C(4)-C(3)-C(2) and C(3)-C(2)-C(1) bond angles increase from 111.9° to 119.0° (which increases angle strain).

Thus, the energy-minimized eclipsed conformation represents a balance between a decrease in steric strain and an increase in angle strain. The calculated strain in the

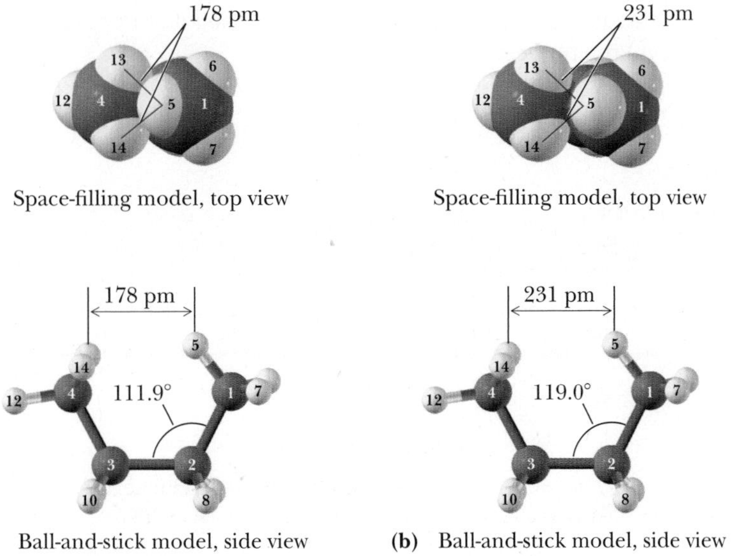
178 pm
231 pm
Space-filling model, top view
Space-filling model, top view
178 pm
111.9°
231 pm
119.0°
(a) Ball-and-stick model, side view
(b) Ball-and-stick model, side view

FIGURE 2.10 Eclipsed conformations of butane. (a) Non-energy-minimized and (b) energy-minimized eclipsed conformations. The calculated difference in energy between the non-energy-minimized and energy-minimized eclipsed conformations is 3.6 kJ (0.86 kcal)/mol.

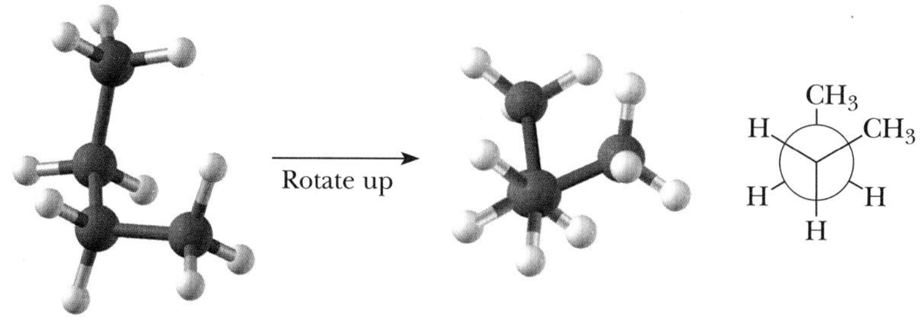

Butane

FIGURE 2.11 One of two equivalent energy-minimized gauche conformations of butane.

energy-minimized eclipsed conformation of butane is 21 kJ (5.0 kcal)/mol, relative to the staggered conformation.

An energy-minimized gauche conformation of butane (Figure 2.11) is approximately 3.8 kJ (0.90 kcal)/mol higher in energy than the anti, staggered conformation. This difference in energy is caused almost entirely by the steric strain (nonbonded interaction strain) between the two methyl groups. At any given instant, a larger number of butane molecules are in the anti conformation than in the gauche conformation and the number of molecules in the eclipsed conformation is vanishingly small. The percentage of the anti conformation present at 20°C is about 70%.

Although the two gauche conformations ($\theta = 60°$ and $300°$) have equal energies, they are not identical. They are related by reflection; that is, one gauche conformation is the reflection of the other, just as your right hand is the reflection of your left hand. The conformations with eclipsed —CH_3 and —H groups ($\theta = 120°$ and $240°$) are also related by reflection. We shall have more to say about objects and their mirror reflections in Chapter 3.

Example 2.8 Newman Projections

Following is the structural formula of 1,2-dichloroethane:

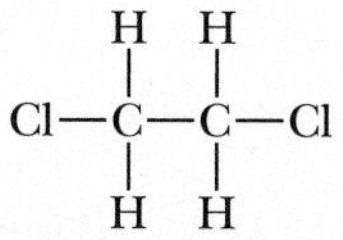

1,2-Dichloroethane

(a) Draw Newman projections for all staggered conformations formed by rotation from 0° to 360° about the carbon-carbon single bond.

(b) Which staggered conformation(s) has (have) the lowest energy? Which has (have) the highest energy?

(c) Which, if any, of these staggered conformations are related by reflection?

Solution

(a) If we take the dihedral angle when the chlorines are eclipsed as a reference point, staggered conformations occur at dihedral angles 60°, 180°, and 300°.

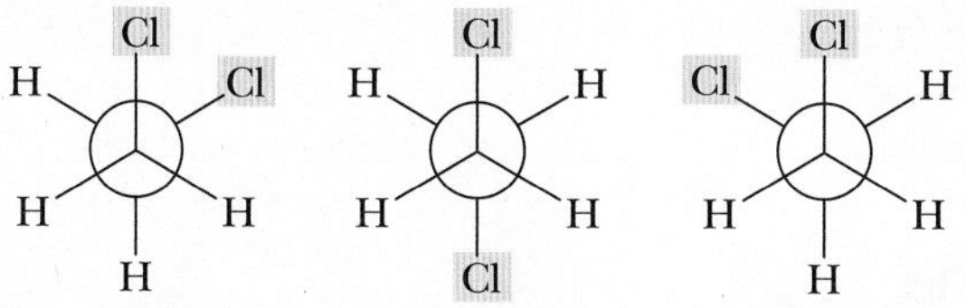

Gauche $\theta = 60°$ | Anti $\theta = 180°$ | Gauche $\theta = 300°$

(b) We predict that the anti conformation (dihedral angle $\theta = 180°$) has the lowest energy. The two gauche conformations (dihedral angle $\theta = 60°$ and 300°) are of higher but equal energy. We are not given data in the problem to calculate the actual energy differences.

(c) The two gauche conformations are related by reflection.

Problem 2.8

For 1,2-dichloroethane:

(a) Draw Newman projections for all eclipsed conformations formed by rotation from 0° to 360° about the carbon-carbon single bond.

(b) Which eclipsed conformation(s) has (have) the lowest energy? Which has (have) the highest energy?

(c) Which, if any, of these eclipsed conformations are related by reflection?

Watch a video explanation

A word of caution. Although we talk about eclipsed along with staggered conformations, this can be misleading. Of all the conformations of butane, the eclipsed conformations are at the highest points on the energy profile. The gauche and anti conformations are in energy troughs. As a result, butane molecules spend their time in the anti and gauche conformations and only fleetingly pass through the eclipsed conformations when interconverting between anti and gauche. A further nuance is that C—H and C—C bonds have vibrational motions at room temperature, and these vibrations occur simultaneously with bond rotations. The reality is that hydrocarbons are exceedingly complex and dynamic structures in solution, undergoing constant interconversions between low-energy conformations through combined vibrations and rotations.

B. Cycloalkanes

The structures and energies of cyclic alkanes are highly dependent on the size of their rings. Of major importance is **small ring strain**, which exists in ring sizes below six carbons and is due to the C—C—C bond angles not being able to achieve the optimal tetrahedral angle of 109.5°. Further, just as alkanes have rapidly interconverting conformations with varying degrees of torsional strain along their C—C single bonds, so do cyclic alkanes of four carbons or more. The interconversions involve rotations along C—C single bonds of the rings. However, the bonds can only rotate so far without breaking the ring; hence, the rotations are limited to certain angles.

Small ring strain

A strain associated with ring sizes below six that arises from nonoptimal bond angles.

i. Cyclopropane

The observed C—C—C bond angles in cyclopropane are 60° (Figure 2.12), a value considerably smaller than the bond angle of 109.5° predicted for sp^3 hybridized carbon atoms. This compression from the optimal bond angle introduces a considerable angle strain. Furthermore, because cyclopropane is planar, there are six pairs of fully eclipsed C—H bonds, which introduce considerable torsional strain. The combined angle and torsional strain energy in cyclopropane is approximately 116 kJ (27.7 kcal)/mol. Because of their extreme degree of intramolecular strain, cyclopropane and its derivatives undergo several ring-opening reactions not seen with larger cycloalkanes.

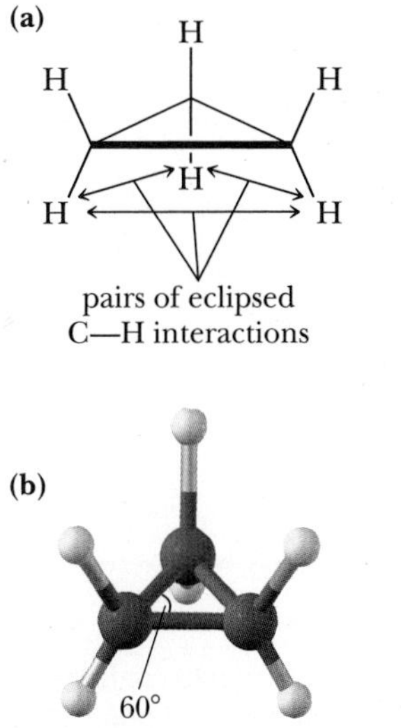

FIGURE 2.12 Cyclopropane. (a) Structural formula and (b) ball-and-stick model.

ii. Cyclobutane

Nonplanar or puckered conformations are favored in all cycloalkanes larger than cyclopropane. If cyclobutane were planar [Figure 2.13(a)], all C—C—C bond angles would be 90° and there would be eight pairs of eclipsed hydrogen interactions (which would maximize torsional strain). Rotations along the C—C bonds can slightly relieve strain. Puckering of the ring [Figure 2.13(b)] alters the strain energy in two ways: (1) it decreases the torsional strain associated with eclipsed interactions, but (2) it increases further the angle strain caused by the compression of C—C—C bond angles. Because the decrease in torsional strain is greater than the increase in angle strain, puckered cyclobutane is more stable than planar cyclobutane. In the conformation of lowest energy, the measured C—C—C bond angles are 88°, and the strain energy in cyclobutane is approximately 110 kJ (26.3 kcal)/mol. Just like butane, cyclobutane is not static, but undergoes interconversion between the puckered conformations.

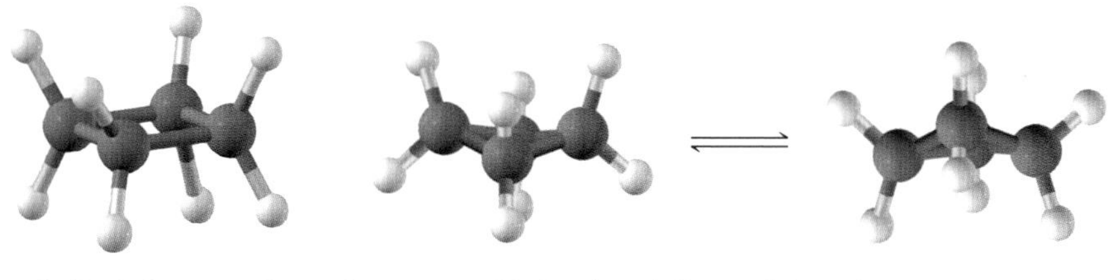

(a) Hypothetical planar conformation (b) Puckered (butterfly) conformations

FIGURE 2.13 Cyclobutane. (a) In the planar conformation, there are eight pairs of eclipsed C—H interactions. (b) The energy is a minimum in the puckered (butterfly) conformation.

iii. Cyclopentane

▶ Five- and six-membered rings are the most stable.

If cyclopentane were to adopt a planar conformation, all C—C—C bond angles would be 108° [Figure 2.14(a)]. This angle differs only slightly from the tetrahedral angle of 109.5°; consequently, there would be little angle strain in this conformation. In a planar conformation, however, there are ten pairs of fully eclipsed C—H bonds creating a torsional strain of approximately 42 kJ (10 kcal)/mol. To relieve at least a part of this torsional strain, the ring twists by rotations along the C—C bonds into the **"envelope" conformation** shown in Figure 2.14(b). In this conformation, four carbon atoms are in a plane and the fifth bends out of the plane, rather like an envelope with its flap bent upward. Cyclopentane exists as a dynamic equilibrium of five equivalent envelope conformations in which each carbon atom alternates as the out-of-plane atom. Note that the average C—C—C bond angle is reduced to 105° (increasing angle strain). In the envelope conformation, the number of eclipsed C—H interactions is also reduced (decreasing torsional strain). Thus, the envelope conformation should be thought of as an energetic compromise that balances a decreased torsional strain with an increased angle strain. The total strain energy in puckered cyclopentane is approximately 27 kJ (6.5 kcal)/mol. ▶

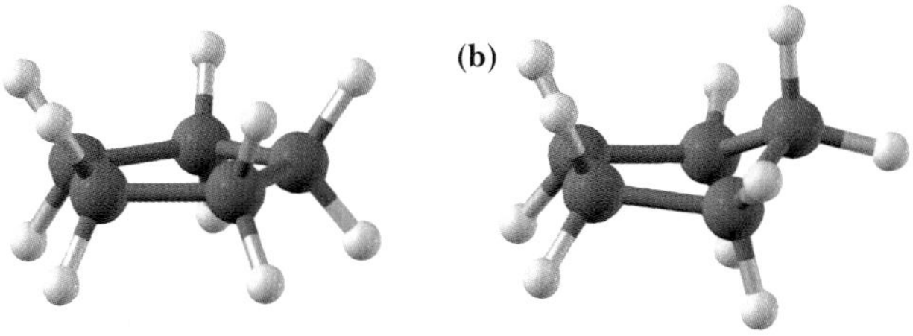

(a) Hypothetical planar conformation (b) Puckered envelope conformation

FIGURE 2.14 Cyclopentane. (a) In the planar conformation, there are ten pairs of eclipsed C—H interactions. (b) The most stable conformation is a puckered envelope conformation.

iv. Cyclohexane

Chair conformation

The most stable conformation of a cyclohexane ring; all bond angles are close to 109.5°, and all bonds on adjacent carbons are staggered.

Cyclohexane adopts a number of puckered conformations that interconvert via C—C rotations, the most stable of which is the **chair conformation** (Figure 2.15). In this conformation, all C—C—C bond angles are 110.9° (minimizing angle strain) and all hydrogens on adjacent carbons are staggered with respect to one another (minimizing torsional strain). In addition, no two atoms are close enough to each other for nonbonded interaction strain to exist. Thus, there is little strain in a chair conformation of cyclohexane.

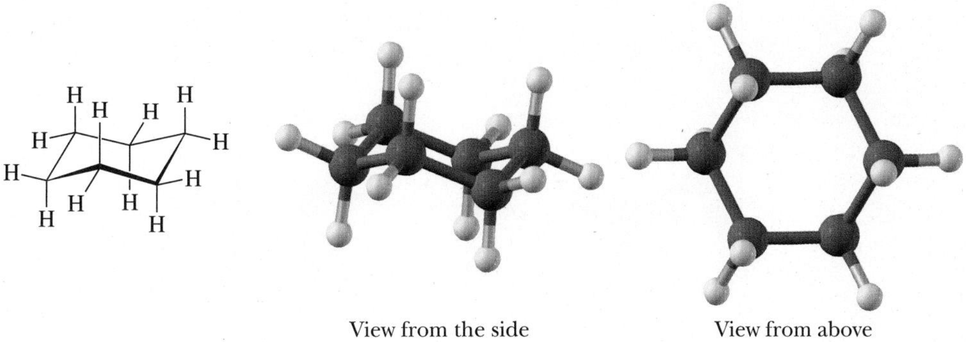

FIGURE 2.15 A chair conformation of cyclohexane.

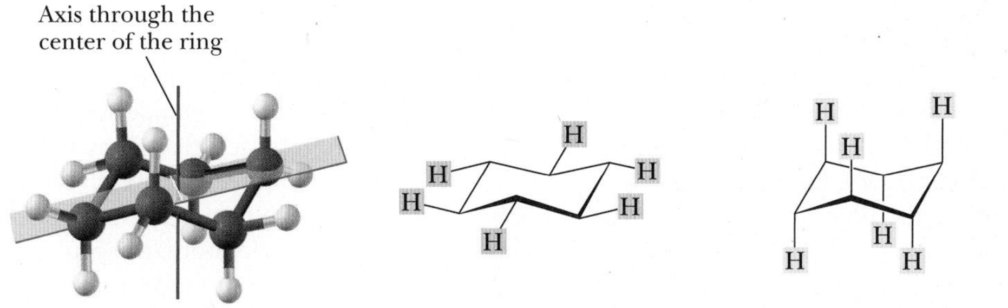

(a) Ball-and-stick model showing all 12 hydrogens (b) The six equatorial C—H bonds shown in red (c) The six axial C—H bonds shown in blue

FIGURE 2.16 A chair conformation of cyclohexane, showing axial and equatorial C—H bonds. The plane of the ring is defined by four carbons; the fifth carbon is above the plane, and the sixth carbon is below it.

The C—H bonds in a chair conformation of cyclohexane are arranged in two different orientations. Six C—H bonds are called **axial bonds**, and the other six are called **equatorial bonds**. One way to visualize the difference between these two types of bonds is to imagine an axis through the center of the chair, perpendicular to the floor (Figure 2.16).

Equatorial bonds are approximately perpendicular to the imaginary axis and form an equator about the ring. Equatorial bonds alternate, first slightly up and then slightly down as you move from one carbon of the ring to the next. Axial bonds are parallel to the imaginary axis. Three axial bonds point straight up; the other three axial bonds point straight down. Axial bonds also alternate, first up and then down as you move from one carbon of the ring to the next. Notice also that if the axial bond on a carbon points upward, the equatorial bond on that carbon points slightly downward. Conversely, if the axial bond on a particular carbon points downward, the equatorial bond on that carbon points slightly upward.

There are many other nonplanar conformations of cyclohexane, two of which, a **boat** and a **twist-boat**, are shown in Figure 2.17.

Axial bond

A bond to a chair conformation of cyclohexane that extends from the ring parallel to the imaginary axis through the center of the ring; a bond that lies roughly perpendicular to the equator of the ring.

Equatorial bond

A bond to a chair conformation of cyclohexane that extends from the ring roughly perpendicular to the imaginary axis through the center of the ring; a bond that lies roughly along the equator of the ring.

Boat conformation

A nonplanar conformation of a cyclohexane ring in which carbons 1 and 4 of the ring are bent toward each other.

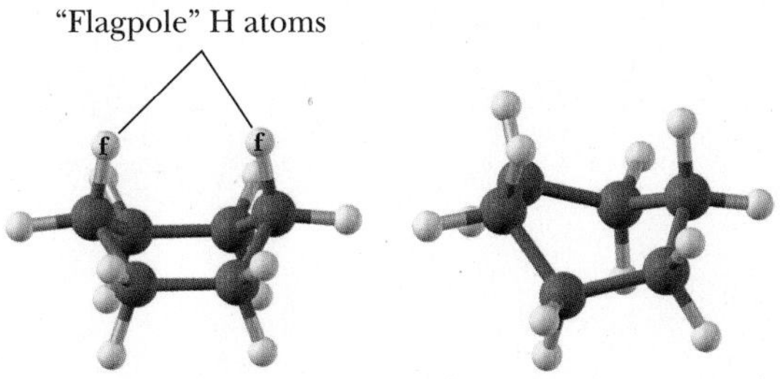

Boat conformation Twist-boat conformation

FIGURE 2.17 Boat and twist-boat conformations of cyclohexane.

You can visualize interconversion of chair and boat conformations by twisting about two carbon-carbon bonds as illustrated in Figure 2.18. A boat conformation is considerably less stable than a chair conformation because of the torsional strain associated with four pairs of eclipsed C—H interactions and the steric strain between the two "flagpole" hydrogens. The difference in energy between chair and boat conformations is approximately 27 kJ (6.5 kcal)/mol.

Twist-boat conformation

A nonplanar conformation of a cyclohexane ring that is twisted from and slightly more stable than a boat conformation.

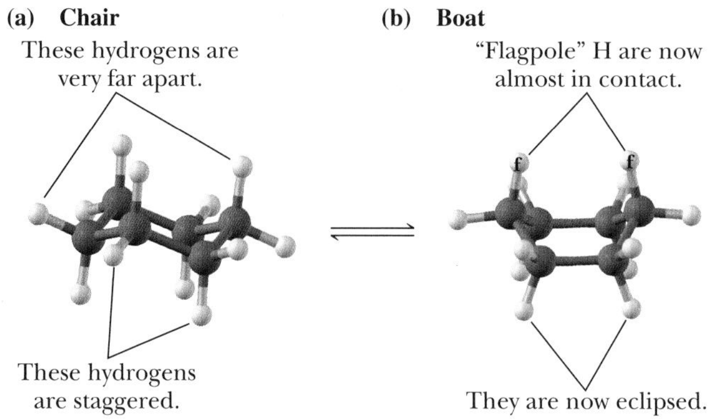
(a) Chair
These hydrogens are very far apart.
These hydrogens are staggered.
(b) Boat
"Flagpole" H are now almost in contact.
f
f
They are now eclipsed.

FIGURE 2.18 Interconversion of (a) a chair conformation to (b) a boat conformation produces one set of flagpole steric interactions and four sets of eclipsed hydrogen interactions.

Some of the strain in the boat conformation can be relieved by a slight twisting of the ring to a twist-boat conformation. It is estimated by computer modeling that a twist-boat conformation is favored over a boat conformation by approximately 6.3 kJ (1.5 kcal)/mol. Figure 2.19 shows an energy diagram for the interconversion between chair, twist-boat, and boat conformations. The large difference in energy between chair and boat or twist-boat conformations means that at room temperature, molecules in the chair conformation make up more than 99.99% of the equilibrium mixture.

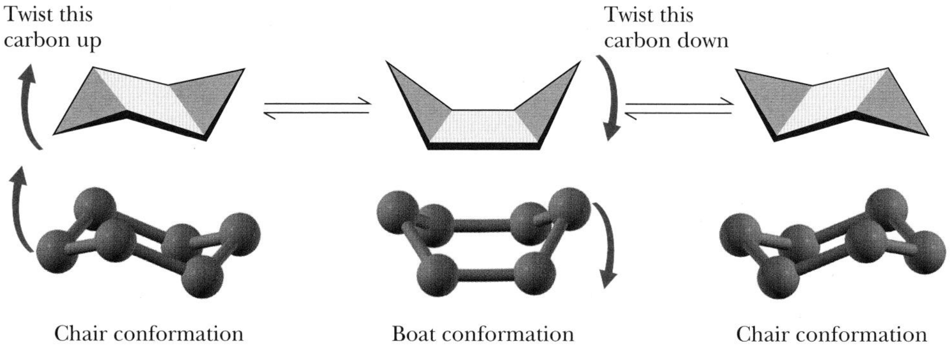
Twist this carbon up
Twist this carbon down
Chair conformation
Boat conformation
Chair conformation

Watch a video explanation

For cyclohexane, the two equivalent chair conformations can be interconverted by twisting one chair first into a boat and then into the alternative chair.

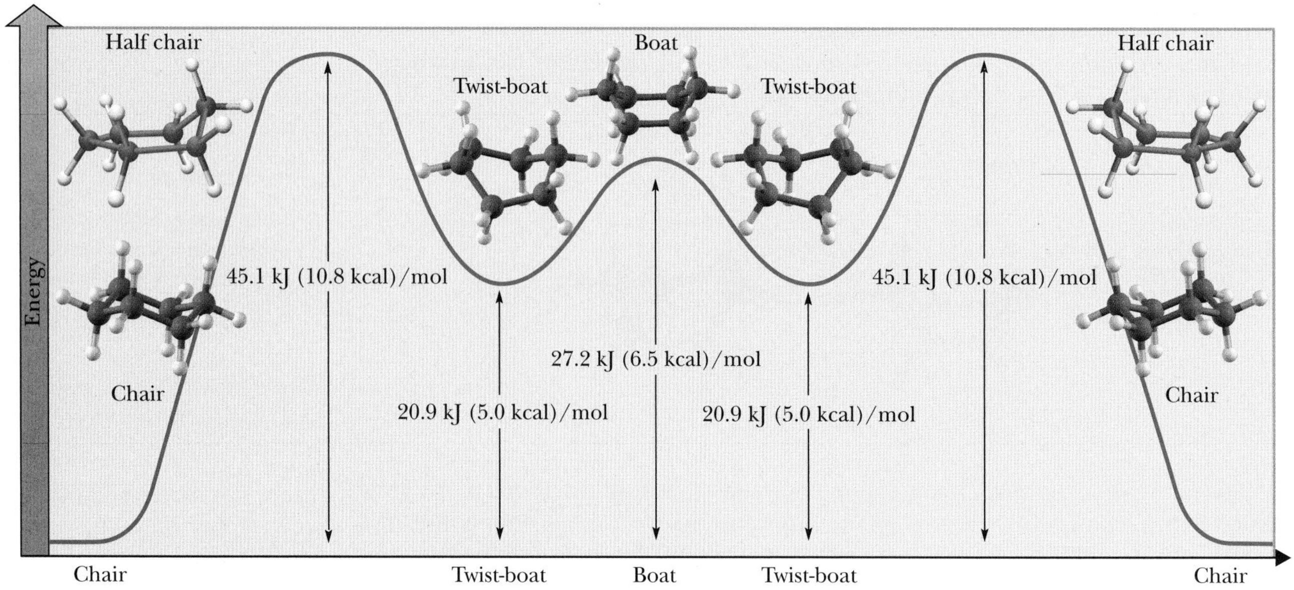
Energy
Half chair
Boat
Twist-boat
Twist-boat
Half chair
45.1 kJ (10.8 kcal)/mol
45.1 kJ (10.8 kcal)/mol
27.2 kJ (6.5 kcal)/mol
20.9 kJ (5.0 kcal)/mol
20.9 kJ (5.0 kcal)/mol
Chair
Chair
Chair
Twist-boat
Boat
Twist-boat
Chair

FIGURE 2.19 Energy diagram for interconversion of chair, twist-boat, and boat conformations of cyclohexane. The chair conformation is most stable because angle, torsional, and steric strain are at a minimum.

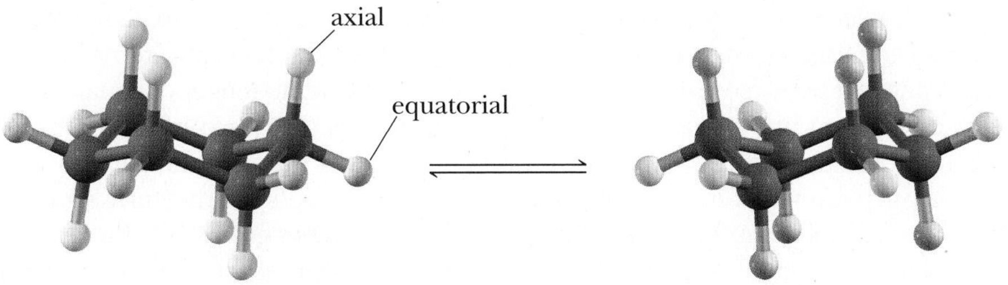

FIGURE 2.20 Interconversion of alternative chair conformations of cyclohexanes. All C—H bonds equatorial in one chair are axial in the alternative chair and vice versa.

When one chair is converted to the other, a change occurs in the relative orientations in space of the hydrogen atoms bonded to each carbon. All hydrogen atoms axial in one chair become equatorial in the other and vice versa (Figure 2.20). The conversion of one chair conformation of cyclohexane to the other occurs rapidly at room temperature.

Example 2.9 Axial Versus Equatorial Groups I

Following is a chair conformation of cyclohexane showing one methyl group and one hydrogen.

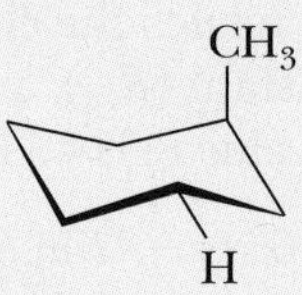

(a) Indicate using a label whether each group is equatorial or axial.

(b) Draw the alternative chair conformation and again label each group as axial or equatorial.

Solution

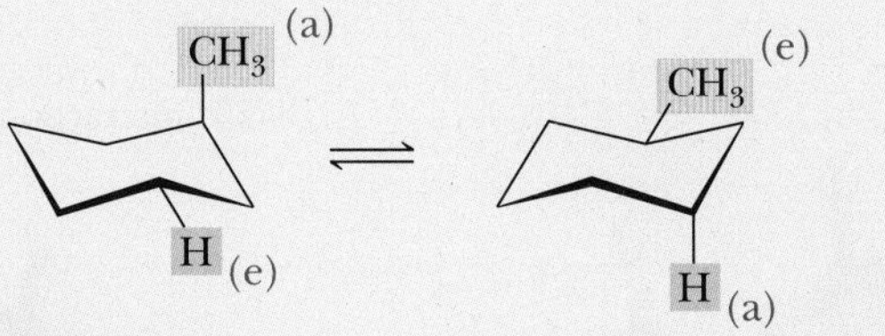

Problem 2.9

Following is a chair conformation of cyclohexane with the carbon atoms numbered 1 through 6.

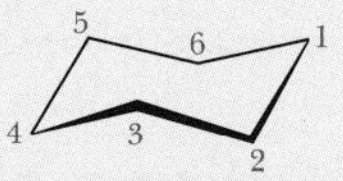

(a) Draw hydrogen atoms that are above the plane of the ring on carbons 1 and 2 and below the plane of the ring on carbon 4.

(b) Which of these hydrogens are equatorial? Which are axial?

(c) Draw the alternative chair conformation. Which hydrogens are equatorial? Which are axial? Which are above the plane of the ring? Which are below it?

If we replace a hydrogen atom of cyclohexane with an alkyl group, the group occupies an equatorial position in one chair and an axial position in the alternative chair. This means that the two chairs are no longer equivalent and no longer of equal stability.

A convenient way to describe the relative stabilities of chair conformations with equatorial and axial substituents is in terms of a type of steric strain called **diaxial (axial-axial) interaction**. The term *diaxial interaction* refers to the steric strain between an axial substituent and an axial hydrogen (or another group) on the same side of a cyclohexane ring. The axial positions on the same side of the ring are extremely close to each other, and any groups larger than hydrogen atoms will introduce steric strain between the larger group and the other two axial hydrogen atoms. Because this type of steric strain originates between groups on carbons 1 and 3 of a cyclohexane ring, it is often called a 1,3-diaxial interaction.

Diaxial interaction

Refers to the steric strain arising from interaction between an axial substituent and an axial hydrogen (or another group) on the same side of a chair conformation of a cyclohexane ring.

Consider methylcyclohexane (Figure 2.21). When —CH_3 is axial, it is parallel to the axial C—H bonds on carbons 3 and 5. Thus, for axial methylcyclohexane, there are two unfavorable methyl-hydrogen diaxial interactions. No such unfavorable interactions exist when the methyl group is in an equatorial position. For methylcyclohexane, the equatorial methyl conformation is favored over the axial methyl conformation by approximately 7.28 kJ (1.74 kcal)/mol.

Given the difference in strain energy between the axial and equatorial conformations of methylcyclohexane, we can calculate the ratio of the two conformations at equilibrium using the equation that relates the change in Gibbs free energy (ΔG^0) for an equilibrium, the equilibrium constant (K_{eq}), and the temperature (T) in kelvins. R, the universal gas constant, has the value 8.314 J (1.987 cal)·K^{-1}·mol^{-1}.

$$\Delta G^0 = -RT \ln K_{eq}$$

Substituting the value of −7.28 kJ/mol (axial methyl → equatorial methyl) for ΔG^0 and solving the equation gives a value of 18.9 for the equilibrium constant at room temperature (25°C = 298 K).

$$\ln K_{eq} = \frac{-(-7280 \text{ J}\cdot\text{mol}^{-1})}{8.314 \text{ J}\cdot\text{K}^{-1}\cdot\text{mol}^{-1} \times 298 \text{ K}} = 2.939$$

$$K_{eq} = \frac{18.9}{1} = \frac{\text{equatorial}}{\text{axial}}$$

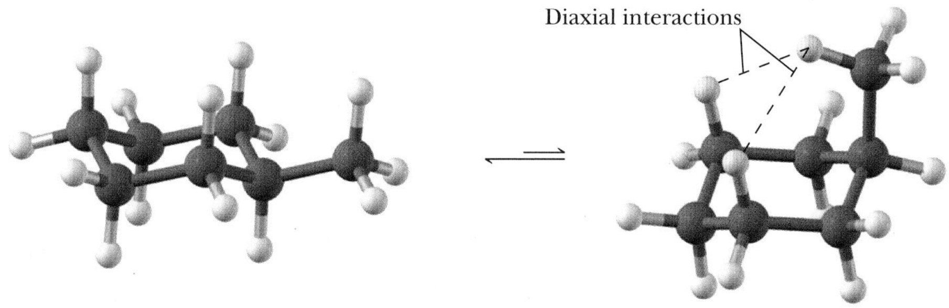

FIGURE 2.21 Two chair conformations of methylcyclohexane. The steric strain introduced by two diaxial interactions makes the axial methyl conformation less stable by approximately 7.28 kJ (1.74 kcal)/mol.

HOW TO

Draw Alternative Chair Conformations of Cyclohexane

You often will be asked to draw chair conformations of cyclohexane because these conformations allow you to identify which substituents are axial and which are equatorial. Although drawing chair conformations takes practice, following a few simple guidelines will make you an expert at drawing even complicated substitution patterns.

Step 1: Start by drawing two sets of parallel lines, each set at a slight angle.

Step 2: Complete each chair by drawing the ends connected to the parallel lines, in each case making one end tip up and the other end tip down.

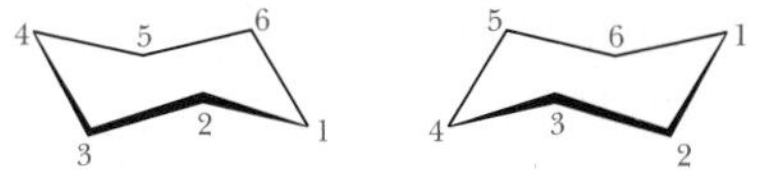

Step 3: Draw the axial bonds as vertical lines drawn on the side of the larger angle (greater than 180°) at each ring atom.

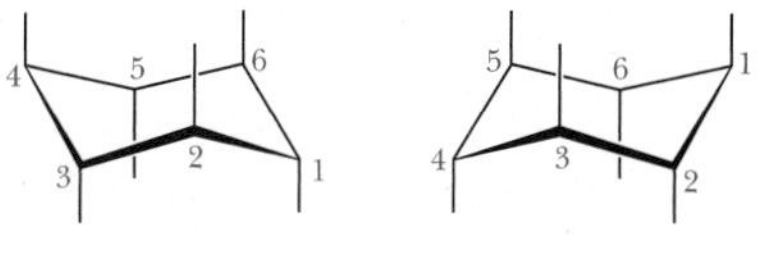

Axial bonds

Step 4: Draw the equatorial bonds using the bonds of the ring as guides for the angles. This is the tricky part.

For the chair conformation on the left, the equatorial bonds on carbons 2 and 5 are parallel to the ring bonds between carbons 3–4 and 1–6 (the two ring bonds drawn in red). The equatorial bonds of carbons 1 and 4 are parallel to the bonds between carbons 2–3 and 5–6 (the two ring bonds drawn in green), and the equatorial bonds of carbons 3 and 6 are parallel to the bonds between carbons 1–2 and 4–5 (the two bonds drawn in purple). Similarly, for the alternative chair on the right, the equatorial bonds on carbons 3 and 6 are parallel to the ring bonds drawn in red, the equatorial bonds of carbons 2 and 5 are parallel to the ring bonds drawn in green, and the equatorial bonds of carbons 1 and 4 are parallel to the ring bonds drawn in purple.

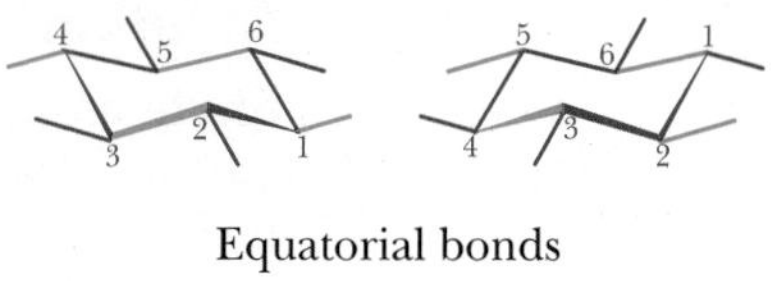

Equatorial bonds

Table 2.4 **ΔG^0 (Axial-Equatorial) for Monosubstituted Cyclohexanes at 25°C**

axial → equatorial

Group	$-\Delta G^0$ kJ/mol	$-\Delta G^0$ kcal/mol	Group	$-\Delta G^0$ kJ/mol	$-\Delta G^0$ kcal/mol
C≡N	0.8	0.19	NH_2	5.9	1.41
F	1.0	0.24	COOH	5.9	1.41
C≡CH	1.7	0.41	$C{=}CH_2$	7.1	1.70
I	1.9	0.45	CH_3	7.28	1.74
Cl	2.2	0.53	CH_2CH_3	7.3	1.75
Br	2.4	0.57	$CH(CH_3)_2$	9.0	2.15
OH	3.9	0.93	$C(CH_3)_3$	21.0	5.00

Thus, at any given instant at room temperature, a much larger number of methylcyclohexane molecules have the methyl group in the equatorial conformation rather than in the axial conformation. The percentage of equatorial is (equatorial/(equatorial + axial) × 100%); that is, (18.9/19.9) × 100% = about 95%.

Table 2.4 shows the difference in free energy between axial and equatorial substituents for several monosubstituted cyclohexanes. Notice that as the size of the alkyl substituent increases, the preference for conformations with the group equatorial increases. With a group as large as *tert*-butyl, the energy of the axial conformer becomes so large that the equatorial conformation is approximately 4000 times more abundant at room temperature than the axial conformation. In fact, a chair with an axial *tert*-butyl group is so unstable that if a *tert*-butyl group is forced into an axial position, the ring adopts a twist-boat conformation.

As shown by the free energy values given in Table 2.4, the preference for the equatorial position among the halogens increases in the order F < I < Cl < Br. Yet, the size of the halogen atoms increases in the order F < Cl < Br < I. This anomaly occurs because the C—I bond is so long that the center of the iodine atom is too far from the axial hydrogen to interact with it strongly.

Example 2.10 Axial Versus Equatorial Groups II

Label all methyl-hydrogen (CH_3/H) diaxial interactions in the following chair conformation of 1,2,4-trimethylcyclohexane.

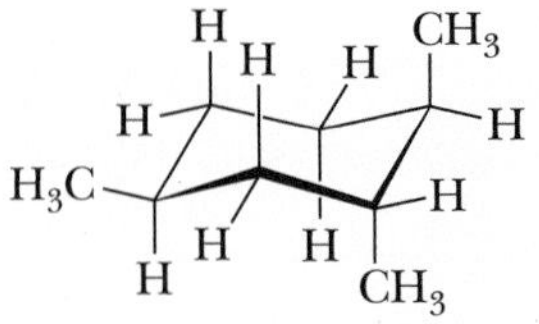

1,2,4-Trimethylcyclohexane

Solution

There are four methyl-hydrogen 1,3-diaxial interactions in this example; each axial methyl group has two sets of 1,3-diaxial interactions with parallel hydrogen atoms on the same side of the ring. The equatorial methyl group has no diaxial interactions.

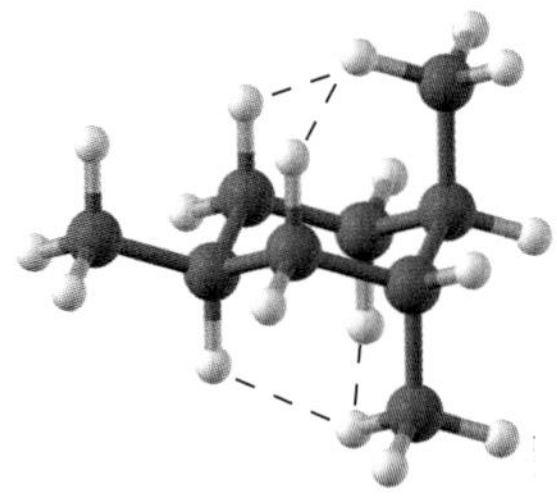

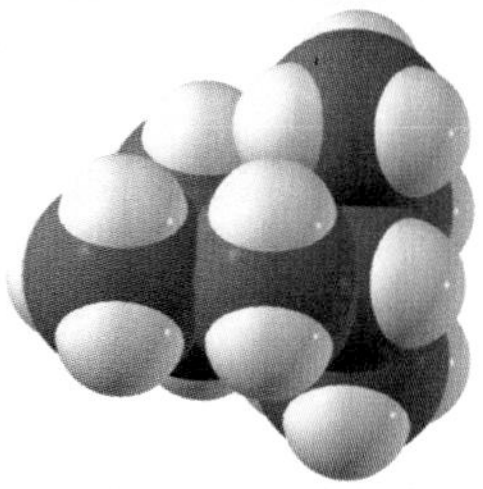

Problem 2.10

Draw the alternative chair conformation for the trisubstituted cyclohexane given in Example 2.10. Label all CH_3/H 1,3-diaxial interactions in this chair conformation.

Example 2.11 Equilibrium Populations of Conformations

Calculate the ratio of the diequatorial to diaxial conformation of this disubstituted cyclohexane at 25°C.

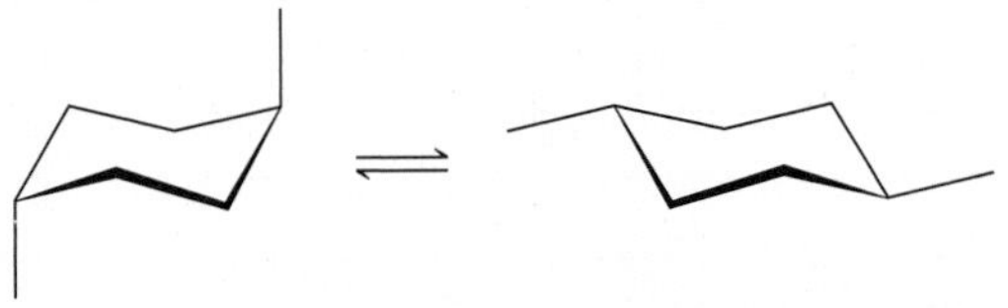

Solution

For these two chair conformations, ΔG^0 (2 axial $CH_3 \rightarrow$ 2 equatorial CH_3) = 2 × (−7.28 kJ/mol) = −14.56 kJ (3.5 kcal)/mol. Substituting this value in the equation $\Delta G^0 = -RT \ln K_{eq}$ gives a ratio of 357:1.

$$\ln K_{eq} = \frac{-(-14{,}560 \text{ J/mol})}{8.314 \text{ J}\cdot \text{K}^{-1}\cdot \text{mol}^{-1} \times 298 \text{ K}} = 5.877$$

$$K_{eq} = \frac{357}{1} = \frac{\text{diequatorial chair conformation}}{\text{diaxial chair conformation}}$$

Thus, at any given instant at room temperature, approximately $357/358 \times 100\% = 99.7\%$ of the molecules of this compound are in the diequatorial chair conformation.

Problem 2.11

Draw a chair conformation of 1,4-dimethylcyclohexane in which one methyl group is equatorial and the other is axial. Draw the alternative chair conformation and calculate the ratio of the two conformations at 25°C.

2.6 *Cis,Trans* Isomerism in Cycloalkanes and Bicycloalkanes

In this section, we introduce the concept of stereoisomerism. **Stereoisomers** are compounds that have (1) the same molecular formula, (2) the same connectivity of their atoms, (3) but a different orientation of their atoms in space. Recall that constitutional isomers, the only other type of isomers we have studied so far, also have the same molecular formula but a different connectivity. The difference between constitutional isomers and stereoisomers is summarized in Figure 2.22.

We begin our study of stereoisomers with the study of *cis,trans* isomerism in cycloalkanes.

Stereoisomers
Compounds that have the same molecular formula, the same connectivity of their atoms, but a different orientation of their atoms in space.

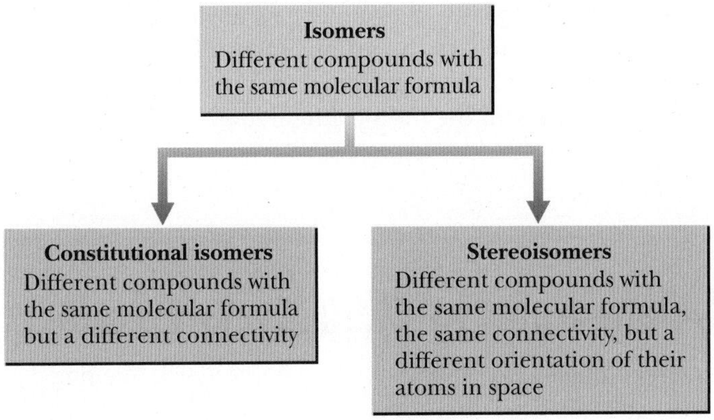

FIGURE 2.22 Relationship between stereoisomers and constitutional isomers.

A. *Cis,Trans* Isomerism in Cycloalkanes

Cycloalkanes with substituents on two or more carbons of the ring show a type of stereoisomerism called ***cis,trans* isomerism**, which we can illustrate by considering 1,2-dimethylcyclopentane. In the following structural formulas, the cyclopentane ring is drawn as a regular pentagon viewed through the plane of the ring. Carbon-carbon bonds of the ring projecting toward you are shown as heavy lines.

***Cis,trans* isomers**
Stereoisomers that have the same connectivity but a different arrangement of their atoms in space as a result of the presence of either a ring or a carbon-carbon double bond.

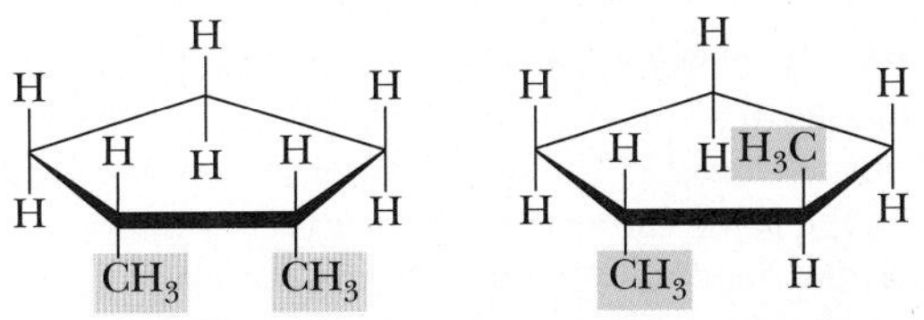

***cis*-1,2-Dimethylcyclopentane** ***trans*-1,2-Dimethylcyclopentane**

Cis

A prefix meaning on the same side.

Trans

A prefix meaning across from.

In one isomer of 1,2-dimethylcyclopentane, the methyl groups are on the same side of the ring; in the other, they are on opposite sides of the ring. The prefix **cis** (Latin: on the same side) indicates that the substituents are on the same side of the ring; the prefix ***trans*** (Latin: across) indicates that they are on opposite sides of the ring. The *cis* isomer cannot be converted to the *trans* isomer and vice versa without breaking and reforming one or more bonds, a process that does not occur at or near room temperature. The *cis* isomer is approximately 7.1 kJ (1.7 kcal)/mol higher in energy (less stable) than the *trans* isomer because of the steric strain of the methyl groups on adjacent carbons in the *cis* isomer.

Alternatively, the cyclopentane ring can be viewed as a regular pentagon seen from above, with the ring in the plane of the page. Substituents on the ring then either project toward you (they project up, above the plane of the page) and are shown by solid wedges or project away from you (they project down, below the plane of the page) and are shown by broken wedges. In the following structural formulas, only the two methyl groups are shown; hydrogen atoms of the ring are not shown.

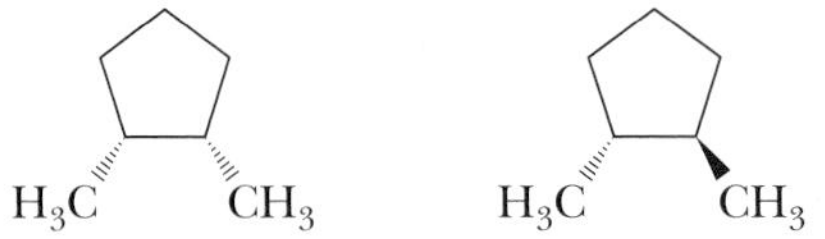

***cis*-1,2-Dimethylcyclopentane** ***trans*-1,2-Dimethylcyclopentane**

Stereocenter

An atom, most commonly carbon, about which exchange of two groups produces a different stereoisomer.

Configuration

Refers to the arrangement of atoms about a stereocenter.

We say that 1,2-dimethylcyclopentane has two stereocenters. A **stereocenter** is an atom, most commonly carbon, about which exchange of two groups produces a different stereoisomer. Both carbons 1 and 2 of 1,2-dimethylcyclopentane, for example, are stereocenters; in this molecule, exchange of H and CH_3 groups at either stereocenter converts a *trans* isomer to a *cis* isomer or vice versa. Alternatively, we refer to the stereoisomers of 1,2-dimethylcyclobutane as having either a *cis* or a *trans* configuration. **Configuration** refers to the arrangement of atoms about a stereocenter. We say, for example, that exchange of groups at either stereocenter in the *cis* configuration gives a stereoisomer with the *trans* configuration.

HOW TO

Convert Planar Cyclohexanes to Chair Cyclohexanes

Following are three different stereorepresentations of *cis*-1,2-dimethylcyclohexane, each with the ring drawn as a planar hexagon.

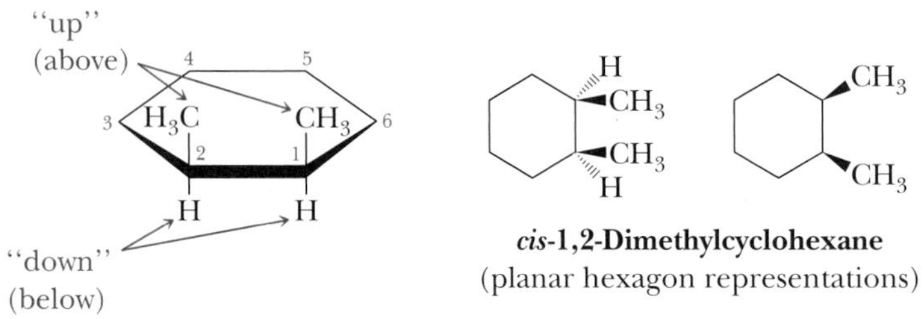

***cis*-1,2-Dimethylcyclohexane**
(planar hexagon representations)

Students often find it difficult to convert substituted cyclohexanes from a planar hexagon representation such as these to a chair conformation. A good rule of thumb is that "up is up and down is down." If a substituent is *up* in a planar hexagon representation, place it *up* on the same carbon of the chair conformation. If a substituent is *down* on a planar hexagon representation, place it *down* on the same carbon of the chair conformation. Note that up or down on a chair conformation may be axial or equatorial depending on which ring carbon you are considering. For *cis*-1,2-dimethylcyclohexane on which both methyl groups are up in the planar hexagon representation, the two methyl groups are also up in a chair conformation. Each of the alternative chair conformations has one methyl group axial and one equatorial. It is generally helpful to draw the hydrogen atoms bonded to the ring carbons bearing substituents to make it absolutely clear which positions are equatorial and which are axial.

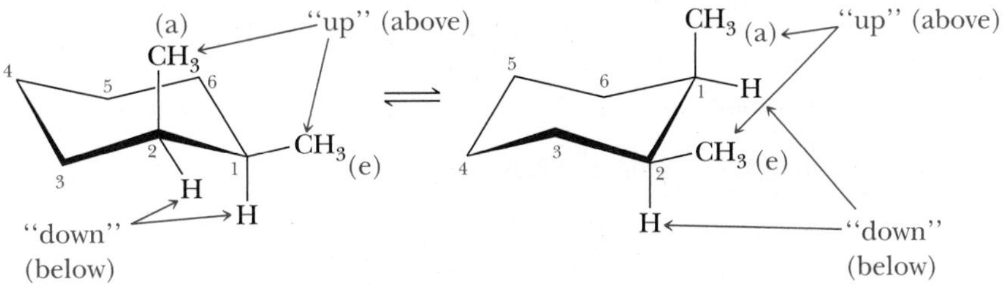

***cis*-1,2-Dimethylcyclohexane**
(alternative chair conformations)

Example 2.12 *Cis* Versus *Trans* Isomerism

Which cycloalkanes show *cis,trans* isomerism? For each that does, draw the *cis* and *trans* isomers.

(a) Methylcyclopentane **(b)** 1,1-Dimethylcyclopentane

(c) 1,3-Dimethylcyclobutane

Solution

(a) Methylcyclopentane does not show *cis,trans* isomerism. It has only one substituent on the ring.

(b) 1,1-Dimethylcyclopentane does not show *cis,trans* isomerism. Because both methyl groups are bonded to the same carbon, only one arrangement is possible for them.

(c) 1,3-Dimethylcyclobutane shows *cis,trans* isomerism. In the following structural formulas, cyclobutane is drawn as a planar ring viewed first from the side and then from above.

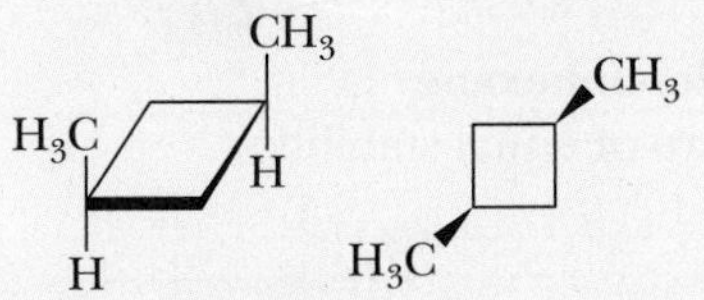

***cis*-1,3-Dimethylcyclobutane**

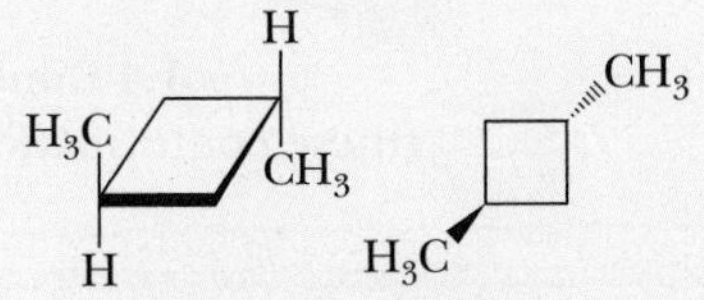

***trans*-1,3-Dimethylcyclobutane**

Problem 2.12

Which cycloalkanes show *cis,trans* isomerism? For each that does, draw both isomers.

(a) 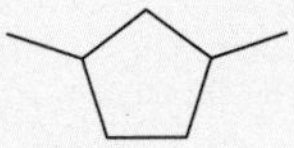 **(b)** 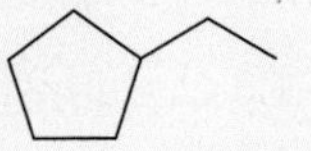 **(c)** 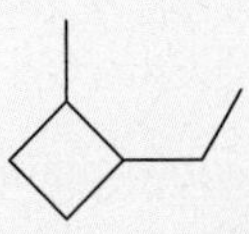

Two *cis,trans* isomers are possible for 1,4-dimethylcyclohexane. For the purposes of determining the number of *cis,trans* isomers in substituted cycloalkanes, it is adequate to draw the cycloalkane ring as a planar polygon as is done here.

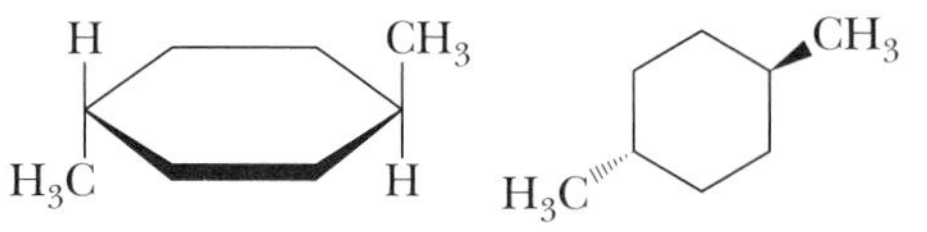

***trans*-1,4-Dimethylcyclohexane**

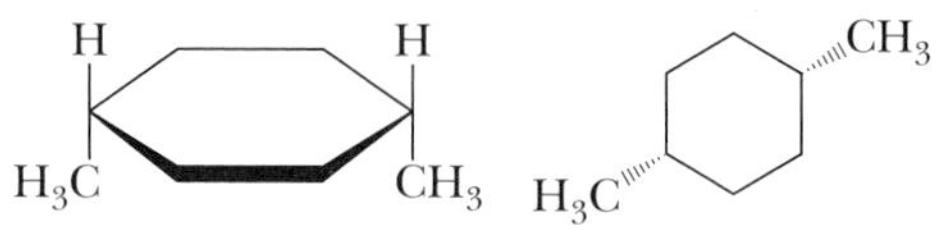

***cis*-1,4-Dimethylcyclohexane**

We can also draw the *cis* and *trans* isomers of 1,4-dimethylcyclohexane as nonplanar chair conformations. In working with alternative chair conformations, it is helpful to remember that all groups axial in one chair become equatorial in the alternative chair and vice versa. In one chair conformation of *trans*-1,4-dimethylcyclohexane, the two methyl groups are axial; in the alternative chair conformation, they are equatorial. Of these chair conformations, the one with both methyl groups equatorial is more stable by approximately 14.6 kJ (3.5 kcal)/mol and makes up the large majority of a sample of *trans*-1,4-dimethylcyclohexane.

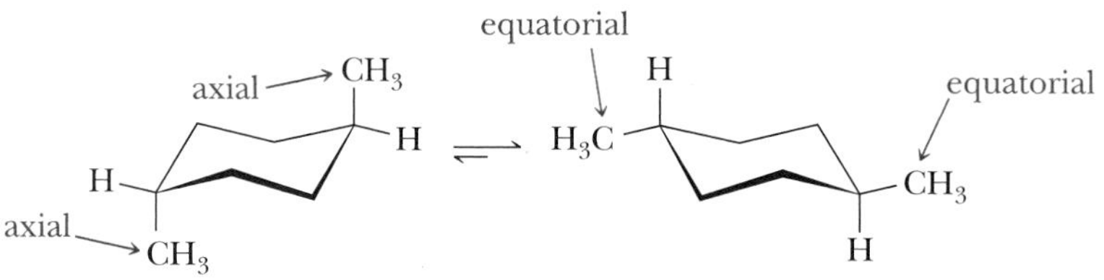

Less stable chair More stable chair

***trans*-1,4-Dimethylcyclohexane**

The alternative chair conformations of *cis*-1,4-dimethylcyclohexane are of equal energy. In one chair, one methyl group is equatorial and the other is axial. In the alternative chair, the orientations in space of the methyl groups are reversed. The result is that a collection of *cis*-1,4-dimethylcyclohexane molecules is composed of rapidly equilibrating alternative chairs in equal proportions.

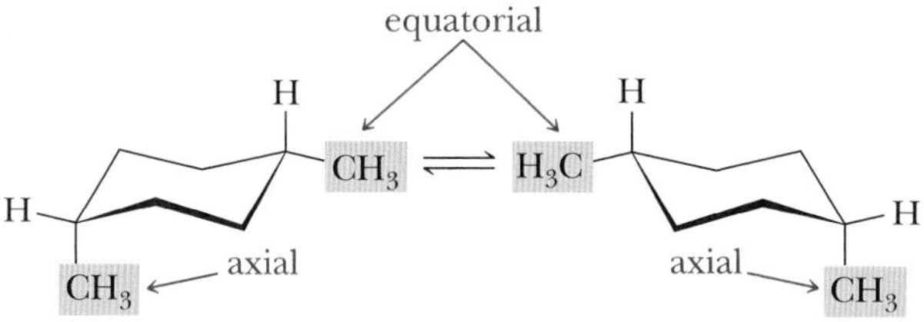

***cis*-1,4-Dimethylcyclohexane**
(these conformations are of equal stability)

Example 2.13 Cyclohexane Rings

Following is a chair conformation of 1,3-dimethylcyclohexane.

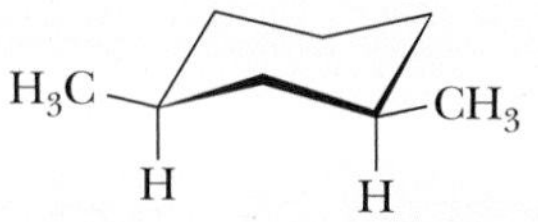

(a) Is this a chair conformation of *cis*-1,3-dimethylcyclohexane or of *trans*-1,3-dimethylcyclohexane?

(b) Draw the alternative chair conformation of this compound. Of the two chair conformations, which is more stable?

(c) Draw a planar hexagon representation of the isomer shown in this example.

Solution

(a) The isomer shown is *cis*-1,3-dimethylcyclohexane; the two methyl groups are on the same side of the ring.

(b)

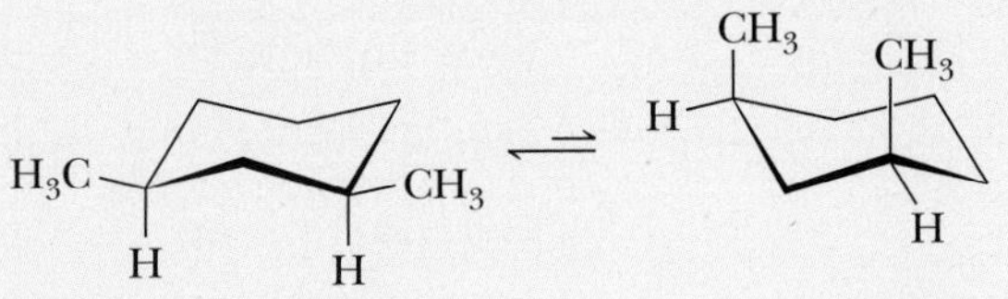

Diequatorial conformation (more stable) — Diaxial conformation (less stable)

(c)

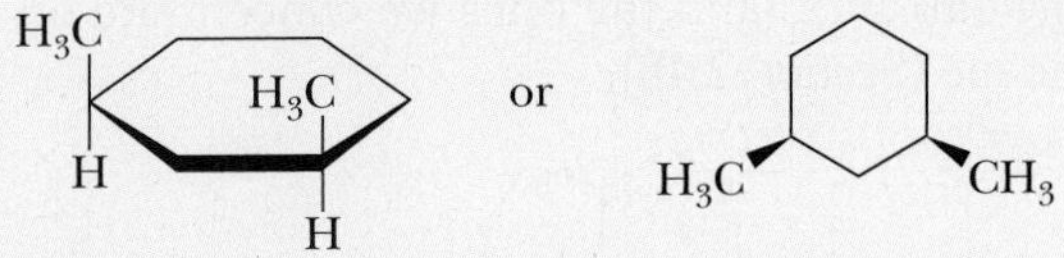

Problem 2.13

Watch a video explanation

Following is a planar hexagon representation for one isomer of 1,2,4-trimethylcyclohexane. Draw the alternative chair conformations of this compound and state which of the two is more stable.

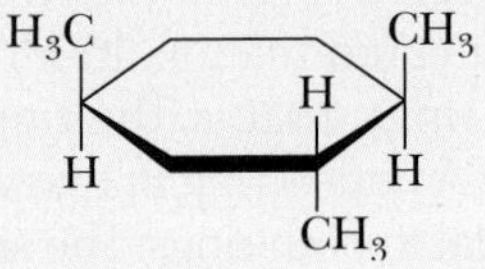

Example 2.14 Substituted Cyclohexane Rings

Here is one *cis,trans* isomer of 2,4-dimethylcyclohexanol. Complete the alternative chair conformations on the right.

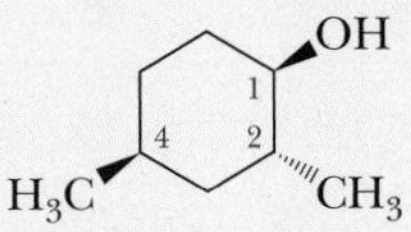

(a)

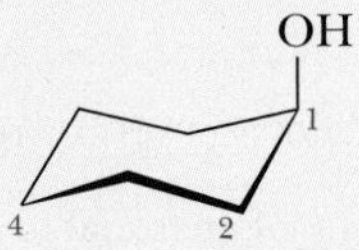

(b)

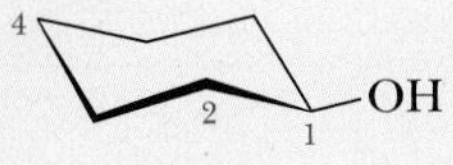

Solution

For (a), the CH_3 group on carbon 2 must be below the plane of the ring, which on this carbon is axial. The CH_3 group on carbon 4 must be above the plane of the ring, which on this carbon is equatorial. (b) The methyl group on carbon 2 is equatorial; the methyl group on carbon 4 is axial.

(a)

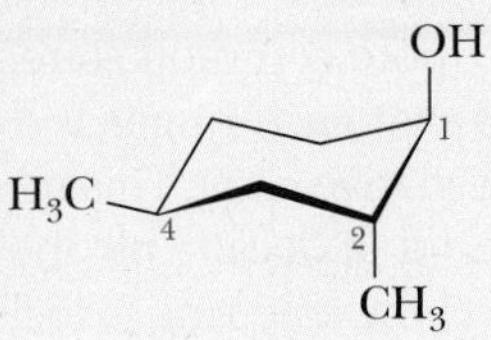

(b)

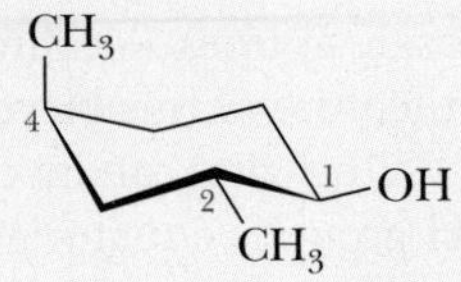

(Continued)

Problem 2.14

Here is one *cis,trans* isomer of 3,5-dimethylcyclohexanol. Complete the alternative chair conformations.

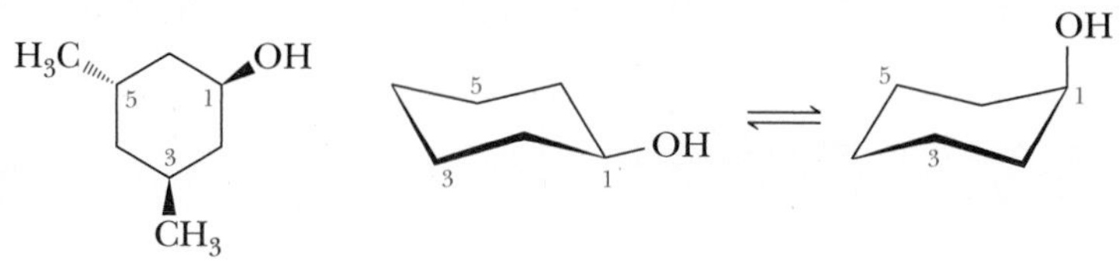

B. *Cis,Trans* Isomerism in Bicycloalkanes

By far, the most common bicycloalkanes (and the ones we concentrate on in this section) are decalin and hydrindane (Section 2.4B).

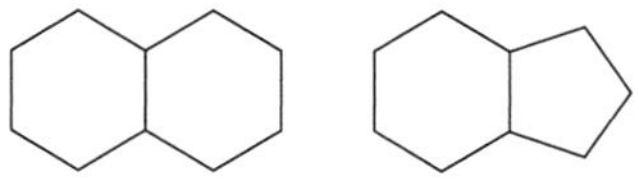

Decalin **Hydrindane**

Two stereoisomers of decalin and hydrindane are possible depending on whether the two hydrogen atoms at the ring junction are *trans* or *cis* to each other. If we draw conformations for the six-membered rings in the two decalins, we see that each ring can exist in its more stable chair conformation. In *trans*-decalin, the hydrogens at the ring junction are axial to both rings; that is, the ring-junction hydrogen above the plane of the rings is axial to ring A and to ring B. Likewise, the ring-junction hydrogen below the plane of the ring is axial to both rings. The situation is different in *cis*-decalin. Each ring-junction hydrogen is axial to one ring but equatorial to the other ring.

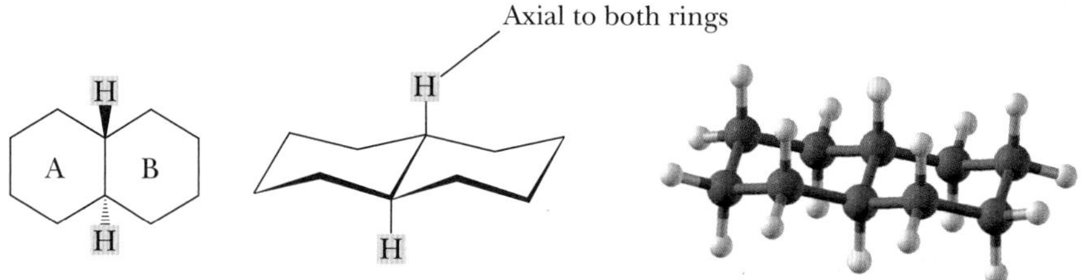

***trans*-Decalin**

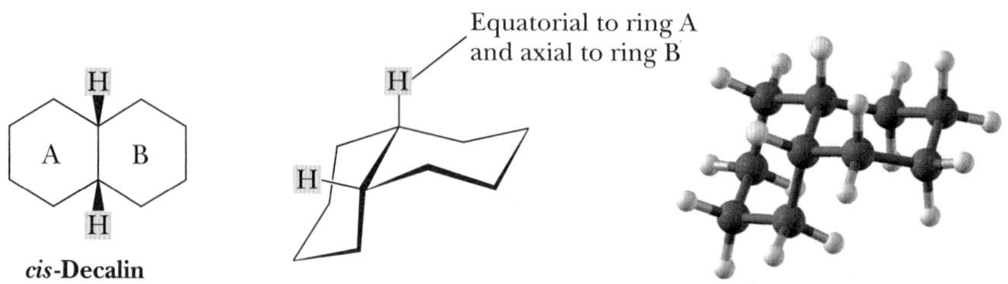

***cis*-Decalin**

Let us look more closely at *trans*-decalin, by far the more common stereoisomer of decalin. An important feature of this bicycloalkane is that each ring is locked into one chair conformation; neither ring can invert to its alternative chair. This means, for example, that if an —OH group is equatorial in a decalinol (a decalin alcohol), it remains equatorial; it cannot become axial because the cyclohexane ring is locked into this one conformation. Likewise, if an —OH group is axial, it remains axial.

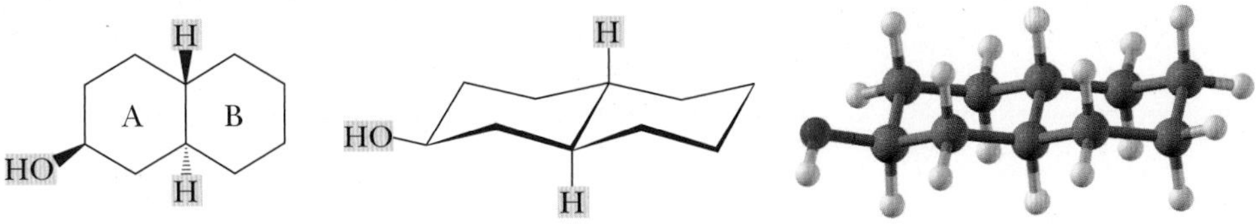

Suppose you are given the structural formula on the left for the decalinol. Can you tell from looking at this structure whether the —OH group is axial or equatorial? You can't tell directly, but you can figure it out. Remember that in *trans*-decalin, the H atoms at the ring junctions are axial to each ring. Remember also that in a chair cyclohexane, axial is up on one carbon, down on the next, up on the next, and so on. Therefore, if you start with the axial group at either ring junction and work your way from one carbon to the next until you come to the carbon bearing the —OH group, you come to the conclusion that the —OH on the structural formula is equatorial to ring A.

A good example of the occurrence of these types of ring systems is in the steroids, all of which contain a carbon skeleton consisting of 3 six-membered rings and 1 five-membered ring connected as shown here. This ring system is present in both animal and plant steroids. Steroids are present in human metabolism as cholesterol, steroid hormones, and bile acids (Section 26.4).

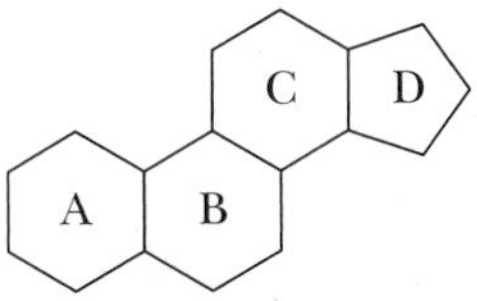

The steroid nucleus

MCAT Practice: *Passage and Questions*

Tetrodotoxin

Tetrodotoxin is one of the most potent toxins known. It is composed of a set of interconnected six-membered rings, each in a chair conformation. All but one of these rings has atoms other than carbon in them. Tetrodotoxin is produced in the liver and ovaries of many species of *Tetraodontidae*, particularly the puffer fish, so called because it inflates itself to an almost spherical spiny ball when it is alarmed. This, however, does not put off everyone. Puffer fish is regarded as a delicacy called "fugu" in Japan. To serve the fish in a public restaurant, the chef must be registered as skilled in removing the toxic organs so as to make the flesh safe to eat.

© Beth Swanson/Shutterstock.com

A puffer fish with its body inflated.

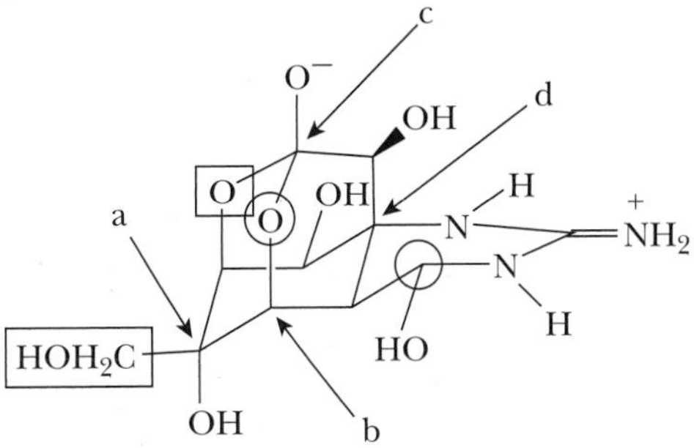

Tetrodotoxin

(*Continued*)

Questions

A. What are the relationships of the boxed atoms and the circled atoms?

1. The boxed atoms are *trans* and the circled atoms are *cis*.
2. The boxed atoms are *cis* and the circled atoms are *trans*.
3. Both sets of atoms are *cis*.
4. Both sets of atoms are *trans*.

B. To what kinds of carbons, 1°, 2°, 3°, or 4°, do the arrows **a**, **b**, **c**, and **d** point?

1. They are all tertiary.
2. Carbons **a** and **c** are tertiary, while **b** and **d** are secondary.
3. Carbon **b** is secondary, while carbons **a**, **b**, and **c** are tertiary.
4. Carbon **c** is primary, carbon **b** is secondary, and carbons **a** and **d** are tertiary.

C. What is the hybridization of the nitrogens within the ring on the right and the nitrogen protruding from the ring?

1. The nitrogens within the ring are sp^3 while the protruding nitrogen is sp^2.
2. They are all sp^3.
3. They are all sp^2.

Following are two stereorepresentations for cholestanol. In the conformational representation on the right, notice that all ring junctions are *trans,* all groups at each ring-junction atom are axial to the ring, and the —OH group on ring A is equatorial.

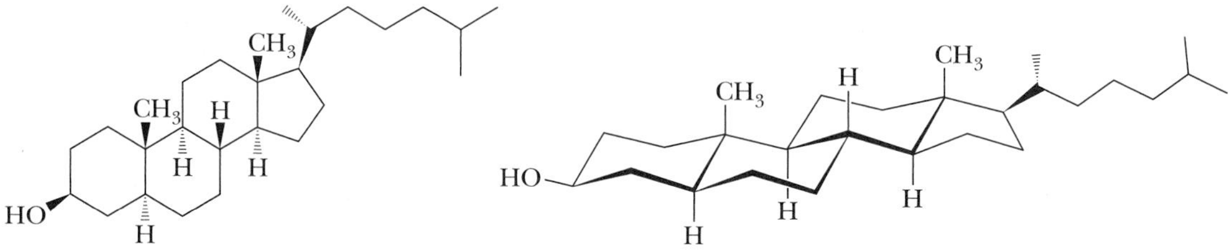

Cholestanol

Another type of bicycloalkane is a six-membered ring in which an added CH_2 group forms a bridge between carbons 1 and 4. You can view and draw this molecule from any number of perspectives. What becomes obvious if you view it from the side, as in (c), is that the one-carbon bridge locks the six-membered ring into a boat conformation. Notice that even though (a) and (b) show the carbon skeleton of the molecule, it is not obvious from them that a locked boat conformation is embedded in the molecule. The lesson here is that it is essential to draw a molecule as a three-dimensional shape to best reveal what you want to show.

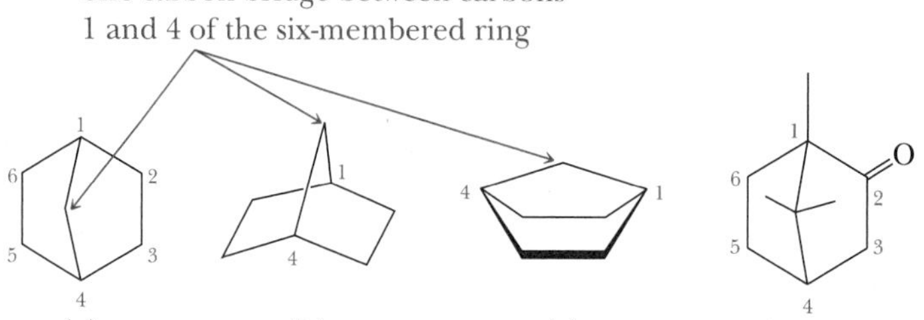

(a) **(b)** **(c)** **Camphor**

An example of a natural product containing this bicyclic skeleton is camphor.

Another example of a carbon skeleton that contains several six-membered rings, all of which are locked into chair conformations, is adamantane (c). To understand how the carbon skeleton of adamantane can be constructed, imagine that you (a) start with a chair cyclohexane, (b) add the three axial bonds on the top side of the ring, and (c) then connect each of the axial bonds to a CH group. You now have adamantane, a compound first isolated from petroleum. Amantadine, a 1° amino derivative of adamantane, is an antiviral agent used to treat influenza A.

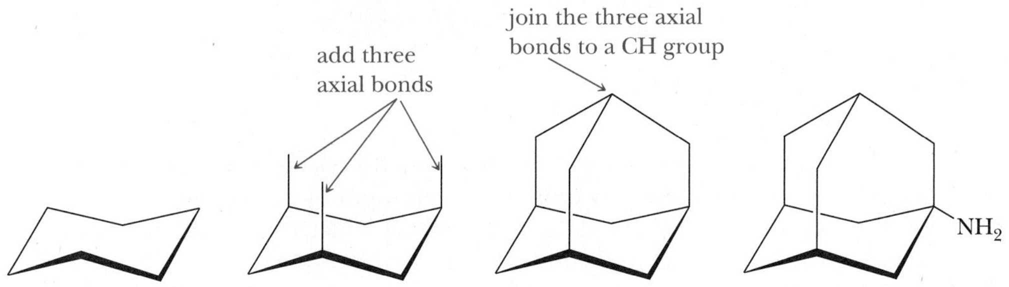

(a) Cyclohexane **(b)** **(c) Adamantane** **Amantadine** (an antiviral agent)

2.7 Physical Properties of Alkanes and Cycloalkanes

The most important property of alkanes and cycloalkanes is their almost complete lack of polarity. As we saw in Section 1.2B, the difference in electronegativity between carbon and hydrogen is $2.5 - 2.1 = 0.4$ on the Pauling scale, and given this small difference, we classify a C—H bond as nonpolar covalent. Therefore, alkanes are nonpolar compounds, and only weak interactions exist between their molecules.

A. Dispersion Forces and Interactions Among Alkane Molecules

Methane is a gas at room temperature and atmospheric pressure. It can be converted to a liquid if cooled to −164°C, and to a solid if further cooled to −182°C. The fact that methane (or any other compound, for that matter) can exist as a liquid or a solid depends upon intermolecular attraction. Although the forces of attraction are generally electrostatic, they vary widely in relative strength. The strongest attractive forces are between ions, as for example between Na^+ and Cl^- in NaCl [787 kJ (188 kcal)/mol]. Dipole-dipole interactions and hydrogen bonding [8–42 kJ (2–10 kcal)/mol] are weaker. We shall have more to say about these intermolecular attractive forces in Chapter 10.

Dispersion forces Very weak intermolecular forces of attraction resulting from the interaction between temporary induced dipoles.

Dispersion forces [0.08 − 8 kJ (0.02 − 2 kcal)/mol] are the weakest intermolecular attractive forces. The existence of dispersion forces accounts for the ability to liquefy low-molecular-weight, nonpolar substances, such as hydrogen (H_2), neon (Ne), and methane (CH_4). To visualize the origin of dispersion forces, think in terms of instantaneous distributions of electron density rather than average distributions. Consider, for example, neon, a gas at room temperature and 1 atm pressure, which can be liquefied when cooled to −246°C. The heat of vaporization tells us that the neon-neon attractive interaction in the liquid state is approximately 0.3 kJ (0.07 kcal)/mol. The intermolecular attractive force is accounted for in the following way. Over time, the distribution of electron density in a neon atom is symmetrical, and there is no dipole moment [Figure 2.23(a)]. However, at any instant, there is a nonzero probability that its electron density will be polarized (shifted) more toward one part of the atom than toward another. This temporary polarization creates a temporary dipole moment, which in turn induces temporary dipole moments in adjacent atoms [Figure 2.23(b)].

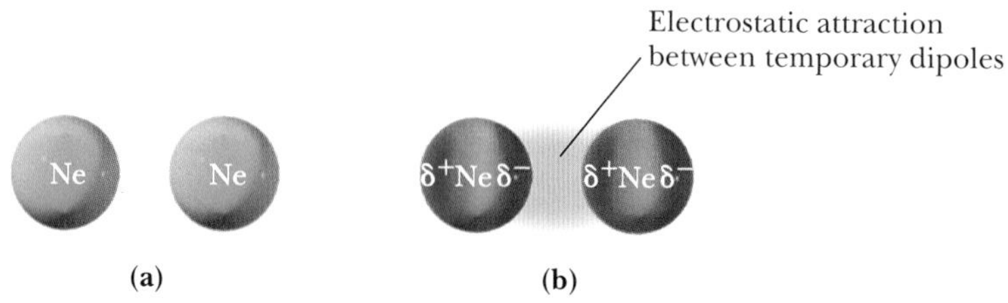

FIGURE 2.23 Dispersion forces. (a) The distribution of electron density averaged over time in a neon atom is symmetrical, and there is no net polarity. (b) Temporary polarization of one neon atom induces temporary polarization in adjacent atoms. Electrostatic attractions between temporary dipoles are called dispersion forces.

The small attraction referred to as dispersion forces is the source of electrostatic attraction between temporary dipoles. The strength of dispersion forces depends on how easily an electron cloud can be polarized. Electrons in smaller atoms and molecules are held closer to their nuclei and, therefore, are not easily polarized. The strength of dispersion forces tends to increase with increasing molecular mass and size. Intermolecular attractive forces between Cl_2 molecules and between Br_2 molecules are estimated to be 2.9 kJ (0.7 kcal)/mol and 4.2 kJ (1.0 kcal)/mol, respectively. Dispersion forces are inversely proportional to the sixth power of the distance between interacting atoms or molecules. For them to be important, the interacting atoms or molecules must be in virtual contact with one another.

B. Boiling Points, Melting Points, and Density

Because interactions between alkane molecules consist only of very weak dispersion forces, the boiling points of alkanes are lower than those of almost any other type of compound of the same molecular weight. As the number of atoms and the molecular weight of alkanes increase, there is more opportunity for dispersion forces between their molecules and boiling points increase. Although, in general, both boiling and melting points of alkanes increase as molecular weight increases (Table 2.5), the increase in melting points is not as regular as that observed for boiling points. In solids, the packing of molecules into ordered patterns of solids changes as molecular size and shape change.

Alkanes containing 1 to 4 carbons are gases at room temperature; those containing 5 to 17 carbons are colorless liquids. High-molecular-weight alkanes (those with

Table 2.5 Physical Properties of Some Unbranched Alkanes

Name	Condensed Structural Formula	Melting Point (°C)	Boiling Point (°C)	Density of Liquid* (g/mL at 0°C)
Methane	CH_4	−182	−164	(a gas)
Ethane	CH_3CH_3	−183	−88	(a gas)
Propane	$CH_3CH_2CH_3$	−190	−42	(a gas)
Butane	$CH_3(CH_2)_2CH_3$	−138	0	(a gas)
Pentane	$CH_3(CH_2)_3CH_3$	−130	36	0.626
Hexane	$CH_3(CH_2)_4CH_3$	−95	69	0.659
Heptane	$CH_3(CH_2)_5CH_3$	−90	98	0.684
Octane	$CH_3(CH_2)_6CH_3$	−57	126	0.703
Nonane	$CH_3(CH_2)_7CH_3$	−51	151	0.718
Decane	$CH_3(CH_2)_8CH_3$	−30	174	0.730

*For comparison, the density of H_2O is 1 g/mL at 4°C

18 or more carbons) are white, waxy solids. Several plant waxes are high-molecular-weight alkanes. The wax found in apple skins, for example, is an unbranched alkane with molecular formula $C_{27}H_{56}$. Paraffin wax, a mixture of high-molecular-weight alkanes, is used for wax candles, in lubricants, and to seal home-canned jams, jellies, and other preserves. Petrolatum, so named because it is derived from petroleum refining, is a liquid mixture of high-molecular-weight alkanes. Sold as mineral oil (typically C_{15} to C_{40}) and Vaseline (C_{25} and up), petrolatum is used as an ointment base in pharmaceuticals and cosmetics and as a lubricant and rust preventative. The average density of the alkanes listed in Table 2.5 is about 0.7 g/mL; that of higher-molecular-weight alkanes is about 0.8 g/mL. All liquid and solid alkanes are less dense than water (1.0 g/mL) and, therefore, float in it.

C. Constitutional Isomers Have Different Physical Properties

Alkanes that are constitutional isomers are different compounds and have different physical and chemical properties. Listed in Table 2.6 are boiling points, melting points, and densities for the five constitutional isomers of C_6H_{14}. The boiling point of each branched-chain isomer of C_6H_{14} is lower than that of hexane itself; the more branching there is, the lower the boiling point. As branching increases, the shape of an alkane molecule becomes more compact and its surface area decreases. As surface area decreases, contact among adjacent molecules decreases, the strength of dispersion forces decreases, and boiling points decrease. Thus, for any group of alkane constitutional isomers, the least branched isomer usually has the highest boiling point and the most branched isomer usually has the lowest boiling point.

Table 2.6 **Physical Properties of the Isomeric Alkanes of Molecular Formula C_6H_{14}**

Name	Boiling Point (°C)	Melting Point (°C)	Density (g/mL)
Hexane	68.7	−95	0.659
2-Methylpentane	60.3	−154	0.653
3-Methylpentane	63.3	−118	0.664
2,3-Dimethylbutane	58.0	−129	0.661
2,2-Dimethylbutane	49.7	−98	0.649

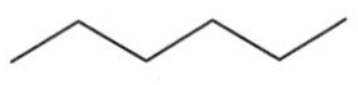

Hexane

2,2-Dimethylbutane

Example 2.15 Alkane Boiling Points

Arrange the alkanes in each set in order of increasing boiling point.

(a) Butane, decane, and hexane

(b) 2-Methylheptane, octane, and 2,2,4-trimethylpentane

Solution

(a) All of these compounds are unbranched alkanes. As the number of carbon atoms in the chain increases, dispersion forces between molecules

(Continued)

increase, as does the amount of energy required to put the molecules into motion. Therefore, the larger the unbranched alkane, the higher the boiling point. Decane has the highest boiling point, and butane has the lowest.

Butane	Hexane	Decane
bp −0.5°C	bp 69°C	bp 174°C

(b) These three alkanes are constitutional isomers with molecular formula C_8H_{18}. Their relative boiling points depend on the degree of branching. 2,2,4-Trimethylpentane, the most highly branched isomer, has the smallest surface area and the lowest boiling point. Octane, the unbranched isomer, has the largest surface area and the highest boiling point.

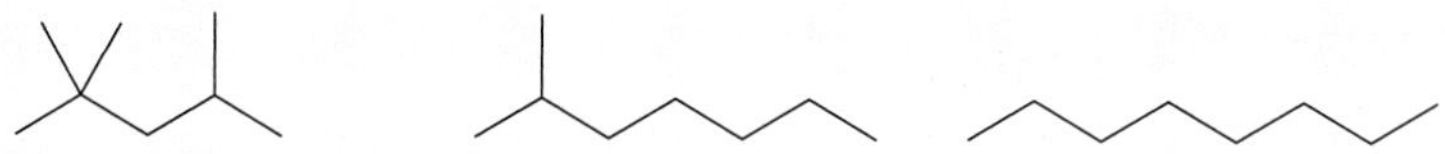

2,2,4-Trimethylpentane	2-Methylheptane	Octane
(bp 99°C)	(bp 118°C)	(bp 126°C)

Problem 2.15

Arrange the alkanes in each set in order of increasing boiling point.

(a) 2-Methylbutane, 2,2-dimethylpropane, and pentane

(b) 3,3-Dimethylheptane, 2,2,4-trimethylhexane, and nonane

2.8 Reactions of Alkanes

Alkanes and cycloalkanes are quite unreactive toward most reagents, a behavior consistent with the fact that they are nonpolar compounds and contain only strong sigma bonds. Under certain conditions, however, alkanes and cycloalkanes do react with O_2 and with the halogens Cl_2 and Br_2. At this point, we present only their combustion with oxygen. We discuss their reaction with halogens in Chapter 8.

A. Oxidation

The oxidation of alkanes by O_2 to give carbon dioxide and water is by far their most economically important reaction. Oxidation of saturated hydrocarbons is the basis for their use as energy sources for heat [natural gas, liquefied petroleum gas (LPG), and fuel oil] and power (gasoline, diesel fuel, and aviation fuel). Following are balanced equations for the complete oxidation of methane (the major component of natural gas) and propane (the major component of LPG).

$$\underset{\textbf{Methane}}{CH_4} + 2O_2 \longrightarrow CO_2 + 2H_2O \qquad \Delta H^0 = -890.4\ \text{kJ}\ (-212.8\ \text{kcal})/\text{mol}$$

$$\underset{\textbf{Propane}}{CH_3CH_2CH_3} + 5O_2 \longrightarrow 3CO_2 + 4H_2O \qquad \Delta H^0 = -2220\ \text{kJ}\ (-530.6\ \text{kcal})/\text{mol}$$

Heat of combustion

Standard heat of combustion is the heat released when 1 mole of a substance in its standard state (gas, liquid, or solid) is oxidized completely to carbon dioxide and water and is given the symbol ΔH^0.

In this and all other hydrocarbon oxidations, the energy of the products is less than that of the reactants, with the difference in energy being given off as the **heat of combustion**. The heat of combustion is the energy of the products minus that of the reactants.

Table 2.7 Heats of Combustion of Four Constitutional Isomers of C_8H_{18}

Hydrocarbon	Line-angle Formula	ΔH^0 [kJ/mol (kcal/mol)]
Octane		−5470.6 (−1307.5)
2-Methylheptane	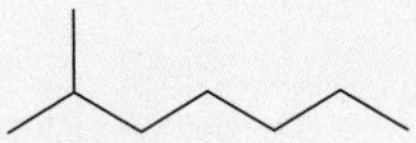	−5465.6 (−1306.3)
2,2-Dimethylhexane	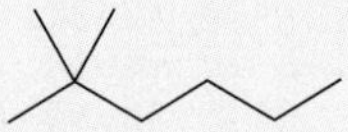	−5458.4 (−1304.6)
2,2,3,3-Tetramethylbutane	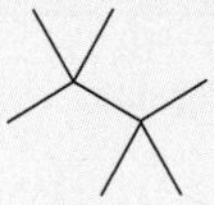	−5451.8 (−1303.0)

B. Heats of Combustion and Relative Stability of Alkanes and Cycloalkanes

One important use of heats of combustion is to give us information about the relative stabilities of isomeric hydrocarbons. To illustrate, consider the heats of combustion of the four constitutional isomers given in Table 2.7. All four compounds undergo combustion according to this equation.

$$C_8H_{18} + \frac{25}{2}O_2 \longrightarrow 8CO_2 + 9H_2O$$

We see that octane has the largest (most negative) heat of combustion. As branching increases, the ΔH^0 decreases (becomes less negative). Of these four isomers, the isomer with four methyl branches has the lowest (least negative) heat of combustion. Therefore, we conclude that branching increases the stability of an alkane.

Figure 2.24 is a graphical analysis of the data in Table 2.7. Because all four compounds give the same products on oxidation, the only differences among them are their relative energies.

As we saw in Section 2.5B, there is considerable strain in small-ring cycloalkanes. We can measure this strain by measuring the heat of combustion versus

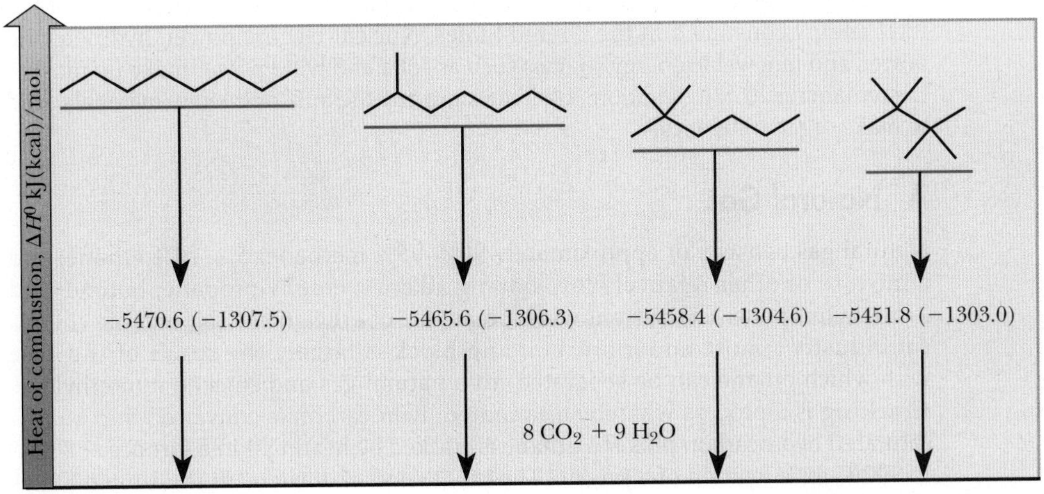

FIGURE 2.24 Heats of combustion in kJ (kcal)/mol of four isomeric alkanes with molecular formula C_8H_{18}.

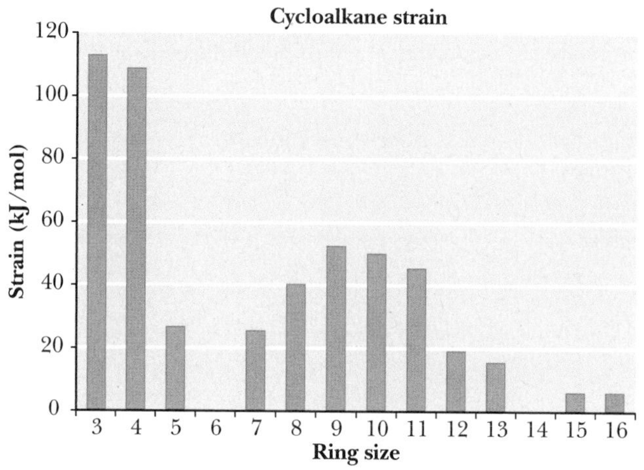

FIGURE 2.25 Strain energy of cycloalkanes as a function of ring size.

the ring size. It has been determined by measurement of the heats of combustion of a series of unbranched alkanes that the average heat of combustion per methylene (CH_2) group is 658.7 kJ (157.4 kcal)/mol. Using this value, we can calculate a predicted heat of combustion for each cycloalkane. Strain energy is the difference between the predicted and actual heats of combustion. These results are displayed graphically in Figure 2.25.

We see that cyclopropane has the largest strain energy of any cycloalkane, which is consistent with the extreme compression of its C—C—C bond angles from 109.5° to 60°. Cyclobutane and cyclopentane each have less strain, and cyclohexane, as expected, has zero strain. What is perhaps surprising is the presence of strain in rings of from 7 to 13 carbon atoms. This strain is primarily the result of torsional and steric strain caused by the fact that these rings are constrained to conformations that cannot achieve ideal bond and torsional angles.

The large amount of strain suffered by the small cycloalkanes means that these structures are unstable relative to the larger cycloalkanes, such as cyclohexane. During the combustion process, all the strain is released because stable entities (CO_2 and H_2O) are formed. The strain in cyclopropane relative to cyclohexane is released as more heat (per carbon) during burning. In other words, cyclopropane burns hotter because considerable energy is stored in its unstable structure.

2.9 Sources and Importance of Alkanes

The three major sources of alkanes throughout the world are the fossil fuels, namely natural gas, petroleum, and coal. Fossil fuels account for approximately 80% of the total energy consumed in the United States. Nuclear electric power, hydroelectric power, and renewable energy sources such as solar and wind power make up most of the remaining 20%. In addition, fossil fuels provide the bulk of the raw materials used to make organic chemicals.

A. Natural Gas

Natural gas consists of approximately 90%–95% methane; 5%–10% ethane; and a mixture of other relatively low-boiling alkanes, chiefly propane, butane, and 2-methylpropane. The current widespread use of ethylene as the organic chemical industry's most important building block is largely the result of the ease with which ethane can be separated from natural gas and cracked into ethylene. **Cracking** is a process whereby a saturated hydrocarbon is converted into an unsaturated hydrocarbon plus H_2. Ethane is cracked by heating it in a furnace at 800°C to 900°C for a fraction of a second. The production of ethylene in the United States in 2010 was 49.8 billion pounds, making it the number one organic compound produced by the U.S. chemical industry, on a weight basis. The bulk of ethylene produced in this manner is used to create organic polymers, as described in Chapter 29.

$$CH_3CH_3 \xrightarrow{800\text{–}900°C} CH_2{=}CH_2 + H_2$$

Ethane **Ethene (Ethylene)**

A petroleum refinery.

B. Petroleum

Petroleum is a thick, viscous liquid mixture of thousands of compounds, most of them hydrocarbons, formed from the decomposition of marine plants and animals. Petroleum and petroleum-derived products fuel automobiles, aircraft, and trains. They provide most of the greases and lubricants required for the machinery of our highly industrialized society. Furthermore, petroleum, along with natural gas, provides nearly 90% of the organic raw materials for the synthesis and manufacture of synthetic fibers, plastics, detergents, drugs, dyes, and a multitude of other products.

From the thousands of different hydrocarbons in the liquid mixture, the task of the petroleum refining industry is to produce usable products with a minimum of waste. The various physical and chemical processes for this purpose fall into two broad categories: separation processes, which separate the complex mixture into various fractions, and reforming processes, which alter the molecular structure of the hydrocarbon components themselves.

The fundamental separation process in refining petroleum is fractional distillation (Figure 2.26). Most crude petroleum that enters a refinery goes to distillation units, where it is heated to temperatures as high as 370°C to 425°C and separated into fractions. Each fraction contains a mixture of hydrocarbons that boils within a particular range.

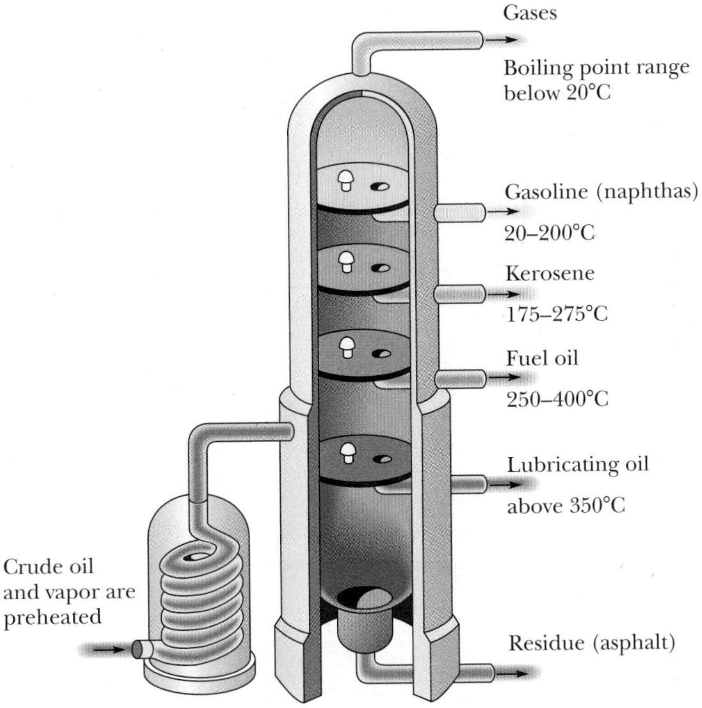

FIGURE 2.26 Fractional distillation of petroleum. The lighter, more volatile fractions are removed from higher up the column; the heavier, less volatile fractions are removed from lower down.

Chemical Connections

Octane Rating: What Those Numbers at the Pump Mean

Gasoline is a complex mixture of C_6 to C_{12} hydrocarbons. The quality of gasoline as a fuel for internal combustion engines is expressed in terms of an octane rating. Engine knocking occurs when a portion of the air-fuel mixture explodes prior to the piston reaching the top of its stroke (usually as a result of heat developed during compression) and independently of ignition by the spark plug. The resulting shockwave of the piston against the cylinder wall reverberates, creating a characteristic metallic "pinging" sound. Two compounds were selected as reference fuels. One of these, 2,2,4-trimethylpentane (isooctane), has very good antiknock properties and was assigned an octane rating of 100. The name *isooctane* is a trivial name; its only relation to 2,2,4-trimethylpentane is that both compounds have eight carbons. Heptane, the other reference compound, has poor antiknock properties and was assigned an octane rating of 0.

Heptane (octane rating 0) **2,2,4-Trimethylpentane** (octane rating 100)

The **octane rating** of a particular gasoline is the percent of isooctane in a mixture of isooctane and heptane that has antiknock properties equivalent to that of the test gasoline. For example, the antiknock properties of 2-methylhexane are the same as those of a mixture of 42% isooctane and 58% heptane; therefore, the octane rating of 2-methylhexane is 42. Octane itself has an octane rating of −20, which means that it produces even more engine knocking than heptane.

The two most common reforming processes are cracking, as illustrated by the thermal conversion of ethane to ethylene (Section 2.9A), and catalytic reforming. Catalytic reforming is illustrated by the conversion of hexane first to cyclohexane and then to benzene.

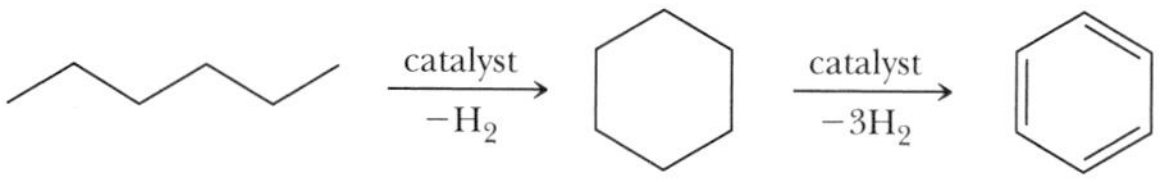

Hexane **Cyclohexane** **Benzene**

C. Coal

To understand how coal can be used as a raw material for the production of organic compounds, it is necessary to discuss synthesis gas. **Synthesis gas** is a mixture of carbon monoxide and hydrogen in varying proportions depending on the means by which it is manufactured. Synthesis gas is prepared by passing steam over hot coal; it is also prepared by partial oxidation of methane with oxygen.

$$\underset{\textbf{Coal}}{C} + H_2O \xrightarrow{\text{heat}} CO + H_2$$

$$\underset{\textbf{Methane}}{CH_4} + \frac{1}{2}O_2 \xrightarrow{\text{catalyst}} CO + 2H_2$$

Two important organic compounds produced today almost exclusively from carbon monoxide and hydrogen are methanol and acetic acid. In the production of methanol, the ratio of carbon monoxide to hydrogen is adjusted to 1:2 and the mixture is passed over a catalyst at elevated temperature and pressure.

$$CO + 2H_2 \xrightarrow{\text{catalyst}} CH_3OH$$

Methanol

Treatment of methanol, in turn, with carbon monoxide over a different catalyst gives acetic acid.

$$\underset{\textbf{Methanol}}{CH_3OH} + CO \xrightarrow{\text{catalyst}} \underset{\textbf{Acetic acid}}{CH_3\overset{O}{\overset{\|}{C}}OH}$$

Because the processes for making methanol and acetic acid directly from carbon monoxide are commercially proven, the decades ahead will likely see the development of routes to other organic chemicals from coal via methanol.

Study Guide

2.1 The Structure of Alkanes

- A **hydrocarbon** is a compound composed only of carbon and hydrogen.
  - *Saturated hydrocarbons (alkanes and cycloalkanes) contain only C—C single bonds. Alkanes have the general formula C_nH_{2n+2}.*

P 2.16–2.18

2.2 Constitutional Isomerism in Alkanes

- **Constitutional isomers** have the same molecular formula but a different connectivity of their atoms. Different constitutional isomers have different physical and chemical properties.

P 2.1, 2.2, 2.19–2.23, 2.25

2.3 Nomenclature of Alkanes and the IUPAC System

- Alkanes are named according to a systematic set of rules developed by the International Union of Pure and Applied Chemistry (IUPAC). To name an alkane:
  - *The main chain is identified, which is the longest carbon chain. The alkyl group substituents are identified and named, each one ending in -yl.*
  - *The main chain is numbered to give the first substituent encountered the lowest number and each substituent is assigned a number from the main chain.*

- *The name is constructed by listing the substituents with their numbers in alphabetical order, followed by the main chain name ending in* -ane.
- *If there is more than one identical substituent, use the prefix di-, tri-, tetra-, penta-, etc. These substituent prefixes are not considered when alphabetizing the substituents in the name.*
- *If a molecule has more than one parent chain of equal length, choose the parent chain with the most substituents.*

P 2.24, 2.26–2.28

- An older common nomenclature for naming molecules is still used for many common alkanes, but only the IUPAC system can name all molecules.
- The IUPAC name of a compound consists of three parts:
  - *A* ***prefix*** *that tells the number of carbon atoms in the parent chain.*
  - *An* ***infix*** *that tells the nature of the carbon-carbon bonds in the parent chain.*
  - *A* ***suffix*** *that tells the class to which the compound belongs.*
  - *A molecule class is determined by the functional groups present, such as an alcohol, aldehyde, ketone, or carboxylic acid.*
- A carbon atom is classified as *primary (1°), secondary (2°), tertiary (3°),* or *quaternary (4°),* depending on the number of carbon atoms bonded to it.

P 2.24, 2.29–2.31

Problem 2.1

2.4 Cycloalkanes

- A saturated hydrocarbon that contains carbon atoms bonded to form a ring is called a **cycloalkane**.
- To name a cycloalkane, name and locate each substituent on the ring and prefix the name of the analogous open-chain alkane with *cyclo-*.
- Five-membered rings and six-membered rings are especially abundant in the biological world.

P 2.29–2.30

2.5 Conformations of Alkanes and Cycloalkanes

- A **conformation** is any three-dimensional arrangement of the atoms of a molecule resulting from rotations about the single bonds.
  - *One convention for showing conformations is the* ***Newman projection***.
  - *A* ***dihedral angle*** *is the angle created by two intersecting planes.*
    - *For ethane,* ***staggered conformations*** *occur at dihedral angles of 60°, 180°, and 300°.* ***Eclipsed conformations*** *occur at dihedral angles of 0°, 120°, and 240°.*
    - *For butane, viewed along the C_2—C_3 bond, the staggered conformation of dihedral angle 180° is called an* ***anti conformation****; the staggered* ***conformations*** *of dihedral angle 60° and 300° are called* ***gauche conformations***.

– *The anti conformation of butane is lower in energy than the gauche conformations by approximately $2.8\ kJ\ (0.9\ kcal)/mol$.*

- **Intramolecular strain** is of three types:
  – ***Torsional strain*** *(also called eclipsed-interaction strain) arises when nonbonded atoms separated by three bonds are forced from a staggered conformation to an eclipsed conformation.*
  – ***Angle strain*** *arises from creation of either abnormally large or abnormally small bond angles.*
  – **Steric strain** *(also called nonbonded interaction or van der Waals strain) arises when nonbonded atoms separated by four or more bonds are forced abnormally close to each other.*

P 2.8, 2.32–2.35

- The relationship between the change in Gibbs free energy, temperature in kelvins, and an equilibrium constant is given by the equation $\Delta G^0 = -RT \ln K_{eq}$.

P 2.37, 2.39, 2.45–2.47

- In all cycloalkanes larger than cyclopropane, nonplanar conformations are favored.
  – *The lowest-energy conformation of cyclopentane is an* ***envelope conformation.***
  – *The lowest-energy conformations of cyclohexane are two interconvertible* ***chair conformations.***
    – *In a chair conformation, six bonds are* ***axial*** *and six bonds are* ***equatorial.***
    – *Bonds axial in one chair are equatorial in the alternative chair.*
    – *Boat and twist-boat conformations are higher in energy than chair conformations.*
    – *Groups larger than H are less stable in the axial position of chair conformations because of 1,3-diaxial steric interactions.*
    – *The more stable chair conformation of a substituted cyclohexane is the one that minimizes diaxial steric interactions and therefore has large groups equatorial.*

P 2.9–2.11, 2.36, 2.38, 2.45, 2.46, 2.48–2.52, 2.61–2.65

▶ Figures 2.9 & 2.19, Problem 2.8

2.6 *Cis,Trans* Isomerism in Cycloalkanes and Bicycloalkanes

- **Stereoisomers** are compounds that have the same molecular formula and the same connectivity, but different orientation of their atoms in space.
  – *A* ***stereocenter*** *is an atom (most commonly a carbon atom) about which exchange of two groups produces a different stereoisomer.*
  – ***Configuration*** *refers to the arrangement of atoms or groups of atoms bonded to a stereocenter.*
  – **Cis,trans** *isomers have the same molecular formula and the same connectivity of their atoms, but the arrangement of their atoms in space cannot be interconverted by rotation about single bonds.*
    – Cis *substituents are on the same side of the ring.*

– Trans *substituents are on opposite sides of the ring.*

– *Most cycloalkanes with substituents on two or more carbons show* cis,trans *isomerism.*

P **2.12–2.14, 2.40–2.44**

Problem 2.13

2.7 Physical Properties of Alkanes and Cycloalkanes

- Alkanes are nonpolar compounds, and the only forces of attraction between their molecules are dispersion forces, which are weak electrostatic interactions between temporary induced dipoles of adjacent atoms or molecules.
  - *Low-molecular-weight alkanes are gases at room temperature and atmospheric pressure.*
  - *Higher-molecular-weight alkanes are liquids. Very-high-molecular-weight alkanes are solids.*
  - *Among a set of alkane constitutional isomers, the least branched isomer generally has the highest boiling point; the most branched isomer generally has the lowest boiling point.*

P **2.15, 2.53–2.55**

2.8 Reactions of Alkanes

- As determined by heats of combustion, strain in cycloalkanes varies with ring size.

KEY REACTIONS

1. Oxidation of Alkanes (Section 2.8A) Oxidation of alkanes to carbon dioxide and water is the basis for their use as energy sources of heat and power.

$$CH_3CH_2CH_3 + 5O_2 \rightarrow 3CO_2 + 4H_2O \qquad \Delta H^0 = -2220 \text{ kJ } (-530.6 \text{ kcal})/\text{mol}$$

- Cyclohexane, which has the most common ring size among organic compounds, is strain-free.

P **2.56–2.60**

2.9 Sources and Importance of Alkanes

- **Natural gas** consists of 90%–95% methane with lesser amounts of ethane and other low-molecular-weight hydrocarbons.
- **Petroleum** is a liquid mixture of literally thousands of different hydrocarbons.
  - *The most important processes in petroleum refining are fractional distillation, catalytic cracking, and catalytic reforming.*
- **Synthesis gas**, a mixture of carbon monoxide and hydrogen, can be derived from natural gas or coal.

Problems

Red numbers indicate applied problems.

2.16 Write a line-angle formula for each condensed structural formula.

(a) 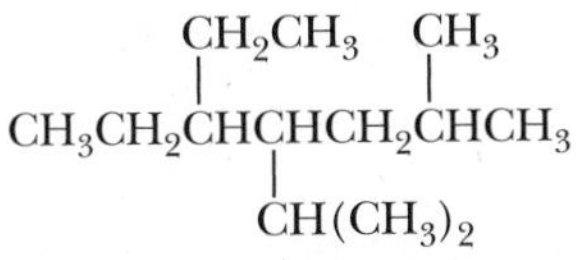 **(b)** 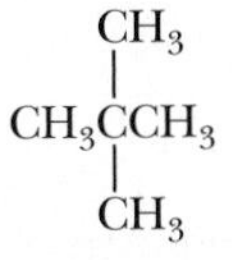 **(c)** $(CH_3)_2CHCH(CH_3)_2$

(d) 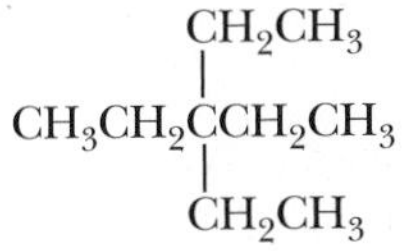 **(e)** $(CH_3)_3CH$ **(f)** $CH_3(CH_2)_3CH(CH_3)_2$

2.17 Write the molecular formula of each alkane.

(a) 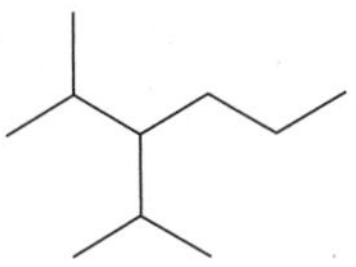 **(b)** 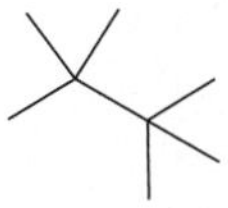 **(c)** 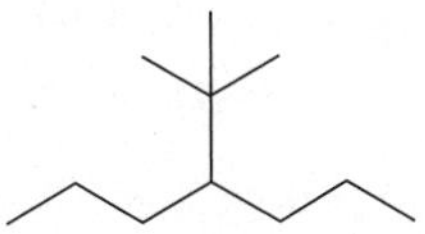

2.18 Using parentheses and subscripts, provide an even more abbreviated formula for each structural formula.

(a) 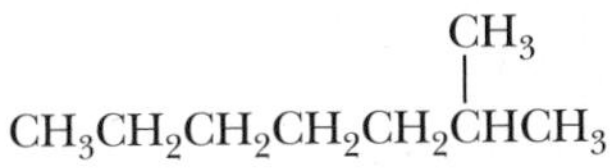 **(b)** 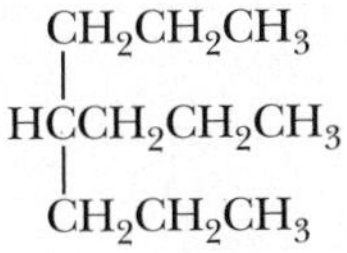 **(c)** 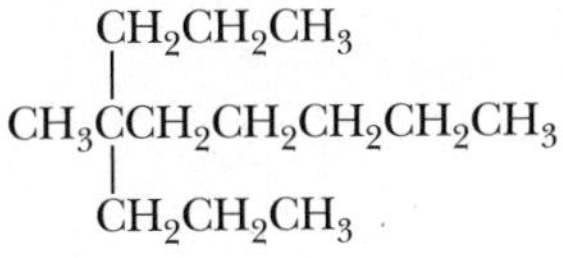

Constitutional Isomerism

2.19 Which statements are true about constitutional isomers?

(a) They have the same molecular formula.
(b) They have the same molecular weight.
(c) They have the same order of attachment of atoms.
(d) They have the same physical properties.

2.20 Indicate whether the compounds in each set are constitutional isomers.

(a) CH_3CH_2OH and CH_3OCH_3 **(b)** 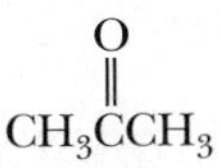 and 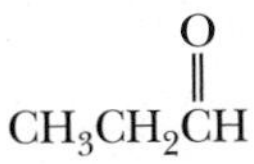

(c) 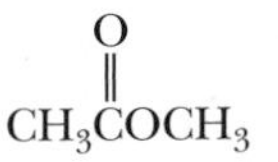 and 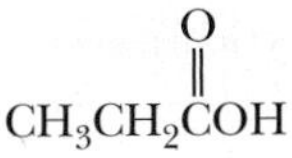 **(d)** 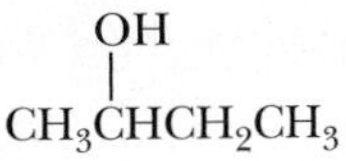 and 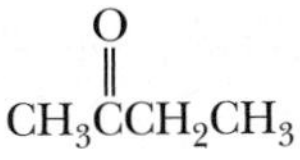

(e) 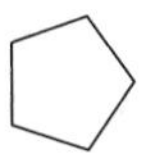 and $CH_3CH_2CH_2CH_2CH_3$ **(f)** 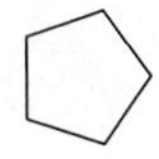 and $CH_2{=}CHCH_2CH_2CH_3$

2.21 Each member of the following set of compounds is an alcohol; that is, each contains an —OH (hydroxyl group, Section 1.3A). Which line-angle formulas represent the same compound? Which represent constitutional isomers?

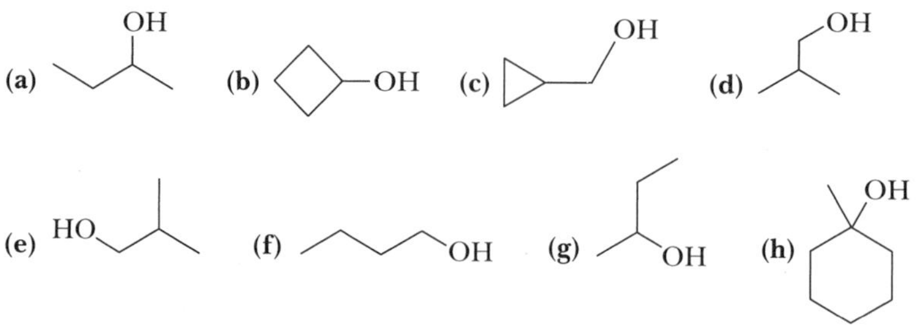

2.22 Each of the following compounds is an amine (Section 1.3B). Which line-angle formulas represent the same compound? Which represent constitutional isomers?

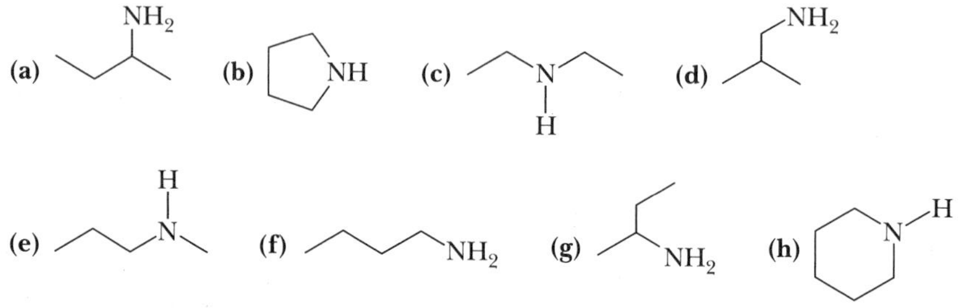

2.23 Each of the following compounds is either an aldehyde or a ketone (Section 1.3C). Which line-angle formulas represent the same compound? Which represent constitutional isomers?

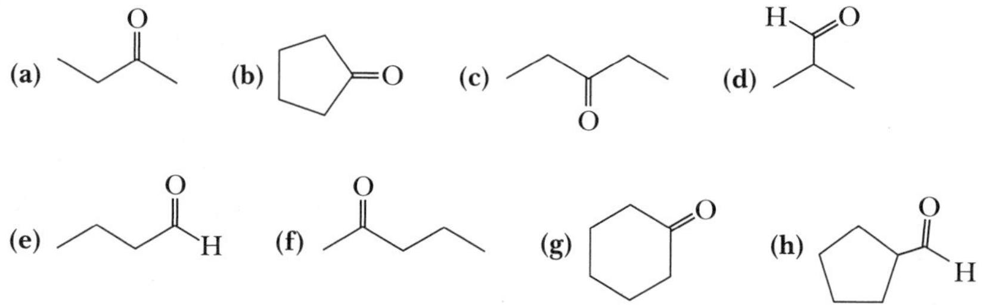

2.24 Draw structural formulas and write IUPAC names for the nine constitutional isomers with the molecular formula C_7H_{16}.

2.25 Draw structural formulas for all of the following.

(a) Alcohols with the molecular formula $C_4H_{10}O$
(b) Aldehydes with the molecular formula C_4H_8O
(c) Ketones with the molecular formula $C_5H_{10}O$
(d) Carboxylic acids with the molecular formula $C_5H_{10}O_2$

Nomenclature of Alkanes and Cycloalkanes

2.26 Write IUPAC names for these alkanes and cycloalkanes.

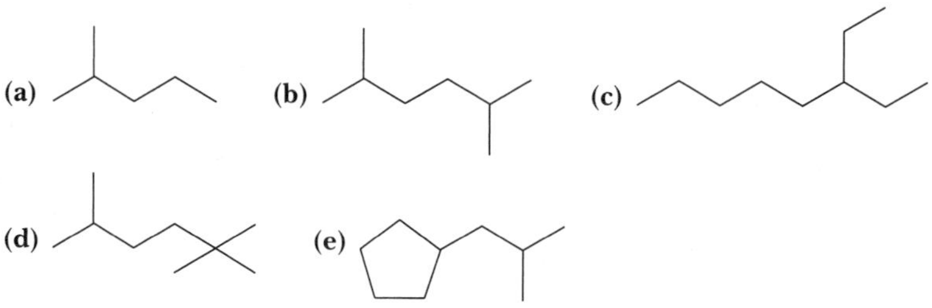

2.27 Write structural formulas and line-angle formulas for the following alkanes and cycloalkanes.

(a) 2,2,4-Trimethylhexane
(b) 2,2-Dimethylpropane
(c) 3-Ethyl-2,4,5-trimethyloctane
(d) 5-Butyl-2,2-dimethylnonane
(e) 4-(1-Methylethyl)octane
(f) 3,3-Dimethylpentane
(g) *trans*-1,3-Dimethylcyclopentane
(h) *cis*-1,2-Diethylcyclobutane

2.28 Explain why each is an incorrect IUPAC name and write the correct IUPAC name for the intended compound.

(a) 1,3-Dimethylbutane
(b) 4-Methylpentane
(c) 2,2-Diethylbutane
(d) 2-Ethyl-3-methylpentane
(e) 2-Propylpentane
(f) 2,2-Diethylheptane
(g) 2,2-Dimethylcyclopropane
(h) 1-Ethyl-5-methylcyclohexane

2.29 For each IUPAC name, draw the corresponding structural formula and line-angle formula.

(a) Ethanol
(b) Butanal
(c) Butanoic acid
(d) Ethanoic acid
(e) Heptanoic acid
(f) Propanoic acid
(g) Octanal
(h) Cyclopentene
(i) Cyclopentanol
(j) Cyclopentanone
(k) Cyclohexanol
(l) Propanone

2.30 Write the IUPAC name for each compound.

(a) $CH_3CH_2\overset{O}{\overset{\|}{C}}CH_3$
(b) $CH_3CH_2\overset{O}{\overset{\|}{C}}H$
(c) $CH_3CH_2CH_2CH_2CH_2\overset{O}{\overset{\|}{C}}OH$

(d) $CH_3\overset{OH}{\overset{|}{C}}HCH_3$
(e) 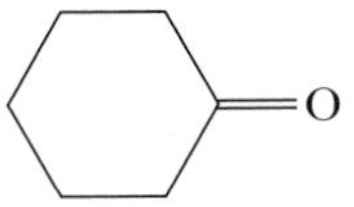
(f) 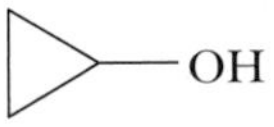

(g) $CH_3CH{=}CH_2$
(h) 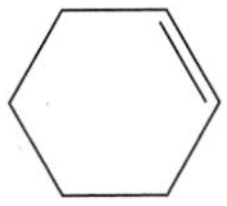

2.31 Assume for the purposes of this problem that to be an alcohol (-ol) or an amine (-amine), the hydroxyl or amino group must be bonded to a tetrahedral (*sp*^{3} hybridized) carbon atom. Write the structural formula of a compound with an unbranched chain of four carbon atoms that is an:

(a) Alkane
(b) Alkene
(c) Alkyne
(d) Alkanol
(e) Alkenol
(f) Alkynol
(g) Alkanamine
(h) Alkenamine
(i) Alkynamine
(j) Alkanal
(k) Alkenal
(l) Alkynal
(m) Alkanone
(n) Alkenone
(o) Alkynone
(p) Alkanoic acid
(q) Alkenoic acid
(r) Alkynoic acid

(*Note:* Only one structural formula is possible for some parts of this problem. For other parts, two or more structural formulas are possible. Where two are more are possible, we will deal with how the IUPAC system distinguishes between them when we come to the chapters on those particular functional groups.)

Conformations of Alkanes and Cycloalkanes

2.32 Torsional strain resulting from eclipsed C—H bonds is approximately 4.2 kJ (1.0 kcal)/mol, and that for eclipsed C—H and C—CH_3 bonds is approximately 6.3 kJ (1.5 kcal)/mol. Given this information, sketch a graph of energy versus dihedral angle for propane.

2.33 How many different staggered conformations are there for 2-methylpropane? How many different eclipsed conformations are there?

2.34 Consider 1-bromopropane, $CH_3CH_2CH_2Br$.

(a) Draw a Newman projection for the conformation in which —CH_3 and —Br are anti (dihedral angle 180°).
(b) Draw Newman projections for the conformations in which —CH_3 and —Br are gauche (dihedral angles 60° and 300°).
(c) Which of these is the lowest energy conformation?
(d) Which of these conformations, if any, are related by reflection?

2.35 Consider 1-bromo-2-methylpropane and draw the following.

(a) The staggered conformation(s) of lowest energy
(b) The staggered conformation(s) of highest energy

2.36 *trans*-1,4-Di-*tert*-butylcyclohexane exists in a normal chair conformation. *cis*-1,4-Di-*tert*-butylcyclohexane, however, adopts a twist-boat conformation. Draw both isomers and explain why the *cis* isomer is more stable in a twist-boat conformation.

2.37 From studies of the dipole moment of 1,2-dichloroethane in the gas phase at room temperature (25°C), it is estimated that the ratio of molecules in the anti conformation to gauche conformation is 7.6 to 1. Calculate the difference in Gibbs free energy between these two conformations.

2.38 Draw structural formulas for the *cis* and *trans* isomers of hydrindane. Show each ring in its most stable conformation. Which of these isomers is more stable?

2.39 Following are the alternative chair conformations for *trans*-1,2-dimethylcyclohexane.

(a) Estimate the difference in free energy between these two conformations.
(b) Given your value in (a), calculate the percent of each chair present in an equilibrium mixture of the two at 25°C.

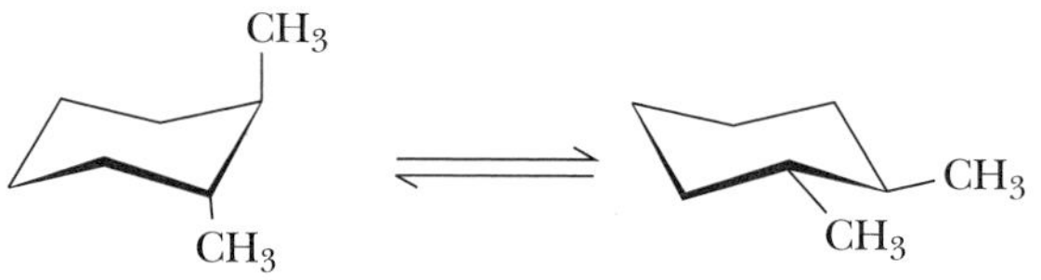

***trans*-1,2-Dimethylcyclohexane**

Cis,Trans Isomerism in Cycloalkanes

2.40 What structural feature of cycloalkanes makes *cis,trans* isomerism in them possible?

2.41 Is *cis,trans* isomerism possible in alkanes?

2.42 Draw line-angle formulas for the *cis* and *trans* isomers of 1,2-dimethylcyclopropane.

2.43 Name and draw structural formulas for all cycloalkanes with molecular formula C_5H_{10}. Include *cis* and *trans* isomers as well as constitutional isomers.

2.44 Using a planar pentagon representation for the cyclopentane ring, draw structural formulas for the *cis* and *trans* isomers of the following.

(a) 1,2-Dimethylcyclopentane **(b)** 1,3-Dimethylcyclopentane

2.45 Gibbs free energy differences between axial-substituted and equatorial-substituted chair conformations of cyclohexane were given in Table 2.4.

(a) Calculate the ratio of equatorial to axial *tert*-butylcyclohexane at 25°C.
(b) Explain why the conformational equilibria for methyl, ethyl, and isopropyl substituents are comparable but the conformational equilibrium for *tert*-butylcyclohexane lies considerably farther toward the equatorial conformation.

2.46 When cyclohexane is substituted by an ethynyl group, —C≡CH, the energy difference between axial and equatorial conformations is only 1.7 kJ (0.41 kcal)/mol. Compare the conformational equilibrium for methylcyclohexane with that for ethynylcyclohexane and account for the difference between the two.

2.47 Calculate the difference in Gibbs free energy in kilojoules per mole between the alternative chair conformations of:

(a) *trans*-4-Methylcyclohexanol **(b)** *cis*-4-Methylcyclohexanol
(c) *trans*-1,4-Dicyanocyclohexane

2.48 Draw the alternative chair conformations for the *cis* and *trans* isomers of 1,2-dimethylcyclohexane, 1,3-dimethylcyclohexane, and 1,4-dimethylcyclohexane.

(a) Indicate by a label whether each methyl group is axial or equatorial.
(b) For which isomer(s) are the alternative chair conformations of equal stability?
(c) For which isomer(s) is one chair conformation more stable than the other?

2.49 Use your answers from Problem 2.48 to complete the table showing correlations between *cis,trans* and axial,equatorial for disubstituted derivatives of cyclohexane.

Position of Substitution	*cis*	*trans*
1,4-	a,e or e,a	e,e or a,a
1,3-	____ or ____	____ or ____
1,2-	____ or ____	____ or ____

2.50 There are four *cis,trans* isomers of 2-isopropyl-5-methylcyclohexanol:

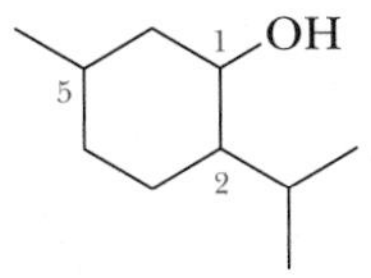

2-Isopropyl-5-methylcyclohexanol

(a) Using a planar hexagon representation for the cyclohexane ring, draw structural formulas for the four *cis,trans* isomers.
(b) Draw the more stable chair conformation for each of your answers in part (a).
(c) Of the four *cis,trans* isomers, which is most stable? (*Hint:* If you answered this part correctly, you picked the isomer found in nature and given the name menthol.)

2.51 Draw alternative chair conformations for each substituted cyclohexane and state which chair is more stable.

(a) 

(b) 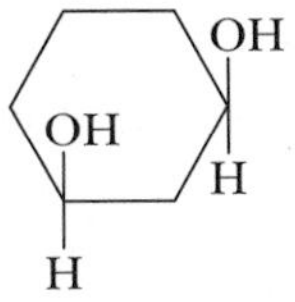

(c) 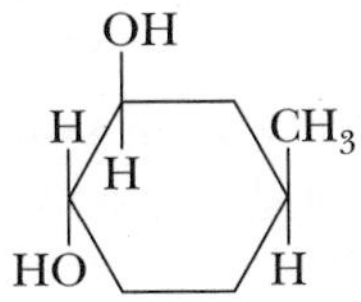

(d) 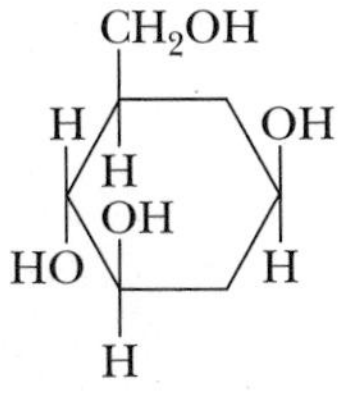

2.52 1,2,3,4,5,6-Hexachlorocyclohexane shows *cis,trans* isomerism. At one time, a crude mixture of these isomers was sold as an insecticide. The insecticidal properties of the mixture arise from one isomer, known as lindane, which is *cis*-1,2,4,5-*trans*-3,6-hexachlorocyclohexane.

(a) Draw a structural formula for 1,2,3,4,5,6-hexachlorocyclohexane disregarding, for the moment, the existence of *cis,trans* isomerism. What is the molecular formula of this compound?
(b) Using a planar hexagon representation for the cyclohexane ring, draw a structural formula for lindane.
(c) Draw a chair conformation for lindane and label which chlorine atoms are axial and which are equatorial.
(d) Draw the alternative chair conformation of lindane and again label which chlorine atoms are axial and which are equatorial.
(e) Which of the alternative chair conformations of lindane is more stable? Explain.

Physical Properties

2.53 In Problem 2.24, you drew structural formulas for all isomeric alkanes with molecular formula C_7H_{16}. Predict which isomer has the lowest boiling point and which has the highest boiling point.

2.54 What generalization can you make about the densities of alkanes relative to the density of water?

2.55 What unbranched alkane has about the same boiling point as water? (Refer to Table 2.5 on the physical properties of alkanes.) Calculate the molecular weight of this alkane and compare it with that of water.

Reactions of Alkanes

2.56 Complete and balance the following combustion reactions. Assume that each hydrocarbon is converted completely to carbon dioxide and water.

(a) Propane + $O_2 \rightarrow$ **(b)** Octane + $O_2 \rightarrow$

(c) Cyclohexane + $O_2 \rightarrow$ **(d)** 2-Methylpentane + $O_2 \rightarrow$

2.57 Following are heats of combustion per mole for methane, propane, and 2,2,4-trimethylpentane. Each is a major source of energy. On a gram-for-gram basis, which of these hydrocarbons is the best source of heat energy?

Hydrocarbon	Component of	ΔH⁰ [kJ (kcal)/mol]
CH_4	Natural gas	−891 (−213)
$CH_3CH_2CH_3$	LPG	−2220 (−531)
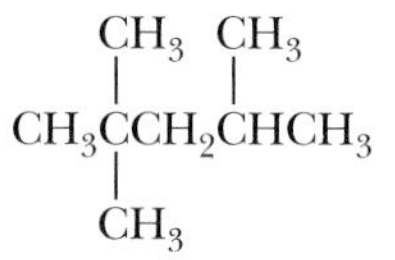	Gasoline	−5452 (−1304)

2.58 Following are structural formulas and heats of combustion of acetaldehyde and ethylene oxide. Which of these compounds is more stable? Explain.

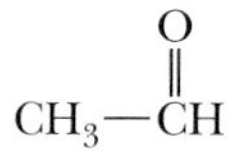

Acetaldehyde
−1164 kJ (−278.8 kcal)/mol

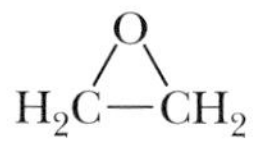

Ethylene oxide
−1264 kJ (−302.1 kcal)/mol

2.59 Without consulting tables, arrange these compounds in order of decreasing (less negative) heat of combustion: hexane, 2-methylpentane, and 2,2-dimethylbutane.

2.60 Which would you predict to have the larger (more negative) heat of combustion, *cis*-1,4-dimethylcyclohexane or *trans*-1,4-dimethylcyclohexane?

Looking Ahead

2.61 Following are structural formulas for 1,4-dioxane and piperidine. 1,4-Dioxane is a widely used solvent for organic compounds. Piperidine is found in small amounts in black pepper (*Piper nigrum*).

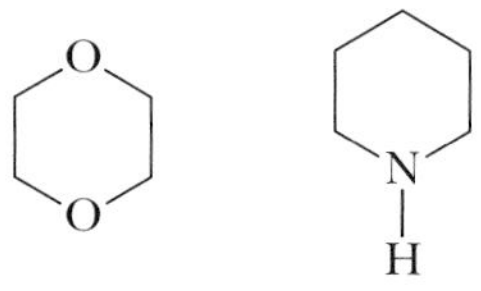

1,4-Dioxane **Piperidine**

(a) Complete the Lewis structure of each compound by showing all unshared electron pairs.
(b) Predict bond angles about each carbon, oxygen, and nitrogen atom.
(c) Describe the most stable conformation of each ring and compare these conformations with the chair conformation of cyclohexane.

2.62 Following is a planar hexagon representation of L-fucose, a sugar component of the determinants of the A, B, O blood group typing. For more on this system of blood typing, see "Chemical Connections: A, B, AB, and O Blood Group Substances" in Chapter 25.

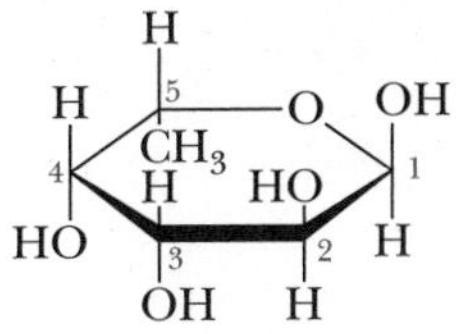

L-Fucose

(a) Draw the alternative chair conformations of L-fucose.
(b) Which of them is more stable? Explain.

2.63 On the left is a stereorepresentation of glucose (we discuss the structure and chemistry of glucose in Chapter 25).

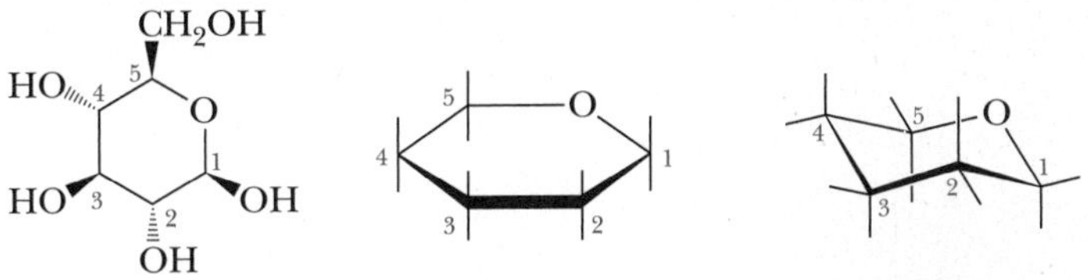

Glucose **(a)** **(b)**

(a) Convert the stereorepresentation on the left to a planar hexagon representation.
(b) Convert the stereorepresentation on the left to a chair conformation. Which substituent groups in the chair conformation are equatorial? Which are axial?

2.64 Following is the structural formula and a ball-and-stick model of cholestanol. The only difference between this compound and cholesterol (Section 26.4) is that cholesterol has a carbon-carbon double bond in ring B.

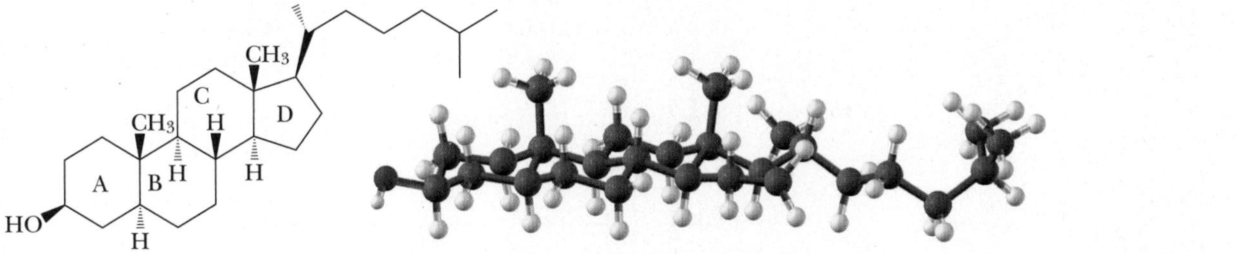

Cholestanol

(a) Describe the conformation of rings A, B, C, and D in cholestanol.
(b) Is the hydroxyl group on ring A axial or equatorial?
(c) Consider the methyl group at the junction of rings A and B. Is it axial or equatorial to ring A? Is it axial or equatorial to ring B?
(d) Is the methyl group at the junction of rings C and D axial or equatorial to ring C?

2.65 Following is the structural formula and a ball-and-stick model of cholic acid (Chapter 26), a component of human bile whose function is to aid in the absorption and digestion of dietary fats.

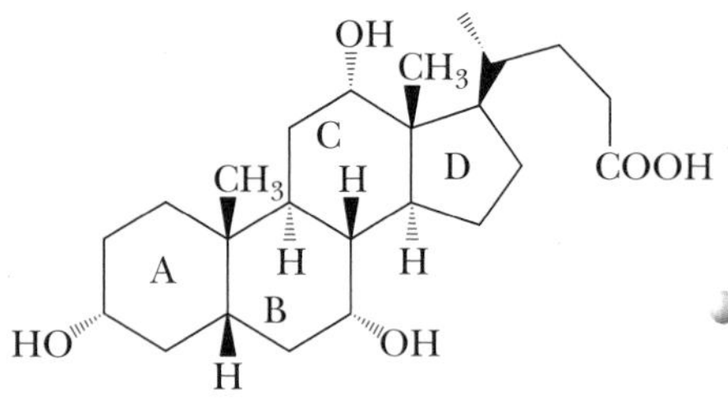

Cholic acid

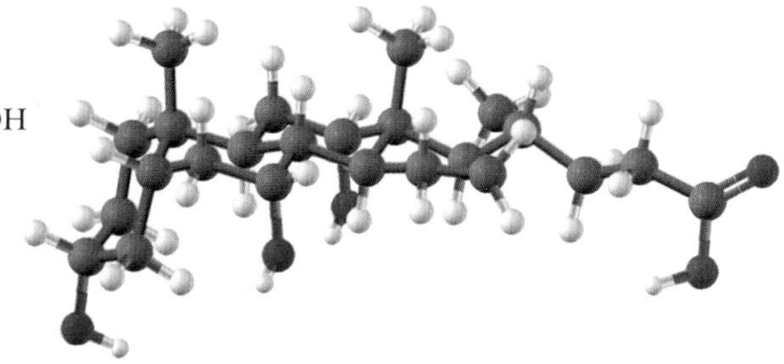

(a) What is the conformation of ring A? of ring B? of ring C? of ring D?

(b) Are the hydroxyl groups on rings A, B, and C axial or equatorial to their respective rings?

(c) Is the methyl group at the junction of rings A and B axial or equatorial to ring A? Is it axial or equatorial to ring B?

(d) Is the hydrogen at the junction of rings A and B axial or equatorial to ring A? Is it axial or equatorial to ring B?

(e) Is the methyl group at the junction of rings C and D axial or equatorial to ring C?

3

Stereoisomerism and Chirality

Tartaric acid (Section 3.4B) is found in grapes and other fruits, both free and as its salts. **Inset**: a model of the *R,R* stereoisomer of tartaric acid. (Bob Nichols/ARS/USDA.)

Outline

3.1 Chirality—The Handedness of Molecules

3.2 Stereoisomerism

HOW TO: Draw Chiral Molecules

3.3 Naming Chiral Centers—The *R,S* System

HOW TO: Assign R *or* S *Configuration to a Chiral Center*

3.4 Acyclic Molecules with Two or More Stereocenters

HOW TO: Quickly Draw and Recognize Enantiomers and Diastereomers

3.5 Cyclic Molecules with Two or More Chiral Centers

3.6 Tying All the Terminology Together

3.7 Optical Activity—How Chirality Is Detected in the Laboratory

3.8 The Significance of Chirality in the Biological World

3.9 Separation of Enantiomers—Resolution

Stereochemistry

The study of three-dimensional arrangements of atoms in molecules.

The study of molecules as three-dimensional objects is called **stereochemistry**. The ability to visualize stereochemical relationships is a survival skill in organic chemistry. The chemistry and properties of the key molecules of biochemistry are critically dependent on stereochemistry. Our goal in this chapter is to expand your awareness of molecules as three-dimensional objects. We suggest you purchase a set of models (or, if you prefer, a computer modeling program such as Spartan). Alternatively, you may have access to a computer lab with a modeling program. Use your models frequently as an aid in visualizing the spatial concepts in this chapter and in later chapters.

3.1 Chirality—The Handedness of Molecules

William H. Brown

The horns of this African gazelle show chirality, due to their twist, and are mirror images of each other.

Chiral

From the Greek, *cheir*, hand; an object that is not superposable on its mirror image; an object that has handedness.

Achiral

An object that lacks chirality; an object that has no handedness.

Plane of symmetry

An imaginary plane passing through an object dividing it so that one half is the mirror image of the other half.

A **mirror image** is the reflection of an object in a mirror. When you look in a mirror, you see a reflection, or mirror image, of yourself. Now suppose your mirror image became a three-dimensional object. We could then ask, "What is the relationship between you and your mirror image?" To clarify what we mean by *relationship,* we might instead ask,"Can your reflection be superposed on (placed on top of) the original'you' in such a way that every detail of the reflection corresponds exactly to the original?" The answer is that you and your mirror image are not superposable if details are included. For example, if you have a ring on the little finger of your right hand, your mirror image has the ring on the little finger of its left hand. If you part your hair on your right side, it will be parted on the left side in your reflection.You and your reflection are different objects.You cannot *exactly* superpose one on the other.

Objects that are not superposable on their mirror images are said to be **chiral** (pronounced ki-ral, to rhyme with spiral; from the Greek: *cheir*, hand). That is, they show handedness. Chirality is encountered in three-dimensional objects of all sorts. Your left hand is chiral, as is your right hand (they are approximately mirror images of each other). A spiral binding on a notebook is chiral. A machine screw with a right-handed thread is chiral. A ship's propeller is chiral. As you examine objects around you, you will undoubtedly conclude that the vast majority of them are chiral as well.

In contrast, when an object and its mirror image are superposable, the object is **achiral**, that is, it lacks chirality. Examples of objects lacking chirality are an undecorated cup, a regular tetrahedron, a cube, and a perfect sphere.

Molecules are objects of the size of several to dozens of Angstroms, yet as discussed in the previous two chapters, they have definite shapes and geometries. These properties make some molecules chiral objects. To figure out if a molecule is chiral, it is simpler to determine whether it is achiral. We look for symmetries in a molecule to see if it is achiral. If these symmetries are missing, we can conclude that the molecule is chiral.

An object or a molecule will be achiral if it has one or more of certain elements of symmetry. The most common such elements in organic compounds are the plane and center of symmetry. As we shall see, any object or molecule with either of these symmetry elements is achiral and can be superposed on its mirror image. A **plane of symmetry** is an imaginary plane passing through an object or a molecule dividing it such that one half is the reflection of the other half. The cube shown in Figure 3.1 has several planes of symmetry. Both the beaker and the compound bromochloromethane

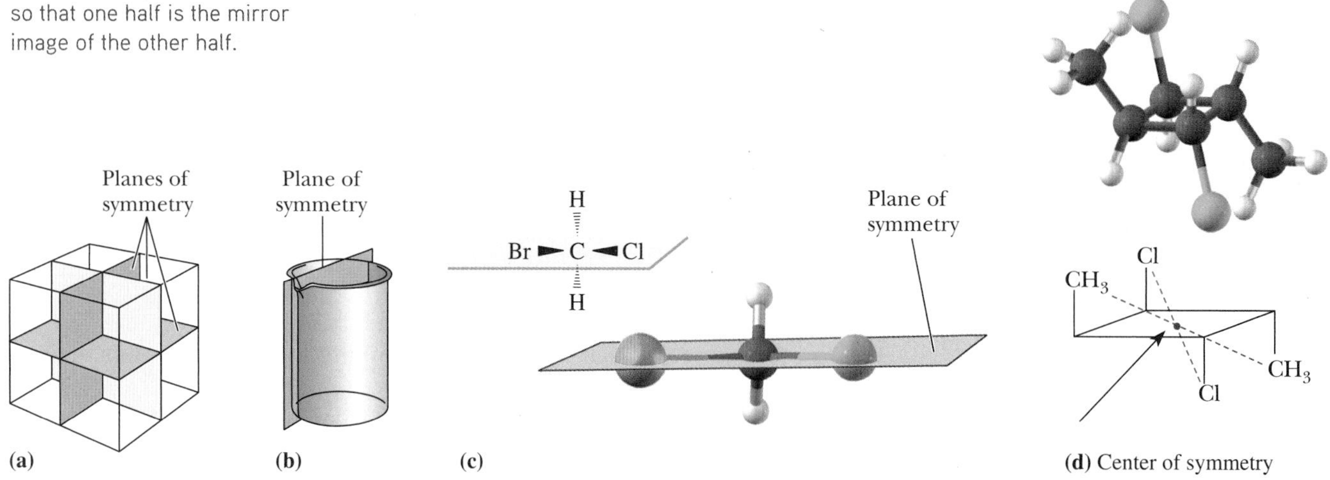

FIGURE 3.1 Symmetry in objects. A cube has several planes of symmetry and a center of symmetry. The beaker and CH_2BrCl each have a single plane of symmetry. The cyclobutane has a center of symmetry.

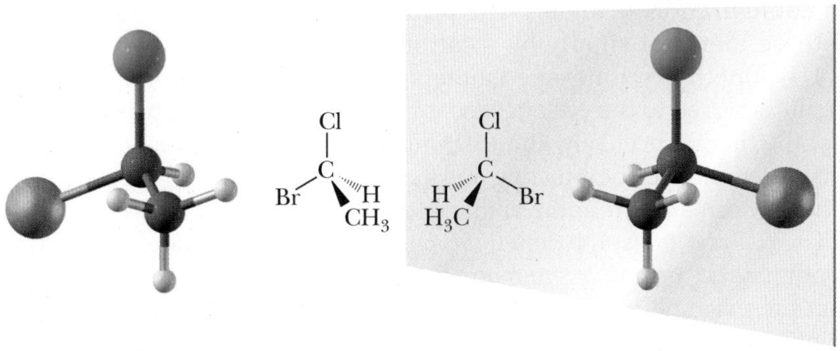

FIGURE 3.2
Stereorepresentations of 1-bromo-1-chloroethane and its mirror image.

have a single plane of symmetry. A **center of symmetry** is a point so situated that identical components of the object or molecule are located equidistant and on opposite sides from the point along any axis passing through that point. The cube shown in Figure 3.1 has a center of symmetry, as does the cyclobutane. Because it has a center of symmetry, the cyclobutane is identical to its mirror image and is achiral.

Objects or molecules that lack both of these symmetry elements are chiral. We can illustrate the chirality of an organic molecule by considering 1-bromo-1-chloroethane. Figure 3.2 shows three-dimensional representations and ball-and-stick models for 1-bromo-1-chloroethane and its mirror image. This molecule has neither a plane nor a center of symmetry. A model of 1-bromo-1-chloroethane can be turned and rotated in any direction in space, but as long as bonds are not broken and rearranged, only two of the four groups bonded to the central carbon of one molecule can be made to coincide with those of its mirror image. Because 1-bromo-1-chloroethane and its mirror image are nonsuperposable, they are chiral objects.

Center of symmetry
A point so situated that identical components of an object are located on opposite sides and equidistant from that point along any axis passing through it.

3.2 Stereoisomerism

Isomers are different compounds with the same molecular formula. Thus far, we have encountered three types of isomers. Constitutional isomers (Section 2.2) have the same molecular formula but a different connectivity of atoms in their molecules. Examples of pairs of constitutional isomers are pentane and 2-methylbutane, and 1-pentene and cyclopentane.

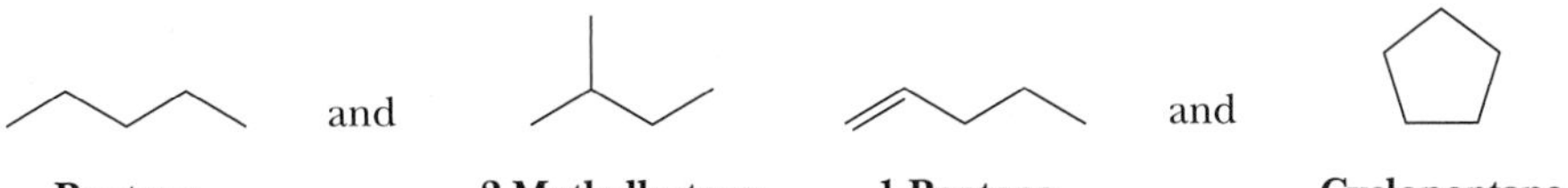

Pentane (C_5H_{12}) and **2-Methylbutane** (C_5H_{12}); **1-Pentene** (C_5H_{10}) and **Cyclopentane** (C_5H_{10})

A second type of isomerism is stereoisomerism. **Stereoisomers** have the same molecular formula and the same connectivity but different orientations of their atoms in space. One example of stereoisomerism we have seen thus far is that of *cis,trans* isomers in cycloalkanes (Section 2.6), which arise because substituents on a ring are locked into one of two orientations in space with respect to one another by the ring. Isomers of this type are called **configurational isomers** because they differ by the configuration of substituents on an atom.

Stereoisomers
Isomers that have the same molecular formula and the same connectivity of their atoms but a different orientation of their atoms in space.

Configurational isomers
Isomers that differ by the configuration of substituents on an atom.

Configurational isomers (*cis,trans* isomers)

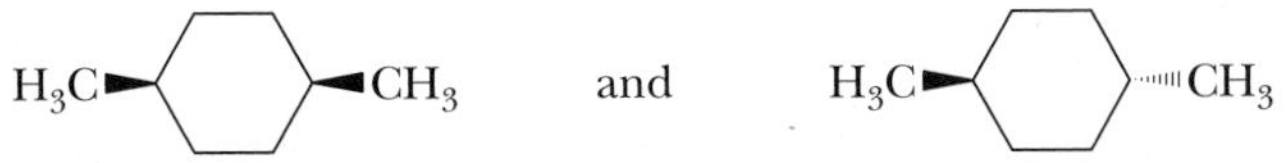

***cis*-1,4-Dimethylcyclohexane** and ***trans*-1,4-Dimethylcyclohexane**

In configurational isomers, the positions of the atoms cannot interchange so as to be identical simply by rotations along single bonds. In the *cis*- and *trans*-1,4-dimethylcyclohexane isomers just presented, the methyl groups are identically attached to the cyclohexane except that they are on the same or opposite sides of the ring, respectively, and their interconversion would require breaking of a bond between the cyclohexane ring and a methyl group.

Chirality can arise from the ability of a molecule to exist as configurational isomers. The *cis*- and *trans*-1,4-dimethylcyclohexane molecules are not chiral because they both possess a plane of symmetry that passes through the methyl groups. However, let's re-examine 1-bromo-1-chloroethane in Figure 3.2. This molecule exists in two configurations that are nonsuperposable mirror images. Hence, these are stereoisomers that are chiral objects, and they are configurational isomers. Such objects, or molecules, are called enantiomers. **Enantiomers** are nonsuperposable mirror images. Note that the terms *chiral* and *achiral* refer to objects, while the term *enantiomer* refers to the relationship between a pair of objects. ▶

▶ **There are two possible arrangements of four different groups around a tetrahedral atom.**

Enantiomers
Stereoisomers that are nonsuperposable mirror images of each other; refers to a relationship between pairs of objects.

Chiral center
A tetrahedral atom, most commonly carbon, that is bonded to four different groups.

Stereocenter
An atom about which exchange of two groups produces a stereoisomer. Chiral centers are one type of stereocenter.

The most common cause of chirality in organic molecules is a tetrahedral atom, most commonly carbon, bonded to four different groups. A carbon atom with four different groups bonded to it lacks the two key symmetry elements and is called a **chiral center**. The carbon atom of 1-bromo-1-chloroethane bearing the —Cl, —H, —CH_3, and —Br groups is a chiral center.

The term **stereocenter** is also used to describe a carbon bonded to four different groups but is broader. A stereocenter is an atom about which exchange of two groups produces a stereoisomer. For example, both carbons in *cis*-2-butene and *trans*-2-butene are stereocenters. The C-2 in 2-butanol is a stereocenter as well because exchange of the OH and H creates a stereoisomer, an enantiomer in this case.

So far, we have seen that configurational isomers can exist as chiral and achiral objects. When the isomers are chiral, they are enantiomers of one another, just as right and left hands are enantiomers. In contrast, *cis*- and *trans*-1,4-dimethylcyclohexane and *trans*- and *cis*-2-butene are configurational isomers, but they are not chiral objects.

How To

Draw Chiral Molecules

It is worthwhile to notice that there are several different ways to represent the three-dimensional structure of chiral molecules on a two-dimensional page. For example, following are four different representations of one enantiomer of 2-butanol.

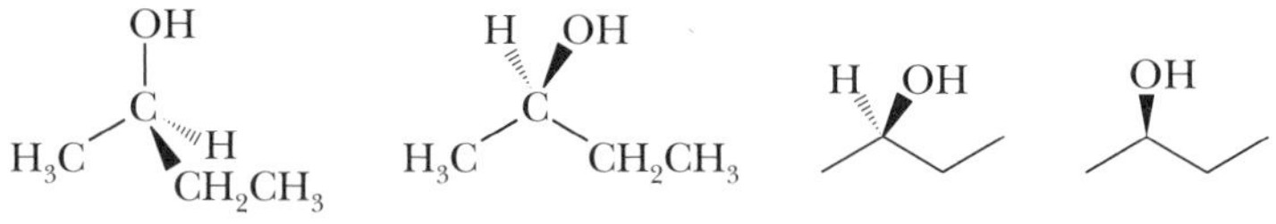

(1) (2) (3) (4)

Representation (1) shows the tetrahedral geometry of the chiral center. We can turn (1) slightly in space and tip it a bit to place the carbon framework in the plane of the paper and give representation (2). In (2), we still have two groups in the plane of the paper, one coming toward us and one going away from us. As discussed in Chapter 2, these are represented by wedged and dashed bonds, respectively. For an even more abbreviated representation, we can turn (2) into the line-angle formula (3). Although we don't normally show hydrogens in a line-angle formula, we do in (3) just to remind ourselves that the fourth group on this chiral center is really there and that it is H. Finally, we can carry the abbreviation a step further and write this enantiomer of 2-butanol as (4). Here, we omit the H on the chiral center. We know it must be there because carbon needs four bonds, and we know it must be behind the plane of the paper because the OH is in front. Clearly, the abbreviated formulas (3) and (4) are the easiest to write, and we will rely on these representations throughout the remainder of the text. When you need to write three-dimensional representations of chiral centers, try to keep the carbon framework in the plane of the paper and the other two atoms or groups of atoms on the chiral center toward and away from you. Often, it is important to draw both enantiomers of a chiral molecule. An easy way to do this is to interchange two of the groups bonded to the chiral center. Using representation (4) as a model, here are two different representations for its enantiomers.

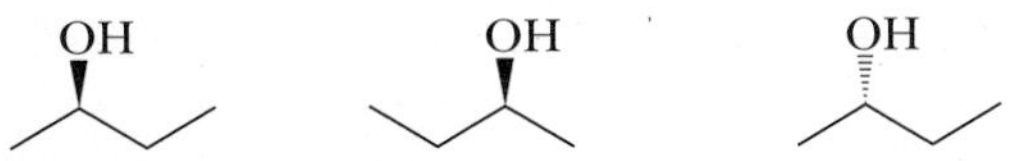

(4) Enantiomer of (4) Enantiomer of (4)

Therefore, they cannot exist as enantiomers. Instead, we call these isomers diastereomers.

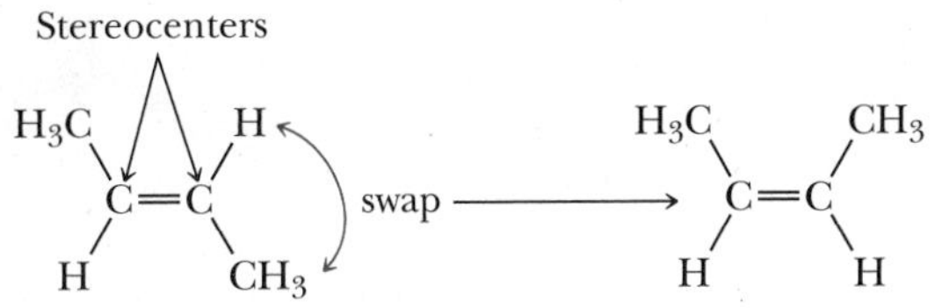

trans-2-Butene *cis*-2-Butene

Watch a video explanation

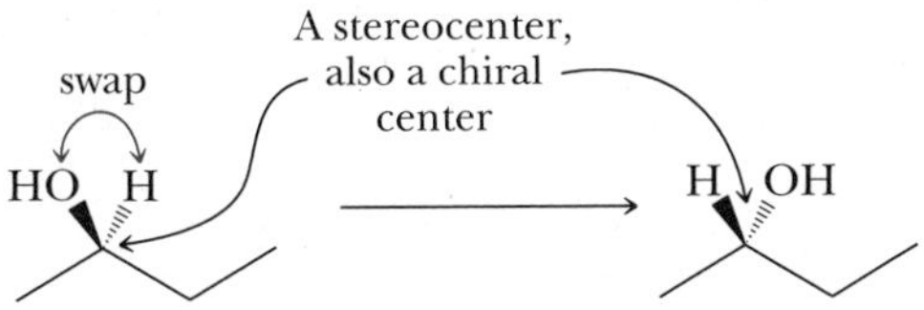

Diastereomers are stereoisomers that are not mirror images of each other. The term *diastereomers* refers to a relationship between a pair of objects. It is important to note that diastereomers can be chiral or achiral objects but that enantiomers must be chiral objects. We will see in Section 3.4 that diastereomers will arise whenever there are two or more stereocenters in a molecule, as is the case with 1,4-dimethylcyclohexane and 2-butene.

Diastereomers
Stereoisomers that are not mirror images of each other; refers to relationships among two or more objects.

Example 3.1 Stereoisomers

Each molecule has one chiral center. Draw stereorepresentations for the enantiomers of each.

(a) **(b)**

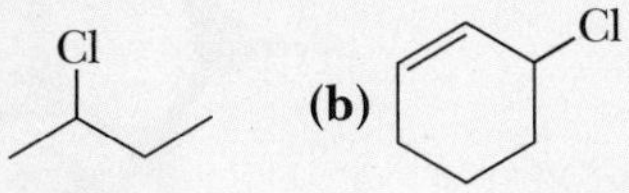

Solution

You will find it helpful to view models of enantiomer pairs from different perspectives, as is done in these representations. As you work with these pairs of enantiomers, notice that each has a tetrahedral carbon atom bonded to four different groups, which makes the molecule chiral.

(Continued)

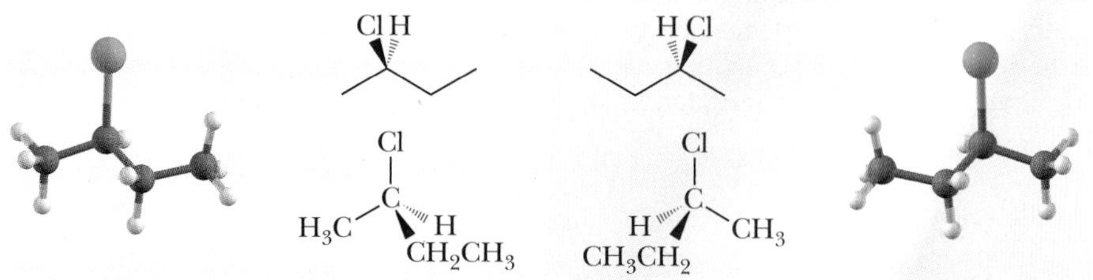

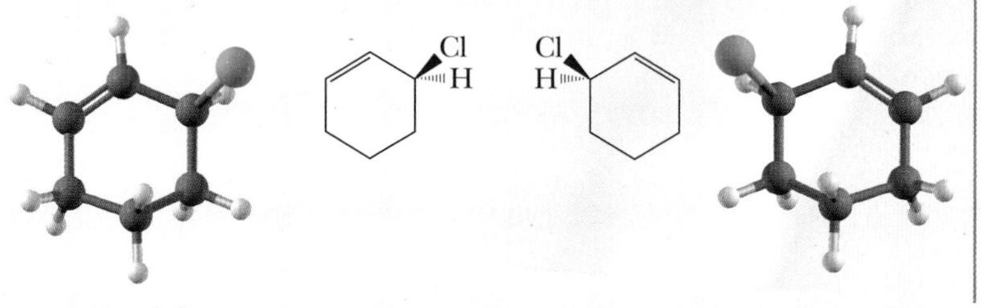

Watch a video explanation

Problem 3.1

Each molecule has one chiral center. Draw stereorepresentations for the enantiomers of each.

(a) 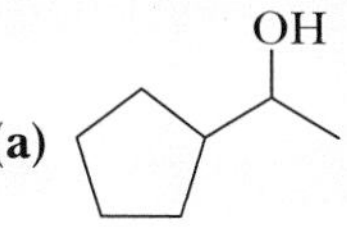

(b) 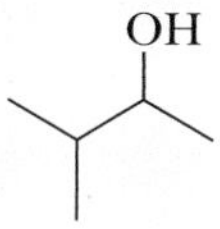

In all the molecules studied so far, chirality arises because of the presence of a tetrahedral carbon chiral center. Chiral centers are not limited to carbon. Following are stereorepresentations of a chiral cation in which the chiral center is nitrogen. We discuss the chirality of nitrogen centers in more detail in Section 23.3. Enantiomers of tetrahedral silicon, phosphorus, and germanium compounds have also been isolated.

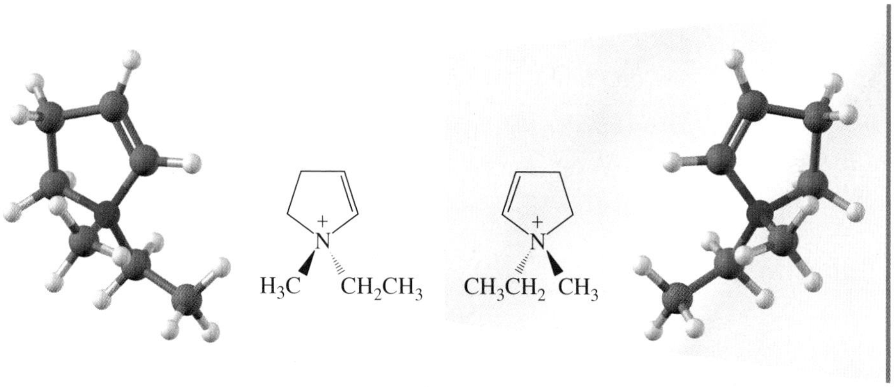

Another form of stereoisomerism is **conformational isomerism**. In Section 2.5, we discussed that conformational changes occur within molecules via rotations along single bonds. For example, recall that butane exists in gauche and anti forms, as shown below using Newman projections. These are not configurational isomers because bonds do not need to be broken in order to interchange atoms; instead, the bonds only need to be rotated. Clearly, they are stereoisomers because the atoms are arranged differently in space.

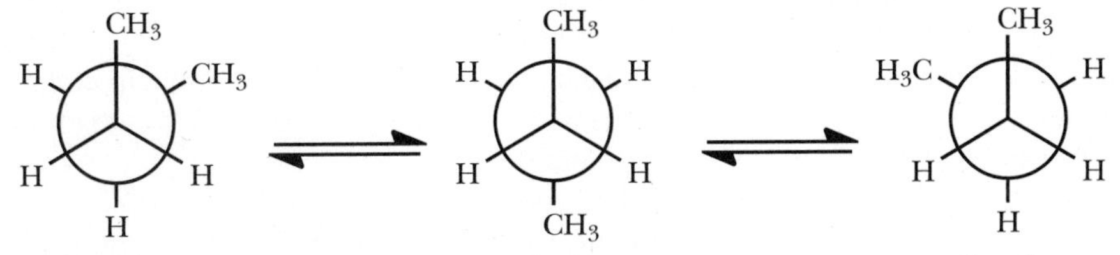

Gauche Anti Gauche

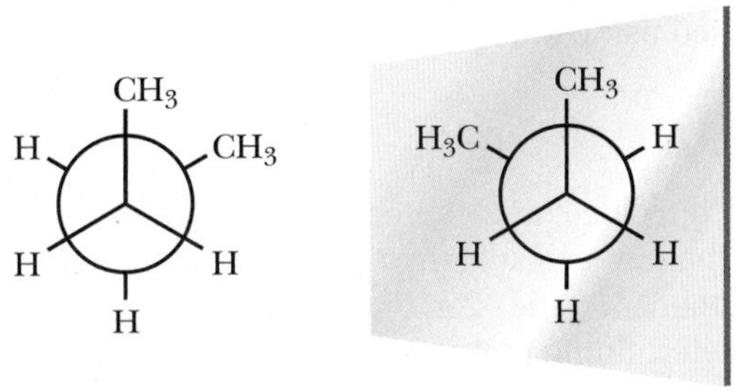

Stereoisomers are either diastereomers or enantiomers; therefore, we also can apply these terms to conformational isomers. The gauche and anti forms of butane are diastereomers because they are not mirror images. The two gauche forms of butane are enantiomers because they are mirror images and not superposable (see the reflection in the mirror given). Hence, these forms of butane are chiral. However, butane is *not* a chiral molecule because these three isomers interconvert very rapidly at room temperature and because they interconvert through the intermediacy of the anti isomer, which is achiral (refer back to Figure 2.9 to see the interconversion). The anti isomer is achiral because there is a plane of symmetry when all four carbons are planar.

This discussion highlights an important point: chirality can be present in molecules without chiral centers. This condition occurs via conformational isomerism, but it is less common than chirality arising from configurational isomerism. When, unlike in butane, the barrier to interconversion is large and the enantiomers cannot interconvert at ambient temperature through a planar form, the molecule will be chiral and the enantiomers can be separated. An example is the following substituted biphenyl. Because of the large groups on the rings, there are substantial nonbonded interactions in the planar conformer and the twisted forms shown have far lower energies. The nonbonded interactions result in a very high barrier to rotation around the carbon-carbon single bond connecting the rings, and rotation is very slow. Although this molecule has no chiral center, the mirror images are not superposable, and the molecule is chiral. Isomers of this sort, which lack a chiral center but do not interconvert because of hindered rotation, are called **atropisomers**.

Atropisomers

Enantiomers that lack a chiral center and differ because of hindered rotation.

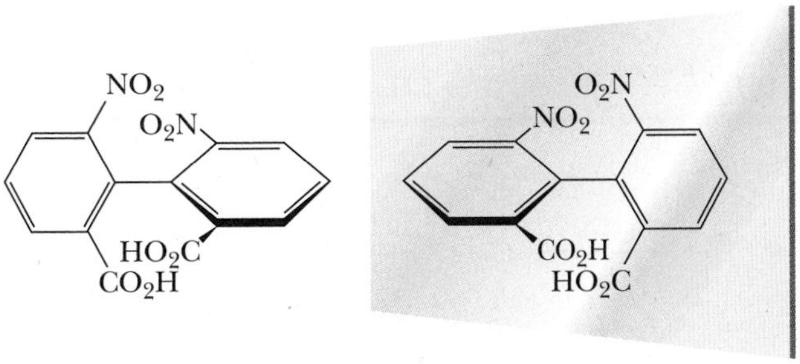

3.3 Naming Chiral Centers—The *R,S* System

Absolute configuration

Which of the two possible isomers an enantiomer is (i.e., whether it is the right- or left-handed isomer).

So far, we have discussed the fact that enantiomers exist. We have not considered the question of which isomer is which (i.e., the **absolute configuration**) or which is the right-handed enantiomer and which is the left. For a given sample of a pure enantiomer, the correct arrangement must be determined by experiment. Experimental determination of absolute configuration can be accomplished by using x-ray analysis of a derivative that has a chiral center with a known absolute configuration. In biological molecules, many absolute configurations were determined by comparison to absolute configurations of the chiral center in glyceraldehyde.

R,S System

A set of rules for specifying absolute configuration about a chiral center; also called the Cahn-Ingold-Prelog system.

R

From the Latin, *rectus*, straight, correct; used in the *R,S* convention to show that the order of priority of groups on a chiral center is clockwise.

S

From the Latin, *sinister*, left; used in the *R,S* convention to show that the order of priority of groups on a chiral center is counterclockwise.

A system for designating the absolute configuration of a chiral center was devised in the late 1950s by R. S. Cahn and C. K. Ingold in England and V. Prelog in Switzerland and is named after them. The system, also called the ***R,S* system**, has been incorporated into the IUPAC rules of nomenclature. The orientation of groups about a chiral center is specified using a set of priority rules.

A. Priority Rules

1. Each atom bonded to the chiral center is assigned a priority. Priority is based on atomic number; the higher the atomic number, the higher the priority. Following are several substituents arranged in order of increasing priority. The atomic number of the atom determining priority is shown in parentheses.

(1)	(6)	(7)	(8)	(16)	(17)	(35)	(53)
—H,	$—CH_3$,	$—NH_2$,	—OH,	—SH,	—Cl,	—Br,	—I

Increasing priority →

2. If priority cannot be assigned on the basis of the atoms bonded directly to the chiral center [because of a tie (i.e., the same first atom on more than one substituent)], look at the next set of atoms and continue until a priority can be assigned. Priority is assigned at the first point of difference. Following is a series of groups arranged in order of increasing priority. The atomic number of the atom on which the assignment of priority is based is shown above it.

(1)	(6)	(7)	(8)	(17)
$—CH_2—H$	$—CH_2—CH_3$	$—CH_2—NH_2$	$—CH_2—OH$	$—CH_2—Cl$

Increasing priority →

If two carbons have substituents of the same priority, priority is assigned to the carbon that has more of the higher priority substituents. Thus, $—CHCl_2 > —CH_2Cl$.

3. Atoms participating in a double or triple bond are considered to be bonded to an equivalent number of similar "phantom" atoms (shown here highlighted) by single bonds; that is, atoms of the double bond are duplicated and atoms of a triple bond are triplicated. The phantom atoms are bonded to no other atoms.

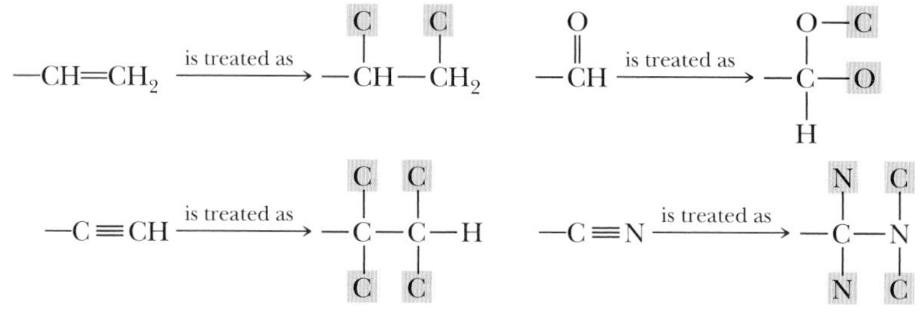

4. *Note:* Priority assignment is made at the *first point of difference* between groups. A common mistake is to assume that larger groups must always have higher priority, but this might not be the case. For example, a —CH_2Cl group has priority over a —$CH_2CH_2CH_2CH_3$ group because the Cl atom is the *first point of difference.*

Example 3.2 Cahn, Ingold, Prelog Priorities

Assign priorities to the groups in each set.

(a) 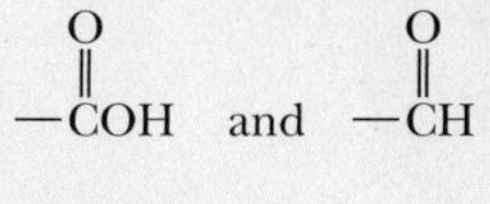

(b) 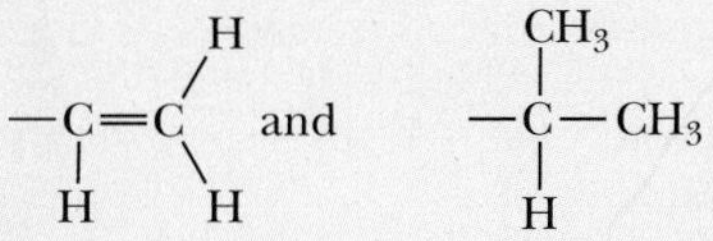

Solution

(a) The first point of difference is O of the —OH in the carboxyl group compared to —H in the aldehyde group. The carboxyl group has a higher priority.

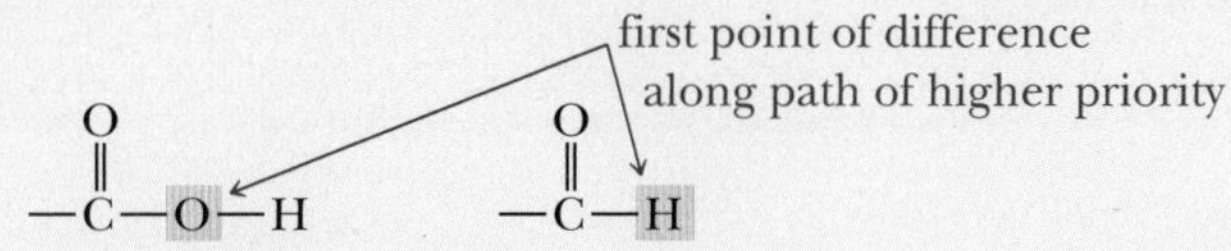

Carboxyl group (higher priority) | Aldehyde group (lower priority)

(b) Carbon 1 in each group has the same pattern of atoms; namely C(C,C,H) (i.e., carbon bonded to two carbons and a hydrogen). For the vinyl group, bonding at carbon 2 is C(C,H,H). For the isopropyl group, at carbon 2, it is C(H,H,H). The vinyl group is higher in priority than is the isopropyl group.

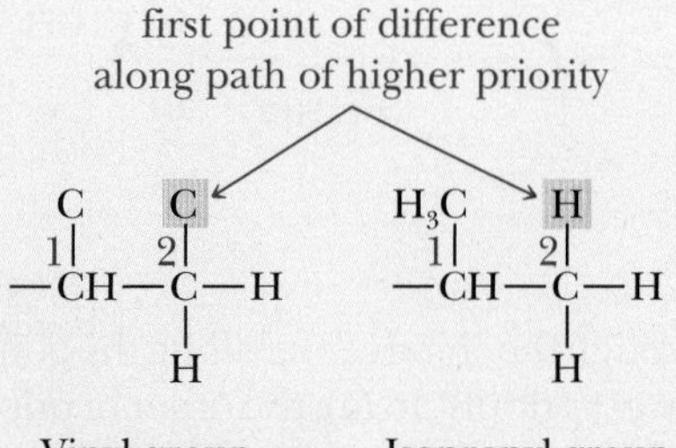

Vinyl group (higher priority) | Isopropyl group (lower priority)

Problem 3.2

Assign priorities to the groups in each set.

(a) —CH_2OH and —CH_2CH_2OH **(b)** —CH_2OH and —$CH{=}CH_2$

(c) —CH_2OH and —$C(CH_3)_3$

HOW TO

Assign *R* or *S* Configuration to a Chiral Center

1. Locate the chiral center, identify its four substituents, and assign a priority from 1 (highest) to 4 (lowest) to each substituent.
2. Orient the molecule in space so that the group of lowest priority (4) is directed away from you, analogously to the steering column of a car. The three groups of higher priority (1–3) then project toward you, as would the spokes of the steering wheel.
3. Read the three groups projecting toward you in order from highest priority (1) to lowest priority (3).
4. If the groups are read in a clockwise direction, the configuration is designated as ***R*** (Latin: *rectus,* straight, correct); if they are read in a counterclockwise direction, the configuration is **S** (Latin: *sinister,* left). You can also visualize this as follows: turning the steering wheel to the right (down the order of priority) equals *R*; turning it to the left equals *S*.

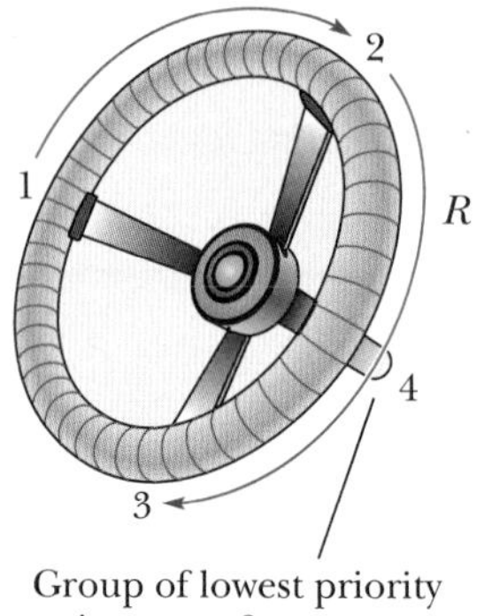

Group of lowest priority points away from you

Example 3.3 *R,S* Configurations

Assign an *R* or *S* configuration to the chiral center in each molecule.

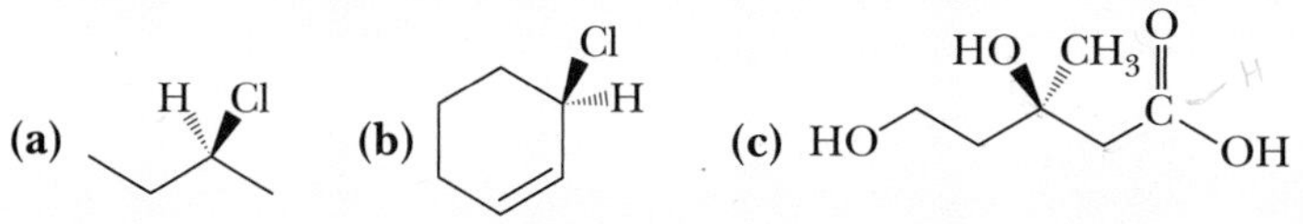

Solution

View each molecule through the chiral center along the bond from the chiral center toward the group of lowest priority. In (a), the order of priority is $Cl > CH_2CH_3 > CH_3 > H$; the configuration is *S*. In (b), the order of priority is $Cl > CH{=}CH > CH_2 > H$; the configuration is *R*. In (c), the order of priority is $OH > CH_2COOH > CH_2CH_2OH > CH_3$; the configuration is *R*.

(a)

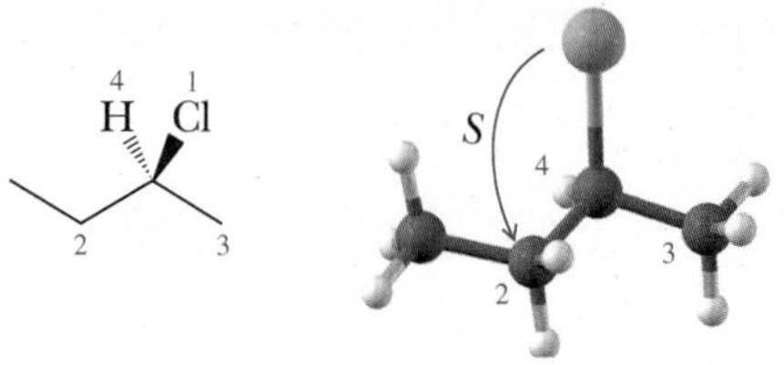

(b)

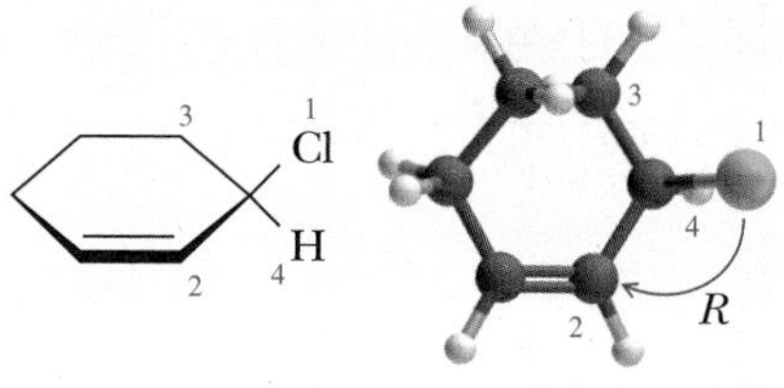

(c)

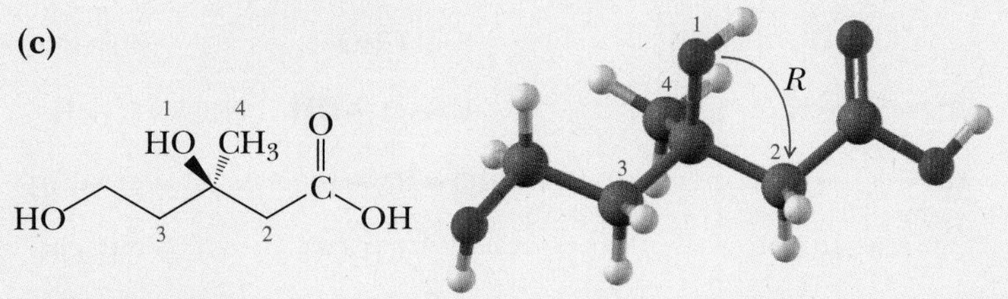

Problem 3.3

Watch a video explanation

Assign an *R* or *S* configuration to the chiral center in each molecule.

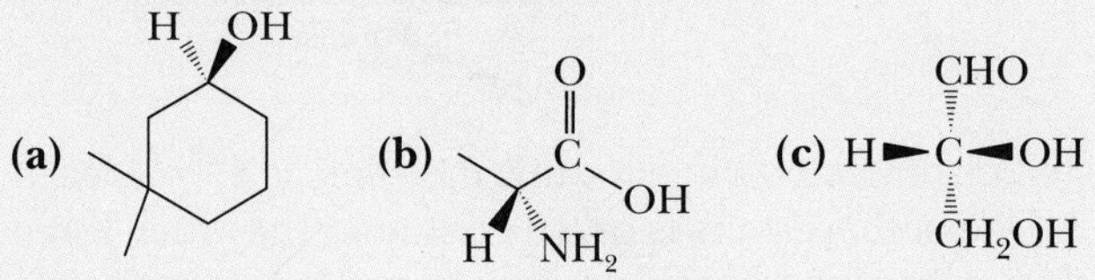

3.4 Acyclic Molecules with Two or More Stereocenters

When two or more stereocenters exist in a molecule, multiple stereoisomers are possible. When stereoisomerism was first introduced in Section 2.6, we noted that both double-bond carbons in *trans*-2-butene are stereocenters because the swapping of a methyl and a hydrogen on either carbon creates the stereoisomer *cis*-2-butene. More double bonds lead to more possible stereoisomers. For example, the molecule 2,4-heptadiene can exist as four stereoisomers. This situation gets even more complicated when the stereocenters are also chiral centers.

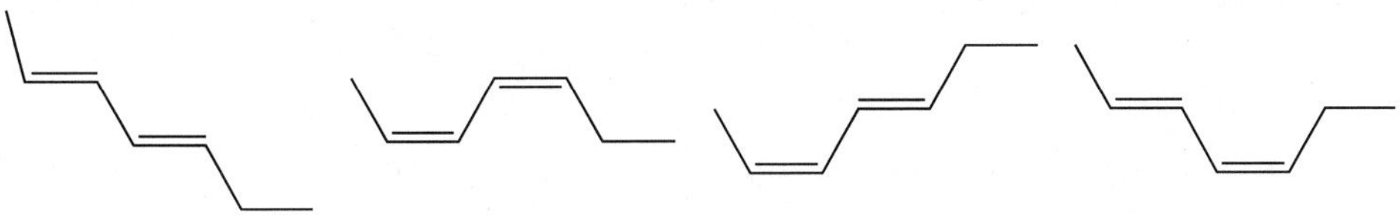

2,4-Heptadiene

A. Enantiomers and Diastereomers

We have now seen several examples of molecules with one chiral center and verified that, for each, two stereoisomers (one pair of enantiomers) are possible. Now let us consider molecules with two or more chiral centers. To generalize, for a molecule with *n* chiral centers, the maximum number of stereoisomers possible is 2^n. We have already seen that, for a molecule with one chiral center, $2^1 = 2$ stereoisomers are possible. For a molecule with two chiral centers, $2^2 = 4$ stereoisomers are possible; for a molecule with three chiral centers, $2^3 = 8$ stereoisomers are possible; and so forth.

Let us begin our study of molecules with multiple chiral centers by considering 2,3,4-trihydroxybutanal, a molecule with two chiral centers, shown here highlighted.

$$HOCH_2{-}\underset{OH}{\underset{|}{CH}}{-}\underset{OH}{\underset{|}{CH}}{-}CHO$$

2,3,4-Trihydroxybutanal

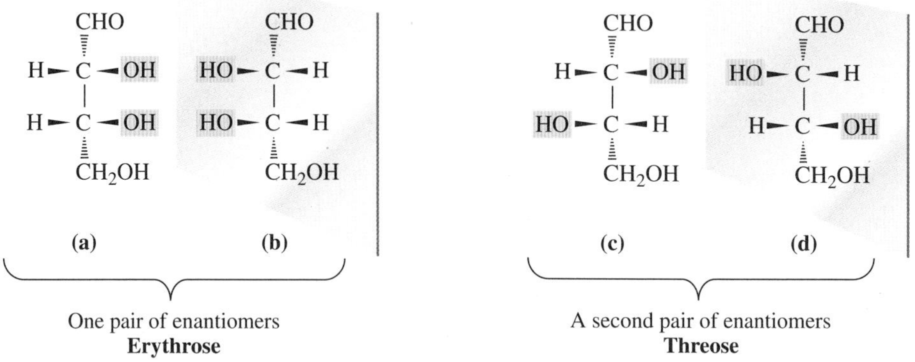

FIGURE 3.3 The four stereoisomers of 2,3,4-trihydroxybutanal, a compound with two chiral centers.

The maximum number of stereoisomers possible for this molecule is $2^2 = 4$, each of which is drawn in Figure 3.3. One of these pairs is called erythrose; the other, threose.

Stereoisomers (a) and (b) are nonsuperposable mirror images and are, therefore, a pair of enantiomers. Stereoisomers (c) and (d) are also nonsuperposable mirror images and are a second pair of enantiomers. One way to describe the four stereoisomers of 2,3,4-trihydroxybutanal is to say that they consist of two pairs of enantiomers. Enantiomers (a) and (b) of 2,3,4-trihydroxybutanal are given the names (2*R*,3*R*)-erythrose and (2*S*,3*S*)-erythrose; enantiomers (c) and (d) are given the names (2*R*,3*S*)-threose and (2*S*,3*R*)-threose. Note that all of the chiral centers in a molecule are reversed in its enantiomer. The molecule with the 2*R*,3*S* configuration is the enantiomer of the molecule with 2*S*,3*R*, and the molecule with 2*S*,3*S* is the enantiomer of the molecule with 2*R*,3*R*. Erythrose and threose belong to the class of compounds called carbohydrates, which we discuss in Chapter 25. Erythrose is found in erythrocytes (red blood cells), hence the derivation of its name.

We have specified the relationship between (a) and (b) and between (c) and (d); each represents a pair of enantiomers. What is the relationship between (a) and (c), between (a) and (d), between (b) and (c), and between (b) and (d)? The answer is that they are diastereomers. Recall that diastereomers are stereoisomers that are not mirror images (enantiomers). As we see in this example, molecules with at least two chiral centers can have diastereomers.

Example 3.4 *R* and *S* Assignments

Following are stereorepresentations for the four stereoisomers of 1,2,3-butanetriol. *R* and *S* configurations are given for the chiral centers in (1) and (4).

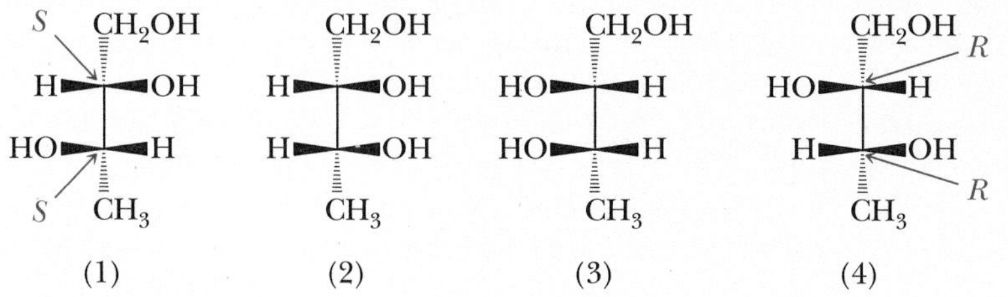

(a) Write the IUPAC names for each compound showing the *R* or *S* configuration of each chiral center.

(b) Which molecules are enantiomers?

(c) Which molecules are diastereomers?

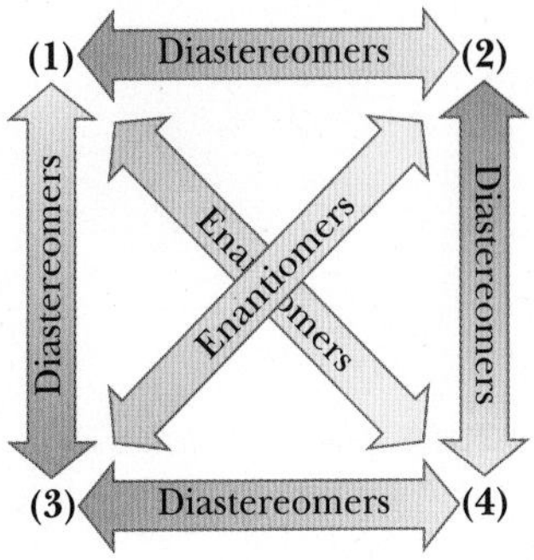

Solution

(a) (1) (2*S*,3*S*)-1,2,3-Butanetriol (2) (2*S*,3*R*)-1,2,3-Butanetriol
(3) (2*R*,3*S*)-1,2,3-Butanetriol (4) (2*R*,3*R*)-1,2,3-Butanetriol

(b) Enantiomers are stereoisomers that are nonsuperposable mirror images. As you see from their configurations, compounds (1) and (4) are one pair of enantiomers and compounds (2) and (3) are a second pair of enantiomers.

(c) Diastereomers are stereoisomers that are not mirror images. Compounds (1) and (2), (1) and (3), (2) and (4), and (3) and (4) are pairs of diastereomers. In the margin is a diagram that shows the relationships among these isomers.

Problem 3.4

Following are stereorepresentations for the four stereoisomers of 3-chloro-2-butanol.

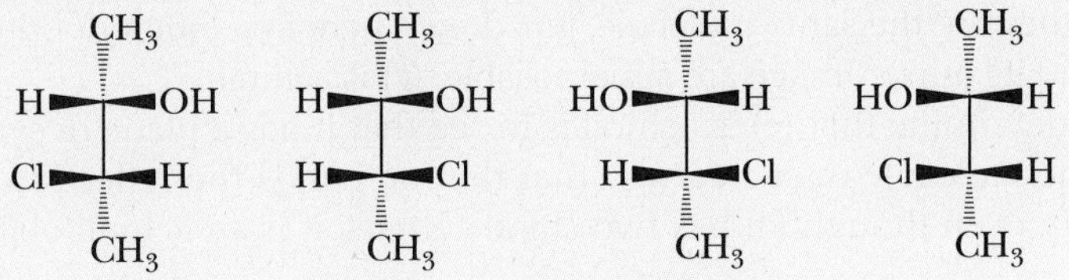

(1) (2) (3) (4)

(a) Assign an *R* or *S* configuration to each chiral center.

(b) Which compounds are enantiomers?

(c) Which compounds are diastereomers?

B. Meso Compounds

Certain molecules containing two or more chiral centers have special symmetry properties that reduce the number of stereoisomers to fewer than the maximum number predicted by the 2^n rule. One such molecule is 2,3-dihydroxybutanedioic acid, more commonly named tartaric acid.

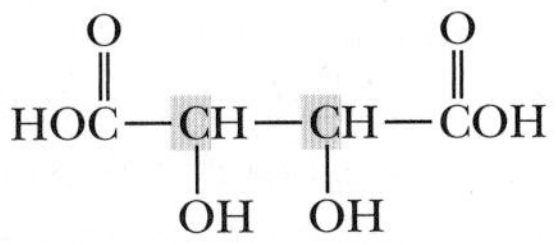

2,3-Dihydroxybutanedioic acid
(Tartaric acid)

Tartaric acid is a colorless crystalline compound. During the fermentation of grape juice, potassium bitartrate (one carboxyl group is present as a potassium salt, —COO^-K^+) deposits as a crust on the sides of wine casks. When collected and purified, it is sold commercially as cream of tartar.

In tartaric acid, carbons 2 and 3 are chiral centers and the maximum number of stereoisomers possible is $2^2 = 4$; these stereorepresentations are drawn in Figure 3.4. Structures (a) and (b) are nonsuperposable mirror images and are, therefore, a pair of enantiomers. Structures (c) and (d) are also mirror images, but they are superposable.

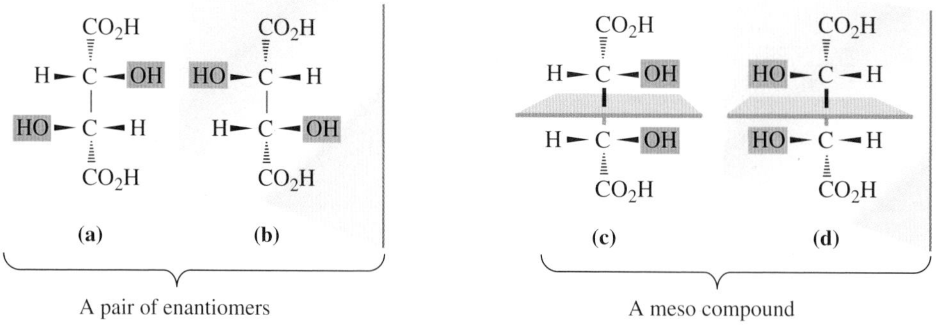

FIGURE 3.4 Stereoisomers of tartaric acid. One pair of enantiomers and one meso compound.

To see this, imagine that you first rotate (d) by 180° in the plane of the paper, lift it out of the plane of the paper, and place it on top of (c). If you do this mental manipulation correctly, you find that (d) is superposable on (c). Therefore, (c) and (d) are not different molecules; they are the same molecule, just drawn here in a different orientation. Because (c) and its mirror image are superposable, (c) is achiral.

Another way to determine that (c) is achiral is to see that it has a plane of symmetry that bisects the molecule in such a way that the top half is the reflection of the bottom half. Thus, even though (c) has two chiral centers, it is an achiral object (Section 3.2).

Meso compound
An achiral compound possessing two or more chiral centers that also has chiral isomers.

The stereoisomer of tartaric acid represented by (c) or (d) is called a **meso compound**. A meso compound contains two or more chiral centers but is achiral. To be a meso compound, a molecule must also have chiral isomers. We can now answer this question: how many stereoisomers are there of tartaric acid? The answer is three: one meso compound and one pair of enantiomers. Note that the meso compound is a diastereomer of each member of the pair of enantiomers.

From this example, we can make this generalization about meso compounds: They have an internal mirror plane (or center of symmetry). Commonly, there are two chiral centers, each with the same four groups: one is *R;* the other, *S*.

Notice that the stereoisomers in Figure 3.4 are shown in only one conformation. Because conformational isomers in acyclic systems interconvert extremely rapidly, we need consider only the most symmetric conformers. Other conformers may be rotated to give the most symmetric one for determination of symmetry properties (see Problem 3.5).

It is sometimes helpful to identify stereochemical relationships among molecules based on assignment of configuration to all the chiral centers. In some cases, this can be easier than searching for and comparing the most symmetric conformations. Two stereoisomers will be enantiomers if all of the chiral centers present are reversed, and the stereoisomers will be diastereomers if only some of the chiral centers are reversed.

Enantiomers have identical physical and chemical properties in an achiral environment. Examples of achiral environments are solvents that have no chiral centers, such as H_2O, CH_3CH_2OH, and CH_2Cl_2. The enantiomers of tartaric acid (Table 3.1), for example, have the same melting point, the same boiling point, the same solubility in water and other common solvents, and the same value of pK_a, and they undergo the same acid-base reactions. The enantiomers of tartaric acid do, however, differ in optical activity (the ability to rotate the plane of plane-polarized light), which we discuss in Section 3.7B. Diastereomers have different physical and chemical properties, even in achiral environments. Meso tartaric acid has different physical properties from those of the enantiomers and can be separated from them by methods such as crystallization.

Table 3.1 Some Physical Properties of the Stereoisomers of Tartaric Acid

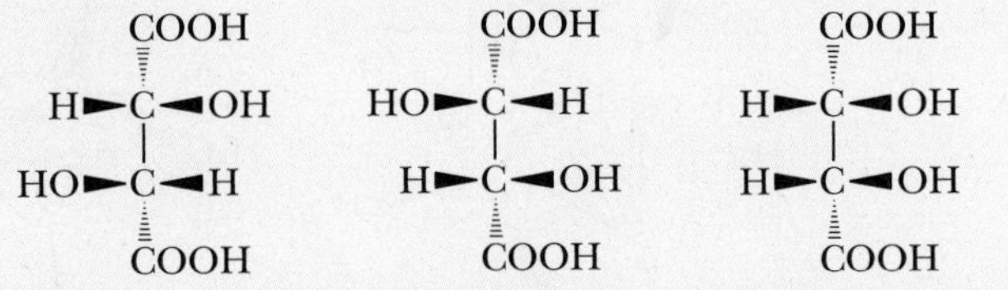

	(*R,R*)-Tartaric Acid	(*S,S*)-Tartaric Acid	Meso Tartaric Acid
Specific rotation*	+12.7	−12.7	0
Melting point (°C)	171–174	171–174	146–148
Density at 20°C (g/cm^3)	1.7598	1.7598	1.660
Solubility in water at 20°C (g/100 mL)	139	139	125
pK_1 (25°C)	2.98	2.98	3.23
pK_2 (25°C)	4.34	4.34	4.82

*Specific rotation is discussed in Section 3.7B.

Example 3.5 Stereoisomerism

Following are stereorepresentations for the three stereoisomers of 2,3-butanediol. The carbons are numbered beginning from the left, as shown in (1).

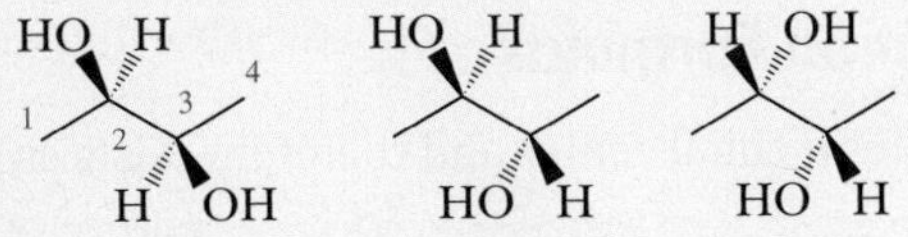

(1) (2) (3)

(a) Assign an *R* or *S* configuration to each chiral center.

(b) Which are enantiomers?

(c) Which is the meso compound?

(d) Which are diastereomers?

Solution

(a) (1) (2*R*,3*R*)-2,3-Butanediol (2) (2*R*,3*S*)-2,3-Butanediol
(3) (2*S*,3*S*)-2,3-Butanediol

(b) Compounds (1) and (3) are enantiomers.

(c) Compound (2) is a meso compound. Note that compound (2) can be drawn in two symmetric conformations, one of which has a plane of symmetry and the other of which has a center of symmetry.

(Continued)

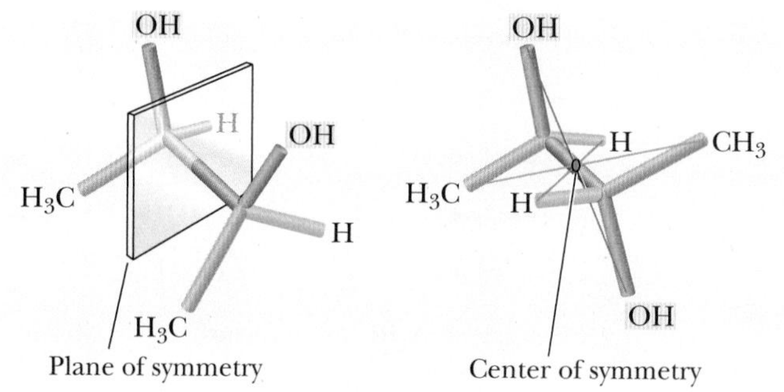
OH
H
OH
H3C
H
H3C
Plane of symmetry
OH
H
CH3
H3C
H
OH
Center of symmetry

(d) (1) and (2) are diastereomers; (2) and (3) are also diastereomers.

Watch a video explanation

Problem 3.5

Following are four Newman projection formulas for tartaric acid.

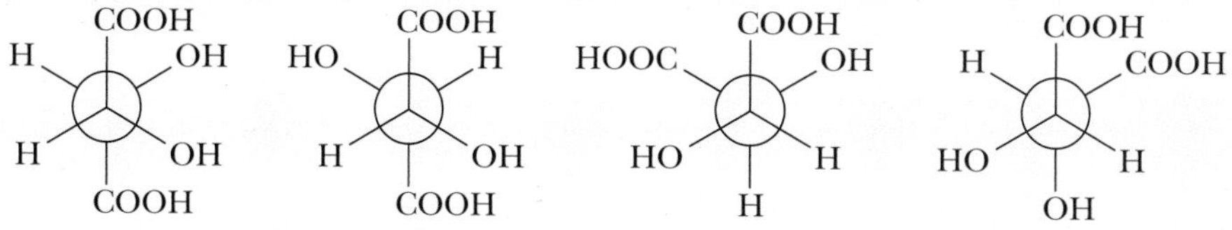
COOH
H
OH
H
OH
COOH
COOH
HO
H
H
OH
COOH
COOH
HOOC
OH
HO
H
H
COOH
H
COOH
HO
H
OH

(1) (2) (3) (4)

(a) Which represent the same compound?

(b) Which represent enantiomers?

(c) Which represent a meso compound?

(d) Which are diastereomers?

C. Fischer Projection Formulas

Glyceraldehyde contains a chiral center and therefore exists as a pair of enantiomers.

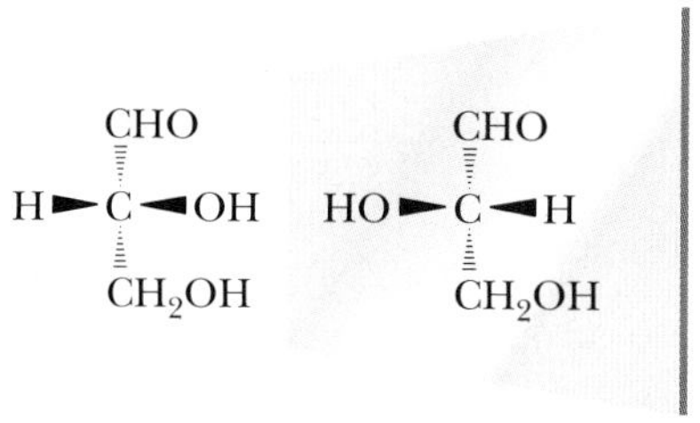
CHO
H
C
OH
CH2OH
CHO
HO
C
H
CH2OH

(*R*)-Glyceraldehyde **(*S*)-Glyceraldehyde**

Fischer projection

A two-dimensional projection of a molecule; in these projections, groups on the right and left are by convention in front, while those at the top and bottom are to the rear.

Chemists use two-dimensional representations called **Fischer projections** to show the configuration of molecules with multiple chiral centers, especially carbohydrates. To write a Fischer projection, draw a three-dimensional representation of the molecule oriented so that the vertical bonds from the chiral center are directed away from you and the horizontal bonds from the chiral center are directed toward you. Then write the molecule as a two-dimensional figure with the chiral center indicated by the point at which the lines cross.

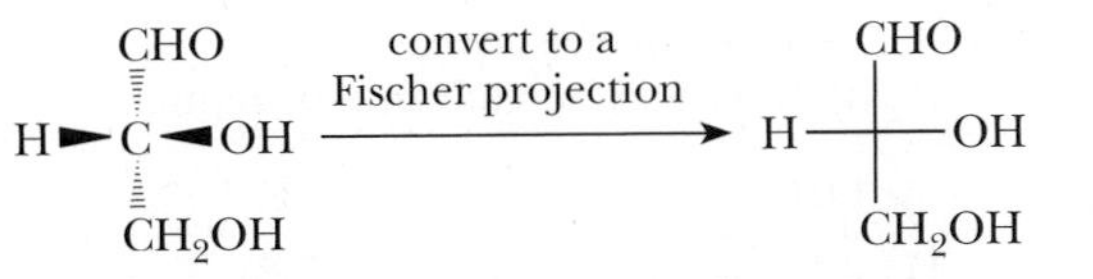
CHO
H
C
OH
CH₂OH
convert to a
Fischer projection
CHO
H
OH
CH₂OH

(*R*)-Glyceraldehyde (three-dimensional representation)

(*R*)-Glyceraldehyde (Fischer projection)

The horizontal segments of this Fischer projection represent bonds directed toward you, and the vertical segments represent bonds directed away from you. The only atom in the plane of the paper is the chiral center. Because a Fischer projection implies that the groups to each side are in front and those at top and bottom are behind the plane of the paper, rotations of these drawings by 90° are not permissible.

Example 3.6 Fischer Projections

Draw a Fischer projection of (2*R*,3*R*)-erythrose (Figure 3.3).

Solution

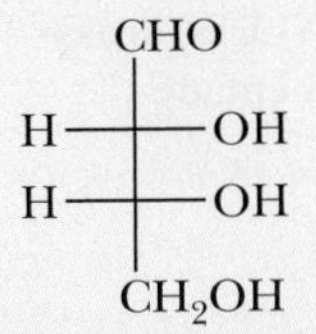
CHO
H
OH
H
OH
CH₂OH

(2*R*, 3*R*)-Erythrose
(2,3,4-Trihydroxybutanal)

Problem 3.6

Give a complete stereochemical name for the following compound, which is a 1,2,3-butanetriol.

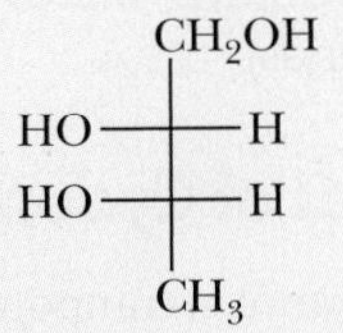
CH₂OH
HO
H
HO
H
CH₃

HOW TO

Quickly Draw and Recognize Enantiomers and Diastereomers

Recall that the definition of a **stereocenter** is an atom where the swapping of any two groups creates a stereoisomer. Combining this recollection with *R* and *S* nomenclature, it is clear that swapping any two groups on a chiral center will switch an *R*-center to an *S*-center, and vice versa. When there is only one chiral center, swapping any two groups creates an enantiomer. However, when more than one chiral center exists, you need to swap any two groups on *all* centers to create an enantiomer; that is, *all R*-centers become *S* and vice versa. Of course, drawing the enantiomer can also be achieved by examining the molecule in a mirror as discussed earlier, but the idea of swapping two groups is fast and easy.

(*Continued*)

swap any two groups to create an enantiomer

swap any two groups on each chiral center to create an enantiomer

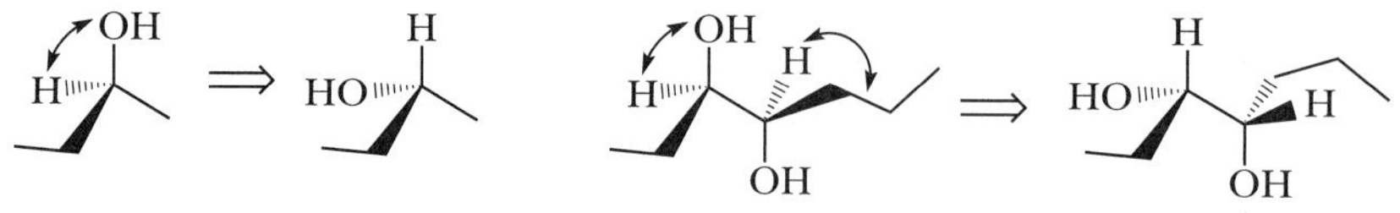
OH
H
H
HO
OH
H
H
OH
H
HO
H
OH

(*S*)-2-Butanol (*R*)-2-Butanol (3*R*,4*R*)-Dihydroxy-heptane (3*S*,4*S*)-Dihydroxy-heptane

When *n* chiral centers exist, creating a diastereomer can be achieved by swapping two groups on any number of the chiral centers less than *n*. In other words, as long as all the centers are not inverted, one would create a diastereomer. The *R*- or *S*- designation only changes on the chiral centers where a swap of groups occurred.

swap any two groups on less than all chiral centers to create a disastereomer

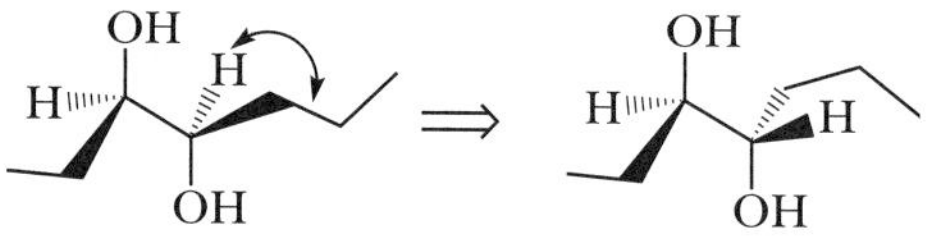
OH
H
H
OH
OH
H
H
OH

(3*R*,4*R*)-Dihydroxy-heptane (3*R*,4*S*)-Dihydroxy-heptane

3.5 Cyclic Molecules with Two or More Chiral Centers

In this section, we concentrate on derivatives of cyclopentane and cyclohexane containing two stereocenters. We can analyze stereoisomerism in cyclic compounds in the same way as in acyclic compounds.

A. Disubstituted Derivatives of Cyclopentane

Let us start with 2-methylcyclopentanol, a compound with two chiral centers. We predict a maximum of $2^2 = 4$ stereoisomers. Both the *cis* isomer and the *trans* isomer are chiral: The *cis* isomer exists as one pair of enantiomers, and the *trans* isomer exists as a second pair of enantiomers. The *cis* and *trans* isomers are stereoisomers that are not mirror images of each other; that is, they are diastereomers.

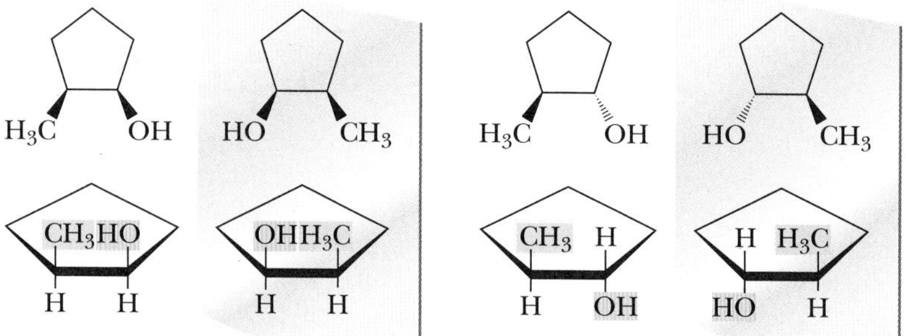
H3C
OH
HO
CH3
H3C
OH
HO
CH3
CH3HO
H
H
OHH3C
H
H
CH3
H
H
OH
H
H3C
HO
H

cis-2-Methylcyclopentanol (a pair of enantiomers)

trans-2-Methylcyclopentanol (a pair of enantiomers)

1,2-Cyclopentanediol also has two chiral centers; therefore, the 2^n rule predicts a maximum of $2^2 = 4$ stereoisomers. As shown in the following stereodrawings, only three stereoisomers exist for this compound. The *cis* isomer is achiral (meso) because it and its mirror image are superposable. An alternative way to identify the *cis* isomer as achiral is to notice that it possesses a plane of symmetry that bisects the molecule into two mirror halves. The *trans* isomer is chiral and exists as a pair of enantiomers.

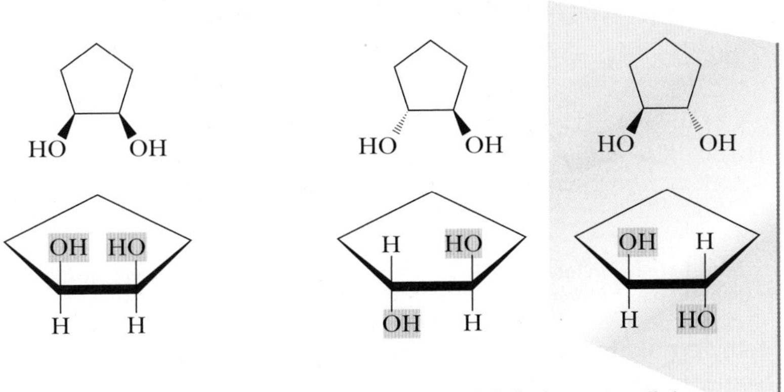

***cis*-1,2-Cyclopentanediol**
(a meso compound)

***trans*-1,2-Cyclopentanediol**
(a pair of enantiomers)

Example 3.7 Stereoisomerism with Rings I

How many stereoisomers exist for 3-methylcyclopentanol?

Solution

There are four stereoisomers of 3-methylcyclopentanol. The *cis* isomer exists as one pair of enantiomers; the *trans* isomer, as a second pair of enantiomers.

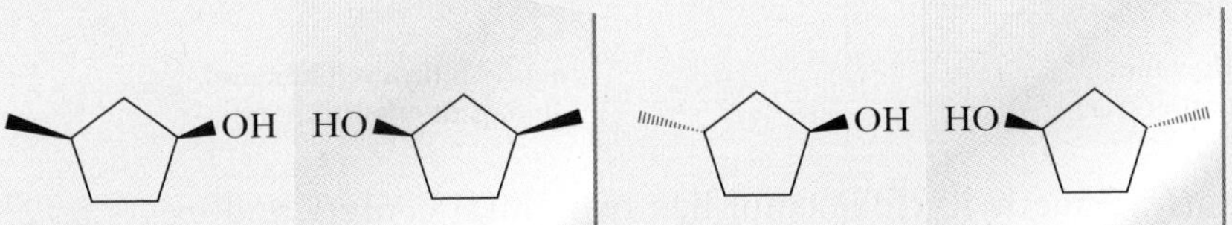

***cis*-3-Methylcyclopentanol**
(a pair of enantiomers)

***trans*-3-Methylcyclopentanol**
(a pair of enantiomers)

Problem 3.7

How many stereoisomers exist for 1,3-cyclopentanediol?

B. Disubstituted Derivatives of Cyclohexane

Cyclohexane derivatives exist in chair conformations. However, when evaluating disubstituted cyclohexane derivatives for symmetry, it is helpful to analyze structures drawn flat. As an example of a disubstituted cyclohexane, let us consider the methylcyclohexanols. 4-Methylcyclohexanol has two stereocenters, but they are not chiral centers because the carbons of the stereocenters do not have four different groups attached. 4-Methylcyclohexanol therefore exists as two diastereomers, a pair of *cis,trans* isomers. Both of these isomers are achiral. In each, a plane of symmetry runs through the —CH_3 and —OH groups and the carbons bonded to them.

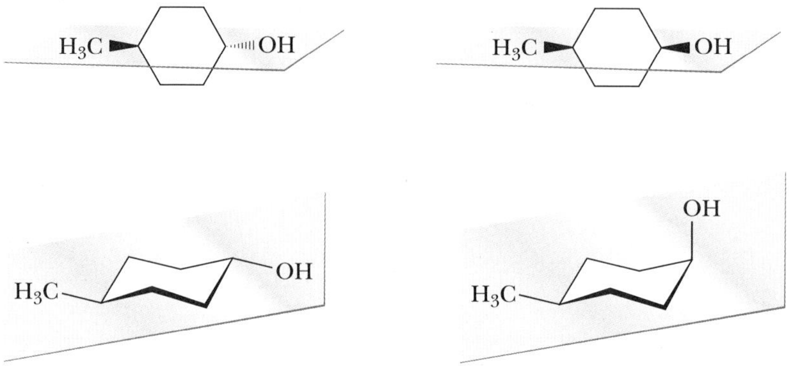

***trans*-4-Methylcyclohexanol**
(achiral, plane of symmetry)

***cis*-4-Methylcyclohexanol**
(achiral, plane of symmetry)

3-Methylcyclohexanol has two chiral centers and exists as $2^2 = 4$ stereoisomers. The *cis* isomer exists as one pair of enantiomers; the *trans* isomer, as a second pair of enantiomers.

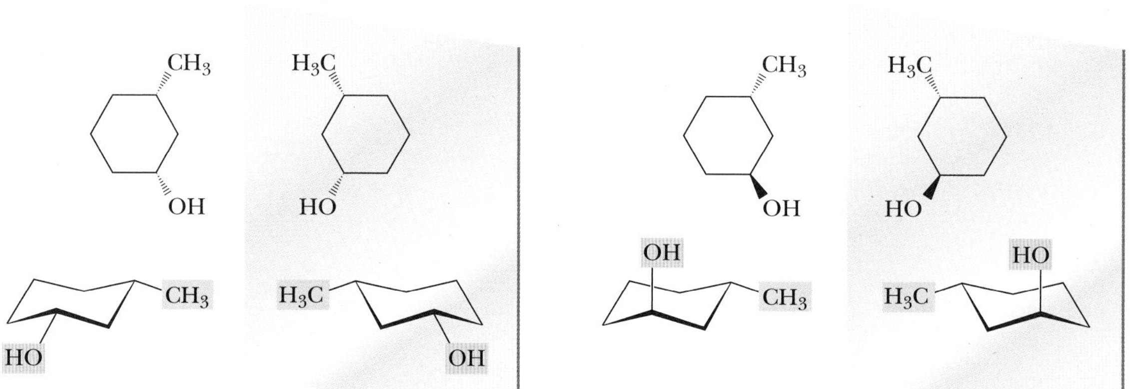

***cis*-3-Methylcyclohexanol**
(a pair of enantiomers)

***trans*-3-Methylcyclohexanol**
(a pair of enantiomers)

Similarly, 2-methylcyclohexanol has two chiral centers and exists as $2^2 = 4$ stereoisomers. The *cis* isomer exists as one pair of enantiomers; the *trans* isomer, as a second pair of enantiomers.

Example 3.8 Stereoisomerism with Rings II

How many stereoisomers exist for 1,3-cyclohexanediol?

Solution

1,3-Cyclohexanediol has two chiral centers and, according to the 2^n rule, has a maximum of $2^2 = 4$ stereoisomers. The *trans* isomer of this compound exists as a pair of enantiomers. The *cis* isomer has a plane of symmetry and is a meso compound. Therefore, although the 2^n rule predicts a maximum of four stereoisomers for 1,3-cyclohexanediol, only three exist: one meso compound and one pair of enantiomers.

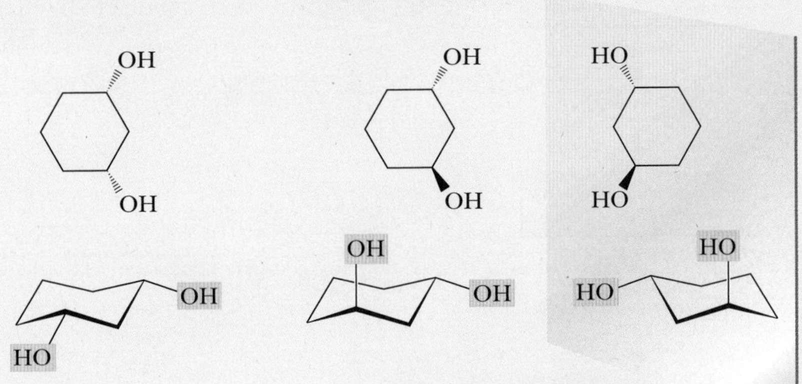

***cis*-1,3-Cyclohexanediol**
(meso)

***trans*-1,3-Cyclohexanediol**
(a pair of enantiomers)

Problem 3.8
How many stereoisomers exist for 1,4-cyclohexanediol?

Examining the planar structures indicates that 1,2-cyclohexanediol has three stereoisomers. The *trans* isomer exists as a pair of enantiomers, while the *cis* isomer is a meso compound because it has a plane of symmetry. Interestingly, the two alternative chair conformations of *cis*-1,2-cyclohexanediol are each chiral and each is the mirror image of the other. However, because they are the same energy and they interconvert rapidly at room temperature, *cis*-1,2-cyclohexanediol is effectively meso as predicted by analysis of the planar structures.

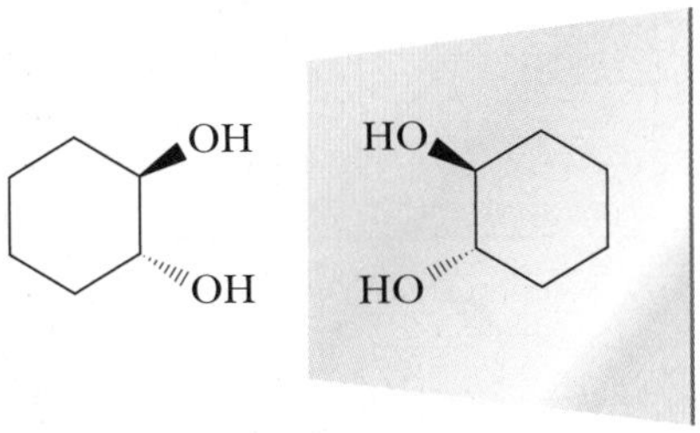

***trans*-1,2-Cyclohexanediol**
(a pair of enantiomers)

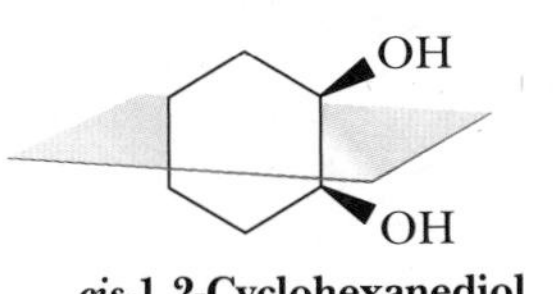

***cis*-1,2-Cyclohexanediol**
(achiral, plane of symmetry)

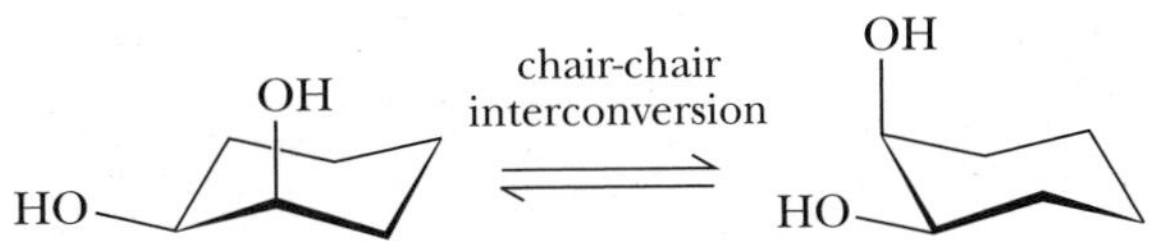
chair-chair
interconversion

***cis*-1,2-Cyclohexanediol**
(chair conformations)

3.6 Tying All the Terminology Together

One of the major challenges in learning stereochemistry is mastering all of the terminology. In this chapter and in the previous chapters, we presented several terms that describe the various kinds of isomers. Figure 3.5 summarizes the different isomers

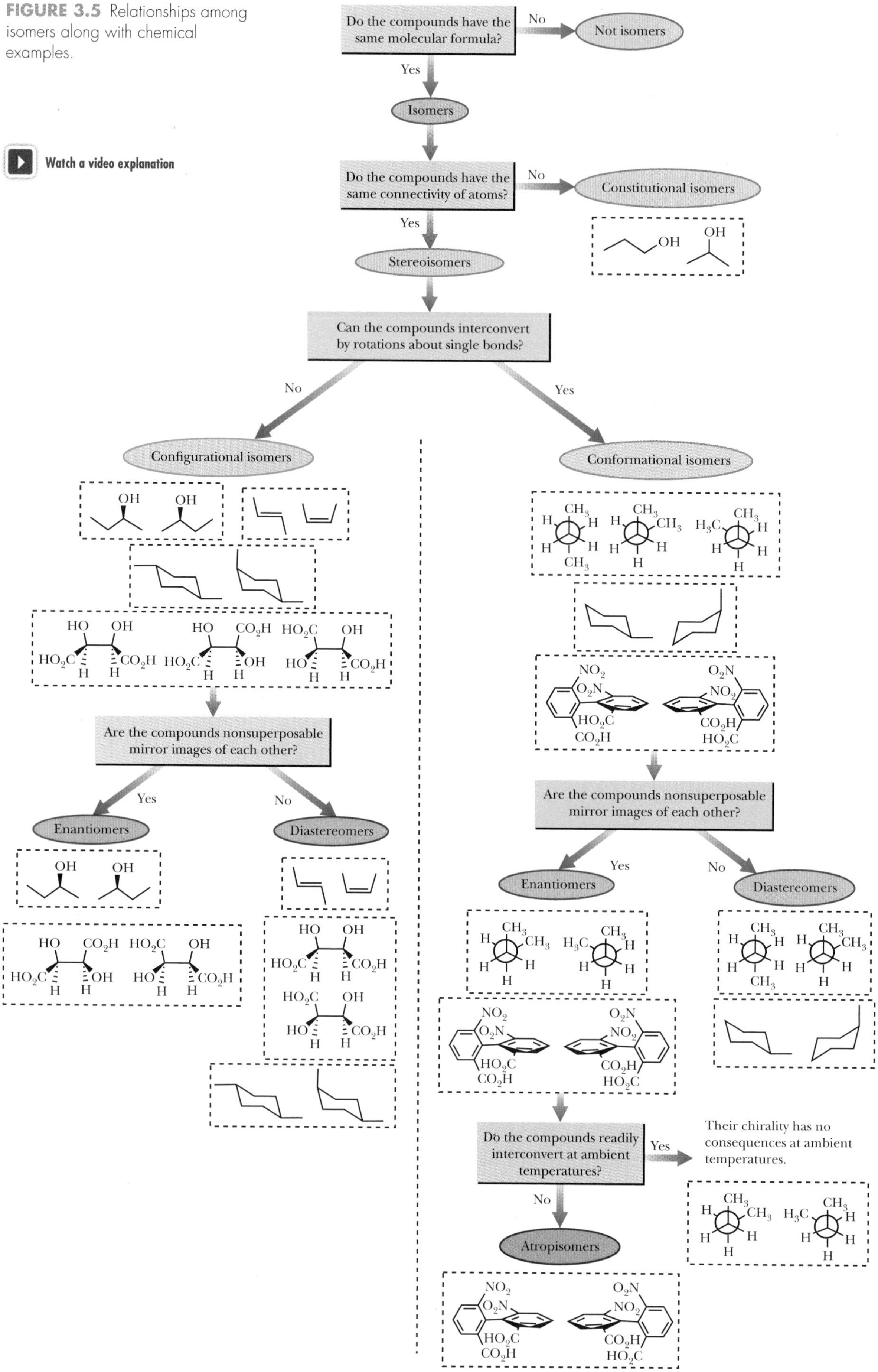

FIGURE 3.5 Relationships among isomers along with chemical examples.

Watch a video explanation

we have discussed so far, as well as the relationships among them. It delineates a step-by-step series of questions to guide you through the decision-making process for stereochemical terminology along with some of the same chemical examples that have been used thus far.

As a review, let's summarize Figure 3.5. Isomers occur when chemicals have the same molecular formula, but stereoisomers occur when the compounds have the same connectivity of their atoms but those atoms are arranged differently in space. Conformational isomers are a form of stereoisomers that interconvert by rotations along single bonds. Sometimes these isomers are nonsuperposable mirror images and are thus enantiomers; if the isomers are not nonsuperposable mirror images, they are diastereomers. If there is restricted rotation such that the enantiomers or diastereomers cannot interconvert at ambient temperatures, then they are called atropisomers. When the enantiomers interconvert rapidly, there are no chemical consequences of their isomeric relationship. So it is generally ignored.

When the stereoisomers are not interconverted by rotations along single bonds, they are referred to as configurational isomers. These isomers are nonsuperposable mirror images and are thus enantiomers, or they are diastereomers.

3.7 Optical Activity—How Chirality Is Detected in the Laboratory

As we have already established, enantiomers are different compounds; thus, we must expect that they differ in some properties. One property that differs between enantiomers is their effect on the rotation of the plane of polarized light. Each member of a pair of enantiomers rotates the plane of polarized light in opposite directions, and for this reason, enantiomers are said to be **optically active**.

Optically active
Refers to a compound that rotates the plane of polarized light.

The phenomenon of optical activity was discovered in 1815 by the French physicist Jean Baptiste Biot. To understand how it is detected in the laboratory, we must first understand something about plane-polarized light and a polarimeter, the device used to detect optical activity.

A. Plane-Polarized Light

Ordinary light consists of waves vibrating in all planes perpendicular to its direction of propagation (Figure 3.6). Certain materials, such as calcite and a Polaroid sheet (a plastic film containing properly oriented crystals of an organic substance embedded in it), selectively transmit light waves vibrating only in parallel planes. Such radiation is said to be **plane polarized**.

Plane-polarized light
Light oscillating in only a single plane.

Plane-polarized light is the vector sum of left and right circularly polarized light that propagates through space as left- and right-handed helices. These two forms of light are enantiomers, and because of their opposite handedness, each component interacts in an

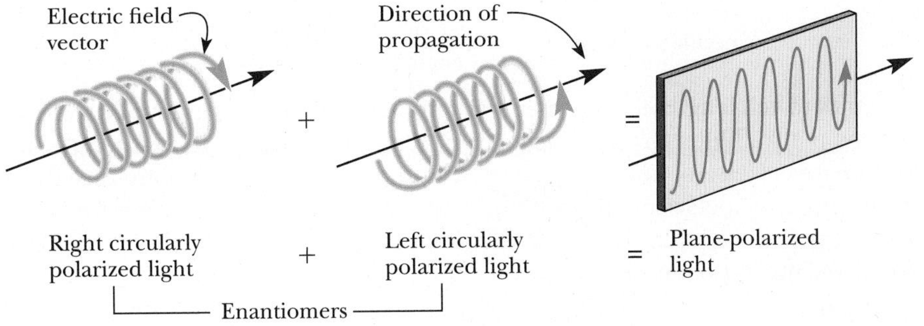

FIGURE 3.6 Plane-polarized light is a mixture of left and right circularly polarized light.

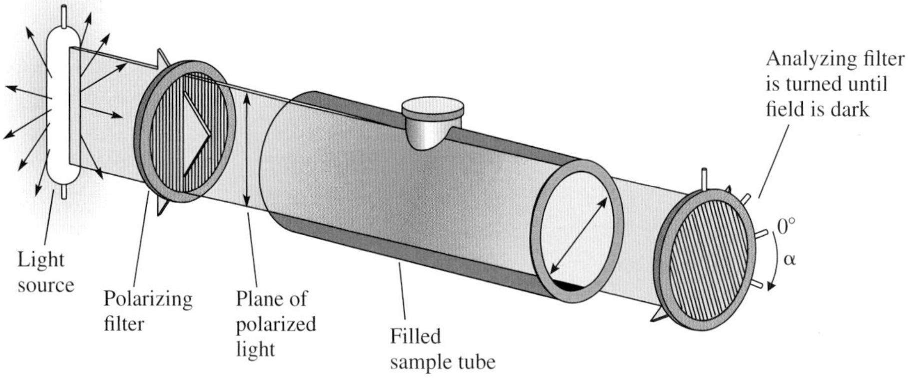

FIGURE 3.7 Schematic diagram of a polarimeter with its sample tube containing a solution of an optically active compound. The analyzing filter has been turned clockwise by α degrees to restore the dark field to the observer.

opposite way with chiral molecules. The result of this interaction is that each member of a pair of enantiomers rotates the plane of polarized light in an opposite direction.

Polarimeter

An instrument for measuring the ability of a compound to rotate the plane of polarized light.

Observed rotation

The number of degrees through which a compound rotates the plane of polarized light.

Dextrorotatory

Refers to a substance that rotates the plane of polarized light to the right.

Levorotatory

Refers to a substance that rotates the plane of polarized light to the left.

Specific rotation

Observed rotation of the plane of polarized light when a sample is placed in a tube 1.0 dm in length and at a concentration of 1 g/mL. For a pure liquid, concentration is expressed in g/mL (density).

B. Polarimeters

A **polarimeter** consists of a monochromatic light source, a polarizing filter and an analyzing filter (each made of calcite or Polaroid film), and a sample tube (Figure 3.7). If the sample tube is empty, the intensity of the light reaching the eye is at its minimum when the polarizing axes of the two filters are at right angles. If the analyzing filter is turned either clockwise or counterclockwise, more light is transmitted.

The ability of molecules to rotate the plane of polarized light can be observed using a polarimeter in the following way. First, a sample tube filled with solvent is placed in the polarimeter. The analyzing filter is adjusted so that the field is dark (the easiest position for the human eye to observe). This position of the analyzing filter is taken as 0°. When a solution of an optically active compound is placed in the sample tube, some light passes through the analyzing filter; the optically active compound has rotated the plane of polarized light from the polarizing filter so that it is now no longer at an angle of 90° to the analyzing filter. The analyzing filter is then rotated to restore darkness in the field of view. The number of degrees, α, through which the analyzing filter must be rotated to restore darkness to the field of view is called the **observed rotation**. If the analyzing filter must be turned to the right (clockwise) to restore darkness, we say that the compound is **dextrorotatory** (Latin: *dexter,* on the right side). If the analyzing filter must be turned to the left (counterclockwise), we say that the compound is **levorotatory** (Latin: *laevus,* on the left side).

The magnitude of the observed rotation for a particular compound depends on its concentration, the length of the sample tube, the temperature, the solvent, and the wavelength of the light used. **Specific rotation, $[\alpha]$**, is defined as the observed rotation at a specific cell length and sample concentration.

$$\text{Specific rotation} = [\alpha]_{\lambda}^{T} = \frac{\text{Observed rotation (degrees)}}{\text{Length (dm)} \times \text{Concentration}}$$

The standard cell length is 1 decimeter (1 dm or 10 cm). Concentration is expressed in grams per milliliter (g/mL). Because specific rotation depends on temperature (T, in degrees Celsius) and wavelength λ of light, these variables are designated, respectively, as superscript and subscript. The light source most commonly used in polarimetry is the sodium D line (λ = 589 nm), the line responsible for the yellow color of sodium-vapor lamps.

In reporting either observed or specific rotation, a dextrorotatory compound is indicated with a plus sign in parentheses, (+), and a levorotatory compound is indicated with a minus sign in parentheses, (−). For any pair of enantiomers, one enantiomer is dextrorotatory and the other is levorotatory. For each member of the pair, the absolute value of the specific rotation is exactly the same, but the sign is opposite. Following are the specific rotations of the enantiomers of 2-butanol at 25°C using the D line of sodium. Note that for molecules with single chiral centers, there is no absolute relationship between *R* and *S* and (+) and (−) rotation. For some molecules, the *R* enantiomer is (+), and for others, the *S* enantiomer is (+). This is why it is not redundant to report both the *R* or *S* and the rotation direction when naming a specific enantiomer. Other nomenclature systems for chirality, such as D and L that will be introduced for sugars and amino acids in Chapters 25 and 26, respectively, have no relationship to (+) and (−). The direction of rotation of plane-polarized light must always be determined experimentally.

A polarimeter is used to measure the rotation of plane-polarized light as it passes through a sample.

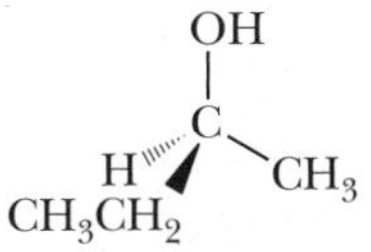

(*S*)-(+)-2-Butanol
$[\alpha]_D^{25}$ +13.52

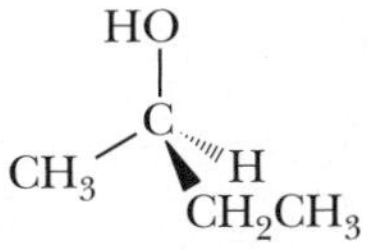

(*R*)-(−)-2-Butanol
$[\alpha]_D^{25}$ −13.52

Example 3.9 Specific Rotation

A solution is prepared by dissolving 400 mg of testosterone, a male sex hormone, in 10.0 mL of ethanol and placing it in a sample tube 10.0 cm in length. The observed rotation of this sample at 25°C using the D line of sodium is +4.36°. Calculate the specific rotation of testosterone.

Solution

The concentration of testosterone is 400 mg/10.0 mL = 0.0400 g/mL. The length of the sample tube is 1.00 dm. Inserting these values in the equation for calculating specific rotation gives

$$\text{Specific rotation} = \frac{\text{Observed rotation (degrees)}}{\text{Length (dm)} \times \text{Concentration (g/mL)}} = \frac{+4.36°}{1.00 \times 0.0400} = +109$$

Problem 3.9

The specific rotation of progesterone, a female sex hormone, is +172. Calculate the observed rotation for a solution prepared by dissolving 300 mg of progesterone in 15.0 mL of dioxane and placing it in a sample tube 10.0 cm long.

C. Racemic Mixtures

An equimolar mixture of two enantiomers is called a **racemic mixture**, a term derived from the name *racemic acid* (Latin: *racemus,* a cluster of grapes). Racemic acid is the name originally given to an equimolar mixture of the enantiomers of tartaric acid (Table 3.1). Because a racemic mixture contains equal numbers of dextrorotatory and levorotatory molecules, its specific rotation is zero. Alternatively, we say that a racemic mixture is optically inactive. A racemic mixture is indicated by adding the prefix (±) to the name of the compound [or sometimes the prefix (*d, l*)].

Racemic mixture
A mixture of equal amounts of two enantiomers.

D. Achiral Molecules

While chiral molecules are well known to rotate plane-polarized light, it is commonly thought that achiral molecules do not. It turns out this is wrong. Optical rotation depends on the orientation of a molecule with respect to the direction of the beam of incident light. Many achiral molecules will rotate a plane of polarized light, but light incident from the opposite direction will be rotated by exactly equal and opposite amounts. Hence, because of the random orientation of molecules in a solution, the light in effect will be incident from all directions on the tumbling molecules, so all of the rotations exactly cancel. For chiral molecules, the rotations from opposite directions do not cancel each other, so in solution, we observe an overall net rotation as described above. The important thing to remember is that in solution, the net optical activity for achiral molecules or racemic mixtures of chiral molecules, will be zero.

E. Optical Purity (Enantiomeric Excess)

Optical purity

The specific rotation of a mixture of enantiomers divided by the specific rotation of the enantiomerically pure substance (expressed as a percent). Optical purity is numerically equal to enantiomeric excess, but experimentally determined.

Enantiomeric excess (ee)

The difference between the percentage of two enantiomers in a mixture.

When dealing with a pair of enantiomers, it is essential to have a means of describing the composition of that mixture and the degree to which one enantiomer is in excess relative to its mirror image. One way of describing the composition of a mixture of enantiomers is by its percent **optical purity**, a property that can be observed directly. Optical purity is the specific rotation of a mixture of enantiomers divided by the specific rotation of the enantiomerically pure substance when they are at the same concentration.

$$\text{Percent optical purity} = \frac{[\alpha]_{\text{sample}}}{[\alpha]_{\text{pure enantiomers}}} \times 100$$

An alternative way to describe the composition of a mixture of enantiomers is by its **enantiomeric excess (ee)**, which is the difference in the number of moles of each enantiomer in a mixture compared to the total number of moles of both. Enantiomeric excess in percent is calculated by taking the difference in the percentage of each enantiomer.

$$\text{Enantiomeric excess (ee)} = \%\text{R} - \%\text{S}$$

For example, if a mixture consists of 75% of the *R* enantiomer and 25% of the *S* enantiomer, then the enantiomeric excess of the *R* enantiomer is 50%. Enantiomeric excess and optical purity are numerically identical.

Example 3.10 Enantiomeric Excess

Figure 3.9 presents a scheme for separation of the enantiomers of mandelic acid. The specific rotation of optically pure (*S*)-(−)-mandelic acid is −158. Suppose that instead of isolating pure (*S*)-(−)-mandelic acid from this scheme, the sample is a mixture of enantiomers with a specific rotation of −134. For this sample, calculate the following:

(a) The enantiomeric excess of this sample of (*S*)-(−)-mandelic acid.

(b) The percentage of (*S*)-(−)-mandelic acid and of (*R*)-(+)-mandelic acid in the sample.

Solution

(a) The enantiomeric excess of (*S*)-(−)-mandelic acid is 84.8%.

$$\text{Enantiomeric excess} = \frac{-134}{-158} \times 100 = 84.8\%$$

(b) This sample is 84.8% (*S*)-(−)-mandelic acid and 15.2% (*R,S*)-mandelic acid. The (*R,S*)-mandelic acid is 7.6% *S* enantiomer and 7.6% *R* enantiomer. The sample, therefore, contains 92.4% of the *S* enantiomer and 7.6% of the *R* enantiomer. We can check these values by calculating the observed rotation of a mixture containing 92.4% (*S*)-(−)-mandelic acid and 7.6% (*R*)-(+)-mandelic acid as follows:

$$\text{Specific rotation} = 0.924 \times (-158) + 0.076 \times (+158) = -146 + 12 = -134$$

which agrees with the experimental specific rotation.

Problem 3.10

One commercial synthesis of naproxen (the active ingredient in Aleve and a score of other over-the-counter and prescription nonsteroidal anti-inflammatory drug preparations) gives the enantiomer shown in 97% enantiomeric excess.

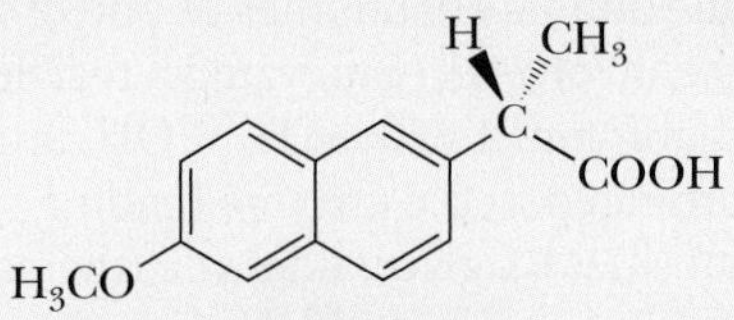

Naproxen
(a nonsteroidal anti-inflammatory drug)

(a) Assign an *R* or *S* configuration to this enantiomer of naproxen.

(b) What are the percentages of *R* and *S* enantiomers in the mixture?

3.8 The Significance of Chirality in the Biological World

Except for inorganic salts and a relatively few low-molecular-weight organic substances, the molecules in living systems, both plant and animal, are chiral. Although these molecules can exist as a number of stereoisomers, almost invariably, only one stereoisomer is found in nature. This occurrence is a consequence of the fact that their natural syntheses are catalyzed by enzymes, which are also chiral. Of course, instances do occur in which more than one stereoisomer is found, but these rarely exist together in the same biological system.

A. Chirality in Enzymes

Let us look more closely at the chirality of enzymes. An illustration is chymotrypsin, an enzyme in the intestines of animals, which catalyzes the hydrolysis of proteins during digestion. Chymotrypsin, like all proteins, is composed of a long molecular chain of amino acids that folds up into the active enzyme. Human chymotrypsin has 268 chiral centers that result from the amino acids; so the maximum number of stereoisomers possible is 2^{268}, a staggeringly large number, almost beyond comprehension.

Fortunately, because each chiral amino acid is only present as a single stereoisomer, only one of the possible stereoisomers of chymotrypsin is produced. Because enzymes are chiral substances and are present as single stereoisomers, most either produce or react only with substances that are single stereoisomers (if chiral).

As an interesting illustration of this enzyme chirality principle, Stephen B. Kent, then at The Scripps Research Institute in La Jolla, California, synthesized the enantiomer of a natural enzyme by using the enantiomers of the amino acids normally found in proteins as building blocks. As expected, this synthetic enzyme enantiomer only catalyzed a reaction with a substrate that was the enantiomer of the chiral substrate utilized by the natural enzyme.

B. How an Enzyme Distinguishes Between a Molecule and Its Enantiomer

Enzymes are chiral catalysts. Some are completely specific for the catalysis of the reaction of only one particular compound, whereas others are less specific and catalyze similar reactions of a family of compounds. An enzyme catalyzes a biological reaction of molecules by first positioning them at a binding site on its surface. These molecules may be held at the binding site by a combination of hydrogen bonds, electrostatic attractions, dispersion forces, or even temporary covalent bonds.

An enzyme with a specific binding site for a molecule with a chiral center can distinguish between a molecule and its enantiomer or one of its diastereomers. Assume, for example, that an enzyme involved in catalyzing a reaction of glyceraldehyde has a binding site with groups that interact with —H, —OH, and —CHO. Assume further that the binding sites are arranged in the enzyme binding site as shown in Figure 3.8. The enzyme can distinguish (*R*)-(+)-glyceraldehyde (the natural or biologically active form) from its enantiomer (*S*)-(−)-glyceraldehyde because the natural enantiomer can be bound to the binding site with three groups interacting with their appropriate binding sites; the other enantiomer can, at best, bind to only two of these sites.

Because interactions between molecules in living systems take place in a chiral environment, it should be no surprise that a molecule and its enantiomer or diastereomers have different physiological properties. The tricarboxylic acid (TCA) cycle, for example, produces and then metabolizes only (*S*)-(+)-malic acid. Because only one enantiomer is produced, both the production and metabolism of (*S*)-(+)-malic acid are said to be enantioselective.

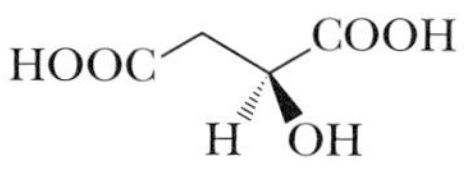

(*S*)-(+)-Malic acid

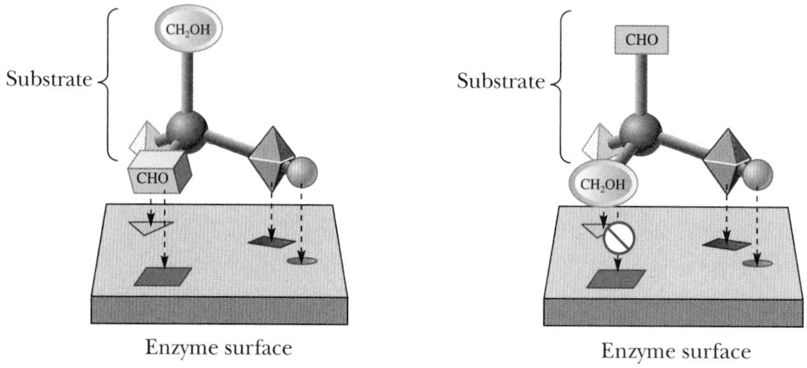

This enantiomer of glyceraldehyde fits the three specific binding sites on the enzyme surface.

This enantiomer of glyceraldehyde does not fit the same binding sites.

FIGURE 3.8 A schematic diagram of an enzyme surface capable of interacting with (*R*)-(+)-glyceraldehyde at three binding sites, but with (*S*)-(−)-glyceraldehyde at only two of these sites.

That interactions between molecules in the biological world are highly enantioselective is not surprising, but just how these interactions are accomplished at the molecular level with such precision and efficiency is a great puzzle that is an active area of scientific research. Scientists in the field of bioorganic chemistry seek to better understand and harness the incredible power of biological molecules, especially enzymes, for use in organic chemistry applications.

Connections to Biological Chemistry

Chiral Drugs

Some of the common drugs used in human medicine (e.g., aspirin) (Section 18.5B) are achiral. Others are chiral and sold as single enantiomers. The penicillin and erythromycin classes of antibiotics and the drug captopril are all chiral drugs. Captopril, which is very effective for the treatment of high blood pressure and congestive heart failure, is manufactured and sold as the *S,S* stereoisomer. A large number of chiral drugs, however, are sold as racemic mixtures. The popular analgesic ibuprofen (the active ingredient in Motrin and many other nonaspirin analgesics) is an example of a drug sold as a racemic mixture.

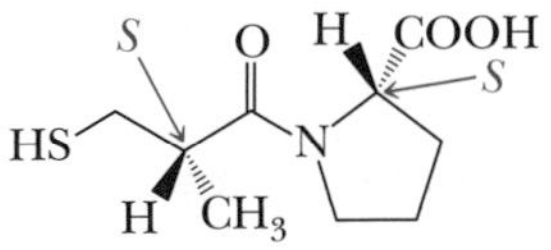

Captopril

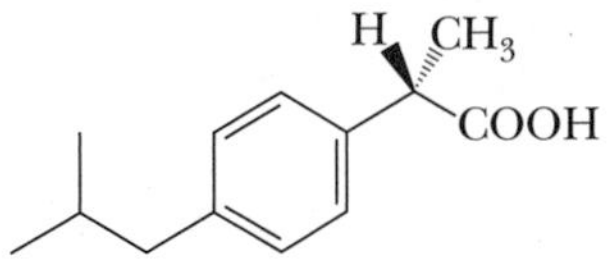

(*S*)-Ibuprofen

For racemic drugs, most often, only one enantiomer exerts the beneficial effect, whereas the other enantiomer either has no effect or may even exert a detrimental effect. For example, only the *S* enantiomer of ibuprofen is biologically active. Interestingly, the body converts the inactive *R* enantiomer to the active *S* enantiomer.

Another good example is the drug dihydroxyphenylalanine used in the treatment of Parkinson's disease. The active drug is dopamine. This compound does not cross the blood-brain barrier to the required site of action in the brain. What is administered, instead, is 3,4-dihydroxyphenylalanine. It crosses the blood-brain barrier and then undergoes decarboxylation catalyzed by the enzyme dopamine decarboxylase. This enzyme is specific for the *S* enantiomer, which is commonly known as L-DOPA. It is essential, therefore, to administer the enantiomerically pure prodrug. Were the prodrug to be administered in a racemic form, there could be a dangerous buildup of the *R* enantiomer, which cannot be metabolized by the enzymes present in the brain.

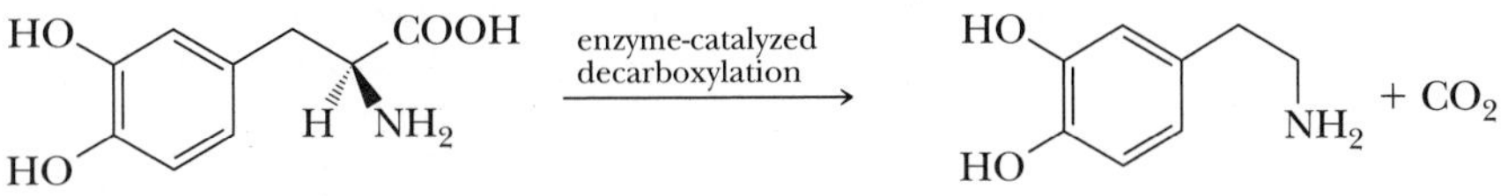

(*S*)-(−)-3,4-Dihydroxyphenylalanine (L-DOPA)
$[\alpha]_D^{13}$ −13.1

Dopamine

MCAT Practice: *Passage and Questions*

Amino Acid Stereochemistry

In organisms ranging from bacteria to humans, there are 20 common amino acids that share structural and stereochemical motifs. In each amino acid there is a central carbon atom bonded to a hydrogen atom, an amine, and a carboxylic acid. At neutral pH, the amines and carboxylic acids exist as ammonium ions and carboxylates, respectively. In addition, for 19 of the 20 amino acids, there is a fourth group on the central carbon other than hydrogen, referred to as a "side chain." The central carbon of these 19 amino acids is therefore a chiral center, and all

(Continued)

19 natural amino acids have the configuration drawn here in a wedge/dash representation as well as a Fischer projection.

Proteins are long chains of amino acids covalently bonded together by amide bonds formed between the carboxyl group of one amino acid to the amine group of another amino acid. Because they are made from pure amino acid stereoisomers, proteins themselves are single stereoisomers despite having several hundred or more chiral centers.

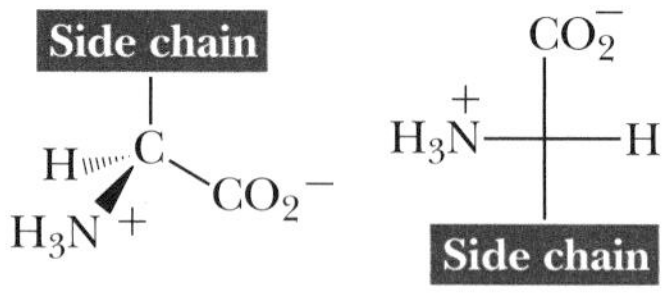

L-amino acids

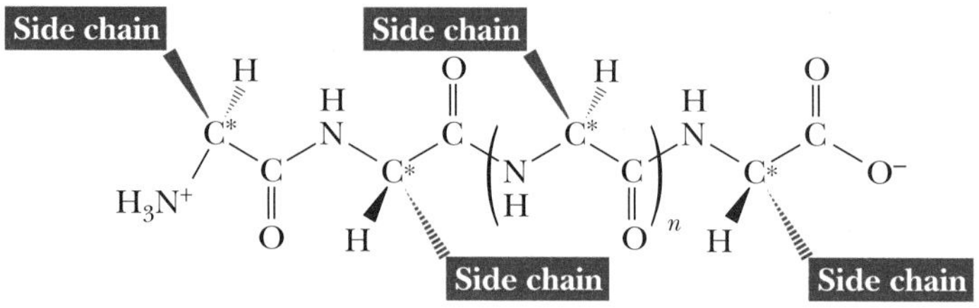

For most proteins, n = 150–750

Questions

A. If the side chain of the amino acid is a methyl group, the compound is referred to as alanine. What is the Cahn-Ingold-Prelog stereochemical descriptor for the chiral center in alanine?

1. It is a *Z*-stereocenter.
2. It is an *S*-stereocenter.
3. It is an *E*-stereocenter.
4. It is an *R*-stereocenter.

B. The Fischer projection is referred to as "L," meaning that the 19 amino acids with a central chiral center are referred to as L-amino acids. The L-descriptor derives from a single stereoisomer of the molecule glyceraldehyde, which rotates plane-polarized light to the left [levorotatory, (−)] when it has an analogous configuration to that drawn above for all the chiral amino acids. The amino acid enantiomers, which are not commonly found in nature, are referred to as D-amino acids (shown below) because they are structurally related to the confirmation of glyceraldehyde that rotates plane-polarized light to the right [dextrorotatory, (+)].

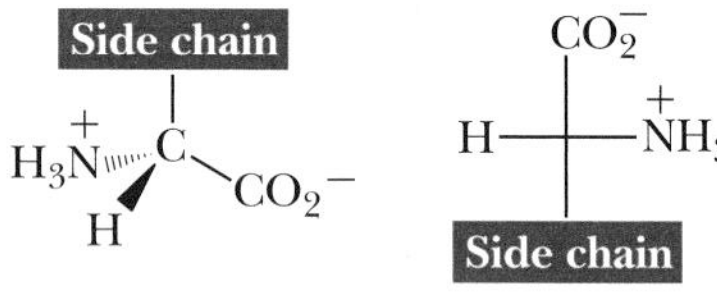

D-amino acids

What direction of rotation of plane-polarized light must the common 19 chiral amino acids show?

1. The 19 L-amino acids will rotate plane-polarized light to the left.
2. The 19 L-amino acids will not rotate plane-polarized to a significant extent because the positive and negative charges balance.
3. The direction of rotation of plane-polarized light will be to the right for *R*-amino acids and to the left for *S*-amino acids.
4. The direction of rotation of plane-polarized light is not known because there is no correlation to nomenclature systems such as L or D, or *R* or S.

C. The amino acids cysteine and serine are shown. What are the Cahn-Ingold-Prelog stereochemical descriptors for these two amino acids?

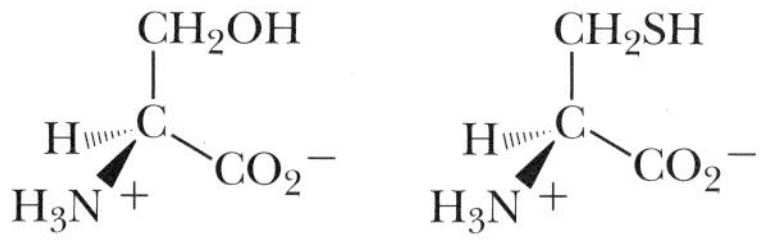

Serine **Cysteine**

1. Serine is *S* while cysteine is *R*.
2. Cysteine is *S* while serine is *R*.
3. Cysteine and serine are both *R*.
4. Cysteine and serine are both *S*.

D. The amino acids isoleucine and threonine are shown. What are the correct stereochemical descriptors for these two amino acids?

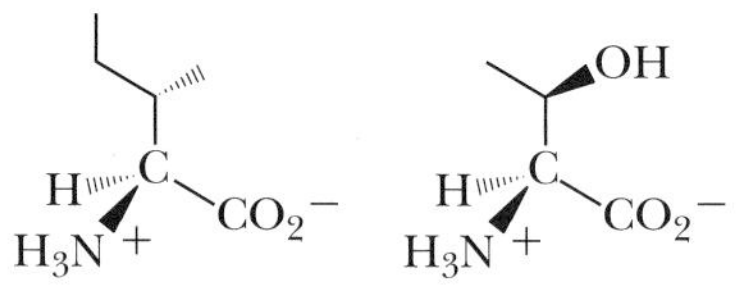

Isoleucine **Threonine**

1. Both amino acids are *S,S*.
2. The central carbon between the amine and carboxylic acid is *R* while the side chain contains an *S* stereocenter for both amino acids.
3. The central carbon between the amine and carboxylic acid is *S* while the side chain contains an *R* stereocenter for both amino acids.

4. The central carbon in both amino acids is *S*, while the side chain in isoleucine is *S* and the side chain in threonine is *R*.

E. As stated, proteins are stereochemically pure because only a single enantiomer of each amino acid building block is used by nature. How many stereoisomers are possible for a chain of only 3 chiral amino acids if both enantiomers of the amino acids are used?

1. 2 **2.** 4 **3.** 8 **4.** 16

F. If racemic mixtures of all the possible enantiomers of the amino acids were used in proteins, which of the following would be correct statements?

1. Proteins would exist as mixtures of diastereomers.
2. Proteins would exist as pairs of enantiomers.
3. Proteins would not show an optical rotation.
4. All of the above.

3.9 Separation of Enantiomers—Resolution

The separation of a racemic mixture into its enantiomers is called **resolution.**

Resolution

Separation of a racemic mixture into its enantiomers.

A. Resolution by Means of Diastereomeric Salts

One general scheme for separating enantiomers requires chemical conversion of a pair of enantiomers into two diastereomers with the aid of an enantiomerically pure chiral resolving agent. This chemical resolution is successful because the diastereomers thus formed are different compounds, have different physical properties, and often can be separated by physical means (most commonly fractional crystallization or column chromatography) and purified. The final step in this scheme for resolution is chemical conversion of the separated diastereomers back to the individual enantiomers and recovery of the chiral resolving agent.

A reaction that lends itself to chemical resolution is salt formation because it is readily reversible.

$$\underset{\text{Carboxylic acid}}{\text{RCOOH}} + \underset{\text{Base}}{:\text{B}} \rightleftharpoons \underset{\text{Salt}}{\text{RCOO}^-\text{HB}^+}$$

Chiral bases available from plants are often used as chiral resolving agents for racemic acids. More commonly, chemists now use commercially available chiral amines such as (*R*,*S*)-1-phenylethanamine.

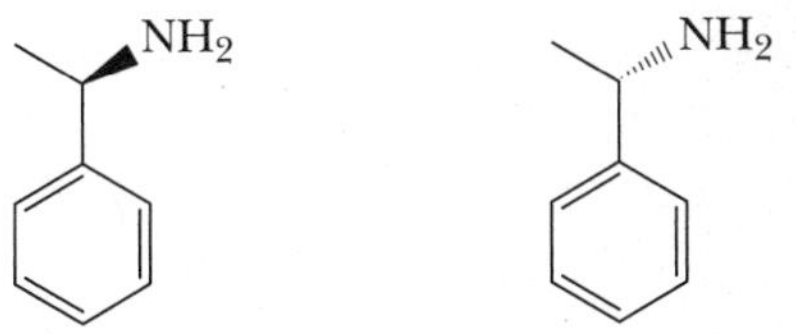

(*R*)-1-Phenylethanamine (*S*)-1-Phenylethanamine

The resolution of mandelic acid by way of its diastereomeric salts with the natural chiral base cinchonine is illustrated in Figure 3.9. Racemic mandelic acid and optically pure (+)-cinchonine (Cin) are dissolved in boiling water, giving a solution of a pair of diastereomeric salts. Diastereomers have different solubilities, and when the solution cools, the less soluble diastereomeric salt crystallizes. This salt is collected and purified by further recrystallization. The filtrates, richer in the more soluble diastereomeric salt, are concentrated to give this salt, which is also purified by further recrystallization. The purified diastereomeric salts are treated with aqueous HCl to precipitate the nearly pure enantiomers of mandelic acid. Cinchonine remains in the aqueous solution as its water-soluble hydrochloride salt.

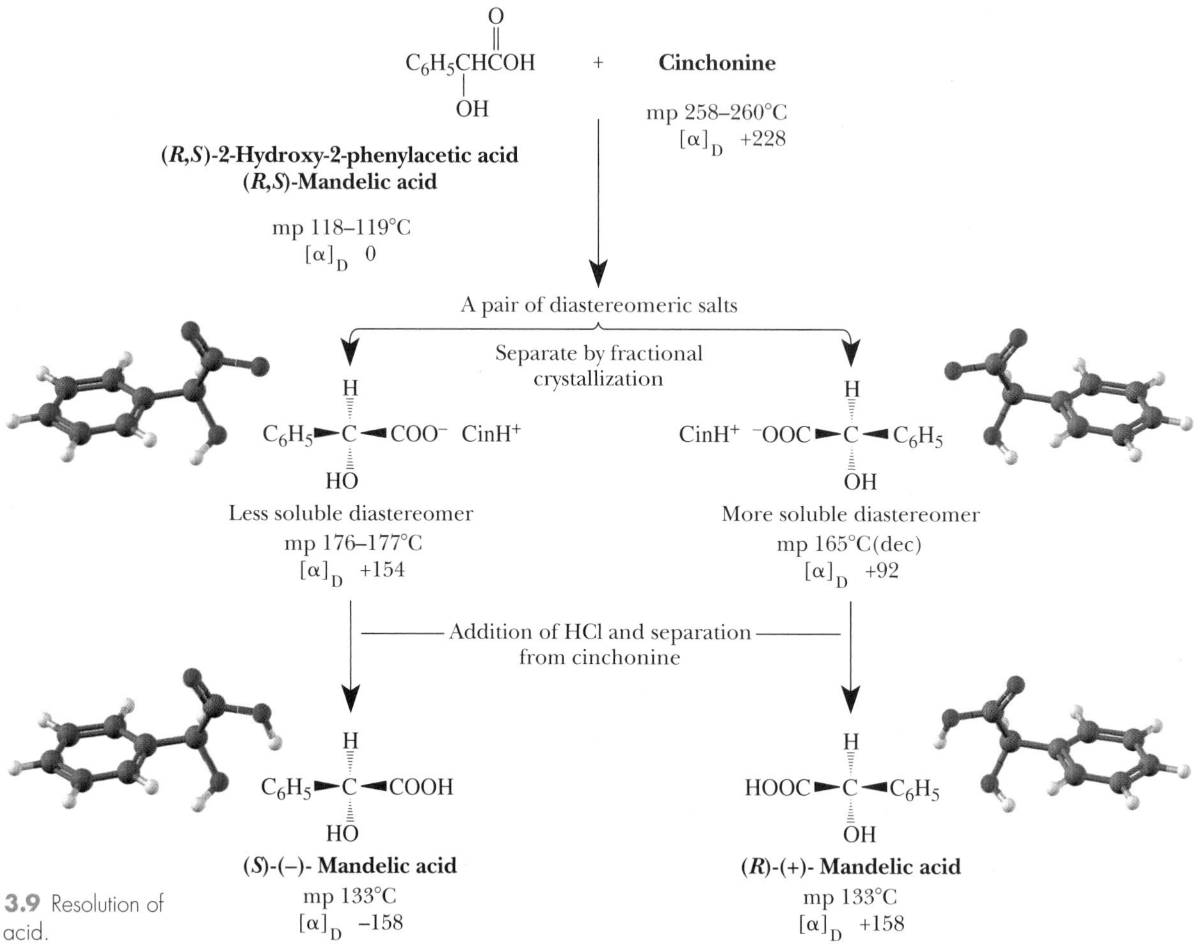

FIGURE 3.9 Resolution of mandelic acid.

Optical rotations and melting points of racemic mandelic acid, cinchonine, the purified diastereomeric salts, and the pure enantiomers of mandelic acid are given in Figure 3.9. Note the following two points: (1) The diastereomeric salts have different specific rotations and different melting points. (2) The enantiomers of mandelic acid have identical melting points and have specific rotations that are identical in magnitude but opposite in sign.

Resolution of a racemic base with a chiral acid is carried out in a similar way. Acids that are commonly used as chiral resolving agents are (+)-tartaric acid, (−)-malic acid, and (+)-camphoric acid (Figure 3.10). These and other naturally occurring chiral resolving agents are produced in plant and animal systems as single enantiomers.

B. Enzymes as Resolving Agents

In their quest for enantiomerically pure compounds, organic chemists have developed several new techniques for the preparation of enantiomerically pure materials.

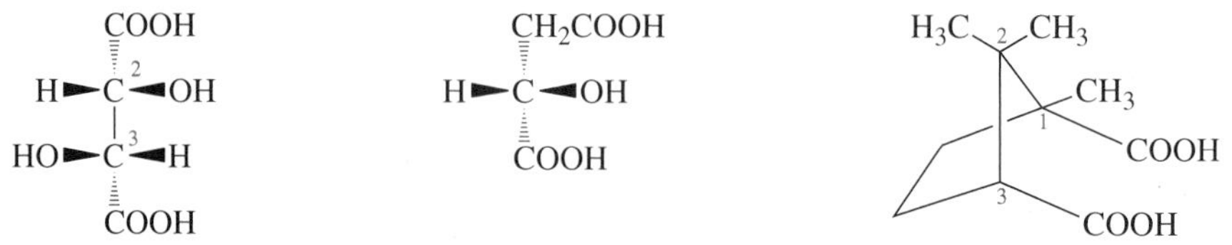

(2*R*,3*R*)-(+)-Tartaric acid $[\alpha]^{20}_{D}$ +12.5

(*S*)-(−)-Malic acid $[\alpha]^{20}_{D}$ −27

(1*S*,3*R*)-(+)-Camphoric acid $[\alpha]^{20}_{D}$ +46.5

FIGURE 3.10 Some carboxylic acids used as chiral resolving agents.

One approach is to use enzymes. Enzymes are themselves chiral, so they can produce single enantiomer products. A class of enzymes used for this purpose is the esterases, which catalyze the hydrolysis of esters to give an alcohol and a carboxylic acid.

The ethyl esters of naproxen crystallize in two enantiomeric crystal forms, one containing the *R* ester and the other containing the *S* ester. Each is insoluble in water. Chemists then use an esterase in alkaline solution to hydrolyze selectively the *S* ester to the (*S*)-carboxylic acid, which goes into solution as the sodium salt. The *R* ester is unaffected by these conditions. Filtering the alkaline solution recovers the crystals of the *R* ester. After crystals of the *R* ester are removed, the alkaline solution is acidified to give enantiomerically pure (*S*)-naproxen.

The recovered *R* ester is racemized (converted to an *R,S* mixture) and treated again with the esterase. Thus, by recycling the *R* ester, all the racemic ester is converted to (*S*)-naproxen.

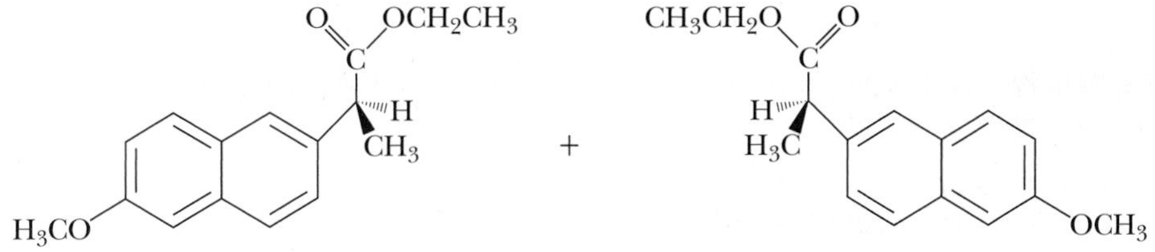

Ethyl ester of (*S*)-naproxen

Ethyl ester of (*R*)-naproxen
(Not affected by the esterase)

1. Esterase NaOH, H_2O
2. HCl, H_2O

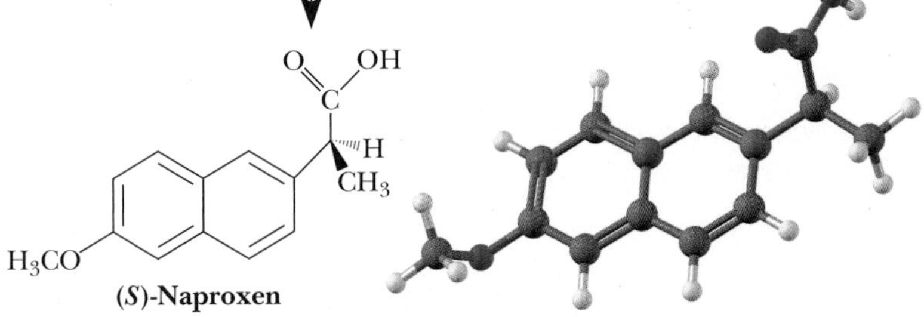

(*S*)-Naproxen

Chemists are no longer limited to using only natural enzymes for resolution of chiral molecules. Using genetic engineering techniques, it is now possible to develop new esterases or other enzymes in the laboratory that react specifically with single enantiomers of important substrates. Scientists have also engineered new properties into enzymes, such as better stability at higher temperatures and enhanced solubility in organic solvent/water mixtures, increasing the practical value of enzymes for organic chemistry.

C. Resolution by Means of Chromatography on a Chiral Substrate

Chromatography is a term used to describe the purification of molecules in which a sample to be purified interacts with a solid material and different components of the sample separate based on their different relative interactions with the solid material. Separation can be accomplished using the sample in either the gas phase (usually for analytical purposes) or the liquid phase (analytical or preparative separations are possible). The solid material is packed into a column and a solvent (or in gas chromatography, a gas) passes down the column, carrying the more weakly bound components of the mixture with it more rapidly than the more tightly bound ones.

Chromatography
A separation method involving passing a vapor or solution mixture through a column packed with a material with different affinities for different components of the mixture.

A common method of resolving enantiomers today is chromatography using a chiral column packing material. Each enantiomer interacts differently with the chiral molecules of the packing material, and the elution time will (in principle at least) be different for the two enantiomers. A wide variety of chiral column packings have been developed for this purpose.

Study Guide

3.1 Chirality—The Handedness of Molecules

- Molecules that are not superposable on their mirror images are said to be **chiral**.
  - *A **mirror image** is the reflection of an object in a mirror.*
  - ***Chirality** is a property of an object such as a molecule as a whole, not of a particular atom.*
  - *An **achiral** object lacks chirality; that is, it is an object with a superposable mirror image. Almost all achiral objects possess at least one plane or center of symmetry.*
    - *A **plane of symmetry** is an imaginary plane passing through an object dividing it such that one half is the reflection of the other half.*
    - *A **center of symmetry** is a point so situated that identical components of the object are located on opposite sides and equidistant from the point along any axis passing through that point.*

P 3.11, 3.12, 3.33, 3.34

3.2 Stereoisomerism

- **Isomers** are molecules that have the same molecular formula but possess different structures because of different connectivity of atoms and/or spatial arrangement of atoms.
- **Constitutional isomers** are molecules that have the same molecular formula but have a different connectivity of atoms.
- **Stereoisomers** are isomers that have the same connectivity of atoms but have a different spatial arrangement of their atoms.
  - ***Conformational isomers** are stereoisomers that interconvert by rotation along single bonds.*
  - ***Configurational isomers** are stereoisomers that do not interconvert at room temperature because their interconversion would require the breaking of individual single or double bonds.*
  - ***Enantiomers** are stereoisomers that are nonsuperposable mirror images.*
  - ***Diastereomers** are stereoisomers that are not mirror images.*
  - ***Atropisomers** are enantiomers or diastereomers that differ because of hindered rotation along a single bond.*
- A **stereocenter** is an atom about which exchange of two groups produces a stereoisomer.
- A **chiral center** is a tetrahedral atom with four different groups bonded to it.
- The term stereocenter is more general that the term chiral center.
  - *A chiral center is a specific type of stereocenter.*
  - *The sp^2 carbon atom of an alkene that exhibits E,Z stereoisomerism is only a stereocenter.*

– *An sp^3 carbon atom with four different groups attached is most commonly described as a chiral center, but it is also a type of stereocenter.*

P **3.1, 3.13–3.15, 3.17, 3.19, 3.36–3.38**

▶ **Stereocenters, problem 3.1**

3.3 Naming Chiral Centers—The *R*,*S* System

- The **absolute configuration** at any chiral center can be specified by the ***R,S*** **system**.
  - *To apply this convention:*
    - *(1) Each atom or group of atoms bonded to the chiral center is assigned a priority and is numbered from highest priority to lowest priority.*
      - *The higher the atomic number of an atom attached to a chiral center, the higher the priority.*
      - *If there is a tie (i.e., the same type of atom is attached to the chiral center), continue out from the chiral center until the first point of difference is reached. This can be tricky, so remember to look for the* **first** *point of difference.*
      - *Atoms participating in double or triple bonds are considered as if they had two or three, respectively, single bonds to their multiple bond partner atom.*
    - *(2) The molecule is oriented in space so that the group of lowest priority is directed away from the observer.*
      - *The lowest priority group is often, but not necessarily always, a hydrogen atom.*
    - *(3) The remaining three groups are read in order from highest priority to lowest priority. If the order of groups is clockwise, the configuration is* ***R*** *(Latin:* rectus, *right, correct). If the order is counterclockwise, the configuration is* ***S*** *(Latin:* sinister, *left).*
      - *If you wish to draw a chiral center with the opposite configuration (i.e., switch from R to S, or vice versa), switch the locations of any two groups bonded to the chiral center.*

P **3.2, 3.3, 3.16, 3.20–3.24, 3.35**

▶ **Problem 3.3**

3.4 Acyclic Molecules with Two or More Stereocenters

- For a molecule with *n* chiral centers, the maximum number of stereoisomers possible is 2^n.
  - *Certain molecules have special symmetry properties that reduce the number of stereoisomers to fewer than that predicted by the 2^n rule.*
  - *Two stereoisomers are enantiomers if all of the chiral centers have the opposite configuration (R versus S or vice versa).*
  - *If even one chiral center is the same between two stereoisomers, they are diastereomers.*

– *A **meso compound** contains two or more chiral centers bonded in such a way that it is achiral. Meso compounds usually have an internal plane of symmetry.*

P **3.4, 3.5, 3.25–3.28, 3.30–3.32**

- **Fischer projections** are two-dimensional projections of chiral molecules that are useful when evaluating stereochemistry.
  - *In Fischer projections, groups on the right and left are by convention in front.*
  - *Groups on the top and bottom are by convention behind.*

P **3.6**

- The type of stereoisomer relationship dictates the relationships of properties between stereoisomers.
  - *Enantiomers have identical physical and chemical properties in achiral environments but different properties in chiral environments, such as in the presence of plane-polarized light. They also have different properties in the presence of chiral reagents and enzymes as chiral catalysts.*
  - *Diastereomers have different physical and chemical properties even in achiral environments.*

▶ **Problem 3.5**

3.5 Cyclic Molecules with Two or More Chiral Centers

- When evaluating symmetry in **ring structures** with chiral centers, such as cyclohexane derivatives, it is helpful to evaluate the planar representations.

P **3.7, 3.8**

3.6 Tying All the Terminology Together

- There are two overarching classes of isomers: constitutional isomers and stereoisomers.
  - *Stereoisomerism can arise either from configurational or conformational isomerism.*
  - *Configurational isomers are either diastereomers or enantiomers (if the compound is chiral).*
  - *Conformational isomers are either diastereomers or enantiomers (if the compound is chiral). Enantiomers or diastereomers that do not readily interconvert by bond rotations are **atropisomers**.*

▶ **Figure 3.5**

3.7 Optical Activity—How Chirality Is Detected in the Laboratory

- Plane-polarized light is used to evaluate stereoisomers.
  - *Light that oscillates in parallel planes is said to be plane-polarized.*
  - *Plane-polarized light contains equal components of left and right circularly polarized light.*
  - *A **polarimeter** is an instrument used to detect and measure the magnitude of optical activity.*
- A compound is said to be **optically active** if it rotates the plane of plane-polarized light.

– ***Observed rotation*** *is the number of degrees the plane of plane-polarized light is rotated.*

– ***Specific rotation*** *is the observed rotation measured in a cell 1 dm long and at a sample concentration of 1 g/mL.*

– *A compound is* ***dextrorotatory*** *if the analyzing filter must be turned clockwise to restore the zero point.*

– *A compound is* ***levorotatory*** *if the analyzing filter must be turned counterclockwise to restore the zero point.*

P 3.9, 3.29

– *Each member of a pair of enantiomers rotates the plane of plane-polarized light an equal number of degrees but opposite in direction.*

– *A* ***racemic mixture*** *is a mixture of equal amounts of two enantiomers and has a specific rotation of zero.*

– *Percent* ***optical purity*** *(identical to* ***enantiomeric excess****) is defined as the specific rotation of a mixture of enantiomers divided by the specific rotation of the pure enantiomer times 100.*

– ***Meso compounds*** *are not chiral, so they are optically inactive.*

P 3.10, 3.39, 3.40

3.8 The Significance of Chirality in the Biological World

- Biological molecules are chiral and are usually found as single enantiomers in living systems.

  – *Enzymes are made of chiral amino acids and they catalyze biological reactions in a chiral way by first positioning the molecule or molecules at binding sites and holding them there by a combination of hydrogen bonds, electrostatic attractions, dispersion forces, and sometimes covalent bonds.*

  – *An enzyme with specific binding sites for three of the four groups on a chiral center can distinguish between a molecule and its enantiomer.*

  – *Because enzymes usually only react with and/or produce a single enantiomer, enzymes are referred to as being stereoselective.*

3.9 Separation of Enantiomers—Resolution

- ***Resolution*** is the experimental process of separating a mixture of enantiomers into the two pure enantiomers.

  – *A common chemical means of resolving organic compounds is to treat the racemic mixture with a chiral resolving agent that converts the mixture of enantiomers into a pair of diastereomers. The diastereomers are separated based on differences in their physical properties; each diastereomer is then converted to a pure stereoisomer, uncontaminated by its enantiomer.*

  – *Enzymes are also used as resolving agents because of their ability to catalyze a reaction of one enantiomer but not that of its mirror image.*

  – ***Chromatography*** *on a chiral substrate is also an effective separation method.*

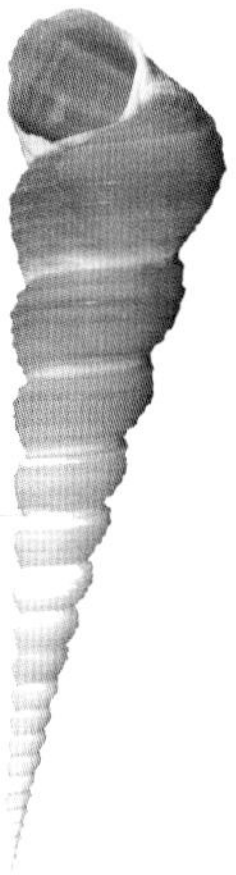

Problems

Red numbers indicate applied problems.

Chirality

3.11 Think about the helical coil of a telephone cord or a spiral binding. Suppose that you view the spiral from one end and find that it is a left-handed twist. If you view the same spiral from the other end, is it a right-handed or left-handed twist?

3.12 The next time you have the opportunity to view a collection of seashells that have a helical twist, study the chirality of their twists. Do you find an equal number of left-handed and right-handed spiral shells or mostly all of the same chirality? What about the handedness of different species of spiral shells?

3.13 One reason we can be sure that sp^3-hybridized carbon atoms are tetrahedral is the number of stereoisomers that can exist for different organic compounds.

(a) How many stereoisomers are possible for $CHCl_3$, CH_2Cl_2, and CHClBrF if the four bonds to carbon have a tetrahedral arrangement?

(b) How many stereoisomers would be possible for each of these compounds if the four bonds to the carbon had a square planar geometry?

Enantiomers

3.14 Which compounds contain chiral centers?

(a) 2-Chloropentane **(b)** 3-Chloropentane

(c) 3-Chloro-1-pentene **(d)** 1,2-Dichloropropane

3.15 Using only C, H, and O, write structural formulas for the lowest-molecular-weight chiral.

(a) Alkane **(b)** Alcohol **(c)** Aldehyde
(d) Ketone **(e)** Carboxylic acid **(f)** Carboxylic ester

3.16 Draw mirror images for these molecules. Are they different from the original molecule?

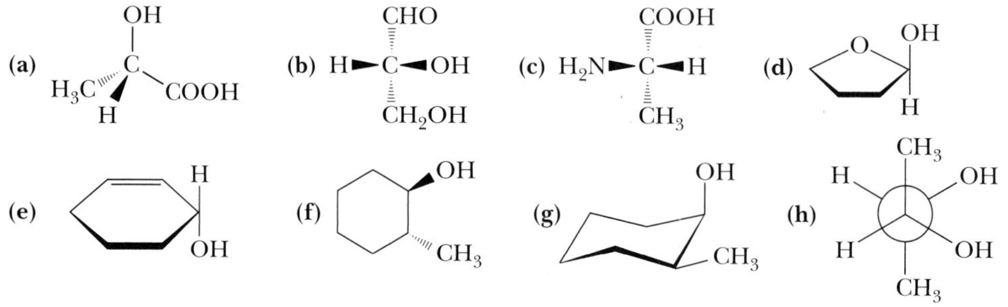

3.17 Following are several stereorepresentations for lactic acid. Use (a) as a reference structure. Which stereorepresentations are identical with (a)? Which are mirror images of (a)?

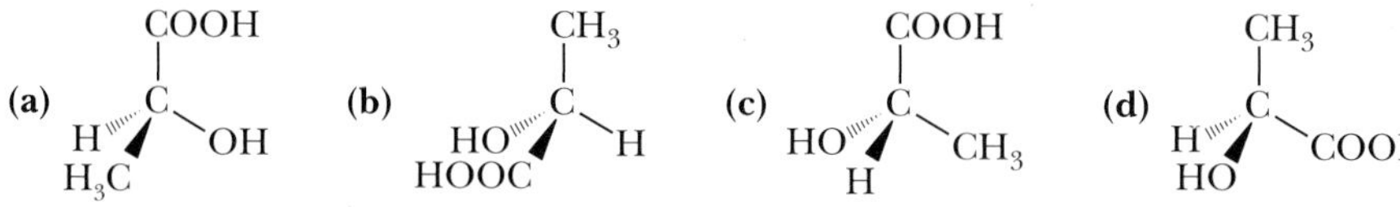

3.18 Mark each chiral center in the following molecules with an asterisk. How many stereoisomers are possible for each molecule?

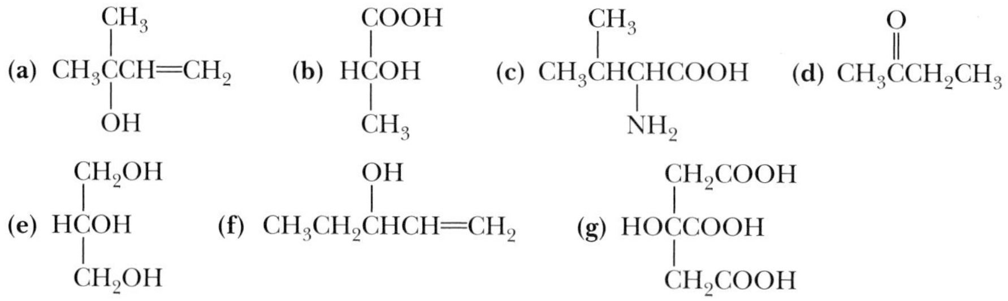

3.19 Show that butane in a gauche conformation is chiral. Do you expect that resolution of butane at room temperature is possible?

Designation of Configuration—The *R*,*S* System

3.20 Assign priorities to the groups in each set.

(a) $—H$ $—CH_3$ $—OH$ $—CH_2OH$
(b) $—CH_2CH{=}CH_2$ $—CH{=}CH_2$ $—CH_3$ $—CH_2COOH$
(c) $—CH_3$ $—H$ $—COO^-$ $—NH_3^+$
(d) $—CH_3$ $—CH_2SH$ $—NH_3^+$ $—CHO$

3.21 Following are structural formulas for the enantiomers of carvone. Each has a distinctive odor characteristic of the source from which it is isolated. Assign an *R* or *S* configuration to the single chiral center in each enantiomer. Why do they smell different when they are so similar in structure?

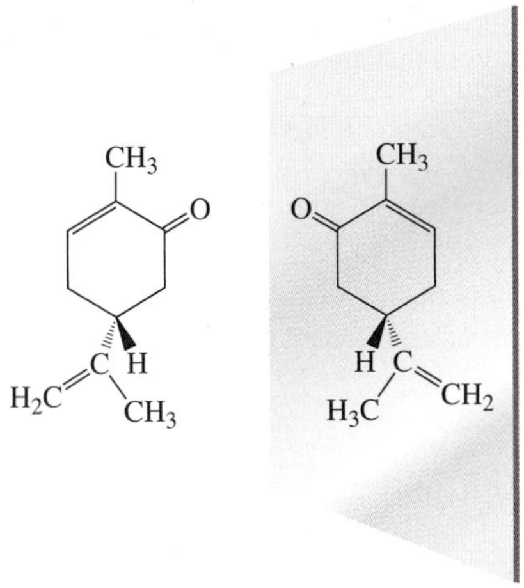

(−)-Carvone | **(+)-Carvone**
Spearmint oil | Caraway and dill seed oil

3.22 Following is a staggered conformation for one of the enantiomers of 2-butanol.

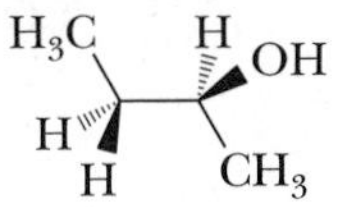

(a) Is this (*R*)-2-butanol or (*S*)-2-butanol?
(b) Viewed along the bond between carbons 2 and 3, draw a Newman projection for this staggered conformation.
(c) Draw a Newman projection for two more staggered conformations of this molecule. Which of your conformations is most stable? Assume that $—OH$ and $—CH_3$ are comparable in size.

3.23 For centuries, Chinese herbal medicine has used extracts of *Ephedra sinica* to treat asthma. *Ephedra* as an "herbal supplement" has been implicated in the deaths of several athletes and has recently been banned as a dietary supplement. Phytochemical investigation of this plant resulted in isolation of ephedrine, a very potent dilator of the air passages of the lungs. Ephedrine also has profound effects on the cardiovascular system. The naturally occurring stereoisomer is levorotatory and has the following structure. Assign an *R* or *S* configuration to each chiral center.

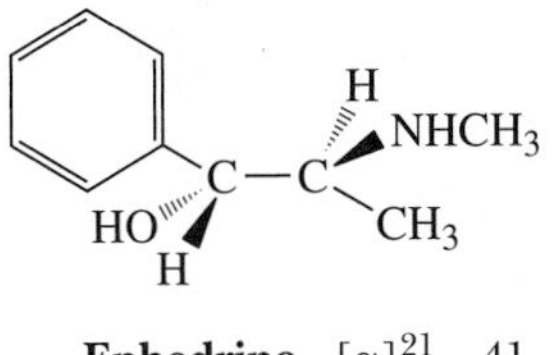

Ephedrine $[\alpha]_D^{21}$ −41

Paolo Koch/Science Source

Ephedra sinica, the source of ephedrine, a potent bronchodilator.

3.24 When oxaloacetic acid and acetyl-coenzyme A (acetyl-CoA) labeled with radioactive carbon-14 in position 2 are incubated with citrate synthase, an enzyme of the tricarboxylic acid cycle, only the following enantiomer of [2-^{14}C]citric acid is formed stereoselectively. Note that citric acid containing only ^{12}C is achiral. Assign an *R* or *S* configuration to this enantiomer of [2-^{14}C]citric acid. (*Note:* Carbon-14 has a higher priority than carbon-12.)

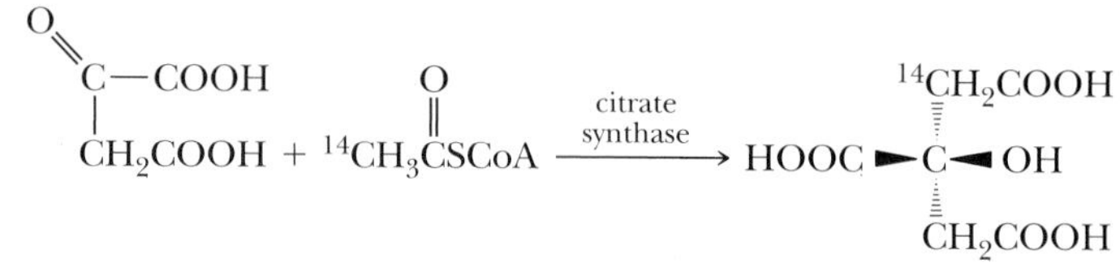

Oxaloacetic acid **Acetyl-CoA** **[2-^{14}C]Citric acid**

Molecules with Two or More Chiral Centers

3.25 Draw stereorepresentations for all stereoisomers of this compound. Label those that are meso compounds and those that are pairs of enantiomers.

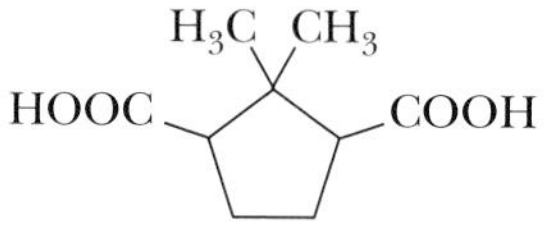

3.26 Mark each chiral center in the following molecules with an asterisk. How many stereoisomers are possible for each molecule?

(a) 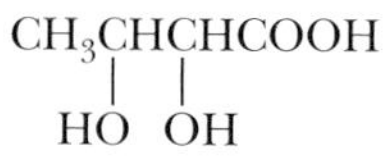

(b) 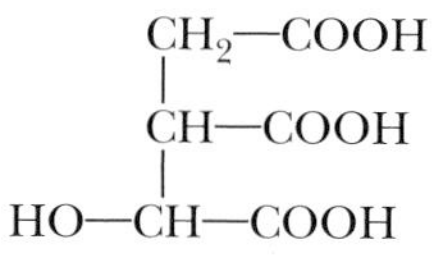

(c) 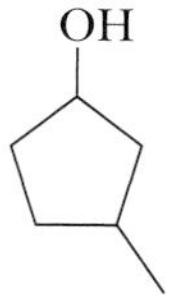

(d) 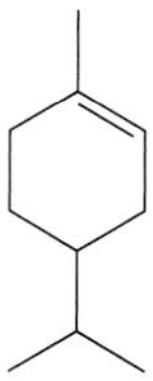

(e) 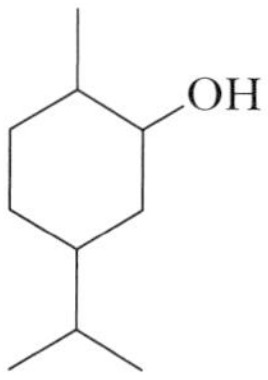

(f) 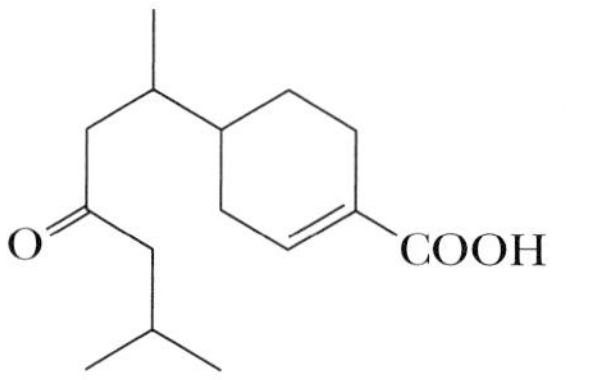

(g) 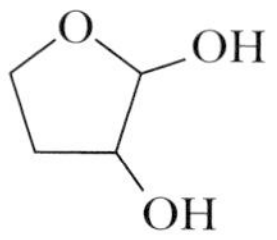

(h) 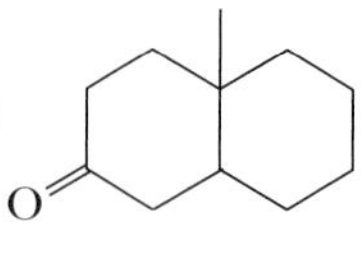

3.27 Label the eight chiral centers in cholesterol. How many stereoisomers are possible for a molecule with this many chiral centers?

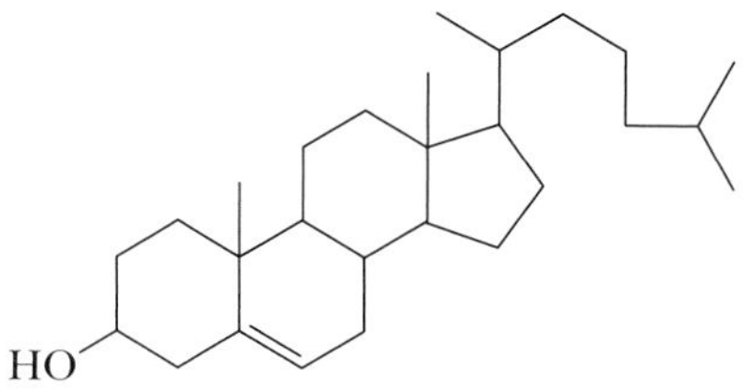

Cholesterol

3.28 Label the four chiral centers in amoxicillin, which belongs to the family of semisynthetic penicillins.

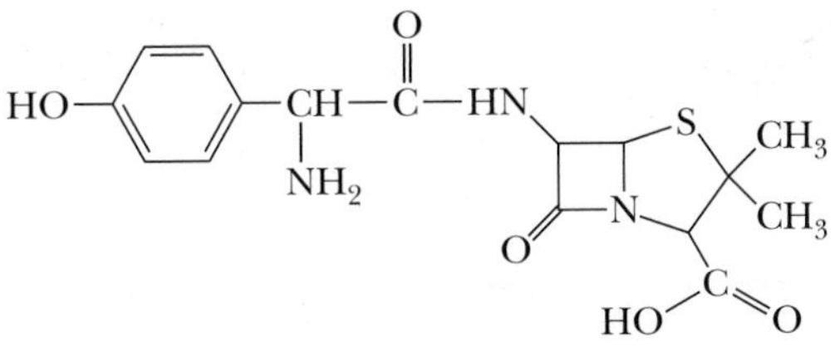

Amoxicillin

3.29 If the optical rotation of a new compound is measured and found to have a specific rotation of +40, how can you tell if the actual rotation is not really +40 plus some multiple of +360? In other words, how can you tell if the rotation is not actually a value such as +400 or +760?

3.30 Are the formulas within each set identical, enantiomers, or diastereomers?

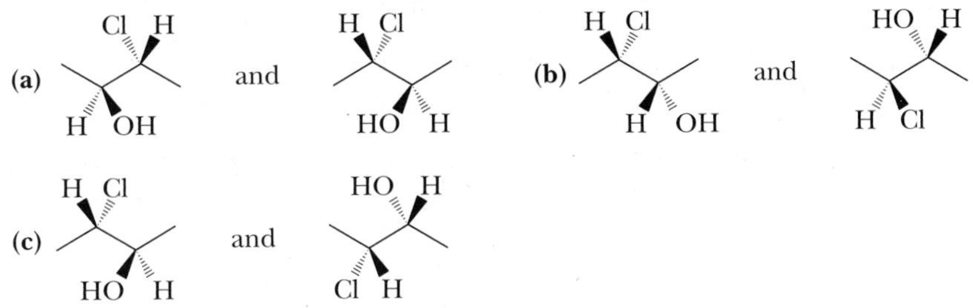

3.31 Which of the following are meso compounds?

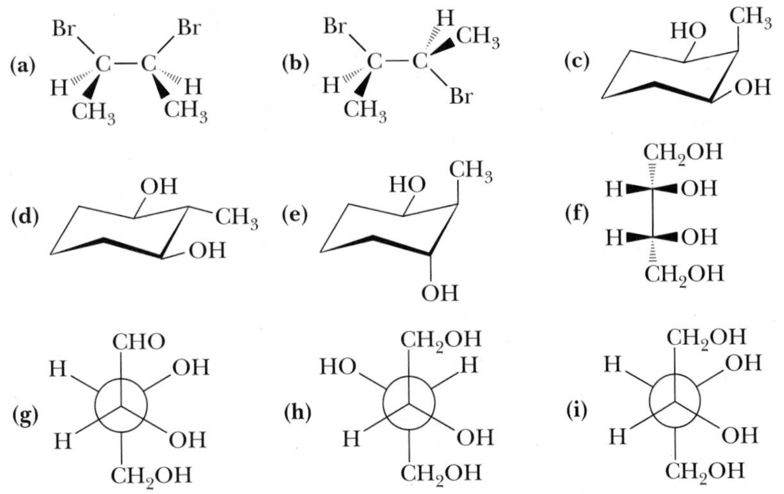

3.32 Vigorous oxidation of the following bicycloalkene breaks the carbon-carbon double bond and converts each carbon of the double bond to a COOH group. Assume that the conditions of oxidation have no effect on the configuration of either the starting bicycloalkene or the resulting dicarboxylic acid. Is the dicarboxylic acid produced from this oxidation one enantiomer, a racemic mixture, or a meso compound?

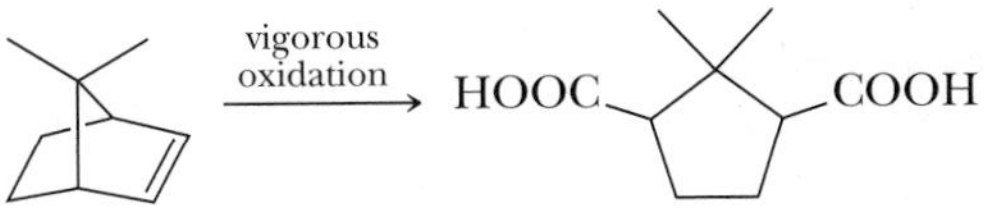

3.33 A long polymer chain, such as polyethylene $(\text{—}CH_2CH_2\text{—})_n$, can potentially exist in solution as a chiral object. Give two examples of chiral structures that a polyethylene chain could adopt.

3.34 Which of the following compounds are chiral? Which, if any, are meso? Which, if any, does not have a possible diastereomer?

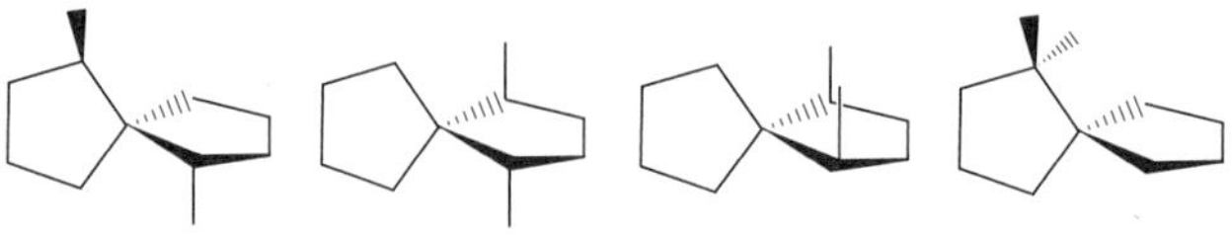

(a) **(b)** **(c)** **(d)**

3.35 Will the following compound show any optical activity if there is restricted rotation along the central C—C bond? What will happen to the optical activity at elevated temperatures as the rotation becomes less restricted?

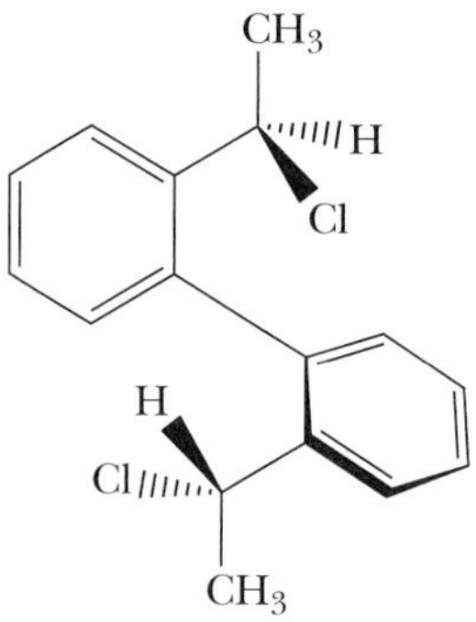
CH_3
H
Cl
H
Cl
CH_3

3.36 Are the following structures chiral as drawn? When placed in a solution at 298 K, which structure will show an optical rotation? Explain.

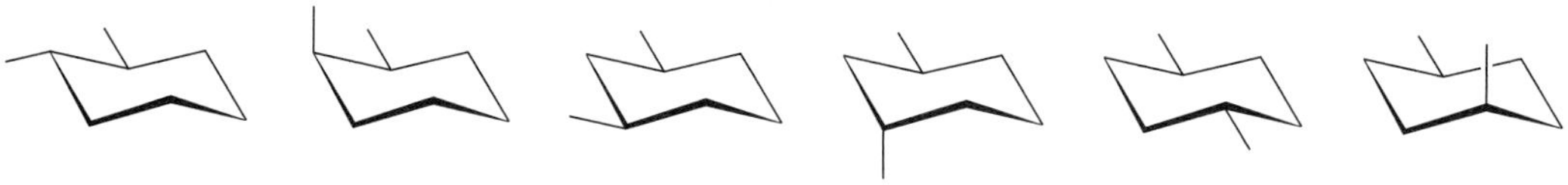

(a) **(b)** **(c)** **(d)** **(e)** **(f)**

3.37 To the following statements, answer true or false and explain your answer.

(a) All chiral centers are also stereocenters.
(b) All stereocenters are also chiral centers.
(c) All chiral molecules are optically active when pure.
(d) All mixtures of chiral molecules are optically active.
(e) To be optically active, a molecule must have a chiral center.
(f) To be meso, a molecule must have at least two chiral centers.

Looking Ahead

3.38 The chiral catalyst (*R*)-BINAP-Ru is used to hydrogenate alkenes to give alkanes (Section 6.7C). The products are produced with high enantiomeric excess. An example is the formation of (*S*)-naproxen, a pain reliever.

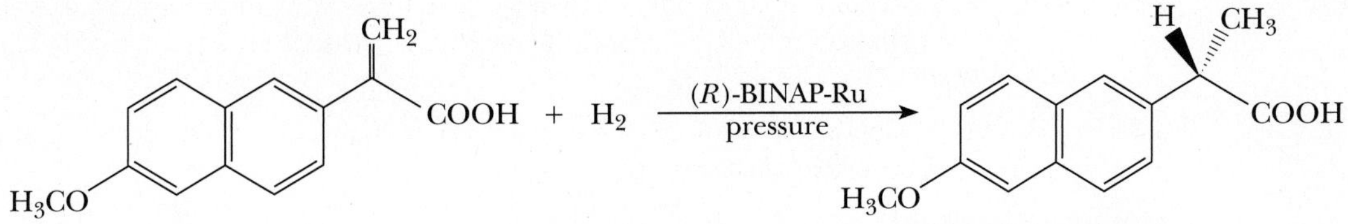

(S)-Naproxen
(ee > 98%)

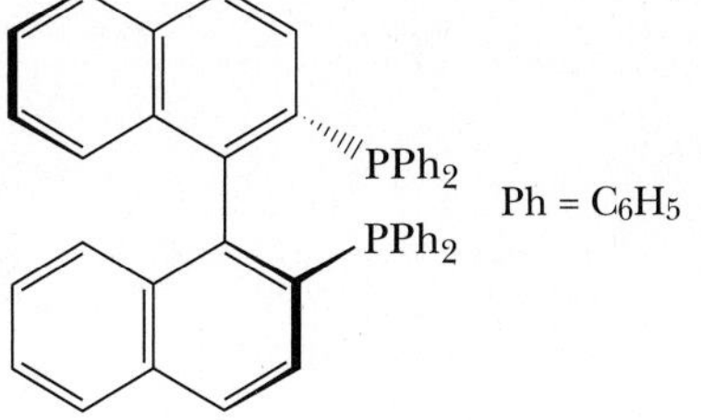

BINAP

(a) What kind of isomers are the enantiomers of BINAP?
(b) How can one enantiomer of naproxen be formed in such high yield?

3.39 In Section 10.5D, the following reactions are discussed. Ts is the tolunesulfonyl group.

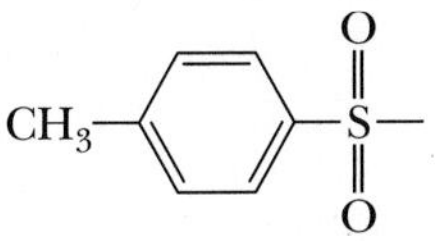

Tolunesulfonyl group (Ts)

(a)

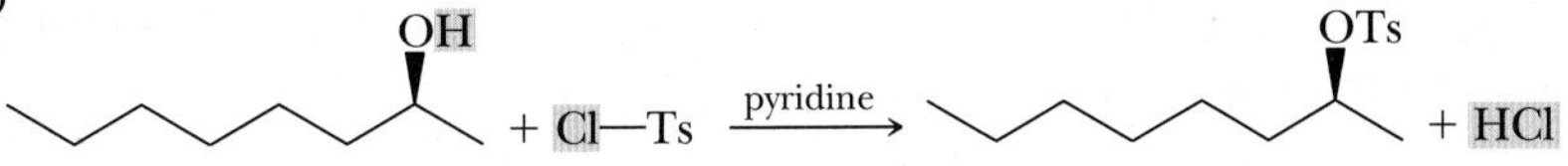

(S)-2-Octanol **(S)-2-Octyl tosylate**

(b)

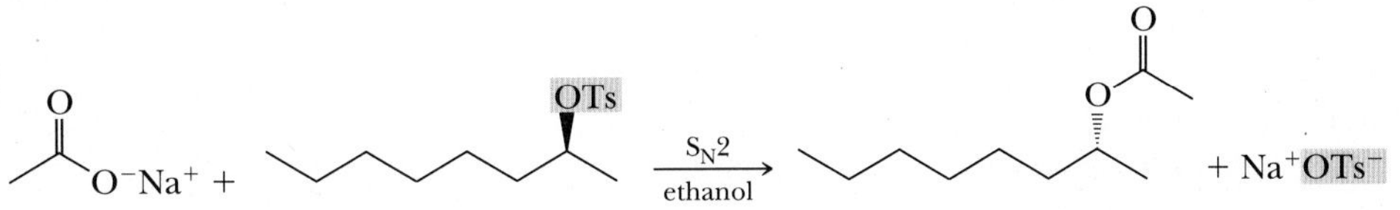

Sodium acetate **(S)-2-Octyl tosylate** **(R)-2-Octyl acetate**

In reaction (a), an *S* compound gives an *S* product. In reaction (b), an *S* compound gives an *R* product. Explain what is probably going on. (*Hint:* The oxygen atom in the starting material and product is the same in one reaction but not in the other.) What might this say about the second reaction?

4

Acids and Bases

Citrus fruits are sources of citric acid. Lemon juice, for example, contains 5%–8% citric acid. **Inset**: a model of citric acid. (Emilio Ereza/ Alamy Stock photo)

Outline

4.1 Arrhenius Acids and Bases

4.2 Brønsted-Lowry Acids and Bases

4.3 Acid Dissociation Constants, pK_a, and the Relative Strengths of Acids and Bases

4.4 The Position of Equilibrium in Acid-Base Reactions

HOW TO: Calculate the Equilibrium Constants for Acid-Base Reactions

4.5 Thermochemistry and Mechanisms of Acid-Base Reactions

4.6 Molecular Structure and Acidity

4.7 Lewis Acids and Base

A great many organic reactions either are acid-base reactions or involve catalysis by an acid or a base at some stage. Of the reactions involving acid catalysis, some use proton-donating acids such as H_2SO_4, H_3O^+, and $CH_3CH_2OH_2^+$. Others use Lewis acids such as BF_3 and $AlCl_3$. It is essential, therefore, that you have a good grasp of the fundamentals of acid-base chemistry. In this chapter and in following chapters, we study the acid-base properties of the major classes of organic compounds.

4.1 Arrhenius Acids and Bases

The first useful definition of acids and bases was put forward by Svante Arrhenius in 1884. According to the original Arrhenius definition, an acid is a substance that dissolves in water to produce H^+ ions. A base is a substance that dissolves in water to produce OH^- ions. Today we know that it is not accurate to refer to an acid as giving rise to simple H^+ ions. An H^+ ion immediately combines with a water molecule to give a hydronium ion, H_3O^+. Hydration of the hydronium ion itself gives the ion $H_5O_2^+$. Modern research suggests that the monohydrated and dihydrated forms (H_3O^+ and $H_5O_2^+$) are the major hydrated forms present in aqueous solution, and they are present in approximately equal concentrations. Throughout this text, we will represent a proton dissolved in aqueous solution as a hydronium ion, H_3O^+, although it is an oversimplification.

$$H^+(aq) + H_2O(l) \longrightarrow \underset{\textbf{Hydronium ion}}{H_3O^+(aq)}$$

$$\underset{\textbf{Hydronium ion}}{H_3O^+(aq)} + H_2O(l) \rightleftharpoons H_5O_2^+(aq)$$

We therefore must modify the Arrhenius acid definition to take into account these interactions of H^+ with water molecules. Apart from this modification, the Arrhenius definitions of an acid and a base are still valid and useful today, as long as we are talking about aqueous solutions.

4.2 Brønsted-Lowry Acids and Bases

In 1923, the Danish chemist Johannes Brønsted and the English chemist Thomas Lowry independently proposed the following definitions: an **acid** is a proton donor, a **base** is a proton acceptor, and an acid-base reaction is a proton-transfer reaction.

Brønsted-Lowry acid A proton donor.

Brønsted-Lowry base A proton acceptor.

A. Conjugate Acid-Base Pairs Differ by a Proton

According to the Brønsted-Lowry definitions, any pair of molecules or ions that can be interconverted by transfer of a proton is called a **conjugate acid-base pair**. When an acid transfers a proton to a base, the acid is converted to its **conjugate base**. When a base accepts a proton, it is converted to its **conjugate acid**. Fundamental to these definitions is the fact that the members of a conjugate acid-base pair differ by a proton.

Conjugate base The species formed when an acid transfers a proton to a base.

Conjugate acid The species formed when a base accepts a proton from an acid.

We can illustrate the relationships among conjugate acid-base pairs by examining the reaction of hydrogen chloride with water to form chloride ion and hydronium ion.

conjugate acid-base pair (HCl / Cl^-); conjugate acid-base pair (H_2O / H_3O^+)

$$HCl(aq) + H_2O(l) \rightleftharpoons Cl^-(aq) + H_3O^+(aq)$$

$HCl(aq)$	$H_2O(l)$	$Cl^-(aq)$	$H_3O^+(aq)$
Hydrogen chloride (acid)	**Water** (base)	**Chloride ion** (conjugate base of HCl)	**Hydronium ion** (conjugate acid of H_2O)

The acid HCl donates a proton and is converted to its conjugate base Cl^-. The base H_2O accepts a proton and is converted to its conjugate acid H_3O^+. The members of each conjugate acid-base pair differ only by a proton.

We can show the transfer of a proton from an acid to a base by using a curved arrow, the same curved arrow symbol used in Section 1.8 to show the relocation of electron pairs among resonance contributing structures. For acid-base reactions, we write the Lewis structure of each reactant and product, showing all valence electrons on reacting atoms. We then use curved arrows to show the change in position of electron pairs during the reaction. The tail of the curved arrow is located at an electron pair, either a lone pair or a bonding pair, as the case may be. The head of the curved arrow shows the new location of the electron pair. For example, a curved arrow originating at a lone pair and pointing to an adjacent atom indicates formation of a new bond, while an arrow originating at a bonding electron pair and pointing toward a previously bonded atom indicates a breaking of that bond (see Things You Should Know III, Reaction Mechanisms, prior to Chapter 6, for a more thorough discussion of the use of curved arrows).

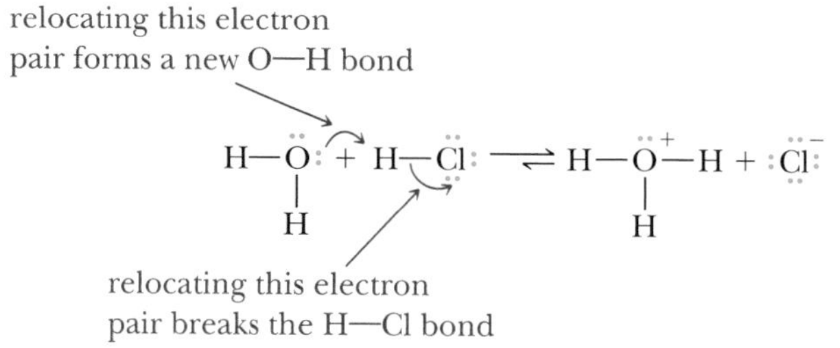
relocating this electron
pair forms a new O—H bond
relocating this electron
pair breaks the H—Cl bond

In this equation, the curved arrow on the left shows that an unshared pair of electrons on oxygen changes position to form a new covalent bond with hydrogen. The curved arrow on the right shows that the H—Cl bond breaks and that its electron pair is given entirely to chlorine to form a chloride ion. Thus, in the reaction of HCl with H_2O, a proton is transferred from HCl to H_2O. In the process, an O—H bond forms and an H—Cl bond breaks. It is important to keep in mind that arrows indicate movement of electrons, not atoms.

The most common mistake when writing curved arrows is to draw arrows that point backward. In the following example, the arrow originates at the H—Cl bond and terminates at an oxygen atom. As drawn, this notation means that the two electrons of the H—Cl bond are moving with the hydrogen; hence, $H:^-$ (hydride anion) is being transferred instead of a H^+ (proton). Things You Should Know IV: Common Mistakes in Arrow Pushing contains more examples of common mistakes.

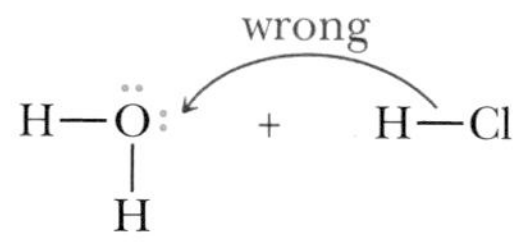
wrong

We have illustrated the application of the Brønsted-Lowry definitions using water as a reactant. The Brønsted-Lowry definitions, however, do not require water as a reactant. Consider the following reaction between acetic acid and ammonia.

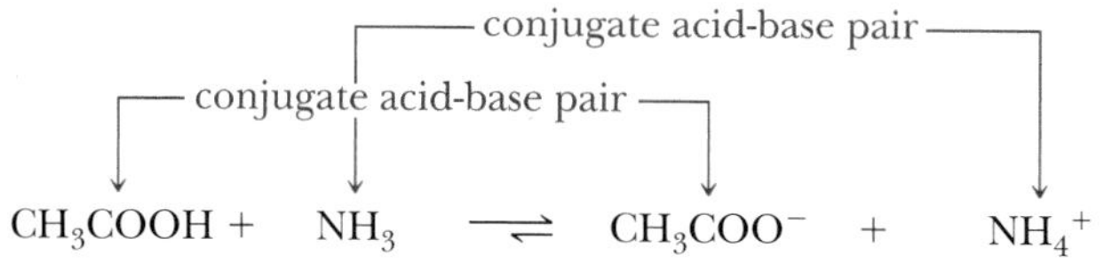
conjugate acid-base pair
conjugate acid-base pair

Acetic acid (acid)	**Ammonia** (base)	**Acetate ion** (conjugate base of acetic acid)	**Ammonium ion** (conjugate acid of ammonia)

The curved arrows below depict the two electron-pair shifts necessary to transfer a proton from an acetic acid molecule to an ammonia molecule.

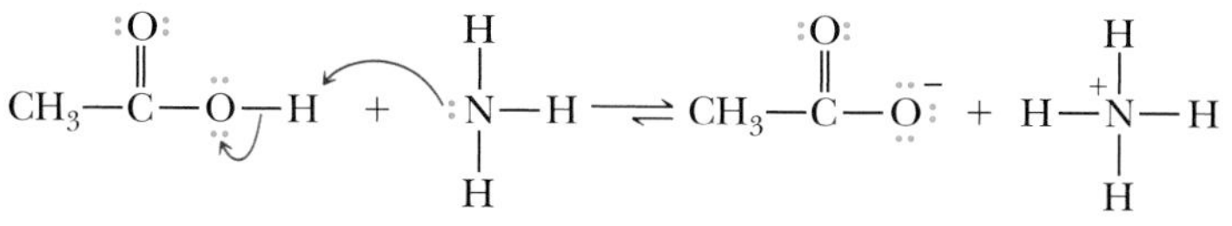

Acetic acid (proton donor)	**Ammonia** (proton acceptor)	**Acetate ion**	**Ammonium ion**

Example 4.1 Acids and Bases

For each conjugate acid-base pair, identify the first species as an acid or a base and the second species as its conjugate base or conjugate acid. In addition, draw Lewis structures for each species, showing all valence electrons and any formal charges.

(a) CH_3OH, $CH_3OH_2^+$ **(b)** H_2CO_3, HCO_3^- **(c)** CH_3NH_2, $CH_3NH_3^+$

Solution

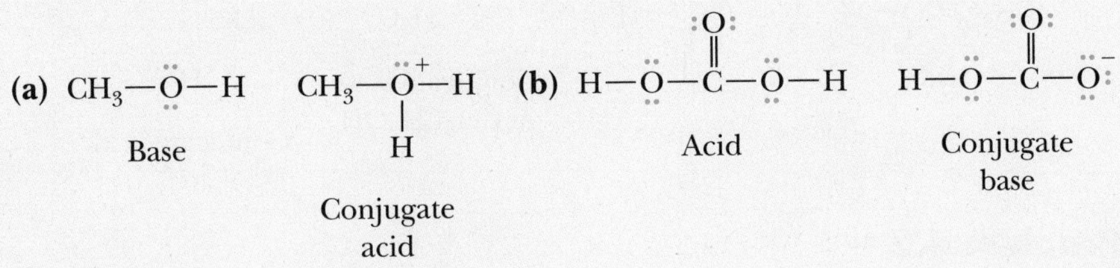

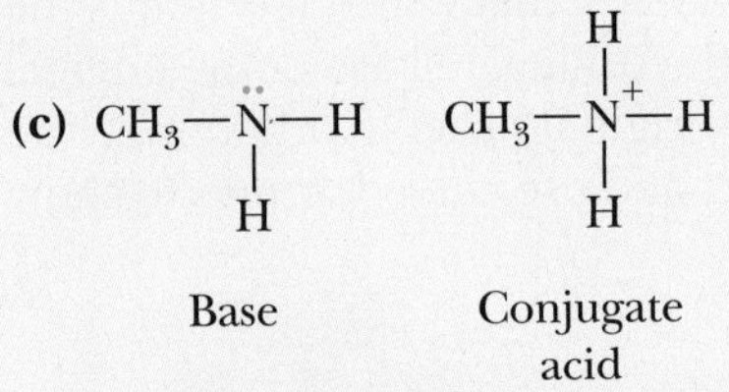

Problem 4.1

For each conjugate acid-base pair, identify the first species as an acid or a base and the second species as its conjugate acid or conjugate base. In addition, draw Lewis structures for each species, showing all valence electrons and any formal charges.

(a) H_2SO_4, HSO_4^- **(b)** NH_3, NH_2^- **(c)** CH_3OH, CH_3O^-

Example 4.2 Acid-Base Reactions

Write these reactions as proton-transfer reactions. Label which reactant is the acid and which is the base, which product is the conjugate base of the original acid, and which is the conjugate acid of the original base. In addition, write Lewis structures for each reactant and product and use curved arrows to show the flow of electrons in each reaction.

(a) $H_2O + NH_4^+ \longrightarrow H_3O^+ + NH_3$

(b) $CH_3CH_2OH + NH_2^- \longrightarrow CH_3CH_2O^- + NH_3$

Solution

(a) Water is the base (proton acceptor), and ammonium ion is the acid (proton donor).

(Continued)

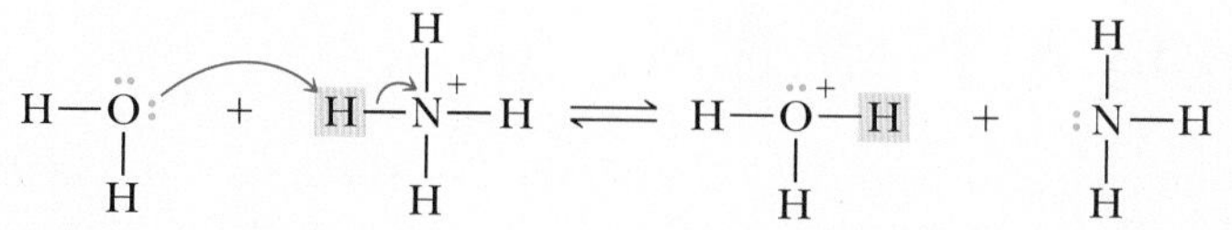

Base | Acid | Conjugate acid of H_2O | Conjugate base of NH_4^+

(b) Ethanol is the acid (proton donor), and amide ion (NH_2^-) is the base (proton acceptor).

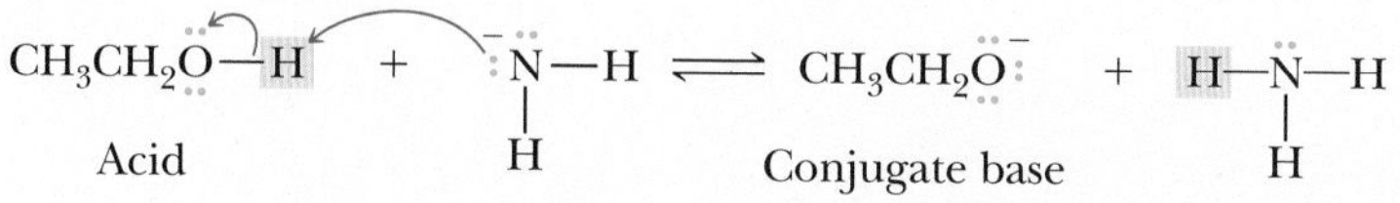

Acid | Base | Conjugate base of CH_3CH_2OH | Conjugate acid of NH_2^-

Watch a video explanation

Problem 4.2

Write these reactions as proton-transfer reactions. Label which reactant is the acid and which is the base, which product is the conjugate base of the original acid, and which is the conjugate acid of the original base. In addition, write Lewis structures for each reactant and product and use curved arrows to show the flow of electrons in each reaction.

(a) $CH_3SH + OH^- \rightleftharpoons CH_3S^- + H_2O$

(b) $CH_2{=}O + HCl \rightleftharpoons CH_2{=}OH^+ + Cl^-$

B. Brønsted-Lowry Bases with Two or More Receptor Sites

Thus far, we have dealt with Brønsted-Lowry bases that have only one site that can act as a proton acceptor in an acid-base reaction. Many organic compounds have two or more such sites. In the following discussion, we restrict our attention to compounds containing a carbonyl group in which the carbonyl carbon is bonded to either an oxygen or a nitrogen. The principle we develop here is an extremely important one that applies to other types of molecules as well. The more stable charged species is the one in which the charge is more delocalized. Relative charge delocalization can often be understood by considering resonance (Section 1.8). Let us consider first the potential sites for proton transfer to an oxygen atom of a carboxylic acid such as acetic acid. Proton transfer to the carbonyl oxygen gives cation A, and proton transfer to the hydroxyl oxygen gives cation B.

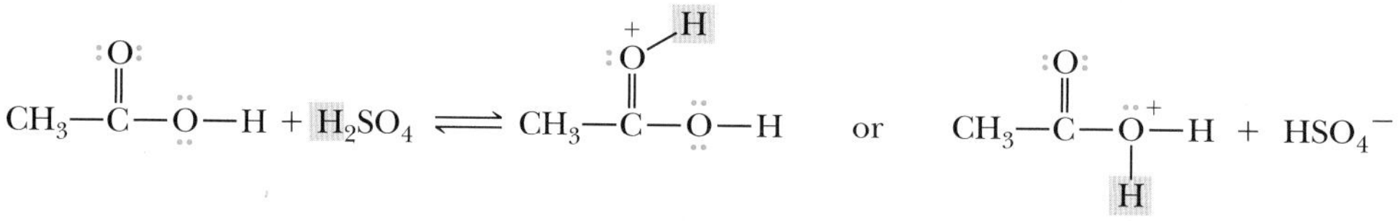

A (protonation on the carbonyl oxygen) | B (protonation on the hydroxyl oxygen)

▶ Delocalization of charge over a larger area is stabilizing.

We now examine each cation and determine which is more stable (lower in energy). For cation A, we can write three contributing structures. Two of these place the positive charge on oxygen, and one places it on carbon. ▶

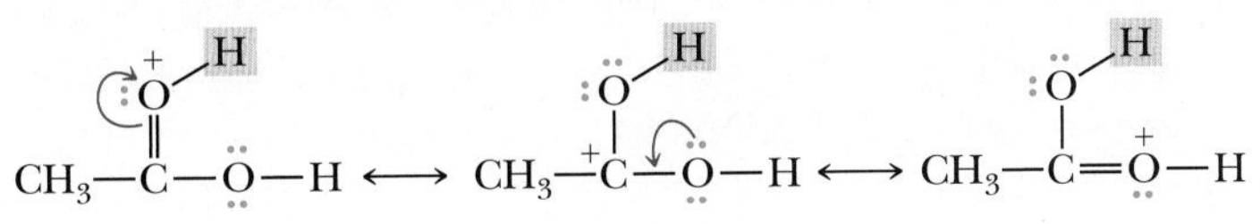

A-1 (C and O have complete octets) | A-2 (C has incomplete octet) | A-3 (C and O have complete octets)

Of these three structures, A-1 and A-3 make the greater contributions to the hybrid because all atoms in each have complete octets; A-2 makes a lesser contribution because its carbonyl carbon has an incomplete octet. Thus, on protonation of the carbonyl oxygen, the positive charge is delocalized over three atoms, with the greater share of it being on the two equivalent oxygen atoms. (The two oxygens were not equivalent before protonation, but they are now.)

Protonation on the hydroxyl oxygen gives cation B, for which we can write two resonance contributing structures.

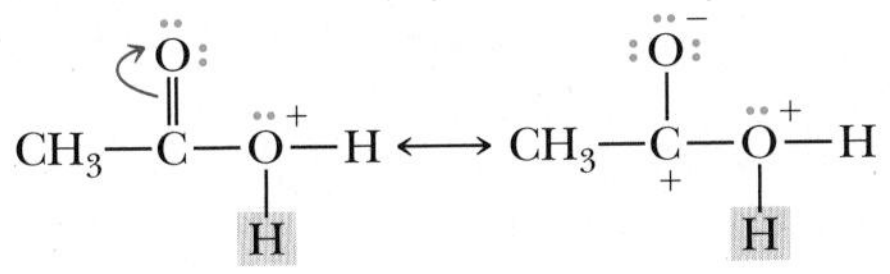

B-1 | B-2 (charge separation and adjacent positive charges)

Of these, B-2 makes, at best, only a minor contribution to the hybrid because of the adjacent positive charges; therefore, the charge on this cation is, in effect, localized on the hydroxyl oxygen.

From this analysis of cations A and B, we see that protonation of a carboxyl group occurs preferentially on the carbonyl oxygen because this cation has greater delocalization of the positive charge.

Example 4.3 Effect of Resonance

The functional group created when the —OH of a carboxyl group is replaced by an NH_2 group is called an amide (Section 1.3F). Draw the structural formula of acetamide, which is derived from acetic acid, and determine whether proton transfer to the amide group from HCl occurs preferentially on the amide oxygen or the amide nitrogen.

Solution

Following is a Lewis structure for acetamide and its two possible protonated forms.

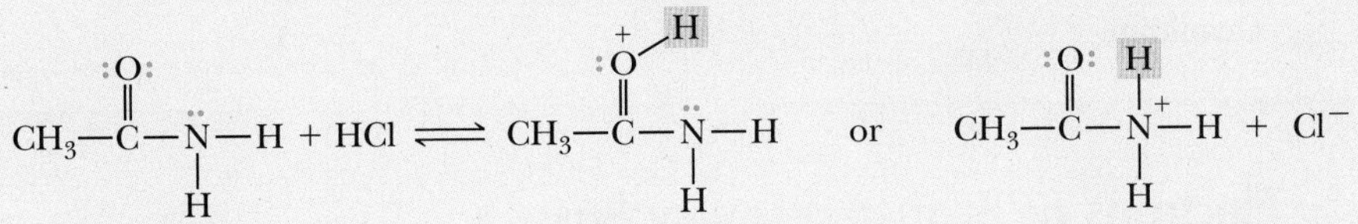

Acetamide (an amide) | A (protonation on the amide oxygen) | B (protonation on the amide nitrogen)

(Continued)

Of the three contributing structures that can be drawn for cation A, structures A-1 and A-3 make greater contributions to the hybrid because all atoms in each have complete octets; of these two contributors, A-3 has the positive charge on the less electronegative atom and, therefore, makes a greater contribution than A-1. The result is that the positive charge in cation A is delocalized over three atoms, the greater share of it being on nitrogen and oxygen.

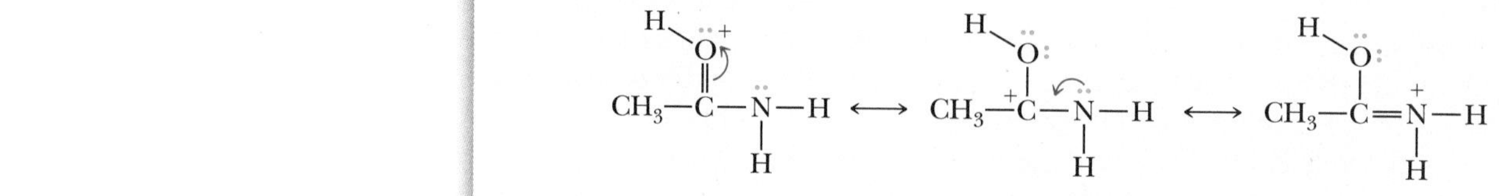

A-1 (complete octets; positive charge on more electronegative atom) | A-2 (incomplete octet for C) | A-3 (complete octets; positive charge on less electronegative atom)

Only two contributing structures can be drawn for cation B. Of these, B-2 requires creation and separation of unlike charges and places positive charges on adjacent atoms. It therefore makes little contribution to the hybrid. Thus, the positive charge in cation B is essentially localized on the amide nitrogen.

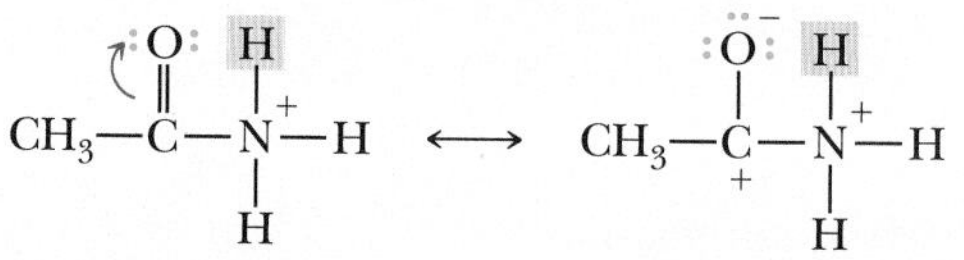

B-1 (protonation on the amide nitrogen) | B-2 (adjacent positive charges; charge separation)

From this analysis, we conclude that as in the acetic acid example, proton transfer to the carbonyl oxygen of the amide group gives the more delocalized and, hence, more stable cation, A.

Problem 4.3

Following is a structural formula for guanidine, the compound by which migratory birds excrete excess metabolic nitrogen. The hydrochloride salt of this compound is a white crystalline powder, freely soluble in water and ethanol.

(a) Write a Lewis structure for guanidine showing all valence electrons.

(b) Does proton transfer to guanidine occur preferentially to one of its —NH_2 groups (cation A) or to its =NH group (cation B)? Explain.

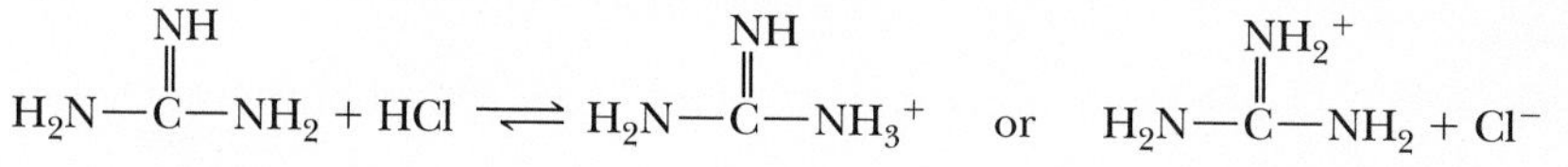

Guanidine | A | B

C. π Electrons as Brønsted-Lowry Bases

Thus far, we have considered proton transfer to atoms having a nonbonding pair of electrons. Proton-transfer reactions also occur with compounds having π electrons (e.g., the π electrons of carbon-carbon double and triple bonds). The π electrons of the carbon-carbon double bond of 2-butene, for example, react with strong acids such as H_2SO_4, H_3PO_4, HCl, HBr, and HI by proton transfer to form a new carbon-hydrogen bond.

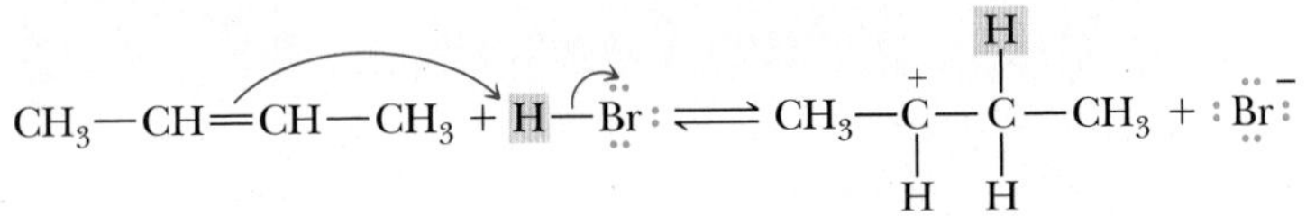

2-Butene

sec-Butyl cation
(a 2° carbocation)

The result of this proton-transfer reaction is formation of a **carbocation**, a species in which one of its carbons has only six electrons in its valence shell and carries a charge of +1. Because the carbon bearing the positive charge in the *sec*-butyl cation has only two other carbons bonded to it, it is classified as a secondary (2°) carbocation. We will study the formation, structure, and reactions of carbocations in detail in Chapter 6.

Example 4.4 Proton Transfer

The acid-base reaction between 2-methyl-2-butene and HI can, in principle, form two carbocations. Write chemical equations for the formation of each carbocation and use curved arrows to show the proton transfer in each reaction.

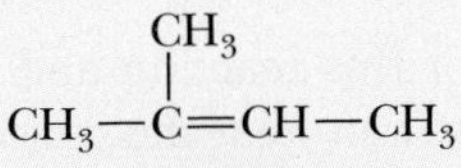

2-Methyl-2-butene

Solution

Proton transfer to carbon 3 of this alkene gives a tertiary (3°) carbocation. Proton transfer to carbon 2 gives a secondary (2°) carbocation.

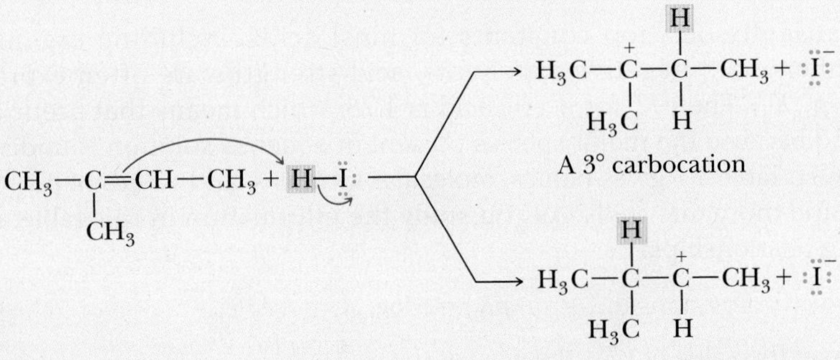

A 3° carbocation

A 2° carbocation

Problem 4.4

Write an equation to show the proton transfer between each alkene or cycloalkene and HCl. Where two carbocations are possible, show each.

(a) $CH_3CH_2CH{=}CHCH_3$ **(b)** 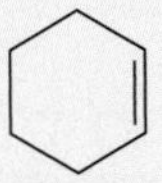

2-Pentene

Cyclohexene

4.3 Acid Dissociation Constants, pK_a, and the Relative Strengths of Acids and Bases

Any quantitative measure of the acidity of organic acids or bases involves measuring the equilibrium concentrations of the various components in an acid-base equilibrium. The strength of an acid is then expressed by an equilibrium constant. The dissociation (ionization) of acetic acid in water is given by the following equation:

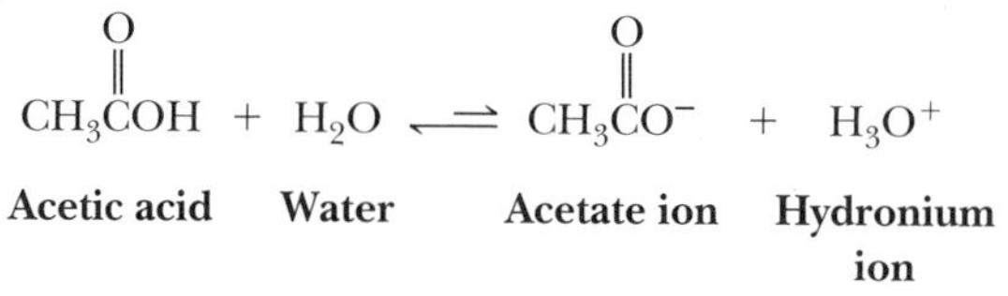

We can write an equilibrium expression for the dissociation of this or any other uncharged acid in a more general form; dissociation of the acid, HA, in water gives an anion, A^-, and the hydronium ion, H_3O^+. The general chemical equation and equilibrium constant for this ionization are

$$HA + H_2O \rightleftharpoons A^- + H_3O^+$$

$$K_{eq} = \frac{[H_3O^+][A^-]}{[HA][H_2O]}$$

Because water is the solvent for this reaction and its concentration changes very little when HA is added to it, we can treat the concentration of water as a constant equal to 1000 g/L, or approximately 55.6 mol/L. We can then combine these two constants (K_{eq} and the concentration of water) to define a new constant called an **acid dissociation constant**, given the symbol $\boldsymbol{K_a}$.

Acid dissociation constant

Equal to the equilibrium constant (K_{eq}) for an acid dissociation reaction multiplied by the concentration of water $[H_2O]$.

$$K_a = K_{eq}[H_2O] = \frac{[H_3O^+][A^-]}{[HA]}$$

Because dissociation constants for most acids, including organic acids, are numbers with negative exponents, acid strengths are often expressed as **pK_a** ($-\log_{10}K_a$). The pK_a for acetic acid is 4.76, which means that acetic acid is a weak acid because the major species present in aqueous solution is undissociated CH_3COOH. Table 4.1 gives names, molecular formulas, and values of pK_a for some organic and inorganic acids. As you study the information in this table, note the following relationships:

$$pK_a = -\log_{10}K_a$$

- The larger the value of pK_a, the weaker the acid.
- The smaller the value of pK_a, the stronger the acid.
- The weaker the acid, the stronger its conjugate base.
- The stronger the acid, the weaker its conjugate base.

Example 4.5 K_a and pK_a

For each value of pK_a, calculate the corresponding value of K_a. Which compound is the stronger acid?

(a) Ethanol, pK_a = 15.9 **(b)** Carbonic acid, pK_a = 6.36

Solution

Recall that pK_a = $-\log_{10}K_a$, so calculating $K_a = 10^{-pK_a}$ values gives the following:

(a) For ethanol, $K_a = 1.3 \times 10^{-16}$ **(b)** For carbonic acid, $K_a = 4.4 \times 10^{-7}$

Because the value of pK_a for carbonic acid is smaller than that for ethanol, carbonic acid is the stronger acid and ethanol is the weaker acid.

Problem 4.5

For each value of K_a, calculate the corresponding value of pK_a. Which compound is the stronger acid?

(a) Acetic acid, $K_a = 1.74 \times 10^{-5}$ **(b)** Chloroacetic acid, $K_a = 1.38 \times 10^{-3}$

Table 4.1 pK_a **Values for Some Organic and Inorganic Acids**

Weaker acid → Stronger acid (down the table); Stronger conjugate base ← Weaker conjugate base (up the table)

Acid	Formula	pK_a	Conjugate Base
Ethane	CH_3CH_3	51	$CH_3CH_2^-$
Ethylene	$CH_2{=}CH_2$	44	$CH_2{=}CH^-$
Ammonia	NH_3	38	NH_2^-
Hydrogen	H_2	35	H^-
Acetylene	$HC{\equiv}CH$	25	$HC{\equiv}C^-$
Ethanol	CH_3CH_2OH	15.9	$CH_3CH_2O^-$
Water	H_2O	15.7	HO^-
Methylammonium ion	$CH_3NH_3^+$	10.64	CH_3NH_2
Bicarbonate ion	HCO_3^-	10.33	CO_3^{2-}
Phenol	C_6H_5OH	9.95	$C_6H_5O^-$
Ammonium ion	NH_4^+	9.24	NH_3
Hydrogen sulfide	H_2S	7.04	HS^-
Pyridinium	$C_5H_5NH^+$	5.2	C_5H_5N
Benzoic acid	C_6H_5COOH	4.19	$C_6H_5COO^-$
Hydrogen fluoride	HF	3.2	F^-
Phosphoric acid	H_3PO_4	2.1	$H_2PO_4^-$
p-Toluenesulfonic acid	$CH_3C_6H_4SO_3H$	0.7	$CH_3C_6H_4SO_3^-$
Nitric acid	HNO_3	−1.5	NO_3^-
Hydronium ion	H_3O^+	−1.74	H_2O
Sulfuric acid	H_2SO_4	−5.2	HSO_4^-
Hydrogen chloride	HCl	−7	Cl^-
Hydrogen bromide	HBr	−8	Br^-
Hydrogen iodide	HI	−9.9	I^-

The pK_a values can also be used to estimate the equilibrium constants (K_a). For example, if the pK_a of an acid is near zero, then the equilibrium constant for the reaction of that acid protonating water is near 1. Negative pK_a values correlate to acids with equilibrium constants greater than 1, while positive pK_a values are for acids with equilibrium constants less than 1. Each single unit difference between pK_a values represents a tenfold increase or decrease in the strength of the acids being compared.

Values of pK_a in aqueous solution in the range 2 to 12 can be measured quite accurately. Values of pK_a smaller than 2 are less accurate because very strong acids, such as HCl, HBr, and HI, are completely ionized in water, and the only acid present in solutions of these acids is H_3O^+. For acids too strong to be measured accurately in water, less basic solvents such as acetic acid or mixtures of water and sulfuric acid are used. Although none of the halogen acids, for example, is completely ionized in acetic acid, HI shows a greater degree of ionization in this solvent than either HBr or HCl; therefore, HI is the strongest acid of the three. Values of pK_a greater than 12 are also less precise. For bases too strong to be measured in aqueous solution, more basic solvents such as liquid ammonia and dimethyl sulfoxide are used. Because different solvent systems are used to measure relative strengths at either end of the acidity scale, pK_a values smaller than 2 and greater than 12 should be used only in a qualitative way when comparing them with values in the middle of the scale.

4.4 The Position of Equilibrium in Acid-Base Reactions

We know from the value of pK_a for an acid whether an aqueous solution of the acid contains more molecules of the undissociated acid or its ions. A negative pK_a value indicates that the majority of molecules of the acid are dissociated in water, while a positive value indicates that most acid molecules remain undissociated in water. HCl, for example, a strong acid with a pK_a of −7, is almost completely dissociated at equilibrium in aqueous solution, and the major species present are H_3O^+ and Cl^-.

$$HCl + H_2O \rightleftharpoons H_3O^+ + Cl^- \qquad pK_a = -7$$

For acetic acid on the other hand, which is a weak acid with a pK_a of 4.76, the major species present at equilibrium in aqueous solution are CH_3COOH molecules.

$$CH_3\overset{O}{\overset{\|}{C}}OH + H_2O \rightleftharpoons CH_3\overset{O}{\overset{\|}{C}}O^- + H_3O^+ \qquad pK_a = 4.76$$

In these acid-base reactions, water is the base. But what if we have a base other than water as the proton acceptor, or what if we have an acid other than hydrogen chloride or acetic acid as the proton donor? How do we determine quantitatively or even qualitatively which species are present at equilibrium; that is, how do we determine where the position of equilibrium lies?

Let us look at an example of the acid-base reaction of acetic acid and ammonia to form acetate ion and ammonium ion.

$$\underset{\textbf{Acetic acid}}{CH_3\overset{O}{\overset{\|}{C}}OH} + \underset{\textbf{Ammonia}}{NH_3} \rightleftharpoons \underset{\textbf{Acetate ion}}{CH_3\overset{O}{\overset{\|}{C}}O^-} + \underset{\textbf{Ammonium ion}}{NH_4^+}$$

Two acids are present in this equilibrium, acetic acid and ammonium ion. Two bases also are present, ammonia and acetate ion. A way to analyze this equilibrium is to view it as a competition between the two bases, ammonia and acetate ion, for a proton. The question we then ask is, "Which of these is the stronger base?" The information we need to answer this question is in Table 4.1. We first determine which conjugate acid is the stronger acid and couple this with the fact that the stronger the acid, the weaker its conjugate base. From Table 4.1, we see that acetic acid, pK_a 4.76, is the stronger acid, which means that CH_3COO^- is the weaker base. Conversely, ammonium ion, pK_a 9.24, is the weaker acid, which means that NH_3 is the stronger base. We can now label the relative strengths of each acid and base.

$$CH_3\overset{\overset{O}{\|}}{C}OH \quad + \quad NH_3 \quad \rightleftharpoons \quad CH_3\overset{\overset{O}{\|}}{C}O^- \quad + \quad NH_4^+$$

Acetic acid	Ammonia	Acetate ion	Ammonium ion
pK_a 4.76 (stronger acid)	(stronger base)	(weaker base)	pK_a 9.24 (weaker acid)

HOW TO

Calculate the Equilibrium Constants for Acid-Base Reactions

Besides using values of pK_a for each acid to estimate the position of equilibrium, we can also use them to calculate an equilibrium constant for the acid-base reaction. A series of mathematical manipulations allows us to do so. For example, consider the following acid-base reaction:

$$\underset{\text{Acid}}{HA} \quad + \quad \underset{\text{Base}}{B} \quad \rightleftharpoons \quad \underset{\text{Conjugate base of HA}}{A^-} \quad + \quad \underset{\text{Conjugate acid of B}}{BH^+}$$

The equilibrium constant for this reaction is

$$K_{eq} = \frac{[A^-][BH^+]}{[HA][B]}$$

Multiplying the right-hand side of the equation by $[H_3O^+]/[H_3O^+]$ gives a new expression, which, upon rearrangement, becomes the K_a of acid HA divided by the K_a of acid BH^+:

$$K_{eq} = \frac{[A^-][BH^+]}{[HA][B]} \times \frac{[H_3O^+]}{[H_3O^+]} = \frac{[A^-][H_3O^+]}{[HA]} \times \frac{[BH^+]}{[B][H_3O^+]} = \frac{K_{HA}}{K_{BH^+}}$$

Upon taking the negative logarithm of each side of this equation, we arrive at this expression:

$$pK_{eq} = pK_{HA} - pK_{BH^+}$$

Thus, if we know the pK_a values of each acid in the equilibrium, we can calculate the equilibrium constant for the acid-base reaction using these steps:

1. Look up the pK_a values of the acid and conjugate acid involved in the reactants and the products, respectively, of an acid-base reaction.
2. Subtract the pK_a of the conjugate acid (product) from the pK_a of the acid (reactant).
3. Because pK_{eq} = $-\log_{10}K_{eq}$, take the antilog of $-$pK_{eq} to arrive at K_{eq}.

Example: Let's calculate the K_{eq} for the reaction of acetic acid with ammonia.

$$\underset{pK_a = 4.76}{CH_3CO_2H} \quad + \quad NH_3 \quad \rightleftharpoons \quad CH_3CO_2^- \quad + \quad \underset{pK_a = 9.24}{NH_4^+}$$

Following Steps 1 through 3, we have pK_{eq} = 4.76 − 9.24 = −4.48, so $K_{eq} = 3.0 \times 10^4$. Because we know that acid-base reactions are favored when the stronger acid reacts with the stronger base to give the weaker acid and the weaker base, we can conclude that the equilibrium for the reaction between acetic acid and ammonia lies to the right. Using the mathematical approach just developed, we can calculate that the preference to the right is 3.0×10^4. This means that if we started with equal amounts of acetic acid and ammonia, the reaction would prefer the products 30,000 to 1.

The most important concept is as follows:

In an acid-base reaction, the position of equilibrium always favors reaction of the stronger acid and stronger base to form the weaker acid and weaker base.

Thus, at equilibrium, the major species present are the weaker acid and weaker base. Therefore, in the reaction between acetic acid and ammonia, the equilibrium lies to the right and the major species present are acetate ion and ammonium ion.

Connections to Biological Chemistry

The Ionization of Functional Groups at Physiological pH

The pH of living cells is generally between 7.0 and 8.5, a range often referred to as physiological pH. At physiological pH, several of the common functional groups found in biological molecules are ionized because they are either acids that are deprotonated or bases that are protonated. A good rule of thumb is that an acid will be substantially deprotonated if its pK_a is two or more units lower than the pH of the solution. A carboxyl group (the functional group of carboxylic acids) is present in all amino acids as well as in the side chains of glutamic acid and aspartic acid. The pK_a values for carboxylic acids are typically between 4 and 5. At pH values of 7 or above, carboxylic acids are essentially fully deprotonated and therefore anionic.

Carboxylic acid pK_a 4–5 — Form present at physiological pH

Another functional group found in biomolecules is the phosphodiester group. This group is found as part of the backbone of nucleic acids such as DNA and RNA. The pK_a values for phosphodiesters are between 1 and 3.

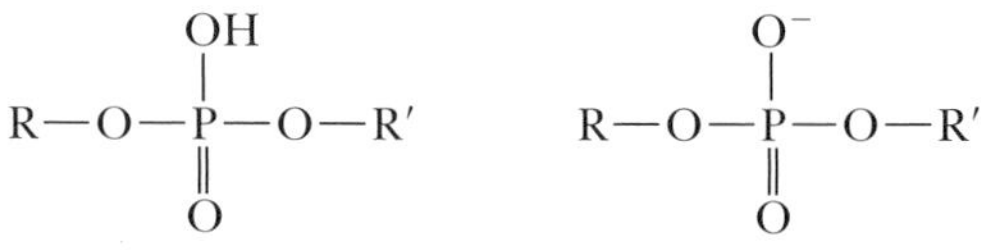

Phosphodiester pK_a 1–3 — Form present at physiological pH

At physiological pH values, the phosphodiester group will be present in its anion form. Therefore, the backbones of nucleic acids (in which there is a repeating pattern of phosphodiester groups) will be polyanionic, a factor that has a major influence on their overall properties.

There are also basic functional groups in biological molecules. A good rule of thumb here is that a base will be substantially protonated if the pK_a of its conjugate acid is two or more units higher than the pH of the solution. Two important examples include amino groups and guanidino groups. The pK_a values for the conjugate acids of amines and guanidines are about 9 to 11 and 13 to 14, respectively. As a result, these groups are protonated and therefore positively charged at physiological pH.

$R{-}NH_2$	$R{-}NH_3^+$
An amine group	pK_a 9–11 Form present at physiological pH

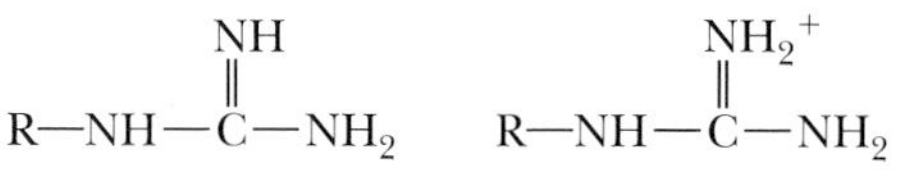

A guanidine group — pK_a 13–14 Form present at physiological pH

An interesting case is the imidazole group, which comprises the side chain of the amino acid histidine. An imidazole group, whose conjugate acid has a pK_a between 6 and 7, will be present at physiological pH as a mixture of protonated and deprotonated forms. This ability to exist in both forms can be significant in situations in which proton transfer reactions are important for the function of the protein containing an imidazole group.

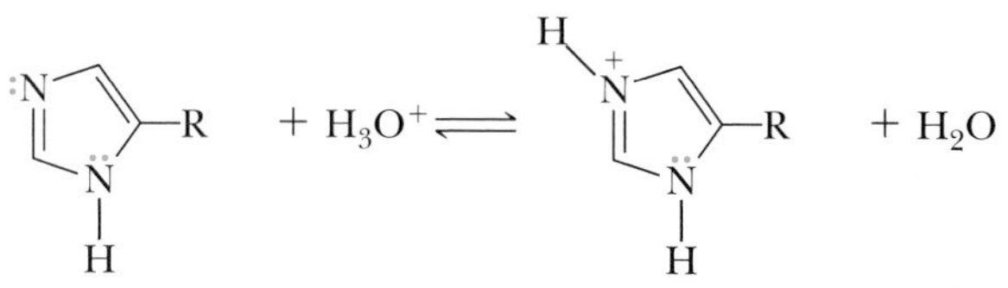

Imidazole group — pK_a 6–7 Conjugate acid of imidazole

Finally, no highlight section on the acid-base properties of biological molecules would be complete without a discussion of amino acids. At physiological pH, both the amino and carboxyl groups are ionized. Free amino acids are found in a number of situations in organisms (e.g., as neurotransmitters in mammals).

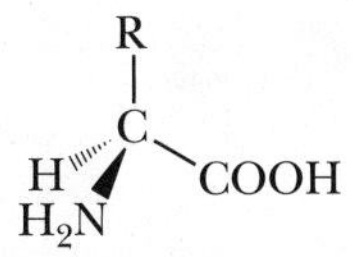

An amino acid contains an amino group and a carboxyl group

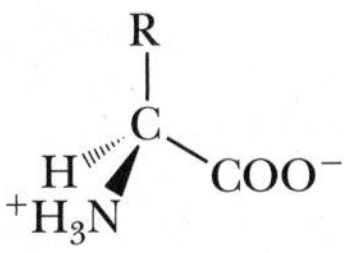

Ionized or zwitterion form present at physiological pH

Example 4.6 Acid-Base Equilibria

Predict the position of equilibrium and calculate the equilibrium constant, K_{eq}, for each acid-base reaction.

(a) $C_6H_5OH + HCO_3^- \rightleftharpoons C_6H_5O^- + H_2CO_3$

Phenol | **Bicarbonate ion** | **Phenoxide ion** | **Carbonic acid**

(b) $HC{\equiv}CH + NH_2^- \rightleftharpoons HC{\equiv}C^- + NH_3$

Acetylene | **Amide ion** | **Acetylide ion** | **Ammonia**

Solution

(a) Carbonic acid is the stronger acid; the position of this equilibrium lies to the left. Phenol does not transfer a proton to bicarbonate ion to form carbonic acid.

$$C_6H_5OH + HCO_3^- \rightleftharpoons C_6H_5O^- + H_2CO_3 \qquad pK_{eq} = 9.95 - 6.36 = 3.59$$

pK_a 9.95 Weaker acid; pK_a 6.36 Stronger acid; $K_{eq} = 10^{-3.59} = 2.57 \times 10^{-4}$

(b) Acetylene is the stronger acid; the position of this equilibrium lies to the right.

$$HC{\equiv}CH + NH_2^- \rightleftharpoons HC{\equiv}C^- + NH_3 \qquad pK_{eq} = 25 - 38 = -13$$

pK_a 25 Stronger acid; pK_a 38 Weaker acid; $K_{eq} = 10^{13}$

Problem 4.6

Watch a video explanation

Predict the position of equilibrium and calculate the equilibrium constant, K_{eq}, for each acid-base reaction.

(a) $CH_3NH_2 + CH_3COOH \rightleftharpoons CH_3NH_3^+ + CH_3COO^-$

Methylamine | **Acetic acid** | **Methylammonium ion** | **Acetate ion**

(b) $CH_3CH_2O^- + NH_3 \rightleftharpoons CH_3CH_2OH + NH_2^-$

Ethoxide ion | **Ammonia** | **Ethanol** | **Amide ion**

4.5 Thermochemistry and Mechanisms of Acid-Base Reactions

The transfer of a proton from an acid to a base is among the simplest of all chemical reactions discussed in this book. It is therefore an excellent reaction to examine as a means of introducing the concepts of thermochemistry and reaction mechanisms.

Reaction mechanism

A sequence of steps occurring during a reaction.

Thermochemistry

The study of the energy of chemical structures.

A **reaction mechanism** describes in detail how a reaction occurs. It describes which bonds are broken and which bonds are formed, as well as the order and relative rates of the various bond-breaking and bond-forming steps. A complete mechanism ideally describes the positions of all atoms and the energy of the entire system during every instant of the reaction. This ideal, however, is rarely ever achieved in practice.

Thermochemistry is the study of the energy of the entire system at every instant of a reaction. It is therefore part of the information gained when deciphering a complete mechanism. Before examining the mechanisms and thermochemistry of acid-base reactions, we must examine what causes chemical reactions to occur and introduce some basic energy principles that are key to chemical reactivity.

A. Thermal Reactions and Transition States

Most chemical reactions occur via collisions. Imagine placing two chemicals, A—H and B^-, in a reaction vessel along with a solvent. Compounds A—H and B^- will move through the solvent by jostling around, hitting and bouncing off individual solvent molecules. On occasion, A—H and B^- will collide with each other. The velocity at which the molecules move through the vessel is proportional to their kinetic energy (*kinetic* meaning motion). Higher-energy collisions occur between molecules possessing larger kinetic energy. During collision, the structures of the molecule contort and flex, and collisions of higher energy lead to larger distortions in structure. Analogously, a head-on collision is more damaging between automobiles moving at 65 mph than at 35 mph. At higher temperatures, the energy of the collisions is greater because the molecules are moving more rapidly. Reactions that result by virtue of the kinetic energy put into a reaction vessel due to temperature are called **thermal reactions**.

During the collisions, the kinetic energy of the reactants is converted into potential energy. The potential energy is stored in the chemical structures in the form of the structural strains, and the energy is released as the molecules again adopt their optimal geometries (see Section 2.5 for a discussion of strain). This situation is analogous to jumping up and down on the branch of a tree. Let's say the most stable geometry of the branch is horizontal. Each jump (collision) strains the branch by bending away from the optimal horizontal geometry. If you jump hard enough, the branch will break. Similarly, if molecules collide with enough force, bonds will break. Further, some collisions cause distortions that lead to rearrangements of bonds.

Transition state

The highest energy point on a reaction coordinate diagram. The chemical structure at this point is commonly called the activated complex.

During a collision process that yields a reaction, a structure forms that is called the **transition state**, also commonly called the activated complex. It has a particular geometry possessing partially broken and partially formed bonds, and it is so strained that it transitions to new structures that are less strained. The structure of a bent tree branch that finally cannot take any more strain and therefore snaps is a good analogy to the transition state in chemical reactions. We will talk more about the transition state in the next section.

B. Reaction Coordinate Diagrams and Thermochemistry

Reaction coordinate diagram

A graph showing the energy changes that occur during a chemical reaction; energy is plotted on the vertical axis, while the reaction progress is plotted along the horizontal axis.

Chemists use **reaction coordinate diagrams** to show the changes in energy for the molecules involved in a chemical reaction. Energy is measured on the vertical axis, and the change in position of the atoms during the reaction is represented on the horizontal axis, called the **reaction coordinate**. The reaction coordinate is not a time axis; rather, it shows how far the reaction has progressed in terms of structural changes. Figure 4.1 shows a reaction coordinate diagram for the reaction of acid H—A and base B^- (reactants) to form H—B and A^- (products). Wells on the plots

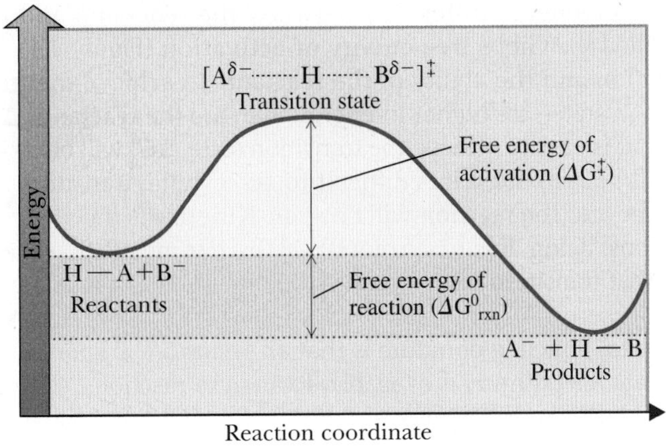

FIGURE 4.1 A reaction coordinate diagram for a one-step reaction between H—A and B^-. The dashed lines in the transition state indicate that the H—A bond is partially broken and the new H—B bond is partially formed. On completion of the reaction, the H—A bond is fully broken and the H—B bond is fully formed. The energy of the reactants is higher than that of the products. The energy axis in this diagram is Gibbs free energy (G^0).

represent stable structures that have lifetimes, while other points along the curve represent unstable structures that cannot be isolated. The transition state exists at the highest energy point on a reaction coordinate diagram. Transition states last less than one picosecond. The example given in Figure 4.1 occurs in one step, meaning that bond breaking in the reactants and bond formation in the products occur simultaneously upon collision. Every point along the reaction coordinate that is not a reactant or product represents a chemical structure that has characteristics of both the reactants and products. Positions along the x-axis close to the reactants represent structures more resembling the reactants, while positions close to the products resemble the products. Any collisions that raise the energy of the reactants (moving toward the products from the reactants in the diagram) but do not raise the energy enough to achieve the transition state simply result in glancing blows. In contrast, most collisions that possess enough energy to achieve the transition state result in passing of the chemical structures over this point. The resulting entities relax to the structure of the products with a release of energy.

Thermodynamics is the study of the relative energies between any two entities on a reaction coordinate diagram that are shown in wells, such as the energy between the reactants and products in Figure 4.1. The energy that chemists plot on the vertical axis of a reaction coordinate diagram can vary. For reactions at constant pressure, the Gibbs free energy (G^0) is often used. The naught superscript (^{0}) indicates standard states (298 K and 1 atm). This kind of energy controls the rates and equilibrium of reactions. For example, a **change in Gibbs free energy** (ΔG^0 or ΔG^0_{rxn}) between the reactants and products is related to the equilibrium constant through this equation:

Gibbs free energy change, ΔG^0

The energy that dictates the position of chemical equilibria and rates of chemical reactions. It is a thermodynamic function of enthalpy, entropy, and temperature given by the equation $\Delta G^0 = \Delta H^0 - T\Delta S^0$. The position of the equilibrium for the reaction favors the product(s) if $\Delta G^0 < 0$ and favors the reactant(s) if $\Delta G^0 > 0$.

$$\Delta G^0 = -RT \ln K_{eq,}$$

where $R = 8.31$ J/K mol and T is the temperature in kelvins.

Example 4.7 Gibbs Free Energy

Calculate the ΔG^0 for a reaction with a K_{eq} of 1.0×10^3 at 298 K.

Solution

$\Delta G^0 = -(8.31 \text{ J/K mol}) \times 298 \text{ K} \times \ln(1.0 \times 10^3) = -17.1 \text{ kJ/mol} \ (-4.09 \text{ kcal/mol})$

Problem 4.7

Calculate K_{eq} for a reaction with $\Delta G^0 = -17.1$ kJ/mol (-4.09 kcal/mol) at 328 K. Compare this value to the 1×10^3 seen at 298 K.

The Gibbs free energy difference between the reactants and the transition state, $\Delta G^{\ddagger}$, is called the Gibbs **free energy of activation** (Figure 4.1). It controls the rate of the reaction and the ability of the reactants to achieve the transition state. Because transition states are higher in energy than are the reactants, $\Delta G^{\ddagger}$ is always a positive value. If it is easy to achieve the transition state, $\Delta G^{\ddagger}$ will be a small value and the reaction will be fast, but if it is difficult to achieve the transition state, $\Delta G^{\ddagger}$ will be relatively large and the reaction will be slow. **Kinetics** is the study of the rates of chemical reactions. Being able to surmount the barrier means that there are obtainable collisions that transition the reactants to the products; that is, a pathway for the reaction exists.

A fundamental principle of nature is that all systems will spontaneously attain a lower energy state if a pathway is available. To analyze reactions, chemists often look at the difference between the energy of the products ($G^0{}_P$) and the energy of the reactants ($G^0{}_R$).

$$\Delta G^0 = G^0{}_P - G^0{}_R$$

If the products are more stable (a lower numerical energy value), the change in energy is negative. A negative ΔG^0 (called an exergonic reaction) is favorable, while a positive ΔG^0 (called an endergonic reaction) is unfavorable. The more negative the value of ΔG^0, the greater the driving force for the reaction to occur. Finally, there are some simple qualitative relationships between Gibbs free energy changes and equilibrium constants for favorable and unfavorable reactions: when ΔG^0 is greater than 0, K_{eq} is less than 1 (unfavorable); and when $\Delta G^0 = 0$, $K_{eq} = 1$; and when ΔG^0 is less than 0, K_{eq} is greater than 1 (favorable).

The change in Gibbs free energy is a function of the change in two terms: enthalpy and entropy, as expressed by the Gibbs-Helmholtz equation:

$$\Delta G^0 = \Delta H^0 - T\Delta S^0$$

Enthalpy (H^0) is the energy contained in chemical bonds and solvation. The difference in enthalpy between reactants and products is called the heat of reaction (ΔH^0). It reflects differences in bond strengths and solvation. If the bonds formed in the product(s) are stronger than the bonds in the reactant(s), heat will be released, the solution will warm, and the reaction is called exothermic (ΔH^0 is negative). Conversely, if the bonds formed in the product(s) are weaker, heat will be absorbed from the solution so that it cools and the reaction is endothermic (ΔH^0 is positive). Although Gibbs free energy (ΔG^0) dictates rates and equilibria, enthalpy is commonly plotted along the *y*-axes of reaction coordinate diagrams. This is because enthalpy reflects the intrinsic stability of the chemical structures involved in the reaction. Structures that have stronger bonds and/or that are better solvated are more stable and thereby have lower enthalpy.

Entropy (S^0) measures disorder versus order, and disorder is favorable. It is a fundamental principle of nature that, given time, everything degrades (the second law of thermodynamics). This is because all systems become more stable as the number of freely moving particles increases and the chaotic movement of the particles increases. Hence, when molecules fragment and/or when molecular vibrations such as bond rotations increase their freedom of movement, entropy becomes more favorable. A favorable change in entropy (ΔS^0) is defined as a positive value and represents an increase in disorder, whereas an increase in order is reflected by a negative value. Because entropy is multiplied by temperature in the Gibbs-Helmholtz equation, and bond cleavage is associated with a positive ΔS^0, high temperature often leads to a negative ΔG^0 and spontaneous bond breaking. This occurs even if the reaction is endothermic by virtue of the breaking of bonds. Cracking of petroleum (Section 2.9B) is an example of a reaction that occurs with an increase in entropy. The relationship among ΔG^0, ΔH^0, and ΔS^0 and the position of equilibrium in chemical reactions are summarized in Table 4.2.

Exergonic reaction

A reaction in which the Gibbs free energy of the products is lower than that of the reactants. The position of equilibrium for an exergonic reaction favors products.

Endergonic reaction

A reaction in which the Gibbs free energy of the products is higher than that of the reactants. The position of equilibrium for an endergonic reaction favors starting materials.

Enthalpy change, ΔH^0

The difference in total bond strengths and solvation between various points under comparison on a reaction coordinate diagram.

Heat of reaction, ΔH^0

The difference in the enthalpy between the reactants and products.

Exothermic reaction

A reaction in which the enthalpy of the products is lower than the enthalpy of the reactants; heat will be released.

Endothermic reaction

A reaction in which the enthalpy of the products is higher than the enthalpy of the reactants; heat will be absorbed.

Table 4.2 **Relationship Among ΔG^0, ΔH^0, ΔS^0**

	$\Delta S^0 < 0$	$\Delta S^0 > 0$
$\Delta H^0 > 0$	$\Delta G^0 > 0$ the position of equilibrium favors reactants	At higher temperatures when $T\Delta S^0 > \Delta H^0$ and $\Delta G^0 < 0$, the position of equilibrium favors products
$\Delta H^0 < 0$	At lower temperatures when $T\Delta S^0 < \Delta H^0$ and $\Delta G^0 < 0$; the position of equilibrium favors products	$\Delta G^0 < 0$; the position of equilibrium favors products

It is important to note that there is commonly a correlation between the Gibbs free energy of reaction (ΔG^0) and the Gibbs free energy of activation ($\Delta G^\ddagger$). Often, more favorable reactions occur more quickly. Yet, there is no strict rule that correlates kinetics with thermodynamics, and many exceptions are seen. ▶

▶ Reactions will occur if the products are more stable than the reactants and the energy barrier is low enough.

C. Mechanism and Thermochemistry of Acid-Base Reactions

We will use the concepts just set forth in many subsequent chapters of this book. Yet, now we can use them to better understand acid-base chemistry. Let's first focus on issues that affect the rate of an acid-base reaction.

For an acid to transfer its proton to a base, the two entities must collide. Moreover, they must collide with a specific geometry. The trajectory of the base (B^-) must be toward the H of H—A, not toward the A group. Further, it makes sense that there would be an optimal trajectory for the approach of H—A and of B^- that will lead to the lowest-energy pathway for passing the proton from the acid to the base. In this case, a linear approach is optimal. The transition state has a linear A—H—B bond angle, and the proton is partially shared between both A and B (look back at Figure 4.1).

This optimal trajectory for the collision of H—A and B^- can be rationalized by examination of an A—H antibonding orbital, and Figure 4.2 shows a specific example of this kind of orbital for an O—H bond of an alcohol (ROH). The orbital has greatest magnitude on H, and it lies along the O—H axis. This orbital will be most efficiently filled with a lone pair of electrons from the base colliding with the H along the H—O axis, which will lead to cleavage of the O—H bond. As depicted by the arrow pushing that describes the mechanism, the lone pair on the base can be considered as the electron pair that forms the new H—B bond, while the electron pair in the O—H bond can be considered as the electrons that create the lone pair on O^-.

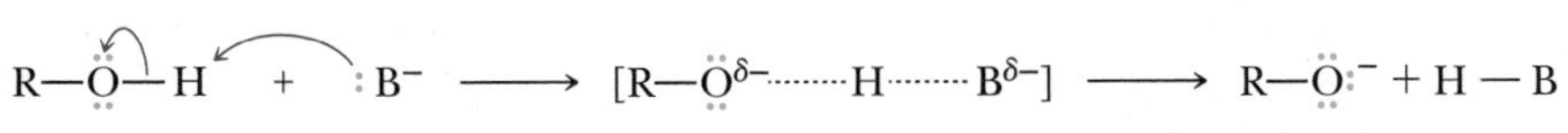
R—O—H + :B⁻ ⟶ [R—Oᵟ⁻·········H·········Bᵟ⁻] ⟶ R—O:⁻ + H—B

The following sections of this chapter focus on predicting relative acidities, which is an analysis of thermodynamics. The focus will be on enthalpy because it measures the intrinsic stabilities of the acids and bases on both sides of the equilibria. We do not consider entropy because an acid and a base exist in both the reactants and the products; therefore, the number of molecules does not change during the reaction. Hence, enthalpy is a good predictor for acid-base reaction equilibria.

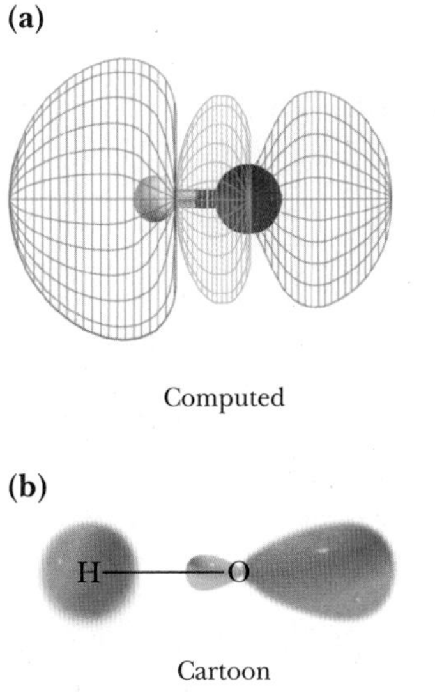
(a)
Computed
(b)
H
O
Cartoon

FIGURE 4.2 Calculated mesh diagram of an antibonding O—H orbital and a cartoon representation. Both pictures display that the orbital is primarily on the hydrogen.

4.6 Molecular Structure and Acidity

Now let us examine in some detail the relationships between molecular structure and acidity. The overriding principle in determining the relative acidities of uncharged organic acids is the stability of the conjugate base anion, A^-, resulting from loss of a proton; the more stable the anion, the greater the acidity of the acid, HA. As we discuss in this section, ways to stabilize A^- include the following:

- Having the negative charge on a more electronegative atom.
- Having the negative charge on a larger atom.
- Delocalizing the negative charge as described by resonance contributing structures.
- Spreading the negative charge onto electron-withdrawing groups by the inductive effect (polarization of sigma bonds).
- Having the negative charge in an orbital with more *s* character.

A. Electronegativity of the Atom Bearing the Negative Charge

Let us consider the relative acidities of the following series of hydrogen acids, all of which are in the same period of the Periodic Table.

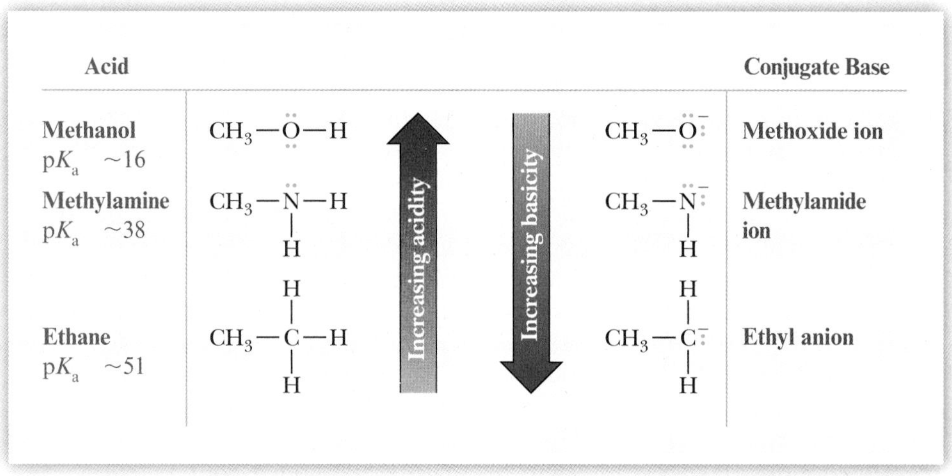

The pK_a value for ethane is given in Table 4.1, but values for methylamine and methanol are not. We can, however, make good guesses about the pK_a values of these acids by reasoning that the nature of the alkyl group bonded to nitrogen or oxygen has only a relatively small effect on the acidity of the hydrogen bonded to the heteroatom (in organic chemistry, an atom other than carbon). Therefore, we estimate that the pK_a of methylamine is approximately the same as that of ammonia (pK_a 38) and that the pK_a of methanol is approximately the same as that of ethanol (pK_a 15.9). As we see, ethane is the weakest acid in this series and ethyl anion is the strongest conjugate base. Conversely, methanol is the strongest acid and methoxide ion is the weakest conjugate base.

The relative acidity within a period of the Periodic Table is related to the electronegativity of the atom in the anion that bears the negative charge. The greater the electronegativity of this atom, the more strongly its electrons are held and the more stable the anion is. Conversely, the smaller the electronegativity of this atom, the less tightly its electrons are held and the less stable the anion is. Oxygen, the most electronegative of the three atoms compared, has the largest electronegativity (3.5 on the Pauling scale), and methanol forms the most stable anion. Carbon, the least electronegative of the three (2.5 on the Pauling scale), forms the least stable anion. Because methanol forms the most stable anion, it is the strongest acid in this series. Ethane is the weakest acid in the series.

It is essential to understand that this argument based on electronegativity applies only to acids within the same period (row) of the Periodic Table. Anions of atoms within the same period have approximately the same size, and their energies of solvation are approximately the same.

B. Size of the Atom Bearing the Negative Charge

To illustrate how the acidity of hydrogen acids varies within a group (column) of the Periodic Table, let us compare the acidities of methanol and methanethiol, CH_3SH. We estimated in the previous section that the pK_a of methanol is around 16, which is fairly close to its measured pK_a value of 15.5. We can estimate the pK_a of methanethiol in the following way. The pK_a of hydrogen sulfide, H_2S, is given in Table 4.1 as 7.04. If we assume that substitution of a methyl group for a hydrogen makes only a slight change in acidity, we might estimate that the pK_a of methanethiol is approximately 7.0. Actually, the pK_a of methanethiol is around 10.4, the difference being a result of the different way in which the methanethiolate anion is solvated versus HS^-. Nevertheless, methanethiol is the stronger acid and methanethiolate ion is the weaker conjugate base.

$$CH_3{-}\ddot{\underset{\cdot\cdot}{S}}{-}H \; + \; CH_3{-}\ddot{\underset{\cdot\cdot}{O}}{:}^- \; \rightleftharpoons \; CH_3{-}\ddot{\underset{\cdot\cdot}{S}}{:}^- \; + \; CH_3{-}\ddot{\underset{\cdot\cdot}{O}}{-}H$$

Methanethiol pK_a 10.4	Methoxide ion	Methanethiolate ion	Methanol pK_a 15.5
(stronger acid)	(stronger base)	(weaker base)	(weaker acid)

The relative acidity of these two hydrogen acids, and, in fact, any set of hydrogen acids within a group of the Periodic Table, is related to the size of the atom bearing the negative charge. Size increases from top to bottom within a group because the valence electrons are in increasingly higher principal energy levels. This means that (1) they are farther from the nucleus and (2) they occupy a larger volume of space. Because sulfur is below oxygen in the Periodic Table, it is larger than oxygen. Accordingly, the negative charge on sulfur in methanethiolate ion is spread over a larger volume of space (more delocalized, Section 4.2B); therefore, the CH_3S^- anion is more stable. The negative charge on oxygen in methoxide ion is confined to a smaller volume of space; therefore, the CH_3O^- anion is less stable.

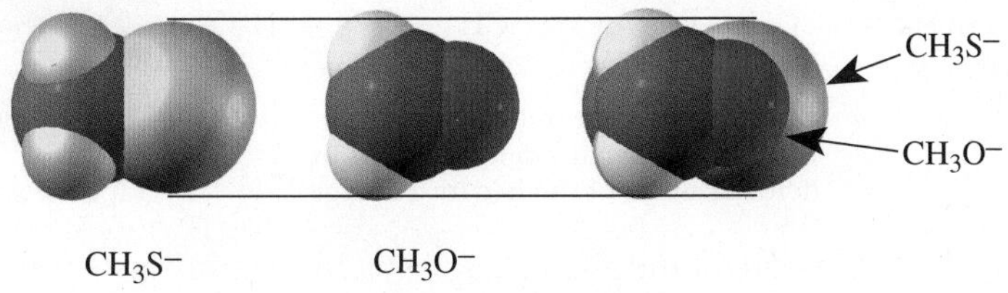

We see this same trend in the strength of the halogen acids, HF, HCl, HBr, and HI, which increase in strength from HF (the weakest) to HI (the strongest). Of their anions, iodide ion is the largest; its charge is delocalized over the largest volume of space and, therefore, is the most stable. HI is the strongest of the halogen acids. Conversely, fluoride ion is the smallest anion; its charge is the most concentrated, and fluoride ion is the least stable. HF is, therefore, the weakest acid of the halogen acids.

This size trend that we use to rationalize that HI is the strongest and HF is the weakest acid of the halogen acids runs contrary to what would be predicted from electronegativity arguments. Based on electronegativity, fluoride has the most stable negative charge because fluorine is the most electronegative element. Therefore, you would predict that HF should be the strongest acid. This is one of many examples in chemistry where there are opposing trends that predict reactivity and one of the trends dominates what we observe experimentally.

We will return to the relative sizes of these ions in Chapter 9, at which point we will discuss the nature of their solvation in polar solvents. We will see that for the ions of the same charge, the smaller the ion, the greater its solvation and that the degree of solvation has a profound effect on their relative reactivities.

C. Delocalization of Charge in the Anion

▶ Delocalization of charge over a larger area is stabilizing.

Carboxylic acids are weak acids. Values of pK_a for most unsubstituted carboxylic acids fall within the range 4 to 5. The pK_a for acetic acid, for example, is 4.76. Values of pK_a for most alcohols, compounds that also contain an —OH group, fall within the range 15 to 18; the pK_a for ethanol, for example, is 15.9. Thus, alcohols are slightly weaker acids compared to water (pK_a = 15.7). ▶

We account for the greater acidity of carboxylic acids compared with alcohols using the resonance model and looking at the relative stabilities of the alkoxide and carboxylate ions. Our guideline is this: the more stable the conjugate base anion, the farther the position of equilibrium is shifted toward the right and the more acidic the compound is.

Here, we take the acid ionization of an alcohol as a reference equilibrium.

$$CH_3CH_2O{-}H + H_2O \rightleftharpoons CH_3CH_2O^- + H_3O^+$$

An alcohol — An alkoxide ion

Elpot of ethoxide ion

In the alkoxide anion, the negative charge is localized on oxygen. In contrast, ionization of a carboxylic acid gives an anion for which we can write two equivalent contributing structures that result in delocalization of the negative charge of the conjugate base anion. Because of this delocalization of negative charge, a carboxylate anion is more stable than an alkoxide anion. Therefore, a carboxylic acid is a stronger acid than is an alcohol.

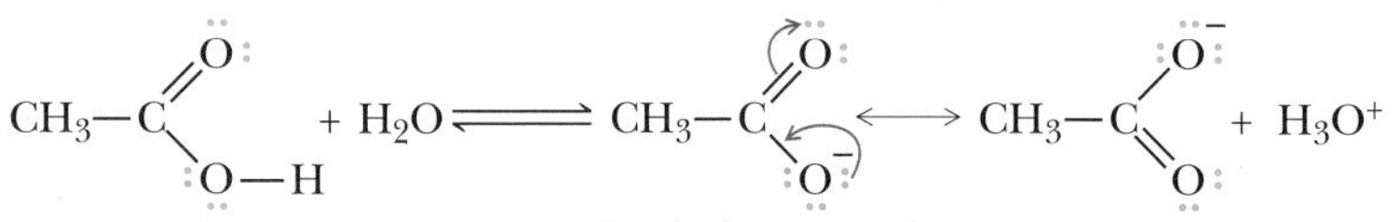

Equivalent contributing structures; the carboxylate anion is stabilized by delocalization of the negative charge

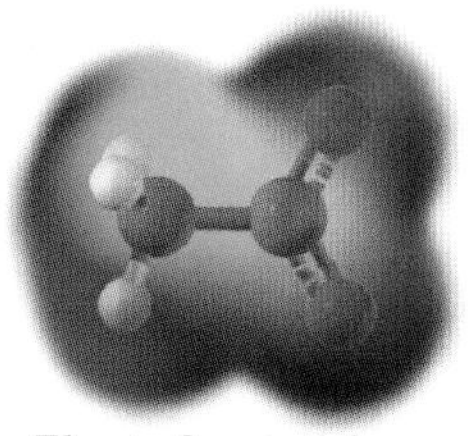

Elpot of acetate ion

D. Inductive Effect and Electrostatic Stabilization of the Anion

Inductive effect

The polarization of the electron density of a covalent bond caused by the electronegativity of a nearby atom.

We see an example of the **inductive effect** in alcohols in the fact that an electronegative substituent adjacent to the carbon bearing the —OH group increases the acidity of the alcohol. Compare, for example, the acidities of ethanol and 2,2,2-trifluoroethanol. The acid dissociation constant for 2,2,2-trifluoroethanol is larger than that of ethanol by more than three orders of magnitude, which means that the 2,2,2-trifluoroethoxide ion is considerably more stable than the ethoxide ion.

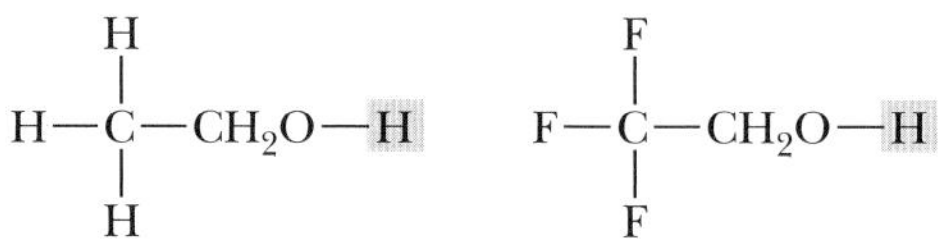

Ethanol pK_a 15.9 — **2,2,2-Trfluoroethanol** pK_a 12.4

We account for the increased stability of the 2,2,2-trifluoroethoxide ion in the following way. Fluorine is more electronegative than carbon (4.0 versus 2.5); therefore, the C—F bond has a significant dipole, indicated in the following figure by symbols to show the partial charges. There is an attractive stabilization by the interaction of the negatively charged oxygen and the partial positive charge on the carbon bearing the fluorines, which results in stabilization of the trifluoroethoxide ion. ▶

▶ **Delocalization of charge over a larger area is stabilizing.**

Another way to think about the relative acidity of these two compounds is to realize that some of the negative charge from the oxygen atom of the anion has been delocalized onto the electronegative fluorine atoms, thereby stabilizing the charged species through delocalization of the charge.

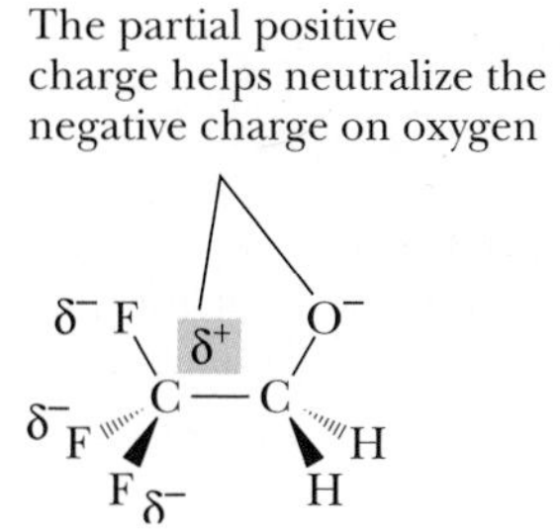
The partial positive charge helps neutralize the negative charge on oxygen
δ⁺
δ⁻ F
δ⁻ F
F δ⁻
C—C
O⁻
H
H

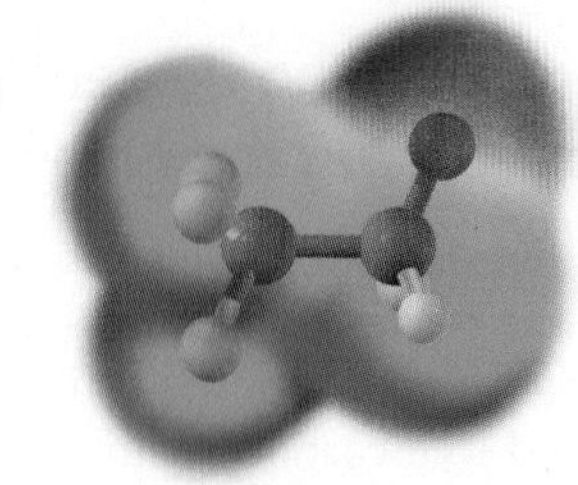
Elpot of trifluoroethoxide ion

Stabilization by the inductive effect falls off rapidly with increasing distance of the electronegative atom(s) from the site of the negative charge. Compare, for example, the pK_a values of alcohols substituted with fluorine on carbons 2 versus 3 versus 4. When fluorine atoms are more than two carbons away from the carbon bearing the —OH group, they have almost no effect on acidity.

$CF_3—CH_2—OH$	$CF_3—CH_2—CH_2—OH$	$CF_3—CH_2—CH_2—CH_2—OH$
2,2,2-Trifluoroethanol (pK_a 12.4)	**3,3,3-Trifluoro-1-propanol** (pK_a 14.6)	**4,4,4-Trifluoro-1-butanol** (pK_a 15.4)

We also see the operation of the inductive effect in the acidity of halogen-substituted carboxylic acids.

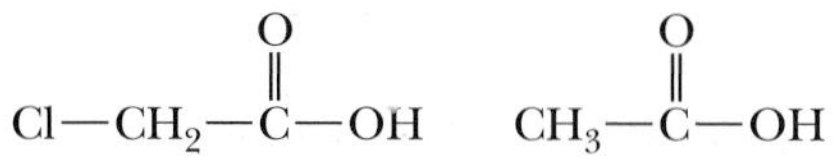

Chloroacetic acid pK_a 2.86 **Acetic acid** pK_a 4.76

The pK_a values of these two acids indicate that chloroacetic acid is approximately two orders of magnitude more acidic than acetic acid. In the case of chloroacetate anion, the negative charge is stabilized by electrostatic interaction between the partial negative charges on the oxygens and the partial positive charge on the carbon bearing the chlorine atom.

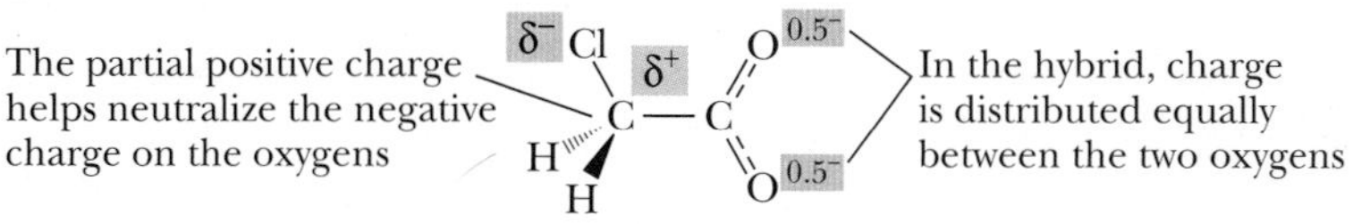
The partial positive charge helps neutralize the negative charge on the oxygens
δ⁻ Cl
δ⁺
C—C
H
H
O 0.5⁻
O 0.5⁻
In the hybrid, charge is distributed equally between the two oxygens

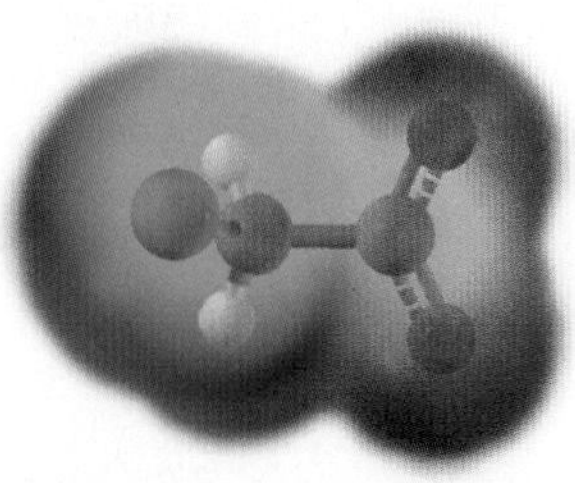
Elpot of chloroacetate ion

Watch a video explanation

As was the case with halogen substitution and the acidity of alcohols, the acid-enhancing effect of halogen substitution in carboxylic acids falls off rapidly with increasing distance between the point of substitution and the carboxyl group.

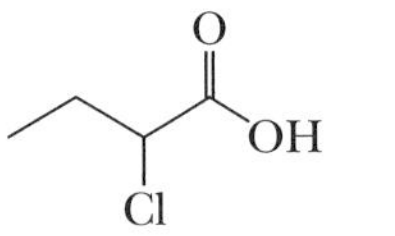

2-Chlorobutanoic acid
pK_a 2.83

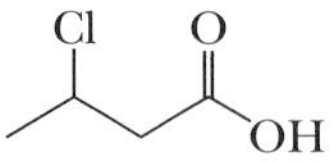

3-Chlorobutanoic acid
pK_a 3.98

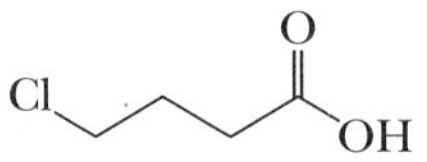

4-Chlorobutanoic acid
pK_a 4.52

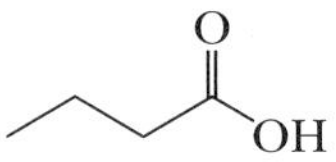

Butanoic acid
pK_a 4.82

E. Hybridization and the Percent *s* Character of the Atom Bearing the Negative Charge

To see the effect of hybridization, we consider the case of two or more conjugate base anions, each with the same charge and same element bearing the charge. The only difference is the hybridization of the atom bearing the negative charge. The acidity of a hydrogen bound to a carbon atom of an alkane, an alkene, and an alkyne is of special importance to us.

One of the major differences between the chemistry of alkynes and that of alkenes and alkanes is that a hydrogen bonded to a triply bonded carbon atom is sufficiently acidic that it can be removed by a strong base, such as sodium amide, $NaNH_2$, or sodium hydride, NaH. Table 4.3 gives pK_a values for an alkyne, alkene, and alkane. Also given for comparison are values for ammonia and water.

We account for the greater acidity of alkynes in the following way. The lone pair of electrons on a carbon anion lies in a hybrid orbital: an *sp* orbital for an alkyne anion, an sp^2 orbital for an alkene anion, and an sp^3 orbital for an alkane anion. An *sp* orbital has 50% *s* character; an sp^2 orbital, 33%; and an sp^3 orbital, 25% (Section 1.7). Electrons in an *s* orbital are lower in energy than those in a *p* orbital; that is, they are held more tightly to the nucleus. Therefore, the more *s* character in a hybrid orbital, the more electronegative the atom and the more acidic a hydrogen bonded to it (and the more stable the anion). Of the three types of compounds, the carbon in an alkyne (*sp* hybridized with 50% *s* character) is the most electronegative. Therefore, an alkyne anion is the most stable of the series and an alkyne is the strongest acid of the series. By similar reasoning, the alkane carbon (sp^3 hybridized with 25% *s* character) is the least electronegative and an alkane is the weakest acid of the series. An alkene anion, with 33% *s* character, is intermediate.

Table 4.3 Acidity of Alkanes, Alkenes, and Alkynes

Weak Acid		Conjugate Base	pK_a
Water	HO—H	HO^-	15.7
Alkyne	HC≡C—H	$HC{\equiv}C^-$	25
Ammonia	H_2N—H	H_2N^-	38
Alkene	CH_2=CH—H	$CH_2{=}CH^-$	44
Alkane	CH_3CH_2—H	$CH_3CH_2^-$	51

Increasing acidity (↑)

MCAT Practice: *Passage and Questions*

Acid-Base Equilibria

Many factors contribute to the acidity of organic compounds. Electronegativity, resonance, induction, hybridization, aromaticity, and atomic size, all play a role. In the following comparisons, you are asked to identify the factor(s) that would be most important to analyze when predicting relative acidity, and then to predict the trend in acidity and pK_a values. For each of the following pairs of compounds answer the following two multiple-choice questions.

(a)

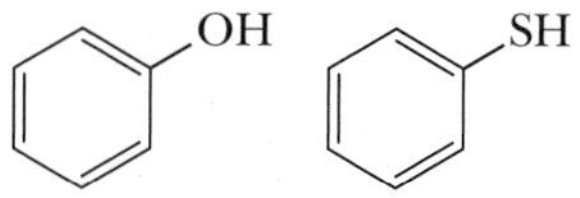

I II

(b)

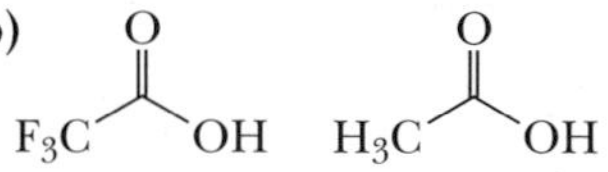

I II

(c)

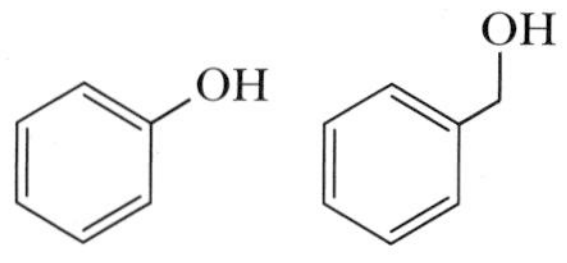

I II

(d)

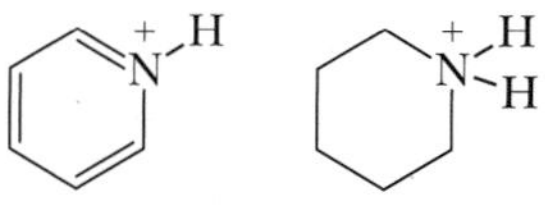

I II

(e)

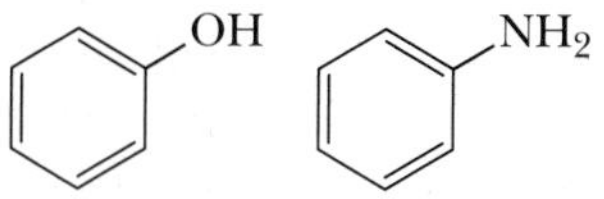

I II

Questions

1. What factor(s) are the most important to consider when predicting the relative acidity of the two compounds?

- **a.** Electronegativity of the atom possessing the hydrogen.
- **b.** Resonance stabilization of the anionic conjugate base.
- **c.** Inductive stabilization of the anionic conjugate base.
- **d.** Hybridization of the atom possessing the hydrogen.
- **e.** The atomic size of the atom possessing the hydrogen.

2. What is the relative trend in acidity and pK_a of the two compounds?

- **a.** Structure I is the most acidic, and Structure I has the highest pK_a.
- **b.** Structure I is the most acidic, and Structure I has the lowest pK_a.
- **c.** Structure II is the most acidic, and Structure II has the highest pK_a.
- **d.** Structure II is the most acidic, and Structure II has the lowest pK_a.

4.7 Lewis Acids and Bases

Gilbert N. Lewis, who proposed that covalent bonds are formed by sharing one or more pairs of electrons (Section 1.2A), further generalized the theory of acids and bases to include a group of substances not included in the Brønsted-Lowry concept. According to the Lewis definition, an acid is a species that can form a new covalent bond by accepting a pair of electrons; a base is a species that can form a new covalent bond by donating a pair of electrons. In the following general equation, the **Lewis acid**, A, accepts a pair of electrons in forming the new covalent bond and acquires a negative formal charge. The **Lewis base**, :B, donates the pair of electrons and acquires a positive formal charge.

Lewis acid
Any molecule or ion that can form a new covalent bond by accepting a pair of electrons.

Lewis base
Any molecule or ion that can form a new covalent bond by donating a pair of electrons.

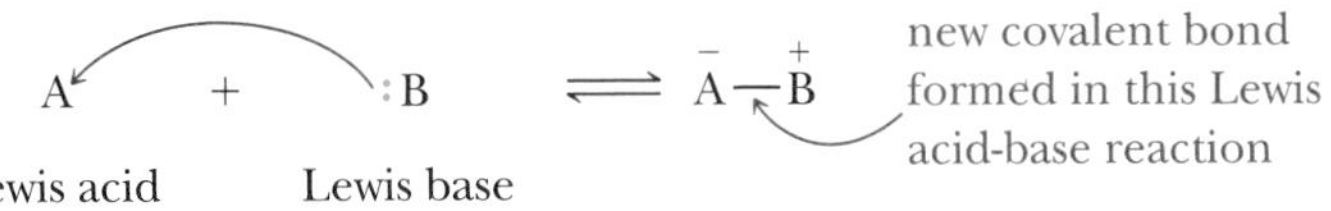

Lewis acid (electron pair acceptor) | Lewis base (electron pair donor)

Note that although we speak of a Lewis base as "donating" a pair of electrons, the term is not fully accurate. Donating in this case does not mean that the electron pair under consideration is removed completely from the valence shell of the base. Rather, donating means that the electron pair becomes shared with another atom to form a covalent bond. Charged species can also be Lewis acids and Lewis bases. In these cases, Lewis acids have a positive charge and Lewis bases have a negative charge. If both reactants are equally but oppositely charged, a Lewis acid-Lewis base reaction will result in the formation of a new covalent bond in which the product has no charge.

$$A^+ \quad + \quad :B^- \rightleftharpoons A{-}B$$

Lewis acid (electron pair acceptor) | Lewis base (electron pair donor) | New covalent bond formed in this Lewis acid-base reaction

An example of this type of a Lewis acid-base reaction is that of a carbocation (a Lewis acid) with bromide ion (a Lewis base). The *sec*-butyl cation, for example, reacts with bromide ion to form 2-bromobutane.

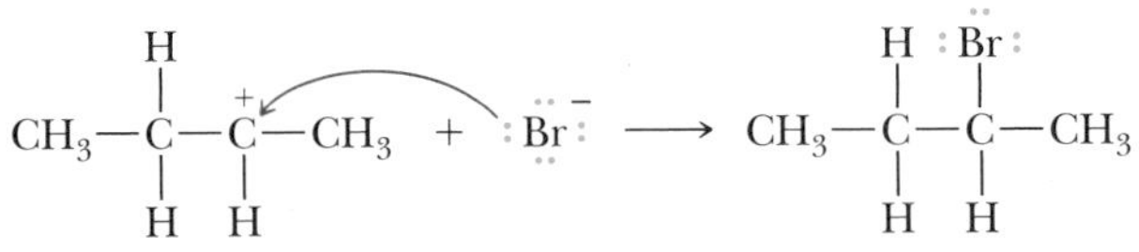

***sec*-Butyl cation** (a carbocation) | **Bromide ion** | **2-Bromobutane** (racemic)

The Lewis concept of acids and bases includes proton-transfer reactions; all Brønsted-Lowry bases (proton acceptors) are also Lewis bases, and all Brønsted-Lowry acids (proton donors) are also Lewis acids. The Lewis model, however, is more general in that it is not restricted to proton-transfer reactions.

Consider the reaction that occurs when boron trifluoride gas is dissolved in diethyl ether.

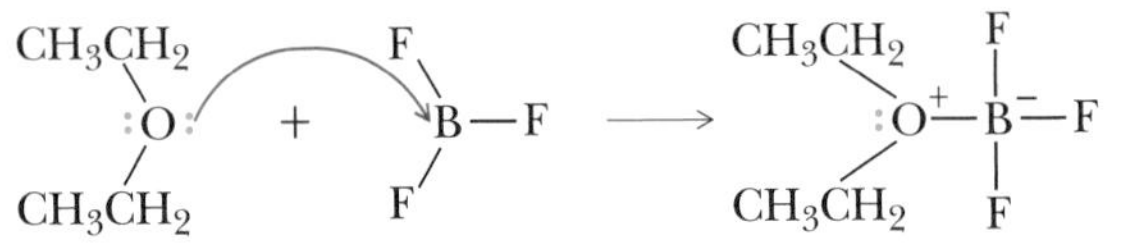

Diethyl ether (a Lewis base) | **Boron trifluoride** (a Lewis acid) | A BF_3-ether complex

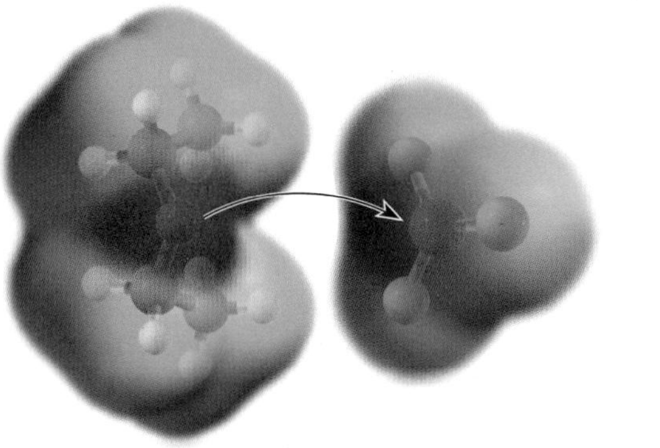

Elpot of diethyl ether | Elpot of BF_3 ▶

▶ In most stable molecules, all the atoms will have filled valence shells.

Boron, a Group 3A element, has three electrons in its valence shell, and after forming single bonds with three fluorine atoms to give BF_3, boron still has only six electrons in its valence shell. Because it has an empty orbital in its valence shell and can accept two electrons into it, boron trifluoride is electron deficient and, therefore, highly reactive as a Lewis acid. In forming the **O—B** bond, the oxygen atom of diethyl ether (a Lewis base) donates an electron pair and boron accepts the electron pair. The reaction between diethyl ether and boron trifluoride is classified as an acid-base reaction according to the Lewis model, but because there is no proton transfer involved, it is not classified as an acid-base reaction by the Brønsted-Lowry model. Said another way, all Brønsted-Lowry acids are **protic acids**; Lewis acids may be protic acids or **aprotic acids**.

Protic acid
An acid that is a proton donor in an acid-base reaction.

Aprotic acid
An acid that is not a proton donor; an acid that is an electron pair acceptor in a Lewis acid-base reaction.

As a final note, we make a tie to terminology that will be used extensively in future chapters. In our analyses of many organic chemical reactions, we will show that high-electron-density regions on molecules or ions react with low-electron-density regions of other molecules or ions, quite often resulting in the formation of a new covalent bond. We call the electron-rich species a **nucleophile** (nucleus-loving), meaning that it is seeking a region of low electron density (such as an atomic nucleus). We call the low-electron-density species an **electrophile** (electron-loving), meaning that it is seeking a region of high electron density. Therefore, nucleophiles are analogous to Lewis bases and electrophiles are analogous to Lewis acids. Although chemists use the terms interchangeably, nucleophile and electrophile are most commonly used in kinetics discussions, while Lewis acid-base terminology is mostly used in thermodynamics discussions. We will talk about these connections in more detail throughout the book and in "Things You Should Know II" before Chapter 6.

Example 4.8 Lewis Acid-Base Reactions

Write an equation for the reaction between each Lewis acid-base pair, showing electron flow by means of curved arrows.

(a) $BF_3 + NH_3 \longrightarrow$ **(b)** $(CH_3)_2CH^+ + Cl^- \longrightarrow$

Solution

(a) BF_3 has an empty orbital in the valence shell of boron and is the Lewis acid. NH_3 has an unshared pair of electrons in the valence shell of nitrogen and is the Lewis base. In this example, each of these atoms takes on a formal charge; the resulting structure, however, has no net charge.

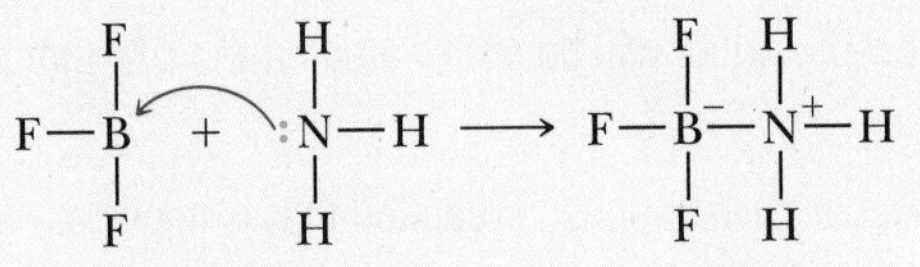

Lewis acid Lewis base

(b) The trivalent carbon atom in the isopropyl cation has an empty orbital in its valence shell and is, therefore, the Lewis acid. Chloride ion is the Lewis base.

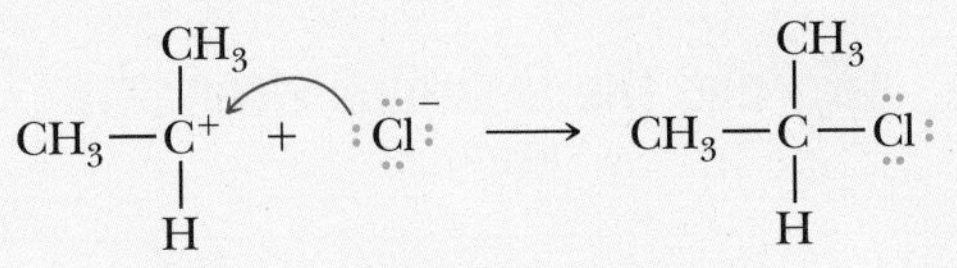

Lewis acid Lewis base

Problem 4.8

Write an equation for the reaction between each Lewis acid-base pair, showing electron flow by means of curved arrows.

(a) $(CH_3CH_2)_3B + OH^- \longrightarrow$ **(b)** $CH_3Cl + AlCl_3 \longrightarrow$

Study Guide

4.1 Arrhenius Acids and Bases

By the Arrhenius definitions, an acid is a substance that dissolves in water to produce H^+ (H_3O^+) ions and a base is a substance that dissolves in water to produce OH^- ions.

4.2 Brønsted-Lowry Acids and Bases

- A **Brønsted-Lowry acid** is a proton donor, and a **Brønsted-Lowry base** is a proton acceptor.

KEY REACTIONS

1. **Proton-Transfer Reaction (Section 4.2)** A proton-transfer reaction involves transfer of a proton from a proton donor (a Brønsted-Lowry acid) to a proton acceptor (a Brønsted-Lowry base).

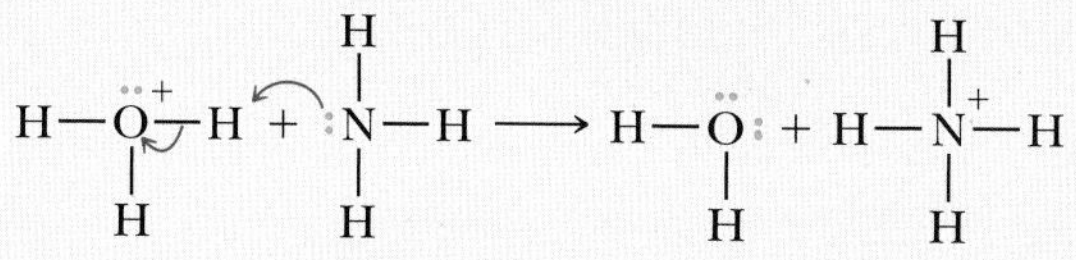

- Any pair of molecules or ions that can be interconverted by transfer of a proton is called a **conjugate acid-base pair**.
  - *Neutralization of an acid by a base is a proton-transfer reaction in which the acid is transformed into its* ***conjugate base*** *and the base is transformed into its* ***conjugate acid****.*
- Brønsted-Lowry bases with two or more receptor sites will be protonated on the site that gives the more delocalized charge.
- π electrons can act as Brønsted-Lowry bases to give cations, a reaction that is important in the chemistry of alkenes.

P **4.1–4.4, 4.9–4.13, 4.29, 4.30, 4.38, 4.40, 4.45, 4.48**

Problem 4.2

4.3 Acid Dissociation Constants, pK_a, and the Relative Strengths of Acids and Bases

- A strong acid or strong base is one that is completely ionized in water, and a weak acid or weak base is one that is only partially ionized in water.
  - *The* ***acid dissociation constant*** K_a *is a quantitative measure of acid strength.*
  - *Acid strengths are generally reported as* **pK_a** *values, which are equal to* $-\log_{10}K_a$.
  - *Stronger acids have larger* K_a *values and therefore smaller* pK_a *values.*

– *Among the most common weak organic acids are carboxylic acids, compounds that contain the* —COOH *(carboxyl) group.*

– *The value of* K_a *for acetic acid, a representative carboxylic acid, is* 1.74×10^{-5} M; *the value of* pK_a *for acetic acid is 4.76.*

P **4.5, 4.14–4.17, 4.32–4.34**

4.4 The Position of Equilibrium in Acid-Base Reactions

- The **position of equilibrium** in an acid-base reaction favors reaction of the stronger acid (lower pK_a value) with the stronger base to form the weaker acid (higher pK_a value) and the weaker base.
  - *For quantitative calculations, the* pK_{eq} *for an acid-base reaction equals the difference of the* pK_a *for the stronger acid* $(pK_{a(HA)})$ *and the protonated form of the stronger base* $(pK_{a(BH^+)})$
  - $pK_{eq} = pK_{a(HA)} - pK_{a(BH^+)}$
  - K_{eq} *for an acid-base reaction is equal to* $10^{(-pK_{eq})}$.

KEY REACTIONS

2. Position of Equilibrium in an Acid-Base Reaction (Section 4.4) The stronger acid reacts with the stronger base to give a weaker acid and a weaker base. K_{eq} for this equilibrium can be calculated from pK_a values for the two acids.

$$CH_3\overset{\overset{O}{\|}}{C}OH + CN^- \rightleftharpoons CH_3\overset{\overset{O}{\|}}{C}O^- + HCN \qquad pK_{eq} = 4.76 - 9.31 = -4.55$$

CH_3COOH: pK_a 4.76 (stronger acid); HCN: pK_a 9.31 (weaker acid); $K_{eq} = 3.55 \times 10^4$

- – *A good rule of thumb is that an acid will be substantially deprotonated if its* pK_a *is two or more units lower than the pH of an aqueous solution. A base will be substantially protonated if the* pK_a *of its conjugate acid is two or more units higher than the pH of an aqueous solution.*
  - – *Carboxylic acids and phosphodiesters are generally deprotonated and anionic at pH 7–8, while amines and guanidinium groups are generally found in their protonated and positively charged forms.*

P **4.6, 4.18–4.25, 4.41, 4.44, 4.46**

▶ **Problem 4.2**

4.5 Thermochemistry and Mechanisms of Acid-Base Reactions

- A **reaction mechanism** describes in detail how a reaction occurs.
  - *Most chemical reactions occur via collisions.*
  - *The reactants must collide with the proper orientation and enough energy to reach the* ***transition state*** *and proceed to products.*
  - *A reaction can be described by a* ***reaction coordinate diagram****, which is a plot of Gibbs free energy versus the progress of the reaction.*

- **Thermochemistry** is the study of energy of the entire system at each moment of a reaction.
  - ***Thermodynamics*** *is the study of the relative energies between any two states in wells on a reaction coordinate diagram.*
    - *If products are more stable than reactants, the overall thermodynamics are favorable for reaction and the step is* ***exergonic****.*
    - *If products are less stable than reactants, the overall thermodynamics are not favorable for reaction and the step is* ***endergonic****.*
  - **Kinetics** *is the study of rates of chemical reactions.*
    - *The lower the* ***Gibbs free energy of activation*** *(energy difference between the transition state and starting state), the faster the rate of the corresponding reaction step and vice versa.*
  - ***Enthalpy*** *is the energy contained within chemical bonds and solvation.*
    - *If bonds formed in a product are stronger than those broken in the starting materials, heat is given off and the reaction is* ***exothermic****. If bonds formed in a product are weaker than those broken in the starting materials, heat is absorbed and the reaction is* ***endothermic****.*
  - ***Entropy*** *measures chaos versus order, and chaos is favorable.*
- For acid-base reactions to occur, the acid and base must collide with a geometry in which the proton to be transferred is between the proton-donor and proton-acceptor atoms, more or less in a linear geometry.

P **4.7, 4.18–4.23**

4.6 Molecular Structure and Acidity

- The acidity of an acid is determined by the stability of the anion formed on deprotonation, according to the rule that more acidic molecules form more stable anions upon deprotonation. Factors that influence the stability of an anion are:
  - *Electronegativity of the atom bearing the negative charge because more electronegative atoms are more stable as anions.*
  - *Size of the atom bearing the negative charge because larger atoms can more easily accommodate a negative charge (it is spread over a larger area).*
  - *Delocalization of charge in the anion, usually described by resonance contributing structures because greater delocalization of charge is stabilizing.*
  - *The inductive effect because adjacent electronegative atoms such as the halogens will attract and therefore delocalize a nearby negative charge.*
  - *The hybridization of the atom bearing the negative charge because the greater the percentage of s character in a hybrid orbital, the more stable the anion.*

P **4.35, 4.36, 4.39, 4.42, 4.43, 4.47**

▶ **The Inductive Effect**

4.7 Lewis Acids and Bases

- A **Lewis acid** is a species that can form a new covalent bond by accepting a pair of electrons; a **Lewis base** is a species that can form a new covalent bond by donating a pair of electrons.

– *All Brønsted-Lowry acids (proton donors) are also Lewis acids, and all Brønsted-Lowry bases (proton acceptors) are also Lewis bases. But the Lewis acid-base model is far more general in that it applies to reactions beyond just proton transfers.*

P 4.8, 4.26–4.29, 4.31, 4.37, 4.40

KEY REACTIONS

3. Lewis Acid-Base Reaction (Section 4.7) A Lewis acid-base reaction involves sharing an electron pair between an electron pair donor (a Lewis base) and an electron pair acceptor (a Lewis acid).

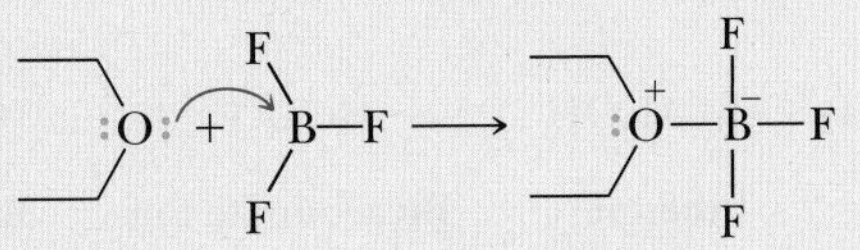

Problems

Red numbers indicate applied problems.

4.9 For each conjugate acid-base pair, identify the first species as an acid or a base and the second species as its conjugate acid or base. In addition, draw Lewis structures for each species, showing all valence electrons and any formal charge.

(a) $HCOOH$ $HCOO^-$ **(b)** NH_4^+ NH_3 **(c)** $CH_3CH_2O^-$ CH_3CH_2OH
(d) HCO_3^- CO_3^{2-} **(e)** $H_2PO_4^-$ HPO_4^{2-} **(f)** CH_3CH_3 $CH_3CH_2^-$
(g) CH_3S^- CH_3SH

4.10 Complete a net ionic equation for each proton-transfer reaction using curved arrows to show the flow of electron pairs in each reaction. In addition, write Lewis structures for all starting materials and products. Label the original acid and its conjugate base; label the original base and its conjugate acid. If you are uncertain about which substance in each equation is the proton donor, refer to Table 4.1 for the relative strengths of proton acids.

(a) $NH_3 + HCl \rightleftharpoons$ **(b)** $CH_3CH_2O^- + HCl \rightleftharpoons$
(c) $HCO_3^- + OH^- \rightleftharpoons$ **(d)** $CH_3COO^- + NH_4^+ \rightleftharpoons$

4.11 Complete a net ionic equation for each proton-transfer reaction using curved arrows to show the flow of electron pairs in each reaction. Label the original acid and its conjugate base; then label the original base and its conjugate acid.

(a) $NH_4^+ + OH^- \rightleftharpoons$ **(b)** $CH_3COO^- + CH_3NH_3^+ \rightleftharpoons$
(c) $CH_3CH_2O^- + NH_4^+ \rightleftharpoons$ **(d)** $CH_3NH_3^+ + OH^- \rightleftharpoons$

4.12 Each molecule or ion can function as a base. Write a structural formula of the conjugate acid formed by reaction of each with HCl.

(a) CH_3CH_2OH **(b)** $H\overset{O}{\overset{\|}{C}}H$
(c) $(CH_3)_2NH$ **(d)** HCO_3^-

4.13 In acetic acid, CH_3COOH, the OH hydrogen is more acidic than the CH_3 hydrogens. Explain.

Quantitative Measure of Acid and Base Strength

4.14 Which has the larger numerical value?

(a) The pK_a of a strong acid or the pK_a of a weak acid

(b) The K_a of a strong acid or the K_a of a weak acid

4.15 In each pair, select the stronger acid.

(a) Pyruvic acid (pK_a 2.49) or lactic acid (pK_a 3.08)
(b) Citric acid (pK_{a1} 3.08) or phosphoric acid (pK_{a1} 2.10)

4.16 Arrange the compounds in each set in order of increasing acid strength. Consult Table 4.1 for pK_a values of each acid.

(a) CH_3CH_2OH — **Ethanol**; $HOC(=O)O^-$ — **Bicarbonate ion**; $C_6H_5C(=O)OH$ — **Benzoic acid**

(b) $HOC(=O)OH$ — **Carbonic acid**; $CH_3C(=O)OH$ — **Acetic acid**; HCl — **Hydrogen chloride**

4.17 Arrange the compounds in each set in order of increasing base strength. Consult Table 4.1 for pK_a values of the conjugate acid of each base.

(a) NH_3 $HOC(=O)O^-$ $CH_3CH_2O^-$ **(b)** OH^- $HOC(=O)O^-$ $CH_3C(=O)O^-$

(c) H_2O NH_3 $CH_3C(=O)O^-$ **(d)** NH_2^- $CH_3C(=O)O^-$ OH^-

Thermodynamics, Kinetics, and Reaction Coordinate Diagrams

4.18 If the ΔG^0 for a reaction is -4.5 kcal/mol at 298 K, what is the K_{eq} for this reaction? What is the change in entropy of this reaction if $\Delta H^0 = -3.2$ kcal/mol?

4.19 Calculate the K_{eq} for the following reactions from the pK_a's given. State whether the reaction is exergonic or endergonic.

(a)

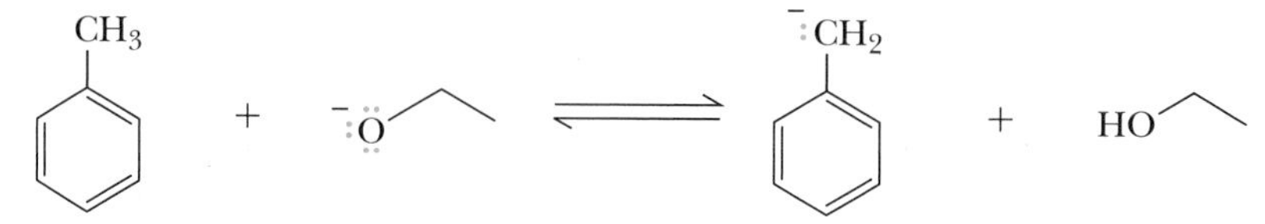

$pK_a = 41.2$ $pK_a = 15.9$

(b)

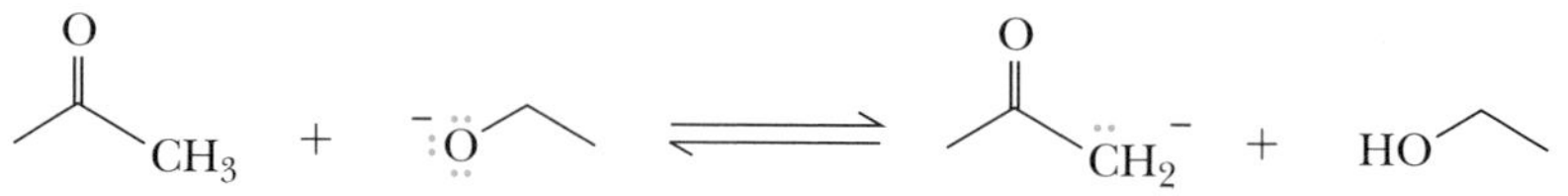

$pK_a = 20.2$ $pK_a = 15.9$

(c)

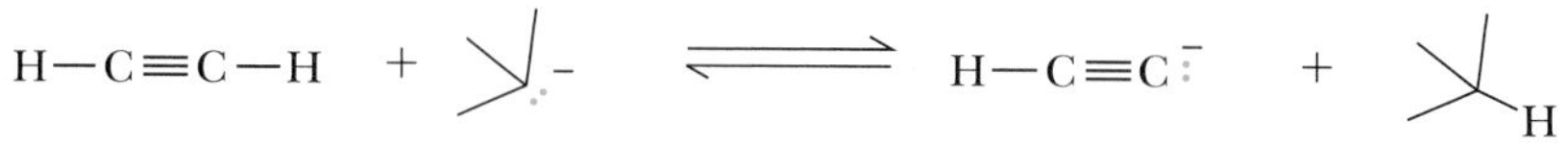

$pK_a = 24$ $pK_a = 71$

4.20 Answer true or false to the following statements about energy diagrams and reactions.

(a) A reaction coordinate diagram is used to visualize the change in the internal energy of chemical structures that occurs during chemical reactions.
(b) Thermodynamics is the study of the energies of structures that are represented by wells on reaction coordinate diagrams.

(c) Kinetics is the study of the rates of chemical reactions.
(d) One part of a reaction mechanism would be the understanding of which bonds break and form during a reaction.
(e) Thermal reactions occur via collisions between molecules, and the more energy in those collisions, the greater the rate of the reactions.
(f) The enthalpy of a reaction is the sole determinant of whether it will occur.
(g) An exergonic reaction will always occur during the life span of the standard human being.

4.21 Answer true or false to the following statements about the mechanism of acid-base reactions.

(a) The acid and base must encounter each other by a collision in order for the proton to transfer.
(b) All collisions between acids and bases result in proton transfer.
(c) During an acid-base reaction the lone pair on the base fills the A-H antibonding sigma orbital.

4.22 In each of the following three reaction coordinate diagrams, state:

(a) Whether the reaction is exothermic or endothermic.
(b) Whether the reaction is the slowest, the fastest, or intermediate in rate.
(c) If all three reactions have the same entropy change between the reactant and product, which reaction has the largest favorable ΔG^0.

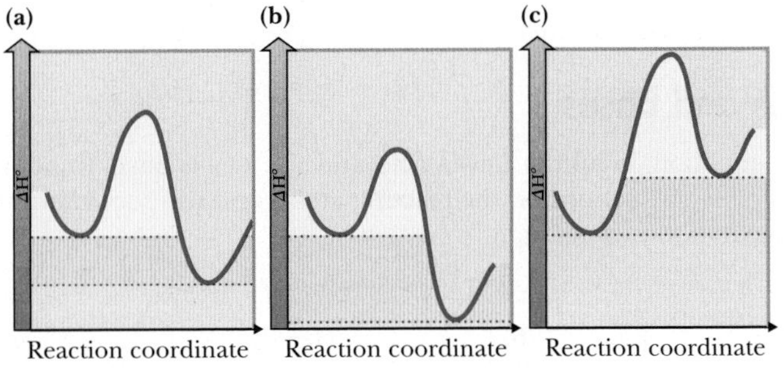

4.23 The acid-base chemistry reaction of barium hydroxide ($Ba(OH)_2$) with ammonium thiocyanate (NH_4SCN) in water creates barium thiocyanate, ammonia, and water. The reaction is highly favorable, but is also so endothermic that the solutions cools to such an extent that a layer of frost forms on the reaction vessel. Explain how an endothermic reaction can be favorable.

Position of Equilibrium in Acid-Base Reactions

4.24 Unless under pressure, carbonic acid (H_2CO_3) in aqueous solution breaks down into carbon dioxide and water and carbon dioxide is evolved as bubbles of gas. Write an equation for the conversion of carbonic acid to carbon dioxide and water.

4.25 Will carbon dioxide be evolved when sodium bicarbonate is added to an aqueous solution of each compound? Explain.

(a) Sulfuric acid **(b)** Ethanol **(c)** Ammonium chloride

4.26 Acetic acid, CH_3COOH, is a weak organic acid, pK_a 4.76. Write an equation for the equilibrium reaction of acetic acid with each base. Which equilibria lie considerably toward the left? Which lie considerably toward the right?

(a) $NaHCO_3$ **(b)** NH_3 **(c)** H_2O **(d)** NaOH

4.27 Benzoic acid, C_6H_5COOH (pK_a 4.19), is only slightly soluble in water, but its sodium salt, $C_6H_5COO^-\ Na^+$, is quite soluble in water. In which solution(s) will benzoic acid dissolve?

(a) Aqueous NaOH **(b)** Aqueous $NaHCO_3$ **(c)** Aqueous Na_2CO_3

4.28 4-Methylphenol, $CH_3C_6H_4OH$ (pK_a 10.26), is only slightly soluble in water, but its sodium salt, $CH_3C_6H_4O^- Na^+$, is quite soluble in water. In which solution(s) will 4-methylphenol dissolve?

(a) Aqueous NaOH **(b)** Aqueous $NaHCO_3$ **(c)** Aqueous Na_2CO_3

4.29 One way to determine the predominant species at equilibrium for an acid-base reaction is to say that the reaction arrow points to the acid with the higher value of pK_a. For example,

$$\underset{pK_a\ 9.24}{NH_4^+} + H_2O \longleftarrow NH_3 + \underset{pK_a\ -1.74}{H_3O^+}$$

$$\underset{pK_a\ 9.24}{NH_4^+} + OH^- \longrightarrow NH_3 + \underset{pK_a\ 15.7}{H_2O}$$

Explain why this rule works.

4.30 Will acetylene react with sodium hydride according to the following equation to form a salt and hydrogen, H_2? Using pK_a values given in Table 4.1, calculate K_{eq} for this equilibrium.

$$\underset{\textbf{Acetylene}}{HC{\equiv}CH} + \underset{\textbf{Sodium hydride}}{Na^+H^-} \rightleftharpoons \underset{\textbf{Sodium acetylide}}{HC{\equiv}C^-Na^+} + \underset{\textbf{Hydrogen}}{H_2}$$

4.31 Using pK_a values given in Table 4.1, predict the position of equilibrium in this acid-base reaction and calculate its K_{eq}.

$$H_3PO_4 + CH_3CH_2OH \rightleftharpoons H_2PO_4^- + CH_3CH_2OH_2^+$$

Lewis Acids and Bases

4.32 For each equation, label the Lewis acid and the Lewis base. In addition, show all unshared pairs of electrons on the reacting atoms and use curved arrows to show the flow of electrons in each reaction.

(a) $F^- + BF_3 \longrightarrow BF_4^-$ **(b)**

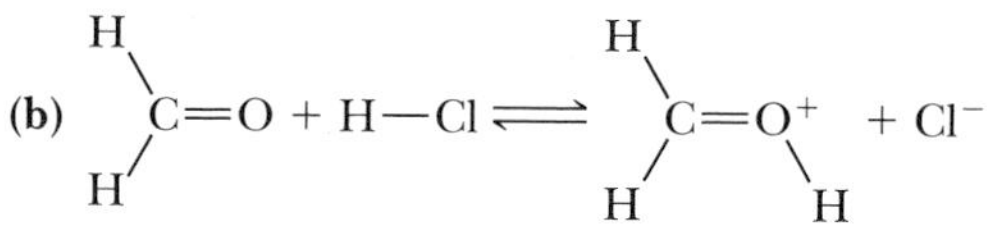

4.33 Complete the equation for the reaction between each Lewis acid-base pair. In each equation, label which starting material is the Lewis acid and which is the Lewis base; use curved arrows to show the flow of electrons in each reaction. In doing this problem, it is essential that you show valence electrons for all atoms participating in each reaction.

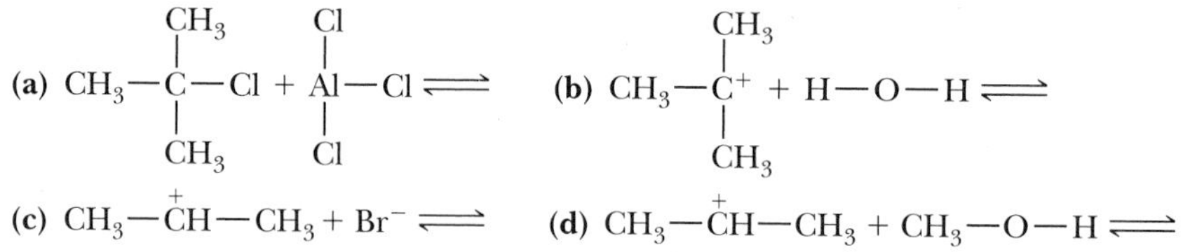

4.34 Each of these reactions can be written as a Lewis acid-Lewis base reaction. Label the Lewis acid and the Lewis base; use curved arrows to show the flow of electrons in each reaction. In doing this problem, it is essential that you show valence electrons for all atoms participating in each reaction.

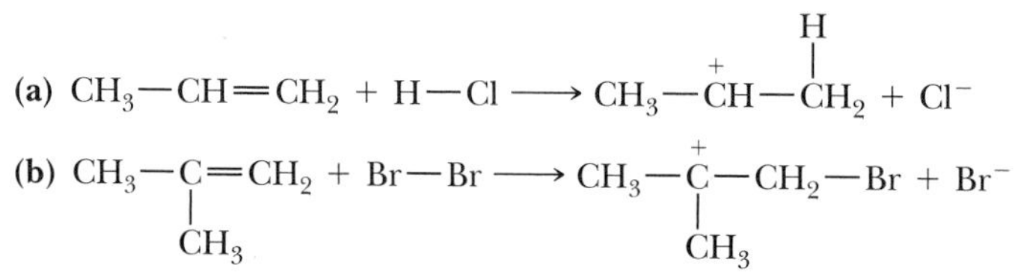

Additional Problems

4.35 The *sec*-butyl cation can react as both a Brønsted-Lowry acid (a proton donor) and a Lewis acid (an electron pair acceptor) in the presence of a water–sulfuric acid mixture. In each case, however, the product is different. The two reactions are as follows:

$$(1)\ CH_3-\overset{+}{C}H-CH_2-CH_3 + H_2O \rightleftharpoons CH_3-\underset{}{\overset{H-\overset{+}{O}-H}{\overset{|}{C}H}}-CH_2-CH_3$$

***sec*-Butyl cation**

$$(2)\ CH_3-\overset{+}{C}H-CH_2-CH_3 + H_2O \rightleftharpoons CH_3-CH{=}CH-CH_3 + H_3O^+$$

***sec*-Butyl cation**

(a) In which reaction(s) does this cation react as a Lewis acid? In which reaction(s) does it react as a Brønsted-Lowry acid?

(b) Write Lewis structures for reactants and products and show by the use of curved arrows how each reaction occurs.

4.36 Write equations for the reaction of each compound with H_2SO_4, a strong protic acid.

(a) CH_3OCH_3 **(b)** $CH_3CH_2SCH_2CH_3$ **(c)** $CH_3CH_2NHCH_2CH_3$

(d) $CH_3\overset{CH_3}{\overset{|}{N}}CH_3$ **(e)** $CH_3\overset{O}{\overset{\|}{C}}CH_3$ **(f)** $CH_3\overset{O}{\overset{\|}{C}}OCH_3$

4.37 Write equations for the reaction of each compound in Problem 4.36 with BF_3, a Lewis acid.

4.38 Label the most acidic hydrogen in each molecule and justify your choice by using appropriate pK_a values.

(a) $HOCH_2CH_2NH_2$ **(b)** $HSCH_2CH_2NH_2$ **(c)** $HOCH_2CH_2C{\equiv}CH$

(d) $HO\overset{O}{\overset{\|}{C}}CH_2CH_2SH$ **(e)** $CH_3\overset{HO}{\overset{|}{C}H}\overset{O}{\overset{\|}{C}}OH$ **(f)** $H_3\overset{+}{N}CH_2CH_2\overset{O}{\overset{\|}{C}}OH$

(g) $H_3\overset{+}{N}CH_2CH_2\overset{O}{\overset{\|}{C}}O^-$ **(h)** $HSCH_2CH_2OH$

4.39 Explain why the hydronium ion, H_3O^+, is the strongest acid that can exist in aqueous solution. What is the strongest base that can exist in aqueous solution?

4.40 What is the strongest base that can exist in liquid ammonia as a solvent?

4.41 For each pair of molecules or ions, select the stronger base and write its Lewis structure.

(a) CH_3S^- or CH_3O^- **(b)** CH_3NH^- or CH_3O^-
(c) CH_3COO^- or OH^- **(d)** $CH_3CH_2O^-$ or H^-
(e) NH_3 or OH^- **(f)** NH_3 or H_2O
(g) CH_3COO^- or HCO_3^- **(h)** HSO_4^- or OH^-
(i) OH^- or Br^-

4.42 Account for the fact that nitroacetic acid, O_2NCH_2COOH (pK_a 1.68), is a considerably stronger acid than acetic acid, CH_3COOH (pK_a 4.76).

4.43 Sodium hydride, NaH, is available commercially as a gray-white powder. It melts at 800°C with decomposition. It reacts explosively with water and ignites spontaneously upon standing in moist air.

(a) Write a Lewis structure for the hydride ion and for sodium hydride. Is your Lewis structure consistent with the fact that this compound is a high-melting solid? Explain.

(b) When sodium hydride is added very slowly to water, it dissolves with the evolution of a gas. The resulting solution is basic to litmus. What is the gas evolved? Why has the solution become basic?

(c) Write an equation for the reaction between sodium hydride and 1-butyne, $CH_3CH_2C{\equiv}CH$. Use curved arrows to show the flow of electrons in this reaction.

4.44 Methyl isocyanate, $CH_3{-}N{=}C{=}O$, is used in the industrial synthesis of a type of pesticide and herbicide known as a carbamate. As a historical note, an industrial accident in Bhopal, India, in 1984 resulted in leakage of an unknown quantity of this chemical into the air. An estimated 200,000 people were exposed to its vapors, and over 2000 of these people died.

(a) Write a Lewis structure for methyl isocyanate and predict its bond angles. What is the hybridization of its carbonyl carbon? Of its nitrogen atom?

(b) Methyl isocyanate reacts with strong acids, such as sulfuric acid, to form a cation. Will this molecule undergo protonation more readily on its oxygen or nitrogen atom? In considering contributing structures to each hybrid, do not consider structures in which more than one atom has an incomplete octet.

4.45 Offer an explanation for the following observations.

(a) H_3O^+ is a stronger acid than NH_4^+.

(b) Nitric acid, HNO_3, is a stronger acid than nitrous acid, HNO_2.

(c) Ethanol and water have approximately the same acidity.

(d) Trifluoroacetic acid, CF_3COOH, is a stronger acid than trichloroacetic acid, CCl_3COOH.

Looking Ahead

4.46 Following is a structural formula for the *tert*-butyl cation. (We discuss the formation, stability, and reactions of cations such as this one in Chapter 6.)

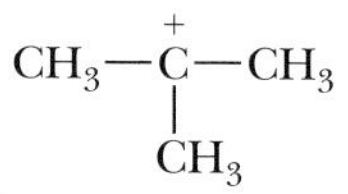
CH3—C—CH3
CH3

***tert*-Butyl cation**
(a carbocation)

(a) Predict all C—C—C bond angles in this cation.

(b) What is the hybridization of the carbon bearing the positive charge?

(c) Write a balanced equation to show its reaction as a Lewis acid with water.

(d) Write a balanced equation to show its reaction as a Brønsted-Lowry acid with water.

4.47 Alcohols (Chapter 10) are weak organic acids, pK_a 15–18. The pK_a of ethanol, CH_3CH_2OH, is 15.9. Write equations for the equilibrium reactions of ethanol with each base. Which equilibria lie considerably toward the right? Which lie considerably toward the left?

(a) $NaHCO_3$ **(b)** $NaOH$ **(c)** $NaNH_2$ **(d)** NH_3

4.48 As we shall see in Chapter 19, hydrogens on a carbon adjacent to a carbonyl group are far more acidic than those not adjacent to a carbonyl group. The anion derived from acetone, for example, is more stable than is the anion derived from ethane. Account for the greater stability of the anion from acetone.

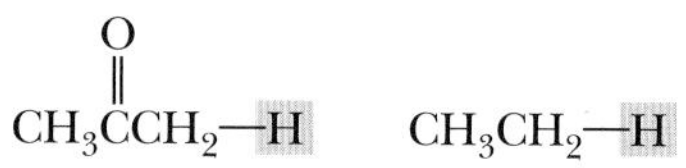
O
CH3CCH2—H
CH3CH2—H

Acetone pK_a 20.2 **Ethane** pK_a 51

4.49 2,4-Pentanedione is a considerably stronger acid than is acetone (Chapter 19). Write a structural formula for the conjugate base of each acid and account for the greater stability of the conjugate base from 2,4-pentanedione.

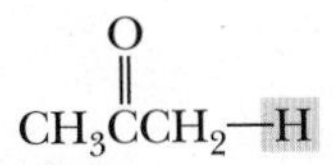

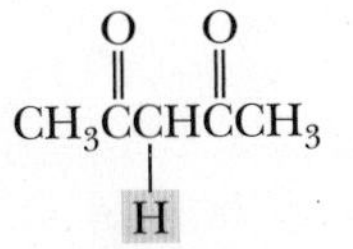

Acetone pK_a 20.2 **2,4-Pentanedione** pK_a 9

4.50 Write an equation for the acid-base reaction between 2,4-pentanedione and sodium ethoxide and calculate its equilibrium constant, K_{eq}. The pK_a of 2,4-pentanedione is 9; that of ethanol is 15.9.

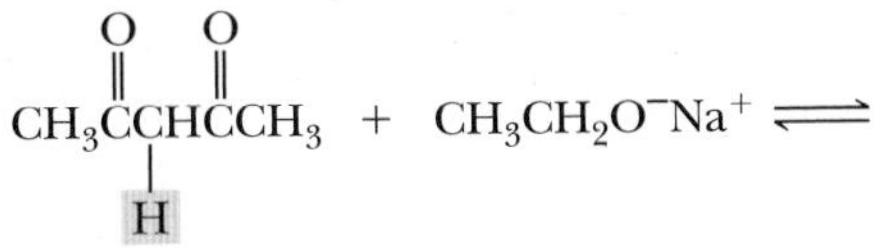

2,4-Pentanedione **Sodium ethoxide**

4.51 An ester is a derivative of a carboxylic acid in which the hydrogen of the carboxyl group is replaced by an alkyl group (Section 1.3E). Draw a structural formula of methyl acetate, which is derived from acetic acid by replacement of the H of its —OH group by a methyl group. Determine whether proton transfer to this compound from HCl occurs preferentially on the oxygen of the C=O group or on the oxygen of the OCH_3 group.

4.52 Alanine is one of the 20 amino acids (it contains both an amino and a carboxyl group) found in proteins (Chapter 27). Is alanine better represented by structural formula A or B? Explain.

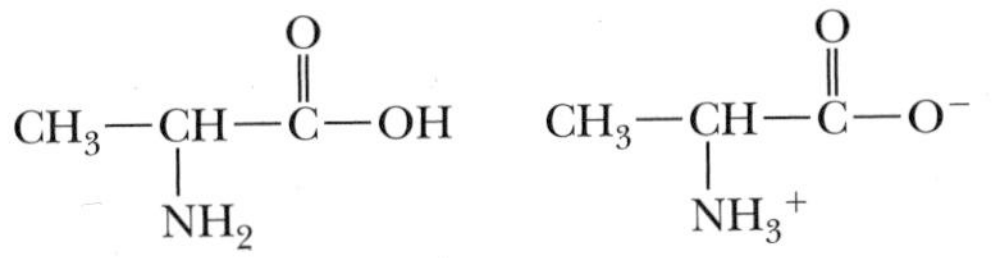

(A) (B)

4.53 Glutamic acid is another of the amino acids found in proteins (Chapter 27). Glutamic acid has two carboxyl groups, one with pK_a 2.10 and the other with pK_a 4.07.

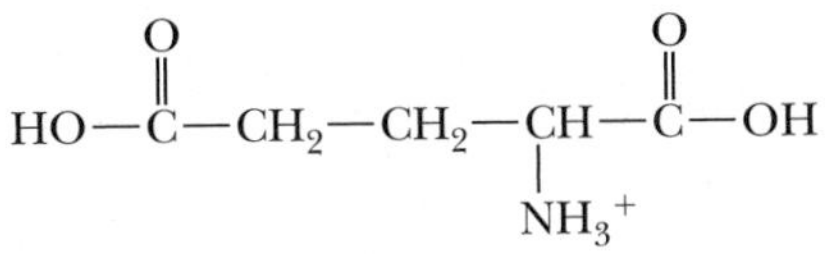

Glutamic acid

(a) Which carboxyl group has which pK_a?
(b) Account for the fact that one carboxyl group is a considerably stronger acid than the other carboxyl group.

4.54 Following is a structural formula for imidazole, a building block of the essential amino acid histidine (Chapter 27). It is also a building block of histamine, a compound all too familiar to people with allergies and takers of antihistamines. When imidazole is dissolved in water, proton transfer to it gives a cation. Is this cation better represented by structure A or B? Explain.

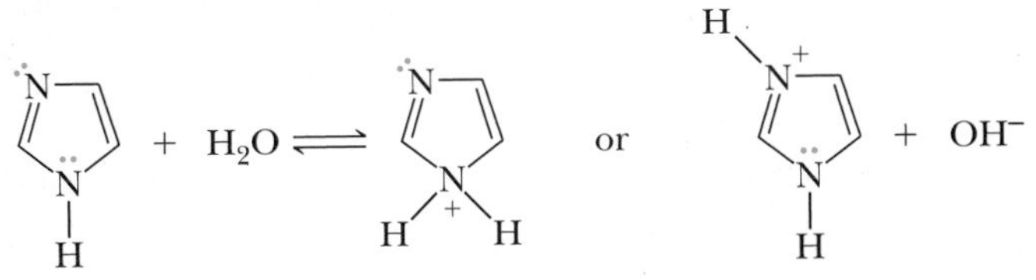

Imidazole (A) (B)

5

Alkenes: Bonding, Nomenclature, and Properties

Outline

5.1 Structure of Alkenes
HOW TO: Calculate the Index of Hydrogen Deficiency
5.2 Nomenclature of Alkenes
5.3 Physical Properties of Alkenes
5.4 Naturally Occurring Alkenes—Terpene Hydrocarbons

Haze in the Blue Ridge Mountains. The aerosolization of hydrocarbons emitted by trees and other plants causes light to scatter and appear as haze. Many naturally occurring hydrocarbons are formed from isoprene units and are oxidized by light (see Section 5.4).
(© Zack Frank/ Shutterstock.com)

An **unsaturated hydrocarbon** contains one or more carbon-carbon double or triple bonds. The term *unsaturation* indicates that fewer hydrogens are bonded to carbon than in an alkane, C_nH_{2n+2}. The three most important classes of unsaturated hydrocarbons are alkenes, alkynes, and arenes. **Alkenes** contain a carbon-carbon double bond and; with one double bond and no rings, have the general formula C_nH_{2n}. Alkynes contain a carbon-carbon triple bond and, with one triple bond and no rings, have the general formula C_nH_{2n-2}. The simplest alkene is ethylene, and the simplest alkyne is acetylene.

Unsaturated hydrocarbon

A hydrocarbon containing one or more carbon-carbon double or triple bonds. The three most important classes of unsaturated hydrocarbons are alkenes, alkynes, and arenes.

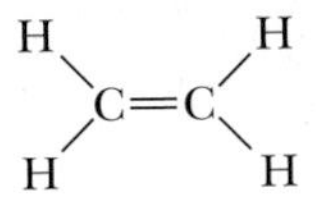

Ethylene
(an alkene)

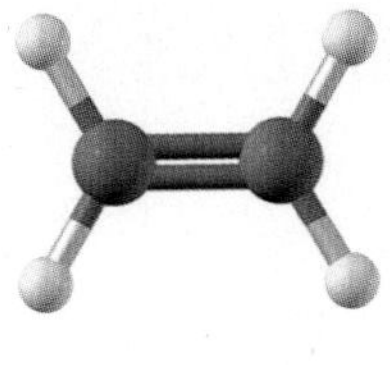

Side view

End view

H—C≡C—H

Acetylene
(an alkyne)

Side view

End view

In this chapter, we study the structure, nomenclature, and physical properties of alkenes. Alkynes are discussed separately in Chapter 7.

Arenes are a third class of unsaturated hydrocarbons. The Lewis structure of benzene, the simplest arene, is shown below.

Arene

A term used to classify benzene and its derivatives.

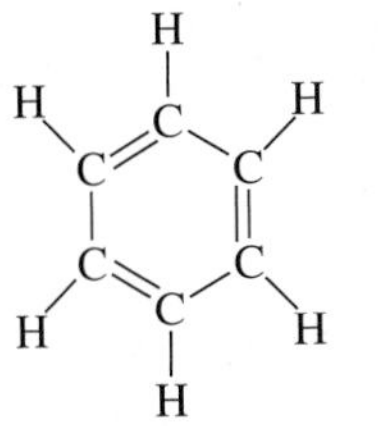

Benzene
(an arene)

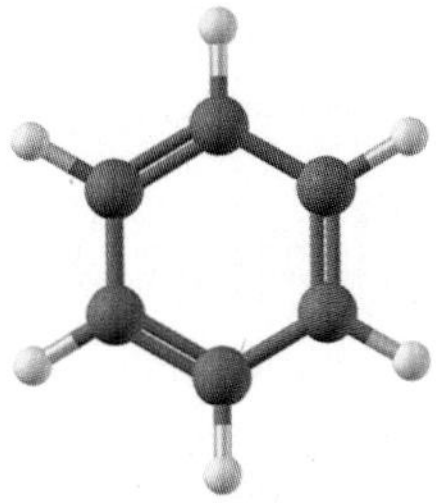

Top view

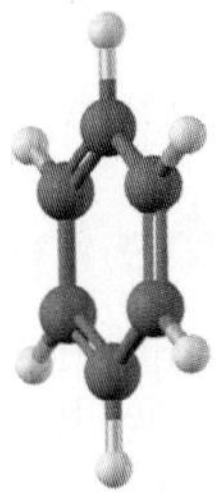

Side view

Just as a group derived by removal of an H from an alkane is called an alkyl group and is given the symbol R— (Section 2.3A), a group derived by removal of an H from an arene is called an **aryl group** and is given the symbol **Ar—**.

When the aryl group substituent on a parent chain is a benzene ring, it is given the special name **phenyl group**. Although ethane becomes ethyl in its substituent form, the derivative name for benzene stems from "phene," a now-obsolete term for benzene. Throughout this text, we represent benzene by a hexagon with three inscribed double bonds. It is also common to represent it by a hexagon with an inscribed circle. The structural formula for the phenyl group and two alternative representations follow.

Aryl group (Ar—)

A group derived from an arene by removal of an H.

Phenyl group

A group derived by removing an H from benzene; abbreviated C_6H_5— or Ph—.

Benzene

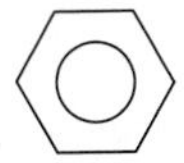

An alternative representation for benzene

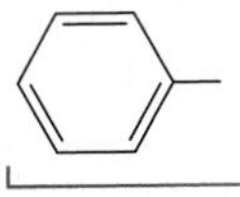

C_6H_5— Ph—

Alternative representations for the phenyl group

The chemistry of benzene and its derivatives is quite different from that of alkenes and alkynes, but even though we do not study the chemistry of arenes until Chapters 21 and 22, we will show structural formulas of compounds containing aryl groups before then. The three double bonds in a six-membered ring create a special stabilization called aromaticity, which lowers the reactivity of benzene relative to other alkenes. What you need to remember at this point is that an aryl group is not chemically reactive under any of the conditions we describe in Chapters 6 through 20.

5.1 Structure of Alkenes

A. Shapes of Alkenes

Using valence-shell electron-pair repulsion (Section 1.4) for a carbon-carbon double bond, we predict a value of 120° for the bond angles about each carbon. The observed H—C—C bond angle in ethylene is 121.1°, close to that predicted. In other alkenes, deviations from the predicted angle of 120° may be somewhat larger because of the strain introduced by nonbonded interactions created by groups bonded to the carbons of the double bond. The C—C—C bond angle in propene, for example, is 123.9°.

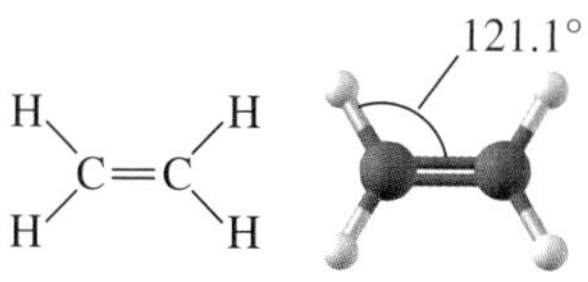

Ethylene

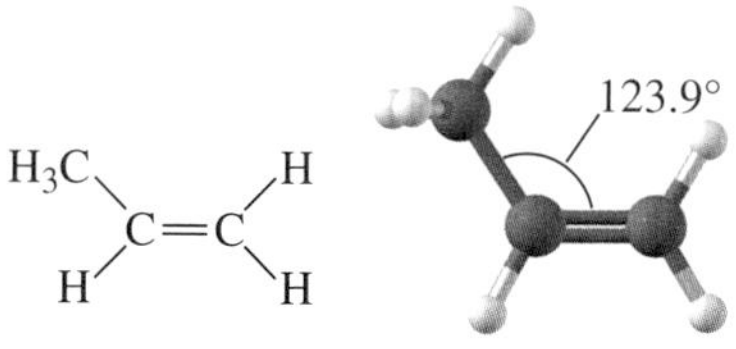

Propene

HOW TO

Calculate the Index of Hydrogen Deficiency

Valuable information about the structural formula of an unknown compound can be obtained by inspecting its molecular formula. In addition to learning the number of atoms of carbon, hydrogen, oxygen, nitrogen, and so forth, in a molecule of the compound, we can also determine what is called its index of hydrogen deficiency. For each ring and π bond, the molecular formula has two fewer hydrogens. The **index of hydrogen deficiency** is the sum of the number of rings and π bonds in a molecule. It is determined by comparing the number of hydrogens in the molecular formula of a compound whose structure is to be determined ($H_{molecule}$) with the number of hydrogens in a reference alkane of the same number of carbon atoms ($H_{reference}$).

$$\text{Index of hydrogen deficiency} = \frac{(H_{reference} - H_{molecule})}{2}$$

1. The molecular formula of a reference acyclic alkane is C_nH_{2n+2} (Section 2.1).
2. To compare the molecular formula for a compound containing elements besides carbon and hydrogen, write the formula of the reference hydrocarbon with the same number of carbon atoms and make the following adjustments to the number of hydrogen atoms in the unknown.

   a. Replace each monovalent atom of a Group 7 element (F, Cl, Br, I) with one hydrogen; halogen substitutes for hydrogen and reduces the number of hydrogens by one per halogen. The general formula of an acyclic monochloroalkane, for example, is $C_nH_{2n+1}Cl$; the general formula of the corresponding acyclic alkane is C_nH_{2n+2}.

   b. No correction is necessary for the addition of divalent atoms of Group 6 elements (O, S, Se). Insertion of a divalent Group 6 element into a hydrocarbon does not change the number of hydrogens.

   c. For each atom of a trivalent Group 5 element (N, P, As) present, add one hydrogen, because insertion of a trivalent Group 5 element adds one hydrogen to the molecular formula. The general molecular formula for an acyclic alkylamine, for example, is $C_nH_{2n+3}N$.

Index of hydrogen deficiency

The sum of the number of rings and π bonds in a molecule.

B. Carbon-Carbon Double Bond Orbitals

In Section 1.7C, we described the formation of a carbon-carbon double bond in terms of the overlap of atomic orbitals. A carbon-carbon double bond consists of one σ bond and one π bond (Figure 5.1). Each carbon of the double bond uses its three sp^2 hybrid orbitals to form σ bonds to three atoms. The unhybridized $2p$ atomic orbitals, which lie perpendicular to the plane created by the axes of the three sp^2 hybrid orbitals, combine to form two π molecular orbitals: one bonding and the other antibonding.

For the unhybridized $2p$ orbitals to be parallel, thus giving maximum overlap, the two carbon atoms of the double bond and the four bonded atoms must lie in a plane.

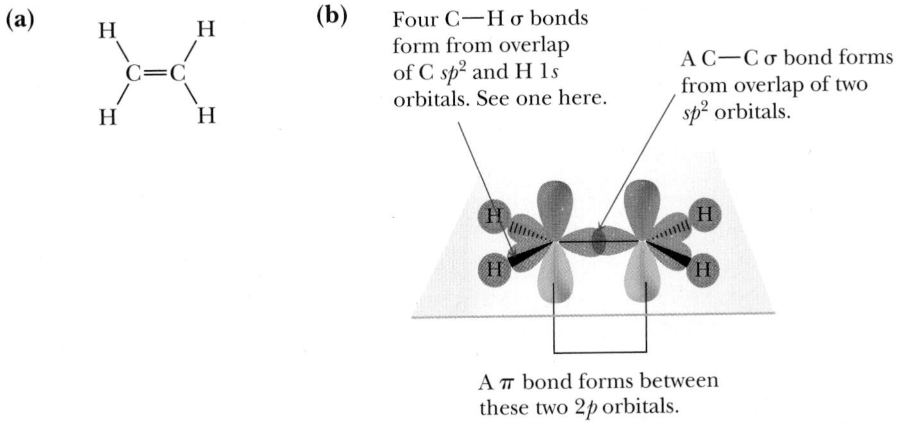
(a)
(b)
Four C—H σ bonds form from overlap of C sp^2 and H $1s$ orbitals. See one here.
A C—C σ bond forms from overlap of two sp^2 orbitals.
A π bond forms between these two $2p$ orbitals.

FIGURE 5.1 Covalent bonding in ethylene. (a) Lewis structure and (b) orbital overlap model showing the σ and π bonds.

It takes approximately 264 kJ (63 kcal)/mol to break the π bond in ethylene [i.e., to rotate one carbon by 90° with respect to the other where zero overlap occurs between $2p$ orbitals on adjacent carbons (Figure 5.2)]. This energy is considerably greater than the thermal energy available at room temperature; consequently, rotation about a carbon-carbon double bond does not occur under normal conditions. You might compare rotation about a carbon-carbon double bond, such as in ethylene, with that about a carbon-carbon single bond, such as in ethane (Section 2.5A). Whereas rotation about the carbon-carbon single bond in ethane is relatively free [rotation barrier approximately 12.5 kJ (3.0 kcal)/mol], rotation about the carbon-carbon double bond in ethylene is severely restricted.

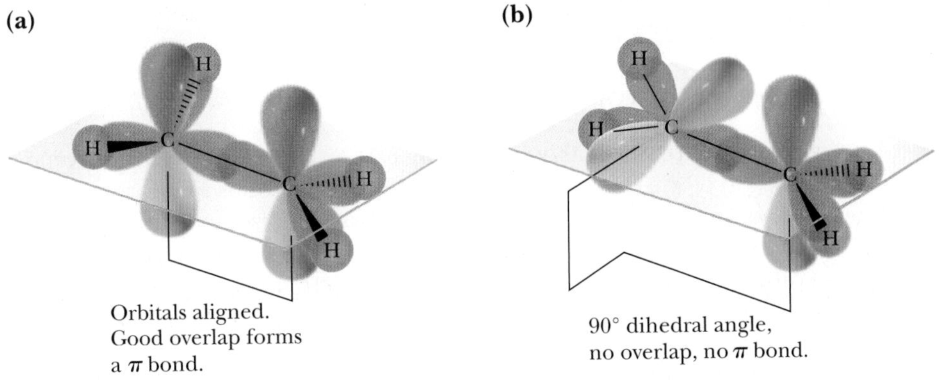
(a)
(b)
Orbitals aligned. Good overlap forms a π bond.
90° dihedral angle, no overlap, no π bond.

FIGURE 5.2 Restricted rotation about a carbon-carbon double bond. (a) Orbital overlap model showing the π bond. (b) The π bond is broken by rotating the plane of one H—C—H group by 90° with respect to the plane of the other H—C—H group.

Example 5.1 Index of Hydrogen Deficiency I

Calculate the index of hydrogen deficiency for 1-hexene, C_6H_{12}, and account for this deficiency by reference to its structural formula.

Solution

The molecular formula of the reference acyclic alkane of six carbon atoms is C_6H_{14}. The index of hydrogen deficiency of 1-hexene $(14 - 12)/2 = 1$ and is accounted for by the one π bond in 1-hexene.

Problem 5.1

Calculate the index of hydrogen deficiency of cyclohexene, C_6H_{10}, and account for this deficiency by reference to its structural formula.

Example 5.2 Index of Hydrogen Deficiency II

Isopentyl acetate, a compound with a banana-like odor, is a component of the alarm pheromone of honeybees. The molecular formula of isopentyl acetate is $C_7H_{14}O_2$. Calculate the index of hydrogen deficiency of this compound.

Solution

The molecular formula of the reference hydrocarbon is C_7H_{16}. Adding oxygens does not require any correction in the number of hydrogens. The index of hydrogen deficiency is $(16 - 14)/2 = 1$, indicating either one ring or one π bond. Following is the structural formula of isopentyl acetate. It contains one π bond—in this case, a carbon-oxygen π bond.

Isopentyl acetate

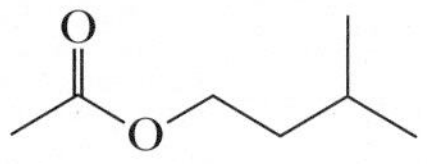

Problem 5.2

The index of hydrogen deficiency of niacin is 5. Account for this index of hydrogen deficiency by reference to the structural formula of niacin.

Nicotinamide (Niacin)

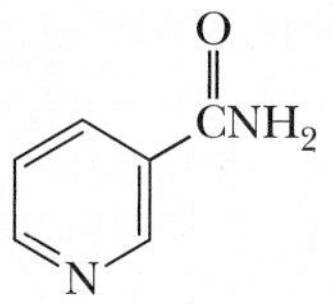

C. *Cis,Trans* Isomerism in Alkenes

***Cis,trans* isomers**

Isomers that have the same order of attachment of their atoms but a different arrangement of their atoms in space owing to the presence of either a ring (Section 2.6) or a carbon-carbon double bond (Section 5.1C).

Because of restricted rotation about a carbon-carbon double bond, any alkene in which each carbon of the double bond has two different groups bonded to it shows ***cis,trans* isomerism**. For example, 2-butene has two stereoisomers. In *cis*-2-butene, the two methyl groups are on one side of the double bond and the two hydrogens are on the other side. In *trans*-2-butene, the two methyl groups are on opposite sides of the double bond. These two compounds cannot be converted into one another at room temperature because of the restricted rotation about the double bond; they are different compounds (diastereomers), with different physical and chemical properties.

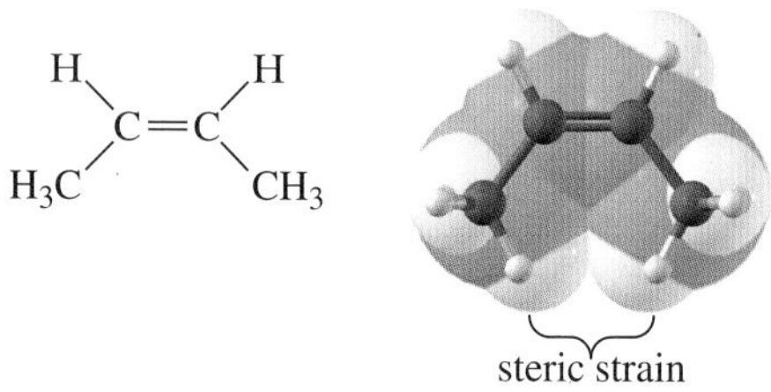

***cis*-2-Butene**
mp −139°C, bp 4°C

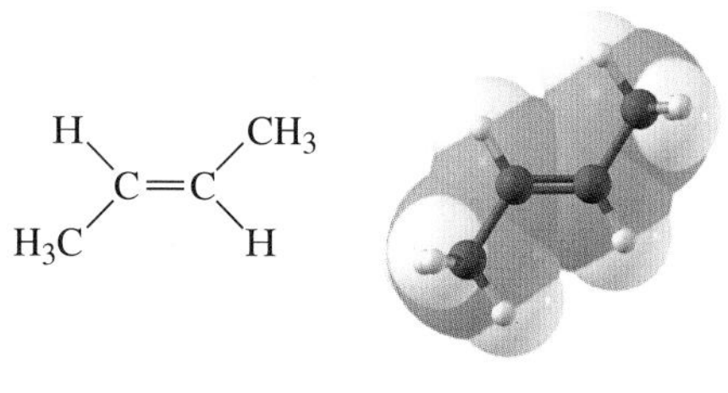

***trans*-2-Butene**
mp −106°C, bp 1°C

Cis alkenes with double bonds in open chains are less stable than their *trans* isomers because of steric strain between alkyl substituents on the same side of the double bond, as can be seen in space-filling models of the *cis* and *trans* isomers of 2-butene. This is the same type of strain that results in the preference for equatorial methylcyclohexane over axial methylcyclohexane (Section 2.5B). *trans*-2-Butene is more stable than the *cis* isomer by about 4.2 kJ (1.0 kcal)/mol because of the sum of steric strain and angle strain that results from the two methyls moving apart. At the energy minimum, the C═C—CH_3 angle for *cis*-2-butene is about 127°.

5.2 Nomenclature of Alkenes

Alkenes are named using the IUPAC system, but as we shall see, some are usually referred to by their common names.

A. IUPAC Names

To form IUPAC names for alkenes, change the *-an-* infix of the parent alkane to *-en-* (Section 2.3C). Hence, $CH_2{=}CH_2$ is named ethene and $CH_3CH{=}CH_2$ is named propene. In higher alkenes, where isomers exist that differ in location of the double bond, a numbering system must be used. According to the IUPAC system:

1. Number the longest carbon chain that contains the double bond in the direction that gives the carbon atoms of the double bond the lowest possible numbers.
2. Indicate the location of the double bond by the number of its first carbon.
3. Name branched or substituted alkenes in a manner similar to alkanes.
4. Number the carbon atoms, locate and name substituent groups, locate the double bond, and name the main chain.

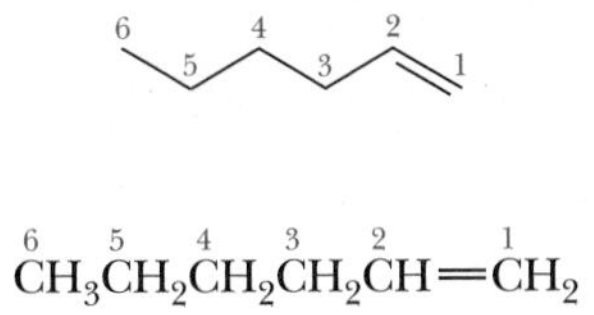

1-Hexene

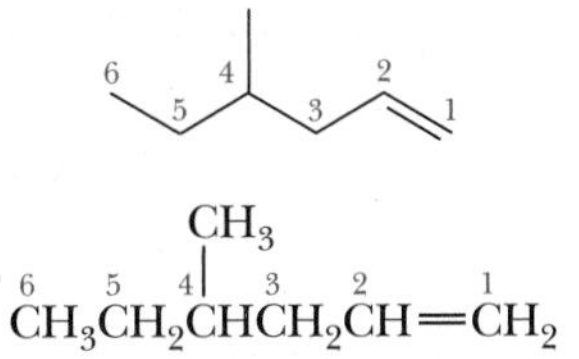

4-Methyl-1-hexene

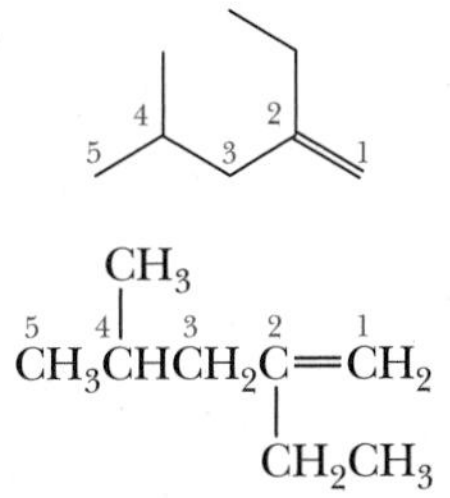

2-Ethyl-4-methyl-1-pentene

Note that there is a chain of six carbon atoms in 2-ethyl-4-methyl-1-pentene. However, because the longest chain that contains the double bond has only five carbons, the parent hydrocarbon is pentane, and the molecule is named as a disubstituted 1-pentene.

Example 5.3 Nomenclature of Alkenes

Write the IUPAC name of each alkene.

(a) 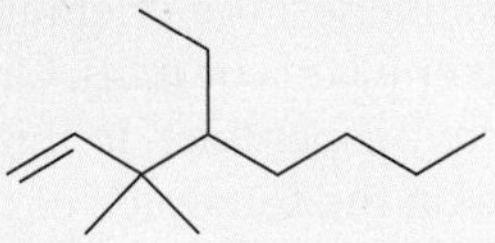 **(b)** 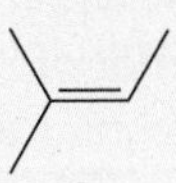

Solution

(a) 4-Ethyl-3,3-dimethyl-1-octene **(b)** 2-Methyl-2-butene

(Continued)

Problem 5.3

Write the IUPAC name of each alkene.

(a) 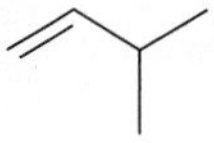 **(b)** 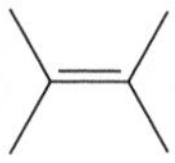

B. Common Names

Some alkenes, particularly those of low molecular weight, are known almost exclusively by their common names, as illustrated by the common names of these alkenes.

	$CH_2{=}CH_2$	$CH_3CH{=}CH_2$	$CH_3C(CH_3){=}CH_2$
IUPAC name:	**Ethene**	**Propene**	**2-Methylpropene**
Common name:	**Ethylene**	**Propylene**	**Isobutylene**

Methylene

A $CH_2{=}$ group.

Vinyl

A $CH_2{=}CH{-}$ group.

Allyl

A $CH_2{=}CHCH_2{-}$ group.

Furthermore, the common names **methylene** (a CH_2 group), **vinyl**, and **allyl** are often used to show the presence of the following alkenyl groups:

Alkenyl Group	IUPAC Name	Common Name	Example	IUPAC Name (Common Name)
$CH_2{=}$	Methylidene	Methylene	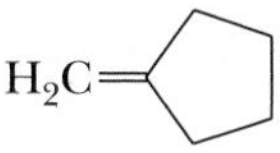	Methylidenecyclopentane (Methylenecyclopentane)
$CH_2{=}CH{-}$	Ethenyl	Vinyl	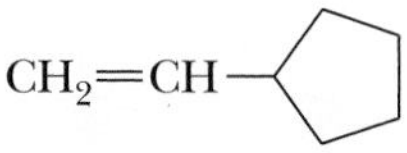	Ethenylcyclopentane (Vinylcyclopentane)
$CH_2{=}CHCH_2{-}$	2-Propenyl	Allyl	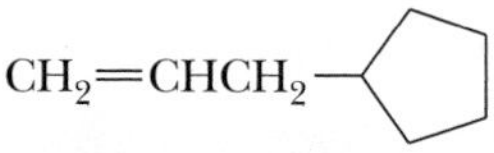	2-Propenylcyclopentane (Allylcyclopentane)

C. Systems for Designating Configuration in Alkenes

The *Cis,Trans* System

cis

In *cis,trans*-alkene nomenclature, it refers to molecules in which the carbon atoms of the main chain are on the same side of the double bond.

trans

In *cis,trans*-alkene nomenclature, it refers to molecules in which the carbon atoms of the main chain are on opposite sides of the double bond.

The most common method for specifying the configuration in alkenes uses the prefixes ***cis*** and ***trans***. There is no doubt which isomers are intended by the names *trans*-3-hexene and *cis*-3-hexene. For more complex alkenes, the orientation of the atoms of the parent chain determines whether the alkene is *cis* or *trans*. On the right is a structural formula for the *cis* isomer of 3,4-dimethyl-2-pentene. In this example, carbon atoms of the main chain (carbons 1 and 4) are on the same side of the double bond; therefore, this alkene is *cis*.

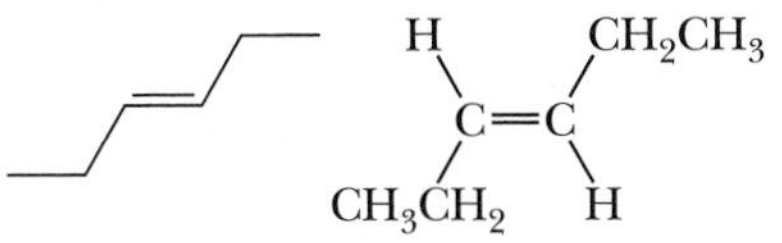

trans-3-Hexene

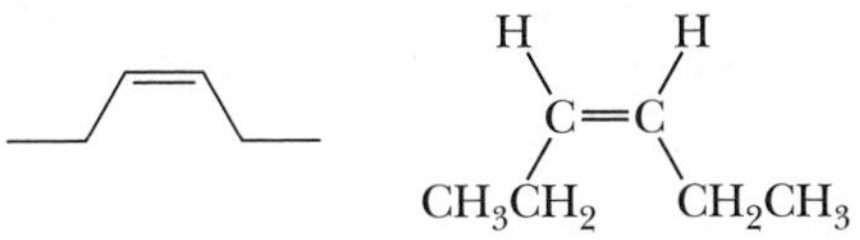

cis-3-Hexene

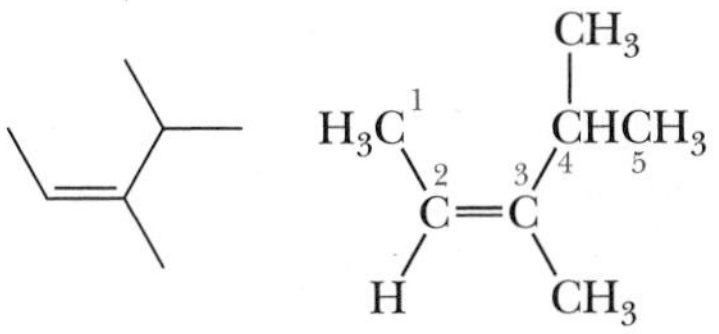

cis-3,4-Dimethyl-2-pentene

Example 5.4 *Trans* Versus *Cis* Alkene Nomenclature

Name each alkene and show the configuration about each double bond using the *cis,trans* system.

(a)

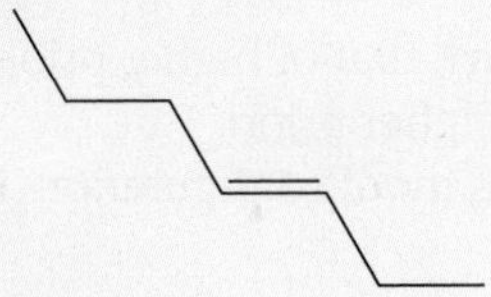

(b)

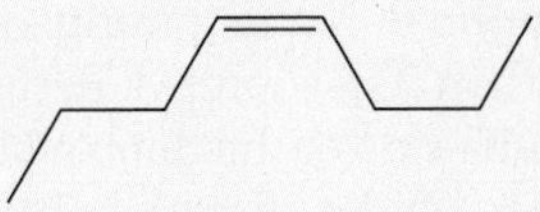

Solution

(a) The chain contains seven carbon atoms and is numbered from the end that gives the lower number to the first carbon of the double bond. Its name is *trans*-3-heptene.

(b) The longest chain contains eight carbon atoms and is numbered from either end of the chain so that the first carbon of the double bond is carbon 4 of the chain. Its name is *cis*-4-octene.

Problem 5.4

Which alkenes show *cis,trans* isomerism? For each alkene that does, draw the *trans* isomer.

(a) 2-Pentene **(b)** 2-Methyl-2-pentene **(c)** 3-Methyl-2-pentene

E,Z system
A system to specify the configuration of groups about a carbon-carbon double bond.

Z
From the German *zusammen*, together. Specifies that groups of higher priority on the carbons of a double bond are on the same side.

E
From the German *entgegen*, opposite. Specifies that groups of higher priority on the carbons of a double bond are on opposite sides.

The *E,Z* System

Because the *cis,trans* system becomes confusing with tri- and tetrasubstituted alkenes and is not detailed enough to name all alkenes, chemists developed the **E,Z system**. This system uses the priority rules of the *R,S* system (Section 3.3) to assign priority to the substituents on each carbon of a double bond. Using these rules, we decide which group on each carbon has the higher priority. If the groups of higher priority are on the same side of the double bond, the configuration of the alkene is *Z* (German: *zusammen,* together). If they are on opposite sides of the double bond, the alkene is *E* (German: *entgegen,* opposite).

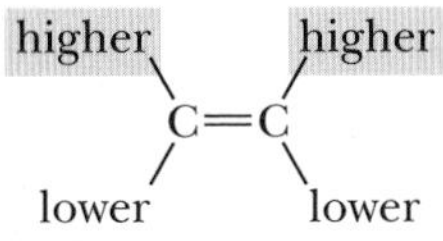

Z (*zusammen*)

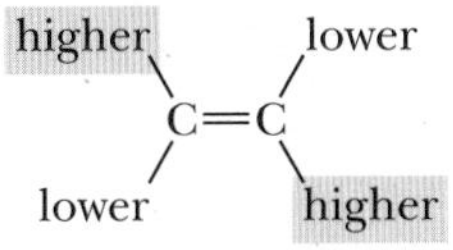

E (*entgegen*)

Watch a video explanation

Throughout this text, we use the *cis,trans* system for alkenes in which each carbon of the C═C bond bears a hydrogen. We use the *E,Z* system in all other cases. The *E,Z* system should always be used if confusion is possible.

Example 5.5 *E,Z* Nomenclature

Name each alkene and specify its configuration by the *E,Z* system.

(a) 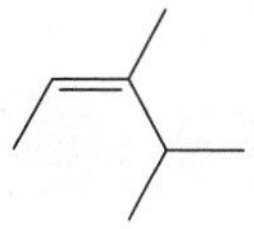

(b) 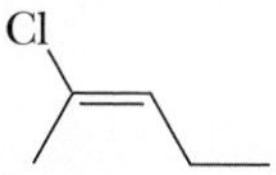

(c) 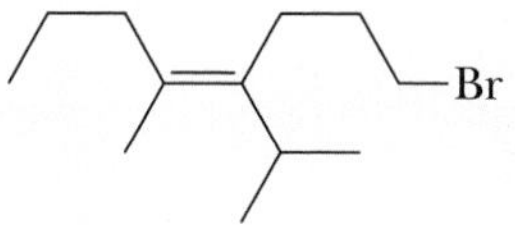

Solution

(a) The group of higher priority on carbon 2 is methyl; that of higher priority on carbon 3 is isopropyl. Because the groups of higher priority are on the same side of the double bond, the alkene has the *Z* configuration. Its name is (*Z*)-3,4-dimethyl-2-pentene.

(b) Groups of higher priority on carbons 2 and 3 are —Cl and —CH_2CH_3. Because these groups are on opposite sides of the double bond, the configuration of this alkene is *E*. Its name is (*E*)-2-chloro-2-pentene.

(c) The groups of higher priority are on opposite sides of the double bond; the configuration is *E*. The name of this bromoalkene is (*E*)-1-bromo-4-isopropyl-5-methyl-4-octene.

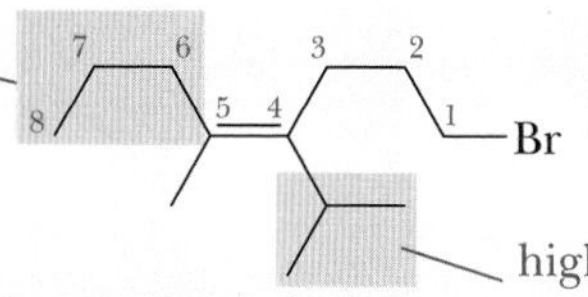

Watch a video explanation

Problem 5.5

Name each alkene and specify its configuration by the *E,Z* system.

(a) 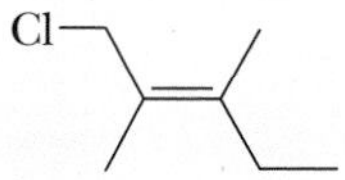

(b) 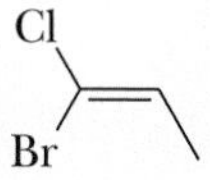

(c) 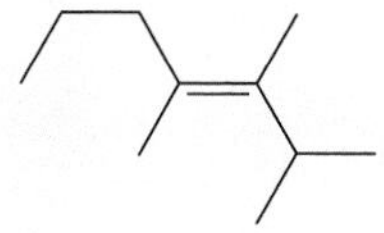

D. Cycloalkenes

In naming **cycloalkenes**, the carbon atoms of the ring double bond are numbered 1 and 2 in the direction that gives the substituent encountered first the smaller number.

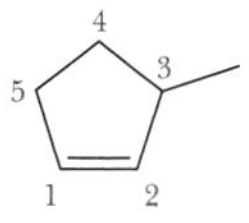

3-Methylcyclopentene

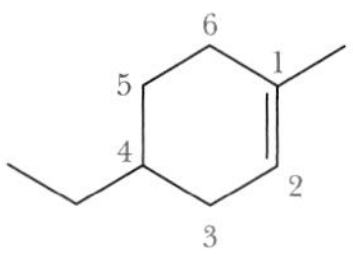

4-Ethyl-1-methylcyclohexene

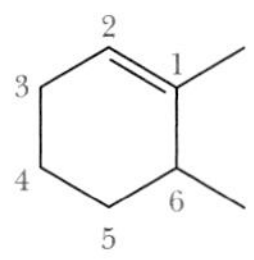

1,6-Dimethylcyclohexene

Example 5.6 Nomenclature for Cyclic Alkenes

Write the IUPAC name of each cycloalkene.

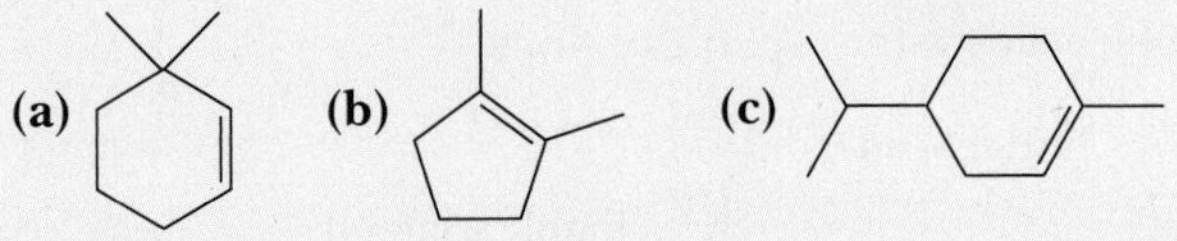

Solution

(a) 3,3-Dimethylcyclohexene **(b)** 1,2-Dimethylcyclopentene

(c) 4-(1-Methylethyl)-1-methylcyclohexene

Problem 5.6

Write the IUPAC name of each cycloalkene.

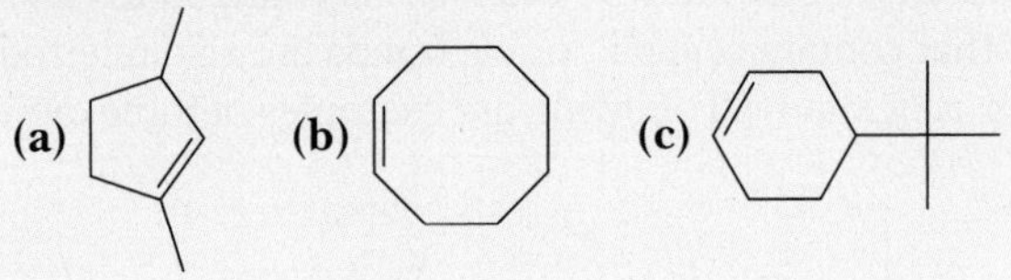

E. *Cis,Trans* Isomerism in Cycloalkenes

Following are structural formulas for four cycloalkenes:

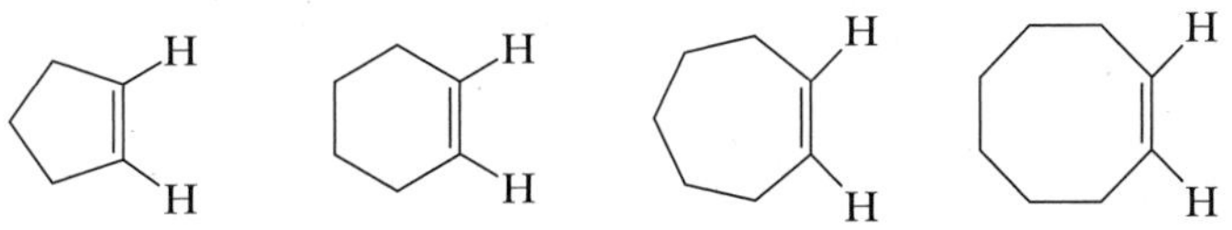

Cyclopentene **Cyclohexene** **Cycloheptene** **Cyclooctene**

In these representations, the configuration about each double bond is *cis*. Is it possible to have a *trans* configuration in these and larger cycloalkenes? To date, *trans*-cyclooctene is the smallest *trans* cycloalkene that has been prepared in pure form and is stable at room temperature. Yet, even in this *trans* cycloalkene, there is considerable angle strain; the double bond's 2*p* orbitals make an angle of 44° to each other. *Cis*-cyclooctene is more stable than its *trans* isomer by 38 kJ (9.1 kcal)/mol. Note that the *trans* isomer is chiral even though it has no chiral center.

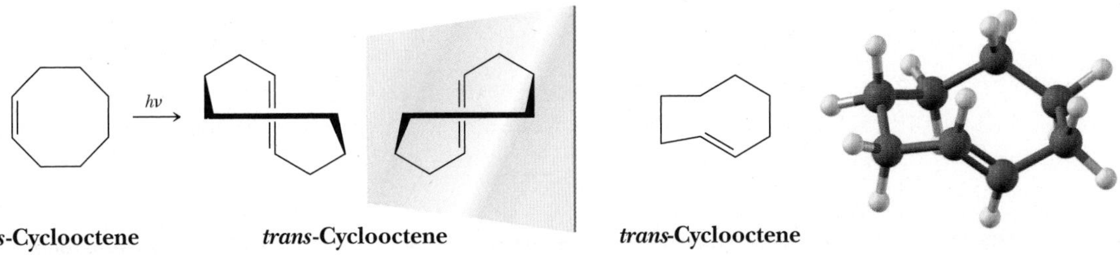

***cis*-Cyclooctene** ***trans*-Cyclooctene** ***trans*-Cyclooctene**

F. Bridgehead Alkenes

Another manner in which alkenes become strained is when one of the carbons in the double bond is placed at a bridgehead carbon of a bicyclic ring system.

A bridgehead carbon is a carbon that is contained in both rings of the hydrocarbon. For example, the carbons shown below with arrows are at bridgeheads. In norbornene, the double bond does not contain a bridgehead carbon, but in the other illustration, the double bond is at a bridgehead. This arrangement of a double bond imparts considerable strain because the alkene cannot be planar while the rest of the carbons in the bicyclic system span the rings.

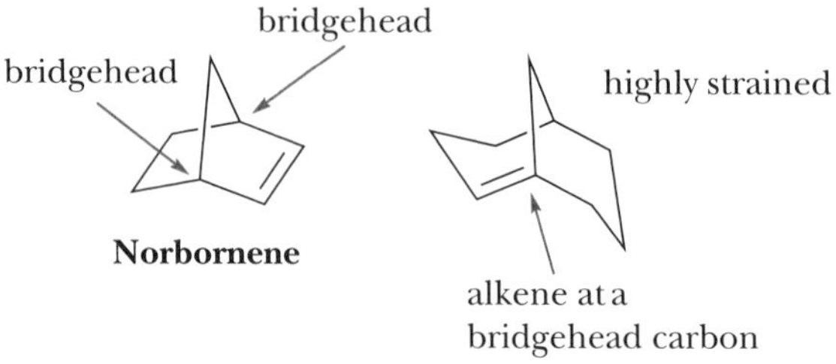

G. Dienes, Trienes, and Polyenes

For alkenes containing two or more double bonds, the infix *-en-* is changed to *-adien-*, *-atrien-*, and so on. Those alkenes that contain several double bonds are also referred to more generally as polyenes (Greek: *poly*, many). Following are examples of three dienes.

1,4-Pentadiene

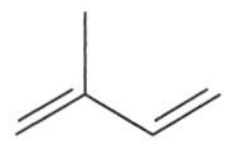
2-Methyl-1,3-butadiene (Isoprene)

1,3-Cyclopentadiene

H. *Cis,Trans* Isomerism in Dienes, Trienes, and Polyenes

Thus far, we have considered *cis,trans* isomerism in alkenes containing only one carbon-carbon double bond. For an alkene with one carbon-carbon double bond that can show *cis,trans* isomerism, two stereoisomers are possible. For an alkene with ***n*** carbon-carbon double bonds, each of which can show *cis,trans* isomerism, $\mathbf{2^n}$ stereoisomers are possible.

Example 5.7 Stereoisomers of Polyenes I

How many stereoisomers are possible for 2,4-heptadiene?

Solution

This molecule has two carbon-carbon double bonds, each of which shows *cis, trans* isomerism. As shown in this table, there are $2^2 = 4$ stereoisomers. Two of these are drawn on the right.

Double Bond C2—C3	Double Bond C4—C5
trans	*trans*
trans	*cis*
cis	*trans*
cis	*cis*

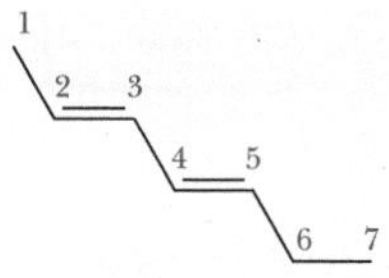

(2*E*,4*E*)-2,4-Heptadiene
***trans,trans*-2,4-Heptadiene**

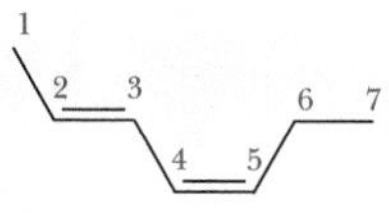

(2*E*,4*Z*)-2,4-Heptadiene
***trans,cis*-2,4-Heptadiene**

Problem 5.7

Draw structural formulas for the other two stereoisomers of 2,4-heptadiene.

Example 5.8 Stereoisomers of Polyenes II

How many stereoisomers are possible for 10,12-hexadecadien-1-ol?

$$CH_3(CH_2)_2CH{=}CHCH{=}CH(CH_2)_8CH_2OH$$

10,12-Hexadecadien-1-ol

Solution

Cis,trans isomerism is possible about both double bonds, so there are four stereoisomers.

Problem 5.8

(10*E*,12*Z*)-10,12-hexadecadien-1-ol is a sex pheromone of the silkworm. Draw a structural formula for this compound.

Paul Chesley/Getty Images

Silkworms spinning cocoons at a silk farm, Japan.

An example of a biologically important compound for which a number of *cis, trans* isomers are possible is vitamin A. There are four carbon-carbon double bonds in the chain of carbon atoms bonded to the substituted cyclohexene ring, and each has the potential for *cis,trans* isomerism. There are $2^4 = 16$ stereoisomers possible for this structural formula. Vitamin A is the all-*E* isomer.

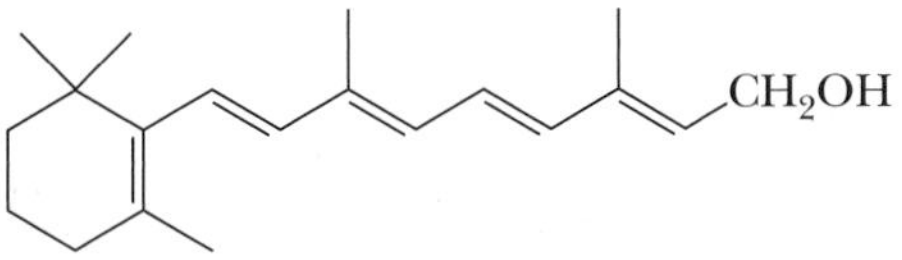

Vitamin A (Retinol)

5.3 Physical Properties of Alkenes

The physical properties of alkenes are similar to those of alkanes. The only attractive forces between alkene molecules, which are nonpolar, are dispersion forces (Section 2.7B). Two, three, and four carbon alkenes are gaseous at room temperature; the larger ones are colorless liquids less dense than water. Alkenes are insoluble in water but soluble in one another, in other nonpolar organic liquids, and in ethanol. Table 5.1 lists physical properties of some alkenes.

Chemical Connections

The Case of the Iowa and New York Strains of the European Corn Borer

Chemicals are the primary means of communication for most animal species. Communication within a species is often specific for one or more configurational isomers. For example, a given species may respond to a *cis* isomer of a chemical but not the *trans* isomer. Alternatively, it might respond to a precise blend of *cis* and *trans* isomers but not to other blends of the same isomers.

Several groups of scientists have studied the components of the sex pheromones of both the Iowa and New York strains of the European corn borer. Females of these closely related species secrete the sex attractant 11-tetradecenyl acetate. Males of the Iowa strain show maximum response to a mixture containing 96% of the *cis* isomer and 4% of the *trans* isomer. When the pure *cis* isomer is

(Continued)

used alone, males are only weakly attracted. Males of the New York strain show an entirely different response. They respond maximally to a mixture containing 3% of the *cis* isomer and 97% of the *trans* isomer.

Response to a narrow range of stereoisomers appears widespread in nature. It has been observed that many insects maintain species isolation for mating and reproduction by the stereochemistry of their pheromones.

Scott Camazine/Science Source

The European corn borer, *Pyrausta nubilalis*.

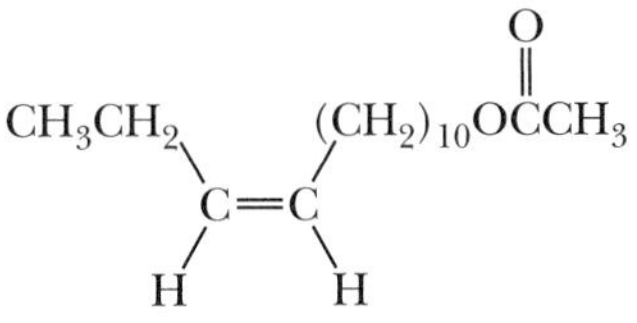

***cis*-11-Tetradecenyl acetate**

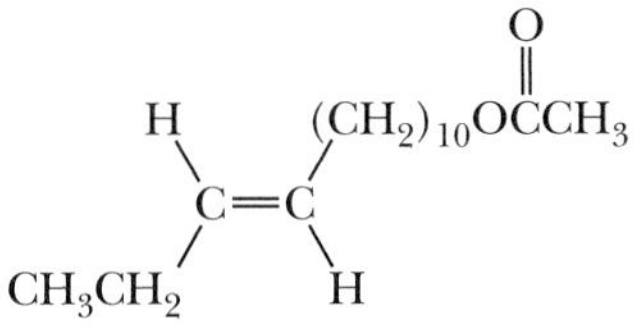

***trans*-11-Tetradecenyl acetate**

Table 5.1 Physical Properties of Some Alkenes

Name	Structural Formula	mp (°C)	bp (°C)
Ethylene	$CH_2{=}CH_2$	−169	−104
Propylene	$CH_3CH{=}CH_2$	−185	−47
1-Butene	$CH_3CH_2CH{=}CH_2$	−185	−6
1-Pentene	$CH_3CH_2CH_2CH{=}CH_2$	−138	30
***cis*-2-Pentene**	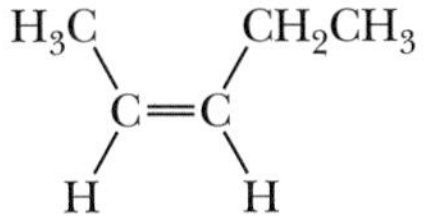	−151	37
***trans*-2-Pentene**	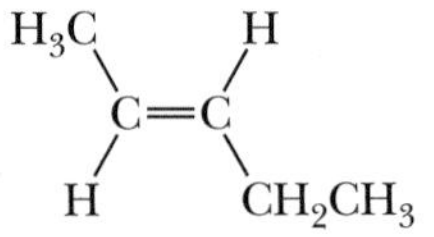	−156	36
2-Methyl-2-butene	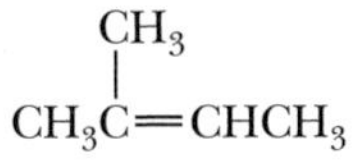	−134	39

5.4 Naturally Occurring Alkenes—Terpene Hydrocarbons

Terpene

A compound whose carbon skeleton can be divided into two or more units identical with the carbon skeleton of isoprene.

A **terpene** is a compound whose carbon skeleton can be divided into two or more units that are identical with the carbon skeleton of isoprene. Carbon 1 of an isoprene unit is called the head; carbon 4 is called the tail. Terpenes are formed by bonding the tail of one isoprene unit to the head of another. This is called the **isoprene rule**.

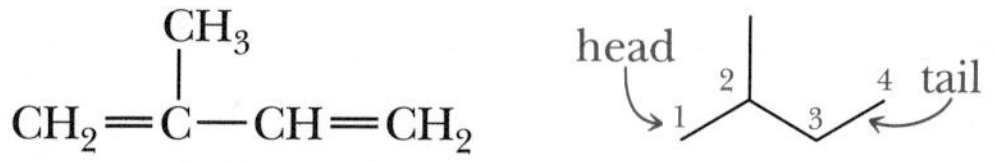
CH3
CH2=C—CH=CH2
head
tail
1
2
3
4

2-Methyl-1,3-butadiene
(Isoprene)

A study of terpenes provides a glimpse of the wondrous diversity that nature can generate from a simple carbon skeleton. Terpenes also illustrate an important principle of the molecular logic of living systems—namely, that in building large molecules, small subunits are bonded together enzymatically by an iterative process and then modified by subsequent precise enzyme-catalyzed reactions. Chemists use the same principles in the laboratory, but their methods do not have the precision and selectivity of the enzyme-catalyzed reactions of living systems.

Probably the terpenes most familiar to you, at least by odor, are components of the so-called essential oils obtained by steam distillation or ether extraction of various parts of plants. Essential oils contain the relatively low-molecular-weight substances that are in large part responsible for characteristic plant fragrances. Many essential oils, particularly those from flowers, are used in perfumes.

One example of a terpene obtained from an essential oil is myrcene, $C_{10}H_{16}$, a component of bayberry wax and oils of bay and verbena. Myrcene is a triene with a parent chain of eight carbon atoms and two one-carbon branches [Figure 5.3(a)].

Head-to-tail bonds between isoprene units are vastly more common in nature than are the alternative head-to-head or tail-to-tail patterns. Figure 5.4 shows structural formulas of five more terpenes, all derived from two isoprene units. Geraniol has the same carbon skeleton as myrcene. In the last four terpenes of Figure 5.4, the carbon atoms present in myrcene and geraniol are cross-linked to give cyclic structures. To help you identify the points of cross-linkage and ring formation, the carbon atoms of the geraniol skeleton are numbered 1 through 8. This numbering pattern is used in the remaining terpenes to show points of cross-linking. In both limonene and menthol, a carbon-carbon bond is present between carbons 1 and 6. In α-pinene, carbon-carbon bonds are present between carbons 1 and 6 and between carbons 4 and 7. In camphor, they are between carbons 1 and 6 and between carbons 3 and 7.

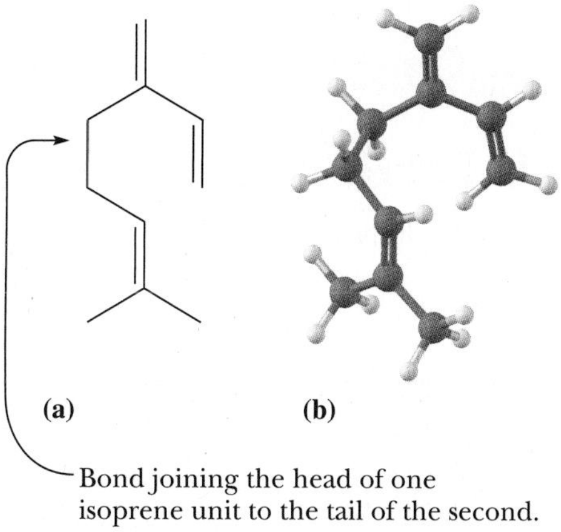
(a)
(b)
Bond joining the head of one
isoprene unit to the tail of the second.

FIGURE 5.3 Myrcene. (a) Structural formula and (b) ball-and-stick model.

FIGURE 5.4 Five terpenes, each divisible into two isoprene units.

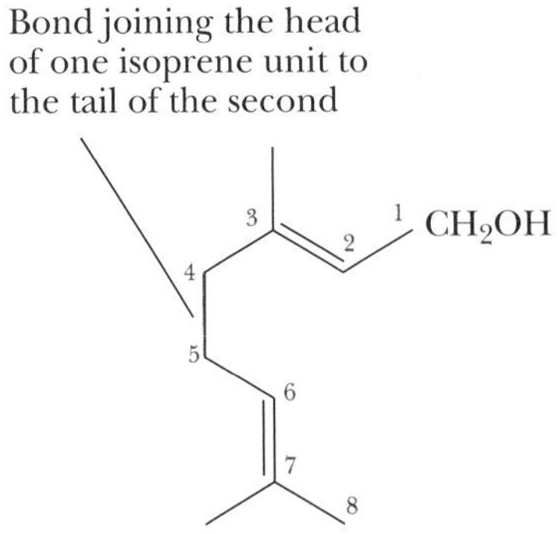

Geraniol
(rose and other flowers)

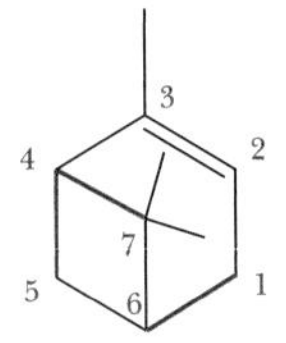

α-Pinene
(turpentine)

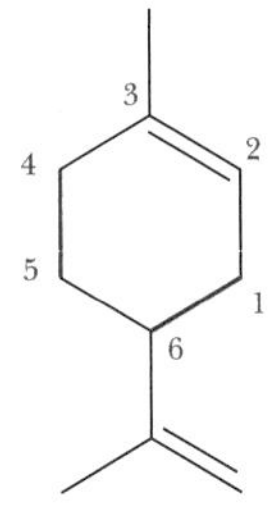

Limonene
(oil of lemon and orange)

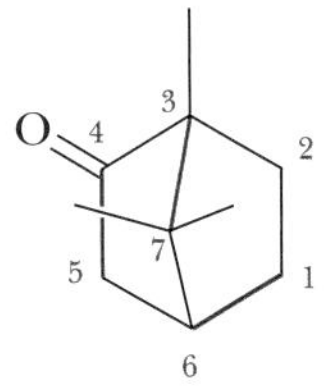

Camphor
(camphor tree)

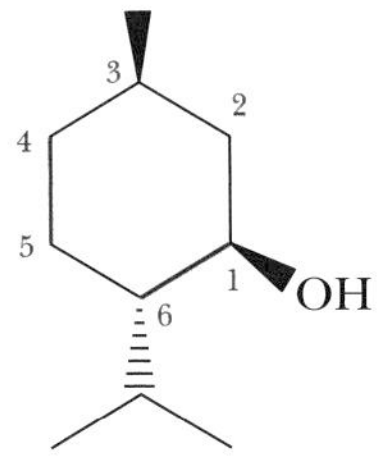

Menthol
(peppermint)

Vitamin A (Section 5.2H), a terpene with the molecular formula $C_{20}H_{30}O$, consists of four isoprene units linked head-to-tail and cross-linked at one point to form a six-membered ring.

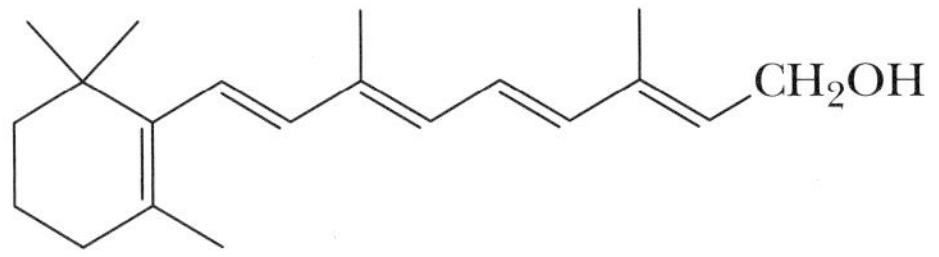

Vitamin A (retinol)

The synthesis of substances in living systems is a fascinating area of research and one of the links between organic chemistry and biochemistry. However tempting it might be to propose that nature synthesizes terpenes by joining together molecules of isoprene, this is not quite the way it is done. We will discuss the creation of new carbon-carbon bonds during terpene biosynthesis in Section 26.4B.

Connections to Biological Chemistry

The Importance of *Cis* Double Bonds in Fats Versus Oils

Fats and oils are very similar in that both are triesters of glycerol, hence the name triglyceride. Hydrolysis of a triglyceride in aqueous base followed by acidification gives glycerol and three carboxylic acids. Because these carboxylic acids can be derived from fats, they are called fatty acids.

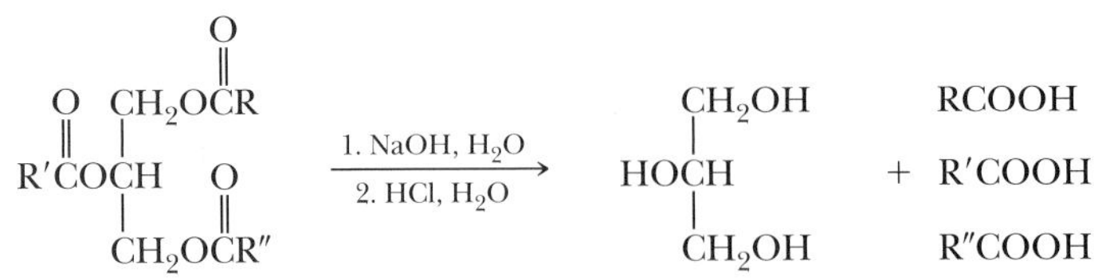

A triglyceride
(a triester of glycerol)

1,2,3-Propanetriol
(Glycerol)

Fatty acids

The most common fatty acids have between 12 and 24 carbon atoms in an unbranched chain.

The main difference between fats and oils is the temperature at which they melt. Fats are solids or semisolids at or near room temperature, while oils are liquids. The different physical properties of fats and oils result from the presence of different fatty acids.

relatively well, and are held together by dispersion forces (Section 2.7A). As a result, both saturated fatty acids and the triglycerides derived from them are solids at room temperature. However, the *cis* double bonds place a considerable "kink" in the chains of monounsaturated and polyunsaturated fatty acid chains limiting their packing, decreasing surface contact and dispersion forces, thereby increasing fluidity.

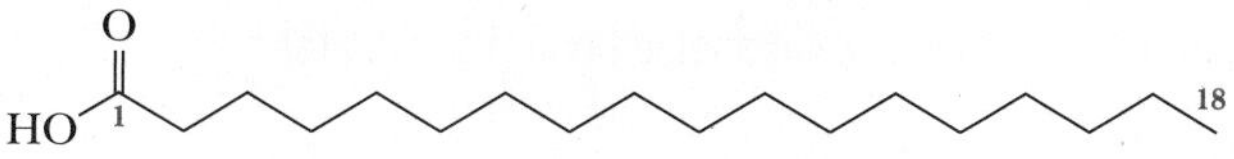

Stearic acid
mp 70 °C
(a saturated C_{18} fatty acid)

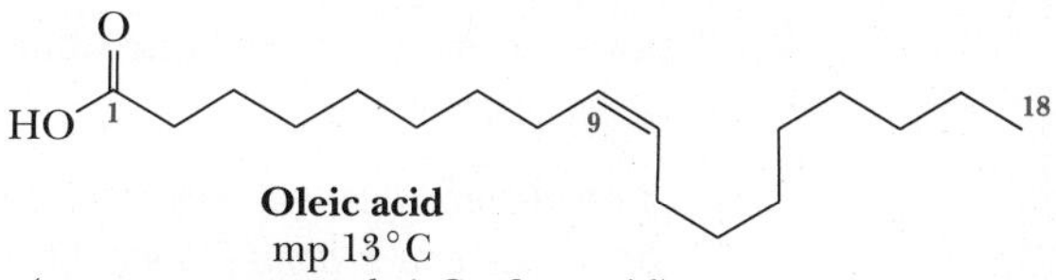

Oleic acid
mp 13 °C
(a monounsaturated *cis* C_{18} fatty acid)

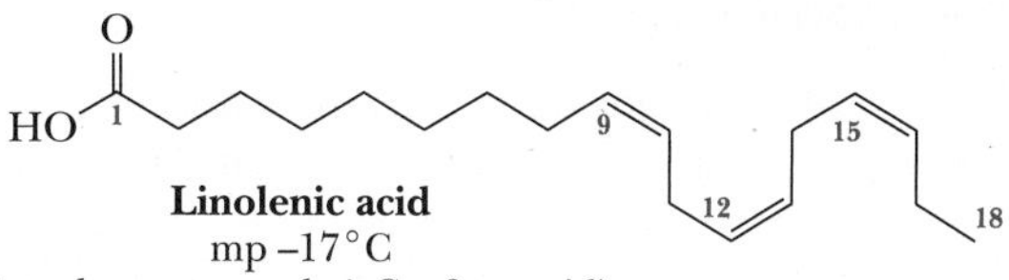

Linolenic acid
mp –17 °C
(a polyunsaturated *cis* C_{18} fatty acid)

Fatty acids with no double bonds are referred to as saturated fatty acids, those with a single double bond are called monounsaturated fatty acids, and those with more than one double bond are called polyunsaturated fatty acids.

The double bonds in almost all naturally occurring fatty acids have *cis* configurations. The triglycerides of

As a result, monounsaturated and polyunsaturated fatty acids and the triglycerides composed of them are liquid at room temperature. Overall, oils have fatty acids with more *cis* double bonds. Fats have fewer *cis* double bonds and more saturated fatty acids. For example, butter has a high content of saturated fats and is a solid at room temperature,

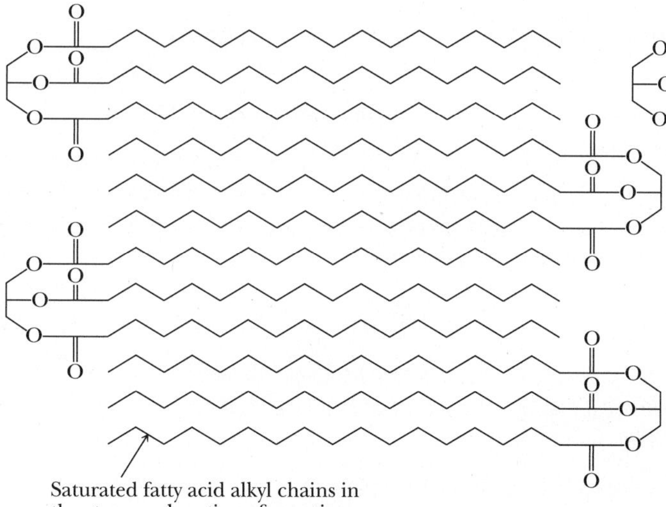

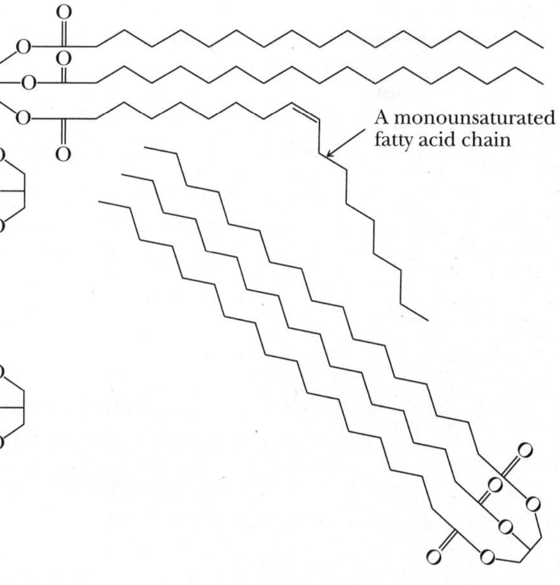

animal fats are richer in saturated fatty acids, whereas the triglycerides of plant oils (e.g., corn, soybean, canola, olive, and palm oils) are richer in unsaturated fatty acids.

The carbon-carbon single bonds of saturated fatty acid alkyl chains exist largely in the staggered, anti conformation, meaning they can pack together

whereas salad oil (from plant oils) has a high content of polyunsaturated fatty acids and is liquid, even at freezing temperatures. Olive oil, which has a high content of the monounsaturated fatty acid oleic acid (hence the name of oleic acid), will solidify in the refrigerator.

Study Guide

5.1 Structure of Alkenes

- An **alkene** is an **unsaturated hydrocarbon** that contains a carbon-carbon double bond. The general formula of an alkene is C_nH_{2n}.
  - *A carbon-carbon double bond consists of one σ bond formed by the overlap of* sp^2 *hybrid orbitals and one π bond formed by the overlap of parallel* 2p *orbitals.*
  - *The strength of the π bond in ethylene is approximately 264 kJ (63 kcal)/mol, which is considerably weaker than the carbon-carbon σ bond.*
  - *The structural feature that makes* **cis,trans** ***isomerism*** *possible in alkenes is lack of rotation about the two carbons of the double bond at or near room temperature.*

P 5.9–5.12, 5.25–5.28

- **Index of hydrogen deficiency** is the sum of the number of π bonds and rings in a molecule.

P 5.1, 5.2, 5.32

5.2 Nomenclature of Alkenes

- According to the IUPAC system, the presence of a carbon-carbon double bond is shown by changing the infix of the parent alkane from *-an-* to *-en-*.
  - *For compounds containing two or more double bonds, the infix* -en- *is changed to* -adien-, -atrien-, *and so on.*
  - *The names* ***methylene***, ***vinyl***, *and* ***allyl*** *are commonly used to show the presence of* $=CH_2$, $-CH=CH_2$, *and* $-CH_2CH=CH_2$ *groups, respectively.*
- Whether an alkene is ***cis*** or ***trans*** is determined by the orientation of the main carbon chain about the double bond. The configuration of a carbon-carbon double bond is specified more precisely by the ***E,Z* system**, which uses the same set of priority rules used for the *R,S* system.

P 5.3

  - *If the two groups of higher priority are on the same side of the double bond, the alkene is designated* **Z** *(German:* zusammen, *together); if they are on opposite sides, the alkene is designated* **E** *(German:* entgegen, *opposite).*
- To date, *trans*-cyclooctene is the smallest *trans* cycloalkene that has been prepared in pure form and is stable at room temperature.

P 5.4–5.8, 5.13–5.24

▶ The *E,Z* System, Problem 5.2

5.3 Physical Properties of Alkenes

- Because alkenes are essentially nonpolar compounds and the only attractive forces between their molecules are **dispersion forces**, their physical properties are similar to those of alkanes.

5.4 Naturally Occurring Alkenes—Terpene Hydrocarbons

- The characteristic structural feature of a **terpene** is a carbon skeleton that can be divided into two or more isoprene units.
- Terpenes illustrate an important principle of the molecular logic of living systems—namely, that in building large molecules, small subunits are strung together by an iterative process and then chemically modified by precise enzyme-catalyzed reactions.

P 5.29–5.31, 5.33–5.36

Problems

Red numbers indicate applied problems.

Structure of Alkenes

5.9 Predict all approximate bond angles about each highlighted carbon atom. To make these predictions, use valence-shell electron-pair repulsion (VSEPR) theory (Section 1.4).

(a) 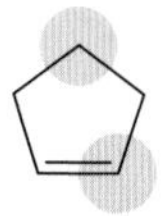

(b) 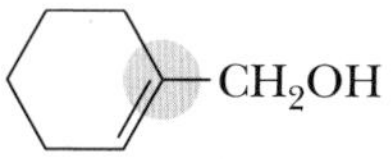

(c) 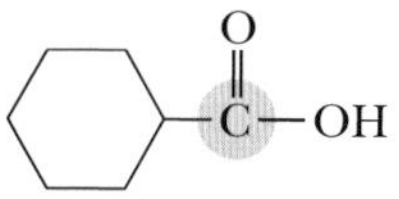

(d) 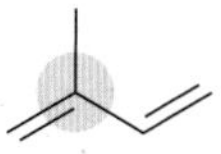

(e) $HC{\equiv}C{-}CH{=}CH_2$

5.10 For each highlighted carbon atom in Problem 5.9, identify which atomic orbitals are used to form each σ bond and which are used to form each π bond.

5.11 The structure of 1,2-propadiene (allene) is shown to the right.
(a) Predict all approximate bond angles in this molecule.
(b) State the orbital hybridization of each carbon.
(c) Explain the three-dimensional geometry of allene in terms of the orbitals used.

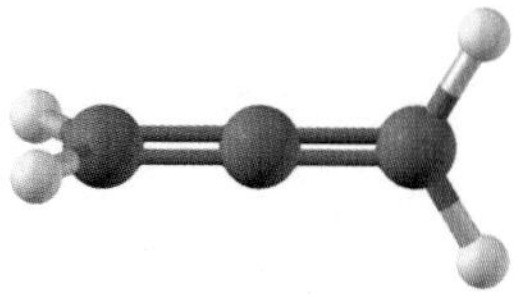

$CH_2{=}C{=}CH_2$

1,2-Propadiene (Allene)

5.12 Following are lengths for a series of C—C single bonds. Propose an explanation for the differences in bond lengths.

Structure	Length of C—C Single Bond (pm)
$CH_3—CH_3$	153.7
$CH_2=CH—CH_3$	151.0
$CH_2=CH—CH=CH_2$	146.5
$HC\equiv C—CH_3$	145.9

Nomenclature of Alkenes

5.13 Draw structural formulas for these alkenes.

(a) *trans*-2-Methyl-3-hexene
(b) 2-Methyl-2-hexene
(c) 2-Methyl-1-butene
(d) 3-Ethyl-3-methyl-1-pentene
(e) 2,3-Dimethyl-2-butene
(f) *cis*-2-Pentene
(g) (*Z*)-1-Chloropropene
(h) 3-Methylcyclohexene
(i) 1-Isopropyl-4-methylcyclohexene
(j) (*E*)-2,6-Dimethyl-2,6-octadiene
(k) 3-Cyclopropyl-1-propene
(l) Cyclopropylethene
(m) 2-Chloropropene
(n) Tetrachloroethylene
(o) 1-Chlorocyclohexene

5.14 Name these alkenes and cycloalkenes.

(a) 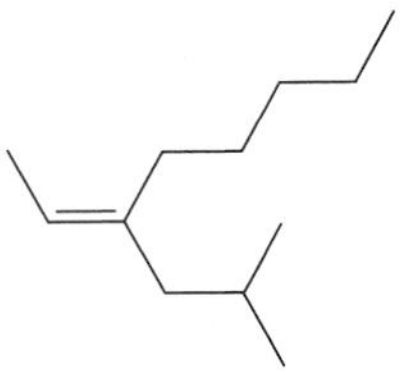

(b) 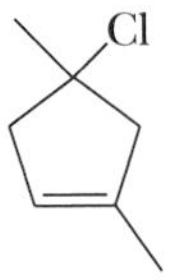

(c) 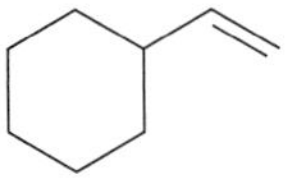

(d) 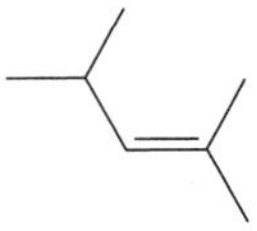

(e) 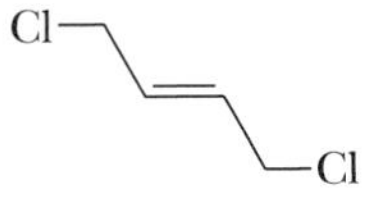

(f) 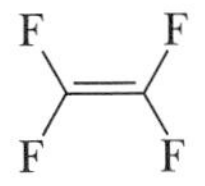

(g) 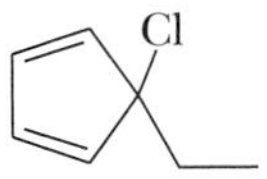

(h) 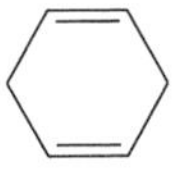

5.15 Arrange the following groups in order of increasing priority.

(a) $—CH_3$ $—H$ $—Br$ $—CH_2CH_3$
(b) $—OCH_3$ $—CH(CH_3)_2$ $—B(CH_2CH_3)_2$ $—H$
(c) $—CH_3$ $—CH_2OH$ $—CH_2NH_2$ $—CH_2Br$

5.16 Assign an *E* or *Z* configuration to these dicarboxylic acids, each of which is an intermediate in the tricarboxylic acid cycle. Under each is its common name.

(a) 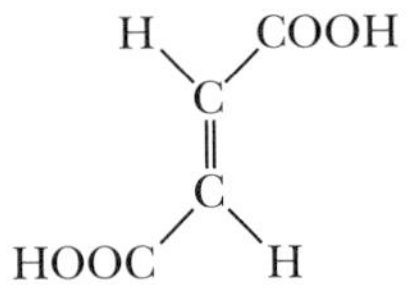

Fumaric acid

(b) 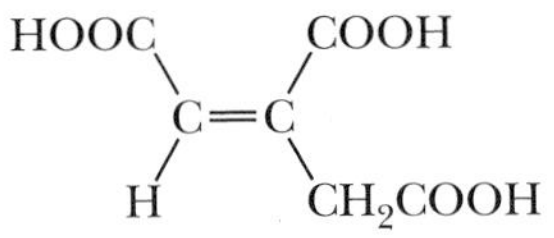

Aconitic acid

5.17 Name and draw structural formulas for all alkenes with the molecular formula C_5H_{10}. As you draw these alkenes, remember that *cis* and *trans* isomers are different compounds and must be counted separately.

5.18 For each molecule that shows *cis,trans* isomerism, draw the *cis* isomer.

(a) 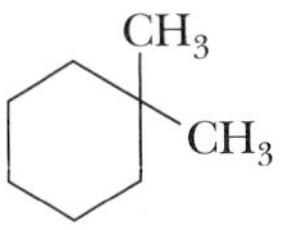

(b) 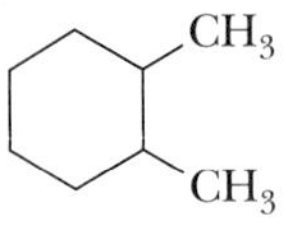

(c) 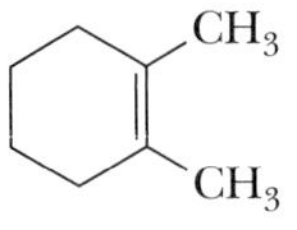

(d) 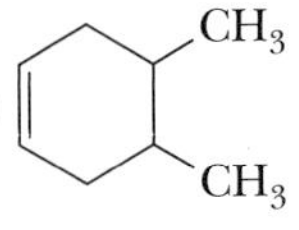

5.19 β-Ocimene, a triene found in the fragrance of cotton blossoms and several essential oils, has the IUPAC name (*Z*)-3,7-dimethyl-1,3,6-octatriene. Draw a structural formula for β-ocimene.

5.20 Draw the structural formula for at least one bromoalkene with the molecular formula C_5H_9Br that shows:

(a) Neither *E,Z* isomerism nor chirality.
(b) *E,Z* isomerism but not chirality.
(c) Chirality but not *E,Z* isomerism.
(d) Both chirality and *E,Z* isomerism.

5.21 Following are structural formulas and common names for four molecules that contain both a carbon-carbon double bond and another functional group. Give each an IUPAC name.

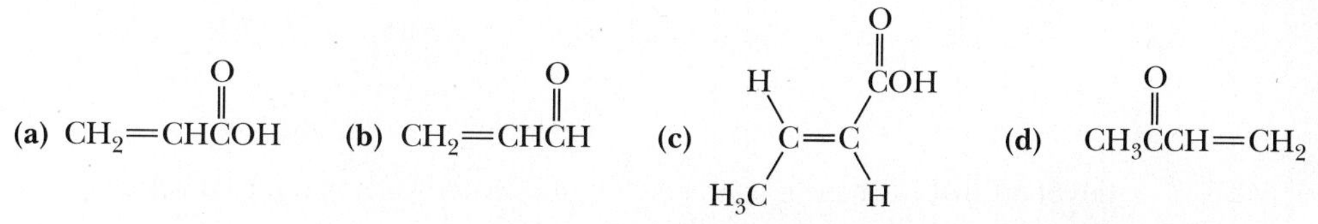

Acrylic acid **Acrolein** **Crotonic acid** **Methyl vinyl ketone**

5.22 *trans*-Cyclooctene has been resolved, and its enantiomers are stable at room temperature. *trans*-Cyclononene has also been resolved, but it racemizes with a half-life of 4 min at 0°C. How can racemization of this cycloalkene take place without breaking any bonds? Why does *trans*-cyclononene racemize under these conditions but *trans*-cyclooctene does not? You will find it especially helpful to examine the molecular models of these cycloalkenes.

5.23 Which alkenes exist as pairs of *cis,trans* isomers? For each that does, draw the *trans* isomer.

(a) $CH_2{=}CHBr$ **(b)** $CH_3CH{=}CHBr$ **(c)** $BrCH{=}CHBr$
(d) $(CH_3)_2C{=}CHCH_3$ **(e)** $(CH_3)_2CHCH{=}CHCH_3$

5.24 Four stereoisomers exist for 3-penten-2-ol.

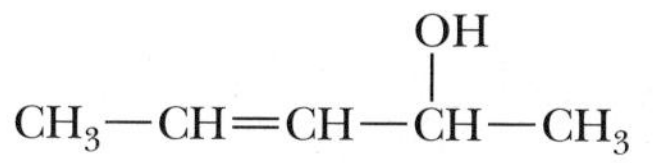

3-Penten-2-ol

(a) Explain how these four stereoisomers arise.
(b) Draw the stereoisomer having the *E* configuration about the carbon-carbon double bond and the *R* configuration at the chiral center.

Molecular Modeling

These problems require molecular modeling programs such as Chem 3D or Spartan to solve. Pre-built models can be found on the book companion site available at **www.cengagebrain.com**.

5.25 Measure the CH_3,CH_3 distance in the energy-minimized model of *cis*-2-butene and the CH_3,H distance in the energy-minimized model of *trans*-2-butene. In which isomer is the nonbonded interaction strain greater?

5.26 Measure the C=C—C bond angles in the energy-minimized models of the *cis* and *trans* isomers of 2,2,5,5-tetramethyl-3-hexene. In which case is the deviation from VSEPR predictions greater?

5.27 Measure the C—C—C and C—C—H bond angles in the energy-minimized model of cyclohexene and compare them with those predicted by VSEPR. Explain any differences.

5.28 Measure the C—C—C and C—C—H bond angles in the energy-minimized models of *cis* and *trans* isomers of cyclooctene. Compare these values with those predicted by VSEPR. In which isomer are deviations from VSEPR predictions greater?

Terpenes

5.29 Show that the structural formula of vitamin A (Section 5.4) can be divided into four isoprene units bonded head-to-tail and cross-linked at one point to form the six-membered ring.

Carotene and lycopene are polyenes occurring in tomatoes and carrots. Carotene is a natural source of vitamin A.

5.30 Following is the structural formula of lycopene, $C_{40}H_{56}$, a deep-red compound that is partially responsible for the red color of ripe fruits, especially tomatoes. Approximately 20 mg of lycopene can be isolated from 1 kg of ripe tomatoes. Lycopene is an important antioxidant that may help prevent oxidative damage in atherosclerosis.

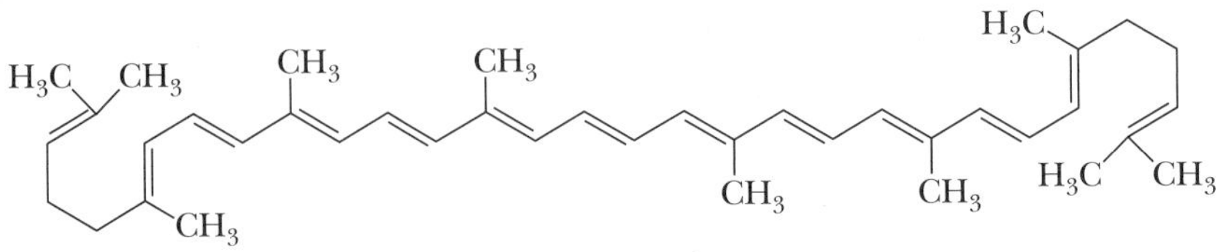
H3C
CH3
CH3
CH3
H3C
CH3
CH3
CH3
H3C
CH3

Lycopene

(a) Show that lycopene is a terpene (i.e., its carbon skeleton can be divided into two sets of four isoprene units with the units in each set joined head-to-tail).

(b) How many of the carbon-carbon double bonds in lycopene have the possibility for *cis,trans* isomerism? Of these, which are *trans* and which are *cis*?

5.31 As you might suspect, β-carotene, $C_{40}H_{56}$, precursor to vitamin A, was first isolated from carrots. Dilute solutions of β-carotene are yellow, hence its use as a food coloring. In plants, it is almost always present in combination with chlorophyll to assist in the harvesting of the energy of sunlight and to protect the plant against reactive species produced in photosynthesis. As tree leaves die in the fall, the green of their chlorophyll molecules is replaced by the yellows and reds of carotene and carotene-related molecules. Compare the carbon skeletons of β-carotene and lycopene. What are the similarities? What are the differences?

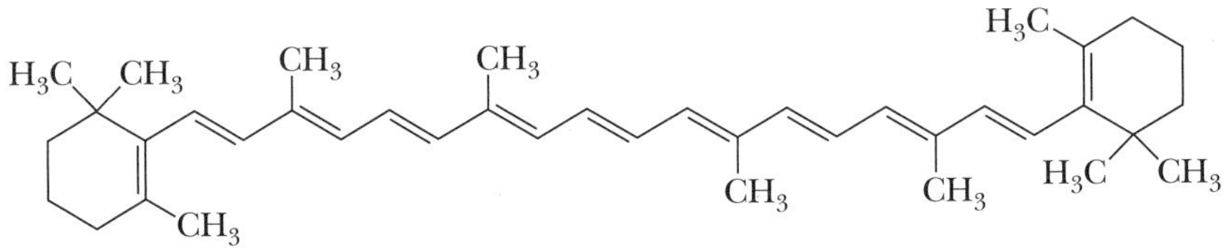
H3C
CH3
CH3
CH3
H3C
CH3
CH3
CH3
H3C
CH3

β-Carotene

5.32 Calculate the index of hydrogen deficiency for β-carotene and lycopene.

5.33 α-Santonin, isolated from the flower heads of certain species of *Artemisia*, is an anthelmintic (meaning against intestinal worms). This terpene is used in oral doses of 60 mg to rid the body of roundworms such as *Ascaris lumbricoides*. It has been estimated that over one-third of the world's population is infested with these slender, threadlike parasites.

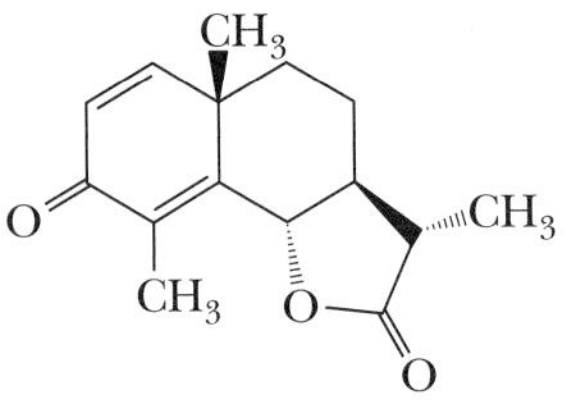
CH3
O
CH3
CH3
O
O

α-Santonin

α-Santonin can be isolated from the flower heads of wormwood, *Artemisia absinthium*. This plant has also been used to make the drink absinthe, popular in nineteenth-century France but now banned for its neurotoxicity.

(a) Locate the three isoprene units in santonin and show how the carbon skeleton of farnesol might be coiled and then cross-linked to give santonin. Two different coiling patterns of the carbon skeleton of farnesol can lead to santonin. Try to find them both.

(b) Label all chiral centers in santonin. How many stereoisomers are possible for this molecule?

(c) Calculate the index of hydrogen deficiency for santonin.

5.34 Pyrethrin II and pyrethrosin are two natural products isolated from plants of the chrysanthemum family. Pyrethrin II is a natural insecticide and is marketed as such.

(a) Label all chiral centers in each molecule and all carbon-carbon double bonds about which there is the possibility for *cis,trans* isomerism.
(b) State the number of stereoisomers possible for each molecule.
(c) Show that the bicyclic ring system of pyrethrosin is composed of three isoprene units.
(d) Calculate the index of hydrogen deficiency for each of these natural products.

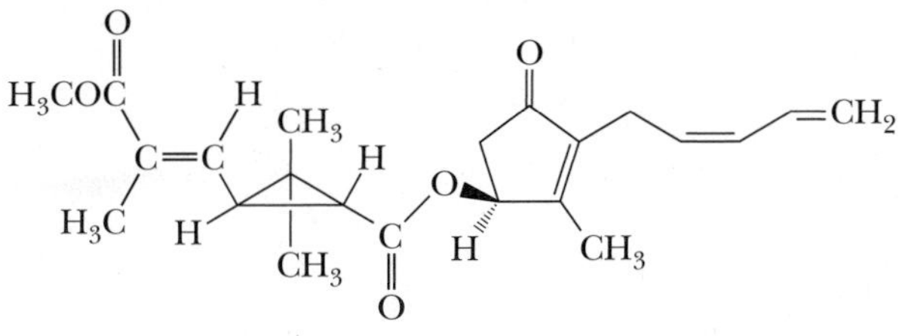

Pyrethrin II

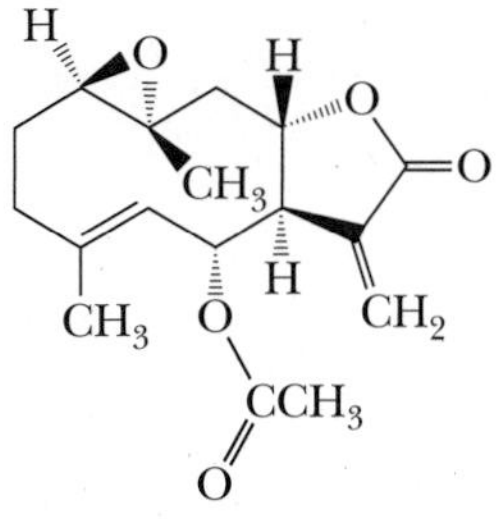

Pyrethrosin

5.35 Limonene is one of the most common inexpensive fragrances. Two isomers of limonene can be isolated from natural sources. They are shown below. The one on the left has the odor of lemons, and the one on the right has the odor of oranges.

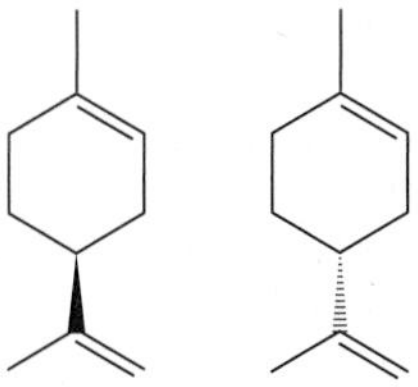

Isomers of Limonene

(a) What kind of isomers are they?
(b) Are *E,Z* isomers possible in limonene?
(c) Why do these two isomers smell different?

5.36 Nepetalactone is the active ingredient of catnip. It is isolated as an oil from the plant *Nepata cataria*. Show that it is a terpene (i.e., that its carbon skeleton can be divided into isoprene units). Is the molecule chiral? How many stereoisomers are possible?

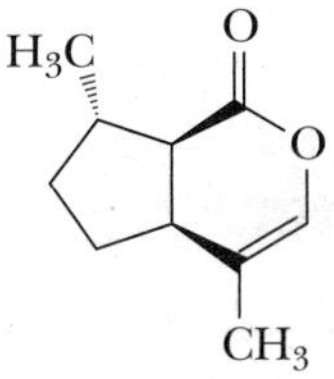

Nepetalactone
(oil of catnip)

Looking Ahead

5.37 Bromine adds to *cis*- and *trans*-2-butene to give different diastereomers of 2,3-dibromobutane. What does this say about the mode of addition of bromine to this alkene?

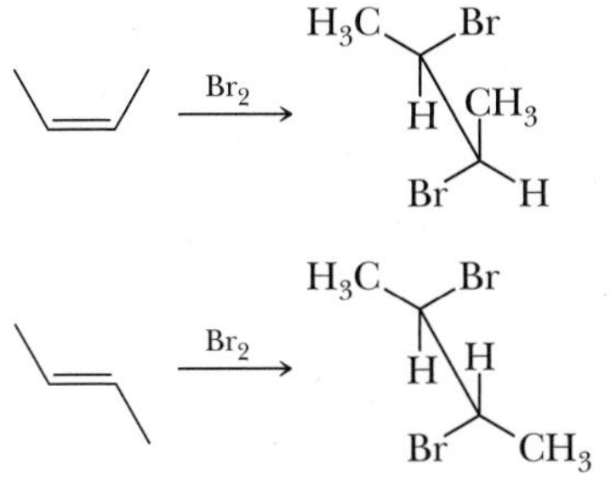

We discuss the addition of bromine to alkenes in Chapter 6.

Things You Should Know II

Nucleophiles and Electrophiles

One of the unifying principles of organic chemistry is that molecules with electron-rich areas, generally lone pairs, or in many cases bonds, have characteristic patterns of reactivity. Similarly, molecules with electron-poor areas or weak bonds also have characteristic patterns of reactivity. Such electron-rich and electron-poor species are described by three different sets of terms. The following section is intended to clarify the confusion students often experience when they initially encounter these somewhat overlapping sets of terms.

In the context of acid-base chemistry, Chapter 4 introduced Brønsted-Lowry and Lewis acid and base definitions. The Brønsted-Lowry definitions are restricted to proton transfers. For other reactions, chemists can refer to the reactants as being Lewis acids and bases (Section 4.7). Recall that a Lewis acid is a species that can accept an electron pair from a Lewis base by virtue of the fact that the Lewis acid has an empty orbital while the Lewis base has the electron pair. An exemplary case is the coordination of ammonia to borane (see below). Importantly, Section 4.5C showed that a Brønsted-Lowry acid, such as H—X, also has an empty orbital (i.e., the antibonding H—X sigma orbital) that can accept a lone pair from a base, thereby breaking the H—X bond leading to proton transfer. Thus, strictly speaking, a proton transfer reaction might *also* be classified as a Lewis acid-base reaction.

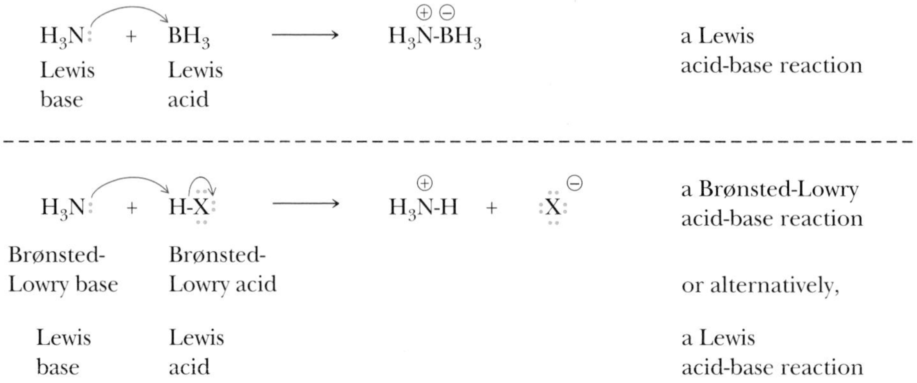

Clearly, then, the Lewis acid-base definition is broader—in fact, it is so broad that many chemists consider most reactions (excluding those involving radicals, Chapter 8) as falling under the definition of a Lewis acid-base interaction. However, in practice, most chemists commonly refer to Brønsted-Lowry acid-base reactions as proton transfers, and this is the terminology we will use throughout this book. In fact, from Chapter 6 onward, the Brønsted-Lowry base will commonly be an organic functional group, such as an alkene, alcohol, or ester. In writing reaction mechanisms, we will use terminology such as "add a proton" or "take a proton away" (see "Things You Should Know III" on Reaction Mechanisms immediately following this section) to describe proton transfers.

While the Lewis terminology is very broad, it is the most common way to describe a situation in which a new bond is created by electron-donation from the base to the acid without breaking other bonds and when the newly formed bond is dative. A **dative bond** (alternatively called a *coordinate covalent bond*, see Section 1.2F) occurs when the two electrons shared between the atoms derive from only one of the two reactants (in this case the Lewis base). The ammonia-borane interaction involves formation of a dative bond because it fits both criteria; the bond is created without breaking other bonds and the new bond is best thought of as having the two electrons coming from only the nitrogen atom of the ammonia.

Looking forward to Chapter 6, addition of a lone pair from a halogen (X = F, Cl, Br, I) to a carbocation might be considered to be a Lewis acid-base reaction. Once again, the bond is created without the breaking of other bonds and is formed solely from the addition of a lone pair from the halogen to the carbocation carbon. All chemists recognize this as a type of Lewis acid-base interaction, as we briefly mentioned in Section 4.7. Yet, a C—X bond is not considered dative. Thus, as we describe below, while organic chemists might recognize this as a Lewis acid-base reaction, we have a different set of more specific terms that are more commonly used because the new bond being formed is not considered to be a dative bond.

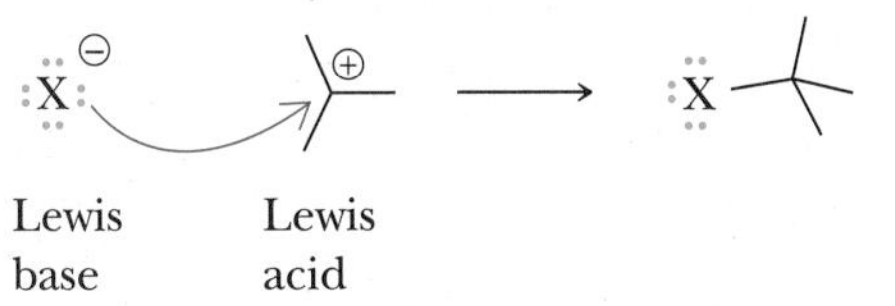

a Lewis acid-base reaction

Before looking at these new terms in detail, we need to explain why the C—X bond is not dative. This can best be seen by considering Lewis dot structures. Recall that Lewis dot structures are generated from neutral groups for any molecule, independent of the chemical reactions that could create that molecule. In constructing a Lewis dot structure for $H_3N—BH_3$ from two neutral groups (NH_3 and BH_3), the two electrons in the N-B bond arise from the nitrogen due to the electron count on ammonia (eight electrons) and borane (only six electrons). Hence, the N—B bond is dative. However, for a C—X bond, let's say in CH_3—Cl, the Lewis dot structure constructed from neutral groups would have seven electrons on $CH_3\cdot$ and seven coming from the chlorine atom (Cl·), and thus the bond is considered to arise from a sharing of one electron from each group. Therefore, the C—Cl bond is considered to be covalent, not dative.

Nucleophiles and Electrophiles: A Comparison to Lewis and Brønsted-Lowry Definitions

When discussing reaction mechanisms in which a new covalent (as opposed to dative) bond is created and the reaction is not a proton transfer, a third set of terms is often used to describe the reactants. The electron-rich species is often referred to as the *nucleophile* and the electron deficient species as the *electrophile*. As we introduced briefly in Chapter 4, nucleophile means, "nucleus loving," or seeking positive charge. Electrophile means, "electron loving," or seeking negative charge. Further, the region of the molecule that is electron rich, commonly a lone pair or a bond, will be referred to as nucleophilic. Analogously, the electron-poor region of a molecule, typically carrying a partial positive charge or a weak bond, will be referred to as electrophilic. Moreover, the relative reactivity of nucleophiles or electrophiles is referred to as the nucleophilicity or electrophilicity of the reactants, respectively. The above terms are commonly evoked for *polar reactions*, those involving reactions between full or partial negative charges with full or partial positive charges. In this book, we describe mechanistic steps that create new covalent bonds that are not proton transfers as "make a new bond between a nucleophile and electrophile" (the next "Things You Should Know III" section discusses this in detail).

The confusing part is that although the three sets of terms are distinct, they are overlapping and can be used somewhat interchangeably by chemists in certain situations. Let's take a look at some examples of how the various terms are used in mechanisms that are coming up in Chapter 6. In example A, shown below, an alkene acts as a Brønsted-Lowry base, and thus we will refer to this as "add a proton," even though some chemists might call it a reaction between Lewis acids and bases or electrophiles and nucleophiles. Because example B does not involve a proton transfer and the bond formed is dative, we will refer to this as a Lewis acid-base reaction. Here, the arrow between the π bond of the alkene and the boron signifies that the bond is dative. Note that some chemists might refer to this as a combination between a nucleophile and an electrophile, but we will not. Next, example C is a repetition of the one given in the discussion of Lewis interactions and dative bonds, but we can now appreciate that it is best described as "make a bond between a nucleophile and an electrophile" because a covalent bond is formed. Nevertheless, some chemists might refer to this as a Lewis acid-base reaction. Lastly, in example D the alkene is again acting as a nucleophile because it is donating electrons to the bromine molecule, which, in turn, must be considered the electrophile. (What makes molecules electrophilic is discussed in Chapter 6 and briefly below.) Once again, because the Lewis definition is so broad, this reaction could be considered a Lewis acid-base combination by some chemists. In this later example, a bond is also breaking to expel a stable ion (bromide).

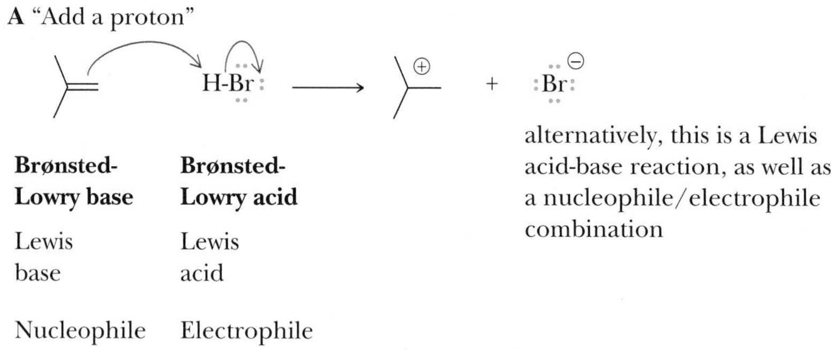

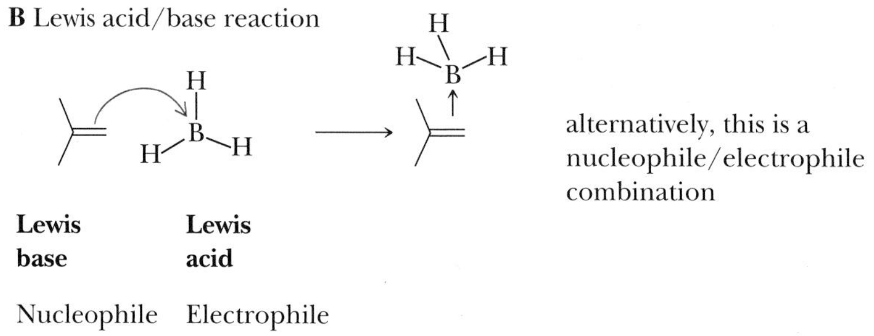

C "Make a bond between a nucleophile and an electrophile"

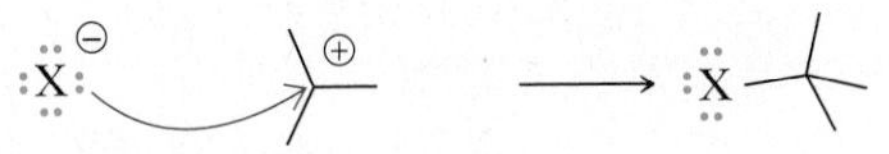

alternatively, this is a
Lewis acid-base reaction

Nucleophile	**Electrophile**
Lewis base	Lewis acid

D "Make a bond between a nucleophile and an electrophile"

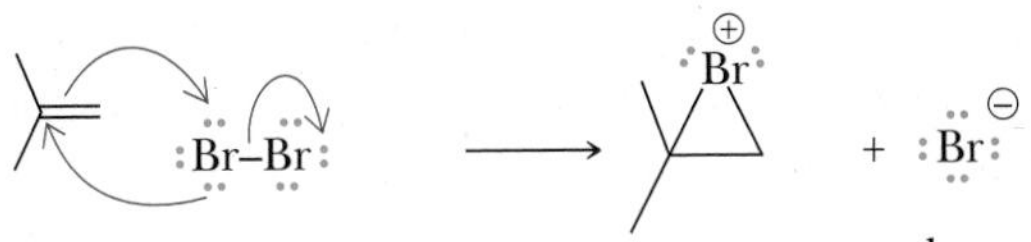

alternatively, this is a
Lewis acid-base reaction

Nucleophile	**Electrophile**
Lewis base	Lewis acid

Although a full or partial negative charge, or alternatively a bond or lone pair, are all possible nucleophilic sites in a molecule, it is sometimes harder to recognize what makes a compound electrophilic. The most obvious is an empty orbital or a full/partial positive charge. Yet, weak bonds are also electrophilic. In the last example, Br_2 is the electrophile due to the relative weak **Br—Br** bond. In actuality, there is a unifying principle in operation here. Weak bonds have low-energy empty antibonding orbitals that can accept electron density from a nucleophile. Hence, these bonds are also electrophilic due to an empty orbital. Filling this orbital leads to breakage of that bond.

In summary, we have classified three sets of related and sometimes overlapping terms to describe chemical reactions in which electron-rich regions of molecules react with electron-poor regions of molecules. Although having three overlapping sets of terms might seem confusing at first, the numerous examples laid out in every chapter of this book will clarify the terminology. As a staring point, it is best to keep the following in mind:

Although all these reactions can in some sense be considered Lewis acid-base combinations, those involving proton transfer are best referred to as Brønsted-Lowry reactions, those involving covalent bond formation are best referred to as combinations of nucleophiles and electrophiles and only those involving dative bond formation are best referred to as Lewis acid-base combinations.

Things You Should Know III

Reaction Mechanisms

Mastering reaction mechanisms is a critical step you must take to begin to understand, as opposed to simply memorize, chemical reactions in organic chemistry. Being able to rationalize reaction outcomes on the basis of a surprisingly small number of fundamental principles will provide you with the ability to group reactions according to common mechanisms and even predict reaction products for unfamiliar reactions. This section teaches you the formal notation used by chemists to describe chemical reaction mechanisms, shows you how to avoid making common mistakes, and presents the four fundamental mechanistic elements found in the majority of chemical reactions. This information is so important to your ultimate success in understanding organic chemistry that it was put in this format for special emphasis and easy reference.

A. Developing a Reaction Mechanism

To develop an understanding of a reaction mechanism, chemists begin by first using a combination of experience and intuition to propose several sets of steps, or mechanisms, each of which might account for the overall chemical transformation. Next, chemists test each proposed mechanism by designing and carrying out experiments that provide experimental observations that will allow them to exclude those mechanisms that are not consistent with the facts. Modern computational methods now allow detailed predictions of mechanistic pathways more accurately than ever before, but the results must still be compared with experiment for confirmation.

A mechanism becomes generally established by excluding reasonable alternatives and by showing that it is consistent with every test that can be devised. This, of course, does not mean that a generally accepted mechanism is a completely accurate description of the chemical events, only that it is the best mechanism chemists have been able to devise. It is important to keep in mind that as new experimental evidence is obtained, it may be necessary to modify a generally accepted mechanism or possibly even discard it in favor of an alternative.

Before we go on to consider reactions and reaction mechanisms, we might ask why chemists go to the trouble of establishing them and why you must spend time learning about them. One reason is very practical: mechanisms provide a framework within which to organize a great deal of descriptive chemistry. For example, with insight into how reagents add to particular alkenes, it is possible to generalize and then predict how the same reagents might add to other alkenes. A second reason lies in the intellectual satisfaction derived from constructing models that accurately reflect the behavior of chemical systems. Finally, to a creative scientist, a mechanism is a tool to be used in the search for new information and new understanding. A mechanism consistent with all that is known about a reaction can be used to make predictions about chemical interactions as yet unexplored, and experiments can be designed to test these predictions. Thus, reaction mechanisms provide a way not only to organize knowledge but also to extend it.

B. The Correct Use of Arrows to Indicate Electron Movement

The ability to write an organic reaction mechanism properly is key to student success in organic chemistry classes. Organic chemists use a technique called **electron pushing** (sometimes called arrow pushing) to depict the flow or movement of electrons during chemical reactions. Arrow pushing helps chemists keep track of the way in which electrons and their associated atoms redistribute as bonds are made and broken. The first essential rule to keep in mind is the following:

Electron pushing
The use of arrows in organic chemistry mechanisms to indicate the flow or movement of electrons.

First Rule: Arrows are used to indicate movement of *electrons*.

A regular arrow (double-sided arrowhead) is used to indicate the movement of two electrons, while a line with a single-sided arrowhead (sometimes called a "fish hook arrow") is used for single electron movement involved with radical reactions that are first described in Chapter 8.

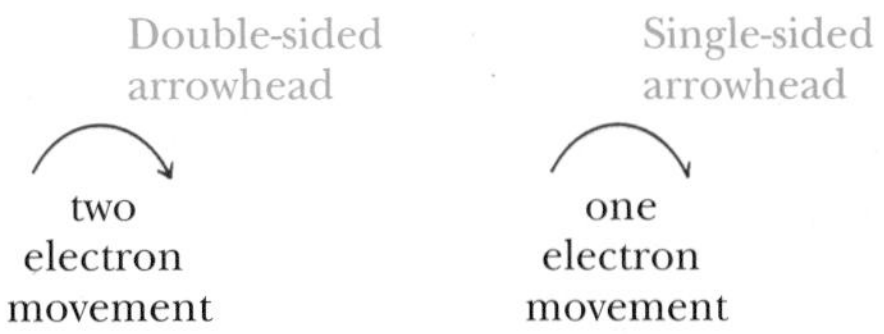

The majority of reactions discussed in this book will involve movement of pairs of electrons. Hence, the mechanisms will be represented by double-sided arrowheads.

Arrow pushing was first introduced in Section 1.8A in the discussion of resonance contributing structures. Recall that in a comparison of two or more contributing structures, an arrow was used to show how two electrons (lines representing bonds or pairs of dots representing lone pairs) could be redistributed within a single chemical structure to create an alternative Lewis line structure representation of the bonding. By convention, arrows are used to keep track of *all* pairs of electrons that are in different locations in the two different contributing Lewis line structures, shown here for the acetate anion and benzene molecule.

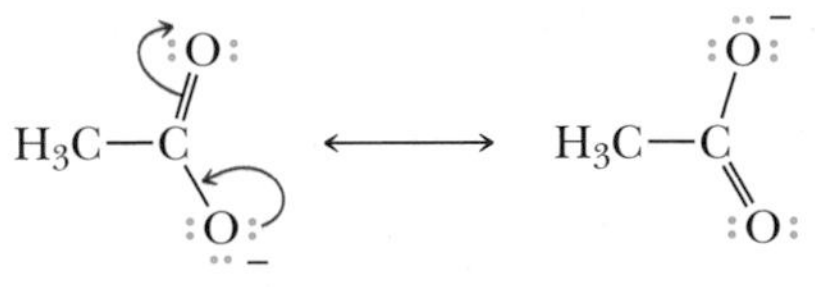

Acetate anion contributing structures

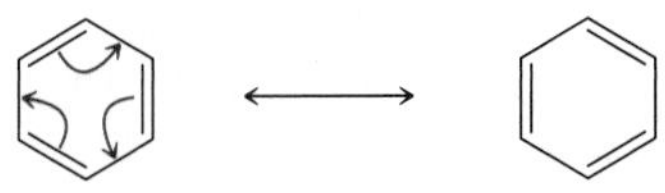

Benzene contributing structures

Keep in mind that in the case of resonance, (1) the atoms do not move between contributing structures and (2) the electrons are not actually moving. The true chemical structure should be thought of as a hybrid of the contributing Lewis line structures. It is worth pointing out that when used with contributing structures, arrows generally indicate only the interconversion of π bonds and lone pairs (acetate anion) or just π bonds (benzene), not the formation or breaking of σ bonds.

In chemical reactions, both electrons and atoms change positions as both π and σ bonds are formed and broken. Arrow pushing is used to keep track of the movement of all electrons involved with each step of the overall transformation. Because electrons are located in orbitals surrounding atoms, when bonds are formed or broken, the movement of electrons between orbitals is necessarily accompanied by the movement of the associated atoms, which leads to the second rule of arrow pushing when depicting chemical reaction mechanisms.

Second Rule: Arrows are never used to indicate the movement of atoms directly. The arrows only show atom movement indirectly as a consequence of electron movement when covalent bonds are made and broken.

We already used arrow pushing to show proton transfer several times in Chapter 4. The example below shows the transfer of a proton from the relatively acidic acetic acid molecule to the relatively basic hydroxide anion. We show this process with one arrow (labeled "a" in the diagram) that starts at a lone pair of electrons on the basic oxygen atom of the hydroxide anion and then points to the acidic H atom of acetic acid to indicate formation of the new bond being made. A second arrow originates at the line representing the breaking O—H bond and points to the O atom to denote creation of a lone pair (arrow "b"). In this reaction, the proton is being transferred between molecules, whereas the arrows indicate movement of the electrons involved.

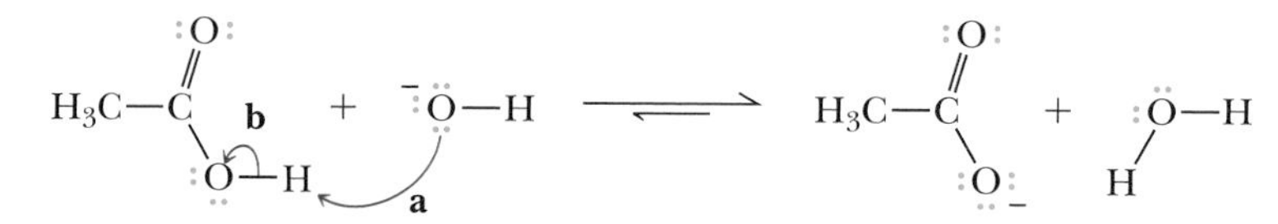

Correct use of arrows to indicate electron movement during a reaction

A common mistake for beginning students is to erroneously write an arrow pointing *from* the H of the acetic acid *to* the O atom of the hydroxide anion. This is wrong, because such an arrow would be indicating the H *atom* movement directly, not *electron* movement. Other common mistakes in arrow pushing are given in Things You Should Know IV.

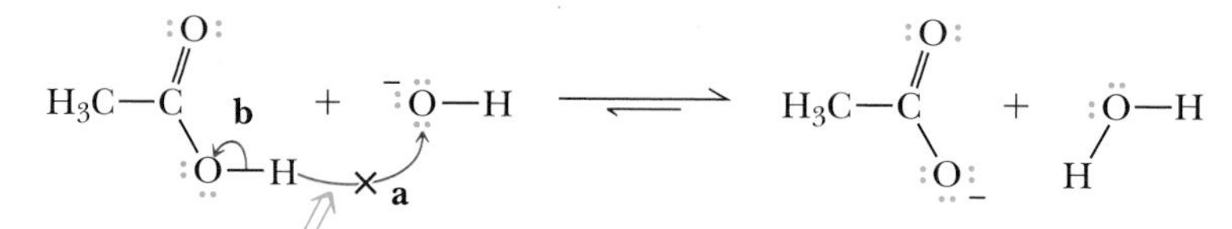

Incorrect arrow because it is pointing in the wrong direction! Never use arrows to indicate atom movement directly

C. Electron Sources and Sinks: How to *Predict* What Will Occur in an Organic Reaction Mechanism

We have now seen all three of the situations illustrated by arrows with double-sided arrowheads, namely the redistribution of π bonds and/or lone pairs, formation of a new σ bond (generally from a lone pair or π bond), and breaking of a σ bond (generally to form a new lone pair or sometimes a new π bond). Often, as in the case of the acetic acid–hydroxide ion reaction, more than one arrow is used in a given mechanism step. Now that you have seen all of the important types of arrows, we can point out their most important common feature.

Third Rule: Arrows always start at an **electron source** and end at an **electron sink**.

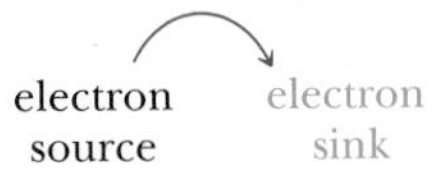

An **electron source** is a bond or a lone pair of electrons. It is most commonly a π bond or a lone pair on an atom of relatively high electron density or a σ bond that must break during a reaction. An **electron sink** is an atom on a molecule or an ion that can accept a new bond or lone pair of electrons.

Electron source

A bond or lone pair of electrons, usually an area of relatively high electron density, that serves as the origin of a mechanism arrow. An electron source characteristically interacts with an electron sink in an organic mechanism.

Electron sink

An atom that accepts a new bond or lone pair of electrons. An electron sink is what an arrow points toward in an organic chemistry mechanism.

Learning to identify the characteristic sources and sinks in different functional groups is the key to learning organic chemistry reaction mechanisms. For example, for arrows that depict the formation of a new σ bond between two molecules, the electron source is often readily identified as being a lone pair on the most electron-rich atom of one molecule or ion and the electron sink is readily identified as the most electron-poor atom of the other molecule or ion. Thus, the prediction of many of the most important electron sources and sinks comes down to lessons concerning the differences in electronegativity between atoms that were presented in Section 1.2, which allow you to identify the most electron-rich and electron-poor atoms in molecules. As an aid to your analysis, the red and blue colors of the various electrostatic surface maps given throughout this book indicate the negative (electron-rich) and positive (electron-poor) regions of molecules. We will have more to say about this reactivity pattern a little later.

This leads us to another commonly encountered type of process that deserves mention. As you will see in many later chapters, making a new bond to an electron sink often requires the simultaneous breaking of one of the bonds present at the sink atom to avoid overfilling its valence orbitals, a situation referred to as hypervalence.

Fourth Rule: Bond breaking will occur to avoid overfilling valence (hypervalence) on an atom serving as an electron sink.

In these cases, the electron source for the arrow is the bond being broken and the sink is an atom able to accommodate the electrons as a lone pair, generally an electronegative atom such as an O atom or a halogen. If an ion is created, that ion is often stabilized by resonance delocalization or other stabilizing interactions.

Returning to the proton transfer reaction between acetic acid and hydroxide, we can now summarize our analysis of this simple one-step mechanism.

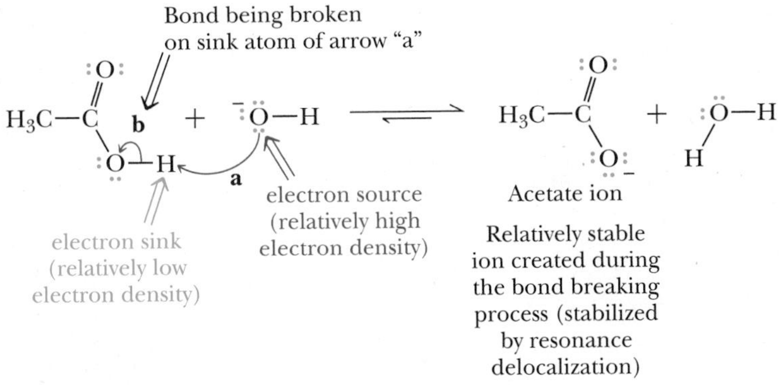

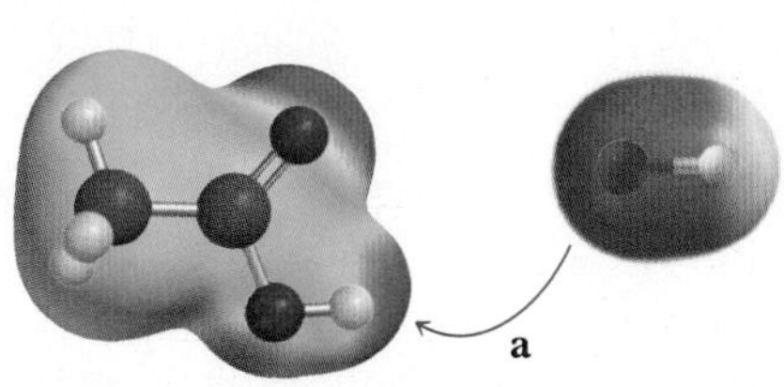

Considering the arrow used to make a new σ bond (arrow "a"), the hydroxide O atom is the electron source (most negatively charged atom) and the acetic acid H atom is the electron sink (atom with highest partial positive charge). In the electrostatic molecular surfaces depicted above, the oxygen in the hydroxide ion has the greatest localized negative charge. This is indicated by the most intense red color. Meanwhile, the acetic acid proton being transferred has the largest positive charge, which is indicated by the most intense blue color. To avoid overfilling the valence of the H atom during the reaction the O—H bond of acetic acid must be broken (arrow"b"). In so doing, the acetate ion is formed. Note that the acetate ion is stabilized by resonance delocalization.

You should now appreciate that the transfer of a proton (a so-called Brønsted acid-base reaction) is really just a special case of the common pattern of reactivity between an electron source (the base) and the proton as an electron sink, combined with the breaking of a bond to satisfy valence and create a relatively stable ion.

The addition or removal of protons during chemical reactions is common and is often referred to as "adding a proton" or "taking a proton away."

Meanwhile, a broader terminology is also applied to reactions in which new σ bonds form between electron-rich and electron-poor regions of molecules. **Nucleophiles** (meaning nucleus seeking) are molecules that have lone pairs or relatively electron-rich π bonds that act as electron sources for arrows making new bonds. **Electrophiles** (meaning electron seeking) are molecules with relatively electron-poor atoms that serve as sinks. Analogously, a molecule (or region of a molecule) that is a source for such an arrow is called nucleophilic, while a molecule or region of a molecule that is a sink for these arrows is referred to as being electrophilic. The term *nucleophile* is analogous to Lewis bases, and the term *electrophile* is analogous to Lewis acids. The choice of terminology depends on context. Nucleophile and electrophile are used more commonly in kinetics discussions, while Lewis acid and Lewis base are used in discussions of reaction thermodynamics.

Nucleophile

From the Greek meaning nucleus loving, seeking a region of low electron density. An electron source that can donate a pair of electrons to form a new covalent bond; alternatively, a Lewis base.

Electrophile

From the Greek meaning electron loving, seeking a region of high electron density. An electron sink that can accept a pair of electrons to form a new covalent bond; alternatively, a Lewis acid.

It is helpful to summarize the appropriate use of key terms associated with arrow pushing and reaction mechanisms. The terms *source* and *sink* are used to identify the start and end, respectively, of each reaction mechanism arrow, which is indicating the change in location of electron pairs. The terms *nucleophile* and *electrophile* (as well as *Lewis base* and *Lewis acid*) are used to describe molecules based on their chemical reactivity and propensity to either donate or receive electrons when they interact. Protons can be thought of as a specific type of electrophile, and for reactions in which a proton is transferred, the nucleophile is called a base.

Example 1 Avoiding Arrow Pushing Mistakes I

The following two sets of reactions [(a) and (b)] show possibilities for arrow pushing in individual reaction steps. Identify which is wrong and explain why. Then using arrow pushing correctly, label which molecule is the nucleophile and which is the electrophile.

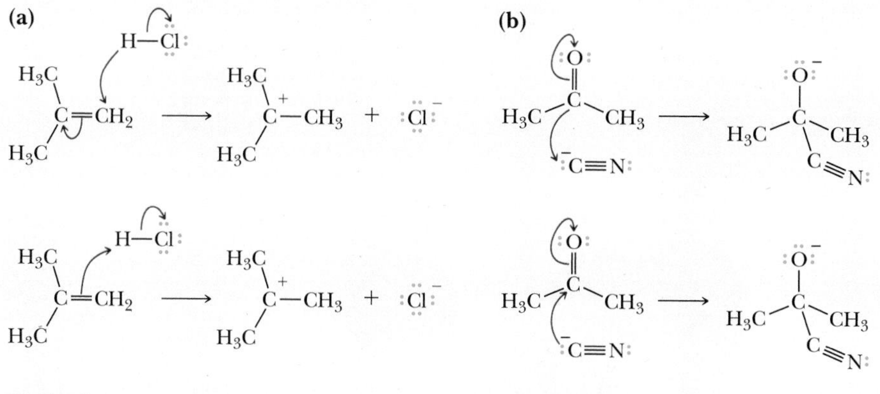

Solution

In each case, the first arrow pushing scenario is wrong. The arrows shown below with stars over them do not start at a source of electrons; rather, they start at positions of relative positive charge, which is incorrect. In the correct arrow pushing, the arrow labeled "a" depicts the interaction of a region of relatively high negative charge (a π-bond or lone pair) with an atom of relatively high partial positive charge on the other reactant. Therefore, the molecule acting as the source for the σ bond-forming arrow "a" is the nucleophile, while the molecule containing the sink atom is the electrophile.

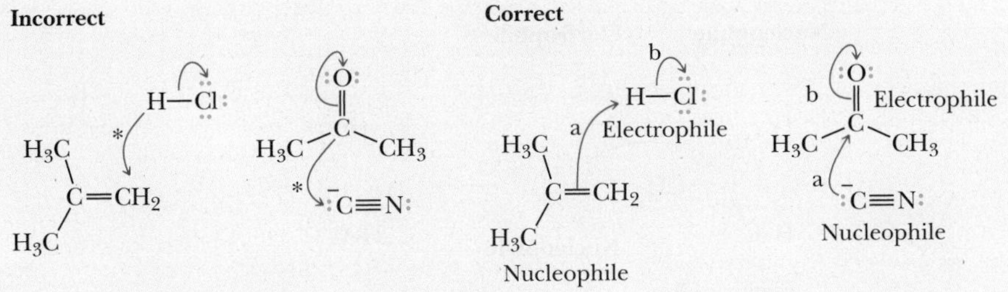

The arrow labeled "b" is needed to satisfy valence and is not considered when defining the nucleophile and electrophile.

Example 2 Avoiding Arrow Pushing Mistakes II

The following two sets of reactions [(a) and (b)] show possibilities for arrow pushing in individual reaction steps. Identify which is wrong and explain why. Then using the correct arrow pushing, label which molecule is the nucleophile and which is the electrophile.

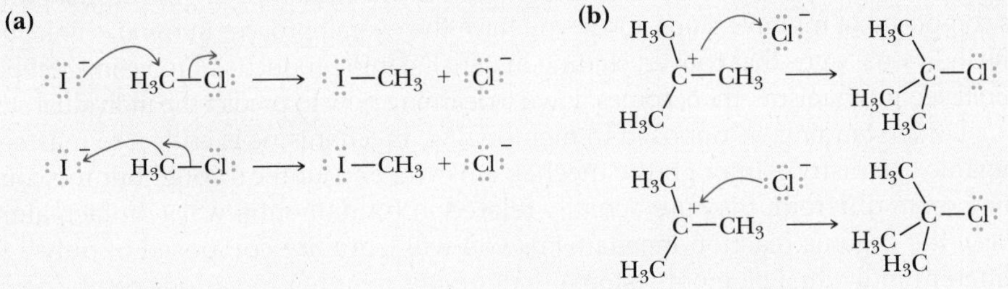

Solution

The incorrect scenarios have arrows pointing the wrong way, namely from electron sinks to electron sources.

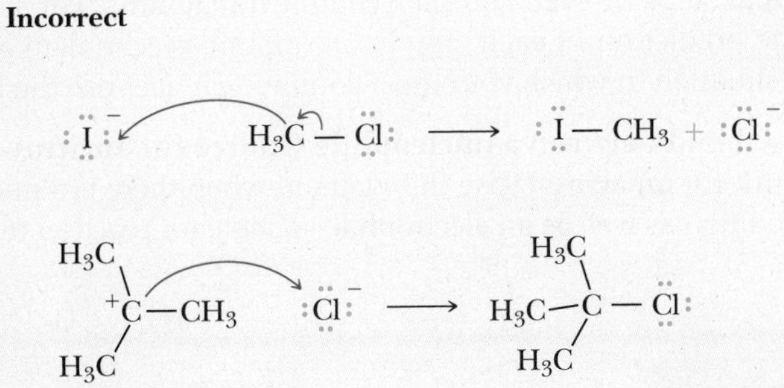

The correct scenarios have arrows pointing from electron sources, areas of relatively high electron density, to electron sinks, atoms of relatively low electron density. In each case, the arrows labeled "a" indicate a lone pair on the

(Continued)

halide anion (nucleophile) reacting with an electrophilic carbon atom, a carbon with a full or partial positive charge. The arrow labeled "b" on methyl chloride is needed to satisfy valence.

Correct

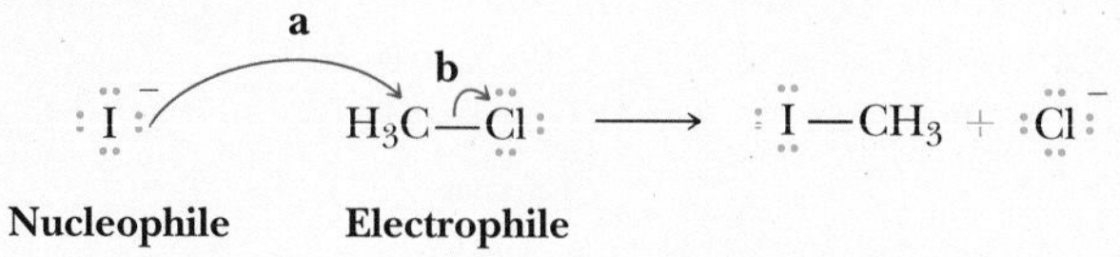
a
b
I
H3C—Cl
I—CH3 + Cl
Nucleophile
Electrophile

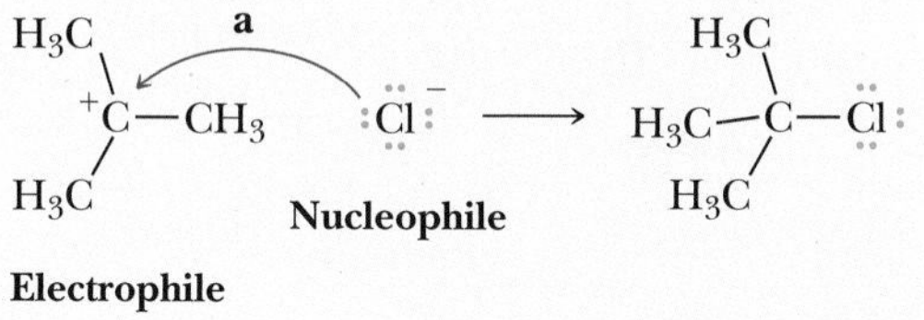
H3C
a
C—CH3
H3C
Cl
Nucleophile
H3C
H3C—C—Cl
H3C
Electrophile

D. Putting it All Together: A Multiple-Choice Situation

In the sections and chapters that follow, many different reaction mechanisms will be described in a stepwise fashion. Each arrow can be classified according to one of the three overall situations we already encountered: (1) redistribution of π bonds and/or lone pairs, (2) formation of a new σ bond from a lone pair or π bond, and (3) breaking of a σ bond to give a new lone pair or π bond. When learning new mechanisms, focus on the overall transformation that takes place. It might be a reaction in which atoms or groups are added (an addition reaction), a reaction in which atoms or groups are removed (an elimination reaction), a reaction in which atoms or groups replace an atom or a group (a substitution reaction), or other processes we will encounter. Often, the overall process is composed of multiple steps. Once you have the overall process in mind, think about the individual steps that convert starting material(s) into product(s). Predicting complete multistep mechanisms, then, comes down to learning how to predict the individual steps.

Understanding, as opposed to memorizing, mechanisms is critical to mastering organic chemistry. Although the mechanisms you encounter throughout the course may seem different, they are actually related in fundamental ways. In fact, almost all of the organic reaction mechanisms you will learn are composed of only a few different individual elements (steps) that are put together in various combinations. Your job is to learn these individual mechanism elements and then understand how to assemble them into the steps of the correct mechanism for an overall reaction.

Fortunately, there are a surprisingly small number of different types of characteristic mechanism elements (patterns of arrows) to be considered when trying to predict individual steps of even complex chemical reactions. For this reason, you should view the prediction of each step in an organic mechanism as essentially a *multiple-choice* situation in which your most common choices are the following four:

1. **Make a new bond between a nucleophile (source for an arrow) and an electrophile (sink for an arrow).** Use this element when there is a nucleophile present in the solution as well as an electrophile suitable for reaction to occur.

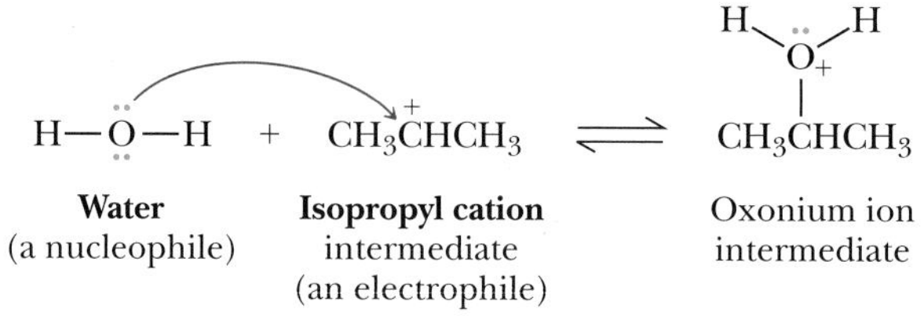
H—O—H
CH3CHCH3
H
H
O
CH3CHCH3
Water
(a nucleophile)
Isopropyl cation
intermediate
(an electrophile)
Oxonium ion
intermediate

2. **Break a bond so that relatively stable molecules or ions are created.** Use this element when there is no suitable nucleophile-electrophile or proton transfer reaction but breaking a bond can create neutral molecules or relatively stable ions or both.

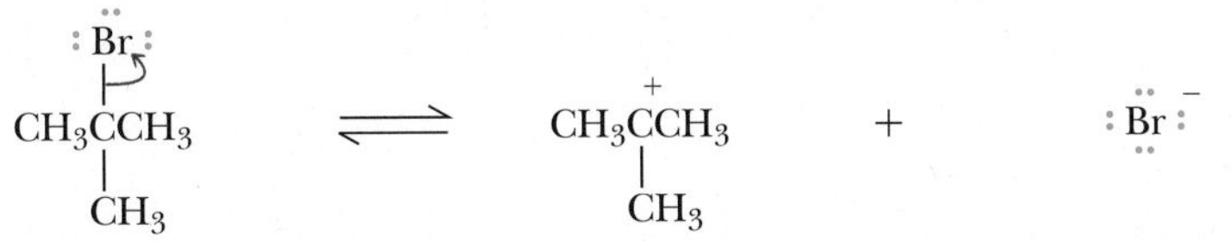

2-Bromo-2-methylpropane | ***tert*-Butyl cation** intermediate (stabilized by inductive and hyperconjugation effects. See Section 6.3A) | **Bromide ion** (relatively stable anion)

3. **Add a proton.** Use this element when there is no suitable nucleophile-electrophile reaction but a strong acid is present or the molecule has a strongly basic functional group.

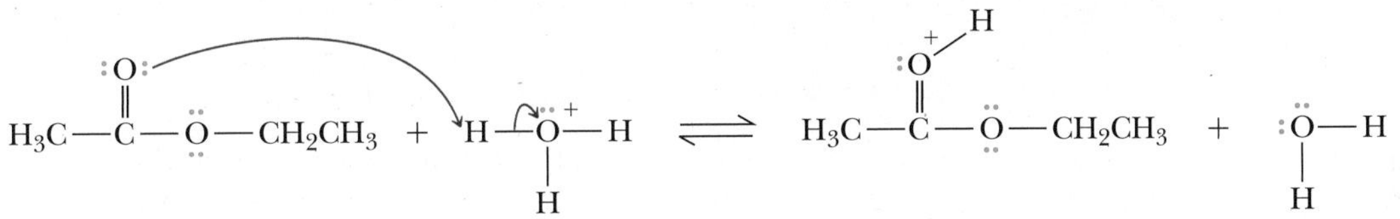

Ethyl acetate (a carboxylic ester) | **Hydronium ion** (a strong acid) | A protonated intermediate | **Water**

4. **Take a proton away.** Use this element when there is no suitable nucleophile-electrophile reaction but the molecule has a strongly acidic proton or a strong base is present.

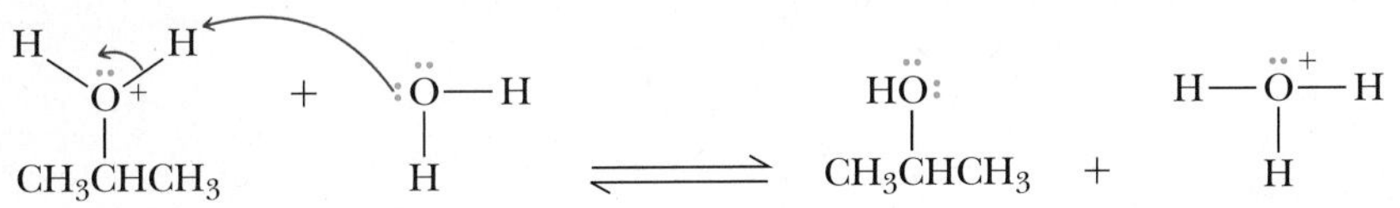

Oxonium ion intermediate (strong acid) | **Water** (can act as a base) | **2-Propanol** | **Hydronium ion**

The situation is even simpler than you might expect because 1 and 2 are the functional reverse of each other, as are 3 and 4. Many times, more than one of the four choices occurs simultaneously in the same mechanistic step.

These four mechanistic elements are the foundation upon which you will build an understanding of organic reaction mechanisms. They are far and away the most common; however, they are not the only mechanistic elements you will see. Once in a while, you will encounter a reaction with a unique mechanistic element specific to the particular molecules involved. For example, Section 6.4 describes a mechanistic element unique to hydroboration reactions. In addition, we will describe other important mechanistic elements, such as *1,2 shifts* (Section 6.3C), that occur in several reactions.

When you are able to predict which of the common mechanistic elements is/are most appropriate for a given step in a mechanism, you will be able to push electrons correctly without relying on memorization. Moreover, you will have taken a major step toward mastering organic chemistry. Throughout the rest of this book, we will label each mechanistic step as one of the common ones when appropriate, emphasizing analogous features even between complex mechanisms.

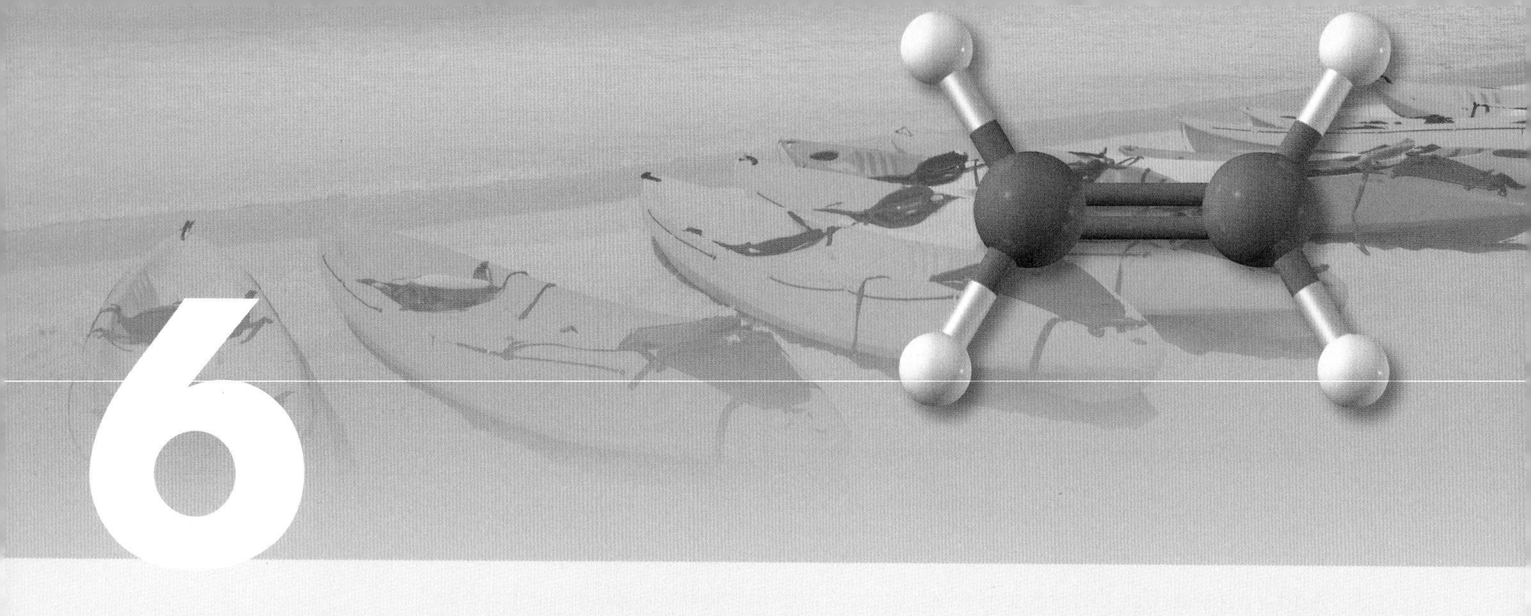

6 Reactions of Alkenes

These kayaks are fabricated from polyethylene. **Inset:** a model of ethylene, the alkene monomer from which polyethylene is derived. (© Pichugin Dmitry/Shutterstock.com)

Outline

6.1 Reactions of Alkenes—An Overview

6.2 Organic Reactions Involving Reactive Intermediates

6.3 Electrophilic Additions

6.4 Hydroboration-Oxidation

6.5 Oxidation

HOW TO: Write a Balanced Half-Reaction

6.6 Reduction

6.7 Molecules Containing Chiral Centers as Reactants or Products

Over 10 million organic compounds have been discovered or made by organic chemists! It would seem to be an almost impossible task to learn the chemical properties of so many compounds. Fortunately, the study of organic compounds is not as formidable a task as you might think. Although organic compounds can undergo a wide variety of chemical reactions, only certain portions of their structures are changed in any particular reaction. As we will see in this chapter, the same functional group, in whatever organic molecule we find it, undergoes the same types of chemical reactions. Therefore, you do not have to study the chemical reactions of even a fraction of the 10 million known organic compounds. Instead, you need only identify a few characteristic types of functional groups and then study the chemical reactions that each undergoes. ▶

▶ Functional groups react the same in different molecules.

6.1 Reactions of Alkenes—An Overview

The most characteristic reaction of alkenes is **addition** to the carbon-carbon double bond in such a way that the π bond is broken and, in its place, σ bonds form to two new atoms or groups of atoms. Table 6.1 gives several examples of reactions at a carbon-carbon double bond along with the descriptive name(s) associated with each. Some of these reactions are treated separately under oxidations (Section 6.5) and reductions (Section 6.6), but are included in this table because they are formally additions.

In the following sections, we study these alkene reactions in considerable detail, paying particular attention to the mechanism by which each occurs.

Another **addition reaction** of alkenes is the formation of chain-growth polymers (Greek: *poly,* many, and *meros,* part). In the presence of certain catalysts called initiators, many alkenes form polymers made by the addition of monomers (Greek: *mono,* one, and *meros,* part) to a growing polymer chain as illustrated by the formation of polyethylene from ethylene.

Addition Reaction
A reaction in which two atoms or ions react with a double bond, forming a compound with the two new groups bonded to the carbon of the original double bond.

$$n\,CH_2{=}CH_2 \xrightarrow{\text{initiator}} \left(CH_2CH_2\right)_n$$

Table 6.1 Characteristic Alkene Addition Reactions

Reaction	Descriptive Name(s)
$C{=}C + HCl\ (HX) \longrightarrow$ H–C–C–Cl (X)	Hydrochlorination (hydrohalogenation)
$C{=}C + H_2O \longrightarrow$ H–C–C–OH	Hydration
$C{=}C + Br_2\ (X_2) \longrightarrow$ (X) Br–C–C–Br (X)	Bromination (halogenation)
$C{=}C + Br_2\ (X_2) \xrightarrow{H_2O}$ HO–C–C–Br (X)	Bromo(halo)hydrin formation
$C{=}C + Hg(OAc)_2 \xrightarrow{H_2O}$ HO–C–C–HgOAc	Oxymercuration
$C{=}C + BH_3 \longrightarrow$ H–C–C–BH_2	Hydroboration
$C{=}C + OsO_4 \longrightarrow$ HO–C–C–OH	Diol formation (oxidation)
$C{=}C + H_2 \longrightarrow$ H–C–C–H	Hydrogenation (reduction)

In alkene polymers of industrial and commercial importance, n is a large number, typically several thousand. We discuss this alkene reaction in Chapter 29.

6.2 Organic Reactions Involving Reactive Intermediates

A. Reaction Coordinate Diagrams

In Chapter 4, we introduced reaction mechanisms and reaction coordinate diagrams. These concepts were explained in the context of acid-base chemistry, which involves the transfer of a proton from one chemical entity to another occurring in a single step. Although many chemical transformations take place in one step, many others involve two or more steps.

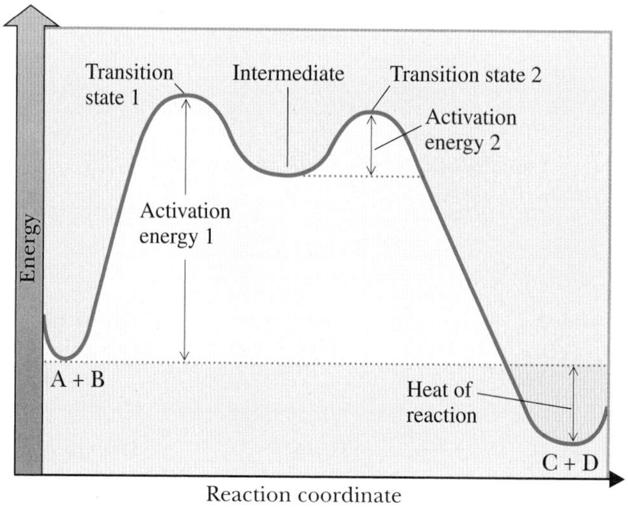

FIGURE 6.1 An energy diagram for a two-step reaction involving formation of an intermediate. The energy of the reactants is higher than that of the products, and energy is released in the conversion of A + B to C + D.

Reactive intermediate

A high-energy species, formed between two successive reaction steps, that lies in an energy minimum between the two transition states.

In a reaction that occurs in two or more steps, each step has its own transition state and activation energy. Figure 6.1 shows an energy diagram for the conversion of reactants to products in two steps. A **reactive intermediate** corresponds to an energy minimum between two transition states—in this case between transition states 1 and 2. Because the energies of the intermediates we describe in this chapter are higher than that of either reactants or products, they are highly reactive and usually cannot be isolated. However, some have significant lifetimes and can be observed directly using special experimental strategies.

The most common reactive intermediates in organic chemistry involve abnormal bonding at a carbon atom. In this chapter, the mechanisms involve *carbocation* intermediates where carbon has three bonds and only six valence electrons and is positively charged (see Section 6.3A). In future chapters, we will encounter radical intermediates where carbon has three bonds and only seven valence electrons. When carbon has three bonds but retains a full octet, it is negatively charged and is referred to as a carbanion. All three intermediates—carbocations, radicals, and carbanions—are significantly higher in energy than either the reactants or the products; therefore, a reaction coordinate diagram analogous to Figure 6.2 applies.

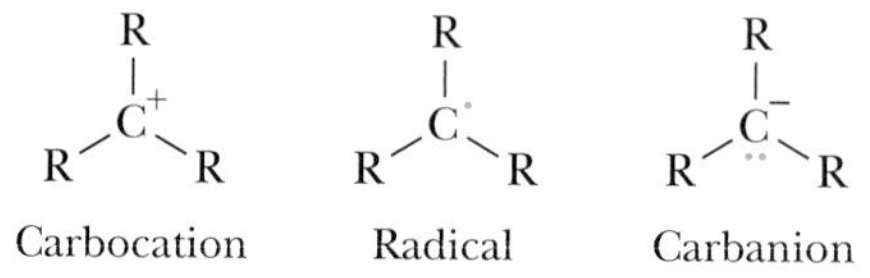

FIGURE 6.2 The reactive intermediates common in organic chemistry transformations.

The slowest step in a multistep reaction, called the **rate-determining step**, is the step that crosses the highest energy barrier. In the two-step sequence shown in Figure 6.1, Step 1 crosses the higher energy barrier and is, therefore, the rate-determining step.

Rate-determining step
The step in a multistep reaction sequence that crosses the highest energy barrier.

Each step of a chemical reaction involves the crossing of a peak on an energy diagram. The peaks represent chemical structures on the energy surfaces that are transitions between stable structures. Reaction intermediates are represented in wells (troughs) on the energy surfaces. The peaks on the surface are referred to as **transition states** (recall Section 4.5A).

It is important to make a distinction between intermediates, such as carbocations, and transition states. *Reaction intermediates* have lifetimes longer than the time it takes for a bond to vibrate, whereas *transition states* have lifetimes only on the order of the lifetime of a bond vibration (bond vibrations occur in the picosecond range, or a trillionth of a second, 10^{-12} s). For most practical purposes, the chemical structures of transition states can be considered as having no measurable lifetime, whereas intermediates have distinct measurable lifetimes even though they may be extremely short.

B. Thermodynamics of Addition Reactions

Equilibrium in a chemical reaction usually favors the side with stronger chemical bonds. Predicting whether a reaction will be thermodynamically favorable is a critical issue in chemistry. In Chapter 4, we noted that the Gibbs free energy controls reaction equilibria, which, in turn, is a function of enthalpy and entropy. Therefore, calculating the reaction enthalpy (ΔH^0) can be used in part to predict the thermodynamics of a reaction, especially when the reaction entropy ($\Delta S°$) is relatively small. ΔH^0 can be calculated by examining the bond strengths in the products relative to the reactants. Bond strengths are reported as **bond dissociation enthalpies** (BDEs), defined as the energy needed to break any bond, X—Y, into its corresponding radicals, X· and Y·, which is an endothermic process. Therefore, the bond enthalpies are the negative of the bond strengths. Reactions in which the new bonds made in the product(s) are stronger than the bonds broken in the starting materials will have a negative value of ΔH^0. To calculate ΔH^0 for the reaction, we subtract the BDEs of the products from the BDEs of the reactants:

Bond dissociation enthalpy
The amount of energy required to break a bond into two radicals in the gas phase at 25°C, A−B ⟶ A· + B·.

$$\Delta H^0_{rxn} = \text{BDEs}_R - \text{BDEs}_P$$

In Problem 6.14 a table of BDEs is given. (An expanded BDE Table is provided in Appendix 3.) Using this table, let's calculate the reaction enthalpy of just one of the addition reactions covered in this chapter—for example, the addition of bromine to an alkene:

▶ Reactions will occur if the products are more stable than the reactants and the energy barrier is low enough.

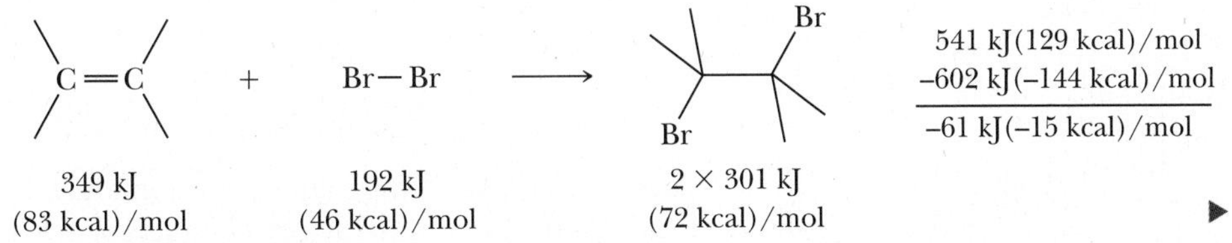
C=C
+
Br—Br
Br
Br
349 kJ
(83 kcal)/mol
192 kJ
(46 kcal)/mol
2 × 301 kJ
(72 kcal)/mol
541 kJ(129 kcal)/mol
−602 kJ(−144 kcal)/mol
−61 kJ(−15 kcal)/mol

This reaction involves cleavage of a C=C π bond and a Br—Br σ bond and the formation of two C—Br σ bonds. We can estimate the BDE of the C=C π bond by taking the difference between the BDE for CH_2=CH_2 (727 kJ/mol) and CH_3—CH_3 (378 kJ/mol). The strength of an π bond between two sp^2 carbons is thereby estimated to be 349 kJ/mol. Other important BDEs for this calculation are Br—Br (192 kJ/mol) and C—Br (301 kJ/mol). Therefore, ΔH^0_{rxn} is

$$\begin{aligned} &\quad\ \ \text{C=C } \pi \text{ bond} \quad \text{Br—Br bond} \quad \text{C—Br bond} \quad \text{C—Br bond} \\ \Delta H^0_{rxn} &= (349 \text{ kJ/mol} + 192 \text{ kJ/mol}) - (301 \text{ kJ/mol} + 301 \text{ kJ/mol}) \\ &= -61 \text{ kJ/mol} \end{aligned}$$

The reaction is therefore exothermic, because stronger bonds are made in the product compared to those broken in the starting materials. If the entropy is favorable or near zero, the reaction will be exergonic. In other words, the reaction is thermodynamically favorable and will spontaneously occur (assuming a relatively low activation energy).

Example 6.1 Calculating Exothermicity/ Endothermicity Values

Using the BDE values from Appendix 3, calculate the ΔH^0_{rxn} value and state if the reaction below is exothermic or endothermic.

$$CH_2{=}CH_2 + H_2O \rightarrow CH_3CH_2OH$$

Solution

The reactants have a C—C π-bond and an H—O bond that are broken, while a C—H and a C—O bond are created in the products. As discussed above, the BDE of a π-bond can be estimated to be 349 kJ/mol (83 kcal/mol). The BDE for a C—H bond, an H—O bond, and a C—O bond are 422 kJ/mol (101 kcal/mol), 497 kJ/mol (119 kcal/mol), and 385 kJ/mol (92 kcal/mol), respectively. Hence, ΔH^0_{rxn} = (349 kJ/mol + 497 kJ/mol) − (422 kJ/mol + 385 kJ/mol) = −39 kJ/mol (9.3 kcal/mol). The reaction is exothermic.

Problem 6.1

Using the BDE values from Appendix 3, calculate the ΔH^0_{rxn} value and state if the reaction below is exothermic or endothermic.

$$CH_2{=}CH_2 + HCl \rightarrow CH_3CH_2Cl$$

6.3 Electrophilic Additions

We begin our introduction to the chemistry of alkenes by examining five common addition reactions. Through experimental observation of selected addition reactions and their mechanisms, you will develop an understanding of how alkenes undergo addition reactions in general.

The overall reaction of electrophilic species adding to π bonds is called **electrophilic addition**. The key step in these mechanisms involves the π electrons of the alkene reacting as a nucleophile or Brønsted-Lowry base with an electrophilic species.

Electrophilic addition

A common type of reaction with alkenes in which an electrophilic species adds to a π bond.

At this point, it is appropriate to ask what makes a molecule or an ion an electrophile? As defined previously, electrophiles are "electron-loving" chemical species. Hence, electrophiles react with nucleophiles, which are molecular species that are electron-rich. The formation of a bond via nucleophilic attack on an electrophile is one of organic chemistry's fundamental reactions. ▶

▶ **Most bond-making steps in reaction mechanisms involve nucleophiles reacting with electrophiles.**

At least three characteristic features can make a molecule or an ion electrophilic. The most common one is a region of low electron density, reflected by a partial or full positive charge. The full or partial positive charge aids reaction with an electron-rich region of a nucleophile through electrostatic attraction. A second characteristic is the lack of an octet on an atom. The electrophiles listed below fit one or both of these criteria. We examine some of their reactions with alkenes in this chapter. Note that Brønsted-Lowry acids can be considered as electrophiles, as illustrated by the two species on the left in the following list.

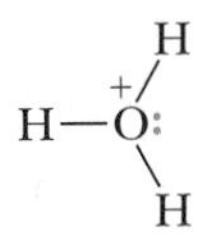

A positive charge with an acidic hydrogen

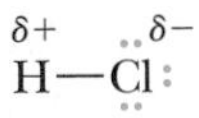

A partial positive charge with an acidic hydrogen

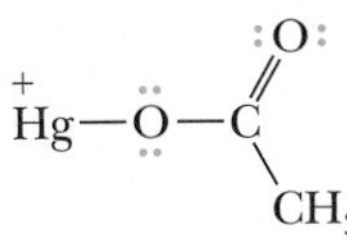

A positive charge with a Lewis acidic mercury atom

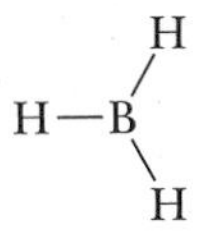

A boron atom lacking an octet

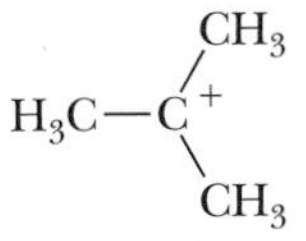

A positive charge on a carbon atom that is lacking an octet

A third characteristic that can make a molecule or an ion electrophilic is a relatively weak bond to an atom that can depart as a stable ion or molecule. These electrophilic species can be considered electron-seeking because their reactions with nucleophiles create stronger bonds and therefore more stable molecules. In such cases, there is often no partial or full positive charge on the electrophilic atom. Molecular halogens (X_2) are good examples. Their bonds are weak relative to the ones they form upon reaction with a nucleophile; this process is further aided by release of a stable halogen anion (X^-). Another example of a molecule that is electrophilic for these reasons is methyl iodide (we examine its reactions with nucleophiles in Chapter 9). There is little to no charge polarity in methyl iodide. However, the carbon is electrophilic because its bond to iodine is weak, and reaction with a nucleophile produces the stable iodide (I^-) anion.

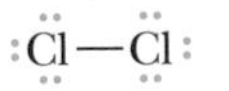

A weak bond between chlorine atoms

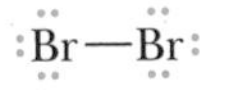

A weak bond between bromine atoms

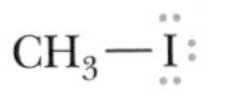

A weak bond between a carbon and iodine atom

A. Addition of Hydrogen Halides

The hydrogen halides HCl, HBr, and HI add to alkenes to give haloalkanes (alkyl halides). These additions may be carried out either with the pure reagents (neat) or in the presence of a polar solvent such as acetic acid. HCl reacts sluggishly compared to the other two acids. Addition of HBr to ethylene gives bromoethane (ethyl bromide):

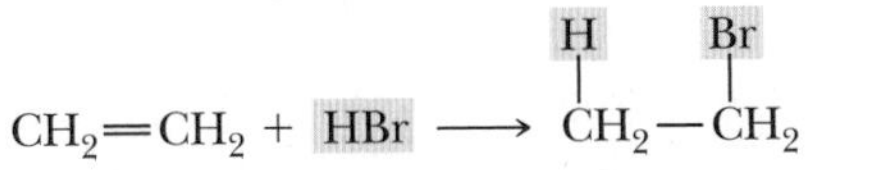

Ethylene **Bromoethane (Ethyl bromide)**

Addition of HBr to propene gives 2-bromopropane (isopropyl bromide); hydrogen adds to carbon 1 of propene, and bromine adds to carbon 2. If the orientation of addition were reversed, 1-bromopropane (propyl bromide) would be formed. The observed result is that 2-bromopropane is formed to the virtual exclusion of 1-bromopropane. We say that addition of HBr to propene is highly regioselective. A **regioselective reaction** is a reaction in which one direction of bond forming or breaking occurs in preference to all other directions of bond forming or breaking.

Regioselective reaction

An addition or substitution reaction in which one of two or more possible products is formed in preference to all others that might be formed.

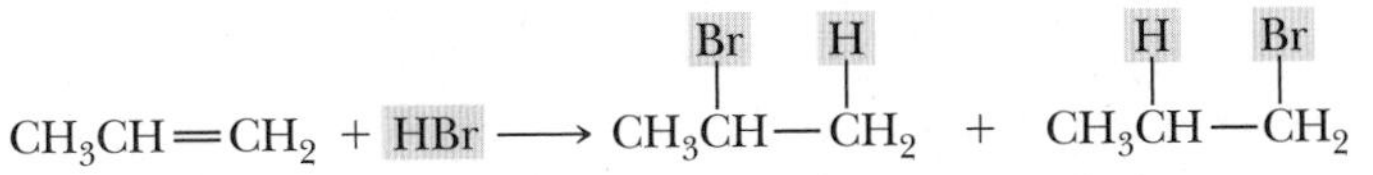

Propene **2-Bromopropane** **1-Bromopropane** (not observed)

This regioselectivity was noted by Vladimir Markovnikov who made the generalization known as **Markovnikov's rule**: in the addition of H—X to an alkene, hydrogen adds to the double-bonded carbon that has the greater number of hydrogens already

Markovnikov's rule

In the addition of HX, H_2O, or ROH to an alkene, hydrogen adds to the carbon of the double bond having the greater number of hydrogens.

bonded to it. Although Markovnikov's rule provides a way to predict the products of many alkene addition reactions, it does not explain why one product predominates over other possible products.

Example 6.2 Addition of HX to an Alkene

Name and draw a structural formula for the product of each alkene addition reaction.

(a) $CH_3C(CH_3){=}CH_2 + HI \longrightarrow$ **(b)** 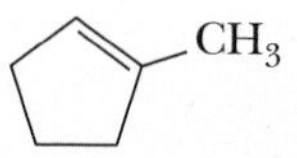 $+ HCl \longrightarrow$

Solution

Using Markovnikov's rule, we predict that 2-iodo-2-methylpropane is the product in (a) and that 1-chloro-1-methylcyclopentane is the product in (b).

(a) $CH_3C(CH_3)(I)CH_3$ **(b)** 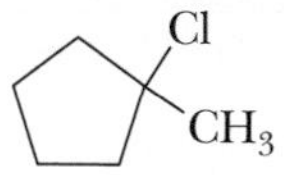

2-Iodo-2-methylpropane **1-Chloro-1-methylcyclopentane**

Problem 6.2

Name and draw a structural formula for the product of each alkene addition reaction.

(a) 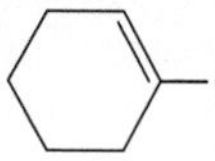 $+ HBr \longrightarrow$ **(b)** 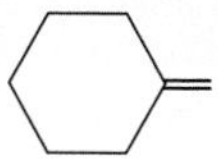 $+ HI \longrightarrow$

▶ A reaction mechanism describes the sequence of steps occurring in a reaction.

Chemists account for the addition of HX to an alkene by arrow pushing and a two-step mechanism, which we illustrate by the reaction of 2-butene with hydrogen bromide to give 2-bromobutane. We will first look at this two-step mechanism in overview and then go back and study each step in detail. ▶

Mechanism 6.1

Electrophilic Addition of HBr to 2-Butene

Step 1: **Add a proton.** Addition begins with the transfer of a proton from H—Br to 2-butene, as shown by the two curved arrows on the left side of the equation. The first curved arrow shows that the π bond of the alkene breaks and that its electron pair forms a new covalent bond with the hydrogen atom of H—Br. The second curved arrow shows that the polar covalent bond in H—Br breaks and that its electron pair moves to bromine to form a bromide ion. The result of this step is the formation of an organic cation and bromide ion.

Watch a video explanation

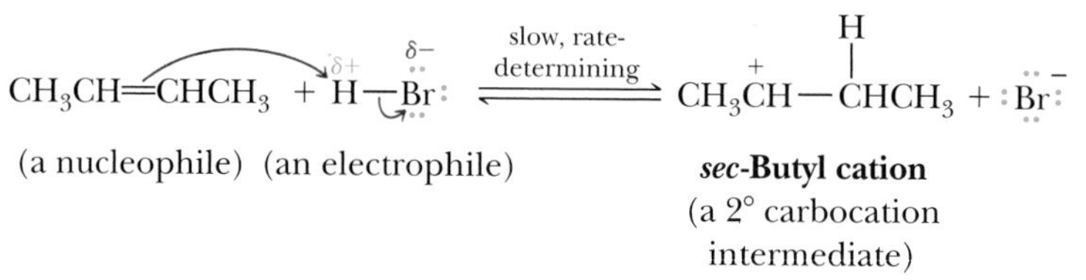
CH3CH=CHCH3 + H—Br:
δ+
δ−
slow, rate-determining
CH3CH—CHCH3 + :Br:−
H
(a nucleophile) (an electrophile)
sec-Butyl cation
(a 2° carbocation intermediate)

Step 2: **Make a new bond between a nucleophile and an electrophile.** Reaction of the *sec*-butyl cation (an electrophile) with bromide ion (a nucleophile) completes the valence shell of carbon and gives 2-bromobutane.

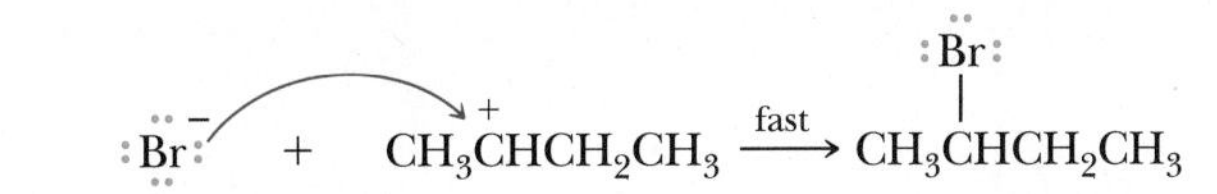

Bromide ion (a nucleophile) **sec-Butyl cation** (an electrophile) **2-Bromobutane** (a racemic mixture)

Now that we have looked at this two-step mechanism in overview, let us go back and look at the individual steps in more detail. Much important organic chemistry is embedded in them.

The reaction commences when the electron-rich π bond of the alkene (the electron source) interacts with the relatively electropositive H atom of H—Br (an electron sink) to form a new bond. The π bond is relatively electron-rich because the π-bonding electron density is above and below the bond axis, not between the positively charged atomic nuclei as is the case with σ-bonding electron density. The first two reactions of alkenes described in this chapter belong to this category, with the alkene π bond acting as a Brønsted-Lowry base. We refer to this step as **addition of a proton**. HBr can be identified as a Brønsted-Lowry acid because it has partial positive charge on the H atom and breaking the H—Br bond leads to a stable bromide anion.

Step 1, the addition of a proton to the alkene, results in formation of a cationic intermediate. One carbon atom in this intermediate has only six electrons in its valence shell and carries a charge of +1. A species containing a positively charged carbon atom is called a **carbocation** (*carbon* + *cation*). Such carbon-containing cations are also called *carbonium ions* and *carbenium ions.* Carbocations are classified as primary (1°), secondary (2°), or tertiary (3°), depending on the number of carbon atoms bonded to the carbon bearing the positive charge. All carbocations are electrophiles as well as Lewis acids (Section 4.7).

Carbocation

A species in which a carbon atom has only six electrons in its valence shell and bears a positive charge.

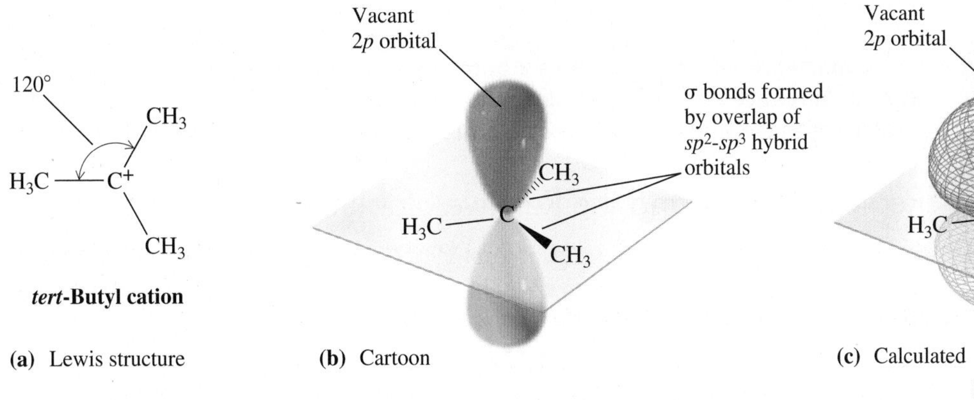

FIGURE 6.3 The structure of the *tert*-butyl cation. (a) Lewis structure, (b) cartoon, and (c) a calculated structure indicating a vacant 2*p* orbital.

In a carbocation, the carbon bearing the positive charge is bonded to three other atoms and, as predicted by VSEPR, the three bonds about the cationic carbon are coplanar and form bond angles of approximately 120°. According to the orbital hybridization model, the electron-deficient carbon of a carbocation uses sp^2 hybrid orbitals to form σ bonds to the three attached groups. The unhybridized 2*p* orbital lies perpendicular to the σ bond framework and contains no electrons. Figure 6.3 shows a Lewis structure, a cartoon, and a calculated structure indicating the vacant 2*p* orbital for the *tert*-butyl cation.

Figure 6.4 shows an energy diagram for the two-step reaction of 2-butene with HBr. The slower, rate-determining step (the one that crosses the higher energy

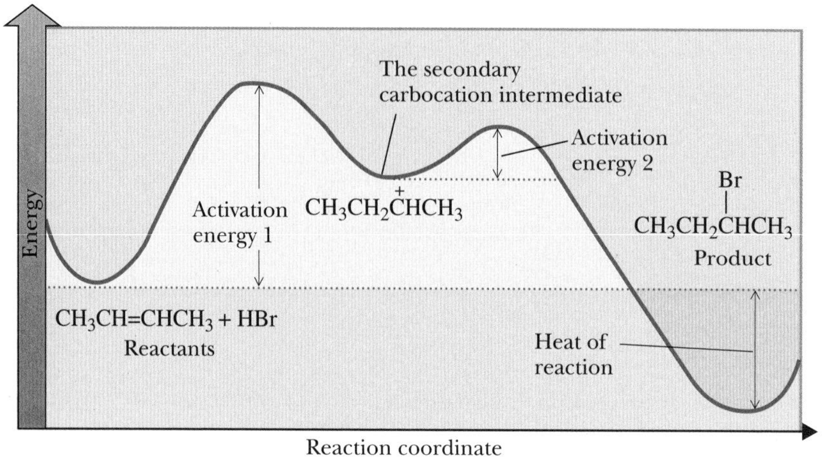
The secondary carbocation intermediate
Activation energy 2
Activation energy 1
CH3CH2CHCH3
Br
CH3CH2CHCH3
Product
Energy
CH3CH=CHCH3 + HBr
Reactants
Heat of reaction
Reaction coordinate

FIGURE 6.4 An energy diagram for the two-step addition of HBr to 2-butene. The reaction is exergonic.

barrier) is Step 1, which leads to the formation of the 2° carbocation intermediate. This carbocation intermediate lies in an energy minimum between the transition states for Steps 1 and 2. The carbocation intermediate (an electrophile) reacts within microseconds, or faster, with bromide ion (a nucleophile) in a nucleophile-electrophile combination to give 2-bromobutane. Note that the energy level for 2-bromobutane (the product) is lower than the energy level for 2-butene and HBr (the reactants). Thus, in this electrophilic addition reaction, energy is released; the reaction is exergonic.

The reaction involves a collision with the appropriate trajectory for transfer of the electrophile to the alkene. As with any collision in everyday life, strain results from physical deformations introduced into the colliding partners. In this case, the collision must occur with enough energy to both contort the structure of the alkene toward that of a carbocation and break the HBr bond. The physical structure of the transition state is the lowest-energy contorted geometry that gives way to the carbocation intermediate.

The second step of the reaction is much easier than the first step because the reactant is now a carbocation. Carbocations are highly reactive electrophiles because of their positive charge and the lack of an octet at carbon, and they commonly have lifetimes of microseconds or less. Because the highly electrophilic cation intermediate is so unstable, the energy necessary in the collision with the nucleophilic bromide to surmount the second barrier is very low. As with a highly unstable object in the macroscopic world (such as a book teetering on the edge of a desk), only a small disturbance is necessary to give the intermediate a pathway to a lower energy state (such as tapping the book to cause it to fall).

Regioselectivity and the Relative Stabilities of Carbocations

Reaction of HX and an alkene can, at least in principle, give two different carbocation intermediates because the proton could be transferred to either of the doubly bonded carbon atoms, as illustrated by the reaction of HBr with propene.

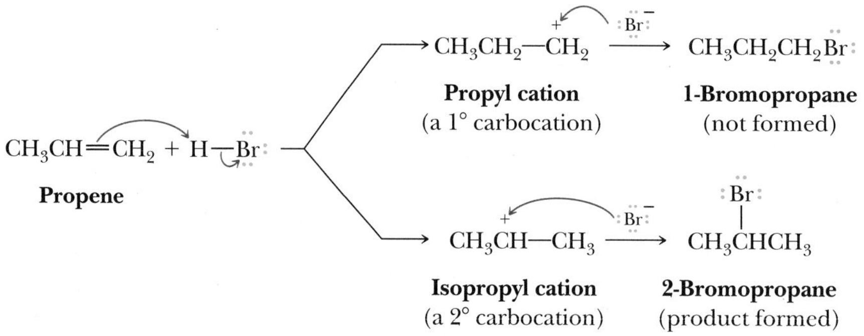
CH3CH=CH2 + H—Br:
Propene
CH3CH2—CH2
Br
Propyl cation
(a 1° carbocation)
CH3CH2CH2Br:
1-Bromopropane
(not formed)
CH3CH—CH3
Isopropyl cation
(a 2° carbocation)
:Br:
CH3CHCH3
2-Bromopropane
(product formed)

The observed product is 2-bromopropane, indicating that the 2° carbocation intermediate is formed in preference to the 1° carbocation intermediate.

Similarly, in the reaction of HBr with 2-methylpropene, proton transfer to the carbon-carbon double bond might form either the isobutyl cation (a 1° carbocation) or the *tert*-butyl cation (a 3° carbocation).

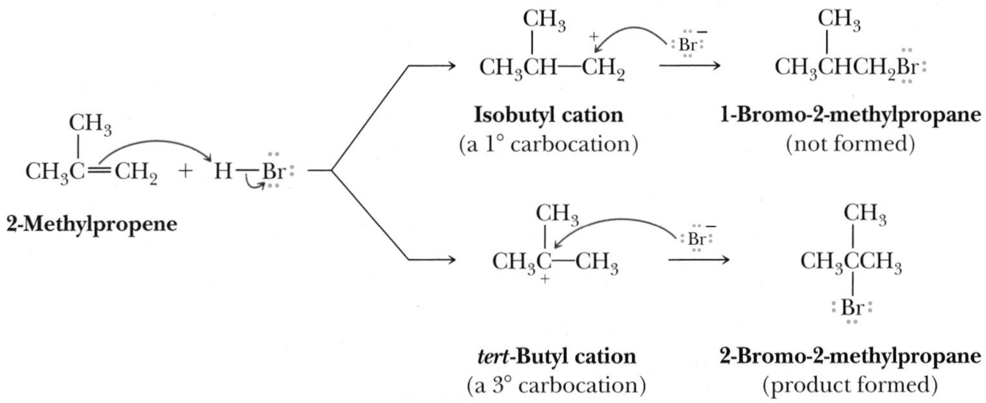
2-Methylpropene
Isobutyl cation
(a 1° carbocation)
1-Bromo-2-methylpropane
(not formed)
tert-Butyl cation
(a 3° carbocation)
2-Bromo-2-methylpropane
(product formed)

The observed product of this reaction is 2-bromo-2-methylpropane, indicating that the 3° carbocation is formed in preference to the 1° carbocation.

From experiments such as these and a great amount of other experimental evidence, we know that a 3° carbocation is more stable and requires a lower activation energy for its formation than a 2° carbocation. A 2° carbocation, in turn, is more stable and requires a lower activation energy for its formation than a 1° carbocation. Methyl and 1° carbocations are so unstable they are rarely observed in solution. It follows, then, that a more stable carbocation intermediate forms faster than a less stable carbocation intermediate. Following is the order of stability of four types of alkyl carbocations.

▶ Delocalization of charge over a larger area is stablizing.

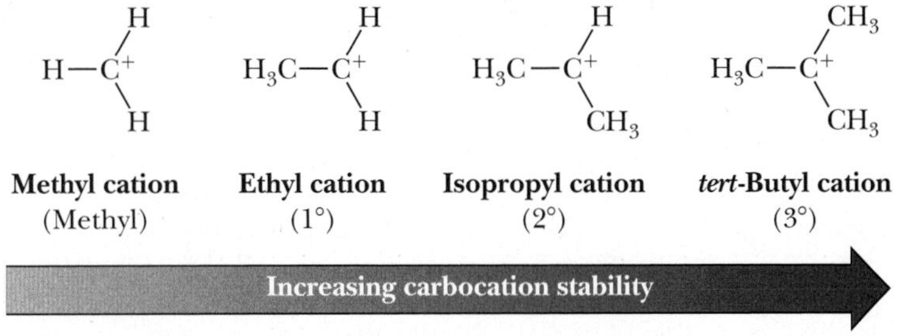
Methyl cation
(Methyl)
Ethyl cation
(1°)
Isopropyl cation
(2°)
tert-Butyl cation
(3°)
Increasing carbocation stability

Now that we know the order of stability of carbocations, how do we account for this order? As we saw during the discussion of anion stability in Section 4.5, a system bearing a charge (either positive or negative) is more stable if the charge is delocalized. We can explain the order of stability of carbocations if we assume that alkyl groups bonded to a positively charged carbon release electron density toward that carbon and thereby help delocalize the positive charge on the cation. We account for the electron-releasing ability of alkyl groups bonded to a cationic carbon by two effects: the inductive effect and hyperconjugation.

The Inductive Effect

The inductive effect operates in the following way. The electron deficiency of the cationic carbon exerts an electron-withdrawing **inductive effect** that polarizes electrons from adjacent sigma bonds toward it. In this way, the positive charge of the cation is not localized on the trivalent carbon but rather is delocalized over nearby atoms. The larger the volume over which the positive charge is delocalized, the greater the stability of the cation. As the number of alkyl groups bonded to the cationic carbon increases, the stability of the cation increases. Figure 6.5 illustrates the electron-withdrawing

Inductive effect The polarization of the electron density of a covalent bond resulting from the electronegativity of a nearby atom.

inductive effect of the positively charged carbon and the resulting delocalization of charge. According to quantum mechanical calculations, the charge on carbon in the methyl cation is approximately +0.645 and the charge on each of the hydrogen atoms is +0.118. Thus, even in the methyl cation, the positive charge is not localized entirely on carbon. Rather, it is delocalized over the volume of space occupied by the entire ion. Delocalization of charge is even more extensive in the *tert*-butyl cation.

(a) H_3C^+ **(b)** $(CH_3)_3C^+$

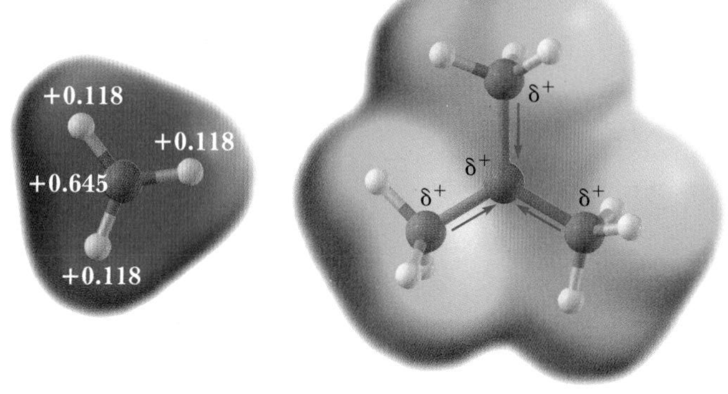

Watch a video explanation

FIGURE 6.5 Electrostatic potential plots showing the distribution of the positive charge in (a) the methyl cation and (b) the *tert*-butyl cation. Electron donation by the methyl groups decreases the positive charge (blue area) on the central carbon of the *tert*-butyl cation.

Hyperconjugation

The second related effect by which alkyl groups stabilize carbocations is hyperconjugation. **Hyperconjugation** involves partial overlapping of the σ-bonding orbital of an adjacent C—H or C—C bond of the alkyl group with the vacant $2p$ orbital of the cationic carbon (Figure 6.6). In other words, some electron density of the alkyl group C—H or C—C bond is mixed into the $2p$ orbital. As shown in Figure 1.25 for analogous mixing, the result is lower and higher energy orbitals, the first filled and the second empty, respectively. In the case of hyperconjugation, the populated orbital remains primarily C—H bonding (Figure 6.6b) and the empty orbital remains primarily $2p$ (Figure 6.6c), but Figure 6.6b and c both show the delocalization that is the hallmark of hyperconjugation. The net result of hyperconjugation is an increase of electron density on the cationic carbon, thereby delocalizing the positive charge onto the adjacent alkyl groups. As more alkyl groups are bonded to a cationic carbon, the hyperconjugation effect becomes stronger and the carbocation becomes more stable.

Hyperconjugation

Interaction of electrons in a sigma-bonding orbital with the vacant $2p$ orbital of an adjacent positively charged carbon.

(a)

Cartoon diagram of the ethyl cation. There is delocalization of C—H bonding electrons into the empty $2p$ orbital of the positively charged carbon. The orbital that results from this delocalization contains two electrons [see **(b)**].

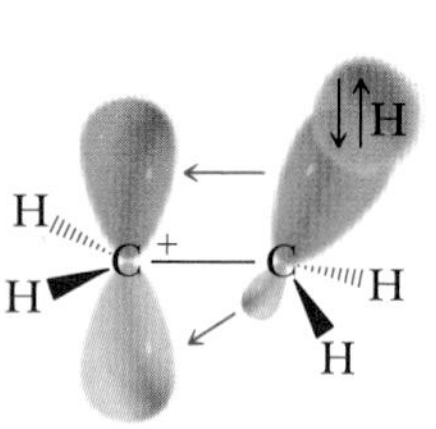

(b)

A calculated orbital populated with two electrons. This orbital is delocalized over the C—H bond and the $2p$ orbital of the positively charged carbon of the ethyl cation.

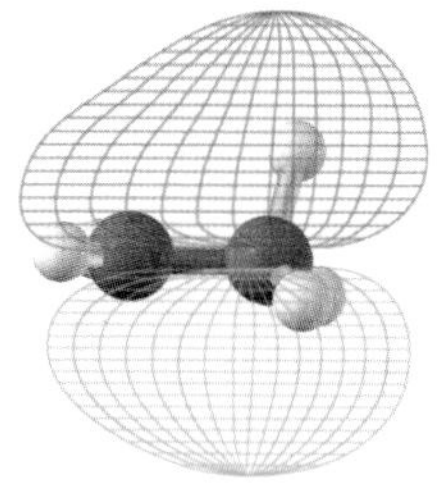

(c)

Calculated empty orbital residing mostly on the positively charged carbon of ethyl cation. Some empty orbital character exists on the methyl group also.

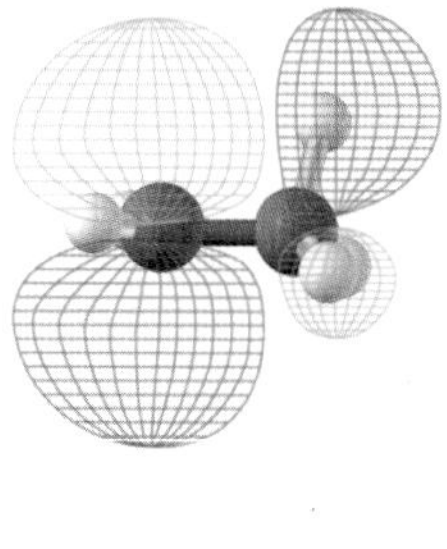

FIGURE 6.6 Hyperconjugation.

Example 6.3 Stability of Carbocations

Arrange these carbocations in order of increasing stability.

(a) 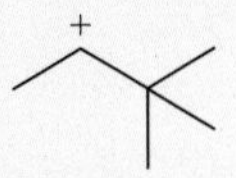 **(b)** 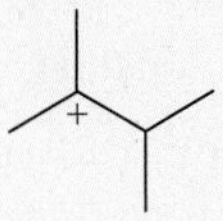 **(c)** 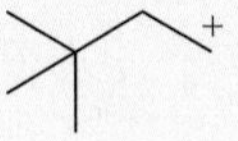

Solution

Carbocation (a) is 2°, (b) is 3°, and (c) is 1°. In order of increasing stability, they are c < a < b.

Problem 6.3

Arrange these carbocations in order of increasing stability.

(a) 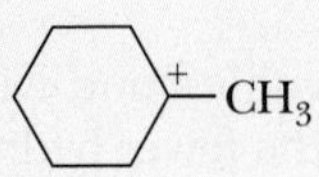 **(b)** 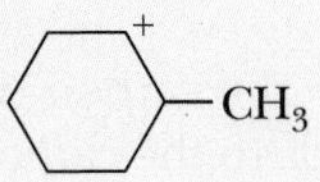 **(c)** 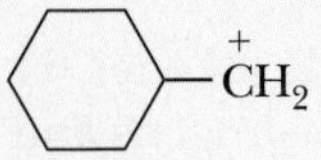

Example 6.4 HX Addition to an Alkene

Propose a mechanism for the addition of HI to methylidenecyclohexane to give 1-iodo-1-methylcyclohexane. Which step in your mechanism is rate-determining?

Solution

Propose a two-step mechanism similar to that proposed for the addition of HBr to propene.

***Step 1:* Add a proton.** A rate-determining addition of a proton from HI to the carbon-carbon double bond gives a 3° carbocation intermediate.

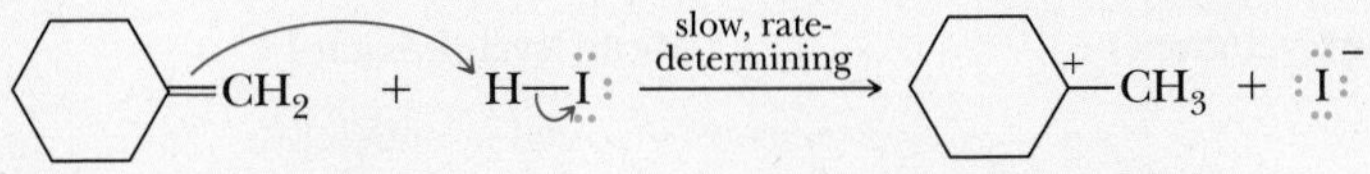

Methylidenecyclohexane (Methylenecyclohexane)

A 3° carbocation intermediate

***Step 2:* Make a new bond between a nucleophile and an electrophile.** Reaction of the 3° carbocation intermediate (an electrophile) with iodide ion (a nucleophile) completes the valence shell of carbon and gives the product.

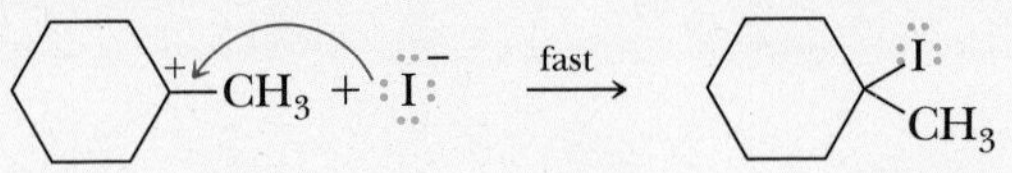

1-Iodo-1-methylcyclohexane

Problem 6.4

Propose a mechanism for addition of HI to 1-methylcyclohexene to give 1-iodo-1-methylcyclohexane. Which step in your mechanism is rate-determining?

Watch a video explanation

B. Addition of Water: Acid-Catalyzed Hydration

Hydration The addition of water.

In the presence of an acid catalyst, most commonly concentrated sulfuric acid, water adds to an alkene to give an alcohol. The addition of water is called **hydration**.

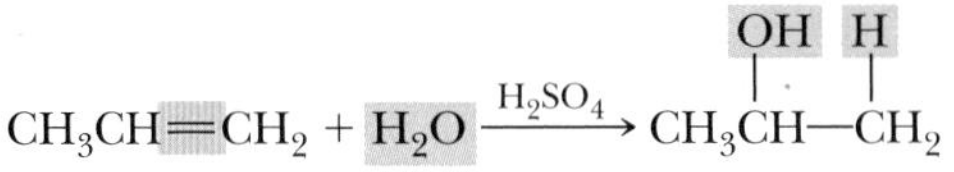

Propene **2-Propanol**

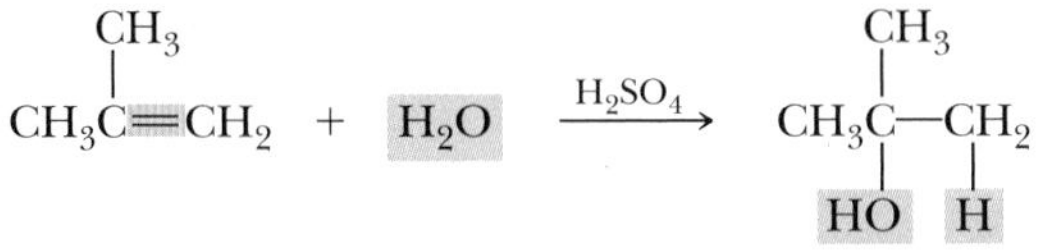

2-Methylpropene **2-Methyl-2-propanol**

In the case of simple alkenes, H adds to the carbon of the double bond with the greater number of hydrogens and OH adds to the carbon with the fewer hydrogens. Thus, H—OH adds to alkenes in accordance with Markovnikov's rule.

Example 6.5 Acid-Catalyzed Hydration of Alkenes I

Draw a structural formula for the product of the acid-catalyzed hydration of 1-methylcyclohexene.

Solution

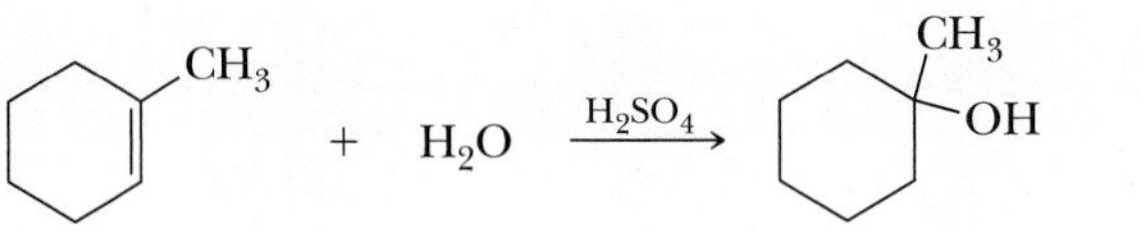

1-Methylcyclohexene **1-Methylcyclohexanol**

Problem 6.5

Draw a structural formula for the product of each alkene hydration reaction.

(a) 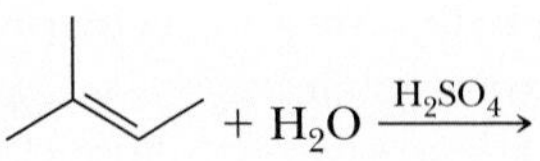 **(b)** 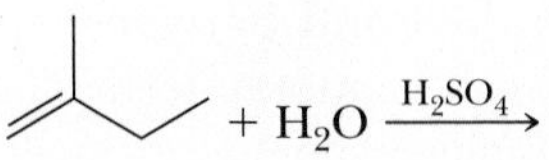

The mechanism for acid-catalyzed hydration of alkenes is quite similar to what we already proposed for addition of HCl, HBr, and HI to alkenes and is illustrated by conversion of propene to 2-propanol. Note that this mechanism is consistent with the fact that acid is a catalyst; an H_3O^+ is consumed in Step 1 but another is generated in Step 3.

Mechanism 6.2

Acid-Catalyzed Hydration of Propene

***Step 1:* Add a proton.** Addition of a proton to propene from H_3O^+ gives a 2° carbocation intermediate.

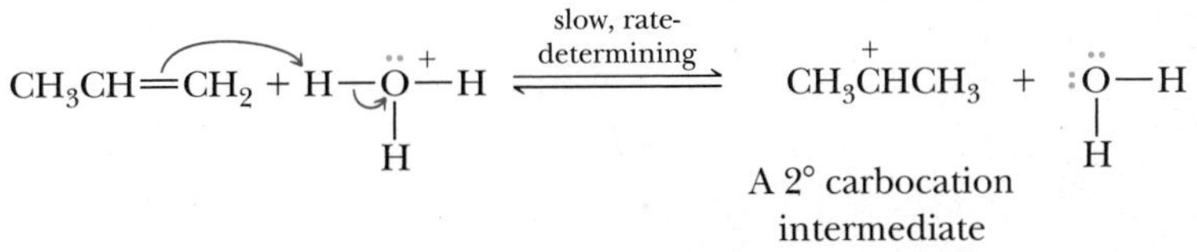

***Step 2:* Make a new bond between a nucleophile and an electrophile.** The 2° carbocation intermediate (an electrophile) completes its valence shell by forming a new covalent bond with an unshared pair of electrons of the oxygen atom of water (a nucleophile) and gives an **oxonium ion**.

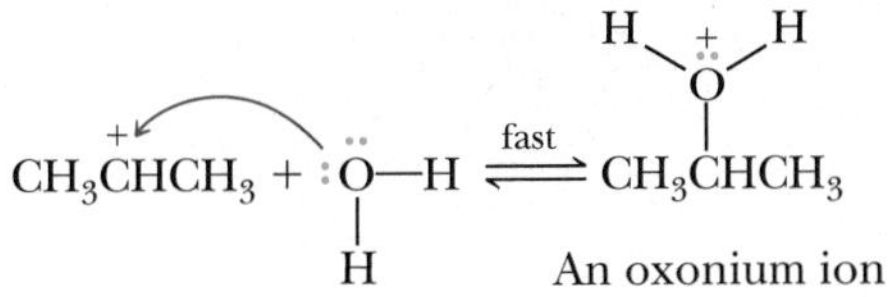

***Step 3:* Take a proton away.** Loss of a proton from the relatively acidic oxonium ion to water gives the alcohol and generates a new acid catalyst.

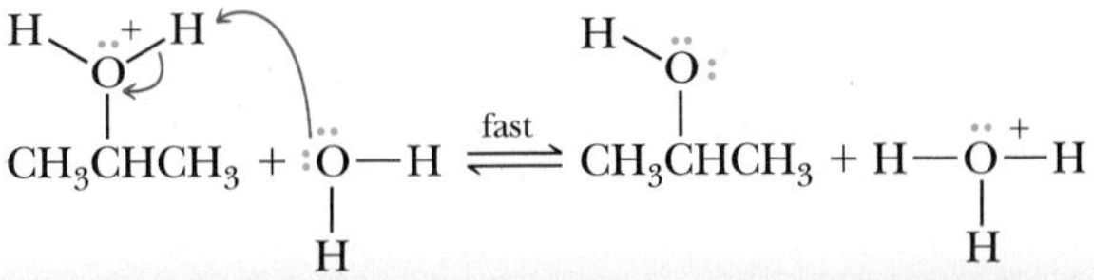

Oxonium ion

An ion in which oxygen bears a positive charge.

Example 6.6 Acid-Catalyzed Hydration of Alkenes II

Propose a mechanism for the acid-catalyzed hydration of methylidenecyclohexane to give 1-methylcyclohexanol. Which step in your mechanism is rate-determining?

Solution

Propose a three-step mechanism similar to that for the acid-catalyzed hydration of propene.

***Step 1:* Add a proton.** Proton transfer from the acid catalyst to the alkene gives a 3° carbocation intermediate. Formation of the 3° carbocation intermediate is rate-determining.

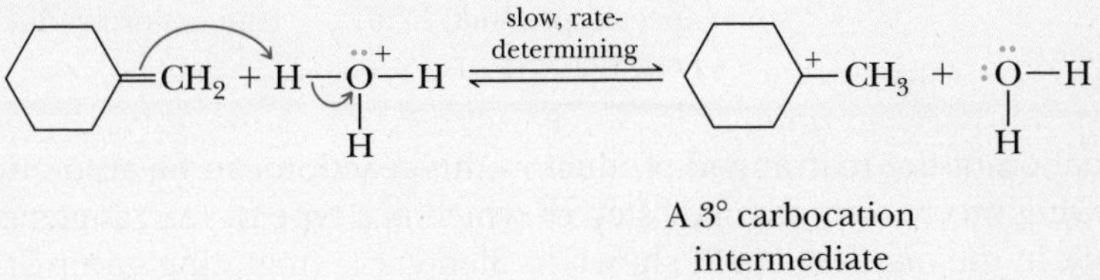

(Continued)

***Step 2:* Make a new bond between a nucleophile and an electrophile.** Reaction of the 3° carbocation intermediate (an electrophile) with water (a nucleophile) completes the valence shell of carbon and gives an oxonium ion.

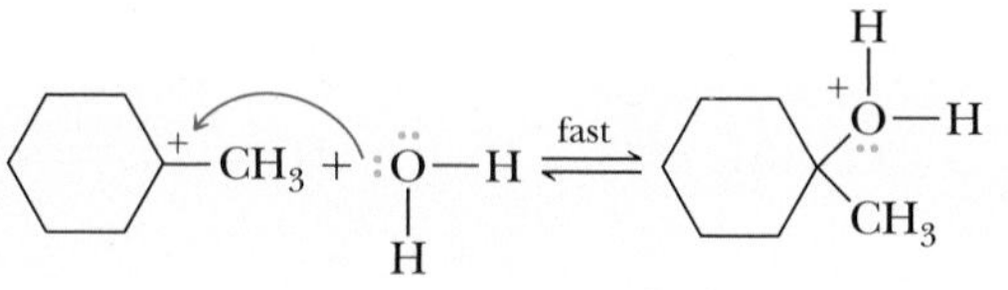

An oxonium ion

***Step 3:* Take a proton away.** Proton transfer from the oxonium ion to water gives the alcohol and generates a new acid catalyst.

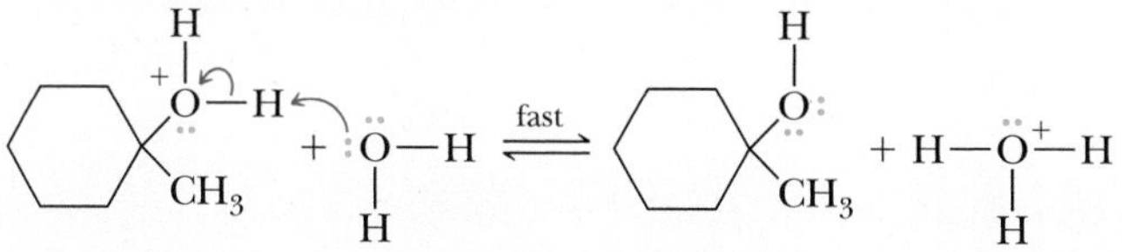

Problem 6.6

Propose a mechanism for the acid-catalyzed hydration of 1-methylcyclohexene to give 1-methylcyclohexanol. Which step in your mechanism is rate-determining?

C. Carbocation Rearrangements

As we have seen in the preceding discussions, the expected product of electrophilic addition to a carbon-carbon double bond involves rupture of the π bond and formation of two new σ bonds in its place. In addition of HCl to 3,3-dimethyl-1-butene, however, only 17% of 2-chloro-3,3-dimethylbutane, the expected product, is formed. The major product is 2-chloro-2,3-dimethylbutane, a compound with a different connectivity of its atoms compared with that in the starting alkene. We say that formation of 2-chloro-2,3-dimethylbutane involves a **rearrangement**. Typically, either an alkyl group or a hydrogen migrates, each with its bonding electrons, from an adjacent atom to the electron-deficient atom. In the rearrangements we examine in this chapter, migration is to an adjacent electron-deficient carbon atom bearing a positive charge.

Rearrangement
A change in connectivity of the atoms in a product compared with the connectivity of the same atoms in the starting material.

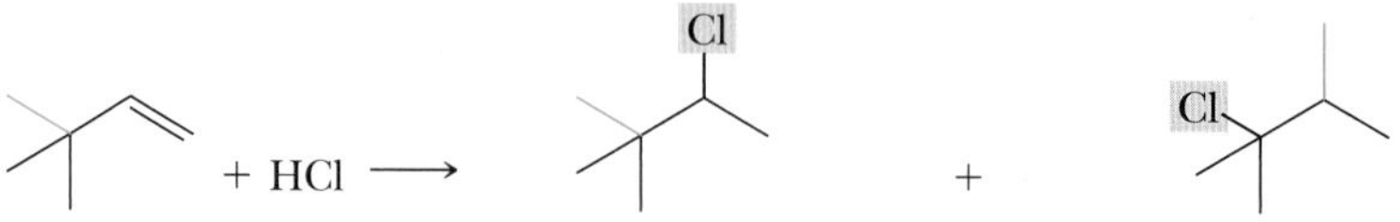

3,3-Dimethyl-1-butene | **2-Chloro-3,3-dimethylbutane** (the expected product; 17%) (a racemic mixture) | **2-Chloro-2,3-dimethylbutane** (the major product; 83%)

1,2 shift
A type of rearrangement in which an atom or a group of atoms moves with its bonding electrons from one atom to an adjacent electron-deficient atom.

Formation of the rearranged product in this reaction can be accounted for by the following mechanism, the key step of which is a type of rearrangement called a **1,2 shift**. In the rearrangement shown in Step 2, the migrating group is a methyl group with its bonding electrons. A 1,2 shift represents a new mechanistic choice in addition to the four choices presented in Things You Should Know III, Reaction Mechanisms. A 1,2 shift is far less common than the other choices.

Mechanism 6.3

Carbocation Rearrangement in the Addition of HCl to an Alkene

Step 1: **Add a proton.** Proton addition to the alkene gives a 2° carbocation intermediate.

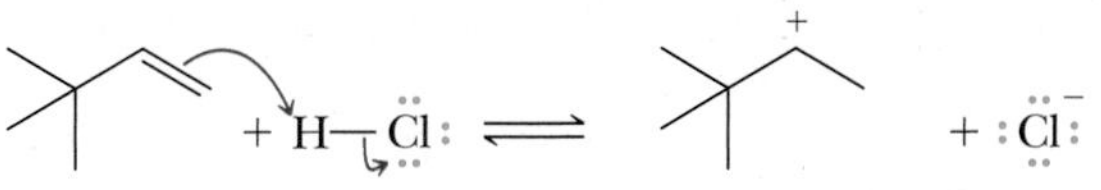

A 2° carbocation intermediate

Step 2: **1,2 Shift.** Migration of a methyl group with its bonding electrons from an adjacent carbon to the positively charged carbon of the 2° carbocation gives a more stable 3° carbocation. In this rearrangement, the major movement is that of the bonding electron pair with the methyl group following.

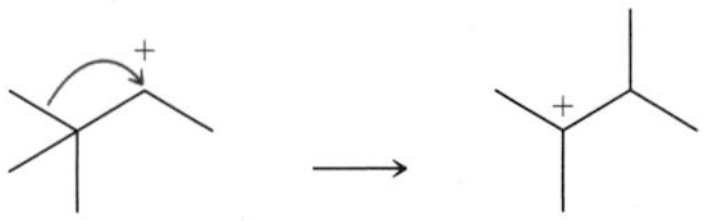

A 2° carbocation intermediate | A 3° carbocation intermediate

Step 3: **Make a new bond between a nucleophile and an electrophile.** Reaction of the 3° carbocation intermediate (an electrophile) with chloride ion (a nucleophile) gives the rearranged product.

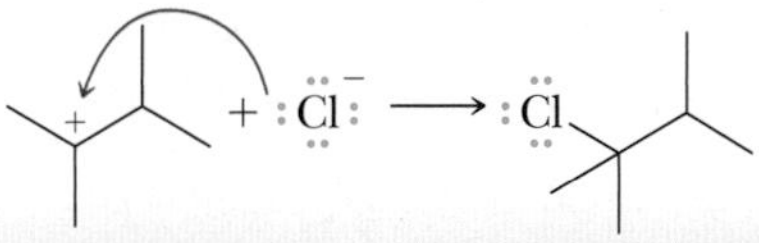

The driving force is that the less stable 2° carbocation is converted to a more stable 3° carbocation. From the study of this and other carbocation rearrangements, we find that 2° carbocations rearrange to more stable 2° or 3° carbocations. They rearrange in the opposite direction only under special circumstances, such as where the relief of ring strain provides added driving force.

Rearrangements also occur in the acid-catalyzed hydration of alkenes, especially when the carbocation formed in the first step can rearrange to a more stable carbocation. For example, acid-catalyzed hydration of 3-methyl-1-butene gives 2-methyl-2-butanol. In this example, the group that migrates is a hydrogen with its bonding pair of electrons, in effect a hydride ion $H:^-$.

$$CH_3\underset{|}{\overset{CH_3}{C}}HCH{=}CH_2 + H_2O \xrightarrow{H_2SO_4} CH_3\overset{CH_3}{\underset{OH}{C}}CH_2CH_3$$

3-Methyl-1-butene | **2-Methyl-2-butanol**

Example 6.7 Acid-Catalyzed Hydration of Alkenes III

Propose a mechanism for the acid-catalyzed hydration of 3-methyl-1-butene to give 2-methyl-2-butanol.

Solution

Following is a four-step mechanism for the formation of 2-methyl-2-butanol.

***Step 1:* Add a proton.** Proton transfer from H_3O^+, the acid catalyst, to the double bond of the alkene gives a 2° carbocation intermediate.

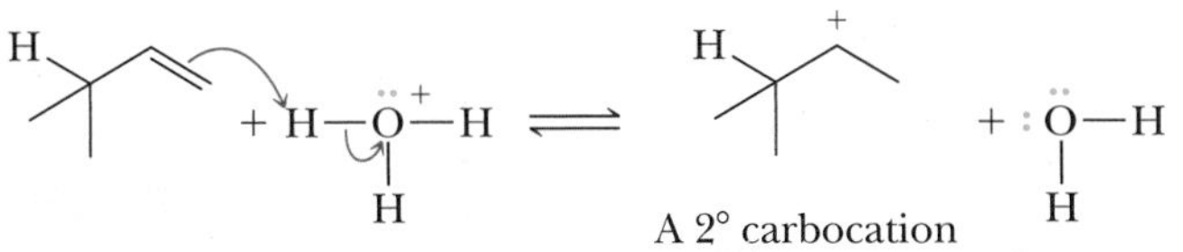

A 2° carbocation intermediate

***Step 2:* 1,2 Shift.** The less stable 2° carbocation rearranges to a more stable 3° carbocation by migration of a hydrogen with its pair of bonding electrons (in effect, a hydride ion).

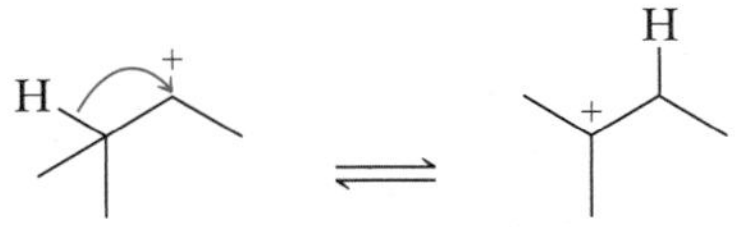

A 2° carbocation intermediate — A 3° carbocation intermediate

***Step 3:* Make a new bond between a nucleophile and an electrophile.** Reaction of the 3° carbocation intermediate (an electrophile) with water (a nucleophile) completes the valence shell of carbon and gives an oxonium ion.

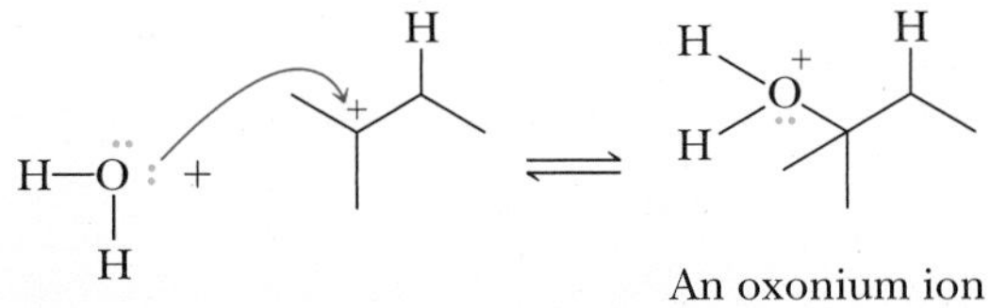

An oxonium ion

***Step 4:* Take a proton away.** Proton transfer from the oxonium ion to water gives the product and generates a new H_3O^+ to continue the hydration reaction.

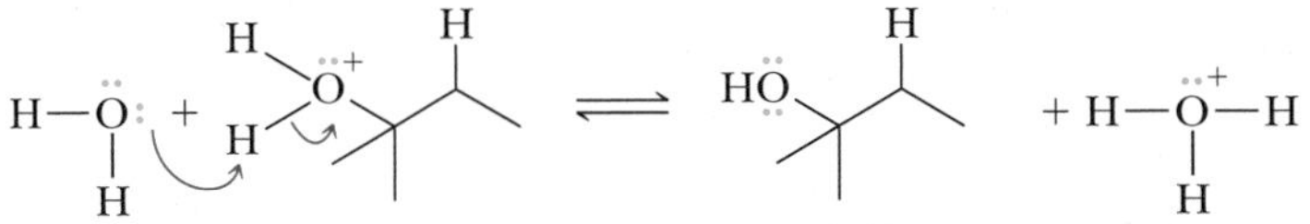

2-Methyl-2-butanol

Problem 6.7

The acid-catalyzed hydration of 3,3-dimethyl-1-butene gives 2,3-dimethyl-2-butanol as the major product. Propose a mechanism for the formation of this alcohol.

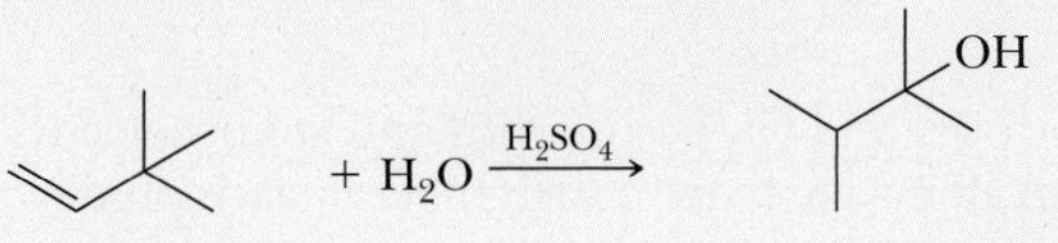

3,3-Dimethyl-1-butene **2,3-Dimethyl-2-butanol**

D. Addition of Bromine and Chlorine

Chlorine, Cl_2, and bromine, Br_2, react with alkenes at room temperature by adding halogen atoms to the two carbon atoms of the double bond with formation of two new carbon-halogen bonds. Fluorine, F_2, adds to alkenes, but because its reactions are very fast and difficult to control, this reaction is not a useful laboratory procedure. Iodine, I_2, also adds, but the reaction is not preparatively useful. Halogenation with bromine or chlorine is generally carried out either with the pure reagents or with them being mixed in an inert solvent such as CH_2Cl_2.

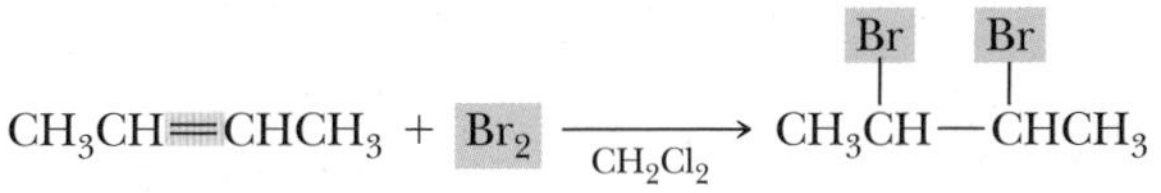

2-Butene **2,3-Dibromobutane**

The addition of bromine or chlorine to a cycloalkene is an example of a **stereoselective reaction**, in that one stereoisomer is formed in preference to all others. In this case, *trans*-dihalocycloalkane is formed as a racemic mixture. The addition of bromine to cyclohexene, for example, gives *trans*-1,2-dibromocyclohexane:

Stereoselective reaction
A reaction in which one stereoisomer is formed in preference to all others. A stereoselective reaction may be enantioselective or diastereoselective, as the case may be.

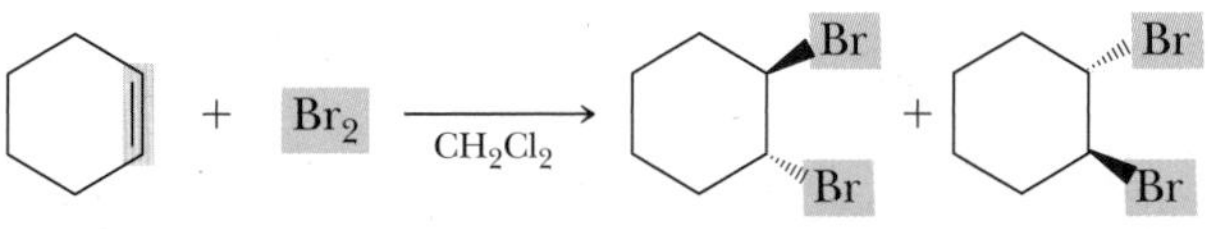

Cyclohexene ***trans*-1,2-Dibromocyclohexane** (a racemic mixture)

At first glance, the two *trans* enantiomers may appear to be the same structure. However, there is no plane of symmetry or center of symmetry, so they are both chiral. You should make molecular models of both enantiomers and convince yourself that they are indeed nonsuperposable mirror images, not just the same molecule viewed from a different perspective.

We discuss the addition stereochemistry of Cl_2 and Br_2 to alkenes in more detail in Section 6.7. For now, it is sufficient to point out that these reactions proceed with anti (from the opposite side or face) addition of halogen atoms; that is, they occur with **anti stereoselectivity**.

Anti stereoselectivity
The addition of atoms or groups of atoms to opposite faces of a carbon-carbon double bond.

Reaction of bromine with an alkene is a particularly useful qualitative test for the presence of a carbon-carbon double bond. If we dissolve bromine in dichloromethane, the solution is red. Both alkenes and dibromoalkanes are colorless. If we now mix a few drops of the bromine solution with an alkene, a dibromoalkane is formed and the red solution becomes colorless.

Example 6.8 Addition of X_2 to an Alkene

Complete these reactions, showing the stereochemistry of the product.

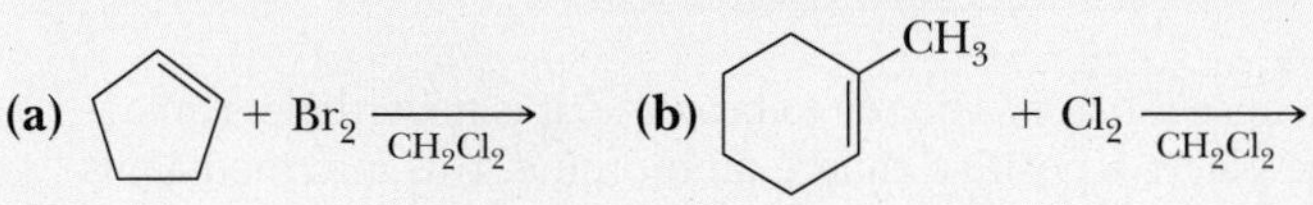

(Continued)

Solution

Addition of both Br_2 and Cl_2 occurs with anti stereoselectivity, which means that the halogen atoms are *trans* to each other in each product.

(a) 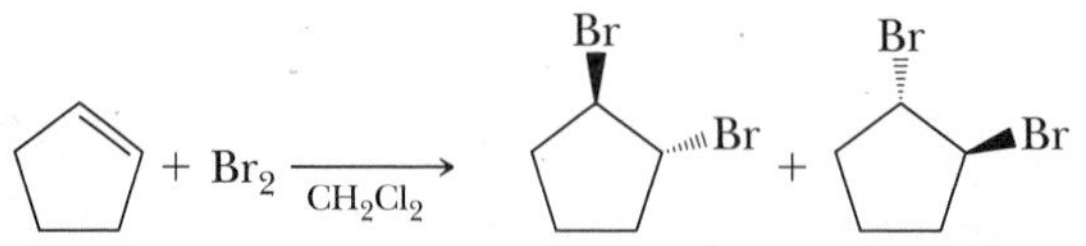

***trans*-1,2-Dibromocyclopentane**
(a racemic mixture)

(b) 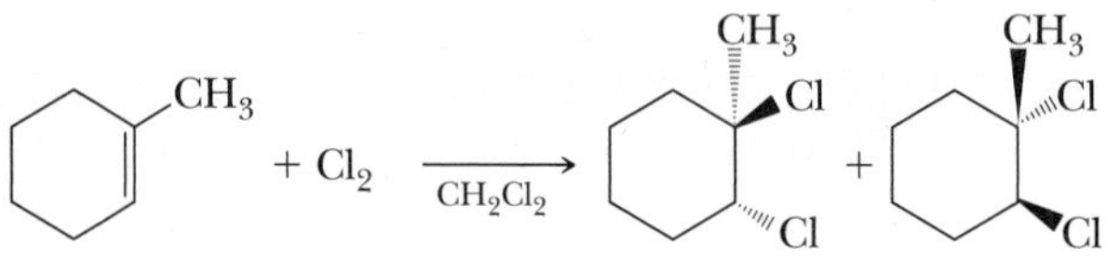

(a racemic mixture)

Problem 6.8

Complete these reactions.

(a) 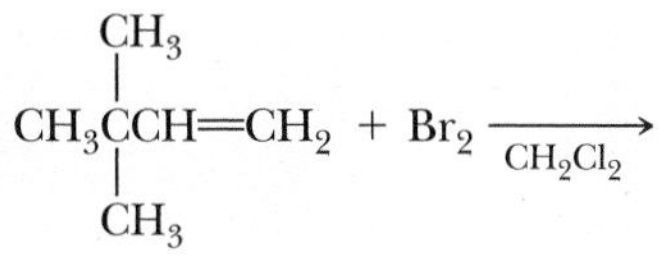

(b) 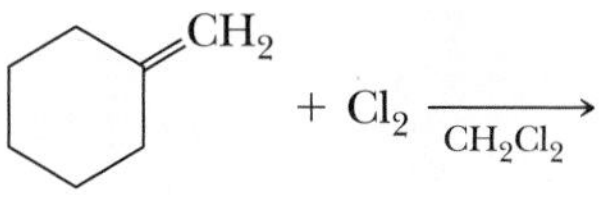

Anti Stereoselectivity and Bridged Halonium Ion Intermediates

We explain the addition of bromine and chlorine to alkenes and its anti stereoselectivity by the following two-step mechanism.

Mechanism 6.4

Addition of Bromine with Anti Stereoselectivity

Step 1: **Make a new bond between a nucleophile (π bond) and an electrophile.** Reaction is initiated by interaction of the π electrons of the alkene with bromine (or chlorine, as the case may be) that acts likes an electrophile to form an intermediate in which bromine bears a positive charge. A bromine atom bearing a positive charge is called a **bromonium ion**, and the cyclic structure of which it is a part is called a **bridged bromonium ion**.

Watch a video explanation

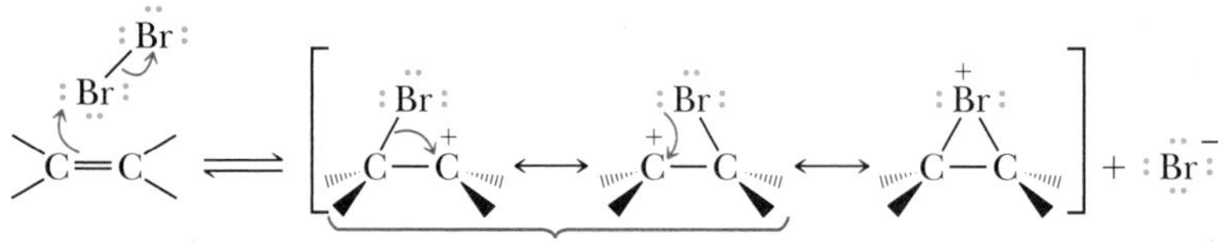

these carbocations are major contributing structures

the bridged bromonium ion retains the geometry

Although a bridged bromonium ion may look odd because it has two bonds to bromine, it is nevertheless an acceptable Lewis structure. Calculation of formal charge places a positive charge on bromine. This intermediate is a hybrid of three resonance-contributing structures. Because of the planarity of the atoms forming the π bond, the bridged bromonium ion can form with equal probability on the top or bottom face of the alkene. In this step, the Br_2 or Cl_2 acts as an electrophile because the halogen-halogen bond is weak and reaction leads to the departure of the stable X^- anion.

Step 2: **Make a new bond between a nucleophile and an electrophile.** Attack of bromide ion (a nucleophile) on carbon from the side opposite the bromonium ion (an electrophile) opens the three-membered ring to give the anti product. The bromines are not only anti but also in the same plane (coplanar). Thus, we call this an anti-coplanar attack. Attack by Br^- can occur at either carbon of the bromonium ion.

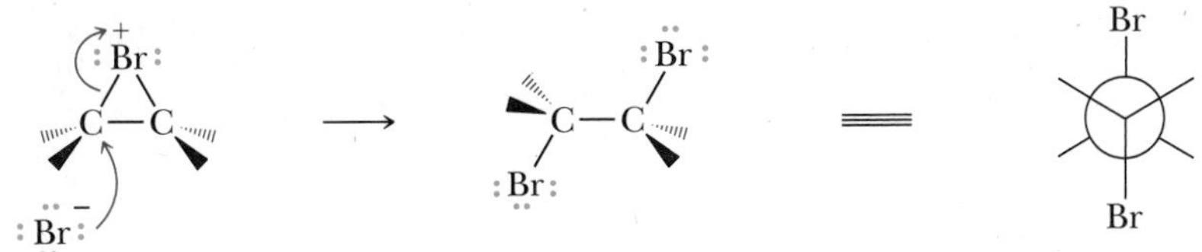

Anti (coplanar) orientation of added bromine atoms

Newman projection of the product

Addition of chlorine or bromine to cyclohexene and its derivatives gives a *trans* diaxial product because only axial positions on adjacent atoms of a cyclohexane ring are anti and coplanar. The initial *trans* diaxial conformation of the product is in equilibrium with the *trans* diequatorial conformation, and in simple derivatives of cyclohexane, the *trans* diequatorial conformation is more stable and predominates. Because the original bromonium ion can form on either face of the double bond with equal probability, both *trans* enantiomers are formed as a racemic mixture.

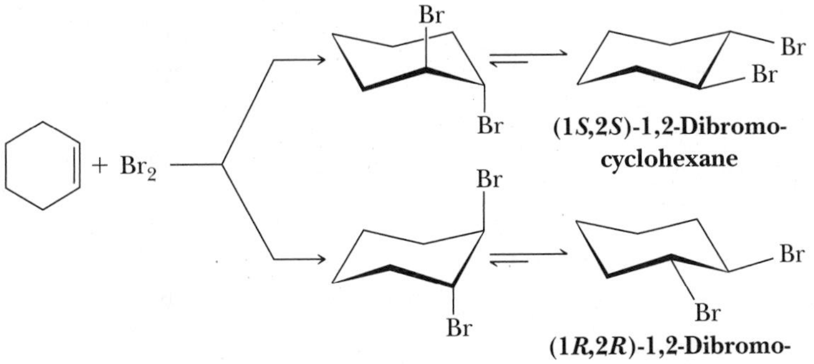

(1*S*,2*S*)-1,2-Dibromo-cyclohexane

(1*R*,2*R*)-1,2-Dibromo-cyclohexane

Trans diaxial conformations (less stable)

Trans diequatorial conformations (more stable)

E. Addition of HOCl and HOBr

Treating an alkene with Br_2 or Cl_2 in the presence of water results in addition of OH and Br, or OH and Cl, to the carbon-carbon double bond to give a **halohydrin.**

Halohydrin

A compound containing a halogen atom and a hydroxyl group on adjacent carbons; those containing Br and OH are bromohydrins, and those containing Cl and OH are chlorohydrins.

$$CH_3CH{=}CH_2 + Cl_2 + H_2O \longrightarrow CH_3\underset{\displaystyle |}{\overset{HO}{}}\!\!CH-\overset{Cl}{}\!CH_2 + HCl$$

Propene → **1-Chloro-2-propanol** (a chlorohydrin) (a racemic mixture)

Addition of HOCl and HOBr is regioselective (halogen adds to the less substituted carbon atom) and anti stereoselective. Both the regioselectivity and anti stereoselectivity are illustrated by the addition of HOBr to 1-methylcyclopentene. Bromine and the hydroxyl group add anti to each other with Br bonding to the less substituted carbon and OH bonding to the more substituted carbon. Dimethyl sulfoxide (DMSO) is used as a cosolvent with water to enhance the solubility of the alkene.

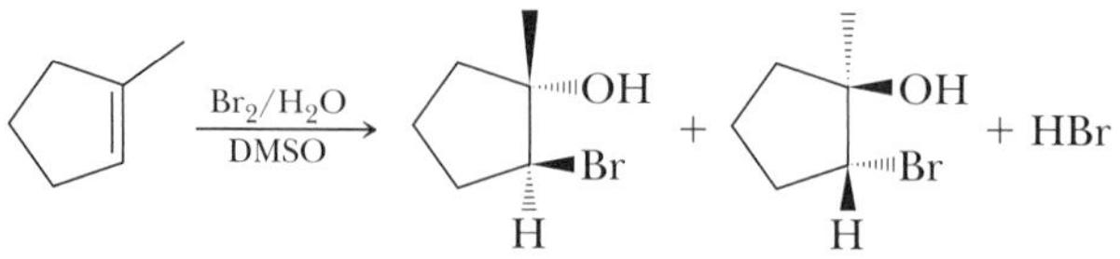

1-Methylcyclopentene **2-Bromo-1-methylcyclopentanol** (a racemic mixture)

To account for the regioselectivity and anti stereoselectivity of halohydrin reactions, chemists propose a three-step mechanism.

Mechanism 6.5

Halohydrin Formation and Its Anti Stereoselectivity

Step 1: **Make a new bond between a nucleophile (π bond) and an electrophile.** Reaction of the π electrons of the alkene with bromine (an electrophile) gives a bridged bromonium ion intermediate. This intermediate has some of the character of a carbocation (to account for the regioselectivity) and some of the character of a halonium ion (to account for the anti stereoselectivity). The secondary carbocation makes a substantial contribution to the structure of the resonance hybrid; the primary carbocation is higher in energy and makes little contribution.

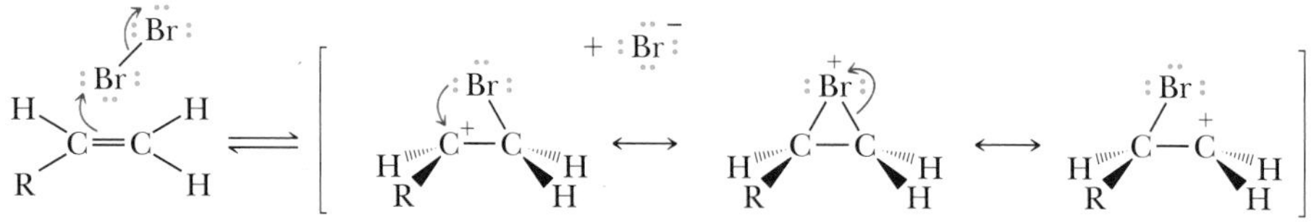

this carbocation is a major contributing structure

the bridged bromonium ion plays an important role

this structure plays little role

Step 2: **Make a new bond between a nucleophile and an electrophile.** Attack of H_2O (a nucleophile) on the more substituted carbon of the bridged bromonium ion (an electrophile) opens the three-membered ring.

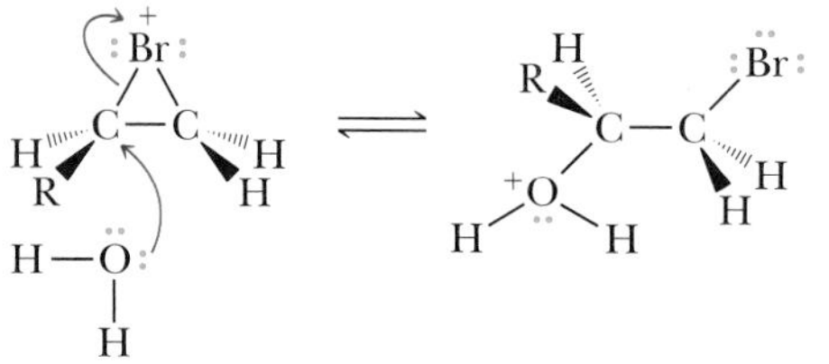

In the case of a bromonium ion derived from a symmetrical alkene, both carbons are attacked by H_2O with equal probability. In the case of unsymmetrical alkenes (as, for example, that derived from 2-methylpropene), there is preferential opening of the cyclic bromonium ion intermediate by attack of H_2O on the more substituted carbon of the alkene. At first glance, this may seem counterintuitive because the more substituted carbon might be considered less accessible to a nucleophile. However, the experimentally observed preferential attack at the more substituted carbon atom can be explained by a combination of two factors working together.

Carbocation Character

As the accompanying electrostatic potential maps show, there is more carbocation character on the more substituted carbon, which directs attack of the nucleophile preferentially to this carbon. Recall that alkyl groups stabilize carbocations, explaining the greater carbocation character at the more substituted carbon.

Activation Energy to Reach the Ring-Opening Transition State

As the accompanying electrostatic potential maps also show, the carbon-halogen bond to the more substituted carbon of the halonium ion is longer than the bond to the less substituted carbon. This difference in bond lengths in the cyclic intermediate state means that the ring-opening transition state can be reached more easily by attack at the more substituted carbon.

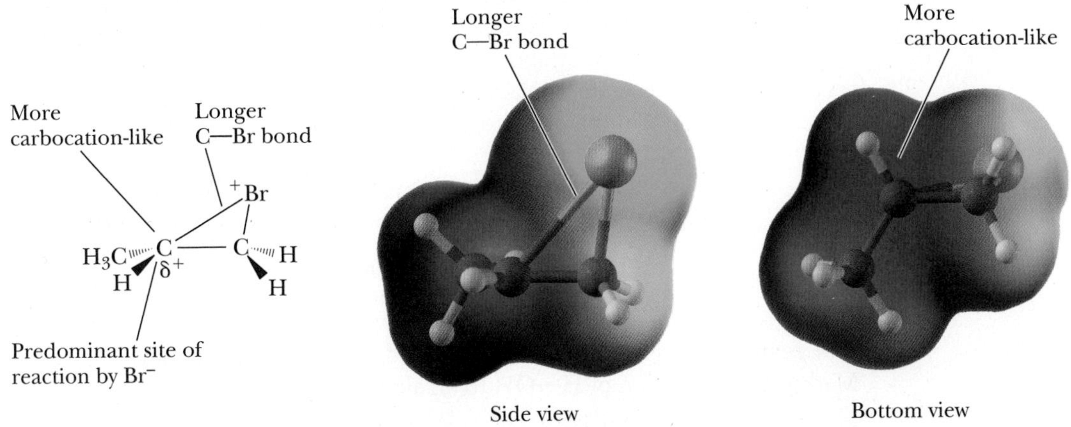

Bridged bromonium ion from propene.

Step 3: **Take a proton away.** Proton transfer to water completes the reaction.

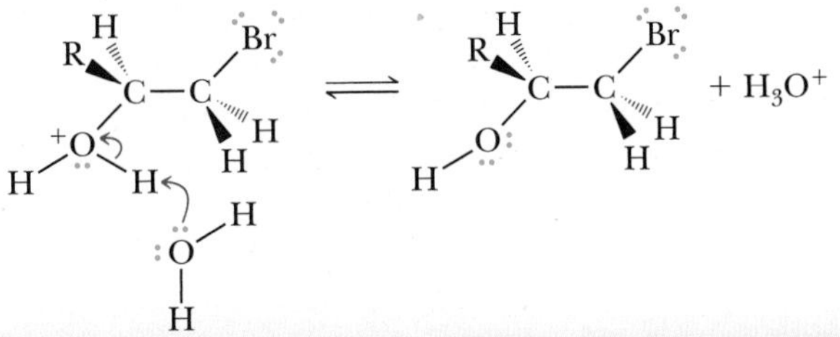

Example 6.9 Bromohydrin Formation

Draw the structure of the bromohydrin formed by treating 2-methylpropene with Br_2/H_2O.

Solution

Addition is regioselective, with —OH adding to the more substituted carbon and —Br adding to the less substituted carbon.

$$\underset{\textbf{2-Methylpropene}}{CH_3-\overset{\overset{\large CH_3}{|}}{C}=CH_2} + Br_2 + H_2O \longrightarrow \underset{\substack{\textbf{1-Bromo-2-methyl-2-propanol}\\ \text{(achiral)}}}{CH_3-\underset{\underset{\large HO}{|}}{\overset{\overset{\large CH_3}{|}}{C}}-\underset{\underset{\large Br}{|}}{CH_2}} + HBr$$

Problem 6.9

Draw the structure of the chlorohydrin formed by treating 1-methylcyclohexene with Cl_2/H_2O.

F. Oxymercuration-Reduction

The hydration of an alkene can be accomplished by treating it with mercury(II) acetate (mercuric acetate) in water followed by reduction of the resulting organomercury compound with sodium borohydride, $NaBH_4$. In the following structural formulas for mercury(II) acetate, the acetate group is written in full in the first formula and abbreviated as AcO in the other formulas.

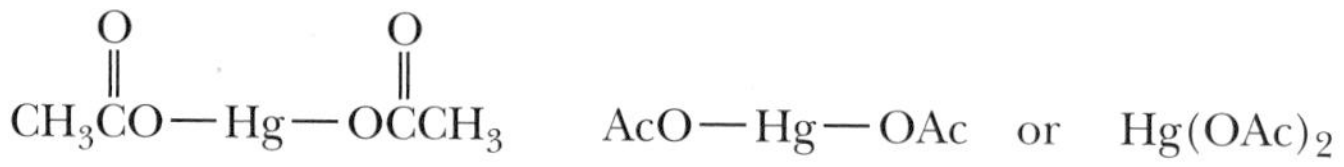
AcO—Hg—OAc
or
Hg(OAc)2

Mercury(II) acetate
(Mercuric acetate)

The result of oxymercuration followed by sodium borohydride reduction is Markovnikov addition of H—OH to an alkene.

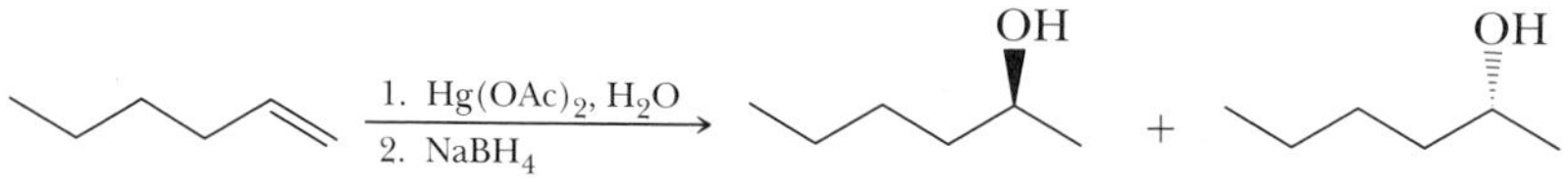
1. Hg(OAc)2, H2O
2. NaBH4
OH
OH
+

1-Hexene **(*S*)-2-Hexanol** **(*R*)-2-Hexanol**
(a racemic mixture)

We discuss this reaction in two stages, first oxymercuration and then reduction.

Oxymercuration-reduction
A method for converting an alkene to an alcohol. The alkene is treated with mercury(II) acetate followed by reduction with sodium borohydride.

Oxymercuration, the addition of mercury(II) to one carbon of the double bond and oxygen to the other, is illustrated by the first step in the two-step conversion of 1-hexene to 2-hexanol. Oxymercuration is regioselective: HgOAc becomes bonded to the less substituted carbon of the alkene, and OH of water becomes bonded to the more substituted carbon.

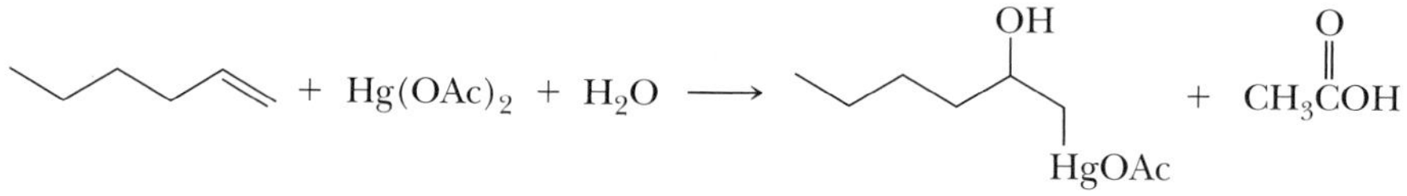
+ Hg(OAc)2 + H2O
OH
HgOAc
+ CH3COH

1-Hexene **Mercury(II) acetate** An organomercury compound (a racemic mixture) **Acetic acid**

Reduction of the organomercury compound by sodium borohydride, $NaBH_4$, replaces HgOAc by H.

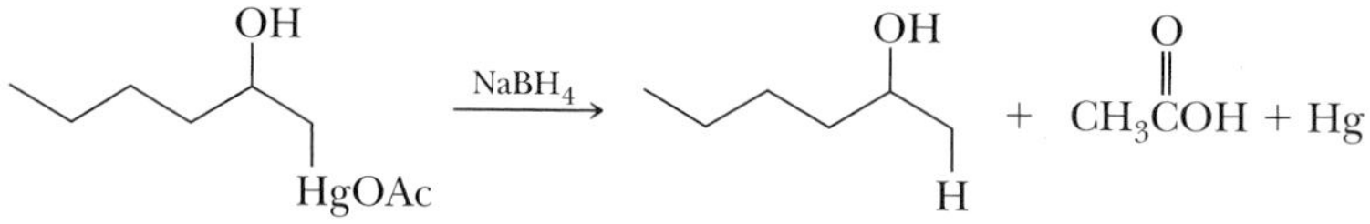
OH
HgOAc
NaBH4
OH
H
+ CH3COH + Hg

(a racemic mixture) **2-Hexanol** (a racemic mixture) **Acetic acid**

Oxymercuration of 3,3-dimethyl-1-butene followed by $NaBH_4$ reduction gives racemic 3,3-dimethyl-2-butanol exclusively and illustrates a very important feature of this reaction sequence: it occurs without rearrangement.

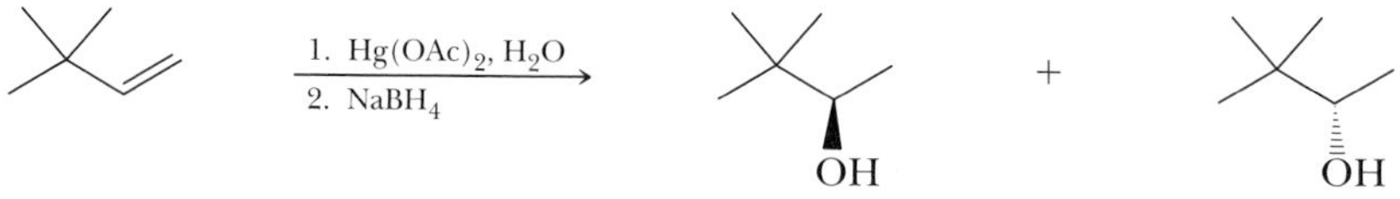
1. Hg(OAc)2, H2O
2. NaBH4
+
OH
OH

3,3-Dimethyl-1-butene **(*R*)-3,3-Dimethyl-2-butanol** **(*S*)-3,3-Dimethyl-2-butanol**
(a racemic mixture)

You might compare the product of oxymercuration-reduction of 3,3-methyl-1-butene with the product formed by acid-catalyzed hydration of the same alkene (Section 6.3C). In the former, no rearrangement occurs. In the latter, the major product is 2,3-dimethyl-2-butanol, a compound formed by rearrangement. The fact that no rearrangement occurs during oxymercuration-reduction indicates that at no time is a free carbocation intermediate formed.

The stereoselectivity of the oxymercuration step is illustrated by the reaction of mercury(II) acetate with cyclopentene.

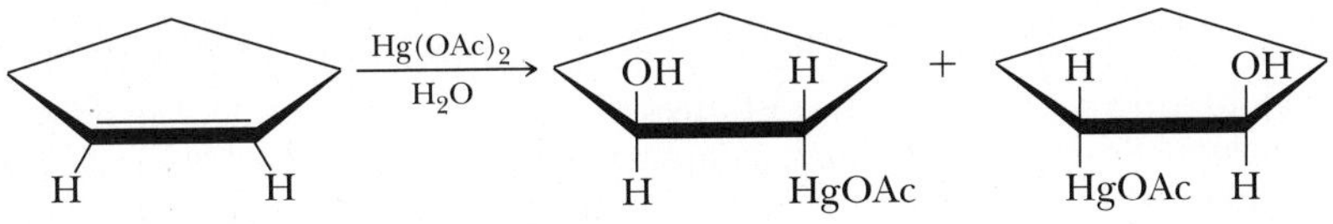

Cyclopentene

(Anti addition of OH and HgOAc)
(racemic)

The fact that the oxymercuration step is both regioselective and anti stereoselective has led chemists to propose the following mechanism, which is closely analogous to that for the addition of Br_2 and Cl_2 to an alkene (Section 6.3D).

Mechanism 6.6

Oxymercuration-Reduction of an Alkene

***Step 1:* Break a bond to give stable molecules or ions.** Dissociation of mercury(II) acetate gives $AcOHg^+$ (an electrophile) and acetate anion.

$$AcO{-}Hg{-}OAc \longrightarrow AcO{-}Hg^+ + AcO^-$$

(an electrophile)

***Step 2:* Make a new bond between a nucleophile (π bond) and an electrophile.** Attack of the alkene π bond onto $AcOHg^+$ (an electrophile) forms a bridged mercurinium ion intermediate.

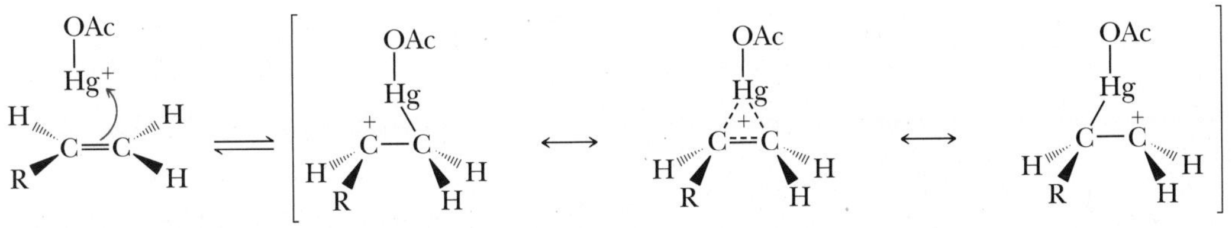

An open carbocation intermediate (a minor contributor)

A bridged mercurinium ion intermediate (the major contributor)

An open carbocation intermediate (a negligible contributor)

This intermediate closely resembles a bridged bromonium ion intermediate (Section 6.3D). However, in the bridged mercurinium ion intermediate, the two π electrons of the carbon-carbon double bond form a ring containing three atoms bonded by two electrons. The open cation structure with the positive charge on the 2° carbon is a minor contributing structure to the resonance hybrid. The open cation contributor with the positive charge on the 1° carbon is a negligible contributor.

The fact that oxymercuration occurs without rearrangement indicates that the intermediate formed in Step 2 is not a true carbocation, but rather a resonance hybrid largely with the character of a bridged mercurinium ion intermediate.

***Step 3:* Make a new bond between a nucleophile and an electrophile.** Anti attack of water (a nucleophile) on the bridged mercurinium ion intermediate (an electrophile) occurs at the more substituted carbon to open the three-membered ring.

(Continued)

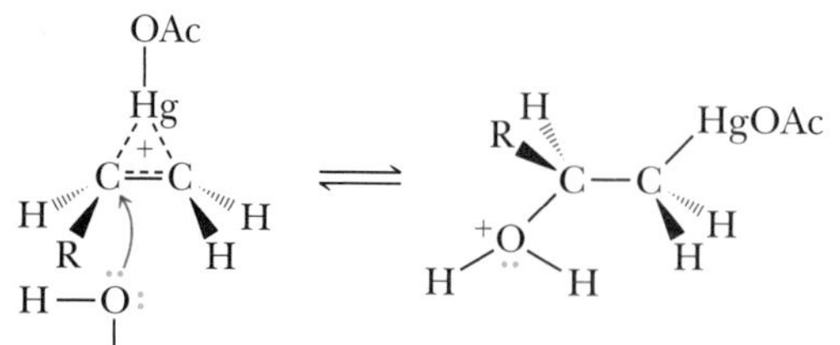
Anti addition of
HgOAc and HOH

We account for the regioselectivity just as we did for the regioselectivity of halohydrin formation in Section 6.3E. Of the two carbons of the mercurinium ion intermediate, the more substituted carbon has a greater degree of partial positive charge and is attacked by the nucleophile, H_2O. In addition, computer modeling indicates that the carbon-mercury bond to the more substituted carbon of the bridged mercurinium ion intermediate is longer than the one to the less substituted carbon, which means that the ring-opening transition state is reached more easily by attack at the more substituted carbon.

Step 4: **Take a proton away.** Transfer of a proton to water completes the onymercuration of the alkene.

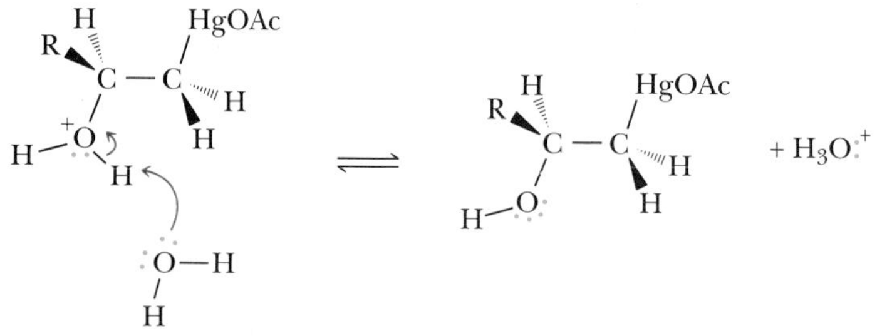

Step 5: Reduction of the C—HgOAc bond to a C—H bond gives the final product and metallic mercury. The mechanism of this step is beyond the scope of this chapter and will not be discussed in detail, which is why we have not drawn arrows to indicate movement of electrons. The mechanism is thought to involve radicals (unpaired electrons). The key from our point of view is that the stereochemistry of any chiral carbon bearing the Hg atom is scrambled in this reduction step. The net result is that even though the OH group and the Hg atom add with anti stereochemistry in the oxymercuration step, following reduction of Hg to H, the products reflect a mixture of syn and anti addition of the H atom and OH group.

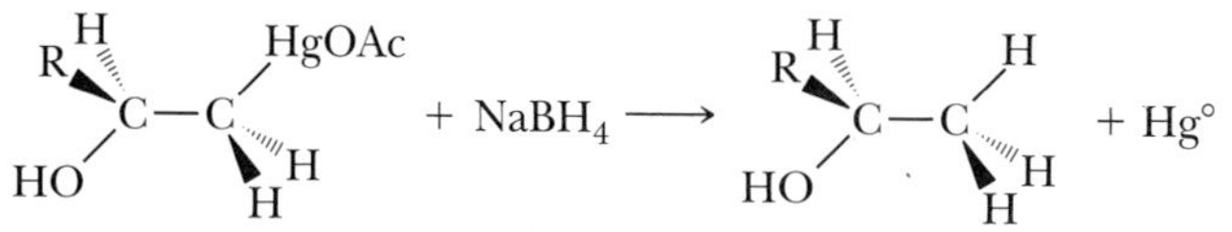

6.4 Hydroboration-Oxidation

The net reaction from hydroboration and subsequent oxidation of an alkene is hydration of a carbon-carbon double bond. Because hydrogen is added to the more substituted carbon of the double bond and —OH to the less substituted carbon, we refer to the regiochemistry of hydroboration and subsequent oxidation as non-Markovnikov hydration:

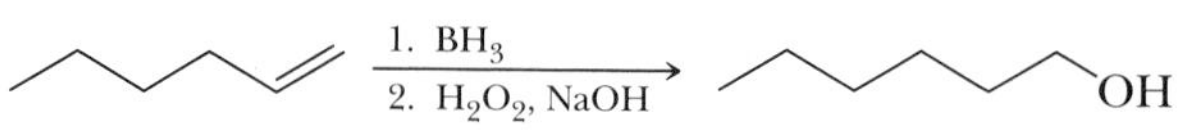
1. BH_3
2. H_2O_2, NaOH
OH

1-Hexene **1-Hexanol**

Hydroboration-oxidation

A method for converting an alkene to an alcohol. The alkene is treated with borane (BH_3) to give a trialkylborane, which is then oxidized with alkaline hydrogen peroxide to give the alcohol.

Hydroboration-oxidation of alkenes is a valuable laboratory method for the regioselective and stereoselective hydration of alkenes. Furthermore, this sequence of reactions occurs without rearrangement.

Hydroboration is the addition of borane, BH_3, to an alkene to form a trialkylborane. The overall reaction occurs in three successive steps. BH_3 reacts with one molecule of alkene to form an alkylborane, then with a second molecule of alkene to form a dialkylborane, and finally with a third molecule of alkene to form a trialkylborane.

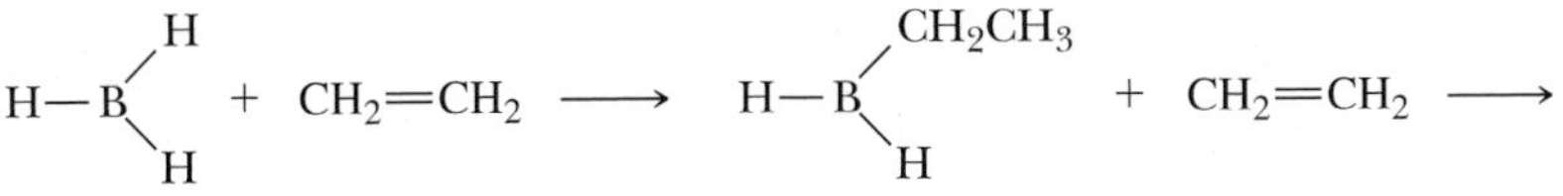

Borane **Ethylborane** (an alkylborane)

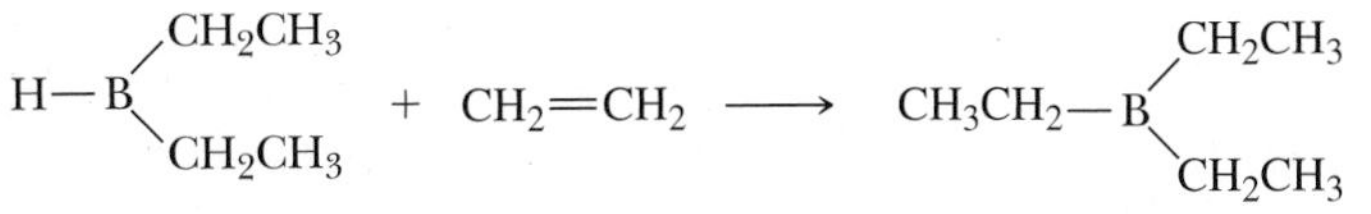

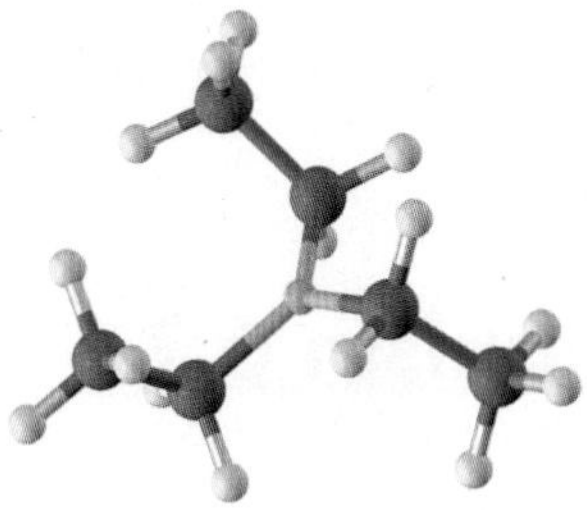

Diethylborane (a dialkylborane) **Triethylborane** (a trialkylborane)

Borane cannot be prepared as a pure compound because it dimerizes to diborane, B_2H_6, a toxic gas that ignites spontaneously in air.

$$2\,BH_3 \rightleftharpoons B_2H_6$$

Borane **Diborane**

However, BH_3 forms stable Lewis acid-base complexes with ethers. Borane is most often used as a commercially available solution of BH_3 in THF.

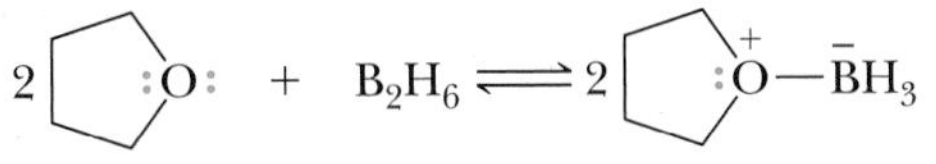

Tetrahydrofuran (THF) **$BH_3 \cdot$ THF**

Boron, atomic number 5, has three electrons in its valence shell. To bond with three other atoms, boron uses sp^2 hybrid orbitals. The unoccupied $2p$ orbital of boron is perpendicular to the plane created by boron and the three other atoms to which it is bonded. An example of a stable, trivalent boron compound is boron trifluoride, BF_3, a planar molecule with F—B—F bond angles of 120° (Section 1.2E). Because of the vacant $2p$ orbital in the valence shell of boron, BH_3, BF_3, and all other trivalent compounds of boron are electrophiles. These compounds of boron resemble carbocations, except that, unlike carbocations, they are electrically neutral. BH_3 is a planar molecule with H—B—H bond angles of 120° (see margin).

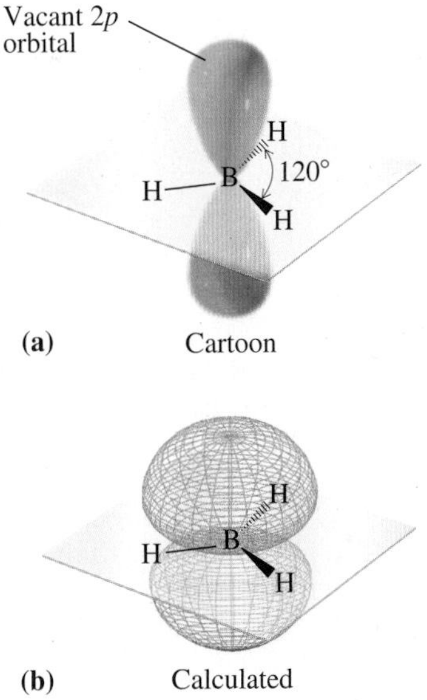

(a) Cartoon and (b) Calculated orbitals and bond angles in BH_3.

Addition of borane to alkenes is regioselective and stereoselective.

- Regioselective: Upon addition of borane to an unsymmetrical alkene, boron becomes bonded predominantly to the less substituted carbon of the double bond **(non-Markovnikov)**.
- Stereoselective: Hydrogen and boron add from the same face of the double bond; that is, the reaction is **syn** (from the same side) **stereoselective**.

Syn stereoselective The addition of atoms or groups of atoms to the same face of a carbon-carbon double bond.

Both the regioselectivity and syn stereoselectivity are illustrated by hydroboration of 1-methylcyclopentene.

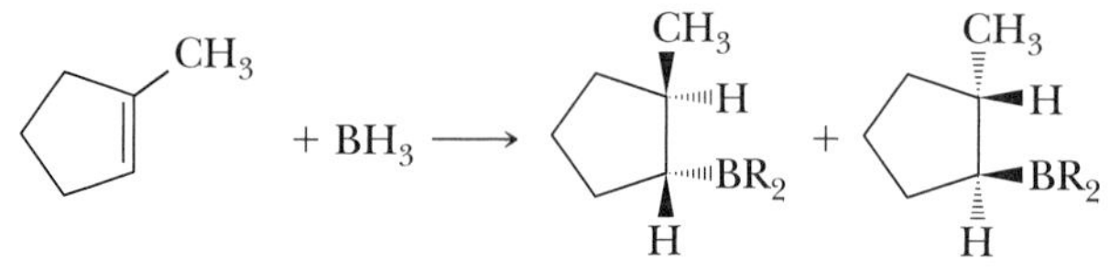

1-Methylcyclopentene

(Syn addition of BH_3)
(R = 2-methylcyclopentyl)
(a racemic mixture)

Note that other textbooks will refer to the regioselectivity of the hydroboration-oxidation reaction as anti-Markovnikov, while we prefer the term non-Markovnikov. We do not favor the use of anti-Markovnikov because it can be confusing to students due to the syn, not anti, stereochemistry of addition observed with this reaction.

Mechanism 6.7

Hydroboration

Step 1: **Lewis Acid-Base Coordination.** The addition of borane to an alkene is initiated by coordination of the vacant $2p$ orbital of boron with the electron pair of the π bond. This coordination is a Lewis acid-base interaction, analogous to the coordination of THF to borane shown above. We designate this coordination with an arrow from the alkene to the boron (see "Things You Should Know II").

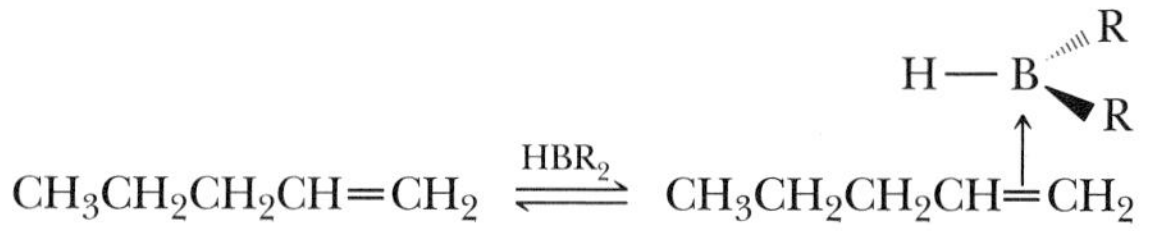

Step 2: **Make a new bond between a nucleophile (π bond) and an electrophile with simultaneous bond formation to H.** After coordination, the boron adds to the less substituted carbon of the alkene, thereby placing the hydrogen on the more substituted carbon, via a cyclic, four-centered transition state. Boron and hydrogen add simultaneously from the same face of the double bond (syn addition).

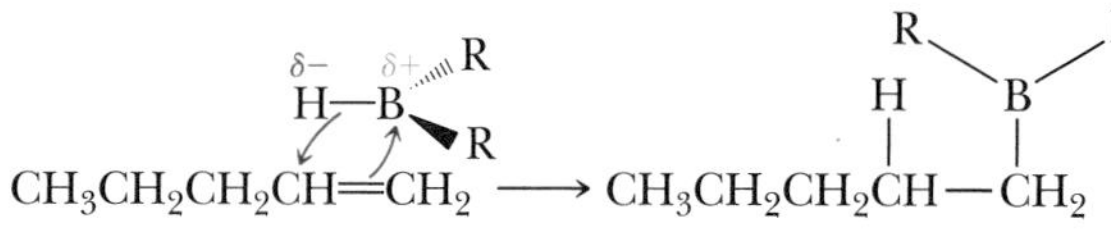

We account for the regioselectivity by a combination of steric and electronic factors. In terms of steric effects, boron, the larger part of the reagent, adds selectively to the less hindered carbon of the double bond, and hydrogen, the smaller part of the reagent, adds to the more hindered carbon. It is believed that the observed regioselectivity is due largely to these steric effects.

Electronic effects probably also influence the regioselectivity. The electronegativity of hydrogen (2.1) is slightly greater than that of boron (2.0); hence, there is a small degree of polarity (approximately 5%) to each B—H bond, with boron bearing a partial positive charge and hydrogen a partial negative charge. It is proposed that there is some degree of carbocation character in the transition state and that the partial positive charge is on the more substituted carbon.

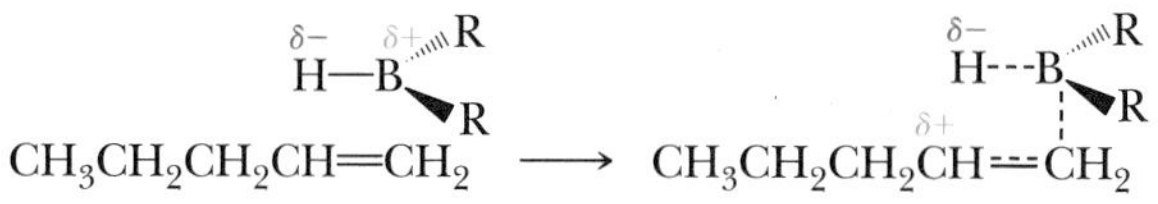

Starting reagents

Transition state
(some carbocation character exists in the transition state)

Trialkylboranes are rarely isolated. Rather, they are converted directly to other products formed by substitution of another atom (H, O, N, C, or halogen) for boron. One of the most important reactions of trialkylboranes is with hydrogen peroxide in aqueous sodium hydroxide. Hydrogen peroxide is an oxidizing agent and, under these conditions, oxidizes a trialkylborane to an alcohol and sodium borate, Na_3BO_3.

Mechanism 6.8

Oxidation of a Trialkylborane by Alkaline Hydrogen Peroxide

Step 1: **Make a new bond between a nucleophile and an electrophile.** Donation of a pair of electrons from a hydroperoxide ion (a nucleophile) to the boron atom of the trialkylborane (an electrophile) gives an intermediate in which boron has a filled valence shell and bears a negative formal charge.

$$HOOH + OH^- \rightleftharpoons HOO^- + HOH$$

Hydrogen peroxide — **Hydroperoxide ion**

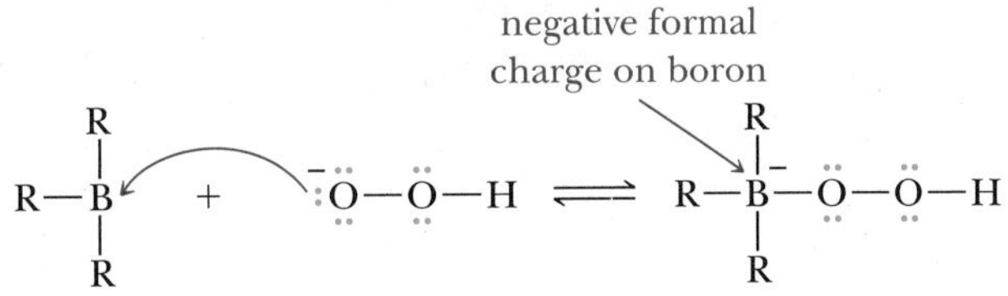

A trialkylborane (an electrophile) — **Hydroperoxide ion** (a nucleophile)

Step 2: **1,2 Shift.** Rearrangement of an R group with its pair of bonding electrons to an adjacent oxygen (a 1,2 shift) results in ejection of hydroxide ion.

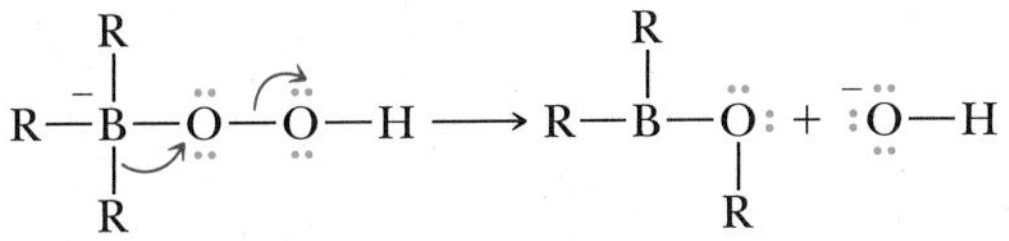

Two more reactions with hydroperoxide ion followed by rearrangements give a trialkylborate.

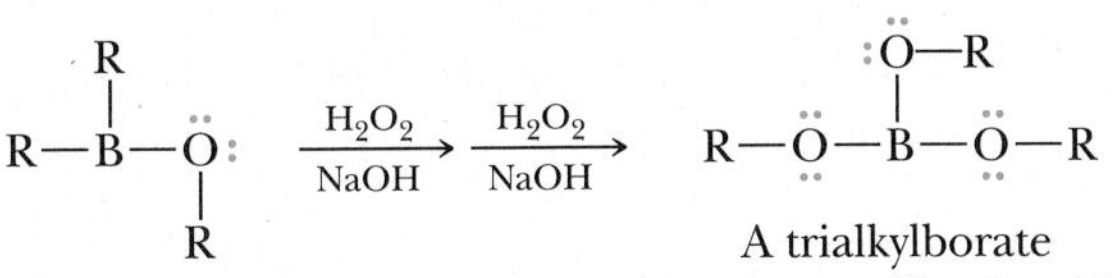

A trialkylborate (a triester of boric acid)

Step 3: Reaction of the trialkylborate with aqueous NaOH gives the alcohol and sodium borate.

$$(RO)_3B + 3\,NaOH \longrightarrow 3\,ROH + Na_3BO_3$$

A trialkylborate — **Sodium borate**

Hydrogen peroxide oxidation of a trialkylborane is stereoselective in that the configuration of the alkyl group is retained; whatever the position of boron in relation to other groups in the trialkylborane, the OH group by which it is replaced occupies the same position. Thus, the net result of hydroboration-oxidation of an alkene is syn stereoselective addition of H and OH to a carbon-carbon double bond, combined with non-Markovnikov regioselectivity.

Example 6.10 Hydroboration-Oxidation

Draw structural formulas for the alcohol formed by hydroboration-oxidation of each alkene.

(a) $CH_3\overset{\displaystyle CH_3}{\overset{|}{C}}{=}CHCH_3$ **(b)** 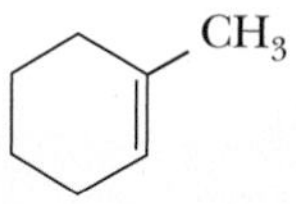

Solution

(a) 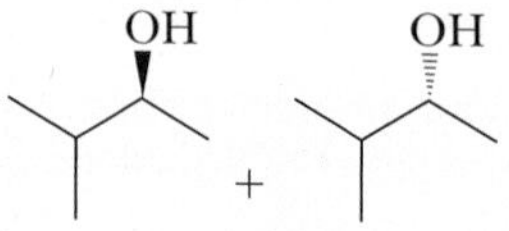

(a racemic mixture)

(b) 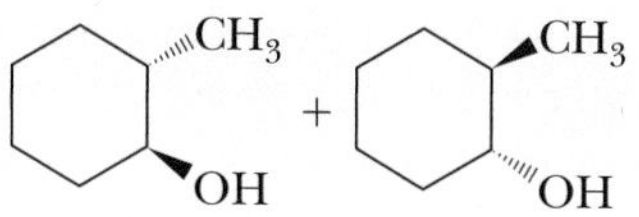

(a racemic mixture)

Watch a video explanation

Problem 6.10

Draw structural formulas for the alkene that gives each alcohol upon hydroboration-oxidation.

(a) 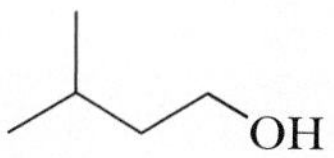 **(b)** 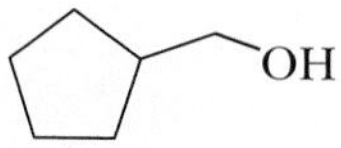

6.5 Oxidation

We begin this section with a review of the definitions of oxidation and reduction. We then consider two common oxidation reactions of alkenes.

Oxidation is the loss of electrons, and **reduction** is the gain of electrons. In the following examples, propene is transformed into two different compounds by reactions we study in this and the following section. The first reaction involves reduction, and the second involves oxidation. These equations, however, do not specify what reagents are necessary to bring about the particular transformation. Each does specify, however, that the carbon atoms of the products are derived from those of propene.

Oxidation
The loss of electrons. Alternatively, the loss of hydrogens, the gain of oxygens, or both.

Reduction
The gain of electrons. Alternatively, the gain of hydrogen, the loss of oxygens, or both.

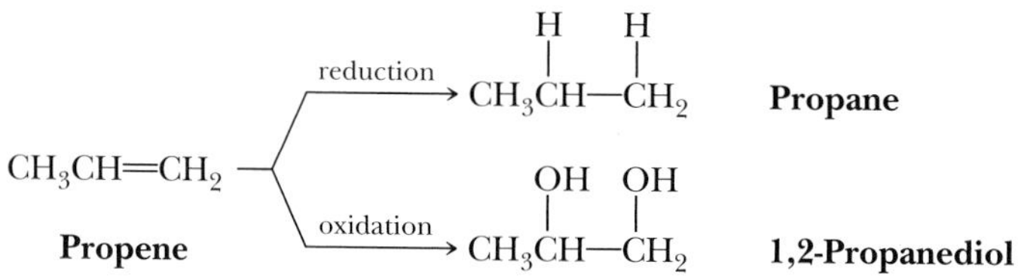

Example 6.11 Half-Reactions

Use a balanced half-reaction to show that each transformation involves an oxidation.

(a) $CH_3CH_2CH_2OH \longrightarrow CH_3CH_2\overset{\displaystyle O}{\overset{\|}{C}}H$ **(b)** $CH_3CH{=}CH_2 \longrightarrow CH_3\overset{\displaystyle O}{\overset{\|}{C}}H + H\overset{\displaystyle O}{\overset{\|}{C}}H$

Solution

Complete a material balance and then a charge balance.

(a) $CH_3CH_2CH_2OH \longrightarrow CH_3CH_2\overset{O}{\overset{\|}{C}}H + 2\,H^+ + 2e^-$

(b) $CH_3CH{=}CH_2 + 2\,H_2O \longrightarrow CH_3\overset{O}{\overset{\|}{C}}H + H\overset{O}{\overset{\|}{C}}H + 4\,H^+ + 4e^-$

The first transformation is a two-electron oxidation; the second is a four-electron oxidation. To bring each about requires an oxidizing agent.

Problem 6.11

Use a balanced half-reaction to show that each transformation involves a reduction.

(a) 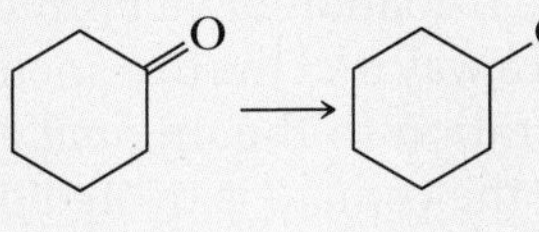

(b) $CH_3CH_2\overset{O}{\overset{\|}{C}}OH \longrightarrow CH_3CH_2CH_2OH$

As an alternative way to recognize oxidation/reduction, recall from your course in general chemistry that oxidation and reduction can be defined in terms of the loss or gain of oxygens or hydrogens. For organic compounds:

oxidation: the addition of O to and/or removal of H from a carbon atom
reduction: the removal of O from and/or addition of H to a carbon atom

Example 6.12 Oxidations or Reductions

Tell which of these transformations are oxidations and which are reductions based on whether there is addition or removal of O or H.

(a) $CH_3CH_2OH \longrightarrow CH_3\overset{O}{\overset{\|}{C}}H$

(b) $CH_3\overset{O}{\overset{\|}{C}}CH_2\overset{O}{\overset{\|}{C}}OH \longrightarrow CH_3\overset{OH}{\overset{|}{C}}HCH_2\overset{O}{\overset{\|}{C}}OH$

(c) $CH_3CH{=}CH_2 \longrightarrow CH_3CH_2CH_3$

Solution

(a) Oxidation; there is a loss of two hydrogens.
(b) Reduction; there is a gain of two hydrogens.
(c) Reduction; there is a gain of two hydrogens.

Problem 6.12

Tell which of these transformations are oxidations and which are reductions based on whether there is addition or removal of O or H.

(a) $CH_3\overset{O}{\overset{\|}{C}}H \longrightarrow CH_3CH_2OH$

(b) 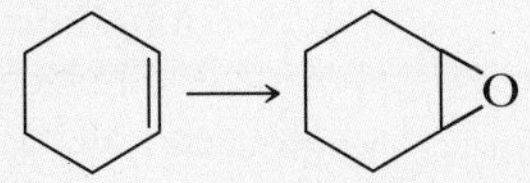

(c) 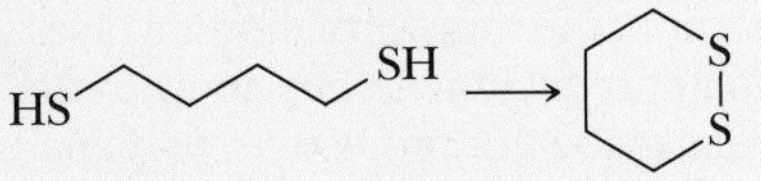

A. OsO_4—Oxidation of an Alkene to a Glycol

Glycol, vicinal diol

A compound with hydroxyl (—OH) groups on adjacent carbons.

Osmium tetroxide, OsO_4, and certain other transition metal oxides are effective oxidizing agents for the conversion of an alkene to a 1,2-diol (a **glycol**, also known as a **vicinal diol**). OsO_4 is highly electrophilic and adds in a single step to the alkene. Oxidation of an alkene by OsO_4 is syn stereoselective; it involves syn addition of an OH group to each carbon of the double bond. For example, oxidation of cyclopentene gives *cis*-1,2-cyclopentanediol. Note that both *cis* and *trans* isomers are possible for this glycol but that only the *cis* glycol forms in this oxidation.

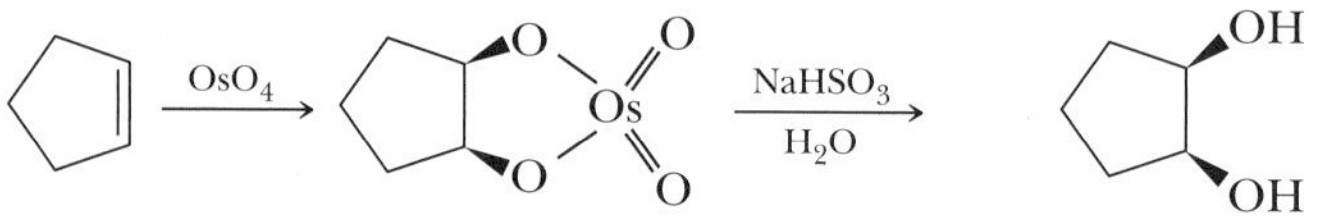

A cyclic osmate

***cis*-1,2-Cyclopentanediol**
(a *cis* glycol)

The syn stereoselectivity of the osmium tetroxide oxidation of an alkene is accounted for by the formation of a cyclic osmate in such a way that the five-membered osmium-containing ring is bonded in a *cis* configuration to the original alkene. Osmates can be isolated and characterized. Usually, the osmate is treated directly with a reducing agent, such as $NaHSO_3$, which cleaves the osmium-oxygen bonds to give a *cis* glycol and reduced forms of osmium.

The drawbacks of OsO_4 are that it is both expensive and highly toxic. One strategy to circumvent the high cost is to use it in catalytic amounts along with a stoichiometric amount of another oxidizing agent whose purpose is to reoxidize the reduced forms of osmium and thus recycle the osmium reagent. Secondary oxidizing agents commonly used for this purpose are hydrogen peroxide and *tert*-butyl hydroperoxide. When this procedure is used, there is no need for a reducing step using $NaHSO_3$.

HOOH

Hydrogen peroxide

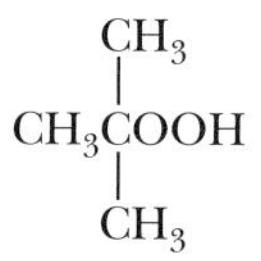

***tert*-Butyl hydroperoxide (*t*-BuOOH)**

HOW TO

Write a Balanced Half-Reaction

One way to tell if these or other transformations involve oxidation, reduction, or neither is to use the method of balanced half-reactions.

To write a balanced half-reaction:

1. Write a half-reaction showing the organic reactant(s) and product(s).
2. Complete a material balance; that is, balance the number of atoms on each side of the half-reaction. To balance the number of oxygens and hydrogens for a reaction that takes place in acid solution, use H_2O for oxygens and H^+ for hydrogens. For a reaction that takes place in basic solution, use H_2O and OH^-.
3. Complete a charge balance; that is, balance the charge on both sides of the half-reaction. To balance the charge, add electrons, e^-, to one side or the other. The equation completed in this step is a balanced half-reaction.

If electrons appear on the right side of the balanced half reaction, the reactant gives up electrons and is oxidized. If electrons appear on the left side of a balanced half reaction, the reactant has gained electrons and is reduced. If no electrons appear in the balanced half reaction, the transformation involves neither oxidation nor reduction. Let us apply these steps to the transformation of propene to propane.

Step 1: Half-reaction $CH_3CH{=}CH_2 \longrightarrow CH_3CH_2CH_3$

Step 2: Material balance $CH_3CH{=}CH_2 + 2\,H^+ \longrightarrow CH_3CH_2CH_3$

Step 3: Balanced half-reaction $CH_3CH{=}CH_2 + 2\,H^+ + 2e^- \longrightarrow CH_3CH_2CH_3$

Because two electrons appear on the left side of the balanced half-reaction (Step 3), conversion of propene to propane is a two-electron reduction. To bring it about requires use of a reducing agent.

A balanced half-reaction for the transformation of propene to 1,2-propanediol requires two electrons on the right side of the equation for a charge balance; this transformation is a two-electron oxidation.

Balanced half-reaction: $$\underset{\textbf{Propene}}{CH_3CH{=}CH_2} + 2\,H_2O \longrightarrow \underset{\textbf{1,2-Propanediol}}{CH_3\overset{OH}{\overset{|}{C}}H{-}\overset{OH}{\overset{|}{C}}H_2} + 2\,H^+ + 2e^-$$

Following is a balanced half-reaction for the transformation of propene to 2-propanol.

Balanced half-reaction: $$\underset{\textbf{Propene}}{CH_3CH{=}CH_2} + H_2O \longrightarrow \underset{\textbf{2-Propanol}}{CH_3\overset{OH}{\overset{|}{C}}HCH_3}$$

Because no electrons are required to achieve an electrical balance in the half-reaction, conversion of propene to 2-propanol is neither oxidation nor reduction; this reaction can be brought about by acid-catalyzed hydration of propene.

It is important to realize that this strategy for recognizing oxidation and reduction is only that, a strategy. In no way does it give any indication of how a particular oxidation or reduction might be carried out in the laboratory. For example, the balanced half-reaction for the transformation of propene to propane requires 2 H^+ and $2e^-$. Yet by far, the most common laboratory procedure for reducing propene to propane does not involve H^+ at all; rather, it involves molecular hydrogen, H_2, and a transition metal catalyst (Section 6.6).

B. Ozone—Cleavage of a Carbon-Carbon Double Bond (Ozonolysis)

Treating an alkene with ozone, O_3, followed by a suitable work-up, cleaves the carbon-carbon double bond and forms two carbonyl (C=O) groups in its place. Once again, as with all reagents discussed in this chapter, O_3 is strongly electrophilic. This reaction is noteworthy because it is one of the few organic reactions that breaks C—C bonds. The alkene is dissolved in an inert solvent, such as CH_2Cl_2, and a stream of ozone is bubbled through the solution. The products isolated from ozonolysis depend on the reaction conditions. Hydrolysis of the reaction mixture with water yields hydrogen peroxide, an oxidizing agent that can bring about further oxidations. To prevent side reactions caused by reactive peroxide intermediates, a weak reducing agent, most commonly dimethyl sulfide, $(CH_3)_2S$, is added during the workup to reduce peroxides to water.

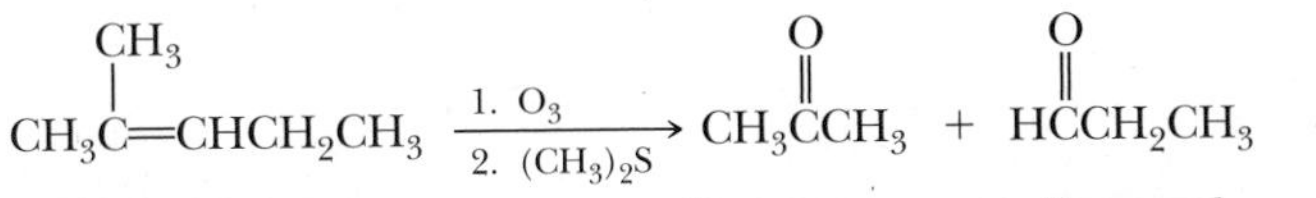
CH3
CH3C=CHCH2CH3
1. O3
2. (CH3)2S
O
CH3CCH3
+
O
HCCH2CH3

2-Methyl-2-pentene **Propanone** (a ketone) **Propanal** (an aldehyde)

The initial product of reaction of an alkene with ozone is an adduct called a molozonide, which rearranges under the conditions of the reaction to an isomeric compound called an ozonide. Low-molecular-weight ozonides are explosive and are rarely isolated. They are treated directly with a weak reducing agent to give the carbonyl-containing products.

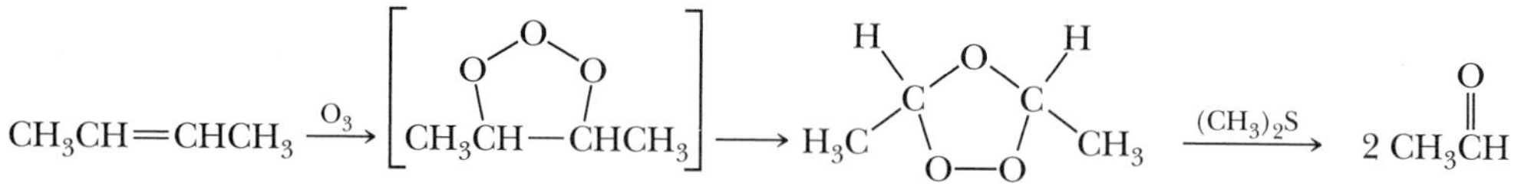

2-Butene A molozonide An ozonide **Acetaldehyde**

To understand how an ozonide is formed, we must first examine the structure of ozone. We can write this molecule as a hybrid of four contributing structures, all of which show separation of unlike charge.

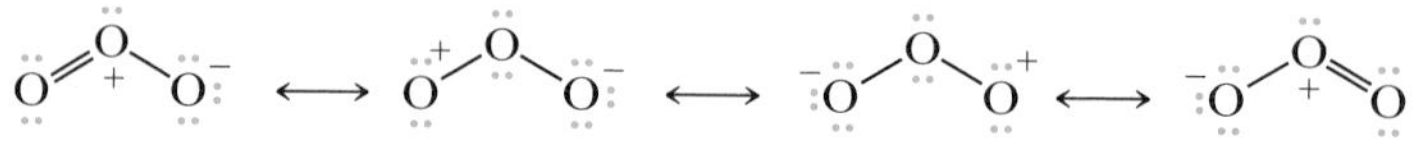

It is not possible to write a Lewis structure for ozone without separation of charges.

Mechanism 6.9

Formation of an Ozonide

Step 1: Ozone reacts with the alkene with both electrophilic and nucleophilic character in a single, simultaneous step.

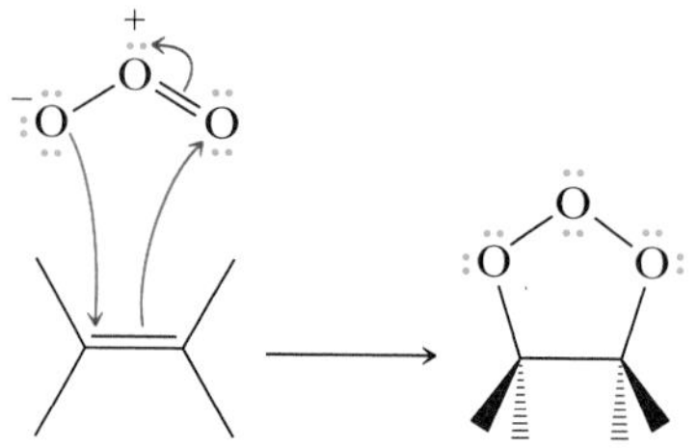

A molozonide

Step 2: **Break a bond to give stable molecules or ions.** Relocating valence electrons in the molozonide results in cleavage of one carbon-carbon and one oxygen-oxygen bond. The resulting fragments then recombine to form an ozonide.

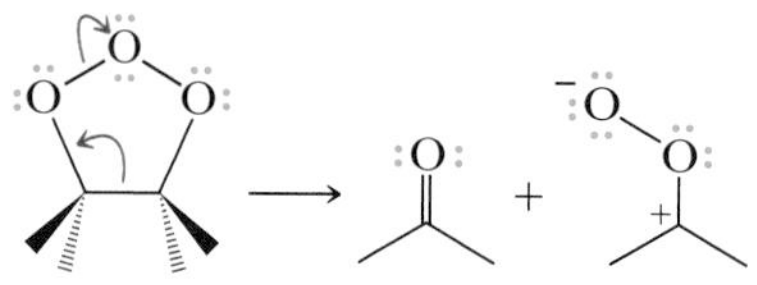

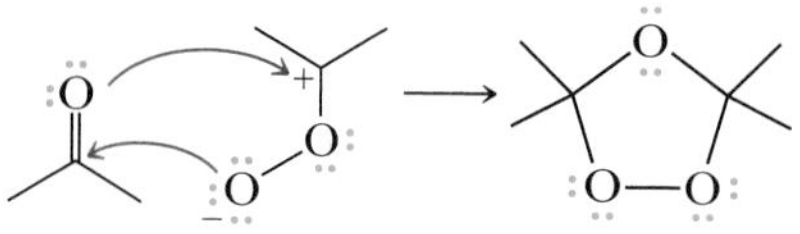

Step 3: Reduction of the ozonide and cleavage results in the final carbonyl fragments.

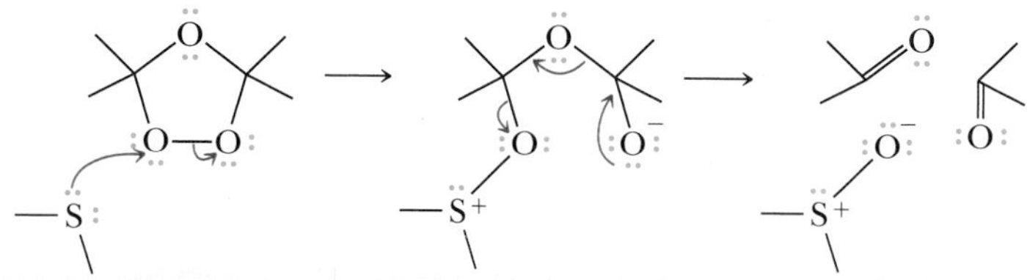

Example 6.13 Ozonolysis

Draw structural formulas for the products of the following ozonolysis reactions and name the new functional groups formed in each oxidation.

(a) 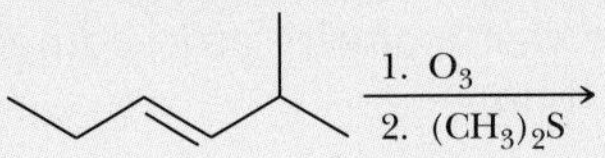

(b) 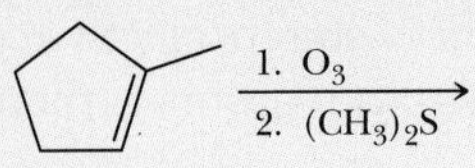

Solution

(a) 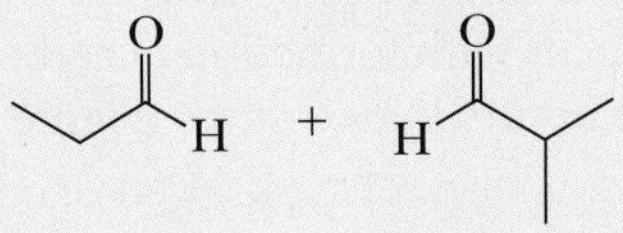

Propanal (an aldehyde) 2-Methylpropanal (an aldehyde)

(b) an aldehyde a ketone

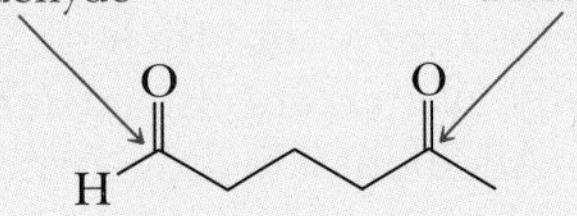

5-Oxohexanal (a ketoaldehyde)

Problem 6.13

Watch a video explanation

What alkene with the molecular formula C_6H_{12}, when treated with ozone and then dimethyl sulfide, gives the following product(s)?

(a) 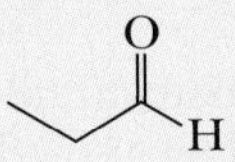

(only product)

(b) 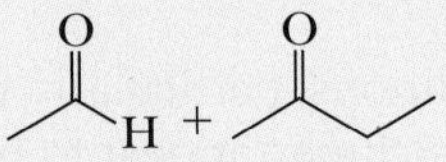

(equal moles of each)

(c) 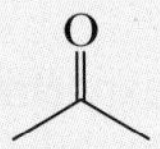

(only product)

6.6 Reduction

Most alkenes are reduced quantitatively by molecular hydrogen, H_2, in the presence of a transition metal catalyst to give an alkane. Yields are usually quantitative or nearly so.

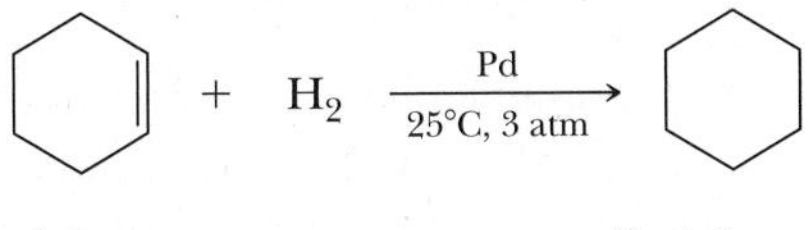

$\Delta H^0 = -119.5$ kJ $(-28.6$ kcal$)$/mol

Cyclohexene Cyclohexane

Although the addition of hydrogen to an alkene is exothermic, reduction is immeasurably slow in the absence of a catalyst. Commonly used transition metal catalysts include platinum, palladium, ruthenium, and nickel. Because the conversion of an alkene to an alkane involves reduction by hydrogen in the presence of a catalyst, the process is called **catalytic reduction** or, alternatively, **catalytic hydrogenation**.

Monosubstituted and disubstituted carbon-carbon double bonds react readily at room temperature under a few atmospheres (atm) pressure of hydrogen. Trisubstituted carbon-carbon double bonds require slightly elevated temperatures and pressures of up to 100 psi (pounds per square inch). Tetrasubstituted carbon-carbon double bonds are difficult to reduce and may require temperatures up to 275°C and hydrogen pressures of 1000 psi. Recall that 1 atm is equal to about 15 psi.

Parr shaker-type hydrogenation apparatus.

The metal catalyst is used as a finely powdered solid or may be supported on some inert material, such as finely powdered charcoal or alumina. The reaction is usually carried out by dissolving the alkene in ethanol or another nonreacting organic solvent, adding the solid catalyst, and then shaking the mixture under hydrogen gas at pressures of from 1 to 50 atm. Alternatively, the metal may be complexed with certain organic molecules and used in the form of a soluble complex (Section 6.7C).

Catalytic reduction is stereoselective, with the vast majority proceeding by syn addition of hydrogens to the carbon-carbon double bond.

A. Mechanism of Catalytic Reduction

The transition metals used in catalytic hydrogenation are able to adsorb large quantities of hydrogen onto their surfaces, probably by forming metal-hydrogen σ bonds. Similarly, alkenes are also adsorbed on metal surfaces with formation of carbon-metal bonds. Addition of hydrogen atoms to the alkene occurs in two steps (Figure 6.7).

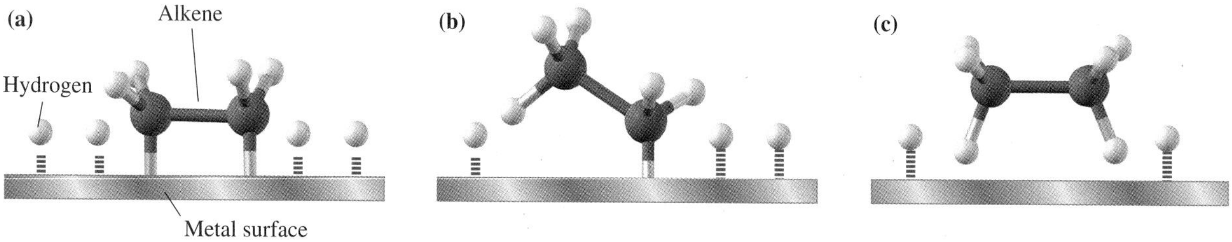

FIGURE 6.7 The addition of hydrogen to an alkene involving a transition metal catalyst. (a) Hydrogen and the alkene are adsorbed on the metal surface, and (b) one hydrogen atom is transferred to the alkene forming a new C—H bond. The other carbon remains adsorbed on the metal surface. (c) A second C—H bond is formed, and the alkane is desorbed.

Under some experimental conditions, particularly with tetrasubstituted double bonds, some percentage of the product may appear to be formed by anti addition of hydrogen. Catalytic reduction of 1,2-dimethylcyclohexene, for example, yields predominantly *cis*-1,2-dimethylcyclohexane. Along with the *cis* isomer are formed lesser amounts of *trans*-1,2-dimethylcyclohexane as a racemic mixture.

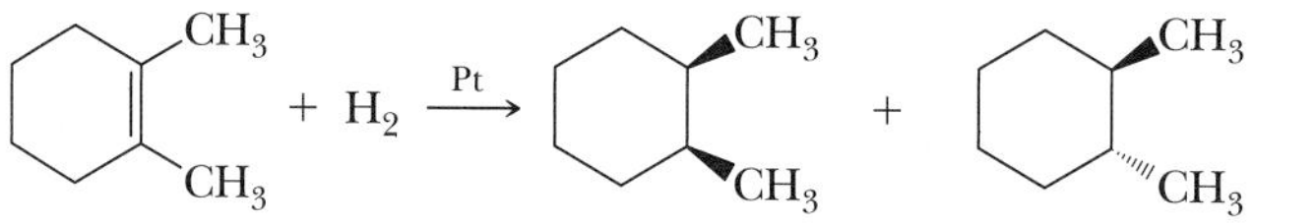

1,2-Dimethylcyclohexene | 70% to 85% ***cis*-1,2-Dimethylcyclohexane** | 30% to 15% ***trans*-1,2-Dimethylcyclohexane** (a racemic mixture)

If addition of hydrogens is syn stereoselective, how do we account for the formation of a *trans* product? It has been found that before a second hydrogen can be delivered from the metal surface to complete the reduction, there is transfer of a hydrogen from a carbon atom adjacent to the original double bond to the metal surface along with a newly added hydrogen. This hydrogen transfer, in effect, reverses the first step and forms a new alkene that is isomeric with the original alkene. As shown in the following equation, 1,2-dimethylcyclohexene undergoes isomerization on the metal surface to 1,6-dimethylcyclohexene. This alkene then leaves the metal surface. When it is later readsorbed and reduced, hydrogens are still added to it with syn stereoselectivity, but not necessarily from the same side as the original hydrogen.

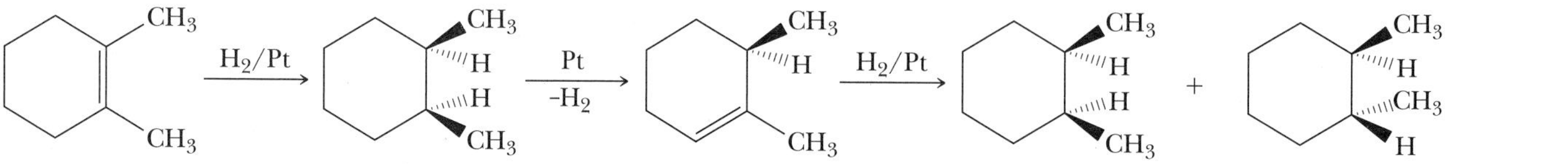

1,2-Dimethylcyclohexene **1,6-Dimethylcyclohexene**

B. Heats of Hydrogenation and the Relative Stabilities of Alkenes

Table 6.2 lists heats of reaction for the catalytic hydrogenation of several alkenes. Three important points are derived from this information.

1. The reduction of an alkene to an alkane is an exothermic process. This observation is consistent with the fact that, during hydrogenation, there is net conversion of a H—H bond and a weaker π bond to stronger σ bonds; that is, one σ bond (H—H) and one π bond (C=C) are broken, and two new σ bonds (C—H) form. For a comparison of the relative strengths of σ and π bonds, refer to Section 1.10.
2. Heats of hydrogenation depend on the degree of substitution of the carbon-carbon double bond; the greater the substitution, the lower the heat of hydrogenation. Compare, for example, heats of hydrogenation of ethylene (no substituents), propene (one substituent), 1-butene (one substituent), and the *cis* and *trans* isomers of 2-butene (two substituents).
3. The heat of hydrogenation of a *trans*-alkene is lower than that of the isomeric *cis*-alkene. Compare, for example, the heats of hydrogenation of *cis*-2-butene and *trans*-2-butene. Because reduction of each alkene gives butane, any difference in heats of hydrogenation must be caused by a difference in relative energy between the two alkenes (Figure 6.8). The alkene with the lower (less negative) value of ΔH^0 is more stable.

We explain the greater stability of *trans*-alkenes relative to *cis*-alkenes in terms of steric strain (Section 2.5A), which can be visualized using space-filling models (Figure 6.9). In *cis*-2-butene, the two $—CH_3$ groups are sufficiently close to each other that steric strain is caused by repulsion between the two methyl groups. This repulsion is reflected in the larger heat of hydrogenation (decreased stability) of

Table 6.2 Heats of Hydrogenation of Several Alkenes

Name	Structural Formula	ΔH^0 [kJ(kcal)/mol]
Ethylene	CH_2CH_2	−137 (−32.8)
Propene	$CH_3CH{=}CH_2$	−126 (−30.1)
1-Butene	$CH_3CH_2CH{=}CH_2$	−127 (−30.3)
***cis*-2-Butene**	H_3C CH_3 C=C H H	−119.7 (−28.6)
***trans*-2-Butene**	H_3C H C=C H CH_3	−115.5 (−27.6)
2-Methyl-2-butene	H_3C CH_3 C=C H_3C H	−113 (−26.9)
2,3-Dimethyl-2-butene	H_3C CH_3 C=C H_3C CH_3	−111 (−26.6)

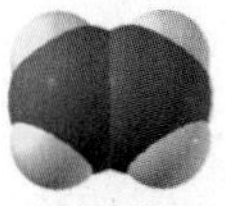

Ethylene

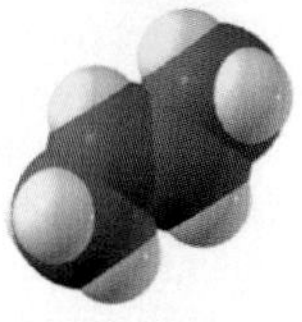

***trans*-2-Butene**

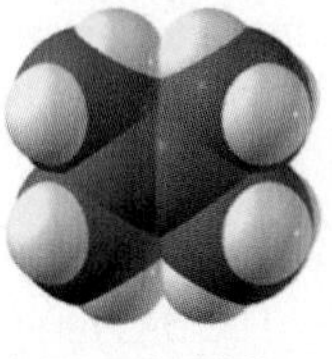

2,3-Dimethyl-2-butene

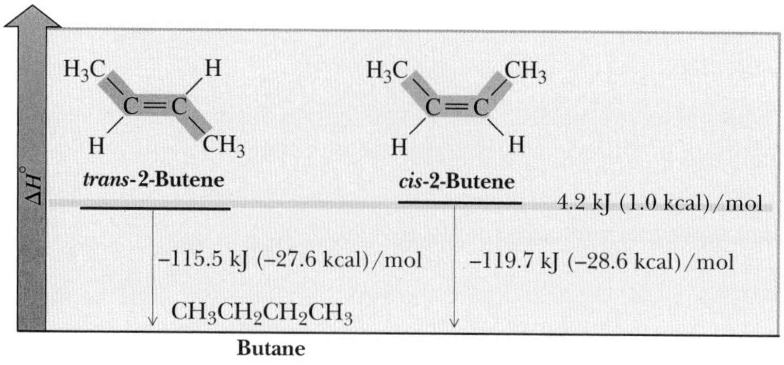

FIGURE 6.8 Heats of hydrogenation of *cis*-2-butene and *trans*-2-butene. Compared to *cis*-2-butene, *trans*-2-butene is more stable by 4.2 kJ (1.0 kcal)/mol.

cis-2-butene compared with *trans*-2-butene [approximately 4.2 kJ (1.0 kcal)/mol]. Thus, hydrogenation allows measurement of the strain energy of *cis*-2-butene directly.

Connections to Biological Chemistry

Trans Fatty Acids: What They Are and How to Avoid Them

Fats and oils were introduced in Chapter 5 (Connections to Biological Chemistry). Fats are added to processed foods to provide a desirable firmness along with a moist texture and pleasant taste. To supply the demand for dietary fats of the appropriate consistency, the *cis* double bonds of vegetable oils are partially hydrogenated by using hydrogen in the presence of a Ni or another transition metal catalyst. The greater the extent of hydrogenation, the higher the melting point of the triglyceride. By controlling the degree of hydrogenation, a triglyceride with a melting point below room temperature can be converted to a semisolid or even a solid product.

Unfortunately, because of the reversible interaction of a carbon-carbon double bond with the Ni catalyst, some of the double bonds remaining in the triglyceride may be isomerized from the less stable *cis* configuration to the more stable *trans* configuration. Recall that a key step in catalytic hydrogenation involves the cleavage of the alkene π bond and the bonding of its two carbons to the surface of the transition metal catalyst (Figure 6.7). This process is reversible, thus allowing equilibration between the *cis* and *trans* configurations.

The oils used for frying in fast-food restaurants are usually partially hydrogenated plant oils and thus contain substantial amounts of *trans* fatty acids that are transferred to the foods fried in them. Other major sources of *trans* fatty acids in the diet include stick margarine, certain commercial bakery products, creme-filled cookies, chips, frozen breakfast foods, and cake mixes.

Recent studies have shown that consuming a significant amount of *trans* fatty acids can lead to serious health problems related to serum cholesterol levels. Low overall serum cholesterol and a decreased ratio of low-density lipoprotein (LDL) cholesterol to high-density lipoprotein (HDL) cholesterol are associated with good overall cardiovascular health. High serum cholesterol and an elevated ratio of LDL cholesterol to HDL cholesterol are linked to a high incidence of cardiovascular disease, especially atherosclerosis. Research has indicated that diets high in either saturated fatty acids or *trans* fatty acids raise the ratio of serum LDL cholesterol to HDL cholesterol and substantially increase the risk of cardiovascular disease.

The Food and Drug Administration requires that processed foods list the content of *trans* fatty acids. A diet low in saturated and *trans* fatty acids is recommended, along with more fish, whole grains, fruits, and vegetables. The recommendation is also for daily exercise, which is tremendously beneficial, regardless of diet.

In addition to the steric reason that makes a *trans*-alkene more stable than a *cis*-alkene, there is an electronic reason for the fact that increased alkyl-group substitution on an alkene imparts stability. Due to the lower energy of 2*s* atomic orbitals relative

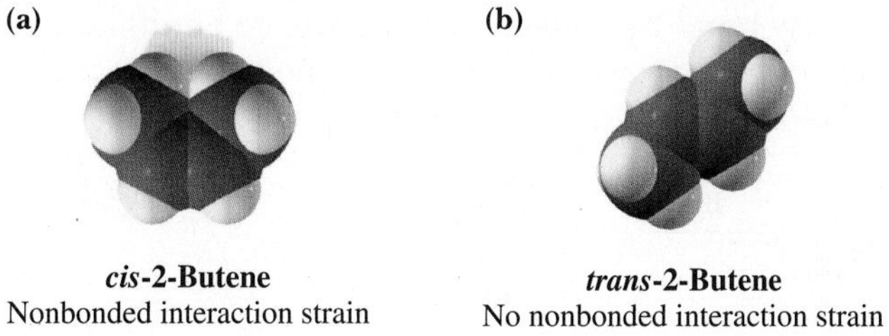
(a)
(b)
cis-2-Butene
Nonbonded interaction strain
trans-2-Butene
No nonbonded interaction strain

FIGURE 6.9 Space-filling models of (a) *cis*-2-butene and (b) *trans*-2-butene. Steric clashing shown as a yellow shading.

to $2p$ atomic orbitals, the increased *s* character of sp^2 hybridized orbitals makes these orbitals more electron withdrawing compared to sp^3 orbitals. Therefore, the carbons involved in double bonds are slightly electron withdrawing. Hence, in a fashion analogous to the electron donation from alkyl groups that stabilizes carbocations, alkyl groups also donate electrons to alkenes, thereby imparting increased stability to the alkenes. The more alkyl groups, the greater the stabilization.

6.7 Molecules Containing Chiral Centers as Reactants or Products

As the structure of an organic compound is altered in the course of a reaction, one or more chiral centers, usually at carbon, may be created, inverted, or destroyed. In Section 6.7A, we consider two alkene addition reactions in which a chiral molecule is created in an achiral environment. In doing so, we will illustrate the point that an optically active compound (i.e., an enantiomerically pure compound or even an enantiomerically enriched compound) can never be produced from achiral starting materials reacting in an achiral environment. Then in Section 6.7B, we consider the reaction of achiral starting materials reacting in a chiral environment—in this case in the presence of a chiral catalyst. We shall see that an enantiomerically pure product may be produced from achiral reagents if the reaction takes place in a chiral environment.

A. Reaction of Achiral Starting Materials in an Achiral Environment

The addition of bromine to 2-butene (Section 6.3D) gives 2,3-dibromobutane, a molecule with two chiral centers. Three stereoisomers are possible for this compound: a meso compound and a pair of enantiomers (Section 3.4). We now ask the following questions:

- What is the stereochemistry of the product? Is it one enantiomer, a pair of enantiomers, the meso compound, or a mixture of all three stereoisomers?
- Is it optically active or optically inactive?

A partial answer is that the stereochemistry of the product formed depends on the configuration of the alkene. Let us first examine the addition of bromine to *cis*-2-butene.

The reaction of bromine with *cis*-2-butene on either face of the planar part of the molecule gives the same bridged bromonium ion intermediate (Figure 6.10). Although this intermediate has two chiral centers, it has a plane of symmetry and is, therefore, meso. Attack of Br^- on this meso intermediate from the side opposite that of the bromonium ion bridge gives a pair of enantiomers. Attack of bromide ion on carbon 2 gives the (2*S*,3*S*) enantiomer; attack on carbon 3 gives the (2*R*,3*R*) enantiomer. Attack of bromide ion occurs at equal rates at each carbon; therefore, the enantiomers are formed in equal amounts, and 2,3-dibromobutane is obtained as a racemic mixture (Figure 6.10). Thus, the product is chiral, but because it is a racemic mixture, it is optically inactive.

***Step 1:* Make a new bond between a nucleophile (π bond) and and an electrophile** Reaction of *cis*-2-butene with bromine forms bridged bromonium ions which are meso and identical.

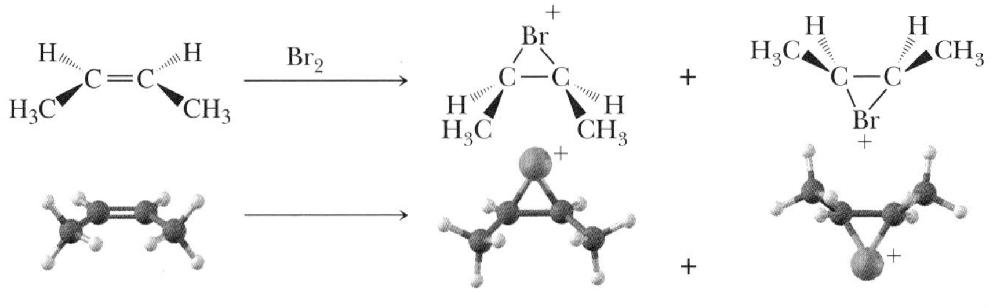

***cis*-2-Butene** (achiral)

(meso bridged bromonium ion intermediate)

***Step 2:* Make a new bond between a nucleophile and an electrophile** Attack of bromide at carbons 2 and 3 occurs with equal probability to give enantiomeric products in a racemic mixture.

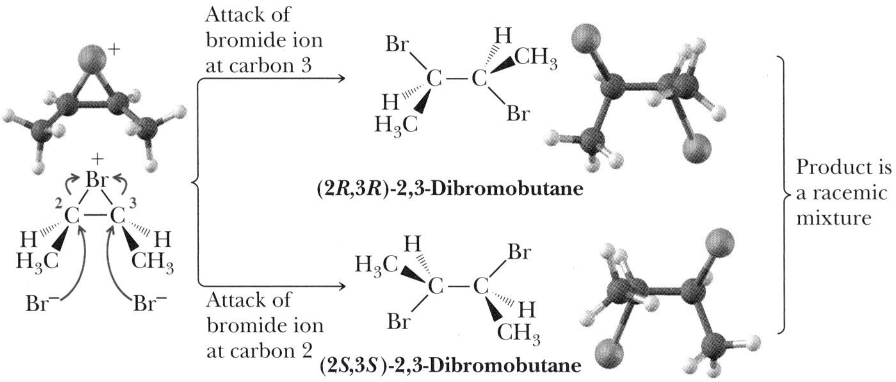

FIGURE 6.10 Anti stereoselective addition of bromine to *cis*-2-butene gives 2,3-dibromobutane as a racemic mixture. The product is chiral, but because it is formed as a racemic mixture, it is optically inactive.

We have described reaction of Br—Br on one side of the carbon-carbon double bond. Reaction on the opposite side followed by opening of the resulting bromonium ion intermediate produces the same two stereoisomers as a racemic mixture.

The addition of Br_2 to *trans*-2-butene leads to two enantiomeric bridged bromonium ion intermediates. Attack by Br^- at either carbon atom of either bromonium ion intermediate gives the meso product, which is achiral and therefore optically inactive (Figure 6.11).

As we have seen in this discussion, the stereochemistry of the product formed by the addition of bromine to 2-butene depends on the stereochemistry of the starting alkene; the addition of halogen to *cis*-2-butene gives a racemic mixture, whereas the addition of halogen to *trans*-2-butene gives a meso product. Accordingly, we say that the addition of Br_2 or Cl_2 to an alkene is stereospecific. A **stereospecific reaction** is a special type of stereoselective reaction in which the stereochemistry of the product depends on the stereochemistry of the starting material.

Stereospecific reaction

A special type of stereoselective reaction in which the stereochemistry of the product is dependent on the stereochemistry of the starting material.

In Section 6.5A, we studied oxidation of alkenes by osmium tetroxide in the presence of peroxides. This oxidation results in syn stereoselective hydroxylation of the alkene to form a glycol. In the case of cycloalkenes, the product is a *cis* glycol. The first step in each oxidation involves formation of a cyclic osmate and is followed immediately by reaction with water to give a glycol.

Syn hydroxylation of *cis*-2-butene gives meso-2,3-butanediol; because the meso compound is achiral, the product is optically inactive. Syn hydroxylation of

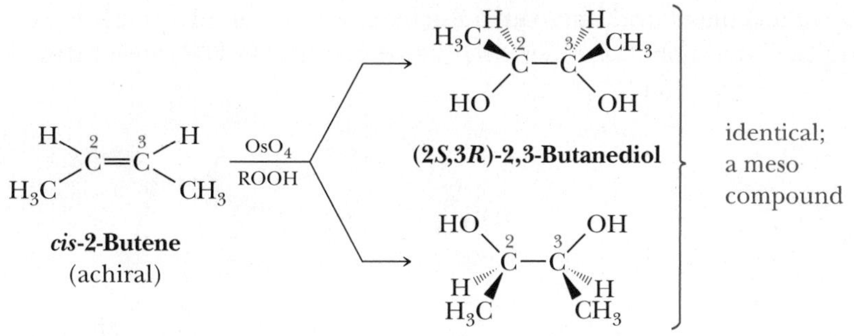

(2*R*,3*S*)-2,3-Butanediol

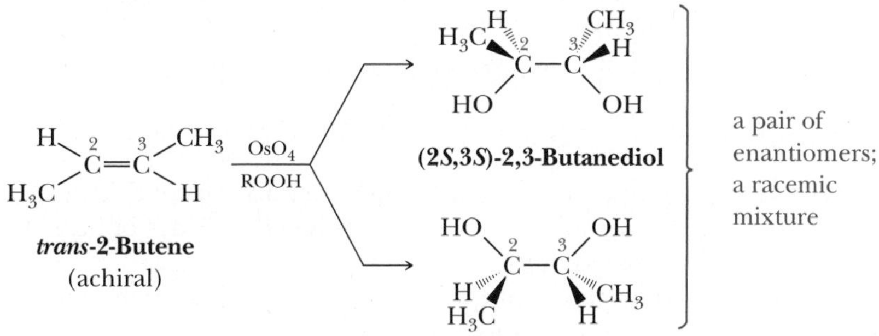

(2*R*,3*R*)-2,3-Butanediol

trans-2-butene gives racemic 2,3-butanediol. Because the diol is formed as a racemic mixture, the product of the oxidation of the *trans*-alkene is also optically inactive. Thus, the osmium tetroxide oxidation of an alkene is stereospecific; the stereochemistry of the product depends on the stereochemistry of the starting alkene.

Notice that the stereochemistry of the product of the osmium tetroxide oxidation of *trans*-2-butene is opposite that formed on the addition of bromine to *trans*-2-butene. Osmium tetroxide oxidation gives the glycol as a pair of enantiomers forming a racemic mixture. Addition of bromine to *trans*-2-butene gives the dibromoalkane as a meso compound. A similar difference is observed between the stereochemical outcomes of these reactions with *cis*-2-butene. The difference in outcomes occurs because bromination of an alkene involves anti addition, whereas oxidation by osmium tetroxide involves syn addition.

In this section, we have seen two examples of reactions in which achiral starting materials give chiral products. In each case, the chiral product is formed as a racemic mixture (which is optically inactive) or as a meso compound (which is also optically inactive). These results illustrate a very important point about the creation of chiral molecules: an optically active product (i.e., an enantiomerically pure compound or even an enantiomerically enriched compound) can never be produced from achiral

Step 1: **Make a new bond between a nucleophile (π bond) and an electrophile**
Reaction of *trans*-2-butene with bromine forms bridged bromonium ions which are enantiomers.

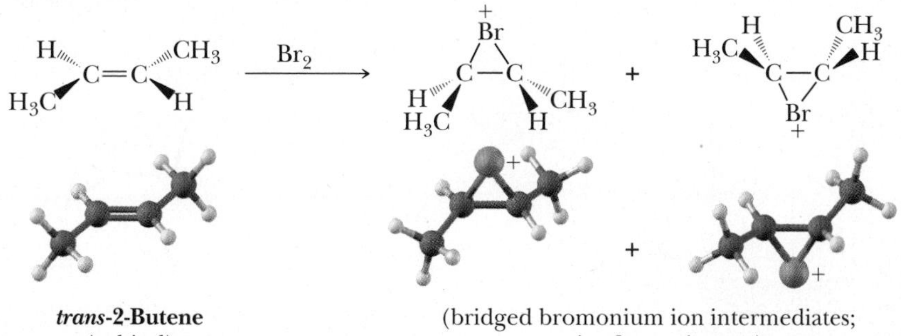

***trans*-2-Butene**
(achiral)

(bridged bromonium ion intermediates;
a pair of enantiomers)

FIGURE 6.11 Anti stereoselective addition of bromine to *trans*-2-butene gives meso-2,3-dibromobutane.

***Step 2:* Make a new bond between a nucleophile and an electrophile** Attack of bromide on either carbon of either enantiomer gives meso-2,3-dibromobutane.

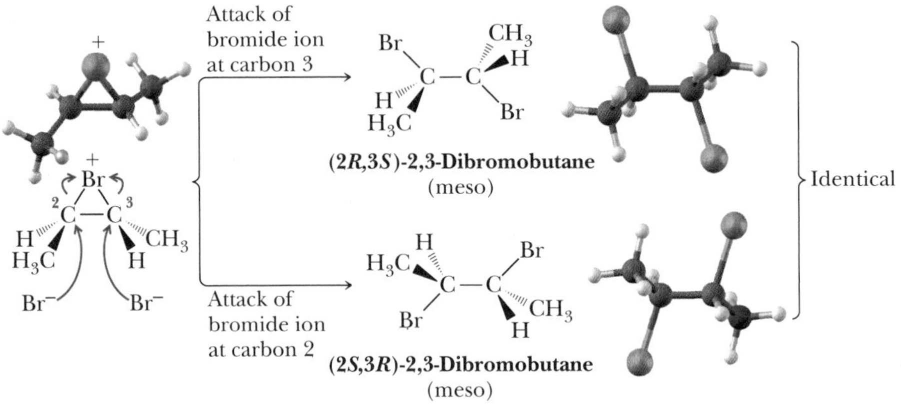

starting materials and achiral reagents reacting under achiral conditions. Although the molecules of the product may be chiral, the product is always optically inactive (either meso or a pair of enantiomers).

We will encounter many reactions throughout the remainder of this course where achiral starting materials are converted into chiral products under achiral reaction conditions. For convenience, we often draw just one of the enantiomeric products, but you must keep in mind that under these experimental conditions, the product will always be racemic and therefore optically inactive.

B. Reaction of a Chiral Starting Material in an Achiral Environment

Let us consider the bromination of (*R*)-4-*tert*-butylcyclohexene. Recall that in derivatives of cyclohexane in which interconversion between one chair conformation and the other is not possible or is severely restricted, the *trans* diaxial product is isolated. If a cyclohexane ring contains a bulky alkyl group, such as *tert*-butyl (Section 2.6B), then the molecule exists overwhelmingly in a conformation in which the *tert*-butyl group is equatorial. Reaction of bromine with enantiomerically pure (*R*)-4-*tert*-butylcyclohexene occurs at both faces of the six-membered ring. Because bromine atoms must add in an axial manner, each bromonium ion intermediate reacts with bromide ion to give the same product. In the favored chair conformation of this product, *tert*-butyl is equatorial, the bromine atoms remain axial, and only a single diastereomer is formed.

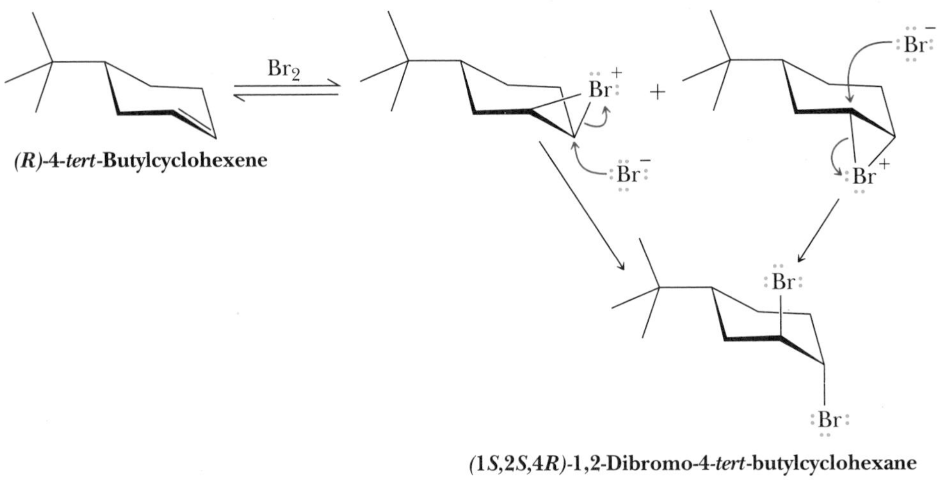

In effect, the presence of the bulky *tert*-butyl group controls the orientation of the two bromine atoms added to the ring.

C. Reaction of Achiral Starting Materials in a Chiral Environment

The reduction of a carbon-carbon double bond can be carried out using hydrogen in the presence of a transition metal catalyst. Because hydrogen atoms are delivered to either face of the double bond with equal probability, if a new chiral center is created, equal amounts of both the *R* and *S* configurations will be produced.

Within the last three decades, chemists have discovered ways to embed transition metal hydrogenation catalysts in chiral molecules with the result that hydrogen can be delivered to only one face of the alkene. In catalytic reductions where a new chiral center is formed, a large enantiomeric excess of one enantiomer may be formed, and the reaction is said to be **enantioselective**. The most widely used of these chiral hydrogenation catalysts involve the chiral ligand 2,2-bis(diphenylphosphanyl)-1,1′-binaphthyl, more commonly known as **BINAP**. **BINAP** has been resolved into its *R* and *S* enantiomers. The fact that **BINAP** can be resolved depends on restricted rotation about the single bond joining the two naphthalene rings. The two enantiomers are atropisomers (Section 3.2).

Enantioselective reaction
A reaction that produces one enantiomer in preference to the other.

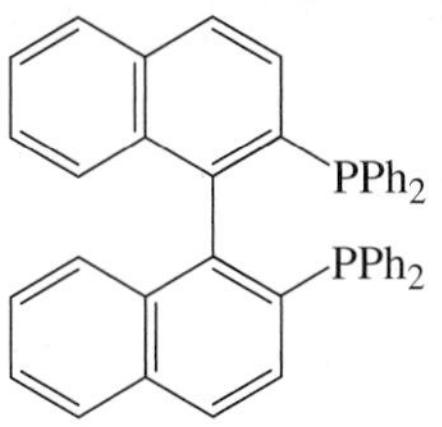

BINAP

(*S*)-(–)-BINAP
$[\alpha]_D^{25}$ –223

(*R*)-(+)-BINAP
$[\alpha]_D^{25}$ +223

Treating either enantiomer of **BINAP** with ruthenium chloride forms a complex in which ruthenium is bound as a complex ion in the chiral environment of the larger **BINAP** molecule. This complex is soluble in dichloromethane, CH_2Cl_2, and can be used as a homogeneous hydrogenation catalyst.

$$(R)\text{-BINAP} + RuCl_3 \longrightarrow (R)\text{-BINAP-RuCl}_3$$

Using (*R*)-**BINAP**-$RuCl_3$ as a hydrogenation catalyst, (*S*)-naproxen is formed in greater than 98% enantiomeric excess. (*S*)-Naproxen is the anti-inflammatory and pain reliever in Aleve and several other over-the-counter medications. Note that in this reduction, neither the benzene rings of the naphthyl group nor the carboxyl group are reduced. We will have more to say about the reduction of these groups in later chapters.

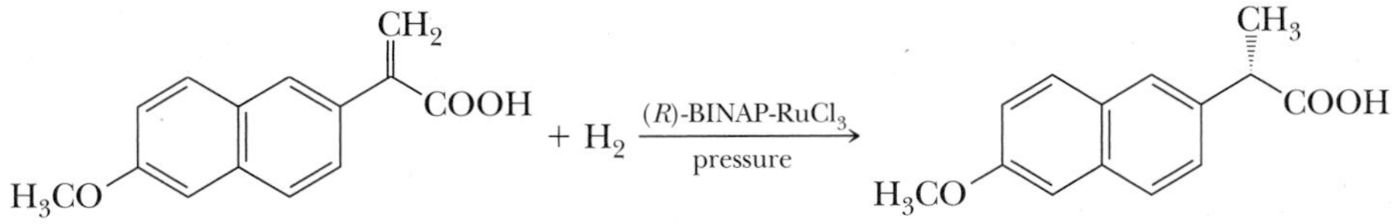

(*S*)-Naproxen
(ee > 98%)

BINAP-$RuCl_3$ complexes are somewhat specific for the types of carbon-carbon double bond reductions they catalyze. To be reduced, the double bond must have some kind of neighboring functional group that serves as a directing group during the reduction. The most common of these directing groups are the carboxyl group of carboxylic acids and esters and the hydroxyl group of alcohols. As shown in the following example, only the carbon-carbon double bond nearer the —OH group is reduced. Geraniol, as the name might suggest, is a natural product isolated from rose and geranium oils. It is also present in citronella and lemon grass oils. With this chiral catalyst, only the *R* enantiomer is formed from the (*S*)-BINAP-$RuCl_3$ complex and only the *S* enantiomer is formed from the (*R*)-BINAP-$RuCl_3$ complex.

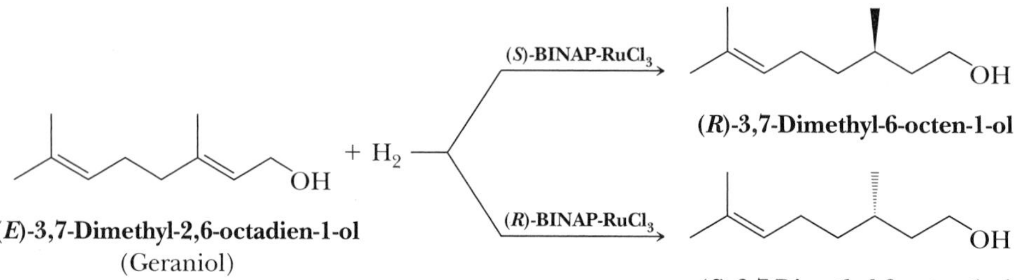

(*E*)-3,7-Dimethyl-2,6-octadien-1-ol (Geraniol)

(*R*)-3,7-Dimethyl-6-octen-1-ol

(*S*)-3,7-Dimethyl-6-octen-1-ol

Study Guide

6.1 Reactions of Alkenes—An Overview

- The most characteristic reaction of alkenes is an **addition reaction** to the carbon-carbon double bond, breaking the π bond and producing two new σ bonds to two new atoms or groups of atoms.

6.2 Organic Reactions Involving Reactive Intermediates

- A **reaction mechanism** is a detailed description of how and why a chemical reaction occurs as it does, which bonds are broken and which new ones are formed, the order in which the various bond-breaking and bond-forming steps take place and their relative rates, the role of the solvent if the reaction takes place in solution, and the role of the catalyst if one is present.
  - *A **reactive intermediate** corresponds to an energy minimum between two transition states.*
  - *In a multistep reaction, the **rate-determining step** is the step that crosses the highest energy barrier.*
- The most common reactive intermediates in organic chemistry lack an octet at carbon or are carbon-based anions.
  - *In particular, common reactive intermediates are **carbocations** (positive charge on a carbon with only six valence electrons), **carbon radicals** (one unpaired electron, so only seven valence electrons on carbon), and **carbanions** (three bonds and one lone pair on carbon).*
- **Bond dissociation enthalpy** (BDE) is the amount of energy required to break a bond into two radicals. BDEs can be used to estimate whether a reaction will be favorable (exothermic) or unfavorable (endothermic).
  - *This analysis assumes that entropy remains relatively constant throughout the reaction.*

- *The change in enthalpy for a reaction (ΔH^0) can be estimated as the difference between the BDEs of bonds broken minus the BDEs of bonds made during the reaction.*
- *A reaction is exothermic if the bonds made in the product(s) are stronger than the bonds broken in the reactant(s).*

P 6.14

6.3 Electrophilic Additions

- **Electrophilic addition** reactions occur when electrophiles add to a π bond.
  - *Electrophiles possess a region of low electron density (full or partial positive charge), or they lack an octet on an atom, and/or they have a relatively weak bond to an atom that can depart as a stable ion or molecule.*
  - *Brønsted-Lowry acids can be considered electrophiles.*
  - *The rate-determining step in electrophilic addition reactions is the making of the first new bond between the alkene π bond and the electrophile.*
- A **carbocation** is created when the electrophile is a Brønsted-Lowry acid and a proton is added.
  - *A carbocation is a positively charged ion that contains a carbon atom with only six electrons in its valence shell.*
  - *Carbocations are planar with bond angles of approximately $120°$ about the positive carbon.*
  - *The order of stability of carbocations is $3° > 2° > 1° >$ methyl.*
  - *Carbocations are stabilized by the electron-releasing* ***inductive effect*** *of alkyl groups bonded to the cationic carbon and by* ***hyperconjugation****. Hyperconjugation is a stabilizing interaction that involves overlap of σ bonding electron density from adjacent alkyl groups with the empty $2p$ orbital of the cationic carbon atom.*
  - *Carbocations can* ***rearrange****.*
    - *The driving force for* ***carbocation rearrangement*** *is conversion to a more stable $2°$ or $3°$ carbocation.*
    - *Rearrangement is by a 1,2 shift in which an atom or a group of atoms with its bonding electrons moves from an adjacent atom to an electron-deficient atom.*
- A **regioselective reaction** is a reaction in which one direction of bond forming or bond breaking occurs in preference to all other directions.
  - *According to* ***Markovnikov's rule,*** *in the addition of HX or H_2O to an unsymmetrical alkene, hydrogen adds to the carbon of the double bond having the greater number of hydrogens and X or OH adds to the more substituted carbon (fewer H atoms).*
  - *Markovnikov's rule can be explained by realizing that two different carbocations can be formed from the reaction between an unsymmetrical alkene and an electrophile. Formation of the more stable carbocation will predominate and will lead to the product predicted by Markovnikov's rule.*
- A **stereoselective reaction** is a reaction in which one stereoisomer is formed in preference to all others that might be formed.
  - *A stereoselective reaction may be* ***enantioselective*** *or* ***diastereoselective****, depending on whether the product stereoisomers are enantiomers or diastereomers, respectively.*

- *Addition of new atoms or groups of atoms from opposite faces (or sides) of a double bond is called* ***anti addition.*** *These reactions often occur with* ***anti stereoselectivity.*** *In cyclic systems, anti addition is equivalent to* trans *coplanar addition.*
- ***Syn addition*** *is the addition of atoms or groups of atoms to the same face (or side) of a double bond.*

• For electrophilic addition reactions of alkenes, keep track of the intermediates (carbocation, halonium ion etc.) and how these predict any observed regiochemistry (Markovnikov or non-Markovnidov) and stereochemistry (anti or syn).

P **6.1–6.8, 6.15–6.36, 6.44, 6.45, 6.49**

▶ **Mechanism 6.1, Figure 6.5, Problem 6.4, Mechanism 6.4**

KEY REACTIONS

1. **Addition of HX (Section 6.3A)** HX is used to convert alkenes to haloalkanes in an electrophilic addition reaction. The two-step mechanism involves initial protonation of the alkene π bond to form a carbocation, which reacts with X^- to give the product haloalkane. The X atom becomes bonded to the more highly substituted atom of the alkene, so it follows Markovnikov regioselectivity (derived from the preference for forming the more stable carbocation intermediate). Carbocation rearrangements are possible.

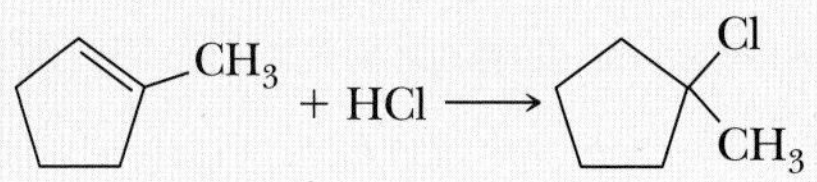

CH3
+ HCl
Cl
CH3

2. **Acid-Catalyzed Hydration (Section 6.3B)** Water, in the presence of an acid catalyst, converts an alkene into an alcohol. The mechanism involves initial protonation of the alkene π bond to give a carbocation, which reacts with the nucleophile water to create a second intermediate, which loses a proton to give the alcohol product. The OH group becomes bonded to the more highly substituted carbon of the alkene, so the reaction displays Markovnikov regioselectivity (derived from the preference for forming the more stable carbocation intermediate). Carbocation rearrangements are possible.

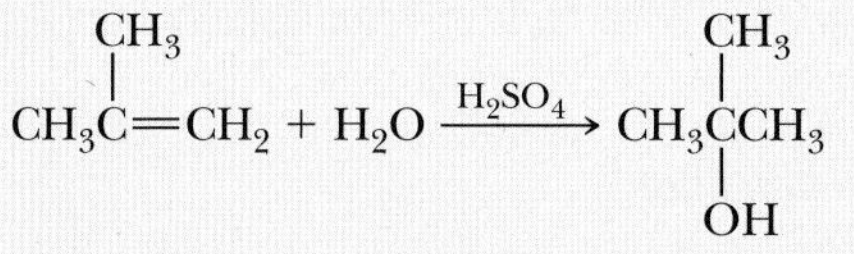

CH3
CH3C=CH2 + H2O
H2SO4
CH3
CH3CCH3
OH

3. **Addition of Bromine and Chlorine (Section 6.3D)** Cl_2 or Br_2 is used to convert an alkene into a vicinal dihalide. The mechanism involves attack by the alkene π bond on one atom of X_2 to give a bridged halonium ion intermediate (a cation) that is, in turn, attacked by X^- from the backside to give the vicinal dihalide. Rearrangements do not occur. The reaction displays anti addition stereoselectivity because of the halonium ion intermediate. The reaction is stereospecific because *Z* alkenes give different products than do *E* alkenes.

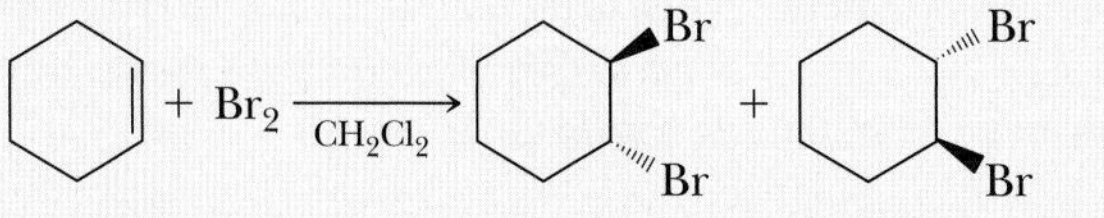

+ Br2
CH2Cl2
Br
Br
+
Br
Br

(a racemic mixture)

4. **Addition of HOCl and HOBr (Section 6.3E)** Cl_2 or Br_2 in the presence of H_2O is used to convert alkenes to halohydrins. The mechanism involves attack by the alkene π bond on one atom of X_2 to give a bridged halonium ion intermediate (a cation) that is, in turn, attacked by H_2O from the backside to give a new intermediate, which loses a proton to give the halohydrin. Rearrangements are not observed and the reaction displays anti addition stereoselectivity because of the halonium ion intermediate. The reaction gives Markovnikov regioselectivity in that the —OH group becomes bonded to the more highly substituted alkene carbon. This regioselectivity is derived from attack by H_2O on the carbon of

the halonium ion intermediate capable of accommodating more carbocation character. The reaction is stereospecific because *Z* alkenes give different products than do isomeric *E* alkenes.

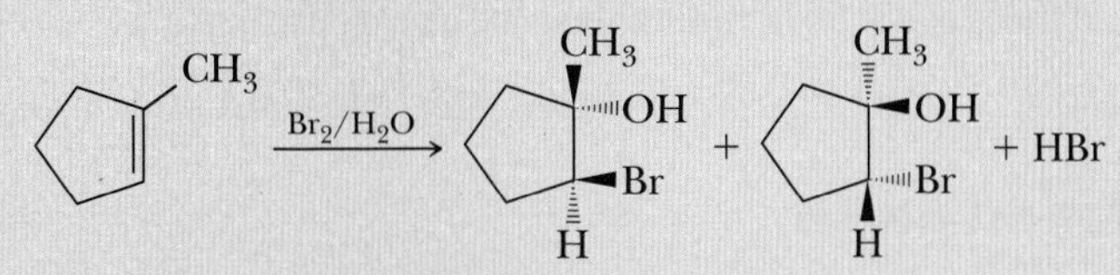

(a racemic mixture)

5. **Oxymercuration-Reduction (Section 6.3F)** Oxymercuration-reduction is used to convert alkenes to alcohols. The mechanism involves reaction of the alkene π bond with an $HgOAc^+$ to give a bridged mercurinium ion intermediate (a cation) that is, in turn, attacked by H_2O from the backside to give a new intermediate, which loses a proton. In a second step, $NaBH_4$ is added to replace the Hg atom with H. The first step is anti stereoselective because HgOAc and OH add from opposite faces of the alkene. However, the $NaBH_4$ reduction step scrambles the stereochemistry as H replaces Hg, so the overall process is scrambled anti and syn addition. The reaction displays Markovnikov regioselectivity in that the —OH group becomes bonded to the more highly substituted alkene carbon. This regioselectivity is derived from attack by H_2O on the carbon of the mercurinium ion intermediate capable of accommodating more carbocation character. Rearrangements are not observed.

$$CH_3CH(CH_3)CH{=}CH_2 \xrightarrow[\text{2. } NaBH_4]{\text{1. } Hg(OAc)_2,\ H_2O} CH_3CH(CH_3)CH(OH)CH_3$$

6.4 Hydroboration-Oxidation

- Non-Markovnikov hydration of an alkene can be achieved by **hydroboration-oxidation,** because the key step in this reaction sequence involves a four-membered transition state featuring syn addition, not a carbocation intermediate.

P 6.9, 6.37, 6.38, 6.44, 6.45

Problem 6.10

KEY REACTIONS

6. **Hydroboration-Oxidation (Section 6.4)** BH_3 followed by basic H_2O_2 converts alkenes into alcohols with non-Markovnikov regioselectivity (the OH adds to the less substituted alkene carbon) and syn stereoselectivity without rearrangement. The reaction is stereospecific because *cis*-alkenes give different products than do *trans*-alkenes. The mechanism involves coordination of the alkene π bond to the vacant 2*p* orbital of borane followed by a four-membered ring transition state, which simultaneously adds H to the more substituted alkene carbon and boron to the less substituted alkene carbon. The basic peroxide replaces the boron with OH.

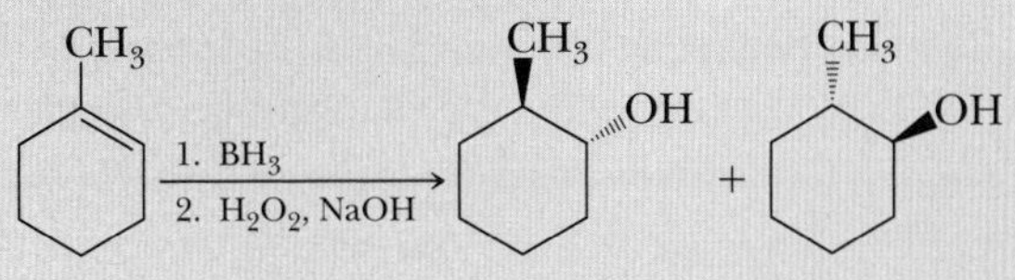

(a racemic mixture)

6.5 Oxidation

- **Oxidation** is the loss of electrons; **reduction** is the gain of electrons.
  - *– Alternatively, oxidation is the gain of oxygen atoms and/or loss of hydrogen atoms.*

P **6.10–6.12, 6.39–6.41, 6.44, 6.45**

▶ **Problem 6.13**

KEY REACTIONS

7. Oxidation to a Vicinal Diol by OsO_4 (Section 6.5A) OsO_4 followed by $NaHSO_3$ converts alkenes into vicinal diols with syn stereoselectivity and without rearrangement. The reaction is stereospecific because *cis*-alkenes give different products than do *trans*-alkenes. The mechanism involves formation of a cyclic osmate ester.

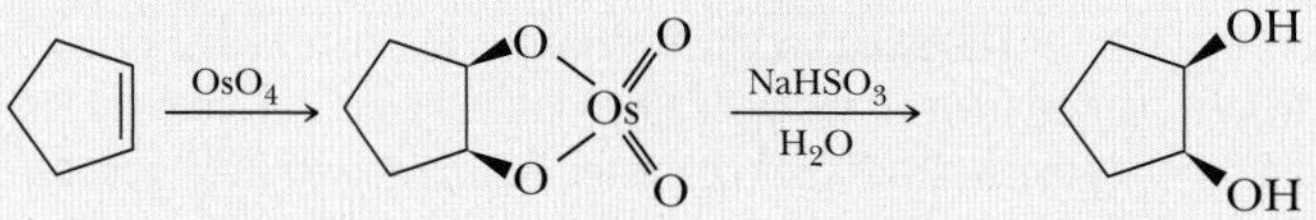

A cyclic osmate

***cis*-1,2-Cyclopentanediol**
(a *cis* glycol)

8. Oxidation by Ozone (Section 6.5B) Ozone, O_3, followed by dimethylsulfide cleaves the carbon-carbon double bond of alkenes and gives two carbonyl groups in its place. The mechanism involves formation of a molozonide intermediate that rearranges to an ozonide intermediate, which is reduced using the dimethyl sulfide to give the carbonyl products.

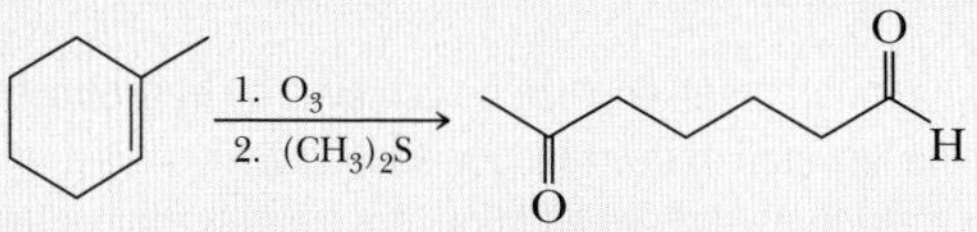

6.6 Reduction

- **Reduction** is the loss of oxygen atoms and/or the gain of hydrogen atoms.
- From **heats of hydrogenation** of a series of alkenes, we conclude that in general:
  - *– The greater the degree of substitution of a carbon-carbon double bond, the more stable the alkene.*
  - *– A* trans*-alkene is more stable than an isomeric* cis*-alkene.*

P **6.42–6.45**

KEY REACTIONS

9. Addition of H_2; Catalytic Reduction (Section 6.6) In the presence of an appropriate transition metal, H_2 reduces alkenes to alkanes with syn stereoselectivity and without rearrangement. The reaction can be stereospecific because *Z* alkenes can give different products than *E* alkenes can.

The mechanism involves cleavage of the alkene π bond through adsorption onto the surface of the transition metal. This step is followed by formation of two carbon-hydrogen bonds from the same face of the alkane molecule as it desorbs.

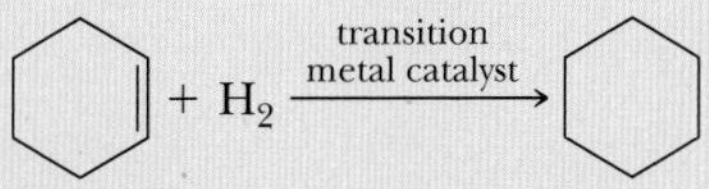

10. **Enantioselective Reduction (Section 6.7)** The most useful chiral hydrogenation catalysts involve a chiral phosphorus-containing ligand complexed with either ruthenium or rhodium.

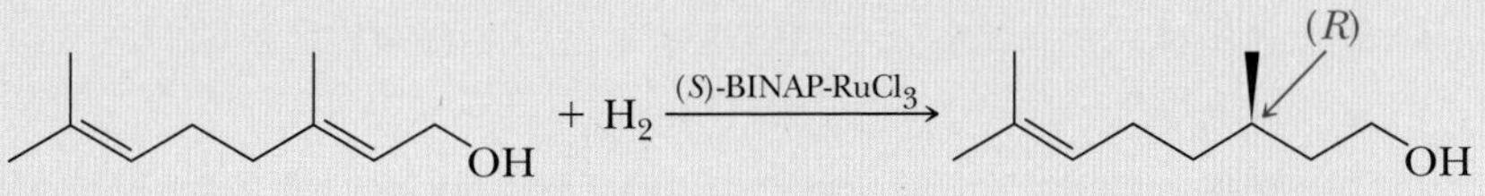

6.7 Molecules Containing Chiral Centers as Reactants or Products

- For reactions in which chiral centers are created, only racemic mixtures are formed from achiral reagents in achiral media.
- **A stereospecific reaction** is a reaction in which the stereochemistry of the product depends on the stereochemistry of the starting material.
  - *All stereospecific reactions are stereoselective, but not all stereoselective reactions are stereospecific, as you will see later.*
  - *Optically active products can never be formed by the reaction of achiral starting materials in an achiral environment.*
  - *Optically active products may be formed, however, by the reaction of achiral starting materials in a chiral environment.*

P 6.46–6.48

Problems

Red numbers indicate applied problems.

Energetics of Chemical Reactions

6.14 Using the table of average bond dissociation enthalpies at 25°C, determine which of the following reactions are energetically favorable at room temperature. Assume that $\Delta S = 0$.

Bond	Bond Dissociation Enthalpy kJ (kcal)/mol	Bond	Bond Dissociation Enthalpy kJ (kcal)/mol
H—H	435 (104)	C—I	238 (57)
O—H	439 (105)	C—Si	301 (72)
C—H(—CH_3)	422 (101)	C=C	727 (174)
C—H(=CH_2)	464 (111)	C=O (aldehyde)	728 (174)
C—H(≡CH)	556 (133)	C=O (CO_2)	803 (192)
N—H	391 (93)	C≡O	1075 (257)
Si—H	318 (76)	N≡N	950 (227)
C—C	378 (90)	C≡C	966 (231)
C—N	355 (85)	O=O	498 (119)
C—O	385 (92)		

(a) $CH_2{=}CH_2 + 2H_2 + N_2 \longrightarrow H_2NCH_2CH_2NH_2$
(b) $CH_2{=}CH_2 + CH_4 \longrightarrow CH_3CH_2CH_3$
(c) $CH_2{=}CH_2 + (CH_3)_3SiH \longrightarrow CH_3CH_2Si(CH_3)_3$
(d) $CH_2{=}CH_2 + CHI_3 \longrightarrow CH_3CH_2CI_3$

(e) $CH_2{=}CH_2 + CO + H_2 \longrightarrow CH_3CH_2\overset{O}{\overset{\|}{C}}H$

(f) 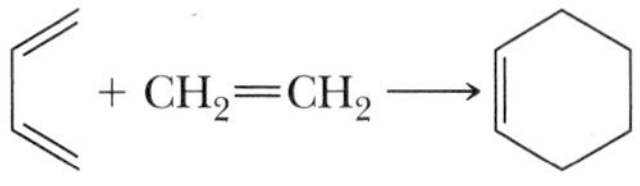

(g) 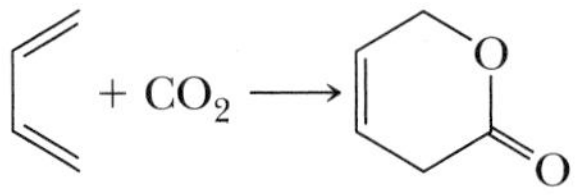

(h) $HC{\equiv}CH + O_2 \longrightarrow H\overset{O}{\overset{\|}{C}}{-}\overset{O}{\overset{\|}{C}}H$

(i) $2CH_4 + O_2 \longrightarrow 2CH_3OH$

Electrophilic Additions

6.15 Draw structural formulas for the isomeric carbocation intermediates formed on treatment of each alkene with HCl. Label each carbocation 1°, 2°, or 3° and state which of the isomeric carbocations forms more readily.

(a) $CH_3CH_2\overset{CH_3}{\overset{|}{C}}{=}CHCH_3$ **(b)** $CH_3CH_2CH{=}CHCH_3$ **(c)** 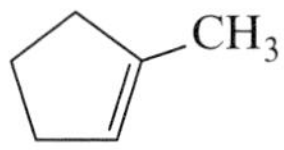 **(d)** 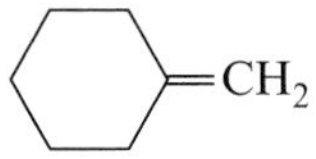

6.16 Arrange the alkenes in each set in order of increasing rate of reaction with HI and explain the basis for your ranking. Draw the structural formula of the major product formed in each case.

(a) $CH_3CH{=}CHCH_3$ and $CH_3\overset{CH_3}{\overset{|}{C}}{=}CHCH_3$ **(b)** 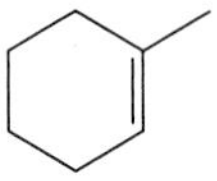 and 

6.17 Predict the organic product(s) of the reaction of 2-butene with each reagent.

(a) H_2O (H_2SO_4) **(b)** Br_2 **(c)** Cl_2
(d) Br_2 in H_2O **(e)** HI **(f)** Cl_2 in H_2O
(g) $Hg(OAc)_2$, H_2O **(h)** product (g) + $NaBH_4$

6.18 Draw a structural formula of an alkene that undergoes acid-catalyzed hydration to give each alcohol as the major product (more than one alkene may give each alcohol as the major product).

(a) 3-Hexanol **(b)** 1-Methylcyclobutanol
(c) 2-Methyl-2-butanol **(d)** 2-Propanol

6.19 Reaction of 2-methyl-2-pentene with each reagent is regioselective. Draw a structural formula for the product of each reaction and account for the observed regioselectivity.

(a) HI **(b)** HBr
(c) H_2O in the presence of H_2SO_4 **(d)** Br_2 in H_2O
(e) $Hg(OAc)_2$ in H_2O

6.20 Account for the regioselectivity and stereoselectivity observed when 1-methylcyclopentene is treated with each reagent.

(a) BH_3 **(b)** Br_2 in H_2O **(c)** $Hg(OAc)_2$ in H_2O

6.21 Draw a structural formula for an alkene with the indicated molecular formula that gives the compound shown as the major product (more than one alkene may give the same compound as the major product).

(a) $C_5H_{10} + H_2O \xrightarrow{H_2SO_4}$ 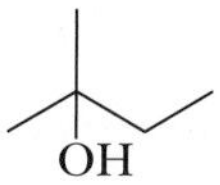

(b) $C_5H_{10} + Br_2 \longrightarrow$ 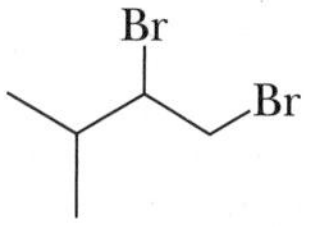

(c) $C_7H_{12} + HCl \longrightarrow$ 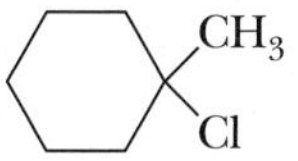

6.22 Account for the fact that addition of HCl to 1-bromopropene gives exclusively 1-bromo-1-chloropropane.

$$CH_3CH{=}CHBr + HCl \longrightarrow CH_3CH_2CHBrCl$$

1-Bromopropene **1-Bromo-1-chloropropane**

6.23 Account for the fact that treating propenoic acid (acrylic acid) with HCl gives only 3-chloropropanoic acid.

$$CH_2{=}CH\overset{\overset{O}{\|}}{C}OH + HCl \longrightarrow ClCH_2CH_2\overset{\overset{O}{\|}}{C}OH \qquad \left[CH_3\overset{\overset{Cl}{|}}{C}H\overset{\overset{O}{\|}}{C}OH \right]$$

Propenoic acid (Acrylic acid) — **3-Chloropropanoic acid** — [**2-Chloropropanoic acid** (this product is not formed)]

6.24 Draw a structural formula for the alkene with the molecular formula C_5H_{10} that reacts with Br_2 to give each product.

(a) 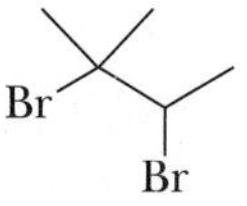

(b) 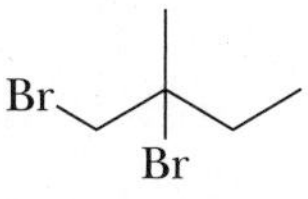

(c) 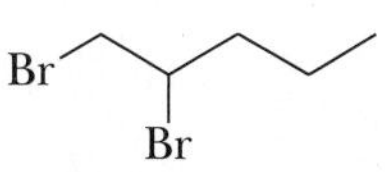

6.25 Draw the alternative chair conformations for the product formed by the addition of bromine to 4-*tert*-butylcyclohexene. The Gibbs free energy differences between equatorial and axial substituents on a cyclohexane ring are 21 kJ (4.9 kcal)/mol for *tert*-butyl and 2.0–2.6 kJ (0.48–0.62 kcal)/mol for bromine. Estimate the relative percentages of the alternative chair conformations you drew in the first part of this problem.

6.26 Draw a structural formula for the cycloalkene with the molecular formula C_6H_{10} that reacts with Cl_2 to give each compound.

(a) 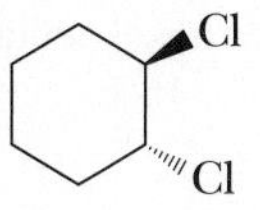

(b) 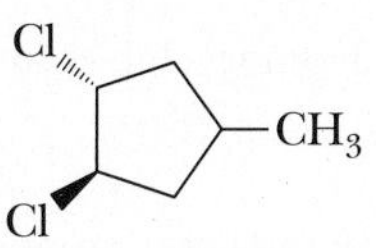

(c) 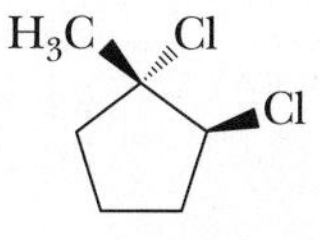

(d) 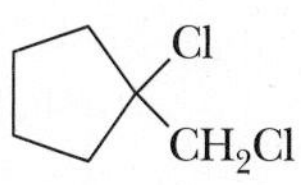

6.27 Reaction of this bicycloalkene with bromine in carbon tetrachloride gives a *trans* dibromide. In both (a) and (b), the bromine atoms are *trans* to each other. However, only one of these products is formed.

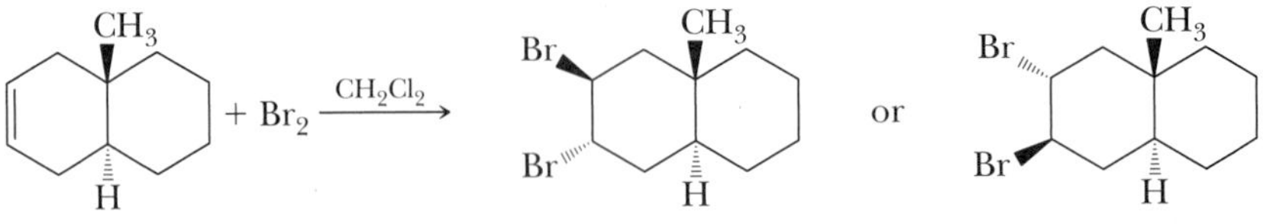

(a) (b)

Which *trans* dibromide is formed? How do you account for the fact that it is formed to the exclusion of the other *trans* dibromide?

6.28 Terpin, prepared commercially by the acid-catalyzed hydration of limonene, is used medicinally as an expectorant for coughs.

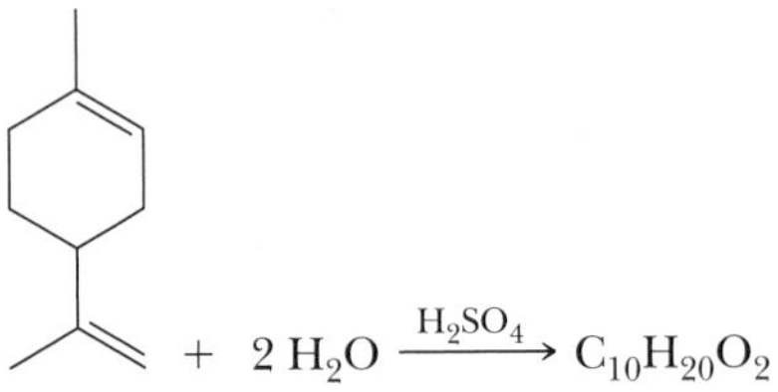

Limonene **Terpin**

(a) Propose a structural formula for terpin and a mechanism for its formation.
(b) How many *cis,trans* isomers are possible for the structural formula you propose?

6.29 Propose a mechanism for this reaction and account for its regioselectivity.

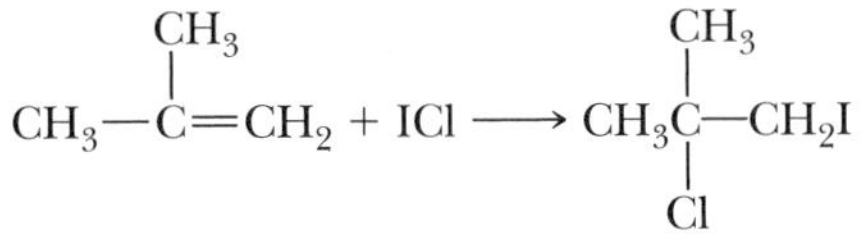

6.30 Treating 2-methylpropene with methanol in the presence of sulfuric acid gives *tert*-butyl methyl ether.

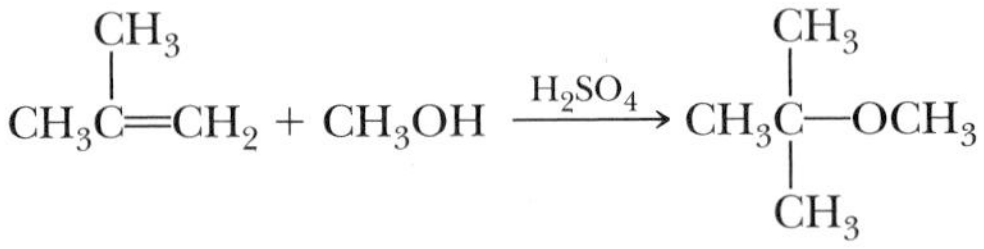

Propose a mechanism for the formation of this ether.

6.31 When 2-pentene is treated with Cl_2 in methanol, three products are formed. Account for the formation of each product (you need not explain their relative percentages).

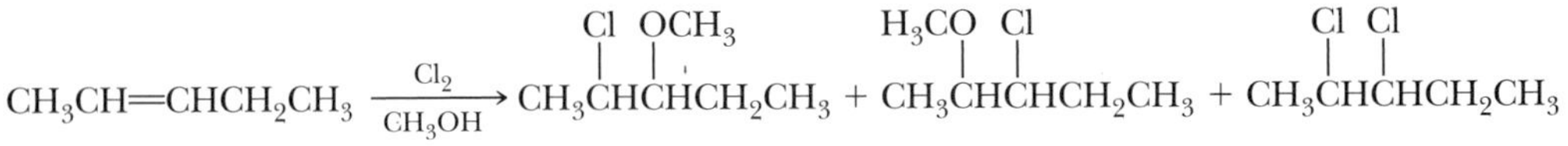

50% 35% 15%

6.32 Treating cyclohexene with HBr in the presence of acetic acid gives bromocyclohexane (85%) and cyclohexyl acetate (15%).

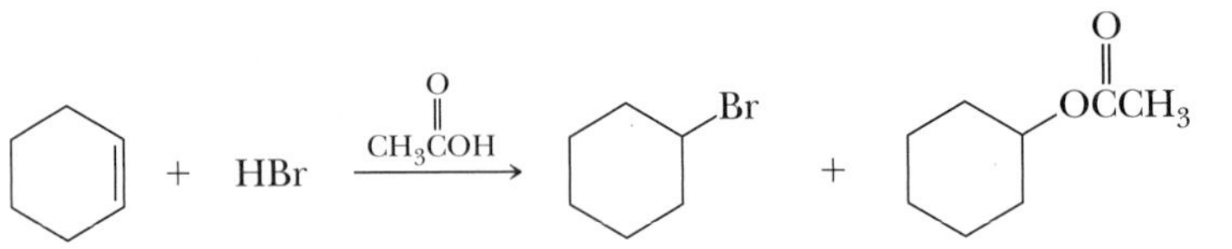

Cyclohexene **Bromocyclohexane** (85%) **Cyclohexyl acetate** (15%)

Propose a mechanism for the formation of the latter product.

6.33 Propose a mechanism for this reaction.

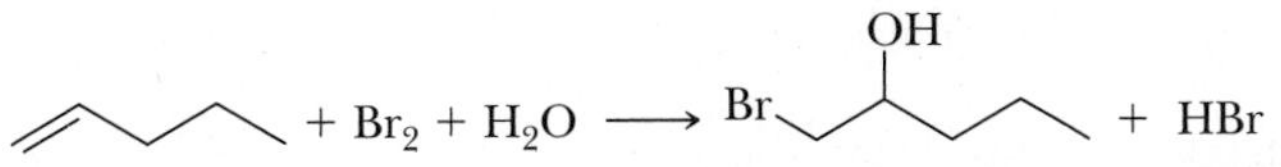

1-Pentene **1-Bromo-2-pentanol** (a racemic mixture)

6.34 Treating 4-penten-1-ol with bromine in water forms a cyclic bromoether.

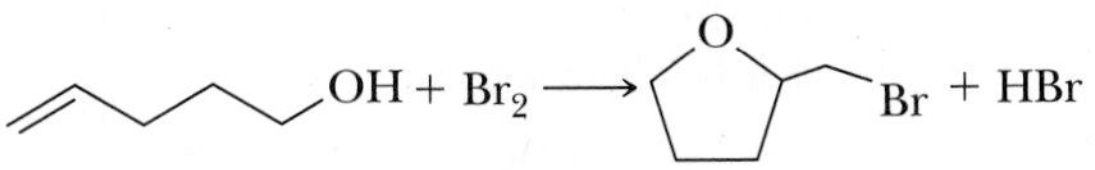

4-Penten-1-ol (a racemic mixture)

Account for the formation of this product rather than a bromohydrin as was formed in Problem 6.33.

6.35 Provide a mechanism for each reaction.

(a) 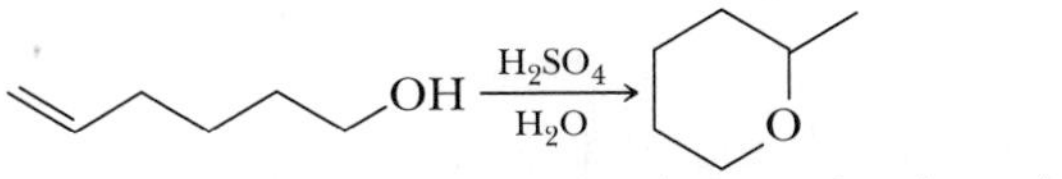

(a racemic mixture)

(b) 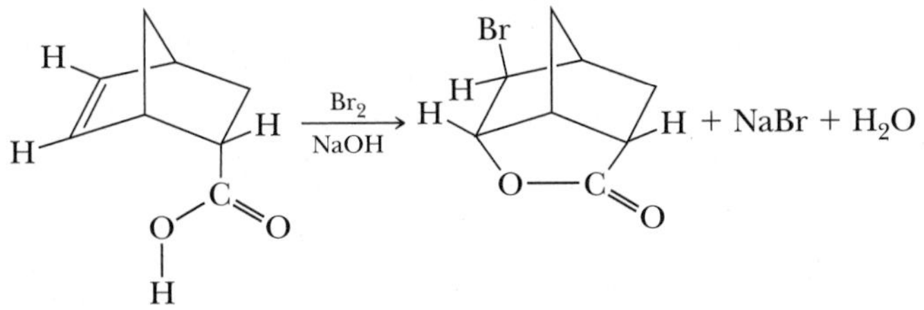

6.36 Treating 1-methyl-1-vinylcyclopentane with HCl gives mainly 1-chloro-1,2-dimethylcyclohexane.

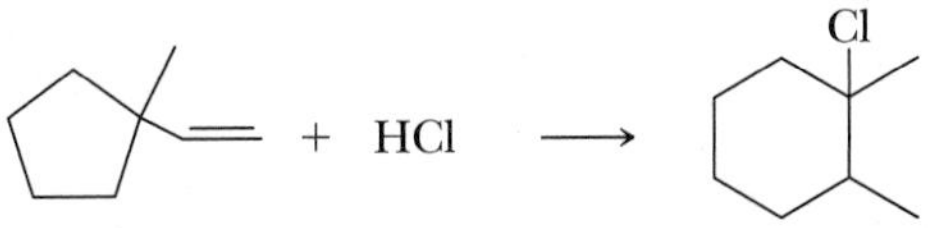

1-Methyl-1-vinylcyclopentane **1-Chloro-1,2-dimethylcyclohexane** (a racemic mixture)

Propose a mechanism for the formation of this product.

Hydroboration

6.37 Draw a structural formula for the alcohol formed by treating each alkene with borane in tetrahydrofuran (THF) followed by hydrogen peroxide in aqueous sodium hydroxide and specify stereochemistry where appropriate.

(a) 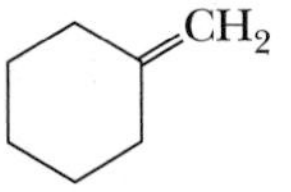 **(b)** 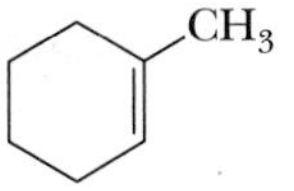 **(c)** 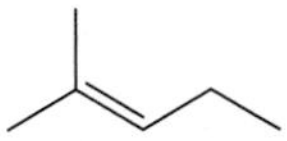

(d)  **(e)** 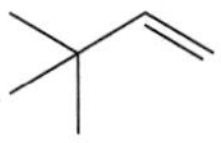

6.38 Reaction of α-pinene with borane followed by treatment of the resulting trialkylborane with alkaline hydrogen peroxide gives the following alcohol.

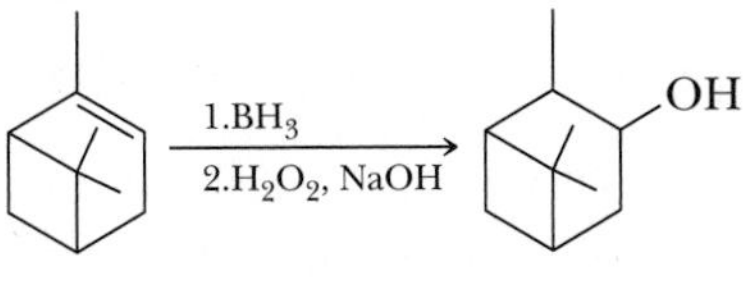

α-Pinene

Of the four possible *cis,trans* isomers, one is formed in over 85% yield.

(a) Draw structural formulas for the four possible *cis,trans* isomers of the bicyclic alcohol.

(b) Which is the structure of the isomer formed in 85% yield? How do you account for its formation? Create a model to help you make this prediction.

Oxidation

6.39 Write structural formulas for the major organic product(s) formed by reaction of 1-methylcyclohexene with each oxidizing agent.

(a) OsO_4/H_2O_2 **(b)** O_3 followed by $(CH_3)_2S$

6.40 Draw the structural formula of the alkene that reacts with ozone followed by dimethyl sulfide to give each product or set of products.

(a) 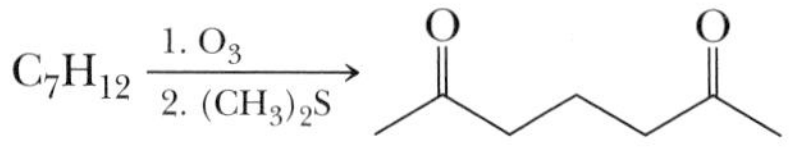

(b) 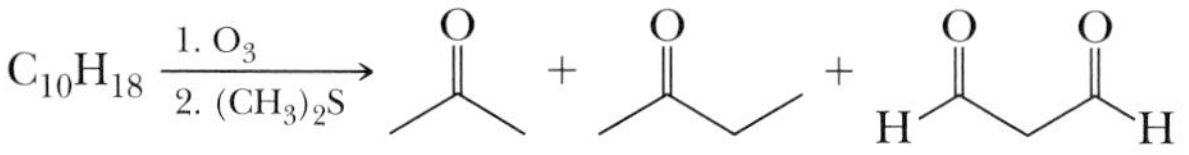

(c) 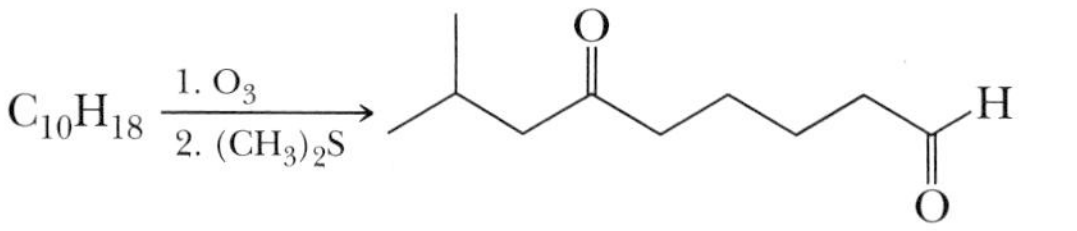

6.41 Consider the following reaction.

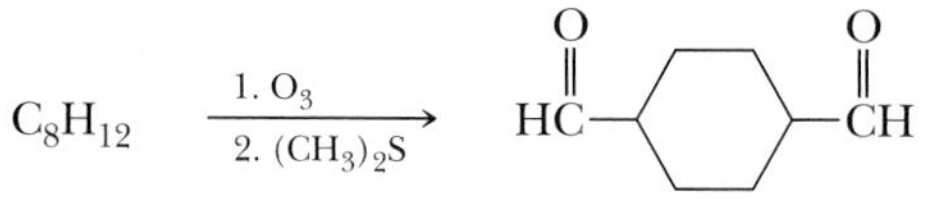

Cyclohexane-1,4-dicarbaldehyde

(a) Draw a structural formula for the compound with the molecular formula C_8H_{12}.

(b) Do you predict the product to be the *cis* isomer, the *trans* isomer, or a mixture of *cis* and *trans* isomers? Explain.

(c) Draw a suitable stereorepresentation for the more stable chair conformation of the dicarbaldehyde formed in this oxidation.

Reduction

6.42 Predict the major organic product(s) of the following reactions and show stereochemistry where appropriate.

(a) 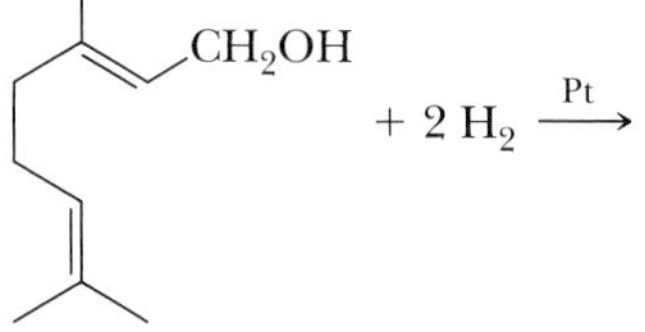

Geraniol

(b) 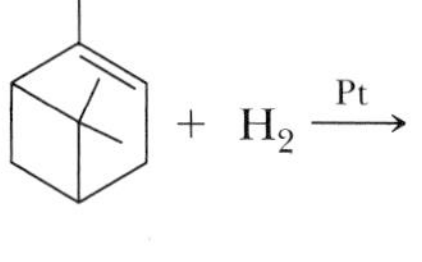

α-Pinene

6.43 The heat of hydrogenation of *cis*-2,2,5,5-tetramethyl-3-hexene is −154 kJ (−36.7 kcal)/mol, while that of the *trans* isomer is only −113 kJ (−26.9 kcal)/mol.

(a) Why is the heat of hydrogenation of the *cis* isomer so much larger (more negative) than that of the *trans* isomer?

(b) If a catalyst could be found that allowed equilibration of the *cis* and *trans* isomers at room temperature (such catalysts do exist), what would be the ratio of *trans* to *cis* isomers?

Synthesis

6.44 Show how to convert ethylene to these compounds.

(a) Ethane
(b) Ethanol
(c) Bromoethane
(d) 2-Chloroethanol
(e) 1,2-Dibromoethane
(f) 1,2-Ethanediol
(g) Chloroethane

6.45 Show how to convert cyclopentene into these compounds.

(a) *trans*-1,2-Dibromocyclopentane
(b) *cis*-1,2-Cyclopentanediol
(c) Cyclopentanol
(d) Iodocyclopentane
(e) Cyclopentane
(f) Pentanedial

Reactions That Produce Chiral Compounds

6.46 State the number and kind of stereoisomers formed when (*R*)-3-methyl-1-pentene is treated with these reagents. Assume that the starting alkene is enantiomerically pure and optically active. Will each product be optically active or inactive?

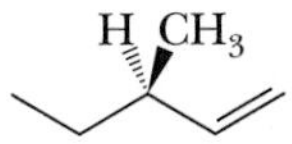

(*R*)-3-Methyl-1-pentene

(a) $Hg(OAc)_2$, H_2O followed by $NaBH_4$
(b) H_2/Pt
(c) BH_3 followed by H_2O_2 in NaOH
(d) Br_2 in CCl_4

6.47 Describe the stereochemistry of the bromohydrin formed in each reaction (each reaction is stereospecific).

(a) *cis*-3-Hexene + Br_2/H_2O
(b) *trans*-3-Hexene + Br_2/H_2O

6.48 In each of these reactions, the organic starting material is achiral. The structural formula of the product is given. For each reaction, determine the following.

(1) How many stereoisomers are possible for the product?
(2) Which of the possible stereoisomers is/are formed in the reaction shown?
(3) Will the product be optically active or optically inactive?

(a) 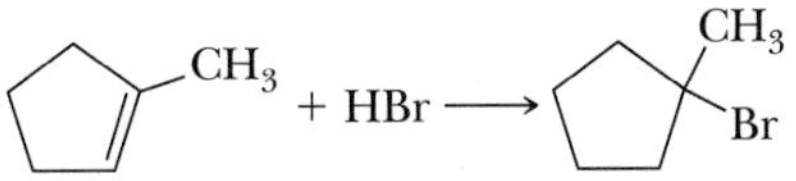

(b) 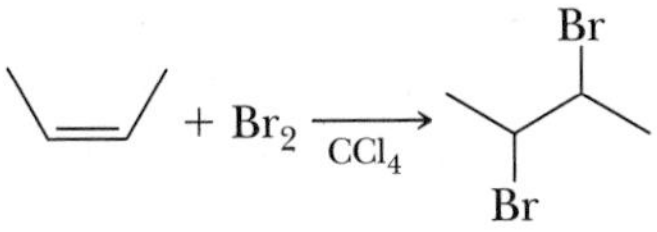

(c) 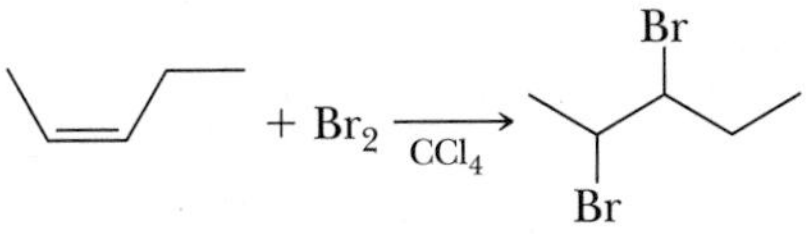

(d) 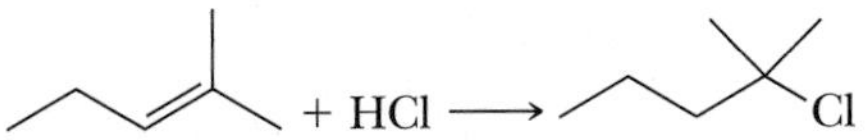

(e) 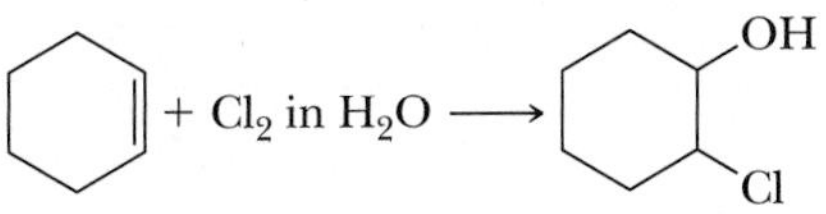

(f) 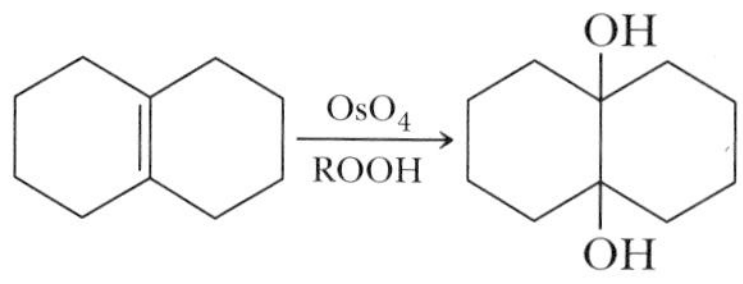
OsO4
ROOH
OH
OH

(g) 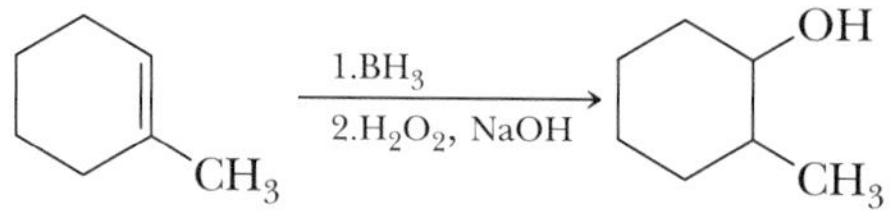
CH3
1.BH3
2.H2O2, NaOH
OH
CH3

Looking Ahead

6.49 The 2-propenyl cation appears to be a primary carbocation, and yet it is considerably more stable than a 1° carbocation such as the 1-propyl cation.

$$CH_2{=}CH{-}CH_2^{+} \qquad\qquad CH_3{-}CH_2{-}CH_2^{+}$$

2-Propenyl cation **1-Propyl cation**

How would you account for the differences in the stability of the two carbocations?

6.50 Treating 1,3-butadiene with 1 mole of HBr gives a mixture of two isomeric products.

$$CH_2{=}CH{-}CH{=}CH_2 + H{-}Br \longrightarrow CH_2{=}CH{-}\underset{}{\overset{\displaystyle Br}{\overset{|}{C}}}H{-}CH_3 + CH_3{-}CH{=}CH{-}CH_2{-}Br$$

1,3-Butadiene **3-Bromo-1-butene** (a racemic mixture) **1-Bromo-2-butene**

Propose a mechanism that accounts for the formation of these two products.

6.51 In this chapter, we studied the mechanism of the acid-catalyzed hydration of an alkene. The reverse of this reaction is the acid-catalyzed dehydration of an alcohol.

$$CH_3{-}\overset{\displaystyle OH}{\overset{|}{C}}H{-}CH_3 \xrightarrow{H_2SO_4} CH_3{-}CH{=}CH_2 + H_2O$$

2-Propanol (Isopropyl alcohol) **Propene**

Propose a mechanism for the acid-catalyzed dehydration of 2-propanol to propene.

6.52 As we have seen in this chapter, carbon-carbon double bonds are electron-rich regions that are attacked by electrophiles (e.g., HBr); they are not attacked by nucleophiles (e.g., diethylamine).

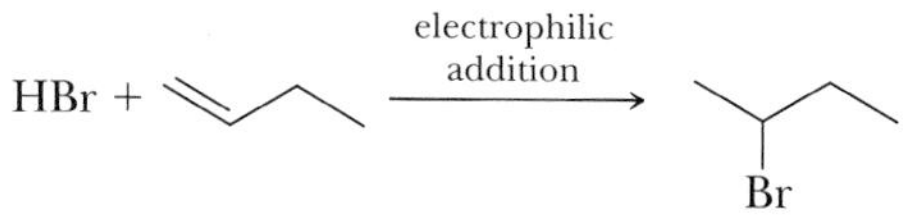
HBr +
electrophilic
addition
Br

(a racemic mixture)

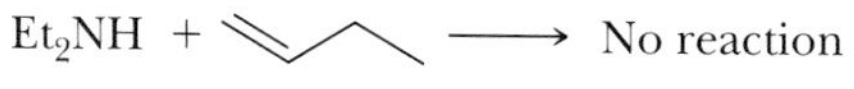
Et2NH +
No reaction

Diethylamine
(a nucleophile)

However, when the carbon-carbon double bond has a carbonyl group adjacent to it, the double bond reacts readily with nucleophiles by nucleophilic addition (Section 19.8).

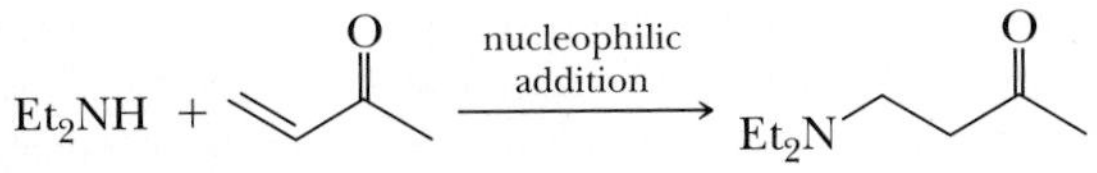

Et_2NH +
nucleophilic addition
Et_2N

Diethylamine
(a nucleophile)

Account for the fact that nucleophiles add to a carbon-carbon double bond adjacent to a carbonyl group and account for the regiochemistry of the reaction.

6.53 Following is an example of a type of reaction known as a Diels-Alder reaction (Chapter 20).

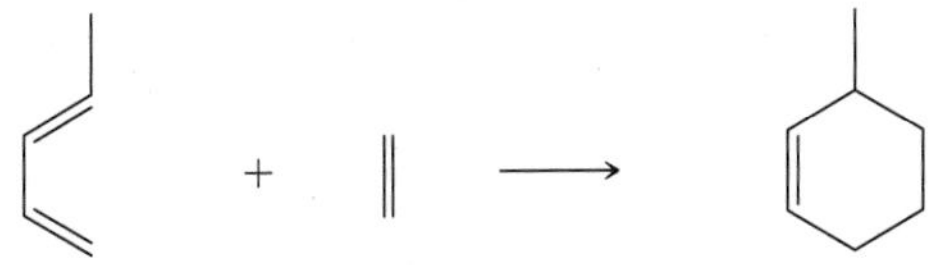

1,3-Pentadiene **Ethylene** **3-Methylcyclohexene**
(a racemic mixture)

The Diels-Alder reaction between a diene and an alkene is quite remarkable in that it is one of the few ways that chemists have to form two new carbon-carbon bonds in a single reaction. Given what you know about the relative strengths of carbon-carbon sigma and pi bonds, would you predict the Diels-Alder reaction to be exothermic or endothermic? Explain your reasoning.

Organic Chemistry Reaction Roadmap

6.54 We now introduce the concept of an organic chemistry "roadmap." An organic chemistry roadmap is a graphical representation of the different reactions that can be used to interconvert functional groups in molecules. Comparing the organic chemistry roadmap to a real roadmap, the functional groups are analogous to cities and the reactions are the roads between them. The power of the organic chemistry roadmap is that it helps students visualize how to interconvert key functional groups for use in multistep syntheses problems. It also will be a useful place for you to keep track of the reactions we encounter in future chapters so that you can see how they complement the reactions from previous chapters.

Reaction Roadmap

To make your own roadmap, take a blank full sheet of paper and write the following functional groups in the orientations shown. Fill the entire sheet of paper and leave plenty of room between functional groups. Most students find it helpful to use a poster-sized sheet of paper filled out in landscape orientation.

vicinal tetrahaloalkanes		aldehydes/ ketones	carboxylic acids
geminal dihaloalkanes	alkynes	alcohols	silyl ethers
allylic halides	vicinal dihaloalkanes	ethers	
thiols	alkenes	vicinal diols	
thioethers	↓ **HX** *Markovnikov regiochemistry* *Mixed stereochemistry of addition* *Carbocation intermediate*		
nitriles			
amines	haloalkanes	epoxides	vicinal aminoalcohols
alkyl azides	alkanes	halohydrins	

Refer to the reactions listed in the Study Guide section of this chapter. Draw arrows between functional groups to account for each reaction. Write the reagents required to bring about each reaction next to the arrow. Next, record any regiochemistry or stereochemistry considerations relevant to the reaction, such as "Markovnikov regiochemistry" or "anti addition stereochemistry." You should also record any key aspects of the mechanism, such as formation of a carbocation intermediate, as a helpful reminder. On the above organic chemistry roadmap template, the information for the first reaction, hydrohalogenation of an alkene, has been added to help you get started. For this initial roadmap, do not write an arrow for reaction 10, enantioselective reduction, because it is of a highly specific nature and a roadmap is intended to organize reactions that are of more general use.

Note that the roadmap template applies to Chapters 6–11, so you will not use all of the functional groups listed until you are finished with Chapter 11. Appendix 10 contains a series of roadmaps for different sections of the book, but you should use those for reference only after you have completed your own.

Cutting with an oxyacetylene torch. **Inset**: a model of acetylene. (Adam Crowley/Getty Images)

7 Alkynes

Outline

7.1 Structure of Alkynes
7.2 Nomenclature of Alkynes
7.3 Physical Properties of Alkynes
7.4 Acidity of 1-Alkynes
7.5 Preparation of Alkynes
7.6 Electrophilic Addition to Alkynes
7.7 Hydration of Alkynes to Aldehydes and Ketones
7.8 Reduction of Alkynes
7.9 Organic Synthesis

In this chapter, we continue our discussion of the chemistry of carbon-carbon π bonds as we now consider the chemistry of alkynes. Because alkenes and alkynes are similar in that the multiple bond in each is a combination of σ and π bonds, both types of functional groups undergo similar chemical reactions. Alkynes undergo electrophilic additions of X_2, HX, and H_2O. They undergo hydroboration-oxidation, and the carbon-carbon triple bond can be reduced first to a double bond and then to a single bond. An important reaction of alkynes that is not typical of alkenes is the conversion of terminal alkynes to their alkali metal salts, which are good nucleophiles and, therefore, valuable building blocks for the construction of larger molecules through the formation of carbon-carbon bonds.

7.1 Structure of Alkynes

Alkyne
An unsaturated hydrocarbon that contains one or more carbon-carbon triple bonds.

The functional group of an **alkyne** is a carbon-carbon triple bond. The simplest alkyne is ethyne, C_2H_2, more commonly named acetylene (Figure 7.1). Acetylene is a linear molecule; all bond angles are 180°. The carbon-carbon bond length in acetylene is 121 pm (1.21 Å) (Figure 7.1).

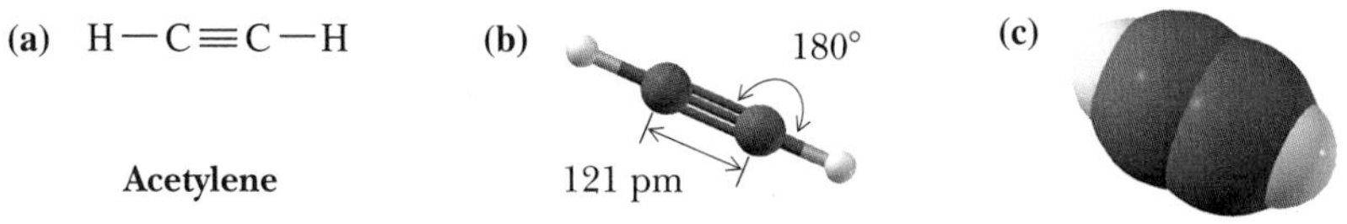
(a) H—C≡C—H
Acetylene
(b)
180°
121 pm
(c)

FIGURE 7.1 Acetylene. (a) Lewis structure, (b) ball-and-stick model, and (c) space-filling model.

By comparison, the length of the carbon-carbon double bond in ethylene is 134 pm (1.34 Å), and that of the carbon-carbon single bond in ethane is 153 pm (1.53 Å). Thus, triple bonds are shorter than double bonds, which, in turn, are shorter than single bonds. The bond dissociation enthalpy of the carbon-carbon triple bond in acetylene [966 kJ (231 kcal)/mol] is considerably larger than that for the carbon-carbon double bond in ethylene [727 kJ (174 kcal)/mol] and the carbon-carbon single bond in ethane [376 kJ (90 kcal)/mol]. The difference in bond dissociation enthalpies between the carbon-carbon triple bond in acetylene and the carbon-carbon double bond in ethylene is only 239 kJ (57 kcal)/mol. This difference indicates that a π bond in an alkyne is weaker than a π bond in an alkene.

A triple bond is described in terms of the overlap of *sp* hybrid orbitals of adjacent carbon atoms to form a σ bond, the overlap of parallel $2p_y$ orbitals to form one π bond, and the overlap of parallel $2p_z$ orbitals to form a second π bond (Figure 1.22). In acetylene, each carbon-hydrogen bond is formed by the overlap of a 1*s* orbital of hydrogen with an *sp* orbital of carbon. Because of the 50% *s*-character of the acetylenic C—H bond, it is unusually strong (see Table 1.11 and related text).

7.2 Nomenclature of Alkynes

A. IUPAC Names

According to the rules of the IUPAC system, the infix *-yn-* is used to show the presence of a carbon-carbon triple bond (Section 2.3). Thus, HC≡CH is named ethyne and CH_3C≡CH is named propyne. The IUPAC system retains the name acetylene; therefore, there are two acceptable names for HC≡CH, ethyne and acetylene. Of these names, acetylene is used more frequently.

There is no need to use a number to locate the position of the triple bond in ethyne and propyne; there is only one possible location for it in each compound. For larger molecules, number the longest carbon chain that contains the triple bond from the end that gives the triply bonded carbons the lower numbers. Show the location of the triple bond by the number of its first carbon. If a hydrocarbon chain contains more than one triple bond, we use the infixes *-adiyn-*, *-atriyn-*, and so forth.

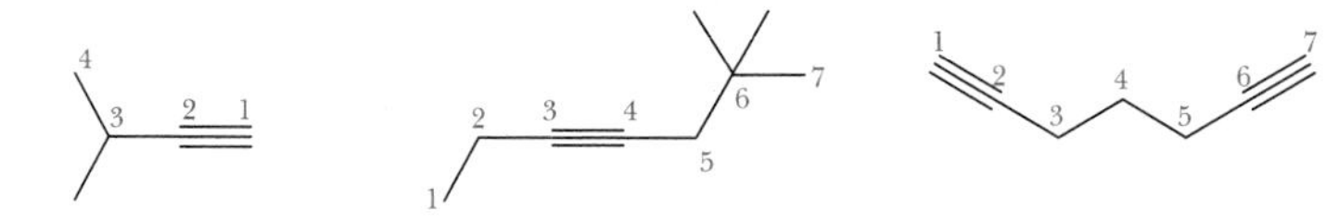
4
3
2
1
2
3
4
6
7
5
1
1
2
3
4
5
6
7

3-Methyl-1-butyne **6,6-Dimethyl-3-heptyne** **1,6-Heptadiyne**

Example 7.1 IUPAC Nomenclature for Alkynes

Write the IUPAC name of each compound.

(a) 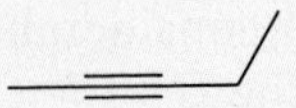 **(b)** 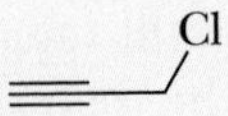 **(c)** 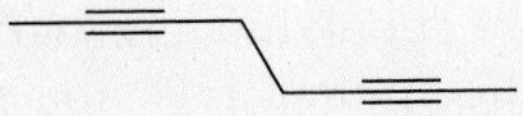

Solution

(a) 2-Pentyne **(b)** 3-Chloropropyne **(c)** 2,6-Octadiyne

Problem 7.1

Write the IUPAC name of each compound.

(a) 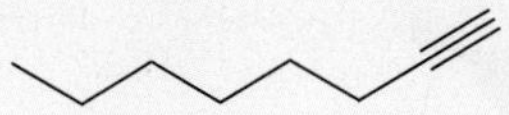 **(b)** 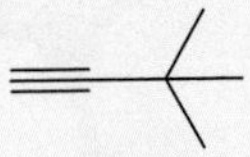 **(c)** 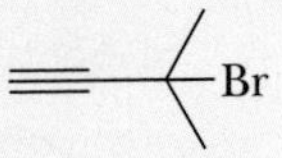

B. Common Names

Common names for alkynes are derived by prefixing the names of substituents on the carbon-carbon triple bond to the word *acetylene.* Note in the third example that when a carbon-carbon double bond (indicated by *-en-*) and a carbon-carbon triple bond (indicated by *-yn-*) are both present in the same molecule, the IUPAC rules specify that the location of the double bond takes precedence in numbering the compound.

	$CH_3C{\equiv}CH$	$CH_3C{\equiv}CCH_3$	$CH_2{=}CHC{\equiv}CH$
IUPAC name:	**Propyne**	**2-Butyne**	**1-Buten-3-yne**
Common name:	**Methylacetylene**	**Dimethylacetylene**	**Vinylacetylene**

Alkynes in which the triple bond is between carbons 1 and 2 are commonly referred to as **terminal alkynes**. Examples of terminal alkynes are propyne and 1-butyne.

Terminal alkyne
An alkyne in which the triple bond is between carbons 1 and 2.

Example 7.2 Common Nomenclature of Alkynes

Write the common name of each alkyne.

(a) 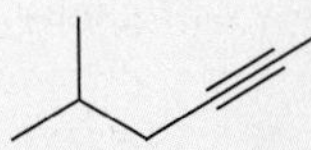 **(b)** 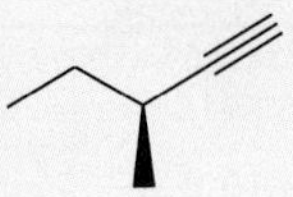 **(c)** 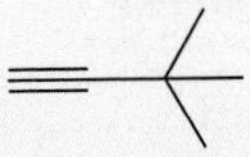

Solution

(a) Isobutylmethylacetylene **(b)** (S)-*sec*-Butylacetylene **(c)** *tert*-Butylacetylene

Problem 7.2

Write the common name of each alkyne.

(a) 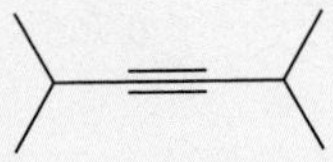 **(b)** 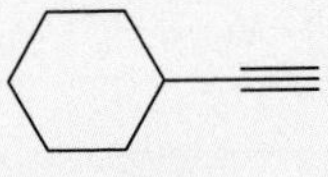 **(c)** 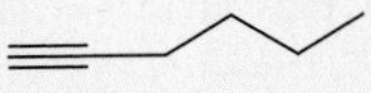

The smallest cycloalkyne that has been isolated is cyclooctyne. This molecule is quite unstable and polymerizes rapidly at room temperature. The C—C≡C bond angle in cyclooctyne is calculated to be approximately 155°, indicating a high degree of angle strain. Cyclononyne (Figure 7.2) has also been prepared and is stable at room temperature. The calculated C—C≡C bond angles in this cycloalkyne are approximately 160°, which still represents a considerable distortion from the optimal 180°. You can see the distortion of the C—C≡C bond angles in the accompanying optimized molecular model. You can also see the degree to which C—C and C—H bonds on adjacent carbons are staggered, thus minimizing torsional strain.

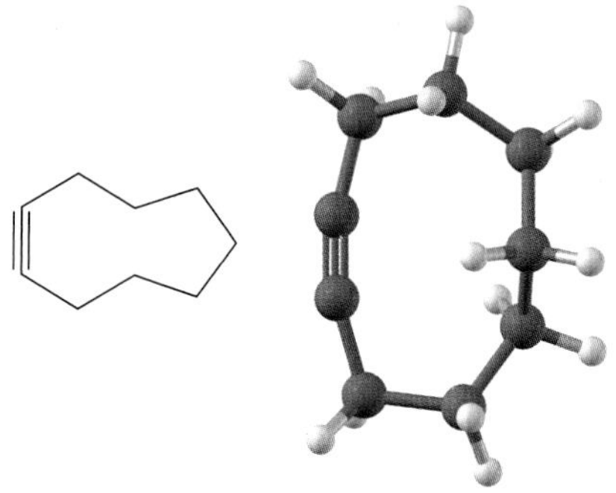

FIGURE 7.2 Structure of cyclononyne.

7.3 Physical Properties of Alkynes

The physical properties of alkynes are similar to those of alkanes and alkenes with analogous carbon skeletons. The lower-molecular-weight alkynes are gases at room temperature. Those that are liquids at room temperature have densities less than 1.0 g/mL (less dense than water). Listed in Table 7.1 are melting points, boiling points, and densities of several low-molecular-weight alkynes. Because alkynes, like alkanes and alkenes, are nonpolar compounds, they are insoluble in water and other polar solvents. They are soluble in each other and in other nonpolar organic solvents.

Table 7.1 Physical Properties of Some Low-Molecular-Weight Alkynes

Name	Formula	Melting Point (°C)	Boiling Point (°C)	Density at 20°C (g/mL)
Ethyne	HC≡CH	−81	−84	(a gas)
Propyne	$CH_3C{\equiv}CH$	−102	−23	(a gas)
1-Butyne	$CH_3CH_2C{\equiv}CH$	−126	8	(a gas)
2-Butyne	$CH_3C{\equiv}CCH_3$	−32	27	0.691
1-Pentyne	$CH_3(CH_2)_2C{\equiv}CH$	−90	40	0.690
1-Hexyne	$CH_3(CH_2)_3C{\equiv}CH$	−132	71	0.716
1-Octyne	$CH_3(CH_2)_5C{\equiv}CH$	−79	125	0.746
1-Decyne	$CH_3(CH_2)_7C{\equiv}CH$	−36	174	0.766

7.4 Acidity of 1-Alkynes

One of the major differences between the chemistry of alkynes and that of alkenes or alkanes is that a hydrogen bonded to a triply bonded carbon atom of a terminal alkyne is sufficiently acidic that it can be removed by a strong base, such as sodium amide $NaNH_2$ (Table 4.1), to give an acetylide anion.

$$H{-}C{\equiv}C{-}H + {:}\ddot{N}H_2^- \rightleftharpoons H{-}C{\equiv}C{:}^- + \ddot{N}H_3 \qquad K_{eq} = 10^{13}$$

pK_a 25 (Stronger acid) — (Stronger base) — (Weaker base) — pK_a 38 (Weaker acid)

Other strong bases commonly used to form acetylide anions are sodium hydride and lithium diisopropylamide (LDA).

$Na^+H{:}^-$ **Sodium hydride**

$[(CH_3)_2CH]_2\ddot{N}{:}^-Li^+$ **Lithium diisopropylamide (LDA)**

Because water is a stronger acid than acetylene, the hydroxide ion is not a strong enough base to convert a terminal alkyne to an alkyne anion.

$$H{-}C{\equiv}C{-}H + {:}\ddot{O}H^- \rightleftharpoons H{-}C{\equiv}C{:}^- + H{-}\ddot{O}H \qquad K_{eq} = 10^{-9.3}$$

pK_a 25 (Weaker acid) — (Weaker base) — (Stronger base) — pK_a 15.7 (Stronger acid)

The pK_a values for alkene and alkane hydrogens are so large (they are so weakly acidic) that neither the commonly used alkali metal hydroxides nor sodium hydride, sodium amide, or lithium diisopropylamide are strong enough bases to remove a proton from alkanes or alkenes.

7.5 Preparation of Alkynes

A. Alkylation of Acetylide Anions with Methyl and 1° Haloalkanes

As we have already seen, an acetylide anion is a strong base. An acetylide anion is also a nucleophile; it has an unshared pair of electrons that it can donate to another atom to form a new covalent bond. In this instance, an acetylide anion donates its unshared pair of electrons to the carbon of a methyl or primary haloalkane, and in so doing, the acetylide nucleophile replaces the halogen atom. This type of reaction is called a nucleophilic substitution. For example, treating sodium acetylide with 1-bromobutane gives 1-hexyne.

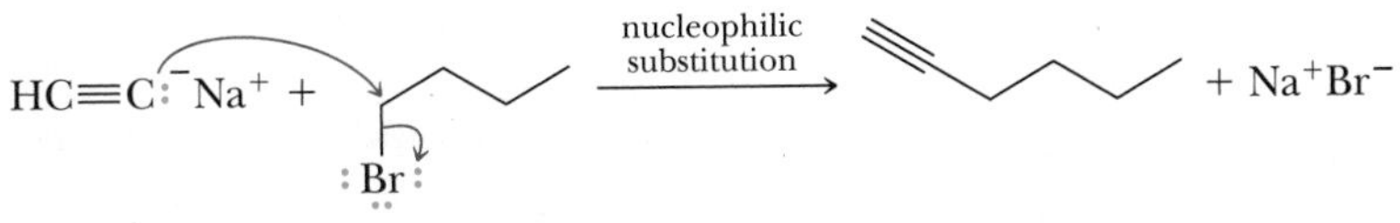

Sodium acetylide — **1-Bromobutane** — **1-Hexyne**

▶ Most bond making steps in reaction mechanisms involve nucleophiles reacting with electrophiles.

Alkylation reaction

Any reaction in which a new carbon-carbon bond to an alkyl group is formed.

Because an alkyl group (in this case a butyl group) is added to a molecule, this type of reaction is also called an **alkylation reaction**. We limit our discussion of nucleophilic substitution in this chapter to the reactions of acetylide anions with methyl and 1° haloalkanes. We discuss the scope and limitations of nucleophilic substitution in more general terms in Chapter 9. For reasons we discuss in Chapter 9, alkylation of acetylide anions is practical only with methyl and primary halides.

Because of the ready availability of acetylene and the ease with which it is converted to a good nucleophile, alkylation of an acetylide anion is the most convenient laboratory method for the synthesis of terminal alkynes, meaning one with the carbon-carbon triple bond at the end of a carbon chain. The process of alkylation can be repeated, and a terminal alkyne can, in turn, be converted to an alkyne with the carbon-carbon triple bond in the middle of a carbon chain. Such an alkyne is usually referred to as an **internal alkyne**. An important feature of this reaction is that new carbon-carbon bonds are made, allowing the construction of larger carbon backbones from smaller ones.

Internal alkyne

A carbon-carbon triple bond in the middle of a carbon chain.

$$CH_3CH_2C{\equiv}C^-Na^+ + CH_3CH_2{-}Br \xrightarrow[\text{substitution}]{\text{nucleophilic}} CH_3CH_2C{\equiv}CCH_2CH_3 + Na^+Br^-$$

Sodium butynide **Bromoethane** **3-Hexyne**

B. Alkynes from Alkenes

To prepare an alkyne from an alkene, the alkene is first treated with one mole of either bromine (Br_2) or chlorine (Cl_2) to give a dihaloalkane (Section 6.3D). Treating the dihaloalkane with two moles of a strong base such as sodium amide ($NaNH_2$) in liquid ammonia [$NH_3(l)$] brings about two successive **dehydrohalogenations**. Recall that addition of HX to an alkene is called hydrohalogenation; removal of HX from a haloalkane is called dehydrohalogenation. The removal of atoms from adjacent carbons to form an alkene is also called an elimination reaction and is discussed fully in Chapter 9. The following example shows the conversion of 2-butene to 2-butyne.

Dehydrohalogenation

The removal of HX from a molecule.

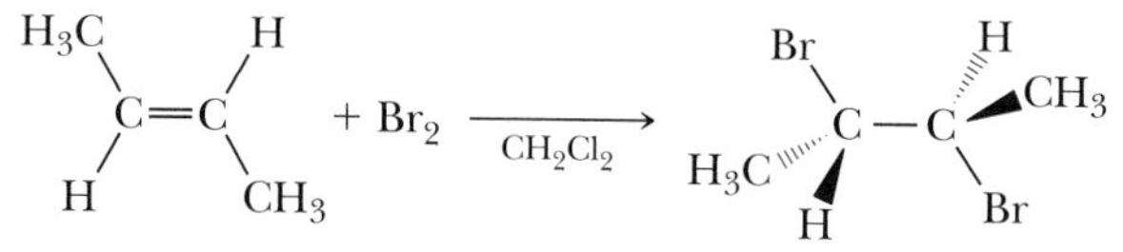

***trans*-2-Butene** **2,3-Dibromobutane** (meso)

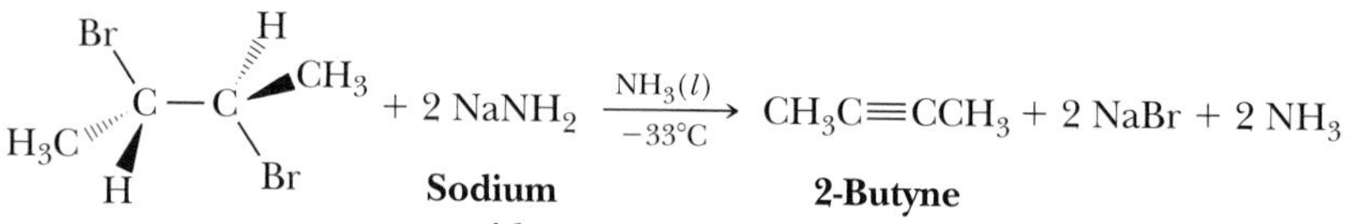

2,3-Dibromobutane (meso) **Sodium amide** **2-Butyne**

Given the ease of converting alkenes to dihaloalkanes and then to alkynes, alkenes are versatile starting materials for the preparation of alkynes.

With a strong base such as sodium amide, both dehydrohalogenations occur readily. However, with weaker bases such as sodium hydroxide or potassium hydroxide in ethanol, it is often possible to stop the reaction after the first dehydrohalogenation and isolate the haloalkene.

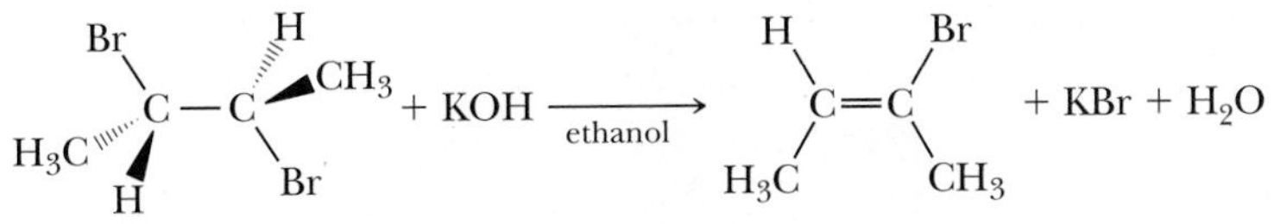

2,3-Dibromobutane (meso) **(*E*)-2-Bromo-2-butene**

In practice, it is much more common to use a stronger base and go directly to the alkyne.

The following equations show the conversion of 1-hexene to 1-hexyne. Note that three moles of sodium amide are used in this sequence. Two moles are required for the double dehydrohalogenation reaction, which gives 1-hexyne. As soon as any 1-hexyne (a weak acid, pK_a 25) forms, it reacts with sodium amide (a strong base) to give an alkyne salt. Thus, a third mole of sodium amide is required to complete the dehydrohalogenation of the remaining bromoalkene. Addition of water (a weak acid) or aqueous acid completes the sequence and gives 1-hexyne.

$$\underset{\textbf{1-Hexene}}{CH_3(CH_2)_3CH{=}CH_2} \xrightarrow{Br_2} \underset{\substack{\textbf{1,2-Dibromohexane}\\ \text{(a racemic mixture)}}}{CH_3(CH_2)_3\overset{Br}{\overset{|}{C}}H\overset{Br}{\overset{|}{C}}H_2} \xrightarrow[-2\,HBr]{2\,NaNH_2} \underset{\textbf{1-Hexyne}}{[CH_3(CH_2)_3C{\equiv}CH]} \xrightarrow{NaNH_2}$$

$$\underset{\textbf{Sodium salt of 1-hexyne}}{CH_3(CH_2)_3C{\equiv}C^-Na^+} \xrightarrow{H_2O} \underset{\textbf{1-Hexyne}}{CH_3(CH_2)_3C{\equiv}CH}$$

In dehydrohalogenation of a haloalkene with at least one hydrogen on each adjacent carbon, a side reaction occurs, namely the formation of an allene.

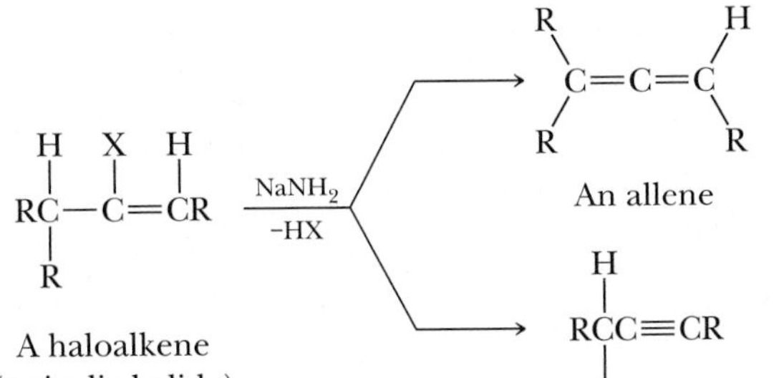

A haloalkene (a vinylic halide) An allene An alkyne

An **allene** has two adjacent carbon-carbon double bonds; that is, it contains a C═C═C functional group. The simplest allene is 1,2-propadiene, commonly named allene. In it, each end carbon is sp^2 hybridized and the middle carbon is sp hybridized. Each carbon-carbon σ bond is formed by the overlap of sp and sp^2 hybrid orbitals. One π bond is formed by the overlap of parallel $2p_y$ orbitals; the other, by the overlap of parallel $2p_z$ orbitals. The two π-bonding molecular orbitals are in planes perpendicular to each other, as are the two H—C—H groups.

Allene

The compound $CH_2{=}C{=}CH_2$. Any compound that contains adjacent carbon-carbon double bonds (i.e., any molecule that contains a C═C═C functional group).

$CH_2{=}C{=}CH_2$

1,2-Propadiene
(Allene)

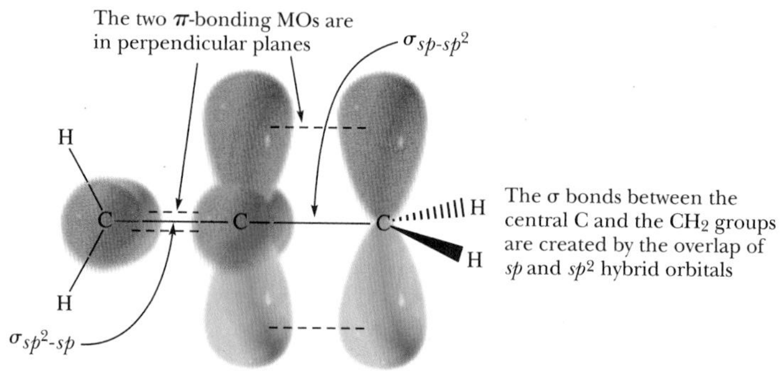
The two π-bonding MOs are in perpendicular planes
σ sp-sp²
H
H
C
C
C
H
H
The σ bonds between the central C and the CH₂ groups are created by the overlap of sp and sp² hybrid orbitals
σ sp²-sp

Cartoon representation of two π bonds at 90 degree angles

Most allenes are less stable than their isomeric alkynes. For example, allene itself is less stable by 6.7 kJ (1.6 kcal)/mol than its constitutional isomer propyne, and 1,2-butadiene is less stable than 2-butyne by 16.7 kJ (4.0 kcal)/mol.

$$CH_2{=}C{=}CH_2 \longrightarrow CH_3C{\equiv}CH \qquad \Delta H^0 = -6.7\text{ kJ } (-1.6\text{ kcal})/\text{mol}$$

$$CH_2{=}C{=}CHCH_3 \longrightarrow CH_3C{\equiv}CCH_3 \qquad \Delta H^0 = -16.7\text{ kJ } (-4.0\text{ kcal})/\text{mol}$$

Because of their lower stability relative to isomeric alkynes, allenes are generally only minor products of alkyne-forming dehydrohalogenation reactions.

Example 7.3 Synthesis of Alkynes

Show how you might convert 1-pentene to 1-pentyne.

Solution

This synthesis can be done in two steps: Treating 1-pentene with one mole of bromine gives 1,2-dibromopentane. Treating this dibromoalkane with three moles of sodium amide followed by H_2O gives 1-pentyne.

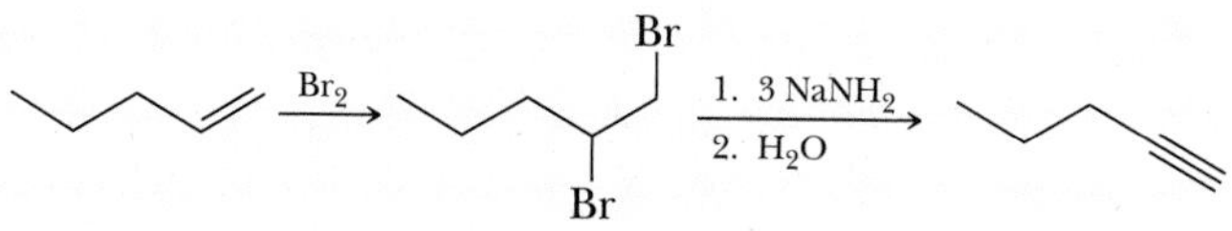
Br₂
Br
Br
1. 3 NaNH₂
2. H₂O

1-Pentene **1,2-Dibromopentane** (a racemic mixture) **1-Pentyne**

Watch a video explanation

Problem 7.3

Draw a structural formula for an alkene and a dichloroalkane with the given molecular formula that yields the indicated alkyne by each reaction sequence.

(a) $C_6H_{12} \xrightarrow{Cl_2} C_6H_{12}Cl_2 \xrightarrow{2\ NaNH_2}$ 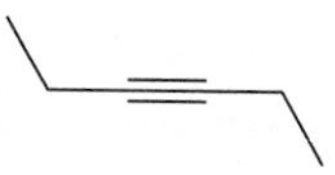

(b) $C_7H_{14} \xrightarrow{Cl_2} C_7H_{14}Cl_2 \xrightarrow{2\ NaNH_2}$ 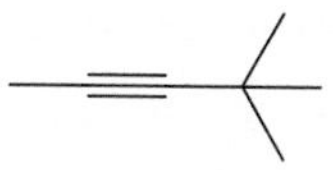

7.6 Electrophilic Addition to Alkynes

Alkynes undergo many of the same electrophilic additions as alkenes. A characteristic reaction is a π bond acting as a nucleophile to make a new bond with an electrophile. In this section, we study the addition of bromine and chlorine as well as the addition of hydrogen halides.

A. Addition of Bromine and Chlorine

Addition of one mole of Br_2 to an alkyne gives a dibromoalkene. Addition of bromine to a triple bond is stereoselective, as illustrated by the reaction of 2-butyne with one mole of Br_2. The major product corresponds to anti addition of the two bromine atoms. Carrying out the bromination in acetic acid with added bromide ion (e.g., LiBr) significantly increases the preference for anti addition.

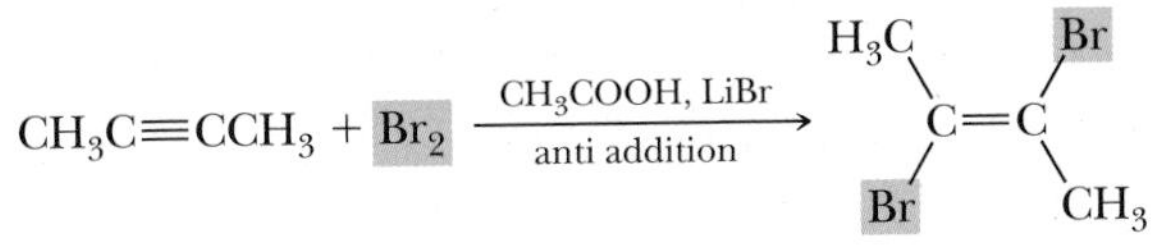

2-Butyne **(*E*)-2,3-Dibromo-2-butene**

Alkynes similarly undergo addition of Cl_2, although less stereoselectively than with Br_2.

Addition of bromine to alkynes follows much the same type of mechanism as it does for addition to alkenes (Section 6.3D), namely formation of a bridged bromonium ion intermediate, which is then attacked by bromide ion from the face opposite that occupied by the positively charged bromine atom.

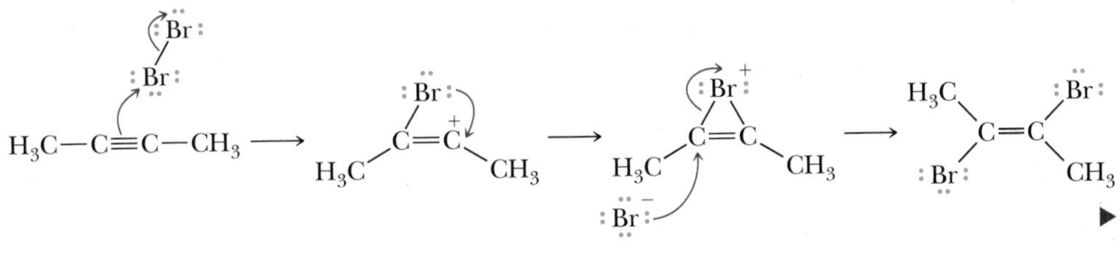

▶ Most bond-making steps in reaction mechanisms involve nucleophiles reacting with electrophiles.

Addition of a second mole of Br_2 gives a tetrabromoalkane.

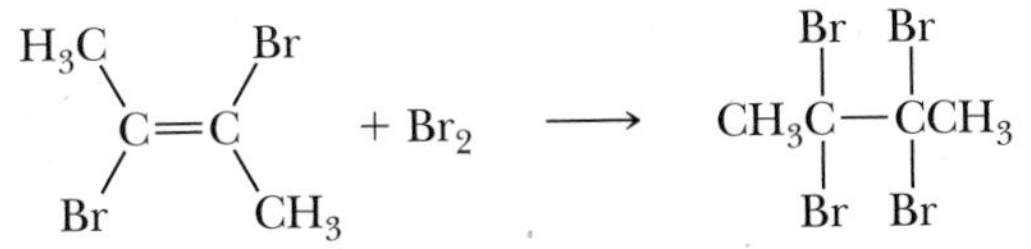

(*E*)-2,3-Dibromo-2-butene **2,2,3,3-Tetrabromobutane**

B. Addition of Hydrogen Halides

Alkynes add either one or two moles of HBr and HCl, depending on the ratios in which the alkyne and halogen acid are mixed.

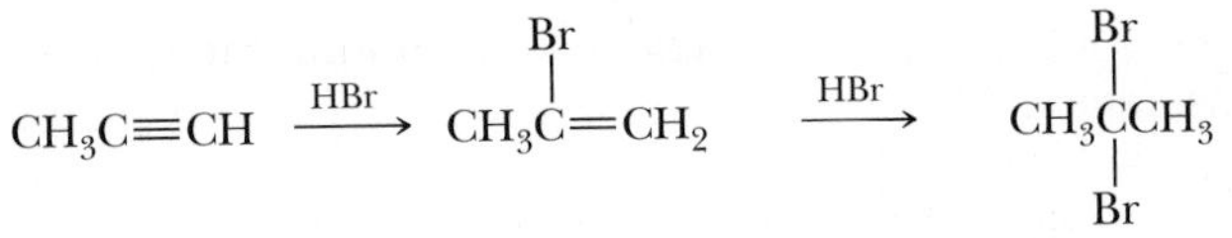

Propyne **2-Bromopropene** **2,2-Dibromopropane**

As shown in this equation, addition of both the first and second moles of HBr is regioselective. Addition of hydrogen halides follows Markovnikov's rule (Section 6.3A); hydrogen adds to the carbon that has the greater number of hydrogens. We can account for this regioselectivity of addition of HX by a two-step mechanism for each addition.

Mechanism 7.1

Addition of HBr to an Alkyne

Step 1: **Make a new bond between a nucleophile (π bond) and an electrophile—add a proton.** Proton transfer from HBr to the alkyne gives a **vinylic carbocation**; the more stable 2° vinylic carbocation is formed in preference to the less stable 1° vinylic carbocation.

$$CH_3C{\equiv}CH + H{-}\ddot{\underset{..}{Br}}: \longrightarrow CH_3\overset{+}{C}{=}CH_2 + :\ddot{\underset{..}{Br}}:^-$$

A 2° vinylic carbocation

Step 2: **Make a new bond between a nucleophile and an electrophile.** Reaction of the vinylic carbocation (an electrophile) with bromide ion (a nucleophile) gives the vinylic bromoalkene.

$$CH_3\overset{+}{C}{=}CH_2 + :\ddot{\underset{..}{Br}}:^- \longrightarrow CH_3\overset{:\ddot{Br}:}{\overset{|}{C}}{=}CH_2$$

2-Bromopropene

Vinylic carbocation

A carbocation in which the positive charge is on one of the carbons of a carbon-carbon double bond.

Alkynes are considerably less reactive toward most electrophilic additions than are alkenes. The major reason for this difference is the instability of the *sp*-hybridized vinylic carbocation intermediate formed from an alkyne compared with the sp^2-hybridized alkyl carbocation formed from an alkene.

In the addition of the second mole of HX, Step 1 is the reaction of the electron pair of the remaining π bond with HBr to form a carbocation. Of the two possible carbocations, the one with the positive charge on the carbon bearing the halogen is favored because of delocalization of the positive charge through resonance.

▶ Delocalization of charge over a larger area is stabilizing.

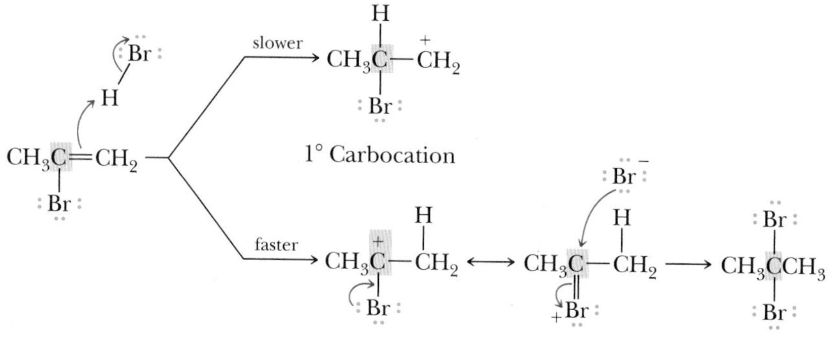

slower
1° Carbocation
faster

Resonance-stabilized 2° carbocation

Addition of one mole of HCl to acetylene gives chloroethene (vinyl chloride), a compound of considerable industrial importance.

$$HC{\equiv}CH + HCl \longrightarrow CH_2{=}CHCl$$

Ethyne (Acetylene) → **Chloroethene (Vinyl chloride)**

Vinyl chloride is the monomer in the production of the polymer poly(vinyl chloride), abbreviated PVC.

$$n\,CH_2{=}CHCl \xrightarrow{\text{catalyst}} \left(CH_2\underset{}{\overset{Cl}{\overset{|}{C}}}H\right)_n$$

Vinyl chloride — **Poly(vinyl chloride) (PVC)**

PVC dominates much of the plumbing and construction market for plastics. Approximately 67% of all pipe, fittings, and conduit, along with 42% of all plastics used in construction at the present time, are fabricated from PVC. We will describe the synthesis of this polymer and its properties in Chapter 29. Our purpose here is to describe the synthesis of vinyl chloride.

At one time, hydrochlorination of acetylene was the major source of vinyl chloride. As the cost of production of acetylene increased, however, manufacturers of vinyl chloride sought other routes to this material. The starting material chosen was ethylene, which can be converted to vinyl chloride in two steps: treating ethylene with chlorine gives 1,2-dichloroethane, which, when heated in the presence of charcoal or other catalyst, loses a molecule of HCl to form vinyl chloride.

$$CH_2{=}CH_2 \xrightarrow{Cl_2} ClCH_2CH_2Cl \xrightarrow[\text{catalyst}]{\text{heat}} CH_2{=}CHCl + HCl$$

Ethylene — **1,2-Dichloroethane** — **Vinyl chloride**

Chlorine atoms of the byproduct are recycled by passing the HCl mixed with air over a copper(II) catalyst, which results in the oxidation of HCl to Cl_2.

$$4\,HCl + O_2 \xrightarrow{CuCl_2} 2\,H_2O + 2\,Cl_2$$

We described the production of vinyl chloride first from acetylene and then from ethylene to illustrate an important point about industrial organic chemistry. The aim is to produce a desired chemical from readily available and inexpensive starting materials by reactions in which byproducts can be recycled. All chemical companies now support this objective to minimize both costs and production of materials that require disposal or can harm the environment.

7.7 Hydration of Alkynes to Aldehydes and Ketones

The elements of H_2O can be added to the carbon-carbon triple bond of an alkyne by the same two reactions used for the hydration of alkenes, namely hydroboration-oxidation and acid-catalyzed hydration. Even though the reagents are similar, the products from hydration of alkenes and alkynes are quite different.

A. Hydroboration-Oxidation

Borane adds readily to an internal alkyne as illustrated by its reaction with 3-hexyne.

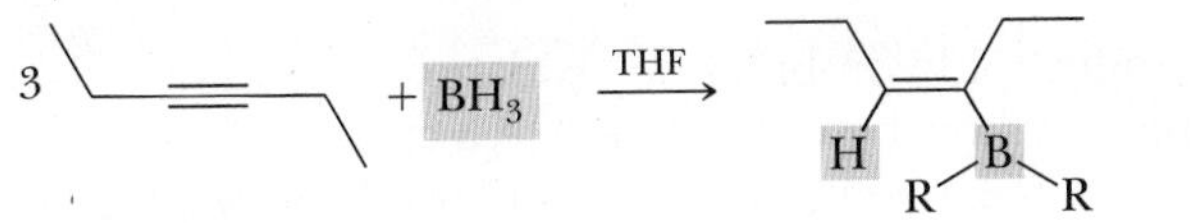

3-Hexyne — A trialkenylborane (R = *cis*-3-hexenyl group)

Notice that the hydroboration of an internal alkyne stops after the addition of one mole of borane. The product is a trialkenylborane (the infix *-enyl-* shows the presence of a carbon-carbon double bond on the carbon bonded to boron). As with hydroboration of alkenes (Section 6.4), hydroboration of alkynes is stereoselective; it involves syn addition of hydrogen and boron.

Terminal alkynes also react regioselectively with borane to form trialkenylboranes. In practice, however, the reaction is difficult to stop at this stage because the alkenyl group reacts further with borane to undergo a second hydroboration.

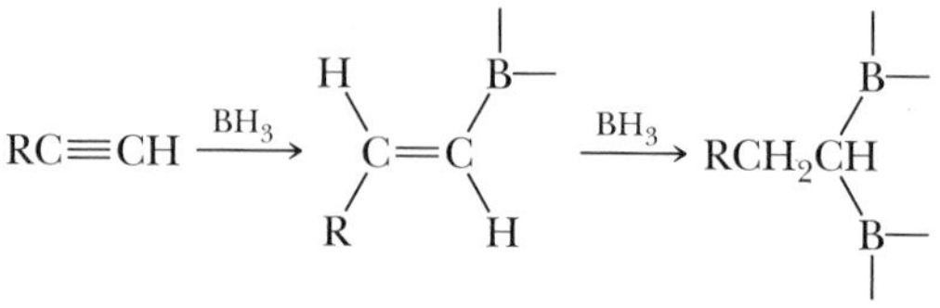

It is possible to prevent the second hydroboration step and, in effect, stop the reaction at the alkenylborane stage by using a sterically hindered disubstituted borane. One of the most widely used of these is di-*sec*-isoamylborane, $(sia)_2BH$, prepared by treating borane with two equivalents of 2-methyl-2-butene (amyl is an older common name for pentyl).

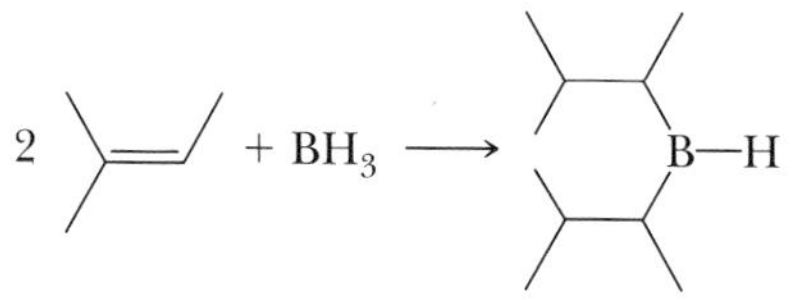

2-Methyl-2-butene **Di-*sec*-isoamylborane $[(sia)_2BH]$**

Reaction of this sterically hindered dialkylborane with a terminal alkyne results in a single hydroboration and formation of an alkenylborane.

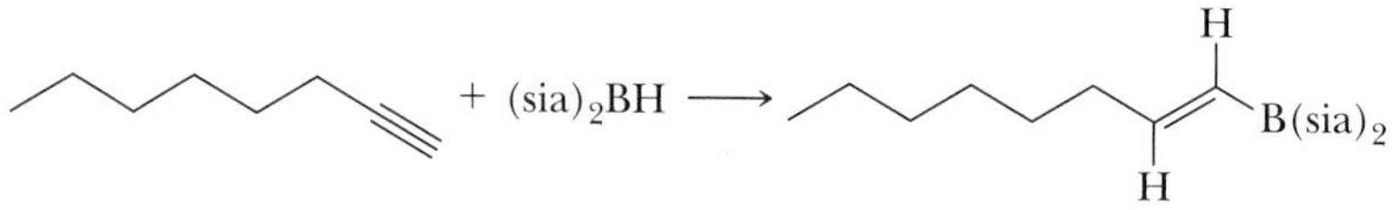

1-Octyne An alkenylborane

As with hydroboration of unsymmetrical alkenes, the addition of $(sia)_2BH$ to a carbon-carbon triple bond of a terminal alkene is regioselective; boron adds to the less substituted carbon.

Treatment of an alkenylborane with hydrogen peroxide in aqueous sodium hydroxide gives a product that corresponds to hydration of an alkyne; that is, it corresponds to addition of H to one carbon of the triple bond and OH to the other as illustrated by the hydroboration-oxidation of 2-butyne.

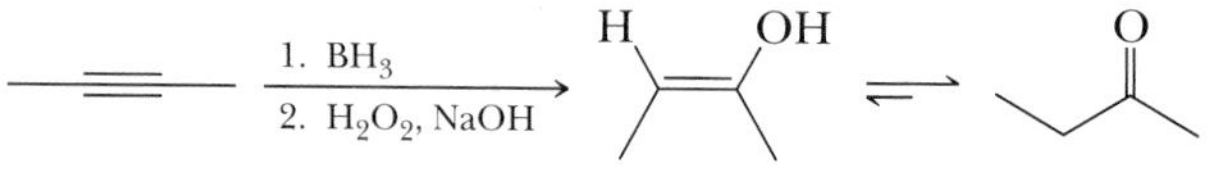

$K_{eq} = 6.7 \times 10^6$

2-Butyne **2-Buten-2-ol** (an enol) **2-Butanone** (a ketone) (for keto-enol tautomerism)

Watch a video explanation

Enol

A compound containing a hydroxyl group bonded to a doubly bonded carbon atom.

The initial product of hydroboration-oxidation of an alkyne is an **enol**, a compound containing a hydroxyl group bonded to a carbon of a carbon-carbon double bond. The name *enol* is derived from the fact that it is both an alkene (*-en-*) and an alcohol (*-ol*). To this point, hydroboration-oxidation of alkynes is identical to that of alkenes (Section 6.4).

Enols are in equilibrium with a constitutional isomer formed by migration of a hydrogen atom from oxygen to carbon and rearrangement of the carbon-carbon double bond to form a carbon-oxygen double bond. As can be seen from the value of K_{eq}, 2-butanone (the keto form) is much more stable than its enol. Keto forms in general are more stable than enol forms because (1) a C═O π bond is generally stronger than a C═C π bond, whereas (2) C—H and O—H σ bonds generally have similar bond strengths.

The keto and enol forms of 2-butanone are said to be tautomers. **Tautomers** are constitutional isomers that are in equilibrium with each other and differ only in the location of a hydrogen atom or another atom and a double bond relative to a heteroatom, most commonly O, N, or S. This type of isomerism is called tautomerism. Because the type of tautomerism we are dealing with in this section involves keto (from ketone) and enol forms, it is commonly called **keto-enol tautomerism**. We discuss keto-enol tautomerism in more detail in Section 16.9.

Tautomers

Constitutional isomers in equilibrium with each other that differ in the location of a hydrogen atom and a double bond relative to a heteroatom, most commonly O, N, or S.

Keto-enol tautomerism

A type of isomerism involving keto (from ketone) and enol tautomers.

Hydroboration of a terminal alkyne using di-*sec*-isoamylborane $(sia)_2BH$ followed by oxidation in alkaline hydrogen peroxide also gives an enol that, in this case, is in equilibrium with the more stable aldehyde. Thus, hydroboration-oxidation of a terminal alkyne gives an aldehyde.

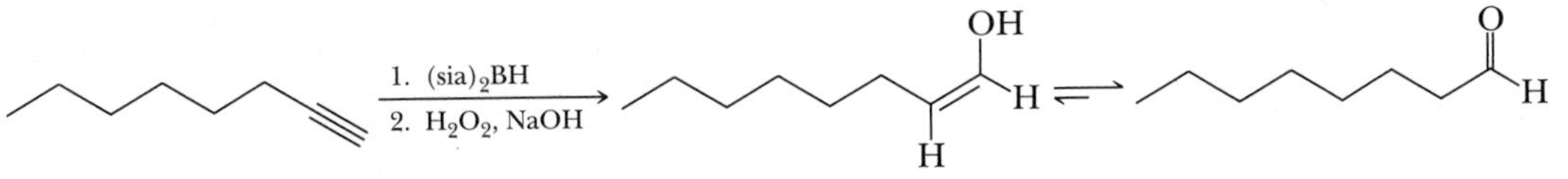
1. $(sia)_2BH$
2. H_2O_2, NaOH
OH
H
H
O
H

1-Octyne An enol **Octanal**

Example 7.4 Hydroboration-Oxidation of Alkynes

Hydroboration-oxidation of 2-pentyne gives a mixture of two ketones, each with the molecular formula $C_5H_{10}O$. Propose structural formulas for these two ketones and for the enol from which each is derived.

Solution

Because each carbon of the triple bond in 2-pentyne has the same degree of substitution, very little regioselectivity occurs during hydroboration. Two enols are formed, and the isomeric ketones are formed from them.

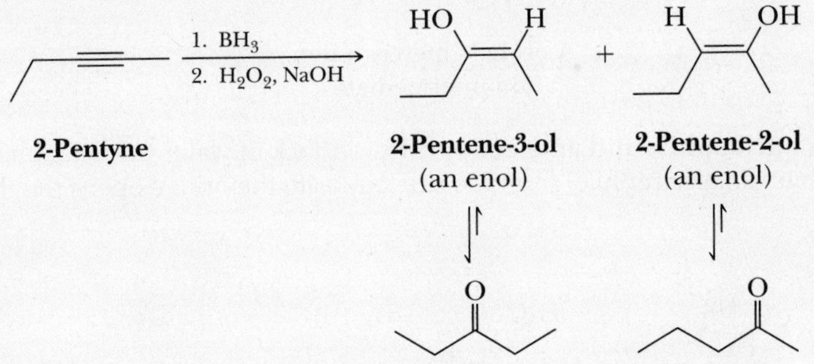
1. BH_3
2. H_2O_2, NaOH
HO
H
H
OH
O
O

2-Pentyne **2-Pentene-3-ol** (an enol) **2-Pentene-2-ol** (an enol)

3-Pentanone **2-Pentanone**

Problem 7.4

Draw a structural formula for a hydrocarbon with the given molecular formula that undergoes hydroboration-oxidation to give the indicated product.

(Continued)

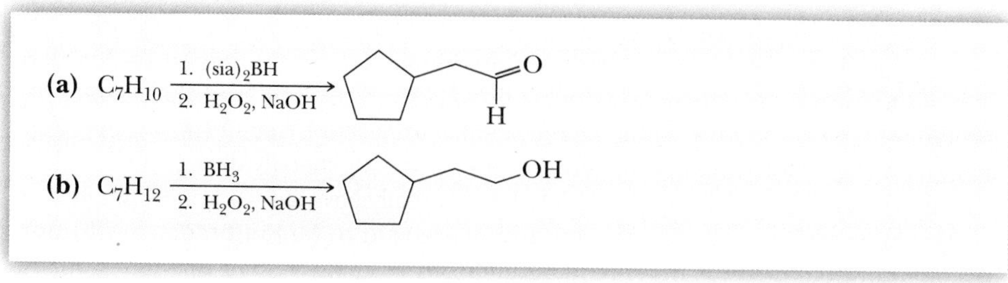

B. Acid-Catalyzed Hydration

In the presence of concentrated sulfuric acid and Hg(II) salts as catalysts, alkynes undergo the addition of water in a reaction analogous to the oxymercuration of alkenes (Section 6.3F). The Hg(II) salts most often used for this purpose are HgO, $HgSO_4$, or $Hg(OAc)_2$. For terminal alkynes, addition of water follows Markovnikov's rule; hydrogen adds to the carbon atom of the triple bond bearing the hydrogen. The resulting enol is in equilibrium with the more stable keto form, so the product isolated is a ketone (an aldehyde in the case of acetylene itself).

$$CH_3C{\equiv}CH + H_2O \xrightarrow[HgSO_4]{H_2SO_4} CH_3\overset{OH}{\overset{|}{C}}{=}CH_2 \rightleftharpoons CH_3\overset{O}{\overset{\|}{C}}CH_3$$

Propyne — **Propen-2-ol** (an enol) — **Propanone (Acetone)**

The mechanism of this reaction is illustrated by the hydration of propyne to give propanone (acetone).

Mechanism 7.2

$HgSO_4/H_2SO_4$ Catalyzed Hydration of an Alkyne

Step 1: **Make a new bond between a nucleophile (π bond) and an electrophile.** Attack of the C—C triple bond on the Hg^{2+} (an electrophile) gives a bridged mercurinium ion intermediate.

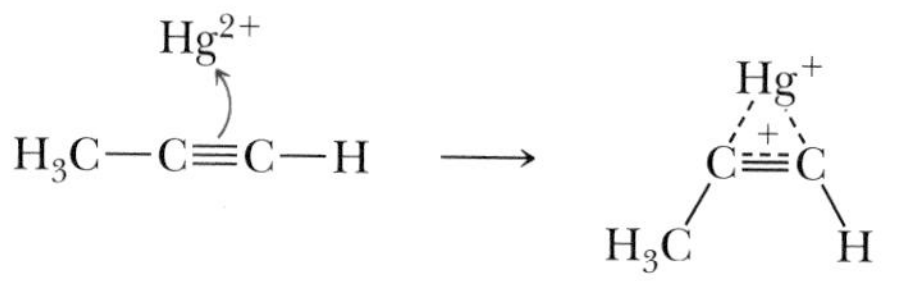

A bridged mercurinium ion intermediate

Step 2: **Make a new bond between a nucleophile and an electrophile.** Attack of water (a nucleophile) on the bridged mercurinium ion intermediate (an electrophile) from the side opposite the bridge opens the three-membered ring.

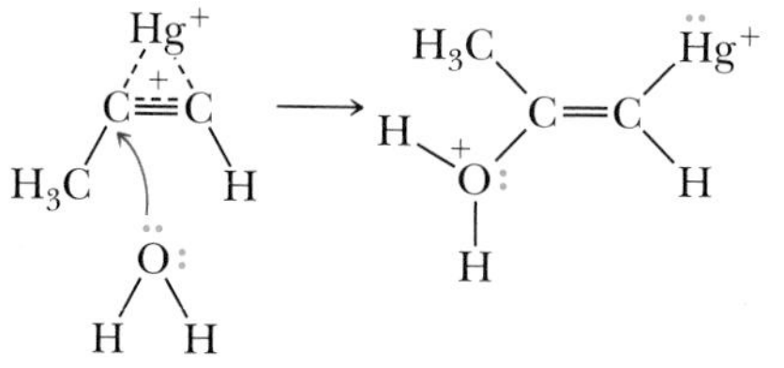

Because the 2° vinylic cation structure makes a greater contribution to the hybrid than does the 1° vinylic cation structure, attack of water occurs preferentially at the more substituted carbon, which accounts for the observed regioselectivity of the reaction.

Step 3: **Take a proton away.** Proton transfer to solvent gives an organomercury enol.

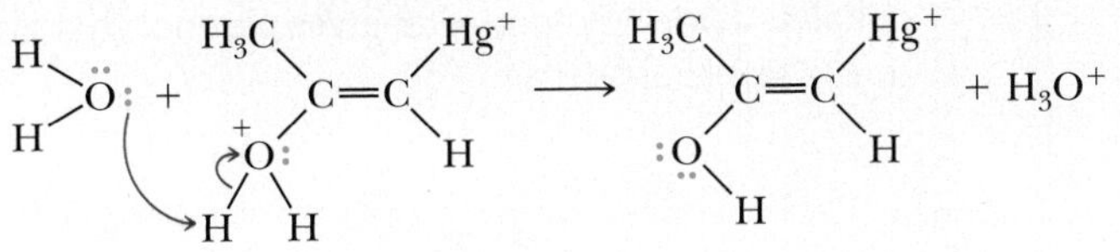

Step 4: **Keto-enol tautomerism.** Tautomerism of the enol gives the keto form.

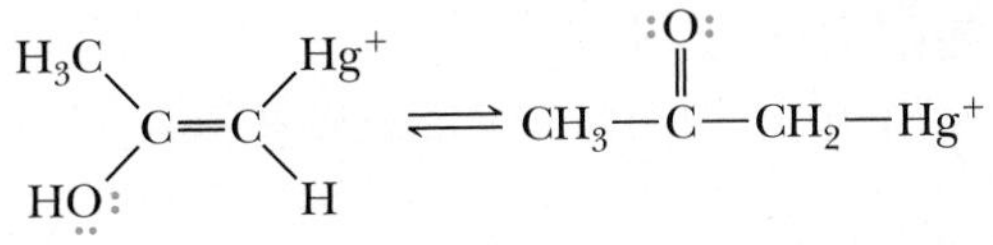

Enol form Keto form

Step 5: **Add a proton.** Proton transfer to the carbonyl group of the ketone gives an oxonium ion.

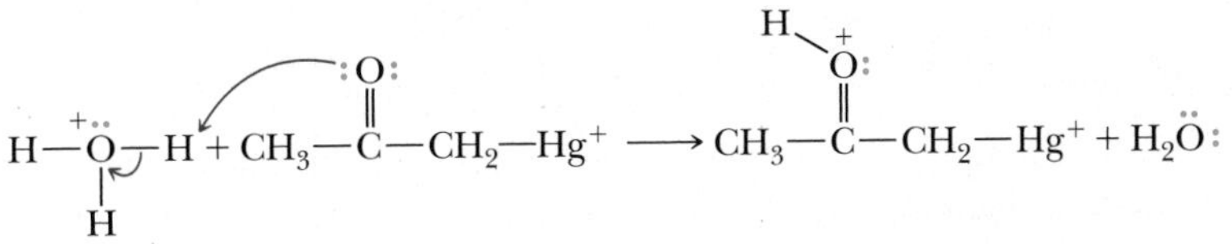

Steps 6 and 7: **Break a bond to give stable molecules or ions followed by keto-enol tautomerism.** Loss of Hg^{2+} from the oxonium ion gives the enol form of the final product. Tautomerism of the enol gives the ketone.

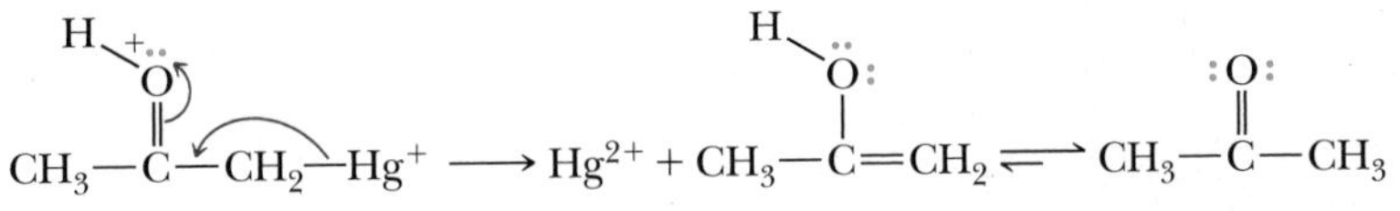

(Enol form) (Keto form)

Example 7.5 Reactions of Alkynes

Show reagents to bring about the following conversions.

(a) 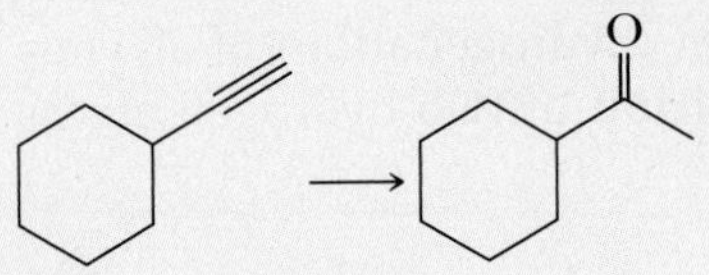 (b) 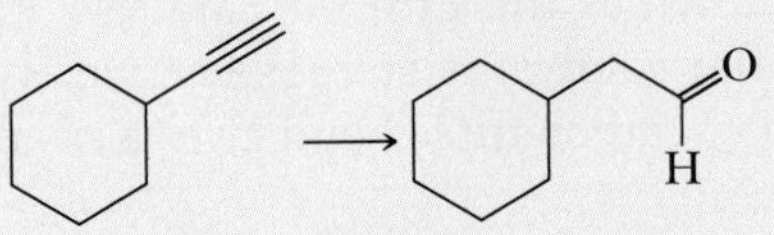

Solution

(a) Hydration of this monosubstituted alkyne using a mercuric ion catalyst gives an enol that is in equilibrium with the more stable keto form.

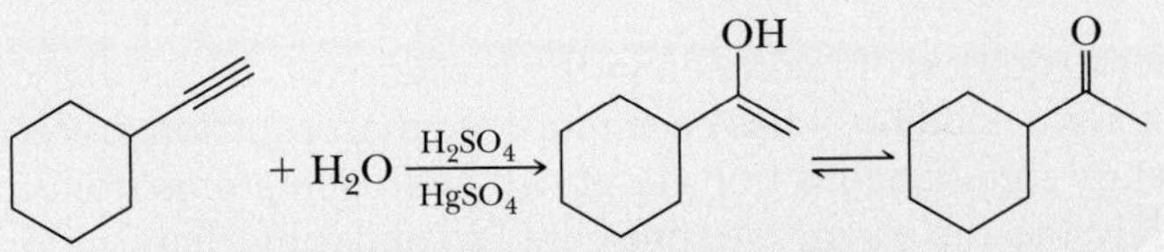

(Continued)

(b) Hydroboration using di-*sec*-isoamylborane followed by treatment with alkaline hydrogen peroxide gives an enol that is in equilibrium with the more stable aldehyde.

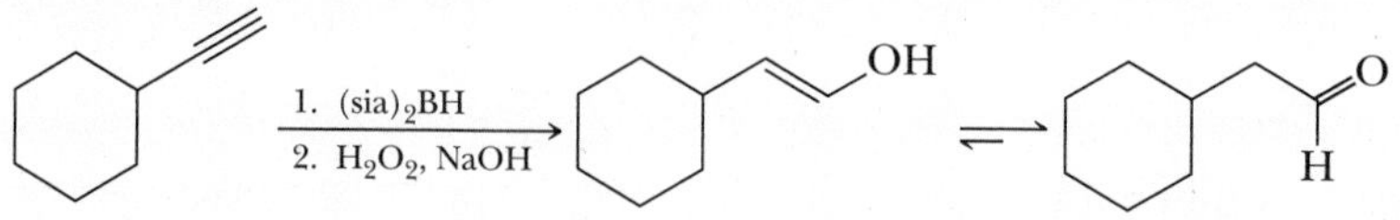

Problem 7.5

Hydration of 2-pentyne gives a mixture of two ketones, each with the molecular formula $C_5H_{10}O$. Propose structural formulas for these two ketones and for the enol from which each is derived.

7.8 Reduction of Alkynes

Three types of reactions are used to convert alkynes to alkenes and alkanes: catalytic reduction, hydroboration-protonolysis, and dissolving-metal reduction.

A. Catalytic Reduction

Treatment of an alkyne with H_2 in the presence of a transition metal catalyst, most commonly palladium, platinum, or nickel, results in the addition of two moles of H_2 to the alkyne and its conversion to an alkane. Catalytic reduction of an alkyne can be brought about at or slightly above room temperature and with moderate pressures of hydrogen gas.

$$\underset{\textbf{2-Butyne}}{CH_3C{\equiv}CCH_3} + 2\,H_2 \xrightarrow[\text{3 atm}]{\text{Pd, Pt, or Ni}} \underset{\textbf{Butane}}{CH_3CH_2CH_2CH_3}$$

Reduction of an alkyne occurs in two stages: (1) addition of one mole of H_2 to form an alkene and (2) addition of the second mole to the alkene to form the alkane. In most cases, it is not possible to stop the reaction at the alkene stage.

However, by careful choice of catalyst, it is possible to stop the reduction after the addition of one mole of hydrogen. The catalyst most commonly used for this purpose consists of finely powdered palladium metal deposited on solid calcium carbonate that has been specially modified with lead salts. This combination is known as the **Lindlar catalyst**. Reduction (hydrogenation) of alkynes over a Lindlar catalyst is stereoselective; **syn addition** of two hydrogen atoms to the carbon-carbon triple bond gives a *cis*-alkene.

Lindlar catalyst
Finely powdered palladium metal deposited on solid calcium carbonate that has been specially modified with lead salts. Its particular use is as a catalyst for the reduction of an alkyne to a *cis*-alkene.

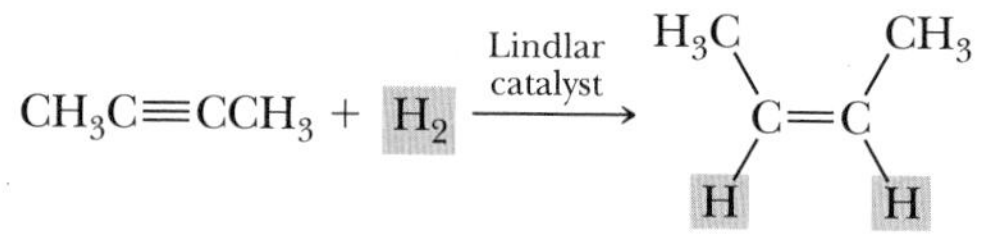

2-Butyne **cis-2-Butene**

Because addition of hydrogen in the presence of the Lindlar catalyst is stereoselective for syn addition, it has been proposed that reduction proceeds by simultaneous or nearly simultaneous transfer of two hydrogen atoms from the surface of the metal catalyst to the alkyne. We presented a similar mechanism in Section 6.6A for the catalytic reduction of an alkene to an alkane.

B. Hydroboration-Protonolysis

As we just saw in Section 7.7A, internal alkynes react with borane to give a trialkenylborane. Treating a trialkenylborane with a carboxylic acid, such as acetic acid, results in stereoselective replacement of boron by hydrogen: a *cis*-alkenyl group bonded to a boron is converted to a *cis*-alkene.

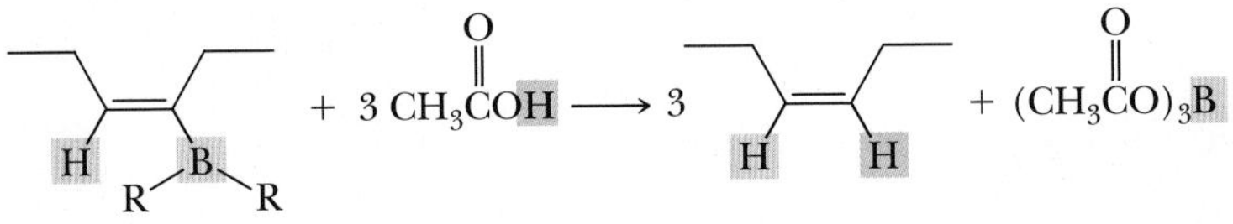

A trialkenylborane **cis-3-Hexene**

The net effect of hydroboration of an internal alkyne followed by treatment with acetic acid is reduction of the alkyne to a *cis*-alkene. Thus, hydroboration-protonolysis and catalytic reduction over a Lindlar catalyst provide alternative schemes for conversion of an alkyne to a *cis*-alkene.

C. Dissolving-Metal Reduction

Alkynes can also be reduced to alkenes by using either sodium or lithium metal in liquid ammonia or in low-molecular-weight primary or secondary amines. The alkali metal is the reducing agent and, in the process, is oxidized to M^+, which dissolves as a metal salt in the solvent for the reaction. Reduction of an alkyne to an alkene by lithium or sodium in liquid ammonia, $NH_3(l)$, is stereoselective; it involves mainly **anti addition** of two hydrogen atoms to the triple bond.

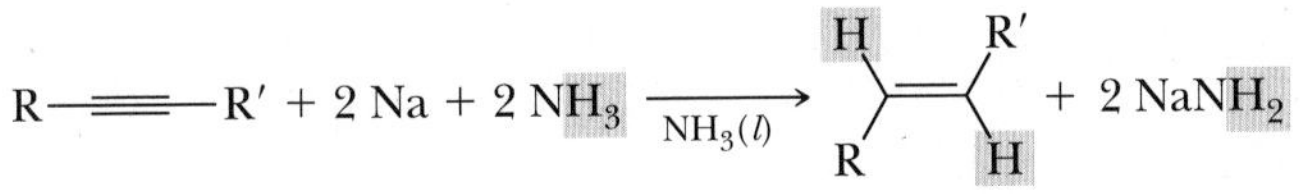

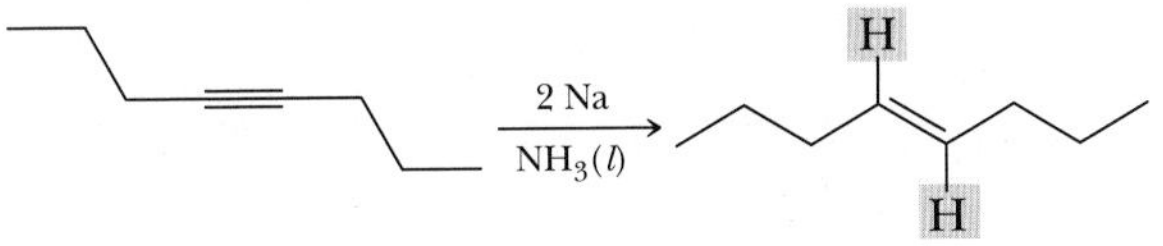

4-Octyne ***trans*-4-Octene**

Thus, by the proper choice of reagents and reaction conditions, it is possible to reduce an alkyne to either a *cis*-alkene (by catalytic reduction or hydroboration-protonolysis) or to a *trans*-alkene (by dissolving-metal reduction).

The stereoselectivity of alkali metal reduction of alkynes to alkenes can be accounted for by the following mechanism. As you study this mechanism, note that it involves two one-electron reductions and two proton-transfer reactions. The stereochemistry of the alkene is determined in Step 3. Adding the four steps and canceling species that appear on both sides of the equation gives the overall equation for the reaction.

Mechanism 7.3

Reduction of an Alkyne by Sodium in Liquid Ammonia

Step 1: A one-electron reduction of the alkyne gives an alkenyl radical anion [i.e., an ion containing an unpaired electron on one carbon and a negative charge on an adjacent carbon (note that we use a single-headed arrow to show the repositioning of single electrons)].

(Continued)

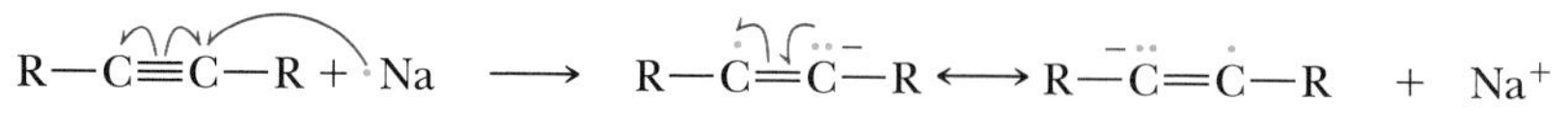

A resonance-stabilized alkenyl radical anion

Step 2: **Add a proton.** The alkenyl radical anion (a very strong base) abstracts a proton from a molecule of ammonia (under these conditions, a weak acid) to give an alkenyl radical.

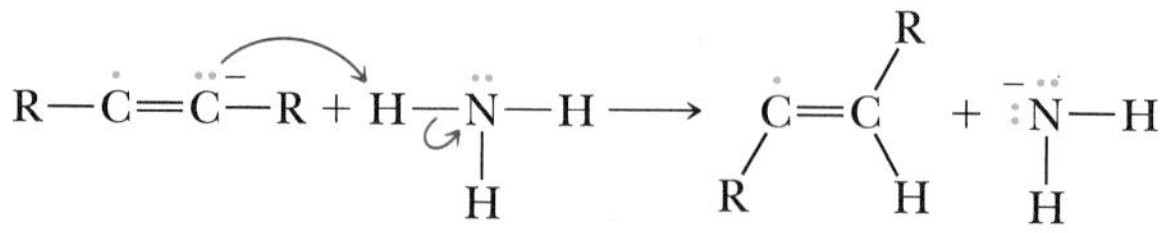

An alkenyl radical **Amide anion**

Step 3: A one-electron reduction of the alkenyl radical gives an alkenyl anion. The *trans*-alkenyl anion is more stable than its *cis* isomer, and the stereochemistry of the final product is determined in this step.

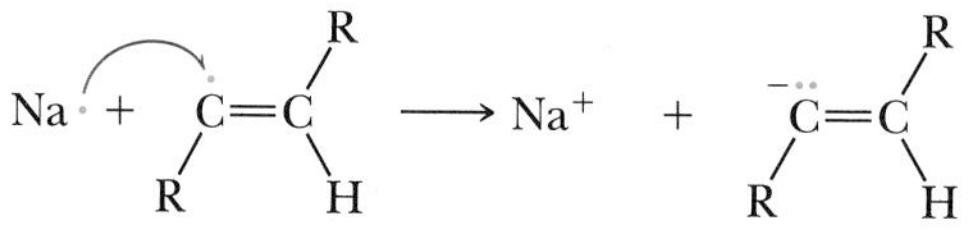

An alkenyl anion

Step 4: **Add a proton.** A second proton-transfer reaction completes the reduction and gives the *trans*-alkene.

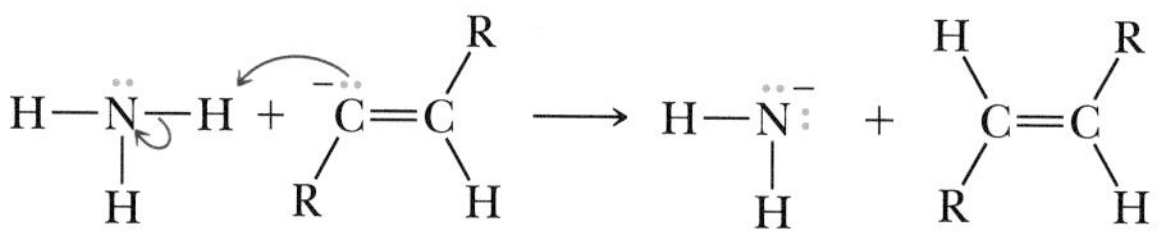

Amide anion A *trans*-alkene

7.9 Organic Synthesis

We have now seen how to prepare both terminal and internal alkynes from acetylene and substituted acetylenes, and we have seen several common reactions of alkynes, including addition (HX, X_2, and H_2O), hydroboration-oxidation, and reduction. Now let us move a step farther to consider what might be called the art of **organic synthesis**.

Organic synthesis
A series of reactions by which a set of organic starting materials is converted to a more complicated structure.

A. Retrosynthetic Analysis

Synthesis is generally the most important objective of organic chemists, applicable to the preparation of compounds for use as pharmaceuticals, agrochemicals, plastics, elastomers, and textile fibers. A successful synthesis must provide the desired product in a maximum yield with a maximum control of stereochemistry at all stages of the synthesis. Furthermore, there is an increasing desire to develop "green" syntheses (i.e., syntheses that do not produce or release byproducts harmful to the environment).

Our goal in this section is to develop an ability to plan a successful synthesis. The best strategy is to work backward from the desired product. First, we analyze the target molecule in the following way.

1. Count the carbon atoms of the carbon skeleton of the target molecule. Determining how to build the carbon skeleton from available starting materials is

often the most challenging part of a synthesis. If you must add carbons, you need to consider what carbon-carbon bond-forming reactions are available to you. At this stage in the course, you have only one such reaction, namely alkylation of acetylide anions with methyl or primary halides (Section 7.5).

2. Analyze the functional groups. What are they, and how can they be changed to facilitate formation of the carbon skeleton? How can they then be changed to give the final set of functional groups in the desired product? Regiospecificity and stereospecificity can be important considerations.
3. Now work backward; surprisingly, it is often easier to identify the last step in a synthesis than trying to start from the beginning. This process is referred to as **retrosynthetic analysis**. The idea is to start with the product structure and work backward, one step at a time until you reach a point at which you can create your synthetic intermediate from the starting materials. If you cannot do this using your first proposed route, try working backward using a different sequence of reactions. Synthesis problems are creative in the sense that they force you to create entirely new ways to connect the reactions you have been taught. To be good at synthesis, you must be very familiar with the reactions, including how to carry out those reactions on new molecules. The best way to keep track of the reactions you are learning is to use an organic chemistry reaction roadmap. You should work all of the end-of-chapter problems under the "Organic Chemistry Reaction Roadmap" heading and become familiar with Appendix 10.

Retrosynthesis
A process of reasoning backward from a target molecule to a suitable set of starting materials.

Reaction Roadmap

Target Molecule: *cis*-3-Hexene

As readily available starting materials, we use acetylene and haloalkanes.

Target molecule: 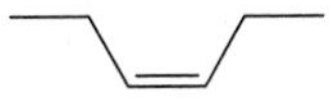

***cis*-3-Hexene**

Analysis We note that there are six carbons in the product and only two in acetylene. We will need to construct the carbon skeleton through carbon-carbon bond formation with haloalkanes totaling four additional carbon atoms. The functional group in the product is a *cis* carbon-carbon double bond, which can be prepared by catalytic reduction of a carbon-carbon triple bond using the Lindlar catalyst (Section 7.8A). We then disconnect the carbon skeleton into possible starting materials, which we can later reconnect by known reactions. In the example here, we disconnect at the two carbon-carbon single bonds adjacent to the triple bond. These bonds can be formed during the synthesis by alkylation of the acetylide dianion using two haloalkanes (Section 7.5A), each with two carbon atoms (e.g., bromoethane). We use an open arrow to symbolize a step in a retrosynthesis.

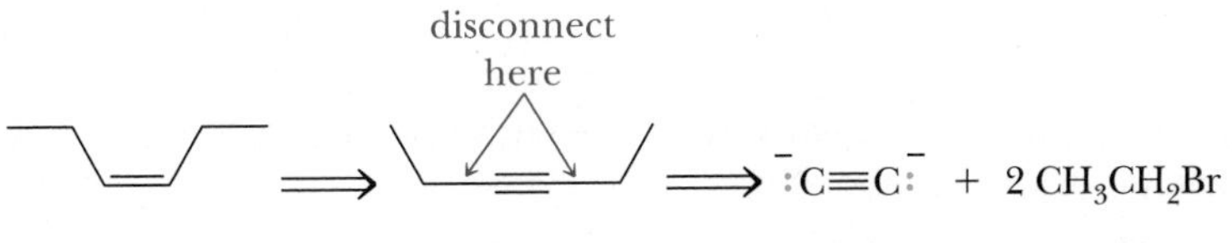
disconnect here
:C≡C:
+ 2 CH3CH2Br

***cis*-3-Hexene** **3-Hexyne** **Acetylide dianion** **Bromoethane**

Synthesis Our starting materials for this synthesis of *cis*-3-hexene are acetylene and bromoethane, both readily available compounds. This synthesis is carried out in five steps as follows.

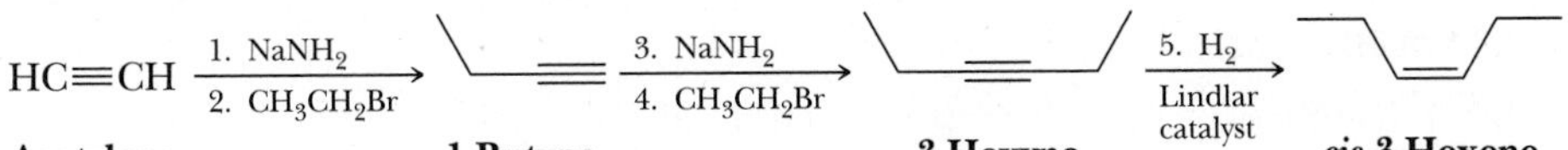
HC≡CH
1. NaNH2
2. CH3CH2Br
3. NaNH2
4. CH3CH2Br
5. H2
Lindlar catalyst

Acetylene **1-Butyne** **3-Hexyne** ***cis*-3-Hexene**

Target Molecule: 2-Heptanone

2-Heptanone is responsible for the "peppery" odor of cheeses of the Roquefort type. As readily available starting materials, we again use acetylene and haloalkanes.

Watch a video explanation

Target molecule: 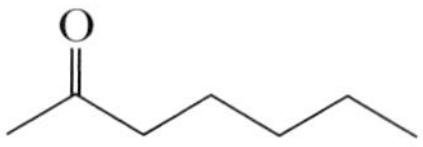

2-Heptanone

Analysis We note that there are seven carbons in the product and only two in acetylene. We will need to construct the carbon skeleton through carbon-carbon bond formation with haloalkanes totaling five carbon atoms. The functional group in the target molecule is a ketone, which we can prepare by hydration of a carbon-carbon triple bond. Hydration of 1-heptyne gives only 2-heptanone, whereas hydration of 2-heptyne gives a mixture of 2-heptanone and 3-heptanone. Therefore, we choose a functional group interconversion via 1-heptyne.

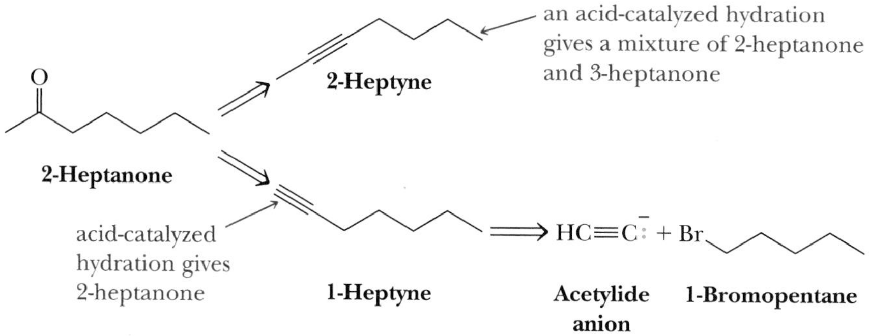
an acid-catalyzed hydration gives a mixture of 2-heptanone and 3-heptanone
2-Heptyne
2-Heptanone
acid-catalyzed hydration gives 2-heptanone
1-Heptyne
HC≡C:⁻ + Br
Acetylide anion
1-Bromopentane

Synthesis This synthesis can be carried out in three steps as follows.

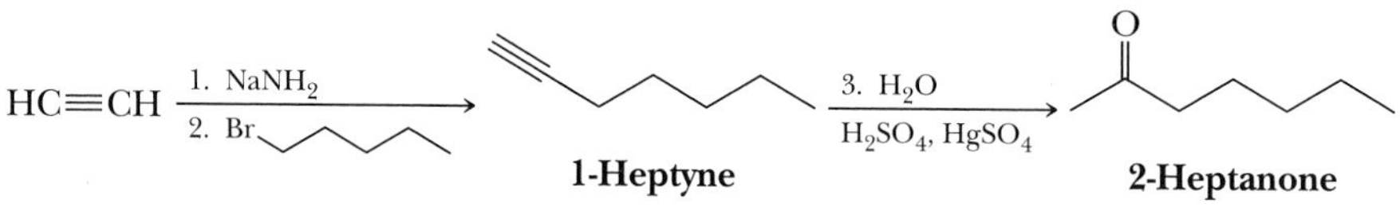
HC≡CH
1. NaNH2
2. Br
1-Heptyne
3. H2O
H2SO4, HgSO4
2-Heptanone

Example 7.6 Synthesis Using Alkynes

How might the scheme for the synthesis of 2-heptanone be modified so that the product is heptanal?

Solution

Steps 1 and 2 are the same and give 1-heptyne. Instead of acid-catalyzed hydration of 1-heptyne, treat the alkyne with $(sia)_2BH$ followed by alkaline hydrogen peroxide (Section 7.7).

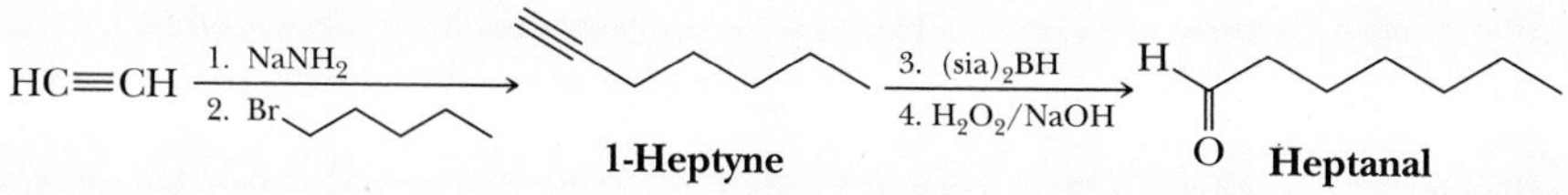
HC≡CH
1. NaNH2
2. Br
1-Heptyne
3. (sia)2BH
4. H2O2/NaOH
H
O
Heptanal

Problem 7.6

Show how the synthetic scheme in Example 7.6 might be modified to give the following.

(a) 1-Heptanol **(b)** 2-Heptanol

B. Reactions of Functional Groups in Complex Molecules

A defining feature of organic chemistry is that a functional group often reacts in the same fashion whether it is in a relatively simple molecule or in a more complex molecule that contains numerous other functional groups. As a result, the synthesis of increasingly complex molecules can be carried out in a rational manner, provided the reactivities of all the functional groups present are taken into account.

Fexofenadine, sold under the trade name Allegra, is a powerful nonsedating antihistamine. An efficient synthesis of fexofenadine published by a team from McGill University in 1994 involved using **HgO** in the conversion of an alkyne to a ketone in acidic solution (Section 7.7B). The ketone function ended up on the carbon adjacent to the benzene ring because positive charge in the intermediate is stabilized at this position by resonance delocalization. This ketone product is converted to fexofenadine through two additional reactions we will not discuss here. Although the McGill synthesis is efficient, the use of toxic **HgO** makes this overall synthesis inappropriate for commercial use, and an alternative approach is used when the product is prepared for sale. Fexofenadine has one chiral center and is sold as a racemic mixture.

▶ Functional groups react the same in different molecules.

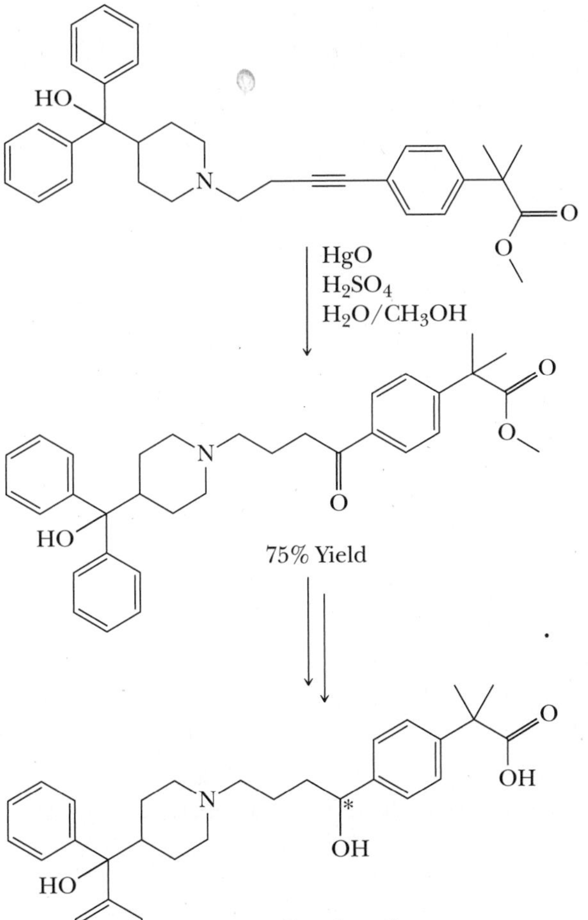
HO
N
O
O
HgO
H_2SO_4
H_2O/CH_3OH
N
O
O
HO
O
75% Yield
N
*
O
OH
HO
OH

Fexofenadine
(racemic)

This example illustrates two important points about how functional group analysis is used to develop systematic syntheses of complex molecules. First, the alkyne was the only functional group to react under the conditions used, and it reacted as expected based on comparison to simpler alkyne molecules such as those discussed in this chapter. You will learn about the reactivities of many functional groups in this course, and in each case, they can be used to predict what happens in complex molecules that contain them. Second, although they do not react directly, other functional groups often have an influence over the regiochemistry or stereochemistry of a reaction. The benzene ring in the present example did not react directly with the HgO reagent but did influence the regiochemistry of the product.

Study Guide

7.1 Structure of Alkynes

- **Alkynes** contain one or more carbon-carbon triple bonds.
  - *The triple bond is a combination of one σ bond formed by the overlap of sp hybrid orbitals and two π bonds formed by the overlap of two sets of parallel 2p orbitals.*

7.2 Nomenclature of Alkynes

- According to IUPAC nomenclature, the infix *-yn-* is used to show the presence of a carbon-carbon triple bond.
  - *We name the main chain as the longest one that contains the triple bond and assign numbers from the end that gives the carbon atoms of the triple bond the lower set of numbers.*
  - *The IUPAC system retains the name acetylene.*
- Common names are derived by adding a prefix for the substituent to the word *acetylene*.

P 7.1, 7.2

7.3 Physical Properties of Alkynes

- The physical properties of alkynes are similar to those of alkanes and alkenes of comparable carbon skeletons.

7.4 Acidity of 1-Alkynes

- The pK_a values of terminal alkynes are approximately 25; they are less acidic than water and alcohols but more acidic than alkanes, alkenes, and ammonia.
  - *The hydrogen atom bonded to a carbon-carbon triple bond is sufficiently acidic so that it can be removed by a strong base, most commonly sodium amide ($NaNH_2$), sodium hydride (NaH), or lithium diisopropylamide (LDA).*

P 7.10

KEY REACTIONS

1. **Acidity of Terminal Alkynes (7.4)** Treatment of terminal alkynes (pK_a 25) with a strong base [most commonly $NaNH_2$, NaH, or lithium diisopropylamide (LDA)] gives an acetylide salt.

$$HC{\equiv}CH + Na^{+}NH_2^{-} \longrightarrow HC{\equiv}C^{-}Na^{+} + NH_3$$

7.5 Preparation of Alkynes

- Alkynes can be prepared through alkylation of acetylide anions with methyl or primary halides.
  - *This reaction is important because it creates a carbon-carbon bond. Starting with acetylene, even unsymmetrical alkynes can be made using two sequential deprotonation–**alkylation reactions**.*
- Alkynes can be made from alkenes through a two-step sequence of halogenation with X_2 followed by double dehydrohalogenation.
  - ***Allenes** are sometimes seen as a byproduct of the double dehydrohalogenation.*

P **7.3, 7.8, 7.9, 7.15, 7.19, 7.22, 7.26**

Problem 7.3

KEY REACTIONS

2. Alkylation of Acetylide Anions (Section 7.5A) Acetylide anions are nucleophiles and will displace halide ion from methyl and 1° haloalkanes.

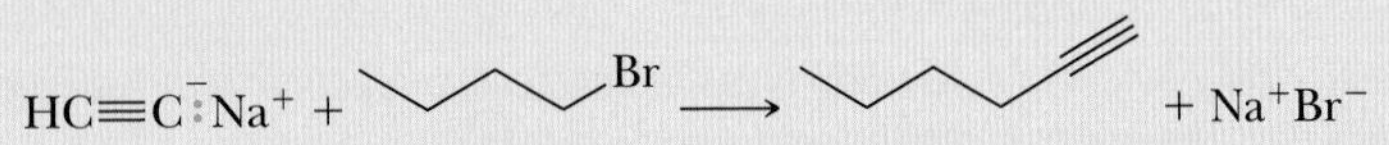

3. Synthesis of an Alkyne from an Alkene (Section 7.5B) Treating an alkene with Br_2 or Cl_2 gives a dihaloalkane. Treating the dihaloalkane with $NaNH_2$ or another strong base results in two successive dehydrohalogenations to give an alkyne.

$$\mathrm{RCH{=}CHR} \xrightarrow{Br_2} \mathrm{RCH(Br){-}CH(Br)R} \xrightarrow[NH_3(l)]{2\,NaNH_2} \mathrm{RC{\equiv}CR}$$

7.6 Electrophilic Addition to Alkynes

- Alkynes add Br_2 to give dibromoalkenes via anti addition stereochemistry.
  - *A second mole of Br_2 can be added to give a tetrabromoalkane.*
  - *Addition of Cl_2 occurs with less stereoselectivity than the addition of Br_2.*
- Alkynes add two moles of HCl or HBr, with both halogens adding to the same carbon.
  - *For terminal alkynes, Markovnikov's rule is followed and the hydrogen ends up on the terminal carbon atom.*

P **7.11, 7.16**

KEY REACTIONS

4. Addition of Br_2 and Cl_2 (Section 7.6A) Addition of one mole of Br_2 or Cl_2 is anti stereoselective; anti addition of halogen to an alkyne gives an (*E*)-dihaloalkene. Addition of a second mole of halogen gives a tetrahaloalkane.

(Continued)

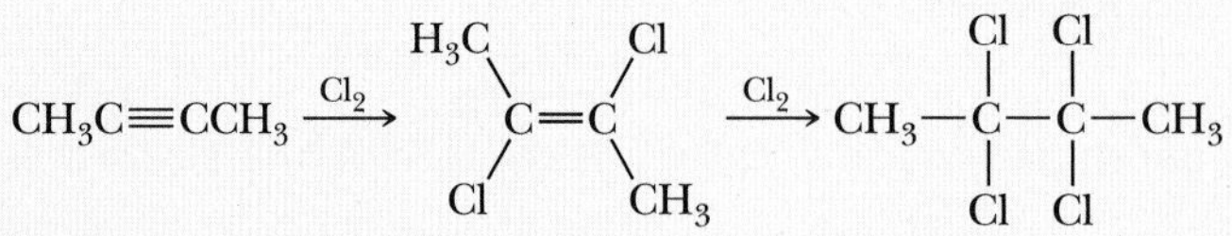

5. **Addition of HX (Section 7.6B)** Addition of HX is regioselective. Reaction by way of a vinylic carbocation intermediate follows Markovnikov's rule. Addition of 2 HX gives a geminal dihaloalkane.

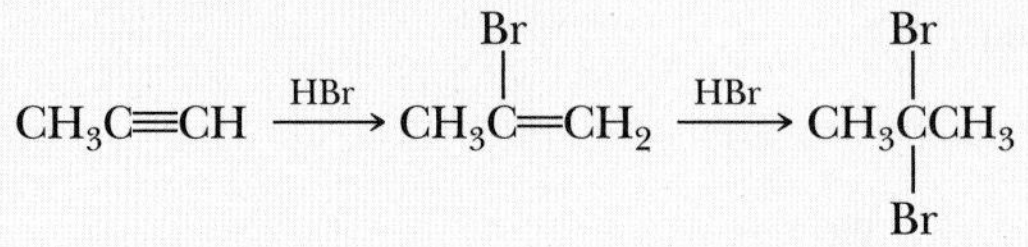

7.7 Hydration of Alkynes to Aldehydes and Ketones

- Alkynes react with borane [BH_3 or $(sia)_2BH$] followed by basic peroxide to give aldehydes or ketones.
  - *The sterically hindered $(sia)_2BH$ gives regioselective reaction with 1-alkynes leading to aldehydes. The reaction initially produces an* ***enol****.*
- The functional group of an enol is an —OH group on a carbon atom of an alkene.
  - *The enol form is in equilibrium with the keto form, namely an aldehyde or ketone, and the equilibrium almost always lies far on the side of the keto form.*
  - ***Tautomers*** *are constitutional isomers that are in equilibrium with each other but differ in the location of a hydrogen and a double bond relative to a heteroatom, most commonly O, N, and S.*
  - ***Keto-enol tautomerism*** *is the most common type of tautomerism we encounter in this course.*
- Alkynes react with water in the presence of H_2SO_4 and Hg(II) salts to give ketones (or acetaldehyde in the case of acetylene) via an enol intermediate.

Hydroboration-oxidation of 2-butyne

KEY REACTIONS

6. **Keto-Enol Tautomerism (Section 7.7A)** In an equilibrium between a keto form and an enol form, the keto form generally predominates.

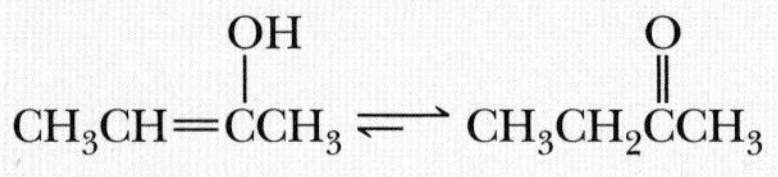

Enol form Keto form

7. **Hydroboration-Oxidation (Section 7.7A)** Hydroboration of an internal alkyne is syn stereoselective. Oxidation of the resulting trialkenylborane by H_2O_2/NaOH gives an enol that is in equilibrium, through keto-enol tautomerism, with a ketone.

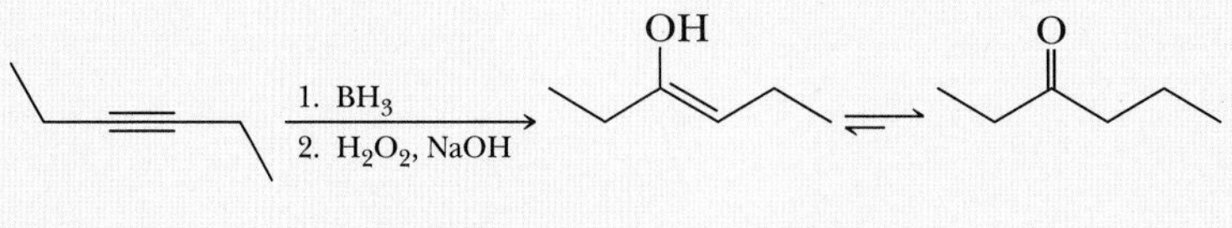

Hydroboration of a terminal alkyne using a hindered dialkylborane followed by oxidation of the resulting trialkenylborane with $H_2O_2/NaOH$ and then keto-enol tautomerism gives an aldehyde.

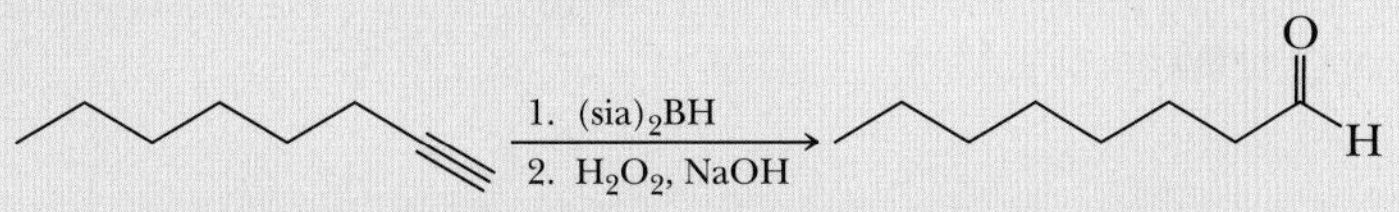
1. (sia)$_2$BH
2. H_2O_2, NaOH
O
H

8. **Acid-Catalyzed Hydration (Section 7.7B)** Acid-catalyzed addition of water in the presence of Hg(II) salts is regioselective. Keto-enol tautomerism of the resulting enol gives a ketone.

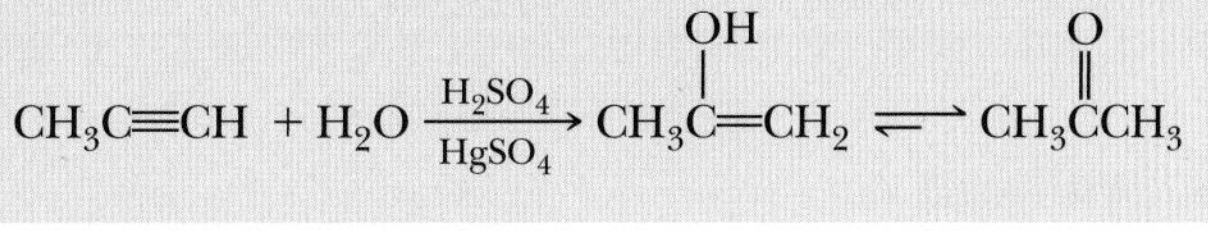
OH
O
$CH_3C{\equiv}CH + H_2O$
H_2SO_4
$HgSO_4$
$CH_3C{=}CH_2$
CH_3CCH_3

7.8 Reduction of Alkynes

- Alkynes react with two moles of H_2 in the presence of a transition metal catalyst to give alkanes in a reaction that does not ordinarily stop at the alkene stage.
- The **Lindlar catalyst** is a deactivated hydrogenation catalyst that stops alkyne hydrogenation at the alkene stage, allowing conversion of alkynes to *cis*-alkenes. Hydroboration followed by an acid workup (instead of basic peroxide) also gives *cis*-alkenes from alkynes. Hydroboration–protonolysis is an alternative way to prepare a *cis*-alkene from an alkyne.
- Dissolving metal reduction gives *trans*-alkenes from alkynes.

P 7.17

KEY REACTIONS

9. **Catalytic Reduction (Section 7.8A)** Reaction of an alkyne with two moles of H_2 under moderate pressure in the presence of a transition metal catalyst at room temperature gives an alkane.

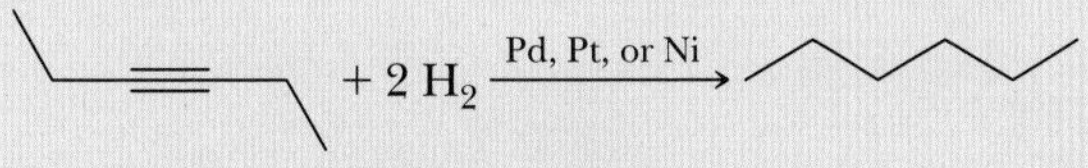
$+ 2 H_2$
Pd, Pt, or Ni

Catalytic reduction of an alkyne in the presence of the Lindlar catalyst is syn stereoselective; Lindlar reduction of an internal alkyne gives a *cis*-alkene.

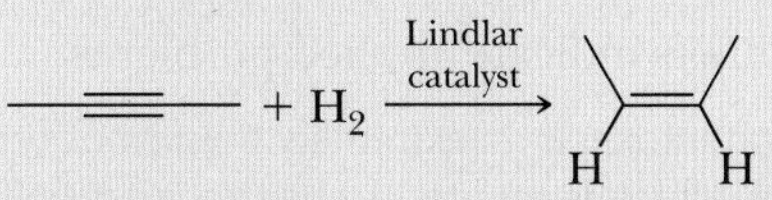
$+ H_2$
Lindlar
catalyst
H
H

10. **Hydroboration-Protonolysis (Section 7.8B)** Hydroboration of an alkyne followed by protonolysis also converts an alkyne to a *cis*-alkene.

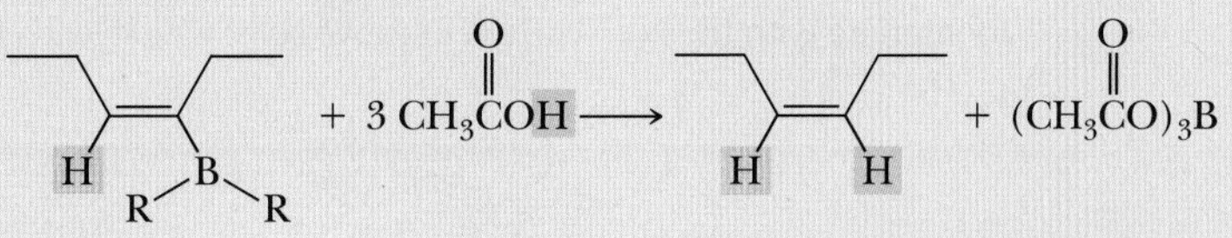
H
B
R
R
O
$+ 3\ CH_3COH$
H
H
O
$+ (CH_3CO)_3B$

(Continued)

11. Reduction Using Na or Li Metal in $NH_3(l)$ (Section 7.8C) Alkali metal reduction is stereoselective: anti addition of hydrogens to an internal alkyne gives a *trans*-alkene. The mechanism involves a radical mechanism with two sequential single-electron transfers from the Na or Li metal, each followed by deprotonation of the amine solvent.

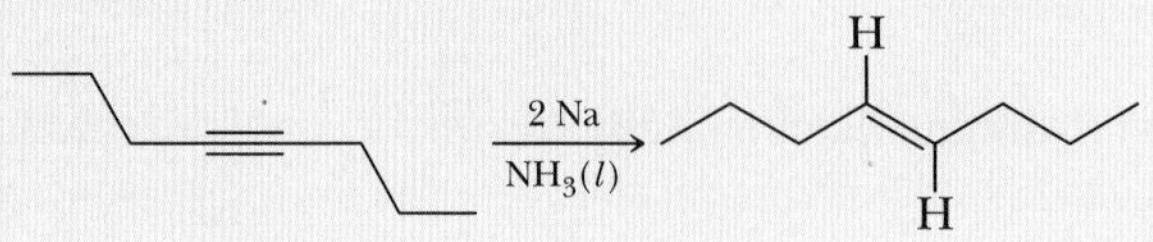

7.9 Organic Synthesis

- When planning an **organic synthesis** involving several reactions, it is best to use the following systematic approach:
  1. *Count the carbons in the products versus the starting material so that you know what fragments must be added or subtracted.*
  2. *Analyze the functional groups.*
  3. *Work backward (**retrosynthesis**).*
- A **functional group** often reacts in the same fashion whether it is in a relatively simple molecule or a more complex molecule that contains numerous other functional groups.
  – *Although they do not react directly, other functional groups often have an influence over the regiochemistry or stereochemistry of a reaction.*

P **7.6, 7.14, 7.20, 7.21–7.34; 7.35**

Organic synthesis of 2-Heptanone

Problems

Red numbers indicate applied problems.

7.7 Enanthotoxin is an extremely poisonous organic compound found in hemlock water dropwart, which is reputed to be the most poisonous plant in England. It is believed that no British plant has been responsible for more fatal accidents. The most poisonous part of the plant is the roots, which resemble small white carrots, giving the plant the name "five finger death." Also poisonous are its leaves, which look like parsley. Enanthotoxin is thought to interfere with the Na^+ current in nerve cells, which leads to convulsions and death.

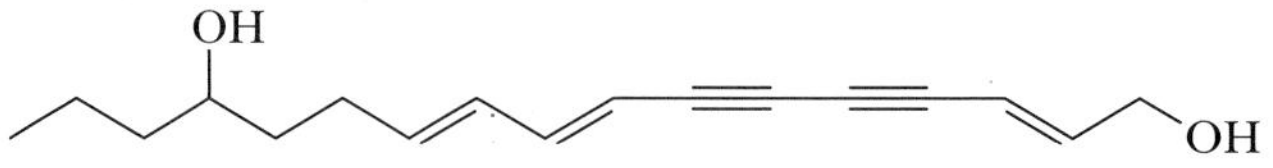

How many stereoisomers are possible for enanthotoxin?

Preparation of Alkynes

7.8 Show how to prepare each alkyne from the given starting material. In part (c), D indicates deuterium. Deuterium-containing reagents such as BD_3, D_2O, and CH_3COOD are available commercially.

(a) 

(b) 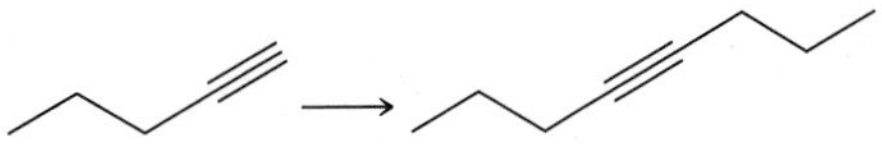

(c) 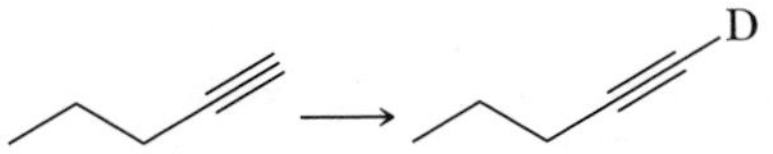

7.9 If a catalyst could be found that would establish an equilibrium between 1,2-butadiene and 2-butyne, what would be the ratio of the more stable isomer to the less stable isomer at 25°C?

$$CH_2{=}C{=}CHCH_3 \rightleftharpoons CH_3C{\equiv}CCH_3 \qquad \Delta G^0 = -16.7 \text{ kJ } (-4.0 \text{ kcal})\text{/mol}$$

Reactions of Alkynes

7.10 Complete each acid-base reaction and predict whether the position of equilibrium lies toward the left or toward the right.

(a) $CH_3C{\equiv}CH + CH_3CH_2O^-Na^+ \underset{CH_3CH_2OH}{\rightleftharpoons}$

(b) $CH_3C{\equiv}CCH_2CH_2OH + Na^+NH_2^- \underset{NH_3(l)}{\rightleftharpoons}$

(c) $CH_3C{\equiv}C^-Na^+ + CH_3\overset{O}{\overset{\|}{C}}OH \rightleftharpoons$

7.11 Draw structural formulas for the major product(s) formed by reaction of 3-hexyne with each of these reagents. (Where you predict no reaction, write NR.)

(a) H_2 (excess)/Pt
(b) H_2/Lindlar catalyst
(c) Na in $NH_3(l)$
(d) BH_3 followed by H_2O_2/NaOH
(e) BH_3 followed by CH_3COOH
(f) BH_3 followed by CH_3COOD
(g) Cl_2 (1 mol)
(h) $NaNH_2$ in $NH_3(l)$
(i) HBr (1 mol)
(j) HBr (2 mol)
(k) H_2O in $H_2SO_4/HgSO_4$

7.12 Draw the structural formula of the enol formed in each alkyne hydration reaction; then draw the structural formula of the carbonyl compound with which each enol is in equilibrium.

(a) $CH_3(CH_2)_5C{\equiv}CH + H_2O \xrightarrow[H_2SO_4]{HgSO_4}$ (an enol) $\longrightarrow$

(b) $CH_3(CH_2)_5C{\equiv}CH \xrightarrow[\text{2. NaOH}/H_2O_2]{\text{1. (sia)}_2BH}$ (an enol) $\longrightarrow$

7.13 Propose a mechanism for this reaction.

$$HC{\equiv}CH + CH_3\overset{O}{\overset{\|}{C}}OH \xrightarrow[HgSO_4]{H_2SO_4} CH_3\overset{O}{\overset{\|}{C}}OCH{=}CH_2$$

Acetylene **Acetic acid** **Vinyl acetate**

Vinyl acetate is the monomer for the production of poly(vinyl acetate), the major use of which is as an adhesive in the construction and packaging industry, but it is also used in the paint and coatings industry.

Syntheses

7.14 Show how to convert 9-octadecynoic acid to the following.

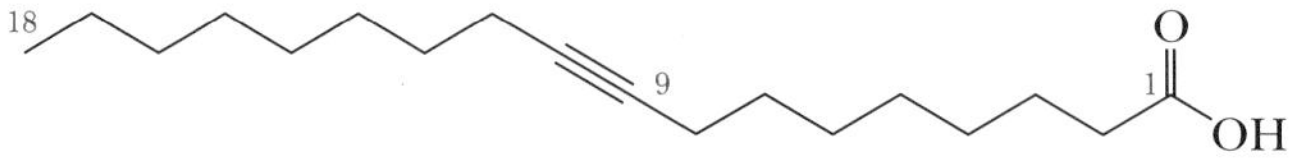

9-Octadecynoic acid

(a) (*E*)-9-Octadecenoic acid (eliadic acid)
(b) (*Z*)-9-Octadecenoic acid (oleic acid)
(c) 9,10-Dihydroxyoctadecanoic acid
(d) Octadecanoic acid (stearic acid)

7.15 For small-scale and consumer welding applications, many hardware stores sell cylinders of MAAP gas, which is a mixture of propyne (methylacetylene) and 1,2-propadiene (allene), with other hydrocarbons. How would you prepare the methylacetylene/allene mixture from propene in the laboratory?

7.16 Show reagents and experimental conditions you might use to convert propyne into each product. (Some of these syntheses can be done in one step; others require two or more steps.)

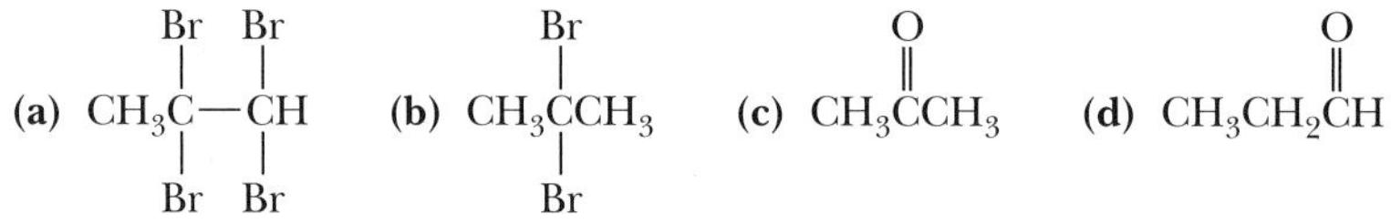

7.17 Show reagents and experimental conditions you might use to convert each starting material into the desired product. (Some of these syntheses can be done in one step; others require two or more steps.)

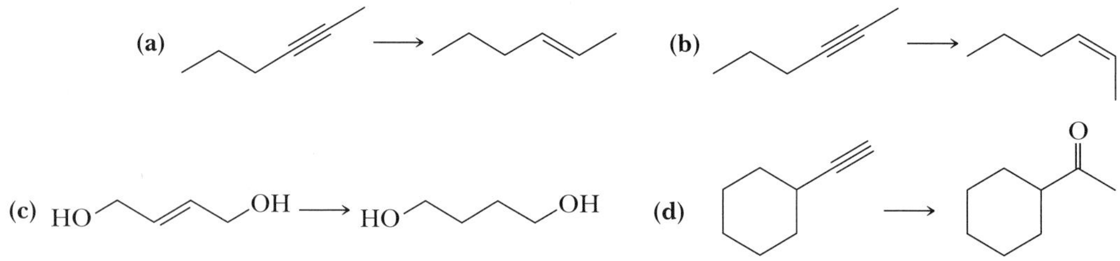

7.18 Show how to convert 1-butyne to each of these compounds.

(a) $CH_3CH_2C{\equiv}C^-Na^+$ **(b)** $CH_3CH_2C{\equiv}CD$

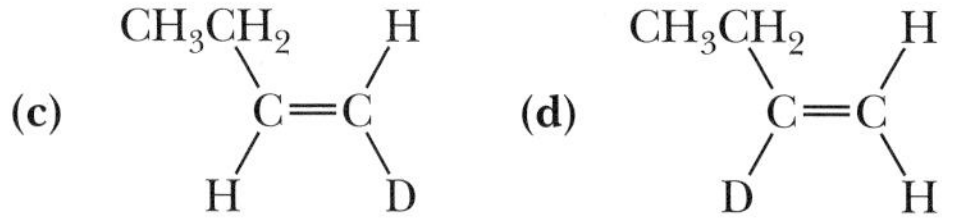

7.19 Rimantadine was among the first antiviral drugs to be licensed in the United States to use against the influenza A virus and to treat established illnesses. It is synthesized from adamantane by the following sequence (we discuss the chemistry of Step 1 in Chapter 8 and the chemistry of Step 5 in Section 16.8A).

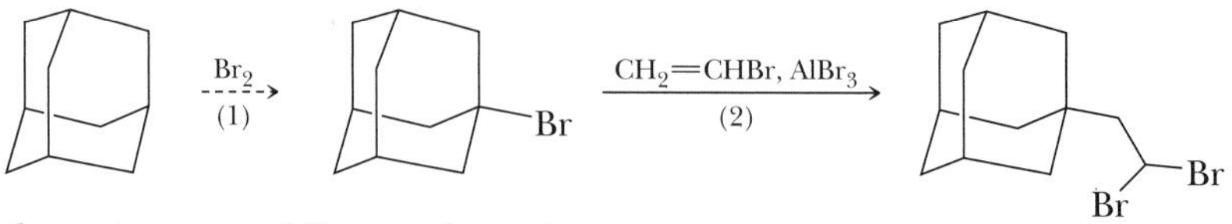

Adamantane **1-Bromoadamantane**

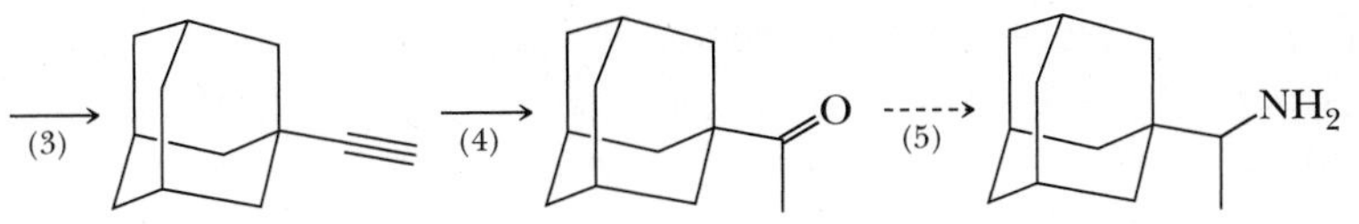

Rimantadine

Rimantadine is thought to exert its antiviral effect by blocking a late stage in the assembly of the virus.

(a) Propose a mechanism for Step 2. *Hint:* As we shall see in Section 22.1C, reaction of a bromoalkane such as 1-bromoadamantane with aluminum bromide (a Lewis acid, Section 4.7) results in the formation of a carbocation and $AlBr_4^-$. Assume that adamantyl cation is formed in Step 2 and proceed from there to describe a mechanism.

(b) Account for the regioselectivity of carbon-carbon bond formation in Step 2.

(c) Describe experimental conditions to bring about Step 3.

(d) Describe experimental conditions to bring about Step 4.

7.20 Show reagents and experimental conditions to bring about the following transformations.

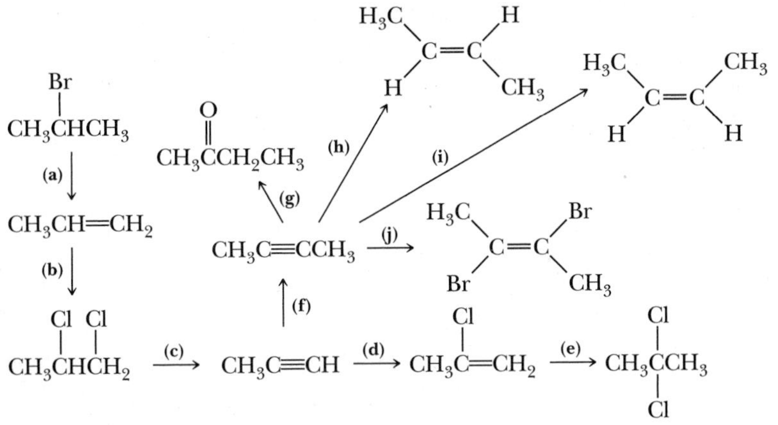

7.21 Show reagents to bring about each conversion.

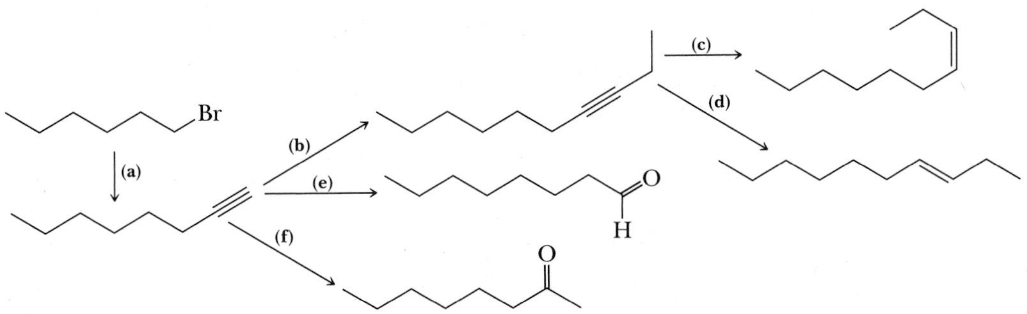

7.22 Propose a synthesis for (*Z*)-9-tricosene (muscalure), the sex pheromone for the common housefly (*Musca domestica*), starting with acetylene and haloalkanes as sources of carbon atoms.

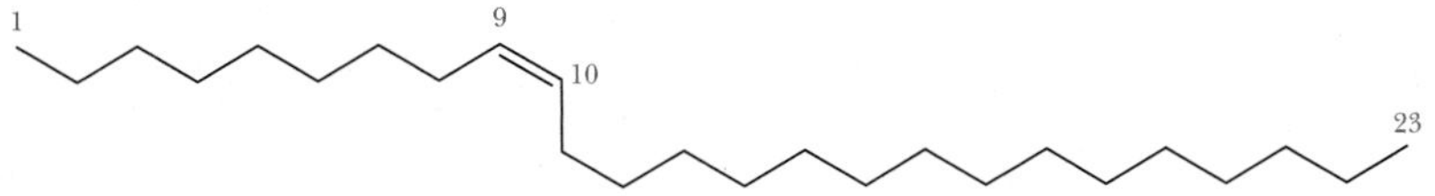

7.23 Propose a synthesis of each compound starting from acetylene and any necessary organic and inorganic reagents.

(a) 4-Octyne	**(b)** 4-Octanone	**(c)** *cis*-4-Octene
(d) *trans*-4-Octene	**(e)** 4-Octanol	**(f)** meso-4,5-Octanediol

7.24 Show how to prepare each compound from 1-heptene.

(a) 1,2-Dichloroheptane **(b)** 1-Heptyne **(c)** 1-Heptanol
(d) 2-Octyne **(e)** *cis*-2-Octene **(f)** *trans*-2-Octene

7.25 Show how to bring about this conversion.

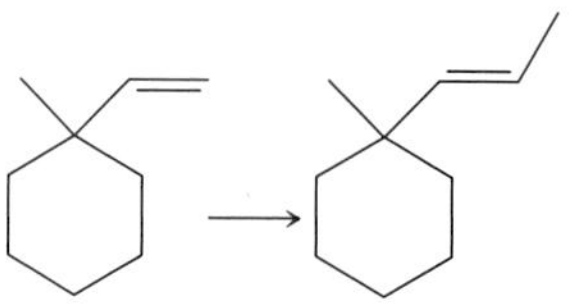

Looking Ahead

7.26 Alkyne anions react with the carbonyl groups of aldehydes and ketones to form alkynyl alcohols, as illustrated by the following sequence.

$$CH_3C{\equiv}C\!:^-Na^+ + H{-}\overset{\overset{O}{\|}}{C}{-}H \longrightarrow [CH_3C{\equiv}C{-}CH_2O^-Na^+] \xrightarrow[H_2O]{HCl} CH_3C{\equiv}C{-}CH_2OH$$

Propose a mechanism for the formation of the bracketed compound, using curved arrows to show the flow of electron pairs in the course of the reaction.

7.27 Following is the structural formula of the tranquilizer meparfynol (Oblivon).

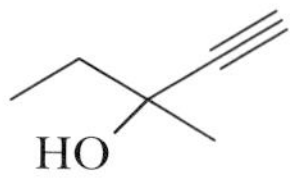

Oblivon

Propose a synthesis for this compound starting with acetylene and a ketone. (Notice the *-yn-* and *-ol* in the chemical name of this compound, indicating that it contains alkyne and hydroxyl functional groups.)

7.28 The standard procedure for synthesizing a compound is the stepwise progress toward a target molecule by forming individual bonds through single reactions. Typically, the product of each reaction is isolated and purified before the next reaction in the sequence is carried out. One of the ways nature avoids this tedious practice of isolation and purification is by the use of a domino sequence in which each new product is built on a preexisting one in stepwise fashion. A great example of a laboratory domino reaction is William S. Johnson's elegant synthesis of the female hormone progesterone. Johnson first constructed the polyunsaturated monocyclic 3° alcohol (A) and then, in an acid-induced domino reaction, formed compound B, which he then converted to progesterone.

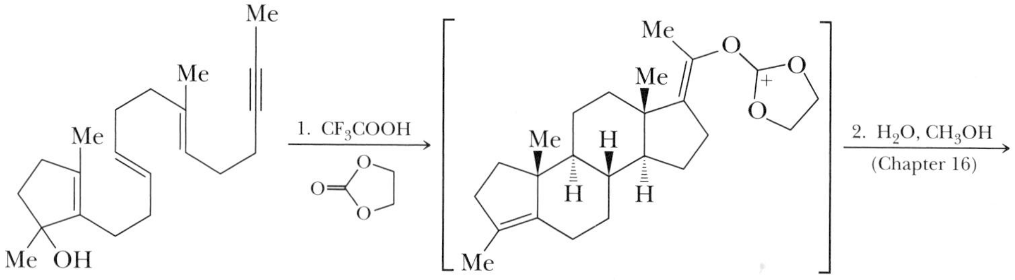

A B

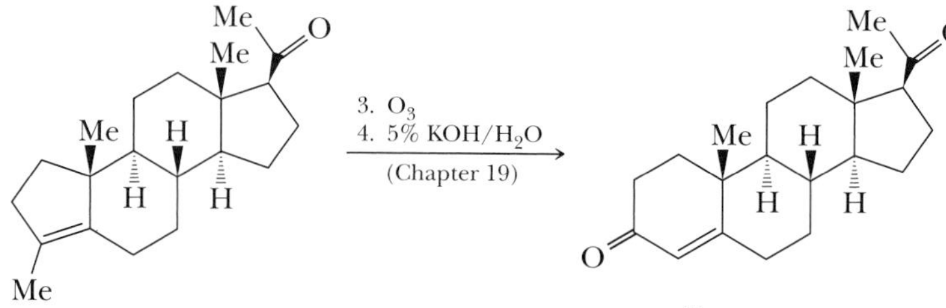

C **Progesterone**

A remarkable feature of this synthesis is that compound A, which has only one stereocenter, gives compound B, which has five stereocenters, each with the same configuration as those in progesterone. We will return to the chemistry of Step 2 in Section 16.7 and to the chemistry of Steps 3 and 4 in Chapter 19. In this problem, we focus on Step 1.

(a) Assume that the domino reaction in Step 1 is initiated by protonation of the 3° alcohol in compound A followed by loss of H_2O to give a 3° carbocation. Show how the series of reactions initiated by the formation of this cation gives compound B.

(b) If you have access to a large enough set of molecular models or to a computer modeling program, build a model of progesterone and describe the conformation of each ring. There are two methyl groups and three hydrogen atoms at the set of ring junctions in progesterone. Which of these five groups occupies an equatorial position? Which occupies an axial position?

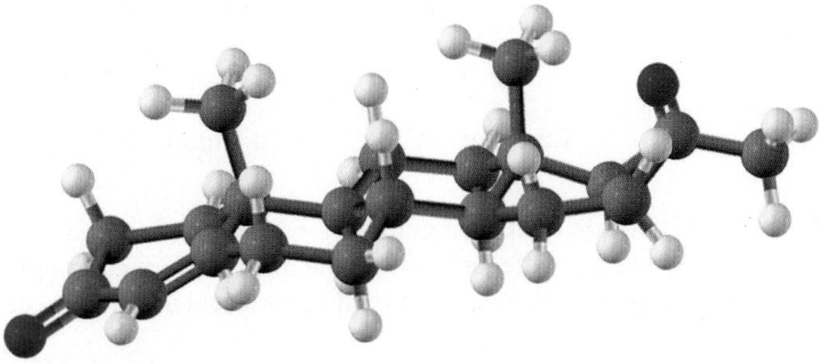

Progesterone

Organic Chemistry Reaction Roadmap

Reaction Roadmap

7.29 Use the reaction roadmap you made for Problem 6.54 and update it to contain the reactions in the Study Guide section of this chapter. Because of their highly specific nature, do not use reactions 1 and 6 from this chapter on your roadmap.

7.30 Write the products of the following sequences of reactions. Refer to your reaction roadmap to see how the combined reactions allow you to "navigate" between the different functional groups. For example, in part (a) below, notice how the reaction sequence results in the conversion of an alkyne into a haloalkane in two steps.

(a) 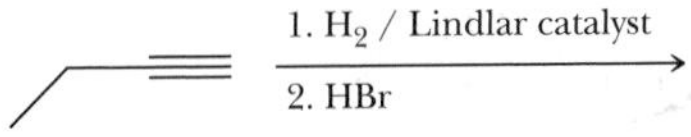
1. H2 / Lindlar catalyst
2. HBr

An alkyne

(b) 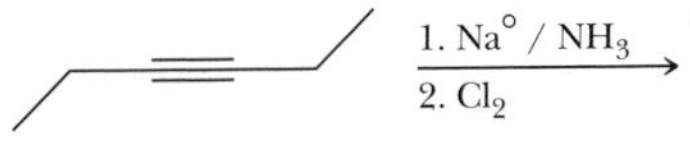
1. Na° / NH3
2. Cl2

An alkyne

(c) 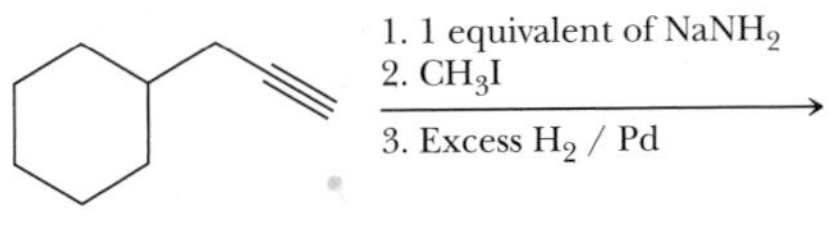
1. 1 equivalent of NaNH2
2. CH3I
3. Excess H2 / Pd

An alkyne

(d) 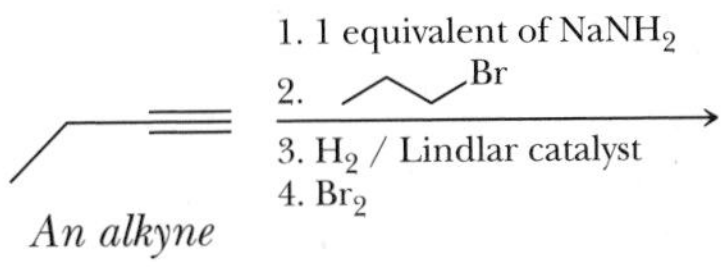
1. 1 equivalent of NaNH2
2. Br
3. H2 / Lindlar catalyst
4. Br2

An alkyne

(e) 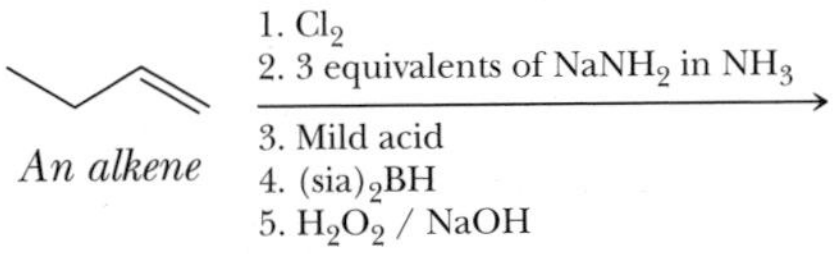
An alkene
1. Cl2
2. 3 equivalents of NaNH2 in NH3
3. Mild acid
4. (sia)2BH
5. H2O2 / NaOH

Multistep Synthesis

Reaction Roadmap

7.31 Using your reaction roadmap as a guide, show how to convert acetylene and bromoethane into 1-butene. All of the carbon atoms of the target molecule must be derived from the given starting materials. Show all intermediate molecules synthesized along the way.

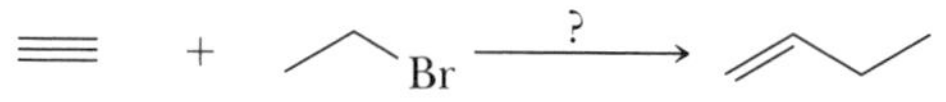

Acetylene **Bromoethane** **1-Butene**

Reaction Roadmap

7.32 Using your reaction roadmap as a guide, show how to convert ethylene into 1-butene. All of the carbon atoms of the target molecule must be derived from ethylene. Show all intermediate molecules synthesized along the way.

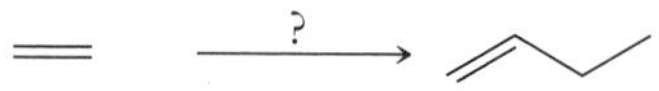

Ethylene **1-Butene**

Reaction Roadmap

7.33 Using your reaction roadmap as a guide, show how to convert 3-hexyne into propanal. All of the carbon atoms of the target molecule must be derived from the starting material as efficiently as possible. Show all intermediate molecules synthesized along the way.

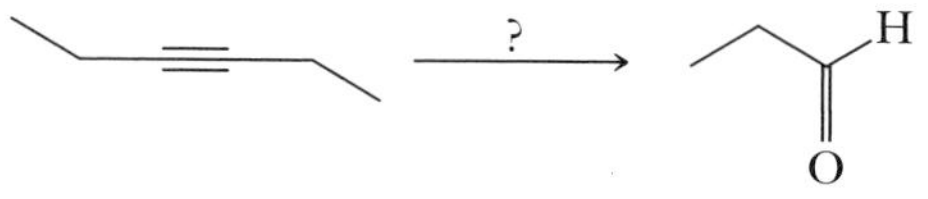

3-Hexyne **Propanal**

Reaction Roadmap

7.34 Using your reaction roadmap as a guide, show how to convert the starting *trans*-alkene to the *cis*-alkene in high yield. Show all intermediate molecules synthesized along the way.

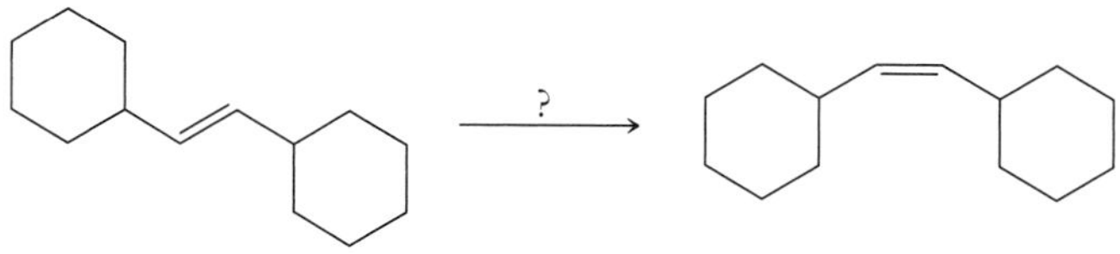

trans-alkene *cis*-alkene

Reactions in Context

7.35 Functional groups such as alkynes react the same in complex molecules as they do in simpler structures. The following examples of alkyne reactions were taken from syntheses carried out in the research group of E. J. Corey at Harvard University. You can assume that the reactions listed involve only the alkyne, not any of the functional groups present in the molecules. Draw the expected products for the following reactions.

(a)

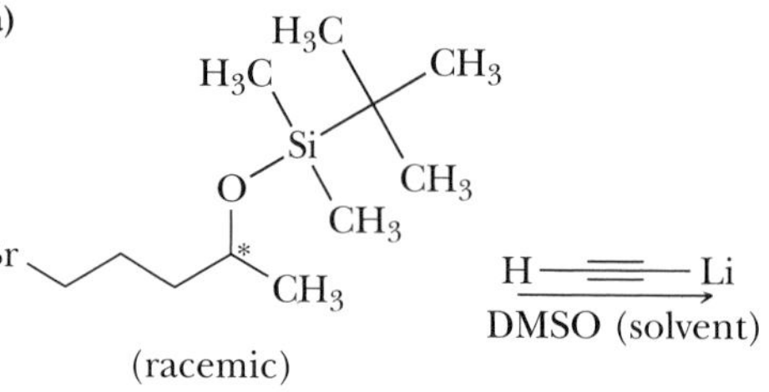

(racemic)

(b)

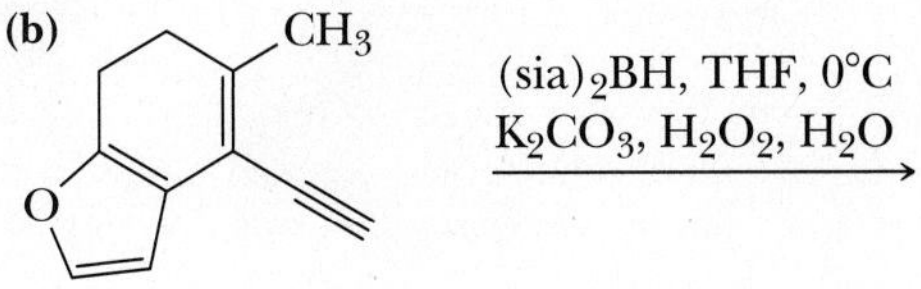

(c)

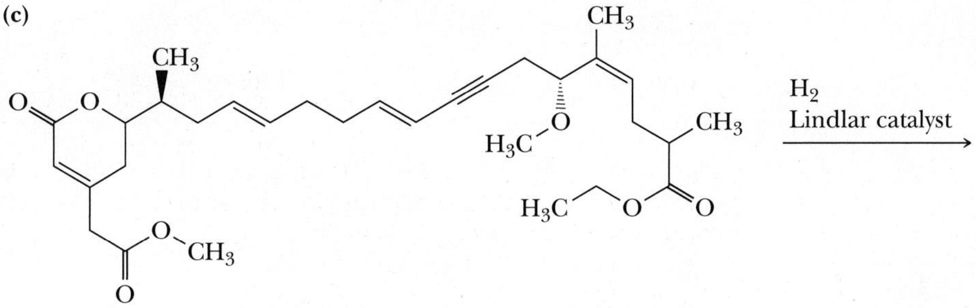

8

Haloalkanes, Halogenation, and Radical Reactions

Many common objects such as these pipes are made of poly(vinyl chloride). **Inset:** a model of chloroethene (vinyl chloride).
(© Christina Richards/ Shutterstock.com)

Outline

8.1 Structure
8.2 Nomenclature
8.3 Physical Properties of Haloalkanes
8.4 Preparation of Haloalkanes by Halogenation of Alkanes
8.5 Mechanism of Halogenation of Alkanes
8.6 Allylic Halogenation
8.7 Radical Autoxidation
8.8 Radical Addition of HBr to Alkenes

Compounds containing a halogen atom covalently bonded to an sp^3 hybridized carbon atom are named haloalkanes, or, in the common system of nomenclature, alkyl halides. Several haloalkanes are important laboratory and industrial solvents. In addition, haloalkanes are invaluable building blocks for organic synthesis because of the variety of ways in which the halogen may be substituted with other groups or eliminated to produce π bonds.

In this chapter, we begin with the structure and physical properties of haloalkanes. We then study radical halogenation of alkanes as a vehicle to introduce an important type of reaction mechanism, namely the mechanism of radical chain reactions. Reactions of oxygen with alkenes and a radical mechanism for HBr addition to alkenes complete the chapter.

8.1 Structure

The general symbol for a **haloalkane** is R—X, where —X may be —F, —Cl, —Br, or —I. If a halogen is bonded to a doubly bonded carbon of an alkene, the compound belongs to a class called **haloalkenes**. If it is bonded to a benzene ring, the compound belongs to a class called **haloarenes**, which have the general symbol Ar—X.

Haloalkane (alkyl halide) A compound containing a halogen atom covalently bonded to an sp^3 hybridized carbon atom. Given the symbol R—X.

Haloalkene (vinylic halide) A compound containing a halogen bonded to one of the carbons of a carbon-carbon double bond.

Haloarene (aryl halide) A compound containing a halogen atom bonded to a benzene ring. Given the symbol Ar—X.

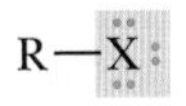

A haloalkane (an alkyl halide)

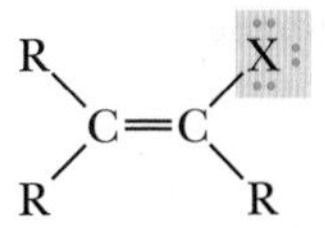

A haloalkene (an alkenyl or vinylic halide)

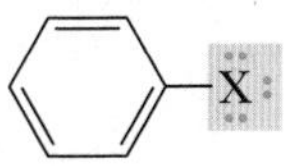

A haloarene (an aryl halide)

8.2 Nomenclature

A. IUPAC System

IUPAC names for haloalkanes are derived by naming the parent alkane according to the rules given in Section 2.3A.

- The parent chain is numbered from the direction that gives the first substituent encountered the lowest number, whether it is a halogen or an alkyl group. If two groups could have the same lowest number from the end of the chain, give the group of lower alphabetical order the lower number. An example is 2-bromo-4-methylpentane.
- Halogen substituents are indicated by the prefixes *fluoro-*, *chloro-*, *bromo-*, and *iodo-* and are listed in alphabetical order with other substituents.
- The location of each halogen atom on the parent chain is given by a number preceding the name of the halogen.
- In haloalkenes, numbering the parent hydrocarbon is determined by the location of the carbon-carbon double bond. Numbering is done in the direction that gives the carbon atoms of the double bond and substituents the lowest set of numbers.

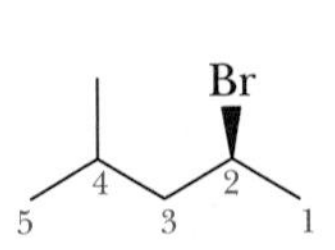

(*S*)-2-Bromo-4-methylpentane

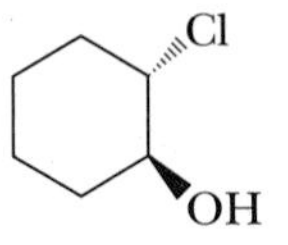

(1*S*,2*S*)-2-Chlorocyclohexanol

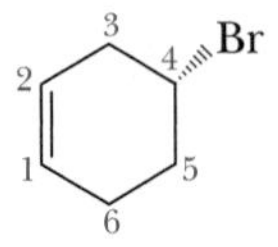

(*R*)-4-Bromocyclohexene

B. Common Names

Common names of haloalkanes and haloalkenes consist of the common name of the alkyl group followed by the name of the halide as a separate word. Hence, the name **alkyl halide** is a common name for this class of compounds. In the following examples, the IUPAC name of the compound is given first, followed by its common name in parentheses.

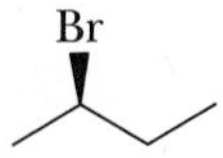

(*R*)-2-Bromobutane ((*R*)-*sec*-Butyl bromide)

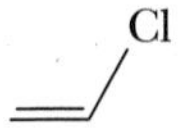

Chloroethene (Vinyl chloride)

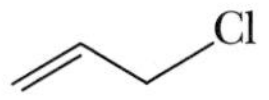

3-Chloropropene (Allyl chloride)

Several polyhaloalkanes are important solvents and are generally referred to by their common names. Dichloromethane (methylene chloride) is the most widely used

Haloform

A compound of the type CHX_3 where X is a halogen.

haloalkane solvent. Compounds of the type CHX_3 are called **haloforms**. The common name for $CHCl_3$, for example, is chloroform. The common name methyl chloroform for the compound CH_3CCl_3 derives from this name. Methyl chloroform and trichloroethylene (trichlor) were once common solvents for industrial cleaning. Because they are somewhat toxic and cause environmental problems, they have been phased out.

CH_2Cl_2	$CHCl_3$	CH_3CCl_3	$CCl_2{=}CHCl$
Dichloromethane (Methylene chloride)	**Trichloromethane (Chloroform)**	**1,1,1-Trichloroethane (Methyl chloroform)**	**Trichloroethylene (Trichlor)**

Hydrocarbons in which all hydrogens are replaced by halogens are commonly called perhaloalkanes or perhaloalkenes. Perchloroethylene, commonly known as perc, is a dry cleaning solvent that is also being phased out.

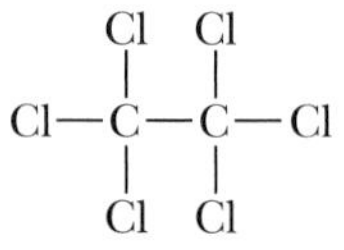

Perchloroethane

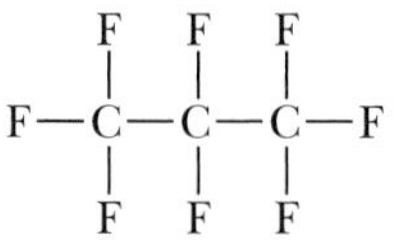

Perfluoropropane

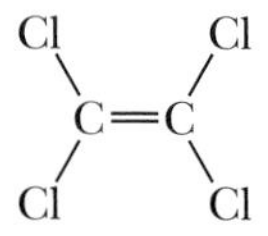

Perchloroethylene

Example 8.1 IUPAC Nomenclature for Haloalkanes

Write the IUPAC name and, where possible, the common name of each compound. Show stereochemistry where relevant.

(a) 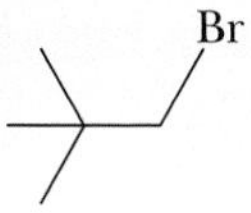

(b) 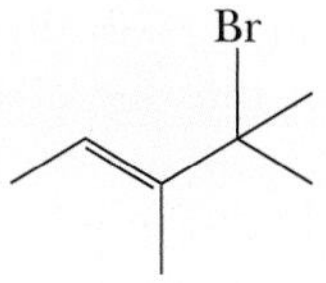

(c) 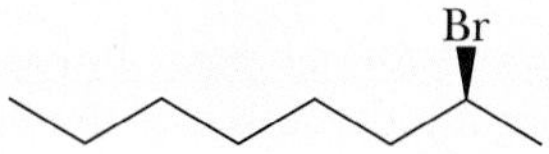

Solution

(a) 1-Bromo-2,2-dimethylpropane. Its common name is neopentyl bromide.

(b) (*E*)-4-Bromo-3,4-dimethyl-2-pentene.

(c) (*S*)-2-Bromooctane.

Problem 8.1

Write the IUPAC name, and where possible, the common name of each compound. Show stereochemistry where relevant.

(a) 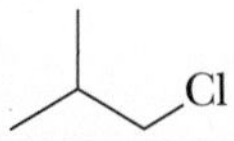

(b) 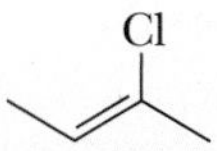

(c) 

(d) 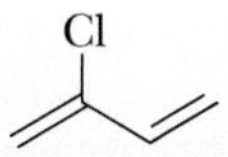

8.3 Physical Properties of Haloalkanes

A. Polarity

Fluorine, chlorine, and bromine are all more electronegative than carbon (Table 1.5); as a result, C—X bonds with these atoms are polarized with a partial negative charge on halogen and a partial positive charge on carbon. Table 8.1 shows that each of the

Table 8.1 Dipole Moments (Gas Phase) of Halomethanes

Halomethane	Electronegativity of Halogen	Carbon-Halogen Bond Length (pm)	Dipole Moment (debyes, D)
CH_3F	4.0	139	1.85
CH_3Cl	3.0	178	1.87
CH_3Br	2.8	193	1.81
CH_3I	2.5	214	1.62

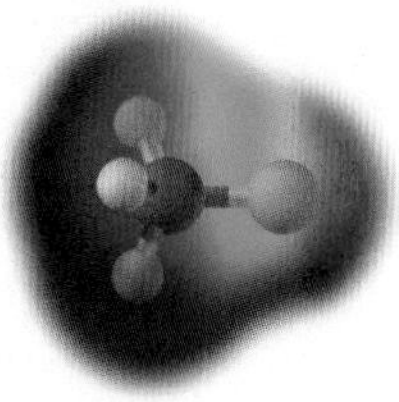

Electrostatic potential map of fluoromethane

halomethanes has a substantial dipole moment. The electrostatic potential map of fluoromethane shows the large charge separation in this compound caused by the dipole.

The magnitude of a dipole moment depends on the size of the partial charges, the distance between them, and the polarizability of the three pairs of unshared electrons on each halogen (see Table 8.1). For the halomethanes, the dipole moment increases as the electronegativity of the halogen and the bond length increase. These two trends run counter to each other, the net effect being that chloromethane has the largest dipole moment of the series. The experimental dipole moments also show that the electronegativity of carbon is less than the standard Pauling value of 2.5 in many compounds (as mentioned in Section 1.2B); otherwise, CH_3I would have no dipole moment.

B. Boiling Point

Haloalkanes are associated in the liquid state by a combination of attractive dipole-dipole, dipole-induced dipole, and induced dipole-induced dipole (dispersion) forces. These forces are grouped together under the term **van der Waals forces**, in honor of J. D. van der Waals, the nineteenth-century Dutch physicist. As atoms or molecules are brought closer and closer, van der Waals attractive forces are overcome by repulsive forces between the electron clouds of adjacent atoms. The energy minimum is where the net attraction is the strongest. Nonbonded interatomic and intermolecular distances at these minima can be measured by X-ray crystallography of solid compounds, and each atom and group of atoms can be assigned an atomic or molecular radius called a **van der Waals radius**. Van der Waals radii for selected atoms and groups of atoms are given in Table 8.2.

van der Waals forces
A group of intermolecular attractive forces including dipole-dipole, dipole-induced dipole, and induced dipole-induced dipole (dispersion) forces.

van der Waals radius
The minimum distance of approach to an atom that does not cause nonbonded interaction strain.

Notice in Table 8.2 that the van der Waals radius of fluorine is only slightly greater than that of hydrogen and that, among the halogens, only iodine has a larger van der Waals radius than methyl.

Table 8.2 van der Waals Radii (pm) for Selected Atoms and Groups of Atoms

H	F	Cl	Br	CH_2	CH_3	I
120	135	180	195	200	200	215

Increasing van der Waals radius →

Boiling points of several low-molecular-weight haloalkanes and the alkanes from which they are derived are given in Table 8.3. Several trends are to be noticed from these data.

Table 8.3 Boiling Points of Some Low-Molecular-Weight Alkanes and Haloalkanes

Alkyl Group	Name	Boiling Point (°C)				
		H	F	Cl	Br	I
CH_3—	**Methyl**	−161	−78	−24	4	43
CH_3CH_2—	**Ethyl**	−89	−37	13	38	72
$CH_3(CH_2)_2$—	**Propyl**	−45	3	46	71	102
$(CH_3)_2CH$—	**Isopropyl**	−45	−11	35	60	89
$CH_3(CH_2)_3$—	**Butyl**	0	32	77	100	130
$CH_3CH_2(CH_3)CH$—	***sec*-Butyl**	0	25	67	90	119
$(CH_3)_2CHCH_2$—	**Isobutyl**	−1	16	68	91	120
$(CH_3)_3C$—	***tert*-Butyl**	−1	12	51	72	98
$CH_3(CH_2)_4$—	**Pentyl**	36	63	108	129	157
$CH_3(CH_2)_5$—	**Hexyl**	69	92	134	155	181

1. As with hydrocarbons, constitutional isomers with branched chains have lower boiling points than their unbranched-chain isomers (Section 2.7C). Compare, for example, the boiling points of unbranched-chain 1-bromobutane (butyl bromide, bp 100°C) with the more branched and compact 2-bromo-2-methylpropane (*tert*-butyl bromide, bp 72°C). Branched-chain constitutional isomers have lower boiling points because they have a more spherical shape and, therefore, decreased surface area, leading to smaller van der Waals forces between their molecules.
2. For an alkane and haloalkane of comparable size and shape, the haloalkane has a higher boiling point. Compare, for example, the boiling points of ethane (bp −89°C) and bromomethane (bp 4°C). Although both molecules are roughly the same size and have roughly the same effective contact area, the boiling point of bromomethane is considerably higher. This difference in boiling points is due to the dipole moment in bromomethane, as well as the greater **polarizability** of bromine compared to methyl. Recall from Section 2.7A that the strength of dispersion forces, the weakest of all intermolecular forces, depends on the polarizability of electrons, which, in turn, depends on how tightly they are held by the nucleus. Unshared electron pairs have a higher polarizability than electrons shared in a covalent bond. In addition, the farther electrons are from the nucleus, the less tightly they are held and the greater their polarizability. Therefore, the larger the halogen, the greater its polarizability.
3. The boiling points of fluoroalkanes are comparable to those of hydrocarbons of similar molecular weight. Compare, for example, the boiling points of hexane (MW 86.2, bp 69°C) and 1-fluoropentane (MW 90.1, bp 63°C) and the boiling points of 2-methylpropane (MW 58.1, bp −1°C) and 2-fluoropropane (MW 62.1, bp −11°C). This low boiling point is attributable to the small size of fluorine, the tightness with which its electrons are held, and their particularly low polarizability.

Polarizability

A measure of the ease of distortion of the distribution of electron density about an atom or a group in response to interaction with other molecules or ions. Fluorine, which has a high electronegativity, holds its electrons tightly and has a very low polarizability. Iodine, which has a lower electronegativity and holds its electrons less tightly, has a very high polarizability.

C. Density

The densities of liquid haloalkanes are greater than those of hydrocarbons of comparable molecular weight because of the halogens' large mass-to-volume ratio. A bromine atom and a methyl group have almost identical van der Waals radii, but bromine has a mass of 79.9 atomic mass units (amu) compared with 15 amu for methyl. Table 8.4 gives densities for some low-molecular-weight haloalkanes that are liquid at 25°C. The densities of all liquid bromoalkanes and iodoalkanes are greater than that of water.

Table 8.4 Densities of Some Low-Molecular-Weight Haloalkanes

Alkyl Group	Name	Density of Liquid (g/mL) at 25°C: Cl	Br	I
CH_3—	Methyl	—	—	2.279
CH_3CH_2—	Ethyl	—	1.460	1.936
$CH_3(CH_2)$—	Propyl	0.891	1.354	1.749
$(CH_3)_2CH$—	Isopropyl	0.862	1.314	1.703
$CH_3(CH_2)_3$—	Butyl	0.886	1.276	1.615
$(CH_3)_3C$—	*tert*-Butyl	0.842	1.221	1.545

Although the densities of liquid chloroalkanes (Table 8.4) are less than that of water, further substitution of chlorine for hydrogen increases the density to the point where di- and polychloroalkanes have a greater density compared to that of water (Table 8.5). These compounds sink in water and form the lower layer when mixed because they are immiscible with water.

Table 8.5 Density of Polyhalomethanes

Haloalkane	X = Cl	Br	I
	Density of Liquid (g/mL) at 25°C		
CH_2X_2	1.327	2.497	3.325
CHX_3	1.483	2.890	4.008*
CX_4	1.594	3.273*	4.23*

* Density of solids

D. Bond Lengths and Bond Strengths

With the exception of C—F bonds, C—X bonds are weaker than C—H bonds as measured by bond dissociation enthalpies (BDEs), which are a measure of bond strengths (Section 6.2B). A table of BDE values for many bonds is given in Appendix 3. Recall that bond dissociation enthalpy is defined as the amount of energy required to break a bond homolytically into two radicals in the gas phase at 25°C.

$$A{-}B \longrightarrow A\cdot + \cdot B$$

A **radical**, sometimes called a free radical, is any chemical species that contains one or more unpaired electrons. Radicals are produced from a molecule by cleavage of a bond in such a way that each atom or fragment participating in the bond retains one electron, a process called **homolytic bond cleavage**. In the more common **heterolytic bond cleavage**, a bond breaks in such a way that one of the species retains both electrons. We use **fishhook arrows** to show the change in position of single electrons and to indicate a homolytic mechanism.

This reaction used to determine BDE is a "virtual" one because it can't be carried out in most cases. Instead, the extensive tables of BDEs are collected from

Radical
Any chemical species that contains one or more unpaired electrons.

Homolytic bond cleavage
Cleavage of a bond so that each fragment retains one electron, producing radicals.

Heterolytic bond cleavage
Cleavage of a bond so that one fragment retains both electrons and the other has none.

Fishhook arrow
A barbed, curved arrow used to show the change in position of a single electron.

Table 8.6 Average Bond Dissociation Enthalpies for C—H and C—X Bonds

Bond	Bond Length (pm)	Bond Dissociation Enthalpy [kJ (kcal)/mol]
C—H	109	414 (99)
C—F	142	464 (111)
C—Cl	178	355 (85)
C—Br	193	309 (78)
C—I	214	228 (57)

thermochemical data on heats of combustion, hydrogenation, and other reactions. The useful thing about these data is that by adding and subtracting them, heats of reaction can be calculated with confidence for reactions that have never been measured.

C—X BDEs are tabulated in Table 8.6. As the size of the halogen atom increases, the C—X bond length increases and its strength decreases. These relationships between bond strength and bond length help us to understand the difference in the ease with which haloalkanes undergo reactions that involve carbon-halogen bond breaking. Fluoroalkanes, for example, with the strongest and shortest C—X bonds, are highly resistant to bond breaking under most conditions. This characteristic inertness is one of the factors that makes perfluoroalkanes such as Teflon such useful materials.

8.4 Preparation of Haloalkanes by Halogenation of Alkanes

As we saw in Sections 6.3A and 6.3D, haloalkanes can be prepared by the addition of HX and X_2 to alkenes. They are also prepared by replacement of the —OH group of alcohols by halogen (Section 10.5). Many of the simpler low-molecular-weight haloalkanes are prepared by the halogenation of alkanes, illustrated here by treating 2-methylpropane with bromine at an elevated temperature.

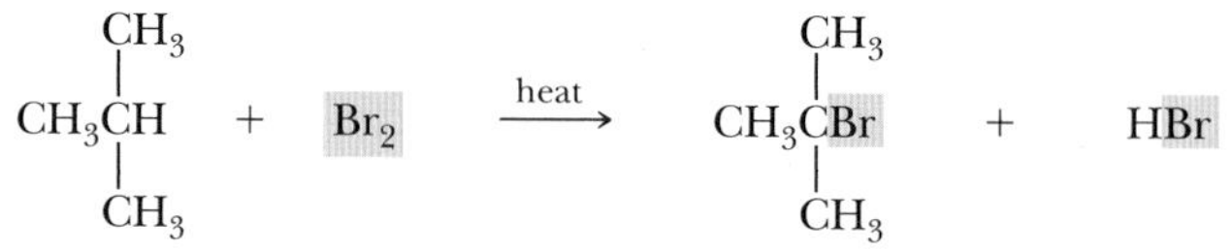
CH3
CH3CH
CH3
+
Br2
heat
CH3
CH3CBr
CH3
+
HBr

2-Methylpropane (Isobutane) **2-Bromo-2-methylpropane (*tert*-Butyl bromide)**

Halogenation of alkanes is common with Br_2 and Cl_2. Fluorine, F_2, is seldom used because its reactions with alkanes are so exothermic that they are difficult to control and can actually cause C—C bond cleavage and even explosions. Iodine, I_2, is seldom used because the reaction is endothermic and the position of equilibrium favors alkane and I_2 rather than iodoalkane and HI.

Substitution A reaction in which an atom or a group of atoms in a compound is replaced by another atom or group of atoms.

If a mixture of methane and chlorine gas is kept in the dark at room temperature, no detectable change occurs. If, however, the mixture is heated or exposed to light, a reaction begins almost at once with the evolution of heat. The products are chloromethane and hydrogen chloride. What occurs is a **substitution** reaction—in this case, substitution of a methane hydrogen atom by a chlorine atom and the production of an equivalent amount of hydrogen chloride.

$$CH_4 + Cl_2 \xrightarrow{\text{heat}} CH_3Cl + HCl$$

Methane → **Chloromethane (Methyl chloride)**

If chloromethane is allowed to react with more chlorine, further chlorination produces a mixture of dichloromethane (methylene chloride), trichloromethane (chloroform), and tetrachloromethane (carbon tetrachloride). Notice that in the last equation, the reagent Cl_2 is placed over the reaction arrow and the equivalent amount of HCl formed is not shown. Placing reagents over reaction arrows and omitting byproducts is commonly done to save space.

$$CH_3Cl + Cl_2 \xrightarrow{\text{heat}} CH_2Cl_2 + HCl$$

Chloromethane (Methyl chloride) → **Dichloromethane (Methylene chloride)**

$$CH_2Cl_2 \xrightarrow[\text{heat}]{Cl_2} CHCl_3 \xrightarrow[\text{heat}]{Cl_2} CCl_4$$

Dichloromethane (Methylene chloride) → **Trichloromethane (Chloroform)** → **Tetrachloromethane (Carbon tetrachloride)**

It is possible to prepare chloromethane or tetrachloromethane in relatively pure form by this reaction. In the case of chloromethane, a large excess of methane is used; for tetrachloromethane, a large excess of chlorine drives the reaction to complete halogenation. The other chlorinated methanes can be separated by distillation of partially chlorinated mixtures.

Treating ethane with bromine gives bromoethane (ethyl bromide).

$$CH_3CH_3 + Br_2 \xrightarrow{\text{heat}} CH_3CH_2Br + HBr$$

Ethane → **Bromoethane (Ethyl bromide)**

In all cases, monosubstituted products are only obtained using an excess of alkane.

A. Regioselectivity

Treating propane with bromine gives a mixture consisting of approximately 92% of 2-bromopropane and 8% of 1-bromopropane.

$$CH_3CH_2CH_3 + Br_2 \xrightarrow[\text{or light}]{\text{heat}} CH_3CH(Br)CH_3 + CH_3CH_2CH_2Br + HBr$$

Propane → **2-Bromopropane** (92%) + **1-Bromopropane** (8%)

Propane contains eight hydrogens—one set of six equivalent primary hydrogens and one set of two secondary hydrogens (Section 2.3D). The hydrogens in each set are equivalent because of rapid bond rotation about C—C single bonds. Substitution of bromine for a primary hydrogen gives 1-bromopropane; substitution of bromine for a secondary hydrogen gives 2-bromopropane. If there were random substitution of any one of the eight hydrogens in propane, we would predict that the isomeric bromopropanes would be formed in the ratio of 6:2, or 75% 1-bromopropane and 25% 2-bromopropane. In fact, in the bromination of propane, substitution of a secondary hydrogen rather than a primary hydrogen is strongly favored. 2-Bromopropane is the major product, and the reaction is highly regioselective.

Product Distribution	$CH_3CH_2CH_2Br$	$CH_3CH(Br)CH_3$
Prediction based on ratio of six 1° H to two 2° H	75%	25%
Experimental observation	8%	92%

Other experiments have shown that substitution at a tertiary hydrogen is favored over both secondary and primary hydrogens. For example, monobromination of 2-methylpentane is very regioselective and gives almost exclusively 2-bromo-2-methylpentane.

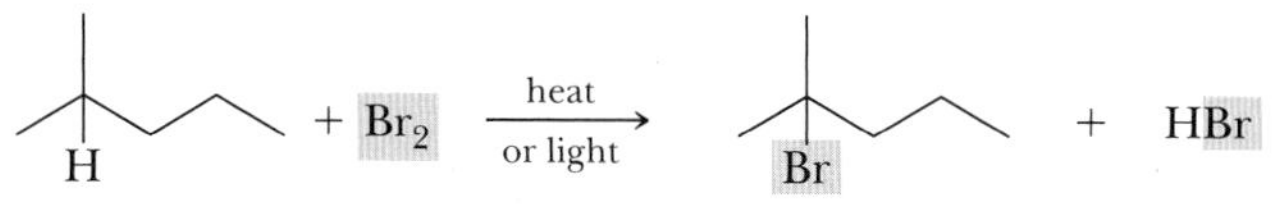

2-Methylpentane **2-Bromo-2-methylpentane**

The reaction of bromine with an alkane occurs in the order 3° > 2° > 1° hydrogen. Chlorination of alkanes is also regioselective, but much less so than bromination. For example, treatment of propane with chlorine gives a mixture of approximately 57% 2-chloropropane and 43% 1-chloropropane.

Propane **2-Chloropropane** (57%) **1-Chloropropane** (43%)

Thus, we can conclude that although both bromine and chlorine are regioselective in hydrogen replacement in the order 3° > 2° > 1°, regioselectivity is far greater for bromination than for chlorination. From data on product distribution, it has been determined that regioselectivity per hydrogen for bromination is approximately 1600:80:1, whereas it is only about 5:4:1 for chlorination. We will discuss reasons for this difference in Section 8.5.

Example 8.2 Free-Radical Halogenation

Name and draw structural formulas for all monobromination products formed by treating 2-methylpropane with Br_2. Predict the major product based on the regioselectivity of the reaction of Br_2 with alkanes.

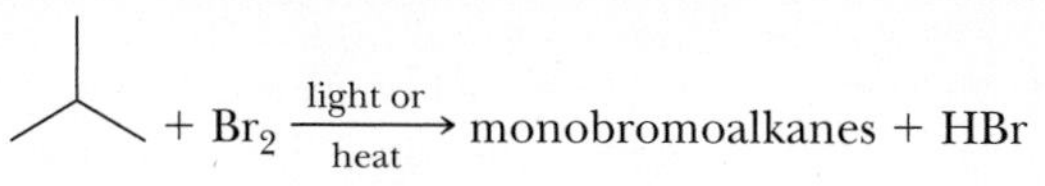

2-Methylpropane

Solution

2-Methylpropane has nine equivalent primary hydrogens and one tertiary hydrogen. Substitution of bromine for a primary hydrogen gives 1-bromo-2-methylpropane; substitution for the tertiary hydrogen gives 2-bromo-2-methylpropane. Given that the regioselectivity per hydrogen of bromination for 3° > 2° > 1° hydrogens is approximately 1600:80:1, it is necessary to correct for the number of hydrogens: nine primary and one tertiary. The result is that 99.4% of the product is 2-bromo-2-methylpropane and 0.6% is 1-bromo-2-methylpropane.

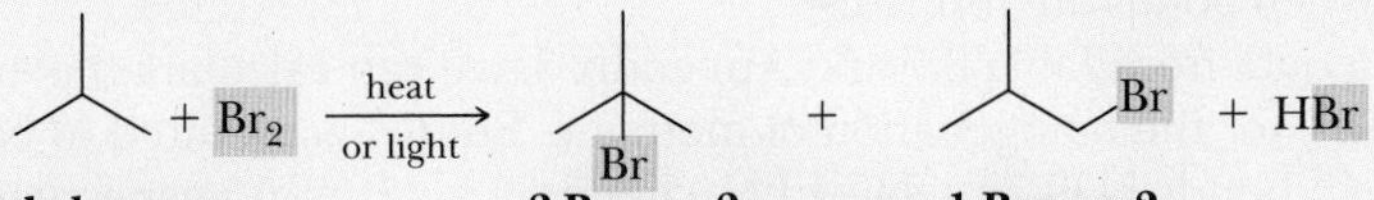

2-Methylpropane **2-Bromo-2-methylpropane** (major product) **1-Bromo-2-methylpropane**

$$\text{Predicted \% 2-bromo-2-methylpropane} = \frac{1 \times 1600}{(1 \times 1600) + (9 \times 1)} \times 100 = 99.4\%$$

Problem 8.2

Name and draw structural formulas for all monochlorination products formed by treatment of 2-methylpropane with Cl_2. Predict the major product based on the regioselectivity of the reaction of Cl_2 with alkanes.

B. Energetics

We can learn a lot about these reactions by careful consideration of the energetics of each step. A selection of C—H BDE values is given in Table 8.7.

Note that the BDE values for saturated hydrocarbons depend on the type of hydrogen being abstracted and are in the order methane > 1° > 2° > 3°. This order can best be understood by comparison to carbocations. In general, interactions that are stabilizing to carbocations are also stabilizing to radicals. This makes sense because radicals have only seven electrons in the valence shell of the carbon bearing the radical. Therefore, like carbocations, the more alkyl groups the carbon radicals have, the more stable they are. Note also that an sp^2 C—H bond is particularly strong, as mentioned in Section 1.10.

Chemists calculate the heat of any reaction (enthalpy change) by comparing the bond strengths in the reactants and products. The heat of a reaction equals the

Table 8.7 Bond Dissociation Enthalpies for Selected C—H Bonds

Hydrocarbon	Radical	Name of Radical	Type of Radical	ΔH^0 [kJ(kcal)/mol]
$CH_2{=}CHCH_2{-}H$	$CH_2{=}CHCH_2\cdot$	Allyl	Allylic	372 (89)
$C_6H_5CH_2{-}H$	$C_6H_5CH_2\cdot$	Benzyl	Benzylic	376 (90)
$(CH_3)_3C{-}H$	$(CH_3)_3C\cdot$	*tert*-Butyl	3°	405 (97)
$(CH_3)_2CH{-}H$	$(CH_3)_2CH\cdot$	Isopropyl	2°	414 (99)
$CH_3CH_2{-}H$	$CH_3CH_2\cdot$	Ethyl	1°	422 (101)
$CH_3{-}H$	$CH_3\cdot$	Methyl	Methyl	439 (105)
$CH_2{=}CH{-}H$	$CH_2{=}CH\cdot$	Vinyl	Vinylic	464 (111)

Radical stability ↑

sum of the BDEs of the bonds broken minus the sum of the BDEs of the bonds formed:

▶ Reactions will occur if the products are more stable than the reactants and the energy barrier is low enough.

$$\Delta H^0 = \Sigma\text{BDEs(broken)} - \Sigma\text{BDEs(formed)}$$

This equation makes sense because energy must be added to break bonds while it is released when bonds are formed. ▶

Using data from Table 8.7 and Appendix 3, we can calculate the enthalpy of reaction, ΔH^0, for the halogenation of methane. Energy is required to break CH_3—H and Cl—Cl bonds [439 and 247 kJ (105 and 59 kcal)/mol, respectively]. Energy is released in making the CH_3—Cl and H—Cl bonds [−351 and −431 kJ (−84 and −103 kcal)/mol, respectively]. Based upon these enthalpies, we calculate that chlorination of methane to form chloromethane and hydrogen chloride liberates 96 kJ (23 kcal)/mol.

	CH_4	+ Cl_2 ⟶	CH_3Cl	+ HCl	$\Delta H^0 = -96$ kJ/mol (−23 kcal/mol)
BDE, kJ/mol	+439	+247	−351	−431	

Example 8.3 Enthalpy of Reactions I

Using the table of bond dissociation enthalpies in Appendix 3, calculate ΔH^0 for bromination of propane to give 2-bromopropane and hydrogen bromide.

Solution

Under each molecule is given the enthalpy for breaking or forming each corresponding bond. The calculated heat of reaction is −71 kJ (−17 kcal)/mol.

	$CH_3CH(H)CH_3$	+ Br—Br ⟶	$CH_3CH(Br)CH_3$	+ H—Br
BDE, kJ/mol	+414	+192	−309	−368

$\Delta H^0 = -71$ kJ/mol (−17 kcal/mol)

Watch a video explanation

Problem 8.3

Using the table of bond dissociation enthalpies in Appendix 3, calculate ΔH^0 for bromination of propane to give 1-bromopropane and hydrogen bromide.

8.5 Mechanism of Halogenation of Alkanes

From detailed studies of the conditions and products for halogenation of alkanes, chemists have concluded that these reactions occur by a type of mechanism called a radical chain mechanism.

A. Formation of Radicals

Following are four reactions that result in homolytic cleavage to give radicals. Note that in the first three cases, the bond being broken is between atoms with at least two lone pairs. Lone pair repulsion is one factor that weakens sigma bonds. BDEs of these reactions from Appendix 3 are shown at the right of each equation.

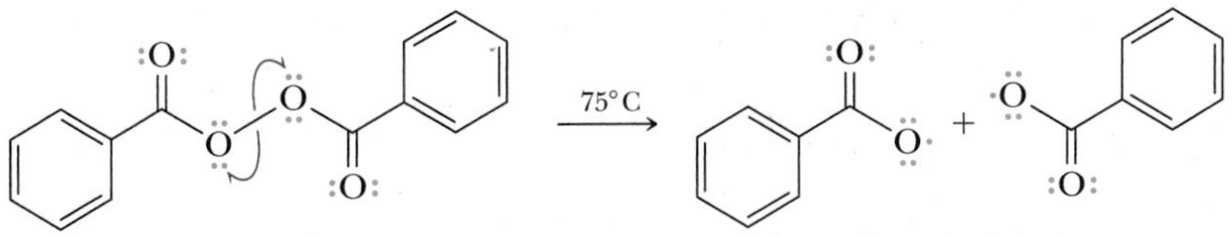

$\Delta H^0 = +130$ kJ/mol (+31 kcal/mol)

Dibenzoyl peroxide **Benzoyl radicals**

$$CH_3CH_2\ddot{O}-\ddot{O}CH_2CH_3 \xrightarrow{80°C} CH_3CH_2\ddot{O}\cdot + \cdot\ddot{O}CH_2CH_3 \qquad \Delta H^0 = +159 \text{ kJ/mol } (+38 \text{ kcal/mol})$$

Diethyl peroxide **Ethoxy radicals**

$$:\ddot{Cl}-\ddot{Cl}: \xrightarrow[\text{light}]{\text{heat or}} :\ddot{Cl}\cdot + \cdot\ddot{Cl}: \qquad \Delta H^0 = +247 \text{ kJ/mol } (+59 \text{ kcal/mol})$$

Chlorine **Chlorine atoms**

$$CH_3-CH_3 \xrightarrow{\text{heat}} CH_3\cdot + \cdot CH_3 \qquad \Delta H^0 = +376 \text{ kJ/mol } (+90 \text{ kcal/mol})$$

Ethane **Methyl radicals**

Energy to cause bond cleavage and generation of radicals can be supplied by either light or heat. The energy of visible and ultraviolet radiation (wavelength from 200 to 700 nm) falls in the range of 585 to 167 kJ (140 to 40 kcal)/mol and is of the same order of magnitude as the bond dissociation enthalpies of halogen-halogen covalent bonds. The bond dissociation enthalpy of Br_2 is 192 kJ (46 kcal)/mol; that for Cl_2 is 247 kJ (59 kcal)/mol. Dissociation of these halogens can also be brought about by heating at temperatures above 350°C.

Oxygen-oxygen single bonds in peroxides (ROOR) and hydroperoxides (ROOH) have dissociation enthalpies in the range of 146 to 209 kJ (35 to 50 kcal)/mol, and compounds containing these bonds are cleaved to radicals at considerably lower temperatures than those required for rupture of carbon-carbon bonds. Diethyl peroxide, for example, begins to dissociate to ethoxy radicals at 80°C. Dissociation of ethane into two methyl radicals occurs only at very high temperature.

B. A Radical Chain Mechanism

To account for the products formed from halogenation of alkanes, chemists propose a radical chain mechanism involving three types of steps: (1) **chain initiation,** (2) **chain propagation,** and (3) **chain termination.** We illustrate radical halogenation of alkanes by the reaction of chlorine with ethane.

Mechanism 8.1

Radical Chlorination of Ethane

Chain initiation involves formation of radicals from nonradical species. Chlorine is homolytically dissociated by heat or light.

Step 1: $$:\ddot{Cl}-\ddot{Cl}: \xrightarrow[\text{or light}]{\text{heat}} :\ddot{Cl}\cdot + \cdot\ddot{Cl}:$$

Chain propagation involves reaction of a radical and a molecule to form a new radical. The chlorine atom formed in Step 1 attacks the alkane, removing a hydrogen atom in another homolytic reaction.

(Continued)

Step 2: $CH_3CH_2{-}H + \cdot\ddot{Cl}: \longrightarrow CH_3CH_2\cdot + H{-}\ddot{Cl}:$

Step 3: $CH_3CH_2\cdot + :\ddot{Cl}{-}\ddot{Cl}: \longrightarrow CH_3CH_2{-}\ddot{Cl}: + \cdot\ddot{Cl}:$

Chain termination involves destruction of radicals. The first three possible chain termination steps involve coupling of radicals to form a new covalent bond. The fourth chain termination step, called disproportionation, involves transfer of a hydrogen atom from the beta position of one radical to another radical and formation of an alkane and an alkene.

Step 4: $CH_3CH_2\cdot + \cdot CH_2CH_3 \longrightarrow CH_3CH_2{-}CH_2CH_3$

Step 5: $CH_3CH_2\cdot + \cdot\ddot{Cl}: \longrightarrow CH_3CH_2{-}\ddot{Cl}:$

Step 6: $:\ddot{Cl}\cdot + \cdot\ddot{Cl}: \longrightarrow :\ddot{Cl}{-}\ddot{Cl}:$

Step 7: $CH_3CH_2\cdot + H{-}CH_2{-}CH_2\cdot \longrightarrow CH_3CH_3 + CH_2{=}CH_2$

Initiation

Chain initiation

A step in a chain reaction characterized by the formation of reactive intermediates (radicals, anions, or cations) from nonradical or noncharged molecules.

The characteristic feature of a **chain initiation** step is formation of radicals from nonradical compounds. In the case of chlorination of ethane, chain initiation is by thermal or light-induced homolysis of the Cl—Cl bond to give two chlorine radicals.

Chain Propagation

Chain propagation

A step in a chain reaction characterized by the reaction of a reactive intermediate and a molecule to give a new reactive intermediate and a new molecule.

The characteristic feature of a **chain propagation** step is reaction of a radical and a molecule to give a new radical. A chlorine atom, also called a chlorine radical, is consumed in Step 2, but an ethyl radical is produced. Similarly, an ethyl radical is consumed in Step 3, but a chlorine radical is produced. Steps 2 and 3 can repeat thousands of times as long as neither radical is removed by a different reaction.

A second characteristic feature of chain propagation steps is that, when added together, they give the observed stoichiometry of the reaction. Adding Steps 2 and 3 and canceling structures that appear on both sides of the equation gives the balanced equation for the radical chlorination of ethane (Figure 8.1).

Steps 2 + 3 $\quad CH_3CH_3 + Cl_2 \longrightarrow CH_3CH_2Cl + HCl$

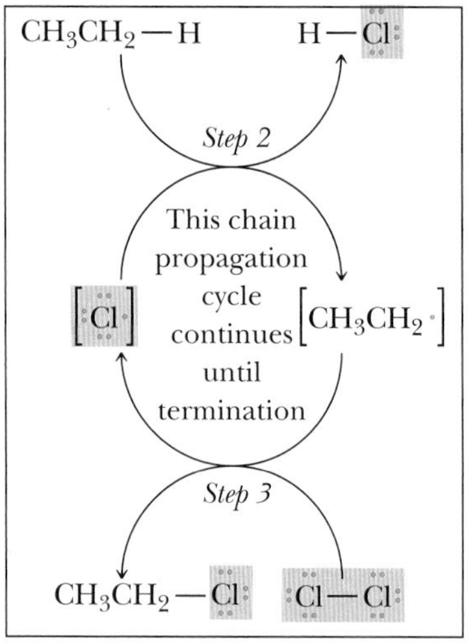

FIGURE 8.1 The cycle of chain propagation steps from the halogenation of ethane radical chain mechanism.

Watch a video explanation

The number of times a cycle of chain propagation steps repeats is called **chain length**. Chain lengths can range from a few to many thousand, depending on the relative rates of reactions and the concentrations of various species.

Chain length
The number of times the cycle of chain propagation steps repeats in a chain reaction.

Chain Termination

A characteristic feature of a **chain termination** step is destruction of radicals. Among the most important chain termination reactions during halogenation of alkanes are radical couplings (illustrated by Steps 4, 5, and 6 in the mechanism for the halogenation of ethane) and disproportionation (illustrated by Step 7). One of these termination steps gives the product ethyl chloride. However, the loss of ethyl and chlorine radicals terminates two chains and therefore stops many dozens to thousands of subsequent product-forming propagation steps.

Chain termination
A step in a chain reaction that involves destruction of reactive intermediates.

Note that chain termination steps are usually relatively rare compared to chain propagation steps in radical chain reactions. This is because at any one time, the concentration of radical species is very low, making a collision between two radicals a relatively rare event. The relatively rare occurrence of chain termination steps explains why there can be chain lengths of many thousands of steps in a radical chain reaction.

A major concern with chlorofluorocarbons is destruction of the ozone layer (mentioned in"Chemical Connections: Freons"in this chapter). The ozone-consuming reactions are thought to occur through a radical chain mechanism. The production of one radical species derived from a CFC can destroy a large number of ozone molecules before any termination steps occur.

The structures, geometries, and relative stabilities of simple alkyl radicals are similar to those of alkyl carbocations. Methyl radical is planar, and all other radicals are nearly so, with bond angles near 120° about the carbon with the unpaired electron. This geometry indicates that carbon is sp^2 hybridized and that the unpaired electron occupies the unhybridized 2*p* orbital. As mentioned, the order of stability of alkyl radicals, like alkyl carbocations, is 3° > 2° > 1° > methyl.

C. Energetics of Chain Propagation Steps

After the radical chain is initiated, the heat of reaction is derived entirely from the heat of reaction of the individual chain propagation steps. In Step 2 of radical chlorination of ethane, for example, energy is required to break the CH_3CH_2—H bond [422 kJ (101 kcal)/mol], but energy is released on formation of the H—Cl bond [431 kJ (103 kcal)/mol]. Similarly, energy is required in Step 3 to break the Cl—Cl bond [247 kJ (59 kcal)/mol], but energy is released on formation of the CH_3CH_2—Cl bond [355 kJ (85 kcal)/mol]. We see that, just as the sum of the chain propagation steps for radical halogenation gives the observed stoichiometry, the sum of the heats of reaction for each propagation step is equal to the observed heat of reaction:

Bond Dissociation Enthalpies for Chain Propagation Steps

	Reaction Step								ΔH^0, kJ/mol (kcal/mol)
Step 2:	CH_3CH_2—H +422	+	·Cl	⟶	CH_3CH_2·	+	H—Cl −431		−9 (−2)
Step 3:	CH_3CH_2·	+	Cl—Cl +247	⟶	CH_3CH_2—Cl −355	+	·Cl		−108 (−26)
Sum:	CH_3CH_2—H	+	Cl—Cl	⟶	CH_3CH_2—Cl	+	H—Cl		−117 (−28)

Chemical Connections

Freons

Of all the fluoroalkanes, **chlorofluorocarbons (CFCs)** manufactured under the trade name **Freons** are the most widely known. CFCs are nontoxic, nonflammable, odorless, and noncorrosive and seemed to be ideal replacements for the hazardous compounds such as ammonia and sulfur dioxide formerly used as heat-transfer agents in refrigeration systems. Among the CFCs most widely used for this purpose were trichlorofluoromethane (CCl_3F, Freon-11) and dichlorodifluoromethane (CCl_2F_2, Freon-12). They are particularly desirable refrigerants because of their low boiling points.

The CFCs found wide use as industrial cleaning solvents to prepare surfaces for coatings, to remove cutting oils and waxes from millings, and to remove protective coatings. CFCs were also used as propellants for aerosol sprays.

Concern about the environmental impact of CFCs arose in the 1970s when it was shown that more than 4.5×10^5 kg/yr of these compounds were being emitted into the atmosphere. Then in 1974, Sherwood Rowland of the University of California, Irvine, and Mario Molina, now at the University of California, San Diego, announced their theory, which has since been amply confirmed, of ozone (O_3) destruction by these compounds. When released into the air, CFCs escape to the lower atmosphere, but because of their inertness, they do not decompose there. Slowly they find their way to the stratosphere. What makes CFCs so dangerous is that a single CFC molecule in the stratosphere can lead to the destruction of many thousands of molecules of ozone through a radical chain mechanism (Section 8.5B). The process is initiated when the sun's unfiltered UV light in the stratosphere causes homolytic cleavage of the CFC to give a Cl radical. The Cl radical reacts with ozone in the first chain propagation step to give O_2 and ClO. The ClO reacts with an O atom (created when UV light hits O_2 or O_3 in the stratosphere) to give O_2 and another Cl radical, the latter of which reacts with a new ozone molecule, thus continuing the radical chain reaction. Not only is an ozone molecule destroyed for each cycle of the chain propagation steps, but also the O atom consumed might have created a new O_3 molecule had it not been intercepted by the ClO.

The problem for humans and all other living things on the planet is that ozone acts as a shield for the earth against short-wavelength ultraviolet radiation from the sun. Scientists believe that an increase in short-wavelength ultraviolet radiation reaching the earth will lead to the destruction of certain crops and agricultural species and even to an increased incidence of skin cancer in light-skinned individuals.

The results of this concern were that in 1987, most countries subscribed to the so-called Montreal Protocol, which set limits on the production and use of ozone-depleting CFCs and urged a complete phaseout of their production by 1996. This phaseout has resulted in enormous costs and is not yet complete in developing countries. The fact that an international agreement on the environment that set limits on the production of any substance could be reached is indeed amazing and bodes well for the health of the planet. Rowland, Molina, and Paul Crutzen, a Dutch chemist at the Max Planck Institute for Chemistry in Germany, were awarded the 1995 Nobel Prize in Chemistry for their work on this topic.

The chemical industry has responded by developing less-ozone-depleting alternatives to CFCs, among which are the hydrofluorocarbons (HFCs) and hydrochlorofluorocarbons (HCFCs). These compounds are much more chemically reactive in the atmosphere than the Freons and are destroyed before reaching the stratosphere. However, they tend to act as "greenhouse gases" and may contribute to global warming. For this reason, they are likely to be replaced in turn.

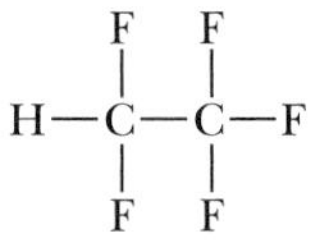

HFC-134a

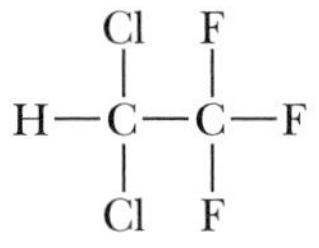

HCFC-123

We must not assume, however, that haloalkanes are introduced into the environment only by human action. It is estimated, for example, that annual production of bromomethane from natural sources is 2.7×10^8 kg, largely from marine algae, giant kelp, and volcanoes. Furthermore, global emission of chloromethane is estimated to be 4.5×10^9 kg/yr, most of it from terrestrial and marine biomass. These haloalkanes, however, have only short atmospheric lifetimes, and only a tiny fraction of them reach the stratosphere. The CFCs are the problem; they have longer atmospheric lifetimes, reach the stratosphere, and do their damage there.

Example 8.4 Enthalpy of Reactions II

Using the table of bond dissociation enthalpies in Appendix 3, calculate ΔH^0 for each propagation step in the radical bromination of propane to give 2-bromopropane and HBr.

Solution

Here are the two chain propagation steps along with bond dissociation enthalpies for the bonds broken and the bonds formed. The first chain propagation step is endothermic, the second is exothermic, and the overall reaction is exothermic by 71 kJ (17 kcal)/mol.

		ΔH^0
Step 2:	$CH_3CH(H)CH_3 + Br\cdot \longrightarrow CH_3\dot{C}HCH_3 + H{-}Br$	+46 kJ/mol (+11 kcal/mol)
Step 3:	$CH_3\dot{C}HCH_3 + Br{-}Br \longrightarrow CH_3CH(Br)CH_3 + Br\cdot$	−117 kJ/mol (−28 kcal/mol)
Sum:	$CH_3CH(H)CH_3 + Br{-}Br \longrightarrow CH_3CH(Br)CH_3 + H{-}Br$	−71 kJ/mol (−17 kcal/mol)

Problem 8.4

Write a pair of chain propagation steps for the radical bromination of propane to give 1-bromopropane. Then calculate ΔH^0 for each propagation step and for the overall reaction.

D. Regioselectivity of Bromination Versus Chlorination: Hammond's Postulate

The regioselectivity in halogenation of alkanes can be accounted for in terms of the relative stabilities of radicals (3° > 2° > 1° > methyl). As we will see, the energy of the transition state reflects the energy of the radicals; more stable radical products are formed with a lower activation energy, making them faster. But how do we account for the greater regioselectivity in bromination of alkanes compared with chlorination of alkanes? To do so, we need to consider **Hammond's postulate**, a refinement of transition state theory proposed in 1955 by George Hammond, then at Iowa State University. According to this postulate:

> The structure of the transition state for an exothermic reaction step is reached relatively early in the reaction, so it resembles the reactants of that step more than the products. Conversely, the structure of the transition state for an endothermic reaction step is reached relatively late, so it resembles the products of that step more than the reactants.

It is important to realize that we cannot observe a transition state directly. Until the advent of modern computational theory, we could only infer its existence, structure, and stability from experiment. Hammond's postulate gives us a reasonable way of deducing something about the structure of a transition state by examining things we can observe: the structure of reactants and products and heats of reaction.

Hammond's postulate

The structure of the transition state for an exothermic step looks more like the reactants of that step than the products. Conversely, the structure of the transition state for an endothermic step looks more like the products of that step than the reactants.

Hammond's postulate applies equally well to multistep reactions. The transition state of any exothermic step in a multistep sequence looks more like the starting material(s) of that step; the transition state of any endothermic step in the sequence looks more like the product(s) of that step. Thus, changes in starting material energy affect the transition state of an exothermic reaction more than changes in product energy. The converse is true for an endothermic reaction. Today we can carry out high-level computations that reveal details of complex transition states and strongly support the validity of Hammond's postulate.

Shown in Figure 8.2 are energy diagrams for a highly exothermic reaction and a highly endothermic reaction, each occurring in one step.

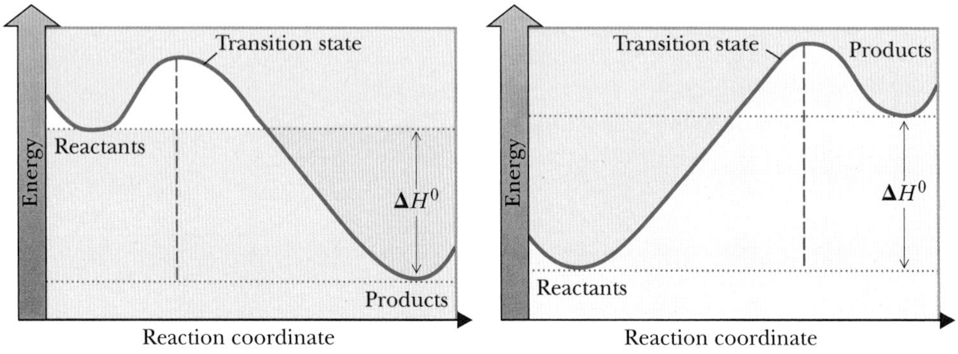

(a) Highly exothermic reaction **(b)** Highly endothermic reaction

FIGURE 8.2 Hammond's postulate. Energy diagrams for two one-step reactions. In the exothermic reaction, the transition state occurs early, and its structure resembles that of the reactants. In the endothermic reaction, the transition state occurs late, and its structure resembles that of the products.

Now let us apply Hammond's postulate to explain the relative regioselectivities of chlorination versus bromination of alkanes. In applying this postulate, we deal with the rate-determining step of the reaction, which, in radical halogenation of alkanes, is the abstraction of a hydrogen atom by a halogen radical. Given in Table 8.8 are

Table 8.8 Enthalpies of Reaction for Hydrogen Abstraction in 2-Methylpropane

Reaction Step	ΔH^0, [kJ, (kcal)/mol]	
Chlorination		
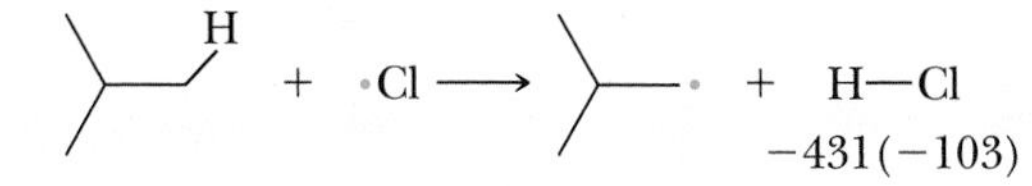 +422(101) Primary radical	−9(−2)	17(4)
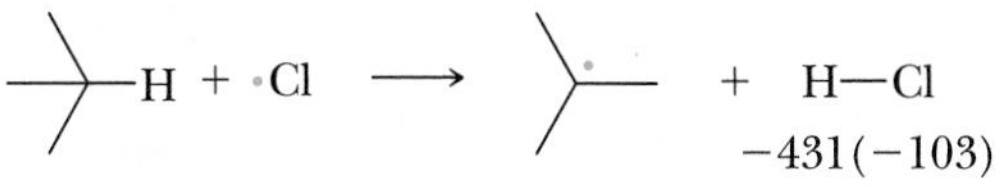 +405(97) Tertiary radical	−26(−6)	
Bromination		
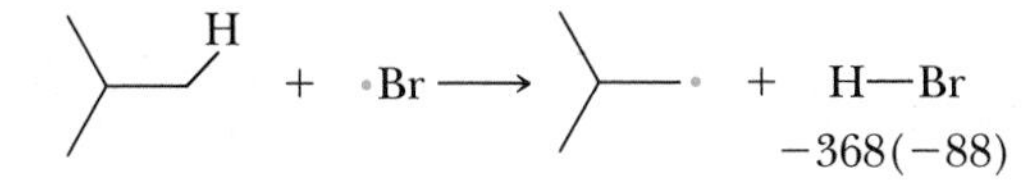 +422(101)	+54(+13)	17(4)
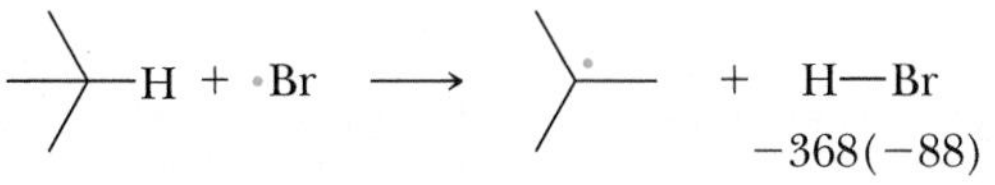 +405(97)	+37(+9)	

heats of reaction, ΔH^0, for the hydrogen abstraction step in chlorination and bromination of the different hydrogens of 2-methylpropane (isobutane). Also given under the formulas of isobutane, HCl, and HBr are bond dissociation enthalpies for the bonds broken (1° and 3° C—H) and formed (H—Cl and H—Br) in each step. Because the 3° radical is more stable than the 1° radical, the BDE for the 3° H is lower than that of a 1° H by about 17 kJ (4 kcal)/mol and the difference in ΔH^0 between 1° and 3° for the two reactions is just this amount.

Abstraction of hydrogen by chlorine is exothermic, which, according to Hammond's postulate, means that the transition state for H abstraction by Cl· is reached early in the course of the reaction [Figure 8.3(a)]. Therefore, the structure of the transition state for this step resembles the reactants, namely the alkane and a chlorine atom, not the product radicals. As a result, there is relatively little radical character on carbon in this transition state, and regioselectively in radical chlorination is only slightly influenced by the relative stabilities of radical intermediates. Products are determined more by whether a chlorine atom happens to collide with a 1°, 2°, or 3° H.

In Section 8.4A, we were given the fact that the selectivity for abstraction of a 3° H compared to a 1° H in chlorination is 5:1; this ratio directly reflects the relative reaction rates of these hydrogens with chlorine atoms. Using this ratio of reaction rates and the relationship between $\Delta G^\ddagger$ and rate constants, we can calculate that the difference in activation energies, $\Delta\Delta G^\ddagger$, for the abstraction of a 3° H versus a 1° H is about 4 kJ (1 kcal)/mol. However, we can calculate from the primary and tertiary C—H bond dissociation enthalpies [Figure 8.3(a)] that $\Delta\Delta H^0$ for the two reactions is about 17 kJ (4 kcal)/mol. Thus, the difference in product stabilities is only slightly reflected in the transition states and the resulting reaction rates.

Contrast this reaction with bromination [Figure 8.3(b)]. For bromination, the selectivity of 3° H to 1° H is 1600:1, which corresponds to $\Delta\Delta G^\ddagger$ of approximately 18 kJ (4.2 kcal)/mol. The $\Delta\Delta H^0$ for the formation of the primary and tertiary radicals is

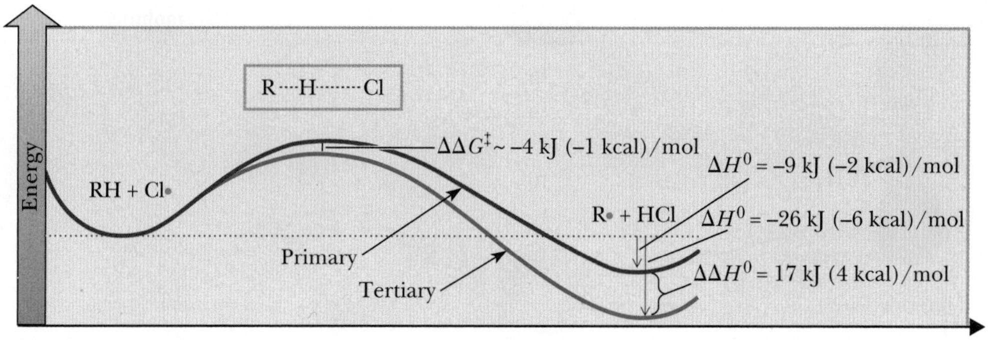

(a) Chlorination

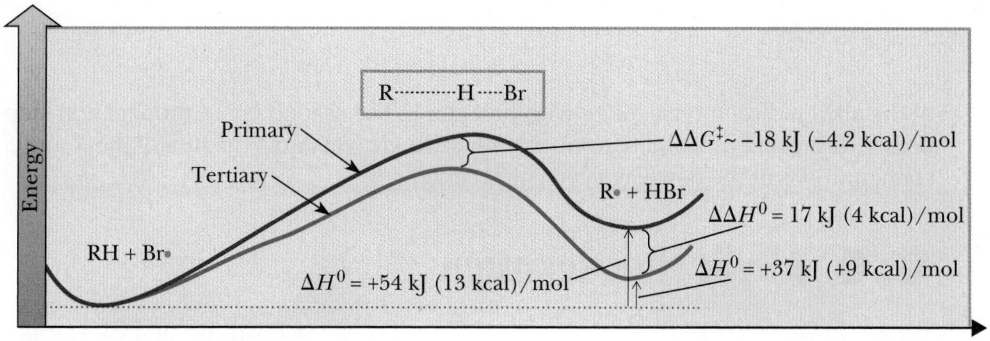

(b) Bromination

FIGURE 8.3 Transition states and energetics for hydrogen abstraction in the radical chlorination and bromination of 2-methylpropane (isobutane). The product is the intermediate radical, R·.

Watch a video explanation

▶ BDE values are ΔH^0 and not ΔG^0. Recall that $\Delta G = \Delta H - T\Delta S$. Because we are dealing with similar reactions, we can assume that entropy differences between them are nearly zero and, therefore, $\Delta\Delta G^0 \approx \Delta\Delta H^0$ and $\Delta\Delta G^\ddagger \approx \Delta\Delta H^\ddagger$, which allows us to make these comparisons.

the same in bromination and in chlorination (it is just the difference in BDE of the C—H bonds). But the rate-determining step for bromination, because it is endothermic, has a transition state more like the product radical, and the transition state reflects nearly all the energy difference of the primary and tertiary radicals. The later transition state, and the correspondingly larger $\Delta\Delta G^\ddagger$ (which causes a large difference in reaction rates), is the reason for the much larger regioselectivity in radical bromination than in radical chlorination.

E. Stereochemistry of Radical Halogenation

When radical halogenation produces a chiral center or takes place at a hydrogen on an existing chiral center, the product is an equal mixture of *R* and *S* enantiomers. Consider, for example, radical bromination of butane, which produces 2-bromobutane.

$$\underset{\textbf{Butane}}{CH_3CH_2CH_2CH_3} + Br_2 \xrightarrow[\text{or light}]{\text{heat}} \underset{\substack{\textbf{2-Bromobutane}\\ \text{(racemic)}}}{CH_3CH_2\overset{\overset{\large Br}{|}}{C}HCH_3} + HBr$$

In this example, both of the starting materials are achiral, and as is true for any reaction of achiral starting materials taking place in an achiral environment that gives a chiral product (Section 6.7A), the product is a racemic mixture (Figure 8.4).

In the case of the *sec*-butyl radical, the carbon bearing the unpaired electron is sp^2 hybridized and the unpaired electron lies in the unhybridized $2p$ orbital. Reaction of the alkyl radical intermediate with halogen in the second chain propagation step occurs with equal probability from either face to give an equal mixture of the *R* and *S* configurations at the newly created chiral center.

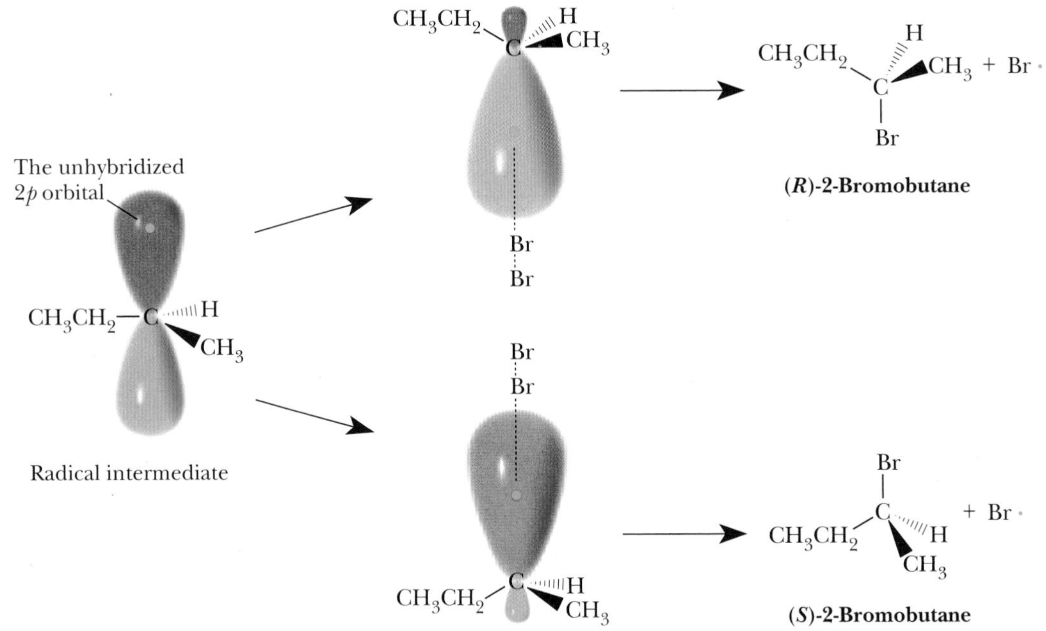

FIGURE 8.4 Radical bromination of butane.

8.6 Allylic Halogenation

We saw in Section 6.3D that propene and other alkenes react with Br_2 and Cl_2 at room temperature by addition to the carbon-carbon double bond. If, however, propene and one of these halogens are allowed to react at a high temperature, an entirely different reaction takes place; namely, substitution of a halogen occurs at the **allylic carbon** (the carbon next to a carbon-carbon double bond). We illustrate **allylic substitution** by the reaction of propene with chlorine at high temperature.

Allylic carbon A carbon adjacent to a carbon-carbon double bond.

Allylic substitution Any reaction in which an atom or a group of atoms is substituted for another atom or group of atoms at an allylic carbon.

$$CH_2{=}CHCH_3 + Cl_2 \xrightarrow{350°C} CH_2{=}CHCH_2Cl + HCl$$

Propene — **3-Chloropropene (Allyl chloride)**

A comparable reaction takes place when propene is treated with bromine at an elevated temperature.

To predict which of the various C—H bonds in propene is most likely to break when a mixture of propene and bromine or chlorine is heated, we need to look at bond dissociation enthalpies. We find that the bond dissociation enthalpy of an allylic C—H bond in propene (Table 8.7) is approximately 92 kJ (22 kcal)/mol less than that of a vinylic C—H bond and 50 kJ (12 kcal)/mol less than a C—H bond of ethane. The allyl radical is even more stable than a 3° radical; this unusual stability also applies to carbocations. The reason the allylic C—H bond is so weak is discussed in Section 8.6B. Note from Table 8.7 that the benzyl radical $C_6H_5CH_2\cdot$ is stabilized in exactly the same way as the allyl radical and for the same reason; benzylic compounds undergo many of the same reactions as allylic compounds (Section 21.5).

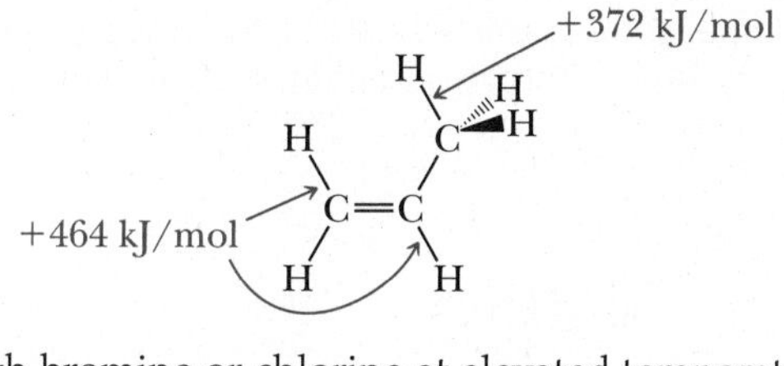

Treating propene with bromine or chlorine at elevated temperatures illustrates a very important point about organic reactions: it is often possible to change the product(s) by changing the mechanism through a change in reaction conditions. Under the high temperatures used in this reaction, the concentration of bromine radicals becomes much higher than at room temperature; this greatly accelerates the substitution reaction, which occurs by the radical halogenation mechanism. At room temperature, there are far fewer radicals and electrophilic addition is observed.

$$CH_2{=}CHCH_3 + Br_2 \xrightarrow{\text{high temp.}} CH_2{=}CHCH_2Br + HBr \quad \text{(allylic substitution)}$$

Propene — **3-Bromopropene**

$$CH_2{=}CHCH_3 + Br_2 \xrightarrow[CH_2Cl_2]{\text{room temp.}} CH_2(Br)CH(Br)CH_3 \quad \text{(electrophilic addition)}$$

Propene — **1,2-Dibromopropane** (racemic)

A very useful way to carry out allylic bromination in the laboratory at or slightly above room temperature is to use the reagent *N*-bromosuccinimide (NBS) in dichloromethane (CH_2Cl_2). Reaction between an alkene and NBS is most commonly initiated by light. This reaction involves a net double substitution: a bromine in NBS and a hydrogen in the alkene exchange places.

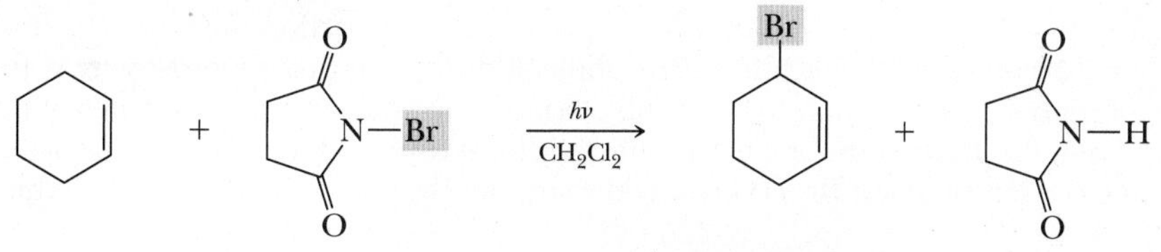

Cyclohexene **N-Bromosuccinimide (NBS)** **3-Bromocyclohexene** (racemic) **Succinimide**

A. Mechanism of Allylic Halogenation

Allylic bromination and chlorination proceed by a radical chain mechanism involving the same type of chain initiation, chain propagation, and chain termination steps involved in the radical halogenation of alkanes.

Mechanism 8.2

Allylic Bromination of Propene using NBS

Chain Initiation Chain initiation involves formation of radicals from NBS by light-induced homolytic cleavage of the N—Br bond in NBS. This step is analogous to the homolytic dissociation of chlorine to chlorine radicals.

Step 1:

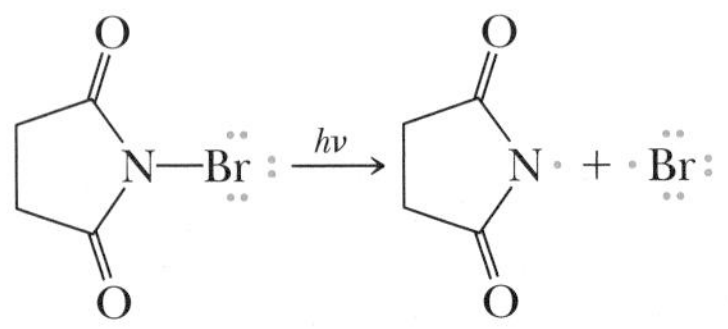

Chain Propagation Chain propagation involves the formation of products. Reaction of a radical and a nonradical gives a new radical. (Both radicals formed in the initiation can abstract hydrogen atoms. We show only the Br· reaction.) In the first propagation step, a bromine atom abstracts an allylic hydrogen (the weakest C—H bond in propene) to produce an allyl radical. The allyl radical, in turn, reacts with a bromine molecule to form allyl bromide and a new bromine atom.

Step 2:

$$CH_2{=}CHCH_2{-}H + \cdot\ddot{Br}: \longrightarrow CH_2{=}CHCH_2\cdot + H{-}\ddot{Br}:$$

Allylic hydrogen — Allyl radical

Step 3:

$$CH_2{=}CHCH_2\cdot + :\ddot{Br}{-}\ddot{Br}: \longrightarrow CH_2{=}CHCH_2{-}\ddot{Br}: + \cdot\ddot{Br}:$$

Note that, as always, this combination of chain propagation steps adds up to the observed stoichiometry. This reaction is exactly like halogenation of alkanes, but is strongly regioselective for the allylic hydrogen because of its weak bond.

Chain Termination—The Destruction of Radicals Propagation of the chain reaction continues until termination steps produce nonradical products and thus stop further reaction.

Step 4:

$$:\ddot{Br}\cdot + \cdot\ddot{Br}: \longrightarrow :\ddot{Br}{-}\ddot{Br}:$$

Step 5:

$$CH_2{=}CHCH_2\cdot + \cdot\ddot{Br}: \longrightarrow CH_2{=}CHCH_2{-}\ddot{Br}:$$

Step 6:

$$CH_2{=}CHCH_2\cdot + \cdot CH_2CH{=}CH_2 \longrightarrow CH_2{=}CHCH_2{-}CH_2CH{=}CH_2$$

The Br_2 necessary for Step 2 is formed by reaction of product **HBr** with **NBS**.

Step 7:

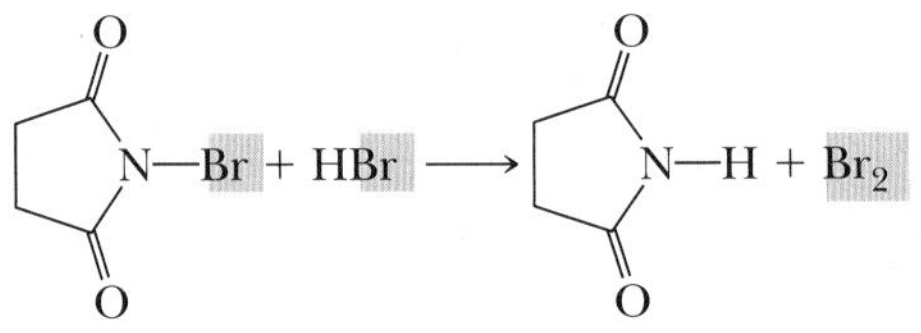

Bromine formed in this step then reacts with an allyl radical to continue the chain propagation reactions (Step 3).

The mechanism we described for allylic bromination by NBS poses the following problem. NBS is the indirect source of Br_2, which then takes part in chain propagation. But if Br_2 is present in the reaction mixture, why does it not react instead with the carbon-carbon double bond by electrophilic addition? In other words, why is the observed reaction allylic substitution rather than addition to the double bond? The answer is that the rates of the chain propagation steps are much faster than the rate of electrophilic addition of bromine to the alkene when radicals are present. Furthermore, the concentration of Br_2 is very low throughout the course of the reaction, which slows the rate of electrophilic addition.

B. Structure of the Allyl Radical

The allyl radical can be represented as a hybrid of two contributing structures. Here, fishhook arrows show the redistribution of single electrons between contributing structures. Note that three π electrons take part in this resonance.

$$CH_2{=}CH{-}\dot{C}H_2 \longleftrightarrow \dot{C}H_2{-}CH{=}CH_2$$

(Equivalent contributing structures)

▶ Delocalization of unpaired electron density over a larger area is stabilizing.

The position of the radical electron in the two contributing structures predicts that radical reactivity will occur at carbons 1 and 3 but not at carbon 2. This result is experimentally observed. ▶

The eight atoms of the allyl radical lie in a plane, and all bond angles are approximately 120°. Each carbon atom is sp^2 hybridized, and the three $2p$ orbitals participating in resonance delocalization of the radical are parallel to one another as shown in Figure 8.5. Like charged systems, in which a delocalized charge is more stable than a localized one, delocalized unpaired electron density leads to more stable structures than localized unpaired electron density.

Because of the radical-stabilizing influence of this electron delocalization, it is reasonable to expect that the BDE of an allylic C—H is significantly weaker than that of a primary C—H. In fact, based on bond dissociation enthalpies, we conclude that an allyl radical is even more stable than a 3° alkyl radical. Note that because of the larger amount of *s* character in its carbon sp^2 hybrid orbital, a vinylic C—H bond is stronger (has a larger bond dissociation enthalpy) than any sp^3 C—H bond and is never abstracted in homolytic reactions.

According to the molecular orbital description, the conjugated system of the allyl radical involves the formation of three molecular orbitals by overlap of three $2p$ atomic orbitals (Figure 8.5). The lowest energy MO has zero nodes, the next MO has one node, and the highest energy MO has two nodes. The molecular orbital of intermediate energy in this case leads to neither net stabilization nor destabilization and is therefore called a nonbonding MO. The lowest π MO is at a lower energy than the isolated $2p$ orbitals.

FIGURE 8.5 Molecular orbital model of covalent bonding in the allyl radical. Combination of three $2p$ atomic orbitals gives three π MOs. The lowest, a π-bonding MO, has zero nodes; the next in energy, a π-nonbonding MO, has one node; and the highest in energy, a π-antibonding MO, has two nodes.

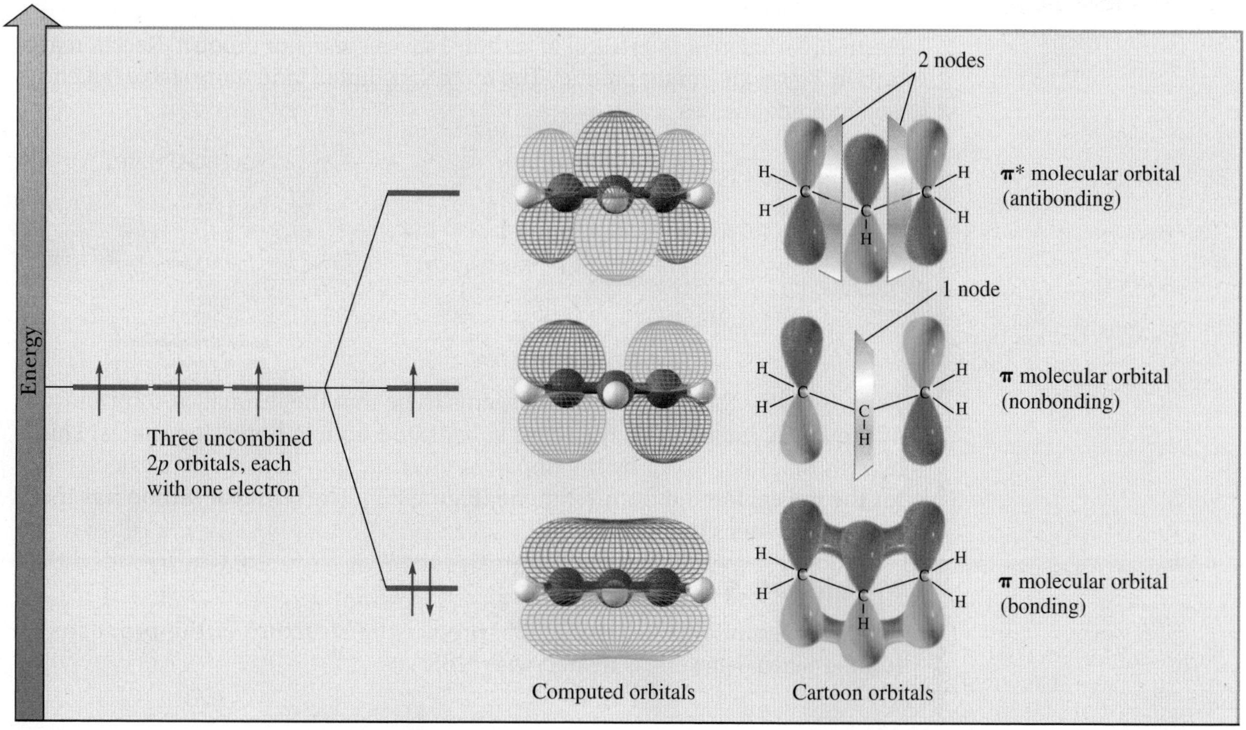

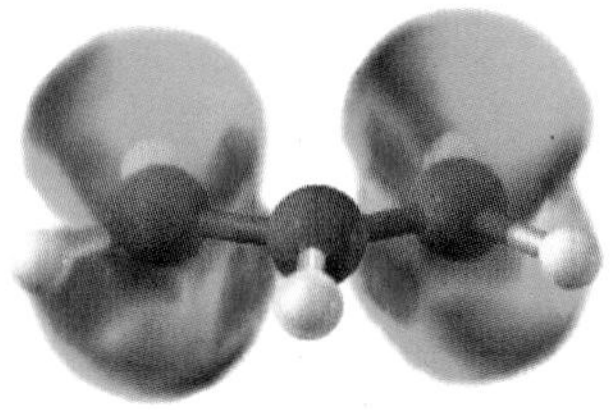

FIGURE 8.6 Unpaired electron spin density map for the allyl radical. Unpaired electron density (green cones) appears only on carbons 1 and 3.

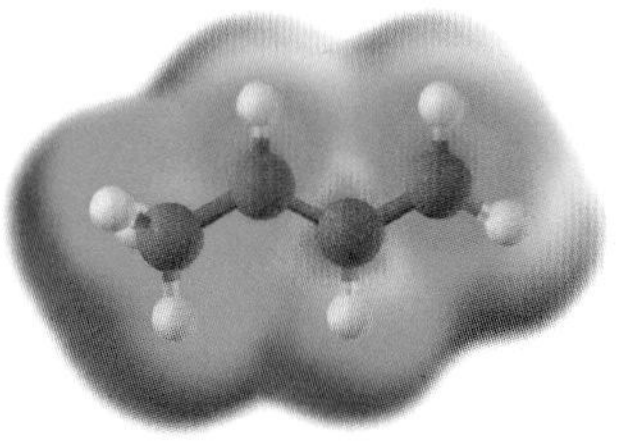

FIGURE 8.7 Unpaired electron spin density map of the radical formed from 1-butene. Spin (blue) is on carbons 1 and 3, but more is on carbon 1.

In the lowest energy (ground) state of the allyl radical, two electrons of the π system lie in the π-bonding MO and the third lies in the π-nonbonding MO; the π-antibonding MO is unoccupied. Because the lowest π MO is at a lower energy than the isolated $2p$ atomic orbitals, putting two electrons in this MO releases considerable energy, which accounts for the stability of the allyl radical.

The lone electron of the allyl radical is associated with the π-nonbonding MO, which places electron density on carbons 1 and 3 only. This localization is shown clearly in the unpaired electron density map in Figure 8.6. Thus, both the resonance model and molecular orbital theory are consistent in predicting radical character on carbons 1 and 3 of the allyl radical but no radical character on carbon 2, consistent with the experimental observation. Importantly, when there is a difference, the reaction will occur to generate the alkene product that is most stable—in other words, with the more highly substituted double bond.

Example 8.5 Allylic Bromination

Account for the fact that allylic bromination of 1-octene by NBS gives these products.

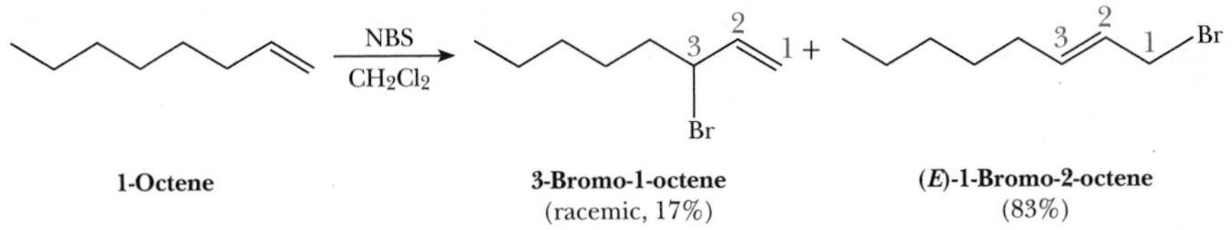
NBS
CH_2Cl_2
3
2
1 +
Br
3
2
1
Br
1-Octene
3-Bromo-1-octene
(racemic, 17%)
(*E*)-1-Bromo-2-octene
(83%)

Solution

The rate-determining step of this radical chain mechanism is hydrogen abstraction from the allylic position on 1-octene to give a 2° allylic radical. This radical is stabilized by delocalization of the two π electrons and the unpaired electron. Reaction of the radical at carbon 1 gives the major product. Reaction at carbon 3 gives the minor product. The more substituted (and more stable) alkene isomer predominates.

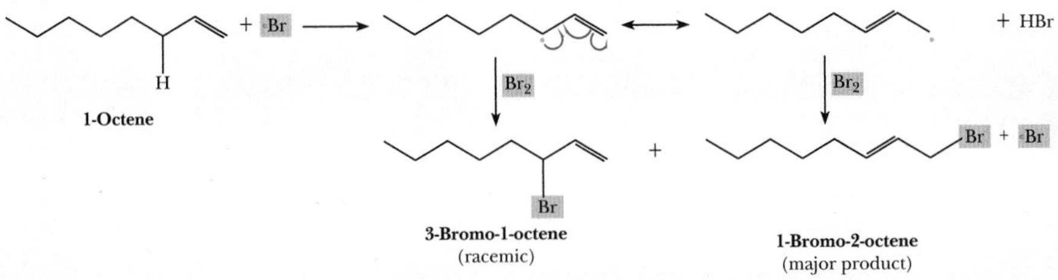
+ ·Br
H
1-Octene
+ HBr
Br_2
Br_2
+
Br
Br + ·Br
3-Bromo-1-octene
(racemic)
1-Bromo-2-octene
(major product)

The reason for this regioselectivity seems to be that the resonance contributor of the allylic radical with the more substituted double bond dominates. This hypothesis is borne out by the unpaired electron spin density, here calculated for the radical formed from 1-butene (Figure 8.7). The terminal carbon has the highest spin density (blue).

Problem 8.5

Given the solution to Example 8.5, predict the structure of the product(s) formed when 3-hexene is treated with NBS.

8.7 Radical Autoxidation

One of the most important destructive reactions for materials, foods, and living systems is called **autoxidation** (i.e., oxidation requiring oxygen and no other reactant). This reaction takes place by a radical chain mechanism very similar to that for allylic bromination. If you open a bottle of cooking oil that has stood for a long time, you will notice the hiss of air entering the bottle because of the negative pressure caused by the consumption of oxygen by autoxidation of the oil.

Autoxidation
Air oxidation of materials such as unsaturated fatty acids.

Cooking oil contains **polyunsaturated fatty acid esters**. (See Section 5.4, "Connections to Biological Chemistry: The Importance of *Cis* Double Bonds in Fats Versus Oils," as well as Section 26.1.) The most common of these compounds have chains of 16 or 18 carbons containing 1,4-diene functional groups. (Both double bonds are *cis*; the nature of R_1 and R_2 need not concern us at this stage.) The hydrogens on the CH_2 group between the double bonds are doubly allylic; that is, they are allylic with respect to both double bonds. As you might expect, the radical formed by abstraction of one of these hydrogens is unusually stable because it is even more delocalized (described by resonance contributing structures) than an allylic radical. An allylic C—H bond is much weaker than a corresponding alkane C—H bond, and the doubly allylic C—H is even weaker.

Autoxidation begins when a radical initiator, X·, which is formed either by light activation of an impurity in the oil or by thermal decomposition of peroxide impurities, abstracts a doubly allylic hydrogen to form a radical. This radical is delocalized through resonance with both double bonds (1 and 2 in the following structure).

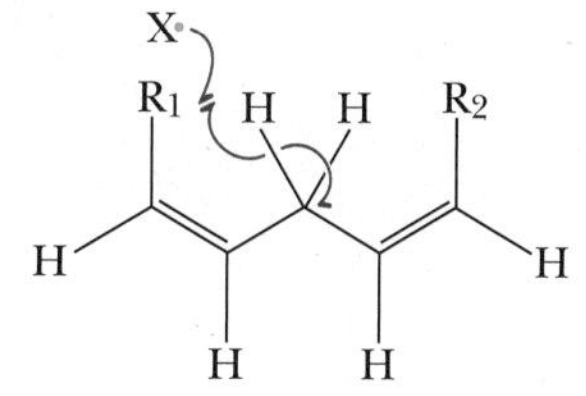

A polyunsaturated fatty acid ester (R—H)

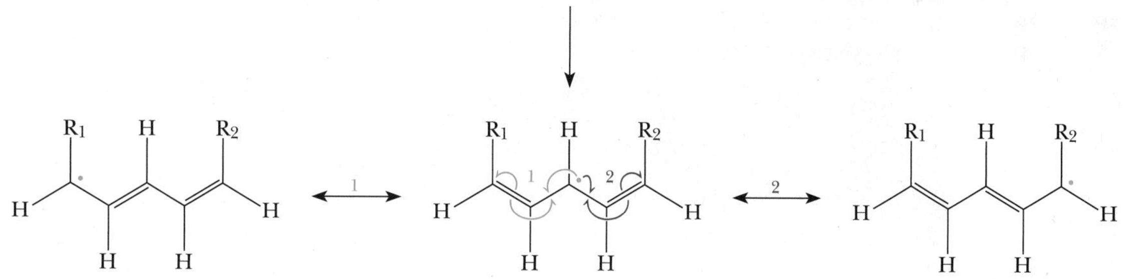

This radical reacts with oxygen, itself a (very unreactive) diradical, to form a peroxy radical, which then reacts with the CH_2 of another 1,4-diene fatty acid ester (R—H) to give a new radical (R·) and a hydroperoxide. Hydroperoxides are formed on both sides by reactions with the resonance hybrid; only one is shown. The new radical reacts again with oxygen, causing a radical chain reaction in which hundreds of molecules of fatty acid ester are oxidized for each initiator radical.

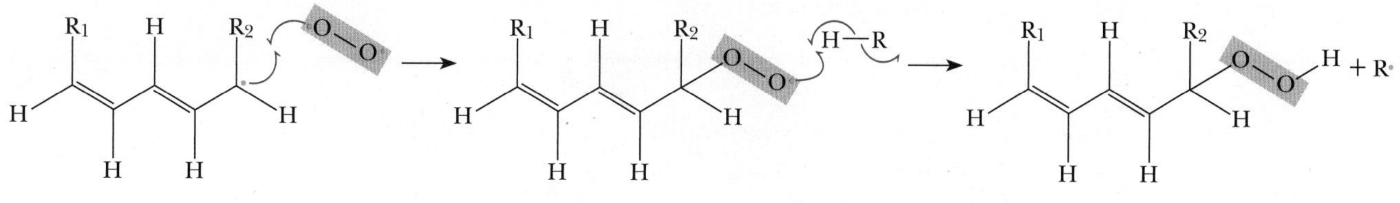

Peroxy radical

A hydroperoxide

The ultimate fate of the peroxide, and some of the peroxy radical as well, is complex. Some autoxidation products degrade to short-chain aldehydes and carboxylic acids with unpleasant"rancid"smells familiar to anyone who has smelled old cooking oil or aged foods that contain polyunsaturated oils. It has been suggested that some products of autoxidation of oils are toxic and/or carcinogenic. Oils lacking the 1,4-diene structure are much less easily oxidized.

Radical inhibitor

A compound such as a phenol that selectively reacts with radicals to remove them from a chain reaction and terminate the chain.

Many natural and unnatural compounds can act to terminate the radical chain reaction and are referred to as **radical inhibitors**. Many are phenols, as are described in the MCAT Practice.

Example 8.6 Radical Autoxidation

What products would you expect from the following reaction? Indicate the major one and specify stereochemistry if relevant.

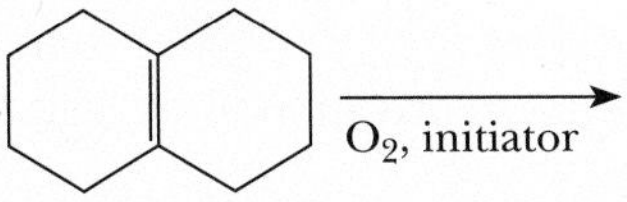
O₂, initiator

Solution

The major product has the more substituted double bond. (Both are racemic.)

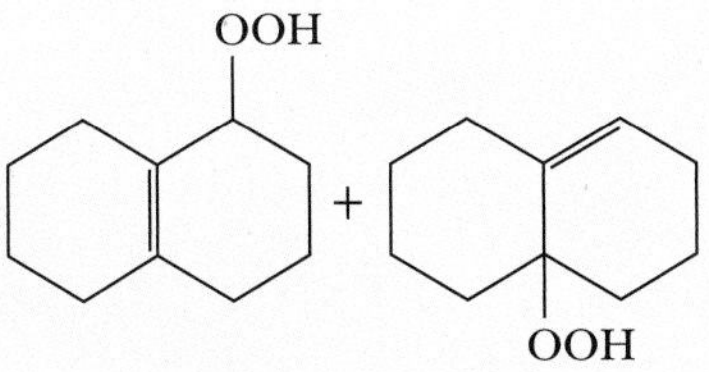
OOH
+
OOH

Major (racemic) Minor (racemic)

Problem 8.6

Show the products of the following reaction and indicate the major one.

$$CH_3(CH_2)_4CH_2CH{=}CH_2 \xrightarrow[O_2,\ \text{initiator}]{}$$

MCAT Practice: *Passage and Questions*

Antioxidants

Many plants contain polyunsaturated fatty acid esters in their leaves or seeds, as do the foods derived from these plants. Such structures are prone to autoxidation that gives hydroperoxides leading to lower molecular weight oxidized products, such as volatile aldehydes, ketones, and hydrocarbons. These products are indicative of the rotting of the food, and result in the unpleasant "rancid" odors familiar to anyone who has smelled old cooking oil.

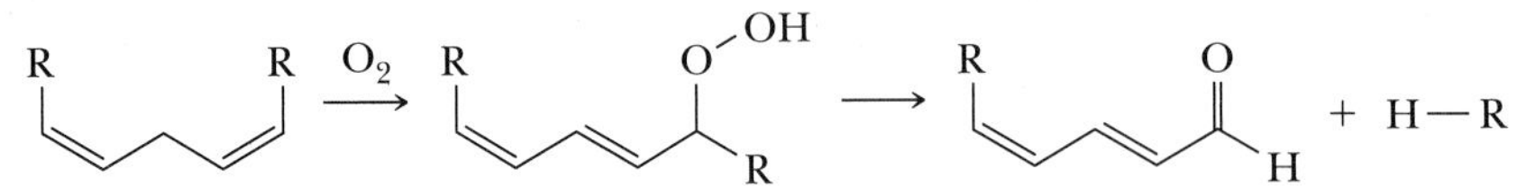
R
R
O₂
R
O
OH
R
R
O
H
+ H—R

Part of a polyunsaturated fatty acid

More volatile species with rancid odors

Questions

A. Linoleic acid is shown below. What makes this fatty acid particularly susceptible to autoxidation?

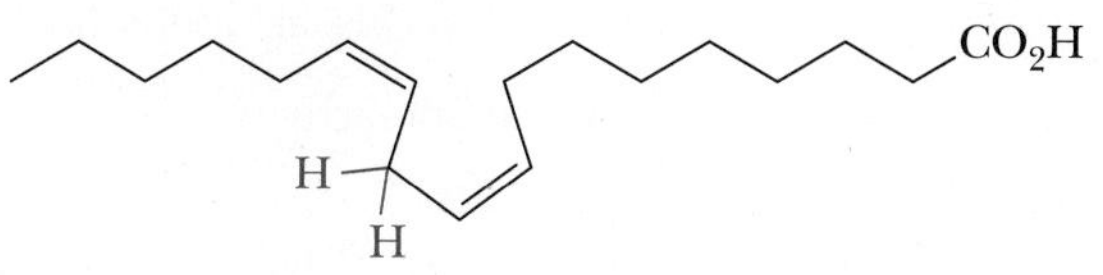

Linoleic acid

1. The red C—H bond has a low bond dissociation energy because it is doubly allylic.
2. The red C—H bond has a high bond dissociation energy because it is doubly allylic.
3. The red C—H bond is the most accessible to reaction with O_2 because it is the least sterically crowded C—H bond.
4. Both 2 and 3.

B. The reaction of the red C—H bond with oxygen leads to a carbon radical. This radical is,

1. stabilized primarily via induction.
2. stabilized because sp^2 carbons are more stable than sp^3 carbons.
3. stabilized primarily via hyperconjugation.
4. stabilized primarily via resonance delocalization.

Nature protects against autoxidation by a variety of agents, one of the most important of which is α-tocopherol (vitamin E). This compound is a phenol, which is defined as an OH group on a phenyl ring. The characteristic of phenols that makes them protective agents against autoxidation is their O—H bond, which is even weaker than the C—H bond from which hydrogen is abstracted by O_2 in the polyunsaturated fatty acid. Vitamin E reacts preferentially with the initial peroxy radical to give a resonance stabilized phenoxy radical, which is less reactive and survives to scavenge another peroxy radical. The resulting peroxide derived from vitamin E is relatively stable.

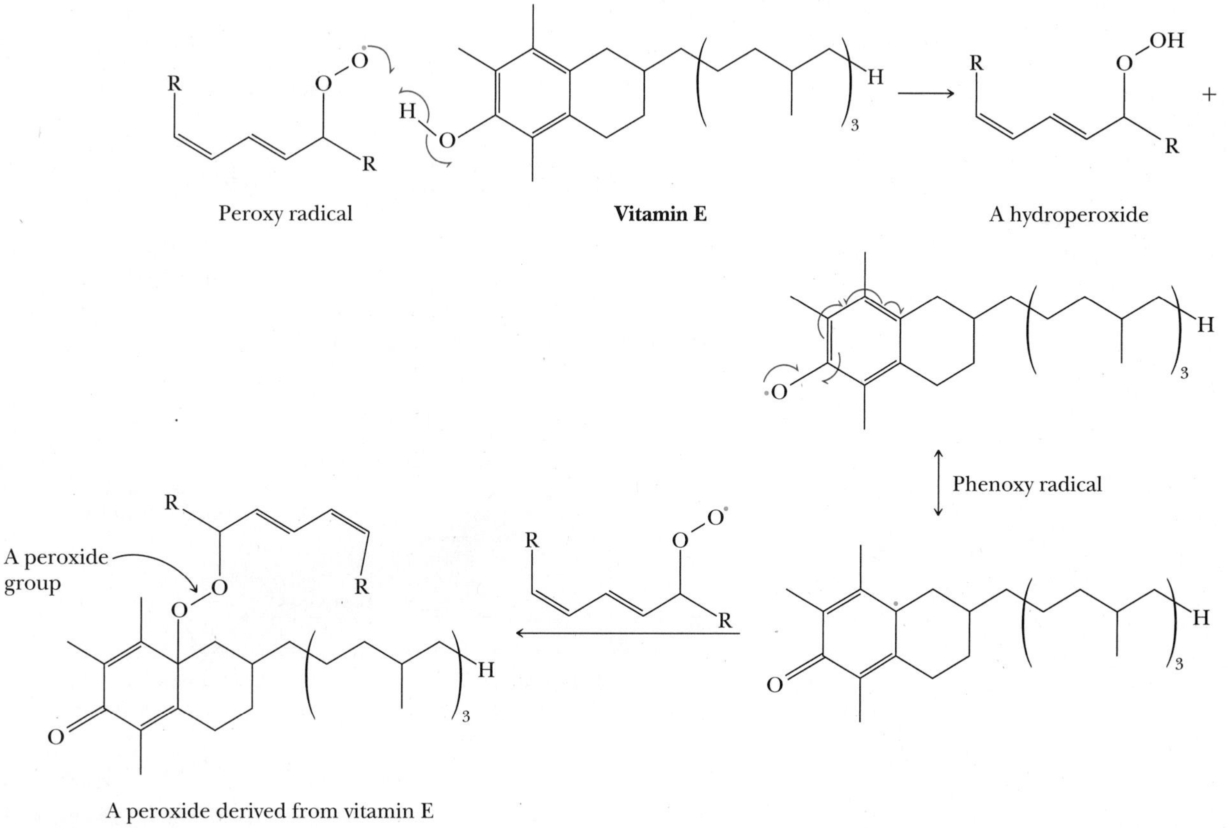

C. The action of each vitamin E molecule removes how many peroxy radicals in the autoxidation process?

1. one
2. two
3. three
4. four

(Continued)

D. The strength of the H—O bond in vitamin E is weaker than the C—H from which the hydrogen is abstracted in the fatty acid primarily because,

1. O—H bonds are always weaker than C—H bonds.
2. The radical created from hydrogen abstraction from O—H is stabilized by resonance delocalization and many of the contributing structures have secondary carbon radical character.
3. The sp^2 hybridization of the phenyl ring changes the hybridization of the oxygen in the O—H bond.
4. The radical created from hydrogen abstraction from O—H is stabilized by resonance delocalization and many of the contributing structures have tertiary carbon radical character.

Because vitamin E is often removed during the processing of foods, similar phenols such as BHT are often added to retard spoilage, and are referred to as preservatives. Upon hydrogen abstraction from BHT, a stable radical is created that does not react further with more peroxy radicals, as does vitamin E.

E. Why is the BHT radical unreactive?

1. It is sterically hindered.
2. It has more contributing structures than does the radical from vitamin E.
3. The O—H bond in BHT is expected to be stronger than the O—H bond in vitamin E due to the inductive effect of the *t*-butyl groups.
4. All of the above.

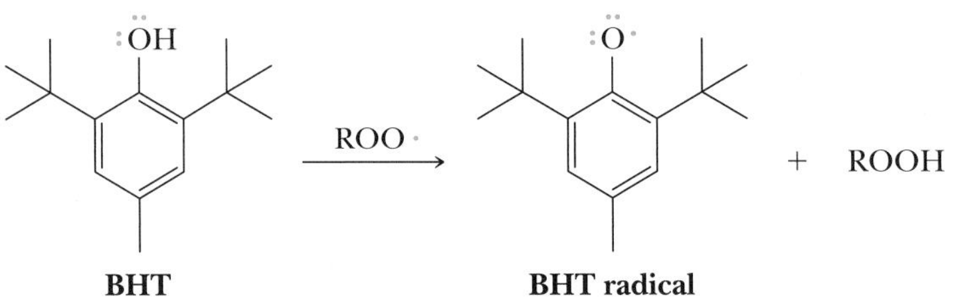

8.8 Radical Addition of HBr to Alkenes

When the addition of hydrogen halides to alkenes was first studied systematically in the 1930s, chemists observed that the addition of HBr sometimes gave Markovnikov addition and sometimes gave non-Markovnikov addition. These two modes of addition of HBr are illustrated for 2-methylpropene (isobutylene).

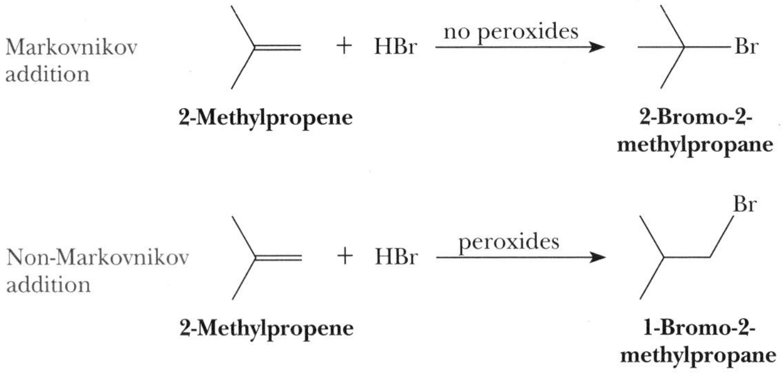

The puzzle was solved in 1933 when it was discovered that non-Markovnikov products were observed only in the presence of peroxides or other sources of radicals. In the absence of radicals, addition of HBr gave only the expected Markovnikov product.

To account for the products of HBr addition to alkenes in the presence of peroxides, chemists proposed a radical chain mechanism like the one for halogenation (Section 8.6A). In the following mechanism, the source of initiating radicals is a dialkyl peroxide, which is frequently present as an impurity in the solvent or alkene.

Mechanism 8.3

Radical Initiated Non-Markovnikov Addition of HBr to Alkenes

Chain Initiation Homolytic cleavage of a dialkyl peroxide is induced by light or heat to give two alkoxy radicals. An alkoxy radical then reacts with **HBr** by hydrogen abstraction to give an alcohol and a bromine radical.

Step 1:

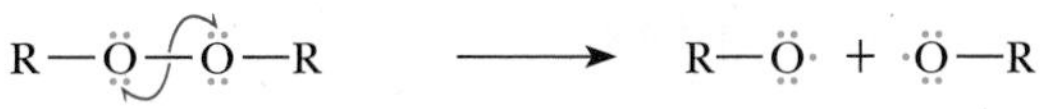

A dialkyl peroxide — Two alkoxy radicals

Step 2:

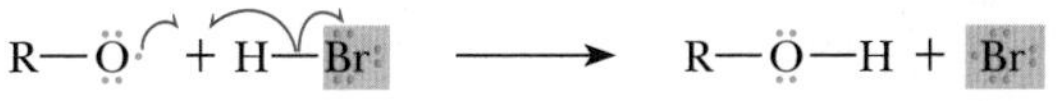

Bromine radical

Chain Propagation A bromine radical adds to the carbon-carbon double bond regioselectively to give the more substituted (and more stable) carbon radical. The carbon radical, in turn, reacts with a molecule of **HBr** to give the bromoalkane and to generate a new bromine radical. Note that in each propagation step, one radical is consumed, but another is formed.

Step 3:

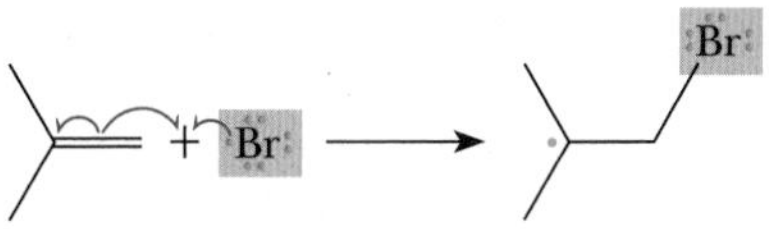

A 3° radical

Step 4:

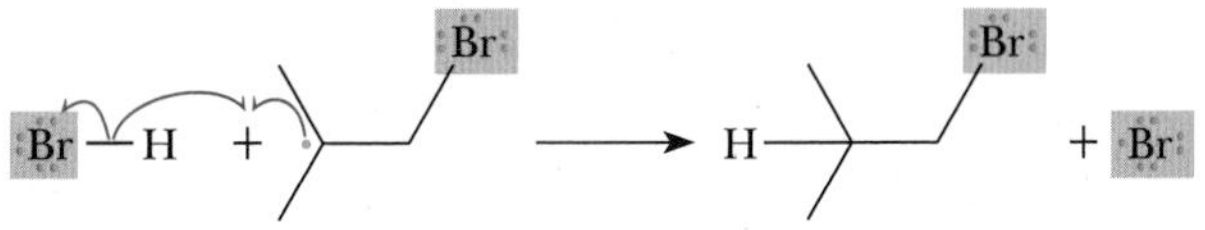

1-Bromo-2-methylpropane

Chain Termination The most important chain termination steps are the combination of a carbon radical with a bromine radical and the combination of two bromine radicals. Each of these steps destroys one or both of the radical intermediates in the chain.

Step 5:

Step 6:

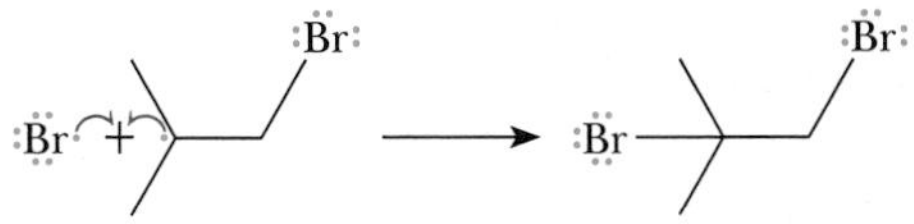

The observed non-Markovnikov regioselectivity of radical addition of **HBr** to an alkene is a combination of a steric factor and an electronic factor. First, a bromine radical attacks the less hindered carbon of the double bond (the steric factor). Second, as mentioned in Section 8.5D, the relative stabilities of radicals parallel those of carbocations (Section 6.3A).

Because the intermediate in Step 3 is a radical, it does not preserve stereochemistry in Step 4 when relevant.

The addition of the bromine atom to the alkene could occur at either C of the double bond, but it is dominated by addition that gives the more stable radical. In the mechanism just described, the two choices are the formation of a primary or tertiary radical. Because a tertiary carbon radical is more stable than a primary radical, the regiochemistry ends up as non-Markovnikov. Recall that the polar addition of HBr to an alkene is regioselective (Section 6.3A), with bromine adding to the more substituted carbon (Markovnikov addition). There is an important similarity between the polar mechanism (Section 6.3A) and the radical mechanism. The regiochemistry of each reaction is dominated by the reactions that proceed through the most stable reactive intermediates, which are in both cases tertiary, as shown below. This pair of alkene additions illustrates how the products of a reaction often can be altered by a change in experimental conditions and a change in mechanism—in this case, a change from a polar mechanism to a radical mechanism. The non-Markovnikov creation of a haloalkane can be very useful in organic synthesis.

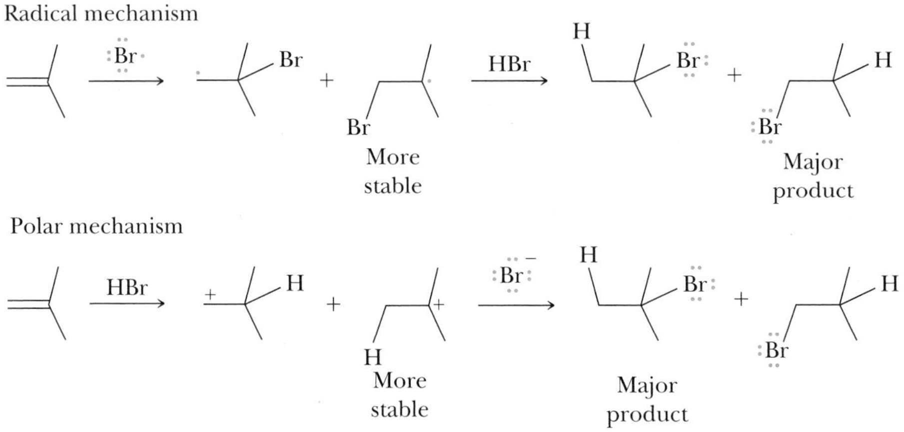

Interestingly, the non-Markovnikov addition occurs only with HBr. Both HCl and HI always add to alkenes according to Markovnikov's rule. The reason for the differences in reactivity between HCl, HBr, and HI is a subtle balance of bond strengths. As just discussed, the mechanism of non-Markovnikov addition involves hydrogen radical abstraction from the H—X bond and addition of a halogen radical to the alkene. The bond strength in HCl is too strong for the abstraction of an H atom, while the bond is sufficiently weak enough in HBr and HI. However, with HI, the addition of iodine atom to the double bond does not occur because the π bond is stronger than a C—I bond. With HCl and HBr, the C—X bond strengths exceed the π bond; therefore, these reactions are favorable. In summary, only HBr can add to alkenes via a radical mechanism, whereas all HX acids can add via mechanisms involving carbocations (Chapter 6).

Example 8.7 Free-Radical Addition of HX to an Alkene

Predict the product of the following reaction:

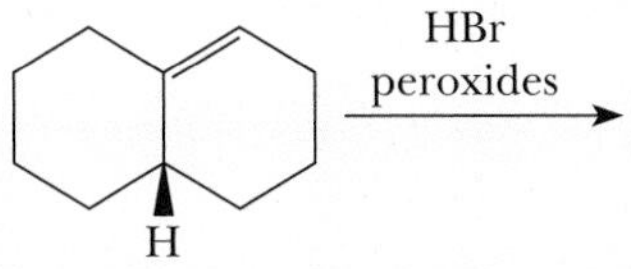

Solution

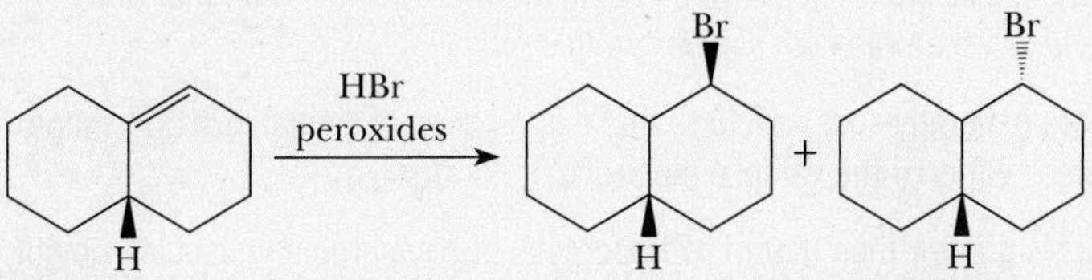
HBr
peroxides
Br
Br
H
H
H

Mixture of *cis* and *trans*; the radical intermediate does not preserve stereochemistry

Problem 8.7

Predict the major product of the following reaction:

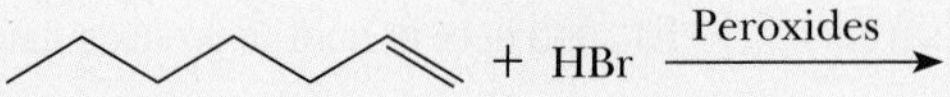
+ HBr
Peroxides

Watch a video explanation

Study Guide

8.1 Structure

- **Haloalkanes** contain a halogen covalently bonded to an sp^3 hybridized carbon.

8.2 Nomenclature

- In the IUPAC system, halogen atoms are named *fluoro-, chloro-, bromo-,* and *iodo-* and are listed in alphabetical order with other substituents. In the common system, they are named **alkyl halides**.
- **Haloalkenes** contain a halogen covalently bonded to an sp^2 hybridized carbon of an alkene. In the common system, they are named alkenyl or vinylic halides.
- **Haloarenes** contain a halogen atom covalently bonded to a benzene ring.

P **8.1, 8.8, 8.9**

8.3 Physical Properties of Haloalkanes

- The **van der Waals radius** of fluorine is only slightly greater than that of hydrogen, and among the other halogens, only iodine has a larger van der Waals radius than methyl.
- Among alkanes and chloro-, bromo-, and iodoalkanes of comparable size and shape, the haloalkanes have the higher boiling points predominantly because of the greater **polarizability** of the unshared electrons of the halogen atom.
    - *Polarizability refers to the distortion of the distribution of electron density around an atom that is interacting with another atom or ion.*

- *The electron density on larger, less electronegative atoms is more polarizable than that of electrons on more electronegative atoms with smaller atomic radii.*

- Boiling points of fluoroalkanes are generally comparable to those of alkanes of similar size and shape because of the uniquely low polarizability of the valence electrons of fluorine.
- The density of liquid haloalkanes is greater than that of hydrocarbons of comparable molecular weight because of the halogen's larger mass-to-volume ratio.

P 8.10–8.12

8.4 Preparation of Haloalkanes by Halogenation of Alkanes

- **Free-radical halogenation** of alkanes uses Cl_2 or Br_2 and light or heat to produce haloalkanes.
  - *Free-radical halogenation is regioselective and replaces* H *atoms with the halogen in the order allylic* $> 3° > 2° > 1° >$ *methyl.*
    - *This order of reactivity can be predicted on the basis of radical intermediate stabilities, which follow the same order. Radical intermediates are analogous to carbocations in the sense that they are stabilized by the same interactions, namely resonance delocalization and hyperconjugation with attached alkyl groups.*
- Regioselectivity of halogenation is greater for bromination than for chlorination.

P 8.2, 8.13–8.15, 8.20, 8.33–8.34

- Bond dissociation enthalpies (BDEs) are the enthalpies of homolytic bond cleavage for a given type of bond.
  - *The overall energetics of a reaction can be calculated by adding all the BDEs of the bonds broken minus the BDEs of the bonds made in the reaction.*
    - *Exothermic reactions (i.e., reactions favored to give products) have overall negative values calculated in this way.*

P 8.3, 8.17, 8.27

Problem 8.3

KEY REACTIONS

1. Chlorination and Bromination of Alkanes (Section 8.4) Chlorination and bromination of alkanes are regioselective in the order $3°\ H > 2°\ H > 1°\ H$. Bromination has a higher regioselectivity than chlorination. The mechanism involves a radical chain process.

$$CH_3CH_2CH_3 + Br_2 \xrightarrow[\text{or light}]{\text{heat}} \underset{92\%}{CH_3\overset{\overset{\large Br}{|}}{C}HCH_3} + \underset{8\%}{CH_3CH_2CH_2{-}Br}$$

8.5 Mechanism of Halogenation of Alkanes

- A **radical chain mechanism** consists of three types of steps: chain initiation, chain propagation, and chain termination.
  - *In* ***chain initiation****, radicals are formed from nonradical compounds.*

  - *In a **chain propagation** step, a radical and a molecule react to give a new radical and a new molecule.*
    - *When summed, chain propagation steps give the observed stoichiometry of the reaction.*
    - ***Chain length** is the number of times a cycle of chain propagation steps repeats.*
  - *In a **chain termination** step, radicals are destroyed, as two radicals collide to make a new bond.*

- Simple alkyl radicals are planar or almost so with bond angles of 120° about the carbon with the unpaired electron.
- Heats of reaction for a radical reaction and for individual chain initiation, propagation, and termination steps can be calculated from bond dissociation energies (BDEs).

P 8.4, 8.16, 8.21, 8.32

- According to **Hammond's postulate**, the structure of the transition state of an exothermic reaction step occurs early and looks more like the reactants of that step than like the products; thus, changes in the reactants have a large effect on the rate.
  - *Conversely, the structure of the transition state of an endothermic reaction step occurs later and looks more like the products of that step than like the reactants, and changes in products have a large effect on the rate. Hammond's postulate accounts for the fact that bromination of an alkane is more regioselective than chlorination.*
    - *For both bromination and chlorination of alkanes, the rate-determining step is hydrogen abstraction to form an alkyl radical. Hydrogen abstraction is endothermic for bromination and exothermic for chlorination.*

P 8.18, 8.19

▶ Figures 8.1 and 8.3

8.6 Allylic Halogenation

- **Allylic substitution** is any reaction in which an atom or a group of atoms is substituted for another atom or group of atoms at a carbon adjacent to a carbon-carbon double bond.
  - ***Allylic halogenation** proceeds by a radical chain mechanism.*
- Because of delocalization of unpaired electron density through resonance, the allyl radical is more stable than even the *tert*-butyl radical.

P 8.5, 8.22–8.26

KEY REACTIONS

2. Allylic Bromination (Section 8.6) These reactions occur at high temperatures (heat is the radical initiator) using the halogens themselves. Bromination using *N*-bromosuccinimide (NBS) is initiated by light. The mechanism involves a radical chain process, with a resonance delocalized allyl radical intermediate.

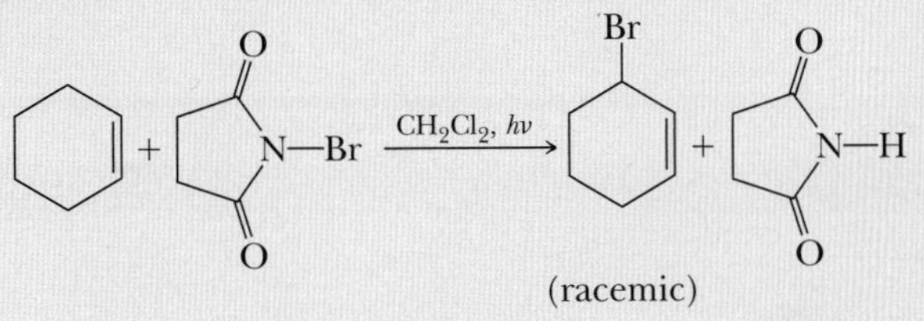

8.7 Radical Autoxidation

- **Autoxidation** of unsaturated compounds is an important process in aging and degradation of materials.

P 8.5, 8.29, 8.31

Problem 8.7

KEY REACTIONS

3. **Autoxidation (Section 8.7)** Autoxidation involves reaction of a CH bond, especially an allylic one, with oxygen under radical initiation conditions. The primary product is a hydroperoxide. The mechanism involves a radical chain process in which resonance-delocalized allylic radical intermediates react with molecular oxygen to give a peroxy radical that continues the radical chain.

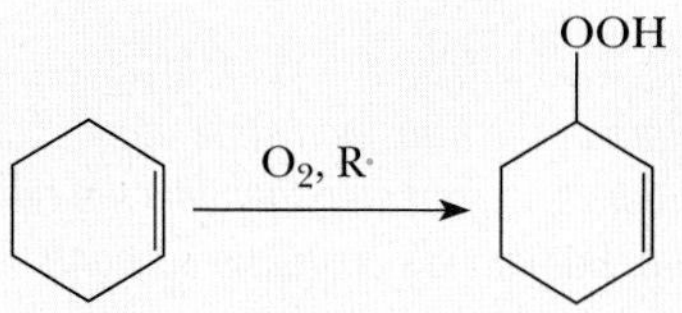

(racemic)

8.8 Radical Addition of HBr to Alkenes

- Addition of HBr to alkenes in the presence of peroxides can lead to non-Markovnikov addition because under these conditions, a free-radical chain mechanism operates.
  - *This reaction is particularly useful for making 1° haloalkanes from terminal alkenes.*

P 8.7, 8.30, 8.32

KEY REACTIONS

4. **HBr Addition to Alkenes Under Radical Conditions** Non-Markovnikov addition of HBr to alkenes occurs by a radical mechanism in the presence of peroxides, in which a Br· reacts with the π bond of the alkene to create a radical intermediate that abstracts H· from H—Br to continue the chain process. The regioselectivity of the products is the opposite of the ordinary Markovnikov addition products that form under polar conditions; hence, it is a useful alternative to polar addition of HBr in organic synthesis.

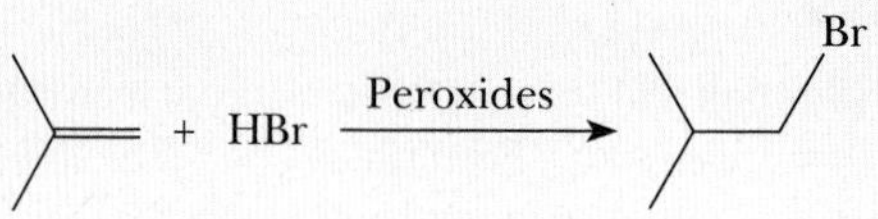

Problems

Red numbers indicate applied problems.

Nomenclature

8.8 Give IUPAC names for the following compounds. Where stereochemistry is shown, include a designation of configuration in your answer.

(a) 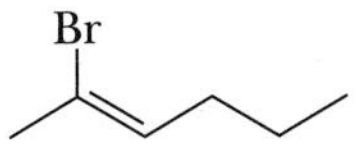

(b) 

(c) 

(d) 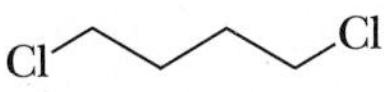

(e) 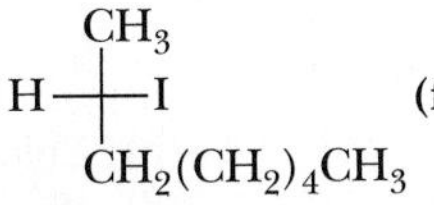

Fischer projection

(f) 

(g) 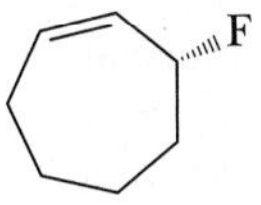

(h) 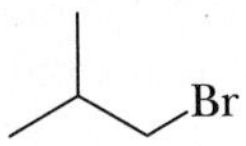

(i) 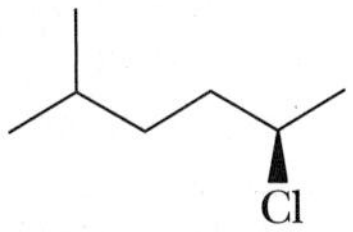

8.9 Draw structural formulas for the following compounds.

(a) 3-Iodo-1-propene
(b) (*R*)-2-Chlorobutane
(c) meso-2,3-Dibromobutane
(d) *trans*-1-Bromo-3-isopropylcyclohexane
(e) 1-Iodo-2,2-dimethylpropane
(f) Bromocyclobutane

Physical Properties

8.10 Water and dichloromethane are insoluble in each other. When each is added to a test tube, two layers form. Which layer is water, and which is dichloromethane?

8.11 The boiling point of methylcyclohexane (C_7H_{14}, MW 98.2) is 101°C. The boiling point of perfluoromethylcyclohexane (C_7F_{14}, MW 350) is 76°C. Account for the fact that although the molecular weight of perfluoromethylcyclohexane is over three times that of methylcyclohexane, its boiling point is lower than that of methylcyclohexane.

8.12 Account for the fact that among the chlorinated derivatives of methane, chloromethane has the largest dipole moment and tetrachloromethane has the smallest dipole moment.

Name	Molecular Formula	Dipole Moment (debyes, D)
Chloromethane	CH_3Cl	1.87
Dichloromethane	CH_2Cl_2	1.60
Trichloromethane	$CHCl_3$	1.01
Tetrachloromethane	CCl_4	0

Halogenation of Alkanes

8.13 Name and draw structural formulas for all possible monohalogenation products that might be formed in the following reactions.

(a) 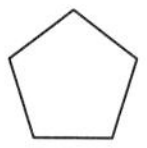 $+ Cl_2 \xrightarrow{\text{light}}$

(b) 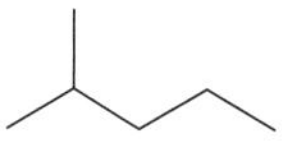 $+ Cl_2 \xrightarrow{\text{light}}$

(c) 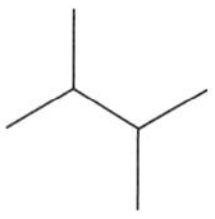 $+ Br_2 \xrightarrow{\text{light}}$

8.14 Which compounds can be prepared in high yield by halogenation of an alkane?

(a) 2-Chloropentane
(b) Chlorocyclopentane
(c) 2-Bromo-2-methylheptane
(d) (*R*)-2-Bromo-3-methylbutane
(e) 2-Bromo-2,4-trimethylpentane
(f) Iodoethane

8.15 There are three constitutional isomers with the molecular formula C_5H_{12}. When treated with chlorine at 300°C, isomer A gives a mixture of four monochlorination products. Under the same conditions, isomer B gives a mixture of three monochlorination products and isomer C gives only one monochlorination product. From this information, assign structural formulas to isomers A, B, and C.

8.16 Following is a balanced equation for bromination of toluene.

$$C_6H_5CH_3 + Br_2 \longrightarrow C_6H_5CH_2Br + HBr$$

Toluene **Benzyl bromide**

(a) Using the values for bond dissociation enthalpies given in Appendix 3, calculate ΔH^0 for this reaction.
(b) Propose a pair of chain propagation steps and show that they add up to the observed reaction.
(c) Calculate ΔH^0 for each chain propagation step.
(d) Which propagation step is rate-determining?

8.17 Write a balanced equation and calculate ΔH^0 for reaction of CH_4 and I_2 to give CH_3I and HI. Explain why this reaction cannot be used as a method of preparation of iodomethane.

8.18 Following are balanced equations for fluorination of propane to produce a mixture of 1-fluoropropane and 2-fluoropropane.

$$CH_3CH_2CH_3 + F_2 \longrightarrow CH_3CH_2CH_2F + HF$$

Propane **1-Fluoropropane**

$$CH_3CH_2CH_3 + F_2 \longrightarrow CH_3\overset{\overset{\large F}{|}}{C}HCH_3 + HF$$

Propane **2-Fluoropropane**

Assume that each product is formed by a radical chain mechanism.

(a) Calculate ΔH^0 for each reaction.
(b) Propose a pair of chain propagation steps for each reaction and calculate ΔH^0 for each step.
(c) Reasoning from Hammond's postulate, predict the regioselectivity of radical fluorination relative to that of radical chlorination and bromination.

8.19 As you demonstrated in Problem 8.18, fluorination of alkanes is highly exothermic. Per Hammond's postulate, assume that the transition state for radical fluorination is almost identical to the starting material. Assuming this fact, estimate the fraction of each monofluoro product formed in the fluorination of 2-methylbutane.

8.20 Cyclobutane reacts with bromine to give bromocyclobutane, but bicyclobutane reacts with bromine to give 1,3-dibromocyclobutane. Account for the differences between the reactions of these two compounds.

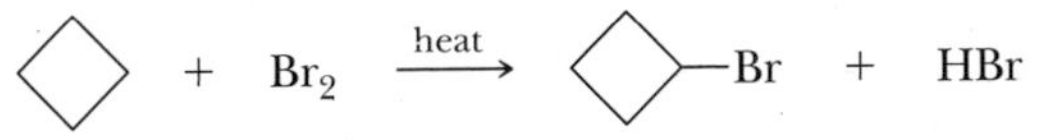

Cyclobutane **Bromocyclobutane**

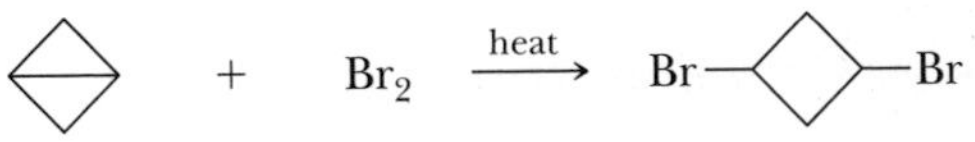

Bicyclobutane **1,3-Dibromocyclobutane**

8.21 The first chain propagation step of all radical halogenation reactions we considered in Section 8.5B was abstraction of hydrogen by the halogen atom to give an alkyl radical and HX, as for example

$$CH_3CH_3 + \cdot\ddot{\underset{..}{Br}}: \longrightarrow CH_3CH_2\cdot + H\ddot{\underset{..}{Br}}:$$

Suppose, instead, that radical halogenation occurs by an alternative pair of chain propagation steps, beginning with this step.

$$CH_3CH_3 + \cdot\ddot{\underset{..}{Br}}: \longrightarrow CH_3CH_2\ddot{\underset{..}{Br}}: + H\cdot$$

(a) Propose a second chain propagation step. Remember that a characteristic of chain propagation steps is that they add to the observed reaction.
(b) Calculate the heat of reaction, ΔH^0, for each propagation step.
(c) Compare the energetics and relative rates of the set of chain propagation steps in Section 8.5B with the set proposed here.

Allylic Halogenation

8.22 Following is a balanced equation for the allylic bromination of propene.

$$CH_2{=}CHCH_3 + Br_2 \longrightarrow CH_2{=}CHCH_2Br + HBr$$

(a) Calculate the heat of reaction, ΔH^0, for this conversion.
(b) Propose a pair of chain propagation steps and show that they add up to the observed stoichiometry.
(c) Calculate the ΔH^0 for each chain propagation step and show that they add up to the observed ΔH^0 for the overall reaction.

8.23 Using the table of bond dissociation enthalpies (Appendix 3), estimate the BDE of each indicated bond in cyclohexene.

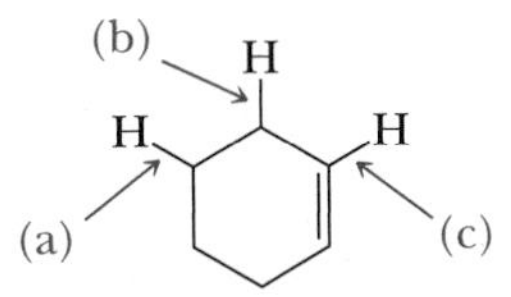

8.24 Propose a series of chain initiation, propagation, and termination steps for this reaction and estimate its heat of reaction.

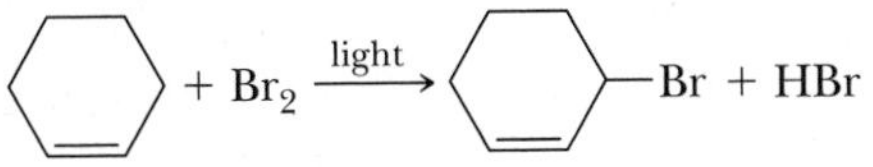

(racemic)

8.25 The major product formed when methylenecyclohexane is treated with NBS in dichloromethane is 1-(bromomethyl)cyclohexene. Account for the formation of this product.

8.26 Draw the structural formula of the products formed when each alkene is treated with one equivalent of NBS in CH_2Cl_2 in the presence of light.

(a) $CH_3CH{=}CHCH_2CH_3$ **(b)**  **(c)** 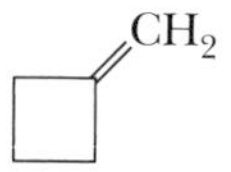

8.27 Calculate the ΔH^0 for the following reaction step. What can you say regarding the possibility of bromination at a vinylic hydrogen?

$$CH_2{=}CH_2 + Br\cdot \longrightarrow CH_2{=}CH\cdot + HBr$$

Reaction Roadmap

Organic Chemistry Reaction Roadmap

8.28 Use the reaction roadmap you made for Problems 6.54 and 7.29 and update it to contain the reactions in the Study Guide section of this chapter. Because of its highly specific nature, do not use reaction 3 of this chapter on your roadmap.

8.29 Write the products of the following sequences of reactions. Refer to your reaction roadmap to see how the combined reactions allow you to "navigate" between the different functional groups.

(a) *An alkyne* → 1. Excess H_2 / Pd; 2. Br_2 and light

(b)  *An alkyne* → 1. H_2 / Lindlar catalyst; 2. NBS and light

(c) 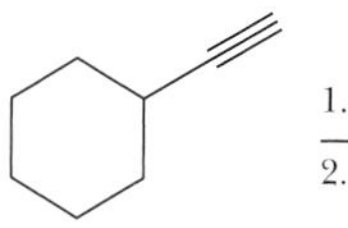 *An alkyne* → 1. H_2 / Lindlar catalyst; 2. NBS and light

(d) 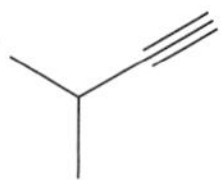 *An alkyne* → 1. 1 equivalent of $NaNH_2$; 2. 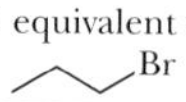; 3. Excess H_2 / Pt; 4. Br_2 and light

(e) 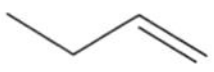 *An alkene* → 1. Cl_2; 2. 3 equivalents of $NaNH_2$ in NH_3; 3. CH_3I; 4. Excess H_2 / Pt; 5. Br_2 and light

Synthesis

Reaction Roadmap

8.30 Using your reaction roadmap as a guide, show reagents and conditions to bring about these conversions, which may require more than one step.

(a) 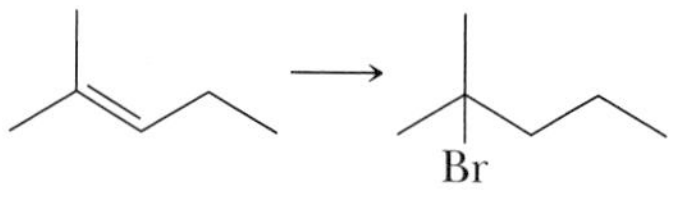

(b) $CH_3CH{=}CHCH_3 \longrightarrow CH_3CH{=}CHCH_2Br$

(c) $CH_3CH{=}CHCH_3 \longrightarrow CH_3CH(Br){-}CH(Br)CH_3$

(racemic)

(d) 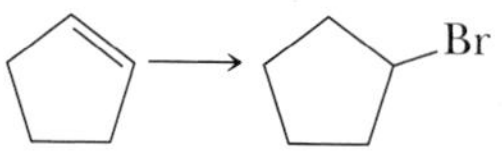

(e) 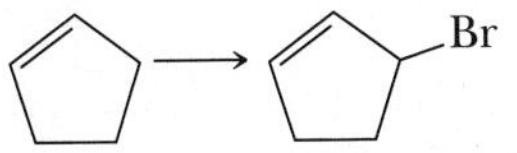
Br

(f) CH_4 ⟶  (Review section 7.5.)

(g) 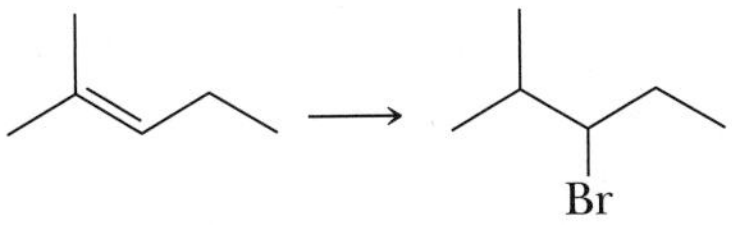
Br

(racemic)

Autoxidation

8.31 Predict the products of the following reactions. Where isomeric products are formed, label the major product.

(a) 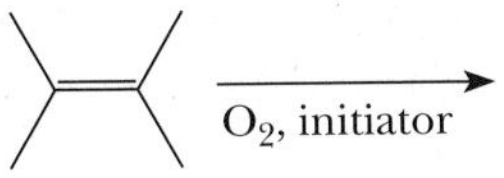
O_2, initiator

(b) 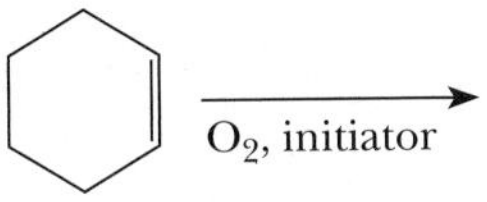
O_2, initiator

8.32 Give the major product of the following reactions.

(a) 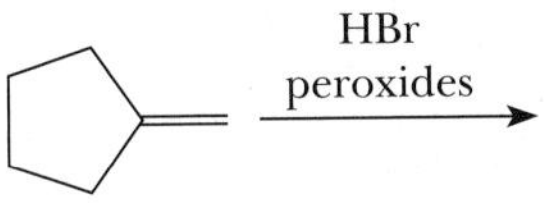
HBr
peroxides

(b) 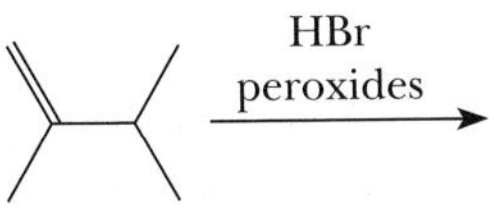
HBr
peroxides

Looking Ahead

8.33 A major use of the compound cumene is in the industrial preparation of phenol and acetone in the two-step synthesis, shown below.

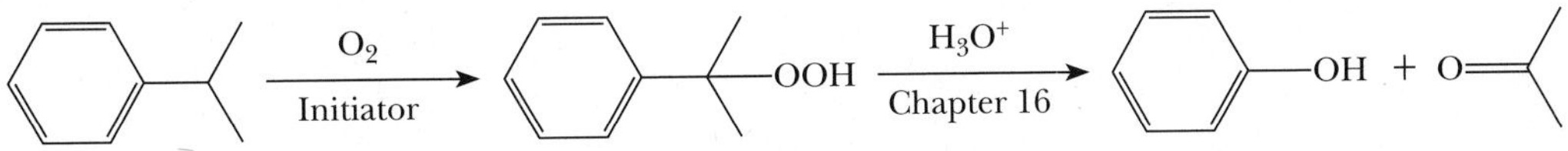
O_2
Initiator
OOH
H_3O^+
Chapter 16
OH
+
O

Cumene **Cumene hydroperoxide** **Phenol** **Acetone**

Write a mechanism for the first step. We will see in Problem 16.65 how to complete the synthesis.

8.34 An important use of radical-chain reactions is in the polymerization of ethylene and substituted ethylene monomers such as propene, vinyl chloride [the synthesis of which was discussed in Section 7.6 along with its use in the synthesis of poly(vinyl chloride), (PVC)], and styrene. The reaction for the formation of PVC, where n is the number of repeating units and is very large, follows.

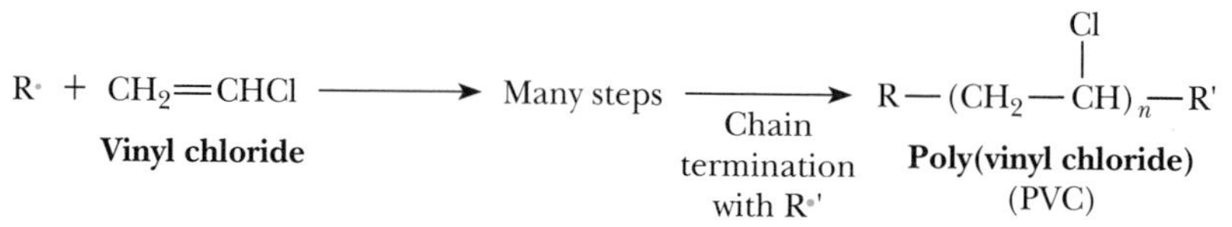
R· + CH₂=CHCl
Vinyl chloride
Many steps
Chain termination with R·'
R—(CH₂—CH)ₙ—R'
Cl
Poly(vinyl chloride)
(PVC)

(a) Give a mechanism for this reaction (see Chapter 29).

(b) Give a similar mechanism for the formation of poly(styrene) from styrene. Which end of the styrene double bond would you expect R· to attack? Why?

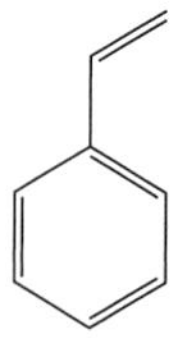

Styrene

Things You Should Know IV

Common Mistakes in Arrow Pushing

We've now had several chapters in which arrow pushing, as presented in "Things You Should Know III," has been used to depict organic reaction mechanisms. Throughout this book, arrow pushing will be used to indicate the flow of electrons in various mechanisms.

Before proceeding further, it seems like a good idea to pause and examine the mistakes that students commonly make when using arrow pushing. This is particularly true before launching into Chapter 9, where many of the most important mechanisms in organic chemistry are introduced: S_N2, S_N1, E2, and E1. The mistakes given here are the ones seen most often by the authors during their cumulative dozens of years of experience in teaching Introductory Organic Chemistry.

Backward Arrows

Reversing the direction of one or more arrows during a chemical step is the most common mistake made by students when writing organic reaction mechanisms. Backward arrow pushing usually derives from a student thinking about the movement of atoms, not the movement of electrons. Hence, to avoid this mistake it is important to remember that arrows depict how electrons move, not where atoms move, within or between chemical structures. Further, one can avoid this mistake by remembering that every arrow must start at an electron source (a bond or lone pair) and terminate at an electron sink (an atom that can accept a new bond or lone pair).

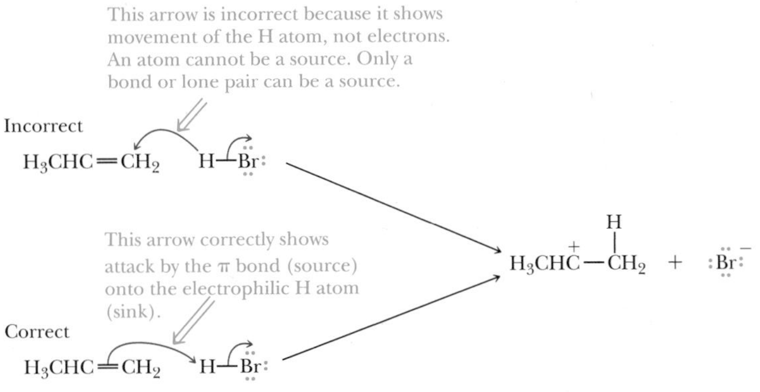

This arrow is incorrect because it shows movement of the H atom, not electrons. An atom cannot be a source. Only a bond or lone pair can be a source.
Incorrect
H3CHC=CH2
H—Br:
This arrow correctly shows attack by the π bond (source) onto the electrophilic H atom (sink).
Correct
H3CHC=CH2
H—Br:
H
H3CHC—CH2
+
:Br:

Not Enough Arrows

A second common mistake in writing arrow-pushing schemes is to not use enough arrows. This usually results from not keeping track of all lone pairs, bonds made, or bonds broken in a mechanism step. In other words, if you analyze exactly the new position of electrons resulting from each arrow, missing arrows will become evident. In the following example we compare two arrow-pushing scenarios, one of which is missing an arrow. In the incorrect scheme, there is no arrow that indicates breaking of the C—H bond of the reactant and formation of the π bond in the alkene product. Note that when an arrow is missing, the result is commonly too many bonds and/or lone pairs on one atom (see the next section on hypervalency) and not enough bonds or lone pairs on another.

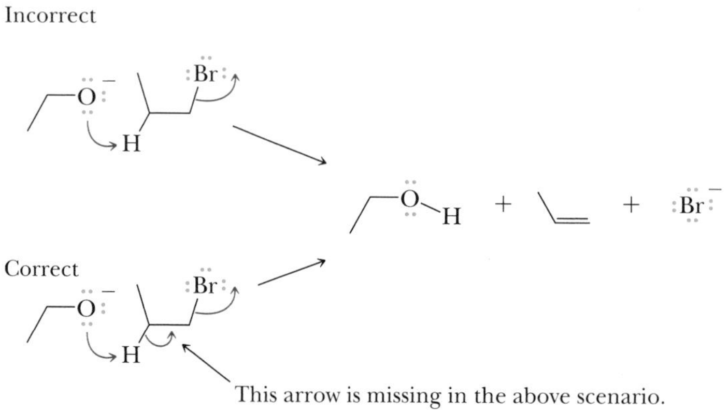

Hypervalency

Another frequent mistake when writing arrow-pushing schemes is to expand the valency of an atom to more electrons than an atom can accommodate, a situation referred to as hypervalency. An overarching principle of organic chemistry is that carbon has eight electrons in its valence shell when present in stable organic molecules (the Octet Rule, Section 1.2). Analogously, many of the other most common elements in organic molecules, such as nitrogen, oxygen, and chlorine, also obey the Octet Rule. There are three common ways in which students incorrectly draw hypervalent atoms: (1) Too many bonds to an atom, (2) Forgetting the presence of hydrogens, and (3) Forgetting the presence of lone pairs.

In the following case an arrow is used to depict a potential resonance contributing structure of nitromethane. However, the result is a nitrogen atom with 10 electrons in its valence shell because there are too many bonds to N. Such mistakes can be avoided by remembering to draw all bonds and lone pairs on an atom so that the total number of electrons in each atom's valence shell is apparent.

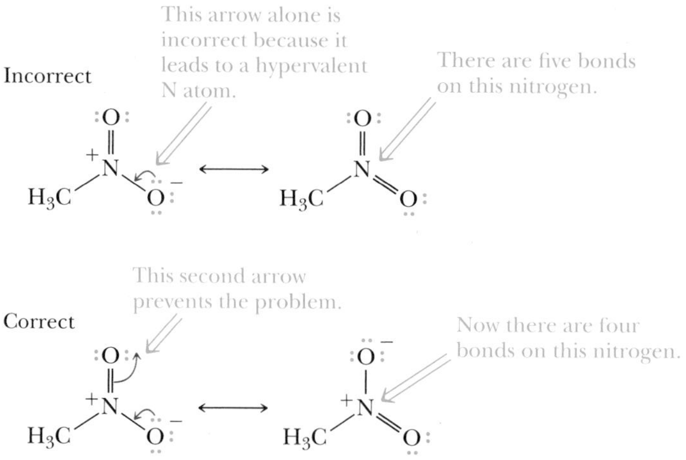

Another common way students mistakenly end up with a hypervalent atom is to forget the presence of hydrogens that are not explicitly written. When using line-angle diagrams to write organic chemical structures, not all the hydrogens are drawn, and hence it is common to forget them during an arrow pushing exercise. The following example shows two proposed resonance contributing structures of an amide anion. The arrow drawn on the molecule to the left is incorrect because it depicts the formation of a new bond to a carbon that already has four bonds. When both bonds to hydrogen are drawn explicitly, as on the structure farthest to the right, it is clear there are now five bonds around the indicated carbon atom. Notice also that the negative charge was lost upon drawing the contributing structures on the right, providing another clear signal that something was wrong because overall charge is always conserved when arrows are drawn correctly.

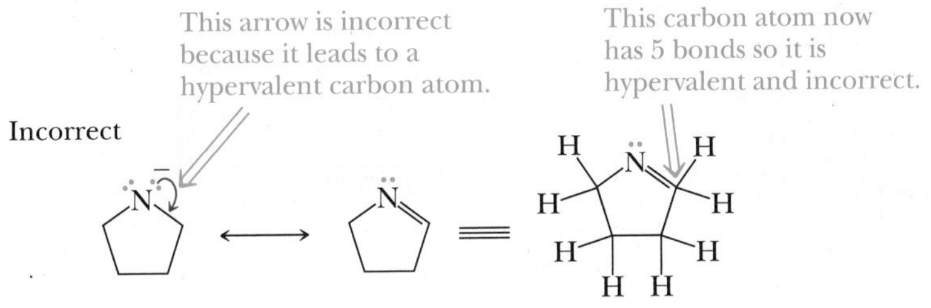

Another common way to make a hypervalency mistake is by forgetting to count all lone pairs of electrons. The following example shows a negatively charged nucleophile incorrectly adding to the formal positive charge on an alkylated ketone. This may look correct because atoms with positive and negative charges are being directly combined, but when counting bonds and lone pairs of electrons, it is found that the oxygen ends up with 10 electrons overall. Hence, this is a mistake.

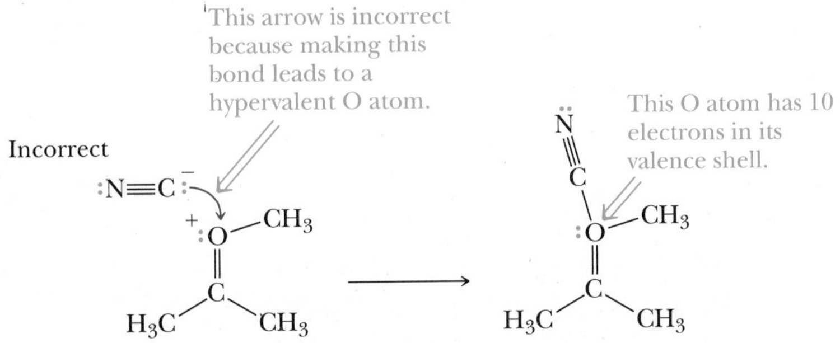

Mixed Media Errors

Acids and bases are catalysts, reactants, products, and intermediates in many organic chemistry transformations. When writing mechanisms for reactions involving acids and bases, there are three general rules that will guide you in depicting the correct mechanism:

1. Do not show the creation of a strong acid for the mechanism of a reaction that is performed in strongly basic media.
2. Do not show the creation of a strong base for the mechanism of a reaction that is performed in strongly acidic media.
3. In strongly acidic media, all the intermediates and products will be either neutral or positively charged, while in strongly basic media, all the products and intermediates will be neutral or negatively charged.

The reason for these rules is that significant amounts of strong acids and bases cannot coexist in the same medium because they would rapidly undergo a proton transfer reaction before anything else would happen in the solution.

An example of a mixed media error is given below. The first equation shows a strong base being created although the reaction is performed under acidic conditions (see conditions over the first equilibrium arrows). Not shown are the three steps that lead to the intermediate. A mistake is made in the arrow pushing because a strong base (methoxide) is generated as the leaving group even though the reaction is run in strong acid. In the correct mechanism, the next step would be protonation of the ether oxygen atom followed by loss of methanol (not shown) to give a carboxylic acid product.

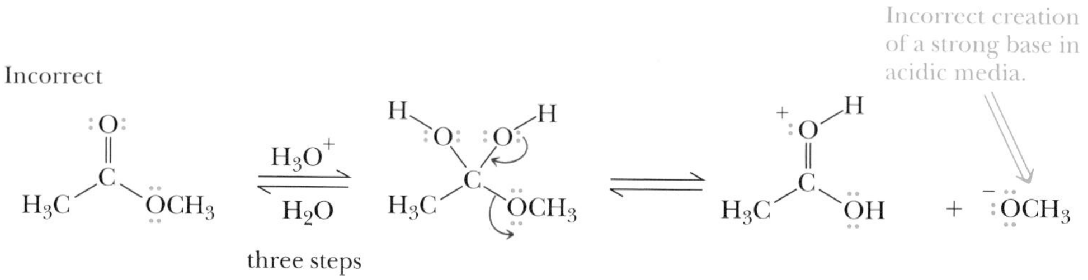
Incorrect
Incorrect creation of a strong base in acidic media.
three steps

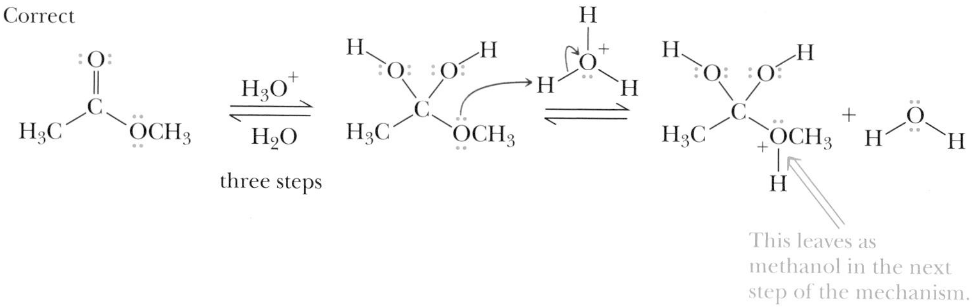
Correct
three steps
This leaves as methanol in the next step of the mechanism.

Failing to Conserve Charge

Overall charge must be conserved in all mechanism steps. Failure to conserve overall charge could be caused by some of the preceding errors (hypervalency, failure to draw arrows, mixed media errors), but we mention it by itself because it is always helpful to check that your arrow pushing is consistent by confirming that overall charge conservation is obeyed. In the example shown below, an arrow is missing, leading to a neutral intermediate even though the overall charge on the left side of the equation was minus one. Notice there are five bonds to carbon on the intermediate (hypervalency), providing another obvious indication that something was incorrect in the mechanism step as drawn.

This side of the equation has an overall negative charge due to the hydroxide anion.

As drawn there is no negative charge on this intermediate and this C atom has 5 bonds.

Incorrect

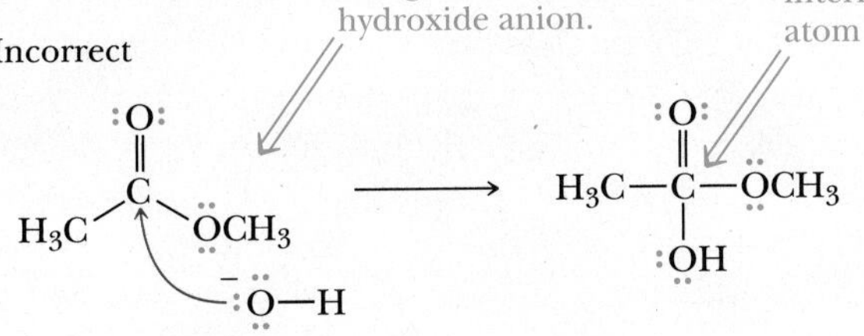

This C atom has only four bonds and there is an overall negative charge on both sides of the equation.

Correct

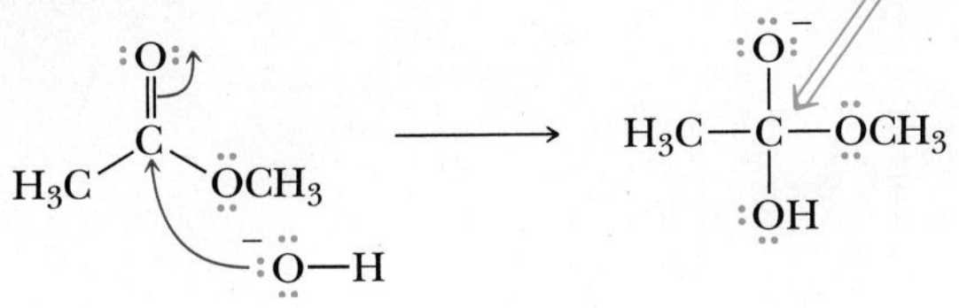

9

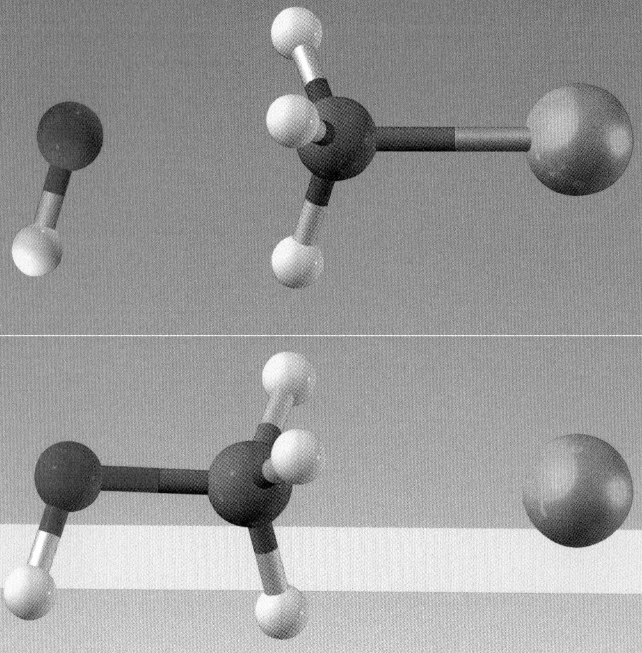

Nucleophilic Substitution and β-Elimination

Hydroxide ion reacts with bromomethane (upper models) to give methanol and bromide ion (lower models) by an S_N2 mechanism (Section 9.3).

Outline

9.1 Nucleophilic Substitution in Haloalkanes
9.2 Mechanisms of Nucleophilic Aliphatic Substitution
9.3 Experimental Evidence for S_N1 and S_N2 Mechanisms
9.4 Analysis of Several Nucleophilic Substitution Reactions
9.5 β-Elimination
9.6 Mechanisms of β-Elimination
9.7 Experimental Evidence for E1 and E2 Mechanisms
9.8 Substitution Versus Elimination
9.9 Analysis of Several Competitions Between Substitutions and Eliminations
9.10 Neighboring Group Participation

Nucleophilic substitution refers to any reaction in which an electron-rich **nucleophile** (meaning nucleus loving) (Nu:) replaces a **leaving group** (Lv). Viewed in the context of the mechanism elements first described in Things You Should Know III prior to Chapter 6, nucleophilic substitution is a combination of making a new bond between a nucleophile and an electrophile and breaking a bond so that relatively stable molecules or ions are created. All nucleophiles are electron sources and can be considered Lewis bases (Section 4.7). With the exception of radical reactions (Chapter 8), essentially every reaction you will study involves a reaction of a Lewis acid (that can also be considered a good electron sink) with a Lewis base. In these reactions, the Lewis base, which is electron-rich, reacts with the Lewis acid, which is electron-poor. The Lewis acid is called an **electrophile** (meaning electron loving). The leaving group (Lv) can be a halide (X) or another electronegative group that can form a stable anion or another stable species. It should be noted that not

Nucleophilic substitution
Any reaction in which one nucleophile is substituted for another at a tetravalent carbon atom.

Nucleophile
From the Greek, meaning nucleus loving. A molecule or an ion that donates a pair of electrons to another atom or ion to form a new covalent bond; a Lewis base.

all nucleophiles and leaving groups covered in this chapter are negatively charged. Further, not all nucleophiles react with all electrophiles; recognizing which do and do not is part of what you should take from this chapter. Here is a general equation for a nucleophilic substitution reaction.

Leaving group (Lv) The group that is displaced in a substitution or that is lost in an elimination.

Electrophile From the Greek, meaning electron loving. A molecule or an ion that accepts a pair of electrons from another atom or molecule in a reaction; a Lewis acid.

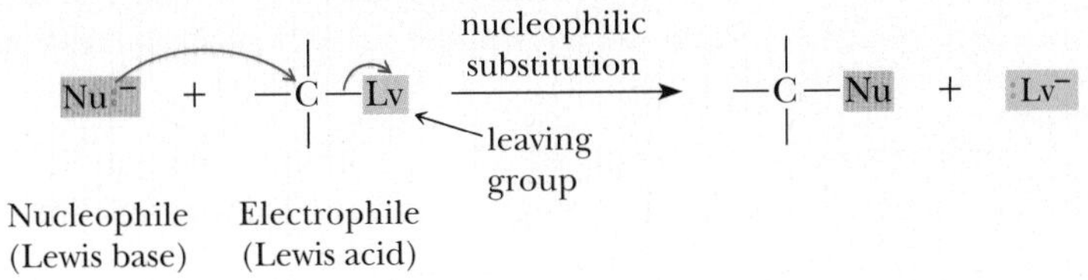

An example of this reaction that you have already studied is the alkylation of terminal alkynes (Section 7.5A). Another is the reaction of hydroxide ion with chloromethane. In this reaction, chloromethane is the electrophile. Because of the electronegativity of chlorine, there is a partial positive charge on the carbon. An electrostatic potential map shows the negative electron density on HO^- (the nucleophile) interacting with the partial positive charge on the methyl group.

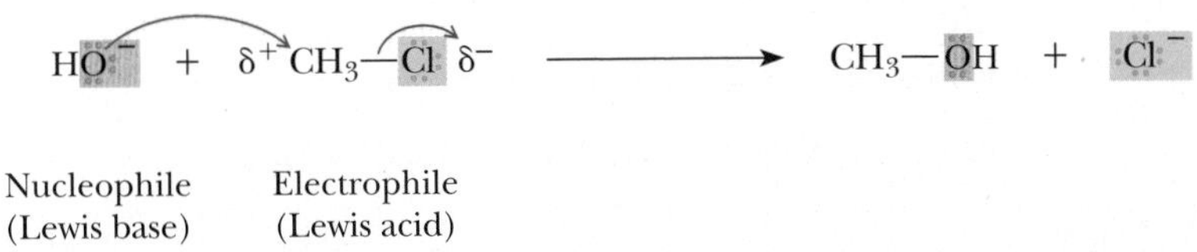

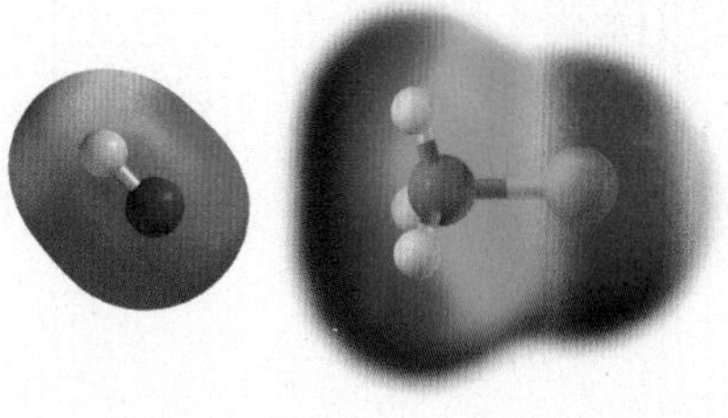

Electrostatic potential map showing the nucleophile (OH^-) reacting at its negative (red) end with the electrophilic carbon (blue) in the reaction of hydroxide with chloromethane.

Nucleophiles are also Brønsted–Lowry bases (Section 4.2), although some are very weak ones. The stronger ones can remove protons as well as attack at carbon centers. A reaction in which a halide and a hydrogen on the neighboring (β) carbon are removed is called a **β-elimination**. Nucleophilic substitution and base-promoted β-elimination are therefore competing reactions. For example, ethoxide ion reacts with bromocyclohexane as a nucleophile to give ethoxycyclohexane (cyclohexyl ethyl ether) and as a Brønsted–Lowry base to give cyclohexene and ethanol.

β-Elimination A reaction in which a molecule, such as HCl, HBr, HI, or HOH, is split out or eliminated from adjacent carbons.

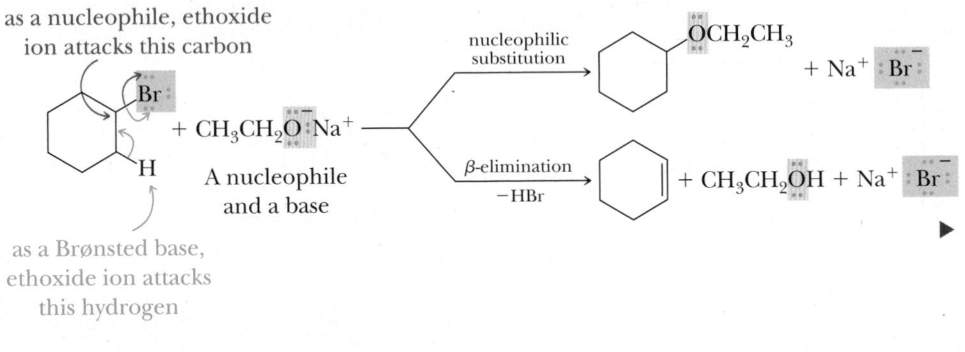

▶ **Most bond-making steps in reaction mechanisms involve nucleophiles reacting with electrophiles.**

In this chapter, we study substitution and β-elimination. By using these reactions, we can convert haloalkanes to compounds with other functional groups, including alcohols, ethers, thiols, sulfides, amines, nitriles, alkenes, and alkynes. Nucleophilic substitution and β-elimination open entirely new areas of organic chemistry and are a major method of interconverting functional groups. One of the most challenging aspects of the study of these reactions is deciding whether substitution or elimination is likely to prevail, and this will be the major focus of the last part of the chapter.

9.1 Nucleophilic Substitution in Haloalkanes

Nucleophilic substitution is one of the most important reactions of haloalkanes and can lead to a wide variety of new functional groups, many of which are illustrated in Table 9.1. Some of these reactions proceed smoothly at room temperature; others occur only at elevated temperatures, as we will see in later sections. As you study the entries in this table, note these points:

1. If the nucleophile is negatively charged, as, for example, OH^- and $HC{\equiv}C^-$, in a substitution reaction, the atom donating the pair of electrons becomes neutral in the product.
2. If the nucleophile is uncharged, as, for example, NH_3 and CH_3OH, in the substitution reaction, the atom donating the pair of electrons becomes positively charged in the initial product.
3. In the middle of the table are two reactions involving $N{\equiv}C^-$ and $HC{\equiv}C^-$ nucleophiles. In these nucleophilic substitution reactions, the products have new carbon-carbon bonds, as we saw for alkynes in Section 7.5A. The formation of new carbon-carbon bonds is important in organic chemistry because it provides a means of extending a molecular carbon skeleton.

Table 9.1 Some Nucleophilic Substitution Reactions

Reaction: $Nu\!:^- + CH_3Br \longrightarrow CH_3Nu + :\!Br^-$

Nucleophile		Product	Class of Compound Formed
$^-\!:\!OH$	$\longrightarrow$	CH_3OH	An alcohol
$^-\!:\!OR$	$\longrightarrow$	CH_3OR	An ether
$^-\!:\!SH$	$\longrightarrow$	CH_3SH	A thiol (a mercaptan)
$^-\!:\!SR$	$\longrightarrow$	CH_3SR	A sulfide (a thioether)
$^-\!:\!C{\equiv}CH$	$\longrightarrow$	$CH_3C{\equiv}CH$	An alkyne
$^-\!:\!C{\equiv}N\!:$	$\longrightarrow$	$CH_3C{\equiv}N\!:$	A nitrile
$:\!I\!:^-$	$\longrightarrow$	$CH_3I\!:$	An alkyl iodide
$^-\!:\!N{=}\overset{+}{N}{=}N\!:^-$	$\longrightarrow$	$CH_3{-}N{=}\overset{+}{N}{=}N\!:^-$	An alkyl azide
$:\!NH_3$	$\longrightarrow$	$CH_3NH_3^+$	An alkylammonium ion
$:\!O(H){-}H$	$\longrightarrow$	$CH_3\overset{+}{O}(H){-}H$	An alcohol (after proton is taken away)
$:\!O(H){-}CH_3$	$\longrightarrow$	$CH_3\overset{+}{O}(H){-}CH_3$	An ether (after proton is taken away)

Example 9.1 Nucleophilic Substitution Products I

Complete these nucleophilic substitution reactions. In each reaction, show all electron pairs on both the nucleophile and the leaving group.

(a) 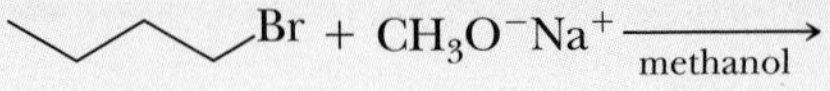

(b) 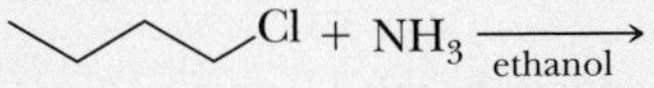

Solution

(a) Methoxide ion is the nucleophile, and bromide is the leaving group.

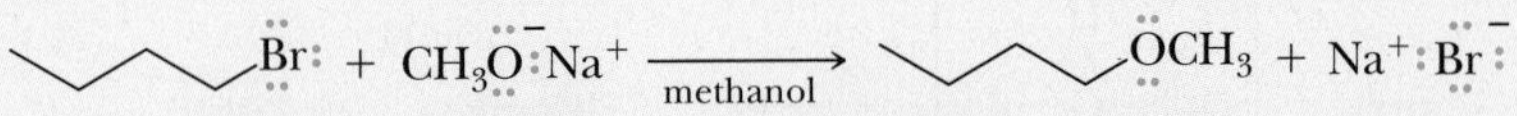

1-Bromobutane **Sodium methoxide** **1-Methoxybutane (Butyl methyl ether)** **Sodium bromide**

(b) Ammonia is the nucleophile, and chloride is the leaving group.

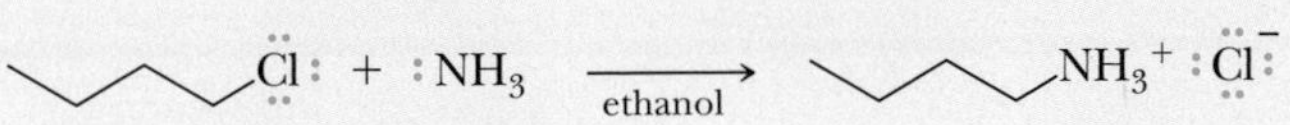

1-Chlorobutane **Ammonia** **Butylammonium chloride**

Problem 9.1

Complete the following nucleophilic substitution reactions. In each reaction, show all electron pairs on both the nucleophile and the leaving group.

(a) 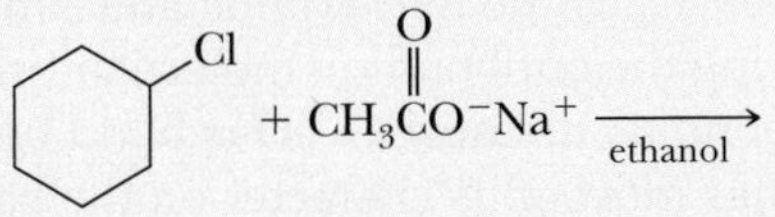

(b) 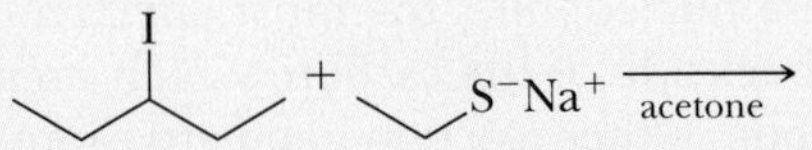

(c) 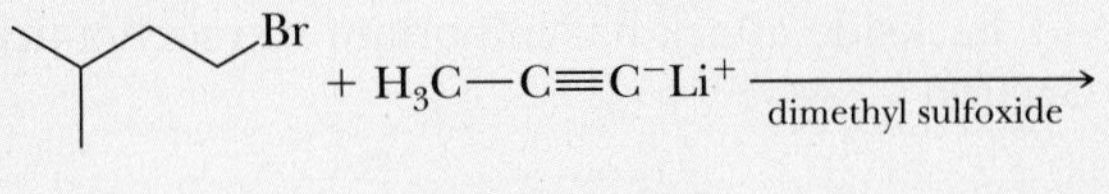

9.2 Mechanisms of Nucleophilic Aliphatic Substitution

On the basis of experimental observations developed over a 70-year period, two limiting mechanisms for nucleophilic substitutions have been proposed, called S_N2 and S_N1. A fundamental difference between them is the timing of bond breaking between carbon and the leaving group and of bond forming between carbon and the nucleophile.

A. S_N2 Mechanism

At one extreme, bond breaking and bond forming occur simultaneously. Thus, departure of the leaving group is assisted by the incoming nucleophile. This

Bimolecular reaction

A reaction in which two species are involved in the rate-determining step.

mechanism is designated $\mathbf{S_N2}$, where S stands for *S*ubstitution, N for *N*ucleophilic, and 2 for a **bimolecular reaction**. This type of substitution reaction is classified as bimolecular because both the haloalkane and nucleophile are involved in the rate-determining step.

Mechanism 9.1

An S_N2 Reaction

Make a new bond between a nucleophile and an electrophile and simultaneously break a bond to give stable molecules or ions. The nucleophile attacks the reactive center from the side opposite the leaving group; that is, an S_N2 **reaction** involves backside attack of the nucleophile. In this diagram, the dashed lines in the transition state represent partially formed or broken bonds.

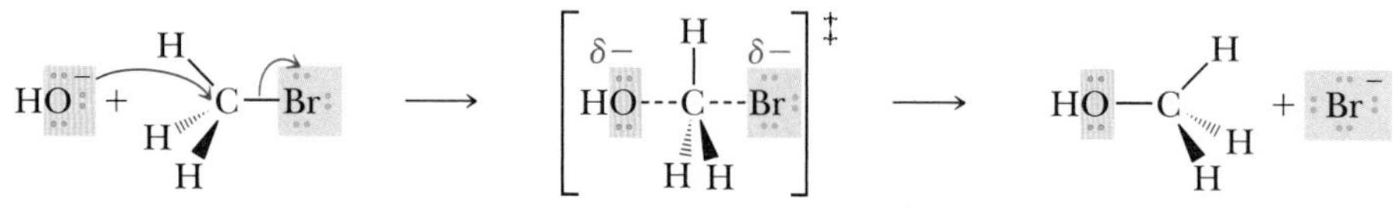

Transition state with simultaneous bond breaking and bond forming

S_N2 reaction

A bimolecular nucleophilic substitution reaction.

As noted in the mechanism just given, the nucleophile attacks from the backside. Backside attack by the nucleophile is facilitated in two ways. First, because of the polarization of the C—Br bond, the carbon atom has a partial positive charge and therefore attracts the electron-rich nucleophile (as shown for methyl chloride at the beginning of the chapter). Second, the electron density of the nucleophile entering from the backside assists in breaking the C—Br bond, thereby helping the bromide leave. The electron density of the nucleophile attacking from the backside can be thought of as populating the antibonding molecular orbital of the C—Br bond, weakening the C—Br bond as the new C—O σ bond becomes stronger. This antibonding C—Br orbital has most of its character on the backside of the C. Therefore, upon collision with a nucleophile, the most effective way to fill this orbital is by collision from the backside of the carbon, which breaks the C—Br bond on the other side of the carbon. Other reaction geometries are higher in energy because they do not produce an efficient orbital overlap that leads to weakening of the C—Lv bond. Backside attack has important stereochemical consequences, as we will see in Section 9.3A.

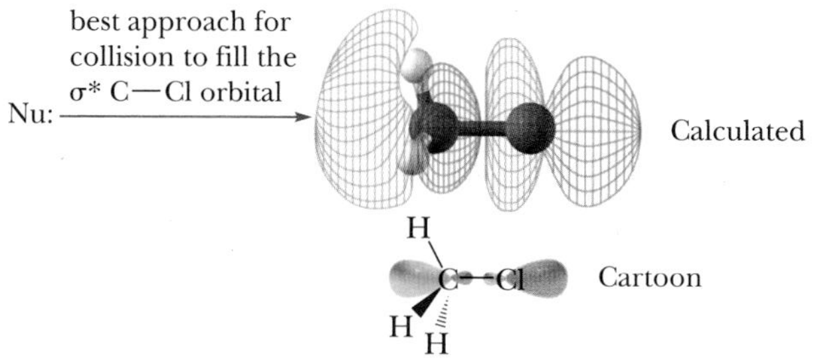
best approach for collision to fill the σ* C—Cl orbital
Nu:
Calculated
Cartoon

Antibonding (σ^*) C—Cl orbital

Figure 9.1 shows an energy diagram for an S_N2 reaction. Because there is a single step in the S_N2 mechanism, the energy diagram has one energy barrier that corresponds to a single transition state. Recall that transition states are fleeting structures

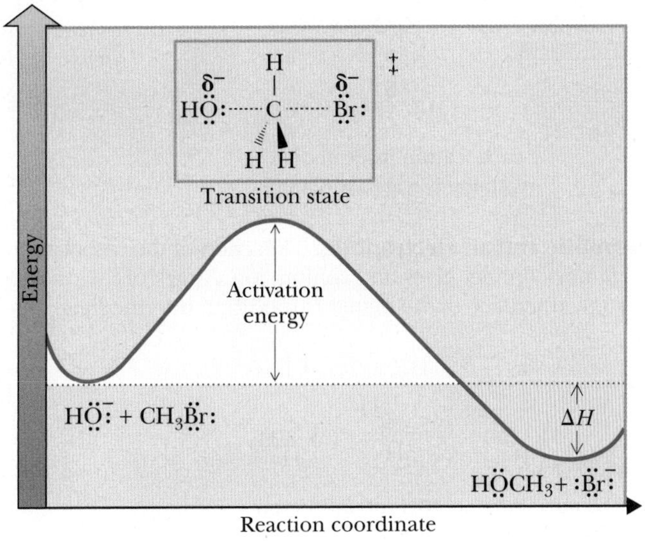

FIGURE 9.1 An energy diagram for an S_N2 reaction. There is one transition state (no reactive intermediate).

with essentially no lifetime. They are not intermediates, but instead represent transitions between two structures that lie in wells (troughs) on energy surfaces, as shown in Figure 9.1.

The collision between the nucleophile and the electrophile must occur with enough energy to surmount the barrier to the reaction. This energy barrier is present because of the distortion from optimal bonding arrangements that occurs at the transition state. At the transition state of an S_N2 reaction, the C is distorted into a trigonal bipyramidal geometry with one bond partially forming and one bond partially breaking. Hence, the transition state structure is considerably strained relative to the reactant and product and is a hybrid structure that is transitioning between the reactant and product. A good way to think about this is that the transition state has higher internal energy due to the structural distortions caused by the collision—like the strain that takes place when two rubber balls collide, deform, and then bounce off each other. The difference is that in chemistry, the collisions can occur with sufficient energy to cause bond breaking and forming, leading to a new structure.

S_N1 reaction A unimolecular nucleophilic substitution reaction.

Unimolecular reaction A reaction in which only one species is involved in the rate-determining step.

Solvolysis A nucleophilic substitution in which the solvent is also the nucleophile.

B. S_N1 Mechanism

The other limiting mechanism is called the **S_N1 reaction**. In this mechanism, bond breaking between carbon and the leaving group is entirely completed before bond forming with the nucleophile begins. In the designation S_N1, 1 stands for a **unimolecular reaction**. This type of substitution is classified as unimolecular because only the haloalkane is involved in the rate-determining step. An S_N1 mechanism is illustrated by the solvolysis of 2-bromo-2-methylpropane (*tert*-butyl bromide) in methanol to form 2-methoxy-2-methylpropane (*tert*-butyl methyl ether) and HBr. In this reaction, the nucleophile (methanol) is also the solvent, hence the name **solvolysis**.

Mechanism 9.2

An S_N1 Reaction

Step 1: **Break a bond to give stable molecules or ions.** Ionization of the C—Lv (Lv=Br) bond forms a carbocation intermediate. Because no nucleophile is assisting the departure of the halide anion, this is the relatively slow, rate-determining step of the reaction.

(Continued)

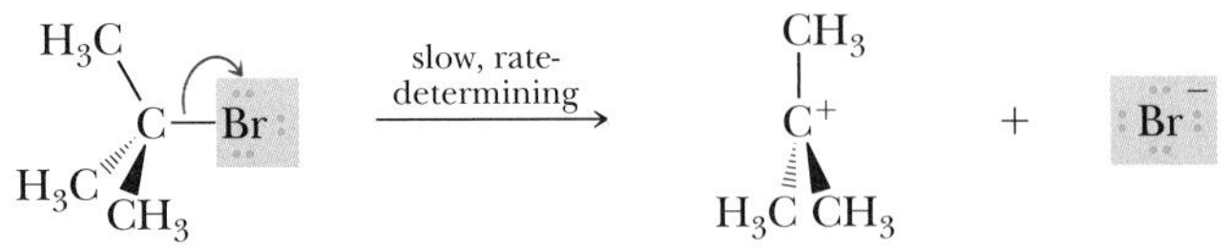

A carbocation intermediate;
its shape is trigonal planar

Step 2: **Make a new bond between a nucleophile and an electrophile.** Reaction of the carbocation intermediate (an electrophile) with methanol (a nucleophile) gives an oxonium ion. Attack of the nucleophile occurs with approximately equal probability from either face of the planar carbocation intermediate.

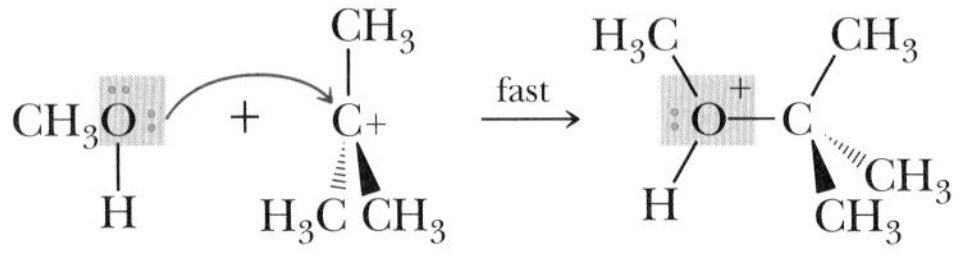

Nucleophile (Lewis base) Electrophile (Lewis acid) An oxonium ion

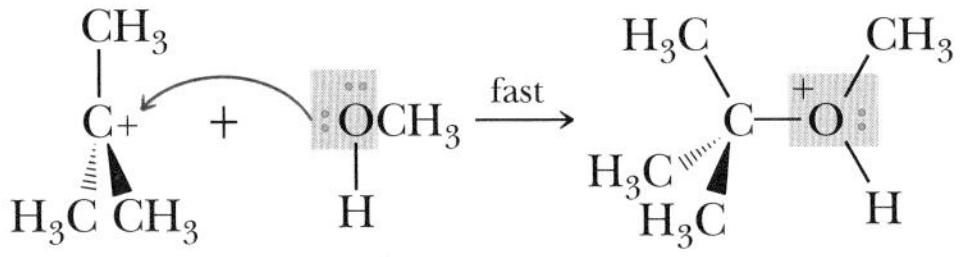

Step 3: **Take a proton away.** Proton transfer from the oxonium ion to methanol completes the reaction and gives *tert*-butyl methyl ether.

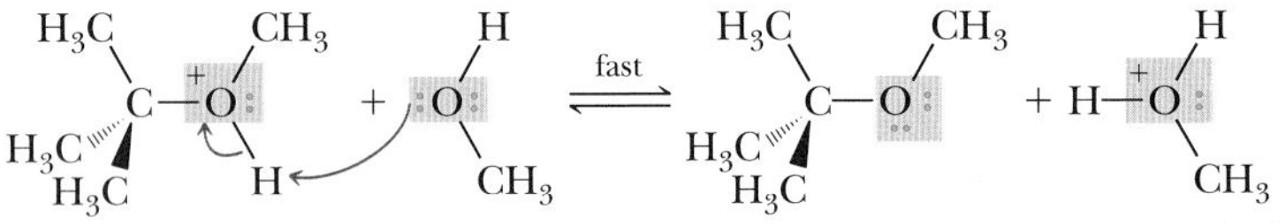

In an S_N1 reaction, the rate-determining step is the cleavage of the C—Lv bond to form a carbocation intermediate, shown as the structure in the second well of Figure 9.2. As we presented in Chapter 6, carbocations are two electrons shy of an octet and quickly react with nucleophiles, such as the solvent methanol. In the example of Figure 9.2, after reaction of the carbocation with methanol, the structure created has a proton on oxygen. A transition state exists on the energy diagram for each individual step. The last step in this three-step mechanism is a proton-transfer reaction following the S_N1 reaction.

As discussed, the rate-determining step of an S_N1 reaction involves unimolecular cleavage of the haloalkane to a carbocation and a halide anion. One can envision this occurring due to collisions with the solvent. Recall that the reactant is dissolved in a solvent and that there is continual thermal motion consisting of translation and tumbling of both the reactant and solvent molecules. The haloalkane is continually being jostled around within the solvent, being hit by the solvent from all directions. When one of these collisions is of high enough energy to distort the haloalkane into a geometry in which the bond to the leaving group is almost completely broken, the transition state for departure of the leaving group can be achieved and the S_N1 mechanism enabled. This is in contrast to the S_N2 mechanism, where a collision with the nucleophile from the backside initiates the reaction.

C. Key Mechanistic Differences Between S_N2 and S_N1 Reactions

Now that we have introduced the two most dominant mechanisms for substitution reactions on haloalkanes (R—Lv), it is worthwhile to point out some of the key differences. First, an S_N2 reaction involves a single step and therefore has no

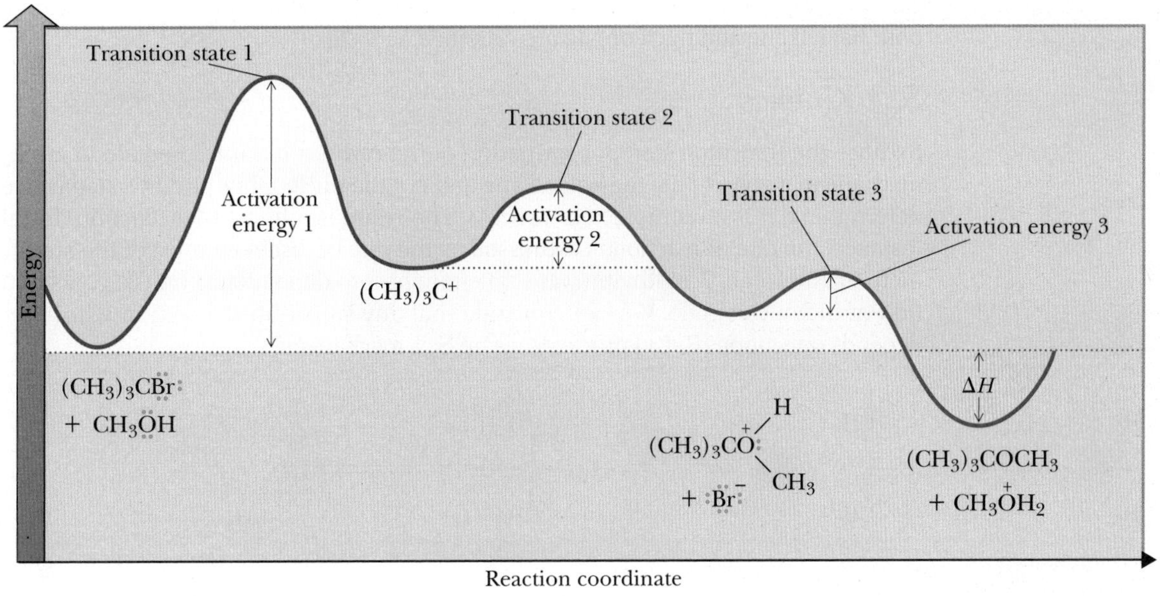

FIGURE 9.2 An energy diagram for an S_N1 reaction. There are three transition states before the final product is created. The first is for formation of the carbocation intermediate, the second is for the reaction of the carbocation with methanol to give an oxonium ion, and the third is for taking off the proton. Step 1 crosses the highest energy barrier and therefore is rate-determining.

intermediates. As with all chemical reactions, however, it has a transition state. In contrast, an S_N1 reaction has two steps (or three steps when a proton is removed as the last step), each with a transition state. Importantly, an intermediate carbocation is formed. The single-step versus the two-step mechanisms for S_N2 and S_N1, respectively, along with the positive charge on the carbocation intermediate involved in an S_N1 reaction are key factors that influence the preference for one mechanism over the other. Let us now examine the experimental evidence on which these two contrasting mechanisms are based and learn what structural features cause one mechanism to dominate over the other.

9.3 Experimental Evidence for S_N1 and S_N2 Mechanisms

We consider the following questions as a means of contrasting the two commonly observed mechanisms.

1. What are the kinetics and stereochemistry of S_N2 and S_N1 mechanisms?
2. What effect does the structure of the haloalkane have on the rate of reaction?
3. What effect does the structure of the leaving group have on the rate of reaction?
4. What is the role of the solvent?
5. What effect does the structure of the nucleophile have on the rate of reaction?
6. Under what conditions are skeletal rearrangements observed?

A. Kinetics and Stereochemistry

Chemists routinely perform two experiments to distinguish S_N2 and S_N1 mechanisms. The first involves performing a kinetic analysis, which means that the rates of reactions are followed as the concentration of reactants is changed. The second experiment is to run the substitution reaction with an alkyl-Lv structure, where the C bonded to the Lv is a chiral center, and then examine the stereochemistry of the products. Because these two approaches are commonly applied for studying substitution reactions, we group them together here.

The kinetic order of nucleophilic substitutions can be studied by measuring the effect on rate of varying the concentrations of haloalkane and nucleophile. Those reactions whose rate is dependent only on the concentration of haloalkane are classified as S_N1; those reactions whose rate is dependent on the concentration of both haloalkane and nucleophile are classified as S_N2.

S_N1 Kinetics

Because the transition state for formation of the carbocation intermediate in an S_N1 mechanism involves only the haloalkane and not the nucleophile and this step is rate-determining, it is a unimolecular process. The result is a first-order reaction. In this instance, the rate of reaction is expressed as the rate of disappearance of the starting material, 2-bromo-2-methylpropane. The rate has no dependence on the concentration of the nucleophile. We can conclude that any substitution reaction whose rate depends only upon [R-Lv] proceeds via an S_N1 mechanism.

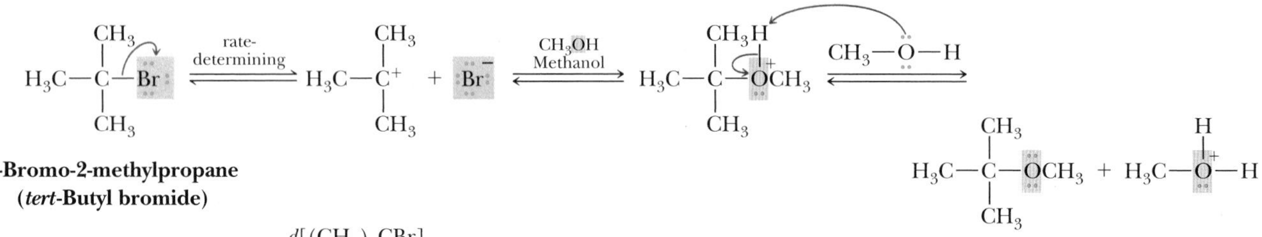

2-Bromo-2-methylpropane (*tert*-Butyl bromide)

2-Methoxy-2-methylpropane

$$\text{Rate} = -\frac{d[(CH_3)_3CBr]}{dt} = k[(CH_3)_3CBr]$$

S_N2 Kinetics

By contrast, there is only one step in the S_N2 mechanism. For the reaction of OH^- and CH_3Br, for example, both species must collide and are present in the transition state; that is, the reaction is bimolecular. The reaction between CH_3Br and NaOH to give CH_3OH and NaBr is second order: it is first order in CH_3Br and first order in OH^-, so doubling the concentration of either increases the rate by a factor of two. Doubling both of the reactants increases the rate by a factor of four. When the rate of a substitution reaction depends on both [R-Lv] and [Nu^-], we conclude that an S_N2 mechanism is occurring.

Bromomethane **Methanol**

$$\text{Rate} = -\frac{d[CH_3Br]}{dt} = k[CH_3Br][OH^-]$$

Example 9.2 Kinetics of S_N1 Reactions

The reaction of *tert*-butyl bromide with azide ion (N_3^-) in methanol is a typical S_N1 reaction. What happens to the rate of the reaction if [N_3^-] is doubled?

Solution

The rate remains the same because the nucleophile concentration does not appear in the rate equation for an S_N1 reaction.

Problem 9.2

The reaction of bromomethane with azide ion (N_3^-) in methanol is a typical S_N2 reaction. What happens to the rate of the reaction if [N_3^-] is doubled?

S_N1 Stereochemistry

Experiments in which nucleophilic substitution takes place at a chiral center provide us with information about the stereochemical course of the reaction. One of the compounds studied to determine the stereochemistry of an S_N1 reaction utilized the following chloroalkane. When either enantiomer of this molecule undergoes nucleophilic substitution by an S_N1 pathway, the product is racemic. The reason is that ionization of this secondary chloride forms an achiral carbocation. Attack of the nucleophile can occur from either side of the planar carbocation carbon, resulting in enantiomeric products. The *R* and *S* enantiomers are formed in equal amounts, and in this case, the product is a racemic mixture.

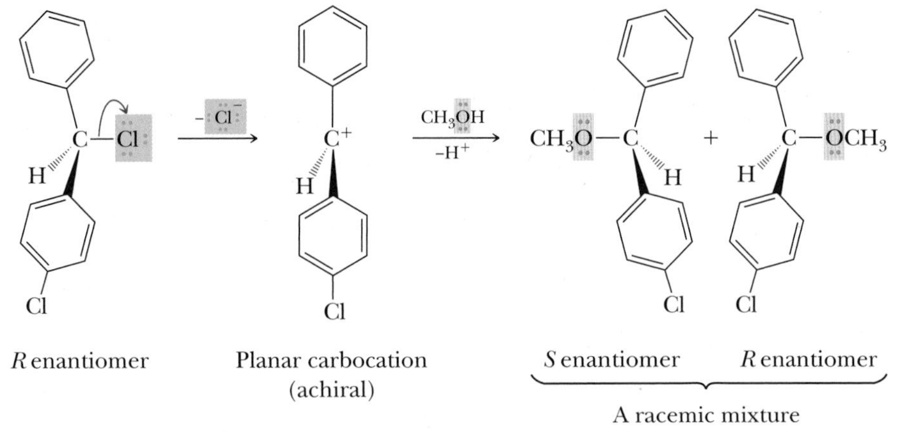
$-:\ddot{Cl}:^-$
CH_3OH
$-H^+$
+
R enantiomer
Planar carbocation (achiral)
S enantiomer
R enantiomer
A racemic mixture

Watch a video explanation

The S_N1 mechanism just described results in complete racemization. Although examples of complete racemization have been observed, it is common to find only partial racemization, with the predominant product being the one with inversion of configuration at the chiral center. Although bond breaking between carbon and the leaving group is complete, the leaving group (chloride ion in this example) remains associated for a short time with the carbocation in an ion pair.

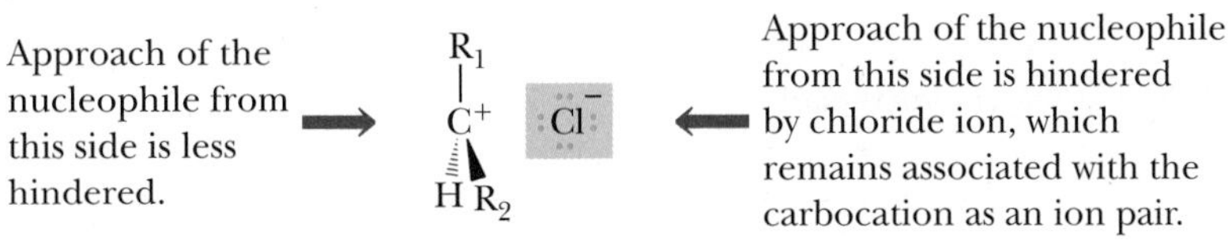
Approach of the nucleophile from this side is less hindered.
Approach of the nucleophile from this side is hindered by chloride ion, which remains associated with the carbocation as an ion pair.

To the extent that the leaving group remains associated with the carbocation as an ion pair, it hinders approach of the nucleophile from that side of the carbocation. The result is that somewhat more than 50% of the product is formed by attack of the nucleophile from the side of the carbocation opposite that of the leaving group. Whenever we observe partial to complete racemization of stereochemistry, we conclude that an S_N1 mechanism is operative.

S_N2 Stereochemistry

Every S_N2 reaction proceeds with backside attack by the nucleophile and therefore inversion of configuration. This was shown in an ingenious experiment designed by the English chemists E. D. Hughes and C. K. Ingold. They studied the exchange reaction between enantiomerically pure 2-iodooctane and iodine-131, a radioactive isotope of iodine. Iodine-127, the naturally occurring isotope of iodine, is stable and does not undergo radioactive decay. Here, acetone is the solvent.

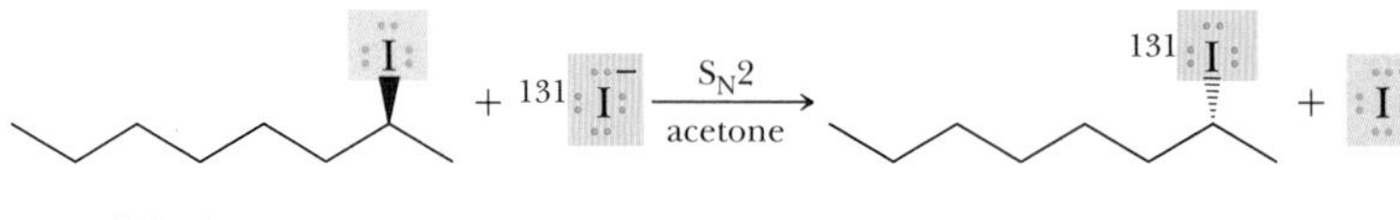

2-Iodooctane

Hughes and Ingold first demonstrated that the reaction is second order: first order in 2-iodooctane and first order in iodide ion. Therefore, the reaction proceeds by an S_N2 mechanism. They observed further that the rate of racemization of enantiomerically pure 2-iodooctane is exactly twice the rate of incorporation of iodine-131. This observation must mean, they reasoned, that each displacement of iodine-127 by iodine-131 proceeds with inversion of configuration, as illustrated in the following equation. Note that the reaction was run to only a low percent conversion in order to minimize further reactions from the product; such a result would invert stereochemistry again.

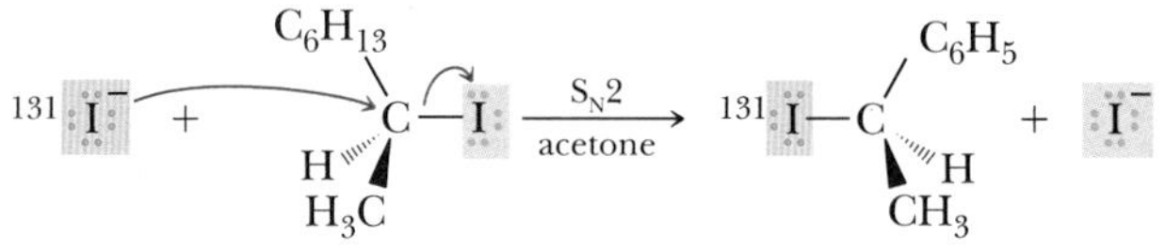

(*S*)-2-Iodooctane **(*R*)-2-Iodooctane**

Substitution with inversion of configuration in one molecule cancels the rotation of one molecule that has not reacted; so for each molecule undergoing inversion, one racemic pair is formed. Inversion of configuration in 50% of the molecules results in 100% racemization. Whenever complete inversion of configuration is found in a substitution reaction, we conclude that an S_N2 mechanism is occurring.

Example 9.3 Kinetics of S_N2 Reactions

Complete these S_N2 reactions, showing the configuration of each product.

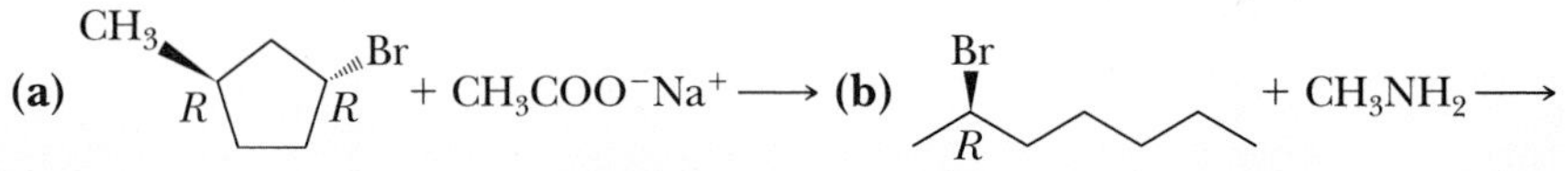

Solution

S_N2 reactions occur with inversion of configuration at the chiral center. In (a), the starting material is the (*R*,*R*) isomer; the product is the (*R*,*S*) isomer. In (b), the starting material is the (*R*) enantiomer; the product is the (*S*) enantiomer.

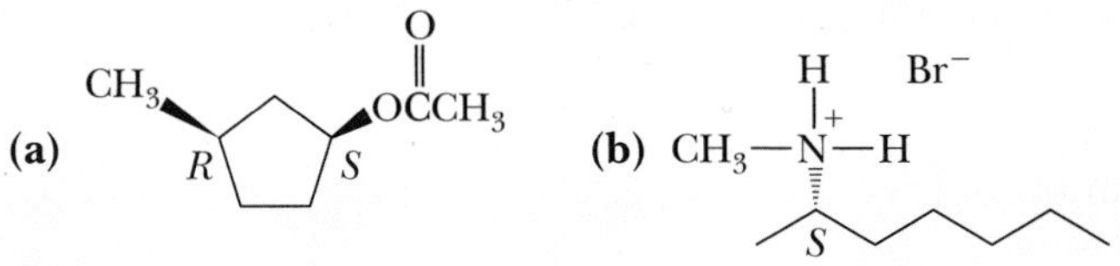

Problem 9.3

Complete these S_N2 reactions, showing the configuration of each product.

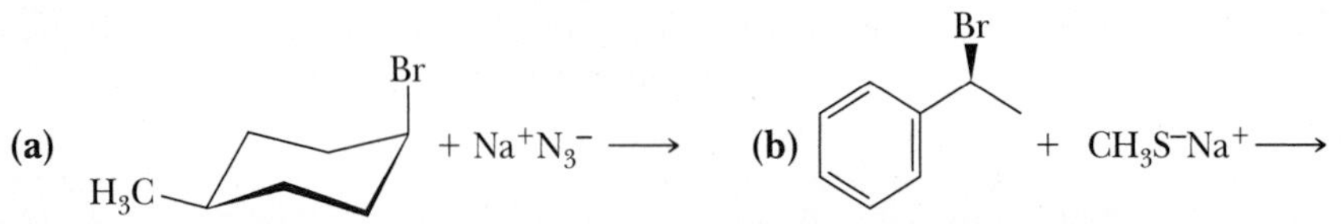

B. Structure of the Alkyl Portion of the Haloalkane

The rates of S_N1 reactions are governed mainly by electronic factors, namely the relative stabilities of carbocation intermediates. The rates of S_N2 reactions, on the other hand, are governed mainly by steric factors, and their transition states are particularly sensitive to *bulky groups* at the site of reaction. The ability of groups, because of their size, to hinder access to a reaction site within a molecule is called **steric hindrance**.

Steric hindrance
The ability of groups, because of their size, to hinder access to a reaction site within a molecule.

S_N1 Considerations

Relative Stabilities of Carbocations Let us first consider the effect of the alkyl group of the haloalkane on S_N1 reactions. As discussed, the rate-determining step of an S_N1 mechanism is formation of a carbocation; therefore, the stability of the resulting carbocation is a dominant consideration.

As we learned in Section 6.3A, 3° carbocations are most stable (lowest activation energy for their formation) due to hyperconjugation, whereas 1° carbocations are least stable (highest activation energy for their formation). In fact, 1° carbocations are so unstable that they rarely are ever formed in solution. Because carbocations are high-energy intermediates, the transition states for their formation are very similar to the carbocation in energy (Hammond's postulate, Section 8.5D). Therefore, 3° haloalkanes are most likely to react by carbocation formation; 2° haloalkanes are the next most likely to react by carbocation formation, while methyl and 1° haloalkanes react in this manner only when they are specially stabilized.

Allylic carbocations, like allylic radicals (Section 8.6), have a double bond next to the electron-deficient carbon. The allyl cation is the simplest allylic carbocation. Because the allyl cation has only one substituent on the carbon bearing the positive charge, it is a primary **allylic** carbocation. Allylic carbocations are considerably more stable than comparably substituted alkyl carbocations because delocalization is associated with the resonance interaction between the positively charged carbon and the adjacent π bond. The allyl cation, for example, can be represented as a hybrid of two equivalent contributing structures. The result is that the positive charge appears only on carbons 1 and 3, as shown in the accompanying electrostatic potential map.

Allylic carbocation
A carbocation in which an allylic carbon bears the positive charge.

Allylic
Next to a carbon-carbon double bond.

$$CH_2{=}CH{-}\overset{+}{C}H_2 \longleftrightarrow \overset{+}{C}H_2{-}CH{=}CH_2$$

Allyl cation
(a hybrid of two equivalent contributing structures)

▶ Delocalization of charge over a larger area is stabilizing.

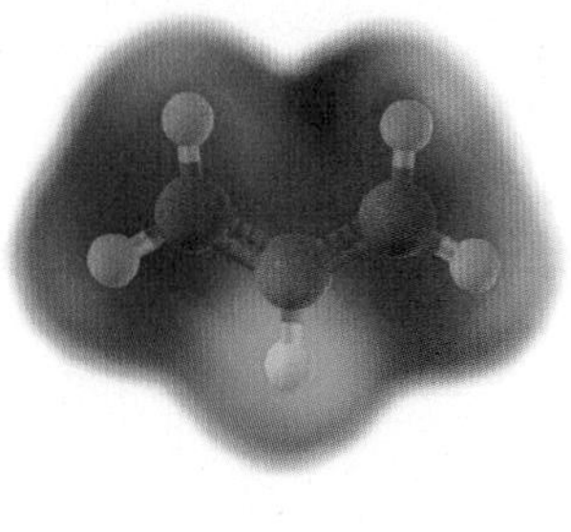

Electrostatic potential map for the allyl cation. The positive charge (blue) is on carbons 1 and 3.

Recall that in general, a distributed charge in a molecule is more stabilizing than a more localized charge. It has been determined experimentally that the double bond of one adjacent vinyl group provides approximately as much stabilization as two alkyl groups. Thus, the allyl cation and 2° isopropyl cation are of comparable stability.

These cations are of comparable stability

$CH_2{=}CH{-}\overset{+}{C}H_2$	$CH_3{-}\overset{+}{C}H{-}CH_3$
1° Allylic cation	2° Alkyl cation

The classification of allylic cations as 1°, 2°, and 3° is determined by the location of the positive charge in the more important contributing structure. Following are examples of 2° and 3° allylic carbocations.

$$CH_2{=}CH{-}\overset{+}{C}H{-}CH_3 \qquad CH_2{=}CH{-}\overset{+}{\underset{\large CH_3}{C}}{-}CH_3$$

2° Allylic cation — 3° Allylic cation

Benzylic carbocation

A carbocation in which a carbon attached to a benzene ring bears the positive charge.

Benzylic carbocations show approximately the same stability as allylic carbocations. Both are stabilized by resonance delocalization of the positive charge due to adjacent π bonds. Benzylic carbocations can be written as $C_6H_5{-}CH_2^{\ +}$.

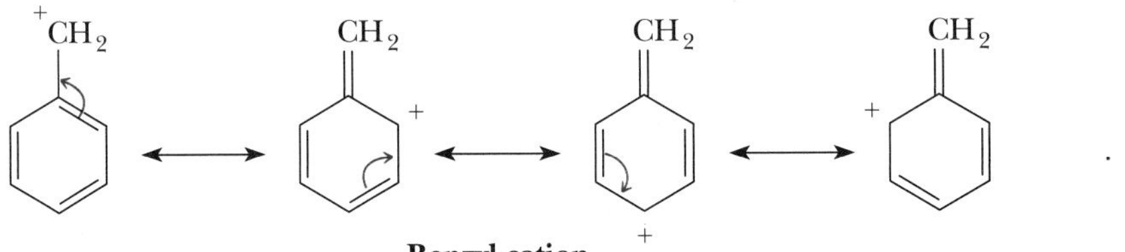

CH2
CH2
CH2
CH2

Benzyl cation
(a benzylic carbocation)

In Section 6.3A, we presented the order of stability of methyl, 1°, 2°, and 3° carbocations. We can now expand this order to include 1°, 2°, and 3° allylic, as well as 1°, 2°, and 3° benzylic carbocations.

$$\text{methyl} < 1°\text{ alkyl} < \left\{\begin{matrix}2°\text{ alkyl}\\1°\text{ allylic}\\1°\text{ benzylic}\end{matrix}\right\} < \left\{\begin{matrix}3°\text{ alkyl}\\2°\text{ allylic}\\2°\text{ benzylic}\end{matrix}\right\} < \left\{\begin{matrix}3°\text{ allylic}\\3°\text{ benzylic}\end{matrix}\right\}$$

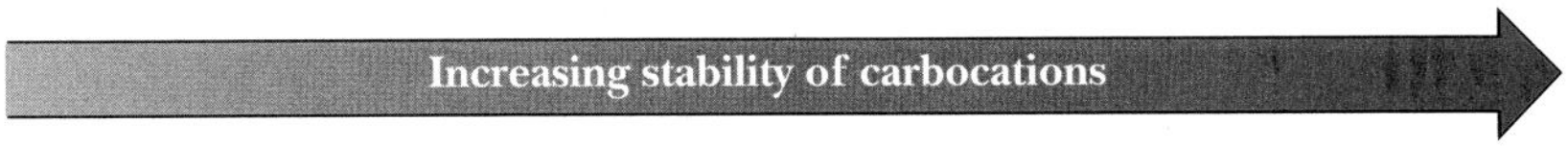

Increasing stability of carbocations

Thus, in summary, S_N1 mechanisms should be considered for allyl and benzylic haloalkanes, even if they are primary haloalkanes.

Finally, we note that S_N1 reactions rarely occur with sp^2 carbons and never occur on sp carbons. The carbocations derived from sp^2 C—X or sp C—X bonds are too unstable to form.

Example 9.4 Carbocation Resonance

Write an additional resonance contributing structure for each carbocation and state which of the two makes the greater contribution to the resonance hybrid. Classify each additional contributing structure as a 1°, 2°, or 3° allylic cation.

(a) 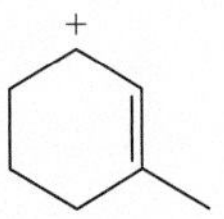 **(b)** 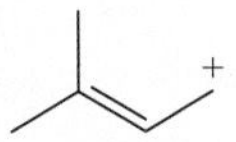

Solution

The additional resonance contributing structure in each case is a 3° allylic cation. The contributing structure having the greater degree of substitution on the positively charged carbon makes the greater contribution to the hybrid.

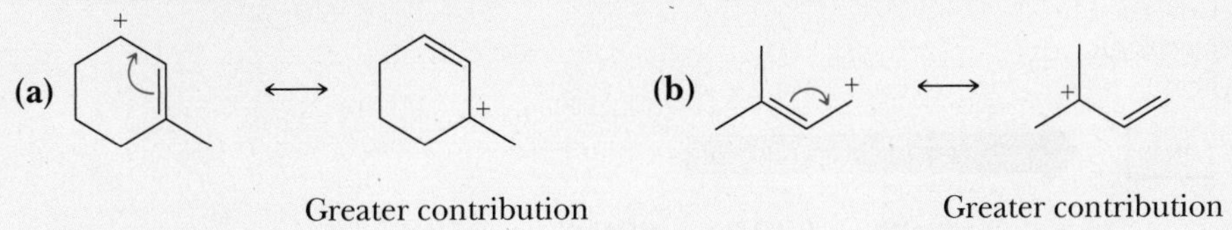

Problem 9.4

Write an additional resonance contributing structure for each carbocation and state which of the two makes the greater contribution to the resonance hybrid.

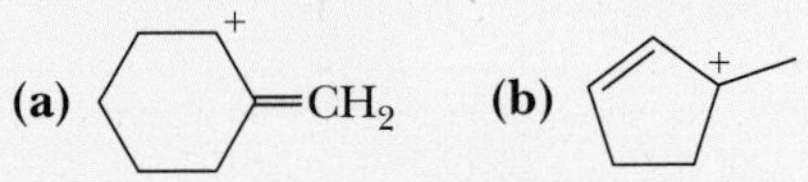

S_N2 Considerations

Steric Hindrance In an S_N2 mechanism the nucleophile must approach the substitution center and begin to form a new covalent bond to it while the leaving group is departing. If we compare the ease of approach to the substitution center of a 1° haloalkane with that of a 3° haloalkane, we see that the approach is considerably easier for bromoethane than for *tert*-butyl bromide. Two hydrogen atoms and one alkyl group screen the backside of the substitution center of a 1° haloalkane. In contrast, three alkyl groups screen the backside of the 3° haloalkane.

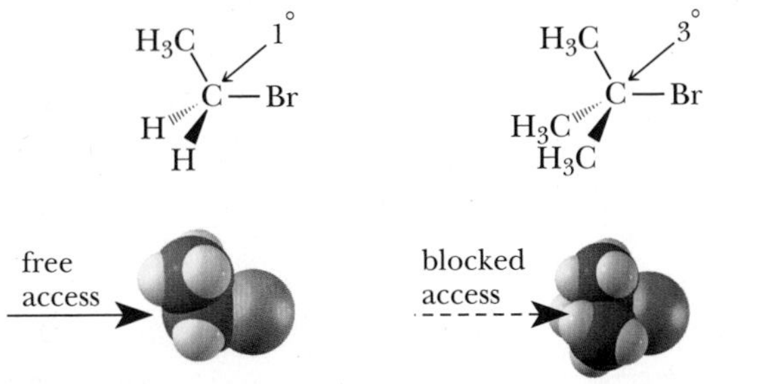

Bromoethane (Ethyl bromide) **2-Bromo-2-methylpropane (*tert*-Butyl bromide)**

Tertiary haloalkanes react by an S_N1 mechanism because 3° carbocation intermediates are relatively stable and tertiary haloalkanes are protected against backside attack. In fact, 3° haloalkanes are never observed to react by an S_N2 mechanism. In contrast, halomethanes and primary haloalkanes are never observed to react by an S_N1 mechanism. They have little crowding around the reaction site and react by an S_N2 mechanism because methyl and primary carbocations are unstable. Secondary haloalkanes may react by either S_N1 or S_N2 mechanisms, depending on the nucleophile and solvent. The competition between electronic and steric factors and their effects on relative rates of nucleophilic substitution reactions of haloalkanes are summarized in Figure 9.3.

We see a similar effect of steric hindrance on S_N2 reactions in molecules with branching at the β-carbon. The carbon bearing the halogen in a haloalkane is called the α-carbon, and the next carbon is called the β-carbon. Table 9.2 shows relative rates of S_N2 reactions on a series of primary bromoalkanes. In these data, the rate of nucleophilic substitution of bromoethane is taken as a reference and is given the value 1.0. As CH_3 branches are added to the β-carbon, the relative rate of reaction decreases. Compare the relative rates of bromoethane (no β-branch) with

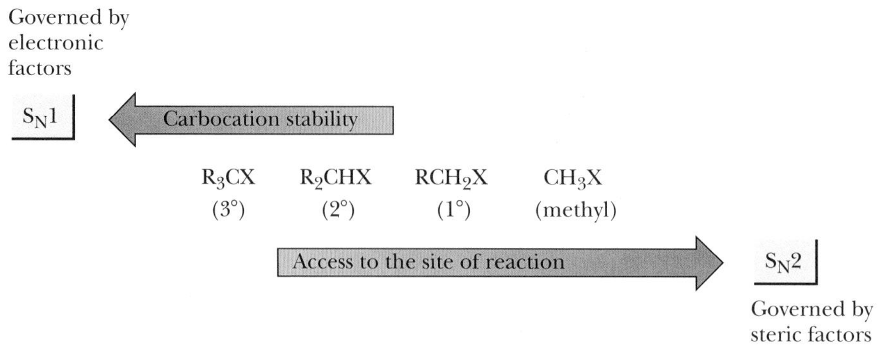

FIGURE 9.3 Effect of steric and electronic factors in competition between S_N1 and S_N2 reactions of haloalkanes. Methyl and primary haloalkanes react only by the S_N2 mechanism; they do not react by S_N1. Tertiary haloalkanes do not react by S_N2; they react only by S_N1. Secondary haloalkanes may be made to react by either S_N1 or S_N2 mechanisms depending on the solvent and the choice of nucleophile, which are topics addressed later in this chapter.

Table 9.2 Effect of β-Branches on the Rate of S_N2 Reactions

Alkyl bromide	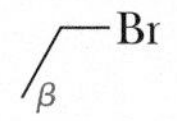	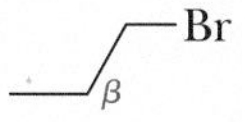	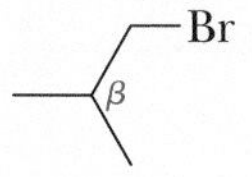	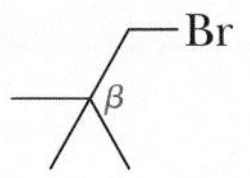
β-Branches	0	1	2	3
Relative rate	1.0	4.1×10^{-1}	1.2×10^{-3}	1.2×10^{-5}

that of 1-bromo-2,2-dimethylpropane (neopentyl bromide), a compound with three β-branches. The rate of S_N2 substitution of this compound is only 10^{-5} that of bromoethane. For all practical purposes, primary halides with three β-branches do not undergo S_N2 reactions.

As shown in Figure 9.4, the carbon of the C—Br bond in bromoethane is unhindered and open to attack by a nucleophile in an S_N2 reaction. On the other hand, three β-methyl groups screen the corresponding carbon in neopentyl bromide. Thus, although the carbon bearing the leaving group is primary, approach of the nucleophile is so hindered that the rate of S_N2 reaction of neopentyl bromide is greatly reduced compared to bromoethane.

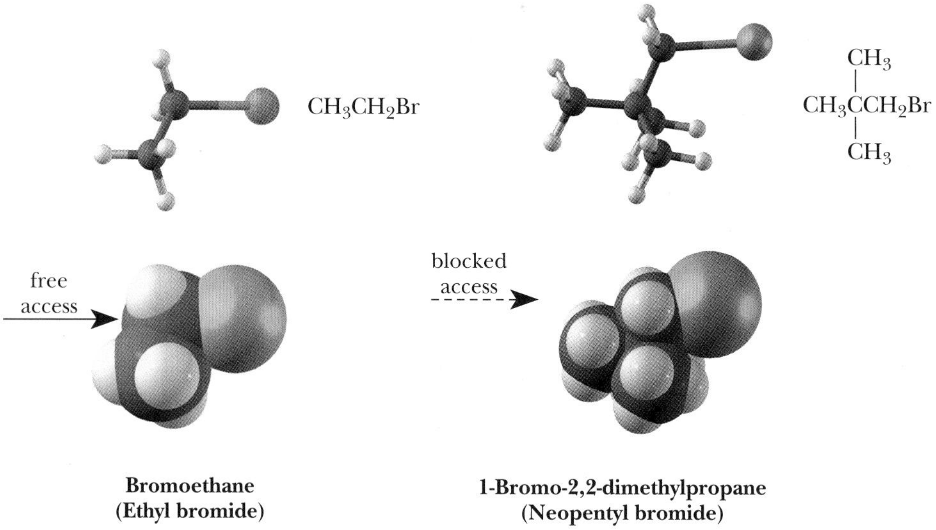

FIGURE 9.4 The effect of β-branching in S_N2 reactions on a primary haloalkane. With bromoethane, attack of the nucleophile is unhindered. With 1-bromo-2,2-dimethylpropane, the three β-branches block approach of the nucleophile to the backside of the C—Br bond, thus drastically reducing the rate of S_N2 reaction of this compound.

Finally, we note that S_N2 reactions never occur on sp^2 or sp hybridized carbons. Hence, you should never consider performing a substitution on a vinyl halide (C═C—X), an aryl halide, or an alkynyl halide (C≡C—X).

C. The Leaving Group

In the transition state for nucleophilic substitution on a haloalkane, the **leaving group** develops a partial negative charge in both S_N1 and S_N2 reactions; therefore, the ability of a group to function as a leaving group is related to how stable it is as an anion. The most stable anions, and therefore the best leaving groups, are the (weak) conjugate bases of strong acids. Thus, we can use the information on the relative strengths of organic and inorganic acids in Table 4.1 to determine which anions are the best leaving groups. This order is shown here.

← Reactivity as a leaving group

rarely function as leaving groups (applies to F^- through NH_2^-)

$$I^- > Br^- > Cl^- \sim H_2O >> F^- > CH_3\overset{\overset{\displaystyle O}{\|}}{C}O^- > HO^- > CH_3O^- > NH_2^-$$

← Stability of anion; strength of conjugate acid

The best leaving groups in this series are the halides, I^-, Br^-, and Cl^-. Hydroxide ion, OH^-, methoxide ion, CH_3O^-, and amide ion, NH_2^-, are such poor leaving groups that they rarely, if ever, are displaced in nucleophilic aliphatic substitution.

D. The Solvent

The solvent plays a role of paramount importance in substitution reactions. It influences the rates of both S_N1 and S_N2 reactions, the relative nucleophilicity (see the next section), and the balance between S_N1 and S_N2 mechanisms for secondary haloalkanes.

Common solvents can be divided into two groups: protic and aprotic. Furthermore, solvents are classified as polar and nonpolar based on their **dielectric constant**. The greater the value of the dielectric constant of a solvent, the better it solvates and thus the smaller the interaction between ions of opposite charge dissolved in it. We say that a solvent is a **polar solvent** if it has a dielectric constant of 15 or greater. A solvent is a **nonpolar solvent** if it has a dielectric constant of less than 5. Solvents with a dielectric constant between 5 and 15 are borderline.

Dielectric constant
A measure of a solvent's ability to insulate opposite charges from one another.

The common **protic solvents** for nucleophilic substitution reactions are water, low-molecular-weight alcohols, and low-molecular-weight carboxylic acids (Table 9.3). Each of these has a partially negatively charged oxygen bonded to a partially positively charged hydrogen atom. Protic solvents solvate ionic substances by electrostatic interactions between anions and the partially positively charged hydrogens of the solvent and between cations and partially negatively charged atoms of the solvent. By our guideline, water, formic acid, methanol, and ethanol are classified as

Protic solvent
A solvent that is a hydrogen-bond donor; the most common protic solvents contain —OH groups. Common protic solvents are water, low-molecular-weight alcohols such as ethanol, and low-molecular-weight carboxylic acids.

Table 9.3 Common Protic Solvents

Solvent	Structure	Dielectric Constant (25°C)
Water	H_2O	79
Formic acid	$HCOOH$	59
Methanol	CH_3OH	33
Ethanol	CH_3CH_2OH	24
Acetic acid	CH_3COOH	6

Table 9.4 Common Aprotic Solvents

Solvent	Structure	Dielectric Constant
Polar		
Dimethyl sulfoxide (DMSO)	$CH_3\overset{O}{\overset{\|}{S}}CH_3$	48.9
Acetonitrile	$CH_3C{\equiv}N$	37.5
N,N-Dimethylformamide (DMF)	$H\overset{O}{\overset{\|}{C}}N(CH_3)_2$	36.7
Acetone	$CH_3\overset{O}{\overset{\|}{C}}CH_3$	20.7
Moderately Polar		
Dichloromethane	CH_2Cl_2	9.1
Tetrahydrofuran (THF)	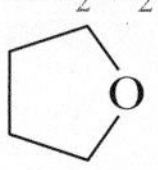	7.6
Nonpolar		
Diethyl ether	$CH_3CH_2OCH_2CH_3$	4.3
Toluene	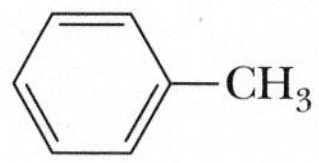	2.3
Hexane	$CH_3(CH_2)_4CH_3$	1.9

Increasing solvent polarity ↑

polar protic solvents. Because of its smaller dielectric constant, acetic acid is classified as a moderately polar protic solvent.

Aprotic solvent A solvent that cannot serve as a hydrogen-bond donor; nowhere in the molecule is there a hydrogen bonded to an atom of high electronegativity. Common aprotic solvents are dichloromethane, diethyl ether, and dimethyl sulfoxide.

The **aprotic solvents** most commonly used for nucleophilic substitution reactions are given in Table 9.4. Of these, dimethyl sulfoxide (DMSO), acetonitrile, *N,N*-dimethylformamide (DMF), and acetone are classified as **polar aprotic solvents**. Dichloromethane and tetrahydrofuran (THF) are moderately polar aprotic solvents. Diethyl ether, toluene, and hexane are classified as nonpolar aprotic solvents.

Effect of Solvent on S_N1 Reactions

Nucleophilic substitution by an S_N1 pathway involves creation and separation of opposite charges in the transition state of the rate-determining step. For this reason, the rate of S_N1 reactions depends on both the ability of the solvent to keep opposite charges separated and its ability to stabilize both positive and negative sites by solvation. The solvents that best solvate charges are polar protic solvents such as H_2O; low-molecular-weight alcohols such as methanol and ethanol; and, to a lesser degree, low-molecular-weight carboxylic acids such as formic acid and acetic acid. As shown in Table 9.5, the rate of solvolysis of 2-chloro-2-methylpropane (*tert*-butyl chloride) increases by a factor of 10^5 when the solvent is changed from ethanol to water because water better solvates the carbocation and chloride anion.

The increased rate of an S_N1 reaction in higher polarity solvents can also be explained by an analysis of reaction coordinate diagrams. Figure 9.5a shows curves for two different solvents. The higher polarity solvent will generally better solvate the reactants, intermediates, and products; therefore, the entire curve is lower in energy with the higher polarity solvent. However, the increased solvation is greatest for the intermediate carbocation, which lowers the barrier to its formation and thereby accelerates the reaction.

Table 9.5 Rates of an S_N1 Reaction as a Function of Solvent

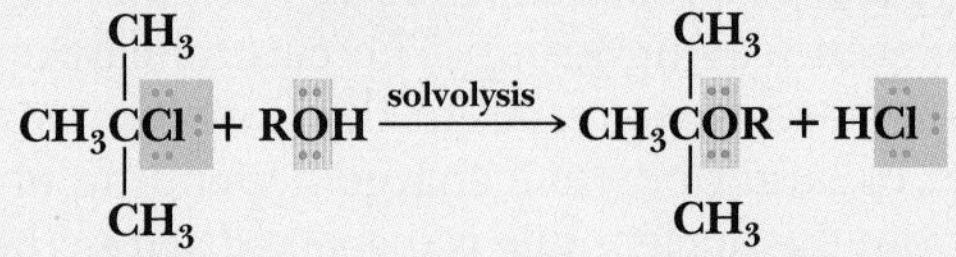

Solvent	$\frac{k_{(solvent)}}{k_{(ethanol)}}$
Water	100,000
80% water: 20% ethanol	14,000
40% water: 60% ethanol	100
Ethanol	1

Increasing polarity of solvent ↑ Increasing rate of solvolysis ↑

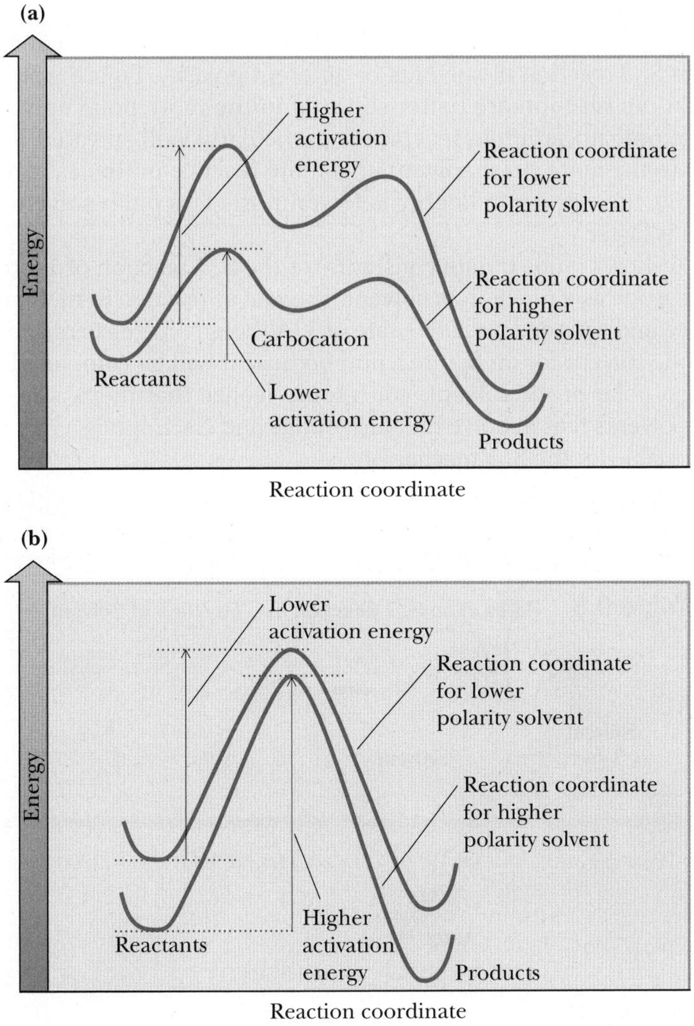

FIGURE 9.5 Energy diagrams for substitution reactions in different polarity solvents. (a) An S_N1 reaction run in two different solvents. The higher polarity solvent better solvates all species but has the greatest difference in solvation for the intermediate carbocation. The reaction is therefore faster in the higher polarity solvent. (b) An S_N2 reaction run in two different solvents. The higher polarity solvent better solvates all the species but has the greatest difference in solvation for the anionic nucleophile reactant. The reaction is therefore faster in the lower polarity solvent.

The Effect of Solvent on S_N2 Reactions

The most common type of S_N2 reaction involves a negatively charged nucleophile and a negatively charged leaving group. The central carbon atom has a partial positive charge in the starting material. In the transition state, however, it may have either a smaller or larger positive charge depending on the conditions, but the negative charge from the nucleophile is dispersed across the adding nucleophile and the departing leaving group.

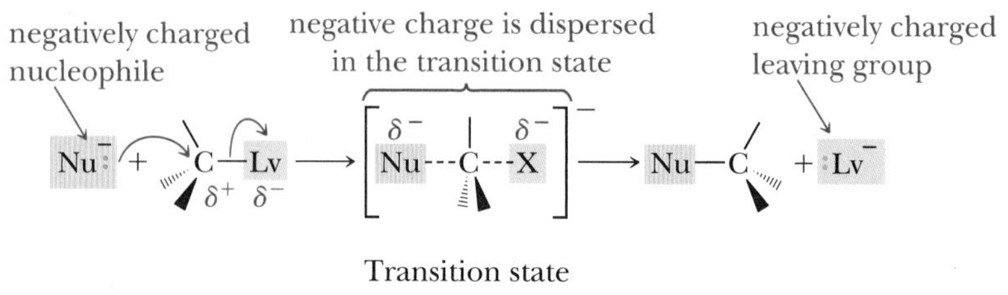

The stronger the solvation of the nucleophile, the greater the energy required to remove the nucleophile from its solvation shell to reach the transition state and hence the lower the rate of the S_N2 reaction.

Polar aprotic solvents can solvate cations very well, but they solvate anions (nucleophiles) relatively poorly, because they cannot donate a hydrogen bond to an anion, as can a protic solvent. For this reason, nucleophiles are freer and more reactive in polar aprotic solvents than in protic solvents; so the rates of S_N2 reactions are dramatically accelerated, often by several orders of magnitude compared to the same reaction in protic solvents.

These effects can be seen pictorially by comparing reaction coordinate diagrams for an S_N2 reaction in solvents of different polarity. Figure 9.5b shows that all species in the reaction are better solvated in the more polar solvent. But because the anionic nucleophile reactant is particularly well solvated (lower starting energy) in the more polar solvent, it is more reactive in the less polar solvent. Consequently, the activation energy is lower in the less polar solvent; therefore, the reaction is faster.

Table 9.6 shows ratios of rate constants for the S_N2 reaction of 1-bromobutane with sodium azide as a function of solvent. The rate of reaction in methanol is taken as a reference and assigned a relative rate of 1. Although chemists may prefer to use a polar aprotic solvent for an S_N2 reaction because it will be completed faster than when using a polar protic solvent, you should realize that the S_N2 mechanism is viable in all solvents that dissolve the nucleophile and electrophile. Only polar protic solvents are viable for the S_N1 mechanism.

Table 9.6 Rates of an S_N2 Reaction as a Function of Solvent

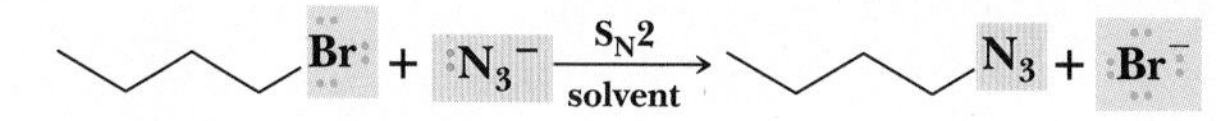

Solvent Type	Solvent		$\frac{k_{(solvent)}}{k_{(methanol)}}$
Polar aprotic	Acetonitrile	$CH_3C{\equiv}N$	5000
	DMF	$(CH_3)_2NCHO$	2800
	DMSO	$(CH_3)_2S{=}O$	1300
Polar protic	Water	H_2O	7
	Methanol	CH_3OH	1

E. Structure of the Nucleophile

Nucleophilicity is a kinetic property measured by relative rates of reaction. Relative nucleophilicities for a series of nucleophiles are established by measuring the rate at which each displaces a leaving group from a haloalkane (e.g., the rate at which each nucleophile displaces bromide ion from bromoethane in ethanol at 25°C). Here is a reaction using ammonia as the nucleophile.

$$CH_3CH_2\ddot{\underset{..}{Br}}: + :NH_3 \xrightarrow[\text{ethanol}]{25°C} CH_3CH_2\overset{+}{N}H_3 + :\ddot{\underset{..}{Br}}:^-$$

Nucleophilicity
A kinetic property measured by the rate at which a nucleophile causes nucleophilic substitution on a reference compound under a standardized set of experimental conditions.

From these studies, we can make correlations between the structure of a nucleophile and its relative nucleophilicity. Listed in Table 9.7 are the types of nucleophiles we deal with most commonly in this text and their nucleophilicity in alcohol or water. The more rapidly a nucleophile reacts with a substrate in an S_N2 reaction, the more nucleophilic it is, by definition.

Table 9.7 Common Nucleophiles and Their Relative Nucleophilicities in Alcohol or Water

	Effectiveness in Nucleophilic Substitution Reactions	Nucleophile
Increasing nucleophilicity ↑	Good	Br^-, I^-
		CH_3S^-, RS^-
		HO^-, CH_3O^-, RO^-, $R—C{\equiv}C^-$
		CN^-, N_3^-, H_2N^-
	Moderate	Cl^-, F^-
		$CH_3\overset{O}{\overset{\Vert}{C}}O^-$, $R\overset{O}{\overset{\Vert}{C}}O^-$
		CH_3SH, RSH, R_2S
		NH_3, RNH_2, R_2NH, R_3N
	Poor	H_2O
		CH_3OH, ROH
		$CH_3\overset{O}{\overset{\Vert}{C}}OH$, $R\overset{O}{\overset{\Vert}{C}}OH$

Because all nucleophiles are Brønsted–Lowry bases as well (see Chapter 4), we also study correlations between nucleophilicity and basicity. **Basicity** and nucleophilicity are often related because they both involve a lone pair of electrons making a bond to another atom. In the case of a base, the lone pair makes a bond to a proton, while with a nucleophile, the lone pair most commonly creates a bond to an electrophilic carbon. In general, sterically unhindered strong bases are good nucleophiles.

Basicity
An equilibrium property measured by the position of equilibrium in an acid-base reaction, such as the acid-base reaction between ammonia and water.

For example, oxygen anions such as hydroxide and methoxide (CH_3O^-) are good nucleophiles because they are also strong bases. Weaker oxygen bases are similarly weaker nucleophiles. As an example, carboxylate anions (RCO_2^-) are classified as moderate nucleophiles. Because of this trend with basicity, we can confidently conclude that anionic atoms are more nucleophilic than their neutral counterparts. Hence, neutral oxygen species such as water, alcohols, and carboxylic acids are weak nucleophiles. As a rough guideline for oxygen and amine nucleophilicities, we consider those nucleophiles that have conjugate acids with pK_a's above 11 to be **strong nucleophiles**, around 11 to be **moderate nucleophiles**, and below 11 to be **weak nucleophiles**. This guideline classifies amines as moderate nucleophiles, as shown in Table 9.7.

But upon inspection of Table 9.7, you might be asking, "Why are the extremely weak bases iodide and bromide anions good nucleophiles?" Also, the table shows that anionic sulfur species, as well as cyanide and azide, are good nucleophiles even though none of these species are particularly strong bases. As we now describe, nucleophilicity is complex and depends on solvent and shape, not just base strength.

Solvation Effects on Nucleophilicities

The solvent in which nucleophilic substitutions are carried out has a marked effect on relative nucleophilicities. For a fuller understanding of the role of the solvent, let us consider nucleophilic substitution reactions carried out in polar aprotic solvents and in polar protic solvents. An organizing principle for substitution reactions is the following:

> *All other factors being equal, the stronger the interaction of the nucleophile with the solvent, the lower its nucleophilicity.*

Conversely, the less the nucleophile interacts with solvent, the greater its nucleophilicity.

The most commonly used polar aprotic solvents (DMSO, acetone, acetonitrile, and DMF) are very effective in solvating cations (in addition to the attraction to the negative end of the dipole, the lone pairs on oxygen and nitrogen act as Lewis bases) but are not nearly as effective in solvating anions. Consider, for example, acetone. Because the negative end of its dipole and the lone pairs on oxygen can come close to the center of positive charge in a cation, acetone is effective in solvating cations. The positive end of its dipole, however, is shielded by the two methyl groups and is therefore less effective in solvating anions. The sodium ion of sodium iodide, for example, is effectively solvated by acetone and DMSO, but the iodide ion is only poorly solvated. Because anions are only poorly solvated in polar aprotic solvents, they are freer and participate readily in nucleophilic substitution reactions. In these solvents, their relative nucleophilicities parallel their relative basicities. The relative nucleophilicities of halide ions in polar aprotic solvents, for example, are $F^- > Cl^- > Br^- > I^-$.

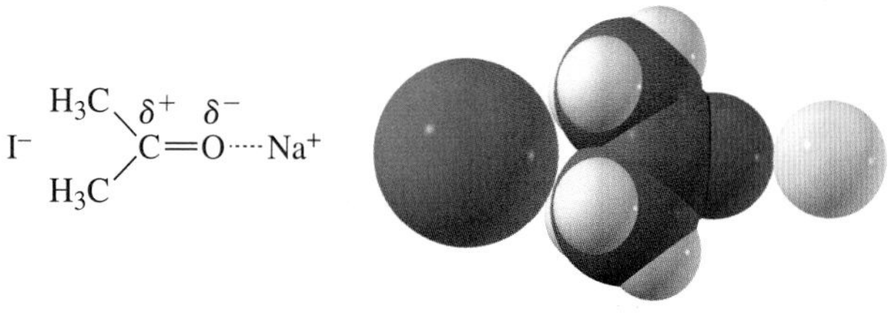

Solvation of NaI in acetone

The relative nucleophilicities of halide ions in polar protic solvents are quite different from those in polar aprotic solvents (Table 9.8).

Table 9.8 Relative Nucleophilicities of Halide Ions in Polar Aprotic and Protic Solvents

Solvent	Increasing nucleophilicity →
Polar aprotic	$I^- < Br^- < Cl^- < F^-$
Polar protic	$F^- < Cl^- < Br^- < I^-$

In polar protic solvents, iodide ion, the least basic of the halide ions, has the greatest nucleophilicity. Conversely, fluoride ion, the most basic of the halide ions, has the smallest nucleophilicity. The reason for this reversal of correlation between nucleophilicity and basicity lies in the degree of solvation of anions in protic solvents compared with aprotic solvents and in polarizability trends.

- In polar aprotic solvents, anions are only weakly solvated and therefore relatively free to participate in nucleophilic substitution reactions and basicity dictates nucleophilicity.
- In polar protic solvents, anions are highly solvated by hydrogen bonding with solvent molecules and therefore are less free to participate in nucleophilic substitution reactions and polarizability dictates nucleophilicity.

The negative charge on the fluoride ion, the smallest of the halide ions, is concentrated in a small volume, and the very tightly held solvent shell formed by a polar protic solvent constitutes a barrier between fluoride ion and substrate. The fluoride ion must be at least partially removed from its tightly held solvation shell before it can participate in nucleophilic substitution. The negative charge on the iodide ion, the largest and most polarizable of the halide ions, is far less concentrated, the solvent shell is less tightly held, and iodide is considerably freer to participate in nucleophilic substitution reactions.

Recall that the order of polarizability is $F^- < Cl^- < Br^- < I^-$, which reflects the greater ability of electron clouds to undergo perturbations during chemical reactions such as S_N2 displacements. Iodide, being the most polarizable, makes it the best nucleophile in a polar protic solvent.

Effect of Shape on Nucleophilicity

As described previously, the nucleophile in an S_N2 reaction attacks the backside of a C—Lv bond. That backside attack can have varying degrees of steric hindrance depending upon the R-group of the haloalkane. Not surprisingly, nucleophiles that are shaped like bullets or spears can better penetrate past the steric hindrance and are generally better nucleophiles. Two prime examples are azide and cyanide, both of which are cylindrically shaped anions. Although neither are particularly basic nor polarizable, both are excellent nucleophiles. In contrast, when an otherwise good nucleophile, such as an alkoxide, is large and bulky, its ability to be a nucleophile diminishes. For example, whereas ethoxide (EtO^-) is an excellent nucleophile, *tert*-butoxide (t-BuO^-) is not a nucleophile.

F. Skeletal Rearrangement

As we saw in Section 6.3C, skeletal rearrangement is typical of reactions involving a carbocation intermediate that can rearrange to a more stable one. Because there is little or no carbocation character at the substitution center, S_N2 reactions are free of rearrangement. In contrast, S_N1 reactions often proceed with rearrangement. An example of an S_N1 reaction involving rearrangement is solvolysis of 2-chloro-3-phenylbutane in methanol, a polar protic solvent and a weak nucleophile. The major substitution product is the ether with a rearranged structure. The chlorine atom in the starting material is on a 2° carbon, but the methoxy group in the product is on the adjacent 3° carbon.

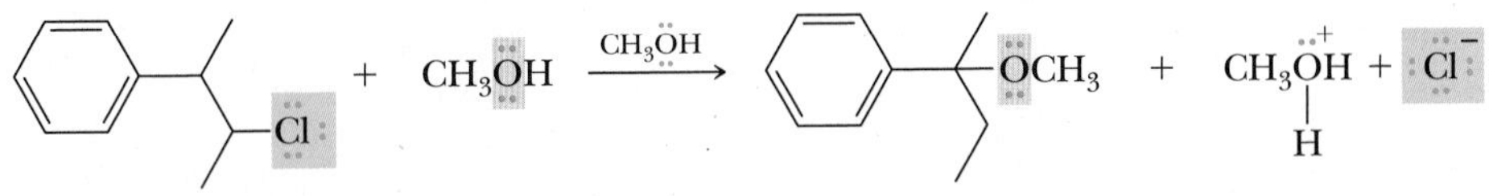

2-Chloro-3-phenylbutane (racemic)

2-Methoxy-2-phenylbutane (racemic)

As shown in the following mechanism, reaction is initiated by heterolytic cleavage of the carbon-chlorine bond to form a 2° carbocation, which rearranges to a considerably more stable 3° carbocation by shift of a hydrogen with its pair of electrons (a hydride ion) from the adjacent benzylic carbon. Note that the rearranged carbocation is not only tertiary (hyperconjugation stabilization) but also benzylic (stabilization by resonance delocalization).

Mechanism 9.3

Rearrangement During Solvolysis of 2-Chloro-3-phenylbutane

Step 1: **Break a bond to give stable molecules or ions.** Ionization of the C—Cl bond gives a 2° carbocation intermediate.

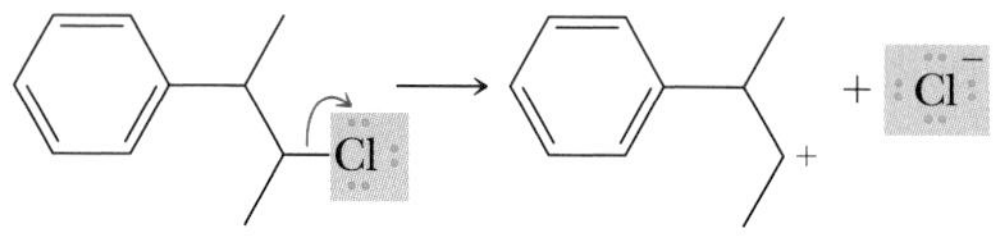

A 2° carbocation

Step 2: **1,2 Shift.** Migration of a hydrogen atom with its bonding electrons (a hydride ion) gives a more stable 3° benzylic carbocation intermediate.

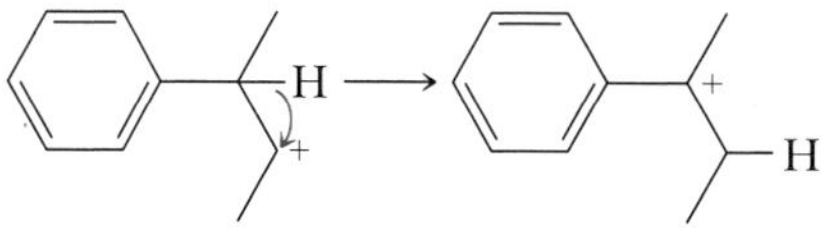

A 3° benzylic carbocation

Step 3: **Make a new bond between a nucleophile and an electrophile.** Reaction of the 3° benzylic carbocation intermediate (an electrophile) with methanol (a nucleophile) forms an oxonium ion.

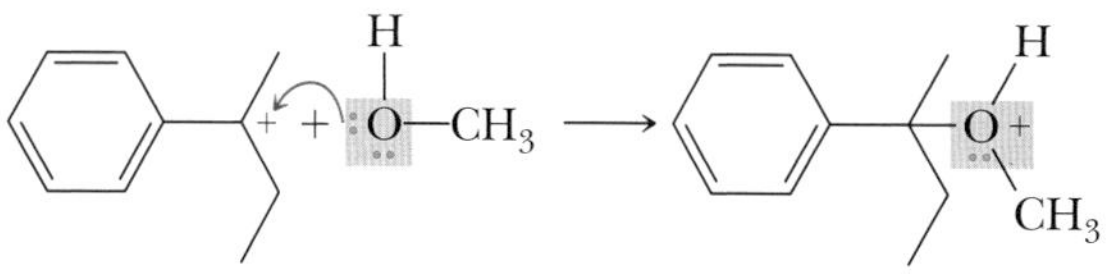

An oxonium ion

Step 4: **Take a proton away.** Proton transfer to solvent (in this case, methanol) gives the final product.

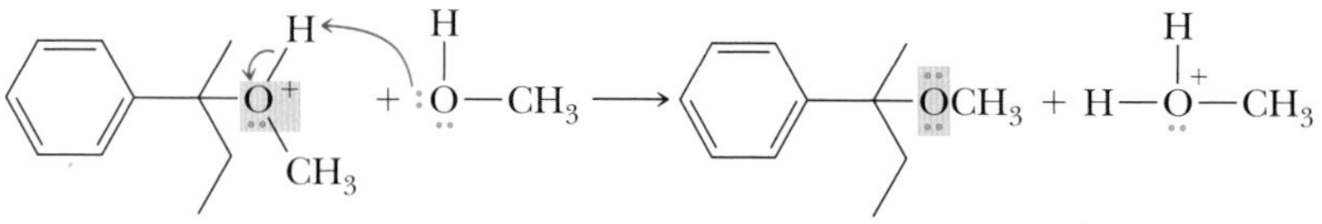

In general, migration of a hydrogen atom or an alkyl group with its bonding electrons occurs when a more stable carbocation can be formed.

Now that we have considered the many factors involved in substitution reactions, we present an overview useful to predict the type of mechanism that dominates under certain reaction conditions (Table 9.9). Examples of how to use this summary are given in Section 9.4.

Table 9.9 Summary of S_N1 Versus S_N2 Reactions of Haloalkanes

Type of Alkyl Halide	S_N2	S_N1
Methyl CH_3X	**S_N2 is favored.**	**S_N1 does not occur.** The methyl cation is so unstable that it is never observed in solution.
Primary RCH_2X	**S_N2 is favored.**	**S_N1 rarely occurs.** Primary cations are so unstable that they are not formed in solution (allylic and benzylic cations are the exceptions).
Secondary R_2CHX	**S_N2 is favored** in aprotic solvents with good nucleophiles.	**S_N1 is favored** in protic solvents with poor nucleophiles. Carbocation rearrangements may occur.
Tertiary R_3CX	**S_N2 does not occur** because of steric hindrance around the reaction center.	**S_N1 is favored** because of the ease of formation of tertiary carbocations.
Substitution at a chiral center	**Inversion of configuration.** The nucleophile attacks the chiral center from the side opposite the leaving group.	**Racemization is favored.** The carbocation intermediate is planar, and attack of the nucleophile occurs with more or less equal probability from either side, although there is often some net inversion of configuration.

9.4 Analysis of Several Nucleophilic Substitution Reactions

Predictions about the mechanism for a particular nucleophilic substitution reaction must be based on considerations of the structure of the haloalkane, the nucleophile, the leaving group, and the solvent. Following are five nucleophilic substitution reactions and an analysis of the factors that favor an S_N1 or S_N2 mechanism for each and the products that result from the mechanism used. Note that in the following examples, we ignore competing elimination because it has not been discussed yet.

Nucleophilic Substitution 1

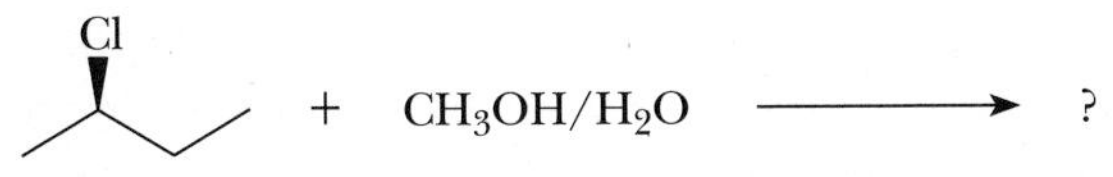

(*R*)-2-Chlorobutane

The mixture of methanol and water is a polar protic solvent and a good ionizing solvent in which to form carbocations. 2-Chlorobutane ionizes in this solvent to form a fairly stable 2° carbocation intermediate. Both water and methanol are poor nucleophiles. From this analysis, we predict that reaction occurs primarily by an S_N1 mechanism. Ionization of the 2° chloroalkane gives a carbocation intermediate, which then reacts with either water or methanol as the nucleophile to give the observed products. Each product is formed as an approximately 50:50 mixture of *R* and *S* enantiomers.

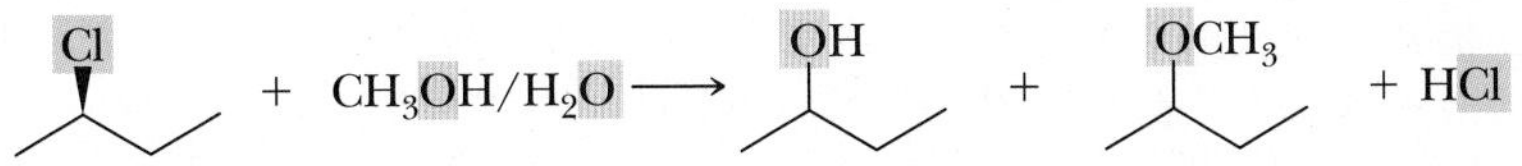

(*R*)-2-Chlorobutane **2-Butanol** (racemic) **2-Methoxybutane** (racemic)

Nucleophilic Substitution 2

This is a primary bromoalkane with two beta branches in the presence of a cyanide ion, a good nucleophile. Dimethyl sulfoxide (DMSO), a polar aprotic solvent, is a particularly good solvent in which to carry out nucleophile-assisted substitution reactions because of its good ability to solvate cations (in this case, Na^+) and its poor ability to solvate anions (in this case, CN^-). From this analysis, we predict that this reaction occurs by an S_N2 mechanism.

Nucleophilic Substitution 3

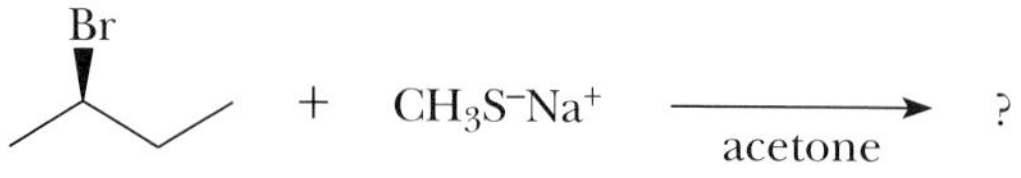

(*R*)-2-Bromobutane

Bromide is a good leaving group, and it is on a 2° carbon. The methylsulfide ion is a good nucleophile. Acetone, a polar aprotic solvent, is a good medium in which to carry out S_N2 reactions but a poor medium in which to carry out S_N1 reactions. From this analysis, we predict that this reaction occurs by an S_N2 mechanism and that the product is the *S* enantiomer.

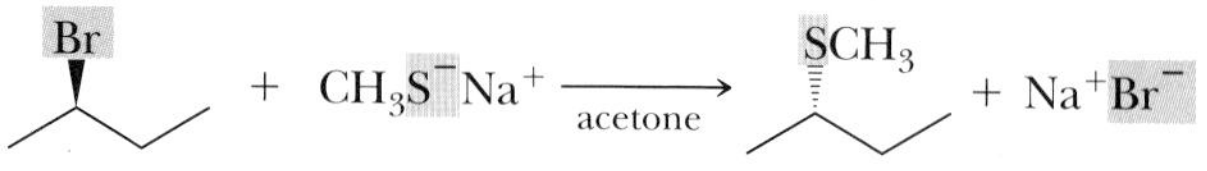

(*R*)-2-Bromobutane **(*S*)-2-Methylsulfanylbutane**

Nucleophilic Substitution 4

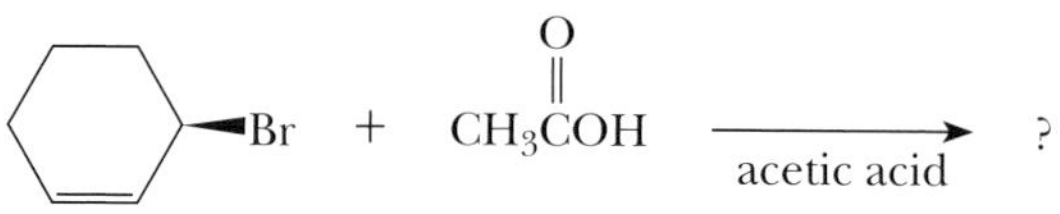

Ionization of the carbon-bromine bond forms a resonance-stabilized 2° allylic carbocation. Acetic acid is a poor nucleophile, which reduces the likelihood of an S_N2 reaction. Further, acetic acid is a moderately polar protic (hydroxylic) solvent that favors S_N1 reaction. From this analysis, we predict that this reaction occurs by an S_N1 mechanism and both enantiomers of the product are observed.

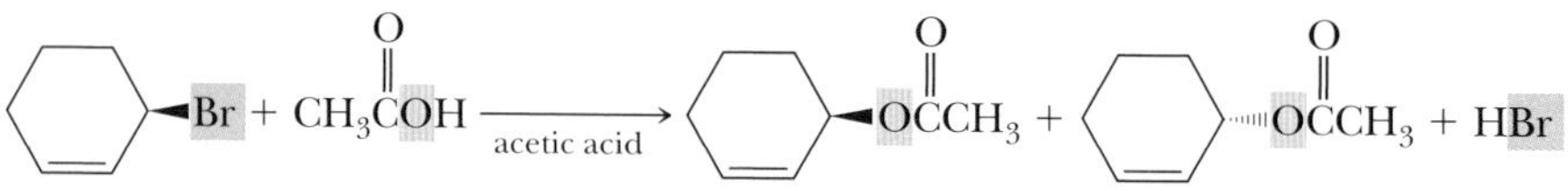

(*R*)-3-Bromocyclohexene **(*R*)-3-Acetoxycyclohexene** **(*S*)-3-Acetoxycyclohexene**

Nucleophilic Substitution 5

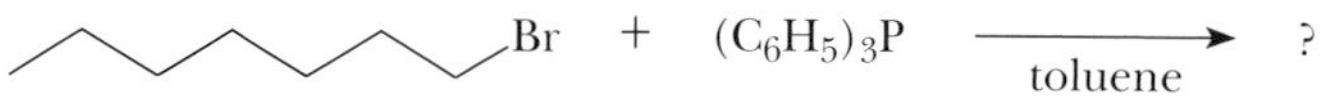

The bromoalkane is primary, and bromide is a good leaving group. Trivalent compounds of phosphorus, a third-row element, are moderate nucleophiles. Toluene is a nonpolar aprotic solvent. Given the combination of a primary halide, a good leaving group, a moderate nucleophile, and a nonpolar aprotic solvent, we predict the reaction occurs by an S_N2 pathway.

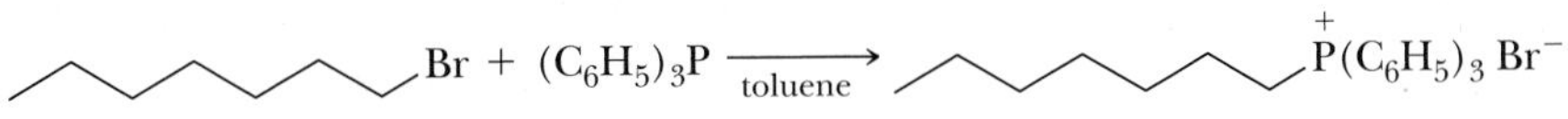

Example 9.5 Nucleophilic Substitution Products II

Write the expected substitution product(s) for each reaction and predict the mechanism by which each product is formed.

(a) 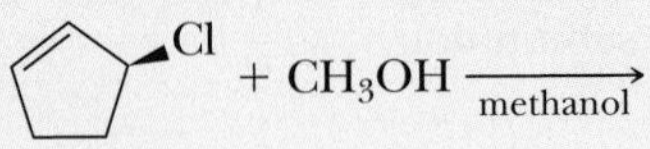

S enantiomer

(b) 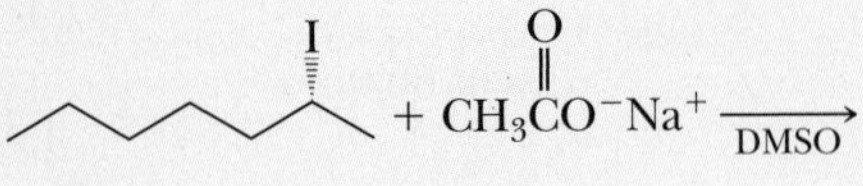

R enantiomer

Solution

(a) This 2° allylic chloride is treated with methanol, a poor nucleophile and a polar protic solvent. Ionization of the carbon-chlorine bond forms a secondary allylic cation that is stabilized by resonance delocalization. Therefore, we predict reaction by an S_N1 mechanism and formation of the product as a roughly racemic mixture.

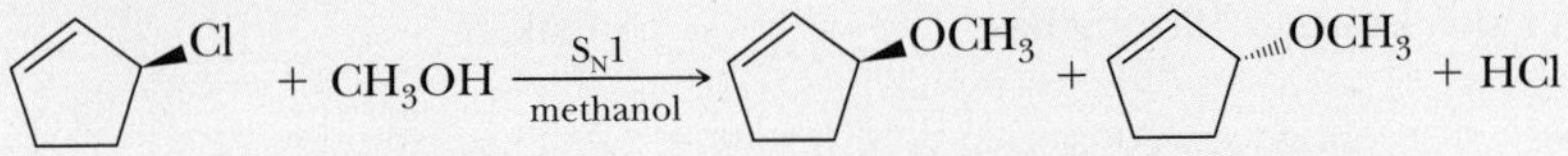

S enantiomer — *S* enantiomer (~50%) — *R* enantiomer (~50%)

(b) Iodide is a good leaving group on a moderately accessible secondary carbon. Acetate ion dissolved in a polar aprotic solvent is a moderate nucleophile. We predict substitution by an S_N2 pathway with inversion of configuration at the chiral center.

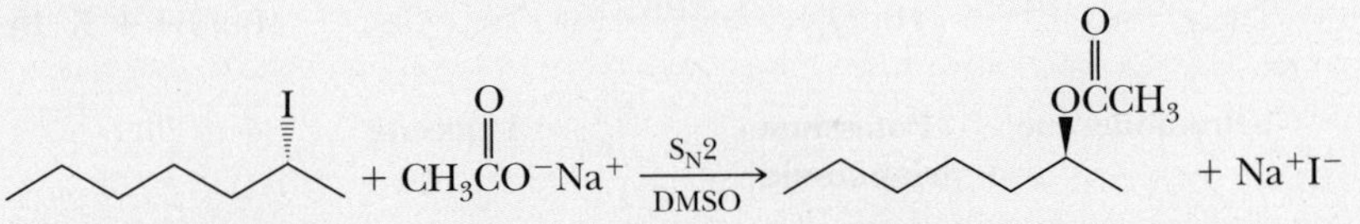

R enantiomer — *S* enantiomer

Problem 9.5

Watch a video explanation

Write the expected substitution product(s) for each reaction and predict the mechanism by which each product is formed.

(a) 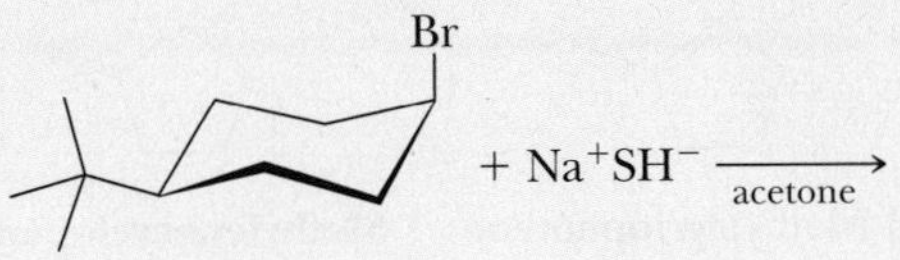

(b) 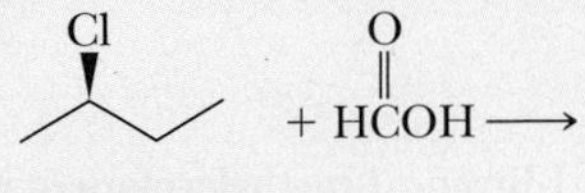

(*R*)-2-Chlorobutane

9.5 β-Elimination

All nucleophiles have an electron pair that can take part in a reaction as a lone pair or sometimes as a π bond. This means that all nucleophiles are also bases, because any pair of electrons can accept a proton. Hence, chemists are routinely confronted with competing reactions that depend on a balance between the basicity and nucleophilicity of the reactants we use. As mentioned in the introduction to this chapter, the **β-elimination reaction** is the competing process to substitution that we observe. Viewed in the context of the mechanistic elements described in Things You Should Know III prior to Chapter 6, β-elimination is the combination of **take a proton away** and **break a bond to give stable molecules or ions**.

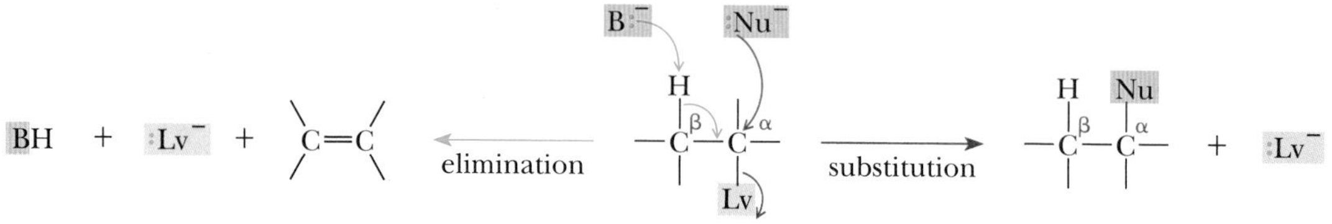

Dehydrohalogenation
Removal of —H and —X from adjacent carbons; a type of β-elimination.

Here, we study a type of β-elimination called **dehydrohalogenation**. In the presence of base, halogen is removed from one carbon of a haloalkane and hydrogen is removed from an adjacent carbon to form an alkene.

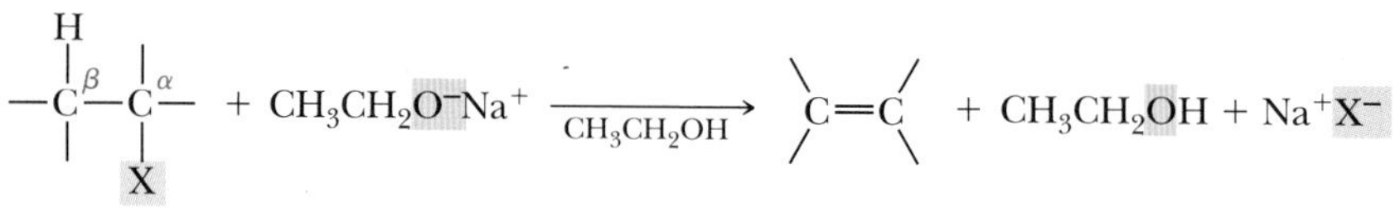

A haloalkane Strong base An alkene

Strong bases promote β-elimination reactions. Strong bases that serve effectively in β-eliminations of haloalkanes are OH^-, OR^-, NH_2^-, and acetylide anions. Following are three examples of base-promoted β-elimination reactions. In the first example, the base is shown as a reactant. In the second and third examples, the base is a reactant but is shown over the reaction arrow. Note that the solvent used is commonly the conjugate acid of the base used in the elimination.

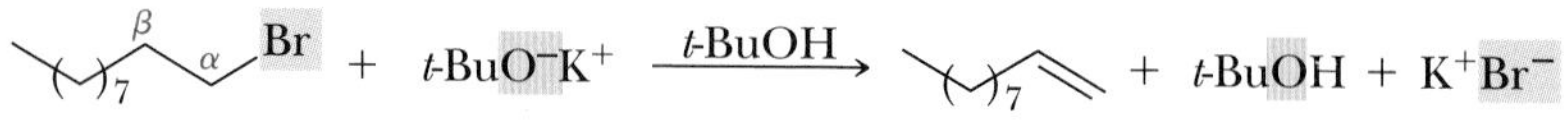

1-Bromodecane **Potassium *tert*-butoxide** **1-Decene**

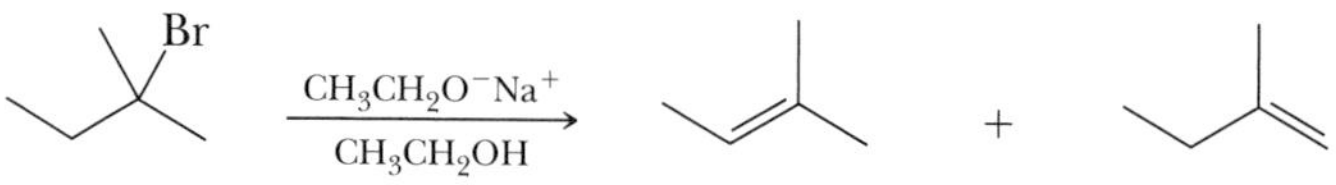

2-Bromo-2-methylbutane **2-Methyl-2-butene** (major product) **2-Methyl-1-butene**

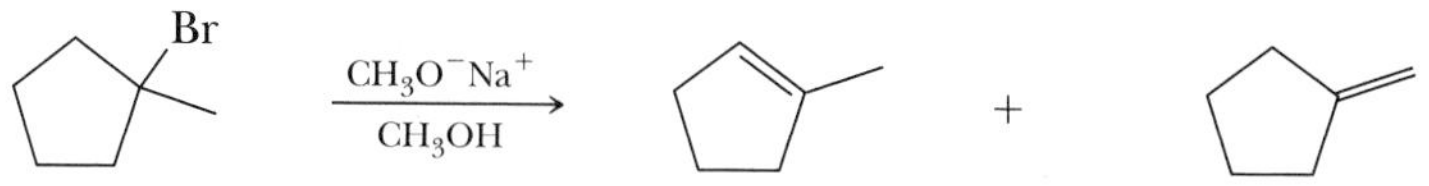

1-Bromo-1-methylcyclopentane **1-Methylcyclopentene** (major product) **Methylenecyclopentane**

In the second and third illustrations, there are two nonequivalent β-carbons, each bearing a hydrogen; therefore, two alkenes are possible. In each case, the major product of these and most other β-elimination reactions is the more substituted (and therefore the more stable) alkene (Section 6.6B). Formation of the more substituted alkene in an elimination is common, but it is not always the outcome. When the more substituted alkene is the dominant product, the reaction is said to follow **Zaitsev's rule** or to undergo **Zaitsev elimination**.

Zaitsev's rule
A rule stating that the major product of a β-elimination reaction is the most stable alkene; that is, it is the alkene with the greatest number of substituents on the carbon-carbon double bond.

Example 9.6 β-Elimination Products

Predict the β-elimination product(s) formed when each bromoalkane is treated with sodium ethoxide in ethanol. If two or more products might be formed, predict which is the major product.

(a) 

(racemic)

(b) 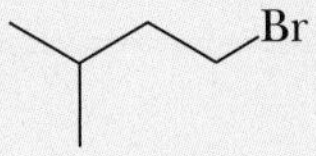

Solution

(a) There are two nonequivalent β-carbons in this bromoalkane, and two alkenes are possible. 2-Methyl-2-butene, the more substituted alkene, is the major product.

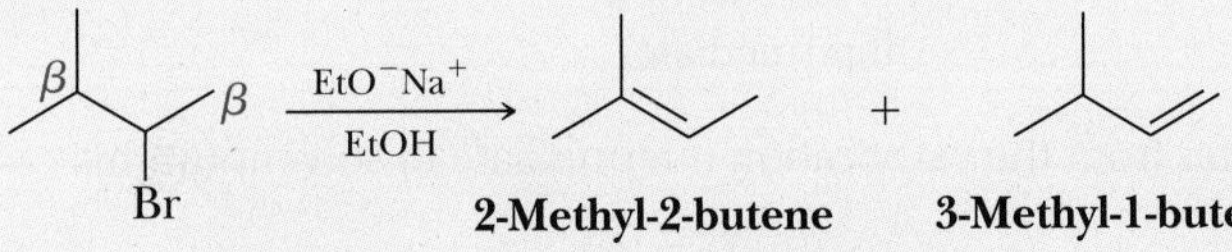

2-Methyl-2-butene (major product) **3-Methyl-1-butene**

(b) There is only one β-carbon in this bromoalkane, and only one alkene is possible.

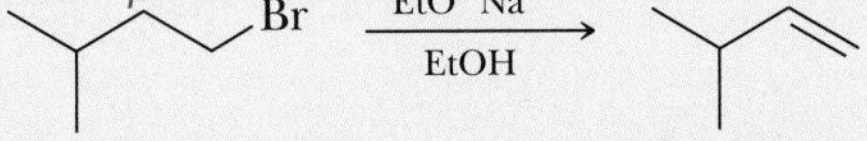

3-Methyl-1-butene

Problem 9.6

Predict the β-elimination product(s) formed when each chloroalkane is treated with sodium ethoxide in ethanol. If two or more products might be formed, predict which is the major product.

(a) 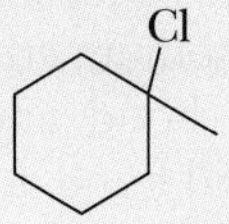

(b) 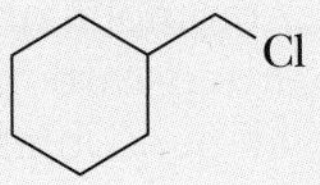

(c) 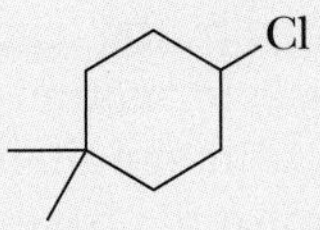

9.6 Mechanisms of β-Elimination

There are two limiting mechanisms for β-eliminations. A fundamental difference between them is the timing of the bond-breaking and bond-forming steps. Recall that we made the same statement about the two limiting mechanisms for nucleophilic substitution reactions (Section 9.3).

A. E1 Mechanism

At one extreme, breaking of the C—Lv bond to give a carbocation is complete before any reaction occurs with the base to lose a hydrogen and form the carbon-carbon double bond. This mechanism is designated an **E1 reaction**, where E stands for *E*limination and 1 stands for unimolecular. One species (in this case, the haloalkane) is involved in the rate-determining step. The mechanism of an E1 reaction is illustrated here by the reaction of 2-bromo-2-methylpropane to form 2-methylpropene.

E1 reaction

A unimolecular β-elimination reaction.

Mechanism 9.4

E1 Reaction of 2-Bromo-2-methylpropane

Step 1: **Break a bond to give stable molecules or ions.** Rate-determining ionization of the C—Br bond gives a carbocation intermediate.

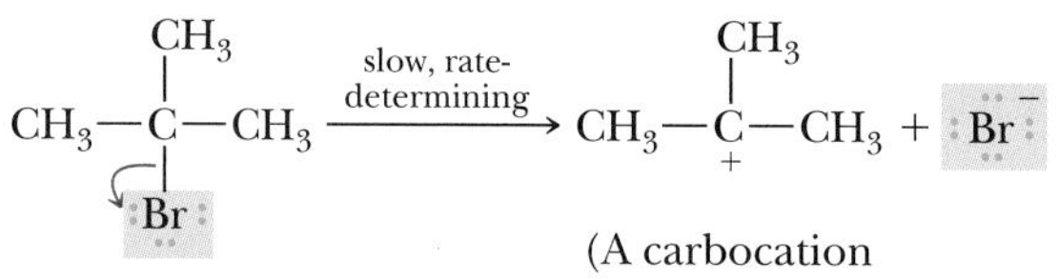

(A carbocation intermediate)

Step 2: **Take a proton away.** Proton transfer from the carbocation intermediate to solvent (in this case, methanol) gives the alkene.

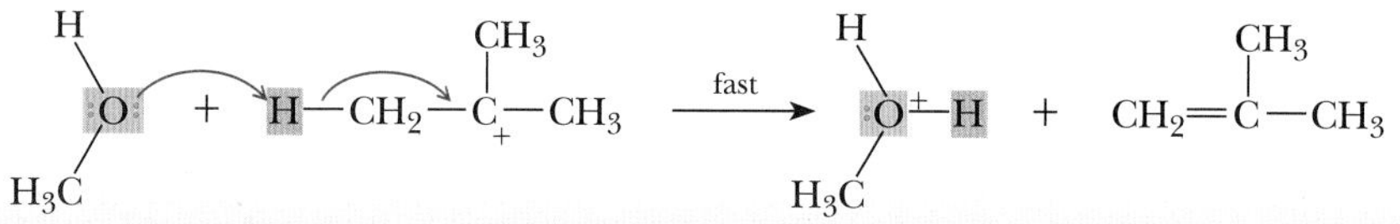

In an E1 mechanism, one transition state exists for the formation of the carbocation in Step 1 and a second exists for the loss of a hydrogen in Step 2 (Figure 9.6). Formation of the carbocation intermediate in Step 1 crosses the higher energy barrier and is the rate-determining step. This reaction competes with S_N1 substitution. E1 and S_N1 almost always occur together.

B. E2 Mechanism

At the other extreme of elimination mechanisms is a concerted process. In an **E2 reaction** (here illustrated by the reaction of 2-bromobutane with sodium ethoxide) proton transfer to the base, formation of the carbon-carbon double bond, and ejection of the bromide ion occur simultaneously; all bond-breaking and bond-forming steps are concerted. Because the base removes a β-hydrogen at the same time that the C—Br bond is broken to form a halide ion, the transition state has considerable double-bond character (Figure 9.7).

E2 reaction

A bimolecular β-elimination reaction.

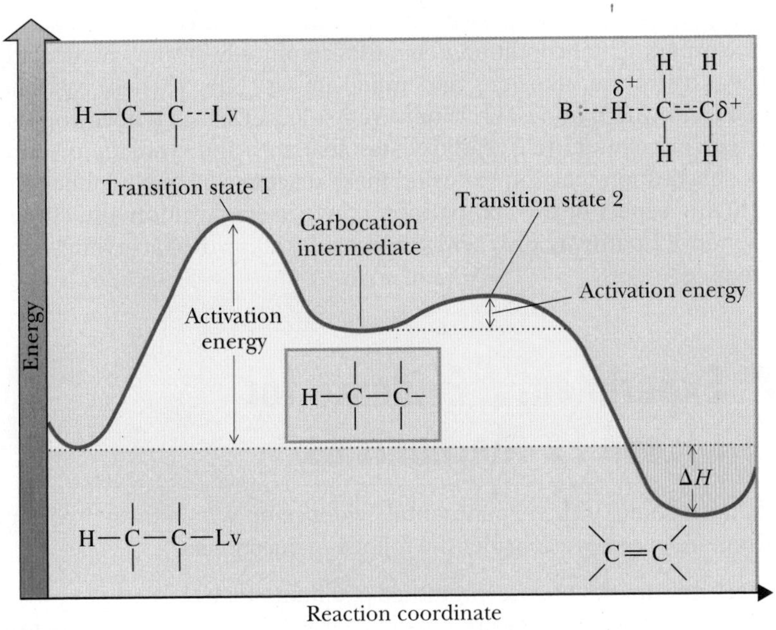

FIGURE 9.6 An energy diagram for an E1 reaction showing two transition states and one carbocation intermediate.

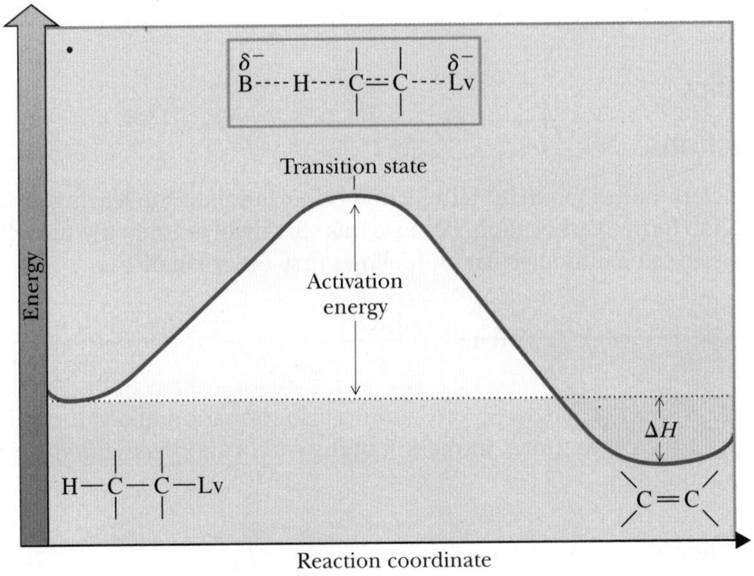

FIGURE 9.7 An energy diagram for an E2 reaction. There is considerable double-bond character in the transition state.

Mechanism 9.5

E2 Reaction of 2-Bromobutane

Take a proton away and simultaneously break a bond to give stable molecules or ions. Bond breaking and bond forming are concerted; that is, they occur simultaneously.

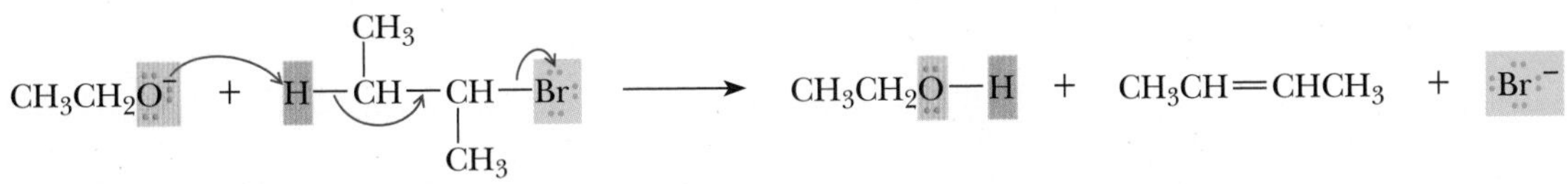

This mechanism is designated E2, where E stands for *E*limination and 2 stands for bimolecular; both the haloalkane and the base are involved in the transition state for the rate-determining step.

Although in principle any base can be made to induce an E2 reaction under appropriate experimental conditions, chemists commonly employ particularly strong bases such as hydroxide, alkoxides, and amide anions (NR_2^-). These bases have conjugate acids with pK_a's above 11. When we use other bases whose conjugate acid pK_a's are near or below 11 (e.g., carboxylates, thiolates, and cyanide), the intention is to effect a substitution reaction via using these reactants as nucleophiles. Therefore, one simplifying aspect of the competition between substitution and elimination is to consider an E2 pathway only when hydroxide, alkoxides, acetylides, and amide anions are used.

9.7 Experimental Evidence for E1 and E2 Mechanisms

As we examine some of the experimental evidence on which these two contrasting mechanisms are based, we consider the following questions:

1. What are the kinetics of base-promoted β-eliminations?
2. Where two or more alkenes are possible, what factors determine the ratio of the possible products?
3. What is the stereoselectivity?

A. Kinetics

E1 Reactions

The rate-determining step in an E1 reaction is ionization of the leaving group (often a halide, X) to form a carbocation. Because this step involves only the haloalkane, the reaction is said to be unimolecular and follows first-order kinetics.

$$\text{Rate} = -\frac{d[\text{RX}]}{dt} = k[\text{RX}]$$

Recall that the first step in an S_N1 reaction is also formation of a carbocation. Thus, for both S_N1 and E1 reactions, formation of the carbocation is the first step and the rate-determining step.

E2 Reactions

Only one step occurs in an E2 mechanism, and the transition state is bimolecular. The reaction is second order: first order in haloalkane and first order in base.

$$\text{Rate} = -\frac{d[\text{RX}]}{dt} = k[\text{RX}][\text{Base}]$$

B. Regioselectivity

E1 Reactions

The major product in E1 reactions is the more stable alkene; [i.e., the alkene with the more highly substituted carbon-carbon double bond (Zaitsev's rule)]. After the carbocation is formed in the rate-determining step of an E1 reaction, it may lose a hydrogen to complete β-elimination or it may rearrange to a more stable carbocation and then lose a hydrogen.

E2 Reactions

For E2 reactions that use strong bases and in which the leaving group is a halide ion, the major regioisomeric product is also that formed following Zaitsev's rule, unless special steric relations apply (Section 9.7C). Double-bond character is so highly developed in the transition state that the relative stability of possible alkenes commonly determines which regioisomer is the major product. Thus, the transition state of lowest energy is commonly that leading to the most highly substituted alkene. For similar reasons, *trans* double bonds predominate over *cis* double bonds in the products when either is possible. Note that E2 elimination at a 2° carbon predominates over S_N2 reaction with the strongly basic alkoxide ions.

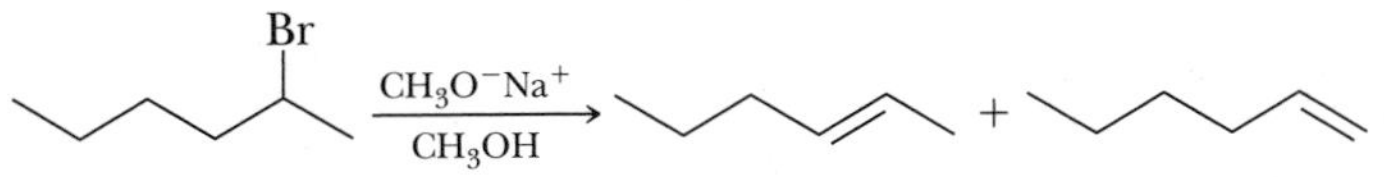

Br
$CH_3O^-Na^+$
CH_3OH
+

2-Bromohexane (racemic) — **2-Hexene** (74%) (*cis* and *trans*) — **1-Hexene** (26%)

With larger, sterically hindered bases such as *tert*-butoxide, however, where isomeric alkenes are possible, the major product is often the less substituted alkene because reaction occurs primarily at the most accessible H atom. Sterically hindered bases such as *tert*-butoxide are also noteworthy because the steric hindrance prevents them from reacting as nucleophiles, even with primary haloalkanes.

C. Stereoselectivity

The stereochemistry of E2 reactions is controlled by a conformational effect. The lowest-energy transition state of an E2 reaction is commonly the one in which the —Lv and —H are oriented anti and coplanar (at a dihedral angle of 180°) to each other. The reason for this preferred geometry is that it allows for proper orbital overlap between the base, the proton being removed, and the departing leaving group. Remembering the anti and coplanar geometry requirement is important because it allows prediction of alkene stereochemistry in E2 reactions, namely whether *E* or *Z* products are produced.

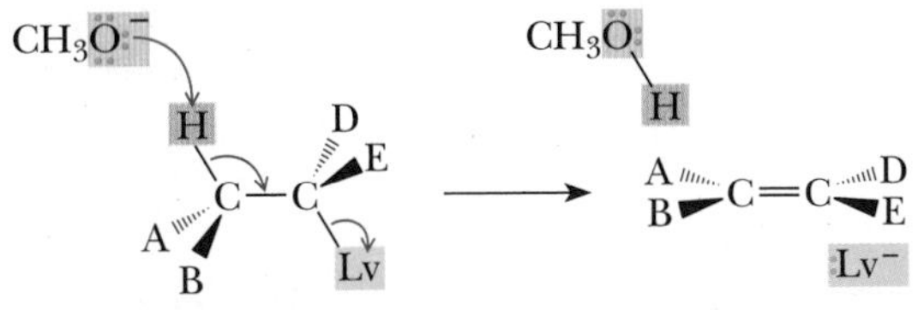

CH_3O^-
H
D
E
C
C
A
B
Lv
CH_3O
H
A
B
C=C
D
E
Lv^-

—H and —Lv are anti and coplanar (dihedral angle 180°)

This is shown more clearly in a Newman projection with a bromide as the leaving group.

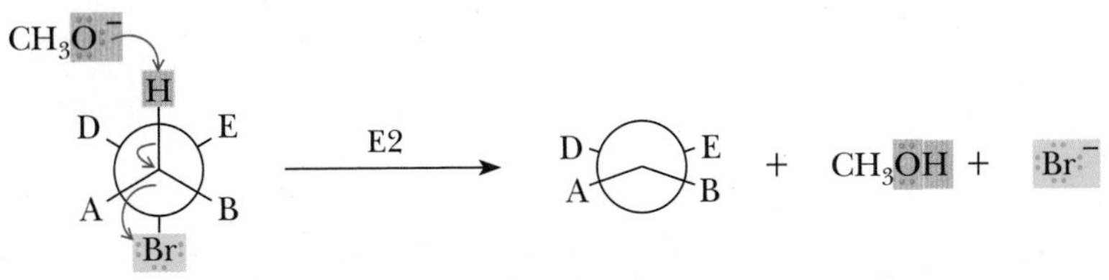

CH_3O^-
H
D
E
A
B
Br
E2
D
E
A
B
+
CH_3OH
+
Br^-

As with the required backside attack associated with an S_N2 reaction, there is an orbital-based reason for the anti and coplanar arrangement of the —H and —Lv involved in an E2 reaction. The following diagram shows a filled C—H σ bonding molecular orbital aligned with the empty C—Lv σ antibonding molecular orbital.

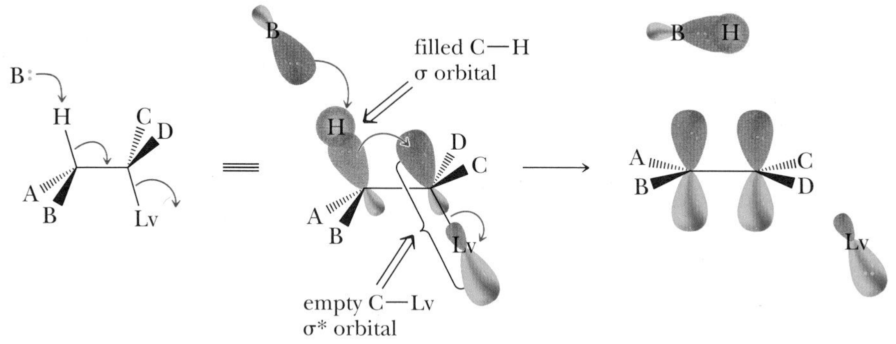
B:
H
C
D
A
B
Lv
B
filled C—H
σ orbital
H
D
C
A
B
Lv
empty C—Lv
σ* orbital
B H
A
B
C
D
Lv

As the strong base removes the proton, we consider the two electrons in the C—H orbital filling the antibonding C—Lv orbital and thereby breaking the C—Lv bond. An anti and coplanar arrangement of the C—H and C—Lv leads to proper phasing in the resulting π bond. When the H and Lv are aligned as shown, a collision of the base along the C—H bond leads to the lowest-energy E2 pathway.

For example, treatment of 1,2-dibromo-1,2-diphenylethane with sodium methoxide in methanol gives 1-bromo-1,2-diphenylethylene. The meso isomer of 1,2-dibromo-1,2-diphenylethane gives *(E)*-1-bromo-1,2-diphenylethylene, whereas the racemic mixture of 1,2-dibromo-1,2-diphenylethane gives *(Z)*-1-bromo-1,2-diphenylethylene. We use the anti coplanar requirement of the transition state to account for the stereospecificity of these E2 β-eliminations.

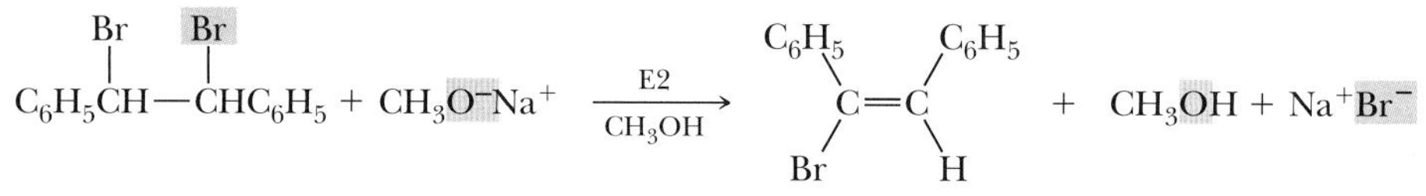
Br Br
C6H5CH—CHC6H5 + CH3O⁻Na⁺
E2
CH3OH
C6H5 C6H5
C=C
Br H
+ CH3OH + Na⁺Br⁻

meso-1,2-Dibromo-1,2-diphenylethane

(*E*)-1-Bromo-1,2-diphenylethylene

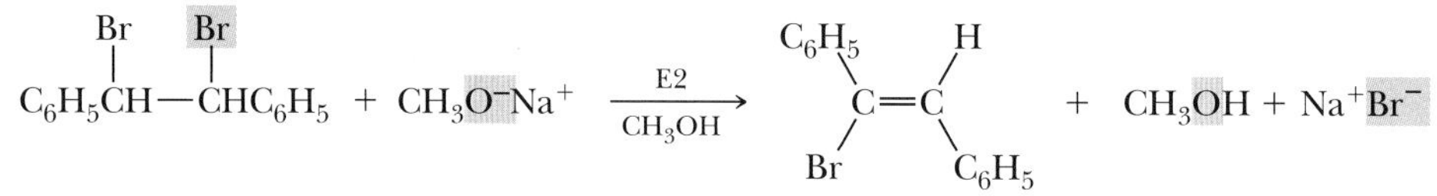
Br Br
C6H5CH—CHC6H5 + CH3O⁻Na⁺
E2
CH3OH
C6H5 H
C=C
Br C6H5
+ CH3OH + Na⁺Br⁻

racemic 1,2 Dibromo-1,2-diphenylethane

(*Z*)-1-Bromo-1,2-diphenylethylene

Because it is preferred for an E2 reaction that —H and —Lv be anti and coplanar, it is important to identify the reactive conformation of a haloalkane starting material. Following is a stereorepresentation of the meso isomer of 1,2-dibromo-1,2-diphenylethane, drawn to show the plane of symmetry.

Mechanism 9.6

E2 Reaction of meso-1,2-Dibromo-1,2-diphenylethane

Take a proton away and simultaneously break a bond to give stable molecules or ions. Clockwise rotation of the left carbon by 60° brings —H and —Br into the required anti and coplanar relationship.

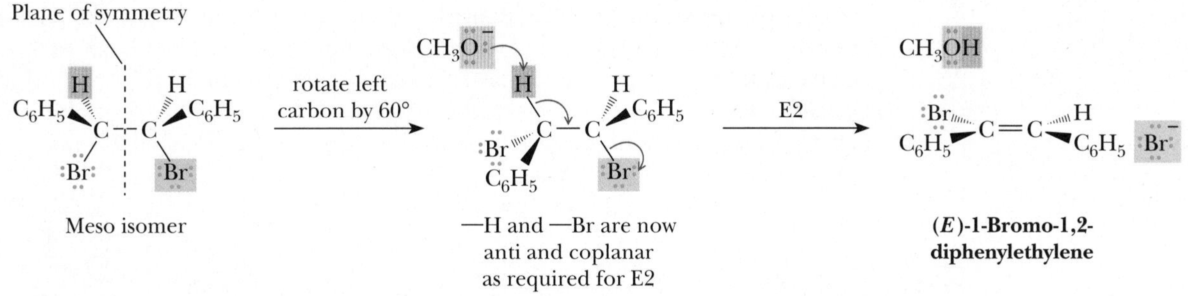
Plane of symmetry
rotate left carbon by 60°
E2
Meso isomer
—H and —Br are now anti and coplanar as required for E2
(E)-1-Bromo-1,2-diphenylethylene

E2 reaction on this conformation gives only the (*E*)-alkene, as shown below in a Newman projection. The —H and —Br are anti and coplanar. All bond-breaking and bond-forming steps are concerted.

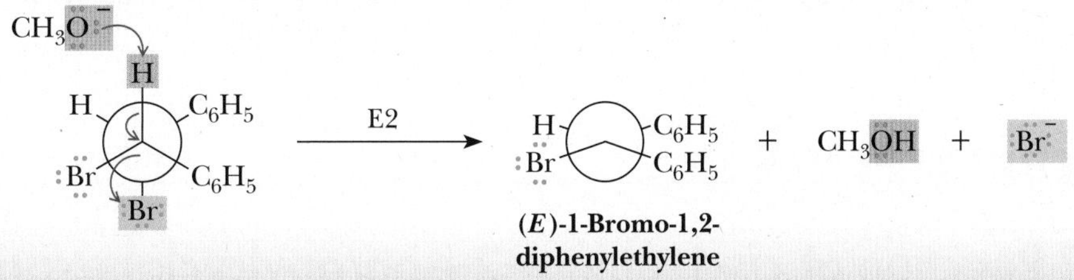
E2
(E)-1-Bromo-1,2-diphenylethylene

E2 reaction of either enantiomer of the racemic mixture of 1,2-dibromo-1,2-diphenylethane gives only the (*Z*)-alkene as predicted by analysis of the proper anti and coplanar conformations.

Mechanism 9.7

E2 Reaction of the Enantiomers of 1,2-Dibromo-1,2-diphenylethane

Take a proton away and simultaneously break a bond to give stable molecules or ions. The —H and —Br are anti and coplanar. All bond-breaking and bond-forming steps are concerted.

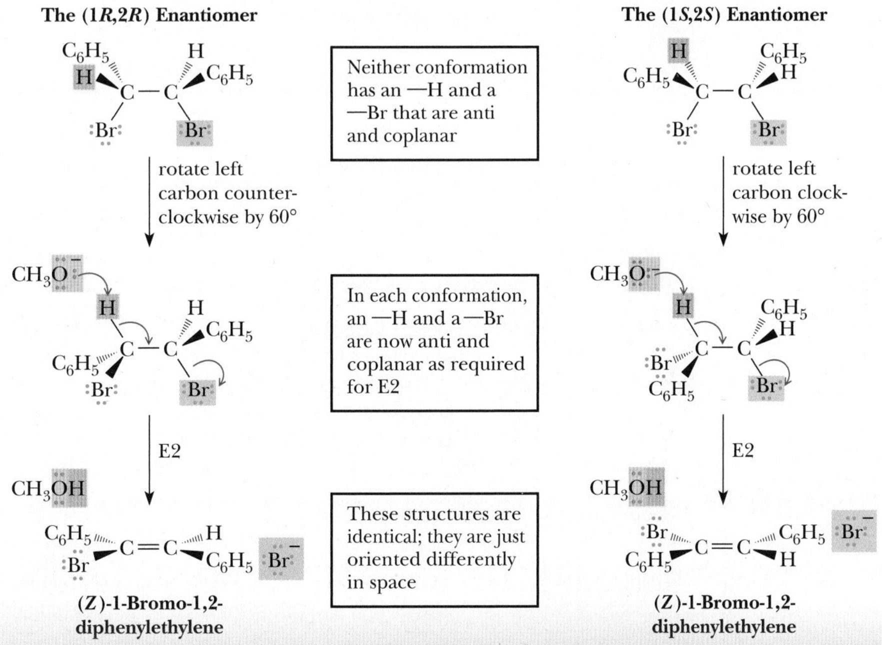
The (1R,2R) Enantiomer
The (1S,2S) Enantiomer
Neither conformation has an —H and a —Br that are anti and coplanar
rotate left carbon counter-clockwise by 60°
rotate left carbon clock-wise by 60°
In each conformation, an —H and a —Br are now anti and coplanar as required for E2
E2
E2
These structures are identical; they are just oriented differently in space
(Z)-1-Bromo-1,2-diphenylethylene
(Z)-1-Bromo-1,2-diphenylethylene

The required anti and coplanar transition state geometry can also be used to predict the regiochemistry of E2 elimination in halocyclohexanes such as chlorocyclohexanes. In these molecules, anti and coplanar correspond to *trans* and diaxial. Consider the E2 reaction of the *cis* isomer of 1-chloro-2-isopropylcyclohexane. The major product is 1-isopropylcyclohexene, the more substituted cycloalkene.

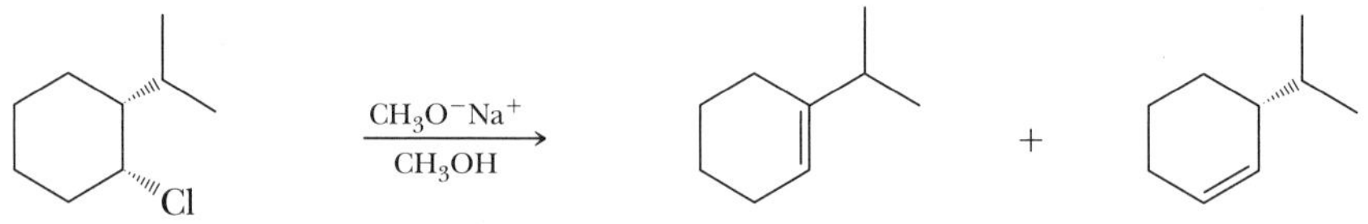

***cis*-1-Chloro-2-isopropylcyclohexane** **1-Isopropylcyclohexene** (major product) **(*R*)-3-Isopropylcyclohexene**

Mechanism 9.8

E2 Reaction of *cis*-1-Chloro-2-isopropylcyclohexane

Take a proton away and simultaneously break a bond to give stable molecules or ions. In the more stable chair conformation of the *cis* isomer, the considerably larger isopropyl group is equatorial and the smaller chlorine is axial. In this chair conformation, —H on carbon 2 and —Cl on carbon 1 are anti and coplanar. Concerted E2 elimination gives 1-isopropylcyclohexene, a trisubstituted alkene, as the major product. Note that —H on carbon 6 and —Cl are also anti and coplanar. Dehydrohalogenation of this combination of —H and —Cl gives 3-isopropylcyclohexene, a disubstituted (and therefore less stable) alkene. The formation of the 1-isomer as the major product is in agreement with Zaitsev's rule. However, Zaitsev's rule can be counteracted by the anti coplanar arrangement of —H and —Lv (Example 9.7).

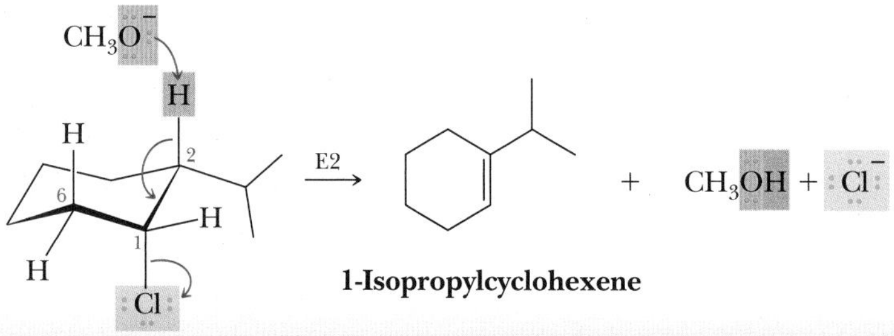

The factors favoring E1 or E2 elimination are summarized in Table 9.10.

Table 9.10 Summary of E1 Versus E2 Reactions for Haloalkanes

Alkyl Halide	E1	E2
Primary RCH_2X	E1 not observed. Primary carbocations are so unstable that they are never observed in solution.	E2 is favored if elimination is observed. Usually requires sterically hindered strong base.
Secondary R_2CHX	Main reaction with weak bases such as H_2O, ROH.	Main reaction with strong bases such as OH^- and OR^-.
Tertiary R_3CX	Main reaction with weak bases such as H_2O, ROH.	Main reaction with strong bases such as OH^- and OR^-.

Example 9.7 Anti and Coplanar Arrangements in E2 Reactions

From *trans*-1-chloro-2-isopropylcyclohexane, only 3-isopropylcyclohexene, the less substituted alkene, is formed. Using conformational analysis, explain why this product is observed. Also, will the E2 reaction with *trans*-1-chloro-2-isopropylcyclohexane or *cis*-1-chloro-2-isopropylcyclohexane occur faster under the same basic conditions?

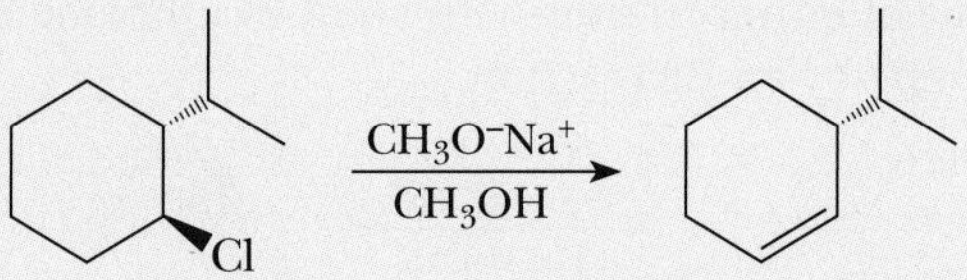

***trans*-1-Chloro-2 isopropylcyclohexane** **(*R*)-3-Isopropylcyclohexene**

Solution

Step 1: In the more stable chair conformation of the *trans* isomer, both isopropyl and chlorine are equatorial. In this conformation, the hydrogen atom on carbon 2 is *cis* to the chlorine atom. One of the hydrogen atoms on carbon 6 is *trans* to —Cl, but it is not anti and coplanar. Therefore, the reaction is not favored from this conformation. In the alternative, less stable chair conformation of the *trans* isomer, both isopropyl and chlorine are axial. In this conformation, the axial hydrogen on carbon 6 is anti and coplanar to chlorine and E2 β-elimination can occur to give 3-isopropylcyclohexene. Thus, even though the diaxial conformation is less stable, the reaction goes through this conformation because it is the only one with an anti-coplanar arrangement of the Cl and a β-H; consequently, the non-Zaitsev product is formed.

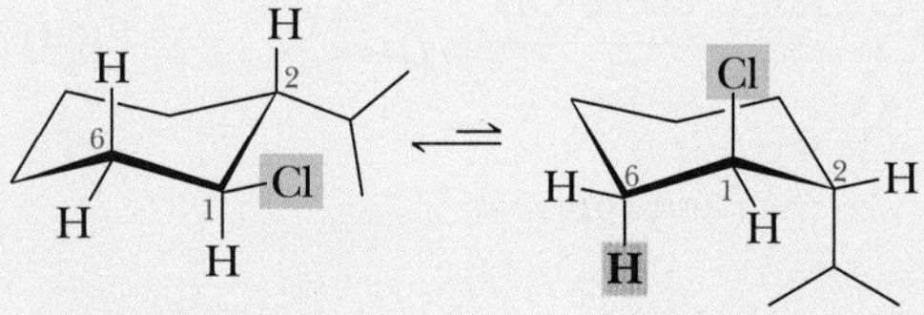

More stable chair (no H is anti and coplanar to Cl)

Less stable chair (H on carbon 6 is anti and coplanar to Cl)

Step 2: E2 reaction can take place now that an —H and a —Cl are anti and coplanar. This reaction doesn't follow the Zaitsev rule because the mechanism of the reaction requires the anti arrangement.

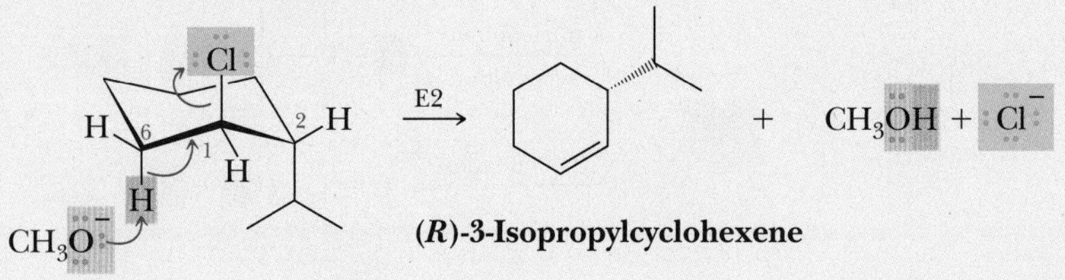

(*R*)-3-Isopropylcyclohexene

The rate at which the *cis* isomer undergoes E2 reaction is considerably greater than the rate for the *trans* isomer. We can account for this observation in the

(*Continued*)

following manner. The more stable chair conformation of the *cis* isomer has —H and —Cl anti and coplanar, and the activation energy for the reaction is that required to reach the E2 transition state. The more stable chair conformation of the *trans* isomer, however, cannot undergo anti elimination. To react, it must first be converted to the less stable chair, and the transition state for elimination is correspondingly higher in energy because of the axial isopropyl group.

Watch a video explanation

Problem 9.7

1-Chloro-4-isopropylcyclohexane exists as two stereoisomers: one *cis* and one *trans*. Treatment of either isomer with sodium ethoxide in ethanol gives 4-isopropylcyclohexene by an E2 reaction.

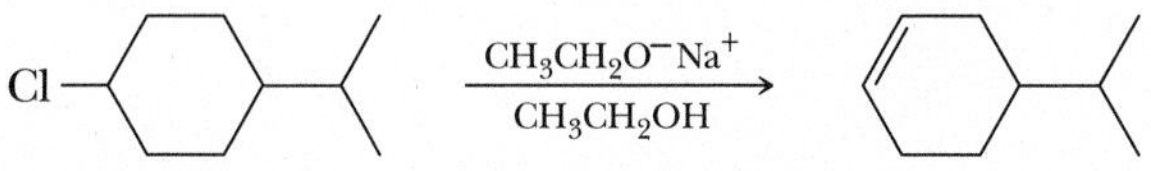

1-Chloro-4-isopropylcyclohexane **4-Isopropylcyclohexene**

The *cis* isomer undergoes E2 reaction several orders of magnitude faster than the *trans* isomer. How do you account for this experimental observation?

Although *much* rarer than an anti and coplanar arrangement of the C—H and C—Lv bonds in an E2 reaction, a syn and coplanar arrangement of these bonds can also lead to E2. Such an arrangement means that the C—H and C—Lv bonds are eclipsed; therefore, only certain constrained ring systems have this geometry. As an example, the following reaction occurs via elimination of H_a rather than H_b because the C—H_a bond is aligned with the C—Cl bond while the C—H_b bond is gauche to the C—Cl bond, that is, the latter two bonds lie at a dihedral angle of 60°.

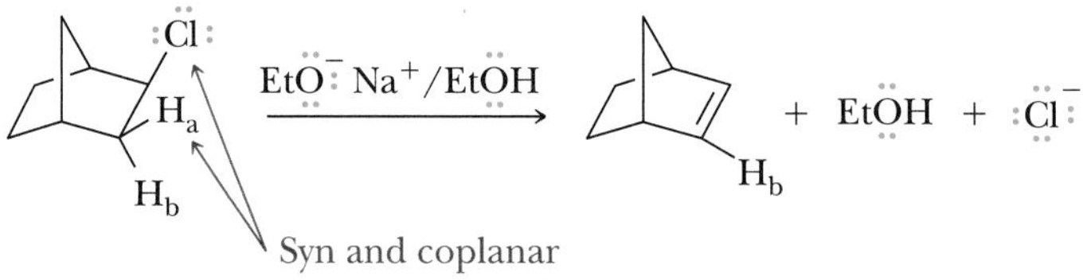

9.8 Substitution Versus Elimination

Nucleophilic substitution and β-elimination often compete with each other, and the ratio of products formed by these reactions depends on the relative rates of the two reactions. In this section, we consider factors that influence this competition.

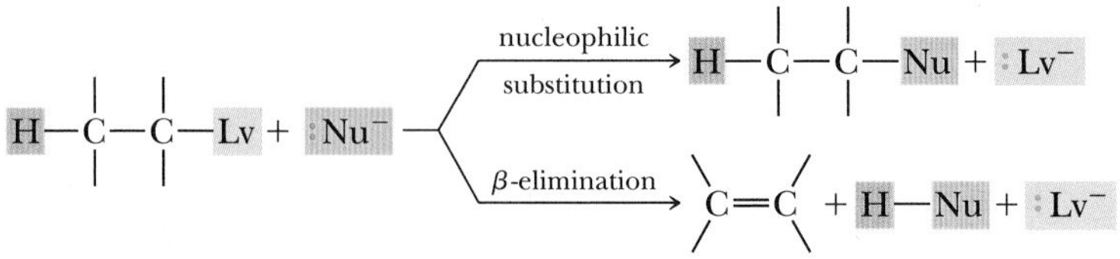

A. S_N1 Versus E1 Reactions

Reactions of secondary and tertiary haloalkanes in polar protic solvents give mixtures of substitution and elimination products. In both reactions, Step 1 is the formation of a carbocation intermediate. This step is then followed by one or more characteristic carbocation reactions: (1) loss of a hydrogen (E1) to give an alkene, (2) reaction with solvent (S_N1) to give a substitution product, or (3) rearrangement followed by reaction (1) or (2). In polar protic solvents, the products formed depend only on the structure of the particular carbocation. For example, *tert*-butyl chloride and *tert*-butyl iodide in 80% aqueous ethanol both react with solvent, giving the same mixture of substitution and elimination products. Because iodide ion is a better leaving group than chloride ion, *tert*-butyl iodide reacts over 100 times faster than *tert*-butyl chloride. Yet, the ratio of products is the same because the intermediate *tert*-butyl cation is the same.

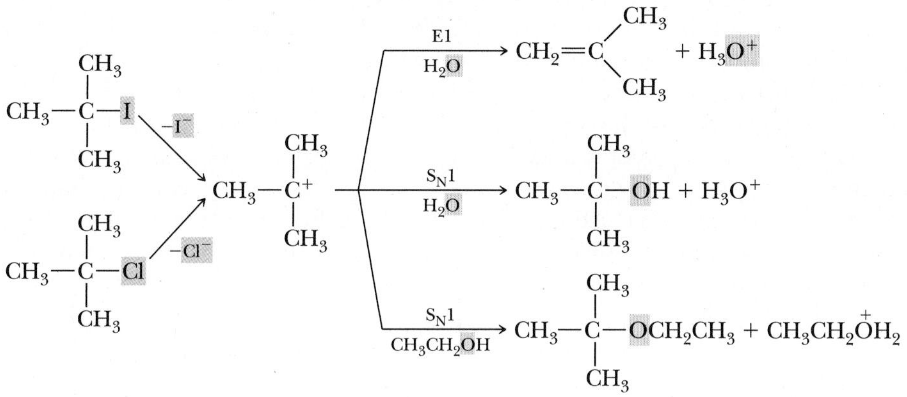

It is difficult to predict the ratio of substitution to elimination products for first-order reactions of haloalkanes. For the majority of cases, however, S_N1 predominates over E1 when weak bases are used.

B. S_N2 Versus E2 Reactions

It is considerably easier to predict the ratio of substitution to elimination products for second-order reactions of haloalkanes with reagents that act both as nucleophiles and bases. The guiding principles are:

1. Branching at the α-carbon or β-carbon(s) increases steric hindrance about the α-carbon and significantly retards S_N2 reactions. Conversely, branching at the α-carbon or β-carbon(s) increases the rate of E2 reactions because of the increased stability of the alkene product.
2. The greater the nucleophilicity of the attacking reagent, the greater the S_N2-to-E2 ratio. Conversely, the greater the basicity of the attacking reagent, the greater the E2-to-S_N2 ratio.

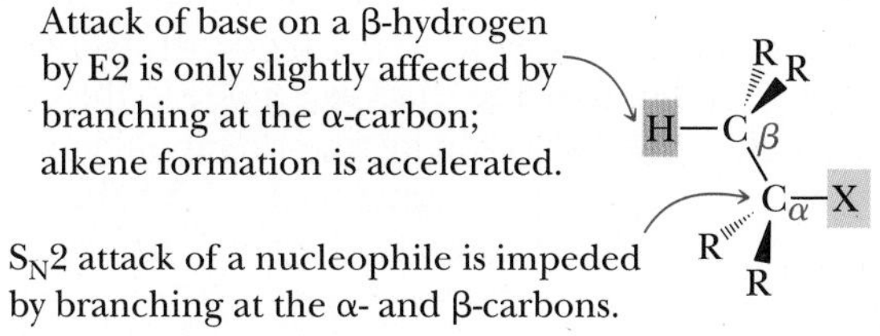
Attack of base on a β-hydrogen by E2 is only slightly affected by branching at the α-carbon; alkene formation is accelerated.
S_N2 attack of a nucleophile is impeded by branching at the α- and β-carbons.

A second point involves a relative comparison of nucleophilicity to basicity. It is often difficult to definitively predict in advance whether nucleophilicity will outcompete basicity, thereby favoring or not favoring S_N2 versus E2. This competition is particularly important with secondary haloalkanes (see below). However, a general guideline is reasonably predictive for secondary haloalkanes. If a nucleophile/base has a conjugate acid with a pK_a below 11 and is a good nucleophile, then an S_N2 reaction will dominate. If the pK_a of the conjugate acid of the nucleophile/base is above 11, the basic character will usually outcompete the nucleophilic character and an E2 reaction will dominate. A particularly good example of this phenomenon is the comparison between a thiolate anion (RS^-) and an alkoxide anion (RO^-). Thiolates are excellent nucleophiles in polar protic media, and the pK_a's of their conjugate acids (thiols, RSH) are in the range of 10 to 12. However, alkoxides are also excellent nucleophiles, but the pK_a's of their conjugate acids (alcohols, ROH) are much higher—in the range of 16 to 18. Hence, for a reaction with the same haloalkane in the same polar protic solvent, the percent of S_N2 will be greater for a thiolate nucleophile/base, while the percent of E2 will be greater for the reaction performed with an alkoxide nucleophile/base.

Temperature is another factor that influences the balance between S_N2 and E2 reactions. In general, higher temperatures result in increasing extents of elimination at the expense of substitution. The reason derives from differences in the number of products compared to the number of reactants. Elimination reactions involve the creation of increasing numbers of molecules because a base and an R—Lv react to give the conjugate acid of the base, an alkene, and the free leaving group. In contrast, substitution reactions do not change the number of molecules because a nucleophile and an R—Lv react to give the substituted product and the free leaving group. The more particles formed in a reaction, the more entropically favored the reaction. Hence, because elimination reactions create more particles than substitution reactions do, they are more entropically favored, which will be reflected in differences in the energies of the transition states for these two reactions. Recall that $\Delta G = \Delta H - T\Delta S$; therefore, entropy effects become more accentuated at higher temperatures because the $T\Delta S$ term becomes increasingly important. Thus, when two or more reactions are in competition, at higher temperatures, the reactions with the more favorable entropies will increase at the expense of those with less favorable entropies. With the competition between S_N2 and E2, the more favorable entropy for elimination results in an increase in elimination at higher temperatures.

S_N2

$$\text{Nu:}^- + \text{R—Lv} \longrightarrow \text{Nu—R} + \text{Lv:}^-$$

E2

$$\text{Base:}^- + \text{R—Lv} \longrightarrow \text{Base—H} + \text{alkene} + \text{Lv:}^-$$

Watch a video explanation

C. Putting It All Together

In this chapter, we examined S_N2, S_N1, E2, and E1 mechanisms and learned how they compete with each other depending upon the alkyl group, the leaving group, the solvent, and the nucleophile. We also examined solvent effects on nucleophilicity. Nature does not always have clear-cut rules, but here we summarize guidelines that chemists use to predict the outcome of reactions between haloalkanes and various nucleophiles and bases.

Figure 9.8 shows a flowchart that allows you to predict the major product of substitution or elimination reactions. Use the chart as a guide to the following discussion. Alternatively, you can follow the discussion by referring to Table 9.11.

We first classify the haloalkane as (a) primary (RCH_2X), (b) secondary (R_2CHX), or (c) tertiary (R_3CX). We consider 1° carbons that are not sterically hindered in Part (a) because this covers most 1° cases. If the 1° carbon is sterically hindered, such as that in the neopentyl group [R = $(CH_3)_3CCH_2X$, recall Section 9.3B], we treat it as if it were a secondary carbon (b) that cannot undergo an elimination reaction. The flowchart does not show what happens if the alkyl group is methyl (CH_3X), because the only possible outcome is an S_N2 reaction irrespective of the structure of the nucleophile, the leaving group, and the solvent (Table 9.11 puts CH_3X at the top). Recall that methyl cations are too unstable to form (S_N1 is ruled out) and there is only one carbon, meaning that elimination to create a double bond is impossible (E1 and E2 are ruled out).

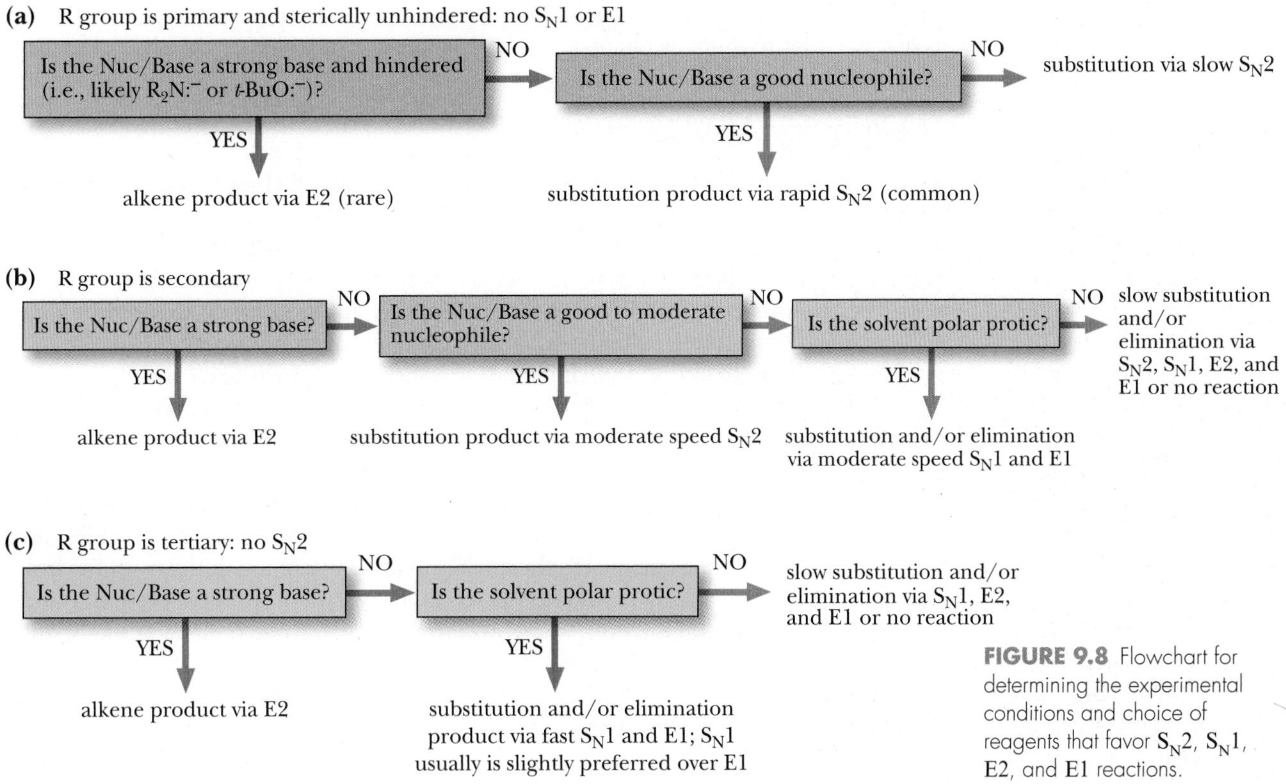
(a) R group is primary and sterically unhindered: no S_N1 or E1
Is the Nuc/Base a strong base and hindered (i.e., likely $R_2N:^-$ or t-BuO:$^-$)?
NO
Is the Nuc/Base a good nucleophile?
NO
substitution via slow S_N2
YES
alkene product via E2 (rare)
YES
substitution product via rapid S_N2 (common)
(b) R group is secondary
Is the Nuc/Base a strong base?
NO
Is the Nuc/Base a good to moderate nucleophile?
NO
Is the solvent polar protic?
NO
slow substitution and/or elimination via S_N2, S_N1, E2, and E1 or no reaction
YES
alkene product via E2
YES
substitution product via moderate speed S_N2
YES
substitution and/or elimination via moderate speed S_N1 and E1
(c) R group is tertiary: no S_N2
Is the Nuc/Base a strong base?
NO
Is the solvent polar protic?
NO
slow substitution and/or elimination via S_N1, E2, and E1 or no reaction
YES
alkene product via E2
YES
substitution and/or elimination product via fast S_N1 and E1; S_N1 usually is slightly preferred over E1

FIGURE 9.8 Flowchart for determining the experimental conditions and choice of reagents that favor S_N2, S_N1, E2, and E1 reactions.

Table 9.11 Summary of Substitution Versus Elimination Reactions of Haloalkanes

Halide	Reaction	Comments
Methyl CH_3X	S_N2	S_N1 reactions of methyl halides are never observed. The methyl cation is so unstable that it is not observed in common solvents.
Primary RCH_2X	S_N2	The main reaction with good nucleophiles/weak bases such as I^- and CH_3COO^-.
	E2	The main reaction with strong, bulky bases such as $(CH_3)_3CO^-$.
	S_N1/E1	Primary cations are rarely formed in solution; therefore, S_N1 and E1 reactions of primary halides are unlikely.
Secondary R_2CHX	S_N2	The main reaction with bases/nucleophiles where pK_a of the conjugate acid is 11 or less, as, for example, I^- and CH_3COO^-.
	E2	The main reaction with bases/nucleophiles where the pK_a of the conjugate acid is 11 or greater, as, for example, OH^- and $CH_3CH_2O^-$.
	S_N1/E1	Common in reactions with weak nucleophiles in polar protic solvents, such as water, methanol, and ethanol.
Tertiary R_3CX	E2	Main reaction with strong bases such as HO^- and RO^-.
	S_N1/E1	Main reactions with poor nucleophiles/weak bases if the solvent is polar protic.
	S_N2	S_N2 reactions of tertiary halides are never observed because of the extreme crowding around the 3° carbon.

We next examine the structure of the nucleophile. Because all nucleophiles are bases, we refer to them as Nuc/Base.

(a) Primary Alkyl Groups:

- Because primary carbocations are too unstable to form, S_N1 or E1 mechanisms are not possible.
- If the Nuc/Base is a strong base and sterically hindered, it will *not* be a good nucleophile and E2 is the major pathway. A common example is the use of *tert*-butoxide ion as the Nuc/Base. Amide anions are exceptions. Although they are not hindered, they are so basic that E2 dominates.
- If the Nuc/Base is a strong base and not sterically hindered, we next consider whether it is a good nucleophile. Examples of strong bases that are also good nucleophiles are hydroxide, acetylide, and methoxide. Weak bases are defined as bases that have conjugate acids with pK_a's below 11. Examples of weak bases that are good nucleophiles are thiolate (RS^-), cyanide (NC^-), iodide (I^-), and azide (N_3^-) anions. Even moderate nucleophiles that are weak bases, such as unhindered amines (NR_3) and phosphines (PR_3), participate in efficient S_N2 reactions. Hence, with any of the basic or weakly basic good-to-moderate nucleophiles, we find products predominantly from S_N2 pathways. However, some accompanying E2 mechanism is likely with strong bases.
- Finally, if the Nuc/Base is neither a good nor moderate nucleophile, we are likely to get S_N2 and E2 in a ratio that is difficult to predict or to get no reaction. Examples are water, alcohols, and carboxylic acids.

(b) Secondary Alkyl Groups:

- If the Nuc/Base is a strong base, whether or not it is hindered, E2 will dominate. Strong bases are defined as bases that have conjugate acids with pK_a's above 11, such as hydroxide, alkoxides, acetylides, and H_2N^-.
- With a weak base that is a good to moderate nucleophile, S_N2 will dominate. Examples are those nucleophiles that have conjugate acids with pK_a's below 11. However, because the alkyl group is secondary, the S_N2 reaction may be sluggish, and S_N1 and E2/E1 elimination pathways may compete to a small extent.
- When the nucleophile is not good, we need to examine the solvent. In a polar protic solvent, often with gentle warming, we can induce S_N1 and E1 pathways. The extent that substitution or elimination occurs is hard to predict.
- Finally, if the solvent is neither polar nor protic and the Nuc/Base is neither a strong base nor a good nucleophile, all four reaction pathways are possible, and it is difficult to predict which will dominate or whether a reaction will occur at all.

(c) Tertiary Alkyl Groups:

- We start by noting that an S_N2 mechanism cannot occur on a tertiary alkyl group.
- If the Nuc/Base is a strong base, E2 will dominate.
- Because S_N2 is not possible, we do not have to consider whether the Nuc/Base is a strong or weak nucleophile. Hence, this question is not relevant to predicting the dominant reaction pathway. Therefore, instead we check the solvent.
- If the solvent is polar and protic, we can induce S_N1 and E1 pathways, often by applying heat. Whether substitution or elimination dominates is hard to predict.
- Finally, if the Nuc/Base is not a strong base and the solvent is not polar and protic, S_N1, E2, and E1 reaction pathways are possible, and it is difficult to predict which will dominate or whether a reaction will occur at all.

9.9 Analysis of Several Competitions Between Substitutions and Eliminations

Now that we have outlined a step-by-step process with which to make predictions about relative extents of S_N2, S_N1, E2, and E1 mechanisms, we'll examine specific examples as we did in Section 9.4. Following are five examples of reactions between a haloalkane and a Nuc/Base in specific solvents, along with an analysis of the factors that favor the various mechanistic pathways. We start with examples that have clear-cut predictions and move to those that are more challenging. In some cases, we also cover predictions for the reactions under slightly different experimental conditions.

Competition 1

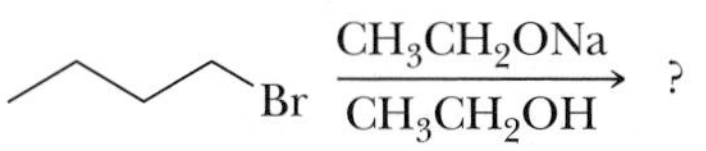
Br
CH_3CH_2ONa
CH_3CH_2OH
?

The haloalkane is primary; therefore, S_N1 and E1 cannot occur. The Nuc/Base is a strong base because the pK_a of its conjugate acid is several units above 11 (pK_a HOEt = 15.9). But due to its basicity and because it is not sterically hindered, ethoxide is also a good nucleophile. Hence, S_N2 will dominate over E2.

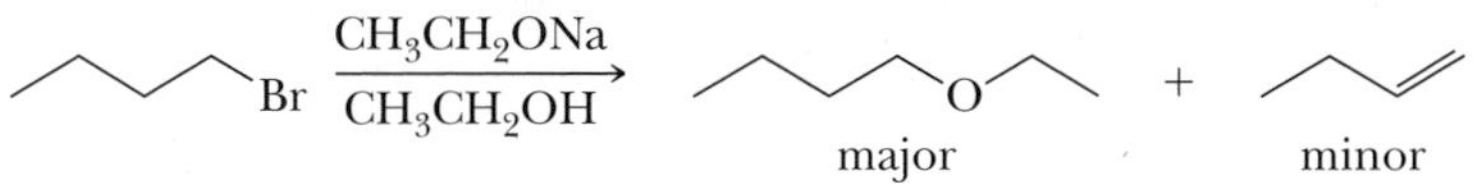
Br
CH_3CH_2ONa
CH_3CH_2OH
O
+
major
minor

Competition 2

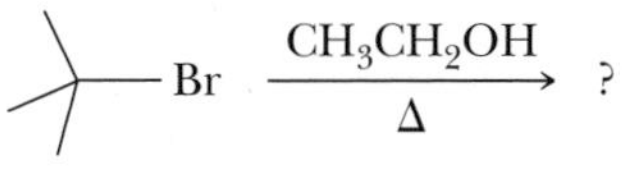
Br
CH_3CH_2OH
Δ
?

The haloalkane is tertiary; therefore, S_N2 cannot occur. The Nuc/Base is a weak base. The solvent is polar protic; therefore, S_N1 and E1 mechanisms will occur.

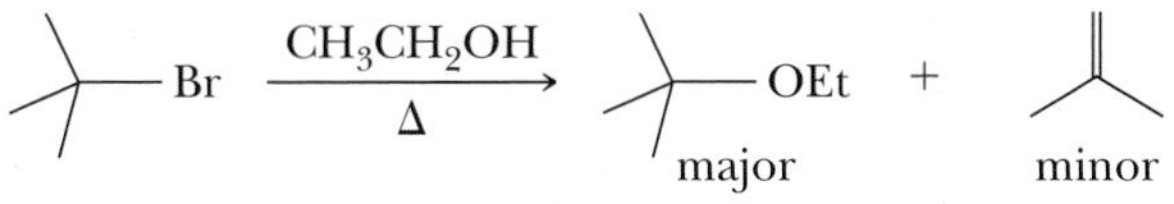
Br
CH_3CH_2OH
Δ
OEt
+
major
minor

If this same reaction were performed with NaOEt in the ethanol, E2 would have been the dominant pathway because ethoxide is a strong base.

Competition 3

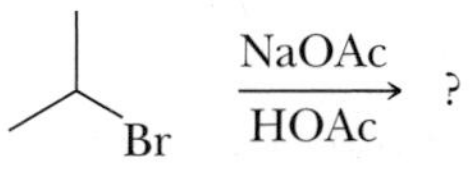
Br
NaOAc
HOAc
?

The haloalkane is secondary. Because the Nuc/Base has a pK_a of its conjugate acid far below 11 (pK_a of HOAc = 4.7), it is a weak base; hence, there will be little to no E2. However, acetate is a moderate nucleophile. Hence, S_N2 is the best prediction.

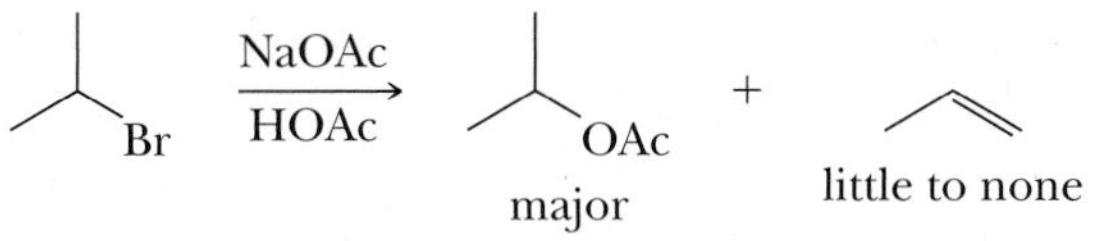
Br
NaOAc
HOAc
OAc
+
major
little to none

Because the solvent is polar protic, there could be a minor extent of $S_N1/E1$. If the sodium acetate were left out of the reaction and it were heated, the prediction would be $S_N1/E1$.

Competition 4

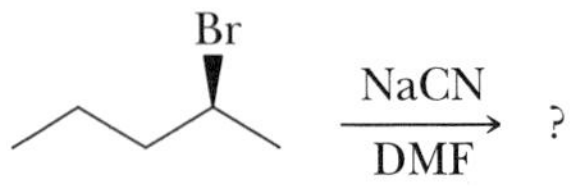

The haloalkane is secondary. The Nuc/Base has a pK_a of its conjugate acid near or slightly below 11 (HCN, pK_a = 9.3) and hence is a moderate to weak base. However, cyanide anion is an excellent nucleophile. Consequently, S_N2 will dominate over E2. Furthermore, the solvent DMF (dimethylformamide) is polar and aprotic and supports S_N2 or E2, but it does not assist S_N1 or E1. Because the reactant is chiral, the S_N2 inverts the configuration.

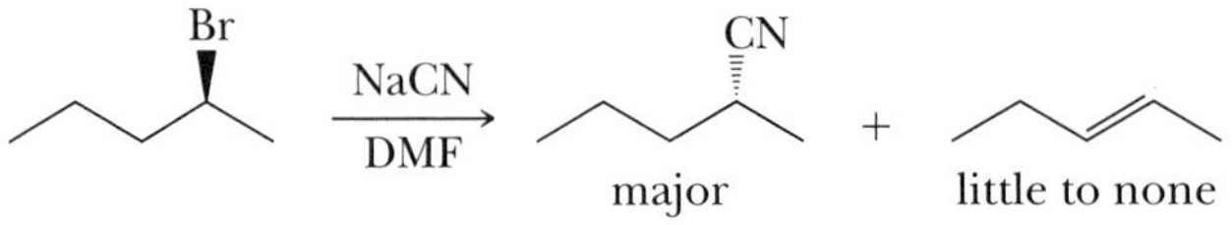

Competition 5

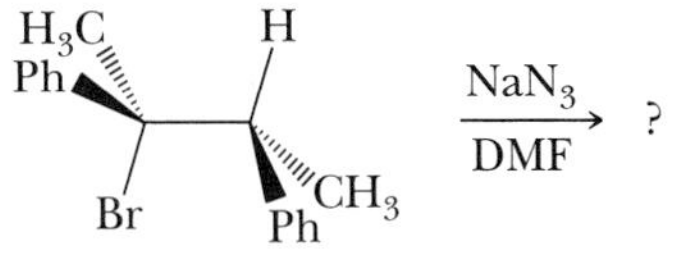

The haloalkane is tertiary; therefore, S_N2 cannot occur. The Nuc/Base is a weak base (pK_a HN_3 = 4.9); therefore, E2 is not obviously favored. However, the solvent is not protic but is simply polar. Therefore, S_N1 and E1 are not going to be favored. This is a case that is difficult to predict using Figure 9.8 or Table 9.11.

However, the lack of a polar protic solvent means that E2 is most likely, even with the weak base. The E2 occurs with an anti and coplanar arrangement of the **Br** and **H** that are eliminated, giving an *E* alkene. Any substitution from an S_N1 pathway would lead to racemization of the chiral center that possessed the leaving group.

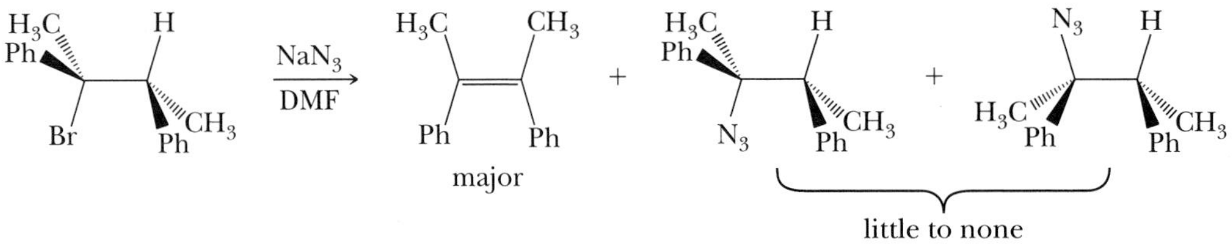

Example 9.8 S_N1 or S_N2, E1 or E2

Predict whether each reaction proceeds predominantly by substitution (S_N1 or S_N2) or elimination (E1 or E2) or whether the two compete. Write structural formulas for the major organic product(s).

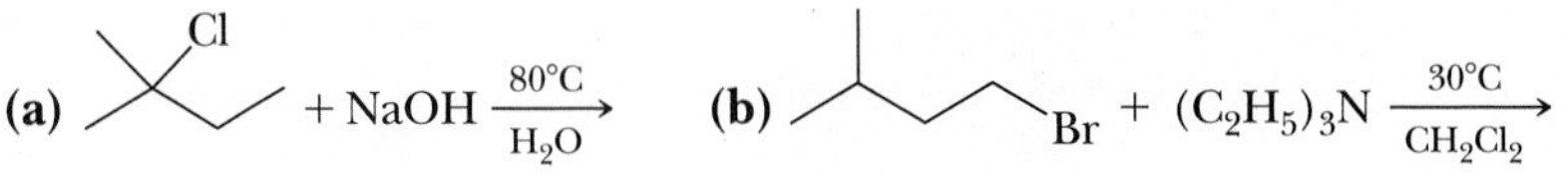

Solution

(a) A 3° haloalkane is heated with a strong base/good nucleophile. Elimination by an E2 reaction predominates to give 2-methyl-2-butene as the major product.

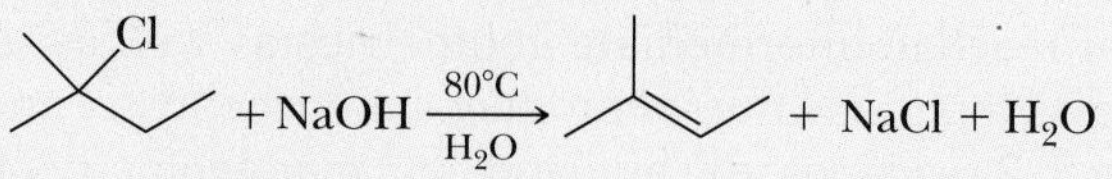

(b) Reaction of a 1° haloalkane with this moderate nucleophile/weak base gives substitution by an S_N2 reaction.

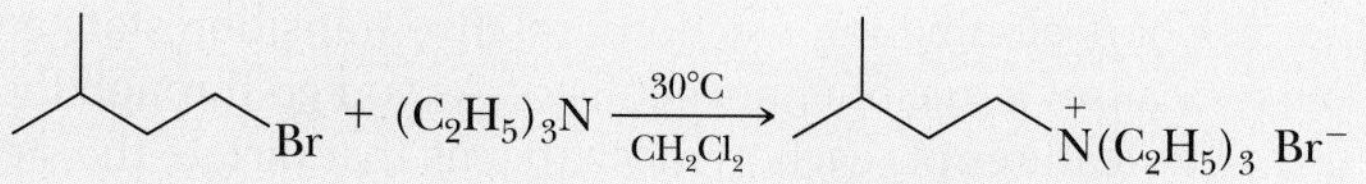

Problem 9.8

Watch a video explanation

Predict whether each reaction proceeds predominantly by substitution (S_N1 or S_N2) or elimination (E1 or E2) or whether the two compete. Write structural formulas for the major organic product(s).

(a)

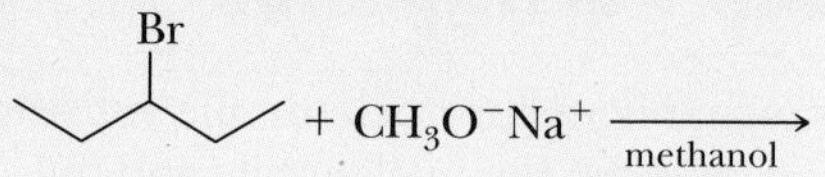

(b)

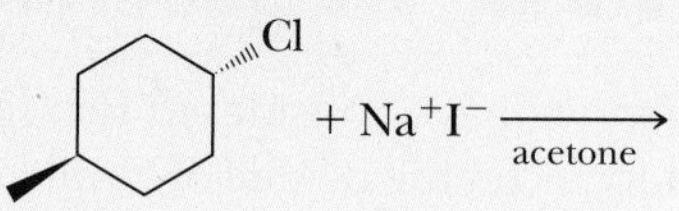

(c)

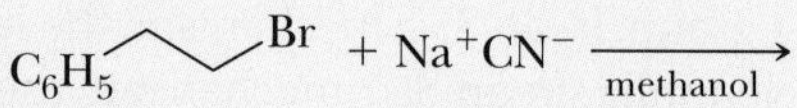

MCAT Practice: *Passage and Questions*

Solvents and Solvation

Choosing the best solvent for a chemical reaction is an extremely important aspect of organic chemistry. When deciding upon a solvent, chemists consider the solubility of the reactants and products, as well as the mechanism of the reaction and the solvation of intermediates. Further, for reactions that need heating to proceed in a reasonable amount of time, the choice of solvent is guided by its boiling point because this sets the temperature at which the reaction refluxes. Lastly, unless the solvent is intentionally used as a reactant, such as in a solvolysis, it must remain inert.

Questions

A. When performing an S_N1 solvolysis, which of the following solvents would be a poor choice for *tert*-butyl iodide ("dried" means that water has been removed)?

1. 80% water, 20% ethanol
2. Pure water
3. Dried acetonitrile
4. Dried acetic acid

B. When attempting to enhance the extent of S_N2 substitution by the nucleophile ethylamine ($EtNH_2$), which of the following solvents would be a poor choice for *sec*-butyl iodide?

1. Pure water
2. Acetonitrile
3. DMSO
4. *tert*-Butyl alcohol

C. When performing an S_N2 reaction using NaCN as the nucleophile reacting with *n*-butyl iodide, which of the following solvents would be the worst choice?

1. DMSO
2. DMF
3. Acetonitrile
4. Toluene

D. The reaction of diethylamine (Et_2NH) and *sec*-butyl iodide requires heating, but to optimize the extent of S_N2 over E2 the reaction cannot be too hot. Which of the following solvents would best represent a compromise solvent in which to reflux this reaction?

1. Diphenyl ether
2. Diethyl ether
3. THF
4. DMSO

An important take-home lesson from this chapter is that understanding key transition state or reactive intermediate geometries as well as relative transition state energies allows the prediction of product stereochemistry and regiochemistry. Backside attack in S_N2 reactions, the anti and coplanar geometry of the H atom and leaving group in E2 reactions, and the presence of carbocation intermediates in S_N1 reactions are important examples of reaction geometries that dictate stereochemistry. Understanding the relative energies of alternative possible transition states is also important. In the case of β-elimination reactions, relative transition state energies provide a rationale for Zaitsev's rule of regiochemistry. As you go through the rest of this book, try to learn key features of reaction mechanisms that dictate the stereochemistry and regiochemistry of reaction products. You should think of mechanisms as more than just electron pushing: they involve three-dimensional molecular interactions with associated relative energies that control the formation of products.

9.10 Neighboring Group Participation

So far, we have considered two limiting mechanisms for nucleophilic substitutions that focus on the degree of covalent bonding between the nucleophile and the substitution center during departure of the leaving group. In an S_N2 mechanism, the leaving group is assisted in its departure by the nucleophile. In an S_N1 mechanism, the leaving group is not assisted in this way. An essential criterion for distinguishing between these two pathways is the order of reaction. Nucleophile-assisted substitutions are second order: first order in RX and first order in nucleophile. Nucleophile-unassisted substitutions are first order: first order in RX and zero order in nucleophile.

Chemists recognize that certain nucleophilic substitutions have the kinetic characteristics of first-order (S_N1) substitution but, in fact, involve two successive displacement reactions. A characteristic feature of a great many of these reactions is the presence of an internal nucleophile (most commonly sulfur, nitrogen, or oxygen) on the carbon atom beta to the leaving group. This neighboring nucleophile participates in the departure of the leaving group to give an intermediate, which then reacts with an external nucleophile to complete the reaction.

The mustard gases are one group of compounds that react by participation of a neighboring group. The characteristic structural feature of a mustard gas is a two-carbon chain, with a halogen on one carbon and a divalent sulfur or trivalent nitrogen on the other carbon (S-C-C-Lv or N-C-C-Lv). An example of a mustard gas is bis(2-chloroethyl)sulfide, a poison gas used extensively in World War I and at one time, at least, manufactured by Iraq. This compound is a deadly vesicant (blistering agent) and quickly causes conjunctivitis and blindness.

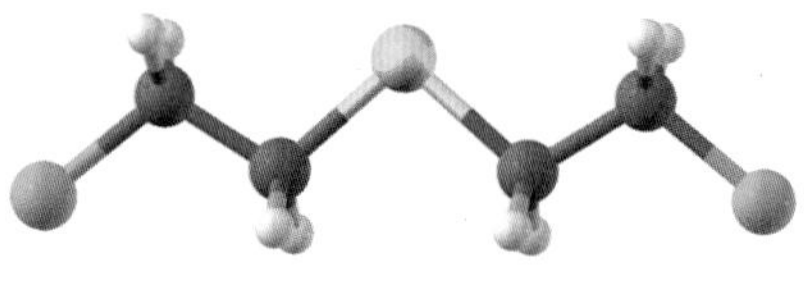

$ClCH_2CH_2SCH_2CH_2Cl$

bis(2-Chloroethyl)sulfide
(a sulfur mustard gas)

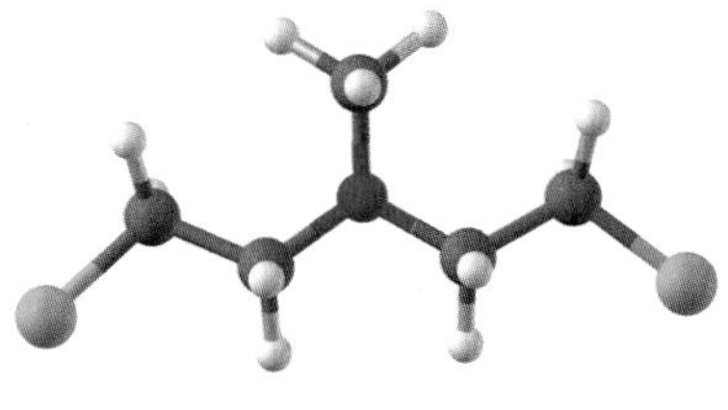

$ClCH_2CH_2{-}N(CH_3){-}CH_2CH_2Cl$

bis(2-Chloroethyl)methylamine
(a nitrogen mustard gas)

Bis(2-chloroethyl)sulfide and bis(2-chloroethyl)methylamine are not gases at all. They are oily liquids with a high vapor pressure, hence the designation "gas." Nitrogen and sulfur mustards react very rapidly with moisture in the air and in the mucous membranes of the eye, nose, and throat to produce HCl, which then burns and blisters these sensitive tissues. What is unusual about the reactivity of the mustard gases is that they react very rapidly with water, a very poor nucleophile.

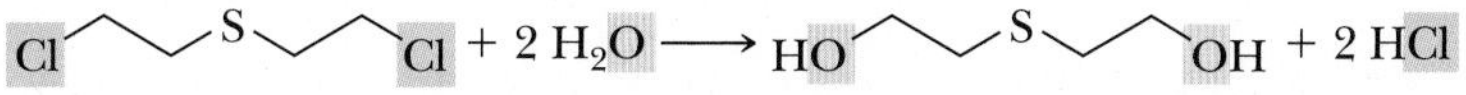

Mustard gases also react rapidly with other nucleophiles, such as those in biological molecules, which makes them particularly dangerous chemicals. Of the two steps in the mechanism of the hydrolysis of a sulfur mustard, the first is the slower and is rate-determining. As a result, the rate of reaction is proportional to the concentration of the sulfur mustard but independent of the concentration of the external nucleophile. Thus, although this reaction has the kinetic characteristics of an S_N1 reaction, it actually involves two successive S_N2 displacement reactions.

Mechanism 9.9

Hydrolysis of a Sulfur Mustard—Participation by a Neighboring Group

Step 1: **Make a new bond between a nucleophile and an electrophile and simultaneously break a bond to give stable molecules or ions.** The reason for the extremely rapid hydrolysis of the sulfur mustards is neighboring group participation by sulfur in the ionization of the carbon-chlorine bond to form a cyclic sulfonium ion. This is the rate-determining step of the reaction; although it is the slowest step, it is much faster than reaction of a typical primary chloroalkane with water. At this point, you should review halogenation of alkenes (Sections 6.3D and 6.3F) and compare the cyclic halonium ions formed there with the cyclic sulfonium ion formed here.

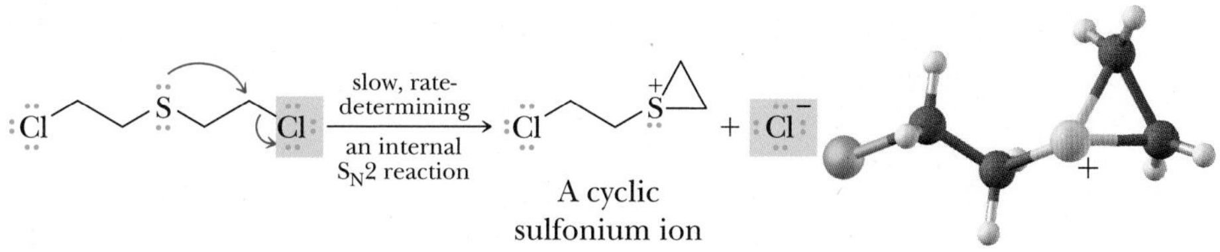

Step 2: **Make a new bond between a nucleophile and an electrophile.** The cyclic sulfonium ion contains a highly strained three-membered ring and reacts rapidly with an external nucleophile to open the ring followed by proton transfer to H_2O to give H_3O^+. In this S_N2 reaction, H_2O is the nucleophile and sulfur is the leaving group.

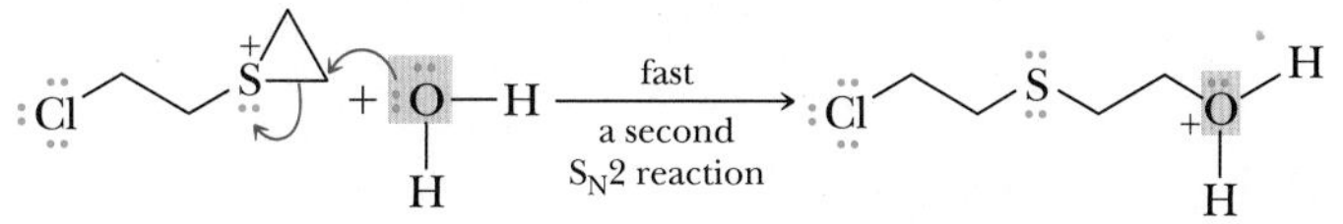

Step 3: **Take a proton away.** Proton transfer to water completes the reaction.

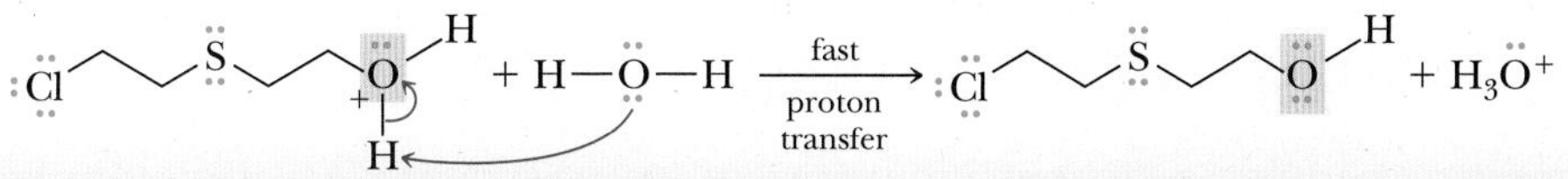

The net effect of these reactions is nucleophilic substitution of Cl by OH.

We continue to use the terms S_N2 and S_N1 to describe nucleophilic substitution reactions. You should realize, however, that these designations do not adequately describe all nucleophilic substitution reactions.

Example 9.9 Hydrolysis of Nitrogen Mustards

Write a mechanism for the hydrolysis of the nitrogen mustard bis(2-chloroethyl)-methylamine.

Solution

Following is a three-step mechanism.

Step 1: **Make a new bond between a nucleophile and an electrophile and simultaneously break a bond to give stable molecules or ions.** This is an internal S_N2 reaction in which ionization of the C—Cl bond is assisted by the neighboring nitrogen atom to form a highly strained three-membered ring.

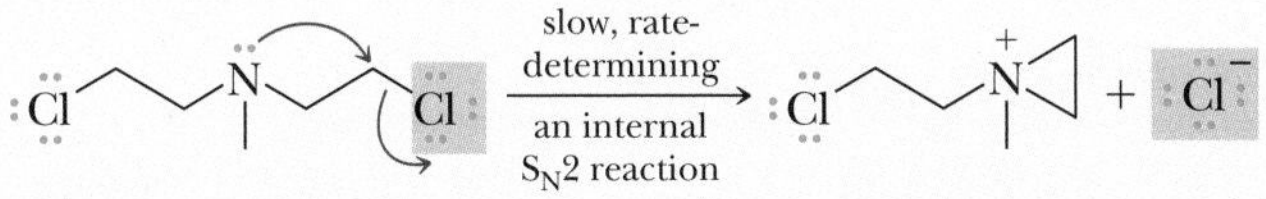

A cyclic ammonium ion

Step 2: **Make a new bond between a nucleophile and an electrophile.** Reaction of the cyclic ammonium ion with water opens the three-membered ring. In this S_N2 reaction, H_2O is the nucleophile and nitrogen is the leaving group.

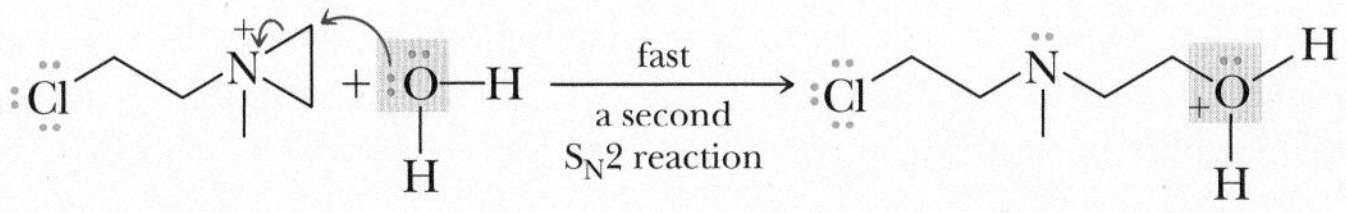

Step 3: **Take a proton away.** Proton transfer to the basic nitrogen completes the reaction.

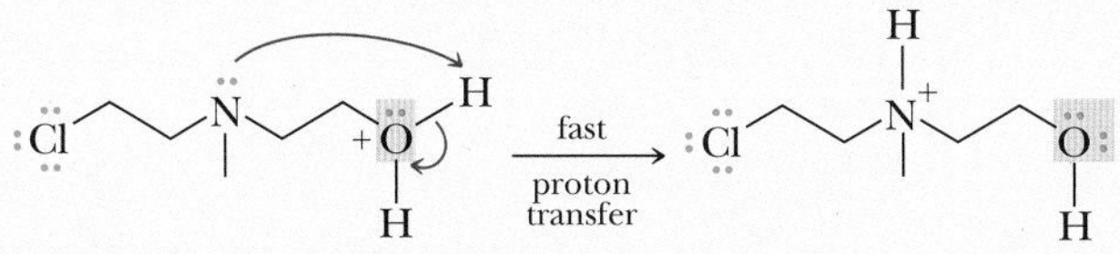

Problem 9.9

Knowing what you do about the stereochemisry of S_N2 reactions, predict the product of hydrolysis of this compound.

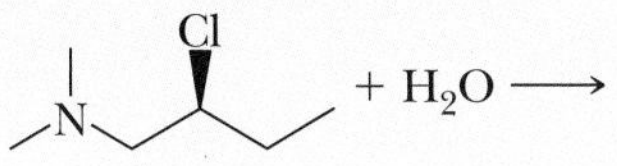

Connections to Biological Chemistry

Mustard Gases and the Treatment of Neoplastic Diseases

Autopsies of soldiers killed by sulfur mustards in World War I revealed, among other things, very low white blood cell counts and defects in bone marrow development. From these observations, it was realized that sulfur mustards have profound effects on rapidly dividing cells. This became a lead observation in the search for less toxic alkylating agents for use in treatment of cancers, which have rapidly dividing cells. Attention turned to the less reactive nitrogen mustards. One of the first compounds tested was

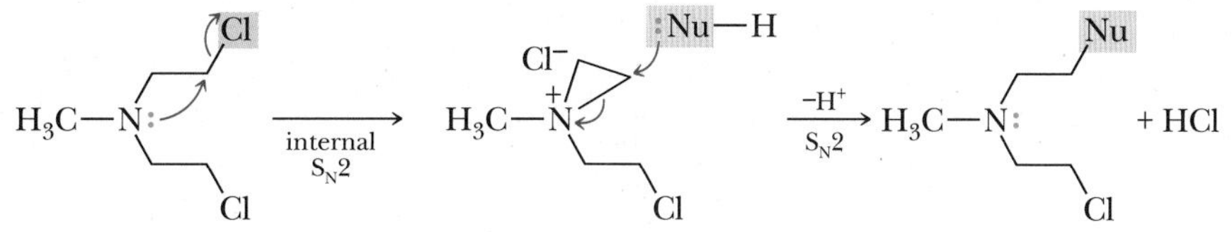

Mechlorethamine

(An aziridinium ion intermediate)

mechlorethamine. As with other mustards, the reaction of mechlorethamine with nucleophiles is rapid because of the formation of an aziridinium ion.

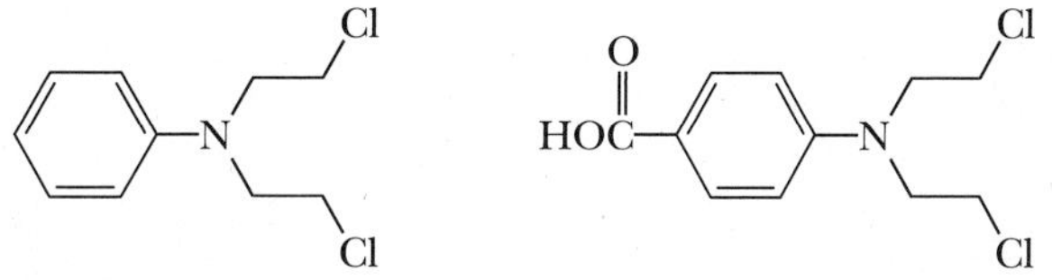

Nucleophilicity of nitrogen is acceptable, but the compound is too insoluble in water for intravenous injection

Solubility in water is acceptable, but nucleophilicity of nitrogen is reduced and compound is unreactive

Mechlorethamine undergoes very rapid reaction with water (hydrolysis) and with other nucleophiles, so much so that within minutes after injection into the body, it has completely reacted. The problem for the chemist, then, was to find a way to decrease the nucleophilicity of nitrogen while maintaining reasonable water solubility. Substitution of phenyl for methyl reduced the

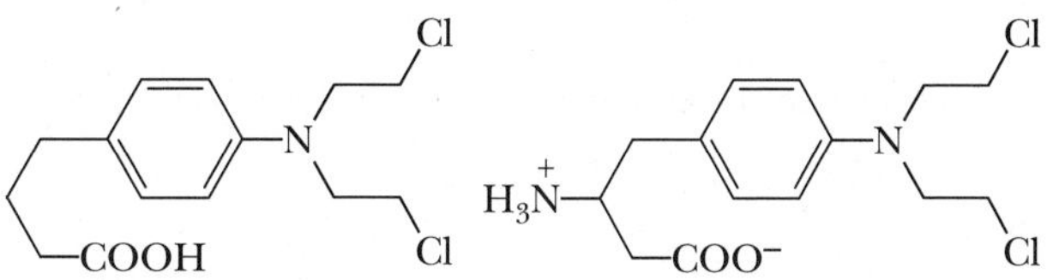

Chlorambucil

Melphalan
(racemic)

nucleophilicity, but the resulting compound was not sufficiently soluble in water for intravenous injection. The solubility problem was solved by adding a carboxyl group. When the carboxyl group was added directly to the aromatic ring, however, the resulting compound was too stable and therefore not biologically active.

Adding a propyl bridge (chlorambucil) or an aminoethyl bridge (melphalan) between the aromatic ring and the carboxyl group solved both the solubility problem and the reactivity problem. Note that melphalan is chiral. It has been demonstrated that the *R* and *S* enantiomers have approximately equal therapeutic potency.

The clinical value of the nitrogen mustards lies in the fact that they undergo reaction with certain nucleophilic sites on the heterocyclic aromatic amine bases in DNA (see Chapter 28). For DNA, the most reactive nucleophilic site is N-7 of guanine. Next in reactivity is N-3 of adenine, followed by N-3 of cytosine.

The nitrogen mustards are bifunctional alkylating agents; one molecule of nitrogen mustard undergoes reaction with two molecules of nucleophile. Guanine alkylation leaves one free reactive alkylating group, which can react with another base, giving cross links that lead to miscoding during DNA replication. The therapeutic value of the nitrogen mustards lies in their ability to disrupt normal base pairing. This prevents replication of the cells, and the rapidly dividing cancer cells are more sensitive than normal cells.

(*Continued*)

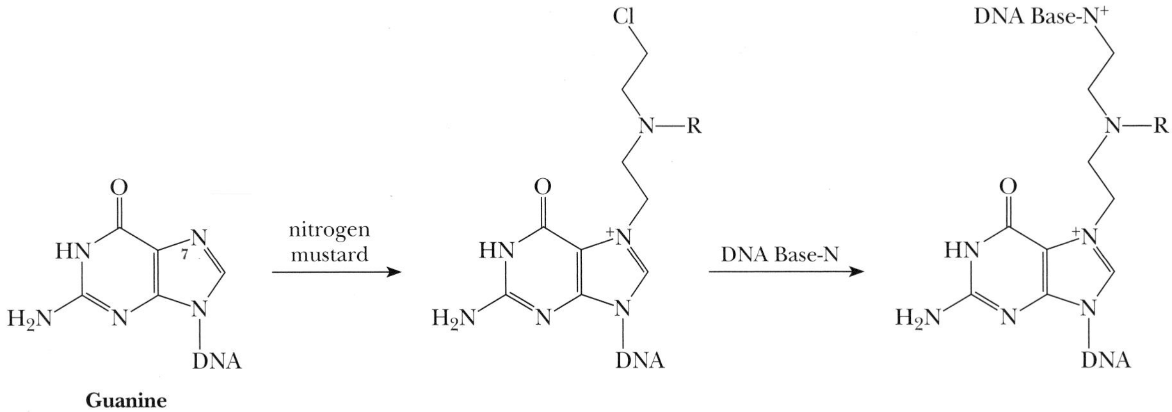

Study Guide

9.1 Nucleophilic Substitution in Haloalkanes

- **Nucleophilic substitution** is any reaction in which a nucleophile replaces another electron-rich group called a **leaving group (Lv)**.
  - *A nucleophile* (Nu:$^-$) *is an electron-rich molecule or ion that donates a pair of electrons to another atom or ion to form a new covalent bond.*

P 9.1

9.2 Mechanisms of Nucleophilic Aliphatic Substitution

- There are two limiting mechanisms for nucleophilic substitution, namely S_N2 and S_N1.
  - *In the S_N2 reaction mechanism, bond forming and bond breaking occur simultaneously.*
    - *S_N2 reactions are bimolecular because both nucleophile and haloalkane concentrations influence reaction rate.*
    - *The nucleophile must approach the carbon-leaving group (C—Lv) bond from the backside in order to populate the C—Lv antibonding orbital and allow reaction.*
  - *In the S_N1 mechanism, the leaving group departs first in the rate-determining step, leaving a carbocation intermediate that reacts with the nucleophile in a second step.*
    - *S_N1 reactions are unimolecular because only the haloalkane concentration influences reaction rate.*

9.3 Experimental Evidence for S_N1 and S_N2 Mechanisms

- The S_N2 reaction can be identified based on kinetics of the reaction and stereochemistry of the products.

– *Because an S_N2 reaction is bimolecular, doubling the concentration of either haloalkane or nucleophile will double the rate of the reaction.*

– *Because backside attack geometry is required, reaction at a chiral center results in inversion of configuration.*

– *The S_N1 reaction can also be identified based on kinetics of the reaction and stereochemistry of the products.*

– *Because an S_N1 reaction is unimolecular, doubling the concentration of only the haloalkane can double the rate of the reaction.*

– *Because a planar and achiral carbocation intermediate is formed that can be attacked with roughly equal probability from either face, reaction at a chiral center results in racemization of stereochemistry.*

  – *The chiral center is often not completely racemized because the leaving group forms an ion pair with the carbocation intermediate, partially blocking one face.*

- The structure of the haloalkane influences the reaction rate and mechanism.

  – *Haloalkanes that can form more stable carbocations react faster if an S_N1 mechanism occurs.*

  – *Because S_N1 reactions involve carbocations, rearrangements (1,2 shifts) can occur if they lead to more stable carbocation intermediates.*

  – *Steric hindrance on the backside of the $C—Lv$ bond of a haloalkane slows down or possibly prevents an S_N2 mechanism.*

- The more stable the anion produced upon reaction, the better the **leaving group ability**.
- **Solvent** properties can have an important influence on reaction mechanisms.

  – ***Protic solvents** are hydrogen-bond donors. The most common protic solvents are those containing $—OH$ groups.*

  – ***Aprotic solvents** cannot serve as hydrogen-bond donors. Common aprotic solvents are acetone, diethyl ether, dimethyl sulfoxide, and* N,N*-dimethylformamide.*

  – ***Polar solvents** interact strongly with ions and polar molecules.*

  – ***Nonpolar solvents** do not interact strongly with ions and polar molecules.*

  – *The **dielectric constant** is the most commonly used measure of solvent polarity.*

  – ***Solvolysis** is a nucleophilic substitution reaction in which the solvent is the nucleophile.*

  – ***Polar protic solvents** accelerate S_N1 reactions by stabilizing the charged carbocation intermediate.*

  – ***Polar aprotic solvents** accelerate S_N2 reactions because they do not interact strongly with the nucleophile.*

- **Nucleophiles** are categorized as good, moderate, or poor.

  – ***Good nucleophiles** are generally anions. **Moderate nucleophiles** are generally neutral, with one or more available lone pairs. **Poor nucleophiles** are generally polar protic solvents.*

  – *All things being equal, the stronger the interaction of a nucleophile with solvent, the lower the nucleophilicity.*

  – *Small nucleophiles with very little steric hindrance are better nucleophiles for S_N2 reactions.*

P 9.2–9.4, 9.10–9.13, 9.15–9.22, 9.24–9.36

Stereochemistry of an S_N1 Reaction

KEY REACTIONS

1. **Nucleophilic Aliphatic Substitution: S_N2 (Section 9.3)** S_N2 reactions occur in one step; departure of the leaving group is assisted by the incoming nucleophile, and both nucleophile and leaving group are involved in the transition state. The nucleophile may be negatively charged as in the first example or neutral as in the second example.

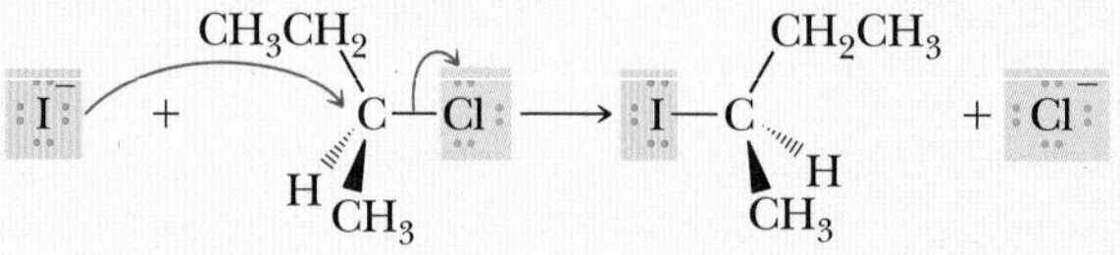

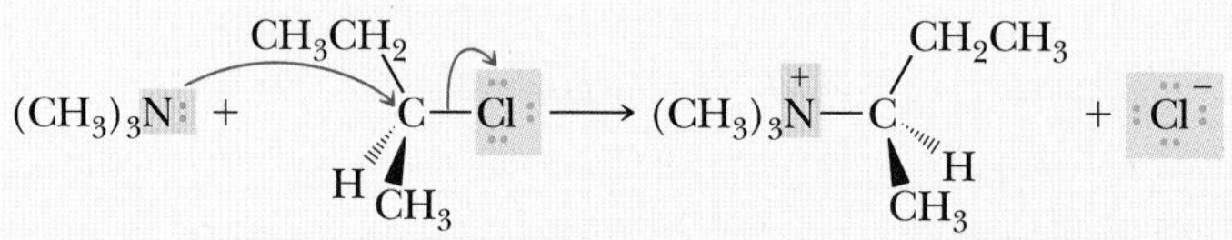

S_N2 reactions result in inversion of configuration at the reaction center. They are accelerated more in polar aprotic solvents than in polar protic solvents. The relative rates of S_N2 reactions are governed by steric factors, namely the degree of crowding around the site of reaction.

2. **Nucleophilic Aliphatic Substitution: S_N1 (Section 9.3)** An S_N1 reaction occurs in two steps. Step 1 is a slow, rate-determining ionization of the C—Lv bond to form a carbocation intermediate followed in Step 2 by rapid reaction of the carbocation intermediate with a nucleophile to complete the substitution. Reaction at a chiral center gives largely racemization, often accompanied with a slight excess of inversion of configuration. Reactions often involve carbocation rearrangements and are accelerated by polar protic solvents. S_N1 reactions are governed by electronic factors, namely the relative stabilities of carbocation intermediates. The following reaction involves an S_N1 reaction with a hydride shift.

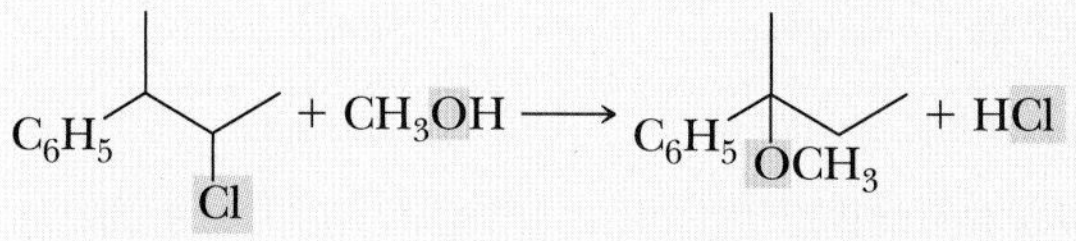

(racemic) (racemic)

9.4 Analysis of Several Nucleophilic Substitution Reactions

- Methyl or primary haloalkanes react through S_N2 mechanisms because of an absence of steric hindrance and lack of carbocation stability.
- Secondary haloalkanes react through an S_N2 mechanism in aprotic solvents with good nucleophiles, but through an S_N1 mechanism in protic solvents with poor nucleophiles.
- Tertiary haloalkanes react through an S_N1 mechanism because the steric hindrance disfavors S_N2 backside attack, and the attached alkyl groups stabilize a carbocation.

P 9.5, 9.14, 9.23

Problem 9.5

9.5 β-Elimination

- A **β-elimination reaction** involves removal of atoms or groups of atoms from adjacent carbon atoms.
  - ***Dehydrohalogenation*** *is a β-elimination reaction that involves loss of an H and a halogen atom from adjacent carbons to create an alkene from a haloalkane.*
  - ***Zaitzev's rule*** *predicts that β-elimination reactions give primarily the more highly substituted alkene. Such reactions are called* ***Zaitsev eliminations****.*

P 9.6

9.6 Mechanisms of β-Elimination

- The two limiting mechanisms for β-elimination reactions are the **E1** and **E2** mechanisms.
  - *In the E1 mechanism, the leaving group departs to give a carbocation; then a proton is taken off an adjacent carbon atom by base to create the product alkene.*
    - *E1 reactions are unimolecular because only the haloalkane concentration influences the rate of the reaction.*
  - *In the E2 mechanism, the halogen departs at the same time that an H atom is removed by base from an adjacent carbon atom to create the product alkene.*
    - *E2 reactions are bimolecular because both the haloalkane and base concentrations influence the rate of the reaction.*

9.7 Experimental Evidence for E1 and E2 Mechanisms

- E2 reactions are stereoselective in that the lowest energy transition state is the state in which the leaving group and H atoms that depart are oriented anti and coplanar.
  - *This anti and coplanar requirement determines whether E or Z alkenes are produced. For cyclohexane derivatives, both the leaving group and departing H atom must be axial.*
- Both E1 and E2 reactions are regioselective, favoring formation of the more stable (Zaitsev) product alkene (as long as Lv and H can be oriented anti and coplanar in the case of E2).
  - *The more stable alkene is generally the more highly substituted alkene.*

P **9.7, 9.37–9.42**

Problem 9.7

KEY REACTIONS

3. β-Elimination: E1 (Sections 9.6, 9.7) An E1 reaction occurs in two steps: slow, rate-determining breaking of the C—Lv bond to form a carbocation intermediate followed by rapid proton transfer to solvent to form an alkene. An E1 reaction is first order in haloalkane and zero order in base. Skeletal rearrangements are common.

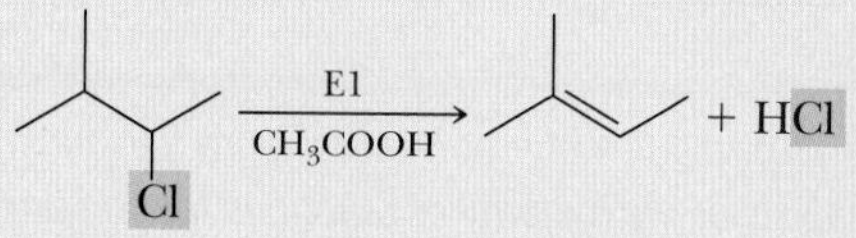

(racemic)

(Continued)

4. β-Elimination: E2 (Sections 9.6, 9.7) An E2 reaction occurs in one step: simultaneous reaction with base to remove a hydrogen, formation of the alkene, and departure of the leaving group. Elimination is stereoselective, requiring an anti and coplanar arrangement of the groups being eliminated.

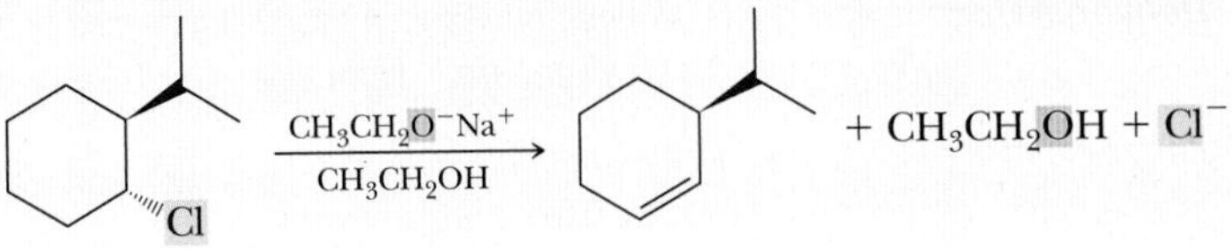

Cl
$CH_3CH_2O^-Na^+$
CH_3CH_2OH
$+ CH_3CH_2OH + Cl^-$

9.8 Substitution Versus Elimination

- When deciding which substitution or elimination mechanism dominates a reaction, analyze the structure of the haloalkane, the choice of the solvent, and the relative base strength of the nucleophile.
- Methyl or primary haloalkanes do not react through E1 or S_N1 mechanisms.
  - *S_N2 is favored for all nucleophiles except for exceptionally strong bases (H_2N^-) or sterically hindered ones (tert-butoxide), which cause E2 to predominate.*
- Secondary haloalkanes can react through any of the mechanisms.
  - *If the nucleophile is a strong base (conjugate acids with pK_a's above 11, such as hydroxide, alkoxides, acetylides, and H_2N^-), E2 predominates.*
  - *Weak bases (conjugate acids with pK_a's below 11) that are good or moderate nucleophiles (see Table 9.7) react predominantly by an S_N2 mechanism.*
  - *Poor nucleophiles (that are polar protic solvents) react through a combination of $S_N1/E1$ pathways, the exact ratio of which is hard to predict.*
- Tertiary haloalkanes cannot react by an S_N2 mechanism.
  - *If the nucleophile is a strong base (conjugate acids with pK_a's above 11, such as hydroxide, alkoxides, acetylides, and H_2N^-), E2 predominates.*
  - *For other nucleophiles in a polar protic solvent, reaction is through a combination of $S_N1/E1$ pathways, the exact ratio of which is hard to predict.*

P **9.8, 9.43–9.62**

▶ **Contrasting S_N2, S_N1, E2 and E1**

9.9 Analysis of Several Competitions Between Substitutions and Eliminations

- Predicting whether substitution or elimination reactions will dominate is a matter of following the logic given in Section 9.8.
  - *Either Table 9.11 or the flowchart given in Figure 9.8 (summarized just above) will lead to a successful analysis of the majority of reactions that organic chemists perform.*

▶ **Problem 9.8**

9.10 Neighboring Group Participation

- Certain nucleophilic displacements that have the kinetic characteristic of S_N1 reactions (first order in haloalkane and zero order in nucleophile) involve two successive S_N2 reactions.
  - *Many such reactions involve participation of a neighboring nucleophile.*
  - *The mustard gases are one group of compounds whose nucleophilic substitution reactions involve neighboring group participation.*

P 9.9

KEY REACTIONS

5. Neighboring Group Participation: (Section 9.10) Neighboring group participation is characterized by first-order kinetics and participation of an internal nucleophile in departure of the leaving group, as in hydrolysis of a sulfur or nitrogen mustard gas. The mechanism for their solvolysis involves two successive nucleophilic displacements.

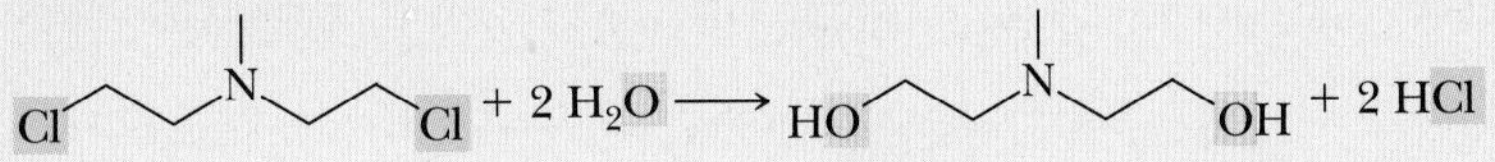

Cl
N
Cl
+ 2 H2O
HO
N
OH
+ 2 HCl

Problems

Red numbers indicate applied problems.

Nucleophilic Aliphatic Substitution

9.10 Draw a structural formula for the most stable carbocation with each molecular formula.

(a) $C_4H_9^+$ **(b)** $C_3H_7^+$ **(c)** $C_5H_{11}^+$ **(d)** $C_3H_7O^+$

9.11 The reaction of 1-bromopropane and sodium hydroxide in ethanol occurs by an S_N2 mechanism. What happens to the rate of this reaction under the following conditions?

(a) The concentration of NaOH is doubled.
(b) The concentrations of both NaOH and 1-bromopropane are doubled.
(c) The volume of the solution in which the reaction is carried out is doubled.

9.12 From each pair, select the stronger nucleophile.

(a) H_2O or OH^- **(b)** CH_3COO^- or OH^-
(c) CH_3SH or CH_3S^- **(d)** Cl^- or I^- in DMSO
(e) Cl^- or I^- in methanol **(f)** CH_3OCH_3 or CH_3SCH_3

9.13 Draw a structural formula for the product of each S_N2 reaction. Where configuration of the starting material is given, show the configuration of the product.

(a) $CH_3CH_2CH_2Cl + CH_3CH_2O^-Na^+ \xrightarrow[\text{ethanol}]{}$ **(b)** $(CH_3)_3N + CH_3I \xrightarrow[\text{acetone}]{}$

(c)  $-CH_2Br + Na^+CN^- \xrightarrow[\text{acetone}]{}$ **(d)** H_3C  $Cl + CH_3S^-Na^+ \xrightarrow[\text{ethanol}]{}$

(e) $CH_3CH_2CH_2Cl + CH_3C{\equiv}C^-Li^+ \xrightarrow[\text{diethyl ether}]{}$

(f) 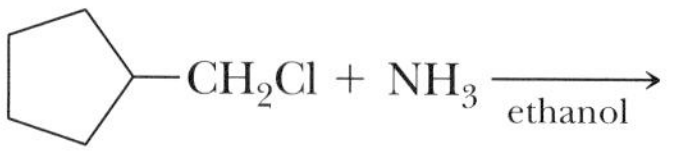

(g) 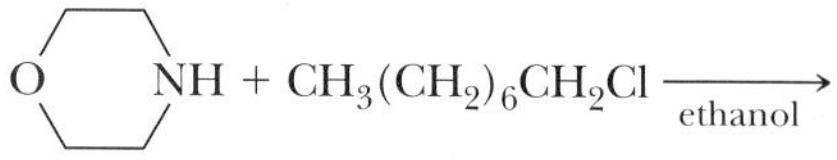

(h) $CH_3CH_2CH_2Br + Na^+CN^- \xrightarrow[\text{acetone}]{}$

9.14 Suppose you are told that each reaction in Problem 9.13 is a substitution reaction but are not told the mechanism. Describe how you could conclude from the structure of the haloalkane, the nucleophile, and the solvent that each reaction is an S_N2 reaction.

9.15 Treatment of 1,3-dichloropropane with potassium cyanide results in the formation of pentanedinitrile. The rate of this reaction is about 1000 times greater in DMSO than in ethanol. Account for this difference in rate.

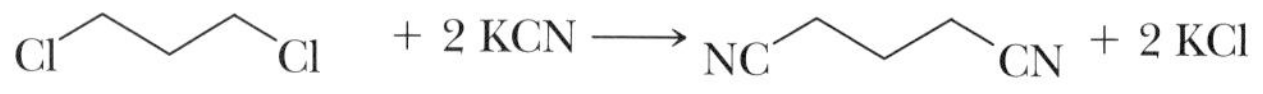

1,3-Dichloropropane **Pentanedinitrile**

9.16 Treatment of 1-aminoadamantane, $C_{10}H_{17}N$, with methyl 2,4-dibromobutanoate in the presence of a nonnucleophilic base, R_3N, involves two successive S_N2 reactions and gives compound A. Propose a structural formula for compound A.

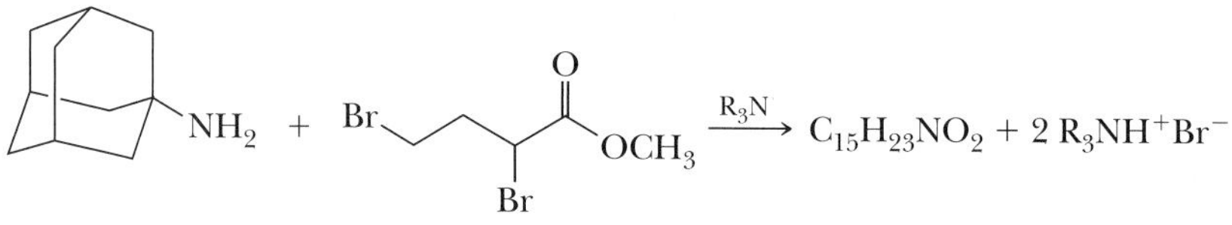

1-Aminoadamantane **Methyl 2,4-dibromobutanoate** **A**

9.17 Select the member of each pair that shows the greater rate of S_N2 reaction with KI in acetone.

(a) 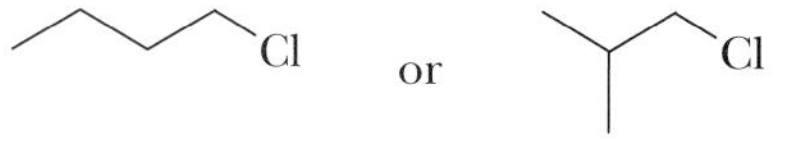

(b) 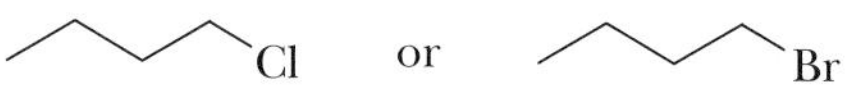

(c) 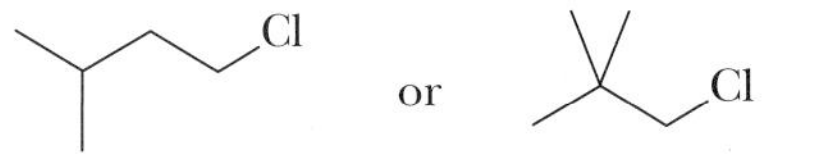

(d) 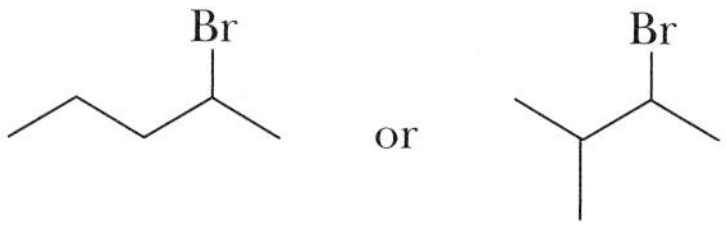

9.18 Select the member of each pair that shows the greater rate of S_N2 reaction with KN_3 in acetone.

(a) 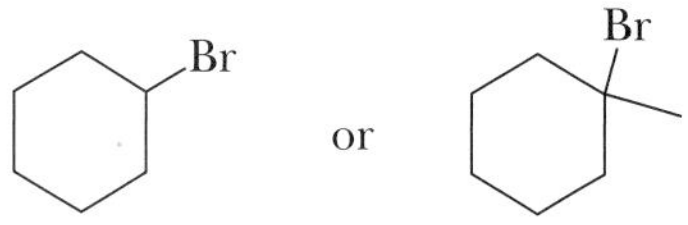

(b) 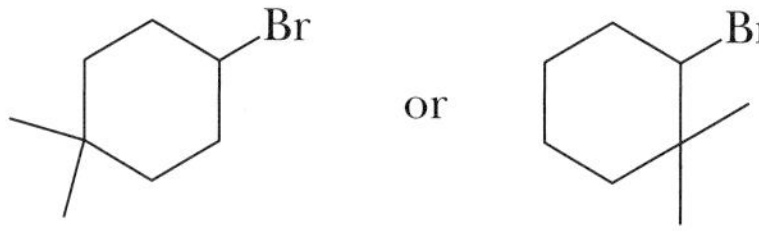

9.19 What hybridization best describes the reacting carbon in the S_N2 transition state?

9.20 Each carbocation is capable of rearranging to a more stable carbocation. Limiting yourself to a single 1,2-shift, suggest a structure for the rearranged carbocation.

(a) 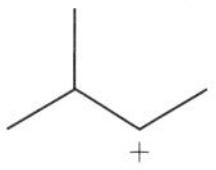

(b) 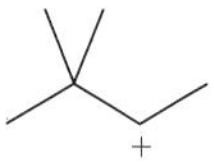

(c) 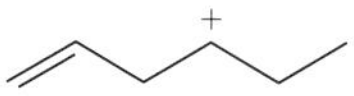

(d) 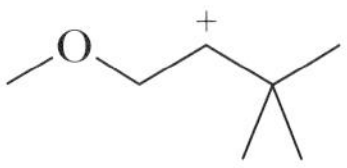

(e) 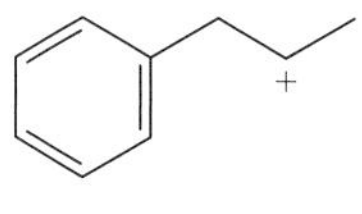

(f) 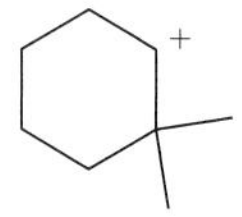

9.21 Attempts to prepare optically active iodides by nucleophilic displacement on optically active bromides using I^- normally produce racemic iodoalkanes. Why are the product iodoalkanes racemic?

9.22 Draw a structural formula for the product of each S_N1 reaction. Where configuration of the starting material is given, show the configuration of the product.

(a) 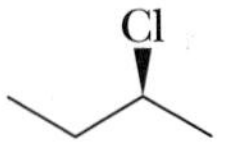 $+ CH_3CH_2OH \xrightarrow{\text{ethanol}}$

S enantiomer

(b)  $+ CH_3OH \xrightarrow{\text{methanol}}$

(c) $CH_3\overset{CH_3}{\underset{CH_3}{C}}Cl + CH_3\overset{O}{C}OH \xrightarrow{\text{acetic acid}}$

(d) 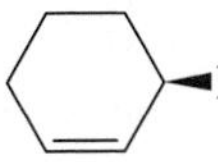 $+ CH_3OH \xrightarrow{\text{methanol}}$

9.23 Suppose you were told that each reaction in Problem 9.22 is a substitution reaction, but you were not told the mechanism. Describe how you could conclude from the structure of the haloalkane or cycloalkene, the nucleophile, and the solvent that each reaction is an S_N1 reaction.

9.24 Alkenyl halides such as vinyl bromide, $CH_2{=}CHBr$, undergo neither S_N1 nor S_N2 reactions. What factors account for this lack of reactivity?

9.25 Select the member of each pair that undergoes S_N1 solvolysis in aqueous ethanol more rapidly.

(a)  or 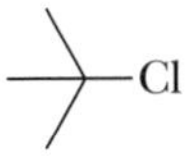

(b) 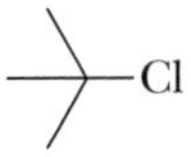 or 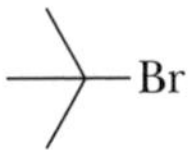

(c) 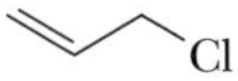 or 

(d) 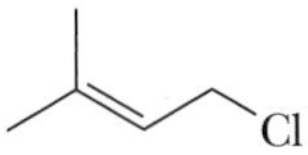 or 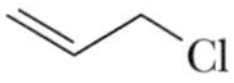

(e) 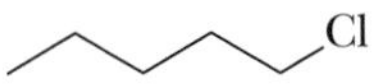 or 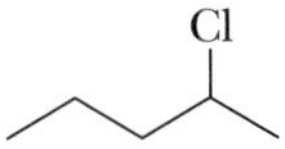

(f) 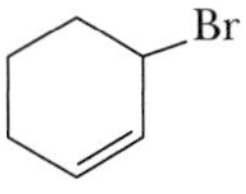 or 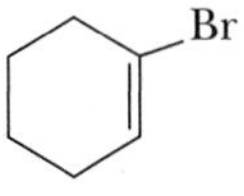

9.26 Account for the following relative rates of solvolysis under experimental conditions favoring S_N1 reaction.

			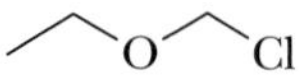
Relative rate of solvolysis (S_N1)	0.2	1	10^9

9.27 Not all tertiary haloalkanes undergo S_N1 reactions readily. For example, the bicyclic compound shown below is very unreactive under S_N1 conditions. What feature of this molecule is responsible for such lack of reactivity? You will find it helpful to examine a model of this compound.

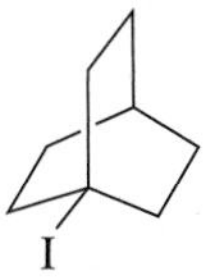

1-Iodobicyclooctane

9.28 Show how you might synthesize the following compounds from a haloalkane and a nucleophile.

(a) 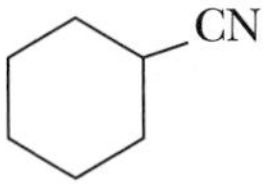

(b) 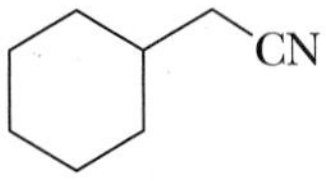

(c) 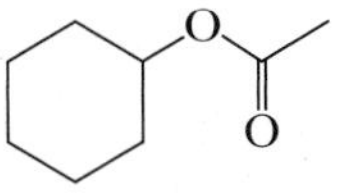

(d) 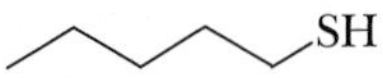

(e) 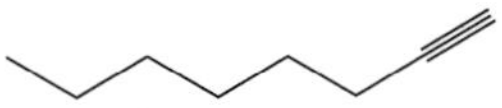

(f) 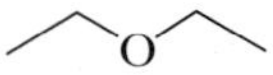

(g) 

9.29 3-Chloro-1-butene reacts with sodium ethoxide in ethanol to produce 3-ethoxy-1-butene. The reaction is second order, first order in 3-chloro-1-butene, and first order in sodium ethoxide. In the absence of sodium ethoxide, 3-chloro-1-butene reacts with ethanol to produce both 3-ethoxy-1-butene and 1-ethoxy-2-butene. Explain these results.

9.30 1-Chloro-2-butene undergoes hydrolysis in warm water to give a mixture of these allylic alcohols. Propose a mechanism for their formation.

$$CH_3CH{=}CHCH_2Cl \xrightarrow{H_2O} CH_3CH{=}CHCH_2OH + CH_3CH(OH)CH{=}CH_2$$

1-Chloro-2-butene — **2-Buten-1-ol** — **3-Buten-2-ol** (racemic)

9.31 The following nucleophilic substitution occurs with rearrangement. Suggest a mechanism for formation of the observed product. If the starting material has the *S* configuration, what is the configuration of the stereocenter in the product?

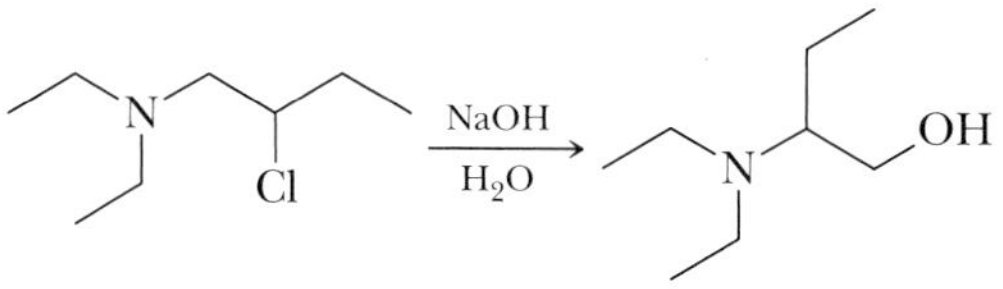

9.32 Propose a mechanism for the formation of these products in the solvolysis of this bromoalkane.

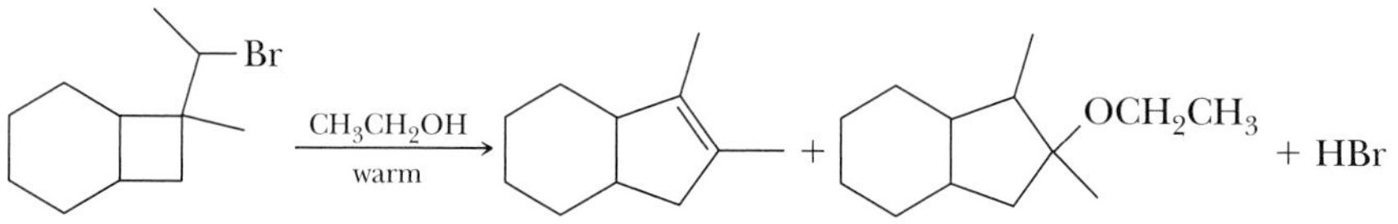

9.33 Solvolysis of the following bicyclic compound in acetic acid gives a mixture of products, two of which are shown. The leaving group is the anion of a sulfonic acid, $ArSO_3H$. A sulfonic acid is a strong acid, and its anion, $ArSO_3^-$, is a weak base and a good leaving group. Propose a mechanism for this reaction.

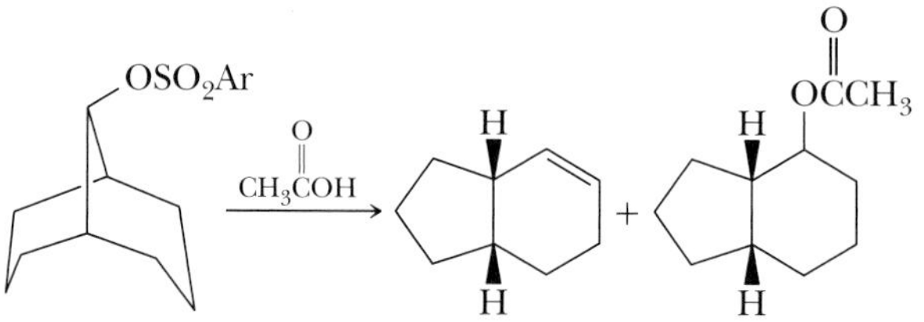

9.34 Which compound in each set undergoes more rapid solvolysis when refluxed in ethanol? Show the major product formed from the more reactive compound.

(a) 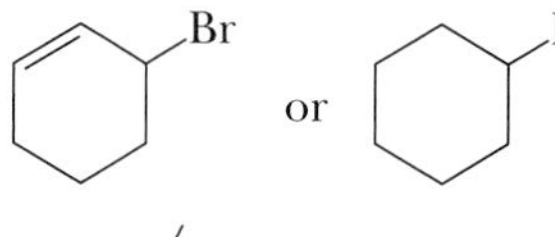

(b) 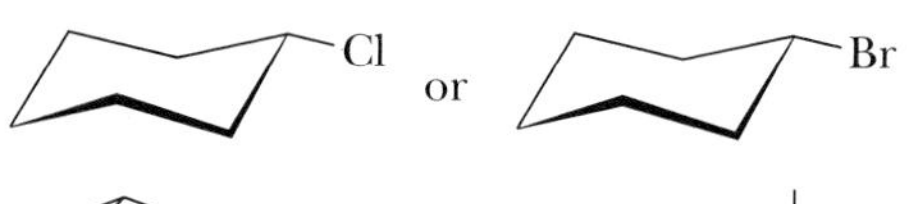

(c) 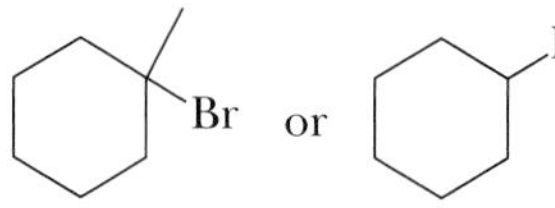

(d) 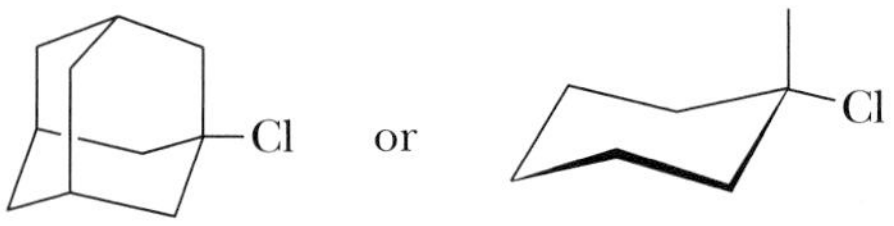

9.35 Account for the relative rates of solvolysis of these compounds in aqueous acetic acid.

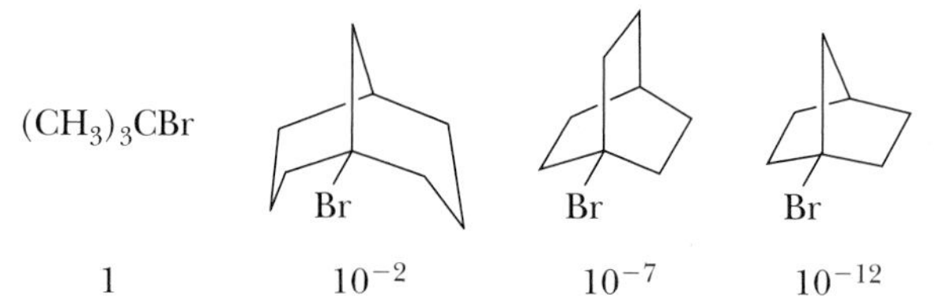

9.36 A comparison of the rates of S_N1 solvolysis of the bicyclic compounds (1) and (2) in acetic acid shows that compound (1) reacts 10^{11} times faster than compound (2). Furthermore, solvolysis of (1) occurs with complete retention of configuration: the nucleophile occupies the same position on the one-carbon bridge as did the leaving $-OSO_2Ar$ group.

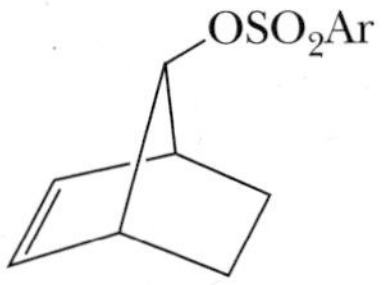
(1)

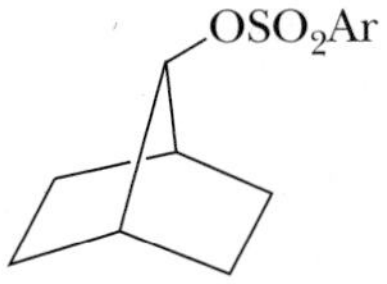
(2)

(a) Draw structural formulas for the products of solvolysis of each compound.
(b) Account for the difference in rate of solvolysis of (1) and (2).
(c) Account for complete retention of configuration in the solvolysis of (1).

β-Eliminations

9.37 Draw structural formulas for the alkene(s) formed by treatment of each haloalkane or halocycloalkane with sodium ethoxide in ethanol. Assume that elimination occurs by an E2 mechanism.

(a) 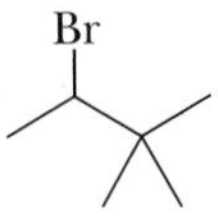 **(b)** 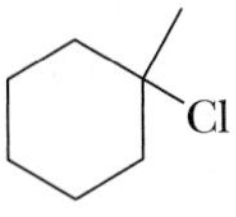 **(c)** 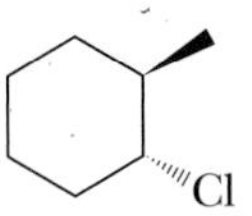 **(d)** 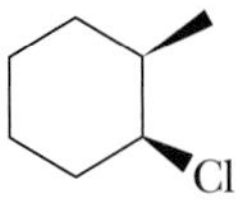

(e) 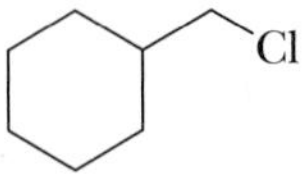 **(f)** 

9.38 Draw structural formulas of all chloroalkanes that undergo dehydrohalogenation when treated with KOH to give each alkene as the major product. For some parts, only one chloroalkane gives the desired alkene as the major product. For other parts, two chloroalkanes may work.

(a) 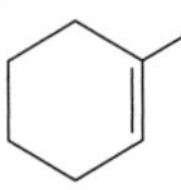 **(b)** 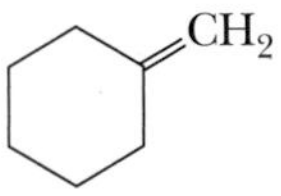 **(c)** 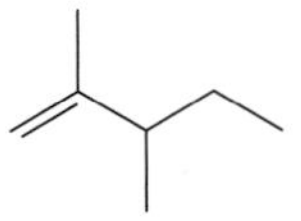 **(d)** 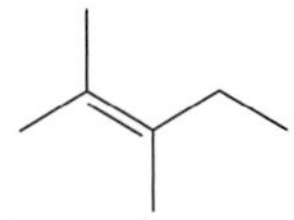 **(e)** 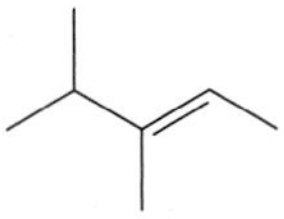

9.39 Following are diastereomers (A) and (B) of 3-bromo-3,4-dimethylhexane. On treatment with sodium ethoxide in ethanol, each gives 3,4-dimethyl-3-hexene as the major product. One diastereomer gives the *E* alkene, and the other gives the *Z* alkene. Which diastereomer gives which alkene? Account for the stereoselectivity of each β-elimination.

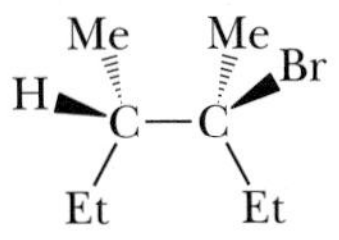
(A)

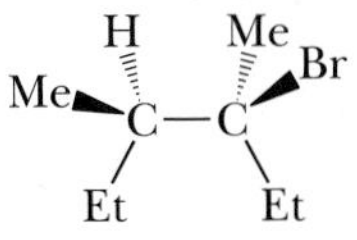
(B)

9.40 Treatment of the following stereoisomer of 1-bromo-1,2-diphenylpropane with sodium ethoxide in ethanol gives a single stereoisomer of 1,2-diphenylpropene. Predict whether the product has the *E* configuration or the *Z* configuration.

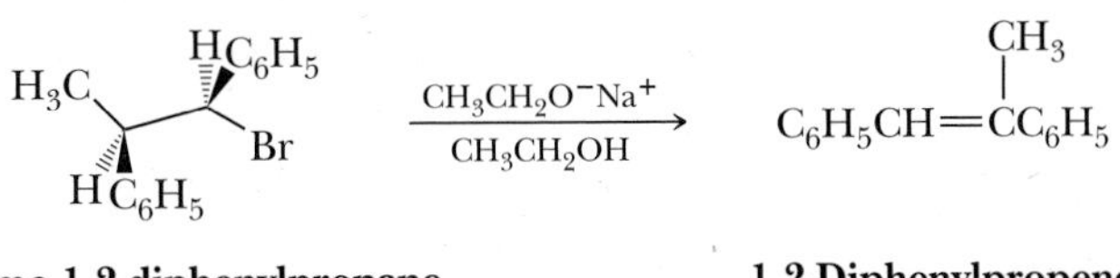

1-Bromo-1,2-diphenylpropane **1,2-Diphenylpropene**

9.41 Elimination of HBr from 2-bromonorbornane gives only 2-norbornene and no 1-norbornene. How do you account for the regioselectivity of this dehydrohalogenation? In answering this question, you will find it helpful to look at molecular models of both 1-norbornene and 2-norbornene and analyze the strain in each.

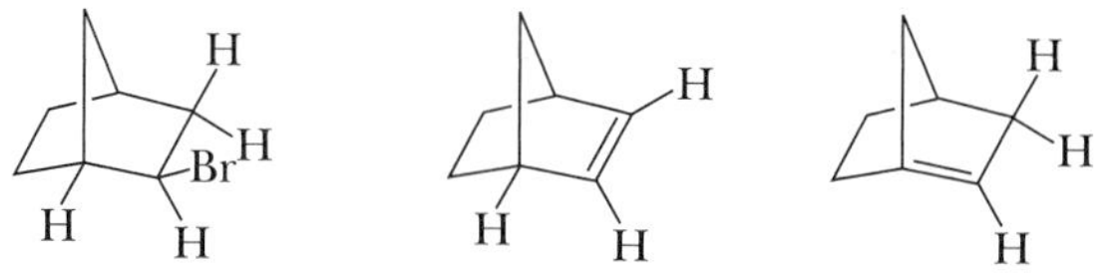

2-Bromonorbornane **2-Norbornene** **1-Norbornene**

9.42 Which isomer of 1-bromo-3-isopropylcyclohexane reacts faster when refluxed with potassium *tert*-butoxide, the *cis* isomer or the *trans* isomer? Draw the structure of the expected product from the faster-reacting compound.

Substitution Versus Elimination

9.43 Consider the following statements in reference to S_N1, S_N2, E1, and E2 reactions of haloalkanes. To which mechanism(s), if any, does each statement apply?

(a) Involves a carbocation intermediate.
(b) Is first order in haloalkane and first order in nucleophile.
(c) Involves inversion of configuration at the site of substitution.
(d) Involves retention of configuration at the site of substitution.
(e) Substitution at a stereocenter gives predominantly a racemic product.
(f) Is first order in haloalkane and zero order in base.
(g) Is first order in haloalkane and first order in base.
(h) Is greatly accelerated in protic solvents of increasing polarity.
(i) Rearrangements are common.
(j) Order of reactivity of haloalkanes is 3° > 2° > 1°.
(k) Order of reactivity of haloalkanes is methyl > 1° > 2° > 3°.

9.44 Arrange these haloalkanes in order of increasing ratio of E2 to S_N2 products observed on reaction of each with sodium ethoxide in ethanol.

(a) CH_3CH_2Br **(b)** $CH_3\overset{\overset{CH_3}{|}}{C}HCH_2Br$ **(c)** $CH_3\underset{\underset{Cl}{|}}{\overset{\overset{CH_3}{|}}{C}}CH_2CH_3$ **(d)** $CH_3\overset{\overset{CH_3}{|}}{C}HCH_2CH_2Br$

9.45 Draw a structural formula for the major organic product of each reaction and specify the most likely mechanism by which each is formed.

(a) 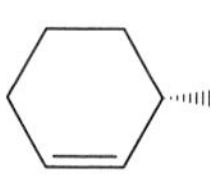 Br + CH_3OH $\xrightarrow[\text{methanol}]{}$ **(b)** $CH_3\underset{\underset{Cl}{|}}{\overset{\overset{CH_3}{|}}{C}}CH_2CH_3$ + NaOH $\xrightarrow[H_2O]{80°C}$

(c) 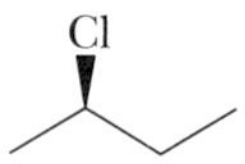 + $CH_3\overset{\overset{O}{||}}{C}O^-Na^+$ $\xrightarrow[\text{DMSO}]{}$

(*R*)-2-Chlorobutane

(d) 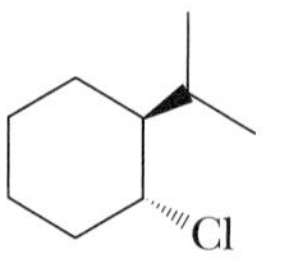 + $CH_3O^-Na^+$ $\xrightarrow[\text{methanol}]{}$ **(e)** 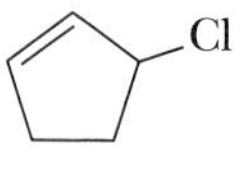 + NaI $\xrightarrow[\text{acetone}]{}$

R enantiomer

(f) $CH_3\overset{\overset{Cl}{|}}{C}HCH_2CH_3$ + $H\overset{\overset{O}{||}}{C}OH$ $\xrightarrow[\text{formic acid}]{}$ **(g)** $CH_3CH_2O^-Na^+$ + $CH_2{=}CHCH_2Cl$ $\xrightarrow[\text{ethanol}]{}$

R enantiomer

9.46 When *cis*-4-chlorocyclohexanol is treated with sodium hydroxide in ethanol, it gives mainly the substitution product *trans*-1,4-cyclohexanediol (1). Under the same reaction conditions, *trans*-4-chlorocyclohexanol gives 3-cyclohexenol (2) and the bicyclic ether (3).

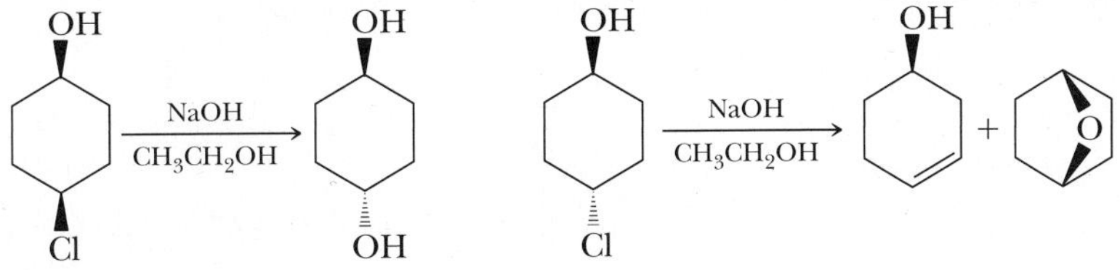

cis-**4-Chloro-cyclohexanol** (1) *trans*-**4-Chloro-cyclohexanol** (2) (3)

(a) Propose a mechanism for formation of product (1), and account for its configuration.
(b) Propose a mechanism for formation of product (2).
(c) Account for the fact that the bicyclic ether (3) is formed from the *trans* isomer but not from the *cis* isomer.

Synthesis

9.47 Show how to convert the given starting material into the desired product. Note that some syntheses require only one step, whereas others require two or more.

(a) 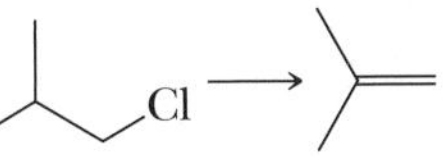

(b) 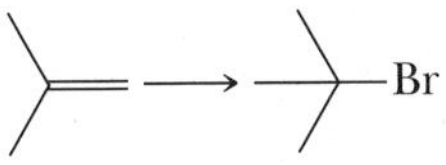

(c) 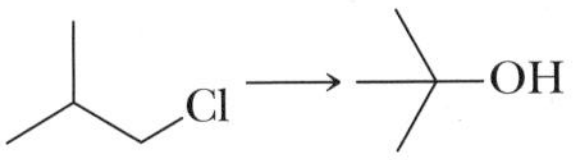

(d) 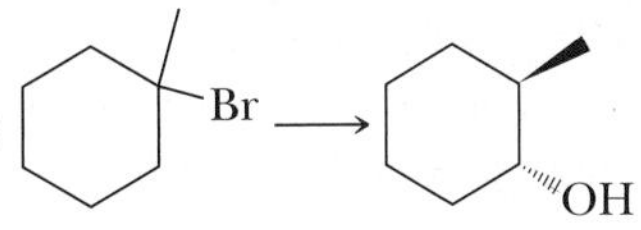

(e) 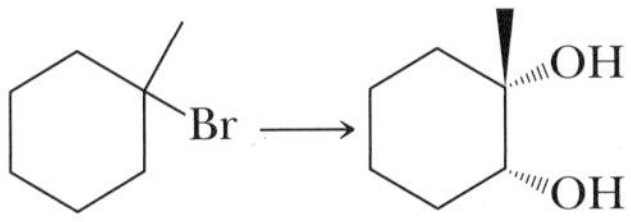

(f) 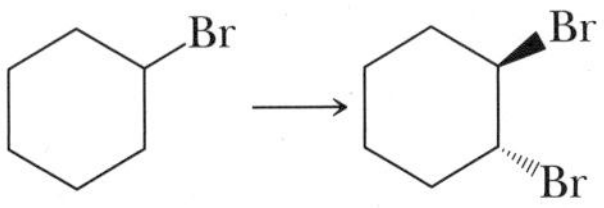

(g) 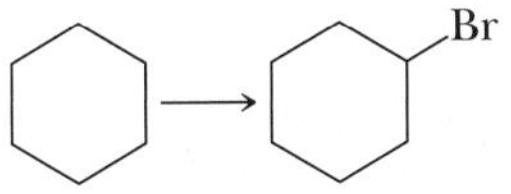

(h) 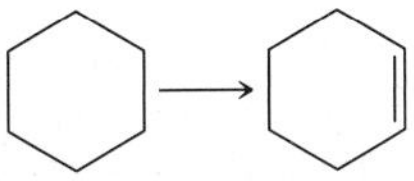

(i) 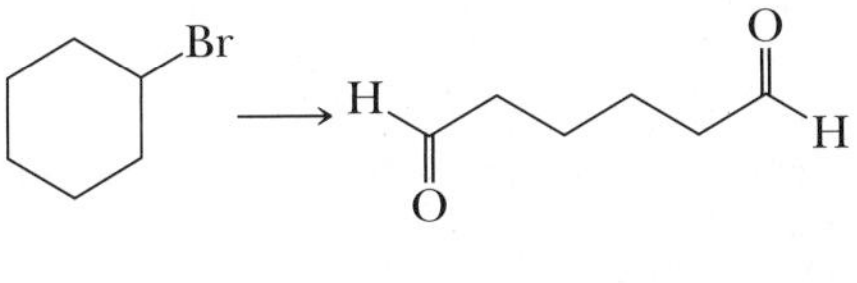

9.48 The Williamson ether synthesis involves treatment of a haloalkane with a metal alkoxide. Following are two reactions intended to give benzyl *tert*-butyl ether. One reaction gives the ether in good yield, and the other reaction does not. Which reaction gives the ether? What is the product of the other reaction, and how do you account for its formation?

(a) 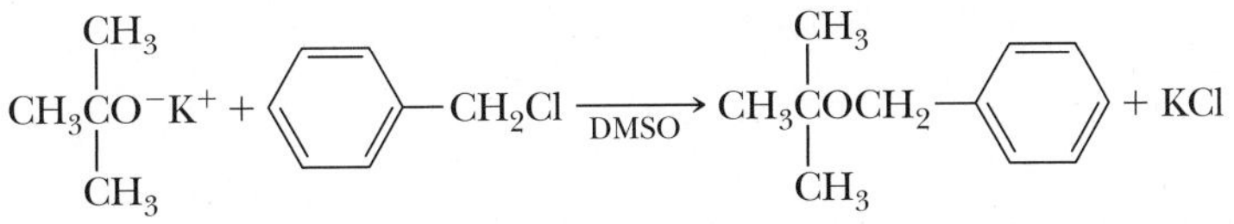

(b) 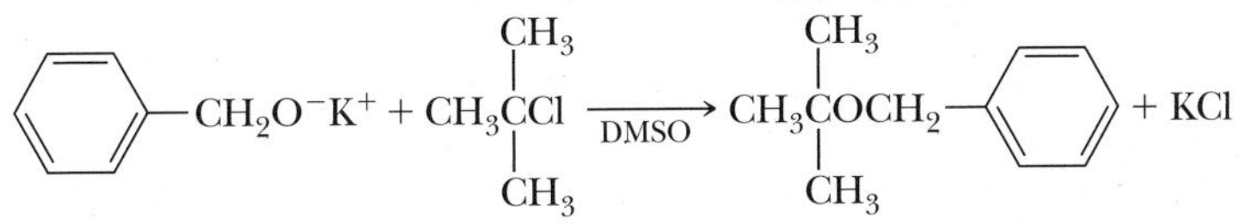

9.49 The following ethers can, in principle, be synthesized by two different combinations of haloalkane or halocycloalkane and metal alkoxide. Show one combination that forms ether bond (1) and another that forms ether bond (2). Which combination gives the higher yield of ether?

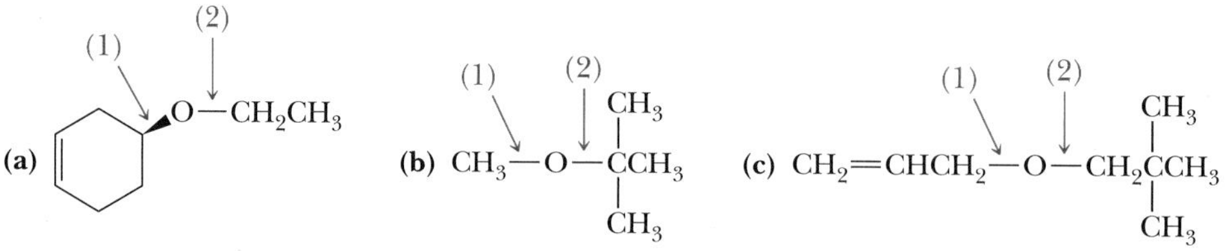

9.50 Propose a mechanism for this reaction.

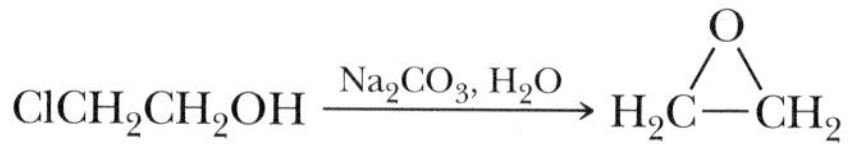

9.51 Each of these compounds can be synthesized by an S_N2 reaction. Suggest a combination of haloalkane and nucleophile that will give each product.

(a) CH_3OCH_3 **(b)** CH_3SH

(c) $CH_3CH_2CH_2PH_2$ **(d)** CH_3CH_2CN

(e) $CH_3SCH_2C(CH_3)_3$ **(f)** $(CH_3)_3NH^+ Cl^-$

(g) $C_6H_5\overset{O}{\overset{\|}{C}}OCH_2C_6H_5$ **(h)** (*R*)-$CH_3\underset{}{\overset{N_3}{\overset{|}{C}}}HCH_2CH_2CH_3$

(i) $CH_2{=}CHCH_2OCH(CH_3)_2$ **(j)** $CH_2{=}CHCH_2OCH_2CH{=}CH_2$

(k) 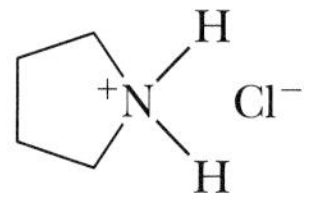

(l) 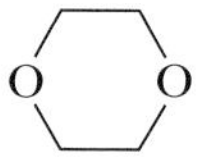

Looking Ahead

9.52 OH^- is a very poor leaving group. However, many alcohols react with alkyl or aryl sulfonyl chlorides to give sulfonate esters.

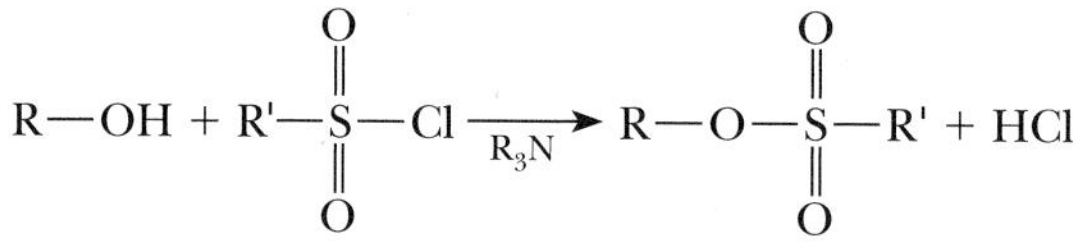

(a) Explain what this change does to the leaving group ability of the substituent.
(b) Suggest the product of the following reaction.

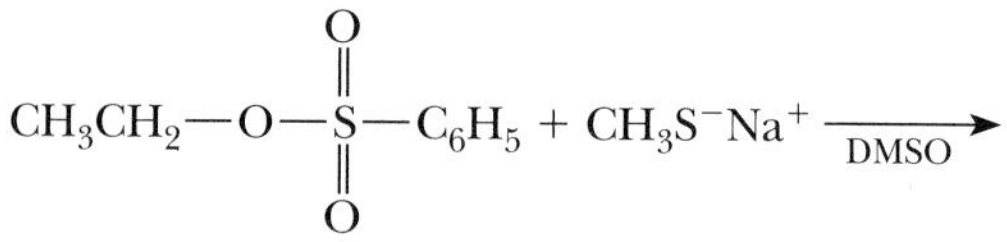

9.53 Suggest a product of the following reaction. HI is a very strong acid.

$$CH_3CH_2OCH_2CH_3 + 2\ HI \longrightarrow$$

Organic Chemistry Reaction Roadmap

Reaction Roadmap

9.54 Use the reaction roadmap you made for Problems 6.54, 7.29, and 8.28 and update it to contain the reactions in the "Study Guide" section as well as Table 9.1 of this chapter. Because of their highly specific nature, do not use reactions 3 and 5 or entry 7 of Table 9.1 on your roadmap.

9.55 Write the products of the following sequences of reactions. Refer to your reaction roadmap to see how the combined reactions allow you to "navigate" between the different functional groups.

(a) 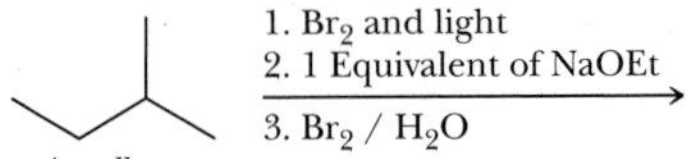
1. Br_2 and light
2. 1 Equivalent of NaOEt
3. Br_2 / H_2O
An alkane

(b) 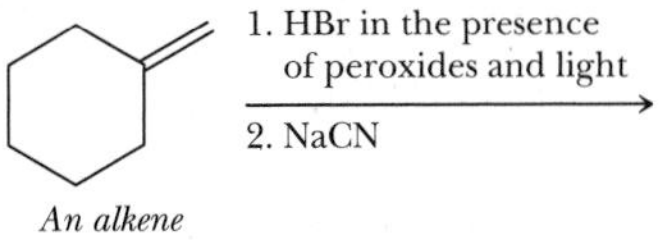
1. HBr in the presence of peroxides and light
2. NaCN
An alkene

(c) 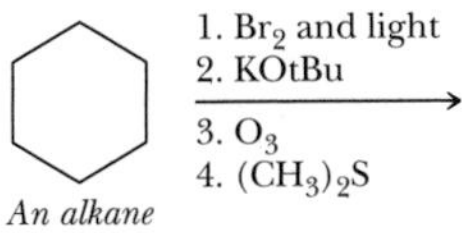
1. Br_2 and light
2. KOtBu
3. O_3
4. $(CH_3)_2S$
An alkane

(d) 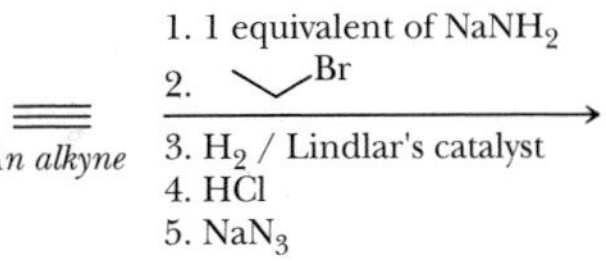
1. 1 equivalent of $NaNH_2$
2. Br
3. H_2 / Lindlar's catalyst
4. HCl
5. NaN_3
An alkyne

Multistep Synthesis Problems

Some reaction sequences are more useful than others in organic synthesis. Among the reactions you have a learned thus far, a particularly useful sequence involves the combination of free radical halogenation of an alkane to give a haloalkane, which is then subjected to an E2 elimination to give an alkene. The alkene is then converted to a variety of possible functional groups. Note that free radical halogenation is the only reaction you have seen that uses an alkane as a starting material.

9.56 Using your reaction roadmap as a guide, show how to convert butane into 2-butyne. Show all reagents and all molecules synthesized along the way.

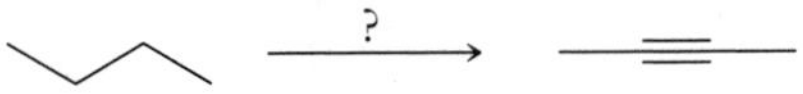
?

Butane **2-Butyne**

9.57 Using your reaction roadmap as a guide, show how to convert 2-methylbutane into racemic 3-bromo-2-methyl-2-butanol. Show all reagents and all molecules synthesized along the way.

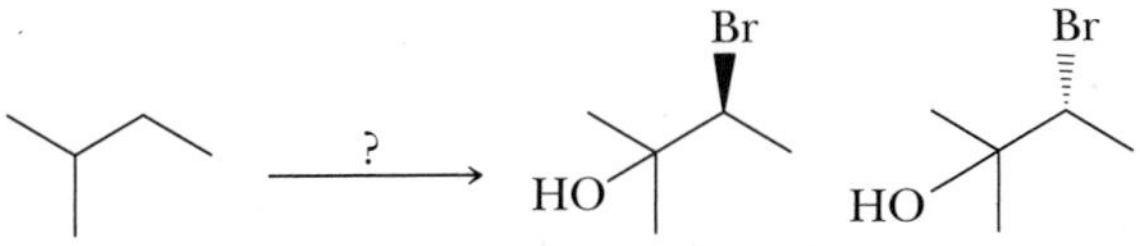
?
Br
HO
Br
HO

2-Methylbutane **3-Bromo-2-methyl-2-butanol** (racemic)

9.58 Using your reaction roadmap as a guide, show how to convert cyclohexane into hexanedial. Show all reagents and all molecules synthesized along the way.

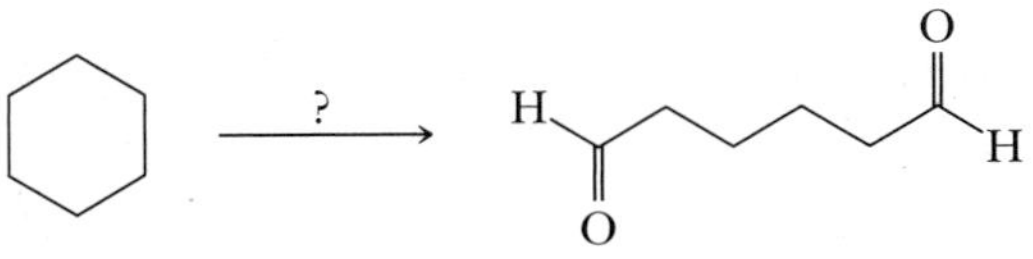
?
H
O
O
H

Cyclohexane **Hexanedial**

9.59 Using your reaction roadmap as a guide, show how to convert cyclohexane into racemic 3-bromocyclohexene. Show all reagents and all molecules synthesized along the way.

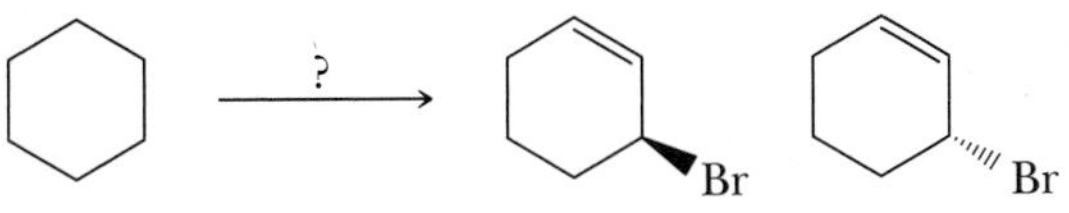
?
Br
Br

Cyclohexane **3-Bromocyclohexene** (racemic)

Reaction Roadmap

9.60 Another important pattern in organic synthesis is the construction of C—C bonds. Using your reaction roadmap as a guide, show how to convert propane into hex-1-en-4-yne. You must use propane as the source of all of the carbon atoms in the hex-1-en-4-yne product. Show all reagents needed and all molecules synthesized along the way.

Propane **Hex-1-en-4-yne**

Reaction Roadmap

9.61 Using your reaction roadmap as a guide, show how to convert propane into butyronitrile. You must use propane and sodium cyanide as the source of all of the carbon atoms in the butyronitrile product. Show all reagents and all molecules synthesized along the way.

Propane **Sodium cyanide** **Butyronitrile**

Reactions in Context

9.62 Fluticasone is a glucocorticoid drug that has been used to treat asthma. In the synthesis of fluticasone, the following transformation is used that involves a limiting amount of sodium iodide. Analyze the structure using the chemistry you learned in this chapter and draw the product of the reaction.

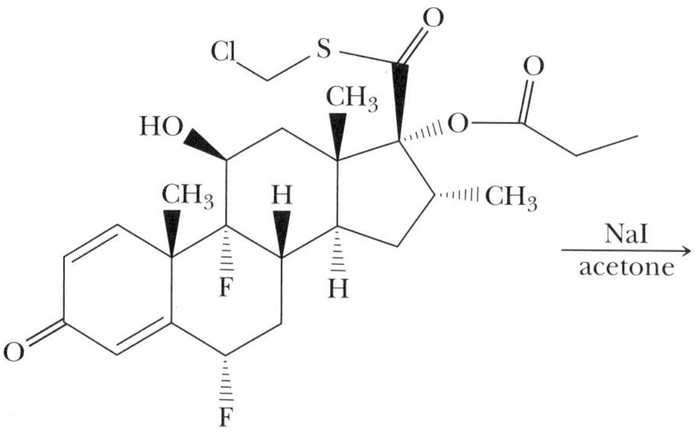

9.63 The following reaction sequence was used in the synthesis of several derivatives of prostaglandin C2. Analyze the structure using the chemistry you learned in this chapter and draw the structures of the synthetic intermediates A and B.

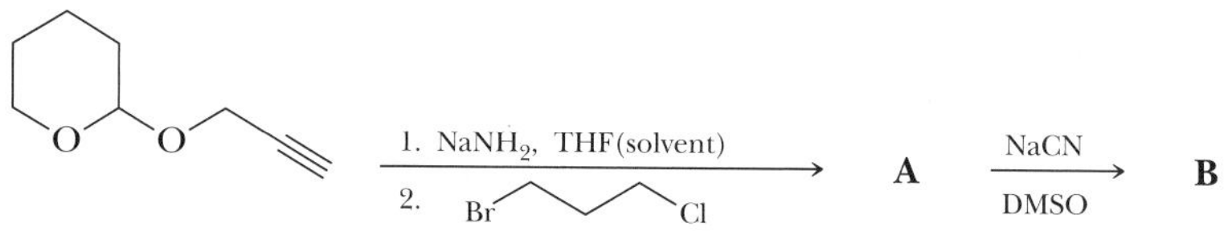

(racemic) (racemic) (racemic)

9.64 The following reaction was used in the synthesis of various prostaglandin derivatives. Analyze the structure using the chemistry you learned in this chapter and draw the product of the reaction.

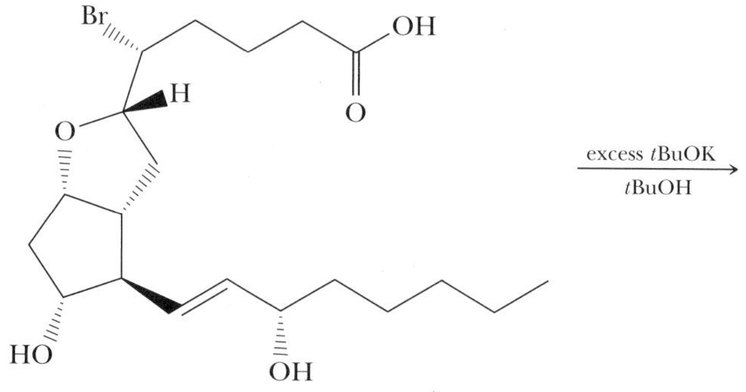

10

Alcohols

Fermentation vats of wine grapes at the Beaulieu Vineyards, California. **Inset:** a model of ethanol. (Ian Shaw/ Alamy Stock Photo)

Outline

10.1 Structure and Nomenclature of Alcohols
10.2 Physical Properties of Alcohols
10.3 Acidity and Basicity of Alcohols
10.4 Reaction of Alcohols with Active Metals
10.5 Conversion of Alcohols to Haloalkanes and Sulfonates
10.6 Acid-Catalyzed Dehydration of Alcohols
10.7 The Pinacol Rearrangement
10.8 Oxidation of Alcohols
10.9 Thiols

In this chapter, we study the physical and chemical properties of alcohols, a class of compounds containing the —OH (hydroxyl) group. We also study thiols, a class of compounds containing the —SH (sulfhydryl) group.

CH_3CH_2OH	CH_3CH_2SH
Ethanol	**Ethanethiol**
(an alcohol)	(a thiol)

Ethanol is the additive in the fuel blend known as E85, the alcohol in alcoholic beverages, and an important industrial solvent. Ethanethiol, like all other low-molecular-weight thiols, has a stench; such smells from skunks, rotten eggs, and sewage are caused by thiols or H_2S.

Alcohols are important because they can be converted into many other types of compounds, including alkenes, haloalkanes, aldehydes, ketones, carboxylic acids, and esters. Not only can alcohols be converted to these compounds, but these compounds can also be converted to alcohols. Thus, alcohols play a central role in the interconversion of organic functional groups.

Hydroxyl groups are found in carbohydrates and certain amino acids. Following are two representations for glucose, the most abundant organic compound in nature. On the left is a Fischer projection showing the configuration of all chiral centers. On the right is a cyclic structure, the predominant form in which this molecule exists in both the solid form and in solution. The amino acid L-serine is one of the 20 amino acid building blocks of proteins.

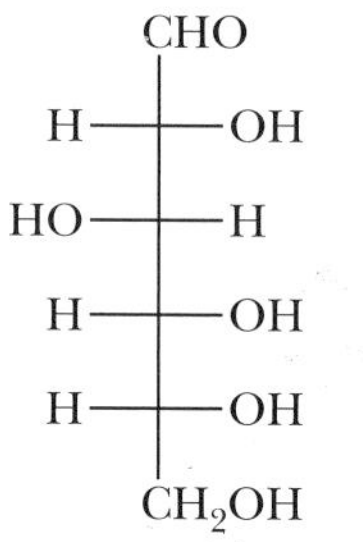

D-Glucose
(Fischer projection)

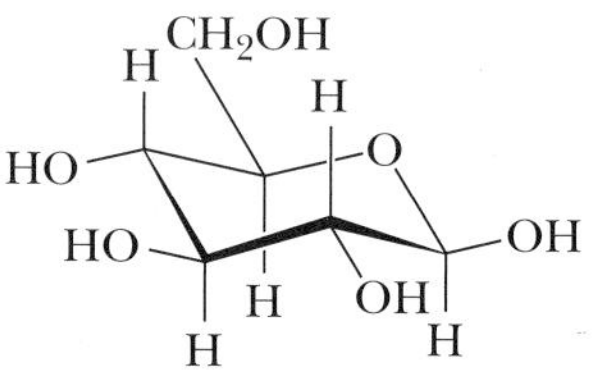

β-D-Glucopyranose

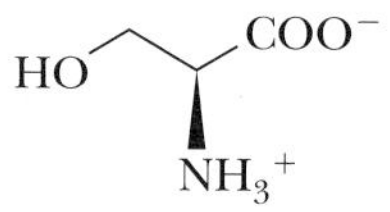

L-Serine

Because sulfur and oxygen are both Group 6 elements, thiols and alcohols undergo many of the same types of reactions. Sulfur, a third-row element, however, can undergo some reactions that are not possible for alcohols. In addition, sulfur's electronegativity and basicity are less than those of oxygen.

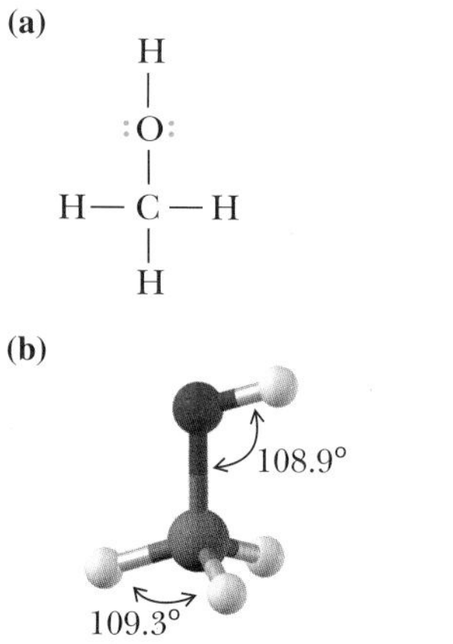

FIGURE 10.1 Methanol, CH_3OH. (a) Lewis structure and (b) ball-and-stick model.

10.1 Structure and Nomenclature of Alcohols

A. Structure

The functional group of an alcohol is an **—OH (hydroxyl) group** (Section 1.3A) bonded to an sp^3 hybridized carbon. The oxygen atom of an alcohol is also sp^3 hybridized. Two sp^3 hybrid orbitals of oxygen form σ bonds to atoms of carbon and hydrogen, and the remaining two sp^3 hybrid orbitals each contain an unshared pair of electrons. Figure 10.1 shows a Lewis structure and a ball-and-stick model of methanol, CH_3OH, the simplest alcohol. The measured C—O—H bond angle in methanol is 108.9°, very close to the perfectly tetrahedral angle of 109.5°.

B. Nomenclature

In the IUPAC system, the longest chain of carbon atoms containing the —OH group is selected as the parent alkane and numbered from the end closer to —OH. To show that the compound is an alcohol, change the suffix *-e* of the parent alkane to *-ol* (Section 2.3) and use a number to show the location of the —OH group. The location of the —OH group takes precedence over alkyl groups and halogen atoms in numbering the parent chain. For cyclic alcohols, numbering begins with the carbon bearing the —OH group. Because the —OH group is understood to be on carbon 1 of the ring, there is no need to give its location a number. In complex alcohols, the number for the hydroxyl group is often placed between the infix and the suffix. Thus, for example, both 2-methyl-1-propanol and 2-methylpropan-1-ol are acceptable names.

Common names for alcohols are derived by naming the alkyl group bonded to —OH and then adding the word *alcohol*. Here are IUPAC names and, in parentheses, common names for several low-molecular-weight alcohols.

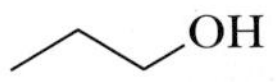

1-Propanol
(Propyl alcohol)

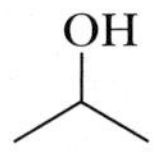

2-Propanol
(Isopropyl alcohol)

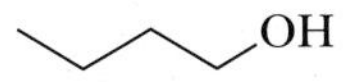

1-Butanol
(Butyl alcohol)

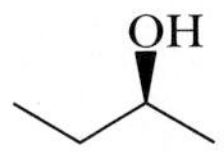

(*S*)-2-Butanol
((*S*)-*sec*-Butyl alcohol)

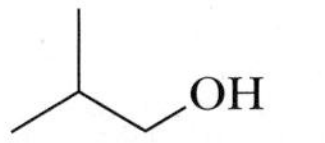

2-Methyl-1-propanol
(Isobutyl alcohol)

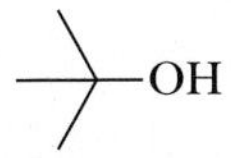

2-Methyl-2-propanol
(*tert*-Butyl alcohol)

Example 10.1 Alcohol Nomenclature I

Write IUPAC names for these alcohols.

(a) 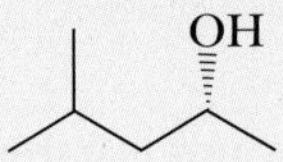 **(b)** 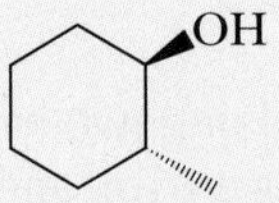

Solution

(a) (*R*)-4-Methyl-2-pentanol.
(b) (1*R*,2*R*)-2-Methylcyclohexanol. Note that the designation of the configuration as *R*,*R* specifies not only the absolute configuration of each chiral center but also the fact that the —CH_3 and —OH groups are *trans* to each other on the ring. The alcohol can also be named *trans*-2-methylcyclohexanol, and while this name specifies that the hydroxyl and methyl groups are *trans* to each other, it does not specify the absolute configuration of either group.

Problem 10.1

Write IUPAC names for these alcohols and include the configuration for (a).

(a) 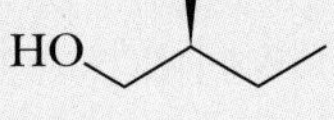 **(b)** 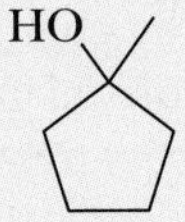

We classify alcohols as **primary (1°)**, **secondary (2°)**, **or tertiary (3°)**, depending on whether the —OH group is on a primary, secondary, or tertiary carbon.

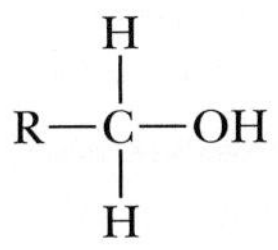

Primary (1°)

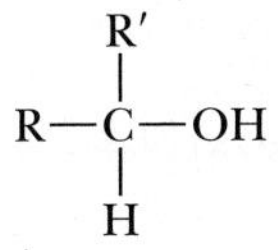

Secondary (2°)

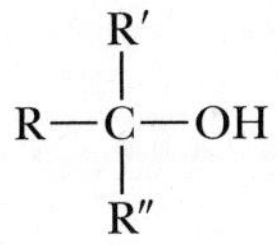

Tertiary (3°)

Example 10.2 Classification of Alcohols

Classify each alcohol as primary, secondary, or tertiary.

(a) 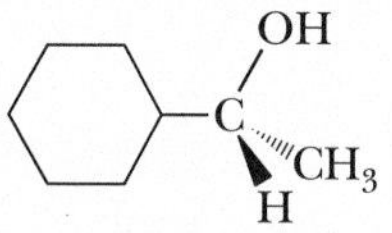

(b) 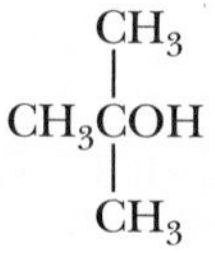

(c) 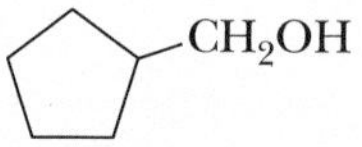

Solution

(a) Secondary (2°) **(b)** Tertiary (3°) **(c)** Primary (1°)

Problem 10.2

Classify each alcohol as primary, secondary, or tertiary.

(a) 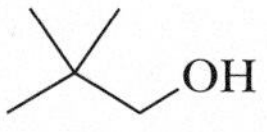

(b) 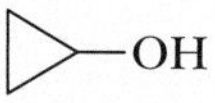

(c) 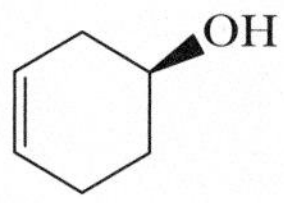

(d) 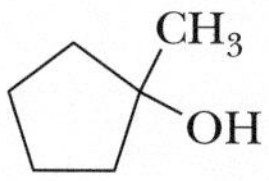

Diol A compound containing two hydroxyl groups.

Triol A compound containing three hydroxyl groups.

In the IUPAC system, a compound containing two hydroxyl groups is named as a **diol**, one containing three hydroxyl groups as a **triol**, and so on. In IUPAC names for diols, triols, and so on, the final *-e* (the suffix) of the parent alkane name is retained, as, for example, in the name 1,2-ethanediol. As with many organic compounds, common names for certain diols and triols have persisted. Compounds containing hydroxyl groups on adjacent carbons are often referred to as **glycols** (Section 6.5). Ethylene glycol and propylene glycol are synthesized from ethylene and propylene, respectively, hence their common names.

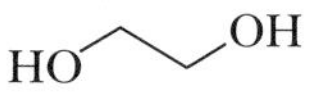

1,2-Ethanediol (Ethylene glycol)

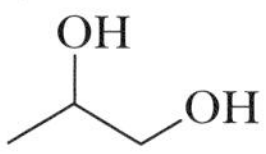

1,2-Propanediol (Propylene glycol)

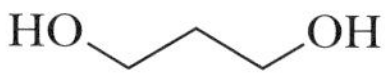

1,3-Propanediol

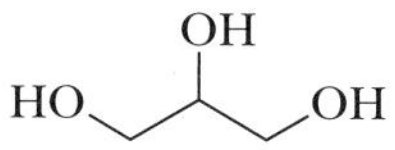

1,2,3-Propanetriol (Glycerol, glycerine)

Compounds containing —OH and C=C groups are often referred to as **unsaturated alcohols** because of the presence of the carbon-carbon double bond. In the IUPAC system, the double bond is shown by changing the infix of the parent alkane from *-an-* to *-en-* (Section 2.3) and the hydroxyl group is shown by changing the suffix of the parent alkane from *-e* to *-ol*. Numbers must be used to show the location of both the carbon-carbon double bond and the hydroxyl group. The parent alkane is numbered to give the —OH group the lowest possible number; that is, the group shown by a suffix (in this case, *-ol*) takes precedence over the group shown by an infix (in this case, *-en-*).

Example 10.3 Alcohol Nomenclature II

Write IUPAC names for these unsaturated alcohols.

(a) $CH_2{=}CHCH_2OH$ **(b)** 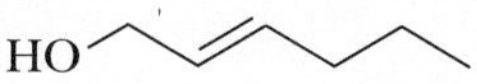

Solution

(a) 2-Propen-1-ol. Its common name is allyl alcohol.

(b) (*E*)-2-Hexen-1-ol (*trans*-2-Hexen-1-ol).

Problem 10.3

Write IUPAC names for these unsaturated alcohols.

(a) 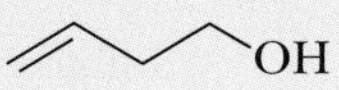

(b) 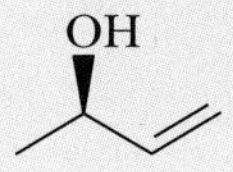

10.2 Physical Properties of Alcohols

Because of the presence of the polar —OH group, alcohols are polar compounds, with partial positive charges on carbon and hydrogen and a partial negative charge on oxygen (Figure 10.2).

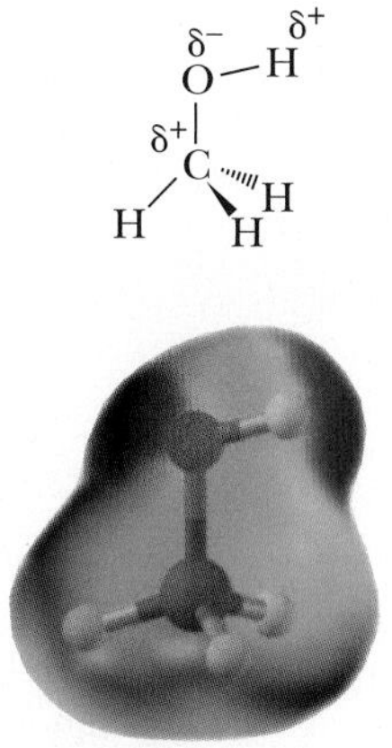

FIGURE 10.2 Polarity of the C—O—H bond in an alcohol.

The attraction between the positive end of one dipole and the negative end of another is called **dipole-dipole interaction**. When the positive end of one of the dipoles is a hydrogen atom bonded to O or N (atoms of high electronegativity) and the negative end of the other dipole is an O or N atom, the attractive interaction between dipoles is particularly strong and is given the special name of **hydrogen bonding**. The length of a hydrogen bond in water is 177 pm, about 80% longer than an O—H covalent bond. The strength of a hydrogen bond in water is approximately 21 kJ (5 kcal)/mol. For comparison, the strength of the O—H covalent bond in water is approximately 498 kJ (118 kcal)/mol. As can be seen by comparing these numbers, an O—H hydrogen bond is considerably weaker than an O—H covalent bond. The presence of a large number of hydrogen bonds in liquid water, however, has an important cumulative effect on the physical properties of water. Because of hydrogen bonding, extra energy is required to separate each water molecule from its neighbors, hence the relatively high boiling point of water.

Dipole-dipole interaction

The attraction between the positive end of one dipole and the negative end of another.

Hydrogen bonding

The attractive interaction between a hydrogen atom bonded to an atom of high electronegativity (most commonly O or N) and a lone pair of electrons on another atom of high electronegativity (again, most commonly O or N).

Similarly, there is extensive hydrogen bonding between alcohol molecules in the pure liquid. Figure 10.3 shows the association of ethanol molecules by hydrogen bonding between the partially negative oxygen atom of one ethanol molecule and the partially positive hydrogen atom of another ethanol molecule.

Table 10.1 lists the boiling points and solubilities in water for several groups of alcohols and hydrocarbons of similar molecular weight. Of the compounds compared

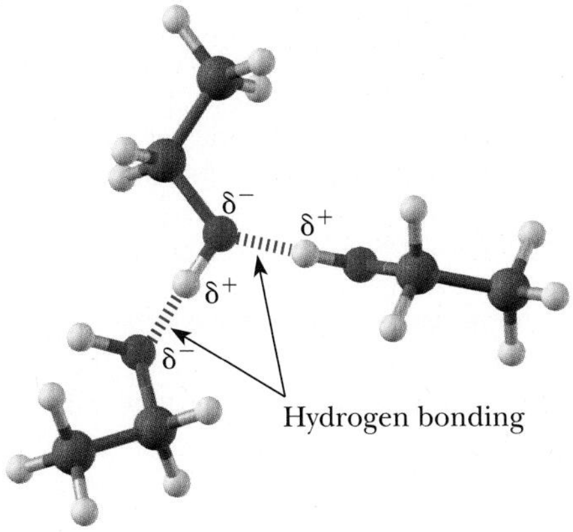

Watch a video explanation

FIGURE 10.3 The association of ethanol molecules in the liquid state by hydrogen bonding. Each O—H can participate in up to three hydrogen bonds (one through hydrogen and two through oxygen). Only two of the three possible hydrogen bonds per molecule are shown.

Table 10.1 Boiling Points and Solubilities in Water of Five Groups of Alcohols and Hydrocarbons of Similar Molecular Weight

Structural Formula	Name	Molecular Weight (g/mol)	Boiling Point (°C)	Solubility in Water
CH_3OH	Methanol	32	65	Infinite
CH_3CH_3	Ethane	30	−89	Insoluble
CH_3CH_2OH	Ethanol	46	78	Infinite
$CH_3CH_2CH_3$	Propane	44	−42	Insoluble
$CH_3CH_2CH_2OH$	1-Propanol	60	97	Infinite
$CH_3CH_2CH_2CH_3$	Butane	58	0	Insoluble
$CH_3CH_2CH_2CH_2OH$	1-Butanol	74	117	8 g/100 g
$CH_3CH_2CH_2CH_2CH_3$	Pentane	72	36	Insoluble
$HOCH_2CH_2CH_2CH_2OH$	1,4-Butanediol	90	230	Infinite
$CH_3CH_2CH_2CH_2CH_2OH$	1-Pentanol	88	138	2.3 g/100 g
$CH_3CH_2CH_2CH_2CH_2CH_3$	Hexane	86	69	Insoluble

in each group, the alcohols have the higher boiling points because more energy is needed to overcome the attractive forces of hydrogen bonding between their polar —OH groups. The presence of additional hydroxyl groups in a molecule further increases the extent of hydrogen bonding, as can be seen by comparing the boiling points of hexane (bp 69°C), 1-pentanol (bp 138°C), and 1,4-butanediol (bp 230°C), all of which have approximately the same molecular weight. Because of increased dispersion forces between larger molecules, boiling points of all types of compounds, including alcohols, increase with increasing molecular weight. Compare, for example, the boiling points of ethanol (bp 78°C), 1-propanol (bp 97°C), 1-butanol (bp 117°C), and 1-pentanol (bp 138°C).

Connections to Biological Chemistry

The Importance of Hydrogen Bonding in Drug-Receptor Interactions

Hydrogen bonds have directionality in that the donor and acceptor groups must be oriented appropriately with respect to each other for hydrogen bonding to occur. Important hydrogen bond donors in biological molecules include —OH groups (proteins, carbohydrates) and —NH groups (proteins, nucleic acids). Important hydrogen bond acceptors are any N or O with a lone pair of electrons, such as C=O groups (proteins, carbohydrates, nucleic acids), —OH groups (proteins, carbohydrates), and COO^- groups (proteins).

With directionality comes the potential for hydrogen bonds to organize molecules at many levels ranging from the folding of biological molecules to the specific binding and recognition between a pharmaceutical and its receptor. The drug atorvastatin (Lipitor) is used to treat high cholesterol. Cholesterol is synthesized in the liver from the two-carbon acetyl group of acetyl coenzyme A (acetyl-CoA). A key intermediate in the sequence of reactions leading to the synthesis of cholesterol is a six-carbon molecule named mevalonate (Section 26.4B). Atorvastatin specifically binds to and blocks the action of HMG-CoA reductase, a key enzyme in the biosynthesis of mevalonate. Atorvastatin binds to this enzyme in preference to the large number of other potential enzyme targets because (1) the drug has a shape complementary to the catalytic cavity (the active site) of HMG-CoA reductase (Figure 1) and (2) it can form at least nine specific hydrogen bonds with functional groups at the active site on the enzyme (Figure 2).

The complementary shape and pattern of hydrogen bonding ensure that atorvastatin binds to HMG-CoA reductase and inhibits its ability to catalyze the formation of mevalonate. The hallmark of this and other effective drugs is their ability to bind strongly with their intended target molecules, while at the same time not interacting with other molecules that could lead to unwanted side effects.

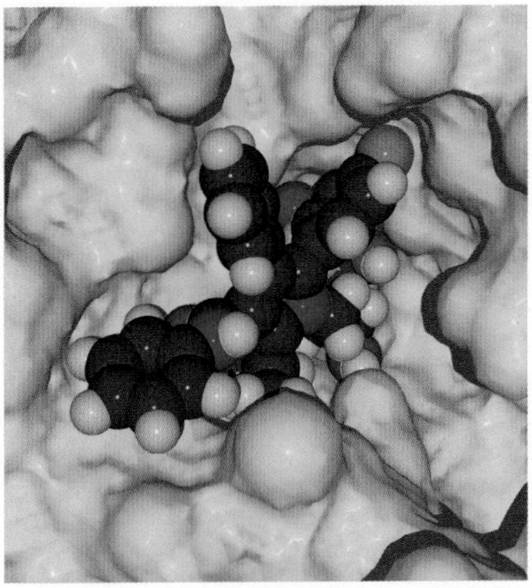

FIGURE 1 A space-filling model of the cholesterol-lowering drug atorvastatin (Lipitor) bound to the active site of its enzyme target HMG-CoA reductase (shown as a yellow surface). The shape of the drug is complementary to the active site of the enzyme.

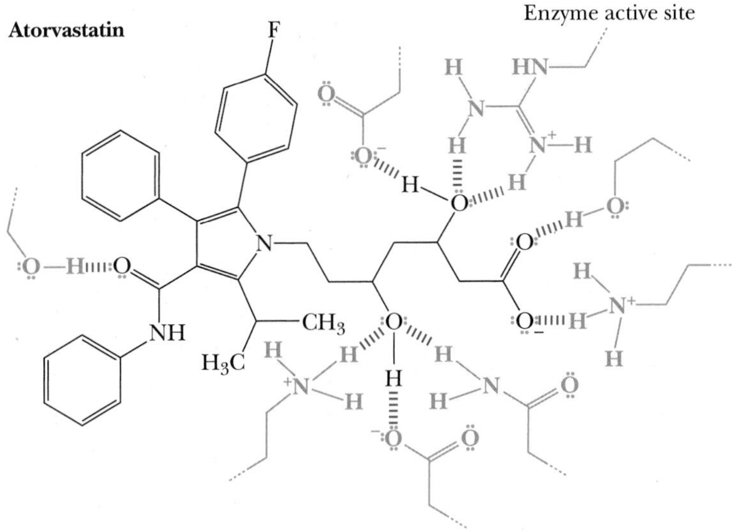

FIGURE 2 Hydrogen bonding (shown in red) between atorvastatin and the functional groups at the active site of the enzyme HMG-CoA reductase. The nine hydrogen bonds (shown in red), many of which involve hydroxyl groups on atorvastatin or the enzyme surface, help to provide the specificity that directs the binding of the drug to its target enzyme.

The effect of hydrogen bonding in alcohols is illustrated dramatically by comparing the boiling points of ethanol (bp 78°C) and its constitutional isomer dimethyl ether (bp −24°C). The difference in boiling point between these two compounds is caused by the presence of a polar O—H group in the alcohol, which is capable of forming intermolecular hydrogen bonds. This hydrogen bonding increases the attractive forces between molecules of ethanol; thus, ethanol has a higher boiling point than dimethyl ether.

CH_3CH_2OH	CH_3OCH_3
Ethanol	**Dimethyl ether**
bp 78°C	bp −24°C

Because alcohols can interact by hydrogen bonding with water, they are more soluble in water than alkanes and alkenes of comparable molecular weight. Methanol, ethanol, and 1-propanol are soluble in water in all proportions. As molecular weight increases, the physical properties of alcohols become more like those of hydrocarbons of comparable molecular weight. Higher-molecular-weight alcohols are much less soluble in water because of the increase in size of the hydrocarbon portion of their molecules.

Example 10.4 Alcohols and Boiling Points

Following are three alcohols with the molecular formula $C_4H_{10}O$. Their boiling points, from lowest to highest, are 82.3°C, 99.5°C, and 117°C. Which alcohol has which boiling point?

(Continued)

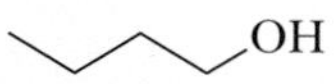 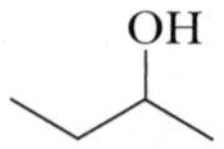 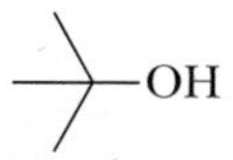

1-Butanol **2-Butanol** **2-Methyl-2-propanol**

Solution

Boiling points of these constitutional isomers depend on the strength of intermolecular hydrogen bonding. The primary —OH group of 1-butanol is most accessible for intermolecular hydrogen bonding; this alcohol has the highest boiling point, 117°C. The tertiary —OH group of 2-methyl-2-propanol is least accessible for intermolecular hydrogen bonding; this alcohol has the lowest boiling point, 82.3°C.

Problem 10.4

Arrange these compounds in order of increasing boiling point.

Heptane **1,5-Pentanediol** **1-Hexanol**

Example 10.5 Solubility of Alcohols

Arrange these compounds in order of increasing solubility in water.

 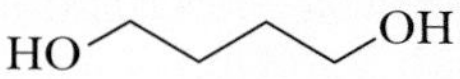 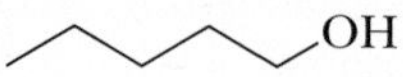

Hexane **1,4-Butanediol** **1-Pentanol**

Solution

Hexane, C_6H_{14}, a nonpolar hydrocarbon, has the lowest solubility in water. Both 1-pentanol and 1,4-butanediol are polar compounds due to the presence of —OH groups, and each interacts with water molecules by hydrogen bonding. Because 1,4-butanediol has more sites within its molecules for hydrogen bonding than 1-pentanol, the diol is more soluble in water than is 1-pentanol. The water solubilities of these compounds are given in Table 10.1.

 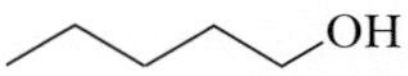 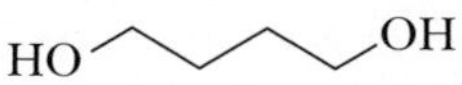

Insoluble 2.3 g/100 mL Infinitely soluble

Problem 10.5

Arrange these compounds in order of increasing solubility in water.

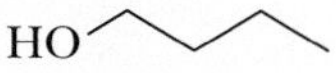  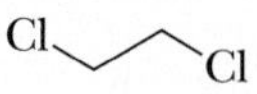

1-Butanol **1-Propanol** **1,2-Dichloroethane**

10.3 Acidity and Basicity of Alcohols

Alcohols can function as both weak acids (proton donors) and weak bases (proton acceptors). Table 10.2 lists the acid ionization constants for several low-molecular-weight alcohols.

In dilute aqueous solution, only methanol (pK_a 15.5) is more acidic than water.

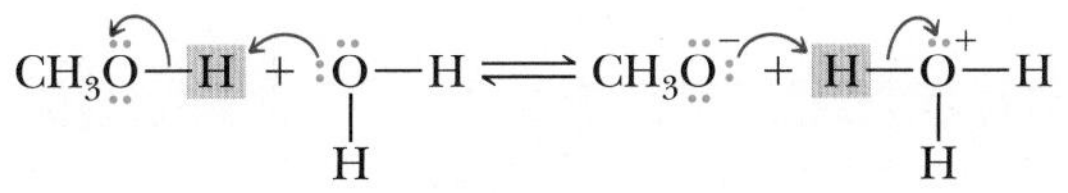

$$K_a = \frac{[CH_3O^-][H_3O^+]}{[CH_3OH]} = 10^{-15.5}$$

$$pK_a = 15.5$$

Ethanol has about the same acidity as water. Higher-molecular-weight, water-soluble alcohols are slightly weaker acids than water. Thus, although alcohols have some acidity, they are not strong enough acids to react with weak bases such as sodium bicarbonate or sodium carbonate. (At this point, it would be wise to review Section 4.4, which discusses the position of equilibrium in acid-base reactions.)

For simple alcohols such as methanol and ethanol, acidity depends primarily on the degree of solvation and stabilization of the alkoxide ion by water molecules. The negatively charged oxygen atoms of the methoxide and ethoxide ions are almost as accessible for solvation as the hydroxide ion is; therefore, these alcohols are about as acidic as water. As the bulk of the alkyl group bonded to oxygen increases, the ability of water molecules to solvate the alkoxide ion decreases. 2-Methyl-2-propanol (*tert*-butyl alcohol) is a weaker acid than either methanol or ethanol, primarily because of the bulk of the *tert*-butyl group, which reduces solvation of the *tert*-butoxide anion by surrounding water molecules.

In the presence of strong acids, the oxygen atom of an alcohol is a base and reacts with an acid by proton transfer to form an oxonium ion.

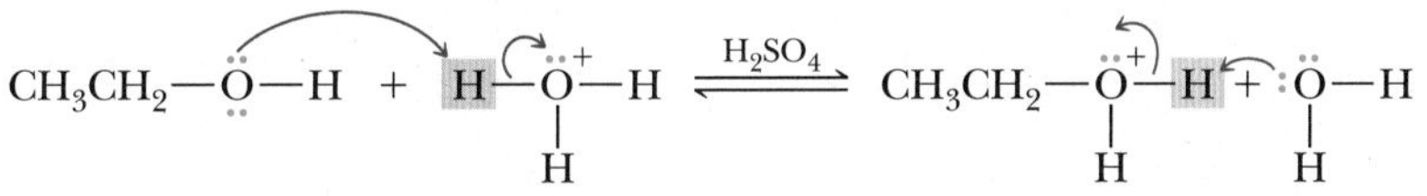

Ethanol — **Hydronium ion** pK_a −1.7 (weaker acid) — **Ethyloxonium ion** pK_a −2.4 (stronger acid)

To summarize, when trying to predict the mechanisms of reactions involving a hydroxyl group, you need to keep in mind that it is both a weak acid and a weak base;

Table 10.2 pK_a Values for Selected Alcohols in Dilute Aqueous Solution*

Compound	Structural Formula	pK_a	
Hydrogen chloride	HCl	−7	Stronger acid
Acetic acid	CH_3COOH	4.8	
Methanol	CH_3OH	15.5	↑
Water	H_2O	15.7	
Ethanol	CH_3CH_2OH	15.9	
2-Propanol	$(CH_3)_2CHOH$	17	Weaker acid
2-Methyl-2-propanol	$(CH_3)_3COH$	18	

*Also given for comparison are pK_a values for water, acetic acid, and hydrogen chloride.

so consider adding a proton or taking a proton away in the initial steps of mechanisms when a strong acid or base is present, respectively. In addition, an important mechanistic theme in many of the reactions of alcohols is that the —OH group, a poor leaving group, reacts with protons or a variety of strong electrophiles to create $—\overset{+}{O}H_2$ or analogous group, a much better leaving group, enabling subsequent substitution or elimination reactions to take place.

10.4 Reaction of Alcohols with Active Metals

Alcohols react with Li, Na, K, and other active metals to liberate hydrogen and form metal alkoxides. In this oxidation/reduction reaction, Na is oxidized to Na^+ and H^+ is reduced to H_2.

$$2\,CH_3OH + 2\,Na \longrightarrow 2\,CH_3O^-Na^+ + H_2$$

Sodium methoxide
(MeO^-Na^+)

To name a metal alkoxide, name the cation first, followed by the name of the anion. The name of the anion is derived from the prefix showing the number of carbon atoms and their arrangement (*meth-*, *eth-*, *isoprop-*, tert-*but-*, and so on) followed by the suffix *-oxide*.

Alkoxide ions are nearly the same or somewhat stronger bases than the hydroxide ion. In addition to sodium methoxide, the following metal salts of alcohols are commonly used in organic reactions requiring a strong base in a nonaqueous solvent, as, for example, sodium ethoxide in ethanol and potassium *tert*-butoxide in 2-methyl-2-propanol (*tert*-butyl alcohol).

$CH_3CH_2O^-Na^+$

Sodium ethoxide
(EtO^-Na^+)

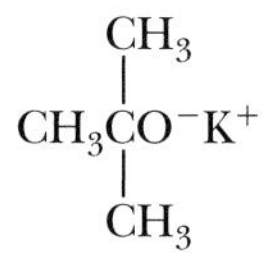

Potassium *tert*-butoxide
(*t*-BuO^-K^+)

Alcohols can also be converted to salts by reaction with bases stronger than alkoxide ions. One such base is sodium hydride, NaH. Hydride ion, $H{:}^-$, the conjugate base of H_2, is an extremely strong base.

$$CH_3CH_2OH + Na^+H^- \longrightarrow CH_3CH_2O^-Na^+ + H_2$$

Ethanol **Sodium hydride** **Sodium ethoxide**

Reactions of sodium hydride with compounds containing acidic hydrogens are irreversible and driven to completion by the formation of H_2, which is given off as a gas.

Example 10.6 Reactions of Alcohols

Write a balanced equation for the reaction of cyclohexanol with sodium metal.

Solution

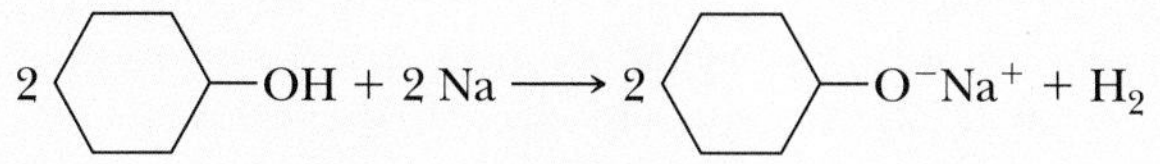

Problem 10.6

Predict the position of equilibrium for this acid-base reaction.

$$CH_3CH_2O^-Na^+ + CH_3\overset{\overset{O}{\|}}{C}OH \rightleftharpoons CH_3CH_2OH + CH_3\overset{\overset{O}{\|}}{C}O^-Na^+$$

Watch a video explanation

10.5 Conversion of Alcohols to Haloalkanes and Sulfonates

Conversion of an alcohol to a haloalkane involves substitution of halogen for —OH at a saturated carbon. The most common reagents for this conversion are the halogen acids (HCl, HBr, and HI) and certain inorganic halides (PBr_3, $SOCl_2$, and $SOBr_2$).

A. Reaction with HCl, HBr, and HI

Tertiary alcohols react rapidly with HCl, HBr, and HI. Mixing a low-molecular-weight, water-soluble tertiary alcohol with concentrated hydrochloric acid for a few minutes at room temperature results in conversion of the alcohol to a chloroalkane.

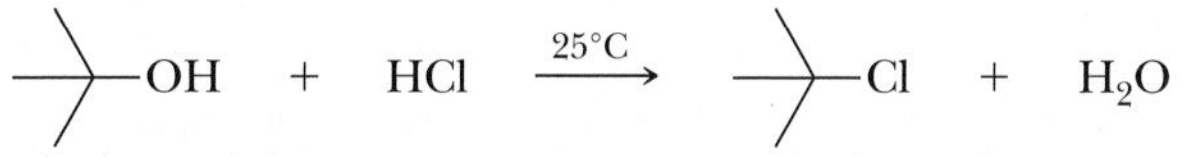

2-Methyl-2-propanol **2-Chloro-2-methylpropane**

Reaction is evident by formation of a water-insoluble chloroalkane that separates from the aqueous layer. Low-molecular-weight, water-soluble primary and secondary alcohols are unreactive under these conditions.

Water-insoluble tertiary alcohols are converted to tertiary halides by bubbling gaseous HX through a solution of the alcohol dissolved in diethyl ether or tetrahydrofuran (THF).

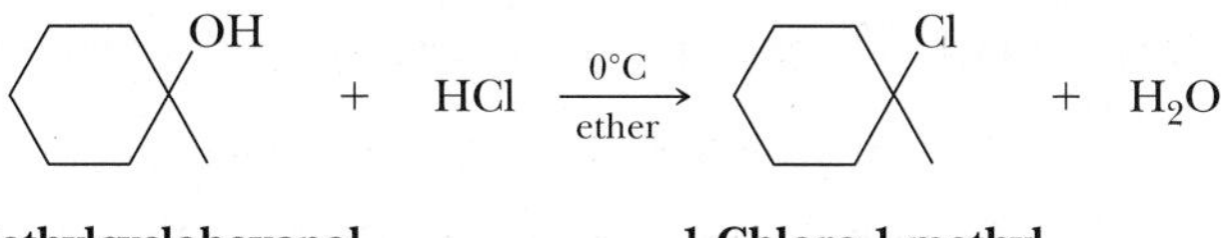

1-Methylcyclohexanol **1-Chloro-1-methyl-cyclohexane**

Water-insoluble primary and secondary alcohols react only slowly under these conditions.

Primary and secondary alcohols are converted to bromoalkanes and iodoalkanes by treatment with hydrobromic and hydroiodic acids. For example, when heated to reflux with concentrated HBr, 1-butanol is converted smoothly to 1-bromobutane.

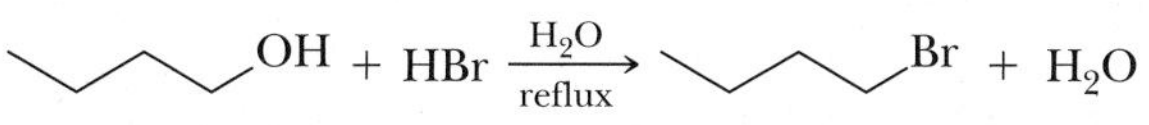

1-Butanol **1-Bromobutane**

Many secondary alcohols give at least some rearranged product, evidence for the formation of carbocation intermediates during their reaction. For example, treating 3-pentanol with HBr gives 3-bromopentane as the major product, along with some 2-bromopentane.

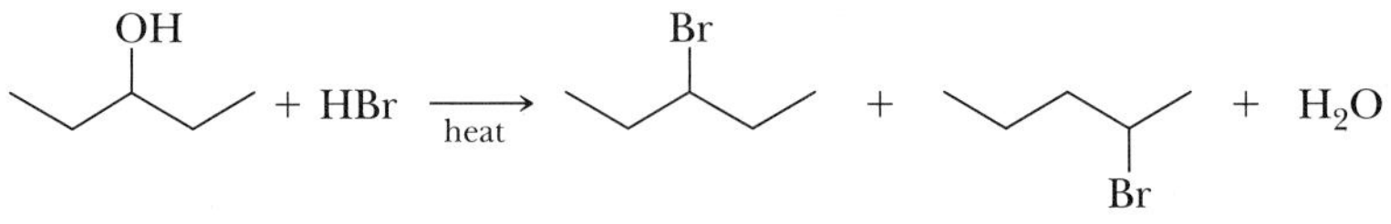

3-Pentanol **3-Bromopentane** (major product) **2-Bromopentane** (a product of rearrangement, racemic)

Primary alcohols with extensive β-branching give large amounts of a product derived from rearrangement. For example, treatment of 2,2-dimethyl-1-propanol (neopentyl alcohol) with HBr gives a rearranged product almost exclusively.

2,2-Dimethyl-1-propanol **2-Bromo-2-methylbutane** (a product of rearrangement)

Based on observations of the relative ease of reaction of alcohols with HX (3° > 2° > 1°) and the occurrence of rearrangements, chemists propose an S_N1 mechanism for the conversion of tertiary and secondary alcohols to haloalkanes by concentrated HX, with the formation of a carbocation intermediate.

Mechanism 10.1

Reaction of a 3° Alcohol with HBr—An S_N1 Reaction

Step 1: **Add a proton.** While we often show HBr as the acid present in solution, the actual acid involved in this reaction is H_3O^+ formed by dissociation of HBr in aqueous solution.

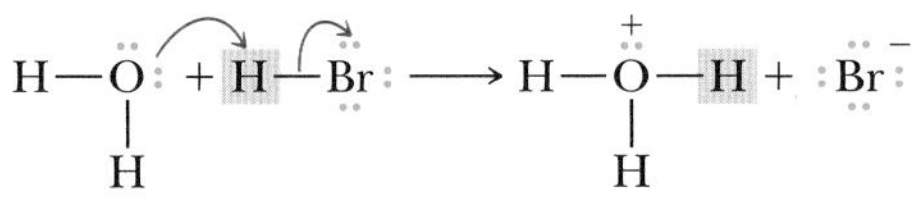

Rapid and reversible proton transfer from H_3O^+ to the —OH group of the alcohol gives an oxonium ion, which converts —OH, a poor leaving group, into, $—\overset{+}{O}H_2$, a better leaving group.

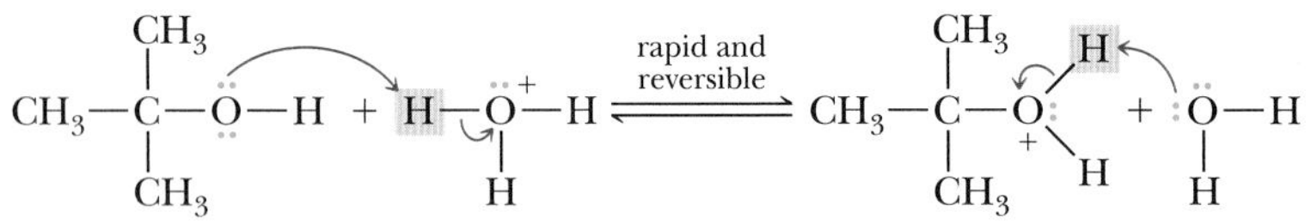

2-Methyl-2-propanol (*tert*-Butyl alcohol) **An oxonium ion**

Step 2: **Break a bond to give stable molecules or ions.** Loss of water gives a 3° carbocation intermediate.

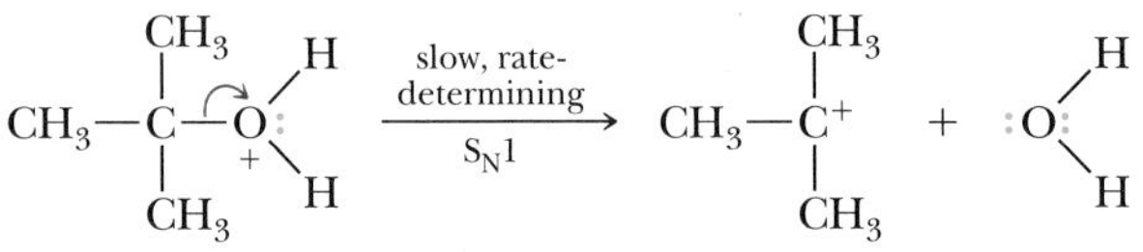

A 3° carbocation intermediate

Step 3: **Make a new bond between a nucleophile and an electrophile.** Reaction of the 3° carbocation (an electrophile) with bromide ion (a nucleophile) gives the haloalkane.

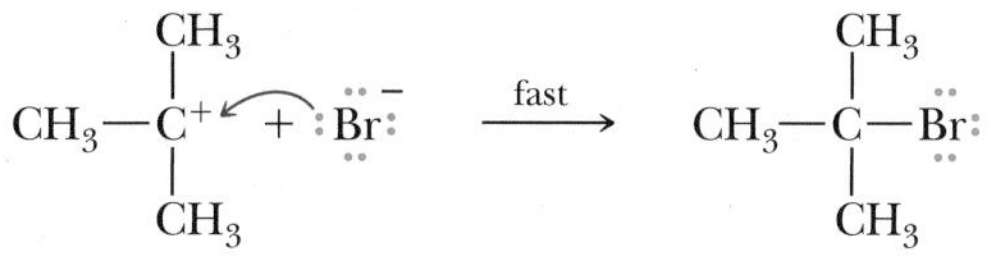

2-Bromo-2-methylpropane
(*tert*-Butyl bromide)

Primary alcohols with no β-branching react with HX by an S_N2 mechanism. In the rate-determining step, halide ion reacts at the carbon bearing the oxonium ion to displace H_2O and form the C—X bond.

Mechanism 10.2

Reaction of a 1° Alcohol with HBr—An S_N2 Reaction

Step 1: **Add a proton.** Rapid and reversible proton transfer gives an oxonium ion, which transforms —OH, a poor leaving group, into —$\overset{+}{O}H_2$, a better leaving group.

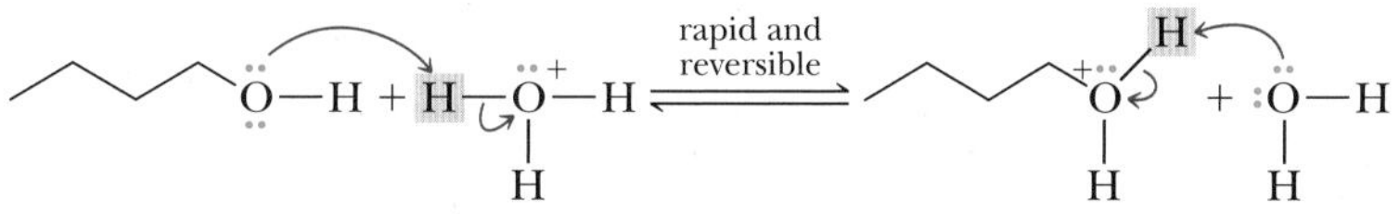

An oxonium ion

Step 2: **Make a new bond between a nucleophile and an electrophile and simultaneously break a bond to give stable molecules or ions.** Nucleophilic displacement of H_2O by Br^- gives the bromoalkane.

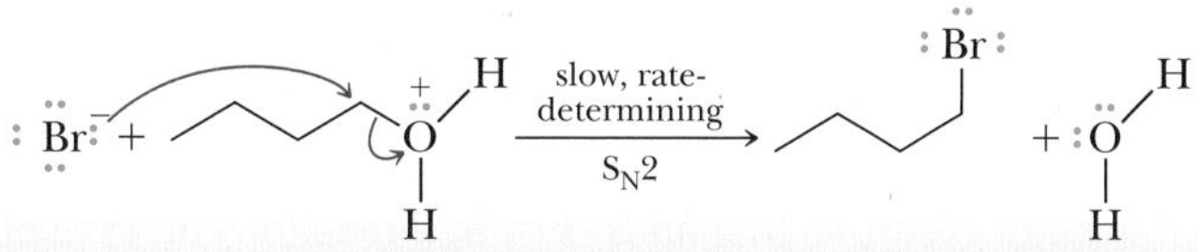

For primary alcohols with extensive β-branching, such as 2,2-dimethyl-1-propanol (neopentyl alcohol), it is difficult, if not impossible, for reaction to occur by direct displacement of H_2O from the primary carbon. Furthermore, formation of a 1° carbocation is also difficult, if not impossible. Instead, primary alcohols with extensive β-branching react by a mechanism involving formation of a 3° carbocation intermediate by simultaneous loss of H_2O and migration of an alkyl group, as illustrated by the conversion of 2,2-dimethyl-1-propanol to 2-chloro-2-methylbutane. Because the rate-determining step of this transformation involves only one reactant, namely the protonated alcohol, it is classified as an S_N1 reaction.

Mechanism 10.3

Rearrangement upon Treatment of Neopentyl Alcohol with HCl

Step 1: **Add a proton.** Rapid and reversible proton transfer gives an oxonium ion. This step converts —OH, a poor leaving group, into —$\overset{+}{O}H_2$, a better leaving group.

(*Continued*)

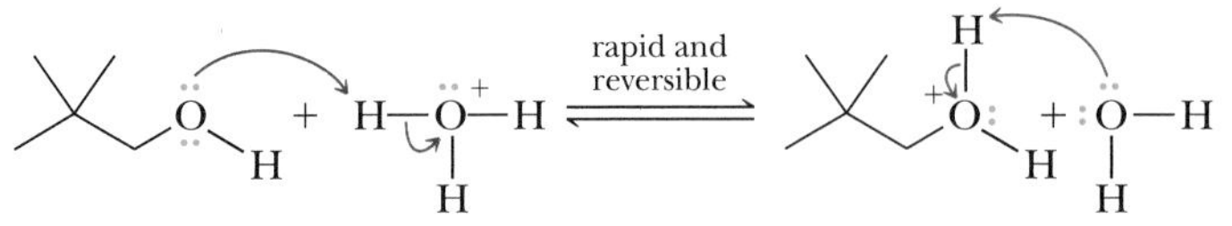

2,2-Dimethyl-1-propanol An oxonium ion

Step 2: **1,2 Shift and simultaneously break a bond to give stable molecules or ions.** Two changes take place simultaneously in this step; the C—O bond breaks, and a methyl group with its pair of bonding electrons migrates to the site occupied by the departing H_2O group. The result of these changes is loss of H_2O and the formation of a 3° carbocation.

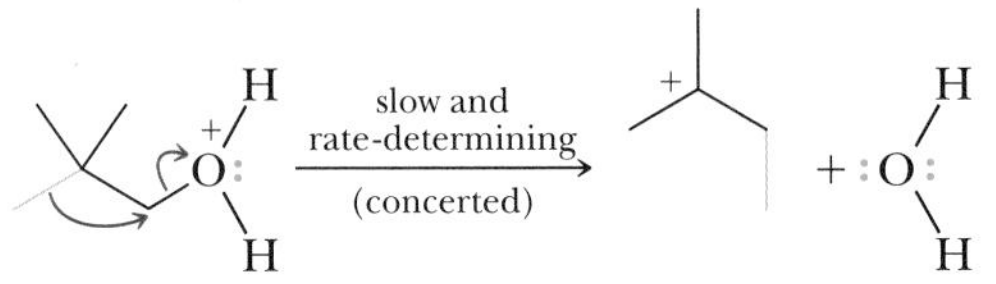

A 3° carbocation intermediate

Step 3: **Make a new bond between a nucleophile and an electrophile.** Reaction of the 3° carbocation (an electrophile) with chloride ion (a nucleophile) gives the 3° haloalkane.

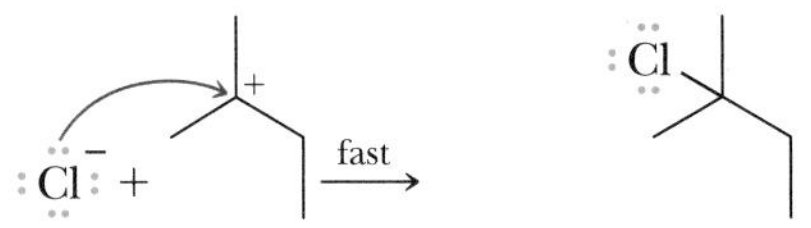

2-Chloro-2-methylbutane

In summary, preparation of haloalkanes by treatment of ROH with HX is most useful for primary and tertiary alcohols. The central theme in all these reactions is that protonation of —OH, a very poor leaving group, transforms it into $—\overset{+}{O}H_2$, a better leaving group, so that an S_N1 or S_N2 reaction can take place with a halide nucleophile. Because of the possibility of rearrangement, this process is less useful for secondary alcohols (except for simple cycloalkanols) and for primary alcohols with extensive branching on the β-carbon.

B. Reaction with Phosphorus Tribromide

An alternative method for the synthesis of bromoalkanes from primary and secondary alcohols is through the use of phosphorus tribromide, PBr_3.

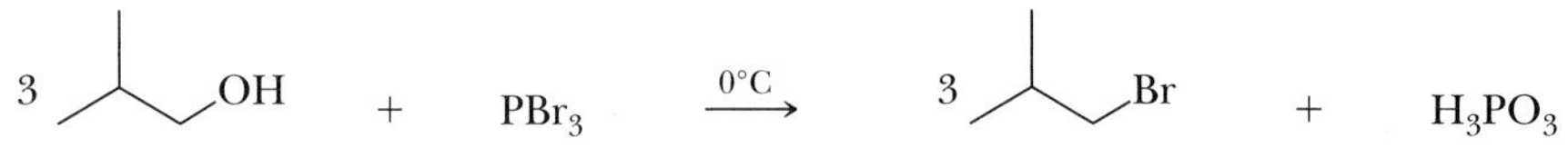

2-Methyl-1-propanol (Isobutyl alcohol) **Phosphorus tribromide** **1-Bromo-2-methylpropane (Isobutyl bromide)** **Phosphorous acid**

This method of preparation of bromoalkanes takes place under milder conditions than treatment with HBr. Although rearrangement sometimes occurs with PBr_3, the extent is considerably less than that with HBr, especially when the reaction mixture is kept at or below 0°C.

Mechanism 10.4

Reaction of a Primary Alcohol with PBr_3

Conversion of an alcohol to a bromoalkane takes place in two steps.

Step 1: **Make a new bond between a nucleophile and an electrophile and simultaneously break a bond to give stable molecules or ions.** Nucleophilic displacement on phosphorus by the oxygen atom of the alcohol gives a protonated dibromophosphite group, which converts —OH, a poor leaving group, into a good leaving group.

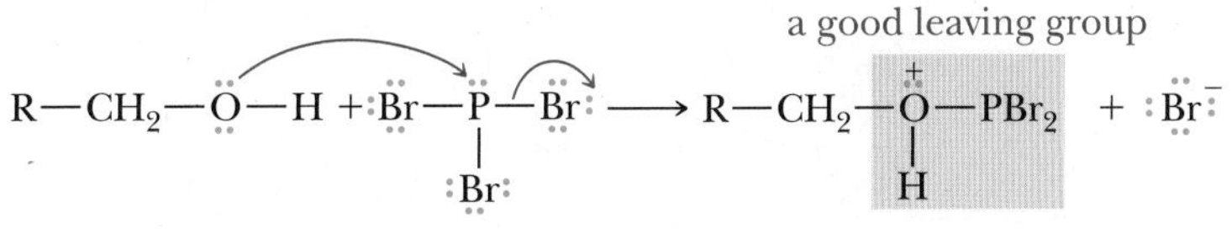

Step 2: **Make a new bond between a nucleophile and an electrophile and simultaneously break a bond to give stable molecules or ions.** Nucleophilic displacement of the protonated dibromophosphite group by bromide ion gives the bromoalkane.

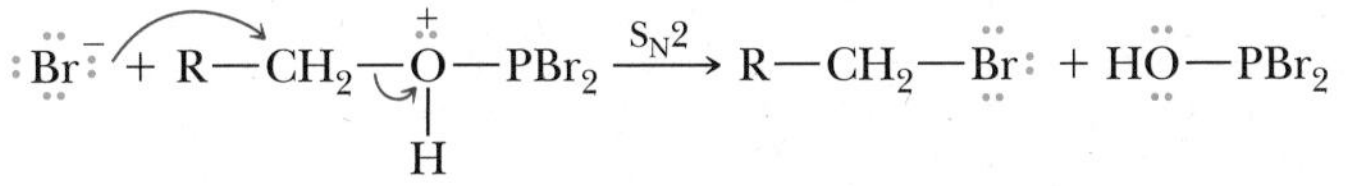

The other two bromine atoms on phosphorus are replaced in similar reactions, giving three moles of RBr and one mole of phosphorous acid, H_3PO_3.

C. Reaction with Thionyl Chloride and Thionyl Bromide

The most widely used reagent for the conversion of primary and secondary alcohols to chloroalkanes is thionyl chloride, $SOCl_2$. Yields are high, and rearrangements are seldom observed. The byproducts of this conversion are HCl and SO_2.

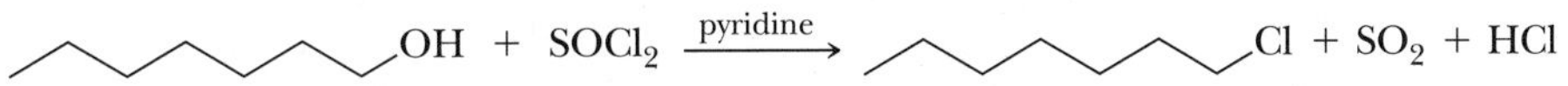

1-Heptanol **Thionyl chloride** **1-Chloroheptane**

Similarly, thionyl bromide, $SOBr_2$ can be used to convert an alcohol to a bromoalkane.

Reactions with these reagents are most commonly carried out in the presence of pyridine (Section 23.1) or a tertiary amine such as triethylamine, Et_3N. The function of the amine (a weak base) is twofold. First, it catalyzes the reaction by forming a small amount of the alkoxide in equilibrium. The alkoxide is more reactive than the alcohol as a nucleophile. In addition, the amine neutralizes the HCl or HBr generated during the reaction and in this way prevents unwanted side reactions.

The 3° amine promotes formation of an alkoxide:

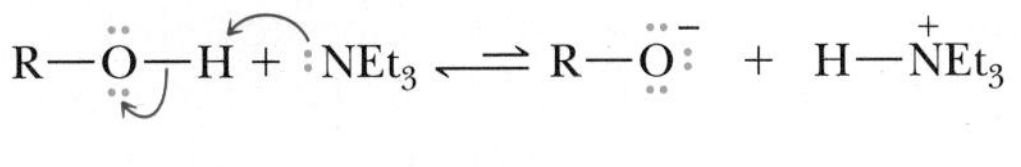

It also neutralizes the HCl formed during the reaction:

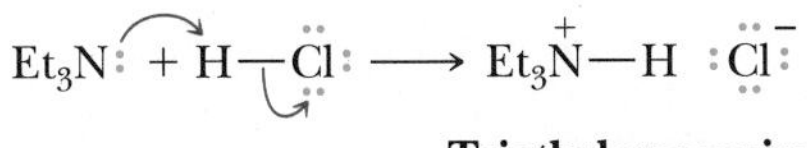

Triethylammonium chloride

A particular value of thionyl halides is that their reaction with alcohols is stereoselective; it occurs with inversion of configuration. Reaction of thionyl chloride with (*S*)-2-octanol, for example, in the presence of a tertiary amine occurs with inversion of configuration and gives (*R*)-2-chlorooctane.

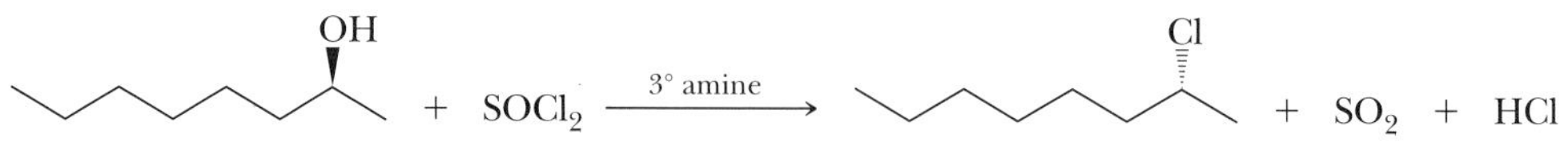

(*S*)-2-Octanol   Thionyl chloride   (*R*)-2-Chlorooctane

A key feature of the reaction of an alcohol with thionyl chloride is the formation of an alkyl chlorosulfite, which converts OH^-, a poor leaving group, into a chlorosulfite that now contains a good leaving group. If the reaction between the alcohol and thionyl chloride is carried out at 0°C or below, the alkyl chlorosulfite can be isolated.

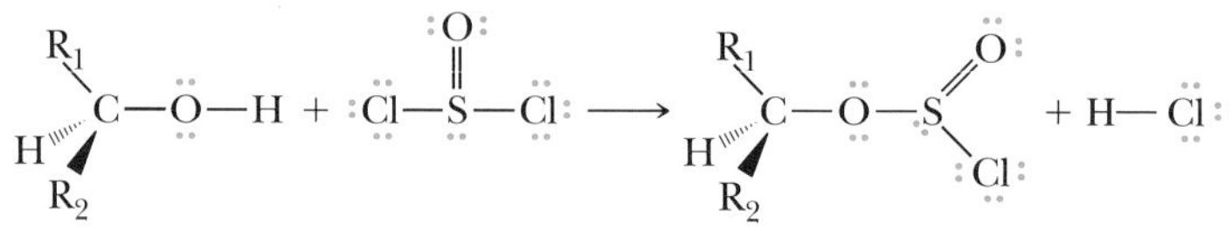

An alkyl chlorosulfite

Nucleophilic displacement of this leaving group by chloride ion gives the product.

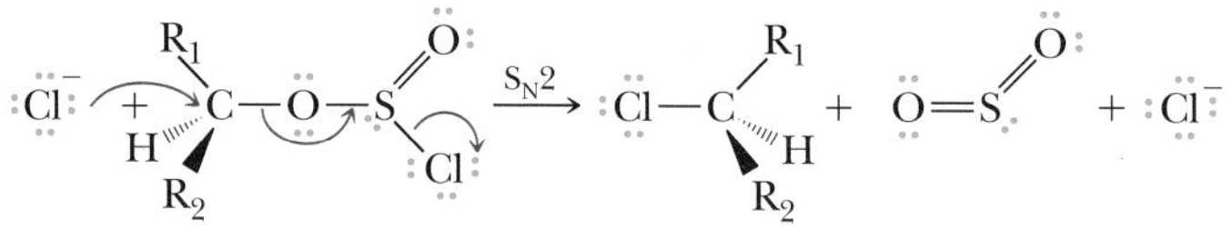

D. Formation of Aryl and Alkyl Sulfonates

As we have just seen, alcohols react with thionyl chloride to form alkyl chlorosulfites. Alcohols also react with compounds called sulfonyl chlorides to form alkylsulfonates. Sulfonyl chlorides are derived from sulfonic acids, which are comparable in strength to sulfuric acid.

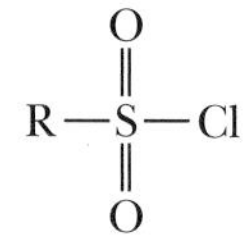

A sulfonyl chloride

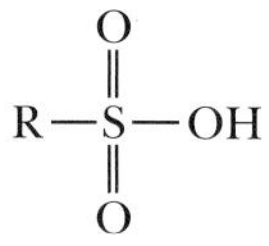

A sulfonic acid
(a very strong acid)

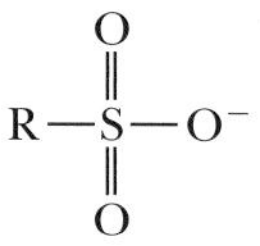

A sulfonate anion
(a very weak base and stable anion;
a very good leaving group)

What is important at this point is that a sulfonate anion is a very weak base and stable anion; therefore, it is a very good leaving group in nucleophilic substitution reactions.

Two of the most commonly used sulfonyl chlorides are *p*-toluenesulfonyl chloride (abbreviated tosyl chloride, TsCl) and methanesulfonyl chloride (abbreviated mesyl chloride, MsCl). Treating ethanol with *p*-toluenesulfonyl chloride in the presence of pyridine gives ethyl *p*-toluenesulfonate (ethyl tosylate). Pyridine is added to catalyze the reaction and to neutralize the HCl formed as a byproduct. Cyclohexanol is converted to cyclohexyl methanesulfonate (cyclohexyl mesylate) by a similar reaction of cyclohexanol with methanesulfonyl chloride.

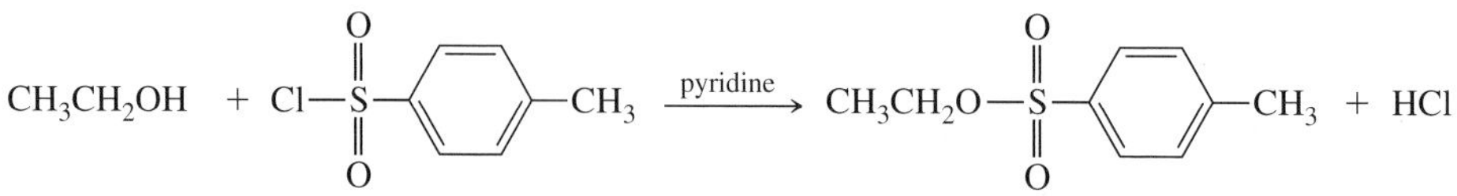

Ethanol   ***p*-Toluenesulfonyl chloride**   **Ethyl *p*-toluenesulfonate (Ethyl tosylate)**

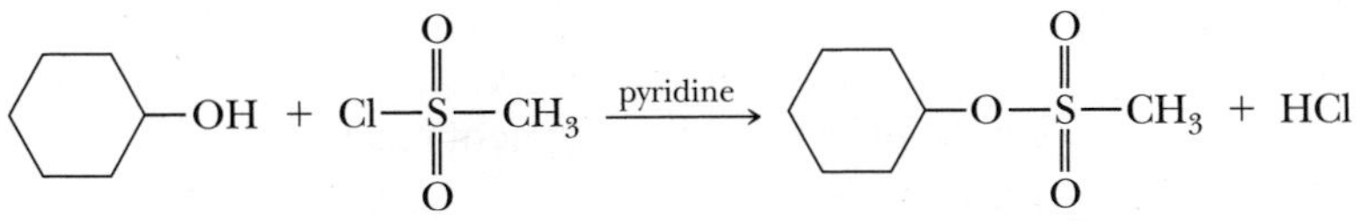

Cyclohexanol **Methanesulfonyl chloride** **Cyclohexyl methanesulfonate (Cyclohexyl mesylate)**

In formation of either a tosylate or a mesylate, the reaction involves breaking the O—H bond of the alcohol; it does not affect the C—O bond in any way. If the carbon bearing the —OH group is a chiral center, sulfonate ester formation takes place with retention of configuration.

A particular advantage of sulfonate esters is that through their use, a hydroxyl group, a very poor leaving group, can be converted to a tosylate or mesylate group, often shown as OTs and OMs, respectively. Both are very good leaving groups readily displaced by nucleophilic substitution.

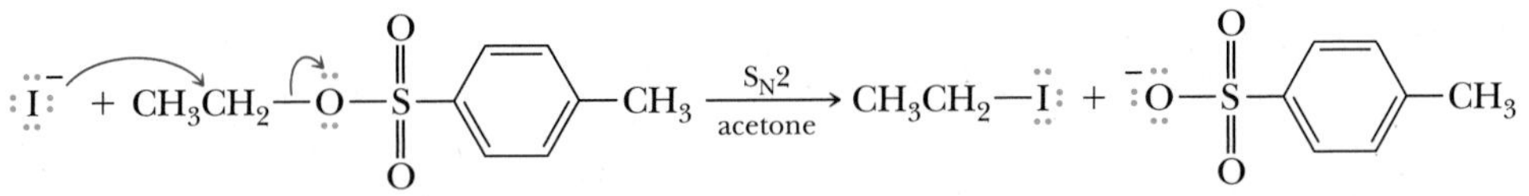

Ethyl *p*-toluenesulfonate **_p_-Toluenesulfonate anion** (a good leaving group)

Following is a two-step sequence for conversion of (*S*)-2-octanol to (*R*)-2-octyl acetate via a tosylate. The first step involves cleavage of the O—H bond and proceeds with retention of configuration at the chiral center. The second step involves S_N2 nucleophilic displacement of tosylate by acetate ion and proceeds with inversion of configuration at the chiral center.

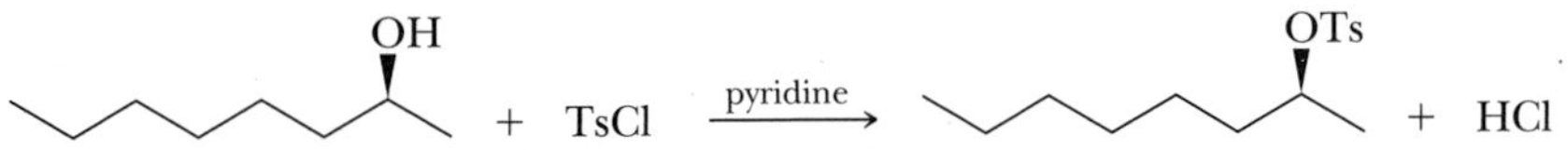

(*S*)-2-Octanol **Tosyl chloride** **(*S*)-2-Octyl tosylate**

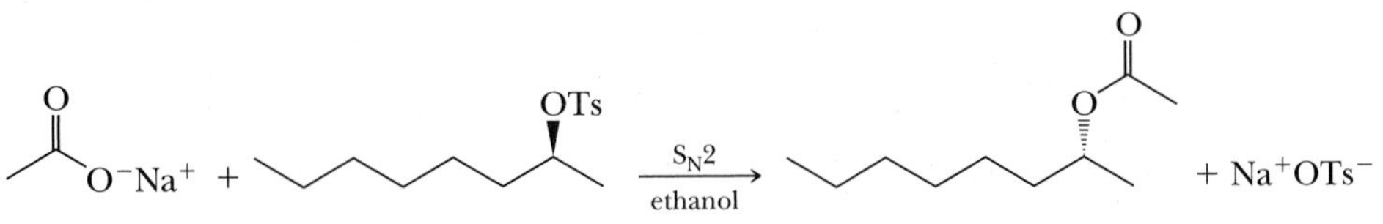

Sodium acetate **(*S*)-2-Octyl tosylate** **(*R*)-2-Octyl acetate**

Example 10.7 Reaction via a Tosylate

Show how to convert *trans*-4-methylcyclohexanol to *cis*-1-iodo-4-methylcyclohexane via a tosylate.

Solution

Treat the alcohol with *p*-toluenesulfonyl chloride in pyridine to form a tosylate with retention of configuration. Then treat the tosylate with sodium iodide in acetone. The S_N2 reaction with inversion of configuration gives the product. Because of the requirement for backside attack by the I^- nucleophile, the tosylate group must be in the axial position to react. Backside attack is not

(Continued)

possible for an equatorial leaving group on a cyclohexane ring. The molecule must undergo a ring flip before the I^- displacement reaction can occur.

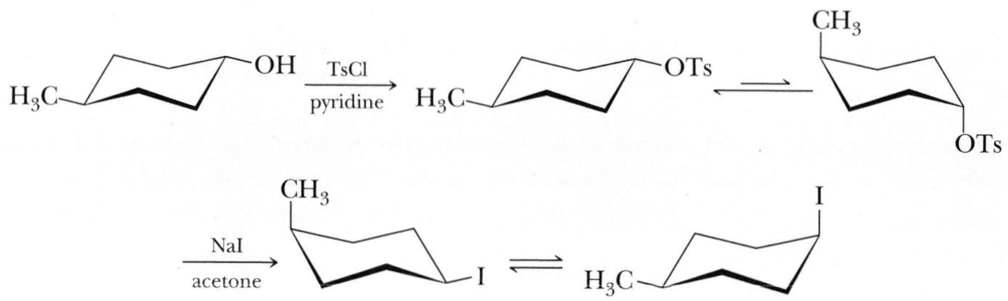

Watch a video explanation

Problem 10.7

Show how to convert (*R*)-2-pentanol to (*S*)-2-pentanethiol via a tosylate.

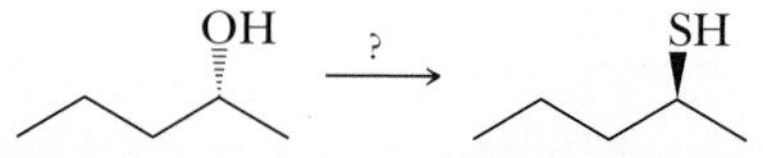

(*R*)-2-Pentanol **(*S*)-2-Pentanethiol**

10.6 Acid-Catalyzed Dehydration of Alcohols

Dehydration Elimination of water.

An alcohol can be converted to an alkene by **dehydration** (i.e., by the elimination of a molecule of water from adjacent carbon atoms). Dehydration is most often brought about by heating the alcohol with either 85% phosphoric acid or concentrated sulfuric acid. Primary alcohols are the most difficult to dehydrate and generally require heating in concentrated sulfuric acid at temperatures as high as 180°C. Secondary alcohols undergo acid-catalyzed dehydration at somewhat lower temperatures. Acid-catalyzed dehydration of tertiary alcohols often requires temperatures only slightly above room temperature.

$$CH_3CH_2OH \xrightarrow[180°C]{H_2SO_4} CH_2{=}CH_2 + H_2O$$

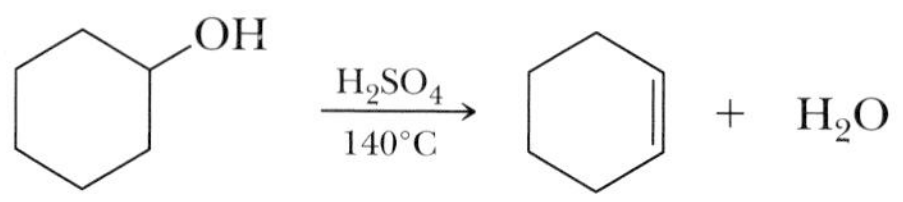

Cyclohexanol **Cyclohexene**

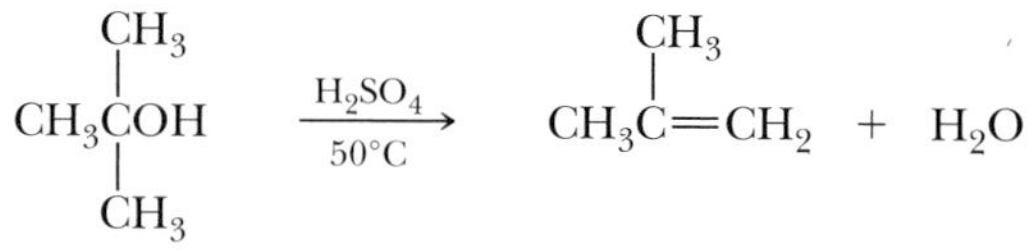

2-Methyl-2-propanol (*tert*-Butyl alcohol) **2-Methylpropene (Isobutylene)**

Thus, the ease of acid-catalyzed dehydration of alcohols is in this order:

1° alcohol < 2° alcohol < 3° alcohol

Ease of dehydration of alcohols →

When isomeric alkenes are obtained in acid-catalyzed dehydration of an alcohol, the alkene having the greater number of substituents on the double bond (the more stable alkene) generally predominates (Zaitsev's rule, Section 9.5).

$$\underset{\substack{\textbf{2-Butanol}\\\text{(racemic)}}}{CH_3CH_2\overset{\overset{\displaystyle OH}{|}}{C}HCH_3} \xrightarrow[\text{heat}]{85\%\ H_3PO_4} \underset{\substack{\textbf{(\textit{E} and \textit{Z})-2-Butene}\\(80\%)}}{CH_3CH{=}CHCH_3} + \underset{\substack{\textbf{1-Butene}\\(20\%)}}{CH_3CH_2CH{=}CH_2} + H_2O$$

Example 10.8 Conversion of Alcohols to Alkenes

Draw structural formulas for the alkenes formed on acid-catalyzed dehydration of each alcohol. Where isomeric alkenes are possible, predict which alkene is the major product.

(a) 3-Methyl-2-butanol (racemic) **(b)** 2-Methylcyclopentanol (racemic)

Solution

(a) Elimination of H_2O from carbons 2-3 gives 2-methyl-2-butene; elimination from carbons 1-2 gives 3-methyl-1-butene. 2-Methyl-2-butene, with three alkyl groups (three methyl groups) on the double bond, is the major product (Zaitsev rule). 3-Methyl-1-butene, with only one alkyl group (an isopropyl group) on the double bond, is the minor product. A small amount of 2-methyl-1-butene is formed by rearrangement.

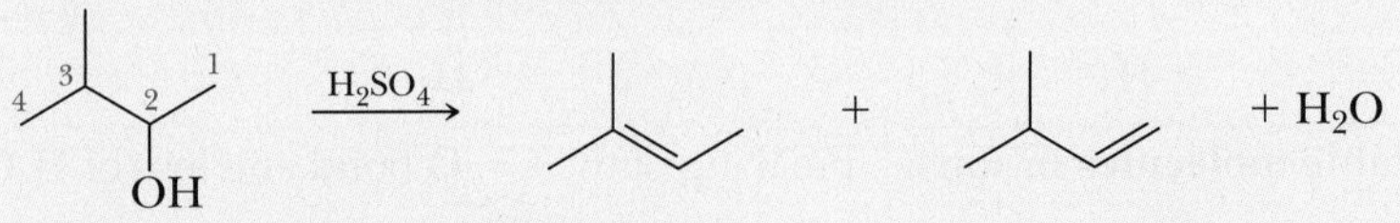

3-Methyl-2-butanol (racemic) **2-Methyl-2-butene** (major product) **3-Methyl-1-butene**

(b) The major product, 1-methylcyclopentene, has three alkyl substituents on the double bond. 3-Methylcyclopentene has only two substituents on the double bond.

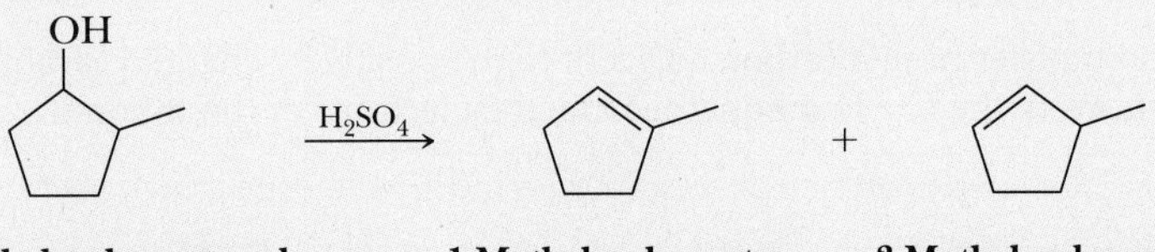

2-Methylcyclopentanol (racemic) **1-Methylcyclopentene** (major product) **3-Methylcyclopentene** (racemic)

Problem 10.8

Draw structural formulas for the alkenes formed by acid-catalyzed dehydration of each alcohol. Where isomeric alkenes are possible, predict which is the major product.

(a) 2-Methyl-2-butanol **(b)** 1-Methylcyclopentanol

Dehydration of primary and secondary alcohols is often accompanied by rearrangement. Acid-catalyzed dehydration of 3,3-dimethyl-2-butanol, for example, gives a mixture of two alkenes, each of which is the result of a rearrangement.

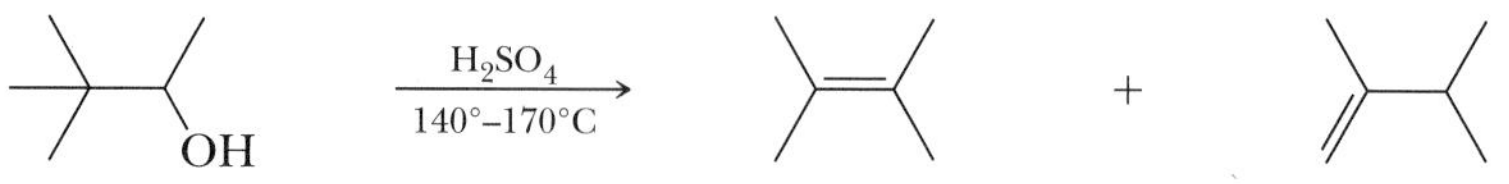
OH
H_2SO_4
140°–170°C
+

3,3-Dimethyl-2-butanol (racemic) **2,3-Dimethyl-2-butene** (80%) **2,3-Dimethyl-1-butene** (20%)

Based on the relative rates of dehydration of alcohols (3° > 2° > 1°) and the prevalence of rearrangement, particularly among primary and secondary alcohols, chemists propose a three-step mechanism for acid-catalyzed dehydration of secondary and tertiary alcohols. This mechanism involves formation of a carbocation in the rate-determining step and therefore is classified as an E1 mechanism.

Mechanism 10.5

Acid-Catalyzed Dehydration of 2-Butanol—An E1 Reaction

Step 1: **Add a proton.** Proton transfer from H_3O^+ to the OH group of the alcohol gives an oxonium ion; —OH, a poor leaving group, is converted to $—\overset{+}{O}H_2$, a better leaving group.

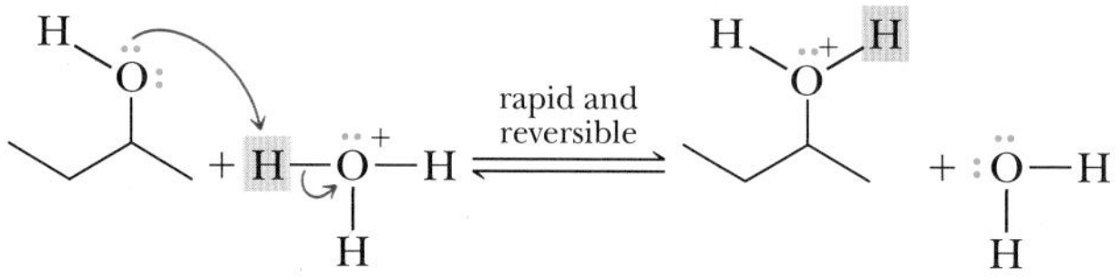
rapid and
reversible

Step 2: **Break a bond to give stable molecules or ions.** Breaking of the C—O bond and loss of H_2O gives a 2° carbocation intermediate.

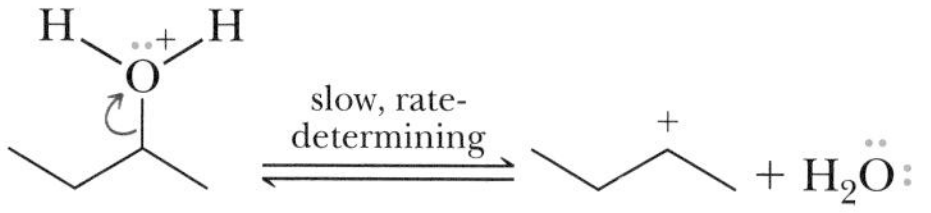
slow, rate-
determining
+ H_2O

A 2° carbocation intermediate

Step 3: **Take a proton away.** Proton transfer from a carbon adjacent to the positively charged carbon to H_2O gives the alkene. In this step, the electrons of the C—H σ bond become the electrons of the π bond.

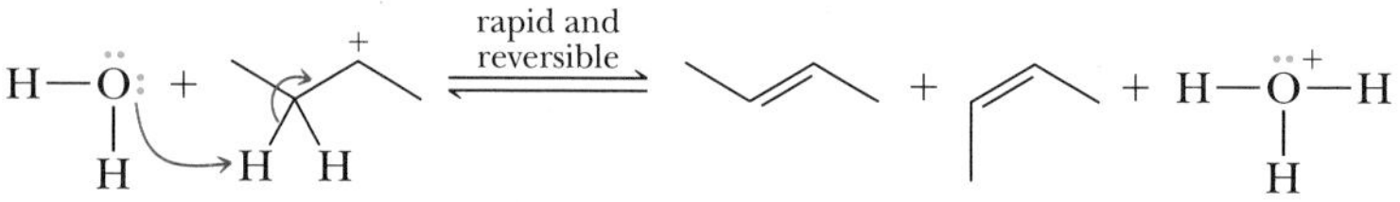
rapid and
reversible

Rearrangement occurs through formation of a carbocation intermediate followed by migration of an atom or a group, with its bonding pair of electrons, from the β-carbon to the carbon bearing the positive charge.

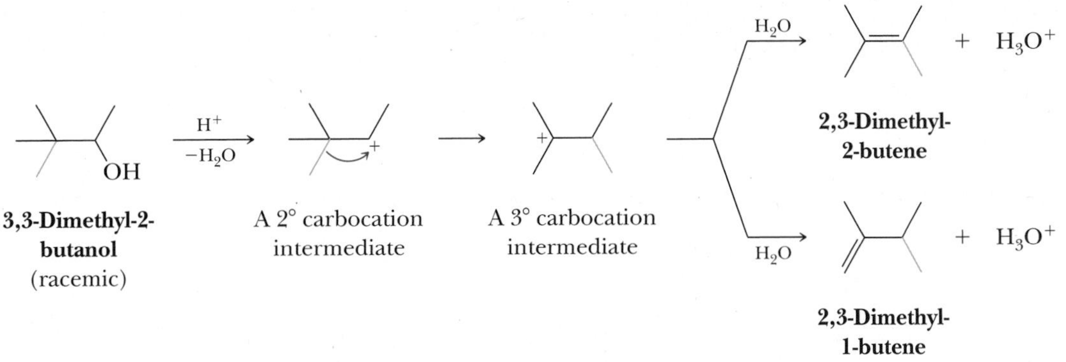

The driving force for rearrangements of this type is conversion of a less stable carbocation to a more stable one. Proton transfer to H_2O then gives the alkenes. As in other cases of acid-catalyzed dehydration of alkenes, the Zaitsev rule applies, and the more substituted alkene predominates.

Primary alcohols with little or no β-branching undergo acid-catalyzed dehydration to give a terminal alkene and rearranged alkenes. Acid-catalyzed dehydration of 1-butanol, for example, gives only 12% of 1-butene. The major product is a mixture of the *trans* and *cis* isomers of 2-butene.

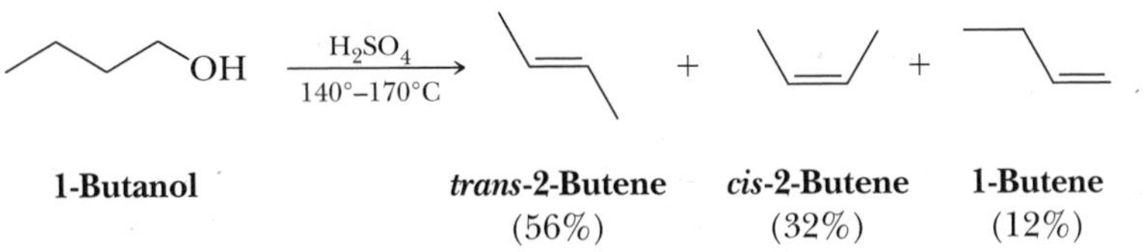

We account for the formation of these products by a combination of E1 and E2 mechanisms.

Mechanism 10.6

Acid-Catalyzed Dehydration of an Unbranched Primary Alcohol

Step 1: **Add a proton.** Proton transfer from H_3O^+ to the OH group gives an oxonium ion.

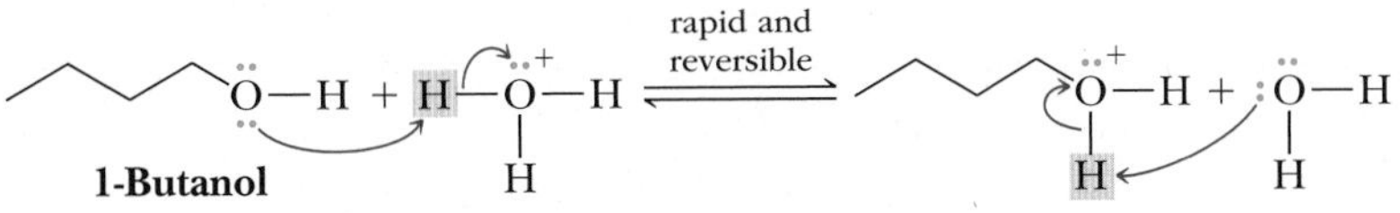

Step 2: **Take a proton away and simultaneously break a bond to give stable molecules or ions.** Simultaneous proton transfer to solvent and loss of H_2O gives the carbon-carbon double bond of the terminal alkene.

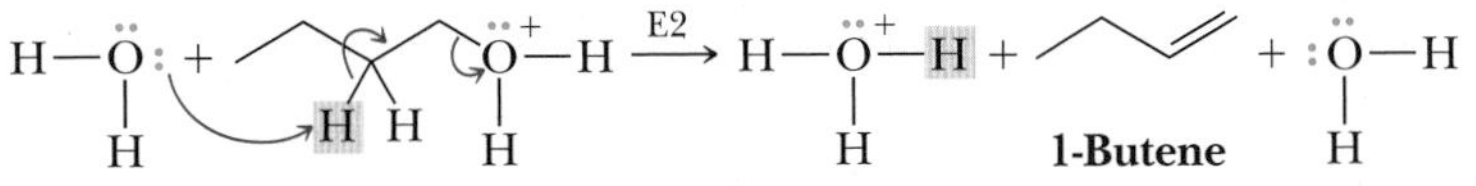

Step 3: **1,2 Shift and simultaneously break a bond to give stable molecules or ions.** Simultaneous shift of a hydride ion from the β-carbon to the α-carbon and loss of H_2O gives a carbocation intermediate.

(Continued)

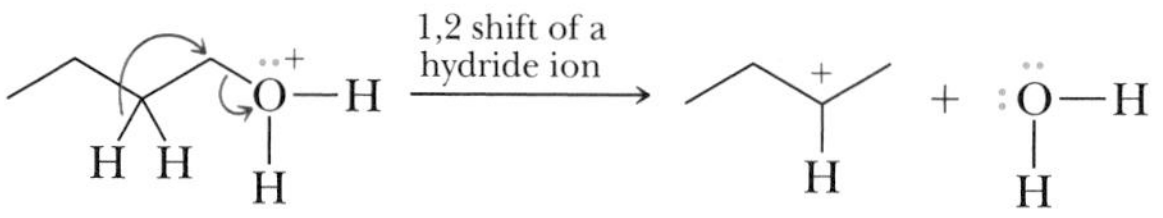

A 2° carbocation

Step 4: **Take a proton away.** Transfer of a proton from a carbon adjacent to the carbocation to solvent gives the rearranged alkenes.

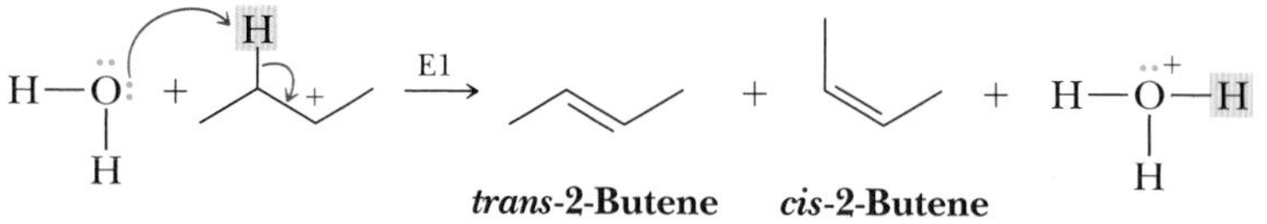

***trans*-2-Butene** ***cis*-2-Butene**

Example 10.9 Acid-Catalyzed Dehydration

Propose a mechanism to account for this acid-catalyzed dehydration.

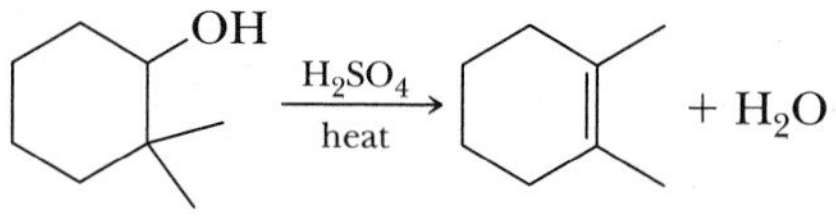

(racemic)

Solution

Proton transfer to the OH group to form an oxonium ion **(add a proton)** followed by loss of H_2O **(break a bond to give stable molecules or ions)** gives a 2° carbocation intermediate. Migration of a methyl group with its bonding pair of electrons from the adjacent carbon to the positively charged carbon gives a more stable 3° carbocation **(1,2 shift)**. Proton transfer from this intermediate to a base (here shown as H_2O) gives the observed product **(take a proton away)**.

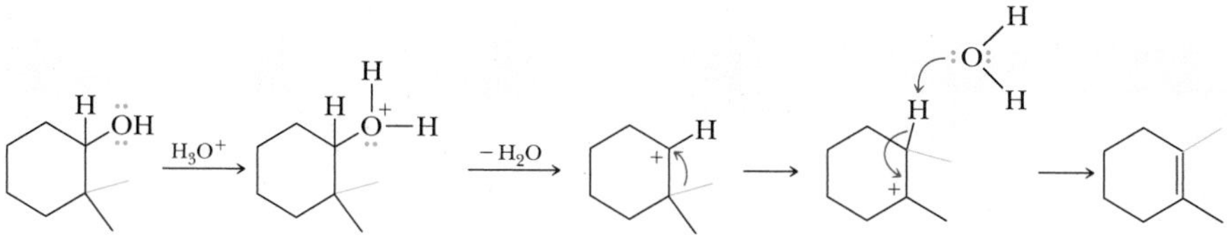

A 2° carbocation intermediate

A 3° carbocation intermediate

Watch a video explanation

Problem 10.9

Propose a mechanism to account for this acid-catalyzed dehydration.

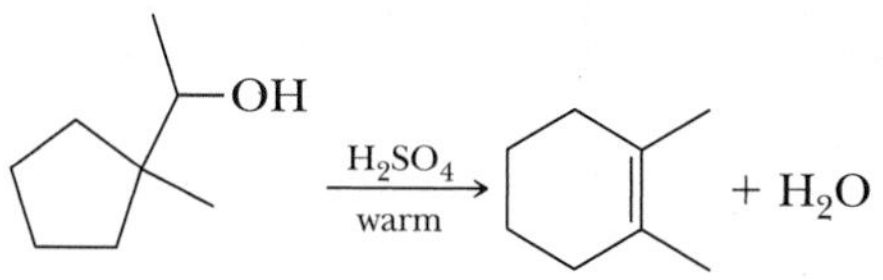

(racemic)

In Section 6.3B, we discussed the acid-catalyzed hydration of alkenes to give alcohols. In the present section, we discussed the acid-catalyzed dehydration of alcohols to give alkenes. In fact, both the alkene hydration and the alcohol dehydration reactions are reversible and represent different directions of the same process. The following equilibrium exists:

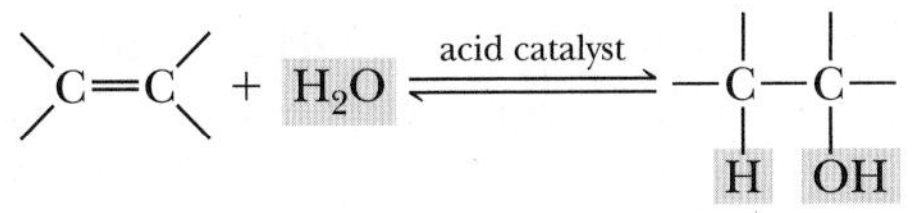

An alkene An alcohol

Large amounts of water (use of dilute aqueous acid) favor alcohol formation, whereas scarcity of water (use of concentrated acid) or experimental conditions where water is removed (heating the reaction mixture above 100°C) favor alkene formation. Thus, depending on experimental conditions, it is possible to use the hydration-dehydration equilibrium to prepare either alcohols or alkenes in high yields.

This hydration-dehydration equilibrium illustrates a very important principle in the study of reaction mechanisms—the **principle of microscopic reversibility**. According to this principle, the sequence of transition states and reactive intermediates (i.e., the mechanism) for any reversible reaction must be the same, but in reverse order, for the reverse reaction as for the forward reaction.

Principle of microscopic reversibility

This principle states that the sequence of transition states and reactive intermediates in the mechanism of any reversible reaction must be the same, but in reverse order, for the reverse reaction as for the forward reaction.

As an illustration of the principle of microscopic reversibility, notice that the mechanism presented in this section for the acid-catalyzed dehydration of 2-butanol to give 2-butene is exactly the reverse of that presented in Section 6.3B for the acid-catalyzed hydration of 2-butene to give 2-butanol.

10.7 The Pinacol Rearrangement

Compounds containing hydroxyl groups on two adjacent carbon atoms are called vicinal *diols*, or alternatively, *glycols*. Such compounds can be synthesized by a variety of methods, including oxidation of alkenes by OsO_4 (Section 6.5A). The products of acid-catalyzed dehydration of glycols are quite different from those of acid-catalyzed dehydration of alcohols. For example, treating 2,3-dimethyl-2,3-butanediol (commonly called pinacol) with concentrated sulfuric acid gives 3,3-dimethyl-2-butanone (commonly called pinacolone).

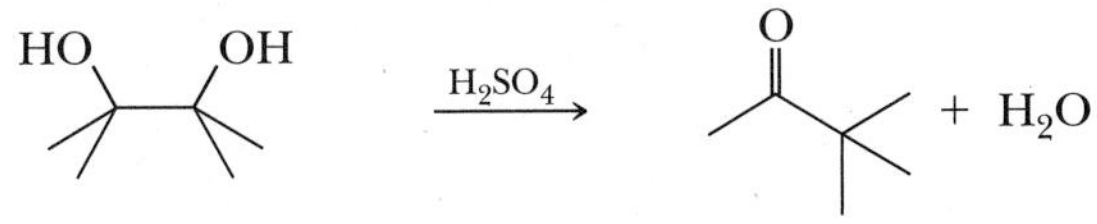

2,3-Dimethyl-2,3-butanediol (Pinacol) **3,3-Dimethyl-2-butanone (Pinacolone)**

Note two features of this reaction: (1) it involves dehydration of a glycol to form a ketone, and (2) it involves migration of a methyl group from one carbon to an adjacent carbon. Acid-catalyzed conversion of pinacol to pinacolone is an example of a type of reaction called the **pinacol rearrangement**. We account for the conversion of pinacol to pinacolone by a four-step mechanism.

Mechanism 10.7

The Pinacol Rearrangement of 2,3-Dimethyl-2,3-butanediol (Pinacol)

Step 1: **Add a proton.** Proton transfer from the acid catalyst to one of the —OH groups gives an oxonium ion, which converts —OH, a poor leaving group, into $—\overset{+}{O}H_2$, a better leaving group.

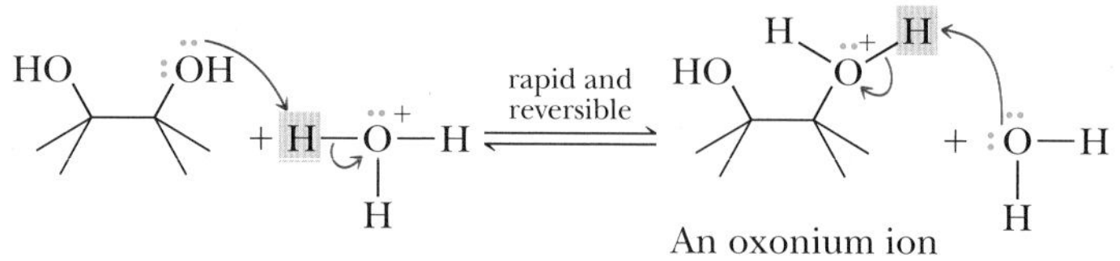

An oxonium ion

Step 2: **Break a bond to give stable molecules or ions.** Loss of H_2O from the oxonium ion gives a 3° carbocation intermediate.

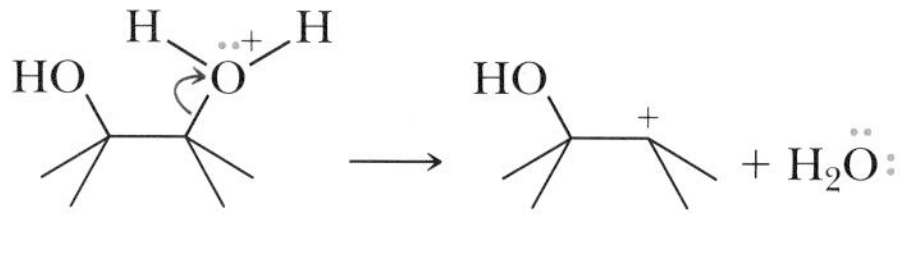
A 3° carbocation intermediate

Step 3: **1,2 Shift.** Migration of a methyl group from the adjacent carbon with its bonding electrons gives a new, more stable resonance-stabilized cation intermediate. Of the two contributing structures we can draw for it, the one on the right makes the greater contribution because, in it, both carbon and oxygen have complete octets of valence electrons (Section 1.8).

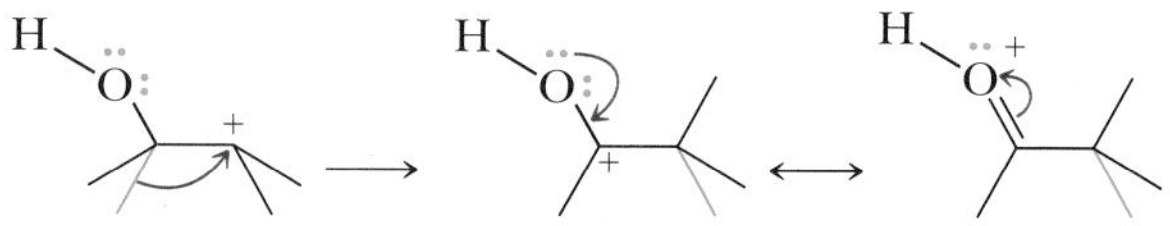
Cation intermediate stabilized by charge delocalization

Step 4: **Take a proton away.** Proton transfer to solvent gives pinacolone.

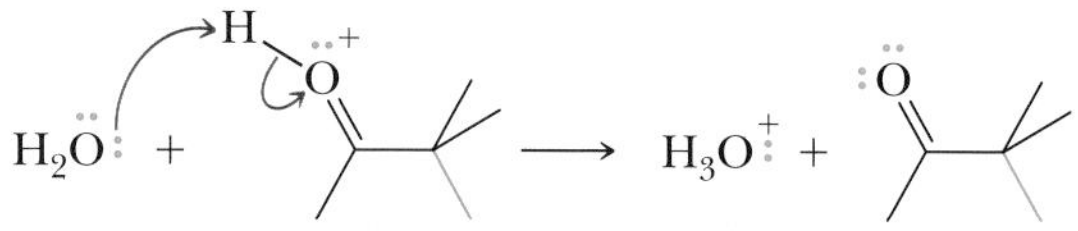

The pinacol rearrangement is general for all glycols. In the rearrangement of pinacol, a symmetrical diol, equivalent carbocations are formed no matter which —OH becomes protonated and leaves. Studies of unsymmetrical vicinal diols reveal that the —OH group that becomes protonated and leaves is the one that gives rise to the more stable carbocation. For example, treatment of 2-methyl-1,2-propanediol with cold concentrated sulfuric acid gives a 3° carbocation. Subsequent migration of hydride ion ($H:^-$) followed by transfer of a proton from the new cation to solvent gives 2-methylpropanal.

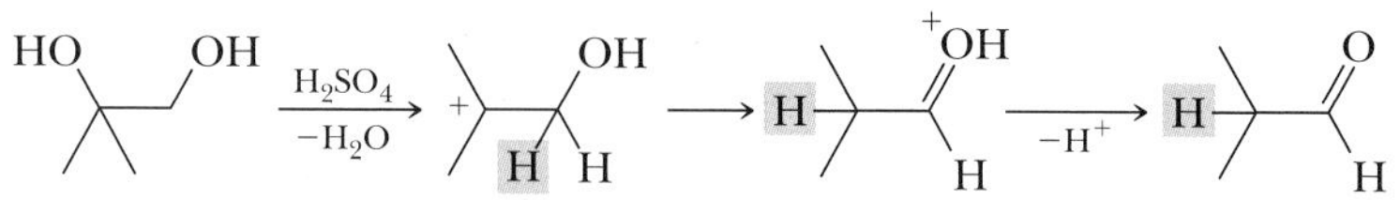

2-Methyl-1,2-propanediol A 3° carbocation intermediate **2-Methylpropanal**

Example 10.10 Diol Reactions

Predict the product formed by treating each vicinal diol with H_2SO_4.

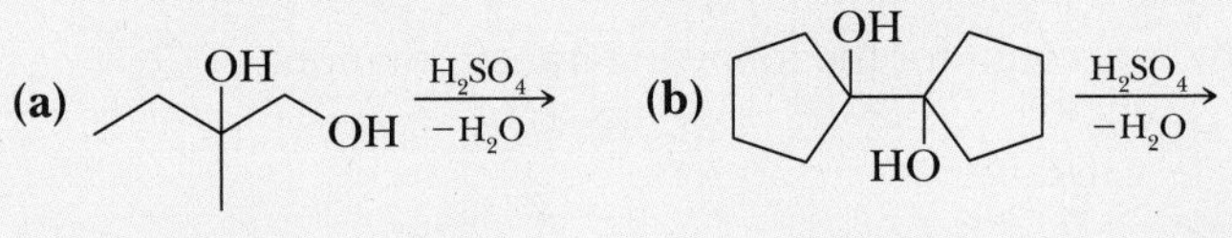

(racemic)

Solution

(a) ***Step 1:*** **Add a proton.** The hydroxyl group is protonated.

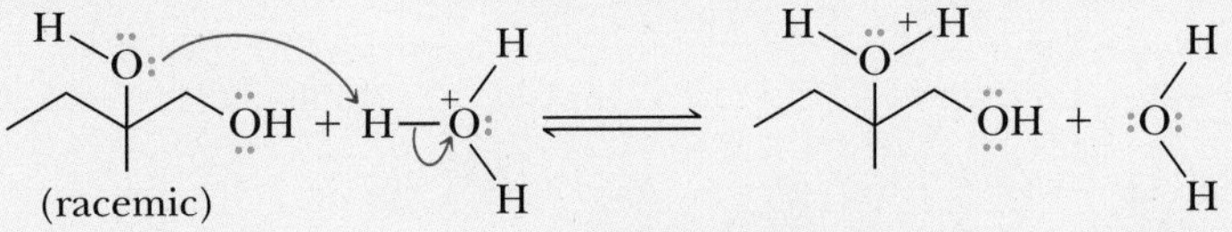

Step 2: **Break a bond to give stable molecules or ions.** Loss of H_2O gives a 3° carbocation intermediate.

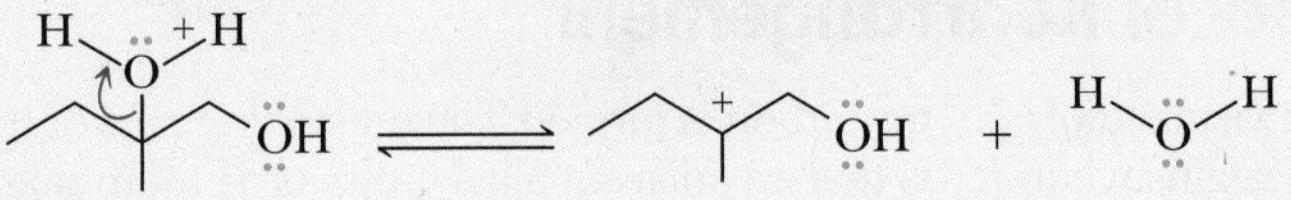

Step 3: **1,2 Shift.** Migration of a hydride ion from the adjacent carbon gives a resonance-stabilized cation intermediate.

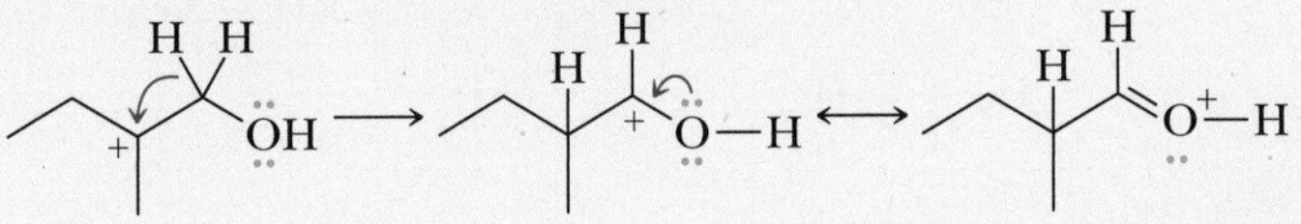

A resonance-stabilized cation intermediate

The contributing structure on the right has filled valence shells on both carbon and oxygen and therefore makes the greater contribution to the hybrid.

Step 4: **Take a proton away.** Proton transfer from the resonance-stabilized cation intermediate to water completes the reaction to give 2-methylbutanal.

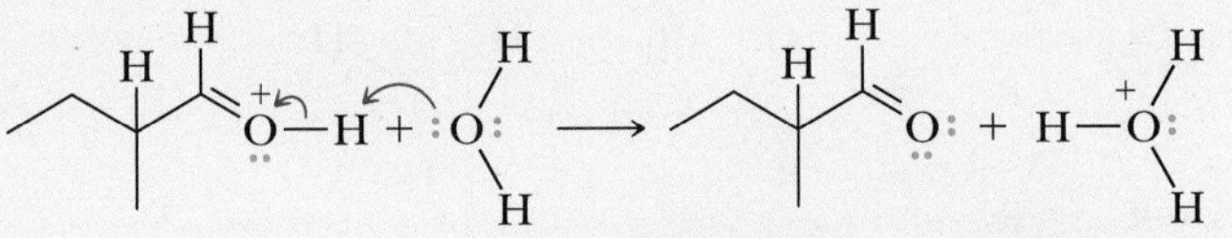

2-Methylbutanal
(racemic)

(b) Here we condense the mechanism. Protonation of either hydroxyl group followed by loss of water gives a 3° carbocation. The group that then migrates is a CH_2 group of the five-membered ring, and the product is a bicyclic ketone.

1,2 Shift

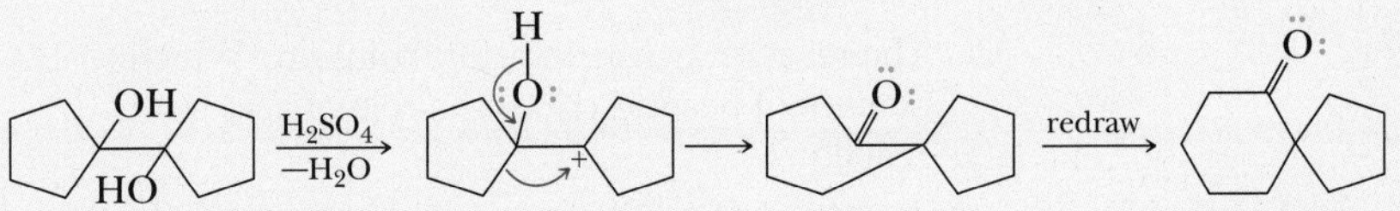

(racemic) A 3° carbocation intermediate **Spiro[4.5]decan-6-one**

(Continued)

The product belongs to the class of compounds called spiro compounds, in which two rings share only one carbon atom.

Problem 10.10

Propose a mechanism to account for the following transformation.

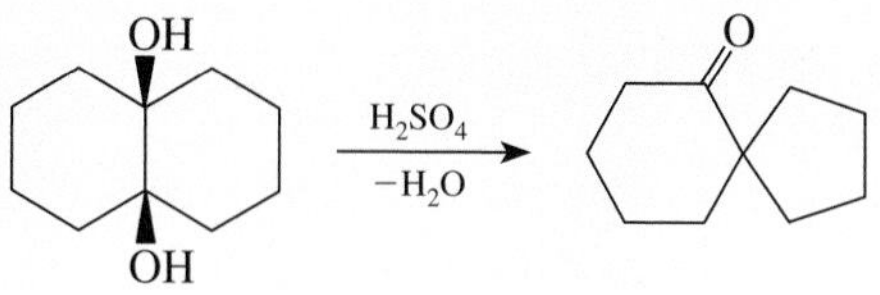

MCAT Practice: *Passage and Questions*

Pinacol Rearrangement

The pinacol rearrangement occurs when a vicinal diol is exposed to acidic conditions. The reaction begins when one of the —OH groups is protonated. Following departure of water, an intermediate with a positive charge on carbon is formed. The interesting part of this reaction is that an adjacent alkyl group or H atom migrates to the positively charged carbon to create a more stable cation. Loss of a proton gives the carbonyl product.

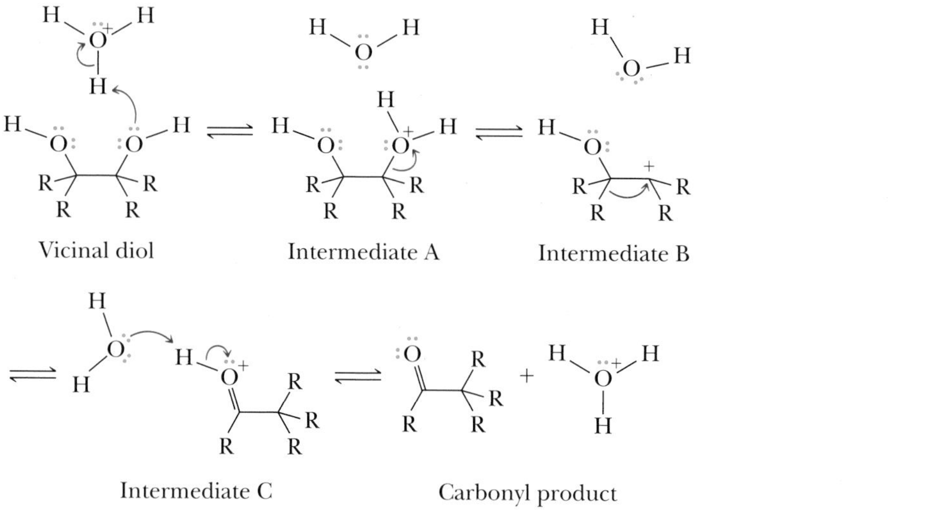

Questions

A. In the pinacol rearrangement, what is the proper name of the intermediate with a positive charge on carbon?

1. Carbonium ion
2. Carbanion
3. Carbrilium ion
4. Carbocation

B. Which of the following intermediates in this reaction is stabilized by resonance delocalization?

1. Intermediate A only
2. Intermediate B only
3. Intermediate C only
4. Intermediates B and C

C. The reaction occurs in acidic solution. Which statement best describes the role of acid in this reaction?

1. The acid is acting in a catalytic fashion.
2. The acid is not involved during the reaction.
3. The reaction consumes one equivalent of acid.
4. The reaction produces one equivalent of acid.

D. Based on your answer to Question C, identify the correct statement.

1. The pH of the solution will drop over time.
2. The pH of the solution will rise over time.
3. The pH of the solution will rise at first and then drop.
4. The pH of the solution will remain constant.

E. Which step in the reaction would you expect to be rate-determining?

1. Vicinal diol to intermediate A.
2. Intermediate A to intermediate B.
3. Intermediate B to intermediate C.
4. Intermediate C to carbonyl product.

F. If there are different R-groups on the two alcohol carbons (as shown here), which statement would be correct about what the chemist is testing by running this reaction?

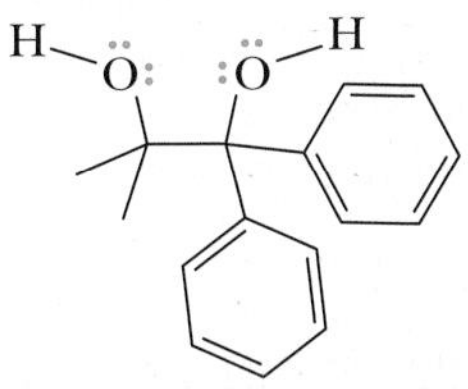

1. The chemist is testing which of the R groups has the greatest propensity to migrate.
2. The chemist is testing which of the R groups makes the attached alcohol most basic.
3. The chemist is testing which of the R groups is best at stabilizing a positive charge on carbon.
4. The chemist is testing whether the acid is a catalyst.

G. If there are two different R groups on each alcohol carbon (as shown here), which statement would be correct about what the chemist is testing by running this reaction?

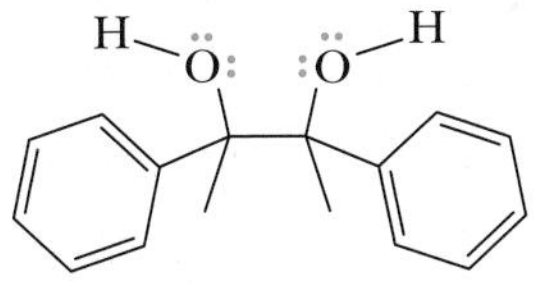

1. The chemist is testing which of the two R groups has the greatest propensity to migrate.
2. The chemist is testing which of the two R groups makes the attached alcohol most basic.
3. The chemist is testing which of the two R groups is best at stabilizing a positive charge on carbon.
4. The chemist is testing whether the acid is a catalyst.

10.8 Oxidation of Alcohols

Oxidation of a primary alcohol gives an aldehyde or a carboxylic acid, depending on experimental conditions. Secondary alcohols are oxidized to ketones. Tertiary alcohols are not oxidized. Following is a series of transformations in which a primary alcohol is oxidized first to an aldehyde and then to a carboxylic acid. The fact that each transformation involves oxidation is indicated by the symbol O in brackets over the reaction arrow.

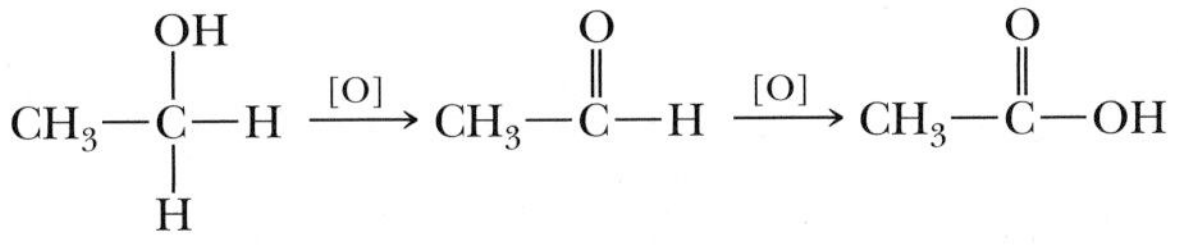

A primary alcohol | An aldehyde | A carboxylic acid

A. Chromic Acid

One reagent used in the laboratory for the oxidation of a primary alcohol to a carboxylic acid is chromic acid, H_2CrO_4. A solution of chromic acid in aqueous sulfuric acid is known as the **Jones reagent**.

Because of the low solubility of most organic compounds in water, their oxidation by chromic acid is commonly carried out by dissolving them in acetone and then adding a stoichiometric amount of Jones reagent to complete the oxidation.

Oxidation of 1-hexanol, for example, using chromic acid in the mixed solvent of aqueous sulfuric acid and acetone gives hexanoic acid in high yield.

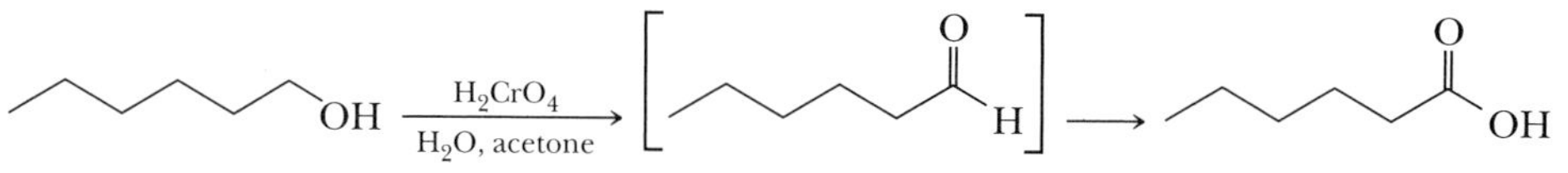
OH
H_2CrO_4
H_2O, acetone
O
H
O
OH

1-Hexanol **Hexanal** (not isolated) **Hexanoic acid**

These experimental conditions are more than sufficient to oxidize the intermediate aldehyde to a carboxylic acid.

Secondary alcohols are oxidized to ketones by chromic acid.

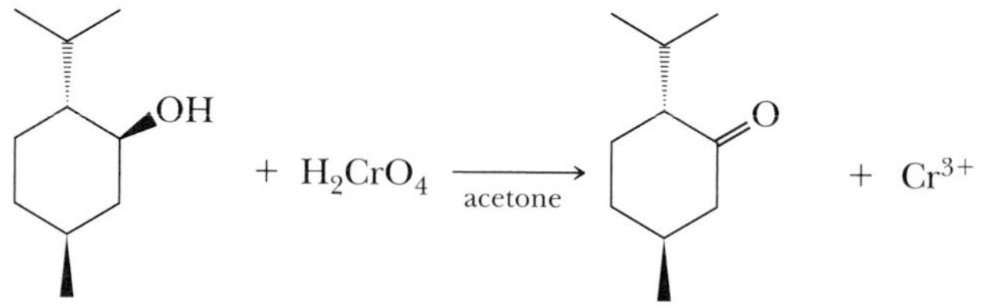
OH
+ H_2CrO_4
acetone
O
+ Cr^{3+}

2-Isopropyl-5-methyl-cyclohexanol (Menthol) **2-Isopropyl-5-methyl-cyclohexanone (Menthone)**

Tertiary alcohols are resistant to oxidation because the carbon bearing the —OH is already bonded to three carbon atoms and therefore cannot form an additional carbon-oxygen bond.

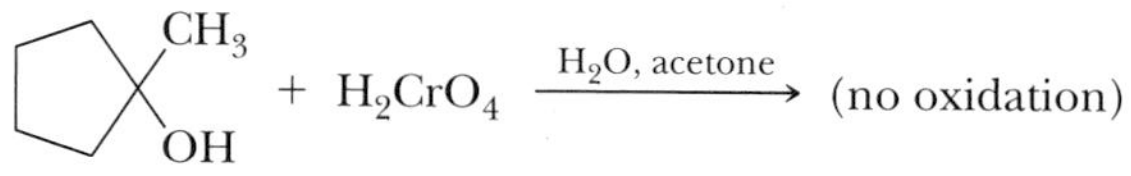
CH_3
OH
+ H_2CrO_4
H_2O, acetone
(no oxidation)

1-Methylcyclopentanol

Thus, the prerequisite for the oxidation of an alcohol to an aldehyde or a ketone is at least one H on the carbon bearing the OH group.

Mechanism 10.8

Chromic Acid Oxidation of an Alcohol

Step 1: Reaction of the alcohol and chromic acid gives an alkyl chromate by a mechanism similar to that for the formation of a carboxylic ester (Section 17.7). There is no change in oxidation state of either carbon or chromium as a result of this step.

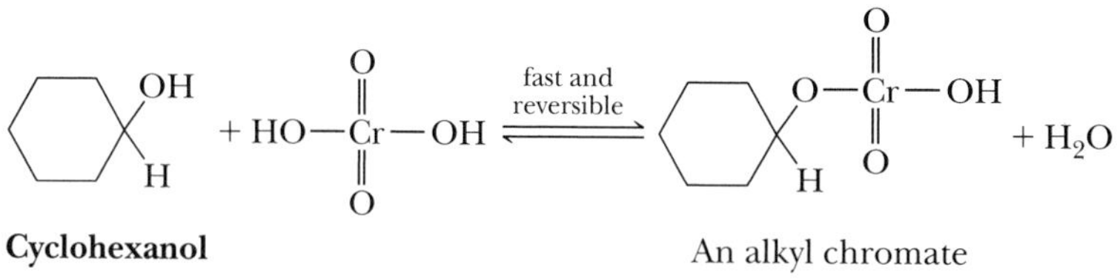
OH
H
+ HO—Cr—OH
fast and reversible
O—Cr—OH
H
+ H_2O

Cyclohexanol An alkyl chromate

Step 2: **Take a proton away and simultaneously break bonds to give stable molecules or ions.** Reaction of the alkyl chromate with a base (here shown as a water molecule) results in cleavage of a C—H bond, formation of the carbonyl group, and reduction of chromium(VI) to chromium(IV).

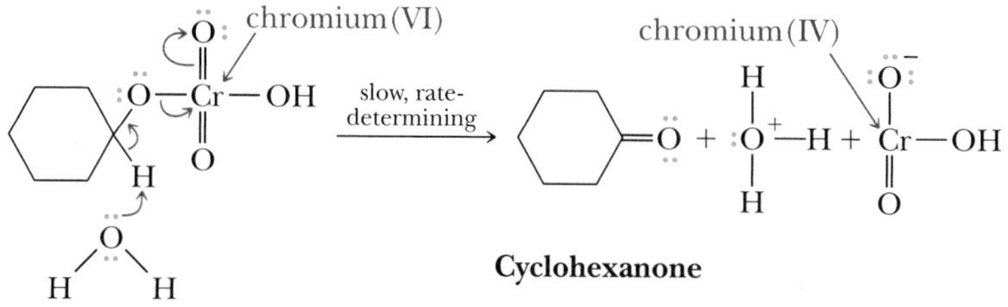
chromium(VI)
chromium(IV)
O—Cr—OH
slow, rate-determining
H
O
H H

Cyclohexanone

This step is the oxidation-reduction step; carbon undergoes a two-electron oxidation and chromium(VI) undergoes a two-electron reduction to chromium(IV). Chromium(IV) then participates in further oxidations by a similar mechanism and eventually is transformed to Cr(III).

We have shown that in aqueous chromic acid, a primary alcohol is oxidized first to an aldehyde and then to a carboxylic acid. In the second step, it is not the aldehyde that is oxidized, but rather the aldehyde hydrate formed by addition of a molecule of water to the aldehyde carbonyl group (hydration). An —OH of the aldehyde hydrate reacts with chromic acid to complete the oxidation of the aldehyde to a carboxylic acid.

The hydration step is critical because it converts the aldehyde carbonyl into two hydroxyl groups. Chromic acid can only react with an —OH to form an alkyl chromate ester, not a carbonyl. The hydration process is catalyzed by acid, and although equilibrium favors the aldehyde, enough hydrate is made to complete the reaction. We will study the hydration of aldehydes and ketones in more detail in Section 16.7.

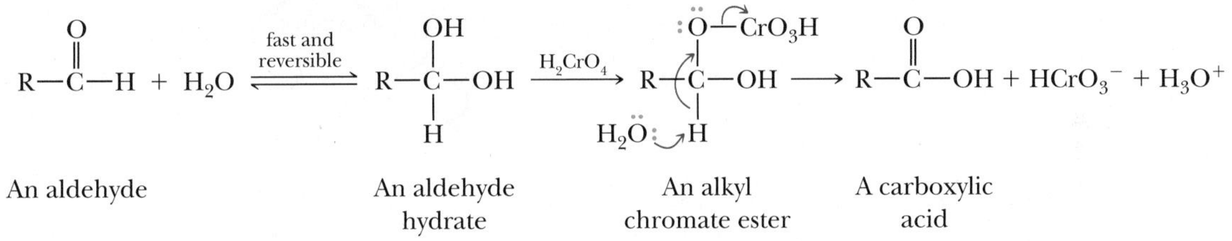

B. Pyridinium Chlorochromate

The form of Cr(VI) most commonly used for oxidation of a primary alcohol to an aldehyde is prepared by dissolving CrO_3 in aqueous HCl and adding pyridine to precipitate **pyridinium chlorochromate (PCC)** as a solid.

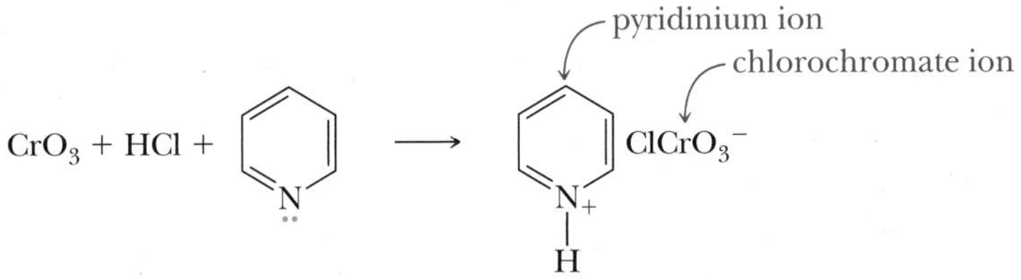

Pyridine **Pyridinium chlorochromate (PCC)**

PCC not only is selective for the oxidation of primary alcohols to aldehydes but also has little effect on carbon-carbon double bonds or other easily oxidized functional groups. In the following example, geraniol, a primary terpene alcohol, is oxidized to geranial without affecting either carbon-carbon double bond.

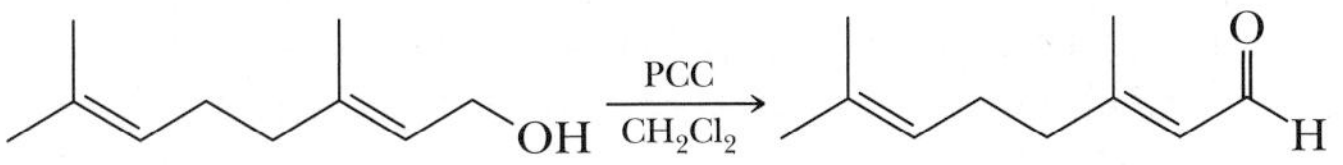

Geraniol **Geranial**

PCC does not oxidize aldehydes further because the PCC reagent is not used in water but rather in an organic solvent, usually CH_2Cl_2. Without water, the product aldehyde is not in equilibrium with the aldehyde hydrate. Recall that only an —OH of an aldehyde hydrate is susceptible to further oxidation by Cr(VI), not an aldehyde carbonyl. Both PCC and H_2CrO_4 can be used for the oxidation of a 2° alcohol to a ketone.

Example 10.11 Predicting Reaction Products I

Draw the product of treating each alcohol with PCC.

(a) 1-Hexanol **(b)** 2-Hexanol **(c)** Cyclohexanol

Solution

1-Hexanol, a primary alcohol, is oxidized to hexanal. 2-Hexanol, a secondary alcohol, is oxidized to 2-hexanone. Cyclohexanol, a secondary alcohol, is oxidized to cyclohexanone.

(a) 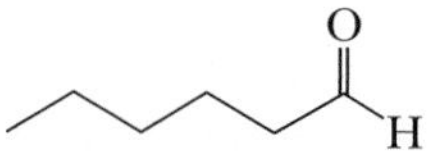
O
H

Hexanal

(b) 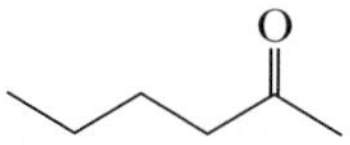
O

2-Hexanone

(c) 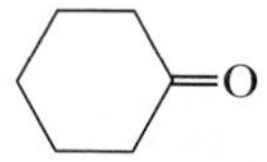
O

Cyclohexanone

Problem 10.11

Draw the product of treating each alcohol in Example 10.11 with chromic acid.

Chemical Connections

Blood Alcohol Screening

Potassium dichromate oxidation of ethanol to acetic acid is the basis for the original breath alcohol screening test used by law enforcement agencies to determine a person's blood alcohol content. The test is based on the difference in color between the dichromate ion (reddish orange) in the reagent and the chromium(III) ion (green) in the product. Thus, color change from reddish orange to green can be used as a measure of the quantity of ethanol present in a sample of a person's breath.

In its simplest form, a breath alcohol screening test consists of a sealed glass tube containing a potassium dichromate-sulfuric acid reagent impregnated on silica gel. To administer the test, the ends of the tube are broken off, a mouthpiece is fitted to one end, and the other end is inserted into the neck of a plastic bag. The person being tested then blows into the mouthpiece until the plastic bag is inflated.

As breath containing ethanol vapor passes through the tube, reddish orange dichromate is reduced to green chromium(III). The concentration of ethanol in the breath is then estimated by measuring how far the green color extends along the length of the tube. When the green color extends beyond the halfway point, the person is judged to have a sufficiently high blood alcohol content to warrant further, more precise testing.

The Breathalyzer, a more accurate testing device, operates on the same principle as the simplified screening test. In a Breathalyzer test, a measured volume of breath is bubbled through a solution of potassium dichromate in aqueous sulfuric acid and the color change is measured spectrophotometrically.

These tests measure alcohol in the breath. The legal definition of being under the influence of alcohol, however, is based on blood alcohol content,

$$\underset{\textbf{Ethanol}}{CH_3CH_2OH} + \underset{\substack{\textbf{Dichromate ion}\\ \text{(reddish orange)}}}{Cr_2O_7^{2-}} \xrightarrow[H_2O]{H_2SO_4} \underset{\textbf{Acetic acid}}{CH_3\overset{O}{\overset{\|}{C}}OH} + \underset{\substack{\textbf{Chromium(III) ion}\\ \text{(green)}}}{Cr^{3+}}$$

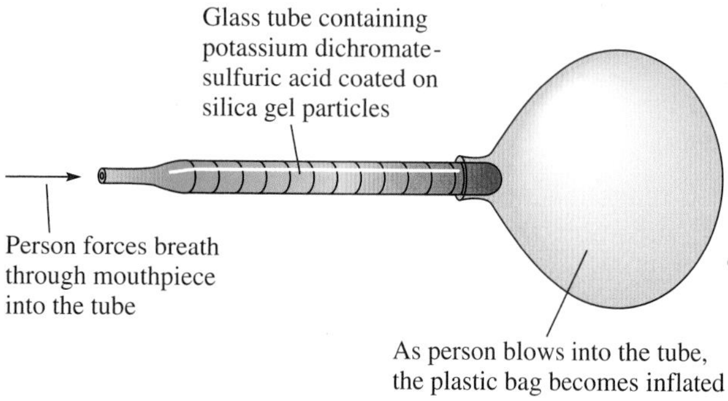
Glass tube containing potassium dichromate-sulfuric acid coated on silica gel particles
Person forces breath through mouthpiece into the tube
As person blows into the tube, the plastic bag becomes inflated

not breath alcohol content. The chemical correlation between these two measurements is that air deep within the lungs is in equilibrium with blood passing through the pulmonary arteries, and an equilibrium is established between blood alcohol and breath alcohol. It has been determined by tests in a person drinking alcohol that 2100 mL of breath contains the same amount of ethanol as 1.00 mL of blood.

C. Swern Oxidation

Due to the toxic nature of chromium compounds, alternatives to the Jones reagent and PCC have been developed. These more modern methods are gradually replacing H_2CrO_4 and PCC.

One such replacement is the Swern oxidation. The oxidizing agent itself is a chlorosulfonium salt, which is generated at −78°C by the reaction of DMSO with oxalyl chloride. (Running this reaction at ambient temperature is explosive!) Slow addition of the alcohol at low temperature, followed by the addition of a tertiary amine such as triethyl amine (Et_3N), yields the product.

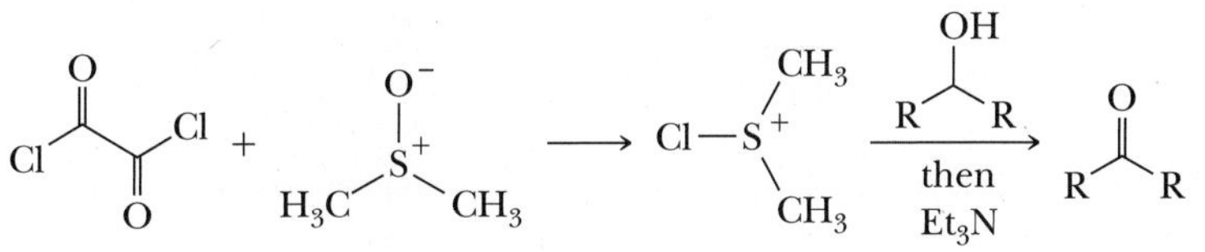

Oxalyl chloride $(COCl)_2$ **Dimethyl sulfoxide** DMSO Chlorosulfonium salt

Primary alcohols are cleanly oxidized to aldehydes (not to carboxylic acids as with Jones reagent), secondary alcohols yield ketones, and tertiary alcohols are again unreactive. Therefore, a Swern oxidation accomplishes the same transformations as PCC. Hence, in Example 10.11, the same products would be obtained with a Swern oxidation as with the application of PCC.

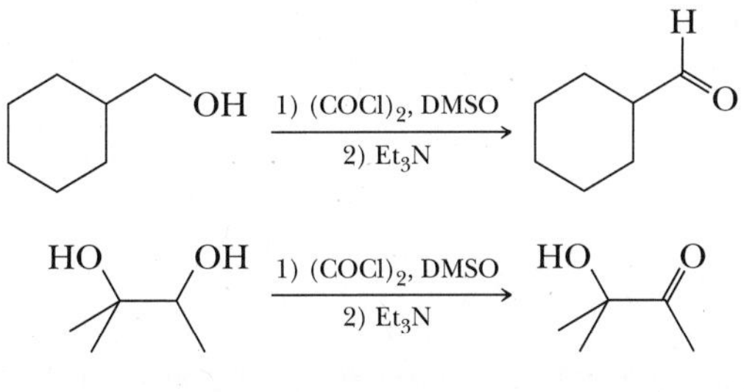

Mechanism 10.9

Swern Oxidation, Starting at the Point of the Chlorosulfonium Ion

Step 1: The reaction of the chlorosulfonium with an alcohol creates an alkylsulfonium ion and HCl. The exact sequence of steps involved in this transformation is still being investigated.

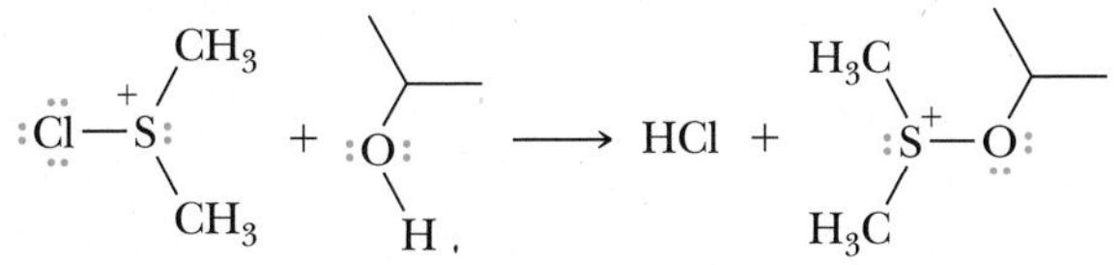

Chlorosulfonium Alkylsulfonium

(Continued)

Step 2: **Take a proton away.** The addition of a tertiary amine, such as triethylamine, leads to a deprotonation of a methyl adjacent to the cationic sulfur, leading to what is referred to as an alkoxysulfonium ylide.

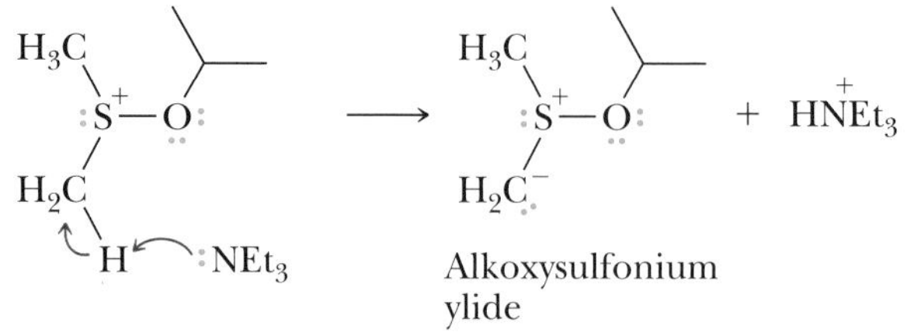

Step 3: **Take a proton away while simultaneously breaking a bond so that stable molecules or ions are created.** The carbanion performs an intramolecular proton transfer that induces cleavage of the alkoxysulfonium ylide into dimethylsulfide and the oxidized product.

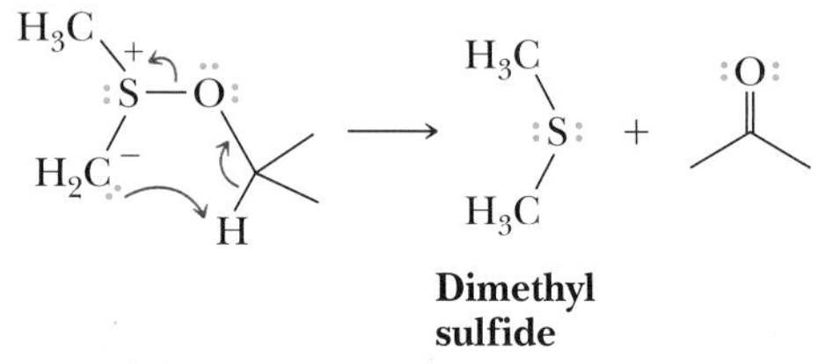

D. Dess-Martin Oxidation

Another more environmentally friendly class of oxidants than chromium species involve hypervalent iodine compounds, such as the reagent commonly referred to as the Dess-Martin periodinane (DMP). Hypervalent means a compound with a greater electron count than predicted by the octet rule. The name DMP derives from the two chemists that developed its reactivity, and its use is called the Dess-Martin oxidation. Once again, primary alcohols are oxidized to aldehydes, while secondary alcohols are oxidized to ketones, and tertiary alcohols are unreactive. Hence, DMP performs the same transformations as PCC and the Swern oxidation (same products as in Example 10.11).

Mechanism 10.10

Dess-Martin Oxidation

Step 1: Reaction of an alcohol with DMP gives a diacetoxyalkoxy periodinane, similar to the formation of an alkyl chromate in the first step of the Jones oxidation ($^{-}$OAc = acetate).

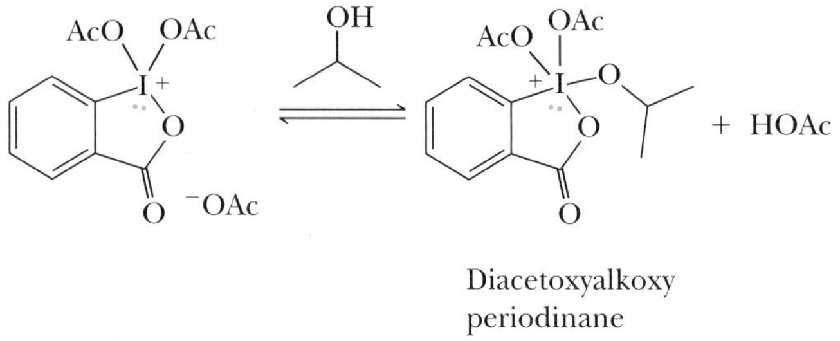

Step 2: **Take a proton away while simultaneously breaking a bond so that stable molecules or ions are created.** Deprotonation of the hydrogen on the coordinated alkoxy group leads to a reduction of the iodine, expulsion of acetate, and the oxidized aldehyde or ketone product.

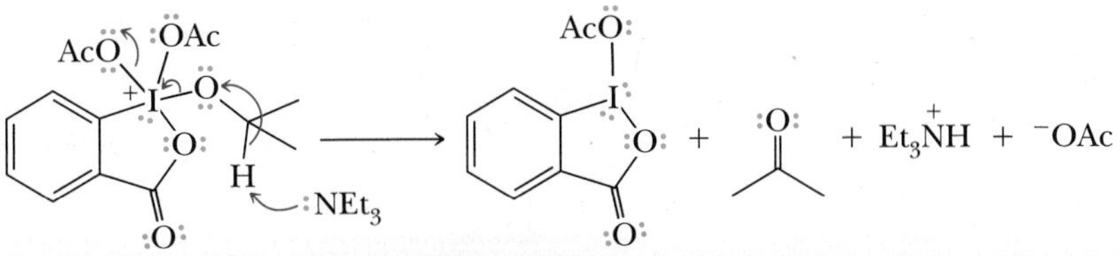

E. Periodic Acid Oxidation of Glycols

Periodic acid, H_5IO_6 (or alternatively, $HIO_4 \cdot 2H_2O$), is a white crystalline solid, mp 122°C. Its major use in organic chemistry is for the cleavage of a glycol to two carbonyl groups. In the process, periodic acid is reduced to iodic acid.

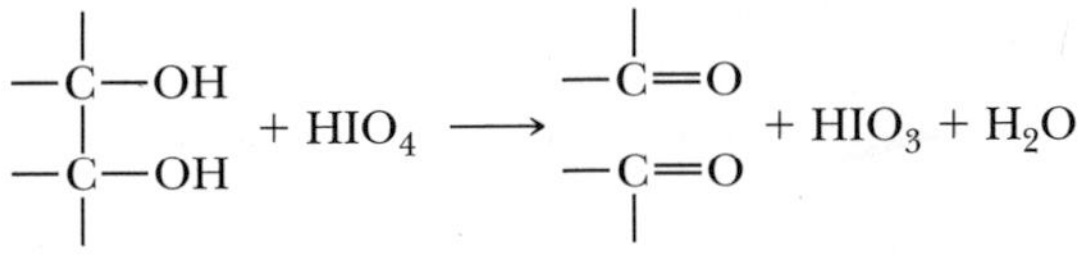

Periodic acid **Iodic acid**

For example, periodic acid oxidizes *cis*-1,2-cyclohexanediol to hexanedial.

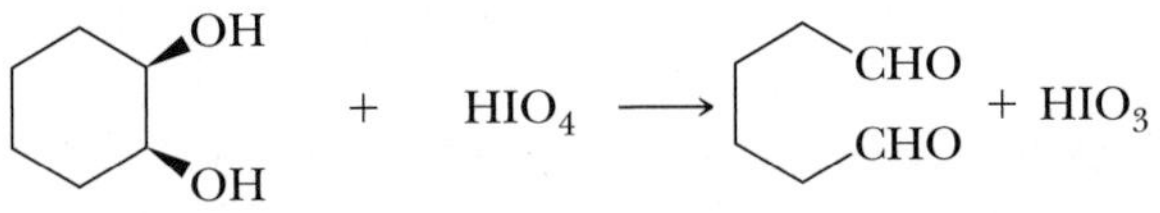

***cis*-1,2-Cyclohexanediol** **Periodic acid** **Hexanedial** **Iodic acid**

Mechanism 10.11

Oxidation of a Glycol by Periodic Acid

Step 1: Reaction of the glycol with periodic acid gives a five-membered cyclic periodate.

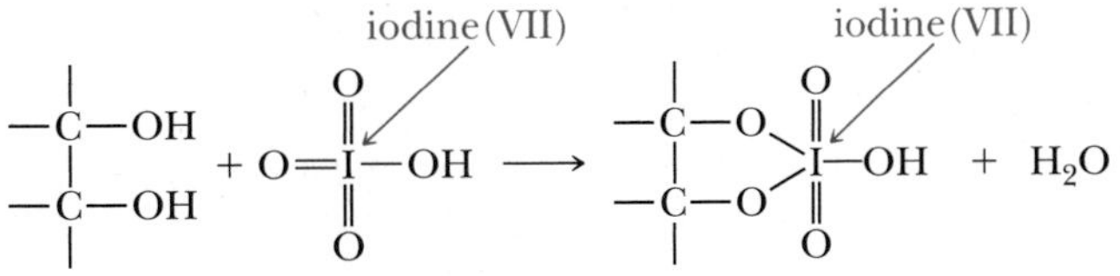
iodine(VII)
iodine(VII)

A cyclic periodate

There is no change in the oxidation state of either iodine or the glycol as a result of formation of the cyclic periodate.

Step 2: **Break bonds to give stable molecules or ions.** Redistribution of valence electrons within the cyclic periodate gives HIO_3 and two carbonyl groups. A result of this electron redistribution is an oxidation of the organic component and a reduction of the iodine-containing component.

(*Continued*)

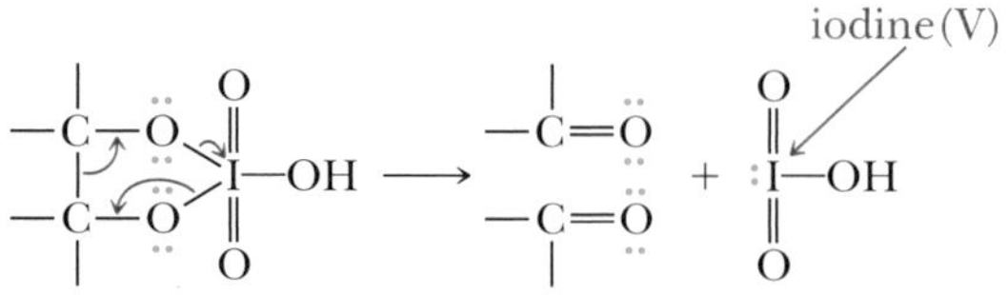
iodine(V)
—OH
+
—OH

HIO_3
(Iodic acid)

This mechanism is consistent with the fact that HIO_4 oxidations are restricted to glycols that can form a five-membered cyclic periodate. Any glycol that cannot form such a cyclic periodate is not oxidized by periodic acid. Following are structural formulas for two isomeric decalindiols. Only the *cis* glycol can form a cyclic periodate with periodic acid, and only the *cis* glycol is oxidized by this reagent.

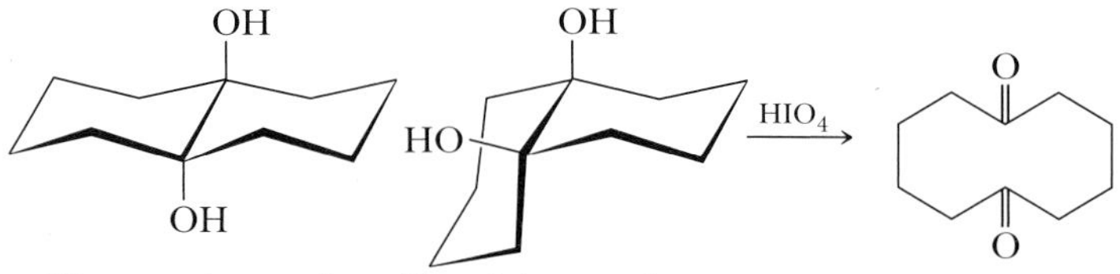
OH
OH
OH
HO
HIO_4
O
O

The *trans* isomer is unreactive toward periodic acid

The *cis* isomer forms a cyclic periodate and is cleaved

Connections to Biological Chemistry

The Oxidation of Alcohols by NAD^+

It should come as no surprise that biological systems do not use agents such as potassium dichromate or the oxides of other transition metals for the oxidation of alcohols to aldehydes and ketones or for the oxidation of aldehydes to carboxylic acids. What biological systems use instead is NAD^+ (AD = a combination of adenine and ribose). The function of the Ad portion of the molecule is to position NAD^+ on the surface of the enzyme in the proper orientation relative to the molecule it is to oxidize.

When NAD^+ functions as a biological oxidizing agent, it is reduced to NADH. In this transformation,

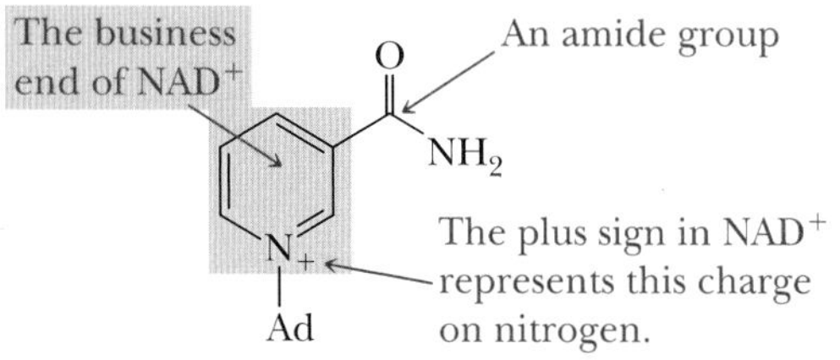
The business end of NAD^+
An amide group
O
NH_2
N+
Ad
The plus sign in NAD^+ represents this charge on nitrogen.

Nicotinamide adenine dinucleotide (NAD^+)

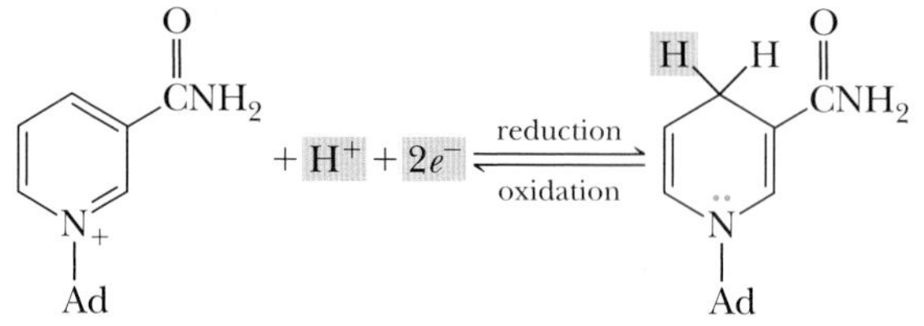
O
CNH_2
N+
Ad
$+ H^+ + 2e^-$
reduction
oxidation
H H
O
CNH_2
N
Ad

NAD^+
(oxidized form)

NADH
(reduced form)

NAD^+ gains one H and loses the positive charge on its nitrogen. The key concept here is that NAD^+ is a two-electron oxidizing agent and in the process undergoes a two-electron reduction to NADH.

NAD^+ serves as an oxidizing agent in a wide variety of enzyme-catalyzed reactions, two of which are shown here. The oxidation of ethanol to acetaldehyde is the first step in the reaction by which the liver detoxifies ethanol. The oxidation of lactate to pyruvate is one step in the process by which the body derives energy from the oxidation of carbohydrates. Lactate is the end product of anaerobic (without oxygen) glycolysis.

$$\underset{\textbf{Ethanol}}{CH_3CH_2OH} + NAD^+ \xrightarrow[\text{dehydrogenase}]{\text{alcohol}} \underset{\textbf{Ethanal (Acetaldehyde)}}{CH_3\overset{\overset{\displaystyle O}{\|}}{C}H} + NADH + H^+$$

$$\underset{\textbf{Lactate}}{CH_3\overset{\overset{\displaystyle OH}{|}}{C}HCOO^-} + NAD^+ \xrightarrow[\text{dehydrogenase}]{\text{lactate}} \underset{\textbf{Pyruvate}}{CH_3\overset{\overset{\displaystyle O}{\|}}{C}COO^-} + NADH + H^+$$

Following is a mechanism for the oxidation of an alcohol by alcohol dehydrogenase and NAD^+.

Mechanism 10.12

Oxidation of an Alcohol by NAD+

Arrow①: A basic group, B^-, on the enzyme removes H^+ from the —OH group.

Arrow②: The H—O σ bond breaks as a C═O bond forms.

Arrow③: Transfer of a hydride ion from the carbon bearing the —OH group to NAD^+ creates the new C—H bond in NADH. This is the oxidation-reduction step: the alcohol is oxidized, and NAD^+ is reduced.

Arrows④–⑤: Electrons within the ring flow to the positively charged nitrogen.

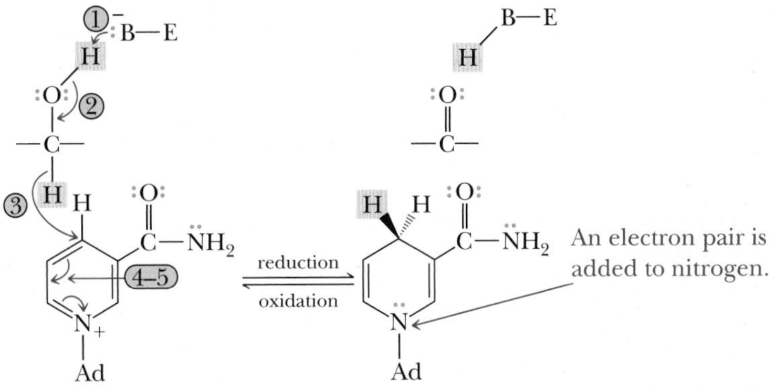
B—E
H
:O:
C
H
H
:O:
C—NH2
4–5
N+
Ad
reduction
oxidation
B—E
H
:O:
C
H
H
:O:
C—NH2
N
Ad
An electron pair is added to nitrogen.

NAD+ **NADH**

Because enzymes are chiral catalysts, NAD^+ oxidation takes place in a chiral environment, with the result that hydride transfer to NAD^+ is stereoselective; some enzymes catalyze the addition of hydride ion to the top face of NAD^+; others, to the bottom face. In the case of alcohol dehydrogenase, the hydride ion is transferred to the top face.

As we will see in Section 16.11, NADH can in turn reverse the process and transfer a hydride ion stereoselectively to the carbonyl group of an aldehyde or a ketone, thus reducing these types of molecules to either primary or secondary alcohols, respectively.

MCAT Practice: *Passage and Questions*

Alcohol Oxidations

The mechanisms of alcohol oxidation by several reagents have many aspects in common. Shown to the right are the key steps in a variety of oxidations, along with electron flow arrows.

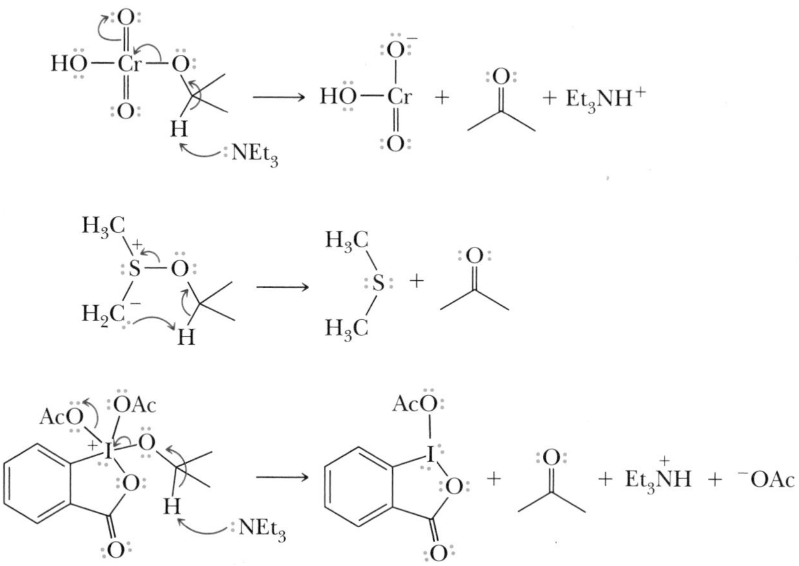

Questions

A. What is the common role of the chromium-containing group, the periodinane, and the dimethylsufide in each of these reactions?

1. All these groups are acting as nucleophiles.
2. All these groups are acting as leaving groups.
3. All these groups are acting as Lewis bases.
4. All these groups are acting as Brønsted-Lowry bases.

B. In each reaction shown, a deprotonation occurs simultaneously with the cleavage of a bond to create two molecules, one of which develops a π bond. What general mechanistic acronym best describes each reaction?

1. E2
2. E1
3. S_N2
4. S_N1

The key step for alcohol oxidation by nature's reagent, nicotinamide adenine dinucleotide (NAD^+), is given in the mechanism above. Use that reaction for the reduction of NAD^+ to answer the following questions.

C. In contrast to the laboratory oxidation mechanisms given above where a base removes a proton from the C—H bond, in nature's alcohol oxidation, the base removes a proton from the O—H bond. How is the hydrogen on the C—H bond is removed with NAD^+?

1. The NAD^+ deprotonates the C—H bond of the alcohol.
2. The NAD^+ performs a hydrogen atom abstraction from the C—H bond.
3. The NAD^+ accepts a hydride from the C—H bond.

D. Why does nature use a reagent as complex as NAD^+ to perform an oxidation?

1. Nature binds the oxidizing agent in an enzyme in proximity to the alcohol being oxidized.
2. Nature cannot use strong bases in an aqueous environment.
3. Nature cannot use toxic heavy metals.
4. All of the above are true.

Example 10.12 Predicting Reaction Products II

What products are formed when each glycol is treated with HIO_4?

(a) 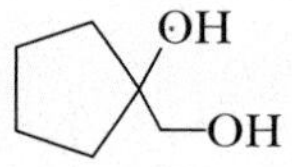

(b) 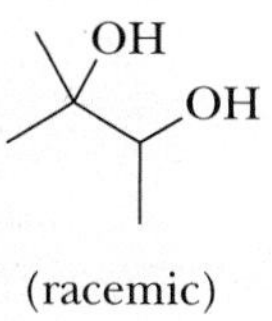

(racemic)

Solution

The bond between the carbons bearing the —OH groups is cleaved, and each —OH group is converted to a carbonyl group.

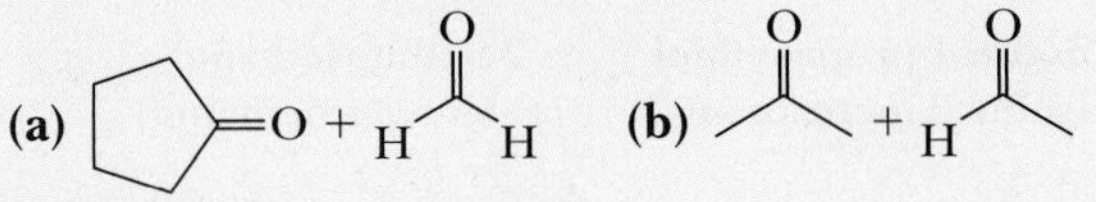

Problem 10.12

α-Hydroxyketones and α-hydroxyaldehydes are also oxidized by treatment with periodic acid.

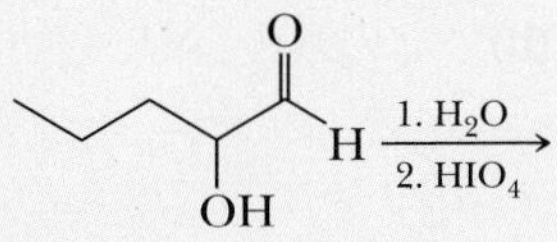

(racemic)

It is not the α-hydroxyketone or aldehyde, however, that undergoes reaction with periodic acid, but the hydrate formed by addition of water to the carbonyl group of the α-hydroxyketone or aldehyde. Write a mechanism for the oxidation of this α-hydroxyaldehyde by HIO_4.

10.9 Thiols

A. Structure

The functional group of a **thiol** is an **—SH (sulfhydryl)** group bonded to an sp^3 hybridized carbon. Figure 10.4 shows a Lewis structure and a ball-and-stick model of methanethiol, CH_3SH, the simplest thiol. The C—S—H bond angle in methanethiol is 100.3°. By way of comparison, the H—S—H bond angle in H_2S is 93.3°. If a sulfur atom were bonded to two other atoms by fully hybridized sp^3 hybrid orbitals, bond angles about sulfur would be approximately 109.5°. If, instead, a sulfur atom were bonded to two other atoms by unhybridized 3*p* orbitals, bond angles would be approximately 90°. The fact that the C—S—H bond angle in methanethiol is 100.3° and the H—S—H bond angle in H_2S is 93.3° indicates that there is considerably more *p*-character (and hence less *s*-character) in the bonding orbitals of divalent sulfur than there is in those of divalent oxygen.

Thiol
A compound containing an —SH (sulfhydryl) group bonded to an sp^3 hybridized carbon.

Mercaptan
A common name for a thiol [i.e., any compound that contains an —SH (sulfhydryl) group].

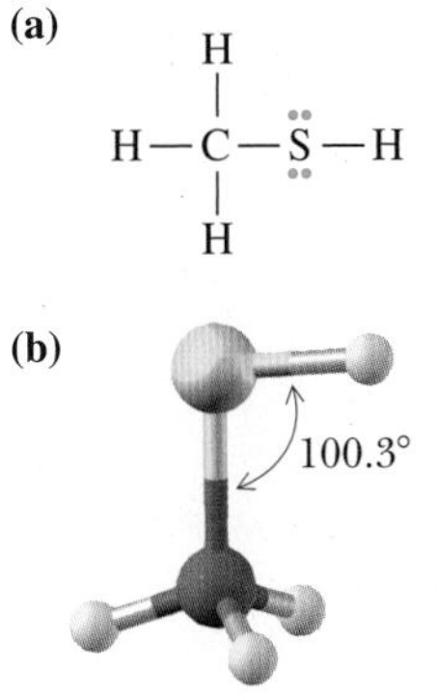

FIGURE 10.4 Methanethiol, CH_3SH. (a) Lewis structure and (b) ball-and-stick model.

B. Nomenclature

In the older literature, thiols are often referred to as **mercaptans**, which literally means mercury capturing. They react with Hg^{2+} in aqueous solution to give sulfide salts as insoluble precipitates. Thiophenol, C_6H_5SH, for example, gives $(C_6H_5S)_2Hg$.

In the IUPAC system, thiols are named by selecting as the parent alkane the longest chain of carbon atoms that contains the —SH group. To show that the compound is a thiol, retain the final *-e* in the name of the parent alkane and add the suffix *-thiol*. The location of the —SH group takes precedence over alkyl groups and halogens in numbering the parent chain.

In the IUPAC system, —OH takes precedence over —SH in both numbering and naming. In compounds containing these two functional groups, an —SH group is indicated by the IUPAC prefix *sulfanyl-*. Alternatively, it may be indicated by the common-name prefix *mercapto-*.

Common names for simple thiols are derived by naming the alkyl group bonded to —SH and adding the word *mercaptan*.

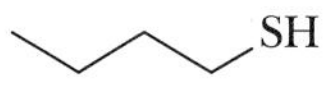

1-Butanethiol (Butyl mercaptan)

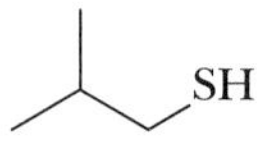

2-Methyl-1-propanethiol (Isobutyl mercaptan)

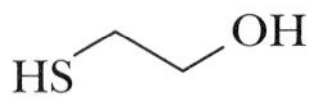

2-Sulfanylethanol (β-Mercaptoethanol)

Example 10.13 Thiol Nomenclature

Write names for these thiols.

(a) 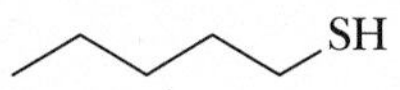 **(b)** 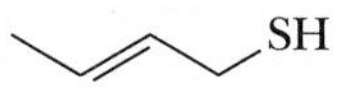

Solution

(a) 1-Pentanethiol (pentyl mercaptan)

(b) (*E*)-2-Butene-1-thiol (*trans*-2-butene-1-thiol)

Problem 10.13

Write IUPAC names for these thiols.

(a) 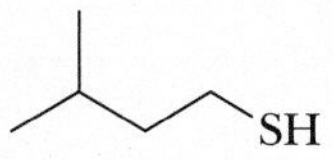 **(b)** 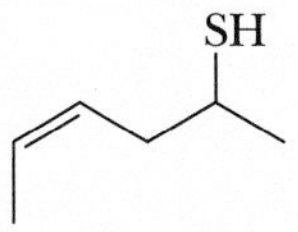

C. Physical Properties

The most outstanding physical characteristic of low-molecular-weight thiols is their stench. The scent of skunks is primarily the result of the following two thiols.

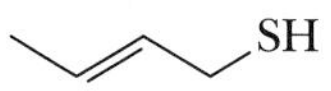

2-Butene-1-thiol

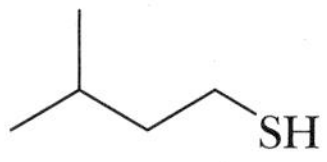

3-Methyl-1-butanethiol (Isopentyl mercaptan)

A blend of low-molecular-weight thiols is added to natural gas as odorants. The most common of these is 2-methyl-2-propanethiol (*tert*-butyl mercaptan) because it is the most resistant to oxidation and has the greatest soil penetration. 2-Propanethiol is also used for this purpose, usually as a blend with 2-methyl-2-propanethiol.

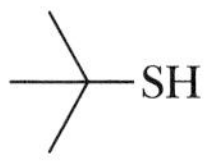

2-Methyl-2-propanethiol (*tert*-Butyl mercaptan)

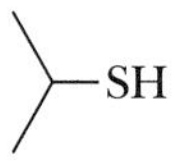

2-Propanethiol (Isopropyl mercaptan)

Because of the very low polarity of the S—H bond, thiols show little association by hydrogen bonding. Consequently, they have lower boiling points and are less soluble in water and other polar solvents than alcohols of comparable molecular weights. Table 10.3 gives names and boiling points for three low-molecular-weight thiols. Shown for comparison are boiling points for alcohols that contain the same number of carbon atoms.

In Section 10.2, we illustrated the importance of hydrogen bonding in alcohols by comparing the boiling points of ethanol (bp 78°C) and its constitutional isomer dimethyl ether (bp −24°C). By comparison, the boiling point of ethanethiol is 35°C and that of its constitutional isomer dimethyl sulfide is 37°C.

CH_3CH_2SH	CH_3SCH_3
Ethanethiol	**Dimethyl sulfide**
bp 35°C	bp 37°C

The fact that the boiling points of these constitutional isomers are almost identical indicates that little or no association by hydrogen bonding occurs between thiol molecules.

Table 10.3 Boiling Points of Thiols and Alcohols of the Same Number of Carbon Atoms

Thiol	bp (°C)	Alcohol	bp (°C)
Methanethiol	6	Methanol	65
Ethanethiol	35	Ethanol	78
1-Butanethiol	98	1-Butanol	117

D. Thiols in Biological Molecules

The thiol group is found in the amino acid L-cysteine. L-Cysteine is important because the thiol groups of pairs of cysteines are oxidized to disulfide bonds (Section 10.9G), which are a major factor in stabilizing the three-dimensional structure of protein molecules. The thiol group of cysteine functions as a nucleophile in certain enzyme mechanisms. In addition, it binds the metal in certain metal-containing enzymes.

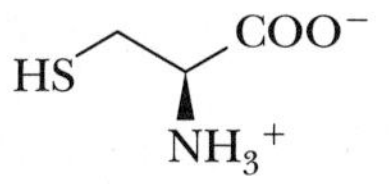

(*R*)-2-Amino-3-mercaptopropanoic acid
(L-Cysteine)

E. Preparation

The most common preparation of thiols, RSH, depends on the high nucleophilicity of the hydrosulfide ion, HS^- (Section 9.3E). Sodium hydrosulfide is prepared by bubbling H_2S through a solution of NaOH in water or aqueous ethanol. Reaction of HS^- with a haloalkane gives a thiol.

$$CH_3(CH_2)_8CH_2I + Na^+SH^- \xrightarrow[\text{ethanol}]{S_N2} CH_3(CH_2)_8CH_2SH + Na^+I^-$$

1-Iododecane **Sodium hydrosulfide** **1-Decanethiol**

In practice, the scope and limitations of this reaction are governed by the limitations of the S_N2 reaction and by competition between substitution and β-elimination (Section 9.8). The reaction is most useful for preparation of thiols from primary haloalkanes. Yields are lower from secondary halides because of the competing β-elimination (E2) reaction. With tertiary halides, β-elimination (E2) predominates, and the alkene formed by dehydrohalogenation is the major product.

In a commercial application of thiol formation by this nucleophilic substitution, the sodium salt of thioglycolic acid is prepared by the reaction of sodium hydrosulfide and sodium iodoacetate.

$$\underset{\textbf{Sodium hydrosulfide}}{Na^+\,SH^-} + \underset{\textbf{Sodium iodoacetate}}{ICH_2\overset{O}{\overset{\|}{C}}O^-\,Na^+} \xrightarrow{S_N2} \underset{\substack{\textbf{Sodium mercaptoacetate}\\ \textbf{(Sodium thioglycolate)}}}{HSCH_2\overset{O}{\overset{\|}{C}}O^-\,Na^+} + Na^+\,I^-$$

The sodium and ammonium salts of thioglycolic acid are used in cold waving of hair. These compounds work by breaking the disulfide bonds of hair proteins, which maintain the overall structure of hair (Section 27.6C). The disulfide bonds are subsequently reformed by oxidation (Section 10.9G) in a second step. The calcium salt of thioglycolic acid is used as a depilatory; that is, it is used to remove body hair.

$$(HSCH_2COO^-)_2Ca^{2+}$$

Calcium mercaptoacetate
(Calcium thioglycolate)

F. Acidity

Hydrogen sulfide is much more acidic than water.

$$H_2O + H_2O \rightleftharpoons HO^- + H_3O^+ \qquad pK_a = 15.7$$

$$H_2S + H_2O \rightleftharpoons HS^- + H_3O^+ \qquad pK_a = 7.0$$

Similarly, thiols are stronger acids than alcohols. Compare, for example, the pK_a values of ethanol and ethanethiol in dilute aqueous solution.

$$CH_3CH_2OH + H_2O \rightleftharpoons CH_3CH_2O^- + H_3O^+ \qquad pK_a = 15.9$$

$$CH_3CH_2SH + H_2O \rightleftharpoons CH_3CH_2S^- + H_3O^+ \qquad pK_a = 10.5$$

The greater acidity of thiols compared to alcohols can be explained by the fact that sulfur (a third-period element) is larger than oxygen (a second-period element). The negative charge on an alkylsulfide ion (RS^-) is delocalized over a larger area and is therefore more stable than the negative charge on an alkoxide ion (RO^-).

Thiols are sufficiently strong acids so that when dissolved in aqueous sodium hydroxide, they are converted completely to alkylsulfide salts.

$$\underset{\substack{pK_a = 10.5\\ \text{(Stronger acid)}}}{CH_3CH_2SH} + \underset{\text{(Stronger base)}}{Na^+OH^-} \longrightarrow \underset{\text{(Weaker base)}}{CH_3CH_2S^-Na^+} + \underset{\substack{pK_a = 15.7\\ \text{(Weaker acid)}}}{H_2O}$$

To name salts of thiols, give the name of the cation first, followed by the name of the alkyl group to which is attached the suffix *-sulfide*. For example, the sodium salt derived from ethanethiol is named sodium ethylsulfide.

G. Oxidation

Many of the chemical properties of thiols stem from the fact that the sulfur atom of a thiol is oxidized easily to several higher oxidation states, the most common of which are shown in the following flow diagram.

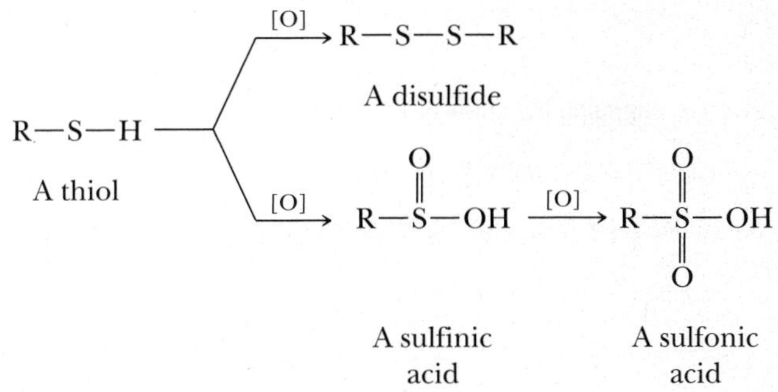

Each oxidation requires a specific oxidizing agent to avoid overoxidation. There are other oxidation states, too, but they are not very stable. Note that as drawn, the valence shell of sulfur appears to contain 8 electrons in a thiol and a disulfide, 10 electrons in a sulfinic acid, and 12 electrons in a sulfonic acid. Recall that although the matter is still being debated, recent calculations indicate that in sulfinic acids and sulfonic acids, it is best to consider each S—O bond to be a single bond, with a negative charge on O and one or two positive charges on S, respectively (Section 1.7C).

The most common oxidation-reduction reaction of sulfur compounds in biological systems is interconversion between a thiol and a disulfide. The functional group of a disulfide is an —S—S— group.

$$2\ \text{RSH} + \text{I}_2 \longrightarrow \text{RSSR} + 2\ \text{HI}$$

A thiol   A disulfide

Thiols are also oxidized to disulfides by molecular oxygen.

$$2\ \text{RSH} + \tfrac{1}{2}\text{O}_2 \longrightarrow \text{RSSR} + \text{H}_2\text{O}$$

A thiol   A disulfide

In fact, thiols are so susceptible to oxidation that they must be protected from contact with air during storage.

The disulfide bond is an important structural feature stabilizing the tertiary structure of many proteins (Section 27.6C).

Recall from Table 9.7 that thiols are moderate nucleophiles and thiolates are good nucleophiles. As such, they rank as better nucleophiles in both their neutral and anionic states as compared to alcohols and alkoxides, respectively. The reason thiols and thiolates are better nucleophiles is that the sulfur atom is more polarizable than oxygen, and increased polarizability enhances nucleophilicity. Also, because the pK_a values for thiols are generally less than 11, thiolate anions react with secondary haloalkanes primarily through an S_N2 substitution mechanism rather than an E2 elimination. The reverse is true for the significantly more basic alkoxides.

Study Guide

10.1 Structure and Nomenclature of Alcohols

- The functional group of an alcohol is an **—OH (hydroxyl) group** bonded to an sp^3 hybridized carbon.
- Alcohols are classified as **1°**, **2°**, or **3°** depending on whether the —OH group is bonded to a **primary**, **secondary**, or **tertiary** carbon, respectively.

- IUPAC names of alcohols are derived by changing the suffix of the parent alkane from *-e* to *-ol*.
  - *The chain is numbered from the direction that gives the carbon bearing the —OH the lower number.*
  - *In compounds containing other functional groups of higher precedence, the presence of —OH is indicated by the prefix* hydroxy-.
- Common names for alcohols are derived by naming the alkyl group bonded to —OH and adding the word *alcohol.*

P **10.1–10.3, 10.14–10.17**

10.2 Physical Properties of Alcohols

- Alcohols are polar compounds with oxygen bearing a partial negative charge and both the carbon and the hydroxyl hydrogen bearing partial positive charges.
- Hydrogen bonds are relatively strong attractive interactions between a hydrogen atom bonded to an atom of high electronegativity, usually O or N, and a lone pair of electrons on a highly electronegative element, again usually O or N.
  - *Because of intermolecular association by* ***hydrogen bonding****, the boiling points of alcohols are higher than those of hydrocarbons of comparable molecular weight.*
  - *Because of increased dispersion forces, the boiling points of alcohols increase with increasing molecular weight.*
  - *Alcohols interact with water by hydrogen bonding and therefore are more soluble in water than hydrocarbons of comparable molecular weight.*

P **10.4–10.5, 10.18–10.24, 10.47, 10.48**

▶ **Figure 10.3**

10.3 Acidity and Basicity of Alcohols

- Alcohols can function as both weak acids (proton donors) and weak bases (proton acceptors).
  - *The pK_a of most alcohols is in the 16–18 range, similar to water. Loss of a proton from an alcohol produces an alkoxide anion, a relatively strong base.*
  - *In the presence of a strong acid, the —OH group can be protonated. Protonation of the —OH group produces the good leaving group —$\overset{+}{O}H_2$, and this is a common mechanistic theme for alcohol reactions.*

P **10.25–10.28, 10.49**

KEY REACTIONS

1. Acidity of Alcohols (Section 10.3) In dilute aqueous solution, methanol and ethanol are comparable in acidity to water; secondary and tertiary alcohols are somewhat weaker acids.

$$CH_3OH + H_2O \rightleftharpoons CH_3O^- + H_3O^+ \qquad pK_a = 15.5$$

10.4 Reaction of Alcohols with Active Metals

- Alcohols react with active metals such as Li, Na, and K to liberate H_2 and form metal alkoxides.

P 10.6, 10.29, 10.30

Problem 10.6

KEY REACTIONS

2. **Reaction with Active Metals (Section 10.4)** Alcohols react with Li, Na, K, and other active metals to form metal alkoxides, which are nearly the same or somewhat stronger bases than the alkali metal hydroxides such as NaOH and KOH.

$$2\,CH_3OH + 2\,Na \longrightarrow 2\,CH_3O^-Na^+ + H_2$$

10.5 Conversion of Alcohols to Haloalkanes and Sulfonates

- A common theme for reactions of alcohols is the conversion of the —OH group into a better leaving group by protonation and/or replacement with a halogen or other group.
- Alcohols can be converted to haloalkanes by reaction with HCl, HBr, and HI or PBr_3, $SOCl_2$ or $SOBr_2$.
  - *The HX reactions work best for tertiary alcohols, where they involve a carbocation intermediate.*
  - *PBr_3, $SOCl_2$, or $SOBr_2$ react well with secondary and primary alcohols. These reactions involve an S_N2 displacement, so inversion of stereochemistry is seen when the OH group is bonded to a chiral center.*
- Alcohols react with sulfonyl chlorides to give alkyl sulfonates. The sulfonate group is a good leaving group analogous to a halogen atom, so alkyl sulfonates can take part in substitution and elimination reactions.

P 10.7, 10.29–10.31, 10.33, 10.35, 10.40, 10.43–10.45, 10.50–10.59

Problem 10.7

KEY REACTIONS

3. **Reaction with HCl, HBr, and HI (Section 10.5A)** Primary alcohols react by an S_N2 mechanism. Strong acid protonates the —OH group, converting it to the good leaving group, $—\overset{+}{O}H_2$, setting up either S_N2 or S_N1 reactions.

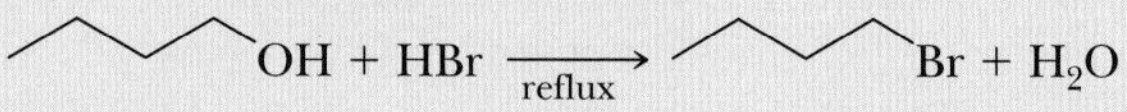

Tertiary alcohols react by an S_N1 mechanism with formation of a carbocation intermediate.

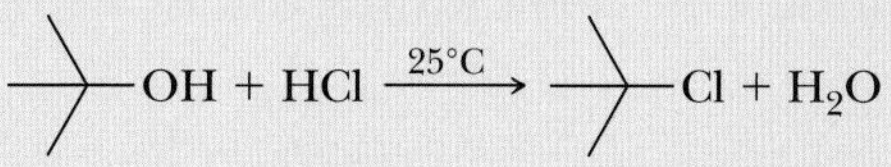

Secondary alcohols may react by an S_N2 or an S_N1 mechanism, depending on the alcohol and experimental conditions. Primary alcohols with extensive β-branching react by an S_N1 mechanism involving formation of a rearranged carbocation.

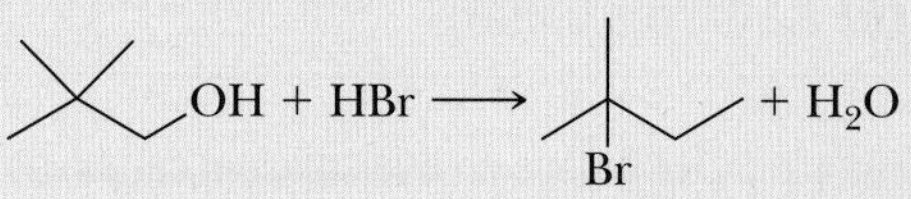

4. **Reaction with PBr_3 (Section 10.5B)** Alcohols react with PBr_3 to initially displace Br^- to give a protonated dibromophosphite intermediate, which is displaced from the backside by Br^- to give the

(*Continued*)

bromoalkane. Although some rearrangement may occur with this reagent, it is less likely than in the reaction of an alcohol with HBr.

$$3\ CH_3CH(OH)CH_3 + PBr_3 \longrightarrow 3\ CH_3CH(Br)CH_3 + H_3PO_3$$

5. **Reaction with $SOCl_2$ and $SOBr_2$ (Section 10.5C)** Alcohols react with $SOCl_2$ to initially give an alkyl chlorosulfite intermediate, which is displaced from the backside by Cl^- to give the chloroalkane. This is often the method of choice for converting a primary or secondary alcohol to an alkyl chloride or alkyl bromide.

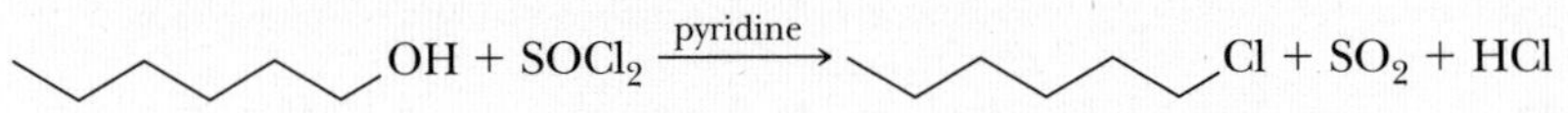

10.6 Acid-Catalyzed Dehydration of Alcohols

- **Dehydration reactions** involve the elimination of water from adjacent carbon atoms.
- Treatment of alcohols with strong acid leads to dehydration to give an alkene.
  - *– Zaitsev's rule is followed; that is, the predominant product is the most stable (usually the most highly substituted) alkene. The mechanism is the reverse of acid-catalyzed hydration of an alkene.*
- According to the **principle of microscopic reversibility**, the sequence of transition states and reactive intermediates (i.e., the mechanism) for any reversible reaction must be the same, but in reverse order, for the reverse reaction as for the forward reaction.

P **10.8, 10.9, 10.29–10.35, 10.37, 10.40, 10.42, 10.45, 10.46, 10.50**

Problem 10.9

KEY REACTIONS

6. **Acid-Catalyzed Dehydration (Section 10.6)** When isomeric alkenes are possible, the major product is generally the more substituted alkene (Zaitsev's rule). Rearrangements are common with secondary alcohols and with primary alcohols with extensive β-branching. The mechanism is the reverse of acid-catalyzed hydration of an alkene, involving initial protonation of the —OH group followed by loss of water to give a carbocation, which loses a proton to give an alkene.

$$\underset{\textbf{2-Butanol}}{CH_3CH_2CH(OH)CH_3} \xrightarrow[\text{heat}]{85\%\ H_3PO_4} \underset{\substack{\textbf{2-Butene}\\(80\%)}}{CH_3CH{=}CHCH_3} + \underset{\substack{\textbf{1-Butene}\\(20\%)}}{CH_3CH_2CH{=}CH_2} + H_2O$$

10.7 The Pinacol Rearrangement

- Some compounds containing OH groups on two adjacent carbon atoms (sometimes called glycols) undergo a characteristic reaction in acid to generate a rearranged aldehyde or ketone product.

P **10.10, 10.36, 10.50**

KEY REACTIONS

7. **Pinacol Rearrangement (Section 10.7)** Dehydration of a glycol involves formation of a carbocation intermediate, rearrangement, and loss of H^+ to give an aldehyde or a ketone.

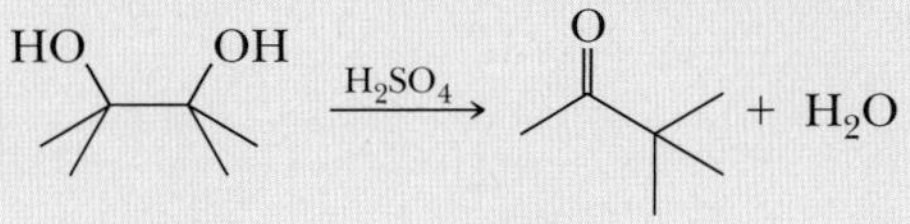

10.8 Oxidation of Alcohols

- Primary alcohols are oxidized by aqueous H_2CrO_4 (called the **Jones reagent**) to give carboxylic acids in a process that involves initial aldehyde formation, followed by conversion to an aldehyde hydrate that is further oxidized to the carboxylic acid.
- Primary alcohols are oxidized to aldehydes using PCC, a reagent that is not used in water, thereby precluding hydrate formation and any further oxidation.
- Primary alcohols are oxidized to aldehydes using DMSO and oxalyl chloride, followed by addition of a tertiary amine, in a reaction known as the **Swern oxidation**.
- Primary alcohols are oxidized to aldehydes using the **Dess-Martin periodinane** (DMP).
- Secondary alcohols can be oxidized to ketones using either PCC or H_2CrO_4.
- Secondary alcohols are oxidized to ketones using DMSO and oxalyl chloride, followed by addition of a tertiary amine, in a reaction known as the **Swern oxidation**.
- Secondary alcohols are oxidized to ketones using the **Dess-Martin periodinane** (DMP).
- Tertiary alcohols are not oxidized because there are already three bonds to carbon preventing any new bonds to oxygen.
- Glycols can be oxidized by **periodic acid** to give two carbonyl species along with cleavage of the C—C bond between the —OH groups.

P **10.11, 10.12, 10.29–10.31, 10.35, 10.40, 10.41, 10.45, 10.50, 10.51, 10.53, 10.54, 10.56**

▶ **Reaction for the hydration of an aldehyde**

KEY REACTIONS

8. **Oxidation of a Primary Alcohol to a Carboxylic Acid (Section 10.8A)** A primary alcohol is oxidized to a carboxylic acid by chromic acid. The mechanism involves initial formation of an alkyl chromate intermediate, followed by reaction with base to remove a proton, generating the carbonyl group of an aldehyde and simultaneously reducing the chromium(VI) to chromium(IV). An initially formed aldehyde adds water, generating an aldehyde hydrate, which is oxidized according to the same mechanism to give the carboxylic acid.

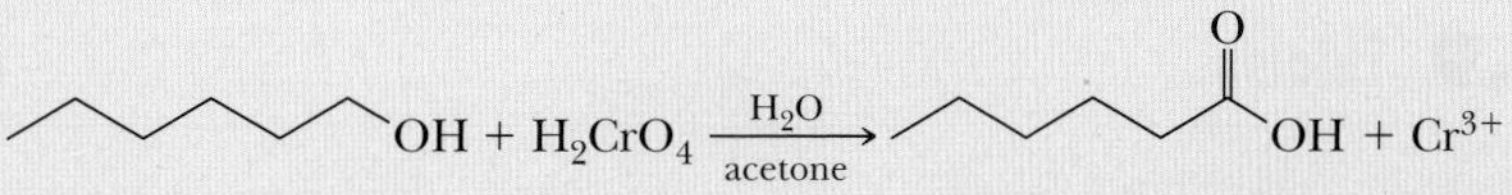

9. **Oxidation of a Secondary Alcohol to a Ketone (Sections 10.8A–10.8D)** A secondary alcohol is oxidized to a ketone by chromic acid, by CO_2Cl_2, DMSO, Et_3N (Swern oxidation), or by DMP (Dess-Martin periodinane).

(Continued)

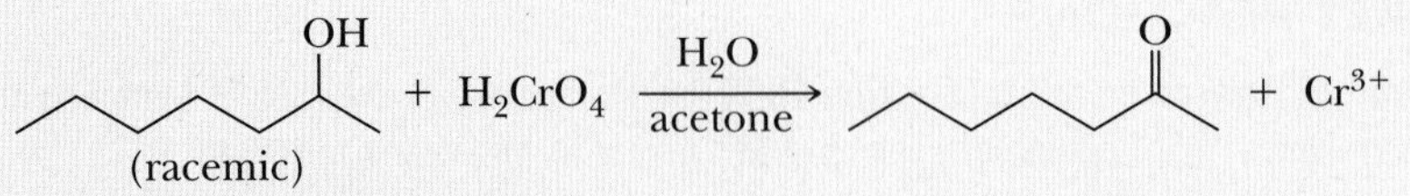

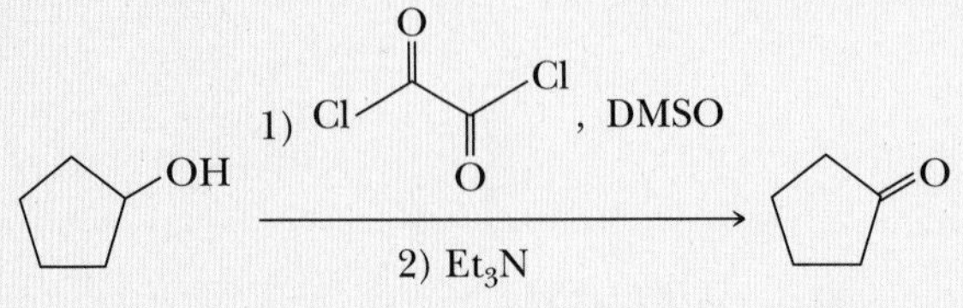

10. **Oxidation of a Primary Alcohol to an Aldehyde (Section 10.8B)** The oxidation of a primary alcohol to an aldehyde can be carried out using pyridinium chlorochromate (PCC). Because there is no water, the aldehyde does not form the hydrate, and the oxidation reaction stops at the aldehyde stage. Alternatively, a Swern or Dess-Martin oxidation can be used.

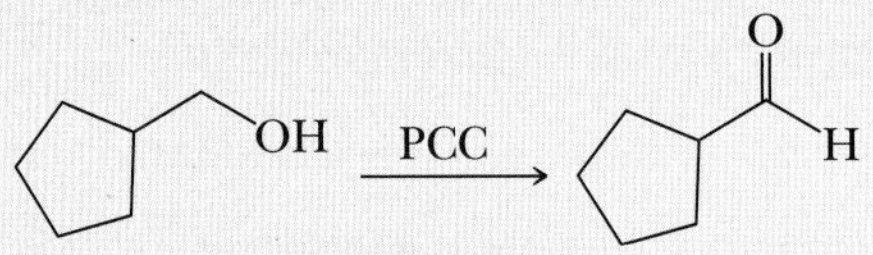

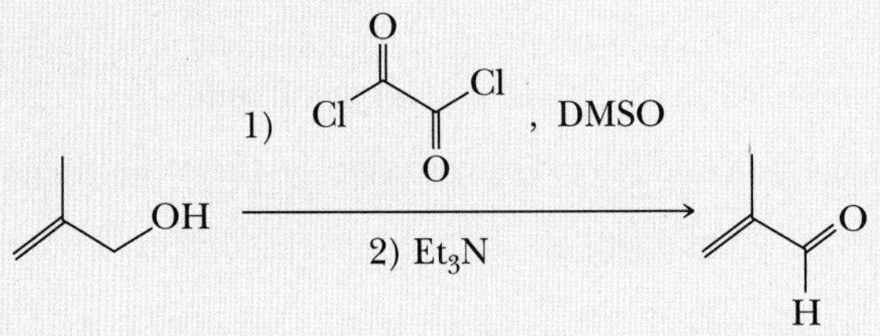

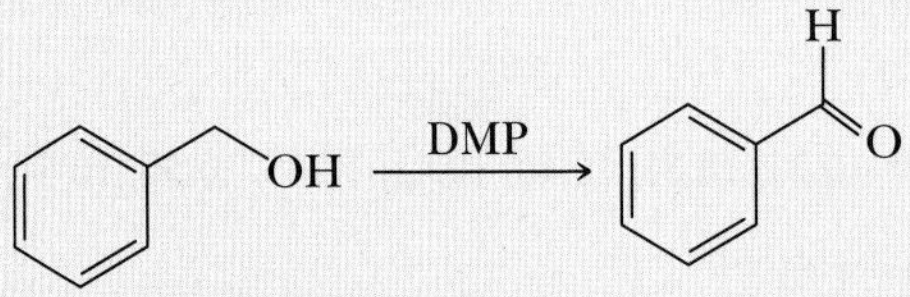

11. **Oxidative Cleavage of a Glycol (Section 10.8E)** HIO_4 reacts with a glycol to form a five-membered cyclic periodate intermediate that undergoes carbon-carbon bond cleavage to form two carbonyl groups.

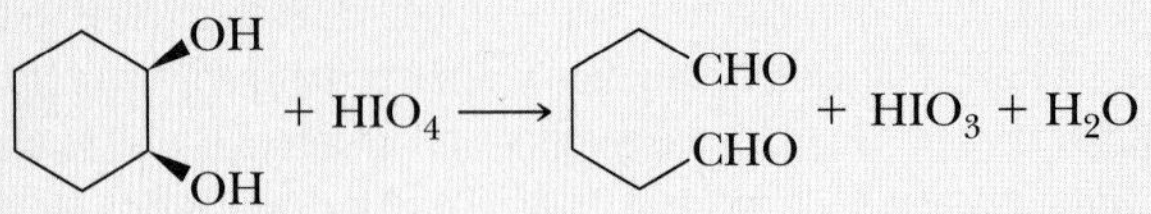

10.9 Thiols

- A **thiol** is a sulfur analog of an alcohol; it contains an —SH **(sulfhydryl) group** in place of an —OH group.

- Thiols are named in the same manner as alcohols, but the suffix *-e* of the parent alkane is retained and *-thiol* added.
  - *In compounds containing other functional groups of higher precedence, the presence of —SH is indicated by the prefix* sufanyl-.
- Common names for thiols are derived by naming the alkyl group bonded to —SH and adding the word **mercaptan**.
- Because the S—H bond is almost nonpolar, the physical properties of thiols are more like those of hydrocarbons of comparable molecular weight.
- Thiols are much more acidic than alcohols.
- Thiols are oxidized to give **disulfides, sulfinic acids,** and **sulfonic acids**.

P **10.13**

KEY REACTIONS

12. Acidity of Thiols (Section 10.9F) Thiols are weak acids, pK_a 10–11, but considerably stronger than alcohols, pK_a 15.5–18.

$$CH_3CH_2SH + H_2O \rightleftharpoons CH_3CH_2S^- + H_3O^+ \qquad pK_a = 10.5$$

13. Oxidation of Thiols to Disulfides (Section 10.9G) Oxidation by weak oxidizing agents such as O_2 and I_2 gives disulfides.

$$2\ RSH + \tfrac{1}{2}\ O_2 \longrightarrow RSSR + H_2O$$

Problems

Red numbers indicate applied problems.

Structure and Nomenclature

10.14 Which are secondary alcohols?

(a) 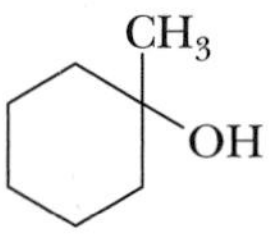

(b) $(CH_3)_3COH$

(c) 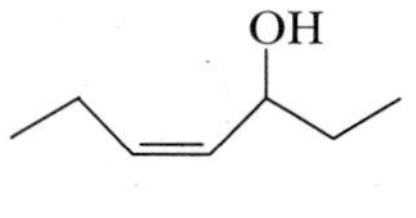

(racemic)

(d) 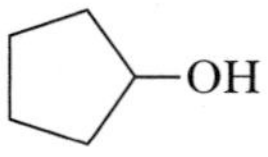

10.15 Name each compound.

(a) 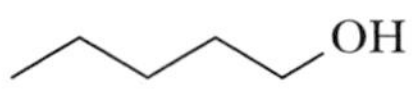

(b) 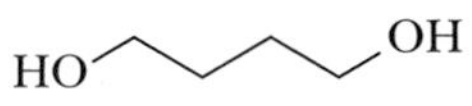

(c) 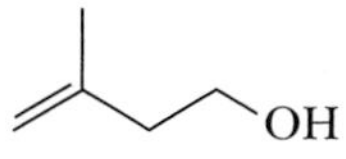

(d) 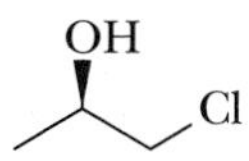

(e) 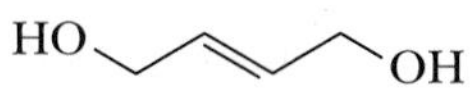

(f) 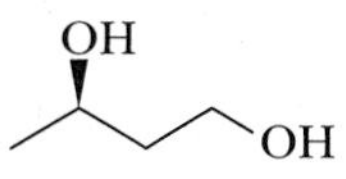

(g) 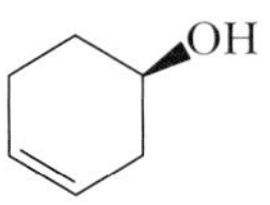 (h) 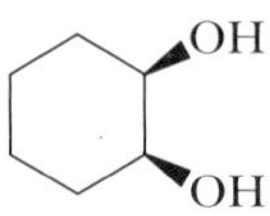 (i) 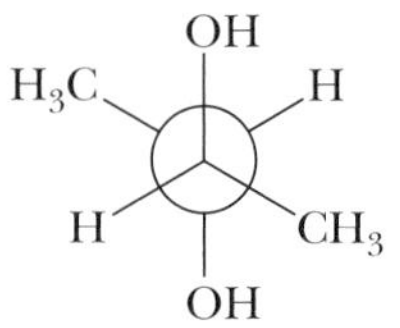

(j) 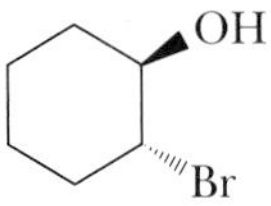 (k) 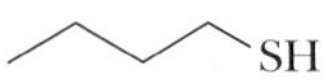 (l) 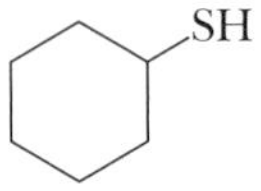

(m) 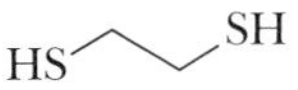 (n) 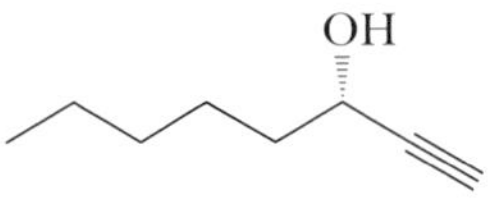 (o) 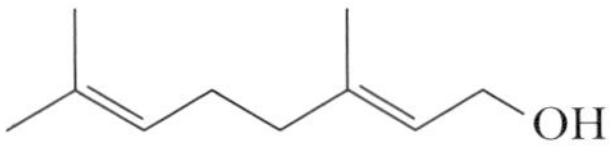

10.16 Write a structural formula for each compound.

(a) Isopropyl alcohol
(b) Propylene glycol
(c) 5-Methyl-2-hexanol
(d) 2-Methyl-2-propyl-1,3-propanediol
(e) 1-Chloro-2-hexanol
(f) *cis*-3-Isobutylcyclohexanol
(g) 2,2-Dimethyl-1-propanol
(h) 2-Mercaptoethanol
(i) Allyl alcohol
(j) *trans*-2-Vinylcyclohexanol
(k) (*Z*)-5-Methyl-2-hexen-1-ol
(l) 2-Propyn-1-ol
(m) 3-Chloro-1,2-propanediol
(n) *cis*-3-Pentene-1-ol

10.17 Name and draw structural formulas for the eight constitutional isomeric alcohols with molecular formula $C_5H_{12}O$. Classify each alcohol as primary, secondary, or tertiary. Which are chiral?

Physical Properties of Alcohols

10.18 Arrange these compounds in order of increasing boiling point (values in °C are −42, 78, 138, and 198)

(a)  **(b)** 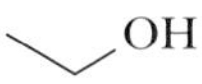 **(c)** 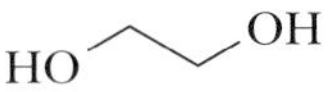 **(d)** 

10.19 Arrange these compounds in order of increasing boiling point (values in °C are −42, −24, 78, and 118).

(a) CH_3CH_2OH **(b)** CH_3OCH_3 **(c)** $CH_3CH_2CH_3$ **(d)** CH_3COOH

10.20 Which compounds can participate in hydrogen bonding with water? State which compounds can act only as hydrogen-bond donors, which can act only as hydrogen-bond acceptors, and which can act as both hydrogen-bond donors and hydrogen-bond acceptors.

(a) 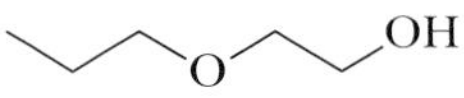 **(b)**  **(c)** 

(d)  **(e)** 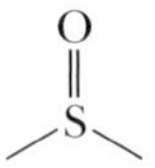 **(f)** 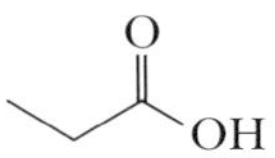

10.21 From each pair of compounds, select the one that is more soluble in water.

(a) CH_2Cl_2 or CH_3OH
(b) $CH_3\overset{\overset{\displaystyle O}{\|}}{C}CH_3$ or $CH_3\overset{\overset{\displaystyle CH_2}{\|}}{C}CH_3$
(c) CH_3CH_2Cl or NaCl
(d) 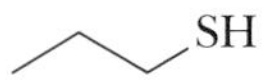 or 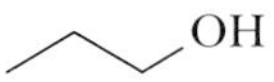
(e) 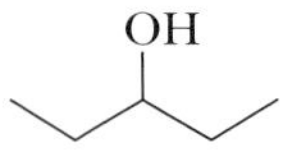 or 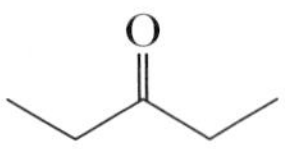

10.22 Arrange the compounds in each set in order of decreasing solubility in water.

(a) Ethanol, butane, diethyl ether
(b) 1-Hexanol, 1,2-hexanediol, hexane

10.23 Each compound given in this problem is a common organic solvent. From each pair of compounds, select the solvent with the greater solubility in water.

(a) CH_2Cl_2 or CH_3CH_2OH **(b)** Et_2O or EtOH

(c)  or 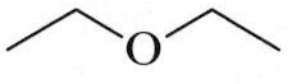 **(d)** 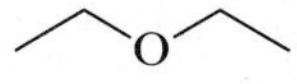 or 

10.24 The decalinols A and B can be equilibrated using aluminum isopropoxide in 2-propanol (isopropyl alcohol) containing a small amount of acetone.

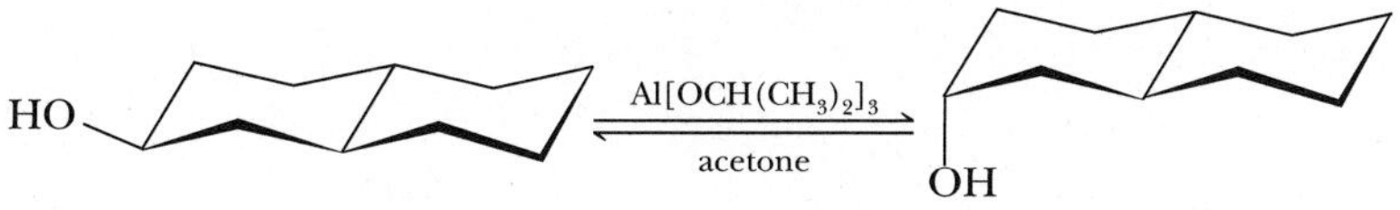

A B

Assuming a value of ΔG^0 (equatorial to axial) for cyclohexanol is 4.0 kJ (0.95 kcal)/mol, calculate the percent of each decalinol in the equilibrium mixture at 25°C.

Acid-Base Reactions of Alcohols

10.25 Complete the following acid-base reactions. Show all valence electrons on the interacting atoms and show by the use of curved arrows the flow of electrons in each reaction.

(a) $CH_3CH_2OH + H_3O^+ \longrightarrow$ **(b)** $CH_3CH_2OCH_2CH_3 + HOS(=O)_2OH \longrightarrow$

(c)  OH + HI $\longrightarrow$ **(d)** 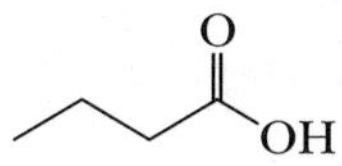 + $HOS(=O)_2OH \longrightarrow$

(e) 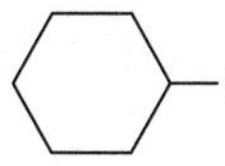 —OH + $BF_3 \longrightarrow$ **(f)** 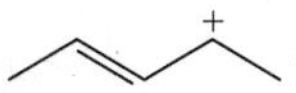 + $H_2O \longrightarrow$

10.26 Select the stronger acid from each pair and explain your reasoning. For each stronger acid, write a structural formula for its conjugate base.

(a) H_2O or H_2CO_3 **(b)** CH_3OH or CH_3COOH
(c) CH_3CH_2OH or $CH_3C{\equiv}CH$ **(d)** CH_3CH_2OH or CH_3CH_2SH

10.27 From each pair, select the stronger base. For each stronger base, write a structural formula of its conjugate acid.

(a) OH^- or CH_3O^- (each in H_2O) **(b)** $CH_3CH_2O^-$ or $CH_3C{\equiv}C^-$
(c) $CH_3CH_2S^-$ or $CH_3CH_2O^-$ **(d)** $CH_3CH_2O^-$ or NH_2^-

10.28 In each equilibrium, label the stronger acid, the stronger base, the weaker acid, and the weaker base. Also estimate the position of each equilibrium.

(a) $CH_3CH_2O^- + CH_3C{\equiv}CH \rightleftharpoons CH_3CH_2OH + CH_3C{\equiv}C^-$
(b) $CH_3CH_2O^- + HCl \rightleftharpoons CH_3CH_2OH + Cl^-$
(c) $CH_3COOH + CH_3CH_2O^- \rightleftharpoons CH_3COO^- + CH_3CH_2OH$

Reactions of Alcohols

10.29 Write equations for the reaction of 1-butanol with each reagent. Where you predict no reaction, write NR.

(a) Na metal **(b)** HBr, heat **(c)** HI, heat
(d) PBr_3 **(e)** $SOCl_2$, pyridine **(f)** $K_2Cr_2O_7$, H_2SO_4, H_2O, heat
(g) HIO_4 **(h)** PCC **(i)** CH_3SO_2Cl, pyridine

(j) CO_2Cl_2 and DMSO, followed by triethylamine **(k)** DMP
(l) triethylamine

10.30 Write equations for the reaction of 2-butanol with each reagent listed in Problem 10.29. Where you predict no reaction, write NR.

10.31 Draw structural formulas for the major organic products of each reaction.

(a) 
OH + H2CrO4

(b) 
OH + SOCl2

(c) 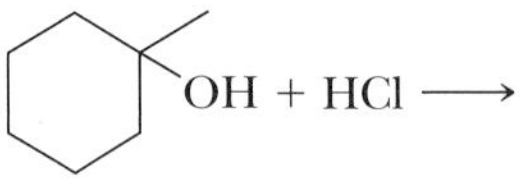
OH + HCl

(d) 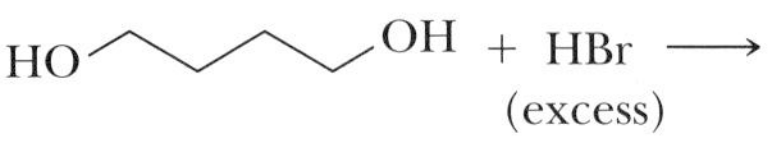
HO
OH + HBr
(excess)

(e) 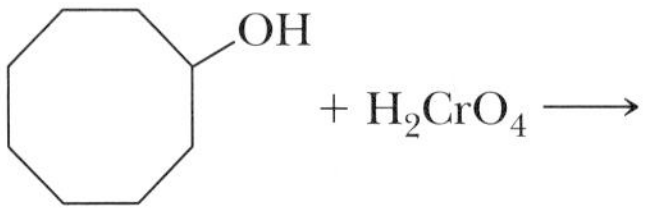
OH
+ H2CrO4

(f) 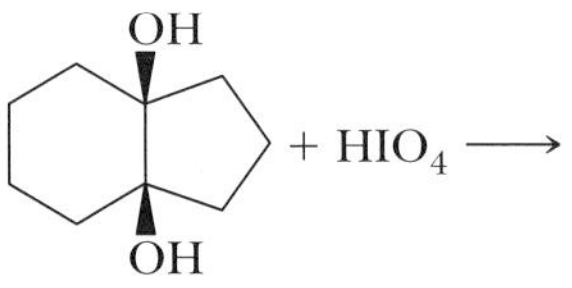
OH
OH
+ HIO4

(g) 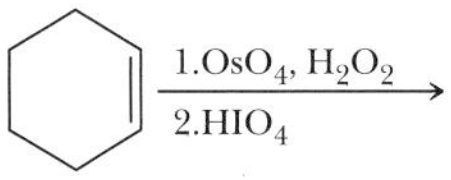
1.OsO4, H2O2
2.HIO4

(h) 
OH + SOCl2

(i) 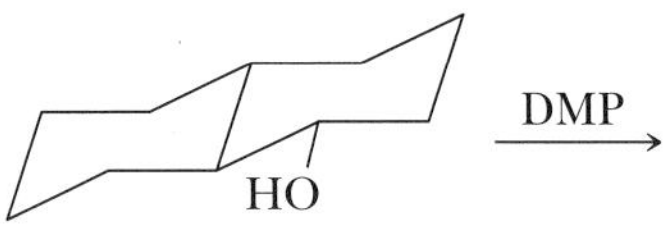
HO
DMP

(j) 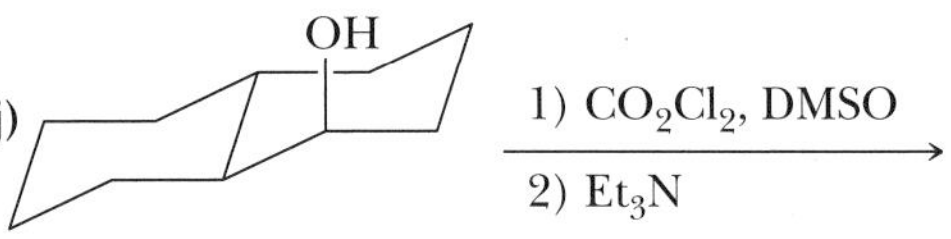
OH
1) CO2Cl2, DMSO
2) Et3N

10.32 When (*R*)-2-butanol is left standing in aqueous acid, it slowly loses its optical activity. Account for this observation.

10.33 Two diastereomeric sets of enantiomers, A/B and C/D, exist for 3-bromo-2-butanol.

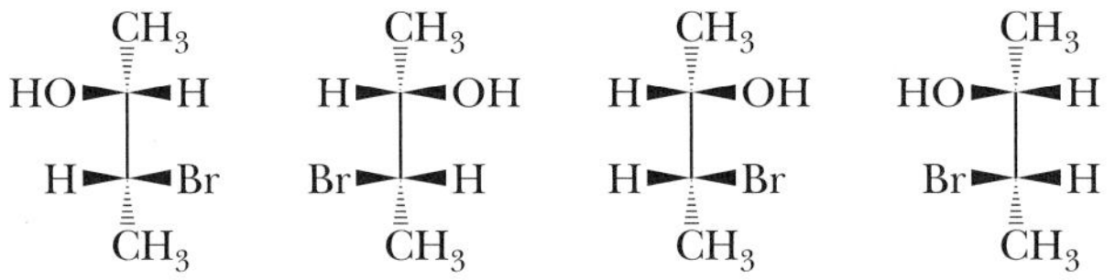
CH3
HO H
H Br
CH3
CH3
H OH
Br H
CH3
CH3
H OH
H Br
CH3
CH3
HO H
Br H
CH3

A B C D

When enantiomer A or B is treated with HBr, only racemic 2,3-dibromobutane is formed; no meso isomer is formed. When enantiomer C or D is treated with HBr, only meso 2,3-dibromobutane is formed; no racemic 2,3-dibromobutane is formed. Account for these observations.

10.34 Acid-catalyzed dehydration of 3-methyl-2-butanol gives three alkenes: 2-methyl-2-butene, 3-methyl-1-butene, and 2-methyl-1-butene. Propose a mechanism to account for the formation of each product.

10.35 Show how you might bring about the following conversions. For any conversion involving more than one step, show each intermediate compound.

(a) 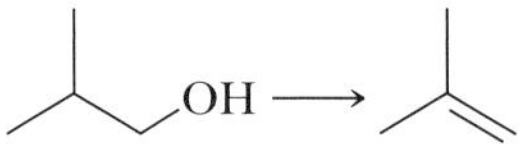
OH

(b) 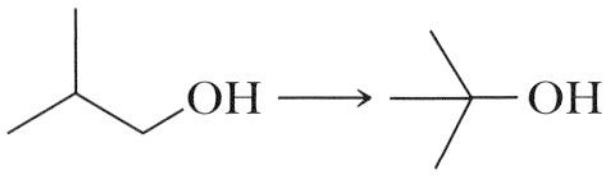
OH
OH

(c) 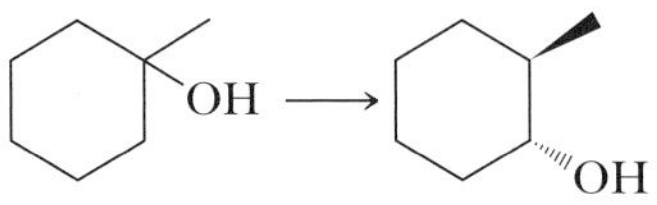
OH
OH

(d) 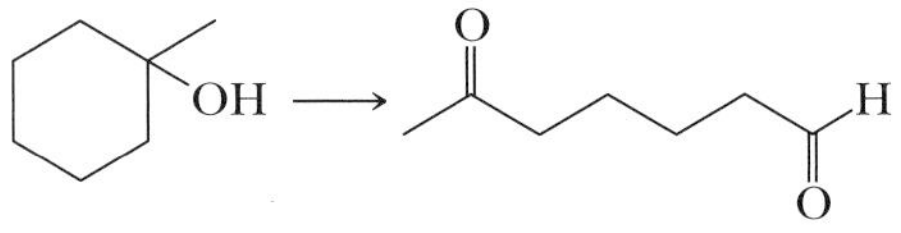
OH
O
H
O

(e) 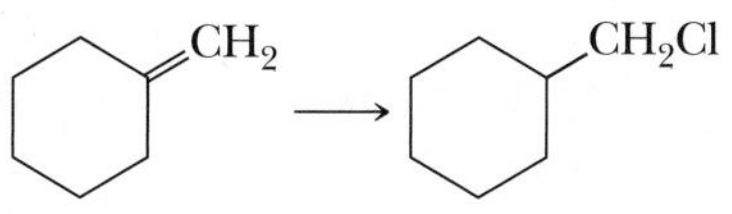

(f) 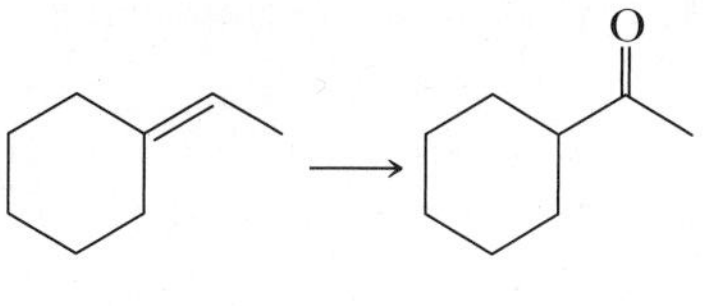

(g) 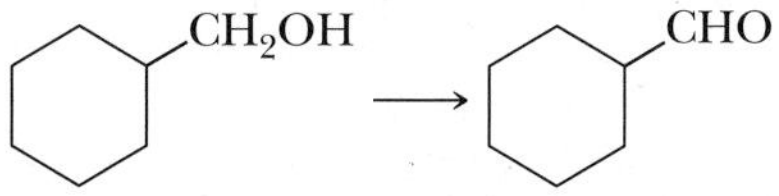

(h) 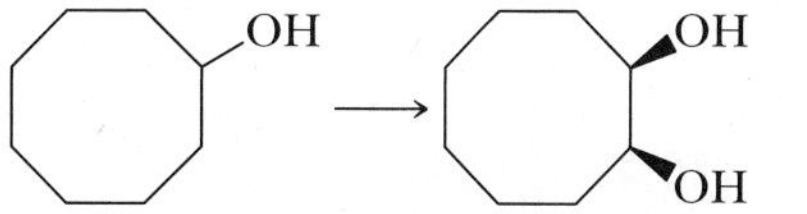

(i) 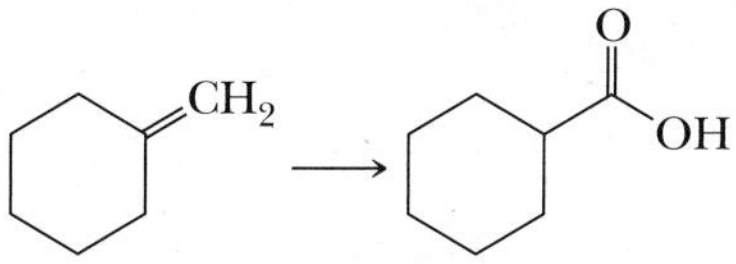

Pinacol Rearrangement

10.36 Propose a mechanism for the following pinacol rearrangement catalyzed by boron trifluoride ethyl etherate.

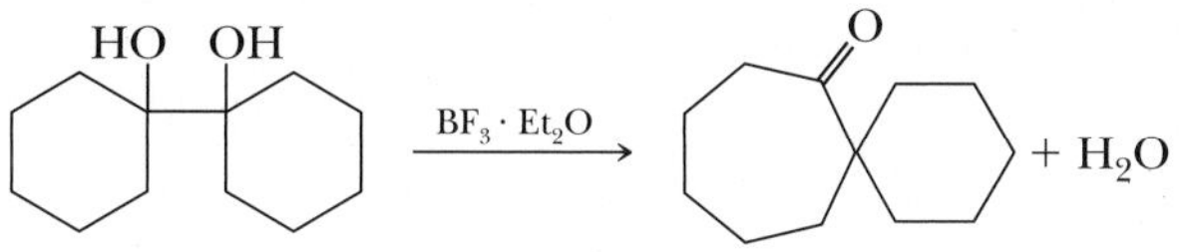

Synthesis

10.37 Alkenes can be hydrated to form alcohols by (1) hydroboration followed by oxidation with alkaline hydrogen peroxide and (2) acid-catalyzed hydration. Compare the product formed from each alkene by sequence (1) with those formed from (2).

(a) Propene **(b)** *cis*-2-Butene **(c)** *trans*-2-Butene
(d) Cyclopentene **(e)** 1-Methylcyclohexene

10.38 Show how each alcohol or diol can be prepared from an alkene.

(a) 2-Pentanol **(b)** 1-Pentanol **(c)** 2-Methyl-2-pentanol
(d) 2-Methyl-2-butanol **(e)** 3-Pentanol **(f)** 3-Ethyl-3-pentanol
(g) 1,2-Hexanediol

10.39 Dihydropyran is synthesized by treating tetrahydrofurfuryl alcohol with an arenesulfonic acid, $ArSO_3H$. Propose a mechanism for this conversion.

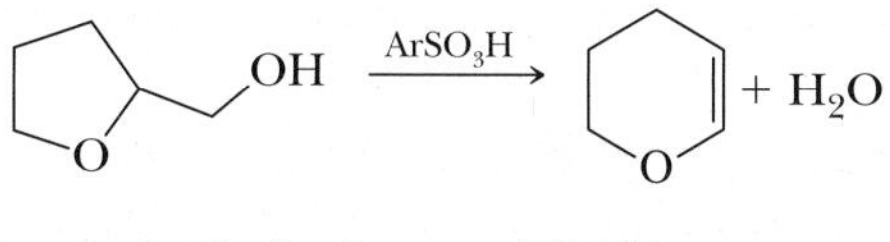

Tetrahydrofurfuryl alcohol **Dihydropyran**

10.40 Show how to convert propene to each of these compounds, using any inorganic reagents as necessary.

(a) Propane **(b)** 1,2-Propanediol **(c)** 1-Propanol
(d) 2-Propanol **(e)** Propanal **(f)** Propanone
(g) Propanoic acid **(h)** 1-Bromo-2-propanol **(i)** 3-Chloropropene
(j) 1,2,3-Trichloropropane **(k)** 1-Chloropropane **(l)** 2-Chloropropane
(m) 2-Propen-1-ol **(n)** Propenal

10.41 **(a)** How many stereoisomers are possible for 4-methyl-1,2-cyclohexanediol?
(b) Which of the possible stereoisomers are formed by oxidation of (*S*)-4-methylcyclohexene with osmium tetroxide?
(c) Is the product formed in part (b) optically active or optically inactive?

10.42 Show how to bring about this conversion in good yield.

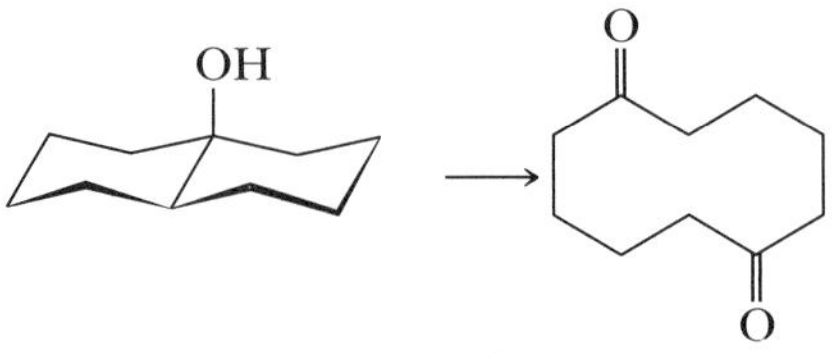

10.43 The tosylate of a primary alcohol normally undergoes an S_N2 reaction with hydroxide ion to give a primary alcohol. Reaction of this tosylate, however, gives a compound of molecular formula $C_7H_{12}O$.

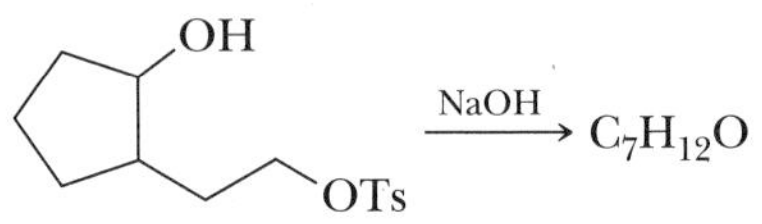

Propose a structural formula for this compound and a mechanism for its formation.

10.44 Chrysanthemic acid occurs as a mixture of esters in flowers of the chrysanthemum (pyrethrum) family. Reduction of chrysanthemic acid to its alcohol (Section 17.6A) followed by conversion of the alcohol to its tosylate gives chrysanthemyl tosylate. Solvolysis (Section 9.2) of the tosylate gives a mixture of artemesia and yomogi alcohols.

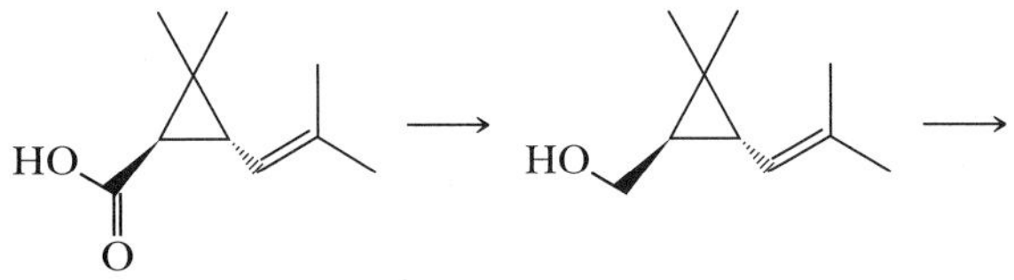

Chrysanthemic acid **Chrysanthemyl alcohol**

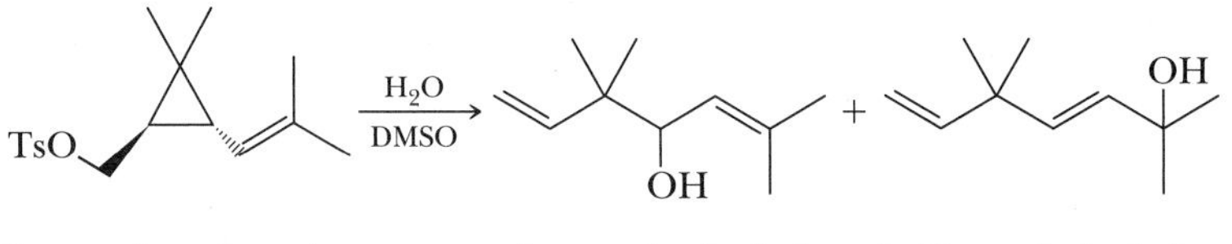

Chrysanthemyl tosylate **Artemesia alcohol** **Yomogi alcohol**

Propose a mechanism for the formation of these alcohols from chrysanthemyl tosylate.

10.45 Show how to convert cyclohexene to each compound in good yield.

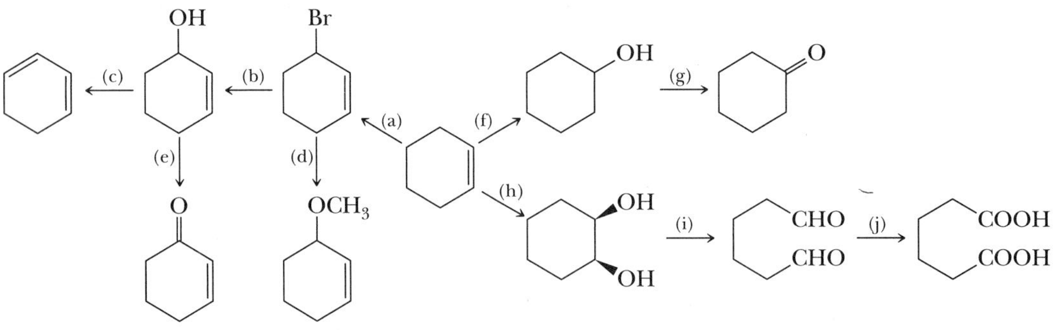

10.46 Hydroboration of the following bicycloalkene followed by oxidation in alkaline hydroperoxide is both stereoselective and regioselective. The product is a single alcohol in better than 95% yield.

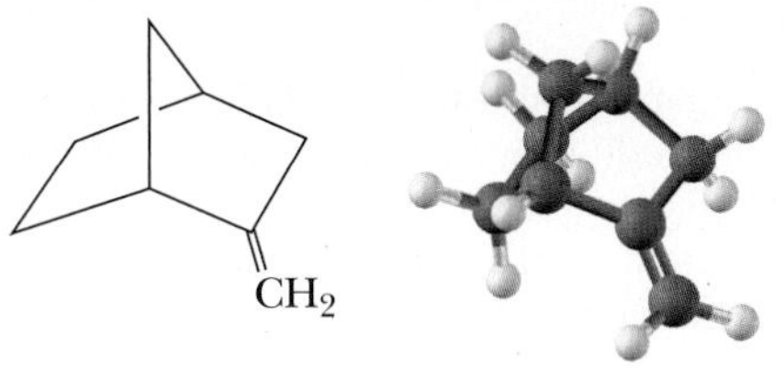
CH2

Propose a structural formula for this alcohol and account for the stereo- and regioselectivity of its formation. *Hint:* Examine a molecular model of this alkene and see if you can determine which face of the double bond is more accessible to hydroboration.

10.47 Ethanol (CH_3CH_2OH) and dimethyl ether (CH_3OCH_3) are constitutional isomers.

(a) Predict which of the two has the higher boiling point.
(b) Predict which of the two is more soluble in water.

Looking Ahead

10.48 Compounds that contain an N—H group associate by hydrogen bonding.

(a) Do you expect this association to be stronger or weaker than that of compounds containing an O—H group?
(b) Based on your answer to part (a), which would you predict to have the higher boiling point, 1-butanol or 1-butanamine?

10.49 Following are structural formulas for phenol and cyclohexanol along with the acid dissociation constants for each.

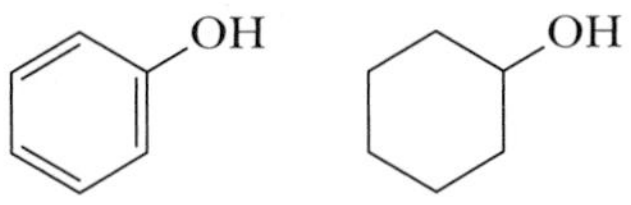
OH
OH

Phenol pK_a 9.96 | **Cyclohexanol** pK_a 18

Propose an explanation for the fact that phenol is a considerably stronger acid than cyclohexanol.

Organic Chemistry Reaction Roadmap

10.50 Use the reaction roadmap you made for Problems 6.54, 7.29, 8.28, and 9.54 and update it to contain the reactions in the "Study Guide" section of this chapter. Because of their highly specific nature, do not use reactions 1, 2, 7, 12, and 13 on your reaction roadmap. But, you should include all the choices of oxidants given in Key Reactions 9 and 10. Reaction Roadmap

10.51 Write the products of the following sequences of reactions. Refer to your reaction roadmap to see how the combined reactions allow you to "navigate" between the different functional groups. Reaction Roadmap

(a)
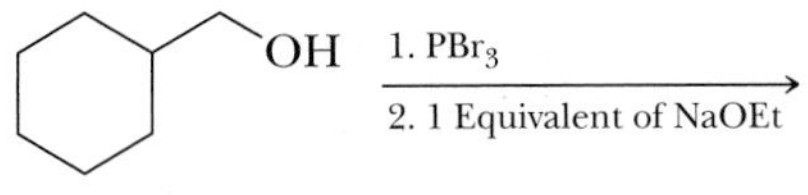
OH
1. PBr3
2. 1 Equivalent of NaOEt

An alcohol

(b)
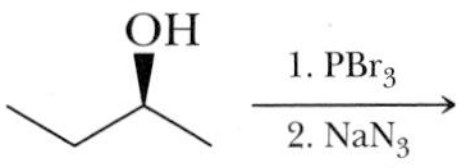
OH
1. PBr3
2. NaN3

An alcohol

(c)
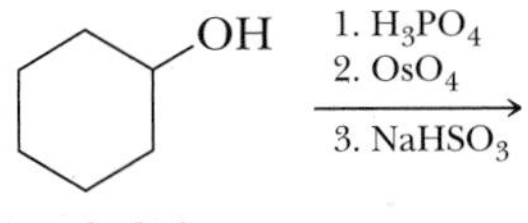
OH
1. H3PO4
2. OsO4
3. NaHSO3

An alcohol

(d)
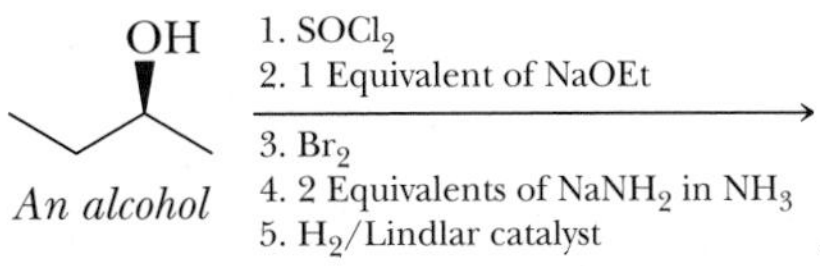
OH
1. SOCl2
2. 1 Equivalent of NaOEt
3. Br2
4. 2 Equivalents of NaNH2 in NH3
5. H2/Lindlar catalyst

An alcohol

(e)
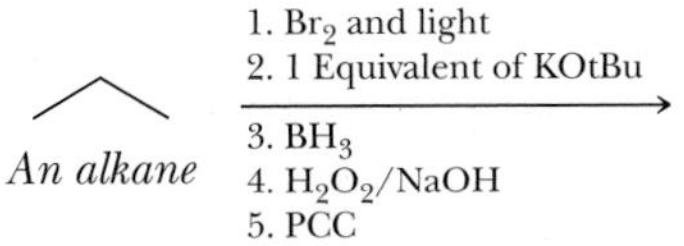
1. Br2 and light
2. 1 Equivalent of KOtBu
3. BH3
4. H2O2/NaOH
5. PCC

An alkane

Mixed Synthesis

Alcohols are important for organic synthesis, especially in situations involving alkenes. The alcohol might be the desired product, or the OH group might be transformed into another functional group via halogenation, oxidation, or perhaps conversion to a sulfonic ester derivative. Formation of an alcohol from an alkene is particularly powerful because conditions can be chosen to produce either the Markovnikov or non-Markovnikov product from an unsymmetrical alkene.

Reaction Roadmap

10.52 Using your reaction roadmap as a guide, show how to convert 4-methyl-1-pentene into 5-methylhexanenitrile. You must use 4-methyl-1-pentene and sodium cyanide as the source of all carbon atoms in the target molecule. Show all reagents needed and all molecules synthesized along the way.

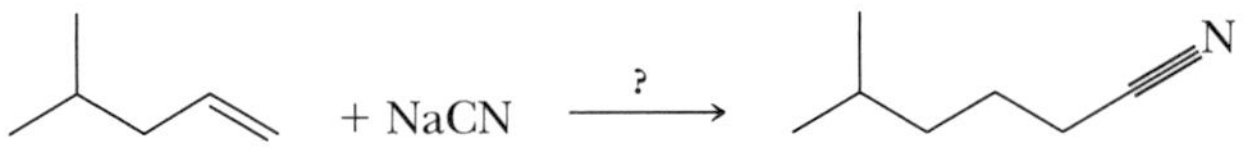

4-Methyl-1-pentene **5-Methylhexanenitrile**

Reaction Roadmap

10.53 Using your reaction roadmap as a guide, show how to convert butane into 2-butanone. Show all reagents and all molecules synthesized along the way.

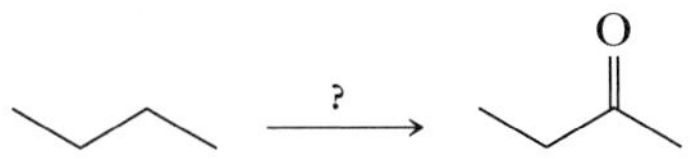

Butane **2-Butanone**

Reaction Roadmap

10.54 Using your reaction roadmap as a guide, show how to convert butane into butanal. Show all reagents needed and all molecules synthesized along the way.

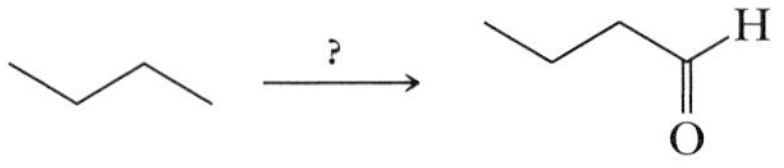

Butane **Butanal**

Reaction Roadmap

10.55 Using your reaction roadmap as a guide, show how to convert 1-propanol into 2-hexyne. You must use 1-propanol as the source of all carbon atoms in the target molecule. Show all reagents needed and all molecules synthesized along the way.

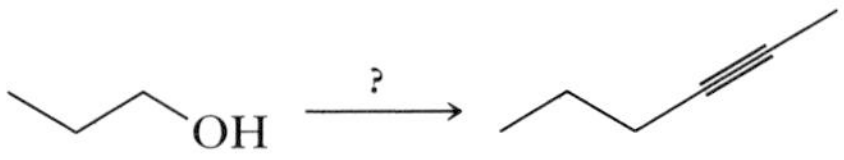

1-Propanol **2-Hexyne**

Reaction Roadmap

10.56 Using your reaction roadmap as a guide, show how to convert 2-methylpentane into 2-methyl-3-pentanone. Show all reagents needed and all molecules synthesized along the way.

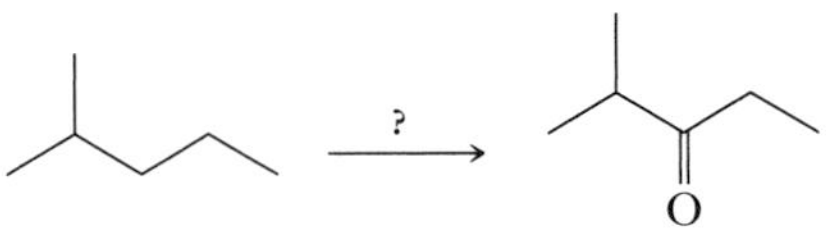

2-Methylpentane **2-Methyl-3-pentanone**

Reactions in Context

10.57 Atorvastatin (Lipitor) is used to decrease patient serum cholesterol levels. It works by inhibiting an enzyme called HMG-CoA reductase. See "Connections to Biological Chemistry" in Section 10.2 for more information about the action of atorvastatin. In one synthesis of atorvastatin that produces the desired single enantiomer of the final product, the following reagents are used. Draw the structures of synthetic intermediates A and B.

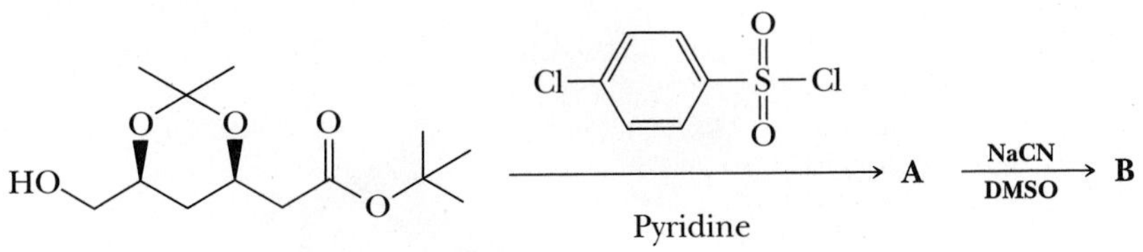

10.58 Paroxetine (Paxil) is an antidepressant that is a member of a family of drugs known as *S*elective *S*erotonin *R*euptake *I*nhibitors (SSRIs). This family of drugs also includes fluoxetine (Prozac) and sertraline (Zoloft). SSRIs work by inhibiting the reuptake of the neurotransmitter serotonin in the synapses of the central nervous system following release of serotonin during excitation of individual nerve cells. Between firings, the serotonin is taken back up by a nerve cell in preparation for firing again. Inhibition of reuptake has the effect of increasing the time serotonin molecules remain in the synapses following excitation, leading to a therapeutic effect. In one synthesis of paroxetine, the following reagents are used. Draw the structures of synthetic intermediates A and B.

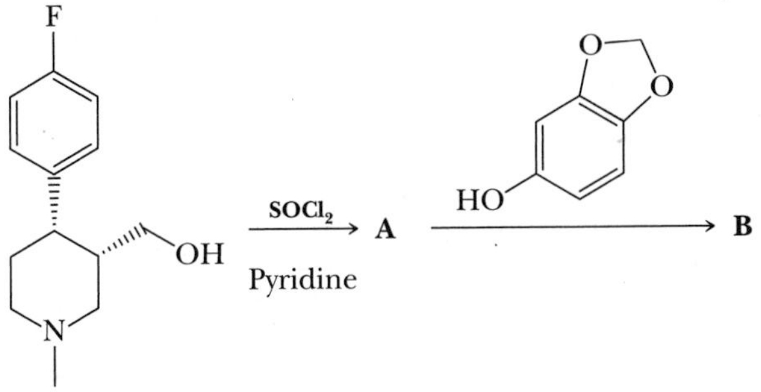

10.59 Tartaric acid is an inexpensive and readily available chiral starting material for the synthesis of chiral molecules. In a well-known prostaglandin synthesis, the (*S*,*S*)-tartaric acid enantiomer was used to prepare the chiral diol in several steps. The chiral diol was isolated as a synthetic intermediate, and the following reagents are used. Draw the structures of synthetic intermediates A and B.

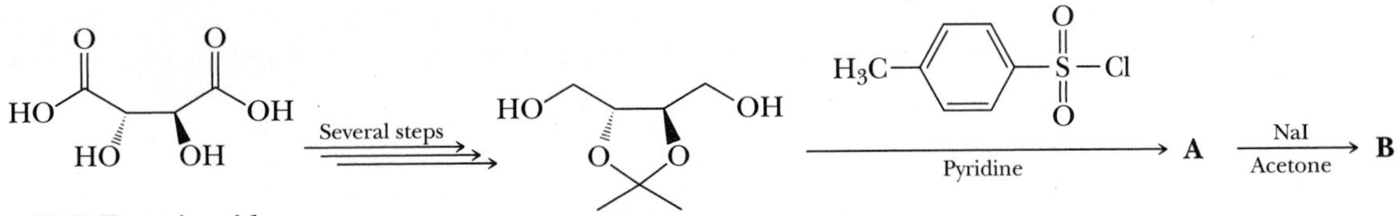

(*S*,*S*)-Tartaric acid

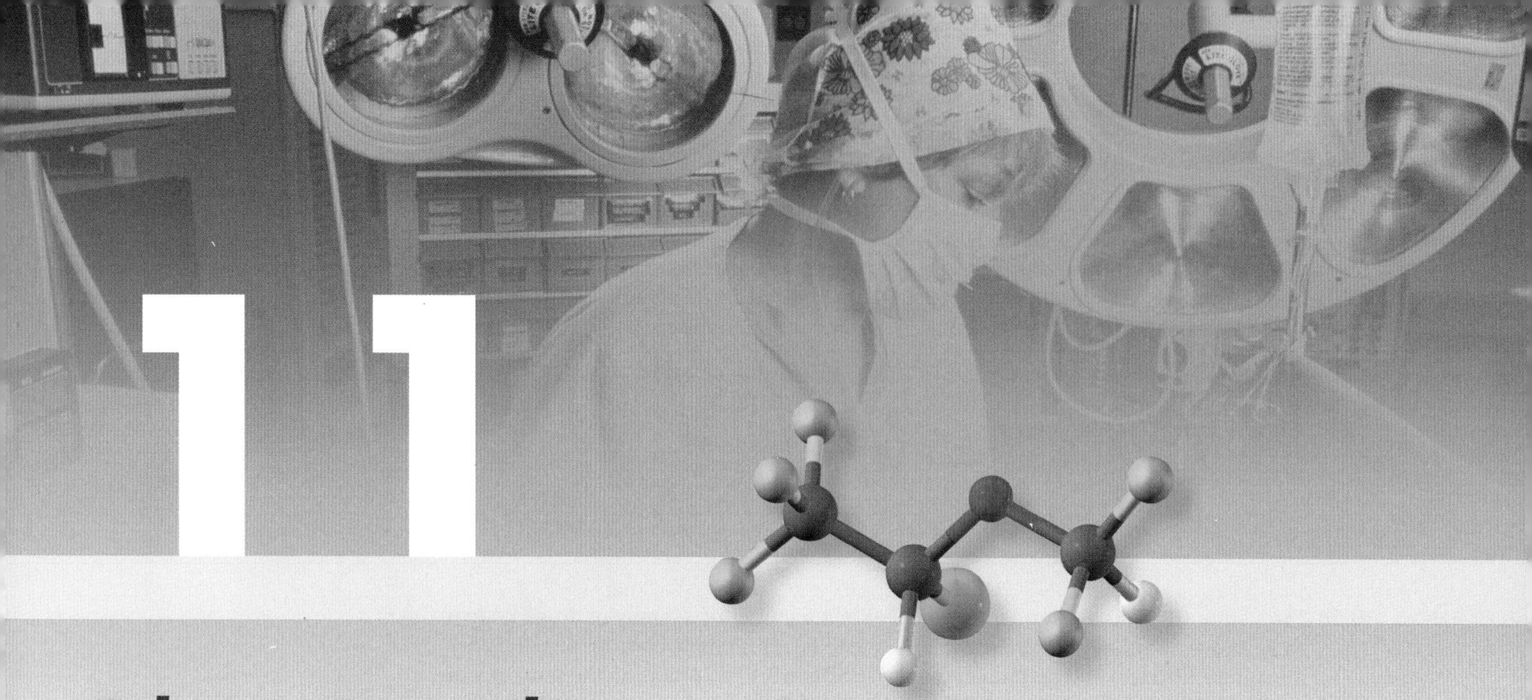

11

Ethers, Epoxides, and Sulfides

Outline

11.1 Structure of Ethers
11.2 Nomenclature of Ethers
11.3 Physical Properties of Ethers
11.4 Preparation of Ethers
11.5 Reactions of Ethers
11.6 Silyl Ethers as Protecting Groups
11.7 Epoxides: Structure and Nomenclature
11.8 Synthesis of Epoxides
11.9 Reactions of Epoxides
11.10 Ethylene Oxide and Epichlorohydrin: Building Blocks in Organic Synthesis
11.11 Crown Ethers
11.12 Sulfides

The discovery that inhaling ethers could make a patient insensitive to pain revolutionized the practice of medicine. **Inset:** A model of isoflurane, $CF_3CHClOCHF_2$, a halogenated ether widely used as an inhalation anesthetic in both human and veterinary medicine. (Alan Levenson/Getty Images)

In this chapter, we discuss the structure, nomenclature, physical properties, and chemical properties of ethers and compare their physical properties with those of isomeric alcohols. Then we study the preparation and chemical properties of a group of cyclic ethers called epoxides. As we shall see, their most important reactions involve nucleophilic substitution. This chapter continues the discussion of S_N1 and S_N2 reaction mechanisms begun in Chapter 9 and continued into Chapter 10.

11.1 Structure of Ethers

The functional group of an **ether** is an atom of oxygen bonded to two carbon atoms. Figure 11.1 shows a Lewis structure and a ball-and-stick model of dimethyl ether, CH_3OCH_3, the simplest ether. In dimethyl ether, two sp^3 hybrid orbitals of oxygen form σ bonds to the two carbon atoms. The other two sp^3 hybrid orbitals of oxygen each contain an unshared pair of electrons. The C—O—C bond angle in dimethyl ether is 110.3°, a value close to the tetrahedral angle of 109.5°.

In still other ethers, the ether oxygen is bonded to sp^2 hybridized carbons. In ethoxyethene (ethyl vinyl ether), for example, the ether oxygen is bonded to one sp^3 and one sp^2 hybridized carbon.

$$CH_3CH_2{-}O{-}CH{=}CH_2$$

Ethoxyethene
(Ethyl vinyl ether)

Ether
A compound containing an oxygen atom bonded to two carbon atoms.

11.2 Nomenclature of Ethers

In the IUPAC system, ethers are named by selecting the longest carbon chain as the parent alkane and naming the —OR group bonded to it as an **alkoxy group**. Common names are derived by listing the alkyl groups bonded to oxygen in alphabetical order and adding the word *ether.* Following are the IUPAC names and, in parentheses, the common names for three low-molecular-weight ethers.

Alkoxy group
An —OR group, where R is an alkyl group.

$CH_3CH_2OCH_2CH_3$

Ethoxyethane
(Diethyl ether)

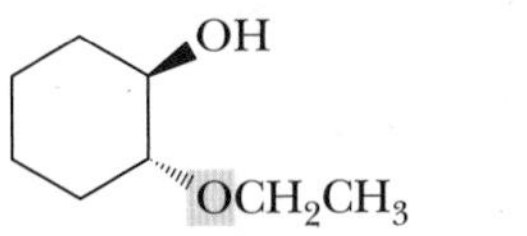

(1*R*,2*R*)-2-Ethoxycyclohexanol
(*trans*-2-Ethoxycyclohexanol)

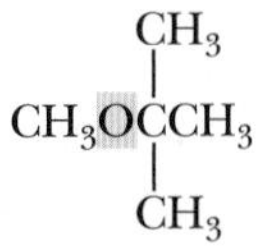

2-Methoxy-2-methylpropane
(*tert*-Butyl methyl ether)

Chemists almost invariably use common names for low-molecular-weight ethers. For example, although ethoxyethane is the IUPAC name for $CH_3CH_2OCH_2CH_3$, it is rarely called that; rather, it is called diethyl ether, ethyl ether, or even more commonly, simply ether. The abbreviation for *tert*-butyl methyl ether is MTBE, after the common name methyl *tert*-butyl ether (incorrectly alphabetized, as you will recognize).

Three other ethers deserve special mention. 2-Methoxyethanol and 2-ethoxyethanol, more commonly known as Methyl Cellosolve and Cellosolve, are good polar protic solvents in which to carry out organic reactions and are used commercially in some paint strippers. *Diethylene glycol dimethyl ether*, more commonly known by its acronym, diglyme, is a common solvent for hydroboration and $NaBH_4$ reductions.

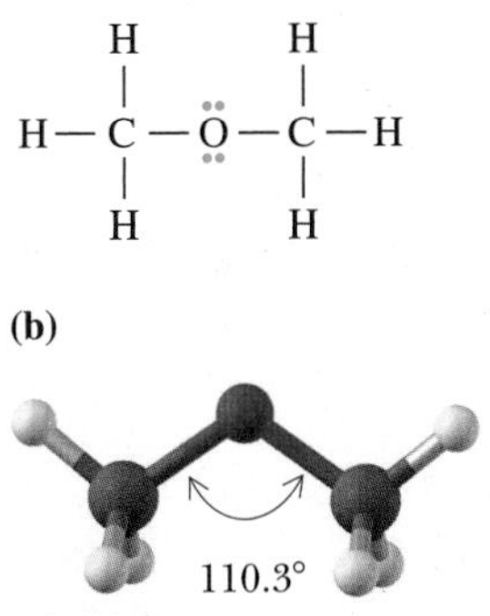

(a)
(b)

FIGURE 11.1 Dimethyl ether, CH_3OCH_3.
(a) Lewis structure and
(b) ball-and-stick model.

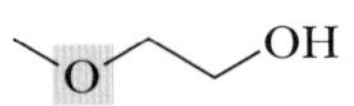

2-Methoxyethanol
(Methyl cellosolve)

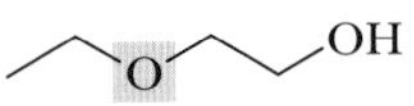

2-Ethoxyethanol
(Cellosolve)

Diethylene glycol dimethyl ether
(Diglyme)

Cyclic ethers are given special names. The presence of an oxygen atom in a saturated ring is indicated by the prefix *ox-*, and ring sizes from three to six are indicated by the endings *-irane*, *-etane*, *-olane*, and *-ane*, respectively. Several of these smaller-ring cyclic ethers are more often referred to by their common names, here shown in parentheses. Numbering of the atoms of the ring begins with the oxygen atom. These compounds and others in which there is a heteroatom (noncarbon atom) in the ring are called **heterocycles**. Heterocycles containing O or N atoms are particularly common in organic chemistry, and several clinically important pharmaceuticals are heterocycles.

Heterocycle
A cyclic compound whose ring contains more than one kind of atom. Oxirane (ethylene oxide), for example, is a heterocycle whose ring contains two carbon atoms and one oxygen atom.

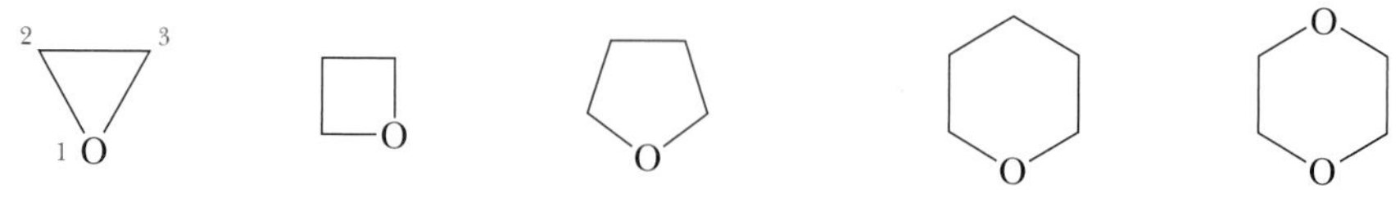

Oxirane (Ethylene oxide) | **Oxetane** | **Oxolane (Tetrahydrofuran)** | **Oxane (Tetrahydropyran)** | **1,4-Dioxane**

Example 11.1 Ether Nomenclature

Write IUPAC and common names for these ethers.

(a) $CH_3\overset{CH_3}{\underset{CH_3}{C}}OCH_2CH_3$ (b) 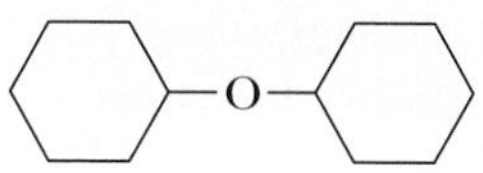 (c) $CH_2{=}CHOCH_3$

Solution

(a) 2-Ethoxy-2-methylpropane. Its common name is *tert*-butyl ethyl ether.

(b) Cyclohexoxycyclohexane. Its common name is dicyclohexyl ether.

(c) Methoxyethene. Its common name is methyl vinyl ether.

Problem 11.1

Write IUPAC and common names for these ethers.

(a) 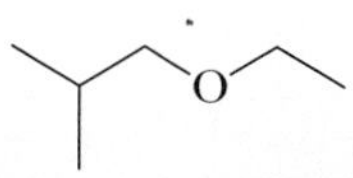 (b) 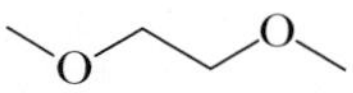 (c) 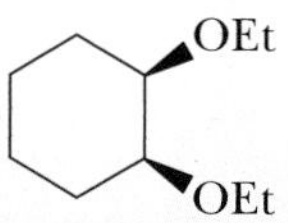

11.3 Physical Properties of Ethers

Ethers are polar molecules in which oxygen bears a partial negative charge and each attached carbon bears a partial positive charge (Figure 11.2). However, only weak dipole-dipole interactions exist between ether molecules in the liquid state. Consequently, boiling points of ethers are much lower than those of alcohols of comparable molecular weight (Table 11.1) and are close to those of hydrocarbons of comparable molecular weight (Table 2.5).

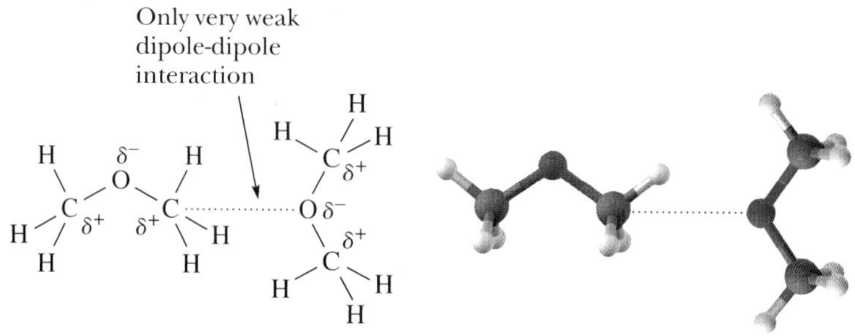

FIGURE 11.2 Although ethers are polar compounds, there are only weak dipole-dipole interactions between their molecules in the liquid state.

Because ethers cannot act as hydrogen bond donors, they are much less soluble in water than alcohols. However, they can act as hydrogen bond acceptors (Figure 11.3), which makes them more water-soluble than hydrocarbons of comparable molecular weight and shape (compare data in Tables 2.5 and 11.1).

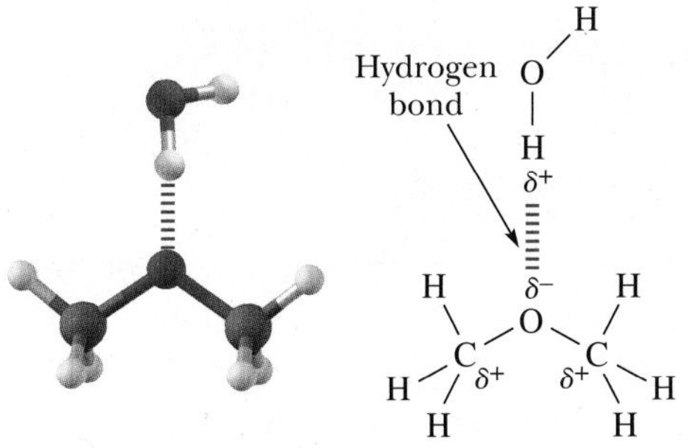

FIGURE 11.3 Ethers are hydrogen bond acceptors only. They are not hydrogen bond donors.

Table 11.1 Boiling Points and Solubilities in Water of Some Ethers and Alcohols of Comparable Molecular Weight

Structural Formula	Name	Molecular Weight	bp (°C)	Solubility in Water
CH_3CH_2OH	Ethanol	46	78	Infinite
CH_3OCH_3	Dimethyl ether	46	−24	7.8 g/100 g
$CH_3CH_2CH_2CH_2OH$	1-Butanol	74	117	7.4 g/100 g
$CH_3CH_2OCH_2CH_3$	Diethyl ether	74	35	8.0 g/100 g
$HOCH_2CH_2CH_2CH_2OH$	1,4-Butanediol	90	230	Infinite
$CH_3CH_2CH_2CH_2CH_2OH$	1-Pentanol	88	138	2.3 g/100 g
$CH_3OCH_2CH_2OCH_3$	Ethylene glycol dimethyl ether	90	84	Infinite
$CH_3CH_2CH_2CH_2OCH_3$	Butyl methyl ether	88	71	Slight

Example 11.2 Ether Water Solubility

Arrange these compounds in order of increasing solubility in water.

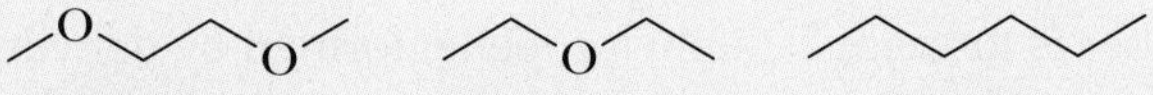

Ethylene glycol dimethyl ether **Diethyl ether** **Hexane**

Solution

Water is a polar solvent. Hexane, a nonpolar hydrocarbon, has the lowest solubility in water. Both diethyl ether and ethylene glycol dimethyl ether are polar compounds because of the presence of the polar C—O—C bond, and each interacts with water as a hydrogen bond acceptor.

Of these three compounds, ethylene glycol dimethyl ether is most soluble in water because it has more sites for hydrogen bonding (a total of four lone pairs on two O atoms) than diethyl ether.

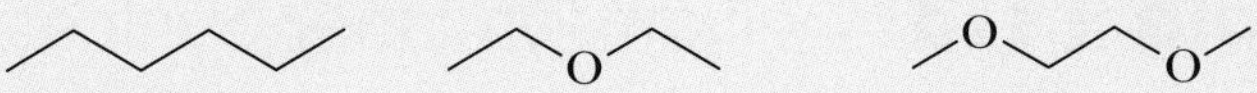

Insoluble 8 g/100 g water Soluble in all proportions

Problem 11.2

Arrange these compounds in order of increasing boiling point.

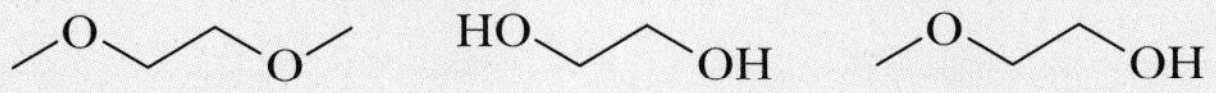

11.4 Preparation of Ethers

A. Williamson Ether Synthesis

Williamson ether synthesis
A general method for the synthesis of dialkyl ethers by an S_N2 reaction between a haloalkane and an alkoxide ion.

The most common general method for the synthesis of ethers, the **Williamson ether synthesis**, involves nucleophilic displacement of a halide ion or another good leaving group by an alkoxide ion.

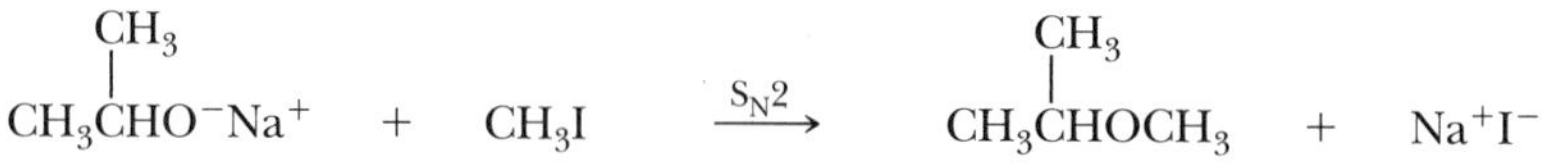

Sodium isopropoxide **Iodomethane (Methyl iodide)** **2-Methoxypropane (Isopropyl methyl ether)**

In planning a Williamson ether synthesis, it is essential to use a combination of reactants that maximizes nucleophilic substitution and minimizes any competing β-elimination (E2, Section 9.6B). Yields of ether are highest when the halide to be displaced is on a methyl or a primary carbon. Yields are low in the displacement from secondary halides (because of competing β-elimination), and the Williamson ether synthesis fails altogether with tertiary halides (because β-elimination by an E2 mechanism is the exclusive reaction). For example, *tert*-butyl methyl ether can be prepared by the reaction of potassium *tert*-butoxide and bromomethane. Note that bromomethane is the only haloalkane with little enough steric hindrance to react with the highly hindered potassium *tert*-butoxide in reasonable yield. Even primary haloalkanes would not react to give a high yield of the corresponding *tert*-butyl ether.

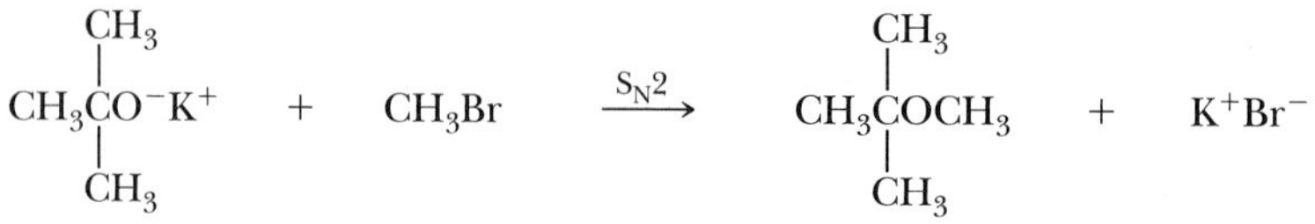

Potassium *tert*-butoxide **Bromomethane (Methyl bromide)** **2-Methoxy-2-methylpropane (*tert*-Butyl methyl ether)**

With the alternative combination of sodium methoxide and 2-bromo-2-methylpropane, no ether is formed; 2-methylpropene, formed by dehydrohalogenation, is the only product.

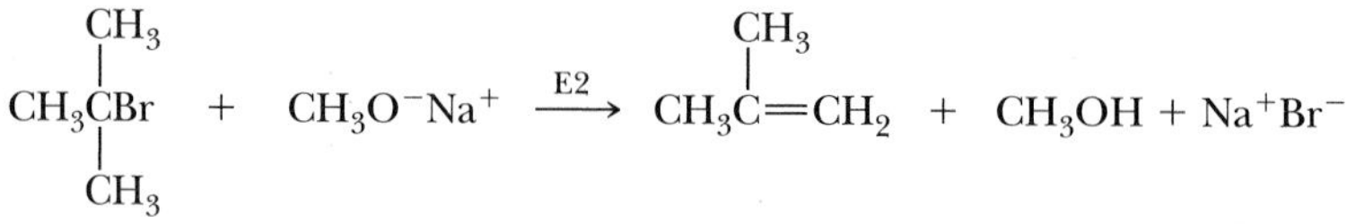

2-Bromo-2-methylpropane **Sodium methoxide** **2-Methylpropene**

Example 11.3 Williamson Ether Synthesis

Show the combination of alcohol and haloalkane that can best be used to prepare each ether by the Williamson ether synthesis.

(a) 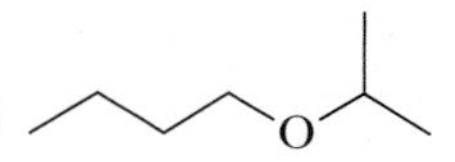 **(b)** 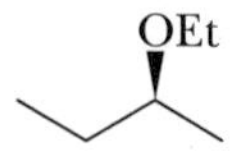

Solution

(a) Treat 2-propanol with sodium metal to form sodium isopropoxide. Then treat this metal alkoxide with 1-bromobutane.

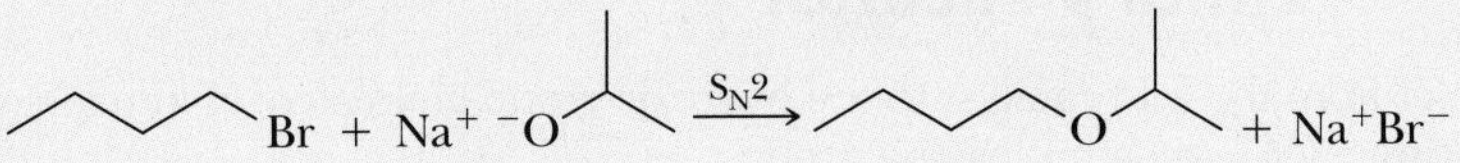

1-Bromobutane **Sodium isopropoxide**

The alternative combination of sodium butoxide and 2-bromopropane gives considerably more elimination product.

(b) Treat (*S*)-2-butanol with sodium metal to form the sodium alkoxide. This reaction involves only the O—H bond and does not affect the chiral center. Then treat this sodium alkoxide with a haloethane [e.g., ethyl iodide (EtI)] to give the desired product.

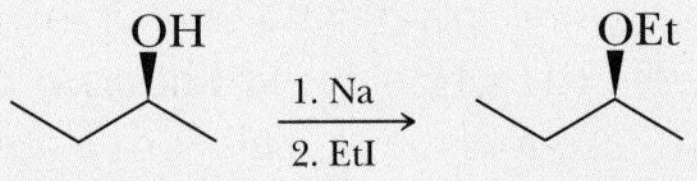

(*S*)-2-Butanol **(*S*)-2-Ethoxyethane**

An alternative synthesis is to convert the (*R*)-2-butanol to its tosylate (Section 10.5D) followed by treatment with sodium ethoxide.

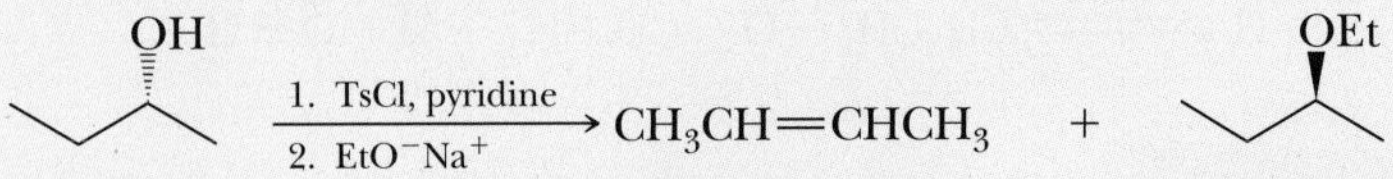

(*R*)-2-Butanol **Product of E2** (major product) **(*S*)-2-Ethoxybutane** (product of S_N2 reaction; a minor product)

This synthesis, however, gives only a low yield of the desired product. Recall from Section 9.8C that when a 2° halide or by analogy a 2° tosylate is treated with a strong base/good nucleophile, E2 is the major reaction.

Problem 11.3

Watch a video explanation

Show how you might use the Williamson ether synthesis to prepare each ether.

(a)

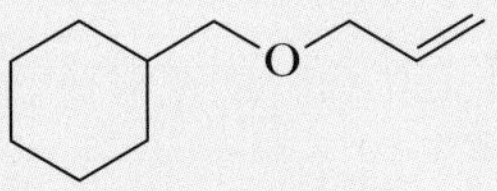

(b)

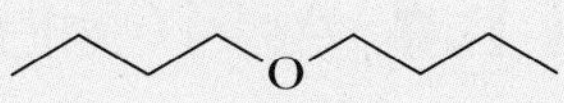

B. Acid-Catalyzed Dehydration of Alcohols

Diethyl ether and several other commercially available ethers are synthesized on an industrial scale by the acid-catalyzed dehydration of primary alcohols. Intermolecular dehydration of ethanol, for example, gives diethyl ether.

$$2\,CH_3CH_2OH \xrightarrow[140°C]{H_2SO_4} CH_3CH_2OCH_2CH_3 + H_2O$$

Ethanol **Diethyl ether**

Mechanism 11.1

Acid-Catalyzed Intermolecular Dehydration of a Primary Alcohol

Step 1: **Add a proton.** Proton transfer from the acid catalyst to the hydroxyl group gives an oxonium ion, which converts —OH, a poor leaving group, into $—\overset{+}{O}H_2$, a better leaving group.

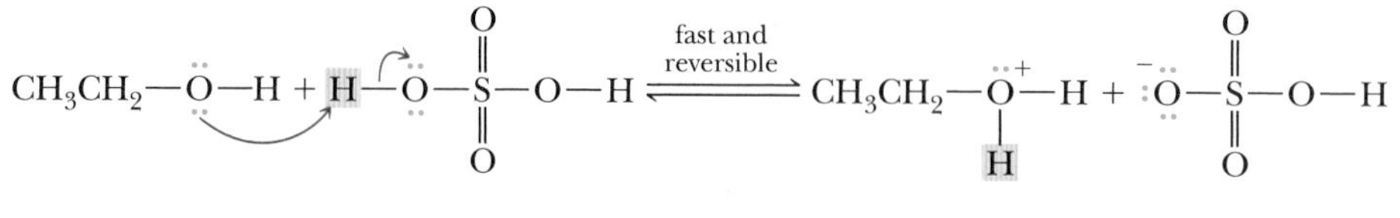

An oxonium ion

Step 2: **Make a new bond between a nucleophile and an electrophile and simultaneously break a bond to give stable molecules or ions.** Nucleophilic displacement of H_2O by the OH group of a second alcohol molecule gives a new oxonium ion.

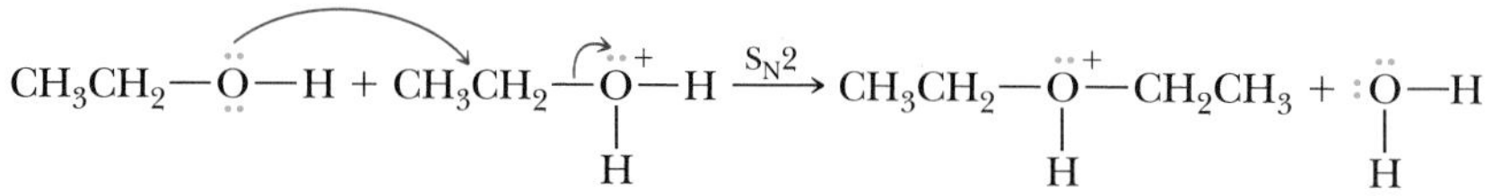

A new oxonium ion

Step 3: **Take a proton away.** Proton transfer from the new oxonium ion to H_2O completes the reaction.

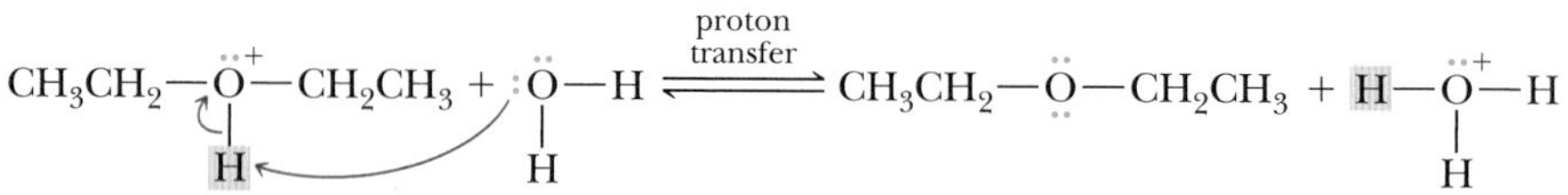

Note that the acid is a catalyst in this reaction. One proton is used in Step 1, but another is generated in Step 3.

Yields of ethers from the acid-catalyzed intermolecular dehydration of alcohols are highest for symmetrical ethers formed from unbranched primary alcohols. Examples of symmetrical ethers formed in good yield by this method are dimethyl ether, diethyl ether, and dibutyl ether. From secondary alcohols, yields of ether are lower because of competition from acid-catalyzed dehydration (Section 10.6). In the case of tertiary alcohols, dehydration to an alkene is the only reaction.

Example 11.4 Acid-Catalyzed Ether Synthesis

Explain why this reaction does not give a good yield of ethyl hexyl ether.

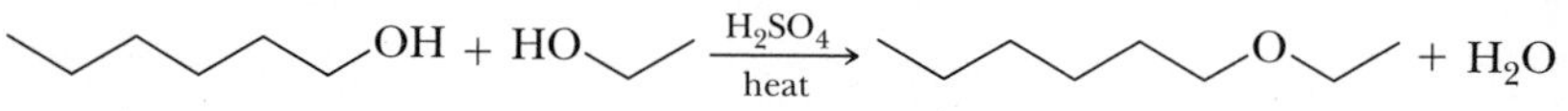

Solution

From this reaction, we expect a mixture of three ethers: diethyl ether, ethyl hexyl ether, and dihexyl ether.

Problem 11.4

Show how ethyl hexyl ether might be prepared by a Williamson ether synthesis.

C. Acid-Catalyzed Addition of Alcohols to Alkenes

Under suitable conditions, alcohols can be added to the carbon-carbon double bond of an alkene to give an ether. The usefulness of this method of ether synthesis is limited to the interaction of alkenes that form stable carbocations and methanol or primary alcohols. An example is the commercial synthesis of *tert*-butyl methyl ether (MTBE). 2-Methylpropene and methanol are passed over an acid catalyst to give the ether.

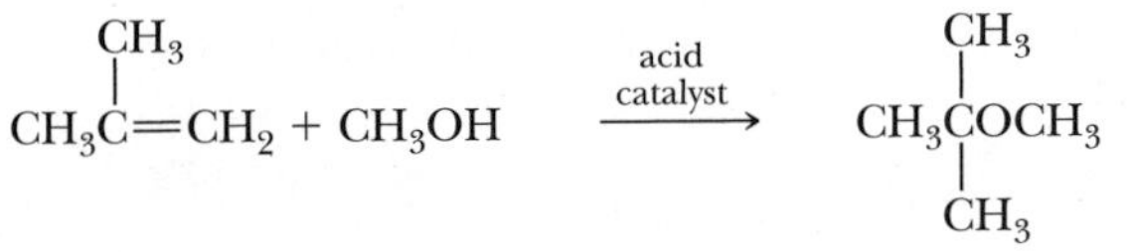

2-Methoxy-2-methylpropane
(*tert*-Butyl methyl ether)

Mechanism 11.2

Acid-Catalyzed Addition of an Alcohol to an Alkene

Step 1: **Add a proton.** Proton transfer from the acid catalyst to the alkene gives a carbocation intermediate.

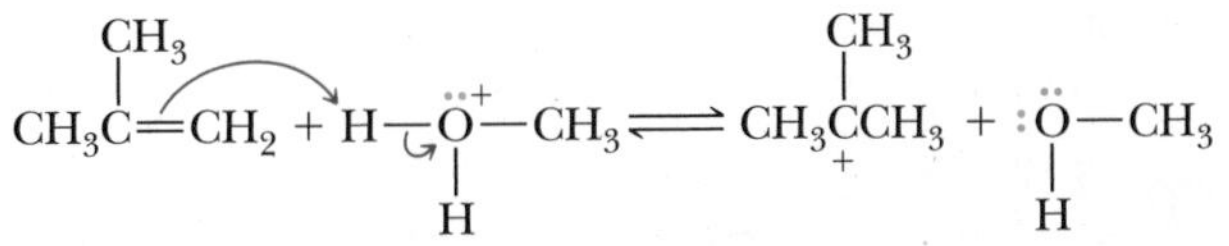

Step 2: **Make a new bond between a nucleophile and an electrophile.** Reaction of the carbocation intermediate (an electrophile and a Lewis acid) with the alcohol (a nucleophile and a Lewis base) gives an oxonium ion.

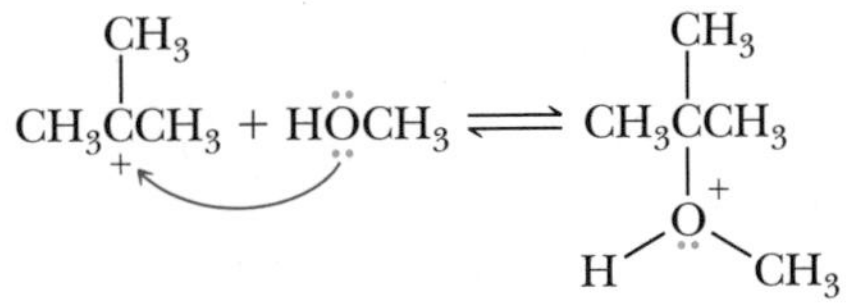

Step 3: **Take a proton away.** Proton transfer to solvent (in this case, methanol) completes the reaction.

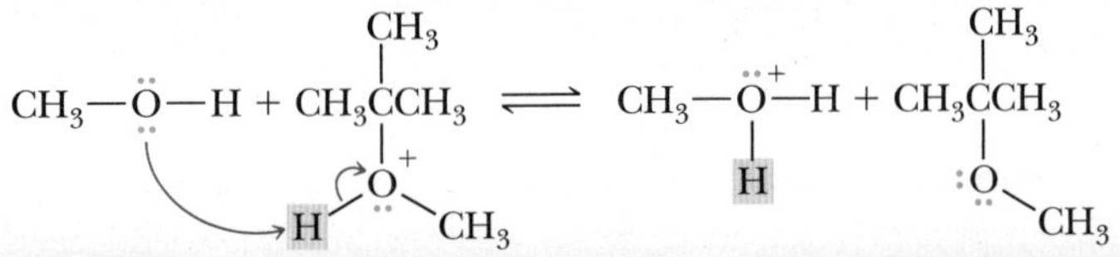

At one time, MTBE was added to gasoline under a mandate from the Environmental Protection Agency to add "oxygenates," which make gasoline burn more smoothly (it raises the octane number) and lower exhaust emissions. As an octane-improving additive, MTBE is superior to ethanol (the additive in ethanol blend fuels such as E10 and E85). A blend of 15% MTBE with gasoline improves octane rating by approximately 5 units. Unfortunately, because MTBE is much more soluble in water than gasoline, it has gotten into the water table in many places—in some cases because of leaky gas station storage tanks. It has been detected in lakes, reservoirs, and water supplies—in some cases at concentrations that exceed limits for both "taste and odor" and human health. Consequently, its use as a gasoline additive was phased out.

11.5 Reactions of Ethers

Ethers resemble hydrocarbons in their resistance to chemical reaction. They do not react with oxidizing agents such as potassium dichromate or potassium permanganate. They are stable toward even very strong bases, and except for tertiary alkyl ethers, they are not affected by most weak acids at moderate temperatures. Because of their good solubilizing properties and general inertness to chemical reaction, ethers are excellent solvents in which to carry out many organic reactions.

A. Acid-Catalyzed Cleavage by Concentrated HX

Cleavage of dialkyl ethers requires both a strong acid and a good nucleophile, hence the use of concentrated aqueous **HI** (57%) or **HBr** (48%). Dibutyl ether, for example, reacts with hot concentrated **HBr** to give two molecules of 1-bromobutane.

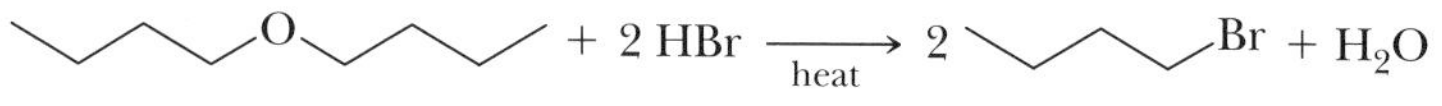
O
+ 2 HBr
heat
2
Br
+ H_2O

Dibutyl ether **1-Bromobutane**

Concentrated HCl (38%) is far less effective in cleaving dialkyl ethers, primarily because Cl^- is a weaker nucleophile in water than either I^- or Br^-.

The mechanism of acid-catalyzed cleavage of dialkyl ethers depends on the nature of the carbons bonded to oxygen. If both carbons are primary, cleavage involves an S_N2 reaction in which a halide ion is the nucleophile. Otherwise, cleavage is by an S_N1 reaction.

Mechanism 11.3

Acid-Catalyzed Cleavage of a Dialkyl Ether

***Step 1:* Add a proton.** Proton transfer from the acid catalyst to the oxygen atom of the ether gives an oxonium ion.

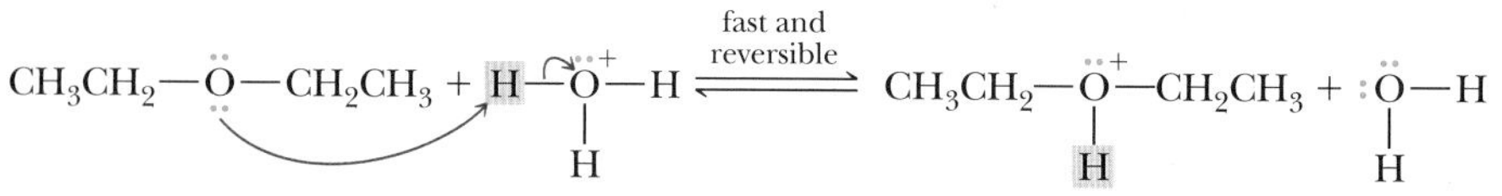
$CH_3CH_2-O-CH_2CH_3$ + H$-O^+-$H
H
fast and reversible
$CH_3CH_2-O^+-CH_2CH_3$ + :O$-$H
H
H

An oxonium ion

***Step 2:* Make a new bond between a nucleophile and an electrophile and simultaneously break a bond to give stable molecules or ions.** Nucleophilic displacement by halide ion on the primary carbon cleaves the C—O bond; the leaving group is CH_3CH_2OH, a weak base and a poor nucleophile.

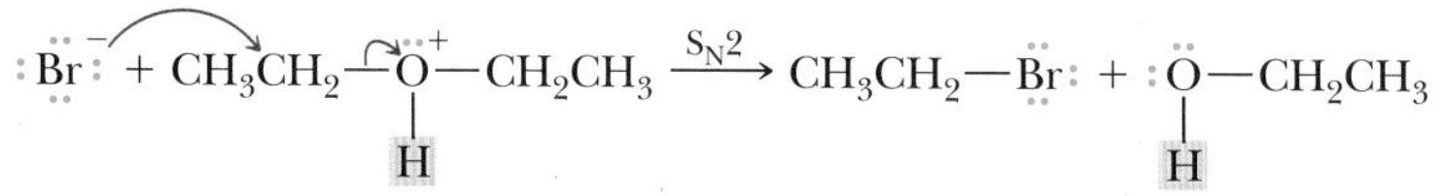
$:Br:^-$ + $CH_3CH_2-O^+-CH_2CH_3$
H
S_N2
$CH_3CH_2-Br:$ + $:O-CH_2CH_3$
H

This cleavage produces one molecule of bromoalkane and one molecule of alcohol. In the presence of excess concentrated **HBr**, the alcohol is converted to a second molecule of bromoalkane by another S_N2 process (Section 9.2).

Tertiary, allylic, and benzylic ethers are particularly susceptible to cleavage by acid, often under quite mild conditions. Tertiary butyl ethers, for example, are cleaved by aqueous HCl at room temperature. Proton transfer from the acid to the oxygen atom of the ether produces an oxonium ion, which then cleaves to produce a particularly stable 3°, allylic, or benzylic carbocation. Reaction of the carbocation with Cl^- completes the reaction.

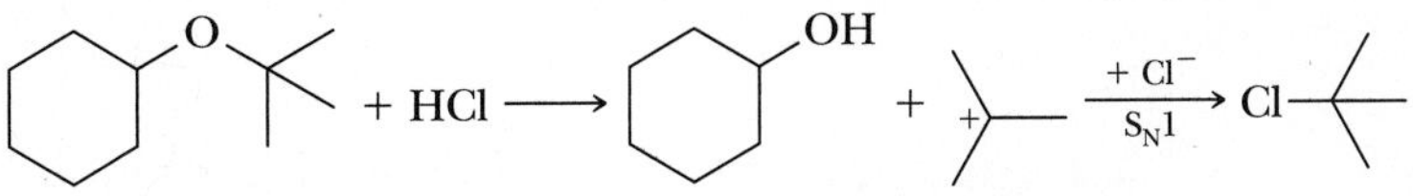

A *tert*-butyl ether   A 3° carbocation intermediate

Example 11.5 Ether Cleavage with HI

Account for the fact that treating most methyl ethers with concentrated HI gives CH_3I and ROH as the initial major products rather than CH_3OH and RI, as illustrated by the following reaction.

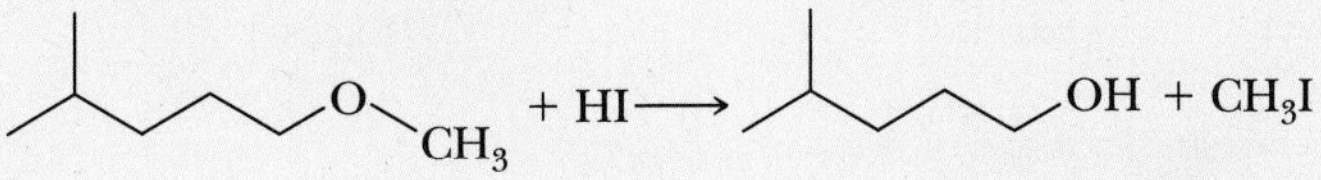

Solution

The first step is protonation of the ether oxygen to give an oxonium ion. Cleavage is by an S_N2 pathway on the less hindered methyl carbon.

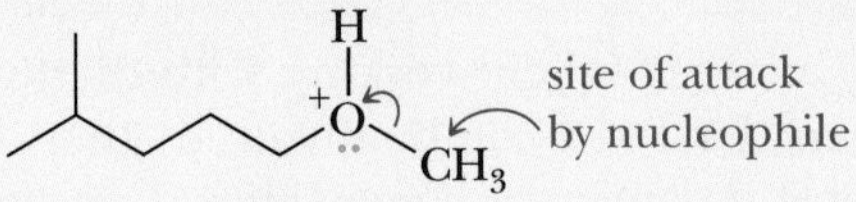

Problem 11.5

Account for the fact that treatment of *tert*-butyl methyl ether with a limited amount of concentrated HI gives methanol and *tert*-butyl iodide rather than methyl iodide and *tert*-butyl alcohol.

Example 11.6 Ether Cleavage with HBr

Draw structural formulas for the major products of each reaction.

(a) 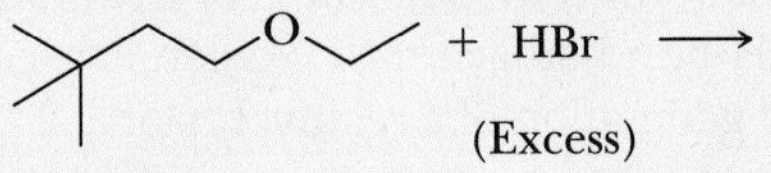

(b) 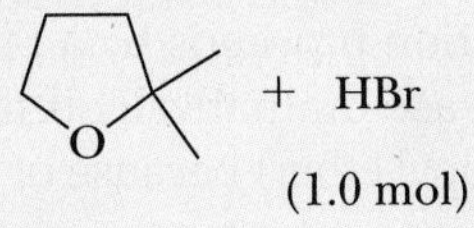

Solution

(a) Cleavage on either side of the ether oxygen by an S_N2 mechanism gives an alcohol and a bromoalkane. Reaction of the alcohol then gives a second molecule of bromoalkane.

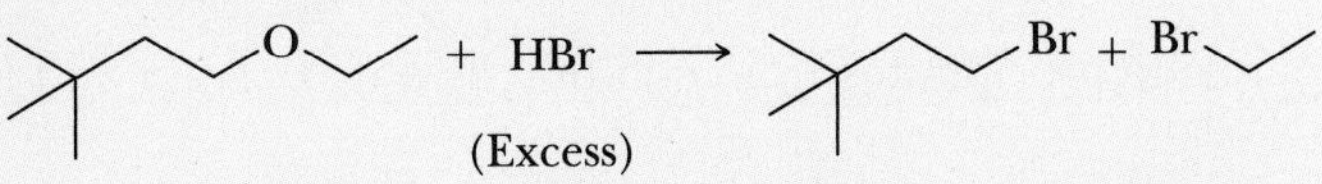

(b) Proton transfer to the ether oxygen followed by cleavage gives a 3° carbocation intermediate, which may then (1) react with bromide ion to give a bromoalcohol or (2) lose a proton to give an unsaturated alcohol.

(Continued)

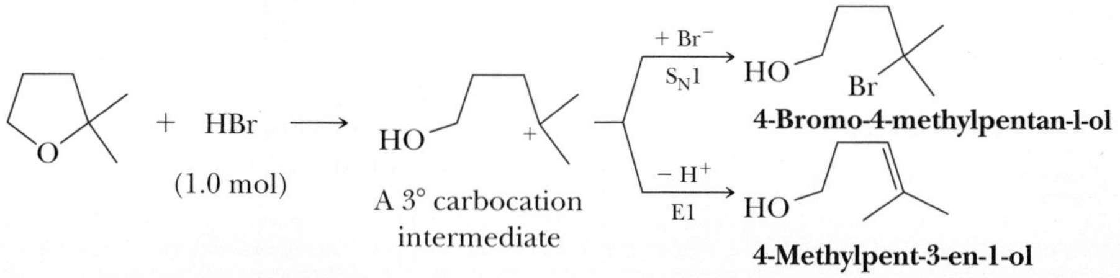
+ HBr
(1.0 mol)
HO
A 3° carbocation intermediate
+ Br⁻
S_N1
HO
Br
4-Bromo-4-methylpentan-l-ol
− H⁺
E1
HO
4-Methylpent-3-en-1-ol

Watch a video explanation

Problem 11.6

Draw structural formulas for the major products of each reaction.

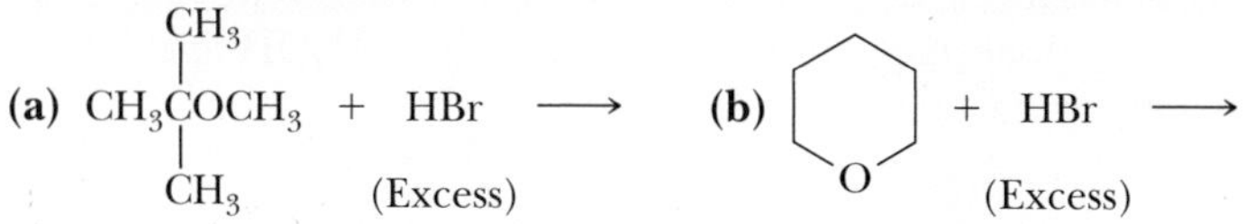
(a) CH_3COCH_3 + HBr (Excess)
CH_3
CH_3
(b) + HBr (Excess)
O

B. Ether Safety Alert: Flammability and Formation of Hydroperoxides

Two hazards must be avoided when working with diethyl ether and other low-molecular-weight ethers. First, the commonly used ethers have low boiling points and are highly flammable, a dangerous combination. Consequently, open flames and electric appliances with sparking contacts must be avoided where ethers are being used (laboratory refrigerators and ovens are frequent causes of ignition). Because diethyl ether is so volatile (its boiling point is 35°C), it should be used in a fume hood to prevent the buildup of vapors and possible explosion. Second, anhydrous ethers react with molecular oxygen at a C—H bond adjacent to the ether oxygen to form **hydroperoxides**, which are dangerous because they are explosive.

Hydroperoxide
A compound containing an —OOH group.

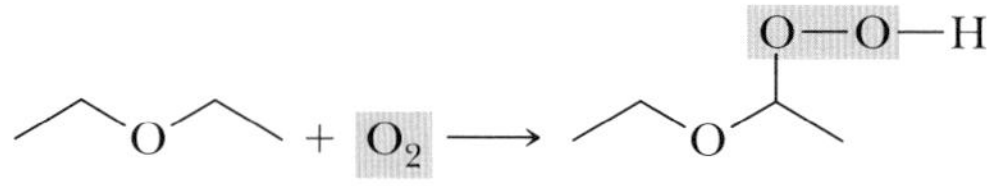
O
+ O_2
O—O—H
O

Diethyl ether A hydroperoxide

Hydroperoxidation proceeds by a radical chain mechanism. Rates of hydroperoxide formation increase dramatically if the C—H bond adjacent to oxygen is secondary (e.g., in diisopropyl ether) because of favored generation of a relatively stable 2° radical intermediate next to oxygen.

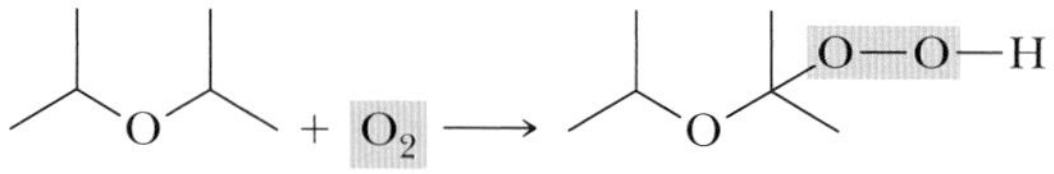
O
+ O_2
O—O—H
O

Diisopropyl ether A hydroperoxide

This hydroperoxide precipitates from solution as a waxy solid and is particularly dangerous.

Hydroperoxides in ethers can be detected by shaking a small amount of the ether with an acidified 10% aqueous solution of potassium iodide, KI, or by using starch iodine paper with a drop of acetic acid. Peroxides oxidize iodide ion to iodine, I_2, which gives a yellow color to the solution. Hydroperoxides can be removed by treating them with a reducing agent. One effective procedure is to shake the hydroperoxide-contaminated ether with a solution of iron(II) sulfate in dilute aqueous sulfuric acid. You should never use ethers past their expiration date, and you should properly dispose of them before then.

11.6 Silyl Ethers as Protecting Groups

When dealing with organic compounds containing two or more functional groups, it is often necessary to protect one functional group (to prevent its reaction) while carrying out a reaction at another functional group. Suppose, for example, that you want to convert 4-pentyn-1-ol to 4-heptyn-1-ol.

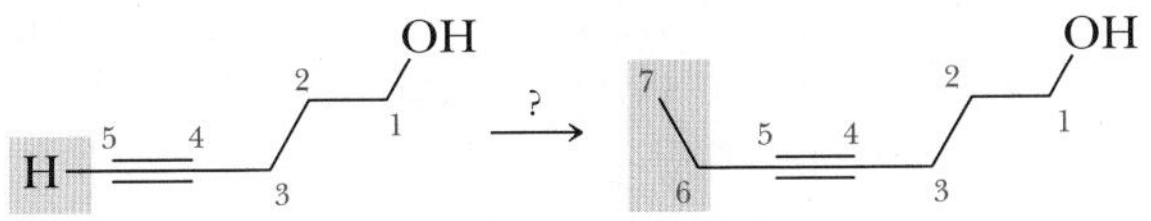

4-Pentyn-1-ol **4-Heptyn-1-ol**

The new carbon-carbon bond can be formed by treating the acetylide anion (Section 7.5) of 4-pentyn-1-ol with bromoethane. 4-Pentyn-1-ol, however, contains two acidic hydrogens, one on the hydroxyl group (pK_a 16–18) and the other on the carbon-carbon triple bond (pK_a 25). Treatment of this compound with one equivalent of $NaNH_2$ forms the alkoxide anion (the —OH group is the stronger acid) rather than the acetylide anion.

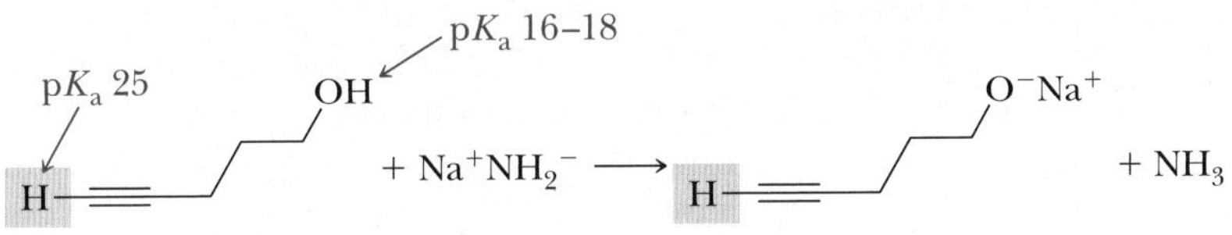

Alkoxide anion

To carry out the synthesis of 4-heptyn-1-ol, we must first protect the —OH group to prevent its reaction with sodium amide. A good **protecting group** is

- Easily added to the sensitive functional group.
- Resistant to the reagents used to transform the unprotected functional group or groups.
- Easily removed to regenerate the original functional group.

Chemists have devised protecting groups for most functional groups, and we will encounter several of them in this text. In this section, we concentrate on the most common type of hydroxyl-protecting group, namely silyl ethers.

Protecting group
An unreactive group reversibly created for the purpose of preventing a functional group from potentially reacting to give an unwanted product or products.

Silicon is in Group 4A of the Periodic Table, immediately below carbon. Like carbon, silicon also forms tetravalent compounds such as the following:

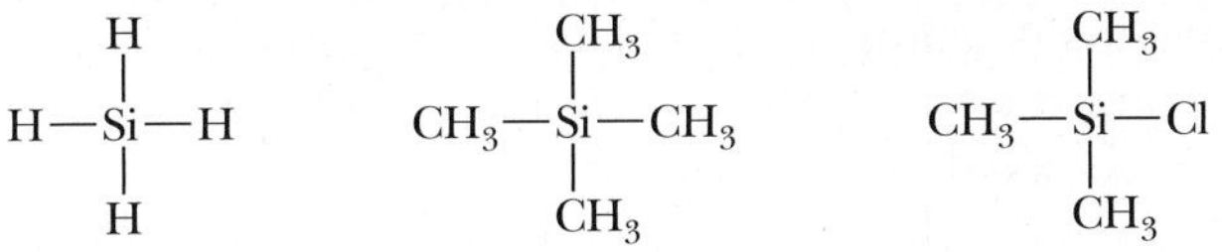

Silane **Tetramethylsilane** **Chlorotrimethylsilane**

An —OH group can be converted to a silyl ether by treating it with a trialkylsilyl chloride in the presence of an amine base. For example, treating an alcohol with chlorotrimethylsilane in the presence of a tertiary amine, such as triethylamine or pyridine, gives a trimethylsilyl ether.

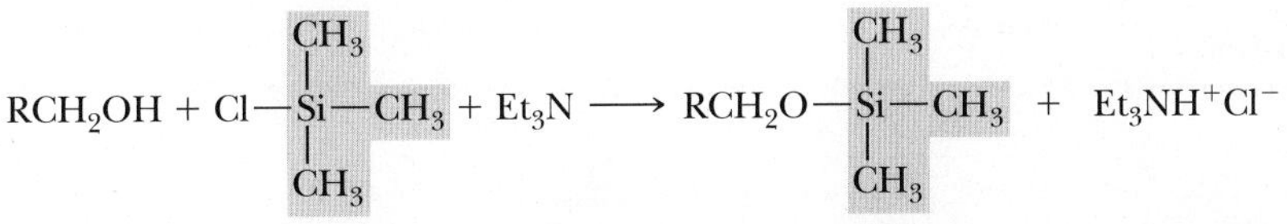

Chlorotri-methylsilane A trimethylsilyl ether **Triethylammonium chloride**

The function of the tertiary amine is to catalyze the reaction by forming some of the more nucleophilic alkoxide ion and to neutralize the HCl formed during the reaction. Tertiary amines are not nucleophilic because the three attached alkyl groups provide steric hindrance. Tertiary amines can react with protons and act as a base but will not react with electrophiles such as chlorotrimethylsilane.

Replacement of one of the methyl groups of the trimethylsilyl (TMS) group by *tert*-butyl gives the *tert*-butyldimethylsilyl (TBS) group, which is considerably more stable than the trimethylsilyl group. In fact, the TMS group is too labile in many situations to be practical. Other common silyl protecting groups are the triethylsilyl (TES) and triisopropylsilyl (TIPS) groups.

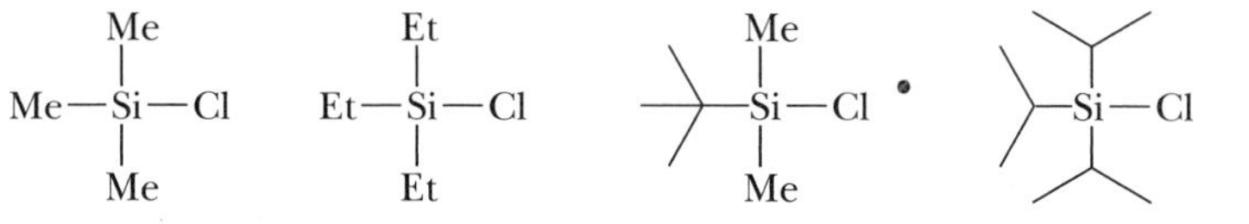

Trimethylsilyl chloride (TMSCl) **Triethylsilyl chloride (TESCl)** ***t*-Butyldimethylsilyl chloride (TBSCl)** **Triisopropylsilyl chloride (TIPSCl)**

Silyl ethers are unaffected by most oxidizing and reducing agents and are stable to most nonaqueous acids and bases. The TBS group is stable in aqueous solution within the pH range 2–12, which makes it one of the most widely used —OH protecting groups. Silyl ether blocking groups are most easily removed by treatment with fluoride ion, generally in the form of tetrabutylammonium fluoride, $Bu_4N^+F^-$.

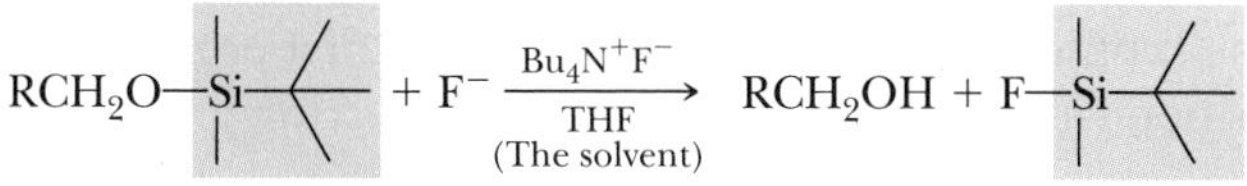

A TBS-protected alcohol

This cleavage of the protecting group depends on the fact that a silicon-fluorine σ bond is considerably stronger (582 kJ/mol) than a silicon-oxygen σ bond (368 kJ/mol). In fact, the Si—F σ bond is one of the strongest σ bonds known. The large difference in bond strengths between Si—O and Si—F bonds drives the silyl ether cleavage reaction to completion. ▶

▶ Reactions will occur if the products are more stable than the reactants and the energy barrier is low enough.

We can use a silyl ether in the following way to convert 4-pentyn-1-ol to 4-heptyn-1-ol. Treating 4-pentyn-1-ol with chlorotrimethylsilane in the presence of triethylamine gives the trimethylsilyl ether. Treatment of the terminal alkyne with sodium amide followed by bromoethane forms the new carbon-carbon bond. Subsequent removal of the TMS protecting group with tetrabutylammonium fluoride gives the desired 4-heptyn-1-ol.

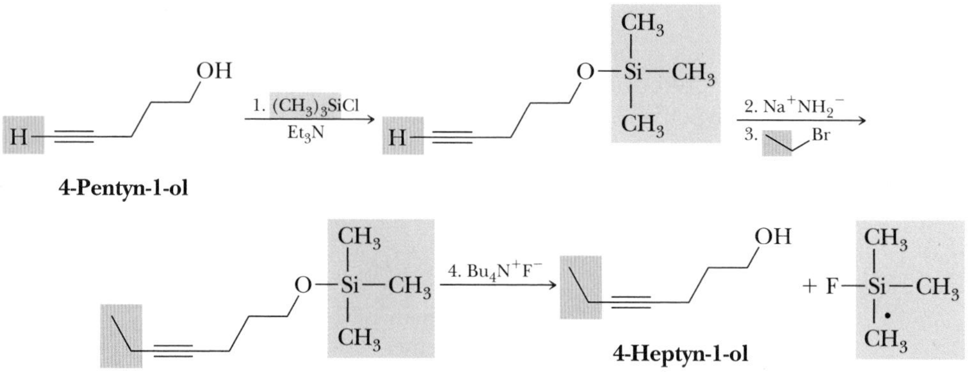

Example 11.7 Carbon and Silicon Bond Polarity

Compare the polarity of the C—Cl bond in $(CH_3)_3C$—Cl with the polarity of the Si—Cl bond in $(CH_3)_3Si$—Cl.

Solution

The difference in electronegativity between carbon and chlorine is $3.0 - 2.5 = 0.5$; that between silicon and chlorine is $3.0 - 1.8 = 1.2$; a Si—Cl bond is more polar than a C—Cl bond.

Problem 11.7

The trimethylsilyl protecting group is easily removed in aqueous solution containing a trace of acid. Propose a mechanism for this reaction. (Note that a TBDMS protecting group is stable under these conditions because of the greater steric crowding around silicon created by the *t*-butyl group.)

11.7 Epoxides: Structure and Nomenclature

Although **epoxides** are technically classed as ethers, we discuss them separately because of their exceptional chemical reactivity compared with other ethers. Simple epoxides are named as derivatives of oxirane, the parent epoxide. Where the epoxide is a part of another ring system, it is named using the prefix *epoxy*.

Epoxide A cyclic ether in which oxygen is one atom of a three-membered ring.

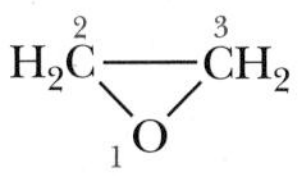
2
3
H_2C
CH_2
1
O

Oxirane (Ethylene oxide)

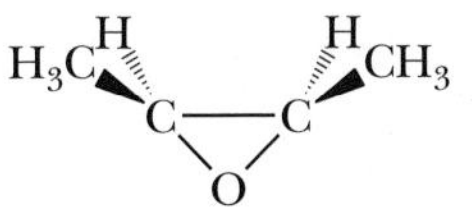
H_3C
H
H
CH_3
C
C
O

***cis*-2,3-Dimethyloxirane (*cis*-2-Butene oxide)**

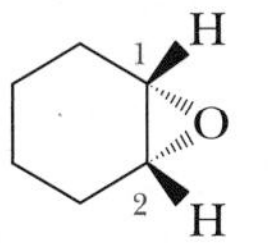
1
H
O
2
H

1,2-Epoxycyclohexane (Cyclohexene oxide)

Common names of epoxides are derived by giving the name of the alkene from which the epoxide is formally derived, followed by the word *oxide;* an example is *cis*-2-butene oxide.

11.8 Synthesis of Epoxides

A. Ethylene Oxide

Ethylene oxide, one of the few epoxides synthesized on an industrial scale, is prepared by passing a mixture of ethylene and air (or oxygen) over a silver catalyst.

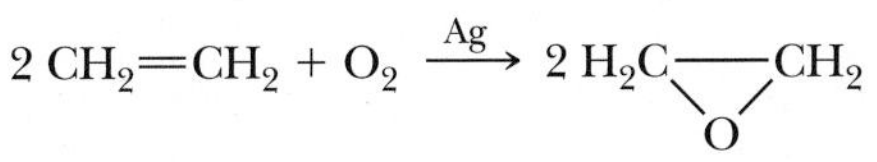
$2\ CH_2{=}CH_2 + O_2$
Ag
$2\ H_2C$
CH_2
O

Oxirane (Ethylene oxide)

This method only works well for the production of oxirane from ethylene. As a result, other methods have been developed for the production of epoxides.

B. Internal Nucleophilic Substitution in Halohydrins

One method for the preparation of epoxides from alkenes involves (1) treating the alkene with chlorine or bromine in water to form a chlorohydrin (or bromohydrin) followed by (2) treating the halohydrin with a base to bring about intramolecular displacement of Cl^-. These steps convert propene first to 1-chloro-2-propanol and then to methyloxirane (propylene oxide).

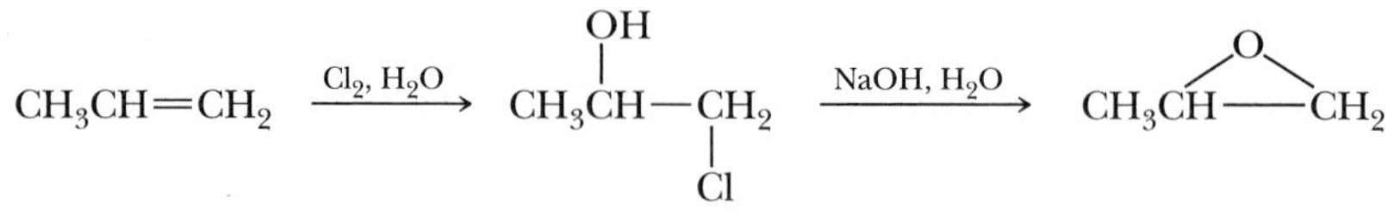

Propene | **1-Chloro-2-propanol** (a chlorohydrin) (racemic) | **Methyloxirane (Propylene oxide)** (racemic)

We studied the reaction of alkenes with chlorine or bromine in water to form halohydrins in Section 6.3E and saw that it is both regioselective and stereoselective (for an alkene that shows *cis-trans* isomerism, it is also stereospecific). Conversion of a halohydrin to an epoxide with base is stereoselective as well and can be viewed as an internal S_N2 reaction. Hydroxide ion or another base abstracts a proton from the halohydrin hydroxyl group to form an alkoxide ion, a good nucleophile, which then displaces halogen on the adjacent carbon. As with all S_N2 reactions, attack of the nucleophile is from the backside of the C—X bond and causes inversion of configuration at the site of substitution.

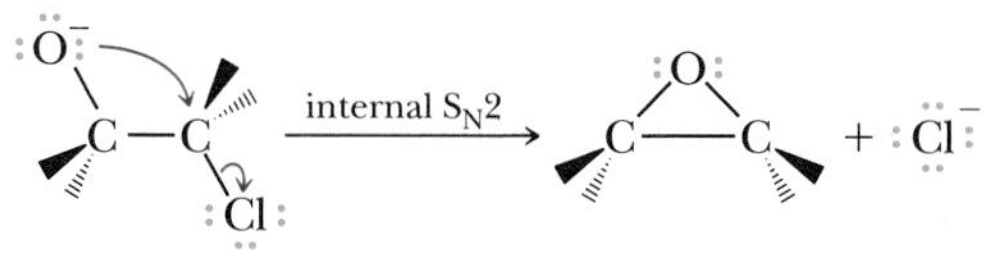

An epoxide

Note that this displacement of halide by the alkoxide ion can also be viewed as an intramolecular variation of the Williamson ether synthesis (Section 11.4A). In this case, the displacing alkoxide and leaving halide ions are on adjacent carbon atoms.

Example 11.8 Stereoselective Oxirane Synthesis

Conversion of an alkene to a halohydrin and internal displacement of a halide ion by an alkoxide ion are both stereoselective. Use this information to demonstrate that the configuration of the alkene is preserved in the epoxide. As an illustration, show that reaction of *cis*-2-butene by this two-step sequence gives *cis*-2,3-dimethyloxirane (*cis*-2-butene oxide).

Solution

Addition of HOCl to an alkene occurs by stereoselective anti addition of —OH and —Cl to the double bond (Section 6.3E). The conformation of this product is also the conformation necessary for stereoselective backside displacement of the halide ion by alkoxide ion. Thus, a *cis*-alkene gives a *cis* disubstituted oxirane, and the transformation from alkene to epoxide is stereospecific.

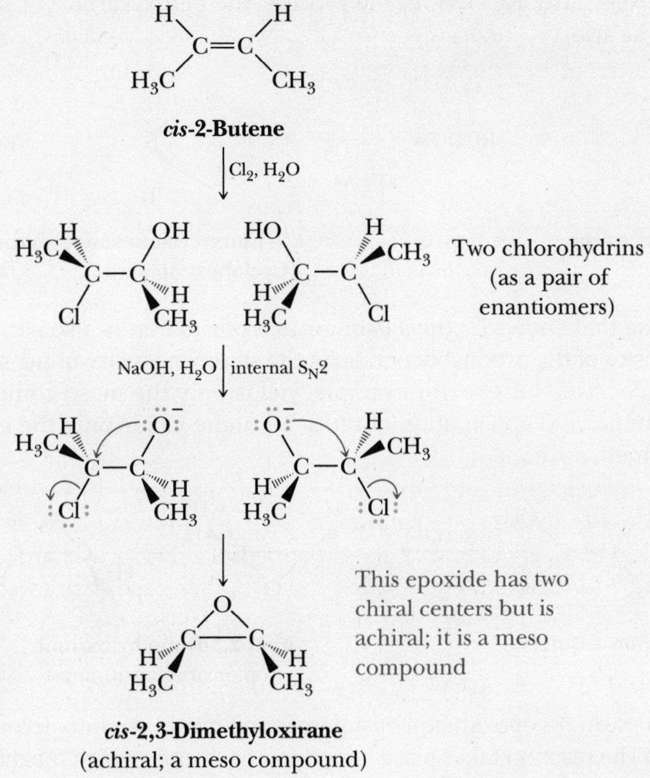
cis-2-Butene
Cl_2, H_2O
Two chlorohydrins (as a pair of enantiomers)
NaOH, H_2O
internal S_N2
This epoxide has two chiral centers but is achiral; it is a meso compound
cis-2,3-Dimethyloxirane (achiral; a meso compound)

Problem 11.8

Watch a video explanation

Consider the possibilities for stereoisomerism in the bromohydrin and epoxide formed from *trans*-2-butene.

(a) How many stereoisomers are possible for the bromohydrin? Which of the possible bromohydrin stereoisomers are formed by treating *trans*-2-butene with bromine in water?

(b) How many stereoisomers are possible for the epoxide? Which of the possible stereoisomers is/are formed in this two-step sequence?

C. Oxidation of Alkenes with Peroxycarboxylic Acids

The most common laboratory method for the synthesis of epoxides from alkenes is oxidation with a peroxycarboxylic acid (a peracid). Three of the most widely used peroxyacids are *meta*-chloroperoxybenzoic acid (MCPBA), the magnesium salt of monoperoxyphthalic acid (MMPP), and peroxyacetic acid.

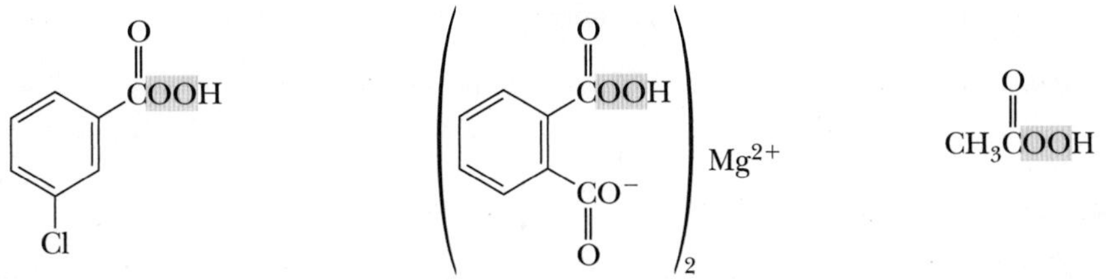

***meta*-Chloroperoxybenzoic acid (MCPBA)** | **Magnesium monoperoxyphthalate (MMPP)** | **Peroxyacetic acid (Peracetic acid)**

Following is a balanced equation for the epoxidation of cyclohexene by a peroxycarboxylic acid, RCO_3H. In the process, the peroxycarboxylic acid is reduced to a carboxylic acid.

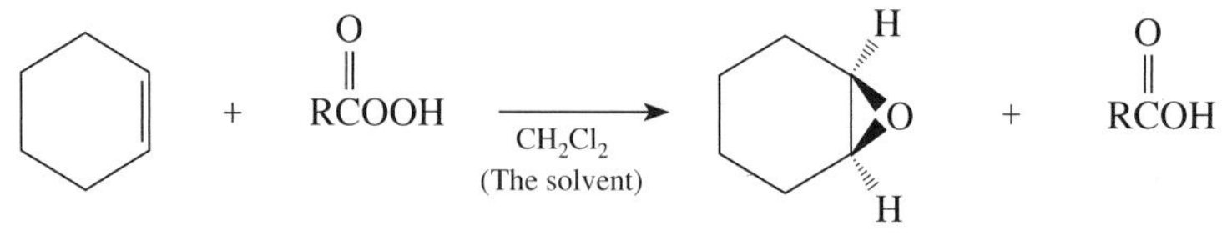

Cyclohexene A peroxy-carboxylic acid **1,2–Epoxycyclohexane (Cyclohexene oxide)** A carboxylic acid

For an alkene that shows *cis,trans* isomerism, epoxidation is also stereospecific: the stereochemistry of the product depends on the stereochemistry of the starting alkene. Epoxidation of *cis*-2-butene, for example, yields only the meso compound *cis*-2,3-dimethyloxirane, and epoxidation of *trans*-2-butene yields only the enantiomers of *trans*-2,3-dimethyloxirane.

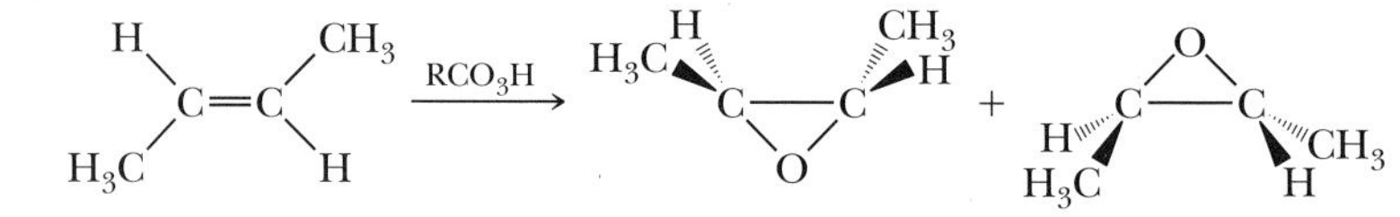

***trans*-2-Butene** ***trans*-2,3-Dimethyloxirane** (a pair of enantiomers)

A mechanism for epoxidation by a peroxyacid must take into account the following facts: (1) The reaction takes place in nonpolar solvents, which means that the reaction cannot involve the formation of ions or any species with significant separation of unlike charges. (2) The reaction is stereospecific, with complete retention of the alkene configuration, which means that even though the π bond of the carbon-carbon double bond is broken, at no time is there free rotation about the remaining σ bond. Following is a mechanism consistent with these observations.

Mechanism 11.4

Epoxidation of an Alkene by RCO_3H

The bond-making and bond-breaking steps are thought to be concerted or nearly so.

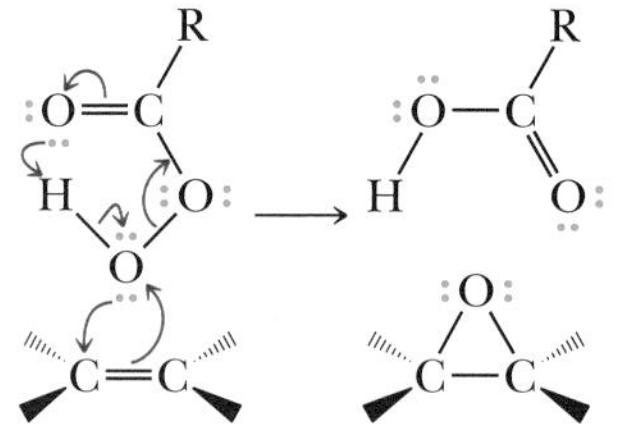

The concerted formation of the two C—O bonds of the epoxide ensures that the reaction is stereospecific (i.e., that *cis*-alkenes give *cis*-epoxides and *trans*-alkenes give *trans*-epoxides).

D. Sharpless Asymmetric Epoxidation

One of the most useful organic reactions discovered in the last several decades is the titanium-catalyzed asymmetric epoxidation of primary allylic alcohols developed by Professor Barry Sharpless, then at Stanford University. The reagent consists of *tert*-butyl hydroperoxide, titanium tetraisopropoxide [$Ti(O\text{-}iPr)_4$], and diethyl tartrate. Recall from

Section 3.4B that tartaric acid has two chiral centers and exists as three stereoisomers: a pair of enantiomers and a meso compound. The form of tartaric acid used in the Sharpless epoxidation is either pure (+)-diethyl tartrate or its enantiomer, (−)-diethyl tartrate. The *tert*-butyl hydroperoxide is the oxidizing agent and must be present in molar amounts. Titanium tetraisopropoxide and diethyl tartrate combine to make the active catalyst and are present in lesser amounts, generally 5–10 mole percent.

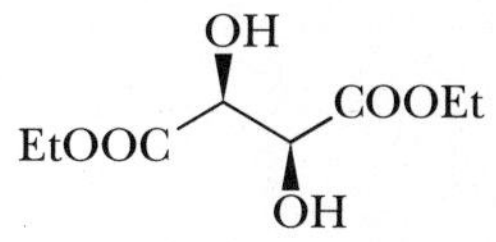

(2*S*,3*S*)-(−)-Diethyl tartrate

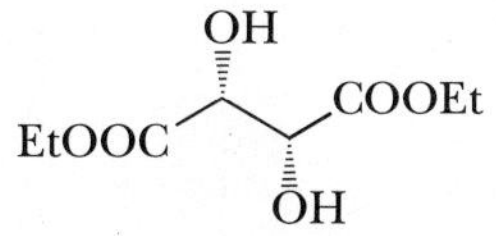

(2*R*,3*R*)-(+)-Diethyl tartrate

What is remarkable about the Sharpless epoxidation is that it is stereospecific based on the diethyl tartrate added; either enantiomer of an epoxide can be produced depending on which enantiomer of diethyl tartrate is used. If the (−)-enantiomer is used, the product is enantiomer A. If the (+)-enantiomer is used, the product is enantiomer B.

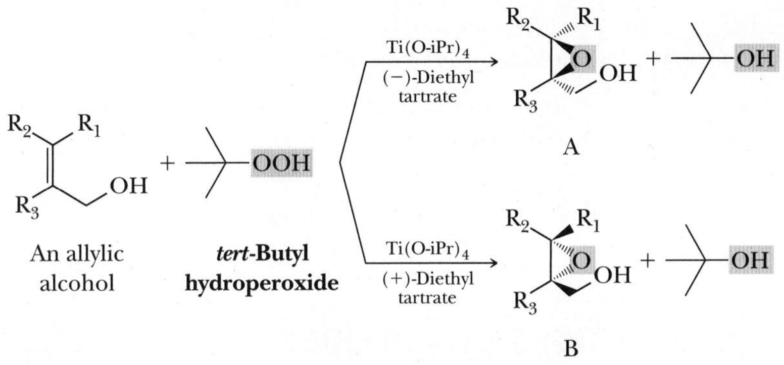

When predicting the stereochemistry of a Sharpless epoxidation product, you will find it helpful to draw the allylic alcohol in the same orientation each time, as shown in the margin, for example.

With (–)-diethyl tartrate, oxygen is delivered to the top face.

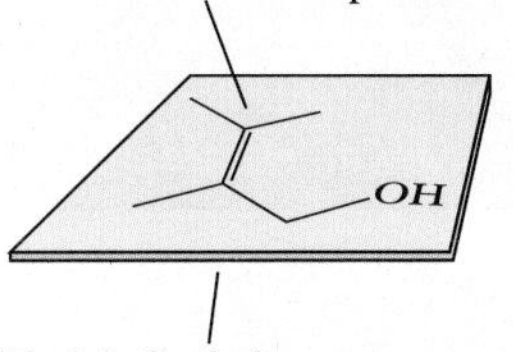

With (+)-diethyl tartrate, oxygen is delivered to the bottom face.

When drawn in this manner, the (−)-tartrate catalyzes delivery of the epoxide oxygen from the top face of the alkene; the (+)-tartrate catalyzes its delivery from the bottom face.

The mechanism of this catalyzed epoxidation has been studied in detail and involves formation of a chiral complex in which the carbonyl oxygen of diethyl tartrate displaces one of the isopropoxide groups on titanium. When the **R—OOH** oxidizing agent is added, it displaces a second isopropoxide group. Finally, the oxygen of the allylic alcohol displaces a third isopropoxide group. Thus, although neither the alkene nor the **ROOH** oxidizing agent is chiral, both are now held in a fixed stereochemical relationship to the other in the chiral environment created by the diethyl tartrate-titanium complex. In this chiral environment, oxygen is delivered to either the top face or the bottom face of the alkene, depending on which enantiomer of diethyl tartrate is present.

For their pioneering work in the field of enantioselective synthesis, Sharpless (along with William Knowles and Ryoji Noyori) received the 2001 Nobel Prize in Chemistry.

Example 11.9 Sharpless Epoxidation

Draw the expected products of Sharpless epoxidation of each allylic alcohol using (+)-diethyl tartrate as the chiral catalyst.

(a)

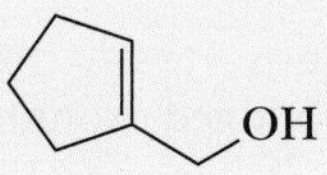

(b)

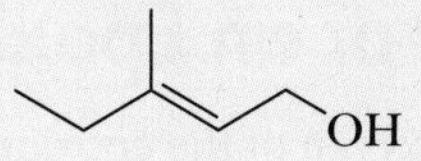

Solution

(a)

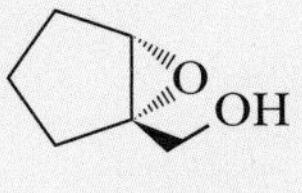

(Continued)

(b) In this solution, the carbon skeleton of the allylic alcohol is first reoriented to match the orientation in the template, the epoxidation is completed, and the carbon skeleton is reoriented to match the original drawing.

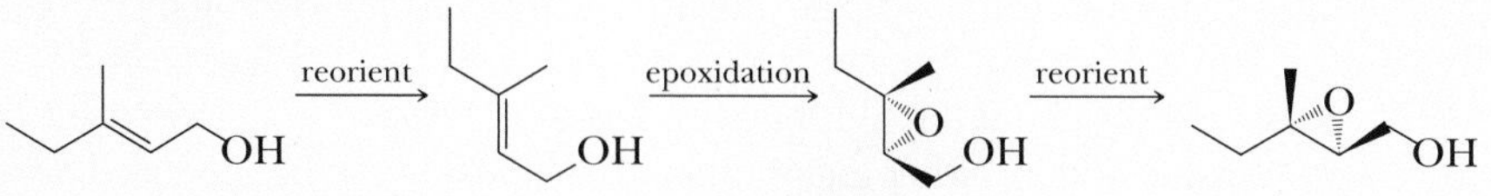

Problem 11.9

Draw the expected products of Sharpless epoxidation of each allylic alcohol using (+)-diethyl tartrate as the chiral catalyst.

(a)
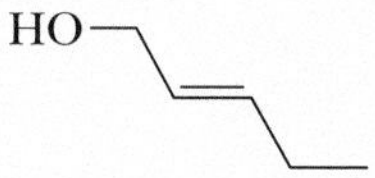

(b)
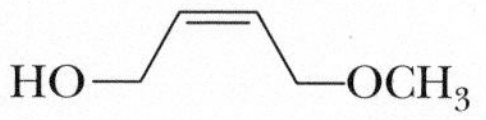

11.9 Reactions of Epoxides

▶ Most bond-making steps in reaction mechanisms involve nucleophiles reacting with electrophiles.

Because of the strain associated with the three-membered ring, epoxides undergo a variety of ring-opening reactions, the characteristic feature of which is nucleophilic substitution at one of the carbons of the epoxide ring with the oxygen atom as the leaving group.

Characteristic reaction of epoxides:

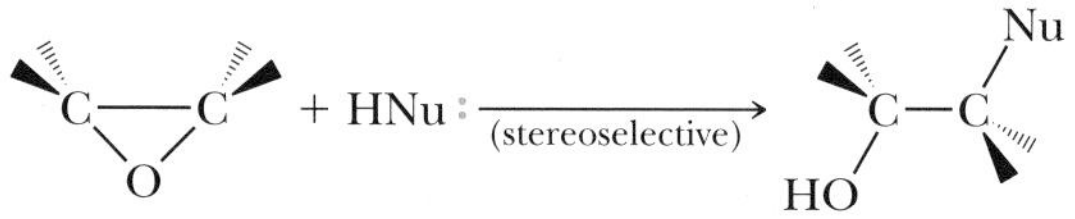

▶

A. Acid-Catalyzed Ring Opening

In the presence of an acid catalyst, such as sulfuric acid, epoxides are hydrolyzed to glycols. As an example, acid-catalyzed hydrolysis of oxirane gives 1,2-ethanediol (ethylene glycol).

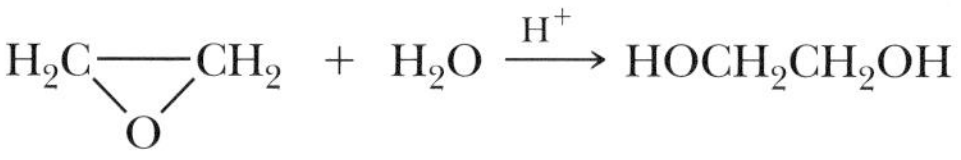

Oxirane (Ethylene oxide) **1,2-Ethanediol (Ethylene glycol)**

Mechanism 11.5

Acid-Catalyzed Hydrolysis of an Epoxide

Step 1: **Add a proton.** Proton transfer from the acid catalyst to oxygen of the epoxide gives a bridged oxonium ion intermediate.

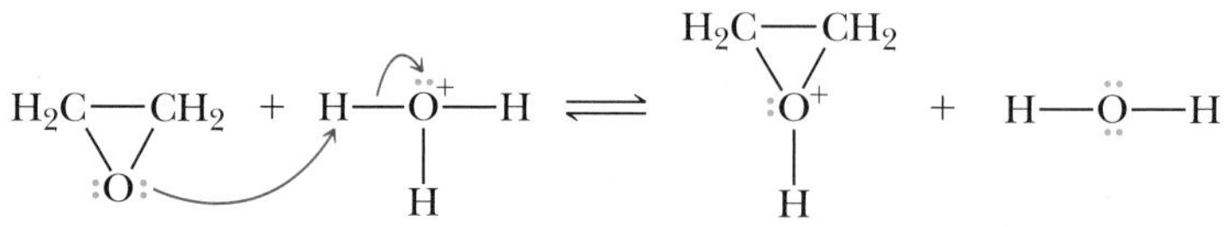

Step 2: **Make a new bond between a nucleophile and an electrophile and simultaneously break a bond to give stable molecules or ions.** Backside attack of H_2O on the protonated epoxide (a bridged oxonium ion) opens the three-membered ring.

Watch a video explanation

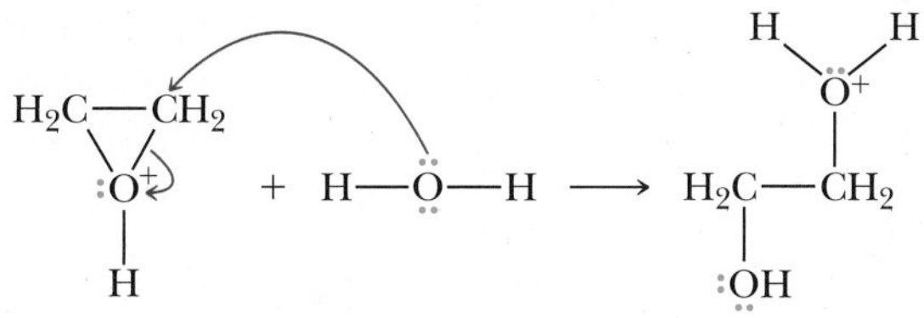

Step 3: **Take a proton away.** Proton transfer to solvent completes the formation of the glycol.

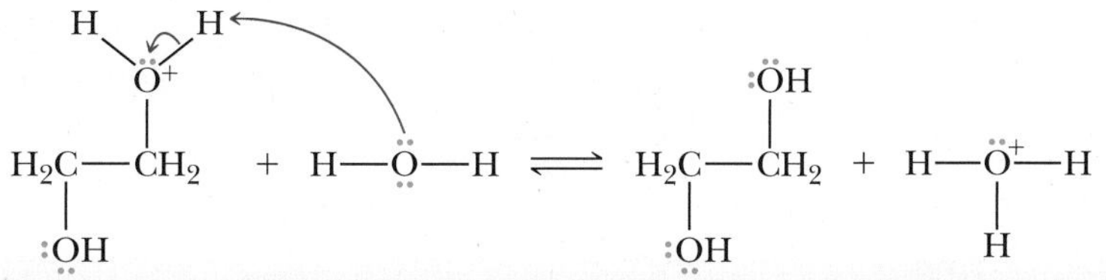

Attack of a nucleophile on a protonated epoxide shows an anti stereoselectivity typical of S_N2 reactions; the nucleophile attacks anti to the leaving hydroxyl group, and the —OH groups in the glycol thus formed are anti. As a result, hydrolysis of an epoxycycloalkane yields a *trans*-1,2-cycloalkanediol.

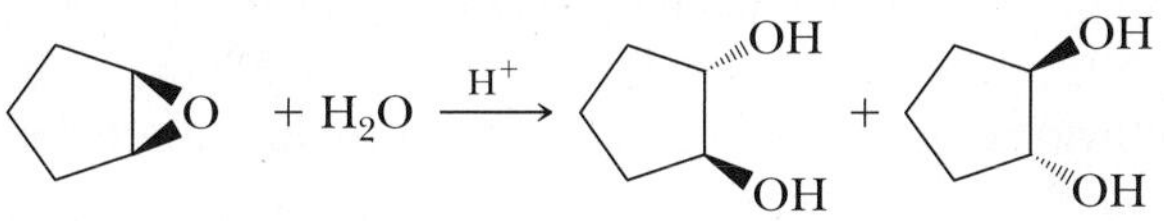

1,2-Epoxycyclopentane (Cyclopentene oxide) (achiral)

***trans*-1,2-Cyclopentanediol** (a racemic mixture)

Note the similarity in ring opening of this bridged oxonium ion intermediate and the bridged halonium ion intermediate in electrophilic addition of halogens or X_2/H_2O to an alkene (Sections 6.3D and 6.3E). In each case, the intermediate is a three-membered ring with a heteroatom bearing a positive charge and attack of the nucleophile is anti to the leaving group.

Because there is some carbocation character developed in the transition state for an acid-catalyzed epoxide ring opening, attack of the nucleophile on unsymmetrical epoxides occurs preferentially at the carbon better able to bear a partial positive charge.

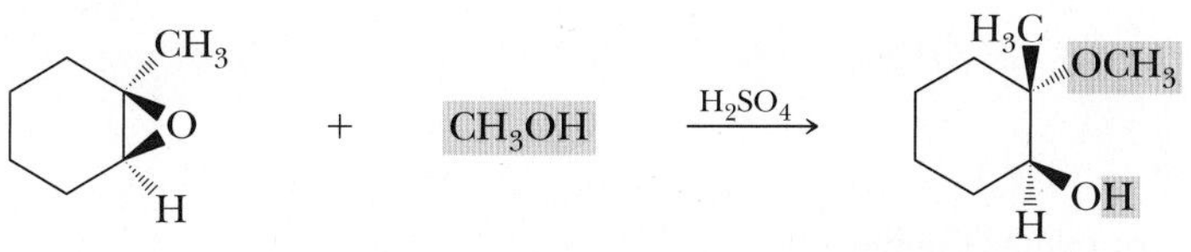

1-Methyl-1,2-epoxycyclohexane

2-Methoxy-2-methylcyclohexanol

The stereochemistry of acid-catalyzed ring openings is S_N2-like in that attack of the nucleophile is from the side opposite the bridged oxonium ion intermediate. The regiochemistry, however, is S_N1-like. Because of the partial carbocation character of the transition state, attack of the nucleophile on the oxonium ion intermediate occurs preferentially at the more substituted carbon. That is, attack occurs at the site better able to bear the partial positive charge that develops on carbon in the transition state in analogy to attack on a bridged bromonium ion.

At this point, let us compare the stereochemistry of the glycol formed by acid-catalyzed hydrolysis of an epoxide with that formed by oxidation of an alkene with osmium tetroxide (Section 6.5A). Each reaction sequence is stereoselective but gives a different stereoisomer. Acid-catalyzed hydrolysis of cyclopentene oxide gives *trans*-1,2-cyclopentanediol; osmium tetroxide oxidation of cyclopentene gives *cis*-1, 2-cyclopentanediol. Thus, a cycloalkene can be converted to either a *cis* glycol or a *trans* glycol by the proper choice of reagents.

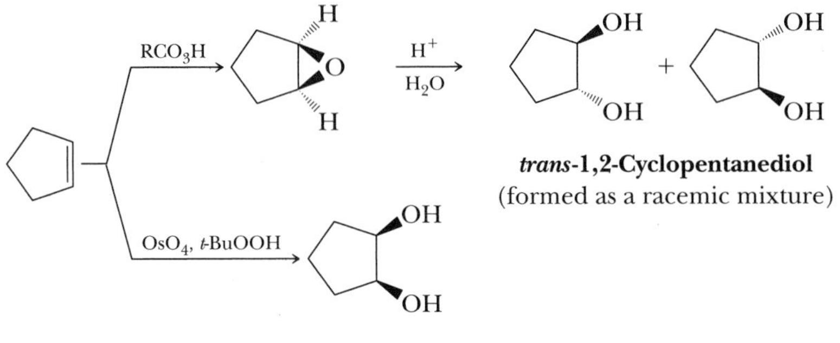
RCO3H
H
O
H
H+
H2O
OH
OH
+
OH
OH
OsO4, t-BuOOH
OH
OH

***trans*-1,2-Cyclopentanediol**
(formed as a racemic mixture)

***cis*-1,2-Cyclopentanediol**
(achiral)

B. Nucleophilic Ring Opening

Ethers are not normally susceptible to reaction with nucleophiles. Epoxides, however, are different. Because of the strain associated with a three-membered ring, epoxides undergo ring-opening reactions with a variety of nucleophiles. Good nucleophiles attack an epoxide ring by an S_N2 mechanism and show an S_N2-like regioselectivity; that is, the nucleophile attacks at the less hindered carbon. Following is an equation for the reaction of methyloxirane (propylene oxide) with sodium methoxide in methanol.

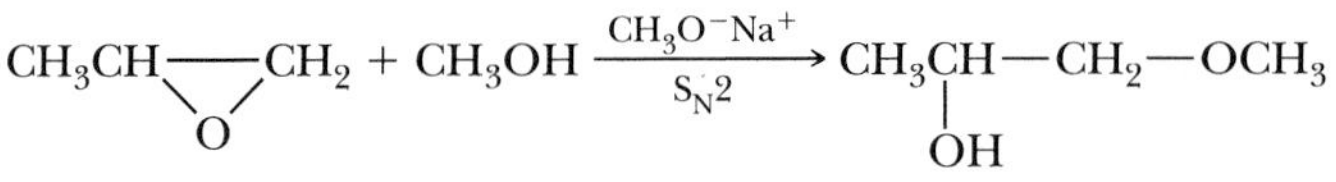
CH3CH—CH2 + CH3OH
O
CH3O−Na+
SN2
CH3CH—CH2—OCH3
OH

Mechanism 11.6

Nucleophilic Opening of an Epoxide Ring

***Step 1:* Make a new bond between a nucleophile and an electrophile and simultaneously break a bond to give stable molecules or ions.** Backside attack of the nucleophile on the less hindered carbon of the highly strained epoxide opens the ring and displaces O^-.

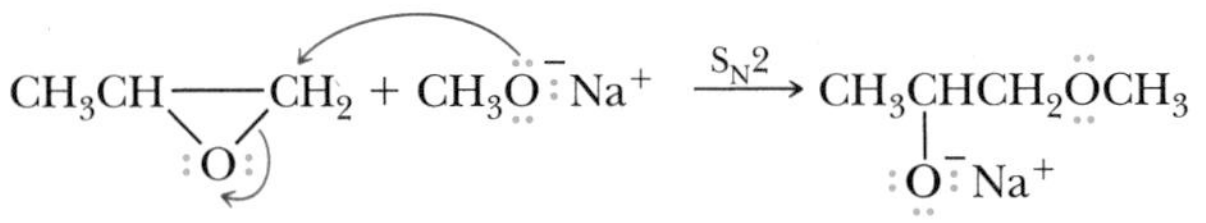
CH3CH—CH2 + CH3O−Na+
O
SN2
CH3CHCH2OCH3
O−Na+

***Step 2:* Add a proton.** Proton transfer completes the reaction.

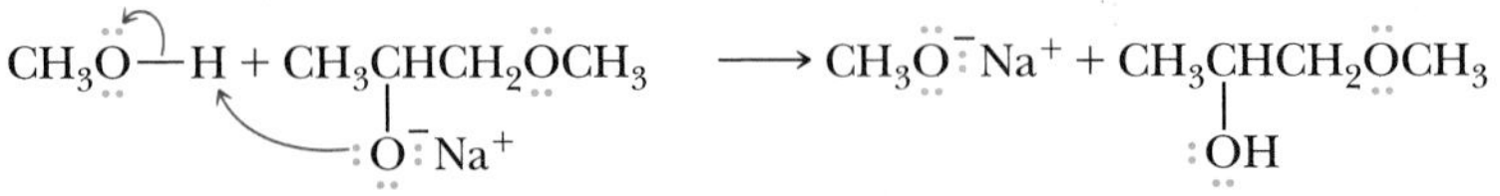
CH3O—H + CH3CHCH2OCH3
O−Na+
CH3O−Na+ + CH3CHCH2OCH3
OH

The nucleophilic ring opening of epoxides is also stereoselective; as expected of an S_N2 reaction, attack of the nucleophile is anti to the leaving group. An illustration is the reaction of cyclohexene oxide with sodium methoxide in methanol to give *trans*-2-methoxycyclohexanol.

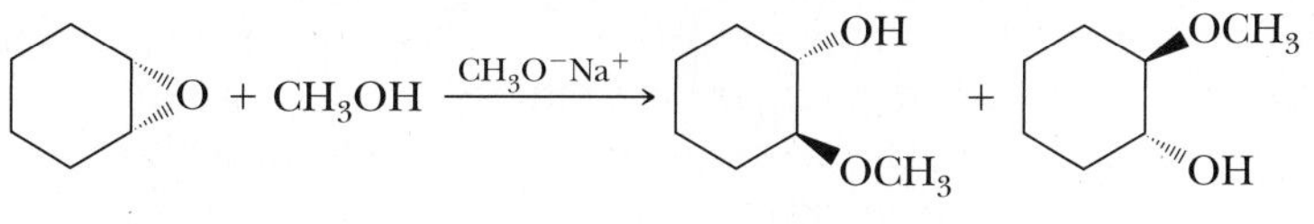
O + CH_3OH
$CH_3O^-Na^+$
OH
OCH_3
+
OCH_3
OH

Cyclohexene oxide **_trans_-2-Methoxycyclohexanol** (a racemic mixture)

The value of epoxides lies in the number of nucleophiles that bring about ring opening and the combinations of functional groups that can be prepared from them. The most important of these ring-opening reactions are summarized in the following chart.

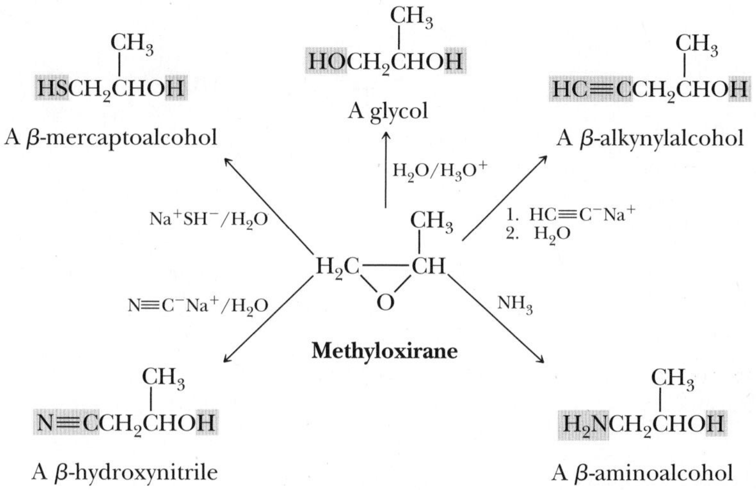
$HSCH_2CH(CH_3)OH$
A β-mercaptoalcohol
$HOCH_2CH(CH_3)OH$
A glycol
$HC{\equiv}CCH_2CH(CH_3)OH$
A β-alkynylalcohol
Na^+SH^-/H_2O
H_2O/H_3O^+
1. $HC{\equiv}C^-Na^+$
2. H_2O
$N{\equiv}C^-Na^+/H_2O$
NH_3
Methyloxirane
$N{\equiv}CCH_2CH(CH_3)OH$
A β-hydroxynitrile
$H_2NCH_2CH(CH_3)OH$
A β-aminoalcohol

The reactions with a terminal alkyne anion or the cyanide anion are particularly noteworthy because a new carbon-carbon bond is formed in each reaction.

Finally, treatment with $LiAlH_4$ reduces an epoxide to an alcohol. Lithium aluminum hydride is similar to sodium borohydride, $NaBH_4$, in that it is a donor of hydride ion, $H:^-$, which is both a strong base and a good nucleophile. In the reduction of a substituted epoxide by $LiAlH_4$, attack of the hydride ion occurs preferentially at the less hindered carbon of the epoxide, an observation consistent with S_N2 reactivity.

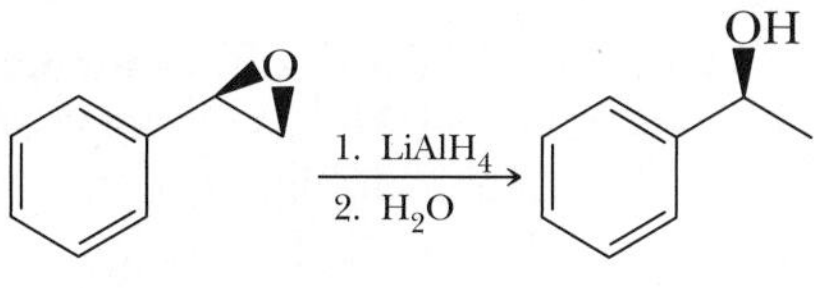
O
1. $LiAlH_4$
2. H_2O
OH

(_R_)-Phenyloxirane ((_R_)-Styrene oxide) **(_S_)-1-Phenylethanol**

MCAT Practice: *Passage and Questions*

Benzo[a]pyrene

Benzo[a]pyrene is a product of incomplete combustion of organic material. It was first recognized as the component of coal tar responsible for various cancers. It is also present in tobacco smoke, automobile exhaust, and charcoal-grilled meats.

The body's metabolism of benzo[a]pyrene converts it into a water-soluble diol epoxide metabolite for easier excretion. However, this metabolite reacts with amine groups in DNA resulting in cancer-causing mutations.

(*Continued*)

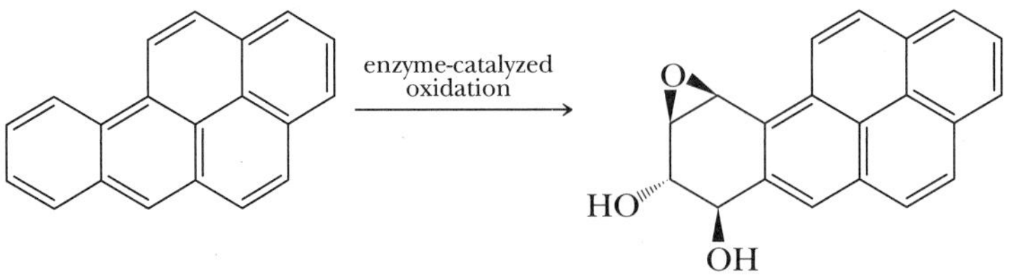

Benzo[a]pyrene A diol epoxide

Reactions of the diol epoxide in water have been studied, and three products are formed. The product distribution is shown below at two different pHs.

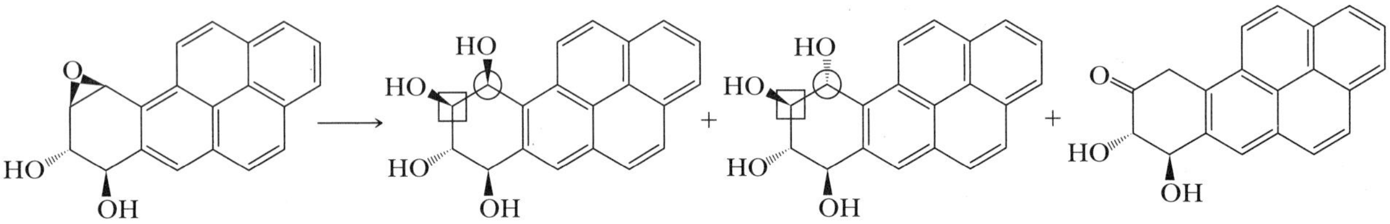

A diol epoxide *cis* tetrol

	cis tetrol		
pH = 3	89	11	0
pH = 12	11	5	84

Questions

A. What is the stereochemical outcome of the reaction at the boxed and circled carbons relative to the reactant?

1. Both carbons have scrambled stereochemistry.
2. Both carbons have undergone an inversion of stereochemistry.
3. The boxed carbon has been inverted while the circled carbon has been scrambled.
4. The circled carbon has been scrambled while the boxed carbon retains configuration.

B. What does the product distribution tell a chemist about the mechanism of the acid-catalyzed epoxide ring opening?

1. That the reaction proceeded via a radical mechanism.
2. That the acid protonated one of the adjacent hydroxyls prior to ring opening.
3. That the reaction completely proceeded via an S_N2 mechanism.
4. That the reaction most likely proceeds via an S_N1 mechanism, with preferential addition of water to one side of a carbocation.

C. What must be occurring in basic condition to generate a carbonyl in the third product?

1. A base removes a proton from an intermediate.
2. A base eliminates water from one of the other two products.
3. Hydroxide replaces OH groups via an S_N2 reaction.
4. Both 2 and 3.

D. Why is the relative amount of *trans*-tetrol to *cis*-tetrol (ratio of 0.45) at pH 12 larger than at pH 3 (ratio of 0.12)?

1. *Trans* arrangements of vicinal diols are more stable at higher pH.
2. Basic conditions led to alkoxides from the alcohols.
3. The mechanism involves increasing amounts of S_N1 ring-opening at higher pH.
4. The mechanism involves increasing amounts of S_N2 ring-opening at higher pH.

11.10 Ethylene Oxide and Epichlorohydrin: Building Blocks in Organic Synthesis

Ethylene oxide is a valuable building block for organic synthesis because each carbon of its two-carbon skeleton has a functional group. Following is a flowchart illustrating some of the functional groups and types of molecules that can be generated from this building block. The key to recognizing a structural unit derived from ethylene oxide is the presence of an Nu—CH_2—CH_2—OH group. In cases where the —OH group is subsequently modified by replacement with another nucleophile, you will find the group Nu—CH_2—CH_2—Nu. The nucleophiles most widely used in modification of the —OH group are ammonia, 1° amines, and 2° amines (Section 1.3B). We have seen all of these reactions before, but not in this form.

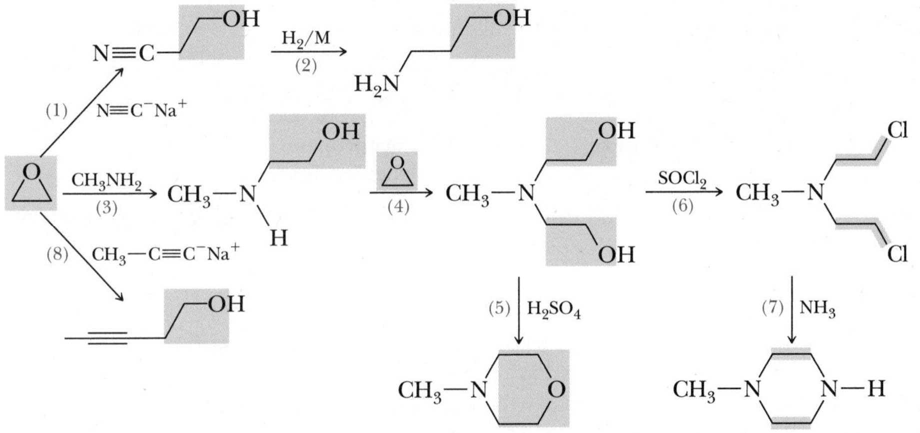

As you study this flowchart, notice the following points:

- Reactions (1) and (8) use carbon nucleophiles to open the three-membered ring and thus form new carbon-carbon bonds.
- Reaction (2) is a catalytic reduction of the carbon-nitrogen triple bond to a 1° amine. Just as a carbon-carbon triple bond can be reduced to a carbon-carbon single bond by hydrogen in the presence of a transition metal catalyst, a carbon-nitrogen triple bond can be similarly reduced.
- Reactions (3) and (4) are openings of the epoxide ring by nitrogen nucleophiles.
- Reaction (5) is an intramolecular acid-catalyzed dehydration of a 1,5-diol to give a cyclic ether.
- Reaction (7) involves two successive S_N2 reactions to form a nitrogen-containing ring.

An example of a compound, part of which is derived from the two-carbon skeleton of ethylene oxide, is the local anesthetic procaine. Here is a retrosynthetic analysis of procaine. ▶

▶ Functional groups react the same in different molecules.

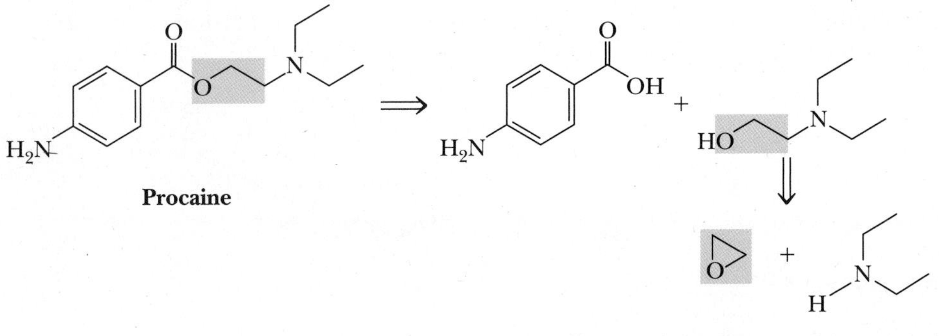

Procaine

Ethylene oxide **Diethylamine**

The hydrochloride salt of procaine is marketed under the trade name Novocaine. We will show how to complete the synthesis of procaine when we study the derivatives of carboxylic acids in Chapter 18.

The epoxide epichlorohydrin is also a valuable synthetic building block because each of its three carbons contains a reactive functional group.

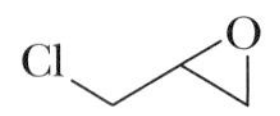

Epichlorohydrin

Epichlorohydrin is an oily liquid (bp 118°C). It is insoluble in water and nonpolar hydrocarbon solvents, but soluble in polar aprotic solvents such as diethyl ether and dichloromethane. Epichlorohydrin is synthesized industrially by the following series of three reactions.

Step 1: Allylic halogenation by a radical chain mechanism (Section 8.6A).

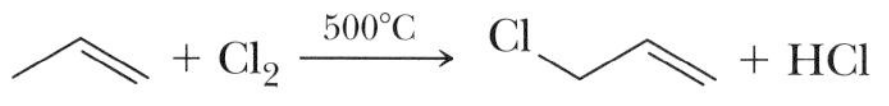

Propene **3-Chloropropene (Allyl chloride)**

Step 2: Treating the haloalkene with chlorine in water gives a chlorohydrin (Section 6.3E).

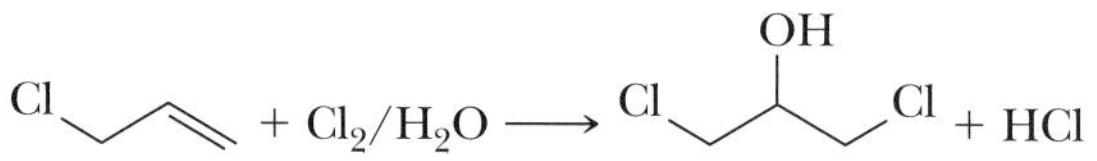

Step 3: Treating the chlorohydrin with calcium hydroxide brings about an internal S_N2 reaction and gives epichlorohydrin.

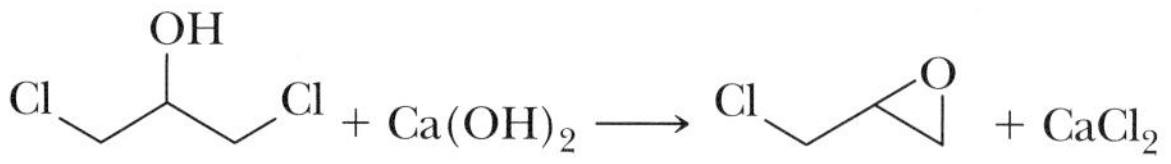

3-Chloro-1,2-epoxypropane (Epichlorohydrin) (racemic)

The characteristic structural feature of a product derived from epichlorohydrin is a three-carbon unit with —OH on the middle carbon and a carbon, nitrogen, oxygen, or sulfur nucleophile bonded to the two end carbons.

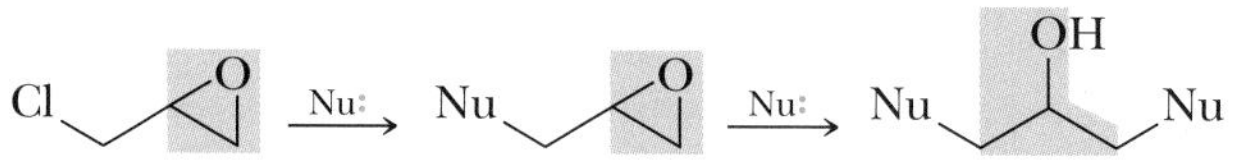

Epichlorohydrin

An example of a compound that contains the three-carbon skeleton of epichlorohydrin is nadolol, a β-adrenergic blocker with vasodilating activity.

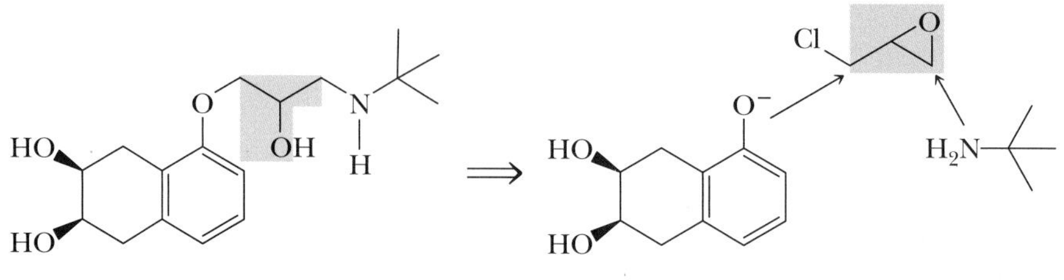

Nadolol (racemic)

A nucleophile derived by removal of the acidic H from an —OH group

The nitrogen nucleophile of a 1° amine

Members of this class of compounds have received enormous clinical attention because of their effectiveness in treating hypertension (high blood pressure), migraine headaches, glaucoma, ischemic heart disease, and certain cardiac arrhythmias. Shown in this retrosynthetic analysis are the two nucleophiles used in the synthesis of nadolol. We will show how to complete the synthesis of nadolol when we study the chemistry of benzene and its derivatives in Chapters 21 and 22.

We are not concerned at this stage with how these nucleophiles are generated or, if two nucleophiles are used, which nucleophile is added first or reacts at which site. Our concern with ethylene oxide and epichlorohydrin at this stage of the course is only that you recognize the structural features in a target molecule that might be derived from these building blocks. Call it pattern recognition if you will. Later, after we study the chemistry of other functional groups, we will discuss in detail the chemistry of how the target molecules are synthesized in the laboratory.

11.11 Crown Ethers

In the early 1960s, Charles Pedersen of DuPont discovered a family of cyclic polyethers derived from ethylene glycol and substituted ethylene glycols. Compounds of this structure are named **crown ethers** because one of their most stable conformations resembles the shape of a crown. These ethers are named by the system devised by Pedersen. The parent name *crown* is preceded by a number describing the size of the ring and followed by a number describing the number of oxygen atoms in the ring, as, for example, 18-crown-6.

Crown ether
A cyclic polyether derived from ethylene glycol and substituted ethylene glycols.

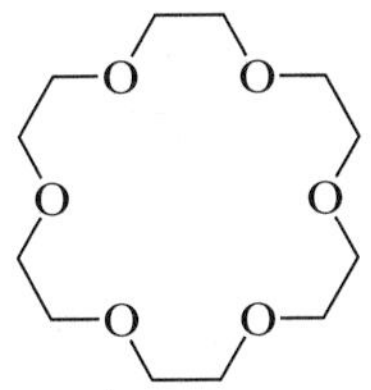

O
O
O
O
O
O

18-Crown-6
(a cyclic hexamer)

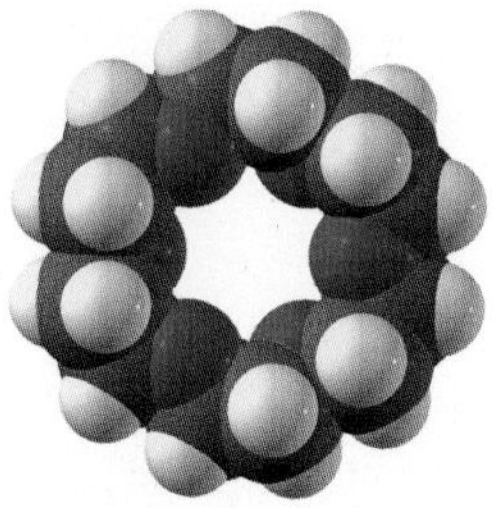

Space-filling model, viewed from above

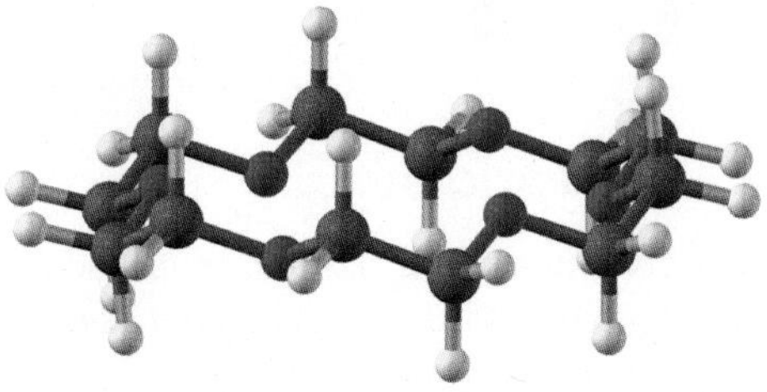

Ball-and-stick model, viewed through an edge

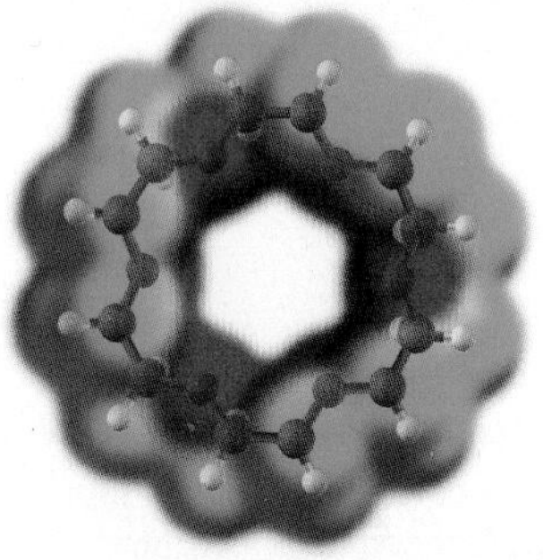

Electrostatic potential map showing the electron-rich interior and the nonpolar exterior

For his work, Pedersen shared the 1987 Nobel Prize in Chemistry with Donald J. Cram of the United States and Jean-Marie Lehn of France.

The most significant structural feature of crown ethers is that the diameter of the cavity created by the repeating oxygen atoms of the ring is comparable to the diameter of alkali metal ions. The diameter of the cavity in 18-crown-6, for example, is approximately the diameter of a potassium ion. When a potassium ion is inserted into the cavity of 18-crown-6, the unshared electron pairs on the six oxygens of the crown ether are close enough to the potassium ion to provide very effective solvation for K^+.

Ion	Diameter (pm)
Li^+	136
Na^+	194
K^+	266
Rb^+	294
Mg^{2+}	164
Ca^{2+}	286

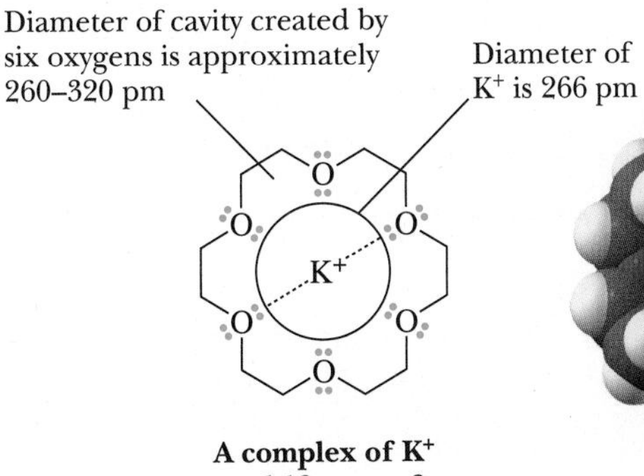

Diameter of cavity created by six oxygens is approximately 260–320 pm
Diameter of K^+ is 266 pm
K^+

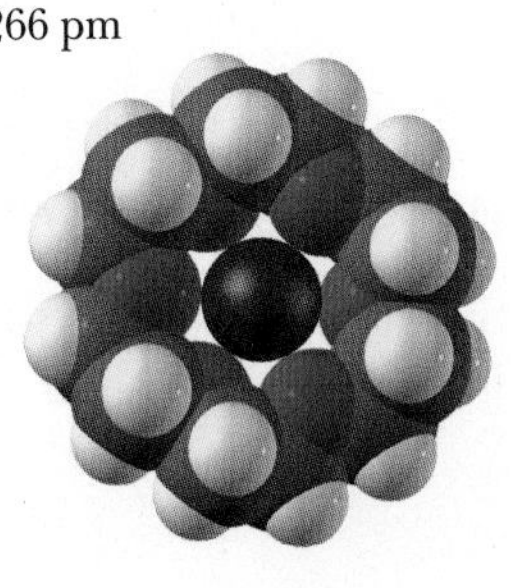

A complex of K^+ and 18-crown-6

18-Crown-6 forms somewhat weaker complexes with rubidium ion (a somewhat larger ion) and with sodium ion (a somewhat smaller ion). It does not coordinate to any appreciable degree with lithium ion (a considerably smaller ion). 12-Crown-4, however, with its smaller cavity, does form a strong complex with lithium ion.

The cavity of a crown ether is a polar region, and the unshared pairs of electrons on the oxygen atoms lining the cavity provide effective solvation for alkali metal ions. The outer surface of the crown is nonpolar and hydrocarbon-like; thus, crown ethers and their alkali metal ion complexes dissolve readily in nonpolar organic solvents.

Crown ethers have proven to be particularly valuable because of their ability to cause inorganic salts to dissolve in nonpolar aprotic organic solvents such as methylene chloride, hexane, and benzene. Potassium permanganate, for example, does not dissolve in benzene. If 18-crown-6 is added to benzene, the solution takes on the purple color characteristic of permanganate ion. The crown-potassium ion complex is soluble in benzene and brings permanganate ion into solution with it. The resulting "purple benzene" is a valuable reagent for the oxidation of water-insoluble organic compounds.

Crown ethers have also proven valuable in nucleophilic displacement reactions. The cations of potassium salts, such as KF, KCN, and KN_3, are very tightly bound within the solvation cavity of 18-crown-6 molecules. The anions, however, are only weakly solvated, and because of the geometry of cation binding within the cavity of the crown, only loose ion pairing occurs between the anion and cation. Thus, in nonpolar aprotic solvents, these anions are without any appreciable solvent shell and, therefore, are highly reactive as nucleophiles. The nucleophilicity of F^-, CN^-, N_3^-, and other anions in nonpolar aprotic solvents containing an 18-crown-6 equals and often exceeds their nucleophilicity in polar aprotic solvents such as DMSO and acetonitrile.

11.12 Sulfides

A. Nomenclature

Sulfide
The sulfur analog of an ether; a molecule containing a sulfur atom bonded to two carbon atoms. Sulfides are also called thioethers.

To derive the IUPAC name of a **sulfide** (also called a **thioether**), select the longest carbon chain as the parent alkane and name the sulfur-containing substituent as an *alkylsulfanyl* group. To derive a common name, list the groups bonded to sulfur and add the word sulfide to show the presence of the —S— group.

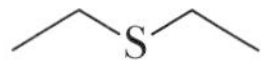

Ethylsulfanylethane
(Diethyl sulfide)

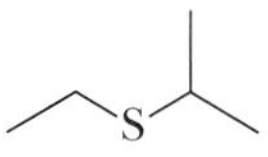

2-Ethylsulfanylpropane
(Ethyl isopropyl sulfide)

Disulfide
A molecule containing an —S—S— group.

The functional group of a **disulfide** is an —S—S— group. IUPAC names of disulfides are derived by selecting the longest carbon chain as the parent alkane and indicating the disulfide-containing substituent as an *alkyldisulfanyl* group. Common names of disulfides are derived by listing the names of the groups bonded to sulfur and adding the word *disulfide.*

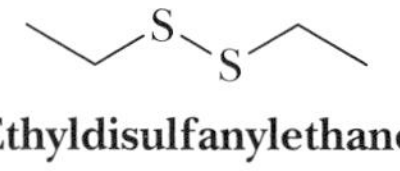

Ethyldisulfanylethane
(Diethyl disulfide)

B. Preparation of Sulfides

Symmetrical sulfides, RSR (also called symmetrical thioethers), are prepared by treating one mole of Na_2S (where S^{2-} is the nucleophile) with two moles of haloalkane.

$$2\,RX + Na_2S \rightarrow \underset{\text{A sulfide}}{RSR} + 2\,NaX$$

This same reaction can also be used to prepare five- and six-membered cyclic sulfides. Treating a 1,4-dihaloalkane with Na_2S gives a five-membered cyclic sulfide; treating a 1,5-dihaloalkane with Na_2S gives a six-membered ring.

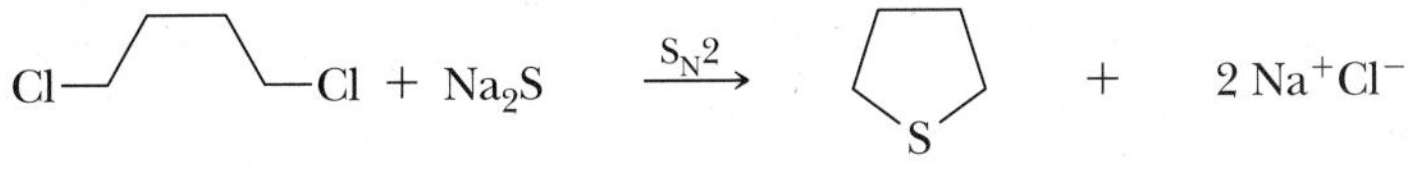

1,4-Dichlorobutane **Thiolane (Tetrahydrothiophene)**

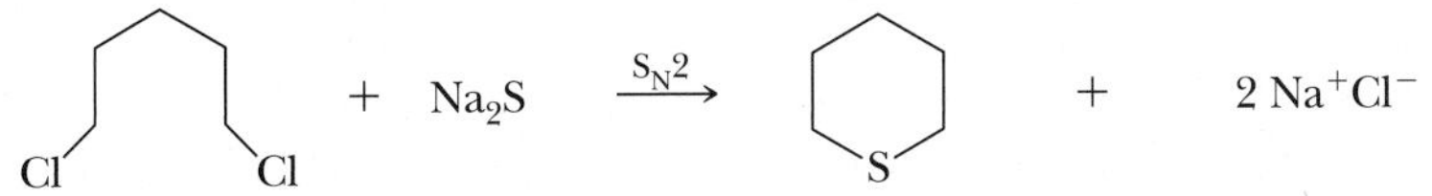

1,5-Dichloropentane **Thiane (Tetrahydrothiopyran)**

Unsymmetrical sulfides, RSR', are prepared by converting a thiol to a sodium salt with either sodium hydroxide or sodium ethoxide and then allowing the salt to react with a haloalkane.

$$CH_3(CH_2)_8CH_2S^-Na^+ + CH_3I \xrightarrow{S_N2} CH_3(CH_2)_8CH_2SCH_3 + Na^+I^-$$

Sodium 1-decanethiolate **1-Methylsulfanyldecane (Decyl methyl sulfide)**

This method of thioether formation is the sulfur analog of the Williamson ether synthesis (Section 11.4A).

Note that all these reactions leading to sulfides (thioethers) are direct applications of nucleophilic substitution reactions (Chapter 9).

C. Oxidation of Sulfides

Many of the properties of sulfides stem from the fact that divalent sulfur is a reducing agent; it is easily oxidized to two higher oxidation states. Treatment of a sulfide with one mole of 30% aqueous hydrogen peroxide at room temperature gives a sulfoxide, as illustrated by oxidation of methyl phenyl sulfide to methyl phenyl sulfoxide. Several other oxidizing agents, including sodium periodate, $NaIO_4$, also bring about the same conversion. Treatment of a sulfoxide with $NaIO_4$ brings about its oxidation to a sulfone.

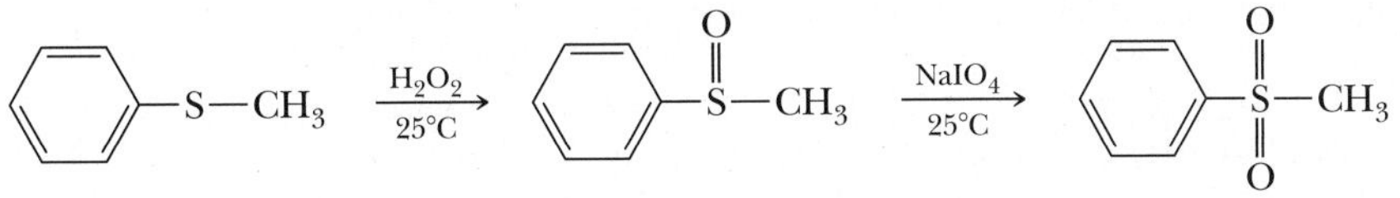

Methyl phenyl sulfide **Methyl phenyl sulfoxide** **Methyl phenyl sulfone**

Dimethyl sulfoxide (DMSO) is manufactured on an industrial scale by air oxidation of dimethyl sulfide in the presence of oxides of nitrogen.

$$2\ CH_3{-}S{-}CH_3 + O_2 \xrightarrow{\text{oxides of nitrogen}} 2\ CH_3{-}\overset{\overset{\large O}{\|}}{S}{-}CH_3$$

Dimethyl sulfide **Dimethyl sulfoxide**

Study Guide

11.1 Structure of Ethers

- An **ether** contains an atom of oxygen bonded to two carbon atoms.

11.2 Nomenclature of Ethers

- In the IUPAC name, the parent chain is named and the —OR group is named as an alkoxy substituent.
- Common names for ethers are derived by naming the two groups bonded to oxygen followed by the word *ether.* Heterocyclic ethers have an oxygen atom as one of the members of a ring.

P 11.1, 11.10, 11.11

11.3 Physical Properties of Ethers

- Ethers are weakly polar compounds and associate by weak dipole-dipole interactions and dispersion forces.
  - *The boiling points of ethers are close to those of hydrocarbons of comparable molecular weight but much lower than those of the corresponding alcohols.*
  - *Because ethers are hydrogen bond acceptors, they are more soluble in water than are hydrocarbons of comparable molecular weight.*

P 11.2, 11.12–11.14

11.4 Preparation of Ethers

- Ethers are prepared using the **Williamson ether synthesis**, which is an S_N2 reaction between an alkoxide and a methyl or primary alkyl halide (secondary or tertiary haloalkanes give too much unwanted E2 elimination).
- Symmetrical ethers can be prepared through dehydration of an alcohol in strong acid.
- Ethers can be prepared through the acid-catalyzed addition of methyl or primary alcohols to alkenes that can form a stable carbocation upon protonation, via a mechanism analogous to acid-catalyzed hydration of an alkene.

P 11.3, 11.4, 11.15, 11.16, 11.41, 11.42, 11.43

Problem 11.3

KEY REACTIONS

1. **Williamson Ether Synthesis (Section 11.4A)** The Williamson ether synthesis is a general method for the synthesis of dialkyl ethers by an S_N2 reaction between a haloalkane and an alkoxide ion.

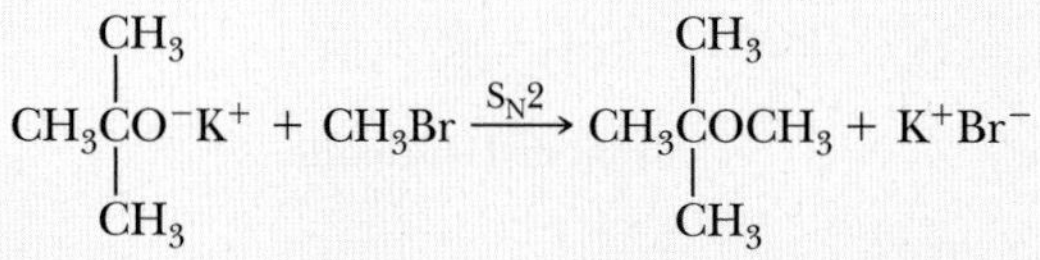

CH3
CH3CO−K+ + CH3Br
SN2
CH3COCH3 + K+Br−
CH3

Yields are highest with methyl, 1° haloalkanes and 1° allylic halides. They are considerably lower with 2° halides because of competition from E2 elimination. The Williamson ether synthesis reaction fails altogether with 3° halides.

2. **Acid-Catalyzed Dehydration of Alcohols (Section 11.4B)** Yields are highest for symmetrical ethers formed from unbranched primary alcohols. The mechanism involves protonation of an —OH group followed by displacement of the H_2O leaving group by a second alcohol molecule acting as a nucleophile followed by loss of a proton to give the ether.

$$2\ CH_3CH_2OH \xrightarrow[140°C]{H_2SO_4} CH_3CH_2OCH_2CH_3 + H_2O$$

3. **Acid-Catalyzed Addition of Alcohols to Alkenes (Section 11.4C)** Proton transfer to the alkene generates a carbocation. Nucleophilic addition of an alcohol to the carbocation followed by proton transfer to the solvent gives the ether.

$$CH_3C(CH_3){=}CH_2 + CH_3OH \xrightarrow{\text{acid catalyst}} CH_3C(CH_3)_2OCH_3$$

11.5 Reactions of Ethers

- Ethers are cleaved by concentrated HX via a mechanism that involves protonation of the ether oxygen followed by S_N1 or S_N2 substitution with X^-, the exact mechanism being determined by the structure of the ether.

KEY REACTIONS

4. **Acid-Catalyzed Cleavage of Dialkyl Ethers (Section 11.5A)** Cleavage of ethers requires both a strong acid and a good nucleophile, hence the use of concentrated HBr and HI.

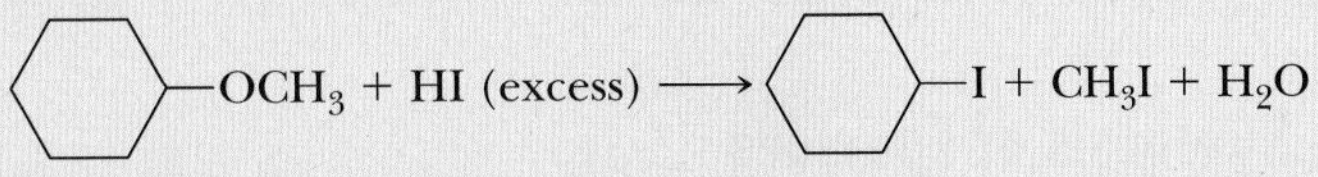

—OCH3 + HI (excess)
—I + CH3I + H2O

The mechanism involves initial protonation of the ether oxygen.

Cleavage of primary and secondary alkyl ethers is by an S_N2 pathway. Cleavage of tertiary alkyl ethers is by an S_N1 pathway.

- When handling ethers, special precautions are necessary.
  - *Common ethers are highly volatile and flammable. They are oxidized in the presence of O_2 to explosive hydroperoxides, so they must be disposed of prior to their expiration date.*

P **11.5, 11.6**

Problem 11.6

11.6 Silyl Ethers as Protecting Groups

- **Protecting groups** are used to temporarily prevent a functional group from reacting.
- Silyl ethers are used to protect alcohol groups.
  - *The silyl ether is prepared from the alcohol and a silyl chloride.*

- *Silyl ethers are stable to oxidations, reductions, nonaqueous acid, and nonaqueous base.*
- *Silyl ethers are removed by treatment with tetrabutylammonium fluoride, a reaction that exploits the extraordinarily strong Si-F bond.*

11.7, 11.38, 11.44

KEY REACTIONS

5. Reaction of Alcohols with Chloro-*tert*-butyldimethylsilane (Section 11.6) The *tert*-butyldimethylsilyl (t-$BuMe_2Si$—) group is used to protect primary and secondary alcohols.

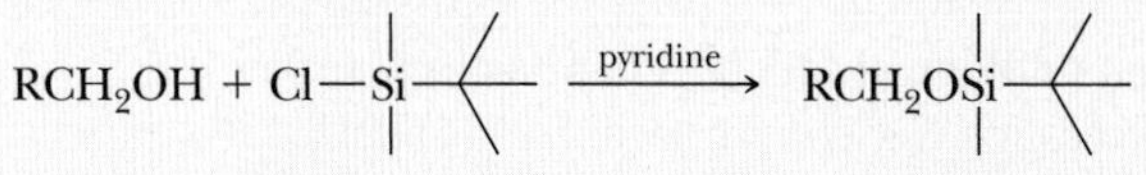
RCH2OH + Cl—Si
pyridine
RCH2OSi

Chloro-*tert*-butyl dimethylsilane A *tert*-butyldimethylsilyl ether

The protecting group is removed by treating the silyl ether with fluoride ion to regenerate the original alcohol.

11.7 Epoxides: Structure and Nomenclature

- **Epoxides** are three-membered ring, cyclic ethers.
  - *In IUPAC nomenclature, epoxides are named as oxirane derivatives, unless they are part of another ring system and given the prefix* epoxy-.
  - *In common nomenclature, epoxides are named from the alkene from which they are derived followed by the word* oxide.

11.10, 11.11

11.8 Synthesis of Epoxides

- Epoxides are synthesized from alkenes using peroxycarboxylic acids (RCO_3H) in a reaction involving a concerted five-membered ring transition state.
- Epoxides can be synthesized from internal nucleophilic substitution of halohydrins, which can be derived from hydrohalogenation of alkenes.
  - *The reaction involves deprotonation of the alcohol followed by backside attack by the alkoxide anion on the adjacent C—X bond.*
- One of the most useful reactions discovered in the last several decades is the **Sharpless epoxidation** of primary allylic alcohols to give single enantiomers of chiral epoxides in a predictable fashion.
  - *This reaction uses a titanium catalyst, a peroxide, and the enantiomers of diethyl tartrate.*
  - *Analyzing the steric environment created by the groups on the alkene allows the predominant product enantiomer to be predicted when either enantiomer of diethyl tartrate is used.*

11.8, 11.9, 11.27–11.29, 11.45

Problem 11.8

KEY REACTIONS

6. **Oxidation of Alkenes with Peroxycarboxylic Acids (Section 11.8C)** Three commonly used peroxycarboxylic acid oxidizing agents are *meta*-chloroperoxybenzoic acid, the magnesium salt of monoperoxyphthalic acid, and peroxyacetic acid. Each reagent oxidizes an alkene to an epoxide. The mechanism is a concerted rearrangement of electrons involving a five-membered ring, the π bond of an alkene, and the terminal O atom of the peroxyacid to give an epoxide and carboxylic acid.

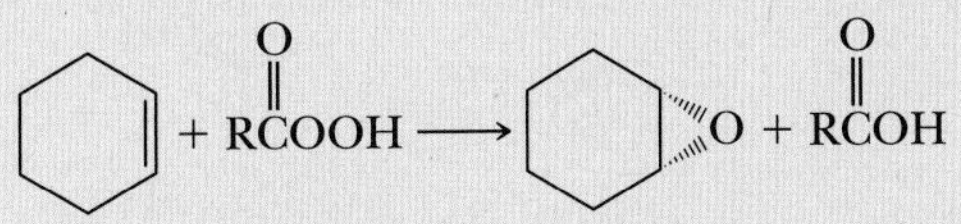

+ RCOOH
O + RCOH

7. **Synthesis of Epoxides from Halohydrins (Section 11.8C)** Formation of the halohydrin and the following intramolecular S_N2 reaction are both stereoselective (the configuration of the alkene is retained in the epoxide) and stereospecific (for alkenes that show *cis,trans* isomerism, the configuration of the epoxide depends on the configuration of the alkene).

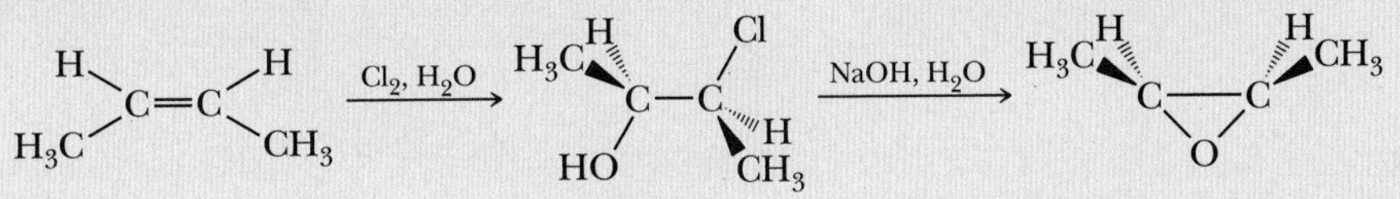

Cl₂, H₂O
NaOH, H₂O

8. **Sharpless Asymmetric Epoxidation (Section 11.8D)** Oxidation of the carbon-carbon double bond of a 1° allylic alcohol by *tert*-butyl hydroperoxide in the presence of a chiral catalyst consisting of either (+)- or (−)-diethyl tartrate and titanium tetraisopropoxide gives an enantiomerically pure epoxide. The enantiomer formed depends on which enantiomer of diethyl tartrate is used in the catalyst.

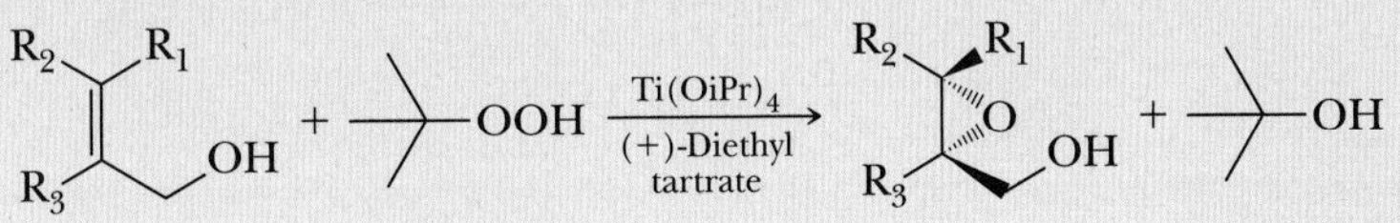

Ti(OiPr)₄
(+)-Diethyl tartrate

A 1° allylic alcohol

***tert*-Butyl hydroperoxide**

11.9 Reactions of Epoxides

- Epoxides undergo acid-catalyzed ring opening to add weak nucleophiles such as water and alcohols to give diols and ether-alcohols, respectively, with anti stereoselectivity.
  - *The mechanism involves protonation of the epoxide oxygen to give a cation intermediate (analogous to a bridged halonium ion) followed by backside nucleophilic attack on the carbon more able to accept a positive charge (in unsymmetrical epoxides).*
- Epoxides react with strong nucleophiles such as hydroxide, alkoxides, and ammonia or amines via an S_N2 mechanism at the less hindered carbon of unsymmetrical epoxides.
  - *The attack is from the back of the C—O bond.*
  - *The observed regiochemistry is often the opposite of that seen with acid-catalyzed ring opening, providing access to both regioisomer products of unsymmetrical epoxides.*

P **11.19–11.26, 11.30–11.38, 11.43**

Mechanism 11.5

KEY REACTIONS

9. **Acid-Catalyzed Hydrolysis of Epoxides (Section 11.9A)** Hydrolysis of an epoxide derived from a cycloalkene gives a *trans* glycol. The reaction involves initial protonation of the epoxide O atom followed by nucleophilic attack of water and then loss of a proton to give the *trans* diol. The reaction also occurs with alcohol nucleophiles, and when there is a difference, reaction of the nucleophile occurs predominantly at the more substituted carbon of the protonated epoxide.

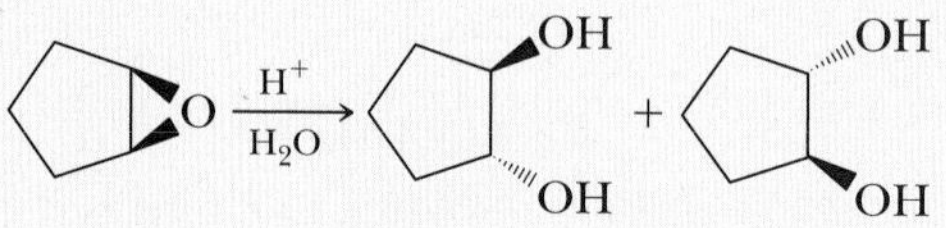
O
H+
H2O
OH
OH
+
OH
OH

(racemic)

10. **Nucleophilic Ring Opening of Epoxides (Section 11.9B)** Attack on the epoxide is regioselective with the nucleophile attacking the less substituted carbon of the epoxide.

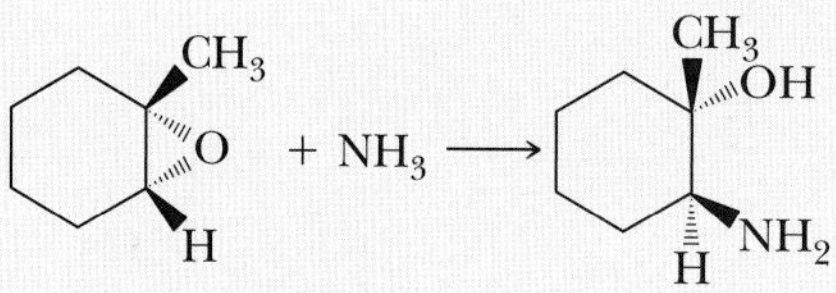
CH3
O
H
+ NH3
CH3
OH
H
NH2

11. **Reduction of an Epoxide to an Alcohol (Section 11.9B)** Regioselective hydride ion transfer from lithium aluminum hydride to the less hindered carbon of the epoxide gives an alcohol.

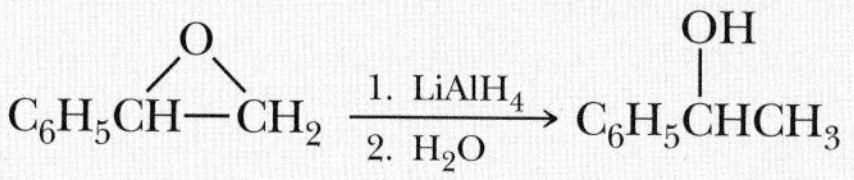
O
C6H5CH—CH2
1. LiAlH4
2. H2O
OH
C6H5CHCH3

(racemic) (racemic)

11.10 Ethylene Oxide and Epichlorohydrin: Building Blocks in Organic Synthesis

- **Ethylene oxide** is a valuable two-carbon building block for synthesis because after nucleophilic attack, both of its carbons contain a functional group.
  - *Ammonia or primary or secondary amines are commonly used nucleophiles, but other sulfur or oxygen nucleophiles can also be used in the reaction.*
  - *In more complex molecules, the ethylene oxide building blocks can be recognized by the* $Nu—CH_2—CH_2—OH$ *pattern or* $Nu—CH_2—CH_2—Nu$ *if the* OH *group is exchanged for another nucleophile.*
- **Epichlorohydrin** is a valuable three-carbon building block for synthesis because after nucleophilic attack, all three of its carbons contain a functional group.
  - *In more complex molecules, epichlorohydrin building blocks can be recognized by the* $Nu—CH_2—CHOH—CH_2—Nu$ *pattern.*

– *Ammonia or primary or secondary amines are commonly used nucleophiles, but other sulfur or oxygen nucleophiles can also be used.*

P **11.19–11.23, 11.39–11.40**

11.11 Crown Ethers

- **Crown ethers** are cyclic polyethers having 12 or more atoms in a ring.
  - *The cavity of a crown ether is a polar region, and the unshared pairs of electrons on the ether oxygens can solvate alkali metal ions.*
  - *The cavity of **18-crown-6**, for example, has approximately the same diameter as the potassium cation, explaining the strong interaction between the two.*
  - *The outer surface of a crown ether is nonpolar and hydrocarbon-like.*
  - *Crown ethers are valuable for their ability to cause ionic compounds to dissolve in nonpolar organic solvents.*

P **11.22**

11.12 Sulfides

- **Sulfides (thioethers)** are named as alkylsulfanylalkanes. Common names for sulfides are derived by naming the two groups bonded to sulfur followed by the word *sulfide.*
- A **disulfide** (R-S-S-R) is named by choosing the longer alkyl group as the parent chain and indicating the disulfide-containing portion as an alkyldisulfanyl group. Common names are derived by naming the two groups bonded to each sulfur followed by the word *disulfide.*
- Symmetrical sulfides are created by reacting Na_2S with two moles of haloalkane. Unsymmetrical sulfides are synthesized by reacting the sodium salt of a thiol with a haloalkane.
- Sulfides can be oxidized with H_2O_2 by addition of one oxygen atom to sulfur to make a **sulfoxide** and can be further oxidized with $NaIO_4$ to add a second oxygen atom to sulfur to make a **sulfone**.

KEY REACTIONS

12. Oxidation of Sulfides (Section 11.12C) Oxidation of a sulfide gives either a sulfoxide or a sulfone, depending on the oxidizing agent and experimental conditions. Air oxidation of dimethyl sulfide is a commercial route to dimethyl sulfoxide, a polar aprotic solvent.

$$CH_3{-}S{-}CH_3 + O_2 \xrightarrow[\text{nitrogen}]{\text{oxides of}} CH_3{-}\overset{\overset{\large O}{\|}}{S}{-}CH_3$$

Problems

Red numbers indicate applied problems.

Structure and Nomenclature

11.10 Write names for these compounds. Where possible, write both IUPAC names and common names.

(a) 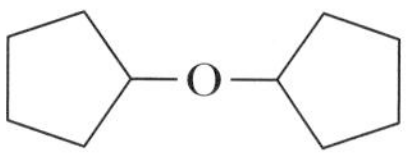 **(b)** 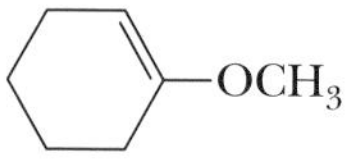 **(c)** 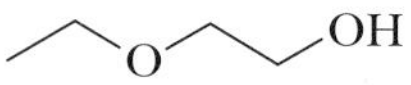

(d) 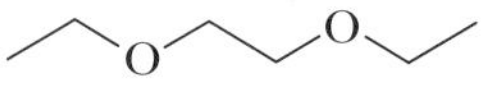 **(e)**  **(f)** 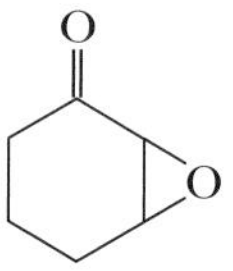

(g) 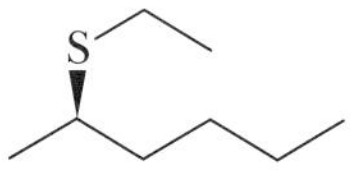 **(h)** 

11.11 Draw structural formulas for these compounds.

(a) 2-(1-Methylethoxy)propane
(b) *trans*-2,3-Diethyloxirane
(c) *trans*-2-Ethoxycyclopentanol
(d) Ethenyloxyethene
(e) Cyclohexene oxide
(f) 3-Cyclopropyloxy-1-propene
(g) (*R*)-2-Methyloxirane
(h) 1,1-Dimethoxycyclohexane

Physical Properties

11.12 Each compound given in this problem is a common organic solvent. From each pair of compounds, select the solvent with the greater solubility in water.

(a) CH_2Cl_2 and EtOH **(b)** Et_2O and EtOH

(c) 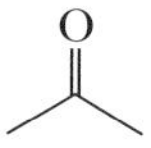 and Et_2O **(d)** Et_2O and 

11.13 Account for the fact that tetrahydrofuran (THF) is very soluble in water, whereas the solubility of diethyl either in water is only 8 g/100 mL water.

11.14 Because of the Lewis base properties of ether oxygen atoms, crown ethers are excellent complexing agents for Na^+, K^+, and NH_4^+. What kind of molecule might serve as a complexing agent for Cl^- or Br^-?

Preparation of Ethers

11.15 Write equations to show a combination of reactants to prepare each ether. Which ethers can be prepared in good yield by a Williamson ether synthesis? If there are any that cannot be prepared by the Williamson method, explain why.

(a) 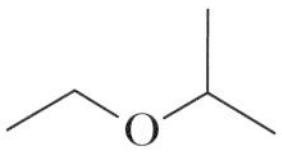 **(b)** 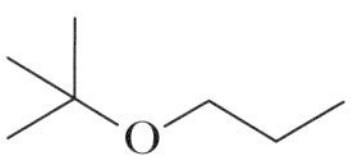 **(c)** 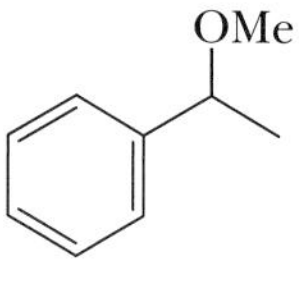

(d) 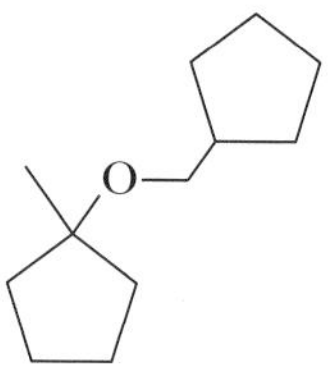 **(e)** 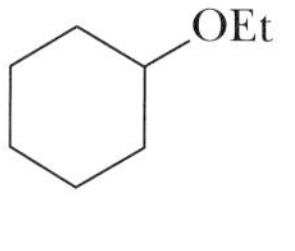 **(f)** 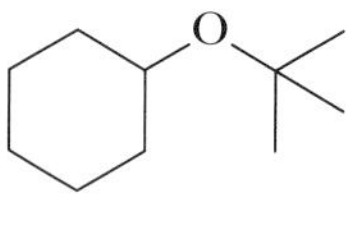

11.16 Propose a mechanism for this reaction.

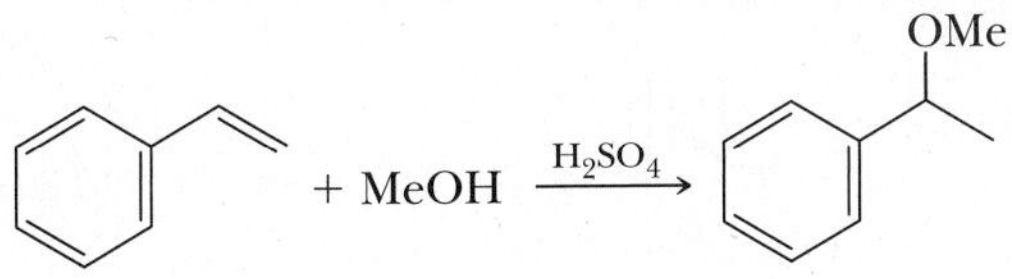
OMe
+ MeOH
H2SO4

(racemic)

Reactions of Ethers

11.17 Draw structural formulas for the products formed when each compound is heated at reflux in concentrated HI.

(a) 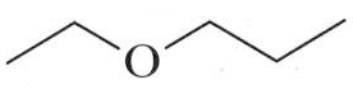 **(b)** 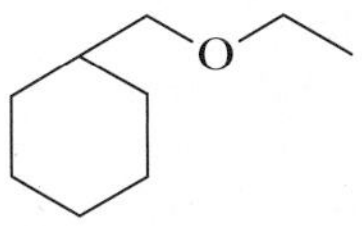 **(c)** 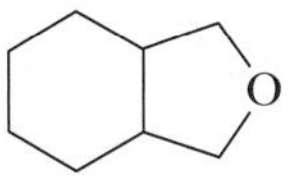 **(d)** 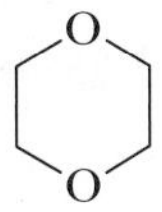

11.18 Following is an equation for the reaction of diisopropyl ether and oxygen to form a hydroperoxide.

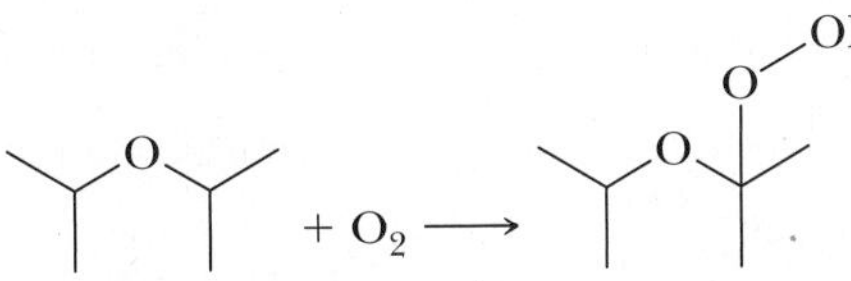
OH
O
O
O
+ O2

Diisopropyl ether A hydroperoxide

Formation of an ether hydroperoxide is a radical chain reaction.

(a) Write a pair of chain propagation steps that accounts for the formation of this ether hydroperoxide. Assume that initiation is by a radical, R·.

(b) Account for the fact that hydroperoxidation of ethers is regioselective (i.e., reaction occurs preferentially at a carbon adjacent to the ether oxygen).

Synthesis and Reactions of Epoxides

11.19 Triethanolamine, $(HOCH_2CH_2)_3N$, is a widely used biological buffer, with maximum buffering capacity at pH 7.8. Propose a synthesis of this compound from ethylene oxide and ammonia.

11.20 Ethylene oxide is the starting material for the synthesis of Cellosolve, an important industrial solvent. Propose a mechanism for this reaction.

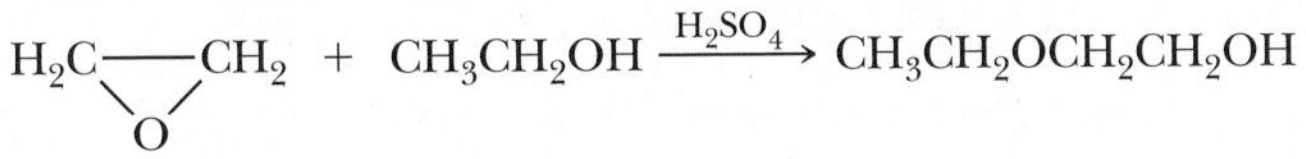
H2C—CH2 + CH3CH2OH
O
H2SO4
CH3CH2OCH2CH2OH

Oxirane (Ethylene oxide) **2-Ethoxyethanol (Cellosolve)**

11.21 Ethylene oxide is the starting material for the synthesis of 1,4-dioxane. Propose a mechanism for each step in this synthesis.

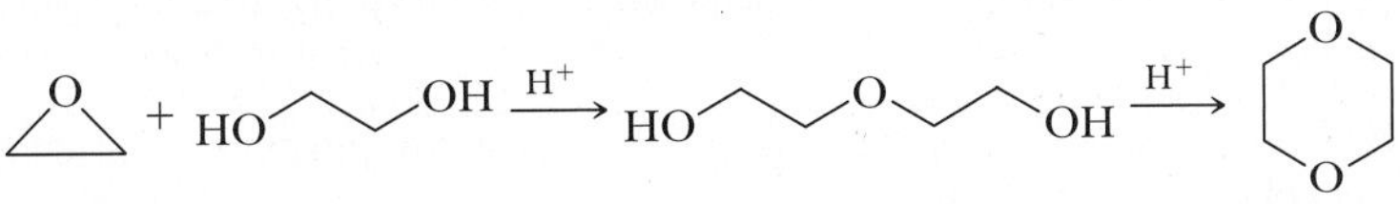
O
+ HO
OH
H+
HO
O
OH
H+
O
O

1,4-Dioxane

11.22 Propose a synthesis for 18-crown-6. If a base is used in your synthesis, does it make a difference whether it is lithium hydroxide or potassium hydroxide? Explain.

11.23 Predict the structural formula of the major product of the reaction of 2,2,3-trimethyloxirane with each set of reagents.

(a) $MeOH/MeO^-Na^+$
(b) $MeOH/H^+$
(c) Me_2NH

11.24 The following equation shows the reaction of *trans*-2,3-diphenyloxirane with hydrogen chloride in benzene to form 2-chloro-1,2-diphenylethanol.

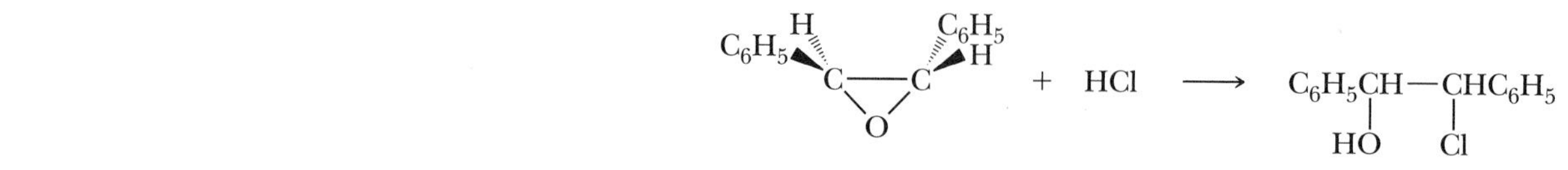

***trans*-2,3-Diphenyloxirane** **2-Chloro-1,2-diphenylethanol**

(a) How many stereoisomers are possible for 2-chloro-1,2-diphenylethanol?

(b) Given that opening of the epoxide ring in this reaction is stereoselective, predict which of the possible stereoisomers of 2-chloro-1,2-diphenylethanol is/are formed in the reaction.

11.25 Propose a mechanism to account for this rearrangement.

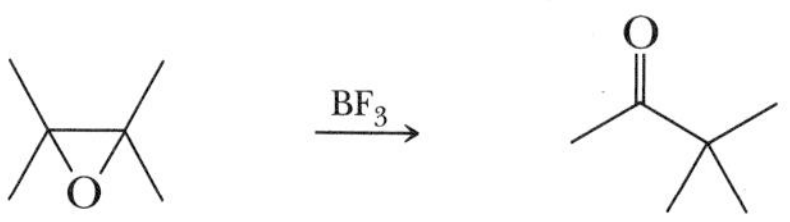

Tetramethyloxirane **3,3-Dimethyl-2-butanone**

11.26 Acid-catalyzed hydrolysis of the following epoxide gives a *trans* diol.

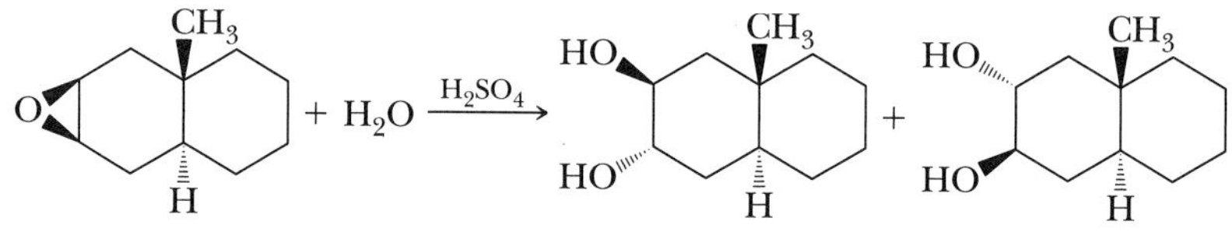

Only this glycol is formed. This glycol is not formed.

Of the two possible *trans* diols, only one is formed. How do you account for this stereoselectivity?

11.27 Following are two reaction sequences for converting 1,2-diphenylethylene into 2,3-diphenyloxirane.

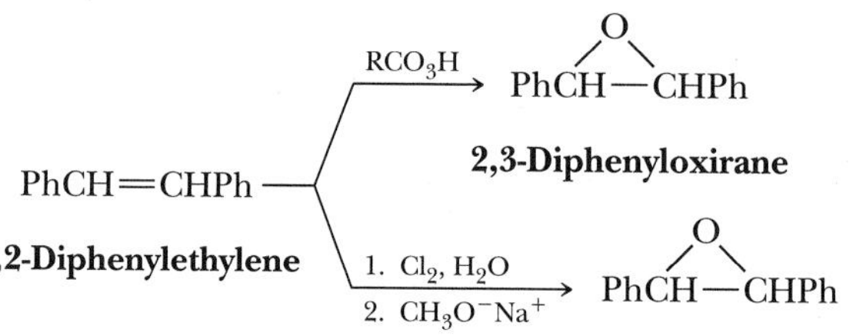

1,2-Diphenylethylene **2,3-Diphenyloxirane** **2,3-Diphenyloxirane**

Suppose that the starting alkene is *trans*-1,2-diphenylethylene.

(a) What is the configuration of the oxirane formed in each sequence?

(b) Will the oxirane formed in either sequence rotate the plane of polarized light? Explain.

11.28 The following enantiomer of a chiral epoxide is an intermediate in the synthesis of the insect pheromone frontalin.

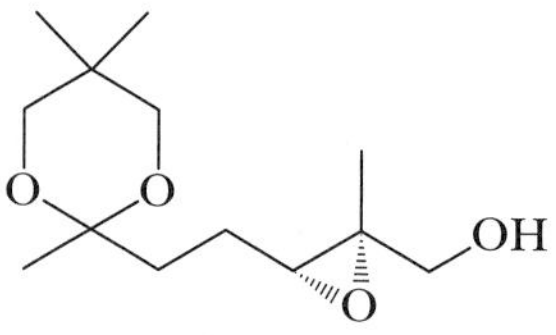

Using the Sharpless epoxidation, show how this enantiomer can be prepared from an allylic alcohol.

11.29 Human white cells produce an enzyme called myeloperoxidase. This enzyme catalyzes the reaction between hydrogen peroxide and chloride ion to produce hypochlorous acid, HOCl, which reacts as if it were Cl^+OH^-. When attacked by white cells, cholesterol gives a chlorohydrin as the major product.

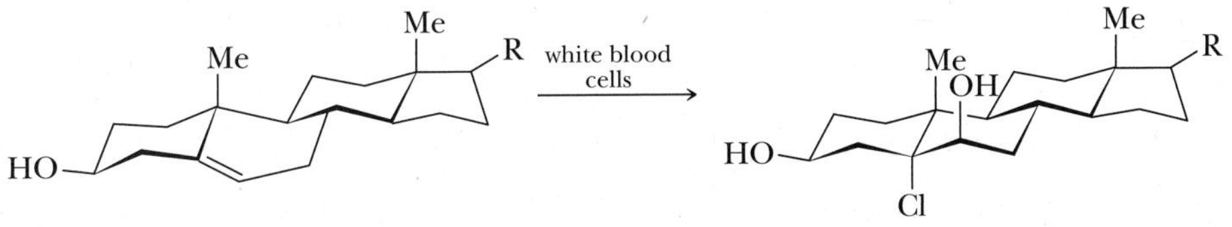

Cholesterol **Cholesterol chlorohydrin**

(a) Propose a mechanism for this reaction. Account for both its regioselectivity and stereoselectivity.

(b) On standing or (much more rapidly) on treatment with base, the chlorohydrin is converted to an epoxide. Show the structure of the epoxide and a mechanism for its formation. This epoxide is believed to be involved in induction of certain cancers.

11.30 Propose a mechanism for the following acid-catalyzed rearrangement.

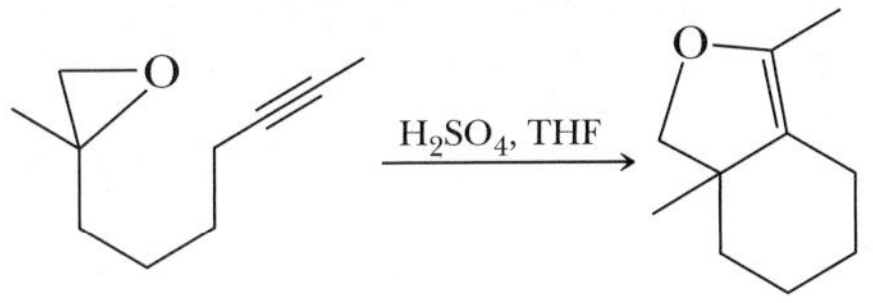

Synthesis

11.31 Show reagents and experimental conditions to synthesize the following compounds from 1-propanol (any derivative of 1-propanol prepared in one part of this problem may be used for the synthesis of another part of the problem).

(a) Propanal
(b) Propanoic acid
(c) Propene
(d) 2-Propanol
(e) 2-Bromopropane
(f) 1-Chloropropane
(g) 1,2-Dibromopropane
(h) Propyne
(i) 2-Propanone
(j) 1-Chloro-2-propanol
(k) Methyloxirane
(l) Dipropyl ether
(m) Isopropyl propyl ether
(n) 1-Mercapto-2-propanol
(o) 1-Amino-2-propanol
(p) 1,2-Propanediol

11.32 Starting with *cis*-3-hexene, show how to prepare the following diols.

(a) Meso 3,4-hexanediol **(b)** Racemic 3,4-hexanediol

11.33 Show reagents to convert cycloheptene to each of the following compounds.

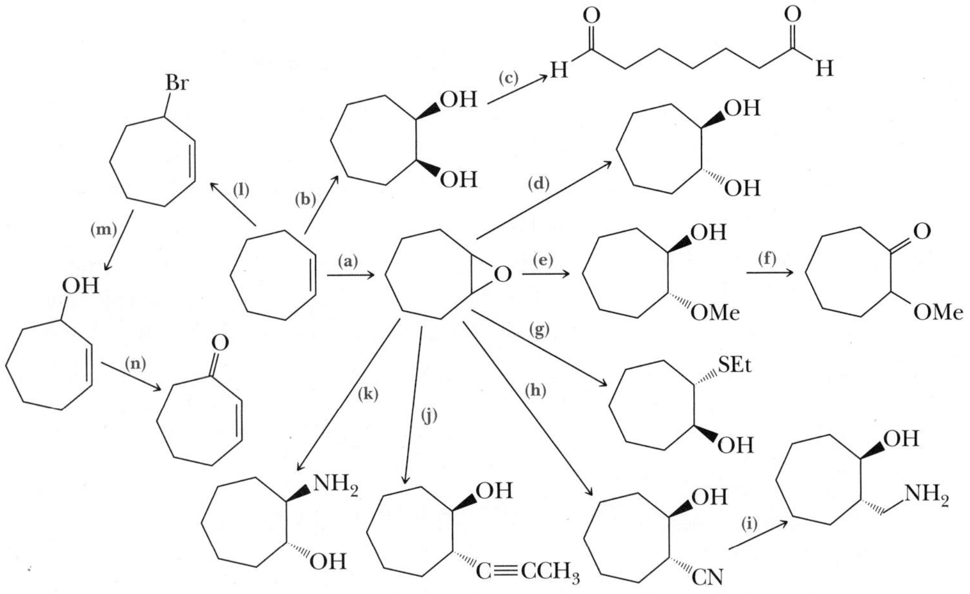

11.34 Show reagents to convert bromocyclopentane to each of the following compounds.

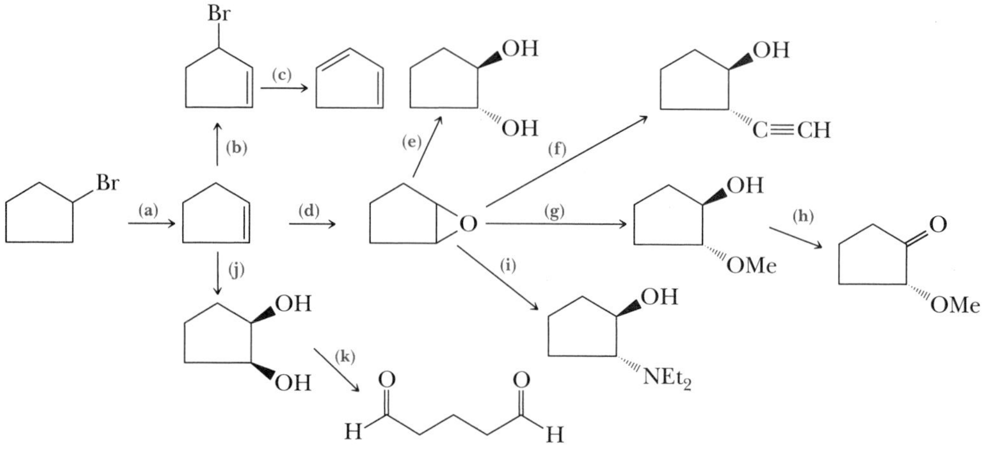

11.35 Given the following retrosynthetic analysis, show how to synthesize the target molecule from styrene and 1-chloro-3-methyl-2-butene.

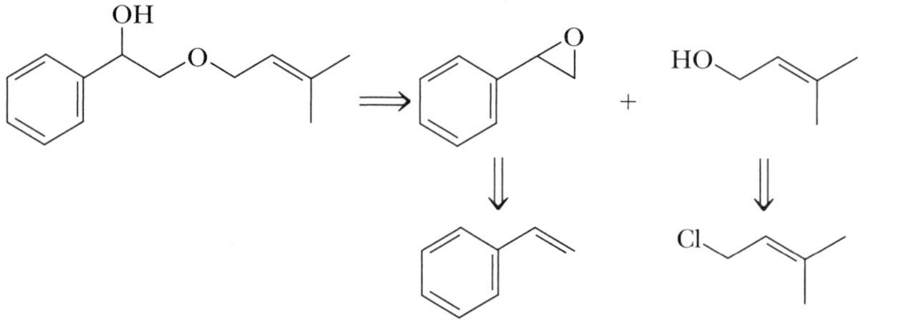

Styrene **1-Chloro-3-methyl-2-butene**

11.36 Starting with acetylene and ethylene oxide as the only sources of carbon atoms, show how to prepare these compounds.

(a) 3-Butyn-1-ol **(b)** 3-Hexyn-1,6-diol **(c)** 1,6-Hexanediol
(d) (*Z*)-3-Hexen-1,6-diol **(e)** (*E*)-3-Hexen-1,6-diol **(f)** Hexanedial

11.37 Following are the steps in the industrial synthesis of glycerin.

$$\underset{\textbf{Propene}}{CH_2{=}CHCH_3} \xrightarrow[\text{heat}]{Cl_2} \textbf{A}\ (C_3H_5Cl) \xrightarrow{NaOH,\ H_2O} \textbf{B}\ (C_3H_6O) \xrightarrow{Cl_2,\ H_2O}$$

$$\textbf{C}\ (C_3H_7ClO_2) \xrightarrow[\text{heat}]{Ca(OH)_2} \textbf{D}\ (C_3H_6O_2) \xrightarrow{H_2O,\ HCl} HOCH_2\overset{\overset{\displaystyle OH}{|}}{C}HCH_2OH$$

1,2,3-Propanetriol (glycerol, glycerin)

Provide structures for all intermediate compounds (**A–D**) and describe the type of mechanism by which each is formed.

11.38 Gossyplure, the sex pheromone of the pink bollworm, is the acetic ester of 7,11-hexadecadien-1-ol. The active pheromone has the *Z* configuration at the C7—C8 double bond and is a mixture of *E*,*Z* isomers at the C11—C12 double bond. Shown here is the *Z*,*E* isomer.

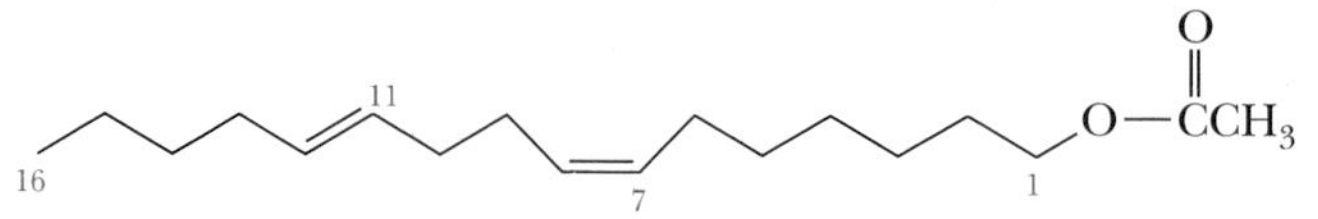

(7*Z*,11*E*)-7,11-Hexadecadienyl acetate

Following is a retrosynthetic analysis for (7*Z*,11*E*)-7,11-hexadecadien-1-ol, which then led to a successful synthesis of gossyplure.

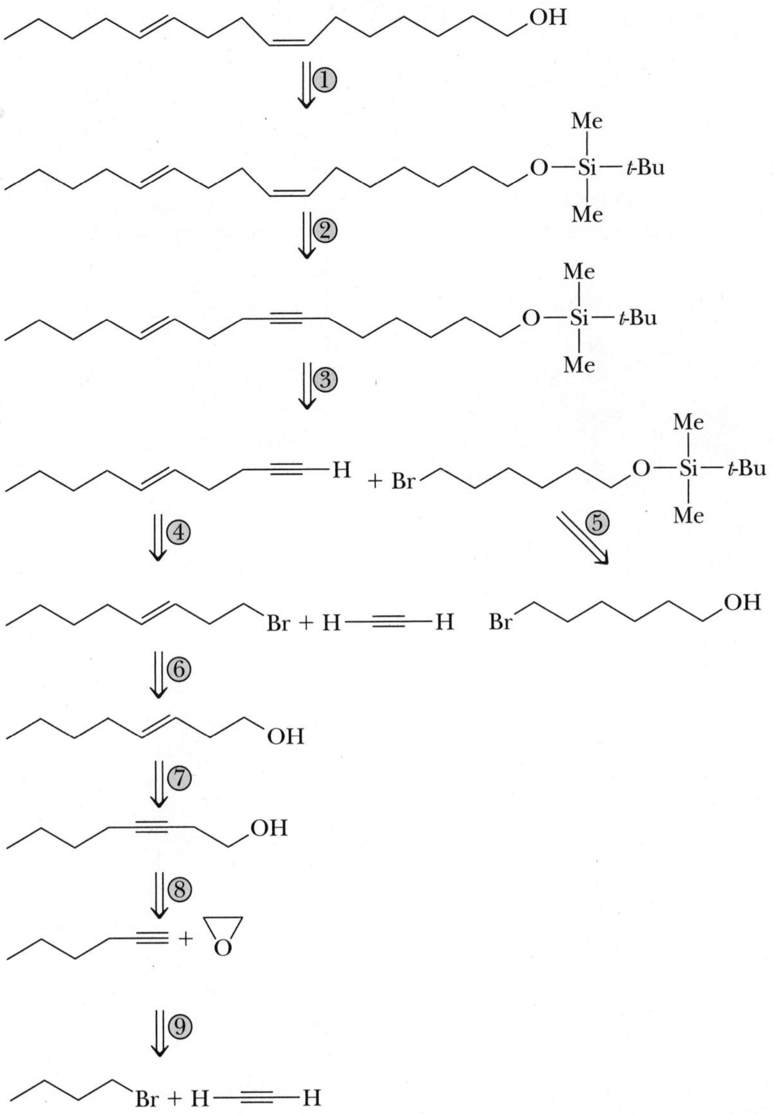

(a) Suggest reagents and experimental conditions for each step in this synthesis.

(b) Why is it necessary to protect the —OH group of 6-bromo-1-hexanol?

(c) How might you modify this synthesis to prepare the 7*Z*,11*Z* isomer of gossyplure?

11.39 Epichlorohydrin (Section 11.10) is a valuable synthetic intermediate because each of its three carbons contains a reactive group. Following is the first step in its synthesis from propene. Propose a mechanism for this step.

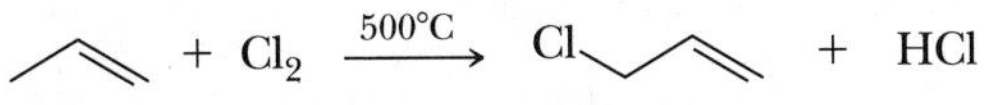

Propene **3-Chloropropene (Allyl chloride)**

11.40 Each of these drugs contains one or more building blocks derived from either ethylene oxide or epichlorohydrin.

(a) 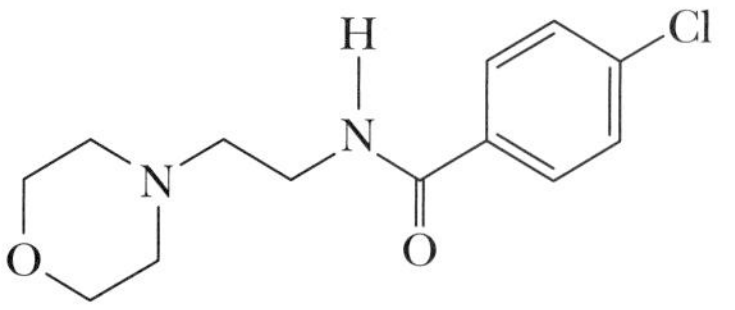

Moclobemide
(an antidepressant)

(b) 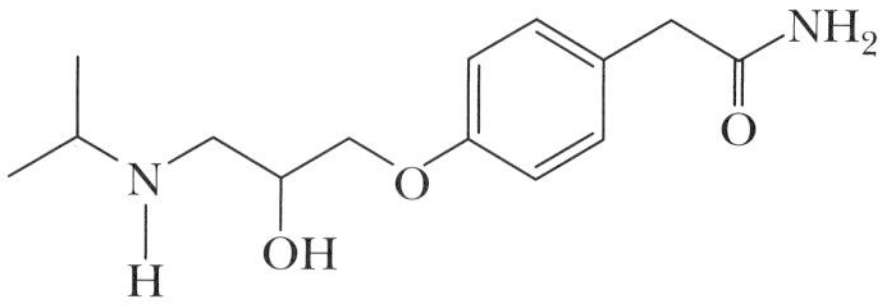

Atenolol
(an antihypertensive)

(c) 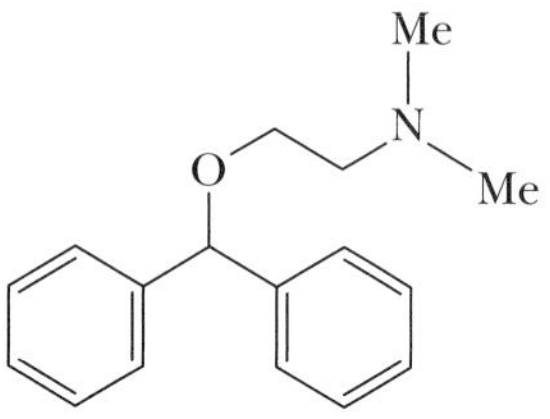

Diphenhydramine
(Benadryl, an antihistamine)

(d) 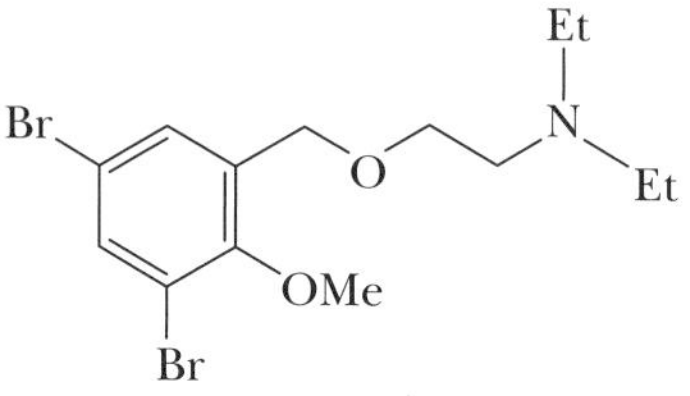

Spasmolytol
(an antispasmodic)

(e) 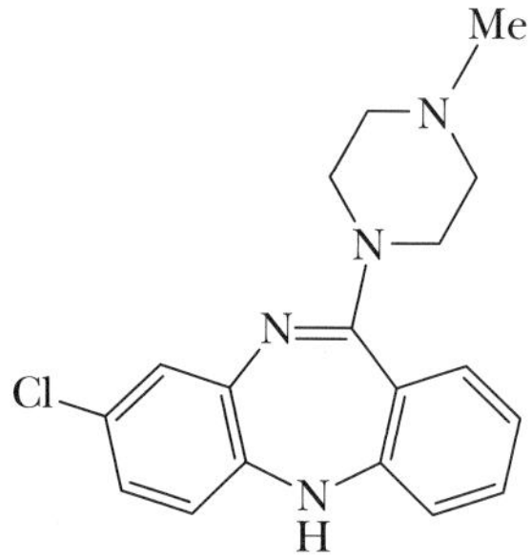

Clozapine
(an antischizophrenic)

(f) 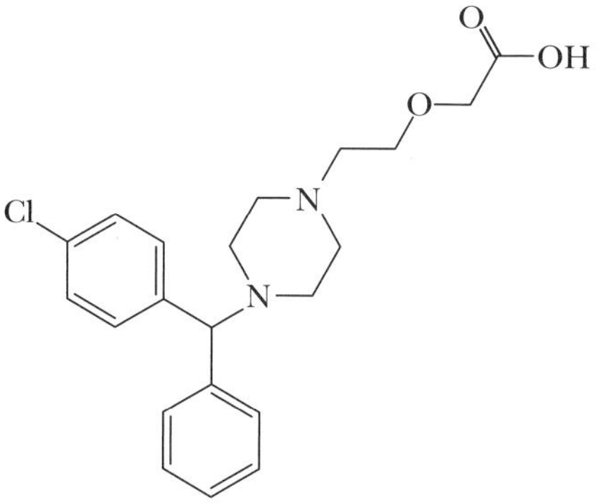

Cetirizine
(Zyrtec, an antihistamine)
(racemic)

Identify the part of each molecule that can be derived from one or the other of these building blocks and propose structural formulas for the nucleophile(s) that can be used along with either ethylene oxide or epichlorohydrin to synthesize each molecule. We will learn about the actual syntheses of each molecule in later chapters.

Looking Ahead

11.41 Aldehydes and ketones react with one molecule of an alcohol to form compounds called hemiacetals, in which there is one hydroxyl group and one ether-like group. Reaction of a hemiacetal with a second molecule of alcohol gives an acetal and a molecule of water. We study this reaction in Chapter 16.

The carbonyl group of an aldehyde or a ketone	A hemiacetal (has an —OH and an —OR group to the same carbon)	An acetal (has two —OR groups to the same carbon)

Draw structural formulas for the hemiacetal and acetal formed from these reagents. The stoichiometry of each reaction is given in the problem.

(a) 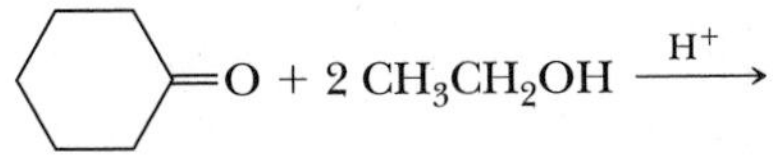

Cyclohexanone **Ethanol**

(b) 

Cyclohexanone **Ethylene glycol**

(c) 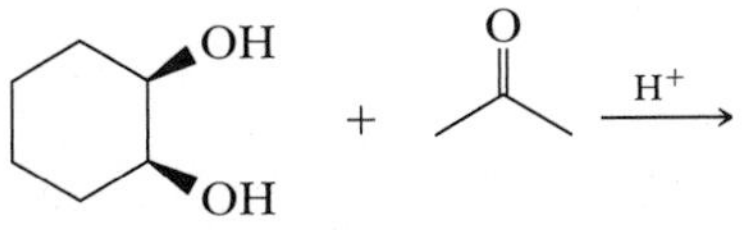

***cis*-1,2-Cyclohexanediol** **Acetone**

Organic Chemistry Reaction Roadmap

11.42 Use the reaction roadmap you made for Problems 6.54, 7.29, 8.28, 9.54, and 10.50 and update it to contain the reactions in the "Study Guide" section of this chapter. Because of their highly specific nature, do not use reactions 2, 8, and 10 on your reaction roadmap.

11.43 Write the products of the following sequences of reactions. Refer to your reaction roadmap to see how the combined reactions allow you to "navigate" between the different functional groups.

(a) 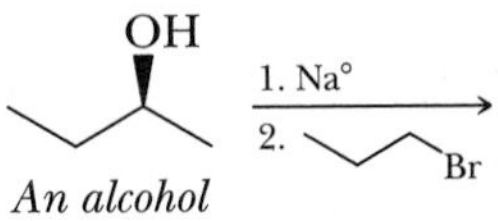

An alcohol

(b) 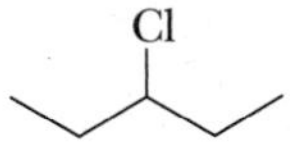

1. 1 Equivalent of KOtBu
2. MCPBA
3. H_3O^+/H_2O

A haloalkane

(c) 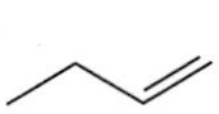

1. BH_3
2. $H_2O_2/NaOH$
3. PBr_3
4. 1 Equivalent of NaOEt

An alkene

(d) 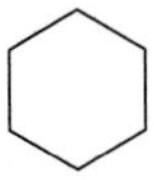

1. Br_2 and light
2. 1 Equivalent of KOtBu
3. MCPBA
4. NH_3

An alkane

Mixed Synthesis

11.44 Using your reaction roadmap as a guide, show how to convert 1-butene into dibutyl ether. You must use 1-butene as the source of all carbon atoms in the dibutyl ether. Show all required reagents and all molecules synthesized along the way.

1-Butene **Dibutyl ether**

Reaction Roadmap

11.45 Using your reaction roadmap as a guide, show how to convert cyclohexanol into racemic *trans*-1,2-cyclohexanediol. Show all required reagents and all molecules synthesized along the way.

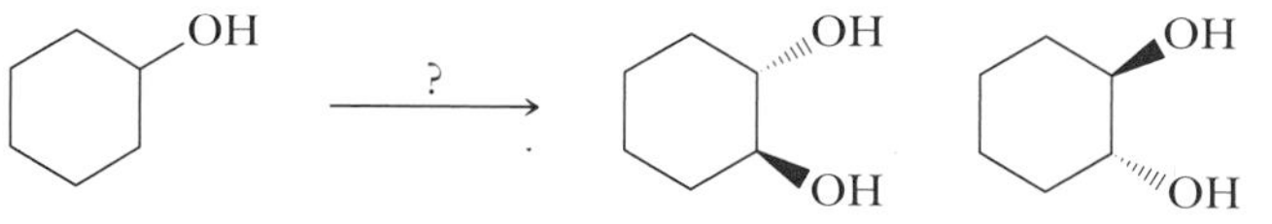

Cyclohexanol **trans-1,2-Cyclohexanediol** (racemic)

Reaction Roadmap

11.46 Using your reaction roadmap as a guide, show how to convert 1-butanol and ethanol into racemic 2-ethoxy-1-butanol. You must use 1-butanol and ethanol as the source of all carbon atoms in the ether product. Show all required reagents and all molecules synthesized along the way.

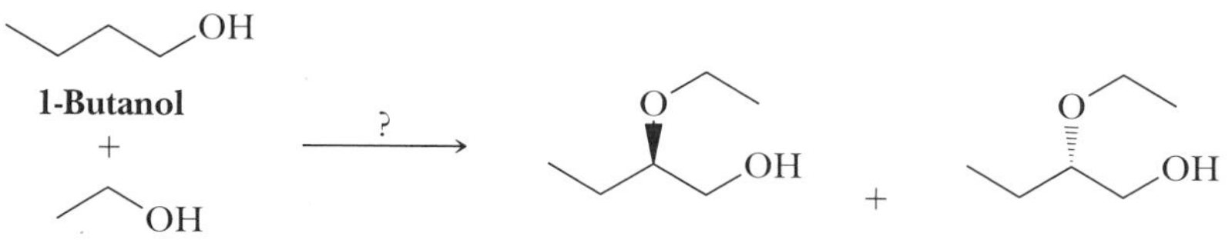

1-Butanol + **Ethanol** **2-Ethoxy-1-butanol** (racemic)

Reactions in Context

11.47 During the synthesis of the antiasthmatic drug montelukast (Singulair), a silyl ether protecting group is used to mask the reactivity of an OH group. The silyl group chosen is the *tert*-butyldimethylsilyl (TBS) group. Draw the product of the following transformation, assuming the TBS-Cl reagent reacts only once with the starting material. Briefly explain your answer.

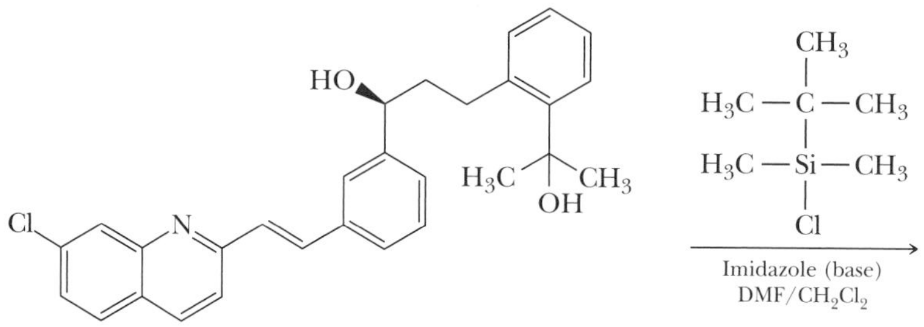

11.48 The Sharpless epoxidation is used when a single enantiomer product is required. Predict the structure of the predominant product of the following transformation.

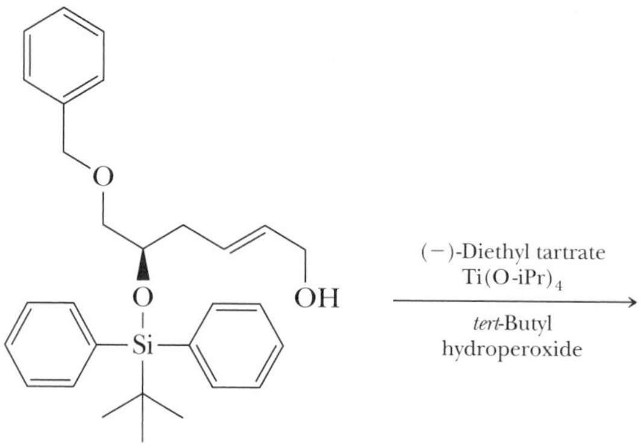

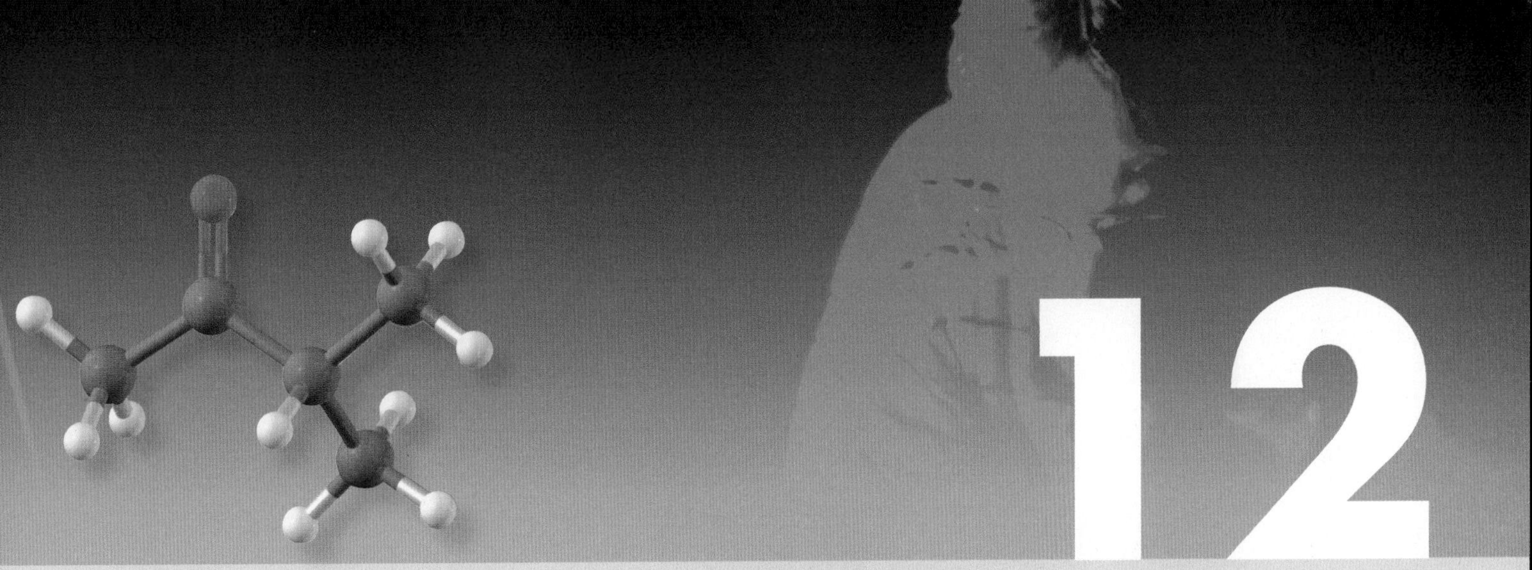

12

Infrared Spectroscopy

A scientist working with a Fourier transform infrared spectrometer. **Inset:** a model of 3-methyl-2-butanone. For an IR spectrum of this compound, see Figure 12.2. (Chris Taylor/CSIRO/Science Source)

Outline

12.1 Electromagnetic Radiation
12.2 Molecular Spectroscopy
12.3 Infrared Spectroscopy
12.4 Interpreting Infrared Spectra
12.5 Solving Infrared Spectral Problems

Determination of molecular structure is one of the central themes of organic chemistry. For this purpose, chemists today rely almost exclusively on instrumental methods, four of which we discuss in this text. We begin in this chapter with infrared (IR) spectroscopy. Then in Chapters 13 and 14, we introduce nuclear magnetic resonance (NMR) spectroscopy and mass spectrometry (MS), respectively. A brief introduction to ultraviolet-visible spectroscopy is contained in Chapter 20 as part of our discussion of conjugated systems.

12.1 Electromagnetic Radiation

Electromagnetic radiation Light and other forms of radiant energy.

Wavelength (λ) The distance between consecutive peaks on a wave.

Frequency The number of full cycles of a wave that pass a given point in a second; it is given the symbol ν (Greek nu) and reported in hertz (Hz), which has the units s^{-1}.

Gamma rays, X-rays, ultraviolet light, visible light, infrared radiation, microwaves, and radio waves are all types of **electromagnetic radiation** that can be described in terms of wavelength and frequency.

Table 12.1 summarizes **wavelengths (λ)**, **frequencies**, and energies of various regions of the electromagnetic spectrum. The wavelengths of visible light fall in the range 400–700 nm. Infrared rays (felt as heat but not visible) fall in the range 0.7–300 μm.

Table 12.1 Wavelengths, Frequencies, and Energies of Some Regions of the Electromagnetic Spectrum

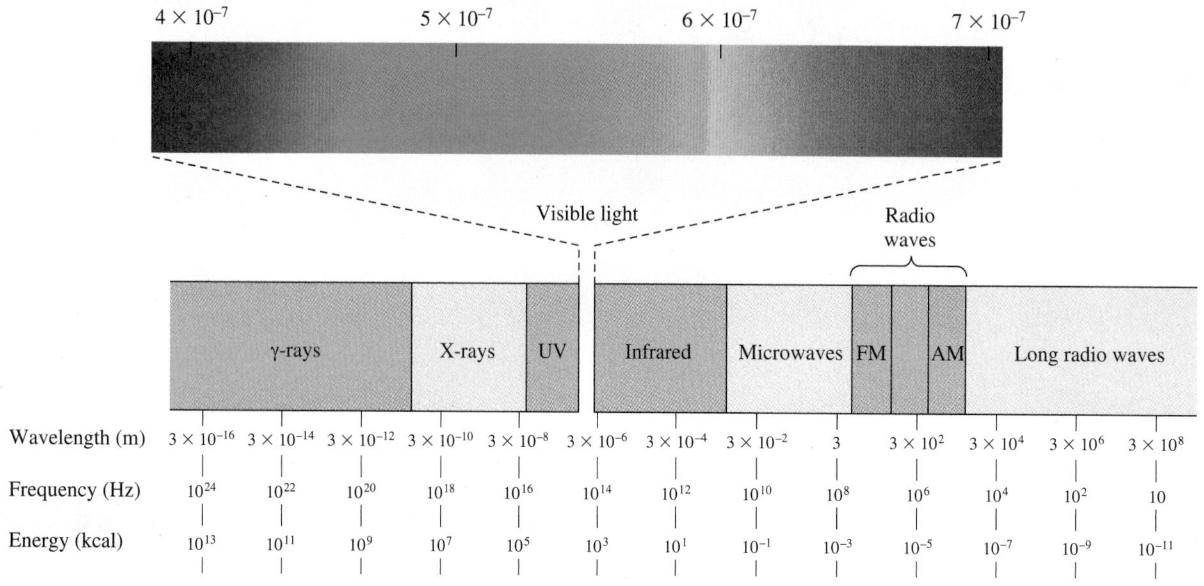
4 × 10⁻⁷
5 × 10⁻⁷
6 × 10⁻⁷
7 × 10⁻⁷
Visible light
Radio waves
γ-rays
X-rays
UV
Infrared
Microwaves
FM
AM
Long radio waves
Wavelength (m)
3 × 10⁻¹⁶ 3 × 10⁻¹⁴ 3 × 10⁻¹² 3 × 10⁻¹⁰ 3 × 10⁻⁸ 3 × 10⁻⁶ 3 × 10⁻⁴ 3 × 10⁻² 3 3 × 10² 3 × 10⁴ 3 × 10⁶ 3 × 10⁸
Frequency (Hz)
10²⁴ 10²² 10²⁰ 10¹⁸ 10¹⁶ 10¹⁴ 10¹² 10¹⁰ 10⁸ 10⁶ 10⁴ 10² 10
Energy (kcal)
10¹³ 10¹¹ 10⁹ 10⁷ 10⁵ 10³ 10¹ 10⁻¹ 10⁻³ 10⁻⁵ 10⁻⁷ 10⁻⁹ 10⁻¹¹

Hertz (Hz)

The unit in which frequency is measured: s^{-1} (read "per second").

Frequency, the number of full cycles of a wave that pass a given point in a second, is given the symbol ν (Greek nu) and is reported in **hertz (Hz)**, which has the units s^{-1}. Wavelength and frequency are inversely proportional, and one can be calculated from the other using the following relationship:

$$\lambda\nu = c$$

where c is the velocity of light, 3.00×10^8 m/s. For example, consider the infrared radiation of wavelength 1.5×10^{-5} m (15 μm). The frequency of this radiation is 2.0×10^{13} Hz.

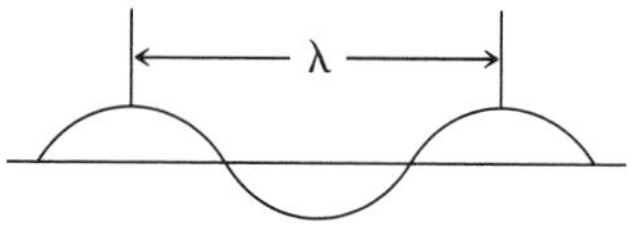
λ

$$\nu = \frac{3 \times 10^8\,\text{m}}{\text{s} \cdot 1.5 \times 10^{-5}\,\text{m}} = 2.0 \times 10^{13}\,\text{Hz}$$

An alternative way to describe electromagnetic radiation is in terms of its properties as a stream of particles called **photons**. The energy in a mole of photons is related to the frequency of the radiation by the equations

$$E = h\nu = h\frac{c}{\lambda}$$

where E is the energy in kJ (kcal)/mol and h is Planck's constant, 3.99×10^{-13} kJ (9.537×10^{-14} kcal) · s · mol^{-1}.

Wavelength is usually expressed in the SI base unit of meters. Other derived units commonly used to express wavelength are given in Table 12.2.

Table 12.2 Common Units Used to Express Wavelength (λ)

Unit	Relation to Meter
Meter (m)	—
Millimeter (mm)	1 mm = 10^{-3} m
Micrometer (μm)	1 μm = 10^{-6} m
Nanometer (nm)	1 nm = 10^{-9} m
Angstrom (Å)	1 Å = 10^{-10} m

12.2 Molecular Spectroscopy

An atom or molecule can be made to undergo a transition from energy state E_1 to a higher energy state E_2 by irradiating the atom or molecule with electromagnetic radiation corresponding to the energy difference between states E_1 and E_2 as illustrated schematically in Figure 12.1.

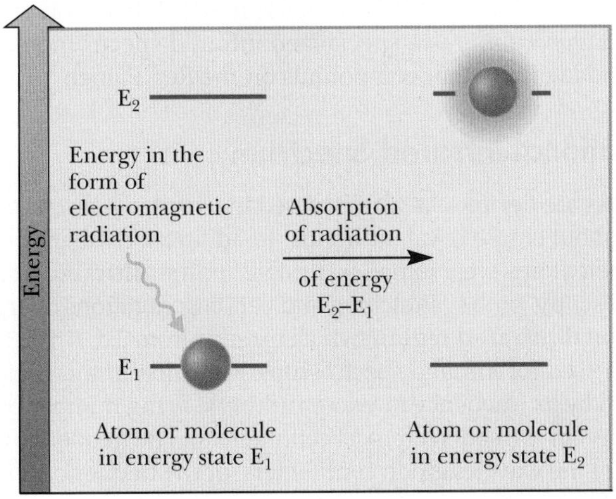

FIGURE 12.1 Absorption of energy in the form of electromagnetic radiation causes an atom or a molecule in energy state E_1 to change to a higher energy state, E_2.

When the atom or molecule returns from state E_2 to state E_1, an equivalent amount of energy is emitted.

Molecular spectroscopy is the experimental process of measuring which frequencies of radiation are absorbed or emitted by a particular substance and then attempting to correlate patterns of energy absorption or emission with details of molecular structure. Table 12.3 summarizes the regions of the electromagnetic spectrum of most interest to us and the relationships of each to changes in atomic and molecular energy levels.

Molecular spectroscopy
The study of which frequencies of radiation are absorbed or emitted by a particular substance and the correlation of these frequencies with details of molecular structure.

Table 12.3 Types of Energy Transitions Resulting from Absorption of Energy from Three Regions of the Electromagnetic Spectrum

Region of Electromagnetic Spectrum	Frequency (hertz)	Type of Spectroscopy	Absorption of Electromagnetic Radiation Results in Transitions Between
Radio frequency	$3 \times 10^7 - 9 \times 10^8$	Nuclear magnetic resonance	Nuclear spin levels
Infrared	$3 \times 10^{11} - 2.5 \times 10^{14}$	Infrared	Vibrational energy levels
Ultraviolet-visible	$2.5 \times 10^{14} - 1.5 \times 10^{15}$	Ultraviolet-visible	Electronic energy levels

12.3 Infrared Spectroscopy

Infrared spectroscopy provides a direct method of detecting certain functional groups in a molecule. In this chapter, we first develop a basic understanding of the theory behind infrared spectroscopy; then we concentrate on the interpretation of spectra and the information they can provide about details of molecular structure. Infrared spectroscopy depends on the absorption of infrared light in the wavelength range 2.5×10^{-6} to 2.5×10^{-5} m by vibrating bonds within molecules. Infrared spectroscopy is useful to organic chemists not only for the determination of molecular structure but also for many other applications. For example, forensic scientists use infrared spectroscopy to identify illegal substances and toxins. The sensors that control automatic windshield wipers in newer cars use infrared detection of water. The Mars rovers *Spirit* and *Opportunity* use specialized infrared spectrometers (Mini-TES) to look for water and identify other compounds on the Red Planet.

A. The Vibrational Infrared Spectrum

Organic molecules are flexible. As we discussed in Chapter 2, atoms and groups of atoms can rotate about single covalent bonds. In addition, covalent bonds can stretch and bend as if their atoms were joined by flexible springs. **Infrared spectroscopy**, also called **IR spectroscopy**, probes stretching and bending vibrations of organic molecules.

Infrared (IR) spectroscopy A spectroscopic technique in which a compound is irradiated with infrared radiation, absorption of which causes covalent bonds to change from a lower vibration state to a higher one. Infrared spectroscopy is particularly valuable for determining the kinds of functional groups present in a molecule.

Vibrational infrared region The portion of the infrared region that extends from 4000 to 400 cm^{-1}.

Wavenumbers, $\bar{\nu}$ The frequency of electromagnetic radiation expressed as the number of waves per centimeter, with units cm^{-1} (read: reciprocal centimeters).

The **vibrational infrared region**, which extends from 2.5×10^{-6} to 2.5×10^{-5} m in wavelength, is used for infrared spectroscopy. Radiation in this region is most commonly referred to by its frequency in **wavenumbers, $\bar{\nu}$**, the number of waves per centimeter, with units cm^{-1} (read: reciprocal centimeters). The frequency in wavenumbers is the reciprocal of the wavelength in centimeters, or the frequency (ν) in hertz divided by c, the speed of light.

$$\bar{\nu} = \frac{1}{\lambda\,(\text{cm})} = \frac{10^{-2}\,(\text{m} \cdot \text{cm}^{-1})}{\lambda\,(\text{m})} = \frac{\nu}{c}$$

When expressed in frequencies, the vibrational region of the infrared spectrum extends from 4000 to 400 cm^{-1}.

$$\bar{\nu} = \frac{10^{-2}\,\text{m} \cdot \text{cm}^{-1}}{2.5 \times 10^{-6}\,\text{m}} = 4000\ \text{cm}^{-1} \qquad \bar{\nu} = \frac{10^{-2}\,\text{m} \cdot \text{cm}^{-1}}{2.5 \times 10^{-5}\,\text{m}} = 400\ \text{cm}^{-1}$$

An advantage of using frequencies is that they are directly proportional to energy; the higher the frequency, the greater the energy of the radiation.

Figure 12.2 is an infrared spectrum of 3-methyl-2-butanone. The horizontal axis at the bottom of the chart paper is calibrated in frequency (wavenumbers, cm^{-1}); that at the top is calibrated in wavelength (micrometers, μm). The frequency scale is often divided into two or more regions. For all spectra reproduced in this text, the scale is divided into three linear regions: 4000–2200 cm^{-1}, 2200–1000 cm^{-1}, and 1000–400 cm^{-1}. The vertical axis measures transmittance (the fraction of light transmitted), with 100% at the top and 0% at the bottom. Thus, the baseline for an infrared spectrum (100% transmittance of radiation through the sample, 0% absorption) is at the top of the chart paper, and absorption of radiation corresponds to a trough or valley. Strange as it may seem, we commonly refer to infrared absorptions as peaks, even though they are conventionally displayed pointing downward.

The spectrum in Figure 12.2 was recorded using a neat sample, which means the pure liquid. A few drops are compressed between transparent salt discs, through which infrared radiation is passed. There are several other ways to prepare gas,

FIGURE 12.2 Infrared spectrum of 3-methyl-2-butanone.

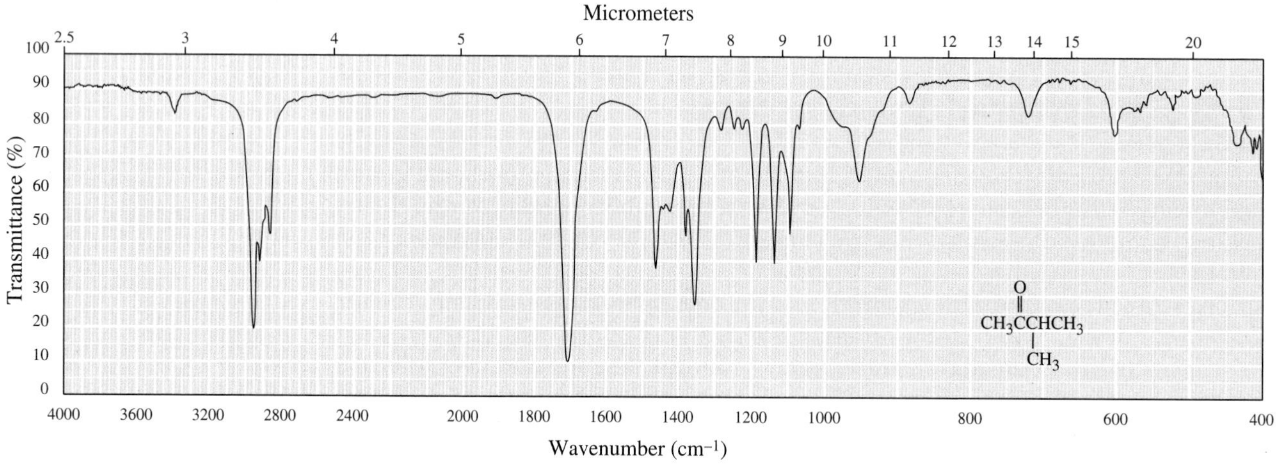

liquid, and solid samples for spectroscopy. NaCl and KBr are often used in sample preparation because, as ionic solids, they have no covalent bonds to absorb infrared radiation. Yet care must be taken to keep the samples dry. NaCl and KBr are salts, and are hygroscopic.

Example 12.1 Unit Conversions

Some infrared spectrophotometers are calibrated to record spectra on an ordinate that is linear in wavelength (μm), whereas others record them on an ordinate that is linear in frequency (cm^{-1}). Carry out the following conversions (note the convenient formula for converting between μm and cm^{-1}: $\bar{\nu}\lambda = 10^4$).

(a) 7.05 μm to cm^{-1} **(b)** 3.35 μm to cm^{-1} **(c)** 3280 cm^{-1} to μm

Solution

(a) 1418 cm^{-1} **(b)** 2985 cm^{-1} **(c)** 3.05 μm

Problem 12.1

Which is higher in energy?

(a) Infrared radiation of 1715 cm^{-1} or of 2800 cm^{-1}?

(b) Radio-frequency radiation of 300 MHz or of 60 MHz?

Watch a video explanation

B. Molecular Vibrations

Atoms joined by covalent bonds are not permanently fixed in one position but rather undergo continual vibrations relative to each other. The energies associated with these vibrations are quantized, which means that within a molecule, only specific vibrational energy levels are allowed. The energies associated with transitions between vibrational energy levels in most covalent molecules correspond to frequencies in the infrared region, 4000–400 cm^{-1}.

For a molecule to absorb this radiation, its vibration must result in a *substantial change in the bond dipole moment.* If two opposite charges are connected by a spring, a change in distance between the charges corresponds to a change in dipole moment. In general, the greater the bond dipole, the greater the change in dipole moment caused by a vibration. Any vibration that leads to a substantial change in dipole moment is said to be **infrared active**. The greater the change is, the more intense the absorption will be. Covalent bonds whose vibration does not result in a change in bond dipole moment, for example, as a result of symmetry in the molecule are said to be infrared inactive. The carbon-carbon double and triple bonds in symmetrically substituted alkenes and alkynes (e.g., 2,3-dimethyl-2-butene and 2-butyne) do not absorb infrared radiation because vibration does not result in a substantial bond dipole change.

Infrared active
Any molecular vibration that leads to a substantial change in dipole moment is observed in an IR spectrum.

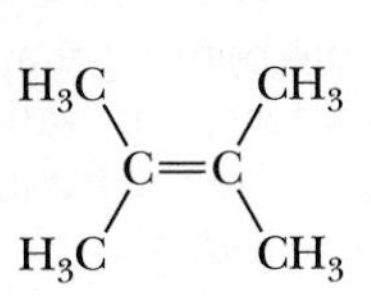

2,3-Dimethyl-2-butene

$H_3C-C\equiv C-CH_3$

2-Butyne

For a nonlinear molecule containing n atoms, $3n - 6$ allowed fundamental vibrations exist. For a molecule as simple as ethanol, CH_3CH_2OH, there are 21 fundamental vibrations, and for hexanoic acid, $CH_3(CH_2)_4COOH$, there are 54. Thus, even for relatively simple molecules, a large number of vibrational energy levels exist and the patterns of energy absorption for these and larger molecules are very complex. For linear molecules with n atoms, there are $3n - 5$ allowed fundamental vibrations.

The simplest vibrational motions in molecules giving rise to absorption of infrared radiation are stretching and bending motions. Figure 12.3 illustrates the fundamental stretching and bending vibrations for a methylene group.

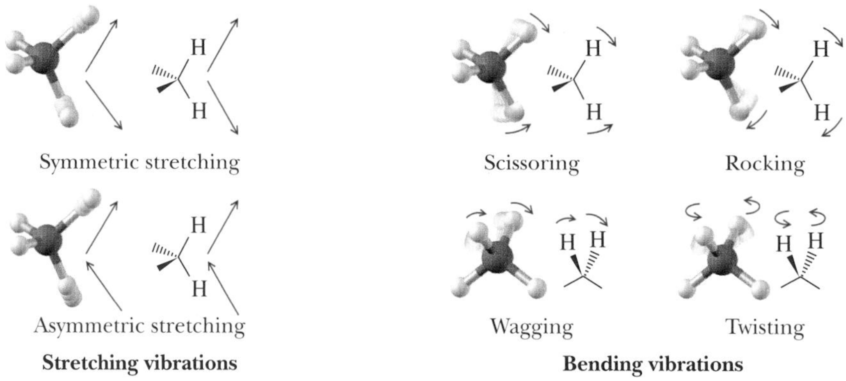
Symmetric stretching
Asymmetric stretching
Stretching vibrations
Scissoring
Rocking
Wagging
Twisting
Bending vibrations

FIGURE 12.3 Fundamental stretching and bending vibrations for a methylene group.

Raman spectroscopy

A vibrational molecular spectroscopy that is complementary to infrared (IR) spectroscopy in that infrared-inactive vibrations are seen in Raman spectra.

A different technique called **Raman spectroscopy** is complementary to infrared spectroscopy in that infrared-inactive vibrations are seen in Raman spectra, while Raman-inactive vibrations are infrared active. A more complete description of Raman spectroscopy is beyond the scope of this text, but it is a very useful technique for studying certain molecules.

C. Characteristic Absorption Patterns

Analysis of the modes of vibration for a molecule is very complex because all the atoms contribute to the vibrational modes. However, we can make useful generalizations about where absorptions due to particular vibrational modes will appear in an infrared spectrum by considering each individual bond and ignoring other bonds in the molecule. As a simplifying assumption, let us consider two covalently bonded atoms as two vibrating masses connected by a spring. The total energy is proportional to the frequency of vibration. The frequency of a stretching vibration is given by the following equation, which is derived from Hooke's law for a vibrating spring.

$$\bar{\nu} = 4.12\sqrt{\frac{K}{\mu}}$$

Here, $\bar{\nu}$ is the frequency of the vibration in wavenumbers (cm^{-1}); K is the force constant of the bond, a measure of the bond's strength, in dynes per centimeter; and μ is the "reduced mass" of the two atoms, $(m_1m_2)/(m_1 + m_2)$, where m is the mass of the atoms.

Force constants for single, double, and triple bonds are approximately 5, 10, and 15×10^5 dynes/cm, respectively, thus approximately in the ratio 1:2:3. Using the value for the force constant for a single bond, we calculate the frequency for the stretching vibration of a single bond between ^{12}C and ^{1}H as follows.

For $^{12}C{-}^{1}H$ stretching:

$$\text{Reduced mass} = 12 \times 1/(12 + 1) = 0.923 \text{ g/atom}$$

and

$$\bar{\nu} = 4.12\sqrt{\frac{K}{\mu}} = 4.12\sqrt{\frac{5 \times 10^5}{0.923}} = 3032 \text{ cm}^{-1}$$

The experimentally determined value for the frequency of an alkyl C—H stretching vibration is approximately 3000 cm^{-1}. Given the simplifying assumptions made in this calculation and the fact that the value of the force constant for a single bond is an average value, the agreement between the calculated value and the experimental value is remarkably good. Although frequencies calculated in this manner can be close to the experimental values, they are generally not accurate enough for precise determination of molecular structure.

Hooke's law predicts that the *position* of the absorption of a stretching vibration in an IR spectrum depends both on the strength of the vibrating bond and on the masses of the atoms connected by the bond. The stronger the bond is and the lighter the atoms are, the higher the frequency of the stretching vibration will be. As we saw earlier, the *intensity* of an absorption depends primarily on the change in dipole of the vibrating bond.

Example 12.2 Calculating IR Wavenumbers

Calculate the stretching frequency in wavenumbers for a carbon-carbon double bond. Assume that each carbon is the most abundant isotope, namely ^{12}C.

Solution

Assume a force constant of 10×10^5 dynes per centimeter for C=C. The calculated frequency is 1682 cm^{-1}, a value close to the experimental value of 1650 cm^{-1}.

$$\bar{\nu} = 4.12 = \sqrt{\frac{10 \times 10^5}{12 \times 12/(12 + 12)}} = 1682 \text{ cm}^{-1}$$

Problem 12.2

Without doing the calculation, which member of each pair do you expect to occur at the higher frequency?

(a) C=O or C=C stretching **(b)** C=O or C—O stretching

(c) C≡C or C=O stretching **(d)** C—H or C—Cl stretching

Detailed interpretation of most infrared spectra is difficult because of the complexity of vibrational modes. In addition to the fundamental vibrational modes we have described, other types of absorptions occur, resulting in so-called overtone and coupling peaks that are usually quite weak.

To one skilled in the interpretation of infrared spectra, the absorption patterns can yield an enormous amount of information about chemical structure. However, we have neither the time nor the need to develop this level of competence. The value of infrared spectra for us is that they can be used to determine the presence or absence of certain functional groups. A carbonyl group, for example, typically shows strong absorption at approximately 1630–1820 cm^{-1}. The position of absorption for a particular carbonyl group depends on whether it is an aldehyde, a ketone, a carboxylic acid, or an ester; if it is in a ring, the position of absorption depends on the size of the ring. In this chapter, we discuss how structural variations, such as ring size or other factors, affect this value.

D. Correlation Tables

Data on absorption patterns of functional groups are collected in tables called **correlation tables.** Table 12.4 lists infrared absorptions for the types of bonds and functional groups we examine most often. A cumulative correlation table can be found in Appendix 6. In these tables, the intensity of a particular absorption is often referred

Watch a video explanation

Table 12.4 Infrared Stretching Frequencies of Selected Functional Groups

Bond	Stretching Frequency (cm^{-1})	Intensity
O—H	3200–3650	Weak to strong (strongest when H-bonded)
N—H	3100–3550	Medium
C—H	2700–3300	Weak to medium
C═C	1600–1680	Weak to medium
C═O	1630–1820	Strong
C—O	1000–1250	Strong

to as strong (s), medium (m), or weak (w). In general, bonds between C and O where the electronegativity difference is largest have the largest dipole moments and tend to give the strongest infrared absorptions.

Organic chemists pay most attention to the region from 3500 to 1500 cm^{-1} because the stretching and bending vibrations for most functional groups are found in this region. Vibrations in the region 1500 to 400 cm^{-1} are more complex and far more difficult to analyze. Because even slight variations in molecular structure and absorption patterns are most obvious in this region, it is often called the **fingerprint region**. If two compounds have even slightly different structures, the differences in their infrared spectra are most clearly discernible in this region.

Fingerprint region

Vibrations in the region 1500 to 400 cm^{-1} of IR spectra are complex and difficult to analyze but are characteristic for different molecules.

Example 12.3 Functional Group Frequencies

What functional group is most likely present if a compound shows IR absorption at these frequencies?

(a) 1705 cm^{-1} **(b)** 2950 cm^{-1}

Solution

(a) A C═O group **(b)** An aliphatic C—H group

Problem 12.3

A compound shows strong, very broad IR absorption in the region 3300–3600 cm^{-1} and strong, sharp absorption at 1715 cm^{-1}. What functional group accounts for both of these absorptions?

Example 12.4 Distinguishing Different Functional Groups

Propanone and 2-propen-1-ol are constitutional isomers. Show how to distinguish between them by IR spectroscopy.

$$CH_3-\overset{\overset{\displaystyle O}{\|}}{C}-CH_3 \qquad CH_2{=}CH-CH_2-OH$$

Propanone (Acetone) **2-Propen-1-ol (Allyl alcohol)**

Solution

Only propanone shows strong absorption in the C=O stretching region, 1630–1820 cm^{-1}. Alternatively, only 2-propen-1-ol shows strong absorption in the O—H stretching region, 3200–3650 cm^{-1}.

Problem 12.4

Propanoic acid and methyl ethanoate are constitutional isomers. Show how to distinguish between them by IR spectroscopy.

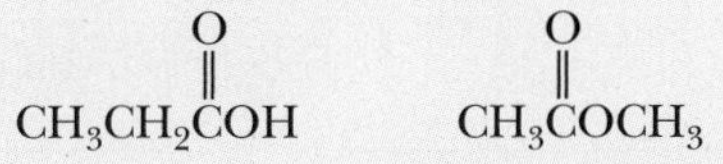

Propanoic acid **Methyl ethanoate (Methyl acetate)**

12.4 Interpreting Infrared Spectra

A. Alkanes

Infrared spectra of alkanes are usually simple with few peaks, the most common of which are given in Table 12.5.

Figure 12.4 is an infrared spectrum of decane. The strong peak with multiple splitting between 2850 and 3000 cm^{-1} is characteristic of alkane C—H stretching; it is strong in this spectrum because there are so many C—H bonds and no other functional groups. The other prominent peaks correspond to methylene bending at 1465 cm^{-1} and methyl bending at 1380 cm^{-1}.

Table 12.5 **Infrared Absorptions of Alkanes, Alkenes, Alkynes, and Arenes**

Hydrocarbon	Vibration	Frequency (cm^{-1})	Intensity
Alkane			
C—H	Stretching	2850–3000	Medium
CH_2	Bending	1450–1475	Medium
CH_3	Bending	1375 and 1450	Weak to medium
C—C	(Not useful for interpretation—too many bands)		
Alkene			
C—H	Stretching	3000–3100	Weak to medium
C=C	Stretching	1600–1680	Weak to medium
Alkyne			
C—H	Stretching	3300	Medium to strong
C≡C	Stretching	2100–2250	Weak
Arene			
C—H	Stretching	3030	Weak to medium
C=C	Stretching	1450–1600	Medium
C—H	Bending	690–900	Strong

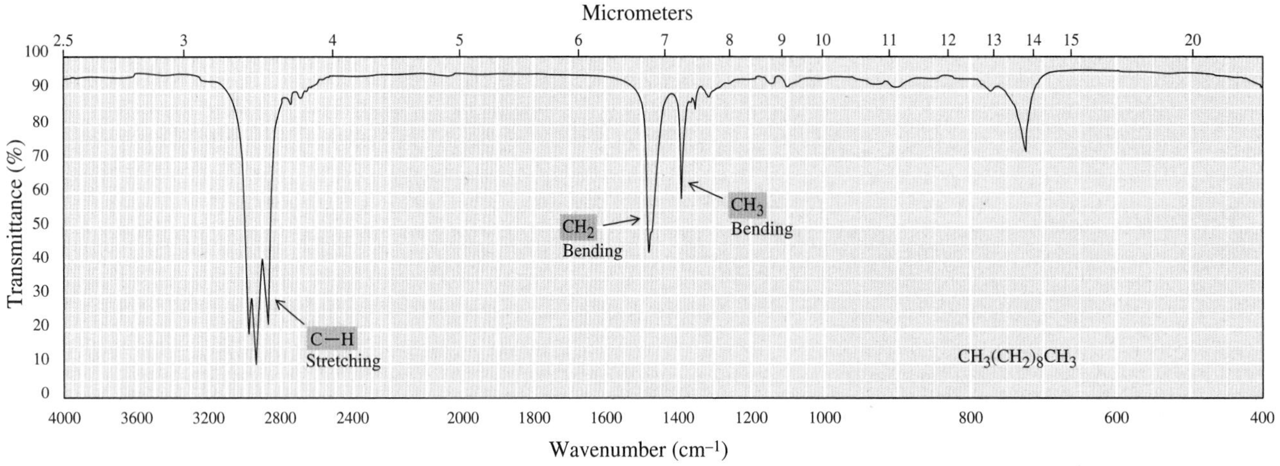

FIGURE 12.4 Infrared spectrum of decane.

B. Alkenes

An easily recognized alkene absorption is the vinylic C—H stretching slightly to the left of 3000 cm^{-1} (i.e., higher frequency). Also characteristic of alkenes is C═C stretching at 1600–1680 cm^{-1}. This vibration, however, is often weak and difficult to observe: the more symmetrical the alkene is, the weaker the absorption will be. Both vinylic C—H stretching and C═C stretching can be seen in the infrared spectrum of cyclohexene (Figure 12.5). Also visible are the aliphatic C—H stretching near 2900 cm^{-1} and methylene bending near 1475 cm^{-1}.

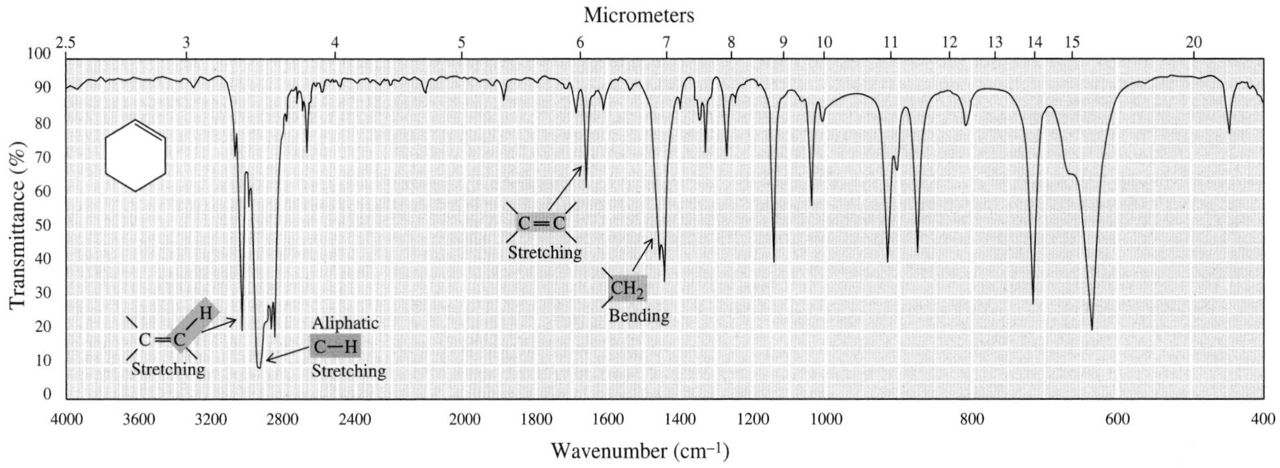

FIGURE 12.5 Infrared spectrum of cyclohexene.

C. Alkynes

Alkyne ≡C—H stretching occurs near 3300 cm^{-1}, at higher frequency than for either alkyl —C—H or vinylic ═C—H stretching. This peak is usually sharp and strong. The (sp-1s) C—H bond is unusually strong and therefore has a higher force constant than alkene (sp^2-1s) C—H bonds, which in turn absorb at higher frequency than the even weaker alkane (sp^3-1s) C—H bonds. Recall that the greater the percent s character of an atom's hybridization, the more tightly held the electrons and, accordingly, the stronger (and shorter) the bonds to atoms such as H. Therefore, the sp-1s C—H bonds (50% s character on carbon) are strongest, and the sp^3-1s C—H bonds (25% s character on carbon) are weakest, with the sp^2-1s C—H bonds in between. Also observed in terminal alkynes is absorption near 2150 cm^{-1} owing to C≡C stretching. Both of these peaks can be seen in the infrared spectrum of 1-octyne

(Figure 12.6). For internal alkynes, the C≡C stretching absorption is often very weak or completely absent (in symmetric alkynes) because stretching of this bond results in little or no change in the bond dipole moment (Section 12.3B).

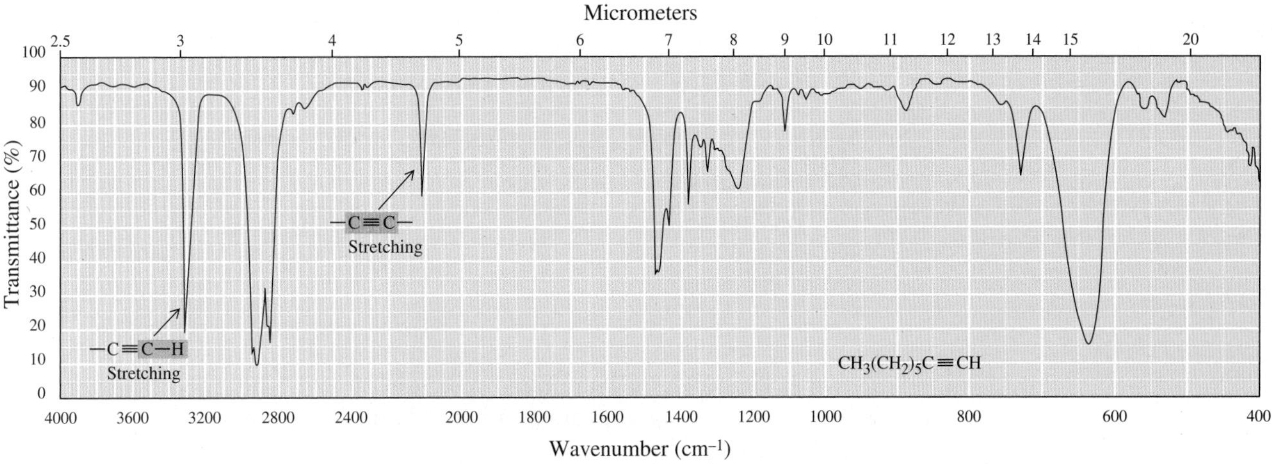

FIGURE 12.6 Infrared spectrum of 1-octyne.

D. Arenes (Benzene and Its Derivatives)

Aromatic rings show a medium to weak peak in the C—H stretching region at approximately 3030 cm^{-1} characteristic of (sp^2-1s) =C—H bonds. In addition, aromatic rings show strong absorption in the region 690–900 cm^{-1} as a result of out-of-plane C—H bending. Finally, these compounds show several absorptions owing to C=C stretching between 1450 and 1600 cm^{-1}. Actually, these are complex vibrational modes of the entire ring. Some modes involve all atoms moving in and out (breathing), whereas others involve some atoms moving in and others moving out. The intensities of these peaks can vary depending on the symmetry of ring substitution patterns. Each of these characteristic absorption patterns can be seen in the infrared spectrum of toluene (Figure 12.7).

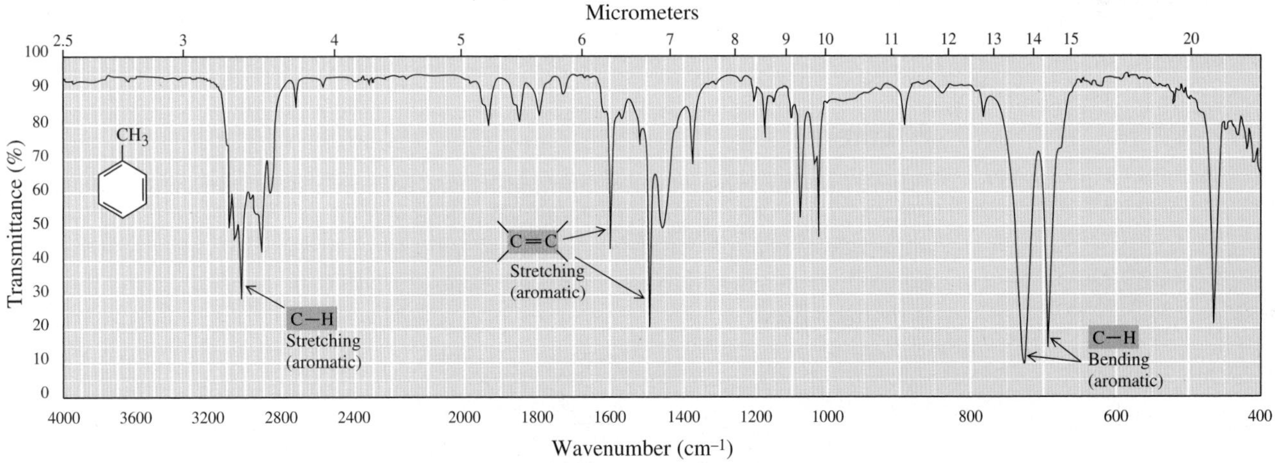

FIGURE 12.7 Infrared spectrum of toluene.

E. Alcohols

Both the position of the O—H stretching absorption and its intensity depend on the extent of hydrogen bonding. Under conditions where there is extensive hydrogen bonding between alcohol molecules (in pure alcohol or in concentrated solutions of the alcohol), the O—H stretching absorption occurs as a broad peak at 3200–3500 cm^{-1}. The variety of hydrogen-bonded states in different molecules leads to this broadening.

The "free" O—H stretch near 3650 cm^{-1} is seen only in very dilute solution in non-hydrogen-bonding solvents. The C—O stretching absorption appears in the range 1000–1250 cm^{-1} (Table 12.6).

Table 12.6 Infrared Absorptions of Alcohols

Bond	Frequency (cm^{-1})	Intensity
O—H (free)	3600–3650	Weak
O—H (hydrogen bonded)	3200–3500	Medium, broad
C—O	1000–1250	Medium

Shown in Figure 12.8 is an infrared spectrum of neat 1-hexanol. The hydrogen-bonded O—H stretching appears as a broad band of strong intensity centered at 3340 cm^{-1}. The C—O stretching appears at 1058 cm^{-1}, a value characteristic of primary alcohols.

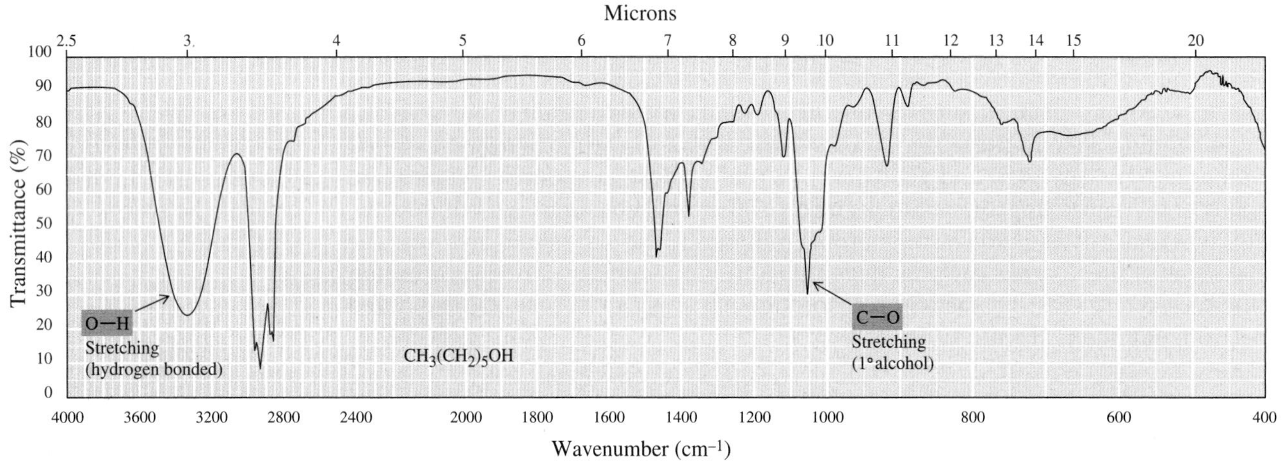

FIGURE 12.8 Infrared spectrum of 1-hexanol (neat, salt plates).

F. Ethers

Ethers have an oxygen atom bonded to two carbon atoms. Either or both of the carbon atoms may be sp^3 hybridized, sp^2 hybridized, or sp hybridized. In the simplest ether, dimethyl ether, both carbons are sp^3 hybridized. In diphenyl ether, both carbons are sp^2 hybridized, and in ethyl vinyl ether, one carbon is sp^3 hybridized and the other is sp^2 hybridized.

$CH_3—O—CH_3$

Methoxymethane (Dimethyl ether)

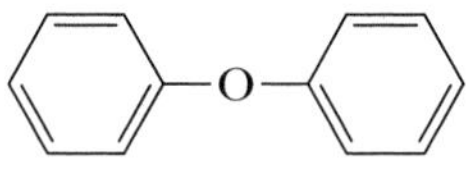

Diphenyl ether

$CH_3CH_2—O—CH{=}CH_2$

Ethoxyethene (Ethyl vinyl ether)

The C—O stretching absorptions of ethers are similar to those observed in alcohols. Dialkyl ethers typically show a single absorption in the region between 1000 and 1250 cm^{-1} as can be seen in the infrared spectrum of dibutyl ether (Figure 12.9).

Aromatic ethers (compounds in which the ether oxygen is bonded to one or two benzene rings) and vinyl ethers (compounds in which the ether oxygen is bonded to one or more sp^2 hybridized carbons of a C=C bond) typically show two

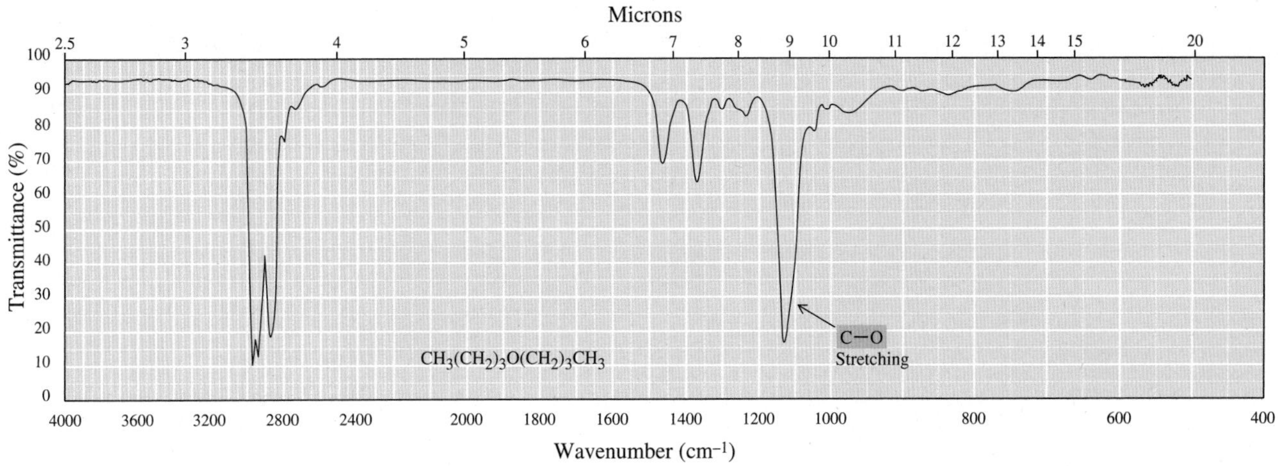

FIGURE 12.9 Infrared spectrum of dibutyl ether.

C—O stretching vibrations, one at either end of the range for C—O stretching. Anisole (Figure 12.10), for example, shows C—O stretching vibrations at 1050 cm^{-1} (*sp*^{3} C—O) and 1250 cm^{-1} (*sp*^{2} C—O). Ethers in which one of the bonds is attached to an *sp*^{2} hybridized carbon typically also have a band in the region between 1200 and 1250 cm^{-1}.

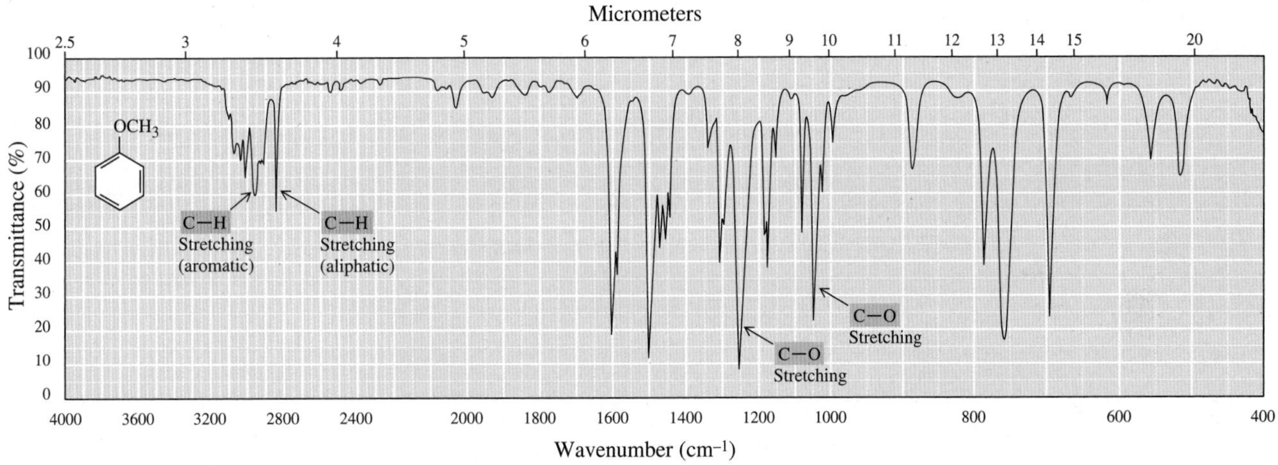

FIGURE 12.10 Infrared spectrum of anisole.

The presence or absence of O—H stretching at 3200 to 3500 cm^{-1} for hydrogen-bonded O—H can be used to distinguish between an ether and an isomeric alcohol. A C—O stretching absorption is also present in esters (see Section 12.4J). In this case, the presence or absence of C═O stretching can be used to distinguish between an ether and an ester.

G. Amines

The most important and readily observed infrared absorptions of primary and secondary amines are the result of N—H stretching vibrations that appear in the region 3300–3500 cm^{-1}. Like O—H bonds, N—H bonds become broader and shift to longer wavelength when they take part in hydrogen bonding. Primary amines have two bands in this region: one caused by symmetric stretching and the other by asymmetric stretching. The two N—H stretching absorptions characteristic of a primary amine can be seen in the IR spectrum of 1-butanamine (Figure 12.11). Secondary amines give only one absorption in this region. Tertiary amines have no N—H bonds and therefore are transparent in this region of the infrared spectrum.

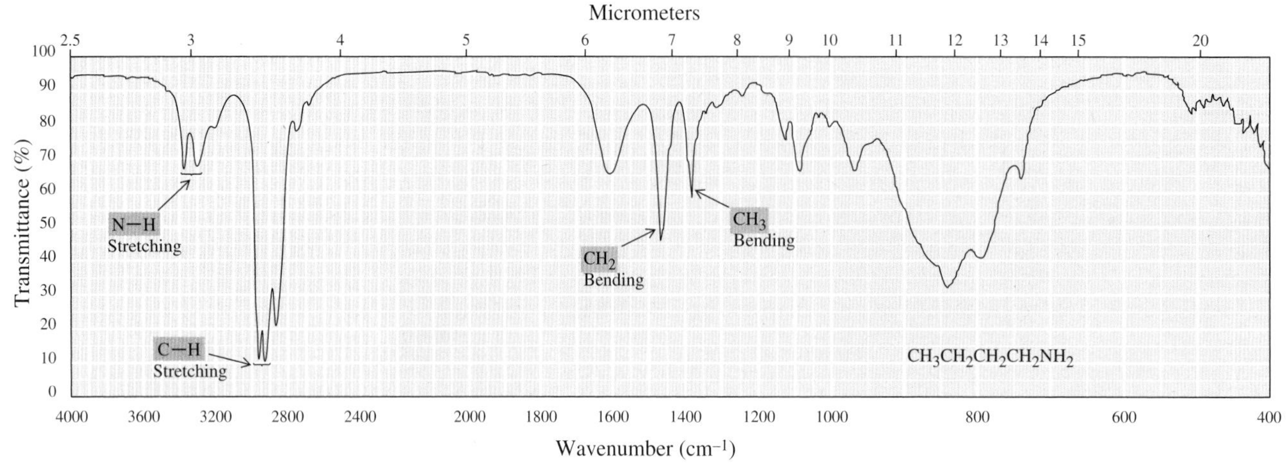
Micrometers
Transmittance (%)
Wavenumber (cm-1)
N−H
Stretching
C−H
Stretching
CH2
Bending
CH3
Bending
CH3CH2CH2CH2NH2

FIGURE 12.11 Infrared spectrum of 1-butanamine, a primary amine.

H. Aldehydes and Ketones

Aldehydes and ketones show characteristic strong infrared absorption between 1630 and 1820 cm^{-1} associated with the stretching vibration of the carbon-oxygen double bond. The stretching vibration for the carbonyl group of menthone occurs at 1705 cm^{-1} (Figure 12.12). The large difference in electronegativity between carbon and oxygen, leading to the correspondingly large dipole moment of the carbonyl group, is the reason for the strong intensity of this IR absorption.

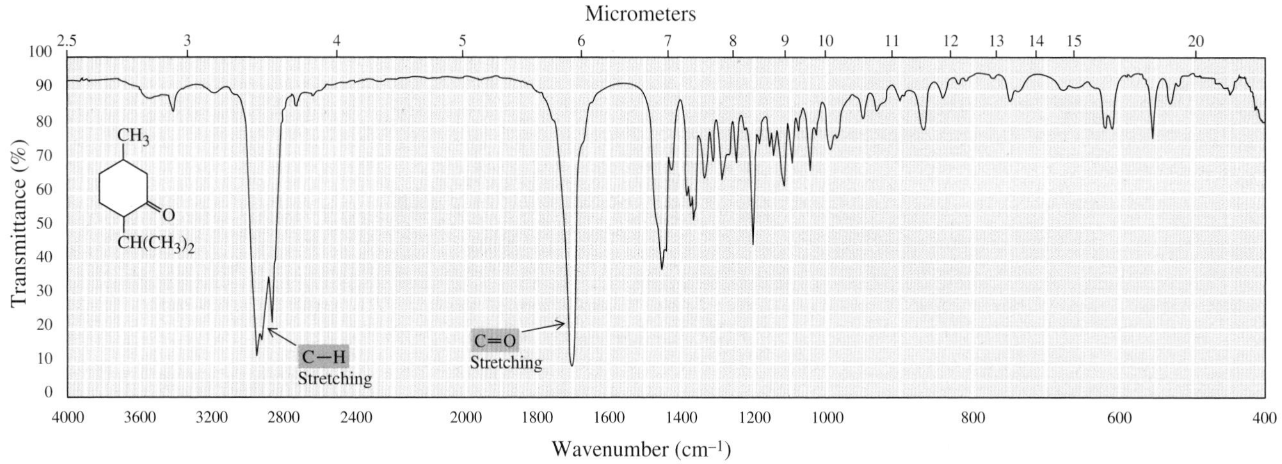
Micrometers
Transmittance (%)
Wavenumber (cm-1)
CH3
O
CH(CH3)2
C−H
Stretching
C=O
Stretching

FIGURE 12.12 Infrared spectrum of menthone.

Because few other bond vibrations absorb energy between 1630 and 1820 cm^{-1}, absorption in this region of the spectrum is a reliable means for confirming the presence of a carbonyl group. Because several different functional groups contain a carbonyl group, it is often not possible to tell from absorption in this region alone whether the carbonyl-containing compound is an aldehyde, a ketone, a carboxylic acid (Section 12.4I), an ester (Section 12.4J), an amide (Section 12.4J), or an anhydride (Section 12.4J). However, other absorptions, such as the C—O stretch in esters, can help distinguish these groups. Aldehydes frequently have a weak but very distinctive absorption at 2720 cm^{-1} caused by the stretching of the C—H of the CHO group.

The position of the C═O stretching vibration is quite sensitive to the molecular environment of the carbonyl group, as illustrated by comparing these cycloalkanones. Cyclohexanone, which has very little angle strain, shows absorption at 1715 cm^{-1}. As ring size decreases and angle strain increases, the C═O absorption shifts to a higher frequency as shown in the following series.

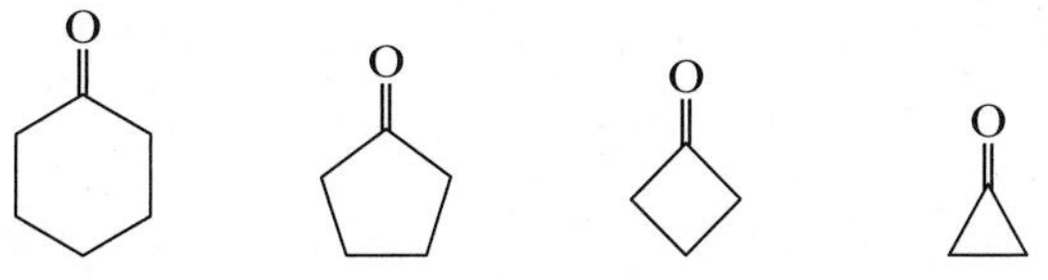

1715 cm^{-1} 1745 cm^{-1} 1780 cm^{-1} 1850 cm^{-1}

The presence of an adjacent carbon-carbon double bond or benzene ring in **conjugation** with the carbonyl group results in a shift of the C═O absorption to a lower frequency, as seen by comparing the carbonyl stretching frequencies of the following molecules. Conjugation occurs when the π electrons of adjacent π bonds interact with each other due to overlap of the unhybridized $2p$ orbitals and is a consequence of the special properties of extended π systems (Chapter 20). Conjugation also moves the C═C stretch to the right (lower frequency) and increases its intensity.

Conjugation

A situation in which two multiple bonds are separated by a single bond. Alternatively, a series of overlapping $2p$ orbitals. 1,3-Butadiene, for example, is a conjugated diene, and 3-butene-2-one is a conjugated ketone.

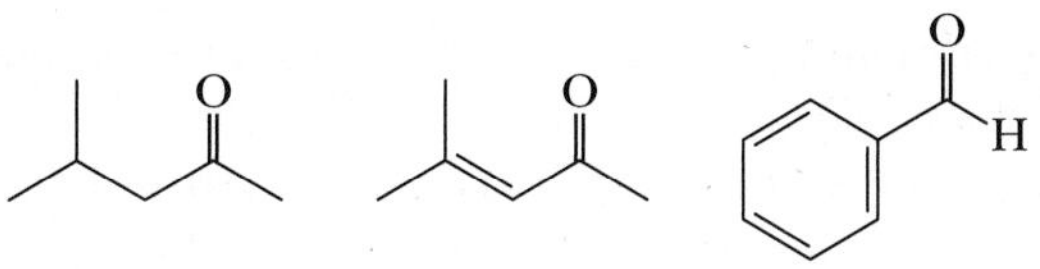

1717 cm^{-1} 1690 cm^{-1} 1700 cm^{-1}

I. Carboxylic Acids

A carboxyl group gives rise to two characteristic absorptions in an infrared spectrum. One of these occurs in the region 1700–1725 cm^{-1} and is associated with the stretching vibration of the carbonyl group. This range of absorption is essentially the same as that for the carbonyl group of aldehydes and ketones, but it is usually broader in the case of the carboxyl carbonyl because of intermolecular hydrogen bonding with the carboxyl O—H of another molecule. The other infrared absorption characteristic of a carboxyl group is a peak between 2500 and 3300 cm^{-1} owing to the stretching vibration of the O—H group, which often overlaps the C—H stretching absorptions. The O—H absorption is generally very broad as a result of hydrogen bonding between molecules of the carboxylic acid. Both C═O and O—H stretches can be seen in the infrared spectrum of pentanoic acid in Figure 12.13.

FIGURE 12.13 Infrared spectrum of pentanoic acid.

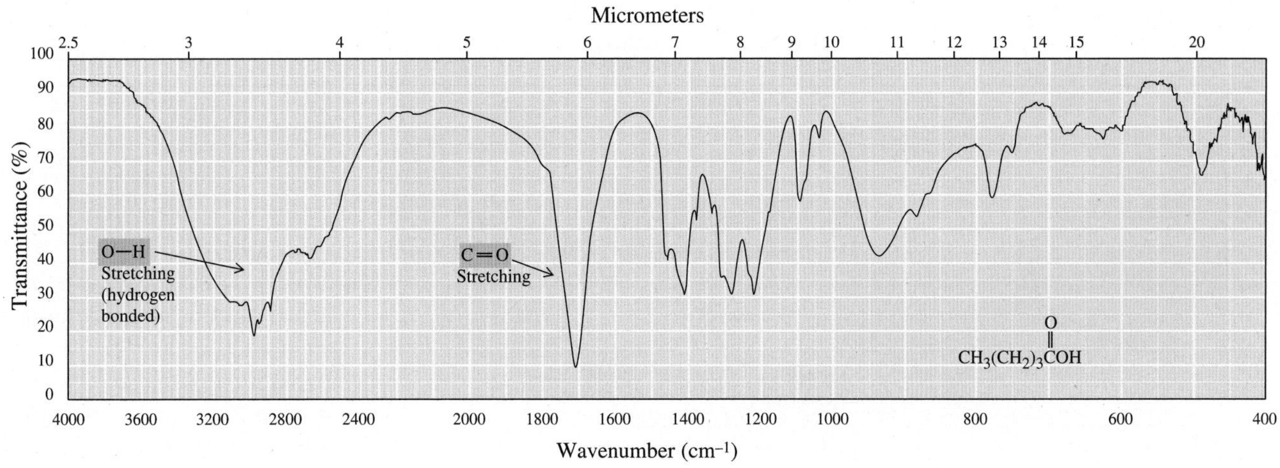

J. Derivatives of Carboxylic Acids

In Chapter 18, you will learn about several important derivatives of carboxylic acids. The most important are amides, esters, anhydrides, and nitriles.

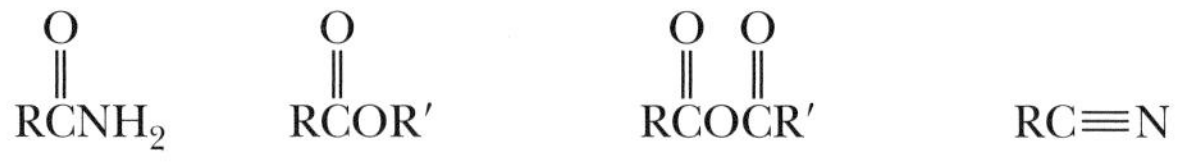

An amide An ester An acid anhydride A nitrile

The functional group of a carboxylic ester, most commonly referred to as simply an ester, is a carbonyl group bonded to an —OR group.

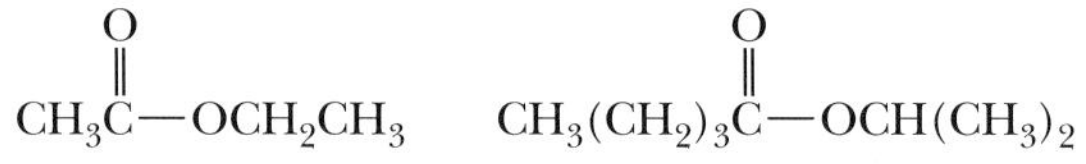

Ethyl ethanoate (Ethyl acetate) **Isopropyl pentanoate**

Note that one of the carbons of the C—O—C group is sp^2 hybridized and the other is sp^3 hybridized.

Esters display strong C═O stretching absorption in the region between 1735 and 1800 cm^{-1}. As in ketones, this band is shifted to higher frequency in smaller rings and to lower frequency by adjacent double bonds. In addition, esters also display strong C—O stretching absorptions in the region 1000–1100 cm^{-1} for the sp^3 C—O stretch and 1200–1250 cm^{-1} for the sp^2 C—O stretch (Figure 12.14). Ethers in which one of the carbons bonded to oxygen is sp^2 hybridized also show this band.

The most important infrared absorption of carboxylic acids and their derivatives is the result of the C═O stretching vibration. Infrared spectroscopic data for these derivatives and the other carbonyl-containing compounds are summarized in Table 12.7.

The carbonyl stretching of amides occurs at 1630–1680 cm^{-1}, at a lower frequency than for other carbonyl compounds. This observation can be explained by a weakened carbonyl C═O bond as described through the three most important resonance-contributing structures of amides, two of which indicate single bond character (see Example 1.19 and "Connections to Biological Chemistry: The Unique Structure of Amide Bonds" in Section 18.1). Primary and secondary amides show N—H stretching in the region 3200–3400 cm^{-1}; primary amides ($RCONH_2$) usually show two N—H absorptions, whereas secondary amides (RCONHR) show only a single N—H absorption (Figure 12.15).

FIGURE 12.14 Infrared spectrum of ethyl butanoate.

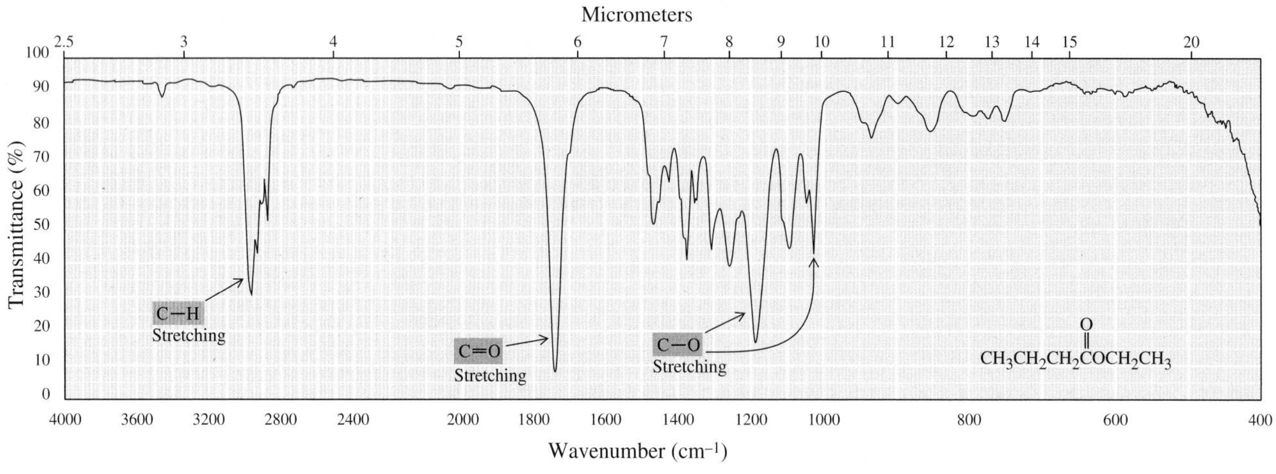
Micrometers
Transmittance (%)
Wavenumber (cm⁻¹)
C–H Stretching
C=O Stretching
C–O Stretching
CH3CH2CH2COCH2CH3

Table 12.7 Infrared Absorptions of Molecules Containing Carbonyl Groups

Carbonyl Group		Vibration	Frequency (cm^{-1})	Intensity
$R\overset{\overset{O}{\parallel}}{C}R'$	Ketones (Section 16.1)			
	C═O	Stretching	1630–1820	Strong
$R\overset{\overset{O}{\parallel}}{C}H$	Aldehydes (Section 16.1)			
	C═O	Stretching	1630–1820	Strong
	C—H	Stretching	2720	Weak
$R\overset{\overset{O}{\parallel}}{C}OH$	Carboxylic Acids (Section 17.1)			
	C═O	Stretching	1700–1725	Strong
	O—H	Stretching	2500–3300	Strong (broad)
$R\overset{\overset{O}{\parallel}}{C}NH_2$	Amides (Section 18.1D)			
	C═O	Stretching	1630–1680	Strong
	N—H	Stretching	3200, 3400	Medium
	(1° amides have two N—H stretches)			
	(2° amides have one N—H stretch)			
$R\overset{\overset{O}{\parallel}}{C}OR'$	Carboxylic Esters (Section 18.1C)			
	C═O	Stretching	1735–1800	Strong
	*sp*² C—O	Stretching	1200–1250	Strong
	*sp*³ C—O	Stretching	1000–1100	Strong
$R\overset{\overset{O}{\parallel}}{C}O\overset{\overset{O}{\parallel}}{C}R$	Acid Anhydrides (Section 18.1B)			
	C═O	Stretching	1740–1760 and 1800–1850	Strong
	C—O	Stretching	900–1300	Strong
RC≡N	Nitriles (Section 18.1E)			
	C≡N	Stretching	2200–2250	Medium

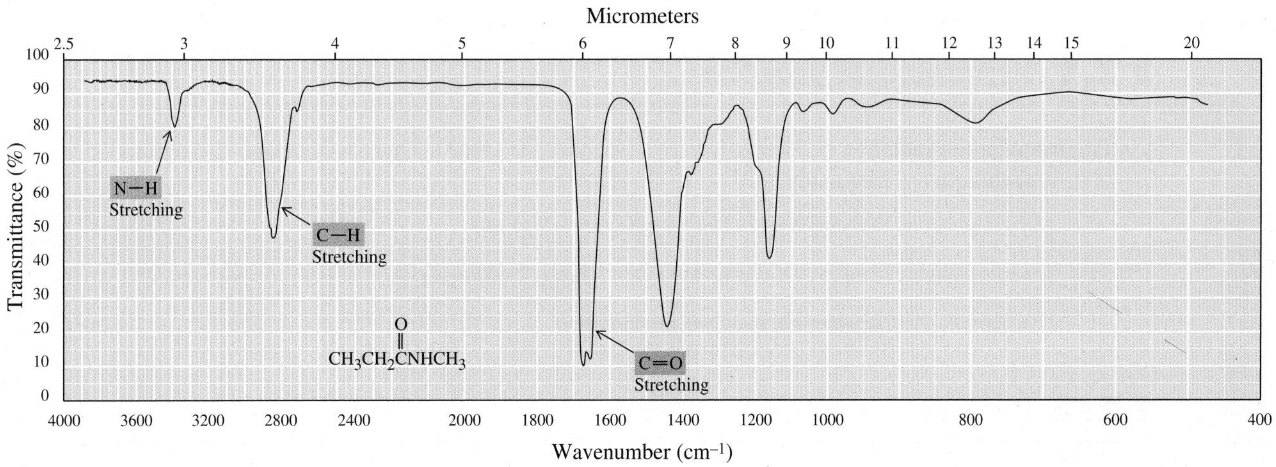

FIGURE 12.15 Infrared spectrum of *N*-methylpropanamide (a secondary amide).

Anhydrides have two carbonyl stretching absorptions, one near 1760 cm^{-1} and the other near 1810 cm^{-1}, the result of symmetric and unsymmetric stretching vibrations of the entire anhydride functional group. In addition, anhydrides display strong C—O stretching absorption in the region 900–1300 cm^{-1}. Nitriles can be distinguished by a medium C≡N stretching absorption at 2200–2250 cm^{-1}.

12.5 Solving Infrared Spectral Problems

The following steps provide a systematic approach to solving IR problems.

Step 1: Check the region around 3000 cm^{-1}. Absorption in this region is caused by C—H stretching. Absorption is generally to the right of 3000 cm^{-1} for sp^3 C—H stretching of alkanes and to the left for the sp^2 C—H stretching of alkenes and aromatic rings.

Step 2: Is there a strong, broad band in the region of 3500 cm^{-1}? If so, the molecule contains an —OH group either of an alcohol or a carboxylic acid. If there is no absorption around 1700 cm^{-1} (caused by a carbonyl group), the functional group is an —OH group of an alcohol. If there is a peak around 1700 cm^{-1}, the functional group may be a carboxyl group. One or two peaks in the 3500 cm^{-1} region at somewhat lower frequency than for —OH may indicate a 2° or 1° amine, respectively.

Step 3: Is there a sharp peak in the region 1630–1820 cm^{-1}? If so, a C=O group is present. This peak will probably be the strongest peak in the spectrum. If no peak is present in this region, no C=O is present. The type of carbonyl-containing functional group can often be determined by looking for the presence or absence of an aldehyde C—H stretch at 2720 cm^{-1}, a carboxyl O—H stretch around 3500 cm^{-1}, and a C—O stretch around 1000–1250 cm^{-1}.

Finally, here is a note of caution on interpreting infrared spectra. Even though it is possible to obtain a great deal of valuable information about a compound from its infrared spectrum, it is often very difficult to determine its structure based solely on this information; so other types of information should be sought.

Study Guide

12.1 Electromagnetic Radiation

- **Electromagnetic radiation** behaves as a wave traveling at the speed of light and is described in terms of its **wavelength** and **frequency**.
  - *The wavelengths of visible light fall in the range 400–700 nm.*
  - *The wavelengths of infrared radiation fall in the range 0.7–300 μm.*
  - *Wavelength and frequency are inversely proportional.*

12.2 Molecular Spectroscopy

- An atom or a molecule can be made to undergo a transition from energy state E_1 to a higher energy state E_2 by irradiating the atom or molecule with electromagnetic radiation corresponding to the energy difference between states E_1 and E_2.

- **Molecular spectroscopy** is the experimental process of measuring which frequencies of radiation are absorbed or emitted by a particular substance and then attempting to correlate patterns of energy absorption or emission with details of molecular structure.

12.3 Infrared Spectroscopy

- **Infrared spectroscopy** provides a direct method for detecting certain functional groups in a molecule.
  - *In infrared spectroscopy, absorption of energy correlates to characteristic stretching and bending vibrations of functional groups in molecules.*
  - *Radiation in the vibrational infrared region is commonly referred to by its frequency in* ***wavenumbers****, or the number of waves per centimeter (cm^{-1}; read: reciprocal centimeters).*
  - *The vibrational infrared spectrum extends from 4000 to 400 cm^{-1}.*
- To be **infrared active**, a bond must be polar and its vibration must cause a substantial change in the dipole moment.
  - *In general, the larger the dipole moment of the bond, the stronger the absorption seen in an infrared spectrum.*
  - *There are $3n-6$ allowed fundamental vibrations for a nonlinear molecule containing n atoms.*
  - *The simplest vibrations that give rise to absorption of infrared radiation are stretching and bending vibrations. Stretching may be symmetrical or asymmetrical.*
  - *The frequency of vibration for an infrared-active bond can be estimated using* ***Hooke's law*** *for the vibration of a simple harmonic oscillator such as a vibrating spring. Hooke's law predicts that the frequency of vibration increases when (1) the bond strength increases and (2) the reduced mass of the vibrating system decreases.*
- A **correlation table** is a list of the absorption patterns of functional groups.
  - *The intensity of a peak is referred to as strong (s), medium (m), or weak (w).*
  - *Bending and stretching vibrations characteristic of most functional groups appear in the region from 3500 to 1500 cm^{-1}.*
  - *The region 1500 to 400 cm^{-1} is called the* ***fingerprint region****, and little useful information concerning functional groups is found in this region.*

P 12.1–12.4, 12.8, 12.9

▶ Problem 12.1 and Table 12.4

12.4 Interpreting Infrared Spectra

- **Alkanes** give infrared spectra that are simple, with few useful peaks, although C—H stretching (2850–3000 cm^{-1}) and bending (1375–1450, 1450–1475 cm^{-1}) peaks are seen.
- The C—H stretch of **alkenes** (3000–3100 cm^{-1}) and terminal **alkynes** (3300 cm^{-1}) are useful for identification of these functional groups.
  - *The C=C stretches (1600–1680 cm^{-1}) for alkenes and the C≡C stretches (2100–2250 cm^{-1}) for alkynes are useful if they are strong enough to see.*
- For **arenes**, C—H stretching (3030 cm^{-1}),—C=C— stretching (1450–1600 cm^{-1}), and C—H bending (690–900 cm^{-1}) are useful peaks.

- For **alcohols**, the O—H (3200–3650 cm^{-1}, depending on hydrogen bonding) and C—O (1000–1250 cm^{-1}) stretches are most useful, while for **ethers**, only the C—O stretches (1000–1250 cm^{-1}) are seen.
- For **amines**, the N—H stretch (3300–3500 cm^{-1}) is useful, with primary amines having two peaks and secondary amines only one.
- For carbonyl species, the strong C═O stretch (1600–1850 cm^{-1}) is characteristic for **aldehydes** (1630–1820 cm^{-1}), **ketones** (1630–1820 cm^{-1}), **carboxylic acids** (1700–1725 cm^{-1}), **amides** (1630–1680 cm^{-1}), **carboxylic esters** (1735–1800 cm^{-1}), and **acid anhydrides** (1740–1760 cm^{-1}).
  - *The presence of other stretches, such as C—O stretches for carboxylic esters (1000–1100 cm^{-1} and 1200–1250 cm^{-1}) and anhydrides (900–1300 cm^{-1}), a C—H stretch (2720 cm^{-1}) for aldehydes, an O—H stretch (2500–3300 cm^{-1}) for carboxylic acids, and an N—H stretch (3200 and 3400 cm^{-1}) for amides, is also used to distinguish these functional groups.*
- For **nitriles**, the C≡N stretch (2200–2250 cm^{-1}) is useful.

P 12.10, 12.11

12.5 Solving Infrared Spectral Problems

- A systematic approach to solving IR problems involves:
  - *Analyzing the 3000 cm^{-1} region to see if there are alkane C—H stretches (to the right of 3000 cm^{-1} or alkene/arene to the left of 3000 cm^{-1}).*
  - *Checking for characteristic broad O—H or N—H peaks (3500 cm^{-1}).*
  - *Checking for strong carbonyl peaks (1630–1820 cm^{-1}), then the C—O (1000–1250 cm^{-1}) or other peaks that might accompany the carbonyl in certain functional groups.*

P 12.5-12.7

Problems

12.5 Following are infrared spectra of methylenecyclopentane and 2,3-dimethyl-2-butene. Assign each compound its correct spectrum.

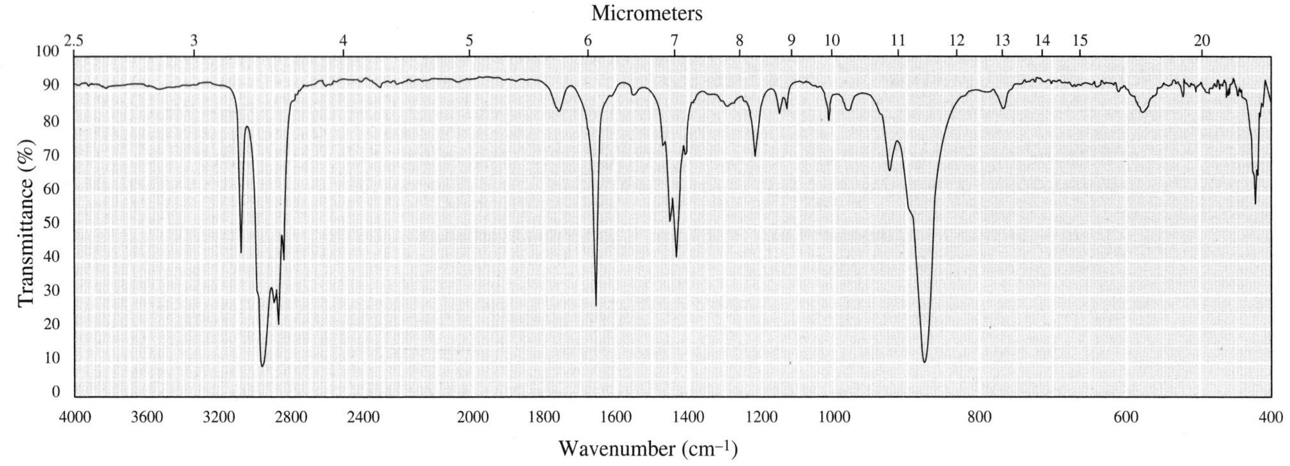

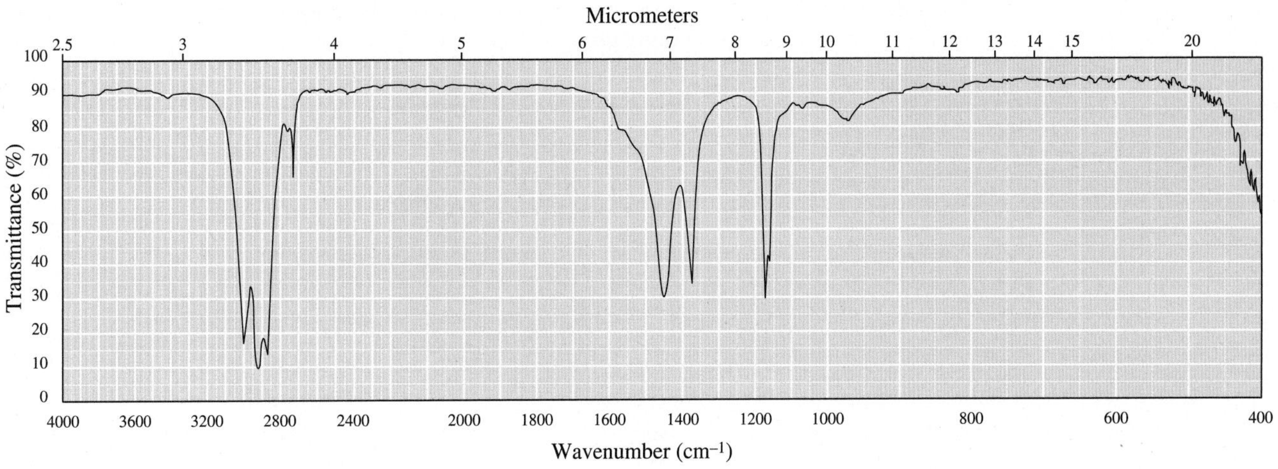

12.6 Following are infrared spectra of nonane and 1-hexanol. Assign each compound its correct spectrum.

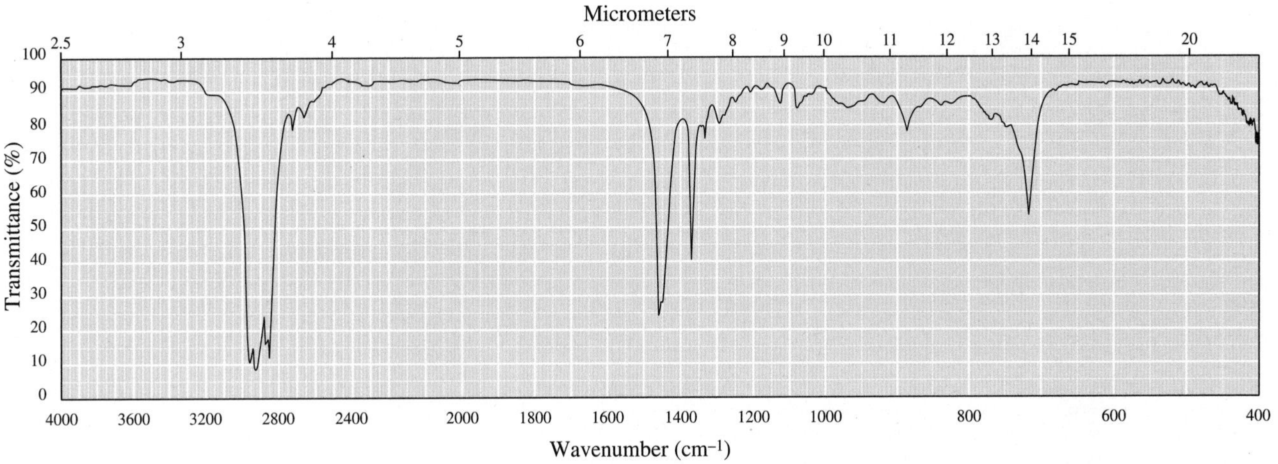

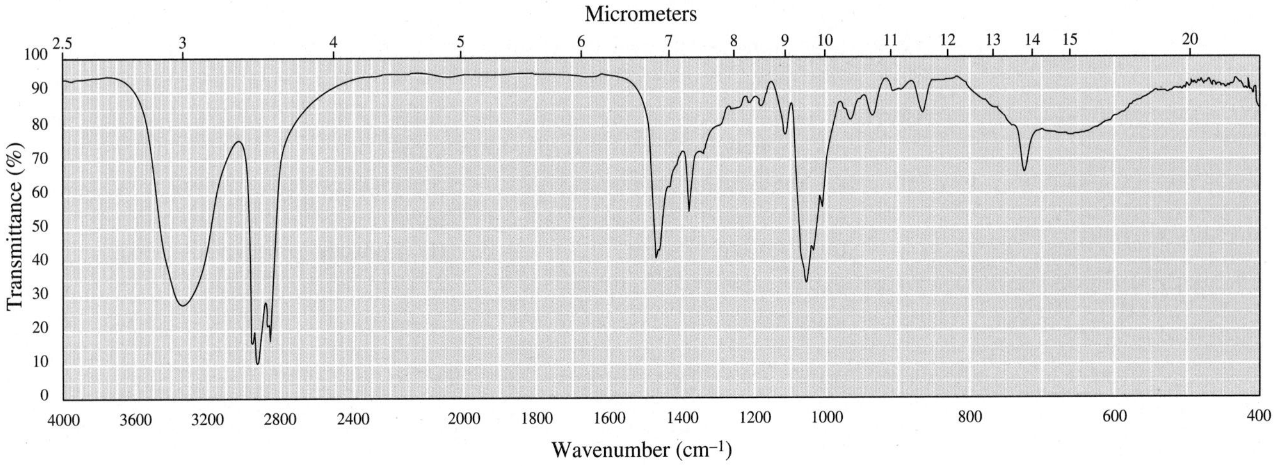

12.7 Following are infrared spectra of 2-methyl-1-butanol and *tert*-butyl methyl ether. Assign each compound its correct spectrum.

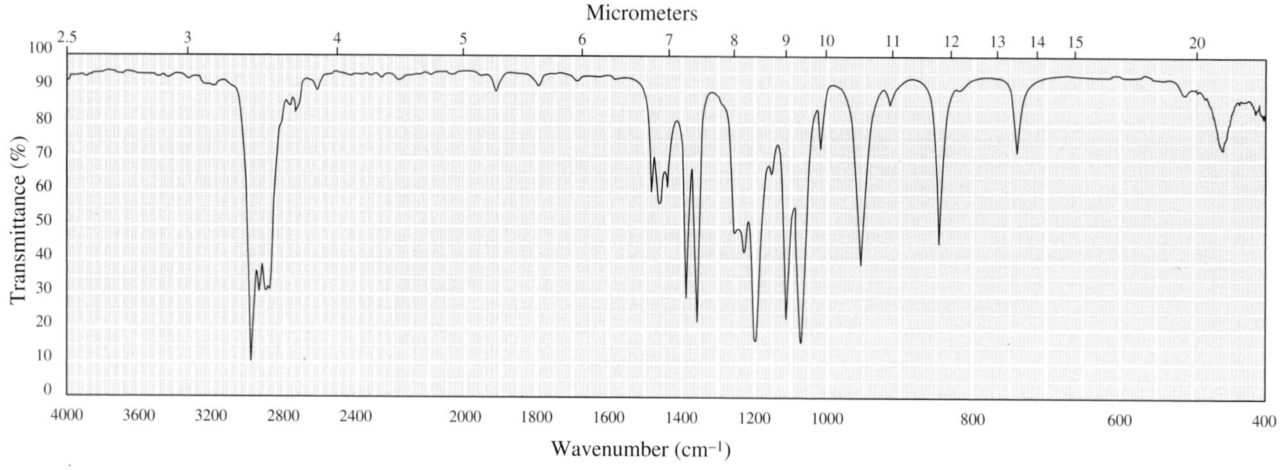

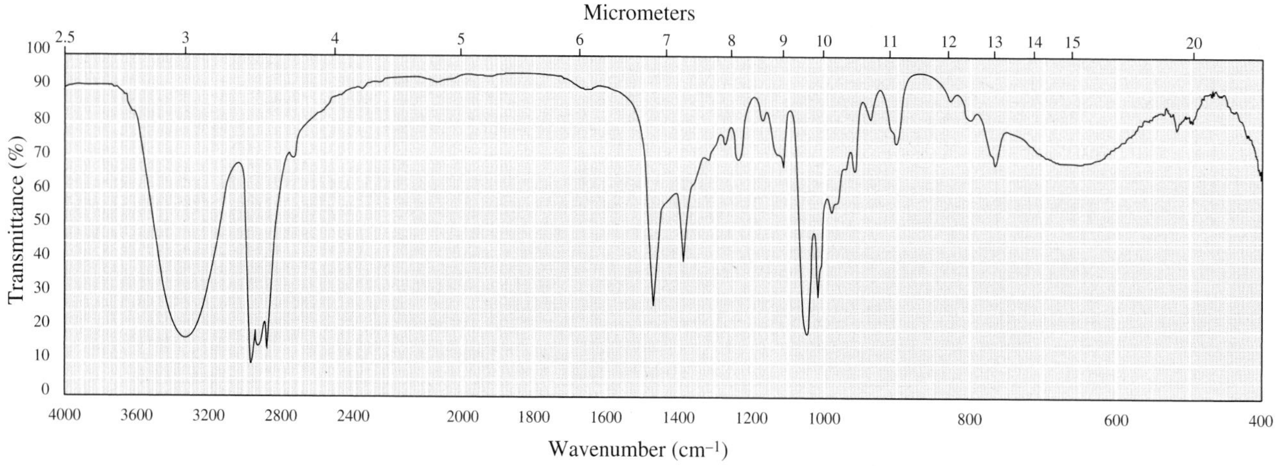

12.8 The IR C≡C stretching absorption in symmetrical alkynes is usually absent. Why?

12.9 Explain the fact that the C—O stretch in ethers and esters occurs at 1000–1100 cm^{-1} when the C is sp^3 hybridized, but at 1250 cm^{-1} when it is sp^2 hybridized.

12.10 A compound has strong infrared absorptions at the following frequencies. Suggest likely functional groups that may be present.

(a) 1735, 1250, and 1100 cm^{-1}
(b) 1745 cm^{-1} but not 1000–1250 cm^{-1}
(c) 1710 and 2500–3400 (broad) cm^{-1}
(d) A single band at about 3300 cm^{-1}
(e) 3600 and 1050 cm^{-1}
(f) 1100 cm^{-1} but not 3300–3650 cm^{-1}

12.11 Show how IR spectroscopy can be used to distinguish between the compounds in each set.

(a) 

and 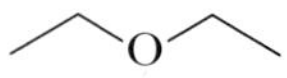

(b) 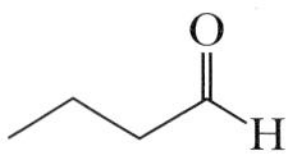

and 

(c) 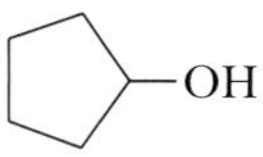

and 

(d) 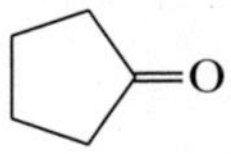 and 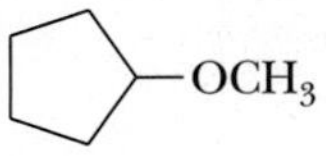

(e) 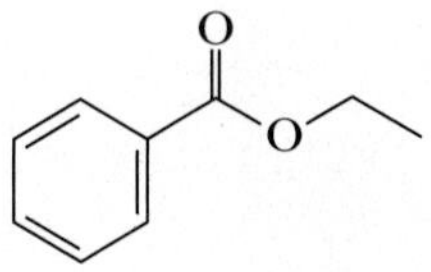 and 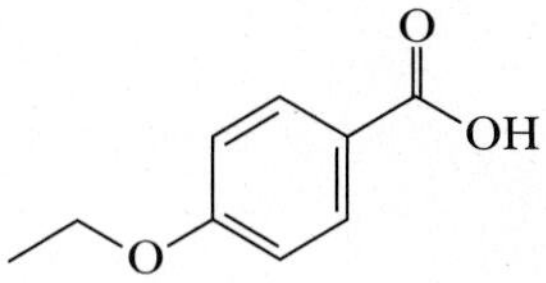

(f)  and 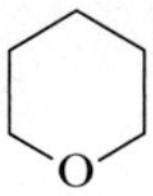

(g) 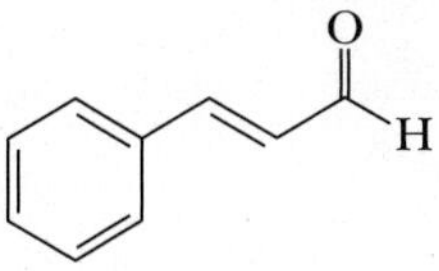 and 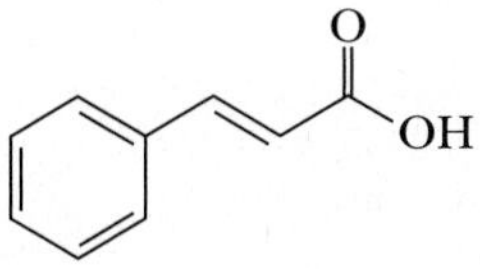

(h)  and 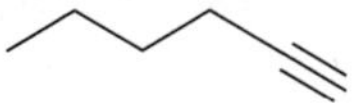

(i) 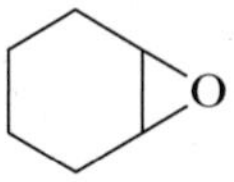 and 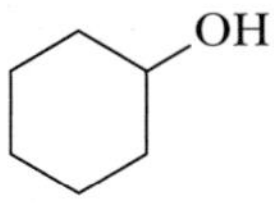

(j) 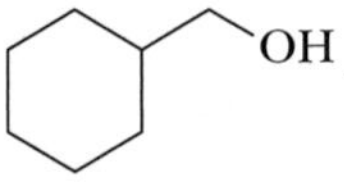 and 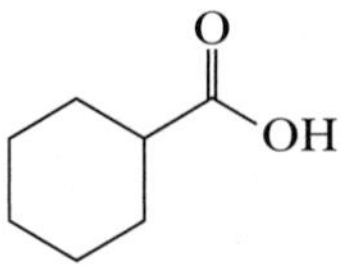

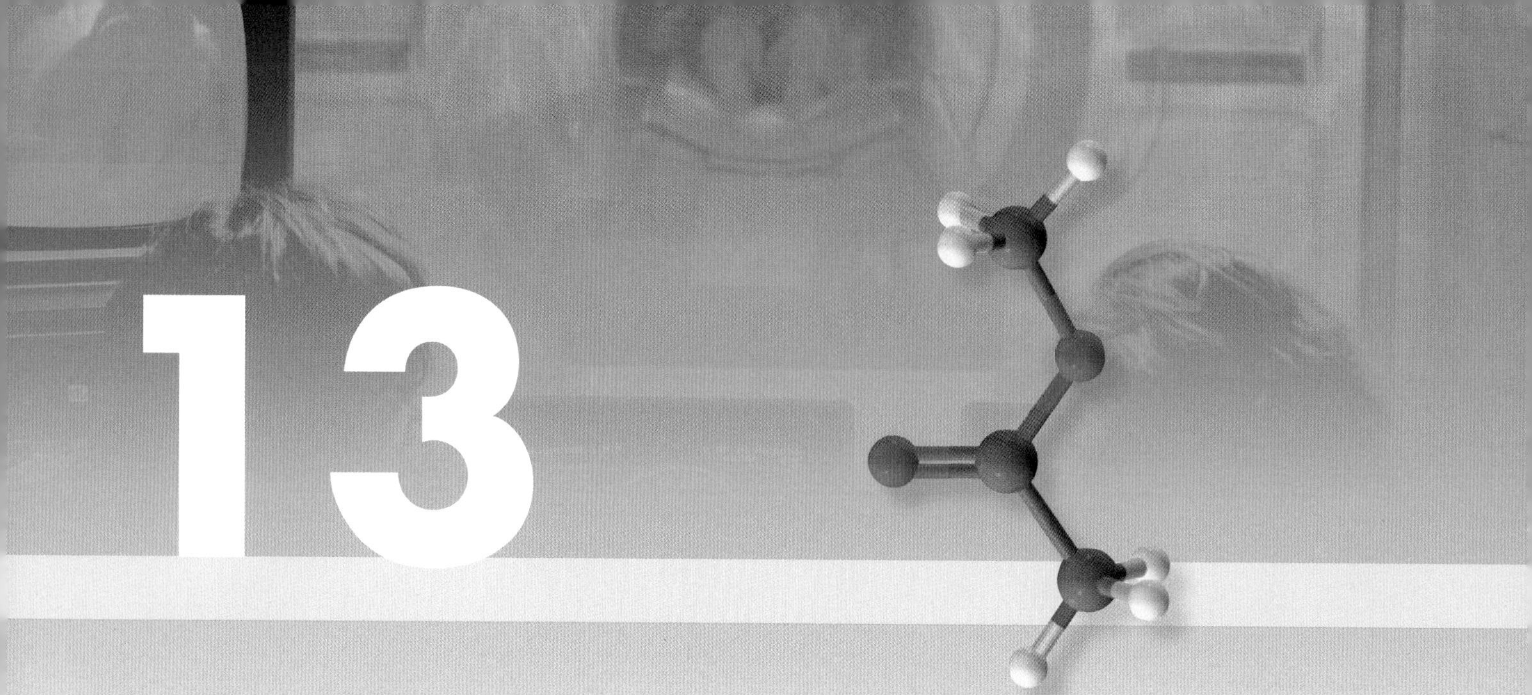

13 Nuclear Magnetic Resonance Spectroscopy

Magnetic resonance imaging is a useful medical diagnostic tool. **Inset:** a model of methyl acetate. For a ^{1}H-NMR spectrum of methyl acetate, see Figure 13.5. (Paul Shambroom/ Science Source)

Outline

13.1 Nuclear Spin States
13.2 Orientation of Nuclear Spins in an Applied Magnetic Field
13.3 Nuclear Magnetic "Resonance"
13.4 An NMR Spectrometer
13.5 Equivalent Hydrogens
13.6 Signal Areas
13.7 Chemical Shift
13.8 Signal Splitting and the $(n + 1)$ Rule
13.9 The Origins of Signal Splitting
13.10 Stereochemistry and Topicity
13.11 ^{13}C-NMR
13.12 Interpretation of NMR Spectra
HOW TO: Solve NMR Spectral Problems

In this chapter, we concentrate on absorption of radio-frequency radiation by nuclei and the resulting transitions between energy levels, better known as **nuclear magnetic resonance (NMR) spectroscopy**. Felix Bloch and Edward Purcell, both of the United States, first detected the phenomenon of nuclear magnetic resonance in 1946. They shared the 1952 Nobel Prize for Physics. Nuclear magnetic resonance (NMR) spectroscopy was developed in the late 1950s, and within a decade, it became the single most important technique available to chemists for the determination of molecular structure. Nuclear magnetic resonance spectroscopy gives us information about the number and types of atoms in a molecule (e.g., about hydrogens using ^{1}H-NMR spectroscopy and about carbons using ^{13}C-NMR spectroscopy).

It can also give us substantial information about the connectivity of the atoms and in many cases can allow determination of the structure of a molecule with no additional information.

13.1 Nuclear Spin States

You are already familiar from general chemistry with the concepts that (1) an electron has a spin quantum number of $\frac{1}{2}$, with allowed values of $+\frac{1}{2}$ and $-\frac{1}{2}$, and that (2) a moving charge has an associated magnetic field. In effect, an electron behaves as if it were a tiny bar magnet that has a magnetic moment. The same effect holds for certain atomic nuclei.

Any atomic nucleus that has an odd mass number, an odd atomic number, or both also has a spin and a resulting nuclear magnetic moment. The allowed nuclear spin states are determined by the spin quantum number, I, of the nucleus. A nucleus with spin quantum number I has $2I + 1$ spin states. Our focus in this chapter is on nuclei of ^{1}H and ^{13}C, isotopes of the two elements most common to organic compounds. Each has a **nuclear spin quantum number** of $\frac{1}{2}$ and therefore has $2(\frac{1}{2}) + 1 = 2$ allowed spin states. Quantum numbers and allowed nuclear spin states for these nuclei and those of other elements common to organic compounds are shown in Table 13.1. Note that ^{12}C, ^{16}O, and ^{32}S each have a spin quantum number of zero and only one allowed nuclear spin state; these nuclei are inactive in NMR spectroscopy.

Table 13.1 Nuclear Spin Quantum Numbers and Allowed Nuclear Spin States for Selected Isotopes of Elements Common to Organic Compounds

Element	^{1}H	^{2}H	^{12}C	^{13}C	^{14}N	^{15}N	^{16}O	^{19}F	^{31}P	^{32}S
Nuclear spin quantum number (I)	$\frac{1}{2}$	1	0	$\frac{1}{2}$	1	$\frac{1}{2}$	0	$\frac{1}{2}$	$\frac{1}{2}$	0
Number of spin states	2	3	1	2	3	2	1	2	2	1

13.2 Orientation of Nuclear Spins in an Applied Magnetic Field

Within a sample containing ^{1}H and ^{13}C atoms, the orientations of the nuclear magnetic moments associated with their nuclear spins are completely random. When placed between the poles of a powerful magnet of field strength B_0, however, interactions between the nuclear spins and the applied magnetic field are quantized, with the result that only certain orientations of nuclear magnetic moments are allowed. For ^{1}H and ^{13}C nuclei, only two orientations are allowed, as illustrated in Figure 13.1. By convention, nuclei with spin designated as $+\frac{1}{2}$ are aligned with the applied magnetic field and are in the lower energy state; nuclei with spin designated as $-\frac{1}{2}$ are aligned against the applied magnetic field and are in the higher energy state.

▶ Note: the SI unit for magnetic field strength is the tesla (T). A unit still in common use, however, is the gauss (G). Values of T and G are related by the equation $1\ T = 10^4\ G$.

The most important NMR physical concept from the point of view of molecular structure determination is that **the difference in energy between nuclear spin states for a given nucleus is proportional to the strength of the magnetic field experienced by that nucleus (Figure 13.2)**. We will come back to this concept several

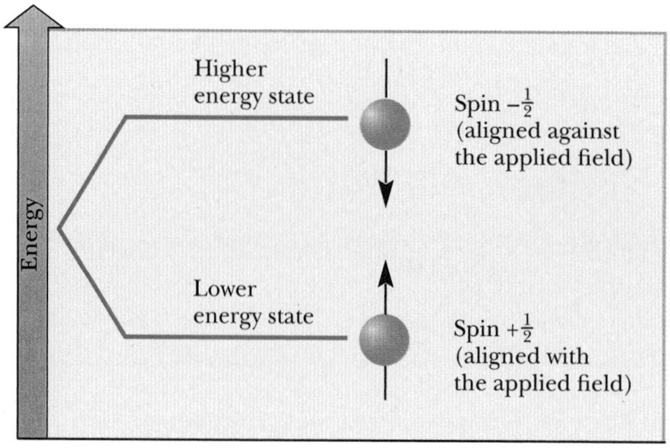

FIGURE 13.1 ^1H and ^{13}C nuclei with spin $+\frac{1}{2}$ are aligned with the applied magnetic field, B_0, and are in the lower spin energy state; those with spin $-\frac{1}{2}$ are aligned against the applied magnetic field and are in the higher spin energy state.

times in this chapter. At an applied field strength of 7.05 T, which is readily available with present-day superconducting electromagnets, the difference in energy between nuclear spin states for ^1H is approximately 0.120 J (0.0286 cal)/mol (corresponding to electromagnetic radiation of 300 MHz). At 7.05 T, the energy difference in nuclear spin states for ^{13}C nuclei is approximately 0.030 J (0.00715 cal)/mol (corresponding to radiation of 75 MHz). Advanced commercial instruments now operate at fields more than three times this value; the operating frequencies are proportional to the field. Sensitivities are more than proportionally higher.

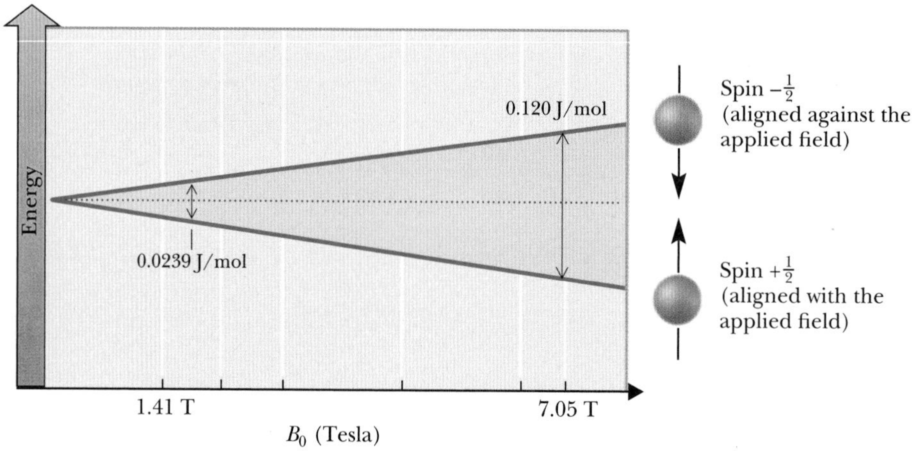

FIGURE 13.2 The energy difference between the allowed nuclear spin states increases linearly with applied field strength. Values shown here are for ^1H nuclei.

To put these values for nuclear spin energy levels in perspective, energies for transitions between vibrational energy levels observed in infrared (IR) spectroscopy are 8 to 63 kJ (2 to 15 kcal)/mol. Those between electronic energy levels in ultraviolet-visible spectroscopy (see Section 20.3) are 167 to 585 kJ (40 to 140 kcal)/mol. Nuclear transitions involve only small energies, on the order of a few hundredths of a calorie per mole.

Example 13.1 Spin-State Population

Calculate the ratio of nuclei in the higher spin state to those in the lower spin state, N_h/N_l, for ^1H at 25°C in an applied field strength of 7.05 T.

Solution

Use the equation given in Section 2.5B for the relationship between the difference in energy states and equilibrium constant. In this problem, this relationship has the form

$$\Delta G^0 = -RT\ln\frac{N_h}{N_l}$$

The difference in energy between the higher and lower nuclear spin states in an applied field of 7.05 T is approximately 0.120 J/mol, and the temperature is 25 + 273 = 298 K. Substituting these values in this equation gives

$$\ln\frac{N_h}{N_l} = \frac{-\Delta G^0}{RT} = \frac{-0.120\ \text{J}\cdot\text{mol}^{-1}}{8.314\ \text{J}\cdot\text{K}^{-1}\cdot\ \text{mol}^{-1} \times 298\ \text{K}} = -4.843 \times 10^{-5}$$

$$\frac{N_h}{N_l} = 0.9999516 = \frac{1.000000}{1.000048}$$

From this calculation, we determine that for every 1,000,000 hydrogen atoms in the higher energy state in this applied field, there are 1,000,048 in the lower energy state. The excess population of the lower energy state under these conditions is only 48 per million. What is important about this number is that the strength of an NMR signal is proportional to the population difference. As you will see, the greater this difference in populations, the stronger the signal because more spins are flipping.

Problem 13.1

Calculate the ratio of nuclei in the higher spin state to those in the lower spin state, N_h/N_l, for ^{13}C at 25°C in an applied field strength of 7.05 T. The difference in energy between the higher and lower nuclear spin states in this applied field is approximately 0.030 J (0.00715 cal)/mol.

13.3 Nuclear Magnetic "Resonance"

As we have seen, when nuclei with spin quantum number $\frac{1}{2}$ are placed in an applied magnetic field, a small majority of nuclear spins are aligned with the applied field in the lower energy state. When nuclei in the lower energy spin state are irradiated with a radio frequency of the appropriate energy, they absorb the energy and nuclear spins flip from the lower energy state to the higher energy state, the only other allowed spin state.

To visualize the mechanism by which a spinning nucleus absorbs energy and the meaning of resonance in this context, think of the nucleus as if it were really spinning. When an applied field of strength B_0 is turned on, the nucleus becomes aligned with the applied field in an allowed spin energy state. The nucleus then begins to **precess** as shown in Figure 13.3(a) and traces out a cone-shaped surface in much the same manner as a spinning top or gyroscope traces out a cone-shaped surface as it precesses in the earth's gravitational field. We can express the rate of precession as a frequency in hertz.

If the precessing nucleus is irradiated with electromagnetic radiation at exactly the precession frequency, then the two frequencies couple, energy is absorbed, and the nuclear spin "flips" from spin state $+\frac{1}{2}$ (with the applied field) to spin state $-\frac{1}{2}$ (against the applied field), as illustrated in Figure 13.3(b). For ^{1}H in an applied magnetic field of 7.05 T, the frequency of precession is approximately 300 MHz. For ^{13}C in the same field, it is approximately 75 MHz. **Resonance** in this context is the absorption of electromagnetic radiation by a precessing nucleus and the resulting flip of its nuclear

Resonance in NMR spectroscopy
The absorption of electromagnetic radiation by a precessing nucleus and the resulting "flip" of its nuclear spin from the lower energy state to the higher energy state.

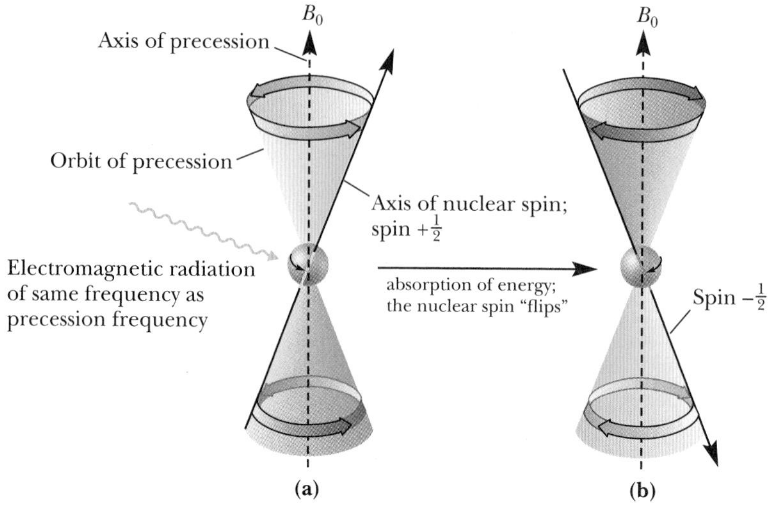

FIGURE 13.3 The origin of nuclear magnetic "resonance." (a) Precession of a spinning nucleus in an applied magnetic field. (b) Absorption of electromagnetic radiation occurs when the frequency of radiation is equal to the frequency of precession.

Signal

A recording in an NMR spectrum of a nuclear magnetic resonance.

Diamagnetic current in NMR

The circulation of electron density in a molecule in an applied magnetic field.

Shielding in NMR

Also called diamagnetic shielding; the term refers to the reduction in magnetic field strength experienced by a nucleus underneath electron density induced to circulate when the molecule is placed in a strong magnetic field.

spin from the lower energy state to the higher energy state. The spectrometer detects this absorption of electromagnetic radiation and records it as a **signal**. The process is quantized so that only electromagnetic radiation of precisely the correct frequency causes a nuclear spin to flip. Electromagnetic radiation of a frequency that is too low or too high is not absorbed.

If we were dealing with ^1H nuclei isolated from all other atoms and electrons, any combination of applied field and electromagnetic radiation that produces a signal for one hydrogen nucleus would produce a signal for all hydrogen nuclei. In other words, hydrogens would be indistinguishable. Hydrogens in an organic molecule, however, are not isolated; they are surrounded by electron density.

A key physical principle for NMR is that circulating electrons induce a magnetic field. This principle, first described by Michael Faraday and his law of electromagnetic induction, also explains electromagnets and electric motors. The direction of electron movement dictates the orientation of the induced magnetic field. Of equal importance to NMR, the converse is also true. An applied magnetic field induces electrons to circulate, and the orientation of the field dictates the direction of circulation. You will learn more details of these relationships in your physics classes. The important point for our purposes is that an applied magnetic field induces the electron density in a molecule to circulate. The spin states of underlying nuclei are, in turn, influenced to a small but measurable degree by the magnetic field created by the induced electron density circulation. The circulation of electron density in a molecule in an applied magnetic field is called a **diamagnetic current**.

It turns out that a molecule's σ-bonding electron density is induced to circulate in a direction that creates a small magnetic field that directly *opposes* the applied magnetic field. As a result of the opposing magnetic fields, the nuclei within the circulating electron density experience a magnetic field that is slightly *smaller* than the actual applied field. In other words, nuclei underneath circulating σ-bonding electron density are *shielded* to a small degree from the applied magnetic field. This nuclear **shielding** is called **diamagnetic shielding**. Although the diamagnetic shielding created by circulating electron density is generally orders of magnitude weaker than the applied fields used in NMR spectroscopy, it is nonetheless significant enough to be measured.

The degree of shielding depends on several factors, which we will take up later. Nevertheless, what we need to establish now is that as the shielding becomes greater, the net magnetic field present at a nucleus becomes smaller so the energy of electromagnetic radiation required to bring that nucleus into resonance (i.e., "flip its spin") also decreases. Energy is proportional to electromagnetic radiation frequency, and resonance frequencies are plotted on an NMR spectrum. Putting all of these ideas together, we see that a nucleus that is more shielded will come into resonance at lower frequency than a nucleus that is less shielded. **Deshielding** is the term commonly used to express the concept of less shielding. The relationship of *greater* shielding leading to resonance at *lower* frequency is fundamental to understanding NMR spectra, but most students find it challenging to remember correctly. Make sure you have a clear understanding of this concept before continuing.

The difference in resonance frequencies caused by differing amounts of shielding is called **chemical shift**. The differences in resonance frequencies among the various hydrogen nuclei within a molecule attributable to shielding/deshielding are generally very small. The difference between the resonance frequencies of hydrogens in chloromethane compared with those in fluoromethane, for example, under an applied field of 7.05 T is only 360 Hz. Considering that the radio-frequency radiation used at this applied field is approximately 300 MHz, the difference in resonance frequencies between these two sets of hydrogens is only slightly greater than 1 part per million (1 ppm) compared with the irradiating frequency.

Chemical shift (δ) The shift in parts per million of an NMR signal relative to the signal of TMS.

$$\frac{360 \text{ Hz}}{300 \times 10^6 \text{ Hz}} = \frac{1.2}{10^6} = 1.2 \text{ ppm}$$

It is customary to measure the resonance frequencies of individual nuclei relative to the resonance frequency of nuclei in a reference compound. The reference compound now universally accepted for ^{1}H-NMR and ^{13}C-NMR spectroscopy is tetramethylsilane (TMS), which is assigned a chemical shift of 0.00 ppm by convention.

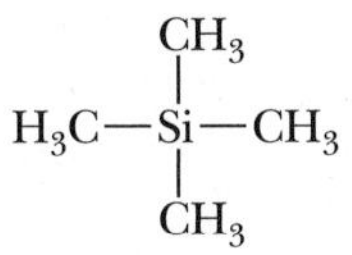

Tetramethylsilane (TMS)

To standardize reporting of NMR data for both ^{1}H and ^{13}C spectra, the chemical shift (δ), in parts per million, is defined as the frequency shift from either the hydrogens or the carbons in TMS divided by the operating frequency of the spectrometer. Thus, by definition, chemical shift is independent of the operating frequency of the spectrometer. On the chart paper used to record NMR spectra, chemical shift values are shown in increasing order to the left of the TMS signal. Having values increase to the left is backward compared to the way you normally see scientific values plotted (with larger values to the right), so make sure you recognize this very unusual feature of NMR spectra.

13.4 An NMR Spectrometer

The essential elements of an NMR spectrometer are a powerful magnet, a radio-frequency generator, a radio-frequency detector, and a sample tube (Figure 13.4).

The sample is dissolved in a solvent, most commonly carbon tetrachloride (CCl_4), deuterochloroform ($CDCl_3$), or deuterium oxide (D_2O), which have no ^{1}H atoms and do not interfere in ^{1}H-NMR spectra. The sample cell is a small glass tube suspended in the magnetic field and set spinning on its long axis to ensure that all parts of the sample experience a homogeneous applied magnetic field. In the simplest form of the ^{1}H-NMR experiment, the absorption of electromagnetic radiation is measured as different ^{1}H nuclei are excited from their $+\frac{1}{2}$ spin states to their $-\frac{1}{2}$ spin states. The frequencies at which the absorptions occur are in the radio-frequency region of the electromagnetic spectrum. The observed absorption frequencies are plotted as peaks relative to the TMS standard on a ppm scale.

Modern Fourier transform NMR (FT-NMR) spectrometers can increase the power of the NMR technique significantly. An FT-NMR spectrometer operates in the following way. The magnetic field is held constant, and the sample is irradiated with a short pulse (approximately 10^{-5} s) of radio-frequency energy that flips the spins of all susceptible

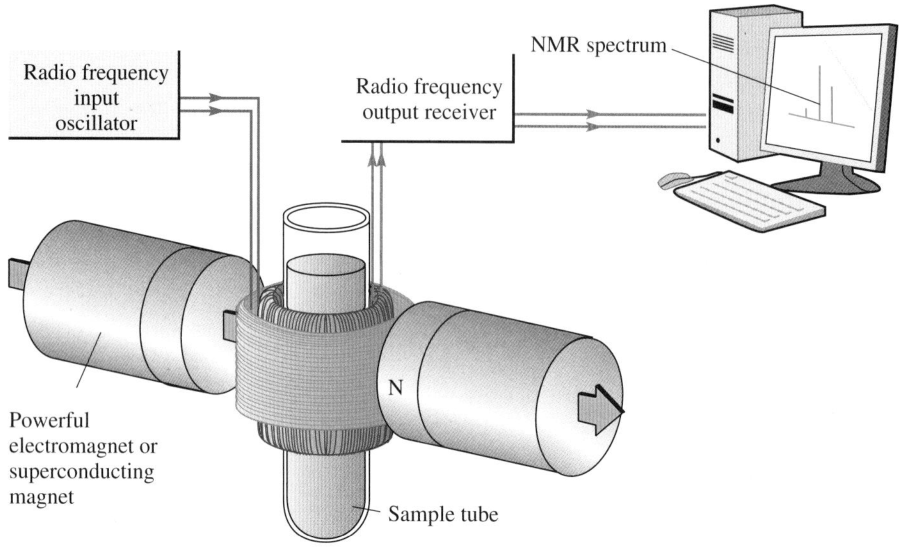

FIGURE 13.4 Schematic diagram of a nuclear magnetic resonance spectrometer.

nuclei simultaneously. As each nucleus returns to its equilibrium state, it emits a sine wave at the frequency of its resonance. The intensity of the sine wave decays with time and falls to zero as nuclei return to their equilibrium state. A computer records this intensity-versus-time information and then uses a mathematical algorithm called a Fourier transform (FT) to convert it to intensity-versus-frequency information. An FT-NMR spectrum can be recorded in less than two seconds. A particular advantage of FT-NMR spectroscopy is that a large number of spectra (as many as several thousand per sample) can be recorded and digitally summed to give a time-averaged spectrum. Instrumental electronic noise is random and partially cancels out when spectra are time-averaged, but sample signals accumulate and become much stronger relative to the electronic noise. The net result is that good NMR spectra can be obtained with very little sample. All NMR spectra shown in this text were recorded and displayed using FT techniques.

Figure 13.5 shows a 300 MHz ^{1}H-NMR spectrum of methyl acetate. The lower axis is δ, in parts per million. The small signal at δ 0 is caused by the hydrogens of the TMS reference, a small amount of which was added to the sample. The remainder of the spectrum consists of two signals: one for the three hydrogens on the methyl adjacent to oxygen and one for the three hydrogens on the methyl adjacent to the carbonyl group. It is not our purpose at the moment to determine which hydrogens give rise to which signal, but only to recognize the form in which an NMR spectrum is recorded and the origin of the calibration marks.

Downfield

A signal of an NMR spectrum that is shifted toward the left (larger chemical shift) on the chart paper.

Here is a note on terminology. If a signal is shifted toward the left on the chart paper (larger chemical shift), we say that it is shifted **downfield**. A downfield shift

FIGURE 13.5 ^{1}H-NMR spectrum of methyl acetate.

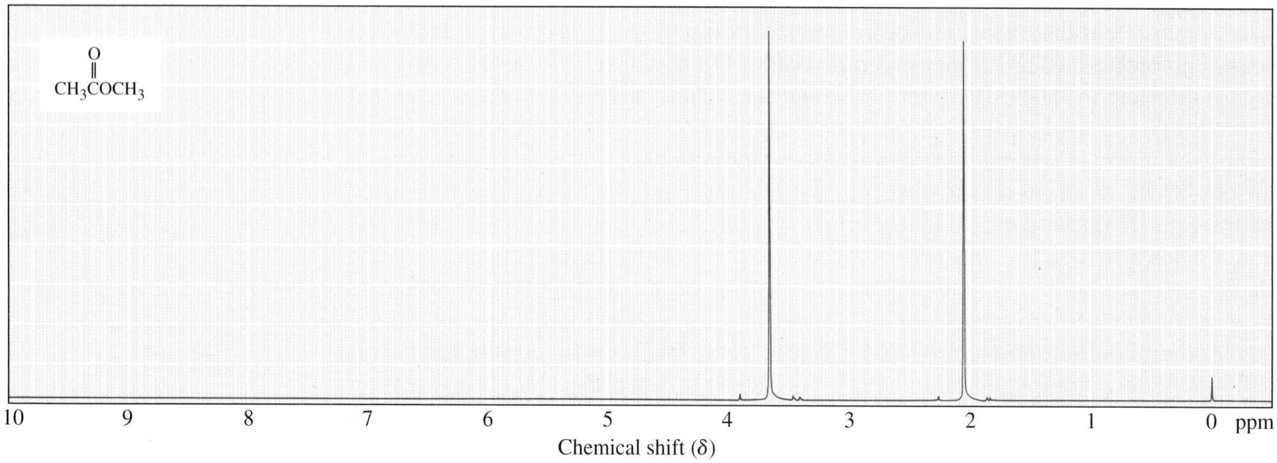

corresponds to decreased shielding around a nucleus (i.e., deshielding). If a signal is shifted toward the right (smaller chemical shift), we say that it is shifted **upfield** and therefore corresponds to increased shielding around a nucleus. Students often remember these relationships through the simple expression "more downfield = more deshielded."

Upfield
A signal of an NMR spectrum that is shifted toward the right (smaller chemical shift) on the chart paper.

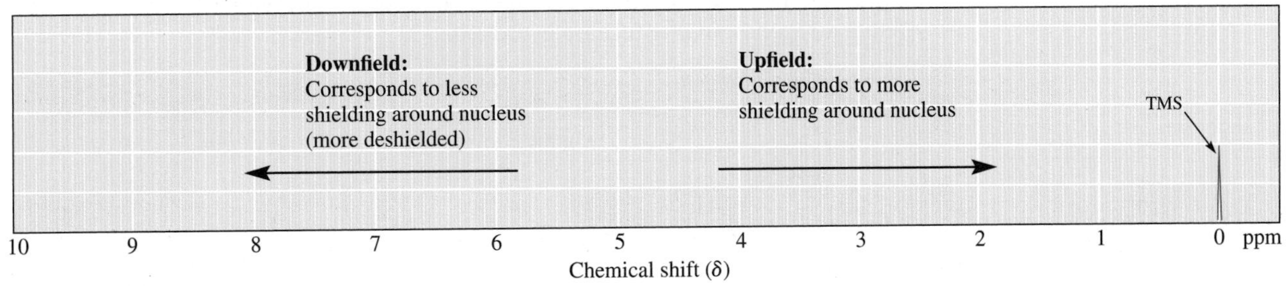

13.5 Equivalent Hydrogens

Given the structural formula of a compound, how do we know how many signals to expect in its ^{1}H-NMR spectrum? The answer is that equivalent hydrogens give the same ^{1}H-NMR signal; nonequivalent hydrogens give different ^{1}H-NMR signals. **Equivalent hydrogens** have the same chemical environment. H atoms are equivalent (in the same chemical environment) if either of the two following conditions exist:

Equivalent hydrogens
Hydrogens that have the same chemical environment.

1. They are bonded to the same sp^3 hybridized carbon atom, and that carbon atom can rotate freely at room temperature. The rapid bond rotation means that, on average, the H atoms bonded to the same carbon atom see the same chemical environment and are therefore equivalent. For example, all three H atoms on a freely rotating —CH_3 group are equivalent, and both H atoms on a freely rotating —CH_2— group are usually equivalent, although see Section 13.10 for the exception that occurs when —CH_2— groups are near a chiral center.
2. They are related by symmetry, namely a plane or point of symmetry in a molecule.

For example, in the 2-chloropropane molecule, all six methyl group H atoms are equivalent. The methyl groups are freely rotating, and they are related by a plane of symmetry as shown (Figure 13.6). A convenient way to determine whether hydrogen atoms are equivalent is to use the replacement test. In your mind, replace each of the hydrogen atoms in question with a test atom (e.g., chlorine). If in each case the same molecule is produced through the replacement, then the original hydrogen atoms are equivalent.

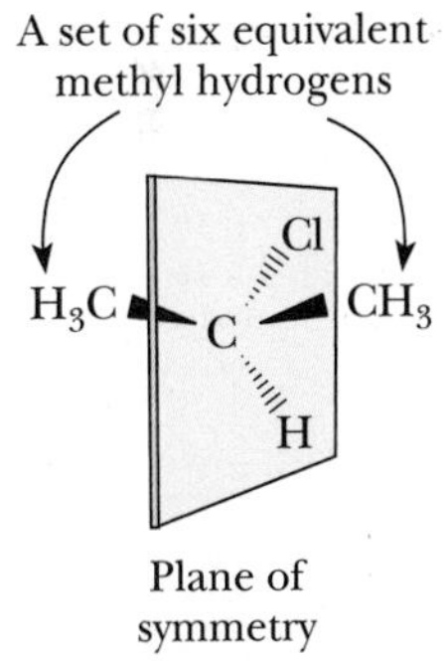

FIGURE 13.6 Structure of 2-chloropropane showing the plane of symmetry responsible for making the two methyl groups (and therefore the six methyl group H atoms) equivalent.

Example 13.2 Equivalent Hydrogens

State the number of types of equivalent hydrogens in each compound and the number of hydrogens in each set.

(a) 2-Methylpropane **(b)** 2-Methylbutane

Solution

(a) 2-Methylpropane contains two sets of equivalent hydrogens: a set of nine equivalent primary hydrogens and one tertiary hydrogen.

(Continued)

(b) 2-Methylbutane contains four sets of equivalent hydrogens. Nine primary hydrogens are in this molecule: one set of three and one set of six. To see that there are two sets, note that replacement by chlorine of any hydrogen in the set of three gives 1-chloro-3-methylbutane. Replacement by chlorine of any hydrogen in the set of six gives 1-chloro-2-methylbutane. In addition, the molecule contains a set of two equivalent secondary hydrogens and one tertiary hydrogen.

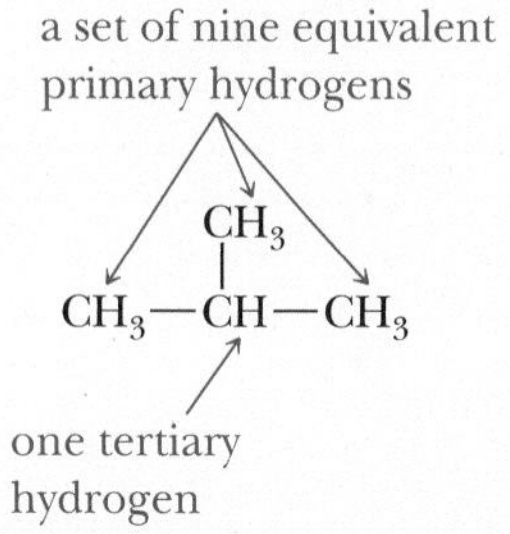

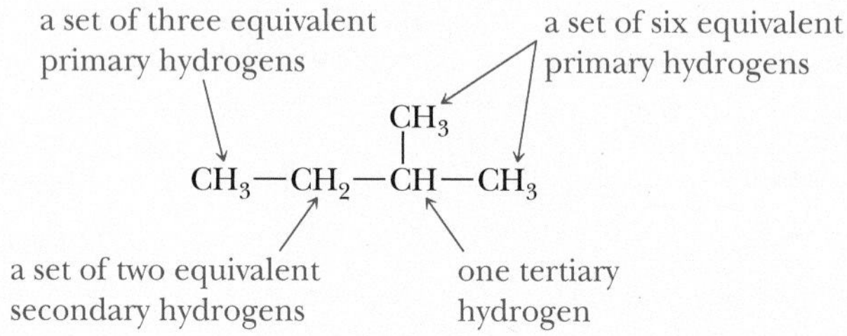

Problem 13.2

State the number of sets of equivalent hydrogens in each compound and the number of hydrogens in each set.

(a) 3-Methylpentane **(b)** 2,2,4-Trimethylpentane

Here are four organic compounds, each of which has one set of equivalent hydrogens and gives one signal in its ^{1}H-NMR spectrum.

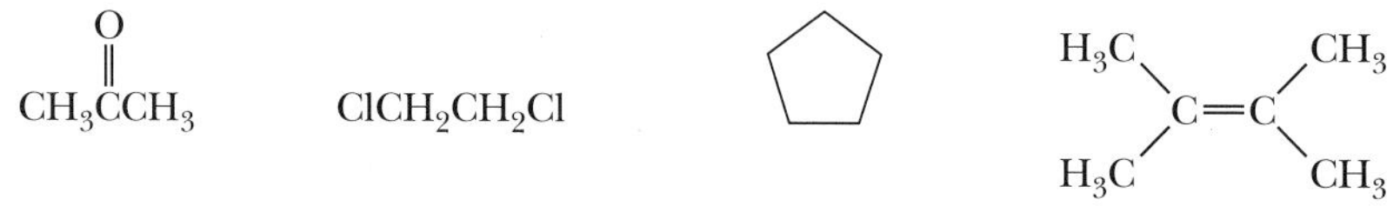

Propanone (Acetone) **1,2-Dichloroethane** **Cyclopentane** **2,3-Dimethyl-2-butene**

Molecules with two or more sets of equivalent hydrogens give rise to a different resonance signal for each set, as illustrated by these four compounds.

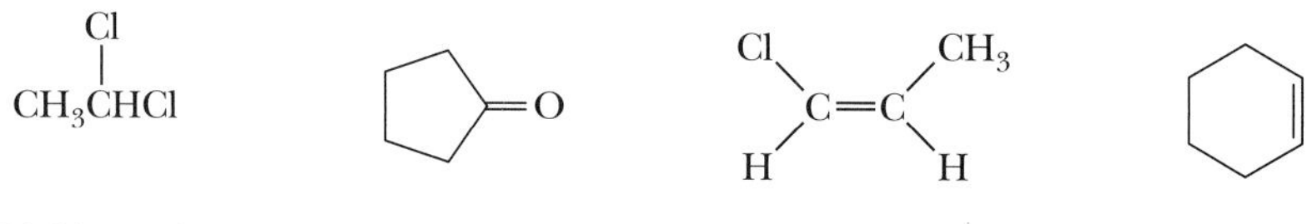

1,1-Dichloroethane (2 signals) **Cyclopentanone** (2 signals) **(Z)-1-Chloropropene** (3 signals) **Cyclohexene** (3 signals)

You should see immediately that valuable information about molecular structure can be obtained simply by counting the number of signals in the ^{1}H-NMR spectrum of a compound. Consider, for example, the two constitutional isomers of molecular formula $C_2H_4Cl_2$. The compound 1,2-dichloroethane has one set of equivalent hydrogens and one signal in its ^{1}H-NMR spectrum. Its constitutional isomer 1,1-dichloroethane has two sets of equivalent hydrogens and two signals in its ^{1}H-NMR spectrum. Thus, simply counting signals allows you to distinguish between these two compounds.

Example 13.3 Structural Prediction I

Each compound gives only one signal in its ^{1}H-NMR spectrum. Propose a structural formula for each compound.

(a) C_2H_6O **(b)** $C_3H_6Cl_2$ **(c)** C_6H_{12} **(d)** C_4H_6

Solution

(a) CH_3OCH_3 **(b)** 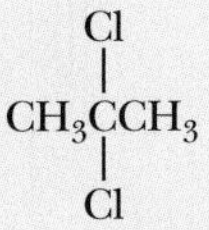
Cl
CH3CCH3
Cl
(c) 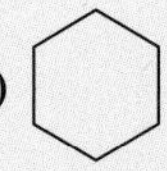 or 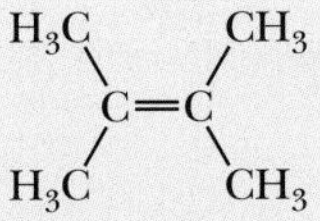
H3C
CH3
C=C
H3C
CH3
(d) $CH_3C{\equiv}CCH_3$

Problem 13.3

Each compound gives only one signal in its ^{1}H-NMR spectrum. Propose a structural formula for each compound.

(a) C_3H_6O **(b)** C_5H_{10} **(c)** C_5H_{12} **(d)** $C_4H_6Cl_4$

13.6 Signal Areas

We have just seen that the number of signals in a ^{1}H-NMR spectrum gives us information about the number of sets of equivalent hydrogens. The relative areas of these signals provide additional information. As a spectrum is being run, the instrument's computer numerically measures the area under each signal. In the spectra shown in this text, this information is displayed in the form of a line of integration superposed on the original spectrum. The vertical rise of the line of integration over each signal is proportional to the area under that signal, which, in turn, is proportional to the number of equivalent hydrogens giving rise to that signal. Figure 13.7 shows an integrated ^{1}H-NMR spectrum of *tert*-butyl acetate, $C_6H_{12}O_2$. The spectrum shows signals at δ 1.44 and 1.95. The integrated signal heights are 23 + 67, or 90 chart divisions, which correspond to 12 hydrogens. From these numbers, we calculate that (23/90) × 12, or 3, hydrogens are in one set and (67/90) × 12, or 9, hydrogens are in the second set. An alternative way of indicating integration is a numerical readout given over or under each signal, and you may see this on many ^{1}H-NMR spectra you encounter.

FIGURE 13.7 ^{1}H-NMR spectrum of *tert*-butyl acetate showing the integration. The total vertical rise of 90 chart divisions corresponds to 12 hydrogens, 9 in one set and 3 in the other.

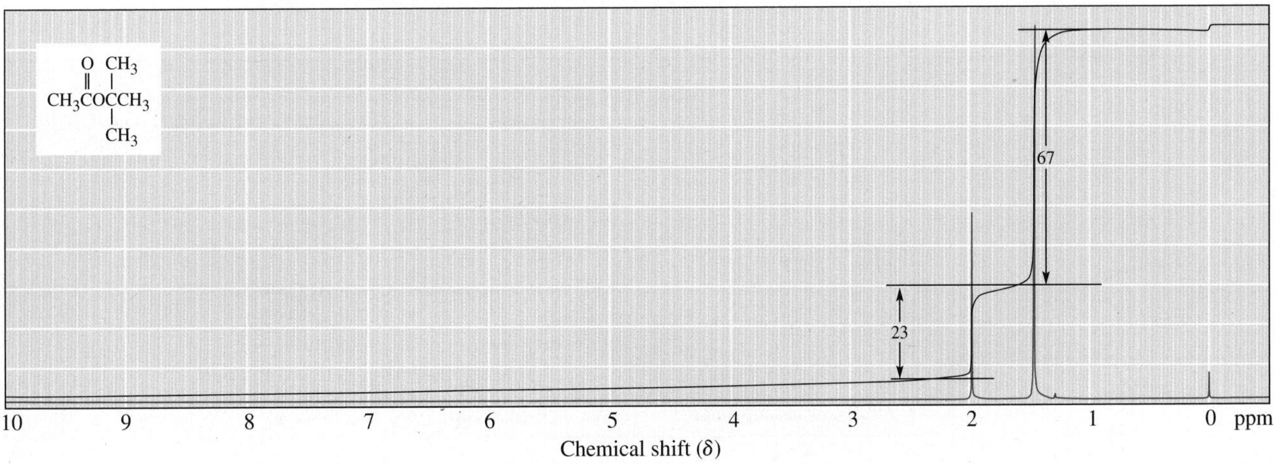
O CH3
CH3COCCH3
CH3
67
23
10
9
8
7
6
5
4
3
2
1
0 ppm
Chemical shift (δ)

Example 13.4 Structural Prediction II

Following is a ^{1}H-NMR spectrum for a compound of molecular formula $C_9H_{10}O_2$. From the integration, calculate the number of hydrogens giving rise to each signal.

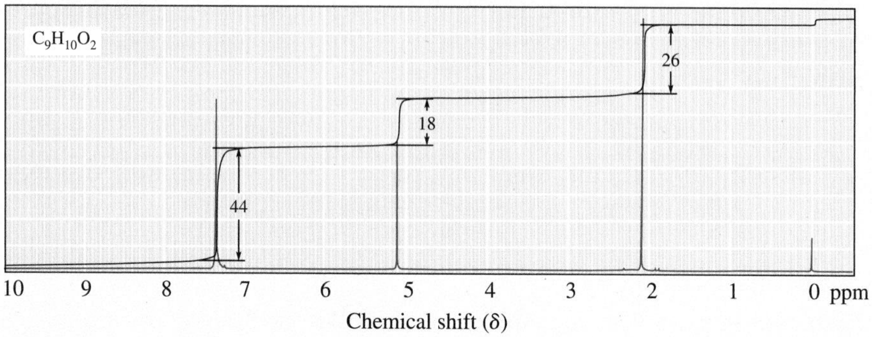

Solution

The total vertical rise in the line of integration is 88 chart divisions and corresponds to 10 hydrogens. From these numbers, we calculate that 44/88 × 10, or 5, of the hydrogens give rise to the signal at δ 7.34. By similar calculations, the signals at δ 5.08 and 2.06 correspond to two hydrogens and three hydrogens, respectively.

Problem 13.4

The line of integration of the two signals in the ^{1}H-NMR spectrum of a ketone with the molecular formula $C_7H_{14}O$ rises 62 and 10 chart divisions, respectively. Calculate the number of hydrogens giving rise to each signal and propose a structural formula for this ketone.

13.7 Chemical Shift

Watch a video explanation

The **chemical shift** for a signal in a ^{1}H-NMR spectrum can give valuable information about the type of hydrogens giving rise to that signal. Hydrogens on methyl groups bonded to sp^3 hybridized carbons, for example, give signals near δ 0.8 to 1.0. Hydrogens on methyl groups bonded to a carbonyl carbon give signals near δ 2.1 to 2.3 (notice the signals near 2.0 ppm in Figures 13.5 and 13.7), and hydrogens on a methyl group bonded to oxygen give signals near δ 3.7 to 3.9 (Figure 13.5). Shown in Figure 13.8 are average chemical shifts for most of the types of hydrogens we deal with in this course. Notice that most of these values fall within a rather narrow range of 0 to 10 δ units (ppm).

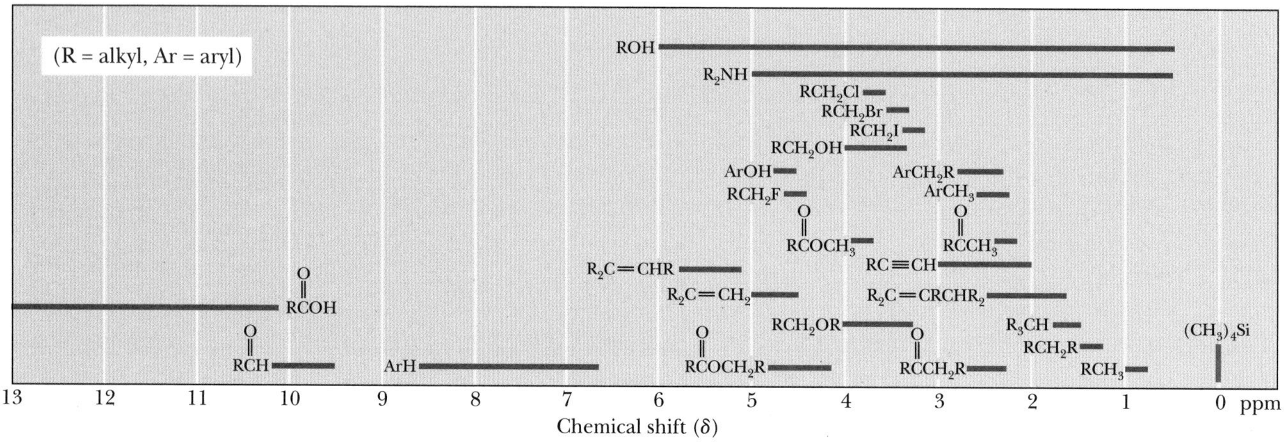

FIGURE 13.8 Average values of chemical shifts of representative types of hydrogens. These values are approximate. Other atoms or groups in the molecules may cause signals to appear outside these ranges.

Example 13.5 Predicting ¹H-NMR Spectra I

Following are structural formulas for two constitutional isomers with the molecular formula $C_6H_{12}O_2$.

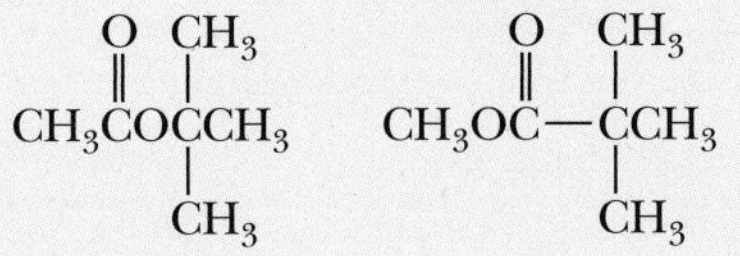

(1) (2)

(a) Predict the number of signals in the ¹H-NMR spectrum of each isomer.

(b) Predict the ratio of areas of the signals in each spectrum.

(c) Show how you can distinguish between these isomers on the basis of chemical shift.

Solution

(a) The ¹H-NMR spectrum of each consists of two signals **(b)** in the ratio 9:3, or 3:1. **(c)** Distinguish between these constitutional isomers by the chemical shift of the single —CH_3 group. The hydrogens of CH_3O are more deshielded (appear farther downfield) than the hydrogens of $CH_3C{=}O$. See Figure 13.8 for approximate values for each chemical shift. Experimentally determined values are shown in red.

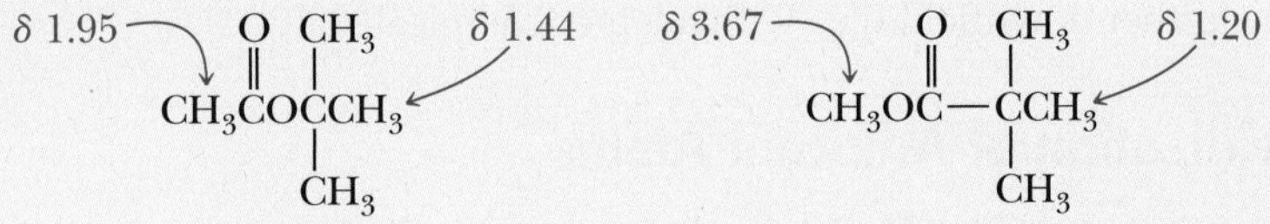

(1) (2)

Please see Figure 13.7 for the spectrum of *tert*-butyl acetate (1).

Problem 13.5

Following are two constitutional isomers with the molecular formula $C_4H_8O_2$.

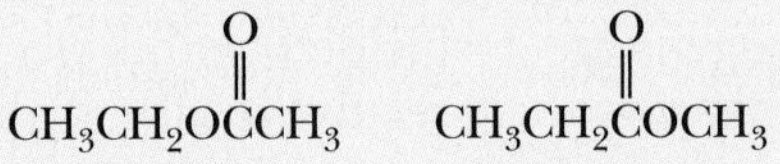

(1) (2)

(a) Predict the number of signals in the ¹H-NMR spectrum of each isomer.

(b) Predict the ratio of areas of the signals in each spectrum.

(c) Show how you can distinguish between these isomers on the basis of chemical shift.

The chemical shift of a particular type of hydrogen depends primarily on the extent of shielding it experiences. Shielding, in turn, depends on three factors: (1) the electronegativity of nearby atoms, (2) the hybridization of the adjacent atoms, and (3) the magnetic induction within an adjacent π bond. Let us consider these factors one at a time.

A. Electronegativity of Nearby Atoms

As illustrated in Table 13.2 for the chemical shift of methyl hydrogens in the series CH_3—X, the greater the electronegativity of X, the greater the chemical shift.

Table 13.2 Dependence of Chemical Shift of CH_3X on the Electronegativity of X

CH_3—X	Electronegativity of X	Chemical Shift (δ) of Methyl Hydrogens
CH_3F	4.0	4.26
CH_3OH	3.5	3.47
CH_3Cl	3.1	3.05
CH_3Br	2.8	2.68
CH_3I	2.5	2.16
$(CH_3)_4C$	2.1	0.86
$(CH_3)_4Si$	1.8	0.00 (by definition)

The effect of an electronegative substituent falls off quickly with distance. The effect of an electronegative substituent two atoms away is only about 10% of that when it is on the adjacent atom. The effect of an electronegative substituent three atoms away is almost negligible.

Electronegativity and chemical shift are related in the following way: the presence of an electronegative atom or group reduces electron density on atoms bonded to it and therefore the shielding. This effect deshields nearby nuclei and causes them to resonate farther downfield (i.e., with a larger chemical shift).

B. Hybridization of Adjacent Atoms

Hydrogens bonded to an sp^3 hybridized carbon typically have signals at δ 0.8 to 1.7. Vinylic hydrogens (those on a carbon of a carbon-carbon double bond) are considerably deshielded and their signals appear at δ 4.6 to 5.7 (Table 13.3). Part of the explanation for the greater deshielding of vinylic hydrogens compared with alkyl hydrogens lies in the hybridization of carbon. Because a σ-bonding orbital of an sp^2 hybridized carbon has more *s*-character than a σ-bonding orbital of an sp^3 hybridized carbon (33% compared with 25%), an sp^2 hybridized carbon atom is more electronegative. Vinylic hydrogens are deshielded by this electronegativity effect, and their nuclei resonate farther downfield relative to alkyl hydrogens. Similarly, signals for acetylene and aldehyde hydrogens also appear farther downfield compared with alkyl hydrogens.

Table 13.3 The Effect of Hybridization on Chemical Shift

Type of Hydrogen (R = alkyl)	Name of Hydrogen	Chemical Shift (δ)
RCH_3, R_2CH_2, R_3CH	Alkyl	0.8–1.7
$R_2C{=}C(R)CHR_2$	Allylic	1.6–2.6
$RC{\equiv}CH$	Acetylenic	2.0–3.0
$R_2C{=}CHR$, $R_2C{=}CH_2$	Vinylic	4.6–5.7
RCHO	Aldehydic	9.5–10.1

Differences in chemical shifts of vinylic and acetylenic hydrogens cannot be accounted for on the basis of the hybridization of carbon alone. If the chemical shift of vinylic hydrogens (δ 4.6–5.7) were caused entirely by the hybridization of carbon, then the chemical shift of acetylenic hydrogens should be even greater than that of vinylic hydrogens. Yet, the chemical shift of acetylenic hydrogens is only δ 2.0 to 3.0. It seems that either the chemical shift of acetylenic hydrogens is abnormally small or the chemical shift of vinylic hydrogens is abnormally large. Theoretical and experimental evidence suggest that the chemical shifts of hydrogens bonded to π-bonded carbons are influenced not only by the relative electronegativities of the sp^2 and sp hybridized carbon atoms but also by magnetic induction from π bonds.

C. Diamagnetic Effects from π Bonds

To understand the influence of π bonds on the chemical shift of an acetylenic hydrogen, imagine that the carbon-carbon triple bond is oriented as shown in Figure 13.9 with respect to the applied field. The applied field induces a circulation of the π electrons, which in turn produces an induced magnetic field. Given the geometry of an alkyne and the cylindrical nature of its π electron cloud, the induced magnetic field is shielding in the vicinity of the acetylenic hydrogen. Therefore, lower frequency electromagnetic radiation is required to make an acetylenic hydrogen nucleus resonate; the local magnetic field induced in the π bonds shifts the signal of an acetylenic hydrogen upfield to a smaller δ value.

The effect of the induced circulation of π electrons on a vinylic hydrogen (Figure 13.10) is opposite that on an acetylenic hydrogen. The direction of the induced magnetic field in the π bond of a carbon-carbon double bond is parallel to the applied field in the region of the vinylic hydrogens. The induced magnetic field deshields vinylic hydrogens and thus shifts their signal downfield to a larger δ value.

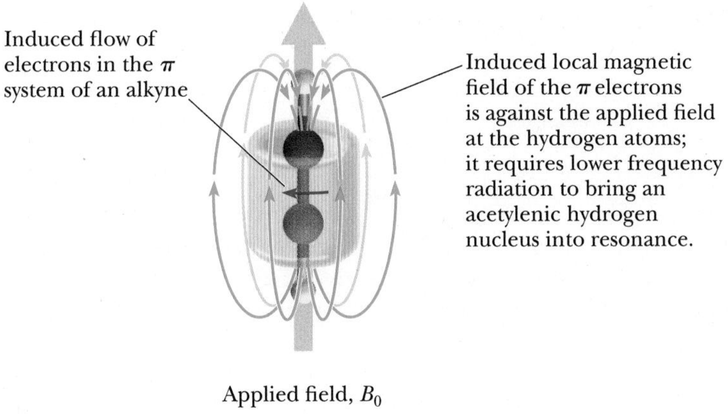
Induced flow of electrons in the π system of an alkyne
Induced local magnetic field of the π electrons is against the applied field at the hydrogen atoms; it requires lower frequency radiation to bring an acetylenic hydrogen nucleus into resonance.
Applied field, B0

FIGURE 13.9 A magnetic field induced in the π bonds of a carbon-carbon triple bond shields an acetylenic hydrogen and shifts its signal upfield.

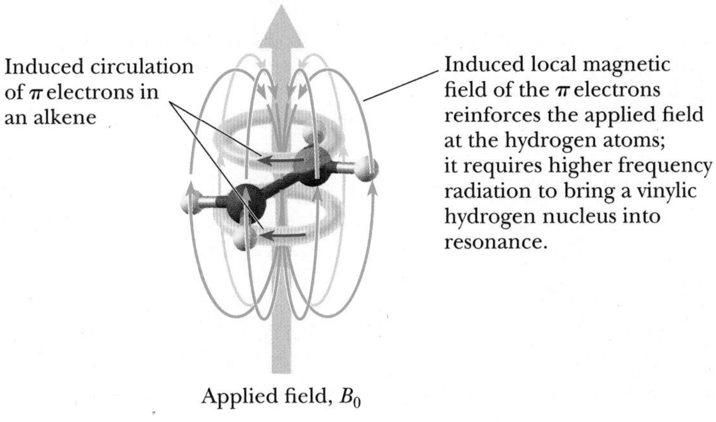
Induced circulation of π electrons in an alkene
Induced local magnetic field of the π electrons reinforces the applied field at the hydrogen atoms; it requires higher frequency radiation to bring a vinylic hydrogen nucleus into resonance.
Applied field, B0

FIGURE 13.10 A magnetic field induced in the π bond of a carbon-carbon double bond deshields vinylic hydrogens and shifts their signals downfield.

The presence of the π electrons in the carbonyl group has a similar effect on the chemical shift of the hydrogen of an aldehyde group.

The bottom line when comparing acetylenic versus vinylic hydrogens is that the hybridization effect, which would predict a larger chemical shift (greater deshielding) for the acetylenic hydrogen signals, is more than overcome by the strongly shielding diamagnetic effect of the cylindrically oriented alkyne π bonds and the deshielding diamagnetic effect of the alkene π bond. The net result is that acetylenic hydrogens have signals that appear less downfield from TMS as compared to the signals for vinylic hydrogens.

The effects of the π electrons in benzene are even more dramatic than in alkenes. All six hydrogens of benzene are equivalent, and its ^{1}H-NMR spectrum is a sharp singlet at δ 7.27. Hydrogens bonded to a substituted benzene ring appear in the region δ 6.5 to 8.5. Few other hydrogens absorb in this region; thus, aryl hydrogens are quite easily identifiable by their distinctive chemical shifts, as much as 2 ppm higher than comparably substituted alkenes.

Ring current

An applied magnetic field causes the π electrons of an aromatic ring to circulate, giving rise to the so-called ring current and an associated magnetic field that opposes the applied field in the middle of the ring but reinforces the applied field on the outside of the ring.

That aryl hydrogens absorb even farther downfield than vinylic hydrogens is accounted for by the existence of a **ring current**, a special property of aromatic rings (Figure 13.11). When the plane of an aromatic ring tumbles in an applied magnetic field, the applied field causes the π electrons to circulate around the ring, giving rise to the so-called ring current. This induced ring current has associated with it a magnetic field that opposes the applied field in the middle of the ring but reinforces the applied field on the outside of the ring. Given the position of aromatic hydrogens relative to the induced ring current, they are deshielded and come into resonance at a larger chemical shift.

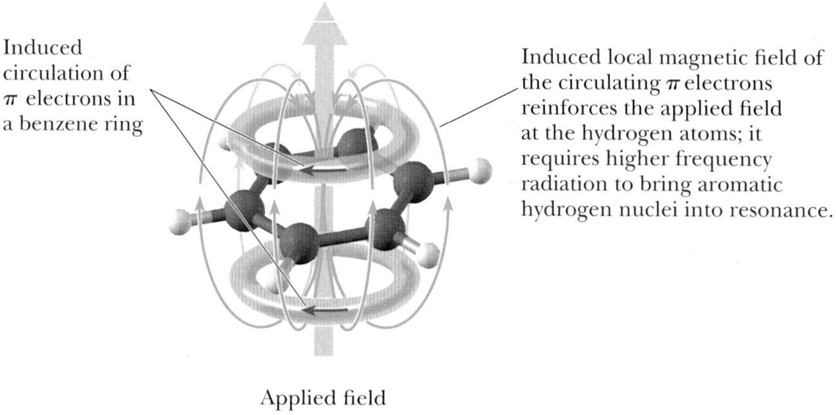
Induced circulation of π electrons in a benzene ring
Induced local magnetic field of the circulating π electrons reinforces the applied field at the hydrogen atoms; it requires higher frequency radiation to bring aromatic hydrogen nuclei into resonance.
Applied field

FIGURE 13.11 The magnetic field induced by circulation of π electrons in an aromatic ring deshields the hydrogens of the aromatic ring and shifts their signal downfield.

13.8 Signal Splitting and the (n + 1) Rule

We have now seen three kinds of information that can be derived from examination of a ^{1}H-NMR spectrum.

1. From the number of signals, we can determine the number of sets of equivalent hydrogens.
2. From integration of signal areas, we can determine the relative numbers of hydrogens giving rise to each signal.
3. From the chemical shift of each signal, we derive information about the types of hydrogens in each set.

A fourth kind of information can be derived from the splitting pattern of each signal. Consider, for example, the ^{1}H-NMR spectrum of 1,1-dichloroethane shown in Figure 13.12. This molecule contains two sets of equivalent hydrogens. According to what we have learned so far, we predict two signals with relative

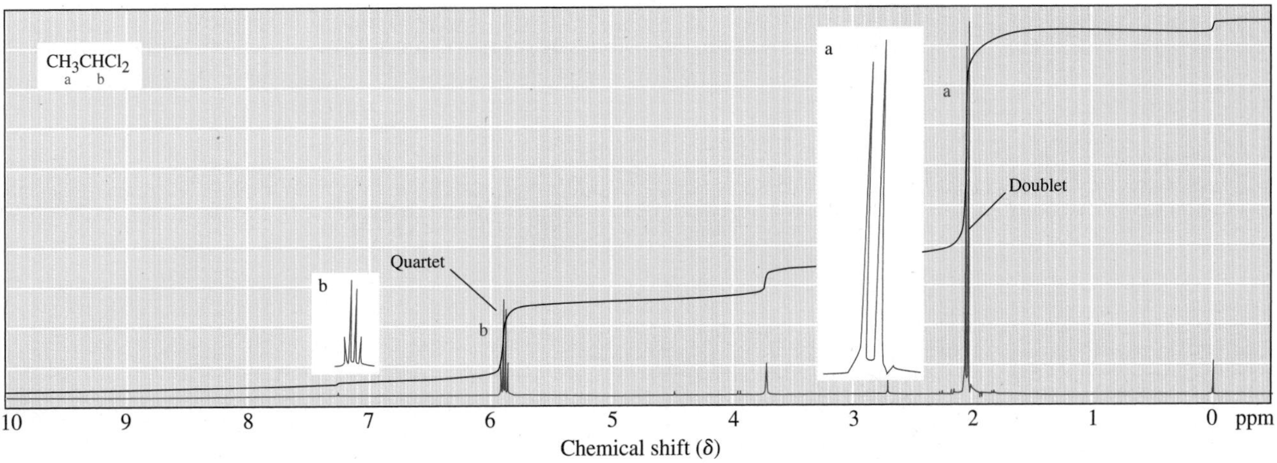

FIGURE 13.12 ^{1}H-NMR spectrum of 1,1-dichloroethane.

areas 3:1 corresponding to the three hydrogens of the $—CH_3$ group and the one hydrogen of the $—CHCl_2$ group, respectively. You see from the spectrum, however, that there are, in fact, six peaks. These peaks are named by how the signal is split: two peaks are a doublet, three peaks are a triplet, and so on. The grouping of two peaks at δ 2.0 is the signal for the three hydrogens of the $—CH_3$ group, and the grouping of four peaks at δ 5.9 is the signal for the single hydrogen of the $—CHCl_2$ group. We say that the CH_3 resonance at δ 2.1 is split into a doublet and that the CH resonance at δ 5.9 is split into a quartet.

In many situations, the degree of **signal splitting** can be predicted on the basis of the **(n + 1) rule**. According to this rule, if a hydrogen has n hydrogens nonequivalent to it but equivalent among themselves on the same or adjacent atom(s), its ^{1}H-NMR signal is split into (n + 1) peaks.

(n + 1) rule

If a hydrogen has n hydrogens nonequivalent to it but equivalent among themselves on the same or adjacent atom(s), its ^{1}H-NMR signal is split into (n + 1) peaks.

Let us apply the (n + 1) rule to the analysis of the spectrum of 1,1-dichloroethane. The three hydrogens of the $—CH_3$ group have one nonequivalent neighbor hydrogen (n = 1); therefore, their signal is split into a doublet. The single hydrogen of the $—CHCl_2$ group has a set of three nonequivalent neighbor hydrogens (n = 3), and its signal is split into a quartet.

For these hydrogens, $n = 1$; their signal is split into (1 + 1), or 2, peaks—a **doublet**.

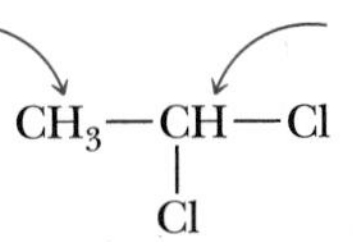

For this hydrogen, $n = 3$; its signal is split into (3 + 1), or 4, peaks—a **quartet**.

Example 13.6 Predicting ^{1}H-NMR Spectra II

Predict the number of signals and the splitting pattern of each signal in the ^{1}H-NMR spectrum of each molecule.

(a) $CH_3\overset{O}{\overset{\|}{C}}CH_2CH_3$ **(b)** $CH_3CH_2\overset{O}{\overset{\|}{C}}CH_2CH_3$ **(c)** $CH_3\overset{O}{\overset{\|}{C}}CH(CH_3)_2$

Solution

The sets of equivalent hydrogens in each molecule are labeled *a*, *b*, and *c*. Molecule (a) has three sets of equivalent hydrogens; its ^{1}H-NMR spectrum shows a singlet, a quartet, and a triplet in the ratio 3:2:3. Molecule (b) has two sets of equivalent hydrogens; its ^{1}H-NMR spectrum shows a triplet and a

(Continued)

quartet in the ratio 3:2. Molecule (c) has three sets of equivalent hydrogens; its ^{1}H-NMR spectrum shows a singlet, a septet, and a doublet in the ratio 3:1:6.

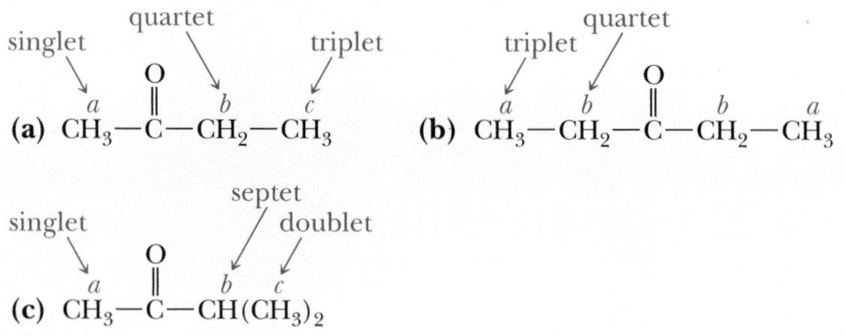

singlet
quartet
triplet
(a)
triplet
quartet
(b)
singlet
septet
doublet
(c)

Problem 13.6

Following are pairs of constitutional isomers. Predict the number of signals in the ^{1}H-NMR spectrum of each isomer and the splitting pattern of each signal.

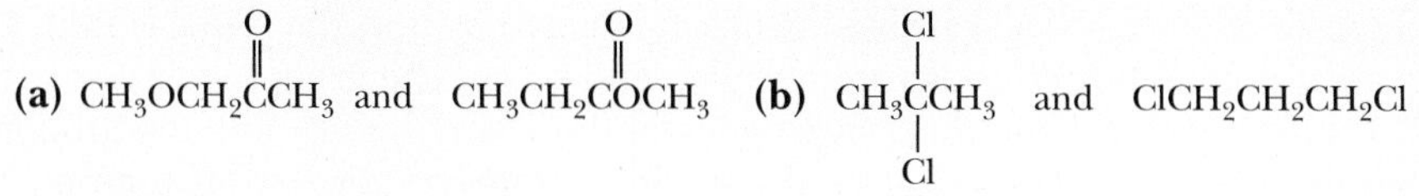

(a)
and
(b)
and

13.9 The Origins of Signal Splitting

^{1}H-NMR signal splitting can be understood by considering **spin-spin coupling** between ^{1}H nuclei, an interaction in which nuclear spins of adjacent atoms influence each other. A common type of spin-spin coupling involves the H atoms on two C atoms that are bonded to each other. These H atoms are three bonds apart and are referred to as **vicinal hydrogens**. Coupling between vicinal hydrogens is called **vicinal coupling**, illustrated in Figure 13.13. A **coupling constant (J)** is the separation on an NMR spectrum between adjacent peaks in a split signal and is a quantitative measure of the influence of adjacent nuclei. The magnitude of a coupling constant is expressed in hertz (Hz); for protons in ^{1}H-NMR spectroscopy, the coupling constant is generally in the range 0 to 18 Hz. The value of J depends only on interactions with other nuclei within a molecule, and so it is independent of the applied field strength.

Nuclear spin and hence the chemical shift of the atom labeled H_a in Figure 13.14 is influenced by the vicinal atom H_b, whose nuclear spin might be aligned with or against an applied magnetic field in a ^{1}H-NMR spectrometer. Because of spin-spin coupling, alignment of the H_b nuclear spin *with* the applied magnetic field results in a slightly different chemical shift of the signal for H_a compared to the situation in which the H_b nuclear spin is aligned *against* the applied magnetic field. Across the population of molecules in a sample, similar numbers of molecules will be having each spin alignment for H_b. Any single molecule gives rise to a single signal for H_a, but the spectrum of the entire sample shows both. The result is that the signal for the H_a atom appears in the spectrum as a **doublet**. In this hypothetical example, the signal for H_b will also be split into a similar doublet owing to H_a because the effect operates in both directions.

The coupling constant for two vicinal hydrogens on adjacent sp^3 hybridized carbon atoms is approximately 7 Hz. For a spectrometer operating at 300 MHz, a coupling constant of 7 Hz corresponds to only 0.023 ppm. Because peaks with this and comparable values of J are so narrowly spaced, splitting patterns from spectra taken

Spin-spin coupling
An interaction in which nuclear spins of adjacent atoms influence each other and lead to the splitting of NMR signals.

Vicinal hydrogens
H atoms on two C atoms that are bonded to each other.

Vicinal coupling
Coupling between nuclei of vicinal H atoms.

Coupling constant (J)
The separation on an NMR spectrum (in hertz) between adjacent peaks in a multiplet and a quantitative measure of the influence of the spin-spin coupling with adjacent nuclei.

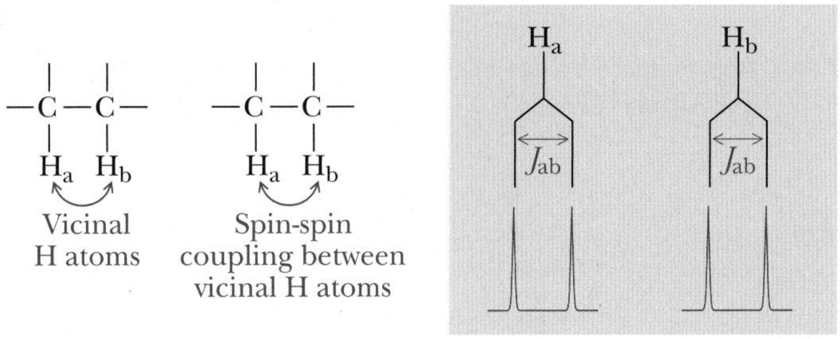

FIGURE 13.13 Vicinal H atoms and the vicinal spin-spin coupling that occurs between them.

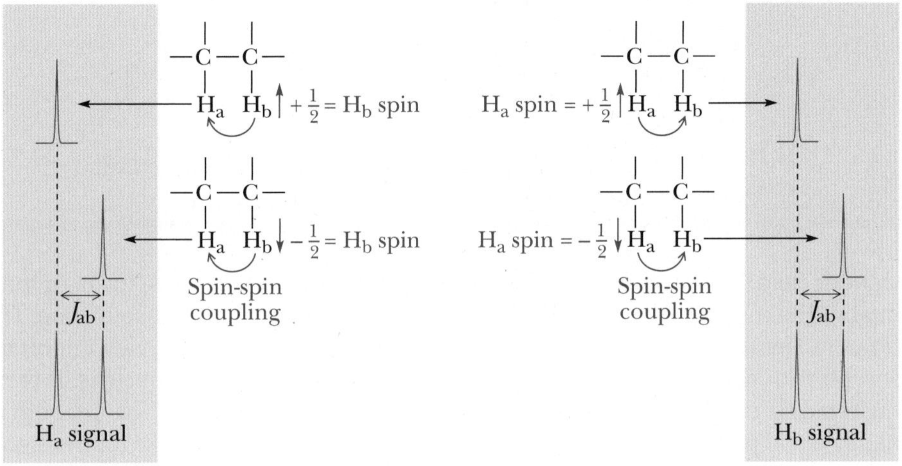

FIGURE 13.14 Illustration of spin-spin coupling that gives rise to signal splitting in ^{1}H-NMR spectra.

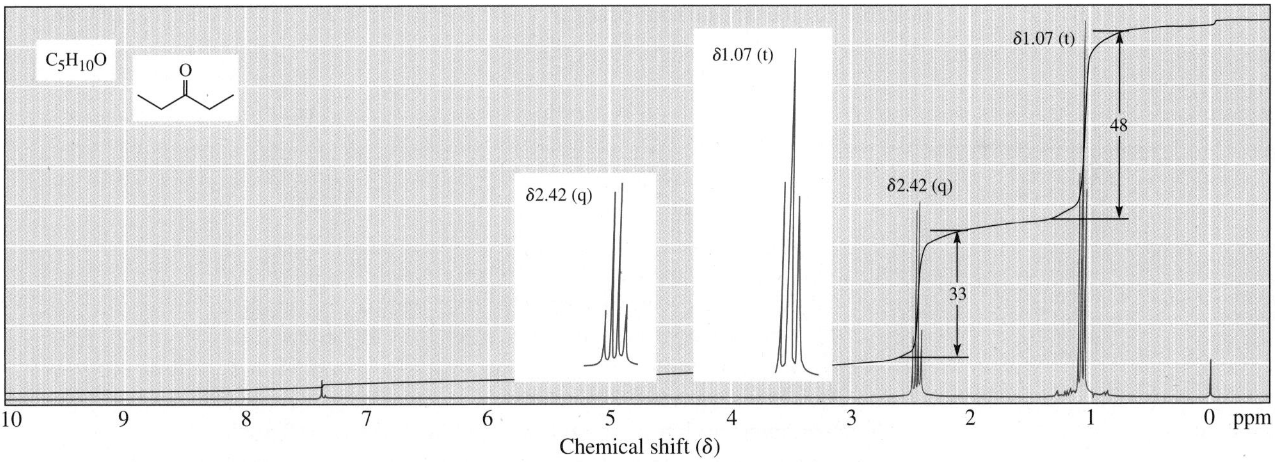

FIGURE 13.15 The quartet-triplet ^{1}H-NMR signals of 3-pentanone showing the original trace and a scale expansion to show the signal splitting pattern more clearly.

at 300 MHz and higher are often very difficult to see by inspection of the spectra. It is, therefore, common practice to retrace certain signals in expanded form so that splitting patterns are easier to observe (Figure 13.15).

Given in Table 13.4 are approximate values for coupling constants for different types of hydrogens.

A. Predicting Peak Intensities

As stated previously, in the general case, n equivalent H atoms will cause signal splitting into $n + 1$ peaks. We can now understand this rule and the relative intensities of these peaks by analyzing all possible spin state combinations. There are $n + 1$ different spin state combinations of n spins aligning with or against an applied

Table 13.4 Approximate Values of *J* for Compounds Containing Alkyl and Alkenyl Groups

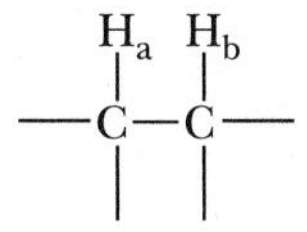 6–8 Hz	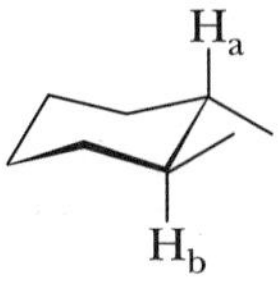 8–14 Hz	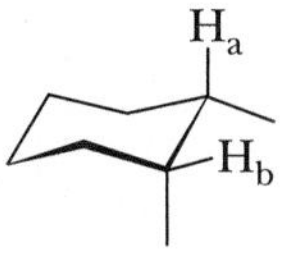 0–5 Hz	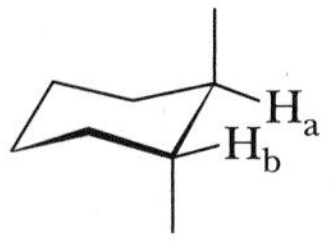 0–5 Hz
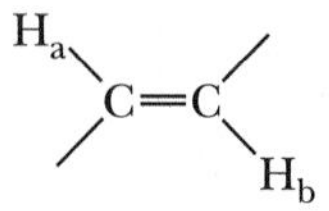 11–18 Hz	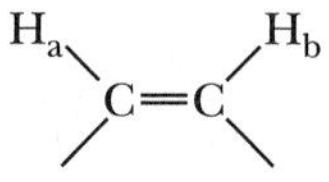 5–10 Hz	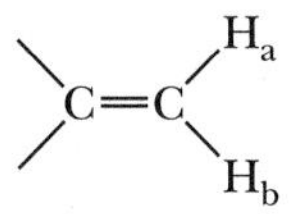 0–5 Hz	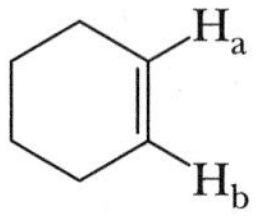 8–11 Hz

magnetic field. The probability of a molecule having a given set of spins is proportional to the number of possible spin alignments giving rise to that spin state. The arrows in Figure 13.16 are particularly helpful in understanding this very important concept, with each arrow representing the spin alignment of a ^{1}H nucleus. If there is just one H_b nucleus to consider, there are only two possibilities (↑ or ↓), both of roughly equal probability, leading to a doublet with a 1:1 ratio of peaks for the signal of H_a (left). Two equivalent H_b nuclei can have three different possible combinations that occur in a 1:2:1 ratio (middle), while three equivalent H_b nuclei can have four possible combinations that occur in a 1:3:3:1 ratio (right).

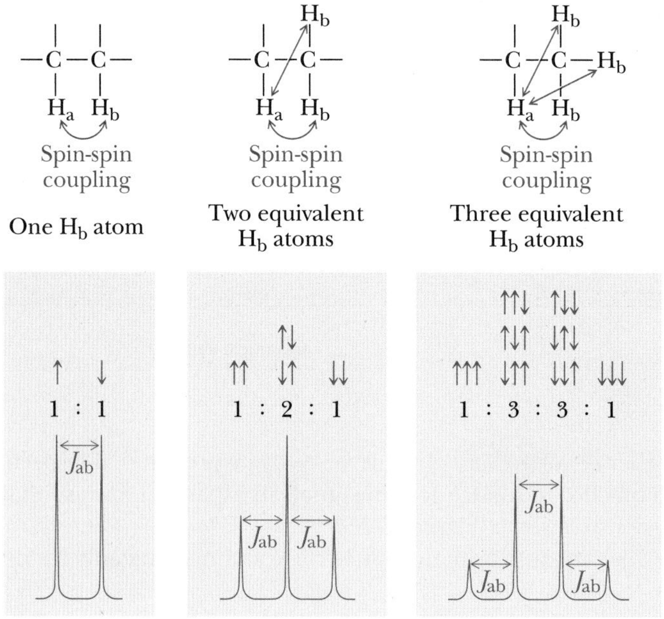

Watch a video explanation

FIGURE 13.16 The origins of signal splitting patterns. Each arrow represents an H_b nuclear spin orientation.

Alternatively, the ratio of peak areas in any multiplet can be derived from a mathematical mnemonic device called **Pascal's triangle** (Figure 13.17). Here is a note of caution in counting the number of peaks in a multiplet: if the signal of a particular hydrogen is of low intensity compared with others in the spectrum, it may not be possible to distinguish some of the smaller side peaks because of electronic noise in the baseline.

B. Physical Basis for the (*n* + 1) Rule

Coupling of nuclear spins is mediated through intervening bonds. The extent of coupling is related to a number of factors, including the number of bonds between the H atoms in question. H atoms with more than three bonds between them generally do not exhibit noticeable coupling, although longer range coupling can be seen in some cases. As described in detail previously, a common type of coupling is between vicinal H atoms.

An important factor in vicinal coupling is the dihedral angle α between the C—H σ bonds and whether it is fixed. As described quantitatively by the Karplus equation, named after Martin Karplus, coupling is maximized when the angle α is 0° and 180° and is minimized when α is 90°.

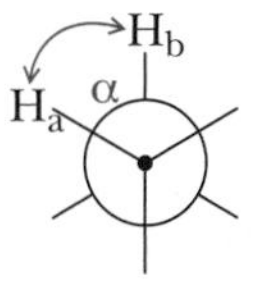

Hb
α
Ha

Bonds that rotate rapidly at room temperature do not have a fixed angle between adjacent C—H bonds, so an average angle and an average coupling are observed. This latter concept is important for the interpretation of ^{1}H-NMR spectra for alkanes and other flexible molecules.

It should be noted here that all the nuclei of adjacent hydrogens couple. It is only when coupling is between nonequivalent hydrogens that signal splitting results; coupling between equivalent hydrogens, whether they are on the same or adjacent carbons, does not produce signal splitting.

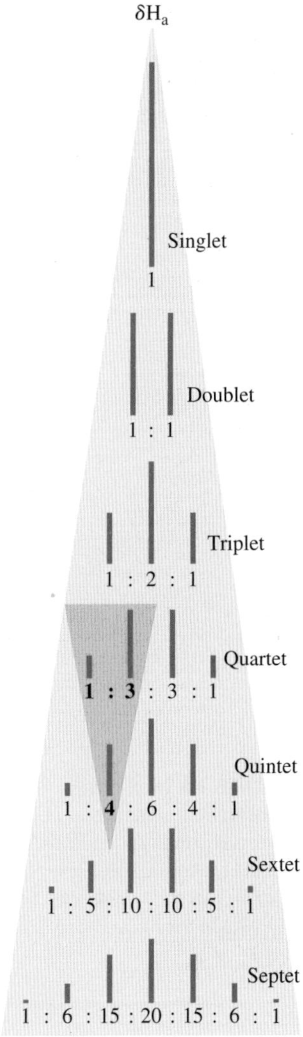

δHa
Singlet
1
Doublet
1 : 1
Triplet
1 : 2 : 1
Quartet
1 : 3 : 3 : 1
Quintet
1 : 4 : 6 : 4 : 1
Sextet
1 : 5 : 10 : 10 : 5 : 1
Septet
1 : 6 : 15 : 20 : 15 : 6 : 1

FIGURE 13.17 Pascal's triangle. As illustrated by the highlighted entries, each entry is the sum of the values immediately above it to the left and the right.

C. More Complex Splitting Patterns

So far, we have concentrated on spin-spin coupling with only one other nonequivalent set of H atoms. However, more complex situations often arise in which the nuclei of a set of H atoms are coupled to the nuclei of more than one set of nonequivalent H atoms in molecules that do not have rapid bond rotation. In these situations, the coupling from adjacent nonequivalent sets of H atom nuclei *combines* to give more complex signal splitting patterns. Use of a tree diagram is helpful in understanding splitting in these cases. In a tree diagram, the different couplings are applied sequentially.

For example, the atom labeled H_b in Figure 13.18 is adjacent to nonequivalent atoms H_a and H_c on either side, so the resulting coupling will give rise to a so-called doublet of doublets (in other words, a signal with four peaks). Here, the signal for H_b is split into a doublet with coupling constant J_{ab} by H_a, and this doublet is split into a doublet of doublets with coupling constant J_{bc} by H_c. If no other H atoms were in the molecule to be considered, then the signal for H_a would be a doublet with coupling constant J_{ab} and the signal for H_c would be a doublet with coupling constant J_{bc}. This analysis assumes that the H_a-H_b and H_b-H_c coupling constants, J_{ab} and J_{bc}, are different. If J_{ab} and J_{bc} are equal, the peaks overlap, a situation discussed in Section 13.9F.

If H_c is a set of two equivalent H atoms and H_a is still a single H atom, then the observed coupling would be a doublet of triplets (in other words, a signal with six peaks). Again, we are assuming that $J_{ab} \neq J_{bc}$. The tree diagram in Figure 13.19 shows the complex pattern that results from this type of splitting. If no other H atoms were in the molecule to be considered, then the signal for H_a would be a doublet with

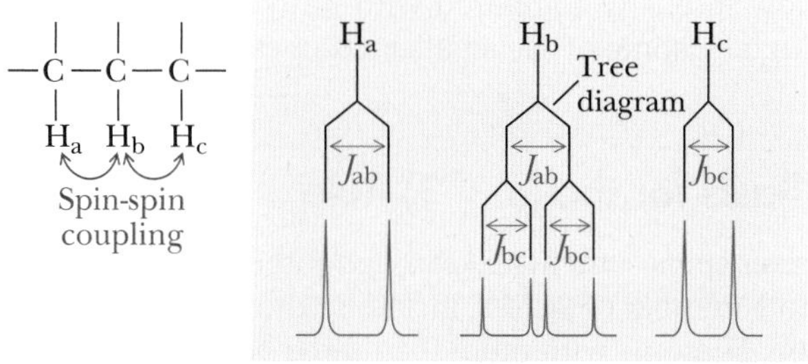

FIGURE 13.18 Coupling that arises when H_b is split by two different nonequivalent H atoms H_a and H_c. This analysis assumes that there is no other coupling in the molecule and that $J_{ab} \neq J_{bc}$.

coupling constant J_{ab} and the signal for H_c would be a doublet with coupling constant J_{bc}. The peaks for H_a and H_b would each integrate to a relative value of one H atom, while the peaks for H_c would integrate to a relative value of two H atoms.

In the general case, a signal will be split into $(n + 1) \times (m + 1)$ peaks for an H atom that is coupled to a set of n H atoms with one coupling constant and to a set of m H atoms with another coupling constant. Note that in a tree diagram, you get the same splitting patterns no matter in which order the two coupling constants are analyzed.

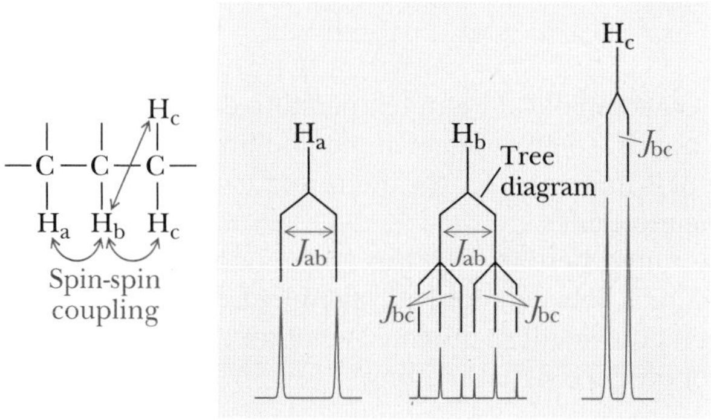

FIGURE 13.19 Complex coupling that arises when H_b is split by H_a and two equivalent atoms H_c. Again, this analysis assumes that there is no other coupling in the molecule and that $J_{ab} \neq J_{bc}$.

D. Bond Rotation

Because the angle between C—H bonds determines the extent of coupling in a molecule, bond rotation is a key parameter. In alkanes and other molecules with relatively free rotation about C—C σ bonds, H atoms bonded to the same C atom in $—CH_2—$ and $—CH_3$ groups are generally equivalent because of the rapid bond rotation. An exception is when a $—CH_2—$ is adjacent to a chiral center, a situation that is discussed in Section 13.10. However, when there is restricted bond rotation, as in alkenes and cyclic structures, H atoms bonded to the same C atom may not be equivalent, especially if the molecule is not symmetrical. Nonequivalent ^1H nuclei on the *same* C atom will couple to each other and cause splitting. This is referred to as **geminal coupling** (Figure 13.20). Geminal coupling constants are generally small, on the order of 0 to 5 Hz.

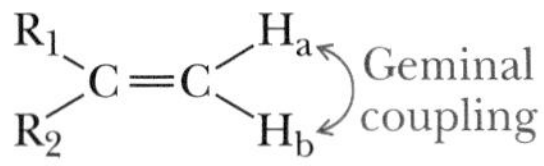

FIGURE 13.20 Geminal coupling that occurs when two H atoms on the same carbon atom are not equivalent. This is most common in unsymmetrical alkenes and cyclic molecules.

Because of the restricted rotation about C═C bonds, the alkenyl (vinylic) H atoms of unsymmetrical alkenes are not equivalent; in other words, they are in unique chemical environments. For example, ethyl propenoate (ethyl acrylate) is an unsymmetrical terminal alkene; therefore, the three alkenyl H atoms are nonequivalent (Figure 13.21). As a result, their nuclei couple with each other. In alkenes, *trans* coupling generally results in larger coupling constants (J_{trans} = 11–18 Hz) compared to *cis* coupling (J_{cis} = 5–10 Hz), with geminal coupling being by far the smallest (J_{gem} = 0–5 Hz). Unless a high-resolution spectrum is taken, the geminal coupling constant is so small that it is often difficult to see in terminal alkenes. In the spectrum of ethyl propenoate, the geminal coupling is only visible upon close inspection of the signals labeled *a* and *c*. You should be able to recognize the characteristic ethyl group pattern of a quartet integrating to two H atoms ($—CH_2—$, H_d) and a triplet integrating to three

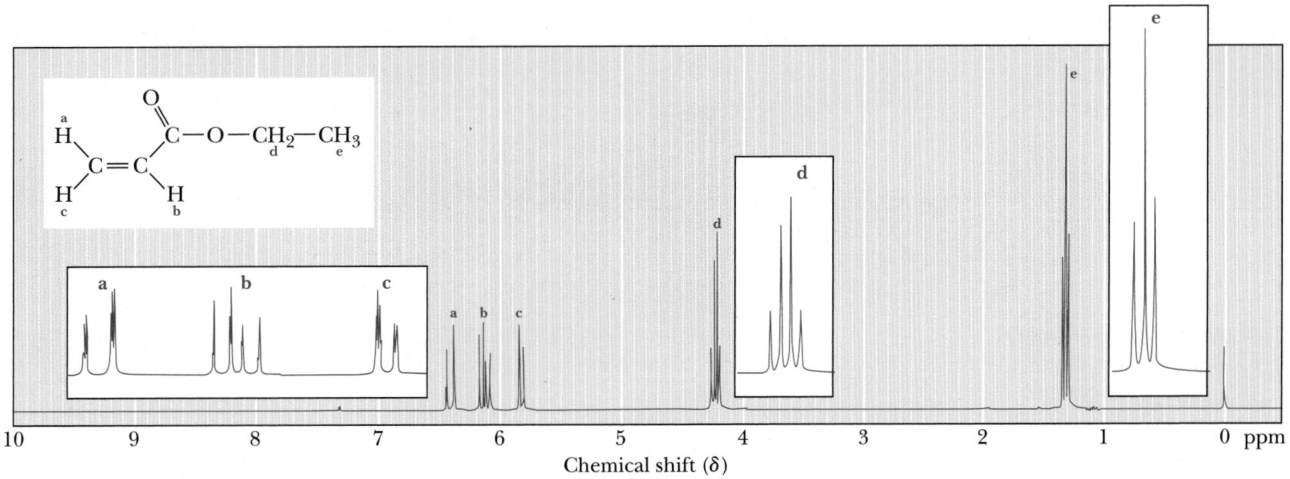

FIGURE 13.21 300 MHz ^{1}H-NMR spectrum of ethyl propenoate.

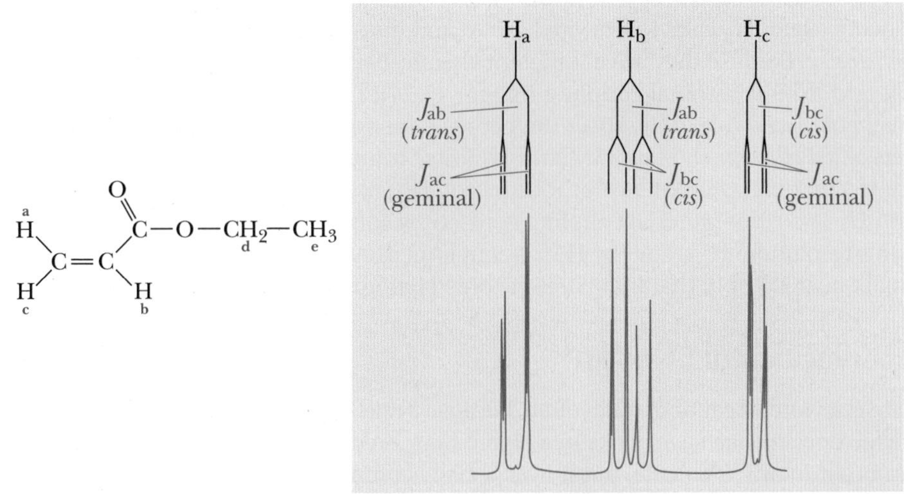

FIGURE 13.22 Tree diagrams for the complex coupling seen for the alkenyl H atoms in the ^{1}H-NMR spectrum of ethyl propenoate.

H atoms (—CH_3, H_e). Tree diagrams are provided in Figure 13.22 to help decipher patterns of the alkenyl signals.

Cyclic structures often exhibit restricted rotation about their C—C σ bonds and can have constrained conformations. The result is that the two H atoms on —CH_2— groups in cyclic molecules can be nonequivalent, leading to complex spin-spin coupling. Substituted epoxides such as 2-methyl-2-vinyloxirane provide a good example (Figure 13.23). The two H atoms on the three-membered epoxide ring are

FIGURE 13.23 300 MHz ^{1}H-NMR spectrum of 2-methyl-2-vinyloxirane. The two H atoms on the oxirane ring are nonequivalent, so they exhibit geminal coupling.

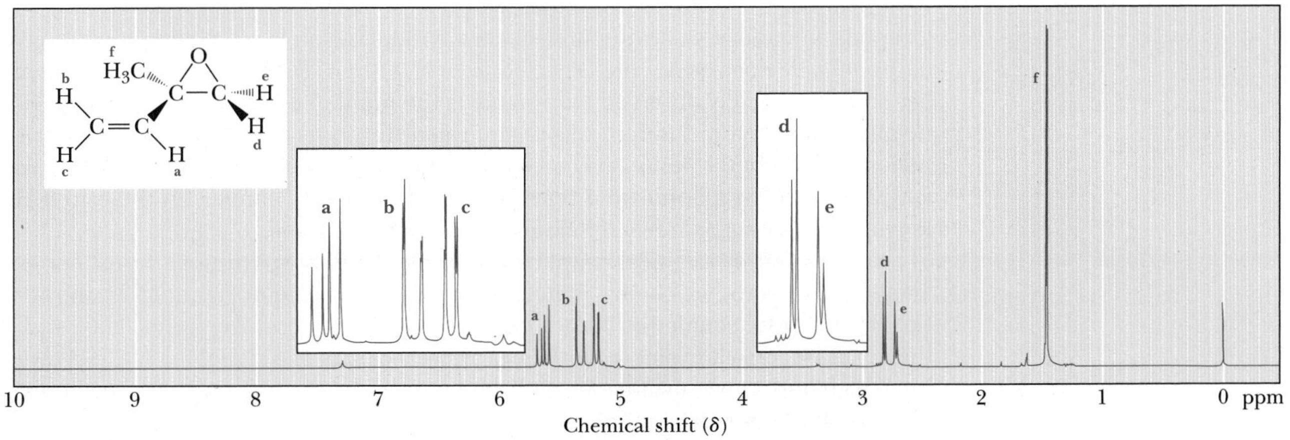

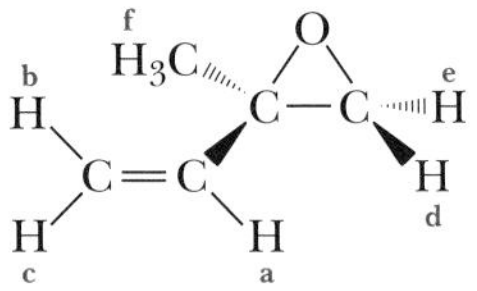
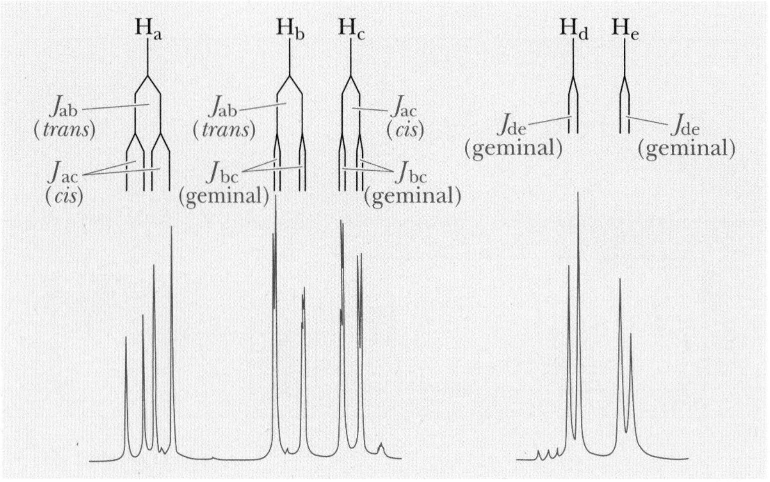

FIGURE 13.24 Tree diagrams that indicate the complex coupling seen in ¹H-NMR signals for the vinyl group and the oxirane ring H atoms of 2-methyl-2-vinyloxirane.

nonequivalent. H_d is *cis* to the vinyl group and *trans* to the methyl group, while H_e is the reverse. Because they are in different chemical environments, they are nonequivalent and exhibit geminal coupling (Figure 13.24). The geminal coupling constant is small but discernible in the spectrum because the signals for both H_d and H_e are doublets. Vinyl H atom H_a is split by both H_b (*trans* coupling) and H_c (*cis* coupling), giving rise to a doublet of doublets, or four peaks. H_b is split by H_a (*trans* coupling) along with H_c; the latter geminal coupling constant is so small that it is barely discernable at this resolution. H_c is split by H_a (*cis* coupling) as well as H_b (geminal coupling). The singlet near 1.5 ppm that integrates to three H atoms is the methyl group labeled *f*.

E. Coincidental Overlap

Here is a word of caution: quite often, because peaks can overlap by coincidence, there are fewer *distinguishable* peaks in a signal than predicted. Coincidental peak overlap can occur in any molecule, but it is especially common with flexible alkyl chains. In addition, some coupling constants are so small that peak splitting is hard to see in a spectrum. Thus, the predicted number of peaks using the $(n + 1) \times (m + 1)$ rule should be considered the maximum that *might* be observed. Detailed analysis using extremely high resolution spectrometers is often required to distinguish all the peaks in a highly split signal. You should note also that the types of splitting patterns we have described are applicable only when the separation between coupled signals is much greater than the coupling constant. When this is not the case, spectra can become much more complex.

F. Complex Coupling in Flexible Molecules

Coupling in molecules having unrestricted bond rotation is often simplified to give only $m + n + 1$ peaks, not the expected $(n + 1) \times (m + 1)$. In other words, the number of peaks actually observed for a signal is the number of adjacent hydrogens + 1, no matter how many different sets of equivalent H atoms this represents. The explanation is that bond rotation averages the coupling constants throughout molecules with freely rotating bonds and tends to make them very similar (in the 6 to 8 Hz range) for H atoms on freely rotating sp^3 hybridized C atoms.

Very similar or identical coupling constants simplify splitting patterns. For example, in the hypothetical unsymmetrical molecule depicted in Figure 13.25, the central H_b atoms are coupled to both H_a atoms as well as to both H_c atoms. If $J_{ab} \neq J_{bc}$, this would lead to a triplet of triplets, or nine peaks, in the signal for H_b. However, if the coupling constants are identical so that $J_{ab} = J_{bc}$, the splitting pattern overlaps considerably to generate only five peaks in the signal for H_b. In general, simplification

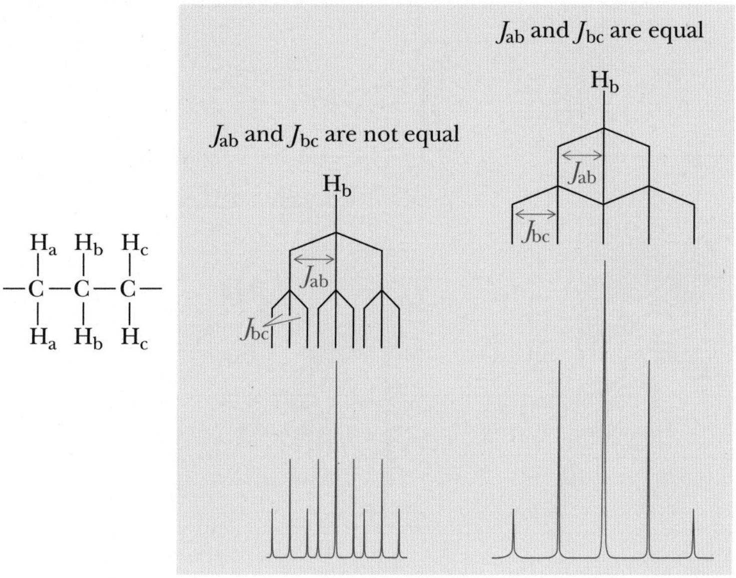

FIGURE 13.25 Simplification of signal splitting that occurs when coupling constants are the same.

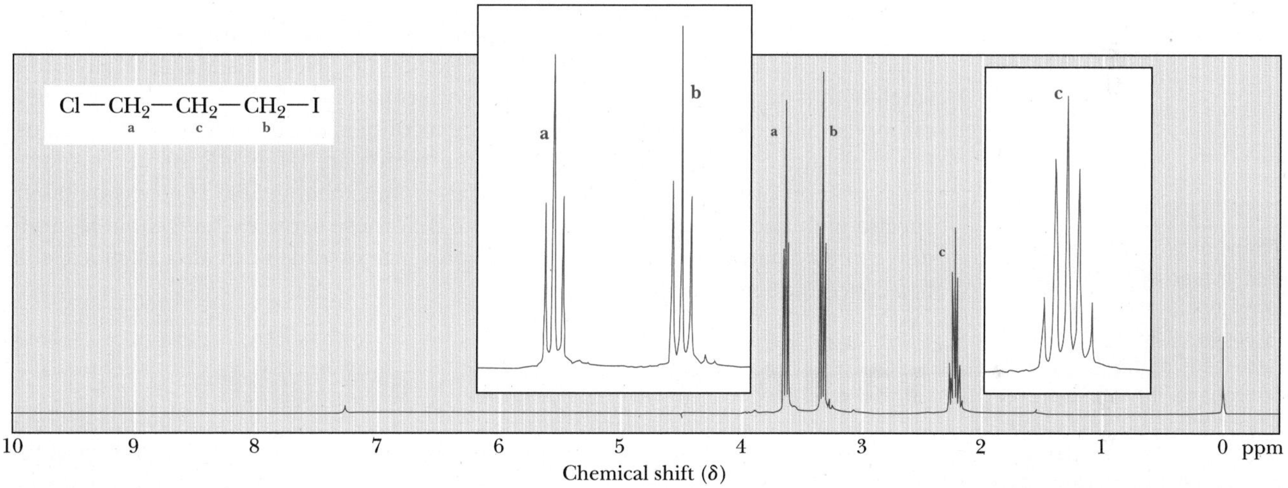

FIGURE 13.26 300 MHz ^{1}H-NMR spectrum of 1-chloro-3-iodopropane.

because of very similar or identical J values gives a number of peaks equal to the number of adjacent H atoms + 1, regardless of patterns of equivalence.

A good example of peak overlap occurs in the spectrum of 1-chloro-3-iodopropane (Figure 13.26). The signal for the H atoms of the central $—CH_2—$ group (labeled *c* on the spectrum) is split by the H atoms on both of the other $—CH_2—$ groups, raising the possibility of splitting into $3 \times 3 = 9$ peaks. However, because the values for J_{ab} and J_{bc} are so similar, only $4 + 1 = 5$ peaks are distinguishable in the spectrum for the H_c signal as a result of peak overlap.

Another common example is the kind of splitting of the signal for the central $—CH_2—$ in a $—CH_2—CH_2—CH_3$ group, such as occurs in the molecule 1-chloropropane, $Cl—CH_2—CH_2—CH_3$ (Figure 13.27). A maximum of $3 \times 4 = 12$ peaks would be possible for the central $—CH_2—$ signal (labeled *b* in Figure 13.27), but because the coupling constants are very similar, only $5 + 1 = 6$ peaks are distinguishable.

G. Fast Exchange

Hydrogen atoms bonded to oxygen or nitrogen atoms can exchange with each other faster than the time it takes to acquire a ^{1}H-NMR spectrum. This process is greatly facilitated by even traces of acid or base in a sample. Important affected functional

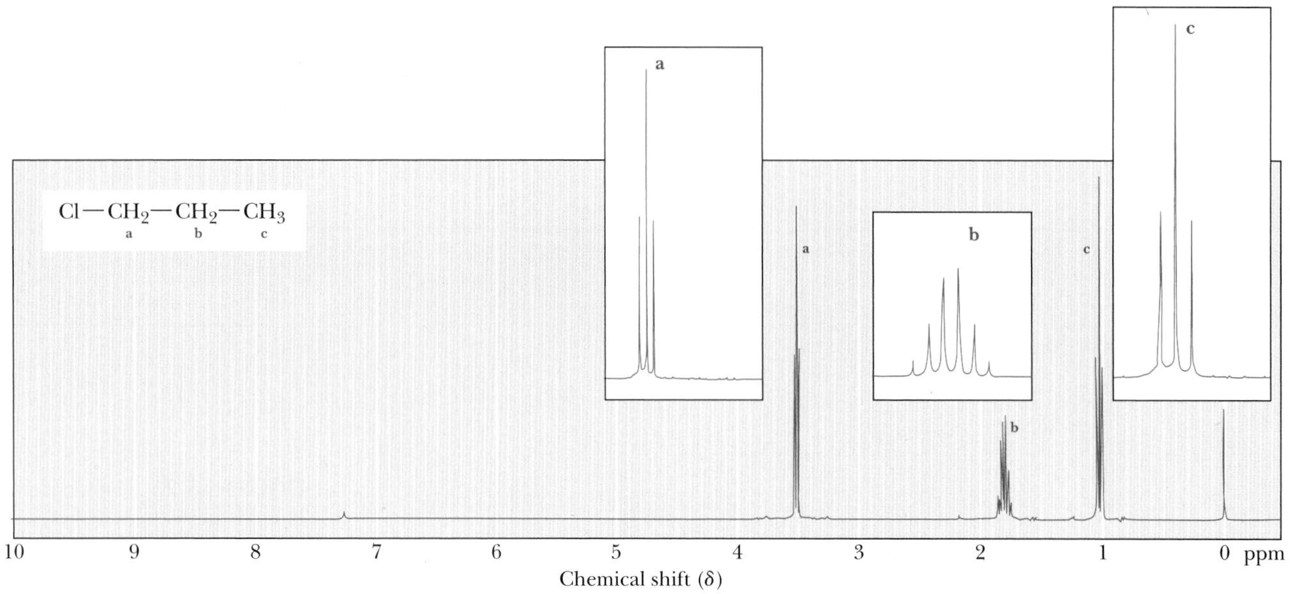

FIGURE 13.27 300 MHz ^{1}H-NMR spectrum of 1-chloropropane.

groups include carboxylic acids, alcohols, amines, and amides. Fast exchange has two important consequences. First, signals for exchanging H atoms are generally broad singlets that do not take part in splitting with other signals. Second, the signal will disappear altogether if D_2O or a deuterated alcohol is added to the sample because the H atoms will be replaced with D atoms, which are ^{1}H-NMR silent. This latter phenomenon can be used to identify signals from exchangeable H atoms by taking spectra with and without added D_2O. Note that these same exchangeable H atoms also can take part in hydrogen bonds, the presence of which can alter chemical shift in a concentration-dependent fashion.

13.10 Stereochemistry and Topicity

The discussion of the number of equivalent hydrogens given at the beginning of this chapter is slightly oversimplified because stereochemistry can affect chemical shift. Depending on the symmetry of the molecule, otherwise equivalent atoms may be **homotopic**, **enantiotopic**, or **diastereotopic**. The simplest way to visualize the topicity of a molecule (i.e., which of the these classes it falls into) is by mentally substituting one of the atoms or groups of atoms by an isotope and then deciding whether the resulting compound would be (a) the same or (b) different from its mirror image or whether (c) diastereomers are possible. Depending on the outcome of the test, the atoms or groups are homotopic, enantiotopic, or diastereotopic, respectively.

Consider the following molecules.

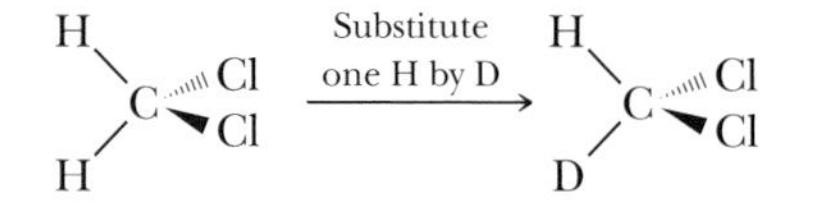

Substitution does not create a chiral center; therefore, hydrogens are homotopic.

Dichloromethane (achiral) achiral

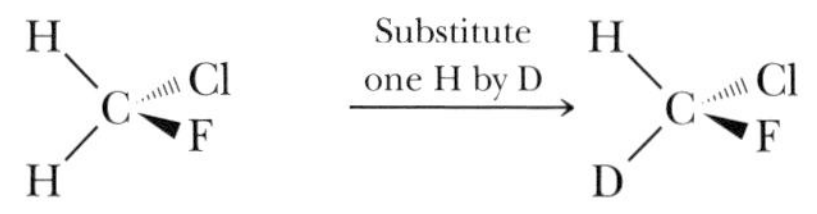

Substitution produces a chiral center; therefore, hydrogens are enantiotopic. Both hydrogens are prochiral—one is pro-*R*-chiral; the other is pro-*S*-chiral.

Chlorofluoromethane (achiral) chiral

Homotopic groups

Atoms or groups on an atom that give an achiral molecule when one of the groups is replaced by an isotope. The hydrogens of the CH_2 group of propane, for example, are homotopic. Replacing either one of them with deuterium gives 2-deuteropropane, which is achiral. Homotopic groups have identical chemical shifts under all conditions.

Enantiotopic groups

Atoms or groups on an atom that give a chiral center when one of the groups is replaced by an isotope. A pair of enantiomers results. The hydrogens of the CH_2 group of ethanol, for example, are enantiotopic. Replacing one of them by deuterium gives (*R*)-1-deuteroethanol; replacing the other gives (*S*)-1-deuteroethanol. Enantiotopic groups have identical chemical shifts in achiral environments but different chemical shifts in chiral environments.

If one hydrogen in dichloromethane is substituted with one deuterium, an achiral compound results. This molecule is identical to its mirror image, and the two hydrogens in dichloromethane are equivalent and homotopic. Homotopic groups have identical chemical shifts in all environments.

If one hydrogen of chlorofluoromethane is substituted with deuterium, the resulting molecule is chiral and not identical to its mirror image. The two hydrogens in this compound are therefore enantiotopic. Enantiotopic hydrogens have identical chemical shifts *except in chiral environments*. In a chiral solvent, for example, the two hydrogens would have different chemical shifts. While the distinction between homotopic and enantiotopic compounds is of little practical consequence in NMR spectroscopy, the two hydrogens in chlorofluoromethane can be distinguished by enzymes, which also provide a chiral environment. The CH_2 hydrogens in this molecule are said to be **prochiral.**

The compound 2-butanol presents a more complex situation.

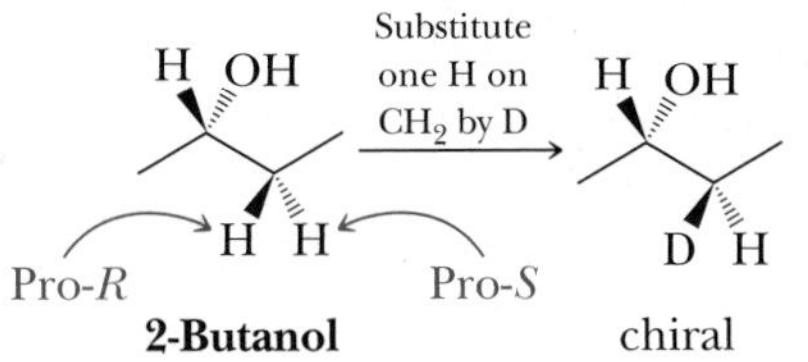

2-Butanol (chiral) chiral

Substitution creates a chiral center that is diastereomeric; therefore, hydrogens are diastereotopic. Both hydrogens are prochiral—one is pro-*R*-chiral; the other is pro-*S*-chiral.

Diastereotopic groups

Atoms or groups on an atom that are bonded to an atom that is bonded to two nonidentical groups, one of which contains a chiral center. When one of the atoms or groups is replaced by an isotope, a new chiral center is created and a set of diastereomers results. The hydrogens of the CH_2 group of 2-butanol, for example, are diastereotopic. Diastereotopic groups have different chemical shifts under all conditions, although the differences are only seen for diastereotopic hydrogens very close to a chiral center.

Molecules such as 3-methyl-2-butanol are even more complicated. If a hydrogen on one of the methyl groups on carbon-3 of 3-methyl-2-butanol is substituted with a deuterium, a new chiral center is created. Because there is already one chiral center, diastereomers are now possible. Thus, the methyl groups on carbon-3 of 3-methyl-2-butanol are diastereotopic. Diastereotopic hydrogens have different chemical shifts under all conditions, which can lead to unexpected complexity in spectra of simple compounds. The ^{1}H-NMR spectrum of 3-methyl-2-butanol is shown in Figure 13.28. The methyl groups on carbon-3 are nonequivalent and give two doublets rather than one doublet of twice the intensity, which would be expected if they were equivalent.

Any molecule with a chiral center near two otherwise identical groups on a carbon with a third substituent has the potential for diastereotopicity. Of course, like any other nonequivalent groups, diastereotopic groups may have very similar or accidentally identical chemical shifts. Generally, the shift differences fall off rapidly with increasing distance from the chiral center.

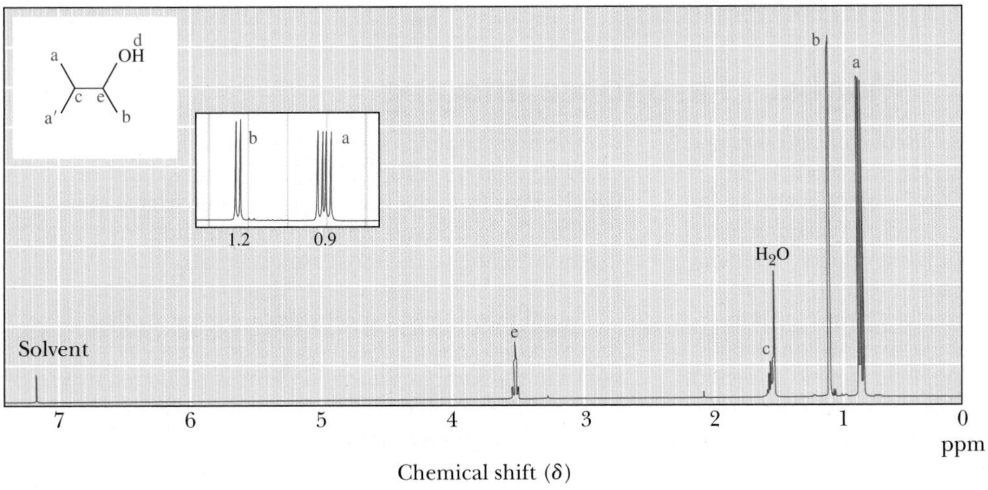

FIGURE 13.28 ^{1}H-NMR spectrum of 3-methyl-2-butanol (500 MHz). The methyl groups on carbon-3 (c) are diastereotopic and therefore nonequivalent. They appear as two doublets.

Example 13.7 Stereochemistry and Chemical Shift

Indicate whether the highlighted hydrogens in the following compounds are homotopic, enantiotopic, or diastereotopic.

(a) 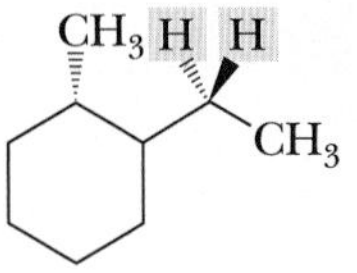

(b) CH_3CH_2Br

(c) 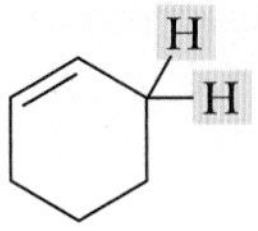

Solution

(a) Diastereotopic (near a chiral center). These hydrogens will have different chemical shifts.

(b) Enantiotopic. These hydrogens will have the same chemical shift except in chiral environments.

(c) Enantiotopic [see part (b)].

Problem 13.7

Following is a ^1H-NMR spectrum of 2-butanol. Explain why the CH_2 protons appear as a complex multiplet rather than as a simple quintet.

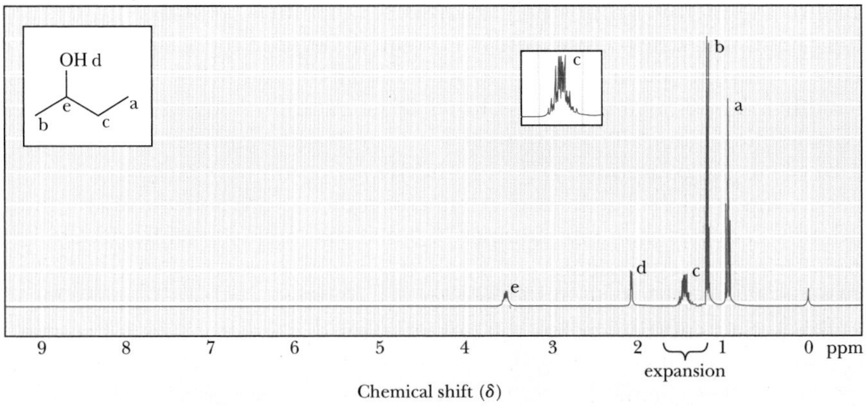

Chemical Connections

Magnetic Resonance Imaging

The NMR phenomenon was discovered and explained by physicists in the 1950s, and by the 1960s, it had become an invaluable analytical tool for chemists. By the early 1970s, scientists realized that imaging parts of the body using NMR could be a valuable addition to diagnostic medicine. Because the term *nuclear magnetic resonance* sounds to many people as if the technique might involve radioactive material, health care personnel call the technique magnetic resonance imaging (MRI). MRI has become so important that in 2003, the Nobel Prize in Medicine or Physiology was awarded to Paul Lauterbur and Peter Mansfield for their discoveries that led to practical MRI.

The body contains several nuclei that, in principle, could be used for MRI. Of these, hydrogens, most of which come from the hydrogens of water, triglycerides (Section 26.1), and membrane phospholipids (Section 26.5) give the most useful signals. Phosphorus MRI is also used in diagnostic medicine.

Recall that in NMR spectroscopy, energy in the form of radio-frequency radiation is absorbed by nuclei in the sample. Relaxation time is a characteristic time at which excited nuclei give up this energy and relax to their ground state.

In 1971, it was discovered that relaxation of water in certain cancerous tumors takes much

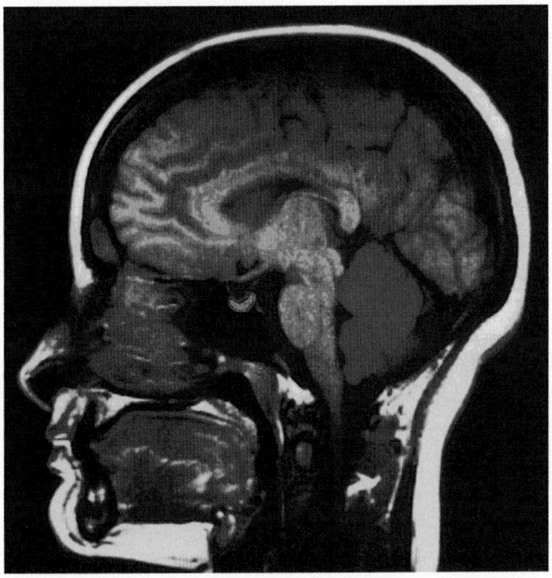
Scott Camazine/Science Source

Computer-enhanced MRI scan of a normal human brain with pituitary gland highlighted.

longer than the relaxation of water in normal cells. Thus, if a relaxation image of the body could be obtained, it might be possible to identify tumors at an early stage. Subsequent work demonstrated that many tumors can be identified this way. Another important application of MRI is in the examination of the brain and spinal cord. White and gray matter are easily distinguished by MRI, which is useful in the study of such diseases as multiple sclerosis. Magnetic resonance imaging and X-ray imaging are, in many cases, complementary. The hard, outer layer of bone is essentially invisible to MRI but shows up extremely well in X-ray images, whereas soft tissue is nearly transparent to X-rays but shows up in MRI.

The key to any medical imaging technique is to know which part of the body gives rise to which signal. In MRI, spatial information is encoded using magnetic field gradients. Recall that a linear relationship exists between the frequency at which a nucleus resonates and the intensity of the magnetic field. In ^{1}H-NMR spectroscopy, we use a homogeneous magnetic field, in which all equivalent hydrogens absorb at the same radio frequency and have the same chemical shift. In MRI, the patient is placed in a magnetic field gradient that can be varied from place to place. Nuclei in the weaker magnetic field gradient absorb at a lower frequency. Nuclei elsewhere in the stronger magnetic field absorb at a higher frequency. In a magnetic field gradient, a correlation exists between the absorption frequency of a nucleus and its position in space. A gradient along a single axis images a plane. Two mutually perpendicular gradients image a line segment, and three mutually perpendicular gradients image a point. In practice, more complicated procedures are used to obtain magnetic resonance images, but they are all based on the idea of magnetic field gradients.

Watch a video explanation

13.11 ^{13}C-NMR

The development of **^{13}C-NMR** spectroscopy lagged behind ^{1}H-NMR spectroscopy primarily because of two problems. One is the particularly low natural abundance of ^{13}C (only 1.1%) and the resulting weak signal. The second problem is that the magnetic moment of ^{13}C is considerably smaller than that of ^{1}H, which causes the population of the higher and lower nuclear spin states to differ by much less than that for ^{1}H. Taken in combination, these two factors mean that ^{13}C-NMR signals in natural samples (those not artificially enriched with carbon-13) are only about 10^{-4} times the strength of ^{1}H-NMR signals. Even though ^{1}H-NMR spectroscopy became a routine analytical tool in the mid-1960s, it was not until 20 years later, with the development of FT-NMR techniques, that ^{13}C-NMR spectroscopy became widely available as a routine analytical tool.

As with ^{1}H-NMR spectra, splitting patterns in ^{13}C-NMR spectra are also explained according to the $(n + 1)$ rule. Because in natural abundance only 1.1% of carbon atoms are ^{13}C, almost all ^{13}C atoms in a molecule have only magnetically inactive ^{12}C next to them; therefore, ^{13}C—^{13}C signal splitting is not normally observed. However, the signal from a ^{13}C nucleus is split by the hydrogens bonded to it. The signal for a ^{13}C atom with three attached hydrogens is split to a quartet, that for an atom of ^{13}C with two attached hydrogens is split to a triplet, and so on. The ^{13}C—H signal splitting provides important information about the number of hydrogen atoms bonded to carbon. The disadvantage of $^{13}C—H$ signal splitting is that coupling constants of between 100 and 250 Hz are common. Coupling constants of this magnitude correspond to 1.33 to 3.33 ppm at 75 MHz, which means that the overlap among signals can be significant and that splitting patterns are often difficult to determine. In addition, there are smaller but significant couplings from hydrogens that are not directly bonded to the carbon, but are separated by two or three bonds. This extensive splitting causes the already weak signals of the ^{13}C to split into many smaller peaks that are easily lost in the noise. For this reason, the most common mode of operation of a ^{13}C-NMR spectrometer is a hydrogen-decoupled mode. (See Problem 13.25 for an interesting problem on the use of coupling constants to determine orbital hybridization.)

In the hydrogen-decoupled mode, the sample is irradiated with two different radio frequencies. The first radio frequency is used to excite the ^{13}C nuclei. The second is a broad spectrum of frequencies that causes all hydrogens in the molecule to undergo rapid transitions among their nuclear spin states. On the time scale of a ^{13}C-NMR spectrum, each hydrogen is in a time average of the two states, with the result that ^{1}H-^{13}C spin-spin interactions are not observed. The term for this process is *spin-spin decoupling*. In a hydrogen-decoupled spectrum, all ^{13}C signals appear as singlets. The hydrogen-decoupled ^{13}C-NMR spectrum of 1-bromobutane (Figure 13.29) consists of four singlets.

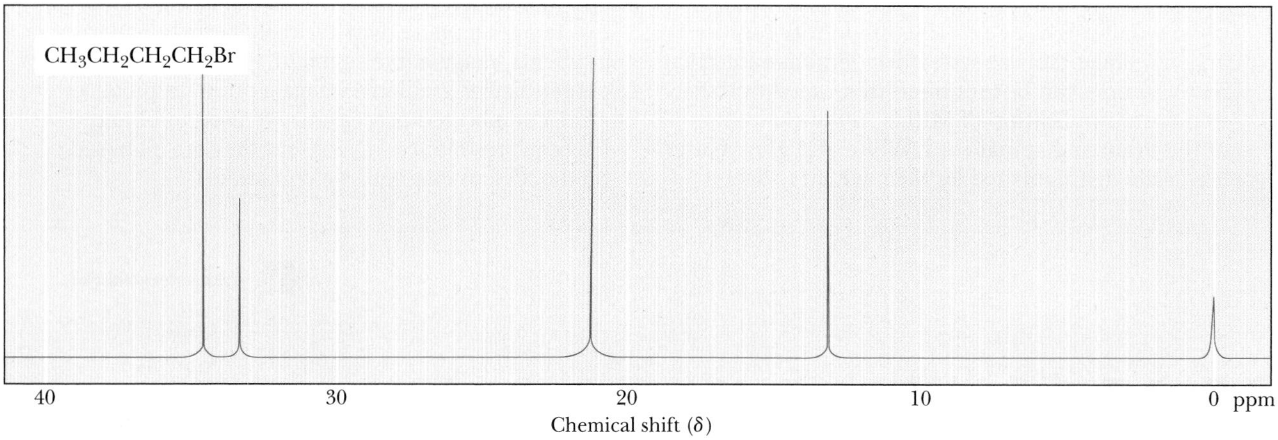

FIGURE 13.29 Hydrogen-decoupled ^{13}C-NMR spectrum of 1-bromobutane.

Figure 13.30 shows approximate chemical shifts for ^{13}C-NMR. Notice how much wider the range of chemical shifts is for ^{13}C-NMR than for ^{1}H-NMR. Most chemical shifts for ^{1}H-NMR fall within a rather narrow range of 0 to 10 ppm; however, those for ^{13}C-NMR cover 0 to 220 ppm. Because of this expanded scale, it is very unusual to find any two nonequivalent carbons in the same molecule with identical chemical shifts. Most commonly, each different type of carbon within a molecule has a distinct signal clearly resolved from all other signals.

Notice further that the chemical shift of carbonyl carbons is quite distinct from those of sp^3 hybridized carbons and of other types of sp^2 hybridized carbons. The presence or absence of a carbonyl carbon is quite easy to recognize in a ^{13}C-NMR spectrum. Note that signals from sp^2 hybridized carbons fall in a distinctive range of 100 to 160 ppm.

A great advantage of ^{13}C-NMR spectroscopy is that it is possible to count the number of types of carbon atoms in a molecule. There is one caution here, however. Because of certain complications, including the long relaxation times of ^{13}C nuclei, it is generally not possible to determine the number of carbons of each type by integration of signal areas.

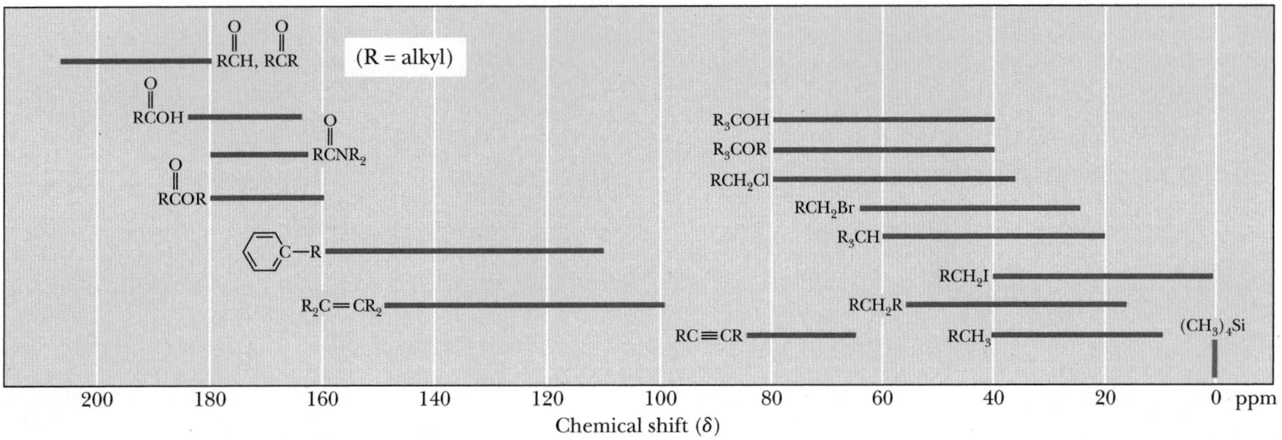

FIGURE 13.30 ^{13}C-NMR chemical shifts of representative groups. These values are approximate. Other atoms or groups in the molecules may cause signals to appear outside of these ranges.

Example 13.8 Predicting ^{13}C-NMR Spectra

Predict the number of signals in a proton-decoupled ^{13}C-NMR spectrum of each compound.

(a) $CH_3\overset{O}{\overset{\|}{C}}OCH_3$ **(b)** $CH_3CH_2CH_2\overset{O}{\overset{\|}{C}}CH_3$ **(c)** $CH_3CH_2\overset{O}{\overset{\|}{C}}CH_2CH_3$

Solution

Following are the number of signals in the proton-decoupled spectrum of each compound, along with the chemical shifts of each signal. The chemical shifts of the carbonyl carbons are quite distinctive (Figure 13.30) and in these examples occur at δ 171.37, 208.85, and 211.97.

(a) Methyl acetate: three signals (δ 171.37, 51.53, and 20.63)

(b) 2-Pentanone: five signals (δ 208.85, 45.68, 29.79, 17.35, and 13.68)

(c) 3-Pentanone: three signals (δ 211.97, 35.45, and 7.92)

Problem 13.8

Explain how to distinguish between the members of each pair of constitutional isomers based on the number of signals in the proton-decoupled ^{13}C-NMR spectrum of each member.

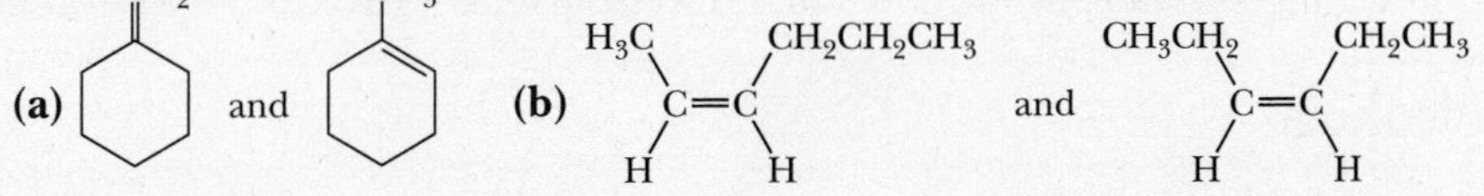

13.12 Interpretation of NMR Spectra

A. Alkanes

All hydrogens in alkanes are in very similar chemical environments; therefore, ^{1}H-NMR chemical shifts of alkane hydrogens fall within a narrow range of δ 0.8 to 1.7. Chemical shifts for alkane carbons in ^{13}C-NMR spectroscopy fall within the considerably wider range of δ 10 to 60.

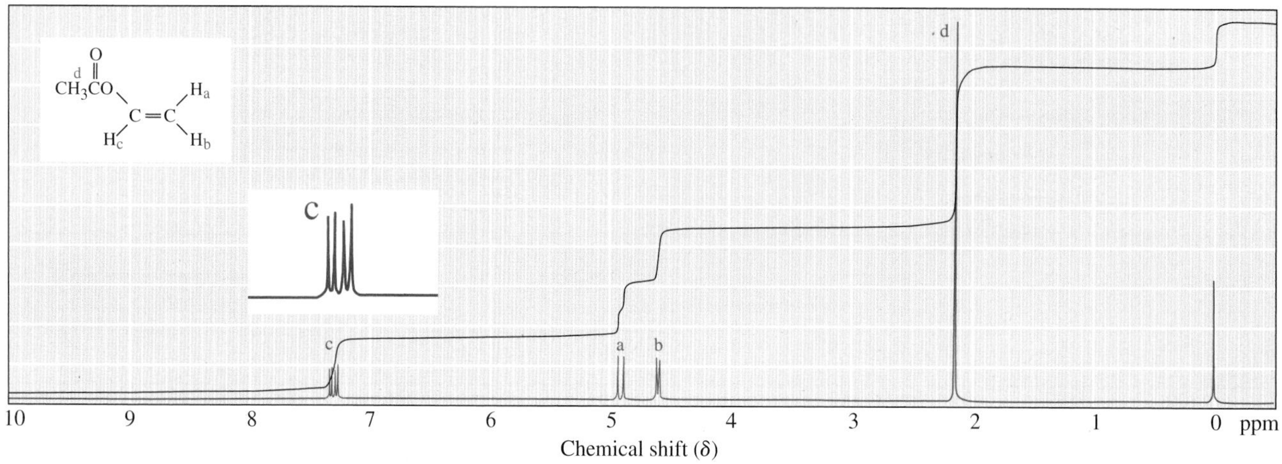

FIGURE 13.31 ^{1}H-NMR spectrum of vinyl acetate.

B. Alkenes

The chemical shifts of vinylic hydrogens are larger than those of alkane hydrogens and typically fall in the range δ 4.6 to 5.7. Vinylic hydrogens are deshielded by the sp^2 hybridized carbons of the double bond (Section 13.7C) and the local magnetic field induced in the π bond of alkenes (Section 13.7C). The splitting pattern observed in the ^{1}H-NMR spectrum of vinyl acetate (Figure 13.31) is typical of monosubstituted alkenes. The singlet at δ 2.12 represents the three hydrogens of the methyl group. The terminal vinylic hydrogens appear at δ 4.58 and δ 4.90. The internal vinylic hydrogen, which normally appears in the range δ 5.0 to 5.7, is shifted farther downfield to δ 7.30 as a result of deshielding by the adjacent electronegative oxygen atom of the ester.

As shown in Table 13.4, coupling constants are generally larger for *trans* vinylic hydrogens (11–18 Hz) than for *cis* vinylic hydrogens (5–10 Hz), and it is often possible to distinguish between *cis*- and *trans*-alkenes by an analysis of their coupling constants. It is also possible to distinguish between vicinal hydrogens and geminal hydrogens (=CH_2), the latter having small coupling constants generally in the 0 to 5 Hz range.

The signal of each vinylic hydrogen in vinyl acetate is predicted to be a doublet of doublets. The signal for H_c, for example, is split to a doublet by coupling with H_a and further split to a doublet of doublets by coupling with H_b. For H_a and H_b, the geminal coupling is so small that it is not visible at this resolution; so their signals appear as doublets. Higher resolution would reveal the geminal coupling.

The sp^2 hybridized carbons of alkenes give ^{13}C-NMR signals in the range δ 100 to 150 ppm (Figure 13.30), which is considerably downfield from sp^3 hybridized carbons.

C. Alcohols

The chemical shift of a hydroxyl hydrogen in a ^{1}H-NMR spectrum is variable and depends on the purity of the sample, the solvent, the concentration, and the temperature. It often appears in the range δ 3.0 to 4.0, but depending on experimental conditions, it may appear as far upfield as δ 0.5. Hydrogens on the carbon bearing the —OH group are deshielded by the electron-withdrawing inductive effect of the oxygen atom, and their signals also typically appear in the range δ 3.4 to 4.0. Shown in Figure 13.32 is the ^{1}H-NMR spectrum of 1-propanol.

Signal splitting between the hydrogen on O—H and its neighbors on the adjacent —CH_2— group is seen in the ^{1}H-NMR spectrum of 1-propanol. However, this splitting is rarely seen. The reason is that most samples of alcohol contain traces of acid, base, or other impurities that catalyze the transfer of the hydroxyl proton from

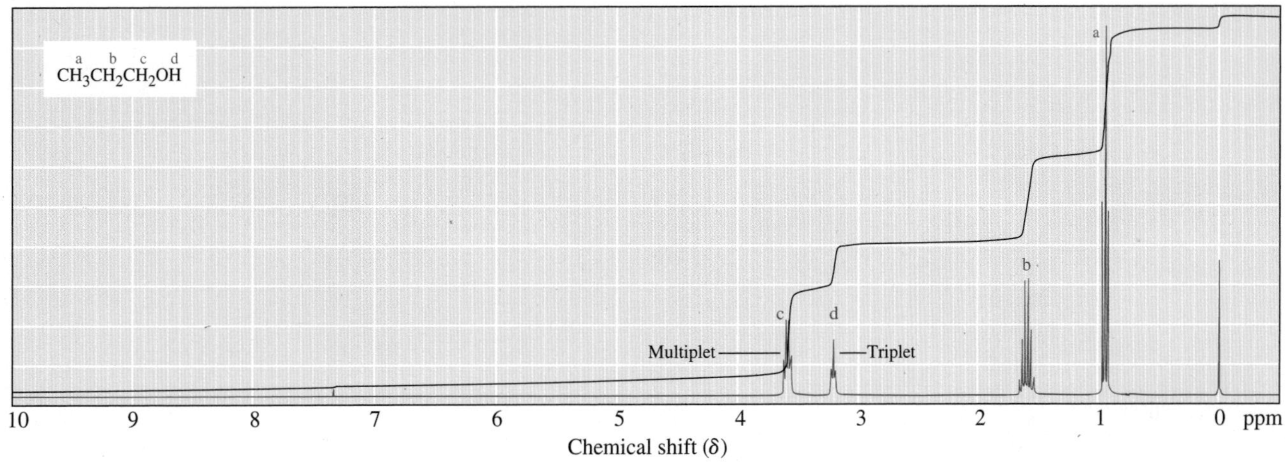

FIGURE 13.32 ^{1}H-NMR spectrum of 1-propanol. The hydroxyl hydrogen appears at δ 3.18 as a narrowly spaced triplet. The signal of hydrogens on carbon 1 of 1-propanol appears as a quartet (labeled in the spectrum as a multiplet) at δ 3.56 (split by the two CH_2 hydrogens and the one OH hydrogen).

the oxygen of one alcohol molecule to the oxygen of another alcohol molecule. This fast exchange decouples the hydroxyl proton from all other nuclei in the molecule (Section 13.9G). For this same reason, the hydroxyl proton does not usually split the signal of any α-hydrogens.

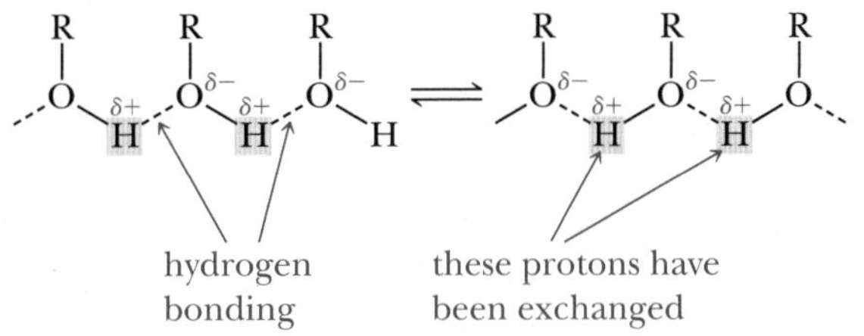

D. Ethers

The most distinctive feature of the ^{1}H-NMR spectra of ethers is the chemical shift of hydrogens on the carbons bonded to the ether oxygen. Signals for this type of hydrogen fall in the range δ 3.3 to 4.0, which corresponds to a downfield shift of approximately 2.4 units compared with their normal position in alkanes. The chemical shifts of H—C—O— hydrogens in ethers are similar to those seen for comparable H—C—OH hydrogens of alcohols.

E. Aldehydes and Ketones

The signal for an aldehyde hydrogen typically appears between δ 9.5 and δ 10.1 in a ^{1}H-NMR spectrum. Because almost nothing else absorbs in this region, it is very useful for identification. Hydrogens on an α-carbon (carbon directly adjacent to carbonyl) of an aldehyde or a ketone appear around δ 2.2 to 2.6. The carbonyl carbons of aldehydes and ketones have characteristic positions in the ^{13}C-NMR between δ 180 and δ 215 (and can be distinguished from carboxylic acid derivatives, which absorb at a higher field).

F. Carboxylic Acids and Esters

Signals for hydrogens on the α-carbon to a carboxyl group in acids and esters appear in a ^{1}H-NMR spectrum in the range δ 2.0 to 2.6. The hydrogen of a carboxyl group gives a very distinctive signal in the range δ 10 to 13, downfield

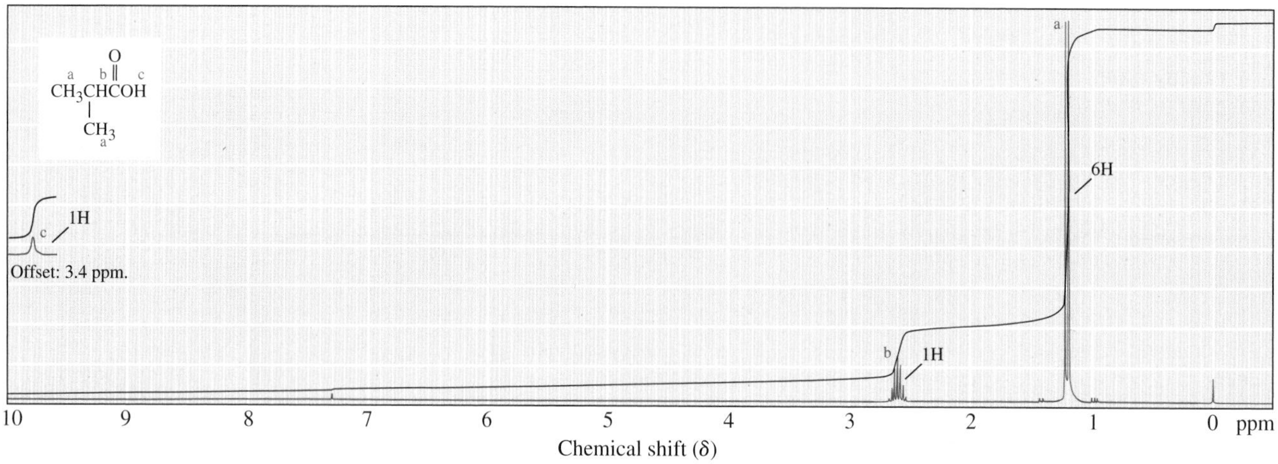

FIGURE 13.33 ^{1}H-NMR spectrum of 2-methylpropanoic acid (isobutyric acid).

of most other types of hydrogens [even farther downfield than that of an aldehyde hydrogen (δ 9.5–10.1)] and serves to distinguish carboxyl hydrogens from most other types of hydrogens. The ^{1}H-NMR signal for the carboxyl hydrogen of 2-methylpropanoic acid, for example, appears at δ 13.2 and is shown at the left in Figure 13.33.

The ^{13}C resonance of the carboxyl carbon in acids and esters appears in the range δ 165 to 185 and at a distinctly higher field than that in ketones. Hydrogens α to an ester oxygen are strongly deshielded and resonate between δ 3.7 and 4.7, more downfield than in alcohols and ethers.

G. Amines

The chemical shifts of amine hydrogens, like those of hydroxyl hydrogens (Section 13.12C), vary between δ 0.5 and δ 5.0, depending on experimental conditions due to hydrogen bonding. As in alcohols, exchange is fast enough that spin-spin splitting between amine hydrogens and hydrogens on adjacent α-carbons is averaged. Thus, amine hydrogens generally appear as broad singlets. Coupling to ^{14}N (beyond the scope of this text) causes these signals to broaden. Hydrogens α to the amine nitrogen appear around δ 2.5 ppm, about 1 ppm higher than for hydrogens α to oxygen in ethers and alcohols.

Carbons bonded to nitrogen appear in the ^{13}C-NMR spectrum approximately 20 ppm higher than in alkanes of comparable structure, but about 20 ppm below carbons attached to oxygen in ethers or alcohols.

H. Final Word

Here is a final word. We have barely scratched the surface of what NMR can do. Something called the Nuclear Overhauser Enhancement (NOE) can determine distances between atoms in molecules that are near each other in three-dimensional space even if more than three bonds separate them. In addition, using modern instruments, spectra that examine multiple parameters simultaneously can be produced to yield immense amounts of information about even very complicated molecules. Such spectra are plotted on more than one axis, so they are referred to as multidimensional spectra. Multidimensional spectra are used to deduce structure and conformation of molecules ranging from small organic molecules to large biological macromolecules such as proteins, DNA, and RNA.

HOW TO

Solve NMR Spectral Problems

One of the first steps in determining molecular structure is establishing the molecular formula. In the past, this task was most commonly done by elemental analysis, combustion analysis to determine percent composition, molecular weight determination, and so forth. More commonly today, molecular weight and molecular formula are determined by mass spectrometry (Chapter 14). In the examples that follow, we assume that the molecular formula of any unknown compound has already been determined, and we proceed from that point using spectral analysis to determine a structural formula.

A. Molecular Formula

Valuable information about the structural formula of an unknown compound can be obtained by inspecting its molecular formula, which gives its index of hydrogen deficiency. Refer to Chapter 5 to review this technique. A molecular formula can often be obtained from the mass spectrum (see Chapter 14).

Watch a video explanation

B. From a ^{1}H-NMR Spectrum to a Structural Formula

The following steps may prove helpful as a systematic approach to solving spectral problems.

Step 1: Molecular formula and index of hydrogen deficiency. Examine the molecular formula, calculate the index of hydrogen deficiency, and deduce what information you can about the presence or absence of rings or π bonds.

Step 2: Number of signals. Count the number of signals to determine the number of sets of equivalent hydrogens present in the compound.

Step 3: Integration. Use the integration and the molecular formula to determine the numbers of hydrogens present in each set.

Step 4: Pattern of chemical shifts. Compare signal chemical shifts with reference tables to determine which functional groups may be present. Keep in mind that these are broad ranges and that signals of each type may be shifted either farther upfield or farther downfield, depending on the details of the molecular structure.

Types of Hydrogens	Descriptive Name	Chemical Shift (δ)
RCH_3 RCH_2R R_3CH	Alkyl hydrogens	0.8–1.7
$R_2C{=}CRCHR_2$	Allylic hydrogens	1.6–2.6
$RCH_2C(=O)R$	Hydrogens on a *sp*^{3} carbon adjacent to a carbonyl group	2.2–2.6
RCH_2OH RCH_2OR	Hydrogens on a carbon adjacent to an *sp*^{3} hybridized oxygen	3.3–4.0
$R_2C{=}CH_2$ $R_2C{=}CHR$	Vinylic hydrogens	4.6–5.7
ArH	Aryl hydrogens	6.5–8.5
$RC(=O)H$	Aldehyde hydrogens	9.5–10.1
$RC(=O)OH$	Carboxyl hydrogens	10–13

(Continued)

***Step 5:* Signal splitting patterns.** Examine splitting patterns for information about the number of nearest nonequivalent hydrogen neighbors.

***Step 6:* Structural formula.** Construct possible molecules from the functional groups present; their relative signal integrations; and any other information you are given, especially the molecular formula and other spectra. Confirm the correct structure by making sure all the available information matches.

Spectral Problem 1: Molecular formula $C_5H_{10}O$

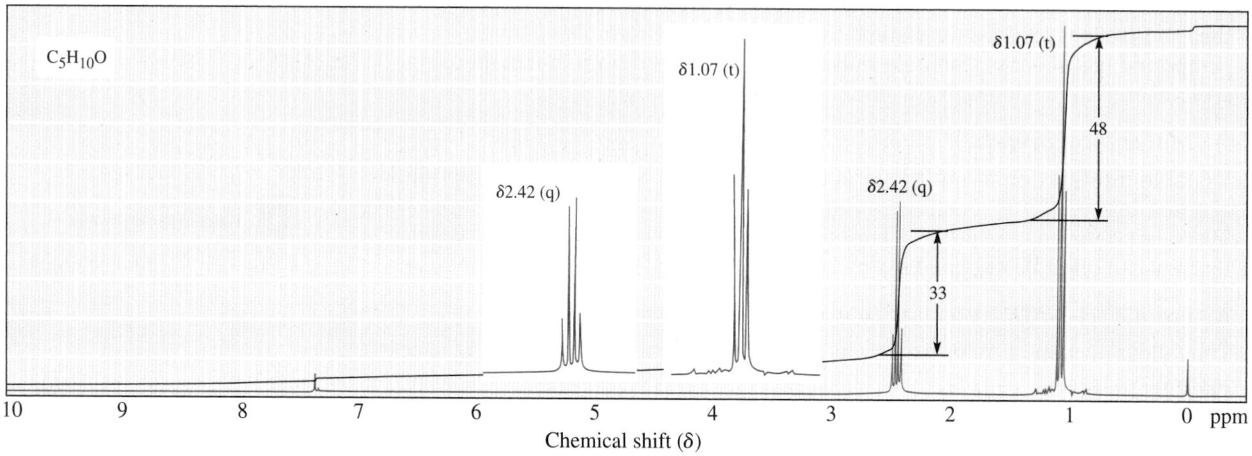

C5H10O
δ1.07 (t)
δ2.42 (q)
δ1.07 (t)
δ2.42 (q)
48
33
10
9
8
7
6
5
4
3
2
1
0 ppm
Chemical shift (δ)

Analysis of Spectral Problem 1

***Step 1:* Molecular formula and index of hydrogen deficiency.** The reference compound is C_5H_{12}; therefore, the index of hydrogen deficiency is 1 and the molecule contains either one ring or one π bond.

***Step 2:* Number of signals.** There are two signals (a triplet and a quartet) and, therefore, two sets of equivalent hydrogens.

***Step 3:* Integration.** From the integration, the hydrogens in each set are in the ratio 3:2. Because there are ten hydrogens, 6H must give rise to the signal at δ 1.07 and 4H must give rise to the signal at δ 2.42.

***Step 4:* Pattern of chemical shifts.** The signal at δ 1.07 is in the alkyl region and, based on its chemical shift, most probably indicates a methyl group. No signal occurs between δ 4.6 and δ 5.7; there are no vinylic hydrogens. If a carbon-carbon double bond is in the molecule, there are no hydrogens on it (i.e., it is tetrasubstituted). The chemical shift of the four protons at δ 2.42 is consistent with two CH_2 groups next to a carbonyl group.

***Step 5:* Signal splitting patterns.** The methyl signal at δ 1.07 is split into a triplet (t); it must have two neighbors, indicating $—CH_2CH_3$. The signal at δ 2.42 is split into a quartet (q); it must have three neighbors. An ethyl group accounts for these two signals. No other signals occur in the spectrum; therefore, there are no other types of hydrogens in the molecule.

***Step 6:* Structural formula.** Put this information together to arrive at the following structural formula. The chemical shift of the methylene group ($-CH_2-$) at δ 2.42 is consistent with an alkyl group adjacent to a carbonyl group.

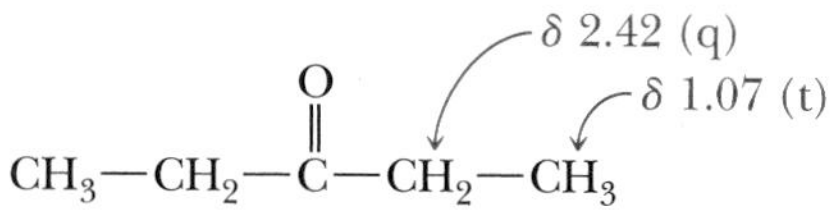

δ 2.42 (q)
δ 1.07 (t)
O
CH3—CH2—C—CH2—CH3

3-Pentanone

Spectral Problem 2: Molecular formula $C_7H_{14}O$

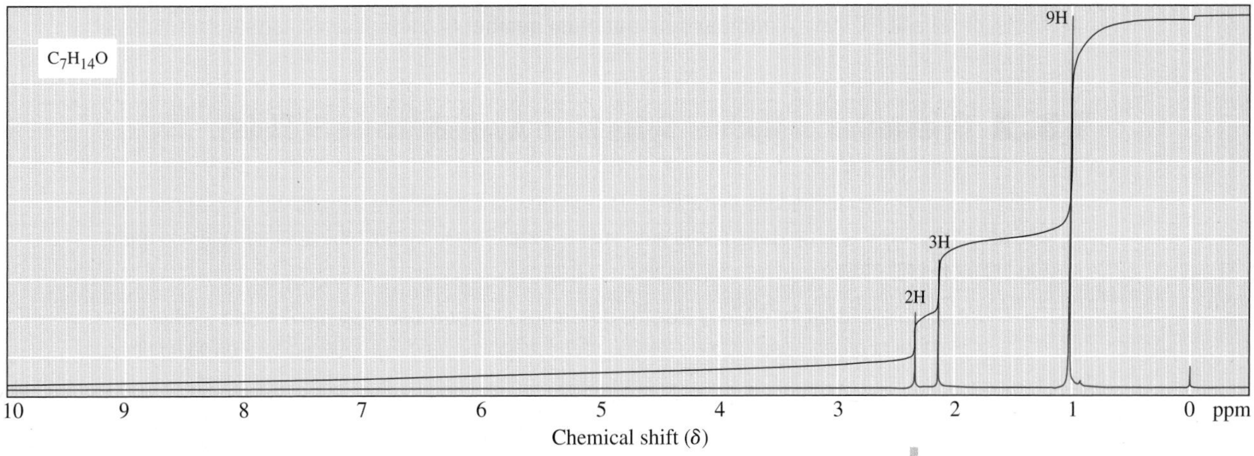

Analysis of Spectral Problem 2

Step 1: ***Molecular formula and index of hydrogen deficiency.*** The index of hydrogen deficiency is 1; the compound contains one ring or one π bond.

Step 2: ***Number of signals.*** There are three signals and, therefore, three sets of equivalent hydrogens.

Step 3: ***Integration.*** Reading from right to left, there are 9, 3, and 2 hydrogens in these signals.

Step 4: ***Pattern of chemical shifts.*** The signal at δ 1.01 is characteristic of a methyl group adjacent to an sp^3 hybridized carbon. The signals at δ 2.11 and δ 2.32 are characteristic of alkyl groups adjacent to a carbonyl group.

Step 5: ***Signal splitting pattern.*** All signals are singlets (s). Therefore, none of the groups has hydrogens on neighboring carbons.

Step 6: ***Structural formula.*** The compound is 4,4-dimethyl-2-pentanone.

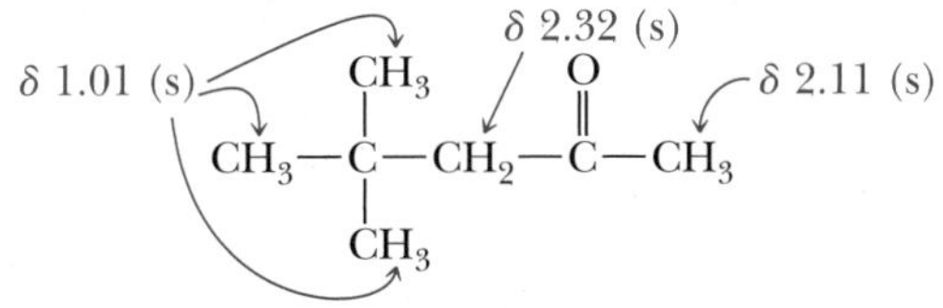

4,4-Dimethyl-2-pentanone

Study Guide

13.1 Nuclear Spin States

- Absorption of electromagnetic energy leads to spectroscopic tools that are important for the determination of structures of organic molecules.
  - *These absorptions can be from electronic: transitions (UV-visible spectra, Section 20.3), vibrations (IR spectra, Chapter 12), and nuclear magnetic spin transitions (NMR spectra at radio frequencies).*

- Nuclei of ^{1}H and ^{13}C have a **nuclear spin quantum number** of $\frac{1}{2}$ and allowed nuclear spin states of $+\frac{1}{2}$ and $-\frac{1}{2}$.

13.2 Orientation of Nuclear Spins in an Applied Magnetic Field

- In the presence of an applied magnetic field, B_0, nuclei with spin $+\frac{1}{2}$ are aligned with the applied field and are in the lower energy state; nuclei with spin $-\frac{1}{2}$ are aligned against the applied field and are in the higher energy state.
  - *The key relationship for NMR is that the difference in energy between the $+\frac{1}{2}$ and $-\frac{1}{2}$ nuclear spin states is proportional to the strength of the magnetic field experienced by a given nucleus.*

P **13.1**

13.3 Nuclear Magnetic "Resonance"

- When placed in a powerful magnetic field, there is a small population bias for the ^{1}H and ^{13}C nuclei to be aligned with the magnetic field.
- A nucleus aligned with the magnetic field **precesses** at a characteristic frequency, analogous to the way a spinning top or gyroscope traces out a cone-shaped surface as it precesses in the Earth's gravitational field.
- **Resonance** is the absorption of electromagnetic radiation by a precessing nucleus and the resulting flip of its nuclear spin from the lower energy spin state to the higher energy spin state.
  - *The electromagnetic radiation must be of precisely the same frequency as the nuclear precession frequency for absorption of energy to occur and thus achieve resonance.*
  - *For typical experiments, this precession frequency corresponds to electromagnetic radiation in the **radio-frequency** range.*
- The experimental conditions required to cause nuclei to resonate are affected by the local chemical and magnetic environments.
  - *Electrons around a hydrogen or carbon create local magnetic fields that shield the nuclei of these atoms from the applied field.*
    - *Any factor that increases the exposure of nuclei to an applied field is said to **deshield** them and shifts their signal downfield to a larger δ value.*
    - *Conversely, any factor that decreases the exposure of nuclei to an applied field is said to **shield** them and shifts their signal upfield to a smaller δ value.*

P **13.10, 13.11**

13.4 An NMR Spectrometer

- An NMR spectrometer records resonance as a **signal** and plots the irradiation frequencies where absorption takes place scaled by the strength of the applied magnetic field using the units of **parts per million (ppm)**.
  - *For ^{1}H-NMR, ppm is a convenient scale because differences in resonance frequencies for different ^{1}H nuclei in molecules are about one-millionth the overall frequency of electromagnetic frequency used to cause the resonance.*

13.5 Equivalent Hydrogens

- **Equivalent hydrogens** within a molecule are in identical chemical environments.
  - *In a molecule, hydrogens are equivalent if they are bonded to the same* sp^3 *carbon atom and there is free bond rotation involving that carbon atom in the molecule.*
    - *A lack of free bond rotation, for example, in small rings or with alkenes, can make hydrogens bonded to the same carbon atom nonequivalent.*
  - *There might be overall symmetry in the molecule making sets of hydrogens equivalent.*

P 13.2, 13.3

13.6 Signal Areas

- The area of a ^{1}H-NMR signal is proportional to the number of equivalent hydrogens giving rise to that signal.

P 13.4

13.7 Chemical Shift

- **Chemical shift,** δ, is defined as the frequency shift from tetramethylsilane (TMS) divided by the operating frequency of the spectrometer.
  - *The resonance signals in* ^{1}*H-NMR spectra are reported according to how far they are shifted from the resonance signal of the 12 equivalent hydrogens in TMS.*
  - *The resonance signals in* ^{13}*C-NMR spectra are reported according to how far they are shifted from the resonance signal of the four equivalent carbons in TMS.*
- The chemical shift of a particular set of equivalent hydrogens depends primarily on three factors:
  - *Nearby electronegative atoms have a deshielding effect.*
  - *In general, the higher the percent of s-character in a hybrid orbital, the greater the deshielding effect on itself and adjacent atoms.*
  - *The induced local magnetic fields in* π *bonds either add to or subtract from the applied field depending on the geometry of the induced electron density circulation relative to the atom in question.*

P 13.6

▶ Figure 13.8

13.8 Signal Splitting and the $(n + 1)$ Rule

- According to the **$(n + 1)$ rule**, if a hydrogen has n hydrogens nonequivalent to it but equivalent among themselves on the same or adjacent atom(s), its ^{1}H-NMR signal will be split into $(n + 1)$ peaks.
  - *Splitting patterns are commonly referred to as singlets (s), doublets (d), triplets (t), quartets (q), quintets, and multiplets (m).*

- *The relative intensities of peaks in a multiplet can be predicted from an analysis of spin combinations for adjacent hydrogens or from the mnemonic device called **Pascal's triangle**.*
- *A **coupling constant (J)** is the distance between adjacent peaks in a multiplet and is reported in hertz (Hz). The value of J depends only on internal fields within a molecule and is independent of the spectrometer field.*

P 13.26

13.9 The Origins of Signal Splitting

- ^{1}H-NMR signals are split because the spin state ($+\frac{1}{2}$ versus $-\frac{1}{2}$) of nuclei of nonequivalent hydrogens no more than three bonds away influence the net magnetic field experienced by a given nucleus, an interaction known as **spin-spin coupling**.
- When a hydrogen nucleus is coupled to more than one set of adjacent hydrogen nuclei, the couplings combine.
  - *In the general case, if a hydrogen nucleus is coupled to a set of n hydrogen nuclei on one side and a set of m hydrogen nuclei on the other, the signal will be split into a maximum of $(n + 1)(m + 1)$ peaks.*
    - *In molecules that are rigid (e.g., alkenes and cyclic molecules) all the $(n + 1) \times (m + 1)$ peaks can often be seen.*
    - *However, because coupling constants can be similar, especially in flexible molecules, this splitting can simplify to a number of observed peaks that is equal to the number of adjacent H atoms + 1, regardless of patterns of equivalence.*

P 13.7, 13.27

▶ Figure 13.16

13.10 Stereochemistry and Topicity

- Groups of atoms in which substitution of one atom by an isotope creates an achiral molecule are called **homotopic**.
  - *Homotopic groups always have identical chemical shifts.*
- Those in which such substitution produces a chiral molecule are **enantiotopic**. Enantiotopic groups have identical chemical shifts, except in a chiral environment.
- Molecules in which substitution produces diastereomers are called **diastereotopic**.
  - *Diastereotopic atoms are nonequivalent in all environments, so they have different chemical shifts. These differences can lead to complex splitting of the signals of diastereotopic H atoms, especially those adjacent to a chiral center.*

▶ Magnetic Resonance Imaging

13.11 ^{13}C-NMR

- **^{13}C-NMR** is like ^{1}H-NMR, except the nuclear spins of ^{13}C nuclei are being analyzed.
  - *^{13}C-NMR spectra are commonly recorded in a hydrogen-decoupled instrumental mode. In this mode, all ^{13}C signals appear as singlets.*

– *The number of different signals in a ^{13}C-NMR spectrum tell you how many nonequivalent carbon atoms are in a molecule.*

– ^{13}C-NMR *chemical shifts tell you what kind of carbon atoms are present.*

P 13.8, 13.27

13.12 Interpretation of NMR Spectra

- Four important types of structural information can be obtained from a ^{1}H-NMR spectrum.
  - *From the number of signals, the number of sets of equivalent hydrogens can be determined.*
  - *From the integration of signal areas, the relative numbers of hydrogens in each set can be determined.*
  - *From the chemical shift of each signal, information about the chemical environment of the hydrogens in each set can be determined.*
  - *From the splitting pattern of each signal, information about the number and chemical equivalency of hydrogens on the same and adjacent carbon atoms (in other words, the connectivities between different groups on the molecule) can be determined.*

P 13.09, 13.12–13.13, 13.15–13.24, 13.28

How To Solve NMR Spectral Problems

Problems

Red numbers indicate applied problems.

13.9 Calculate the index of hydrogen deficiency of these compounds.

(a) Aspirin, $C_9H_8O_4$
(b) Ascorbic acid (vitamin C), $C_6H_8O_6$
(c) Pyridine, C_5H_5N
(d) Urea, CH_4N_2O
(e) Cholesterol, $C_{27}H_{46}O$
(f) Dopamine, $C_8H_{11}NO_2$

Interpretation of ^{1}H-NMR and ^{13}C-NMR Spectra

13.10 Complete the following table. Which nucleus requires the least energy to flip its spin at this applied field? Which nucleus requires the most energy?

Nucleus	Applied Field (tesla, T)	Radio Frequency (MHz)	Energy (J/mol)
^{1}H	7.05	300	______
^{13}C	7.05	75.5	______
^{19}F	7.05	282	______

13.11 The natural abundance of ^{13}C is only 1.1%. Furthermore, its sensitivity in NMR spectroscopy (a measure of the energy difference between a spin aligned with or against an applied magnetic field) is only 1.6% that of ^{1}H. What are the relative signal intensities expected for the ^{1}H-NMR and ^{13}C-NMR spectra of the same sample of $Si(CH_3)_4$?

13.12 Following are structural formulas for three constitutional isomers with the molecular formula $C_7H_{16}O$ and three sets of ^{13}C-NMR spectral data. Assign each constitutional isomer its correct spectral data.

(a) $CH_3CH_2CH_2CH_2CH_2CH_2CH_2OH$

(b) 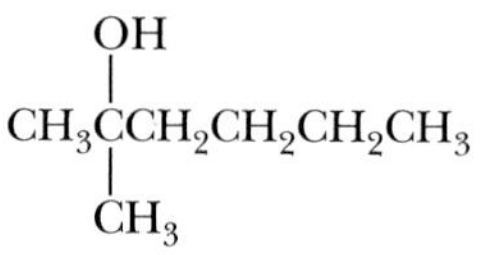

(c) 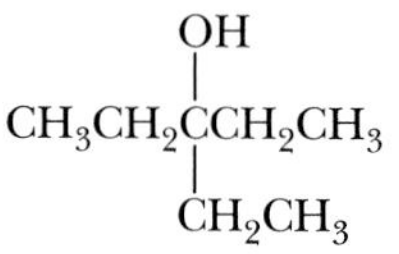

Spectrum 1	Spectrum 2	Spectrum 3
74.66	70.97	62.93
30.54	43.74	32.79
7.73	29.21	31.86
	26.60	29.14
	23.27	25.75
	14.09	22.63
		14.08

13.13 Following are structural formulas for the *cis* isomers of 1,2-, 1,3-, and 1,4-dimethylcyclohexane and three sets of ^{13}C-NMR spectral data. Assign each constitutional isomer its correct spectral data.

(a) 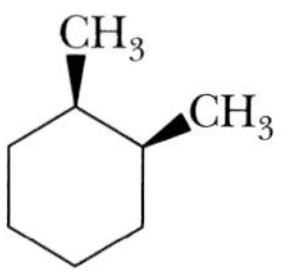

(b) 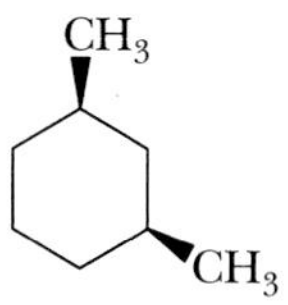

(c) 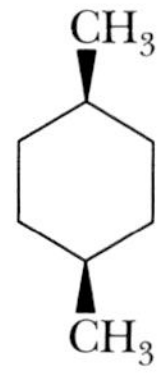

Spectrum 1	Spectrum 2	Spectrum 3
31.35	34.20	44.60
30.67	31.30	35.14
20.85	23.56	32.88
	15.97	26.54
		23.01

13.14 Following are structural formulas, dipole moments, and ^{1}H-NMR chemical shifts for acetonitrile, fluoromethane, and chloromethane.

$CH_3C{\equiv}N$	CH_3F	CH_3Cl
Acetonitrile	**Fluoromethane**	**Chloromethane**
3.92 D	1.85 D	1.87 D
δ 1.97	δ 4.26	δ 3.05

(a) How do you account for the fact that the dipole moments of fluoromethane and chloromethane are almost identical even though fluorine is considerably more electronegative than chlorine?

(b) How do you account for the fact that the dipole moment of acetonitrile is considerably greater than that of either fluoromethane or chloromethane?

(c) How do you account for the fact that the chemical shift of the methyl hydrogens in acetonitrile is considerably less than that for either fluoromethane or chloromethane?

13.15 Following are three compounds with the molecular formula $C_4H_8O_2$ and three ^{1}H-NMR spectra. Assign each compound its correct spectrum and assign all signals to their corresponding hydrogens.

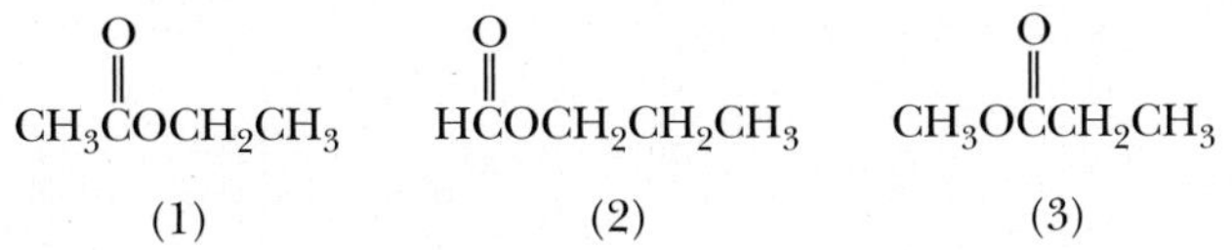

(1) (2) (3)

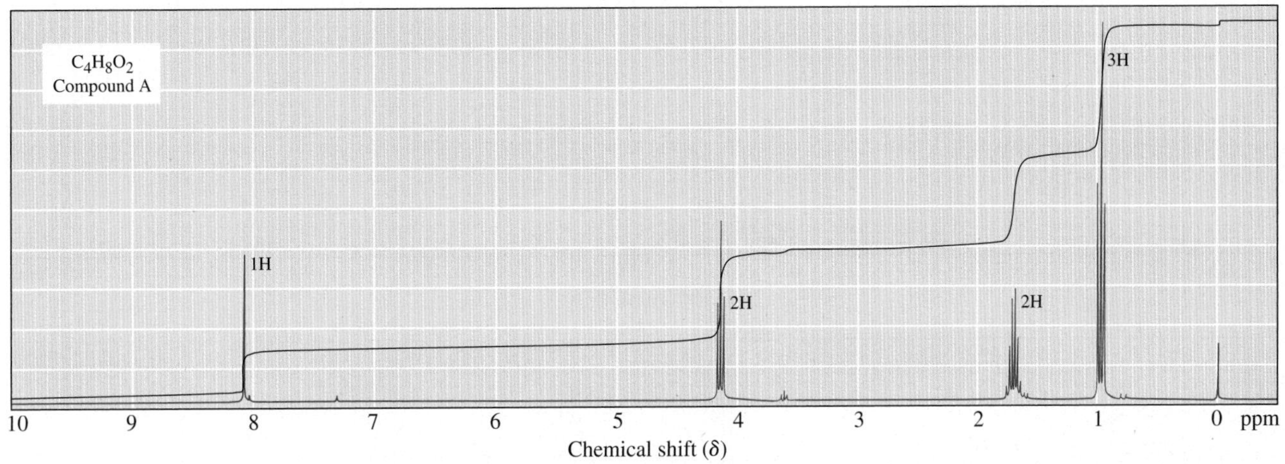

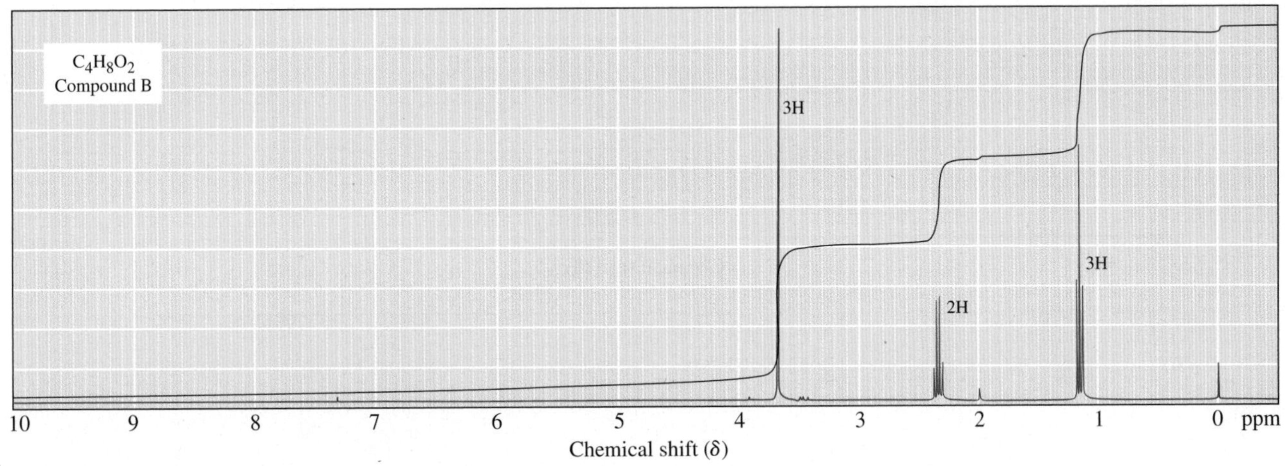

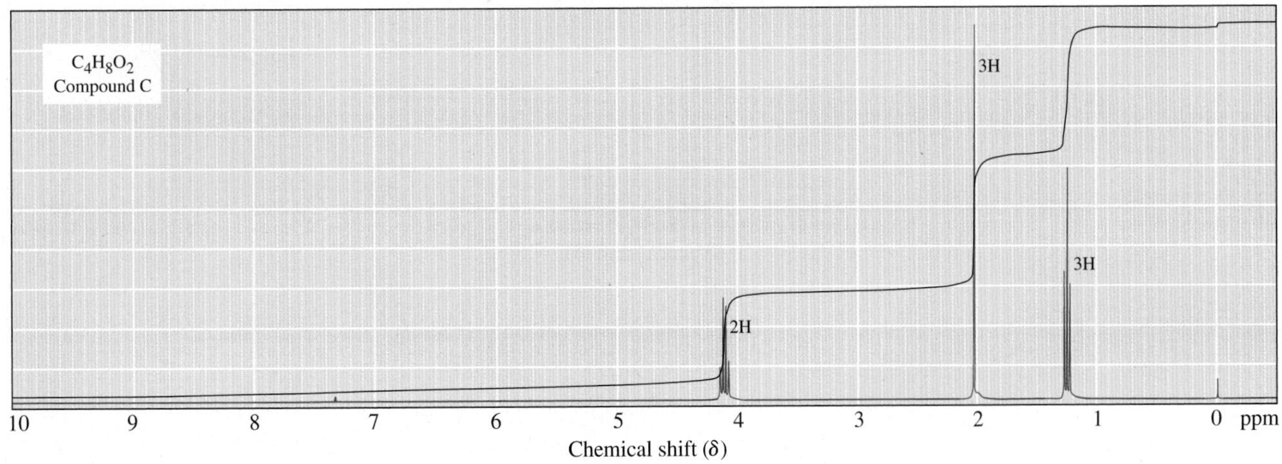

13.16 Following are ^{1}H-NMR spectra for compounds D, E, and F, each with molecular formula C_6H_{12}. Each readily decolorizes a solution of Br_2 in CCl_4. Propose structural formulas for compounds D, E, and F and account for the observed patterns of signal splitting.

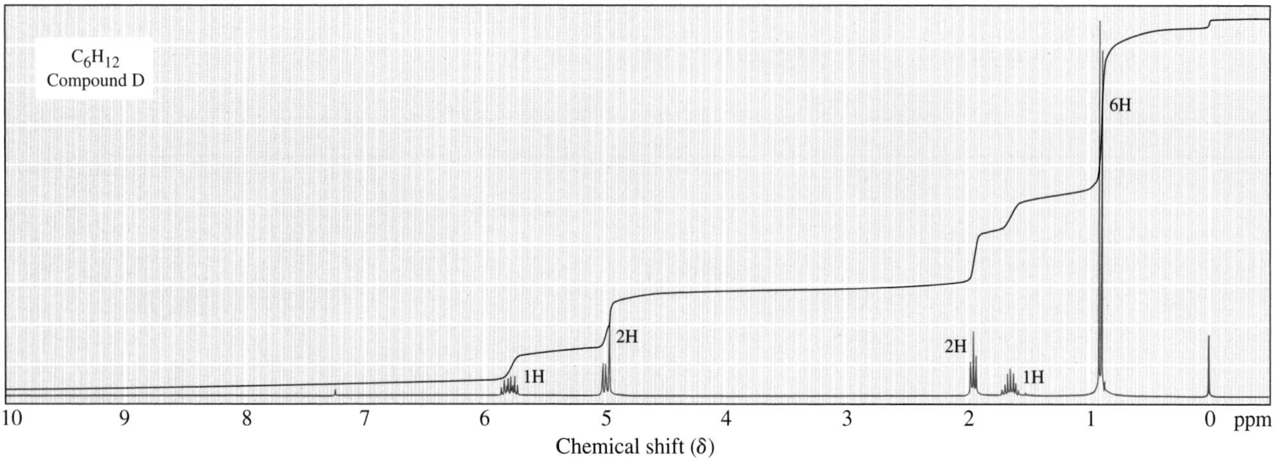

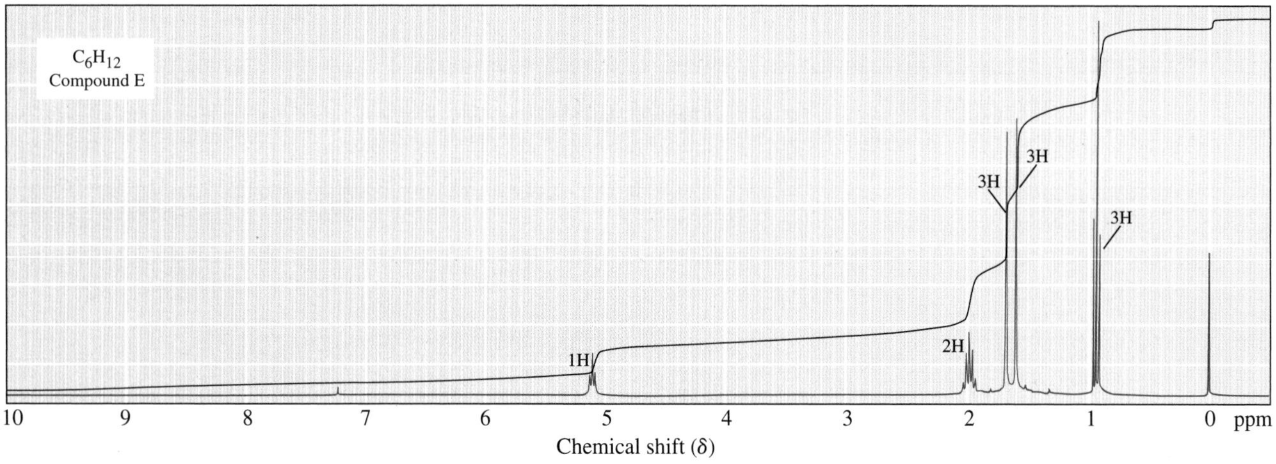

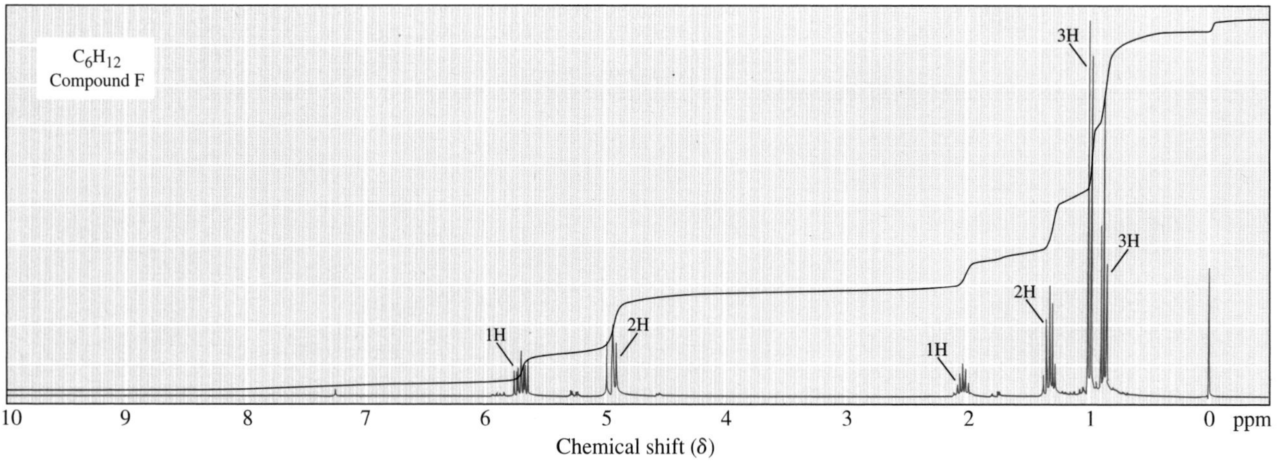

13.17 Following are ^{1}H-NMR spectra for compounds G, H, and I, each with the molecular formula $C_5H_{12}O$. Each is a liquid at room temperature, is slightly soluble in water, and reacts with sodium metal with the evolution of a gas.

(a) Propose structural formulas of compounds G, H, and I.
(b) Explain why there are four lines between δ 0.86 and 0.90 for compound G.
(c) Explain why the 2H multiplets at δ 1.5 and 3.5 for compound H are so complex.

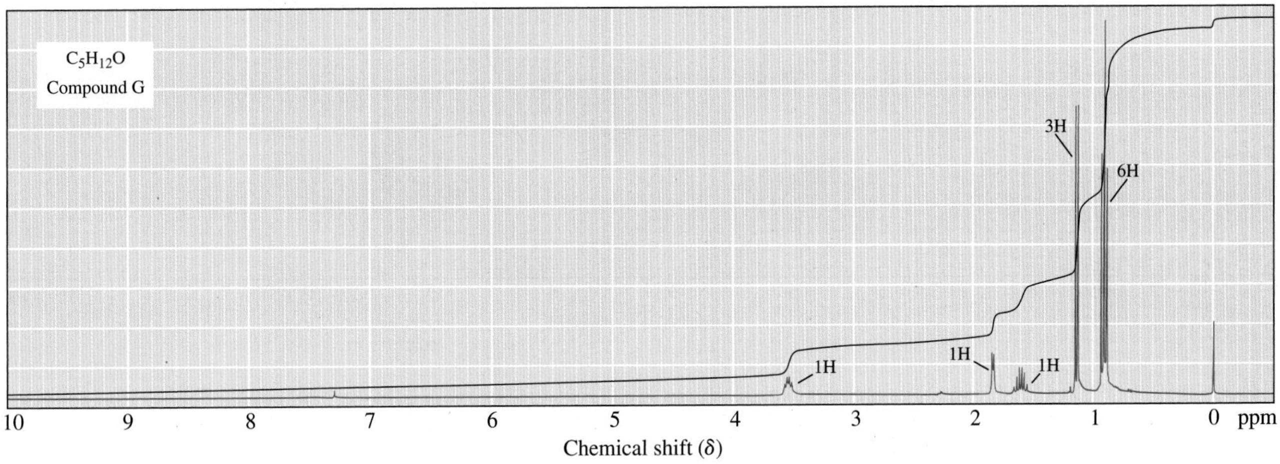

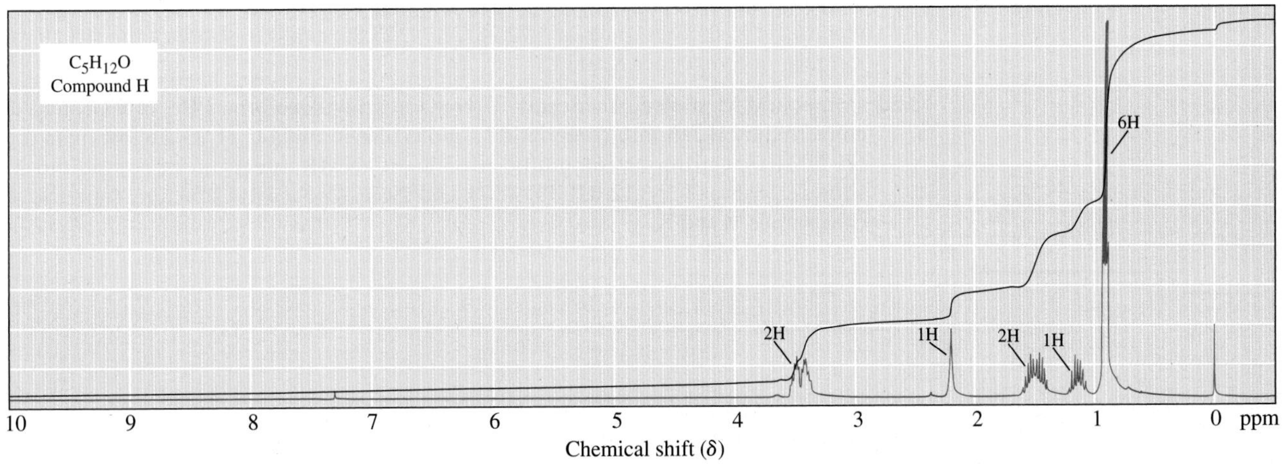

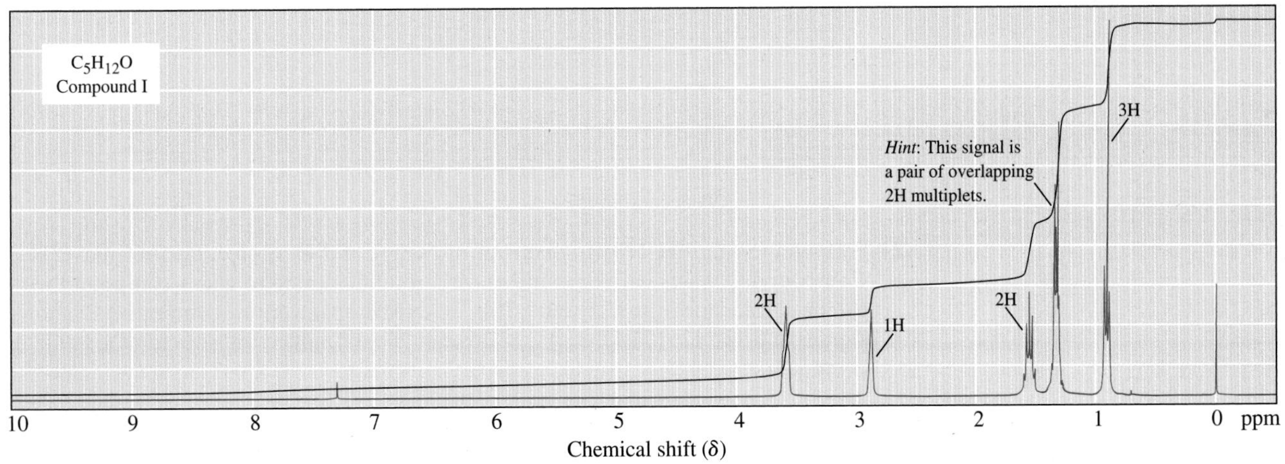

13.18 Propose a structural formula for compound J, molecular formula C_3H_6O, consistent with the following ^{1}H-NMR spectrum.

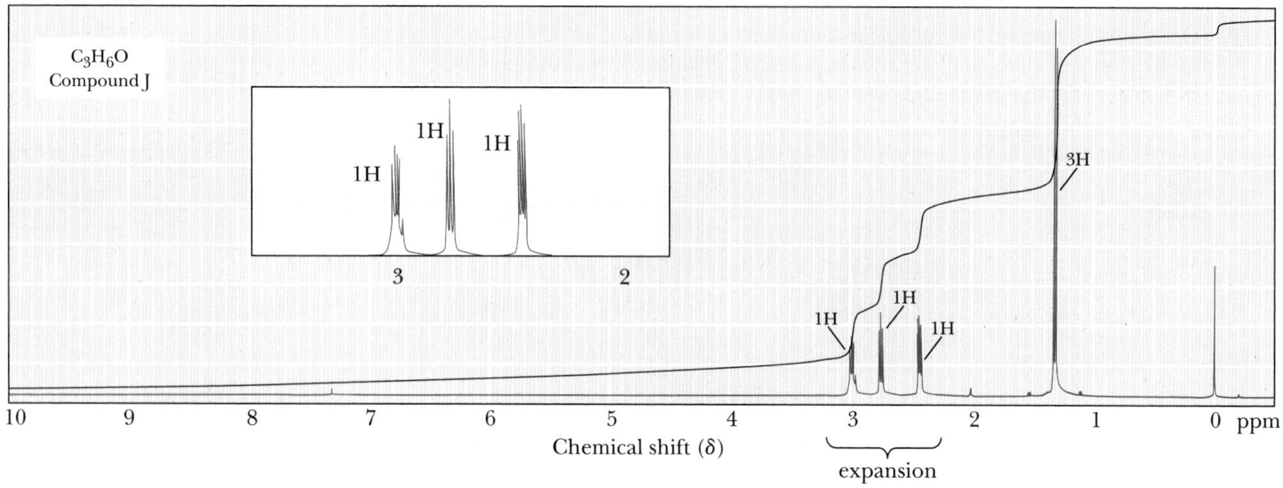

13.19 Compound K, molecular formula $C_6H_{14}O$, readily undergoes acid-catalyzed dehydration when warmed with phosphoric acid to give compound L, molecular formula C_6H_{12}, as the major organic product. The ^{1}H-NMR spectrum of compound K shows signals at δ 0.90 (t, 6H), 1.12 (s, 3H), 1.38 (s, 1H), and 1.48 (q, 4H). The ^{13}C-NMR spectrum of compound K shows signals at δ 72.98, 33.72, 25.85, and 8.16. Deduce the structural formulas of compounds K and L.

13.20 Compound M, molecular formula $C_5H_{10}O$, readily decolorizes Br_2 in CCl_4 and is converted by H_2/Ni into compound N, molecular formula $C_5H_{12}O$. Following is the ^{1}H-NMR spectrum of compound M. The ^{13}C-NMR spectrum of compound M shows signals at δ 146.12, 110.75, 71.05, and 29.38. Deduce the structural formulas of compounds M and N.

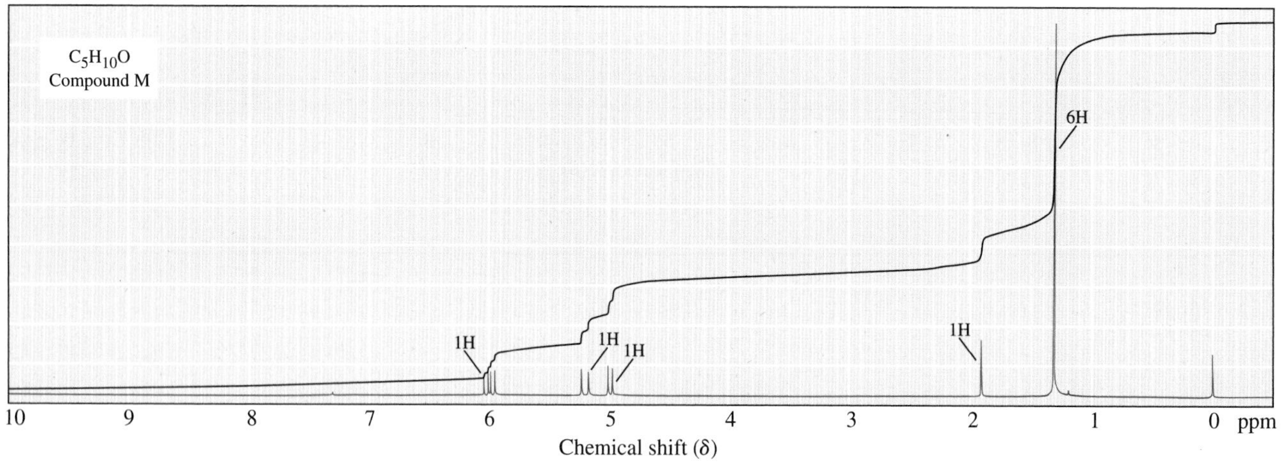

13.21 Following is the ^{1}H-NMR spectrum of compound O, molecular formula C_7H_{12}. Compound O reacts with bromine in carbon tetrachloride to give a compound with the molecular formula $C_7H_{12}Br_2$. The ^{13}C-NMR spectrum of compound O shows signals at δ 150.12, 106.43, 35.44, 28.36, and 26.36. Deduce the structural formula of compound O.

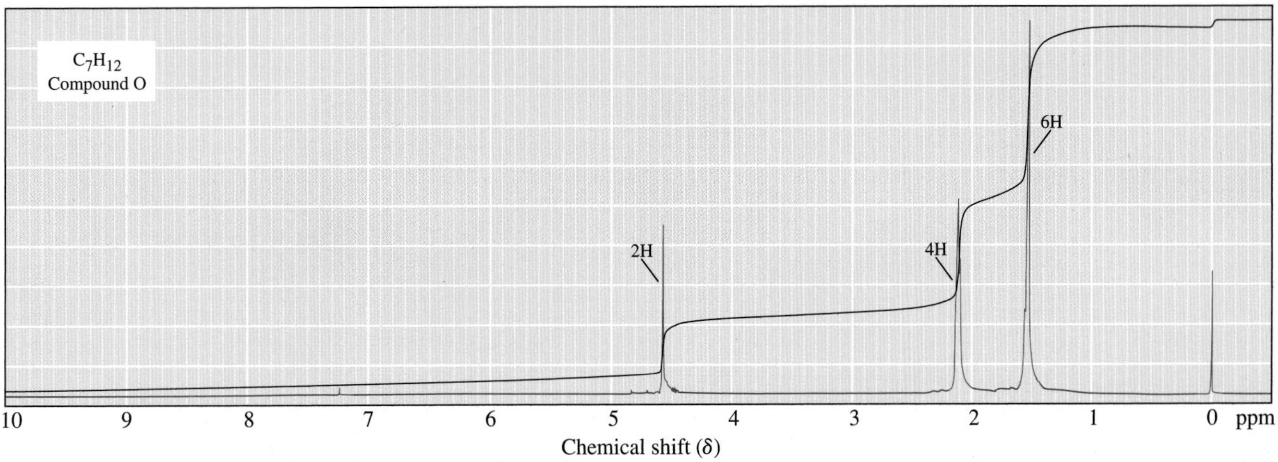

13.22 Treatment of compound P with BH_3 followed by H_2O_2/NaOH gives compound Q. Following are ^{1}H-NMR spectra for compounds P and Q along with ^{13}C-NMR spectral data. From this information, deduce structural formulas for compounds P and Q.

$$\underset{(P)}{C_7H_{12}} \xrightarrow[\text{2. } H_2O_2\text{, NaOH}]{\text{1. } BH_3} \underset{(Q)}{C_7H_{14}O}$$

^{13}C-NMR	
(P)	**(Q)**
132.38	72.71
32.12	37.59
29.14	28.13
27.45	22.68

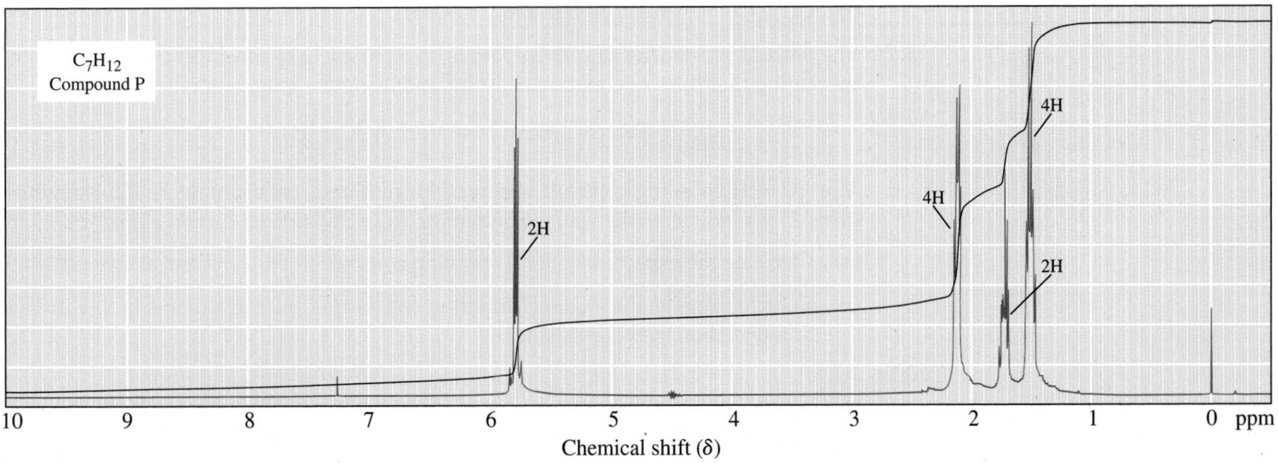

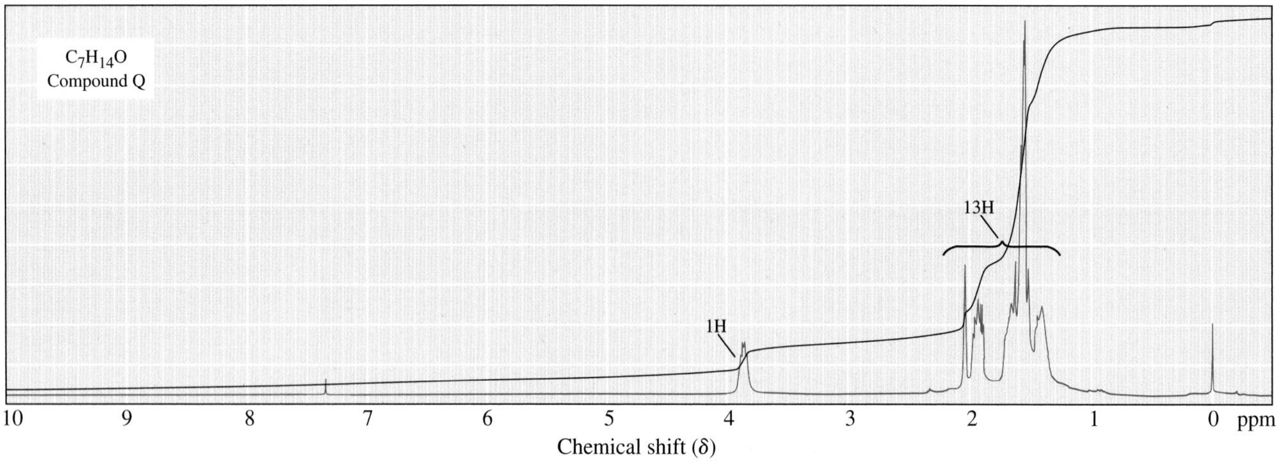

13.23 The ^{1}H-NMR spectrum of compound R, $C_6H_{14}O$, consists of two signals: δ 1.1 (doublet) and δ 3.6 (septet) in the ratio 6:1. Propose a structural formula for compound R consistent with this information.

13.24 Write structural formulas for the following compounds.

(a) $C_2H_4Br_2$:	δ 2.5 (d, 3H) and 5.9 (q, 1H)
(b) $C_4H_8Cl_2$:	δ 1.60 (d, 3H), 2.15 (m, 2H), 3.72 (t, 2H), and 4.27 (m, 1H)
(c) $C_5H_8Br_4$:	δ 3.6 (s, 8H)
(d) C_4H_8O:	δ 1.0 (t, 3H), 2.1 (s, 3H), and 2.4 (quartet, 2H)
(e) $C_4H_8O_2$:	δ 1.2 (t, 3H), 2.1 (s, 3H), and 4.1 (quartet, 2H); contains an ester
(f) $C_4H_8O_2$:	δ 1.2 (t, 3H), 2.3 (quartet, 2H), and 3.6 (s, 3H); contains an ester
(g) C_4H_9Br:	δ 1.1 (d, 6H), 1.9 (m, 1H), and 3.4 (d, 2H)
(h) $C_6H_{12}O_2$:	δ 1.5 (s, 9H) and 2.0 (s, 3H)
(i) $C_7H_{14}O$:	δ 0.9 (t, 6H), 1.6 (sextet, 4H), and 2.4 (t, 4H)
(j) $C_5H_{10}O_2$:	δ 1.2 (d, 6H), 2.0 (s, 3H), and 5.0 (septet, 1H)
(k) $C_5H_{11}Br$:	δ 1.1 (s, 9H) and 3.2 (s, 2H)
(l) $C_7H_{15}Cl$	δ 1.1 (s, 9H) and 1.6 (s, 6H)

13.25 The percent *s*-character of carbon participating in a C—H bond can be established by measuring the ^{13}C—^{1}H coupling constant and using the following relationship:

$$\text{Percent } s\text{-character} = 0.2\, J(^{13}\text{C—}^{1}\text{H})$$

The ^{13}C—^{1}H coupling constant observed for methane, for example, is 125 Hz, which gives 25% *s*-character, the value expected for an sp^3 hybridized carbon atom.

(a) Calculate the expected ^{13}C—^{1}H coupling constant in ethylene and acetylene.

(b) In cyclopropane, the ^{13}C—^{1}H coupling constant is 160 Hz. What is the hybridization of carbon in cyclopropane?

13.26 Ascaridole is a natural product that has been used to treat intestinal worms. Explain why the two methyls on the isopropyl group in ascaridole appear in its ^{1}H-NMR spectrum as four lines of equal intensity, with two sets of two each separated by 7 Hz.

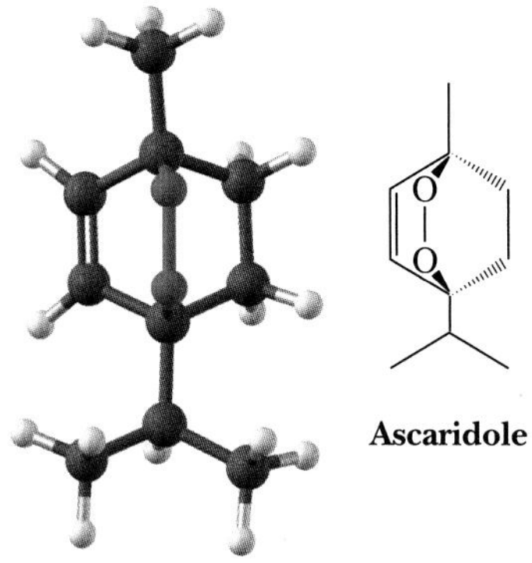

Ascaridole

13.27 The ^{13}C-NMR spectrum of 3-methyl-2-butanol shows signals at δ 17.88 (CH_3), 18.16 (CH_3), 20.01 (CH_3), 35.04 (carbon-3), and 72.75 (carbon-2). Account for the fact that each methyl group in this molecule gives a different signal.

13.28 Sketch the NMR spectrum you would expect from a partial molecule with the following parameters.

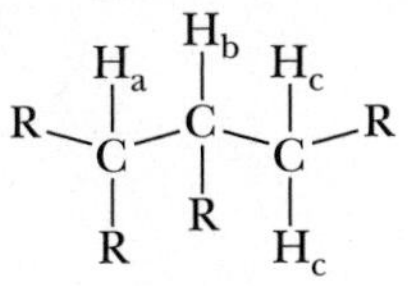

H_a = 1.0 ppm
H_b = 3.0 ppm
H_c = 6.0 ppm
J_{ab} = 5.0 Hz
J_{bc} = 8.0 Hz
J_{ac} = 1.0 Hz

14 Mass Spectrometry

Outline

14.1 A Mass Spectrometer
14.2 Features of a Mass Spectrum
14.3 Interpreting Mass Spectra
14.4 Mass Spectrometry in the Organic Synthesis Laboratory and Other Applications

Some mass spectrometers can be quite sophisticated, such as the one shown above. For a partial mass spectrum of the neurotransmitter dopamine, see Figure 14.2. **Inset:** a model of dopamine (Courtesy of DOE Photos)

Mass spectrometry is an analytical technique for measuring the mass-to-charge ratio (*m*/*z*) of ions, originally positive ions, now both positive and negative. The principles of mass spectrometry were first recognized in 1898. In 1911, J. J. Thomson recorded the first mass spectrum, that of neon, and discovered that this element can be separated into a more abundant isotope, ^{20}Ne, and a less abundant isotope, ^{22}Ne. Using improved instrumentation, F. W. Aston showed that most of the naturally occurring elements are mixtures of isotopes. It was found, for example, that approximately 75% of chlorine atoms in nature are ^{35}Cl and 25% are ^{37}Cl. Mass spectrometry did not come into common use until the 1950s, when commercial instruments that offered high resolution, reliability, and relatively inexpensive maintenance became available. Today mass spectrometry is our most valuable analytical tool for the determination of accurate molecular masses. Furthermore, extensive information about the molecular formula and structure of a compound can be obtained from analysis of its mass spectrum. Mass spectrometry is becoming increasingly important in biochemistry as well; sequencing of proteins using this technique alone allows protein structures to be determined on a virtually single-cell scale.

14.1 A Mass Spectrometer

A mass spectrometer (Figure 14.1) is designed to do three things:

1. Convert neutral atoms or molecules into a beam of positive or negative ions.
2. Separate the ions on the basis of their mass-to-charge (m/z) ratio.
3. Measure the relative abundance of each type of ion.

From this information, we can determine both the molecular mass and the molecular formula of an unknown compound. In addition, we can obtain valuable clues about the molecular structure of the compound.

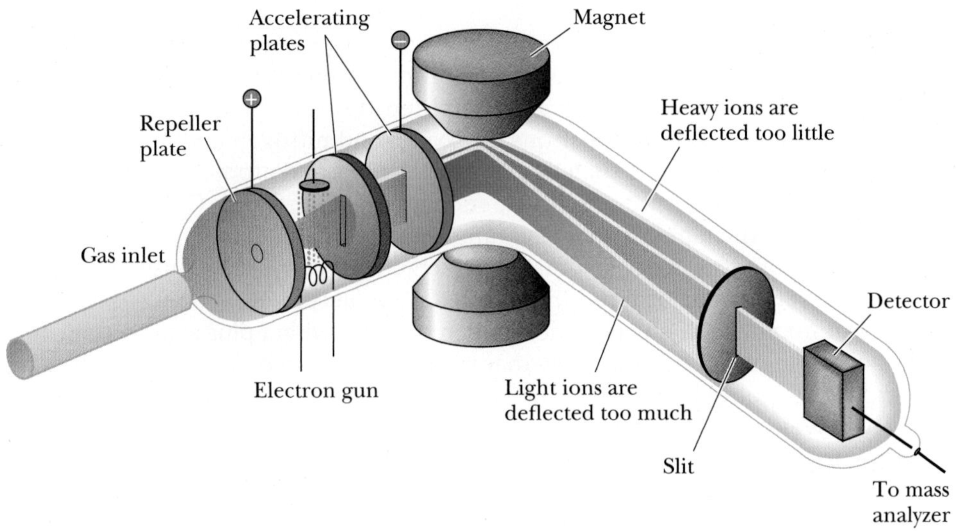

FIGURE 14.1 Schematic diagram of an electron ionization mass spectrometer (EI-MS).

There are many types of mass spectrometers; we have space in this text to describe in detail only the simplest. In the first-generation spectrometers, a vaporized sample in an evacuated ionization chamber is bombarded with high-energy electrons that cause electrons to be stripped from molecules of the sample, giving positively charged ions. Increasingly, radical anions (in which an extra electron has been added to a molecule) are studied; these are beyond the scope of this text. Positive ions are accelerated by a series of negatively charged accelerator plates into an analyzing chamber inside a magnetic (electric in some spectrometers) field perpendicular to the direction of the ion beam. The magnetic field causes the ion beam to curve. The radius of curvature of each ion depends on the charge on the ion (z), its mass (m), the accelerating voltage, and the strength of the magnetic field. A **mass spectrum** is a plot of relative ion abundance versus m/z ratio.

Mass spectrum
A plot of the relative abundance of ions versus their mass-to-charge (m/z) ratio.

Samples of gases and volatile liquids can be introduced directly into the ionization chamber. Because the interior of a mass spectrometer is kept at a high vacuum, volatile liquids and even some solids are vaporized. For less volatile liquids and solids, the sample may be placed on the tip of a heated probe that is then inserted directly into the ionization chamber. Another extremely useful method for introducing a sample into the ionization chamber is to link a gas chromatograph (GC) or liquid chromatograph (LC) directly to the mass spectrometer. These machines can separate complex mixtures of molecules into pure fractions. Each fraction eluted from the chromatograph enters directly into the ionization chamber of the mass spectrometer, enabling mass determination of the individual components.

Once in the ionization chamber, molecules of the sample are bombarded with a stream of high-energy electrons emitted from a hot filament and are then accelerated by an electric field to energies of approximately 70 eV [1 eV = 96.49 kJ (23.05 kcal)/mol]. Collisions between molecules of the sample and these high-energy electrons result in

Molecular ion ($M^{\dot{+}}$)

The radical cation formed by removal of a single electron from a parent molecule in a mass spectrometer.

Radical cation

A species formed when a neutral molecule loses one electron; it contains both an odd number of electrons and a positive charge.

loss of electrons from sample molecules to form positive ions. A **molecular ion, $M^{\dot{+}}$**, is the species formed by removal of a single electron from a molecule. A molecular ion belongs to a class of ions called **radical cations**. When methane, for example, is bombarded with high-energy electrons, an electron is dislodged from a molecule to give a molecular ion at *m*/*z* 16.

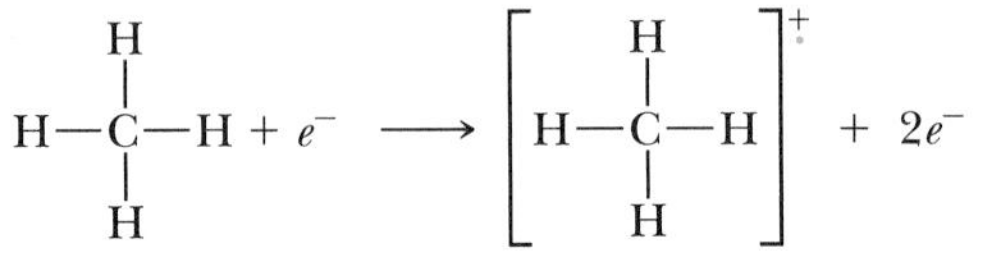

Molecular ion
(a radical cation)

Ionization potential (IP)

The minimum energy required to remove an electron from an atom or a molecule to a distance where there is no electrostatic interaction between the resulting ion and electron.

Which electron is lost in forming the molecular ion is determined by the **ionization potential** of the atom or molecule. Ionization potentials for most organic molecules are between 8 and 15 eV. The potentials are at the lower end of this range both for nonbonding electrons of oxygen and nitrogen and for π electrons in unsaturated compounds such as alkenes, alkynes, and aromatic hydrocarbons. Ionization potentials for σ electrons, such as those of C—C, C—H, and C—O σ bonds, are at the higher end of the range.

For our purposes, it doesn't matter which electron is lost because, in general, the radical cation is delocalized throughout the molecule. Therefore, we write the molecular formula of the parent molecule in brackets with a plus sign to show that it is a cation and with a dot to show that it has an odd number of electrons. See, for example, the molecular ion for ethyl isopropyl ether, shown on the left. At times, however, we will find it useful to depict the radical cation localized in a certain position to better understand its reactions, as in the formula on the right.

$$[CH_3CH_2OCH(CH_3)_2]^{\dot{+}} \qquad CH_3CH_2\overset{\cdot +}{\underset{\cdot\cdot}{O}}CH(CH_3)_2$$

As we shall see in Section 14.2D, a molecular ion can undergo fragmentation to form a variety of smaller cations (which themselves may undergo further fragmentation), as well as radicals and smaller molecules. Only charged fragments are detected.

After molecular ions and their fragments form, a positively charged repeller plate directs the ions toward a series of negatively charged accelerator plates, producing a rapidly traveling ion beam. The ion beam is then focused by one or more slits and passes into a mass analyzer, where it enters a magnetic field perpendicular to the direction of the ion beam. The magnetic field causes the ion beam to curve. Cations with larger values of *m*/*z* are deflected less than those with smaller *m*/*z* values. By varying either the accelerating voltage or the strength of the magnetic field, cations of the same *m*/*z* ratio can be focused on a detector, where the ion current is recorded. Modern detectors are capable of detecting single ions and of scanning a desired mass-to-charge region in a few tenths of a second or less.

Base peak

The peak caused by the most abundant ion in a mass spectrum; the most intense peak. It is assigned an arbitrary intensity of 100.

A mass spectrum is a plot of the relative abundance of each cation versus mass-to-charge ratio. The peak resulting from the most abundant cation is called the **base peak** and is assigned an arbitrary intensity of 100. The relative abundances of all other cations in a mass spectrum are reported as percentages of the base peak. Figure 14.2 shows a partial mass spectrum of dopamine, a neurotransmitter in the brain's caudate nucleus, a center involved with coordination and integration of fine muscle movement. A deficiency of dopamine is an underlying biochemical defect in Parkinson's disease.

As can be seen in Table 14.1, the number of peaks recorded depends on the sensitivity of the detector. If we record all peaks with intensity equal to or greater than 0.5% of the base peak, as in Figure 14.2, we find 45 peaks for dopamine. If we record all peaks with intensity equal to or greater than 0.05% of the base peak, we find 120 peaks.

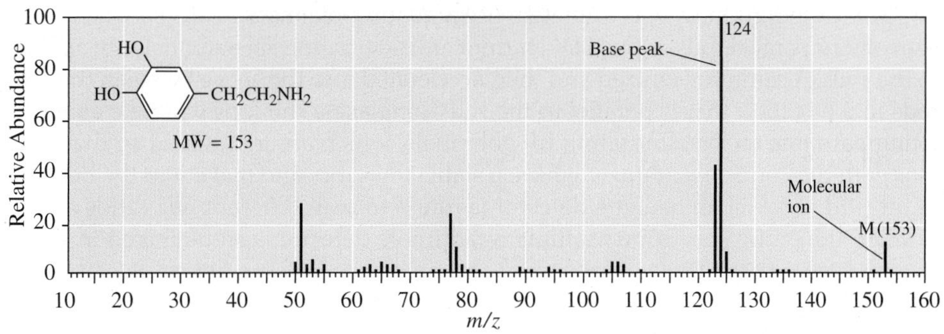

FIGURE 14.2 A partial mass spectrum of dopamine showing all peaks with intensity equal to or greater than 0.5% of the base peak.

Table 14.1 Number of Peaks Recorded in a Mass Spectrum of Dopamine

Peak Intensity Relative to Base Peak (%)	Number of Peaks Recorded
>5	8
>1	31
>0.5	45
>0.05	120

The technique we have described is called **electron ionization mass spectrometry (EI-MS)**. This technique was the first one developed and for a time was the one most widely used. It is limited, however, to relatively low-molecular-weight compounds that are vaporized easily in the evacuated ionization chamber. In recent years, a revolution in ionization techniques has extended the use of mass spectrometry to very high molecular-weight compounds and others that cannot be vaporized directly. Among the new techniques is fast-atom bombardment (FAB), which uses high-energy particles, such as xenon atoms accelerated to keV energies, to bombard a dispersion of a compound in a nonvolatile matrix, producing ions of the compound and expelling them into the gas phase. A second technique is matrix-assisted laser desorption ionization mass spectrometry (MALDI), which uses photons from an energetic laser for the same purpose. A third technique is chemical ionization (CI), which uses gas-phase acid-base reactions to produce ions. CI is particularly useful for identifying the molecular mass of a base (Brønsted-Lowry or Lewis) as its conjugate acid, MH^+. In addition, electrospray ionization mass spectrometry (ESI-MS) has become increasingly popular.

In ESI-MS, a solution of the analyte is introduced directly (e.g., from a liquid chromatograph) through a charged capillary into a high-vacuum chamber. After entry into the vacuum chamber, the analyte exists in small charged droplets that rapidly evaporate due to the vacuum, leaving the charge concentrated on the analyte molecules. The *m/z* ratios of the charged molecules are then determined. ESI is a relatively gentle method of generating charged molecular ions in a vacuum. ESI-MS is therefore particularly effective for the ionization of biological macromolecules, large molecules such as polysaccharides (Chapter 25), proteins (Chapter 27), and nucleotides (Chapter 28), allowing determination of their molecular mass and major fragments without the complications caused by the unavoidable overfragmentation seen with harsher ionization methods. MALDI is also a relatively gentle ionization technique that is now commonly used for the ionization of biological macromolecules.

Many modern mass spectrometers use a quadrupole mass analyzer, which consists of four parallel rods. A fixed DC current and alternating RF potentials are applied to the rods. A sample is ionized and then accelerated into the space between the four rods in a direction that is parallel to the rods' long axes. The ions will take an oscillating path due to the alternating RF potentials, which are modulated so that only ions with the correct *m/z* ratio will have a stable trajectory and make it to the detector at the end of the chamber. The detector is tuned to scan different *m/z* ratios and to develop the entire spectrum. Multiple quadrupole detectors can be linked in series to gain even greater levels of sensitivity and precision. In these linked detectors, individual peaks isolated from the first detector are injected into the second in order to observe fragments of the fragments. Identification of fragments is greatly improved by this arrangement, which is often called tandem mass spectrometry.

14.2 Features of a Mass Spectrum

To understand a mass spectrum, we need to understand the relationships between mass spectra and resolution, the presence of isotopes, and the fragmentation of molecules and molecular ions in both the ionization chamber and the analyzing chamber.

A. Resolution

Resolution
In mass spectrometry, a measure of how well a mass spectrometer separates ions of different mass.

Low-resolution mass spectrometry
Instrumentation that is capable of separating only ions that differ in mass by 1 or more amu.

High-resolution mass spectrometry
Instrumentation that is capable of separating ions that differ in mass by as little as 0.0001 amu.

An important operating characteristic of a mass spectrometer is its **resolution** (i.e., how well it separates ions of different mass). **Low-resolution mass spectrometry** refers to instruments capable of distinguishing among ions of different nominal mass [i.e., ions that differ by one or more atomic mass units (amu)]. **High-resolution mass spectrometry** refers to instruments capable of distinguishing among ions that differ in precise mass by as little as 0.0001 amu.

To illustrate, compounds with the molecular formulas C_3H_6O and C_3H_8O have nominal masses of 58 and 60, respectively, and can be resolved by low-resolution mass spectrometry. The compounds C_3H_8O and $C_2H_4O_2$, however, have the same nominal mass of 60 and cannot be distinguished by low-resolution mass spectrometry. If we calculate the precise mass of each compound using the data in Table 14.2, we see that they differ by 0.03642 amu and can be distinguished by high-resolution mass spectrometry. Observation of a molecular ion with a mass of 60.058 or 60.021 would establish the identities of C_3H_8O and $C_2H_4O_2$, respectively.

Molecular Formula	Nominal Mass	Precise Mass
C_3H_8O	60	60.05754
$C_2H_4O_2$	60	60.02112

B. The Presence of Isotopes

In the mass spectrum of dopamine (Figure 14.2), the molecular ion appears at *m/z* 153. If you look more closely at this mass spectrum, you will see a small peak at *m/z* 154, from an ion 1 amu heavier than the molecular ion of dopamine. This peak is actually the sum of four separate peaks, each of amu 154 and each corresponding to the presence in the ion of a single heavier isotope of H, C, N, or O in dopamine. Because this peak corresponds to an ion 1 amu heavier than the molecular ion, it is called an M + 1 peak. We are concerned in this section primarily with M + 1 and M + 2 peaks.

Virtually all the elements common to organic compounds, including H, C, N, O, S, Cl, and Br, are mixtures of isotopes. Exceptions are fluorine, phosphorus, and

Table 14.2 **Precise Masses and Natural Abundances of Isotopes Relative to 100 Atoms of the Most Abundant Isotope**

Element	Atomic Weight	Isotope	Precise Mass (amu)	Relative Abundance
Hydrogen	1.0079	^{1}H	1.00783	100
		^{2}H	2.01410	0.016
Carbon	12.011	^{12}C	12.0000	100
		^{13}C	13.0034	1.11
Nitrogen	14.007	^{14}N	14.0031	100
		^{15}N	15.0001	0.38
Oxygen	15.999	^{16}O	15.9949	100
		^{17}O	16.9991	0.04
		^{18}O	17.9992	0.20
Sulfur	32.066	^{32}S	31.9721	100
		^{33}S	32.9715	0.78
		^{34}S	33.9679	4.40
Chlorine	35.453	^{35}Cl	34.9689	100
		^{37}Cl	36.9659	31.98
Bromine	79.904	^{79}Br	78.9183	100
		^{81}Br	80.9163	98.0

iodine, which occur in nature exclusively as ^{19}F, ^{31}P, and ^{127}I. Table 14.2 shows average atomic weights for the elements most common to organic compounds along with the masses and relative abundances in nature of the stable isotopes of each. In this table, the relative abundances are tabulated according to the number of atoms of the heavier isotope per 100 atoms of the most abundant isotope. Naturally occurring carbon, for example, is 98.90% ^{12}C and 1.10% ^{13}C. Thus, there are 1.11 atoms of carbon-13 in nature for every 100 atoms of carbon-12.

$$1.10 \times \frac{100}{98.90} = 1.11 \text{ atoms } ^{13}\text{C per 100 atoms } ^{12}\text{C}$$

Example 14.1 Exact Mass Calculation

Calculate the precise mass of each ion to five significant figures. Unless otherwise indicated, use the mass of the most abundant isotope of each element.

(a) $[CH_2Cl_2]^{\dot{+}}$ **(b)** $[^{13}CH_2Cl_2]^{\dot{+}}$ **(c)** $[CH_2Cl^{37}Cl]^{\dot{+}}$

Solution

(a) $12.0000 + 2(1.00783) + 2(34.9689) = 83.953$ **(b)** 84.957 **(c)** 85.951

Problem 14.1

Calculate the nominal mass of each ion. Unless otherwise indicated, use the mass of the most abundant isotope of each element.

(a) $[CH_3Br]^{\dot{+}}$ **(b)** $[CH_3{}^{81}Br]^{\dot{+}}$ **(c)** $[^{13}CH_3Br]^{\dot{+}}$

C. Relative Abundance of M, M + 2, and M + 1 Peaks

The most common elements giving rise to significant M + 2 peaks are chlorine, bromine, and oxygen. Chlorine in nature is 75.77% ^{35}Cl and 24.23% ^{37}Cl. Thus, a ratio of M to M + 2 peaks of approximately 3:1 indicates the presence of a single chlorine atom in the compound. Similarly, bromine in nature is 50.5% ^{79}Br and 49.5% ^{81}Br; a ratio of M to M + 2 of approximately 1:1 indicates the presence of a single bromine atom in the compound. The contribution of ^{18}O is only 0.2%, but it makes the major contribution to the M + 2 peak in compounds containing only C, H, N, and O. Sulfur is the only other element common to organic compounds that gives a significant M + 2 peak.

Let us use pentane, C_5H_{12}, to illustrate the relationship between M and M + 1 peaks. Pentane has a nominal mass of 72, and its molecular ion appears at *m*/*z* 72. In any sample of pentane, there is a probability that there will be a molecule in which one of the atoms of carbon is ^{13}C, the heavier isotope of carbon. This molecule has a nominal mass of 73, and its molecular ion will appear at *m*/*z* 73. Similarly, there is a probability that there will be a molecule in which one of the atoms of hydrogen is the heavier isotope of hydrogen, namely deuterium, ^{2}H. The probability of each of these isotope substitutions occurring is related to the natural abundance of each isotope in the following way:

$$\%(M + 1) = \Sigma\ (\%\ \text{abundance of heavier isotope} \times \text{number of atoms in the formula})$$

Using this formula, we calculate that the relative intensity of the M + 1 peak for pentane is

$$(M + 1) = (1.11 \times 5C + 0.016 \times 12H) = 5.55 + 0.19 = 5.74\%\ \text{of molecular ion peak}$$

Notice that the M + 1 peak for pentane is almost entirely from ^{13}C. The same is true for other compounds containing only C and H. Because M + 1 peaks are relatively low in intensity compared to the molecular ion peak and often difficult to measure with any precision, they are not useful for accurate determinations of molecular formulas. M + 1 and M + 2 peaks, however, can be useful for getting a rough idea of the number of carbons, oxygens, sulfurs, and halogens. For example, the spectrum of chloroethane has peaks at *m*/*z* 64 and 66 (corresponding to $C_2H_5{}^{35}Cl$ and $C_2H_5{}^{37}Cl$, respectively) in a characteristic 3:1 ratio. Figures 14.3 and 14.4 illustrate this for a chloro and a bromo compound.

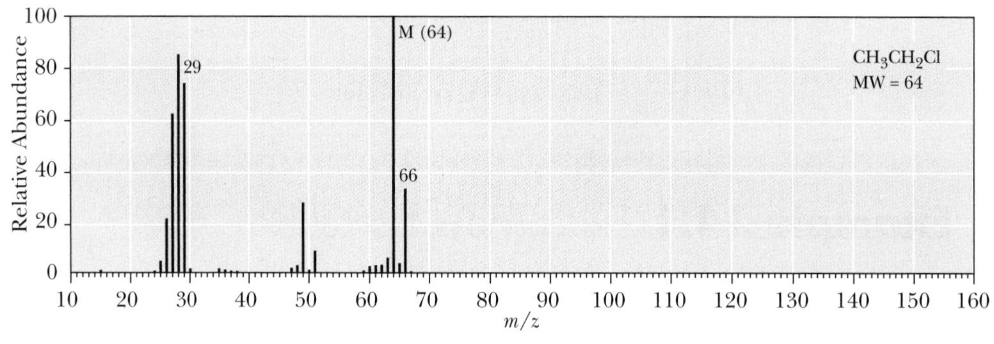

FIGURE 14.3 Mass spectrum of chloroethane.

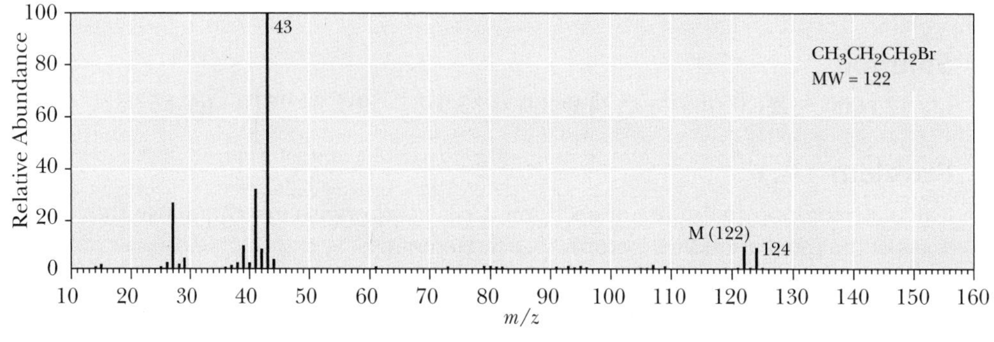

FIGURE 14.4 Mass spectrum of 1-bromopropane.

In contrast, 1-bromopropane has peaks at 122 and 124 for $C_3H_7{}^{79}Br$ and $C_3H_7{}^{81}Br$, respectively, in about a 1:1 ratio. This ratio of M and M + 2 peaks is characteristic of monobrominated compounds. The M and M + 2 peaks in chlorides and bromides in their ratios of 3:1 and 1:1, respectively, are very distinctive and allow almost immediate identification of monochloro and monobromo compounds.

D. Fragmentation of Molecular Ions

To attain high efficiency of molecular ion formation and to give reproducible mass spectra, it is common to use electrons with energies of 70 eV [approximately 6750 kJ (1600 kcal)/mol] in EI. This energy is sufficient not only to dislodge one or more electrons from a molecule but also to cause extensive fragmentation because it is well in excess of bond dissociation enthalpies in organic molecules. These fragments may be unstable as well and, in turn, break apart into even smaller fragments.

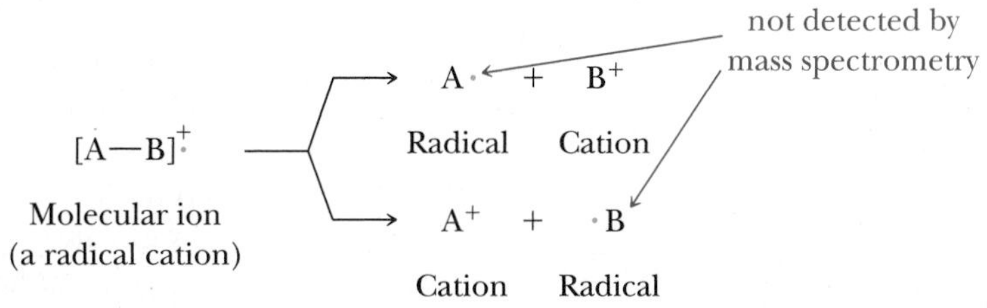

The molecular ions for some compounds have a sufficiently long lifetime in the analyzing chamber that they are observed in the mass spectrum, sometimes as the base (most intense) peak. Molecular ions of other compounds have a shorter lifetime and are present in low abundance or not at all. As a result, the mass spectrum of a compound ionized with one of the harsher ionization methods such as EI or CI typically (but not always) consists of a peak for the molecular ion and a series of peaks for fragment ions. The fragmentation pattern and relative abundances of ions are unique for each compound under a given set of ionizing conditions and are characteristic of that compound. Fragmentation patterns give us valuable information about molecular structure.

A great deal of the chemistry of ion fragmentation can be understood in terms of the formation and relative stabilities of carbocations in solution. Where fragmentation occurs forming new carbocations, the mode of fragmentation that gives the most stable carbocation is favored. Thus, the probability of fragmentation to form a new carbocation increases in the following familiar order.

$CH_3{}^+ < 1° < 2° \cong 1°$ allylic/benzylic $< 3° \cong 2°$ allylic/benzylic $< 3°$ allylic/benzylic

▶ Delocalization of charge over a larger area is stabilizing.

Molecular rearrangements are also characteristic of certain types of functional groups. ▶

14.3 Interpreting Mass Spectra

Chemists often use mass spectra primarily for the determination of molecular mass and molecular formula. Very rarely do they attempt a full interpretation of a mass spectrum, which can be very time consuming, difficult, and dependent on the experimental details of the ionization. The mass spectrum of dopamine (Figure 14.2), for example, contains at least 45 peaks with intensity equal to or greater than 0.5% of the intensity of the base peak. We have neither the need nor the time to attempt to interpret this level of complexity. Rather, we concentrate in this section on the fragmentation mechanisms giving rise to major peaks.

As we now look at typical mass spectra of the classes of organic compounds we have seen so far, keep the following two points in mind. They provide valuable information about the molecular composition of an unknown compound.

1. The only elements giving rise to significant M + 2 peaks are ^{18}O (0.2%), ^{34}S (4.40%), ^{37}Cl (32%), and ^{81}Br (98%). If no large M + 2 peak is present, then these elements are absent.
2. Is the mass of the molecular ion odd or even? According to the **nitrogen rule**, if a compound has an odd number of nitrogen atoms, its molecular ion will appear at an odd *m/z* value. Conversely, if a compound has an even number of nitrogen atoms (including zero), its molecular ion will appear at an even *m/z* value. This rule is most helpful when there is an odd number of nitrogens. You may need additional experimental information to establish the presence of an even number of nitrogens.

Nitrogen rule

A rule stating the molecular ion of a compound with an odd number of nitrogen atoms has an odd *m/z* ratio; if there are zero or an even number of nitrogen atoms, the molecular ion has an even *m/z* ratio.

A. Alkanes

Two rules will help you interpret the mass spectra of alkanes. (1) Fragmentation tends to occur toward the middle of unbranched chains rather than at the ends. (2) The differences in energy among allylic, benzylic, tertiary, secondary, primary, and methyl carbocations in the gas phase are much greater than the differences among comparable radicals. Therefore, where alternative modes of fragmentation are possible, the more stable carbocation tends to form in preference to the more stable radical.

Unbranched alkanes fragment to form a series of cations differing by 14 amu (a CH_2 group), with each fragment formed by a one-bond cleavage having an odd mass number. The mass spectrum of octane (Figure 14.5), for example, shows a peak for the molecular ion (*m/z* 114), as well as peaks for $C_6H_{13}^+$ (*m/z* 85), $C_5H_{11}^+$ (*m/z* 71), $C_4H_9^+$ (*m/z* 57), $C_3H_7^+$ (*m/z* 43), and $C_2H_5^+$ (*m/z* 29). These correspond to loss of ethyl, propyl, and butyl. Fragmentation of the CH_2—CH_3 bond is not observed; there is no peak corresponding to a methyl cation (*m/z* 15), nor is there one corresponding to a heptyl cation (loss of methyl, *m/z* 99). In mass spectrometry, fragmentations are shown by lines through the bond that is cleaved with an angled part toward the fragment that bears the charge.

Fragmentation of branched-chain alkanes leads preferentially to the formation of secondary and tertiary carbocations, and because these cations are more easily formed than methyl and primary carbocations, extensive fragmentation is likely. For this reason, the molecular ion of branched-chain hydrocarbons is often very weak or absent entirely from the spectrum. The molecular ion corresponding to *m/z* 114 is not observed, for example, in the mass spectrum of the highly branched 2,2,4-trimethylpentane (Figure 14.6). The base peak for this hydrocarbon is at *m/z* 57, which corresponds to the *tert*-butyl cation ($C_4H_9^+$). Other prominent peaks are at *m/z* 43 (isopropyl cation) and *m/z* 41 (allyl cation [CH_2=$CHCH_2^+$]).

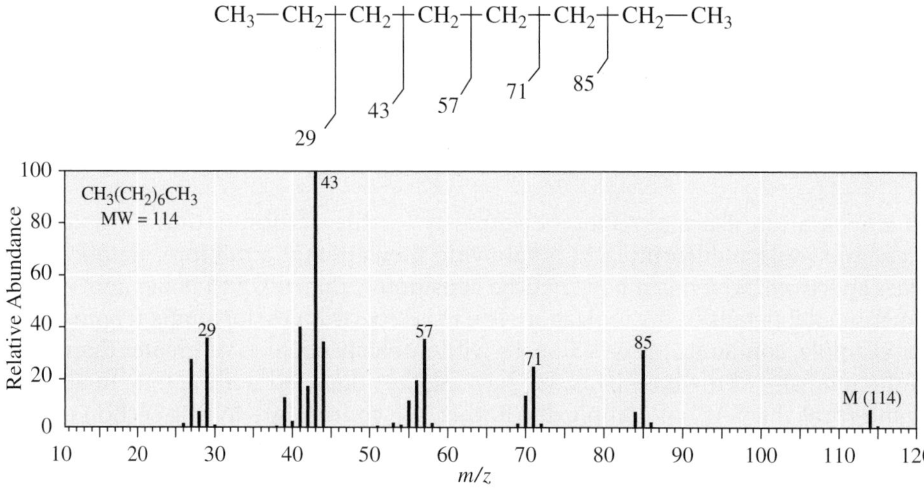

FIGURE 14.5 Mass spectrum of octane.

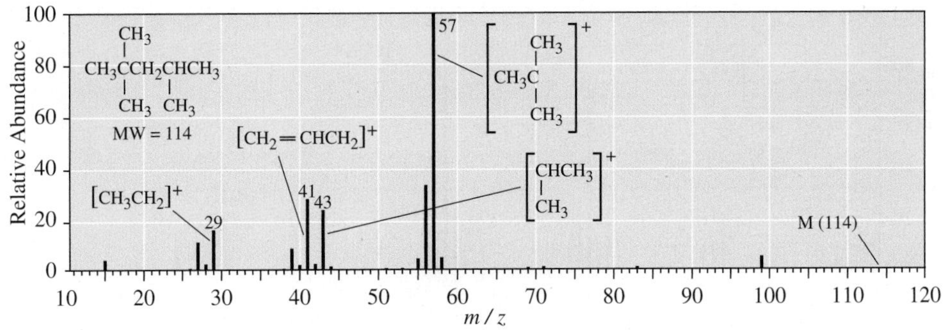

FIGURE 14.6 Mass spectrum of 2,2,4-trimethylpentane. The peak for the molecular ion is of such low intensity that it does not appear in this spectrum.

Sometimes peaks that seem to defy the rules of chemical logic we have encountered so far occur in a mass spectrum. For example, the prominent peak at *m/z* 29 in the mass spectrum of 2,2,4-trimethylpentane (Figure 14.6) is consistent with the ethyl cation, $CH_3CH_2^+$. There is, however, no ethyl group in the parent molecule. This cation must be formed by some combination of fragmentation and rearrangement beyond anything that we have seen up to this point; such rearrangements are common at the high energies of electron-impact mass spectra. Exploration of this point is beyond the scope of this text.

The most common fragmentation patterns of cycloalkanes are loss of side chains and loss of ethylene, $CH_2{=}CH_2$. The peak at *m/z* 69 in the mass spectrum of methylcyclopentane (Figure 14.7) is the result of the loss of the one-carbon side chain to give the cyclopentyl cation, $C_5H_9^+$. The base peak at *m/z* 56 is caused by the loss of ethylene and corresponds to a cation of molecular formula $C_4H_8^+$. Note here that one-carbon cleavages of alkanes and cycloalkanes give fragments with odd mass numbers; two-bond cleavages give fragments with even mass numbers.

Example 14.2 Predicting Structure from Mass Spectra

The base peak at *m/z* 56 in the mass spectrum of methylcyclopentane corresponds to loss of ethylene to give a radical cation with the molecular formula $C_4H_8^{+\cdot}$. Propose a structural formula for this radical cation and show how it might be formed.

Solution

Following is a structural formula for a molecular ion that might be formed in the ionizing chamber. In it, a single electron has been dislodged from a carbon-carbon single bond to give a 1° radical and a 2° carbocation. Rearrangement of bonding electrons in this radical cation gives ethylene and a new radical cation.

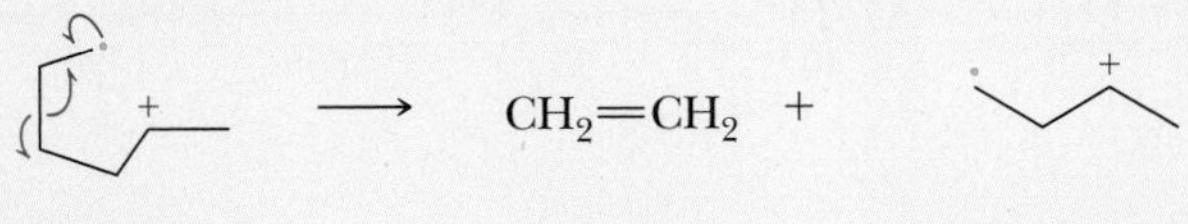

Molecular ion (a radical cation, *m/z* 84) | **Ethylene** | A new radical cation (a 2° carbocation, *m/z* 56)

Problem 14.2

Propose a structural formula for the cation at *m/z* 41 observed in the mass spectrum of methylcyclopentane.

Watch a video explanation

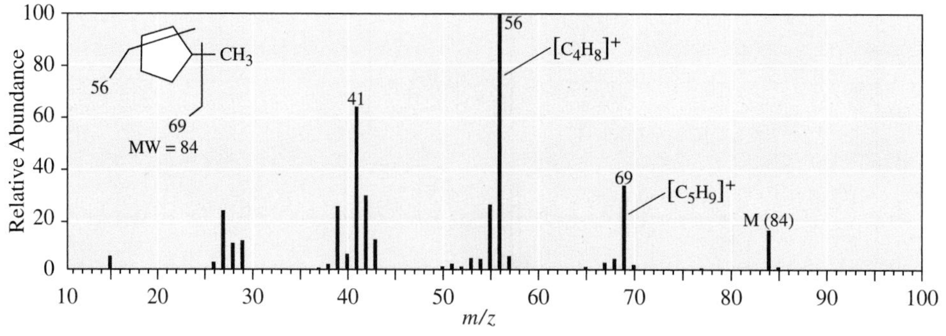

FIGURE 14.7 Mass spectrum of methylcyclopentane.

B. Alkenes

Alkenes characteristically show a strong molecular ion peak, most probably formed by removal of one π electron from the double bond. Furthermore, they cleave readily to form resonance-stabilized allylic cations, such as the allyl cation seen at *m/z* 41 in the mass spectrum of 1-butene (Figure 14.8).

Cyclohexenes undergo fragmentation to give a 1,3-diene and an alkene in a process that is the reverse of a Diels-Alder reaction (Section 20.5). The terpene limonene, a disubstituted cyclohexene, for example, fragments by a reverse Diels-Alder reaction to give two molecules of 2-methyl-1,3-butadiene (isoprene): one formed as a neutral diene and the other formed as a diene radical cation. Note here that the two-bond cleavage of this hydrocarbon gives fragments with even mass numbers.

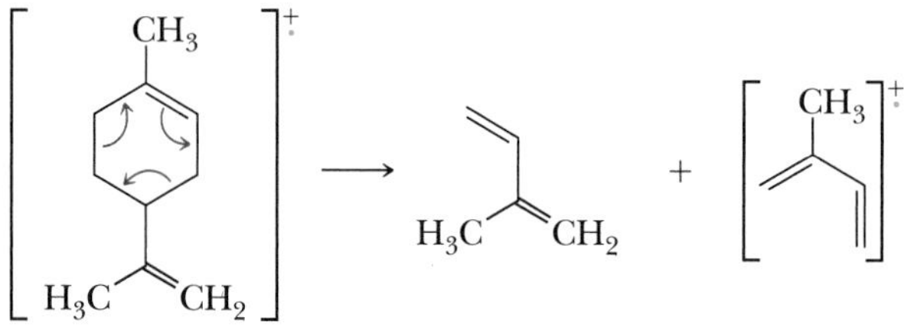

Limonene *m/z* 136 — A neutral diene — A radical cation *m/z* 68

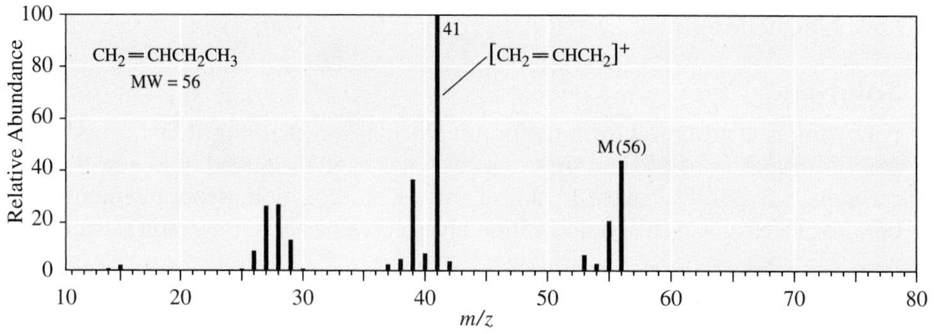

FIGURE 14.8 Mass spectrum of 1-butene.

C. Alkynes

As with alkenes, alkynes show a strong peak for the molecular ion. Their fragmentation patterns are also similar to those of alkenes. One of the most prominent peaks in the mass spectrum of most alkynes is from the delocalization-stabilized 3-propynyl (propargyl) cation (*m/z* 39) or a substituted propargyl cation.

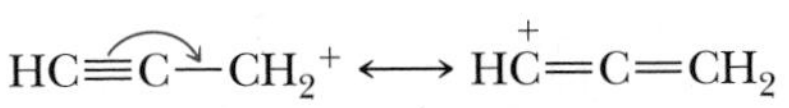

3-Propynyl cation
(Propargyl cation)

Both the molecular ion, *m*/*z* 68, and the propargyl cation, *m*/*z* 39, are seen in the mass spectrum of 1-pentyne (Figure 14.9). Also seen is the ethyl cation, *m*/*z* 29.

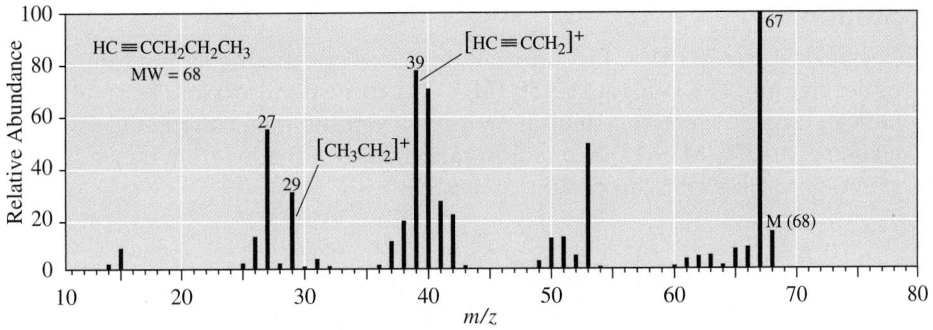

FIGURE 14.9 Mass spectrum of 1-pentyne.

D. Alcohols

The intensity of the molecular ion from primary and secondary alcohols is normally quite low, and there usually is no molecular ion detectable for tertiary alcohols. One of the most common fragmentation patterns for alcohols is loss of a molecule of water to give a peak corresponding to the molecular ion minus 18 (M − 18). Another common pattern is loss of an alkyl group from the carbon bearing the —OH group to form a delocalization stabilized oxonium ion and an alkyl radical. The oxonium ion is particularly stable because of delocalization of charge. ▶

▶ Delocalization of charge over a larger area is stabilizing.

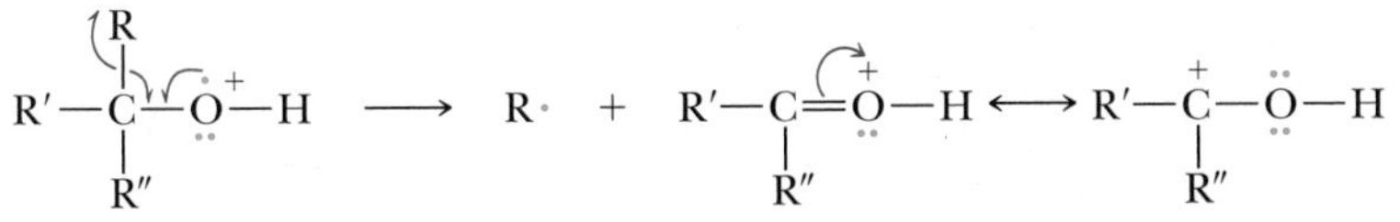

Molecular ion (a radical cation) — A radical — A resonance-stabilized oxonium ion

Each of these patterns is found in the mass spectrum of 1-butanol (Figure 14.10). The molecular ion appears at *m*/*z* 74. The prominent peak at *m*/*z* 56 corresponds to loss of a molecule of water from the molecular ion (M – 18). The base peak at *m*/*z* 31 corresponds to cleavage of a propyl group (M − 43) from the carbon bearing the —OH group. The propyl cation is visible at *m*/*z* 43.

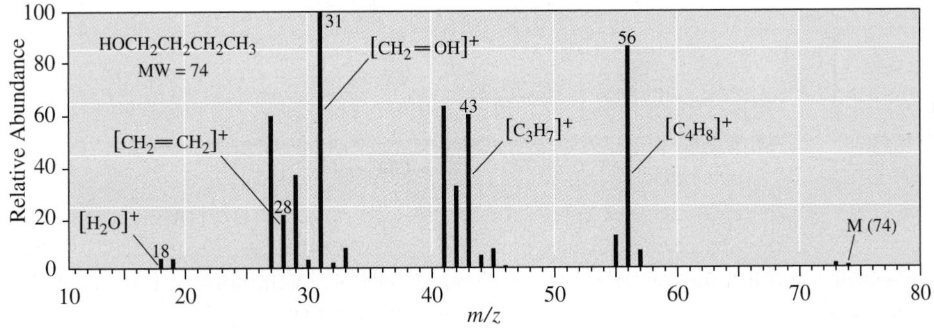

FIGURE 14.10 Mass spectrum of 1-butanol.

Example 14.3 Interpreting Peaks in Mass Spectra

A low-resolution mass spectrum of 2-methyl-2-butanol (MW 88) shows 16 peaks. The molecular ion is absent. Account for the formation of peaks at *m/z* 73, 70, 59, and 55 and propose a structural formula for each cation.

Solution

The peak at *m/z* 73 (M − 15) corresponds to loss of a methyl radical from the molecular ion. The peak at *m/z* 59 (M − 29) corresponds to loss of an ethyl radical. Loss of water as a neutral molecule from the molecular ion gives an alkene at *m/z* 70 (M − 18) as a radical cation. Loss of methyl from this radical cation gives an allylic carbocation at *m/z* 55.

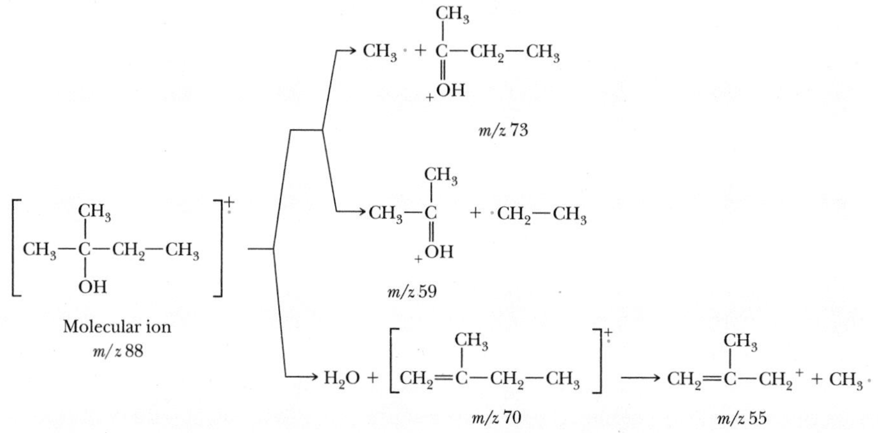

Problem 14.3

The low-resolution mass spectrum of 2-pentanol shows 15 peaks. Account for the formation of the peaks at *m/z* 73, 70, 55, 45, 43, and 41.

E. Aldehydes and Ketones

A characteristic fragmentation pattern of aliphatic aldehydes and ketones is cleavage of one of the bonds to the carbonyl group (α-cleavage). α-Cleavage of 2-octanone, for example, gives carbonyl-containing ions at *m/z* 43 and 113. α-Cleavage of the aldehyde proton gives an M − 1 peak, which is often quite distinct and provides a useful way to distinguish between an aldehyde and a ketone.

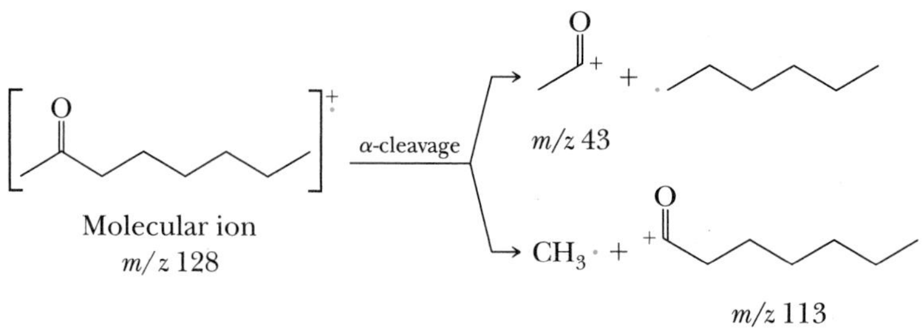

Aldehydes and ketones with a sufficiently long carbon chain show a fragmentation called a McLafferty rearrangement. In a **McLafferty rearrangement** of an aldehyde or a ketone, the carbonyl oxygen abstracts a hydrogen five atoms away to give an alkene and a new radical cation. Because McLafferty rearrangements involve cleavage of two bonds, molecular ions at even *m*/*z* give fragments at even *m*/*z*.

Mechanism 14.1

McLafferty Rearrangement of a Ketone

Reaction occurs through a six-membered ring transition state.

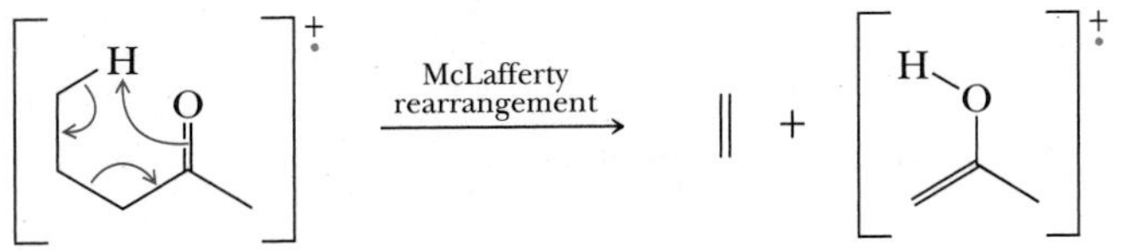

Molecular ion
m/*z* 86

m/*z* 58

McLafferty rearrangement of 2-octanone, for example, gives 1-pentene and a radical cation at *m*/*z* 58, which is the enol of acetone (Section 16.9). ▶

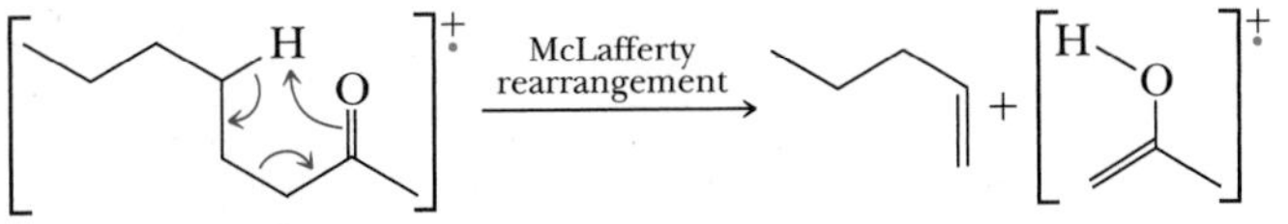

Molecular ion
m/*z* 128

m/*z* 58

▶ Five- and six-membered rings are the most stable.

The results of both α-cleavage and McLafferty rearrangement can be seen in the mass spectrum of 2-octanone (Figure 14.11).

F. Carboxylic Acids, Esters, and Amides

The molecular ion peak from a carboxylic acid is generally observed, although it is often very weak. The most common fragmentation patterns are α-cleavage of the carboxyl group to give the ion $[COOH]^+$ at *m*/*z* 45 and McLafferty rearrangement. The base peak is very often the result of the McLafferty rearrangement product.

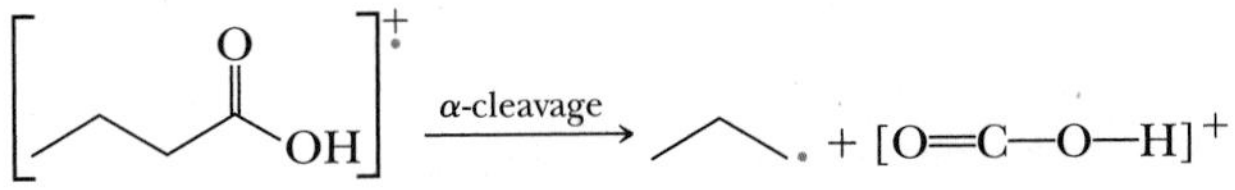

Molecular ion
m/*z* 88

m/*z* 45

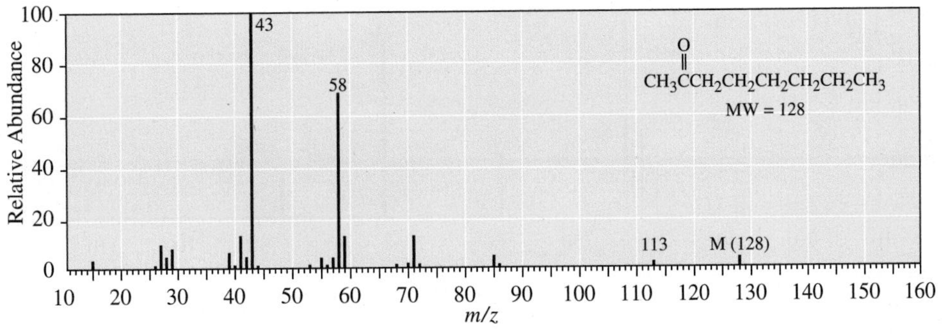

FIGURE 14.11 Mass spectrum of 2-octanone. Ions at *m*/*z* 43 and 113 result from α-cleavage. The ion at *m*/*z* 58 results from McLafferty rearrangement.

Mechanism 14.2

McLafferty Rearrangement of a Carboxylic Acid

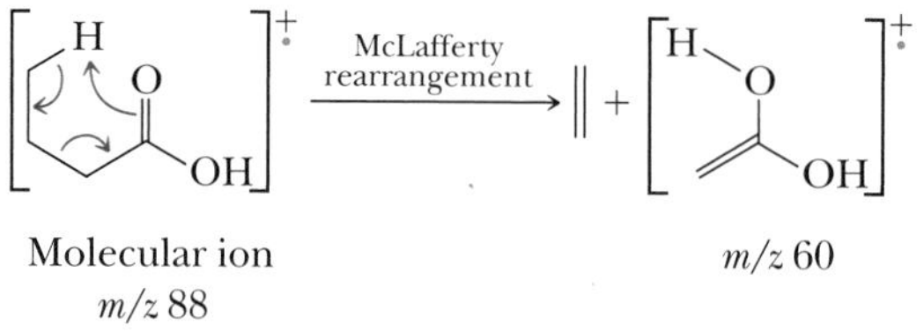

Each of these patterns is seen in the mass spectrum of butanoic acid (Figure 14.12).

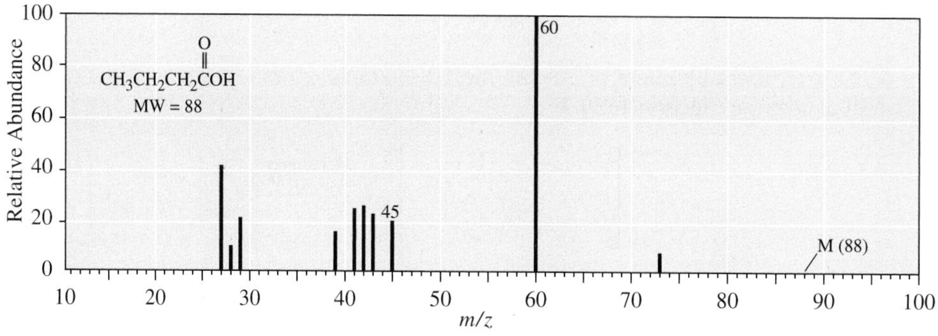

FIGURE 14.12 Mass spectrum of butanoic acid. Common fragmentation patterns of carboxylic acids are α-cleavage to give the ion $[COOH]^+$ at *m/z* 45 and McLafferty rearrangement.

Esters and amides also generally show discernible molecular ion peaks. Like carboxylic acids, their most characteristic fragmentation patterns are α-cleavage and McLafferty rearrangement, both of which can be seen in the mass spectrum of methyl butanoate (Figure 14.13). Peaks at *m/z* 71 and 59 are the result of α-cleavage.

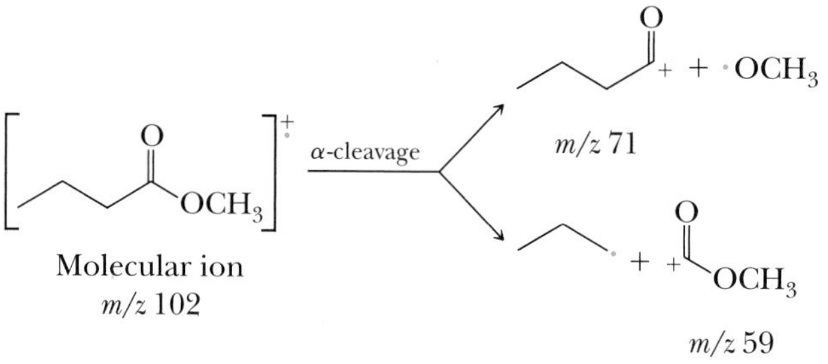

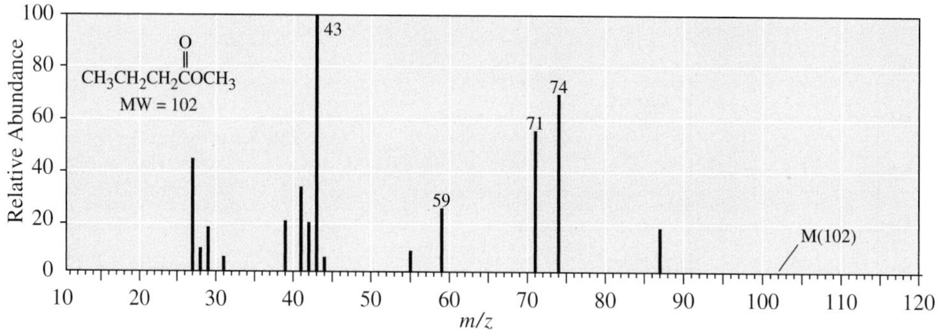

FIGURE 14.13 Mass spectrum of methyl butanoate. Characteristic fragmentation patterns of esters are α-cleavage and McLafferty rearrangement.

Connections to Biological Chemistry

Mass Spectrometry of Biological Macromolecules

As a result of enormous instrumental advances made during the last few years, mass spectrometry is becoming the method of choice for determining the structures of biological macromolecules, in particular, proteins and DNA. The Nobel Prize in Chemistry in 2002 was awarded to John B. Fenn and Koichi Tanaka for the development of the electrospray ionization (ESI) and MALDI methods that have revolutionized structure determination of these molecules. The key advantage of using mass spectrometry for analysis of proteins and nucleic acids is that the entire process can be automated so that information about a large number of molecules can be obtained rapidly. A growing list of DNA sequences at the genome level is now available, and attention is increasingly being turned toward acquiring sequence information at the protein level with high throughput. Mass spectrometry is playing a major role in this effort.

For example, it is now possible to obtain complete amino acid sequences on polypeptides [proteins and biological polyamides (see Chapter 27)] of substantial length. The mass spectrometer cleaves polypeptides into fragments of varying length. Although many cleavage modes are possible, the main cleavage is at peptide (amide) bonds. Both fragments (from the *N*-terminal and *C*-terminal part) can usually be identified. A mass analyzer determines the mass of each fragment. Because each amino acid has a slightly different mass (except for two isomeric amino acids leucine and isoleucine), the exact amino acid composition of the fragment can be determined. Powerful computers align overlapping fragments and determine the exact sequence.

For larger proteins, the mass spectrometric method is usually preceded by enzymatic cleavage into fragments. However, in the most advanced systems, the individual fragments do not need to be separated. They are injected into a tandem mass spectrometer (called an MS-MS). The first segment of the instrument separates the individual fragments by mass, and the second segment fragments each peptide and sequences it separately. The MS-MS can also be used to separate and then further fragment the primary ions obtained in the first stage to obtain sequence information directly. Enormous proteins can be sequenced on a picomole or lower sample size. This method is particularly helpful because it can be used on mixtures (and ultimately, it is believed, on whole cell contents). Similar techniques are also now used for determining the sequence of DNA, the genetic material. Further discussion about applications to macromolecules may be found in Chapters 27 and 28.

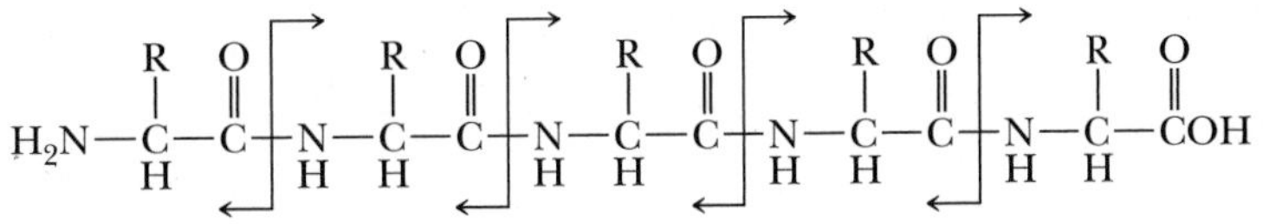

The peak at *m/z* 74 is the result of McLafferty rearrangement.

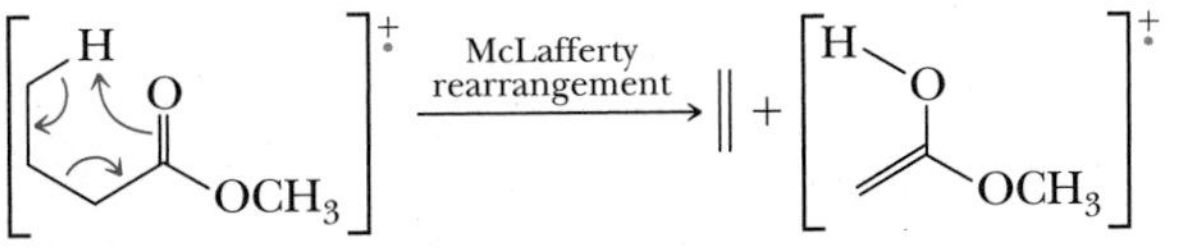
McLafferty
rearrangement
Molecular ion
m/z 102
m/z 74

G. Aromatic Hydrocarbons

The mass spectra of most aromatic hydrocarbons show an intense molecular ion peak. The mass spectrum of toluene (Figure 14.14), for example, shows a large molecular ion peak at *m/z* 92.

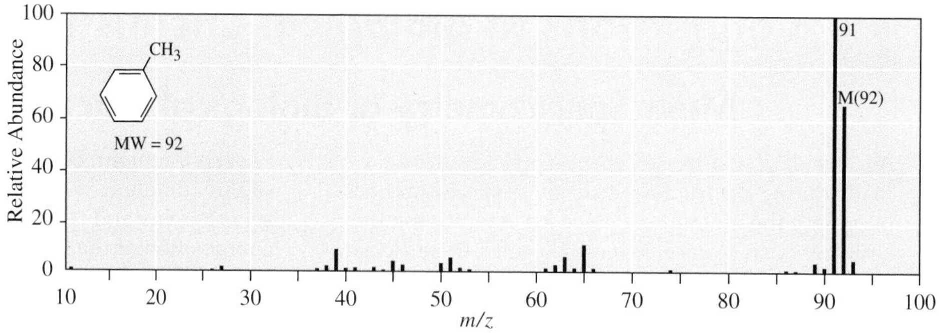

FIGURE 14.14 Mass spectrum of toluene. Prominent are the intense molecular ion peak at *m/z* 92 and the tropylium cation at *m/z* 91.

The mass spectra of toluene and most other alkylbenzenes show a fragment ion at *m/z* 91. Although it might seem that the most likely structure for this ion is that of the benzyl cation, experimental evidence suggests a molecular rearrangement to form the more stable tropylium ion (Section 21.2E). In the tropylium ion, an aromatic cation, the positive charge is delocalized equally over all seven carbon atoms of the cycloheptatrienyl ring.

▶ **Delocalization of charge over a larger area is stabilizing.**

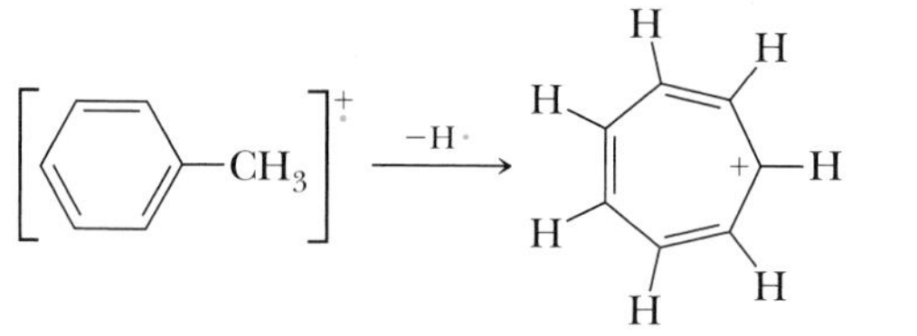

Toluene radical cation **Tropylium cation** (*m/z* 91) ▶

H. Amines

Of the compounds containing C, H, N, O, and the halogens, only those containing an odd number of nitrogen atoms have a molecular ion at odd *m/z* ratio. Thus, mass spectrometry can be a particularly valuable tool for identifying amines. The molecular ion for aliphatic amines, however, is often very weak. The most characteristic fragmentation of amines (and the one that often gives the base peak) is β-cleavage. Where alternative possibilities for β-cleavage exist, it is generally the largest R group that is lost. In contrast to nitrogen-free molecules, single bond fragments from compounds that have one nitrogen or an odd number of nitrogens give compounds with even mass. The most prominent peak in the mass spectrum of 3-methyl-1-butanamine (Figure 14.15) is from $[CH_2{=}NH_2]^+$, *m/z* 30, from β-cleavage. β-Cleavage is also characteristic of secondary and tertiary amines. Complex rearrangement and

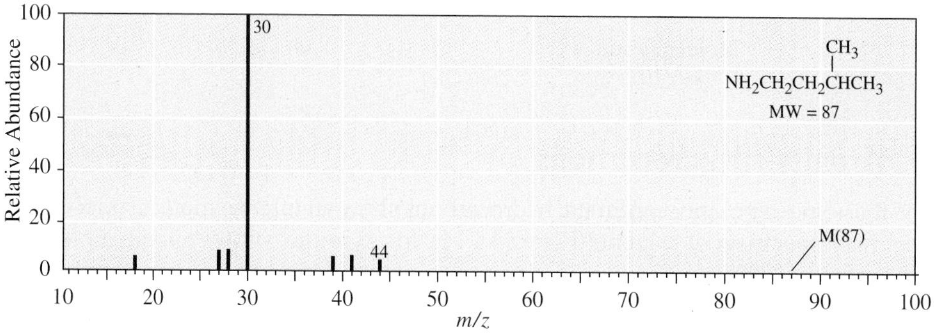

FIGURE 14.15 Mass spectrum of 3-methyl-1-butanamine (isopentylamine). The most characteristic fragmentation pattern of aliphatic amines is β-cleavage.

fragmentation processes give the *m*/*z* 30 peak as a major fragment even from secondary and tertiary amines.

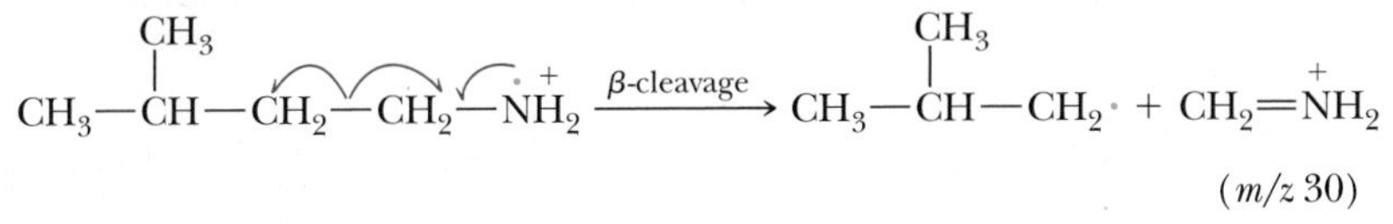

14.4 Mass Spectrometry in the Organic Synthesis Laboratory and Other Applications

Mass spectrometry, especially when interfaced with separation methods such as gas chromatography (GC-MS) and liquid chromatography (LC-MS), represents a powerful and rapid method of identifying compounds in reaction product mixtures. As a result, GC-MS and LC-MS are taking on increased importance as the primary method of routine molecule identification in the organic synthesis laboratory in both industrial and academic settings.

Mass spectrometry is also becoming increasingly important for other practical applications. Some devices used to screen luggage in airports use mass spectrometry to identify traces of known explosives. Drug testing of athletes and advanced forensic science uses GC-MS to identify traces of pharmaceuticals or illicit drugs in blood samples. Looking toward the future, with the advent of powerful new-generation quadrupole mass spectrometers that are increasingly small and inexpensive, there will be a dramatic increase in the use of mass spectrometry in the identification of molecules in many aspects of modern life, not just in the research laboratory.

Study Guide

14.1 A Mass Spectrometer

- A **mass spectrum** is a plot of relative ion abundance versus mass-to-charge (*m*/*z*) ratio. The **base peak** is the most intense peak in a mass spectrum.
  - *Numerous methods of ionization exist, including electron impact (EI), fast-atom bombardment (FAB), chemical ionization (CI), matrix-assisted laser desorption ionization (MALDI), and electrospray ionization (ESI).*
- A **molecular ion**, $M^{+\cdot}$, is a radical cation derived from the parent molecule by loss of one electron.

14.2 Features of a Mass Spectrum

- **Low-resolution mass spectrometry** distinguishes among ions that differ in nominal mass (i.e., ions that differ by at least: 1 amu).

- **High-resolution mass spectrometry** distinguishes among ions that differ by as little as 0.0001 amu.
- $M + 1$ and higher peaks in a mass spectrum are caused by heavier isotopes.
  - *The abundance of these higher mass-to-charge peaks relative to the molecular ion peak provides information about the elemental composition of the molecular ion.*
    - *The presence of a single chlorine atom, for example, is indicated by M and $M + 2$ peaks in a ratio of 3:1.*
- The mass spectrum of a compound typically consists of a peak for the molecular ion and a series of peaks for fragment ions.
  - *The fragmentation pattern and relative abundances of ions are unique for each compound and are characteristic of that compound.*
    - *Fragments formed by cleavage of one bond have odd mass if they contain no nitrogen; those formed from the cleavage of two bonds have even mass.*
  - *Many of the observed fragmentation patterns can be understood in terms of the relative stability of carbocations.*
    - *Where alternative modes of fragmentation are possible, the more stable carbocation tends to be formed in preference to the more stable radical.*

P **14.1, 14.5, 14.8, 14.9, 14.11, 14.12, 14.19–14.23**

14.3 Interpreting Mass Spectra

- Different functional groups have characteristic fragmentation patterns in mass spectra.
  - ***Alkane** mass spectra tend to show fragmentation that occurs toward the middle of unbranched chains, forming a series of fragments differing by 14 amu.*
    - *Branched chains give fragments at the branches to generate secondary and tertiary cations.*
  - ***Alkenes** give a strong molecular ion peak and cleave to form allyl cations.*
  - ***Alkynes** give a strong molecular ion peak and cleave to form the propargyl cation.*
  - ***Alcohols** give generally weak molecular ions (secondary and tertiary alcohols especially) and often cleave to lose water, or they lose an alkyl group bonded to the carbon bonded to the OH group.*
  - ***Aldehydes and ketones** undergo two kinds of characteristic cleavage.*
    - ***α-Cleavage** involves cleavage of one of the bonds to the carbonyl group.*
    - ***McLafferty rearrangement** involves a six-membered ring transition state to give an alkene and a new radical cation.*
  - ***Carboxylic acids, esters, and amides** undergo the same two kinds of characteristic cleavage, α-cleavage and McLafferty rearrangement.*
  - ***Aromatic hydrocarbons** give strong molecular ions, and alkyl benzenes often fragment to give the tropylium cation.*
- According to the **nitrogen rule**, if a compound has an odd number of nitrogen atoms, its molecular ion will have an odd m/z value.

14.2–14.4, 14.6, 14.7, 14.10, 14.13–14.18, 14.24–14.37

Problem 14.2

14.4 Mass Spectrometry in the Organic Synthesis Laboratory and Other Applications

- Mass spectrometry is increasingly used in the forensic laboratory.
- Mass spectral techniques can be used to determine the amino acid sequences of proteins and nucleotide sequences of DNA.

Problems

Red numbers indicate applied problems.

14.4 Draw acceptable Lewis structures for the molecular ion (radical cation) formed from the following molecules when each is bombarded by high-energy electrons in a mass spectrometer.

(a)  **(b)** 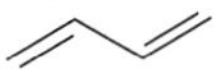 **(c)** 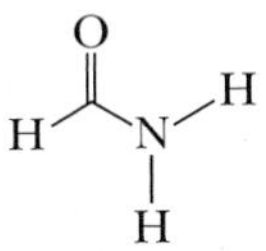 **(d)** 

14.5 The molecular ion for compounds containing only C, H, and O always has an even mass-to-charge value. Why? What can you say about the mass-to-charge ratio of ions that arise from fragmentation of one bond in the molecular ion? from fragmentation of two bonds in the molecular ion?

14.6 For which compounds containing a heteroatom (an atom other than carbon or hydrogen) does the molecular ion have an even-numbered mass? For which does it have an odd-numbered mass?

(a) A chloroalkane with the molecular formula $C_nH_{2n+1}Cl$
(b) A bromoalkane with the molecular formula $C_nH_{2n+1}Br$
(c) An alcohol with the molecular formula $C_nH_{2n+1}OH$
(d) A primary amine with the molecular formula $C_nH_{2n-1}NH_2$
(e) A thiol with the molecular formula $C_nH_{2n+1}SH$

14.7 The so-called nitrogen rule states that if a compound has an odd number of nitrogen atoms, the value of *m*/*z* for its molecular ion will be an odd number. Why?

14.8 Both $C_6H_{10}O$ and C_7H_{14} have the same nominal mass, namely 98. Show how these compounds can be distinguished by the *m*/*z* ratio of their molecular ions in high-resolution mass spectrometry.

14.9 Show how the compounds with the molecular formulas C_6H_9N and C_5H_5NO can be distinguished by the *m*/*z* ratio of their molecular ions in high-resolution mass spectrometry.

14.10 What rule would you expect for the *m*/*z* values of fragment ions resulting from the cleavage of one bond in a compound with an odd number of nitrogen atoms?

14.11 Determine the probability of the following in a natural sample of ethane.

(a) One carbon in an ethane molecule is ^{13}C.
(b) Both carbons in an ethane molecule are ^{13}C.
(c) Two hydrogens in an ethane molecule are replaced by deuterium atoms.

14.12 The molecular ions of both $C_5H_{10}S$ and $C_6H_{14}O$ appear at *m/z* 102 in low-resolution mass spectrometry. Show how determination of the correct molecular formula can be made from the appearance and relative intensity of the M + 2 peak of each compound.

14.13 In Section 14.3, we saw several examples of fragmentation of molecular ions to give resonance-stabilized cations. Make a list of these resonance-stabilized cations and write important contributing structures of each. Estimate the relative importance of the contributing structures in each set.

14.14 Carboxylic acids often give a strong fragment ion at *m/z* (M – 17). What is the likely structure of this cation? Show by drawing contributing structures that it is stabilized by resonance.

14.15 For primary amines with no branching on the carbon bearing the nitrogen, the base peak occurs at *m/z* 30. What cation does this peak represent? How is it formed? Show by drawing contributing structures that this cation is stabilized by resonance.

14.16 The base peak in the mass spectrum of propanone (acetone) occurs at *m/z* 43. What cation does this peak represent?

14.17 A characteristic peak in the mass spectrum of most aldehydes occurs at *m/z* 29. What cation does this peak represent? (No, it is not an ethyl cation, $CH_3CH_2^+$.)

14.18 Predict the relative intensities of the M and M + 2 peaks for the following.

(a) CH_3CH_2Cl **(b)** CH_3CH_2Br **(c)** $BrCH_2CH_2Br$ **(d)** CH_3CH_2SH

14.19 The mass spectrum of compound A shows the molecular ion at *m/z* 85, an M + 1 peak at *m/z* 86 of approximately 6% abundance relative to M, and an M + 2 peak at *m/z* 87 of less than 0.1% abundance relative to M.

(a) Propose a molecular formula for compound A.
(b) Draw at least ten possible structural formulas for this molecular formula.

14.20 The mass spectrum of compound B, a colorless liquid, shows these peaks in its mass spectrum. Determine the molecular formula of compound B and propose a structural formula for it.

m/z	Relative Abundance
43	100 (base)
78	23.6 (M)
79	1.00
80	7.55
81	0.25

14.21 Write molecular formulas for the five possible molecular ions of *m/z* 88 containing the elements C, H, N, and O.

14.22 Write molecular formulas for the five possible molecular ions of *m/z* 100 containing only the elements C, H, N, and O.

14.23 The molecular ion in the mass spectrum of 2-methyl-1-pentene appears at *m/z* 84. Propose structural formulas for the prominent peaks at *m/z* 69, 55, 41, and 29.

14.24 Following is the mass spectrum of 1,2-dichloroethane.

(a) Account for the appearance of an M + 2 peak with approximately two-thirds the intensity of the molecular ion peak.

(b) Predict the intensity of the M + 4 peak.
(c) Propose structural formulas for the cations of *m/z* 64, 63, 62, 51, 49, 27, and 26.

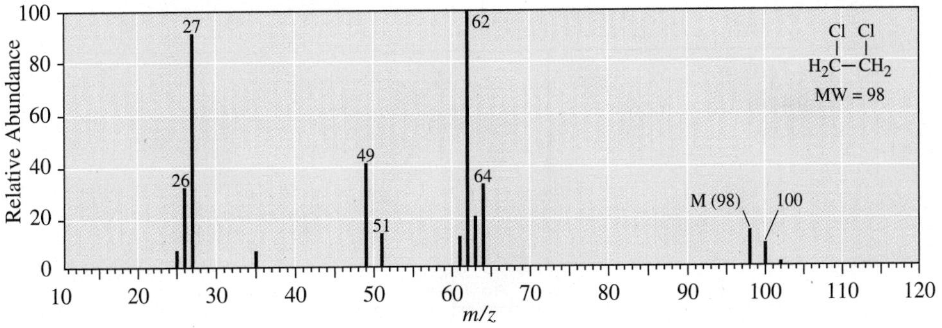

14.25 Following is the mass spectrum of 1-bromobutane.

(a) Account for the appearance of the M + 2 peak of approximately 95% of the intensity of the molecular ion peak.
(b) Propose structural formulas for the cations of *m/z* 57, 41, and 29.

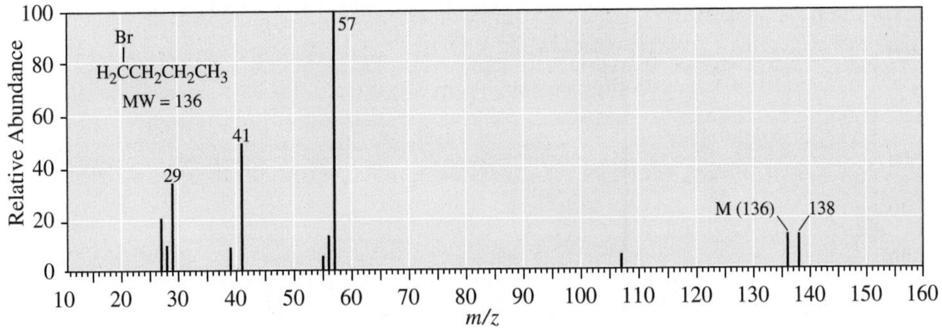

14.26 Following is the mass spectrum of bromocyclopentane. The molecular ion *m/z* 148 is of such low intensity that it does not appear in this spectrum. Assign structural formulas for the cations of *m/z* 69 and 41.

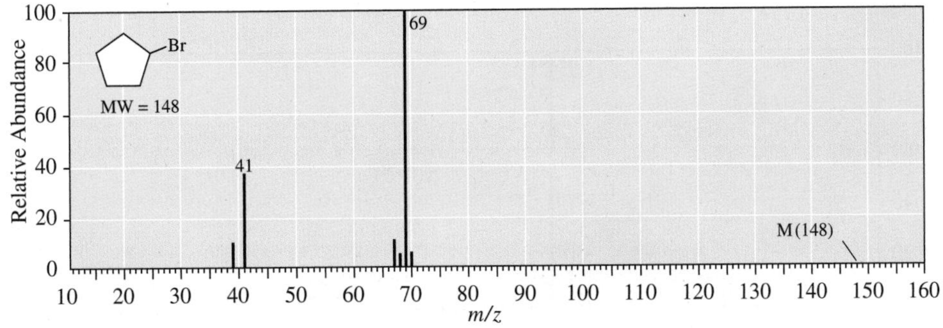

14.27 Following is the mass spectrum of an unknown compound. The two highest peaks are at *m/z* 120 and 122. Suggest a structure for this compound. (Data from http://webbook.nist.gov/chemistry/.)

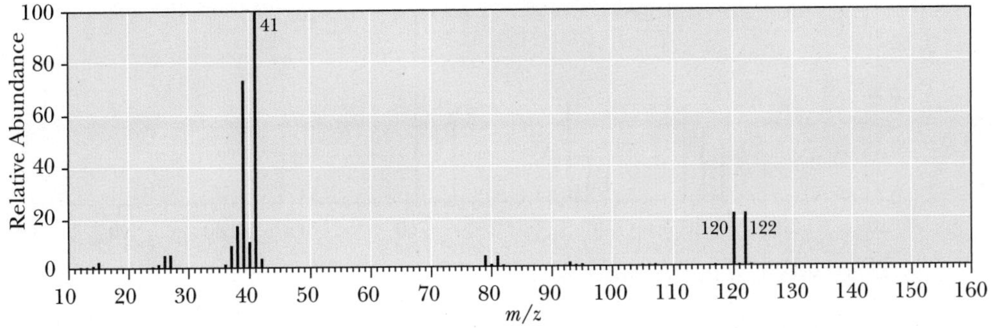

14.28 Following is the mass spectrum of 3-methyl-2-butanol. The molecular ion *m/z* 88 does not appear in this spectrum. Propose structural formulas for the cations of *m/z* 45, 43, and 41.

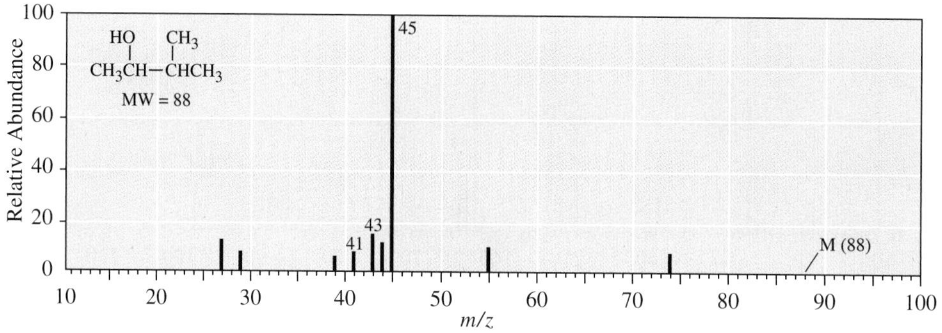

14.29 The following is the mass spectrum of compound C, C_3H_8O. Compound C is infinitely soluble in water, undergoes reaction with sodium metal with the evolution of a gas, and undergoes reaction with thionyl chloride to give a water-insoluble chloroalkane. Propose a structural formula for compound C and write equations for each of its reactions.

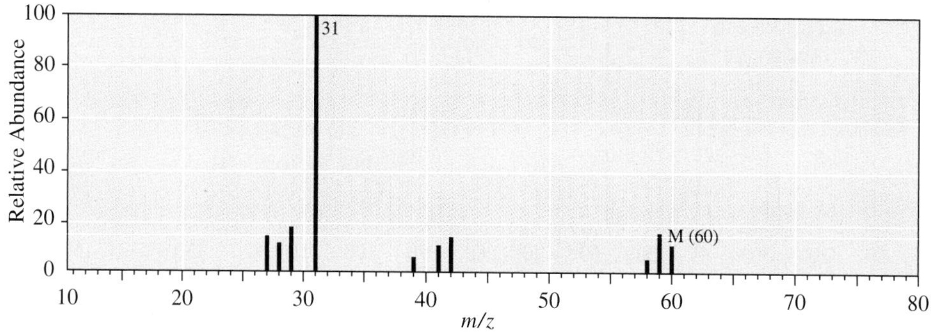

14.30 Following are mass spectra for the constitutional isomers 2-pentanol and 2-methyl-2-butanol. Assign each isomer its correct spectrum.

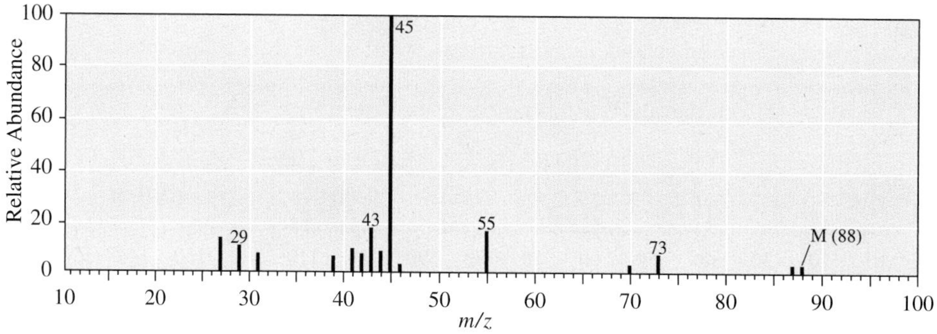

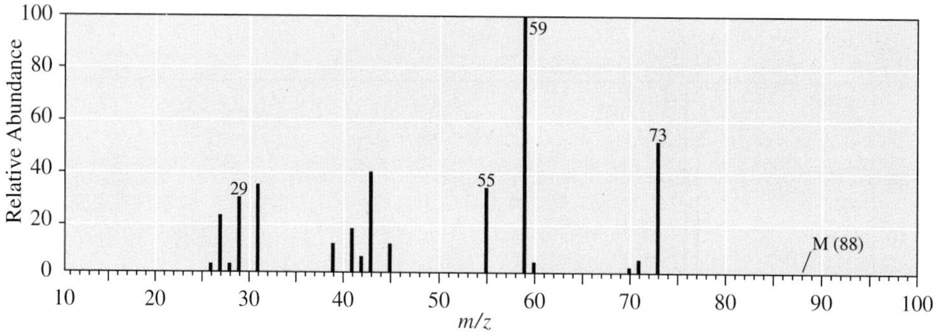

14.31 2-Methylpentanal and 4-methyl-2-pentanone are constitutional isomers with the molecular formula $C_6H_{12}O$. Each shows a molecular ion peak in its mass spectrum at *m/z* 100. Spectrum A shows significant peaks at *m/z* 85, 58, 57, 43, and 42. Spectrum B shows significant peaks at *m/z* 71, 58, 57, 43, and 29. Assign each compound its correct spectrum.

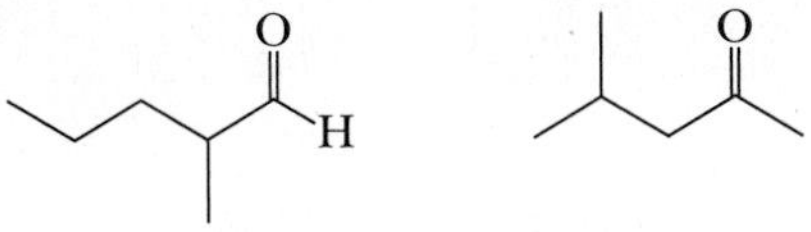

2-Methylpentanal **4-Methyl-2-pentanone**

14.32 Account for the presence of peaks at *m/z* 135 and 107 in the mass spectrum of 4-methoxybenzoic acid (*p*-anisic acid).

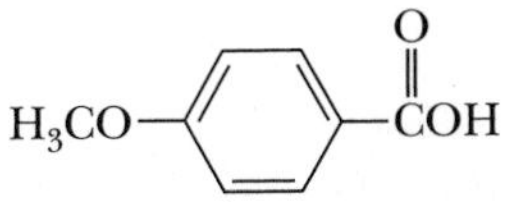

4-Methoxybenzoic acid

14.33 Account for the presence of the following peaks in the mass spectrum of hexanoic acid, $CH_3(CH_2)_4COOH$.

(a) *m/z* 60
(b) A series of peaks differing by 14 amu at *m/z* 45, 59, 73, and 87
(c) A series of peaks differing by 14 amu at *m/z* 29, 43, 57, and 71

14.34 All methyl esters of long-chain aliphatic acids (e.g., methyl tetradecanoate, $C_{13}H_{27}COOCH_3$) show significant fragment ions at *m/z* 74, 59, and 31. What are the structures of these ions? How are they formed?

14.35 Propylbenzene, $C_6H_5CH_2CH_2CH_3$, and isopropyl benzene, $C_6H_5CH(CH_3)_2$, are constitutional isomers with the molecular formula C_9H_{12}. One of these compounds shows prominent peaks in its mass spectrum at *m/z* 120 and 105. The other shows prominent peaks at *m/z* 120 and 91. Which compound has which spectrum?

14.36 Account for the formation of the base peaks in these mass spectra.

(a) Isobutylamine, *m/z* 30 **(b)** Diethylamine, *m/z* 58

14.37 Because of the sensitivity of mass spectrometry, it is often used to detect the presence of drugs in blood, urine, or other biological fluids. Tetrahydrocannabinol (nominal mass 314), a component of marijuana, exhibits two strong fragment ions at *m/z* 246 and 231 (the base peak). What is the likely structure of each ion?

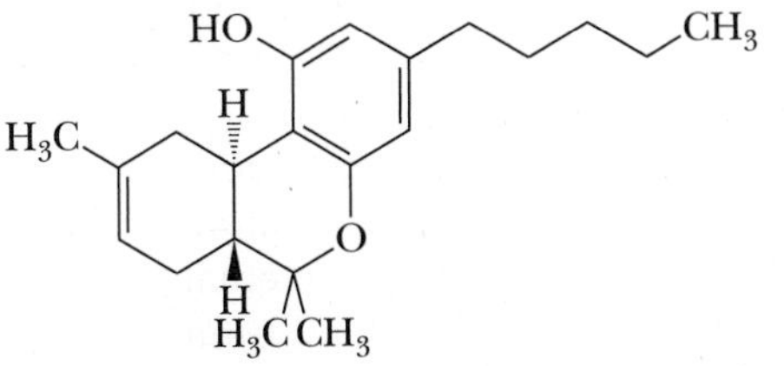

Tetrahydrocannabinol

14.38 Electrospray mass spectrometry is a recently developed technique for looking at large molecules with a mass spectrometer. In this technique, molecular ions, each associated with one or more H^+ ions, are prepared under mild conditions in the mass spectrometer. As an example, a protein (P) with a molecular mass of 11,812 gives clusters of the type $(P + 8H)^{8+}$, $(P + 7H)^{7+}$, and $(P + 6H)^{6+}$. At what mass-to-charge values do these three clusters appear in the mass spectrum?

15

An Introduction to Organometallic Compounds

A common housefly (*Musca domestica*). Commercially available traps with the attractant muscalure (synthesized via a route using an organometallic compound) combined with a poison are efficient at luring and killing houseflies (Section 15.2B). (© Mirko Graul/Shutterstock.com)

Outline

15.1 Organomagnesium and Organolithium Compounds
15.2 Lithium Diorganocopper (Gilman) Reagents
15.3 Carbenes and Carbenoids

In this chapter, we undertake our first discussion of a broad class of organic compounds called **organometallic compounds**, compounds that contain a carbon-metal bond. In recent years, there has been an enormous explosion in our understanding of their chemistry, particularly as stereospecific (and often enantioselective) reagents for synthetic chemistry. We have already seen one example in the Sharpless enantioselective epoxidation of alkenes (Section 11.8D).

This chapter cannot possibly cover the wealth of organometallic reagents and catalysts that have been developed for synthetic organic chemistry, particularly during the last decade. We focus, therefore, on transformations that are fundamental to synthetic chemistry. Organomagnesium, lithium, and copper reagents have been selected because of their historical importance and their continued use in modern organic synthesis. These reagents are particularly important in the reactions of carbonyl compounds—the focus of the next several chapters. Several more recent reactions of organometallic compounds are discussed in the chapter on C—C bond formation and organic synthesis (Chapter 24).

Organometallic compound
A compound that contains a carbon-metal bond.

15.1 Organomagnesium and Organolithium Compounds

We begin with organomagnesium and organolithium compounds and concentrate on their formation and basicity. We discuss their use in organic synthesis in more detail in later chapters, particularly in Chapters 16 and 18.

RMgX	RLi
An organomagnesium compound (a Grignard reagent)	An organolithium compound

A. Formation and Structure

Organomagnesium compounds are among the most readily available, easily prepared, and easily handled organometallics. They are commonly named Grignard reagents after the French chemist Victor Grignard (1871–1935), who was awarded the 1912 Nobel Prize in Chemistry for their discovery and application to organic synthesis.

Grignard reagents are typically prepared by the slow addition of an alkyl, aryl, or alkenyl (vinylic) halide to a stirred suspension of a slight excess of magnesium metal in an ether solvent, most commonly diethyl ether or tetrahydrofuran (THF). Organic iodides and bromides generally react very rapidly under these conditions, whereas most organic chlorides react more slowly. Bromides are the most common starting materials for preparation of Grignard reagents. It is common to use the higher-boiling THF (bp 67°C) to prepare Grignard reagents from the less reactive organic halides. Generally, there is an induction period at the beginning of the reaction caused by the presence of traces of moisture and a thin oxide coating on the surface of the magnesium. When reaction starts, it is exothermic, and the remaining organic halide is added at a rate sufficient to maintain a gentle reflux of the ether.

Butylmagnesium bromide, for example, is prepared by treating 1-bromobutane in diethyl ether with magnesium metal. Aryl Grignard reagents, such as phenylmagnesium bromide, are prepared in the same manner. These reactions are referred to as oxidative additions because they result in an increase in the formal oxidation state of magnesium by two [i.e., from Mg(0) to Mg(II)].

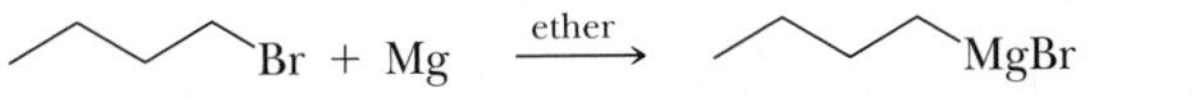

1-Bromobutane → **Butylmagnesium bromide** (an alkyl Grignard reagent)

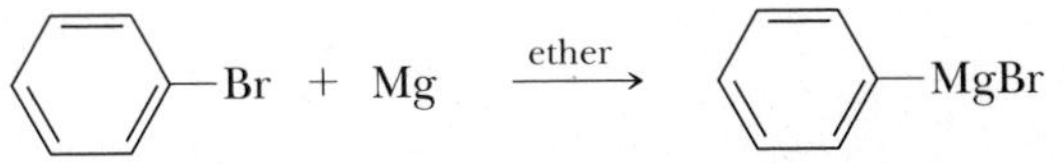

Bromobenzene → **Phenylmagnesium bromide** (an aryl Grignard reagent)

Although the equation for formation of Grignard reagents looks simple, the mechanism is considerably more complicated and involves radicals. We have no need in this course to discuss the mechanism for their formation. However, we note that for many Grignard reagents, there is an equilibrium between monoalkyl and dialkyl magnesium complexes as shown.

$$2\,RMgX \rightleftharpoons R_2Mg + MgX_2$$

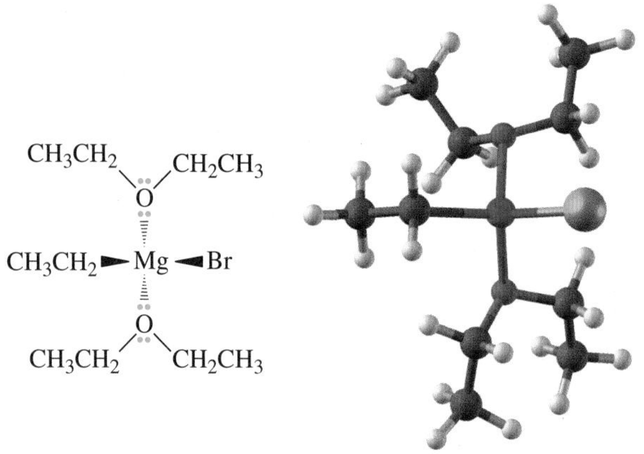
CH3CH2
CH2CH3
O
CH3CH2
Mg
Br
O
CH3CH2
CH2CH3

Ethylmagnesium bromide dietherate

FIGURE 15.1 Ethylmagnesium bromide dietherate, a Grignard reagent solvated by ether.

Grignard reagents form on the surface of the metal and dissolve as coordination complexes solvated by ether. In this ether-soluble complex, magnesium acts as a Lewis acid and the ether acts as a Lewis base (Figure 15.1).

Organolithium reagents are prepared by treating an alkyl, aryl, or alkenyl halide with two equivalents of lithium metal, as illustrated by the preparation of butyllithium. In this reaction, a solution of 1-chlorobutane in pentane is added to lithium wire at −10°C.

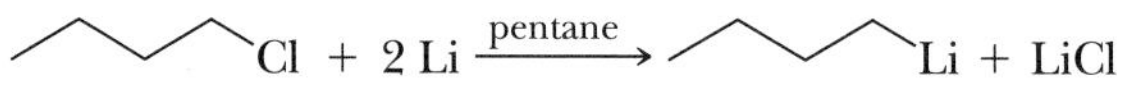
Cl + 2 Li
pentane
Li + LiCl

1-Chlorobutane **Butyllithium**

Organolithium compounds are very reactive as nucleophiles in carbonyl addition reactions even at very low temperatures. They are powerful and effective bases and are widely used in modern synthetic chemistry. As they react rapidly with atmospheric oxygen and moisture, they must be used under an inert atmosphere of N_2 or Ar. This decreases their convenience.

Carbanion

An ion in which carbon has an unshared pair of electrons and bears a negative charge.

The carbon-metal bonds in Grignard and organolithium reagents are best described as polar covalent, with carbon bearing a partial negative charge and the metal bearing a partial positive charge. In their reactions, Grignard and organolithium reagents behave as **carbanions**, which are ions in which carbon has an unshared pair of electrons and bears a negative charge. Shown in Table 15.1 are

Table 15.1 Percent Ionic Character of Some C—M Bonds

$\delta-$ $\delta+$ C—M Bond	Difference in Electronegativity	Percent Ionic Character*
C—Li	2.5 − 1.0 = 1.5	60
C—Mg	2.5 − 1.2 = 1.3	52
C—Al	2.5 − 1.5 = 1.0	40
C—Zn	2.5 − 1.6 = 0.9	36
C—Sn	2.5 − 1.8 = 0.7	28
C—Cu	2.5 − 1.9 = 0.6	24
C—Hg	2.5 − 1.9 = 0.6	24

*Percent ionic character $= \frac{E_C - E_M}{E_C} \times 100$

electronegativity differences (Pauling scale, Table 1.5) between carbon and various metals. From this difference, we can estimate the percent ionic character of each carbon-metal bond.

Organolithium and organomagnesium bonds have the highest partial ionic character, whereas those of organocopper and organomercury compounds are lower. These compounds do not behave as salts. Organolithium reagents, for example, which have the highest percent ionic character, dissolve in nonpolar hydrocarbon solvents such as pentane because they self-assemble into well-ordered aggregates, $(RLi)_x$, that present a nonpolar surface to the surrounding solvent.

An important aspect of the metals listed in Table 15.1 is that they all have electronegativities that are considerably lower than carbon. This means that the polarity of the carbon-metal bond places a partial negative charge on carbon and a partial positive charge on the metal. The partial negative charge makes the carbon atom both basic and nucleophilic. When Grignard and organolithium reagents are prepared, the carbon atom bearing the halogen is transformed from an electrophilic center (partial positive charge) in the haloalkane, alkene, or arene to a nucleophilic center (partial negative charge) in the organometallic compound. In the structural formula of butylmagnesium bromide on the right, the carbon-magnesium bond is shown as ionic to emphasize its nucleophilic character.

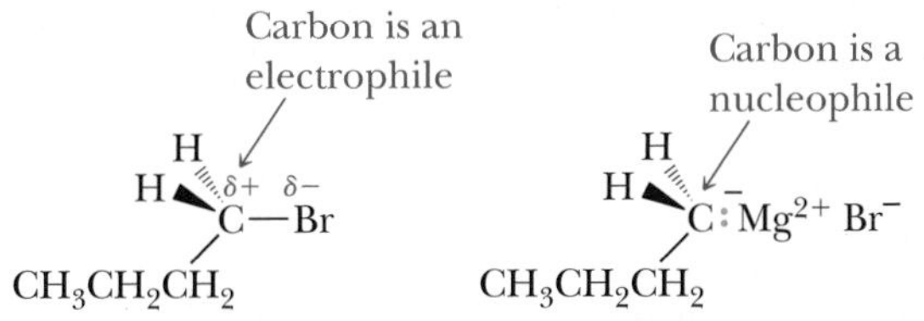
Carbon is an electrophile
Carbon is a nucleophile
H
H
δ+ δ−
C—Br
$CH_3CH_2CH_2$
H
H
C: Mg^{2+} Br^-
$CH_3CH_2CH_2$

As nucleophiles, these compounds react with the electrophilic carbon atom of the carbonyl groups of aldehydes and ketones (Chapter 16) and of carboxylic esters and acid chlorides (Chapter 18). Herein lies the value of organomagnesium and organolithium reagents in synthetic organic chemistry—as carbon-centered nucleophiles, they enable the formation of new carbon-carbon bonds.

B. Reaction with Proton Acids

Both Grignard and organolithium compounds are very strong bases and react readily with any acid (proton donor) stronger than the alkane from which they are derived. Ethylmagnesium bromide, for example, reacts instantly with water, which donates a proton to give ethane and magnesium salts. This reaction is an example of a much stronger acid and a much stronger base reacting to give a weaker acid and a weaker base (Section 4.4). Ethane is evolved from the reaction mixture as a gas.

$$\overset{\delta-}{CH_3CH_2}\!-\!\overset{\delta+}{MgBr} + \overset{\delta+}{H}\!-\!\overset{\delta-}{OH} \longrightarrow CH_3CH_2\!-\!H + Mg^{2+} + OH^- + Br^- \qquad pK_{eq} = -35,\; K_{eq} = 10^{35}$$

	pK_a 15.7	pK_a 51	
Stronger base	Stronger acid	Weaker acid	Weaker base

Following are several classes of proton donors that react readily with Grignard and organolithium reagents. Because they react so readily with these types of compounds, Grignard and organolithium compounds cannot be prepared from any organohalogen that also contains one of these functional groups. Nor can they be prepared from any organohalogen compound that also contains a nitro or carbonyl group because they also react with these groups.

R_2NH	$RC{\equiv}CH$	ROH	HOH	ArOH	RSH	RCOOH
pK_a 38–40 1° and 2° amines	pK_a 25 Terminal alkynes	pK_a 16–18 Alcohols	pK_a 15.7 Water	pK_a 9–10 Phenols	pK_a 8–9 Thiols	pK_a 4–5 Carboxylic acids

Example 15.1 Grignard Reagent as a Strong Base

Write an equation for the acid-base reaction between ethylmagnesium iodide and an alcohol. Use curved arrows to show the flow of electrons in this reaction. In addition, show by using appropriate pK_a values that this reaction is an example of a stronger acid and stronger base reacting to give a weaker acid and weaker base.

Solution

The alcohol is the stronger acid, and the partially negatively charged ethyl group is the stronger base.

$$CH_3CH_2{-}MgI + H{-}\ddot{O}R \longrightarrow CH_3CH_2{-}H + MgI^+ \,{}^-\!:\!\ddot{O}R$$

A Grignard reagent (stronger base)	An alcohol pK_a 16–18 (stronger acid)	An alkane pK_a 51 (weaker acid)	A magnesium alkoxide (weaker base)

Problem 15.1

Explain how these Grignard reagents would react with molecules of their own kind to "self-destruct."

(a)

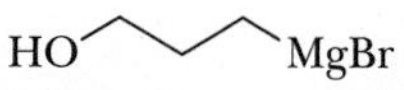

(b)

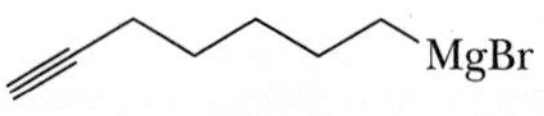

C. Reaction with Oxiranes

▶ Most bond-making steps in reaction mechanisms involve nucleophiles reacting with electrophiles.

As we saw in Section 11.9, the oxirane ring is so strained that it undergoes ring-opening reactions with a variety of nucleophiles. We can now add Grignard and organolithium reagents to the list of reactive nucleophiles. Butylmagnesium bromide, for example, reacts with oxirane (ethylene oxide) to give a magnesium alkoxide, which, upon treatment with aqueous acid, gives 1-hexanol.

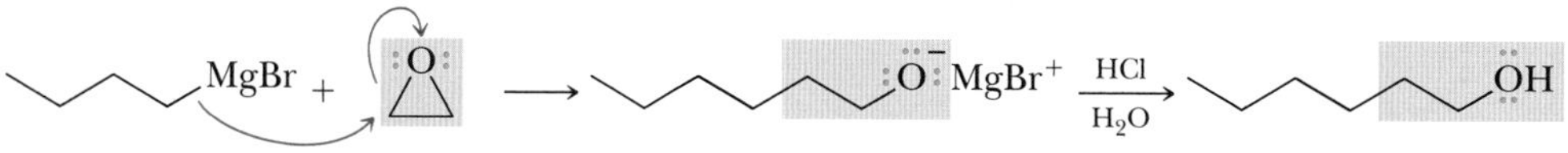

Butylmagnesium bromide **Ethylene oxide** **A magnesium alkoxide** **1-Hexanol** ▶

As illustrated in this example, the product of treatment of a Grignard reagent with oxirane followed by protonation of the alkoxide is a primary alcohol with a carbon chain two carbons longer than the original chain. In reaction of a substituted oxirane, the major product corresponds to attack of the Grignard reagent on the less hindered carbon of the three-membered ring in an S_N2-like reaction. Treatment of racemic methyloxirane (propylene oxide) with phenylmagnesium bromide, for example, followed by workup in aqueous acid gives racemic 1-phenyl-2-propanol. The reaction does not work well if one or more of the oxirane carbons is quaternary.

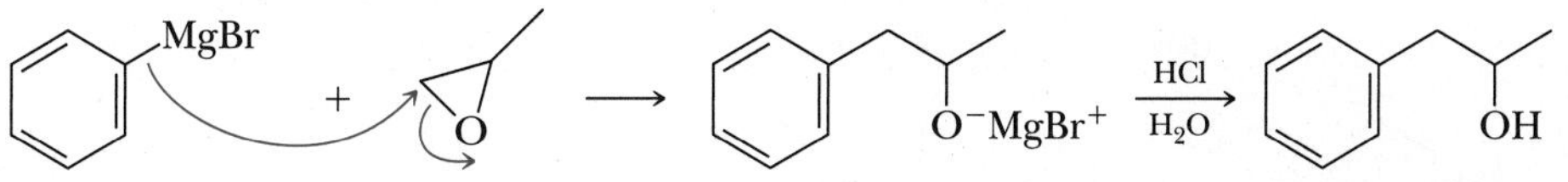

Phenylmagnesium bromide | **Methyloxirane (Propylene oxide)** (racemic) | A magnesium alkoxide (racemic) | **1-Phenyl-2-propanol** (racemic)

Watch a video explanation

Example 15.2 Preparation of an Alcohol

Show how to prepare each alcohol from an organohalogen compound and an oxirane.

(a)
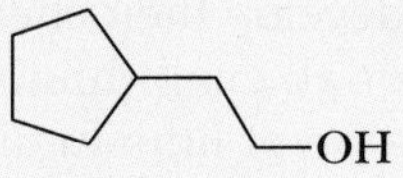

(b)
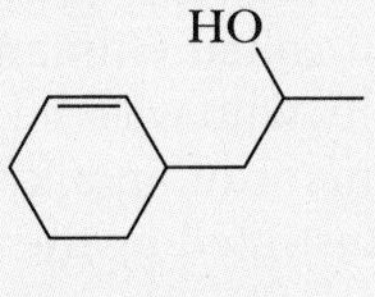

(racemic mixture of two pairs of enantiomers)

Solution

Shown is a retrosynthetic analysis for each compound followed by a synthesis.

(a)
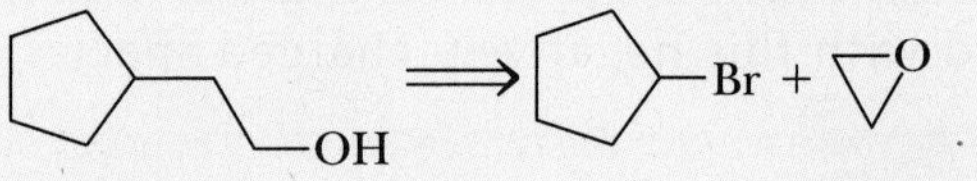

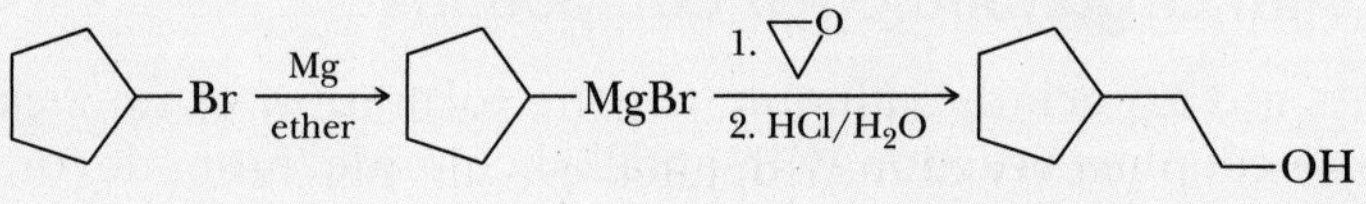

(b)
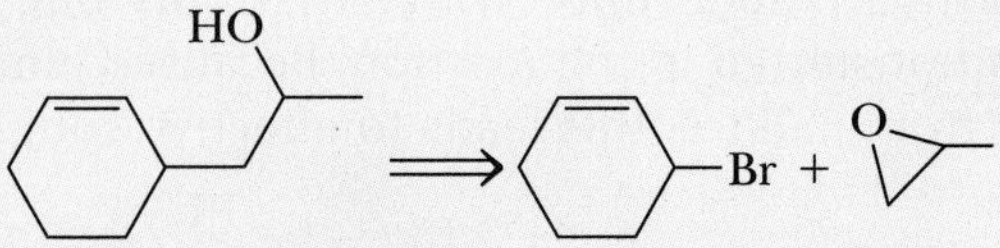

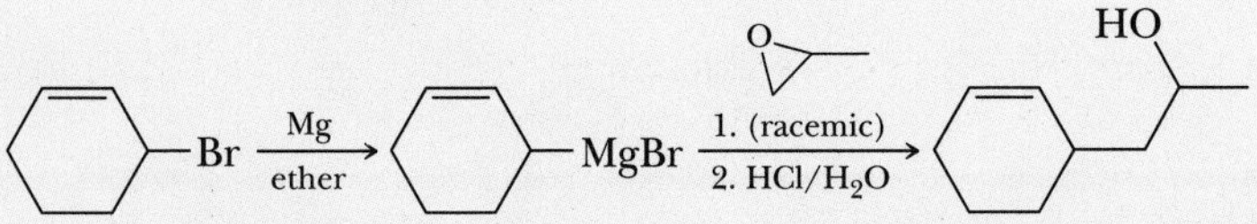

(a racemic mixture of two pairs of enantiomers)

(Continued)

Watch a video explanation

Problem 15.2

Recalling the reactions of alcohols from Chapter 10, show how to synthesize each compound from an organohalogen compound and an oxirane, followed by a transformation of the resulting hydroxyl group to the desired oxygen-containing functional group.

(a) 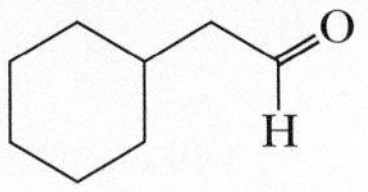
O
H

(b) 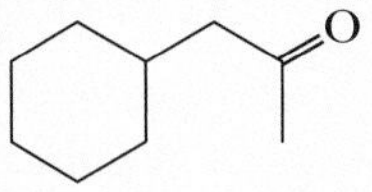
O

15.2 Lithium Diorganocopper (Gilman) Reagents

A. Formation and Structure

An important use of organolithium reagents (Section 15.1) is in the preparation of diorganocopper reagents, often called Gilman reagents after Henry Gilman (1893–1986) of Iowa State University who was the first to develop their chemistry. They are easily prepared by treatment of an alkyl, aryl, or alkenyllithium compound with copper(I) iodide, as illustrated by the preparation of lithium dibutylcopper from butyllithium.

$$2\ CH_3CH_2CH_2CH_2Li + CuI \xrightarrow[\text{or THF}]{\text{diethyl ether}} (CH_3CH_2CH_2CH_2)_2CuLi + LiI$$

Butyllithium — **Copper**(I) **iodide** — **Lithium dibutylcopper** (a Gilman reagent)

Gilman reagents consist of two organic groups associated with a copper(I) ion giving a negatively charged species, which is the source of the carbon nucleophile. Lithium ion is associated with this negatively charged species as the counterion.

B. Coupling with Organohalogen Compounds

Gilman reagents are especially valuable for the formation of new carbon-carbon bonds by a coupling reaction with an alkyl chloride, bromide, or iodide (alkyl fluorides are unreactive under these conditions) as illustrated by the following preparation of 2-methyl-1-dodecene. Notice that only one of the Gilman-reagent alkyl groups is transferred in the reaction. Because Gilman reagents are ultimately prepared from halides, this leads to effective coupling of two halides.

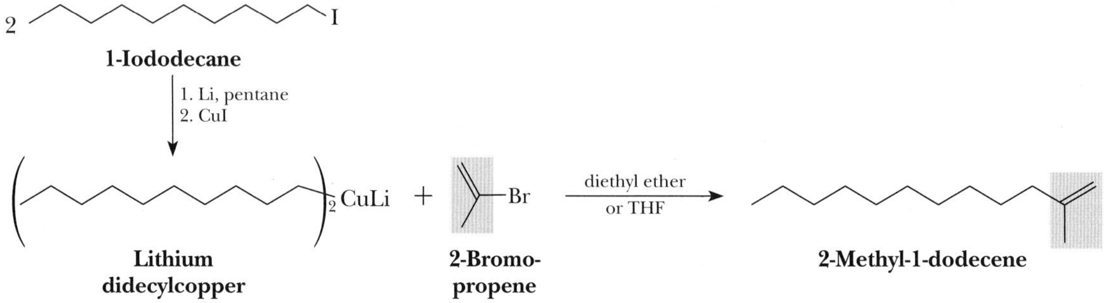
2
I
1-Iododecane
1. Li, pentane
2. CuI
2 CuLi
+
Br
diethyl ether
or THF
Lithium didecylcopper
2-Bromo-propene
2-Methyl-1-dodecene

This example illustrates the coupling of a Gilman reagent, a nucleophile, with a vinylic halide, an electrophile. Vinylic halides are normally quite unreactive toward nucleophilic displacement. Thus, the lithium diorganocopper reaction shown here is unique.

Gilman reagents giving the best yields of coupling products are those prepared from methyl, primary alkyl, allylic, vinylic, and aryl halides via the corresponding organolithium compounds. Yields are lower with secondary and tertiary haloalkanes.

Coupling with a vinylic halide is stereospecific; the configuration of the carbon-carbon double bond is preserved, as illustrated by the synthesis of *trans*-5-tridecene.

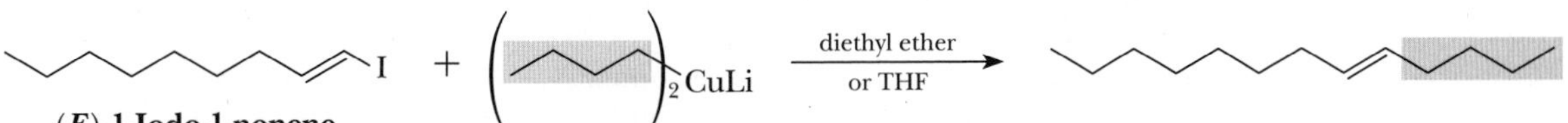

(*E*)-1-Iodo-1-nonene **Lithium dibutylcopper** **(*E*)-5-Tridecene**

A variation on the preparation of Gilman reagents is to use a Grignard reagent in the presence of a catalytic amount of **Cu(I)**. Zoecon Corporation has developed a synthesis of 150 **kg** batches of the housefly sex attractant muscalure by treating (Z)-1-bromo-9-octadecene with pentylmagnesium bromide in the presence of catalytic amounts of **Cu(I)**. The starting bromoalkene is easily prepared from the readily available (Z)-9-octadecenoic acid (oleic acid, Section 26.1). Yields of muscalure are nearly quantitative.

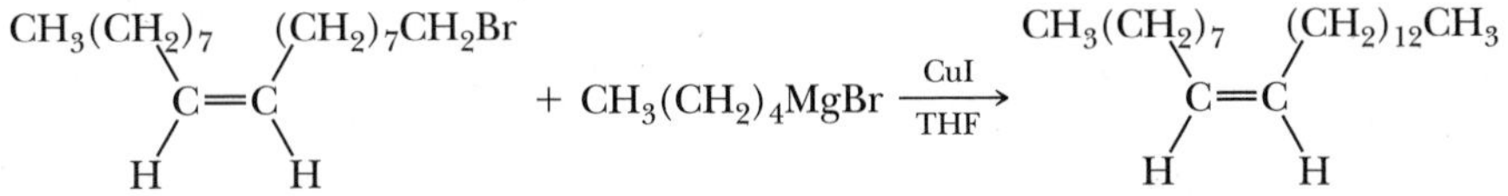

(Z)-1-Bromo-9-octadecene **(Z)-9-Tricosene (Muscalure)**

The mechanism of these coupling reactions is not fully understood and is the subject of active investigation.

Example 15.3 Reactions with Lithium Diorganocopper Reagents

Show how to bring about each conversion using a lithium diorganocopper reagent.

(a) 1-Bromocyclohexene to 1-methylcyclohexene

(b) 1-Bromo-2-methylpropane to 2,5-dimethylhexane using the bromoalkane as the only source of carbon

Solution

(a) Treat 1-bromocyclohexene with lithium dimethylcopper.

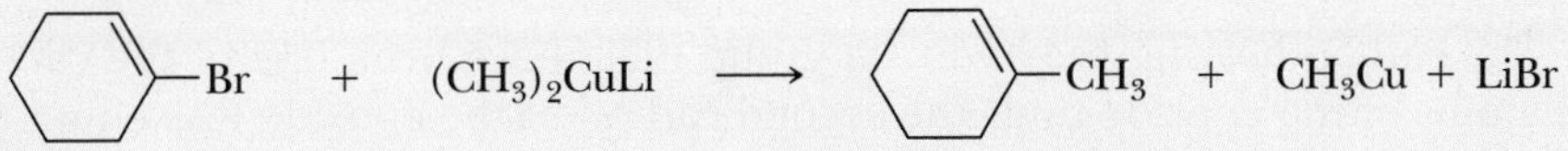

1-Bromocyclohexene **Lithium dimethylcopper** **1-Methylcyclohexene**

(Continued)

(b) Treat 1-bromo-2-methylpropane with lithium diisobutylcopper, itself prepared from 1-bromo-2-methylpropane.

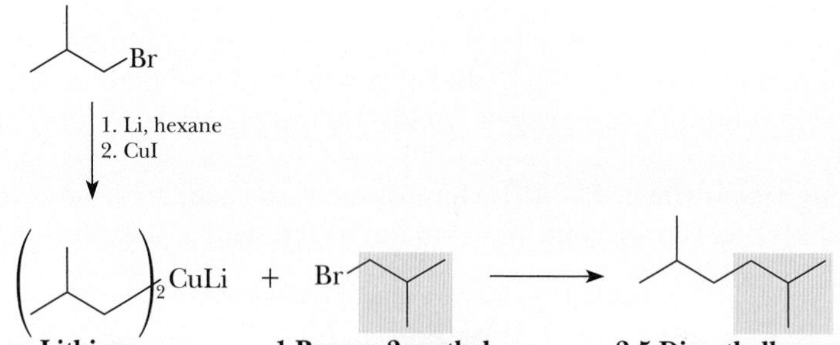

Lithium diisobutylcopper | **1-Bromo-2-methylpropane (Isobutyl bromide)** | **2,5-Dimethylhexane**

Watch a video explanation

Problem 15.3

Show how to bring about each conversion using a lithium diorganocopper reagent.

(a) 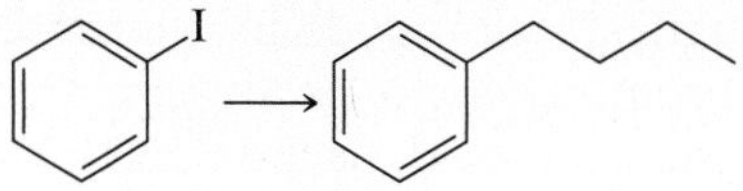

(b) 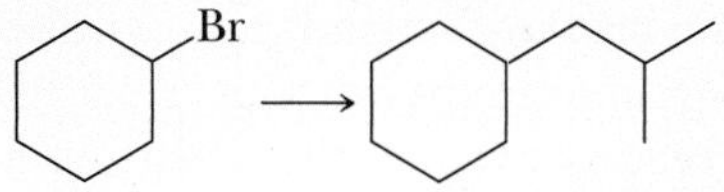

C. Reaction with Oxiranes

The reaction of epoxides with Gilman reagents is an important method for the formation of new carbon-carbon bonds. Like organolithium compounds and Grignard reagents, these compounds bring about regioselective ring opening of substituted epoxides at the less substituted carbon to give alcohols. Treatment of racemic styrene oxide with lithium divinylcopper, for example, followed by workup in aqueous acid gives racemic 1-phenyl-3-buten-1-ol.

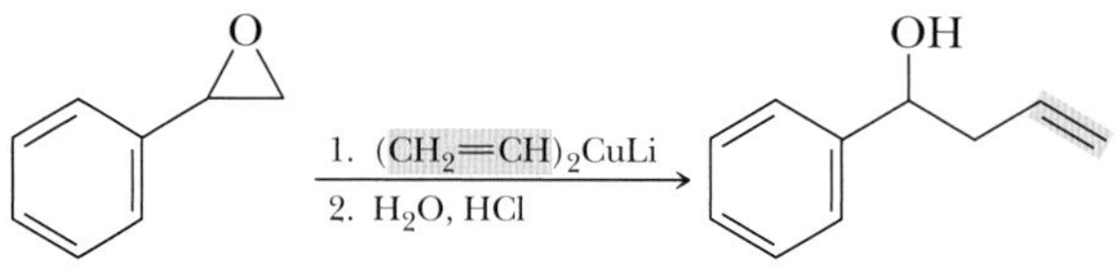

Styrene oxide (racemic) | **1-Phenyl-3-buten-1-ol** (racemic)

Example 15.4 Oxirane Reactions

Show two combinations of epoxide and Gilman reagent that can be used to prepare racemic 1-phenyl-3-hexanol.

Solution

The carbon bearing the hydroxyl group must have been one of the carbon atoms of the epoxide ring. The second carbon of the epoxide was either the one to the right of the carbon now bearing the —OH or the one to the left of it. In these solutions, the phenyl group is written C_6H_5—. Either route would be satisfactory.

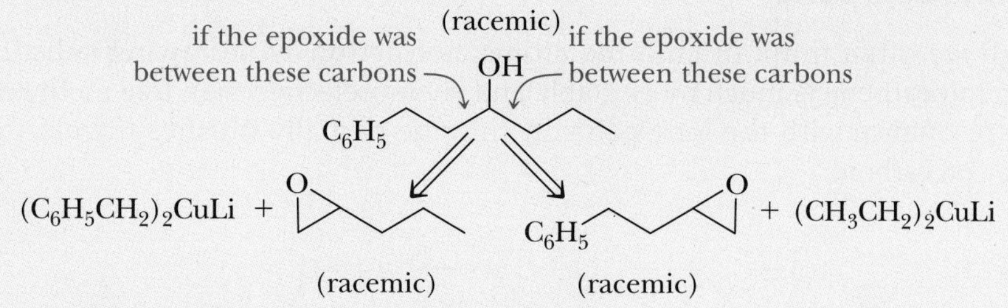

Problem 15.4

Show how to prepare each Gilman reagent in Example 15.4 from an appropriate alkyl halide and each epoxide from an appropriate alkene.

15.3 Carbenes and Carbenoids

A **carbene**, R_2C:, is a neutral molecule in which a carbon atom is surrounded by only six valence electrons. Because they are electron deficient, carbenes are highly reactive and behave as electrophiles. As we will see, one of their most important types of reactions is with alkenes (nucleophiles) to give cyclopropanes.

Carbene A neutral molecule that contains a carbon atom surrounded by only six valence electrons (R_2C:).

A. Methylene

The simplest carbene is methylene, CH_2, prepared by **photolysis** (cleavage by light) or **thermolysis** (cleavage by heating) of diazomethane, CH_2N_2, an explosive, toxic gas.

Photolysis Cleavage by light.

Thermolysis Cleavage by heating.

Methylene
(the simplest carbene)

In the lowest electronic state of most carbenes, carbon is sp^2 hybridized with the unshared pair of electrons occupying the third sp^2 orbital. The unhybridized $2p$ orbital lies perpendicular to the plane created by the three sp^2 orbitals. Note that this orbital description of methylene is very much like that of a carbocation (Section 6.3A). In both species, carbon is sp^2 hybridized with a vacant $2p$ orbital. Methylene in this electronic state combines features of a carbocation and a carbanion in that it has both a vacant p orbital and a lone pair.

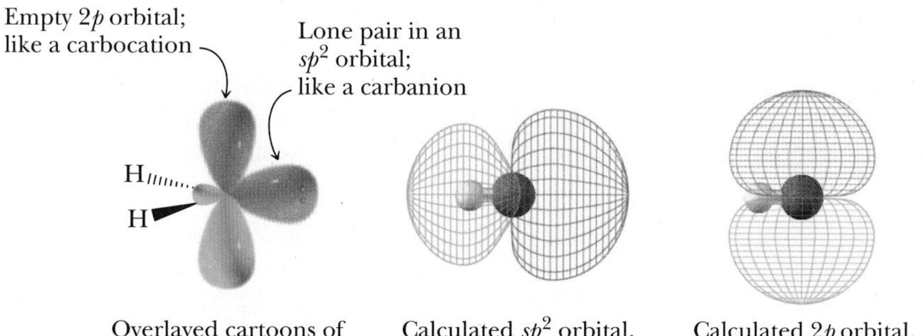

Overlayed cartoons of the sp^2 and $2p$ orbitals of methylene

Calculated sp^2 orbital, containing a lone pair

Calculated $2p$ orbital, empty

Methylene generated in this manner reacts with all C—H and C=C bonds and is so nonselective that it is of little synthetic use.

B. Dichlorocarbene

Although we often think of chlorine atoms as electron-withdrawing substituents, dichlorocarbene is much more stable and chemoselective than free methylene because resonance with the lone pairs on chlorine partially satisfies the electron deficiency on carbon.

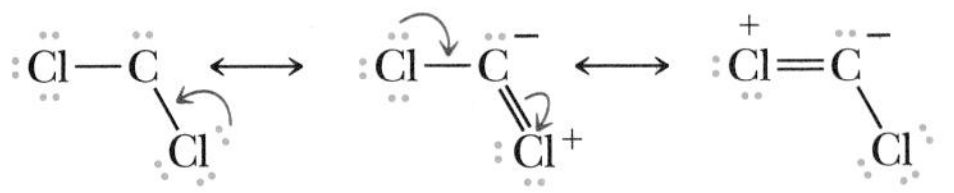

Dichlorocarbene can be prepared by treating chloroform with potassium *tert*-butoxide, removing the elements of HCl. The resulting carbene reacts cleanly with alkenes to give dichlorocyclopropanes. Addition of a dihalocarbene to an alkene shows syn stereoselectivity.

$$CHCl_3 + (CH_3)_3CO^-K^+ \longrightarrow Cl_2C: + (CH_3)_3COH + K^+Cl^-$$

Trichloromethane (Chloroform) | **Potassium *tert*-butoxide** | **Dichlorocarbene** | ***tert*-Butyl alcohol**

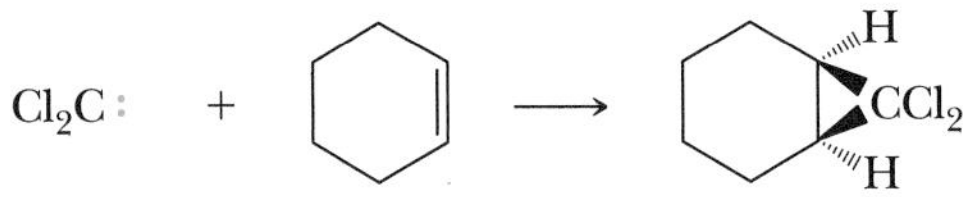

Dichlorocarbene | A dichlorocyclopropane

Reaction of a *cis* alkene with a dihalocarbene gives only a *cis* dihalocyclopropane as illustrated by the reaction of *cis*-3-hexene with dichlorocarbene. Similarly, reaction of a *trans* alkene gives only a *trans* dihalocyclopropane.

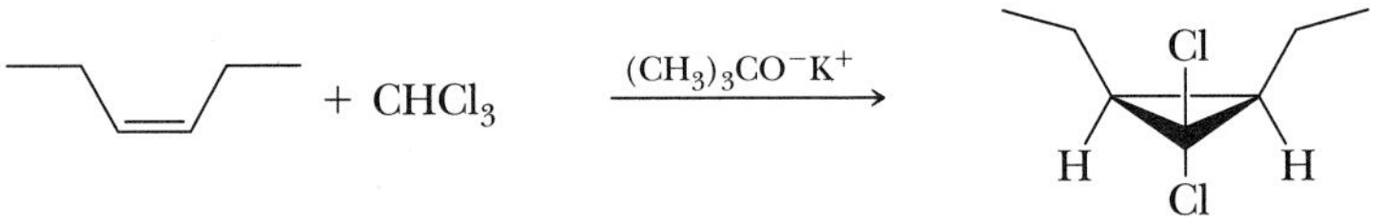

(Z)-3-Hexene | ***cis*-1,1-Dichloro-2,3-diethylcyclopropane**

Mechanism 15.1

Formation of Dichlorocarbene and Its Reaction with Cyclohexene

Taken together, Steps 1 and 2 result in α-elimination of H and Cl; that is, both atoms are eliminated from the same carbon. We have seen many examples of β-elimination, where hydrogen and a leaving group are eliminated from neighboring carbons. There are very few examples of α-elimination, and they are possible only where no β-hydrogen exists.

***Step 1:* Take a proton away.** Treatment of chloroform, which is somewhat acidic because of its three electron-withdrawing chlorine atoms, with potassium *tert*-butoxide gives the trichloromethide anion.

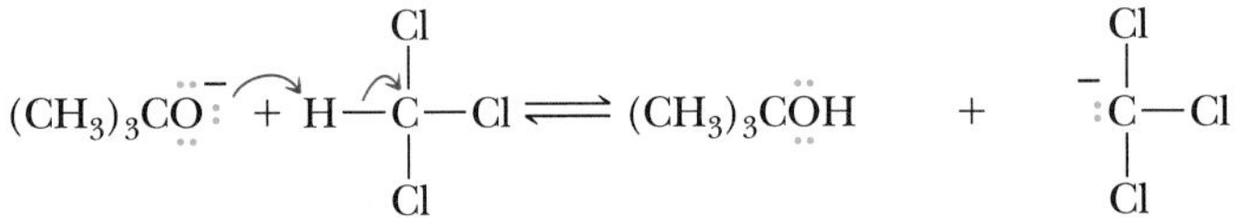

Trichloromethide anion

Step 2: **Break a bond to give stable molecules or ions.** Loss of Cl^- from CCl_3^- gives dichlorocarbene.

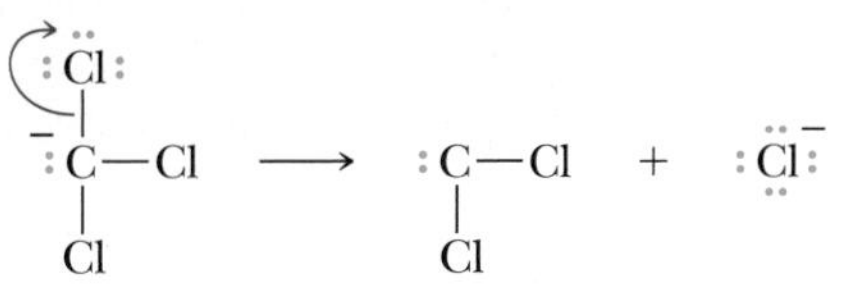
Cl
C
Cl
Cl
C
Cl
Cl
+
Cl

Dichlorocarbene

Step 3: **Make a bond between a nucleophile (π bond) and an electrophile.** Syn addition of dichlorocarbene to cyclohexene gives a dichlorocyclopropane. Compare this step to the electrophilic addition of bromine and chlorine to an alkene first discussed in Section 6.3D.

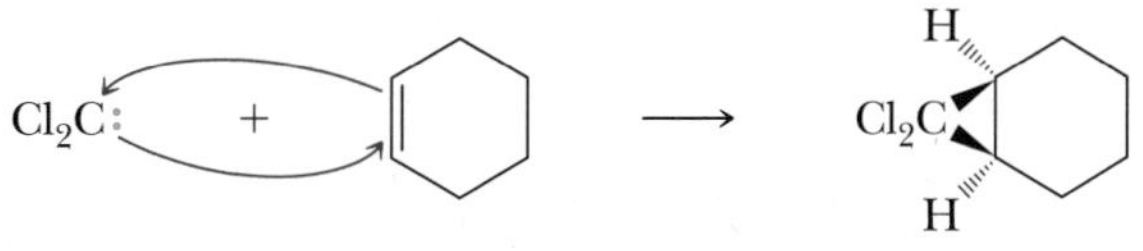
Cl_2C
+
H
Cl_2C
H

Dichlorocarbene (an electrophile) | **Cyclohexene** (a nucleophile) | A dichlorocyclopropane

Example 15.5 Formation of a Dibromocarbene

Predict the product from the following reaction.

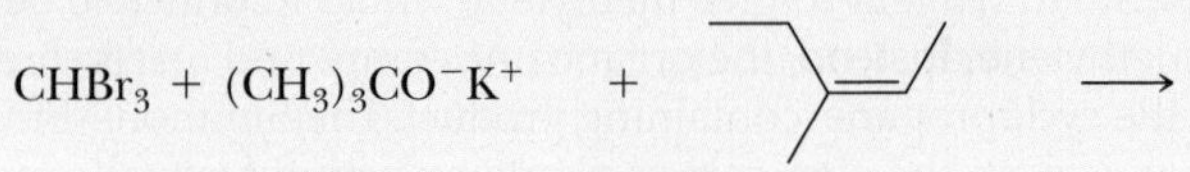
$CHBr_3$ + $(CH_3)_3CO^-K^+$ +

(*Z*)-3-Methyl-2-pentene

Solution

Bromoform gives dibromocarbene, which reacts stereospecifically with the alkene to give a dibromocyclopropane.

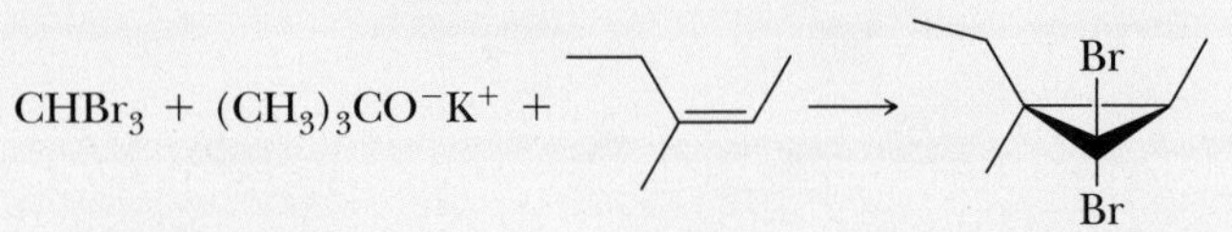
$CHBr_3$ + $(CH_3)_3CO^-K^+$ +
Br
Br

(racemic)

Problem 15.5

Predict the product of the following reaction.

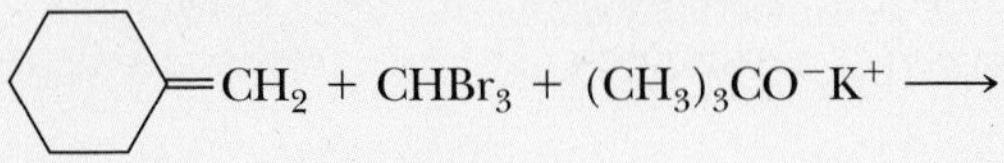
$=CH_2$ + $CHBr_3$ + $(CH_3)_3CO^-K^+$

C. The Simmons-Smith Reaction

Although methylene prepared from diazomethane itself is not synthetically useful, addition of methylene to an alkene can be accomplished using a reaction first reported by the American chemists Howard Simmons and Ronald Smith. The Simmons-Smith reaction uses diiodomethane and zinc dust activated by a small amount of copper (a so-called "zinc-copper couple") to produce iodomethylzinc iodide, in a reaction reminiscent of a Grignard reaction. Even though we show the **Simmons-Smith reagent** here as ICH_2ZnI, its structure is considerably more complex and not fully understood.

$$\underset{\textbf{Diiodomethane}}{CH_2I_2} + \underset{\textbf{Zinc-copper couple}}{Zn(Cu)} \xrightarrow{\text{diethyl ether}} \underset{\substack{\textbf{Iodomethylzinc iodide}\\ \text{(Simmons-Smith reagent)}}}{ICH_2ZnI}$$

This organozinc compound reacts with a wide variety of alkenes to give cyclopropanes.

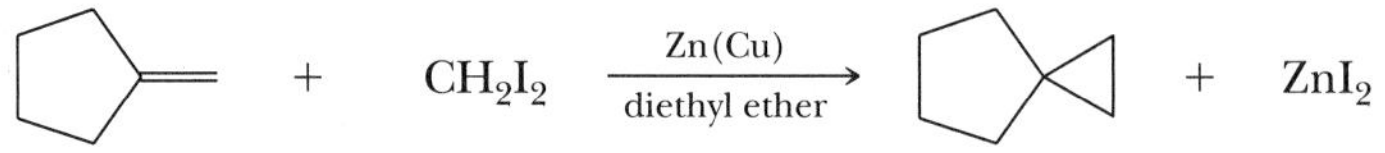

Methylenecyclopentane **Spiro[4.2]heptane**

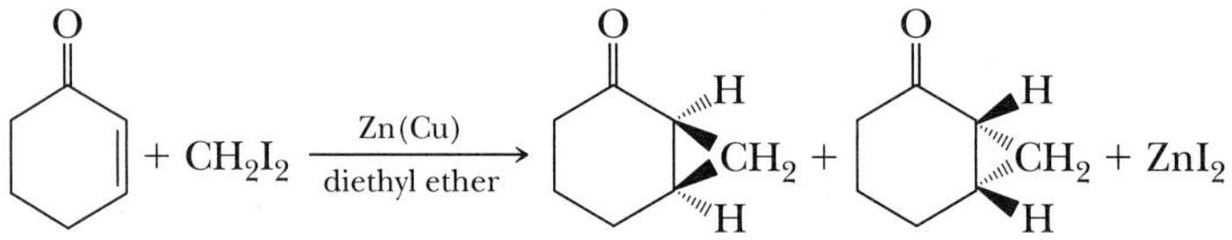

(racemic)

Mechanism 15.2

The Simmons-Smith Reaction with an Alkene

Although an α-elimination from the Simmons-Smith reagent to give methylene would in principle be possible, the reagent is much more selective than free methylene. Instead, the organozinc compound reacts directly with the alkene by a concerted mechanism to give the cyclopropane-containing product. The Simmons-Smith reagent is an example of a **carbenoid**, a compound that delivers the elements of a carbene without actually producing a free carbene.

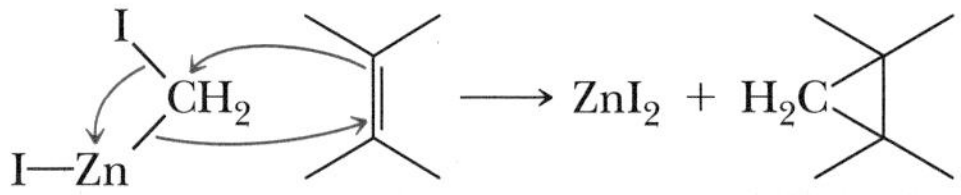

Carbenoid

A compound that delivers the elements of a carbene without actually producing a free carbene.

Example 15.6 Simmons-Smith Reagents

Draw a structural formula for the product of treating each alkene with the Simmons-Smith reagent.

(a) 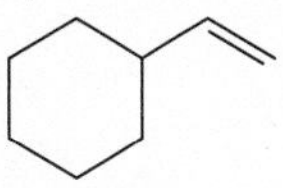 **(b)** 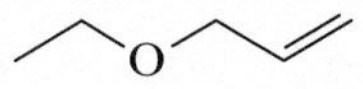

Solution

Reaction at each carbon-carbon double bond forms a cyclopropane ring.

(a) 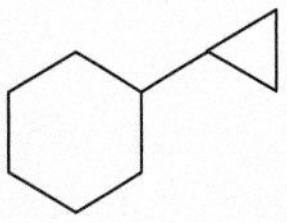 **(b)** 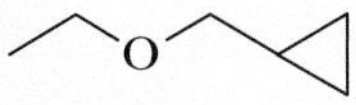

Problem 15.6

Show how the following compound could be prepared from any compound containing ten carbons or fewer.

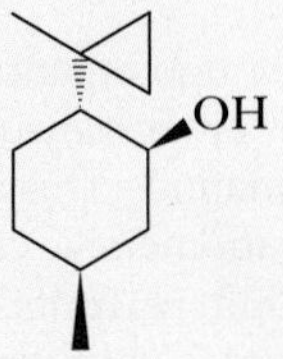

MCAT Practice: *Passage and Questions*

Inorganic Coordination Compounds

Inorganic coordination compounds contain a metal atom or ion bonding to a surrounding array of molecules that are referred to as ligands. If a ligand has carbon directly bonding to the metal, the complex is then referred to as an organometallic compound. Organometallic chemistry is the study of chemical compounds containing bonds between carbon and metals. The carbon-metal bond (C—M) ranges in character between highly ionic to covalent. Primarily ionic bonds are encountered with electropositive metals, such as Group 1 or 2 metals. When examining the periodic table, the C—M bonds generally become increasingly covalent when moving from the left to the right in a row. Other factors can also influence the extent of ionic or covalent bonding, such as the stability of charges on the carbon atom.

Questions

A. Many examples of inorganic coordination compounds exist in nature. For instance, hemoglobin contains an iron coordinated to a porphyrin ligand while chlorophyll has magnesium in the center of a chlorin ring. One of the very few organometallic compounds is methylcobalamin ($R = CH_3$), which possesses a cobalt to methyl bond as well as a corrin ring. Which of the following statements is false?

1. The porphyrin, chlorin, and corrin rings all contain four nitrogens that coordinate the metal but with varying degrees of single and double bond alternation around the rings.
2. The oxidation state (charge on the metal) depicted in these structures corresponds to the number of nitrogen atoms that bond to the metals but are not also doubly bonded to a carbon.

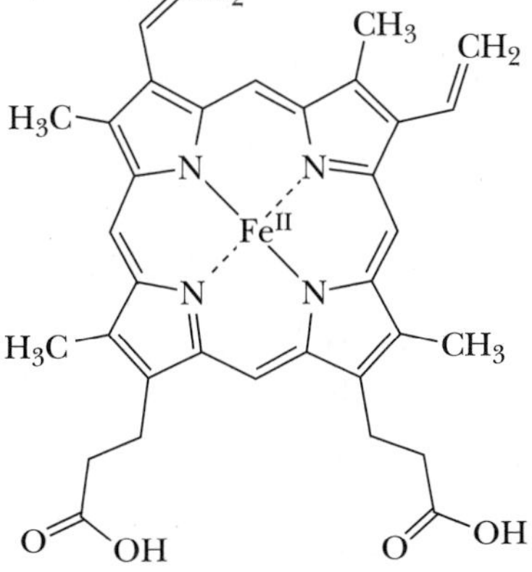

Hemoglobin

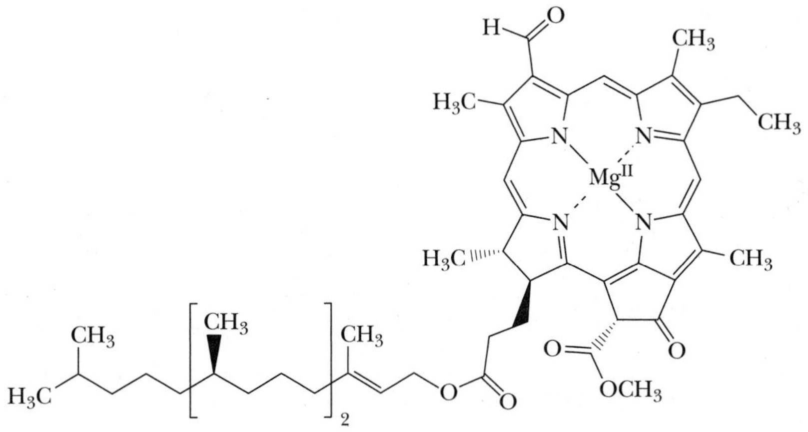

Chlorophyll

(Continued)

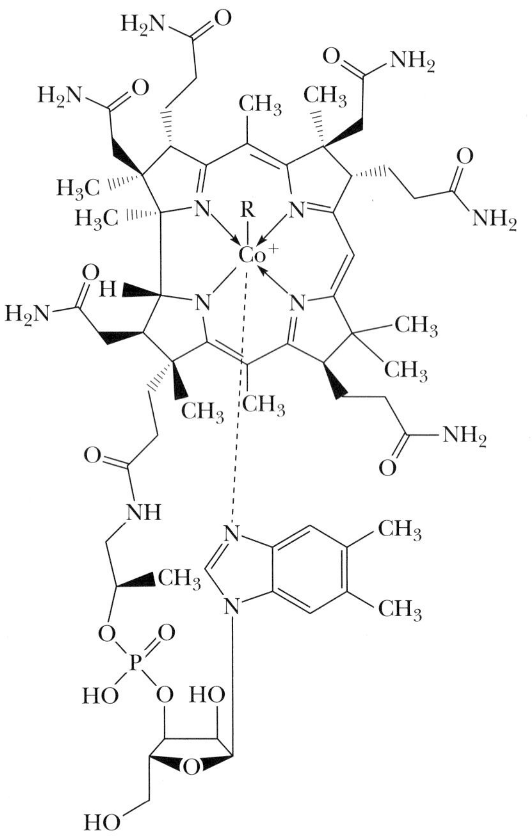

Methylcobalamin

3. One expects the porphyrin and chlorin rings to be nearly planar, while the corrin ring should be puckered.
4. The total charge on all three structures should be neutral at pH 7.3.

B. Methylcobalamin is involved in the biosynthesis of the amino acid methionine. Given what you know about the stability and reactivity of organic intermediates in aqueous media, how would you expect the methyl of methylcobalamin to react in an enzyme active site?

1. As a carbocation.
2. As a radical.
3. As a carbanion.

C. Although organometallic species are rare in nature, they are widely used in organic chemistry transformations. This is especially true of Group 1 and 2 organometallic compounds, such as Grignard and lithium reagents. Which two of the following statements is *not* a reason that Grignard reagents would be incompatible with biochemistry?

1. Magnesium is too rare an element for biology to exploit it.
2. They would react with any ketone or aldehyde containing natural products.
3. Due to the magnesium Lewis basicity, they would rapidly decompose by protonation from water.
4. Due to their alkyl group Brønsted basicity, they would rapidly decompose by protonation from water.

D. Which of the following trends reflects increasing extent of covalent bonding to carbon?

1. Ca—C, Fe—C, Zn—C, Ge—C
2. C—C, N—C, O—C, F—C
3. H—C, Ti—C, Pd—C, Br—C
4. C=C, C=N, C=O

E. Which two of the following trends would you predict increases ionic bonding character of a C—M bond?

1. Increasing *s*-character in the hybridization of the carbon in the C—M bond.
2. Decreasing *s*-character in the hybridization of the carbon in the C—M bond.
3. Electron donating groups on the carbon in the C—M bond.
4. Electron withdrawing groups on the carbon in the C—M bond.

Study Guide

15.1 Organomagnesium and Organolithium Compounds

- An **organometallic compound** is one that contains a carbon-metal bond.
  - *The key feature of many of these reagents is that the carbon of the carbon-metal bond carries a partial negative charge.*

  - *The partial negative charge on carbon makes it basic and nucleophilic; the latter property can be exploited in organic synthesis in the construction of carbon-carbon bonds.*
- Organomagnesium compounds are named **Grignard reagents** after their discoverer, Victor Grignard.
  - *Grignard reagents are prepared by reacting alkyl, aryl, or alkenyl halides (chlorides, bromides, and iodides, not fluorides) with a slight excess of magnesium metal in an ether solvent.*
  - *The carbon-magnesium bond of Grignard reagents is polar covalent, with a partial negative charge on carbon, making it nucleophilc and basic.*
- **Organolithium reagents** are prepared by reaction of an alkyl, aryl, or alkenyl halide with two equivalents of lithium metal.
  - *The carbon-lithium bond in organolithium compounds is polar covalent, with a partial negative charge on carbon, making it nucleophilic and basic.*
- Grignard reagents and organolithium compounds react as **carbon nucleophiles** with a wide range of electrophilic functional groups, including epoxides (and many carbonyl-containing species discussed later in the book).
  - *For unsymmetrical epoxides, reaction occurs at the less hindered carbon.*
  - *A new carbon-carbon bond is formed, making these reactions very useful for synthesis.*
  - *The initial product formed is an alkoxide salt that is converted to an alcohol product following an acidic aqueous workup.*
- Grignard and organolithium reagents are **very basic**, so they will deprotonate functional groups such as amines, terminal alkynes, alcohols, thiols, and carboxylic acids.

P 15.1, 15.2, 15.11, 15.12, 15.17–15.19, 15.20, 15.23–15.25

▶ Treatment of racemic methyloxirane with phenylmagnesium bromide and Problem 15.2.

KEY REACTIONS

1. Formation of Organomagnesium (Grignard) and Organolithium Compounds (Section 15.1A) Organomagnesium compounds are prepared by treating an alkyl, aryl, or alkenyl (vinylic) halide with magnesium in diethyl ether or THF. Organolithium compounds are prepared by treating an alkyl, aryl, or alkenyl halide with lithium in pentane or another hydrocarbon solvent.

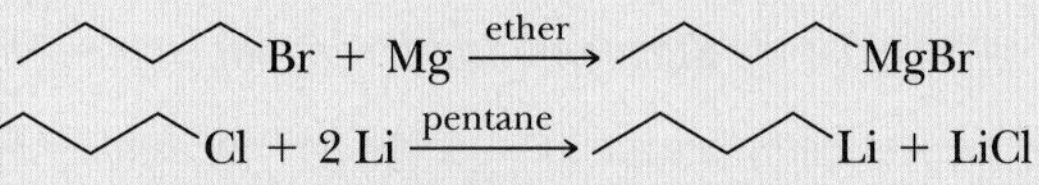

2. Reaction of RMgX and RLi with Proton Donors (Section 15.1B) Both organomagnesium and organolithium compounds are strong bases and react with any proton donor stronger than the alkane from which the organolithium or magnesium compound is derived. Water or other proton donors must be completely excluded during their preparation and use.

$$CH_3CH_2MgBr + H_2O \longrightarrow CH_3CH_2{-}H + Mg(OH)Br$$

3. Reaction of a Grignard Reagent with an Epoxide (Section 15.1C) Treatment of a Grignard reagent with an epoxide followed by protonation of the magnesium alkoxide salt in aqueous acid gives an alcohol with its carbon chain extended by two carbon atoms.

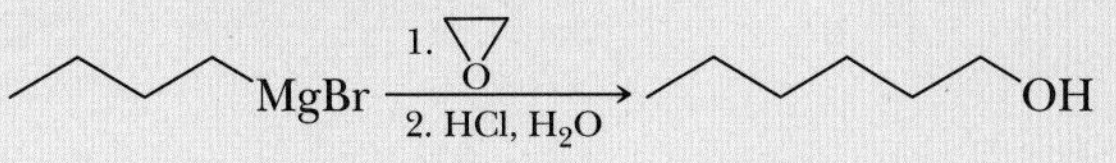

15.2 Lithium Diorganocopper (Gilman) Reagents

- Two equivalents of organolithium compounds react with one equivalent of a Cu(I) salt, usually CuI, to give useful reagents called **Gilman reagents**.
  - *Gilman reagents react with haloalkanes and haloalkenes to form new carbon-carbon bonds.*
  - *Gilman reagents also react with oxiranes to give alcohol products with carbon-carbon bond formation following an acidic aqueous workup.*

P 15.3, 15.4, 15.7–15.11, 15.19, 15.20, 15.21–15.24

Problem 15.3

KEY REACTIONS

4. Formation of Gilman Reagents (Section 15.2A) Lithium diorganocopper (Gilman) reagents are prepared by treating an organolithium compound with copper(I) iodide.

$$2\ CH_3CH_2CH_2CH_2Li + CuI \xrightarrow[\text{or THF}]{\text{ether}} (CH_3CH_2CH_2CH_2)_2CuLi + LiI$$

5. Treatment of a Gilman Reagent with an Alkyl, Aryl, or Alkenyl Halide (Section 15.2B) Coupling of a Gilman reagent with an alkyl, alkenyl, or aryl halide results in formation of a new carbon-carbon bond.

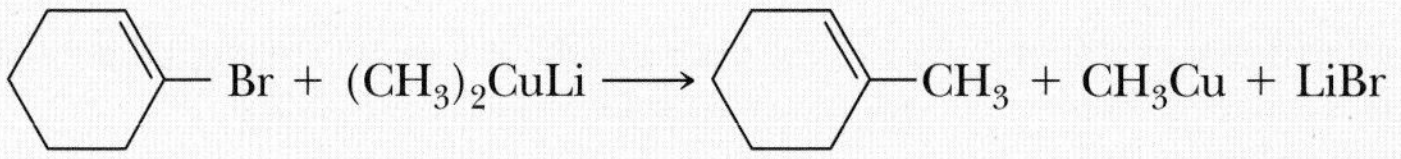

15.3 Carbenes and Carbenoids

- **Carbenes** are neutral molecules that contain a carbon with only six valence electrons, their organometallic-complexed equivalents are called **carbenoids**.
  - *Carbenes are sp^2 hybridized, with an empty $2p$ orbital and a lone pair in one sp^2 hybrid orbital.*
  - *Carbenes are prepared from **photolysis** or **thermolysis** of diazo species such as diazomethane.*
  - *Dichlorocarbene is prepared from the reaction of chloroform with a strong base.*
  - *A useful carbenoid is the **Simmons-Smith reagent** prepared from diiodomethane and* Zn(Cu).
  - *Carbenes such as methylene are too reactive and not used for synthesis.*
  - *Dichlorocarbene and the Simmons-Smith reagent react stereospecifically with alkenes to give cyclopropanes.*

P 15.5, 15.6, 15.12–15.16, 15.23

KEY REACTIONS

6. Reaction of Dichloro- or Dibromocarbene with an Alkene (Section 15.3B) The dihalocarbene is generated by treatment of $CHCl_3$ or $CHBr_3$ with a strong base such as potassium *tert*-butoxide. Addition of the dihalocarbene to an alkene shows syn stereospecificity.

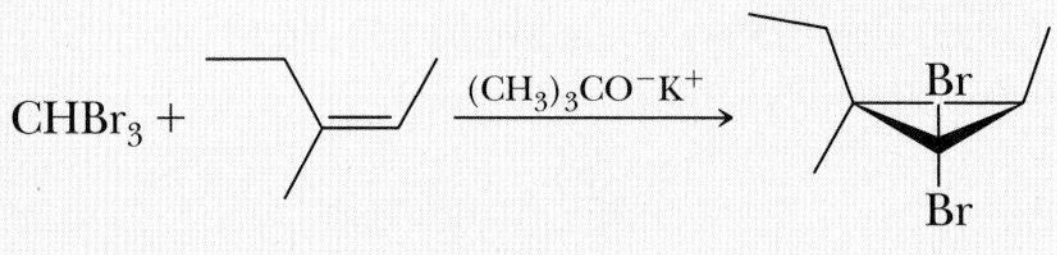

(racemic)

7. The Simmons-Smith Reaction (Section 15.3C) Treatment of CH_2I_2 with a zinc-copper couple generates an organozinc compound, known as the Simmons-Smith reagent, which reacts with alkenes to give cyclopropanes.

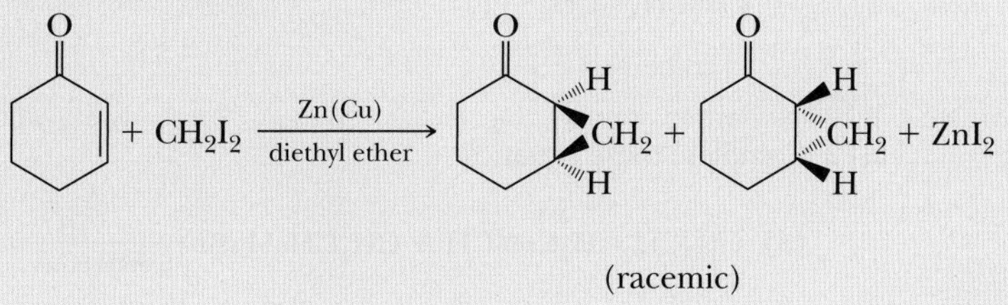

Problems

Red numbers indicate applied problems.

15.7 Complete these reactions involving lithium diorganocopper (Gilman) reagents.

(a) 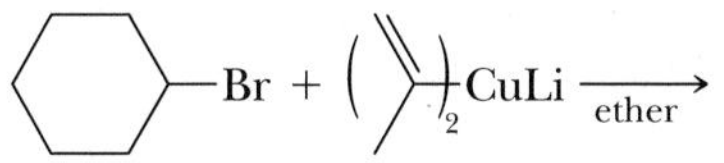

(b) 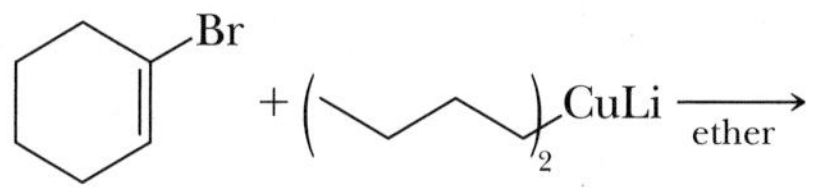

(c) 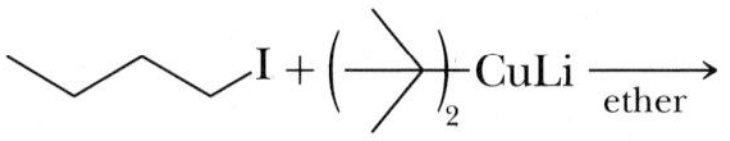

(d) 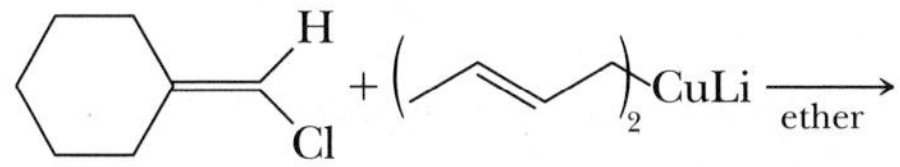

15.8 Show how to convert 1-bromopentane to each of these compounds using a lithium diorganocopper (Gilman) reagent. Write an equation, showing structural formulas, for each synthesis.

(a) Nonane **(b)** 3-Methyloctane **(c)** 2,2-Dimethylheptane
(d) 1-Heptene **(e)** 1-Octene

15.9 In Problem 15.8, you used a series of lithium diorganocopper (Gilman) reagents. Show how to prepare each Gilman reagent from an appropriate alkyl or vinylic halide.

15.10 Show how to prepare each compound from the given starting compound through the use of a lithium diorganocopper (Gilman) reagent.

(a) 4-Methylcyclopentene from 4-bromocyclopentene
(b) (*Z*)-2-Undecene from (*Z*)-1-bromopropene
(c) 1-Butylcyclohexene from 1-iodocyclohexene
(d) 1-Decene from 1-iodooctane
(e) 1,8-Nonadiene from 1,5-dibromopentane

15.11 The following is a retrosynthetic scheme for the preparation of *trans*-2-allylcyclohexanol. Show reagents to bring about the synthesis of this compound from cyclohexane.

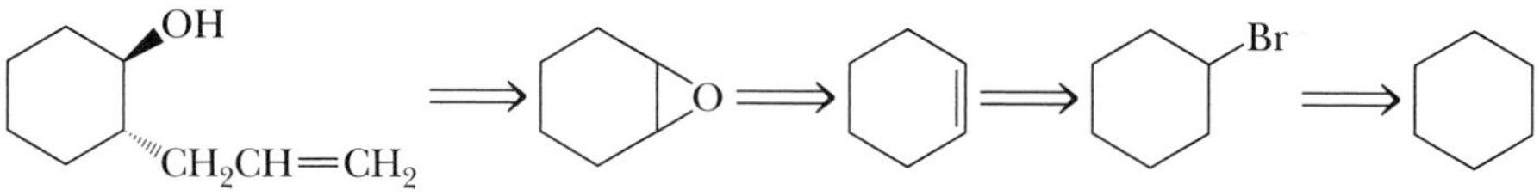

(racemic)

15.12 Complete these equations.

(a) $CH_3CH_2CH_2C{\equiv}CH + CH_3CH_2MgBr \xrightarrow[\text{diethyl ether}]{}$

(b) 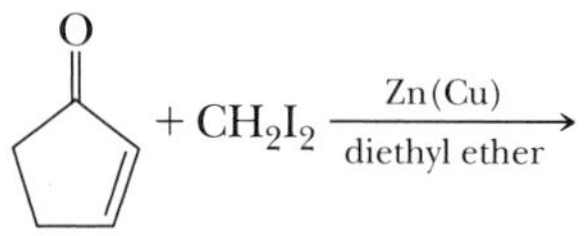

(c) 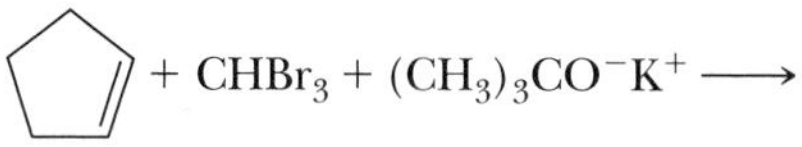

(d) 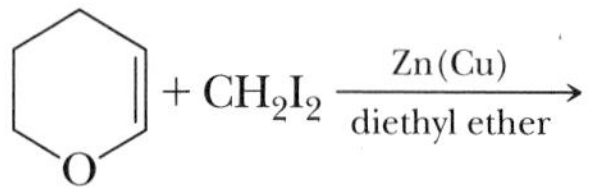

(e) 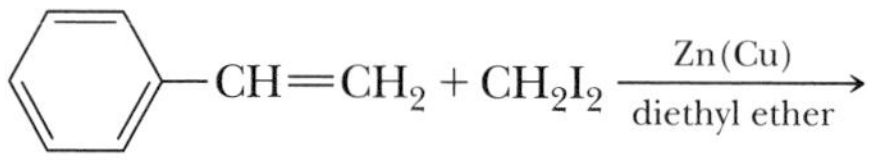

15.13 Reaction of the following cycloalkene with the Simmons-Smith reagent is stereospecific and gives only the isomer shown. Suggest a reason for this stereospecificity.

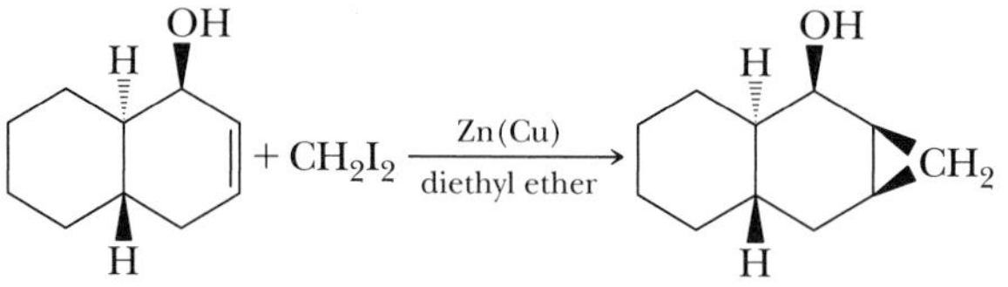

15.14 Show how the following compound can be prepared in good yield.

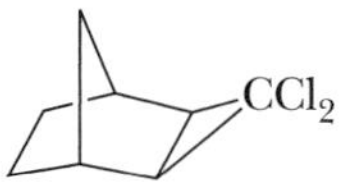

15.15 Show the product of the following reaction (do not concern yourself with which side of the ring is attacked).

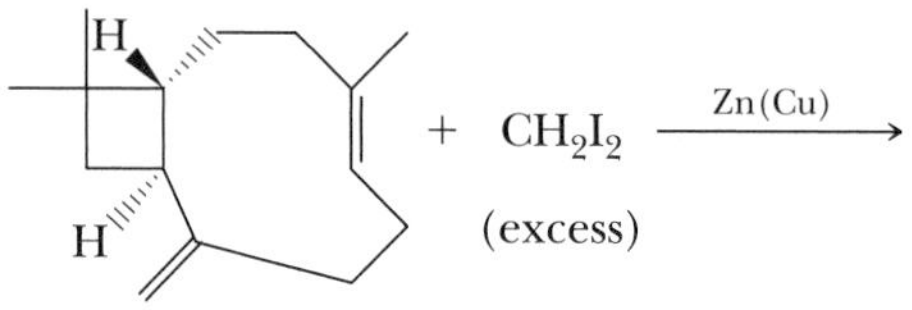

Caryophyllene

15.16 Show how spiro[2.2]pentane can be prepared in one step from organic compounds containing three carbons or fewer and any necessary inorganic reagents or solvents.

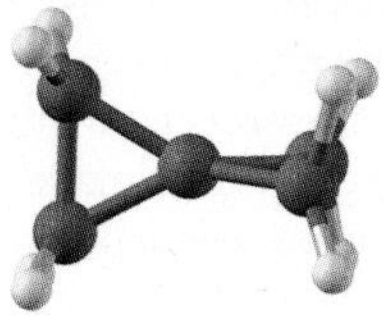

Spiro[2.2]pentane

Looking Ahead

15.17 One of the most important uses for Grignard reagents is their addition to carbonyl compounds to give new carbon-carbon bonds (Section 16.5). In this reaction, the carbon of the organometallic compound acts as a nucleophile to add to the positive carbon of the carbonyl.

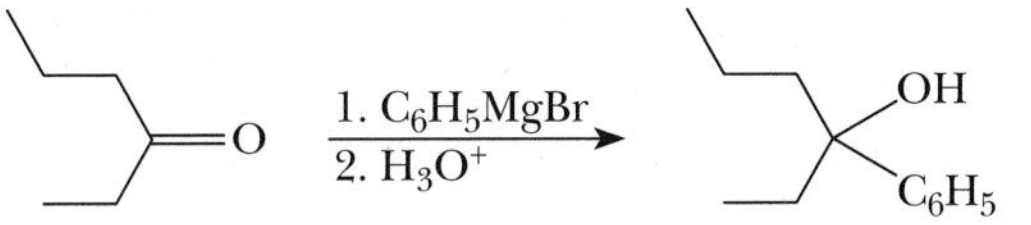
O
1. C_6H_5MgBr
2. H_3O^+
OH
C_6H_5

(racemic)

(a) Give a mechanism for the first step of the reaction.
(b) Explain the function of the acid in the second step.

15.18 Organolithium compounds react with carbonyl compounds in a way that is similar to that of Grignard reactions. Suggest a product of the following reaction.

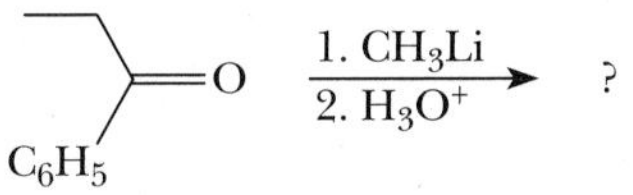
O
C_6H_5
1. CH_3Li
2. H_3O^+
?

Organic Chemistry Reaction Roadmap

15.19 We now continue the introduction of organic chemistry reaction roadmaps. Because of the many new functional groups that will be introduced in coming chapters, we recommend that you make a new roadmap to accommodate the reactions in Chapters 15–18.

To make your own reaction roadmap for Chapters 15–18, take a blank sheet of paper and write the following functional groups in the orientations shown. Fill the entire sheet of paper and leave plenty of room between functional groups. Most students find it helpful to use a poster-sized piece of paper filled out in landscape orientation.

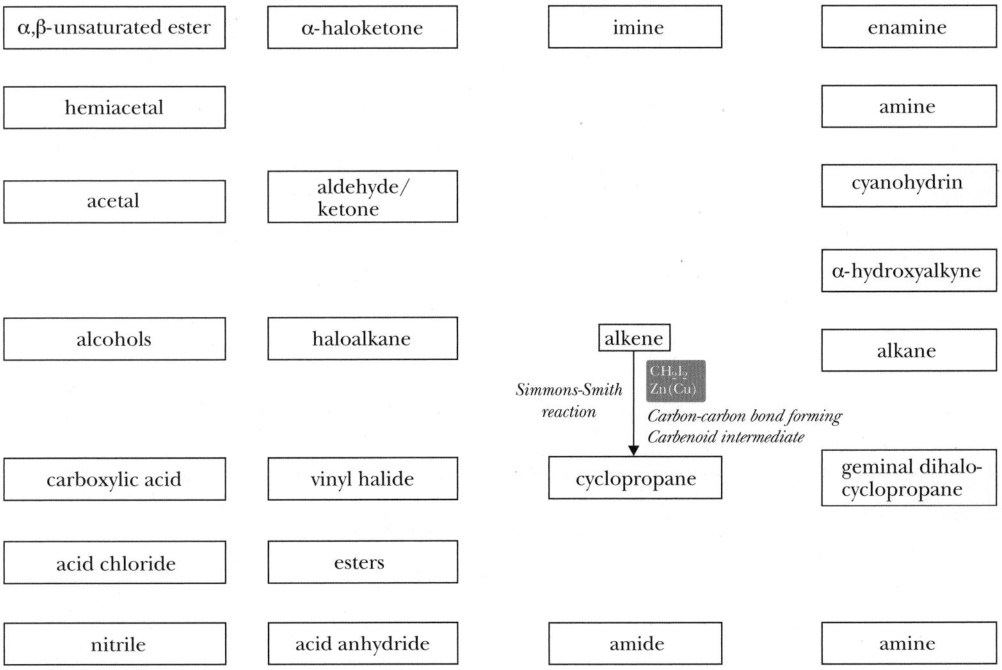
α,β-unsaturated ester
α-haloketone
imine
enamine
hemiacetal
amine
acetal
aldehyde/ ketone
cyanohydrin
α-hydroxyalkyne
alcohols
haloalkane
alkene
alkane
CH_2I_2
Zn(Cu)
Simmons-Smith reaction
Carbon-carbon bond forming
Carbenoid intermediate
carboxylic acid
vinyl halide
cyclopropane
geminal dihalo-cyclopropane
acid chloride
esters
nitrile
acid anhydride
amide
amine

Refer to the "Study Guide" section of this chapter. Draw arrows between functional groups to account for each reaction. Write the reagents required to bring about each reaction next to the arrow. Then record any regiochemistry or stereochemistry considerations relevant to the reaction. You should also record any key aspects of the mechanism, such as formation of a carbocation intermediate, as a helpful reminder. It is important to keep track of all reactions that make carbon-carbon bonds, because these will help you build large molecules from smaller fragments.

On the above organic chemistry reaction roadmap template, the information for the Simmons-Smith reaction, the seventh reaction in the "Study Guide" section has been added to help you get started. For this reaction roadmap, do not write an arrow for reactions 1, 2, and 4 explicitly, because these are considered reagents, which are prepared immediately prior to use. A reaction roadmap is used to indicate interconversion of molecules with more stable functional groups. Appendix 10 contains a series of roadmaps for different sections of the book, but you should use those for reference only after you have completed your own.

Reaction Roadmap

15.20 Write the products of the following sequences of reactions. Refer to your reaction roadmaps to see how the combined reactions allow you to "navigate" between the different functional groups. Note that you will need both your old Chapters 6–11 roadmap and your new Chapter 15 roadmap for these.

(a)

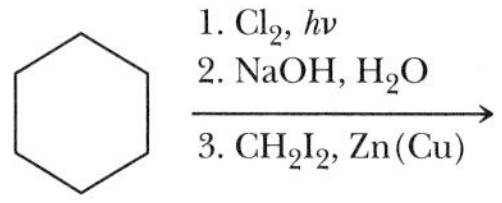
1. Cl_2, *hv*
2. NaOH, H_2O
3. CH_2I_2, Zn(Cu)

An alkane

(b)

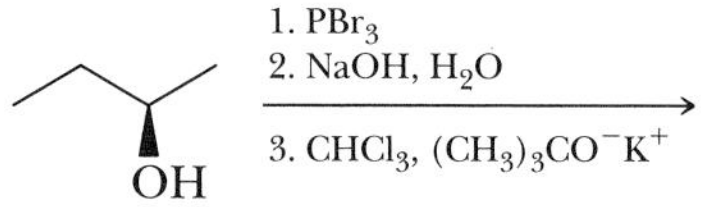
OH
1. PBr_3
2. NaOH, H_2O
3. $CHCl_3$, $(CH_3)_3CO^-K^+$

An alcohol

(c)

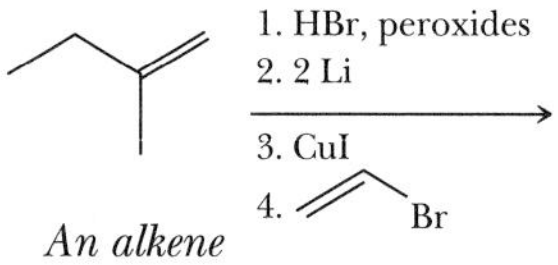
1. HBr, peroxides
2. 2 Li
3. CuI
4.
Br

An alkene

(d)

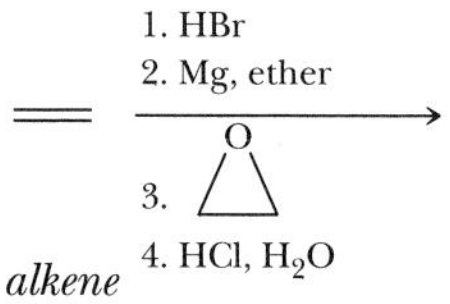
1. HBr
2. Mg, ether
3.
O
4. HCl, H_2O

An alkene

Synthesis

Reaction Roadmap

15.21 Using your old and new reaction roadmaps as a guide, show how 1-bromobutane can be converted into either of the two products shown by a suitable choice of reagents. Give reagents and conditions for each reaction.

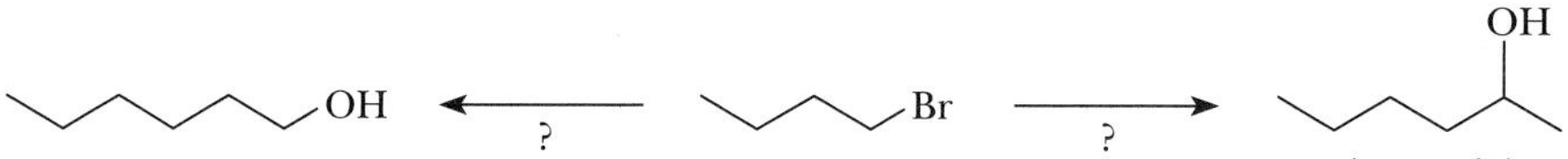
OH
?
Br
?
OH

(racemic)

Reaction Roadmap

15.22 Using your old and new reaction roadmaps as a guide, show how to convert 1-bromo-3-methylbutane into 2,7-dimethyloctane. You must use 1-bromo-3-methylbutane as the source of all carbon atoms in the target molecule. Show all reagents and all molecules synthesized along the way.

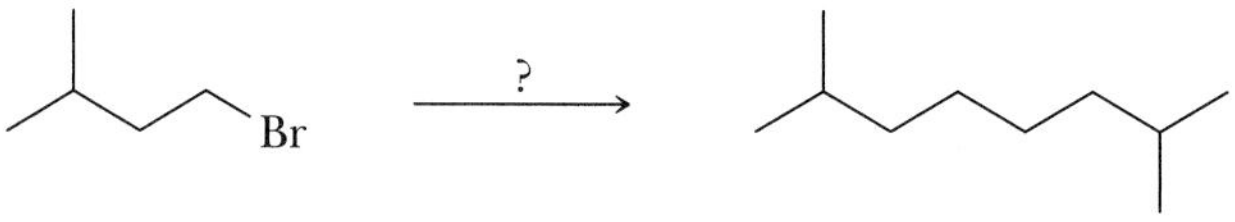
Br
?

1-Bromo-3-methylbutane **2,7-Dimethyloctane**

Reaction Roadmap

15.23 Using your old and new reaction roadmaps as a guide, show how to convert ethane into 1-butanol. You must use ethane as the source of all carbon atoms in the target molecule. Show all reagents and all molecules synthesized along the way.

?
OH

Ethane **1-Butanol**

15.24 Using your old and new reaction roadmaps as a guide, show how to convert 1-propanol and diiodomethane into racemic *trans*-1-methyl-2-propylcyclopropane. You must use 1-propanol and diiodomethane as the source of all carbon atoms in the target molecule. Show all molecules synthesized along the way.

Reaction Roadmap

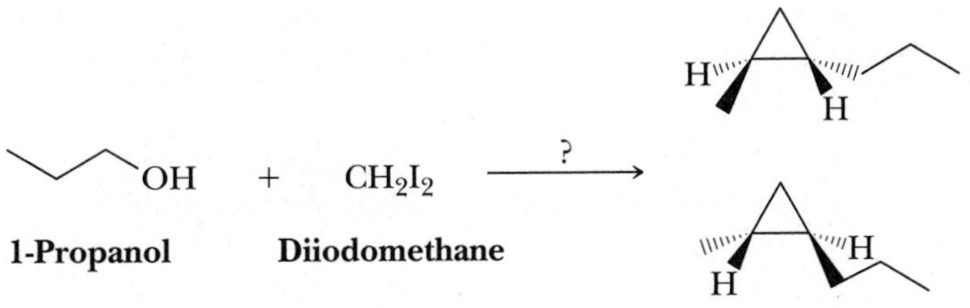
OH
+
CH_2I_2
?
H
H
H
H
1-Propanol
Diiodomethane

***trans*-1-Methyl-2-propylcyclopropane**
(racemic)

Reactions in Context

15.25 The synthesis of carbohydrates can be particularly difficult because of the large number of chiral centers and OH functional groups present. Epoxides can be useful synthetic intermediates in carbohydrate syntheses. Draw the product of the following reactions of a Gilman reagent with each epoxide.

(a)

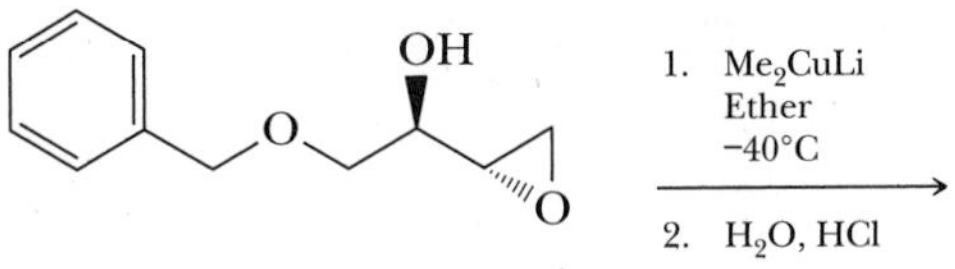
OH
O
O
1. Me_2CuLi
Ether
−40°C
2. H_2O, HCl

(b)

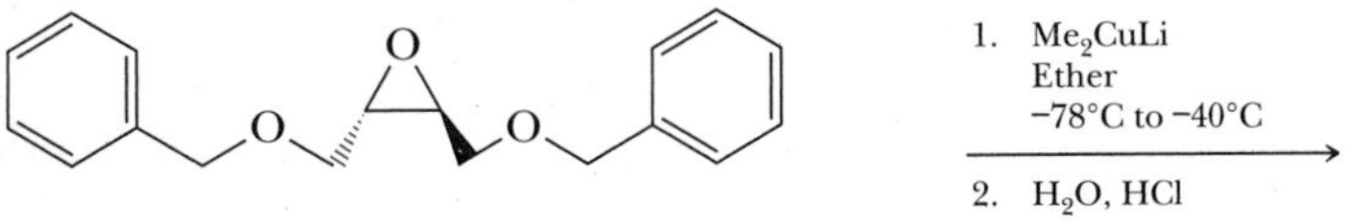
O
O
O
1. Me_2CuLi
Ether
−78°C to −40°C
2. H_2O, HCl

15.26 Gilman reagents are versatile reagents for making new carbon-carbon bonds. Complete the following reactions that use Gilman reagents.

(a)

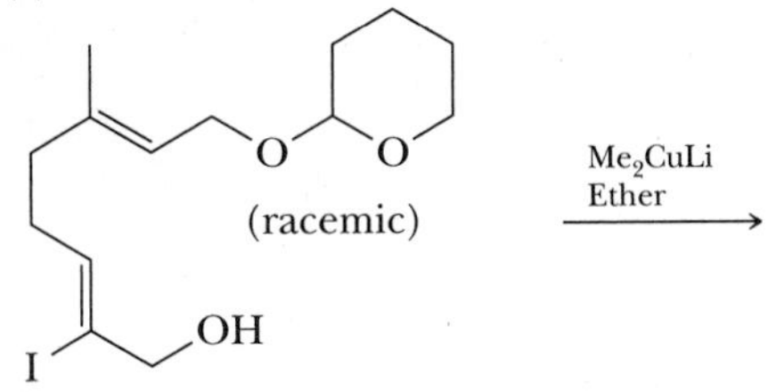
O
O
(racemic)
OH
I
Me_2CuLi
Ether

(b)

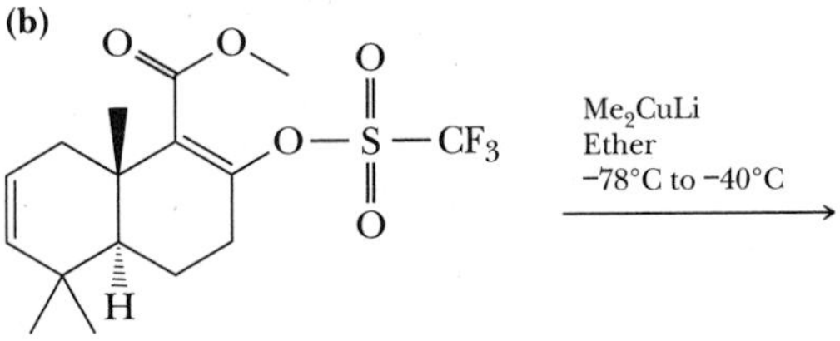
O
O
O
O
S
CF_3
O
H
Me_2CuLi
Ether
−78°C to −40°C

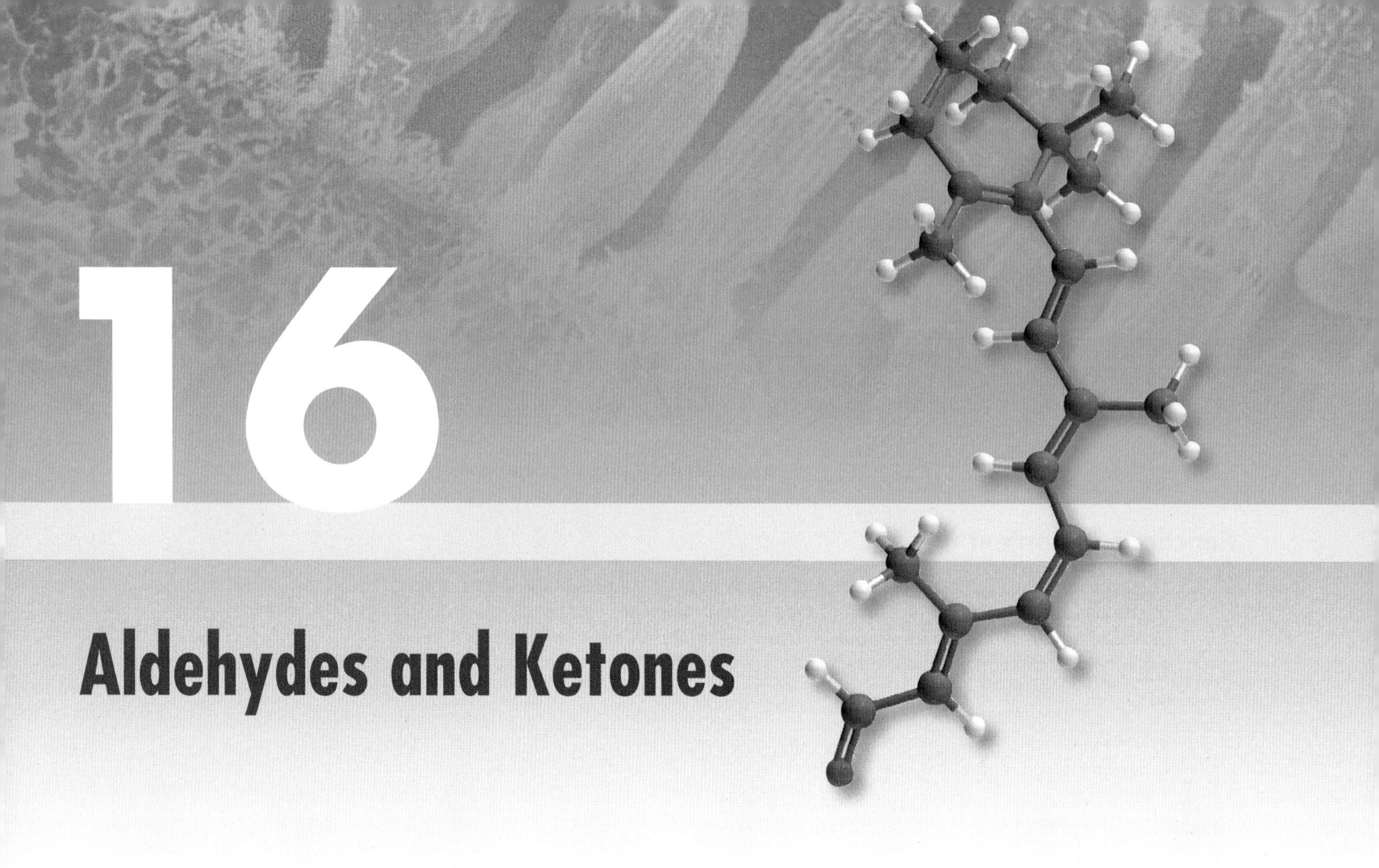

16

Aldehydes and Ketones

Rod cells in the human eye. **Inset:** a model of 11-*cis*-retinal, an oxidized form of vitamin A. For the reaction of 11-*cis*-retinal with opsin to form visual purple, see Section 16.8A. (Omikron/Science Source)

Outline

16.1 Structure and Bonding
16.2 Nomenclature
16.3 Physical Properties
16.4 Reactions
16.5 Addition of Carbon Nucleophiles
16.6 The Wittig Reaction
16.7 Addition of Oxygen Nucleophiles
16.8 Addition of Nitrogen Nucleophiles
16.9 Keto-Enol Tautomerism
16.10 Oxidation
16.11 Reduction
HOW TO: Retrosynthetically Dissect an Amine into the Proper Starting Materials for a Reductive Amination
16.12 Reactions at an α-Carbon

In this and several of the following chapters, we study the physical and chemical properties of compounds containing the carbonyl group, C═O. Because the carbonyl group is the functional group of aldehydes, ketones, carboxylic acids, and carboxyl derivatives, it is one of the most important functional groups in organic chemistry. The chemical properties of this functional group are straightforward, and an understanding of its few characteristic reaction themes leads very quickly to an understanding of a wide variety of organic reactions.

16.1 Structure and Bonding

The functional group of an **aldehyde** is a carbonyl group bonded to a hydrogen atom and a carbon atom (Section 1.3C). In methanal (always called formaldehyde), the simplest aldehyde, the carbonyl group is bonded to two hydrogen atoms. In other aldehydes, it is bonded to one hydrogen atom and one carbon atom. Following are Lewis structures for formaldehyde and ethanal (which is always called acetaldehyde).

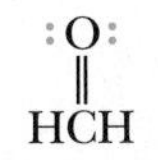

Methanal (Formaldehyde)

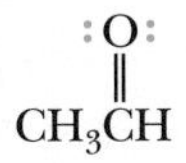

Ethanal (Acetaldehyde)

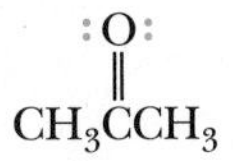

Propanone (Acetone)

The functional group of a **ketone** is a carbonyl group bonded to two carbon atoms (Section 1.3C). The simplest ketone is propanone, which is always called acetone.

According to valence bond theory, the carbon-oxygen double bond consists of one σ bond formed by the overlap of sp^2 hybrid orbitals of carbon and oxygen and one π bond formed by the overlap of parallel $2p$ orbitals. The two nonbonding pairs of electrons on oxygen lie in the remaining sp^2 hybrid orbitals (Figure 1.25).

16.2 Nomenclature

A. IUPAC Nomenclature

IUPAC names for aldehydes and ketones follow the familiar pattern of selecting as the parent alkane the longest chain of carbon atoms that contains the functional group. We show the aldehyde group by changing the suffix *-e* of the parent alkane to *-al* (Section 2.3). Because the carbonyl group of an aldehyde can only appear at the end of a parent chain and because numbering must start with it as carbon-1, its position is unambiguous; there is no need to use a number to locate it.

For unsaturated aldehydes, the presence of a carbon-carbon double or triple bond is indicated by the infix *-en-* or *-yn-*. As with other molecules with both an infix and a suffix, the location of the group corresponding to the suffix determines the numbering pattern.

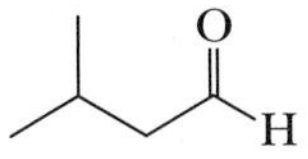

3-Methylbutanal

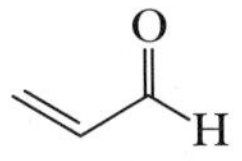

2-Propenal (Acrolein)

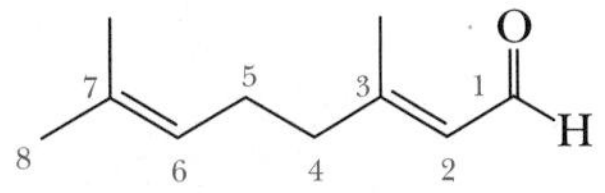

(2*E*)-3,7-Dimethyl-2,6-octadienal (Geranial)

For cyclic molecules in which —CHO is bonded directly to the ring, the molecule is named by adding the suffix *-carbaldehyde* to the name of the ring. The atom of the ring to which the aldehyde group is bonded is numbered 1.

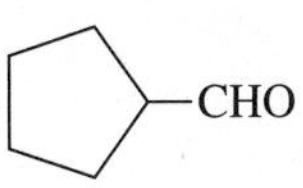

Cyclopentane-carbaldehyde

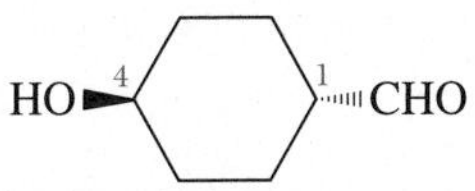

***trans*-4-Hydroxycyclo-hexanecarbaldehyde**

Among the aldehydes for which the IUPAC system retains common names are benzaldehyde and cinnamaldehyde, as well as formaldehyde and acetaldehyde. Note here

Benzaldehyde is found in the kernels of bitter almonds. Cinnamaldehyde is found in Ceylon and Chinese cinnamon oils.

the alternative ways the phenyl group can be written. In benzaldehyde, it is written as a line-angle formula; in cinnamaldehyde, it is written C_6H_5—.

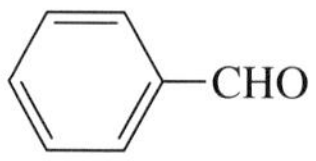

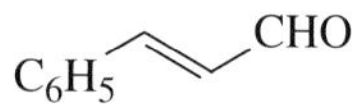

Benzaldehyde ***trans*-3-Phenyl-2-propenal (Cinnamaldehyde)**

In the IUPAC system, ketones are named by selecting as the parent alkane the longest chain that contains the carbonyl group and then indicating its presence by changing the suffix from *-e* to *-one* (Section 2.3). The parent chain is numbered from the direction that gives the carbonyl carbon the smaller number. The IUPAC system retains the common names acetone, acetophenone, and benzophenone.

Propanone (Acetone)

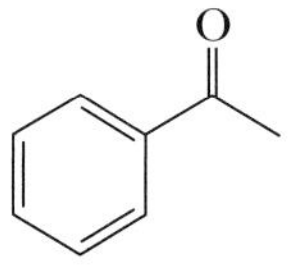

Acetophenone

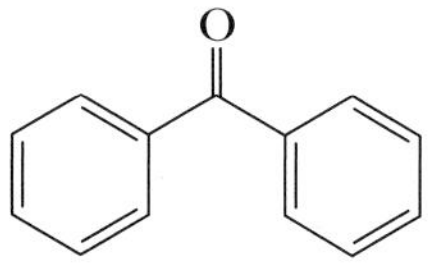

Benzophenone

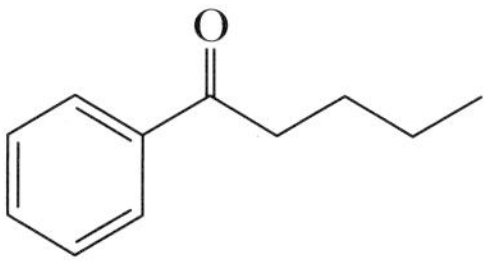

1-Phenyl-1-pentanone

Example 16.1 Naming Aldehydes and Ketones I

Write IUPAC names for each compound. Specify the configuration of all chiral centers in (a) and (c).

(a)
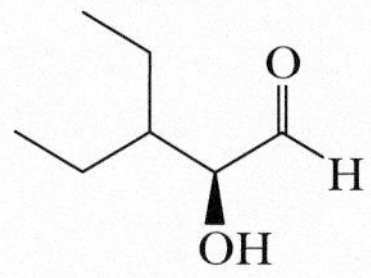

(b)
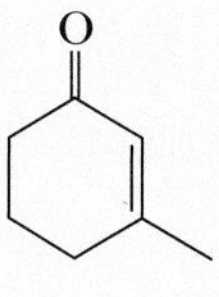

(c)
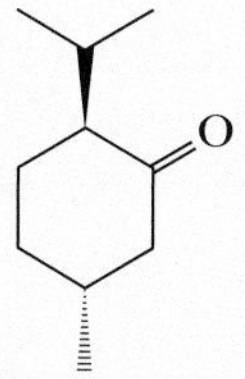

Solution

(a) The parent chain is pentane. The name is (*S*)-3-ethyl-2-hydroxypentanal.

(b) Number the six-membered ring beginning with the carbonyl carbon. The IUPAC name is 3-methyl-2-cyclohexenone.

(c) The name (2*S*,5*R*)-2-isopropyl-5-methylcyclohexanone provides a complete description of the configuration of each chiral center as well as the *trans* relationship between the isopropyl and methyl groups. The common name of this compound is menthone.

Problem 16.1

Write the IUPAC name for each compound. Specify the configuration in (c).

(a)
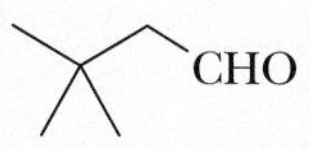

(b)
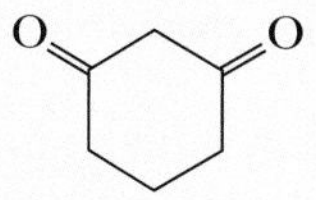

(c)
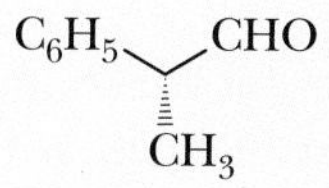

Example 16.2 Naming Aldehydes and Ketones II

Write structural formulas for all ketones with the molecular formula $C_6H_{12}O$ and give each its IUPAC name. Which of these ketones are chiral?

Solution

Following are line-angle formulas and IUPAC names for the six ketones with this molecular formula. Only 3-methyl-2-pentanone is chiral; the *R* enantiomer is drawn here.

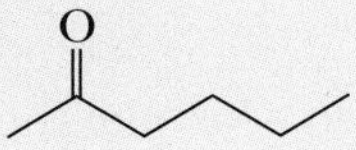

2-Hexanone

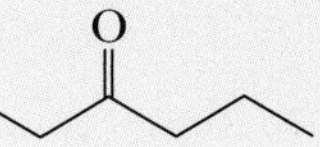

3-Hexanone

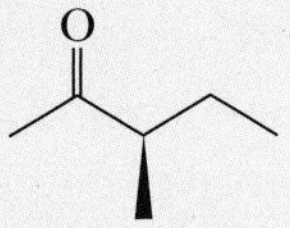

(*R*)-3-Methyl-2-pentanone

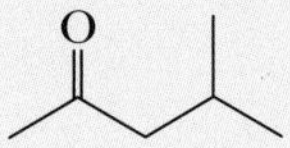

4-Methyl-2-pentanone

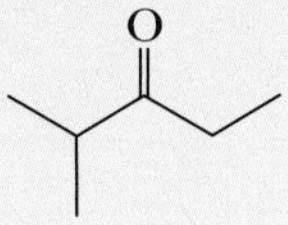

2-Methyl-3-pentanone

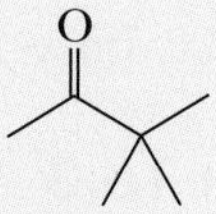

3,3-Dimethyl-2-butanone

Problem 16.2

Write structural formulas for all aldehydes with the molecular formula $C_6H_{12}O$ and give each its IUPAC name. Which of these aldehydes are chiral?

B. IUPAC Names for More Complex Aldehydes and Ketones

In naming compounds that contain more than one functional group that might be indicated by a suffix, the IUPAC system has established an **order of precedence of functions**. Table 16.1 gives the order of precedence for the functional groups we have studied so far.

Order of precedence of functions
A ranking of functional groups in order of priority for the purposes of IUPAC nomenclature.

Table 16.1 Increasing Order of Precedence of Six Functional Groups

(Increasing precedence ↑)

Functional Group	Suffix if Higher Priority	Prefix if Lower Priority	Example When the Functional Group Has Lower Priority	
Carboxyl	-oic acid	—		
Aldehyde	-al	oxo-	3-Oxopropanoic acid	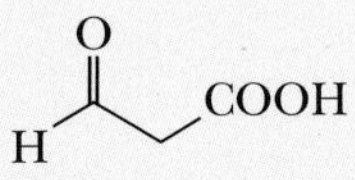
Ketone	-one	oxo-	3-Oxobutanoic acid	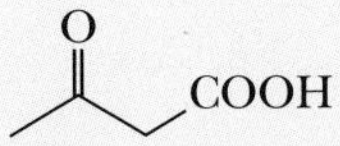
Alcohol	-ol	hydroxy-	4-Hydroxybutanoic acid	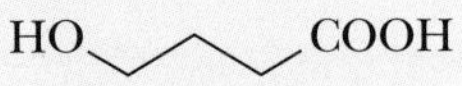
Amino	-amine	amino-	3-Aminobutanoic acid	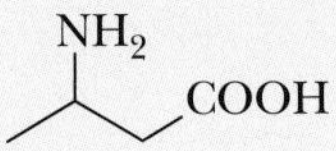
Sulfhydryl	-thiol	mercapto-	2-Mercaptoethanol	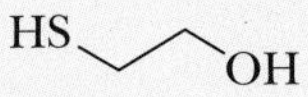

Example 16.3 Naming Aldehydes and Ketones III

Write the IUPAC name for each compound, being certain to specify configuration where appropriate.

(a) 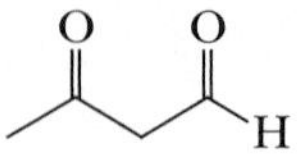 **(b)** 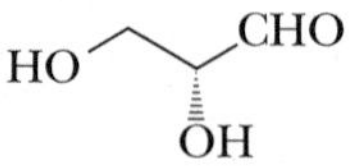 **(c)** 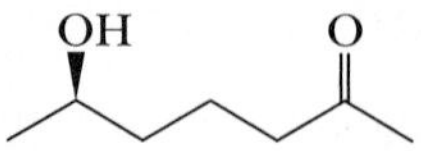

Solution

(a) 3-Oxobutanal. An aldehyde has higher precedence than a ketone. The presence of the carbonyl group of the ketone is indicated by the prefix *oxo-* (Table 16.1).

(b) (*R*)-2,3-Dihydroxypropanal. Its common name is glyceraldehyde. Glyceraldehyde is the simplest carbohydrate (Section 25.1).

(c) (*R*)-6-Hydroxy-2-heptanone.

Problem 16.3

Write the IUPAC name for each compound.

(a) 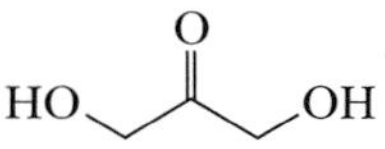 **(b)** 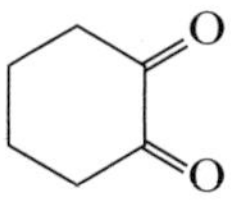 **(c)** 

C. Common Names

The common name for an aldehyde is derived from the common name of the corresponding carboxylic acid by dropping the word *acid* and changing the suffix *-ic* or *-oic* to *-aldehyde*. Because we have not yet studied common names for carboxylic acids, we are not in a position to discuss common names for aldehydes. We can illustrate how they are derived, however, by reference to a few common names with which you are familiar. The name *formaldehyde* is derived from formic acid; the name *acetaldehyde* is derived from acetic acid.

$\overset{O}{\overset{\|}{HCH}}$	$\overset{O}{\overset{\|}{HCOH}}$	$\overset{O}{\overset{\|}{CH_3CH}}$	$\overset{O}{\overset{\|}{CH_3COH}}$
Formaldehyde	**Formic acid**	**Acetaldehyde**	**Acetic acid**

Common names for ketones are derived by naming the two alkyl or aryl groups bonded to the carbonyl group as separate words followed by the word *ketone*.

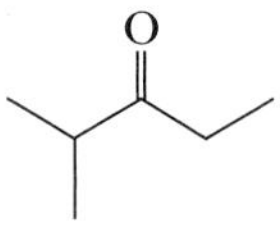

Ethyl isopropyl ketone

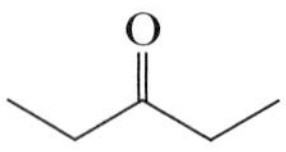

Diethyl ketone

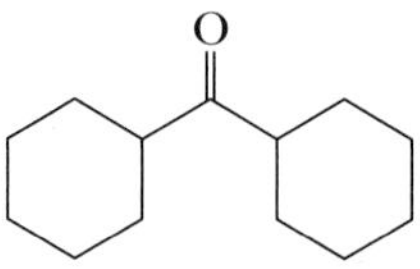

Dicyclohexyl ketone

16.3 Physical Properties

Oxygen is more electronegative than carbon (3.5 compared with 2.5); therefore, a carbon-oxygen double bond is polar, with oxygen bearing a partial negative charge and carbon bearing a partial positive charge. In addition, the resonance structure shown on the right emphasizes the reactivity of the carbonyl oxygen as a Lewis base and the carbonyl carbon as a Lewis acid. The bond dipole moment of a carbonyl group is 2.3 D (Table 1.7).

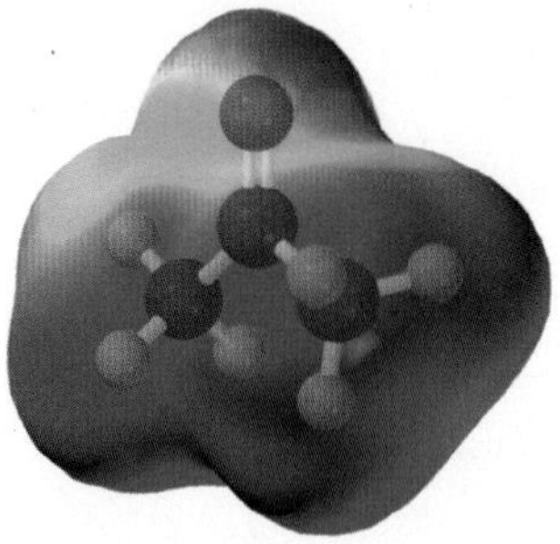

An electrostatic potential map for acetone. Note the large negative charge on oxygen (red) and the positive charges (blue) on the three carbons.

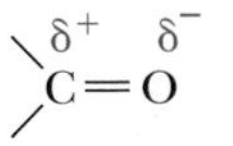

Polarity of a carbonyl group

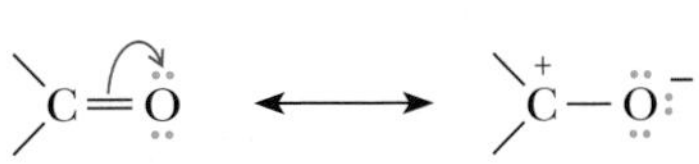

More important contributing structure

Aldehydes and ketones are polar compounds that engage in dipole-dipole interactions in pure liquid. They have higher boiling points than nonpolar compounds of comparable molecular weight (Table 16.2).

Table 16.2 Boiling Points of Six Compounds of Comparable Molecular Weight

Name	Structural Formula	Molecular Weight (g/mol)	Boiling Point (°C)
Diethyl ether	$CH_3CH_2OCH_2CH_3$	74	34
Pentane	$CH_3CH_2CH_2CH_2CH_3$	72	36
Butanal	$CH_3CH_2CH_2CHO$	72	76
2-Butanone	$CH_3CH_2COCH_3$	72	80
1-Butanol	$CH_3CH_2CH_2CH_2OH$	74	117
Propanoic acid	CH_3CH_2COOH	74	141

Of the compounds listed in Table 16.2, diethyl ether and pentane have the lowest boiling points. Diethyl ether is a polar molecule, but because of steric hindrance, only weak dipole-dipole interactions exist between its molecules (Section 11.3). Both butanal and 2-butanone are polar molecules, and because of the intermolecular attraction between their carbonyl groups, their boiling points are higher than those of pentane and diethyl ether. Alcohols (Section 10.2) and carboxylic acids (Section 17.3) are polar molecules that can associate by hydrogen bonding, the strongest intermolecular interaction. The boiling points of 1-butanol and propanoic acid are significantly higher than those of butanal and 2-butanone, compounds whose molecules cannot associate by hydrogen bonding.

The oxygen atoms of the carbonyl groups of aldehydes and ketones act as hydrogen bond acceptors with water; therefore, low-molecular-weight aldehydes and ketones are more soluble in water than are nonpolar compounds of comparable molecular weight. Listed in Table 16.3 are boiling points and solubilities in water for several low-molecular-weight aldehydes and ketones.

Table 16.3 Physical Properties of Selected Aldehydes and Ketones

IUPAC Name	Common Name	Structural Formula	Boiling Point (°C)	Solubility (g/100 g water)
Methanal	Formaldehyde	HCHO	−21	Infinite
Ethanal	Acetaldehyde	CH_3CHO	20	Infinite
Propanal	Propionaldehyde	CH_3CH_2CHO	49	16
Butanal	Butyraldehyde	$CH_3CH_2CH_2CHO$	76	7
Hexanal	Caproaldehyde	$CH_3(CH_2)_4CHO$	129	Slight
Propanone	Acetone	CH_3COCH_3	56	Infinite
2-Butanone	Ethyl methyl ketone	$CH_3COCH_2CH_3$	80	26
3-Pentanone	Diethyl ketone	$CH_3CH_2COCH_2CH_3$	101	5

16.4 Reactions

Nucleophilic acyl addition

A characteristic reaction mechanism of carbonyl-containing compounds such as aldehydes and ketones in which a nucleophile makes a new bond to the electrophilic carbonyl carbon atom.

One of the most common reaction themes of a carbonyl group is addition of a nucleophile to form a tetrahedral carbonyl addition compound. The carbonyl carbon atom is highly electrophilic due to a significant partial positive charge as well as to the ability to accommodate a new bond through the conversion of the π bond to a lone pair on oxygen of the tetrahedral carbonyl addition compound. These reactions are often referred to as **nucleophilic acyl additions**. In the following general reaction, the nucleophilic reagent is written as Nu:$^-$ to emphasize the presence of its unshared pair of electrons. Notice how the π bond of the carbonyl group breaks as the nucleophile attacks, changing the carbon atom hybridization state, all the while maintaining four bonds to carbon in the tetrahedral carbonyl addition compound.

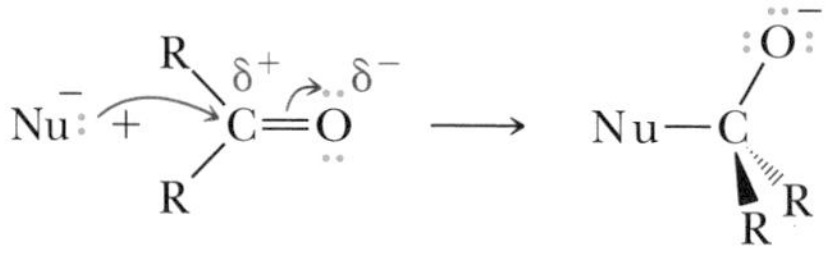

Tetrahedral carbonyl addition compound

A second common reaction theme of a carbonyl group is reaction with a proton or another Lewis acid to form a resonance-stabilized cation. Protonation increases the electron deficiency of the carbonyl carbon and makes it more reactive toward nucleophiles. The reaction is followed by removal of a proton to give a tetrahedral carbonyl addition compound.

▶ Functional groups react the same in different molecules.

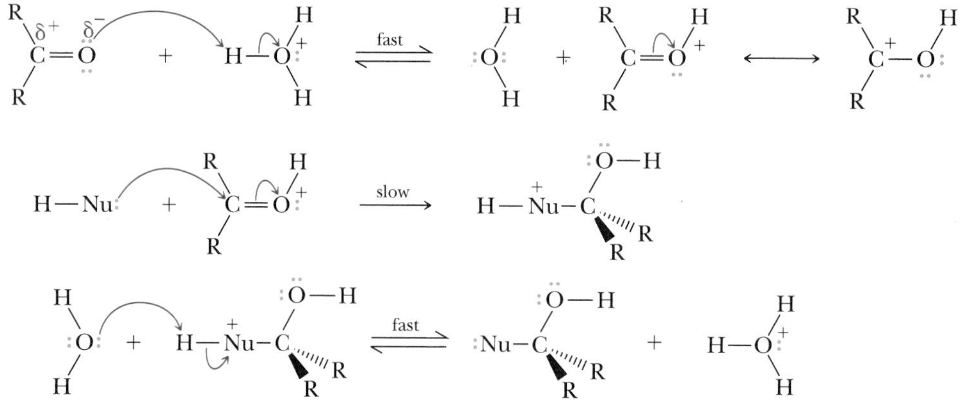

Tetrahedral carbonyl addition compound

Both reactivities can be predicted based on the large bond dipole of the carbonyl group. Using acetaldehyde as an example, Lewis acids such as protons are attracted to the partial negative charge (red color of the electrostatic potential map) of the carbonyl oxygen, while electron-rich nucleophiles are attracted to the partial positive charge (blue color of the electrostatic potential map) of the carbonyl carbon.

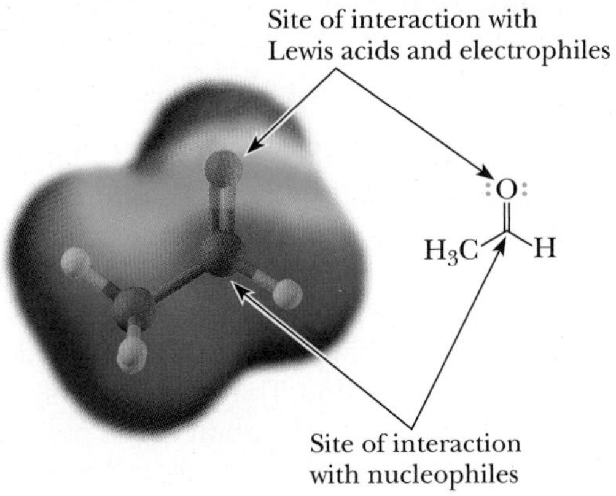

Carbonyl compounds such as aldehydes and ketones undergo a wide variety of important reactions with most involving nucleophilic acyl addition. The mechanisms for these reactions are similar because almost all of the steps can be described as one of the four mechanistic elements first mentioned in Things You Should Know III, Reaction Mechanisms. When choosing among the four to begin each mechanism, look to see whether an acid or a strong base is present. If a nucleophile but no acid or base is present, the reaction mechanism will begin with the nucleophile attacking the electrophilic carbonyl carbon atom (make a new bond between a nucleophile and an electrophile). The π bond of the carbonyl breaks as the nucleophile adds. If an acid is present, the proton will add to the carbonyl oxygen atom (add a proton), a process that makes the carbonyl group even more electrophilic. In these cases, the next step will be attack of a nucleophile on the carbonyl carbon atom. In the presence of strong base, carbonyl compounds undergo a third characteristic reaction (take a proton away) to form a very important species called an enolate anion. (This is the focus of Chapter 19.)

Often, the tetrahedral product of nucleophile addition to a carbonyl is a new chiral center. If none of the starting materials are chiral, the nucleophile will approach the carbonyl from either side with equal probability, resulting in a carbonyl addition product that consists of a racemic mixture.

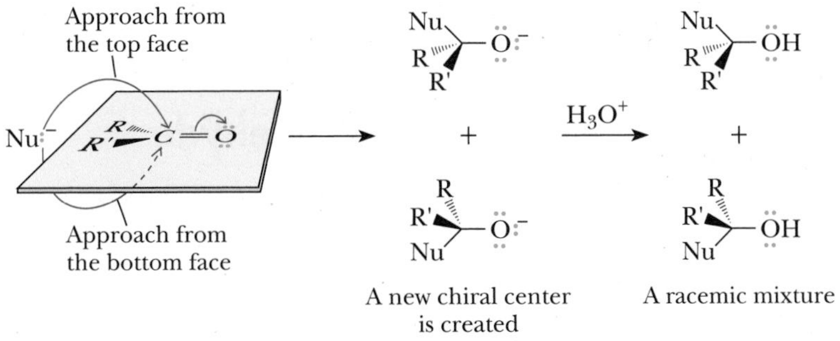

16.5 Addition of Carbon Nucleophiles

In this section, we examine nucleophilic acyl addition reactions of aldehydes and ketones with the following types of carbon nucleophiles.

> ▶ Most bond-making steps in reaction mechanisms involve nucleophiles reacting with electrophiles.

RMgX	RLi	$RC{\equiv}C:^-$	$^-:C{\equiv}N:$
A Grignard reagent	An organolithium reagent	An anion of a terminal alkyne	Cyanide ion

▶

From the perspective of the organic chemist, addition of a carbon nucleophile is the most important type of carbonyl addition reaction because a new carbon-carbon bond is formed in the process. Each of these reactions follows the same mechanistic two-step pattern of **making a bond between the carbon nucleophile and the electrophilic carbonyl carbon atom** to give the tetrahedral carbonyl addition compound, followed by **adding a proton** to give an —OH group in the product.

A. Addition of Grignard Reagents

The special value of Grignard reagents (Section 15.1) is that they provide excellent ways to form new carbon-carbon bonds. Given the difference in electronegativity between carbon and magnesium (2.5 − 1.2 = 1.3), the carbon-magnesium bond of a Grignard reagent is polar covalent with carbon bearing a partial negative charge and magnesium bearing a partial positive charge. A Grignard reagent is therefore a good nucleophile and adds to the carbonyl group of an aldehyde or a ketone to form a tetrahedral carbonyl addition compound. The reaction is assisted by the attraction between the partial negative charge on the carbon of the organometallic compound and the partial positive charge on the carbonyl carbon. The alkoxide ions formed in these reactions are strong bases (Section 10.4) and, when treated with an aqueous acid such as HCl or aqueous NH_4Cl during workup, form alcohols. In the following examples, the magnesium-oxygen bond is written $—O^-[MgBr]^+$ to emphasize its ionic character.

Caution: new chiral centers are often created in Grignard reactions with aldehydes or ketones. When neither the aldehyde/ketone nor the Grignard reagent is chiral but the product has a new chiral center, a racemic mixture is formed.

Addition to Formaldehyde Gives a Primary Alcohol

Treatment of a Grignard reagent with formaldehyde followed by protonation in aqueous acid gives a 1° alcohol.

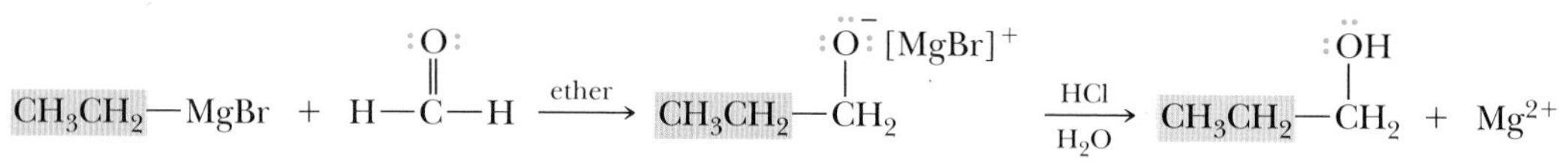

Formaldehyde — A magnesium alkoxide — **1-Propanol** (a 1° alcohol)

Mechanism 16.1

Grignard Reagent Reacting with Formaldehyde

Step 1: **Make a new bond between a nucleophile and an electrophile.** Nucleophilic addition of a Grignard reagent to the electrophilic carbonyl carbon atom gives a tetrahedral carbonyl addition compound.

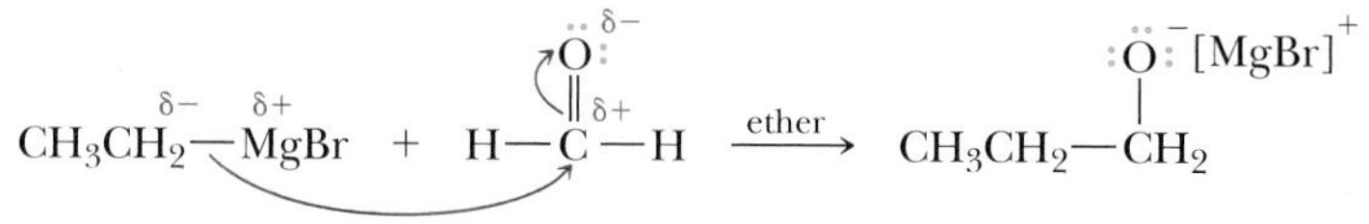

Step 2: **Add a proton.** In a second step, the chemist adds a dilute acid solution to protonate the alkoxide function of the tetrahedral carbonyl addition compound to give the primary alcohol product.

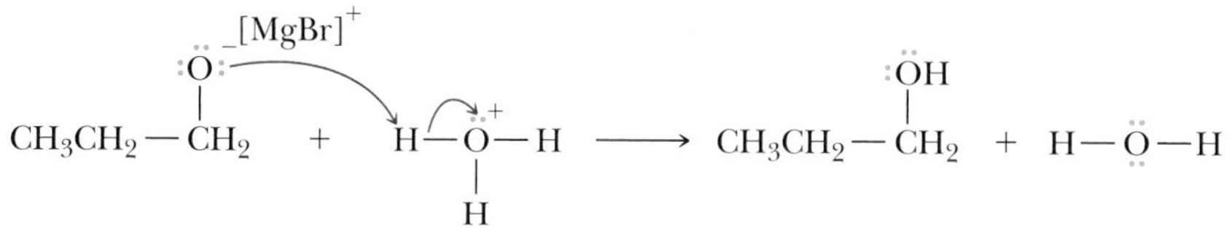

It is important to remember that this second step requires the chemist to add the acid after the tetrahedral carbonyl addition compound forms. If the acid were added with the Grignard reagent in a single step, the acid would immediately protonate the Grignard reagent before any further reaction could take place.

Addition to an Aldehyde (Except Formaldehyde) Gives a Secondary Alcohol

Treatment of a Grignard reagent with any other aldehyde followed by protonation in aqueous acid gives a 2° alcohol.

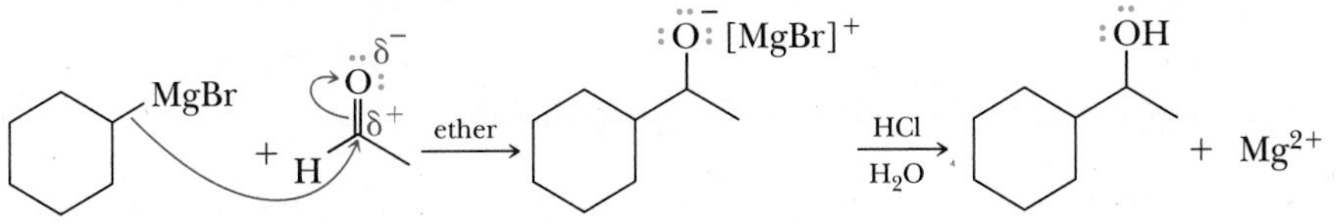

Acetaldehyde (an aldehyde) — A magnesium alkoxide — **1-Cyclohexylethanol** (a 2° alcohol; racemic)

In this example, the product is chiral and is formed as a racemic mixture.

Addition to a Ketone Gives a Tertiary Alcohol

Treatment of a Grignard reagent with a ketone followed by protonation in aqueous acid gives a 3° alcohol.

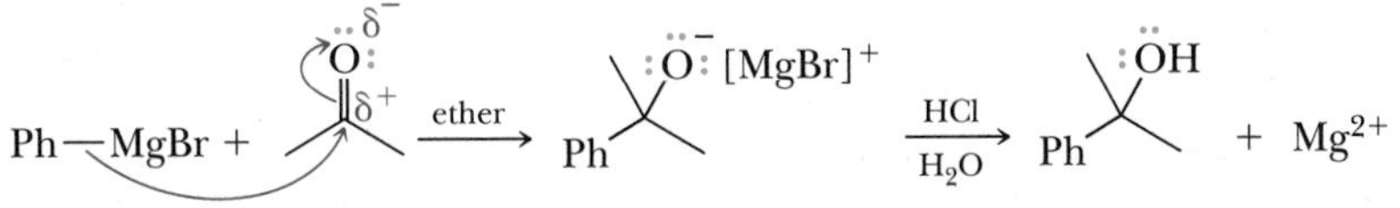

Phenylmagnesium bromide — **Acetone** (a ketone) — A magnesium alkoxide — **2-Phenyl-2-propanol** (a 3° alcohol)

Example 16.4 Reactions of Grignard Reagents

Racemic 2-phenyl-2-butanol can be synthesized by three different combinations of a Grignard reagent and a ketone. Show each combination.

Solution

In each solution, curved arrows show formation of the new carbon-carbon bond and the alkoxide ion. The new carbon-carbon bond formed by each set of reagents is labeled in the final product.

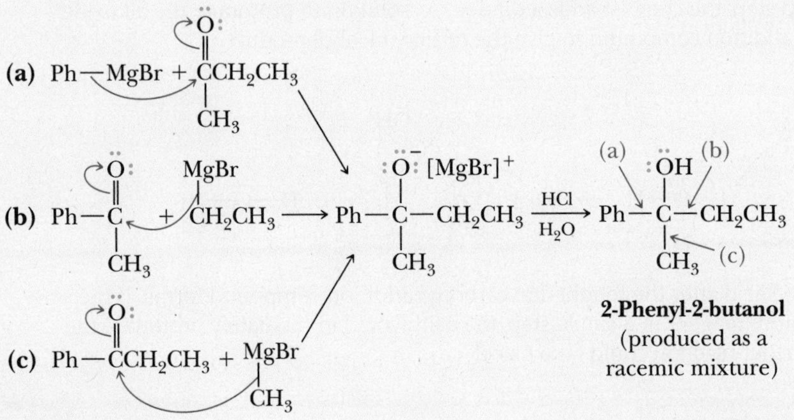

2-Phenyl-2-butanol (produced as a racemic mixture)

(Continued)

Watch a video explanation

Problem 16.4

Show how these four products can be synthesized from the same Grignard reagent.

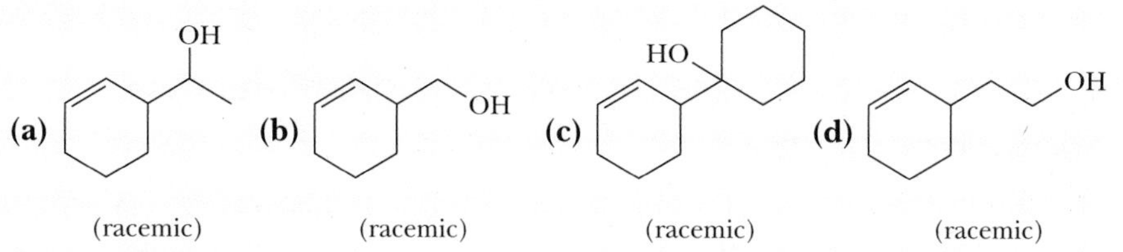

B. Addition of Organolithium Compounds

Organolithium compounds have greater negative charge character on carbon, so they are generally more reactive in nucleophilic acyl addition reactions than organomagnesium compounds, and typically give higher yields of products. They are more troublesome to use, however, because they must be prepared and used under an atmosphere of nitrogen or another inert gas. The following synthesis illustrates the use of an organolithium compound to form a sterically hindered tertiary alcohol.

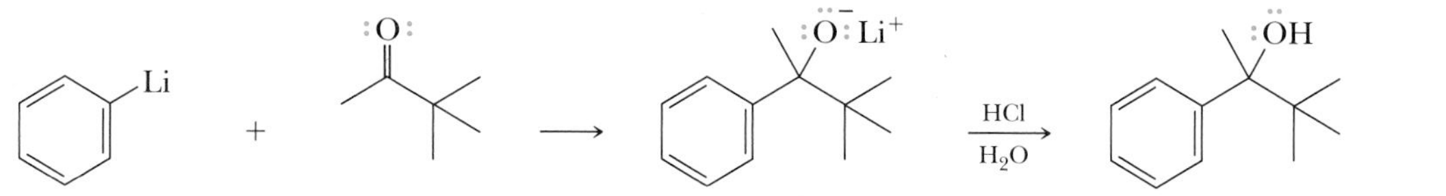

Phenyllithium **3,3-Dimethyl-2-butanone** A lithium alkoxide (racemic) **3,3-Dimethyl-2-phenyl-2-butanol** (racemic)

Mechanism 16.2

Organolithium Reagent Reacting with a Ketone

Step 1: **Make a new bond between a nucleophile and an electrophile.** Nucleophilic addition of an organolithium reagent to the electrophilic carbonyl carbon atom gives a tetrahedral carbonyl addition compound.

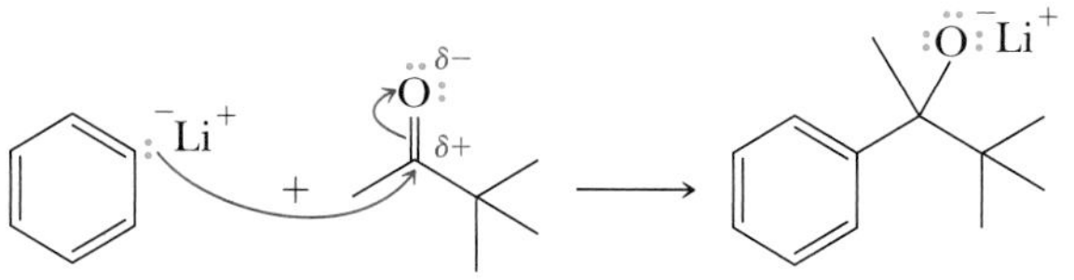

Step 2: **Add a proton.** In a second step, the chemist adds a dilute acid solution to protonate the alkoxide function of the tetrahedral carbonyl addition compound to give the tertiary alcohol product.

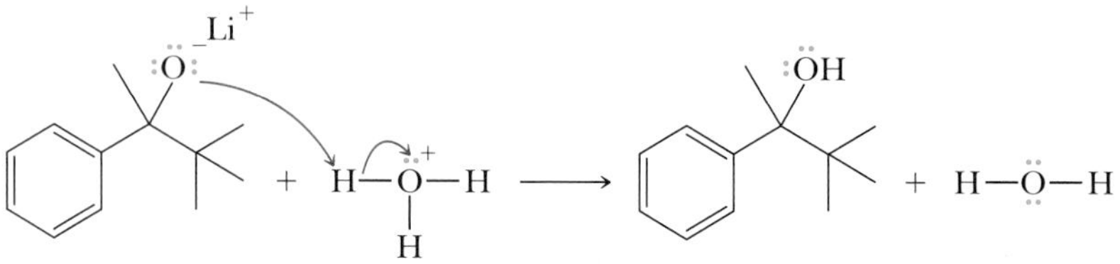

Once again, the chemist must add the acid after the tetrahedral carbonyl addition compound forms. If the acid were added with the organolithium reagent in a single step, the acid would immediately protonate the organolithium reagent before any further reaction could take place.

C. Addition of Anions of Terminal Alkynes

The anion of a terminal alkyne is a nucleophile (Section 7.5) and adds to the carbonyl group of an aldehyde or a ketone to form a tetrahedral carbonyl addition compound. In the following example, addition of sodium acetylide to cyclohexanone followed by hydrolysis in aqueous acid gives 1-ethynylcyclohexanol.

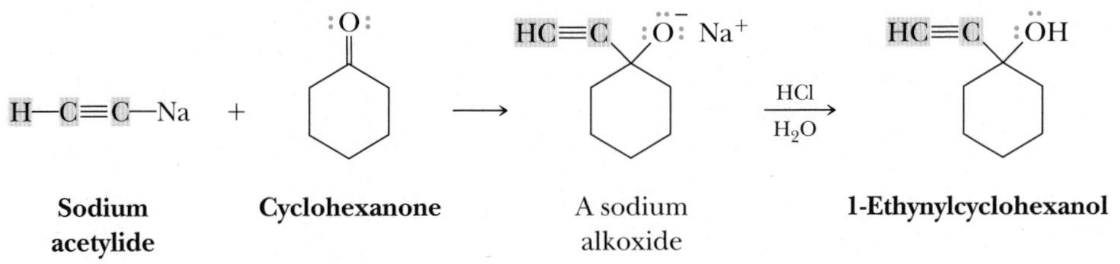

Mechanism 16.3

Alkyne Anion Reacting with a Ketone

Step 1: **Make a new bond between a nucleophile and an electrophile.** Nucleophilic addition of an alkyne anion to the electrophilic carbonyl carbon atom gives a tetrahedral carbonyl addition compound.

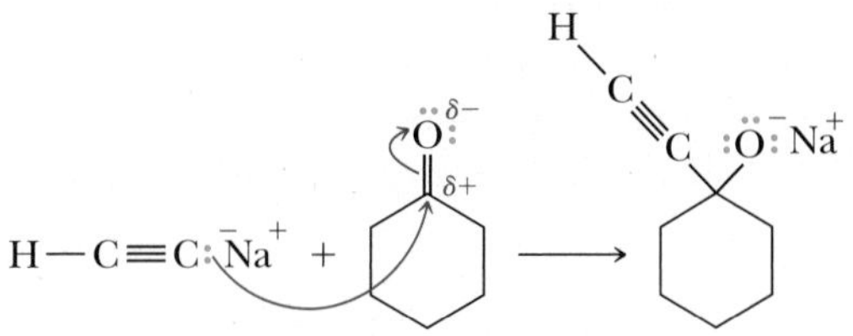

Step 2: **Add a proton.** In a second step, the chemist adds a dilute acid solution to protonate the alkoxide function of the tetrahedral carbonyl addition compound to give the tertiary alcohol product.

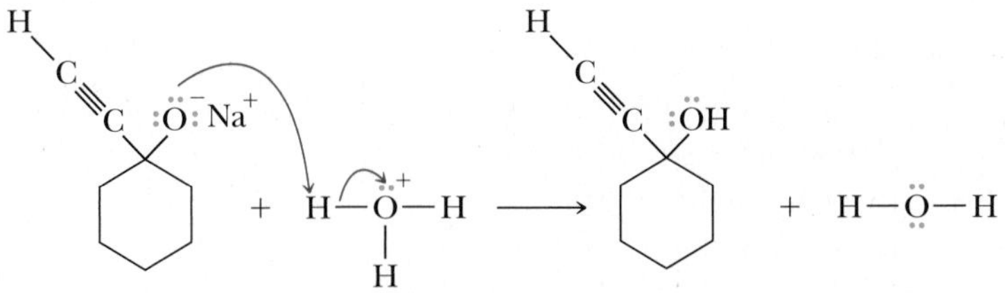

Once again, it is important to remember that this second step requires the chemist to add the acid after the tetrahedral carbonyl addition compound forms. If the acid were added with the alkyne anion in a single step, the acid would immediately protonate the alkyne anion before any further reaction could take place.

These addition compounds (alkynyl alcohols) contain both a hydroxyl group and a carbon-carbon triple bond, each of which can be further modified.

Acid-catalyzed hydration (Section 7.7B) of 1-ethynylcyclohexanol gives an α-hydroxyketone. Alternatively, hydroboration followed by oxidation with alkaline hydrogen peroxide (Section 7.7A) gives a β-hydroxyaldehyde.

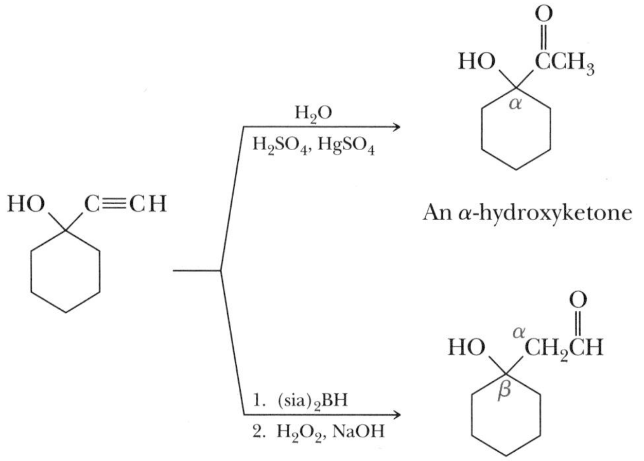

An α-hydroxyketone

A β-hydroxyaldehyde

This example illustrates two of the most valuable reactions of alkynes: (1) addition of the anion of a terminal alkyne to the carbonyl group of an aldehyde or a ketone gives an alkynyl alcohol, and (2) hydration of a terminal alkyne gives either an aldehyde or a ketone, depending on the alkyne and the method of hydration.

D. Addition of Hydrogen Cyanide

Cyanohydrin

A molecule containing an —OH group and a —CN group bonded to the same carbon.

Hydrogen cyanide, HCN, adds to the carbonyl group of an aldehyde or a ketone to form a tetrahedral carbonyl addition compound called a **cyanohydrin**. For example, HCN adds to acetaldehyde to form acetaldehyde cyanohydrin in 75% yield. We study the naming of compounds containing the nitrile group in Section 18.1E.

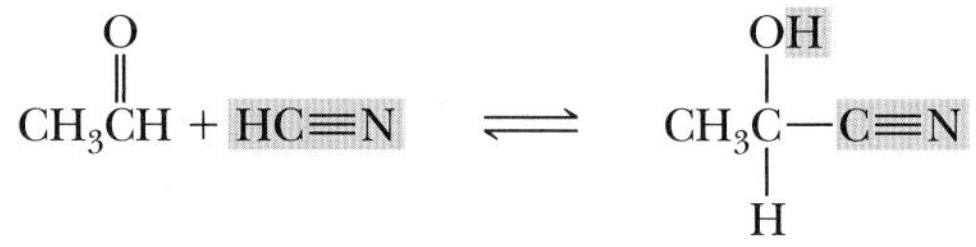

2-Hydroxypropanenitrile
(Acetaldehyde cyanohydrin;
produced as a racemic mixture)

Addition of hydrogen cyanide proceeds by way of cyanide ion. Because HCN is a weak acid, pK_a 9.31, the concentration of cyanide ion in aqueous HCN is too low for cyanohydrin formation to proceed at a reasonable rate. For this reason, cyanohydrin formation is generally carried out by dissolving NaCN or KCN in water and adjusting the pH of the solution to approximately 10.0, giving a solution in which HCN and CN^- are present in comparable concentrations.

Mechanism 16.4

Formation of a Cyanohydrin

Step 1: **Make a new bond between a nucleophile and an electrophile.** Nucleophilic addition of cyanide ion to the carbonyl carbon gives a tetrahedral carbonyl addition compound.

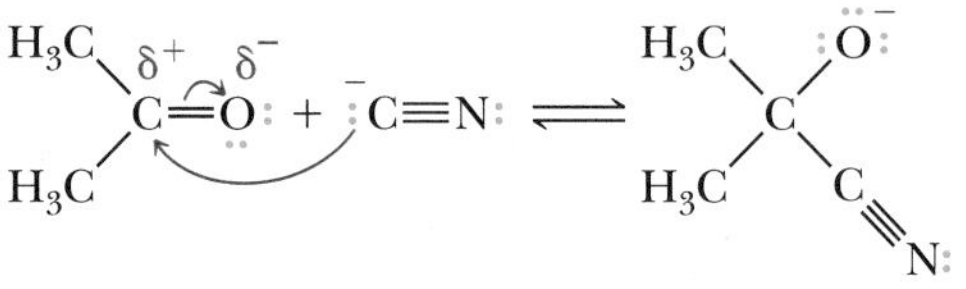

Step 2: **Add a proton.** Proton transfer from HCN gives the cyanohydrin and generates a new cyanide ion.

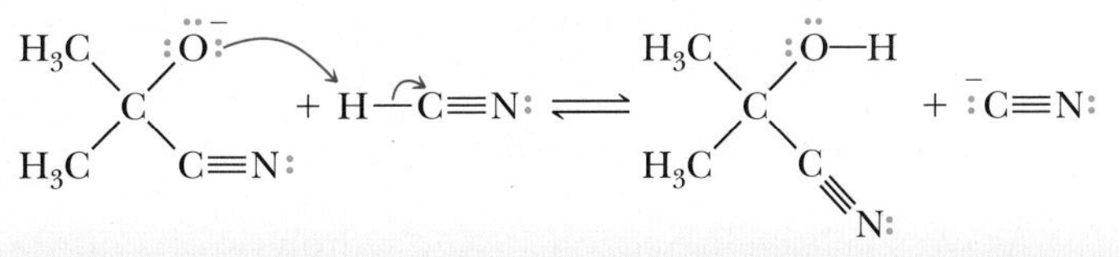

For aldehydes and most aliphatic ketones, the position of equilibrium favors cyanohydrin formation. For many aryl ketones (ketones in which the carbonyl carbon is bonded to a benzene ring) and sterically hindered aliphatic ketones, however, the position of equilibrium favors starting materials; cyanohydrin formation is not a useful reaction for these types of compounds. The following synthesis of ibuprofen, for example, failed because the cyanohydrin was formed only in low yield.

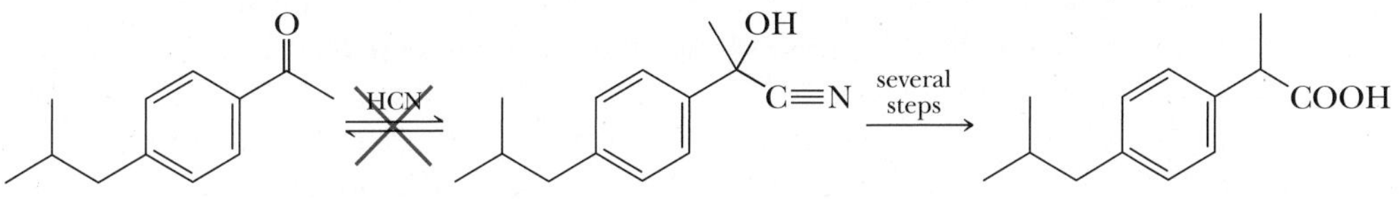

4-Isobutylacetophenone A cyanohydrin **Ibuprofen** (racemic)

Benzaldehyde cyanohydrin (mandelonitrile) provides an interesting example of a chemical defense mechanism in the biological world. This substance is synthesized by millipedes (*Apheloria corrugata*) and stored in special glands. When a millipede is threatened, the cyanohydrin is released from the storage gland and undergoes enzyme-catalyzed reversal of cyanohydrin formation to produce HCN, which is then released to ward off predators. The quantity of HCN emitted by a single millipede is sufficient to kill a small mouse. Mandelonitrile is also found in bitter almonds and peach pits. Its function there is unknown, as is how millipedes survive exposure to hydrogen cyanide.

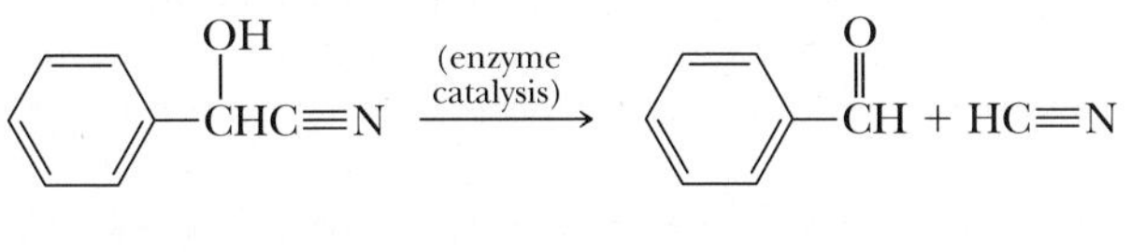

Benzaldehyde cyanohydrin (Mandelonitrile) **Benzaldehyde**

The value of cyanohydrins as synthetic intermediates lies in the new functional groups into which they can be converted. First, the secondary or tertiary hydroxyl group of the cyanohydrin may undergo acid-catalyzed dehydration to form an unsaturated nitrile. For example, acid-catalyzed dehydration of acetaldehyde cyanohydrin gives acrylonitrile, the monomer from which polyacrylonitrile (Orlon, Table 29.1) is made.

$$CH_3CH(OH)C{\equiv}N \xrightarrow{\text{acid catalyst}} CH_2{=}CHC{\equiv}N + H_2O$$

2-Hydroxypropanenitrile (Acetaldehyde cyanohydrin) **Propenenitrile (Acrylonitrile)**

Second, a nitrile is reduced to a primary amine by hydrogen in the presence of nickel or another transition metal catalyst (Section 18.10C). Catalytic reduction of benzaldehyde cyanohydrin, for example, gives 2-amino-1-phenylethanol.

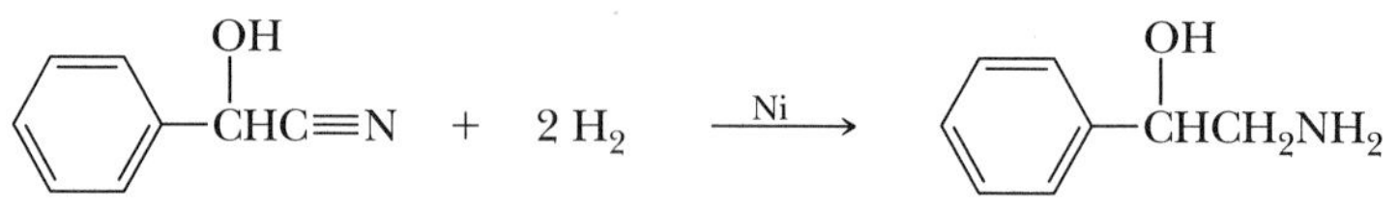

Benzaldehyde cyanohydrin **2-Amino-1-phenylethanol**

As we shall see in Section 18.4E, hydrolysis of a nitrile in the presence of an acid catalyst gives a carboxylic acid. Thus, even though nitriles are little used themselves, they are valuable intermediates for the synthesis of other functional groups.

16.6 The Wittig Reaction

Ylide
A neutral molecule with positive and negative charges on adjacent atoms.

In 1954, Georg Wittig reported a method for the synthesis of alkenes from aldehydes and ketones using compounds called phosphonium **ylides**. For his pioneering study and development of this reaction into a major synthetic tool, Professor Wittig shared the 1979 Nobel Prize in Chemistry. (The other recipient was Herbert C. Brown for his studies of hydroboration and the chemistry of organoboron compounds.) A Wittig synthesis is illustrated by the conversion of cyclohexanone to methylenecyclohexane. In this reaction, a C═O double bond is converted to a C═C double bond. A noteworthy aspect of this reaction is that a strong thermodynamic driving force is provided by formation of the very strong P—O bonding interaction in the phosphine oxide product.

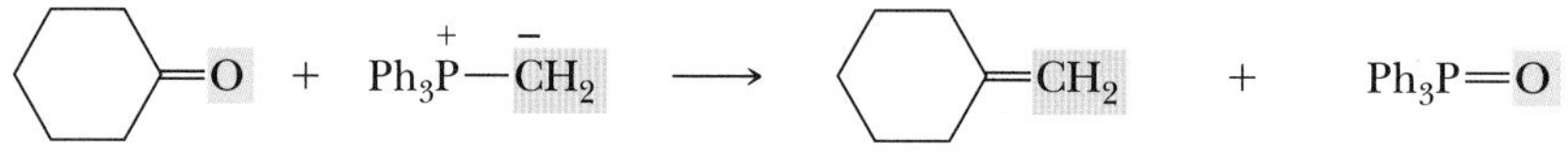

Cyclohexanone A phosphonium ylide **Methylenecyclohexane** **Triphenylphosphine oxide**

We study the **Wittig reaction** in two stages: first, the formation and structure of phosphonium ylides and, second, the reaction of a phosphonium ylide with the carbonyl group of an aldehyde or a ketone to give an alkene. The Wittig reaction is especially valuable as a synthetic tool because it takes place under mild conditions and because the location of the carbon-carbon bond is unambiguously determined. The only disadvantage of the Wittig reaction is that it is subject to steric hindrance. Yields are generally highest with aldehydes that have the least hindered carbonyl group and are lower with ketones in which the carbonyl group is more hindered.

Phosphorus is the second element in Group 5A of the Periodic Table and, like nitrogen, has five electrons in its valence shell. Examples of trivalent phosphorus compounds are phosphine, PH_3, and triphenylphosphine, Ph_3P. Phosphine is a highly toxic, flammable gas. Triphenylphosphine is a colorless, odorless solid. Because phosphorus is below nitrogen in the Periodic Table, phosphines are weaker bases than amines and good nucleophiles (Section 9.3E). Treatment of a phosphine with a methyl, primary, or secondary alkyl halide gives a phosphonium salt by an S_N2 pathway.

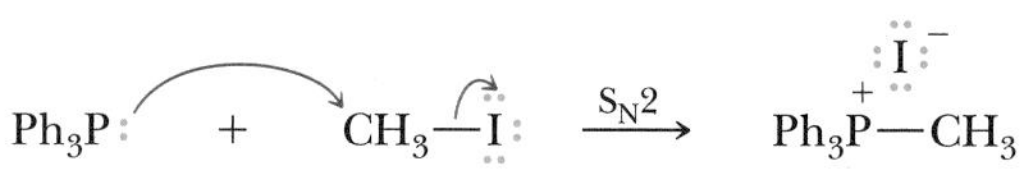

Triphenylphosphine **Methyltriphenylphosphonium iodide** (an alkyltriphenylphosphonium salt)

Because phosphines are also weak bases, treatment of a tertiary halide with a phosphine gives largely an alkene by an E2 pathway.

α-Hydrogen atoms on the alkyl group of an alkyltriphenylphosphonium ion are weakly acidic and can be removed by reaction with a very strong base, typically butyllithium (BuLi), sodium hydride (NaH), or sodium amide ($NaNH_2$).

$$CH_3CH_2CH_2CH_2^{-}\,Li^+ + H{-}CH_2{-}\overset{+}{P}Ph_3\,I^- \longrightarrow CH_3CH_2CH_2CH_3 + {}^{-}\!:CH_2{-}\overset{+}{P}Ph_3 + LiI$$

Butyllithium An alkyltriphenylphosphonium iodide **Butane** A phosphonium ylide

The product of removal of a proton from an alkyltriphenylphosphonium ion is a phosphonium ylide. The important feature of a phosphonium ylide is that the deprotonated carbon atom bears considerable partial negative charge, making it a strong carbon-based nucleophile, analogous to species such as Grignard and organolithium reagents. Just like Grignard and organolithium reagents, the deprotonated carbon of phosphonium ylides readily reacts with the electrophilic carbon atom of aldehyde and ketone carbonyl groups.

Mechanism 16.5

The Wittig Reaction

Step 1: **Make a new bond between a nucleophile and an electrophile.** Reaction of a nucleophilic phosphonium ylide with the electrophilic carbonyl carbon of an aldehyde or a ketone gives a dipolar intermediate called a **betaine**.

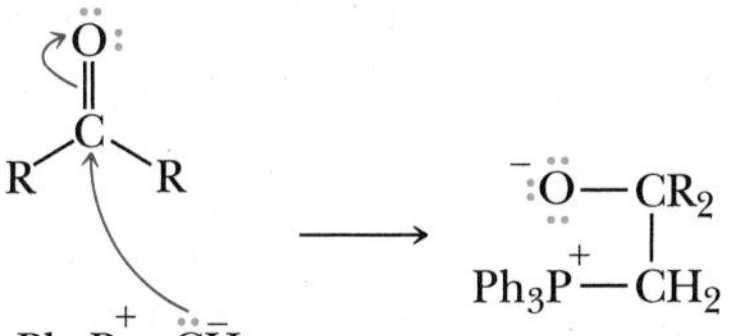

A betaine

Step 2: **The betaine collapses to a four-membered oxaphosphetane ring.** The name for this four-membered ring system is derived by combining the following: *oxa-* shows that it contains oxygen, *-phosph-* shows that it contains phosphorus, *-et-* shows that it is a four-membered ring, and *-ane* shows only carbon-carbon single bonds in the ring. Oxaphosphetanes can be isolated at low temperature.

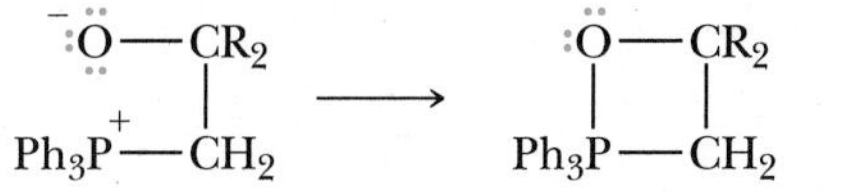

An oxaphosphetane

Step 3: **Break bonds to give stable molecules or ions.** Decomposition of the oxaphosphetane gives triphenylphosphine oxide and an alkene.

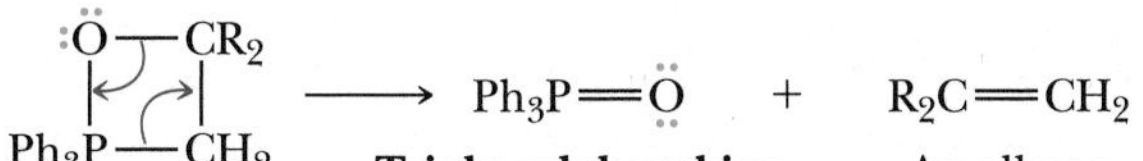

Triphenylphosphine oxide An alkene

The driving force for a Wittig reaction is the formation of the very strong phosphorus-oxygen bond in triphenylphosphine oxide.

The Wittig reaction is effective with a wide variety of aldehydes and ketones and with ylides derived from a wide variety of primary, secondary, and allylic halides as shown by the following examples.

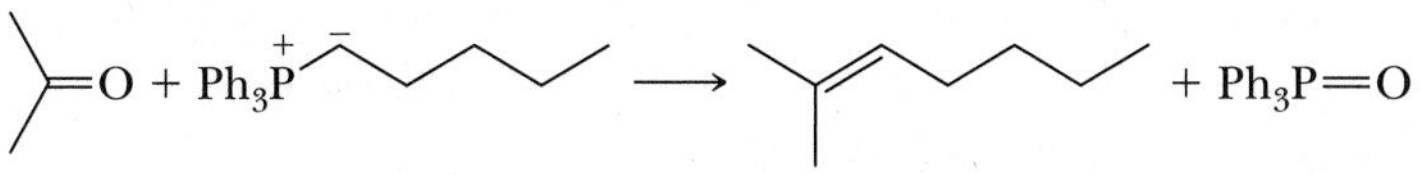

Acetone **2-Methyl-2-heptene**

Betaine

A neutral molecule with nonadjacent positive and negative charges. An example of a betaine is the intermediate formed by addition of a Wittig reagent to an aldehyde or a ketone.

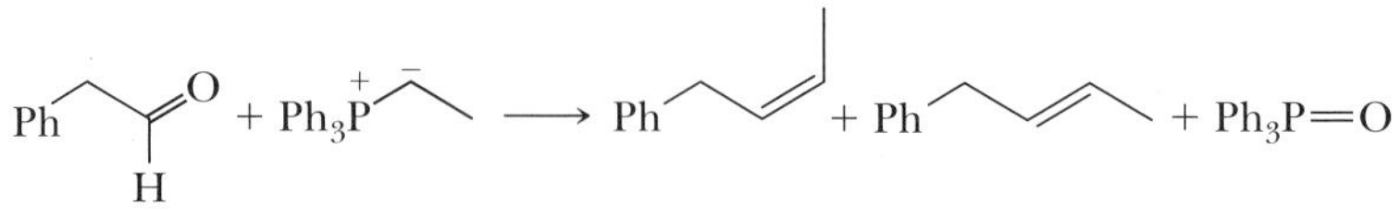

Phenylacetaldehyde **(Z)-1-Phenyl-2-butene** (87%) **(E)-1-Phenyl-2-butene** (13%)

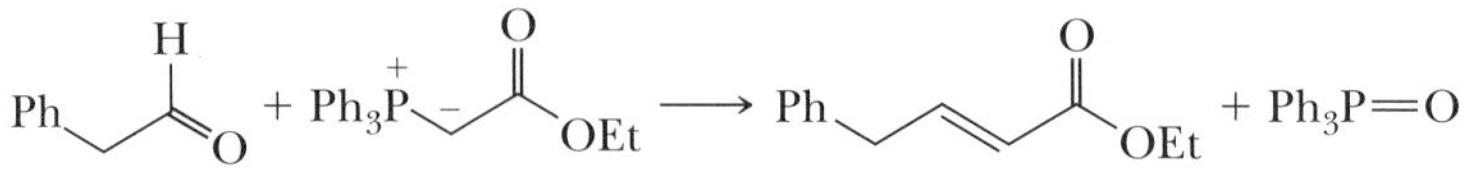

Phenylacetaldehyde **Ethyl (E)-4-phenyl-2-butenoate** (only the *E* isomer is formed)

As illustrated by the second and third examples, some Wittig reactions are *Z* selective, while others are *E* selective. As a general rule, those Wittig reagents with anion-stabilizing substituents, such as a carbonyl group, adjacent to the negative charge are *E* selective. We refer to these ylides as being stabilized. We refer to ylides without an adjacent anion-stabilizing group as being unstabilized; unstabilized ylides are *Z* selective. We can write the following resonance contributing structures for a carbonyl-stabilized ylide.

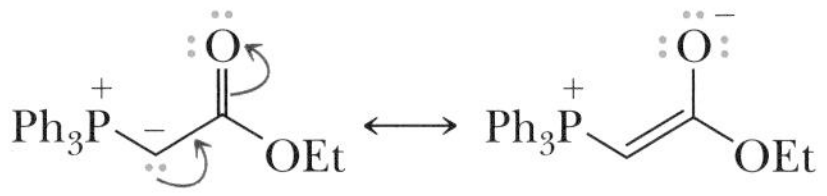

Resonance contributing structures for an ylide stabilized by an adjacent carbonyl group

Stabilization of the ylide through resonance decreases its reactivity, allowing an equilibrium to be established during the product-determining step that favors the more stable *E* isomer.

Because the Wittig reaction is so useful for the preparation of alkenes, chemists have explored several variations of it. One of the most useful of these, known as the Horner-Emmons-Wadsworth modification, uses a phosphonate ester derived from an α-haloester or a ketone to generate the Wittig carbanion.

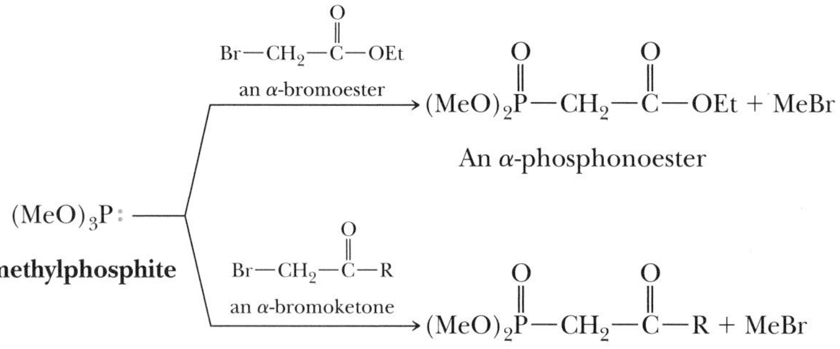

Trimethylphosphite An α-phosphonoester An α-phosphonoketone

The α-phosphonoesters or ketones used in this variation of the Wittig reaction are formed by two successive S_N2 reactions. Trimethylphosphite is an excellent nucleophile and readily displaces bromine from an α-bromoester or α-bromoketone by an S_N2 reaction. Bromide ion then is the nucleophile in the second S_N2 reaction that generates the α-phosphonoester.

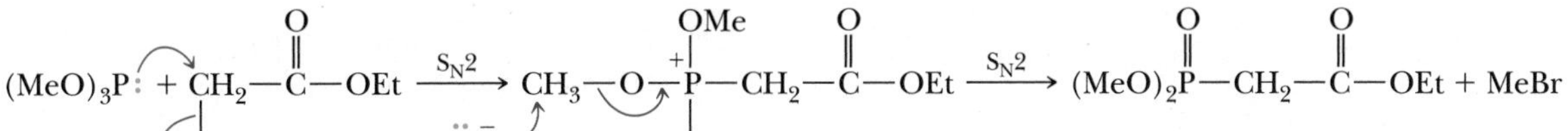

An α-phosphonoester

Treatment of a phosphonoester with a strong base followed by an aldehyde or a ketone gives an alkene, in this case either an α,β-unsaturated ester, a ketone, or an aldehyde. A particular advantage of using a phosphonate-stabilized carbanion as the Wittig reagent is that the resulting alkene is either entirely or almost entirely the *E* isomer; that is, phosphonate-stabilized Wittig reagents are almost exclusively *E* selective. Another advantage of phosphonate-stabilized ylides is that the by-product, dimethylphosphate anion, is water-soluble and therefore easily separated from the desired organic product.

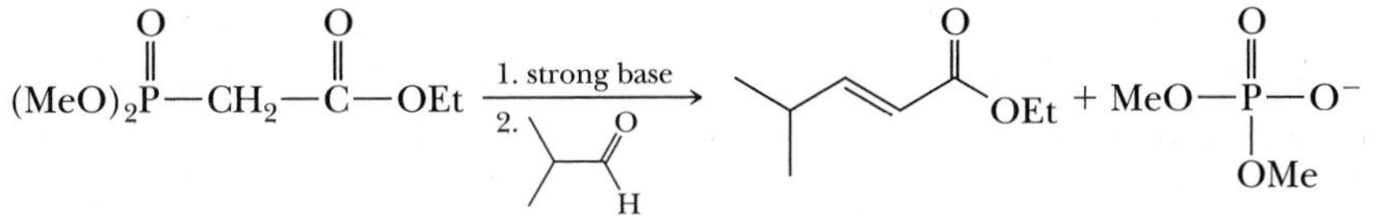

Only the *E* isomer is formed **Dimethylphosphate anion**

Example 16.5 Reactions of Wittig Reagents

Show how this alkene can be synthesized by a Wittig reaction.

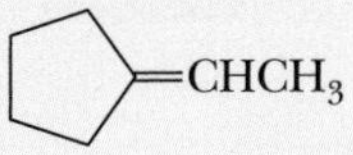

Solution

Starting materials are either cyclopentanone and the triphenylphosphonium ylide derived from bromoethane or acetaldehyde and the triphenylphosphonium ylide derived from chlorocyclopentane. Either route is satisfactory.

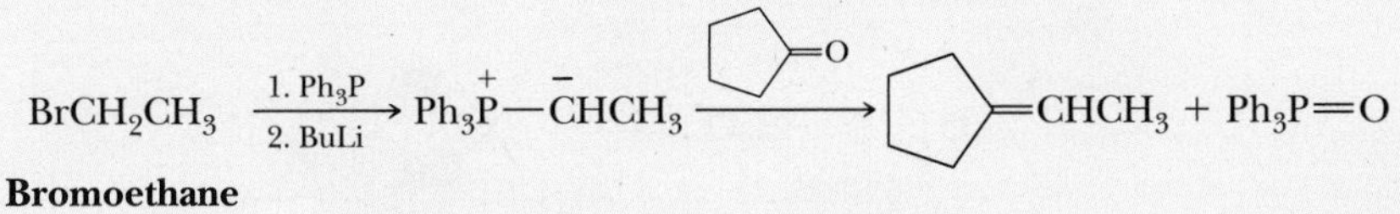

Bromoethane

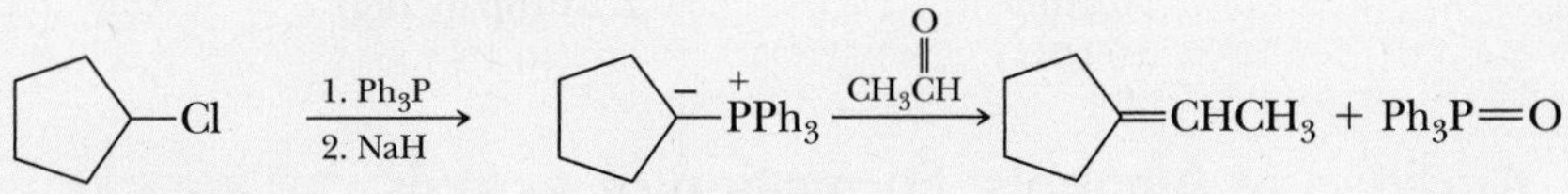

Chlorocyclopentane

Problem 16.5

Show how each alkene can be synthesized by a Wittig reaction (there are two routes to each).

(a) 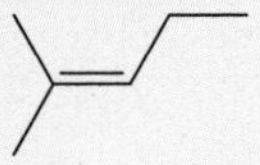 **(b)** 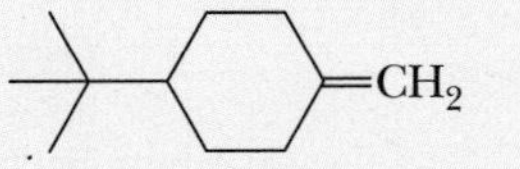

Watch a video explanation

16.7 Addition of Oxygen Nucleophiles

A. Addition of Water: Formation of Carbonyl Hydrates

Nucleophilic acyl addition of water (hydration) to a carbonyl group of an aldehyde or a ketone forms a geminal diol, commonly abbreviated gem-diol.

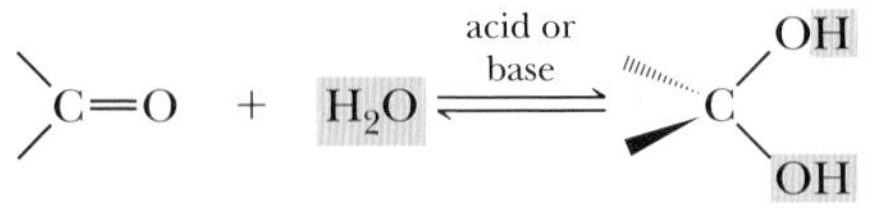

Carbonyl group of an aldehyde or ketone — A hydrate (a gem-diol)

A gem-diol is commonly referred to as the hydrate of the corresponding aldehyde or ketone. These compounds are unstable and are rarely isolated. This reaction is catalyzed by acids and bases. The mechanism is identical to that for the addition of alcohols, which is discussed next.

Hydration of an aldehyde or a ketone is readily reversible, and the diol can eliminate water to regenerate the aldehyde or ketone. In most cases, equilibrium strongly favors the carbonyl group. For a few simple aldehydes, however, the hydrate is favored. For example, when formaldehyde is dissolved in water at 20°C, the position of equilibrium is such that it is more than 99% hydrated.

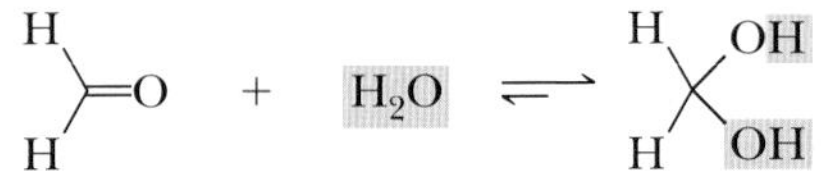

Formaldehyde — **Formaldehyde hydrate** (>99%)

A 37% solution of formaldehyde in water, called formalin, is commonly used to preserve biological specimens.

In contrast, an aqueous solution of acetone consists of less than 0.1% of the hydrate at equilibrium.

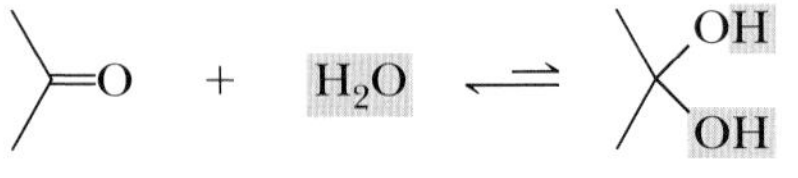

Acetone (99.9%) — **2,2-Propanediol** (0.1%)

B. Addition of Alcohols: Formation of Acetals

Hemiacetal

A molecule containing an —OH and an —OR or —OAr group bonded to the same carbon.

Alcohols add to aldehydes and ketones in the same manner as described for water. Nucleophilic acyl addition of one molecule of alcohol to the carbonyl group of an aldehyde or a ketone forms a **hemiacetal** (a half-acetal).

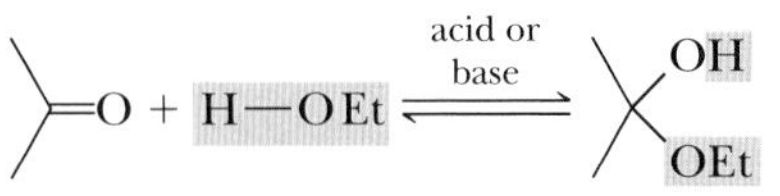

A hemiacetal

The functional group of a hemiacetal is a carbon bonded to an —OH group and an —OR group.

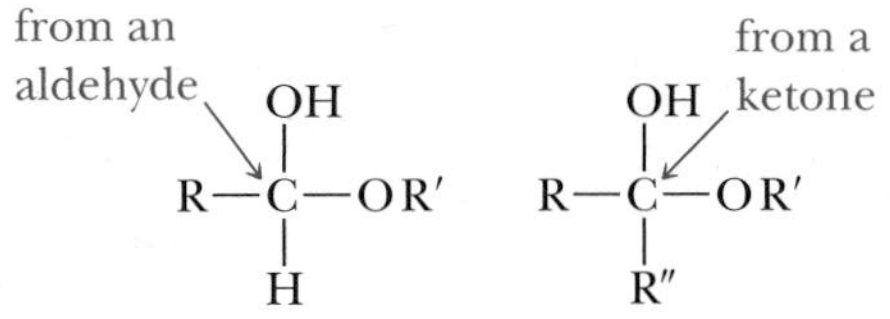

Hemiacetals

Hemiacetals are generally unstable and are only minor components of an equilibrium mixture, except in one very important type of compound. When a hydroxyl group is part of the same molecule that contains the carbonyl group and a five- or six-membered ring can form, the compound exists almost entirely in the cyclic hemiacetal form. Recall that five- and six-membered rings have relatively little ring strain (Section 6.6B). In the following example, (*S*)-4-hydroxypentanal already has a chiral center and a new chiral center is created upon formation of the hemiacetal.

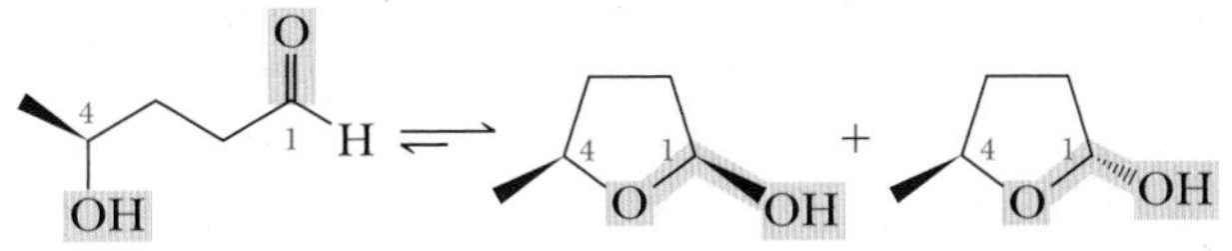

(*S*)-4-Hydroxypentanal

Cyclic hemiacetals
(major forms present at equilibrium)

Because of the original chiral center, the product hemiacetals in this example are diastereomers, not enantiomers, and are not necessarily produced in equal amounts. ▶

▶ **Five- and six-membered rings are the most stable.**

Simple carbohydrates, all of which are polyhydroxyaldehydes or polyhydroxyketones, exist in solution predominantly as cyclic hemiacetals. Because carbohydrates have several hydroxyl groups, they could potentially form rings of different sizes. Generally, only five- and six-membered cyclic hemiacetals (the most strain-free types of rings) are produced to an appreciable extent. The new chiral center created in carbohydrate cyclic hemiacetals can have either an *R* or *S* configuration. The carbon atom at the new chiral center of a carbohydrate cyclic hemiacetal is given the special name of **anomeric carbon** and corresponds to the carbonyl carbon atom in the open chain form. The two different cyclic hemiacetals are called **anomers**, and the configuration of each is designated as α or β depending on whether the hemiacetal —OH group is on the same side of the ring as the terminal —CH_2OH substituent (β anomer) or on the opposite side (α anomer). Note that owing to the presence of multiple chiral centers, the anomers are diastereomers, not enantiomers.

D-Glucose, the most important carbohydrate in mammalian metabolism, exists as a six-membered cyclic hemiacetal form as both α and β anomers.

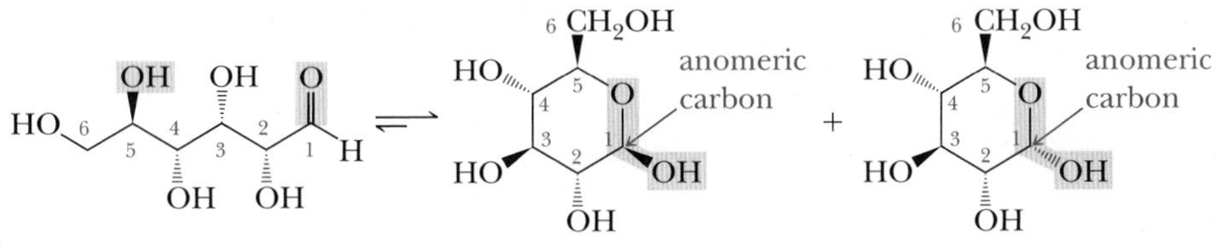

D-Glucose
(open chain form)

β Anomer of D-glucose
cyclic hemiacetal
(predominates at equilibrium)

α Anomer of D-glucose
cyclic hemiacetal

At equilibrium, the β anomer of D-glucose predominates, because the —OH group of the anomeric carbon is in the more stable equatorial position of the more stable chair conformation. In α-D-glucose, the —OH group on the anomeric carbon is axial. When remembering the names of D-glucose anomers, some students find it helpful to remember the phrase *alpha is axial.*

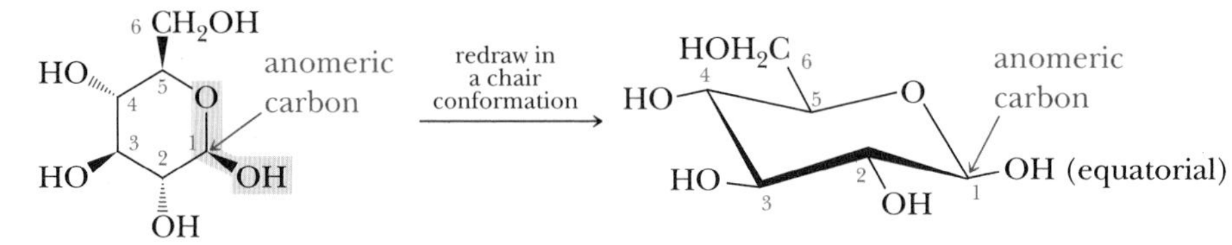

β Anomer of D-glucose
cyclic hemiacetal

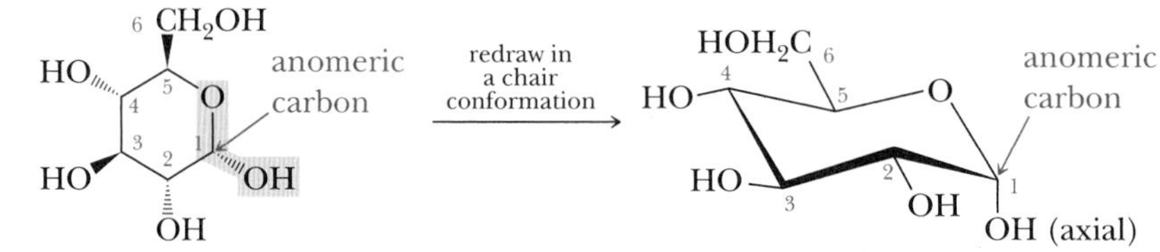

α Anomer of D-glucose
cyclic hemiacetal

We discuss the chemistry of carbohydrate cyclic hemiacetals in more detail in Chapter 25.

Formation of hemiacetals is catalyzed by bases such as hydroxide or alkoxide. The function of the catalyst is to remove a proton from the alcohol, making it a better nucleophile.

Mechanism 16.6

Base-Catalyzed Formation of a Hemiacetal

Step 1: **Take a proton away.** Proton transfer from HOR to the base gives the alkoxide.

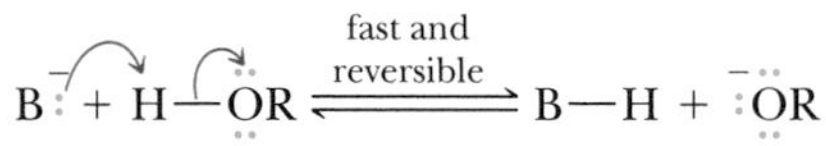

Step 2: **Make a new bond between a nucleophile and an electrophile.** Attack of RO^- on the carbonyl carbon gives a tetrahedral carbonyl addition compound.

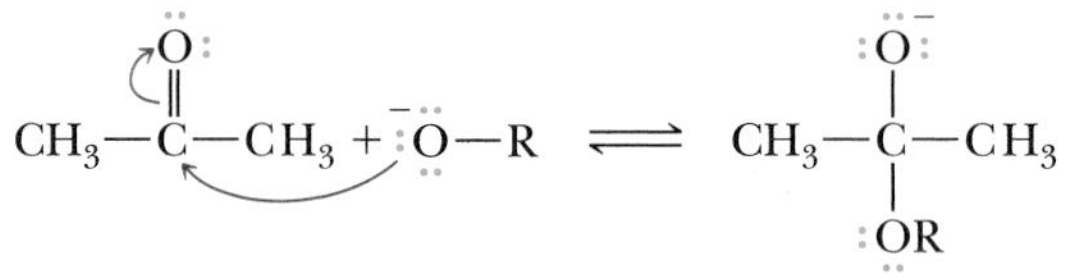

A tetrahedral carbonyl
addition compound

Step 3: **Add a proton.** Proton transfer from the alcohol to the O^- gives the hemiacetal and regenerates the base catalyst.

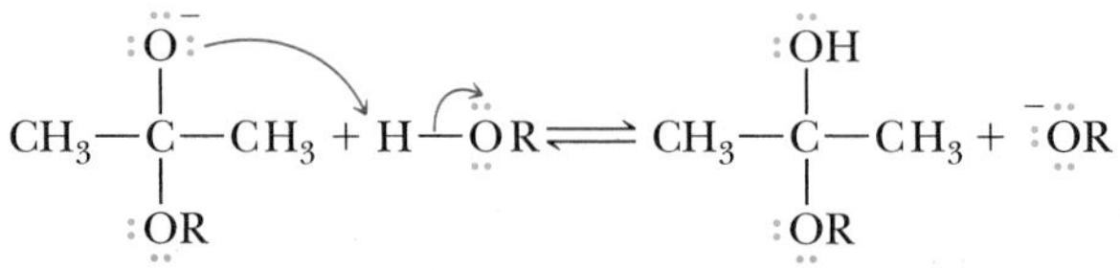

Formation of hemiacetals can also be catalyzed by acid, most commonly sulfuric acid, *p*-toluenesulfonic acid, or hydrogen chloride.

Mechanism 16.7

Acid-Catalyzed Formation of a Hemiacetal

Step 1: **Add a proton.** Using a strong acid such as HCl or sulfuric acid sets up an equilibrium in which the protonated alcohol is preferred. The protonated alcohol is therefore the proton source for subsequent steps in the hemiacetal formation reaction.

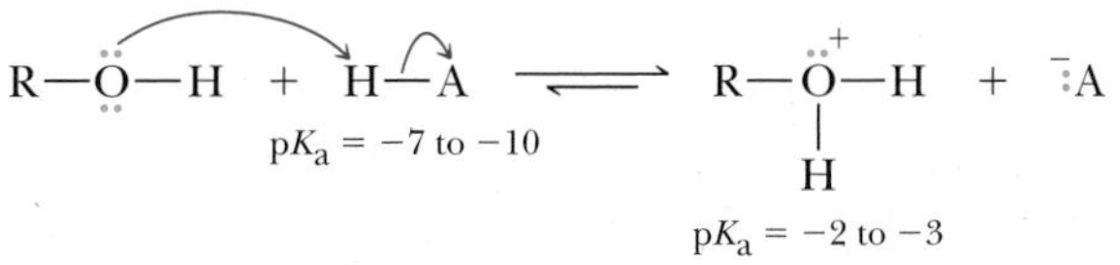

Step 2: **Add a proton.** Proton transfer from the protonated alcohol to the carbonyl oxygen gives the conjugate acid of the aldehyde or ketone.

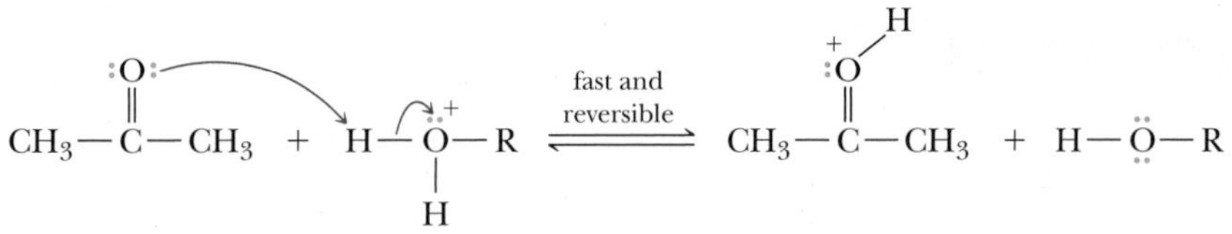

In this way, the acid catalyst functions to protonate the carbonyl oxygen and thus renders the carbonyl carbon more electrophilic and more susceptible to attack by the weakly nucleophilic oxygen atom of the alcohol.

Step 3: **Make a new bond between a nucleophile and an electrophile.** Attack of ROH on the carbonyl carbon gives a tetrahedral carbonyl addition compound.

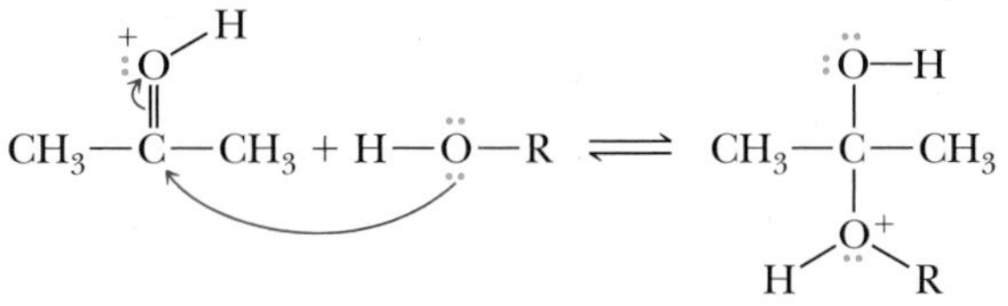

A tetrahedral carbonyl addition compound

Step 4: **Take a proton away.** Proton transfer from the oxonium ion to an alcohol molecule gives the hemiacetal and regenerates the acid catalyst.

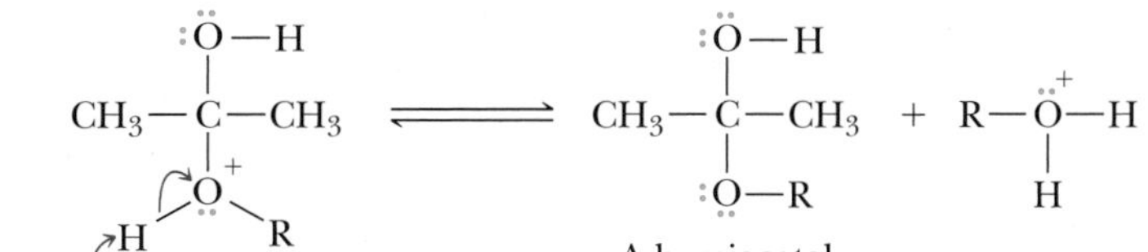

A hemiacetal

Hemiacetals are often not stable relative to starting materials, but they react further with alcohols to form **acetals** and a molecule of water. Acetals are considerably more stable than hemiacetals and can be isolated in good yield under the proper conditions.

Acetal A molecule containing two —OR or —OAr groups bonded to the same carbon.

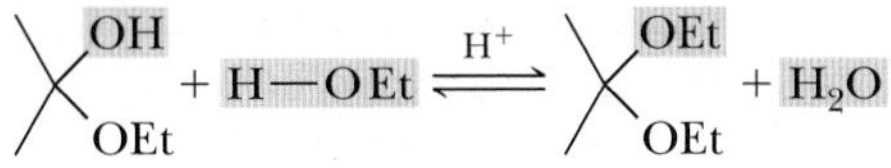

A hemiacetal A diethyl acetal

The formation of acetals and its reverse is catalyzed by acids, not by bases, because the OH group cannot be displaced directly by nucleophiles.

The functional group of an acetal is a carbon bonded to two —OR groups.

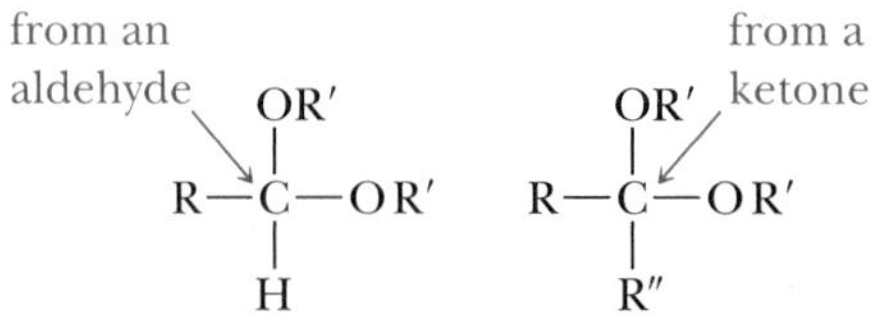

Acetals

The mechanism for the acid-catalyzed conversion of a hemiacetal to an acetal is divided into four steps. As you study this mechanism, note that acid is a true catalyst in this reaction. The protonated alcohol is used to add a proton in Step 1, but another protonated alcohol is generated in Step 4. The latter steps of this mechanism are very similar to those for hemiacetal formation.

Mechanism 16.8

Acid-Catalyzed Formation of an Acetal

Step 1: **Add a proton.** Proton transfer from the protonated alcohol to the hemiacetal **OH** group gives an oxonium ion.

Watch a video explanation

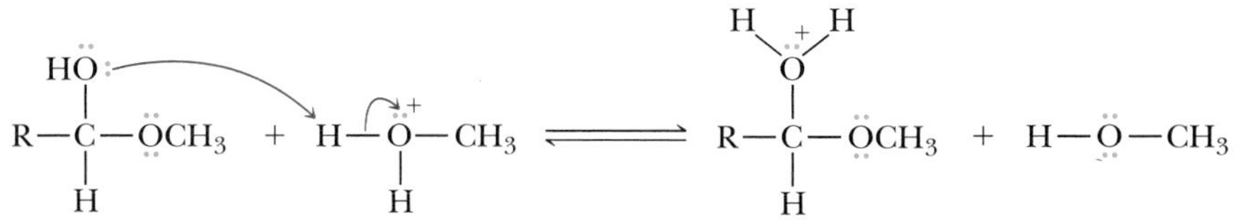

A hemiacetal | An oxonium ion

Step 2: **Break a bond to give stable molecules or ions.** Loss of water gives a new, resonance-stabilized cation.

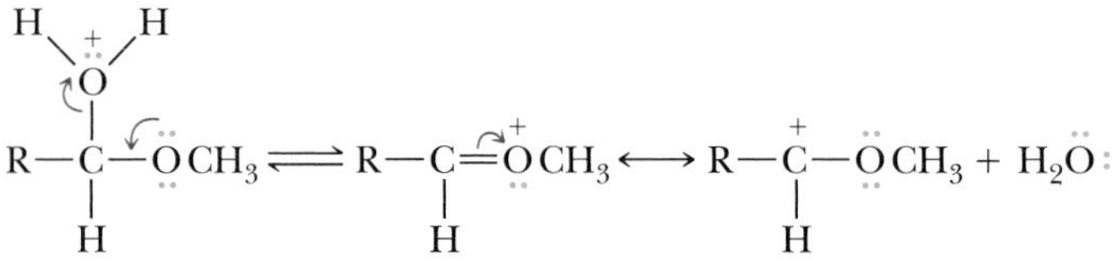

A resonance-stabilized cation

Step 3: **Make a new bond between a nucleophile and an electrophile.** Reaction of the resonance-stabilized cation (an electrophile) with methanol (a nucleophile) gives the conjugate acid of the acetal.

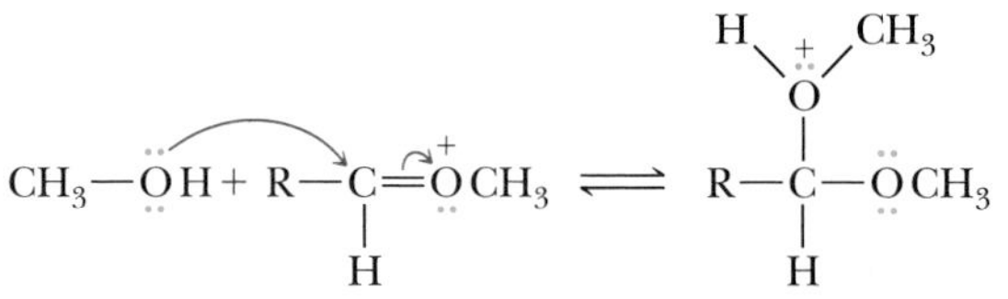

A protonated acetal

Step 4: **Take a proton away.** Proton transfer from the protonated acetal to alcohol gives the acetal and generates a new molecule of the acid catalyst.

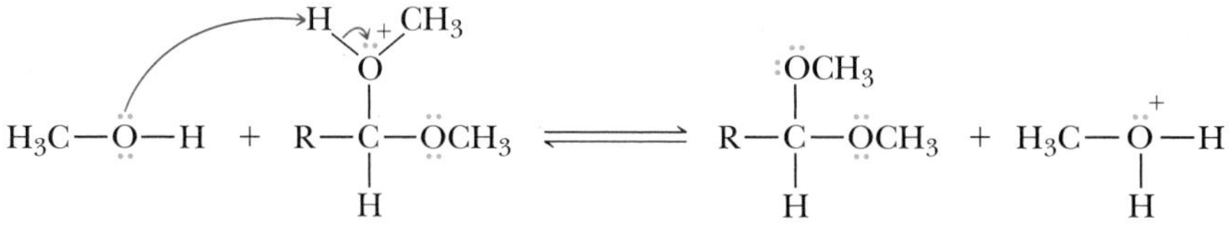

An acetal

Formation of acetals is often carried out using the alcohol as the solvent and dissolving either dry **HCl** (hydrogen chloride gas) or *p*-toluenesulfonic acid in the alcohol. Because the alcohol is both a reactant and solvent, it is present in large molar excess, which forces the equilibrium to the right and favors acetal formation. Note that this reaction is completely reversible. Addition of excess water to an acetal causes hydrolysis to the ketone.

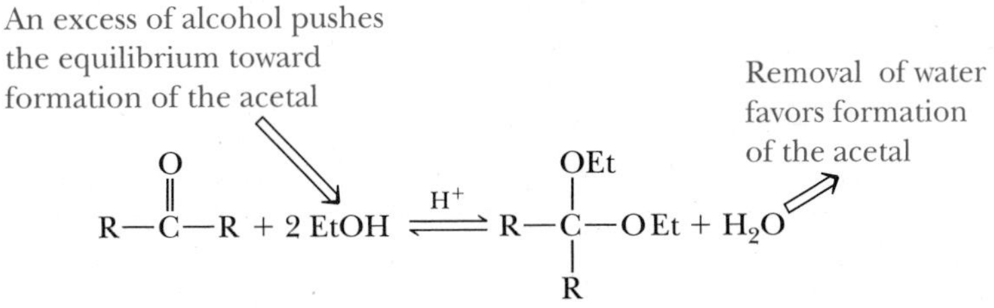

A diethyl acetal

In another experimental technique to force the equilibrium to the right, water is removed from the reaction vessel as an **azeotrope** by distillation using a **Dean-Stark trap** (Figure 16.1). In this method for preparing an acetal, the aldehyde or ketone, alcohol, acid catalyst, and benzene are brought to reflux. The component in this mixture with the lowest boiling point is an azeotrope, bp 69°C, consisting of 91% benzene and 9% water. This vapor is condensed and collected in a side trap, where it separates into two layers. At room temperature, the composition of the upper, less dense layer is 99.94% benzene and 0.06% water. The composition of the lower, more dense layer is almost the reverse, 0.07% benzene and 99.93% water. As reflux continues, benzene from the top layer is returned to the refluxing mixture and water is drawn off at the bottom through a stopcock. A Dean-Stark trap "pumps" water out of the reaction mixture, thus forcing the equilibrium to the right. This same apparatus is used in many other reactions where water needs to be removed, such as formation of enamines (Section 19.5A).

Azeotrope

A liquid mixture of constant composition with a boiling point that is different from that of any of its components.

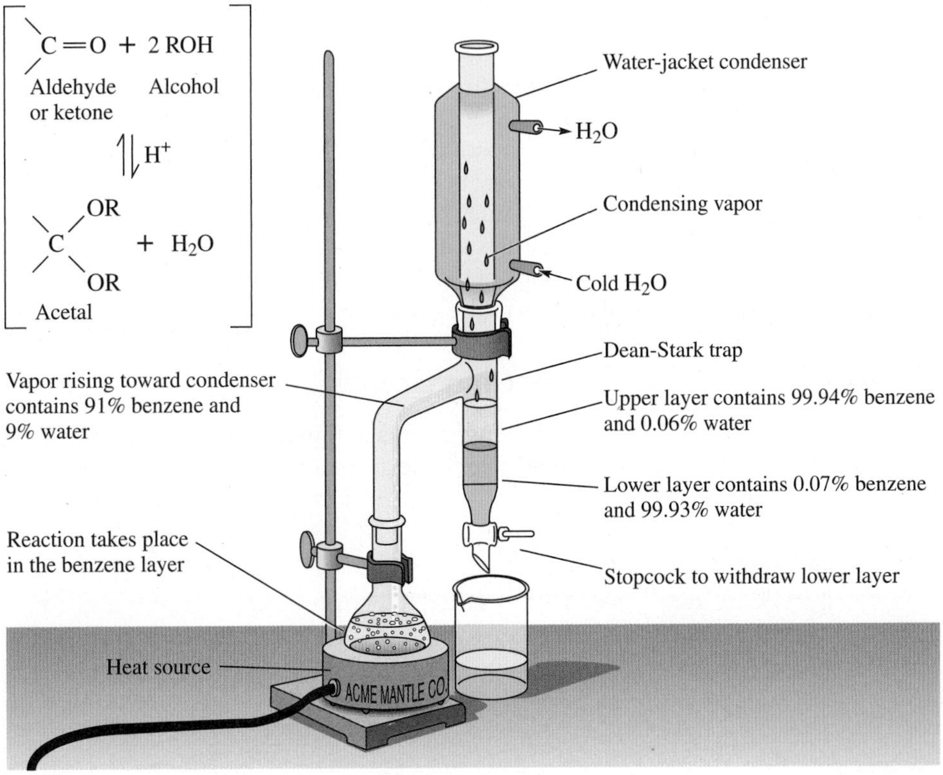

FIGURE 16.1 A Dean-Stark trap for removing water by azeotropic distillation with benzene. Toluene or xylene can be used if a higher reaction temperature is desired.

Example 16.6 Formation of an Acetal

Show the reaction of the carbonyl group of each aldehyde or ketone with one molecule of alcohol to give a hemiacetal and then with a second molecule of alcohol to give an acetal.

(a) 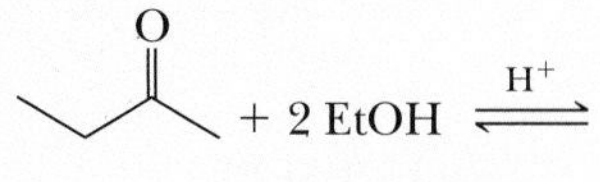 **(b)** 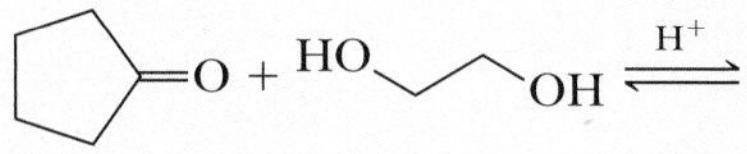

Note that in part (b), ethylene glycol is a diol, and one molecule of it provides both —OH groups.

Solution

Given are structural formulas of the hemiacetal and then the acetal.

(a) 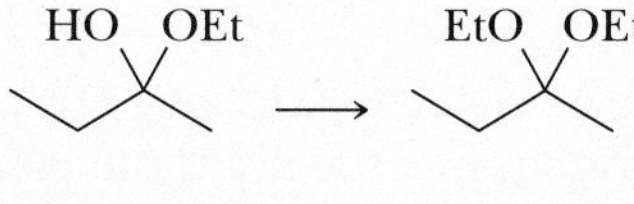

Hemiacetal   Acetal

(b) 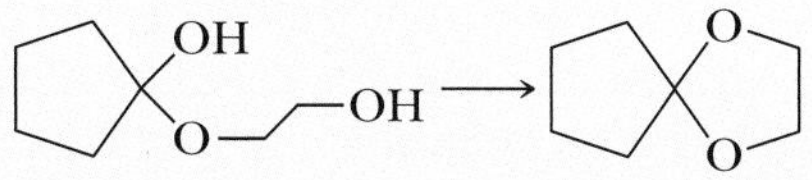

Hemiacetal   Cyclic acetal

Problem 16.6

Hydrolysis of an acetal in aqueous acid gives an aldehyde or a ketone and two molecules of alcohol or one molecule of a diol. Draw the structural formulas for the products of hydrolysis of the following acetals in aqueous acid.

(a) 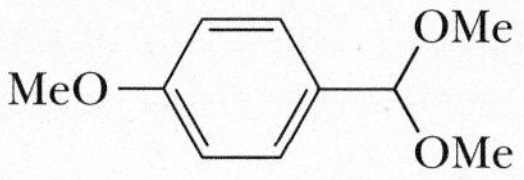 **(b)** 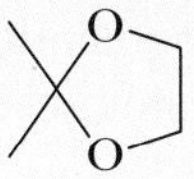 **(c)** 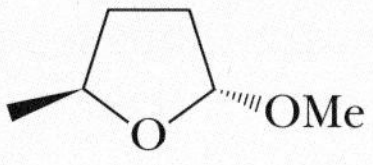

Like ethers (Section 11.5), acetals are unreactive to bases, hydride-reducing agents such as $LiAlH_4$ and $NaBH_4$, Grignard and other organometallic reagents, oxidizing agents (except, of course, those involving the use of aqueous acid), and catalytic reduction. This lack of reactivity is because acetals have no sp^2 hybridized electrophilic carbon atom to react with nucleophiles. Because of their lack of reactivity toward these reagents and ready hydrolysis in aqueous acid, acetals are often used to reversibly "protect" the carbonyl groups of aldehydes and ketones while reactions are carried out on other functional groups in the molecule.

C. Acetals as Carbonyl-Protecting Groups

The use of acetals as carbonyl-protecting groups is illustrated by the synthesis of 5-hydroxy-5-phenylpentanal from benzaldehyde and 4-bromobutanal.

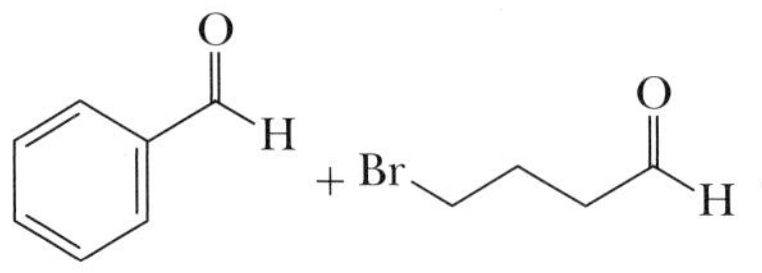

Benzaldehyde   **4-Bromobutanal**

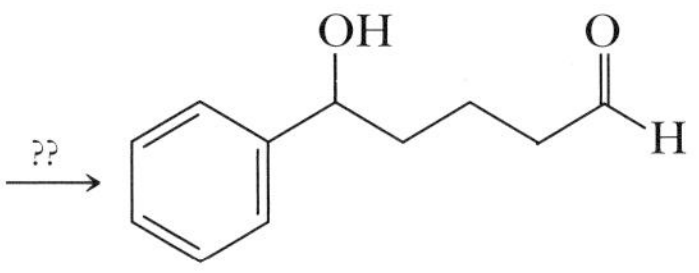

5-Hydroxy-5-phenylpentanal
(racemic)

One obvious way to form a new carbon-carbon bond between these two molecules is to treat benzaldehyde with the Grignard reagent from 4-bromobutanal. However, during preparation, a Grignard reagent formed from 4-bromobutanal reacts immediately with the carbonyl group of another molecule of 4-bromobutanal. As a result, no significant amount of the Grignard reagent is made. A way to avoid this problem is to protect the carbonyl group of the bromoaldehyde by conversion to an acetal. Cyclic acetals are often used because they are particularly easy to prepare.

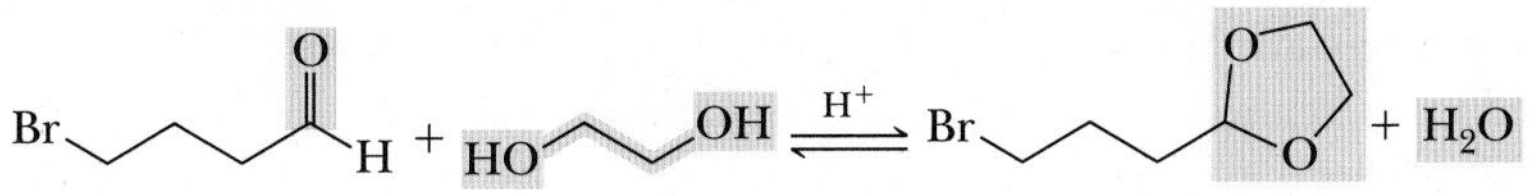

A cyclic acetal

Treatment of the protected bromoaldehyde with magnesium in diethyl ether followed by addition of benzaldehyde gives a chiral magnesium alkoxide as a racemic mixture.

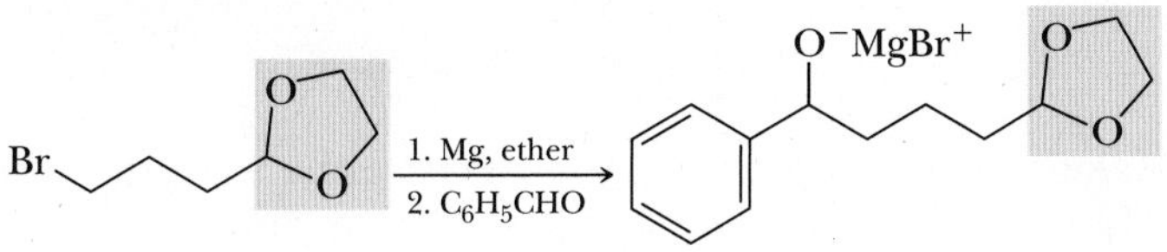

A cyclic acetal

A chiral magnesium alkoxide (produced as a racemic mixture)

Treatment of the magnesium alkoxide with aqueous acid accomplishes two things. First, protonation of the alkoxide anion gives the desired 2° hydroxyl group; second, hydrolysis of the cyclic acetal regenerates the carbonyl group of the aldehyde.

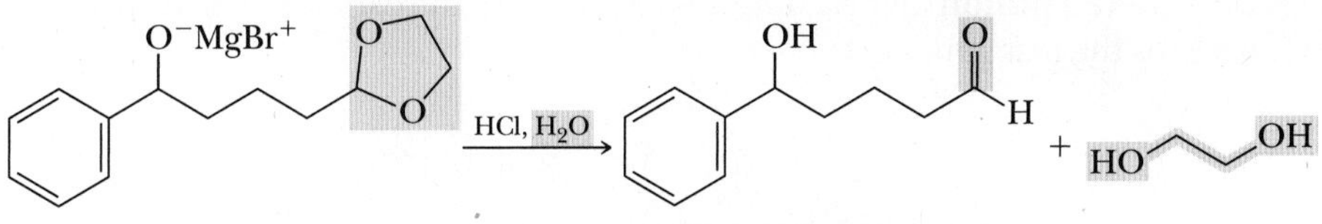

5-Hydroxy-5-phenylpentanal
(formed as a racemic mixture)

D. Tetrahydropyranyl Ethers: Protecting an Alcohol as an Acetal

We just saw in Section 16.7C that an aldehyde or a ketone can be protected by conversion to an acetal. A similar strategy can be used to protect a primary or secondary alcohol. Treatment of the alcohol with dihydropyran in the presence of an acid catalyst, commonly anhydrous HCl or a sulfonic acid, RSO_3H, converts the alcohol into a **tetrahydropyranyl** (THP) **ether**.

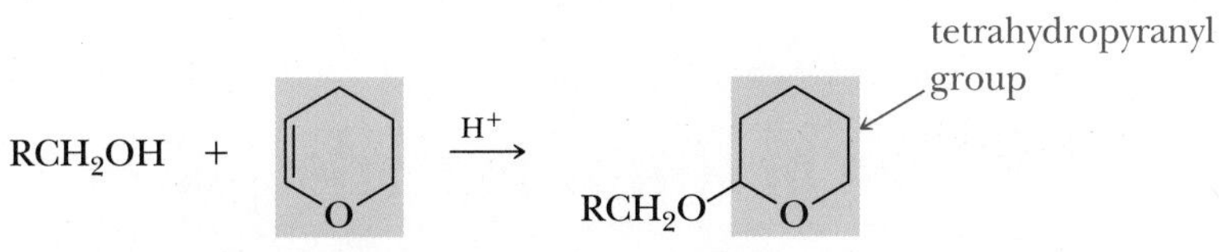

Dihydropyran

A tetrahydropyranyl ether

Because the THP group is an acetal, it is stable in neutral and basic solutions and to most oxidizing and reducing agents. It is removed easily by treatment with dilute aqueous acid to regenerate the original primary or secondary alcohol.

Example 16.7 Formation of a THP Ether

Write a mechanism for the formation of a THP ether from a primary alcohol RCH_2OH catalyzed by a sulfonic acid $ArSO_3H$.

Solution

***Step 1:* Make a new bond between a nucleophile (π bond) and an electrophile—add a proton.** Dihydropyran (a vinyl or enol ether) is weakly basic and is protonated to give a resonance-stabilized cation.

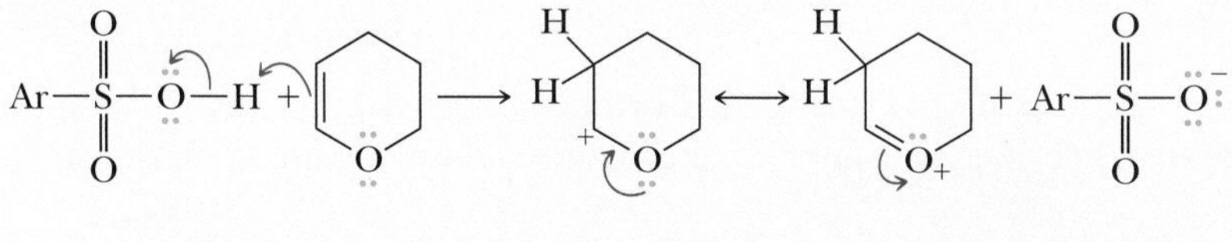

A resonance-stabilized cation

***Step 2:* Make a new bond between a nucleophile and an electrophile.** Reaction of the resonance-stabilized cation (an electrophile) with the alcohol (a nucleophile) gives an oxonium ion.

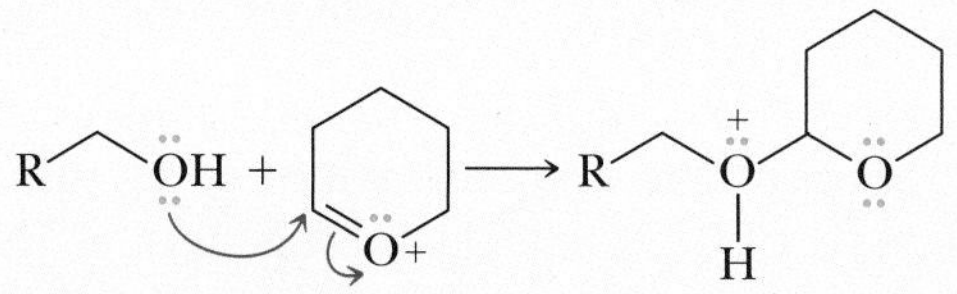

***Step 3:* Take a proton away.** Proton transfer from the oxonium ion to $ArSO_3^-$ completes the reaction.

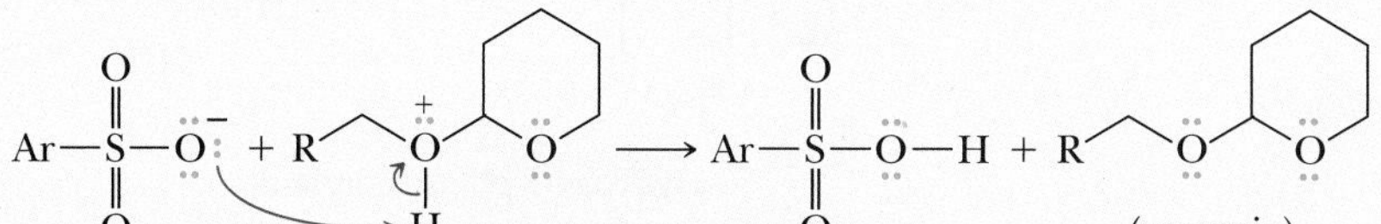

(racemic)

Problem 16.7

Write a mechanism for the acid-catalyzed hydrolysis of a THP ether to regenerate the original alcohol. Into what compound is the THP group converted?

16.8 Addition of Nitrogen Nucleophiles

Imine

A compound containing a carbon-nitrogen double bond, $R_2C{=}NR$; also called a Schiff base.

Schiff base

An alternative name for an imine.

A. Ammonia and Its Derivatives

Ammonia, primary aliphatic amines (RNH_2), and primary aromatic amines ($ArNH_2$) react with the carbonyl group of aldehydes and ketones to give an **imine**, often referred to as a **Schiff base**. Imines are usually unstable unless the C=N group is part of an extended system of conjugation (e.g., rhodopsin) and are generally not isolated.

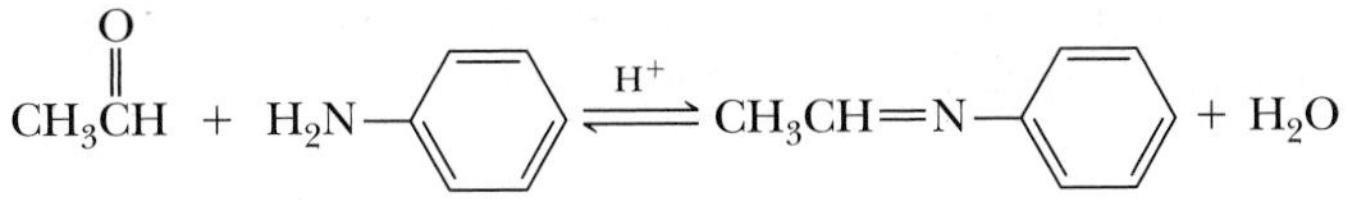

Acetaldehyde **Aniline** An imine (a Schiff base)

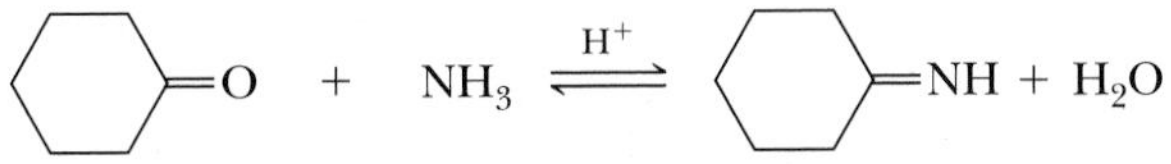

Cyclohexanone **Ammonia** An imine

Mechanism 16.9

Formation of an Imine from an Aldehyde or a Ketone

Step 1: **Make a new bond between a nucleophile and an electrophile.** Nucleophilic addition of a primary amine or ammonia gives a tetrahedral carbonyl addition compound with a negative charge on oxygen and a positive charge on nitrogen. Although this reaction is commonly run in acid as with the formation of an acetal, an alkylamine (not an arylamine, such as aniline) is nucleophilic enough to add directly to the carbonyl group without prior protonation.

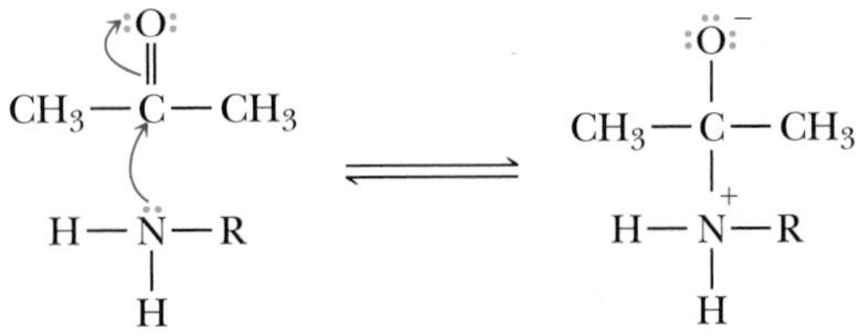

Step 2: **Take a proton away.** Another molecule of amine removes a proton to give a neutral nitrogen atom.

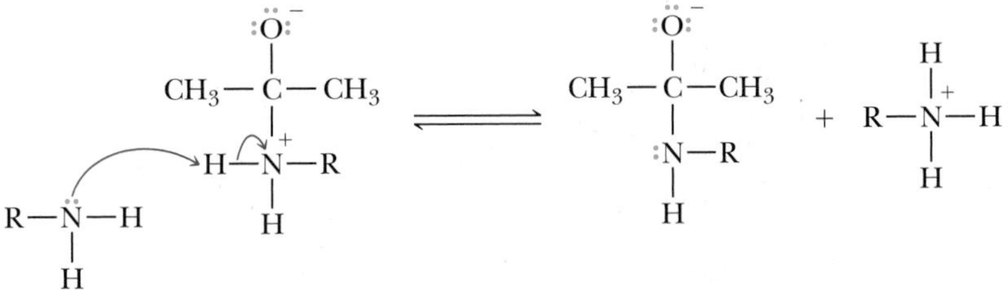

Step 3: **Add a proton.** A proton is added to the oxygen atom to give the neutral aminal intermediate.

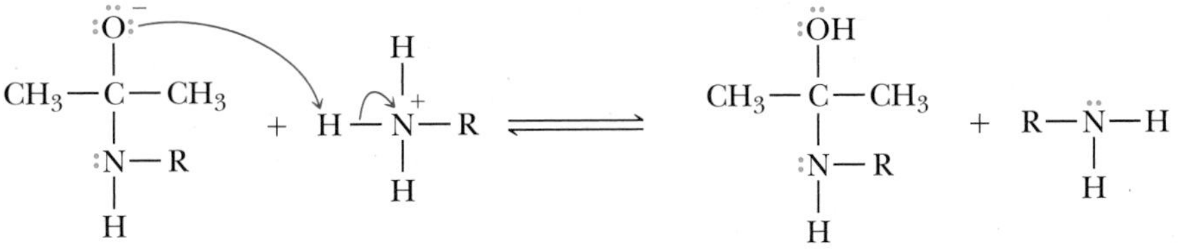

Aminal intermediate

Steps 2 and 3 above are expected to occur more or less at the same time and could take place in either order. The net result of these two steps is the transfer of a proton from one atom on an intermediate to a different site on the same intermediate, although it is unlikely the same proton is transferred because of the other species involved in these steps.

(*Continued*)

Step 4: **Add a proton.** The aminal intermediate is protonated on the hydroxyl group. This is the step that involves the acid. The hydroxyl group is protonated to generate a better leaving group, water (see next step).

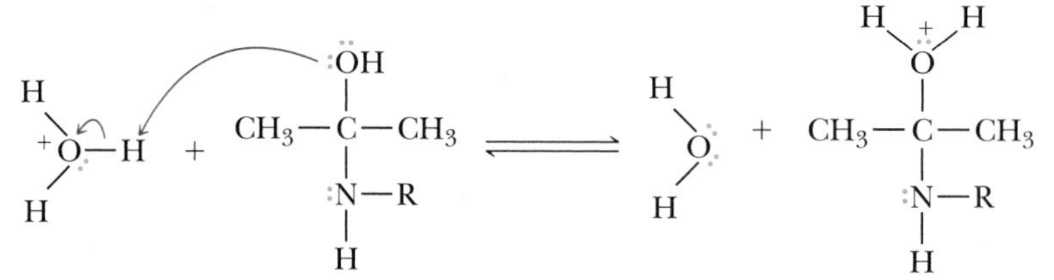

Step 5: **Break a bond to give stable molecules or ions.** The —O^+H_2 group leaves as water to generate an iminium ion, which is stabilized by resonance delocalization.

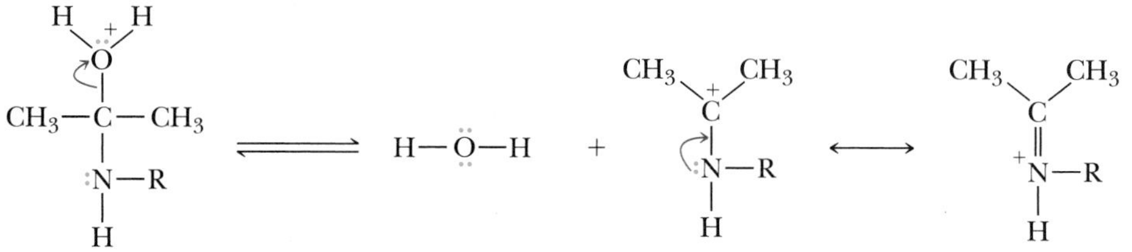

Iminium ion that is stabilized by resonance delocalization

Step 6: **Take a proton away.** The iminium ion is deprotonated to give the imine product.

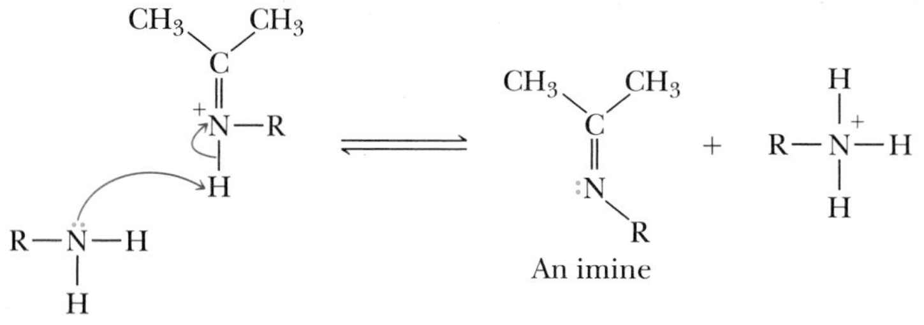

An imine

Imine formation is sensitive to the amount of acid. Acid is required to protonate the aminal intermediate in Step 4, but too much acid would protonate all of the amine and thus prevent it from acting as a nucleophile in Step 1. It turns out that a pH of about 4 is optimum for this process to occur.

As one example of the importance of imines in biological systems, vitamin A aldehyde (retinal) is bound to the protein opsin in the human retina in the form of an imine. The primary amino group of opsin for this reaction is provided by the side chain of the amino acid lysine (Section 27.1). The imine is called rhodopsin or visual purple.

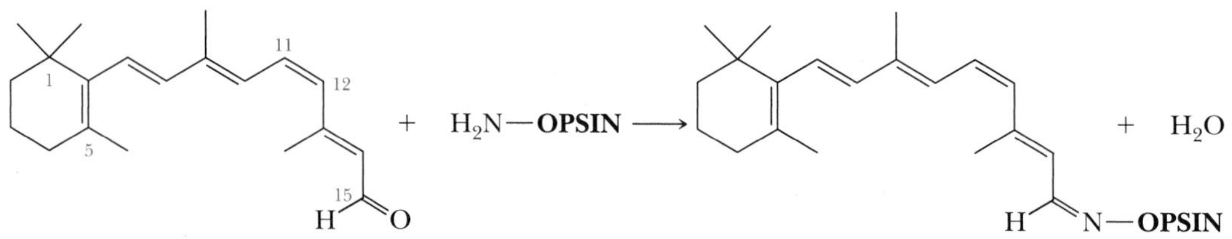

11-*cis*-Retinal **Rhodopsin** (Visual purple)

Absorption of photons by rhodopsin causes a *cis* to *trans* isomerization of the double bond at carbon 11, and the resulting change in molecular shape leads to creation of a nerve impulse that forms the basis of mammalian vision.

Example 16.8 Imine Chemistry

Write a structural formula for the imine formed in each reaction.

(a) 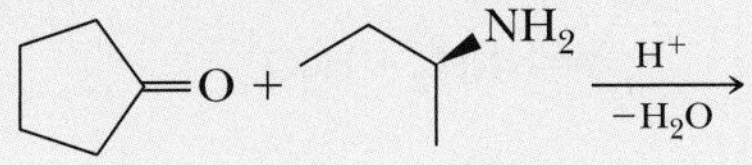

(b) 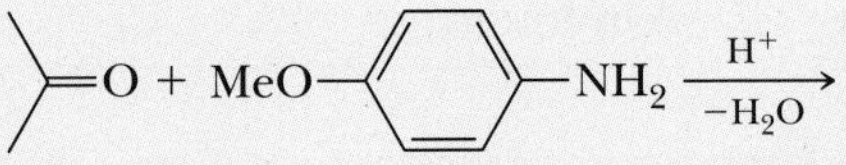

Solution

Given is a structural formula for each imine.

(a) 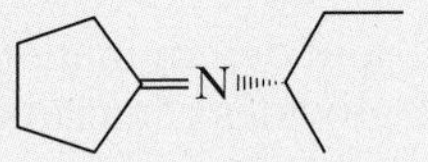 **(b)** 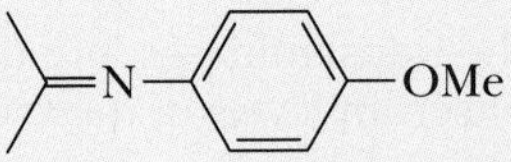

Problem 16.8

Acid-catalyzed hydrolysis of an imine gives an amine and an aldehyde or a ketone. When one equivalent of acid is used, the amine is converted to an ammonium salt. Write structural formulas for the products of hydrolysis of the following imines using one equivalent of HCl.

(a) 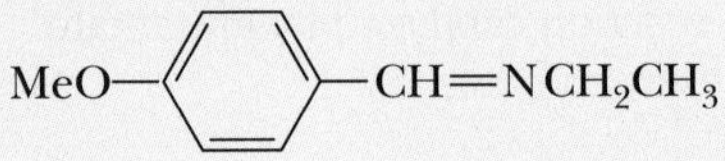 **(b)** 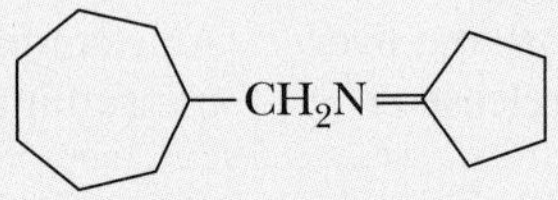

Secondary amines react with aldehydes and ketones to form enamines. The name **enamine** is derived from *-en-* to indicate the presence of a carbon-carbon double bond and *-amine* to indicate the presence of an amino group. An example is enamine formation between cyclohexanone and piperidine, a cyclic secondary amine. Water is removed by a Dean-Stark trap (Figure 16.1), which forces the equilibrium to the right.

Enamine
An unsaturated compound derived by the reaction of an aldehyde or a ketone and a secondary amine followed by loss of H_2O; $R_2C{=}CR{-}NR_2$.

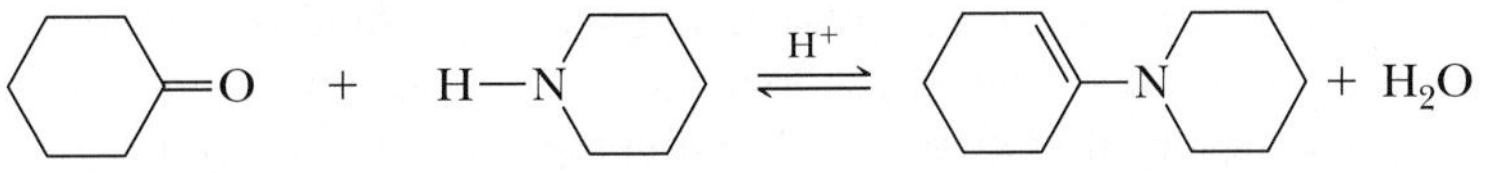

Cyclohexanone | **Piperidine** (a secondary amine) | An enamine

Briefly, the mechanism for formation of an enamine is very similar to that for the formation of an imine. In the first step, nucleophilic addition of the secondary amine to the carbonyl carbon of the aldehyde or ketone followed by proton transfer from nitrogen to oxygen gives a tetrahedral carbonyl addition compound. Acid-catalyzed dehydration gives the enamine. At this stage, enamine formation differs from imine formation. The nitrogen has no proton to lose. Instead, a proton is lost from the α-carbon of the ketone or aldehyde portion of the molecule in an elimination reaction.

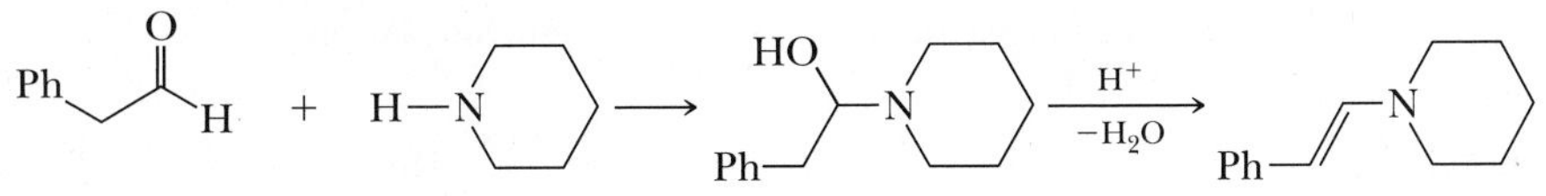

Phenylacetaldehyde | **Piperidine** | A tetrahedral carbonyl addition compound | An enamine (a mixture of *E,Z* isomers)

We will return to the chemistry of enamines and their use in synthesis in Section 19.5.

B. Hydrazine and Related Compounds

Aldehydes and ketones react with **hydrazine** to form compounds called hydrazones, as illustrated by treating cyclopentanone with hydrazine.

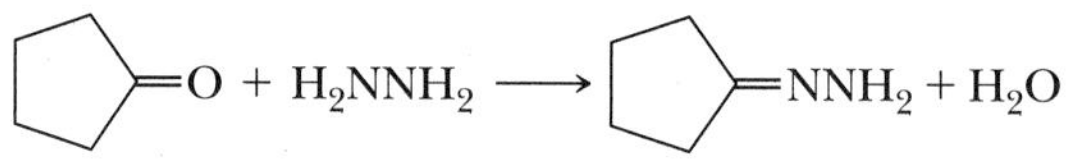
=O + H₂NNH₂ ⟶ =NNH₂ + H₂O

Hydrazine A hydrazone

MCAT Practice: *Passage and Questions*

Pyridoxine (Vitamin B_6): A Carrier of Amino Groups

The transamination reaction involves a shuffling of amino groups and carbonyls and is the last reaction in the biosynthesis of amino acids. In transamination, a donor amino group is transferred from an α-amino acid to an acceptor α-ketoacid via imine intermediates. In the process, the α-ketoacid is transformed to a new α-amino acid.

Transaminations are catalyzed by a specific group of enzymes called **transaminases**. While transaminases are found in all cells, their concentrations are particularly high in heart and liver tissues. Damage to these organs leads to release of the transaminases into the blood, and determination of serum levels of these enzymes provides clinicians with information about the extent of heart or liver damage.

Transaminases serve two vital roles. First, they provide a means of adjusting the relative proportions of amino acids to meet the particular needs of an organism. Second, they collect the nitrogen atoms of all amino acids into glutamate, because glutamate is the central source of nitrogen atoms for biosynthesis. An example of the reaction catalyzed by glutamate transaminase is presented here.

Glutamate transaminase:

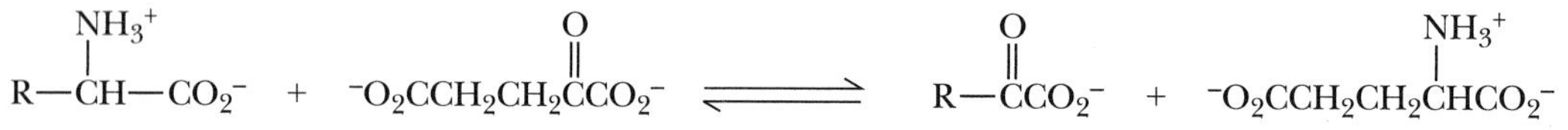

A donor amino acid — **α-ketoglutarate** — A new ketoacid — **L-Glutamate**

The reversible formation of imines from amino groups and carbonyls is the key mechanistic step in transamination reactions. To create imines, all transaminases require pyridoxal phosphate (PLP), a coenzyme derived from pyridoxine (vitamin B_6). This coenzyme undergoes reversible transformations of its aldehyde to a primary amine (pyridoxamine phosphate, PMP). PLP incorporates a phosphate to bind to the enzyme, a pyridinium ring, and an aldehyde that first reacts with the amine of an amino acid.

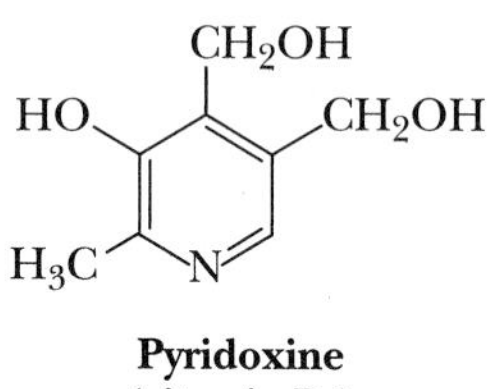

Pyridoxine (vitamin B_6)

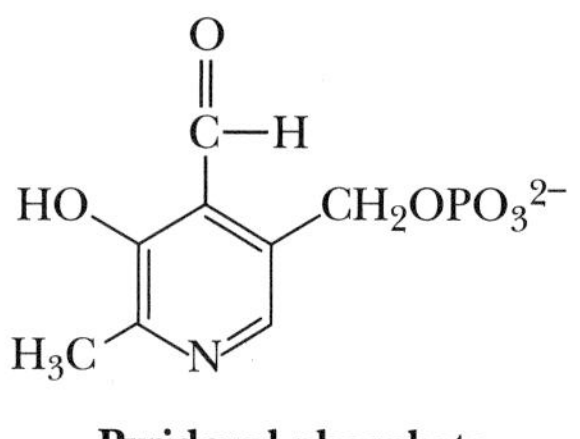

Pyridoxal phosphate (PLP)

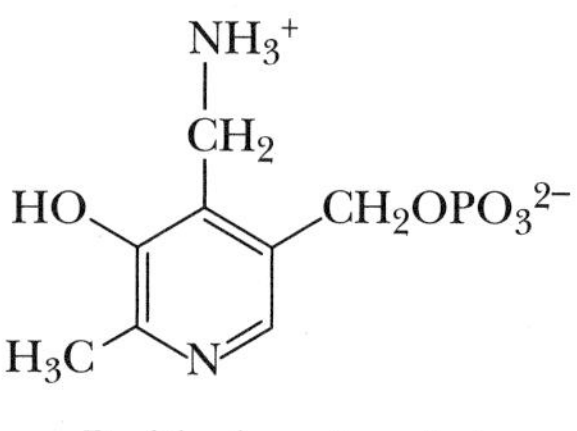

Pyridoxine phosphate (PMP)

The mechanism of transamination is given in detail with the proper electron flow arrows. A series of proton transfers, nucleophilic additions, and leaving group departures are involved that take the amino acid to the ketoacid while simultaneously converting PLP to PMP. A single residue from the transaminase enzyme (HB-Enz) is shown as facilitating proton transfers, although in the real reaction, a series of such groups may be involved.

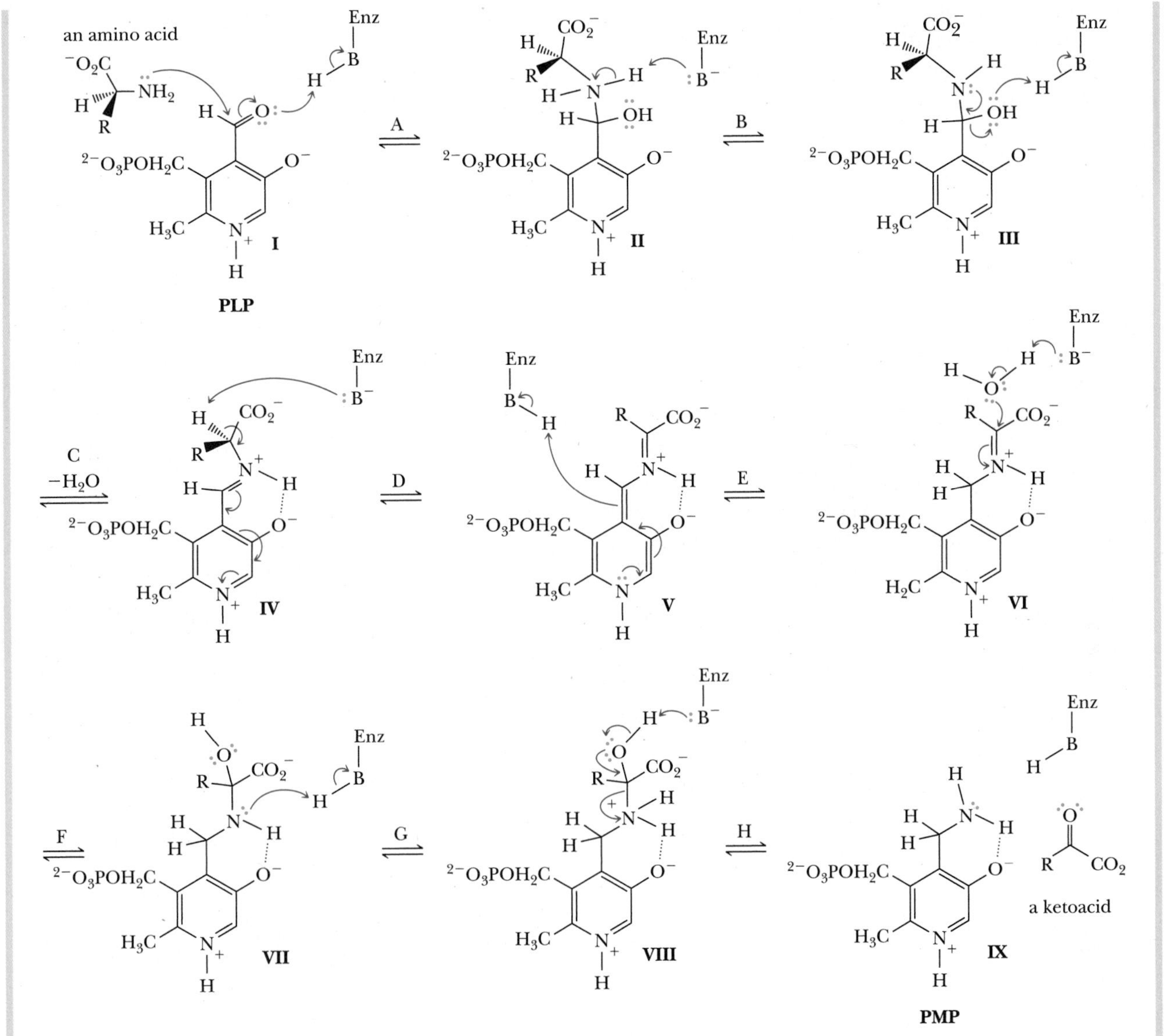

Questions

A. Which structures are imine or iminium intermediates?

1. I, III, and V
2. VII and VIII
3. IV and VII
4. IV, V, and VI

B. Which structures would be classified as tetrahedral intermediates derived from addition reactions?

1. II, III, VII, and VIII
2. IV, V, and VI
3. II, IV, and VIII
4. II and VI

C. What is the role of the pyridinium group in Step D?

1. It acts as an electron-donating group (source) of electrons to enhance deprotonation of the amino acid by the enzyme.
2. It acts as an electron-withdrawing group (sink) to facilitate deprotonation of the amino acid.
3. It plays the role of a Lewis base to coordinate the amino acid.
4. Both 2 and 3 are true.

D. In which steps do nucleophilic additions occur?

1. Steps A, F, and G
2. Steps C, D, and E
3. Steps A and F
4. Steps A, D, and F

E. Which sequence of three consecutive steps constitutes a hydrolysis of an imine/iminium?

1. D, E, and F
2. C, D, and E
3. E, F, and G
4. F, G, and H

(*Continued*)

F. To complete the shuffling of amino groups and carbonyls, the PMP must be converted back to PLP. How would you predict that this occurs?

1. A separate reaction oxidizes the amine group of PMP to an aldehyde.
2. A separate reaction hydrolyzes the amine of PMP to an aldehyde.
3. A different ketoacid reacts with the amine of PMP, and the entire sequence runs backward.
4. Either 2 or 3 could occur.

Table 16.4 lists several other derivatives of ammonia and hydrazine that react with aldehydes and ketones to give imines.

The chief value of the nitrogen nucleophiles listed in Table 16.4 is that most aldehydes and ketones react with them to give crystalline solids with sharp melting points. Historically, these derivatives often provided a convenient way to identify liquid aldehydes or ketones. Now aldehydes and ketones are more readily identified by IR and NMR spectroscopy.

Table 16.4 Derivatives of Ammonia and Hydrazine Used for Forming Imines

Reagent, $H_2N—R$	Name of Reagent	Name of Derivative Formed
$H_2N—OH$	Hydroxylamine	Oxime
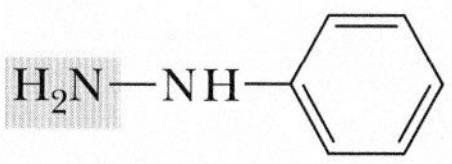	Phenylhydrazine	Phenylhydrazone
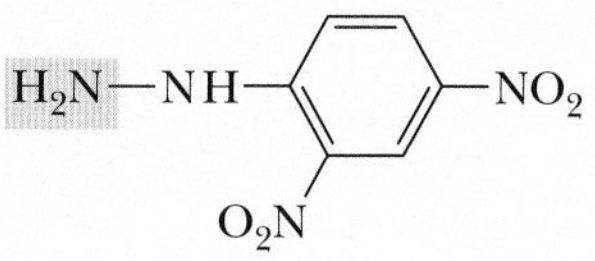	2,4-Dinitrophenylhydrazine	2,4-Dinitrophenylhydrazone
$H_2N—NHC(=O)NH_2$	Semicarbazide	Semicarbazone

16.9 Keto-Enol Tautomerism

A. Acidity of α-Hydrogens

α-Carbon
A carbon atom adjacent to a carbonyl group.

α-Hydrogen
A hydrogen on a carbon adjacent to a carbonyl group.

A carbon atom adjacent to a carbonyl group is called an **α-carbon**, and hydrogen atoms bonded to it are called **α-hydrogens**.

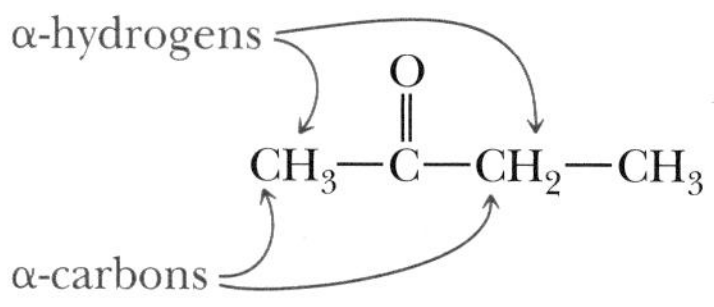
α-hydrogens
O
CH3—C—CH2—CH3
α-carbons

Because carbon and hydrogen have comparable electronegativities, a C—H bond normally has little polarity. In addition, carbon does not have a high electronegativity (compare it, for example, with oxygen, which has an electronegativity of 3.5), so that an anion based on carbon is relatively unstable. As a result, a hydrogen bonded to carbon usually shows very low acidity. The situation is different, however, for hydrogens alpha to a carbonyl group. α-Hydrogens are more acidic than acetylenic, vinylic, and alkane hydrogens but less acidic than —OH hydrogens of alcohols (Table 16.5).

The greater acidity of α-hydrogens arises because the negative charge on the resulting **enolate anion** is delocalized by resonance, thus stabilizing it relative to an alkane, alkene, or alkyne anion.

Enolate anion
An anion derived by loss of a hydrogen from a carbon alpha to a carbonyl group; the anion of an enol.

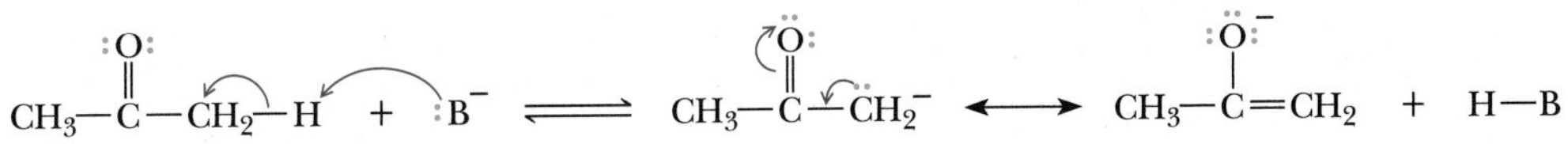

Resonance-stabilized enolate anion

An electrostatic potential map of an enolate anion.

An enolate anion is also stabilized by the electron-withdrawing inductive effect of the electronegative oxygen. Recall that we used these same factors in Section 4.6C to explain the greater acidity of carboxylic acids compared with alcohols. ▶

▶ Delocalization of charge over a larger area is stabilizing.

Example 16.9 Enolate-Forming Reactions

Predict the position of the following equilibrium.

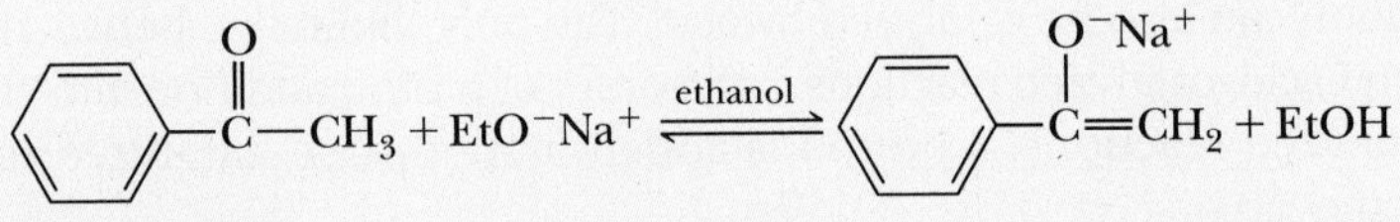

Acetophenone

Solution

The pK_a of ethanol is approximately 16 (Table 4.1). Assume that the pK_a of acetophenone is approximately equal to that of acetone (i.e., about 20). Ethanol is the stronger acid; therefore, the equilibrium lies to the left.

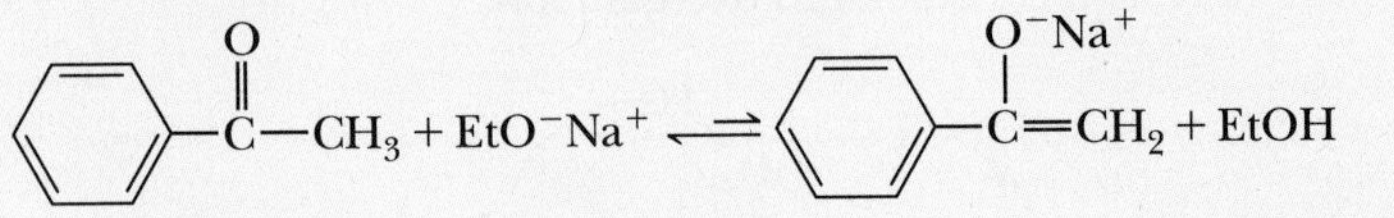

Weaker acid
pK_a 20

Stronger acid
pK_a 16

Problem 16.9

Predict the position of the following equilibrium.

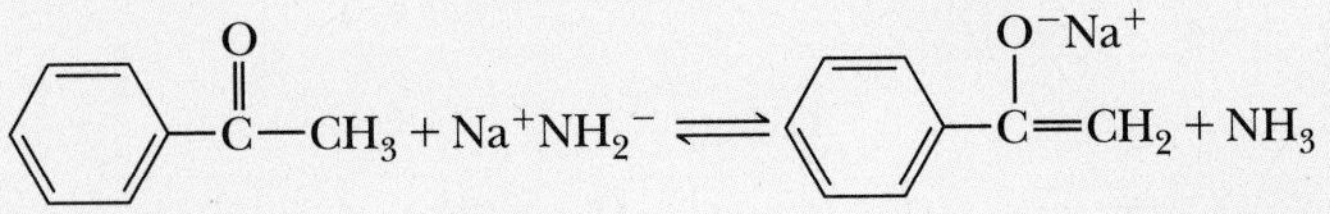

Acetophenone

Table 16.5

Type of Bond	pK_a
CH_3CH_2O—H	16
$CH_3C(=O)CH_2$—H	20
$CH_3C{\equiv}C$—H	25
$CH_2{=}CH$—H	44
CH_3CH_2—H	51

When an enolate anion reacts with a proton donor, it may do so either on oxygen or on the α-carbon. Protonation of the enolate anion on the α-carbon gives the original molecule in what is called the keto form. Protonation on oxygen gives an enol form. In this way, the keto form of an aldehyde or a ketone can be converted into the enol catalyzed by base.

Mechanism 16.10

Base-Catalyzed Equilibration of Keto and Enol Tautomers

Step 1: **Take a proton away.** A base, B^-, removes an α-hydrogen to give an enolate anion intermediate that is stabilized by resonance delocalization of the negative charge.

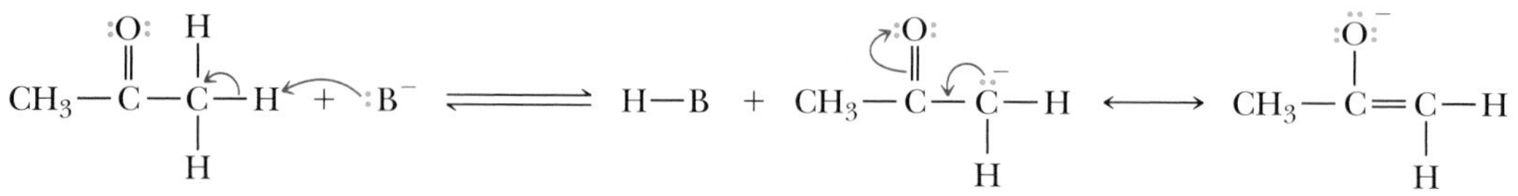

Keto form

An enolate anion intermediate that is stabilized by resonance delocalization

Step 2: **Add a proton.** A proton is added to the enolate oxygen anion to give the enol.

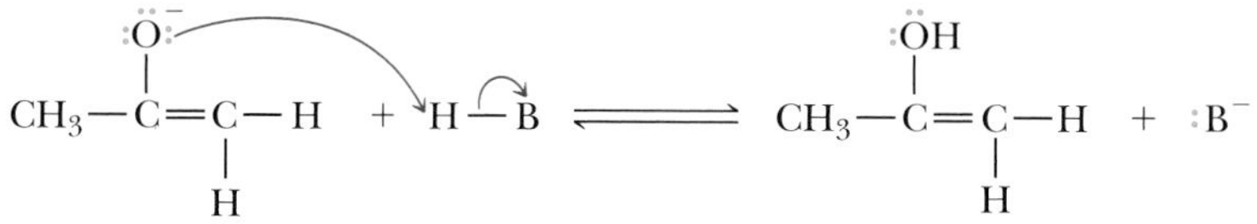

Enol form

Enol formation can also be catalyzed by acid. The only difference between the base-catalyzed and acid-catalyzed reactions is the order of proton addition and elimination. In acid-catalyzed reactions, a proton is added first; in base-catalyzed reactions, a proton is removed first.

Mechanism 16.11

Acid-Catalyzed Equilibration of Keto and Enol Tautomers

Step 1: **Add a proton.** Rapid and reversible proton transfer from the acid catalyst, H—A, to the carbonyl oxygen gives the conjugate acid of the ketone as a resonance-stabilized oxonium ion.

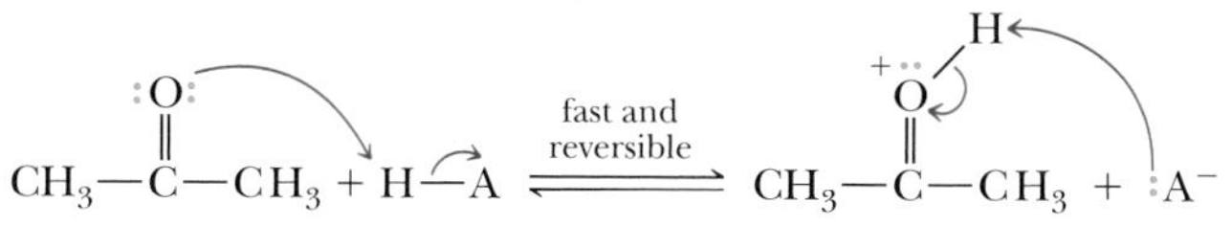

Keto form

The conjugate acid of the ketone

Step 2: **Take a proton away.** Proton transfer from the α-carbon to the base, A^-, gives the enol and generates a new molecule of the acid catalyst.

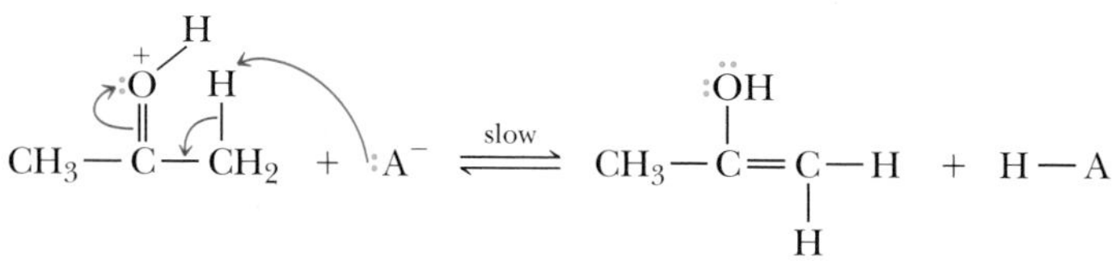

Enol form

B. The Position of Equilibrium in Keto-Enol Tautomerism

Aldehydes and ketones with at least one α-hydrogen are in equilibrium with their enol forms. We first encountered this type of equilibrium in our study of the hydroboration-oxidation and acid-catalyzed hydration of alkynes in Section 7.7. As we see in Table 16.6, the position of keto-enol equilibrium for simple aldehydes and ketones lies far on the side of the keto form, primarily because carbon-hydrogen single bonds are about as strong as oxygen-hydrogen single bonds but a carbon-oxygen double bond is stronger than a carbon-carbon double bond.

Table 16.6 The Position of Keto-Enol Equilibrium for Some Simple Aldehydes and Ketones*

Keto Form		Enol Form	% Enol at Equilibrium
$CH_3CH{=}O$ (O over CH)	⇌	$CH_2{=}CH(OH)$	6×10^{-5}
$CH_3C(=O)CH_3$	⇌	$CH_3C(OH){=}CH_2$	6×10^{-7}
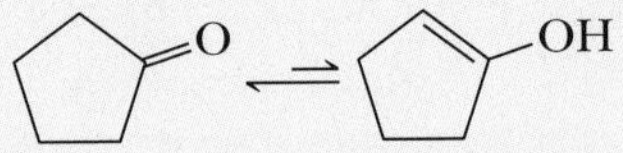			1×10^{-6}
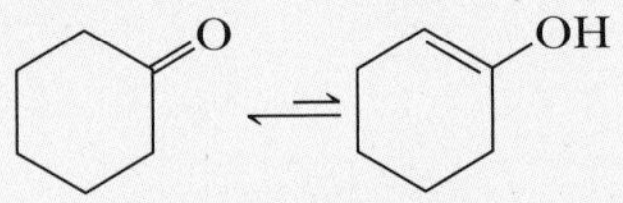			4×10^{-5}

*Data from J. March, *Advanced Organic Chemistry*, 4th ed., Wiley Interscience, New York, 1992, p. 70.

▶ Reaction will occur if the products are more stable than reactants and the energy barrier is low enough.

For certain types of molecules, the enol form may be the major form and, in some cases, the only form present at equilibrium. For β-diketones such as 1,3-cyclohexanedione and 2,4-pentanedione, where an α-carbon lies between two carbonyl groups, the position of equilibrium shifts in favor of the enol form. These enols are stabilized by **conjugation** of the systems of the carbon-carbon double bond and the carbonyl group. The enol of 2,4-pentanedione, an open-chain β-diketone, is further stabilized by intramolecular hydrogen bonding.

Conjugation

A situation that occurs when the electrons of adjacent π bonds interact with each other (i.e., when two double bonds are separated by one single bond).

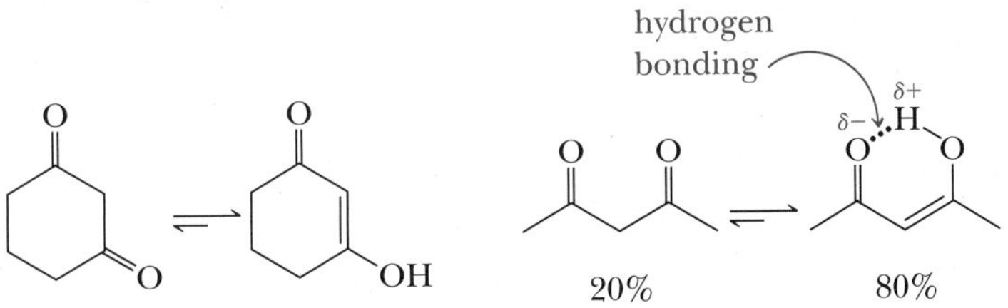

1,3-Cyclohexanedione **2,4-Pentanedione (Acetylacetone)**

Example 16.10 Keto-Enol Equilibrium

Write two enol forms for each compound. Which enol of each has the larger concentration at equilibrium?

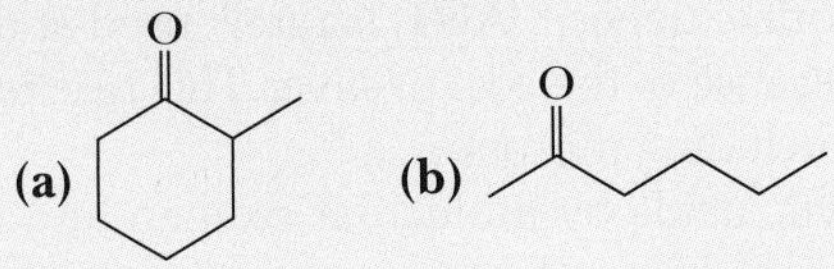

(Continued)

Solution

In each case, the major enol form has the more substituted (and more stable) double bond.

(a) 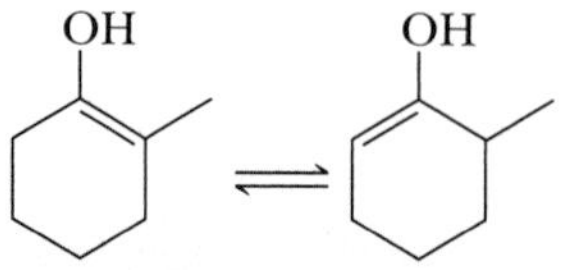 (b) 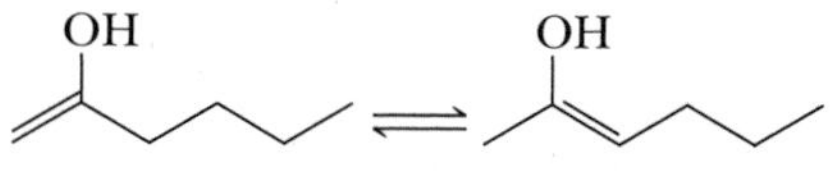

Major enol Major enol

Problem 16.10

Draw a structural formula for the keto form of each enol.

(a) 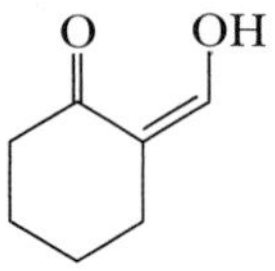 (b) 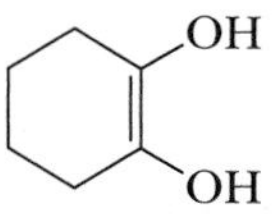 (c) 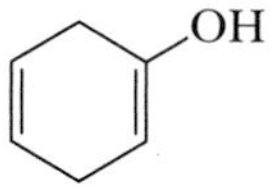

16.10 Oxidation

A. Oxidation of Aldehydes

Aldehydes are one of the most easily oxidized of all functional groups. Oxidation by chromic acid is illustrated by the conversion of hexanal to hexanoic acid (for the mechanism of this oxidation, review Section 10.8A).

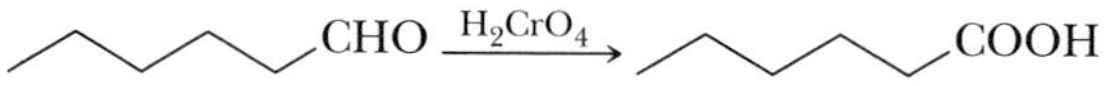

Hexanal **Hexanoic acid**

Aldehydes are also oxidized to carboxylic acids by Ag(I) ion. One laboratory procedure is to shake a solution of the aldehyde in aqueous ethanol or tetrahydrofuran with a slurry of Ag_2O.

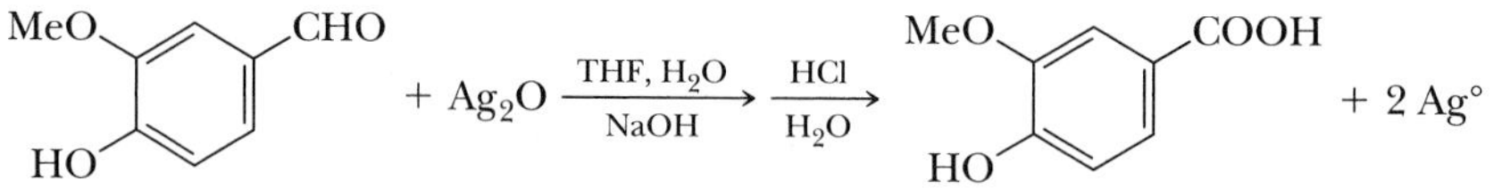

Vanillin **Vanillic acid**

Tollens' reagent A solution prepared by dissolving Ag_2O in aqueous ammonia; used for selective oxidation of an aldehyde to a carboxylic acid.

Tollens' reagent, another form of Ag(I), is prepared by dissolving silver nitrate in water, adding sodium hydroxide to precipitate Ag(I) as Ag_2O, and then adding aqueous ammonia to redissolve silver(I) as the silver-ammonia complex ion.

$$Ag^+NO_3^- + 2\,NH_3 \xrightleftharpoons{NH_3,\ H_2O} Ag(NH_3)_2{}^+NO_3^-$$

When Tollens' reagent is added to an aldehyde, the aldehyde is oxidized to a carboxylic anion and Ag(I) is reduced to metallic silver. If this reaction is carried out properly, silver precipitates as a smooth, mirrorlike deposit. Ag(I) is rarely used at the present time for the oxidation of aldehydes because of the cost of silver. This reaction, however, is still used for silvering glassware, including mirrors.

Aldehydes are also oxidized to carboxylic acids by molecular oxygen and by hydrogen peroxide.

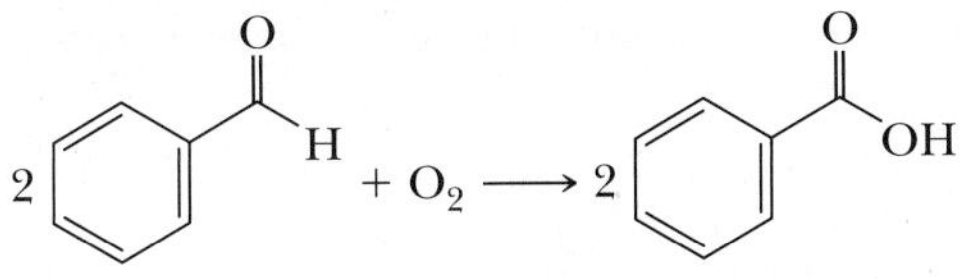

Benzaldehyde Benzoic acid

Reaction with oxygen is a radical chain reaction (Section 8.7). Molecular oxygen is the least expensive and most readily available of all oxidizing agents. On an industrial scale, air oxidation of organic compounds, including aldehydes, is very common.

Another inexpensive method for oxidizing an aldehyde to a carboxylic acid is the **Pinnick oxidation**. This reaction uses sodium chlorite ($NaClO_2$) with a phosphate buffer (NaH_2PO_4) to create chlorous acid, along with a structure to scavenge the hypochlorous acid (HOCl) by-product (see the Mechanism box below). While many scavengers are used, the most common is 2-methyl-2-butene. This is a very selective procedure. It will not oxidize alcohols, so they do not have to be protected.

A silvered mirror has been deposited in the inside of this flask by the reaction of an aldehyde with Tollens' reagent.

Mechanism 16.12

Pinnick Oxidation

Step 1: **Equilibrium between $H_2PO_4^-$ and ClO_2^- creates chlorous acid.**

$$H_2PO_4^- + ClO_2^- \rightleftharpoons HPO_4^{2-} + HClO_2$$

Chlorous acid

Step 2: **Add a proton.**

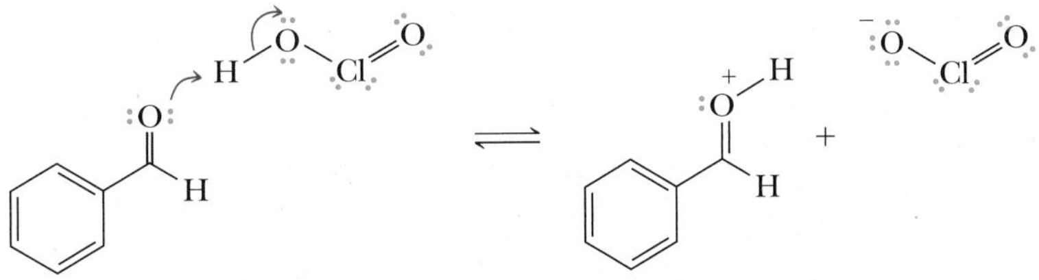

Step 3: **Make a new bond between a nucleophile and an electrophile.**

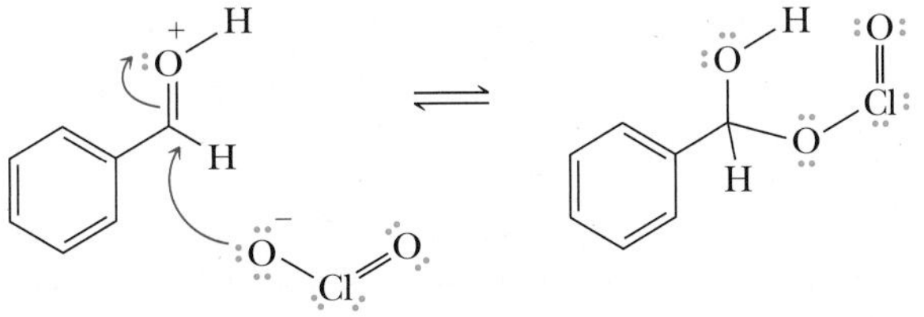

Step 4: **Take a proton away and simultaneously break bonds to give stable molecules or ions.**

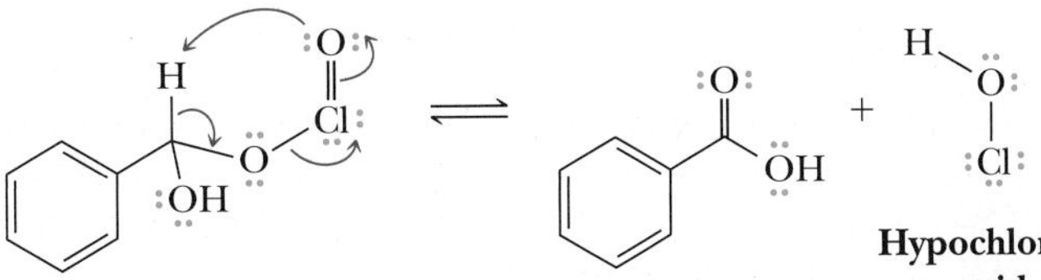

Hypochlorous acid

(Continued)

Step 5: **Scavenge the hypochlorous acid (as an example, using electrophilic addition to an alkene, Section 6.3).**

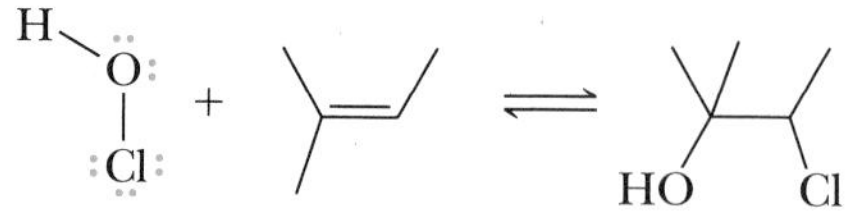

2-Methyl-2-butene

Example 16.11 Using Oxidation Reactions

Draw a structural formula for the product formed by treating each compound with $NaClO_2$ and NaH_2PO_4, along with 2-methyl-2-butene.

(a) Pentanal **(b)** Cyclopentanecarbaldehyde

Solution

The aldehyde group in each compound is oxidized to a carboxyl group.

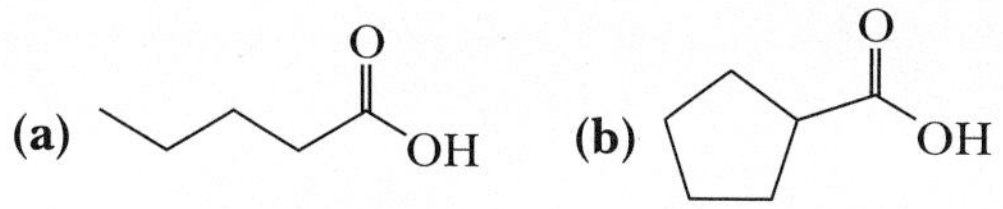

Pentanoic acid **Cyclopentane carboxylic acid**

Problem 16.11

Complete the equations for these oxidations.

(a) Hexanal + $H_2O_2 \longrightarrow$ **(b)** 3-Phenylpropanal + Tollens' reagent $\longrightarrow$

B. Oxidation of Ketones

Ketones are oxidized only under rather special conditions. They are not normally oxidized by chromic acid or potassium permanganate. Rather, chromic acid is used routinely to oxidize secondary alcohols to ketones in good yield (Section 10.8A).

Ketones undergo oxidative cleavage, via their enol form, when treated with potassium dichromate, potassium permanganate, and other strong oxidants at higher temperatures and higher concentrations of acid or base. The carbon-carbon double bond of the enol is cleaved to form two carboxyl or ketone groups, depending on the substitution pattern of the original ketone. An important industrial application of this reaction is oxidation by nitric acid of cyclohexanone to hexanedioic (adipic) acid, one of the two monomers required for the synthesis of the polymer nylon 66 (Section 29.5A).

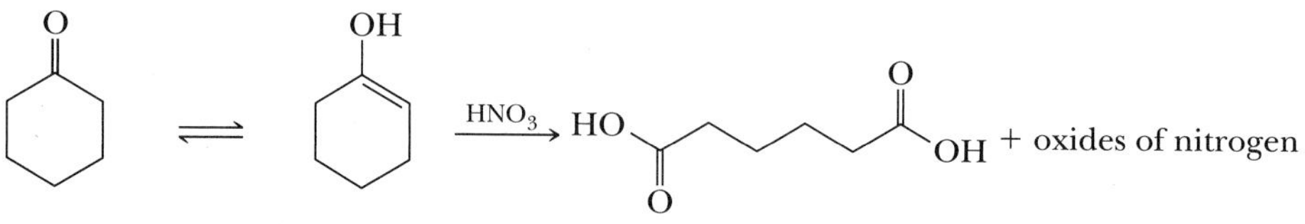

Cyclohexanone (keto form) **Cyclohexanone** (enol form) **Hexanedioic acid (Adipic acid)**

As the oxidation of cyclohexanone shows, this reaction is most useful for oxidation of symmetrical cycloalkanones, which yield a single product. Most other ketones give mixtures of products because the enol can form on either side of the carbonyl group.

16.11 Reduction

Aldehydes are reduced to primary alcohols, and ketones are reduced to secondary alcohols. In addition, both aldehyde and ketone carbonyl groups can be reduced to methylene groups ($-CH_2-$).

Aldehydes	Can Be Reduced to	Ketones	Can Be Reduced to
$R\overset{O}{\overset{\|}{C}}H$	RCH_2OH RCH_3	$R\overset{O}{\overset{\|}{C}}R'$	$R\overset{OH}{\overset{\|}{C}}HR'$ RCH_2R'

A. Metal Hydride Reductions

By far the most common laboratory reagents for reduction of the carbonyl group of an aldehyde or a ketone to a hydroxyl group are sodium borohydride, lithium aluminum hydride (LAH), and their derivatives. These compounds behave as sources of **hydride ion**, a powerful nucleophile, that takes part in a nucleophilic acyl addition reaction.

Hydride ion

A hydrogen atom with two electrons in its valence shell; $H:^-$.

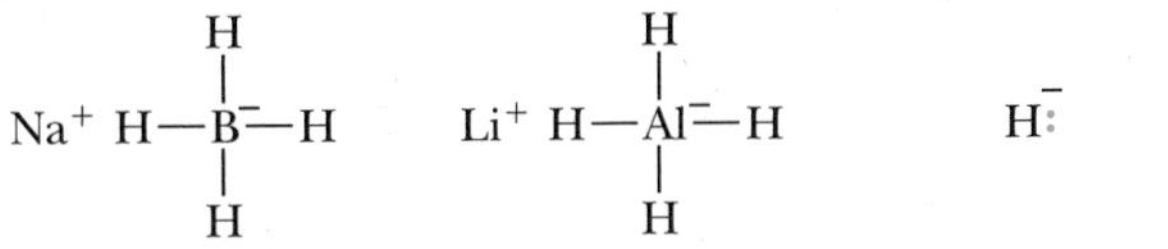

Sodium borohydride **Lithium aluminum hydride (LAH)** **Hydride ion**

Lithium aluminum hydride is a very powerful reducing agent; it reduces not only the carbonyl groups of aldehydes and ketones rapidly but also those of carboxylic acids (Section 17.6A) and their functional derivatives (Section 18.10). Sodium borohydride is a less reactive and therefore much more selective reagent, reducing only aldehydes and ketones rapidly. Neither reagent reduces alkenes or alkynes to alkanes.

Reductions using sodium borohydride are most commonly carried out in aqueous methanol, in pure methanol, or in ethanol. The initial product of reduction is a tetraalkyl borate, which, upon warming with water, is converted to an alcohol and sodium borate salts. One mole of sodium borohydride reduces four moles of aldehyde or ketone. ▶

$$4\,R\overset{O}{\overset{\|}{C}}H + NaBH_4 \xrightarrow{\text{methanol}} (RCH_2O)_4B^-Na^+ \xrightarrow{H_2O} 4\,RCH_2OH + \text{borate salts}$$

A tetraalkyl borate

▶ Most bond-making steps in reaction mechanisms involve nucleophiles reacting with electrophiles.

Unlike sodium borohydride, lithium aluminum hydride reacts violently with water, alcohols, and other protic solvents to liberate hydrogen gas and form metal hydroxides and alkoxides. Therefore, reductions of aldehydes and ketones using this reagent must be carried out in aprotic solvents, most commonly diethyl ether or tetrahydrofuran. The stoichiometry for lithium aluminum hydride reductions is the same as that for sodium borohydride reductions: one mole of lithium aluminum hydride per four moles of aldehyde or ketone. Because of the formation of gelatinous aluminum salts, aqueous acid or base workup is usually used to dissolve them.

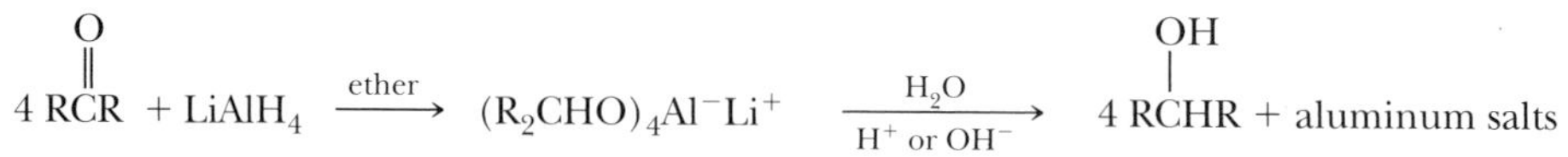
4 RCR + LiAlH4
ether
(R2CHO)4Al−Li+
H2O
H+ or OH−
4 RCHR + aluminum salts

A tetraalkyl aluminate

Mechanism 16.13

Sodium Borohydride Reduction of an Aldehyde or a Ketone

Step 1: **Make a new bond between a nucleophile and an electrophile.** Nucleophilic addition of a hydride to the electrophilic carbonyl carbon atom gives a tetrahedral carbonyl addition compound.

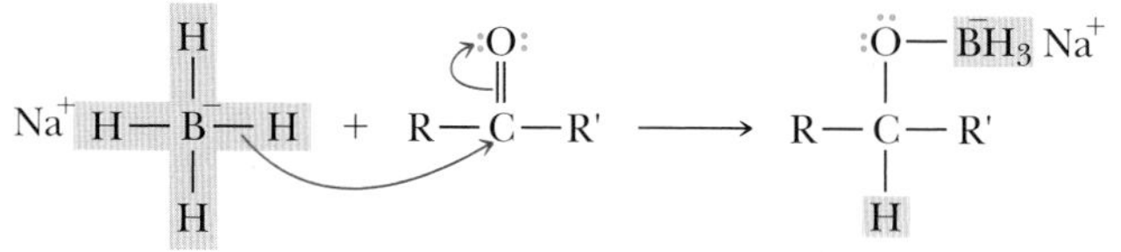

Step 2: **Add a proton.** In a second step, the chemist opens the flask and adds water to give the alcohol product.

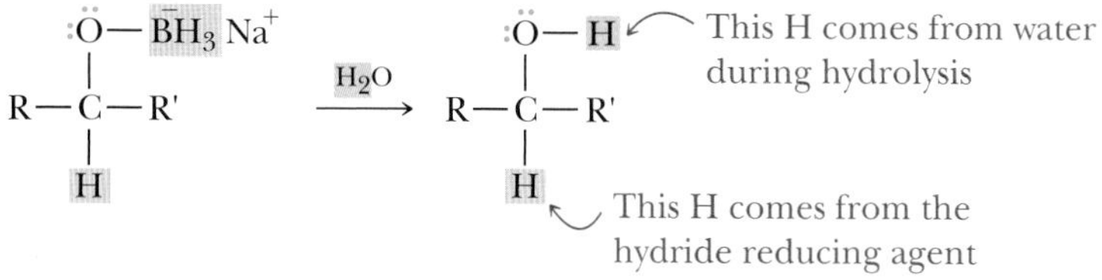
H2O
This H comes from water during hydrolysis
This H comes from the hydride reducing agent

On the alcohol product, the hydrogen atom bonded to carbon comes from the hydride reducing agent and the hydrogen atom bonded to oxygen comes from water during hydrolysis of the metal alkoxide salt.

B. Catalytic Reduction

The carbonyl group of an aldehyde or a ketone is reduced to a hydroxyl group by hydrogen in the presence of a transition metal catalyst, most commonly finely divided platinum or nickel. Reductions are generally carried out at temperatures from 25° to 100°C and at pressures of hydrogen from 1 to 5 atm. Under such conditions, cyclohexanone is reduced to cyclohexanol.

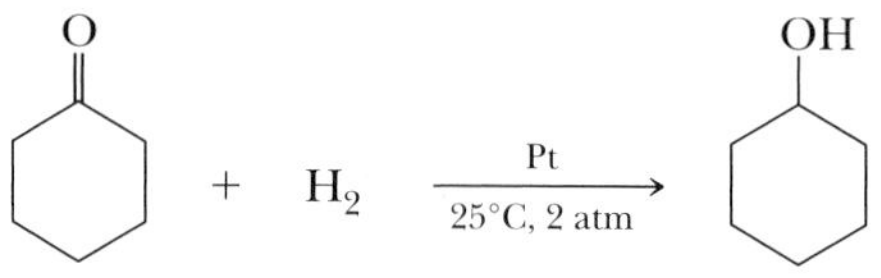
H2
Pt
25°C, 2 atm

Cyclohexanone **Cyclohexanol**

Catalytic reduction of aldehydes and ketones is simple to carry out, yields are generally very high, and isolation of the final product is very easy. A disadvantage is that some other functional groups are also reduced under these conditions (e.g., carbon-carbon double and triple bonds).

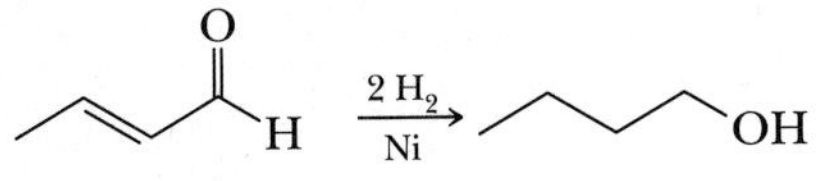

***trans*-2-Butenal (Crotonaldehyde)** **1-Butanol**

C. Selective Reduction

Using metal hydride reductions, it is possible to reduce an aldehyde or ketone carbonyl group without reducing any C═C bonds that might be present in the same molecule. The metal hydride reducing agents, such as $LiAlH_4$ and $NaBH_4$, are nucleophilic reagents. Functional groups that are not electrophiles, such as C═C bonds in alkenes, do not react.

The following equations illustrate selective reduction of a carbonyl group in the presence of a carbon-carbon double bond and, alternatively, selective reduction of a carbon-carbon double bond in the presence of a carbonyl group using rhodium on powdered charcoal as a catalyst.

Selective reduction of a carbonyl group:

$$RCH{=}CH\overset{\overset{\displaystyle O}{\|}}{C}R' \xrightarrow[2.\ H_2O]{1.\ LiAlH_4} RCH{=}CH\overset{\overset{\displaystyle OH}{|}}{C}HR'$$

Selective reduction of a carbon-carbon double bond:

$$RCH{=}CH\overset{\overset{\displaystyle O}{\|}}{C}R' + H_2 \xrightarrow{Rh} RCH_2CH_2\overset{\overset{\displaystyle O}{\|}}{C}R'$$

Example 16.12 Aldehyde and Ketone Reduction Reactions

Complete these reductions.

(a) 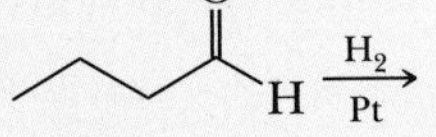

(b) 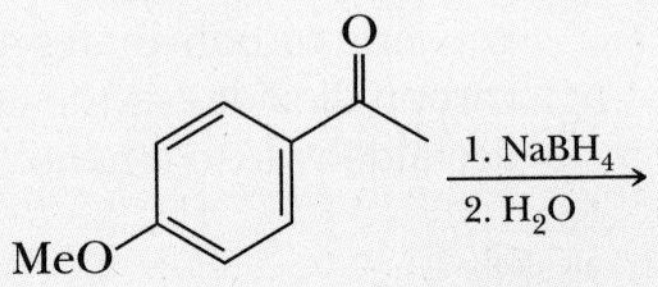

Solution

The carbonyl group of the aldehyde in (a) is reduced to a primary alcohol, and the carbonyl group of the ketone in (b) is reduced to a secondary alcohol.

(a) 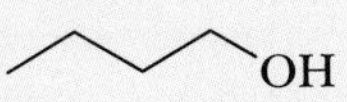

(b) 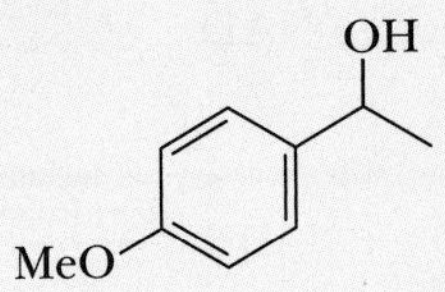

(racemic)

(Continued)

Problem 16.12

What aldehyde or ketone gives these alcohols upon reduction with $NaBH_4$?

(a) 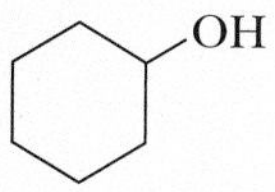

(b) 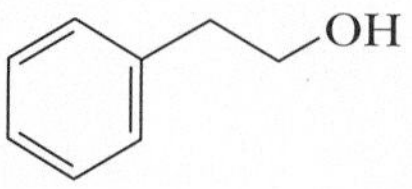

(c) 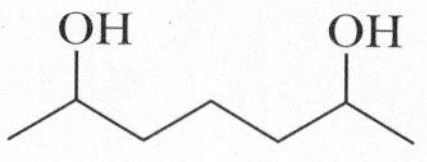

(mixture of three stereoisomers)

D. Reductive Amination

Reductive amination A method of preparing substituted amines by treating an aldehyde or a ketone with an amine in the presence of or followed by a reducing agent.

One of the chief values of imines is that the carbon-nitrogen double bond can be reduced by hydrogen in the presence of a nickel or another transition metal catalyst to a carbon-nitrogen single bond. By this two-step reaction, called **reductive amination**, a primary amine is converted to a secondary amine by way of an imine as illustrated by the conversion of cyclohexylamine to dicyclohexylamine.

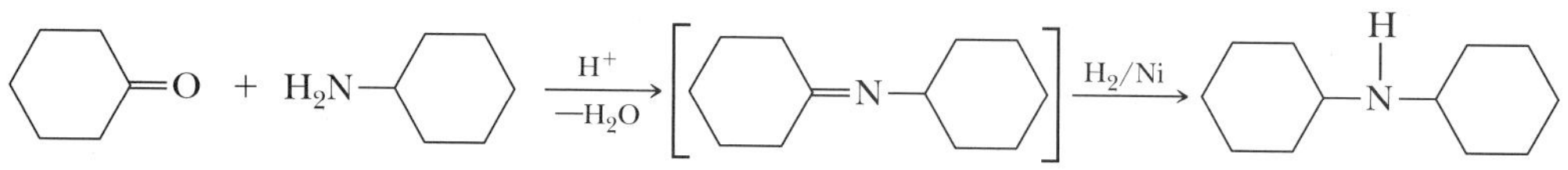

Cyclohexanone **Cyclohexylamine** (an imine intermediate) **Dicyclohexylamine**

It is possible to carry out reductive amination in a single step by using a reducing agent that is not powerful enough to reduce the starting aldehyde or ketone, but is strong enough to reduce the more easily reduced imine that is formed. The reducing agent usually used for this purpose is sodium cyanoborohydride, $NaBH_3CN$.

Connections to Biological Chemistry

NADH: The Biological Equivalent of a Hydride Reducing Agent

For the reduction of aldehydes and ketones to alcohols, biological systems use NADH, a reagent whose results are equivalent to laboratory hydride reducing agents. As an example, the final step in alcoholic fermentation—the process by which yeast converts carbohydrates such as glucose to ethanol and carbon dioxide—is the enzyme-catalyzed reduction of acetaldehyde to ethanol.

$$CH_3-\overset{\overset{\large O}{\|}}{C}-H + NADH + H_3O^+ \xrightarrow[\text{dehydrogenase}]{\text{alcohol}} CH_3CH_2OH + NAD^+ + H_2O$$

Acetaldehyde **Ethanol**

Alcoholic fermentation is the basis for the brewing of beers and the fermentation of grape sugar in wine making.

As another example, the end product of glycolysis is pyruvate and the reduced coenzyme NADH. In the absence of an adequate supply of oxygen (anaerobic metabolism) to reoxidize NADH to NAD^+ and thereby allow glycolysis to continue, cells use the reduction of pyruvate to lactate as a way to regenerate NAD^+:

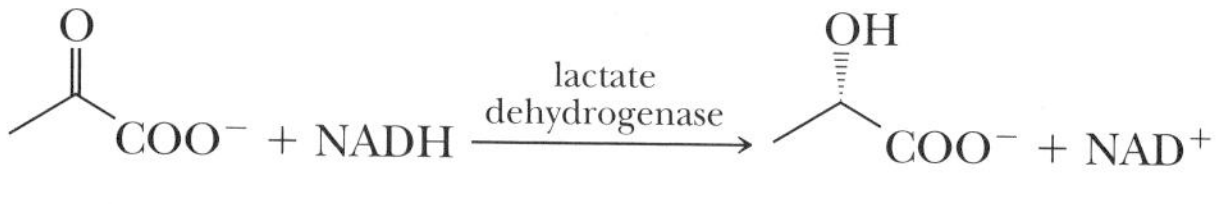

Pyruvate **(*S*)-Lactate**

Anyone who exercises to the point of consuming all available oxygen knows the pain and fatigue associated with the buildup of lactate in muscles. With rest and a renewed supply of oxygen, the concentration of lactate decreases rapidly and muscle pain is relieved. Lactate production by anaerobic organisms during fermentation is responsible for the taste of sour milk and the characteristic taste and fragrance of sauerkraut (fermented cabbage).

In still other aldehyde and ketone reductions, biological systems use NADPH as a reducing agent. This molecule, which is a phosphate ester of NADH, functions in the same manner as NADH as a biological reducing agent.

The mechanism for NADH reduction of the carbonyl group of an aldehyde or a ketone follows. First, the carbonyl-containing compound and NADH are positioned on the surface of the enzyme catalyst in a highly specific relationship to each other. Then follows a redistribution of valence electrons, one part of which is the transfer of a hydrogen atom with its pair of electrons (in effect a hydride ion) from NADH to the carbonyl compound.

Arrows 1 and 2: Electrons within the ring flow from nitrogen.

Arrow 3: Transfer of a hydride ion from the —CH_2— of the six-membered ring to the carbonyl carbon creates the new C—H bond to the carbonyl carbon.

Arrow 4: The C=O π-bond breaks as the new C—H bond forms.

Arrow 5: An acidic group, —BH, on the surface of the enzyme transfers a proton to the newly formed alkoxide ion to complete formation of the hydroxyl group of the product.

The enzyme-catalyzed reduction of pyruvate is completely stereoselective; in muscle tissue, only the *S* enantiomer of lactate is produced. This stereoselectivity arises because the reduction takes place in a chiral environment created by the enzyme. At the actual reduction step, both pyruvate and NADH are positioned precisely on the chiral surface of the enzyme with the result that the hydride ion from NADH can be delivered only to one face of pyruvate, in this case producing only the *S* enantiomer.

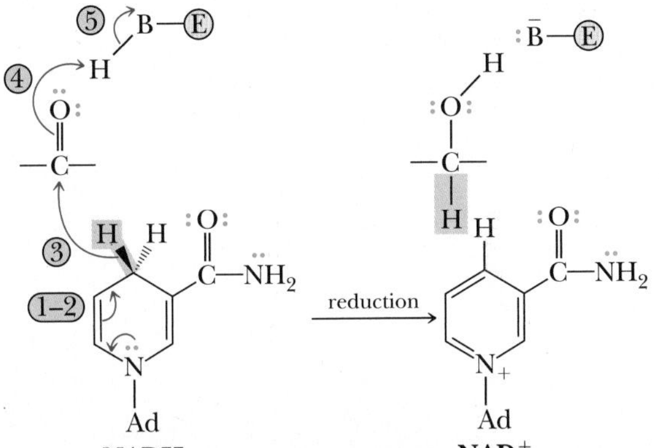

NADH **NAD⁺**

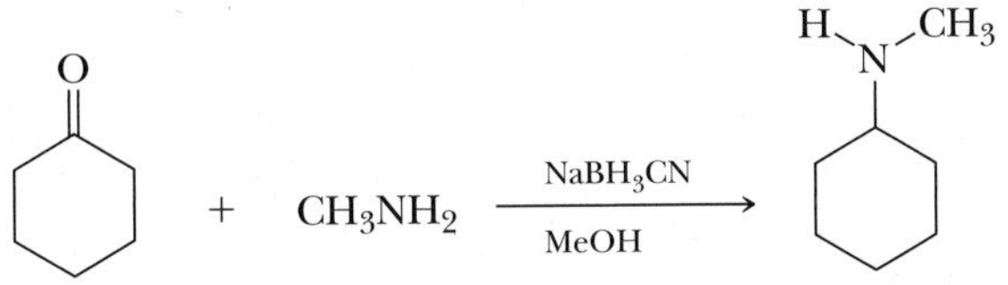

Cyclohexanone **N-Methylcyclohexylamine**

HOW TO

Retrosynthetically Dissect an Amine into the Proper Starting Materials for a Reductive Amination

With a few exceptions, any amine can be envisioned to arise from a reductive amination. Each alkyl carbon attached to the nitrogen can, in a retrosynthetic fashion, be disconnected from the amine and oxidized to the level of a ketone or an aldehyde, depending on whether that carbon has two other carbons or one other carbon attached, respectively. A primary amine has only one possibility, while a secondary amine has two and a tertiary amine has three potential synthetic precursors, as shown below.

(Continued)

Primary amines

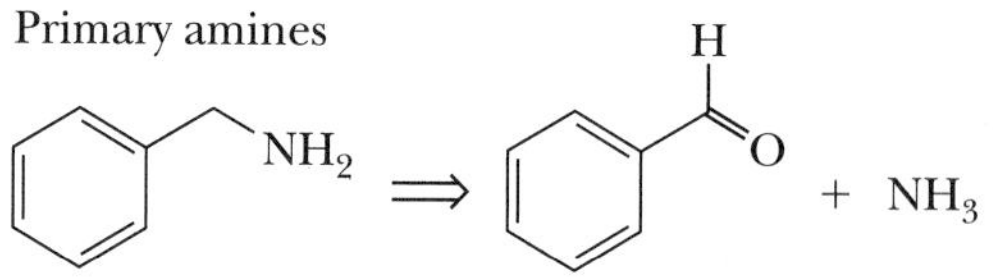

Secondary amines

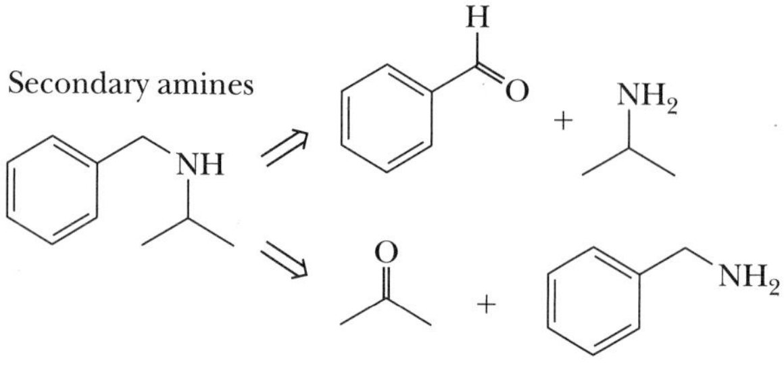

Tertiary amines

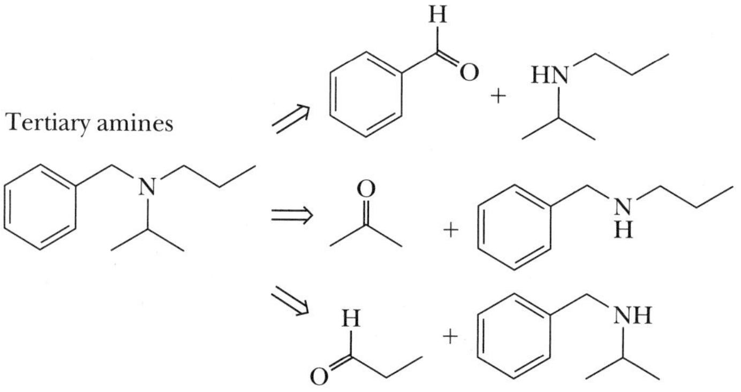

The exceptions include amines with a carbon attached that cannot be at the oxidation level of a ketone or an aldehyde, such as an amine with a directly attached phenyl or tertiary carbon. Two mistakes are shown here where five bonds to carbon would be necessary if one considered a reductive amination.

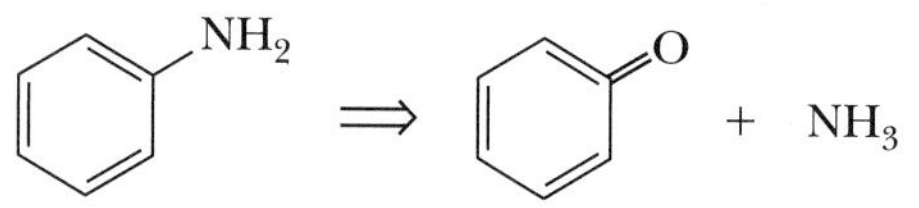

WRONG!

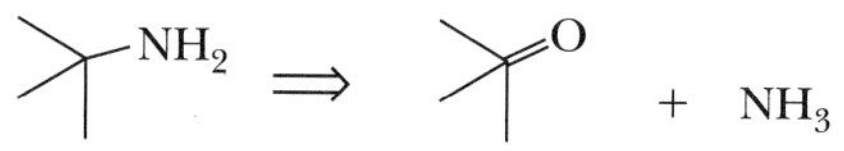

WRONG!

E. Reduction of a Carbonyl Group to a Methylene Group

Several methods are available for reducing the carbonyl group of an aldehyde or a ketone to a methylene group (—CH_2—). One of the first discovered was refluxing the aldehyde or ketone with amalgamated zinc (zinc with a surface layer of mercury) in concentrated HCl.

Clemmensen reduction
Reduction of the C═O group of an aldehyde or a ketone to a CH_2 group using Zn(Hg) and HCl.

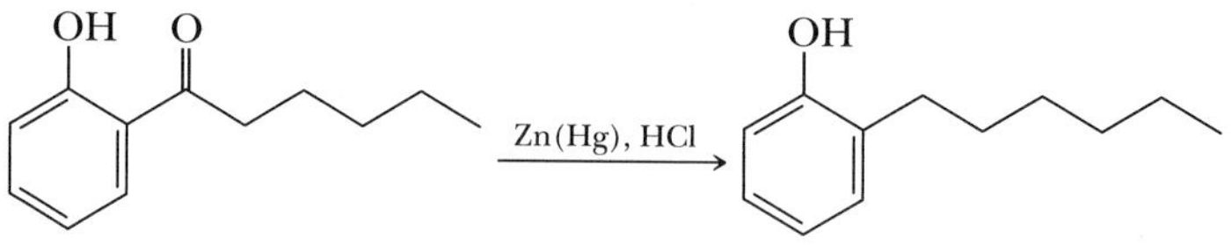

This reaction is known as the **Clemmensen reduction** after the German chemist, E. Clemmensen, who developed it in 1912. The mechanism of Clemmensen reduction, although not well understood, involves transfer of electrons from the Zn to reduce the carbonyl group.

Because the Clemmensen reduction requires the use of concentrated HCl, it cannot be used to reduce a carbonyl group in a molecule that also contains acid-sensitive groups, such as a tertiary alcohol that might undergo dehydration, or an acetal that is hydrolyzed so that its resulting carbonyl group is also reduced.

The **Wolff-Kishner reduction**, discovered independently by N. Kishner in 1911 and L. Wolff in 1912 and reported within months of Clemmensen's discovery, is an alternative method for reduction of a carbonyl group to a methylene group. In this reduction, a mixture of the aldehyde or ketone, hydrazine, and concentrated potassium hydroxide is heated at reflux in a high-boiling solvent such as diethylene glycol (bp 245°C).

Wolff-Kishner reduction
Reduction of the C=O group of an aldehyde or a ketone to a CH_2 group using hydrazine and base.

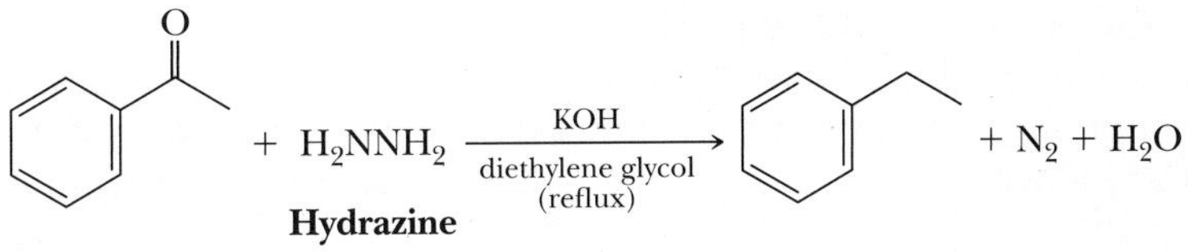
O
+ H2NNH2
Hydrazine
KOH
diethylene glycol
(reflux)
+ N2 + H2O

Mechanism 16.14

Wolff-Kishner Reduction

Step 1: Reaction of the carbonyl group of the aldehyde or ketone with hydrazine gives a hydrazone (Section 16.8B).

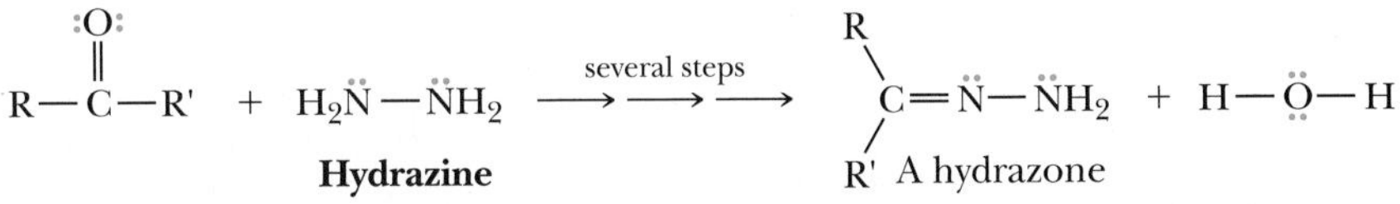
several steps
Hydrazine
A hydrazone

Step 2: **Take a proton away.** Hydroxide removes a proton on the terminal nitrogen on the hydrazone.

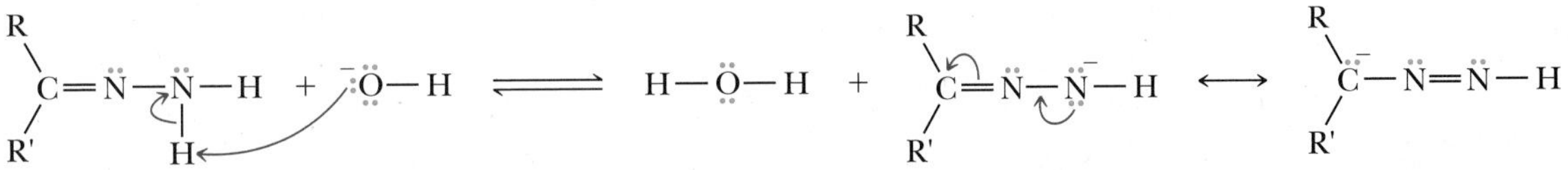

Step 3: **Add a proton.** Protonation takes place on the carbon atom to complete the tautomerism process (compare keto-enol tautomerism, Section 16.9).

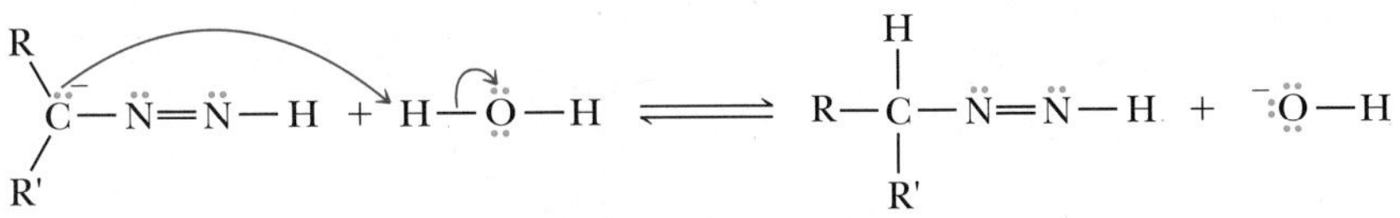

Step 4: **Take a proton away.** Hydroxide again removes a proton on the terminal nitrogen.

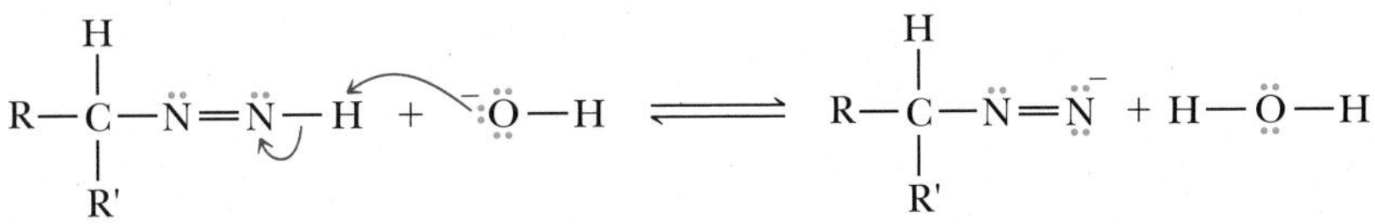

Step 5: **Break a bond to give stable molecules or ions.** Loss of the stable molecule nitrogen drives formation of a highly basic carbanion.

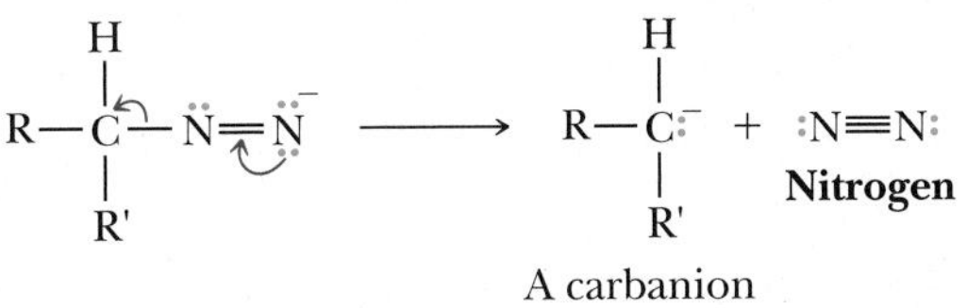
Nitrogen
A carbanion

(Continued)

Step 6: **Add a proton.** Protonation of the highly basic carbanion gives the final reduced product in which the carbonyl O atom has been replaced with two H atoms.

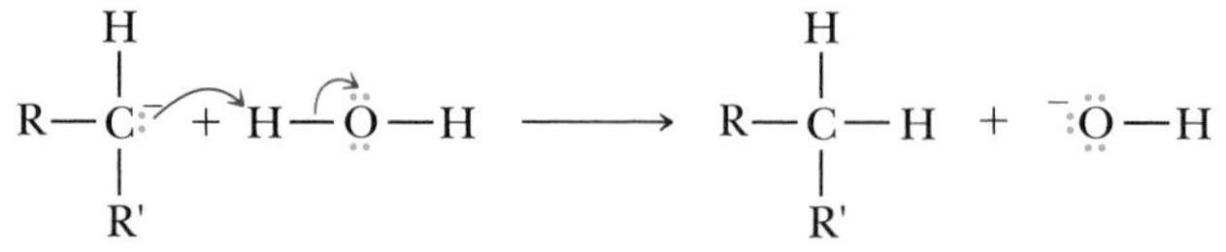

Each of the reductions has its special conditions, advantages, and disadvantages. The Clemmensen reduction cannot be used in the presence of groups sensitive to concentrated acid; the Wolff-Kishner reduction cannot be used in the presence of groups sensitive to concentrated base. However, the carbonyl group of almost any aldehyde or ketone can be reduced to a methylene group by one of these methods.

Example 16.13 Reduction Reactions

Complete the following reactions.

(a)

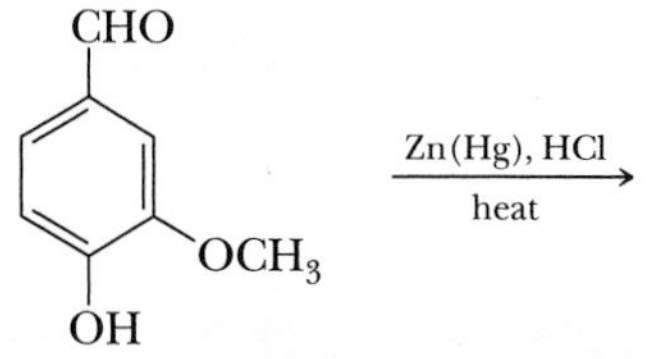

Vanillin
(from vanilla beans)

(b)

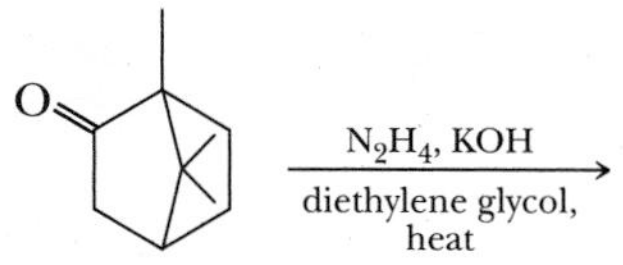

Camphor

Solution

Reaction (a) is a Clemmensen reduction, and reaction (b) is a Wolff-Kishner reduction.

(a)

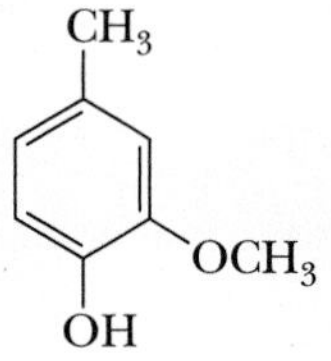

(b)

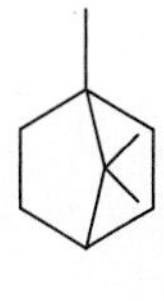

Problem 16.13

Complete the following reactions.

(a)

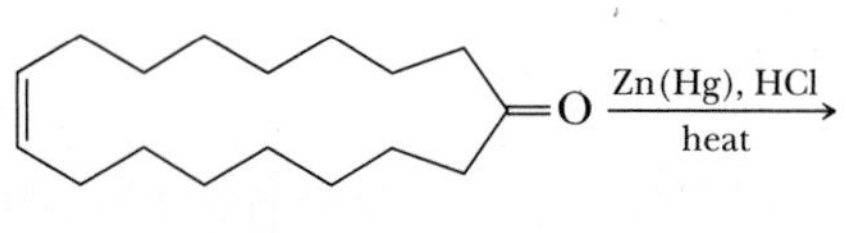

Civetone
(from the civet cat; used in perfumery)

(b)

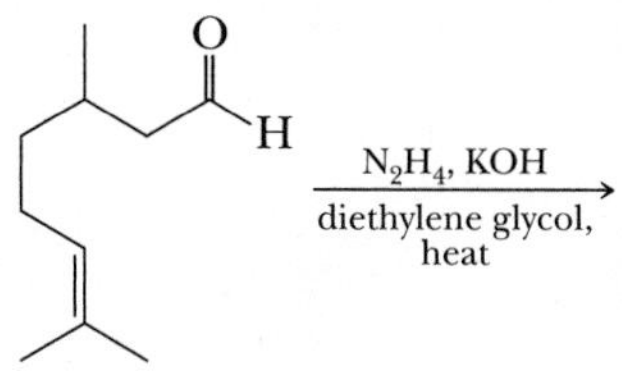

Citronellal
(from citronella and lemon grass oils)

16.12 Reactions at an α-Carbon

A. Racemization

When enantiomerically pure (either *R* or *S*) 3-phenyl-2-butanone is dissolved in ethanol, no change occurs in the optical activity of the solution over time. If, however, a trace of either acid (e.g., aqueous or gaseous HCl) or base (e.g., sodium ethoxide) is added, the optical activity of the solution begins to decrease gradually and eventually drops to zero. When 3-phenyl-2-butanone is isolated from this solution, it is found to be a racemic mixture. Furthermore, the rate of racemization is proportional to the concentration of acid or base. These observations can be explained by a rate-determining acid- or base-catalyzed formation of an achiral enol intermediate. Tautomerism of the achiral enol to the chiral keto form generates the *R* and *S* enantiomers with equal probability.

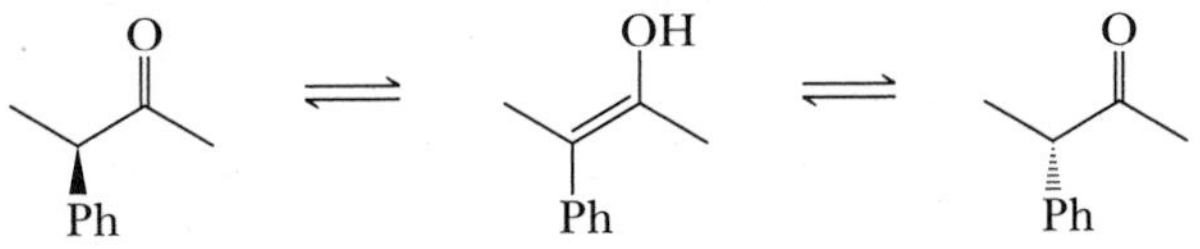

(*R*)-3-Phenyl-2-butanone An achiral enol **(*S*)-3-Phenyl-2-butanone**

Racemization by this mechanism occurs only at α-carbon chiral centers with at least one α-hydrogen.

B. Deuterium Exchange

When an aldehyde or a ketone with one or more α-hydrogens is dissolved in an aqueous solution that is enriched with D_2O and contains catalytic amounts of either D^+ or OD^-, exchange of α-hydrogens occurs at a rate that is proportional to the concentration of the acid or base catalyst. We account for incorporation of deuterium by proposing a rate-determining acid- or base-catalyzed enolization followed by incorporation of deuterium as the enol form converts to the keto form.

$$CH_3\overset{\overset{\displaystyle O}{\|}}{C}CH_3 + 6\,D_2O \xrightleftharpoons{D^+ \text{ or } OD^-} CD_3\overset{\overset{\displaystyle O}{\|}}{C}CD_3 + 6\,HOD$$

Acetone **Acetone-d_6**

Deuterium exchange has two values. First, by observing changes in hydrogen ratios before and after deuterium exchange, it is possible to determine the number of exchangeable α-hydrogens in a molecule. Second, exchange of α-hydrogens is a convenient way to introduce an isotopic label into molecules.

In naming compounds, the presence of deuterium is shown by the symbol "*d*" and the number of deuterium atoms is shown by a subscript following it. In addition to acetone-d_6, more than 225 deuterium-labeled compounds are available commercially in isotopic enrichments of up to 99.8 atom % D. Among these are the following:

$CDCl_3$	$CD_3\overset{\overset{\displaystyle O}{\|}}{C}OD$	$CH_3\overset{\overset{\displaystyle O}{\|}}{C}OD$	$NaBD_4$	CH_3CH_2OD
Chloroform-*d*	**Acetic-d_3 acid-*d***	**Acetic acid-*d***	**Sodium borodeuteride**	**Ethanol-*d***

Deuterated solvents, such as $CDCl_3$, acetone-d_6, and benzene-d_6, are used as solvents in ^{1}H-NMR spectroscopy because they lack protons that might otherwise obscure the spectrum of the compound of interest.

C. α-Halogenation

Aldehydes and ketones with at least one α-hydrogen react at the α-carbon with bromine and chlorine to form α-haloaldehydes and α-haloketones as illustrated by bromination of acetophenone.

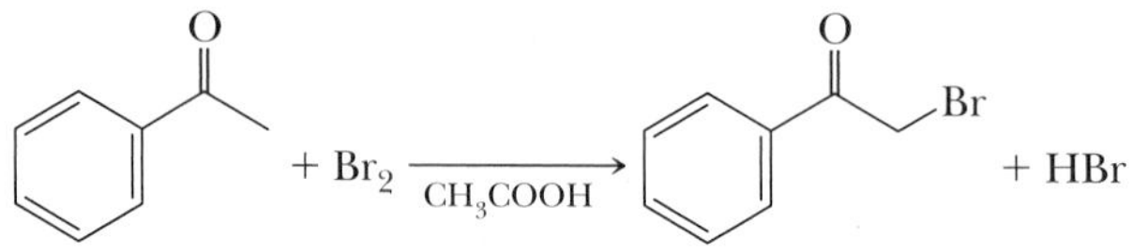

Acetophenone **α-Bromoacetophenone**

Bromination or chlorination at an α-carbon is catalyzed by both acid and base. For acid-catalyzed halogenation, acid generated by the reaction catalyzes further reaction. The slow step of acid-catalyzed halogenation is formation of an enol. This is followed by rapid reaction of the double bond with halogen to give the α-haloketone.

Mechanism 16.15

Acid-Catalyzed α-Halogenation of a Ketone

Step 1: **Keto-enol tautomerism.** Acid-catalyzed keto-enol tautomerism gives the enol.

Watch a video explanation

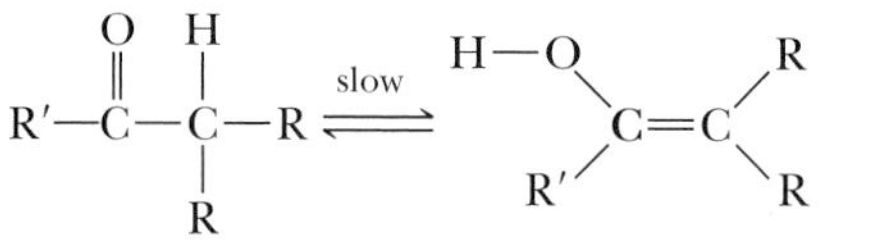

Step 2: **Make a new bond between a π bond and an electrophile.** Nucleophilic attack of the enol on the electrophilic halogen molecule, X_2, gives the conjugate acid of an α-haloketone.

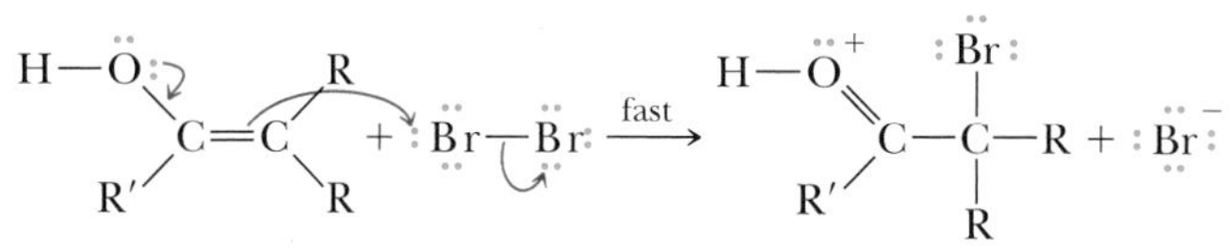

Step 3: **Take a proton away.** Proton transfer to a base (in this case, a molecule of acetic acid) gives the α-haloketone.

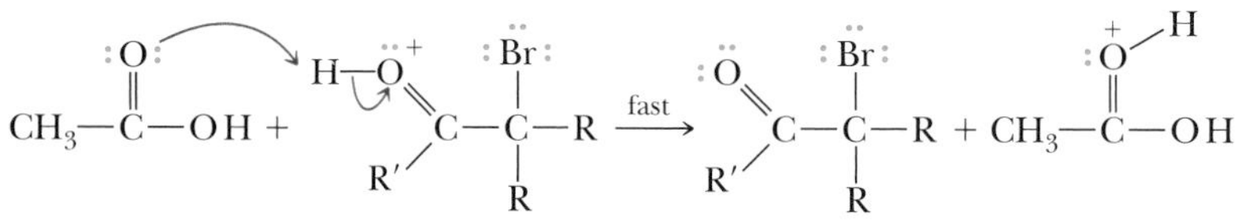

An α-bromoketone

The slow step in base-promoted α-halogenation is removal of an α-hydrogen by base to form an enolate anion, which then reacts with halogen by nucleophilic displacement to form the final product. This procedure for α-halogenation produces HX as a byproduct, and in order to keep the solution basic, it is necessary to add slightly more than one mole of base per mole of aldehyde or ketone. Because base is a reactant required in equimolar amounts, we say that this reaction is base-promoted rather than base-catalyzed.

Mechanism 16.16

Base-Promoted α-Halogenation of a Ketone

Step 1: **Take a proton away.** Proton transfer from the α-carbon to the base gives a resonance-stabilized enolate anion.

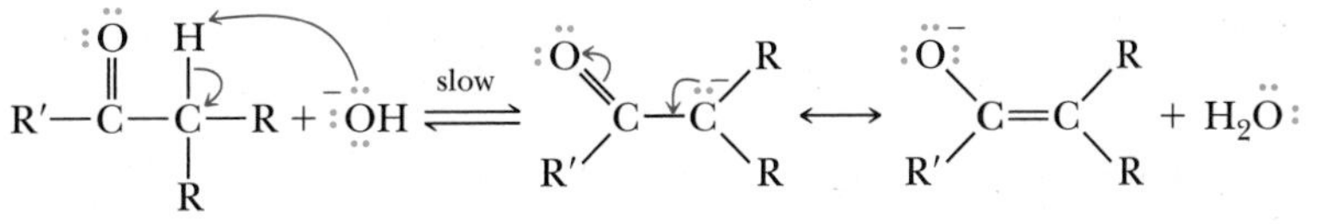

Resonance-stabilized enolate anion

Step 2: **Make a new bond between a nucleophile (π bond) and an electrophile.** Nucleophilic attack of the enolate anion on halogen gives an α-bromoketone.

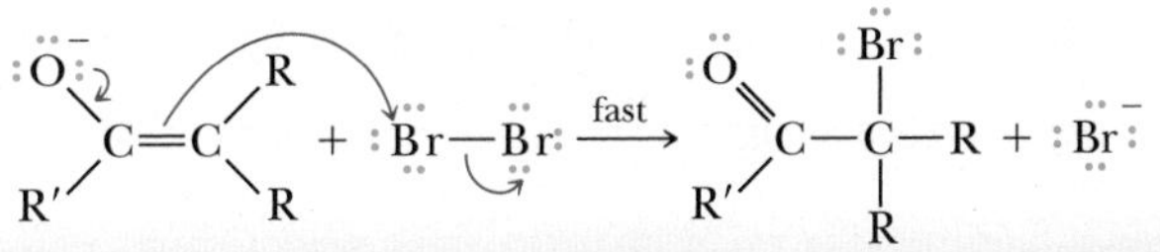

A major difference exists between acid-catalyzed and base-promoted α-halogenation. In principle, both can lead to polyhalogenation. In practice, the rate of acid-catalyzed introduction of a second halogen is considerably lower than the rate of the first halogenation because the electronegative α-halogen decreases the basicity of the carbonyl oxygen toward protonation. Thus, it is generally possible to stop acid-catalyzed halogenation at a single substitution. For base-promoted halogenation, each successive halogenation is more rapid than the previous one because introduction of an electronegative halogen atom on an α-carbon further increases the acidity of remaining α-hydrogens; thus, each successive α-hydrogen is removed more rapidly than the previous one. For this reason, base-promoted halogenation is generally not a useful synthetic reaction.

Study Guide

16.1 Structure and Bonding

- An **aldehyde** contains a carbonyl group bonded to a hydrogen atom and a carbon atom.
- A **ketone** contains a carbonyl group bonded to two carbon atoms.

16.2 Nomenclature

- An aldehyde is named by changing *-e* of the parent alkane to *-al.*
- A ketone is named by changing *-e* of the parent alkane to *-one* and using a number to locate the carbonyl group.

- In naming compounds that contain more than one functional group, the IUPAC system has established an **order of precedence of functions**.
  - *A selection of functional groups ranked from highest to lowest in order of precedence: carboxylic acid (highest), aldehyde, ketone, alcohol, amine, thiol (lowest).*
  - *If the carbonyl group of an aldehyde or a ketone is lower in precedence than other functional groups in the molecule, it is indicated by the infix* -oxo-.

P **16.1–16.3, 16.14, 16.15**

16.3 Physical Properties

- Aldehydes and ketones are polar compounds that engage in dipole-dipole interactions in pure liquid.
  - *They have higher boiling points than nonpolar compounds of comparable molecular weight.*

P **16.16, 16.17**

16.4 Reactions

- An important structural feature of a carbonyl group is the strong dipole moment in which there is a partial negative charge on oxygen and a partial positive charge on carbon.
  - *Lewis acids as well as protons react with carbonyl groups at the oxygen atom.*
  - *Nucleophiles react with carbonyl groups at the carbon atom, breaking the $C{=}O$ π bond in the process.*
  - *One of the most common reaction themes of aldehydes and ketones is addition of a nucleophile to the carbonyl carbon to form a tetrahedral carbonyl addition compound.*
  - *Often, a new chiral center is created by this reaction.*
  - *When none of the starting materials is chiral, a racemic mixture is formed.*
  - *Many of these reactions form new carbon-carbon bonds, making this a very important class of reactions in organic synthesis.*

16.5 Addition of Carbon Nucleophiles

- **Grignard reagents** add to formaldehyde, aldehydes, and ketones to give primary, secondary, and tertiary alcohols, respectively.
- **Organolithium** and **terminal alkyne anions** add to the carbonyl group of aldehydes and ketones to give alcohols according to a mechanism similar to addition of Grignard reagents.
- **Hydrogen cyanide (HCN)** adds to aldehydes and ketones to give cyanohydrins.

P **16.4, 16.18–16.22, 16.51, 16.52, 16.59, 16.62, 16.68, 16.69, 16.72–16.74, 16.78, 16.81**

Problem 16.4

KEY REACTIONS

1. **Reaction with Grignard Reagents (Section 16.5A)** Treating formaldehyde with a Grignard reagent followed by hydrolysis gives a primary alcohol. Similar treatment of any other aldehyde gives a secondary alcohol. Treatment of a ketone gives a tertiary alcohol. The mechanism involves an initial attack of the nucleophilic carbon of the Grignard reagent at the electrophilic carbon of the carbonyl to give the alkoxide Mg salt. The flask is opened and the acid is added to protonate the alkoxide and produce the alcohol.

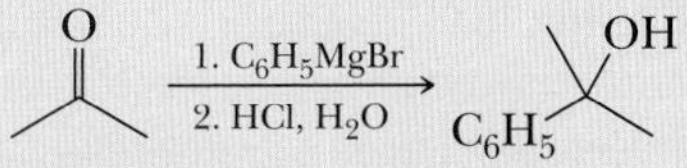

2. **Reaction with Organolithium Reagents (Section 16.5B)** Reactions of aldehydes and ketones with organolithium reagents are similar to those with Grignard reagents.

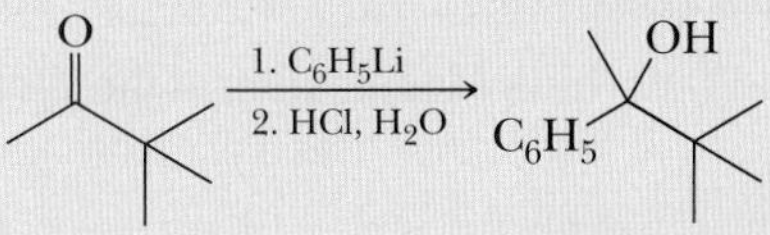

(produced as a racemic mixture)

The mechanism involves an initial attack of the nucleophilic carbon of the organolithium reagent at the electrophilic carbon of the carbonyl to give the alkoxide Li salt. The flask is opened and the acid is added to protonate the alkoxide, thereby producing the alcohol product. Because organolithium reagents are such strong bases, deprotonation of α-hydrogens often competes with nucleophilic attack of aldehydes and ketones.

3. **Reaction with Anions of Terminal Alkynes (Section 16.5C)** Treating an aldehyde or a ketone with the alkali metal salt of a terminal alkyne followed by hydrolysis gives an α-alkynylalcohol.

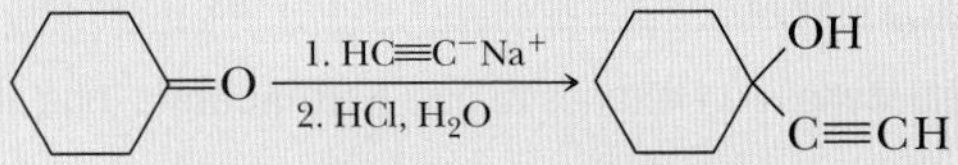

4. **Reaction with HCN to Form Cyanohydrins (Section 16.5D)** For aldehydes and most sterically unhindered aliphatic ketones, equilibrium favors formation of the cyanohydrin. For aryl ketones, equilibrium favors starting materials and little cyanohydrin is obtained. The mechanism involves addition of the cyanide anion to the carbonyl carbon followed by protonation of the resulting alkoxide.

$$C_6H_5\overset{\overset{\displaystyle O}{\|}}{C}H + HC{\equiv}N \xrightarrow{NaCN} C_6H_5\overset{\overset{\displaystyle OH}{|}}{C}HC{\equiv}N$$

(formed as a racemic mixture)

16.6 The Wittig Reaction

- **Wittig reactions** involve addition of phosphonium **ylides** (deprotonated phosphonium salts) with aldehyde and ketone carbonyl groups to give alkenes.
  - E *and* Z *products are generally both formed.*

– *Wittig reagents with anion-stabilizing groups (such as carbonyls) adjacent to the negatively charged carbon are more E selective.*

P 16.5, 16.23–16.28, 16.51, 16.52, 16.53, 16.60, 16.61, 16.72, 16.73, 16.79

Problem 16.5

KEY REACTIONS

5. The Wittig Reaction (Section 16.6) Treating an aldehyde or a ketone with a triphenylphosphonium ylide gives a betaine intermediate, which rearranges to an oxaphosphetane intermediate, which in turn fragments to give triphenylphosphine oxide and an alkene.

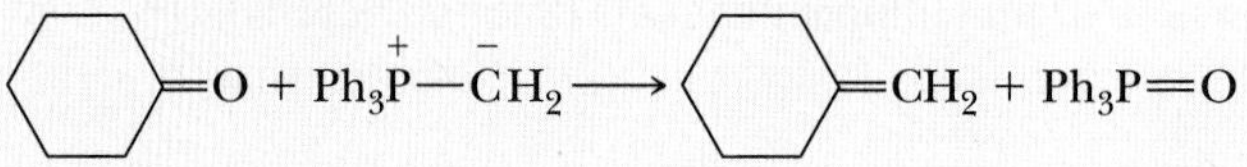

6. The Horner-Emmons-Wadsworth Modification of the Wittig Reaction (Section 16.6) This modification of the original Wittig reaction uses a phosphonate ester derived from an α-haloester, an aldehyde, or a ketone to generate the Wittig carbanion and shows very high E selectivity.

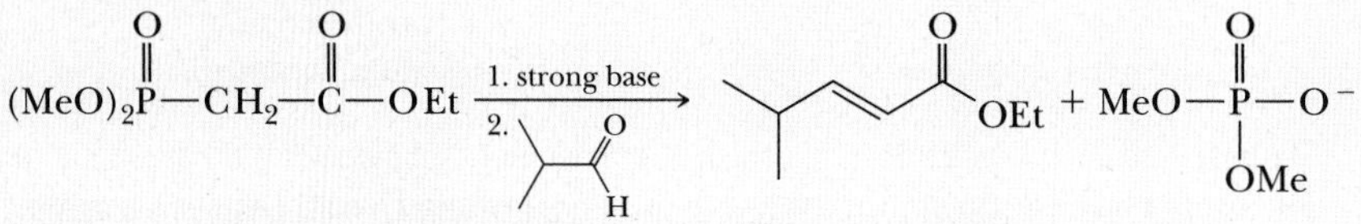

16.7 Addition of Oxygen Nucleophiles

- Water adds to aldehydes and ketones to give hydrates, which are geminal diols.
  – *The reaction is only favorable for simple aldehydes, especially formaldehyde.*
- Alcohols add to aldehydes and ketones to give **hemiacetals** (one alcohol added), then **acetals** (two alcohols added).
  – *Hemiacetals are only stable when five- and six-membered rings are formed from a carbonyl and OH group on the same molecule.*
  – *Hemiacetal formation can be catalyzed by either acid or base.*
  – *Acetal formation can be catalyzed only by acid.*

P 16.6, 16.29–16.37

  – *The overall process of acetal formation from alcohols and an aldehyde or a ketone is acid catalyzed and reversible.*
  – *The relative ratio of alcohol to water in the reaction determines the ratio of carbonyl to acetal species present at equilibrium.*
  – *Water is removed from reactions to favor acetal formation using a* ***Dean-Stark trap****.*
- Carbohydrates are predominantly found in the cyclic hemiacetal form.
  – *The* ***anomeric carbon*** *is the carbon bonded to two oxygen atoms.*
  – *The anomeric carbon can be formed as either the* α *or* β ***anomer****.*

P 16.67, 16.70, 16.71

- Acetals, usually as five- or six-membered rings, are often used as carbonyl-**protecting groups**.
  - *A protecting group reversibly masks the reactivity of a molecule so that unwanted side reactions are prevented.*
  - *Grignard reagents can be prepared from molecules containing carbonyl groups as long as the carbonyl group is protected as a cyclic acetal.*
  - *Cyclic acetal protecting groups are removed by adding excess aqueous acid.*
  - ***Tetrahydropyranyl ethers**, which are cyclic hemiacetals, can be used as a protecting group for alcohols.*

16.7, 16.63, 16.64

Mechanism 16.8

KEY REACTIONS

7. Hydration (Section 16.7A) The degree of hydration is greater for aldehydes than for ketones.

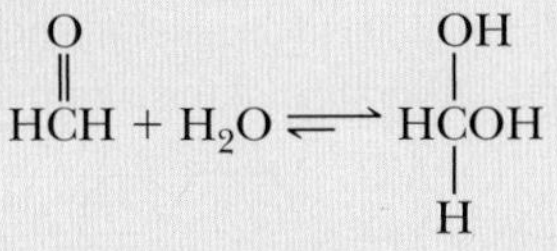

(>99%)

8. Addition of Alcohols to Form Hemiacetals (Section 16.7B) Hemiacetals are only minor components of an equilibrium mixture of aldehyde or ketone and alcohol, except where the —OH and the C—O are parts of the same molecule and a five- or six-membered ring can form. The reaction is catalyzed by acid or base. The mechanism involves protonation of the carbonyl oxygen atom that facilitates attack by an alcohol at the carbonyl carbon followed by loss of a proton to give the hemiacetal.

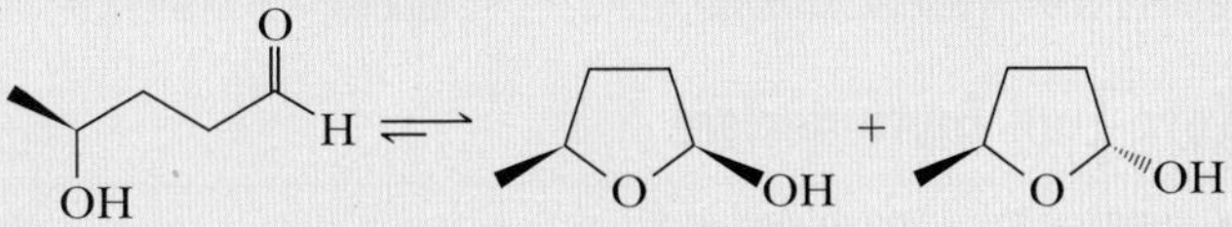

9. Addition of Alcohols to Form Acetals (Section 16.7B) Formation of acetals is catalyzed by acid. Acetals are stable to water and aqueous base but are hydrolyzed in aqueous acid. Acetals are valuable as carbonyl-protecting groups. The mechanism for conversion of a hemiacetal to an acetal involves protonation of the OH group of the hemiacetal followed by loss of water to give a resonance-stabilized cation, which is attacked by the second molecule of alcohol.

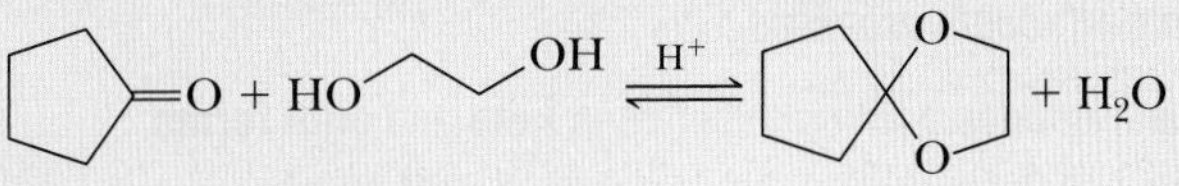

16.8 Addition of Nitrogen Nucleophiles

- Ammonia and primary amines add to aldehydes and ketones to give **imines**, sometimes called **Schiff bases**, that have carbon-nitrogen double bonds.
- Secondary amines, especially cyclic secondary amines such as piperidine, form **enamines** with aldehydes and ketones.

– *Enamines have a carbon-nitrogen single bond with an adjacent carbon-carbon double bond.*

- **Hydrazine** and related compounds react with aldehydes and ketones to give analogous products with carbon-nitrogen double bonds.

P **16.8, 16.38–16.40, 16.43, 16.44**

KEY REACTIONS

10. Addition of Ammonia and Its Derivatives: Formation of Imines (Section 16.8A) Addition of ammonia or a primary amine to the carbonyl group of an aldehyde or a ketone forms a tetrahedral carbonyl addition compound. Loss of water from this intermediate gives an imine. The mechanism for imine formation involves an initial attack of the nucleophilic nitrogen atom on the carbonyl carbon atom followed by proton transfer to the OH, creating an H_2O group that then departs.

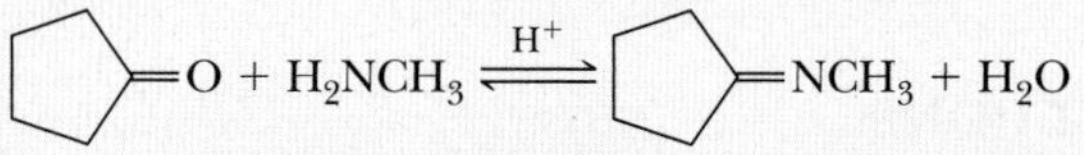
O + H2NCH3
H+
NCH3 + H2O

11. Addition of Secondary Amines: Formation of Enamines (Section 16.8A) Addition of a secondary amine to the carbonyl group of an aldehyde or a ketone forms a tetrahedral carbonyl addition intermediate. Acid-catalyzed dehydration of this intermediate gives an enamine.

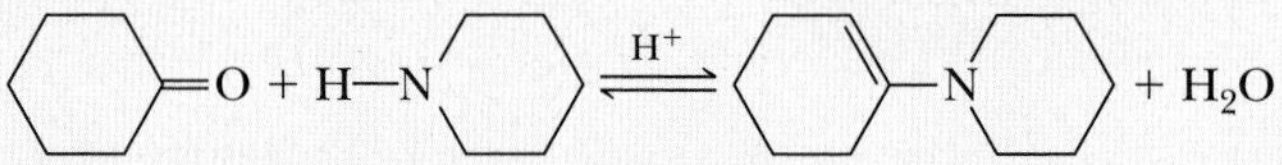
O + H—N
H+
N
+ H2O

12. Addition of Hydrazine and Its Derivatives (Section 16.8B) Treating an aldehyde or a ketone with hydrazine gives a hydrazone.

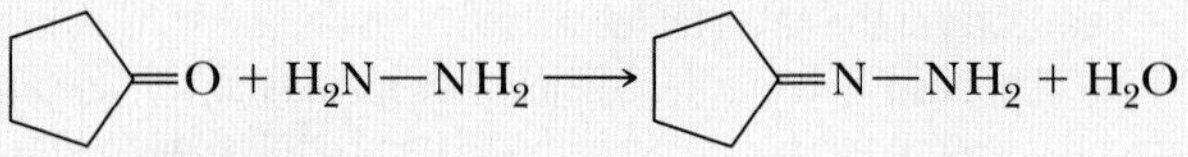
O + H2N—NH2
N—NH2 + H2O

Derivatives of hydrazine react similarly.

16.9 Keto-Enol Tautomerism

- The carbon atom adjacent to a carbonyl group is called an **α-carbon**, and a hydrogen bonded to it is called an **α-hydrogen**.

  – *The pK_a of an α-hydrogen of an aldehyde or a ketone is approximately 20, which makes it less acidic than alcohols but more acidic than terminal alkynes.*

  – *The α-hydrogen is acidic because the deprotonated **enolate anion** is stabilized by delocalization of charge through resonance.*

- Aldehydes and ketones equilibrate between **keto** and **enol forms**.

  Keto-enol equilibration is catalyzed by both acid and base.

  – *Base catalysis of keto-enol equilibration involves an enolate anion intermediate.*

  – *The keto form is favored for most aldehydes and ketones.*

  – *Molecules that favor the enol form at equilibrium generally have conjugated enols.*

P **16.9, 16.10, 16.41, 16.42, 16.57, 16.58, 16.65**

KEY REACTIONS

13. Keto-Enol Tautomerism (Section 16.9B) The keto form predominates at equilibrium, except for those aldehydes and ketones in which the enol is stabilized by resonance or hydrogen bonding.

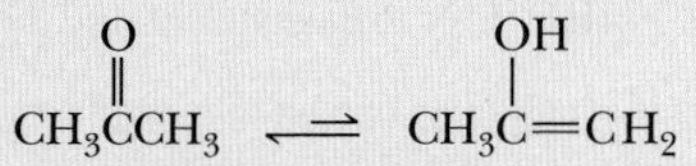

Keto form (approx 99.9%) Enol form

16.10 Oxidation

- Aldehydes are easily oxidized to carboxylic acids using a variety of reagents, including chromic acid, silver salts, peroxides, and molecular oxygen, O_2. Aldehydes are also selectively oxidized, even in the presence of alcohols, to carboxylic acids by treatment with sodium chlorite ($NaClO_2$) and NaH_2PO_4 with added 2-methyl-2-butene.
- Ketones are not easily oxidized, requiring extremely strong oxidizing agents as well as heat.

P **16.11, 16.43, 16.44, 16.72, 16.73**

KEY REACTIONS

14. Oxidation of an Aldehyde to a Carboxylic Acid (Section 16.10A) The aldehyde group is among the most easily oxidized functional groups. Oxidizing agents include H_2CrO_4, $KMnO_4$, Ag_2O, Tollens' reagent, H_2O_2, O_2, and the Pinnick conditions ($NaClO_2$, NaH_2PO_4, and 2-methyl-2-butene).

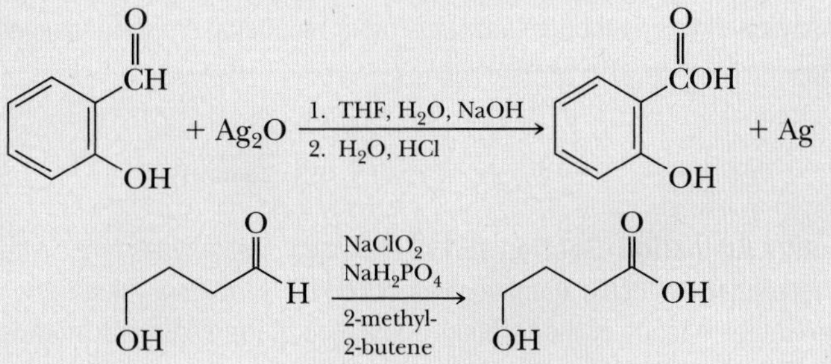

16.11 Reduction

- Aldehydes and ketones are reduced to primary and secondary alcohols, respectively.
  - ***Metal hydride*** *reducing agents such as $LiAlH_4$ and $NaBH_4$ are effective for reducing aldehydes and ketones.*
  - *Metal hydride reducing agents do not reduce carbon-carbon double bonds.*
  - *Hydrogenation using H_2 and a transition metal catalyst can be used to reduce aldehyde or ketone carbonyls, although carbon-carbon double bonds in a molecule may also be reduced.*
  - *Carbon-carbon double bonds are easier to reduce than carbonyls using hydrogenation, so conditions can often be found in which only a carbon-carbon double bond is reduced in the presence of an aldehyde or ketone carbonyl group.*

- *Conversely, by using metal hydride reagents, carbonyls can be reduced without reducing carbon-carbon double bonds.*

- In **reductive amination**, ketones or aldehydes react with amines in the presence of an appropriate reducing agent such as $NaBH_3CN$ to give substituted amines.
- Aldehyde and ketone carbonyl groups can be reduced to methyl or methylene groups, respectively, using two different complementary reactions.
  - *The* ***Clemmensen reduction*** *uses amalgamated zinc, Zn(Hg), in strong acid.*
  - *The* ***Wolff-Kishner reduction*** *uses hydrazine and base.*
  - *The Clemmensen reduction is used when the molecule is stable to strong acid, and the Wolff-Kishner reduction is used when the molecule is stable to strong base.*

P 16.12, 16.13, 16.43, 16.44, 16.51, 16.52, 16.59

KEY REACTIONS

15. Metal Hydride Reduction (Section 16.11A) Both $LiAlH_4$ and $NaBH_4$ are selective in that neither reduces isolated carbon-carbon double or triple bonds. The mechanism for metal hydride reducing agents involves initial nucleophilic attack by an H^- equivalent followed by protonation of the alkoxide ion to give the alcohol.

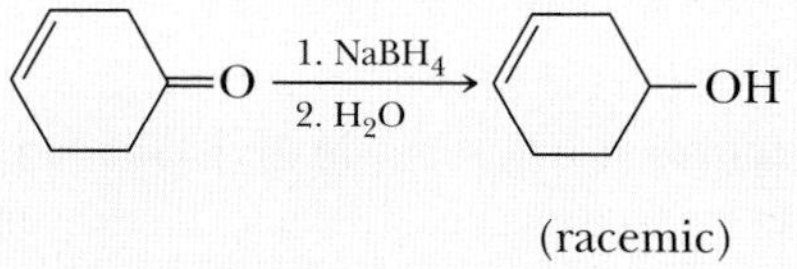

16. Catalytic Reduction (Section 16.11B) Catalytic reduction of the carbonyl group of an aldehyde or a ketone to a hydroxyl group is simple to carry out, and yields of the alcohols are high. A disadvantage of this method is that some other functional groups, including carbon-carbon double and triple bonds, may also be reduced.

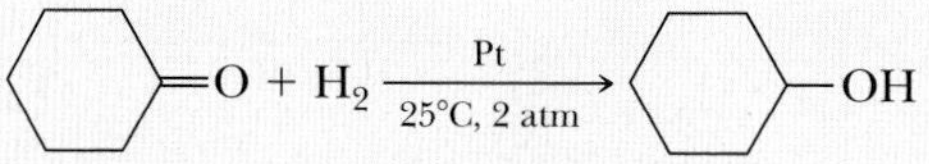

17. Reductive Amination (Section 16.11D) Ketones or aldehydes react with amines in the presence of an appropriate reducing agent such as $NaBH_3CN$ to give substituted amines. The mechanism involves initial reaction of the carbonyl and amine to form an imine intermediate that is reduced by the $NaBH_3CN$ to give the substituted amine.

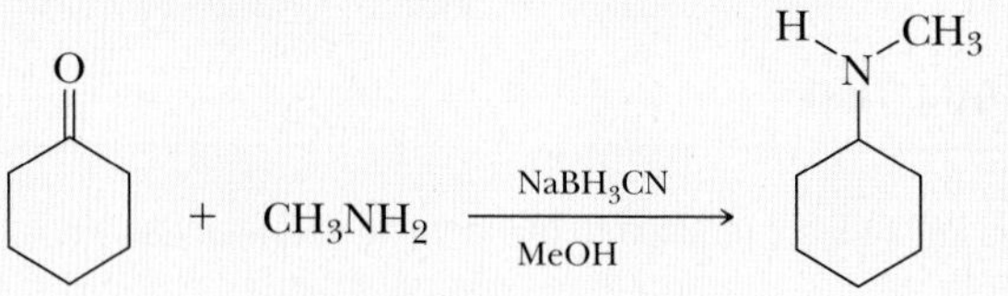

18. Clemmensen Reduction (Section 16.11E) Reduction of the carbonyl group of an aldehyde or a ketone using amalgamated zinc in the presence of concentrated hydrochloric acid gives a methylene group.

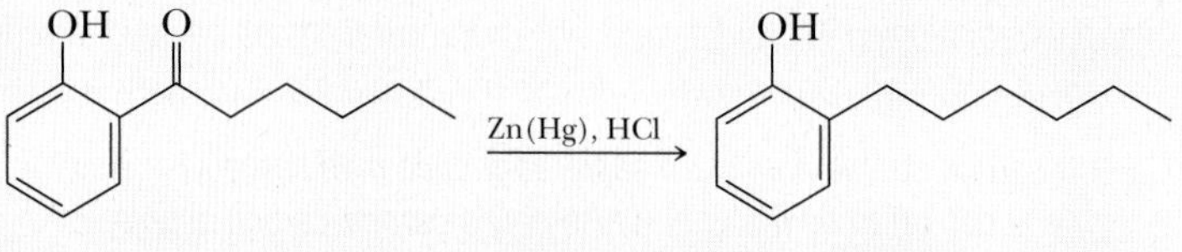

19. Wolff-Kishner Reduction (Section 16.11E) Formation of a hydrazone followed by treatment with base (commonly KOH in diethylene glycol or potassium *tert*-butoxide in dimethyl sulfoxide) reduces the carbonyl group of an aldehyde or a ketone to a methylene group. The mechanism involves initial hydrazone formation followed by base-catalyzed tautomerization, deprotonation, loss of N_2, and reprotonation of the carbanion intermediate to give the fully reduced product.

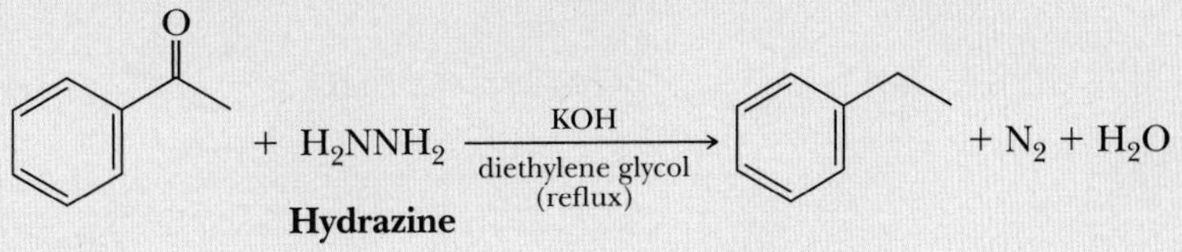
O
+ H2NNH2
KOH
diethylene glycol
(reflux)
Hydrazine
+ N2 + H2O

16.12 Reactions at an α-Carbon

- The α-carbon of aldehydes and ketones has special reactivity derived from the keto-enol equilibrium (referred to as **tautomerization**) that occurs in acid or base.
  - *α-Carbon chiral centers of aldehydes and ketones racemize in the presence of acid or base through formation of an achiral enol during keto-enol tautomerization.*
  - *Deuterium can be exchanged for α-hydrogens catalyzed by acid or base through formation of the enol during keto-enol tautomerization.*
- Aldehydes or ketones with at least one α-hydrogen react with halogens in acid or base to form α-haloaldehydes and α-haloketones.

P **16.45–16.50, 16.66**

▶ **Mechanism 16.15**

KEY REACTIONS

20. Deuterium Exchange at an α-Carbon (Section 16.12B) Acid- or base-catalyzed deuterium exchange at an α-carbon involves formation of an enol or enolate anion intermediate.

$$CH_3\overset{\overset{O}{\|}}{C}CH_3 + 6\,D_2O \xrightleftharpoons{DCl} CD_3\overset{\overset{O}{\|}}{C}CD_3 + 6\,HOD$$

21. Halogenation at an α-Carbon (Section 16.12C) The rate-determining step in acid-catalyzed α-halogenation is the formation of an enol. In base-promoted α-halogenation, it is formation of an enolate anion. Acid-catalyzed α-halogenation involves reaction of the halogen with the enol form at the carbonyl compound. Base-promoted α-halogenation involves reaction of the enolate anion intermediate with the halogen.

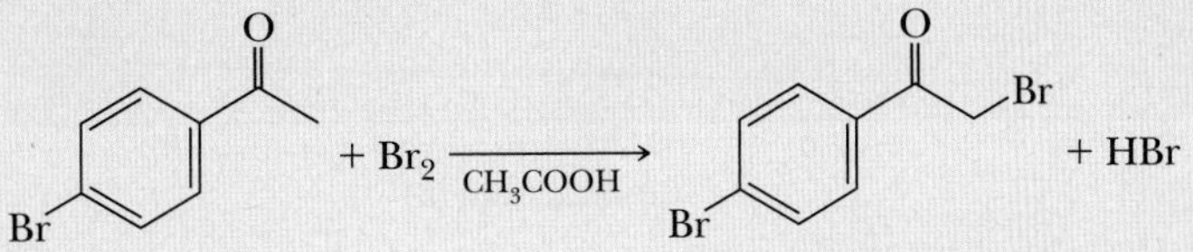
O
Br
+ Br2
CH3COOH
O
Br
Br
+ HBr

Problems

Red numbers indicate applied problems.

Structure and Nomenclature

16.14 Name each compound, showing stereochemistry where relevant.

(a) 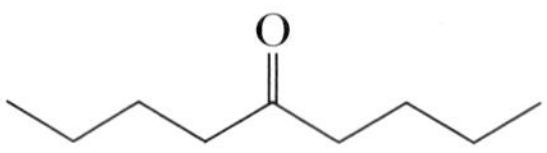

(b) 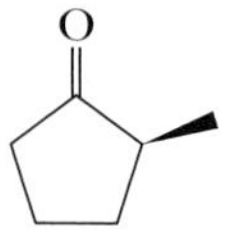

(c) 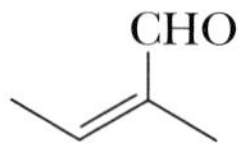

(d) 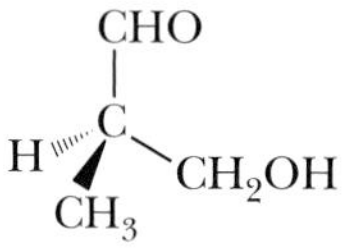

(e) 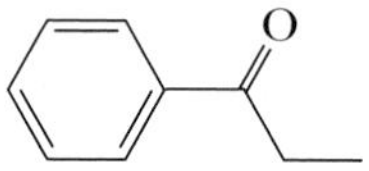

(f) 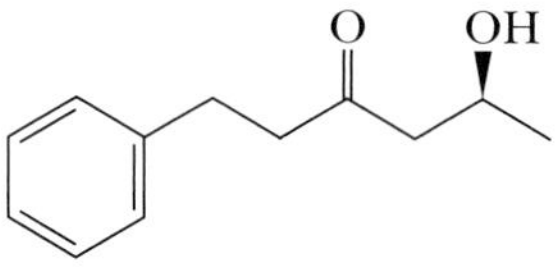

(g) 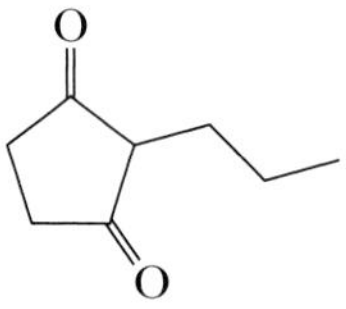

(h) 

(i) 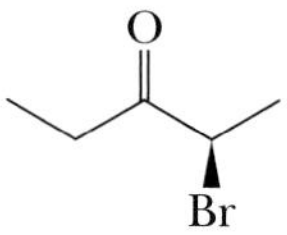

16.15 Draw a structural formula for each compound.

(a) 1-Chloro-2-propanone
(b) 3-Hydroxybutanal
(c) 4-Hydroxy-4-methyl-2-pentanone
(d) 3-Methyl-3-phenylbutanal
(e) 1,3-Cyclohexanedione
(f) 3-Methyl-3-buten-2-one
(g) 5-Oxohexanal
(h) 2,2-Dimethylcyclohexanecarbaldehyde
(i) 3-Oxobutanoic acid

16.16 The infrared spectrum of compound A, $C_6H_{12}O$, shows a strong, sharp peak at 1724 cm^{-1}. From this information and its ^1H-NMR spectrum, deduce the structure of compound A.

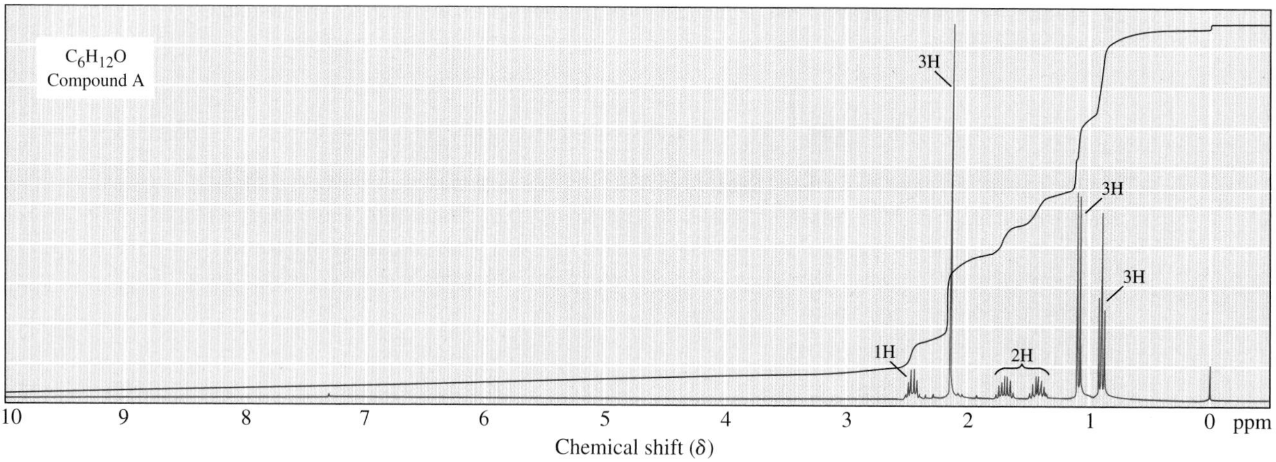

16.17 Following are ^1H-NMR spectra for compounds B ($C_6H_{12}O_2$) and C ($C_6H_{10}O$). Upon warming in dilute acid, compound B is converted to compound C. Deduce the structural formulas for compounds B and C.

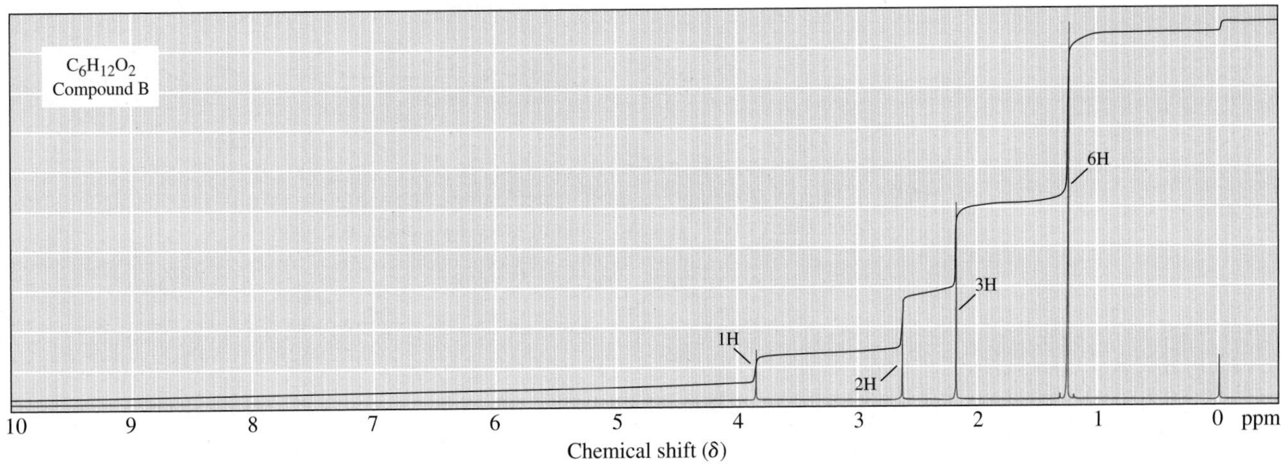

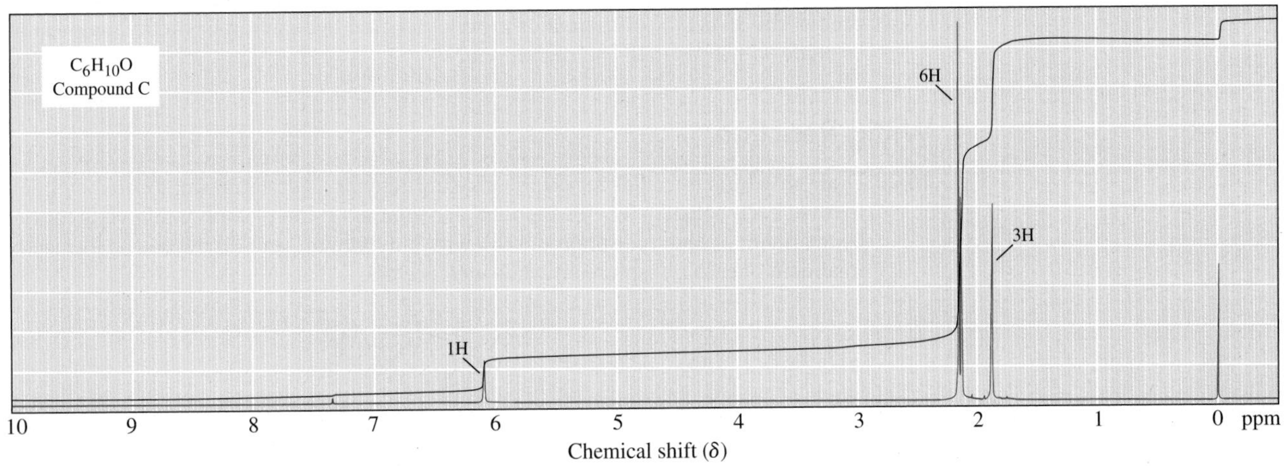

Addition of Carbon Nucleophiles

16.18 Draw structural formulas for the product formed by treating each compound with propylmagnesium bromide followed by aqueous HCl.

(a) CH_2O **(b)**  **(c)** 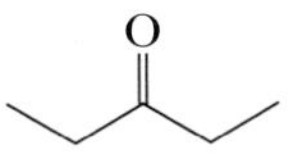 **(d)** 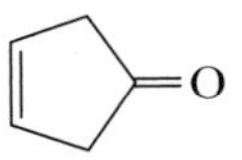

16.19 Suggest a synthesis for the following alcohols starting from an aldehyde or a ketone and an appropriate Grignard reagent. Below each target molecule is the number of combinations of Grignard reagent and aldehyde or ketone that might be used.

(a) 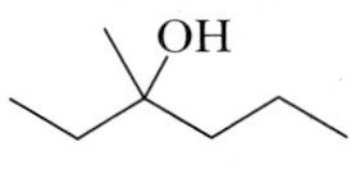
3 Combinations (racemic)

(b) 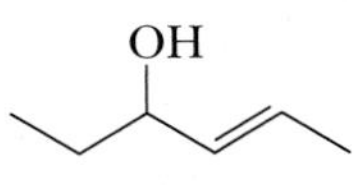
2 Combinations (racemic)

(c) 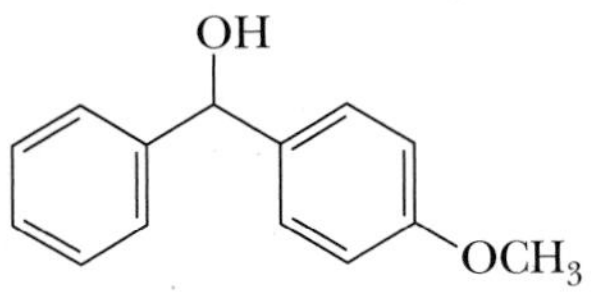
2 Combinations (racemic)

16.20 Show how to synthesize the following alcohol using 1-bromopropane, propanal, and ethylene oxide as the only sources of carbon atoms.

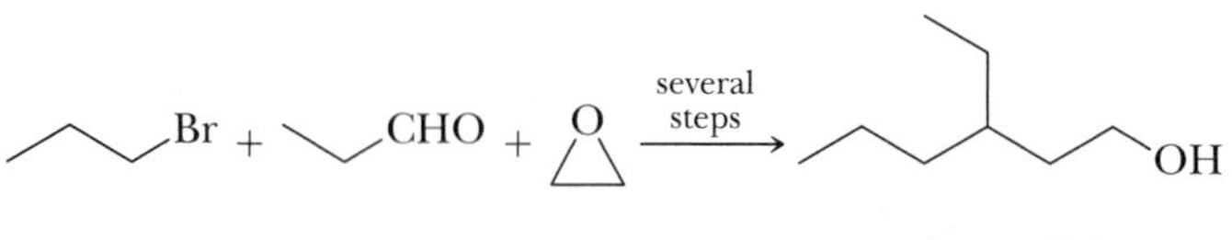

(racemic)

16.21 1-Phenyl-2-butanol is used in perfumery. Show how to synthesize this alcohol from bromobenzene, 1-butene, and any necessary inorganic reagents.

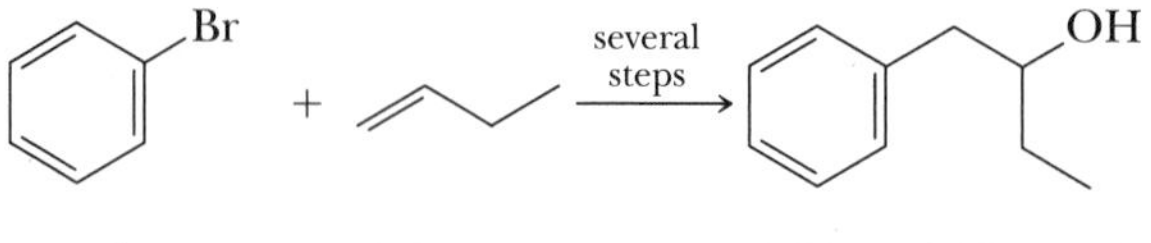

Bromobenzene **1-Butene** **1-Phenyl-2-butanol**
(racemic)

16.22 With organolithium and organomagnesium compounds, approach to the carbonyl carbon from the less hindered direction is generally preferred. Assuming this is the case, predict the structure of the major product formed by reaction of methylmagnesium bromide with 4-*tert*-butylcyclohexanone.

Wittig Reaction

16.23 Draw structural formulas for (1) the alkyltriphenylphosphonium salt formed by treatment of each haloalkane with triphenylphosphine, (2) the phosphonium ylide formed by treatment of each phosphonium salt with butyllithium, and (3) the alkene formed by treatment of each phosphonium ylide with acetone.

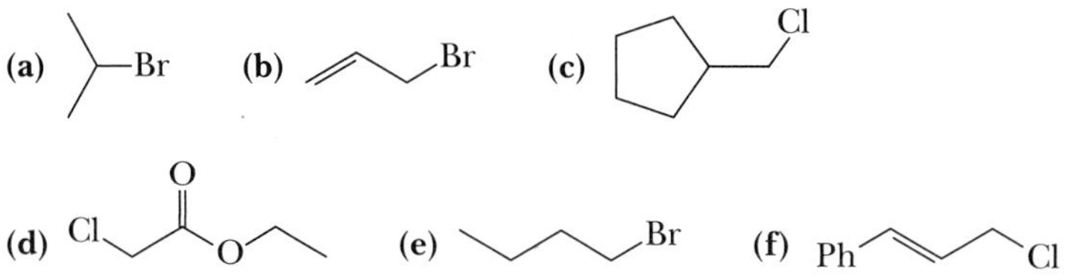

16.24 Show how to bring about the following conversions using a Wittig reaction.

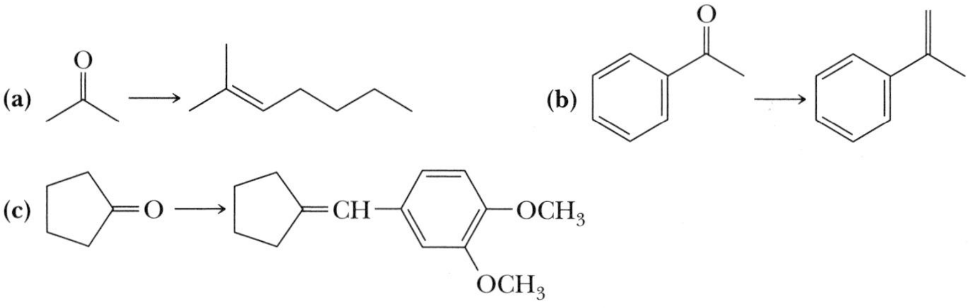

16.25 The Wittig reaction can be used for the synthesis of conjugated dienes, as, for example, 1-phenyl-1,3-pentadiene.

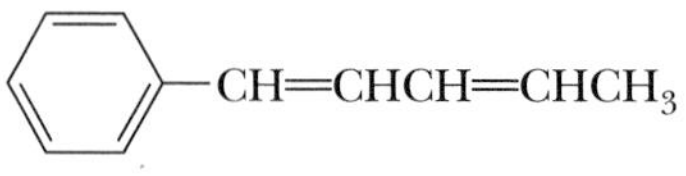

1-Phenyl-1,3-pentadiene

Propose two sets of reagents that might be combined in a Wittig reaction to give this conjugated diene.

16.26 Wittig reactions with the following α-chloroethers can be used for the synthesis of aldehydes and ketones.

$ClCH_2OCH_3$ (A) $\qquad$ $ClCH(CH_3)OCH_3$ (B)

(a) Draw the structure of the triphenylphosphonium salt and Wittig reagent formed from each chloroether.

(b) Draw the structural formula of the product formed by treating each Wittig reagent with cyclopentanone. Note that the functional group is an enol ether or, alternatively, a vinyl ether.

(c) Draw the structural formula of the product formed on acid-catalyzed hydrolysis of each enol ether from part (b).

16.27 It is possible to generate sulfur ylides in a manner similar to that used to produce phosphonium ylides. For example, treating a sulfonium salt with a strong base gives the sulfur ylide.

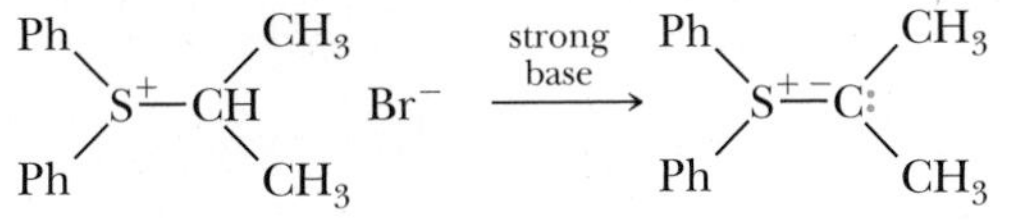

A sulfonium bromide salt $\qquad$ A sulfur ylide

Sulfur ylides react with ketones to give epoxides. Suggest a mechanism for this reaction.

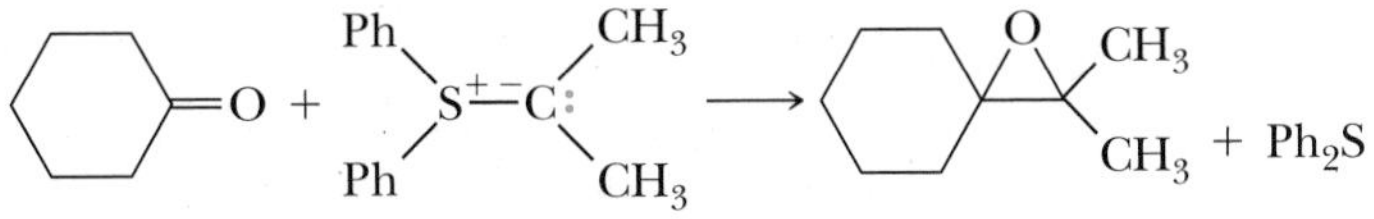

16.28 Propose a structural formula for compound D and for the product, $C_9H_{14}O$, formed in this reaction sequence.

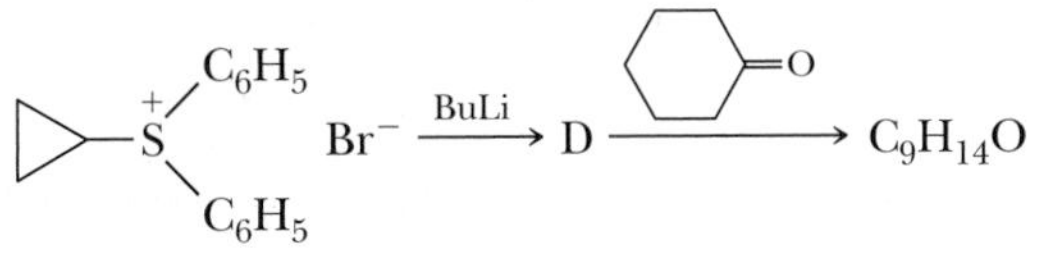

Addition of Oxygen Nucleophiles

16.29 5-Hydroxyhexanal forms a six-membered cyclic hemiacetal, which predominates at equilibrium in aqueous solution.

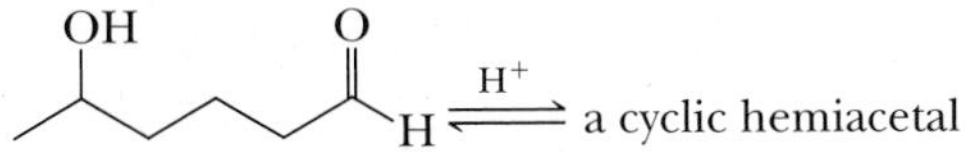

5-Hydroxyhexanal

(a) Draw a structural formula for this cyclic hemiacetal.

(b) How many stereoisomers are possible for 5-hydroxyhexanal?

(c) How many stereoisomers are possible for this cyclic hemiacetal?

(d) Draw alternative chair conformations for each stereoisomer and label groups axial or equatorial. Also predict which of the alternative chair conformations for each stereoisomer is more stable.

16.30 Draw structural formulas for the hemiacetal and then the acetal formed from each pair of reactants in the presence of an acid catalyst.

(a) 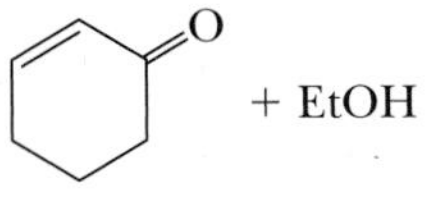

(b) 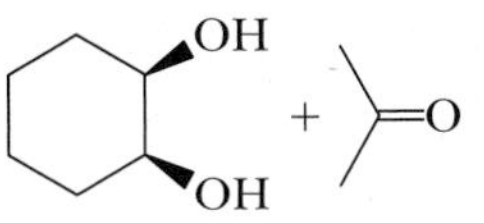

(c) 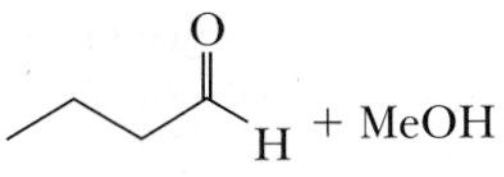

16.31 Draw structural formulas for the products of hydrolysis of the following acetals in aqueous HCl.

(a) 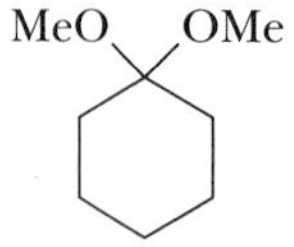 **(b)** 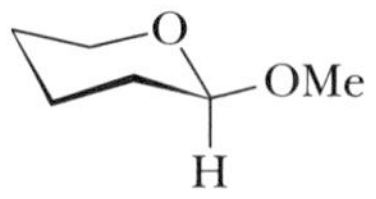 **(c)** 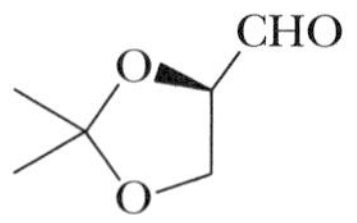

16.32 Propose a mechanism to account for the formation of a cyclic acetal from 4-hydroxypentanal and one equivalent of methanol. If the carbonyl oxygen of 4-hydroxypentanal is enriched with oxygen-18, do you predict that the oxygen label appears in the cyclic acetal or in the water?

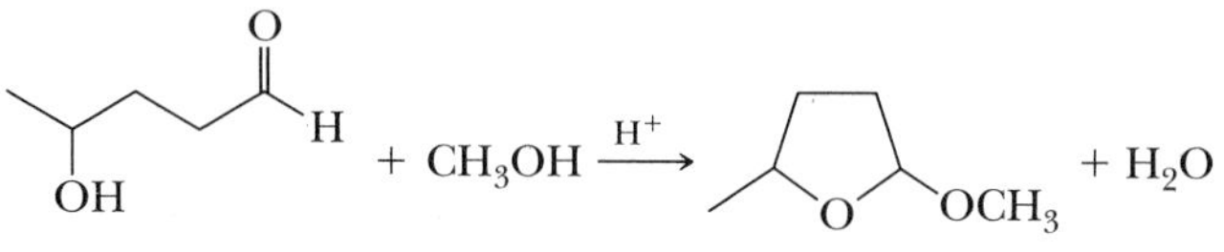

4-Hydroxypentanal A cyclic acetal (racemic)

16.33 Propose a mechanism for this acid-catalyzed hydrolysis.

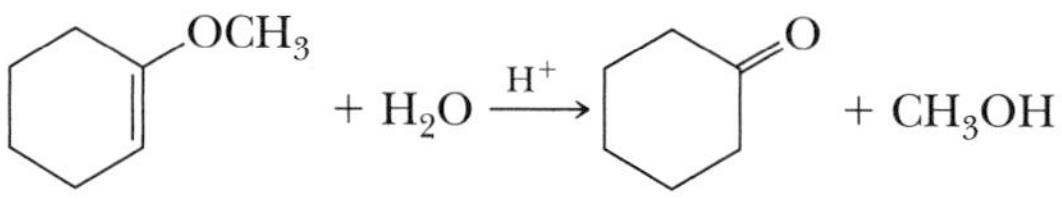

16.34 In Section 11.5, we saw that ethers, such as diethyl ether and tetrahydrofuran, are quite resistant to the action of dilute acids and require hot concentrated HI or HBr for cleavage. However, acetals in which two ether groups are linked to the same carbon undergo hydrolysis readily, even in dilute aqueous acid. How do you account for this marked difference in chemical reactivity toward dilute aqueous acid between ethers and acetals?

16.35 Show how to bring about the following conversion.

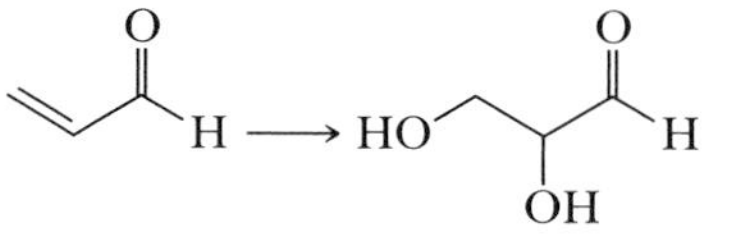

(racemic)

16.36 A primary or secondary alcohol can be protected by conversion to its tetrahydropyranyl ether. Why is formation of THP ethers by this reaction limited to primary and secondary alcohols?

16.37 Which of these molecules will cyclize to give the insect pheromone frontalin?

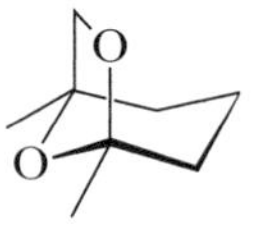

Frontalin

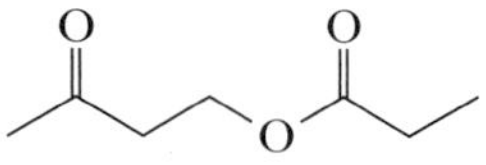

(A)

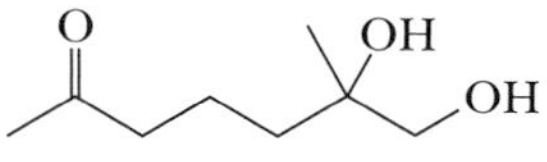

(B)

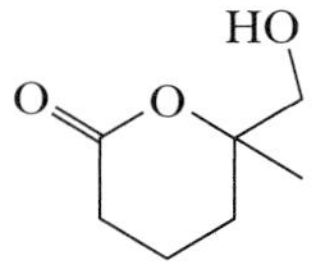

(C)

Addition of Nitrogen Nucleophiles

16.38 Draw a structural formula for the product of each acid-catalyzed reaction.

(a) Phenylacetaldehyde + hydrazine ⟶
(b) Cyclopentanone + semicarbazide ⟶
(c) Acetophenone + 2,4-dinitrophenylhydrazine ⟶
(d) Benzaldehyde + hydroxylamine ⟶

16.39 Following are structural formulas for amphetamine and methamphetamine.

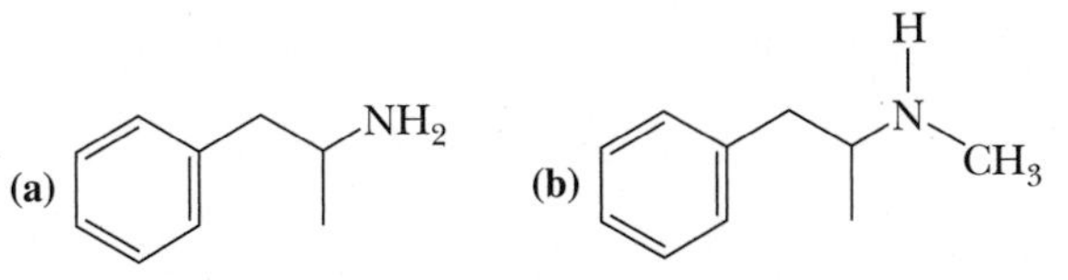

Amphetamine (racemic) **Methamphetamine** (racemic)

The major central nervous system effects of amphetamine and amphetamine-like drugs are locomotor stimulation, euphoria and excitement, stereotyped behavior, and anorexia. Show how each drug can be synthesized by reductive amination of an appropriate aldehyde or ketone and amine. For the structural formulas of several more anorexics, see Problems 23.64 and 23.65.

16.40 Following is the final step in the synthesis of the antiviral drug Rimantadine (Problem 7.19).

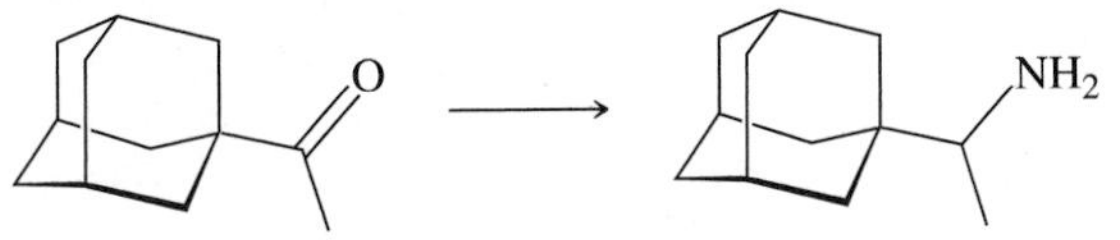

(a) Describe experimental conditions to bring about this conversion.
(b) Is Rimantadine chiral? How many stereoisomers are possible for it?

Keto-Enol Tautomerism

16.41 The following molecule belongs to a class of compounds called enediols; each carbon of the double bond carries an —OH group. Draw structural formulas for the α-hydroxyketone and the α-hydroxyaldehyde with which this enediol is in equilibrium.

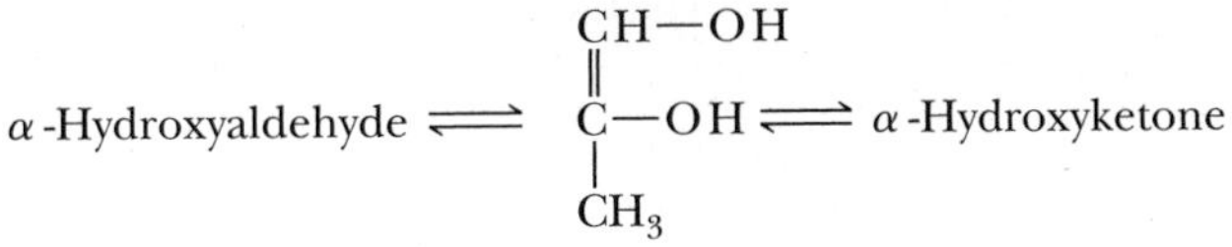

An enediol

16.42 When *cis*-2-decalone is dissolved in ether containing a trace of HCl, an equilibrium is established with *trans*-2-decalone. The latter ketone predominates in the equilibrium mixture.

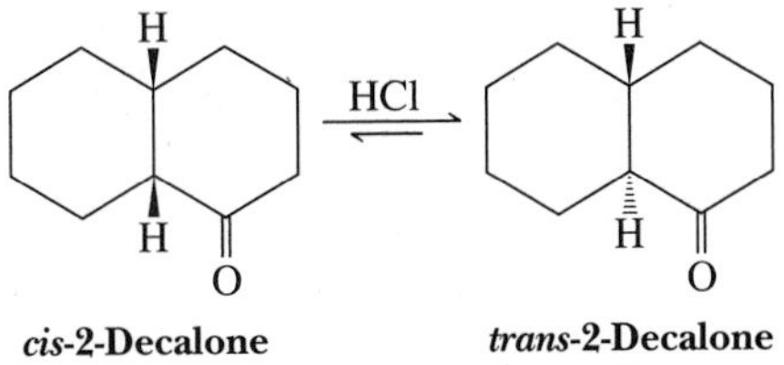

***cis*-2-Decalone** ***trans*-2-Decalone**

Propose a mechanism for this isomerization and account for the fact that the *trans* isomer predominates at equilibrium.

Oxidation/Reduction of Aldehydes and Ketones

16.43 Draw a structural formula for the product formed by treating butanal with each reagent.

(a) $LiAlH_4$ followed by H_2O
(b) $NaBH_4$ in CH_3OH/H_2O
(c) H_2/Pt
(d) $Ag(NH_3)_2^+$ in NH_3/H_2O
(e) H_2CrO_4, heat
(f) $HOCH_2CH_2OH$, HCl
(g) Zn(Hg)/HCl
(h) N_2H_4, KOH at 250°C
(i) $C_6H_5NH_2$
(j) $C_6H_5NHNH_2$
(k) $NaClO_2$, NaH_2PO_4, 2-methyl-2-butene

16.44 Draw a structural formula for the product of the reaction of acetophenone with each reagent given in Problem 16.43.

Reactions at an α-Carbon

16.45 The following bicyclic ketone has two α-carbons and three α-hydrogens. When this molecule is treated with D_2O in the presence of an acid catalyst, only two of the three α-hydrogens exchange with deuterium. The α-hydrogen at the bridgehead does not exchange.

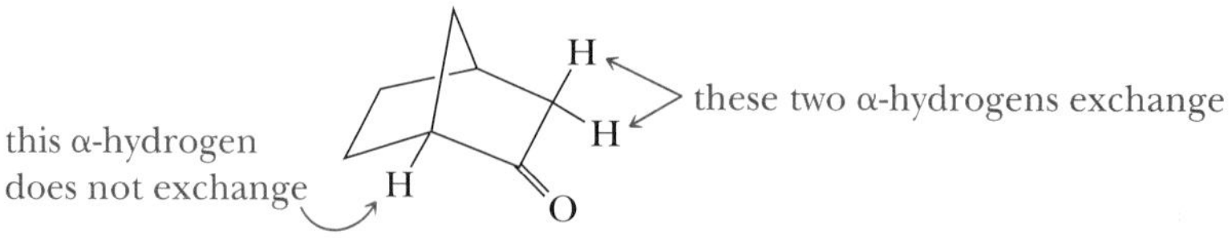

How do you account for the fact that two α-hydrogens do exchange but the third does not? You will find it helpful to build models of the enols by which exchange of α-hydrogens occurs.

16.46 Propose a mechanism for this reaction.

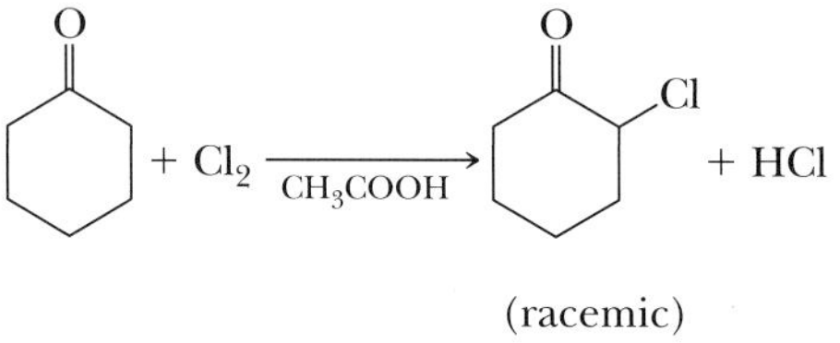

16.47 The base-promoted rearrangement of an α-haloketone to a carboxylic acid, known as the Favorskii rearrangement, is illustrated by the conversion of 2-chlorocyclohexanone to cyclopentanecarboxylic acid.

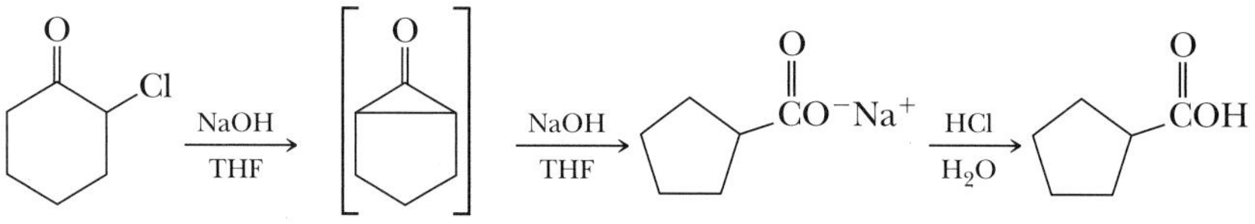

A proposed intermediate

It is proposed that NaOH first converts the α-haloketone to the substituted cyclopropanone shown in brackets and then to the sodium salt of cyclopentanecarboxylic acid.

(a) Propose a mechanism for base-promoted conversion of 2-chlorocyclohexanone to the proposed intermediate.

(b) Propose a mechanism for base-promoted conversion of the proposed intermediate to sodium cyclopentanecarboxylate.

16.48 If the Favorskii rearrangement of 2-chlorocyclohexanone is carried out using sodium ethoxide in ethanol, the product is ethyl cyclopentanecarboxylate.

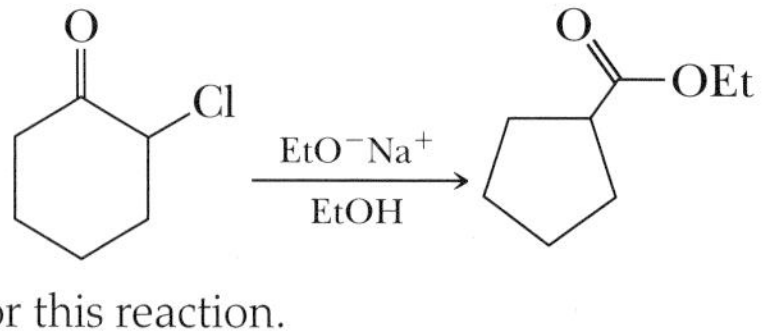

Propose a mechanism for this reaction.

16.49 (*R*)-Pulegone, readily available from pennyroyal oil, is an important enantiopure building block for organic syntheses. Propose a mechanism for each step in this transformation of pulegone.

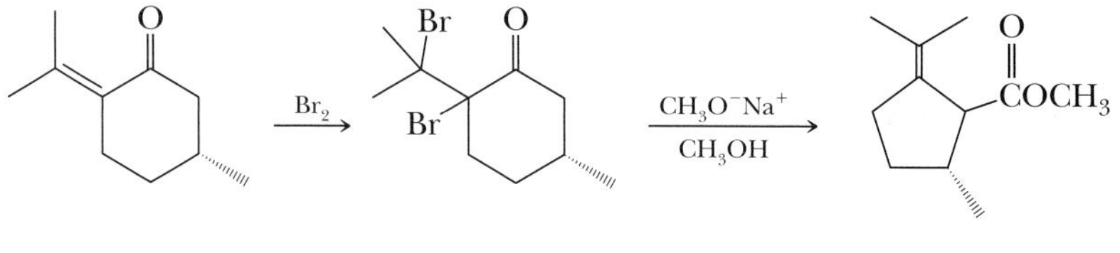

(*R*)-(+)-Pulegone

16.50 (*R*)-Pulegone is converted to (*R*)-citronellic acid by addition of HCl followed by treatment with NaOH.

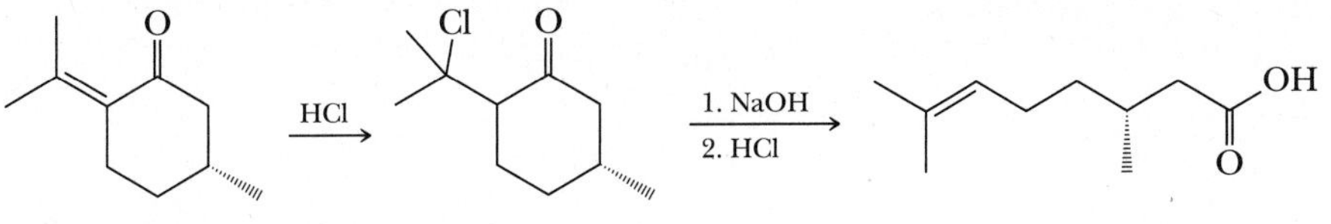

(*R*)-Pulegone **(*R*)-Citronellic acid**

Propose a mechanism for each step in this transformation and account for the regioselectivity of HCl addition.

Synthesis

16.51 Starting with cyclohexanone, show how to prepare these compounds. In addition to the given starting material, use any other organic or inorganic reagents as necessary.

(a) Cyclohexanol
(b) Cyclohexene
(c) *cis*-1,2-Cyclohexanediol
(d) 1-Methylcyclohexanol
(e) 1-Methylcyclohexene
(f) 1-Phenylcyclohexanol
(g) 1-Phenylcyclohexene
(h) Cyclohexene oxide
(i) *trans*-1,2-Cyclohexanediol

16.52 Show how to convert cyclopentanone to these compounds. In addition to cyclopentanone, use any other organic or inorganic reagents as necessary.

(a) 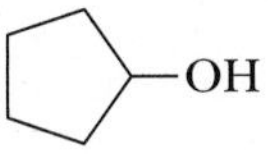
(b) 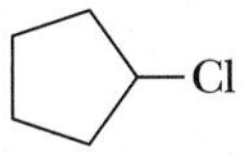
(c) 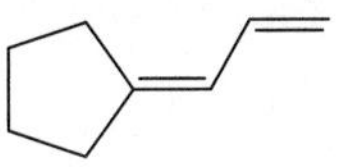
(d) 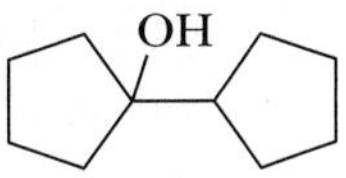

16.53 Disparlure is a sex attractant of the gypsy moth *(Porthetria dispar)*. It has been synthesized in the laboratory from the following *Z* alkene.

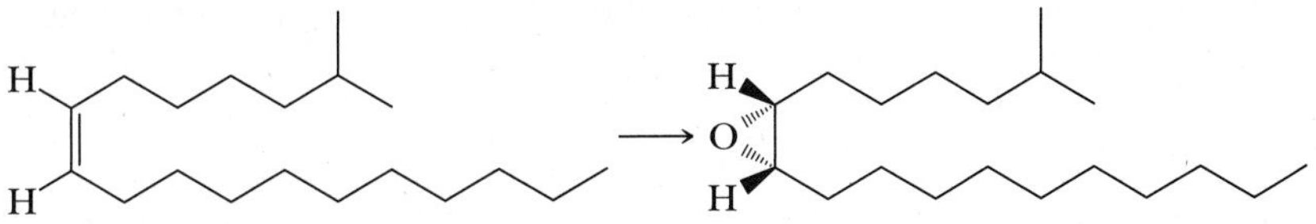

(*Z*)-2-Methyl-7-octadecene **Disparlure**

(a) Propose two sets of reagents that might be combined in a Wittig reaction to give the indicated *Z* alkene.
(b) How might the *Z* alkene be converted to disparlure?
(c) How many stereoisomers are possible for disparlure? How many stereoisomers are formed in the sequence you chose?

16.54 Propose structural formulas for compounds A, B, and C in the following conversion. Also show how to prepare compound C by a Wittig reaction.

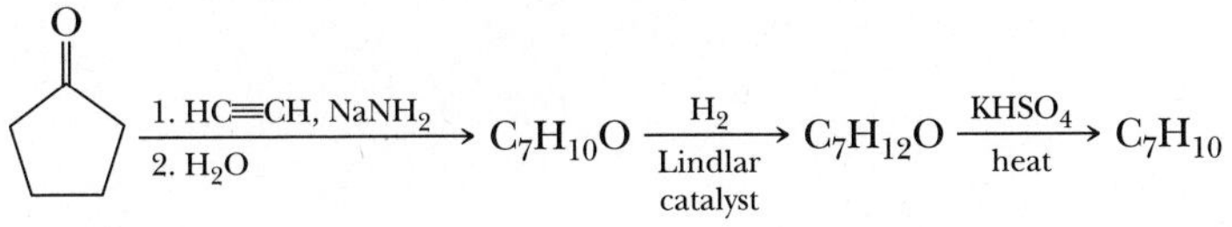

(A) (B) (C)

16.55 Following is a retrosynthetic scheme for the synthesis of *cis*-3-penten-2-ol.

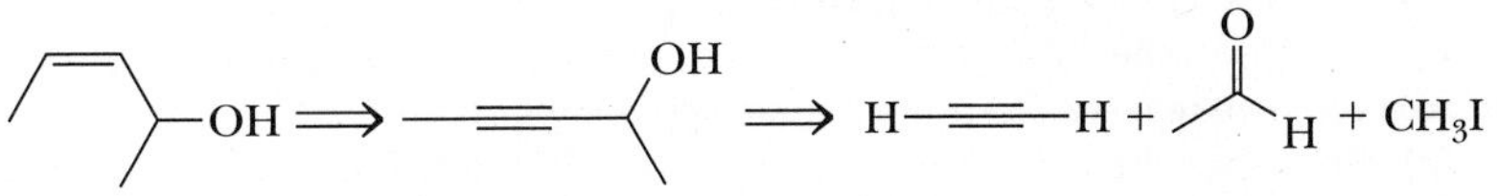

(racemic) (racemic) **Acetylene** **Acetaldehyde** **Iodomethane**

Write a synthesis for this compound from acetylene, acetaldehyde, and iodomethane.

16.56 Following is the structural formula of Surfynol, a defoaming surfactant. Describe the synthesis of this compound from acetylene and a ketone. How many stereoisomers are possible for Surfynol?

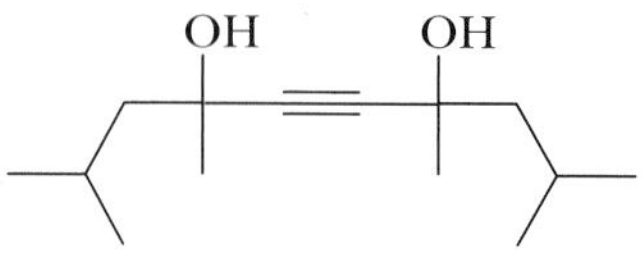

Surfynol

16.57 Propose a mechanism for this isomerization.

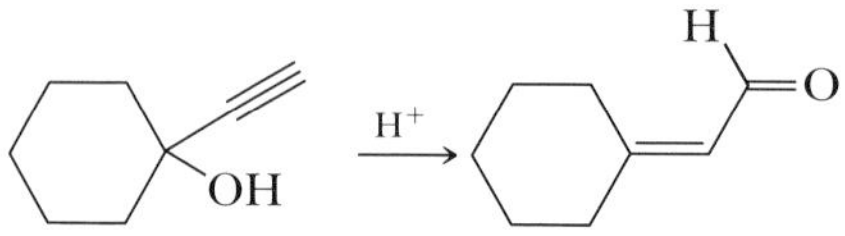

16.58 Propose a mechanism for this isomerization.

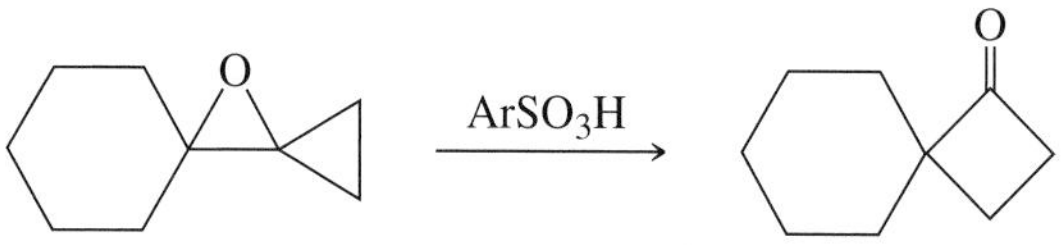

16.59 Starting with acetylene and 1-bromobutane as the only sources of carbon atoms, show how to synthesize the following.

(a) meso-5,6-Decanediol
(b) racemic 5,6-Decanediol
(c) 5-Decanone
(d) 5,6-Epoxydecane
(e) 5-Decanol
(f) Decane
(g) 6-Methyl-5-decanol
(h) 6-Methyl-5-decanone

16.60 Following are the final steps in one industrial synthesis of vitamin A acetate.

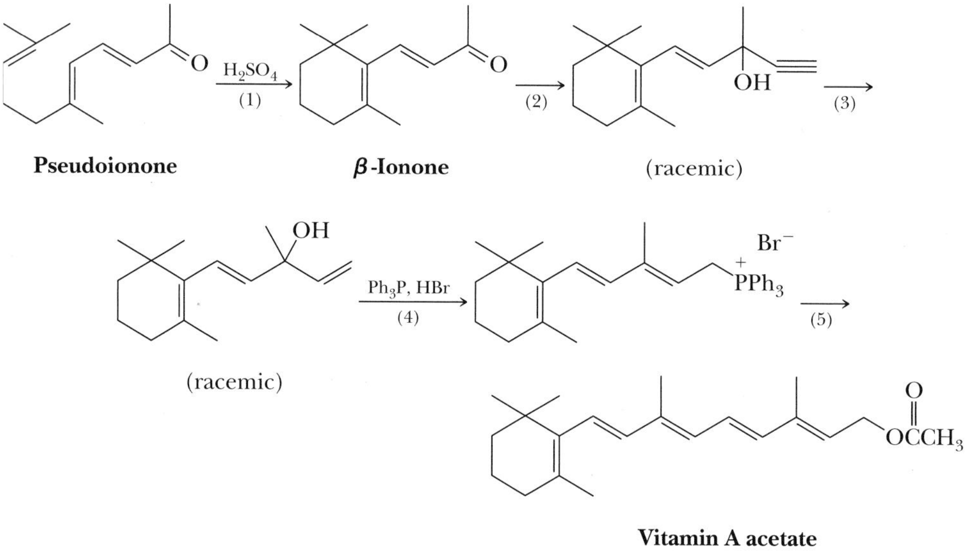

(a) Propose a mechanism for the acid-catalyzed cyclization in Step 1.
(b) Propose reagents to bring about Step 2.
(c) Propose a mechanism for formation of the phosphonium salt in Step 4.
(d) Propose reagents to bring about Step 3.
(e) Show how Step 5 can be completed by a Wittig reaction.

16.61 Following is the structural formula of the principal sex pheromone of the Douglas fir tussock moth (*Orgyia pseudotsugata*), a severe defoliant of the fir trees of western North America.

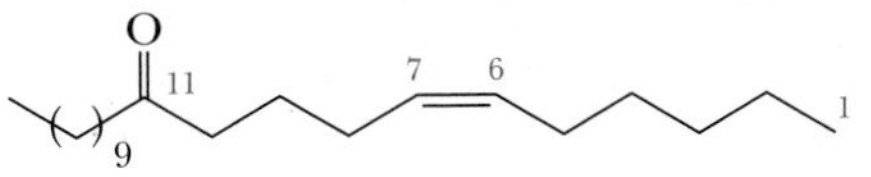

(Z)-6-Heneicosene-11-one

Several syntheses of this compound have been reported, starting materials for three of which are given here.

(a) 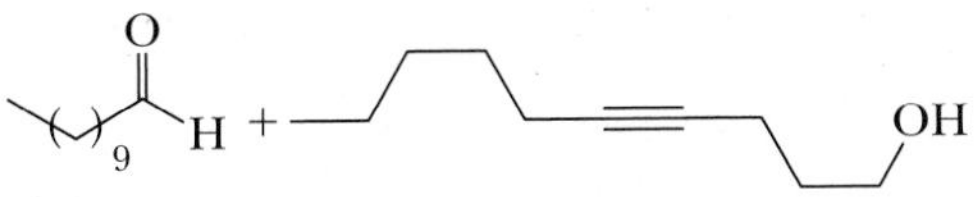

(b) 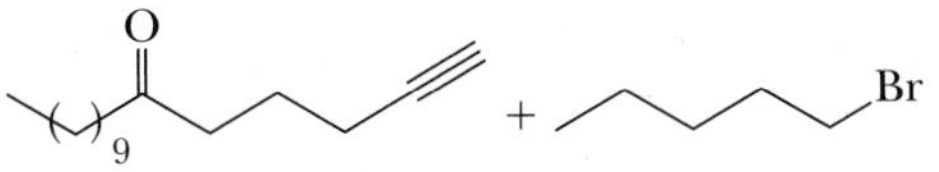

(c) 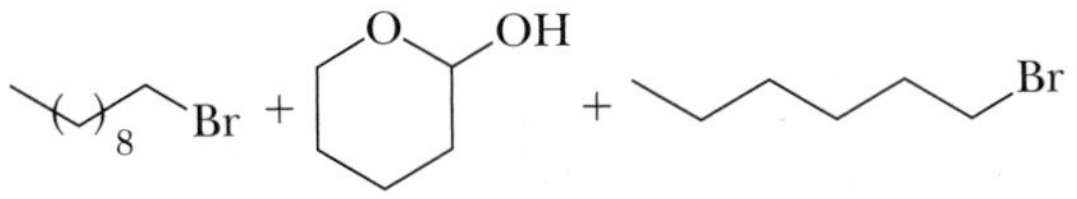

Show a series of steps by which each set of starting materials could be converted into the target molecule.

16.62 Both (*S*)-citronellal and isopulegol are naturally occurring terpenes (Section 5.4). When (*S*)-citronellal is treated with tin(IV) chloride (a Lewis acid) followed by neutralization with aqueous ammonium chloride, isopulegol is obtained in 85% yield.

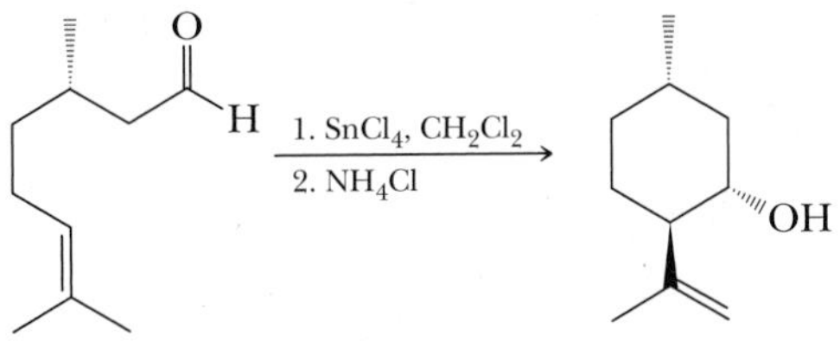

(*S*)-Citronellal ($C_{10}H_{18}O$) **Isopulegol** ($C_{10}H_{18}O$)

(a) Show that both compounds are terpenes.
(b) Propose a mechanism for the conversion of (*S*)-citronellal to isopulegol.
(c) How many stereocenters are present in isopulegol? How many stereoisomers are possible for a molecule with this number of stereocenters?
(d) Isopulegol is formed as a single stereoisomer. Account for the fact that only a single stereoisomer is formed.

16.63 At some point during the synthesis of a target molecule, it may be necessary to protect an —OH group (i.e., to prevent its reacting). In addition to the trimethylsilyl, *tert*-butyldimethylsilyl, and other trialkylsilyl groups described in Section 11.6, and the tetrahydropyranyl group described in Section 16.7D, the ethoxyethyl group may also be used as a protecting group.

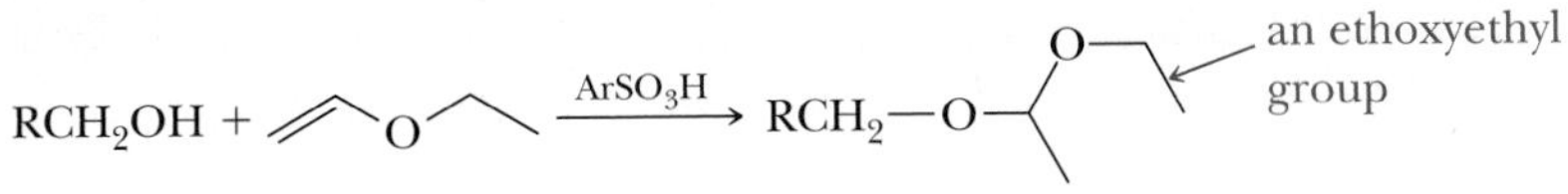

Ethyl vinyl ether

(a) Propose a mechanism for the acid-catalyzed formation of the ethoxyethyl protecting group.

(b) Suggest an experimental procedure whereby this protecting group can be removed to regenerate the unprotected alcohol.

16.64 Both 1,2-diols and 1,3-diols can be protected by treatment with 2-methoxypropene according to the following reaction.

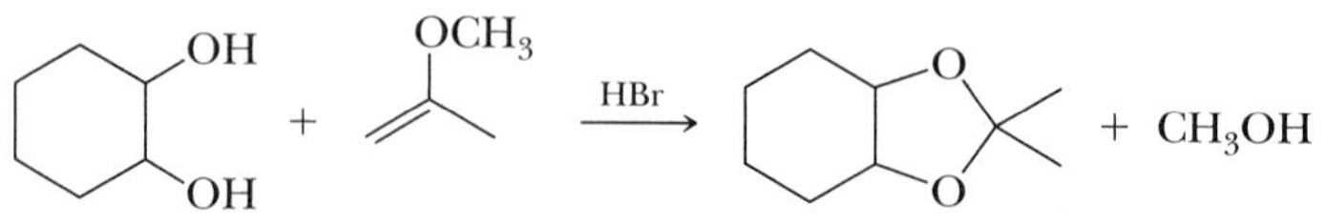

2-Methoxypropene A protected 1,2-diol

(a) Propose a mechanism for the formation of this protected diol.

(b) Suggest an experimental procedure by which this protecting group can be removed to regenerate the unprotected diol.

Looking Ahead

16.65 All rearrangements we have discussed so far have involved generation of an electron-deficient carbon followed by a 1,2-shift of an atom or a group of atoms from an adjacent atom to the electron-deficient carbon. Rearrangements by a 1,2-shift can also occur following the generation of an electron-deficient oxygen. Propose a mechanism for the acid-catalyzed rearrangement of cumene hydroperoxide to phenol and acetone.

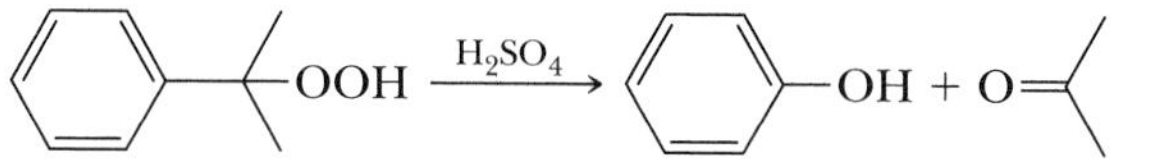

Cumene hydroperoxide **Phenol** **Acetone**

16.66 In dilute aqueous base, (*R*)-glyceraldehyde is converted into an equilibrium mixture of (*R*,*S*)-glyceraldehyde and dihydroxyacetone. Propose a mechanism for this isomerization.

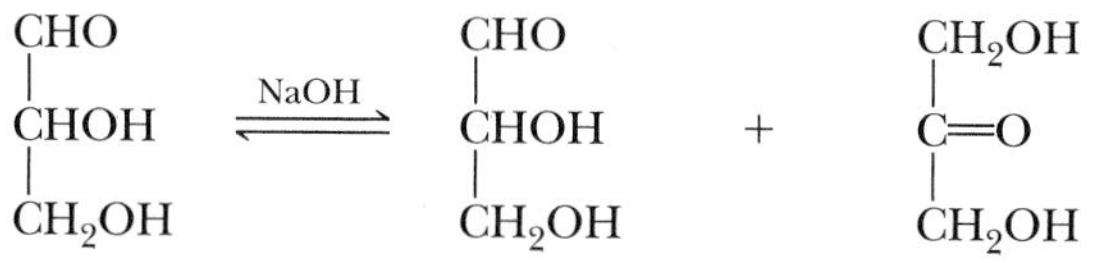

(*R*)-Glyceraldehyde **(*R*,*S*)-Glyceraldehyde** **Dihydroxyacetone**

16.67 Treatment of β-D-glucose with methanol in the presence of an acid catalyst converts it into a mixture of two compounds called methyl glucosides (Section 25.3A).

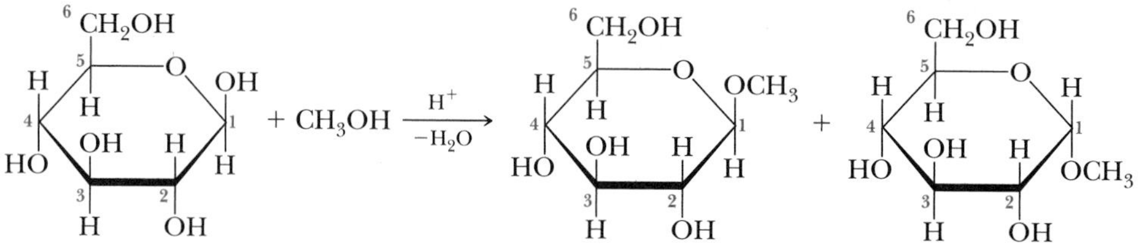

β-D-Glucose **Methyl β-D-glucoside** **Methyl α-D-glucoside**

In these representations, the six-membered rings are drawn as planar hexagons.

(a) Propose a mechanism for this conversion and account for the fact that only the —OH on carbon 1 is transformed into an $—OCH_3$ group.

(b) Draw the more stable chair conformation for each product.

(c) Which of the two products has the chair conformation of greater stability? Explain.

16.68 Treating a Grignard reagent with carbon dioxide followed by aqueous HCl gives a carboxylic acid.

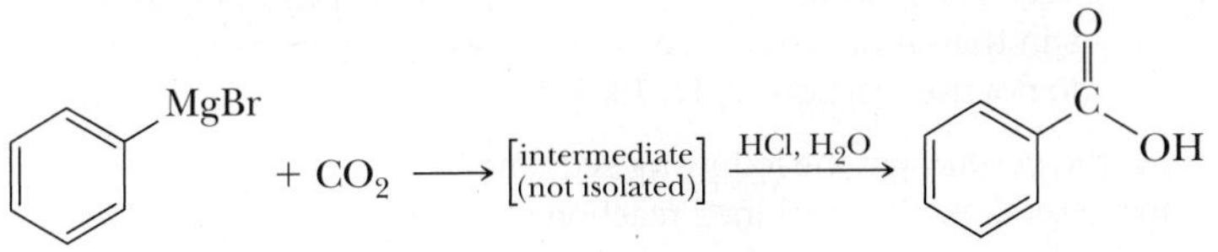

Propose a structural formula for the bracketed intermediate and a mechanism for its formation.

16.69 As we saw in Chapter 6, carbon-carbon double bonds are attacked by electrophiles but not by nucleophiles. An exception to this generalization is the reactivity of α,β-unsaturated aldehydes and ketones toward nucleophiles. Even though an isolated carbon-carbon double bond does not react with 2° amines such as dimethylamine, 3-buten-2-one reacts readily by regioselective addition.

$$Et_2NH + CH_2{=}CH{-}CH_2{-}CH_3 \longrightarrow \text{no reaction}$$

$$Et_2NH + CH_2{=}CH{-}\overset{\overset{\displaystyle O}{\|}}{C}{-}CH_3 \longrightarrow Et_2N{-}CH_2{-}CH_2{-}\overset{\overset{\displaystyle O}{\|}}{C}{-}CH_3$$

Diethylamine **3-Buten-2-one (Methyl vinyl ketone)**

Account for the addition of nucleophiles to the carbon-carbon double bond of an α, β-unsaturated aldehyde or ketone and the regioselectivity of the addition.

16.70 Ribose, a carbohydrate with the formula shown, forms a cyclic hemiacetal, which, in principle, could contain either a four-membered, five-membered, or six-membered ring. When D-ribose is treated with methanol in the presence of an acid catalyst, two cyclic acetals, A and B, are formed, both with molecular formula $C_6H_{12}O_5$. These are separated, and each is treated with sodium periodate (Section 10.8C) followed by dilute aqueous acid. Both A and B yield the same three products in the same ratios.

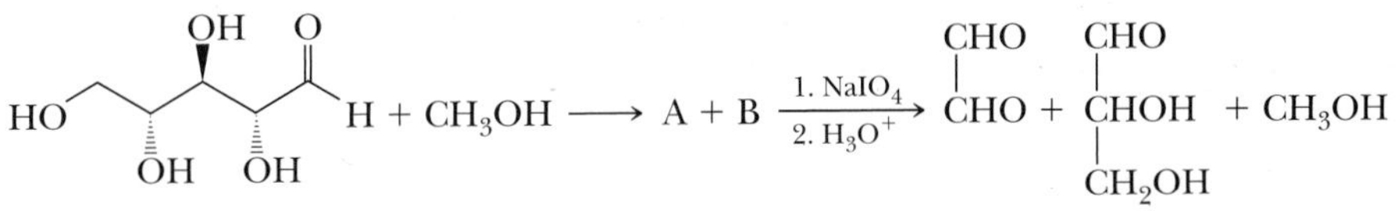

D-Ribose ($C_5H_{10}O_5$)

Isomeric cyclic acetals with molecular formula $C_6H_{12}O_5$

From this information, deduce whether the cyclic hemiacetal formed by D-ribose is four-membered, five-membered, or six-membered.

16.71 The favorite nuclide used in positron emission tomography (PET scan) to follow glucose metabolism is fluorine-18, which decays by positron emission to oxygen-18 and has a half-life of 110 minutes. Fluorine-18 is administered in the form of fludeoxyglucose F-18; the product of this molecule's decay is glucose.

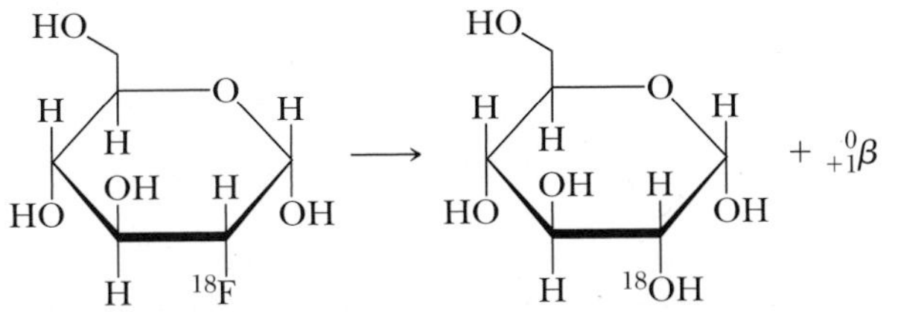

Fludeoxyglucose F-18 **D-Glucose** **A positron**

Draw the alternative chair conformations for fludeoxyglucose F-18 and select the more stable of the two.

Organic Chemistry Reaction Roadmap

Reaction Roadmap

16.72 Use the reaction roadmap you made for Problem 15.19 and update it to contain the reactions in the "Study Guide" section of this chapter. Because of their highly specific nature, do not use reactions 7, 12, 13, 14, and 26 on your roadmap.

Reaction Roadmap

16.73 Write the products of the following sequences of reactions. Refer to your reaction roadmaps to see how the combined reactions allow you to "navigate" between the different functional groups. Note that you will need both your old Chapters 6–11 roadmap and your new Chapters 15–16 roadmap for these.

(a)

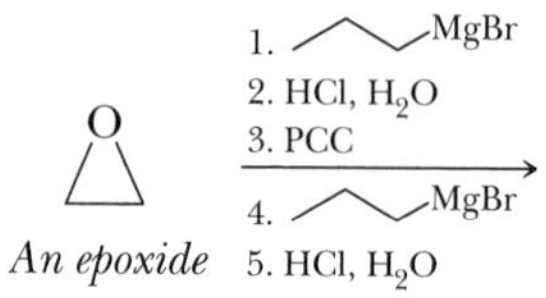
O
An epoxide
1. MgBr
2. HCl, H2O
3. PCC
4. MgBr
5. HCl, H2O

(b)

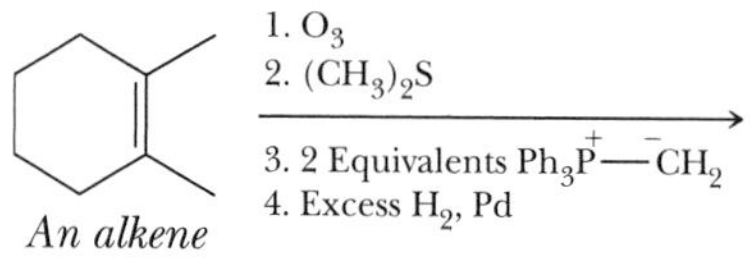
An alkene
1. O3
2. (CH3)2S
3. 2 Equivalents Ph3P—CH2
4. Excess H2, Pd

(c) $CH_3—C≡C—H$

An alkyne

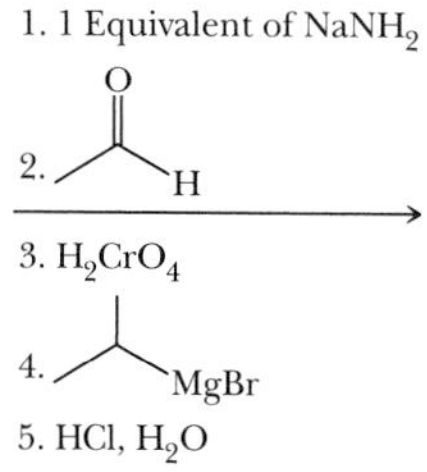
1. 1 Equivalent of NaNH2
O
2. H
3. H2CrO4
4. MgBr
5. HCl, H2O

(d)

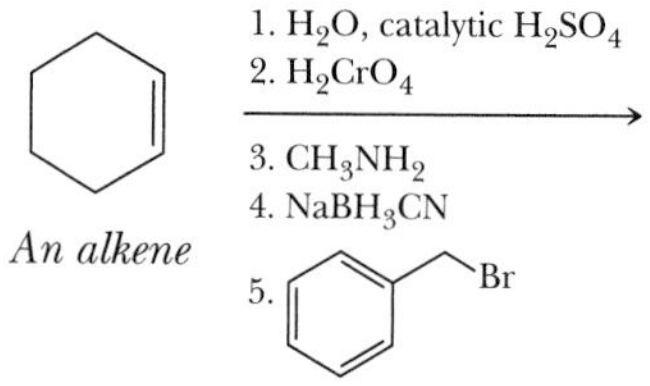
An alkene
1. H2O, catalytic H2SO4
2. H2CrO4
3. CH3NH2
4. NaBH3CN
5. Br

(e)

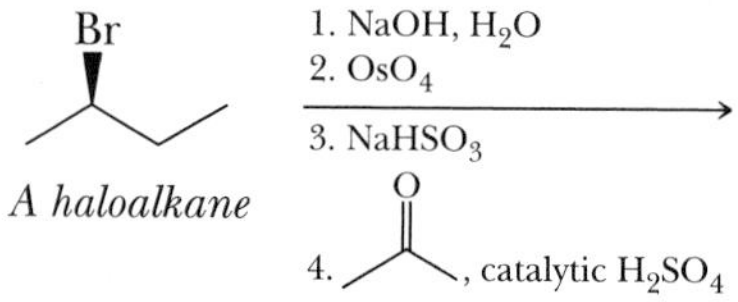
Br
A haloalkane
1. NaOH, H2O
2. OsO4
3. NaHSO3
O
4. , catalytic H2SO4

(f)

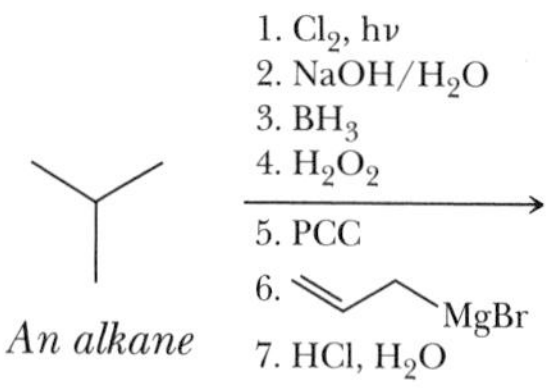
An alkane
1. Cl2, hν
2. NaOH/H2O
3. BH3
4. H2O2
5. PCC
6. MgBr
7. HCl, H2O

Synthesis

Reaction Roadmap

16.74 Using your reaction roadmaps as a guide, show how to convert 1-butanol into racemic 4-octanol. You must use 1-butanol as the source of all carbon atoms in the target molecule. Show all reagents and all molecules synthesized along the way.

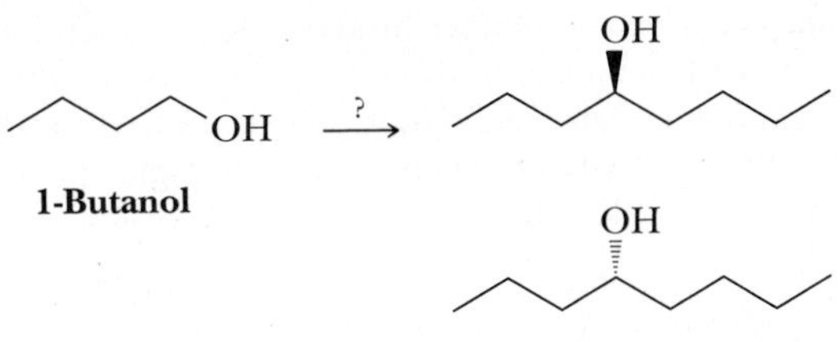

1-Butanol

4-Octanol
(racemic)

16.75 Using your reaction roadmaps as a guide, show how to convert ethylene into 2-methyl-1,3-dioxolane. You must use ethylene as the source of all carbon atoms in the target molecule. Show all reagents and all molecules synthesized along the way.

Reaction Roadmap

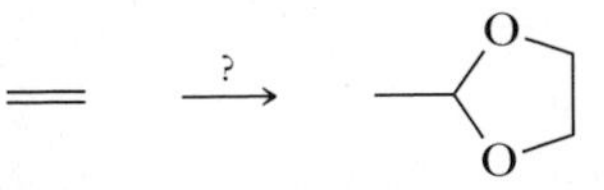

Ethylene **2-Methyl-1,3-dioxolane**

16.76 Using your reaction roadmaps as a guide, show how to convert (2-bromoethyl)benzene into 2-chloro-1-phenylethanone. Show all reagents and all molecules synthesized along the way.

Reaction Roadmap

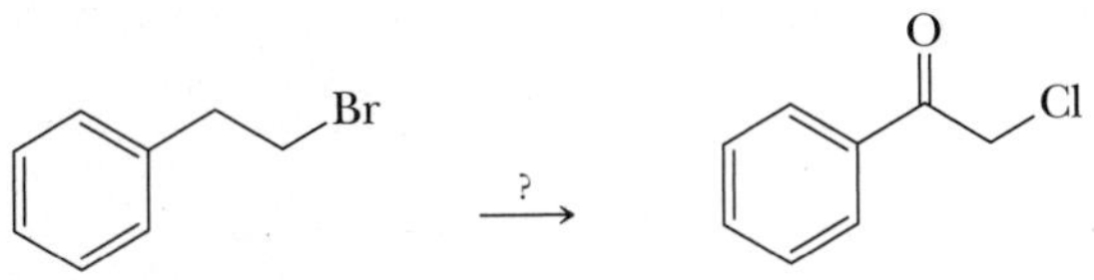

(2-Bromoethyl)benzene **2-Chloro-1-phenylethanone**

16.77 Using your reaction roadmaps as a guide, show how to convert acetaldehyde into racemic 3-hydroxybutanal. You must use acetaldehyde as the source of all carbon atoms in the target molecule. Show all reagents and all molecules synthesized along the way.

Reaction Roadmap

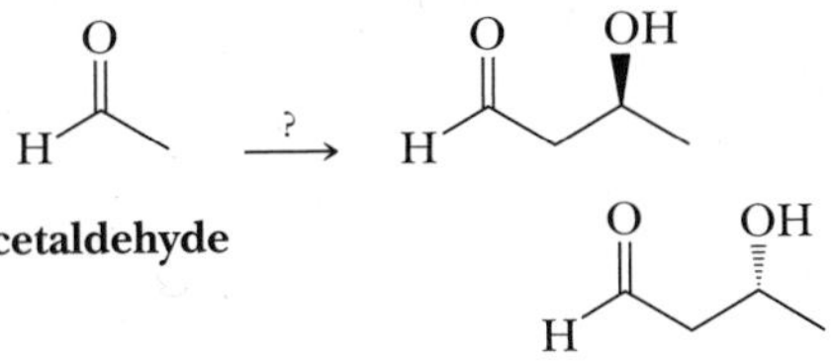

Acetaldehyde

3-Hydroxybutanal
(racemic)

Reactions in Context

16.78 Cetrizine is a nonsedating antihistamine. The first step in a synthesis of cetrizine involves the following Grignard reaction.

(a) Draw the product of this Grignard reaction.

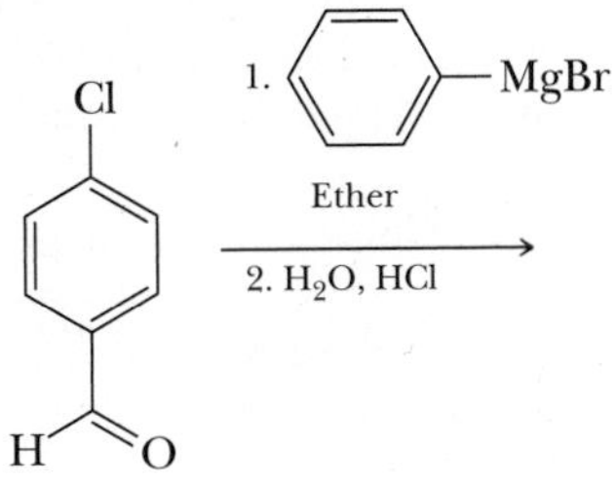

(b) Cetrizine is chiral. Like many chiral drugs, one enantiomer of cetrizine is more active than the other enantiomer or the racemic mixture. The levorotatory (*S*) enantiomer of cetrizine is more active, and syntheses have been developed that produce only the desired enantiomer in high yield. Label the chiral center of cetrizine with an asterisk.

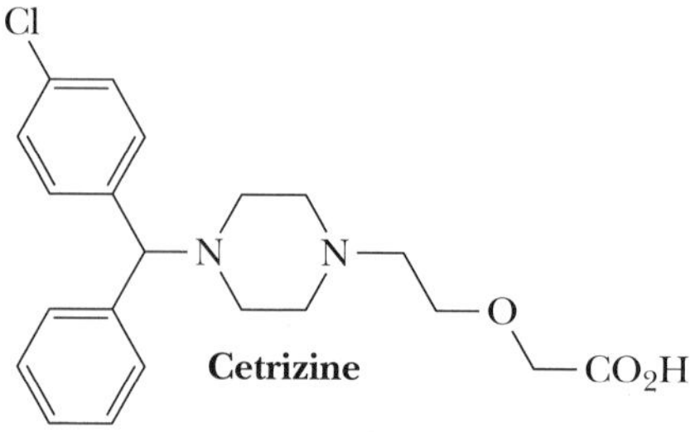

Cetrizine

16.79 Wittig reactions are widely used in drug synthesis. Write the predominant product of the following Wittig reaction used in the first step of a synthesis of elitriptan, which is used to treat migraine headaches.

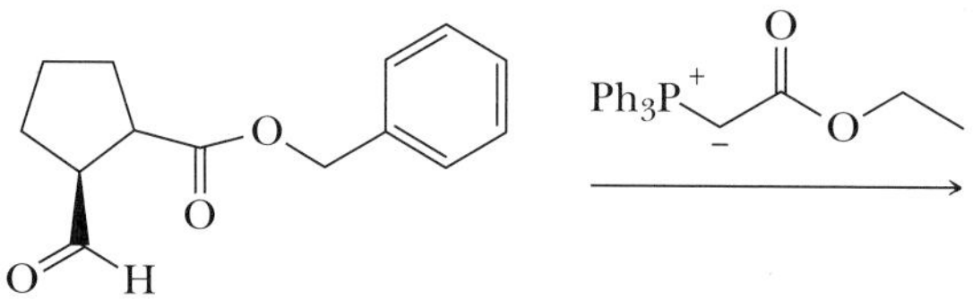

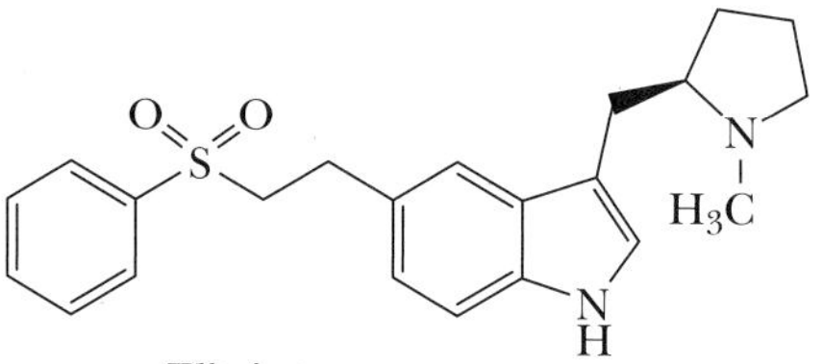

Elitriptan

16.80 Complete the following Wittig reactions.

(a)

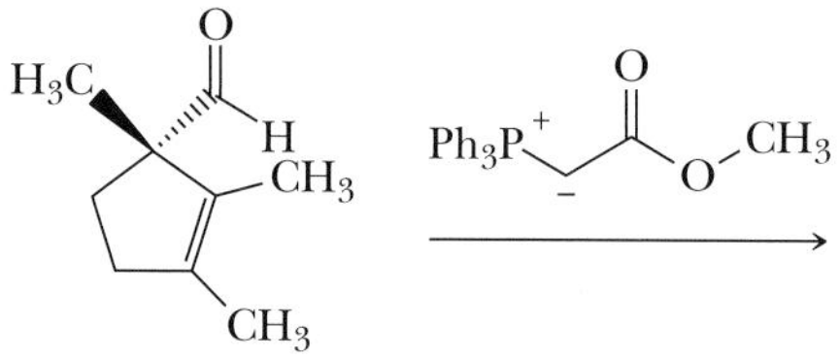

(b)

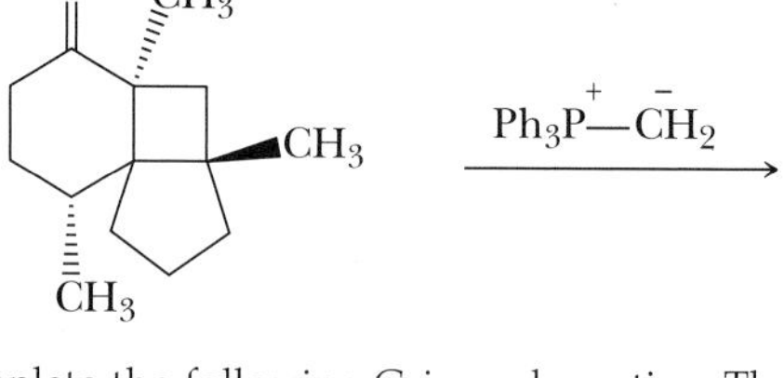

16.81 Complete the following Grignard reaction. The starting material is chiral and present as a single enantiomer. Using models, predict which product enantiomer predominates and include that stereochemical prediction in your answer.

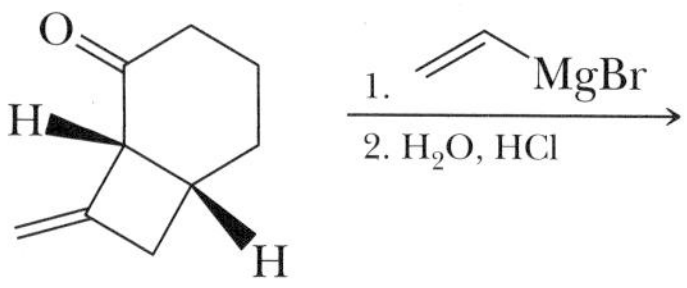

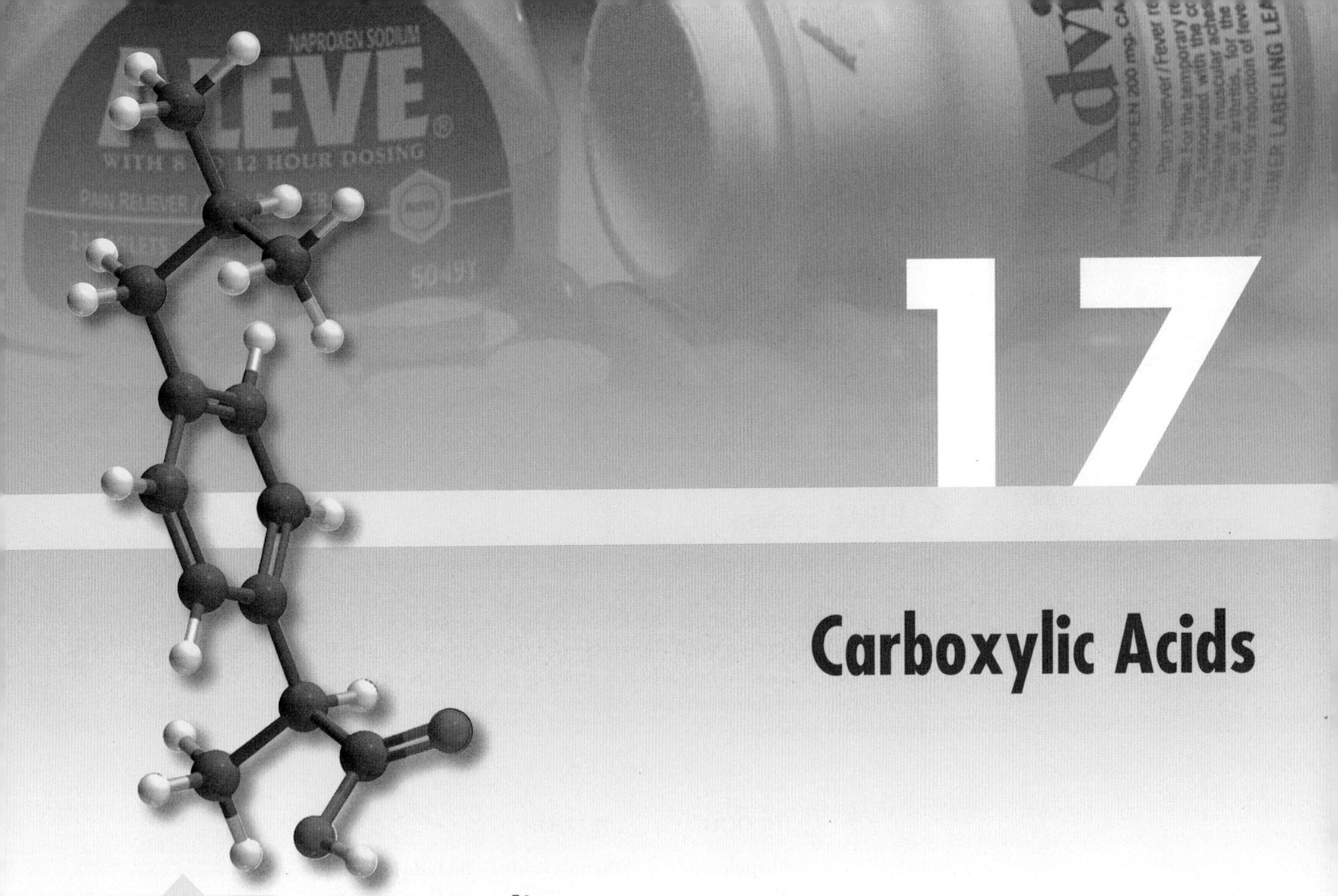

The active ingredients in these two nonprescription pain relievers are derivatives of arylpropanoic acids. **Inset:** a model of (*S*)-ibuprofen. See "Chemical Connections: From Willow Bark to Aspirin and Beyond." (© Charles D. Winters/Cengage Learning)

17

Carboxylic Acids

Outline

17.1 Structure
17.2 Nomenclature
17.3 Physical Properties
17.4 Acidity
17.5 Preparation of Carboxylic Acids
17.6 Reduction
17.7 Esterification
17.8 Conversion to Acid Chlorides
17.9 Decarboxylation

The most important chemical property of carboxylic acids, another class of organic compounds containing the carbonyl group, is their acidity. Furthermore, carboxylic acids form numerous important derivatives, including esters, amides, anhydrides, and acid halides. In this chapter, we study carboxylic acids, and in Chapter 18, we study their derivatives.

17.1 Structure

The functional group of a carboxylic acid is the **carboxyl group** (Section 1.3D), so named because it is made up of a *carb*onyl group and a hydr*oxyl* group. Following is a Lewis structure of the carboxyl group as well as three alternative representations for it.

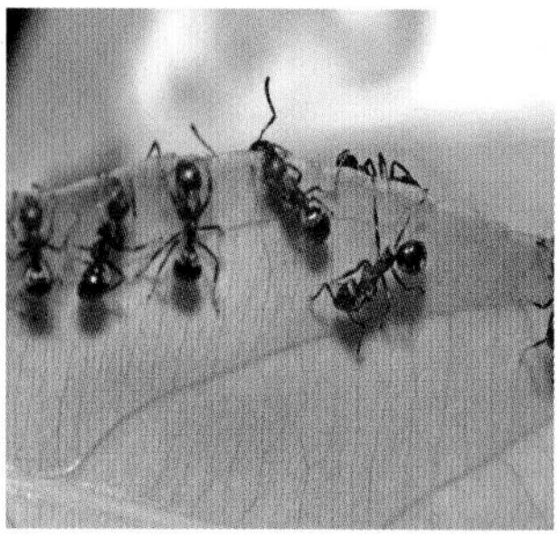

Formic acid was first obtained in 1670 from the destructive distillation of ants, whose Latin genus is *Formica*. It is one of the components of the venom injected by stinging ants.

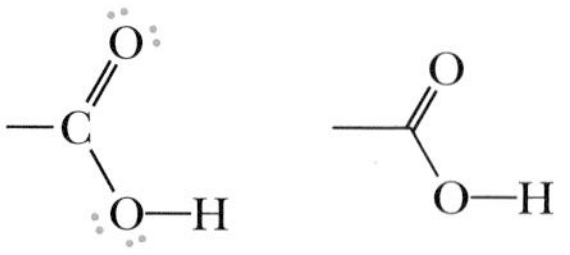

$-COOH$ $-CO_2H$

Alternative representations of a carboxyl group

The general formula for an aliphatic carboxylic acid is RCOOH; the general formula for an aromatic carboxylic acid is ArCOOH.

17.2 Nomenclature

A. IUPAC System

The IUPAC name of a carboxylic acid is derived from that of the longest carbon chain that contains the carboxyl group by dropping the final *-e* from the name of the parent alkane and adding the suffix *-oic* followed by the word *acid* (Section 2.3C). The chain is numbered beginning with the carbon of the carboxyl group. Because the carboxyl carbon is understood to be carbon 1, there is no need to give it a number. The IUPAC system retains the common names formic acid and acetic acid, which are always used to refer to these acids.

HCOOH

Methanoic acid (Formic acid)

CH_3COOH

Ethanoic acid (Acetic acid)

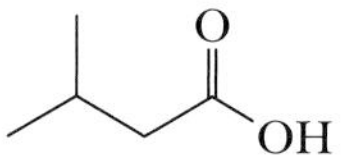

3-Methylbutanoic acid (Isovaleric acid)

If the carboxylic acid contains a carbon-carbon double or triple bond, change the infix from *-an-* to *-en-* or *-yn-*, as the case may be, to indicate the presence of the multiple bond and show the location of the multiple bond by a number.

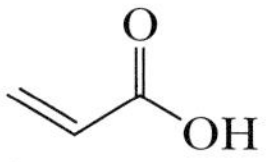

Propenoic acid (Acrylic acid)

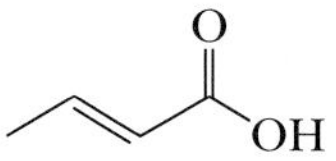

***trans*-2-Butenoic acid (Crotonic acid)**

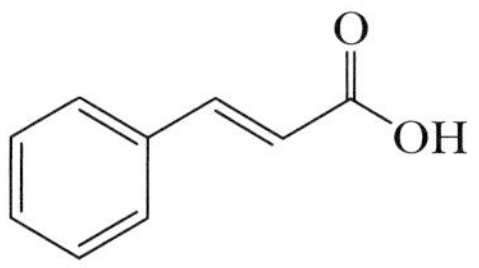

***trans*-3-Phenylpropenoic acid (Cinnamic acid)**

In the IUPAC system, a carboxyl group takes precedence over most other functional groups (Table 16.1), including hydroxyl groups, amino groups, and the carbonyl groups of aldehydes and ketones. As illustrated in the following examples, an —OH group is indicated by the prefix *hydroxy-*; an $-NH_2$ group, by *amino-*; and the C=O group of an aldehyde or ketone, by *oxo-*.

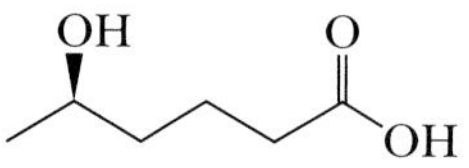

(*R*)-5-Hydroxyhexanoic acid

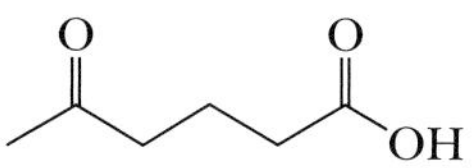

5-Oxohexanoic acid

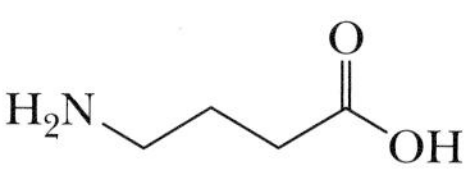

4-Aminobutanoic acid

Dicarboxylic acids are named by adding the suffix *-dioic acid* to the name of the carbon chain that contains both carboxyl groups. The numbers of the carboxyl carbons are not indicated because they can be only at the ends of the parent chain. Following are IUPAC and common names for several important aliphatic dicarboxylic acids.

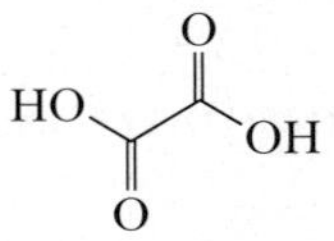

Ethanedioic acid
(Oxalic acid)

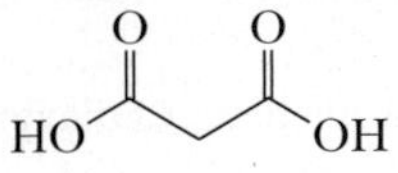

Propanedioic acid
(Malonic acid)

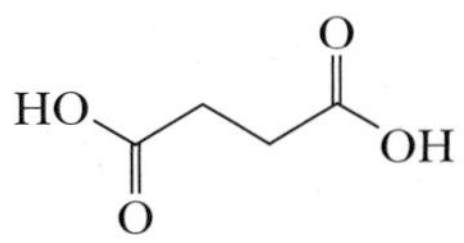

Butanedioic acid
(Succinic acid)

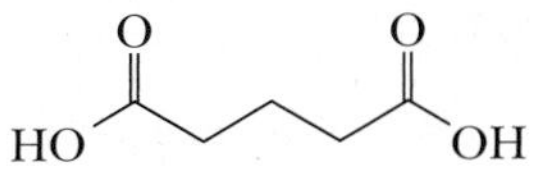

Pentanedioic acid
(Glutaric acid)

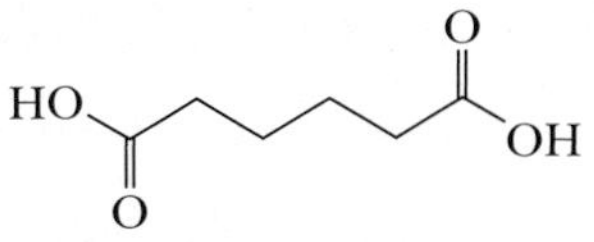

Hexanedioic acid
(Adipic acid)

The name *oxalic acid* is derived from one of its sources in the biological world, namely plants of the genus *Oxalis*, one of which is rhubarb. Adipic acid is one of the two monomers required for the synthesis of the polymer nylon 66 (Section 29.5A). A mnemonic phrase for remembering the common names for the dicarboxylic acids oxalic through adipic is *Oh my, such good apples.*

A carboxylic acid containing a carboxyl group bonded to a cycloalkane ring is named by giving the name of the ring and adding the suffix *-carboxylic acid*. The atoms of the ring are numbered beginning with the carbon bearing the —COOH group.

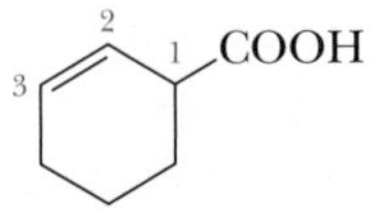

2-Cyclohexenecarboxylic acid

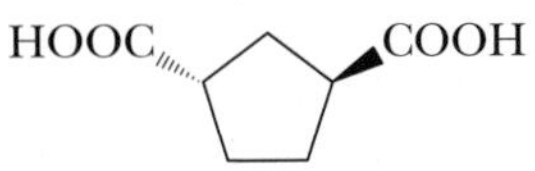

***trans*-1,3-Cyclopentanedicarboxylic acid**

The simplest aromatic carboxylic acid is benzoic acid. Derivatives are named by using numbers to show the location of substituents relative to the carboxyl group. Certain aromatic carboxylic acids have common names by which they are usually known. For example, 2-hydroxybenzoic acid is more often called salicylic acid, a name derived from the fact that this aromatic carboxylic acid was first isolated from the bark of the willow, a tree of the genus *Salix*.

Benzoic acid

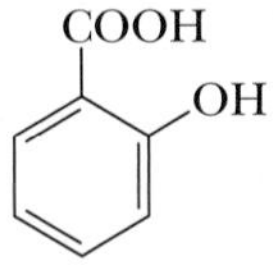

2-Hydroxybenzoic acid
(Salicylic acid)

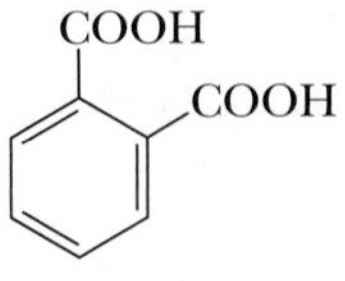

1,2-Benzenedicarboxylic acid
(Phthalic acid)

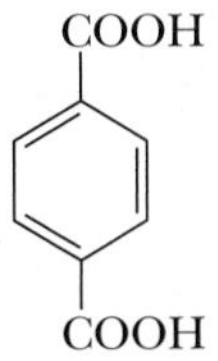

1,4-Benzenedicarboxylic acid
(Terephthalic acid)

Aromatic dicarboxylic acids are named by adding the words *dicarboxylic acid* to *benzene* (e.g., 1,2-benzenedicarboxylic acid and 1,4-benzenedicarboxylic acid). Each is usually known by its common name: phthalic acid and terephthalic acid, respectively. Terephthalic acid is one of the two organic components required for the synthesis of the textile fiber known as Dacron polyester, or Dacron (Section 29.5B).

B. Common Names

Aliphatic carboxylic acids, many of which were known long before the development of structural theory and IUPAC nomenclature, are named according to their source or

Table 17.1 Several Aliphatic Carboxylic Acids—Their Common Names and Derivations

Structure	IUPAC Name	Common Name	Derivation
HCOOH	**Methanoic acid**	**Formic acid**	Latin: *formica*, ant
CH_3COOH	**Ethanoic acid**	**Acetic acid**	Latin: *acetum*, vinegar
CH_3CH_2COOH	**Propanoic acid**	**Propionic acid**	Greek: *propion*, first fat
$CH_3(CH_2)_2COOH$	**Butanoic acid**	**Butyric acid**	Latin: *butyrum*, butter
$CH_3(CH_2)_3COOH$	**Pentanoic acid**	**Valeric acid**	Latin: *valeriana*, a flowering plant
$CH_3(CH_2)_4COOH$	**Hexanoic acid**	**Caproic acid**	Latin: *caper*, goat
$CH_3(CH_2)_6COOH$	**Octanoic acid**	**Caprylic acid**	Latin: *caper*, goat
$CH_3(CH_2)_8COOH$	**Decanoic acid**	**Capric acid**	Latin: *caper*, goat
$CH_3(CH_2)_{10}COOH$	**Dodecanoic acid**	**Lauric acid**	Latin: *laurus*, laurel

for some characteristic property. Table 17.1 lists several of the unbranched aliphatic carboxylic acids found in the biological world along with their common name and Latin or Greek derivation. Those of 16, 18, and 20 carbon atoms are particularly abundant in fats and oils (Section 26.1) and in the phospholipid components of biological membranes (Section 26.5).

When common names are used, the Greek letters α, β, γ, δ, and so forth, are often added as a prefix to locate substituents. The α-position in a carboxylic acid is next to the carboxyl group; an α-substituent in a common name is equivalent to a 2-substituent in an IUPAC name. GABA is an inhibitory neurotransmitter in the central nervous system of humans. Alanine is one of the 20 protein-derived amino acids.

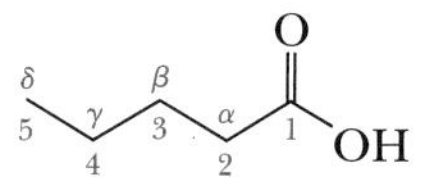

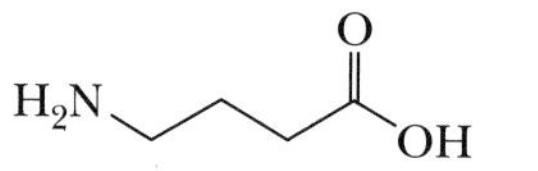

4-Aminobutanoic acid
(γ-Aminobutyric acid, GABA)

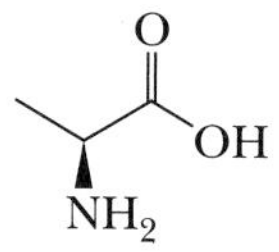

(*S*)-2-Aminopropanoic acid
[(*S*)-α-Aminopropionic acid;
L-Alanine]

In common names, the presence of a ketone carbonyl in a substituted carboxylic acid is indicated by the prefix *keto*-, illustrated by the common name β-ketobutyric acid. This substituted carboxylic acid is also named acetoacetic acid. In deriving this name, 3-oxobutanoic acid is regarded as a substituted acetic acid. In the common nomenclature, the substituent is named an **aceto group**, CH_3CO—.

Aceto group

A CH_3CO— group; also called an acetyl group.

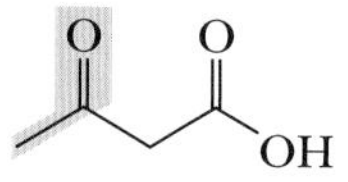

3-Oxobutanoic acid
(β-Ketobutyric acid;
Acetoacetic acid)

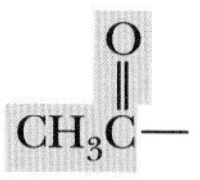

Acetyl group
(an aceto group)

Example 17.1 Naming Carboxylic Acids

Write the IUPAC name for each carboxylic acid.

(a) 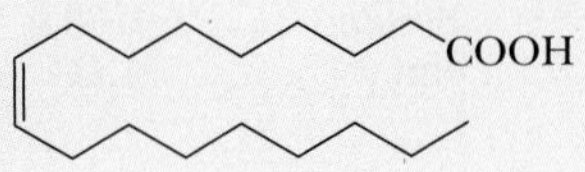

(b) 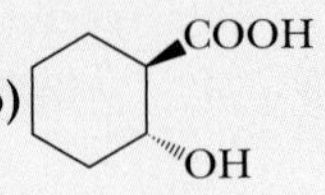

(racemic)

(c) 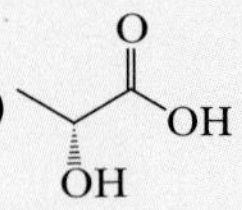

(d) 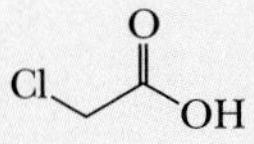

Solution

(a) (*Z*)-9-Octadecenoic acid (oleic acid)

(b) *trans*-2-Hydroxycyclohexanecarboxylic acid

(c) (*R*)-2-Hydroxypropanoic acid [(*R*)-lactic acid]

(d) Chloroacetic acid

Problem 17.1

Each of these carboxylic acids has a well-recognized common name. A derivative of glyceric acid is an intermediate in glycolysis. Maleic acid is an intermediate in the tricarboxylic acid (TCA) cycle. Mevalonic acid is an intermediate in the biosynthesis of steroids. Write the IUPAC name for each compound. Make sure you specify configuration.

(a) 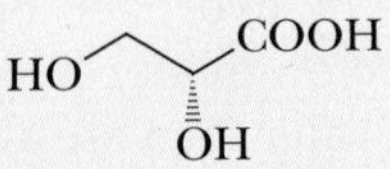

Glyceric acid

(b) 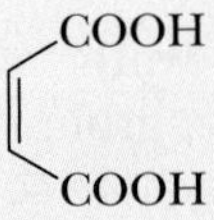

Maleic acid

(c) 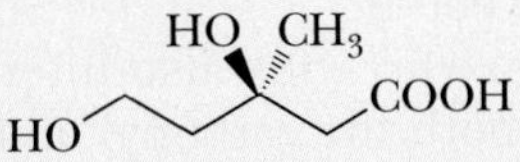

Mevalonic acid

17.3 Physical Properties

In the liquid and solid states, carboxylic acids are associated by hydrogen bonding into dimers, as shown here for acetic acid in the liquid state.

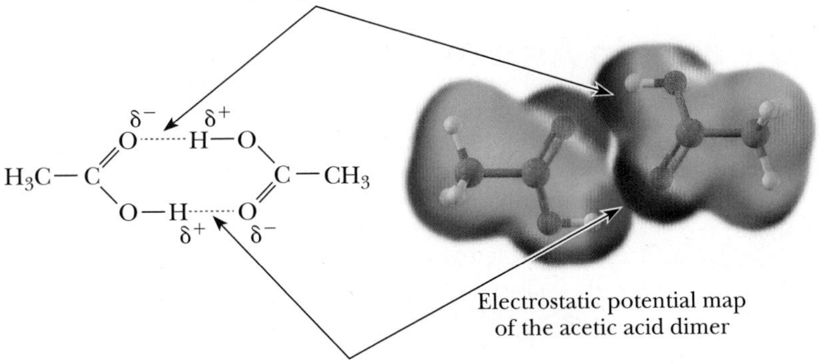
Hydrogen bond
Electrostatic potential map
of the acetic acid dimer
Hydrogen bond

Carboxylic acids have significantly higher boiling points than other types of organic compounds of comparable molecular weight, such as alcohols, aldehydes, and ketones. For example, butanoic acid (Table 17.2) has a higher boiling point than either 1-pentanol or pentanal. The higher boiling points of carboxylic acids result

Table 17.2 Boiling Points and Solubilities in Water of Selected Carboxylic Acids, Alcohols, and Aldehydes of Comparable Molecular Weight

Structure	Name	Molecular Weight (g/mol)	Boiling Point (°C)	Solubility (g/100 g H_2O)
CH_3COOH	**Acetic acid**	60.1	118	Infinite
$CH_3CH_2CH_2OH$	**1-Propanol**	60.1	97	Infinite
CH_3CH_2CHO	**Propanal**	58.1	48	16.0
$CH_3(CH_2)_2COOH$	**Butanoic acid**	88.1	163	Infinite
$CH_3(CH_2)_3CH_2OH$	**1-Pentanol**	88.1	137	2.3
$CH_3(CH_2)_3CHO$	**Pentanal**	86.1	103	Slight
$CH_3(CH_2)_4COOH$	**Hexanoic acid**	116.2	205	1.0
$CH_3(CH_2)_5CH_2OH$	**1-Heptanol**	116.2	176	0.2
$CH_3(CH_2)_5CHO$	**Heptanal**	114.1	153	0.1

from their polarity and from the fact that they form very strong intermolecular hydrogen bonds.

Carboxylic acids also interact with water molecules by hydrogen bonding through both the carbonyl and hydroxyl groups. Because of greater hydrogen-bonding interactions, carboxylic acids are more soluble in water than are alcohols, ethers, aldehydes, and ketones of comparable molecular weight. The solubility of a carboxylic acid in water decreases as its molecular weight increases. We account for this trend in the following way. A carboxylic acid consists of two regions of distinctly different polarity: a polar **hydrophilic** carboxyl group and, except for formic acid, a nonpolar **hydrophobic** hydrocarbon chain. The hydrophilic carboxyl group increases water solubility; the hydrophobic hydrocarbon chain decreases water solubility.

Hydrophilic
From the Greek, meaning water loving.

Hydrophobic
From the Greek, meaning water fearing.

Chemical Connections

From Willow Bark to Aspirin and Beyond

The first drug developed for widespread use was aspirin, one of today's most common pain relievers. Americans alone consume approximately 80 billion tablets of aspirin a year! The story of the development of this modern pain reliever goes back more than 2000 years. In 400 BCE, the Greek physician Hippocrates recommended chewing bark of the willow tree to alleviate the pain of childbirth and to treat eye infections.

The active component of willow bark was found to be salicin, a compound composed of salicyl alcohol bonded to a unit of β-D-glucose (Section 25.1). Hydrolysis of salicin in aqueous acid gives salicyl alcohol, which can be oxidized to salicylic acid. Salicylic acid proved to be an even more effective reliever of pain, fever, and inflammation than salicin, without its extremely bitter taste. Unfortunately, salicylic acid causes severe irritation of the mucous membrane lining of the stomach.

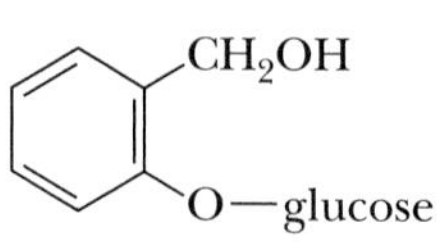

Salicin

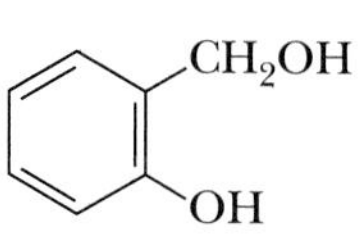

Salicyl alcohol

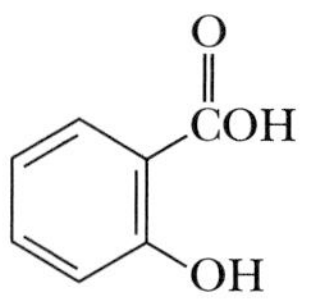

Salicylic acid

In the search for less irritating but still effective derivatives of salicylic acid, chemists at the Bayer division of I. G. Farben in Germany in 1883 prepared acetylsalicylic acid and gave it the name aspirin.

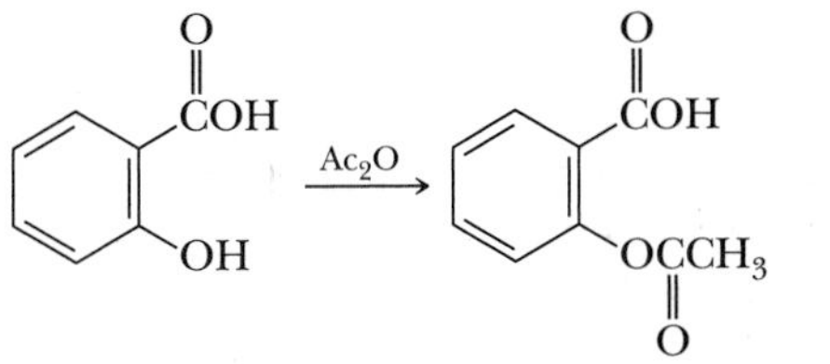

Salicylic acid **Acetylsalicylic acid (Aspirin)**

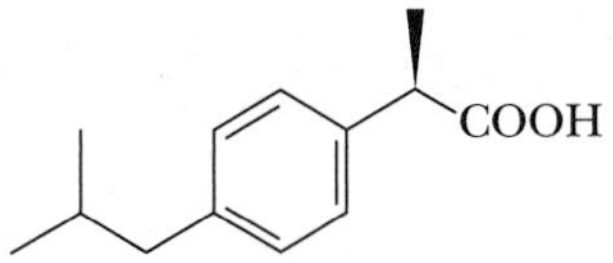

(*R*)-Ibuprofen

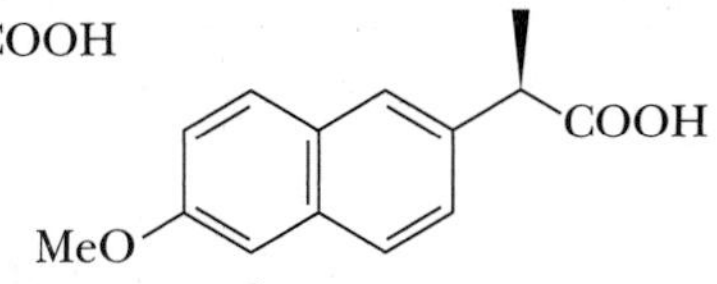

(*R*)-Naproxen

Aspirin proved to be less irritating to the stomach than salicylic acid and more effective in relieving the pain and inflammation of rheumatoid arthritis. Aspirin, however, is still irritating to the stomach, and frequent use can cause duodenal ulcers in susceptible people.

In the 1960s, in a search for even more effective and less irritating analgesics and anti-inflammatory drugs, chemists at the Boots Pure Drug Company in England synthesized a series of compounds related in structure to salicylic acid. Among them, they discovered an even more potent compound, which they named ibuprofen. Soon thereafter, Syntex Corporation in the United States developed naproxen, the active ingredient in Aleve. Each compound has one chiral center and can exist as a pair of enantiomers. For each drug, the physiologically active form is the *S* enantiomer.

In the 1960s, it was discovered that aspirin acts by inhibiting cyclooxygenase (COX), a key enzyme in the conversion of arachidonic acid to prostaglandins (Section 26.3). With this discovery, it became clear why only one enantiomer of ibuprofen and naproxen is active: only the *S* enantiomer of each has the correct handedness to bind to COX and inhibit its activity.

Recently, it was recognized that there are actually two cyclooxygenases; one is more important for the inflammation pathway, and the other affects the stability of blood vessels. Aspirin and other nonsteroidal anti-inflammatory drugs (NSAIDs) inhibit both, which is why they can cause gastrointestinal bleeding. Next-generation drugs (e.g., Celebrex) have been developed that inhibit only the inflammatory enzyme pathway and are remarkably effective for suppression of inflammation (e.g., in arthritis) without the gastrointestinal side effects. However, people taking drugs such as Celebrex have been found to have an increased risk of heart attack or stroke.

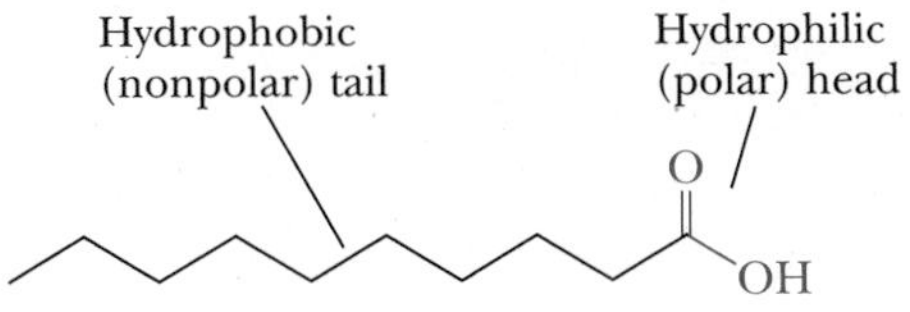
Hydrophobic
(nonpolar) tail
Hydrophilic
(polar) head

Decanoic acid
(0.2 g/100 mL H_2O)

Electrostatic potential map of decanoic acid

The first four aliphatic carboxylic acids (formic, acetic, propanoic, and butanoic acids) are infinitely soluble in water because the hydrophobic character of the hydrocarbon chain is more than counterbalanced by the hydrophilic character of the carboxyl group. As the size of the hydrocarbon chain increases relative to the size of the hydrophilic group, water solubility decreases. The solubility of hexanoic acid is 1.0 g/100 g H_2O, while that of decanoic acid is only 0.2 g/100 g H_2O.

One other physical property of carboxylic acids must be mentioned. The liquid carboxylic acids from propanoic acid to decanoic acid have extremely foul odors, about as bad as those of thiols, although different. Butanoic acid is found in stale perspiration, is a major component of "locker room odor," and provides the characteristic odor of regurgitated milk. Pentanoic acid smells even worse, and goats, which secrete C_6, C_8, and C_{10} acids, are famous for their unpleasant odors.

17.4 Acidity

A. Acid Ionization Constants

Carboxylic acids are weak acids. Values of K_a for most unsubstituted aliphatic and aromatic carboxylic acids fall within the range 10^{-4} to 10^{-5}. The value of K_a for acetic acid, for example, is 1.74×10^{-5}. Its pK_a is 4.76.

$$CH_3COOH + H_2O \rightleftharpoons CH_3COO^- + H_3O^+ \quad K_a = \frac{[CH_3COO^-][H_3O^+]}{[CH_3COOH]} = 1.74 \times 10^{-5}$$

$$pK_a = 4.76$$

As we discussed in Sections 4.6C and 4.6D, the greater acidity of carboxylic acids (pK_a 4–5) compared with alcohols (pK_a 16–18) is because of delocalization of the negative charge of the carboxylate anion through resonance and because of the electron-withdrawing inductive effect of the carbonyl group. There is no comparable resonance or inductive stabilization of alkoxide ions.

We saw in Section 4.6D that substitution at the α-carbon of an atom or a group of atoms of higher electronegativity than carbon further increases the acidity of carboxylic acids by the inductive effect. Compare, for example, the acidity of acetic acid (pK_a 4.76) and chloroacetic acid (pK_a 2.86). To see the effects of multiple halogen substitution, compare the values of pK_a for acetic acid with its mono-, di-, and trichloro derivatives. A single chlorine substituent increases acid strength by nearly 100. Trichloroacetic acid, the strongest of the three acids, is a stronger acid than H_3PO_4.

Formula:	CH_3COOH	$ClCH_2COOH$	$Cl_2CHCOOH$	Cl_3CCOOH
Name:	**Acetic acid**	**Chloroacetic acid**	**Dichloroacetic acid**	**Trichloroacetic acid**
pK_a:	4.76	2.86	1.48	0.70

Increasing acid strength →

The trend found in the above chart for the acidity of the substituted carboxylic acids is best understood by an analysis of the stability of the anionic conjugate bases that are created. The electron withdrawing ability of the electronegative chlorine atoms stabilizes the nearby anionic charge by electron withdrawal through σ-bonds. The strength of this inductive effect is proportional to the number of chlorine atoms present, explaining why trichloroacetic acid is the strongest acid in the series.

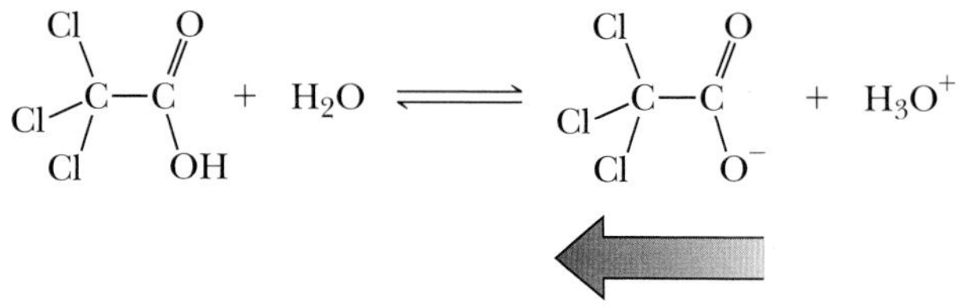

Inductive effect spreads the negative charge onto the chlorine atoms ▶

▶ Delocalization of charge over a larger area is stabilizing.

We also see an example of the inductive effect in a comparison of the relative acidities of benzoic acid and acetic acid. Because of the stronger electron-withdrawing inductive effect of the sp^2 hybridized carbon of the benzene ring compared with the sp^3 hybridized carbon of the methyl group, benzoic acid is a stronger acid than acetic acid; its K_a is approximately four times that of acetic acid.

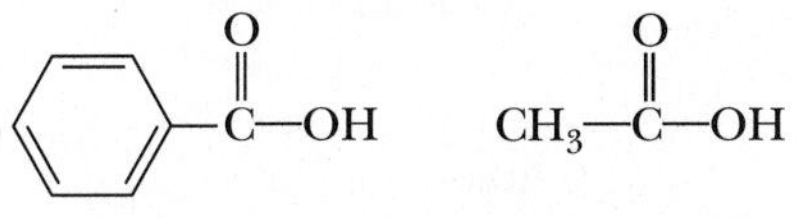

Benzoic acid pK_a 4.19 | **Acetic acid** pK_a 4.76

Example 17.2 Predicting Acid Strength

Which is the stronger acid in each pair?

(a) 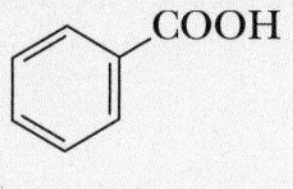 or 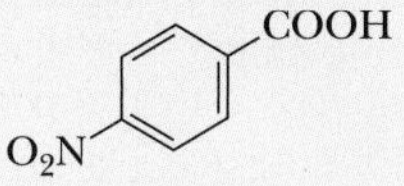

Benzoic acid | **4-Nitrobenzoic acid**

(b) 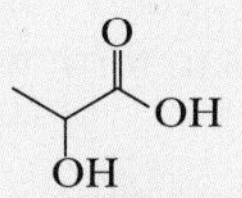 or 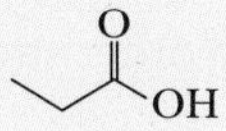

2-Hydroxypropanoic acid (Lactic acid) | **Propanoic acid**

Solution

(a) 4-Nitrobenzoic acid (pK_a 3.42) is a considerably stronger acid than benzoic acid (pK_a 4.19) because of the electron-withdrawing inductive effect of the nitro group. This inductive effect makes the anionic conjugate base of 4-nitrobenzoic acid more stable than that of benzoic acid and hence leads to the greater acidity.

(b) 2-Hydroxypropanoic acid (pK_a 3.08) is a stronger acid than propanoic acid (pK_a 4.87) because of the electron-withdrawing inductive effect of the adjacent hydroxyl oxygen. This effect stabilizes the anionic conjugate base of 2-hydroxypropionic acid.

Problem 17.2

Which is the stronger acid in each pair?

(a) CH_3COOH or CH_3SO_3H

Acetic acid | **Methanesulfonic acid**

(b) 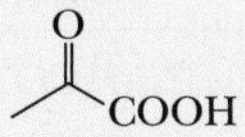 or 

2-Oxopropanoic acid (Pyruvic acid) | **Propanoic acid**

One final point about carboxylic acids: when a carboxylic acid dissolves in an aqueous solution, the form of the carboxylic acid present depends on the pH of the solution in which it is dissolved. Consider typical carboxylic acids, which have pK_a values in the range of 4.0–5.0. When the pH of the solution is equal to the pK_a of the carboxylic acid (i.e., the pH of the solution is in the range 4.0–5.0), the acid and its anion (its conjugate base) are present in equal concentrations. If the pH of the solution is adjusted to 2.0 or lower by the addition of a strong acid, the carboxylic acid then is present in solution almost entirely as RCOOH. If, on the other hand, the pH of the solution is adjusted to 7.0 or higher, the carboxylic acid is present almost entirely as its anion. Thus, even in a neutral solution (pH 7.0), a carboxylic acid is present predominantly as its anion.

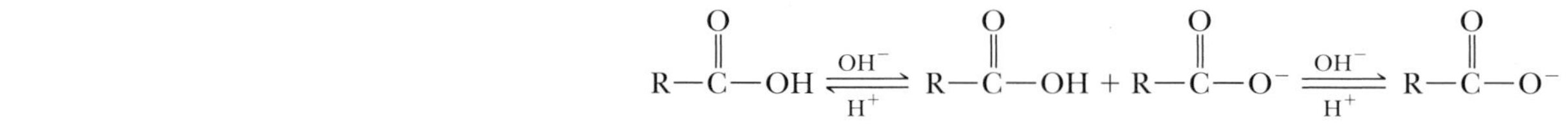

Predominant species when the pH of the solution is 2.0 or less	Present in equal concentrations when the pH of the solution is equal to the pK_a of the acid	Predominant species when the pH of the solution is 7.0 or greater

A carboxylate anion has a negative charge; so in biological systems, molecules with substantial numbers of carboxylate anions have considerable negative charge. Because of the hydrophilic character of carboxylate anions, molecules with a large number of them are highly water-soluble.

B. Reaction with Bases

All carboxylic acids, whether soluble or insoluble in water, react with NaOH, KOH, and other strong bases to form water-soluble salts.

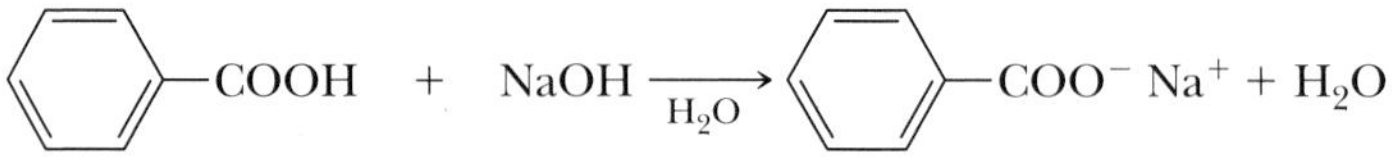

Benzoic acid (slightly soluble in water) — **Sodium benzoate** (60 g/100 mL water)

Carboxylic acids also form water-soluble salts with ammonia and amines.

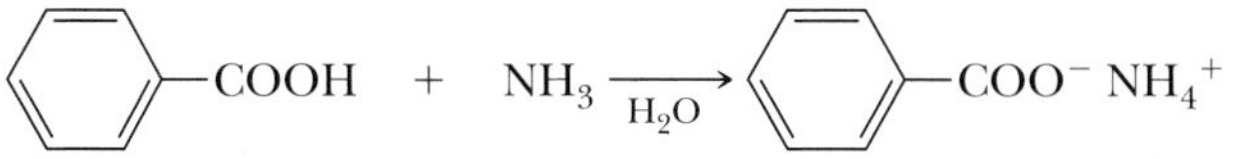

Benzoic acid (slightly soluble in water) — **Ammonium benzoate** (20 g/100 mL water)

Carboxylic acids react with sodium bicarbonate and sodium carbonate to form water-soluble sodium salts and carbonic acid (a weaker acid). Carbonic acid, in turn, decomposes to give water and carbon dioxide, which evolves as a gas.

$$CH_3COOH + NaHCO_3 \longrightarrow CH_3COO^-Na^+ + H_2CO_3$$
$$H_2CO_3 \longrightarrow CO_2 + H_2O$$
$$\overline{CH_3COOH + NaHCO_3 \longrightarrow CH_3COO^-Na^+ + CO_2 + H_2O}$$

Salts of carboxylic acids are named in the same manner as the salts of inorganic acids; the cation is named before the anion. The name of the anion is derived from the name of the carboxylic acid by dropping the suffix *-ic acid* and adding the suffix *-ate*.

Example 17.3 Salts of Carboxylic Acids

Complete each acid-base reaction and name the carboxylic salt formed.

(a) 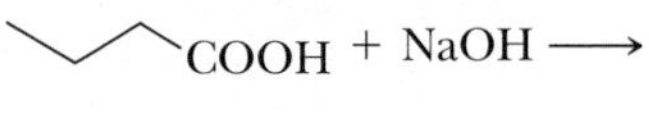

(b) 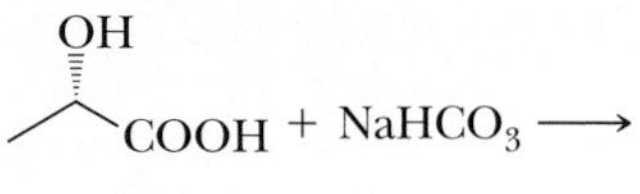

(S)-Lactic acid

Solution

Each carboxylic acid is converted to its sodium salt. In (b), carbonic acid is formed; it decomposes to carbon dioxide and water.

(a) 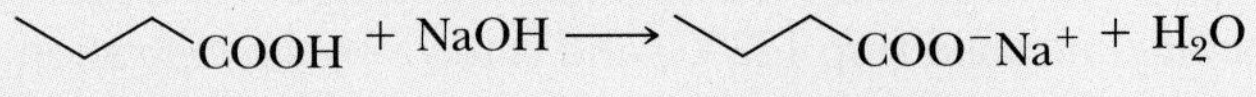

Butanoic acid **Sodium butanoate**

(b) 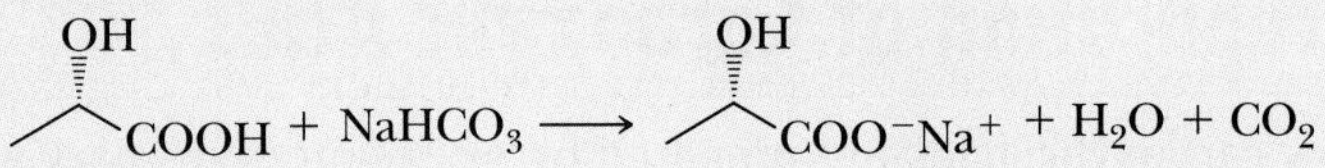

(*S*)-Lactic acid **Sodium (*S*)-lactate**

Problem 17.3

Write equations for the reaction of each acid in Example 17.3 with ammonia and name the carboxylic salt formed.

A consequence of the water solubility of carboxylic acid salts is that water-insoluble carboxylic acids can be converted to water-soluble ammonium or alkali metal salts and extracted into aqueous solution. The salt, in turn, can be transformed back to the free carboxylic acid by addition of HCl, H_2SO_4, or another strong acid. These reactions allow an easy separation of carboxylic acids from water-insoluble nonacidic compounds.

Shown in Figure 17.1 is a flowchart for the separation of benzoic acid, a water-insoluble carboxylic acid, from benzyl alcohol, a nonacidic compound. First, the

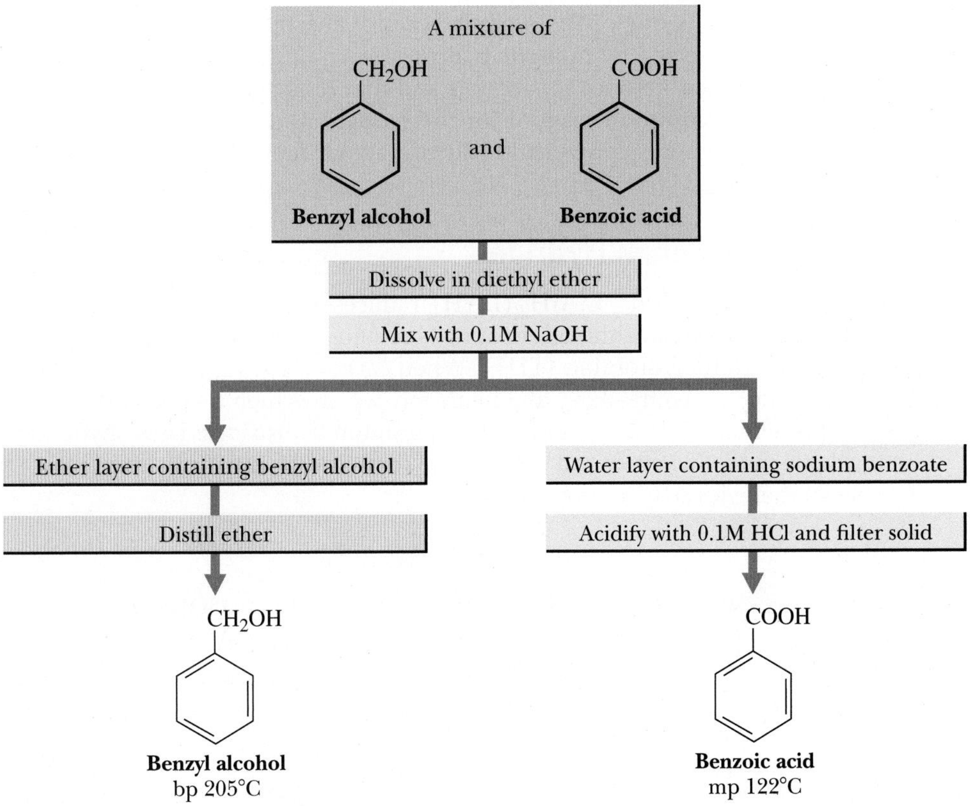

FIGURE 17.1 Flowchart for the separation of benzoic acid from benzyl alcohol.

mixture of benzoic acid and benzyl alcohol is dissolved in diethyl ether. When the ether solution is shaken with aqueous NaOH or another strong base, benzoic acid is converted to its water-soluble sodium salt. Then the ether and aqueous phases are separated. The ether solution is distilled, yielding first diethyl ether (bp 35°C) and then benzyl alcohol (bp 205°C). The aqueous solution is acidified with HCl, and benzoic acid precipitates as a crystalline solid (mp 122°C) and is recovered by filtration.

17.5 Preparation of Carboxylic Acids

We have seen how carboxylic acids are prepared by oxidation of primary alcohols (Section 10.8) and aldehydes (Section 16.10A). We mention an additional method involving a Grignard reagent.

Treating a Grignard reagent with carbon dioxide gives the magnesium salt of a carboxylic acid, which, upon protonation with aqueous acid, gives a carboxylic acid.

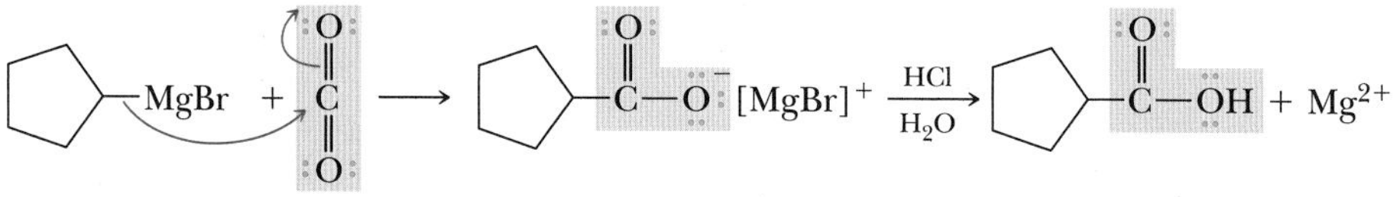

Carbon dioxide A magnesium carboxylate **Cyclopentanecarboxylic acid**

Thus, carbonation of a Grignard reagent is a convenient way to convert an alkyl or aryl halide to a carboxylic acid.

17.6 Reduction

The carboxyl group is one of the organic functional groups most resistant to reduction. It is not affected by catalytic hydrogenation under conditions that easily reduce aldehydes and ketones to alcohols and that reduce alkenes and alkynes to alkanes. The most common reagent for the reduction of carboxylic acids to primary alcohols is the very powerful reducing agent, lithium aluminum hydride (Section 16.11A).

A. Lithium Aluminum Hydride

Lithium aluminum hydride, $LiAlH_4$ (LAH), reduces a carboxylic acid to a primary alcohol in excellent yield, although heating is required. LAH is usually dissolved in diethyl ether or tetrahydrofuran (THF). When carboxylic acids react with $LiAlH_4$, the initial product is a tetraalkoxy aluminate ion, which is then treated with water to give the primary alcohol and lithium and aluminum hydroxides. These hydroxides are insoluble in diethyl ether and THF and are removed by filtration. Evaporation of the solvent then yields the primary alcohol.

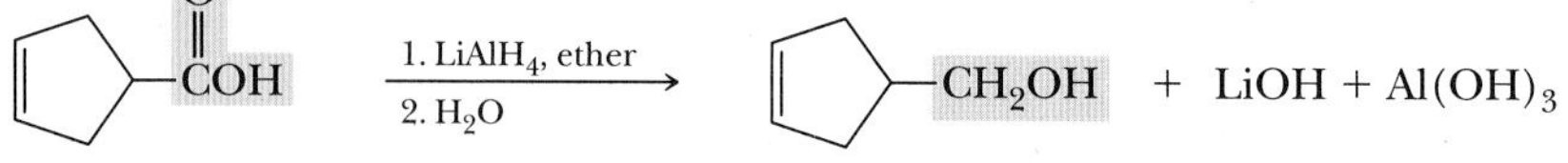

3-Cyclopentenecarboxylic acid **4-Hydroxymethylcyclopentene**

Alkenes are generally not affected by metal hydride reducing agents. These reagents function as hydride ion donors (i.e., as nucleophiles), and alkenes are not attacked by nucleophiles.

Chemical Connections

Industrial Synthesis of Acetic Acid—Transition Metal Catalysis

Yearly production of acetic acid in the United States is approximately 10^7 kg, a volume that ranks it at the top of the list of organic chemicals manufactured by the US chemical industry. The first industrial synthesis of acetic acid was commercialized in 1916 in Canada and Germany, using acetylene as a feedstock. The process involved two stages: (1) hydration of acetylene to acetaldehyde followed by (2) oxidation of acetaldehyde to acetic acid by molecular oxygen, catalyzed by cobalt(III) acetate.

As an alternative feedstock, chemists turned to ethylene, already available in huge quantities from the refining of natural gas and petroleum. The process of producing acetic acid from ethylene depends on the fact that in the presence of catalytic amounts of Pd^{2+} and Cu^{2+} salts, ethylene is oxidized by molecular oxygen to acetaldehyde.

$$\underset{\textbf{Acetylene}}{HC\equiv CH} + H_2O \xrightarrow[HgSO_4]{H_2SO_4} \underset{\text{Enol of acetaldehyde}}{\left[CH_2{=}CH{-}OH\right]} \longrightarrow \underset{\textbf{Acetaldehyde}}{CH_3\overset{O}{\overset{\|}{C}}H} \xrightarrow[Co^{3+}]{O_2} \underset{\textbf{Acetic acid}}{CH_3\overset{O}{\overset{\|}{C}}OH}$$

The technology of producing acetic acid from acetylene is simple, yields are high, and these factors made this procedure the major route to acetic acid for over 50 years. Acetylene was prepared by the reaction of calcium carbide with water. Calcium carbide, in turn, was prepared by heating calcium oxide (from limestone, $CaCO_3$) with coke (from coal) to between 2000 and 2500°C in an electric furnace.

$$\underset{\textbf{Calcium oxide}}{CaO} + 2\,C \xrightarrow{2500^\circ C} \underset{\textbf{Calcium carbide}}{CaC_2} \xrightarrow{2\,H_2O} \underset{\textbf{Acetylene}}{HC\equiv CH} + Ca(OH)_2$$

This preparation of calcium carbide requires enormous amounts of energy; so as the cost of energy rose, acetylene ceased to be an economical feedstock from which to manufacture acetic acid.

$$2\,CH_2{=}CH_2 + O_2 \xrightarrow[\text{(Wacker process)}]{Pd^{2+},\,Cu^{2+}} 2\,CH_3\overset{O}{\overset{\|}{C}}H$$

The first chemical plant to use ethylene oxidation for the manufacture of acetaldehyde was built in Germany by Wacker Chemie in 1959, and the process itself became known as the **Wacker process**.

In another approach to the synthesis of acetic acid, chemists turned to a route based on carbon monoxide, a readily available raw material. The carbonylation of methanol is exothermic. The challenge was to find a catalytic system that would bring about this reaction.

In 1973, the Monsanto Company in the United States developed a process for the carbonylation of methanol in the presence of small amounts of soluble rhodium(III) salts, HI, and H_2O.

$$CH_3OH + CO \xrightarrow[HI,\,H_2O]{Rh(CO)_2I_2^-} CH_3\overset{O}{\overset{\|}{C}}OH \qquad \Delta H^0 = -138\text{ kJ}(33\text{ kcal})/\text{mol}$$

In the reduction of a carboxyl group by lithium aluminum hydride, the first hydride ion reacts with the carboxyl hydrogen to give H_2. The resulting carboxylate anion reacts with a hydride ion at the carbonyl carbon atom, with the assistance of Li^+ and Al complexes acting as Lewis acids. Following hydride reaction, departure of an oxygen atom as an aluminum oxide species produces an intermediate aldehyde. The aldehyde immediately reacts under the reaction conditions (Section 16.11A) to give a tetraalkoxy aluminate ion that hydrolyzes in an aqueous workup to yield the final product alcohol. Following are balanced equations for treatment of a carboxylic acid with $LiAlH_4$ to form a tetraalkoxy aluminate ion, followed by its hydrolysis in water.

$$4\,\mathrm{R\overset{\overset{\displaystyle O}{\|}}{C}OH} + 3\,\mathrm{LiAlH_4} \longrightarrow \left[4\,\mathrm{R\overset{\overset{\displaystyle O}{\|}}{C}H}\right] \longrightarrow [(\mathrm{RCH_2O})_4\mathrm{Al}]\mathrm{Li} + 2\,\mathrm{LiAlO_2} + 4\,\mathrm{H_2}$$

$$\xrightarrow{4\,\mathrm{H_2O}} 4\,\mathrm{RCH_2OH} + \mathrm{LiOH} + \mathrm{Al(OH)_3}$$

In the reduction of a carboxyl group, two hydrogens from $LiAlH_4$ are delivered to the carbonyl group. The hydrogen on the hydroxyl group of the product is provided by water or by aqueous acid during workup. The mechanism of lithium aluminum hydride reduction of carboxyl derivatives is presented in Section 18.10.

B. Selective Reduction of Other Functional Groups

Because carboxyl groups are not affected by the conditions of catalytic hydrogenation, which normally reduce aldehydes, ketones, alkenes, and alkynes, it is possible to selectively reduce these functional groups to alcohols or alkanes in the presence of carboxyl groups.

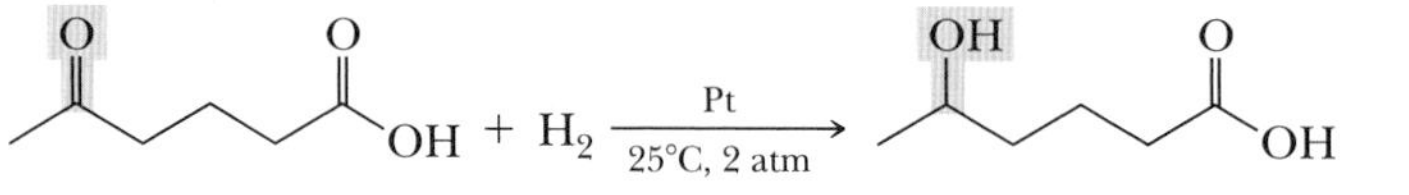

5-Oxohexanoic acid **5-Hydroxyhexanoic acid** (racemic)

We saw in Section 16.11A that both $LiAlH_4$ and $NaBH_4$ reduce aldehydes and ketones to alcohols. Only $LiAlH_4$, however, reduces carboxyl groups. Thus, it is possible to reduce an aldehyde or ketone carbonyl group selectively in the presence of a carboxyl group by using the less reactive $NaBH_4$ as the reducing agent. An example is the selective reduction of the following ketoacid to a hydroxyacid.

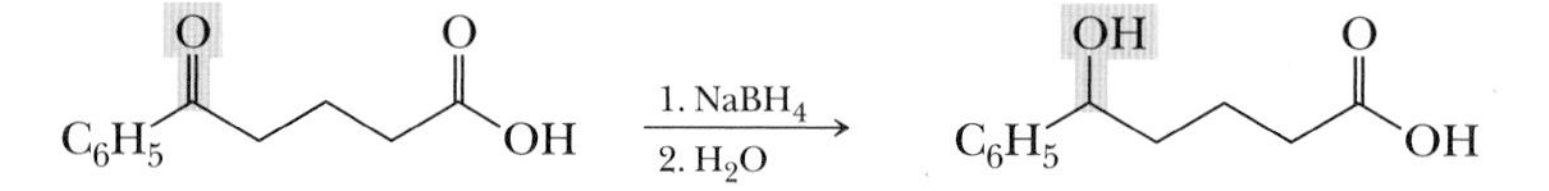

5-Oxo-5-phenylpentanoic acid **5-Hydroxy-5-phenylpentanoic acid** (racemic)

17.7 Esterification

A. Fischer Esterification

Fischer esterification
The process of forming an ester by heating at reflux a carboxylic acid and an alcohol in the presence of an acid catalyst, commonly H_2SO_4, $ArSO_3H$, or HCl.

Esters can be prepared by treating a carboxylic acid with an alcohol in the presence of an acid catalyst, most commonly H_2SO_4, $ArSO_3H$, or gaseous HCl. Conversion of a carboxylic acid and an alcohol to an ester is given the special name **Fischer esterification** after the German chemist, Emil Fischer (1852–1919). An example of Fischer esterification is the treatment of acetic acid with ethanol in the presence of concentrated sulfuric acid, which gives ethyl acetate and water.

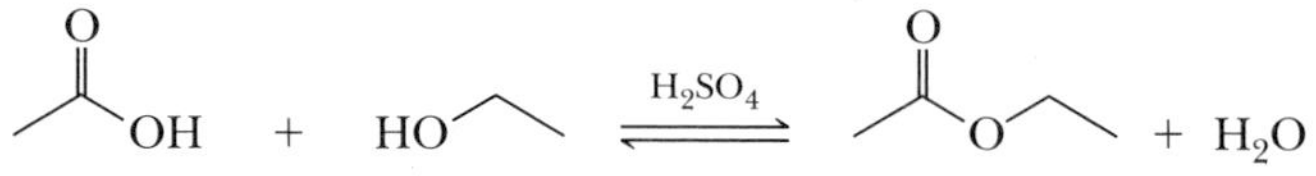

Ethanoic acid (Acetic acid) **Ethanol (Ethyl alcohol)** **Ethyl ethanoate (Ethyl acetate)**

Acid-catalyzed esterification is reversible, and generally, the quantities of both carboxylic acid and alcohol remaining at equilibrium are appreciable. If, for example, 60.1 g (1.00 mol) of acetic acid and 60.1 g (1.00 mol) of 1-propanol are heated under reflux in the presence of a few drops of concentrated sulfuric acid until equilibrium is reached, the reaction mixture contains approximately 0.67 mol each of propyl acetate and water and 0.33 mol each of acetic acid and 1-propanol. Thus, at equilibrium, about 67% of the carboxylic acid and alcohol are converted to the desired ester. The mechanism of Fischer esterification involves nucleophilic acyl substitution, which is a major focus of the next chapter. Consequently, the mechanism of this reaction will be covered in Chapter 18.

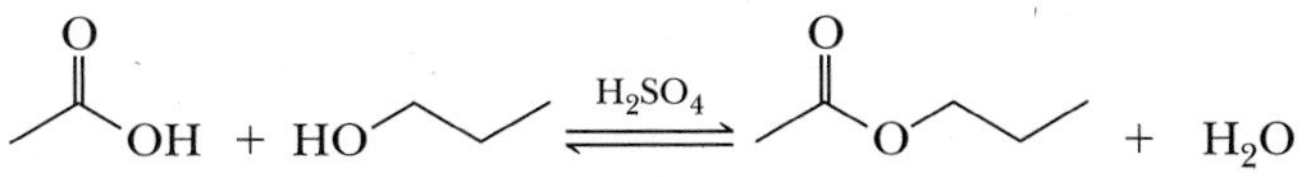

	Acetic acid	1-Propanol	Propyl acetate	
Initial:	1.00 mol	1.00 mol	0.00 mol	0.00 mol
Equilibrium:	0.33 mol	0.33 mol	0.67 mol	0.67 mol

By control of reaction conditions, it is possible to use Fischer esterification to prepare esters in high yields. If the alcohol is inexpensive compared with the carboxylic acid, a large excess of it can be used to drive the equilibrium to the right and achieve a high conversion of carboxylic acid to its ester. Alternatively, water can be removed by azeotropic distillation and a Dean-Stark trap (Figure 16.1).

Example 17.4 Using the Fischer Esterification Reaction

Complete the equation for each Fischer esterification reaction.

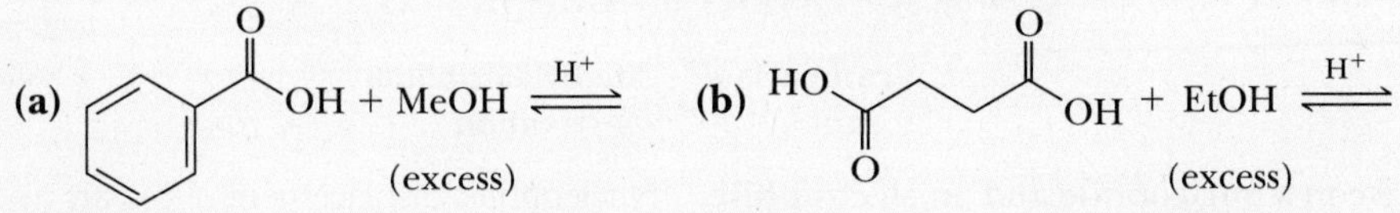

Solution

Following is the structural formula for the ester produced in each reaction.

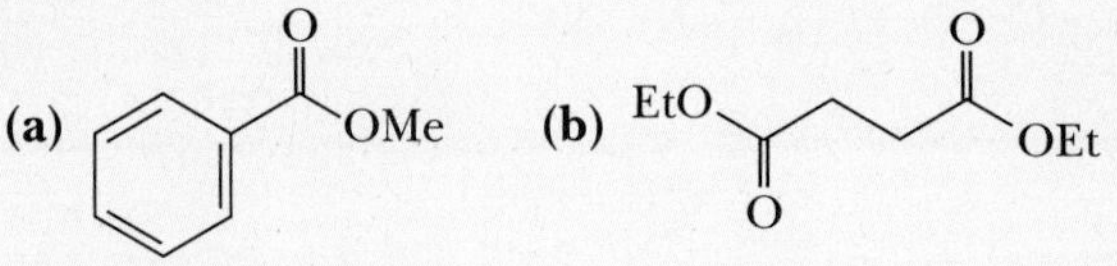

Problem 17.4

Complete the equation for each Fischer esterification.

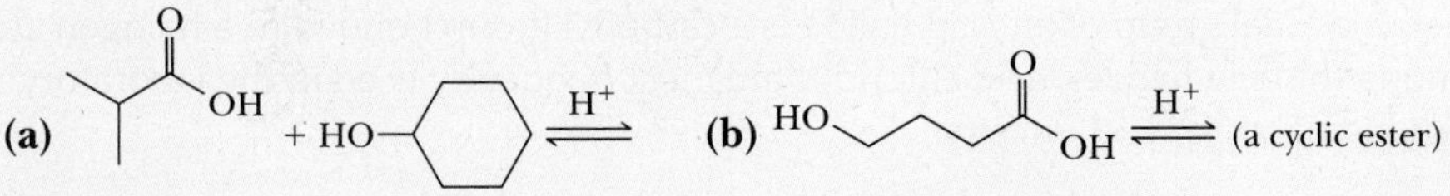

B. Formation of Methyl Esters Using Diazomethane

Treating a carboxylic acid with diazomethane, usually in ether solution, converts the carboxylic acid under mild conditions and in very high yield to its methyl ester.

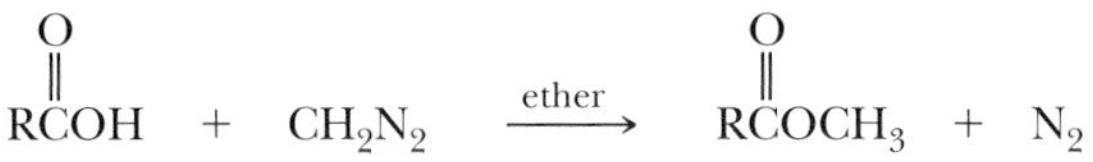

Diazomethane A methyl ester

Diazomethane, a potentially explosive, toxic **yellow gas**, is best represented as a hybrid of two resonance contributing structures.

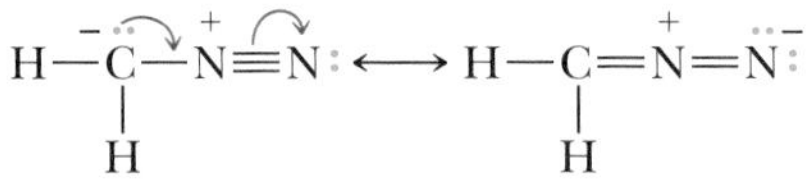

Diazomethane
(a resonance hybrid of two important contributing structures)

Because of the hazards associated with the use of diazomethane, it is used only where other means of preparation of methyl esters are too harsh, and even then, it is used only in small quantities.

Mechanism 17.1

Formation of a Methyl Ester Using Diazomethane

The reaction of a carboxylic acid with diazomethane occurs in two steps.

Step 1: **Take a proton away.** Proton transfer from the carboxyl group to diazomethane gives a carboxylate anion and methyldiazonium cation.

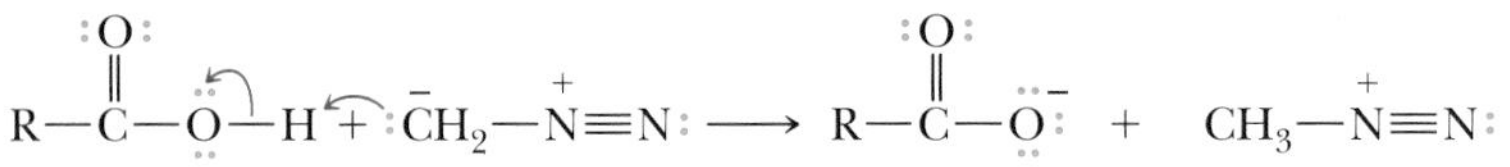

A carboxylate anion **Methyldiazonium cation**

Step 2: **Make a new bond between a nucleophile and an electrophile.** Nucleophilic displacement of N_2, an extraordinarily good leaving group, gives the methyl ester.

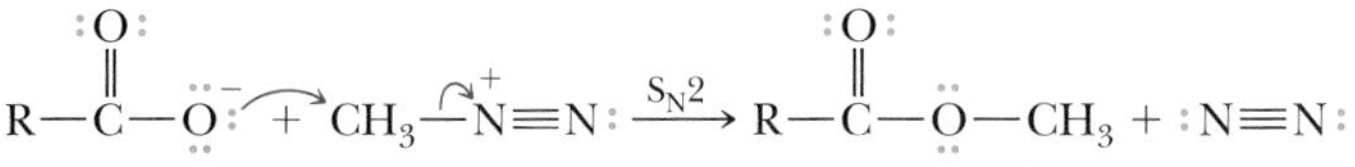

17.8 Conversion to Acid Chlorides

The functional group of an acid halide is a carbonyl group bonded to a halogen atom. Among the acid halides, acid chlorides are most frequently used in the laboratory and in industrial organic chemistry.

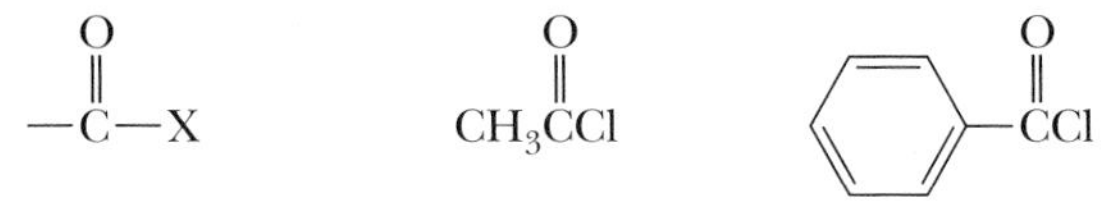

Functional group of an acid halide **Acetyl chloride** **Benzoyl chloride**

We study the nomenclature, structure, and characteristic reactions of acid halides in Chapter 18. Here, we are concerned only with their synthesis from carboxylic acids.

Chemical Connections

Esters as Flavoring Agents

Flavoring agents are the largest class of food additives. At the present time, over a thousand synthetic and natural flavors are available. The majority of these are concentrates or extracts from the material whose flavor is desired. These flavoring agents are often complex mixtures of tens to hundreds of compounds. A number of flavoring agents, many of them esters, however, are synthesized industrially. Many of these synthetic flavoring agents are major components of the natural flavors, and adding only one or a few of them is sufficient to make ice cream, soft drinks, and candy taste naturally flavored.

Structure	Name	Flavor
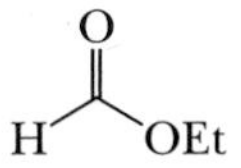	**Ethyl formate**	Rum
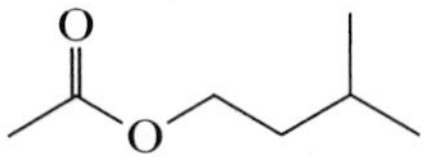	**(3-Methyl)butyl acetate (Isopentyl acetate)**	Banana
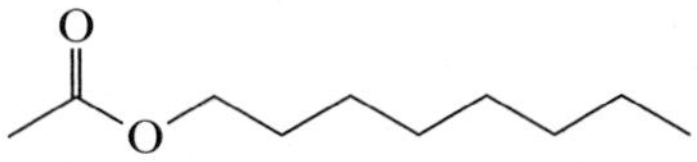	**Octyl acetate**	Orange
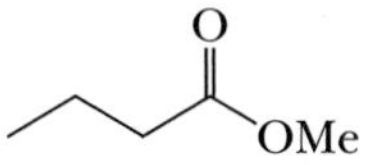	**Methyl butanoate**	Apple
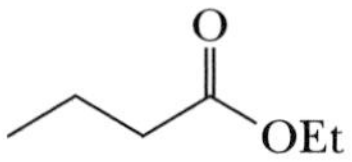	**Ethyl butanoate**	Pineapple
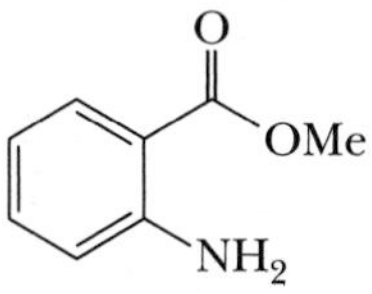	**Methyl 2-aminobenzoate (Methyl anthranilate)**	Grape

Acid chlorides are most often prepared by treating a carboxylic acid with thionyl chloride, the same reagent used to convert an alcohol to a chloroalkane (Section 10.5C).

OH
+
$SOCl_2$
Cl + SO_2 + HCl

Butanoic acid **Thionyl chloride** **Butanoyl chloride**

The mechanism for the reaction of thionyl chloride with a carboxylic acid to form an acid chloride is similar to that presented in Section 10.5C for the conversion of an alcohol to a chloroalkane and involves initial chlorosulfite formation, followed by nucleophilic attack of chloride ion on the carbonyl carbon to give a tetrahedral carbonyl addition intermediate, which decomposes to give the acid chloride, SO_2, and chloride ion.

Example 17.5 Making Acid Chlorides

Complete the equation for each reaction.

(a) 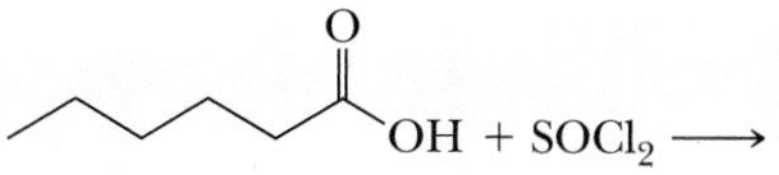
O
OH + SOCl2

(b) 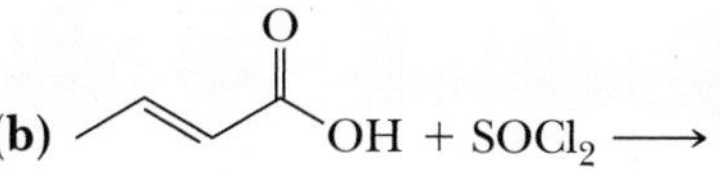
O
OH + SOCl2

Solution

Following are the products of each reaction.

(a) 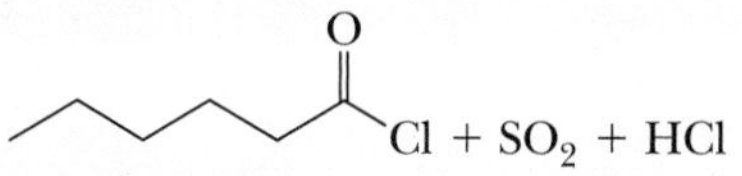
O
Cl + SO2 + HCl

(b) 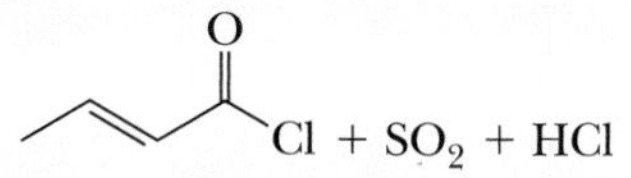
O
Cl + SO2 + HCl

Problem 17.5

Complete the equation for each reaction.

(a) 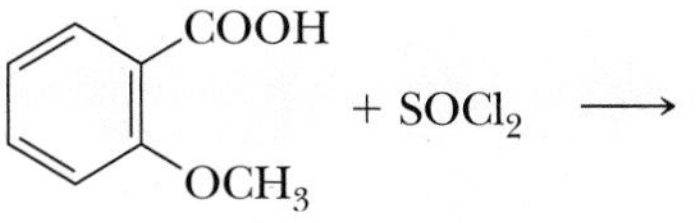
COOH
OCH3
+ SOCl2

(b) 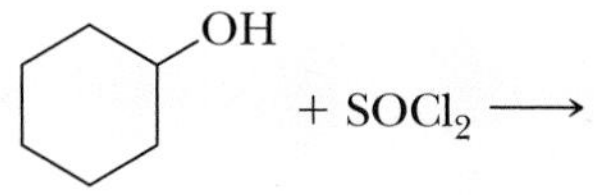
OH
+ SOCl2

MCAT Practice: *Passage and Questions*

Permethrin and Bifenthrin

Pyrethrin is a natural insecticide obtained from the powdered flower heads of several species of *Chrysanthemum*. The active substances in pyrethrum, principally pyrethrins I and II, are contact poisons for insects and cold-blooded vertebrates. Although powders made from *Chrysanthemum* extracts have found widespread use, the active substances in them are destroyed rapidly in the environment. In an effort to develop synthetic compounds as effective as the natural insecticides but with greater biostability, chemists prepared a series of esters related in structure. Among the synthetic pyrethrenoids now in common use in household and agricultural products are permethrin and bifenthrin.

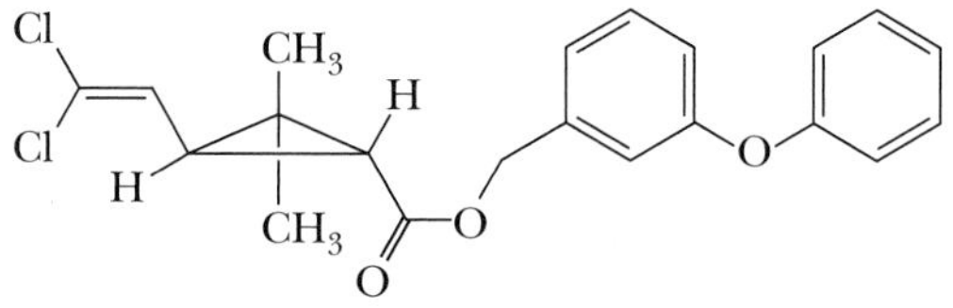
Cl
Cl
CH3
H
H
CH3
O
O
O

Permethrin

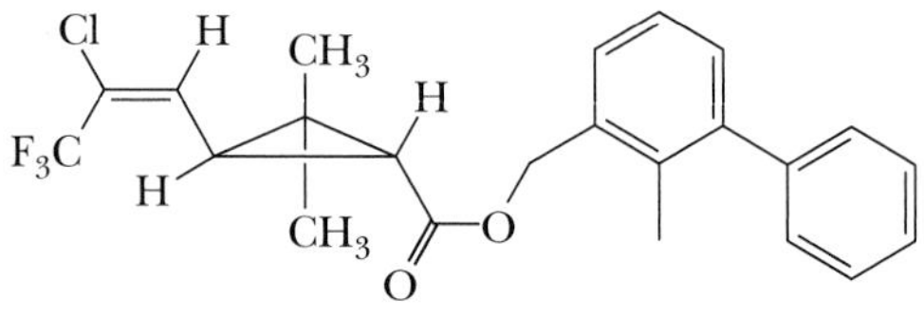
Cl
H
CH3
H
F3C
H
CH3
O
O

Bifenthrin

Questions

A. What is the stereochemical descriptor for the alkenes in permethrin and bifenthrin?

1. *Z* for permethrin and *E* for bifenthrin.
2. *E* for permethrin and *Z* for bifenthrin.
3. Neither alkene needs a stereochemical descriptor.
4. It is not appropriate to apply a descriptor to permethrin, but is it for bifenthrin. That descriptor is *E*.
5. It is not appropriate to apply a descriptor to bifenthrin, but it is for permethrin. That descriptor is *Z*.

B. What is the stereochemical descriptor for the substitution pattern of the cyclopropane rings in permethrin and bifenthrin?

1. Permethrin and bifenthrin are *trans*.
2. Permethrin and bifenthrin are *cis*.
3. Permethrin is *cis*, and bifenthrin is *trans*.
4. Permethrin is *trans*, and bifenthrin is *cis*.

C. Creation of the ester linkage in these two compounds is the last reaction in their syntheses. What reaction conditions could conceivably be used to create the esters?

1. Treatment of the carboxylic acid portion of the molecules with the benzylic alcohol portion under basic conditions.
2. Treatment of the carboxylic acid portion of the molecules with the benzylic alcohol portion under acidic conditions in water.
3. Treatment of the carboxylic acid portion of the molecule with thionyl chloride ($SOCl_2$) followed by addition of the benzylic alcohols corresponding in structure to the right sides of the molecules.
4. Both 2 and 3.
5. All of the above.

D. As discussed above, the natural products pyrethrins I and II (not shown) are destroyed rapidly in the environment. One of the key changes in the structures of permethrin and bifenthrin relative to the pyrethrins was the substitution of naturally occurring methyl groups on the alkene with electron withdrawing groups (EWGs) such as chlorine and trifluoromethyl. What reactions of the alkene would the change of methyl groups to EWGs retard?

1. Oxidation of the double bond.
2. Electrophilic addition reactions.
3. Nucleophilic addition reactions.
4. Both 1 and 2.
5. All of the above.

17.9 Decarboxylation

A. β-Ketoacids

Decarboxylation is the loss of CO_2 from the carboxyl group of a molecule. Almost any carboxylic acid, heated to a very high temperature, undergoes thermal decarboxylation.

Decarboxylation
Loss of CO_2 from a carboxyl group.

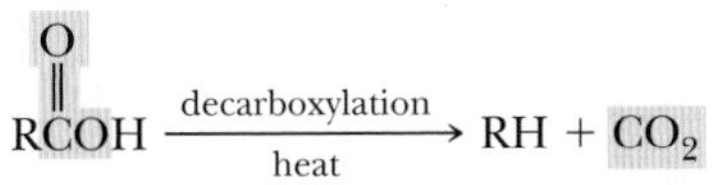

Most carboxylic acids, however, are quite resistant to moderate heat and melt or even boil without decarboxylation. Exceptions are carboxylic acids that have a carbonyl group β to the carboxyl group. This type of carboxylic acid undergoes decarboxylation quite readily on mild heating. For example, warming 3-oxobutanoic acid brings about its decarboxylation to give acetone and carbon dioxide.

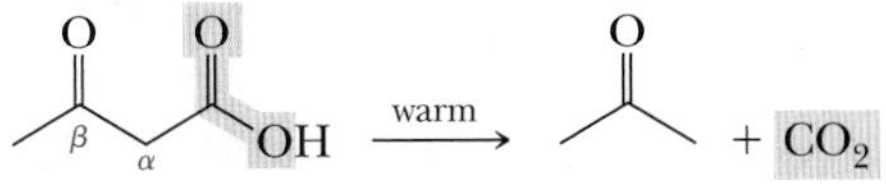

3-Oxobutanoic acid (Acetoacetic acid) **Acetone**

Decarboxylation upon moderate heating is a unique property of 3-oxocarboxylic acids (β-ketoacids) and is not observed with other classes of ketoacids.

Mechanism 17.2

Decarboxylation of a β-Ketocarboxylic Acid

Step 1: Redistribution of six electrons in a cyclic six-membered transition state gives carbon dioxide and an enol.

Step 2: Keto-enol tautomerism (Section 16.9B) of the enol gives the more stable keto form of the product.

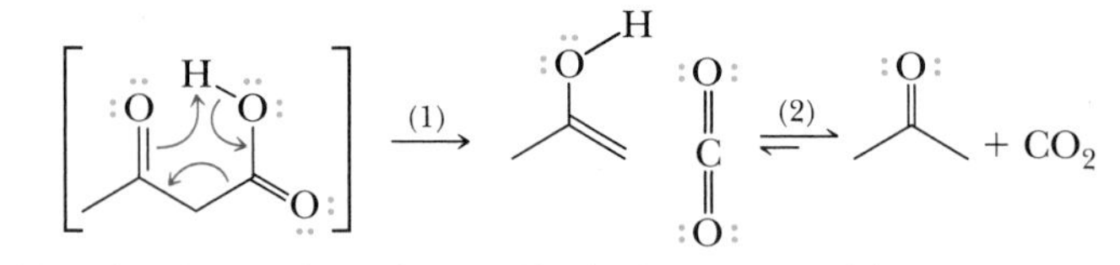

(A cyclic six-membered transition state) — Enol of a ketone — A ketone

A hydrogen bond between the carboxyl hydrogen atom and the β-carbonyl oxygen promotes the reaction by favoring a conformation on the path to the six-membered ring transition state. Through this conformational stabilization, the molecules have a much higher probability of undergoing reaction and the reaction occurs rapidly at moderate temperatures.

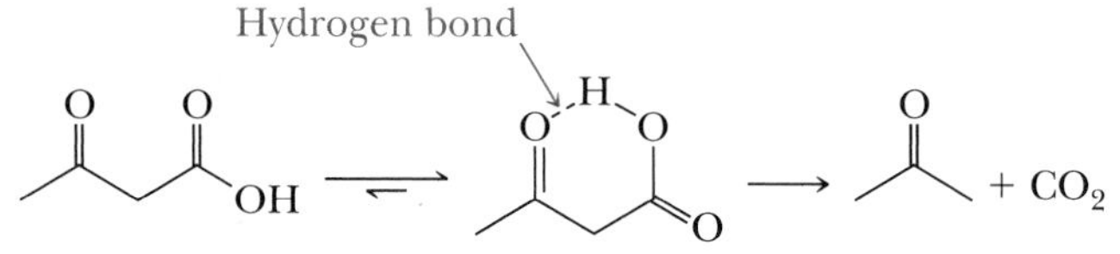

Reactive conformation

▶ Five- and six-membered rings are the most stable.

An important example of decarboxylation of a β-ketoacid in the biological world occurs during the oxidation of foodstuffs in the tricarboxylic acid (TCA) cycle. One of the intermediates in this cycle is oxalosuccinic acid, which undergoes spontaneous decarboxylation to produce α-ketoglutaric acid. Only one of the three carboxyl groups of oxalosuccinic acid has a carbonyl group in the β-position to it, and this carboxyl group is lost as CO_2.

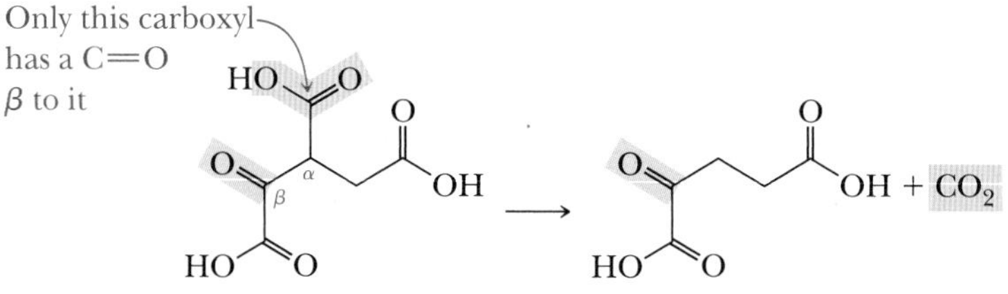

Oxalosuccinic acid — **α-Ketoglutaric acid**

B. Malonic Acid and Substituted Malonic Acids

The presence of a ketone or aldehyde carbonyl group β to the carboxyl group is sufficient to facilitate decarboxylation. In the more general reaction, decarboxylation is facilitated by the presence of any carbonyl group at the β-position, including that of a carboxyl group or an ester. Malonic acid and substituted malonic acids, for example, undergo thermal decarboxylation, as illustrated by the decarboxylation of malonic acid when it is heated slightly above its melting point of 135–137°C.

$$\mathrm{HO\overset{O}{\overset{\|}{C}}CH_2\overset{O}{\overset{\|}{C}}OH} \xrightarrow{140°\text{–}150°C} \mathrm{CH_3\overset{O}{\overset{\|}{C}}OH} + CO_2$$

Propanedioic acid
(Malonic acid)

Connections to Biological Chemistry

Ketone Bodies and Diabetes Mellitus

3-Oxobutanoic acid (acetoacetic acid) and its reduction product, 3-hydroxybutanoic acid, are synthesized in the liver from acetyl-CoA, a product of the metabolism of fatty acids and certain amino acids. 3-Hydroxybutanoic acid and 3-oxobutanoic acid are known collectively as ketone bodies.

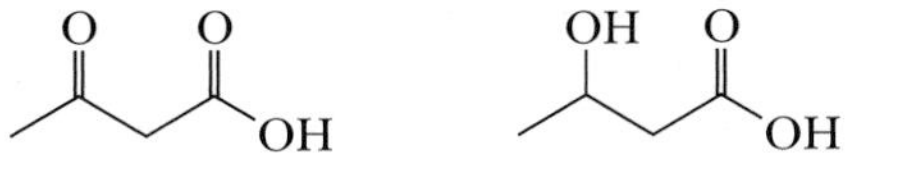

3-Oxobutanoic acid (Acetoacetic acid) **3-Hydroxybutanoic acid (β-Hydroxybutyric acid)**

The concentration of ketone bodies in the blood of healthy, well-fed humans is approximately 0.01 mmol/L. However, in people suffering from starvation or diabetes mellitus, the concentration of ketone bodies may increase to as much as 500 times normal. Under these conditions, the concentration of acetoacetic acid increases to the point that it undergoes spontaneous decarboxylation to form acetone and carbon dioxide. Acetone is not metabolized by humans and is excreted through the kidneys and the lungs. The odor of acetone is responsible for the characteristic "sweet smell" of the breath of severely diabetic patients.

The mechanism of decarboxylation of malonic acids is very similar to what we just saw for the decarboxylation of β-ketoacids. In this case, formation of a cyclic six-membered transition state involving rearrangement of three electron pairs gives the enol form of a carboxylic acid, which is tautomerized to the carboxylic acid.

Mechanism 17.3

Decarboxylation of a β-Dicarboxylic Acid

Step 1: Redistribution of six electrons in a cyclic six-membered transition state gives carbon dioxide and the enol form of a carboxyl group.

Step 2: Keto-enol tautomerism (Section 16.9B) of the enol gives the more stable keto form of the carboxyl group.

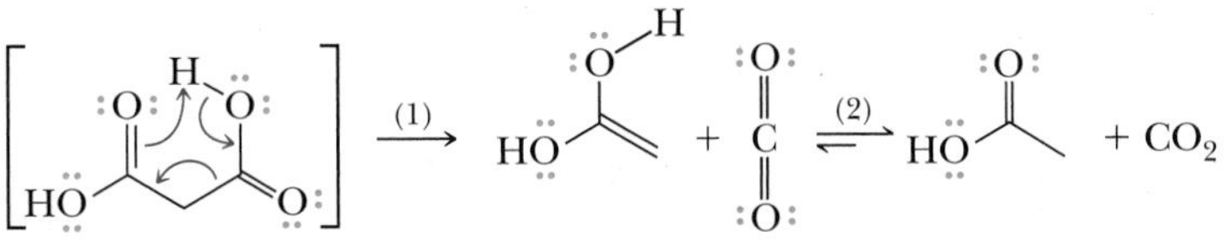

A cyclic six-membered transition state   Enol of a carboxylic acid   A carboxylic acid

Example 17.6 Decarboxylation Reactions

Each of these carboxylic acids undergoes thermal decarboxylation.

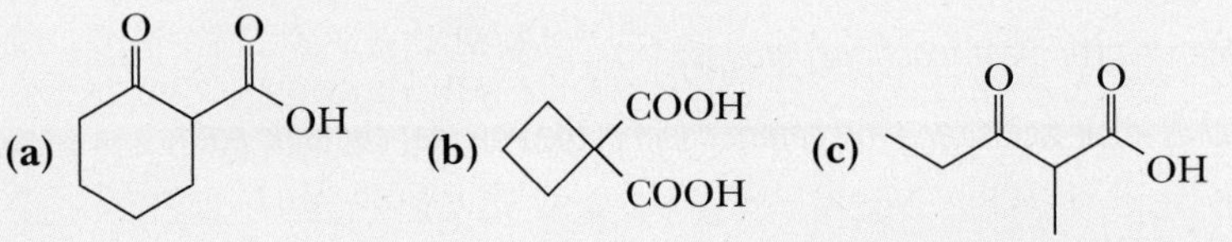

Draw a structural formula for the enol intermediate and final product formed in each reaction.

(Continued)

Solution

Following is a structural formula for the enol intermediate and the final product of each decarboxylation.

(a) 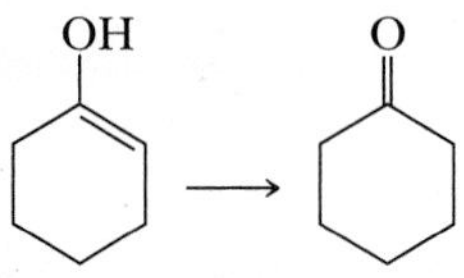

(b) 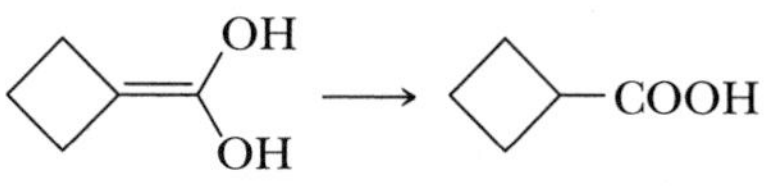

(c) 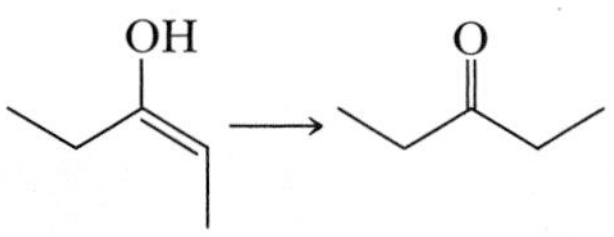

Problem 17.6

Account for the observation that the following β-ketoacid can be heated for extended periods at temperatures above its melting point without noticeable decarboxylation.

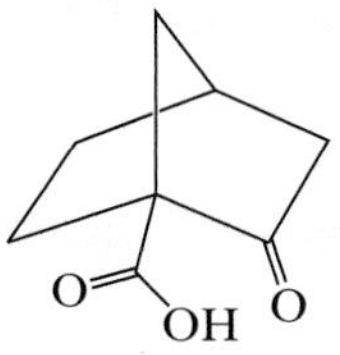

Study Guide

17.1 Structure

- A **carboxylic acid** (—COOH) contains a carbonyl group bonded to an —OH group.

17.2 Nomenclature

- IUPAC names of carboxylic acids are derived from the parent alkane by dropping the suffix *-e* and adding *-oic acid.*
  - *The carbon with the carboxyl group is understood to be carbon 1, so there is no need to give it a number.*
  - *The carboxyl group takes precedence over most functional groups.*
  - **Dicarboxylic acids** *are named as* -dioic acids, *and the parent chain is the one that contains both carboxyl groups.*

P **17.1, 17.7–17.14**

17.3 Physical Properties

- A carboxyl group is polar, and in the liquid and solid states, carboxylic acids are associated by hydrogen bonding into dimers.
  - *Carboxylic acids have higher boiling points and are more soluble in water than alcohols, aldehydes, ketones, and ethers of comparable molecular weight.*
- A carboxylic acid consists of two regions of distinctly different polarity: a polar **hydrophilic** carboxyl group, which increases solubility in water, and a nonpolar **hydrophobic** hydrocarbon chain, which decreases solubility in water.
  - *The low-molecular-weight carboxylic acids are infinitely soluble in water because the hydrophilic carboxyl group more than counterbalances the hydrophobic hydrocarbon chain.*
  - *As the size of the carbon chain increases, however, the hydrophobic group becomes dominant and solubility in water decreases.*

P 17.15–17.17

17.4 Acidity

- Values of pK_a for aliphatic carboxylic acids are in the range 4.0–5.0.
  - *The greater acidity of carboxylic acids compared with alcohols is explained by charge delocalization through resonance in a carboxylate anion relative to an alkoxide ion and the electron-withdrawing inductive effect of the carbonyl group.*
  - *Electron-withdrawing substituents near the carboxyl group increase its acidity via the inductive effect.*

P 17.2, 17.3, 17.25–17.32

KEY REACTIONS

1. **Acidity of Carboxylic Acids (Section 17.4A)** Values of pK_a for most unsubstituted aliphatic and aromatic carboxylic acids are within the range pK_a 4–5.

$$CH_3\overset{O}{\overset{\|}{C}}OH + H_2O \rightleftharpoons CH_3\overset{O}{\overset{\|}{C}}O^- + H_3O^+ \qquad K_a = 1.74 \times 10^{-5}$$

   The presence of electron-withdrawing groups near the carboxyl group decreases its pK_a (increases its acidity).

2. **Reaction of Carboxylic Acids with Bases (Section 17.4B)** Carboxylic acids form water-soluble salts with alkali metal hydroxides, carbonates, and bicarbonates, as well as with ammonia and aliphatic and aromatic amines.

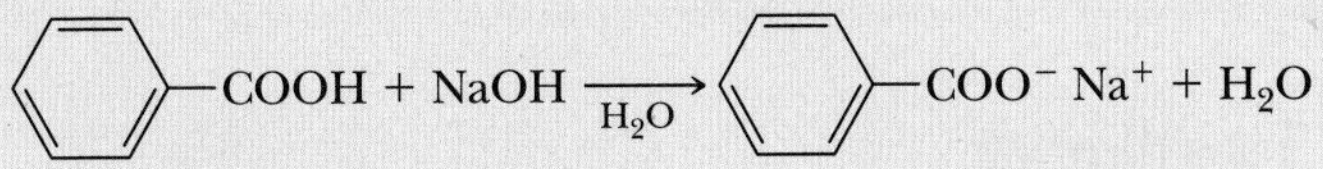

—COOH + NaOH
H_2O
—COO^- Na^+ + H_2O

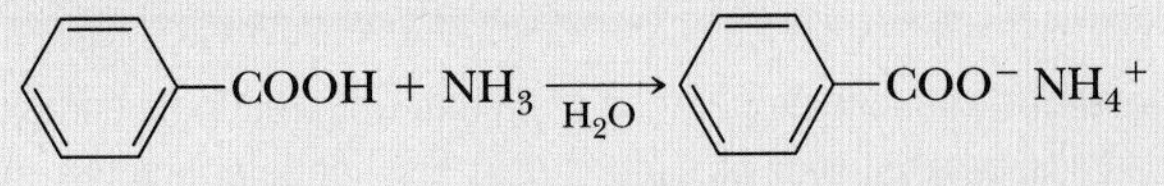

—COOH + NH_3
H_2O
—COO^- NH_4^+

17.5 Preparation of Carboxylic Acids

- Carboxylic acids can be prepared by oxidation of primary alcohols and aldehydes.
- Treating a Grignard reagent with carbon dioxide (CO_2) gives the magnesium salt of a carboxylic acid, which, upon protonation with aqueous acid, gives a carboxylic acid.

P 17.18–17.24, 17.42, 17.48

KEY REACTIONS

3. Carbonation of a Grignard reagent (Section 17.5) Adding CO_2 to a Grignard reagent followed by acidification provides a useful route to carboxylic acids.

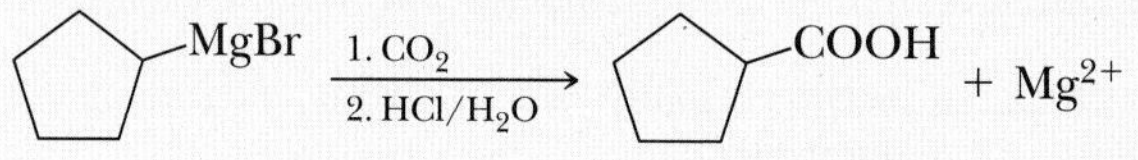

Cyclopentane-carboxylic acid

4. Industrial Preparation of Acetic Acid by the Carbonylation of Methanol (Section 17.6)

$$CO + 2\,H_2 \xrightarrow{\text{catalyst}} CH_3OH \xrightarrow[\text{rhodium(III)}\\ \text{HI, } H_2O]{CO} CH_3\overset{\overset{\displaystyle O}{\|}}{C}OH$$

17.6 Reduction

- Lithium aluminum hydride ($LiAlH_4$) reduces a carboxylic acid to a primary alcohol, although heating is required.
  - *Other reducing agents such as catalytic hydrogenation and* $NaBH_4$ *cannot reduce a carboxylic acid, so these can be used to reduce other functional groups without affecting a carboxyl group in the same molecule.*

P 17.32, 17.33, 17.34, 17.51

KEY REACTIONS

5. Reduction by Lithium Aluminum Hydride (Section 17.6A) Lithium aluminum hydride reduces a carboxyl group to a primary alcohol.

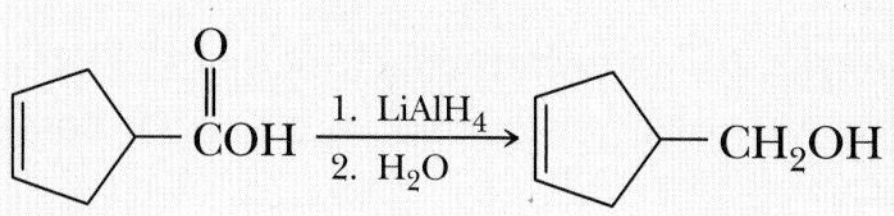

17.7 Esterification

- **Fischer esterification** is the preparation of an ester by treating a carboxylic acid with an alcohol in the presence of an acid catalyst such as sulfuric acid. Fischer esterification is reversible.
- Treating a carboxylic acid with diazomethane (CH_2N_2) gives a methyl ester in high yield.

P **17.4, 17.32, 17.35–17.39, 17.43–17.45**

KEY REACTIONS

6. Fischer Esterification (Section 17.7A) An ester can be prepared by treating a carboxylic acid with an alcohol in the presence of an acid catalyst.

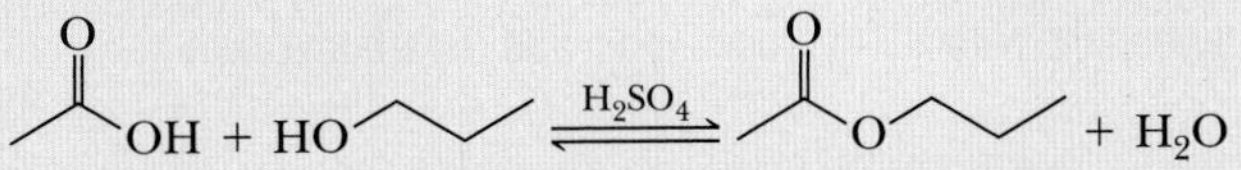

Fischer esterification is reversible. To achieve high yields of ester, it is necessary to force the equilibrium to the right. One way to accomplish this is to use an excess of alcohol; another is to remove water by azeotropic distillation using a Dean-Stark trap.

7. Reaction with Diazomethane (Section 17.7B) Diazomethane is used to form methyl esters from carboxylic acids. The mechanism involves protonation of the diazomethane carbon atom by the carboxylic acid to make a methyldiazonium cation, followed by attack of the resulting carboxylate on the methyldiazonium cation to give the methyl ester and N_2.

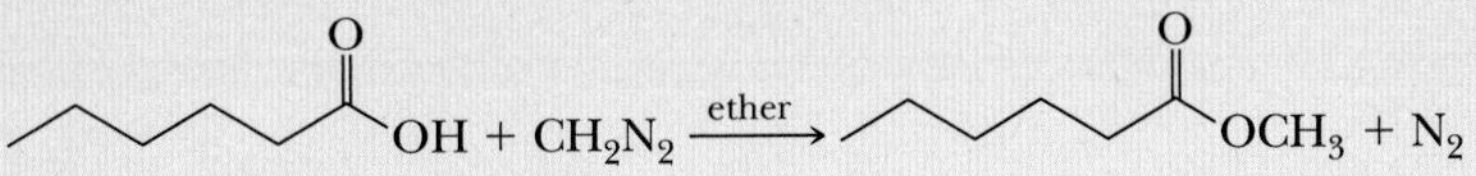

Because diazomethane is explosive and poisonous, it is used only when other means of preparing methyl esters are not suitable.

17.8 Conversion to Acid Chlorides

- Acid chlorides are prepared from a carboxyl group by treatment with thionyl chloride ($SOCl_2$).

P **17.5, 17.32, 17.47, 17.49**

KEY REACTIONS

8. Conversion to Acid Halides (Section 17.8) Acid chlorides, the most common and widely used of the acid halides, are prepared by treating a carboxylic acid with thionyl chloride. The mechanism, similar to that of the conversion of alcohols to chloroalkanes, involves initial chlorosulfite formation, followed by nucleophilic attack of chloride ion on the carbonyl carbon to give a tetrahedral carbonyl addition intermediate, which decomposes to give the acid chloride, SO_2, and chloride ion.

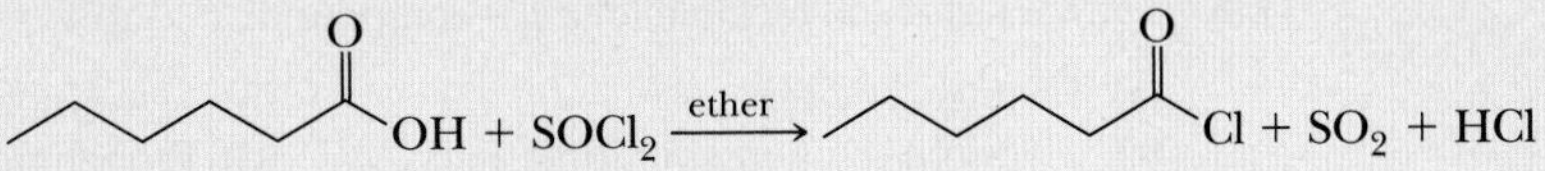

17.9 Decarboxylation

- Carboxylic acids with a carbonyl β to the carboxyl group undergo **decarboxylation** (loss of CO_2) upon heating.
  - *The reaction is important for β-keto acids as well as malonic acid derivatives.*
  - *The mechanism involves the redistribution of six electrons in a six-membered ring transition state.*

P 17.6, 17.40, 17.41

KEY REACTIONS

9. Decarboxylation of β-Ketoacids (Section 17.9A) β-Ketoacids decarboxylate upon heating. The mechanism involves redistribution of electrons in a six-membered transition state to give CO_2 and the enol of a ketone, which tautomerizes to give a ketone. The reaction is facilitated by a hydrogen bond between the carboxyl hydrogen atom and β-carbonyl oxygen.

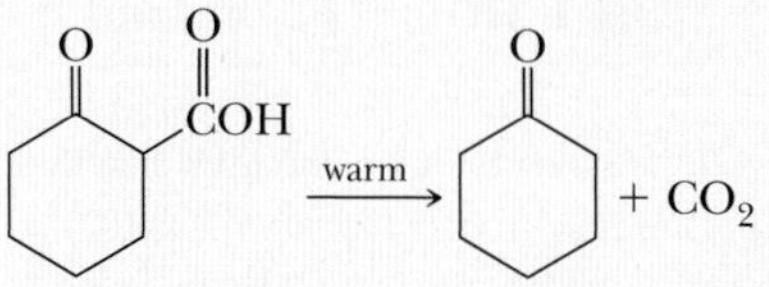

10. Decarboxylation of β-Dicarboxylic Acids (Section 17.9B) The mechanism of decarboxylation of a β-dicarboxylic acid is similar to that for decarboxylation of a β-ketoacid.

$$\mathrm{HO\overset{O}{\overset{\|}{C}}CH_2\overset{O}{\overset{\|}{C}}OH \xrightarrow{\text{heat}} CH_3\overset{O}{\overset{\|}{C}}OH + CO_2}$$

Problems

Red numbers indicate applied problems.

Structure and Nomenclature

17.7 Write the IUPAC name of each compound, showing stereochemistry where relevant.

(a) 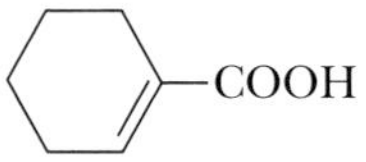

(b) 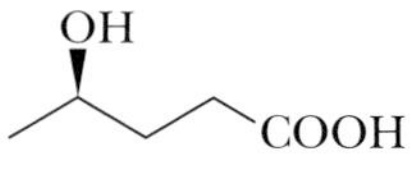

(c) 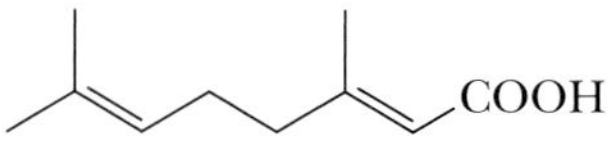

(d) 

(e) 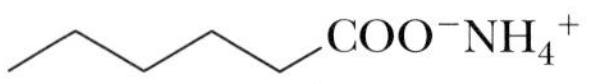

(f) 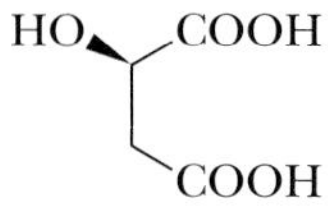

17.8 Draw a structural formula for each compound.

(a) Phenylacetic acid
(b) 4-Aminobutanoic acid
(c) 3-Chloro-4-phenylbutanoic acid
(d) Propenoic acid (acrylic acid)
(e) (Z)-3-Hexenedioic acid
(f) 2-Pentynoic acid
(g) Potassium phenylacetate
(h) Sodium oxalate
(i) 2-Oxocyclohexanecarboxylic acid
(j) 2,2-Dimethylpropanoic acid

17.9 Megatomoic acid, the sex attractant of the female black carpet beetle, has the following structure.

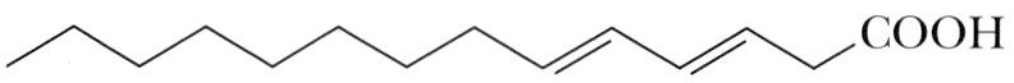

(a) What is its IUPAC name?
(b) State the number of stereoisomers possible for this compound.

17.10 Draw a structural formula for each salt.

(a) Sodium benzoate
(b) Lithium acetate
(c) Ammonium acetate
(d) Disodium adipate
(e) Sodium salicylate
(f) Calcium butanoate

17.11 The monopotassium salt of oxalic acid is present in certain leafy vegetables, including rhubarb. Both oxalic acid and its salts are poisonous in high concentrations. Draw the structural formula of monopotassium oxalate.

17.12 Potassium sorbate is added as a preservative to certain foods to prevent bacteria and molds from causing food spoilage and to extend the foods' shelf life. The IUPAC name of potassium sorbate is potassium (2*E*,4*E*)-2,4-hexadienoate. Draw a structural formula for potassium sorbate.

17.13 Zinc 10-undecenoate, the zinc salt of 10-undecenoic acid, is used to treat certain fungal infections, particularly *Tinea pedis* (athlete's foot). Draw a structural formula for this zinc salt.

17.14 On a cyclohexane ring, an axial carboxyl group has a conformational energy of 5.9 kJ (1.4 kcal)/mol relative to an equatorial carboxyl group. Consider the equilibrium for the alternative chair conformations of *trans*-1,4-cyclohexanedicarboxylic acid. Draw the less stable chair conformation on the left of the equilibrium arrows and the more stable chair on the right. Calculate ΔG^0 for the equilibrium as written and calculate the ratio of the more stable chair to the less stable chair at 25°C.

Physical Properties

17.15 Arrange the compounds in each set in the order of increasing boiling point.

(a) $CH_3(CH_2)_5COOH$ $CH_3(CH_2)_6CHO$ $CH_3(CH_2)_6CH_2OH$

(b) 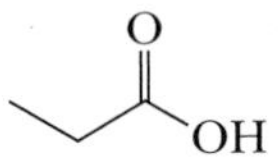 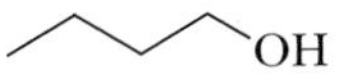 

17.16 Acetic acid has a boiling point of 118°C, whereas its methyl ester has a boiling point of 57°C. Account for the fact that the boiling point of acetic acid is higher than that of its methyl ester even though acetic acid has a lower molecular weight.

17.17 Given here are ^{1}H-NMR and ^{13}C-NMR spectral data for nine compounds. Each compound shows strong absorption between 1720 and 1700 cm^{-1} and strong, broad absorption over the region 2500–3300 cm^{-1}. Propose a structural formula for each compound. Refer to Appendices 4, 5, and 6 for spectral correlation tables.

(a) $C_5H_{10}O_2$

^{1}H-NMR	^{13}C-NMR
0.94 (t, 3H)	180.71
1.39 (m, 2H)	33.89
1.62 (m, 2H)	26.76
2.35 (t, 2H)	22.21
12.0 (s, 1H)	13.69

(b) $C_6H_{12}O_2$

^{1}H-NMR	^{13}C-NMR
1.08 (s, 9H)	179.29
2.23 (s, 2H)	47.82
12.1 (s, 1H)	30.62
	29.57

(c) $C_5H_8O_4$

^{1}H-NMR	^{13}C-NMR
0.93 (t, 3H)	170.94
1.80 (m, 2H)	53.28
3.10 (t, 1H)	21.90
12.7 (s, 2H)	11.81

(d) $C_5H_8O_4$

^{1}H-NMR	^{13}C-NMR
1.29 (s, 6H)	174.01
12.8 (s, 2H)	48.77
	22.56

(e) $C_4H_6O_2$

^{1}H-NMR	^{13}C-NMR
1.91 (d, 3H)	172.26
5.86 (d, 1H)	147.53
7.10 (m, 1H)	122.24
12.4 (s, 1H)	18.11

(f) $C_3H_4Cl_2O_2$

^{1}H-NMR	^{13}C-NMR
2.34 (s, 3H)	171.82
11.3 (s, 1H)	79.36
	34.02

(g) $C_5H_8Cl_2O_2$

^{1}H-NMR	^{13}C-NMR
1.42 (s, 6H)	180.15
6.10 (s, 1H)	77.78
12.4 (s, 1H)	51.88
	20.71

(h) $C_5H_9BrO_2$

^{1}H-NMR	^{13}C-NMR
0.97 (t, 3H)	176.36
1.50 (m, 2H)	45.08
2.05 (m, 2H)	36.49
4.25 (t, 1H)	20.48
12.1 (s, 1H)	13.24

(i) $C_4H_8O_3$

^{1}H-NMR	^{13}C-NMR
2.62 (t, 2H)	177.33
3.38 (s, 3H)	67.55
3.68 (s, 2H)	58.72
11.5 (s, 1H)	34.75

Preparation of Carboxylic Acids

17.18 Complete each reaction.

(a) 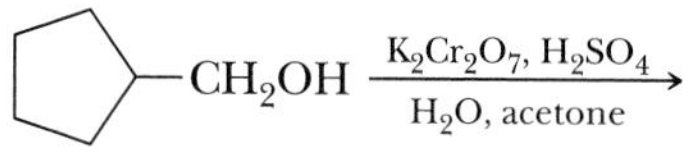
CH2OH
K2Cr2O7, H2SO4
H2O, acetone

(b) 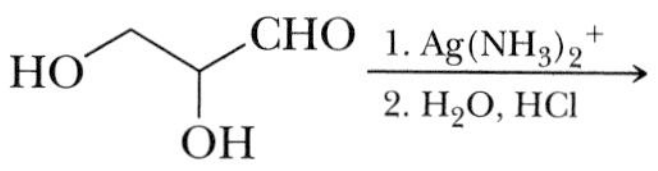
HO
CHO
OH
1. Ag(NH3)2+
2. H2O, HCl

(c) 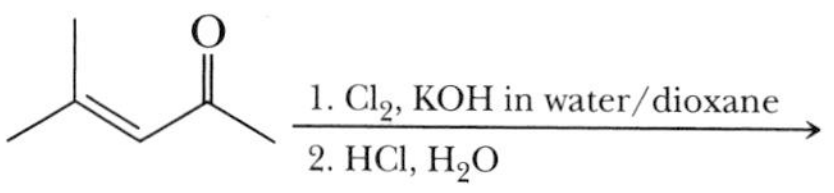
O
1. Cl2, KOH in water/dioxane
2. HCl, H2O

(d) 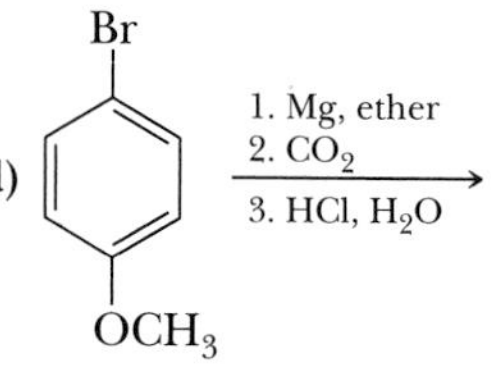
Br
OCH3
1. Mg, ether
2. CO2
3. HCl, H2O

17.19 Show how to bring about each conversion in good yield.

(a) 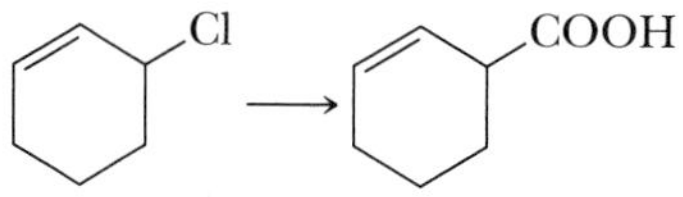
Cl
COOH

(b) 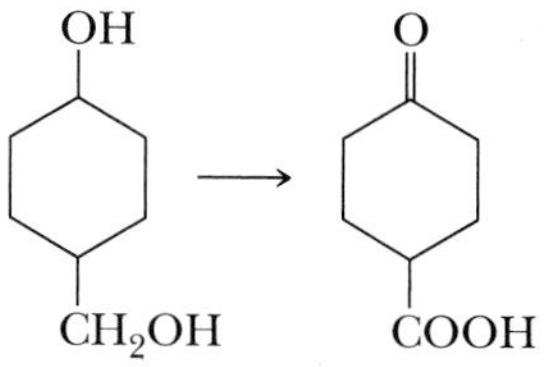
OH
CH2OH
O
COOH

(c) 
C6H5
OH
C6H5
COOH

17.20 Show how to prepare pentanoic acid from each compound.

(a) 1-Pentanol (b) Pentanal (c) 1-Pentene
(d) 1-Butanol (e) 1-Bromopropane (f) 1-Hexene

17.21 Draw the structural formula of a compound with the given molecular formula that, upon oxidation by potassium dichromate in aqueous sulfuric acid, gives the carboxylic acid or dicarboxylic acid shown.

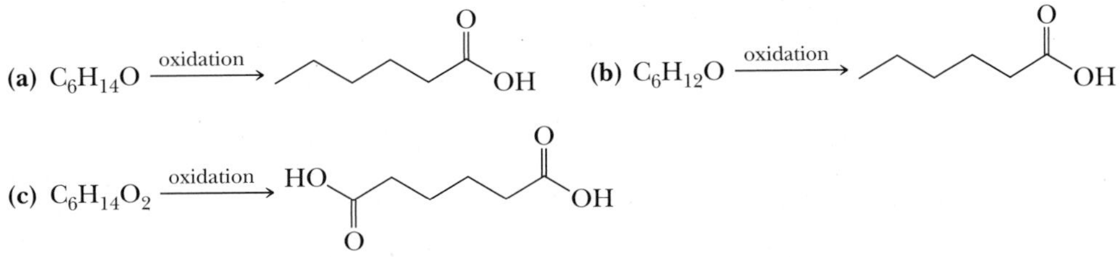
(a) C6H14O
oxidation
OH
O
(b) C6H12O
oxidation
OH
O
(c) C6H14O2
oxidation
HO
OH
O
O

17.22 Show the reagents and experimental conditions necessary to bring about each conversion in good yield.

(a) 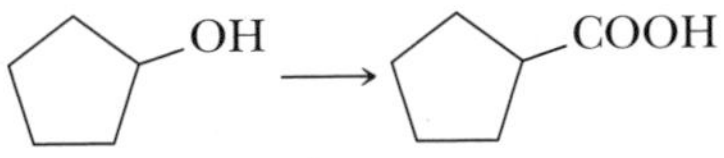
OH
COOH

(b) $CH_3\underset{CH_3}{\overset{CH_3}{|}}COH \longrightarrow CH_3\underset{CH_3}{\overset{CH_3}{|}}CCOOH$

(c) $CH_3\underset{CH_3}{\overset{CH_3}{|}}COH \longrightarrow CH_3\overset{CH_3}{|}CHCOOH$

(d) $CH_3\underset{CH_3}{\overset{CH_3}{|}}COH \longrightarrow CH_3\overset{CH_3}{|}CHCH_2COOH$

(e) $CH_3CH{=}CHCH_3 \longrightarrow CH_3CH{=}CHCH_2COOH$

17.23 Succinic acid can be synthesized by the following series of reactions from acetylene. Show the reagents and experimental conditions necessary to carry out this synthesis.

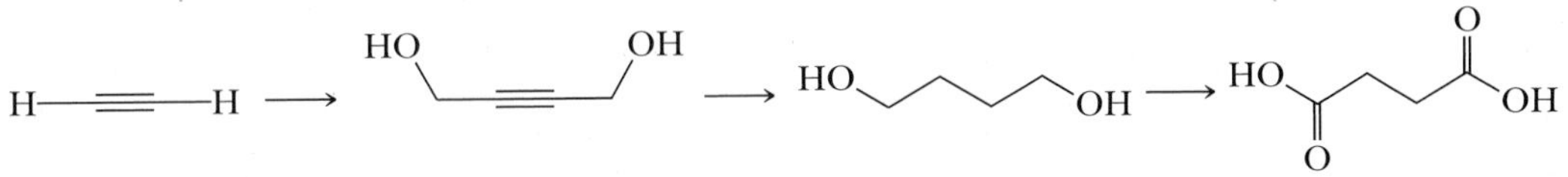
H
H
HO
OH
HO
OH
HO
OH
O
O

Acetylene **2-Butyne-1,4-diol** **1,4-Butanediol** **Butanedioic acid (Succinic acid)**

17.24 The reaction of an α-diketone with concentrated sodium or potassium hydroxide to give the salt of an α-hydroxyacid is given the general name benzil-benzilic acid rearrangement. It is illustrated by the conversion of benzil to sodium benzilate and then to benzilic acid. Propose a mechanism for this rearrangement.

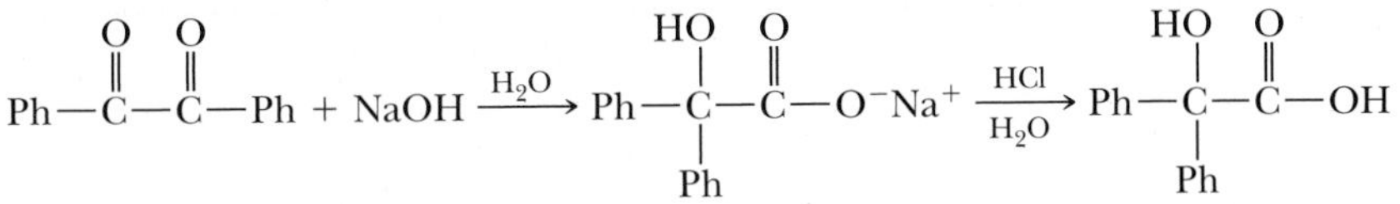
O O
Ph—C—C—Ph + NaOH
H2O
HO O
Ph—C—C—O−Na+
Ph
HCl
H2O
HO O
Ph—C—C—OH
Ph

Benzil (an α-diketone) **Sodium benzilate** **Benzilic acid**

Acidity of Carboxylic Acids

17.25 Select the stronger acid in each set.

(a) Phenol (pK_a 9.95) and benzoic acid (pK_a 4.19)
(b) Lactic acid (K_a 8.4×10^{-4}) and ascorbic acid (K_a 7.9×10^{-5})

17.26 In each set, assign the acid its appropriate pK_a.

(a) 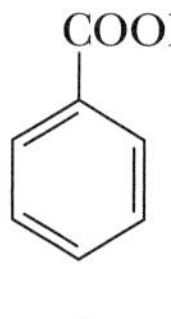 and 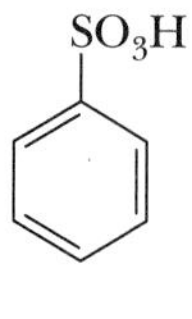 (pK_a 4.19 and 0.70)

(b) 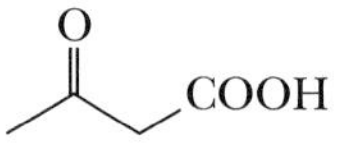 and 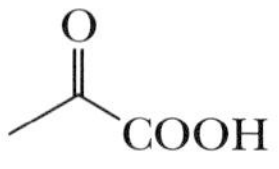 (pK_a 3.58 and 2.49)

(c) CH_3CH_2COOH and $N{\equiv}CCH_2COOH$ (pK_a 4.78 and 2.45)

17.27 Low-molecular-weight dicarboxylic acids normally exhibit two different pK_a values. Ionization of the first carboxyl group is easier than the second. This effect diminishes with molecular size, and for adipic acid and longer chain dicarboxylic acids, the two acid ionization constants differ by about one pK unit.

Dicarboxylic Acid	Structural Formula	pK_{a1}	pK_{a2}
Oxalic	HOOCCOOH	1.23	4.19
Malonic	$HOOCCH_2COOH$	2.83	5.69
Succinic	$HOOC(CH_2)_2COOH$	4.16	5.61
Glutaric	$HOOC(CH_2)_3COOH$	4.31	5.41
Adipic	$HOOC(CH_2)_4COOH$	4.43	5.41

Why do the two pK_a values differ more for the shorter chain dicarboxylic acids than for the longer chain dicarboxylic acids?

17.28 Complete the following acid-base reactions.

(a) 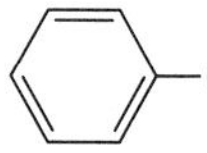 $-CH_2COOH + NaOH \longrightarrow$ (b) $CH_3CH{=}CHCH_2COOH + NaHCO_3 \longrightarrow$

(c) 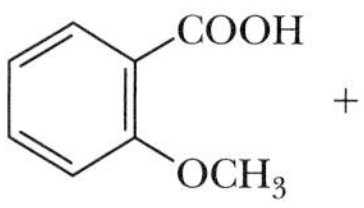 $+ NaHCO_3 \longrightarrow$ (d) $CH_3CH(OH)COOH + H_2NCH_2CH_2OH \longrightarrow$

(e) $CH_3CH{=}CHCH_2COO^-\ Na^+ + HCl \longrightarrow$
(f) $CH_3CH_2CH_2CH_2Li + CH_3COOH \longrightarrow$
(g) $CH_3CH_2CH_2CH_2MgBr + CH_3CH_2OH \longrightarrow$

17.29 The normal pH range for blood plasma is 7.35–7.45. Under these conditions, would you expect the carboxyl group of lactic acid (pK_a 3.08) to exist primarily as a carboxyl group or as a carboxylic anion? Explain.

17.30 The K_{a1} of ascorbic acid is 7.94×10^{-5}. Would you expect ascorbic acid dissolved in blood plasma (pH 7.35–7.45) to exist primarily as ascorbic acid or as ascorbate anion? Explain.

17.31 Excess ascorbic acid is excreted in the urine, the pH of which is normally in the range 4.8–8.4. What form of ascorbic acid would you expect to be present in urine of pH 8.4—free ascorbic acid or ascorbate anion? Explain.

Reactions of Carboxylic Acids

17.32 Give the expected organic product when phenylacetic acid, $PhCH_2COOH$, is treated with each reagent.

(a) $SOCl_2$
(b) $NaHCO_3$, H_2O
(c) NaOH, H_2O
(d) CH_3MgBr (one equivalent)

(e) $LiAlH_4$ followed by H_2O **(f)** CH_2N_2

(g) $CH_3OH + H_2SO_4$ (catalyst)

17.33 Show how to convert *trans*-3-phenyl-2-propenoic acid (cinnamic acid) to each compound.

(a) 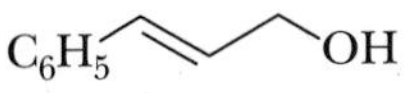 **(b)** 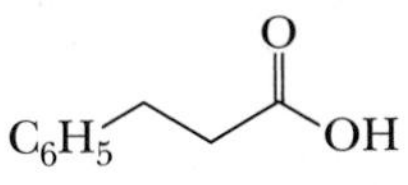 **(c)** 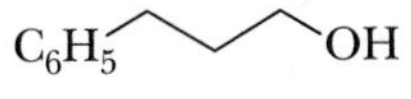

17.34 Show how to convert 3-oxobutanoic acid (acetoacetic acid) to each compound.

(a) 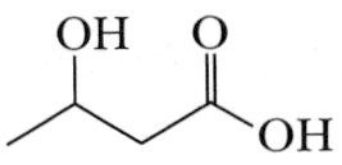
(racemic)

(b) 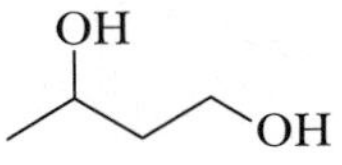
(racemic)

(c) 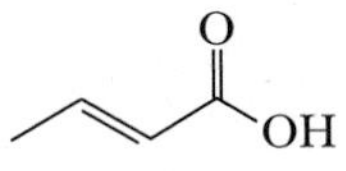

17.35 Complete these examples of Fischer esterification. Assume that the alcohol is present in excess.

(a) 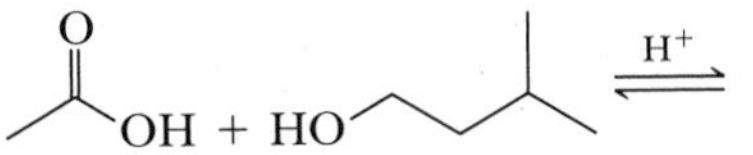

(b) 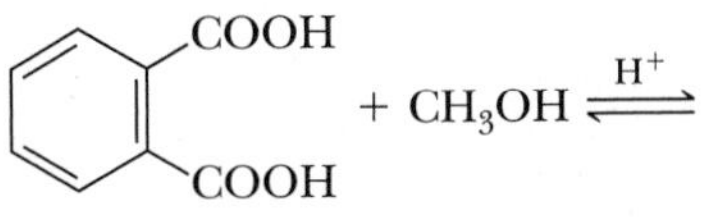

(c) 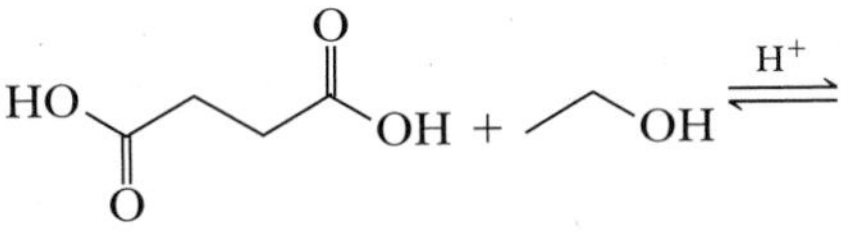

17.36 Benzocaine, a topical anesthetic, is prepared by treatment of 4-aminobenzoic acid with ethanol in the presence of an acid catalyst followed by neutralization. Draw a structural formula for benzocaine.

17.37 Name the carboxylic acid and alcohol from which each ester is derived.

(a) 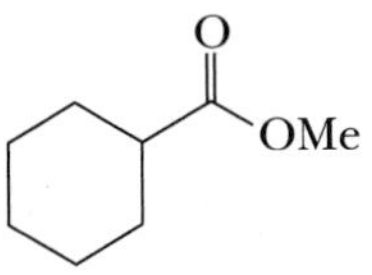

(b) 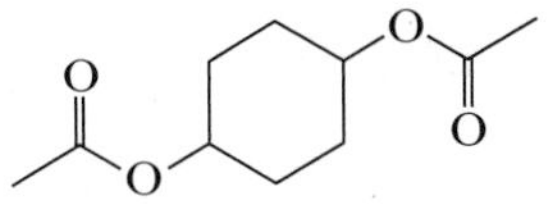

(c) 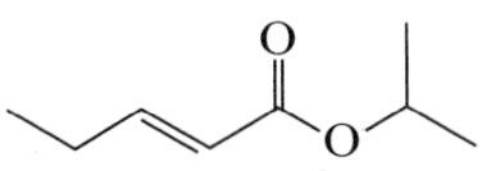

(d) 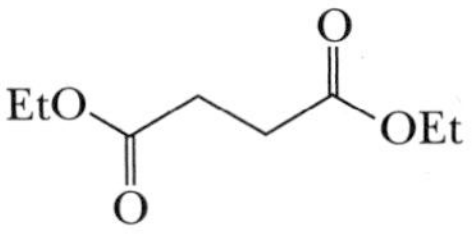

17.38 When 4-hydroxybutanoic acid is treated with an acid catalyst, it forms a lactone (a cyclic ester). Draw the structural formula of this lactone and propose a mechanism for its formation.

17.39 Fischer esterification cannot be used to prepare *tert*-butyl esters. Instead, carboxylic acids are treated with 2-methylpropene in the presence of an acid catalyst to generate them.

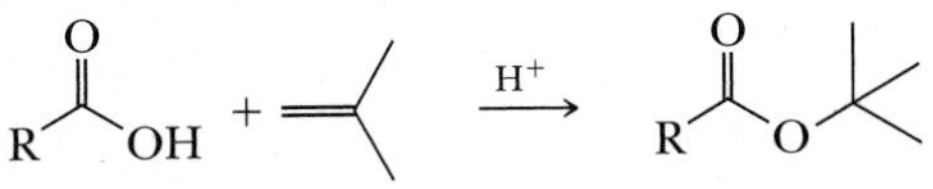

2-Methylpropene (Isobutylene) A *tert*-butyl ester

(a) Why does the Fischer esterification fail for the synthesis of *tert*-butyl esters?

(b) Propose a mechanism for the 2-methylpropene method.

17.40 Draw the product formed on thermal decarboxylation of each compound.

(a) 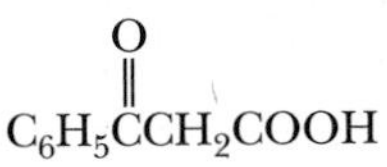 **(b)** 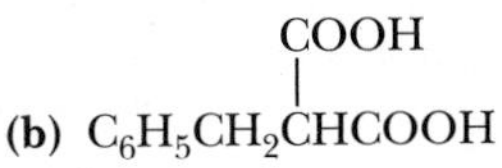 **(c)** 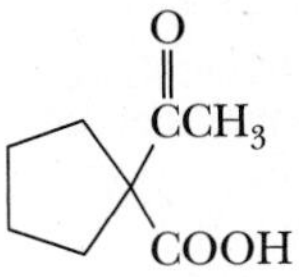

17.41 When heated, carboxylic salts in which there is a good leaving group on the carbon beta to the carboxylate group undergo decarboxylation/elimination to give an alkene.

(a) 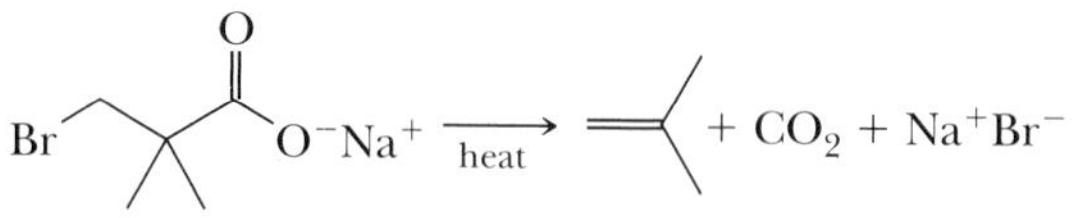

(b) 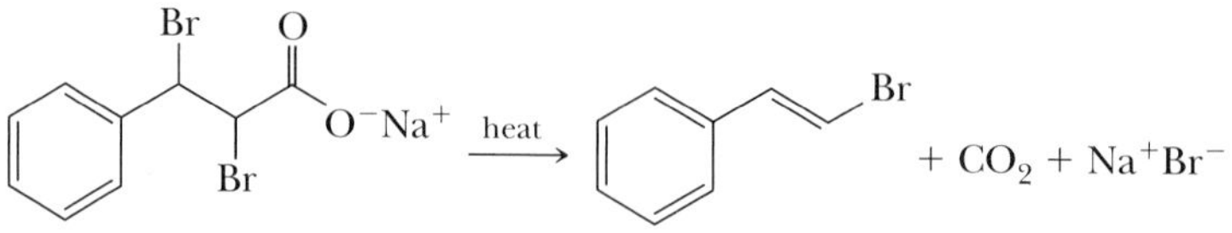

Propose a mechanism for this type of decarboxylation/elimination. Compare the mechanism of these decarboxylations with the mechanism for decarboxylation of β-ketoacids. In what way(s) are the mechanisms similar?

17.42 Show how cyclohexanecarboxylic acid could be synthesized from cyclohexane in good yield.

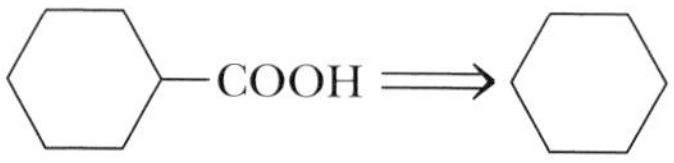

Looking Ahead

17.43 In Section 17.7B, we suggested that the mechanism of Fischer esterification of carboxylic acids is a model for the reactions of functional derivatives of carboxylic acids. One of these reactions is that of an acid chloride with water (Section 18.4A). Suggest a mechanism for this reaction.

$$R-\overset{\overset{\displaystyle O}{\|}}{C}-Cl + H_2O \longrightarrow R-\overset{\overset{\displaystyle O}{\|}}{C}-OH + HCl$$

17.44 We have studied Fischer esterification, in which a carboxylic acid is reacted with an alcohol in the presence of an acid catalyst to form an ester. Suppose that you start instead with a dicarboxylic acid such as terephthalic acid and a diol such as ethylene glycol. Show how Fischer esterification in this case can lead to a macromolecule with a molecular weight several thousand times that of the starting materials.

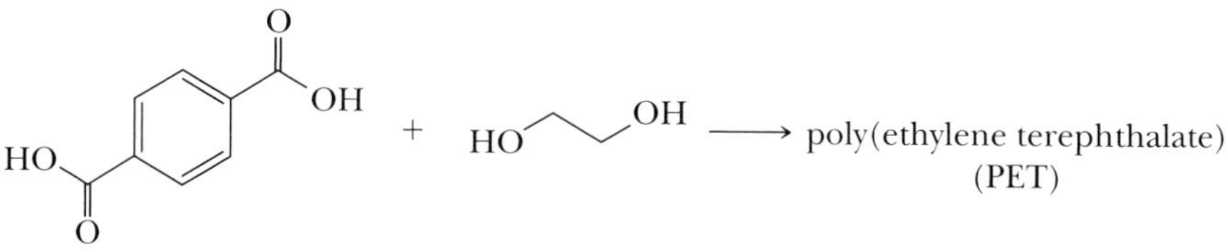

1,4-Benzenedicarboxylic acid (Terephthalic acid) **1,2-Ethanediol (Ethylene glycol)**

As we shall see in Section 29.5B, the material produced in this reaction is a high-molecular-weight polymer, which can be fabricated into Mylar films and into the textile fiber known as Dacron polyester.

Organic Chemistry Reaction Roadmap

17.45 Use the reaction roadmap you made for Problems 15.19 and 16.72 and update it to contain the reactions in the "Study Guide" section of this chapter. Because of their highly specific nature, do not use reactions 1, 2, 4, 9, and 10 on your roadmap.

17.46 Write the products of the following sequences of reactions. Refer to your reaction roadmaps to see how the combined reactions allow you to "navigate" between the different functional groups. Note that you will need both your old Chapters 6–11 roadmap and your new Chapters 15–17 roadmap for these.

(a) 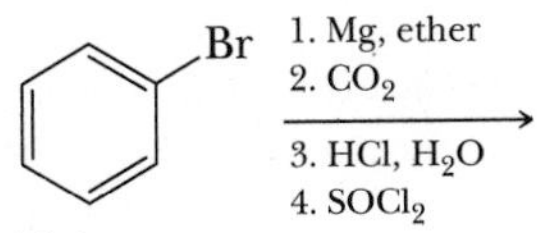
Br
1. Mg, ether
2. CO2
3. HCl, H2O
4. SOCl2
A haloarene

(b) 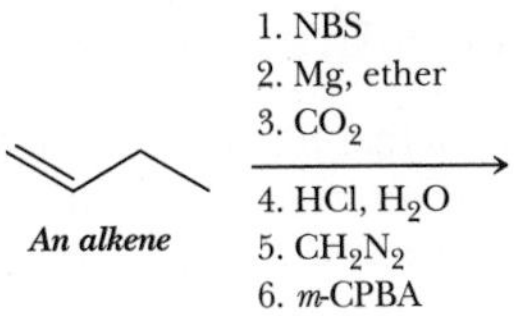
1. NBS
2. Mg, ether
3. CO2
4. HCl, H2O
5. CH2N2
6. m-CPBA
An alkene

(c) 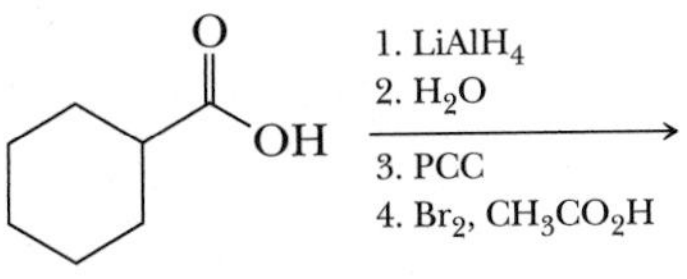
O
OH
1. LiAlH4
2. H2O
3. PCC
4. Br2, CH3CO2H
A carboxylic acid

(d) 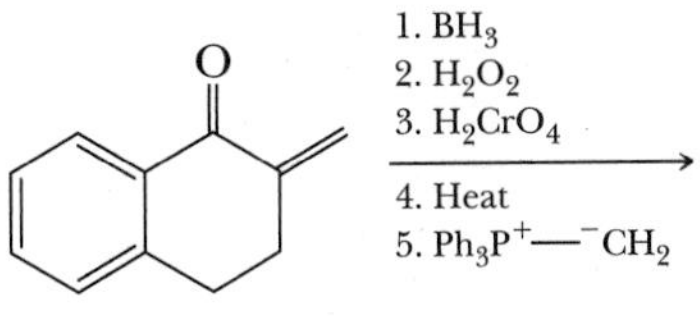
O
1. BH3
2. H2O2
3. H2CrO4
4. Heat
5. Ph3P+—−CH2
An alkene

Synthesis

17.47 Using your reaction roadmaps as a guide, show how to convert propane into propyl propanoate. You must use propane as the source of all carbon atoms in the target molecule. Show all reagents needed and all molecules synthesized along the way. Reaction Roadmap

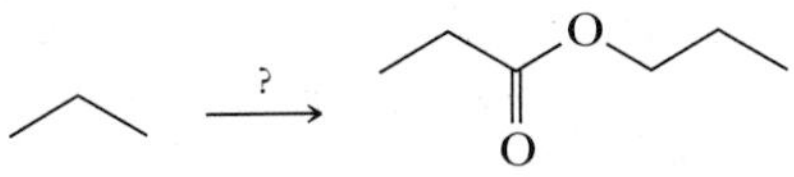
?
O
O

Propane **Propyl propanoate**

17.48 Using your reaction roadmaps as a guide, show how to convert 4-methyl-1-pentene and carbon dioxide into 5-methylhexanoic acid. You must use 4-methyl-1-pentene and carbon dioxide as the source of all carbon atoms in the target molecule. Show all reagents and all molecules synthesized along the way. Reaction Roadmap

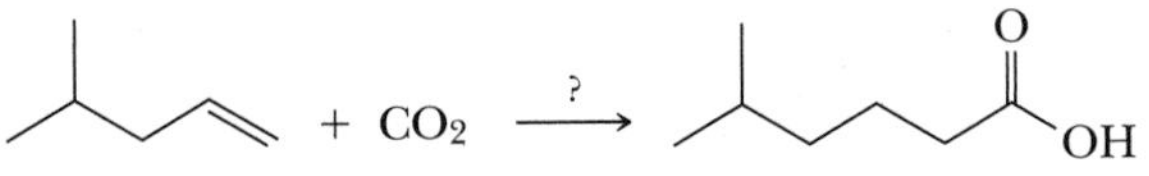
+ CO2
?
O
OH

4-Methyl-1-pentene **5-Methylhexanoic acid**

17.49 Using your reaction roadmaps as a guide, show how to convert cyclohexane into adipoyl dichloride. Show all reagents and all molecules synthesized along the way. Reaction Roadmap

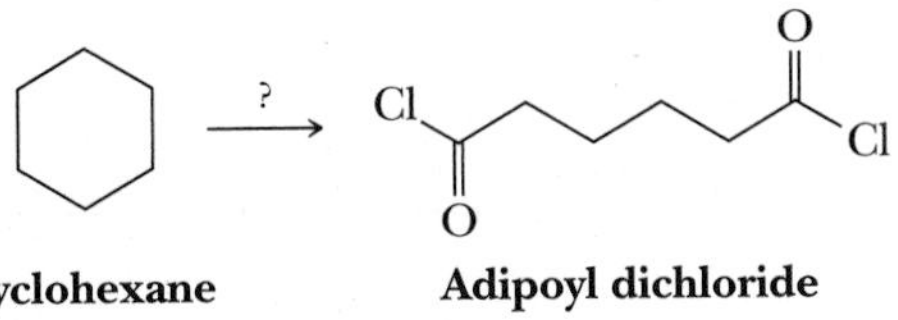
?
Cl
O
O
Cl

Cyclohexane **Adipoyl dichloride**

17.50 Using your reaction roadmaps as a guide, show how to convert 5-chloro-2-pentanone and carbon dioxide into racemic tetrahydro-6-methyl-2-pyranone. You must use 5-chloro-2-pentanone and carbon dioxide as the source of all carbon atoms in the racemic target molecule. Show all reagents and all molecules synthesized along the way. Reaction Roadmap

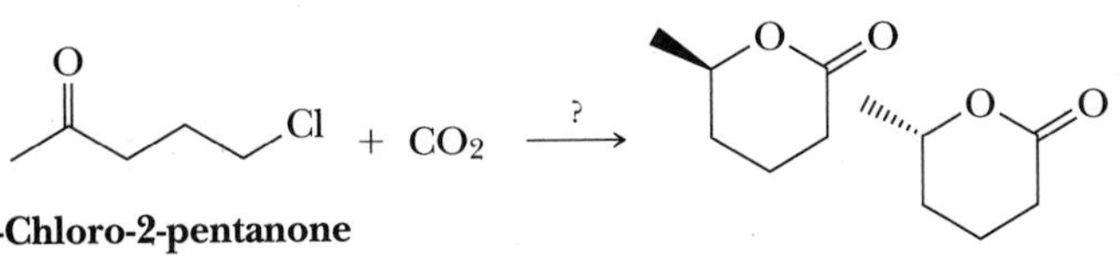
O
Cl
+ CO2
?
O
O
O
O

5-Chloro-2-pentanone

Tetrahydro-6-methyl-2-pyranone
(racemic)

Reactions in Context

17.51 Diazomethane, CH_2N_2, is used in the organic chemistry laboratory despite its danger because it produces very high yields and is selective for reaction with carboxylic acids. Write the products of the following reactions.

(a)

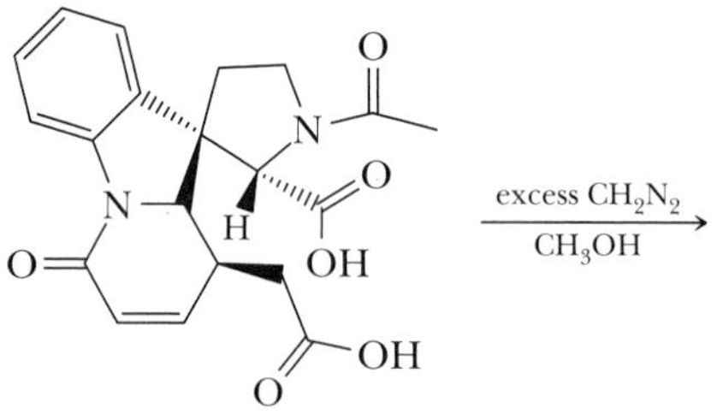
excess CH_2N_2
CH_3OH

(b)

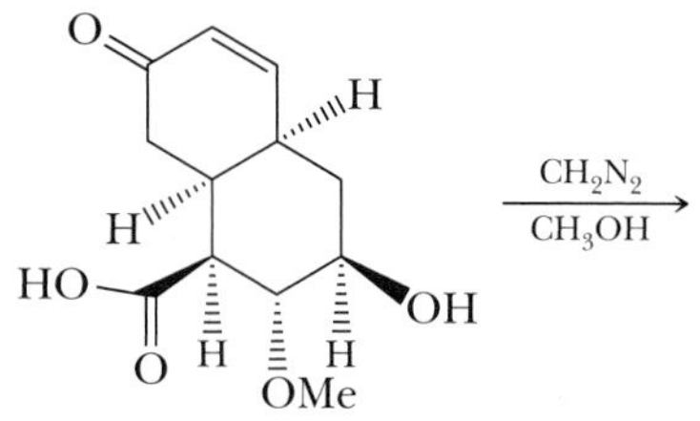
CH_2N_2
CH_3OH

(c)

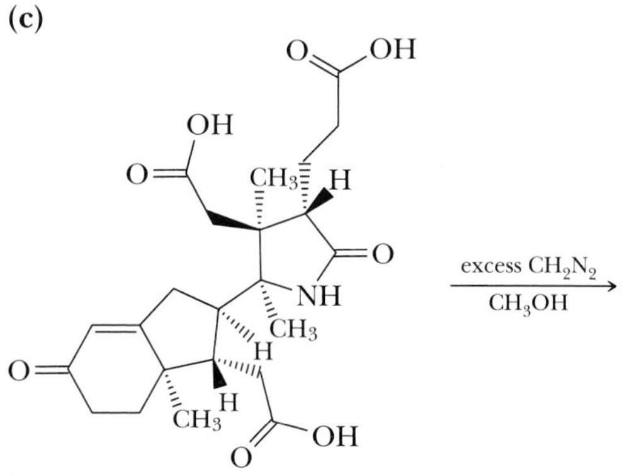
excess CH_2N_2
CH_3OH

17.52 Complete the following Fischer esterification reactions.

(a)

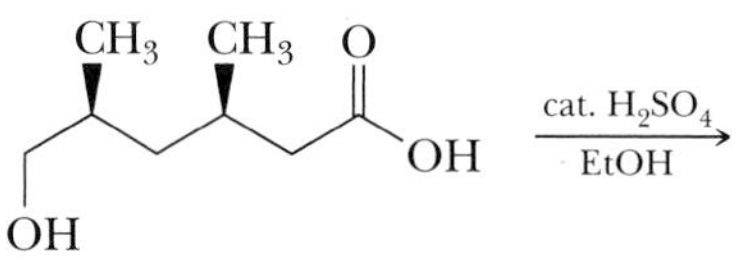
cat. H_2SO_4
EtOH

(b)

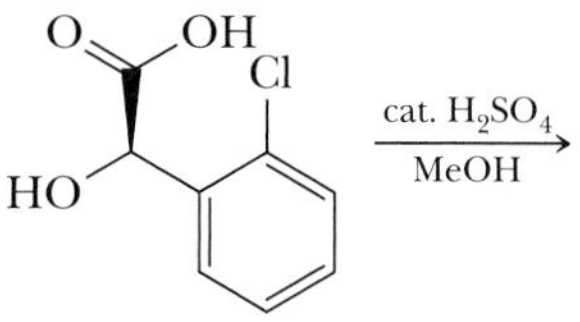
cat. H_2SO_4
MeOH

17.53 So far, you have seen a number of reducing agents used in reactions. Functional groups react differently with each of these reagents. With this in mind, complete the following reaction, which is the last step in a synthesis of fexofenadine, a nonsedating antihistamine sold under the trade name Allegra.

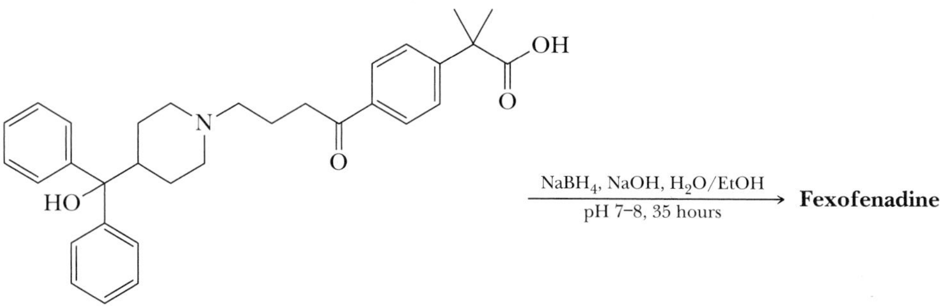
$NaBH_4$, NaOH, H_2O/EtOH
pH 7–8, 35 hours
Fexofenadine

Things You Should Know V

Carboxylic Acid Derivative Reaction Mechanisms

The reaction mechanisms of carboxylic acid derivatives may look intimidating because they involve four to seven individual steps. In addition, the mechanisms are similar, often resembling subtle variations on a theme, making it difficult if not impossible to memorize each one. The best approach to mastering these mechanisms is to develop an intuitive understanding so that you can accurately predict each mechanistic step rather than resort to memorization. This primer gives you the background you need to develop this mechanistic intuition before actively studying carboxylic acid derivatives.

In the next chapter, we will see that acid chlorides react with water, carboxylic acids, alcohols, and amines. Anhydrides undergo reactions with water, alcohols, and amines. Esters undergo reactions with water and amines, and finally, amides undergo reactions with water. Considering that this is a list of ten reactions, each of which can be performed with the addition of acid or base (the acid chloride and anhydride reactions don't require acid or base), there are nearly 20 different reactions for interconversions of carboxylic acids and carboxylic acid derivatives. By combining the following four most common mechanistic elements you have seen throughout this book, you will be able to predict and write each mechanism without resorting to memorization.

1. Make a new bond between a nucleophile and an electrophile.
2. Break a bond to give stable molecules or ions.
3. Add a proton.
4. Take a proton away.

Because the mechanisms for many of the reactions discussed in this chapter are relatively long, these steps may be used repetitively. To put each step together in the proper sequence, we recommend examining each reaction with regard to the following three principles.

I. Figure out which bonds must break and form throughout the mechanism.
II. Avoid **mixed media errors**. In other words, when writing a mechanism for a reaction occurring in strongly basic media (contains hydroxide or alkoxides),

do not create any intermediates that are strong acids (such as protonated and therefore positively charged carbonyls or alcohols). Similarly, when writing a mechanism for a reaction occurring in strongly acidic media (contains hydronium ions or protonated alcohols ROH_2^+), do not create any intermediates that are highly basic (hydroxide, alkoxides, or amide anions). (See Things You Should Know IV for more discussion of mixed media errors.) A good rule of thumb is that in strong acid, every intermediate is either positively charged or neutral, while in strong base, every intermediate is either negatively charged or neutral.

III. Analyze each intermediate in your mechanism to determine whether the next most likely step is a nucleophilic addition, a leaving group departure, a proton addition, or a proton takeaway.

Fischer Esterification Revisited

Let's put this logic together to construct the mechanism for the Fischer esterification reaction we saw in the last chapter. Examination of the overall reaction shows that the **OH** group of the carboxylic acid has been replaced with **OR′**; thus, an **OH** group has to depart as a stable molecule or ion, and an **OR′** group must be a nucleophile at some point during the mechanism (Principle I). Given this fact, we start considering possible steps to write, thinking of each step almost as a multiple-choice situation among the four mechanistic elements.

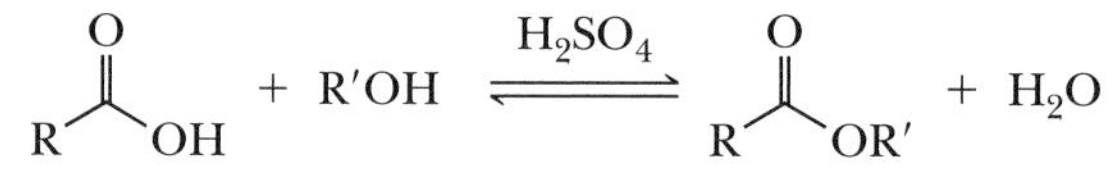

Step 1: If the alcohol was added directly to the carbonyl (Make a bond), we would create an anionic oxygen on the ester carbonyl. Because the reaction is carried out in acid and the anionic oxygen is basic, this constitutes a mixed media error, and therefore is incorrect (Principle II). The **OH** group cannot depart from an sp^2 carbon (Break a bond) because it would leave as hydroxide and we are in acidic media (Principle II). There are no protons that can be removed (Take a proton away; Principle III). Hence, by process of elimination, the first step must be protonation of the carbonyl oxygen to make structure II. Therefore, **Add a proton**.

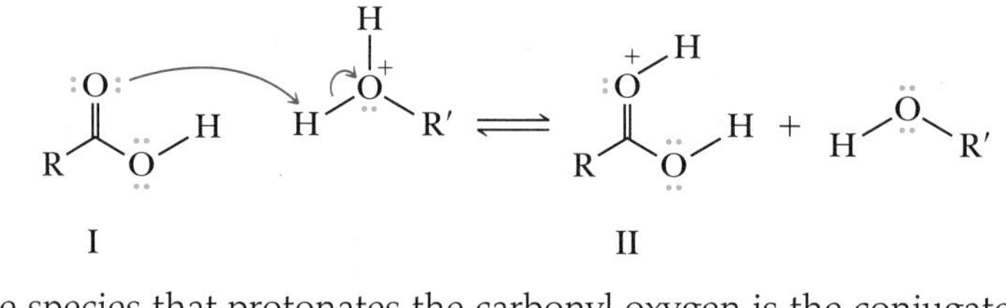

(**Note:** The species that protonates the carbonyl oxygen is the conjugate acid of the alcohol solvent. This is the acid created by adding sulfuric acid to an alcohol.)

Step 2: Structure II still has no leaving group that can depart (Break a bond) given the acid media. We cannot protonate a second time (Add a proton) because doing so would create a dication, and if we take off the proton (Take a proton away), that simply leads back to I. So by process of elimination, we predict that there must be nucleophilic addition to give III. Therefore, **Make a bond between a nucleophile and an electrophile**.

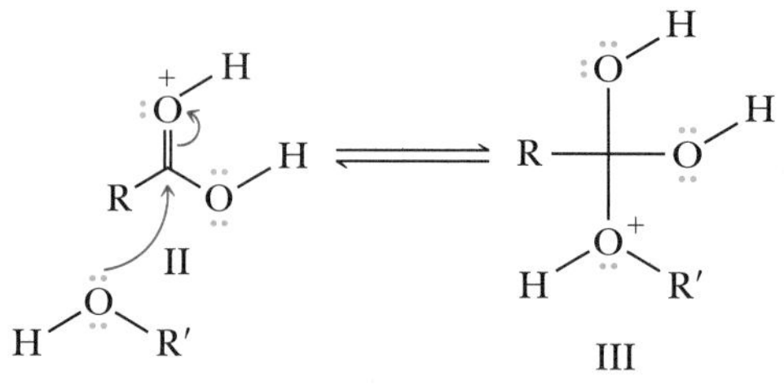

Step 3: From Structure III, the alcohol could depart as a leaving group (Break a bond), but that simply regenerates II. No nucleophilic attack is possible on III (Make a bond) because the carbon is tertiary (cannot undergo S_N2 attack), and we should not put on another proton (Add a proton) because doing so would again create a dication. Thus, again by process of elimination, we conclude that we must take a proton off to give IV. Therefore, **Take a proton away**.

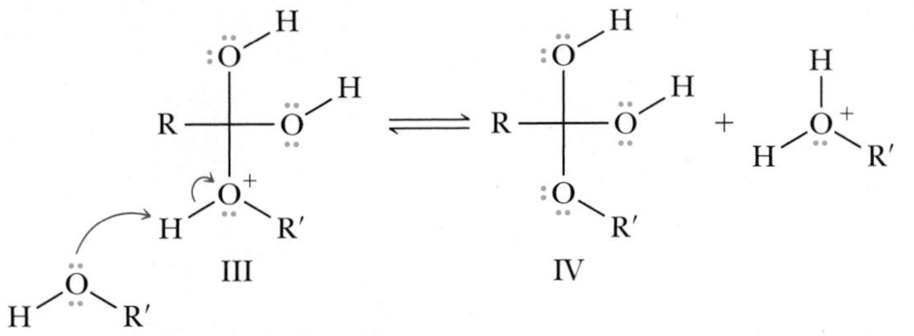

Step 4: There are no acidic protons on IV that can be removed (Take a proton away). There are no electrophilic sites on IV that could be attacked by a nucleophile (Make a bond). A leaving group cannot depart directly from IV (Break a bond) because it would be either a hydroxide or an alkoxide and we are in acidic media (Principle II). Hence, the leaving group must be protonated first, giving V. Therefore, **Add a proton**.

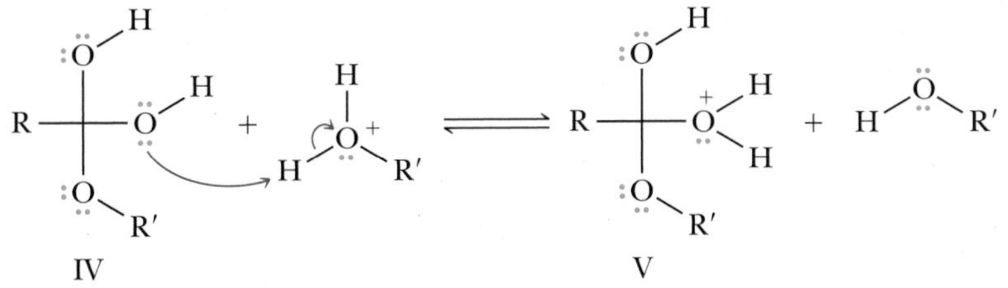

Step 5: Protonation in Step 4 creates the great leaving group water. Therefore, **Break a bond to give stable molecules or ions**.

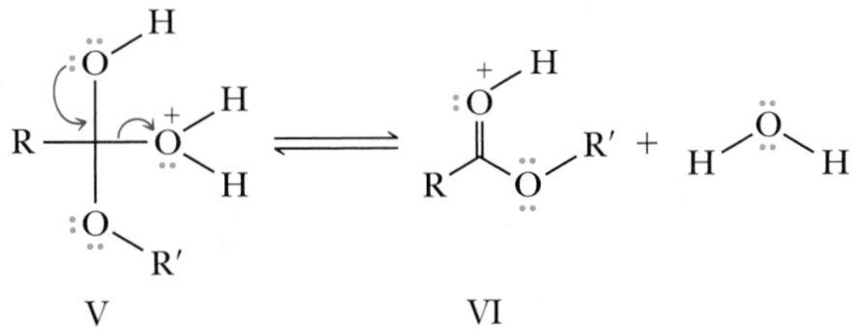

Step 6: Finally, Structure VI just needs to lose a proton to give the product ester VII. Therefore, **Take a proton away**.

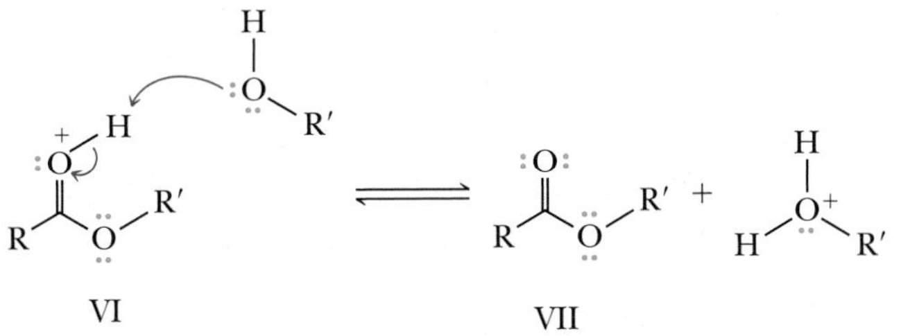

Using the three principles of logic and the four possible mechanistic steps presented here, you should be able to write a reasonable mechanism for all the carboxylic acid and carboxylic acid derivative interconversions discussed in the next chapter, as well as many other mechanisms in other chapters.

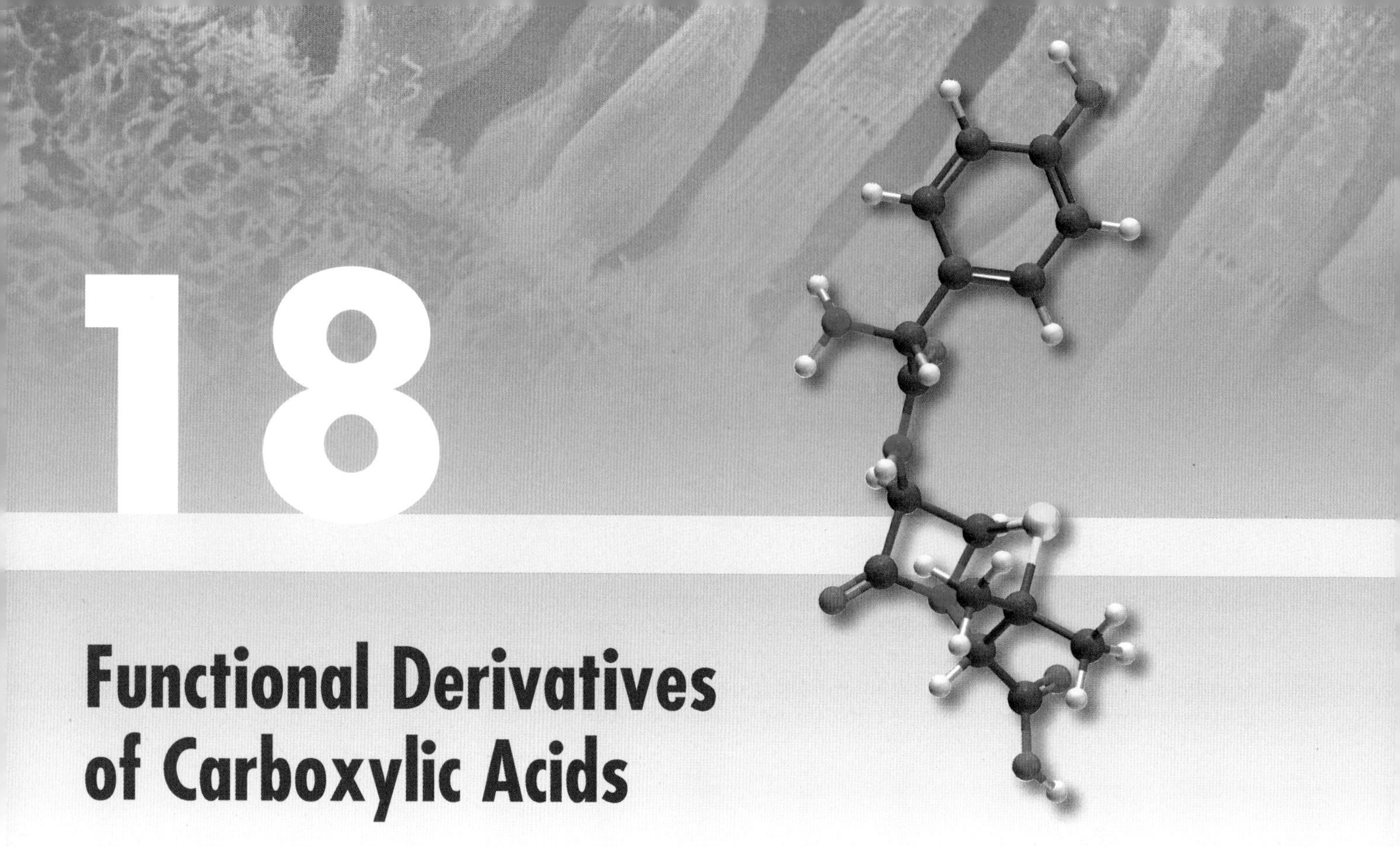

18

Functional Derivatives of Carboxylic Acids

Colored scanning electron micrograph of *Penicillium s.* fungus. The stalklike objects are condiophores to which are attached numerous round condia. The condia are the fruiting bodies of the fungus. **Inset:** a model of amoxicillin. (SCIMAT/ Science Source)

Outline

18.1 Structure and Nomenclature
18.2 Acidity of Amides, Imides, and Sulfonamides
18.3 Characteristic Reactions
18.4 Reaction with Water: Hydrolysis
18.5 Reaction with Alcohols
18.6 Reactions with Ammonia and Amines
18.7 Reaction of Acid Chlorides with Salts of Carboxylic Acids
18.8 Interconversion of Functional Derivatives
18.9 Reactions with Organometallic Compounds
18.10 Reduction

In this chapter, we study five classes of organic compounds, each related to the carboxyl group: acid halides, acid anhydrides, esters, amides, and nitriles.

Under the general formula of each functional group is an illustration to show you how the group is formally related to a carboxylic acid. Formal loss of —OH from a carboxyl group and H— from H—Cl, for example, gives an acid chloride. Similarly, loss of —OH from a carboxyl group and H— from ammonia gives an amide. For illustrative purposes, we show each of these reactions as a formal loss of water. However, as we will see in this chapter, some actual mechanisms do not involve a step in which an H_2O molecule is lost.

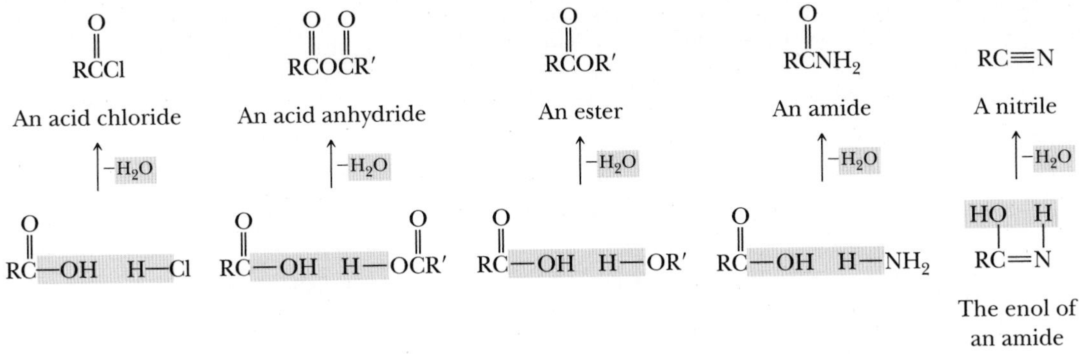
RCCl
An acid chloride
$-H_2O$
RC—OH H—Cl
RCOCR′
An acid anhydride
$-H_2O$
RC—OH H—OCR′
RCOR′
An ester
$-H_2O$
RC—OH H—OR′
$RCNH_2$
An amide
$-H_2O$
$RC—OH H—NH_2$
RC≡N
A nitrile
$-H_2O$
HO H
RC=N
The enol of
an amide

18.1 Structure and Nomenclature

A. Acid Halides

The functional group of an **acid halide** (acyl halide) is an **acyl group** **(RCO—)** bonded to a halogen atom. Acid chlorides are the most common acid halides.

Acyl group An RCO— or ArCO— group.

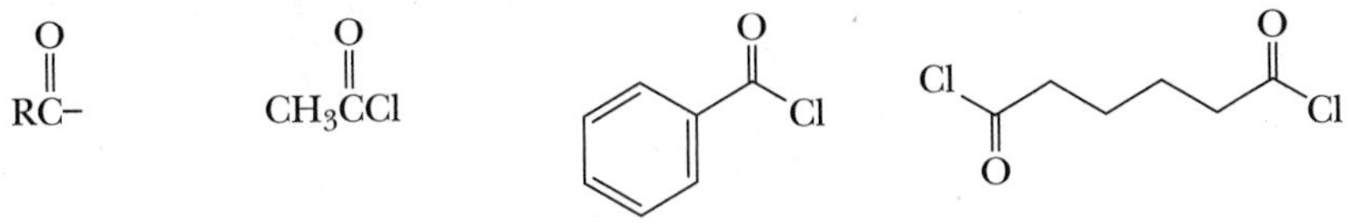

An acyl group | **Ethanoyl chloride (Acetyl chloride)** | **Benzoyl chloride** | **Hexanedioyl chloride (Adipoyl dichloride)**

Acid halides are named by changing the suffix *-ic acid* in the name of the parent carboxylic acid to *-yl halide*.

Similarly, replacement of —OH in a sulfonic acid by chlorine gives a derivative called a **sulfonyl chloride**. Following are structural formulas for two sulfonic acids and the acid chloride derived from each.

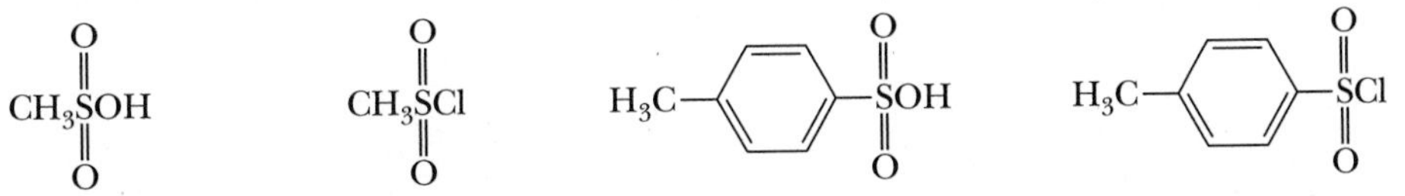

Methanesulfonic acid | **Methanesulfonyl chloride (MsCl)** | ***p*-Toluenesulfonic acid** | ***p*-Toluenesulfonyl chloride (Tosyl chloride, TsCl)**

B. Acid Anhydrides

Carboxylic Anhydrides

The functional group of a **carboxylic anhydride** is two acyl groups bonded to an oxygen atom. These compounds are called **acid anhydrides** because they are formally derived from two carboxylic acids by the loss of water. An anhydride may be symmetrical (two identical acyl groups), or it may be mixed (two different acyl groups). Anhydrides are named by replacing the word *acid* in the name of the parent carboxylic acid with the word *anhydride*.

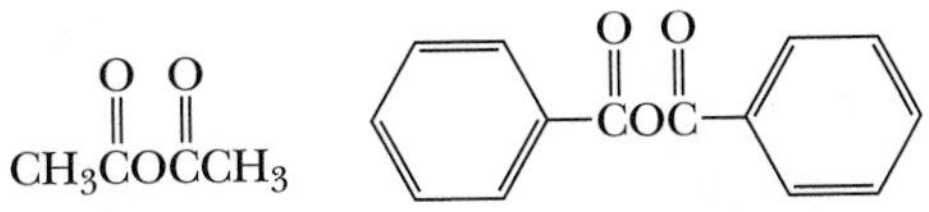

Acetic anhydride | **Benzoic anhydride**

Cyclic anhydrides are named from the dicarboxylic acids from which they are derived. Here are the cyclic anhydrides derived from succinic acid, maleic acid, and phthalic acid.

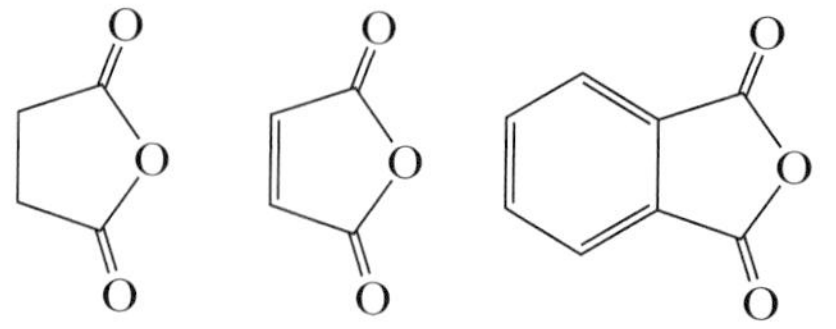

Succinic anhydride **Maleic anhydride** **Phthalic anhydride**

Phosphoric Anhydrides

Because of the special importance of anhydrides of phosphoric acid in biological chemistry, we include them here to show their similarity with the anhydrides of carboxylic acids. The functional group of a **phosphoric anhydride** is two phosphoryl groups bonded to an oxygen atom. Here are structural formulas for two anhydrides of phosphoric acid and the ions derived by ionization of each acidic hydrogen.

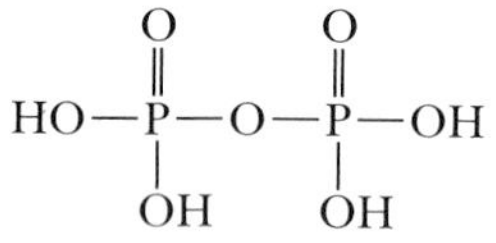

Diphosphoric acid (Pyrophosphoric acid)

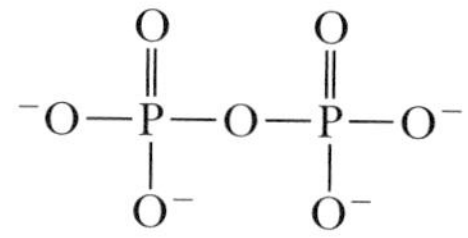

Diphosphate ion (Pyrophosphate ion)

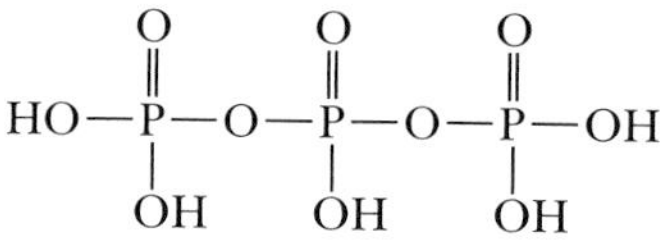

Triphosphoric acid

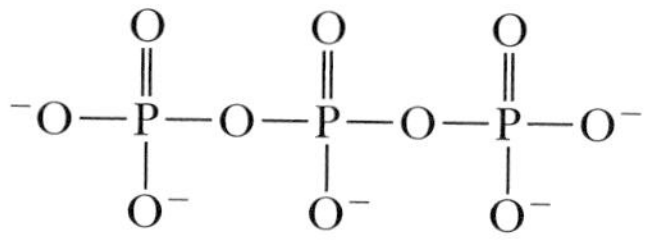

Triphosphate ion

C. Esters

Esters of Carboxylic Acids

The functional group of a **carboxylic ester** is an acyl group bonded to —OR or —OAr. Both IUPAC and common names of esters are derived from the names of the parent carboxylic acids. The alkyl or aryl group bonded to oxygen is named first, followed by the name of the acid in which the suffix *-ic acid* is replaced by the suffix *-ate*.

Ethyl ethanoate (Ethyl acetate)

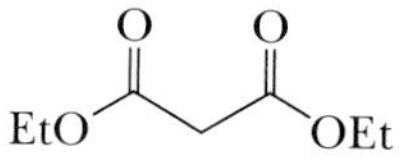

Diethyl propanedioate (Diethyl malonate)

Lactones: Cyclic Esters

Lactone
A cyclic ester.

Cyclic esters are called **lactones**. The IUPAC system has developed a set of rules for naming these compounds. Nonetheless, the simplest lactones are still named by dropping the suffix *-ic acid* or *-oic acid* from the name of the parent carboxylic acid and adding the suffix *-olactone*. The location of the oxygen atom in the ring is indicated by a number if the IUPAC name of the acid is used or by the Greek letter α, β, γ, δ, ε, and so forth, if the common name of the acid is used.

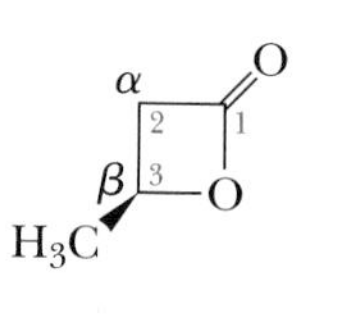

(*S*)-3-Butanolactone ((*S*)-β-Butyrolactone)

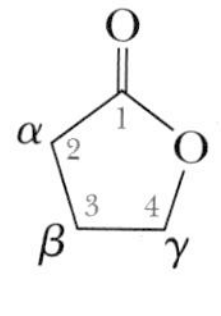

4-Butanolactone (γ-Butyrolactone)

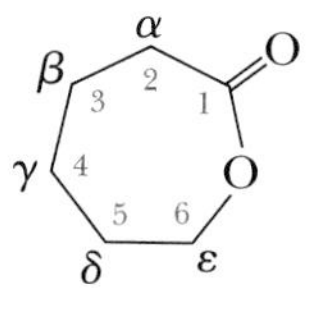

6-Hexanolactone (ε-Caprolactone)

Chemical Connections

From Cocaine to Procaine and Beyond

Cocaine is an alkaloid present in the leaves of the South American coca plant *Erythroxylon coca.* It was first isolated in 1880, and soon thereafter its property as a local anesthetic was discovered. Cocaine was introduced into medicine and dentistry in 1884 by two young Viennese physicians, Sigmund Freud and Karl Koller. Unfortunately, the use of cocaine can create a dependence, as Freud himself observed when he used it to wean a colleague from morphine and thereby produced one of the first documented cases of cocaine addiction.

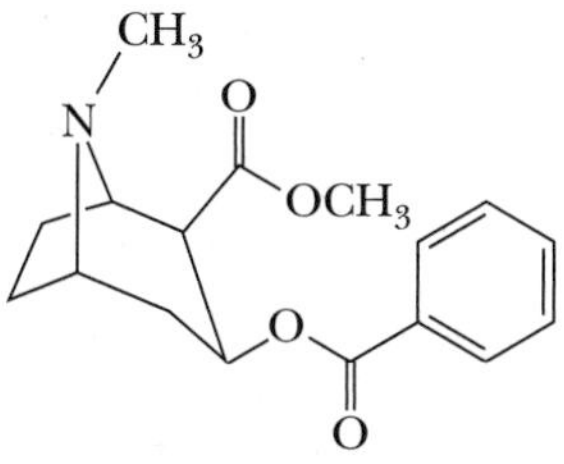

Cocaine

After determining cocaine's structure, chemists could ask, "How is the structure of cocaine related to its anesthetic effects? Can the anesthetic effects be separated from the habituation effect?" If these questions could be answered, it might be possible to prepare synthetic drugs with the structural features essential for the anesthetic activity but without those giving rise to the undesirable effects. Chemists focused on three structural features of cocaine: its benzoic ester, its basic nitrogen atom, and something of its carbon skeleton. This search resulted in 1905 in the synthesis of procaine, which almost immediately replaced cocaine in dentistry and surgery. Lidocaine was introduced in 1948 and today is one of the most widely used local anesthetics. More recently, other members of the "caine" family of local anesthetics have been introduced (e.g., etidocaine). All of these local anesthetics are administered as their water-soluble hydrochloride salts.

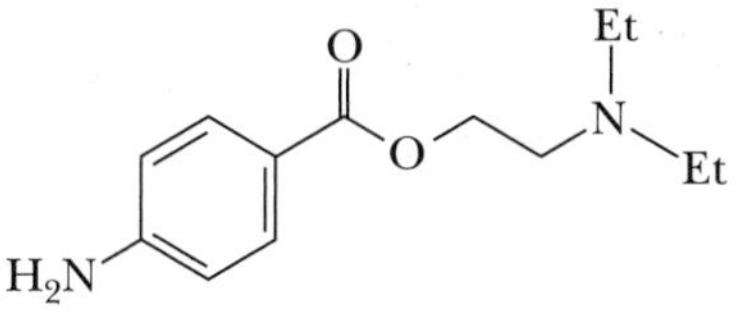

Procaine
(Novocain)

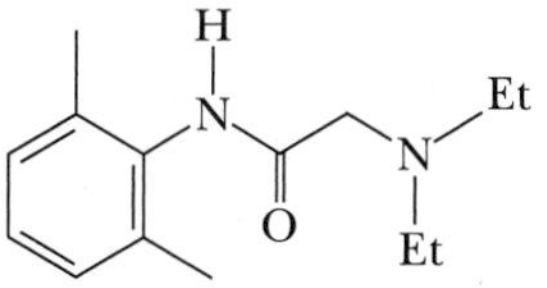

Lidocaine
(Xylocaine)

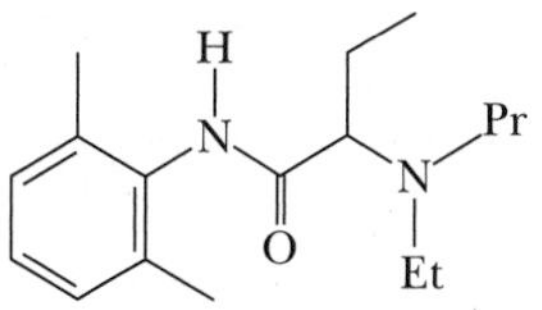

Etidocaine
(Duranest; racemic)

Cocaine reduces fatigue, permits greater physical endurance, and gives a feeling of tremendous confidence and power. In some of the Sherlock Holmes stories, the great detective injects himself with a 7% solution of cocaine to overcome boredom.

Thus, seizing on clues provided by nature, chemists have been able to synthesize drugs far more suitable for a specific function than anything known to be produced by nature itself.

Esters of Phosphoric Acid

Phosphoric acid has three —OH groups and forms mono-, di-, and triesters, which are named by giving the name(s) of the alkyl or aryl group(s) bonded to oxygen followed by the word *phosphate,* as, for example, dimethyl phosphate. In more complex phosphoric esters, it is common to name the organic molecule and then indicate the presence of the phosphoric ester using either the word *phosphate* or the prefix *phospho-*.

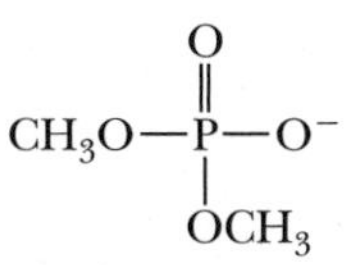

Dimethyl phosphate

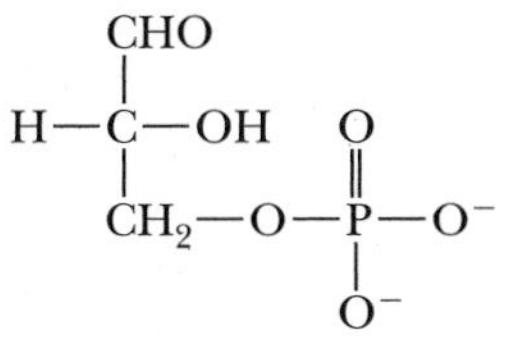

Glyceraldehyde 3-phosphate

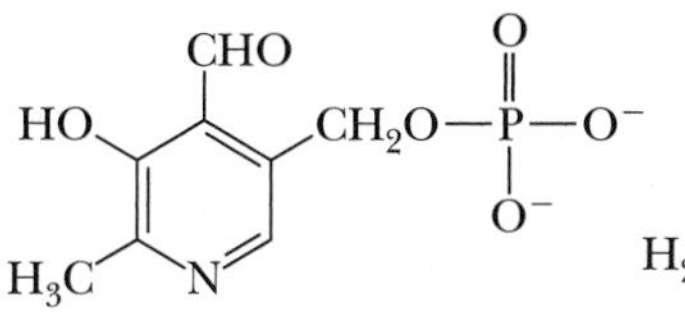

Pyridoxal 5-phosphate

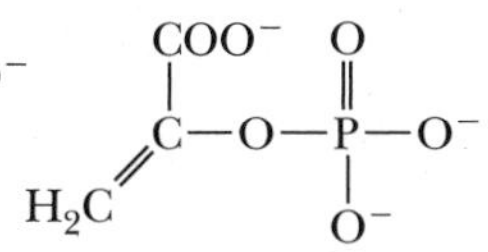

Phosphoenol-pyruvate

Chemical Connections

From Moldy Clover to a Blood Thinner

In 1933, a disgruntled farmer delivered a pail of unclotted blood to the laboratory of Dr. Karl Link at the University of Wisconsin and tales of cows bleeding to death from minor cuts. Over the next couple of years, Link and his collaborators discovered that when cows are fed moldy clover, their blood clotting is inhibited, and they bleed to death from minor cuts and scratches. From the moldy clover, they isolated the anticoagulant dicoumarol, a substance that delays or prevents blood clotting. Dicoumarol exerts its anticoagulation effect by interfering with vitamin K activity. Within a few years after its discovery, dicoumarol became widely used to treat victims of heart attack and others at risk for developing blood clots.

Dicoumarol is a derivative of coumarin, a lactone that gives sweet clover its pleasant smell. Coumarin, which does not interfere with blood clotting, is converted to dicoumarol as sweet clover becomes moldy.

In a search for even more potent anticoagulants, Link developed warfarin (named for the Wisconsin Alumni Research Foundation), now used primarily as a rat poison. When rats consume it, their blood fails to clot, and they bleed to death. Warfarin is also used as a blood anticoagulant in humans. The *S* enantiomer shown here is more active than the *R* enantiomer. The commercial product is sold as a racemic mixture. The synthesis of racemic warfarin is described in Problem 19.59.

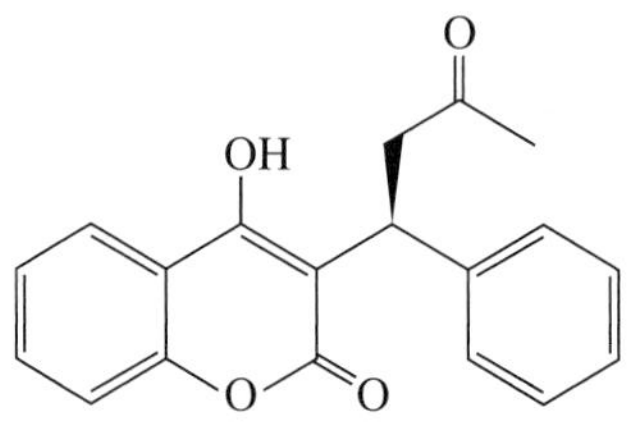

(*S*)-Warfarin
(a synthetic anticoagulant)

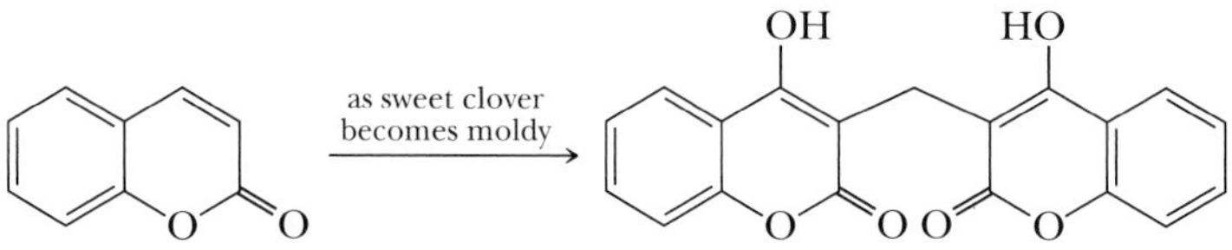

Coumarin
(from sweet clover)

Dicoumarol
(an anticoagulant)

Glyceraldehyde 3-phosphate is an intermediate in glycolysis, the metabolic pathway by which glucose is converted to pyruvate. Pyridoxal phosphate is one of the metabolically active forms of vitamin B_6. Each of these esters is shown as it is ionized at pH 7.4, the pH of blood plasma; the two hydroxyl groups of these phosphoryl groups are ionized, giving each a charge of −2. The molecular backbones of both DNA and RNA contain phosphoric diesters in each repeating unit.

D. Amides and Imides

The functional group of an **amide** is an acyl group bonded to a nitrogen atom. Amides are named by dropping the suffix *-oic acid* from the IUPAC name of the parent acid (or *-ic acid* from its common name) and adding *-amide.* If the nitrogen atom of an amide is bonded to an alkyl or aryl group, the group is named, and its location on nitrogen is indicated by *N*-. Two identical alkyl or aryl groups on nitrogen are indicated by *N,N*-di-. *N,N*-Dimethylformamide (DMF) is a widely used polar aprotic solvent (Section 9.3D).

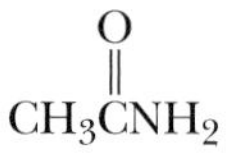

Acetamide
(a 1° amide)

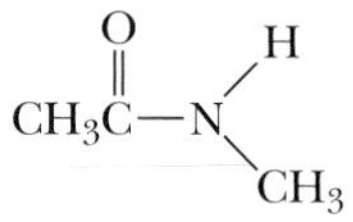

***N*-Methylacetamide**
(a 2° amide)

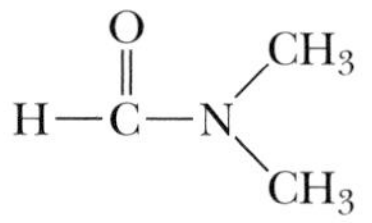

***N,N*-Dimethyl-formamide (DMF)**
(a 3° amide)

Cyclic amides are given the special name **lactam**. Their names are derived in a manner similar to those of lactones, with the difference that the suffix *-lactone* is replaced by *-lactam.*

Lactam
A cyclic amide.

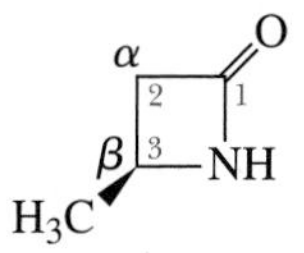

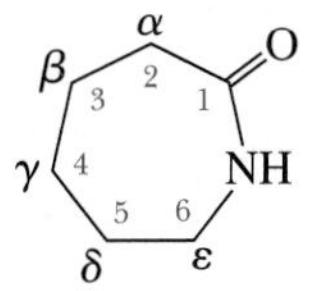

(*S*)-3-Butanolactam ((*S*)-β-Butyrolactam) **6-Hexanolactam (ε-Caprolactam)**

The functional group of an **imide** is two acyl groups bonded to nitrogen. Both succinimide and phthalimide are cyclic imides.

Imide
A functional group in which two acyl groups, RCO— or ArCO—, are bonded to a nitrogen atom.

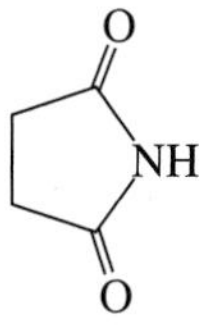

Succinimide

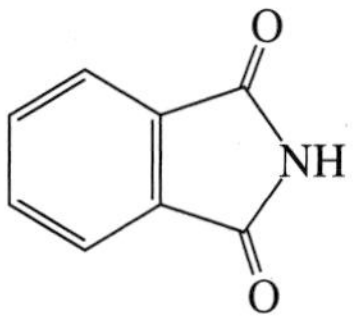

Phthalimide

Example 18.1 Naming Carboxylic Acid Derivatives

Write the IUPAC name for each compound.

(a) 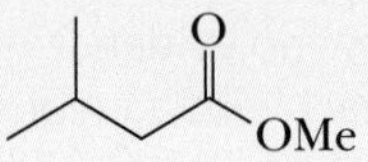 **(b)** 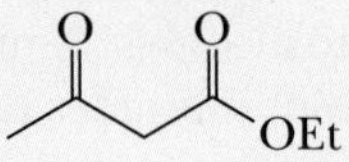

(c) 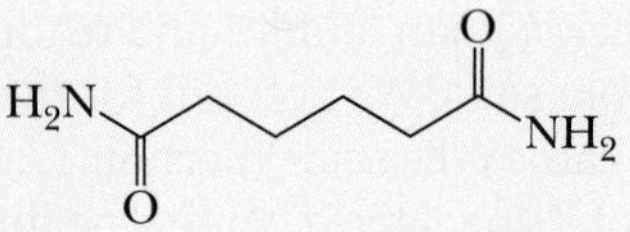 **(d)** 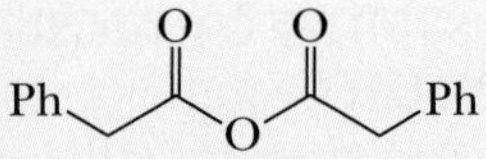

Solution

Given first is the IUPAC name and then, in parentheses, the common name.

(a) Methyl 3-methylbutanoate (methyl isovalerate, from isovaleric acid)

(b) Ethyl 3-oxobutanoate (ethyl β-ketobutyrate, from β-ketobutyric acid)

(c) Hexanediamide (adipamide, from adipic acid)

(d) Phenylethanoic anhydride (phenylacetic anhydride, from phenylacetic acid)

Problem 18.1

Draw a structural formula for each compound.

(a) *N*-Cyclohexylacetamide

(b) 1-Methylpropyl methanoate

(c) Cyclobutyl butanoate

(d) *N*-(1-Methylheptyl)succinimide

(e) Diethyl adipate

(f) 2-Aminopropanamide

E. Nitriles

Nitrile

A compound containing a —C≡N (cyano) group bonded to a carbon atom.

The functional group of a **nitrile** is a cyano (C≡N) group bonded to a carbon atom. IUPAC names follow the pattern alkanenitrile (e.g., ethanenitrile). Common names are derived by dropping the suffix *-ic* or *-oic acid* from the name of the parent carboxylic acid and adding the suffix *-onitrile.*

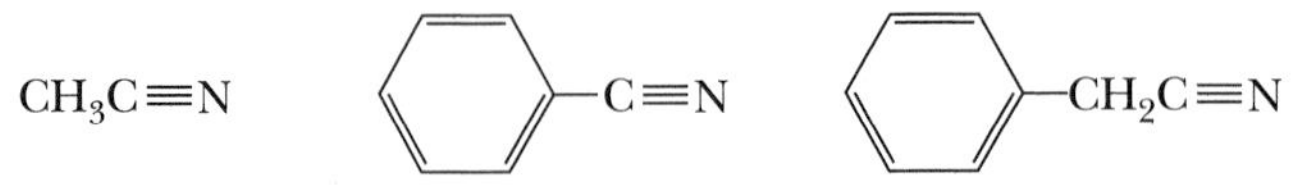

Ethanenitrile (Acetonitrile) **Benzonitrile** **Phenylethanenitrile (Phenylacetonitrile)**

18.2 Acidity of Amides, Imides, and Sulfonamides

Following are structural formulas of a primary amide, a sulfonamide, and two cyclic imides, along with pK_a values for each.

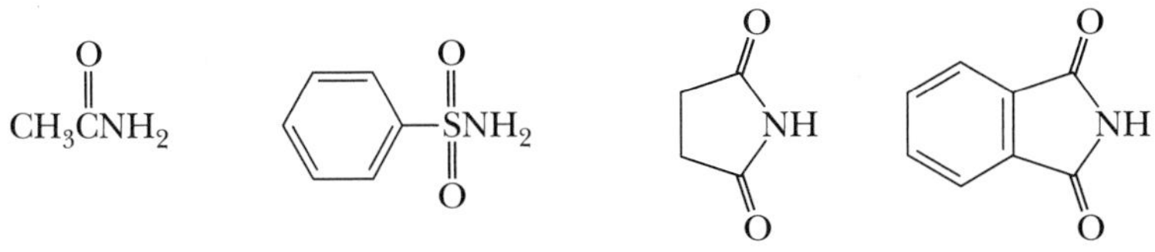

Acetamide	Benzenesulfonamide	Succinimide	Phthalimide
pK_a 15–17	pK_a 10	pK_a 9.7	pK_a 8.3

Values of pK_a for amides of carboxylic acids are in the range of 15–17, which means that they are comparable in acidity to alcohols. Amides show no evidence of acidity in aqueous solution; that is, water-insoluble amides do not react with aqueous solutions of NaOH or other alkali metal hydroxides to form water-soluble salts.

Imides (pK_a 8–10) are considerably more acidic than amides and readily dissolve in 5% aqueous NaOH by forming water-soluble salts. We account for the acidity of imides in the same manner as for the acidity of carboxylic acids (Section 17.4); namely, the imide anion is stabilized by delocalization of its negative charge. The more important contributing structures for the anion formed by ionization of an imide delocalize the negative charge on nitrogen and the two carbonyl oxygens. ▶

▶ Delocalization of charge over a larger area is stabilizing.

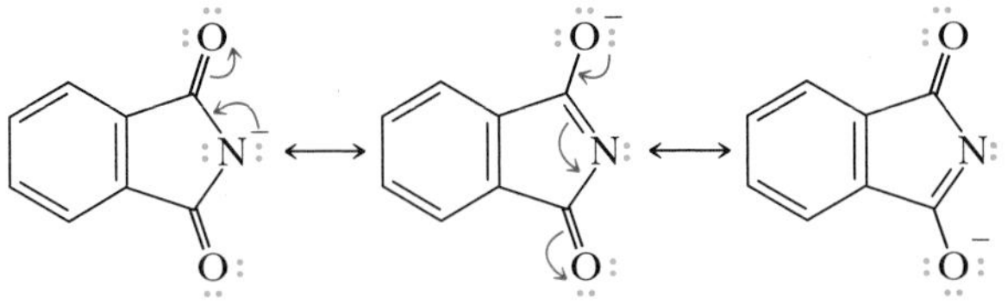

A resonance-stabilized anion

Sulfonamides derived from ammonia and primary amines are also sufficiently acidic to dissolve in aqueous solutions of NaOH or other alkali metal hydroxides by forming water-soluble salts. The pK_a of benzenesulfonamide is approximately 10. We account for the acidity of sulfonamides in the same manner as for imides, namely the resonance stabilization of the resulting anion.

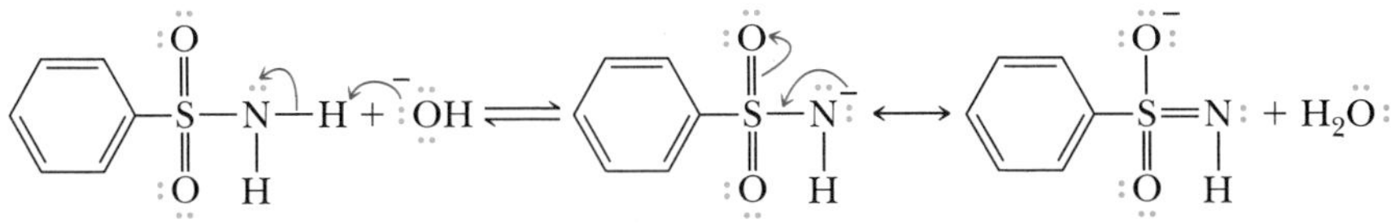

Benzenesulfonamide A resonance-stabilized anion

Example 18.2 Properties of Phthalimide

Phthalimide is insoluble in water. Will phthalimide dissolve in aqueous NaOH?

Solution

Phthalimide is the stronger acid, and NaOH is the stronger base. The position of equilibrium, therefore, lies to the right. Phthalimide dissolves in aqueous NaOH by forming a water-soluble sodium salt.

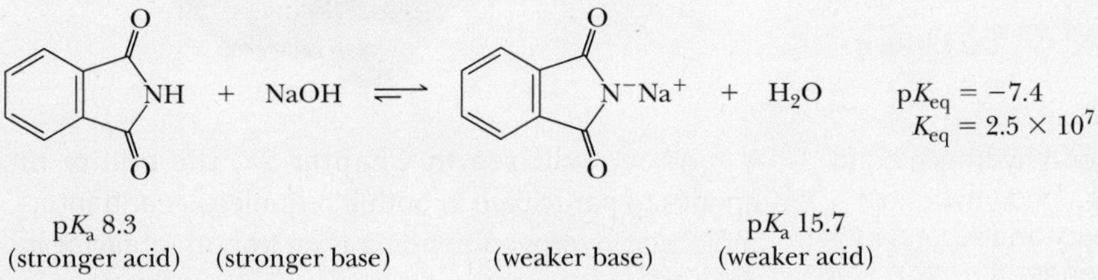

O
NH
+ NaOH
O
N⁻Na⁺
+ H_2O
O
O
pK_{eq} = −7.4
K_{eq} = 2.5 × 10^7
pK_a 8.3
(stronger acid)
(stronger base)
(weaker base)
pK_a 15.7
(weaker acid)

Problem 18.2

Will phthalimide dissolve in aqueous sodium bicarbonate?

Connections to Biological Chemistry

The Unique Structure of Amide Bonds

Amides have structural characteristics that are unique among carboxylic acid derivatives. In the late 1930s, Linus Pauling discovered that the bond angles about the nitrogen atom of an amide bond in proteins are close to 120°; the amide nitrogen is trigonal planar and sp^2 hybridized. We know that amides are best represented as a hybrid of three resonance contributing structures (see Section 1.9C).

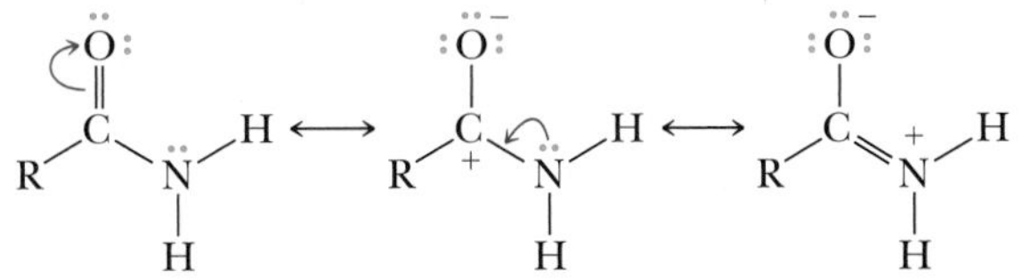

R
C
O
N
H
H
+

This contributing structure places a double bond between C and N

The fact that the six atoms of an amide bond are coplanar with bond angles of 120° means that the contributing structure on the right makes a significant contribution to the hybrid and that the hybrid looks very much like this third structure. Inclusion of the third contributing structure explains why the amide nitrogen is sp^2 hybridized and therefore trigonal planar.

Also, the presence of a partial double bond (π bond) in the resonance hybrid indicates the presence of a restricted bond rotation about the C—N bond. The measured C—N bond rotation barrier in amides is approximately 63–84 kJ (15–20 kcal)/mol, large enough so that at room temperature, rotation about the C—N bond is restricted. In addition, because the lone pair on nitrogen is delocalized into the bond, it is not as available for interacting with protons and other Lewis acids. Thus, amide nitrogens are not basic. In fact, in acid solution, amides are protonated on the carbonyl oxygen atom rather than on the nitrogen (review Example 4.2). Finally, delocalization of the nitrogen lone pair reduces the electrophilic character (partial positive charge) on the carbonyl carbon, thus reducing the susceptibility of amides to nucleophilic attack.

(Continued)

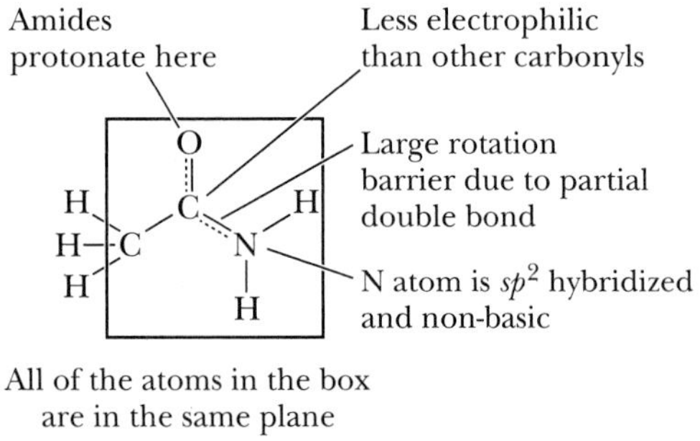
Amides
protonate here
Less electrophilic
than other carbonyls
Large rotation
barrier due to partial
double bond
N atom is sp^2 hybridized
and non-basic
All of the atoms in the box
are in the same plane

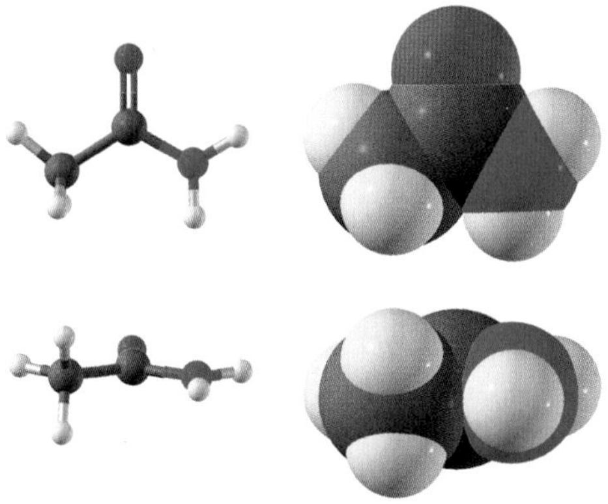

The amide —NH group is a good hydrogen bond donor, while the amide carbonyl is a good hydrogen bond acceptor, allowing both primary and secondary amides to form strong hydrogen bonds.

As we will see in Chapter 27, the ability of amides to participate in both intermolecular and intramolecular hydrogen bonding is an important factor in determining the three-dimensional structure of polypeptides and proteins.

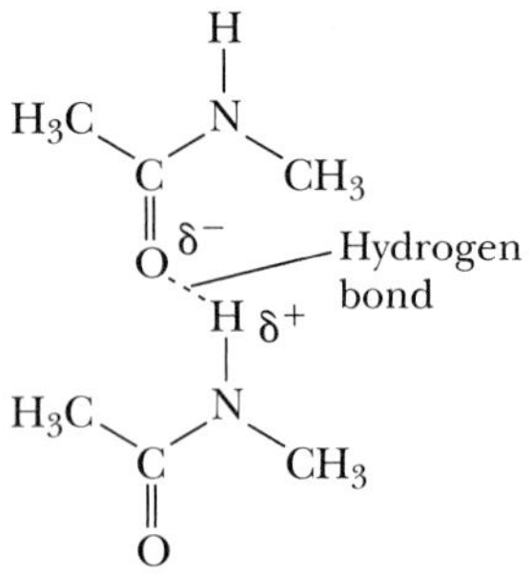
Hydrogen
bond

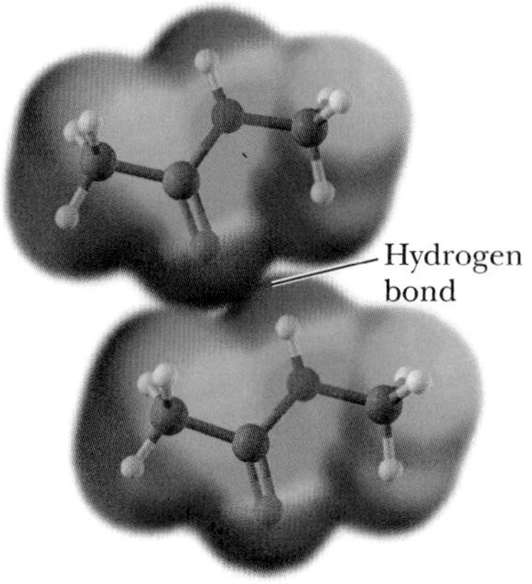
Hydrogen
bond

18.3 Characteristic Reactions

In this and subsequent sections, we examine the interconversions of various carboxylic acid derivatives. All these reactions begin with formation of a tetrahedral carbonyl addition intermediate (make a new bond between a nucleophile and an electrophile).

A. Nucleophilic Acyl Addition

▶ Most bond-making steps in reaction mechanisms involve nucleophiles reacting with electrophiles.

The first step of this reaction is exactly analogous to the addition of alcohols to aldehydes and ketones (Section 16.7B). This reaction can be carried out under basic conditions, in which a negatively charged nucleophile adds directly to the carbonyl carbon. The tetrahedral carbonyl addition intermediate formed then adds a proton from a proton donor, HA. The result of this reaction is nucleophilic acyl addition. ▶

Nucleophilic acyl addition (basic conditions):

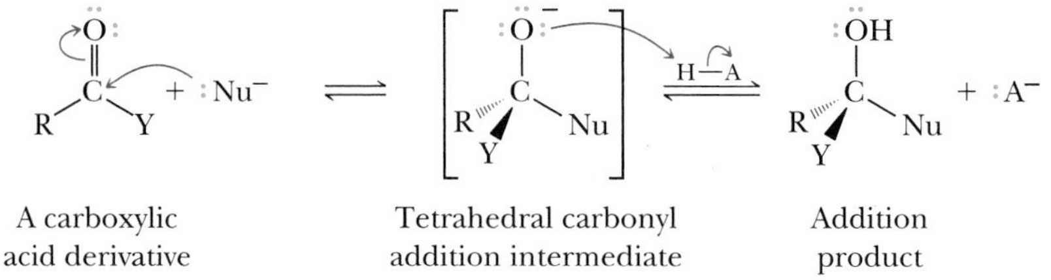
A carboxylic
acid derivative
Tetrahedral carbonyl
addition intermediate
Addition
product

As with aldehydes and ketones, this reaction can also be catalyzed by acid, in which case protonation (add a proton) of the carbonyl oxygen precedes the attack of the nucleophile.

Nucleophilic acyl addition (acidic conditions):

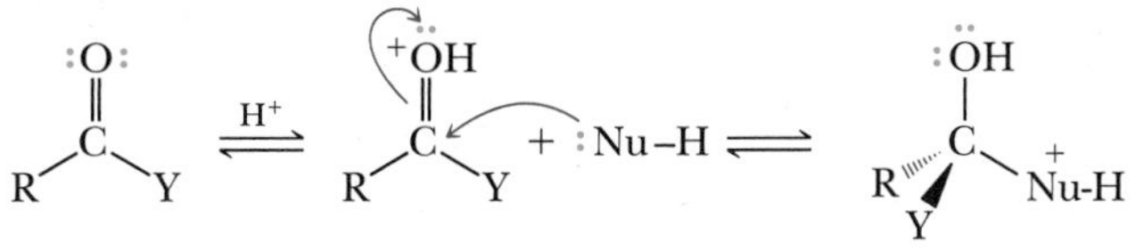

A carboxylic acid derivative

Tetrahedral carbonyl addition intermediate

B. Nucleophilic Acyl Substitution

For functional derivatives of carboxylic acids, the fate of the tetrahedral carbonyl addition intermediate is quite different from that of aldehydes and ketones; the intermediate collapses to expel the leaving group (Lv) and regenerate the carbonyl group (break a bond to give stable molecules or ions). The result of this addition-elimination sequence is **nucleophilic acyl substitution**.

Nucleophilic acyl substitution

A reaction in which a nucleophile bonded to the carbon of an acyl group is replaced by another nucleophile.

Nucleophilic acyl substitution (basic conditions):

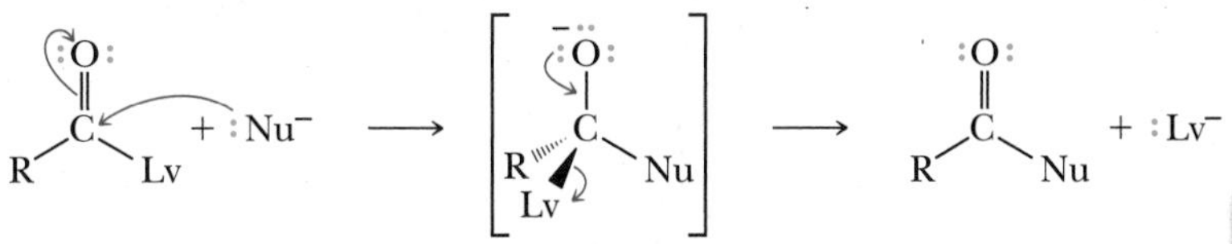

Tetrahedral carbonyl addition intermediate

Substitution product

Watch a video explanation

The major difference between nucleophilic acyl addition and nucleophilic acyl substitution is that aldehydes and ketones do not have a group that can leave as a relatively stable anion. They undergo only nucleophilic addition. The four carboxylic acid derivatives we study in this chapter have a leaving group, Lv, that can leave as a relatively stable anion or as a neutral species. Neutral molecules commonly serve as nucleophiles in this reaction, mainly when it is carried out under acid-catalyzed conditions. When these reactions are catalyzed by acid, protonation precedes nucleophilic attack; similarly, protonation precedes leaving group departure. We will see this sequence numerous times in this chapter.

Nucleophilic acyl substitution (acidic conditions):

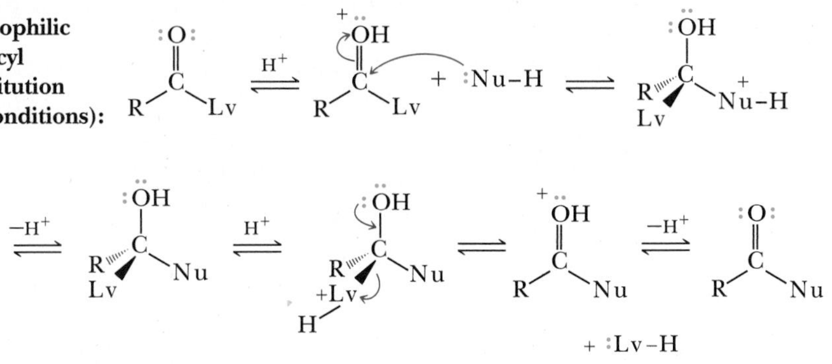

C. Relative Reactivity

The four carboxylic acid derivatives that are the focus of this chapter have the relative reactivity toward nucleophilic acyl substitution as follows. The differences in this trend are dramatic. For example, at common ambient temperatures and neutral pH, acid halides will react with water within seconds to minutes, while anhydrides will do so over minutes to hours. Esters, however, do not react with water at appreciable rates under these conditions, taking many years to hydrolyze; amides take centuries to react. Hence, acid halides and acid anhydrides are so reactive that they are not found in nature, whereas esters and amides are universally present.

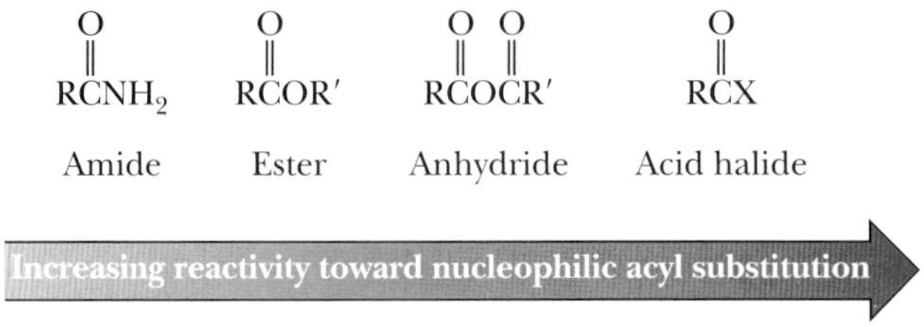

Two effects lead to this trend. One is relative leaving group ability. We show below the leaving groups as anions in order to illustrate an important point: the weaker the base (i.e. the more stable the anion), the better the leaving group (Figure 18.1). The weakest base in the series and the best leaving group is the halide ion; acid halides are most reactive toward nucleophilic acyl substitution. The strongest base and the poorest leaving group is the amide ion; amides are least reactive toward nucleophilic acyl substitution.

FIGURE 18.1 Anion leaving group ability and basicity.

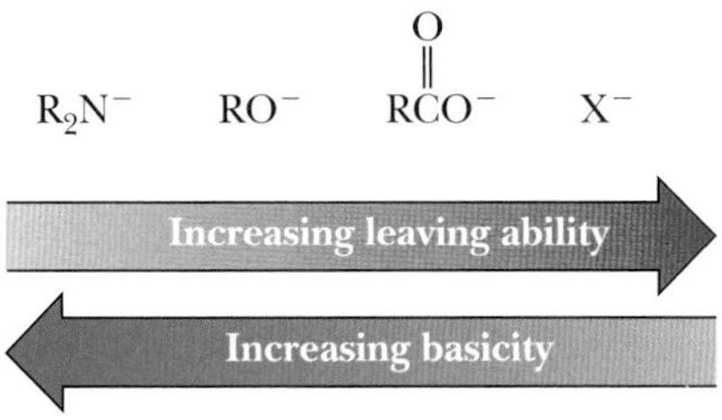

The second effect derives from the relative resonance stabilization of the carboxylic acid derivatives. As follows, each derivative can be written with contributing structures that will be stabilizing to some extent. The second contributing structure that we show for each carboxylic acid derivative has a positive charge on the carbonyl carbon. This structure reflects the electrophilicity of these carbons. However, for each derivative, it is the other contributing structures that reflect the relative resonance stabilization of the derivatives.

Let's start with an analysis of the acid chloride. The third contributing structure for an acid chloride has a carbon-to-chlorine double bond whose π bond is weak due to poor orbital overlap between the differently sized p orbitals on these two atoms. Further, there is a positive charge on the electronegative chlorine atom. Both of these factors make this a poor contributing structure for the acid chloride. An acid anhydride has five contributing structures; the last two shown place positive charges on the central oxygen. However, these positive charges are adjacent to an electron-withdrawing carbonyl group. Hence, these two contributing structures are not very reasonable depictions of an acid anhydride. But the analogous contributing structure for an ester places the positively charged oxygen near an electron-donating alkyl group, which stabilizes this charge. Accordingly, this contributing structure is a reasonable depiction of an ester; it is stabilizing, and it lowers the susceptibility of the carbonyl carbon to nucleophilic attack. Finally, the third contributing structure for an amide has a positive charge on the less electronegative nitrogen (relative to oxygen as with an ester), making this an even more reasonable structure and thereby increasingly stabilizing. In fact, the C=N double bond character of an amide is significant. This increased stability makes the amide least susceptible to nucleophilic attack.

Acid chloride contributing structures

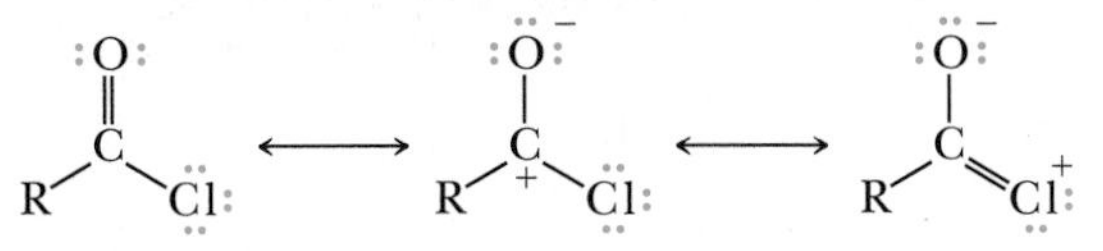

Acid anhydride contributing structures

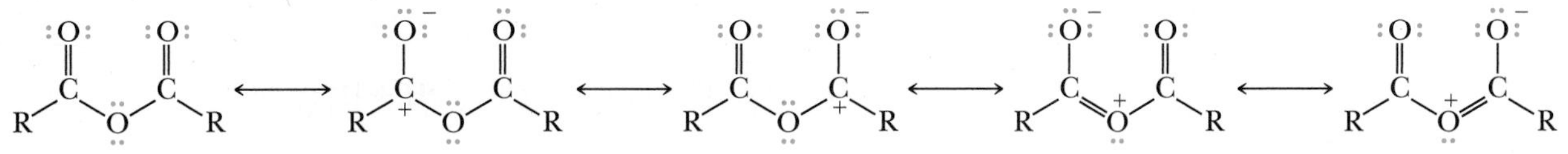

Ester contributing structures

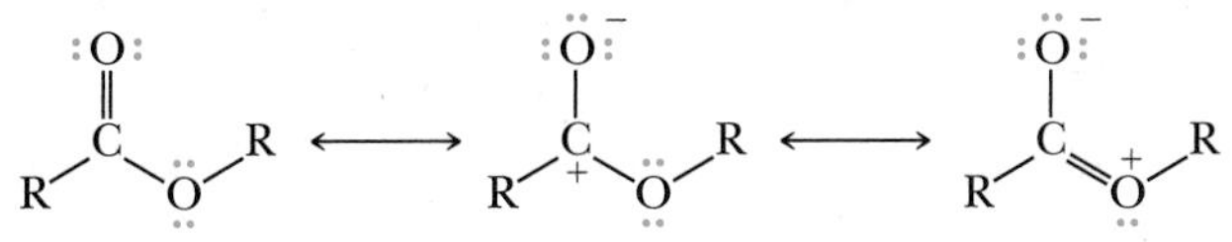

Amide contributing structures

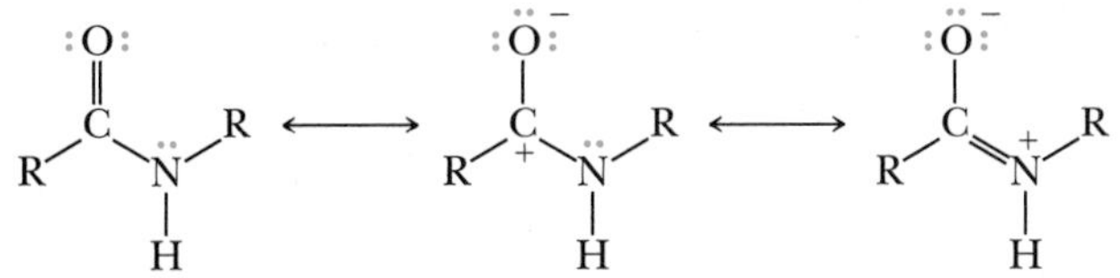

Taken together, the combined effects of leaving group ability and susceptibility to nucleophilic attack reinforce each other, thereby resulting in the order of reactivity given below.

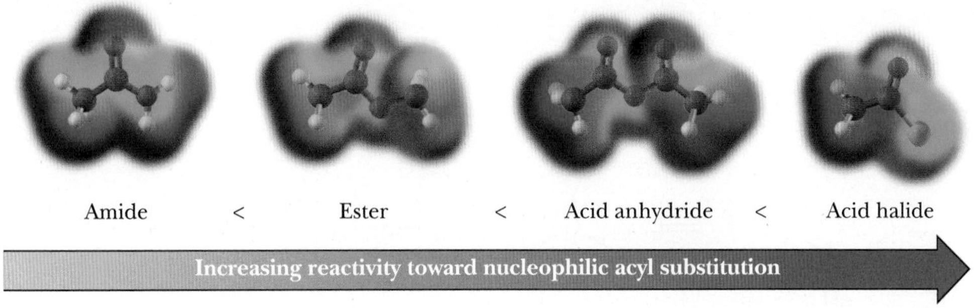

D. Catalysis

The reactivity of acid halides and acid anhydrides is high enough that the common nucleophiles used to interconvert the carboxylic acid derivatives will react directly with these species without any catalysis. However, esters and amides are so stable that some form of acid or base catalysis is required. Acid catalysis is used to increase the electrophilicity of the carboxylic acid derivatives and to facilitate leaving group departure. Placing a proton on the carbonyl oxygen creates significantly more positive charge on the carbonyl carbon, making it more susceptible to nucleophilic attack. In addition, placing a proton on the leaving group makes it more readily depart as a stable molecule.

Base is used to increase nucleophilicity by converting a neutral nucleophile to an anionic nucleophile (e.g., ethanol to sodium ethoxide). In addition, under basic conditions, the tetrahedral addition intermediates are negatively charged and therefore more apt to expel a negatively charged leaving group. We will see detailed mechanisms involving both acid and base in this chapter. Each of the mechanisms presented have the accompanying electron flow arrows. Before beginning to learn the mechanisms and arrow pushing, you may want to refer to "Things You Should Know IV, Common Mistakes in Arrow Pushing" that lists common errors to avoid when writing mechanisms.

18.4 Reaction with Water: Hydrolysis

A. Acid Chlorides

Low-molecular-weight acid chlorides react very rapidly with water to form carboxylic acids and HCl.

$$\underset{\textbf{Acetyl chloride}}{CH_3\overset{O}{\overset{\|}{C}}Cl} + H_2O \longrightarrow CH_3\overset{O}{\overset{\|}{C}}OH + HCl$$

Higher molecular-weight acid halides are less soluble and, consequently, react less rapidly with water.

Mechanism 18.1

Hydrolysis of an Acid Chloride

Acid chlorides are so reactive that hydrolysis does not require acid or base catalysis; therefore, the steps in the mechanism do not involve putting on or taking off protons prior to the nucleophilic attack and/or the leaving group departure.

Step 1: **Make a new bond between a nucleophile and an electrophile.** Water attacks the carbonyl carbon directly to give a tetrahedral addition intermediate.

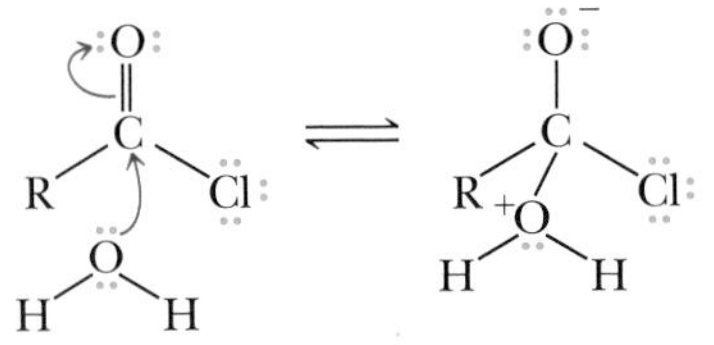

Step 2: **Take a proton away.** Removal of a proton is rapid.

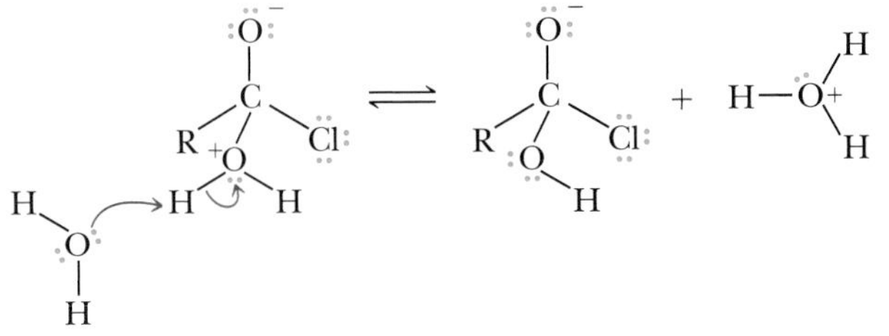

Step 3: **Break a bond to give stable molecules or ions.** Expulsion of the chloride anion leaving group yields the carboxylic acid product.

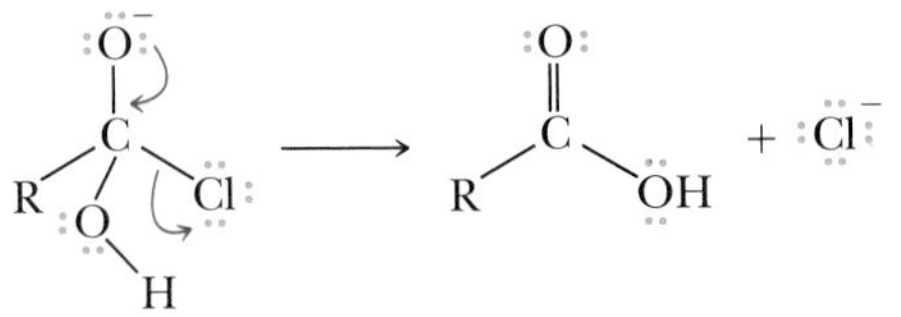

This reaction creates the very strong acid HCl (H_3O^+ and Cl^-). Chemists commonly add a weak base, such as pyridine, to neutralize the acid that is created.

B. Acid Anhydrides

Anhydrides are generally less reactive than acid chlorides. However, the lower molecular-weight anhydrides also react readily with water to form two molecules of carboxylic acid.

$$CH_3\overset{O}{\overset{\|}{C}}O\overset{O}{\overset{\|}{C}}CH_3 + H_2O \longrightarrow CH_3\overset{O}{\overset{\|}{C}}OH + HO\overset{O}{\overset{\|}{C}}CH_3$$

Acetic anhydride

As with the hydrolysis of acid chlorides, the hydrolysis of acid anhydrides will occur without an added acid or base catalyst (although sometimes acid is used); therefore, the mechanism is similar to that given above. The acid-catalyzed mechanism is analogous to that with esters, discussed in the next section.

C. Esters

Esters are hydrolyzed very slowly at neutral pH, even when heated to reflux. Hydrolysis becomes considerably more rapid, however, when esters are heated to reflux in aqueous acid or base. The mechanism of acid-catalyzed hydrolysis highlights the logic and key steps involved in many of the mechanisms discussed in this chapter.

Watch a video explanation

Mechanism 18.2

Acid-Catalyzed Ester Hydrolysis

Step 1: **Add a proton.** The reaction begins with protonation, which increases the electrophilicity of the ester carbonyl carbon.

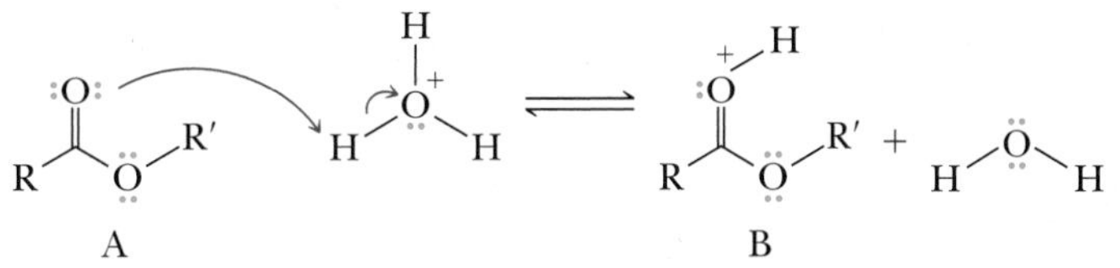

Step 2: **Make a new bond between a nucleophile and an electrophile.** Water adds to the carbonyl carbon atom.

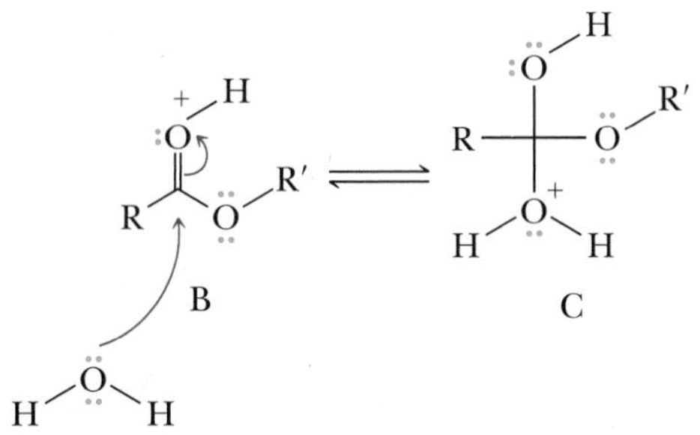

Step 3: **Take a proton away.** Deprotonation gives a tetrahedral addition intermediate.

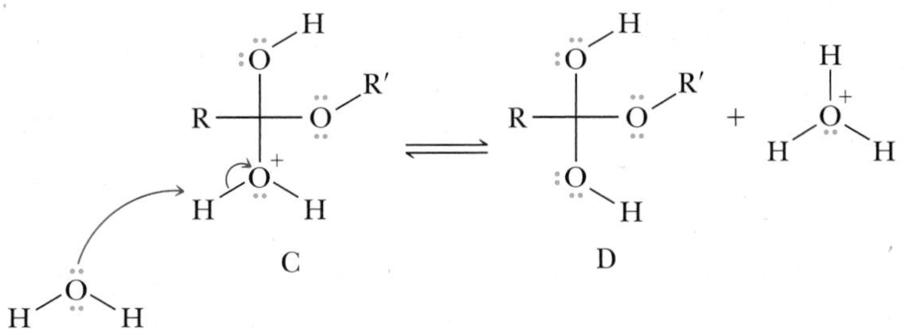

(Continued)

Step 4: **Add a proton.** Placing a proton on —OR′ converts it to —$\overset{+}{O}$HR′; this process will set the stage for the departure of the much better leaving group R′OH in the next step.

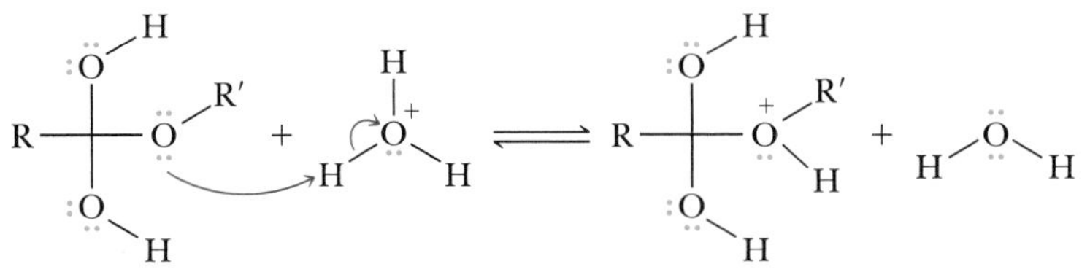

D E

Step 5: **Break a bond to give stable molecules or ions.** R′OH departs as a leaving group.

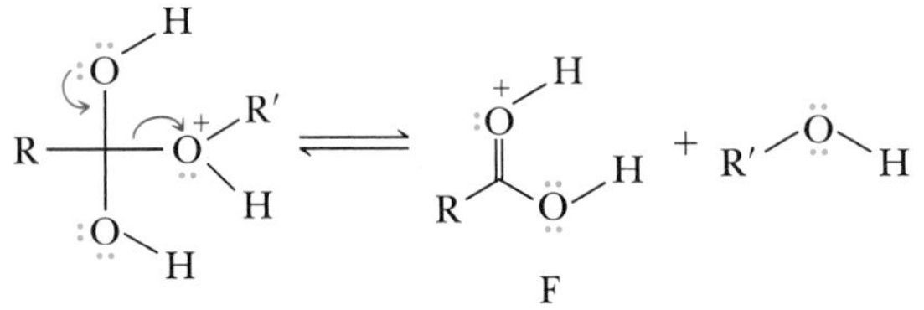

E F

Step 6: **Take a proton away.** A final deprotonation gives the ester product and regenerates the acid catalyst.

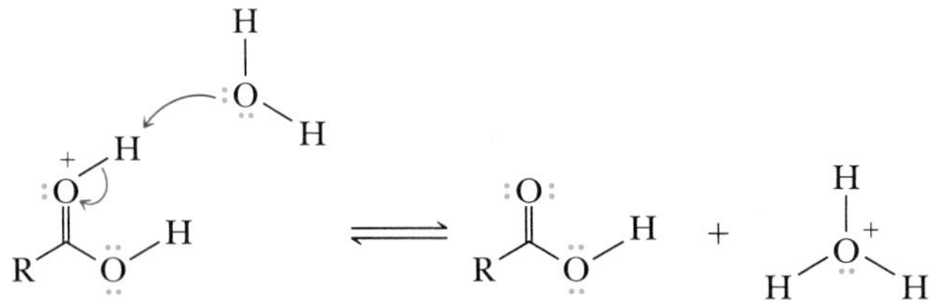

F G

Microscopic Reversibility

We have now discussed Fischer esterification (formation of an ester in an acidic solution of an alcohol) and the hydrolysis of an ester in acidic water. When we discussed Fischer esterification, we pointed out that it is an equilibrium reaction. Ester hydrolysis in aqueous acid is also an equilibrium reaction. The two reactions proceed via the same nucleophilic addition/elimination mechanism, except that they are the reverse of each other. As first introduced in Section 10.6, the principle of microscopic reversibility states that for any reversible reaction, the sequence of intermediates and transition states must be the same but in reverse order for the backward versus forward reaction. In general, the reverse of protonation (Add a proton) is deprotonation (Take a proton away). The reverse of nucleophilic attack (Make a bond between a nucleophile and an electrophile) is leaving group departure (Break a bond to give stable molecules or ions).

With Fischer esterification and ester hydrolysis, we can see the principle of microscopic reversibility by comparing the mechanism for Fischer esterification. (Things You Should Know V, Carboxylic Acid Derivative Reactions Mechanisms, and the ester hydrolysis mechanism box immediately above.) First, note that they both have six overall steps. Now let's examine the corresponding steps. In the following analysis, we will compare within parentheses the structures lettered with Roman numerals in the Fischer esterification mechanism listed in "Things You Should Know V" to the capital letters in the ester hydrolysis Mechanism box.

The esterification starts with a protonation of the carbonyl oxygen, and the hydrolysis ends with a deprotonation of a carbonyl oxygen (I = G). The second step of esterification is nucleophilic attack on the carbonyl carbon, and the second to last step of hydrolysis is leaving group departure to create a carbonyl (II = F). The third step of esterification is to remove a proton of the nucleophile that was added, and the third to last step of hydrolysis is to protonate what will be the leaving group (III = E). The fourth step of esterification is to protonate what will be the leaving group, and the third step of hydrolysis is to deprotonate what was the nucleophile (IV = D). It is important to note at this point that the third step of esterification creates the same neutrally charged tetrahedral intermediate via deprotonation to which the third step of hydrolysis supplies a proton. The fifth step of esterification is leaving group departure, and the second step of hydrolysis is nucleophilic attack (V = C). The last step of esterification is the deprotonation of the carbonyl oxygen, and the first step of hydrolysis is to add a proton to the carbonyl oxygen (VI = B). By using the principle of microscopic reversibility, you should be able to write the mechanism of any reverse reaction once you know and understand the forward reaction.

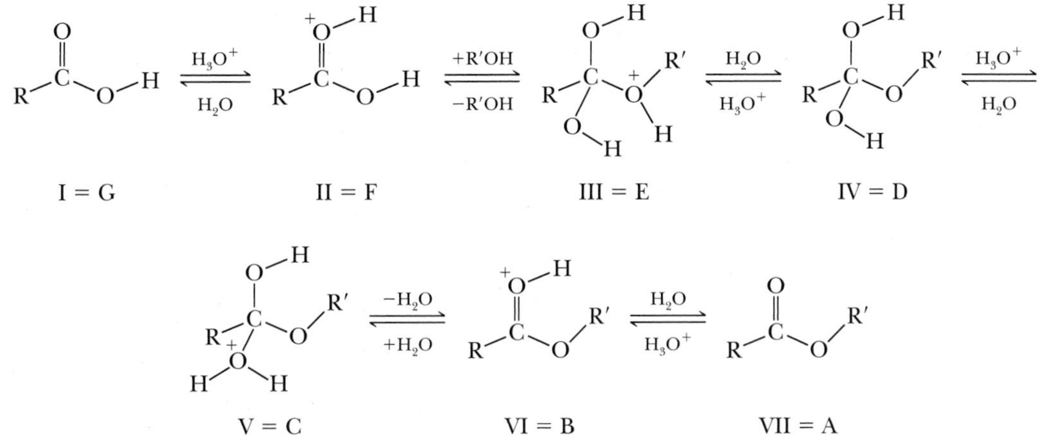
I = G
II = F
III = E
IV = D
V = C
VI = B
VII = A

Saponification

Hydrolysis of esters may also be carried out using hot aqueous base, such as aqueous NaOH.

$$\mathrm{R\overset{O}{\overset{\|}{C}}OCH_3} + \mathrm{NaOH} \xrightarrow{\mathrm{H_2O}} \mathrm{R\overset{O}{\overset{\|}{C}}O^-Na^+} + \mathrm{CH_3OH}$$

Hydrolysis of esters in aqueous base is often called **saponification**, a reference to the use of this reaction in the manufacture of soaps (Section 26.2A) through hydrolysis of triglyceride ester groups. Although the carbonyl carbon of an ester is not strongly electrophilic, hydroxide ion is a good nucleophile and adds to the carbonyl carbon to form a tetrahedral carbonyl addition intermediate, which in turn collapses to give a carboxylic acid and an alkoxide ion. The carboxylic acid reacts with the alkoxide ion or other base present to form a carboxylate anion. Thus, each mole of ester hydrolyzed requires one mole of base.

Saponification
Hydrolysis of an ester in aqueous NaOH or KOH to an alcohol and the sodium or potassium salt of a carboxylic acid.

Mechanism 18.3

Hydrolysis of an Ester in Aqueous Base (Saponification)

Step 1: **Make a new bond between a nucleophile and electrophile.** Addition of hydroxide ion to the carbonyl carbon of the ester gives a tetrahedral carbonyl addition intermediate.

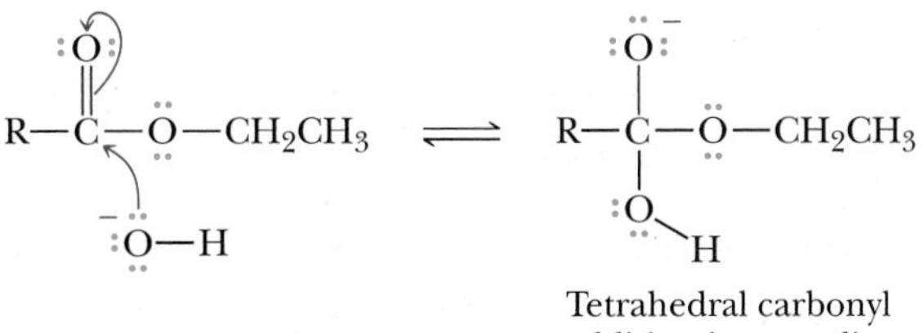

Tetrahedral carbonyl addition intermediate

(Continued)

Step 2: **Break a bond to give stable molecules or ions.** Collapse of this intermediate gives a carboxylic acid and an alkoxide ion.

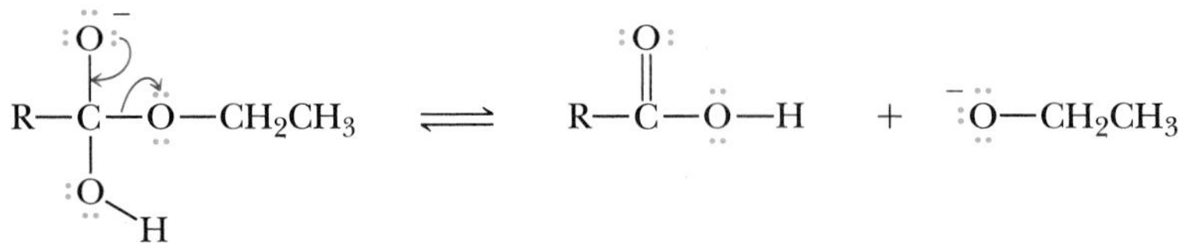

Step 3: **Take a proton away.** Proton transfer between the carboxyl group and the alkoxide ion gives the carboxylate anion. This strongly exothermic acid-base reaction drives the whole reaction to completion.

There are two major differences between hydrolysis of esters in aqueous acid and aqueous base.

1. For hydrolysis of an ester in aqueous acid, acid is required in only catalytic amounts. For hydrolysis in aqueous base, base is required in stoichiometric amounts because it is a reactant, not a catalyst.
2. Hydrolysis of an ester in aqueous acid is reversible, but hydrolysis in aqueous base is irreversible because a carboxylate anion (weakly electrophilic, if at all) is not attacked by ROH (a weak nucleophile).

Other acid derivatives react with base in an identical manner to esters.

Example 18.3 Ester Hydrolysis Reactions

Complete and balance equations for the hydrolysis of each ester in aqueous sodium hydroxide. Show all products as they are ionized under these conditions.

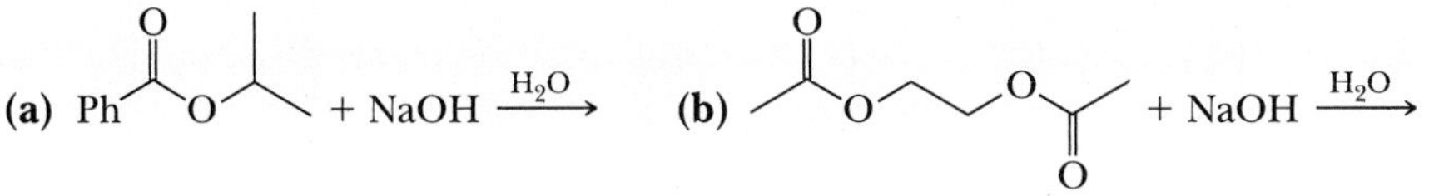

Solution

The products of hydrolysis of (a) are benzoic acid and 2-propanol. In aqueous NaOH, benzoic acid is converted to its sodium salt. Therefore, one mole of NaOH is required for hydrolysis of one mole of this ester. Compound (b) is a diester of ethylene glycol. Two moles of NaOH are required for its hydrolysis.

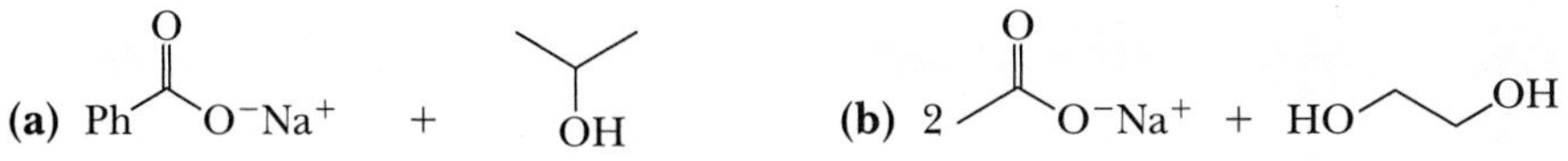

Sodium benzoate	2-Propanol (Isopropyl alcohol)	Sodium acetate	1,2-Ethanediol (Ethylene glycol)

Problem 18.3

Complete and balance equations for the hydrolysis of each ester in aqueous solution; show each product as it is ionized under the indicated experimental conditions.

(a) 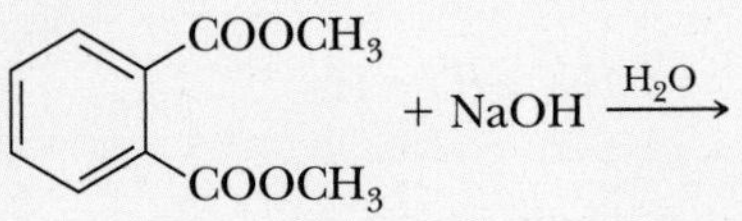

(b) 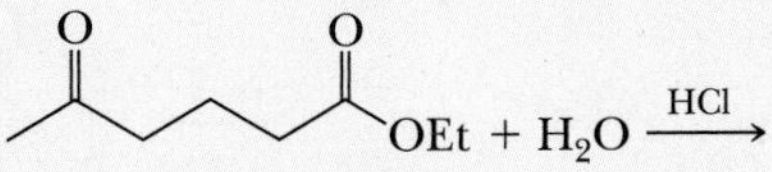

D. Amides

Compared to esters, amides require considerably more vigorous conditions for hydrolysis in both acid and base. Amides undergo hydrolysis in hot aqueous acid to give a carboxylic acid and an ammonium ion. Hydrolysis is driven to completion by the acid-base reaction between ammonia or the amine and acid to form an ammonium salt. One mole of acid is required per mole of amide.

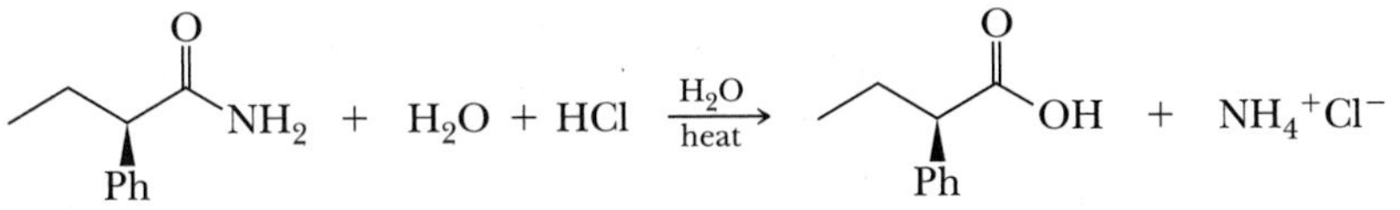

(*R*)-2-Phenylbutanamide **(*R*)-2-Phenylbutanoic acid**

In aqueous base, the products of amide hydrolysis are a carboxylate salt and ammonia or an amine. Hydrolysis in aqueous base is driven to completion by the acid-base reaction between the resulting carboxylic acid and base to form a salt. One mole of base is required per mole of amide.

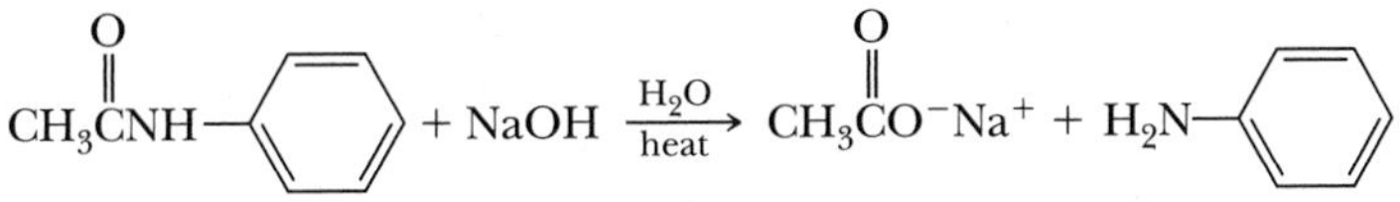

N-Phenylethanamide (N-Phenylacetamide, Acetanilide) **Sodium acetate** **Aniline**

The steps in the mechanism for the hydrolysis of amides in aqueous acid are similar to those for the hydrolysis of esters in aqueous acid.

Chemical Connections

Mechanistic Alternatives for Ester Hydrolysis: S_N2 and S_N1 Possibilities

S_N2

Although an addition/elimination sequence involving the formation of a tetrahedral carbonyl addition intermediate is the most common mechanism for the hydrolysis of esters, alternative pathways are followed in special cases. One such case occurs with methyl esters in basic conditions. Recall that S_N2 reactions are most favorable with CH_3Lv (where Lv = leaving

(Continued)

group) relative to 1°, 2°, and 3° alkyl groups. With methyl esters, an S_N2 mechanism has a lower energy transition state than those involved in the addition/elimination sequence; therefore, the S_N2 pathway dominates.

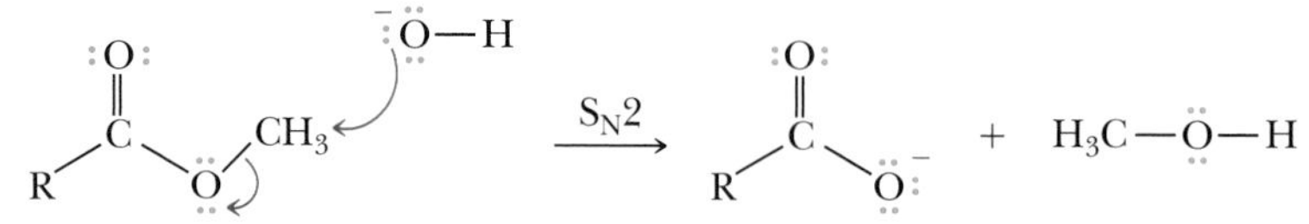

Make a new bond between a nucleophile and an electrophile and simultaneously break a bond to give stable molecules and ions

S_N1

Another special case occurs in acidic media when the alkyl group bonded to the oxygen can form an especially stable carbocation. In these cases, protonation of the carbonyl oxygen is followed by cleavage of the O—C bond to give a carboxylic acid and a carbocation. Benzyl and *tert*-butyl esters readily undergo this type of ester hydrolysis in acid. The carbocation is then trapped by water to create an alcohol. This is an S_N1 reaction in which the leaving group is a carboxylic acid.

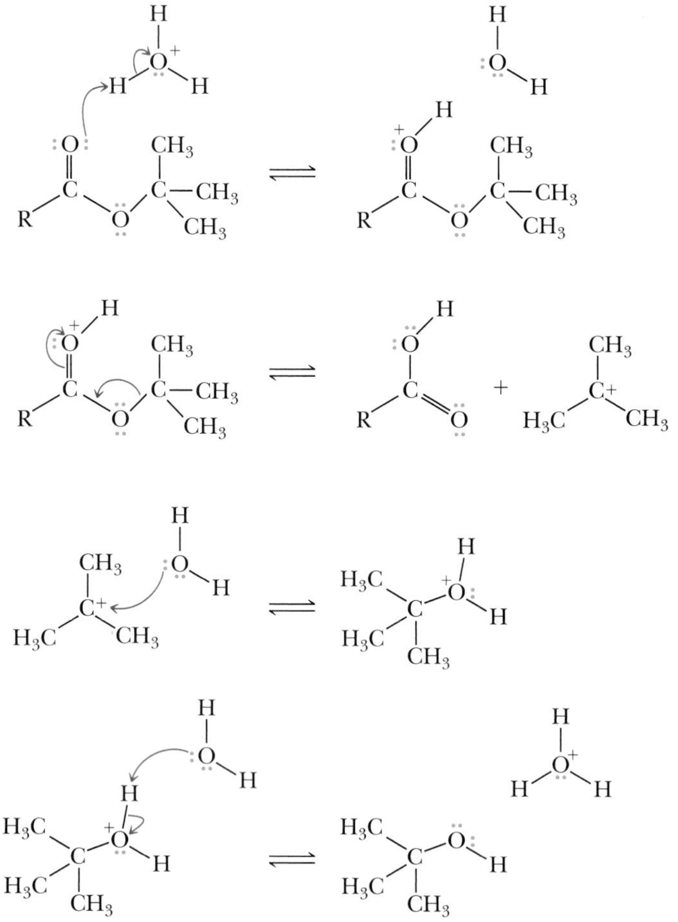

Step 1: Add a proton

Step 2: Break a bond to give stable molecules or ions

Step 3: Make a bond between a nucleophile and an electrophile

Step 4: Take a proton away

Mechanism 18.4

Hydrolysis of an Amide in Aqueous Acid

Step 1: **Add a proton.** Protonation of the carbonyl oxygen gives a resonance-stabilized cation intermediate.

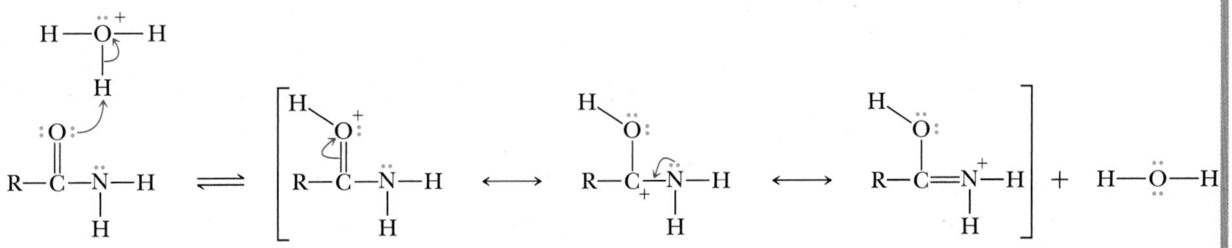

Cation stabilized by resonance delocalization

The role of the proton in this step is to protonate the carbonyl oxygen to increase the electrophilic character of the carbonyl carbon.

Step 2: **Make a new bond between a nucleophile and an electrophile.** Addition of water to the carbonyl carbon.

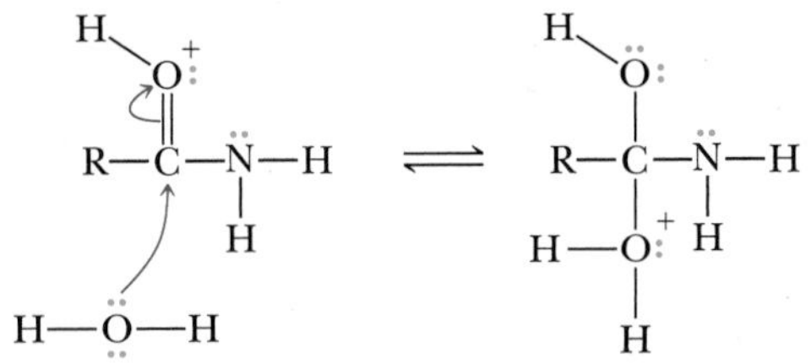

Step 3: **Take a proton away/add a proton.** Proton transfer between the O and N atoms gives a carbonyl addition intermediate. This would occur in two steps, where a solvent molecule accepts the acidic proton on the O atom and that a hydronium ion donates the proton to the N atom, although the exact timing of these events may be different for different molecules in the flask.

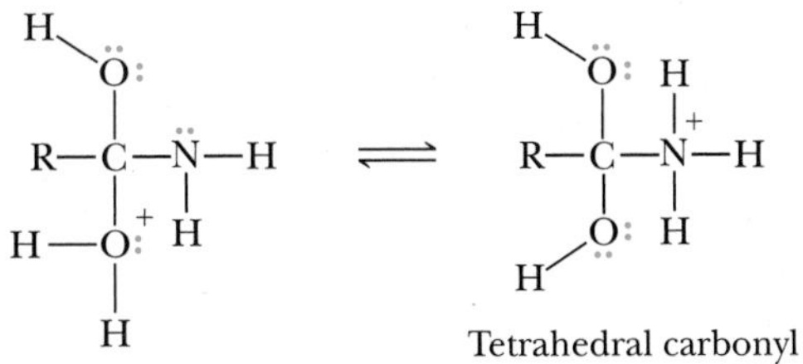

Tetrahedral carbonyl addition intermediate

Step 4: **Break a bond to make stable molecules or ions.** Note that the leaving group in this step is a neutral amine (a weaker base), a far better leaving group than an amide ion (a much stronger base).

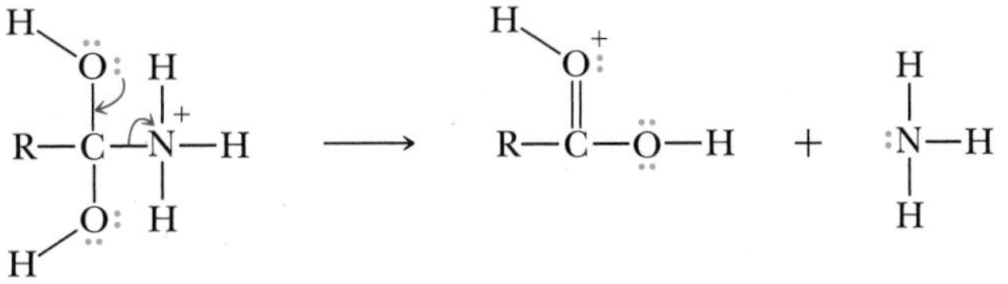

(Continued)

Step 5: **Take a proton away.** Proton transfer between the very acidic protonated carbonyl and relatively basic amine gives the carboxylic acid and ammonium ion products.

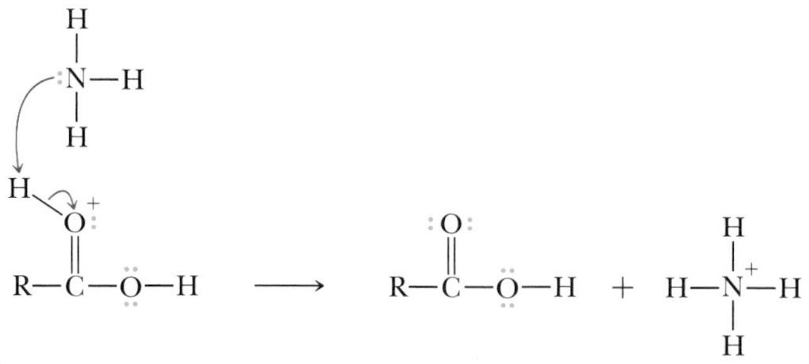

The mechanism for the hydrolysis of amides in aqueous base is more complex than that for the hydrolysis of esters in aqueous base because the amide anion is such a poor leaving group.

Mechanism 18.5

Hydrolysis of an Amide in Aqueous Base

Step 1: **Make a new bond between a nucleophile and an electrophile.** Addition of hydroxide ion to the carbonyl carbon gives a tetrahedral carbonyl addition intermediate.

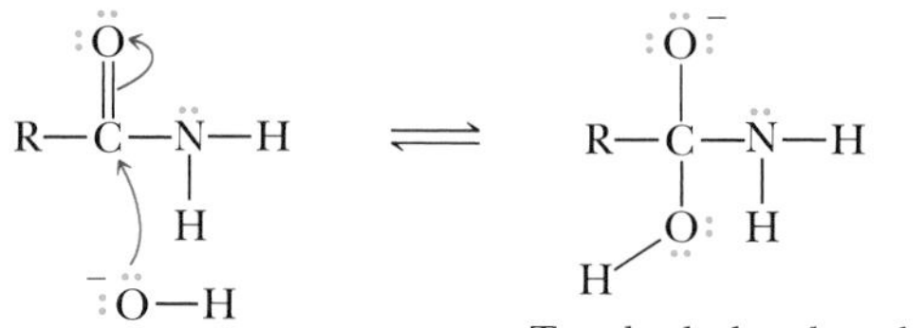

Tetrahedral carbonyl addition intermediate

Step 2: **Take a proton away.** The accepted mechanism involves the creation of a dianionic tetrahedral intermediate, which has enough negative charge to expel the amide anion.

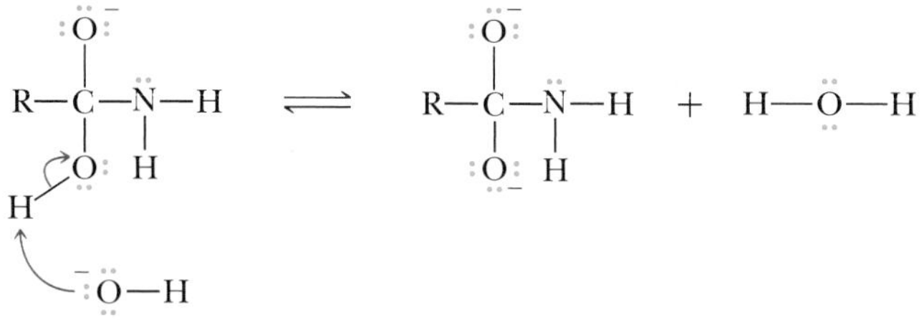

Step 3: **Break a bond to give stable molecules or ions/add a proton.** The amide anion has little to no lifetime in water because it is so basic; therefore, it will be instantly protonated by water upon its formation [or potentially during its expulsion (as shown here)].

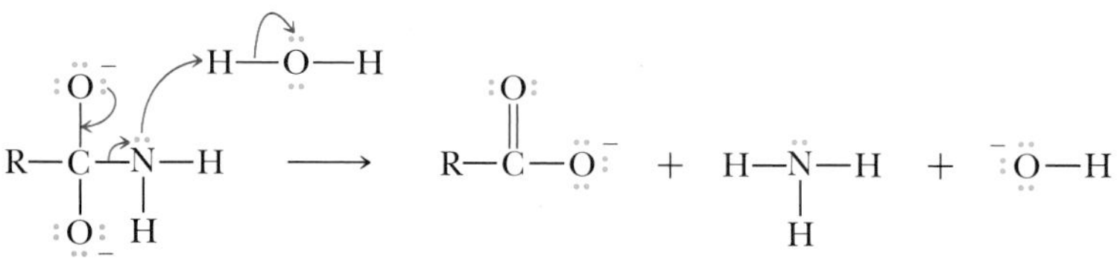

Example 18.4 Amide Hydrolysis Reactions

Write equations for the hydrolysis of these amides in concentrated aqueous HCl. Show all products as they exist in aqueous HCl and the number of moles of HCl required for hydrolysis of each amide.

(a) $CH_3\overset{\overset{O}{\|}}{C}N(CH_3)_2$ **(b)** 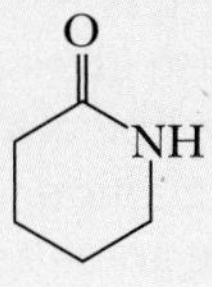
O
NH

Solution

(a) Hydrolysis of *N,N*-dimethylacetamide gives acetic acid and dimethylamine. Dimethylamine, a base, is protonated by HCl to form dimethylammonium ion and is shown in the balanced equation as dimethylammonium chloride. One mole of HCl is required per mole of amide.

$$CH_3\overset{\overset{O}{\|}}{C}N(CH_3)_2 + H_2O + HCl \xrightarrow{\text{heat}} CH_3\overset{\overset{O}{\|}}{C}OH + (CH_3)_2NH_2^+Cl^-$$

(b) Hydrolysis of this δ-lactam gives the protonated form of 5-aminopentanoic acid. One mole of HCl is required per mole of amide.

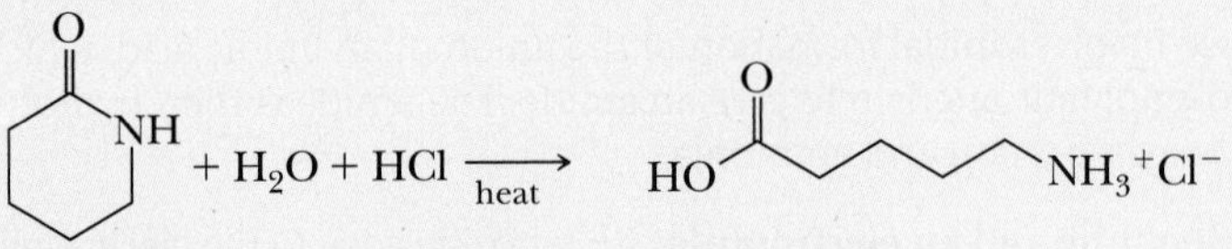
O
NH
+ H2O + HCl
heat
O
HO
NH3+Cl−

Problem 18.4

Complete equations for the hydrolysis of the amides in Example 18.4 in concentrated aqueous NaOH. Show all products as they exist in aqueous NaOH and the number of moles of NaOH required for hydrolysis of each amide.

E. Nitriles

The cyano group of a nitrile is hydrolyzed in aqueous acid to a carboxyl group and ammonium ion as shown in the following equation.

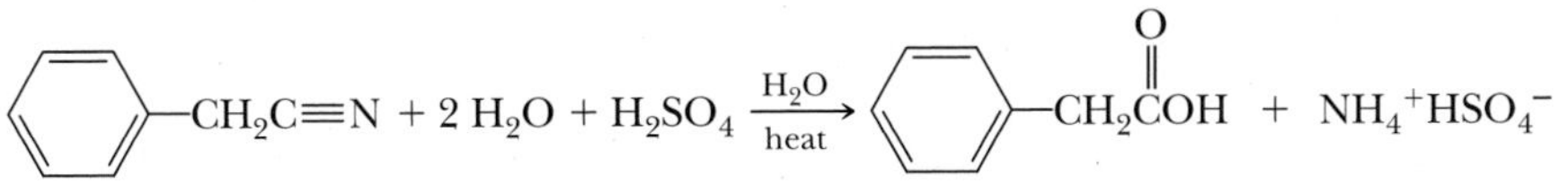
CH2C≡N + 2 H2O + H2SO4
H2O
heat
O
CH2COH + NH4+HSO4−

Phenylacetonitrile **Phenylacetic acid** **Ammonium hydrogen sulfate**

In hydrolysis of a cyano group in aqueous acid, protonation of the nitrogen atom gives a cation that reacts with water to give an imidic acid (the enol of an amide). Keto-enol tautomerism of the imidic acid gives an amide. The amide is then hydrolyzed, as already described, to a carboxylic acid and an ammonium ion.

$$R{-}C{\equiv}N + H_2O \xrightarrow{H^+} R{-}\overset{\overset{OH}{|}}{C}{=}NH \rightleftharpoons R{-}\overset{\overset{O}{\|}}{C}{-}NH_2$$

An imidic acid (enol of an amide) An amide

The reaction conditions required for acid-catalyzed hydrolysis of a cyano group are typically more vigorous than those required for hydrolysis of an amide, and in the presence of excess water, a cyano group is hydrolyzed first to an amide and then to a carboxylic acid. It is possible to stop at the amide by using sulfuric acid as a catalyst and one mole of water per mole of nitrile. Selective hydrolysis of a nitrile to an amide, however, is not a good method for the preparation of amides. They are better prepared from acid chlorides, acid anhydrides, or esters.

Hydrolysis of a cyano group in aqueous base gives a carboxylate anion and ammonia. The reaction is driven to completion by the acid-base reaction between the carboxylic acid and base to form a carboxylate anion. Acidification of the reaction mixture during workup converts the carboxylate anion to the carboxylic acid.

$$CH_3(CH_2)_9C{\equiv}N + H_2O + NaOH \xrightarrow[\text{heat}]{H_2O} CH_3(CH_2)_9\overset{\overset{\displaystyle O}{\|}}{C}O^-Na^+ + NH_3 \xrightarrow[H_2O]{HCl} CH_3(CH_2)_9\overset{\overset{\displaystyle O}{\|}}{C}OH + NaCl + NH_4Cl$$

Undecanenitrile **Sodium undecanoate** **Undecanoic acid**

Mechanism 18.6

Hydrolysis of a Cyano Group to an Amide in Aqueous Base

Hydrolysis of a cyano group in aqueous base involves initial formation of the anion of an imidic acid, which, after proton transfer from water, undergoes keto-enol tautomerism to give an amide. The amide is then hydrolyzed by aqueous base, as we saw earlier, to the carboxylate anion and ammonia.

Step 1: **Make a new bond between a nucleophile and an electrophile.** Hydroxide adds to the electrophilic C atom of the cyano group.

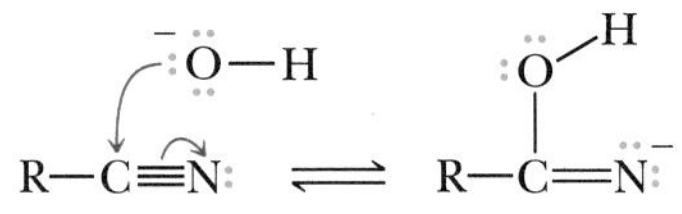

A nitrile

Step 2: **Add a proton.** Proton transfer from water gives an imidic acid.

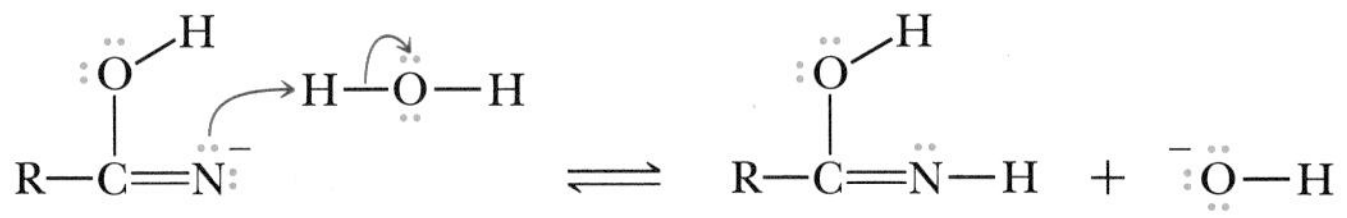

An imidic acid

Step 3: **Keto-enol tautomerism.** Tautomerism of the imidic acid gives the amide.

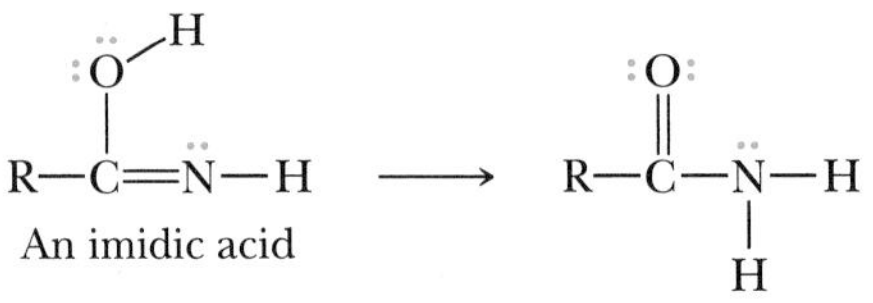

An imidic acid

An amide

The acid-catalyzed reaction proceeds similarly; the only difference is in the order of proton transfers.

Hydrolysis of nitriles provides a valuable way to synthesize carboxylic acids from primary or secondary haloalkanes. To do so, add one carbon in the form of a cyano group (Table 8.1) to the carbon chain and convert this to a carboxyl group.

$$CH_3(CH_2)_8CH_2Cl \xrightarrow[\text{ethanol, water}]{KCN} CH_3(CH_2)_9C{\equiv}N \xrightarrow[\text{heat}]{H_2SO_4,\ H_2O} CH_3(CH_2)_9\overset{O}{\overset{\|}{C}}OH$$

1-Chlorodecane **Undecanenitrile** **Undecanoic acid**

Hydrolysis of cyanohydrins, which are obtained by the addition of HCN to an aldehyde or a ketone (Section 16.5D), provides a valuable way to create α-hydroxycarboxylic acids, as illustrated by the synthesis of mandelic acid.

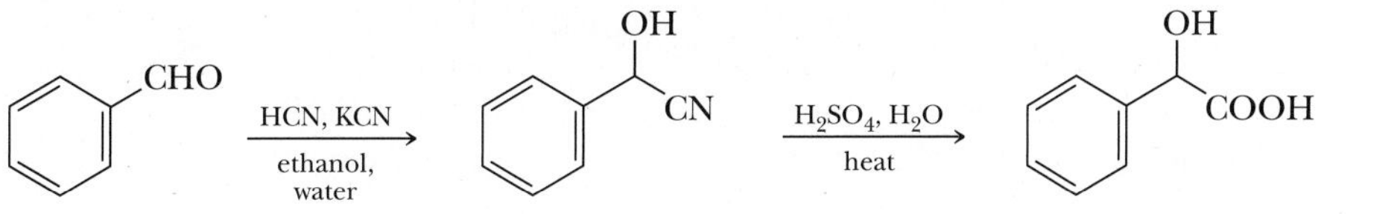

Benzaldehyde **Benzaldehyde cyanohydrin (Mandelonitrile)** (racemic) **2-Hydroxyphenylacetic acid (Mandelic acid)** (racemic)

Example 18.5 Nitrile Hydrolysis Reactions

Show how to bring about the following conversions using as one step the hydrolysis of a cyano group.

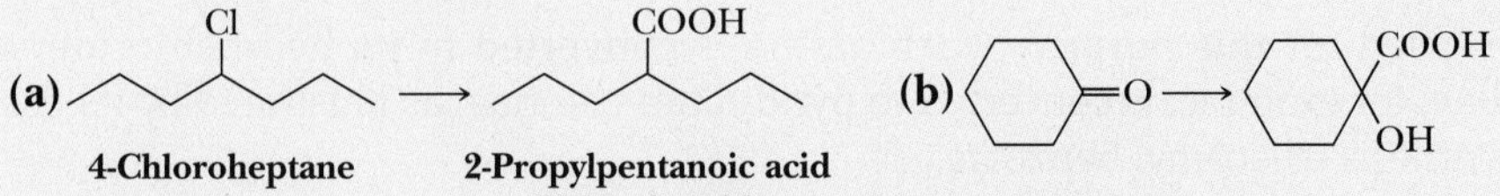

4-Chloroheptane **2-Propylpentanoic acid (Valproic acid)**

Solution

(a) Treatment of 4-chloroheptane with KCN in aqueous ethanol gives a nitrile. Hydrolysis of the cyano group in aqueous sulfuric acid gives the product.

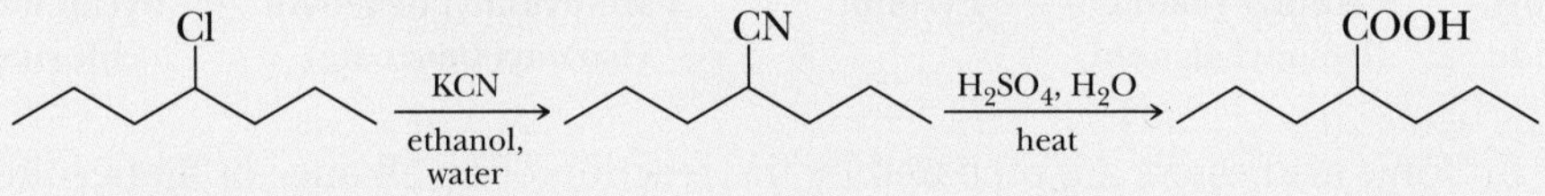

This synthesis can also be accomplished by conversion of the chloroalkane to a Grignard reagent followed by carbonation and hydrolysis in aqueous acid.

(b) Treatment of cyclohexanone with HCN/KCN in aqueous ethanol gives a cyanohydrin. Hydrolysis of the cyano group in concentrated sulfuric acid gives the carboxyl group of the product.

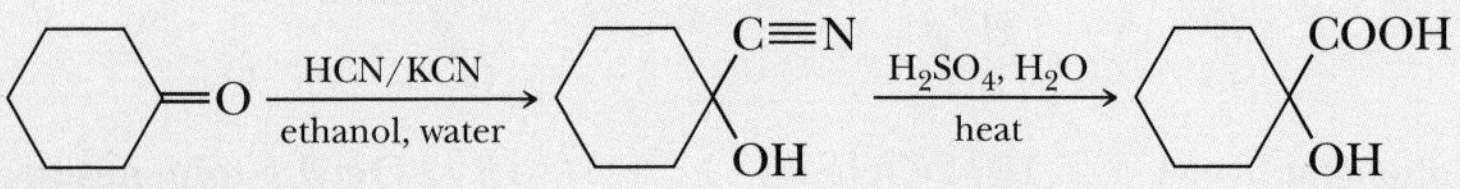

Problem 18.5

Synthesis of nitriles by nucleophilic displacement of halide from a haloalkane (also called an alkyl halide) is practical only with primary and secondary haloalkanes. It fails with tertiary haloalkanes. Why? What is the major product of the following reaction?

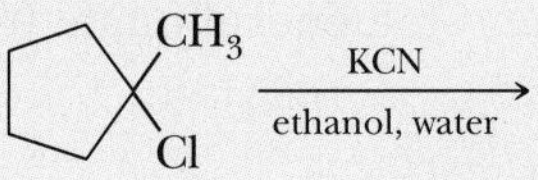

18.5 Reaction with Alcohols

A. Acid Halides

An acid halide reacts with an alcohol to give an ester.

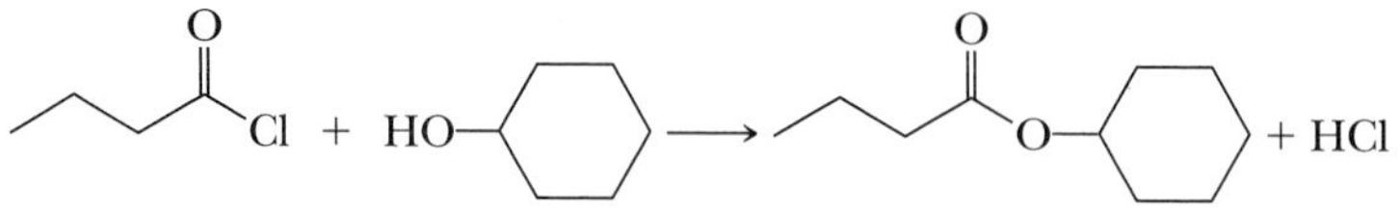

Butanoyl chloride **Cyclohexanol** **Cyclohexyl butanoate**

Because acid halides are so reactive toward even weak nucleophiles such as alcohols, no catalyst is necessary for these reactions.

In cases in which the alcohol or resulting ester is sensitive to acid, the reaction can be carried out in the presence of a tertiary amine to neutralize the HCl as it is formed. The amines most commonly used for this purpose are pyridine and triethylamine.

Pyridine **Triethylamine**

When used for this purpose, each amine is converted to its hydrochloride salt. Pyridine, for example, is converted to pyridinium chloride, as illustrated by its use in the synthesis of isoamyl benzoate.

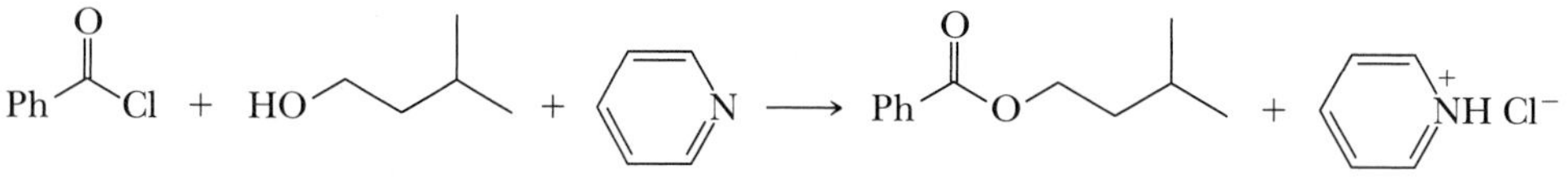

Benzoyl chloride **3-Methyl-1-butanol (Isoamyl alcohol)** **Pyridine** **3-Methylbutyl benzoate (Isoamyl benzoate)** **Pyridinium chloride**

Sulfonic acid esters are prepared by the reaction of an alkane- or arenesulfonyl chloride with an alcohol or phenol. Two of the most common sulfonyl chlorides are *p*-toluenesulfonyl chloride, abbreviated TsCl, and methanesulfonyl chloride, abbreviated MsCl (Section 18.1A).

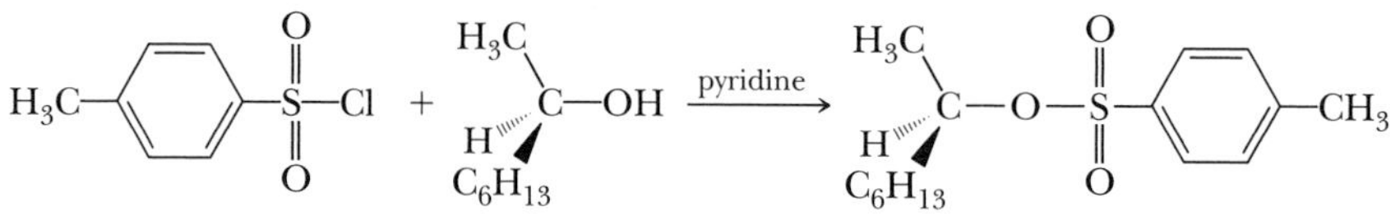

***p*-Toluenesulfonyl chloride (Tosyl chloride; TsCl)** **(*R*)-2-Octanol** **(*R*)-2-Octyl *p*-toluenesulfonate ((*R*)-2-Octyl tosylate)**

As discussed in Section 10.5D, a special value of *p*-toluenesulfonic (tosylate) and methanesulfonic (mesylate) esters is that in forming them, an —OH is converted from a poor leaving group (hydroxide ion) in nucleophilic displacement to an excellent leaving group, the *p*-toluenesulfonate (tosylate) or methanesulfonate (mesylate) anions.

B. Acid Anhydrides

Acid anhydrides react with alcohols to give one mole of ester and one mole of a carboxylic acid.

$$\underset{\textbf{Acetic anhydride}}{CH_3\overset{O}{\overset{\|}{C}}O\overset{O}{\overset{\|}{C}}CH_3} + \underset{\textbf{Ethanol}}{HOCH_2CH_3} \longrightarrow \underset{\textbf{Ethyl acetate}}{CH_3\overset{O}{\overset{\|}{C}}OCH_2CH_3} + \underset{\textbf{Acetic acid}}{CH_3\overset{O}{\overset{\|}{C}}OH}$$

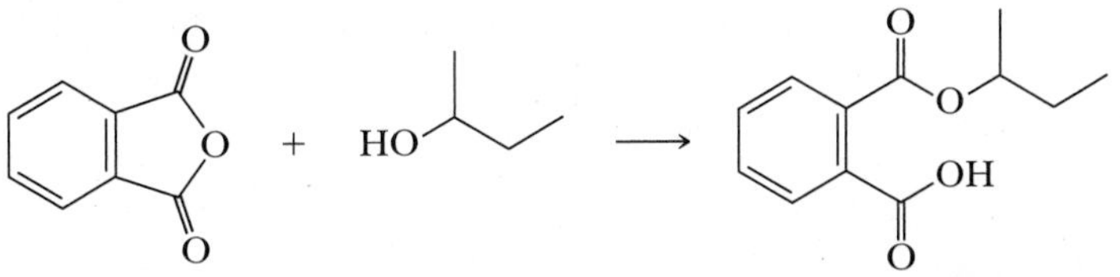

Phthalic anhydride | **2-Butanol (*sec*-Butyl alcohol)** (racemic) | **1-Methylpropyl hydrogen phthalate (*sec*-Butyl hydrogen phthalate)** (racemic)

Thus, the reaction of an alcohol with an anhydride is a useful method for the synthesis of esters. This reaction is catalyzed by acids and by tertiary amines.

Aspirin is synthesized on an industrial scale by the reaction of acetic anhydride and salicylic acid.

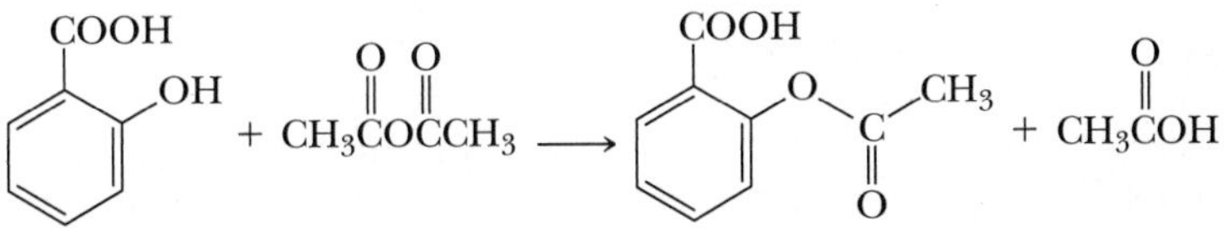

2-Hydroxybenzoic acid (Salicylic acid) | **Acetic anhydride** | **Acetylsalicylic acid (Aspirin)** | **Acetic acid**

C. Esters

Esters react with alcohols in an acid-catalyzed reaction called **transesterification**. For example, it is possible to convert methyl acrylate to butyl acrylate by heating the methyl ester with 1-butanol in the presence of an acid catalyst.

Transesterification
Exchange of the —OR or —OAr group of an ester for another —OR or —OAr group.

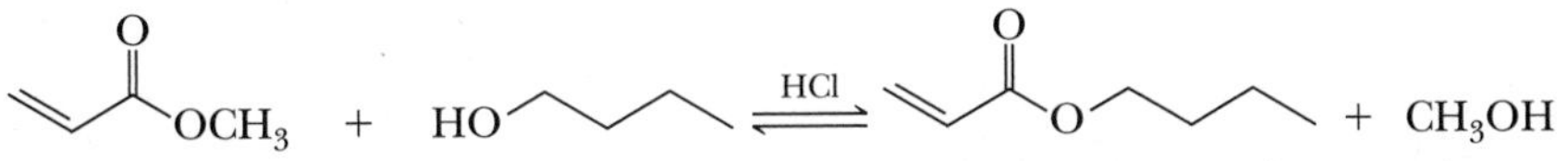

Methyl propenoate (Methyl acrylate) (bp 81°C) | **1-Butanol** (bp 117°C) | **Butyl propenoate (Butyl acrylate)** (bp 147°C) | **Methanol** (bp 65°C)

The acids most commonly used for transesterification are HCl as a gas bubbled into the reaction medium and *p*-toluenesulfonic acid.

Transesterification is an equilibrium reaction that can be driven in either direction by control of experimental conditions. For example, in the reaction of methyl acrylate with 1-butanol, transesterification is carried out at a temperature slightly above the boiling point of methanol (the lowest boiling component in the mixture). Methanol distills from the reaction mixture, thus shifting the position of equilibrium in favor of butyl acrylate. Conversely, reaction of butyl acrylate with a large excess of methanol shifts the equilibrium to favor formation of methyl acrylate.

Example 18.6 Transesterification Reactions

Complete the following transesterification reactions (the stoichiometry of each is given in the problem).

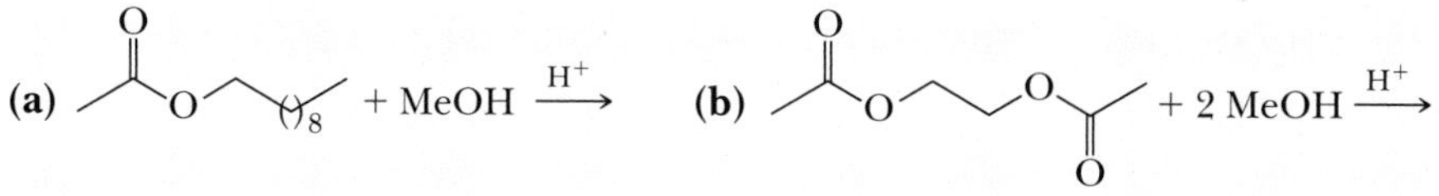

Solution

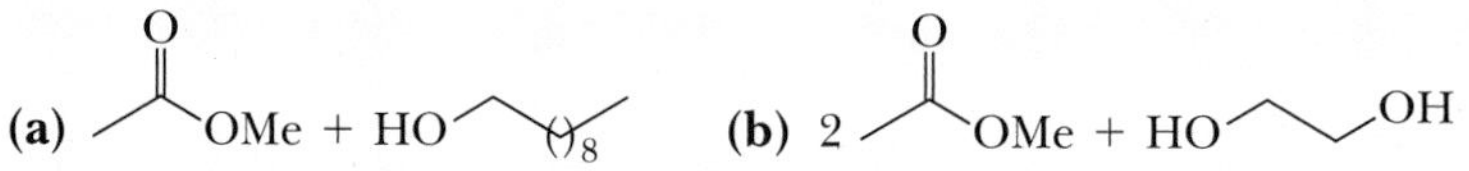

Problem 18.6

Complete the following transesterification reaction (the stoichiometry is given in the equation).

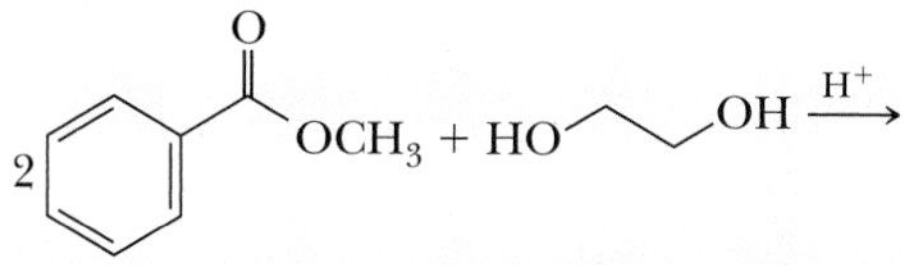

D. Amides

Amides, the least reactive of the functional derivatives of carboxylic acids, do not react with alcohols. Thus, the reaction of an amide with an alcohol cannot be used to prepare an ester.

18.6 Reactions with Ammonia and Amines

A. Acid Halides

Acid halides react readily with ammonia and 1° and 2° amines to form amides. For complete conversion of an acid halide to an amide, two equivalents of ammonia or amine are used, one to form the amide and one to neutralize the hydrogen halide formed.

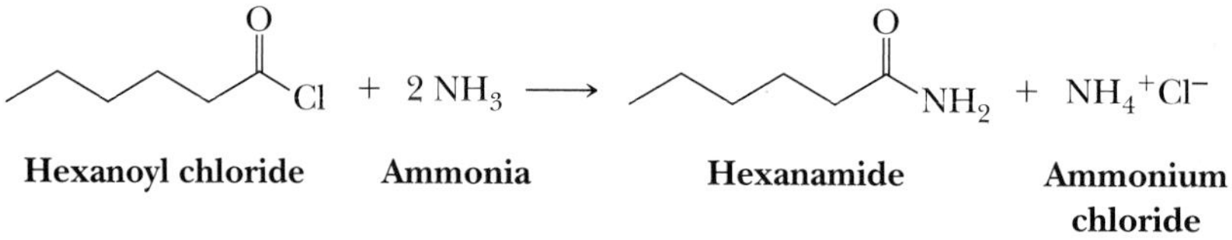

Mechanism 18.7

Reaction of an Acid Chloride and Ammonia

Step 1: **Make a new bond between a nucleophile and an electrophile.** Ammonia adds to the carbonyl carbon.

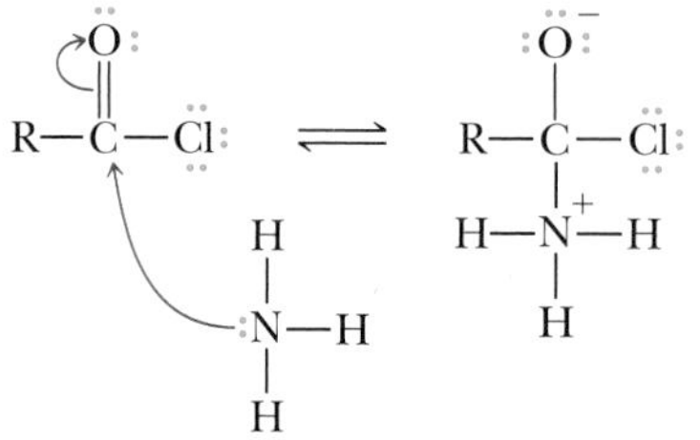

***Step 2:* Take a proton away.** Proton transfer gives a tetrahedral carbonyl addition intermediate.

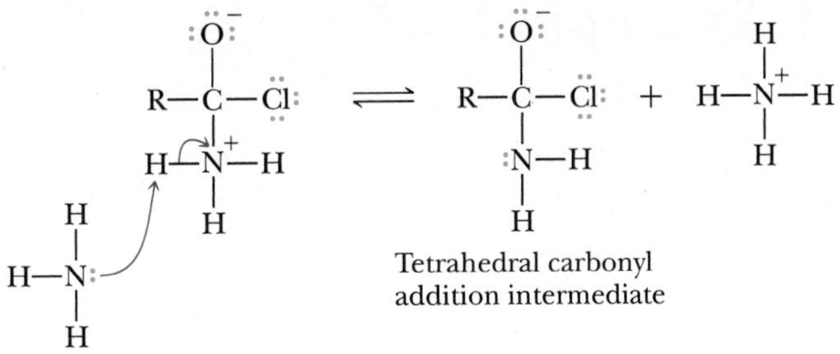

Tetrahedral carbonyl addition intermediate

***Step 3:* Break a bond to give stable molecules or ions.** The tetrahedral carbonyl addition intermediate then expels the chloride as a leaving group.

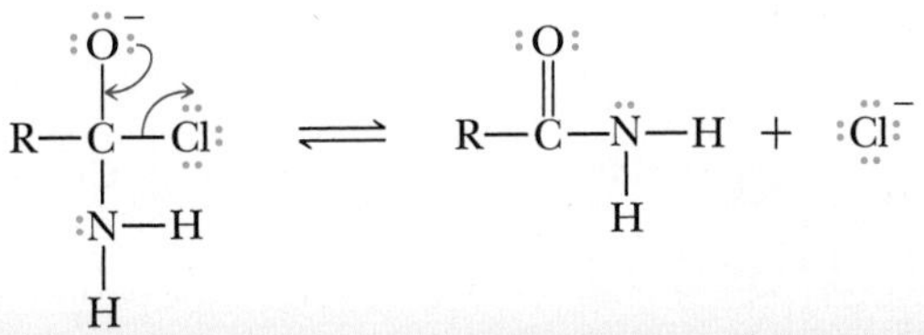

B. Acid Anhydrides

Acid anhydrides react with ammonia and 1° and 2° amines to form amides. As with acid halides, two moles of amine are required, one mole to form the amide and one mole to neutralize the carboxylic acid by product.

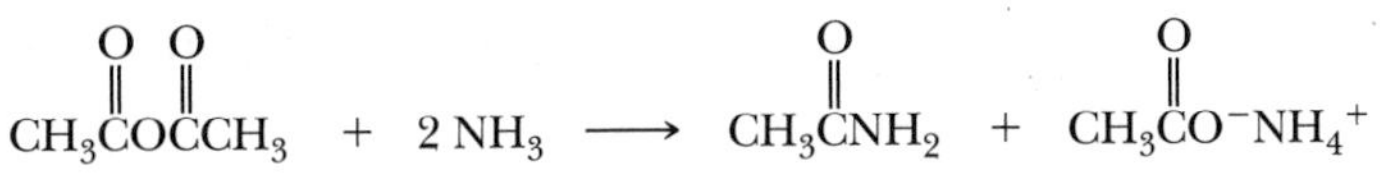

Acetic anhydride **Ammonia** **Ethanamide (Acetamide)** **Ammonium acetate**

Alternatively, if the amine used to make the amide is expensive, a non-nucleophilic tertiary amine such as triethylamine may be used to neutralize the carboxylic acid.

C. Esters

Esters react with ammonia and with 1° and 2° amines to form amides.

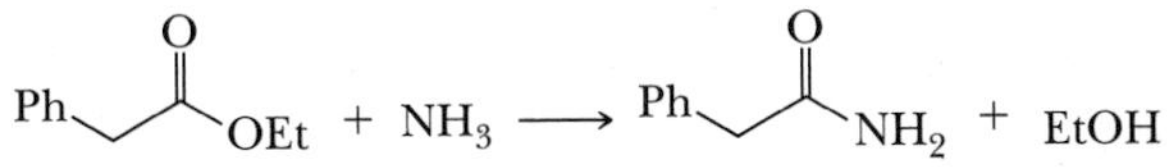

Ethyl phenylacetate **Phenylacetamide** **Ethanol**

Because an alkoxide anion is a poor leaving group compared with either a halide or a carboxylate ion, esters are less reactive toward ammonia, 1° amines, and 2° amines than are acid halides or acid anhydrides. The reaction often requires heating or high concentrations of amine or both.

D. Amides

Amides do not react with ammonia or primary or secondary amines.

Example 18.7 Amide Formation Reactions

Complete the following reactions (the stoichiometry of each reaction is given in the equation).

(a) 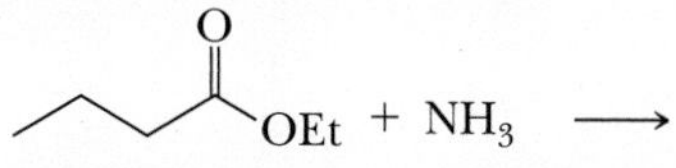

Ethyl butanoate

(b) 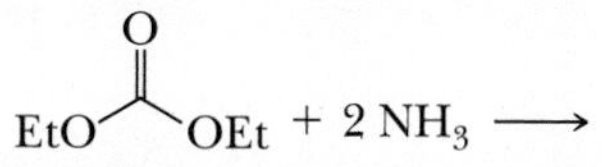

Diethyl carbonate

Solution

(a) 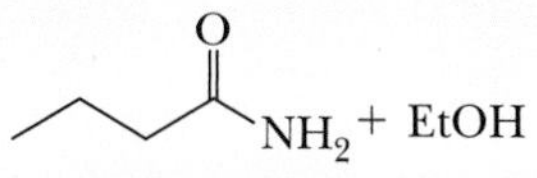

Butanamide

(b) 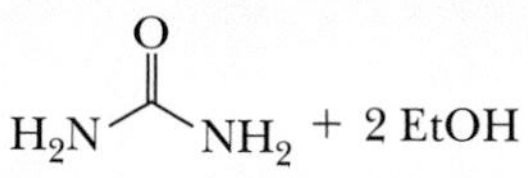

Urea

Problem 18.7

Complete and balance equations for the following reactions (the stoichiometry of each reaction is given in the equation).

(a) 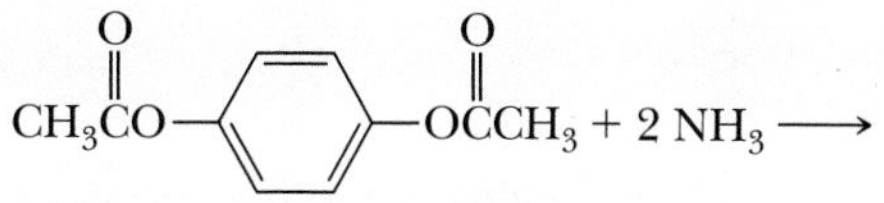

(b) 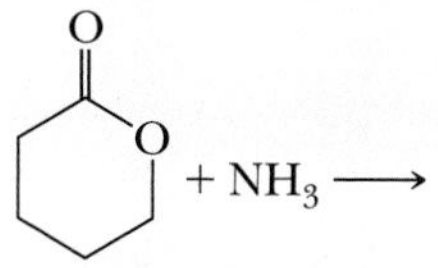

18.7 Reaction of Acid Chlorides with Salts of Carboxylic Acids

Acid chlorides react with salts of carboxylic acids to give anhydrides. Most commonly used are the sodium or potassium salts.

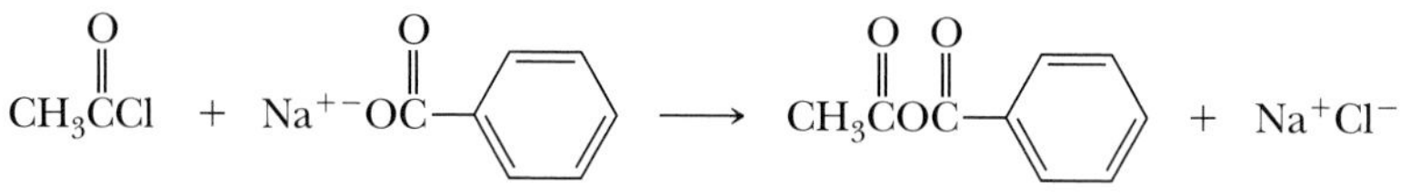

Acetyl chloride **Sodium benzoate** **Acetic benzoic anhydride**

Reaction of an acid halide with a carboxylate anion of a carboxylic acid is a particularly useful method for synthesis of mixed anhydrides.

18.8 Interconversion of Functional Derivatives

We have seen throughout the past several sections that acid chlorides are most reactive toward nucleophilic acyl substitution, followed by acid anhydrides and esters; the least reactive are amides. Carboxylate anions are negatively charged and therefore repel nucleophiles; the resonance in these species is quite stabilizing. Both of these factors make carboxylate anions essentially inert to nucleophilic acyl substitution (hence, we have not examined them to this point in the chapter). Another useful way to think about the reactions of the functional derivatives of carboxylic acids is summarized in Figure 18.2.

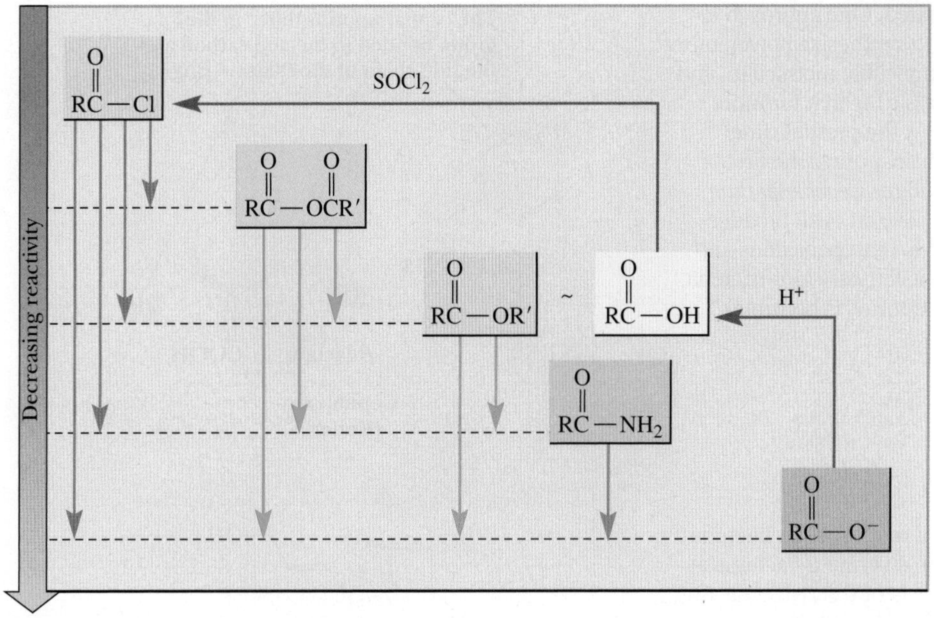

FIGURE 18.2 Reactivities of carboxyl derivatives toward nucleophilic acyl substitution. A more reactive derivative may be converted to a less reactive derivative by treatment with an appropriate reagent. Treatment of a carboxylic acid with thionyl chloride (the acid chloride of sulfurous acid) converts it to the more reactive acid chloride. Carboxylic acids are about as reactive as esters under acidic conditions, but they are converted to unreactive carboxylates under basic conditions.

Watch a video explanation

Any functional group lower in this figure can be prepared from any functional group above it by treatment with an appropriate oxygen or nitrogen nucleophile. An acid chloride, for example, can be converted to an acid anhydride, an ester, an amide, or a carboxylic acid. Acid anhydrides, esters, and amides, however, do not react with chloride ion to give acid chlorides.

MCAT Practice: *Passage and Questions*

β-Lactam Antibiotics

The penicillins were discovered in 1928 by the Scottish bacteriologist Sir Alexander Fleming. Subsequently, due to the brilliant experimental work of the Australian pathologist, Sir Howard Florey, and German chemist, Ernst Chain, who fled Nazi Germany, penicillin G was introduced into the practice of medicine in 1943. Arguably, the discovery of antibiotics has saved more lives worldwide than any other discovery in chemistry and medicine. For their pioneering work in developing one of the most effective antibiotics of all time, Fleming, Florey, and Chain were awarded the 1945 Nobel Prize in Medicine or Physiology.

The structural feature common to all penicillins is a β-lactam ring fused to a five-membered thiazolidine ring. The penicillins owe their antibacterial activity to a common mechanism that inhibits the biosynthesis of a vital part of bacterial cell walls.

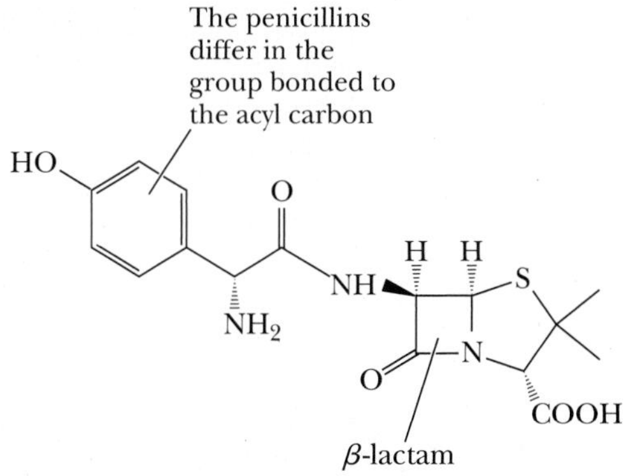

Amoxicillin
(a β-lactam antibiotic)

Soon after the penicillins were introduced into practice, penicillin-resistant strains of bacteria began to

(Continued)

appear and have since proliferated. One approach to combating resistant strains is to synthesize newer, more effective penicillins, such as ampicillin, methicillin, and amoxicillin. Another approach is to search for more effective β-lactam antibiotics. At the present time, the most effective are the cephalosporins, the first of which was isolated from the fungus *Cephalosporium acremonium*. These antibiotics have an even broader spectrum of antibacterial activity than penicillins and are effective against many penicillin-resistant bacteria, although resistance to cephalosporins is becoming widespread.

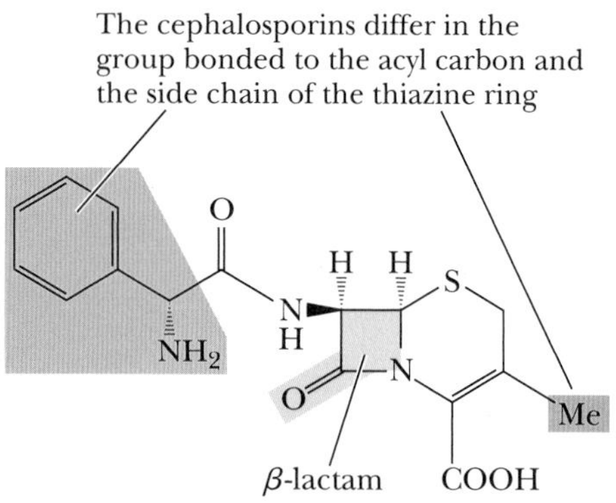
The cephalosporins differ in the group bonded to the acyl carbon and the side chain of the thiazine ring
O
H H
S
N
H
NH2
N
O
Me
β-lactam
COOH

Cephalexin (Keflex)

Questions

A. A common mechanism of resistance in bacteria involves expression of the enzyme β-lactamase that catalyzes the hydrolysis of the β-lactam ring common to all penicillins and cephalosporins, thereby rendering the antibiotics inactive. The first step in the hydrolysis is delivery of water as a nucleophile to the carbonyl carbon of the amide in the β-lactam ring. Due to the high stability of an amide functional group how might you expect an enzyme to catalyze this hydrolysis?

1. Electrophilic activation of the amide carbonyl by coordination of the oxygen to a positively charged metal, or electrophilic activation by hydrogen bonding the oxygen to an acid or hydrogen bond donor.
2. Electrophilic activation of the amide by placing the negative end of bond dipoles near the oxygen of the carbonyl.
3. Nucleophilic activation of the amide carbonyl by coordination of the oxygen to a positively charged metal, or nucleophilic activation by hydrogen bonding the oxygen to an acid or hydrogen bond donor.
4. Nucleophilic activation of the amide by placing the negative end of bond dipoles near the oxygen of the carbonyl.

B. Several compounds have been found to inhibit β-lactamase, and drugs based on these compounds can be taken in combination with penicillins and cephalosporins to restore their effectiveness when resistance is known to be a problem. The commonly prescribed formulation called Augmentin is a combination of the β-lactamase inhibitor shown below with amoxicillin (shown above). It is used to treat childhood ear infections when resistance is suspected, and many kids know it as the white liquid that tastes like bananas. Which of the statements below are true statements?

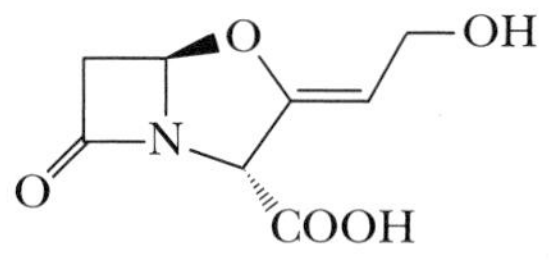
O
OH
N
O
COOH

A β-lactamase inhibitor

1. The stereochemistry of the fusion between the four-and five-membered rings in the inhibitor and amoxicillin are different.
2. The inhibitor possesses enol ether and allylic alcohol functional groups while the antibiotic possesses a phenol and a secondary amide functional group.
3. Neither the inhibitor nor the antibiotic contains strained rings.
4. Both 1 and 2 are true.

C. The formation of the amide that is not part of the β-lactam ring is commonly near the end the synthesis of the penicillins and cephalosporins. Answer true or false about each of the following statements.

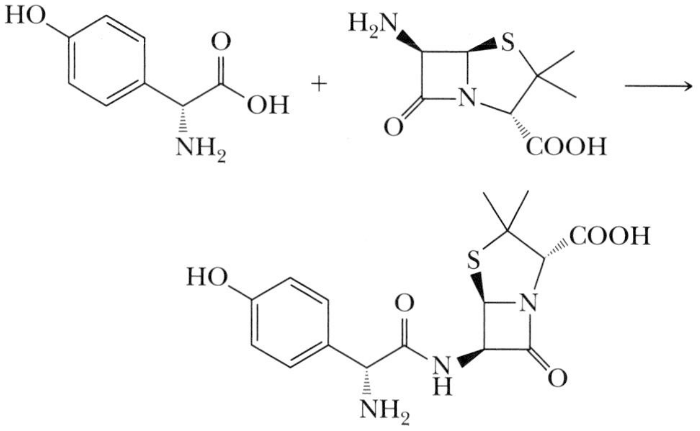
HO
O
OH
NH2
+
H2N
S
N
O
COOH
COOH
S
HO
O
N
N
H
O
NH2

1. An acid chloride could be used as the functional derivative of the carboxylic acid in the first reactant to couple with the amine on the β-lactam ring without protection

of the amine, but the phenol would need protecting.

2. An anhydride could be used as the functional derivative of the carboxylic acid in the first reactant to couple with the amine on the β-lactam ring without protection of the phenol, but the amine would need protecting.
3. Irrespective of which functional derivative of the carboxylic acid in the first reactant is generated before the above reaction is performed, the carboxylic acid in the second reaction does not need protecting.
4. The thioether is not nucleophilic enough to disturb the amide formation step, and therefore does not need to be protected.

D. The following sequence of steps is used to create the carboxylic acids that can be used as the first reactant, where X can be a wide variety of groups. Answer true or false about each of the following statements.

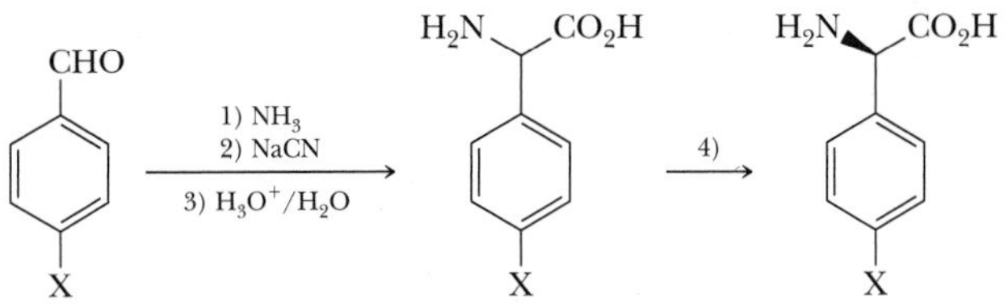

1. Step 1 would best be performed at a pH below the pK_a of ammonium.
2. Step 1 would create an imine that undergoes nucleophilic attack by cyanide in Step 2.
3. The hydrolysis of the nitrile created in Step 2 would be better performed in base rather than the acid shown.
4. Step 4 is a resolution of a racemic mixture that results from Steps 1–3 and does not require a chiral reagent.

18.9 Reactions with Organometallic Compounds

A. Grignard Reagents

Treating a formic ester with two moles of a Grignard reagent followed by hydrolysis of the magnesium alkoxide salt in aqueous acid gives a secondary alcohol.

$$\underset{\text{An ester of formic acid}}{HC(=O)OCH_3} + 2\,RMgX \longrightarrow \text{magnesium alkoxide salt} \xrightarrow{H_2O,\ HCl} \underset{\text{A 2° alcohol}}{HC(OH)(R)-R} + CH_3OH$$

Treating an ester other than a formate with a Grignard reagent gives a tertiary alcohol in which two of the groups bonded to the carbon bearing the —OH group are the same.

$$\underset{\text{An ester of any acid other than formic acid}}{CH_3C(=O)OCH_3} + 2\,RMgX \longrightarrow \text{magnesium alkoxide salt} \xrightarrow{H_2O,\ HCl} \underset{\text{A 3° alcohol}}{CH_3C(OH)(R)-R} + CH_3OH$$

Reaction of an ester with a Grignard reagent involves formation of two successive tetrahedral carbonyl addition intermediates.

Mechanism 18.8

Reaction of an Ester with a Grignard Reagent

Step 1: **Make a new bond between a nucleophile and an electrophile.** The reaction begins with addition of one mole of Grignard reagent to the carbonyl carbon to form a tetrahedral carbonyl addition intermediate.

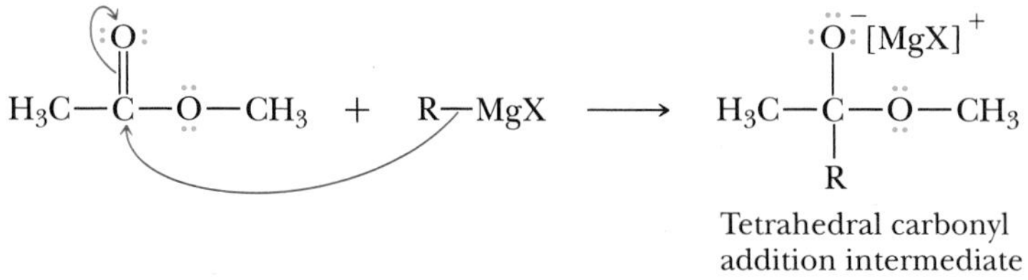

Step 2: **Break a bond to give stable molecules or ions.** Because an alkoxide ion is a moderately good leaving group from a tetrahedral carbonyl addition intermediate, this intermediate collapses to give a ketone and a magnesium alkoxide salt. To this point in the mechanism, we have nucleophilic acyl substitution.

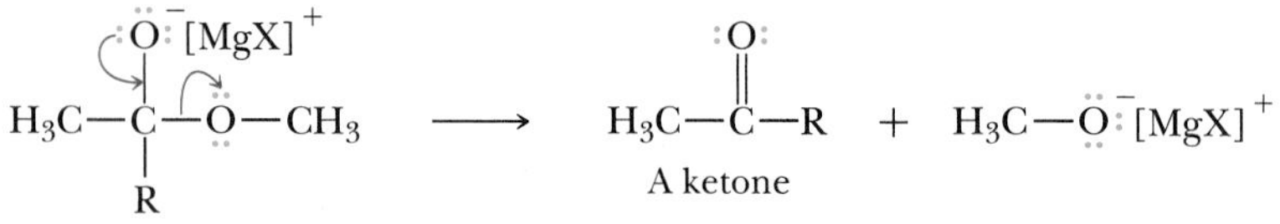

Step 3: **Make a new bond between a nucleophile and an electrophile.** The ketone reacts with a second mole of Grignard reagent to form a second tetrahedral carbonyl addition compound.

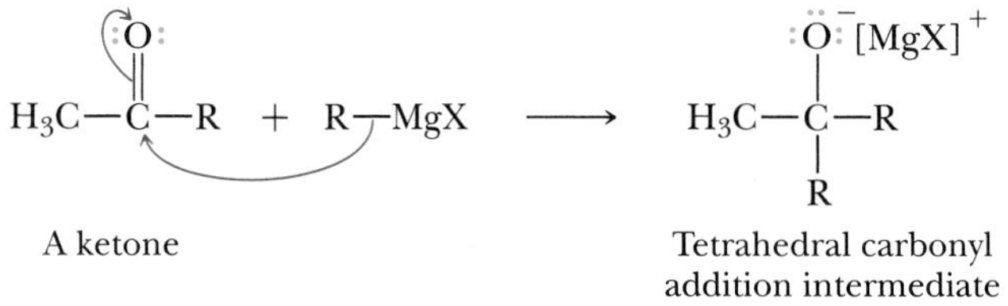

Step 4: **Add a proton.** The chemist adds aqueous acid to the reaction, and the resulting hydrolysis gives a tertiary alcohol. These last two steps constitute nucleophilic acyl addition.

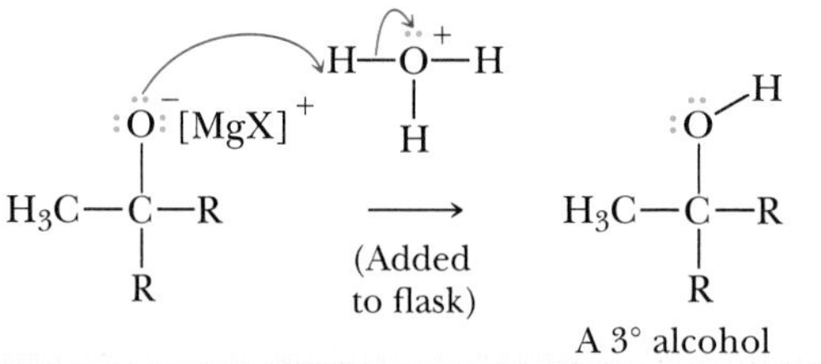

It is important to realize that it is not possible to use RMgX and an ester to prepare a ketone; the intermediate ketone is more reactive than the ester and reacts immediately with the Grignard reagent to give a tertiary alcohol.

Example 18.8 Reacting Esters with Grignard Reagents

Complete each Grignard reaction.

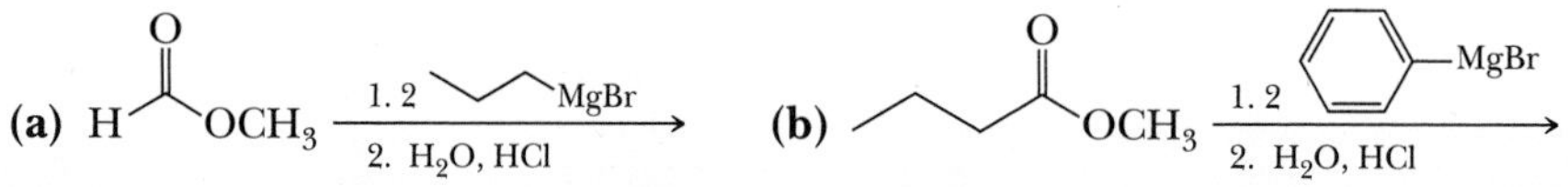

Solution

Sequence (a) gives a secondary alcohol, and sequence (b) gives a tertiary alcohol.

(a) 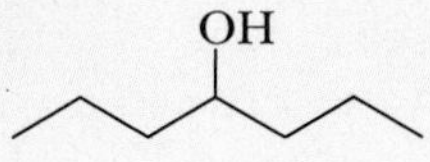
OH

(b) 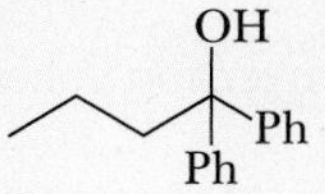
OH
Ph
Ph

Problem 18.8

Show how to prepare each alcohol by treating an ester with a Grignard reagent.

(a) 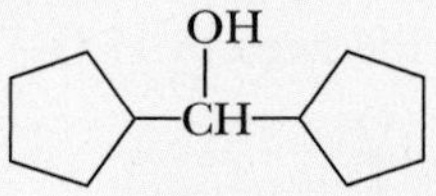
OH
CH

(b) 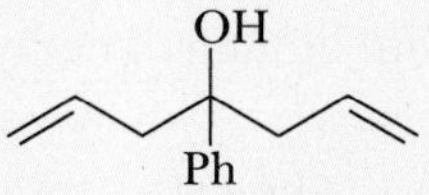
OH
Ph

B. Organolithium Compounds

Organolithium compounds are even more powerful nucleophiles than Grignard reagents and react with esters to give the same types of secondary and tertiary alcohols as shown for Grignard reagents, often in higher yields.

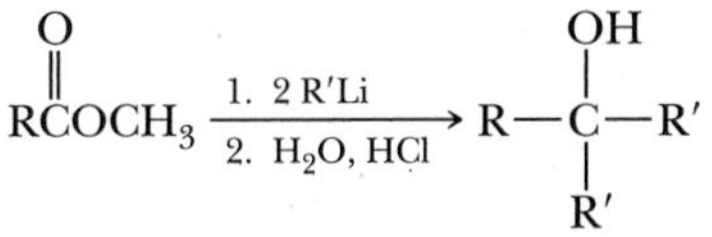
O
RCOCH3
1. 2 R'Li
2. H2O, HCl
OH
R—C—R'
R'

C. Lithium Diorganocuprates

Acid chlorides react readily with lithium diorganocopper (Gilman) reagents (Section 15.2) to give ketones, as illustrated by the conversion of pentanoyl chloride to 2-hexanone. The reaction is carried out at −78°C in either diethyl ether or tetrahydrofuran. Following hydrolysis in aqueous acid, the ketone is isolated in good yield.

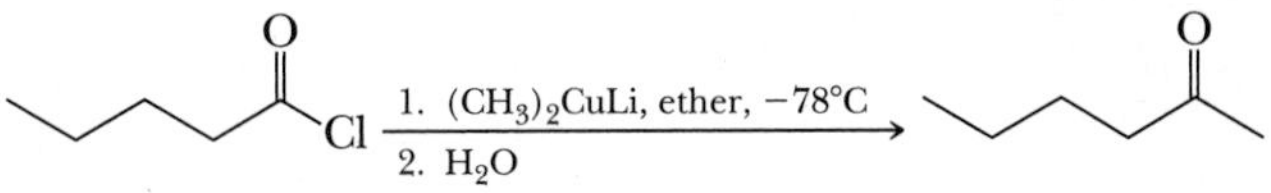
O
Cl
1. (CH3)2CuLi, ether, −78°C
2. H2O
O

Pentanoyl chloride **2-Hexanone**

Notice that under these conditions, the ketone does not react further. This contrasts with the reaction of an ester with a Grignard reagent or an organolithium compound, where the intermediate ketone reacts with a second mole of the organometallic compound to give an alcohol. The reason for this difference in reactivity is that the tetrahedral carbonyl addition intermediate in a diorganocuprate reaction is stable at −78°C; it survives until the workup causes it to decompose to the ketone, at which point the Gilman reagent has been destroyed.

R_2CuLi reagents react readily only with the very reactive acid chlorides; they do not react with aldehydes, ketones, esters, amides, acid anhydrides, or nitriles. The following compound contains both an acid chloride and an ester group. When treated with lithium dimethylcopper, only the acid chloride reacts.

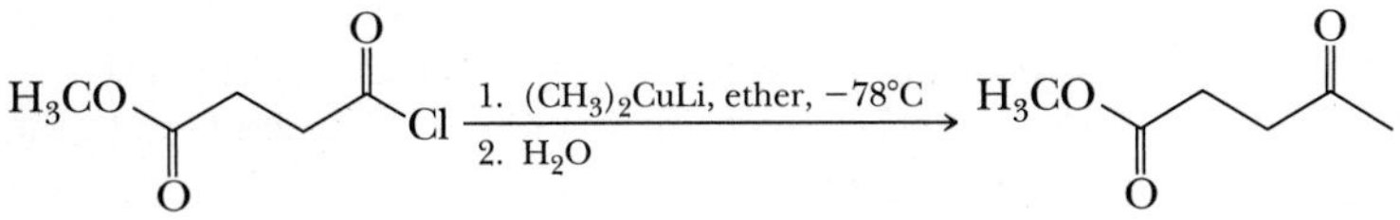
H3CO
O
O
Cl
1. (CH3)2CuLi, ether, −78°C
2. H2O
H3CO
O
O

Example 18.9 Reactions of Carboxylic Acid Derivatives I

Show how to bring about each conversion in good yield.

(a) 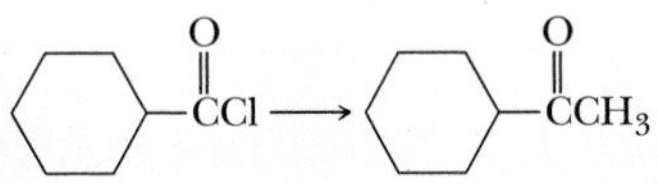 **(b)** 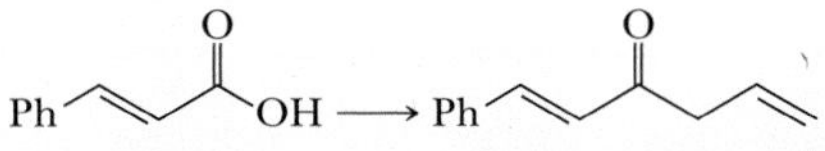

Solution

(a) Treat the acid chloride with lithium dimethylcopper followed by H_2O.

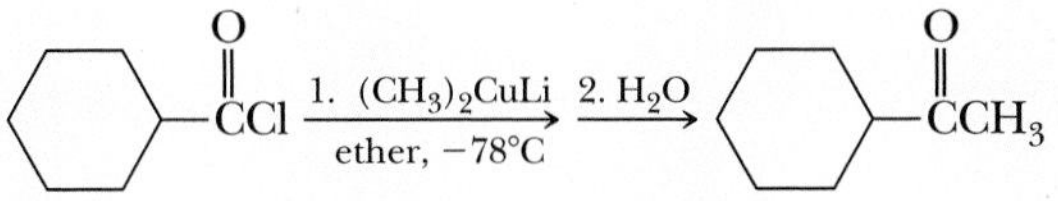

(b) Treat the carboxylic acid with thionyl chloride to form the acid chloride, followed by treatment with lithium diallylcopper and then aqueous acid.

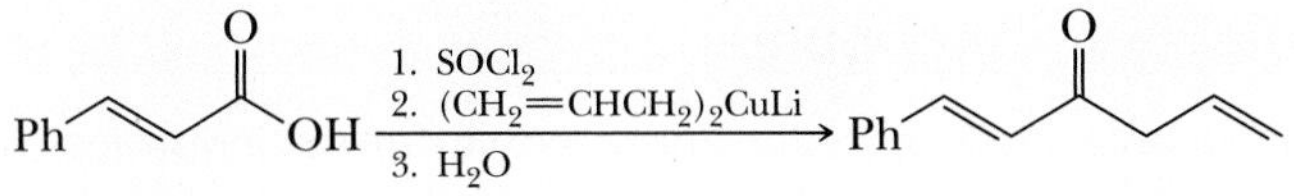

Problem 18.9

Show how to bring about each conversion in good yield.

(a) 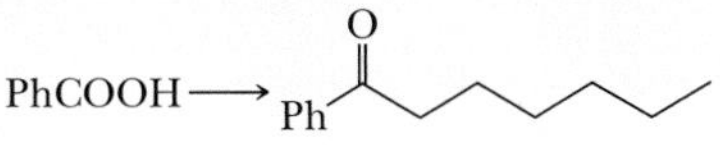 **(b)** 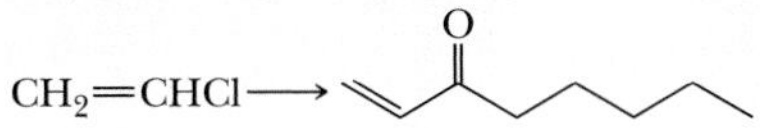

18.10 Reduction

Most reductions of carbonyl compounds, including aldehydes and ketones, are now accomplished by transfer of hydride ions from boron or aluminum hydrides. We have already seen the use of sodium borohydride to reduce the carbonyl group of aldehydes and ketones to hydroxyl groups (Section 16.11A) and the use of lithium aluminum hydride to reduce not only aldehyde and ketone carbonyl groups but also carboxyl groups to hydroxyl groups (Section 17.6A).

A. Esters

Lithium aluminum hydride reduces an ester to two alcohols; the alcohol derived from the acyl group is primary and is usually the objective of the reduction.

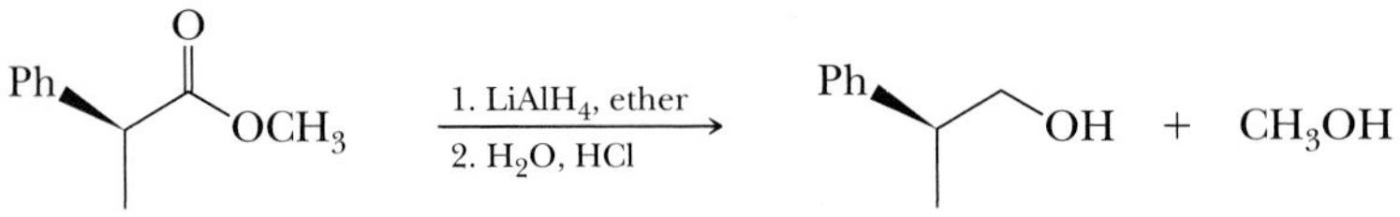

Methyl (*S*)-2-phenylpropanoate **(*S*)-2-Phenyl-1-propanol** **Methanol**

Mechanism 18.9

Reduction of an Ester by Lithium Aluminum Hydride

As you study this mechanism, note that Steps 1 and 3 are closely analogous to the reaction of Grignard reagents with an ester, with the exception that a hydride ion rather than a carbanion is being donated to the carbonyl carbon.

Step 1: **Make a new bond between a nucleophile and an electrophile.** Nucleophilic addition of hydride ion to the carbonyl carbon gives a tetrahedral carbonyl addition intermediate. The hydride ion is not free but is donated by the AlH_4^- ion. The resulting alkoxide ion forms a Lewis acid-base adduct with the AlH_3.

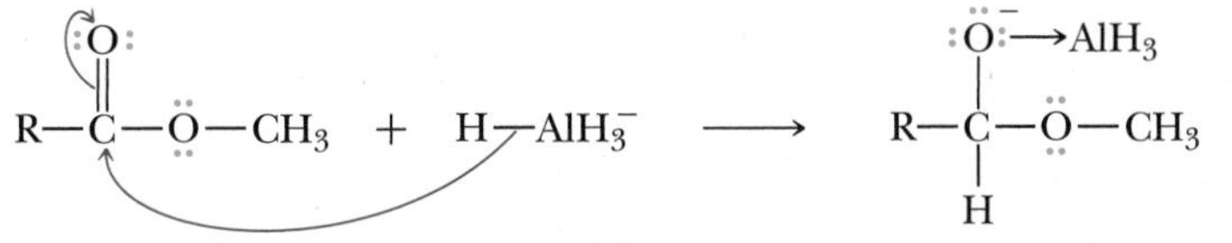

Tetrahedral carbonyl addition intermediate

Step 2: **Break a bond to give stable molecules or ions.** Collapse of this intermediate by loss of alkoxide ion gives a new carbonyl-containing compound.

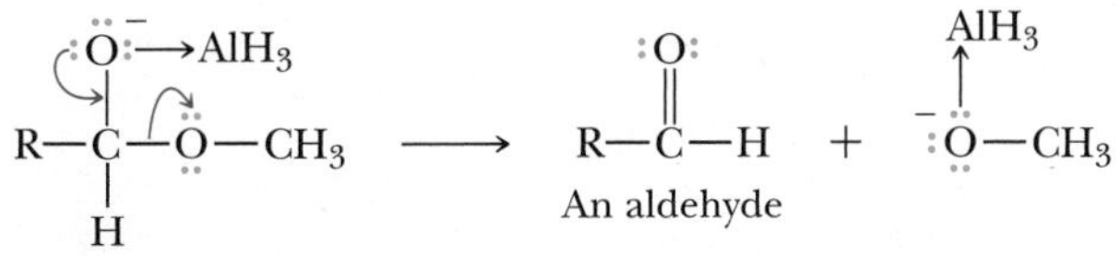

Step 3: **Make a new bond between a nucleophile and an electrophile.** Nucleophilic addition of a second hydride ion to the newly formed carbonyl group gives an alkoxide ion.

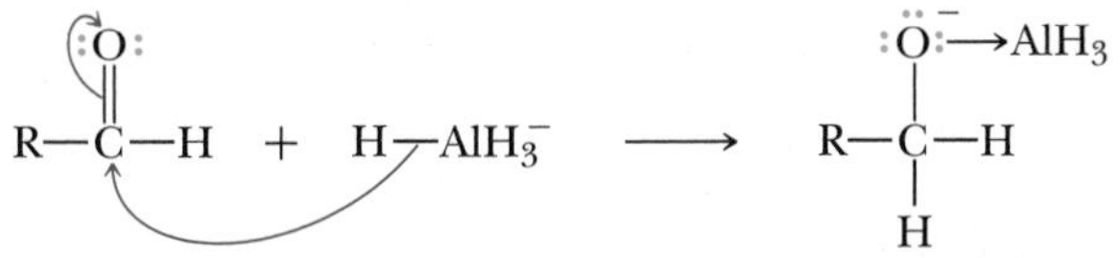

Tetrahedral carbonyl addition intermediate

Step 4: **Add a proton.** The chemist adds water to the reaction, and the resulting hydrolysis gives a primary alcohol. Any Al—H bonds are converted to aluminum hydroxides.

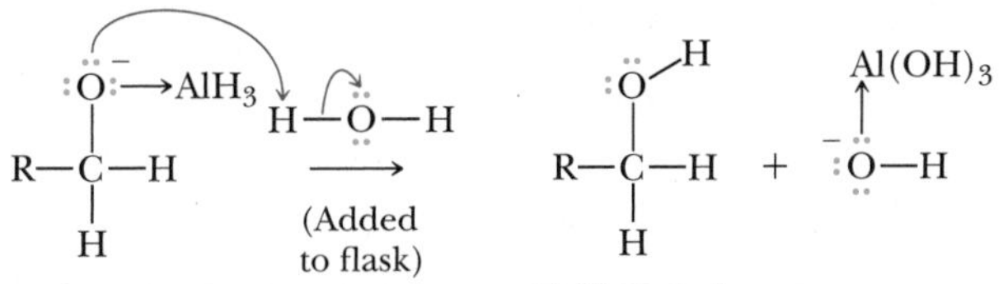

A 1° alcohol

Sodium borohydride is not normally used to reduce esters because the reaction is very slow. Because of this lower reactivity of sodium borohydride toward esters, it is possible to reduce the carbonyl group of an aldehyde or a ketone to a hydroxyl group with this reagent without reducing an ester or carboxyl group in the same molecule.

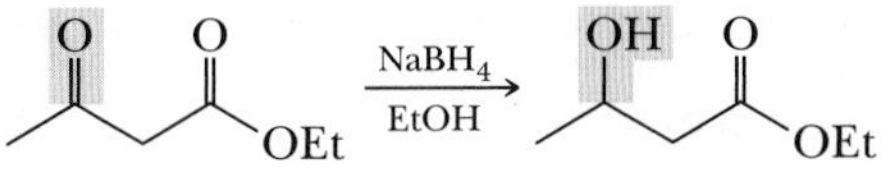

(racemic)

Reduction of an ester to a primary alcohol can be viewed as two successive hydride ion transfers, as shown in the mechanism we just presented. Chemists wondered if it might be possible to modify the structure of the reducing agent to reduce an ester to an aldehyde and no further. A useful modified hydride-reducing agent developed for this purpose is diisobutylaluminum hydride (DIBALH).

$[(CH_3)_2CHCH_2]_2AlH$

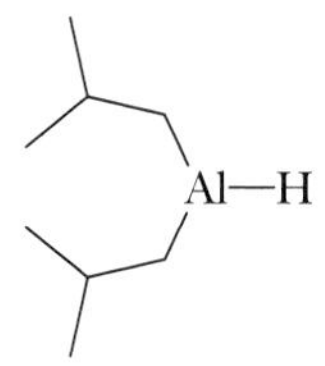

Diisobutylaluminum hydride (DIBALH)

DIBALH reductions are typically carried out in toluene or hexane at −78°C (dry ice/acetone temperature) followed by warming to room temperature and adding aqueous acid to hydrolyze the aluminum salts and liberate the aldehyde. Reduction of esters using DIBALH has become a valuable method for the synthesis of aldehydes, as illustrated by the conversion of methyl hexanoate to hexanal.

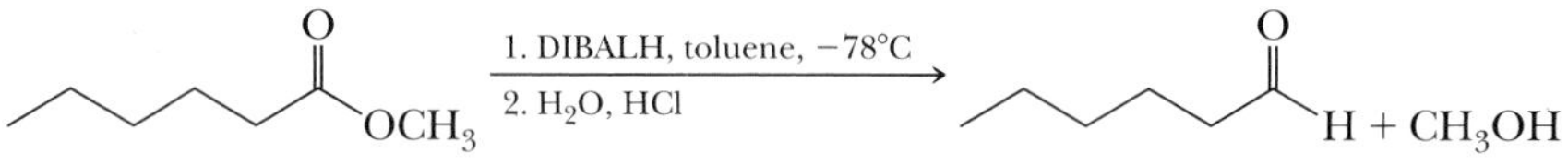

Methyl hexanoate **Hexanal**

If a DIBALH reduction of an ester is carried out at room temperature, the ester is reduced to a primary alcohol. At low temperature, the tetrahedral carbonyl addition intermediate does not eliminate alkoxide ion, and the more reactive aldehyde is not formed until after workup, when the hydride ion has been destroyed. Thus, temperature control is critical for the selective reduction of an ester to an aldehyde.

B. Amides

Lithium aluminum hydride reduction of amides can be used to prepare 1°, 2°, or 3° amines, depending on the degree of substitution of the amide.

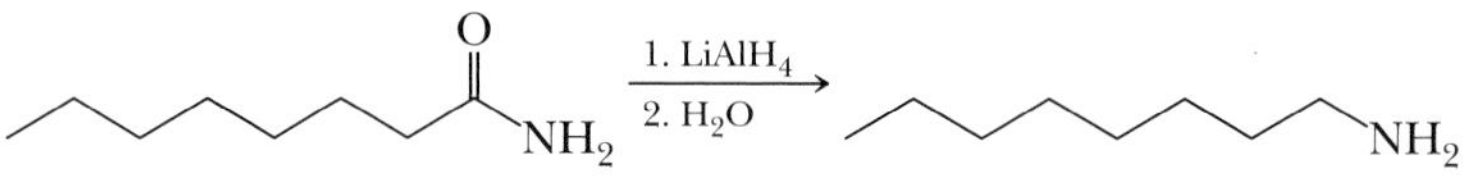

Octanamide **1-Octanamine**

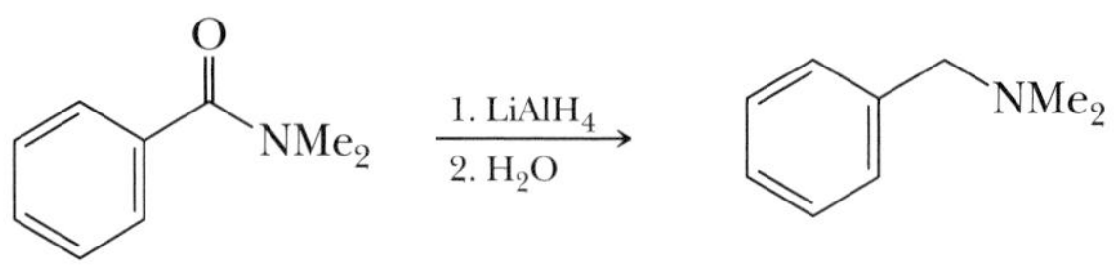

***N,N*-Dimethylbenzamide** ***N,N*-Dimethylbenzylamine**

The mechanism for the reduction of an amide to an amine is shown here divided into four steps.

Mechanism 18.10

Reduction of an Amide by Lithium Aluminum Hydride

Step 1: **Make a new bond between a nucleophile and an electrophile.** Hydride ion adds to the carbonyl carbon. The resulting alkoxide forms a Lewis acid-base adduct.

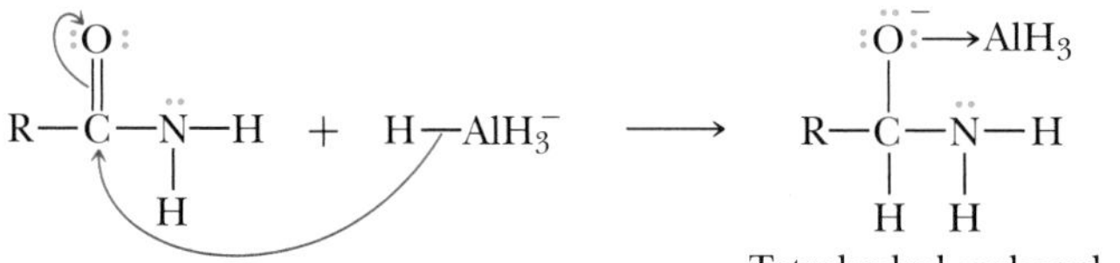

Tetrahedral carbonyl addition intermediate

Step 2: **Break a bond to give stable molecules or ions.** Rearrangement of electron pairs ejects H_3AlO^{2-} and generates an iminium ion. Because aluminum hydroxides are somewhat acidic, H_3AlO^{2-} is a reasonably good leaving group.

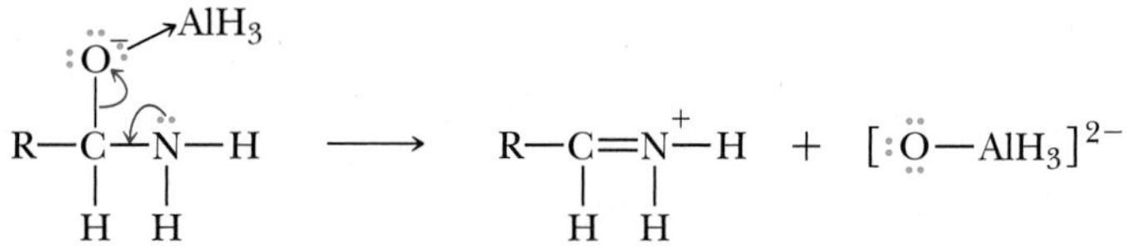

An iminium ion

Step 3: **Make a new bond between a nucleophile and an electrophile.** In the final step, the iminium ion adds a second hydride ion to complete the reduction.

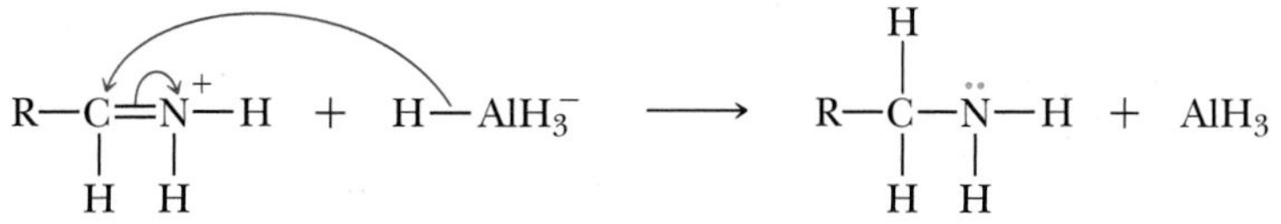

A 1° amine

Example 18.10 Reactions of Carboxylic Acid Derivatives II

Show how to bring about each conversion.

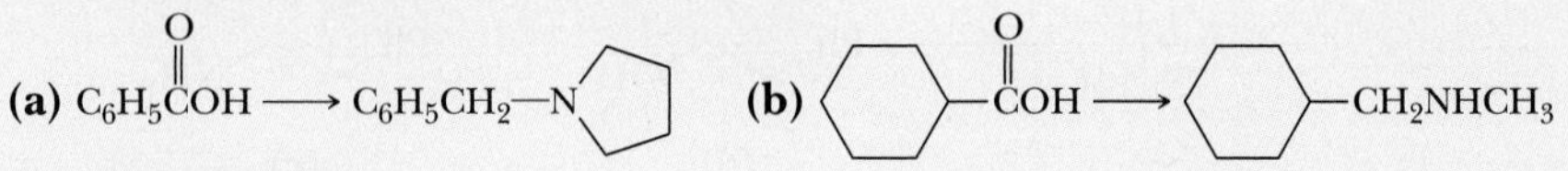

Solution

The key in each part is to convert the carboxylic acid to an amide and then to reduce the amide with $LiAlH_4$. The amide can be prepared by treating the carboxylic acid with $SOCl_2$ to give the acid chloride (Section 17.8) and then treating the acid chloride with an amine (Section 18.6A). Alternatively, the carboxylic acid can be converted to an ethyl ester by Fischer esterification, and the ester can then be treated with an amine to give the amide. Solution (a) uses the acid chloride route, and solution (b) uses the ester route.

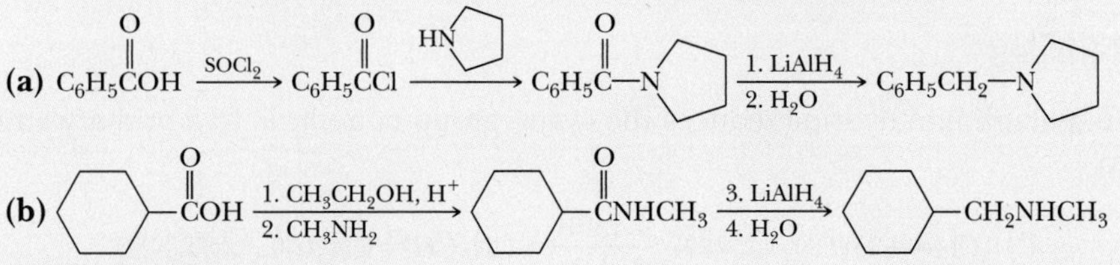

Problem 18.10

Show how to convert hexanoic acid to each amine.

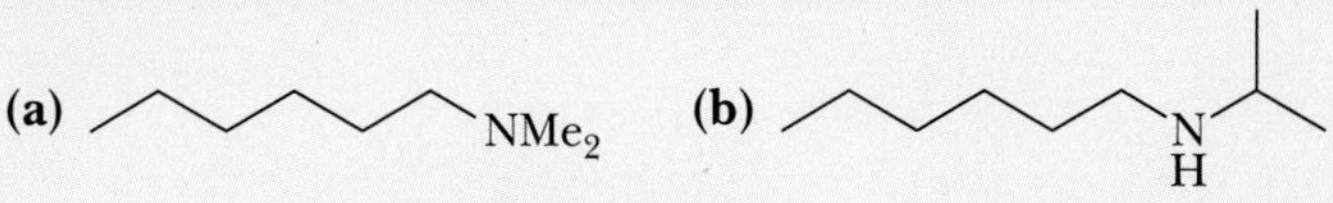

Example 18.11 Reactions of Carboxylic Acid Derivatives III

Show how to convert phenylacetic acid to each compound.

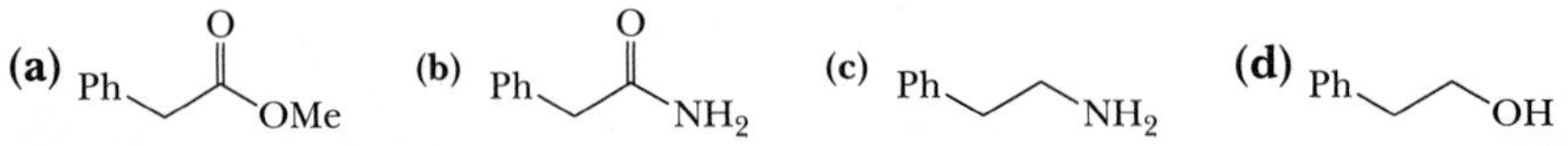

Solution

Prepare methyl ester (a) by Fischer esterification of phenylacetic acid with methanol. Then treat this ester with ammonia to prepare amide (b). Alternatively, treat phenylacetic acid with thionyl chloride (Section 17.8) to give an acid chloride. Then treat this acid chloride with ammonia to give amide (b). Reduction of the amide (b) by $LiAlH_4$ gives the primary amine (c). Similar reduction of either phenylacetic acid or ester (a) gives alcohol (d).

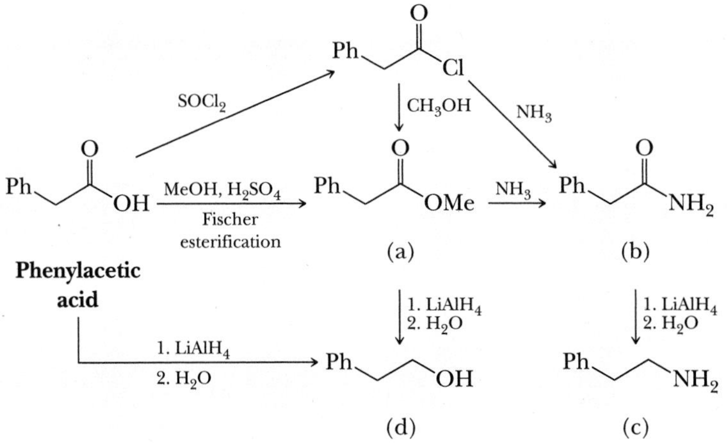

Problem 18.11

Show how to convert (*R*)-2-phenylpropanoic acid to each compound.

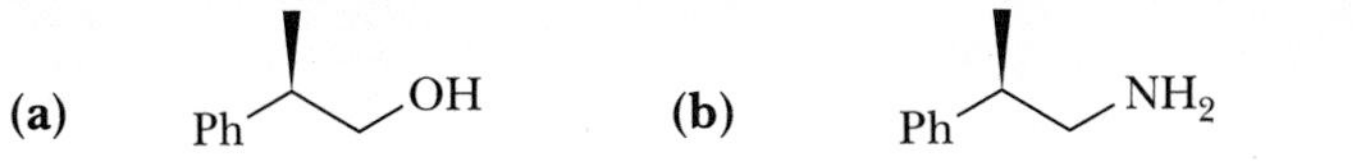

(*R*)-2-Phenyl-1-propanol **(*R*)-2-Phenyl-1-propanamine**

C. Nitriles

Lithium aluminum hydride reduces the cyano group of a nitrile to a primary amino group.

$$CH_3CH{=}CH(CH_2)_4C{\equiv}N \xrightarrow[2.\ H_2O]{1.\ LiAlH_4} CH_3CH{=}CH(CH_2)_4CH_2NH_2$$

6-Octenenitrile **6-Octen-1-amine**

Reduction of cyano groups is useful for the preparation of primary amines only.

Study Guide

18.1 Structure and Nomenclature

- An **acyl group** is $R{-}\overset{\overset{\displaystyle O}{\|}}{C}{-}$.
- The most important carboxylic acid derivatives are acid halides, acid anhydrides, esters, amides, and nitriles.
- **Acid halides** have an acyl group bonded to a halogen, usually Cl.
  - *Acid halides are named by replacing the* -ic acid *of the parent acid with* -yl halide.
- An **acid anhydride** has two acyl groups bonded to an O atom.
  - *Acid anhydrides are named by replacing the word* acid *of the parent acid with* anhydride.
- **Esters** have an acyl group bonded to an —OR group.
  - *The alkyl or aryl group bonded to the O atom is named first, followed by the name of the parent acid in which* -ic acid *is replaced by* -ate.
  - ***Lactones** are cyclic esters in which the acyl carbon and O atom are part of the ring.*
- **Amides** have an acyl group bonded to an N atom.
  - *If the N atom has one or two groups attached, the groups are named after the designator* N-*; then* -ic acid *of the parent acid is replaced by* -amide.
  - ***Lactams** are cyclic amides in which the acyl carbon and N atom are part of the ring.*
  - *In amides, the N atom is* sp^2 *hybridized, owing to a third significant resonance contributing structure that places a double bond between the acyl C and N atoms.*
  - *The acyl C—N bond has partial double bond character and a correspondingly large rotation barrier.*
  - *The amide acyl C atom is less electrophilic than that of the other carboxylic acid derivatives.*
  - *The amide N atom is not basic.*
- **Nitriles** have a cyano group bonded to carbon.
  - *Nitriles are named by replacing* -oic acid *of the parent acid with* -onitrile.

P 18.1, 18.12–18.18

18.2 Acidity of Amides, Imides, and Sulfonamides

- **Imides** have two acyl groups attached to the same N atom.
- Imides are considerably more acidic than amides due to delocalization of the negative charge of the *N*-deprotonated anion over both carbonyls.

KEY REACTIONS

1. Acidity of Imides (Section 18.2) Imides (pK_a 8–10) dissolve in aqueous NaOH by forming water-soluble sodium salts.

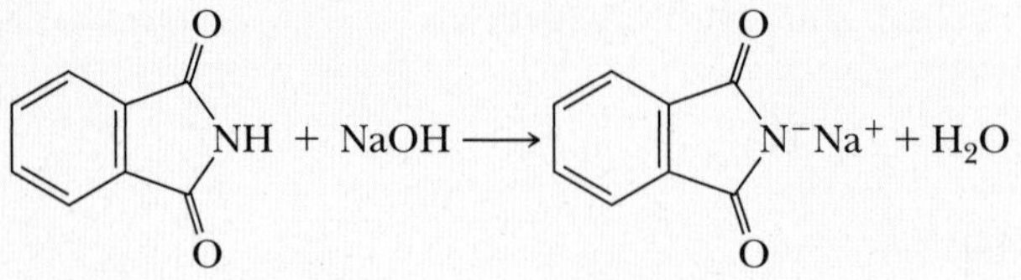

Imides are more acidic than amides because the imide anion is stabilized by delocalization of the negative charge onto the two carbonyl oxygens.

2. Acidity of Sulfonamides (Section 18.2) Sulfonamides (pK_a 9–10) dissolve in aqueous NaOH by forming water-soluble salts. The sulfonamide anion is stabilized by delocalization of the negative charge onto the two O atoms.

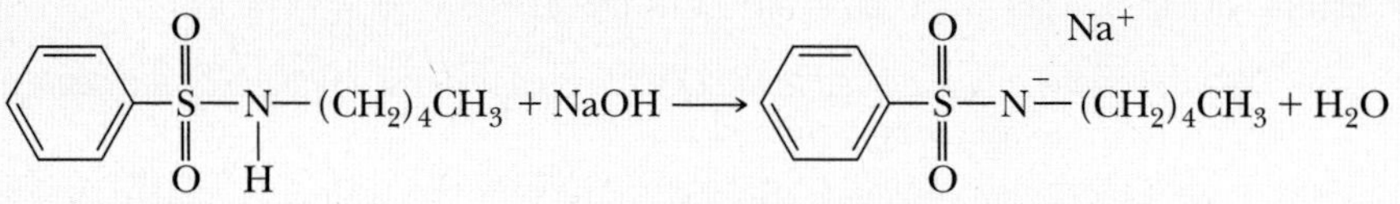

18.3 Characteristic Reactions

- The characteristic reaction of carboxylic acid derivatives is **nucleophilic acyl substitution**.
  - *A strong nucleophile adds directly to the electrophilic acyl carbon, breaking the C=O π bond, thereby creating a tetrahedral carbonyl addition intermediate.*
  - *The reaction with weaker nucleophiles can be catalyzed by acid, in which case the acyl O atom is protonated first.*
  - *The tetrahedral intermediate collapses by losing a leaving group and reforming the C=O π bond.*
  - *More stable anions are better anionic leaving groups.*
  - *Leaving group ability increases in the order $H_2N^- < RO^- < R{-}C({=}O){-}O^- < Cl^-$.*
  - *Stability imparted by resonance increases in the order of acid chloride < acid anhydride < ester < amide.*
  - *The carboxylic acid derivatives increase in reactivity in the order of amide < ester < acid anhydride < acid chloride.*

P 18.19

Acyl substitution mechanisms

18.4 Reaction with Water: Hydrolysis

- Acid chlorides and acid anhydrides react spontaneously with water to give a carboxylic acid and HCl or two molecules of carboxylic acid, respectively.

– *The reaction is catalyzed by acid, but the reaction will occur without added acid because the acid produced in the reaction catalyzes the process.*

- Acid or base is needed for the hydrolysis of esters and amides but is not required for acid halides and acid anhydrides.
  – *The role of an acid is to increase the electrophilicity of the carbonyl and to protonate the leaving group in order to facilitate its departure.*
  – *HO^- is a strong enough nucleophile that it can attack the ester or amide carbonyl directly.*
  – *One HO^- molecule is required so that the process is stoichiometric, not catalytic, in base.*
- Esters react with water in the presence of an acid catalyst to produce a carboxylic acid and alcohol. The mechanism is exactly the reverse of Fischer esterification, explaining why Fischer esterification-acid catalyzed ester hydrolysis is a reversible (equilibrium) process.
- The principle of microscopic reversibility states that the sequence of intermediates and transition states is the same, except in reverse, for the forward and reverse pathways of equilibrium reactions.
- Esters hydrolyze in base in a process called **saponification** because it has been used to hydrolyze triglyceride ester groups in soap manufacturing.
  – *A carboxylate ion is produced along with a molecule of alcohol.*
- Amides hydrolyze in either acid or base, and in both cases, the reaction is stoichiometric, not catalytic, and requires more vigorous conditions than those required to hydrolyze even esters.
- Nitriles hydrolyze in strong acid to give a carboxylic acid and ammonium ion. They hydrolyze in strong base to give a carboxylate ion and an amine.

18.21–18.24, 18.27

Mechanism 18.2

KEY REACTIONS

3. Hydrolysis of an Acid Chloride (Section 18.4A) Low-molecular-weight acid chlorides react vigorously with water.

$$CH_3\overset{O}{\overset{\|}{C}}Cl + H_2O \longrightarrow CH_3\overset{O}{\overset{\|}{C}}OH + HCl$$

Higher molecular-weight acid chlorides react less rapidly.

4. Hydrolysis of an Acid Anhydride (Section 18.4B) Acid anhydrides react readily with water.

$$CH_3\overset{O}{\overset{\|}{C}}O\overset{O}{\overset{\|}{C}}CH_3 + H_2O \longrightarrow CH_3\overset{O}{\overset{\|}{C}}OH + HO\overset{O}{\overset{\|}{C}}CH_3$$

The mechanism involves protonation of the acyl oxygen, attack by water to create the tetrahedral addition intermediate, transfer of a proton to the acyl oxygen of the leaving group, and departure of the leaving carboxylic acid.

5. Hydrolysis of an Ester (Section 18.4C) Esters are hydrolyzed only in the presence of acid or base. Acid is a catalyst. Base is required in an equimolar amount. In acid, the mechanism involves protonation of the acyl oxygen, attack by water to create the tetrahedral addition intermediate, transfer of a proton to the oxygen of the —OR group, and departure of the leaving alcohol.

(Continued)

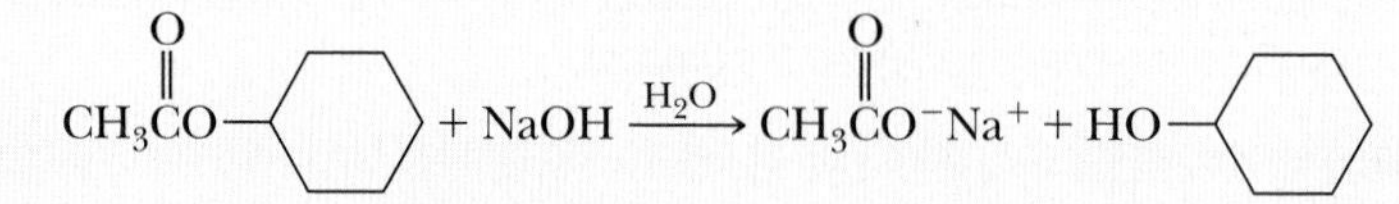

In base, the mechanism involves direct addition of the strong nucleophile HO^- to give the tetrahedral addition intermediate, followed by collapse to give a carboxylic acid and an alkoxide, which transfers a proton from the relatively acidic carboxylic acid to the relatively basic alkoxide to give the carboxylate ion and alcohol as final products.

6. **Hydrolysis of an Amide (Section 18.4D)** Either acid or base is required in an amount equivalent to that of the amide. In acid, the mechanism is similar to that for esters, except the departing amine is basic and reacts with a proton to give an ammonium ion product. This last step consumes a proton, explaining why the process is not catalytic in acid.

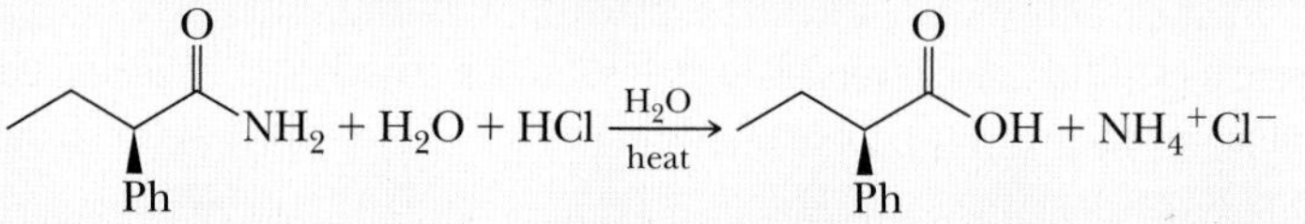

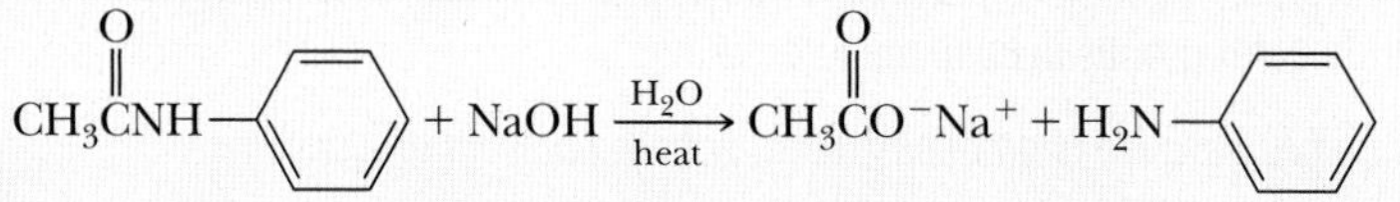

In base, the mechanism for amide hydrolysis is more complex than with esters, involving initial attack by HO^- and deprotonation to give a dianionic tetrahedral intermediate, followed by loss of an amide ion, which is immediately protonated to give the amine product.

7. **Hydrolysis of a Nitrile (Section 18.4E)** Either acid or base is required in an amount equivalent to that of the nitrile. In acid, the mechanism involves an initial protonation of the nitrile N atom, followed by attack by water to give an imidic acid that tautomerizes to give an amide, and the rest proceeds the same as for amide hydrolysis in acid.

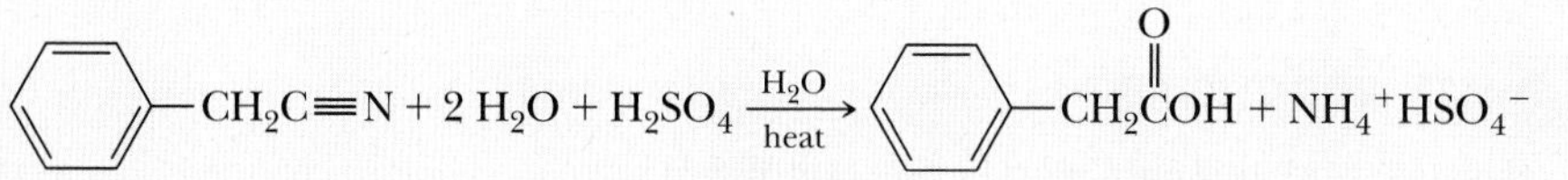

$$CH_3(CH_2)_9C{\equiv}N + H_2O + NaOH \xrightarrow[\text{heat}]{H_2O} CH_3(CH_2)_9\overset{\overset{\displaystyle O}{\|}}{C}O^-Na^+ + NH_3$$

In base, the mechanism involves an initial attack of HO^- on the nitrile C atom to form the anion of an imidic acid, which acquires a proton to give an imidic ion intermediate that tautomerizes to an amide, and the rest proceeds the same as for amide hydrolysis in base.

18.5 Reaction with Alcohols

- Acid chlorides react with alcohols to give esters and HCl. This reaction is analogous to the formation of sulfonic esters discussed in Section 10.5D. When the product ester is acid-sensitive, a base such as a tertiary amine is used to neutralize the HCl as it is formed.
- Acid anhydrides react with alcohols to give one molecule of ester and one molecule of carboxylic acid.
- Esters react with alcohols in an acid-catalyzed reaction called **transesterification**, an equilibrium process in which one ester —OR group is exchanged for another.
- Amides are not reactive enough to react with alcohols.

18.6, 18.19, 18.20, 18.27–18.30, 18.36–18.38, 18.40, 18.42, 18.48, 18.53, 18.56, 18.60–18.62, 18.68, 18.69

KEY REACTIONS

8. **Reaction of an Acid Chloride with an Alcohol (Section 18.5A)** Treating an acid chloride with an alcohol gives an ester plus HCl.

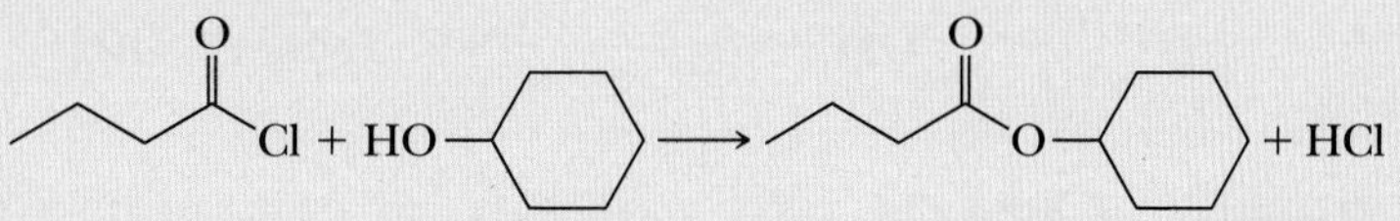

Preparation of an acid-sensitive ester is carried out using an equimolar amount of triethylamine or pyridine to neutralize the HCl.

9. **Reaction of an Acid Anhydride with an Alcohol (Section 18.5B)** Treating an acid anhydride with an alcohol gives one mole of ester and one mole of carboxylic acid.

$$CH_3\overset{O}{\overset{\|}{C}}O\overset{O}{\overset{\|}{C}}CH_3 + HOEt \longrightarrow CH_3\overset{O}{\overset{\|}{C}}OEt + CH_3\overset{O}{\overset{\|}{C}}OH$$

10. **Reaction of an Ester with an Alcohol: Transesterification (Section 18.5C)** Transesterification requires an acid catalyst and an excess of alcohol to drive the reaction to completion.

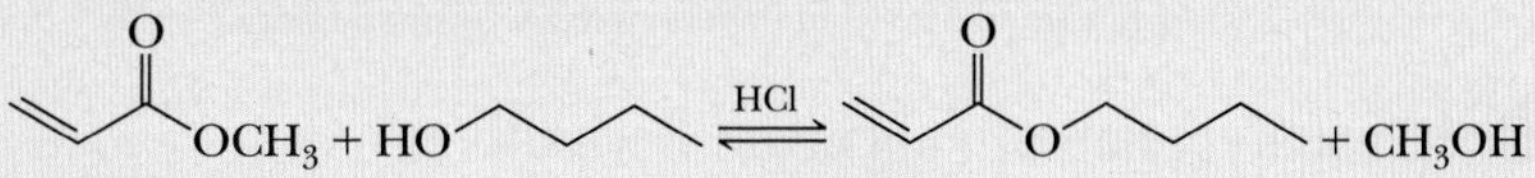

18.6 Reactions with Ammonia and Amines

- Acid chlorides react with two equivalents of ammonia and 1° and 2° amines to form an amide and one equivalent of an ammonium chloride.
- Acid anhydrides react with two equivalents of ammonia and 1° and 2° amines to form an amide and one equivalent of an ammonium carboxylate salt.
- Esters react slowly with ammonia and 1° and 2° amines to form an amide and an alcohol.

KEY REACTIONS

11. **Reaction of an Acid Chloride with Ammonia or an Amine (Section 18.6A)** Reaction requires two moles of ammonia or amine, one to form the amide and one to neutralize the HCl byproduct. The mechanism involves nucleophilic addition of ammonia or the amine to the carbonyl carbon, followed by a proton transfer to give a tetrahedral addition intermediate that can directly eliminate chloride and lose a proton to give products.

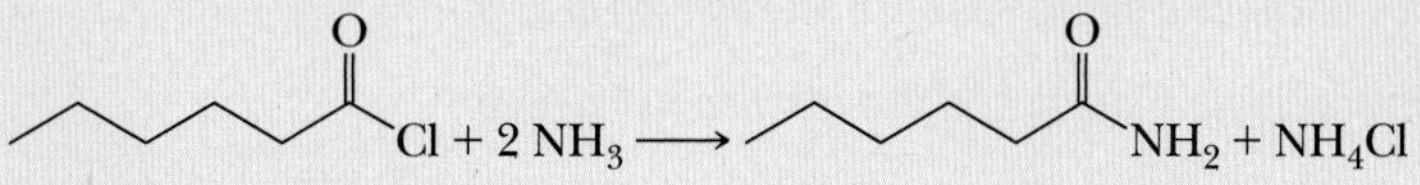

12. **Reaction of an Acid Anhydride with Ammonia or an Amine (Section 18.6B)** Reaction requires two moles of ammonia or amine, one to form the amide and one to neutralize the carboxylic acid byproduct. The mechanism is analogous to that of the acid chloride reaction with ammonia or amines.

$$CH_3\overset{O}{\overset{\|}{C}}O\overset{O}{\overset{\|}{C}}CH_3 + 2\,NH_3 \longrightarrow CH_3\overset{O}{\overset{\|}{C}}NH_2 + CH_3\overset{O}{\overset{\|}{C}}O^-NH_4^+$$

(Continued)

13. **Reaction of an Ester with Ammonia or an Amine (Section 18.6C)** Treating an ester with ammonia or a primary or secondary amine gives an amide. The mechanism involves nucleophilic addition of the ammonia or amine to the carbonyl carbon, followed by a proton transfer to give a tetrahedral addition intermediate that can directly eliminate alkoxide and lose a proton to the alkoxide to give products.

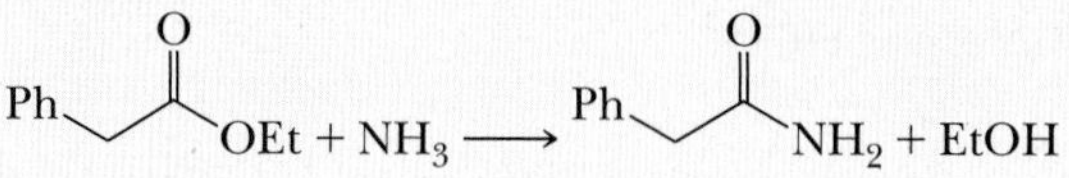

18.7 Reaction of Acid Chlorides with Salts of Carboxylic Acids

- Acid chlorides react with carboxylate anions to give acid anhydrides and a chloride salt.

KEY REACTIONS

14. **Reaction of an Acid Chloride with a Carboxylic Acid Salt (Section 18.7)** Treating an acid chloride with the salt of a carboxylic acid is a valuable method for synthesizing mixed anhydrides.

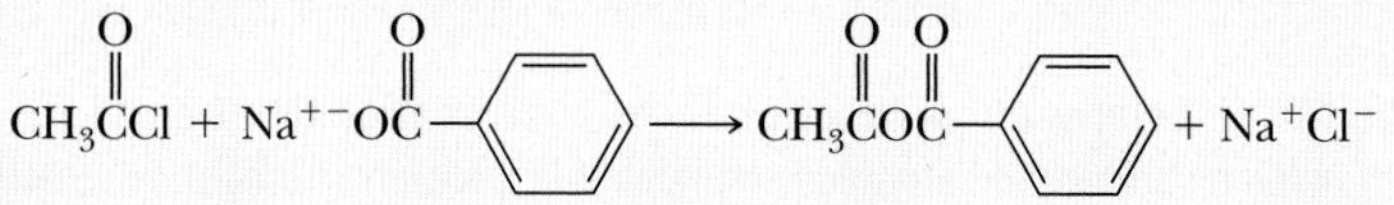

18.8 Interconversion of Functional Derivatives

- The general rule is that you can make any less reactive carboxylic acid derivative from any more reactive carboxylic acid derivative and the appropriate oxygen or nitrogen nucleophile.
  - *Acid chlorides can be used to make any of the other carboxylic acid derivatives.*
  - *Because acid chlorides can be made from carboxylic acids using $SOCl_2$ and all carboxylic acid derivatives can be hydrolyzed, it is possible to interconvert any of the carboxylic acid derivatives.*
- The mechanisms shown in the chapter are combinations of the following four steps:
  - *Putting on a proton (Add a proton)*
  - *Taking off a proton (Take a proton away)*
  - *Attack of a nucleophile on an sp^2 carbon to give a tetrahedral addition intermediate. (Make a new bond between a nucleophile and an electrophile)*
  - *Departure of a leaving group from an sp^3 carbon atom. (Break a bond to give stable molecules or ions)*
- Do not mix media in reaction mechanisms (Things You Should Know IV, Common Mistakes in Arrow Pushing).
  - *Do not create a strong base in a reaction run in acid.*
  - *Do not create a strong acid in a reaction run in base.*

P **18.6, 18.19, 18.20, 18.27–18.30, 18.36–18.38, 18.40, 18.42, 18.48, 18.53, 18.56, 18.60–18.62, 18.68, 18.69**

Figure 18.2

18.9 Reactions with Organometallic Compounds

- Esters add two molecules of a Grignard reagent, and following aqueous acid workup, they give an alcohol product in which two groups bonded to the carbon bearing the —OH group are the same.

– *Formate esters give a secondary alcohol; all other esters give a tertiary alcohol.*

– *Organolithium reagents can be used in place of the Grignard reagents to carry out the same reaction.*

- Acid chlorides react with Gilman reagents to give ketones.

P 18.8, 18.9, 18.19, 18.20, 18.32, 18.41

KEY REACTIONS

15. Reaction of an Ester with a Grignard Reagent (Section 18.9A) Treating a formic ester with a Grignard reagent followed by hydrolysis gives a secondary alcohol. Treating any other ester with a Grignard reagent gives a tertiary alcohol. The mechanism involves nucleophilic attack of the Grignard reagent on the carbonyl carbon atom to give a tetrahedral addition intermediate, which collapses with the departure of alkoxide ion to give a ketone intermediate (an aldehyde in the case of formate esters), which reacts with a second mole of Grignard reagent.

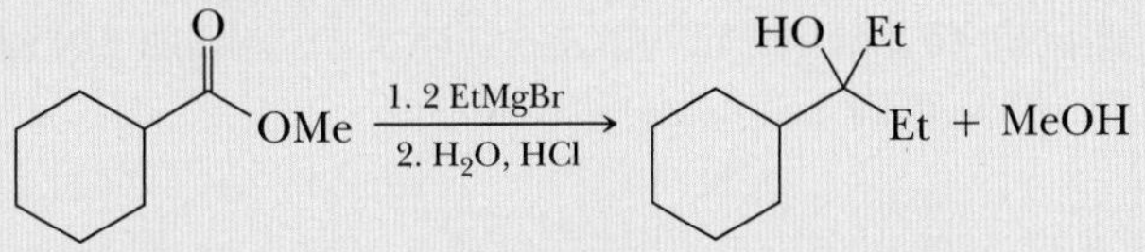
O
OMe
1. 2 EtMgBr
2. H2O, HCl
HO Et
Et + MeOH

16. Reaction of an Acid Chloride with a Lithium Diorganocuprate (Section 18.9C) Acid chlorides react readily with lithium diorganocuprates at $-78°C$ to give ketones.

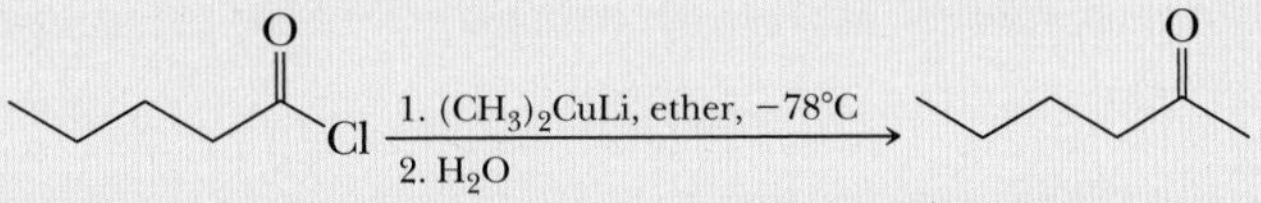
O
Cl
1. (CH3)2CuLi, ether, −78°C
2. H2O
O

18.10 Reduction

- Esters react with lithium aluminum hydride followed by aqueous acid to form two alcohols.

  – *Sodium borohydride reacts very slowly with esters and is not used for this purpose.*

- Diisobutylaluminum hydride (DIBALH) reacts with esters at low temperatures to give aldehydes. At higher temperatures, it reacts all the way to give alcohols.
- Amides react with lithium aluminum hydride to give amines.
- Lithium aluminum hydride reduces the cyano group of nitriles to a primary (1°) amine.

P 18.10, 18.11, 18.19, 18.20, 18.23–18.27, 18.32, 18.33, 18.43, 18.48, 18.52–18.54

KEY REACTIONS

17. Reduction of an Ester (Section 18.10A) Reduction of an ester by lithium aluminum hydride gives two alcohols. The mechanism involves initial nucleophilic attack by a hydride ion onto the carbonyl carbon to give a tetrahedral addition intermediate, which collapses through the loss of alkoxide to give an aldehyde, which reacts with a second hydride to give the product alcohol.

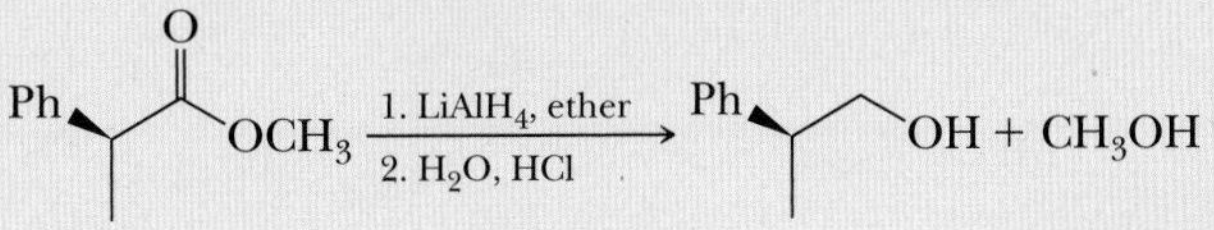
O
Ph
OCH3
1. LiAlH4, ether
2. H2O, HCl
Ph
OH + CH3OH

(Continued)

Reduction by diisobutylaluminum hydride (DIBALH) at low temperature gives an aldehyde and an alcohol.

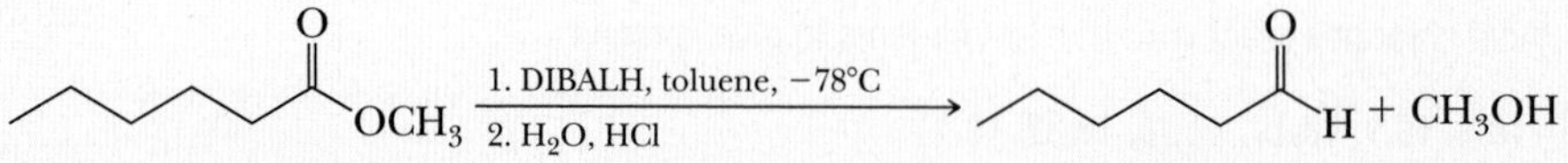

18. Reduction of an Amide (Section 18.10B) Reduction of an amide by lithium aluminum hydride gives an amine. The mechanism involves initial nucleophilic reaction of a hydride equivalent with the carbonyl carbon to give a tetrahedral addition intermediate, followed by a Lewis acid-base reaction to give an oxygen-aluminum bond that rearranges electron pairs to eject an Al-O species to give an electrophilic iminium ion, which reacts with another equivalent of hydride to give the amine product.

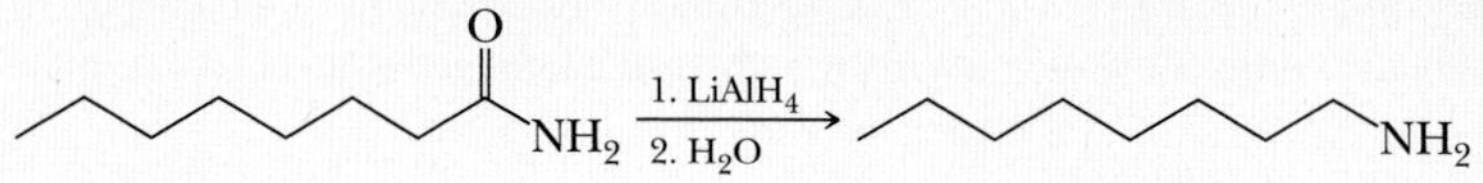

19. Reduction of a Nitrile (Section 18.10C) Reduction of a cyano group by lithium aluminum hydride gives a primary amino group.

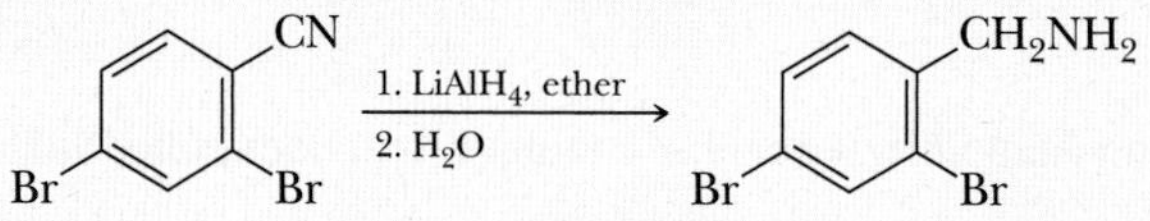

Problems

Red numbers indicate applied problems.

Structure and Nomenclature

18.12 Draw a structural formula for each compound.

(a) Dimethyl carbonate
(b) Benzonitrile
(c) Isopropyl 3-methylhexanoate
(d) Diethyl oxalate
(e) Ethyl (*Z*)-2-pentenoate
(f) Butanoic anhydride
(g) Dodecanamide
(h) Ethyl 3-hydroxybutanoate
(i) Octanoyl chloride
(j) Diethyl *cis*-1,2-cyclohexanedicarboxylate
(k) Methanesulfonyl chloride
(l) *p*-Toluenesulfonyl chloride

18.13 Write the IUPAC name for each compound.

(a) 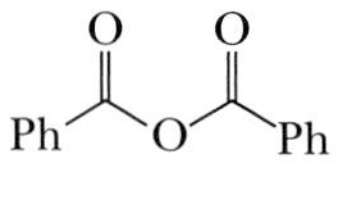

(b) 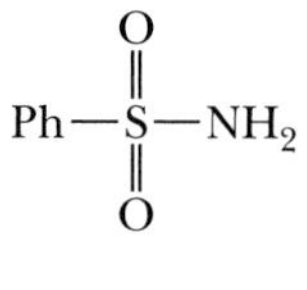

(c) 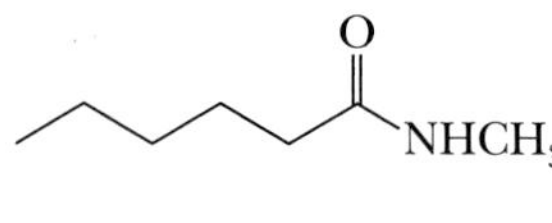

(d) 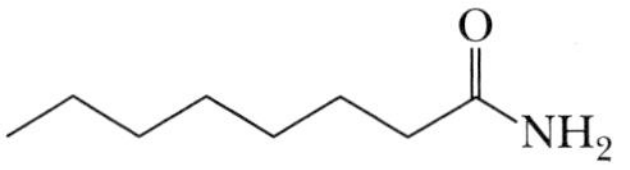

(e) 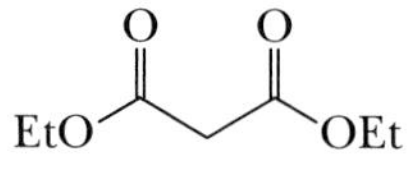

(f) 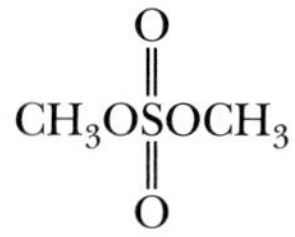

(g) 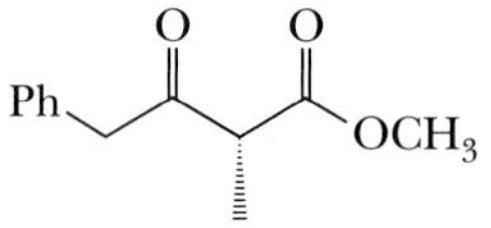

(h) 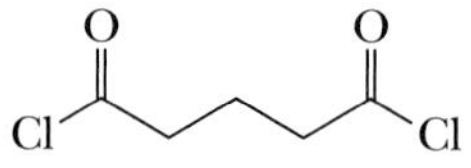

(i) $CH_3(CH_2)_5CN$

Physical Properties

18.14 Both the melting point and boiling point of acetamide are higher than those of its *N,N*-dimethyl derivative. How do you account for these differences?

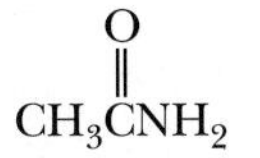

Acetamide
mp 82.3°C, bp 221.2°C

***N,N*-Dimethylacetamide**
mp −20°C, bp 165°C

Spectroscopy

18.15 Each hydrogen of a primary amide typically has a separate ^{1}H-NMR resonance, as illustrated by the separate signals for the two amide hydrogens of propanamide, which fall at δ 6.22 and δ 6.58. Furthermore, each methyl group of *N,N*-dimethylformamide has a separate resonance (δ 3.88 and δ 3.98). How do you account for these observations?

18.16 Propose a structural formula for compound A, $C_7H_{14}O_2$, consistent with its ^{1}H-NMR and IR spectra.

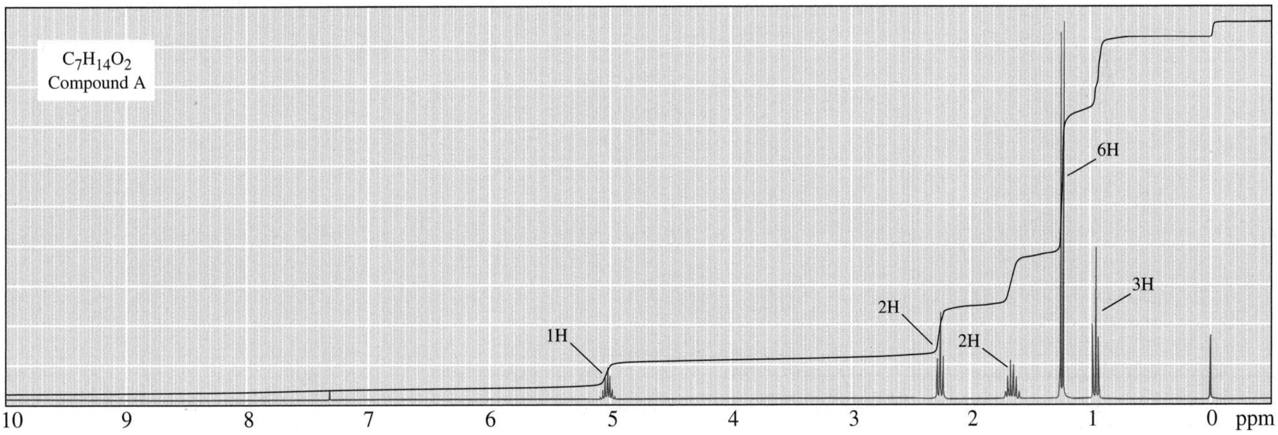

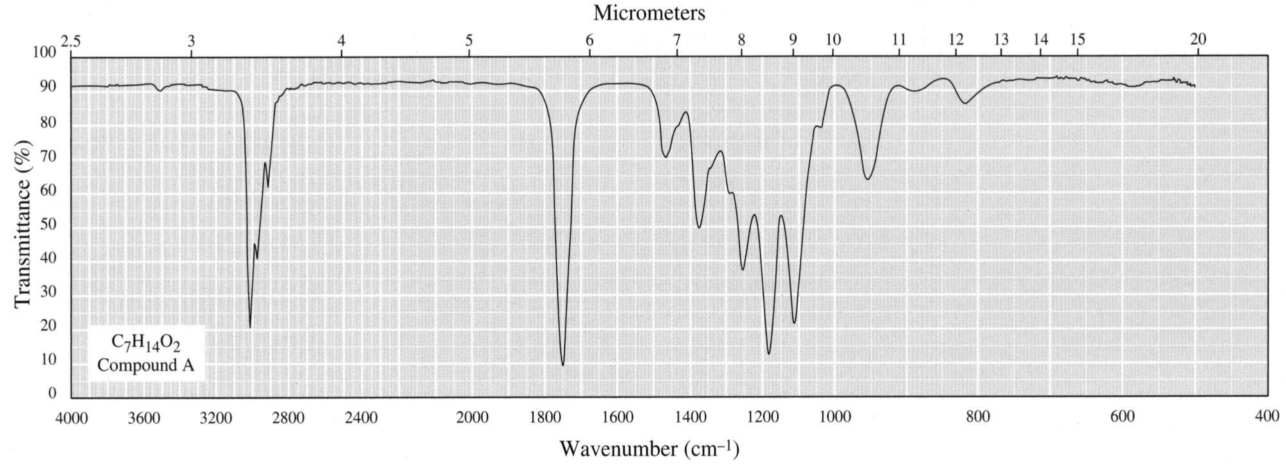

18.17 Propose a structural formula for compound B, $C_6H_{13}NO$, consistent with its ^{1}H-NMR and IR spectra.

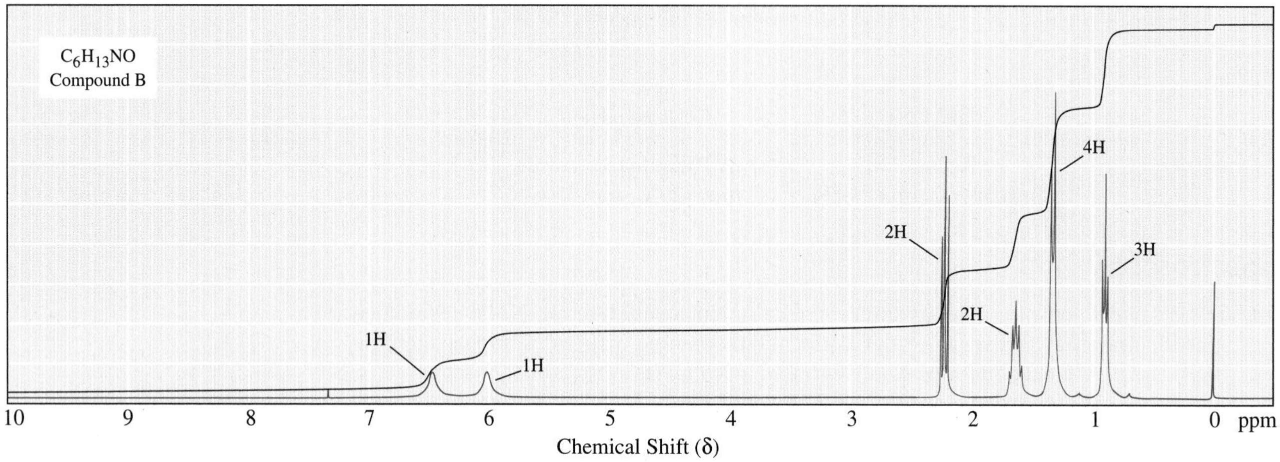

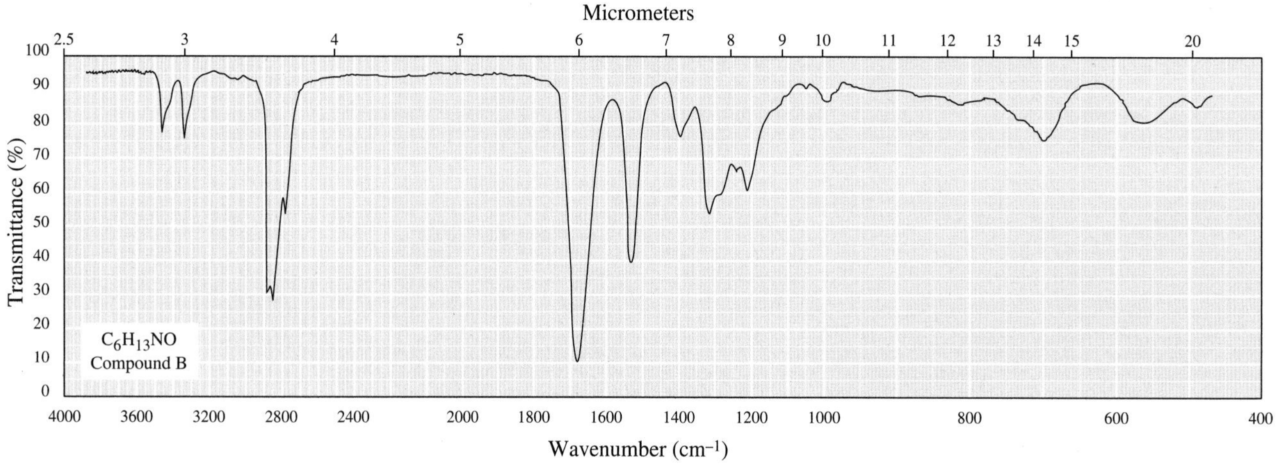

18.18 Propose a structural formula for each compound consistent with its ^{1}H-NMR and ^{13}C-NMR spectra.

(a) $C_5H_{10}O_2$

^{1}H-NMR	^{13}C-NMR
0.96 (d, 6H)	161.11
1.96 (m, 1H)	70.01
3.95 (d, 2H)	27.71
8.08 (s, 1H)	19.00

(b) $C_7H_{14}O_2$

^{1}H-NMR	^{13}C-NMR
0.92 (d, 6H)	171.15
1.52 (m, 2H)	63.12
1.70 (m, 1H)	37.31
2.09 (s, 3H)	25.05
4.10 (t, 2H)	22.45
	21.06

(c) $C_6H_{12}O_2$

^{1}H-NMR	^{13}C-NMR
1.18 (d, 6H)	177.16
1.26 (t, 3H)	60.17
2.51 (m, 1H)	34.04
4.13 (q, 2H)	19.01
	14.25

(d) $C_7H_{12}O_4$

^{1}H-NMR	^{13}C-NMR
1.28 (t, 6H)	166.52
3.36 (s, 2H)	61.43
4.21 (q, 4H)	41.69
	14.07

(e) $C_4H_7ClO_2$

^{1}H-NMR	^{13}C-NMR
1.68 (d, 3H)	170.51
3.80 (s, 3H)	52.92
4.42 (q, 1H)	52.32
	21.52

(f) $C_4H_6O_2$

^{1}H-NMR	^{13}C-NMR
2.29 (m, 2H)	177.81
2.50 (t, 2H)	68.58
4.36 (t, 2H)	27.79
	22.17

Reactions

18.19 Draw a structural formula for the principal product formed when benzoyl chloride is treated with each reagent.

(a) 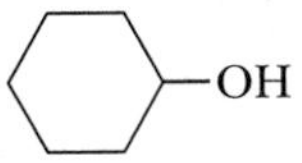

(b) 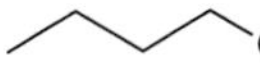OH, pyridine

(c) 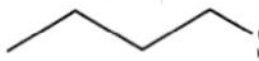SH, pyridine

(d) NH_2 (two equivalents)

(e) 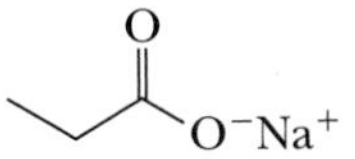

(f) $(CH_3)_2CuLi$, then H_3O^+

(g) 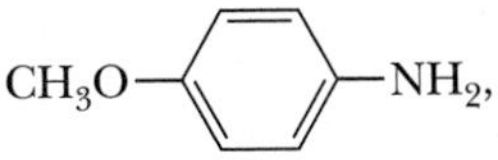, pyridine

(h) C_6H_5MgBr (two equivalents), then H_3O^+

18.20 Draw a structural formula of the principal product formed when ethyl benzoate is treated with each reagent.

(a) H_2O, NaOH, heat
(b) H_2O, H_2SO_4, heat
(c) $CH_3CH_2CH_2CH_2NH_2$
(d) DIBALH (−78°C), then H_2O
(e) $LiAlH_4$, then H_2O
(f) C_6H_5MgBr (two equivalents), then HCl/H_2O

18.21 The mechanism for hydrolysis of an ester in aqueous acid involves formation of a tetrahedral carbonyl addition intermediate. Evidence in support of this mechanism comes from an experiment designed by Myron Bender. He first prepared ethyl benzoate enriched with oxygen-18 in the carbonyl oxygen and then carried out acid-catalyzed hydrolysis of the ester in water containing no enrichment in oxygen-18. If he stopped the experiment after only partial hydrolysis and isolated the remaining ester, the recovered ethyl benzoate lost a portion of its enrichment in oxygen-18. In other words, some exchange had occurred between oxygen-18 of the ester and oxygen-16 of water. Show how this observation bears on the formation of a tetrahedral carbonyl addition intermediate during acid-catalyzed ester hydrolysis.

18.22 Predict the distribution of oxygen-18 in the products obtained from hydrolysis of ethyl benzoate labeled in the ethoxy oxygen under the following conditions.

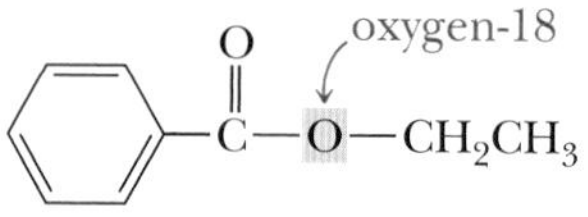

(a) In aqueous NaOH **(b)** In aqueous HCl
(c) What distribution would you predict if the reaction were done with the *tert*-butyl ester in HCl?

18.23 Draw a structural formula for the principal product formed when benzamide is treated with each reagent.

(a) H_2O, HCl, heat **(b)** NaOH, H_2O, heat **(c)** $LiAlH_4$, then H_2O

18.24 Draw a structural formula of the principal product formed when benzonitrile is treated with each reagent.

(a) H_2O (one equivalent), H_2SO_4, heat **(b)** H_2O (excess), H_2SO_4, heat
(c) NaOH, H_2O, heat **(d)** $LiAlH_4$, then H_2O

18.25 Show the product expected when the following unsaturated δ-ketoester is treated with each reagent.

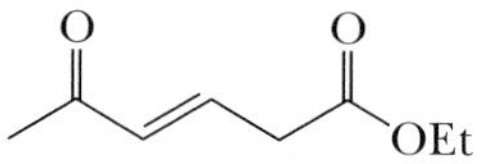

(a) $\xrightarrow[\text{Pd, EtOH}]{H_2\text{ (1 mol)}}$ **(b)** $\xrightarrow[CH_3OH]{NaBH_4}$ **(c)** $\xrightarrow[\text{2. } H_2O]{\text{1. } LiAlH_4\text{, THF}}$ **(d)** $\xrightarrow[\text{2. } H_2O]{\text{1. DIBALH, } -78°}$

18.26 The reagent diisobutylaluminum hydride (DIBALH) reduces esters to aldehydes. When nitriles are treated with DIBALH followed by mild acid hydrolysis, the product is also an aldehyde. Propose a mechanism for this reduction.

18.27 Show the product of treating this anhydride with each reagent.

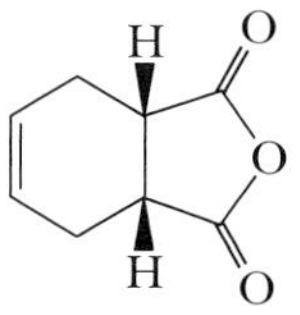

(a) $\xrightarrow[\text{heat}]{H_2O\text{, HCl}}$ **(b)** $\xrightarrow[\text{heat}]{H_2O\text{, NaOH}}$ **(c)** $\xrightarrow[\text{2. } H_2O]{\text{1. } LiAlH_4}$

(d) $\xrightarrow{CH_3OH}$ **(e)** $\xrightarrow{NH_3\text{ (2 mol)}}$

18.28 The analgesic acetaminophen is synthesized by treating 4-aminophenol with one equivalent of acetic anhydride. Draw a structural formula for acetaminophen.

18.29 Treating choline with acetic anhydride gives acetylcholine, a neurotransmitter. Write an equation for the formation of acetylcholine.

$$(CH_3)_3\overset{+}{N}CH_2CH_2OH$$

Choline

18.30 Nicotinic acid, more commonly named niacin, is one of the B vitamins. Show how nicotinic acid can be converted to (a) ethyl nicotinate and then to (b) nicotinamide.

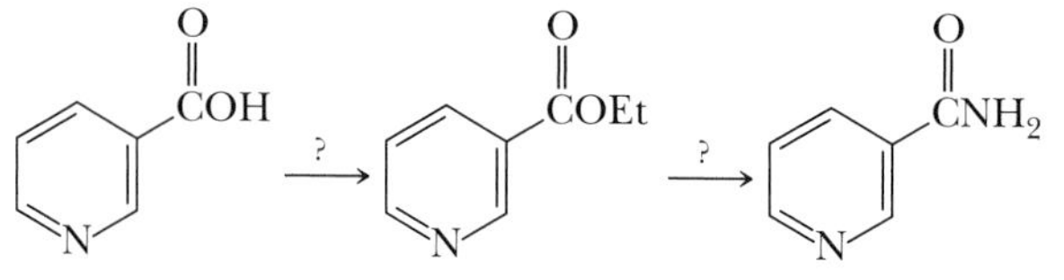

Nicotinic acid (Niacin) **Ethyl nicotinate** **Nicotinamide**

18.31 Complete each reaction.

(a) 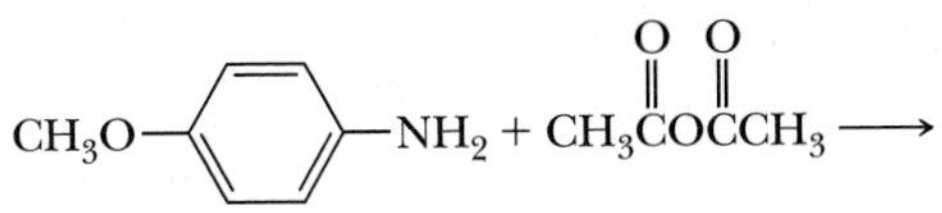

(b) 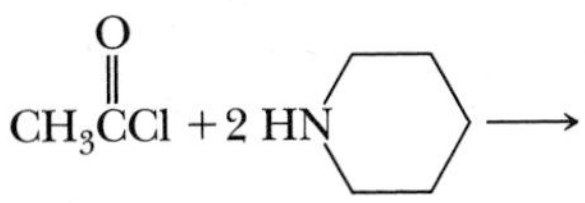

(c) 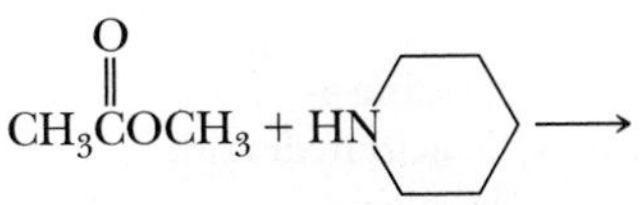

(d) 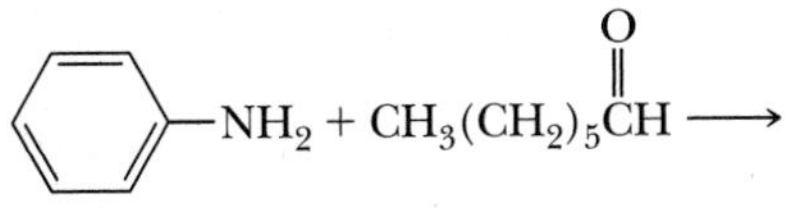

18.32 Show the product of treating γ-butyrolactone with each reagent.

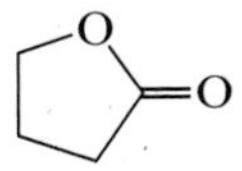

(a) $\xrightarrow{NH_3}$

(b) $\xrightarrow[2.\ H_2O]{1.\ LiAlH_4}$

(c) $\xrightarrow[2.\ H_2O,\ HCl]{1.\ 2\ PhMgBr,\ ether}$

(d) $\xrightarrow[heat]{NaOH/H_2O}$

(e) $\xrightarrow[2.\ H_2O,\ HCl]{1.\ 2\ CH_3Li,\ ether}$

(f) $\xrightarrow[2.\ H_2O,\ HCl]{1.\ DIBALH,\ ether,\ -78°C}$

18.33 Show the product of treating the following γ-lactam with each reagent.

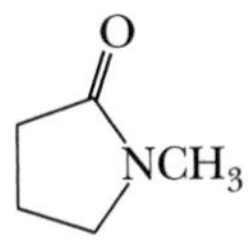

(a) $\xrightarrow[heat]{H_2O,\ HCl}$

(b) $\xrightarrow[heat]{H_2O,\ NaOH}$

(c) $\xrightarrow[2.\ H_2O]{1.\ LiAlH_4}$

18.34 Draw structural formulas for the products of complete hydrolysis of meprobamate, phenobarbital, and pentobarbital in hot aqueous acid.

(a) 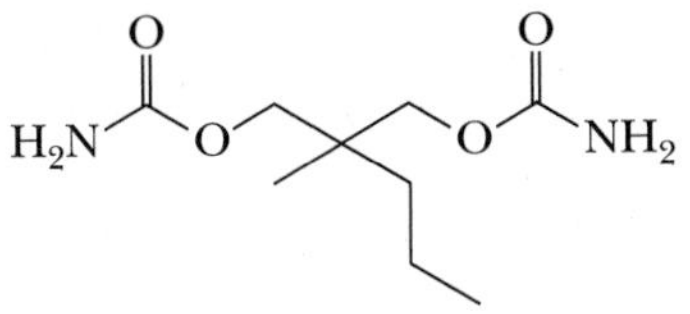

Meprobamate
(racemic)

(b) 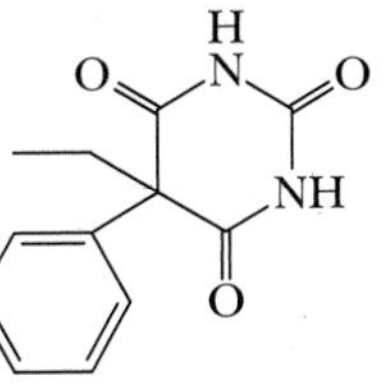

Phenobarbital
(racemic)

(c) 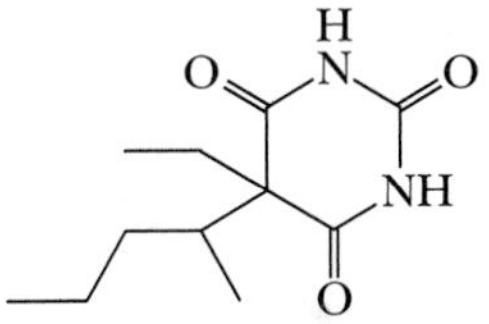

Pentobarbital
(racemic)

Meprobamate is a tranquilizer prescribed under 58 different trade names, including Equanil and Miltown. Phenobarbital is a long-acting sedative, hypnotic, and anticonvulsant. Luminal is one of over a dozen names under which it is prescribed. Pentobarbital is a short-acting sedative, hypnotic, and anticonvulsant. Nembutal is one of several trade names under which it is prescribed.

Synthesis

18.35 *N,N*-Diethyl-*m*-toluamide (DEET) is the active ingredient in several common insect repellents. Propose a synthesis for DEET from 3-methylbenzoic acid.

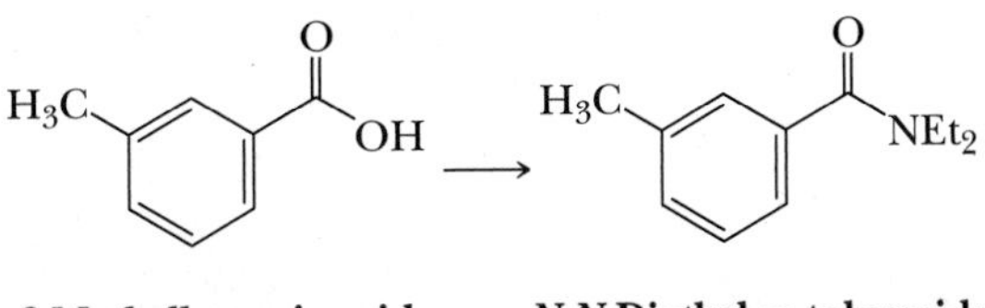

3-Methylbenzoic acid (*m*-Toluic acid) → ***N,N*-Diethyl-*m*-toluamide (DEET)**

18.36 Isoniazid, a drug used to treat tuberculosis, is prepared from pyridine-4-carboxylic acid. How might this synthesis be carried out?

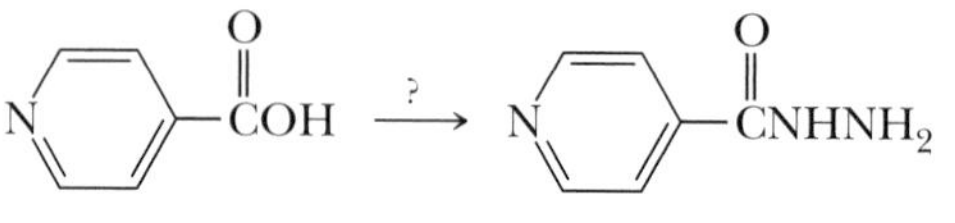

Pyridine-4-carboxylic acid **Pyridine-4-carboxylic acid hydrazide (Isoniazid)**

18.37 Show how to convert phenylacetylene to allyl phenylacetate.

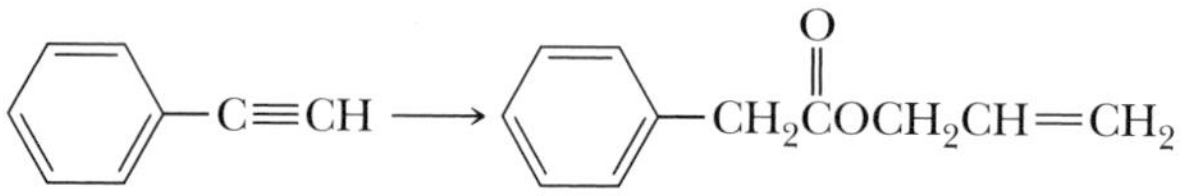

Phenylacetylene **Allyl phenylacetate**

18.38 A step in a synthesis of PGE_1 (prostaglandin E_1, alprostadil) is the reaction of a trisubstituted cyclohexene with bromine to form a bromolactone. Propose a mechanism for formation of this bromolactone and account for the observed stereochemistry of each substituent on the cyclohexane ring.

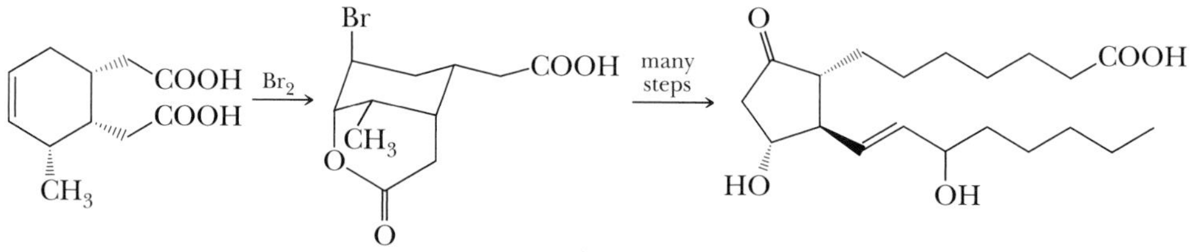

A bromolactone **PGE_1 (Alprostadil)**

Alprostadil is used as a temporary therapy for infants born with congenital heart defects that restrict pulmonary blood flow. It brings about dilation of the ductus arteriosus, which in turn increases blood flow in the lungs and blood oxygenation.

18.39 Barbiturates are prepared by treating a derivative of diethyl malonate with urea in the presence of sodium ethoxide as a catalyst. Following is an equation for the preparation of barbital, a long-duration hypnotic and sedative, from diethyl diethylmalonate and urea.

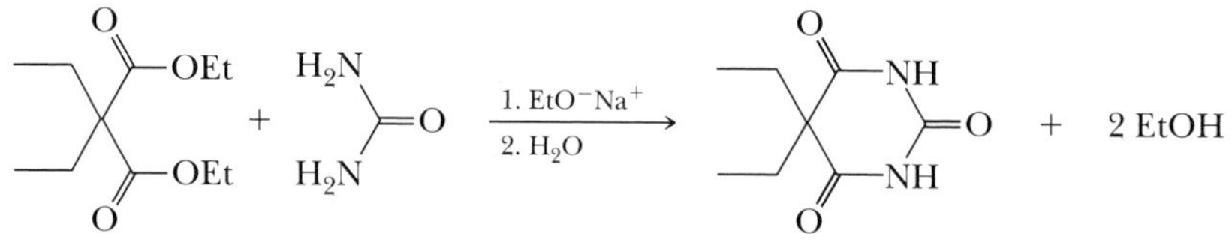

Diethyl diethylmalonate **Urea** **5,5-Diethylbarbituric acid (Barbital)**

Barbital is prescribed under one of a dozen or more trade names.

(a) Propose a mechanism for this reaction.

(b) The pK_a of barbital is 7.4. Which is the most acidic hydrogen in this molecule? How do you account for its acidity?

18.40 The following compound is one of a group of β-chloroamines, many of which have antitumor activity. Describe a synthesis of this compound from anthranilic acid and ethylene oxide.

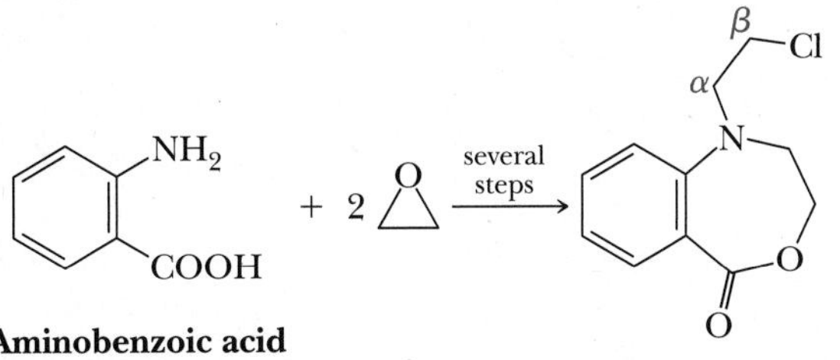

2-Aminobenzoic acid (Anthranilic acid)

A β-chloramine

18.41 Show how to synthesize 5-nonanone from 1-bromobutane as the only organic starting material.

18.42 Procaine (its hydrochloride is marketed as Novocain) was one of the first local anesthetics for infiltration and regional anesthesia. See "Chemical Connections: From Cocaine to Procaine and Beyond." According to the following retrosynthetic scheme, procaine can be synthesized from 4-aminobenzoic acid, ethylene oxide, and diethylamine as sources of carbon atoms.

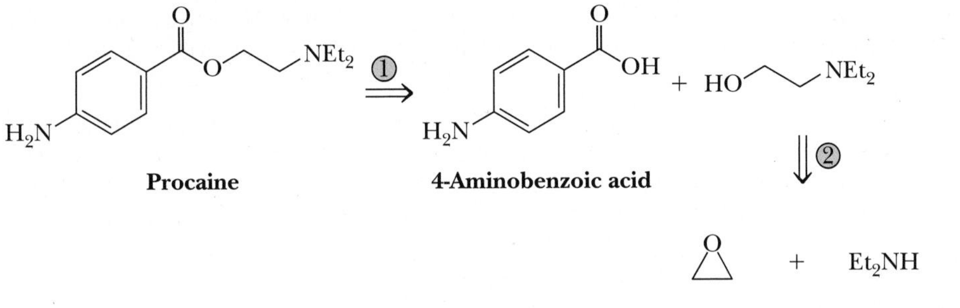

Provide reagents and experimental conditions to carry out the synthesis of procaine from these three compounds.

18.43 The following sequence of steps converts (*R*)-2-octanol to (*S*)-2-octanol.

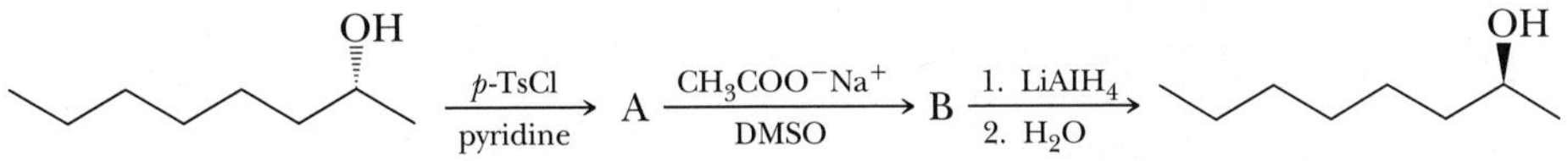

(*R*)-2-Octanol **(*S*)-2-Octanol**

Propose structural formulas for intermediates A and B, specify the configuration of each, and account for the inversion of configuration in this sequence.

18.44 Reaction of a primary or secondary amine with diethyl carbonate under controlled conditions gives a carbamic ester.

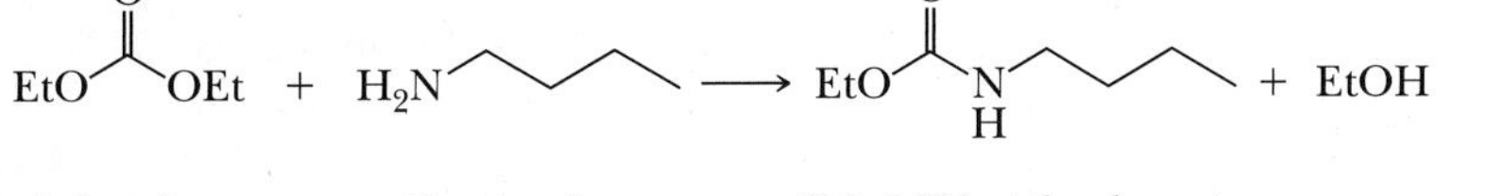

Diethyl carbonate **Butylamine** **Ethyl *N*-butylcarbamate**

Propose a mechanism for this reaction.

18.45 Several sulfonylureas, a class of compounds containing $RSO_2NHCONHR$, are useful drugs as orally active replacements for injected insulin in patients with adult-onset diabetes. These drugs decrease blood glucose concentrations by stimulating β cells of the pancreas to release insulin and by increasing the sensitivity of insulin receptors in peripheral tissues to insulin stimulation. Tolbutamide is synthesized by the reaction of the sodium salt of *p*-toluenesulfonamide and ethyl *N*-butylcarbamate (see Problem 18.44 for the synthesis of this carbamic ester). Propose a mechanism for this step.

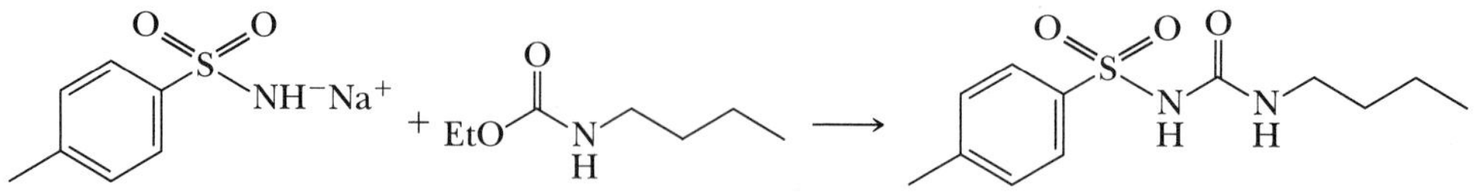

Sodium salt of *p*-toluenesulfonamide — A carbamic ester — **Tolbutamide (Orinase)**

18.46 Following are structural formulas for two more widely used sulfonylurea hypoglycemic agents.

(a)

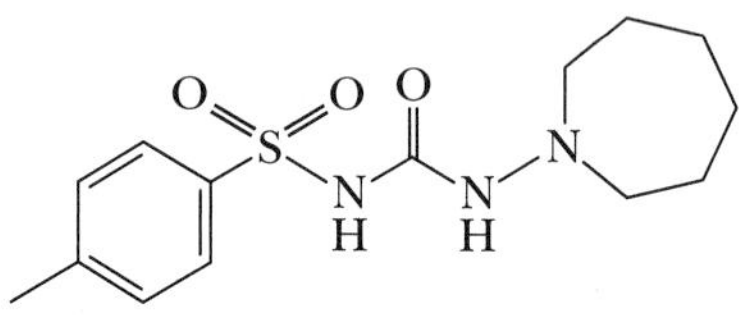

Tolazamide (Tolinase)

(b)

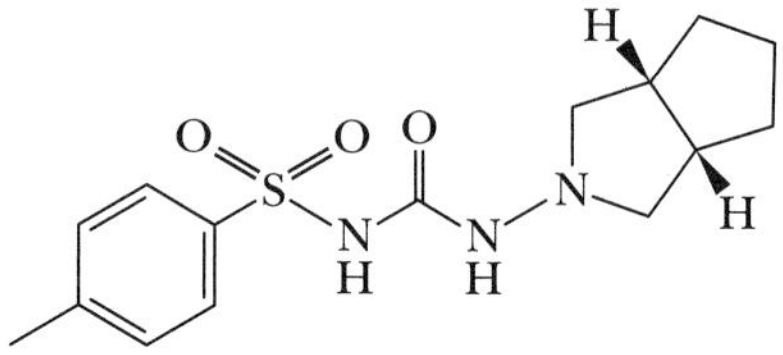

Gliclazide (Diamicron)

Show how each might be synthesized by converting an appropriate amine to a carbamic ester and then treating the carbamate with the sodium salt of a substituted benzenesulfonamide.

18.47 Amantadine is effective in preventing infections caused by the influenza A virus and in treating established illnesses. It is thought to block a late stage in the assembly of the virus. Amantadine is synthesized by treating 1-bromoadamantane with acetonitrile in sulfuric acid to give *N*-adamantylacetamide, which is then converted to amantadine.

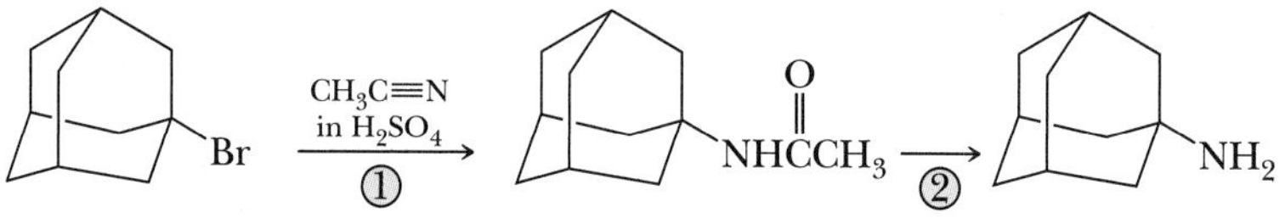

1-Bromoadamantane — **Amantadine**

(a) Propose a mechanism for the transformation in Step 1.
(b) Describe experimental conditions to bring about Step 2.

18.48 In a series of seven steps, (*S*)-malic acid is converted to the bromoepoxide shown on the right in 50% overall yield. This synthesis is enantioselective—of the stereoisomers possible for the bromoepoxide, only one is formed.

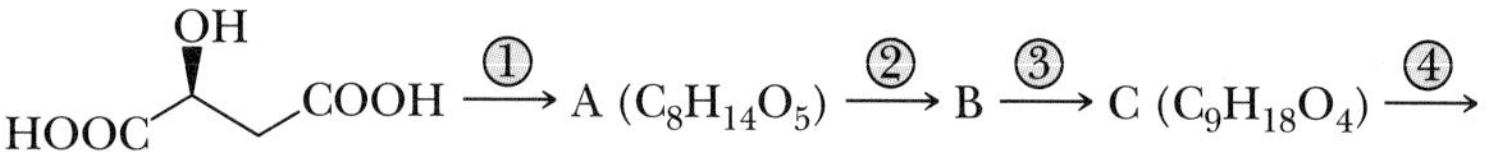

(*S*)-Malic acid

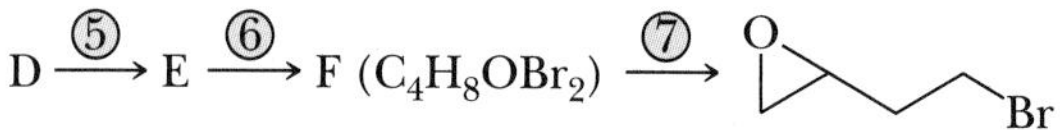

A bromoepoxide

Steps/reagents:
1. CH_3CH_2OH, H^+
2. 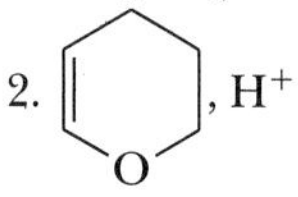, H^+
3. $LiAlH_4$, then H_2O
4. TsCl, pyridine
5. NaBr, DMSO
6. H_2O, CH_3COOH
7. KOH

In thinking about the chemistry of these steps, you will want to review the use of dihydropyran as an —OH protecting group (Section 16.7D) and the use of the *p*-toluenesulfonyl chloride to convert the —OH, a poor leaving group, into a tosylate, a good leaving group (Section 10.5D).

(a) Propose structural formulas for intermediates A through F and specify the configuration at each chiral center.

(b) What is the configuration of the chiral center in the bromoepoxide? How do you account for the stereoselectivity of this seven-step conversion?

18.49 Following is a retrosynthetic analysis for the synthesis of the herbicide (*S*)-Metolachlor from 2-ethyl-6-methylaniline, chloroacetic acid, acetone, and methanol.

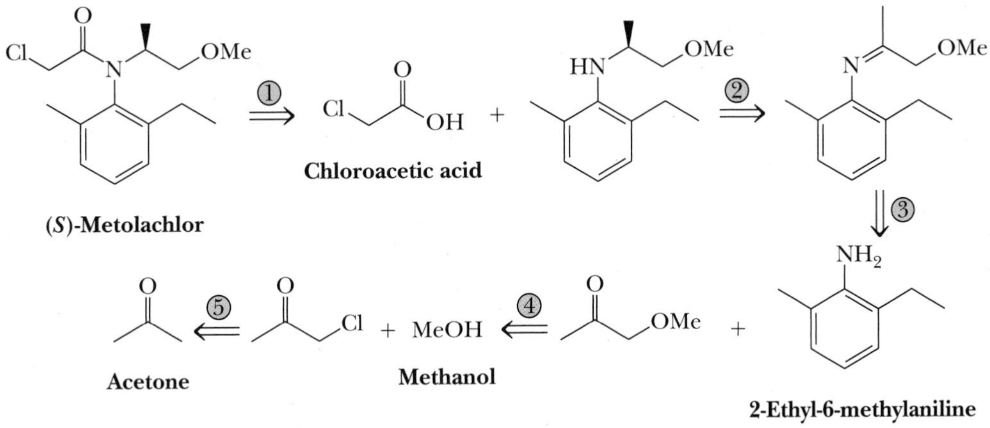

(S)-Metolachlor
Chloroacetic acid
Acetone
Methanol
2-Ethyl-6-methylaniline

Show reagents and experimental conditions for the synthesis of Metolachlor from these four organic starting materials. Your synthesis will most likely give a racemic mixture. The chiral catalyst used by Novartis for reduction in Step 2 gives 80% enantiomeric excess of the *S* enantiomer.

18.50 Following is a retrosynthetic analysis for the anthelmintic (against worms) diethylcarbamazine.

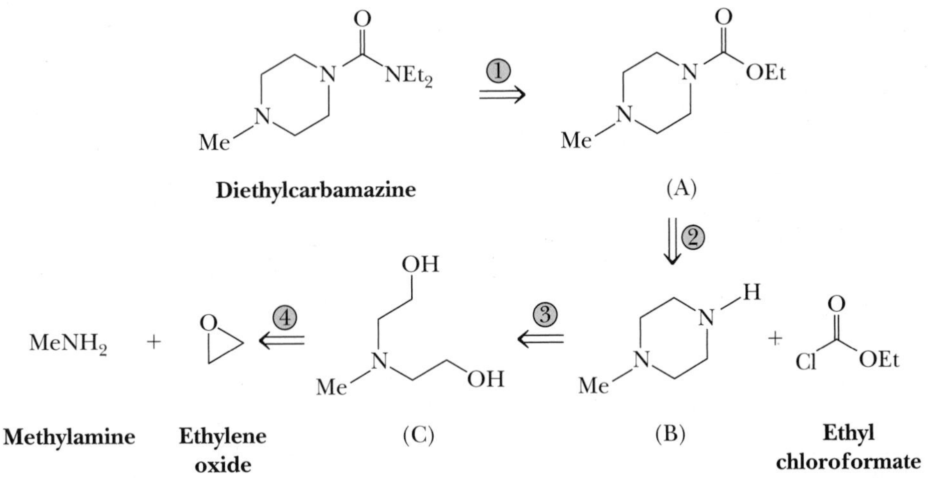

Diethylcarbamazine
(A)
(B)
(C)
Methylamine
Ethylene oxide
Ethyl chloroformate

Diethylcarbamazine is used chiefly against nematodes, small cylindrical or slender threadlike worms such as the common roundworm, which are parasitic in animals and plants. Given this retrosynthetic analysis, propose a synthesis of diethylcarbamazine from the three named starting materials.

18.51 Given this retrosynthetic analysis, propose a synthesis for the antidepressant moclobemide.

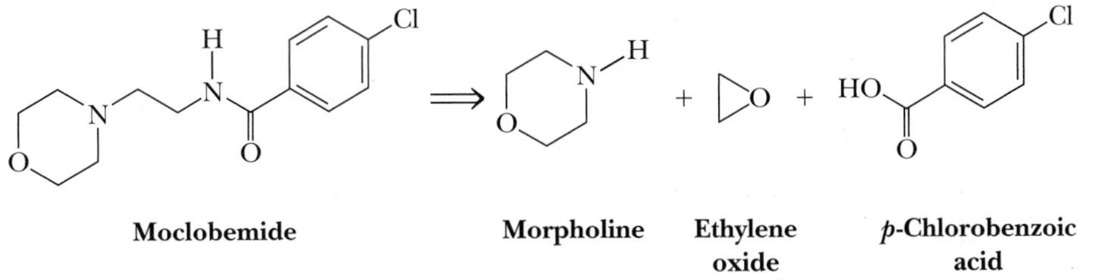

Moclobemide
Morpholine
Ethylene oxide
p-Chlorobenzoic acid

18.52 Propose a synthesis for diphenhydramine starting from benzophenone, benzoic acid, and 2-(*N,N*-dimethylamino)ethanol.

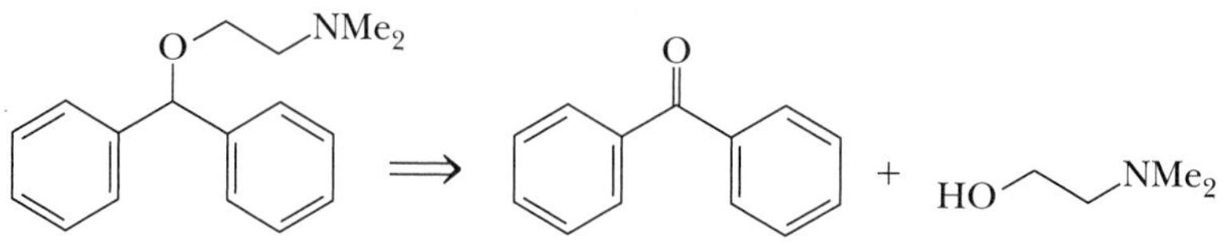

Diphenhydramine **Benzophenone** **2-(*N,N*-Dimethylamino)-ethanol**

The hydrochloride salt of diphenhydramine, best known by its trade name of Benadryl, is an antihistamine.

18.53 Propose a synthesis of the topical anesthetic cyclomethycaine from 4-hydroxybenzoic acid, 2-methylpiperidine, and any other necessary reagents.

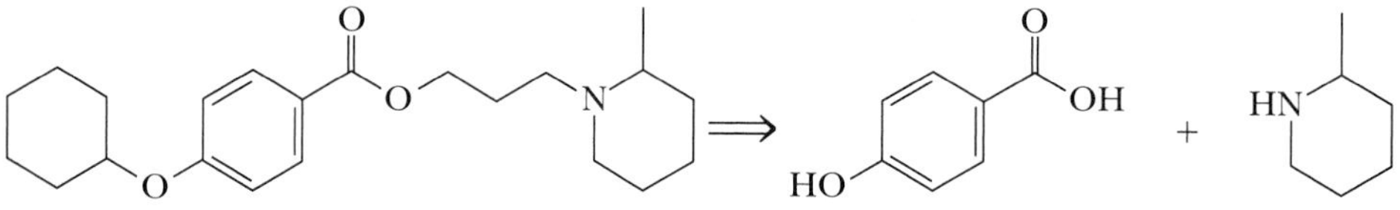

Cyclomethycaine **4-Hydroxybenzoic acid** **2-Methylpiperidine**

18.54 Following is an outline of a synthesis of bombykol, the sex attractant of the male silkworm moth. Of the four stereoisomers possible for this conjugated diene, the 10-*trans*-12-*cis* isomer shown here is over 10^6 times more potent as a sex attractant than any of the other three possible stereoisomers.

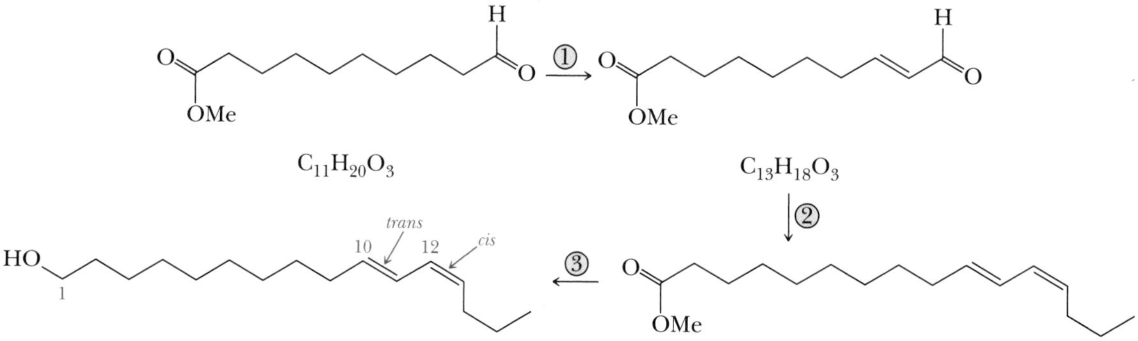

$C_{11}H_{20}O_3$ $C_{13}H_{18}O_3$ $C_{17}H_{30}O_2$

Bombykol
$C_{16}H_{30}O$

Show how this synthesis might be accomplished and explain how your proposed synthesis is stereoselective for the 10-*trans*-12-*cis* isomer.

18.55 In Problem 7.28, we saw this step in Johnson's synthesis of the steroid hormone progesterone.

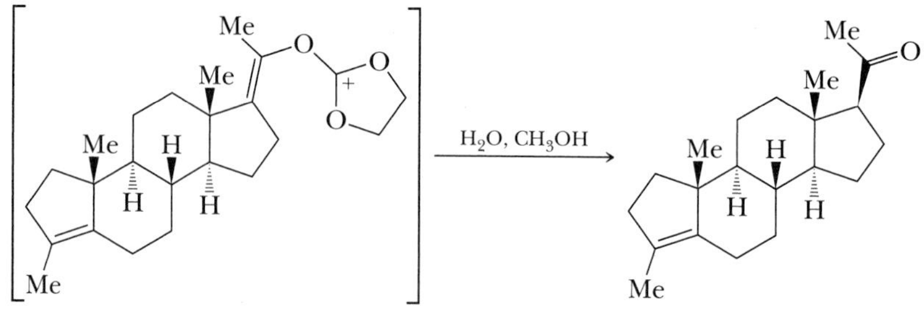

B C

Propose a mechanism for this step in the synthesis.

Mechanisms

18.56 Using the principles for writing mechanisms and the four common mechanistic steps, write mechanisms showing all electron flow arrows for the following reactions:

(a) Hydrolysis of *N,N*-dimethylacetamide in acidic water.
(b) Hydrolysis of acetic anhydride in basic water.
(c) Esterification of acetic acid in acidic ethanol.
(d) The reaction of dimethylamine in water with acetic anhydride to create *N,N*-dimethylacetamide.
(e) Partial hydrolysis of acetonitrile in acidic water to create acetamide.

18.57 The following statements are true experimental observations. Explain the reason behind each observation.

(a) The reaction of acetic acid with ammonia in water does not give any amide products.
(b) The reaction of acetyl chloride with water causes the pH to decrease.
(c) The hydrolysis of an amide at neutral pH takes seven years at room temperature, while the hydrolysis of an acid chloride takes a few minutes.

Looking Ahead

18.58 We have seen two methods for converting a carboxylic acid and an amine into an amide. Suppose that you start instead with a dicarboxylic acid such as hexanedioic acid and a diamine such as 1,6-hexanediamine. Show how amide formation in this case can lead to a polymer (a macromolecule of molecular weight several thousands times that of the starting materials).

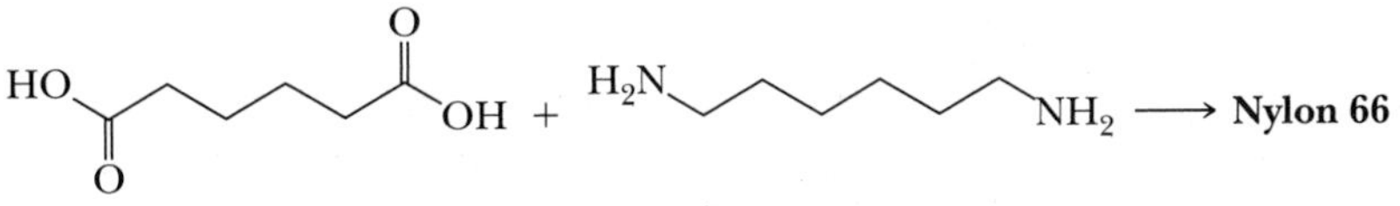

Hexanedioic acid (Adipic acid) **1,6-Hexanediamine (Hexamethylenediamine)**

As we shall see in Section 29.5A, the material produced in this reaction is the high-molecular-weight polymer nylon 66, so named because it is synthesized from two 6-carbon starting materials.

18.59 Using the same reasoning as in Problem 18.58, show how amide formation between this combination of dicarboxylic acid and diamine will also lead to a polymer, in this case Kevlar.

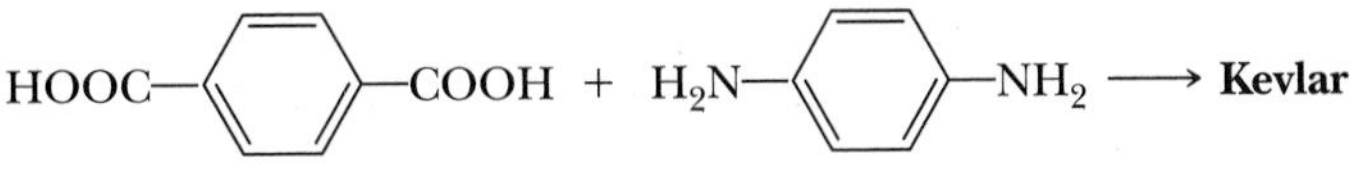

1,4-Benzenedicarboxylic acid (Terephthalic acid) **1,4-Benzenediamine**

18.60 A urethane is a molecule in which a carbonyl group is part of an ester and an amide (it is an amide in one direction and an ester in the other direction). Propose a mechanism for the reaction of an isocyanate with an alcohol to form a urethane.

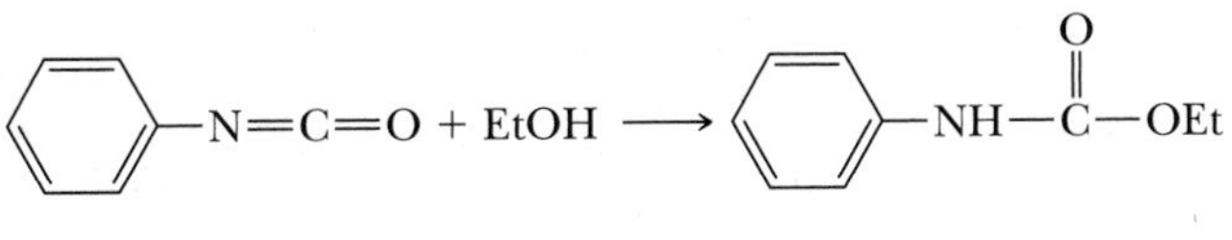

Phenylisocyanate **Ethanol** A urethane

18.61 Suppose that you start with a diisocyanate and a diol. Show how their reaction can lead to a polymer called a polyurethane (Section 29.5D).

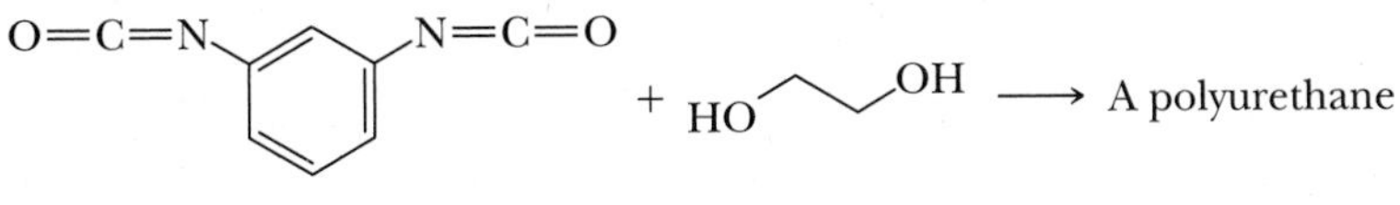

A diisocyanate **Ethylene glycol**

Organic Chemistry Reaction Roadmap

Reaction Roadmap

18.62 Use the reaction roadmap roadmap you made for Problems 15.19, 16.72, and 17.45 and update it to contain the reactions in the "Study Guide" section of this chapter. Because of their highly specific nature, do not use reactions 1, 2, and 10 on your roadmap.

Reaction Roadmap

18.63 Write the products of the following sequences of reactions. Refer to your reaction roadmaps to see how the combined reactions allow you to "navigate" between the different functional groups. Note that you will need both your old Chapters 6–11 roadmap and your new Chapters 15–18 roadmap for these.

(a)
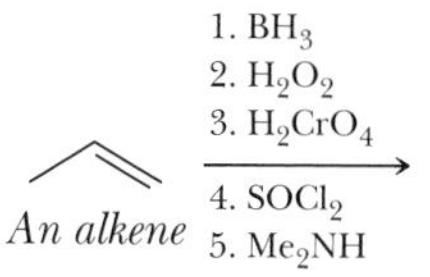
An alkene
1. BH3
2. H2O2
3. H2CrO4
4. SOCl2
5. Me2NH

(b)
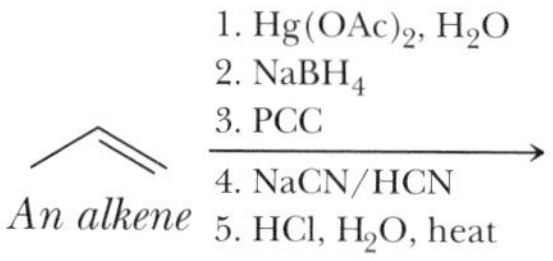
An alkene
1. Hg(OAc)2, H2O
2. NaBH4
3. PCC
4. NaCN/HCN
5. HCl, H2O, heat

(c)
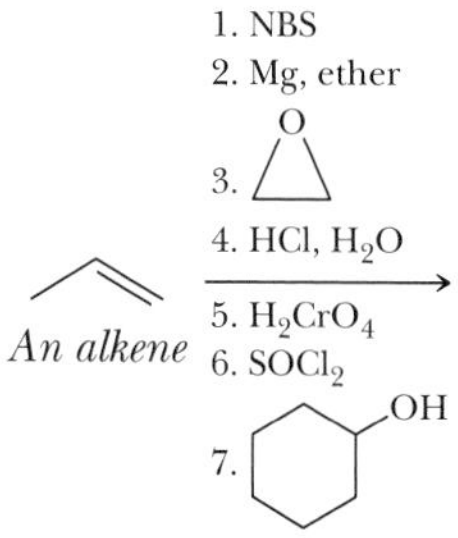
An alkene
1. NBS
2. Mg, ether
3.
4. HCl, H2O
5. H2CrO4
6. SOCl2
7.
OH

(d)
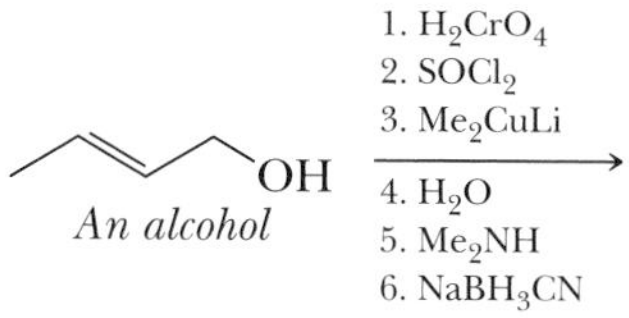
An alcohol
OH
1. H2CrO4
2. SOCl2
3. Me2CuLi
4. H2O
5. Me2NH
6. NaBH3CN

(e)
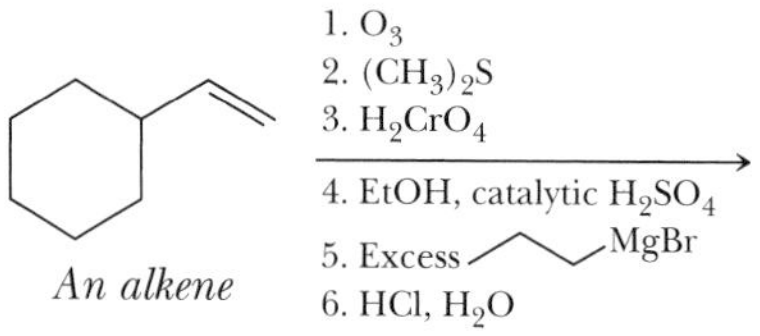
An alkene
1. O3
2. (CH3)2S
3. H2CrO4
4. EtOH, catalytic H2SO4
5. Excess
MgBr
6. HCl, H2O

Synthesis

Reaction Roadmap

18.64 Using your reaction roadmap as a guide, show how to convert (*E*)-3-hexene into propyl propionate. You must use (*E*)-3-hexene as the source of all carbon atoms in the target molecule. Show all reagents and all molecules synthesized along the way.

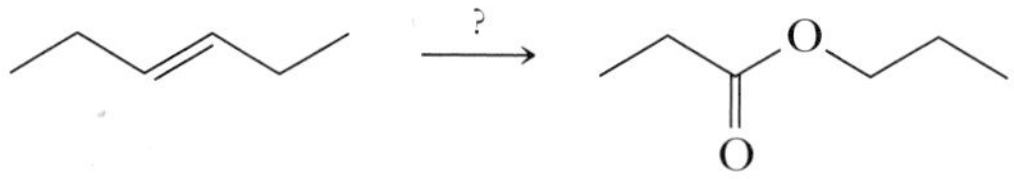

(*E*)-3-Hexene **Propyl propanoate**

Reaction Roadmap

18.65 Using your reaction roadmap as a guide, show how to convert 1-bromopentane and sodium cyanide into *N*-hexylhexanamide. You must use 1-bromopentane and sodium cyanide as the source of all carbon atoms in the target molecule. Show all reagents and all molecules synthesized along the way.

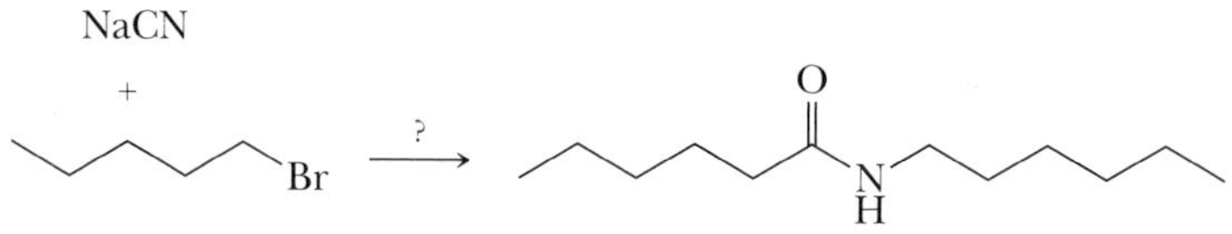
NaCN
+
Br
?
O
N
H

1-Bromopentane ***N*-Hexylhexanamide**

18.66 Using your reaction roadmap as a guide, show how to convert 1-bromopropane and carbon dioxide into 4-propyl-4-heptanol. You must use 1-bromopropane and carbon dioxide as the source of all carbon atoms in the target molecule. Show all reagents and all molecules synthesized along the way.

Reaction Roadmap

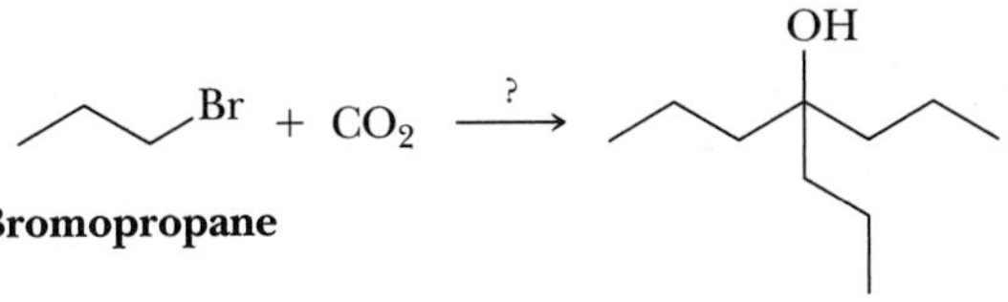

1-Bromopropane

4-Propyl-4-heptanol

18.67 Using your reaction roadmap as a guide, show how to convert 1-bromopropane and carbon dioxide into 4-heptanone. You must use 1-bromopropane and carbon dioxide as the source of all carbon atoms in the target molecule. Show all reagents and all molecules synthesized along the way.

Reaction Roadmap

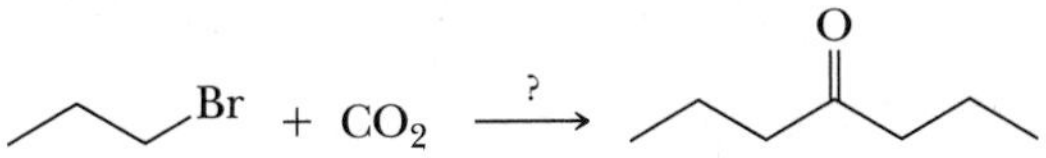

1-Bromopropane **4-Heptanone**

Reactions in Context

18.68 Minoxidil is a molecule that causes hair growth in some people. It was originally synthesized as a vasodilator for the treatment of hypertension (high blood pressure). Most of the patients taking the drug for hypertension were seen to grow body hair. Due to other side effects, its oral use was stopped, but it became popular as a topical cream to promote hair growth.

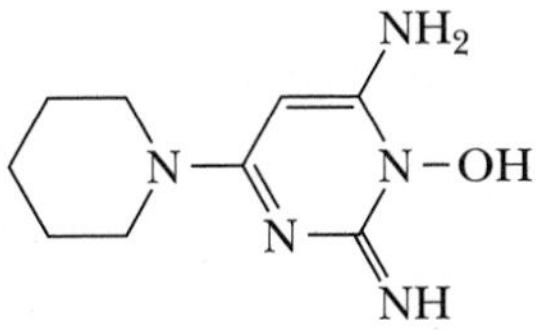

Minoxidil

The first key reaction in one synthesis of minoxidil follows. Draw the product of this reaction.

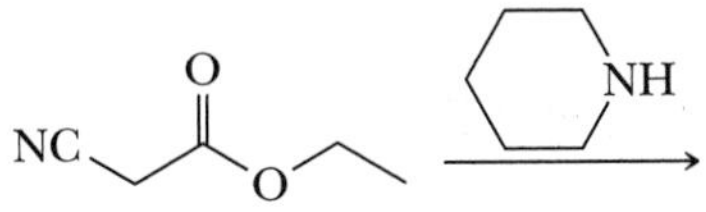

18.69 Chloroformates have the functional group R–O–C(=O)–Cl, in which R is often a *tert*-butyl or benzyl group. A chloroformate is used in the following synthesis of the antibacterial drug linezolid (Zyvox).

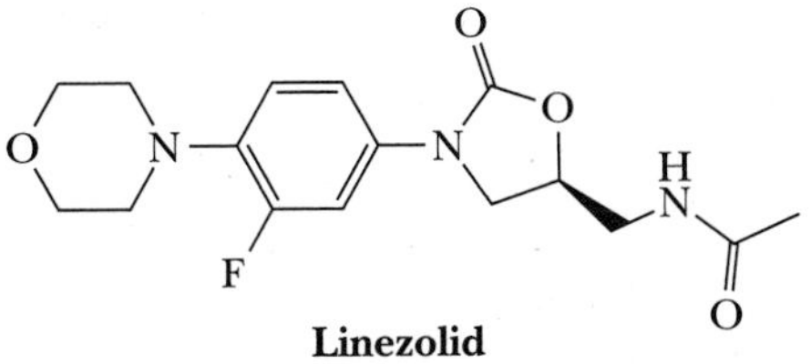

Linezolid (Zyvox®)

Based on your knowledge of carboxylic acid derivatives, predict the product of the following transformation used in a synthesis of linezolid (Zyvox). The new functional group created is called a carbamate. Carbamates are often used as protecting groups for amine groups during complex syntheses.

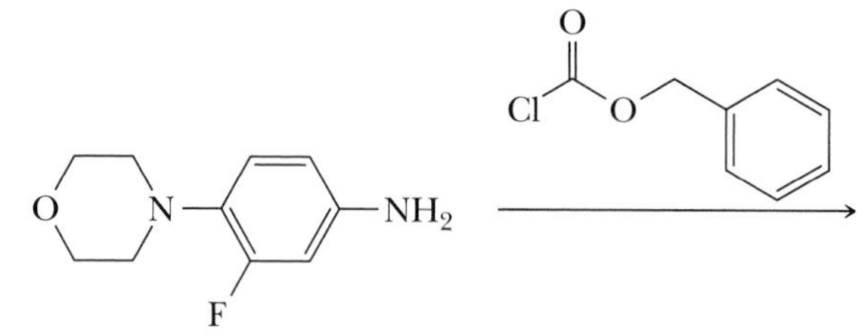

18.70 Acid anhydrides are often used in place of acid chlorides because a less acidic carboxylic acid, not the much stronger acid HCl, is the byproduct of the reaction. In the following reaction of a carbohydrate derivative, acetic anhydride is used to obtain the product in 99% yield as a single stereoisomer. Note that the stereochemistry of the starting anomeric carbon is not indicated. Draw the product of the following transformation in a chair form and show the single stereoisomer product of this transformation, which is also the most stable possible chair species.

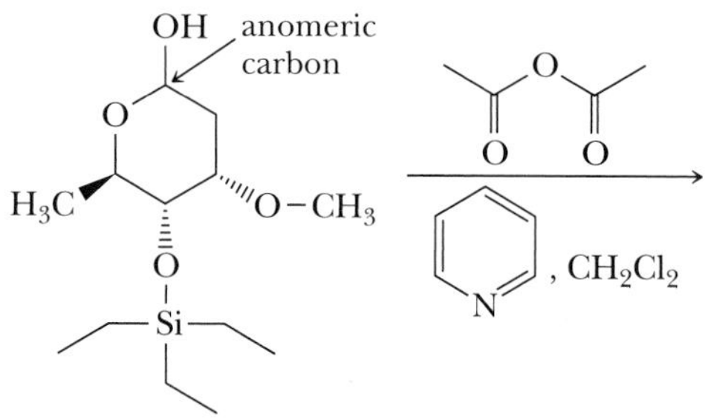

18.71 The benzyl ether group (—OBn) is often used as a protecting group for OH groups during the synthesis of complex molecules. The following structure has a number of benzyl ether groups used for this purpose. Draw the product of the following transformation. What role does the diisopropylethylamine play in this reaction?

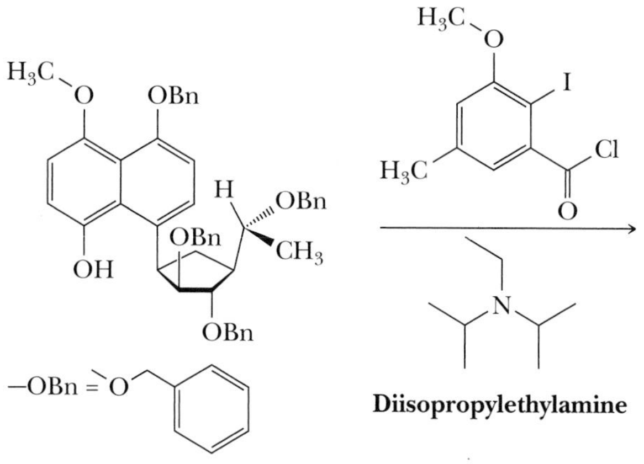

19

Enolate Anions and Enamines

Crystals of tamoxifen (Problems 19.42, 21.63, and 21.64) viewed under polarizing light. **Inset:** a model of tamoxifen.

Outline

19.1 Formation and Reactions of Enolate Anions: An Overview
19.2 Aldol Reaction
19.3 Claisen and Dieckmann Condensations
19.4 Claisen and Aldol Condensations in the Biological World
19.5 Enamines
19.6 Acetoacetic Ester Synthesis
19.7 Malonic Ester Synthesis
19.8 Conjugate Addition to α,β-Unsaturated Carbonyl Compounds
19.9 Crossed Enolate Reactions Using LDA

In this chapter, we continue the chemistry of carbonyl compounds. In Chapters 16 through 18, we concentrated on the carbonyl group itself and on nucleophilic additions to the carbonyl carbon to form tetrahedral carbonyl addition intermediates and on products derived from protonation or collapse of these intermediates. In this chapter, we expand on the chemistry of carbonyl-containing compounds and consider the consequences of the acidity of α-hydrogens and the formation of enolate anions.

19.1 Formation and Reactions of Enolate Anions: An Overview

On the next page is the resonance-stabilized **enolate anion** (Section 16.9A) formed by treating acetaldehyde with base. This anion is best represented as a hybrid of two contributing structures. Of these, the structure with the negative charge on the more electronegative oxygen atom makes the greater contribution to the hybrid. Note that although the majority of the negative charge is on the carbonyl oxygen, there is still a significant partial negative charge on the alpha carbon.

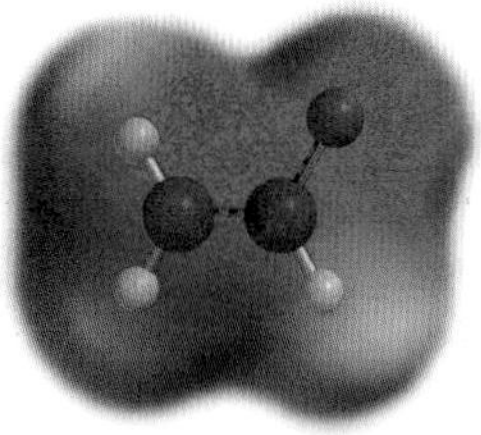

The majority of the negative charge in the hybrid is on oxygen

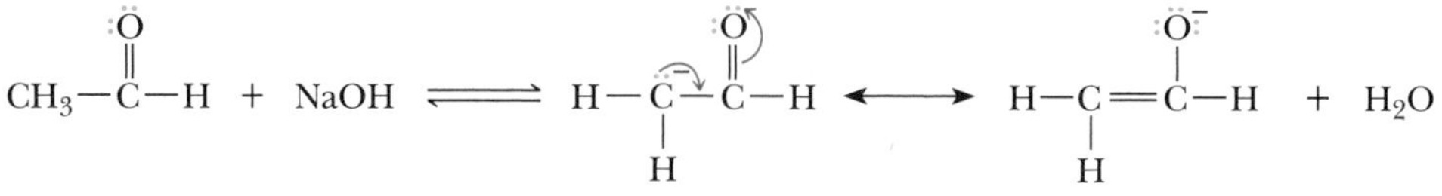

An enolate anion

▶ Delocalization of charge over a larger area is stabilizing.

Enolate anions are important synthetic reagents because they react at carbon to create new carbon-carbon bonds in two types of reactions. First, they can function as nucleophiles in S_N2 reactions as shown in this general reaction.

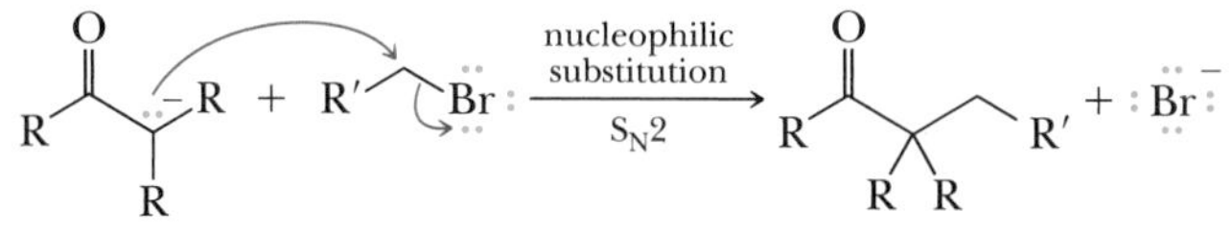

An enolate anion A 1° haloalkane or sulfonate

Second, they function as nucleophiles in carbonyl addition reactions. Here, we show nucleophilic acyl addition of an enolate anion to the carbonyl carbon of an aldehyde.

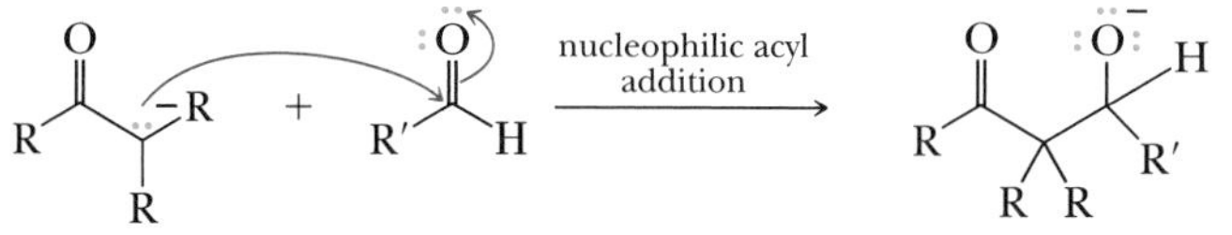

An enolate anion An aldehyde A tetrahedral carbonyl addition intermediate

Enolate anions also add in this manner to the carbonyl groups of ketones and esters.

As shown by the charge distribution on the electrostatic potential map, the majority of negative charge of an enolate anion is on the carbonyl oxygen. If reaction were to occur at the carbonyl oxygen, the product would be a vinyl ether, whereas reaction at the α-carbon leads to alkylation.

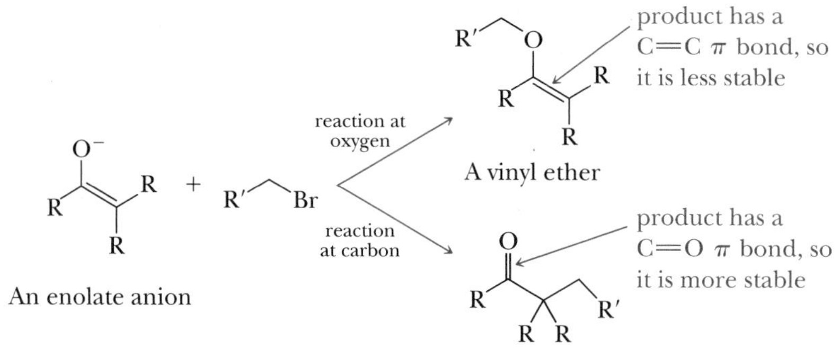

Despite this charge distribution, enolate anions react primarily at carbon for two reasons. First, there is always a counterion such as the Li^+ or Na^+ ion associated with the enolate anion. These counterions are more tightly associated with the oxygen atom than the alpha carbon. As a result, the counterion to some degree blocks the approaching electrophile, thus reducing the likelihood of a productive collision with the oxygen. In fact, enolates are thought to exist in solution as larger aggregates containing several counterions associated with several enolate oxygen atoms, and the solvent, effectively amplifying this effect.

An example of an aggregate of lithium enolates with the solvent THF

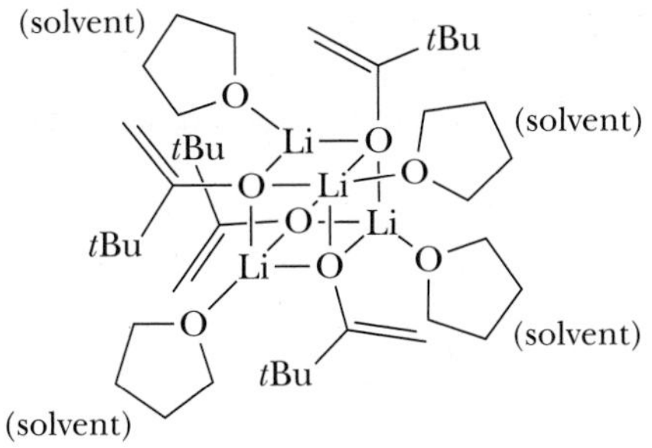

The second reason enolates react at carbon is based on product thermodynamics. We have already seen that, other factors being equal, reactions at equilibrium will favor products with stronger bonds. If an enolate anion were to react at the alpha carbon, the product would contain a C=O π bond. If it were to react at the carbonyl oxygen, the product would contain a C=C π bond. In general, C=O bonds are stronger than C=C bonds. (Recall, for example, the relative percentages of keto and enol forms present at equilibrium for simple aldehydes and ketones, Section 16.9B.) Thus, enolate anions react primarily at the alpha carbon to form new carbon-carbon bonds.

19.2 Aldol Reaction

Unquestionably, the most important reaction of enolate anions derived from aldehydes and ketones is their nucleophilic addition to the carbonyl group of another molecule of the same or different compound, as illustrated by the following reactions.

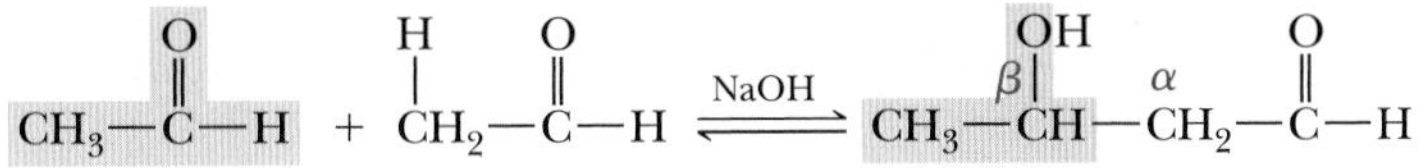

Ethanal (Acetaldehyde) **Ethanal (Acetaldehyde)** **3-Hydroxybutanal** (a β-hydroxyaldehyde; formed as a racemic mixture)

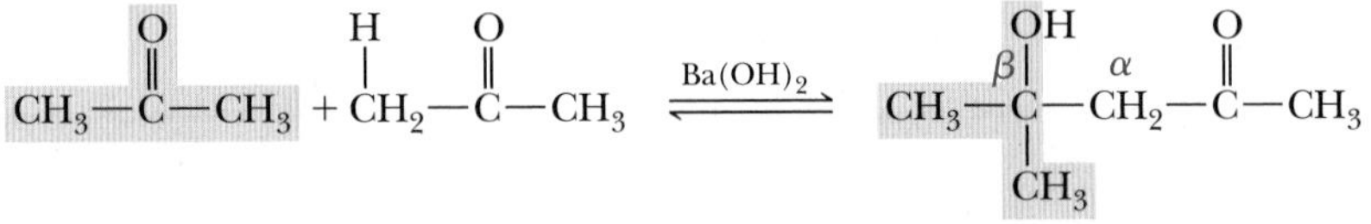

Propanone (Acetone) **Propanone (Acetone)** **4-Hydroxy-4-methyl-2-pentanone** (a β-hydroxyketone)

The common name of the product derived from the reaction of acetaldehyde in base is aldol because it is both an *alde*hyde and an alcoh*ol*. Aldol is also the generic name given to any product formed in this type of reaction. The product of an **aldol reaction** is a β-hydroxyaldehyde or a β-hydroxyketone.

A. Mechanisms

Although such reactions may be catalyzed by either acid or base, base catalysis is more common. The key step in a base-catalyzed aldol reaction is nucleophilic addition of the enolate anion of one carbonyl-containing molecule to the carbonyl group of another to form a tetrahedral carbonyl addition intermediate. This mechanism is illustrated by the aldol reaction between two molecules of acetaldehyde. Notice in this three-step mechanism that OH^- is a catalyst; an OH^- is used in Step 1, but another OH^- is generated in Step 3.

Watch a video explanation

Mechanism 19.1

Base-Catalyzed Aldol Reaction

Step 1: **Take a proton away.** Removal of an α-hydrogen by base gives a resonance-stabilized enolate anion.

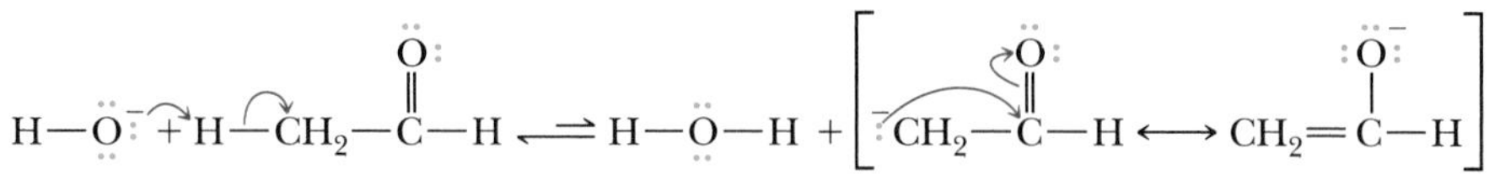

pK_a 20 (weaker acid) pK_a 15.7 (stronger acid) An enolate anion

Given the relative acidities of the two acids in this equilibrium, the position of this equilibrium lies considerably to the left.

Step 2: **Make a new bond between a nucleophile and an electrophile.** Nucleophilic addition of the enolate anion to the carbonyl carbon of another aldehyde (or ketone) gives a tetrahedral carbonyl addition intermediate.

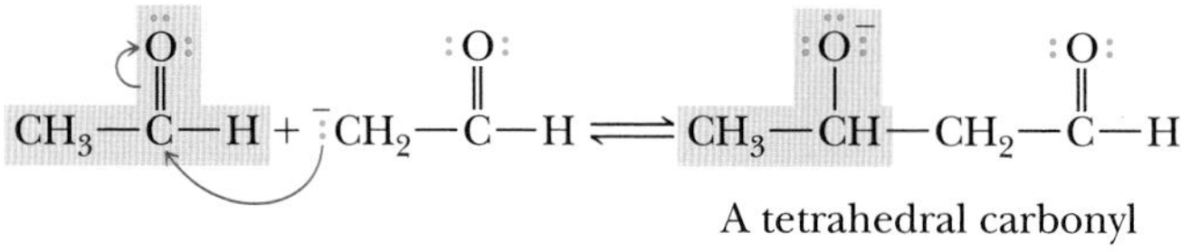

A tetrahedral carbonyl addition intermediate

Step 3: **Add a proton.** Reaction of the tetrahedral carbonyl addition intermediate with a proton donor gives the aldol product as a racemic mixture and generates a new base catalyst.

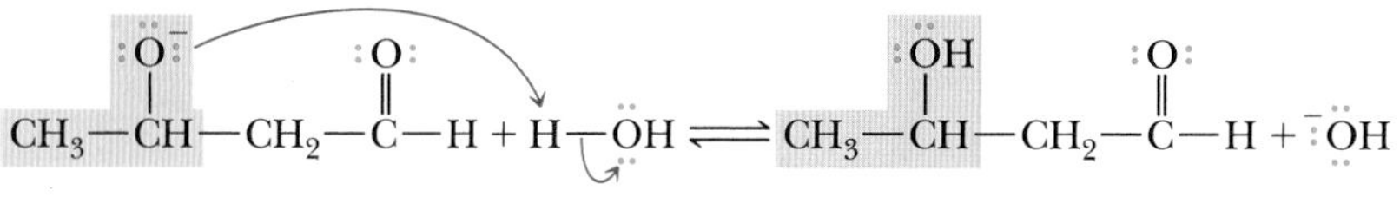

(racemic)

Mechanism 19.2

Acid-Catalyzed Aldol Reaction

The mechanism of an acid-catalyzed aldol reaction involves four steps, the first two of which are preparation of the aldehyde or ketone for formation of the new carbon-carbon bond. The key step is attack of the enol of one molecule on the protonated carbonyl group of a second molecule.

Step 1: **Keto-enol tautomerism.** Keto and enol forms of the aldehyde or ketone undergo acid-catalyzed equilibration (Section 16.9B).

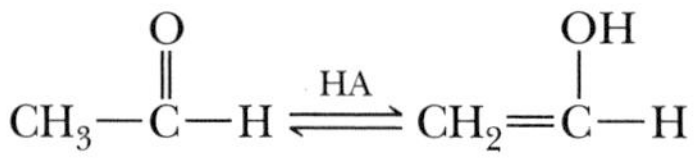

Step 2: **Add a proton.** Proton transfer from the acid, HA, to the carbonyl oxygen of a second molecule of aldehyde or ketone gives an oxonium ion.

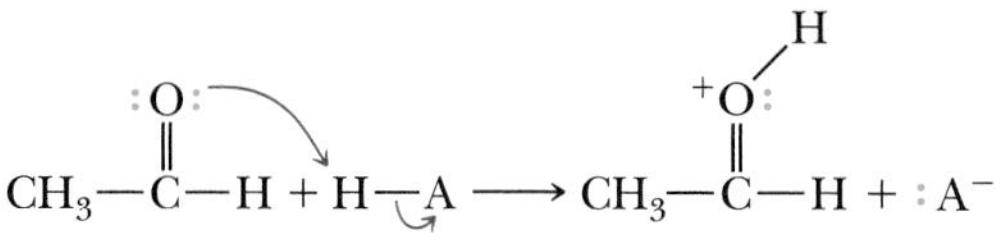

Step 3: **Make a new bond between a nucleophile and an electrophile.** Attack by the enol of one molecule on the protonated carbonyl group of another molecule forms the new carbon-carbon bond.

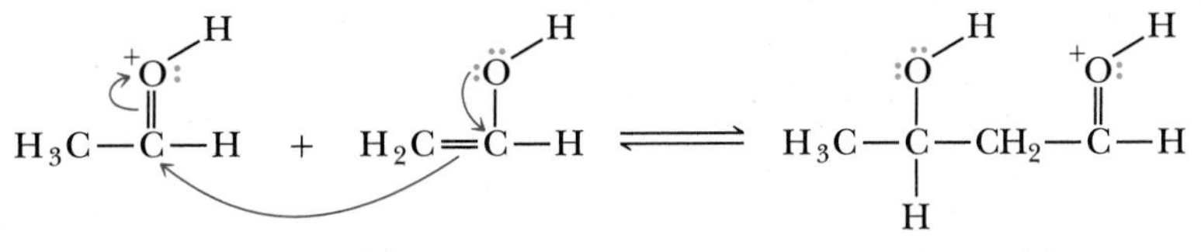

(racemic)

Step 4: **Take a proton away.** Proton transfer to A^- regenerates the acid catalyst and gives the racemic aldol product.

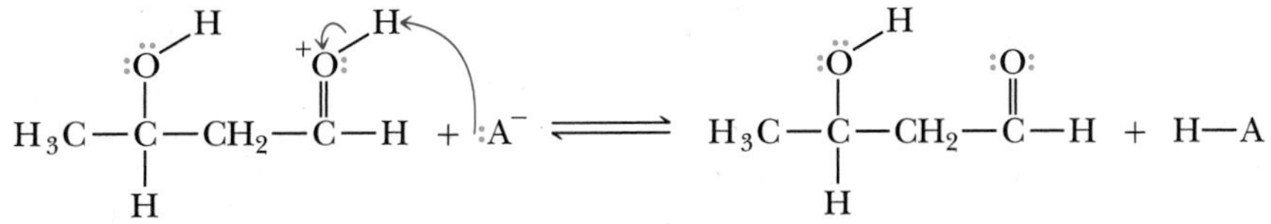

(racemic) (racemic)

You might compare the mechanisms of the acid- and base-catalyzed aldol reactions. Under base catalysis, the carbon-carbon bond-forming step involves attack of an enolate anion (a nucleophile) on the uncharged carbonyl carbon (an electrophile) of a second molecule of the aldehyde or ketone. Under acid catalysis, it involves attack of the enol (a nucleophile) of one molecule on the protonated carbonyl group (an electrophile) of the second molecule.

It is quite common to create chiral products during aldol reactions, as well as in the other enolate reactions we discuss in this chapter. The products will be formed as racemic mixtures unless one of the reactants is chiral and present as a single enantiomer. In cases when two chiral centers are created in the reaction, four stereoisomers are produced as two 1:1 mixtures of enantiomers. A great deal of work has gone into learning how to carry out aldol and other enolate reactions that give predominantly a single enantiomer product, but that work is beyond the scope of this text.

Example 19.1 Aldol Reaction

Draw the product of the base-catalyzed aldol reaction of each compound.

(a) Butanal **(b)** Cyclohexanone

Solution

The aldol product is formed by nucleophilic addition of the α-carbon of one molecule to the carbonyl carbon of another.

(a) 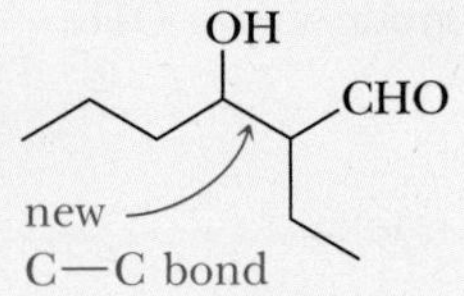
OH
CHO
new
C—C bond

(four stereoisomers; two racemic mixtures)

(b) 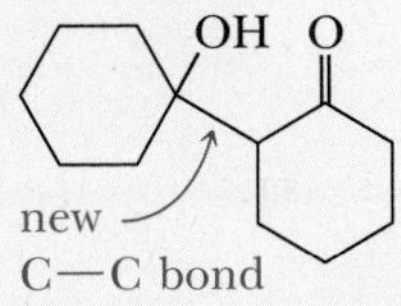
OH O
new
C—C bond

(racemic)

(Continued)

Watch a video explanation

Problem 19.1

Draw the product of the base-catalyzed aldol reaction of each compound.

(a) Phenylacetaldehyde **(b)** Cyclopentanone

β-Hydroxyaldehydes and β-hydroxyketones are very easily dehydrated, and often the conditions necessary to bring about an aldol reaction are sufficient to cause dehydration. Dehydration is particularly facile in the case of acid-catalyzed aldol reactions. Alternatively, dehydration can be brought about by warming the aldol product in dilute acid. The major product from dehydration of an aldol is one in which the carbon-carbon double bond is conjugated with the carbonyl group; that is, the product is an α,β-unsaturated aldehyde or ketone.

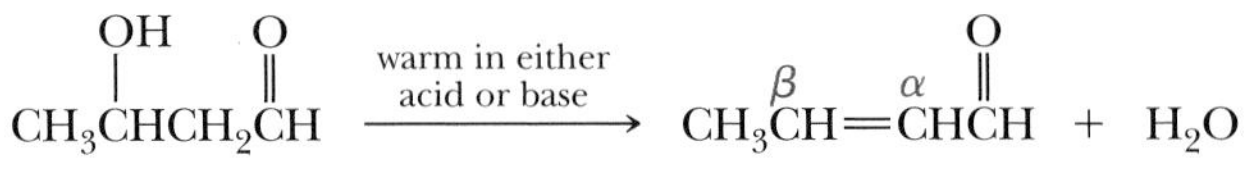

OH
O
CH_3CHCH_2CH
warm in either
acid or base
O
β
α
$CH_3CH{=}CHCH$
+ H_2O

An α,β-unsaturated aldehyde

Aldol reactions are readily reversible, especially when catalyzed by base, and generally, little aldol product is present at equilibrium, especially for aldol reactions of ketones. Equilibrium constants for dehydration, however, are generally large so that if reaction conditions are sufficiently vigorous to bring about dehydration, good yields of product can be obtained. The alkene generally forms as an *E,Z* mixture, although in some cases, such as that below, the *E* alkene will be substantially more stable and thus the predominant product.

Mechanism 19.3

Acid-Catalyzed Dehydration of an Aldol Product

Step 1: **Keto-enol tautomerism.** The acid catalyzes interconversion of the keto and enol forms of the aldehyde.

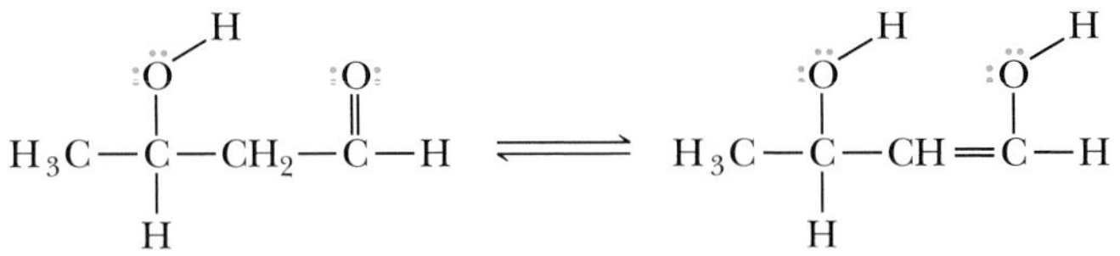

(racemic) The enol of the aldehyde (racemic)

Step 2: **Add a proton.** Proton transfer from the acid catalyst (here shown as H_3O^+) to the enol gives an oxonium ion intermediate. This step has the effect of converting the —OH, a poor leaving group, into $—\overset{+}{O}H_2$, a better leaving group.

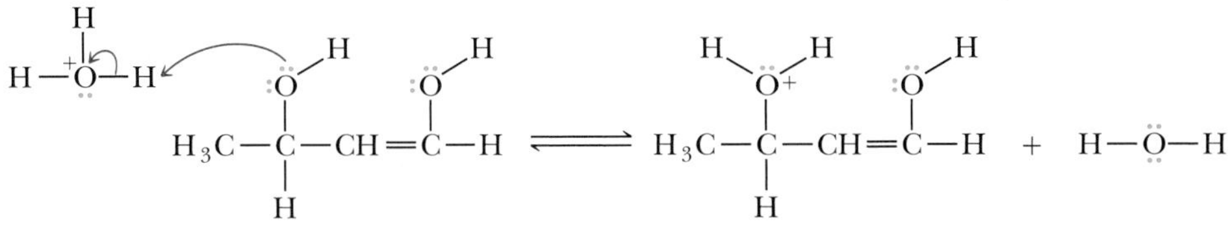

The enol of the aldehyde (racemic) (racemic)

Step 3: **Break a bond to give stable molecules or ions.** Loss of water from the oxonium ion intermediate gives the conjugate acid of the final product. Notice that the double bond forms to give predominantly the more stable *E* alkene in this case.

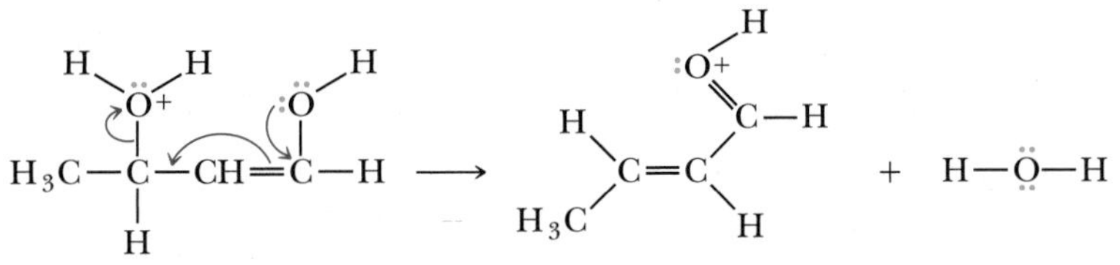

Step 4: **Take a proton away.** Proton transfer from the conjugate acid of the final product to water completes the reaction.

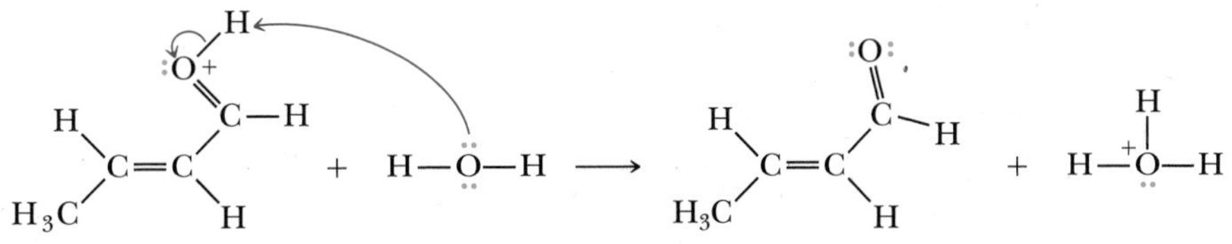

An α,β-unsaturated aldehyde

Example 19.2 Dehydration of Aldol Products

Draw the product of dehydration of each aldol product in Example 19.1.

Solution

Loss of H_2O from aldol product (a) gives an α,β-unsaturated aldehyde; loss of H_2O from (b) gives an α,β-unsaturated ketone.

(a) 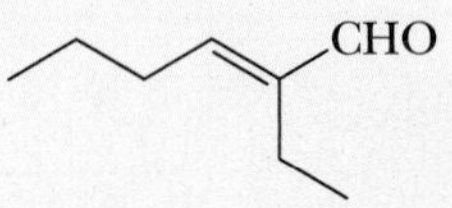

(b) 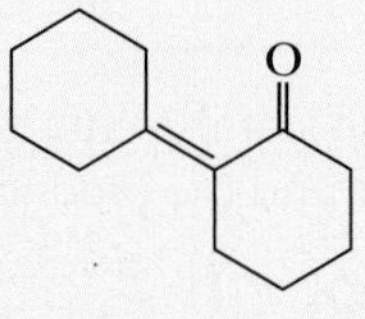

Problem 19.2

Draw the product of dehydration of each aldol product in Problem 19.1.

B. Crossed and Intramolecular Aldol Reactions

The reactants in the key step of an aldol reaction are an enolate anion and an enolate anion acceptor. In self-reactions, both roles are played by one kind of molecule. **Crossed aldol reactions** are also possible, as, for example, the crossed aldol reaction between acetone and formaldehyde. Formaldehyde cannot provide an enolate anion because it has no α-hydrogen, but it can function as a particularly good enolate anion acceptor because its carbonyl group is unhindered. Acetone forms an enolate anion, but its carbonyl group, which is bonded to two alkyl groups, is less reactive than that of formaldehyde. Consequently, the crossed aldol reaction between acetone and formaldehyde gives 4-hydroxy-2-butanone.

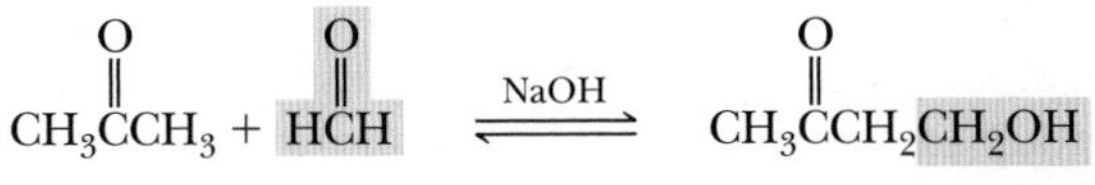

4-Hydroxy-2-butanone

As this example illustrates, for a crossed aldol reaction to be successful, one of the two reactants should have no α-hydrogens so that an enolate anion does not form. It also helps if the compound with no α-hydrogens has the more reactive carbonyl (e.g., an aldehyde). If these requirements are not met, a complex mixture of products results. Following are examples of aldehydes that have no α-hydrogens and can be used in crossed aldol reactions.

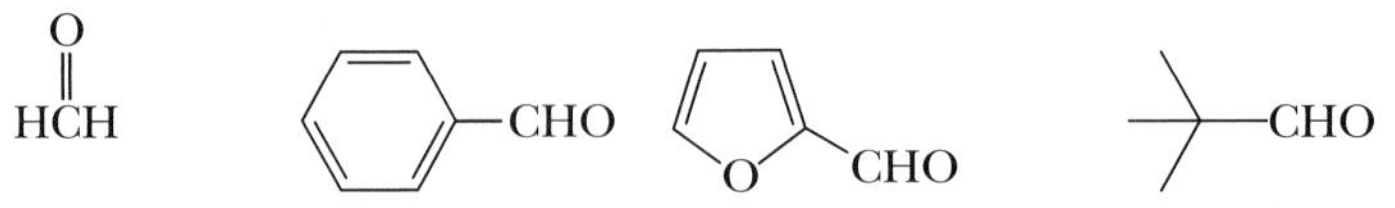

Formaldehyde **Benzaldehyde** **Furfural** **2,2-Dimethylpropanal**

Example 19.3 Crossed Aldol Reactions

Draw a structural formula for the product of the base-catalyzed crossed aldol reaction between furfural and cyclohexanone and for the product formed by its dehydration.

Solution

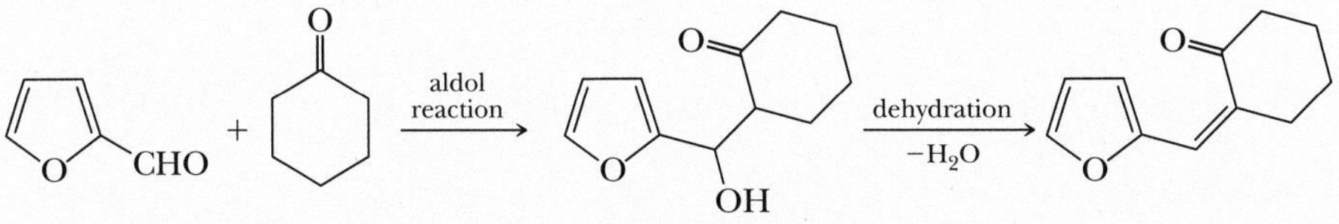

Furfural **Cyclohexanone** Aldol product (racemic)

Problem 19.3

Draw the product of the base-catalyzed crossed aldol reaction between benzaldehyde and 3-pentanone and the product formed by its dehydration.

Nitro groups can be introduced into aliphatic compounds by way of an aldol reaction between the anion of a nitroalkane and an aldehyde or a ketone. The α-hydrogens of nitroalkanes are sufficiently acidic that they are removed by bases such as aqueous NaOH and KOH. The pK_a of nitromethane, for example, is 10.2. The acidity of the α-hydrogen of a nitroalkane is caused by the stabilization of the resulting anion by delocalization of its negative charge into the nitro group.

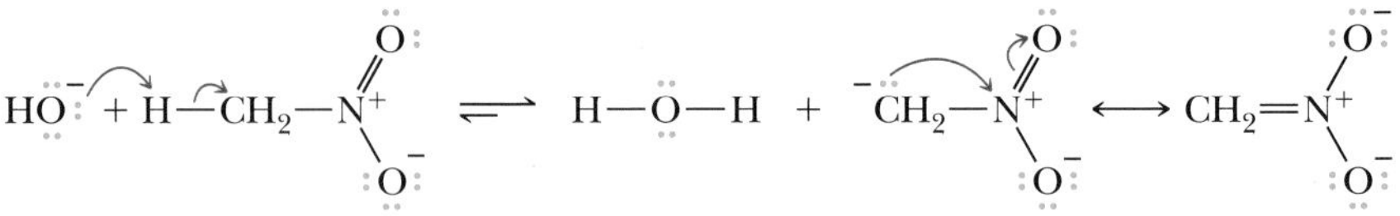

Nitromethane pK_a 10.2 (stronger acid)

Water pK_a 15.7 (weaker acid)

Anion stabilized by charge delocalization

Following is an aldol reaction between nitromethane and cyclohexanone. Reduction of the nitro group in the aldol product thus formed is a convenient synthetic route to β-aminoalcohols.

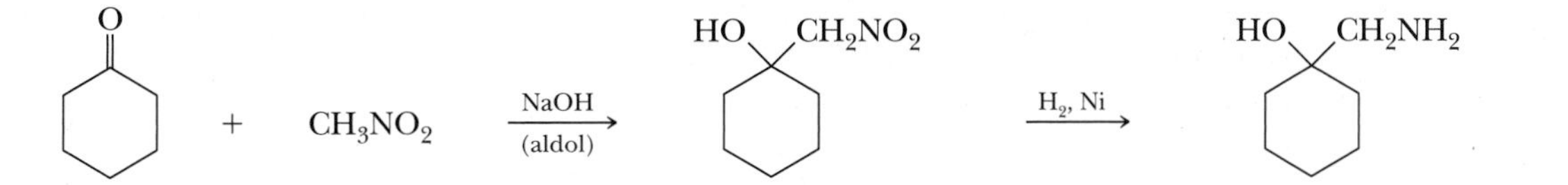

Cyclohexanone **Nitromethane** **1-(Nitromethyl)cyclohexanol** **1-(Aminomethyl)cyclohexanol**

When both the enolate anion and the carbonyl group to which it adds are in the same molecule, the aldol reaction results in the formation of a ring. This type of **intramolecular aldol reaction** is particularly useful for the formation of five- and six-membered rings. Because they are the most stable rings, five- and six-membered rings form much more readily than four-, seven-, or larger membered rings. Intramolecular aldol reaction of 2,7-octanedione via an enolate anion at α_3, for example, gives a five-membered ring. Intramolecular aldol reaction of this same compound via an enolate anion at α_1 would give a seven-membered ring. In the case of 2,7-octadienone, the five-membered ring forms in preference to the seven-membered ring.

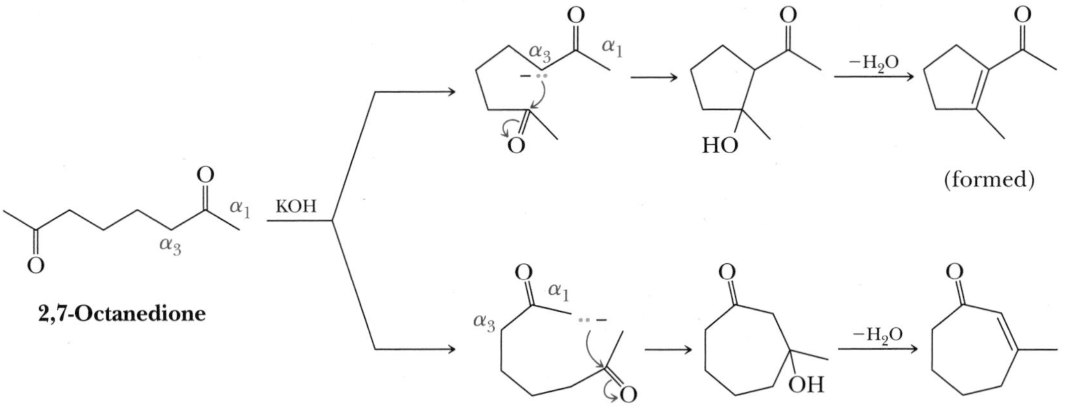

2,7-Octanedione

(formed)

(not formed)

In general, smaller rings form faster than larger rings because the reacting groups are closer together. However, the formation of three- and four-membered rings is disfavored because of the strain in them.

Following is another example in which either a four-membered ring (via an enolate anion at α_3) or a six-membered ring (via an enolate anion at α_1) could be formed. Because of the greater stability of six-membered rings compared to four-membered rings, the six-membered ring is formed exclusively in this intramolecular aldol reaction.

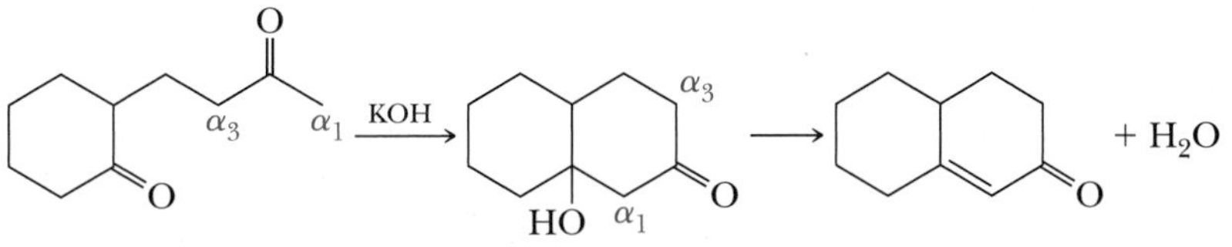

(racemic) (four stereoisomers as two racemic mixtures) (racemic)

C. Retrosynthetic Analysis

The functional groups created by the aldol reaction are a β-hydroxy carbonyl or an α,β-unsaturated carbonyl. Whenever you encounter these patterns in a target molecule, you should consider using an aldol reaction for its construction. Using retrosynthetic analysis (Section 7.9), the aldol product can be dissected into the proper reactants. Keep in mind that if the substituents R_1, R_2, and R_3 are too large, a reaction could fail for steric reasons. In addition, aldol reactions to give the β-hydroxy carbonyl product are generally not favorable for two ketones.

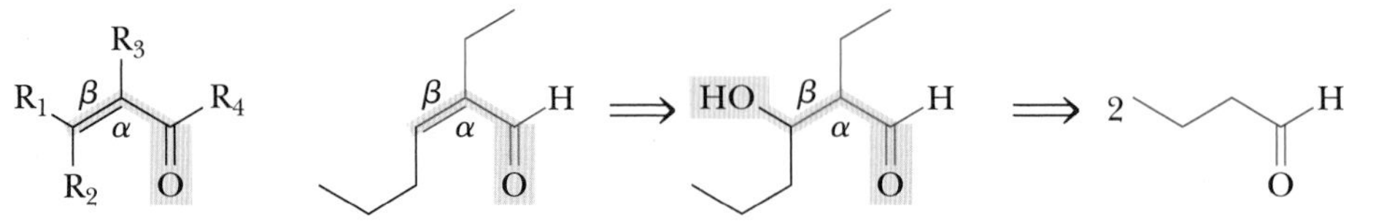

α,β-Unsaturated carbonyl
R_1–R_4 = alkyl group, aryl group, H atom

target product

(racemic)

reactants

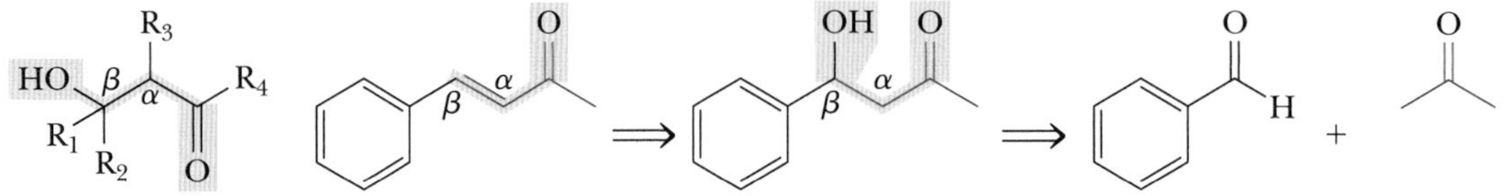

β-Hydroxy carbonyl
Aldol reaction product recognition elements

target product

(racemic)

reactants

19.3 Claisen and Dieckmann Condensations

A. Claisen Condensation

In Chapter 18, we described reactions of esters, all of which take place at the carbonyl carbon and involve nucleophilic acyl substitution by an addition/elimination. In this section, we examine a second type of reaction characteristic of esters, namely one that involves both formation of an enolate anion and its participation in nucleophilic acyl substitution. One of the first of these reactions discovered is the **Claisen condensation**, named after the German chemist Ludwig Claisen (1851–1930). A Claisen condensation is illustrated by the reaction of two molecules of ethyl acetate in the presence of sodium ethoxide followed by acidification to give ethyl acetoacetate.

$$2\ CH_3\overset{O}{\overset{\|}{C}}OEt \xrightarrow[2.\ H_2O,\ HCl]{1.\ EtO^-Na^+} CH_3\overset{O}{\overset{\|}{C}}CH_2\overset{O}{\overset{\|}{C}}OEt + EtOH$$

Ethyl ethanoate (Ethyl acetate) | **Ethyl 3-oxobutanoate (Ethyl acetoacetate)** | **Ethanol**

As this example illustrates, the product of a Claisen condensation is a **β-ketoester**.

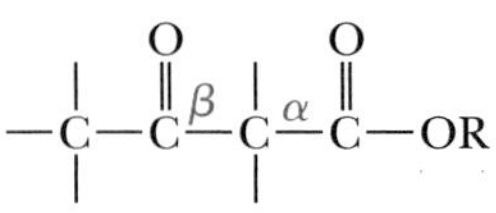

A β-ketoester

Like aldol reactions, Claisen condensations require a base. Aqueous bases such as NaOH, however, cannot be used in Claisen condensations because they would bring about the hydrolysis of the ester. Rather, the bases most commonly used in Claisen condensations are nonaqueous bases such as sodium ethoxide in ethanol and sodium methoxide in methanol.

Claisen condensation of two molecules of ethyl propanoate gives the following β-ketoester as a racemic mixture.

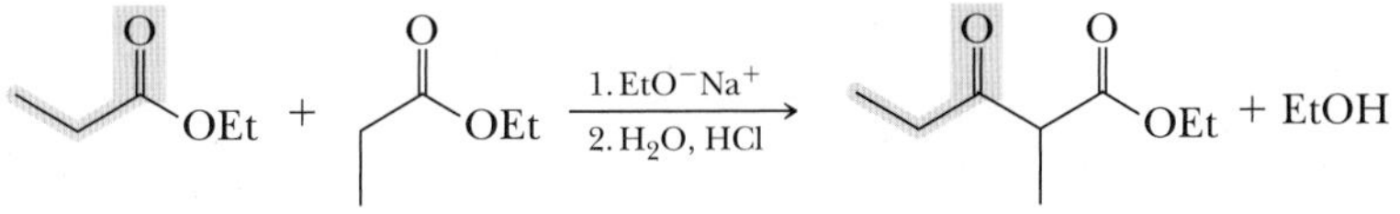

Ethyl propanoate **Ethyl propanoate** **Ethyl 2-methyl-3-oxopentanoate** (racemic)

The first steps of a Claisen condensation bear a close resemblance to the first steps of an aldol reaction (Section 19.2). The carbon-carbon bond-forming step in each reaction involves nucleophilic addition of an enolate anion to the carbonyl group of another molecule.

Watch a video explanation

Mechanism 19.4

Claisen Condensation

Step 1: **Take a proton away.** Removal of an α-hydrogen by base gives a resonance-stabilized enolate anion.

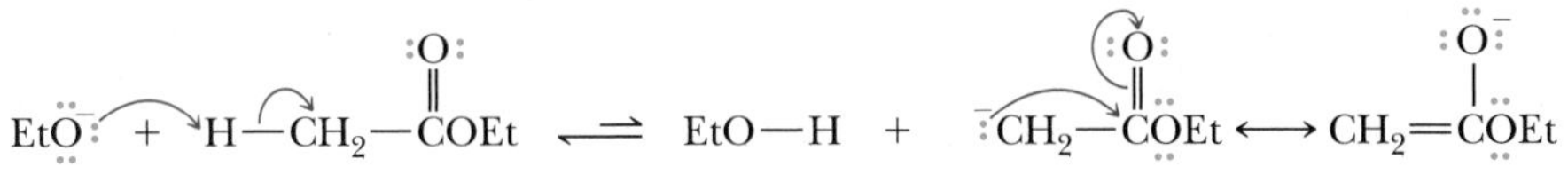

pK_a 22 (weaker acid) — pK_a 15.9 (stronger acid) — Resonance-stabilized enolate anion (stronger base)

Because the α-hydrogen of an ester is the weaker acid and ethoxide ion is the weaker base, the position of equilibrium for this step lies very much toward the left; the concentration of enolate anion is very low compared with that of ethoxide ion and ester. Thus, there is an excess of ester to react with the small amount of enolate anion that forms.

Step 2: **Make a new bond between a nucleophile and an electrophile.** Attack of the enolate anion of one ester on the carbonyl carbon of another ester gives a tetrahedral carbonyl addition intermediate.

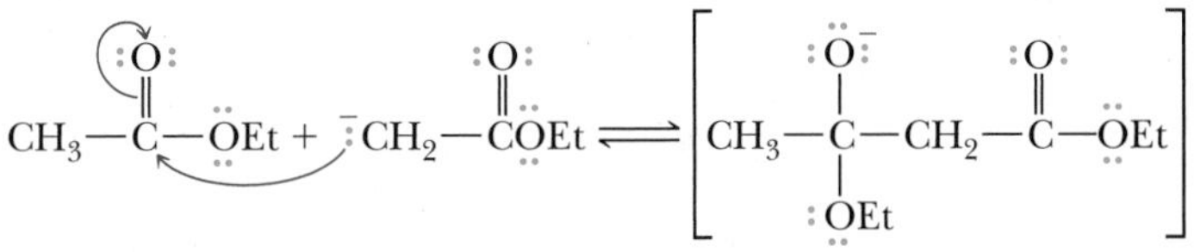

A tetrahedral carbonyl addition intermediate

Unlike similar intermediates in aldol reactions, this intermediate (a hemiacetal anion) has an ethoxy leaving group.

Step 3: **Break a bond to give stable molecules or ions.** Collapse of the tetrahedral carbonyl addition intermediate and ejection of ethoxide ion gives a β-ketoester.

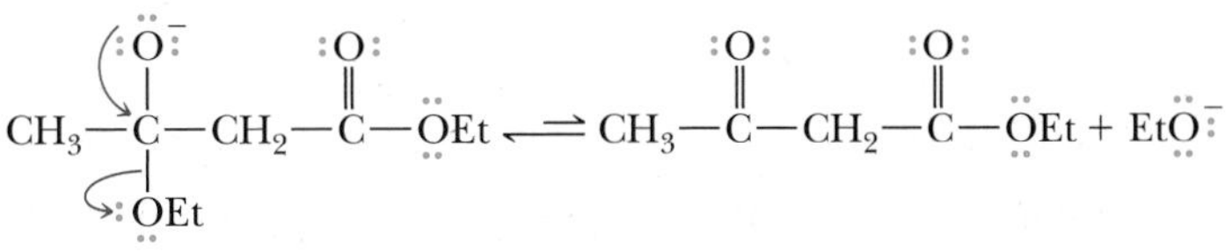

Step 4: **Take a proton away.** The position of equilibrium for Steps 1–3 lies very much on the side of starting materials. The overall condensation, however, is driven to completion by the acid-base reaction between the

(Continued)

β-ketoester (the stronger acid) and ethoxide ion (the stronger base) to give ethanol (the weaker acid) and the anion of the β-ketoester (the weaker base).

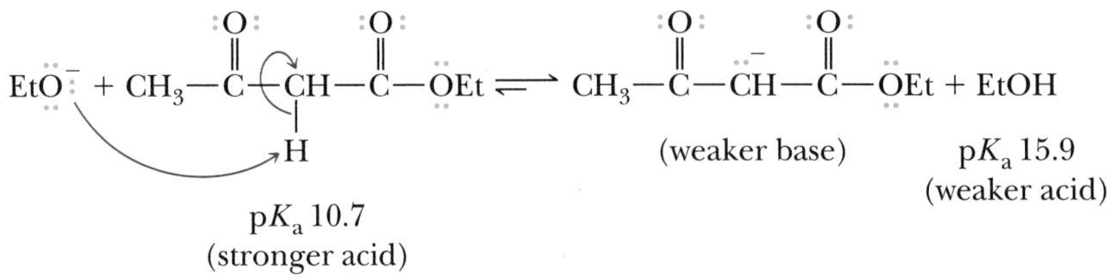
EtO
CH3
C
CH
OEt
H
pKa 10.7
(stronger acid)
(weaker base)
EtOH
pKa 15.9
(weaker acid)

Overall, the Claisen condensation involves consumption of a stronger base (in this case, ethoxide) and creation of a weaker base, the resonance-stabilized enolate anion of the β-ketoester. One molecule of the original base is consumed for every two molecules of ester that react. This is in contrast to an aldol reaction, in which base is catalytic (not consumed).

Step 5: **Add a proton.** The chemist then opens the flask and adds mild acid to protonate the enolate anion giving the β-ketoester.

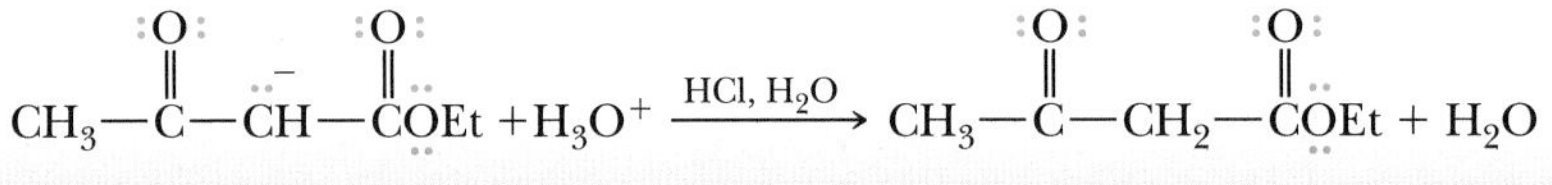
CH3
C
CH
COEt
H3O+
HCl, H2O
CH2
H2O

From an analysis of this mechanism, we see that the structural feature required for a successful Claisen condensation is an ester with two α-hydrogens, one to form the initial enolate anion and the second to form the enolate anion of the resulting β-ketoester.

Example 19.4 Claisen Condensation

Show the product of the Claisen condensation of ethyl butanoate in the presence of sodium ethoxide followed by acidification with aqueous HCl.

Solution

The new bond formed in a Claisen condensation is between the carbonyl group of one ester molecule and the α-carbon of another.

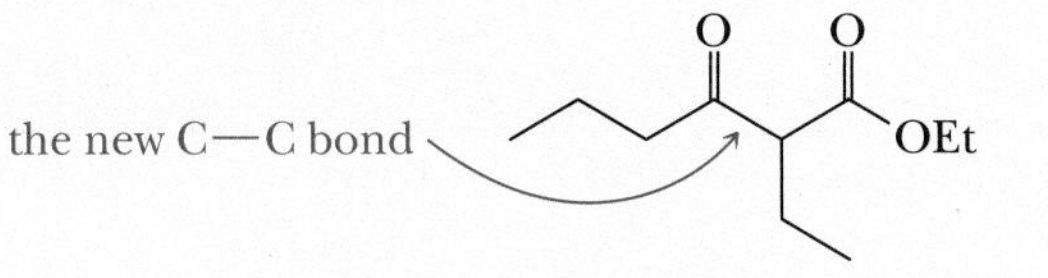
the new C—C bond
O
O
OEt

Ethyl 2-ethyl-3-oxohexanoate
(racemic)

Problem 19.4

Show the product of Claisen condensation of ethyl 3-methylbutanoate in the presence of sodium ethoxide followed by acidification with aqueous HCl.

B. Dieckmann Condensation

An intramolecular Claisen condensation of a dicarboxylic ester to give a five- or six-membered ring is given the special name of **Dieckmann condensation**. In the presence of one equivalent of sodium ethoxide, for example, diethyl hexanedioate (diethyl adipate) undergoes an intramolecular condensation to form a five-membered ring.

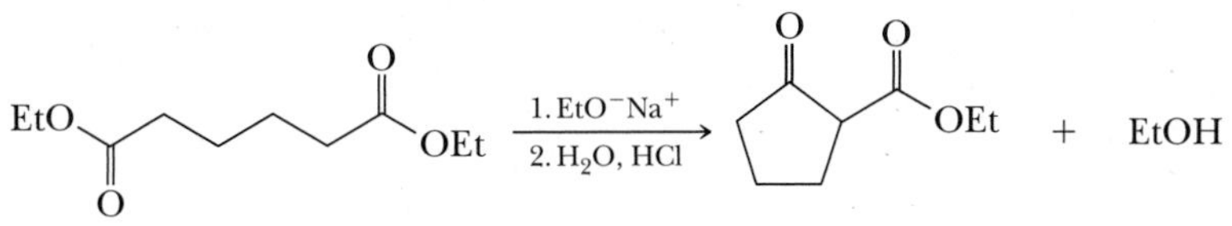

Diethyl hexanedioate (Diethyl adipate)

Ethyl 2-oxocyclopentanecarboxylate (racemic)

The mechanism of a Dieckmann condensation is identical to the mechanism we described for the Claisen condensation. An anion formed at the α-carbon of one ester group adds to the carbonyl of the other ester group to form a tetrahedral carbonyl addition intermediate (Make a bond between a nucleophile and an electrophile). This intermediate ejects ethoxide ion to regenerate the carbonyl group (Break a bond to give stable molecules or ions). Cyclization is followed by formation of the conjugate base of the β-ketoester, as in the Claisen condensation (Take a proton away). The β-ketoester is isolated after acidification with aqueous acid (Add a proton).

C. Crossed Claisen Condensations

In a **crossed Claisen condensation** between two different esters, each with two α-hydrogens, a mixture of four β-ketoesters is possible; therefore, crossed Claisen condensations of this type are not synthetically useful. Such condensations are useful, however, if appreciable differences in reactivity exist between the two esters, as, for example, when one of the esters has no α-hydrogens and can function only as an enolate anion acceptor. Following are four examples of esters without α-hydrogens.

Ethyl formate **Diethyl carbonate** **Diethyl ethanedioate (Diethyl oxalate)** **Ethyl benzoate**

Crossed Claisen condensations of this type are usually carried out by using the ester with no α-hydrogens in excess. In the following illustration, methyl benzoate is used in excess.

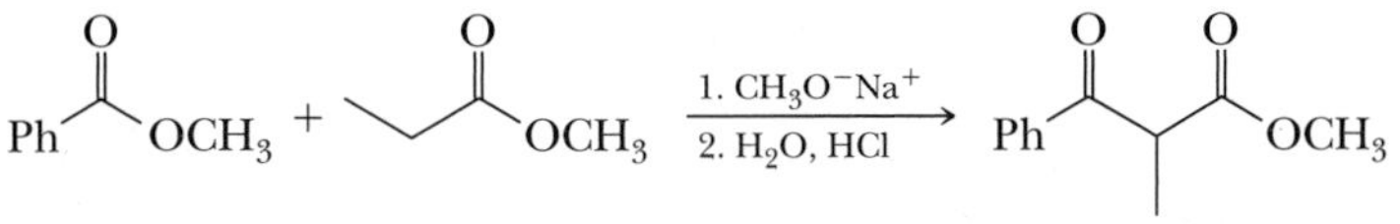

Methyl benzoate **Methyl propanoate** **Methyl 2-methyl-3-oxo-3-phenylpropanoate** (racemic)

Example 19.5 Crossed Claisen Condensation

Complete the equation for this crossed Claisen condensation.

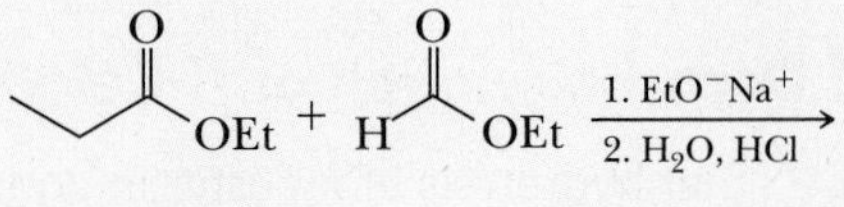

(Continued)

Solution

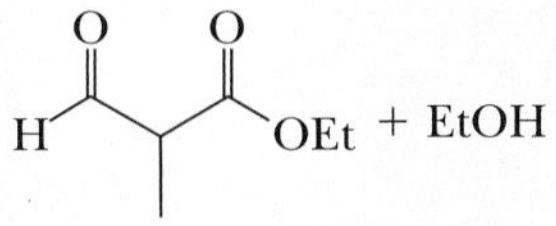

(racemic)

Problem 19.5

Complete the equation for this crossed Claisen condensation.

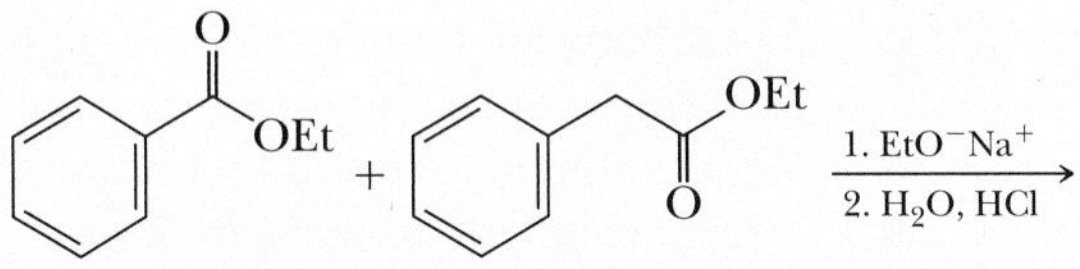

(excess)

D. Retrosynthetic Analysis

The functional group pattern created by a Claisen, mixed Claisen, or Dieckmann condensation is a β-ketoester. Using retrosynthetic analysis, a desired β-ketoester target molecule is dissected into the corresponding ester molecules. For the Dieckmann condensation, the β-ketoester is contained within a ring, generally a five- or six-membered ring, which is derived from the corresponding diester.

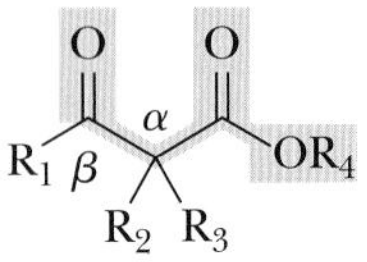

***β*-Ketoester**

Claisen condensation product recognition element

R_1–R_3 = alkyl group, aryl group, H atom

R_4 = alkyl group or aryl group

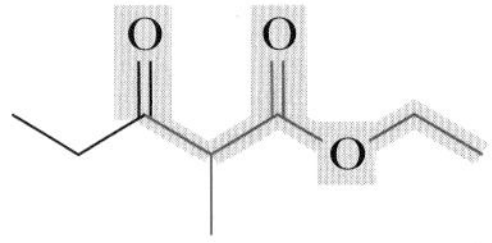

target product (racemic)

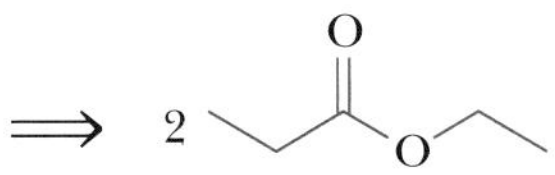

an ester

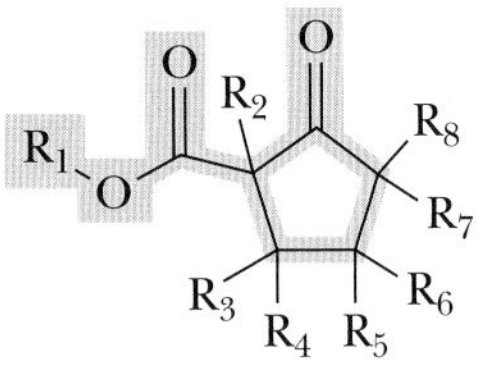

Cyclic β-ketoester

Dieckmann condensation product recognition element

R_1 = alkyl group or aryl group

R_2–R_8 = alkyl group, aryl group, H atom

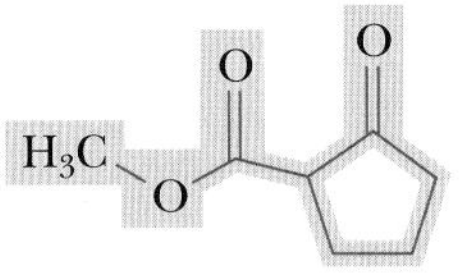

target product (racemic)

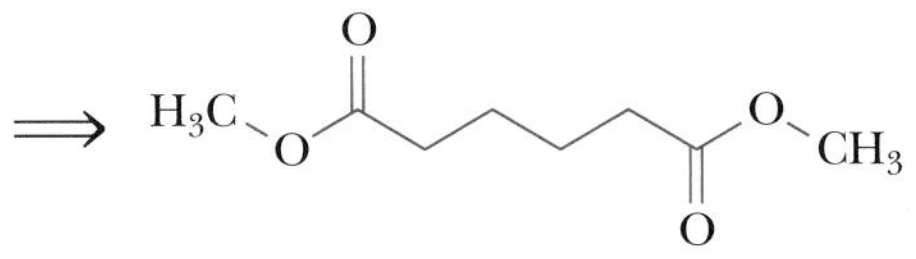

a diester

E. Hydrolysis and Decarboxylation of β-Ketoesters

Recall from Section 18.4C that hydrolysis of an ester in aqueous sodium hydroxide (saponification) followed by acidification of the reaction mixture with aqueous HCl converts an ester to a carboxylic acid. Also recall from Section 17.9 that β-ketoacids and β-dicarboxylic acids (substituted malonic acids) readily undergo decarboxylation (lose CO_2) when heated. Both the Claisen and Dieckmann condensations yield esters of β-ketoacids. The following equations illustrate the results of a Claisen condensation followed by hydrolysis of the ester, acidification, and decarboxylation.

Claisen condensation:

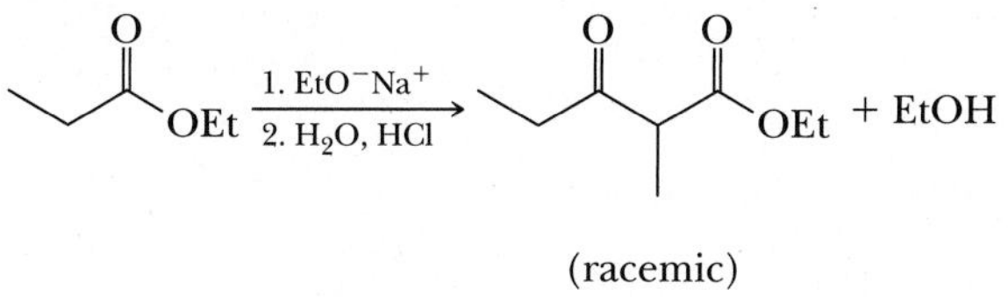

Saponification followed by acidification:

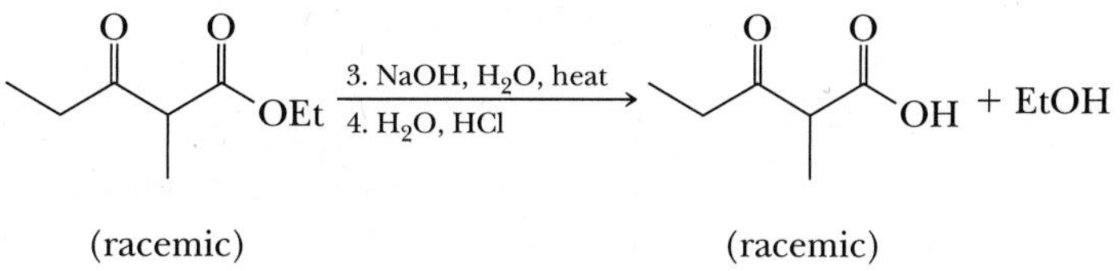

Decarboxylation:

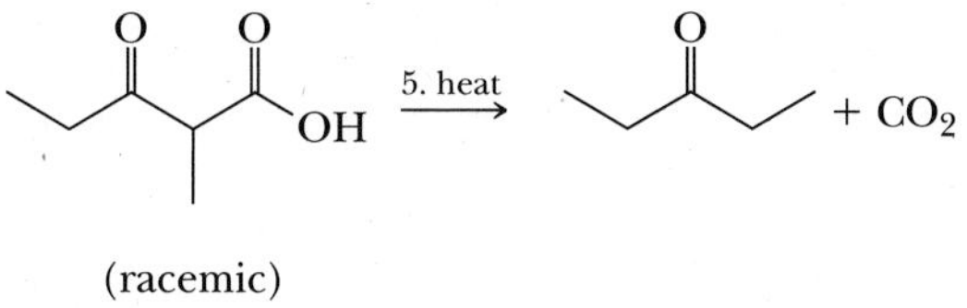

The result of this five-step sequence is reaction between two molecules of ester (one furnishing a carbonyl group and the other furnishing an enolate anion) to give a ketone and carbon dioxide. In the general reaction, both ester molecules are the same and the product is a symmetrical ketone.

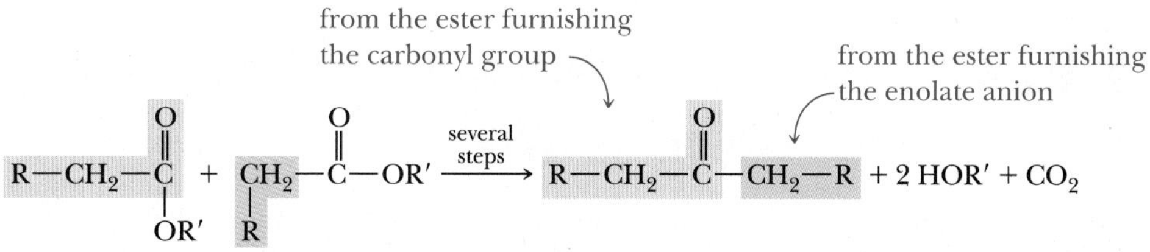

The same sequence of reactions starting with a crossed Claisen condensation gives an unsymmetrical ketone.

Example 19.6 Claisen, Hydrolysis, Decarboxylation

Each compound or set of compounds undergoes a Claisen or Dieckmann condensation followed by acidification, saponification, acidification, and thermal decarboxylation.

(a) $PhCOEt + CH_3COEt$ (each with C=O) **(b)**

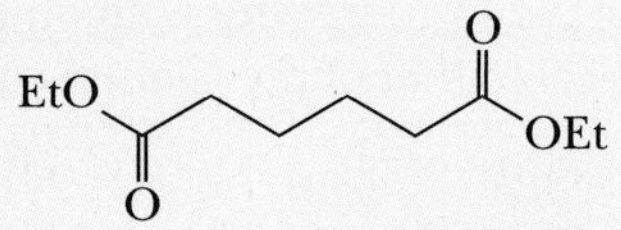

Draw a structural formula of the product isolated after completion of this reaction sequence.

(Continued)

Solution

Steps 1 and 2 bring about a crossed Claisen or Dieckmann condensation to give a β-ketoester. Steps 3 and 4 bring about hydrolysis of the β-ketoester to give a β-ketoacid, and Step 5 brings about decarboxylation to give a ketone.

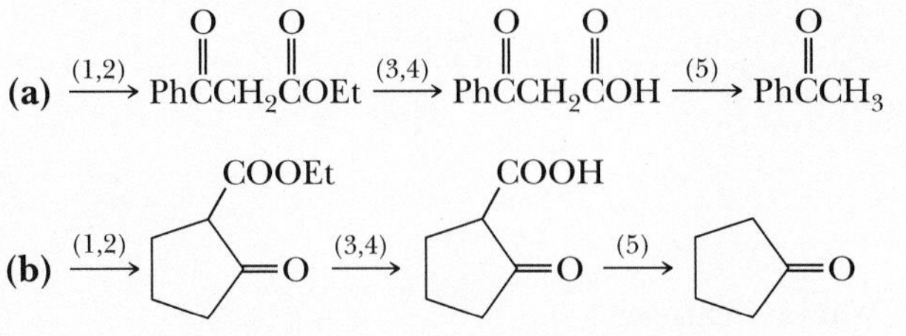
(a)
(1,2)
PhCCH2COEt
(3,4)
PhCCH2COH
(5)
PhCCH3
(b)
COOEt
COOH

(racemic) (racemic)

Problem 19.6

Show how to convert benzoic acid to 3-methyl-1-phenyl-1-butanone (isobutyl phenyl ketone) by the following synthetic strategies, each of which uses a different type of reaction to form the new carbon-carbon bond to the carbonyl group of benzoic acid.

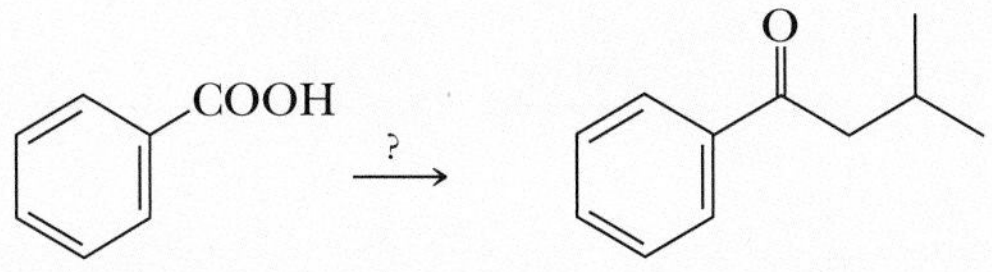
COOH
?
O

Benzoic acid **3-Methyl-1-phenyl-1-butanone**

(a) A lithium diorganocopper (Gilman) reagent **(b)** A Claisen condensation

19.4 Claisen and Aldol Condensations in the Biological World

Carbonyl condensations are among the most widely used reactions in the biological world for the assembly of new carbon-carbon bonds. One source of carbon atoms for the synthesis of biomolecules is **acetyl-CoA**, a thioester of acetic acid and the thiol group of coenzyme A (Problem 25.34). In this section, we examine the series of reactions by which the carbon skeleton of acetic acid is converted to isopentenyl pyrophosphate, a key intermediate in the synthesis of terpenes, cholesterol, steroid hormones, and bile acids. Note that in the discussion that follows, we will not be concerned with the mechanism by which each of these enzyme-catalyzed reactions occurs. Rather, our concern is in recognizing the types of reactions that take place.

In a Claisen condensation catalyzed by the enzyme thiolase, acetyl-CoA is converted to its enolate anion, which then adds to the carbonyl group of a second molecule of acetyl-CoA to give a tetrahedral carbonyl addition intermediate. Collapse of this intermediate by elimination of coenzyme A anion ($CoAS^-$) gives acetoacetyl-CoA. Subsequent proton transfer to coenzyme A anion gives coenzyme A. The mechanism of this reaction is identical to that of the Claisen condensation (Section 19.3A).

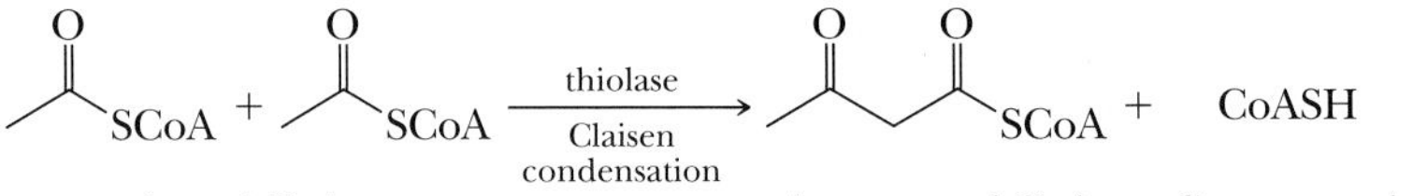
O
SCoA
+
O
SCoA
thiolase
Claisen condensation
O
O
SCoA
+
CoASH

Acetyl-CoA **Acetoacetyl-CoA** **Coenzyme A**

Enzyme-catalyzed aldol reaction with a third molecule of acetyl-CoA on the ketone carbonyl of acetoacetyl-CoA gives (*S*)-3-hydroxy-3-methylglutaryl-CoA.

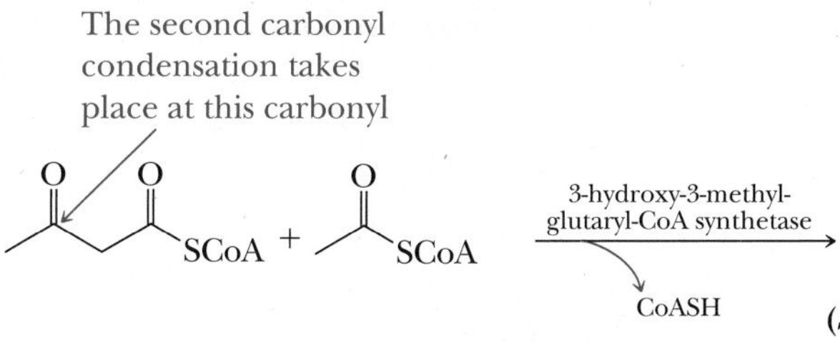

(*S*)-3-Hydroxy-3-methylglutaryl-CoA

Note three features of this second carbonyl-condensation reaction.

- The reaction is completely enantioselective; only the *S* enantiomer is formed. Condensation takes place in a chiral environment created by the enzyme, 3-hydroxy-3-methylglutaryl-CoA synthetase, which induces the formation of one enantiomer of the product to the exclusion of the other.

Chemical Connections

Drugs That Lower Plasma Levels of Cholesterol

Coronary artery disease is the leading cause of death in the United States and other Western countries, where about one-half of all deaths can be attributed to atherosclerosis. Atherosclerosis results from the buildup of fatty deposits called plaque on the inner walls of arteries. A major component of plaque is cholesterol derived from low-density lipoproteins (LDLs), which circulate in blood plasma. Because more than one-half of total body cholesterol in humans is synthesized in the liver from acetyl-CoA, intensive efforts have been directed toward finding ways to inhibit this synthesis. The rate-determining step in cholesterol biosynthesis is reduction of (*S*)-3-hydroxy-3-methylglutaryl CoA to (*R*)-mevalonic acid. This reduction is catalyzed by the enzyme HMG-CoA reductase and requires two moles of NADPH per mole of HMG-CoA.

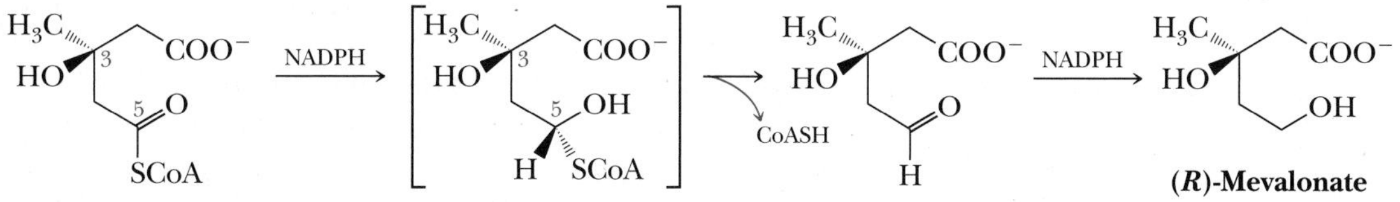

(*S*)-3-Hydroxy-3-methyl-glutaryl-CoA (HMG-CoA)

A hemithioacetal intermediate formed by the first NADPH reduction

(*R*)-Mevalonate

Beginning in the early 1970s, researchers at the Sankyo Company in Tokyo screened more than 8000 strains of microorganisms and in 1976 announced the isolation of mevastatin, a potent inhibitor of HMG-CoA reductase, from culture broths of the fungus *Penicillium citrinum*. The same compound was isolated by researchers at Beecham Pharmaceuticals in England from cultures of *Penicillium brevicompactum*. Soon thereafter, a second, more active compound called lovastatin was isolated at the Sankyo Company from the fungus *Monascus ruber* and at Merck Sharpe & Dohme from *Aspergillus terreus*. Both mold metabolites are extremely effective in lowering plasma concentrations of LDL. The active form of each is the 5-hydroxycarboxylic acid formed by hydrolysis of the δ-lactone.

Soon thereafter, Merck developed a synthesis for simvastatin (Zocor), which came on the market in the late 1980s and is still used worldwide for the control of plasma cholesterol levels. These drugs and several synthetic modifications now available inhibit HMG-CoA reductase by forming an enzyme-inhibitor complex that prevents further catalytic action of the enzyme. It is reasoned that the 3,5-dihydroxycarboxylic acid part of each drug binds tightly to the enzyme because it mimics the hemithioacetal intermediate formed by the first reduction of HMG-CoA.

(*Continued*)

Systematic studies have shown the importance of each part of the drug for effectiveness. It has been found, for example, that the carboxylate anion (—COO^-) is essential, as are both the 3-OH and 5-OH groups. Insertion of a bridging unit other than —CH_2—CH_2— between carbon 5 and the two fused six-membered rings reduces potency, as does almost any modification of the six-membered rings and their pattern of substitution.

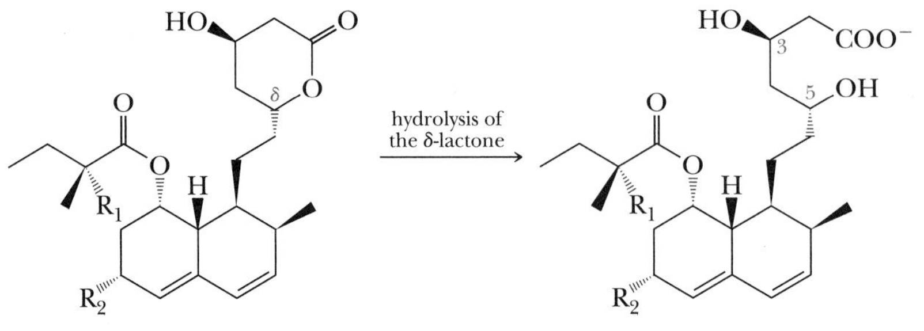

$R_1 = R_2 =$ H, mevastatin
$R_1 =$ H, $R_2 = CH_3$, lovastatin (Mevacor)
$R_1 = R_2 = CH_3$, simvastatin (Zocor)

The active form of each drug

- Hydrolysis of the thioester group of acetyl-CoA is coupled with the aldol reaction.
- The carboxyl group is shown as it is ionized at pH 7.4, the approximate pH of blood plasma and many cellular fluids.

Enzyme-catalyzed reduction by NADPH (a phosphorylated form of NADH) of the thioester group of 3-hydroxy-3-methylglutaryl-CoA to a primary alcohol gives mevalonic acid, shown here as its anion.

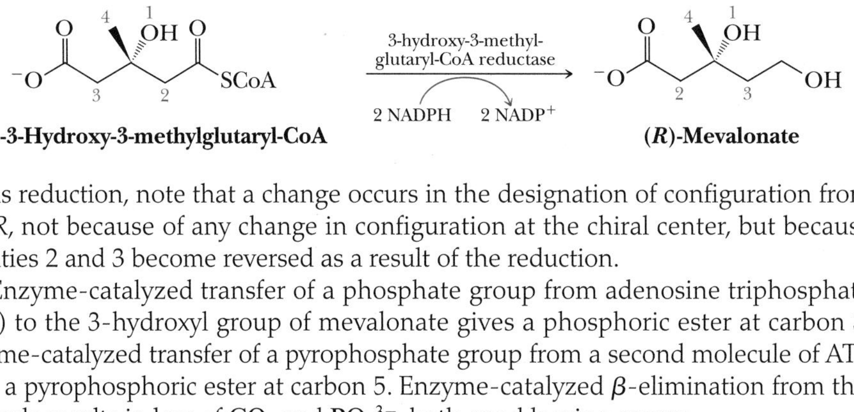

(*S*)-3-Hydroxy-3-methylglutaryl-CoA **(*R*)-Mevalonate**

In this reduction, note that a change occurs in the designation of configuration from *S* to *R*, not because of any change in configuration at the chiral center, but because priorities 2 and 3 become reversed as a result of the reduction.

Enzyme-catalyzed transfer of a phosphate group from adenosine triphosphate (ATP) to the 3-hydroxyl group of mevalonate gives a phosphoric ester at carbon 3. Enzyme-catalyzed transfer of a pyrophosphate group from a second molecule of ATP gives a pyrophosphoric ester at carbon 5. Enzyme-catalyzed β-elimination from this molecule results in loss of CO_2 and PO_4^{3-}, both good leaving groups.

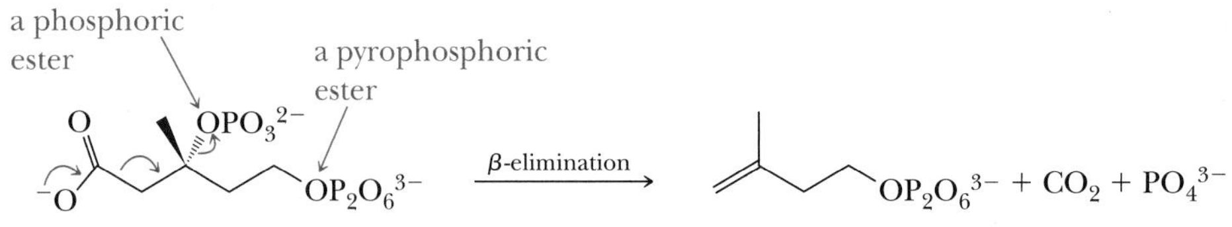

(*R*)-3-Phospho-5-pyrophosphomevalonate **Isopentenyl pyrophosphate**

Isopentenyl pyrophosphate has the carbon skeleton of isoprene, the unit into which terpenes can be divided (Section 5.4). This compound is, in fact, a key intermediate in the synthesis of terpenes, as well as of cholesterol and steroid hormones. We shall return to the chemistry of isopentenyl pyrophosphate in Section 26.4B and discuss its conversion to cholesterol and terpenes.

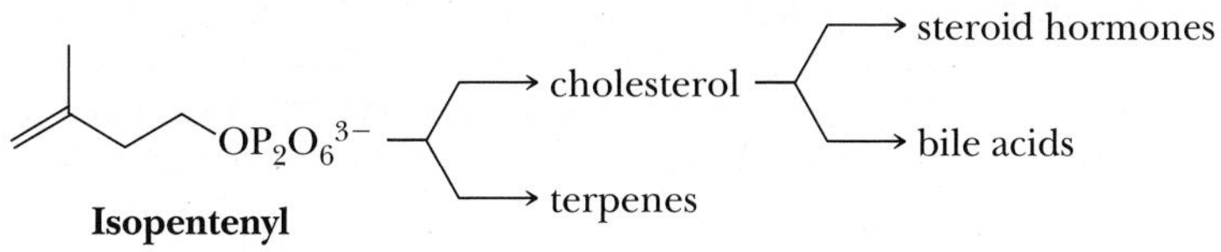

$OP_2O_6^{3-}$
cholesterol
steroid hormones
bile acids
terpenes

Isopentenyl pyrophosphate

19.5 Enamines

Enamines are formed by the reaction of a secondary amine with an aldehyde or a ketone (Section 16.8A). The secondary amines most commonly used for this purpose are pyrrolidine and morpholine (Figure 19.1).

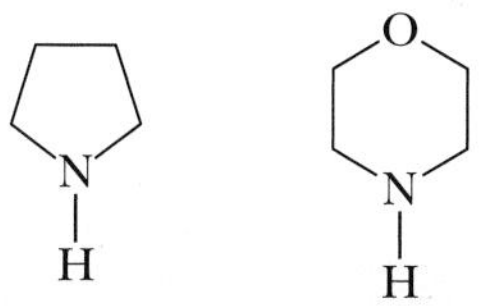

Pyrrolidine **Morpholine**

FIGURE 19.1 Secondary amines used in the formation of enamines.

Example 19.7 Reaction of Enamines

Draw structural formulas for the aminoalcohol and enamine formed in the following reactions.

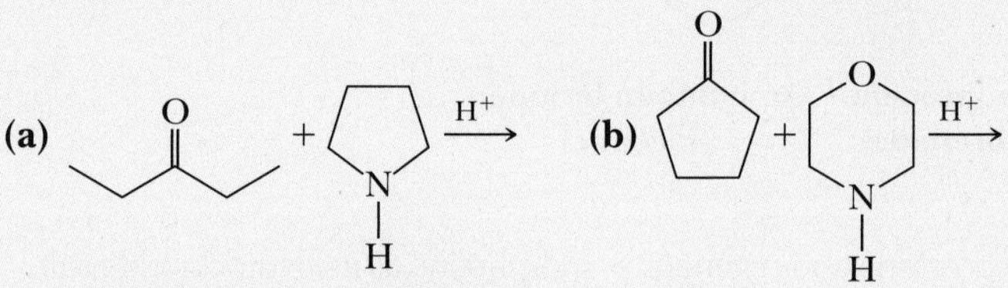

(a)
H^+
(b)
H^+

Solution

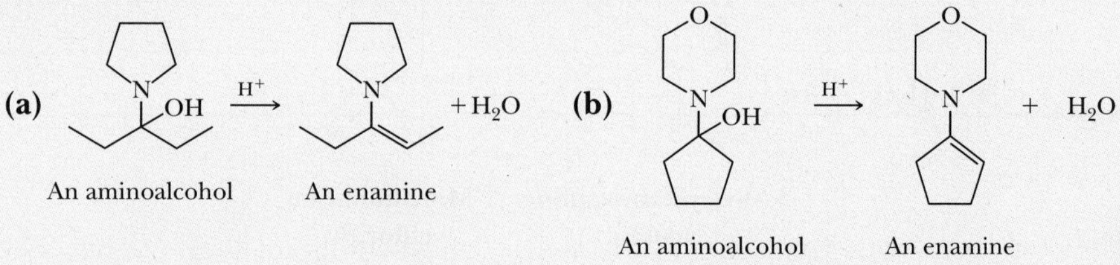

(a)
OH
H^+
$+H_2O$
An aminoalcohol
An enamine
(b)
OH
H^+
$+ H_2O$
An aminoalcohol
An enamine

Problem 19.7

Following are structural formulas for two enamines.

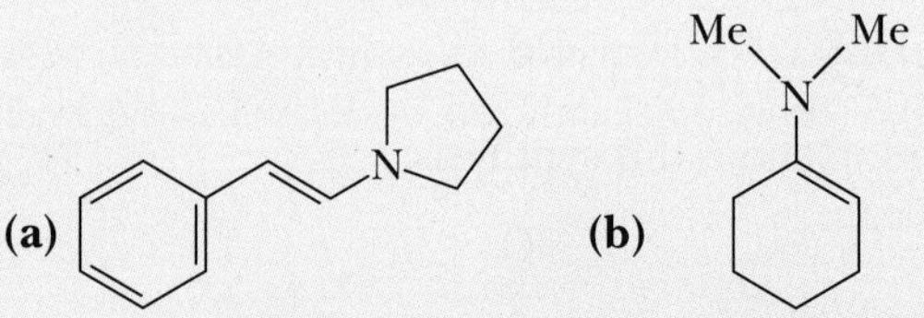

(a)
(b)
Me
Me

Draw structural formulas for the secondary amine and carbonyl compound from which each enamine is derived.

The particular value of enamines in synthetic organic chemistry is the fact that the β-carbon of an enamine is a nucleophile by virtue of the conjugation of the carbon-carbon double bond with the electron pair on nitrogen. Enamines resemble enols and enolate anions in their reactions.

The use of enamines as synthetic intermediates for the alkylation and acylation at the α-carbon of aldehydes and ketones was pioneered by Gilbert Stork of Columbia University. This use of enamines is called the Stork enamine reaction.

A. Alkylation of Enamines

Enamines readily undergo S_N2 reactions with methyl and primary haloalkanes, α-haloketones, and α-haloesters. Enamines are superior to enolate anions for these reactions because they are less basic and consequently give higher ratios of substitution to elimination products. In addition, they also give more alkylation on carbon than do enolate anions.

Mechanism 19.5

Alkylation of an Enamine

Step 1: **Make a new bond between a nucleophile and an electrophile.** Treatment of the enamine with one equivalent of an alkylating agent gives an iminium halide.

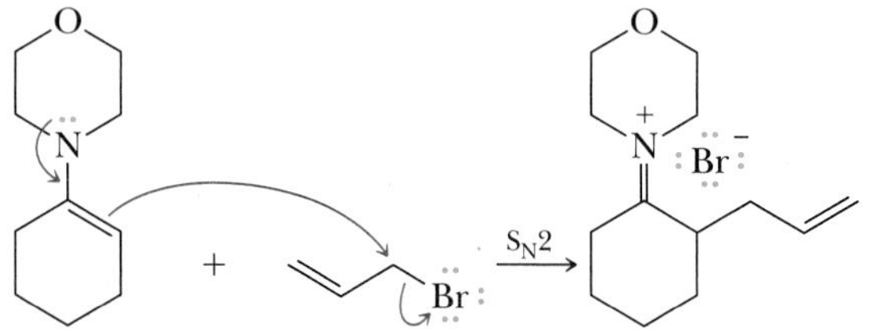

The morpholine enamine of cyclohexanone | **3-Bromopropene (Allyl bromide)** | An iminium bromide (racemic)

Step 2: Hydrolysis of the iminium salt gives the alkylated ketone and regenerates morpholine as its hydrochloride salt.

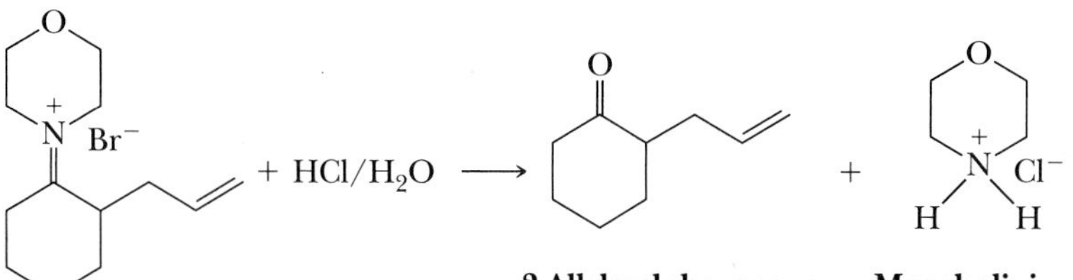

(racemic) | **2-Allylcyclohexanone** (racemic) | **Morpholinium chloride**

Example 19.8 Using Enamines I

Show how to use an enamine to bring about this synthesis.

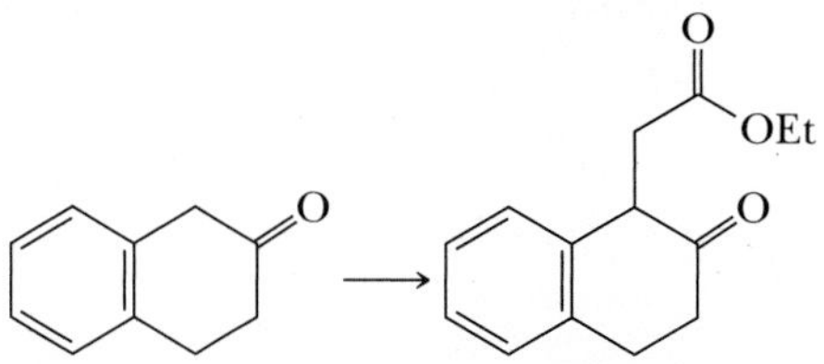

(racemic)

Solution

Prepare an enamine by treating the ketone with either morpholine or pyrrolidine. The intermediate aminoalcohol can undergo dehydration in two directions. The direction shown here is favored because of the stabilization gained by conjugation of the carbon-carbon double bond of the enamine with the aromatic ring. Treatment of the enamine with ethyl 2-chloroacetate followed by hydrolysis of the iminium chloride in aqueous hydrochloric acid gives the product.

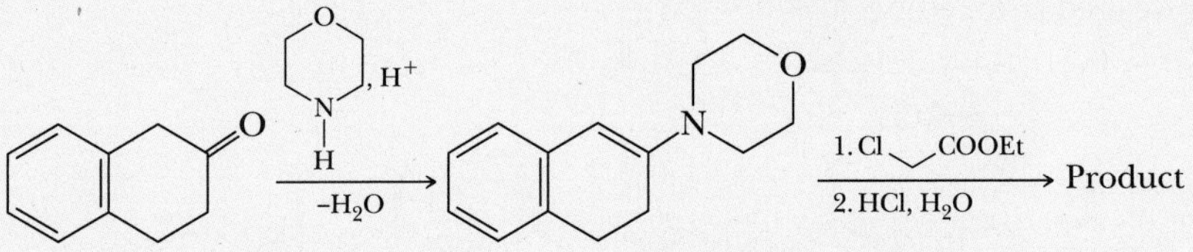

Problem 19.8

Write a mechanism for the hydrolysis of the following iminium chloride in aqueous HCl.

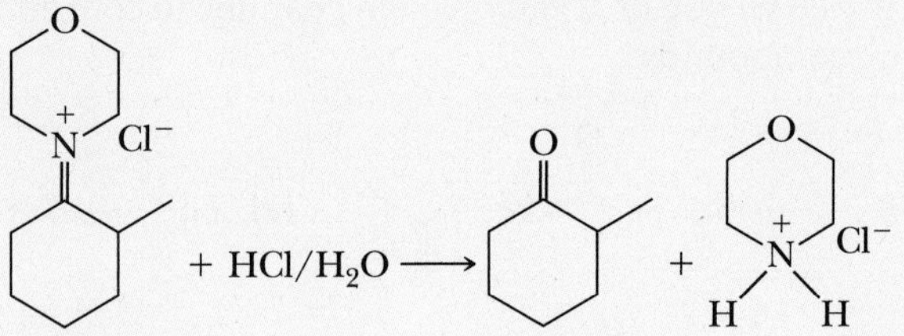

(racemic) (racemic)

B. Acylation of Enamines

Enamines undergo acylation when treated with acid chlorides and acid anhydrides. The reaction is a nucleophilic acyl substitution as illustrated by the conversion of cyclohexanone, via its pyrrolidine enamine, to 2-acetylcyclohexanone.

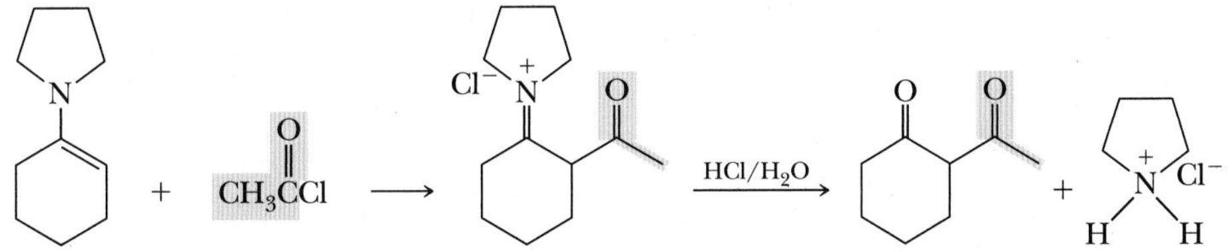

The pyrrolidine enamine of cyclohexanone — **Acetyl chloride** — An iminium chloride (racemic) — **2-Acetylcyclohexanone** (racemic)

Thus, we can attach an acyl group to the α-carbon of an aldehyde or a ketone using its enamine as an intermediate. The process of introducing an acyl group onto an organic molecule is called **acylation**.

Acylation The process of introducing an acyl group, RCO— or ArCO—, onto an organic molecule.

Example 19.9 Using Enamines II

Show how to use an enamine to bring about this synthesis.

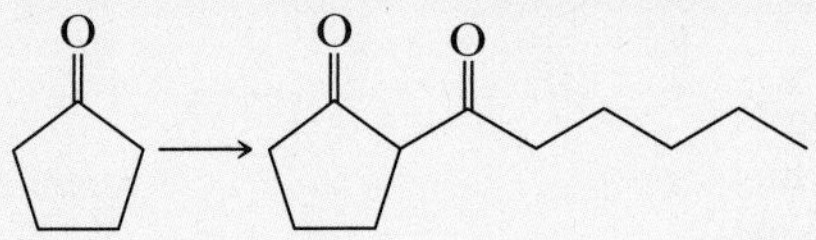

(racemic)

(Continued)

Solution

Treating cyclopentanone with pyrrolidine gives an enamine. Treating the enamine with hexanoyl chloride followed by hydrolysis in aqueous HCl gives the desired β-diketone.

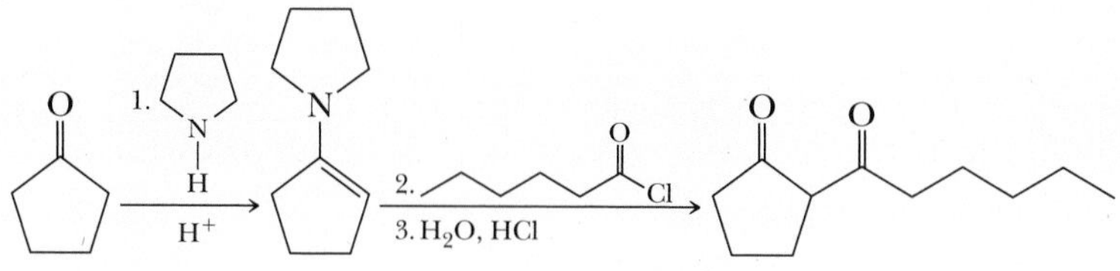

(racemic)

Watch a video explanation

Problem 19.9

Show how to use alkylation or acylation of an enamine to convert acetophenone to the following compounds.

(a) 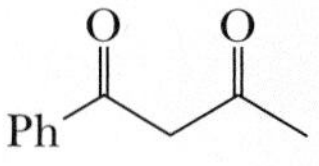

(b) 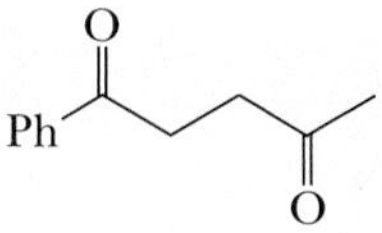

(c) 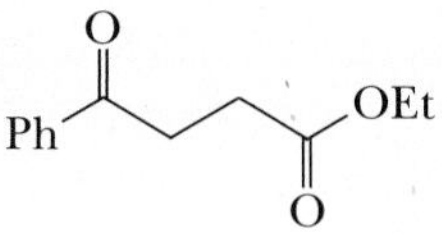

C. Retrosynthetic Analysis

You learned that enamines can undergo both alkylation and acylation reactions. The characteristic functional group created by enamine alkylation is a carbonyl compound with an α-carbon possessing a group that can be derived from an S_N2 reaction (methyl, primary, possibly secondary, but not tertiary halide). The characteristic functional group created by enamine acylation is a β-dicarbonyl (aldehyde or ketone). Thus, in a retrosynthetic analysis, a desired alkylation product can be derived from a carbonyl reactant with an α-hydrogen, a secondary amine (usually pyrrolidine), and the new group R bonded to a leaving group (Lv). A desired acylation product can be derived from a carbonyl reactant with an α-hydrogen, a secondary amine (again usually pyrrolidine), and the appropriate acid chloride.

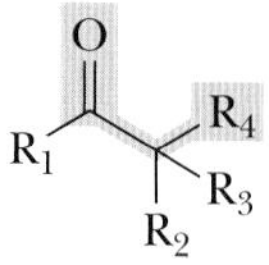

Alkylated α-carbon

Enamine alkylation reaction product recognition element

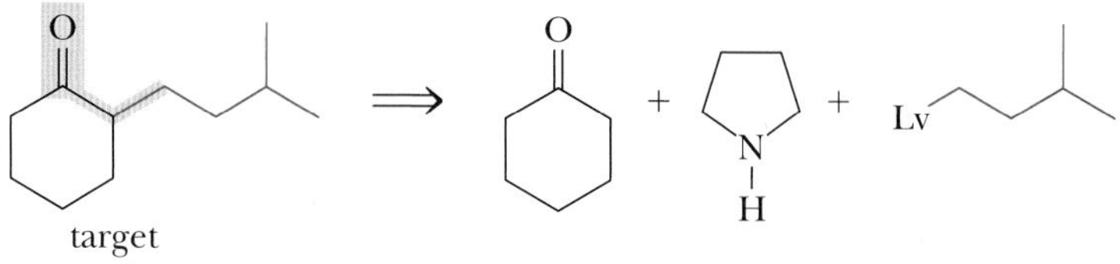

target product (racemic)

R_1–R_3 = alkyl group, aryl group, H atom
R_4 = methyl, primary or possibly secondary alkyl group

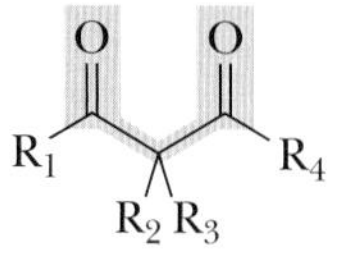

β-Dicarbonyl

Enamine acylation reaction product recognition element

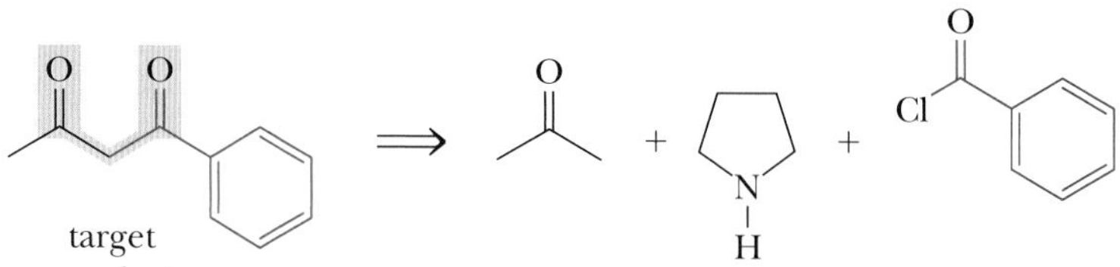

target product (racemic)

R_1–R_4 = alkyl group or aryl group or H atom

19.6 Acetoacetic Ester Synthesis

Acetoacetic ester and other β-ketoesters are versatile starting materials for the formation of new carbon-carbon bonds because of:

1. The acidity of α-hydrogens (pK_a 10–11) between the two carbonyl groups.
2. The nucleophilicity of the enolate anion resulting from loss of an α-hydrogen.
3. The ability of the product to undergo decarboxylation after hydrolysis of the ester.

The **acetoacetic ester synthesis** is useful for the preparation of monosubstituted and disubstituted acetones of the following types.

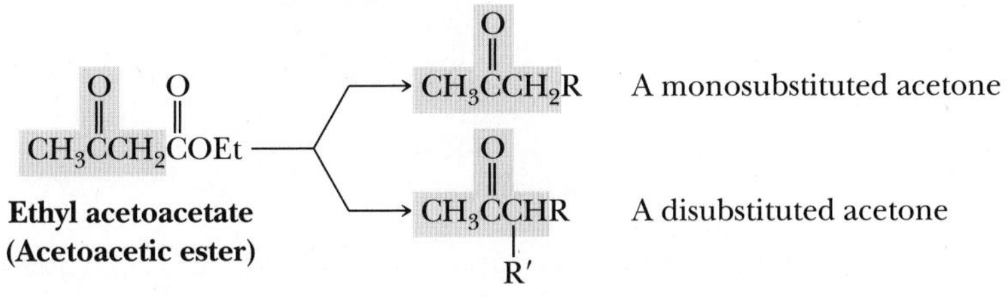

We have already seen the chemistry of the individual steps in this synthesis, but we have not put them together in this particular sequence. Let us illustrate the acetoacetic ester synthesis by choosing 5-hexen-2-one as a target molecule. The three carbons shown in color are provided by ethyl acetoacetate. The remaining three carbons represent the —R group of a substituted acetone.

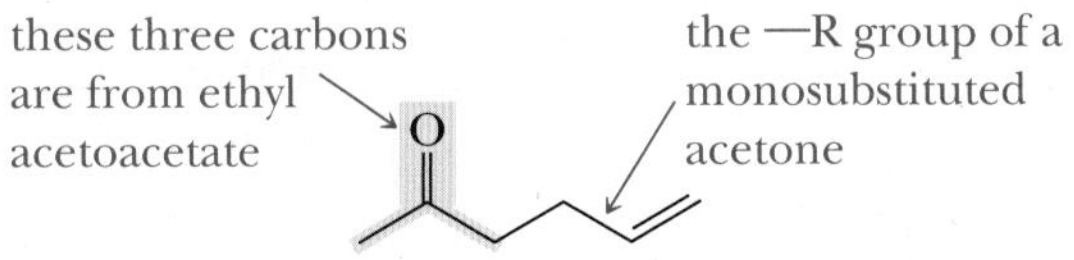

5-Hexen-2-one

A. Five Sequential Reactions

1. The methylene hydrogens of ethyl acetoacetate (pK_a 10.7) are considerably more acidic than the hydroxyl group of ethanol (pK_a 15.9); therefore, ethyl acetoacetate is converted completely to its anion by sodium ethoxide or other alkali metal alkoxides.

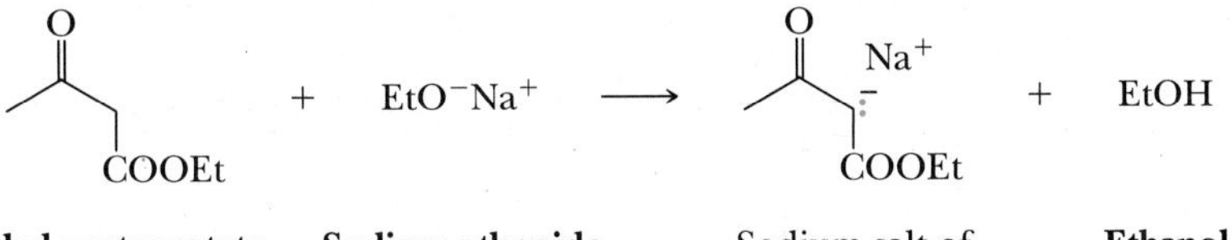

Ethyl acetoacetate	Sodium ethoxide	Sodium salt of ethyl acetoacetate	Ethanol
pK_a 10.7 (stronger acid)	(stronger base)	(weaker base)	pK_a 15.9 (weaker acid)

2. The enolate anion of ethyl acetoacetate is a nucleophile and reacts by an S_N2 pathway with methyl and primary haloalkanes, α-haloketones, and α-haloesters. Secondary haloalkanes give lower yields, and tertiary haloalkanes undergo E2 elimination exclusively. In the following example, the anion of ethyl acetoacetate is alkylated with allyl bromide.

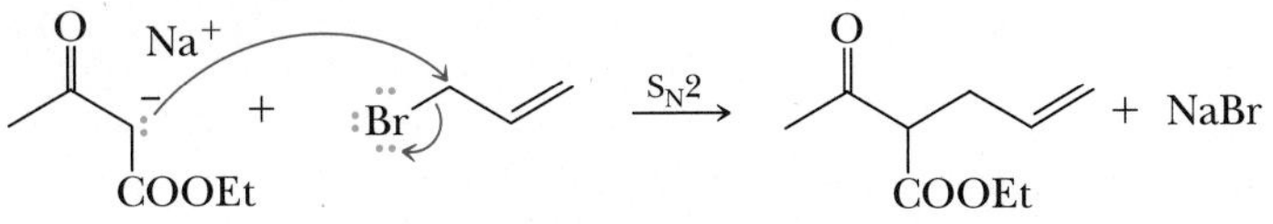

3-Bromopropene (Allyl bromide)

(racemic)

3, 4. Hydrolysis of the alkylated acetoacetic ester in aqueous NaOH followed by acidification with aqueous HCl (Section 18.4C) gives a β-ketoacid.

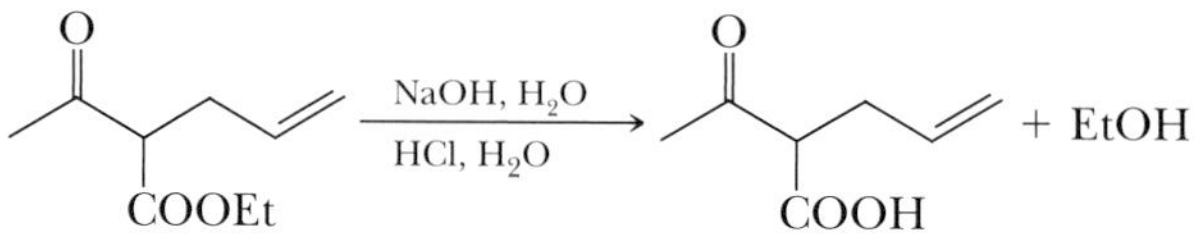

(racemic) (racemic)

5. Heating the β-ketoacid brings about decarboxylation (Section 17.9A) to give 5-hexen-2-one.

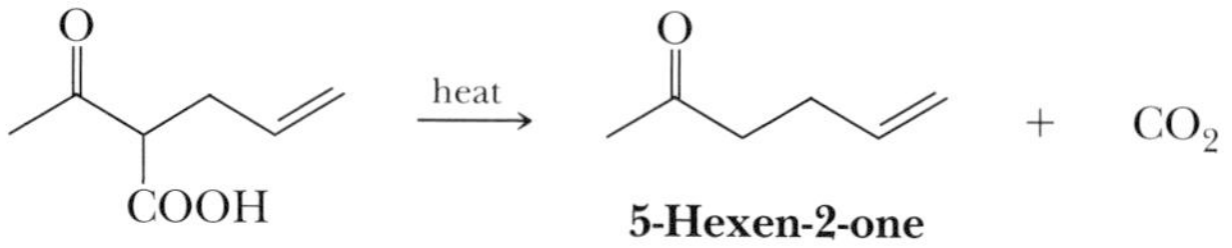

(racemic)

5-Hexen-2-one
(a monosubstituted acetone)

A disubstituted acetone can be prepared by interrupting this sequence after Step 2, treating the monosubstituted acetoacetic ester with a second equivalent of base, carrying out a second alkylation, and then proceeding with Steps 3–5.

1′. Treatment with a second equivalent of base gives a second enolate anion.

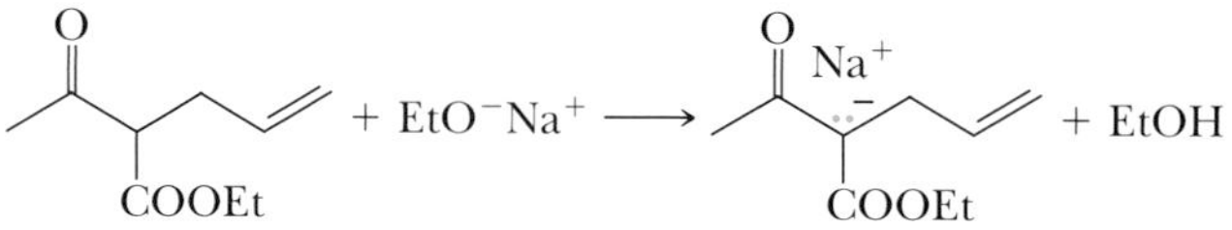

(racemic)

2′. Treatment of this enolate anion with a haloalkane completes the second alkylation. This haloalkane should be methyl or primary for best yields due to steric considerations.

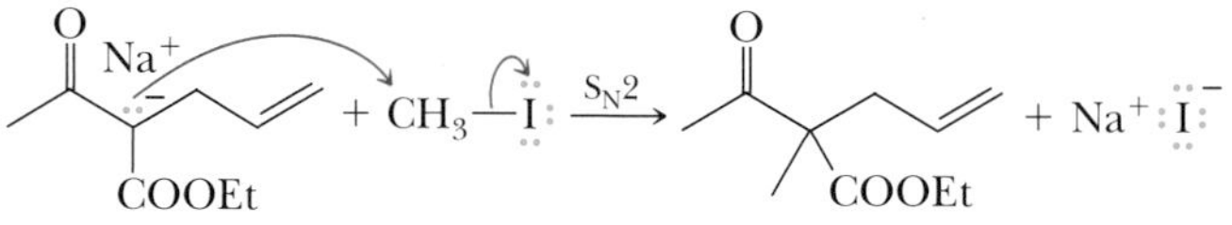

(racemic)

3′, 4′, 5′. Hydrolysis of the ester in aqueous base followed by acidification and heating gives the ketone.

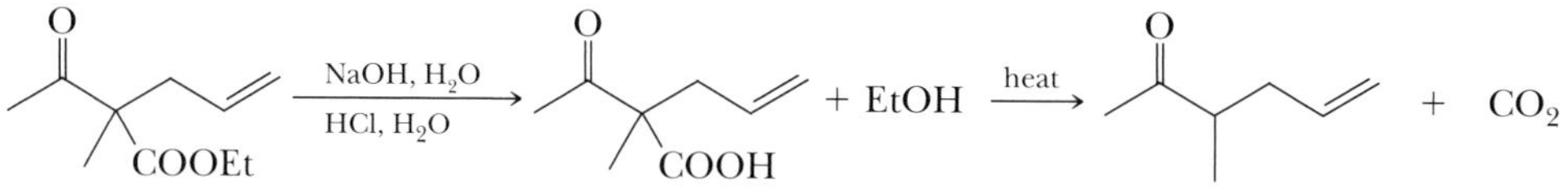

(racemic) (racemic)

3-Methyl-5-hexen-2-one
(a disubstituted acetone; racemic)

B. Retrosynthetic Analysis

The sequence of five steps used in the acetoacetic ester synthesis creates a carbonyl compound with an α-carbon possessing one or two groups, generally methyl, primary, or possibly secondary alkyl groups, which are derived from S_N2 substitution. This is a convenient and versatile method for making complex ketones. Using retrosynthetic analysis, one can derive a desired complex ketone from the appropriate β-ketoester and alkylating agent(s) (R-Lv). Note that other β-ketoesters besides ethyl acetoacetate can be used.

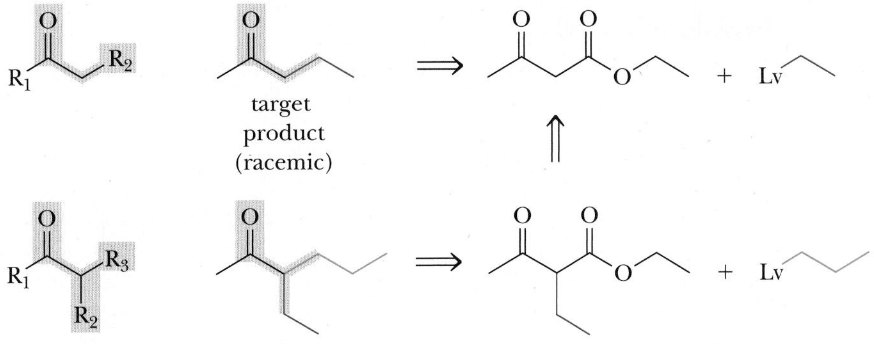

target
product
(racemic)
Lv
target
product
(racemic)
Lv
(racemic)

Alkylated ketone
Acetoacetic ester synthesis
product recognition elements

R_1 = alkyl group, aryl group, H atom
R_2, R_3 = methyl, primary or possibly secondary alkyl group

Example 19.10 Acetoacetic Ester Synthesis

Show how the acetoacetic ester synthesis can be used to prepare this ketone.

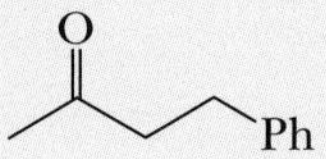

4-Phenyl-2-butanone

Solution

Determine which three carbons of the product originate from ethyl acetoacetate. Then establish the location on the carbon chain of the —COOH lost in decarboxylation and verify the bond formed in the alkylation step. On the basis of this analysis, determine that the starting materials are ethyl acetoacetate and a benzyl halide.

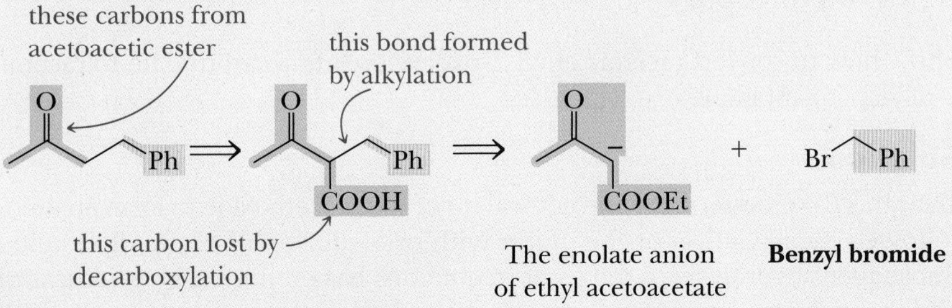

these carbons from
acetoacetic ester
this bond formed
by alkylation
Ph
Ph
COOH
COOEt
Br
Ph
this carbon lost by
decarboxylation
The enolate anion
of ethyl acetoacetate
Benzyl bromide

Now combine these reagents in the following way to prepare the desired ketone.

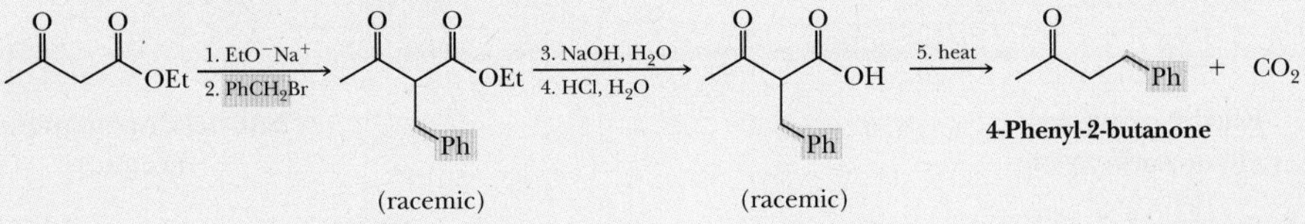

OEt
1. EtO⁻Na⁺
2. PhCH2Br
OEt
3. NaOH, H2O
4. HCl, H2O
OH
5. heat
Ph
+ CO2
Ph
Ph
(racemic)
(racemic)
4-Phenyl-2-butanone

(Continued)

Problem 19.10

Show how the acetoacetic ester synthesis can be used to prepare these compounds.

(a)

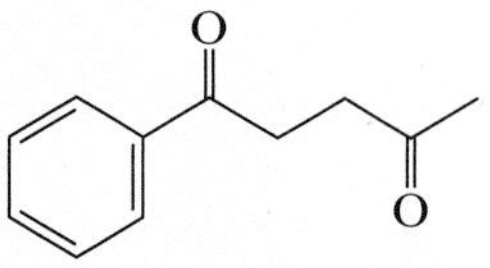

(b)

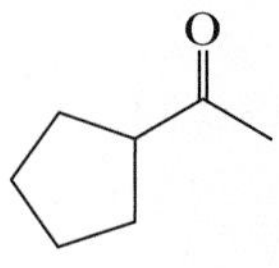

(c)

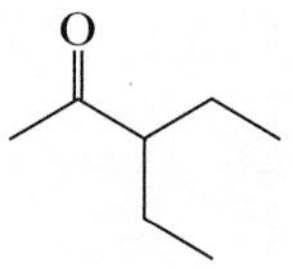

C. Variants

We have described what is commonly known as the acetoacetic ester synthesis and have illustrated the use of ethyl acetoacetate as the starting reagent. This same synthetic strategy is applicable to any β-ketoester, as, for example, those that are available by the Claisen (Section 19.3A) and Dieckmann (Section 19.3B) condensations. For example, following are structural formulas for two β-ketoesters available from Dieckmann and Claisen condensations that can be made to undergo (1) formation of an enolate anion, (2) alkylation or acylation, (3) hydrolysis followed by (4) acidification, and finally (5) decarboxylation just as we have shown for ethyl acetoacetate.

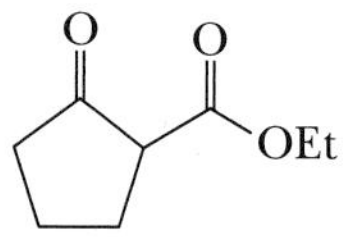
OEt

Ethyl 2-oxocyclopentanecarboxylate
(racemic)

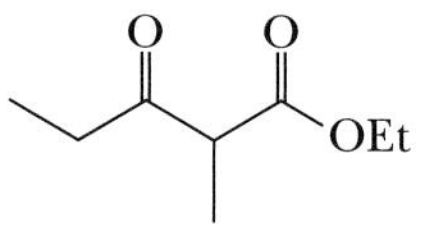
OEt

Ethyl 2-methyl-3-oxopentanoate
(racemic)

Example 19.11 Variants of Acetoacetic Ester Synthesis

Show how to convert racemic ethyl 2-oxocyclopentanecarboxylate to racemic 2-allylcyclopentanone.

Solution

Treat this β-ketoester with one equivalent of sodium ethoxide to form an anion followed by alkylation of the anion with one equivalent of an allyl halide. Subsequent hydrolysis of the ester in aqueous base followed by acidification and thermal decarboxylation gives the desired product.

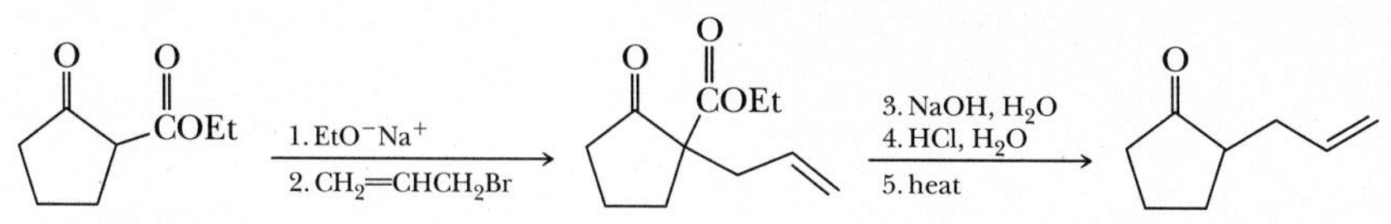
COEt
1. EtO⁻Na⁺
2. CH₂=CHCH₂Br
COEt
3. NaOH, H₂O
4. HCl, H₂O
5. heat

Ethyl 2-oxocyclo-pentanecarboxylate
(racemic)

2-Allylcyclopentanone
(racemic)

Problem 19.11

Show how to convert racemic ethyl 2-oxocyclopentanecarboxylate to this compound.

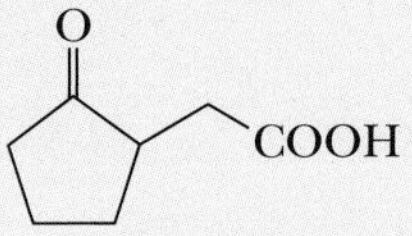

(racemic)

19.7 Malonic Ester Synthesis

The factors that make malonic esters and other β-diesters versatile starting materials for formation of new carbon-carbon bonds are the same as those we have already seen for the acetoacetic ester synthesis, namely:

1. The acidity of α-hydrogens (pK_a 13–14) between the two carbonyl groups.
2. The nucleophilicity of the enolate anion resulting from loss of an α-hydrogen.
3. The ability of the product to undergo decarboxylation after hydrolysis of the ester.

The **malonic ester synthesis** is useful for the preparation of monosubstituted and disubstituted acetic acids of the following types.

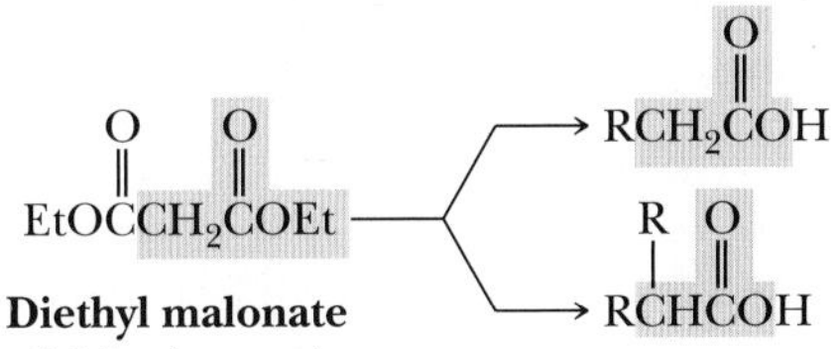

Diethyl malonate (Malonic ester)

A monosubstituted acetic acid

A disubstituted acetic acid

A. Five Sequential Reactions

As with the acetoacetic ester synthesis, we have already encountered all the important chemistry of the malonic ester synthesis, although not in this particular pattern. Let us illustrate this synthesis by choosing 5-methoxypentanoic acid as a target molecule. The two carbons shown in color are provided by diethyl malonate. The remaining three carbons and the methoxy group represent the —R group of a monosubstituted acetic acid.

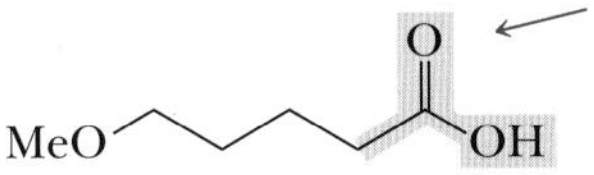

5-Methoxypentanoic acid

1. The α-hydrogens of diethyl malonate (pK_a 13.3) are more acidic than ethanol (pK_a 15.9); therefore, diethyl malonate is converted completely to its anion by sodium ethoxide or some other alkali metal alkoxide.

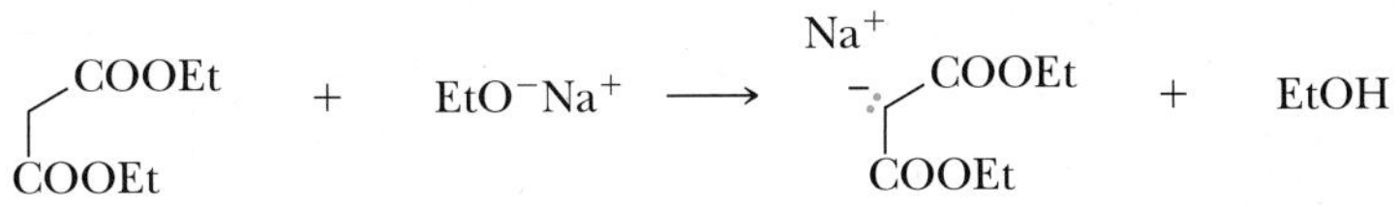

Diethyl malonate	Sodium ethoxide	Sodium salt of diethyl malonate	Ethanol
pK_a 13.3 (stronger acid)	(stronger base)	(weaker base)	pK_a 15.9 (weaker acid)

2. The enolate anion of diethyl malonate is a nucleophile and reacts by an S_N2 pathway with methyl and primary haloalkanes, α-haloketones, and α-haloesters. In the following example, the anion of diethyl malonate is alkylated with 1-bromo-3-methoxypropane.

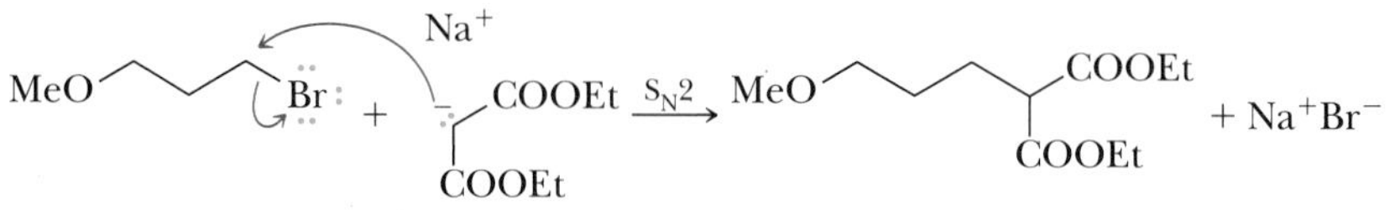

3, 4. Hydrolysis of the alkylated malonic ester in aqueous NaOH followed by acidification with aqueous HCl gives a β-dicarboxylic acid.

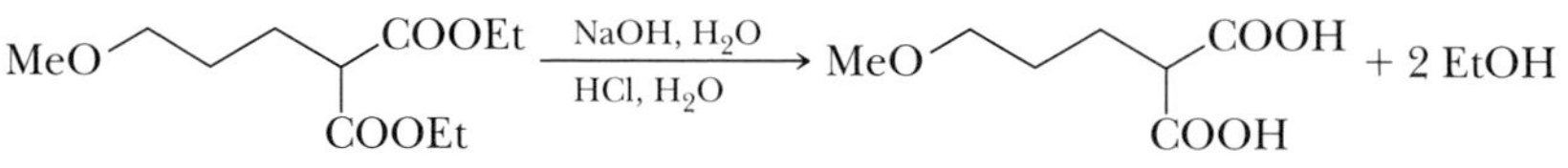

5. Heating the β-dicarboxylic acid slightly above its melting point brings about decarboxylation and gives 5-methoxypentanoic acid.

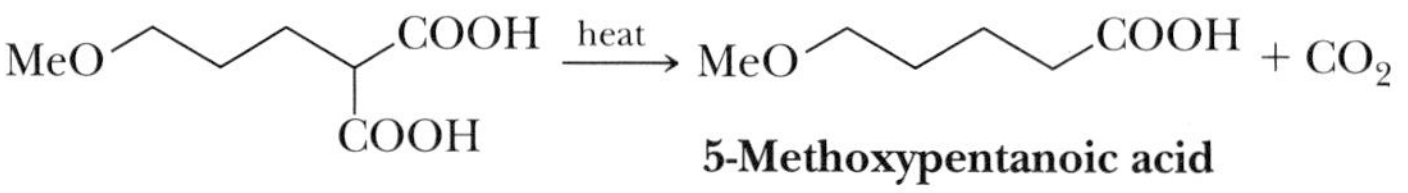

5-Methoxypentanoic acid

A disubstituted acetic acid can be prepared by interrupting the previous sequence after Step 2, treating the monosubstituted diethyl malonate with a second equivalent of base, carrying out a second alkylation, and then proceeding with Steps 3–5.

B. Retrosynthetic Analysis

The characteristic functional group produced by the malonic acid synthesis is a carboxylic acid in which the α-carbon has one or two methyl, primary, or possibly secondary alkyl groups. Using retrosynthetic analysis derives the desired carboxylic acid from a diester of malonic acid, and one or two appropriate methyl, primary, or possibly secondary alkyl groups with a leaving group attached, sufficient to allow an S_N2 reaction.

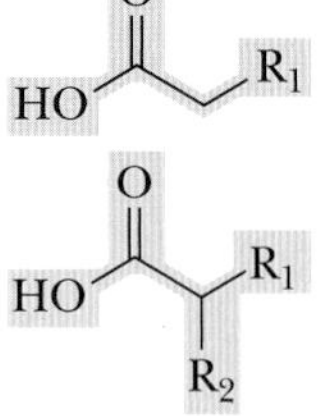

Alkylated carboxylic acid
Malonic ester synthesis product recognition elements

R_1, R_2 = methyl, primary or possibly secondary alkyl group

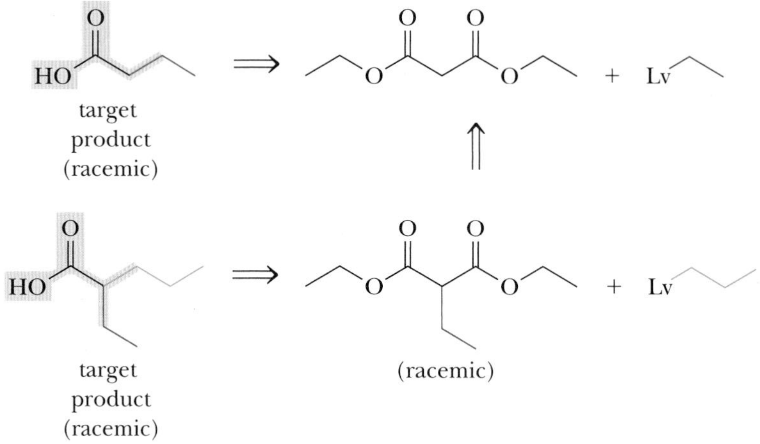

Example 19.12 Malonic Ester Synthesis

Show how the malonic ester synthesis can be used to prepare 3-phenylpropanoic acid.

Solution

Determine which two carbons of the product originate from diethyl malonate, the location on the carbon chain of the —COOH lost in decarboxylation, and the bond formed in the alkylation step. On the basis of this analysis, determine that the starting materials are diethyl malonate and a benzyl halide.

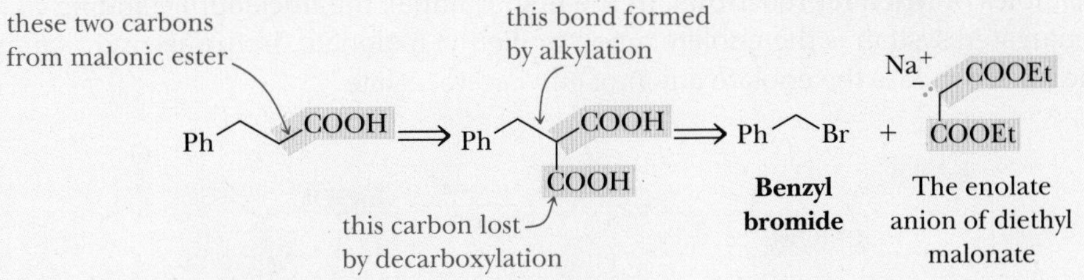

Now combine these reagents in the following way to get the desired product.

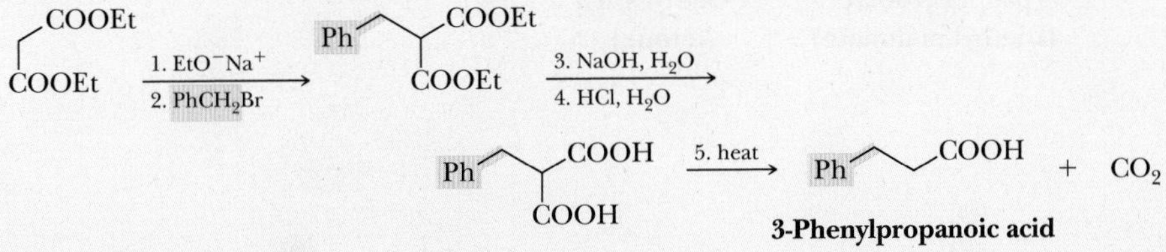

Watch a video explanation

Problem 19.12

Show how the malonic ester synthesis can be used to prepare the following substituted acetic acids.

(a) 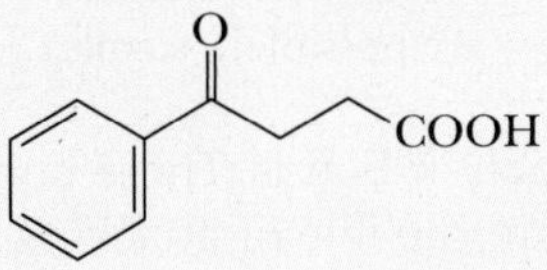

(b) 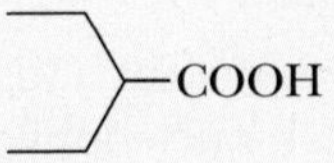

(c) 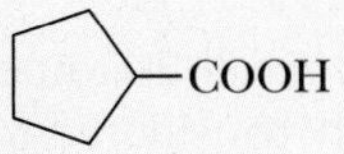

19.8 Conjugate Addition to α,β-Unsaturated Carbonyl Compounds

Thus far, we have used a variety of carbon nucleophiles to form new carbon-carbon bonds.

1. Anions of terminal alkynes (Section 7.5) and the cyanide ion.
2. Organomagnesium (Grignard) reagents, organolithium reagents, and lithium diorganocopper (Gilman) reagents.
3. Enolate anions derived from aldehydes and ketones (aldol reactions), esters (Claisen and Dieckmann condensations), β-diesters (malonic ester syntheses), and β-ketoesters (acetoacetic ester syntheses).
4. Enamines (which are synthetically equivalent to enolate anions).

These species can be used to form new carbon-carbon bonds by two synthetic strategies: (1) substitution by the carbon nucleophile in an S_N2 reaction and (2) addition of the carbon nucleophile to a carbonyl carbon.

Conjugate addition

Addition of a nucleophile to the β-carbon of an α,β-unsaturated carbonyl compound.

Conjugate addition, as it is known, presents a third synthetic strategy: addition of a carbon nucleophile to an electrophilic carbon-carbon double or triple bond conjugated with a carbonyl or another electron-withdrawing group. In this section, we study two types of conjugate additions to electrophilic double bonds: addition of enolate anions (the Michael reaction) and addition of lithium diorganocopper (Gilman) reagents.

A. Michael Addition of Enolate Anions

Nucleophilic addition of enolate anions to α,β-unsaturated carbonyl compounds was first reported in 1887 by the American chemist Arthur Michael. Following are two examples of **Michael reactions**. In the first example, the nucleophile adding to the conjugated system is the enolate anion of diethyl malonate. In the second example, the nucleophile is the enolate anion of ethyl acetoacetate.

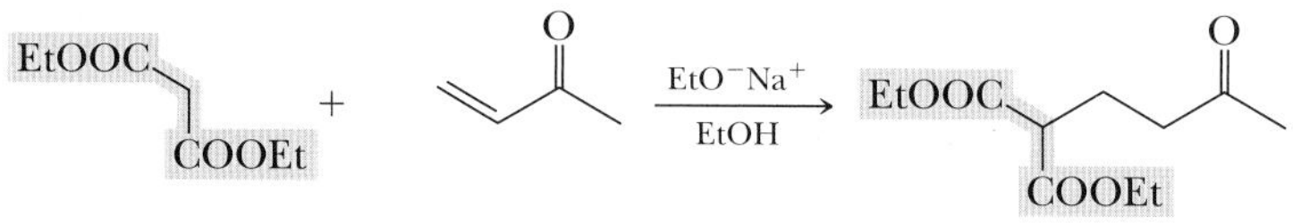

Diethyl propanedioate (Diethyl malonate) **3-Buten-2-one (Methyl vinyl ketone)**

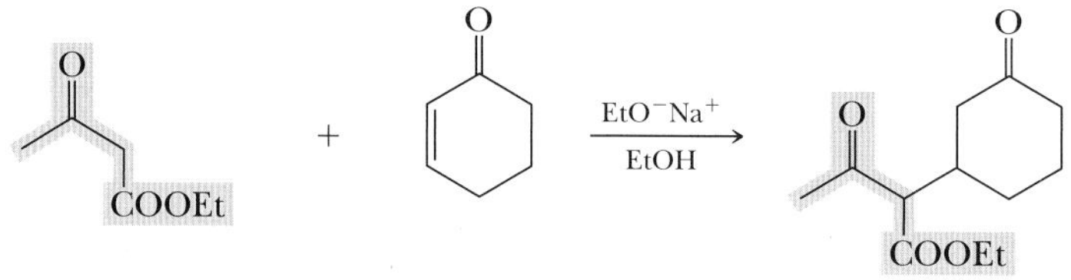

Ethyl 3-oxobutanoate (Ethyl acetoacetate) **2-Cyclohexenone** (four stereoisomers are possible; racemic)

You will recall that nucleophiles do not add to ordinary π bonds. These bonds are generally considered to be weak nucleophiles that are capable of attacking strong electrophiles (Section 6.3). What activates a carbon-carbon double bond for nucleophilic attack in a Michael reaction is the presence of the adjacent carbonyl group. One important resonance structure of the α,β-unsaturated carbonyl compounds puts positive charge at the end (in this case, the β-carbon) of the double bond, making it resemble a carbonyl group in its reactivity. Thus, nucleophiles can add to this type of double bond, which we call "activated" for this reason.

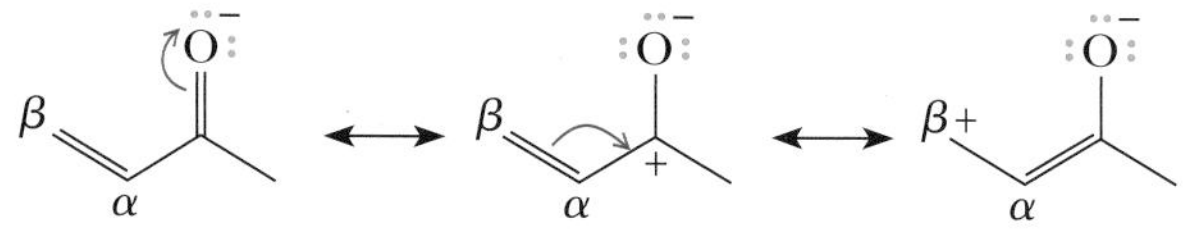

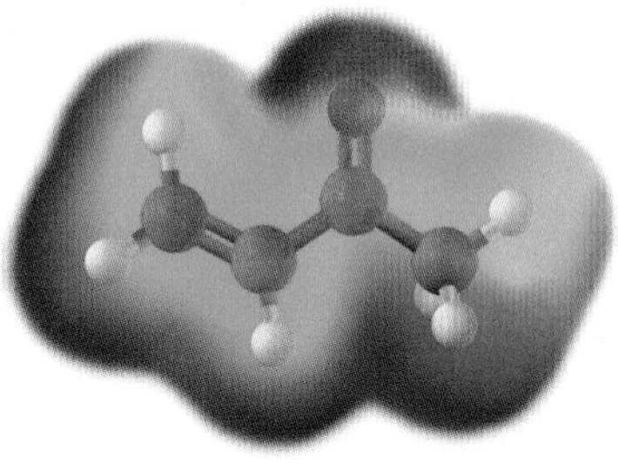

Elpot of α,β-unsaturated carbonyl

Although the major fraction of the partial positive charge (blue) of an α,β-unsaturated aldehyde or ketone is on the carbonyl carbon, there is nevertheless a significant partial positive charge on the beta carbon.

Note that aldol, Claisen, and Dieckmann condensations all give primary products with oxygens in a 1,3 relationship. The Michael reaction with enolate anions gives products with oxygens in a 1,5 relationship. These relationships are a consequence of the polarization of the reagents. In aldol, Claisen, and Dieckmann condensations, the carbonyl carbon is positive and the α-position is negative.

Aldol reaction

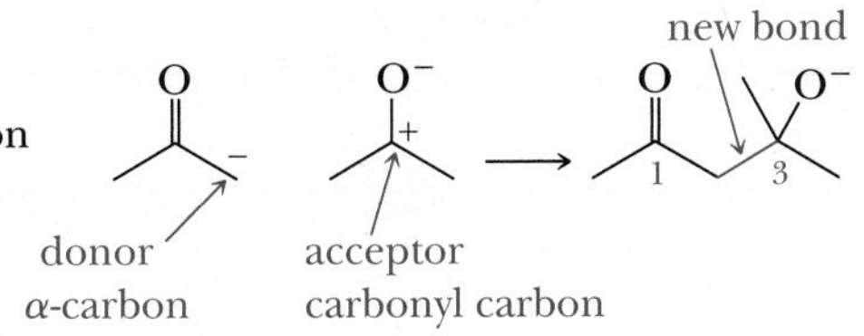

In a Michael reaction, the positive polarization of the carbonyl carbon is transmitted two carbons farther by the double bond.

Michael reaction

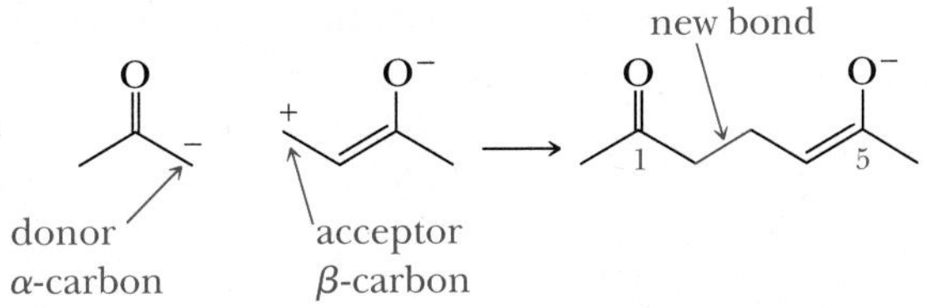

The Michael reaction takes place with a wide variety of α,β-unsaturated carbonyl compounds as well as with α,β-unsaturated nitriles and nitro compounds. The most commonly used types of nucleophiles in Michael reactions are summarized in Table 19.1. The bases most commonly used to generate the nucleophile are metal alkoxides, pyridine, and piperidine. It is important to realize that other nucleophiles can undergo similar additions to the beta carbon of unsaturated carbonyl compounds (e.g., amines, alcohols, and water).

We can write the following general mechanism for a Michael reaction. Note that in Step 3, the base, B^-, is regenerated, in accord with the experimental observation that a Michael reaction requires only a catalytic amount of base rather than a molar equivalent, provided there are no additional acidic H atoms on the product.

Table 19.1 Combinations of Reagents for Effective Michael Reactions

These Types of α,β-Unsaturated Compounds Are Nucleophile Acceptors in Michael Reactions		These Types of Compounds Provide Effective Nucleophiles for Michael Reactions	
$CH_2{=}CHCH$ (C=O)	Aldehyde	$CH_3CCH_2CCH_3$ (both C=O)	β-Diketone
$CH_2{=}CHCCH_3$ (C=O)	Ketone	CH_3CCH_2COEt (both C=O)	β-Ketoester
$CH_2{=}CHCOEt$ (C=O)	Ester	CH_3CCH_2CN (C=O)	β-Ketonitrile
$CH_2{=}CHCNH_2$ (C=O)	Amide	$EtOCCH_2COEt$ (both C=O)	β-Diester
$CH_2{=}CHC{\equiv}N$	Nitrile	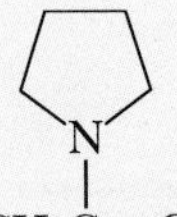 $CH_3C{=}CH_2$	Enamine
$CH_2{=}CHNO_2$	Nitro compound	NH_3, RNH_2, R_2NH	Amine

Mechanism 19.6

Michael Reaction—Conjugate Addition of Enolate Anions

Step 1: **Take a proton away.** Treating H—Nu with base gives the nucleophile, Nu:⁻.

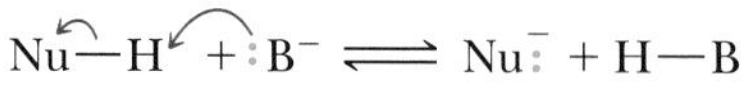

Base

Step 2: **Make a new bond between a nucleophile and an electrophile.** Nucleophilic addition of Nu:⁻ to the β-carbon of the conjugated system gives a resonance-stabilized enolate anion.

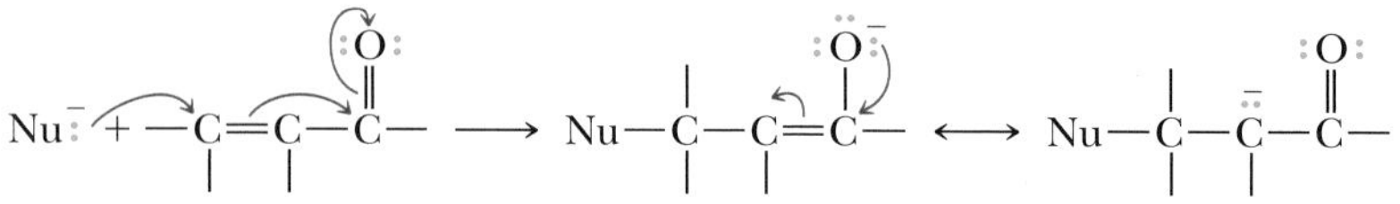

Resonance-stabilized enolate anion

The enolate corresponds to 1,4-addition to the conjugated system of the α,β-unsaturated carbonyl compound. Because this intermediate is formed, the Michael reaction is classified as a 1,4- or conjugate addition.

Step 3: **Add a proton.** Proton transfer from H—B gives the ketone.

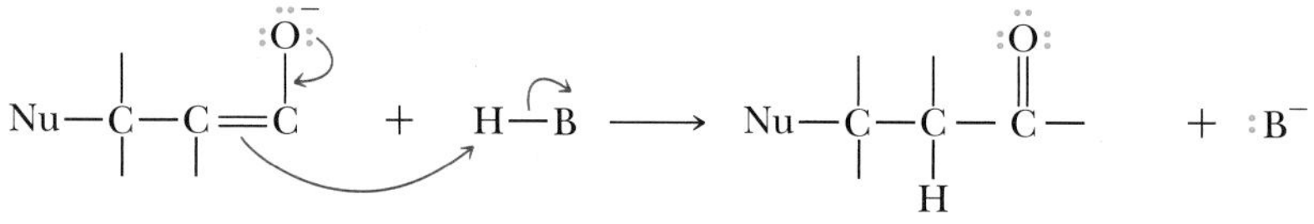

A β-substituted ketone
(a product of 1,4-addition)

Example 19.13 Michael Addition

Draw a structural formula for the product formed by treating each set of reactants with sodium ethoxide in ethanol under conditions of the Michael reaction.

(a) 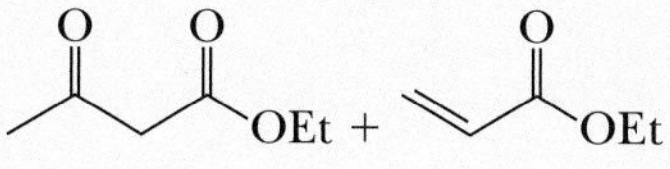

(b) 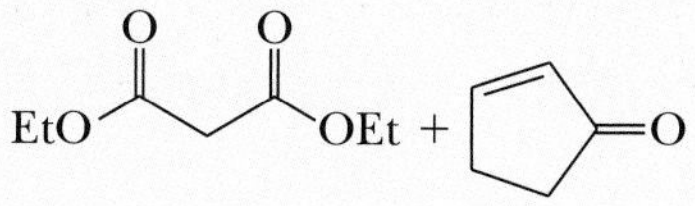

Solution

(a) 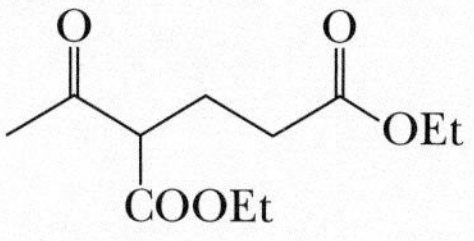

(racemic)

(b) 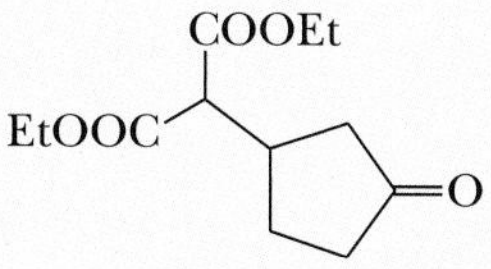

(racemic)

Problem 19.13

Show the product formed from each Michael product in the solution to Example 19.13 after (1) hydrolysis in aqueous NaOH, (2) acidification, and (3) thermal decarboxylation of each β-ketoacid or β-dicarboxylic acid. These reactions illustrate the usefulness of the Michael reaction for the synthesis of 1,5-dicarbonyl compounds.

As noted in Table 19.1, enamines also participate in Michael reactions as illustrated by the addition of the enamine of cyclohexanone to acrylonitrile, $CH_2{=}CHCN$.

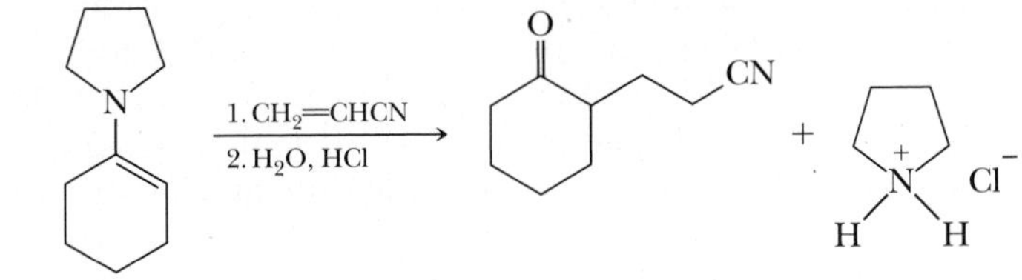

Pyrrolidine enamine of cyclohexanone

(racemic)

A final word about addition of nucleophiles to α,β-unsaturated carbonyl compounds. The Michael reaction is an example of 1,4-addition (conjugate addition) to an α,β-unsaturated carbonyl compound. In general, resonance-stabilized enolate anions and enamines are weak bases, react slowly, and give 1,4-addition products. Organolithium and organomagnesium compounds, on the other hand, are strong bases, react rapidly, and give primarily 1,2-addition products; that is, they give products formed by addition to the carbonyl carbon.

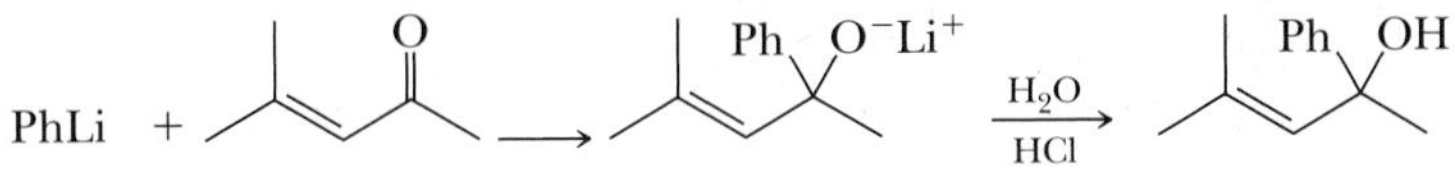

Phenyllithium **4-Methyl-3-penten-2-one** **4-Methyl-2-phenyl-3-penten-2-ol** (racemic)

Why do the nucleophiles listed in Table 19.1 react with conjugated carbonyl compounds by 1,4-addition rather than 1,2-addition? The answer has to do with **kinetic control** versus **thermodynamic control** of product formation. It has been shown that 1,2-addition of nucleophiles to the carbonyl carbon of α,β-unsaturated carbonyl compounds is faster than conjugate addition. If formation of the 1,2-addition product is irreversible, then that is the product observed. If, however, formation of the 1,2-addition product is reversible, then an equilibrium is established between the more rapidly formed but less stable 1,2-addition product and the more slowly formed but more stable 1,4-addition product. As mentioned at the beginning of the chapter, a carbon-oxygen double bond is stronger than a carbon-carbon double bond. Thus, under conditions of thermodynamic (equilibrium) control, the more stable 1,4-Michael addition product is formed.

Kinetic control

Experimental conditions under which the composition of the product mixture is determined by the relative rates of formation of each product.

Thermodynamic control

Experimental conditions that permit the establishment of equilibrium between two or more products of a reaction. The composition of the product mixture is determined by the relative stabilities of the products.

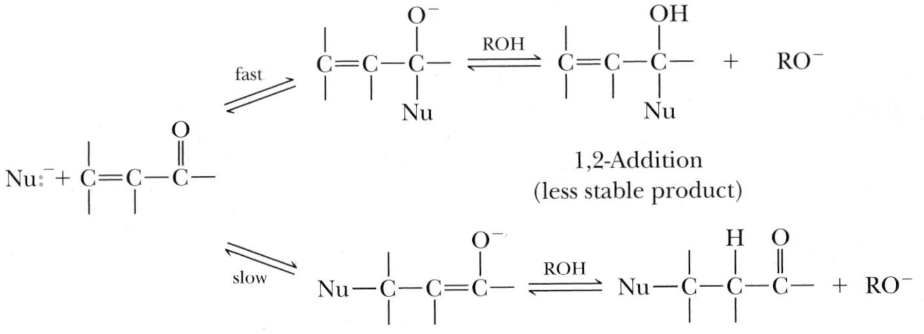

1,2-Addition (less stable product)

1,4-Addition (more stable product)

B. Retrosynthetic Analysis

The Michael addition of the nucleophiles listed in Table 19.1 to α,β-unsaturated carbonyls leads to a large variety of structures involving ketones, nitriles, and carboxylic acids, with a δ-ketone. As one example of a Michael addition retrosynthetic analysis, we examine δ-diketones. Such structures can be derived from a β-ketoester and an α,β-unsaturated carbonyl. Dissecting the product, we find that the α carbon must be derived from a β-ketoester and the β- and γ-carbons must come from the alkene of the Michael acceptor. Note in this example that the ester group is removed by a hydrolysis/decarboxylation sequence. Alternatively, the α-carbon may be nucleophilic by virtue of coming from an enamine, in which case a hydrogen is present at this position in the starting ketone. A chemist's ability to construct complex molecules using more than a single route makes organic synthesis interesting and challenging.

Watch a video explanation

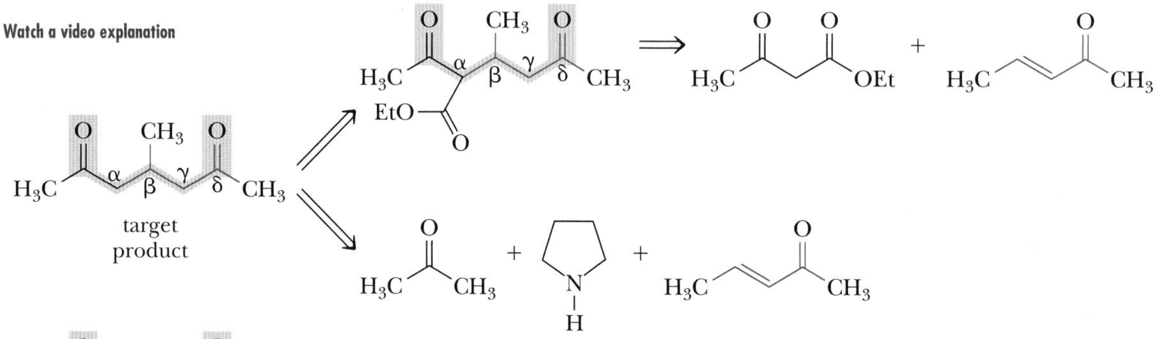
target
product

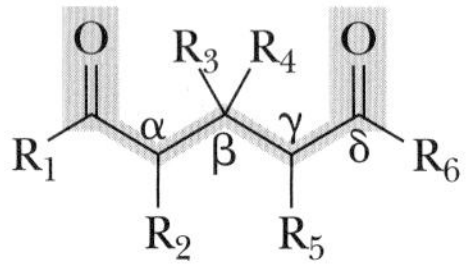

δ-Diketone

Michael addition product recognition element

R_1–R_6 = alkyl group, aryl group, or H atom

Example 19.14 Retrosynthetic Analysis I

Show how the series of reactions in Example 19.13 and Problem 19.13 (Michael reaction, hydrolysis, acidification, and thermal decarboxylation) can be used to prepare 2,6-heptanedione.

Solution

As shown in the following retrosynthetic analysis, this molecule can be constructed from the carbon skeletons of ethyl acetoacetate and methyl vinyl ketone.

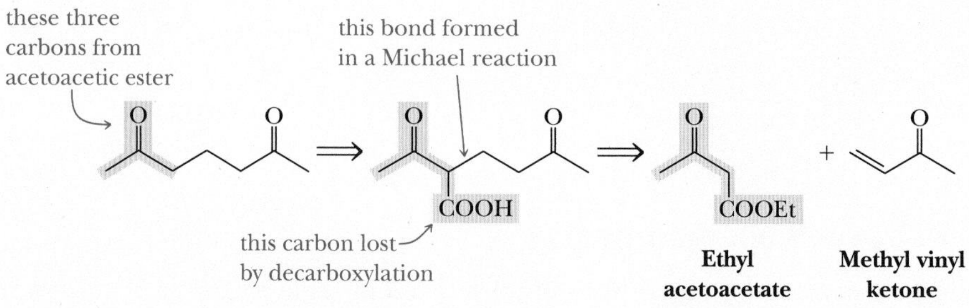
these three
carbons from
acetoacetic ester
this bond formed
in a Michael reaction
this carbon lost
by decarboxylation
Ethyl
acetoacetate
Methyl vinyl
ketone

Following are the steps in their conversion to 2,6-heptanedione.

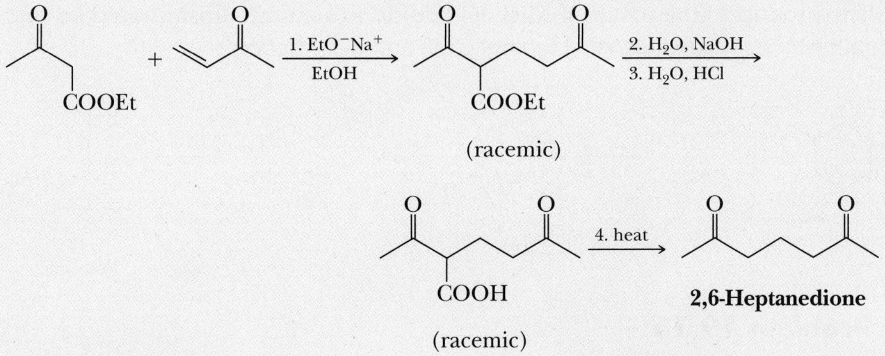
1. EtO⁻Na⁺
EtOH
COOEt
COOEt
(racemic)
2. H2O, NaOH
3. H2O, HCl
4. heat
COOH
(racemic)
2,6-Heptanedione

Problem 19.14

Show how the sequence of Michael reaction, hydrolysis, acidification, and thermal decarboxylation can be used to prepare pentanedioic acid (glutaric acid).

C. Robinson Annulation

Michael reaction with an α,β-unsaturated ketone followed by an intramolecular aldol reaction has proven to be a valuable method for the synthesis of 2-cyclohexenones. An especially important example of a Michael-aldol sequence is the **Robinson annulation**, in which treatment of a cyclic ketone, β-ketoester, or β-diketone with an α,β-unsaturated ketone in the presence of a base catalyst forms a cyclohexenone ring fused to the original ring. When the following racemic β-ketoester, for example, is treated with methyl vinyl ketone in the presence of sodium ethoxide in ethanol, the Michael adduct forms and then, in the presence of sodium ethoxide, undergoes a base-catalyzed intramolecular aldol reaction followed by dehydration to give a racemic substituted cyclohexenone.

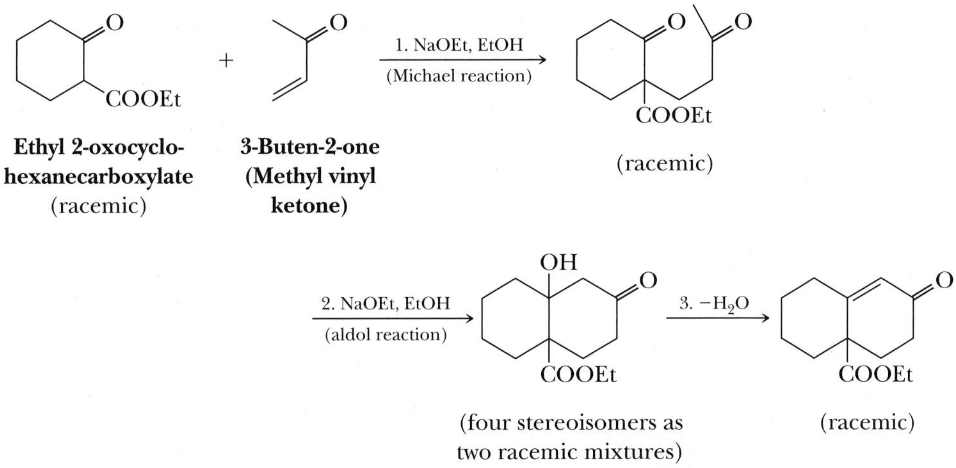
COOEt
Ethyl 2-oxocyclo-hexanecarboxylate
(racemic)
3-Buten-2-one (Methyl vinyl ketone)
1. NaOEt, EtOH
(Michael reaction)
COOEt
(racemic)
2. NaOEt, EtOH
(aldol reaction)
OH
COOEt
(four stereoisomers as two racemic mixtures)
3. −H2O
COOEt
(racemic)

Example 19.15 Robinson Annulation

Draw structural formulas for the lettered compounds (A) and (B) in the following synthetic sequence.

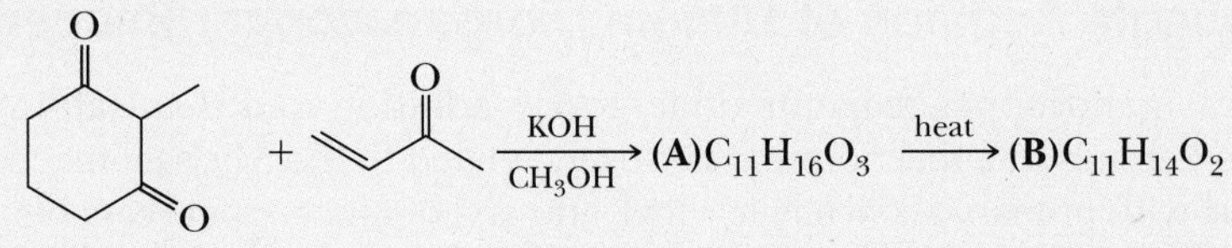
KOH
CH3OH
(A)C11H16O3
heat
(B)C11H14O2

(Continued)

Solution

The product is the result of Michael addition to an α,β-unsaturated ketone followed by base-catalyzed aldol reaction and dehydration.

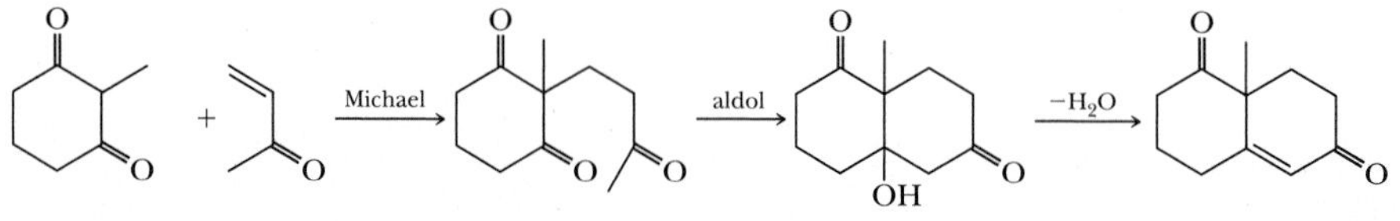

(A) **(B)** (racemic)

Watch a video explanation

Problem 19.15

Show how to bring about the following conversion.

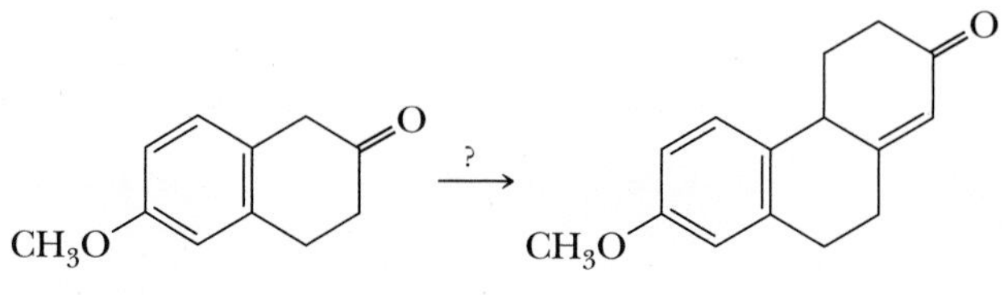

D. Retrosynthetic Analysis

The functional group present in the product of a Robinson annulation is a six-membered ring containing an α,β-unsaturated ketone. Often, there is a carbonyl group on a carbon adjacent to the ring, two carbons away from the β-carbon of the α,β-unsaturated ketone. Retrosynthetic analysis of the Robinson annulation is a combination of those discussed for the Michael addition and the aldol reaction. The carbon-carbon double bond in the ring is derived from the aldol reaction, while another new bond is created by the Michael reaction. The challenge is to choose the proper enolate-forming β-dicarbonyl and α,β-unsaturated ketone derivatives based on these bond connections.

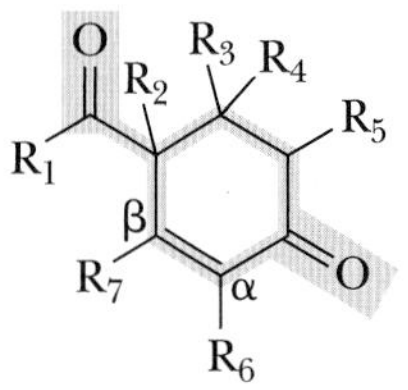

Cyclic α,β-unsaturated carbonyl
Robinson annulation
product recognition element

R_1–R_7 = alkyl group, aryl group, or H atom

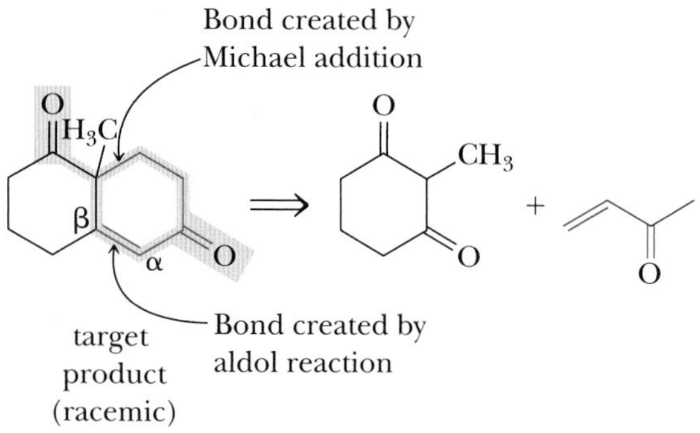

E. Conjugate Addition of Lithium Diorganocopper Reagents

Lithium diorganocopper reagents undergo 1,4-addition to α,β-unsaturated aldehydes and ketones in a reaction that is closely related to the Michael reaction. Yields are highest with primary alkyl, allylic, vinyl, and aryl organocopper reagents.

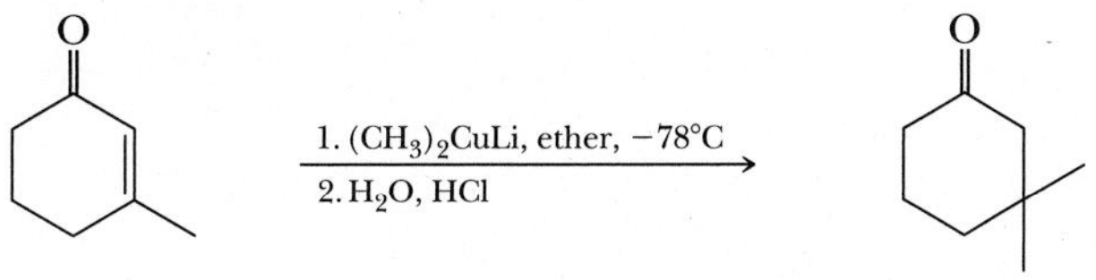
1. $(CH_3)_2CuLi$, ether, −78°C
2. H_2O, HCl

3-Methyl-2-cyclohexenone **3,3-Dimethylcyclohexanone**

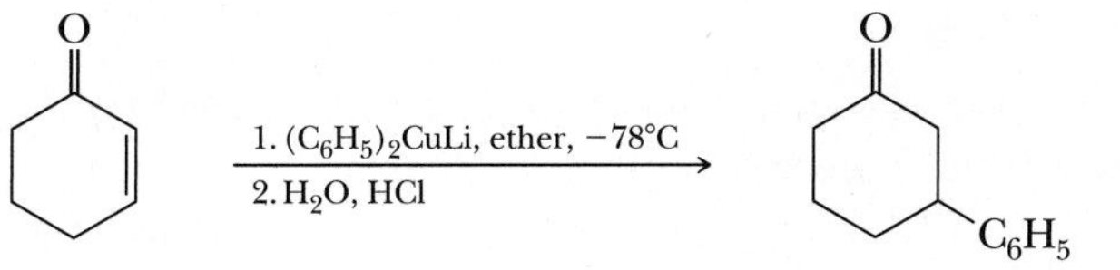
1. $(C_6H_5)_2CuLi$, ether, −78°C
2. H_2O, HCl
C_6H_5

2-Cyclohexenone **3-Phenylcyclohexanone** (racemic)

Grignard reagents give primarily addition reactions to the carbonyl group of α,β-unsaturated carbonyls (1,2-addition). In fact, lithium diorganocopper reagents are unique among organometallic compounds in that they give almost exclusively 1,4-addition, which makes them very valuable reagents in synthetic organic chemistry. The mechanism of conjugate addition of lithium diorganocopper reagents is not fully understood.

Example 19.16 Retrosynthetic Analysis II

Propose two syntheses of 4-octanone, each involving conjugate addition of a lithium diorganocopper reagent.

Solution

A lithium diorganocopper reagent adds to the beta carbon of an α,β-unsaturated aldehyde or ketone. Therefore, locate each carbon beta to the carbonyl group in this target molecule and disconnect at those points.

Synthesis 1:

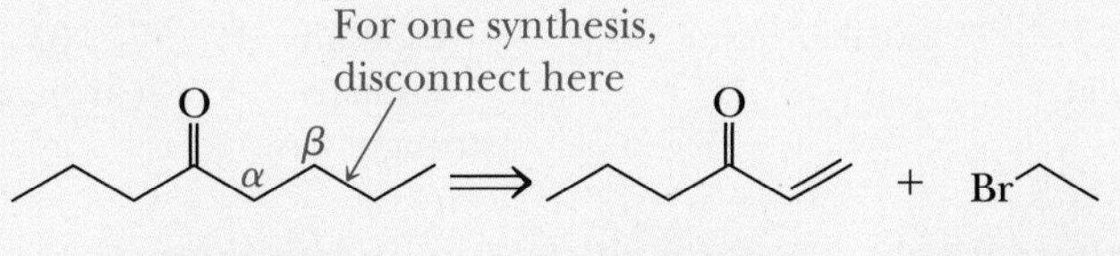
For one synthesis, disconnect here
+ Br

4-Octanone **1-Hexen-3-one** **Bromoethane**

For this synthesis, add lithium diethylcopper to 1-hexen-3-one.

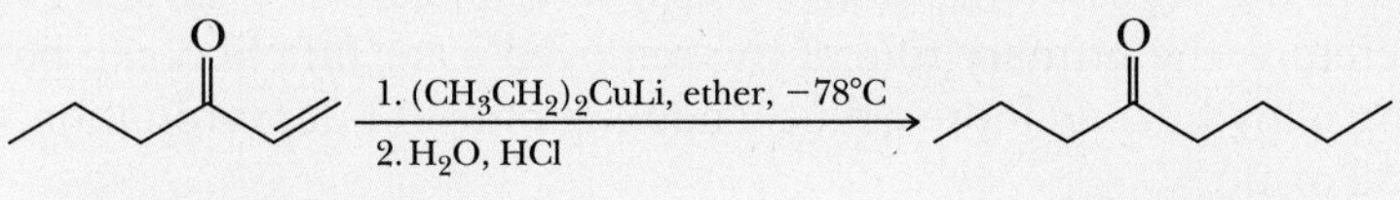
1. $(CH_3CH_2)_2CuLi$, ether, −78°C
2. H_2O, HCl

1-Hexen-3-one **4-Octanone**

Synthesis 2:

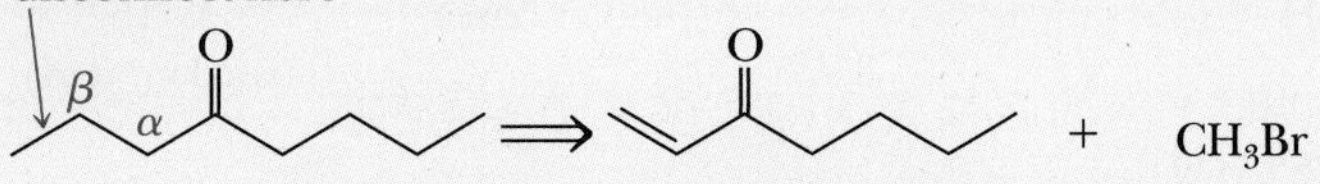
For a second synthesis, disconnect here
+ CH_3Br

4-Octanone **1-Hepten-3-one** **Bromomethane**

(Continued)

For this synthesis, add lithium dimethylcopper to 1-hepten-3-one.

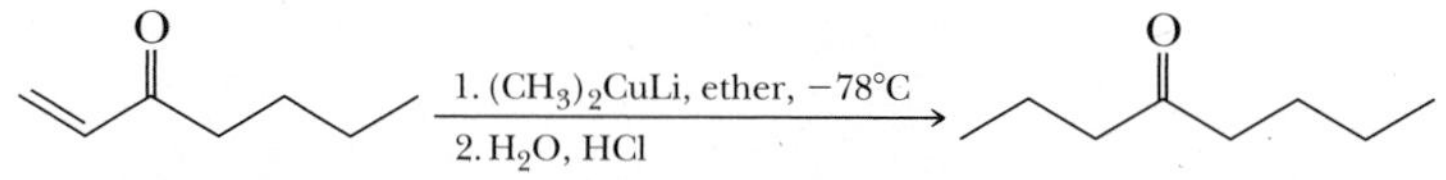
O
1. (CH3)2CuLi, ether, −78°C
2. H2O, HCl
O

1-Hepten-3-one **4-Octanone**

Problem 19.16

Propose two syntheses of 4-phenyl-2-pentanone, each involving conjugate addition of a lithium diorganocopper reagent.

19.9 Crossed Enolate Reactions Using LDA

A. Acid-Base Considerations

As discussed in this chapter, enolate anions are formed when a carbonyl compound containing an α-hydrogen is treated with a base such as hydroxide or an alkoxide. We noted earlier that α-hydrogens normally are considerably less acidic than water or alcohols, so the position of equilibrium in this acid-base reaction greatly favors the reactants rather than enolate products.

$$\underset{\substack{\text{p}K_a\ 20\\ \text{(weaker acid)}}}{CH_3\overset{O}{\overset{\|}{C}}CH_3} + \underset{\text{(weaker base)}}{NaOH} \rightleftharpoons \underset{\substack{\text{A sodium}\\ \text{enolate}\\ \text{(stronger base)}}}{CH_2{=}\overset{O^-Na^+}{\overset{|}{C}}CH_3} + \underset{\substack{\text{p}K_a\ 15.7\\ \text{(stronger acid)}}}{H_2O} \qquad K_{eq} = 5 \times 10^{-5}$$

$$\underset{\substack{\text{p}K_a\ 23\\ \text{(weaker acid)}}}{CH_3\overset{O}{\overset{\|}{C}}OC_2H_5} + \underset{\text{(weaker base)}}{C_2H_5O^-Na^+} \rightleftharpoons \underset{\substack{\text{A sodium}\\ \text{enolate}\\ \text{(stronger base)}}}{CH_2{=}\overset{O^-Na^+}{\overset{|}{C}}OC_2H_5} + \underset{\substack{\text{p}K_a\ 16\\ \text{(stronger acid)}}}{C_2H_5OH} \qquad K_{eq} = 10^{-7}$$

If, alternatively, a second electron-withdrawing group such as a carbonyl is present, as in ethyl acetoacetate or diethyl malonate, the pK_a is shifted so that the equilibrium lies largely toward the enolate products. During the acetoacetic ester and malonic diester sequence of five steps, an ester carbonyl carbon is removed as CO_2. Therefore, the primary role of the removed ester function can be thought of as making α-hydrogens more acidic, enabling efficient deprotonation with a base such as alkoxide.

$$EtO^- + \underset{\substack{\text{p}K_a\ 10.7\\ \text{(stronger acid)}}}{CH_3{-}\overset{O}{\overset{\|}{C}}{-}\underset{H}{\underset{|}{C}}H{-}\overset{O}{\overset{\|}{C}}{-}OEt} \rightleftharpoons \underset{\text{(weaker base)}}{CH_3{-}\overset{O}{\overset{\|}{C}}{-}\overset{-}{C}H{-}\overset{O}{\overset{\|}{C}}{-}OEt} + \underset{\substack{\text{p}K_a\ 15.9\\ \text{(weaker acid)}}}{EtOH} \qquad K_{eq} = 1.6 \times 10^5$$

With a substantially stronger base, the formation of an enolate anion from an aldehyde, a ketone, or an ester can be driven to completion without an additional electron-withdrawing group. A commonly used base for this purpose is lithium diisopropylamide (LDA).

$$[(CH_3)_2CH]_2N^-Li^+$$

Lithium diisopropylamide

LDA is prepared by dissolving diisopropylamine in tetrahydrofuran and treating this solution with butyllithium.

$$[(CH_3)_2CH]_2NH + CH_3(CH_2)_3Li \longrightarrow [(CH_3)_2CH]_2N^-Li^+ + CH_3(CH_2)_2CH_3 \quad K_{eq} = 10^{10}$$

Diisopropylamine	Butyllithium	Lithium diisopropylamide	Butane
pK_a 40 (stronger acid)	(stronger base)	(weaker base)	pK_a 50 (weaker acid)

Although LDA is an extremely strong base, it is a poor nucleophile because of steric crowding around the nitrogen, which prevents its addition to carbonyl groups. LDA is, therefore, ideal for generating enolate anions from carbonyl-containing compounds. Using a molar equivalent of LDA can completely convert an aldehyde, a ketone, or an ester to the corresponding lithium enolate.

$$CH_3\overset{O}{\overset{\|}{C}}OC_2H_5 + [(CH_3)_2CH]_2N^-Li^+ \longrightarrow CH_2{=}\overset{O^-Li^+}{\overset{|}{C}}OC_2H_5 + [(CH_3)_2CH]_2NH \quad K_{eq} = 10^{17}$$

pK_a 23 (stronger acid)	LDA (stronger base)	A lithium enolate (weaker base)	pK_a 40 (weaker acid)

B. Base Stoichiometry

Let's now consider the stoichiometry of the added base needed when performing an enolate reaction. In an aldol reaction carried out with an aldehyde and hydroxide, the amount of hydroxide has a minimal effect on yield. The hydroxide catalyst produces very little enolate in the equilibrium established for the first deprotonation step (about one part in a thousand). The addition of more hydroxide leads to slightly more enolate intermediate (up to a few parts per thousand), speeding up the overall aldol reaction but not changing the outcome. The key is that even with an excess of hydroxide, unreacted aldehyde persists. Therefore, using more than an equivalent of hydroxide would not substantially change aldol reaction product yields.

Yet the amount of LDA used is critical. The proper amount of LDA to use in an aldol reaction is 0.5 equivalents, which is enough to generate 50% enolate, which then reacts with the remaining 50% aldehyde for an efficient and rapid aldol reaction. If alternatively you used one full equivalent of LDA, all the aldehyde would be deprotonated and exist as the enolate, leaving no further aldehyde in solution to react. In this case, no aldol product would form.

C. Crossed Enolate Reactions Using LDA

Using LDA allows one to direct crossed aldol and crossed Claisen reactions. For example, consider how to perform the following reaction.

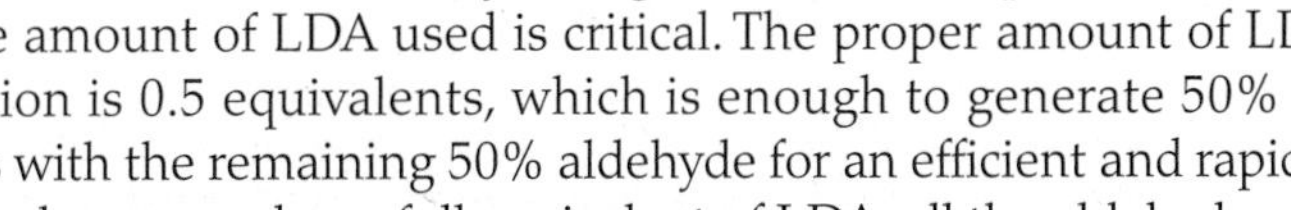

Phenylacetaldehyde **Acetone** **4-Hydroxy-5-phenyl-2-pentanone** (racemic)

This crossed aldol reaction between acetone and an aldehyde may be carried out successfully by treating acetone with one equivalent of LDA to convert it completely to its enolate anion. The preformed enolate is then treated with the aldehyde followed by workup in water to give the crossed aldol condensation product.

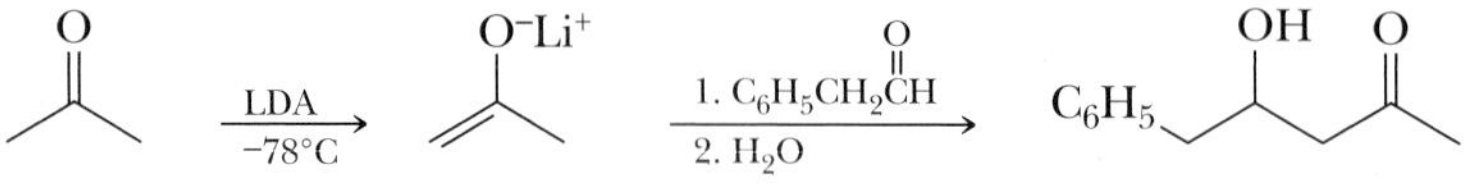

A lithium enolate

4-Hydroxy-5-phenyl-2-pentanone (racemic)

Example 19.17 Using LDA

Show how to prepare 5-hydroxy-4-methyl-3-hexanone using a crossed aldol reaction.

Solution

First, recognize that the two carbonyl-containing compounds to be joined in the aldol reaction are 3-pentanone and acetaldehyde. Treat the symmetrical ketone with LDA to form its lithium enolate. Treatment of this enolate anion with acetaldehyde followed by aqueous workup gives the desired aldol product.

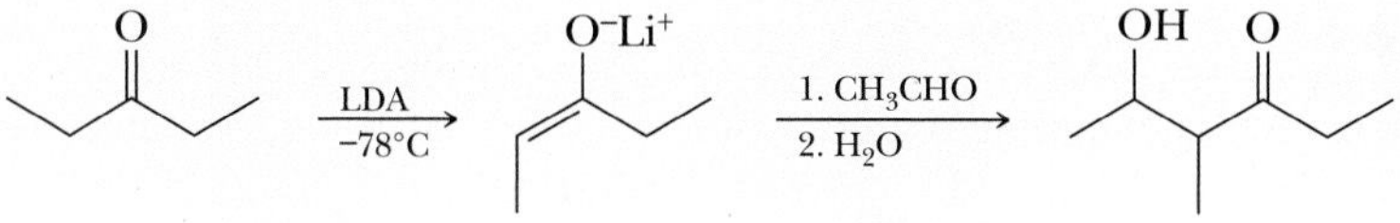

Problem 19.17

Show how you might prepare the following compounds using directed aldol reactions.

(a) 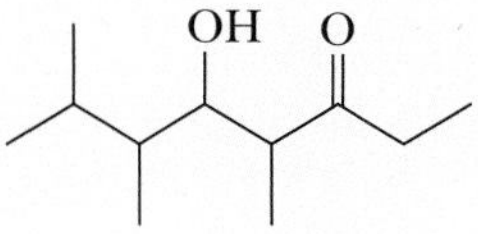

(b) 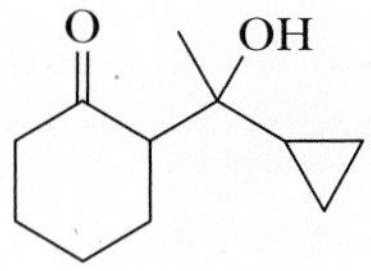

This chapter introduced the use of enolates and/or enamines as nucleophiles in several reactions, including aldol reactions, Claisen condensations and Michael additions, alkylations, and acylations. We can also use LDA to generate the enolate anions and perform the same reactions, as shown here for cyclohexanone and a few specific electrophiles. Similar reactions are possible for aldehydes and esters with α-hydrogens. The synthetic versatility of this approach has made LDA a very popular and important reagent in modern synthetic organic chemistry.

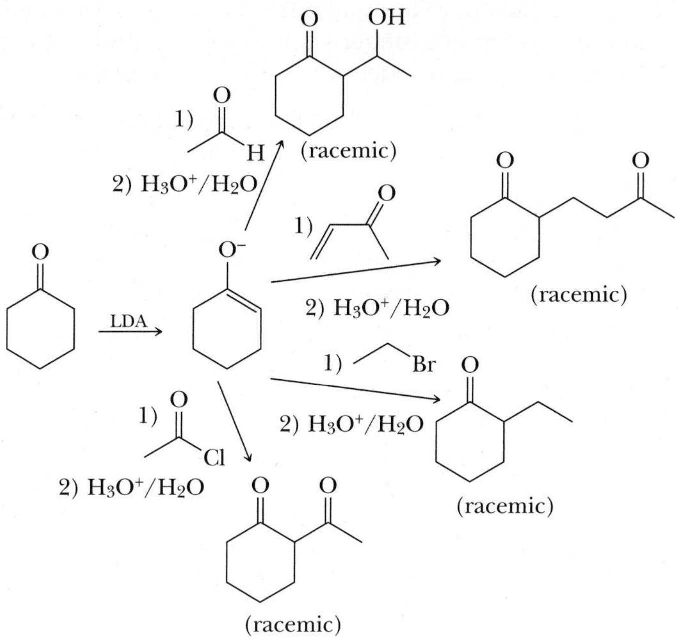

D. Kinetic versus Thermodynamic Enolates

For a ketone with two sets of nonequivalent α-hydrogens, the following questions arise: is formation of an enolate anion regioselective, and if so, what factors determine the degree of regioselectivity? It has been determined experimentally that a high degree of regioselectivity often exists and that its occurrence depends on experimental conditions. When 2-methylcyclohexanone, for example, is added to a slight excess of LDA, the ketone is converted to its lithium enolate, which consists almost entirely of the salt of the less substituted enolate anion.

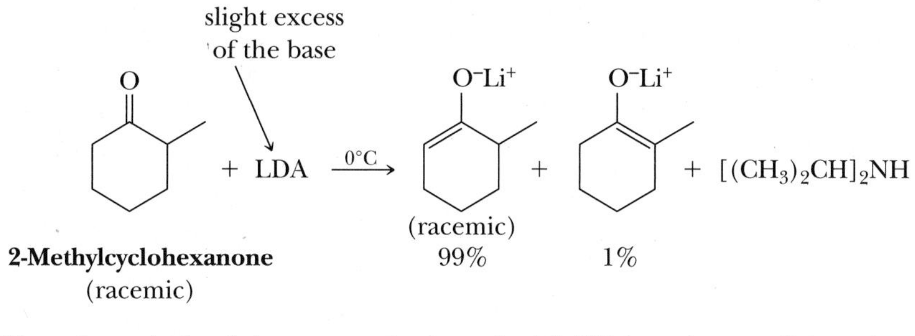

Watch a video explanation

When 2-methylcyclohexanone is treated with LDA under conditions in which the ketone is in slight excess, the composition of the product is quite different; it is richer in the more substituted enolate anion.

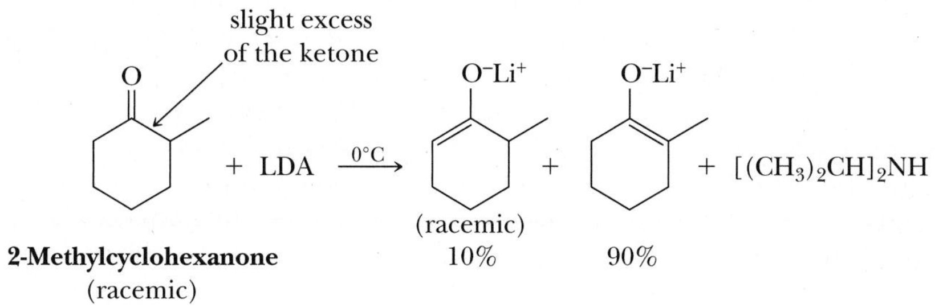

The most important factor determining the composition of an enolate anion mixture is whether the reaction is under kinetic (rate) control or thermodynamic (equilibrium) control. In a reaction under **thermodynamic control**:

Thermodynamic control
Experimental conditions that permit the establishment of equilibrium between two or more products of a reaction. The composition of the product mixture is determined by the relative stabilities of the products.

- The reaction conditions permit the equilibration of alternative products, and
- The composition of the product mixture is determined by the relative stabilities of the alternative products.

Equilibrium among enolate anions is established when the ketone is in slight excess, a condition under which it is possible for proton-transfer reactions to occur between an enolate and an α-hydrogen of an unreacted ketone. Thus, equilibrium is established between alternative enolate anions.

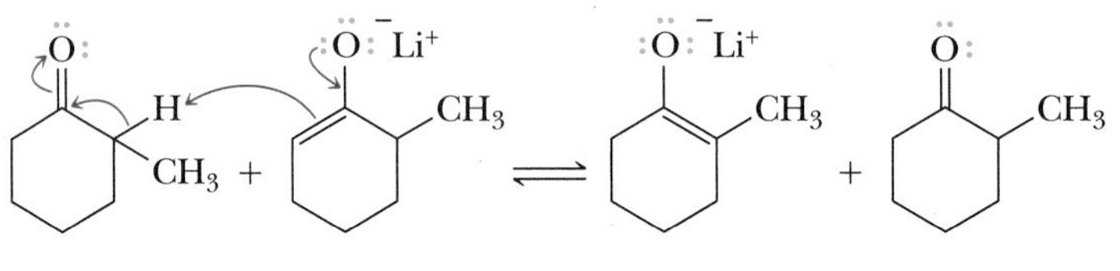

(racemic) Less stable enolate anion (racemic) More stable enolate anion (racemic)

Under these conditions, the more stable enolate anion predominates. The factors that determine the relative stabilities of enolate anions are the same as those that determine the relative stabilities of alkenes; the more substituted the double bond of the enolate anion, the greater its stability. Thus, the composition of the enolate anion mixture formed under conditions of thermodynamic control reflects the relative stabilities of the individual enolate anions.

In a reaction under **kinetic control**, the composition of the product mixture is determined by the relative rates of formation of each product. No equilibrium among possible alternative structures is set up. In the case of formation of enolate anions, kinetic control refers to the relative rates of removal of the alternative α-hydrogens. The less hindered α-hydrogen is removed more rapidly; thus, the major product is the less substituted enolate anion. Because a slight excess of base is used, there is no ketone to serve as a proton donor and the less stable enolate anion cannot equilibrate with a more stable one.

Kinetic control
Experimental conditions under which the composition of the product mixture is determined by the relative rates of formation of each product.

MCAT Practice: *Passage and Questions*

Ibuprofen: The Evolution of an Industrial Synthesis

A major consideration in any industrial synthesis is atom economy; it is most efficient to use only reagents whose atoms appear in the final product. An example of the evolution of syntheses with increasingly improved atom economy is the synthesis of ibuprofen.

Synthesis I: One of the first industrial syntheses of ibuprofen used the following sequence to introduce a methyl group on the carboxyl side chain of 4-isobutylphenylacetic acid.

Even though synthesis I gives ibuprofen in good yield, it wastes carbons; only 13 of the 18 carbons in intermediate C appear in ibuprofen.

Synthesis II: An alternative route with greater atom economy starts with 4-isobutylacetophenone. (below)

Although synthesis II is more atom-efficient than synthesis I, it uses a cyano group, neither atom of which appears in the final product.

Synthesis III: The ultimate in atom economy is the synthesis of ibuprofen as shown at the bottom of the page. The single carbon itself added via carbon monoxide appears in the product.

Synthesis I

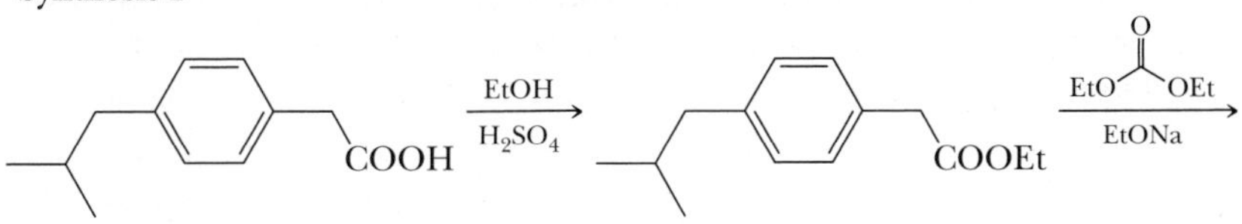

4-Isobutylphenylacetic acid A

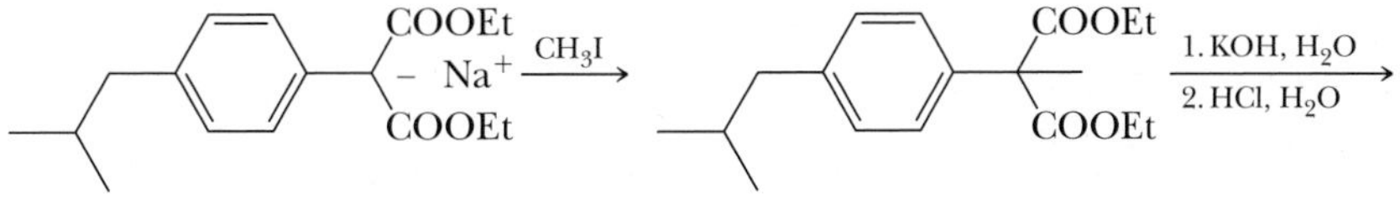

B C

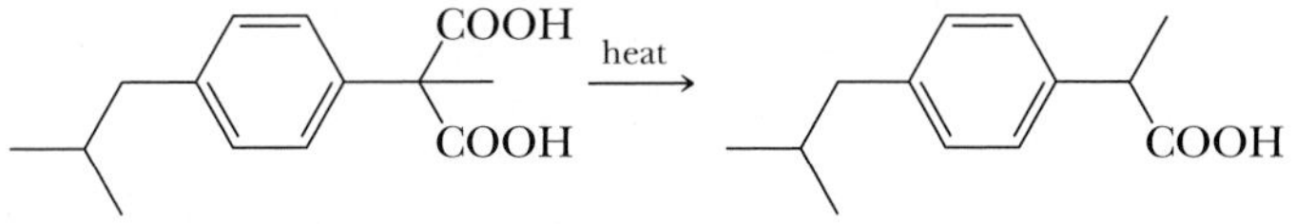

D **Ibuprofen** (racemic)

Synthesis II

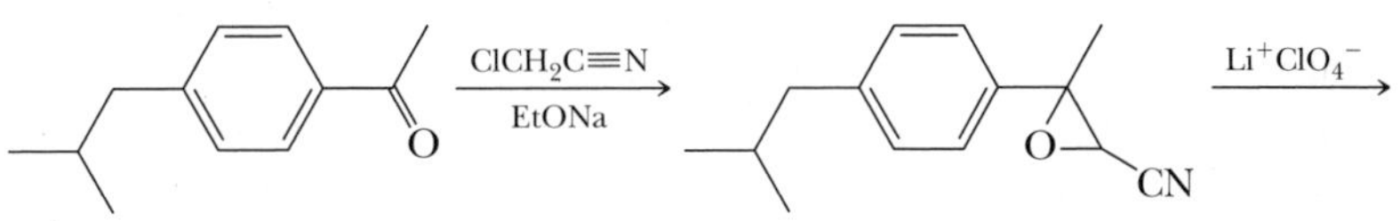

4-Isobutylacetophenone E (racemic)

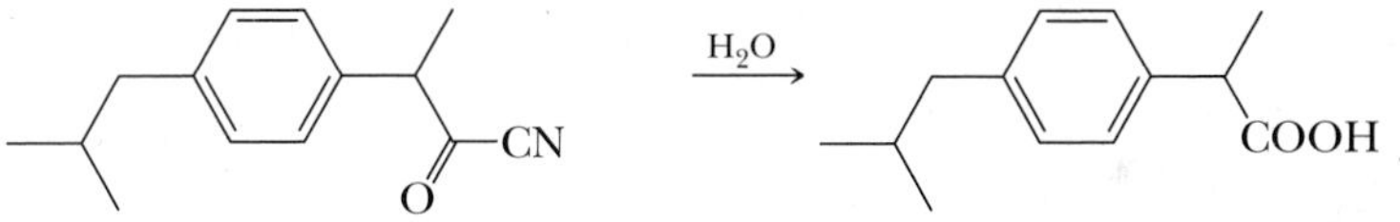

F (racemic) **Ibuprofen** (racemic)

Synthesis III

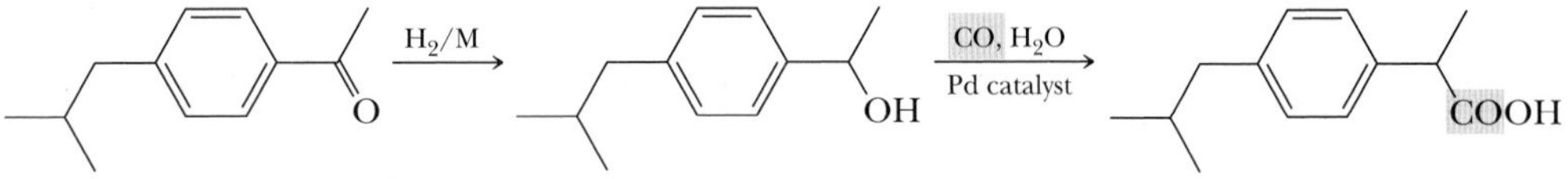

4-Isobutylacetophenone G (racemic) **Ibuprofen** (racemic)

Questions

A. The first reaction in Synthesis I is an example of a(n):

1. Reduction of a carboxylic acid.
2. Oxidation of a carboxylic acid.
3. Michael addition.
4. Fischer esterification.

B. The second reaction in Synthesis I that converts structure A to structure B is an example of a(n):

1. Aldol reaction.
2. Dieckmann condensation.
3. Crossed Claisen condensation.
4. Enamine alkylation.

C. The sequence of reactions that convert structure B to ibuprofen in Synthesis I is analogous to a(n)

1. Michael reaction.
2. Acetoacetic ester synthesis.
3. Malonic ester synthesis.
4. Robinson annulation.

D. The first reaction in Synthesis II must involve the following mechanistic steps:

1. Deprotonation of the α-carbon of the ketone by ethoxide to make an enolate whose carbon does an S_N2 displacement of chloride from chloroacetonitrile, followed by intramolecular

(Continued)

nucleophilic formation of the epoxide ring in structure E.

2. Deprotonation of the α-carbon of the ketone by ethoxide to make an enolate whose oxygen does an S_N2 displacement of chloride from chloroacetonitrile, followed by intramolecular electrophilic formation of the epoxide ring in structure E.
3. Deprotonation of chloroacetonitrile by ethoxide to form a carbanion that deprotonates the α-carbon of the ketone, causing nucleophilic addition to the nitrile carbon and epoxide ring formation in structure E.
4. Deprotonation of chloroacetonitrile by ethoxide to form a carbanion that nucleophilically adds to the ketone carbon, creating an alkoxide that does an intramolecular S_N2, thereby displacing chloride to form the epoxide ring in structure E.

E. The conversion of structure E to structure F in synthesis II using lithium perchlorate must involve:

1. Deprotonation of the α-carbon to the nitrile by perchlorate, followed by an intramolecular rearrangement.
2. The use of lithium as a Lewis acid to promote epoxide ring opening that forms a benzylic carbocation that subsequently undergoes deprotonation at the α-carbon to the nitrile.
3. Epoxide ring opening by water promoted by the lithium perchlorate, followed by dehydration.
4. Nucleophilic epoxide ring opening by perchlorate, followed by elimination of perchloric acid.

F. The last step in Synthesis II occurs upon addition of water. The mechanism would be referred to as

1. An addition.
2. An addition/elimination.
3. Either S_N2 or S_N1.
4. A radical chain.

G. The first step in Synthesis III does not involve the following:

1. Metal-mediated reduction of a ketone.
2. Metal-mediated hydrolysis of a ketone.
3. Metal mediated hydrogenation of a ketone.
4. Metal C—H bond formation.

H. Intermediate G in Synthesis III is produced as a racemic mixture because

1. Chiral metal catalysts have not yet been developed by chemists.
2. Although 4-isobutylacetophenone is chiral, the stereochemical preferences for a particular enantiomeric product must coincidentally be near zero.
3. None of the reactants are chiral.
4. Answers 1 and 3 are correct.

I. Which statement is *not* true about the last step in Synthesis III?

1. It involves a relatively stable benzylic carbocation intermediate.
2. It involves organometallic intermediates.
3. The Pd catalyst speeds up the reaction.
4. The use of carbon monoxide must be done safely because it is a deadly gas.

J. The last reaction is exothermic because

1. The water incorporation creates strong bonds in the product.
2. Pd-catalyzed reactions routinely make stable products.
3. The π bond in CO and the C—O σ bond in the reactants that are broken are weaker than the σ C—C and C—O bonds being formed.
4. The CO bond strength does not matter. It is due to the simple fact that bonds in alcohols are stronger than bonds in carboxylic acids.

Study Guide

19.1 Formation and Reactions of Enolate Anions: An Overview

- The carbon atom immediately adjacent to a carbonyl group is referred to as the **α-carbon**, and hydrogen atoms bonded to the α-carbon are referred to as **α-hydrogens**.
  - *The α position of carbonyl-containing compounds such as aldehydes, ketones, and esters is relatively acidic, having pK_a values in the 21–25 range.*
  - *This relative acidity for a carbon-bound H atom is largely due to the stability of the resonance-stabilized anion, called an **enolate anion**, which is produced upon deprotonation.*

- Enolate anions are best represented as a hybrid of two contributing structures.
  - *One contributing structure places the negative charge on the α-carbon atom and contains a carbonyl π bond.*
  - *The other contributing structure places the negative charge on the oxygen atom and contains a C═C π bond.*
  - *The contributing structure with the negative charge on the O atom is the major contributor to the hybrid because O is more electronegative than C, so O is better able to accommodate the negative charge.*
- Despite their charge distribution, enolate anions react as carbon nucleophiles that create new carbon-carbon bonds, making enolate anions important for organic synthesis.
  - *Enolate anions react with haloalkanes in S_N2 reactions.*
  - *Enolate anions react with aldehydes, ketones, and esters in carbonyl addition reactions.*
    - *In reactions with electrophiles, enolate anions react primarily at C rather than O for at least two reasons.*
    - *Reaction at C gives more stable products that contain a relatively strong C═O π bond. Reaction at O would give products with a weaker C═C π bond.*
    - *The enolate O atom, having most of the negative charge of the resonance hybrid, is more tightly associated with a counterion such as Na^+ or Li^+. The counterion has a shielding effect that inhibits reactions with electrophiles, an effect that is amplified by the aggregate structure of enolate anions in solution.*

19.2 Aldol Reaction

- In the **aldol reaction**, enolate anions derived from aldehydes or ketones react with a second molecule of aldehyde or ketone to give a carbonyl addition reaction and create a new carbon-carbon bond.
  - *The mechanism of the base-catalyzed aldol reactions involves initial deprotonation of the α-hydrogen in base to create an enolate anion, which is a strong nucleophile that attacks another aldehyde or ketone molecule to give a carbonyl addition intermediate, which, in turn, reacts with water to create a β-hydroxy aldehyde or ketone product and regenerates the original base.*
    - *Because the base is regenerated at the end of the reaction, the aldol reaction is considered to be base catalyzed.*
- The aldol reaction can also be catalyzed by acid.
  - *The mechanism of the acid-catalyzed aldol reaction involves an initial acid-catalyzed keto-enol tautomerization to provide the enol form; protonation of a second molecule on the carbonyl oxygen creates an electrophilic oxonium ion that is then attacked by the nucleophilic enol, followed by loss of a proton to give the β-hydroxy aldehyde or ketone product.*
    - *Because the proton is regenerated at the end of the reaction, the reaction is considered to be acid catalyzed.*
  - *In both the acid- and base-catalyzed aldol reactions, one or two new chiral centers are often created, leading to racemic mixtures unless a starting aldehyde, ketone, or catalyst is chiral and present as a single enantiomer.*
- Aldol reactions are readily reversible, especially in base.
  - *Equilibrium in aldol reactions favors products in the case of aldehydes, but for ketones, often little product is made.*

- The β-hydroxy aldehyde or ketone products of aldol reactions are easily dehydrated and lose H_2O to give an α,β-unsaturated aldehyde or ketone.
  - *Dehydration can occur under the conditions of the aldol reaction, or sometimes heating in acid is used, in which case the carbonyl tautomerizes to the enol form, the other (non-enol) OH group is protonated, and H_2O departs along with a proton to give the α,β-unsaturated aldehyde or ketone.*
- **Crossed aldol reactions** that give high yields of a desired product are generally not possible in the presence of two different aldehydes or ketones.
  - *Crossed aldol reactions* do not *generally give high yields of a desired product using catalytic hydroxide and two different aldehydes or ketones, because a mixture of products results.*
  - *High yields of a single desired product are possible in base if the more reactive carbonyl (usually an aldehyde) has no α-hydrogens, so that only the less reactive carbonyl species (usually a ketone) can form the enolate anion.*
- **Nitro groups** can be added to organic molecules through an aldol reaction between the anion of a nitroalkane and an aldehyde or ketone carbonyl.
- **Intramolecular aldol reactions** can be used to create five- or six-membered rings from dicarbonyl compounds (either aldehydes or ketones), which form in preference to smaller or larger rings that may be possible.
- To become skilled at retrosynthetic analysis, it is important to recognize that the α,β-unsaturated carbonyl and β-hydroxy carbonyl functional groups are the characteristic products of an aldol reaction.

P 19.1–19.3, 19.18–19.28, 19.53–19.56, 19.58, 19.59, 19.62, 19.64, 19.70, 19.72, 19.74, 19.77

Mechanism 19.1 and Problem 19.1

KEY REACTIONS

1. Aldol Reaction (Section 19.2) The aldol reaction involves nucleophilic addition of the enolate anion of one aldehyde or ketone to the carbonyl group of another aldehyde or ketone. The product of an aldol reaction is a β-hydroxyaldehyde or a β-hydroxyketone. An aldol reaction can be base catalyzed or acid catalyzed. If base is regenerated at the end of the reaction, it is base catalyzed, and if acid is regenerated, it is acid catalyzed. In both reactions, one or two new chiral centers are often created, leading to racemic products unless a starting aldehyde, ketone, or catalyst is chiral and present as a single enantiomer.

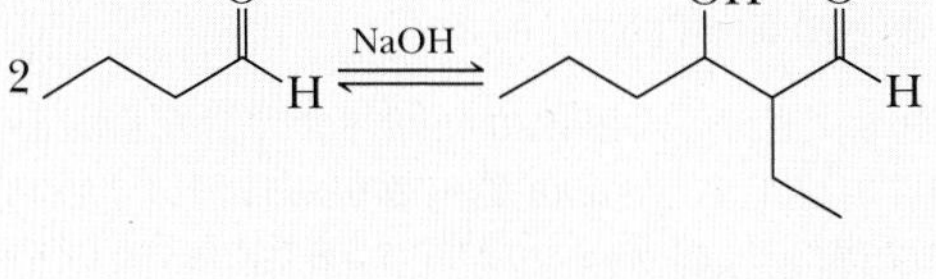

(a racemic mixture of four stereoisomers)

2. Dehydration of the Product of an Aldol Reaction (Section 19.2) Dehydration of the β-hydroxyaldehyde or ketone from an aldol reaction occurs very readily under acidic or basic conditions and gives an α,β-unsaturated aldehyde or ketone.

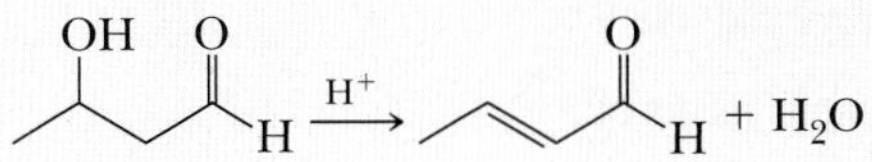

(racemic)

19.3 Claisen and Dieckmann Condensations

- The **Claisen condensation** involves two ester molecules reacting in base to give a β-ketoester product.
  - *The Claisen condensation mechanism involves reaction of one ester molecule with base to form an enolate, which reacts as a nucleophile with another molecule of ester to give a tetrahedral carbonyl addition intermediate, in which the —OR group is lost to give a β-ketoester, which is deprotonated at the α position by the RO^-.*
    - *The base used in a Claisen condensation is RO^-, with R chosen to match the alkoxy groups on the ester starting material.*
    - *Depending on relative acid-base strengths, when RO^- is used as the base, the position of equilibrium for the initial enolate-forming step is far to the side of the starting ester; so the small amount of enolate anion formed will have plenty of ester to react with.*
    - *The reaction is not catalytic in base, because the deprotonated β-ketoester product (β-ketoester $pK_a = 10-11$) is substantially less basic than the starting RO^-.*
    - *An amount of base equal to one-half equivalent compared to the amount of starting ester is the minimum amount that must be used.*
  - *To complete the reaction, the chemist adds dilute acid to generate the neutral β-ketoester product.*
- A **Dieckmann condensation** is an intramolecular Claisen condensation of a diester.
  - *Five- or six-membered ring products are favored.*
  - *One equivalent of base relative to the amount of starting diester is used.*
- Crossed Claisen reactions can be used to give high yields of a desired product from two different esters if one ester has no α-hydrogens (cannot form an enolate anion) and is used in excess.
- For retrosynthetic analysis, the β-ketoester functional group is the characteristic product of a Claisen or Dieckmann condensation.
- The product of Claisen and Dieckmann condensation reactions can be treated with aqueous base (saponification) followed by acidification to convert the β-ketoester group into a β-ketoacid that is then heated to cause decarboxylation to give a ketone product and CO_2.
  - *The general case of a Claisen condensation followed by saponification, acidification, and decarboxylation gives a symmetrical ketone product.*

P **19.4–19.6, 19.29–19.38, 19.53, 19.73–19.78**

▶ **Mechanism 19.4**

KEY REACTIONS

3. Claisen Condensation (Section 19.3A) The product of a Claisen condensation is a β-ketoester. Condensation occurs by nucleophilic acyl substitution in which the attacking nucleophile is the enolate anion of an ester. The Claisen condensation mechanism involves reaction of one ester molecule with base to form an enolate anion, which reacts as a nucleophile with another molecule of ester to give a tetrahedral carbonyl addition intermediate, in which the RO^- group is lost to give a β-ketoester, which is deprotonated at the α position by the RO^-.

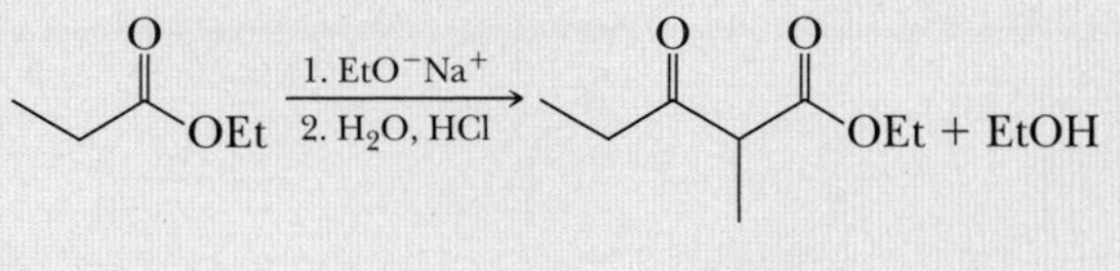
O
OEt
1. EtO⁻Na⁺
2. H₂O, HCl
O O
OEt + EtOH

(racemic)

(Continued)

4. Dieckmann Condensation (Section 19.3B) An intramolecular Claisen condensation is called a Dieckmann condensation.

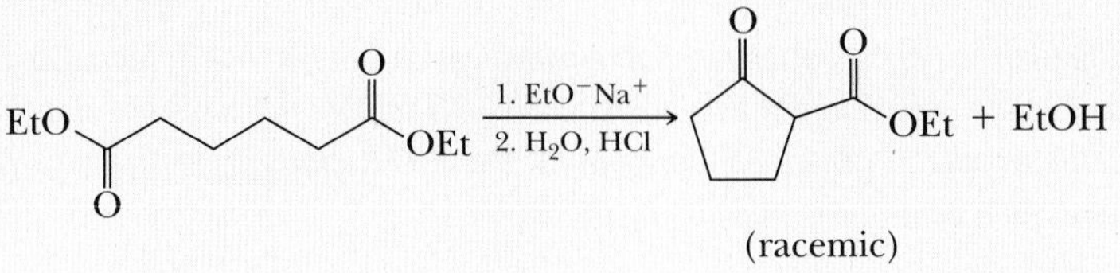
EtO
OEt
1. EtO⁻Na⁺
2. H₂O, HCl
OEt + EtOH
(racemic)

19.4 Claisen and Aldol Condensations in the Biological World

- Biological molecules are created from simple building blocks through enzyme-catalyzed reactions that often resemble the organic transformations presented in this chapter.
  - *Claisen condensations and aldol reactions are common, and* ***acetyl-CoA*** *is a common starting material for these reactions.*

19.5 Enamines

- **Enamines**, compounds with a C═C π bond adjacent to a C—N bond, are formed through reaction of an aldehyde or a ketone with a secondary amine, most commonly pyrrolidine or morpholine.
  - *Enamines are important for synthesis because the β-carbon is a nucleophile by virtue of conjugation of the C═C π bond with the electron pair on nitrogen.*
  - *Enamines resemble enols and enolate anions in their reactions, yet harsh conditions (i.e., strong acid or base) are not required.*
    - *Enamines can be alkylated on the β-carbon in an S_N2 reaction with methyl and primary haloalkanes.*
    - *Enamines can be acylated on the β-carbon by treatment with acid chlorides and acid anhydrides.*
    - *Following the alkylation or acylation reaction, aqueous acid is used to convert the enamine back into a carbonyl group.*
- For retrosynthetic analysis, an unsymmetrical aldehyde/ketone and a β-dicarbonyl are characteristic products of enamine alkylation and acylation, respectively.

P **19.7–19.9, 19.39–19.43, 19.57, 19.76**

Problem 19.9

KEY REACTIONS

5. Alkylation of an Enamine Followed by Hydrolysis (Section 19.5A) Enamines are reactive nucleophiles with methyl and primary haloalkanes, α-haloketones, and α-haloesters.

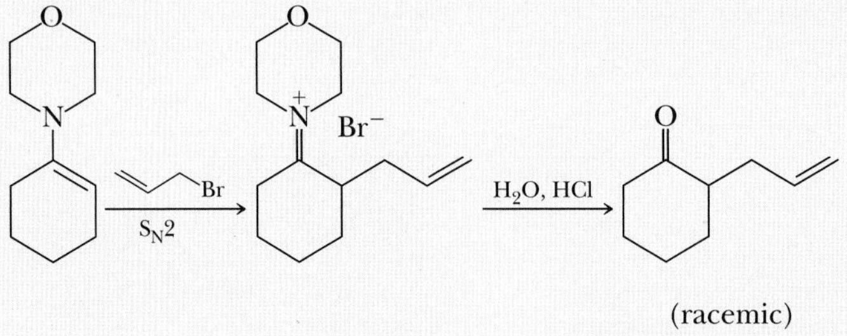
O
N
Br
S$_N$2
O
N
Br⁻
H₂O, HCl
O
(racemic)

6. Acylation of an Enamine Followed by Hydrolysis (Section 19.5B)

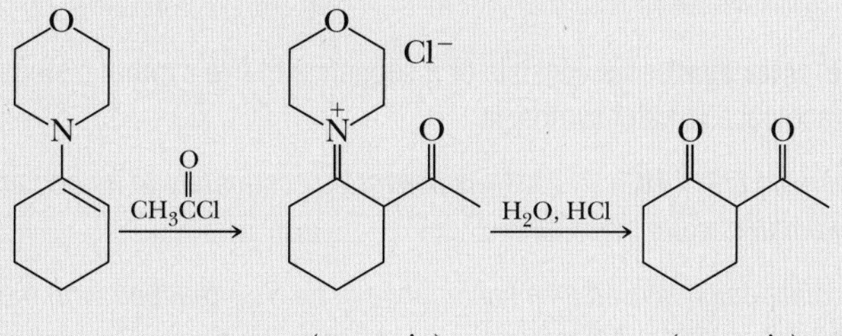
O
N
O
CH_3CCl
O
N
+
Cl^-
O
H_2O, HCl
O
O
(racemic)
(racemic)

19.6 Acetoacetic Ester Synthesis

- α-Hydrogens between two carbonyl groups are especially easy to remove (pK_a values of 10–14) because the resulting anion is stabilized through delocalization with both adjacent carbonyl groups.
- The **acetoacetic ester synthesis** consists of a sequence of five synthetic steps.
  - *Ethyl acetoacetate ($pK_a = 10.7$) is converted completely to its enolate anion using one equivalent of a base such as a sodium alkoxide.*
  - *The enolate anion is used as a nucleophile in an S_N2 reaction with methyl or primary haloalkanes, α-haloketones, and α-haloesters.*
    - *These first two steps can be repeated if a doubly alkylated product is desired.*
  - *The alkylated acetoacetic ester is hydrolyzed using HO^-.*
  - *Acidification gives the alkylated acetoacetic acid.*
  - *Heating causes decarboxylation to give the alkylated ketone product.*
- The same sequence of reactions can be used with other β-ketoacids, not only ethyl acetoacetate.
  - *β-Ketoesters result from Claisen condensation reactions, so the product of a Claisen reaction can be easily manipulated further using the synthetic sequence described here.*
- For retrosynthetic analysis, a methyl ketone is a characteristic product of the acetoacetic ester synthesis starting with ethyl acetoacetate, and other complex ketones can be derived from other β-ketoester starting materials.

P **19.10, 19.11, 19.46, 19.53, 19.73**

KEY REACTIONS

7. Acetoacetic Ester Synthesis (Section 19.6) This sequence is useful for the synthesis of monosubstituted and disubstituted acetones.

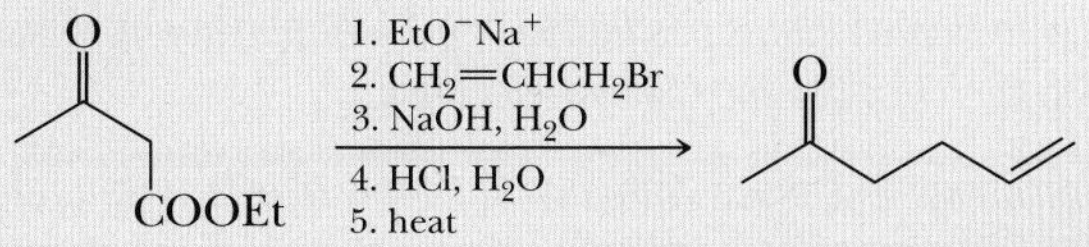
O
COOEt
1. EtO^-Na^+
2. $CH_2{=}CHCH_2Br$
3. NaOH, H_2O
4. HCl, H_2O
5. heat
O

Ethyl acetoacetate A monosubstituted acetone

19.7 Malonic Ester Synthesis

- The **malonic ester synthesis** consists of a sequence of five synthetic steps that are analogous to those of the acetoacetic ester synthesis.
  - *Diethyl malonate ($pK_a = 13.3$) is converted completely to its enolate anion using one equivalent of sodium ethoxide.*
  - *The enolate anion is used as a nucleophile in an S_N2 reaction with methyl or primary haloalkanes, α-haloketones, and α-haloesters.*
    - *These first two steps can be repeated if a doubly alkylated product is desired, subject to steric constraints.*
  - *The alkylated acetoacetic ester is hydrolyzed using HO^-.*
  - *Acidification gives the alkylated acetoacetic acid.*
  - *Heating causes decarboxylation to give the alkylated carboxylic acid product.*
- For retrosynthetic analysis, a carboxylic acid with one or two alkyl groups bonded to the α-carbon is a characteristic product of the acetoacetic ester synthesis starting with diethyl malonate.

P **19.12, 19.44–19.48, 19.53, 19.61–19.64, 19.69**

Problem 19.12

KEY REACTIONS

8. Malonic Ester Synthesis (Section 19.7) This sequence is useful for the synthesis of monosubstituted and disubstituted acetic acids.

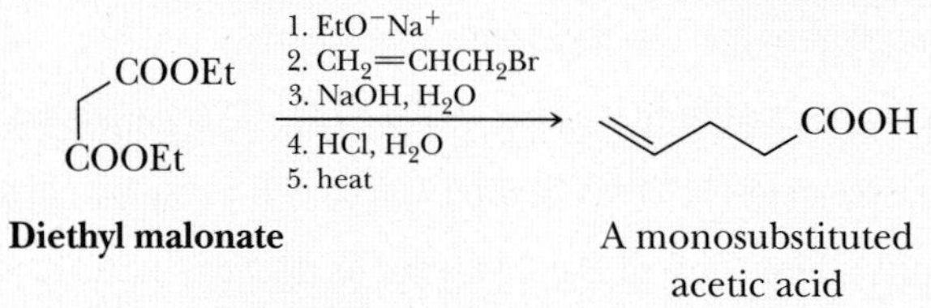
COOEt
COOEt
1. EtO⁻Na⁺
2. CH₂=CHCH₂Br
3. NaOH, H₂O
4. HCl, H₂O
5. heat
COOH
Diethyl malonate
A monosubstituted acetic acid

19.8 Conjugate Addition to α,β-Unsaturated Carbonyl Compounds

- In a **Michael reaction,** the β-carbon of α,β-unsaturated carbonyl species is electrophilic and able to react with certain nucleophiles, especially enolate anions.
  - *One important resonance-contributing structure places positive charge at the β-carbon of α,β-unsaturated carbonyl species, helping to explain reaction with nucleophiles at this position.*
  - *Reaction of enolate anions with α,β-unsaturated carbonyl species in Michael reactions gives products with oxygen atoms positioned in a 1,5 arrangement, which is distinct from the 1,3 positioning seen in aldol and Claisen reactions.*
  - *Michael reactions take place with a wide variety of α,β-unsaturated carbonyl compounds (aldehydes, ketones, esters, and amides) as well as α,β-unsaturated nitro and nitrile compounds.*
    - *These electrophiles in Michael reactions are often referred to as Michael acceptors.*

- *The mechanism of the Michael reaction involves initial formation of an enolate anion in base and attack of the enolate nucleophile onto the β-carbon of the Michael acceptor to create a new resonance-stabilized enolate anion intermediate that is protonated on oxygen to create an enol and regenerate the base; then tautomerization to the keto form completes the reaction.*
  - *The base is catalytic in the Michael reaction.*
  - *This type of addition mechanism is referred to as conjugate addition, or alternatively, 1,4-addition.*

- A **Robinson annulation** is a Michael reaction with an α,β-unsaturated ketone followed by an intramolecular aldol reaction creating a cyclic product.
  - *Six-membered rings are produced in high yields this way.*
- Gilman reagents undergo conjugate addition with α,β-unsaturated carbonyl compounds in a reaction that is closely related to the Michael reaction.

P 19.13–19.16, 19.49, 19.50, 19.53, 19.54–19.56, 19.59, 19.60, 19.64, 19.68, 19.70, 19.74, 19.80

Retrosynthetic analysis of the Michael reaction and Problem 19.15

KEY REACTIONS

9. **Michael Reaction (Section 19.8A)** A Michael reaction involves the addition of a weakly basic nucleophile to a carbon-carbon double bond made electrophilic by conjugation with the carbonyl group of an aldehyde, a ketone, or an ester or with a nitro or cyano group. The mechanism of the Michael reaction involves initial formation of an enolate anion in base and attack of the enolate nucleophile at the β-carbon of the Michael acceptor to create a new resonance-stabilized enolate anion intermediate that is protonated on oxygen to create an enol and regenerate the base; then tautomerization to the keto form completes the reaction. The base is catalytic in the Michael reaction.

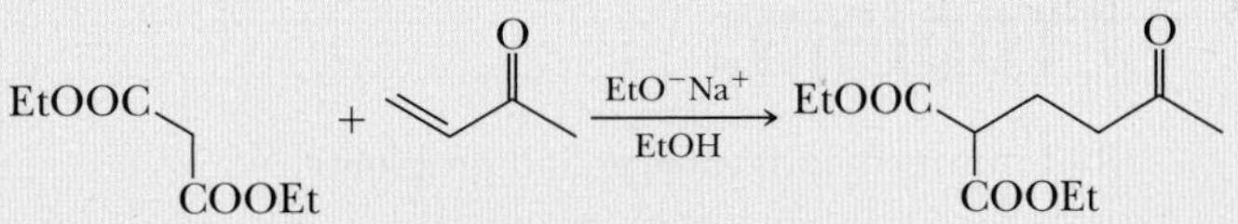

10. **Robinson Annulation (Section 19.8C)** A Robinson annulation comprises a Michael reaction followed by an intramolecular aldol reaction and dehydration to form a substituted 2-cyclohexenone.

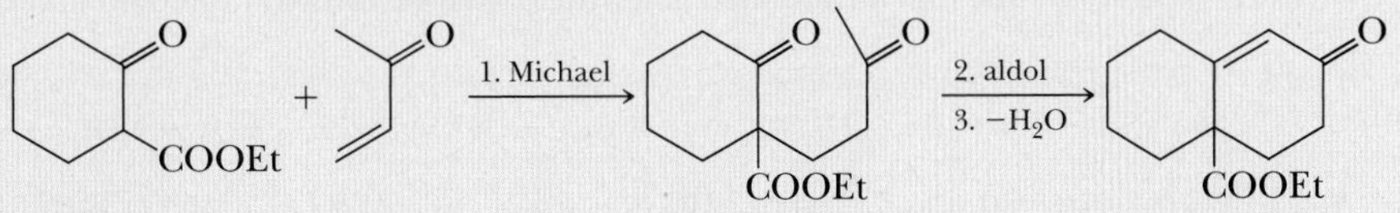

(racemic)

11. **Conjugate Addition of Lithium Diorganocopper Reagents (Section 19.8E)** In a reaction closely related to the Michael reaction, lithium diorganocopper reagents undergo conjugate addition to the electrophilic double bond of α,β-unsaturated aldehydes and ketones.

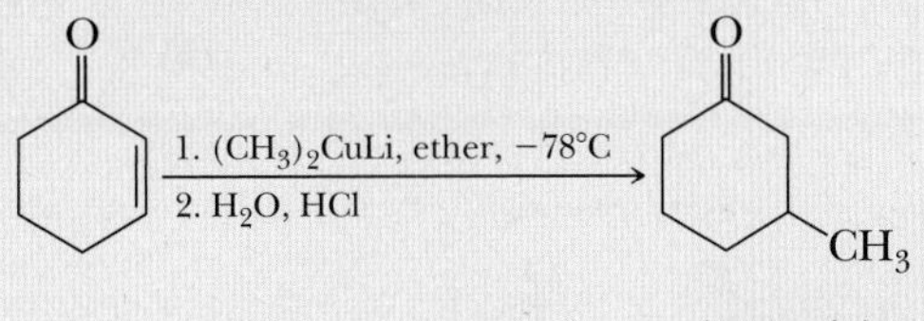

(racemic)

19.9 Crossed Enolate Reactions Using LDA

- **Lithium diisopropylamide (LDA)** is an extremely strong base that is not nucleophilic due to steric hindrance.
  - *One molar equivalent of LDA converts aldehydes, ketones, and esters completely to their enolate anions.*
  - *Preformed enolate anions using LDA can be used to carry out a wide variety of crossed enolate reactions, including aldol reactions, Claisen condensations, Michael additions, alkylations, and acylations.*
  - *If a slight excess of carbonyl species is used, an equilibrium is set up in which the more stable of the alternative enolate anions predominates, a situation known as* ***thermodynamic control****.*
    - *The most highly substituted of the possible enolate anions predominates.*
  - *If a slight excess of LDA is used, no equilibrium is established among the alternative enolate anions and the predominant enolate is the one that forms more rapidly (the removed α-hydrogen is more accessible).*
    - *This situation is referred to as* ***kinetic control****.*
    - *The less substituted of the possible enolate anions generally predominate.*

P **19.17, 19.51–19.53, 19.65–19.67, 19.71, 19.78**

Kinetic vs. thermodynamic enolates

Problems

Red numbers indicate applied problems.

The Aldol Reaction

19.18 Draw a structural formula for the product of the aldol reaction of each compound and for the α,β-unsaturated aldehyde or ketone formed from dehydration of each aldol product.

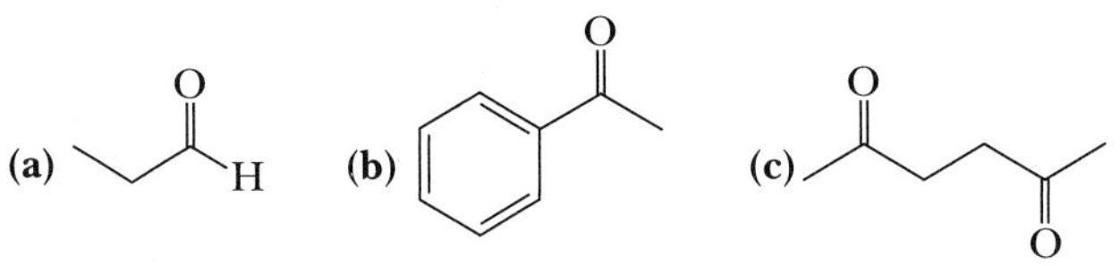

19.19 Draw a structural formula for the product of each crossed aldol reaction and for the compound formed by dehydration of each aldol product.

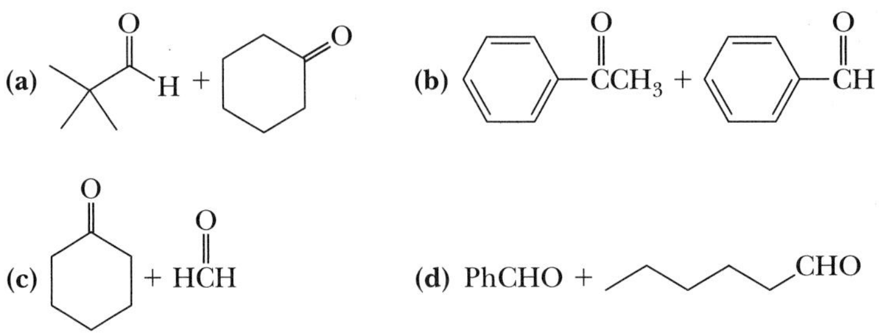

19.20 When a 1:1 mixture of acetone and 2-butanone is treated with base, six aldol products are possible. Draw a structural formula for each product.

19.21 Show how to prepare each α,β-unsaturated ketone by an aldol reaction followed by dehydration of the aldol product.

(a) 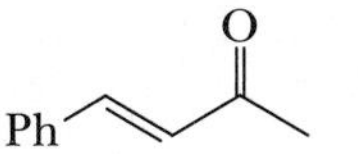 **(b)** 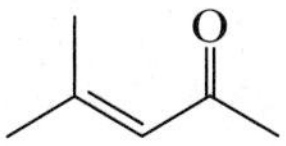

19.22 Show how to prepare each α,β-unsaturated aldehyde by an aldol reaction followed by dehydration of the aldol product.

(a) 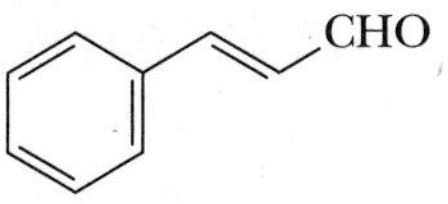 **(b)** 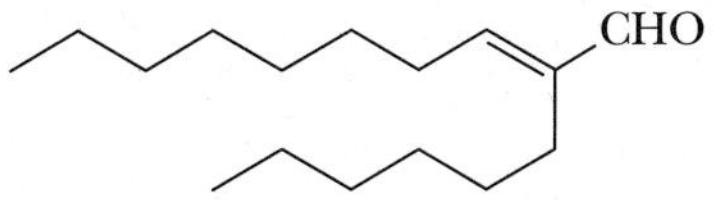

19.23 When treated with base, the following compound undergoes an intramolecular aldol reaction to give a product containing a ring (yield 78%).

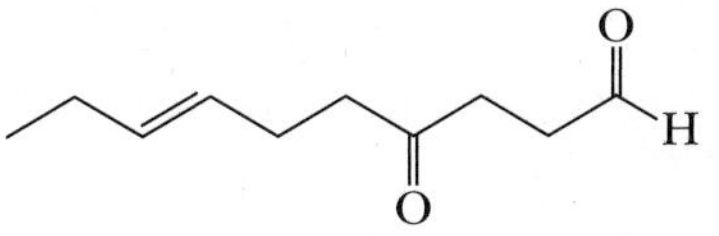

$\xrightarrow{\text{base}}$ $C_{10}H_{14}O + H_2O$

Propose a structural formula for this product.

19.24 Cyclohexene can be converted to 1-cyclopentenecarbaldehyde by the following series of reactions.

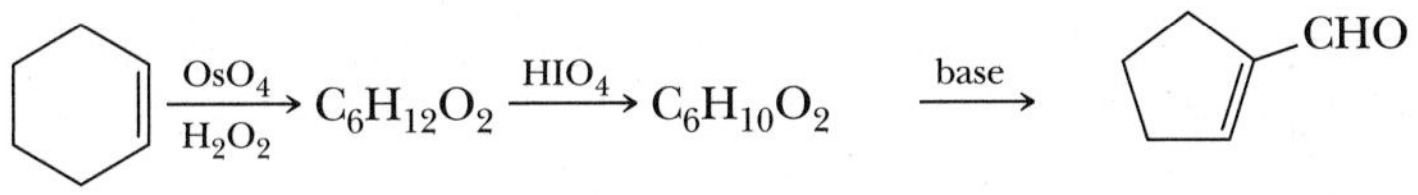

1-Cyclopentenecarbaldehyde

Propose a structural formula for each intermediate compound.

19.25 Propose a structural formula for each lettered compound.

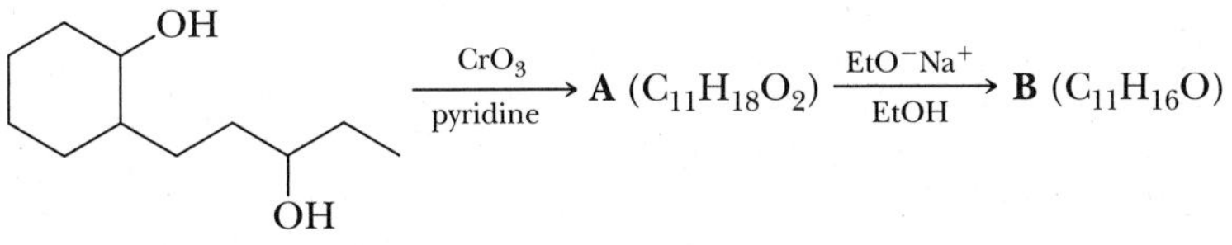

(racemic)

19.26 How might you bring about the following conversion?

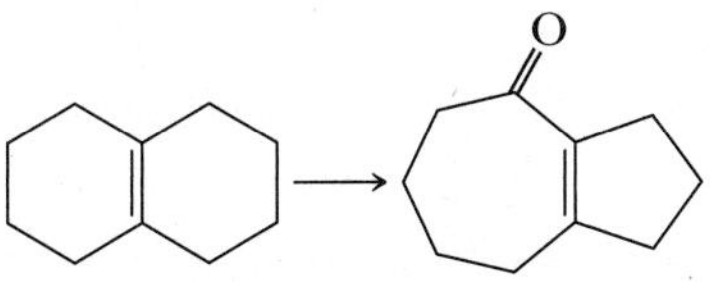

19.27 Pulegone, $C_{10}H_{16}O$, a compound from oil of pennyroyal, has a pleasant odor midway between peppermint and camphor. Treatment of pulegone with steam produces acetone and 3-methylcyclohexanone.

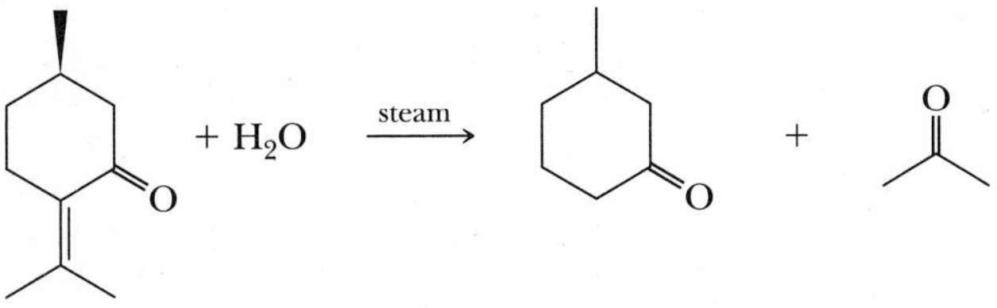

Pulegone **3-Methylcyclohexanone** **Acetone**

(a) Natural pulegone has the configuration shown. Assign an *R* or *S* configuration to its chiral center.

(b) Propose a mechanism for the steam hydrolysis of pulegone to the compounds shown.

(c) In what way does this steam hydrolysis affect the configuration of the chiral center in pulegone? Assign an *R* or *S* configuration to the 3-methylcyclohexanone formed in this reaction.

19.28 Propose a mechanism for this acid-catalyzed aldol reaction and the dehydration of the resulting aldol product.

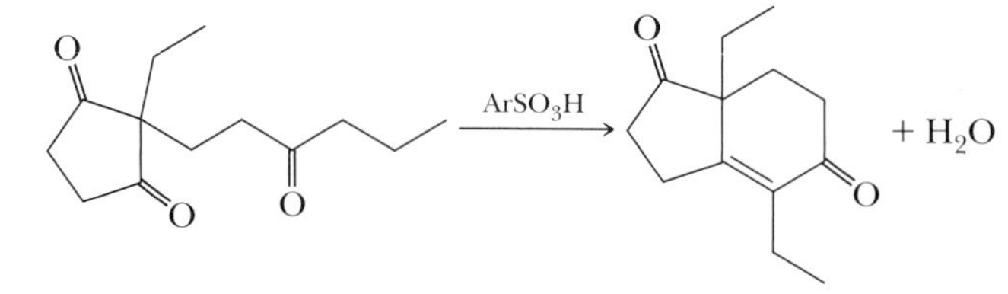

(racemic)

The Claisen Condensation

19.29 Show the product of Claisen condensation of each ester.

(a) Ethyl phenylacetate in the presence of sodium ethoxide

(b) Methyl hexanoate in the presence of sodium methoxide

19.30 When a 1:1 mixture of ethyl propanoate and ethyl butanoate is treated with sodium ethoxide, four Claisen condensation products are possible. Draw a structural formula for each product.

19.31 Draw structural formulas for the β-ketoesters formed by Claisen condensation of ethyl propanoate with each ester.

(a) $EtO\overset{O}{\overset{\|}{C}}-\overset{O}{\overset{\|}{C}}OEt$ **(b)** $Ph\overset{O}{\overset{\|}{C}}OEt$ **(c)** $H\overset{O}{\overset{\|}{C}}OEt$

19.32 Draw a structural formula for the product of saponification, acidification, and decarboxylation of each β-ketoester formed in Problem 19.31.

19.33 The Claisen condensation can be used as one step in the synthesis of ketones, as illustrated by this reaction sequence.

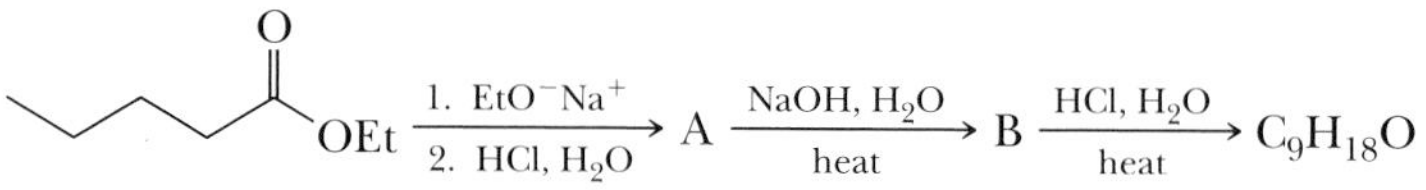

Propose structural formulas for compounds A and B and the ketone formed in this sequence.

19.34 Propose a synthesis for each ketone, using as one step in the sequence a Claisen condensation and the reaction sequence illustrated in Problem 19.33.

(a) 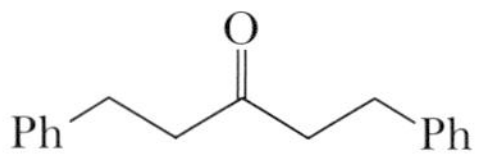

(b) 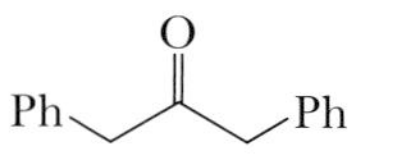

(c) 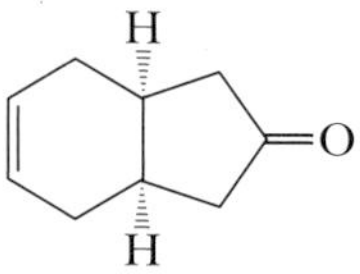

19.35 Propose a mechanism for the following conversion.

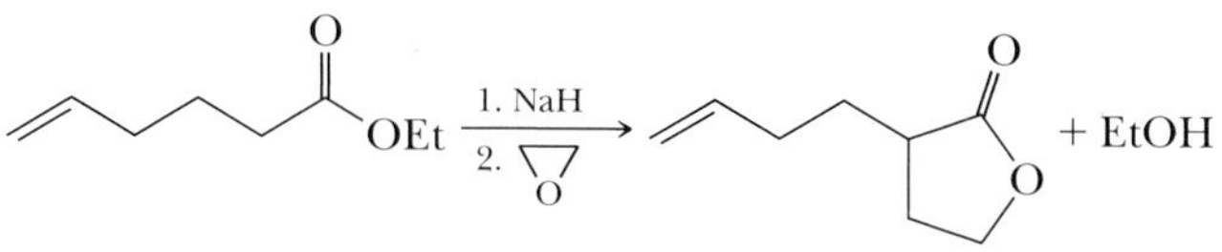

(racemic)

19.36 Claisen condensation between diethyl phthalate and ethyl acetate followed by saponification, acidification, and decarboxylation forms a diketone, $C_9H_6O_2$.

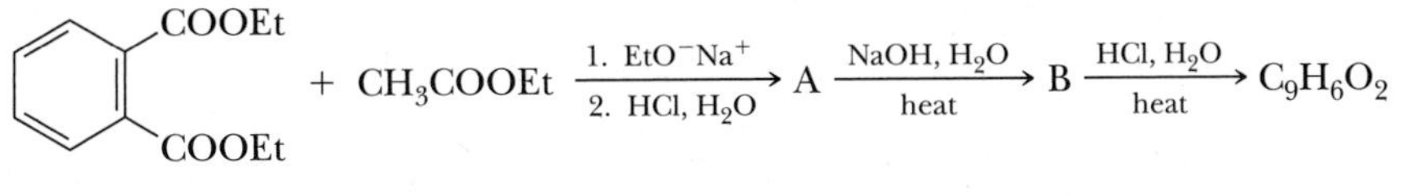

Diethyl phthalate **Ethyl acetate**

Propose structural formulas for compounds A and B and the diketone.

19.37 In 1887, the Russian chemist Sergei Reformatsky at the University of Kiev discovered that treatment of an α-haloester with zinc metal in the presence of an aldehyde or a ketone followed by hydrolysis in aqueous acid results in formation of a β-hydroxyester. This reaction is similar to a Grignard reaction in that a key intermediate is an organometallic compound, in this case, a zinc salt of an ester enolate anion. Grignard reagents, however, are so reactive that they undergo self-condensation with the ester.

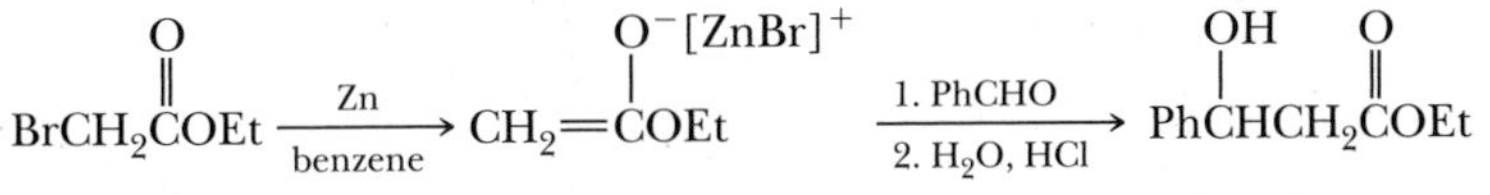

Zinc salt of an enolate anion

A β-hydroxyester (racemic)

Show how a Reformatsky reaction can be used to synthesize these compounds from an aldehyde or a ketone and an α-haloester.

(a) 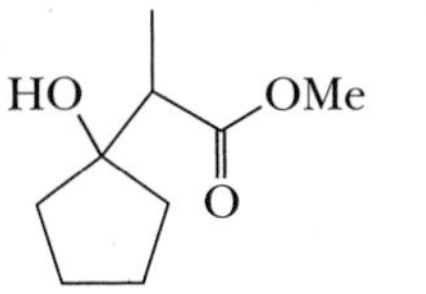

(four stereoisomers as two racemic mixtures)

(b) 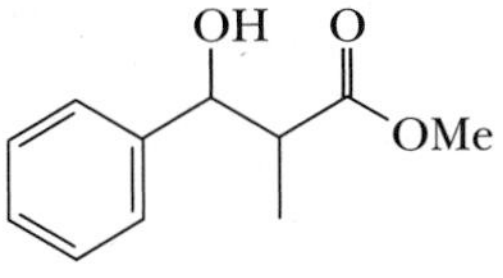

(four stereoisomers as two racemic mixtures)

(c) 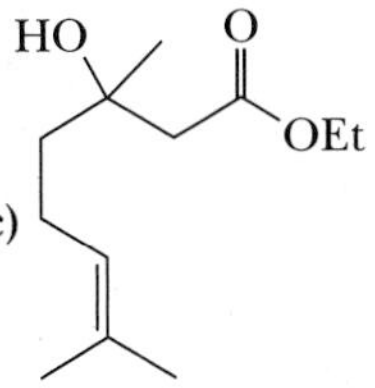

(racemic)

19.38 Many types of carbonyl condensation reactions have acquired specialized names, after the nineteenth-century organic chemists who first studied them. Propose mechanisms for the following named condensations.

(a) Perkin condensation: Condensation of an aromatic aldehyde with an acid anhydride

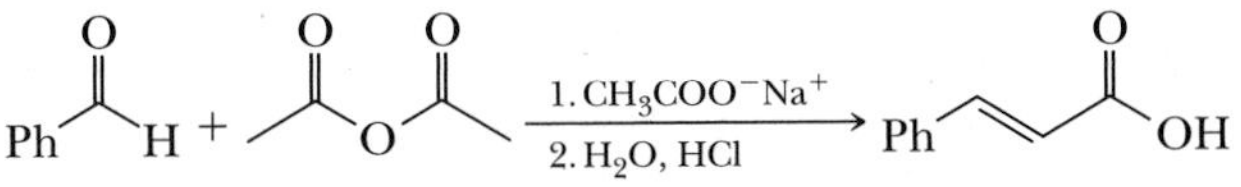

Cinnamic acid

(b) Darzens condensation: Condensation of an α-haloester with a ketone or an aromatic aldehyde

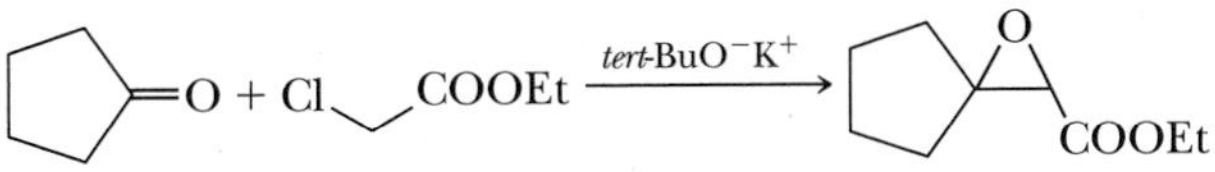

Enamines

19.39 When 2-methylcyclohexanone is treated with pyrrolidine, two isomeric enamines are formed.

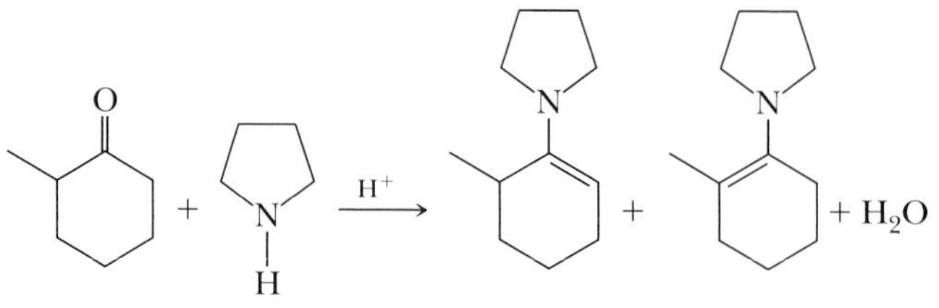

A (85%) B (15%)

Why is enamine A with the less substituted double bond the thermodynamically favored product? (You will find it helpful to examine the models of these two enamines.)

19.40 Enamines normally react with methyl iodide to give two products: one arising from alkylation at nitrogen and the second arising from alkylation at carbon. For example,

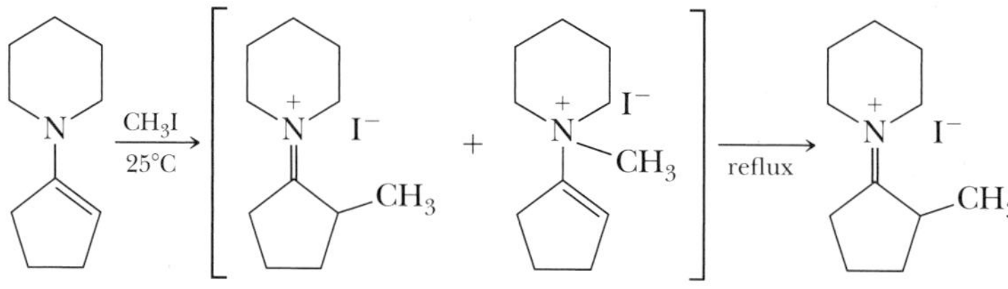

Product of *C*-alkylation | Product of *N*-alkylation | (racemic)

Heating the mixture of *C*-alkylation and *N*-alkylation products gives only the product from *C*-alkylation. Propose a mechanism for this isomerization.

19.41 Propose a mechanism for the following conversion.

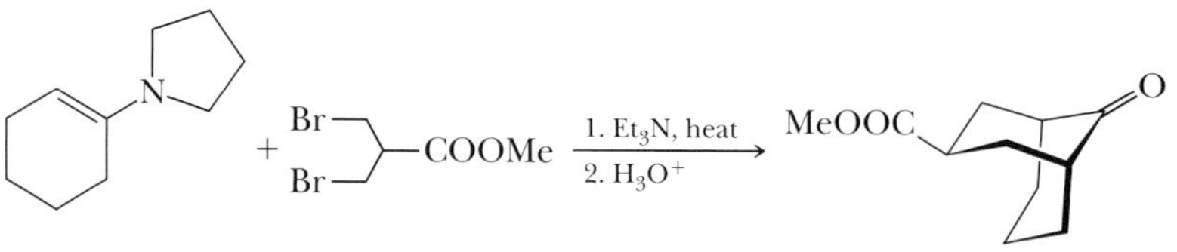

19.42 The following intermediate was needed for the synthesis of tamoxifen, a widely used antiestrogen drug for treating estrogen-dependent cancers such as breast and ovarian cancer.

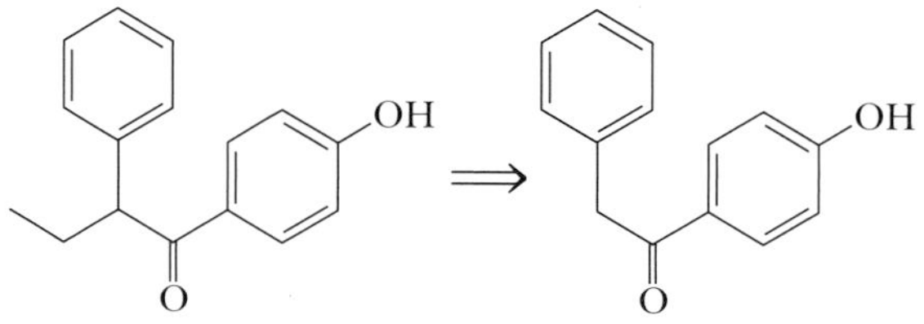

Needed for the synthesis of tamoxifen (A)

Propose a synthesis for this intermediate from compound A.

19.43 Propose a mechanism for the following reaction.

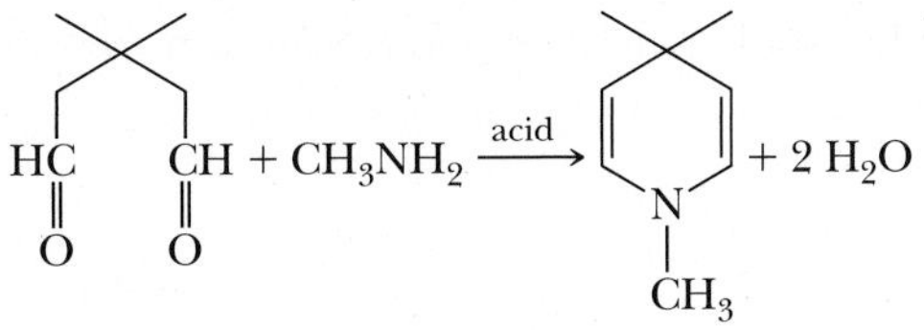

Acetoacetic Ester and Malonic Ester Syntheses

19.44 Propose syntheses of the following derivatives of diethyl malonate, each of which is a starting material for synthesis of a barbiturate.

(a) 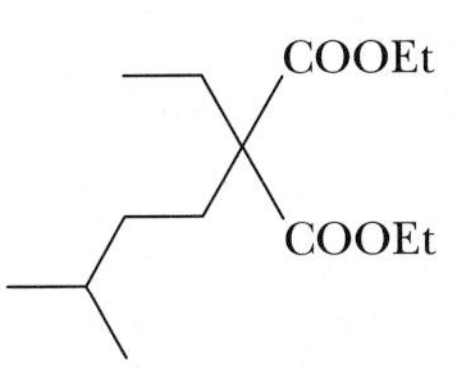

Needed for the synthesis of amobarbital

(b) 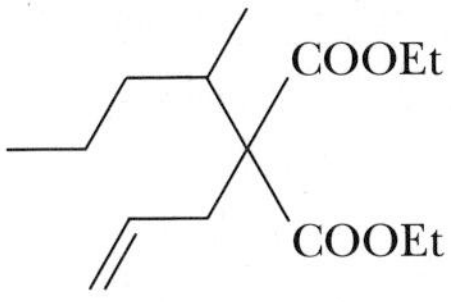

Needed for the synthesis of secobarbital

19.45 2-Propylpentanoic acid (valproic acid) is an effective drug for treatment of several types of epilepsy, particularly absence seizures, which are generalized epileptic seizures characterized by brief and abrupt loss of consciousness. Propose a synthesis of valproic acid starting with diethyl malonate.

19.46 Show how to synthesize the following compounds using either the malonic ester synthesis or the acetoacetic ester synthesis.

(a) 4-Phenyl-2-butanone
(b) 2-Methylhexanoic acid
(c) 3-Ethyl-2-pentanone
(d) 2-Propyl-1,3-propanediol
(e) 4-Oxopentanoic acid
(f) 3-Benzyl-5-hexene-2-one
(g) Cyclopropanecarboxylic acid
(h) Cyclobutyl methyl ketone

19.47 Propose a mechanism for formation of 2-carbethoxy-4-butanolactone and 4-butanolactone (γ-butyrolactone) in the following sequence of reactions.

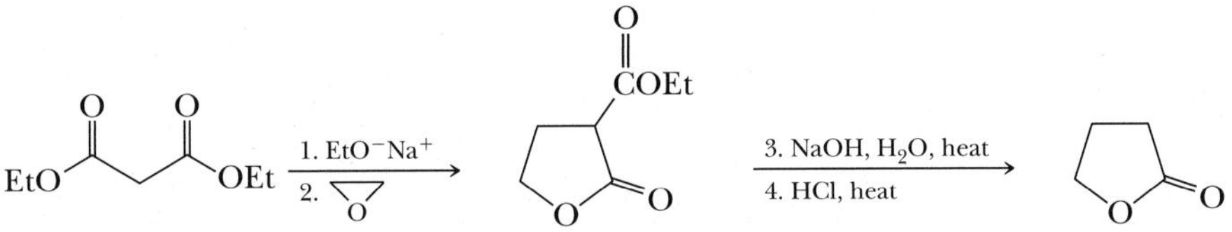

2-Carbethoxy-4-butanolactone

4-Butanolactone (γ-Butyrolactone)

19.48 Show how the scheme for formation of 4-butanolactone in Problem 19.47 can be used to synthesize lactones (a) and (b), each of which has a peach odor and is used in perfumery. As sources of carbon atoms for these syntheses, use diethyl malonate, ethylene oxide, 1-bromoheptane, and 1-nonene.

(a) 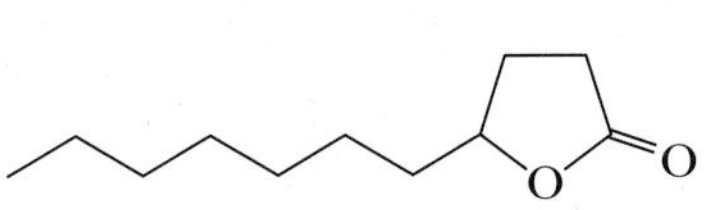

(racemic)

(b) 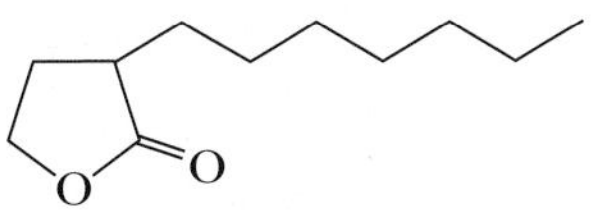

(racemic)

Michael Reactions

19.49 The following synthetic route is used to prepare an intermediate in the total synthesis of the anticholinergic drug benzilonium bromide.

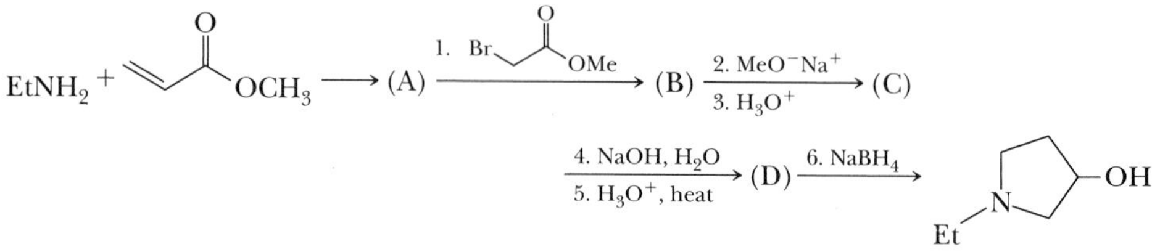

(racemic)

Propose structural formulas for intermediates A, B, C, and D.

19.50 Propose a mechanism for formation of the bracketed intermediate and for the bicyclic ketone formed in the following reaction sequence.

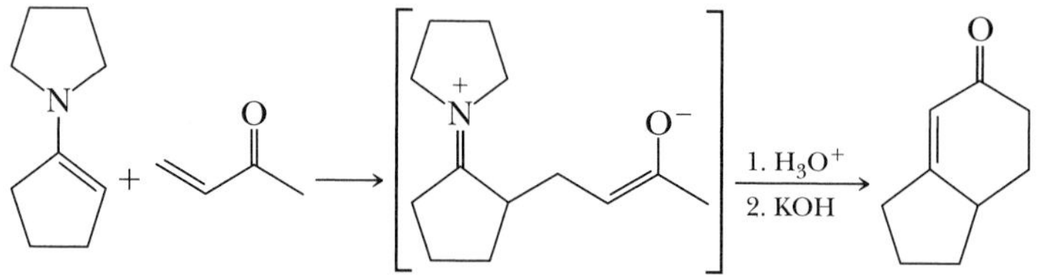

An intermediate (not isolated) (racemic)

Directed Aldol and Alkylation

19.51 Discuss the different experimental conditions used to give the major and minor product distributions shown.

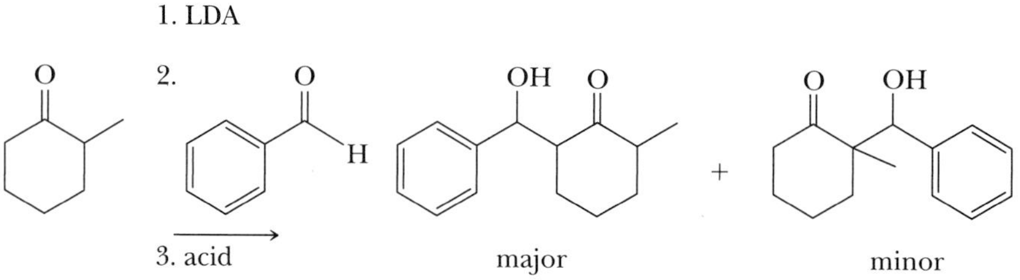

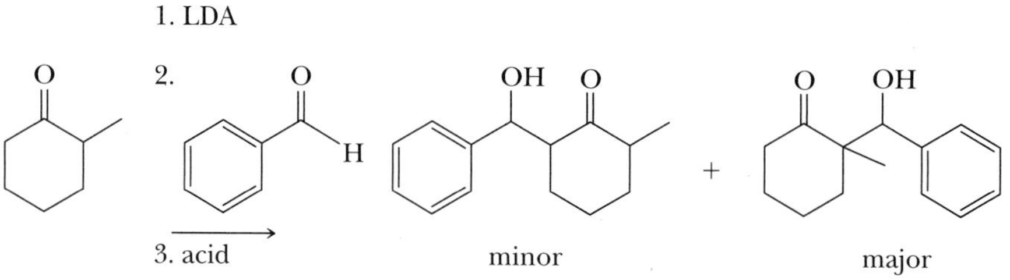

19.52 Why does the following reaction give the product shown as the major product when 0.95 equivalent of LDA relative to ketone is used in the first step?

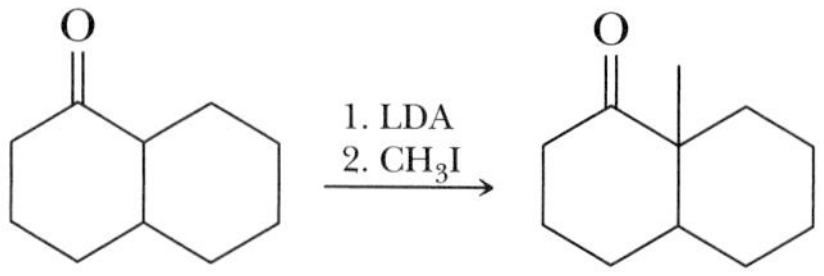

Retrosynthetic Analysis

19.53 Using one of the reactions in this chapter, give the correct starting material (A–L) needed to produce each structure (a–f). Name the type of reaction used.

(a) 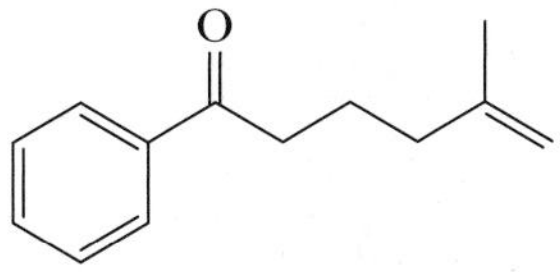 ⟹

(b) 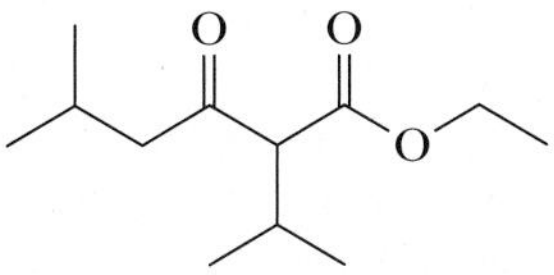 ⟹

(c) 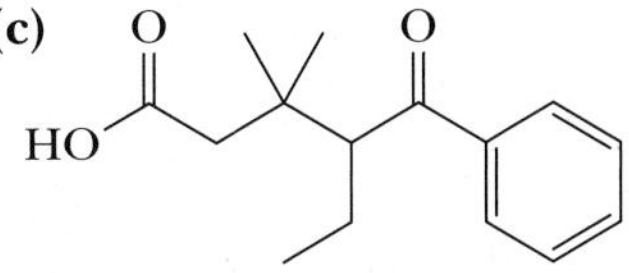 ⟹

(d) 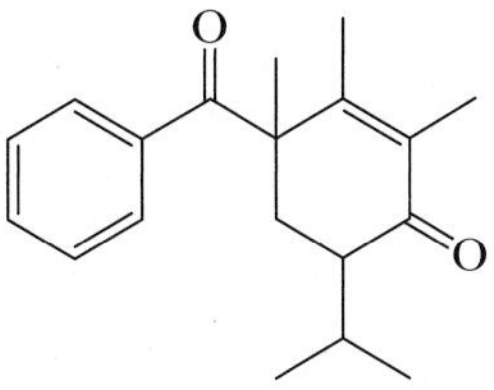 ⟹

(e) 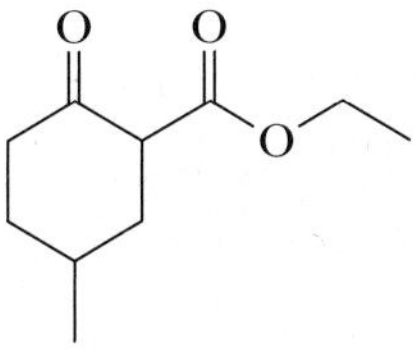 ⟹

(f) 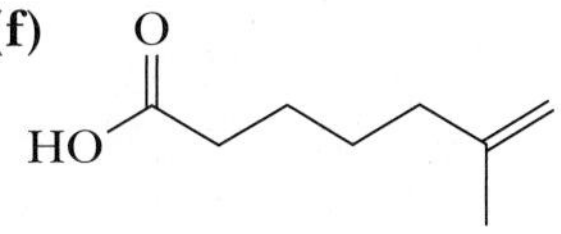 ⟹

A

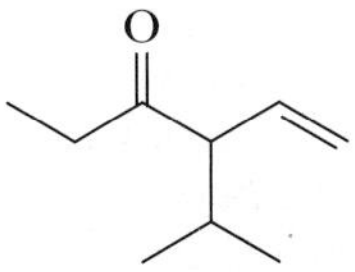

B

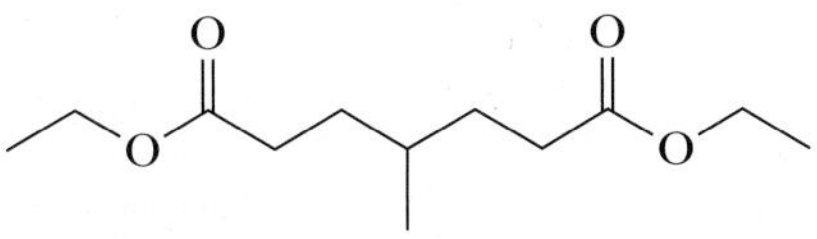

C

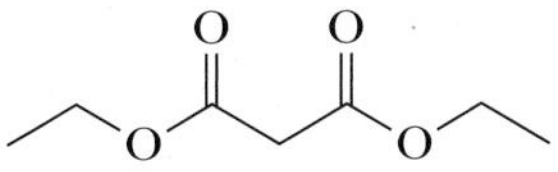

D

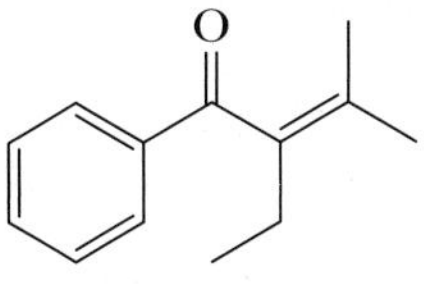

E

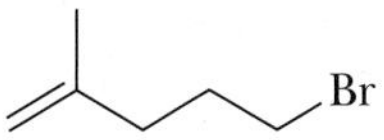

F

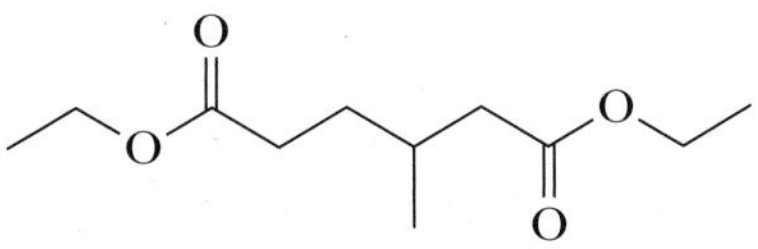

G

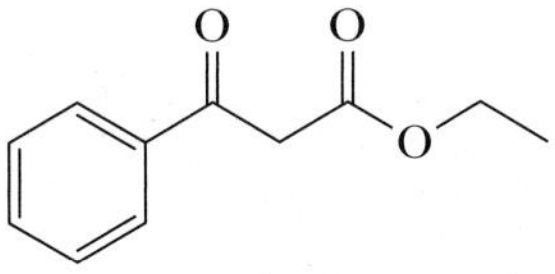

H

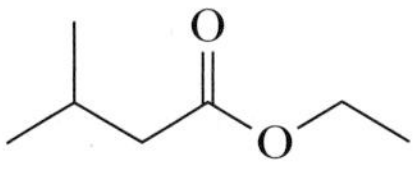

I

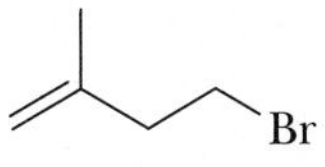

J

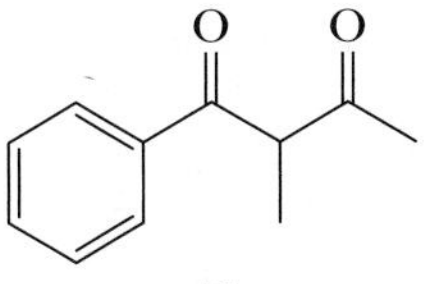

K

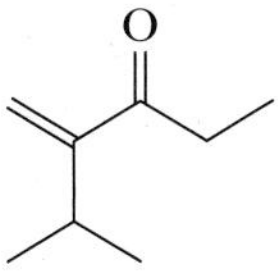

L

Synthesis

19.54 Show experimental conditions by which to carry out the following synthesis.

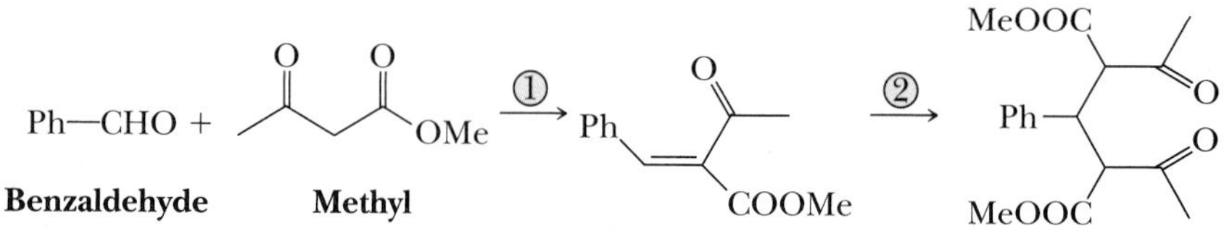

Benzaldehyde **Methyl acetoacetate**

19.55 Nifedipine (Procardia and Adalat) belongs to a class of drugs called calcium channel blockers and is effective in the treatment of various types of angina, including that induced by exercise. Show how nifedipine can be synthesized from 2-nitrobenzaldehyde, methyl acetoacetate, and ammonia. (*Hint:* Review the chemistry of your answers to Problems 19.43 and 19.54 and then combine that chemistry to solve this problem.)

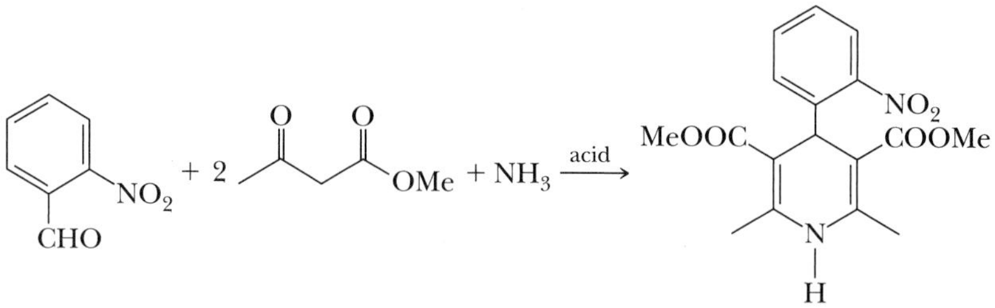

Nifedipine

19.56 The compound 3,5,5-trimethyl-2-cyclohexenone can be synthesized using acetone and ethyl acetoacetate as sources of carbon atoms. New carbon-carbon bonds in this synthesis are formed by a combination of aldol reactions and Michael reactions. Show reagents and conditions by which this synthesis might be accomplished.

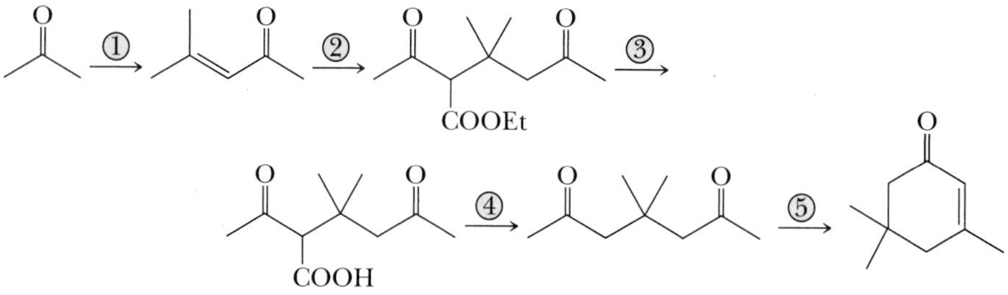

3,5,5-Trimethyl-2-cyclohexenone

19.57 The following β-diketone can be synthesized from cyclopentanone and an acid chloride using an enamine reaction.

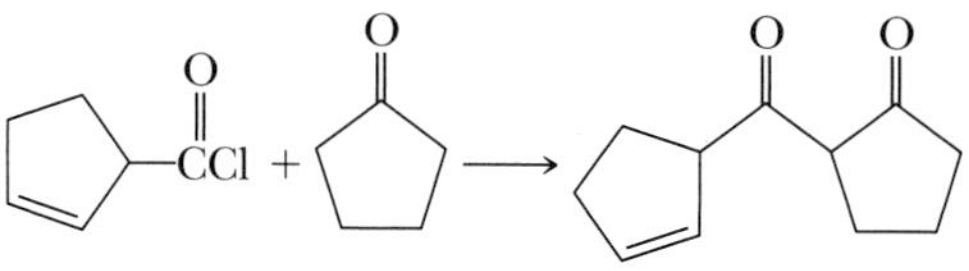

(racemic)

(a) Propose a synthesis of the starting acid chloride from cyclopentene.

(b) Show the steps in the synthesis of the β-diketone using a morpholine enamine.

19.58 Oxanamide is a mild sedative belonging to a class of molecules called oxanamides. As seen in this retrosynthetic scheme, the source of carbon atoms for the synthesis of oxanamide is butanal.

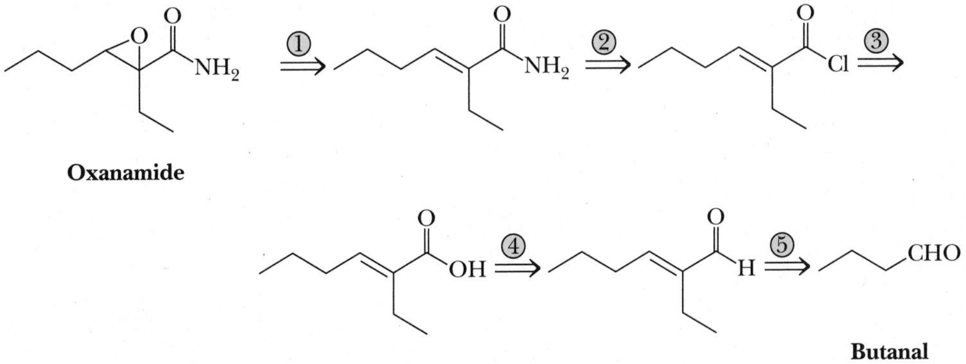

Oxanamide
Butanal

(a) Show reagents and experimental conditions by which oxanamide can be synthesized from butanal.

(b) How many chiral centers are in oxanamide? How many stereoisomers are possible for this compound?

19.59 The widely used anticoagulant warfarin (see "Chemical Connections: From Moldy Clover to a Blood Thinner" in Chapter 18) is synthesized from 4-hydroxycoumarin, benzaldehyde, and acetone as shown in this retrosynthesis. Show how warfarin is synthesized from these reagents.

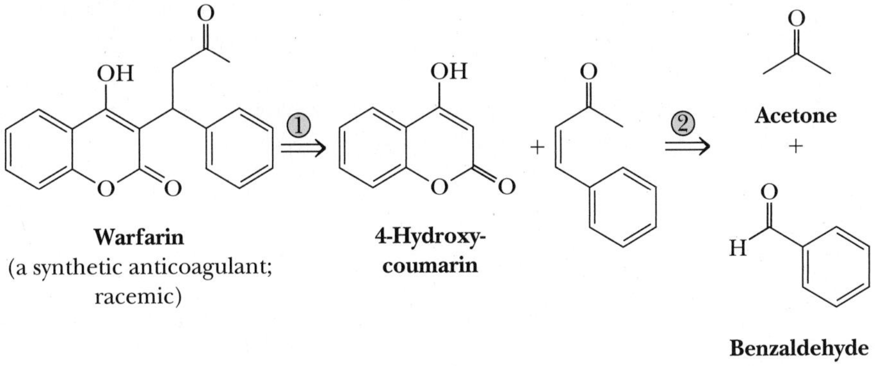

Warfarin
(a synthetic anticoagulant;
racemic)
4-Hydroxy-
coumarin
Acetone
Benzaldehyde

19.60 Following is a retrosynthetic analysis for an intermediate in the industrial synthesis of vitamin A.

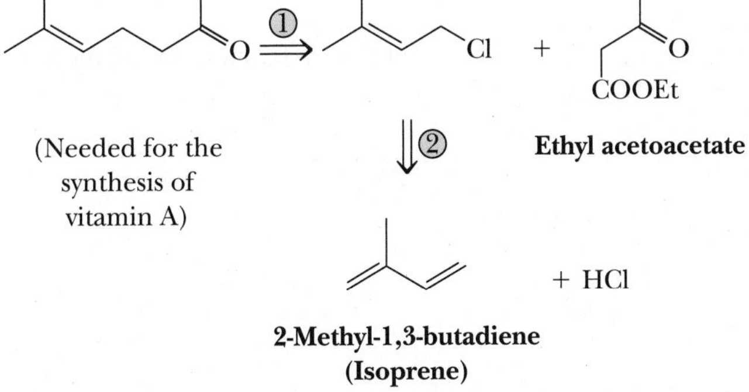

(Needed for the
synthesis of
vitamin A)
Ethyl acetoacetate
+ HCl
2-Methyl-1,3-butadiene
(Isoprene)

(a) Addition of one mole of HCl to isoprene gives 4-chloro-2-methyl-2-butene as the major product. Propose a mechanism for this addition and account for its regioselectivity.

(b) Propose a synthesis of the vitamin A precursor from this allylic chloride and ethyl acetoacetate.

19.61 Following are the steps in one of the several published syntheses of frontalin, a pheromone of the western pine beetle.

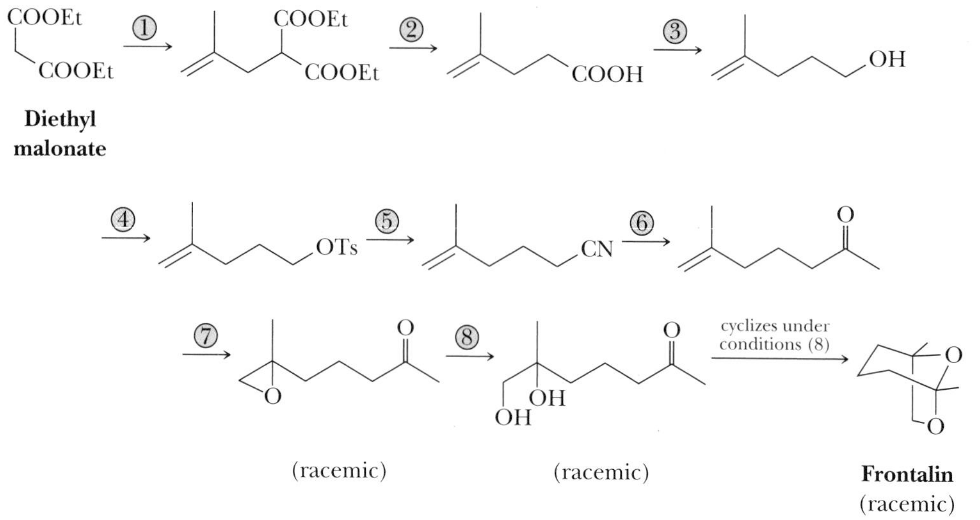

(a) Propose reagents for Steps 1–8.

(b) Propose a mechanism for the cyclization of the ketodiol from Step 8 to frontalin.

19.62 2-Ethyl-1-hexanol was needed for the synthesis of the sunscreen octyl *p*-methylcinnamate. Show how this alcohol could be synthesized (a) by an aldol condensation of butanal and (b) by a malonic ester synthesis starting with diethyl malonate.

19.63 Gabapentin, an anticonvulsant used in the treatment of epilepsy, is structurally related to the neurotransmitter 4-aminobutanoic acid (GABA).

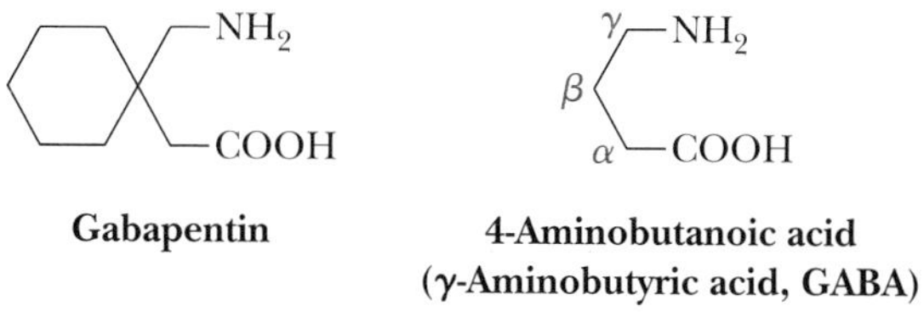

Gabapentin was designed specifically to be more lipophilic than GABA and therefore more likely to cross the blood-brain barrier, the lipidlike protective membrane that surrounds the capillary system in the brain and prevents hydrophilic (water-loving) compounds from entering the brain by passive diffusion. Given the following retrosynthetic analysis, propose a synthesis for gabapentin.

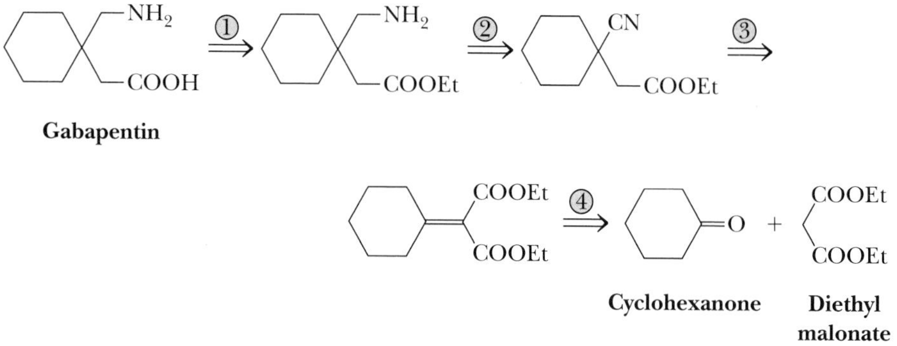

19.64 The following three derivatives of succinimide are anticonvulsants that have found use in the treatment of epilepsy, particularly petit mal seizures.

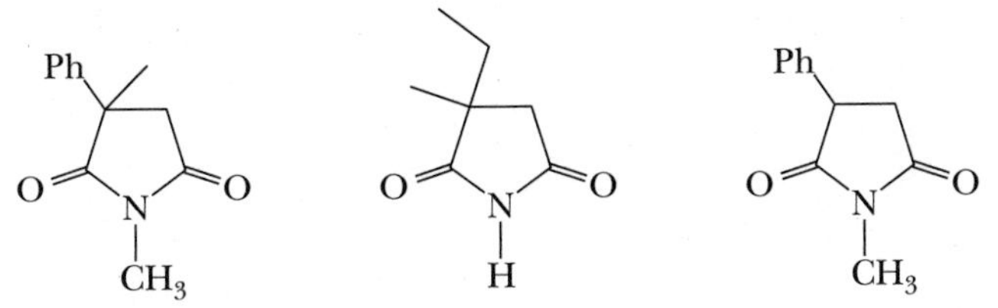

Methsuximide (racemic) **Ethosuximide** (racemic) **Phensuximide** (racemic)

Following is a synthesis of phensuximide.

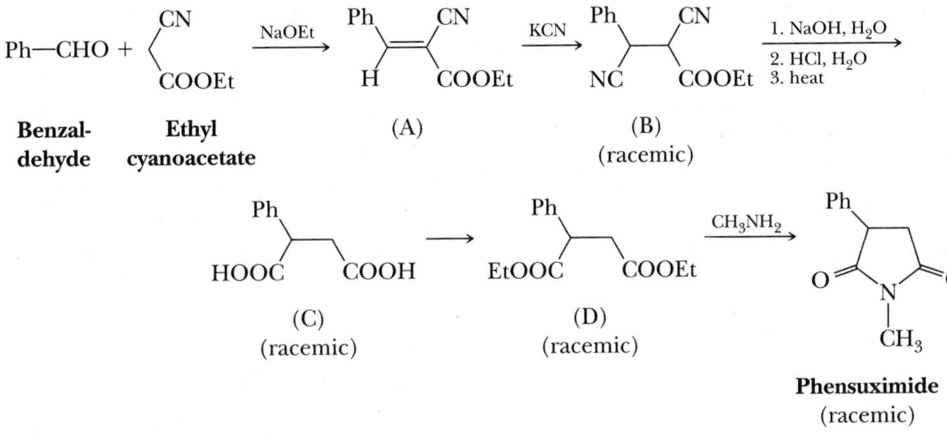

(a) Propose a mechanism for the formation of (A).

(b) What (person's) name is given to this type of reaction involved in the conversion of (A) to (B)?

(c) Describe the chemistry involved in the conversion of (B) to (C). You need not present detailed mechanisms. Rather, state what is accomplished by treating (B) with NaOH and then with HCl followed by heating.

(d) Propose experimental conditions for the conversion of (C) to (D).

(e) Propose a mechanism for the conversion of (D) to phensuximide.

(f) Show how this same synthetic strategy can be used to prepare ethosuximide and methsuximide.

(g) Of these three anticonvulsants, one is considerably more acidic than the other two. Which is the most acidic compound? Estimate its pK_a and account for its acidity. How does its acidity compare with that of phenol? with that of acetic acid?

19.65 The analgesic meperidine (Demerol) was developed in the search for analgesics without the addictive effects of morphine. As shown in these structural formulas, it represents a simplification of morphine's structure.

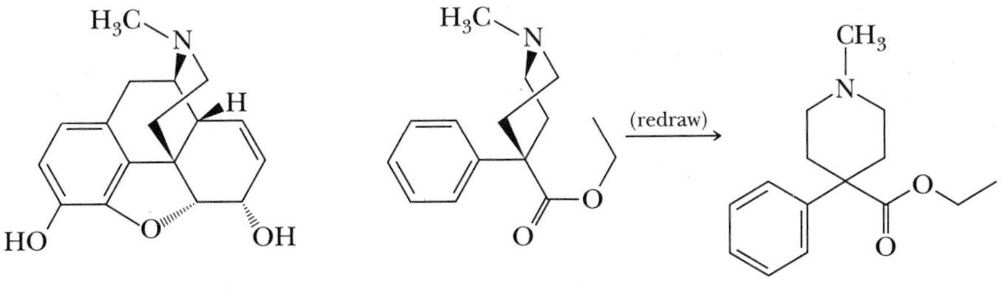

Morphine **Meperidine**

Meperidine is prepared by treating phenylacetonitrile with one mole of bis(*N*-2-chloroethyl)methylamine (a nitrogen mustard) in the presence of two moles of sodium hydride to give (A). Refluxing (A) with concentrated sodium hydroxide followed by neutralization of the reaction mixture with dilute HCl gives (B). Treating (B) with ethanol in the presence of one equivalent of HCl gives meperidine as its hydrochloride salt.

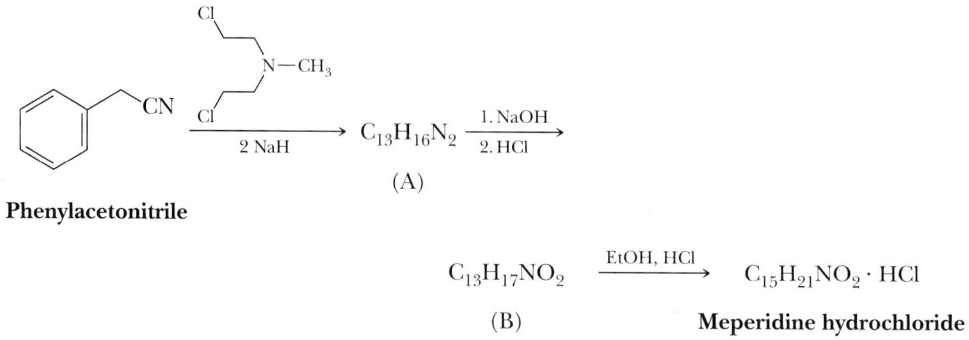

(a) Propose structural formulas for (A) and (B).
(b) Propose a mechanism for the formation of (A).

19.66 Verapamil, a coronary artery vasodilator, is used in the treatment of angina caused by insufficient blood flow to cardiac muscle. Even though its effect on coronary vasculature tone was recognized over 30 years ago, only recently has its role as a calcium channel blocker become understood. Following is a retrosynthetic analysis leading to a convergent synthesis; it is convergent because (A) and (B) are made separately and then combined (i.e., the route converges) to give the final product. Convergent syntheses are generally much more efficient than those in which the skeleton is built up stepwise.

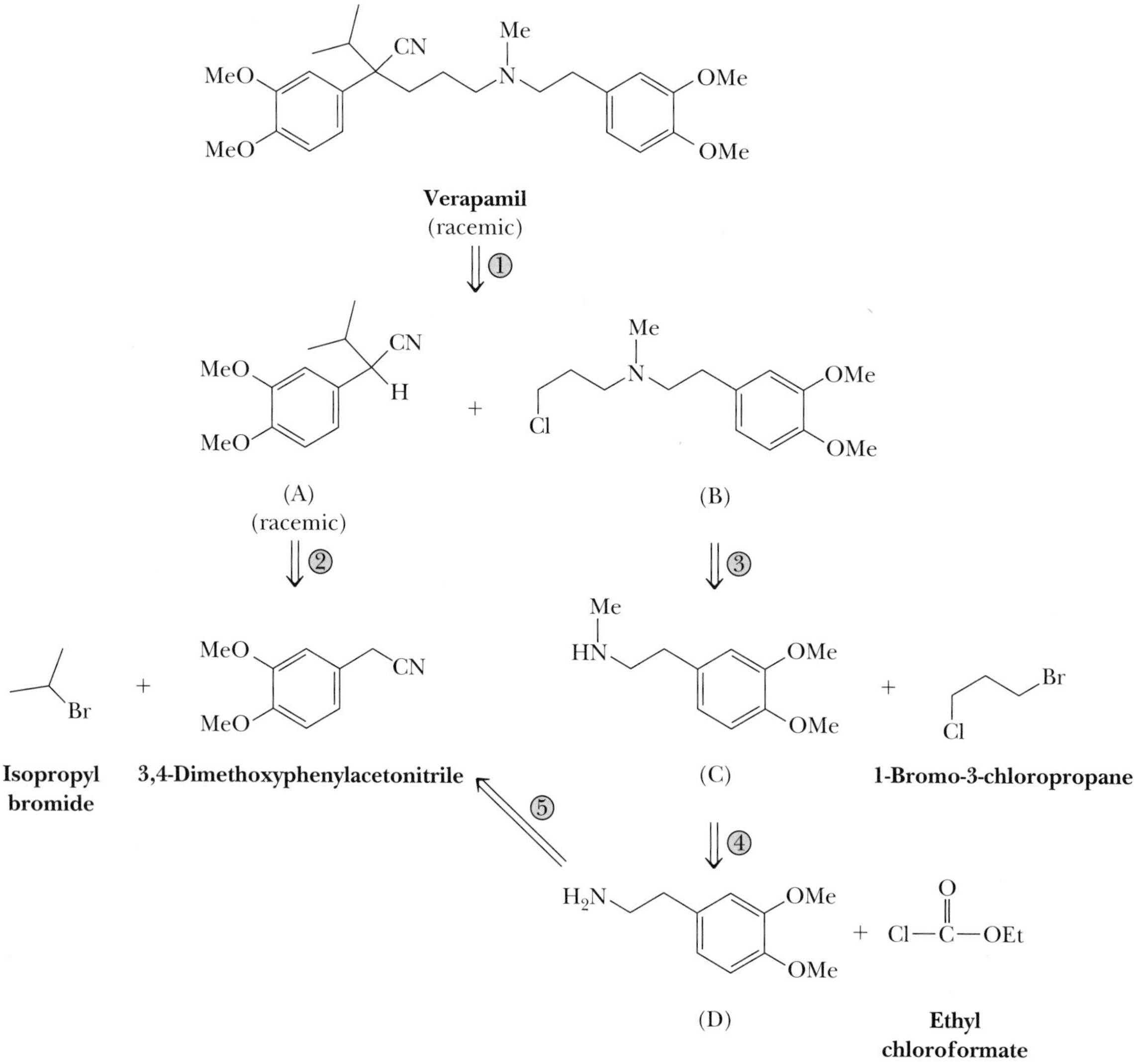

(a) Given this retrosynthetic analysis, propose a synthesis for verapamil from the four named starting materials.

(b) Two steps are required to convert (D) to (C). The first is treatment of (D) with ethyl chloroformate. What is the product of this first step? What reagent can be used to convert this product to (C)?

(c) How do you account for the regioselectivity of the nucleophilic displacement involved in converting (C) to (B)?

19.67 Based on this retrosynthetic analysis, propose a synthesis of the anticoagulant (a substance that inhibits blood clotting) diphenadione.

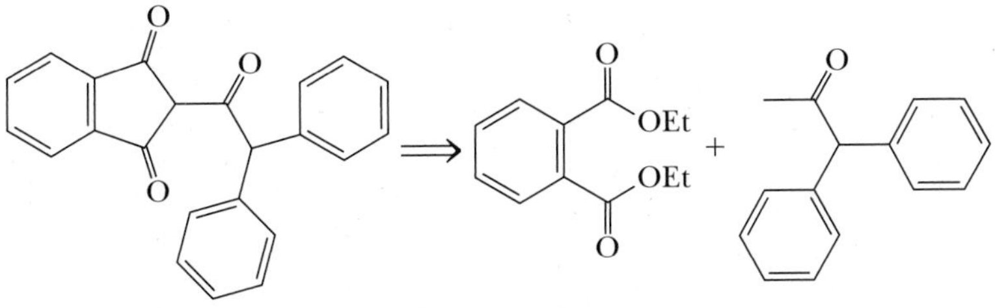

Diphenadione **Diethyl phthalate**

Because of its anticoagulant activity for blood, this compound is used as a rodenticide. For the story of the discovery of the anticoagulant dicoumarin, see "Chemical Connections: From Moldy Clover to a Blood Thinner" in Chapter 18.

19.68 Following are two possible retrosynthetic analyses for the anticholinergic drug cycrimine. Fill in the details of each potential synthesis.

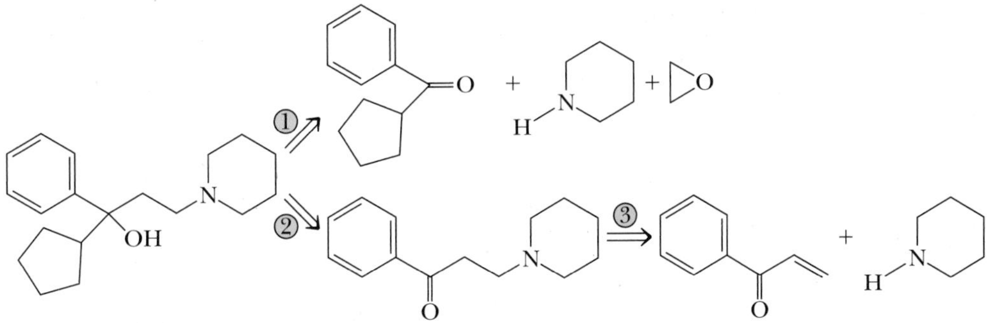

Cycrimine
(racemic)

19.69 Show how the tranquilizer valnoctamide can be synthesized using diethyl malonate as the source of the carboxamide group.

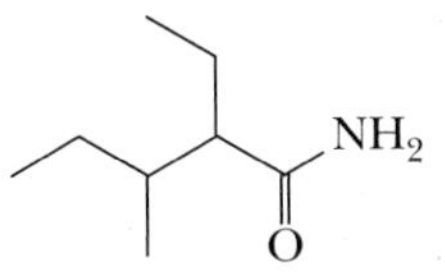

2-Ethyl-3-methylpentanamide
(**Valnoctamide**; racemic)

19.70 In Problem 7.28, we saw this two-step sequence in Johnson's synthesis of the steroid hormone progesterone. Propose a structural formula for the intermediate formed in Step 3 and a mechanism for its conversion in Step 4 to progesterone.

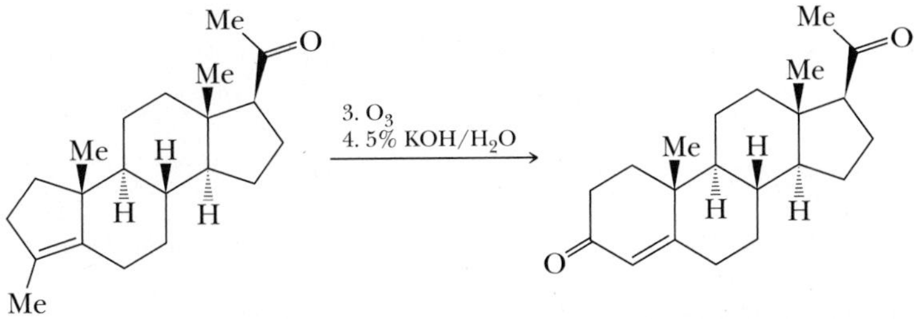

C **Progesterone**

19.71 Monensin, a polyether antibiotic, was isolated from a strain of *Streptomyces cinnamonensis* in 1967, and its structure was determined shortly thereafter.

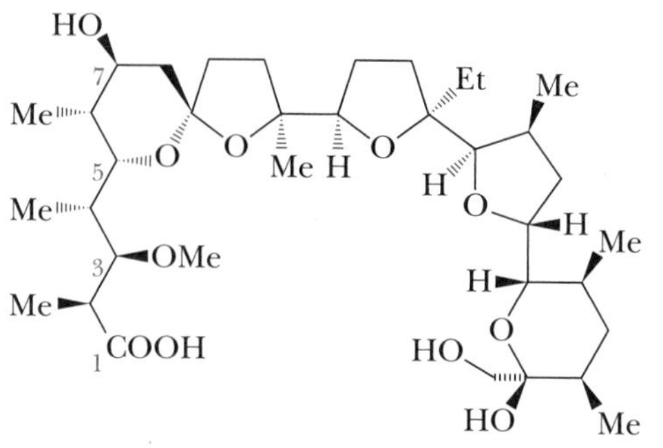

Monensin

This molecule exhibits a broad-spectrum anticoccidial activity that, since its introduction in 1971, has been used as a treatment for coccidial infections in poultry and as an additive in cattle feed. In the synthesis of monensin, Y. Kishi chose to create the molecule in sections and then join them to create the target molecule. Following is an outline of the steps by which he created the seven-carbon-chain building block on the left side of the molecule.

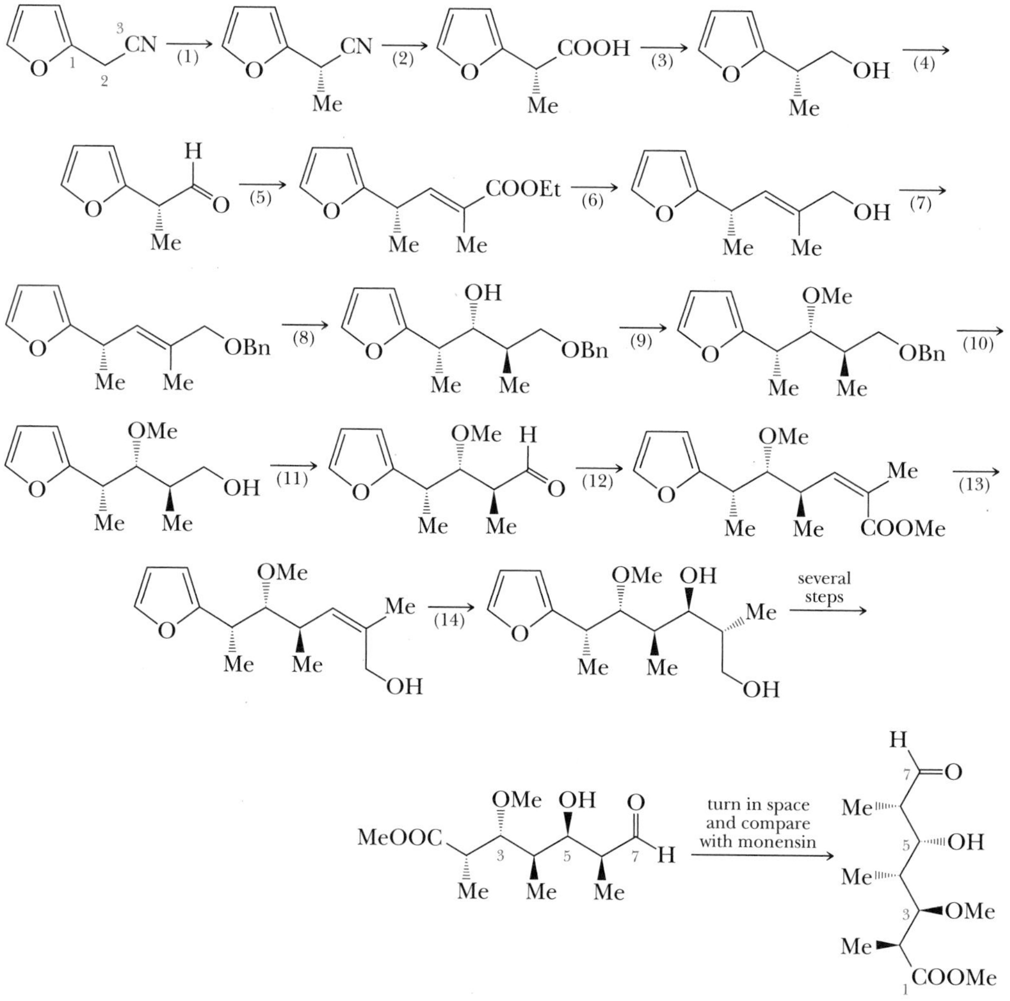

Propose a reagent or reagents for Steps 1–14. Note that this fragment contains five chiral centers. You do not have to predict or rationalize the stereochemistry of each step, but only propose a reagent or type of reagent to bring about each step.

Organic Chemistry Reaction Roadmap

19.72 We now continue the use of organic chemistry reaction roadmaps. Because of the new and unique nature of the carbon-carbon bond forming reactions presented, we recommend that you make a new reaction roadmap only for Chapter 19.

To make your own reaction roadmap for Chapters 19, take a blank sheet of paper and write the following functional groups in the orientations shown. Note that product structures are used here along with names because the products of these reactions generally have more complex patterns of functional groups. Highlighted bonds are used to indicate the key combinations of functional groups that are characteristic for the given reaction. Fill the entire sheet of paper and leave plenty of room between functional groups. Most students find it helpful to use a poster-sized sheet of paper filled out in landscape orientation.

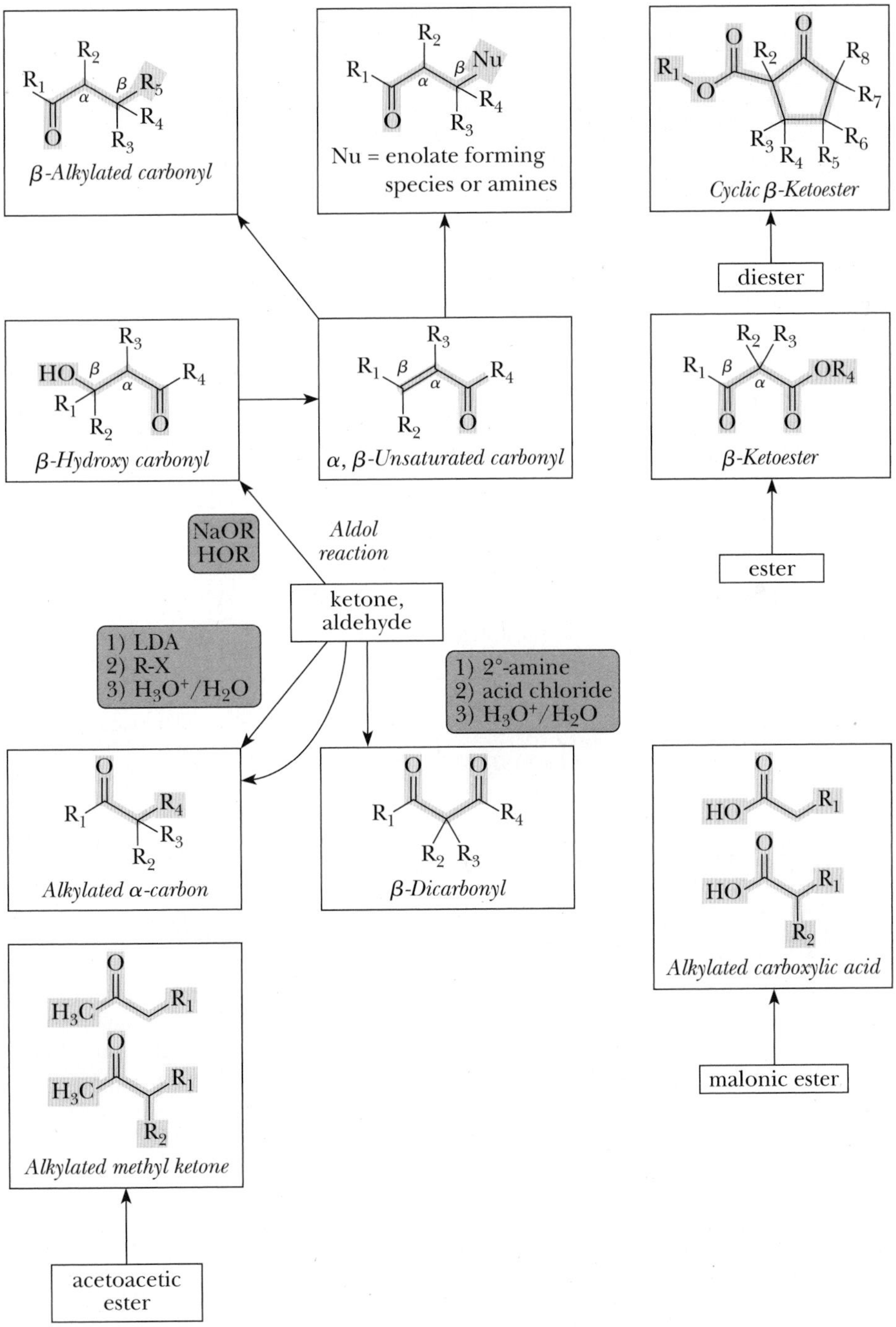

As before, refer to the "Study Guide" section of this chapter. Write the reagents required to bring about each reaction next to the arrows shown. Next, record any regiochemistry or stereochemistry considerations relevant to the reaction. You should also record any key aspects of the mechanism, such as formation of an important intermediate, as a helpful reminder. You may want to keep track of all reactions that make carbon-carbon bonds, because these help you build large molecules from smaller fragments. This especially applies to the reactions in Chapter 19.

On the above organic chemistry roadmap template, the information for the aldol reaction has been added to help you get started. Appendix 10 contains a series of reaction roadmaps for different sections of the book, but you should use those for reference only after you have completed your own.

Reaction Roadmap

19.73 Write the products of the following sequences of reactions. Refer to your reaction roadmaps to see how the combined reactions allow you to "navigate" between the different functional groups. Note that you will need your old Chapters 6–11 and Chapters 15–18 roadmaps along with your new Chapter 19 roadmap for these.

(a)

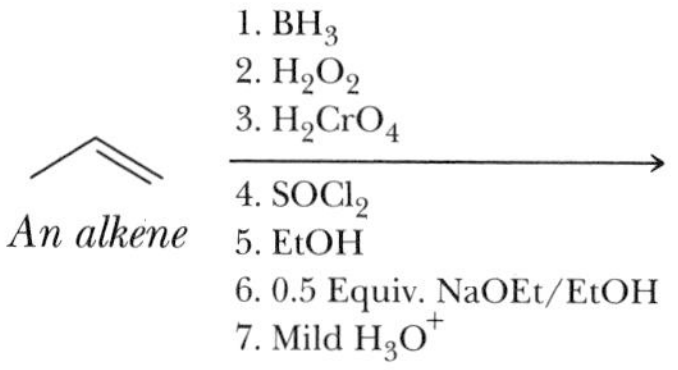
An alkene
1. BH_3
2. H_2O_2
3. H_2CrO_4
4. $SOCl_2$
5. EtOH
6. 0.5 Equiv. NaOEt/EtOH
7. Mild H_3O^+

(b)

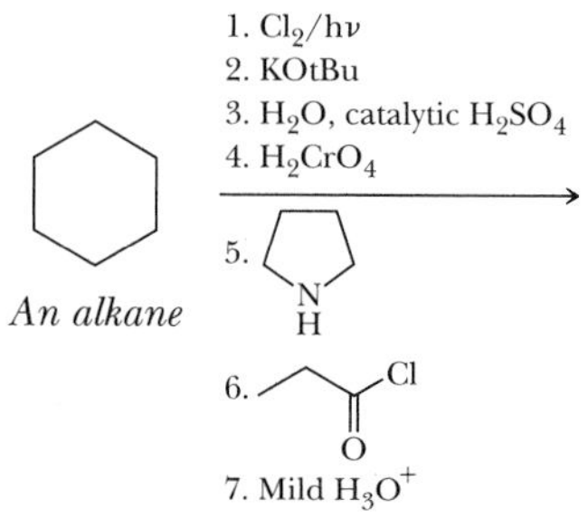
An alkane
1. $Cl_2/h\nu$
2. KOtBu
3. H_2O, catalytic H_2SO_4
4. H_2CrO_4
5.
N
H
6.
Cl
O
7. Mild H_3O^+

(c)

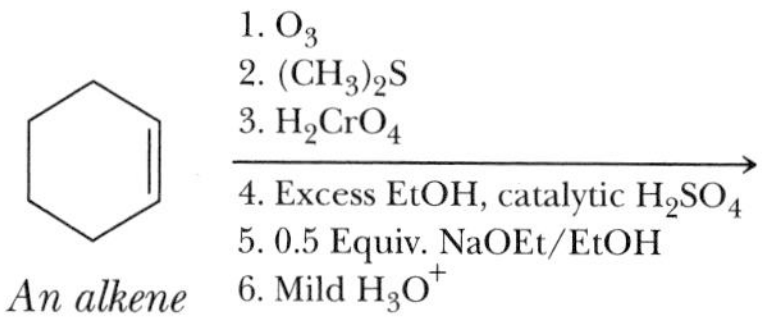
An alkene
1. O_3
2. $(CH_3)_2S$
3. H_2CrO_4
4. Excess EtOH, catalytic H_2SO_4
5. 0.5 Equiv. NaOEt/EtOH
6. Mild H_3O^+

(d)

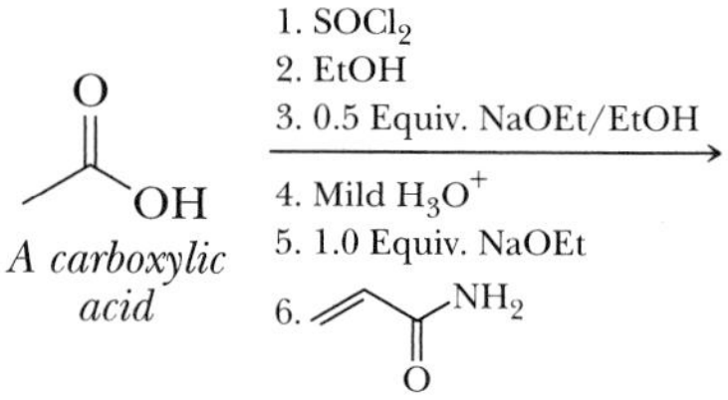
O
OH
A carboxylic acid
1. $SOCl_2$
2. EtOH
3. 0.5 Equiv. NaOEt/EtOH
4. Mild H_3O^+
5. 1.0 Equiv. NaOEt
6.
NH_2
O

(e)

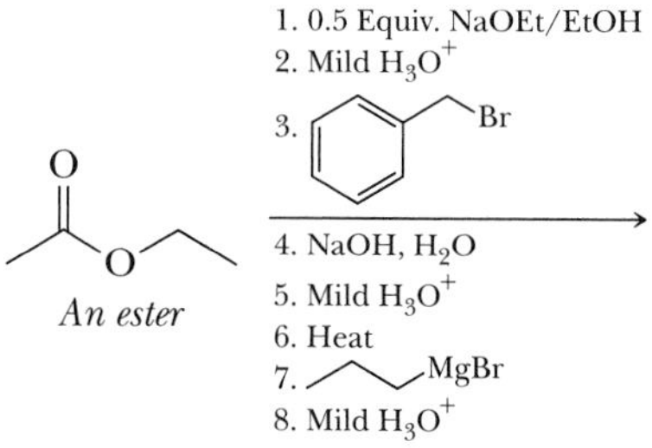
O
O
An ester
1. 0.5 Equiv. NaOEt/EtOH
2. Mild H_3O^+
3.
Br
4. NaOH, H_2O
5. Mild H_3O^+
6. Heat
7.
MgBr
8. Mild H_3O^+

(f)

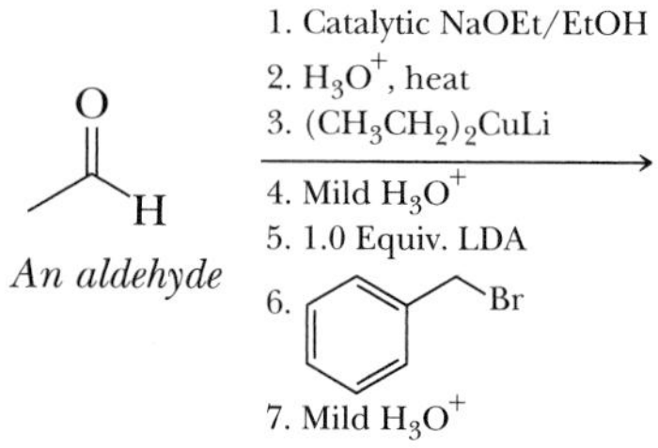
O
H
An aldehyde
1. Catalytic NaOEt/EtOH
2. H_3O^+, heat
3. $(CH_3CH_2)_2CuLi$
4. Mild H_3O^+
5. 1.0 Equiv. LDA
6.
Br
7. Mild H_3O^+

Multi-Step Synthesis

Reaction Roadmap

19.74 Using your reaction roadmaps as a guide, show how to convert (2-methylpropyl)benzene into 4-phenyl-3-buten-2-one. You must use (2-methylpropyl)benzene as the source of all carbon atoms in the target molecule. Show all reagents and all molecules synthesized along the way.

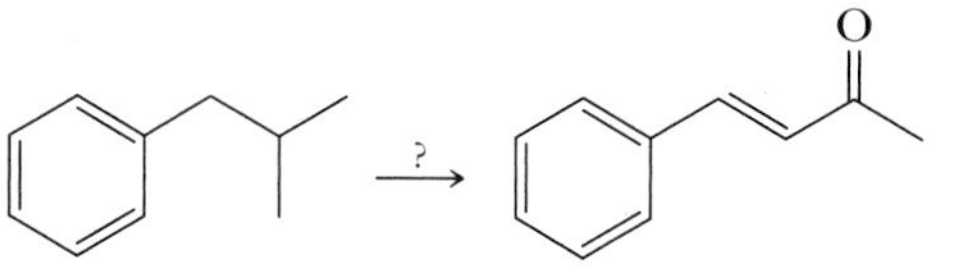
?
O

(2-Methylpropyl)benzene **4-Phenyl-3-buten-2-one**
(*E,Z* mixture)

19.75 Using your reaction roadmaps as a guide, show how to convert ethanol into 2-pentanone. You must use ethanol as the source of all carbon atoms in the target molecule. Show all reagents and all molecules synthesized along the way.

Reaction Roadmap

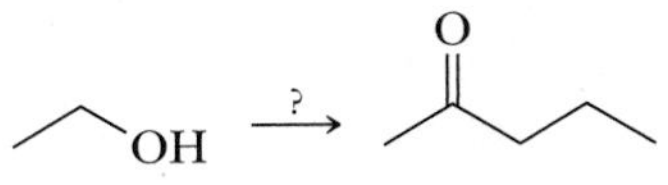

Ethanol **2-Pentanone**

19.76 Using your reaction roadmaps as a guide, show how to convert ethanol, formaldehyde, and acetone into racemic ethyl 2-acetyl-5-oxohexanoate. You must use ethanol, formaldehyde, and acetone as the source of all carbon atoms in the target molecule. Show all reagents and all molecules synthesized along the way.

Reaction Roadmap

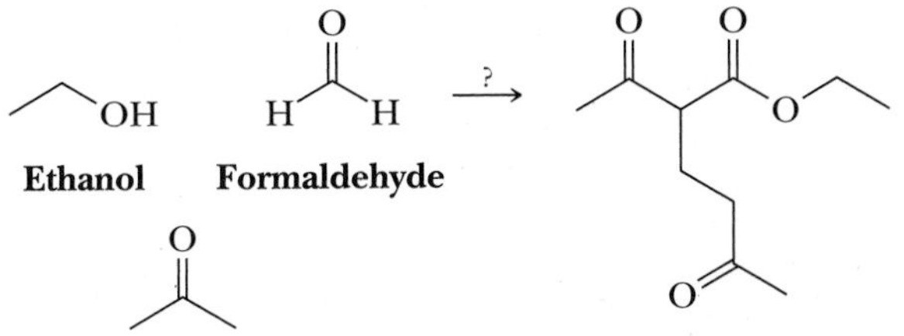

Ethanol **Formaldehyde** **Acetone** **Ethyl 2-acetyl-5-oxohexanoate** (racemic)

19.77 Using your reaction roadmaps as a guide, show how to convert cyclohexane and ethanol into racemic ethyl 2-oxocyclopentanecarboxylate. You must use ethanol and cyclohexane as the source of all carbon atoms in the target molecule. Show all reagents and all molecules synthesized along the way.

Reaction Roadmap

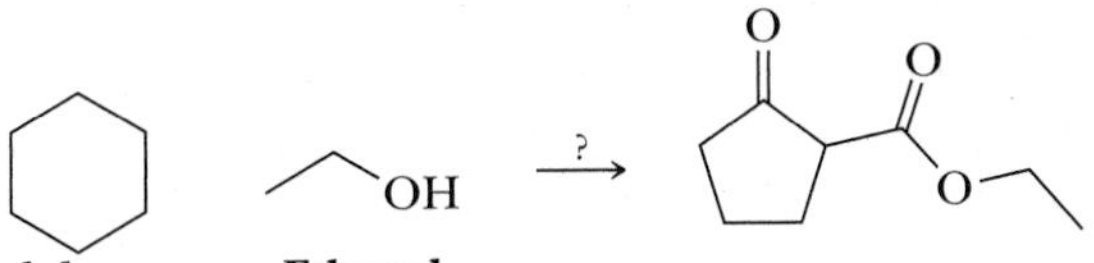

Cyclohexane **Ethanol** **Ethyl 2-oxocyclopentanecarboxylate** (racemic)

19.78 Using your reaction roadmaps as a guide, show how to convert cyclohexane and ethanol into racemic 2-acetylcyclohexanone. You must use ethanol and cyclohexane as the source of all carbon atoms in the target molecule. Show all reagents and all molecules synthesized along the way.

Reaction Roadmap

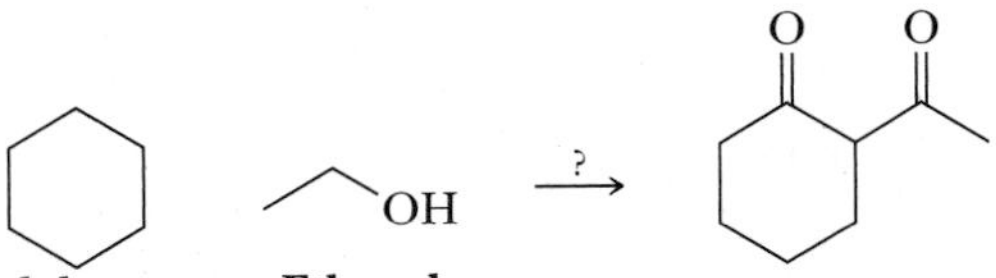

Cyclohexane **Ethanol** **2-Acetylcyclohexanone** (racemic)

19.79 Using your reaction roadmaps as a guide, show how to convert 2-oxepanone and ethanol into 1-cyclopentenecarbaldehyde. You must use 2-oxepanone as the source of all carbon atoms in the target molecule. Show all reagents and all molecules synthesized along the way.

Reaction Roadmap

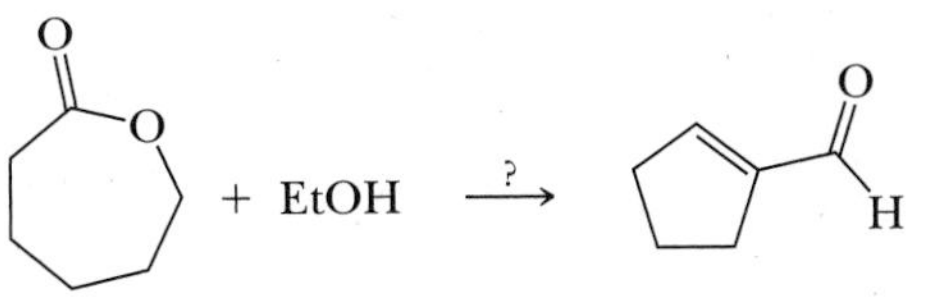

2-Oxepanone **1-Cyclopentenecarbaldehyde**

Reactions in Context

19.80 Atorvastatin (Lipitor) is a popular treatment for high cholesterol. See "The Importance of Hydrogen Bonding in Drug-Receptor Interactions" in Chapter 10 for more information about atorvastatin. One synthesis of atorvastatin involves the following enolate reaction. Draw the predominant product of this reaction, which gives an overall yield of 90%.

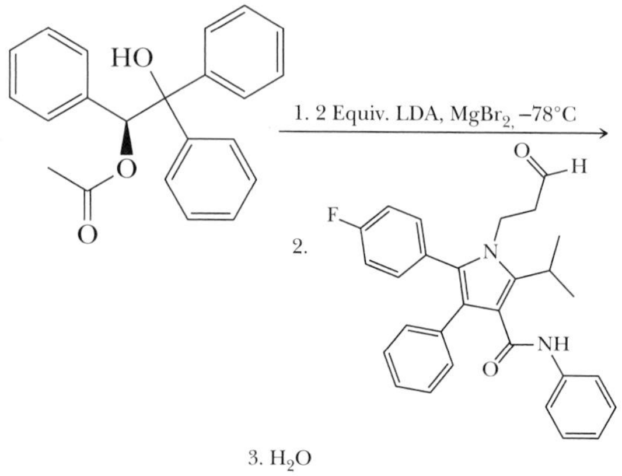

There are a couple of noteworthy aspects to this reaction. First, $MgBr_2$ is added to exchange with Li and make Mg enolates. This is helpful for controlling stereochemistry. Notice that the starting material is chiral and that a single enantiomer is used. The product of this reaction is a 97:3 (94% ee) mixture of two enantiomers, not a racemic mixture. You don't have to be able to deduce which enantiomer is the predominant product, but be aware that being able to control the stereochemical outcome of a reaction by using a single enantiomer of a chiral starting material can save time and resources in the large-scale synthesis of chiral drugs.

19.81 E. J. Corey used the following reaction in a synthesis of thromboxane B_2. Predict the major product of the reaction. There are two possible products here. State why you think the pathway that creates the predominant product is favored under the conditions of the reaction.

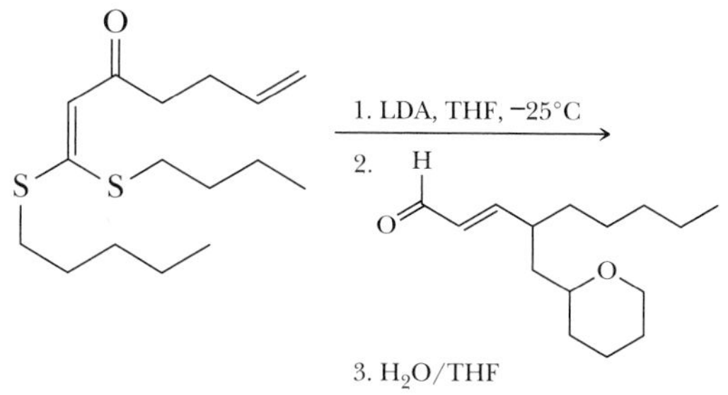

19.82 The following molecule undergoes an intramolecular reaction in the presence of pyrrolidinium acetate, the protonated form of pyrrolidine. Draw the product of this reaction, assuming that a dehydration reaction takes place.

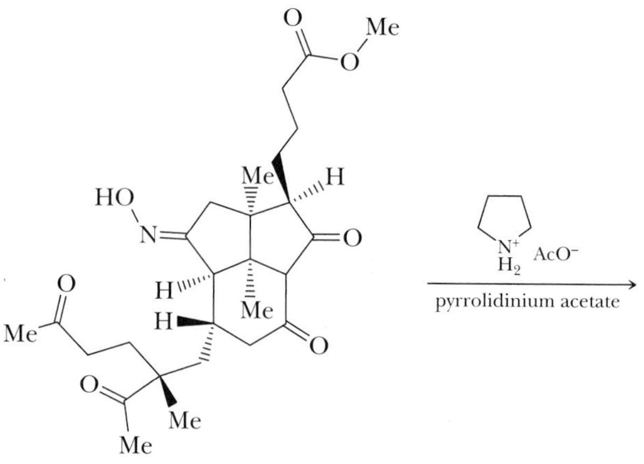

19.83 Organocuprates predominantly react to give 1,4-addition products with α,β-unsaturated carbonyl species, while Grignard reagents often add to the carbonyl, in a process referred to as 1,2-addition. To increase the yield of 1,4-addition products, CuI is added to convert an easily prepared Grignard reagent into a organocuprate reagent *in situ* (during the reaction). Predict the major product and stereochemistry of the following reaction, assuming that the more stable chair product predominates.

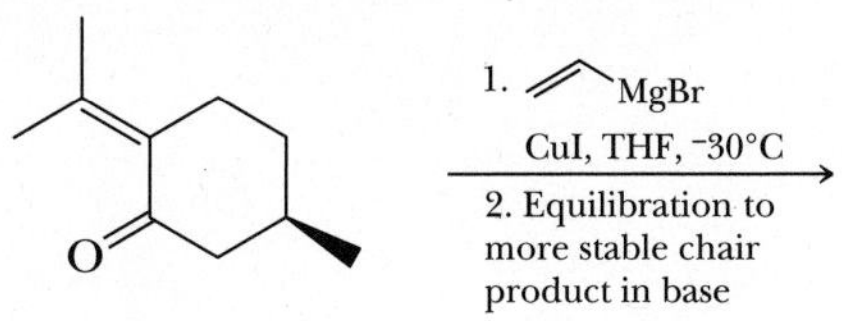
O
1. MgBr
CuI, THF, ⁻30°C
2. Equilibration to
more stable chair
product in base

20

Dienes, Conjugated Systems, and Pericyclic Reactions

Turmeric flower. Curcumin, an orange-yellow powder isolated from the spice turmeric and responsible for much of the color of curry, has recently been found to retard tumor growth. See "Chemical Connections Curry and Cancer." **Inset:** a model of curcumin.

Outline

20.1 Stability of Conjugated Dienes
20.2 Electrophilic Addition to Conjugated Dienes
20.3 UV-Visible Spectroscopy
20.4 Pericyclic Reaction Theory
20.5 The Diels-Alder Reaction
20.6 Sigmatropic Shifts

In Chapters 5 and 6, we discussed the structure and characteristic reactions of alkenes. We limited this discussion to molecules containing isolated double bonds. In this chapter, we extend our study of molecules with π bonds to include molecules that contain two or more adjacent double bonds. Such compounds are called **conjugated**. The important feature of conjugated systems is that all the adjacent $2p$ orbitals combine with each other. The result is that the π electrons are not localized between just two carbon atoms. Instead, they are best thought of as being *delocalized* throughout the entire conjugated π orbital systems.

20.1 Stability of Conjugated Dienes

Dienes are compounds that contain two carbon-carbon double bonds. Dienes can be divided into three groups: unconjugated, conjugated, and cumulated. An **unconjugated diene** is one in which the double bonds are separated by two or more single bonds. A **conjugated diene** is one in which the double bonds are separated by one single bond. A **cumulated diene** is one in which two double bonds share an *sp* hybridized carbon. Because of the geometry of this carbon, the $2p$ orbitals of the two double bonds do not overlap in a cumulated diene and are not conjugated.

Conjugated diene
A conjugated diene or carbonyl is one in which the double bonds are separated by one single bond.

Cumulated diene
A cumulated diene is one in which two double bonds share an *sp* hybridized carbon.

	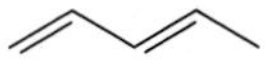	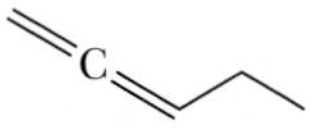
1,4-Pentadiene (an unconjugated diene)	**1,3-Pentadiene** (a conjugated diene)	**1,2-Pentadiene** (a cumulated diene)

Example 20.1 Conjugation

Which of these molecules contain conjugated double bonds?

(a) 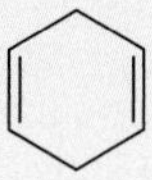 **(b)** 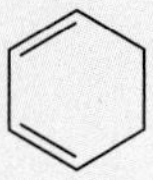 **(c)**  **(d)** 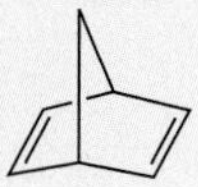

Solution

Compounds (b) and (c) contain conjugated double bonds. The double bonds in compounds (a) and (d) are unconjugated.

Problem 20.1

Which of these terpenes (Section 5.4) contains conjugated double bonds?

(a) 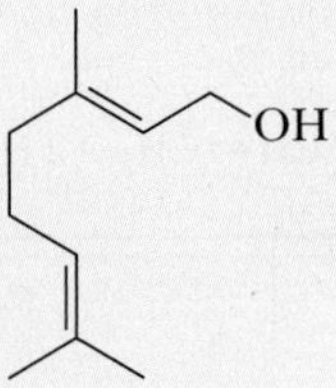

Geraniol

(b) 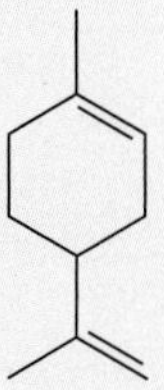

Limonene

(c) 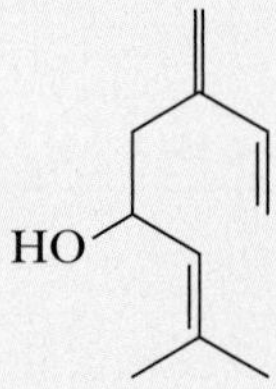

An aggregating pheromone of bark beetles

Given in Table 20.1 are heats of hydrogenation for several alkenes and conjugated dienes. By using these data, we can compare the relative stabilities of conjugated and unconjugated dienes.

Table 20.1 Heats of Hydrogenation of Several Alkenes and Conjugated Dienes

Name	Structural Formula	ΔH^0 kJ (kcal)/mol
1-Butene	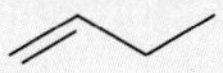	−127 (−30.3)
1-Pentene	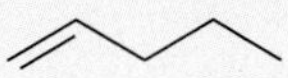	−126 (−30.1)
***cis*-2-Butene**	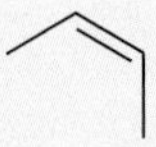	−120 (−28.6)
***trans*-2-Butene**	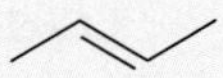	−115 (−27.6)
1,3-Butadiene	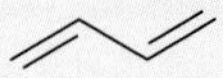	−237 (−56.5)
***trans*-1,3-Pentadiene**	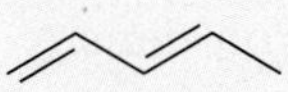	−226 (−54.1)
1,4-Pentadiene		−254 (−60.8)

The simplest conjugated diene is 1,3-butadiene, but because this molecule has only four carbon atoms, it has no unconjugated constitutional isomer. However, we can estimate the effect of conjugation of two double bonds in this molecule in the following way. The heat of hydrogenation of 1-butene is −127 kJ (−30.3 kcal)/mol. A molecule of 1,3-butadiene has two terminal double bonds, each with the same degree of substitution as the one double bond in 1-butene; therefore, we might predict that the heat of hydrogenation of 1,3-butadiene should be 2(−127 kJ/mol) or −254 kJ (−60.6 kcal)/mol. However, the observed heat of hydrogenation of 1,3-butadiene is −237 kJ (−56.5 kcal)/mol, a value 17 kJ (4.1 kcal)/mol less than estimated.

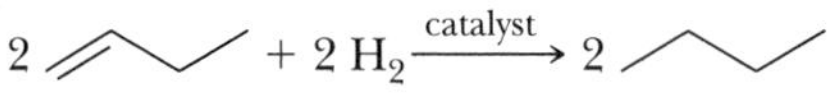

$$\Delta H^0 = 2(-127 \text{ kJ/mol}) = -254 \text{ kJ/mol}$$

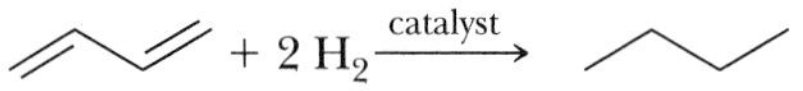

$$\Delta H^0 = -237 \text{ kJ/mol}$$

Both reactions are exothermic and give the same product; the more stable compound (lower in enthalpy) releases less heat upon hydrogenation. The conclusion is that conjugation of two double bonds in 1,3-butadiene gives an extra stability to the molecule of approximately 17 kJ (4.1 kcal)/mol. These energy relationships are displayed graphically in Figure 20.1.

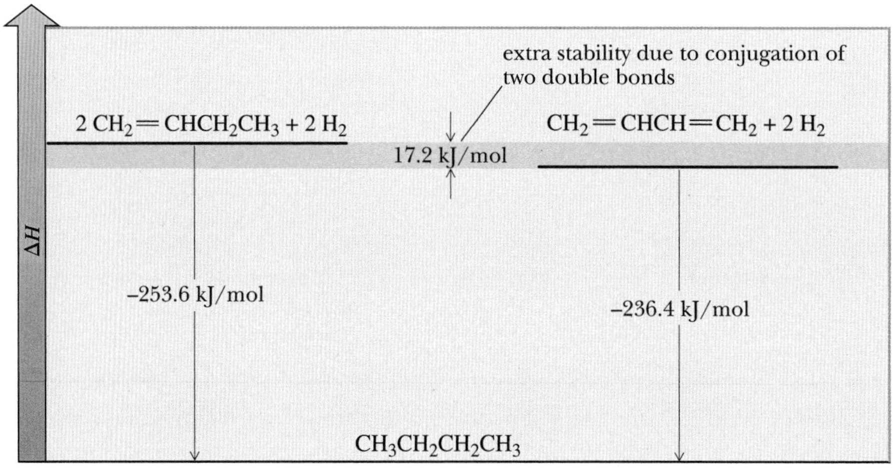

FIGURE 20.1 Conjugation of double bonds in butadiene gives the molecule an additional stability of approximately 17 kJ (4.1 kcal)/mol.

Calculations of this type for other conjugated and unconjugated dienes give similar results: conjugated dienes are more stable than isomeric unconjugated dienes by approximately 14.5–17 kJ (3.5–4.1 kcal)/mol. The effects of conjugation on stability are even more general. Compounds containing conjugated double bonds, not just those in dienes, are more stable than isomeric compounds containing unconjugated double bonds. For example, 2-cyclohexenone is more stable than its isomer 3-cyclohexenone.

▶ Delocalization of π electron density over a larger area is stabilizing.

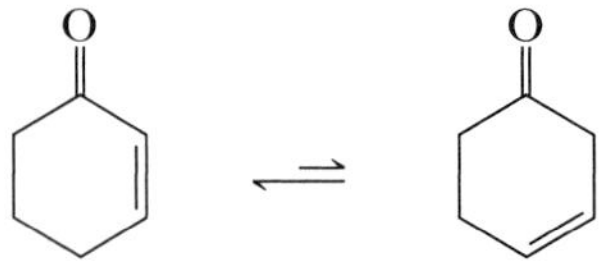

2-Cyclohexenone (more stable) **3-Cyclohexenone** (less stable) ▶

The additional stability of conjugated dienes relative to unconjugated dienes arises from delocalization of electron density in the conjugated diene. In two unconjugated double bonds, each pair of π electrons is localized between two carbons. In a conjugated diene, however, the four π electrons are delocalized over the set of four parallel $2p$ orbitals. As we have seen many times before, delocalization leads to increased stability.

According to the molecular orbital model, the conjugated system of a diene is described as a set of four π molecular orbitals arising from combination of four 2*p* atomic orbitals. The key idea here is that in conjugated systems, the adjacent 2*p* orbitals overlap in space, even between the 2*p* orbitals on C2 and C3 in butadiene. As a result, they all combine to produce π molecular orbitals that cover all the atoms of the conjugated system, in this case, the four carbon atoms. These MOs have zero, one, two, and three nodes, respectively, as illustrated in Figure 20.2. In the ground state, all four π electrons lie in π-bonding MOs. Because the lowest two MOs are at lower energies than that of

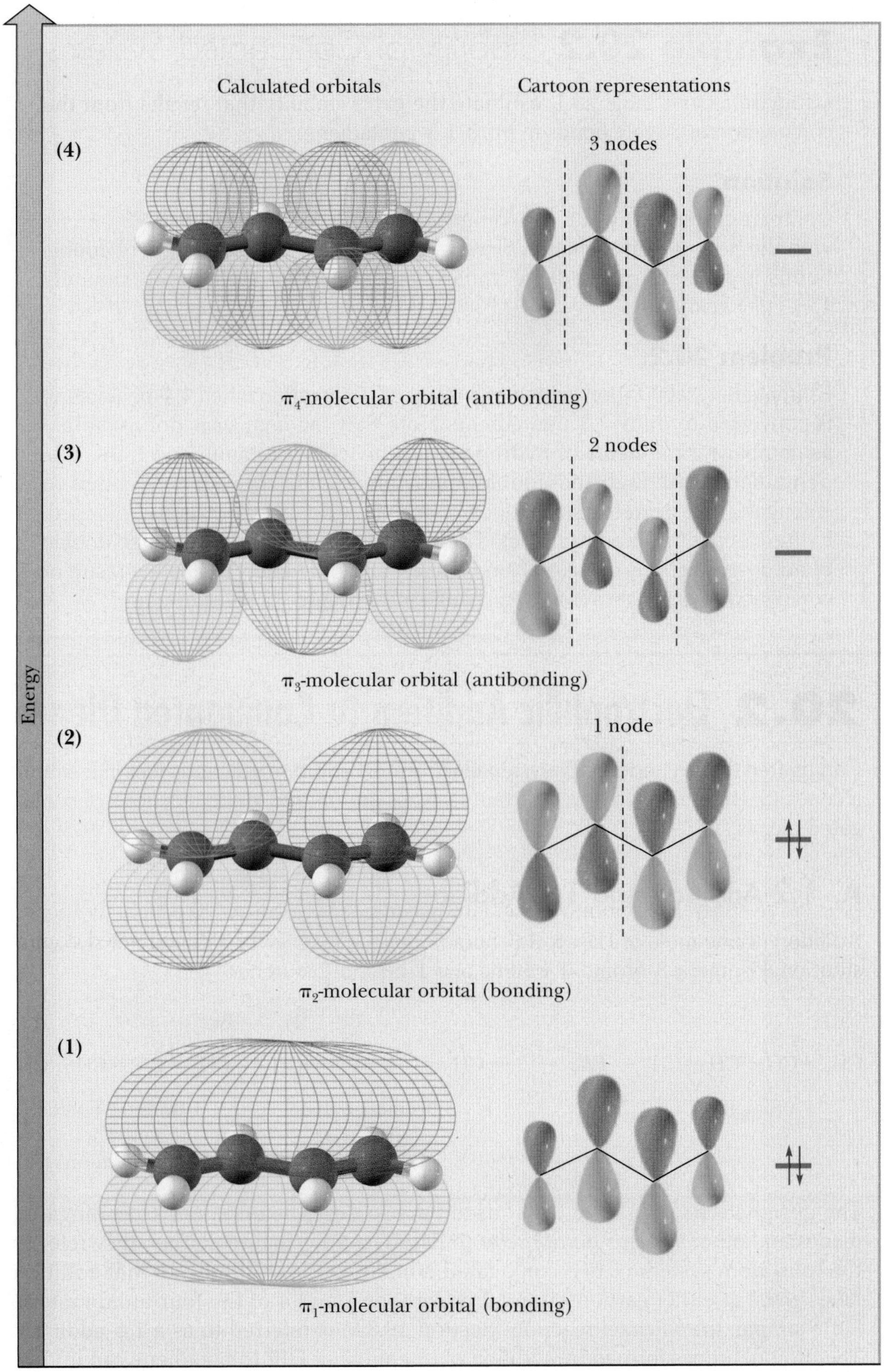
Energy
Calculated orbitals
Cartoon representations
(4)
3 nodes
π_4-molecular orbital (antibonding)
(3)
2 nodes
π_3-molecular orbital (antibonding)
(2)
1 node
π_2-molecular orbital (bonding)
(1)
π_1-molecular orbital (bonding)

Watch a video explanation

FIGURE 20.2 Structure of 1,3-butadiene—molecular orbital model. Combination of four parallel 2*p* atomic orbitals gives two π-bonding MOs and two π-antibonding MOs. In the ground state, each π-bonding MO is filled with two spin-paired electrons. The π-antibonding MOs are unoccupied.

two isolated π bonds, the net heat given off by filling these orbitals is more than would be the case for two isolated π bonds. Note that the electrons in these filled MOs are delocalized over the entire π orbital system. This π electron delocalization is the hallmark of conjugated systems and can be used to explain the spectroscopy and reactivity of conjugated molecules. Finally, it is worth pointing out that in order for maximal orbital overlap to occur, the $2p$ orbitals must be parallel, restricting the four sp^2 hybridized atoms of conjugated systems to a planar geometry.

Example 20.2 Stability of Conjugated Systems

Using data from Table 20.1, estimate the extra stability that results from the conjugation of double bonds in *trans*-1,3-pentadiene.

Solution

Compare the sum of heats of hydrogenation of 1-pentene and *trans*-2-butene with the heat of hydrogenation of *trans*-1,3-pentadiene. Conjugation of double bonds in *trans*-1,3-pentadiene imparts an added stability of approximately 15 kJ (3.6 kcal)/mol.

Problem 20.2

Estimate the stabilization gained as a result of conjugation when 1,4-pentadiene is converted to *trans*-1,3-pentadiene. Note that the answer is not as simple as comparing the heats of hydrogenation of 1,4-pentadiene and *trans*-1,3-pentadiene. Although the double bonds are moved from unconjugated to conjugated, the degree of substitution of one of the double bonds is also changed, in this case, from a monosubstituted double bond to a *trans* disubstituted double bond. To answer this question, you must separate the effect that is the result of conjugation from that caused by a change in the degree of substitution.

20.2 Electrophilic Addition to Conjugated Dienes

Conjugated dienes undergo two-step electrophilic addition reactions just like simple alkenes (Section 6.3). However, certain features are unique to the reactions of conjugated dienes.

A. 1,2-Addition and 1,4-Addition

Addition of one mole of HBr to 1,3-butadiene at −78°C gives a mixture of two constitutional isomers, 3-bromo-1-butene and 1-bromo-2-butene.

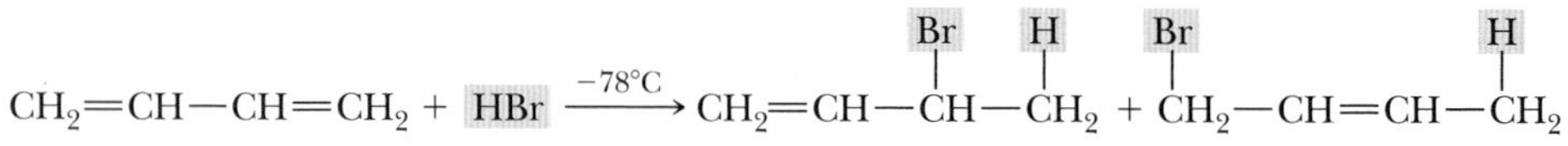
Br
H
Br
H
CH2=CH—CH=CH2 + HBr
−78°C
CH2=CH—CH—CH2 + CH2—CH=CH—CH2

1,3-Butadiene | **3-Bromo-1-butene** (90%) (1,2-addition) | **1-Bromo-2-butene** (10%) (1,4-addition)

The designations "1,2-" and "1,4-" used here to describe additions to conjugated dienes and other systems do not refer to IUPAC nomenclature. Rather, they refer to the four-atom system of two conjugated double bonds and indicate that addition takes place at either carbons 1 and 2 or carbons 1 and 4 of the four-atom system. For example, the Michael reaction (Section 19.8A) is referred to as a 1,4-addition. The bromobutenes formed by addition of one mole of HBr to butadiene can in turn undergo addition of a second mole of HBr to give a mixture of dibromobutanes. Our concern at this point is only with the products of the first reaction.

Addition of one mole of Br_2 at −15°C also gives a mixture of 1,2-addition and 1,4-addition products.

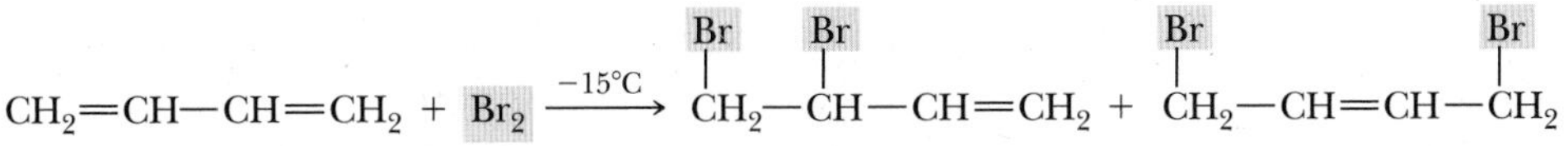

Br
Br
Br
Br
−15°C
CH2=CH—CH=CH2 + Br2
CH2—CH—CH=CH2 + CH2—CH=CH—CH2

1,3-Butadiene | **3,4-Dibromo-1-butene** (54%, racemic) (1,2-addition) | **1,4-Dibromo-2-butene** (46%) (1,4-addition)

We can account for the formation of isomeric products in the addition of HBr and Br_2 by the following mechanism.

Mechanism 20.1

1,2- and 1,4-Addition to a Conjugated Diene

Step 1: **Add a proton.** Electrophilic addition is initiated by the reaction of a terminal carbon of one of the double bonds with HBr to give an allylic carbocation intermediate (Section 9.3B), which can best be represented as a resonance hybrid of two contributing structures. Formation of this stabilized cation is the rate-determining step.

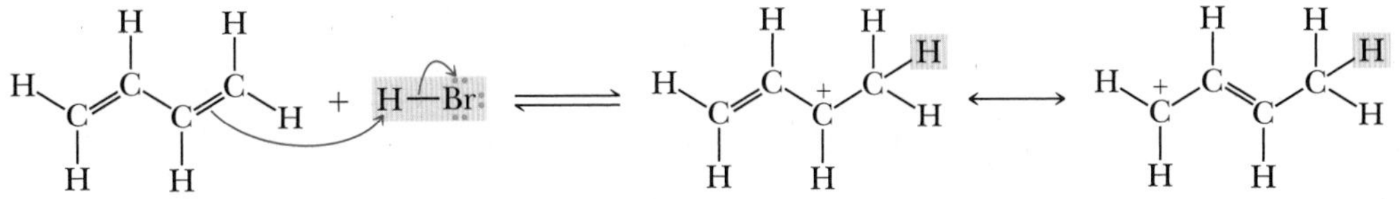

Allylic carbocation stabilized by resonance ▶

Step 2: **Make a new bond between a nucleophile and an electrophile.** Reaction of bromide at one of the carbons bearing partial positive charge gives the 1,2-addition product; reaction at the other gives the 1,4-addition product.

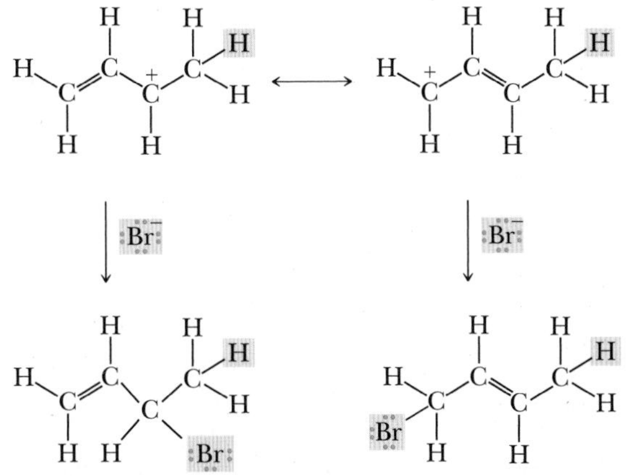

(1,2-addition, racemic) (1,4-addition)

▶ Delocalization of π electron density over a larger area is stabilizing.

Example 20.3 Addition to Conjugated Dienes

Addition of one mole of HBr to 2,4-hexadiene gives a mixture of 4-bromo-2-hexene and 2-bromo-3-hexene. No 5-bromo-2-hexene is formed. Account for the formation of the first two bromoalkenes and for the fact that the third bromoalkene is not formed.

(Continued)

Solution

2,4-Hexadiene is a conjugated diene, and you can expect products from both 1,2-addition and 1,4-addition. Reaction of the diene with HBr at C2 of the diene in Step 1, the rate-determining step, gives a resonance-stabilized 2° allylic carbocation intermediate. Reaction of this intermediate in Step 2 at one of the carbons bearing a partial positive charge gives 4-bromo-2-hexene, a 1,2-addition product; reaction at the other gives 2-bromo-3-hexene, a 1,4-addition product.

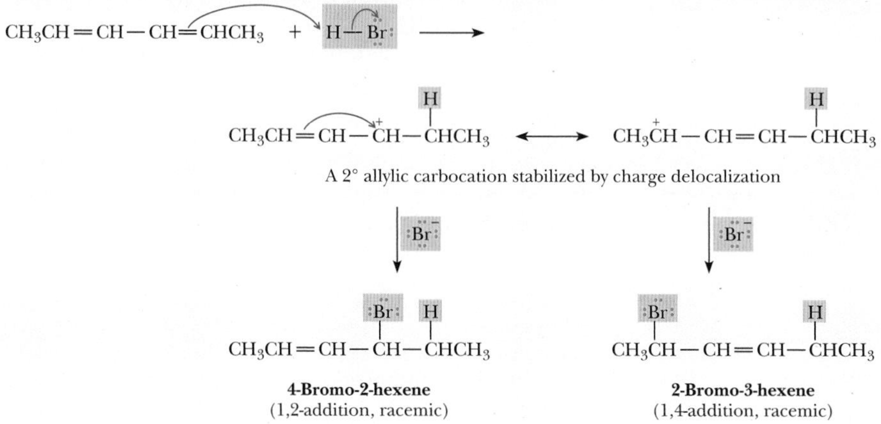

Formation of 5-bromo-2-hexene requires reaction of the diene with HBr to give a secondary, nonallylic carbocation by protonation at C3. The activation energy for formation of this less stable 2° carbocation is considerably greater than that for formation of the resonance-stabilized 2° allylic carbocation; therefore, formation of this carbocation and the resulting 5-bromo-2-hexene does not compete effectively with formation of the observed products.

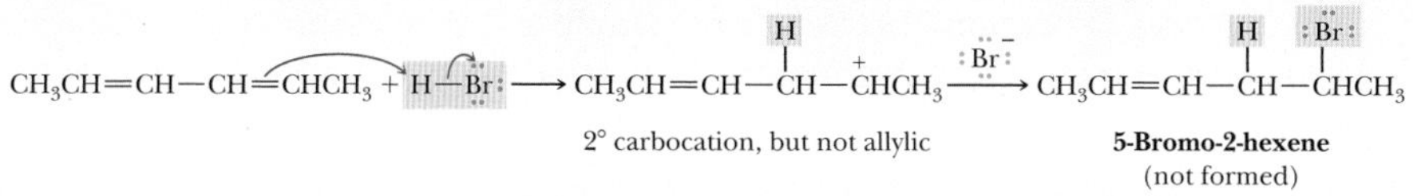

Problem 20.3

Predict the product(s) formed by addition of one mole of Br_2 to 2,4-hexadiene.

B. Kinetic Versus Thermodynamic Control of Electrophilic Addition

The regiochemistry of addition of HBr to conjugated dienes displays an unusual dependence on the kinetics and thermodynamics of the reaction. Before examining this dependence, let's review the regiochemistry of addition of HBr to standard alkenes.

As discussed in Section 6.3, the electrophilic addition of H—X to an alkene obeys Markovnikov's rule: hydrogen adds to the double-bonded carbon that has the greatest number of hydrogens already bonded to it (see the propene example below and Figure 6.4). The kinetic barrier for the rate-determining step in the formation of the Markovnikov product is lower because the carbocation formed is secondary and more stable than the alternative primary carbocation. The product resulting from the secondary carbocation is therefore formed faster. We noted in Section 19.9D that when the distribution of products is determined by the relative rates of formation of each, that reaction is under **kinetic (rate) control**.

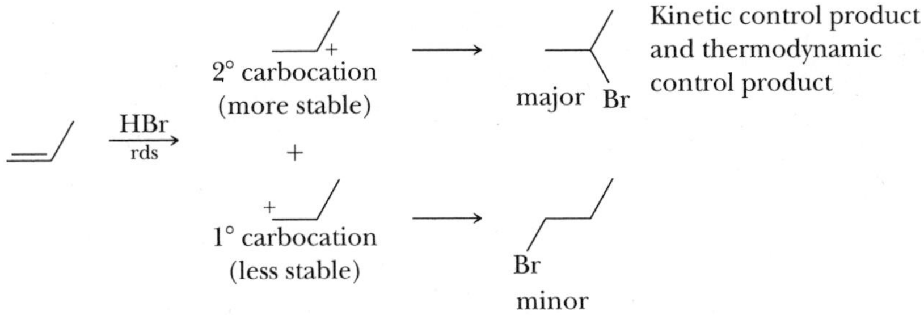

With electrophilic addition to standard alkenes such as propene, the product predicted by Markovnikov's rule is also more stable. For reactions under **thermodynamic (equilibrium) control**, the distribution of products is determined by the relative stability of each. Thus, kinetically controlled and thermodynamically controlled electrophilic additions of H—X to standard alkenes results in the same dominant product. This is the case with many reactions: the product formed fastest is also most stable. Yet, many other reactions do not behave this way. Below we will see that the addition of HBr to conjugated dienes exemplifies reactions in which kinetic and thermodynamic control produce different dominant products.

Whether a reaction is under kinetic or thermodynamic control can be manipulated by changing the experimental conditions. A common approach to switching between kinetic or thermodynamic control is to change the temperature chosen for a reaction. In general, at lower temperatures, little to no equilibrium is established between reactants and products, and the reactions must therefore be under kinetic control. At higher temperatures, reactions become increasingly reversible, and equilibrium can be established between reactants and products, leading to thermodynamic control. The electrophilic addition of HBr to standard alkenes such as propene would give the same dominant product at all temperatures, although the extent of domination will vary somewhat with temperature. We can now return to the addition of HBr to conjugated dienes.

We saw in Section 20.2A that electrophilic addition to conjugated dienes gives a mixture of 1,2-addition and 1,4-addition products. Following are some additional experimental observations about the products of electrophilic additions to 1,3-butadiene.

1. For addition of HBr at −78°C and addition of Br_2 at −15°C, the 1,2-addition products predominate over the 1,4-addition products. Generally, at lower temperatures, the 1,2-addition products predominate over 1,4-addition products.
2. For addition of HBr and Br_2 at higher temperatures (generally, 40–60°C), the 1,4-addition products predominate.
3. If the products of low temperature addition are allowed to remain in solution and then are warmed to a higher temperature, the composition of the product changes over time and becomes identical to that obtained when the reaction is carried out at higher temperature. Thus, under these higher temperature conditions, an equilibrium is established between 1,2- and 1,4-addition products in which 1,4-addition products predominate.

These experimental observations can be explained by considering kinetic versus thermodynamic products, which dominate at lower and higher temperatures, respectively. At the lower temperatures, no equilibrium is established between the 1,2- and 1,4-addition products. Because the 1,2-addition products dominate under these conditions, they must be the products formed by kinetic control (i.e., the 1,2-addition is faster than 1,4-addition). Alternatively, at the higher temperatures, the 1,4-addition products dominate; therefore, we can conclude that an equilibrium is established between the 1,2- and 1,4-addition products and that the 1,4-addition products are thermodynamically more stable than 1,2-addition products.

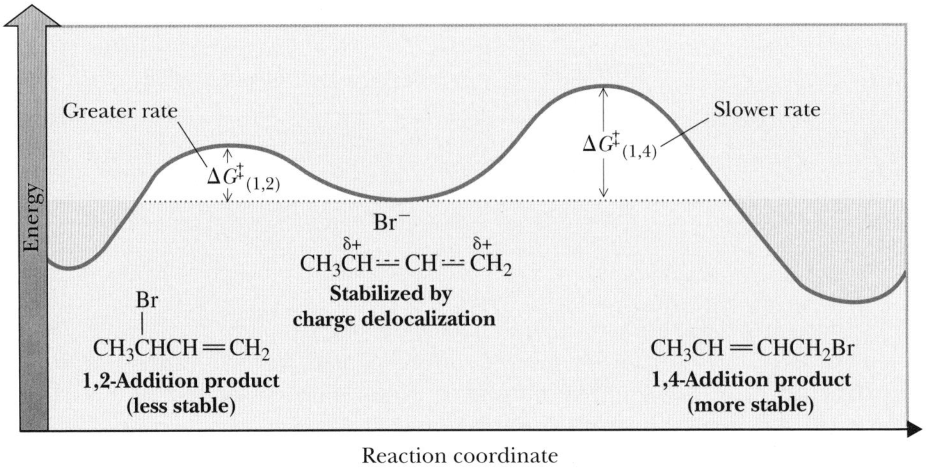

FIGURE 20.3 Kinetic versus thermodynamic control. A plot of Gibbs free energy versus reaction coordinate for Step 2 in the electrophilic addition of **HBr** to 1,3-butadiene. The resonance-stabilized allylic carbocation intermediate reacts with bromide ion by way of the transition state on the left to give the 1,2-addition product. It reacts with bromide ion by way of the alternative transition state on the right to give the 1,4-addition product.

Watch a video explanation

Relationships between kinetic and thermodynamic control for electrophilic addition of HBr to 1,3-butadiene are illustrated graphically in Figure 20.3. The structure shown in the Gibbs free energy well in the center of Figure 20.3 is the resonance-stabilized allylic cation intermediate formed by proton transfer from **HBr** to C1 of 1,3-butadiene. The dashed lines in this intermediate show the partial double bond character between C2 and C3 and between C3 and C4 in the resonance hybrid. To the left of this intermediate is the activation energy for its reaction with bromide ion to form the less stable 1,2-addition product; to the right is the activation energy for its reaction with bromide ion to form the more stable 1,4-addition product. As shown in Figure 20.3, the activation energy for 1,2-addition is less than that for 1,4-addition; therefore, the 1,2-addition product is favored under kinetic control. The 1,4-addition product is more stable and is favored when the reaction is under thermodynamic control.

To complete our discussion of electrophilic addition to conjugated dienes and of kinetic versus thermodynamic control, we need to ask the following questions.

1. Why is the 1,2-addition product (the less stable product) formed more rapidly at lower temperatures? First, we need to look at the resonance-stabilized allylic carbocation intermediate and determine which Lewis structure makes the greater contribution to the hybrid. We must consider the degree of substitution of both the positive carbon and the carbon-carbon double bond in each contributing structure.

$$CH_2{=}CH{-}\overset{+}{C}H{-}CH_3 \longleftrightarrow \overset{+}{C}H_2{-}CH{=}CH{-}CH_3$$

Less substituted double bond, Secondary carbocation ⟷ More substituted double bond, Primary carbocation

A secondary carbocation is more stable than a primary carbocation. If the degree of substitution of the carbon bearing the positive charge were the more important factor, the Lewis structure on the left would make the greater contribution to the hybrid. However, a more substituted double bond is more stable than a less substituted double bond (Section 6.6B). If the degree of substitution of the carbon-carbon double bond were the more important factor, the Lewis structure on the right would make the greater contribution to the hybrid.

We know from other experimental evidence that the location of the positive charge in the allylic carbocation is more important than the location of the double bond. Therefore, in the hybrid, the greater fraction of positive charge is on the secondary carbon. Reaction with bromide ion occurs more rapidly at this carbon, giving 1,2-addition, simply because it has a greater density of positive charge. The electrostatic potential map shows that the positive charge (blue) is more intense on the secondary carbon.

2. Is the 1,2-addition product also formed more rapidly at higher temperatures even though the 1,4-addition product predominates under these conditions? The answer is yes. The factors affecting the structure of a resonance-stabilized allylic carbocation intermediate and the reaction of this intermediate with a nucleophile are not greatly affected by changes in temperature.
3. Why is the 1,4-addition product the thermodynamically more stable product? The answer to this question has to do with the relative degree of substitution of double bonds. In general, the greater the degree of substitution of a carbon-carbon double bond, the greater the stability of the compound or ion containing it. Following are pairs of 1,2- and 1,4-addition products. In each case, the more stable alkene is the 1,4-addition product.

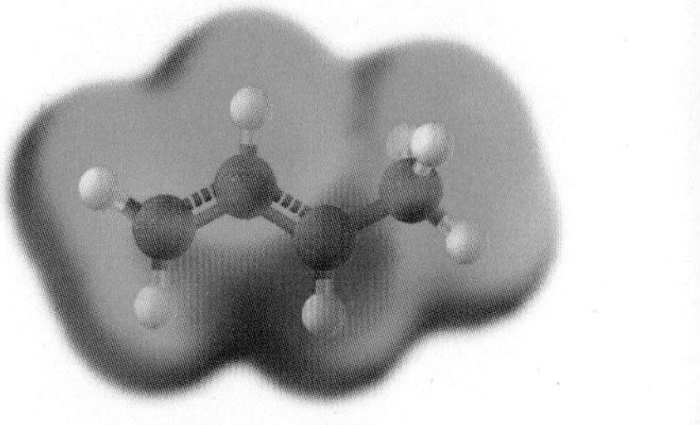

Electrostatic potential map of the allylic carbocation formed by protonating 1,3-butadiene.

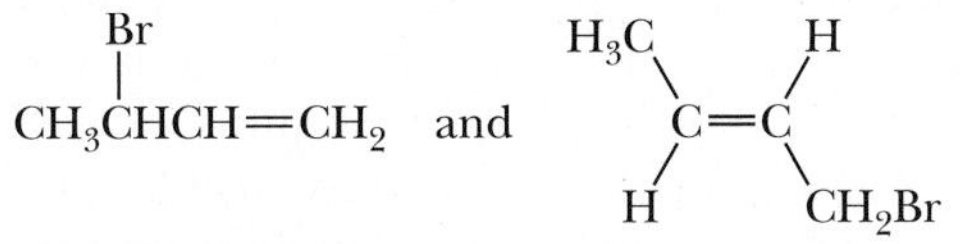

3-Bromo-1-butene (less stable alkene) | **(*E*)-1-Bromo-2-butene** (more stable alkene)

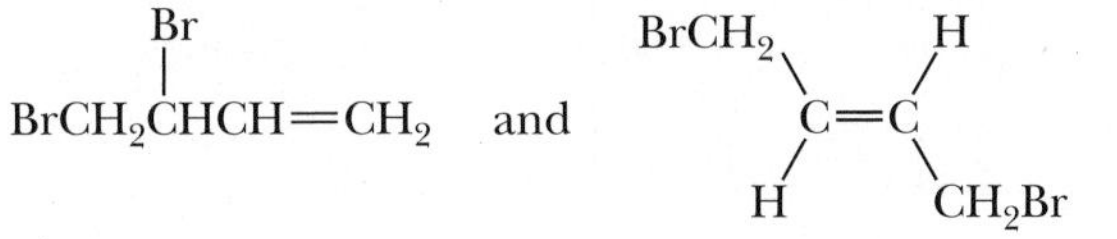

3,4-Dibromo-1-butene (less stable alkene) | **(*E*)-1,4-Dibromo-2-butene** (more stable alkene)

However, there are cases where the 1,2-addition product is more stable and would be the product of thermodynamic control. For example, addition of bromine to 1,4-dimethyl-1,3-cyclohexadiene under conditions of thermodynamic control gives 3,4-dibromo-1,4-dimethylcyclohexene because its trisubstituted double bond is more stable than the disubstituted double bond of the 1,4-addition product.

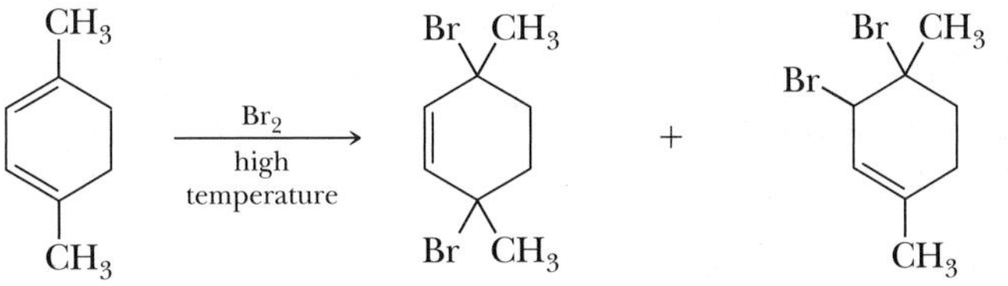

1,4-Addition product (less stable) | 1,2-Addition product (more stable)

4. What is the mechanism by which the thermodynamically less stable product is converted to the thermodynamically more stable product at higher temperatures? At higher temperatures used for electrophilic addition of **HBr** and **Br_2** to conjugated dienes, collisions between the 1,2- and 1,4-addition products with the solvent are energetic enough to reform the resonance-stabilized allylic carbocation intermediate via ionization of the C—**Br** bonds. When the 1,2-addition product reverts to this allylic carbocation, re-addition of bromide can give the more stable 1,4-addition product. At lower temperature, however, the increase in potential energy of the products upon collisions is not sufficient to overcome the activation energy for C—**Br** bond ionization; therefore, the reactions are not reversible.

In summary, although the thermodynamically most stable product is often the most rapidly formed product, such is not always the case. Whether the thermodynamically more stable product is formed at a greater rate—from a reactant or a common intermediate—very much depends on the particular reactants, the reaction mechanism, and the reaction conditions.

20.3 UV-Visible Spectroscopy

An important property of conjugated systems is that they absorb energy in the ultraviolet-visible region of the spectrum as a result of electronic transitions (Table 12.3). In this section, we study the information this absorption gives us about the conjugation of carbon-carbon and carbon-oxygen double bonds and their substitution.

A. Introduction

The region of the electromagnetic spectrum covered by most ultraviolet spectrophotometers is from 200 to 400 nm, a region commonly referred to as the **near ultraviolet**. Wavelengths shorter than 200 nm require special instrumentation and are not used routinely. The region covered by most visible spectrophotometers runs from 400 nm (violet) to 700 nm (red), with extensions into the (near) IR region to 800 or 1000 nm available on many instruments.

Example 20.4 UV-Vis Radiation

Calculate the energy of radiation at either end of the near-ultraviolet spectrum [i.e., at 200 nm and 400 nm (review Section 12.1)].

Solution

Use the relationship $E = hc/\lambda$. Make sure you express the dimension of length in consistent units.

$$E = \frac{hc}{\lambda} = 3.99 \times 10^{-13}\,\frac{\text{kJ} \times \text{s}}{\text{mol}} \times 3.00 \times 10^{8}\,\frac{\text{m}}{\text{s}} \times \frac{1}{200 \times 10^{-9}\,\text{m}}$$
$$= 598\ \text{kJ}\ (143\ \text{kcal})/\text{mol}$$

By a similar calculation, the energy of radiation of wavelength 400 nm is found to be 299 kJ (71.5 kcal)/mol.

Problem 20.4

Wavelengths in ultraviolet-visible spectroscopy are commonly expressed in nanometers; wavelengths in infrared spectroscopy are sometimes expressed in micrometers. Carry out the following conversions.

(a) 2.5 μm to nanometers **(b)** 200 nm to micrometers

Wavelengths and corresponding energies for near-ultraviolet and visible radiation are summarized in Table 20.2.

Table 20.2 Wavelengths and Energies of Near Ultraviolet and Visible Radiation

Region of Spectrum	Wavelength (nm)	Energy	
		kJ/mol	kcal/mol
Near Ultraviolet	200–400	299–598	71.5–143
Visible	400–700	171–299	40.9–71.5

Ultraviolet and visible spectral data are recorded as plots of **absorbance (A)** on the vertical axis versus wavelength on the horizontal axis.

Absorbance (A)
A quantitative measure of the extent to which a compound absorbs radiation of a particular wavelength. $A = \log(I_0/I)$ where I_0 is the incident radiation and I is the transmitted radiation.

$$\text{Absorbance } (A) = \log \frac{I_0}{I}$$

where I_0 is the intensity of radiation incident on the sample and I is the intensity of the radiation transmitted through the sample. The quantity $(I/I_0) \times 100$ is called **percent transmittance**; many spectrophotometers read in this scale.

Typically, UV-visible spectra consist of a small number of broad absorption bands, sometimes just one. Figure 20.4 is an ultraviolet absorption spectrum of 2,5-dimethyl-2,4-hexadiene. Absorption of ultraviolet radiation by this conjugated diene begins at wavelengths below 200 nm and continues to almost 270 nm, with maximum absorption at 242 nm. This spectrum is reported as a single absorption peak using the notation λ_{max} 242 nm.

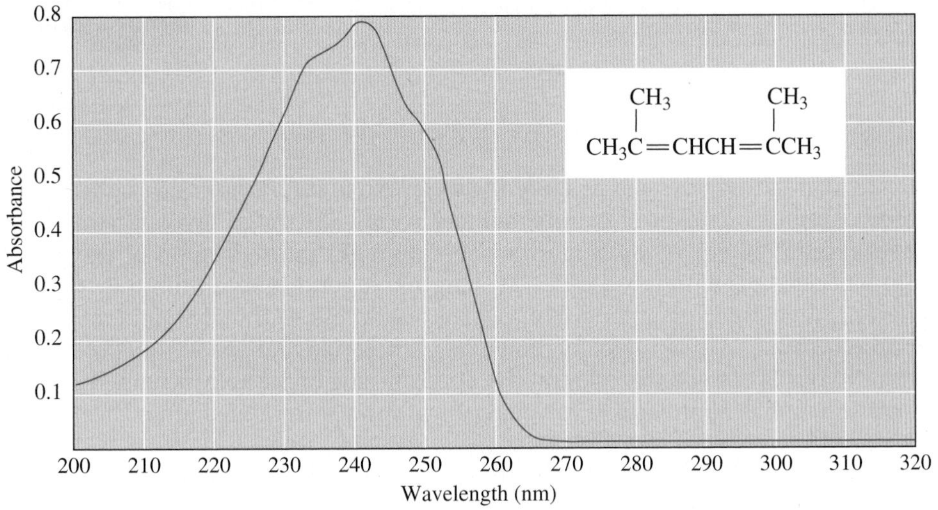

FIGURE 20.4 Ultraviolet spectrum of 2,5-dimethyl-2,4-hexadiene (in methanol).

The extent of absorption of ultraviolet-visible radiation is proportional to the number of molecules capable of undergoing the observed electronic transition; therefore, ultraviolet-visible spectroscopy can be used for quantitative analysis of samples. The relationship between absorbance, concentration, and length of the sample cell (cuvette) is known as the **Beer-Lambert law**. The proportionality constant in this equation is given the name **molar absorptivity** (ε) or extinction coefficient.

Molar absorptivity (ε)
The absorbance of a 1 M solution of a compound.

$$\text{Beer-Lambert Law:} \qquad A = \varepsilon\, c\, l$$

where A is the **absorbance** (unitless), ε is the molar absorptivity (in per moles per liter per centimeter, $M^{-1}\text{cm}^{-1}$), c is the concentration of solute (in moles per liter, M), and l is the length of the sample cell, or cuvette (in centimeters, cm).

The molar absorptivity is a characteristic property of a compound and is not affected by its concentration or the length of the light path. Values range from zero to $10^6\ M^{-1}\text{cm}^{-1}$. Values above $10^4\ M^{-1}\text{cm}^{-1}$ correspond to high-intensity absorptions; values below $10^4\ M^{-1}\text{cm}^{-1}$, to low-intensity absorptions. The molar absorptivity of 2,5-dimethyl-2,4-hexadiene, for example, is 13,100 $M^{-1}\text{cm}^{-1}$, a high-intensity absorption.

An interesting aspect of absorption by molecules in the visible region is that a sample will appear to our eyes as the combination of reflected wavelengths. White light is composed of all wavelengths of light in the visible region (400–740 nm), present in approximately equal intensity. Individual wavelengths of light have individual colors, as indicated on the spectrum (Figure 20.5a). Light of the given wavelength appears as the color indicated to our eyes. For example, monochromatic 400 nm light appears violet and monochromatic 700 nm light appears red.

Absorption by a substance removes the absorbed wavelengths from white light, leaving the remaining wavelengths to be reflected, the combination of which determines the color our eyes see. Figure 20.5b shows the approximate color that a substance would appear if a single wavelength were absorbed. For example, if a molecule absorbs strongly only at 500 nm (lighter blue light), it appears red to our eyes because the remaining reflected colors combine to appear red. Similarly, a molecule that strongly absorbs around 600 nm (orange light) appears blue, because orange is removed from the reflected light, and the remaining reflected wavelengths combine to appear blue.

Watch a video explanation

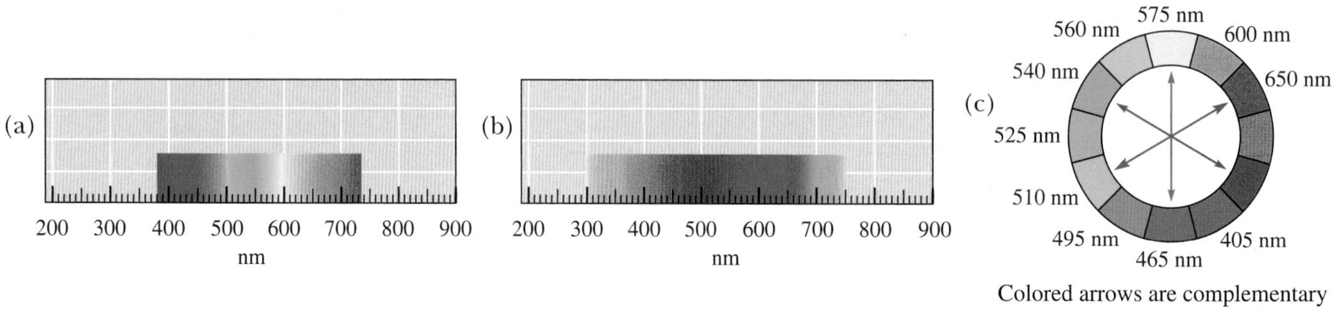
(a)
200 300 400 500 600 700 800 900
nm
(b)
200 300 400 500 600 700 800 900
nm
(c)
575 nm
560 nm
600 nm
540 nm
650 nm
525 nm
510 nm
495 nm
465 nm
405 nm
Colored arrows are complementary

FIGURE 20.5 (a) Visible light color-wavelength correlation. (b) Approximate color of substance (reflected light) if a single wavelength (i.e., the wavelength listed on the numerical scale of the *x*-axis) is absorbed. (c) Complementary colors on a color wheel.

The correlation between absorbed wavelength and reflection can be approximated using the concept of complementary colors illustrated by an artist's color wheel (Figure 20.5c). A molecule that absorbs light in one region of the spectrum will reflect the nonabsorbed wavelengths. A good rule of thumb is that the reflected wavelengths combine to appear more or less as the complement of the absorbed color. To a first approximation then, a molecule that absorbs one color will appear to our eyes to be the color on the opposite side of the color wheel. The color wheel shown has approximate wavelengths of monochromatic light indicated for reference.

More complicated absorptions, with two or more strong absorptions by a single molecule, lead to a more complex interpretation of reflected color, but the concept is the same. A substance will appear to our eyes as the combination of reflected (not absorbed) wavelengths.

Example 20.5 Beer-Lambert Law

The molar absorptivity of 2,5-dimethyl-2,4-hexadiene in methanol is 13,100 $M^{-1}cm^{-1}$. What concentration of this diene in methanol is required to give an absorbance of 1.6? Assume a light path of 1.00 cm. Calculate concentration in these units.

(a) Moles per liter **(b)** Milligrams per milliliter

Solution

Solve the Beer-Lambert equation for concentration and substitute appropriate values for length, absorbance, and molar absorptivity.

(a) $$c = \frac{A}{l \times \varepsilon} = \frac{1.6}{1.00\ \text{cm} \times 13{,}100\ \text{L} \cdot \text{mol}^{-1} \cdot \text{cm}^{-1}} = 1.22 \times 10^{-4}\ \text{mol/L}$$

(b) The molecular weight of 2,5-dimethyl-2,4-hexadiene is 110 g/mol. The concentration of the sample in milligrams per milliliter is

$$1.22 \times 10^{-4}\ \frac{\text{mol}}{\text{L}} \times \frac{110\ \text{g}}{\text{mol}} \times \frac{1\ \text{L}}{1000\ \text{mL}} \times \frac{1000\ \text{mg}}{\text{g}} = 1.34 \times 10^{-2}\ \text{mg/mL}$$

Problem 20.5

The visible spectrum of β-carotene ($C_{40}H_{56}$, MW 536.89, the orange pigment in carrots) dissolved in hexane shows intense absorption maxima at 463 nm and 494 nm, both in the blue-green region. Because light of these wavelengths is absorbed by β-carotene, we perceive the color of this compound as that of the complement to blue-green, namely red-orange.

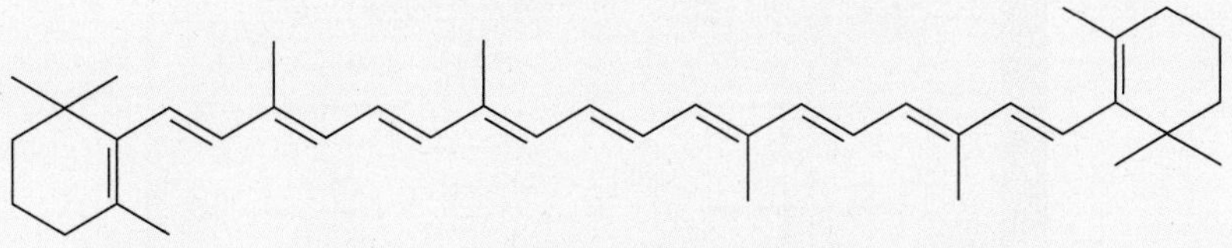

β-Carotene

λ_{max} 463 nm (log ε 5.10); 494 nm (log ε 4.77)

Calculate the concentration in milligrams per milliliter of β-carotene that gives an absorbance of 1.8 at 463 nm.

B. The Origin of Transitions Between Electronic Energy Levels

Absorption of electromagnetic radiation in the ultraviolet-visible region results in promotion of electrons from a lower energy, occupied MO to a higher energy, unoccupied MO. The energy of this radiation is generally insufficient to affect electrons in the much lower energy, σ-bonding molecular orbitals. It is, however, sufficient to cause an electron in a nonbonding (lone pair) or π orbital to be promoted to an antibonding π^* orbital (called an $n \rightarrow \pi^*$ and $\pi \rightarrow \pi^*$ transition, respectively). Conjugated π systems have particularly noteworthy $\pi \rightarrow \pi^*$ transitions. Three examples of conjugated systems follow.

$CH_2{=}CH{-}CH{=}CH_2$

1,3-Butadiene

$CH_2{=}CH{-}C({=}O){-}CH_3$

3-Buten-2-one

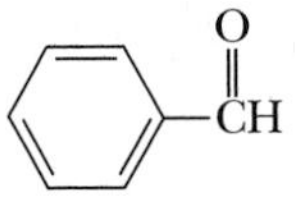

Benzaldehyde

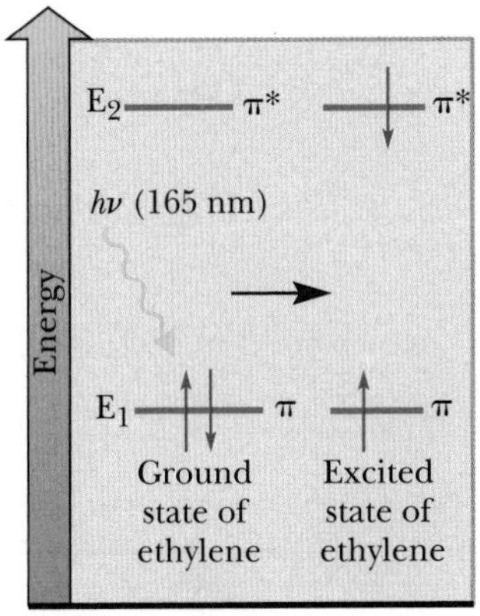

FIGURE 20.6 A $\pi \rightarrow \pi^*$ transition in excitation of ethylene. Absorption of ultraviolet radiation causes a transition of an electron from a π-bonding MO in the ground state to a π-antibonding MO in the excited state. There is no change in electron spin.

As an example of a $\pi \rightarrow \pi^*$ transition, consider ethylene. The double bond in ethylene consists of one σ bond formed by combination of sp^2 orbitals and one π bond formed by combination of $2p$ orbitals. The relative energies of the π-bonding and π-antibonding molecular orbitals are shown schematically in Figure 20.6. The $\pi \rightarrow \pi^*$ transitions for simple, unconjugated alkenes occur below 200 nm (at 165 nm for ethylene). Because these transitions occur at extremely short wavelengths, they are not observed in conventional ultraviolet spectroscopy and therefore are not useful to us for determining molecular structure.

For 1,3-butadiene, the difference in energy between the highest occupied π molecular orbital and the lowest unoccupied π-antibonding molecular orbital is less than it is for ethylene with the result that a $\pi \rightarrow \pi^*$ transition for 1,3-butadiene (Figure 20.7) takes less energy (occurs at longer wavelength) than that for ethylene. This transition for 1,3-butadiene occurs at 217 nm.

Electronic excitation in molecules is accompanied by changes in vibrational or rotational energy levels. The energy levels for these excitations are considerably smaller than the energy differences between electronic excitations. These transitions are superposed on the electronic excitations, which results in a large number of absorption peaks so closely spaced that the spectrophotometer cannot resolve them. For this reason, UV-visible absorption peaks usually are much broader than IR absorption peaks.

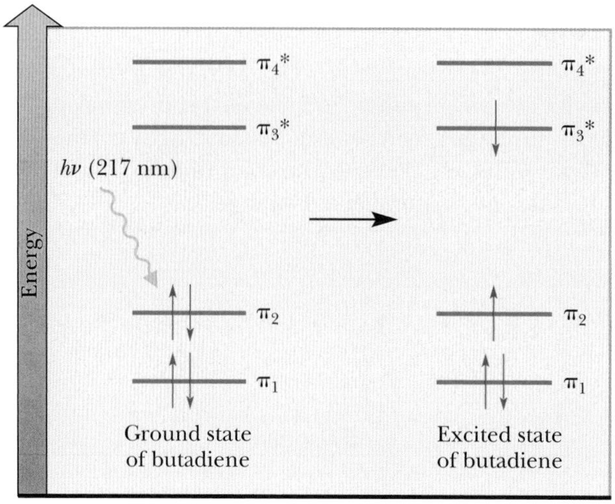

FIGURE 20.7 Electronic excitation of 1,3-butadiene; a $\pi \rightarrow \pi^*$ transition.

Simple aldehydes and ketones show only weak absorption in the ultraviolet region of the spectrum owing to an n to π^* electronic transition of the carbonyl group. If, however, the carbonyl group is conjugated with one or more carbon-carbon double bonds, intense absorption ($\varepsilon = 8{,}000 - 20{,}000\ M^{-1}\text{cm}^{-1}$) occurs as a result of a π to π^* transition; as with polyenes, the position of absorption is shifted to longer wavelengths and the molar absorptivity, ε, of the absorption maximum increases sharply. For the α,β-unsaturated ketone 3-penten-2-one, for example, λ_{max} is 224 nm (log ε 4.10).

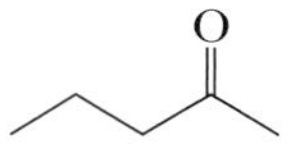

2-Pentanone
λ_{max} 180 nm (ε 900)

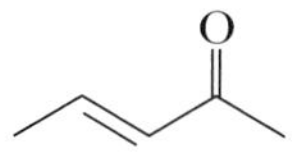

3-Penten-2-one
λ_{max} 224 nm (ε 12,590)

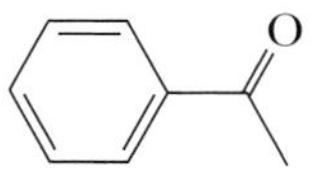

Acetophenone
λ_{max} 246 nm (ε 9,800)

The greater the extent of conjugation of unsaturated systems with the carbonyl group, the more the absorption maximum is shifted toward the visible region of the spectrum.

Like the carbonyl groups of simple aldehydes and ketones, the carboxyl group shows only weak absorption in the ultraviolet spectrum unless it is conjugated with a carbon-carbon double bond or an aromatic ring.

The important point is that conjugation decreases the energy gap between filled and unfilled π orbitals. Therefore, in general, the greater the number of double bonds in conjugation, the longer the wavelength of ultraviolet radiation absorbed. Shown in Table 20.3 are wavelengths and energies required for $\pi \rightarrow \pi^*$ transitions in several conjugated alkenes.

Table 20.3 Wavelengths and Energies Required for $\pi \rightarrow \pi^*$ Transitions of Ethylene and Three Conjugated Polyenes

Name	Structural Formula	λ_{max} (nm)	Energy [kJ (kcal)/mol]
Ethylene	$CH_2{=}CH_2$	165	724 (173)
1,3-Butadiene	$CH_2{=}CHCH{=}CH_2$	217	552 (132)
(3*E*)-1,3,5-Hexatriene	$CH_2{=}CHCH{=}CHCH{=}CH_2$	268	448 (107)
(3*E*,5*E*)-1,3,5,7-Octatetraene	$CH_2{=}CH(CH{=}CH)_2CH{=}CH_2$	290	385 (92)

20.4 Pericyclic Reaction Theory

To this point in this book, there have been only a few cases for which we analyzed the orbitals of reactants in order to understand the reaction mechanisms, as, for example, S_N2 (Section 9.2A) and E2 (Section 9.7C). Even though we have drawn on orbital analysis infrequently, the interactions of orbitals actually dictate all chemical reactions. In fact, there is a class of reactions called *pericyclic* for which an analysis of orbitals is critical for even a rudimentary understanding of the mechanisms. **Pericyclic reactions** occur in a single step with a closed loop of orbitals; that is, we can draw orbitals interacting at the transition states of the reactions in a cyclic ring. Further, because the reactions occur in a single step, there are no radical or ionic intermediates. The examples given below will make this characteristic clear. Because many of these reactions involve conjugated and nonconjugated dienes, we discuss them in this chapter. One of the hallmarks of pericyclic reactions is precise control of the stereochemistry of the reactions, and the examples given below will highlight this feature.

Pericyclic reactions are routinely classified as "allowed" or "forbidden" with a particular structure for the transition state. In practice, this classification means that one geometry for the reaction has a low energy transition state (allowed) or that a different geometry has a very high energy transition state (forbidden). To determine whether a reaction is allowed or forbidden, a handful of approaches exist. We will examine one approach: frontier molecular orbital theory.

Pericyclic reaction

A reaction that takes place in a single step without intermediates and that involves a cyclic distribution of orbitals.

▶ **Although we cover only one approach to the understanding of pericyclic reactions, the various approaches developed over the years have given important different contributions to our understanding. R. B. Woodward (Harvard University), Roald Hoffmann (then at Harvard, now at Cornell University), Kenichi Fukui (Kyoto University), and Howard Zimmerman (University of Wisconsin) provided the key insights into pericyclic reaction mechanisms. Hoffman and Fukui were awarded the Nobel Prize for this work in 1981 (after the death of Woodward).**

Chemical Connections

Curry and Cancer

Curcumin is a natural dye from the root of *Curcuma longa* L. In pure form, it is an orange-yellow crystalline powder that is isolated from the spice turmeric, one of the major ingredients of curry. Its color is a result of the highly conjugated system in curcumin (it is probable that the molecule is actually enolized as shown). It has been known for some time that curcumin retards the growth of new cancers by inhibiting the formation of blood vessels that are necessary for the cancers to grow (angiogenesis). Recently, Korean biochemists have shown that curcumin acts by inhibiting an enzyme that is important to angiogenesis. So curry may be good for you.

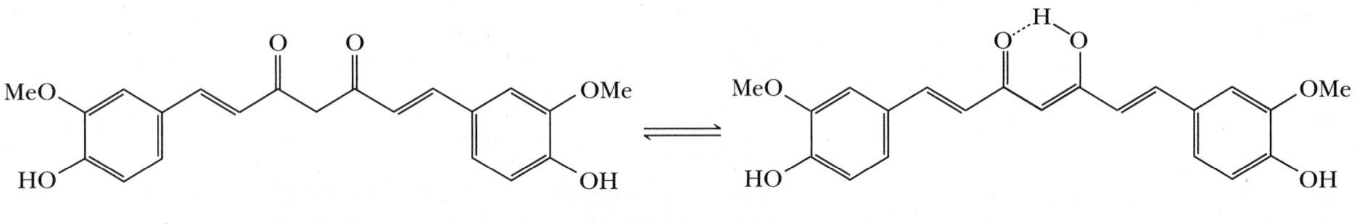

Curcumin
(Natural Yellow-3, Turmeric yellow)

A. Frontier Molecular Orbital Theory (FMOT)

There are five classes of pericyclic reactions, only two of which are covered in this book, cycloadditions and sigmatropic shifts. The most common, **cycloaddition** involves the reaction of a conjugated diene with an alkene, although we also examine the reaction of an alkene with an alkene. Hence, we are examining what are commonly called 4 + 2 and 2 + 2 cycloadditions, respectively, to keep track of the number of π electrons involved in the reaction.

Cycloaddition reaction

A reaction in which two reactants add together in a single step to form a cyclic product.

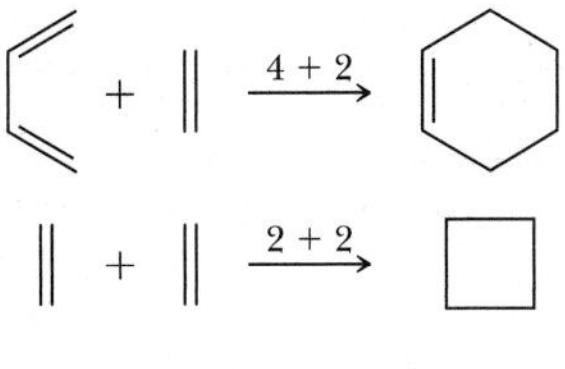

HOMO

Highest occupied molecular orbital.

LUMO

Lowest unoccupied molecular orbital.

To analyze whether these reactions are allowed or forbidden, chemists focus on the frontier molecular orbitals of the reactants. The frontier molecular orbitals consist of the **highest occupied molecular orbitals (HOMOs)** and the **lowest unoccupied molecular orbitals (LUMOs)**. The terms *highest* and *lowest* refer to the energy of the orbitals, and the terms *occupied* and *unoccupied* refer to whether the orbital is populated with two electrons or is empty. For example, the HOMO and LUMO of butadiene would be orbitals 2 and 3 in Figure 20.2, respectively. The HOMO and LUMO of ethylene are simply the π and π^* orbitals given in Figure 1.21.

After identifying the frontier molecular orbitals of the reactants, chemists predict a reaction geometry. The goal is to decide if the predicted geometry for a reaction is allowed or forbidden. In the analysis, we examine the HOMO of one reactant and the LUMO of the other. It does not matter which reactant is assigned the HOMO or LUMO in the analysis because the answer will be the same. What is important is how one reactant's HOMO interacts (contacts) with the other reactant's LUMO in the proposed reaction geometry. When the phasing of the orbitals that are undergoing contact matches (zero phase changes), the reaction is allowed. In fact, if there are any even number of phase changes, the reaction is allowed. However, if there is one (or any odd number of) contacts between the orbitals in which the phasing does not match, the reaction is forbidden.

Suprafacial

When the same face of a π bond performs a reaction.

Antarafacial

When the opposite faces of a π bond perform a reaction.

To demonstrate the frontier molecular orbital analysis, let's analyze a collision geometry for butadiene and ethylene that is called suprafacial for each reactant. A **suprafacial** interaction occurs when the same face (or side) of the π system of an individual reactant undergoes collision (an alternative interaction is referred to as **antarafacial**, shown in Example 20.6). In Figure 20.8(a), the bottoms of the *p* orbitals on carbons 1 and 4 are undergoing reaction (hence the same face) and the tops of the *p* orbitals of ethylene are reacting (hence again the same face). Therefore, each reactant is interacting in a suprafacial manner. Suprafacial collision geometries are shown for both the 4 + 2 and 2 + 2 reactions in Figure 20.8.

Note that the 4 + 2 reaction is allowed because when the HOMO and LUMO on the reactants contact each other, the phasing matches; red on red and blue on blue. The 2 + 2 reaction is forbidden because there is one red on blue contact.

The frontier molecular orbital analysis led to the conclusion that butadiene will react with ethylene to give cyclohexene if both butadiene and ethylene collide in a suprafacial manner. However, ethylene will not react with ethylene in an analogous manner. We can extend these conclusions to the reaction of any conjugated diene with any alkene and to any alkene with another alkene. Presented below are several examples and practical considerations of the allowed 4 + 2 reaction (the Diels-Alder reaction). The suprafacial approach of both reactants has important consequences on the stereochemistry of the Diels-Alder reaction.

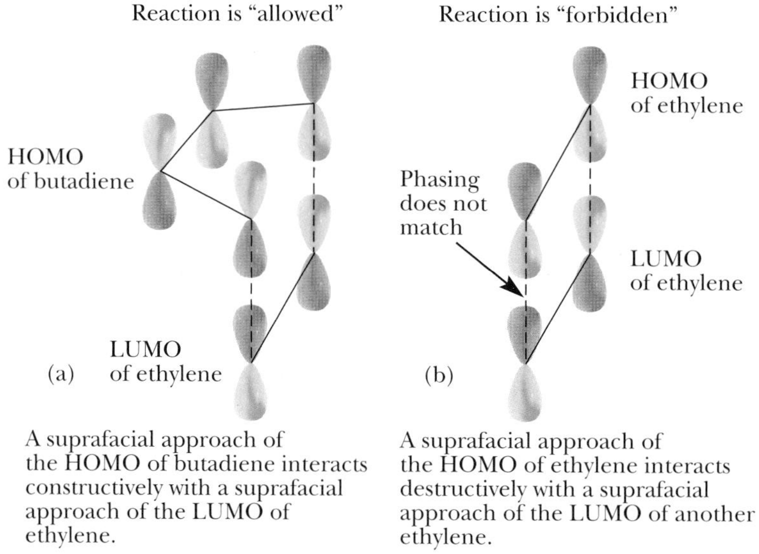

FIGURE 20.8 (a) A frontier molecular orbital analysis for the 4 + 2 cycloaddition of butadiene with ethylene. The phasing is found to match between the HOMO of butadiene and the LUMO of ethylene when they collide face to face, and hence, the reaction is allowed. (b) A frontier molecular orbital analysis for the 2 + 2 cycloaddition of two ethylene molecules. Because the phasing does not match in one spot when the HOMO and LUMO are combined, the reaction is termed forbidden.

Example 20.6 FMOT

Let's examine an alternative collision geometry for the 4 + 2 reaction. In an antarafacial interaction, one reactant collides in such a manner that the opposite faces of the *p* orbitals in the π bonds interact. Show that the 4 + 2 cycloaddition reaction is forbidden when butadiene interacts in an antarafacial manner.

Solution

Shown below is the reaction of butadiene with ethylene, except that we have the ethylene collide with the butadiene at a skewed angle. Recall that molecules in a reaction flask are constantly colliding with the solvent and each other in all possible orientations, but only certain orientations lead to chemical reactions. In the antarafacial geometry for the butadiene, there is one contact between the butadiene and ethylene in which the phasing does not match. Hence, this geometry for a collision between the two reactants will not lead to product and is considered forbidden.

An antarafacial approach of the HOMO of butadiene interacts destructively with a suprafacial approach of the LUMO of ethylene.

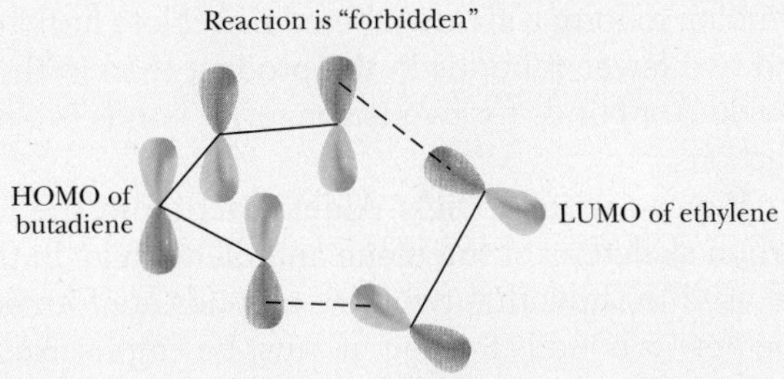

Problem 20.6

The 2 + 2 cycloaddition with one suprafacial and one antarafacial interaction is allowed. Show this conclusion via a frontier molecular orbital analysis. Although the reaction is allowed, it is seldom seen. Can you think of a reason not based on an orbital analysis of why this reaction is difficult?

20.5 The Diels-Alder Reaction

In 1928, Otto Diels and Kurt Alder in Germany discovered a unique reaction of conjugated dienes: they undergo cycloaddition reactions with certain types of carbon-carbon double and triple bonds. For their discovery and subsequent studies of this reaction, Diels and Alder were jointly awarded the 1950 Nobel Prize in Chemistry.

The compound with the double or triple bond that reacts with the diene in a Diels-Alder reaction is given the special name **dienophile** (diene-loving), and the product of a Diels-Alder reaction is given the special name **Diels-Alder adduct**. The designation **cycloaddition** refers to the fact that two reactants add together to give a cyclic product. Following are two examples of Diels-Alder reactions: one with a compound containing a carbon-carbon double bond and the other containing a carbon-carbon triple bond.

Dienophile

A compound containing a double or triple bond (consisting of one or two C, N, or O atoms) that can react with a conjugated diene to give a Diels-Alder adduct.

Diels-Alder adduct

A cyclohexene resulting from the cycloaddition reaction of a diene and a dienophile.

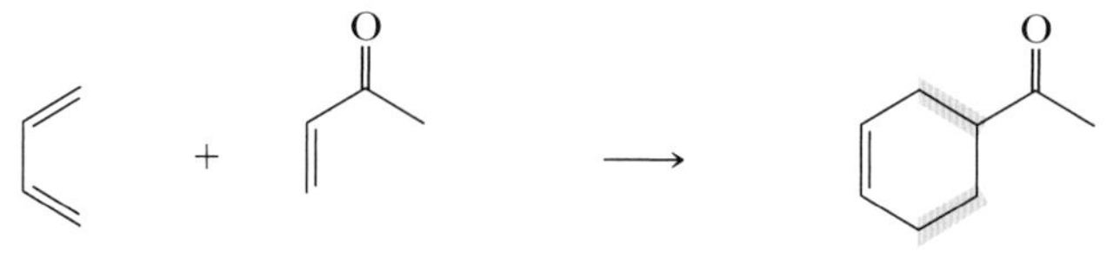

1,3-Butadiene (a diene) **3-Buten-2-one** (a dienophile) **3-Cyclohexenyl methyl ketone** (a Diels-Alder adduct) (racemic mixture)

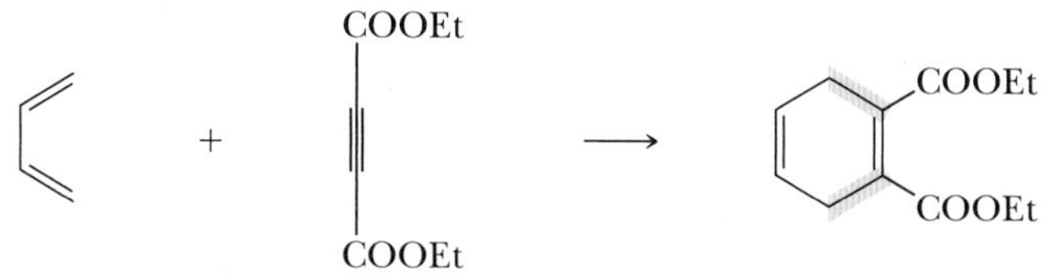

1,3-Butadiene (a diene) **Diethyl 2-butynedioate** (a dienophile) **Diethyl 1,4-cyclohexadiene-1,2-dicarboxylate** (a Diels-Alder adduct)

Note that the four carbon atoms of the diene and two carbon atoms of the dienophile combine to form a six-membered ring. Note further that there are two more σ bonds and two fewer π bonds in the product than in the reactants. This exchange of two (weaker) π bonds for two (stronger) σ bonds is a major driving force in Diels-Alder reactions.

We can write a Diels-Alder reaction in the following way, showing only the carbon skeletons of the diene and dienophile. In this representation, curved arrows are used to show that two new σ bonds are formed, three π bonds are broken, and one new π bond is formed. It must be emphasized here that in this particular case, the curved arrows in this diagram are not meant to show a mechanism. Rather, they are intended to show which bonds are broken, which new bonds are formed, and how many electrons are involved (six in this case). The real mechanism is pericyclic and is a 4 + 2 cycloaddition, as was presented in Section 20.4.

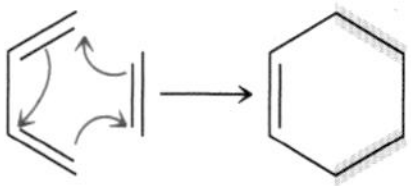

Diels-Alder reaction

The special values of the reaction discovered by Diels and Alder are that (1) it is one of the simplest reactions that can be used to form six-membered rings; (2) it is one of few reactions that can be used to form two new carbon-carbon bonds at the same time; and, as we will see later in this section, (3) it is completely stereospecific and quite regioselective. For these reasons, the Diels-Alder reaction has proved to be enormously valuable in synthetic organic chemistry.

Example 20.7 The Diels-Alder Reaction I

Draw a structural formula for the Diels-Alder adduct formed by reaction of each diene and dienophile pair.

(a) 1,3-Butadiene and propenal
(b) 2,3-Dimethyl-1,3-butadiene and 3-buten-2-one

Solution

First, draw the diene and dienophile so that each molecule is properly aligned to form a six-membered ring. Then complete the reaction to form the six-membered ring Diels-Alder adduct.

(a)

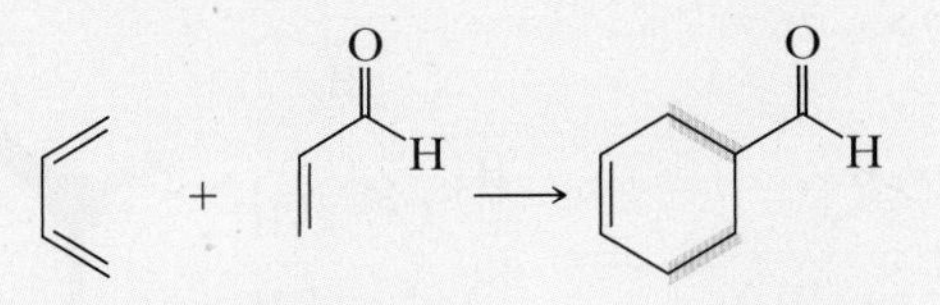

1,3-Butadiene Propenal (Acrolein) (racemic mixture)

(b)

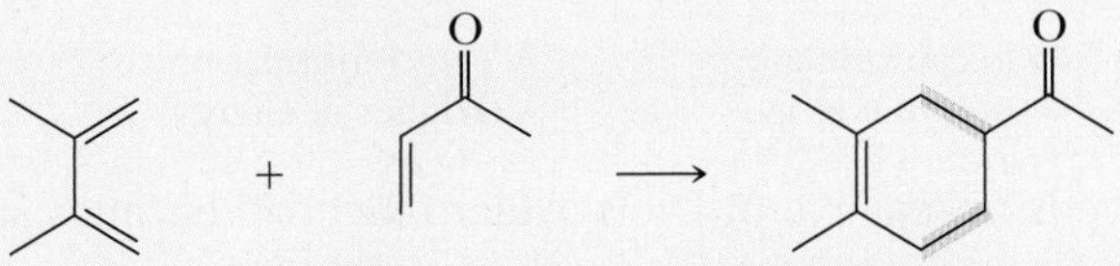

2,3-Dimethyl-1,3-butadiene 3-Buten-2-one (Methyl vinyl ketone) (racemic mixture)

Problem 20.7

What combination of diene and dienophile undergoes Diels-Alder reaction to give each adduct?

(a)

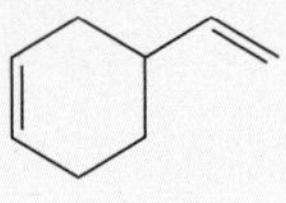

(racemic)

(b)

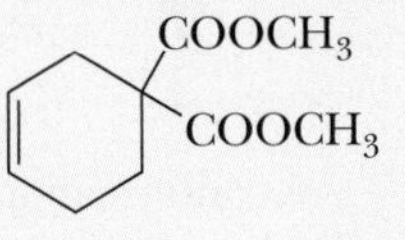

(c)

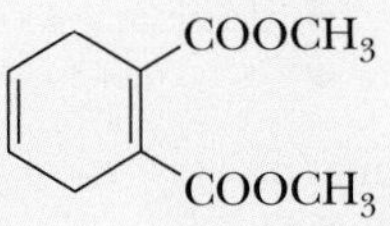

Watch a video explanation

Now let us look more closely at the scope and limitations, stereochemistry, and mechanism of Diels-Alder reactions.

A. Diene Must Be Able to Assume an *s-Cis* Conformation

We can illustrate the significance of conformation of the diene by reference to 1,3-butadiene. For maximum stability of a conjugated diene, overlap of the four unhybridized 2*p* orbitals making up the π system must be complete, a condition that occurs only when all four carbon atoms of the diene lie in the same plane. If the carbon skeleton of 1,3-butadiene is planar, the six atoms bonded to the skeleton of the diene are also contained in the same plane. Bond rotation is somewhat restricted around the central single bond due to conjugation; if the atoms are not coplanar, conjugation is imperfect or broken completely. There are two planar conformations of 1,3-butadiene, called the ***s-trans*** **conformation** and the ***s-cis*** **conformation** where the designation *s* refers to the carbon-carbon single bond of the diene. Of these, the *s-trans* conformation is slightly lower in energy and therefore is slightly more stable.

Although *s-trans*-1,3-butadiene is the more stable conformation, *s-cis*-1,3-butadiene is the reactive conformation in Diels-Alder reactions. In the *s-cis* conformation, carbon atoms 1 and 4 of the conjugated system are close enough to react with the carbon-carbon double or triple bond of the dienophile and to form a six-membered ring. In the *s-trans* conformation, they are too far apart for this to happen.

The energy barrier for interconversion of the *s-trans* and *s-cis* conformations for 1,3-butadiene is low, approximately 11.7 kJ (2.8 kcal)/mol; consequently, 1,3-butadiene can still be a reactive diene in Diels-Alder reactions.

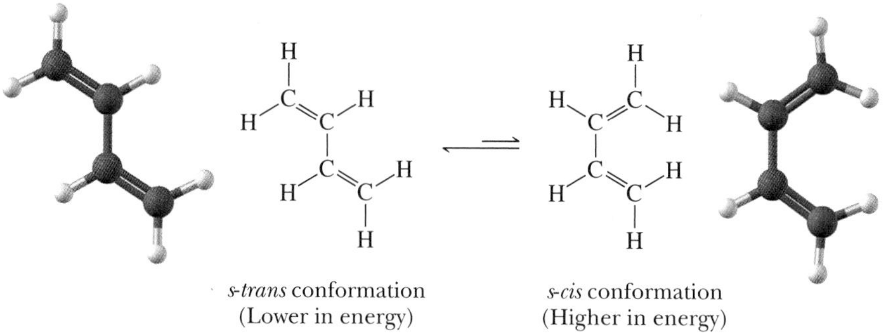

(2Z,4Z)-2,4-Hexadiene is unreactive in Diels-Alder reactions because steric hindrance prevents it from assuming the required *s-cis* conformation.

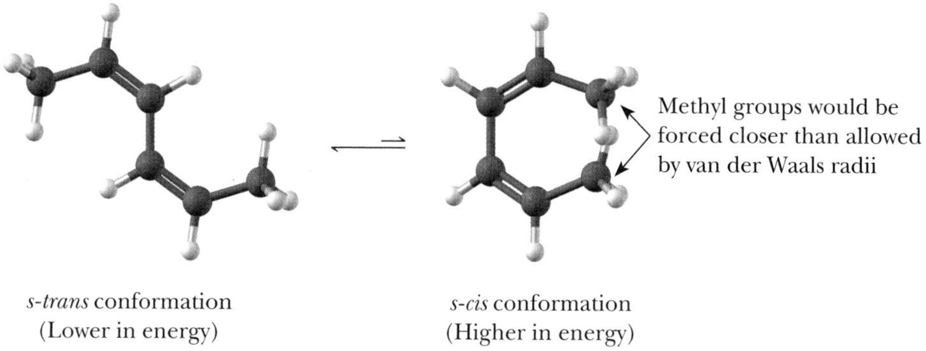
Methyl groups would be forced closer than allowed by van der Waals radii
s-trans conformation
(Lower in energy)
s-cis conformation
(Higher in energy)

(2Z,4Z)-2,4-Hexadiene

Example 20.8 The Diels-Alder Reaction II

Which molecules can function as dienes in Diels-Alder reactions?

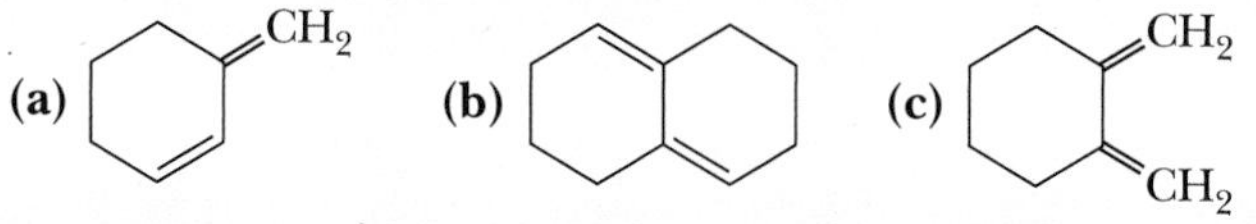

Solution

The dienes in both (a) and (b) are fixed in the *s-trans* conformation and therefore are not capable of participating in Diels-Alder reactions. The diene in (c) is fixed in the *s-cis* conformation and therefore has the proper orientation to participate in Diels-Alder reactions.

Problem 20.8

Which molecules can function as dienes in Diels-Alder reactions?

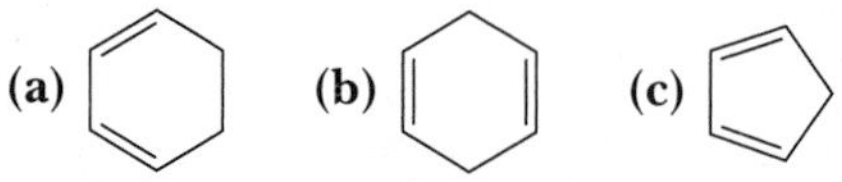

B. The Effect of Substituents on Rate

The simplest example of a Diels-Alder reaction is that between 1,3-butadiene and ethylene, both gases at room temperature. Although this reaction does occur, it is very slow and takes place only when the reactants are heated at 200°C under pressure.

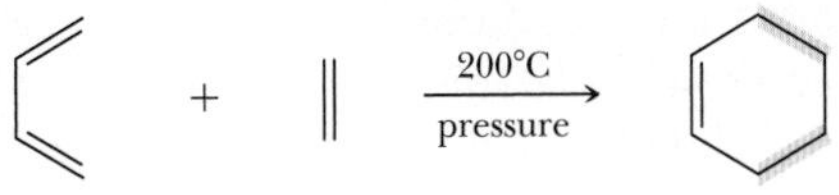

1,3-Butadiene **Ethylene** **Cyclohexene**

Diels-Alder reactions are facilitated by a combination of electron-withdrawing substituents on one of the reactants and electron-releasing substituents on the other. Most commonly, the dienophile is electron deficient and the diene is electron rich. For example, placing a carbonyl group (electron withdrawing because of the partial positive charge on its carbon) on the dienophile facilitates the reaction. To illustrate, 1,3-butadiene and 3-buten-2-one form a Diels-Alder adduct when heated at 140°C.

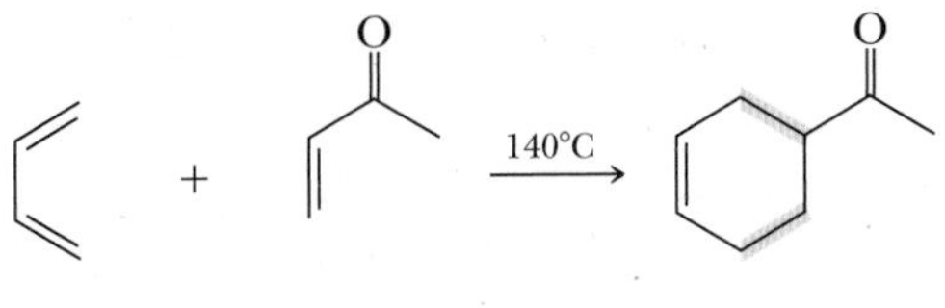

1,3-Butadiene **3-Buten-2-one** (racemic)

Placing electron-releasing methyl groups on the diene further facilitates reaction; 2,3-dimethyl-1,3-butadiene and 3-buten-2-one form a Diels-Alder adduct at 30°C.

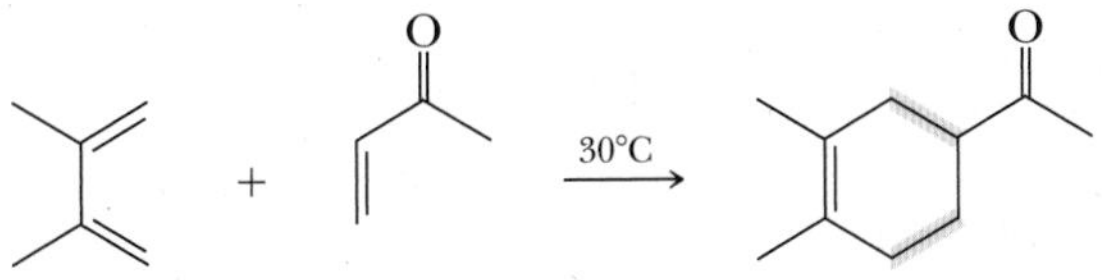

2,3-Dimethyl-1,3-butadiene **3-Buten-2-one** (racemic)

Several of the electron-releasing and electron-withdrawing groups most commonly encountered in Diels-Alder reactions are given in Table 20.4. Note that the ester group can be either electron donating or electron withdrawing depending on whether the oxygen or the carbonyl is attached to the double bond.

Table 20.4 Electron-Releasing and Electron-Withdrawing Groups

Electron-Releasing Groups	Electron-Withdrawing Groups
$-CH_3$	$-\overset{O}{\overset{\|}{C}}H$ (aldehyde)
$-CH_2CH_3$	$-\overset{O}{\overset{\|}{C}}R$ (ketone)
$-CH(CH_3)_2$	$-\overset{O}{\overset{\|}{C}}OH$ (carboxyl)
$-C(CH_3)_3$	$-\overset{O}{\overset{\|}{C}}OR$ (ester)
$-R$ (other alkyl groups)	$-NO_2$ (nitro)
$-OR$ (ether)	$-C\equiv N$ (cyano)
$-O\overset{O}{\overset{\|}{C}}R$ (ester)	

C. Diels-Alder Reactions Can Be Used to Form Bicyclic Systems

Conjugated cyclic dienes, in which the double bonds are of necessity held in an *s-cis* conformation, are highly reactive in Diels-Alder reactions. Two particularly useful dienes for this purpose are cyclopentadiene and 1,3-cyclohexadiene. In fact, cyclopentadiene is reactive both as a diene and as a dienophile, and upon standing at room temperature, it forms a Diels-Alder self-adduct known by the common name dicyclopentadiene. When dicyclopentadiene is heated to 170°C, a reverse Diels-Alder reaction takes place and cyclopentadiene is reformed.

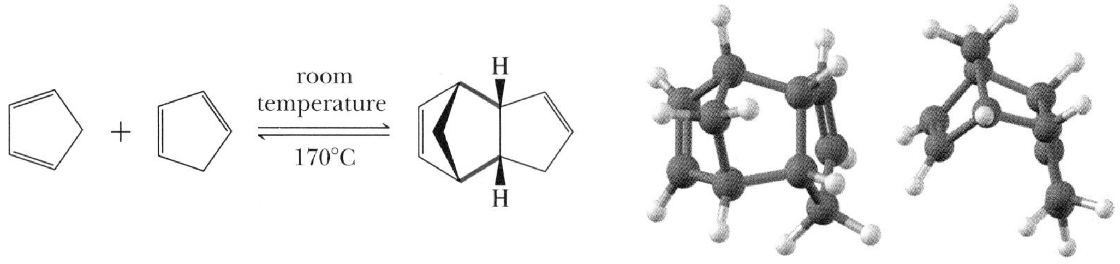

Diene Dienophile **Dicyclopentadiene** (endo form) From top From side

The terms *endo* and *exo* are used for bicyclic Diels-Alder adducts to describe the orientation of substituents of the dienophile in relation to the two-carbon diene-derived bridge. **Exo** (Greek, outside) substituents are on the opposite side from the diene-derived bridge; **endo** (Greek, within) substituents are on the same side.

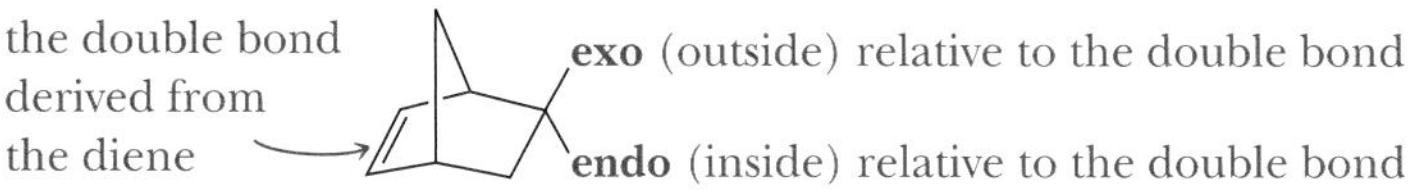

For Diels-Alder reactions under kinetic control, the endo orientation of the dienophile is favored (Figure 20.9). Treatment of cyclopentadiene with methyl propenoate (methyl acrylate) gives the endo adduct exclusively. The exo adduct is not formed. Diels-Alder reactions are not always so stereoselective.

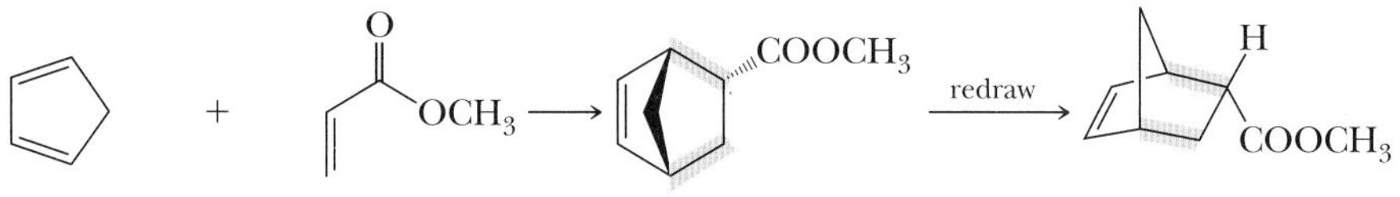

Cyclopentadiene **Methyl propenoate** The endo adduct(racemic)

D. The Configuration of the Dienophile Is Retained

The reaction is completely stereospecific at the dienophile. If the dienophile is a *cis* isomer, then the substituents *cis* to each other in the dienophile are *cis* in the Diels-Alder adduct. Conversely, if the dienophile is a *trans* isomer, substituents that are *trans* in the dienophile are *trans* in the adduct.

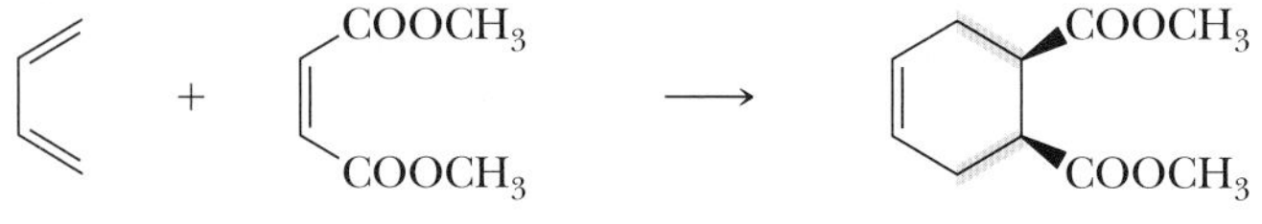

Dimethyl *cis*-2-butenedioate (a *cis* dienophile) **Dimethyl *cis*-4-cyclohexene-1,2-dicarboxylate**

Dimethyl *trans*-2-butenedioate (a *trans* dienophile) **Dimethyl *trans*-4-cyclohexene-1,2-dicarboxylate** (racemic)

E. The Configuration at the Diene Is Retained

The reaction is also completely stereospecific at the diene. Groups on the 1 and 4 positions of the diene retain their relative orientation.

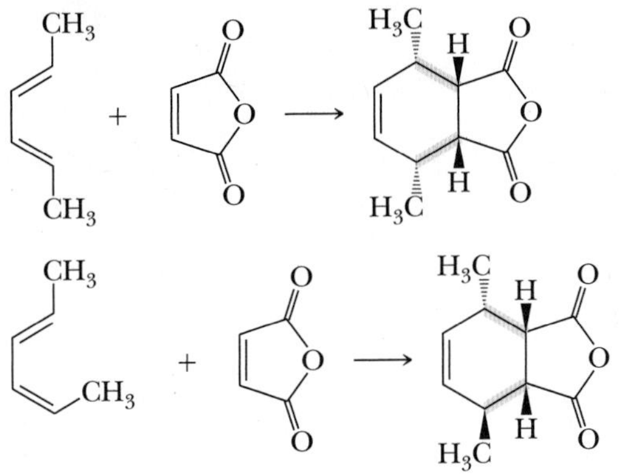

(racemic)

A picture of the transition state will help clarify the reason for this. Bonds being formed in the transition state are shown as dashed red lines; bonds being broken are shown as dashed blue lines. The groups that are inside on the diene end up on the opposite side from the dienophile.

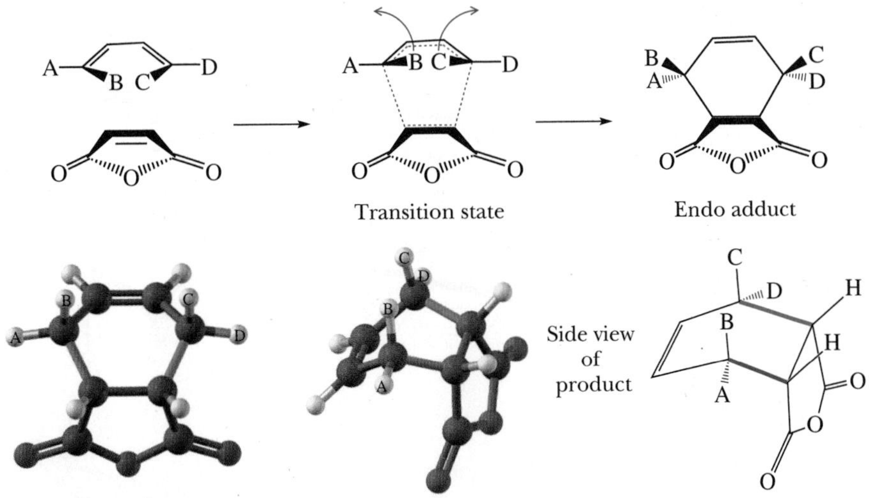

Front view
of
product

Example 20.9 Stereochemistry of the Diels-Alder Reaction

Complete the following Diels-Alder reaction, showing the stereochemistry of the product.

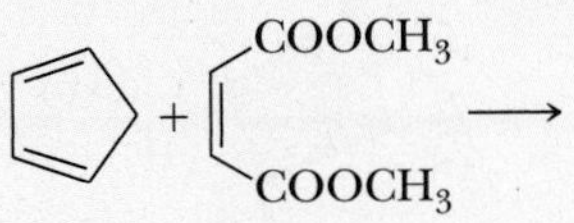

Solution

Reaction of cyclopentadiene with this dienophile forms a disubstituted bicyclic product. The two ester groups are *cis* in the dienophile, and given the stereoselectivity of the Diels-Alder reaction, they are *cis* and endo in the product.

(Continued)

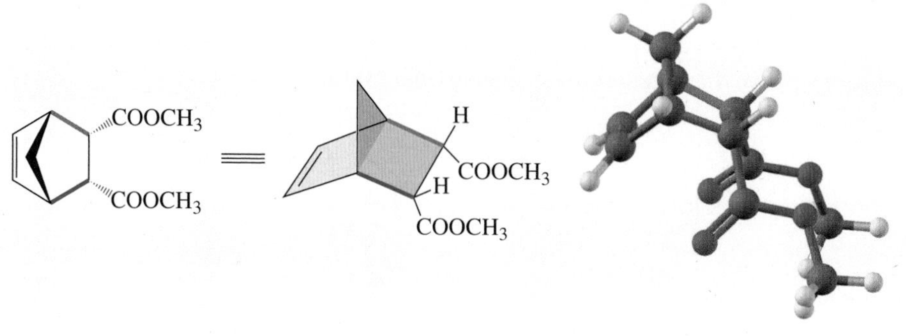

Problem 20.9

What diene and dienophile might you use to prepare the following racemic Diels-Alder adduct?

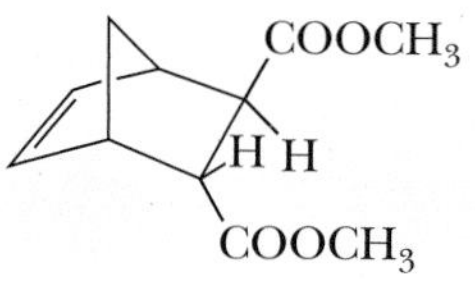

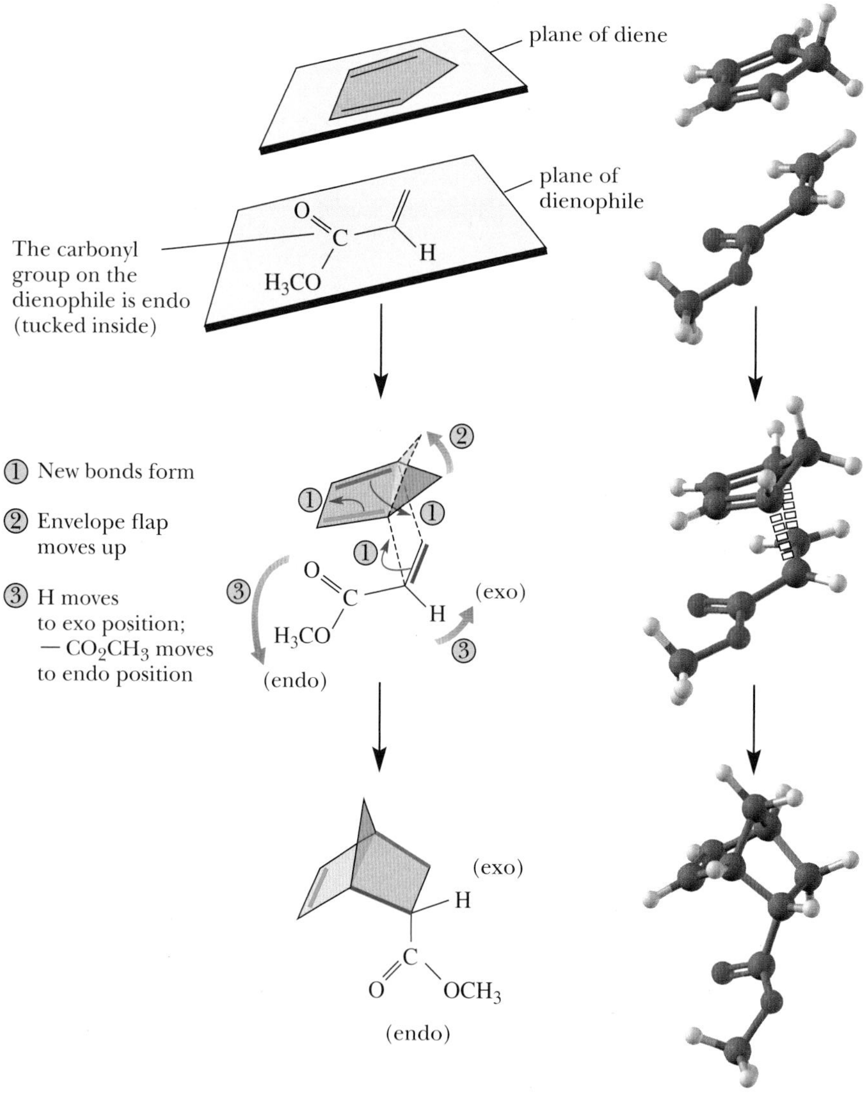

FIGURE 20.9 Mechanism of the Diels-Alder reaction. The diene and dienophile approach each other in parallel planes, one above the other, with the substituents on the dienophile endo to the diene. There is overlap of the π orbitals of each molecule and syn addition of each molecule to the other. As (1) new σ bonds form in the transition state, (2) the $—CH_2—$ on the diene rotates upward and (3) the hydrogen atom of the dienophile becomes exo and the ester group becomes endo.

F. Exploiting the Stereochemistry of the Diels-Alder Reaction

As we have mentioned repeatedly throughout the text, the synthesis of chiral products from achiral starting materials in an achiral environment invariably leads to a racemic mixture of products. Nature achieves the synthesis of single enantiomers by using enzymes that create a chiral environment in which reaction takes place. Enzymes, in fact, show such high enantiomeric and diastereomeric selectivity that the result of an enzyme-catalyzed reaction is generally only a single one of all possible stereoisomers. Chemists have developed chiral catalysts that produce chiral products. However, these catalysts are often far less stereoselective than nature's enzyme catalysts, although great progress has been made in this field in recent years. How then do chemists achieve the synthesis of single enantiomers uncontaminated by their mirror images?

One strategy they use is resolution (Section 3.9) to separate enantiomers and recover each in pure form. The most common methods for resolution depend on (1) the different physical properties of diastereomeric salts, (2) the use of enzymes as resolving agents, and (3) chromatography on a chiral substrate. While resolution is effective in preparing pure enantiomers, half of all product prepared to the point of resolution, namely the unwanted enantiomer, is lost in the process. Thus, this strategy for the preparation of single enantiomers wastes starting materials and reagents.

We illustrate an alternative strategy, namely **asymmetric induction**, by E. J. Corey's preparation of a key intermediate in his synthesis of prostaglandins. In asymmetric induction, the reactive functional group of an achiral molecule is placed in a chiral environment by reacting it with a **chiral auxiliary**. The strategy is that the chiral auxiliary then exerts control over the stereoselectivity of the desired reaction. The chiral auxiliary chosen by Corey was 8-phenylmenthol. This molecule has three chiral centers and can exist as a mixture of $2^3 = 8$ possible stereoisomers. It was prepared in enantiomerically pure form from naturally occurring, enantiomerically pure menthol.

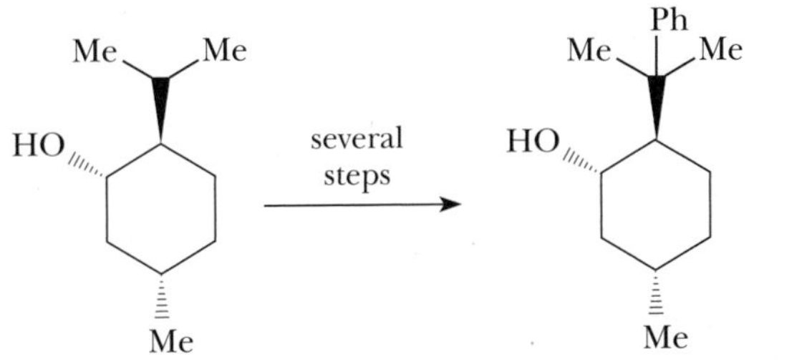

Menthol
(enantiomerically pure)

8-Phenylmenthol
(a chiral auxiliary)

The initial step in Corey's prostaglandin synthesis was a Diels-Alder reaction between a substituted cyclopentadiene and the double bond of an acrylate ester. By binding the achiral acrylate reactant to enantiomerically pure 8-phenylmenthol, Corey placed the carbon-carbon double bond of the dienophile in a chiral environment. The result was that the diene approached the carbon-carbon double bond of the acrylate preferentially from one direction.

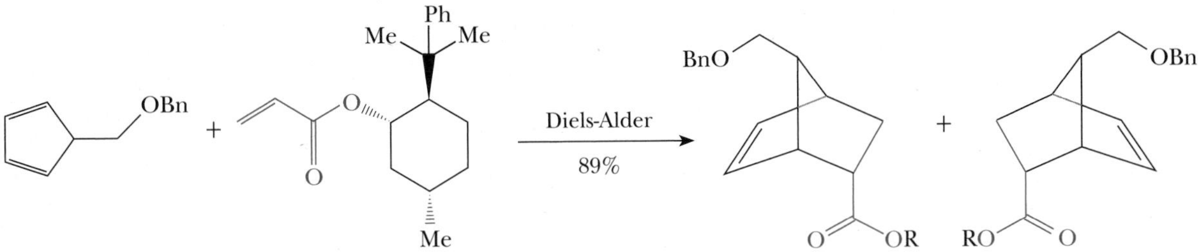

Achiral Enantiomerically pure 97% 3%

A remarkable feature of this reaction is that it creates three chiral centers. Two of the chiral centers, namely those at the two ring junctions, are established by the

Diels-Alder reaction. The third, namely the endo position of the ester group, is also established by the Diels-Alder reaction. Without the chiral auxiliary 8-phenylmenthyl group, two of the eight possible stereoisomers would be produced, namely the pair of enantiomers shown. Although both enantiomers of the bicyclic products were formed in Corey's scheme, they were formed in the ratio of 97:3 and the desired enantiomer could be separated in pure form. In subsequent steps, the 8-phenylmenthyl ester was hydrolyzed and the pure enantiomer was converted to the so-called Corey lactone and then to enantiomerically pure prostaglandin F_{2a}.

G. A Word of Caution About Electron Pushing

Earlier we used curved arrows to show the flow of electrons that takes place in the process of bond breaking and bond forming in the Diels-Alder reaction. As discussed, these reactions involve a four-carbon diene and a two-carbon dienophile and are termed [4 + 2] cycloadditions. We can write similar electron-pushing mechanisms for the dimerization of ethylene by a [2 + 2] cycloaddition to form cyclobutane and for the dimerization of butadiene by a [4 + 4] cycloaddition to form 1,5-cyclooctadiene.

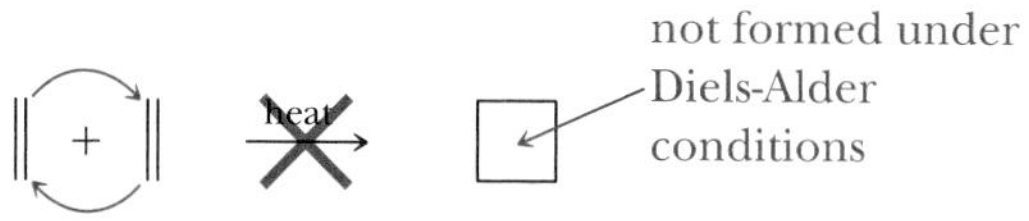

Two molecules of ethylene

Cyclobutane

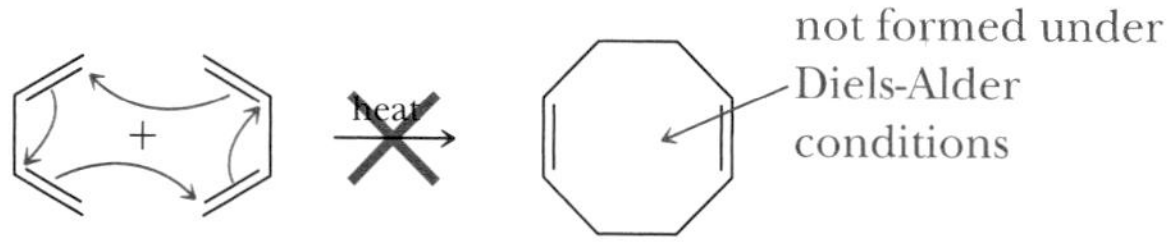

Two molecules of butadiene

1,5-Cyclooctadiene

Although [2 + 2] and [4 + 4] cycloadditions bear a formal relationship to the Diels-Alder reaction, neither, in fact, takes place under the thermal conditions required for Diels-Alder reactions (see Section 20.4) because they are forbidden as determined by the frontier molecular orbital analysis.

20.6 Sigmatropic Shifts

Sigmatropic shift

A reaction in which a σ bond migrates across the face of one or more π bonds.

The second class of pericyclic reactions that we examine is that of **sigmatropic shifts**. These reactions consist of the movement of a σ bond across the face of one or more π bonds. Although many examples of these reactions are known, we are only going to analyze what is known as a [3,3]-shift. The numbering system for the nomenclature of the shift derives from assigning the number 1 to the ends of the σ bond that is shifting and then naming the reaction to denote the number of atoms to which the σ bond migrates. There are two common versions of this reaction, known as the Claisen and Cope rearrangements.

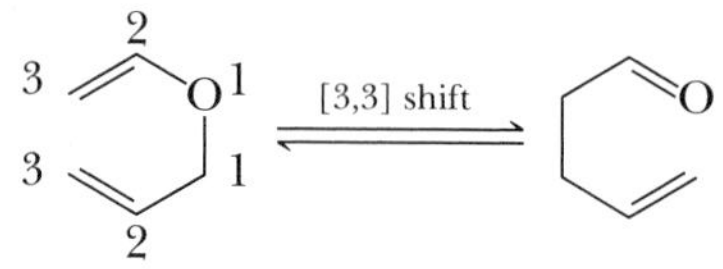

The Claisen rearrangement

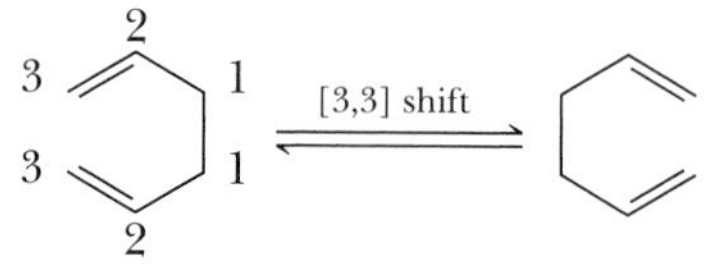

The Cope rearrangement

To derive the frontier molecular orbital analysis for any [3,3]-shift, we'll use 1,5-hexatriene as the model, just as we used butadiene and ethylene as models for the frontier molecular orbital analysis of all Diels-Alder reactions. As always, in a frontier molecular orbital analysis, we first identify a proposed geometry for the reaction. Let's propose a chairlike transition state in which the carbons on the ends of the chain react from the top of one π bond and the bottom of the other.

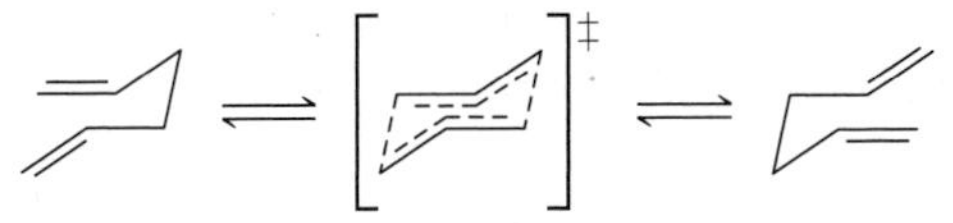

The next step of a frontier molecular orbital analysis involves identifying a **HOMO** and a **LUMO** and checking to see if the **HOMO** and **LUMO** can interact with matched phasings (Figure 20.10). In this case, we assign the σ bond that is migrating to be the **HOMO**; it is thus drawn as the overlap of two sp^3 hybrid orbitals (see Figure 1.18). The **LUMO** is assigned to be a molecular orbital that is a mixture of the two alkenes when their ends are in close proximity [Figure 20.10(a)] and in a trajectory to react in a manner consistent with a chairlike geometry of the transition state. The molecular orbitals that result from the mixture of two separate alkenes are analogous to those found in butadiene. Hence, the **LUMO** for the [3,3]-shift is phased identical to the **LUMO** of butadiene [compare the **LUMO** indicated in Figure 20.10(a) to orbital 3 in Figure 20.2]. However, there is one important difference. In orbital 3 of Figure 20.2, the two central p orbitals are in phase when parallel. But in our analysis of the Cope reaction, the top of one of the central p orbitals is placed in phase with the bottom of the other because this is the interaction geometry we are analyzing.

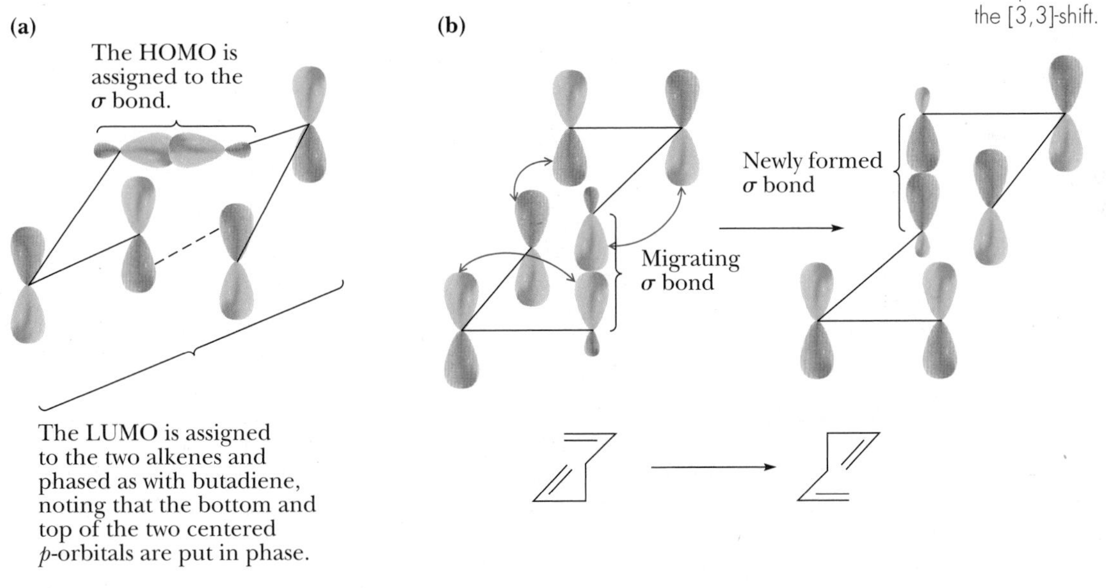
(a)
The HOMO is assigned to the σ bond.
The LUMO is assigned to the two alkenes and phased as with butadiene, noting that the bottom and top of the two centered p-orbitals are put in phase.
(b)
Migrating σ bond
Newly formed σ bond

FIGURE 20.10 (a) Proper phasing of orbitals for the frontier molecular orbital analysis of a [3,3]-sigmatropic shift; note the phasing interaction between the terminal carbons when the top of one π bond interacts with the bottom of the other in-phase. (b) Reorientation showing how the chair conformation leads to in-phase interactions throughout the [3,3]-shift.

We can now check to see if the **HOMO** and **LUMO** phases match. In Figure 20.10(b), we redraw the chair with the σ bond vertically for clarity. The arrows show matched phasing that leads to a σ bond and two π bonds in the product that are all in phase and bonding. Hence, the reaction is allowed.

In summary, the frontier molecular orbital approach finds that there is an allowed geometry for reaction with the ends of the π bonds of a 1,5-diene as in a Cope rearrangement (or analogously a Claisen rearrangement) with a chairlike transition state. Interestingly, a boatlike transition state is also allowed, although it is conformationally less stable (see Problem 20.48).

A. The Claisen Rearrangement

One example of the Claisen rearrangement transforms allyl phenyl ethers to *o*-allylphenols. Heating allyl phenyl ether, for example, the simplest member of this class of compounds, at 200–250°C results in a Claisen rearrangement to form *o*-allylphenol. In this rearrangement, an allyl group migrates from a phenolic oxygen to a carbon atom ortho to it. Carbon-14 labeling, here shown in color, has demonstrated that during a Claisen rearrangement, carbon 3 of the allyl group becomes bonded to the ring carbon ortho to the phenolic oxygen.

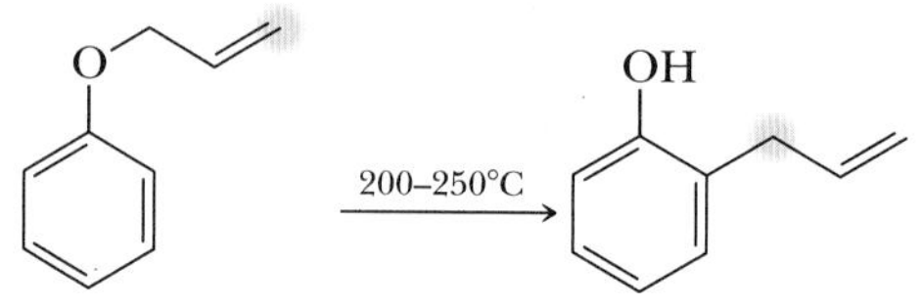

Allyl phenyl ether **2-Allylphenol**

The mechanism of a Claisen rearrangement involves a concerted redistribution of six electrons in a cyclic transition state, as described above. The product of this rearrangement is a substituted cyclohexadienone, which undergoes keto-enol tautomerism to reform the aromatic ring. A new carbon-carbon bond is formed in the process.

Mechanism 20.2

The Claisen Rearrangement

Step 1: **Sigmatropic shift.** Redistribution of six electrons in a cyclic transition state gives a cyclohexadienone intermediate. Dashed red lines indicate bonds being formed in the transition state, and dashed blue lines indicate bonds being broken.

Step 2: **Keto-enol tautomerism.** Keto-enol tautomerism restores the aromatic character of the ring.

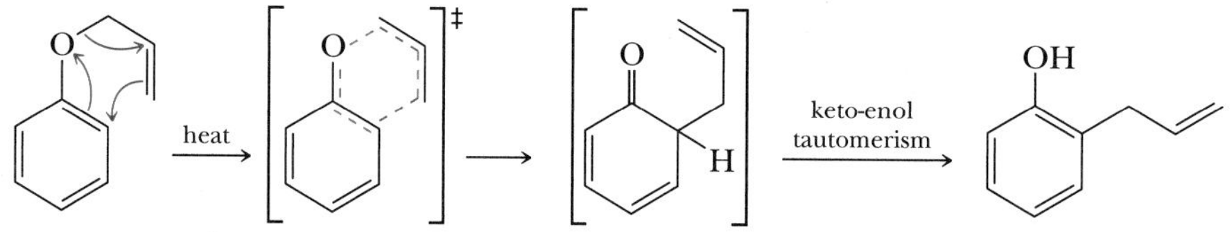

Allyl phenyl ether — Transition state — A cyclohexadienone intermediate — **2-Allylphenol**

Thus, we see that the transition state for the Claisen rearrangement bears a close resemblance to that for the Diels-Alder reaction. Both involve a concerted redistribution of six electrons in a cyclic transition state.

Example 20.10 The Claisen Rearrangement

Predict the product of Claisen rearrangement of *trans*-2-butenyl phenyl ether.

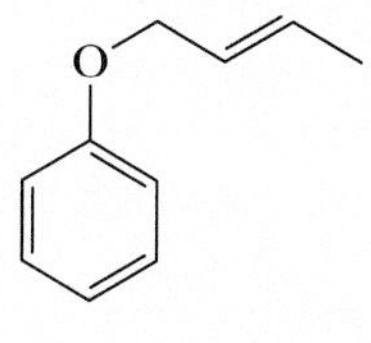

***trans*-2-Butenyl phenyl ether**

Solution

In the six-membered transition state for this rearrangement, carbon 3 of the allyl group becomes bonded to the ortho position of the ring.

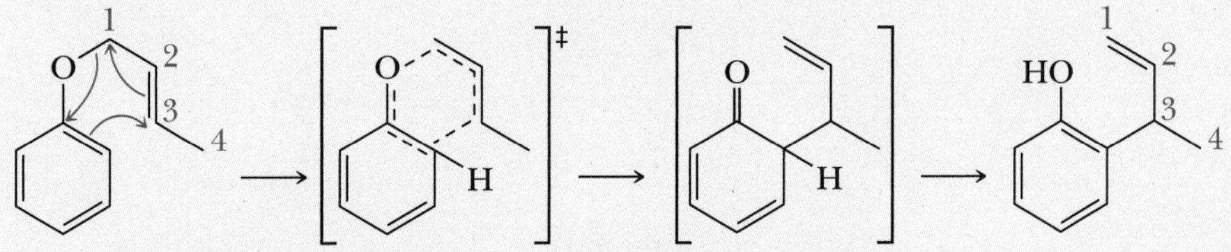

Transition state — A cyclohexadienone intermediate — (racemic)

Problem 20.10

Show how to synthesize allyl phenyl ether and 2-butenyl phenyl ether from phenol and appropriate alkenyl halides.

B. The Cope Rearrangement

The Cope rearrangement of 1,5-dienes also takes place via a cyclic six-electron transition state. In this example, the product is an equilibrium mixture of isomeric dienes. The favored product is the diene on the right, which contains the more highly substituted double bonds.

Mechanism 20.3

The Cope Rearrangement

Pericyclic reaction. Redistribution of six electrons in a cyclic transition state converts a 1,5-diene to an isomeric 1,5-diene.

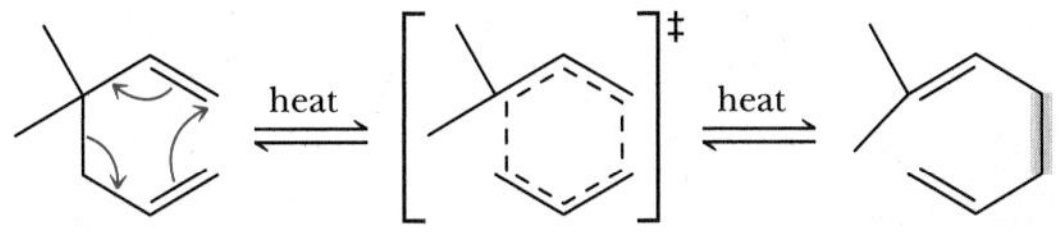

3,3-Dimethyl-1,5-hexadiene — Transition state — **6-Methyl-1,5-heptadiene**

Example 20.11 The Cope Rearrangement

Propose a mechanism for the following Cope rearrangement.

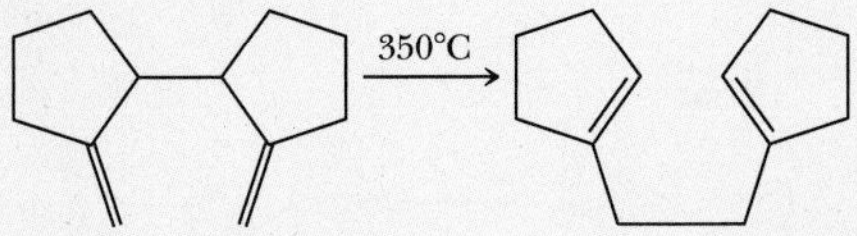

(Continued)

Solution

Redistribution of six electrons in a cyclic transition state gives the observed product.

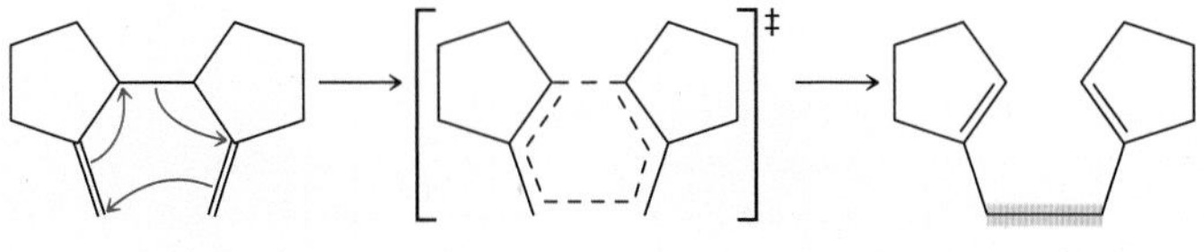

Transition state

Problem 20.11

Propose a mechanism for the following Cope rearrangement.

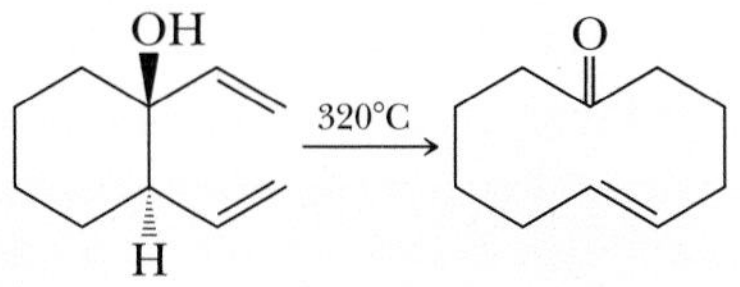

C. Stereochemistry of the Cope Rearrangement

As discussed during the previous frontier molecular orbital theory analysis of [3,3]-sigmatropic shifts, a chairlike transition state is allowed for these reactions. As you will show in Problem 20.48, a boatlike transition state is also allowed by frontier molecular orbital theory. However, chair conformations are more favorable than boat conformations for six-membered cyclic rings (look back at Section 2.5A). This preference influences the stereochemistry of these shifts, as we now show with an example.

Example 20.12 Stereochemistry, the Cope Rearrangement

Upon heating the *meso* version of 3,4-dimethyl-1,5-hexadiene, three products with differing alkene stereochemistry from the Cope rearrangement are possible, but only two are found, with one being highly preferred. Show all three possible products and predict the preference in the distribution.

Solution

The products and preference for the *cis-trans* alkenes can be explained by redrawing the reactant in chair- and boatlike conformations. These drawings reveal that the preferred product arises from a chairlike transition state.

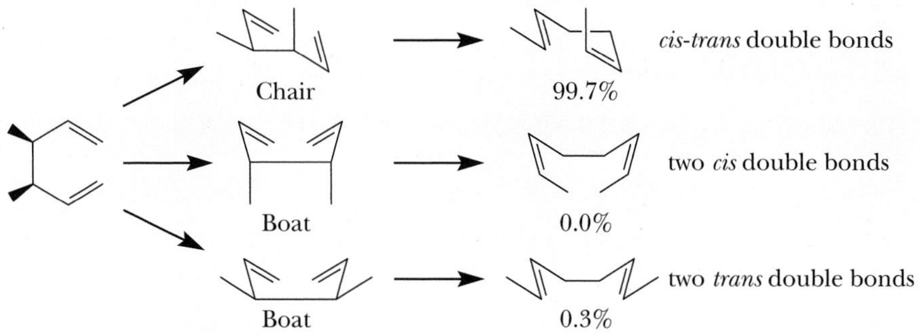

Problem 20.12

Upon heating a racemic mixture of *d,l*-3,4-dimethyl-1,5-hexadiene, three products are possible, and all three are observed. The ratios are 90, 9, and nearly 1 percent. Predict which percentages correspond to which products and explain the ratio by showing the chair and boat conformations that lead to the products.

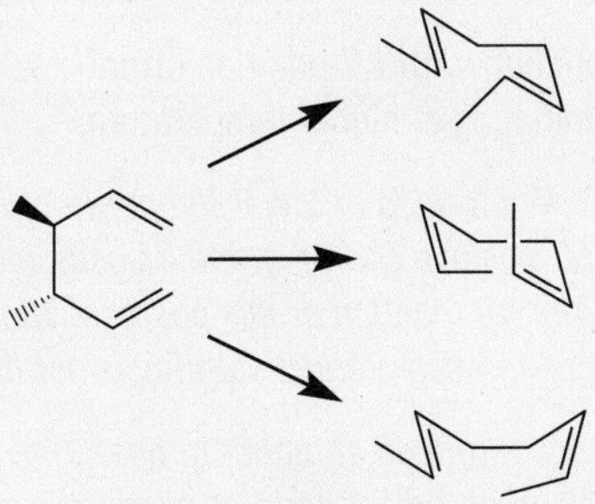
two *trans* double bonds
two *cis* double bonds
cis-trans double bonds

Watch a video explanation

Study Guide

20.1 Stability of Conjugated Dienes

- A **conjugated diene** is one in which the double bonds are separated by only one single bond so that the $2p$ orbitals of the adjacent π bonds overlap.
  - *An* ***unconjugated diene*** *is one in which the double bonds are separated by two or more single bonds.*
  - *A* ***cumulated diene*** *is one in which the two double bonds share an sp hybridized carbon. In a cumulated diene, the 2p orbitals of the π bonds do not overlap; so they are not conjugated.*
- The two conjugated double bonds in conjugated dienes are 14.5–17 kJ (3.5–4.1 kcal)/mol more stable than isomeric nonconjugated dienes, an observation that extends to all conjugated double bonds, not just dienes.
  - *The increased stability of conjugated double bonds results from delocalization of the four π electrons over the set of four parallel 2p orbitals.*
  - *According to molecular orbital theory, two conjugated double bonds are derived from four π molecular orbitals because the four parallel 2p orbitals overlap in space, even the 2p orbitals on either side of the single bond between the conjugated double bonds.*
  - *The lowest two π molecular orbitals have zero and one node, respectively, are bonding orbitals, and are filled with two electrons each.*
  - *Each of these lowest two filled π molecular orbitals is at an energy that is lower than isolated π bonds, accounting for the "extra" stability of conjugated π systems.*
  - *The lowest filled π molecular orbital has large lobes extending over all four atoms, illustrating the delocalization of electron density in conjugated π systems.*
  - *In order for maximal overlap to occur, the 2p orbitals must be parallel; so the sp^2 atoms of the conjugated systems must be coplanar.*

P **20.1, 20.2, 20.6, 20.14, 20.15**

Figure 20.2

20.2 Electrophilic Addition to Conjugated Dienes

- Conjugated dienes undergo both 1,2- and 1,4-addition reactions with electrophiles, often giving mixtures of both kinds of products.
- The ratio of 1,2-addition to 1,4-addition is temperature dependent, with 1,2-addition often predominating at lower temperature and 1,4-addition predominating at higher temperature.
  - *1,2-Addition to butadiene leads to the predominant product formed at lower temperature under **kinetic (rate) control**, because there is usually greater positive charge at the 2 position of the allylic cation intermediate, lowering the activation barrier for reaction at this position. The lower temperature prevents equilibration between products, so relative product stability is not important.*
  - *1,4-Addition to butadiene leads to the predominant product formed at higher temperature under **thermodynamic control**, because the double bond of the 1,4-addition product is more substituted and therefore of lower energy. The higher temperature allows equilibration of products so that product distribution depends on relative product stability.*
  - *Note that the details of a conjugated diene structure will determine relative stabilities of 1,2- and 1,4-addition products, so the preceding statements concerning kinetic and thermodynamic product ratios of butadiene should be considered guidelines only and each new molecule needs to be carefully analyzed.*

P **20.3, 20.16–20.22**

▶ **Figure 20.3**

KEY REACTIONS

1. **Electrophilic Addition to Conjugated Dienes (Section 20.2)** The ratio of 1,2- to 1,4-addition products depends on whether the reaction is under kinetic control or thermodynamic control. When a conjugated diene reacts with HBr, initial protonation of one of the double bonds gives a resonance-stabilized allylic cation; reaction of bromide with one of the carbons of this intermediate bearing the partial positive charge gives the 1,2-addition product, and reaction at the other gives the 1,4-addition product.

$$CH_2{=}CHCH{=}CH_2 + HBr \longrightarrow CH_3CH(Br)CH{=}CH_2 + CH_3CH{=}CHCH_2Br$$

Products at −78°C (kinetic control):	90%	10%
Products at 40°C (thermodynamic control):	15%	85%

20.3 UV-Visible Spectroscopy

- The ultraviolet region of the electromagnetic spectrum has wavelengths extending between 200 nm and 400 nm, and the visible region has wavelengths extending between 400 nm and 700 nm.
- Ultraviolet and visible spectral data are plotted as **absorbance (A)** versus wavelength, where absorbance is calculated as the log base 10 of the ratio of (I_0/I) where I_0 is the intensity of light at a given wavelength irradiating a sample and I is the light transmitted through the sample.
  - *The quantity* $(I/I_0) \times 100$ *is called **percent transmittance**.*
  - *The relationship between absorbance, concentration, and length of the sample cell (cuvette) is known as the **Beer-Lambert Law**,* $A = \varepsilon cl$ *where* A *is absorbance,* ε *is the molar absorptivity*

(also called extinction coefficient) of the molecules in the sample having the units per moles per liter per centimeter $(M^{-1}cm^{-1})$, c *is concentration in moles per liter* (M), *and* l *is the length of the cuvette in centimeters (cm).*

- *The molar absorptivity (extinction coefficient) as a function of wavelength is characteristic for a molecule and is based on the functional groups within the molecule. If the molar absorptivity is known for a given molecule, its concentration in solution can be calculated using the Beer-Lambert law.*

- Absorption by molecules removes the absorbed wavelengths from white light, and a sample will appear to our eyes as the combination of reflected wavelengths.
  - *Wavelengths not absorbed are reflected.*
  - *The color of combined reflected wavelengths can be roughly approximated as the complement of the absorbed color as illustrated using an artist's standard color wheel.*
- Absorption of electromagnetic radiation in the ultraviolet-visible region results in promotion of an electron from a lower energy, occupied molecular orbital to a higher energy, unoccupied molecular orbital.
  - *The amount of energy in the ultraviolet-visible region is appropriate to excite nonbonding (lone pair) or* π *(bonding) electrons to* π^* *(antibonding) orbitals in a process known as an* $n \rightarrow \pi^*$ *or* $\pi \rightarrow \pi^*$ *transition, respectively.*
  - σ *bonding electrons are usually too low in energy to take part in ultraviolet or visible light absorption.*
  - *The* $\pi \rightarrow \pi^*$ *transition for unconjugated alkenes is usually too high in energy (wavelength too short) to observe in the near ultraviolet spectrum.*
  - *Conjugated* π *systems have* $\pi \rightarrow \pi^*$ *transitions that can be seen in the ultraviolet or even visible absorption region, because conjugation decreases the energy difference between filled and unfilled* π *orbitals.*
    - *The greater the number of conjugated* π *bonds, the smaller the* $\pi \rightarrow \pi^*$ *energy gap, so the longer the wavelength of absorbed light.*
    - *Carbonyl groups can take part in conjugation along with* C=C *double bonds.*

P **20.4, 20.5, 20.23–20.27**

The Origins of Color

20.4 Pericyclic Reaction Theory

- **Pericyclic reactions** occur in a single step involving a transition state that has a closed loop of orbitals.
- Although several methods exist to understand these reactions, **frontier molecular orbital theory** is the most common and easiest approach. In this approach, one follows a sequence of steps in order to predict whether the reaction is allowed or forbidden.
  - *First, a reaction geometry is proposed.*
  - *Second, the* $HOMO$ *and* $LUMO$ *of the reacting partners are written.*
  - *Third, the interaction between the* $HOMO$ *and* $LUMO$ *of the partners is examined in order to reveal whether an even number of phase changes (most commonly zero) or an odd number (most commonly one) exist at the points of interaction.*

- *If the number of phase changes is even, the reaction is allowed, and if the number is odd, the reaction is forbidden and is generally not observed.*

- Frontier molecular orbital theory predicts that **4 + 2 cycloadditions** in which the reactants interact in a suprafacial manner are allowed but that analogous **2 + 2 cycloadditions** are forbidden.

P **20.11, 20.28, 20.29**

20.5 The Diels-Alder Reaction

- Conjugated dienes react with certain types of molecules possessing double or triple bonds to form two new σ bonds and a ring structure in a reaction called the Diels-Alder reaction, an example of a 4 + 2 cycloaddition reaction.
  - *The compound with the double or triple bond that reacts with the diene is called a* ***dienophile****, and the cyclic product is usually called the* ***Diels-Alder adduct****.*
    - *Three π bonds are broken and two stronger new σ bonds along with a new π bond are formed in the reaction, providing the driving force.*
- The Diels-Alder reaction is facilitated by having electron-withdrawing groups such as carbonyls on one reactant (usually the dienophile) and electron-releasing groups on the other (usually the diene).
- The diene must be in the ***s-cis*** **conformation** to react, and dienes such as cyclopentadiene that are constrained to be in this conformation are particularly reactive.
  - *When cyclic dienes are used, a bicyclic Diels-Alder adduct is produced.*
  - *The terms* ***exo*** *and* ***endo*** *are used with bicyclic Diels-Alder adducts. Exo substituents are on the opposite side of the newly formed ring from the diene-derived two-carbon bridge, and endo substituents are on the same side. For reactions that give kinetic products (not at equilibrium), the endo orientation of the dienophile is preferred.*
- The configuration of the dienophile (i.e., *E* or *Z*) is retained in the Diels-Alder reaction, as is the relative orientation of groups on the diene, indicating a highly concerted reaction mechanism.
- The mechanism of the reaction is concerted in that there is a single six-membered ring transition state in which the three π bonds are breaking at the same time as the two new σ bonds and one new π bond are being created.
- The Diels-Alder reaction has high stereoselectivity. One way to create enantiomerically pure target molecules is to use a chiral auxiliary, which is a chiral molecule available as a single enantiomer that is bonded to the starting material. The use of a chiral auxiliary can influence the resulting stereochemistry of a Diels-Alder reaction, producing a desired enantiomer in excess. The chiral auxiliary is then removed.
- The arrow pushing in a cycloaddition reaction does not accurately reflect whether a particular reaction—2 + 2, 4 + 2, 4 + 4, etc.—will be allowed or forbidden. One must rely on the frontier molecular orbital approach (or other such approaches) to properly understand the mechanism of these reactions.

P **20.12–20.14, 20.30–20.47**

▶ **Problem 20.7**

KEY REACTIONS

2. The Diels-Alder Reaction: A Pericyclic Reaction (Section 20.5) A Diels-Alder reaction takes place in a single step, without intermediates, and involves a redistribution of six π electrons in a cyclic transition state. The configuration of the diene and dienophile is preserved. Formation of the endo adduct is favored.

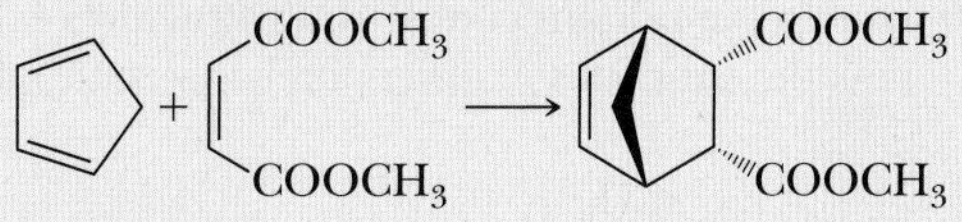

20.6 Sigmatropic Shifts

- **Sigmatropic shifts** involve the migration of a σ bond across one or more π systems.
- One of the most common shifts is called a **[3,3]-shift** that involves the migration of a σ bond across two flanking π bonds.
- Frontier molecular orbital analysis shows that the reaction is allowed in a geometry that creates a chairlike transition state, although boatlike transition states can also occur.
- One example of the Claisen rearrangement transforms allyl phenyl ethers to *o*-allylphenols through the redistribution of six electrons in a cyclic transition state.
- The Cope rearrangement of 1,5-dienes produces an equilibrium mixture of isomeric 1,5-dienes through the redistribution of six electrons in a cyclic transition state.
- By analyzing the possible chair- and boatlike transition states for the Cope rearrangement and by taking into account the lower energy of a chair conformation, one can predict the preferential stereochemistry of the products.

P **20.15, 20.16, 20.48–20.54**

▶ **Problem 20.12**

KEY REACTIONS

3. The Claisen Rearrangement: A Pericyclic Reaction (Section 20.6A) The Claisen rearrangement transforms an allyl phenyl ether to an ortho-substituted phenol. The reaction takes place in a single step and involves the redistribution of six electrons in a cyclic transition state.

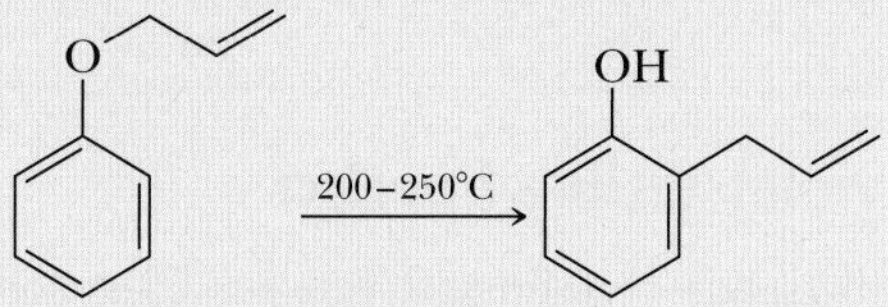

4. The Cope Rearrangement: A Pericyclic Reaction (Section 20.6B) The Cope rearrangement converts a 1,5-diene to give an isomeric 1,5-diene. The reaction takes place in a single step and involves the redistribution of six electrons in a cyclic transition state.

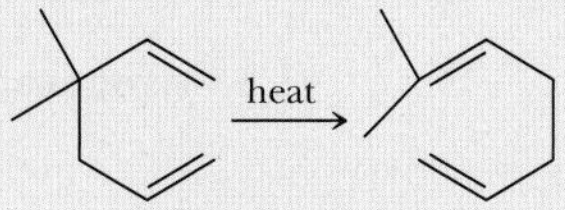

Problems

Red numbers indicate applied problems.

Structure and Stability

20.14 If an electron is added to 1,3-butadiene, into which molecular orbital does it go? If an electron is removed from 1,3-butadiene, from which molecular orbital is it taken?

20.15 Draw all important contributing structures for the following allylic carbocations; then rank the structures in order of relative contributions to each resonance hybrid.

(a) 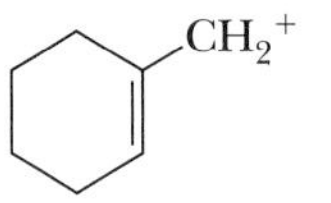

(b) $CH_2{=}CHCH{=}CHCH_2^+$

(c) $CH_3\overset{CH_3}{\underset{+}{C}}CH{=}CH_2$

Electrophilic Addition to Conjugated Dienes

20.16 Predict the structure of the major product formed by 1,2-addition of HCl to 2-methyl-1,3-butadiene (isoprene).

20.17 Predict the major product formed by 1,4-addition of HCl to isoprene.

20.18 Predict the structure of the major 1,2-addition product formed by reaction of one mole of Br_2 with isoprene. Also predict the structure of the major 1,4-addition product formed under these conditions.

20.19 Which of the two molecules shown do you expect to be the major product formed by 1,2-addition of HCl to cyclopentadiene? Explain.

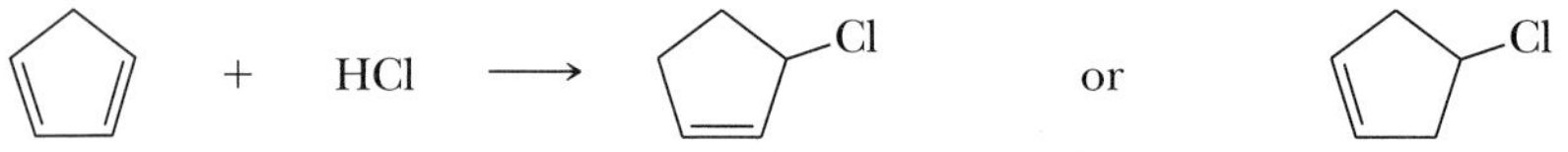

Cyclopentadiene **3-Chlorocyclopentene** (racemic) **4-Chlorocyclopentene** (racemic)

20.20 Predict the major product formed by 1,4-addition of HCl to cyclopentadiene.

20.21 Draw structural formulas for the two constitutional isomers with the molecular formula $C_5H_6Br_2$ formed by adding one mole of Br_2 to cyclopentadiene.

20.22 What are the expected kinetic and thermodynamic products from addition of one mole of Br_2 to the following dienes?

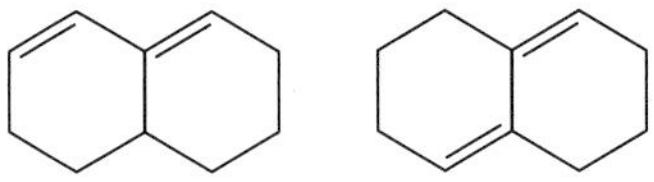

(a) **(b)**

Ultraviolet-Visible Spectra

20.23 Show how to distinguish between 1,3-cyclohexadiene and 1,4-cyclohexadiene by ultraviolet spectroscopy.

1,3-Cyclohexadiene **1,4-Cyclohexadiene**

20.24 Pyridine exhibits a UV transition of the type $n \rightarrow \pi^*$ at 270 nm. In this transition, one of the unshared electrons on nitrogen is promoted from a nonbonding MO to a π^* (antibonding) MO. What is the effect on this UV peak if pyridine is protonated?

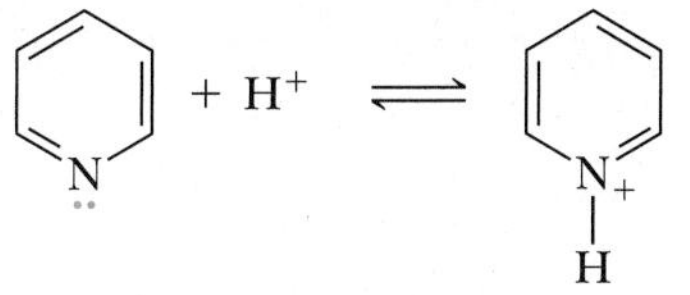

Pyridine **Pyridinium ion**

20.25 The weight of proteins or nucleic acids in solution is commonly determined by UV spectroscopy using the Beer-Lambert law. For example, the ε of double-stranded DNA at 260 nm is 6670 $M^{-1}cm^{-1}$. The formula weight of the repeating unit in DNA (650 Daltons on average) can be used as the molecular weight. What is the weight of DNA in 2.0 mL of aqueous buffer if the absorbance, measured in a 1-cm cuvette, is 0.75?

20.26 A sample of adenosine triphosphate (ATP) (MW 507, ε = 14,700 $M^{-1}cm^{-1}$ at 257 nm) is dissolved in 5.0 mL of buffer. A 250-μL aliquot is removed and placed in a 1 cm cuvette with sufficient buffer to give a total volume of 2.0 mL. The absorbance of the sample at 257 nm is 1.15. Calculate the weight of ATP in the original 5.0 mL sample.

20.27 The following equilibrium was discussed in Section 20.1.

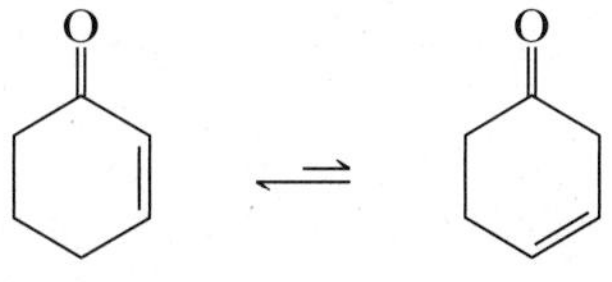

2-Cyclohexenone (more stable) **3-Cyclohexenone** (less stable)

(a) Give a mechanism for this reaction under either acidic or basic conditions.
(b) Explain the position of the equilibrium.

Frontier Molecular Orbital Theory

20.28 Write the frontier molecular orbital analysis for the cycloaddition of butadiene with butadiene when both interact in a suprafacial manner. Is this reaction allowed?

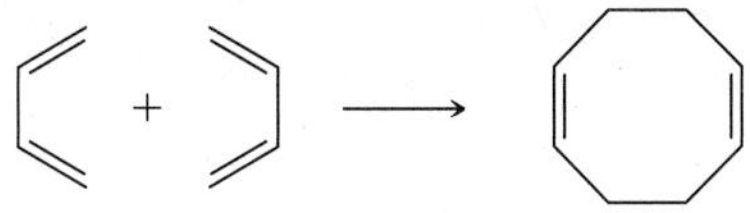

20.29 Write the frontier molecular orbital analysis for a [3,3]-sigmatropic shift in the analogous fashion as presented in the chapter except that you are using a geometry that would lead to a boatlike conformation for the transition state. As a hint, you should find that the reaction is allowed. However, why would this geometry be less favorable?

Diels-Alder Reaction

20.30 Draw structural formulas for the products of reaction of cyclopentadiene with each dienophile.

(a) $CH_2{=}CHCl$ **(b)** $CH_2{=}CH\overset{O}{\overset{\|}{C}}OCH_3$ **(c)** $HC{\equiv}CH$ **(d)** $CH_3O\overset{O}{\overset{\|}{C}}C{\equiv}C\overset{O}{\overset{\|}{C}}OCH_3$

20.31 Propose structural formulas for compounds A and B and specify the configuration of compound B.

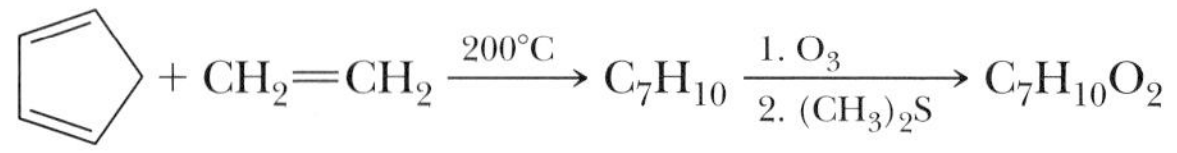
+ $CH_2{=}CH_2$
200°C
C_7H_{10}
1. O_3
2. $(CH_3)_2S$
$C_7H_{10}O_2$

(A) (B)

20.32 Under certain conditions, 1,3-butadiene can function as both a diene and a dienophile. Draw a structural formula for the Diels-Alder adduct formed by reaction of 1,3-butadiene with itself.

20.33 1,3-Butadiene is a gas at room temperature that requires a gas-handling apparatus to use in a Diels-Alder reaction. Butadiene sulfone is a convenient substitute for gaseous 1,3-butadiene. This sulfone is a solid at room temperature (mp 66°C), and when heated above its boiling point of 110°C, it decomposes by a reverse Diels-Alder reaction to give *cis*-1,3-butadiene and sulfur dioxide. Draw Lewis structures for butadiene sulfone and SO_2; then show by curved arrows the path of this reaction, which resembles a reverse Diels-Alder reaction.

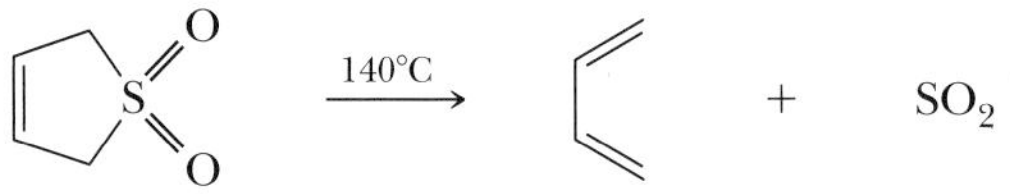
O
S
O
140°C
+
SO_2

Butadiene sulfone **1,3-Butadiene** **Sulfur dioxide**

20.34 The following triene undergoes an intramolecular Diels-Alder reaction to give the product shown. Show how the carbon skeleton of the triene must be coiled to give this product and show by curved arrows the redistribution of electron pairs that takes place to give the product.

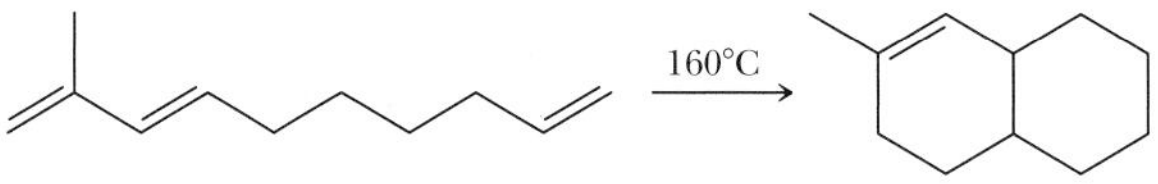
160°C

20.35 The following triene undergoes an intramolecular Diels-Alder reaction to give a bicyclic product. Propose a structural formula for the product. Account for the observation that the Diels-Alder reaction given in this problem takes place under milder conditions (at lower temperature) than the analogous Diels-Alder reaction shown in Problem 20.34.

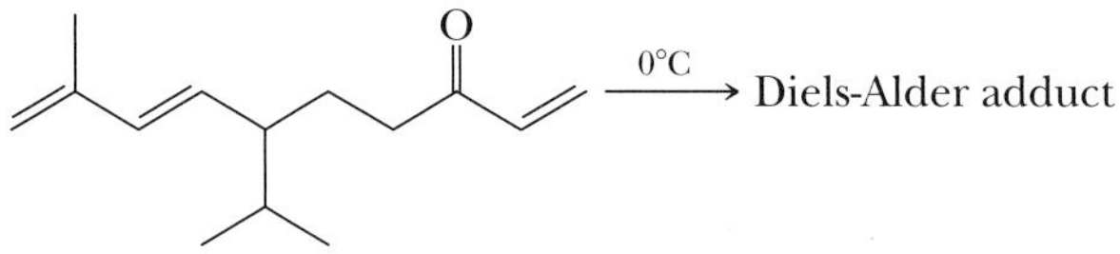
O
0°C
Diels-Alder adduct

20.36 The following compound undergoes an intramolecular Diels-Alder reaction to give a tricyclic product. Propose a structural formula for the product.

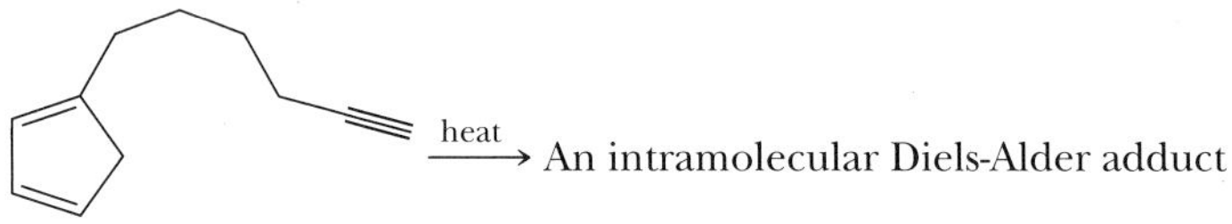
heat
An intramolecular Diels-Alder adduct

20.37 Draw a structural formula for the product of this Diels-Alder reaction, including the stereochemistry of the product.

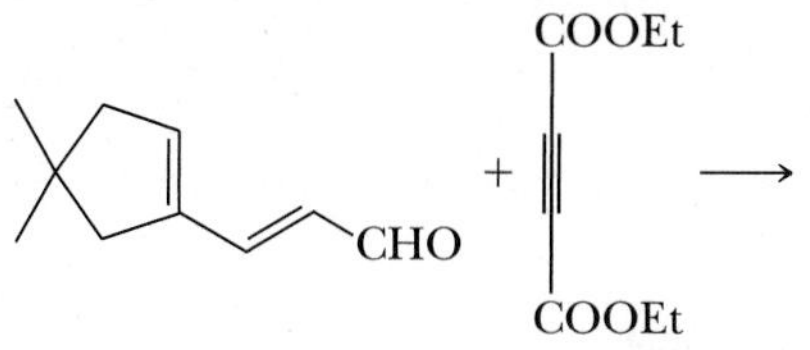

20.38 Following is a retrosynthetic analysis for the dicarboxylic acid shown on the left.

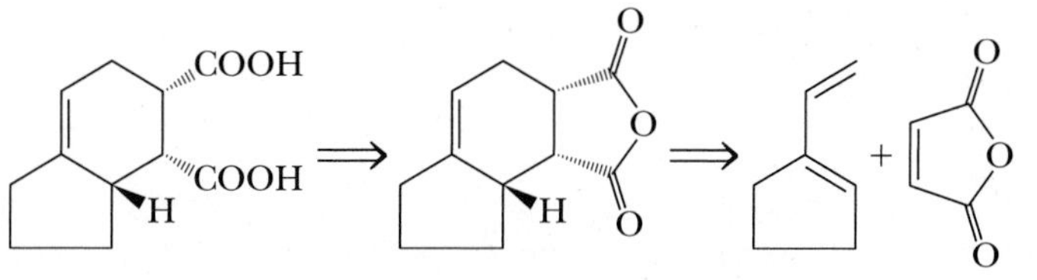

(a) Propose a synthesis of the diene from cyclopentanone and acetylene.
(b) Rationalize the stereochemistry of the target dicarboxylic acid.

20.39 One of the published syntheses of warburganal begins with the following Diels-Alder reaction. Propose a structure for compound **A**.

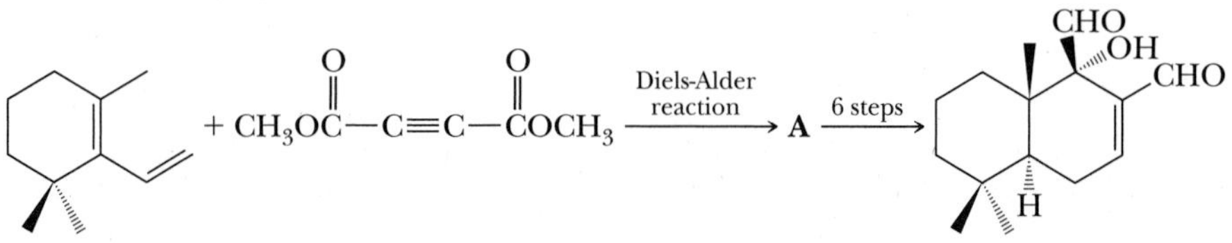

Warburganal
(racemic)

20.40 The Diels-Alder reaction is not limited to making six-membered rings with only carbon atoms. Predict the products of the following reactions that produce rings with atoms other than carbon in them.

(a)

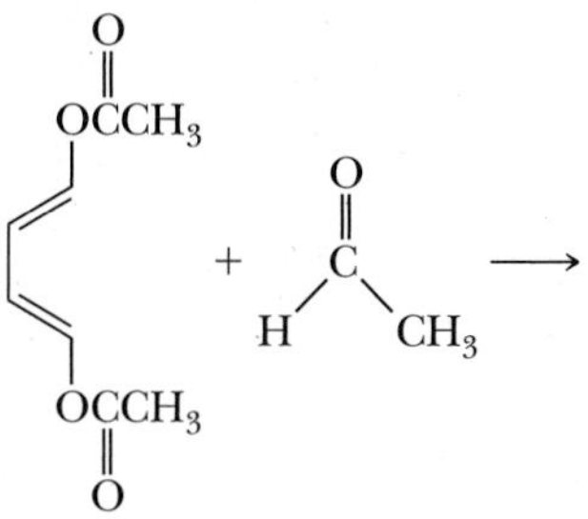

(b)

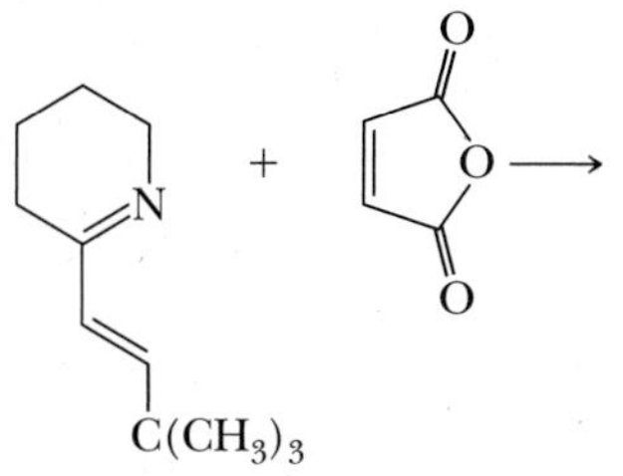

(c)

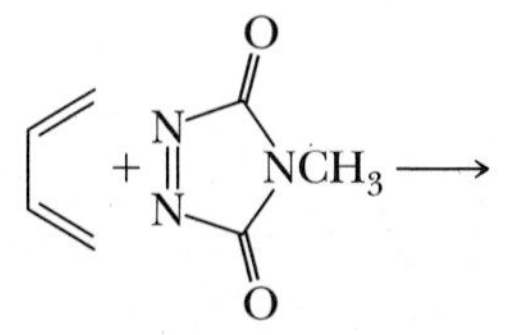

(d)

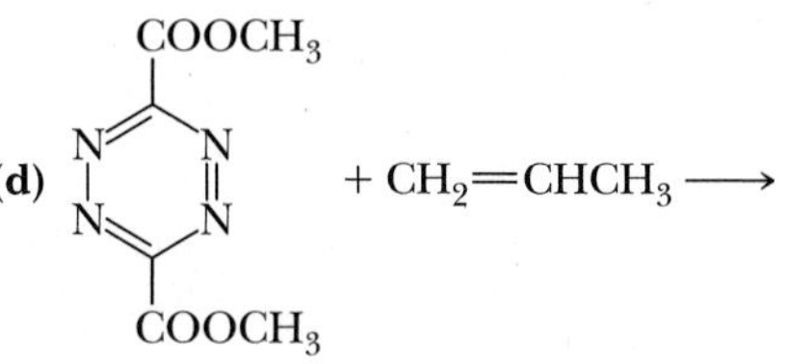

(e)

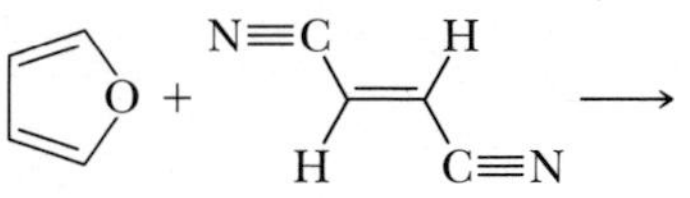

20.41 The first step in a synthesis of dodecahedrane involves a Diels-Alder reaction between the cyclopentadiene derivative (1) and dimethyl acetylenedicarboxylate (2). Show how these two molecules react to form the dodecahedrane synthetic intermediate (3).

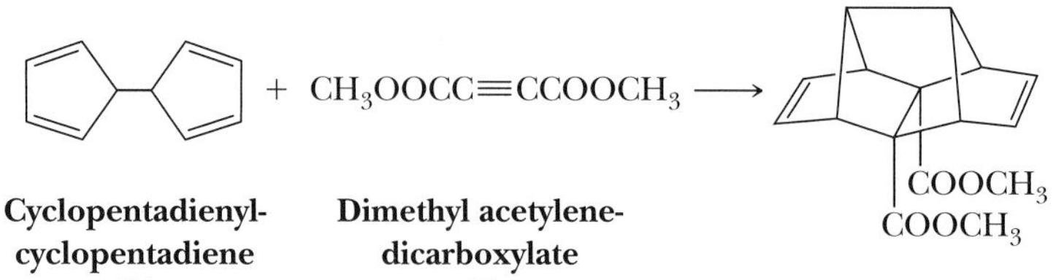

Cyclopentadienyl-cyclopentadiene (1) **Dimethyl acetylene-dicarboxylate** (2) (3)

20.42 Bicyclo-2,5-heptadiene can be prepared in two steps from cyclopentadiene and vinyl chloride. Provide a mechanism for each step.

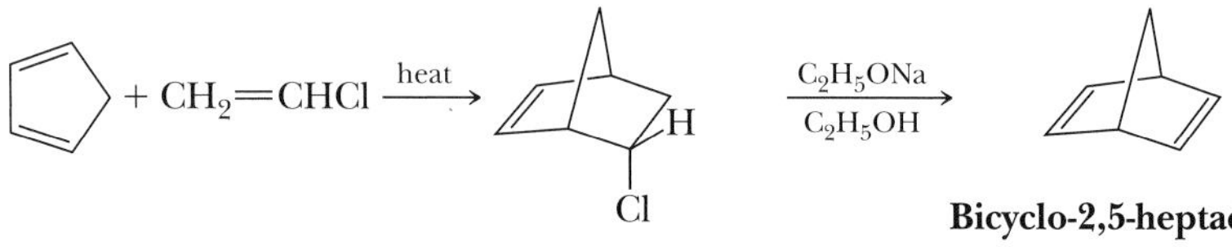

Bicyclo-2,5-heptadiene

20.43 Treatment of anthranilic acid with nitrous acid gives an intermediate, **A**, that contains a diazonium ion and a carboxylate group. When this intermediate is heated in the presence of furan, a tricyclic compound is formed. Propose a structural formula for compound A and a mechanism for the formation of the tricyclic product.

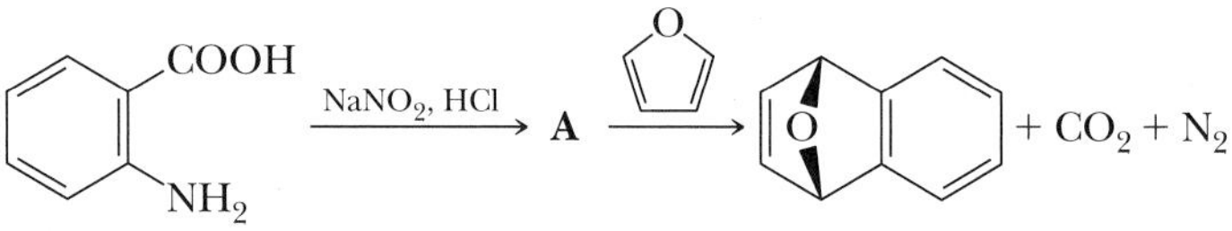

Anthranilic acid

20.44 All attempts to synthesize cyclopentadienone yield only a Diels-Alder adduct. Cycloheptatrienone, however, has been prepared by several methods and is stable. *Hint:* Consider important resonance contributing structures.

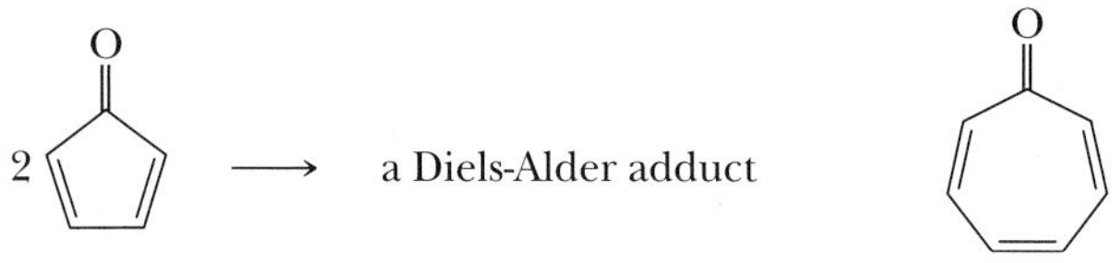

Cyclopentadienone **Cycloheptatrienone**

(a) Draw a structural formula for the Diels-Alder adduct formed by cyclopentadienone.
(b) How do you account for the marked difference in stability of these two ketones?

20.45 Following is a retrosynthetic scheme for the synthesis of the tricyclic diene on the left. Show how to accomplish this synthesis from 2-bromopropane, cyclopentadiene, and 2-cyclohexenone.

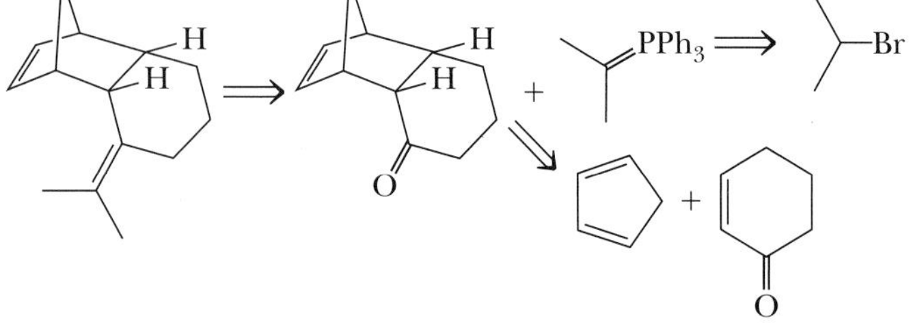

20.46 Show the product of the following reaction. Include stereochemistry.

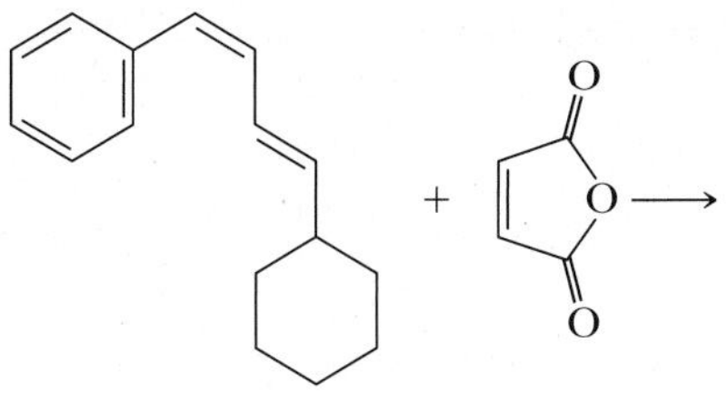

20.47 Following is a synthesis for the antifungal agent tolciclate.

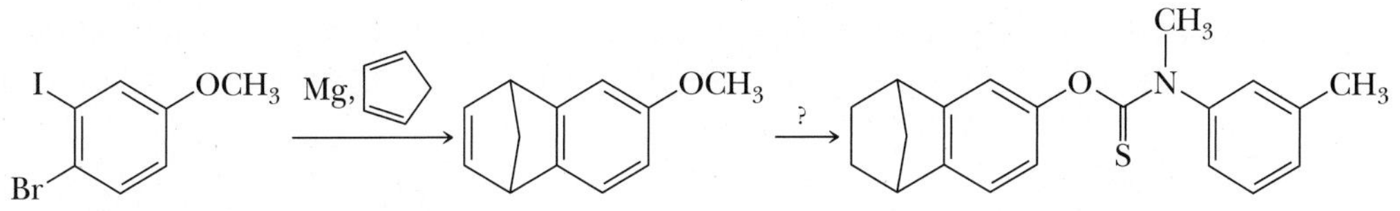

4-Bromo-3-iodoanisole **(A)** **Tolciclate**

(a) Propose a mechanism for formation of **(A)**.

(b) Show how **(A)** can be converted to tolciclate. Use 3-methyl-*N*-methylaniline as the source of the amine nitrogen and thiophosgene, $Cl_2C{=}S$, as the source of the C=S group.

Sigmatropic Shifts

20.48 We showed in Figure 20.10 that a chairlike transition state for a [3,3]-sigmatropic shift is allowed via frontier molecular orbital theory.

(a) Write analogous pictures for a boatlike reaction geometry showing that this is also allowed.

(b) Why are products from this reaction geometry formed to a much lower extent than those that proceed via a chairlike transition state?

20.49 Write the products of the following Cope rearrangements; pay particular attention to the stereochemistry in the products. Predict which is preferred.

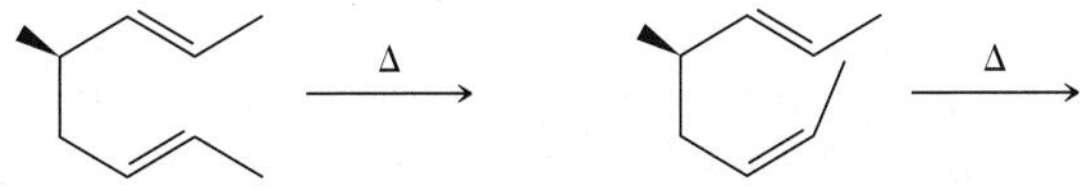

20.50 Predict whether the following reaction will give an achiral product or an equal mixture of two enantiomeric products. Explain your answer by drawing a chairlike transition state geometry for the reaction.

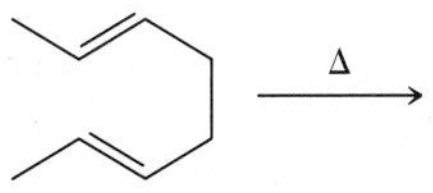

20.51 What reaction presented in this chapter is occurring in the following equation? Explain the resulting stereochemistry of the reaction.

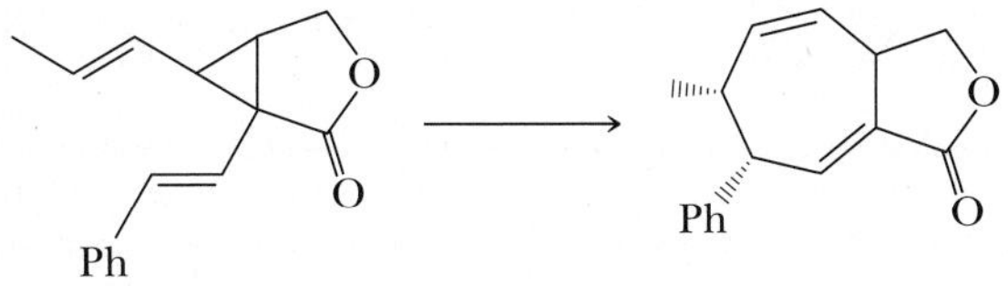

20.52 Claisen rearrangement of an allyl phenyl ether with substituent groups in both ortho positions leads to the formation of a para-substituted product. Propose a mechanism for the following rearrangement.

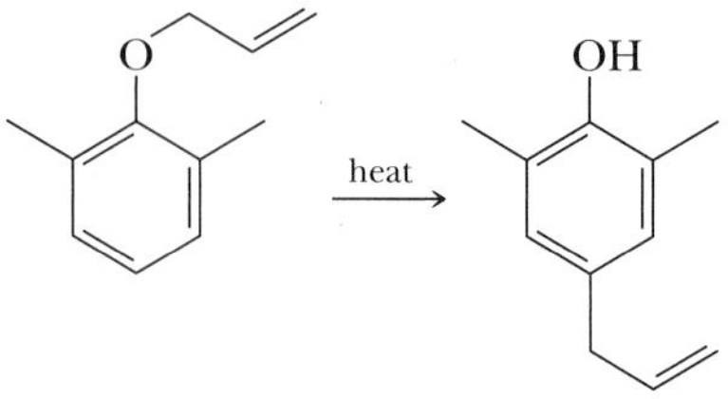

20.53 Following are three examples of Cope rearrangements of 1,5-dienes. Show that each product can be formed in a single step by a mechanism involving redistribution of six electrons in a cyclic transition state.

(a)

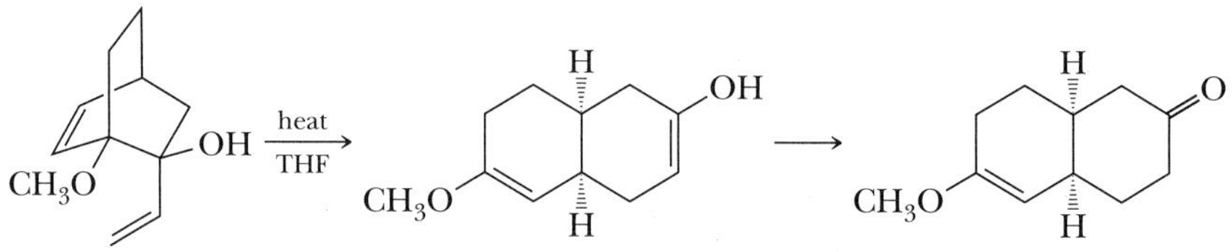

(b)

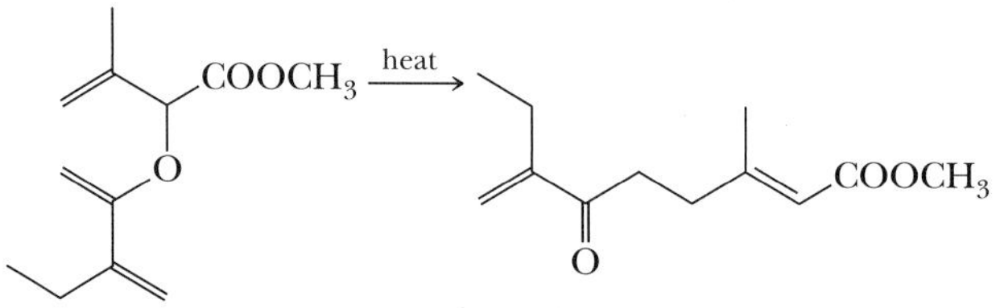

(c)

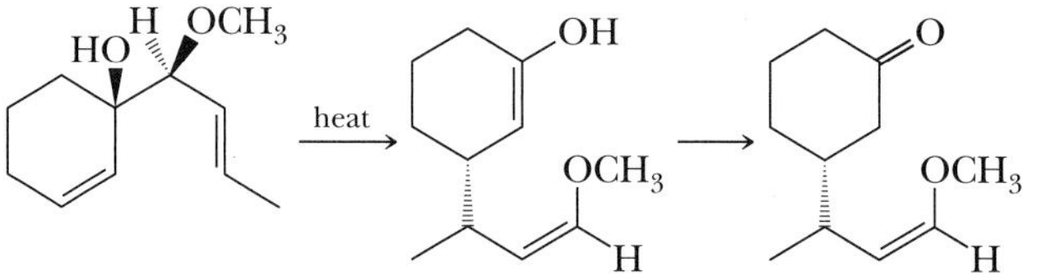

20.54 The following transformation is an example of the Carroll reaction, named after the English chemist M. F. Carroll, who first reported it. Propose a mechanism for this reaction.

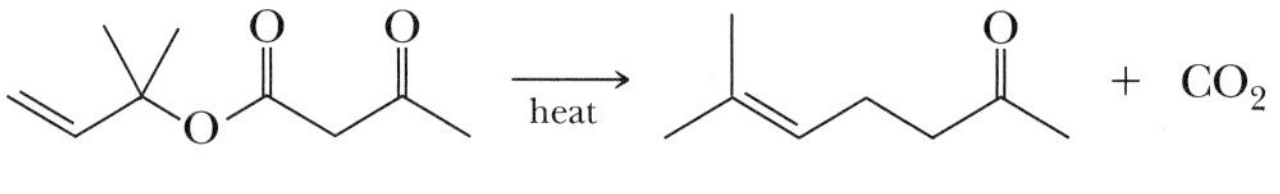

6-Methyl-5-hepten-2-one

Organic Chemistry Reaction Roadmap

Reaction Roadmap

20.55 We now continue the use of organic chemistry reaction roadmaps. Because of the unique nature of the new reactions presented, we recommend that you make a new roadmap only for Chapters 20–23.

To make your own roadmap for Chapters 20–23, take a blank sheet of paper and write the following functional groups in the orientations shown. Fill the entire sheet of paper and leave plenty of room between functional groups. Most students find it helpful to use a poster-sized sheet of paper filled out in landscape orientation.

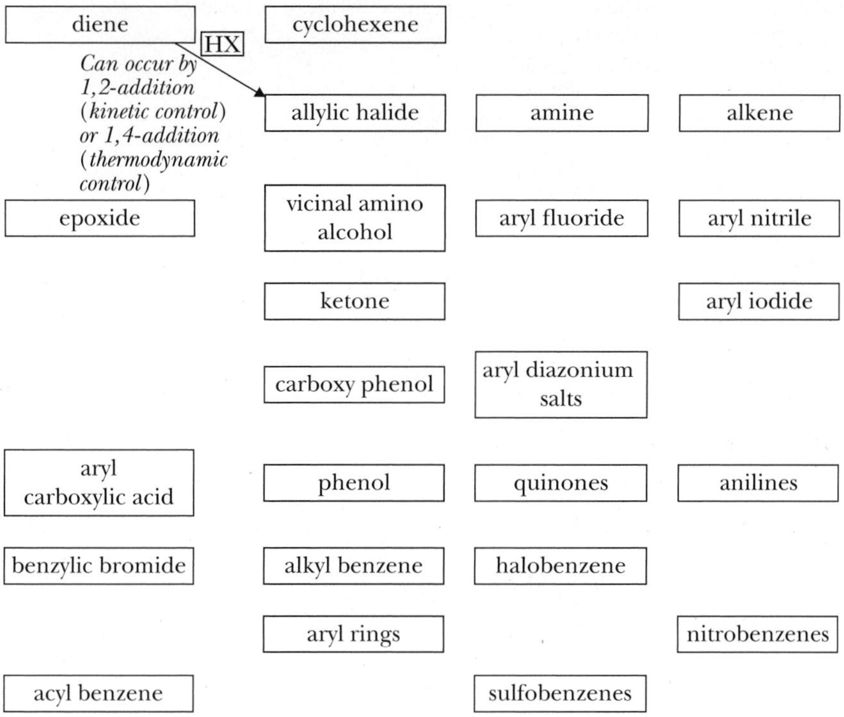

As before, refer to the "Study Guide" section of this chapter. Write the reagents required to bring about each reaction next to the arrows shown. Next, record any regiochemistry or stereochemistry considerations relevant to the reaction. You should also record any key aspects of the mechanism, such as formation of an important intermediate, as a helpful reminder. You may want to keep track of all reactions that make carbon-carbon bonds, because these help you build large molecules from smaller fragments.

On the above organic chemistry roadmap template, the information for the reaction addition of HX to dienes has been added to help you get started. Appendix 10 contains a series of roadmaps for different sections of the book, but you should use those for reference only after you have completed your own.

20.56 Write the products of the following sequences of reactions. Refer to your reaction roadmaps to see how the combined reactions allow you to "navigate" between the different functional groups. Note that you will need your old Chapters 6–11, Chapters 15–18, and Chapter 19 roadmaps along with your new Chapter 20 reaction roadmap for these.

Reaction Roadmap

(a)

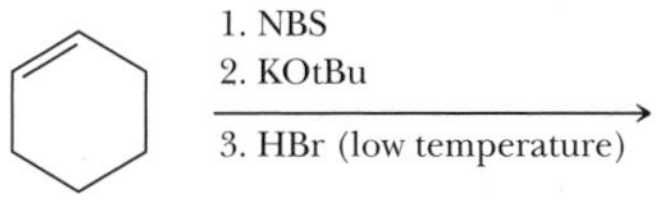

An alkene

(b)

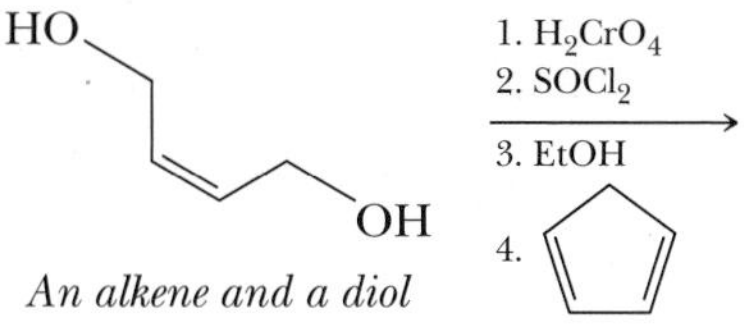

An alkene and a diol

(c)

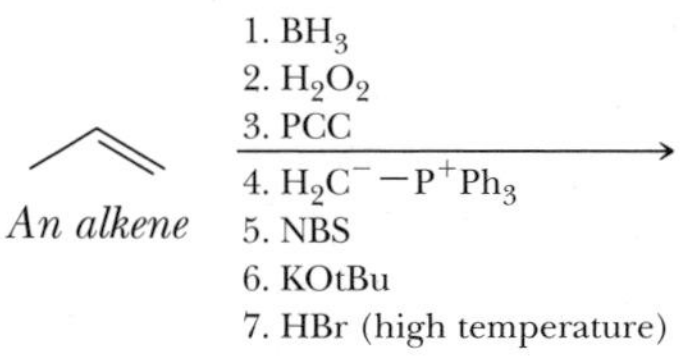

An alkene

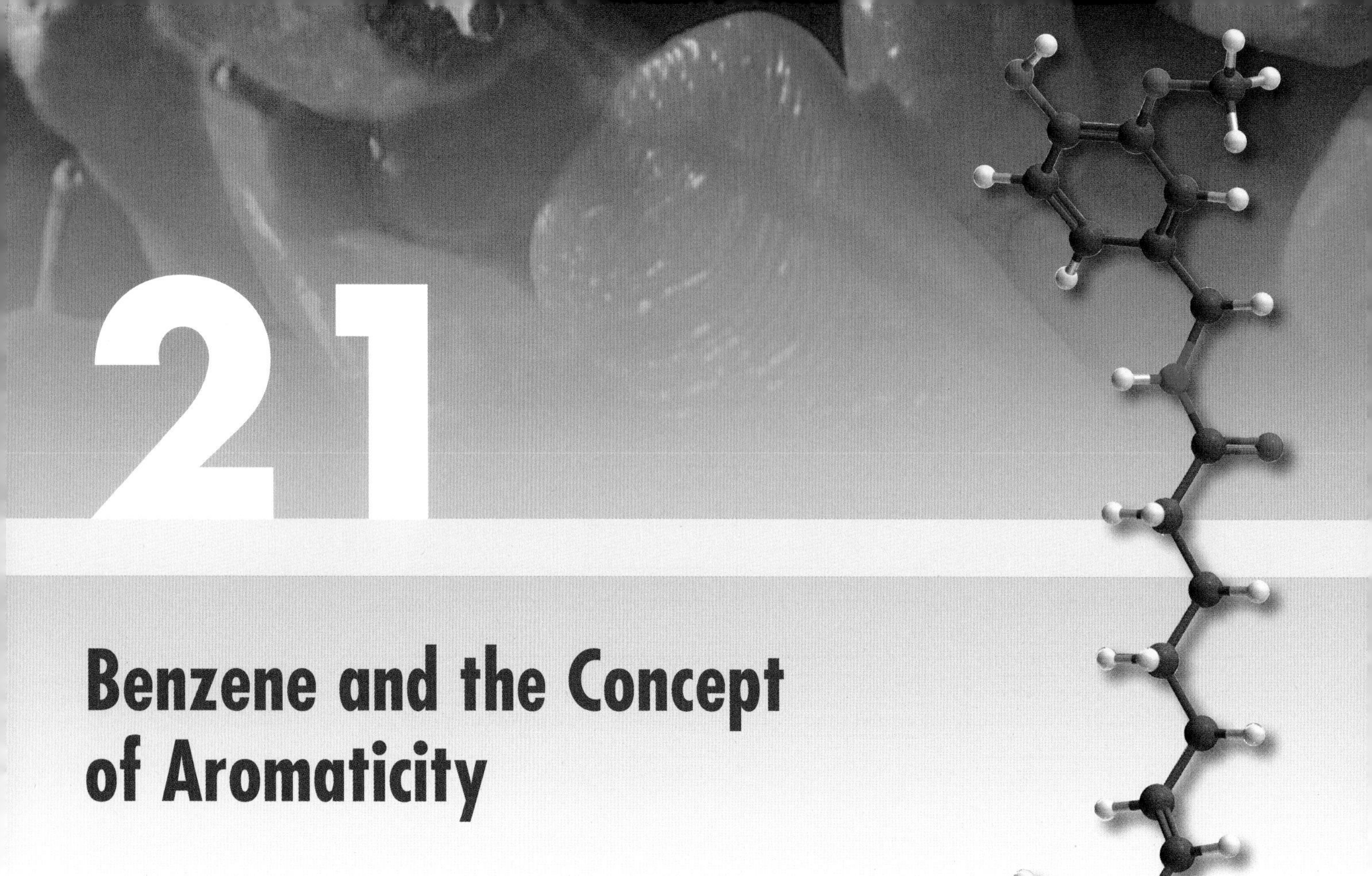

21

Benzene and the Concept of Aromaticity

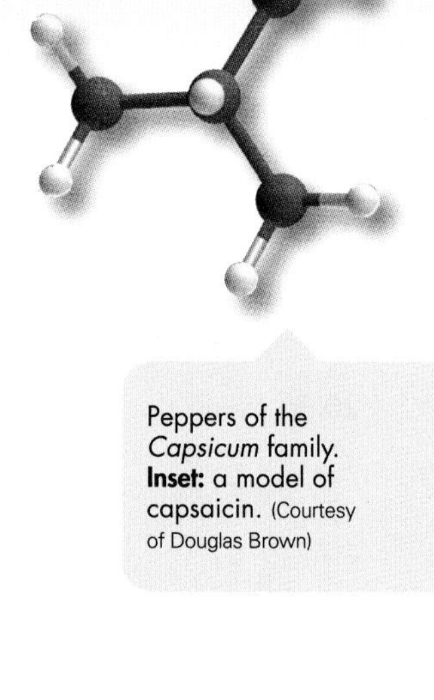

Peppers of the *Capsicum* family. **Inset:** a model of capsaicin. (Courtesy of Douglas Brown)

Outline

21.1 The Structure of Benzene
21.2 The Concept of Aromaticity
HOW TO: Recognize Aromatic Compounds: Criteria and Caveats
21.3 Nomenclature
21.4 Phenols
21.5 Reaction at a Benzylic Position

Benzene, a colorless compound with a melting point of 6°C and a boiling point of 80°C, was first isolated by Michael Faraday in 1825 from the oily residue that collected in the illuminating gas lines of London. Benzene's molecular formula, C_6H_6, suggests a high degree of unsaturation. Compared with the corresponding alkane of molecular formula C_6H_{14}, benzene's index of hydrogen deficiency is four, which can be met by an appropriate combination of rings, double bonds, and triple bonds. For example, a compound of molecular formula C_6H_6 might have four double bonds or three double bonds and one ring or two double bonds and two rings or one triple bond and two rings and so on.

Considering benzene's high degree of unsaturation, it might be expected to show many of the reactions characteristic of alkenes and alkynes. Yet, benzene is remarkably unreactive. It does not undergo the addition, oxidation, and reduction reactions characteristic of alkenes and alkynes. For example, benzene does not react with bromine, hydrogen bromide, or other reagents that usually add to carbon-carbon double and triple bonds. It is not oxidized by chromic acid under conditions that readily oxidize alkenes and alkynes. When benzene reacts, it does so by substitution, in which a hydrogen atom is replaced by another atom or group of atoms.

As noted in Chapter 5, the term **aromatic** was originally used to classify benzene and its derivatives because many of them have distinctive odors. The term *aromatic*, as it is now used, refers instead to the fact that these compounds are highly unsaturated and unexpectedly stable toward reagents that attack alkenes and alkynes. The term **arene** is used to describe aromatic hydrocarbons, by analogy with alkane, alkene, and alkyne. Benzene is the parent arene. Just as a group derived by removal of an H from an alkane is called an alkyl group and given the symbol R–, a group derived by removal of an H from an arene is called an **aryl group** and given the symbol Ar–.

Aromatic compound
A term used initially to classify benzene and its derivatives. More accurately, it is used to classify any compound that meets the Hückel criteria for aromaticity (Section 21.2A).

21.1 The Structure of Benzene

Let us put ourselves in the mid-nineteenth century and examine the evidence on which chemists attempted to build a model for the structure of benzene. First, because the molecular formula of benzene is C_6H_6, it seemed clear that the molecule must be highly unsaturated. Yet, benzene does not show the chemical properties of alkenes, the only unsaturated hydrocarbons known at that time. Benzene does undergo chemical reactions, but its characteristic reaction is substitution rather than addition. When benzene is treated with bromine in the presence of ferric chloride, for example, only one compound with the molecular formula C_6H_5Br is formed.

$$\underset{\textbf{Benzene}}{C_6H_6} + Br_2 \xrightarrow{FeCl_3} \underset{\textbf{Bromobenzene}}{C_6H_5Br} + HBr$$

Chemists concluded, therefore, that all six hydrogens of benzene must be equivalent. When bromobenzene is treated with bromine in the presence of ferric chloride as a catalyst, three isomeric dibromobenzenes are formed.

$$\underset{\textbf{Bromobenzene}}{C_6H_5Br} + Br_2 \xrightarrow{FeCl_3} \underset{\text{Three isomeric dibromobenzenes}}{C_6H_4Br_2} + HBr$$

For chemists in the mid-nineteenth century, the problem was to incorporate these observations, along with the accepted tetravalence of carbon, into a structural formula for benzene. Before we examine these proposals, we should note that the problem of the structure of benzene and other aromatic hydrocarbons has occupied the efforts of chemists for over a century. Only since the 1930s has a general understanding of this problem been realized.

A. Kekulé's Model of Benzene

The first structure for benzene was proposed by August Kekulé in 1865 and consisted of a six-membered ring with one hydrogen bonded to each carbon. Although Kekulé's original structural formula provided for the equivalency of the C—H and C—C bonds, it was inadequate because all the carbon atoms were trivalent. To maintain the tetravalence of carbon, Kekulé proposed in 1872 that the ring contains three double bonds that shift back and forth so rapidly that the two forms cannot be separated. Kekulé regarded this interconversion as an equilibrium, each structure in which has become known as a **Kekulé structure** (Figure 21.1).

Now, more than 135 years after the time of Kekulé, we are apt to misunderstand what scientists in his time did and did not know. For example, it is a given to us that covalent bonds consist of one or more pairs of shared electrons. We must remember, however, that it was not until 1897 that J. J. Thomson, professor of physics at the Cavendish Laboratory of Cambridge University, discovered the electron. Thomson was awarded the 1906 Nobel Prize in Physics. The fact that the electron played any

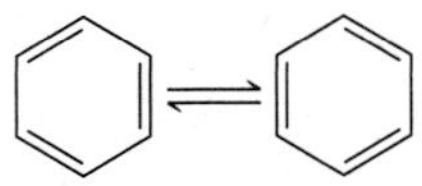

FIGURE 21.1 Kekulé structures for benzene.

role in chemical bonding did not become clear for another 30 years. Thus, at the time Kekulé made his proposal for the structure of benzene, the existence of electrons and their role in chemical bonding was completely unknown.

Kekulé's proposal accounted nicely for the fact that bromination of benzene gives only one bromobenzene and that bromination of bromobenzene gives three isomeric dibromobenzenes.

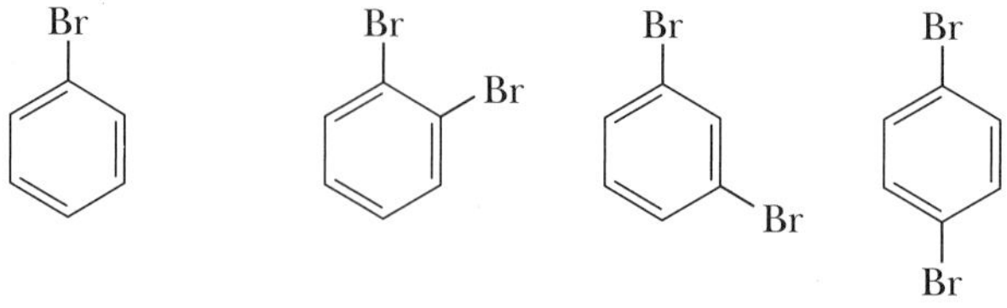

Bromobenzene Three isomeric dibromobenzenes

Although his proposal was consistent with many experimental observations, it did not totally solve the problem and was contested for years. The major objection was that it did not account for the unusual chemical behavior of benzene. If benzene contains three double bonds, Kekulé's critics argued, why doesn't it show reactions typical of alkenes? Why, for example, doesn't benzene add three moles of bromine to form 1,2,3,4,5,6-hexabromocyclohexane? We now understand the surprising unreactivity of benzene on the basis of two complementary descriptions, the molecular orbital model and the resonance model.

B. The Molecular Orbital Model of Benzene

The carbon skeleton of benzene forms a regular hexagon with C—C—C and H—C—C bond angles of 120°. For this type of bonding, carbon uses sp^2 hybrid orbitals. Each carbon forms σ bonds to two adjacent carbons by overlap of sp^2-sp^2 hybrid orbitals and one σ bond to hydrogen by overlap of sp^2-1s orbitals. As determined experimentally, all carbon-carbon bonds are 139 pm in length, a value almost midway between the length of a single bond between sp^3 hybridized carbons (154 pm) and a double bond between sp^2 hybridized carbons (133 pm).

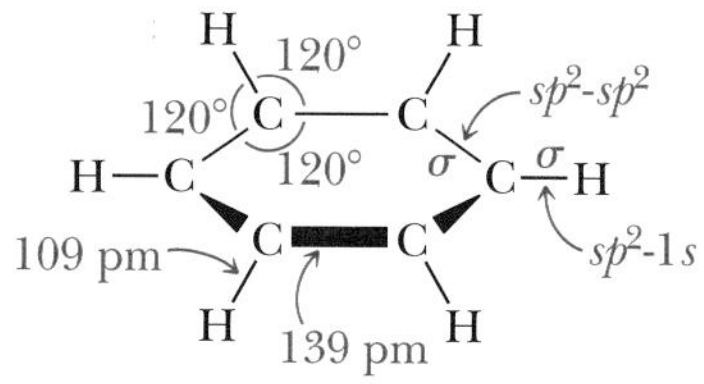

Each carbon also has a single unhybridized 2p orbital that is perpendicular to the plane of the ring and contains one electron. According to molecular orbital theory, the combination of these six parallel 2p atomic orbitals gives a set of six π MOs, three π-bonding MOs, and three π-antibonding MOs. Figure 21.2 shows these six molecular orbitals and their relative energies. Note that π_2 and π_3 MOs are degenerate (they have the same energy) bonding orbitals. Similarly, $\pi_4{}^*$ and $\pi_5{}^*$ are a degenerate pair of π-antibonding MOs.

In the ground-state electron configuration of benzene, the six electrons of the π system occupy the three bonding MOs (Figure 21.2). According to molecular orbital calculations, the great stability of benzene results from the fact that these three bonding MOs are much lower in energy when compared with the six uncombined 2p atomic orbitals.

The π orbitals of benzene are shown in Figure 21.3. It is common to represent the π system of benzene as one torus (a doughnut-shaped region) above the plane of the ring and a second torus below it, shown in Figure 21.3 as π_1.

This picture is useful because it emphasizes the delocalization of electron density of the π system and the equivalence among all six carbon atoms. However, this is not

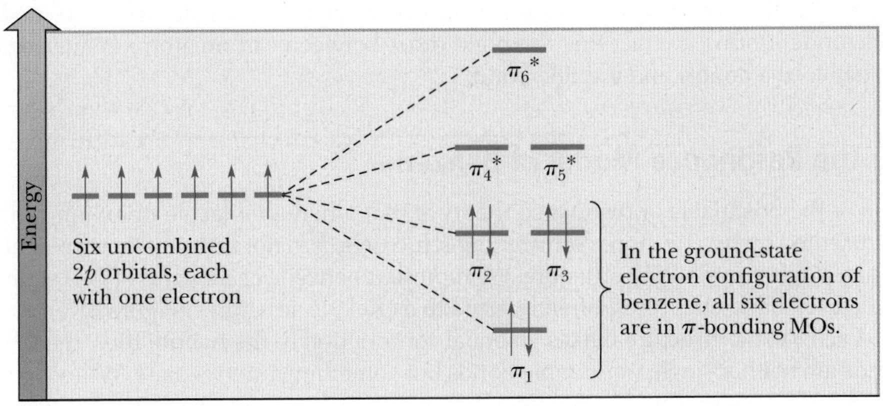

FIGURE 21.2 The molecular orbital representation of the π bonding in benzene.

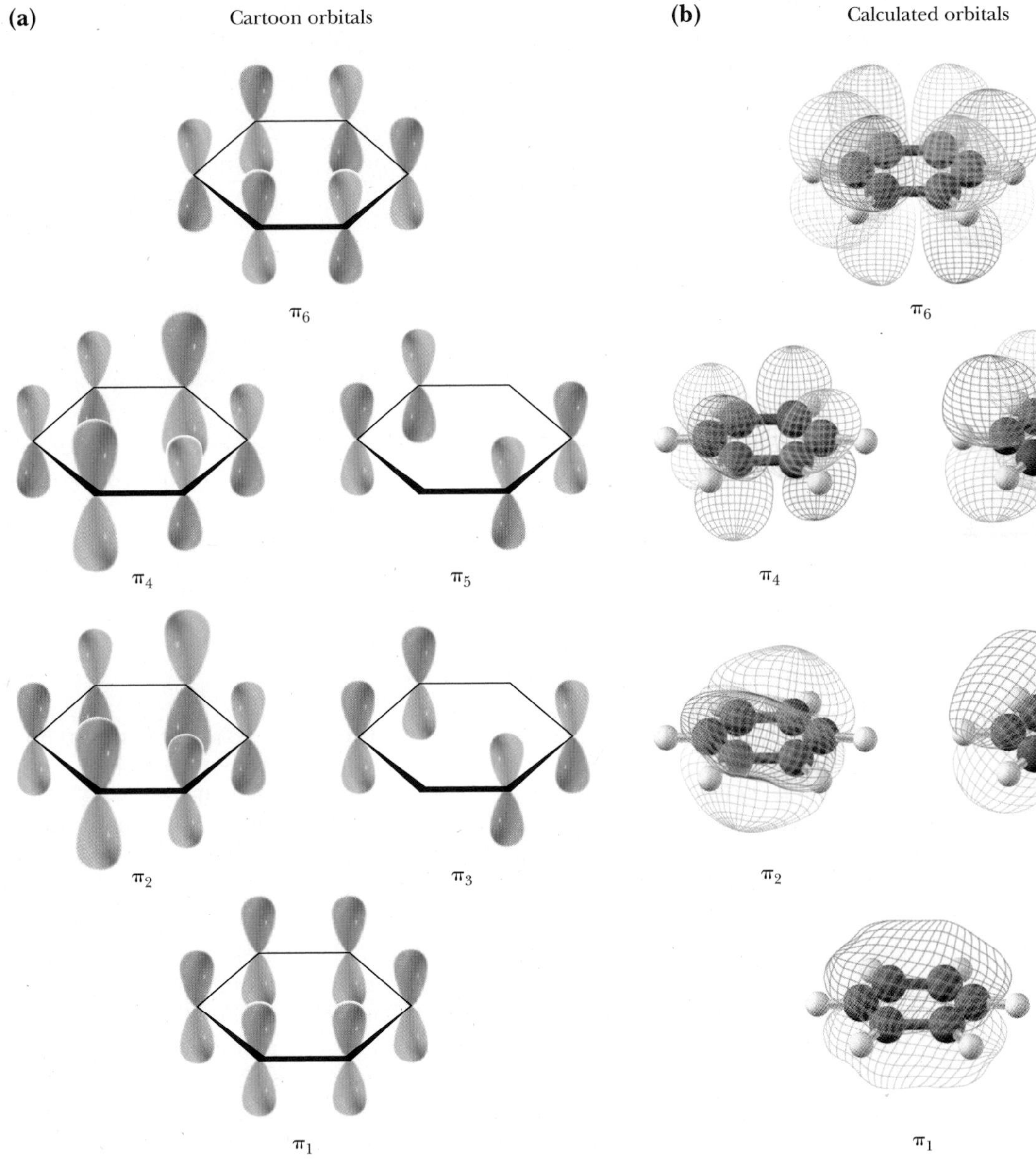

FIGURE 21.3 Orbitals for the π system of benzene. (a) Cartoon representations of the six calculated orbitals that chemists routinely draw. These pictures accentuate the fact that various combinations of parallel 2p orbitals lead to the π system of benzene. (b) Calculated orbitals. The three lowest in energy are occupied with electrons (see Figure 21.2). The lowest of these orbitals is the image most chemists use for the π system of benzene: a torus of electron density above and below the ring.

the whole story. The other two filled molecular orbitals (π_2 and π_3) have two nodes each, underscoring the fact that the bond order between carbon atoms is intermediate between a double and a single bond.

C. The Resonance Model of Benzene

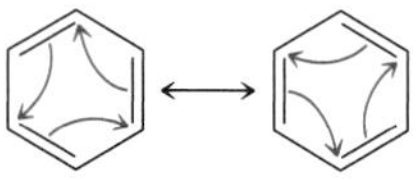

FIGURE 21.4 Benzene as a hybrid of two equivalent contributing structures.

One of the postulates of resonance theory is that when a molecule or an ion can be represented by two or more contributing structures, it is not adequately represented by any single contributing structure. We represent benzene as a hybrid of two equivalent contributing structures, often referred to as Kekulé structures (Figure 21.4).

Each Kekulé structure makes an equal contribution to the hybrid; thus, the C—C bonds are neither single nor double bonds but something intermediate. We recognize that neither of these contributing structures exists (they are merely alternative ways to pair 2*p* orbitals with no reason to prefer one or the other) and that the actual structure is a superposition of both. Nevertheless, chemists continue to use a single contributing structure to represent this molecule because it is as close as they can come to an accurate structure within the limitations of classical valence bond structures and the tetravalence of carbon.

Resonance energy
The difference in energy between a resonance hybrid and the most stable of its hypothetical contributing structures in which electrons are localized on particular atoms and in particular bonds.

One way to estimate the **resonance energy** of benzene is to compare the heats of hydrogenation of cyclohexene and benzene. Cyclohexene is readily reduced to cyclohexane by hydrogen in the presence of a transition metal catalyst (Section 6.6A).

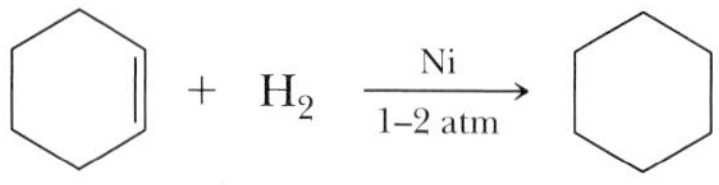

+ H2
Ni
1–2 atm

$\Delta H^0 = -119.7$ kJ (-28.6 kcal)/mol

Cyclohexene **Cyclohexane**

▶ Delocalization of π electron density over a large area is stabilizing.

Benzene is reduced very slowly under these conditions to cyclohexane. It is reduced more rapidly when heated and under a pressure of several hundred atmospheres of hydrogen. ▶

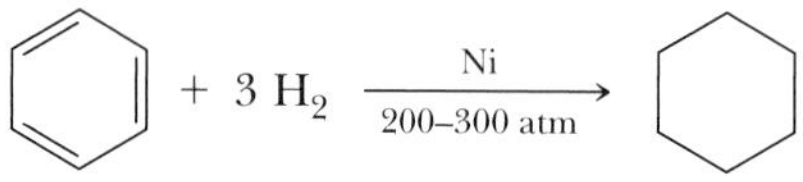

+ 3 H2
Ni
200–300 atm

$\Delta H^0 = -208$ kJ (-49.8 kcal)/mol

Benzene **Cyclohexane**

Catalytic hydrogenation of an alkene is an exothermic reaction (Section 6.6B). The heat of hydrogenation per double bond varies somewhat with the degree of substitution of the particular alkene; for cyclohexene, $\Delta H^0 = -119.7$ kJ (-28.6 kcal)/mol. If we consider benzene to be 1,3,5-cyclohexatriene, a hypothetical unsaturated compound with alternating single and double bonds, we calculate that $\Delta H^0 = 3(-119.7 \text{ kJ/mol}) = -359$ kJ (-85.8 kcal)/mol. The ΔH^0 for reduction of benzene to cyclohexane is -208 kJ (-49.8 kcal)/mol, considerably less than that calculated for 1,3,5-cyclohexatriene. The difference between these values, 151 kJ (36.0 kcal)/mol, is the **resonance energy of benzene**. Note that the product of both reductions is cyclohexane and that both reductions are exothermic. Therefore, the lower heat of hydrogenation for benzene confirms that it is more stable than 1,3,5-cyclohexatriene. These experimental results are shown graphically in Figure 21.5.

Several other experimental determinations of the resonance energy of benzene have been performed using different model compounds, and although these determinations differ somewhat in their results, they all agree that the resonance stabilization of benzene is large. Following are resonance energies for several other aromatic hydrocarbons.

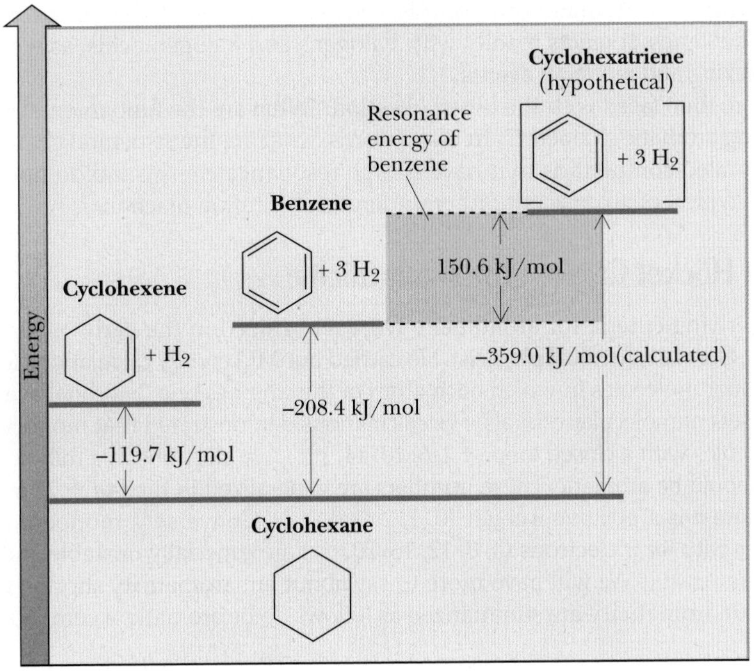

FIGURE 21.5 The resonance energy of benzene as determined by comparison of the heats of hydrogenation of cyclohexene, benzene, and the hypothetical compound 1,3,5-cyclohexatriene.

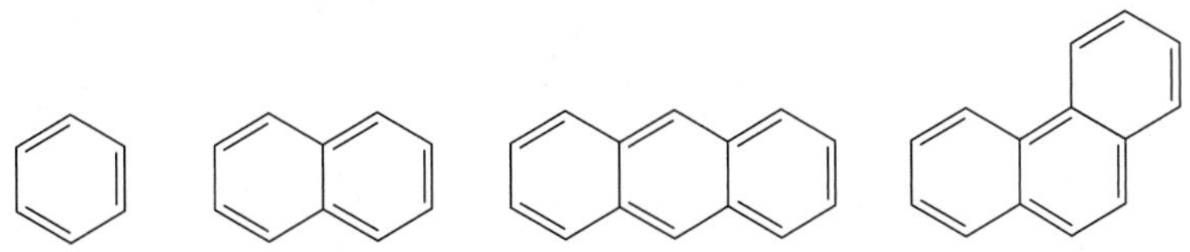

	Benzene	Naphthalene	Anthracene	Phenanthrene
Resonance energy [kJ (kcal)/mol]	151 (36)	255 (61)	347 (83)	381 (91)

21.2 The Concept of Aromaticity

The molecular orbital and resonance theories are powerful tools with which chemists can understand the unusual stability of benzene and its derivatives. According to resonance theory, benzene is best represented as a hybrid of two equivalent contributing structures. By analogy, cyclobutadiene and cyclooctatetraene can also be represented as hybrids of two equivalent contributing structures. Is either of these compounds aromatic?

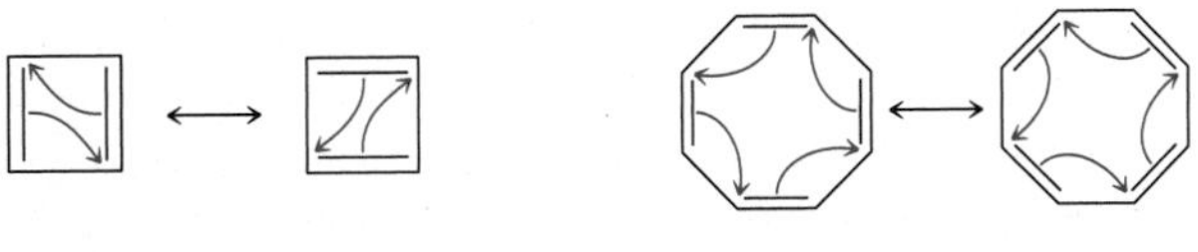

Cyclobutadiene as a hybrid of two equivalent contributing structures

Cyclooctatetraene as a hybrid of two equivalent contributing structures

The answer for both compounds is no. Repeated attempts to isolate cyclobutadiene have all failed. It was not until 1965 that it was finally synthesized, and even then, it could only be detected if trapped at 4K (−269°C). Cyclobutadiene is a highly unstable compound and does not show any of the chemical and physical properties we associate with aromatic compounds. Cyclooctatetraene has chemical properties typical of alkenes. It reacts readily with halogens and halogen acids, as well as with mild oxidizing and reducing agents.

We are then faced with the broad question: "What are the fundamental principles underlying aromatic character?" In other words, what are the structural characteristics of unsaturated compounds that have a large resonance energy and do not undergo reactions typical of alkenes but rather undergo substitution reactions?

Watch a video explanation

A. The Hückel Criteria for Aromaticity

The underlying criteria for aromaticity were recognized in the early 1930s by Erich Hückel, a German chemical physicist. He carried out **MO** energy calculations for monocyclic, planar molecules in which each atom of the ring has one $2p$ orbital available for forming sets of molecular orbitals. His calculations demonstrated that monocyclic, planar molecules with a closed loop of 2, 6, 10, 14, 18, . . . π electrons in a fully conjugated system should be aromatic. These numbers are generalized in the **$(4n + 2)$ π electron rule**, where n is a positive integer (0, 1, 2, 3, 4, . . .). Conversely, monocyclic, planar molecules with $4n$ π electrons (4, 8, 12, 16, 20, . . .) are especially unstable and are said to be antiaromatic. We will have more to say about antiaromaticity shortly. **Hückel's criteria for aromaticity** are summarized as follows. To be aromatic, a compound must:

Hückel criteria for aromaticity

To be aromatic, a monocyclic compound must have one $2p$ orbital on each atom of the ring, be planar or nearly so, and have $(4n + 2)$ π electrons in the cyclic arrangement of $2p$ orbitals.

1. Be cyclic.
2. Have one $2p$ orbital on each atom of the ring.
3. Be planar or nearly planar so that there is continuous or nearly continuous overlap of all $2p$ orbitals of the ring.
4. Have a closed loop of $(4n + 2)$ π electrons in the cyclic arrangement of $2p$ orbitals.

To appreciate the reasons for aromaticity and antiaromaticity, we must examine **MO** energy diagrams for the molecules and ions we will consider in this and the following section. The relative energies of the π MOs for planar, monocyclic, fully conjugated systems can be constructed quite easily using the **Frost circle**, or inscribed polygon method. To construct such a diagram, draw a circle and then inscribe in it a polygon of the same number of sides as the ring in question. Inscribe the polygon in such a way that one of its vertices is at the bottom of the circle. The relative energies of the MOs in the ring are then given by the points where the vertices touch the circle. Those MOs below the horizontal line through the center of the circle are bonding MOs. Those on the horizontal line are nonbonding MOs, and those above the line are antibonding MOs.

Frost circle

A graphic method for determining the relative energies of π MOs for planar, fully conjugated, monocyclic compounds.

Figure 21.6 shows Frost circles describing the MOs of monocyclic, planar, and fully conjugated four-, five-, and six-membered rings. This apparently coincidental method works because it reproduces geometrically the mathematical solutions to the wave equation.

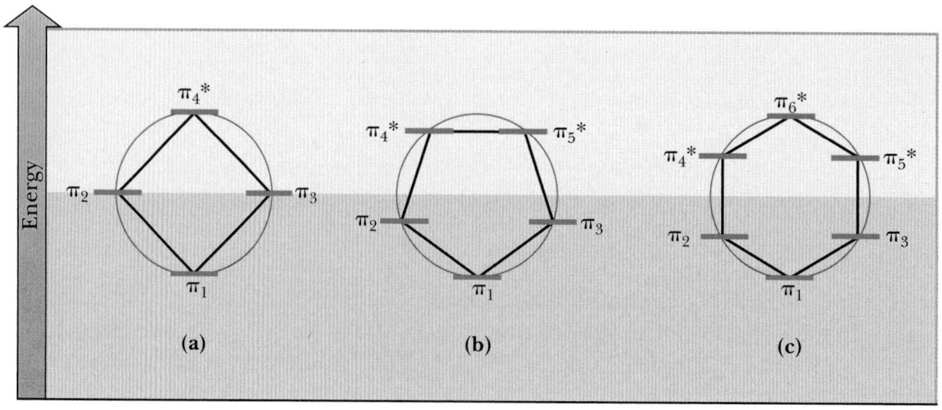

FIGURE 21.6 Frost circles showing the number and relative energies of the π MOs for planar, fully conjugated four-, five-, and six-membered rings.

Example 21.1 Frost Circles

Construct a Frost circle for a planar seven-membered ring with one 2*p* orbital on each atom of the ring and show the relative energies of its seven π molecular orbitals. Which are bonding **MOs**, which are antibonding, and which are nonbonding?

Solution

Of the seven π molecular orbitals, three are bonding and four are antibonding.

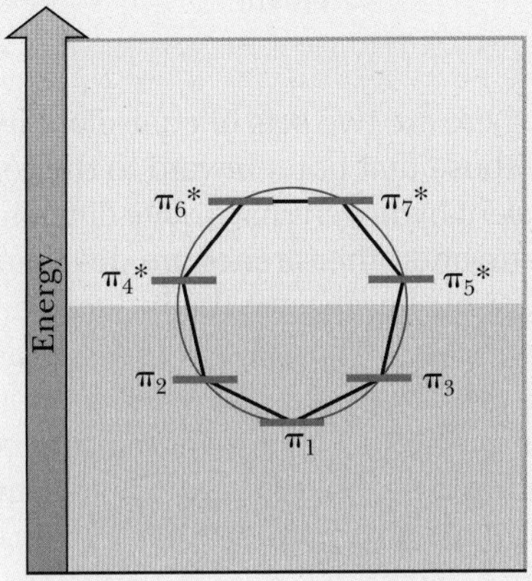

Problem 21.1

Construct a Frost circle for a planar eight-membered ring with one 2*p* orbital on each atom of the ring and show the relative energies of its eight π molecular orbitals. Which are bonding **MOs**, which are antibonding, and which are nonbonding?

B. Aromatic Hydrocarbons

Cyclobutadiene, benzene, and cyclooctatetraene are the first members of a family of molecules called annulenes. An **annulene** is a cyclic hydrocarbon with a continuous alternation of single and double bonds. The name of an annulene is derived by showing the number of atoms in the ring in brackets followed by the word *annulene*. Named as annulenes, cyclobutadiene, benzene, and cyclooctatetraene are [4]annulene, [6]annulene, and [8]annulene, respectively. These compounds, however, are rarely named as annulenes.

Annulene
A cyclic hydrocarbon with a continuous alternation of single and double bonds.

Beginning in the 1960s, Franz Sondheimer and his colleagues, first at the Weizmann Institute in Israel and later at the University of London, synthesized a number of larger annulenes, primarily to test the validity of Hückel's criteria for aromaticity. They found, for example, that both [14]annulene and [18]annulene are aromatic, as predicted by Hückel. [18]Annulene has a resonance energy of approximately 418 kJ (100 kcal)/mol. Notice that for these annulenes to achieve planarity, several of the carbon-carbon double bonds in each must have the *trans* configuration.

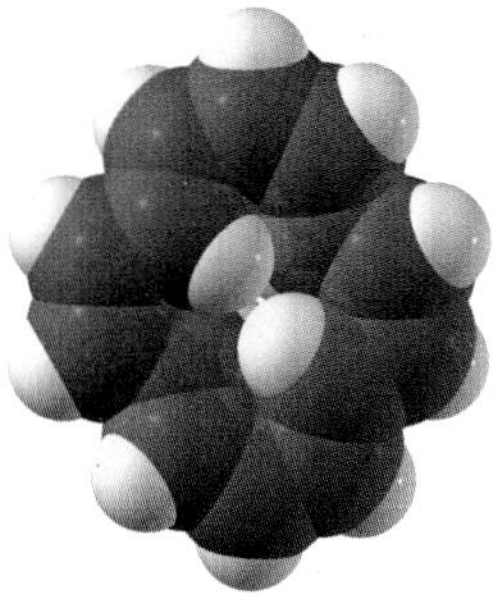

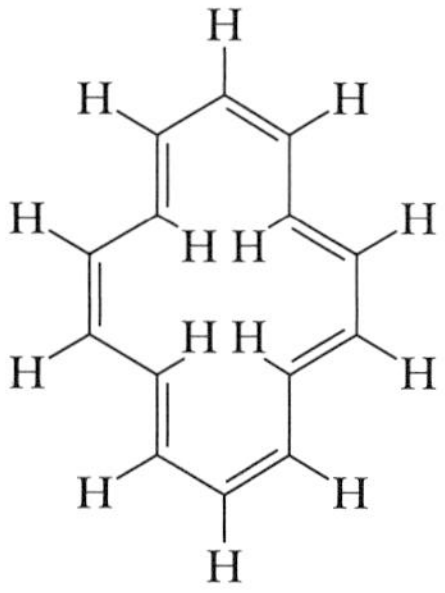

[14]Annulene
(aromatic)

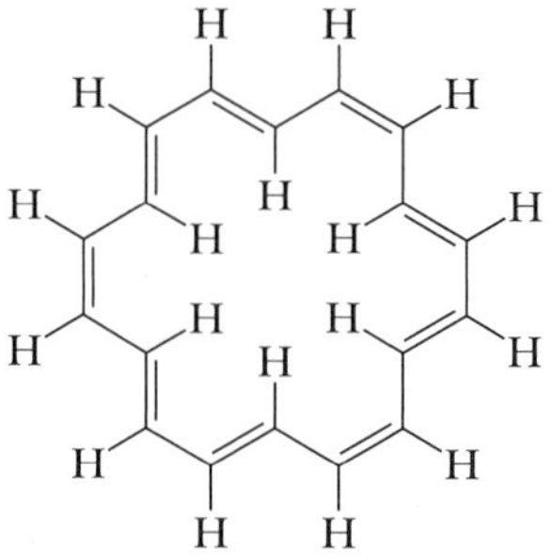

[18]Annulene
(aromatic)

In these larger annulenes, there are two sets of equivalent hydrogens: those that point outward from the ring and those that point inward to the center of the ring. The fact is that these two sets of equivalent hydrogens have quite different ^{1}H-NMR chemical shifts.

The protons on benzene and other arenes are deshielded and appear far downfield (usually around 7–8 ppm) because of the induced ring current that occurs in aromatic molecules (Section 13.7C). The effect of induced ring current is characteristic not only of benzene and its derivatives but also of all compounds that meet the Hückel criteria for aromaticity. This concept of a circulating ring current and of an induced magnetic field predicts that hydrogen atoms outside the ring should come into resonance with a downfield shift. It also predicts that a hydrogen atom inside the ring should come into resonance farther upfield. Of course, no hydrogens are inside the benzene ring, but with larger aromatic annulenes (as, e.g. [18]annulene), there are both "inside" hydrogens and "outside" hydrogens. The degree of the upfield chemical shift of the inside hydrogens of [18]annulene is remarkable. They come into resonance at δ −3.00 [i.e., at 3.00 δ units upfield (to the right) of the TMS standard].

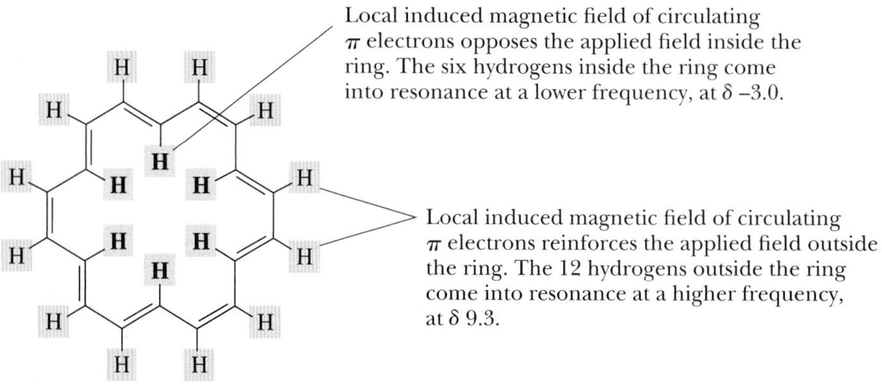
Local induced magnetic field of circulating π electrons opposes the applied field inside the ring. The six hydrogens inside the ring come into resonance at a lower frequency, at δ −3.0.
Local induced magnetic field of circulating π electrons reinforces the applied field outside the ring. The 12 hydrogens outside the ring come into resonance at a higher frequency, at δ 9.3.

[18]Annulene
(aromatic)

Example 21.2 Chemical Shifts

Which hydrogens have a larger chemical shift, the six hydrogens of benzene or the eight hydrogens of cyclooctatetraene? Explain.

Solution

Benzene is an aromatic compound; its six equivalent hydrogens appear as a sharp singlet at δ 7.27. Cyclooctatetraene does not meet the Hückel criteria for aromaticity because it has $4n$ π electrons and is nonplanar. Therefore, the eight equivalent hydrogens of the cyclooctatetraene ring appear as a singlet at δ 5.8 in the region of vinylic hydrogens (δ 4.6–δ 5.7).

Problem 21.2

Which compound gives a signal in the ^{1}H-NMR spectrum with a larger chemical shift, furan or cyclopentadiene? Explain.

According to Hückel's criteria, [10]annulene should be aromatic; it is cyclic, has one $2p$ orbital on each carbon of the ring, and has $4(2) + 2 = 10$ electrons in its π system. It has been found, however, that this molecule shows reactions typical of alkenes and therefore is classified as nonaromatic. The reason for its lack of aromaticity lies in the fact that the ten-membered ring is too small to accommodate the two hydrogens that point inward toward the center of the ring. Nonbonded interaction between these two hydrogens forces the ring into a nonplanar conformation in which the overlap of all ten $2p$ orbitals is no longer continuous. Therefore, because [10]annulene is not planar, it is not aromatic.

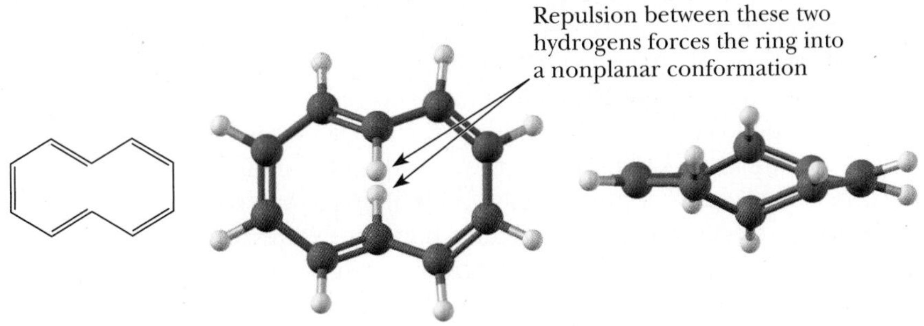

[10]Annulene Viewed from above Viewed through an edge

What is remarkable is that if the two hydrogen atoms facing inward toward the center of the ring in [10]annulene are replaced by a CH_2 group, the ring is now able to assume a conformation close enough to planar that it becomes aromatic.

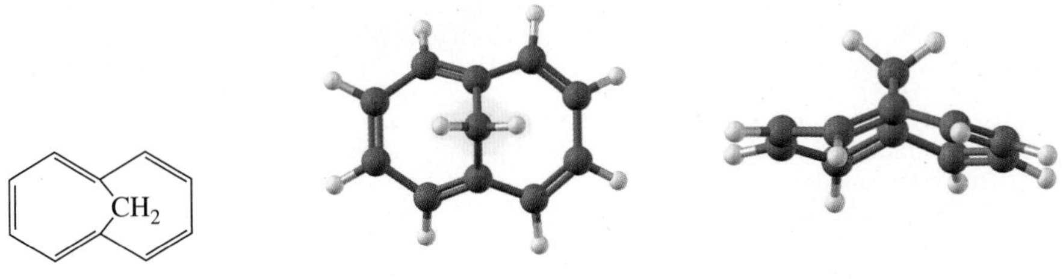

Bridged [10]annulene Viewed from above Viewed through an edge

C. Antiaromatic Hydrocarbons

According to the Hückel criteria, monocyclic, planar molecules with $4n$ π electrons (4, 8, 12, 16, 20, . . .) are especially unstable and are said to be **antiaromatic**. By these criteria, cyclobutadiene with 4 π electrons is antiaromatic. Using the Frost circle energy diagram from Figure 21.6, we can construct a molecular orbital energy diagram for cyclobutadiene (Figure 21.7).

In the ground-state electron configuration of cyclobutadiene, two π electrons fill the π_1-bonding MO. The third and fourth π electrons are unpaired and lie in the π_2- and π_3-nonbonding MOs. The existence of these two unpaired electrons in

Antiaromatic compound

A monocyclic compound that is planar or nearly so, has one $2p$ orbital on each atom of the ring, and has $4n$ π electrons in the cyclic arrangement of overlapping $2p$ orbitals, where n is an integer. Antiaromatic compounds are especially unstable.

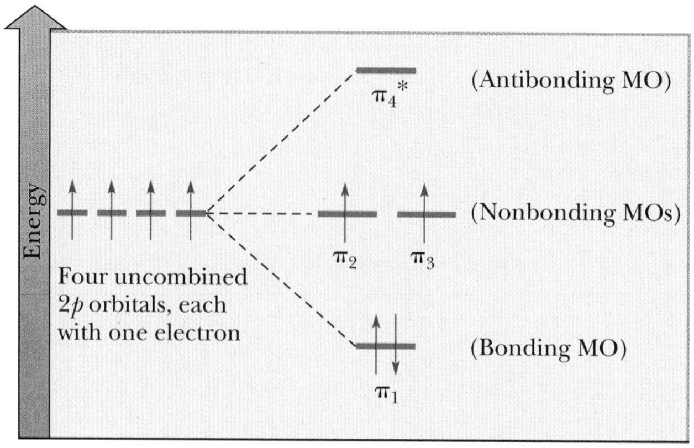

FIGURE 21.7 Molecular orbital energy diagram for cyclobutadiene. In the ground state, two electrons are in the low-lying π_1-bonding **MO**. The remaining two electrons are unpaired and occupy the degenerate π_2- and π_3-nonbonding **MOs**.

planar cyclobutadiene makes this molecule highly unstable and reactive compared to butadiene, a noncyclic molecule containing two conjugated double bonds. It has been found that cyclobutadiene is not planar, but slightly puckered with two shorter bonds and two longer bonds, which makes the two degenerate orbitals no longer equivalent; nevertheless, it retains some apparent diradical character.

Cyclooctatetraene shows reactions typical of alkenes and is classified as nonaromatic. X-ray studies show clearly that the most stable conformation of the molecule is a nonplanar "tub" conformation with two distinct types of carbon-carbon bonds: four longer carbon-carbon single bonds and four shorter carbon-carbon double bonds. The four single bonds are equal in length to the single bonds between sp^2 hybridized carbons (approximately 146 pm), and the four double bonds are equal in length to double bonds in alkenes (approximately 133 pm). In the tub conformation, the overlap of 2*p* orbitals on carbons forming double bonds is excellent, but almost no overlap occurs between 2*p* orbitals at the ends of carbon-carbon single bonds because these 2*p* orbitals are not parallel. Thus, the π system in cyclooctatetraene is not conjugated despite having continuous sp^2 hybridized carbon atoms.

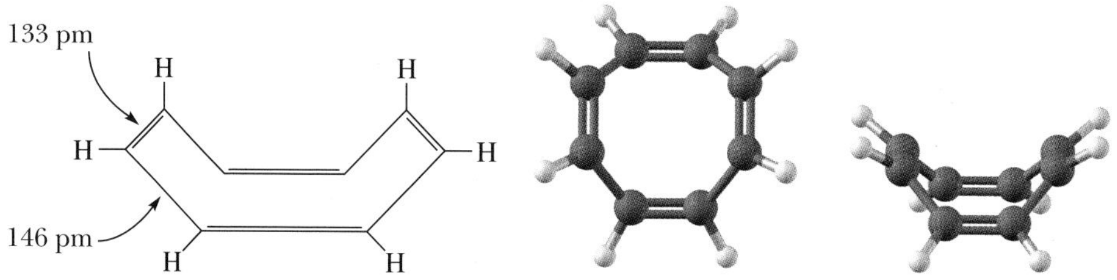

1,3,5,7-Cyclooctatetraene
(tub conformation with alternating single and double bonds)

Viewed from above

Viewed through an edge

To appreciate why planar cyclooctatetraene would be classified as antiaromatic, we need to examine the **MO** energy diagram for an eight-membered ring containing eight π electrons in a cyclic, fully conjugated ring. You constructed a Frost circle for this ring in your answer to Problem 21.1. Note that the most stable conformation of cyclooctatetraene is not planar, but if it were planar, the Frost circle you constructed would be its **MO** energy diagram. The molecular orbital energy diagram for planar cyclooctatetraene is shown in Figure 21.8. In the ground state, six π electrons fill the three low-lying π_1-, π_2-, and π_3-bonding **MOs**. The remaining two π electrons are unpaired and lie in the degenerate π_4- and π_5-nonbonding **MOs**. Because of these two unpaired electrons, planar cyclooctatetraene, if it existed, would be classified as antiaromatic. Cyclooctatetraene, however, is large enough to pucker into a nonplanar conformation and become nonaromatic.

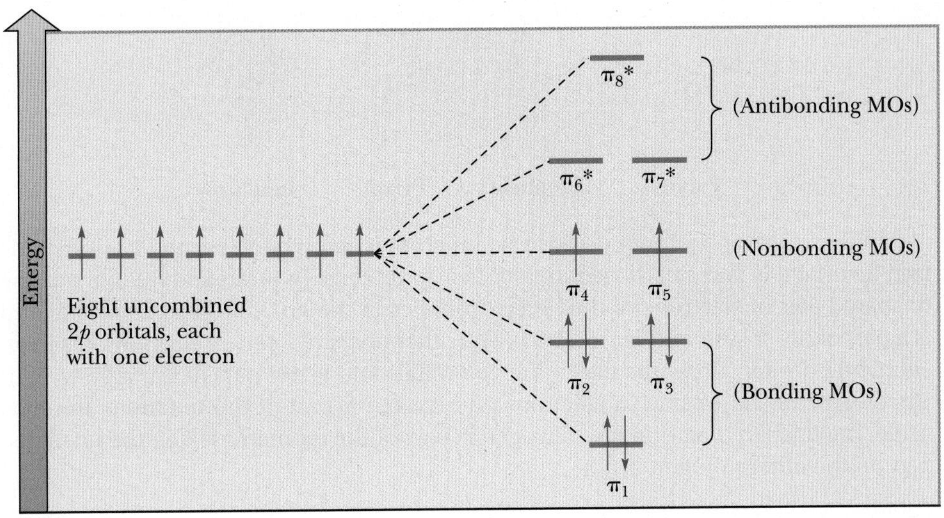

FIGURE 21.8 Molecular orbital energy diagram for a planar conformation of cyclooctatetraene. Three pairs of electrons fill the three low-lying π-bonding molecular orbitals. Two electrons are unpaired in degenerate π-nonbonding molecular orbitals.

If [16]annulene were planar, it too would be antiaromatic. The size of the ring, however, is large enough that it can pucker into a nonplanar conformation in which the double bonds are no longer fully conjugated. [16]Annulene, therefore, is nonaromatic.

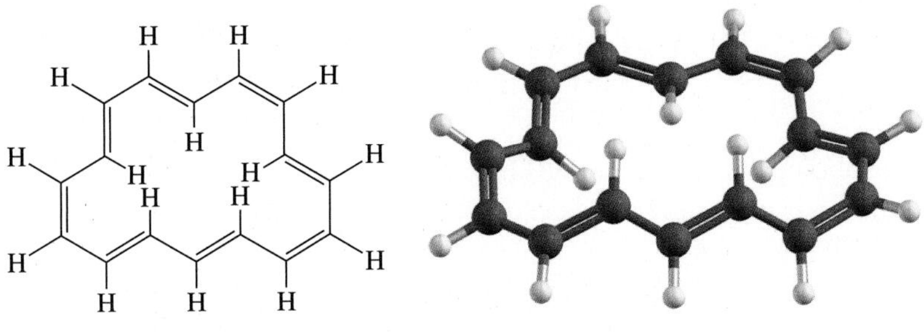

[16]Annulene
(nonaromatic)

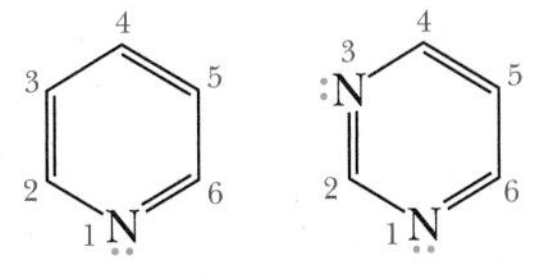

Pyridine **Pyrimidine**

FIGURE 21.9 Two heterocyclic aromatic compounds.

D. Heterocyclic Aromatic Compounds

Aromatic character is not limited to hydrocarbons; it is found in **heterocyclic compounds** as well. Pyridine and pyrimidine are heterocyclic analogs of benzene. In pyridine, one CH group of benzene is replaced by nitrogen, and in pyrimidine, two CH groups are replaced by nitrogens (Figure 21.9).

Each molecule meets the Hückel criteria for aromaticity. Each is monocyclic and planar, each has one 2*p* orbital on each atom of the ring, and each has six electrons in the π system. In pyridine, nitrogen is sp^2 hybridized and its unshared pair of electrons occupies an sp^2 orbital in the plane of the ring and is perpendicular to the 2*p* orbitals of the π system; thus, the unshared pair on the nitrogen of pyridine is not a part of the π system (Figure 21.10).

In pyrimidine, neither unshared pair of electrons of nitrogen is part of the π system. The resonance energy of pyridine is estimated to be 134 kJ (32 kcal)/mol, slightly less than that of benzene. The resonance energy of pyrimidine is estimated to be 108 kJ (26 kcal)/mol.

The five-membered ring heterocyclic compounds furan, thiophene, pyrrole, and imidazole are also aromatic.

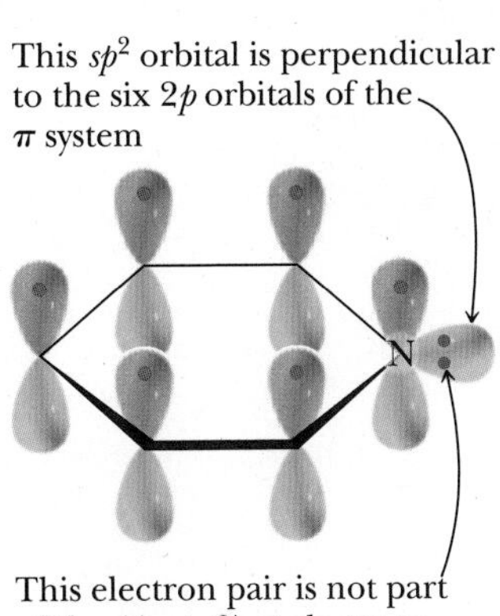

FIGURE 21.10 Pyridine.

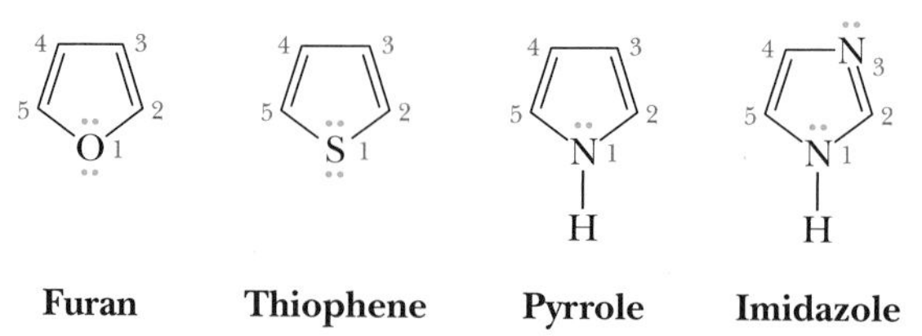
Furan
Thiophene
Pyrrole
Imidazole

In these planar compounds, each heteroatom is sp^2 hybridized, and its unhybridized $2p$ orbital is part of a closed loop of five $2p$ orbitals. In furan and thiophene, one unshared pair of electrons of the heteroatom lies in the unhybridized $2p$ orbital and is a part of the π system (Figure 21.11). The other unshared pair of electrons lies in an sp^2 hybrid orbital perpendicular to the $2p$ orbitals and is not part of the π system. In pyrrole, the unshared pair of electrons on nitrogen is part of the π system. In imidazole, the unshared pair on one nitrogen is part of the aromatic sextet; the unshared pair on the other nitrogen is not.

Watch a video explanation

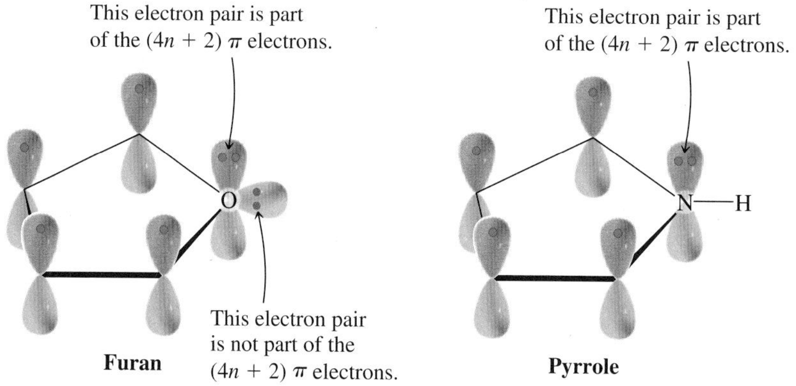
This electron pair is part of the $(4n + 2)$ π electrons.
This electron pair is part of the $(4n + 2)$ π electrons.
This electron pair is not part of the $(4n + 2)$ π electrons.
Furan
Pyrrole

FIGURE 21.11 Origin of the $(4n + 2)$ π electrons in furan and pyrrole. The estimated resonance energy of furan is 67 kJ (16 kcal)/mol, and that of pyrrole is 88 kJ (21 kcal)/mol.

Nature abounds with compounds having a heterocyclic ring fused to one or more other rings. Two such compounds especially important in the biological world are indole and purine.

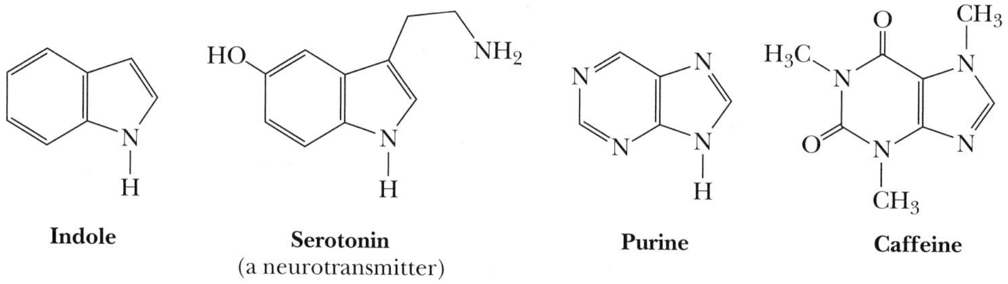
Indole
Serotonin
(a neurotransmitter)
Purine
Caffeine

Indole contains a pyrrole ring fused with a benzene ring. Compounds derived from indole include the essential amino acid L-tryptophan (Section 27.1C) and the neurotransmitter serotonin. Purine contains a six-membered pyrimidine ring fused with a five-membered imidazole ring. Caffeine is a trimethyl derivative of an oxidized purine. Compounds derived from purine and pyrimidine are building blocks of deoxyribonucleic acids (DNA) and ribonucleic acids (RNA, Chapter 28).

E. Aromatic Hydrocarbon Ions

Any neutral monocyclic unsaturated hydrocarbon with an odd number of carbons in the ring must of necessity have at least one CH_2 group in the ring and therefore cannot be aromatic. Examples of such hydrocarbons are cyclopropene, cyclopentadiene, and cycloheptatriene.

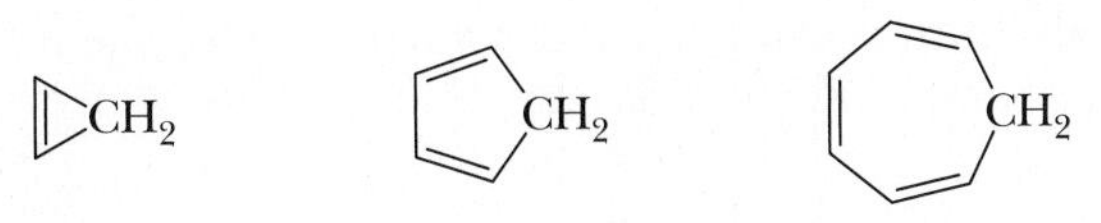

Cyclopropene **Cyclopentadiene** **Cycloheptatriene**

Cyclopropene has the correct number of π electrons to be aromatic, namely $4(0) + 2 = 2$, but it does not have a continuous closed loop of $2p$ orbitals. If, however, the CH_2 group becomes a CH^+ group in which the carbon atom is sp^2 hybridized and has a vacant $2p$ orbital, thus still containing only two electrons, then the overlap of orbitals is continuous, and according to molecular orbital theory, the **cyclopropenyl cation** should be aromatic. The cyclopropenyl cation can be drawn as a resonance hybrid of three equivalent contributing structures. The fact that we can draw three equivalent contributing structures is not the key to the aromaticity of this cation; the key is that it meets the Hückel criteria of aromaticity.

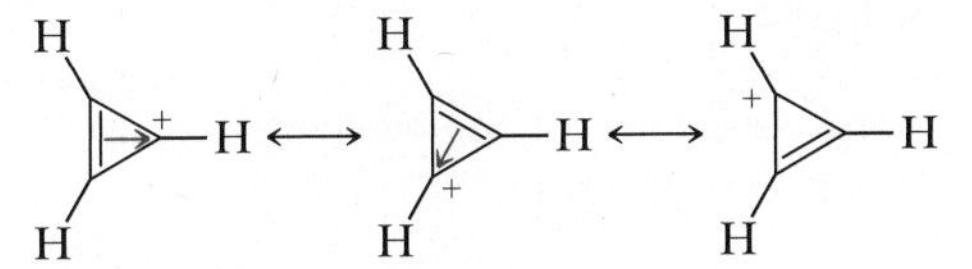

Cyclopropenyl cation
(a hybrid of three equivalent contributing structures)

As an example of the aromatic stabilization of this cation, 3-chlorocyclopropene reacts readily with antimony(V) chloride to form a stable salt.

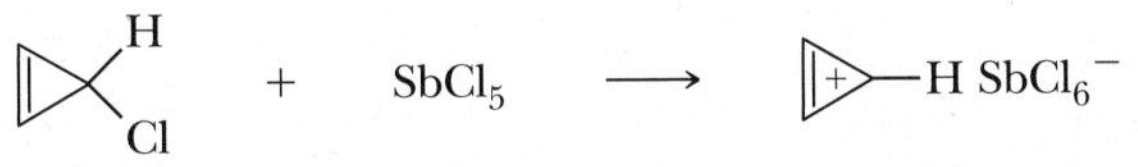

3-Chloro-cyclopropene **Antimony(V) chloride** (a Lewis acid) **Cyclopropenyl hexachloroantimonate**

This chemical behavior is to be contrasted with that of 5-chloro-1,3-cyclopentadiene, which cannot be made to form a stable salt.

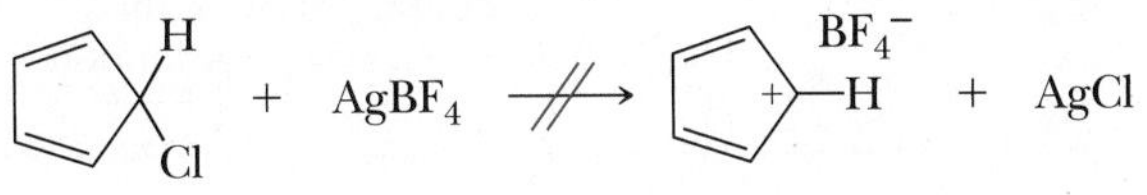

5-Chloro-1,3-cyclopentadiene **Cyclopentadienyl tetrafluoroborate**

In fact, a cyclic, planar, conjugated cyclopentadienyl cation has four π electrons, and if it were to be synthesized, it would be antiaromatic. Note that it is possible to draw five equivalent contributing structures for the cyclopentadienyl cation. Yet, this cation is not aromatic because it has only $4n$ π electrons rather than the required $(4n + 2)$ π electrons.

To form an aromatic ion from cyclopentadiene, it is necessary to convert the CH_2 group to a CH^- group in which the carbon becomes sp^2 hybridized and has two electrons in its unhybridized $2p$ orbital. The resulting **cyclopentadienyl anion** is aromatic. Its aromatic character may also be represented by an inscribed circle with a minus sign (Figure 21.12).

Evidence of the stability of this anion is the fact that the pK_a of cyclopentadiene is approximately 16.0, which makes it one of the most acidic hydrocarbons known. The acidity of cyclopentadiene is comparable to that of water (pK_a 15.7) and ethanol (pK_a 15.9). Consequently, when cyclopentadiene is treated with aqueous sodium

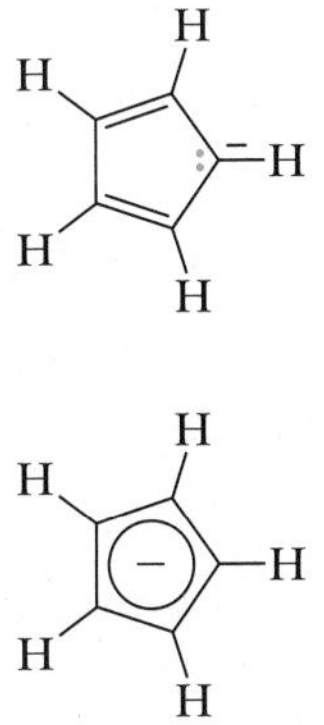

Cyclopentadienyl anion

FIGURE 21.12 Cyclopentadienyl anion (aromatic).

hydroxide, an equilibrium is established in which some of the hydrocarbon is converted to its aromatic anion. K_{eq} for this equilibrium is approximately 0.5.

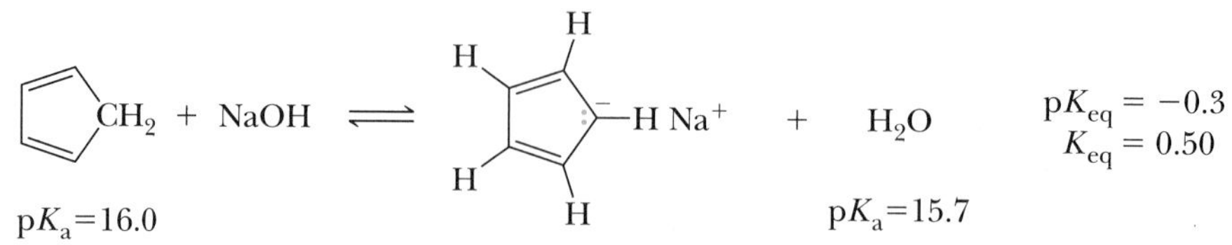
CH2 + NaOH
H
H
H
H
H Na+
+ H2O
pKeq = −0.3
Keq = 0.50
pKa=16.0
pKa=15.7

Cyclopentadiene is converted completely to its anion by treatment with sodium amide.

Example 21.3 MO Energy Diagrams I

Construct an **MO** energy diagram for the cyclopentadienyl anion and describe its ground-state electron configuration.

Solution

Refer to the Frost circle shown in Figure 21.6 for a planar, fully conjugated five-membered ring. The six π electrons occupy the π_1, π_2, and π_3 molecular orbitals, all of which are bonding **MOs**.

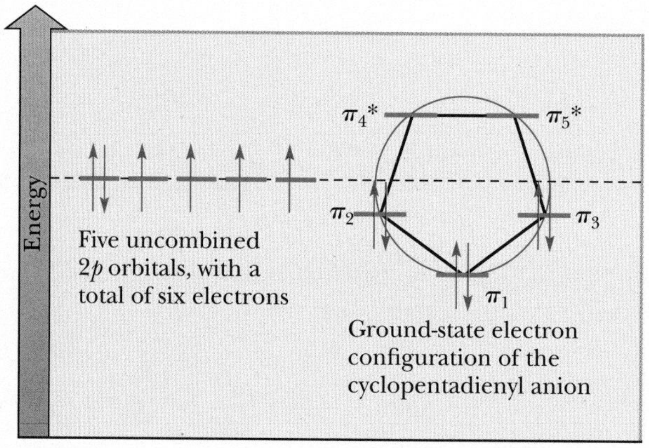
Energy
π4*
π5*
π2
π3
π1
Five uncombined
2p orbitals, with a
total of six electrons
Ground-state electron
configuration of the
cyclopentadienyl anion

Problem 21.3

Describe the ground-state electron configuration of the cyclopentadienyl cation and radical. Assuming each species is planar, would you expect it to be aromatic or antiaromatic?

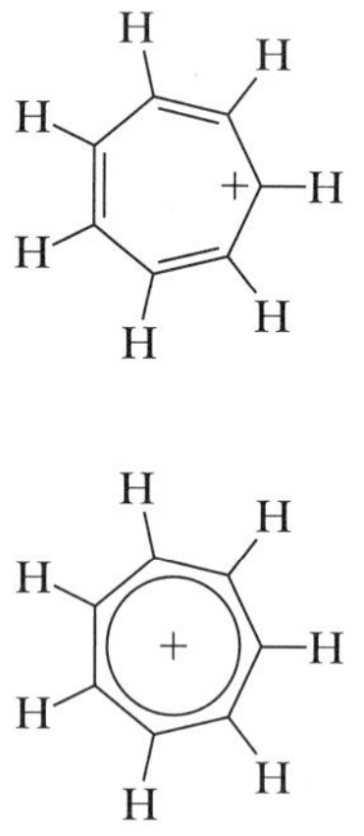

Cycloheptatrienyl cation

FIGURE 21.13
Cycloheptatrienyl cation (tropylium ion) (aromatic).

Cycloheptatriene forms an aromatic cation by conversion of its CH_2 group to a CH^+ group with this sp^2 hybridized carbon having a vacant $2p$ atomic orbital. The **cycloheptatrienyl (tropylium) cation** is planar and has six π electrons in seven $2p$ orbitals, one from each atom of the ring. It can be drawn as a resonance hybrid of seven equivalent contributing structures (Figure 21.13).

Example 21.4 MO Energy Diagrams II

Construct an MO energy diagram for the cycloheptatrienyl cation and describe its ground-state electron configuration.

Solution

Refer to the Frost circle constructed in the answer to Example 21.1. In the ground-state electron configuration of the cycloheptatrienyl cation, the six π electrons occupy the π_1, π_2, and π_3 molecular orbitals, all of which are bonding.

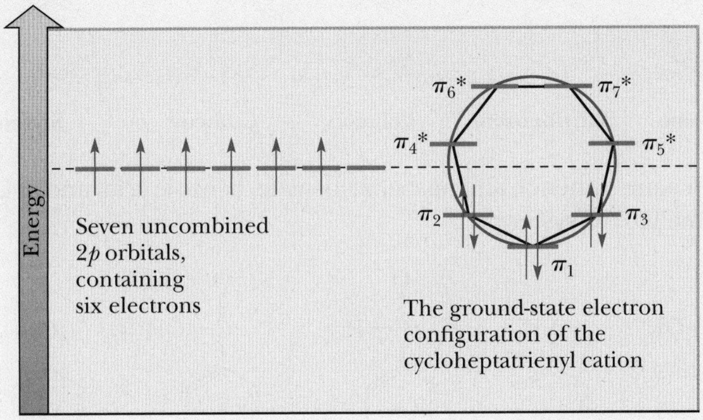

Problem 21.4

Describe the ground-state electron configuration of the cycloheptatrienyl radical and anion. Assuming each species is planar, would you expect them to be aromatic or antiaromatic?

HOW TO

Recognize Aromatic Compounds: Criteria and Caveats

It is worthwhile to recap how to recognize aromaticity. After all, it has been described in the context of hydrocarbons, heterocycles, cyclic cations, and cyclic anions. The Hückel criteria for aromaticity can be summarized as follows: Look for $4n + 2$ electrons where those electrons are in a cyclic array of parallel *p* orbitals; that is, the molecule is planar or nearly planar. Benzene (C_6H_6) is the paradigmatic hydrocarbon example, but other planar hydrocarbons that simply increase the number of electrons in *p* orbitals by a factor of 4*n* are also aromatic {i.e., bridged-[10]annulene ($C_{10}H_{10}$, page 947) and [14]annulene ($C_{14}H_{14}$, page 946)}.

Also remember that the number of *p* orbitals does not matter; rather, it is the number of electrons in the *p* orbitals that is of prime importance. For example, cyclopropenyl cation (page 951) and cycloheptatrienyl cation (page 952) are both aromatic although they have three and seven parallel *p* orbitals, respectively. Furthermore, cyclopentadienyl anion is aromatic, although there are five parallel *p* orbitals (page 951).

Finally, when examining heterocyclic rings, it is of prime importance to delineate whether lone pairs of electrons are in *p* orbitals that are parallel with the other *p* orbitals or whether the lone pair is orthogonal to the cyclic array of *p* orbitals. The lone pair should be used in the electron count only if it is parallel to the other *p* orbitals. For example, in pyridine (page 949), the lone pair is not counted, but in pyrrole (page 950), the lone pair is counted. Keeping the primary Hückel criteria in mind along with the caveats just discussed should allow you to recognize aromaticity in more complex scenarios.

21.3 Nomenclature

A. Monosubstituted Benzenes

Monosubstituted alkylbenzenes are named as derivatives of benzene, as, for example, ethylbenzene. The IUPAC system retains common names for several of the simpler monosubstituted alkylbenzenes. Examples are toluene (rather than methylbenzene), cumene (rather than isopropylbenzene), and styrene (rather than vinylbenzene).

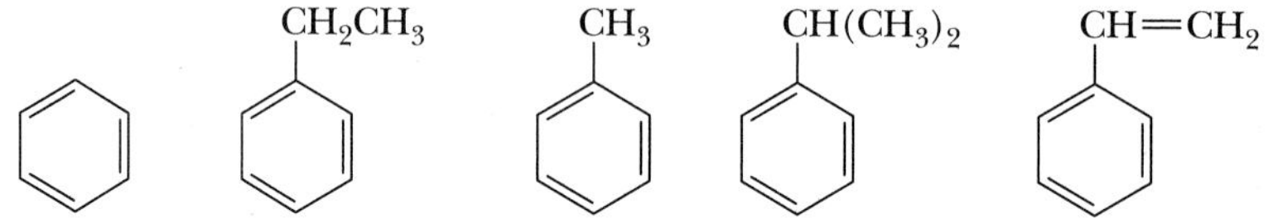

Benzene **Ethylbenzene** **Toluene** **Cumene** **Styrene**

The common names phenol, aniline, benzaldehyde, benzoic acid, and anisole are also retained by the IUPAC system.

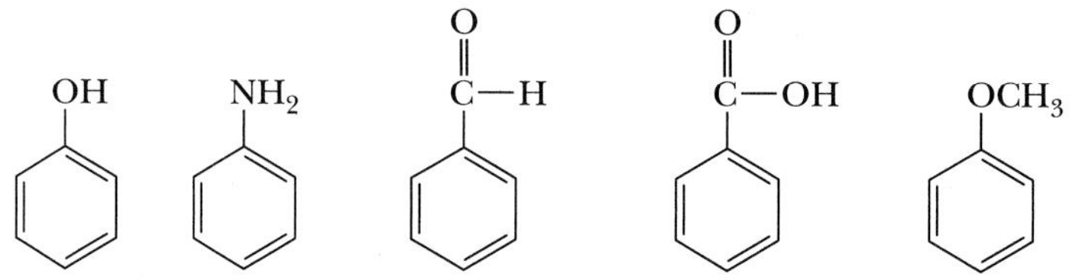

Phenol **Aniline** **Benzaldehyde** **Benzoic acid** **Anisole**

Benzyl group, $C_6H_5CH_2$—

The group derived from toluene by removing a hydrogen from its methyl group.

As we noted in the introduction to Chapter 5, the substituent group derived by loss of an H from benzene is a **phenyl group**, abbreviated Ph–; that derived by loss of an H from the methyl group of toluene is a **benzyl group**, abbreviated Bn–.

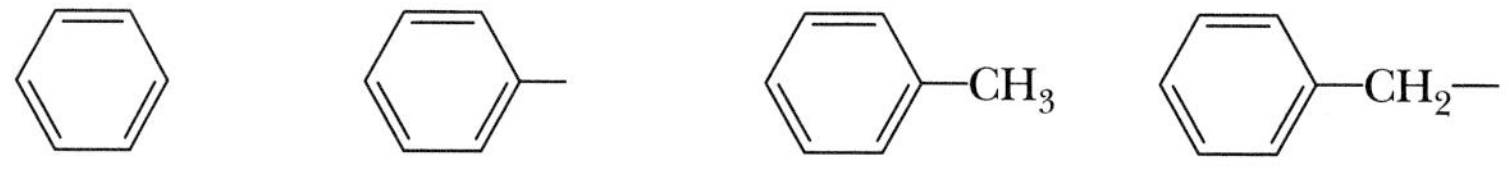

Benzene **Phenyl group, Ph–** **Toluene** **Benzyl group, Bn–**

In molecules containing other functional groups, the phenyl group and its derivatives are named as substituents.

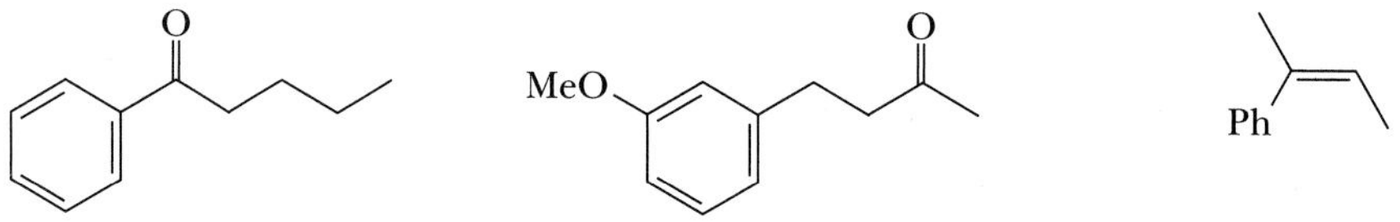

1-Phenyl-1-pentanone **4-(3-Methoxyphenyl)-2-butanone** **(Z)-2-Phenyl-2-butene**

Ortho (*o*)

Refers to groups occupying 1,2-positions on a benzene ring.

Meta (*m*)

Refers to groups occupying 1,3-positions on a benzene ring.

Para (*p*)

Refers to groups occupying 1,4-positions on a benzene ring.

B. Disubstituted Benzenes

When two substituents occur on a benzene ring, three constitutional isomers are possible. The substituents may be located by numbering the atoms of the ring or by using the locators ortho, meta, and para. 1,2- is equivalent to **ortho** (Greek, straight or correct), 1,3- is equivalent to **meta** (Greek, in the middle, between), and 1,4- is equivalent to **para** (Greek, beyond).

When one of the two substituents on the ring imparts a special name to the compound, as, for example, toluene, cumene, phenol, and aniline, then the compound is named as a derivative of that parent molecule. The special substituent is assumed to occupy ring position number 1. The IUPAC system retains the common name xylene for the three isomeric dimethylbenzenes.

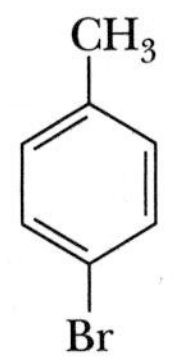
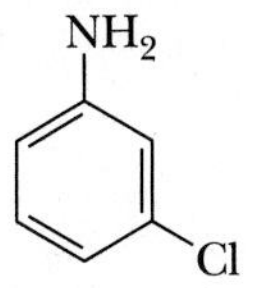
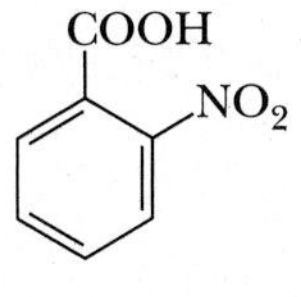
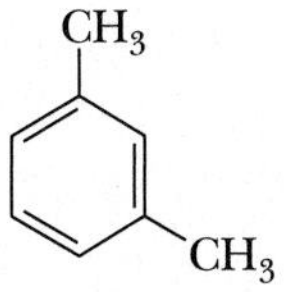

4-Bromotoluene (*p*-Bromotoluene)	**3-Chloroaniline (*m*-Chloroaniline)**	**2-Nitrobenzoic acid (*o*-Nitrobenzoic acid)**	***m*-Xylene**

Where neither group imparts a special name, then the two substituents are located and listed in alphabetical order before the ending *-benzene*. The carbon of the benzene ring with the substituent of lower alphabetical ranking is numbered C-1.

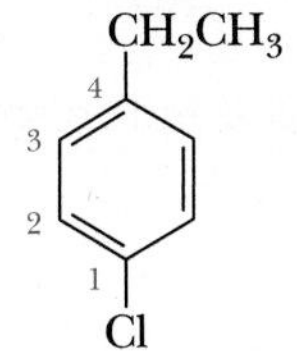
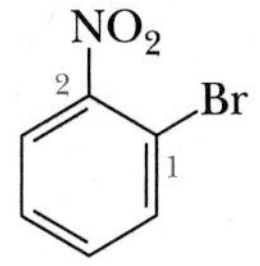
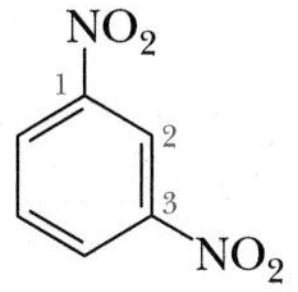

1-Chloro-4-ethylbenzene (*p*-Chloroethylbenzene)	**1-Bromo-2-nitrobenzene (*o*-Bromonitrobenzene)**	**1,3-Dinitrobenzene (*m*-Dinitrobenzene)**

C. Polysubstituted Benzenes

When three or more substituents are present on a ring, their locations are specified by numerals. If one of the substituents imparts a special name, then the compound is named as a derivative of that parent molecule. If none of the substituents imparts a special name, the substituents are numbered to give the smallest set of numbers and are listed in alphabetical order before the ending *-benzene*. In the following examples, the first compound is a derivative of toluene and the second is a derivative of phenol. Because there is no special name for the third compound, its three substituents are listed in alphabetical order and the atoms of the ring are numbered using the lowest possible set of numbers.

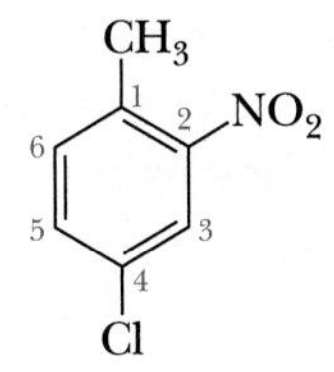
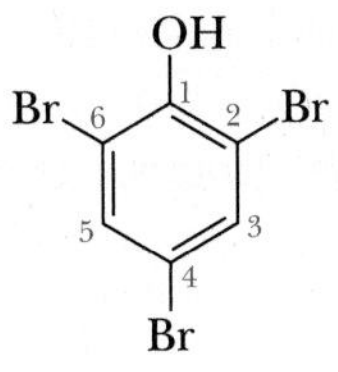
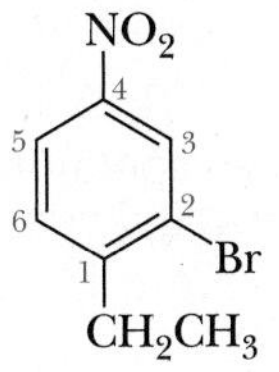

4-Chloro-2-nitrotoluene	**2,4,6-Tribromophenol**	**2-Bromo-1-ethyl-4-nitrobenzene**

Example 21.5 Benzene Nomenclature

Write names for these compounds.

(a) 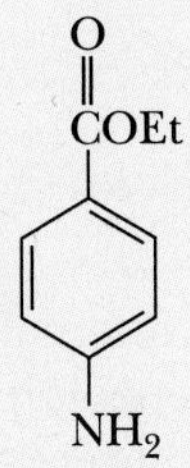
(b) 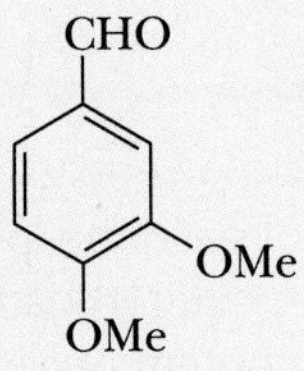
(c) 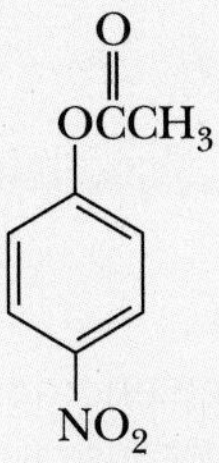
(d) 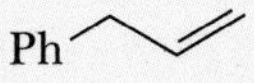

(Continued)

Solution

(a) Ethyl 4-aminobenzoate (ethyl *p*-aminobenzoate)
(b) 3,4-Dimethoxybenzaldehyde
(c) 4-Nitrophenyl ethanoate (*p*-nitrophenyl acetate)
(d) 3-Phenylpropene (allyl benzene)

Problem 21.5

Write names for these compounds.

(a) 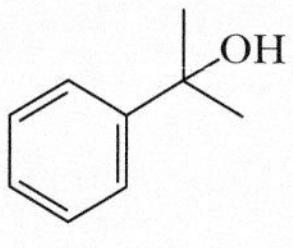

(b) 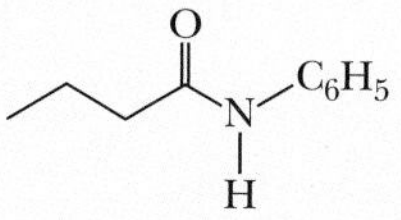

(c) 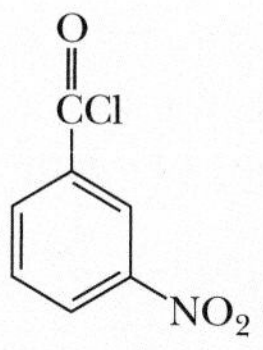

(d) 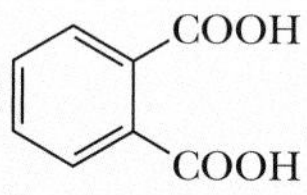

Polynuclear aromatic hydrocarbon (PAH)
A hydrocarbon containing two or more fused benzene rings.

Polynuclear aromatic hydrocarbons (PAHs) contain two or more benzene rings, each pair of which shares two ring carbon atoms. Naphthalene, anthracene, and phenanthrene, the most common PAHs, and substances derived from them are found in coal tar and high-boiling petroleum residues.

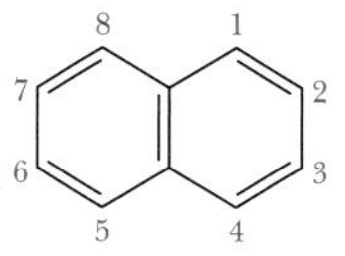

Naphthalene

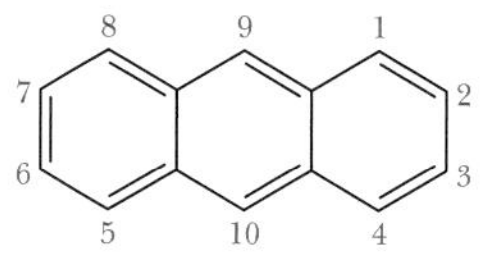

Anthracene

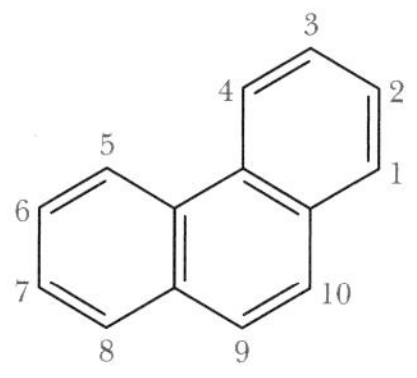

Phenanthrene

At one time, naphthalene was used as a moth repellent and an insecticide in preserving woolens and furs, but its use has decreased owing to the introduction of chlorinated hydrocarbons such as *p*-dichlorobenzene. In numbering PAHs, carbon atoms common to two or more rings are not numbered because they have no replaceable hydrogens.

Also found in petroleum and coal tar are lesser amounts of the following PAHs.

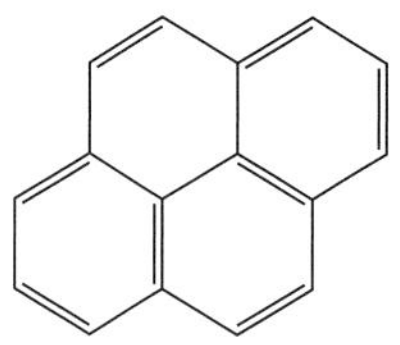
Pyrene

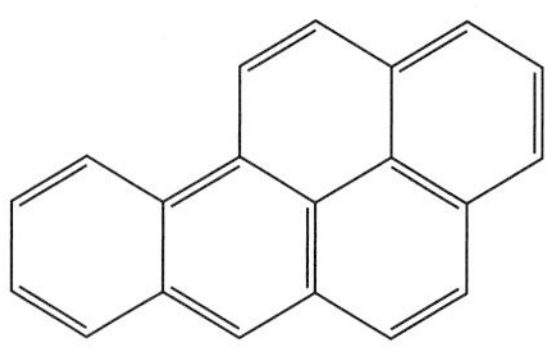
Benzo[a]pyrene

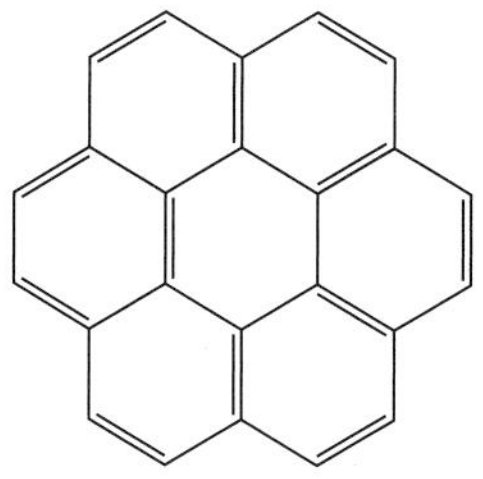
Coronene

These compounds can be found in the exhausts of gasoline-powered internal combustion engines (e.g., automobile engines) and in cigarette smoke. Benzo[a]pyrene has attracted particular interest because it is a very potent carcinogen (cancer-causing substance) and mutagen.

21.4 Phenols

A. Structure and Nomenclature

The functional group of a **phenol** is a hydroxyl group bonded directly to a benzene ring. Substituted phenols are named as derivatives of phenol, as benzenols, or by common names.

Phenol

A compound that contains an —OH bonded to a benzene ring; a benzenol.

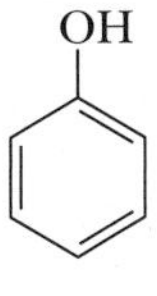

Phenol

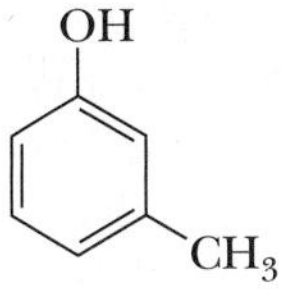

3-Methylphenol (*m*-Cresol)

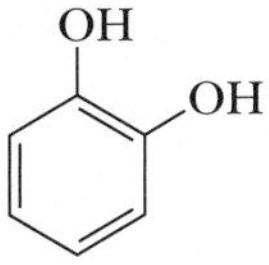

1,2-Benzenediol (Catechol)

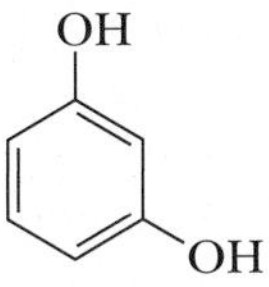

1,3-Benzenediol (Resorcinol)

1,4-Benzenediol (Hydroquinone)

Thymol is a constituent of garden thyme, *Thymus vulgaris.*

Phenols are widely distributed in nature. Phenol itself and the isomeric cresols (*o*-, *m*-, and *p*-cresol) are found in coal tar and petroleum. Thymol and vanillin are important constituents of thyme and vanilla beans, respectively.

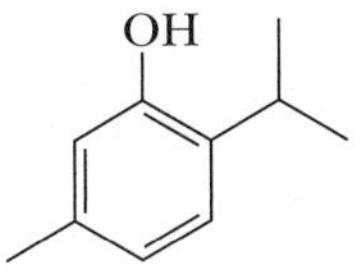

2-Isopropyl-5-methylphenol (Thymol)

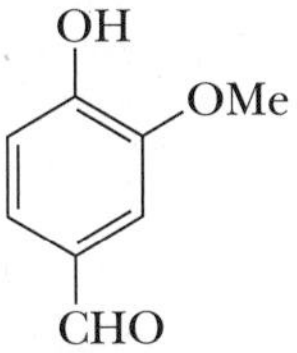

4-Hydroxy-3-methoxybenzaldehyde (Vanillin)

Phenol, or carbolic acid as it was once called, is a low-melting solid that is soluble in water. In sufficiently high concentrations, it is harmful to all kinds of cells. In dilute solutions, it has some antiseptic properties and was introduced into the practice of surgery by Joseph Lister who demonstrated his technique of aseptic surgery in the surgical theater of the University of Glasgow School of Medicine in 1865. Phenol's medical use is now limited. It has been replaced by antiseptics that are more powerful and have fewer undesirable side effects. Among these is hexylresorcinol, which is widely used in nonprescription preparations as a mild antiseptic and disinfectant. Eugenol, which can be isolated from the flower buds (cloves) of *Eugenia aromatica,* is used as a dental antiseptic and analgesic. Urushiol is the main component of the irritating oil of poison ivy.

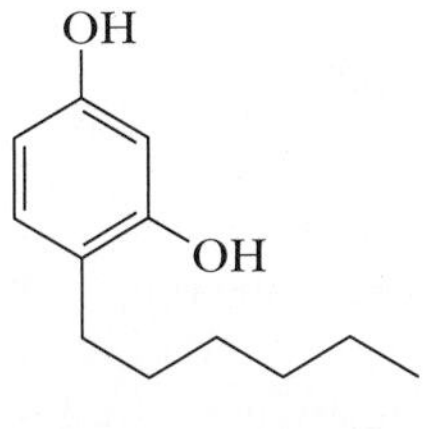

Hexylresorcinol

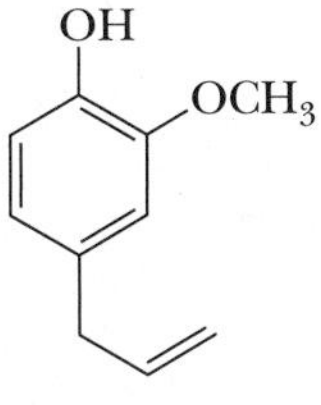

Eugenol

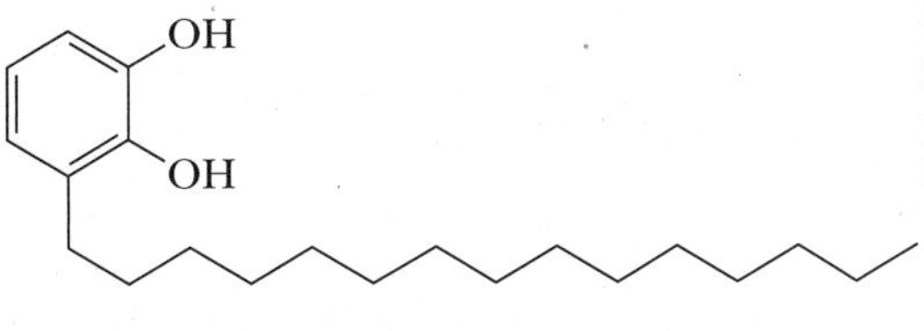

Urushiol

West Indian vanilla, *Vanilla pompona.*

B. Acidity of Phenols

Phenols and alcohols both contain a hydroxyl group. Phenols, however, are grouped as a separate class of compounds because their chemical properties are quite different from those of alcohols. One of the most important of these differences is the fact that

© Cengage Learning/Charles D. Winters

Poison ivy

phenols are significantly more acidic than alcohols. The acid ionization constant of phenol is 10^6 times larger than that of ethanol.

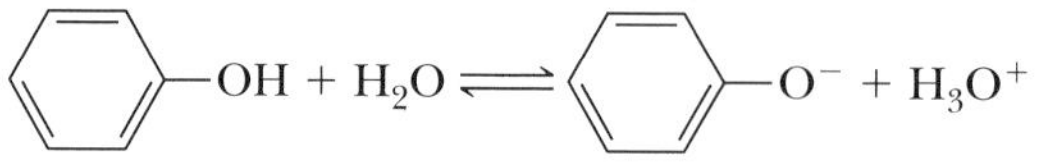

$K_a = 1.1 \times 10^{-10}$ $pK_a = 9.95$

$$CH_3CH_2OH + H_2O \rightleftharpoons CH_3CH_2O^- + H_3O^+ \quad K_a = 1.3 \times 10^{-16} \quad pK_a = 15.9$$

Another way to compare the relative acid strengths of ethanol and phenol is to look at the hydrogen ion concentration and pH of a 0.1 *M* aqueous solution of each (Table 21.1). For comparison, the hydrogen ion concentration and pH of 0.1 *M* HCl are also included.

In aqueous solution, alcohols are neutral substances and the hydrogen ion concentration of 0.1 *M* ethanol is the same as that of pure water. A 0.1 *M* solution of phenol is slightly acidic and has a pH of 5.4. By contrast, 0.1 *M* HCl, a strong acid (completely ionized in aqueous solution), has a pH of 1.0.

Table 21.1 Relative Acidities of 0.1 *M* Solutions of Ethanol, Phenol, and HCl

Acid Ionization Equation	$[H^+]$	pH
$CH_3CH_2OH + H_2O \rightleftharpoons CH_3CH_2O^- + H_3O^+$	1×10^{-7}	7.0
$C_6H_5OH + H_2O \rightleftharpoons C_6H_5O^- + H_3O^+$	3.3×10^{-6}	5.4
$HCl + H_2O \rightleftharpoons Cl^- + H_3O^+$	0.1	1.0

The greater acidity of phenol is a result of the greater stability of the phenoxide ion compared with an alkoxide ion. The negative charge on the phenoxide ion is delocalized by resonance. The two contributing structures on the left place the negative charge on oxygen. The three contributing structures on the right place it on the ortho and para positions of the ring. Taken together, these contributing structures delocalize the negative charge of the phenoxide ion over four atoms. There is no possibility for delocalization of charge in an alkoxide ion.

▶ Delocalization of charge over a larger area is stabilizing.

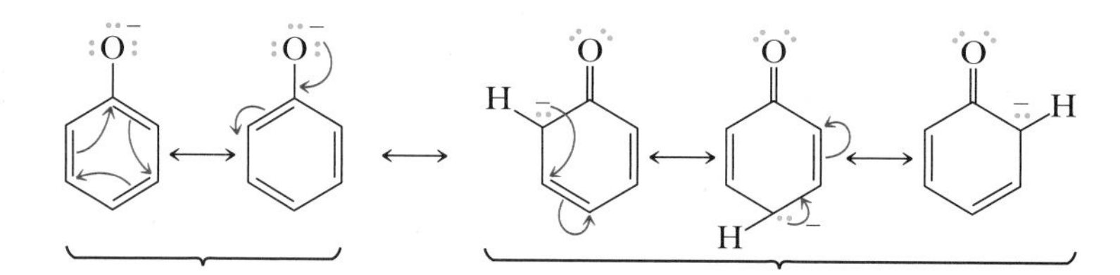

These two Kekulé structures are equivalent

These three contributing structures delocalize the negative charge onto carbon atoms of the ring

Although the charge-delocalization resonance model elucidates why phenol is a stronger acid than ethanol, it does not provide a quantitative means of predicting how much stronger it is. To find out how these acids compare, we must determine their pK_a values experimentally.

Ring substituents, particularly halogens and nitro groups, markedly affect the acidities of phenols through a combination of induction and resonance. Both *m*-cresol and *p*-cresol are weaker acids than phenol itself, and *m*-chlorophenol and *p*-chlorophenol are stronger.

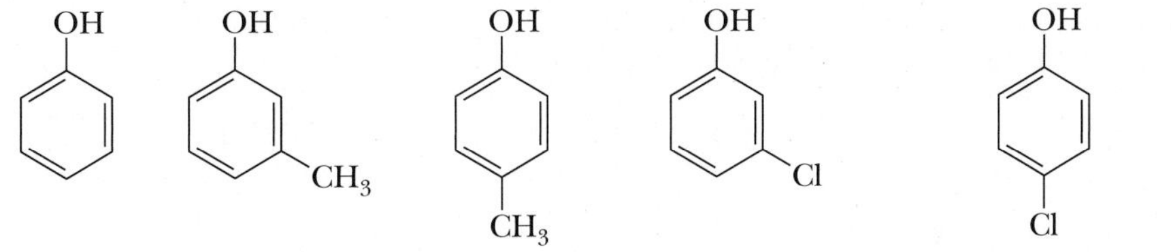

Phenol	*m*-Cresol	*p*-Cresol	*m*-Chlorophenol	*p*-Chlorophenol
pK_a 9.95	pK_a 10.01	pK_a 10.17	pK_a 8.85	pK_a 9.18

The acid-weakening effect of alkyl-substituted phenols can be explained in the following way. The sp^2 hybridized carbon of an aromatic ring is more electronegative than the sp^3 hybridized atom of an alkyl substituent. Alkyl substituents are electron releasing toward the aromatic ring. Because they are electron releasing, they destabilize phenoxide ion-contributing structures and in effect reduce the acidity of substituted phenols.

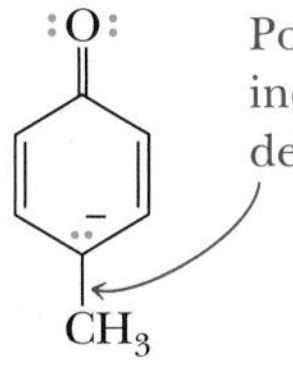

Polarization of this C—C bond by the electron-releasing inductive effect of the sp^3 carbon of the methyl group destabilizes this contributing structure

The inductive effect of the halogens is opposite to that of alkyl substituents. The more highly electronegative halogens withdraw electron density from the aromatic ring and provide greater stability to the halophenoxide ion, compared to phenoxide ion itself. Fluorine, the most electronegative halogen, has the greatest acid-strengthening effect in halophenols; the effect is less for chlorophenols and still less for bromophenols.

Both inductive and resonance effects are observed in nitrophenols.

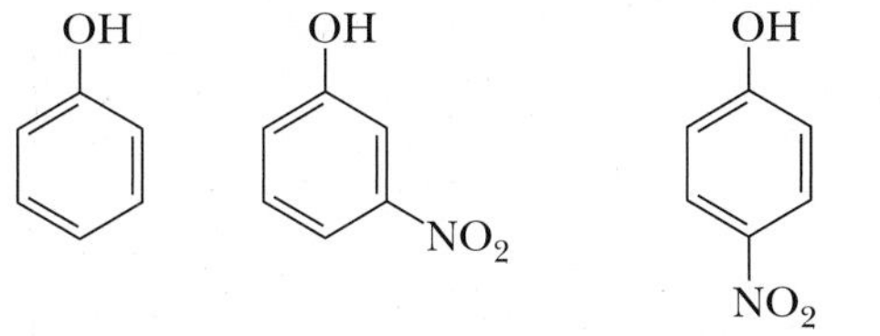

Phenol	*m*-Nitrophenol	*p*-Nitrophenol
pK_a 9.95	pK_a 8.28	pK_a 7.15

Compared to phenol, both *m*-nitrophenol and *p*-nitrophenol are stronger acids. The acid-strengthening effect of the nitro group is greater in the para position, even though it is farther away from the —OH group. Part of the acid-strengthening property of the nitro group stems from its electron-withdrawing inductive effect. In addition, nitro substitution in the ortho or para position increases acidity because the negative charge of the phenoxide ion is delocalized onto an oxygen of the nitro group, as shown in the contributing structure on the right.

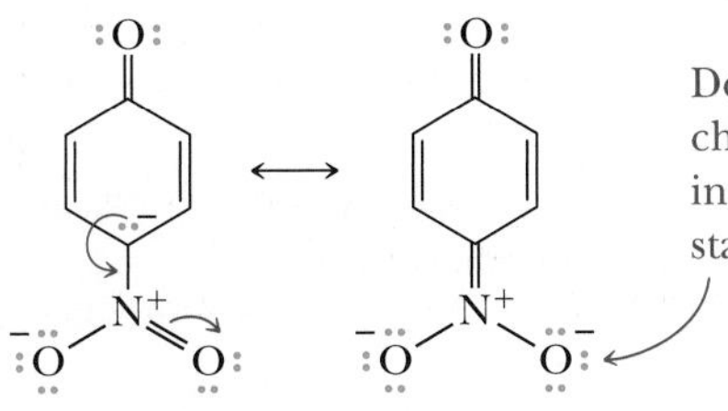

Delocalization of negative charge onto oxygen further increases the resonance stabilization of phenoxide ion

The combined inductive and resonance acid-strengthening effects of the nitro group are such that 2,4,6-trinitrophenol (picric acid) is a stronger acid than phosphoric acid or the hydrogen sulfate ion.

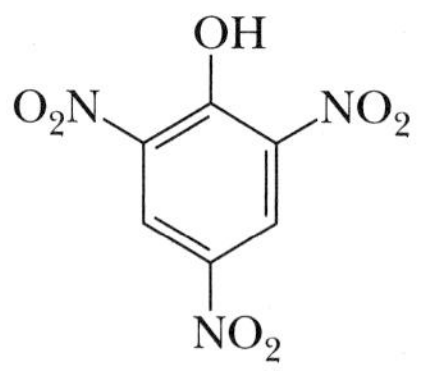

2,4,6-Trinitrophenol (Picric acid) pK_a 0.38

H_3PO_4 **Phosphoric acid** pK_a 2.1

HSO_4^- **Hydrogen sulfate ion** pK_a 1.92

Example 21.6 Acidity of Phenols

Arrange these compounds in order of increasing acidity: 2,4-dinitrophenol, phenol, and benzyl alcohol.

Solution

Benzyl alcohol, a primary alcohol, has a pK_a of approximately 16–18 (Section 10.3). The pK_a of phenol is 9.95. Nitro groups are electron withdrawing and increase the acidity of the phenolic —**OH** group. In order of increasing acidity, they are as follows.

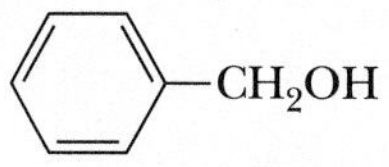

Benzyl alcohol
pK_a 16–18

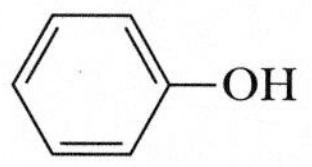

Phenol
pK_a 9.95

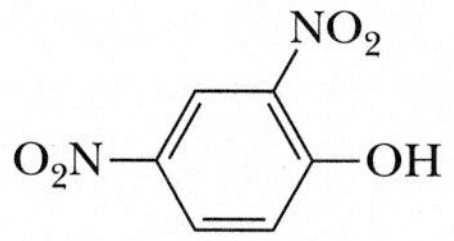

2,4-Dinitrophenol
pK_a 3.96

Watch a video explanation

Problem 21.6

Arrange these compounds in order of increasing acidity: 2,4-dichlorophenol, phenol, cyclohexanol.

C. Acid-Base Reactions of Phenols

Phenols are weak acids and react with strong bases such as **NaOH** to form water-soluble salts.

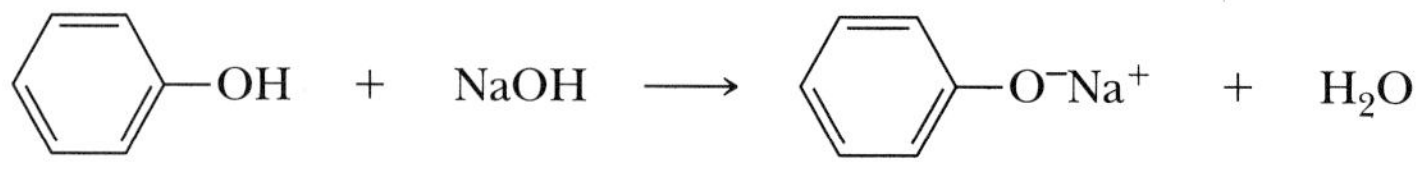

Phenol pK_a 9.95 (stronger acid) — **Sodium hydroxide** — **Sodium phenoxide** — **Water** pK_a 15.7 (weaker acid)

Most phenols do not react with weaker bases such as sodium bicarbonate; they do not dissolve in aqueous sodium bicarbonate. A review of Section 4.4 could be of value.

Carbonic acid is a stronger acid than phenol; consequently, the equilibrium for the reaction of phenol and bicarbonate ion lies far to the left.

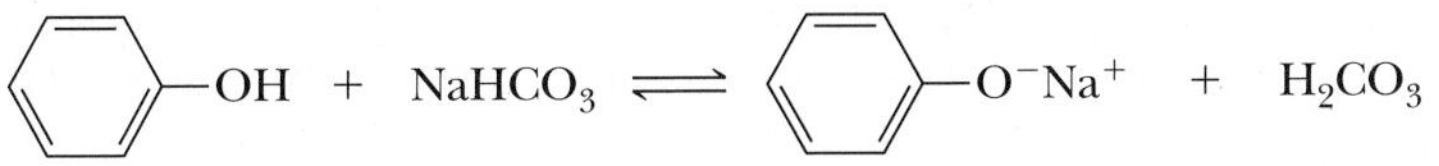

Phenol	Sodium bicarbonate	Sodium phenoxide	Carbonic acid
pK_a 9.95 (weaker acid)			pK_a 6.36 (stronger acid)

Phenols do, however, form water-soluble salts with sodium carbonate, a stronger base than sodium bicarbonate.

The fact that phenols are weakly acidic whereas alcohols are neutral provides a convenient way to separate water-insoluble phenols from water-insoluble alcohols. Suppose we want to separate 4-methylphenol (*p*-cresol) from cyclohexanol. Each is only slightly soluble in water; therefore, they cannot be separated on this basis. They can be separated, however, on the basis of their differences in acidity. First, the mixture of the two is dissolved in diethyl ether or some other water-immiscible solvent. Next, the ether solution is placed in a separatory funnel and shaken with dilute aqueous NaOH. Under these conditions, 4-methylphenol reacts with NaOH and is converted to a water-soluble phenoxide salt. The upper layer in the separatory funnel is now diethyl ether (density 0.74 g/cm^3) containing only dissolved cyclohexanol. The lower aqueous layer contains the dissolved phenoxide salt. The layers are separated, and removal of the ether (bp 35°C) by distillation leaves pure cyclohexanol (bp 161°C). Acidification of the aqueous phase with 0.1 *M* HCl or another strong acid converts the phenoxide salt to 4-methylphenol, which is more soluble in ether than in water and can be extracted with ether and recovered in pure form. These experimental steps are summarized in Figure 21.14.

MCAT Practice: *Passage and Questions*

Capsaicin, "Some Like It Hot"*

Capsaicin is a natural product that is formed in various species of peppers. It was first isolated in 1876, and its structure was determined in 1919. It causes the burning sensation and tearing of the eyes in foods spiced with chili peppers. The human tongue can detect as little as one drop of capsaicin in 5 L of water.

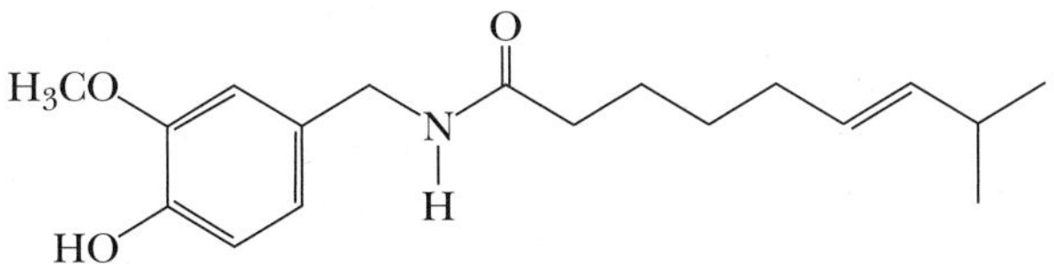

Capsaicin
(from various types of peppers)

© Bizroug/Shutterstock.com

Red chili peppers being dried.

Questions

A. Besides a phenol, what other functional group(s) does capsaicin have in its structure?

1. An ether and an alkene
2. An amide
3. An alkene and a carboxylic acid
4. Both 1 and 2

*See the 1959 movie of this title with Marilyn Monroe, Tony Curtis, and Jack Lemmon.

(Continued)

B. Relative to the pK_a of phenol itself, the pK_a of the phenol in capsaicin is

1. Higher due to the electron-withdrawing nature of the neighboring $—OCH_3$ group.
2. Higher due to the electron-donating nature of the neighboring $—OCH_3$ group.
3. Lower due to the electron-withdrawing nature of the neighboring $—OCH_3$ group.
4. Lower due to the electron-donating nature of the neighboring $—OCH_3$ group.

C. Capsaicin activates an ion-channel protein that modulates the transport of cations across cell membranes in response to heat or abrasion. Hence, the spiciness of capsaicin is akin to a burning sensation although there is no direct tissue damage. If the amide is hydrolyzed, the individual components of capsaicin no longer elicit the burning sensation. Speculate as to what kinds of interactions capsaicin may make with the biological receptor.

1. The nine-carbon tail on the right would dangle in water, and the phenyl ring and associated functional groups would reside within the protein binding site.
2. The nine-carbon tail on the right would interact with the protein via the hydrophobic effect, and the phenol and other functional groups would make hydrogen bonds with the protein.
3. The phenyl ring and attached functional groups would dangle in water, and the nine-carbon chain on the right would bind to the protein with extensive hydrogen bonding.

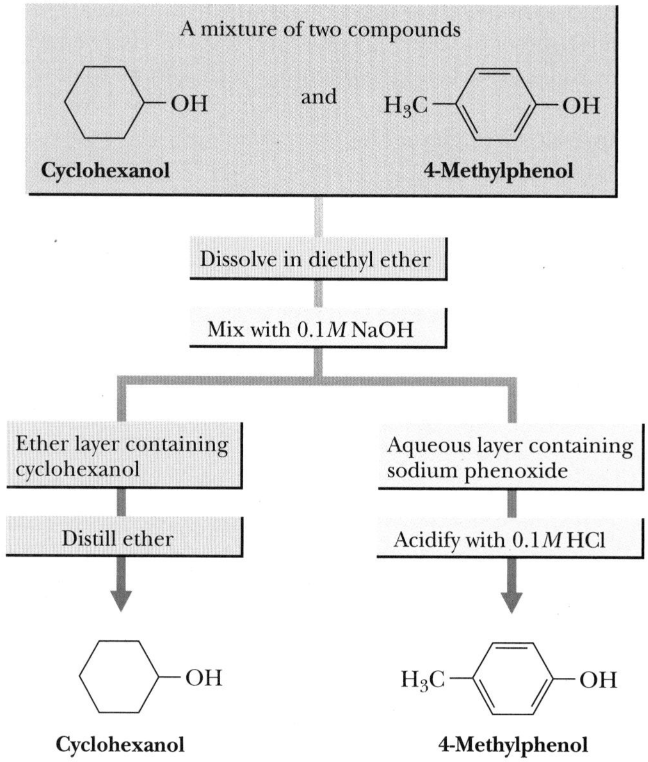

FIGURE 21.14 Experimental steps to separate water-insoluble phenols from water-insoluble alcohols.

D. Preparation of Alkyl-Aryl Ethers

Alkyl-aryl ethers can be prepared from a phenoxide salt and a haloalkane (the Williamson synthesis, Section 11.4A). They cannot be prepared from a haloarene and alkoxide salt, however, because haloarenes are unreactive under the conditions of Williamson synthesis; they do not undergo nucleophilic displacement by either an S_N1 or S_N2 mechanism.

Alkyl-aryl ethers are often synthesized by carefully controlling solubility. Both the haloalkane and phenol are dissolved in dichloromethane; then the solution is mixed with an aqueous solution of sodium hydroxide. Phenol, a poor nucleophile, reacts with sodium hydroxide in the aqueous phase to form the phenoxide ion, a good nucleophile. Alkyl-aryl ethers can be synthesized by treating the sodium salt of a phenol with a haloalkane. The following example illustrates the Williamson synthesis of alkyl-aryl ethers. The $Bu_4N^+Br^-$ is used to facilitate reaction between the polar phenoxide salt and the hydrophobic haloalkane in the mixed solvent.

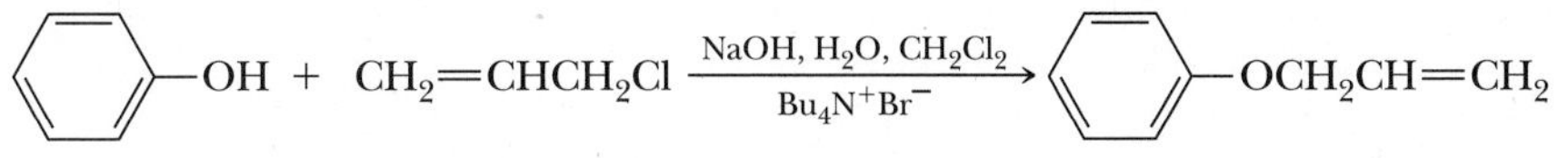

Phenol **3-Chloropropene (Allyl chloride)** **Phenyl 2-propenyl ether (Allyl phenyl ether)**

The synthesis of anisole illustrates the use of dimethyl sulfate as a methylating agent.

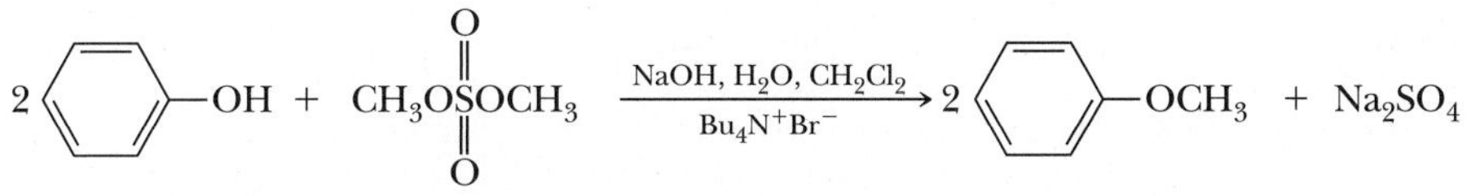

Phenol **Dimethyl sulfate** **Methyl phenyl ether (Anisole)**

An alkyl-aryl ether, ArOR, is cleaved by hydrohalic acids, HX, to form a haloalkane and a phenol.

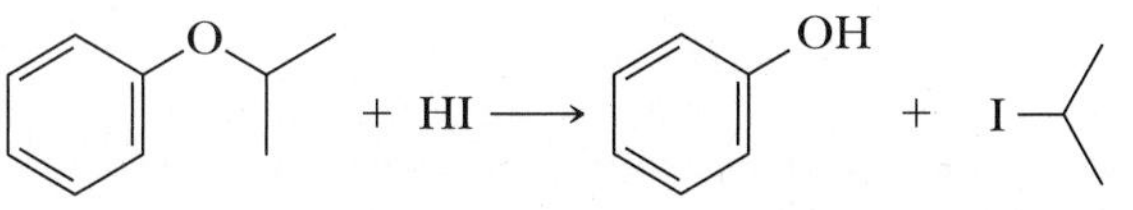

2-Phenoxypropane (Isopropyl phenyl ether) **Phenol** **2-Iodopropane (Isopropyl iodide)**

This cleavage illustrates the fact that nucleophilic substitution is not likely to occur at an aromatic carbon and that phenols, unlike alcohols, are not converted to haloarenes by treatment with concentrated HCl, HBr, or HI.

E. Kolbe Carboxylation: Synthesis of Salicylic Acid

Phenoxide ions react with carbon dioxide to give a carboxylic acid salt as shown by the industrial synthesis of salicylic acid, the starting material for the production of aspirin (Section 18.5B). Phenol is dissolved in aqueous NaOH, and this solution is then saturated with CO_2 under pressure to give sodium salicylate.

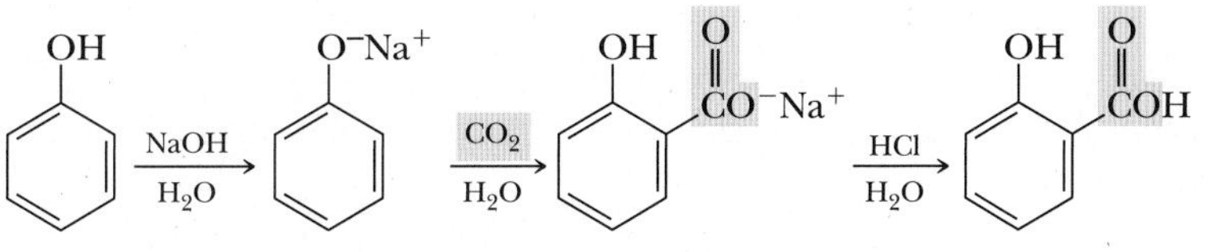

Phenol **Sodium phenoxide** **Sodium salicylate** **Salicylic acid**

This process is referred to as high-pressure Kolbe carboxylation of sodium phenoxide. Upon acidification of the alkaline solution, salicylic acid is isolated as a solid, mp 157–159°C.

The importance of salicylic acid in industrial organic chemistry is demonstrated by the fact that over 6×10^6 kg of aspirin are synthesized in the United States each year.

Mechanism 21.1

Kolbe Carboxylation of Phenol

Step 1: **Make a new bond between a nucleophile (arene ring) and an electrophile.** The phenoxide ion reacts like an enolate anion; it is a strong nucleophile. Nucleophilic attack of the phenoxide anion on a carbonyl group of carbon dioxide gives a substituted cyclohexadienone intermediate.

Sodium phenoxide

Cyclohexadienone intermediate

Step 2: **Keto-enol tautomerism.** Keto-enol tautomerism of the cyclohexadienone intermediate gives the product salicylate anion. Note that in this case, the enol, owing to its aromatic character, is the more stable of the two tautomers.

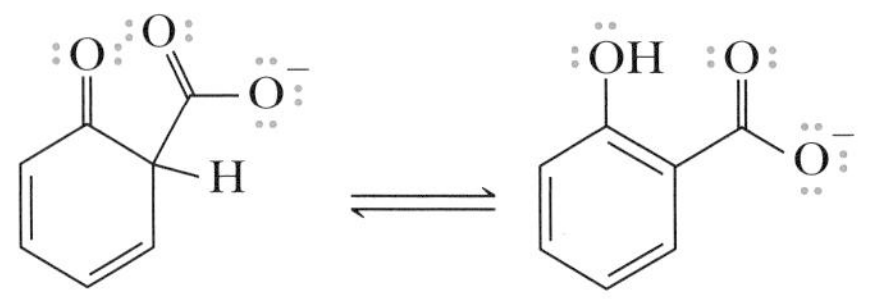

Salicylate anion

F. Oxidation to Quinones

Because of the presence of the electron-donating —OH group on the ring, phenols are susceptible to oxidation by a variety of strong oxidizing agents. For example, oxidation of phenol itself by potassium dichromate gives 1,4-benzoquinone (*p*-quinone).

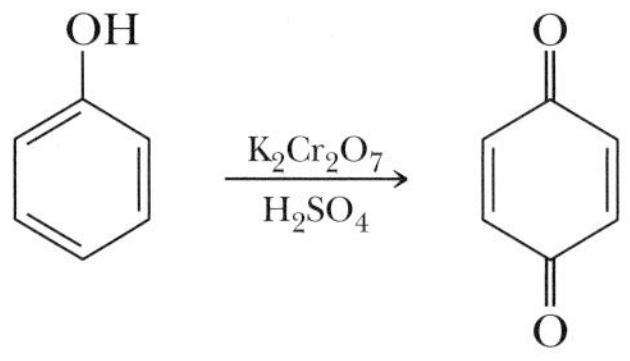

Phenol **1,4-Benzoquinone (*p*-Quinone)**

By definition, a quinone is a cyclohexadienedione. Those with carbonyl groups ortho to each other are called *o*-quinones; those with carbonyl groups para to each other are called *p*-quinones.

Quinones can also be obtained by oxidation of 1,2-benzenediol (catechol) or 1,4-benzenediol (hydroquinone).

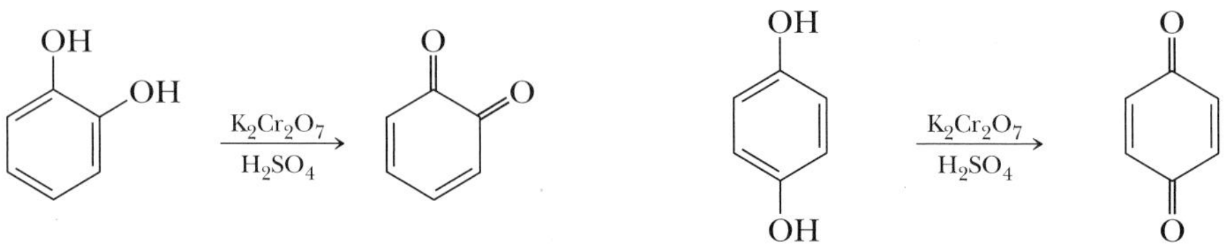

1,2-Benzenediol (Catechol) **1,2-Benzoquinone (*o*-Quinone)** **1,4-Benzenediol (Hydroquinone)** **1,4-Benzoquinone (*p*-Quinone)**

Perhaps the most important chemical property of quinones is that they are readily reduced to benzenediols. For example, *p*-quinone is readily reduced to hydroquinone by sodium dithionite in neutral or alkaline solution. There are many other ways to carry out this reduction. The point is that it can be done very easily, as can the corresponding oxidation of a hydroquinone.

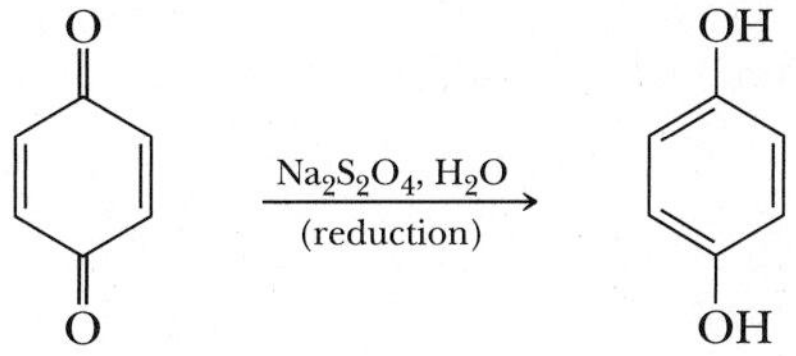

1,4-Benzoquinone (*p*-Quinone) **1,4-Benzenediol (Hydroquinone)**

There are many examples in both chemistry and biology in which the reversible oxidation/reduction of hydroquinones or quinones is important. One such example is coenzyme Q, alternatively known as ubiquinone. The name of this important biomolecule is derived from the Latin *ubique* (everywhere) + quinone.

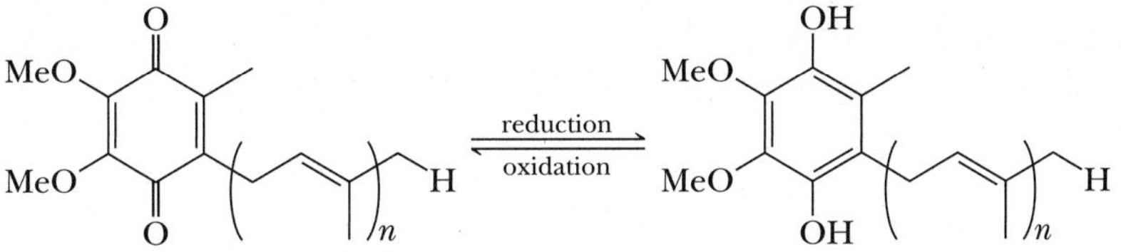

Coenzyme Q (oxidized form) **Coenzyme Q** (reduced form)

Coenzyme Q, a carrier of electrons in the respiratory chain, contains a long hydrocarbon chain of between 6 and 10 isoprene units that serves to anchor it firmly in the nonpolar environment of the mitochondrial inner membrane. The oxidized form of coenzyme Q is a two-electron oxidizing agent. In subsequent steps of the respiratory chain, the reduced form of coenzyme Q transfers the two electrons to another link until they are eventually delivered to a molecule of oxygen, which is, in turn, reduced to water.

Another quinone important in biological systems is vitamin K_2. This compound was discovered in 1935 as a result of a study of newly hatched chicks with a fatal disease in which their blood was slow to clot. It was later discovered that the delayed clotting time of blood was caused by a deficiency of prothrombin. We now know that a prothrombin deficiency is, in turn, caused by a deficiency in vitamin K_2, which is essential to the synthesis of prothrombin in the liver. The natural form of vitamin K_2 has a chain of five to eight isoprene units bonded to a 1,4-naphthoquinone ring. Figure 21.15 shows seven isoprene units in the side chain.

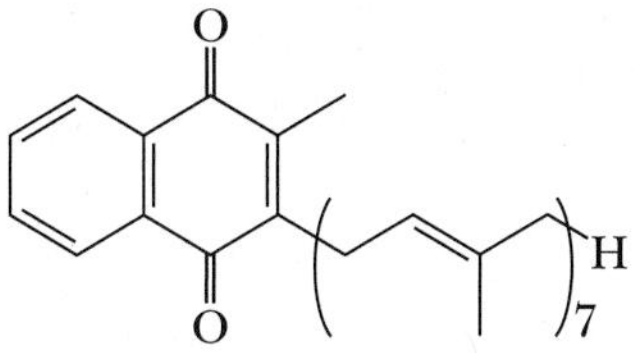

Vitamin K_2

FIGURE 21.15 Vitamin K_2 with seven isoprene units.

The natural vitamins of the K family have for the most part been replaced by synthetic preparations in food supplements. Menadione, one such synthetic material with vitamin K activity, has only hydrogen in place of the long alkyl side chain. Menadione is prepared by chromic acid oxidation of 2-methylnaphthalene under mild conditions.

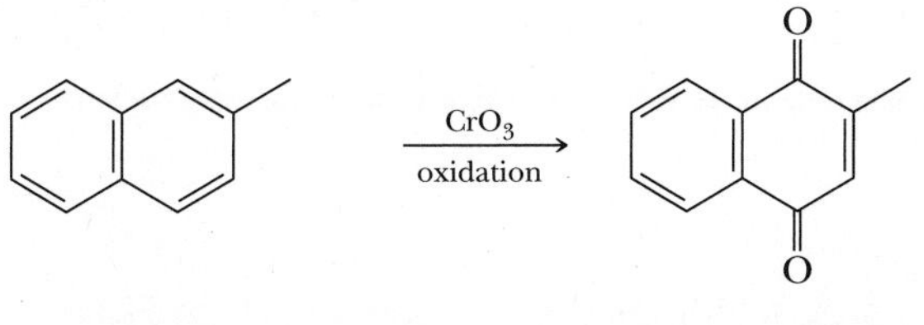

2-Methylnaphthalene **2-Methyl-1,4-naphthoquinone (Menadione)**

A commercial process that uses a quinone is black-and-white photography. Black-and-white film is coated with an emulsion containing silver bromide or silver iodide crystals, which become activated by exposure to light. The activated silver ions are reduced in the developing stage to metallic silver by hydroquinone, which at the same time is oxidized to quinone. Following is an equation showing the relationship between these species.

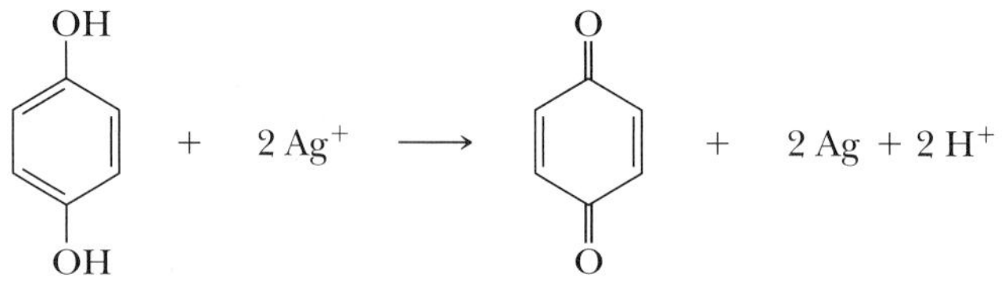
OH
OH
+ 2 Ag+
O
O
+ 2 Ag + 2 H+

1,4-Benzenediol (Hydroquinone) **1,4-Benzoquinone (*p*-Quinone)**

All silver halide not activated by light and then reduced by interaction with hydroquinone is removed in the fixing process, and the result is a black image (a negative) left by deposited metallic silver where the film has been struck by light. Other compounds are now used to reduce "light-activated" silver bromide, but the result is the same—a deposit of metallic silver in response to exposure of film to light.

21.5 Reactions at a Benzylic Position

Benzylic position An sp^3 hybridized carbon bonded to a benzene ring.

In this section, we study two reactions of substituted aromatic hydrocarbons that occur preferentially at the **benzylic position**.

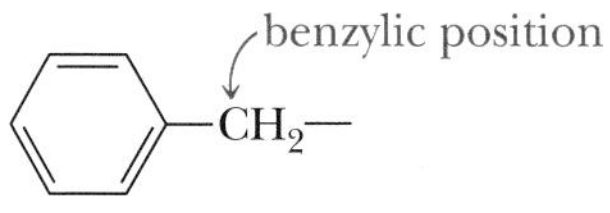
benzylic position
CH2—

Benzyl group

▶ **Delocalization of charge over a larger area is stabilizing.**

Reactions involving alkyl side chains of aromatic compounds occur preferentially at the benzylic position for two reasons. First, the benzene ring is especially resistant to reaction with many of the reagents that normally attack alkanes. Second, benzylic cations and benzylic radicals are easily formed because of resonance stabilization of these intermediates. A benzylic cation or radical is a hybrid of five contributing structures: two Kekulé structures and three that delocalize the positive charge (or the lone electron) onto carbons of the aromatic ring. Following are contributing structures for a benzylic cation. Similar contributing structures can be written for a benzylic radical and anion. Benzylic contributing structures are closely analogous to allylic structures in stabilizing cations, radicals, and anions. ▶

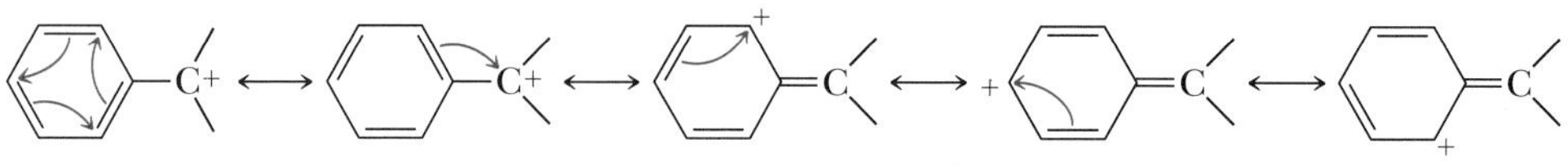

The benzyl cation as a hybrid of five contributing structures

A. Oxidation

Benzene is unaffected by strong oxidizing agents such as H_2CrO_4 and $KMnO_4$. However, when toluene is treated with these oxidizing agents under vigorous conditions, the side-chain methyl group is oxidized to a carboxyl group to give benzoic acid.

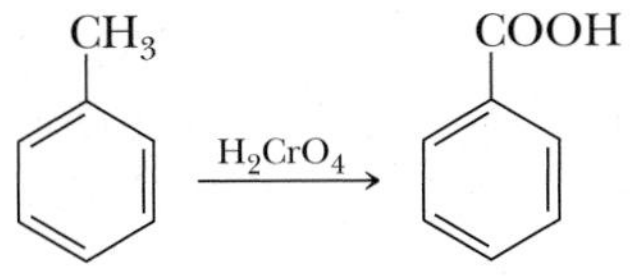

Toluene **Benzoic acid**

Halogen and nitro substituents on an aromatic ring are unaffected by these oxidations. 2-Chloro-4-nitrotoluene, for example, is oxidized to 2-chloro-4-nitrobenzoic acid.

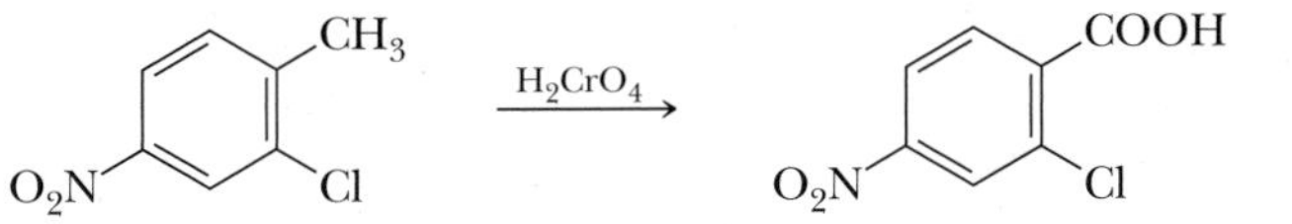

2-Chloro-4-nitrotoluene **2-Chloro-4-nitrobenzoic acid**

Ethyl and isopropyl side chains are also oxidized to carboxyl groups. The side chain of *tert*-butylbenzene, however, is not oxidized.

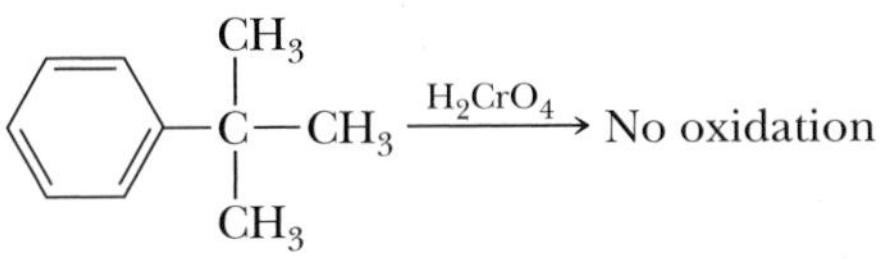

***tert*-Butylbenzene**

From these observations, we conclude that if a benzylic hydrogen exists, then the benzylic carbon is oxidized to a carboxyl group and all other carbons of the side chain are removed as CO_2. If no benzylic hydrogen exists, as in the case of *tert*-butylbenzene, no oxidation of the side chain occurs.

If more than one alkyl side chain exists, each is oxidized to —COOH. Oxidation of *m*-xylene gives 1,3-benzenedicarboxylic acid, more commonly named isophthalic acid.

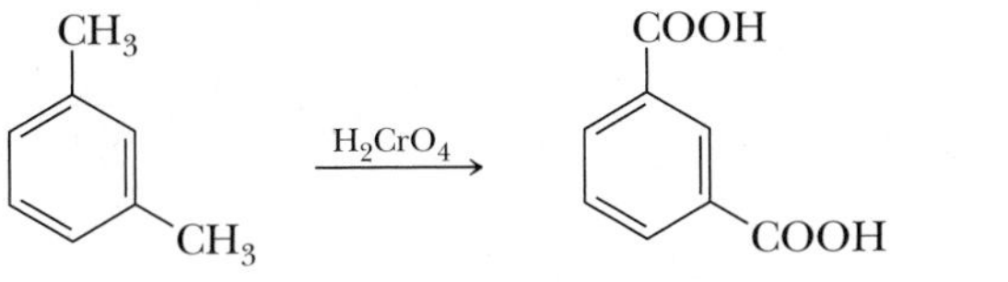

***m*-Xylene** **1,3-Benzenedicarboxylic acid (Isophthalic acid)**

Example 21.7 Benzylic Oxidation

Draw a structural formula for the product of vigorous oxidation of 1,4-dimethylbenzene (*p*-xylene) by H_2CrO_4.

Solution

Both alkyl groups are oxidized to —COOH groups. The product is terephthalic acid, one of two monomers required for the synthesis of Dacron polyester and Mylar (Section 29.5B).

(Continued)

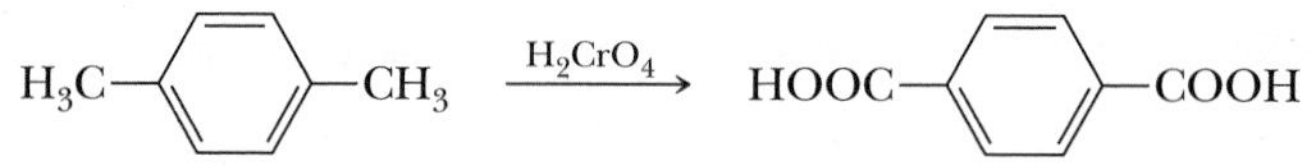

1,4-Dimethylbenzene (*p*-Xylene)

1,4-Benzenedicarboxylic acid (Terephthalic acid)

Problem 21.7

Predict the products resulting from vigorous oxidation of each compound by H_2CrO_4.

(a)

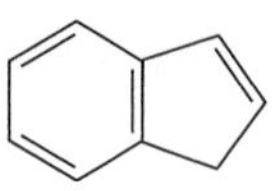

(b)

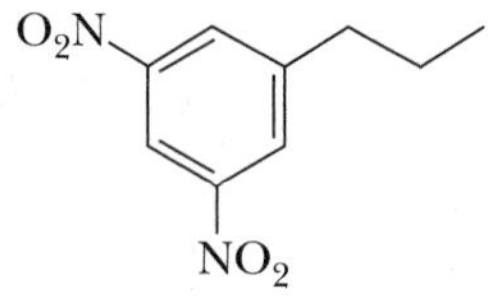

Studying these side-chain oxidations and formulating mechanisms for them is difficult. Available evidence supports the formation of unstable intermediates that are either benzyl radicals or benzylic carbocations.

Naphthalene is oxidized to phthalic acid by molecular oxygen in the presence of a vanadium(V) oxide (vanadium pentoxide) catalyst.

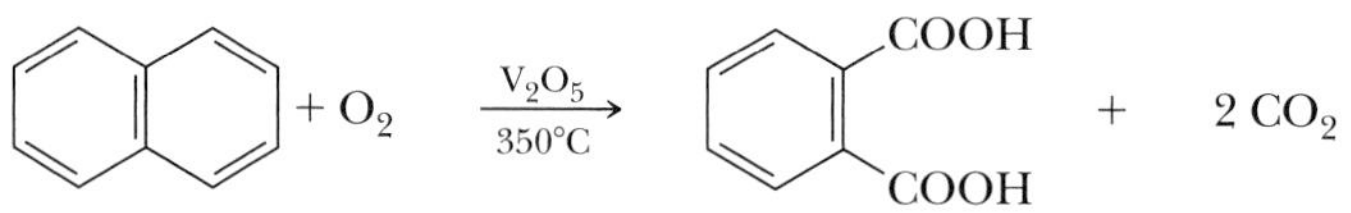

Naphthalene

1,2-Benzenedicarboxylic acid (Phthalic acid)

This conversion, the basis for an industrial synthesis of this aromatic dicarboxylic acid, illustrates the ease of oxidation of condensed benzene rings compared with benzene itself.

B. Halogenation

Reaction of toluene with chlorine in the presence of heat or light results in formation of chloromethylbenzene and HCl.

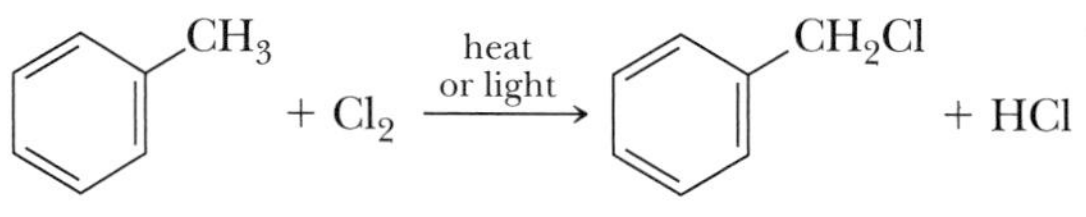

Toluene

Chloromethylbenzene (Benzyl chloride)

Bromination is easily accomplished by using *N*-bromosuccinimide (NBS) in the presence of a peroxide catalyst.

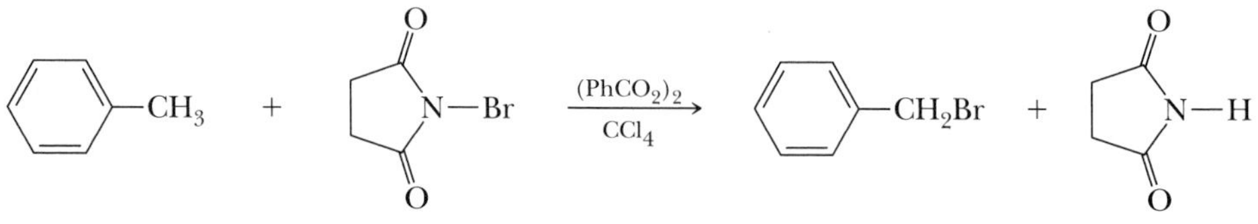

Toluene

***N*-Bromosuccinimide (NBS)**

Bromomethylbenzene (Benzyl bromide)

Succinimide

Halogenation of a larger alkyl side chain is highly regioselective, as illustrated by the halogenation of ethylbenzene. When treated with NBS, the only monobromo organic product formed is 1-bromo-1-phenylethane. This regioselectivity is dictated by the resonance stabilization of the benzylic radical intermediate. The mechanism of radical bromination at a benzylic position is identical to that for allylic bromination (Section 8.6A).

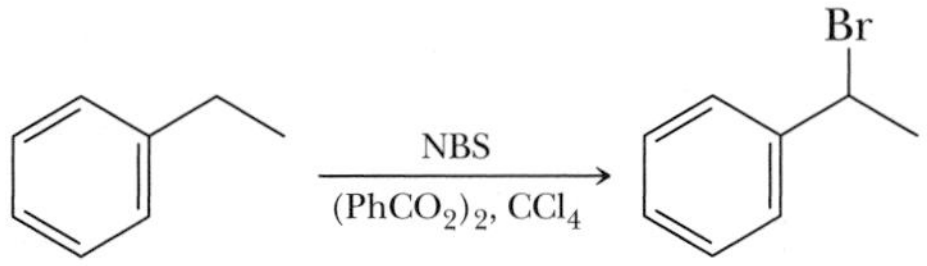

Ethylbenzene **1-Bromo-1-phenylethane** (racemic)

When ethylbenzene is treated with chlorine under radical reaction conditions, two products are formed in the ratio of 9:1.

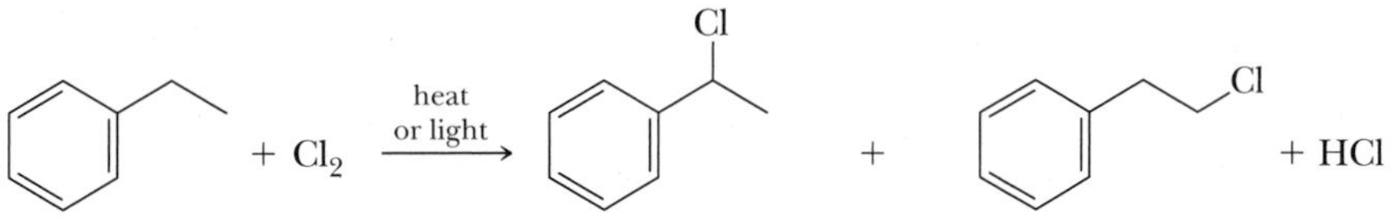

Ethylbenzene **1-Chloro-1-phenylethane** (90%) (racemic) **1-Chloro-2-phenylethane** (10%)

The chlorination of alkyl side chains is also regioselective but not to the same high degree as bromination. Recall that we observed this same pattern in the regioselectivities of bromination and chlorination of alkanes (Section 8.4A).

Combining the information on product distribution for bromination and chlorination of hydrocarbons, we conclude that the order of stability of radicals is

$$\text{methyl} < 1° < 2° < 3° < \text{allylic} \cong \text{benzylic}$$

This order reflects the C—H bond dissociation enthalpies (BDE) for formation of these radicals (Appendix 3).

C. Hydrogenolysis of Benzyl Ethers

Among ethers, benzylic ethers are unique in that they are cleaved under the conditions of catalytic hydrogenation as illustrated by the **hydrogenolysis** of benzyl hexyl ether. In this illustration, the benzyl group is converted to toluene, and the alkyl group is converted to an alcohol.

Hydrogenolysis
Cleavage of a single bond by H_2, most commonly accomplished by treating a compound with H_2 in the presence of a transition metal catalyst.

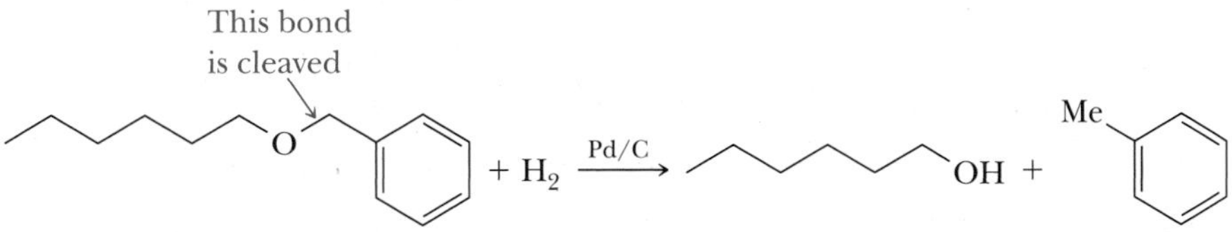

Benzyl hexyl ether **1-Hexanol** **Toluene**

Hydrogenolysis is the cleavage of a single bond by H_2. In the hydrogenolysis of a benzylic ether, the single bond between the benzylic carbon and its attached oxygen is cleaved and replaced by a carbon-hydrogen bond.

Benzyl ethers are formed by treatment of an alcohol or a phenol with benzyl chloride in the presence of a base such as triethylamine or pyridine. The particular value of benzylic ethers is that they can serve as protecting groups for the —OH groups of alcohols and phenols.

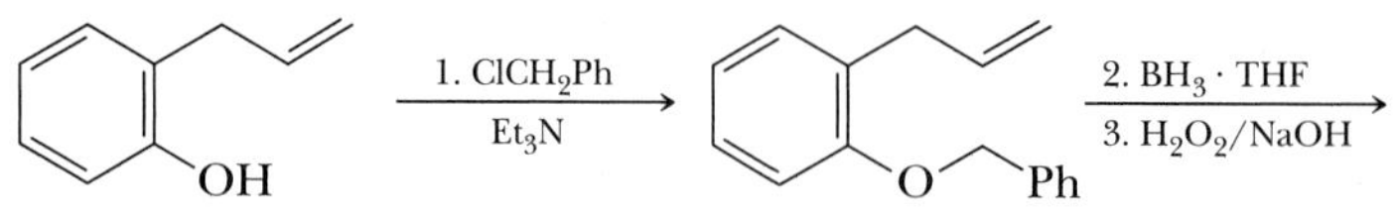

2-(2-Propenyl)phenol
(2-Allylphenol)

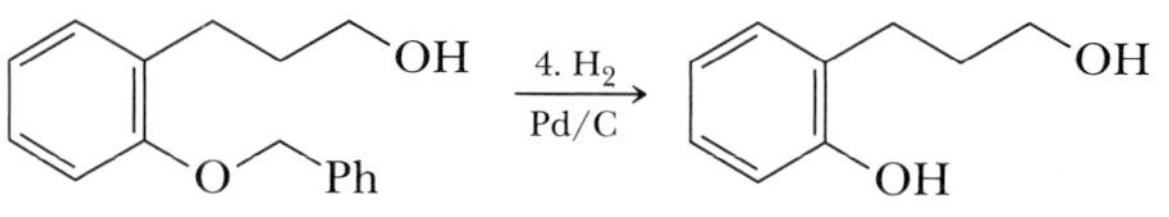

2-(3-Hydroxypropyl)phenol

Study Guide

21.1 The Structure of Benzene

- Benzene and its derivatives are classified as **aromatic**.
  - *Aromatic molecules display a remarkable stability that makes them unreactive toward reagents that attack other unsaturated species such as alkenes and alkynes.*
  - ***Aromaticity** is the term used to describe this special stability of benzene and its derivatives, and the term **arene** refers to aromatic hydrocarbons.*
- August Kekulé proposed that benzene is composed of six carbon atoms in a ring, with one hydrogen atom bonded to each carbon.
  - *The six carbon atoms of the ring are equivalent, and the carbon-carbon bond lengths are all intermediate between a single and double bond.*
  - *It is not accurate to think of benzene as having alternating single and double bonds that are static, because this would incorrectly predict alternating longer and shorter carbon-carbon bonds.*
- According to molecular orbital theory, benzene's structure is described as follows:
  - *Each carbon atom of the ring is* sp^2 *hybridized.*
  - *Each carbon atom of the ring makes* σ *bonds by* sp^2-sp^2 *overlap with the two adjacent carbon atoms and* sp^2-$1s$ *overlap with a hydrogen atom.*
  - *Each carbon atom also has a single unhybridized* $2p$ *orbital containing one electron.*
  - *The six* $2p$ *orbitals overlap to form six* π *molecular orbitals.*
    - *These molecular orbitals are arranged in a 1:2:2:1 pattern with respect to energy.*
    - *The six* π *electrons fill the three* π*-bonding molecular orbitals, all of which are at lower energy than the six isolated* $2p$ *orbitals, explaining why the benzene* π *system is so unreactive.*

  - *The lowest-lying filled molecular orbital has two torus-shaped lobes, one above and one below the plane of the ring, emphasizing the delocalized nature of the π system.*
  - *The other two filled molecular orbitals have one node apiece, underscoring the fact that the bond order between carbon atoms is intermediate between a double and single bond.*

- Benzene is best represented as a resonance hybrid composed of two resonance forms in which the locations of the double bonds are reversed. For simplicity, benzene is often represented as a single contributing structure or as a hexagon with a circle drawn on the inside.
- **Resonance energy** is the difference in energy between a resonance hybrid and the most stable hypothetical contributing structure in which electron density is localized on particular atoms and bonds.
  - *The resonance energy for benzene is large, namely* 36.0 kcal/mol, *meaning that the π system of benzene is extremely stable and less reactive than would be expected for an alkene under conditions such as catalytic hydrogenation.*

P 21.12–21.14

21.2 The Concept of Aromaticity

- Not all cyclic hydrocarbons with alternating double bonds possess the aromatic characteristics of benzene. Erich Hückel used molecular orbital calculations to define a set of requirements for aromaticity.
  - *The molecule must be cyclic.*
  - *There must be a* 2p *orbital on each atom of the ring (there cannot be any* sp^3 *hybridized atoms in the ring).*
  - *The molecule must be planar, or nearly so, to allow overlap of the 2p orbitals.*
  - *There must be* $(4n + 2)$ *π electrons in the aromatic π system, where n is a positive integer (0, 1, 2, 3, 4, 5, . . .) for a total of 2, 6, 10, 14, . . . π electrons.*
- Some hydrocarbons are considered antiaromatic, because they are much *less* stable (more reactive) than an acyclic analog with the same number of π electrons.
  - *An antiaromatic hydrocarbon has* 4n, *not* $(4n + 2)$, *π electrons.*
  - *Antiaromatic hydrocarbons must also be cyclic, have a* 2p *orbital on each atom of the ring, and be planar to display full antiaromatic properties.*
  - *This instability of antiaromatic hydrocarbons such as cyclobutadiene can be explained by using molecular orbital theory.*
    - *For cyclobutadiene, the four* 2p *orbitals of the π system form four molecular orbitals in a 1:2:1 pattern.*
    - *The four π electrons fill these orbitals to give one filled bonding π molecular orbital and two half-filled degenerate nonbonding molecular orbitals.*
    - *The presence of the two unpaired electrons makes cyclobutadiene reactive and unstable relative to aromatic hydrocarbons.*
  - *To avoid antiaromaticity, larger potentially antiaromatic structures such as cyclooctatetraene adopt a nonplanar geometry with alternating double and single bonds. As a result, two*

different carbon-carbon bond lengths are observed corresponding to single and double bonds, respectively.

- To predict the pattern of molecular orbitals found on a molecular orbital energy diagram, it is helpful to use the inscribed polygon method **(Frost circles)**.
  - *The shape of the polygon being analyzed (e.g., a hexagon for benzene) is drawn in a ring with one vertex down, and the relative energies of the molecular orbitals are indicated by the vertices of the polygon that touch the circle.*
  - *A horizontal line is drawn through the center of the figure.*
  - *Bonding molecular orbitals are below the line, nonbonding molecular orbitals (if any) are on the line, and antibonding molecular orbitals are above the line.*
- An **annulene** is a planar, cyclic hydrocarbon that contains continuously overlapping $2p$ orbitals. It is named by adding the number of atoms in the ring in brackets in front of the word *annulene.*
  - *Cyclobutadiene and benzene are annulenes, namely [4]annulene and [6]annulene, respectively.*
  - *Annulenes can be much larger, such as [14]annulene and [18]annulene, which are both aromatic because they are planar and have a Hückel number of* $(4n + 2)$ π *electrons.*
  - *[10]Annulene is not aromatic because the relatively small ring cannot adopt a planar geometry.*
- A **heterocyclic compound** contains more than one kind of atom in a ring. *Certain heterocycles can be aromatic if the Hückel criteria are met.*
  - *Nature is filled with aromatic heterocycles, such as indoles, purines, and pyrimidines.*
- An important parameter to keep track of in aromatic heterocycles is whether lone pairs of electrons are part of the aromatic π system.
  - *In pyridine (C_5H_5N), the lone pair of electrons on nitrogen is in an* sp^2 *hybrid orbital and is perpendicular to the six* $2p$ *orbitals of the aromatic six* π *electron system. This lone pair is not part of the aromatic* π *system and is free to take part in interactions with other species.*
  - *In pyrrole (C_4H_5N), the lone pair of electrons on nitrogen is in a* $2p$ *orbital and is part of the* π *system to allow for a total of six* π *electrons and aromaticity. This lone pair of electrons is not as available to take part in interactions with other species.*
- Because charged ring systems can satisfy the Hückel criteria for aromaticity, they are highly stabilized compared to other nonaromatic cations or anions. Examples include the cyclopropenyl cation, cycloheptatrienyl cation, and cyclopentadienyl anion.

P **21.1–21.4, 21.15–21.18**

▶ **The Hückel Criteria for Aromaticity and Figure 21.11**

21.3 Nomenclature

- The IUPAC system retains certain common names for several of the simpler benzene derivatives, including toluene, cumene, styrene, xylene, phenol, aniline, benzoic acid, and anisole.
  - *These common names are used as the parent name if their characteristic functional groups are present on a benzene derivative.*
- In molecules with other functional groups, the benzene ring is named as a substituent on a parent chain.

– *The C_6H_5 group is given the name* ***phenyl*** *and abbreviated Ph.*

– *The $C_6H_5CH_2$ group has characteristic reactivity, so it is given the name* ***benzyl*** *and is abbreviated Bn.*

- For benzene rings with two substituents, the three possible constitutional isomers are named **ortho** (1,2-substitution), **meta** (1,3-substitution), and **para** (1,4-substitution), which are abbreviated as ***o, m,*** and ***p,*** respectively.

  – *It is also acceptable to name these species with numbers as locators (such as 1,2- or 1,3-).*

  – *When one of the substituents has a special name (e.g., if NH_2 is present, the molecule is an aniline), the molecule is named after that parent molecule and the key group is assigned the number 1.*

  – *If none of the groups impart a special name, the substituents are listed in alphabetical order followed by the word* benzene. *For example, 1-chloro-4-ethylbenzene and* p*-chloroethylbenzene are acceptable names for the same molecule.*

- **Polynuclear aromatic hydrocarbons (PAHs)** contain more than one benzene ring, each pair of which shares two carbon atoms.

  – *Naphthalene contains two benzene rings fused together, and anthracene contains three benzene rings fused together in a linear fashion. Other common PAHs include phenanthrene, pyrene, coronene, and benzo*[a]*pyrene.*

P 21.15, 21.8–21.11

21.4 Phenols

- The characteristic feature of **phenols** is a hydroxyl group bonded to a benzene ring.

- Phenols (pK_a around 10) are more acidic than simple alcohols, because the negative charge of the phenoxide anion is highly delocalized into the aromatic ring as indicated by resonance contributing structures.

  – *Substituents on the ring that lead to a further stabilization of the phenoxide anion increase acidity of a phenol, and substituents that destabilize the phenoxide anion decrease acidity of a phenol.*

  – *According to the inductive effect, electron-withdrawing groups (more electronegative than* sp^2 *hybridized carbon) such as halogens stabilize a phenoxide anion by absorbing some of the negative charge, thus making a phenol more acidic, and electron-releasing groups (less electronegative than* sp^2 *hybridized carbon) such as alkyl groups destabilize a phenoxide anion by pushing even more electron density into the ring, making a phenol less acidic.*

  – *According to the resonance effect, certain groups, such as nitro groups, make phenols more acidic (especially at the ortho and para positions) because they can stabilize phenoxide anions through additional resonance delocalization of the negative charge.*

- Phenols react with strong bases to create water-soluble salts, a procedure that is useful for their isolation from mixtures.

- Phenoxides can be used in a Williamson ether synthesis by reaction with haloalkanes (S_N2 process) to create alkyl-aryl ethers.

- In the Kolbe reaction, phenoxide ion reacts with CO_2 to produce salicylic acid through a mechanism in which the phenoxide ion reacts like an enolate anion attacking the electrophilic carbon of CO_2, followed by loss of a proton and keto-enol tautomerization to give the salicylate anion.
- Phenols, especially benzenediols **(hydroquinones)**, are easily oxidized to **quinones**, and quinones are easily reduced back to benzenediols, forming a reversible oxidation/reduction couple that is the basis for important biological processes.

P 21.6, 21.7, 21.32–21.40

Problem 21.6

KEY REACTIONS

1. Acidity of Phenols (Section 21.4B) Phenols are weak acids, pK_a approximately 10. Ring substituents may increase or decrease acidity by a combination of resonance and inductive effects.

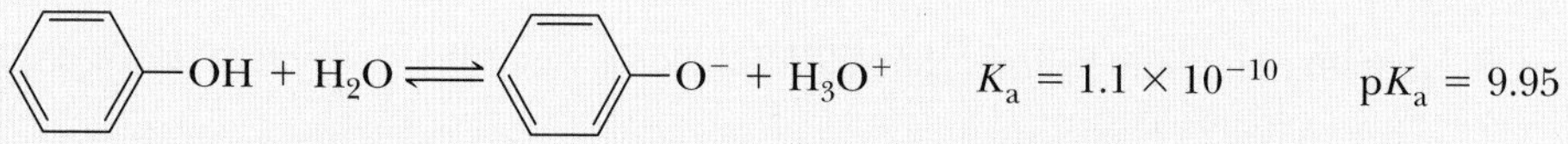

2. Reaction of Phenols with Strong Bases (Section 21.4C) Water-insoluble phenols react quantitatively with strong bases to form water-soluble salts.

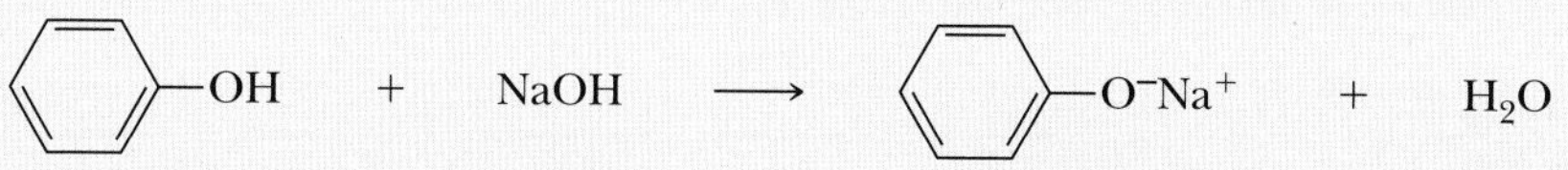

Phenol	Sodium hydroxide	Sodium phenoxide	Water
pK_a 9.95 (stronger acid)			pK_a 15.7 (weaker acid)

3. Kolbe Synthesis: Carboxylation of Phenols (Section 21.4E) Nucleophilic addition of a phenoxide ion to carbon dioxide gives a substituted cyclohexadienone, which then undergoes keto-enol tautomerism to regenerate the aromatic ring.

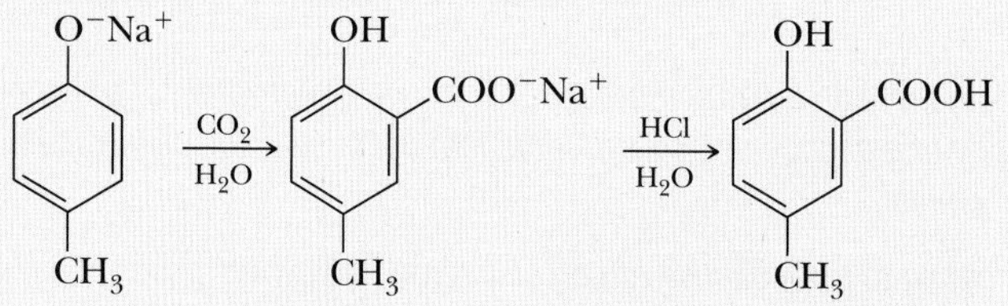

4. Oxidation of Phenols to Quinones (Section 21.4F) Oxidation by H_2CrO_4 gives 1,2-quinones (*o*-quinones) or 1,4-quinones (*p*-quinones), depending on the structure of the particular phenol.

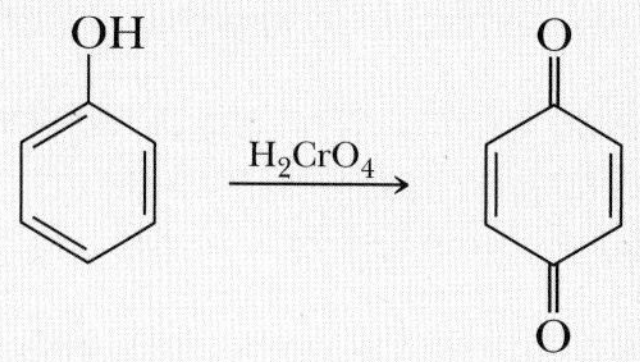

Phenol **1,4-Benzoquinone (*p*-Quinone)**

21.5 Reactions at a Benzylic Position

- Several characteristic reactions occur at the benzylic position of aromatic hydrocarbons because benzylic cation and radical intermediates are stabilized through delocalization into the aromatic ring.
- Compounds with at least one benzylic hydrogen react with $KMnO_4$ in aqueous base or H_2CrO_4 ($Na_2Cr_2O_7$ in aqueous sulfuric acid) to produce benzoic acids. Other groups bonded to the benzylic carbon atom are removed in the process.
- Benzylic hydrogens can be replaced by bromine or chlorine in the presence of light or heat.
  - *In compounds such as toluene, more than one of the benzylic hydrogens can be replaced when excess halogen is used.*
  - *In molecules with alkyl groups larger than methyl bonded to the benzylic carbon, the reaction is selective for replacement of the benzylic hydrogens because the benzylic radical is more stable than the other possible radicals.*
  - *Bromine is more selective than chlorine in these reactions, and NBS is often used as the bromination reagent.*
- Benzylic ethers are unique among ethers in that they can be cleaved using catalytic hydrogenation (H_2 and Pd) to give toluene and an alcohol, a process referred to as **hydrogenolysis**.
  - *Hydrogenolysis of benzyl ethers makes them useful as an alcohol protecting group.*

P 21.41–21.62

KEY REACTIONS

5. Oxidation at a Benzylic Position (Section 21.5A) A benzylic carbon bonded to at least one hydrogen is oxidized to a carboxyl group.

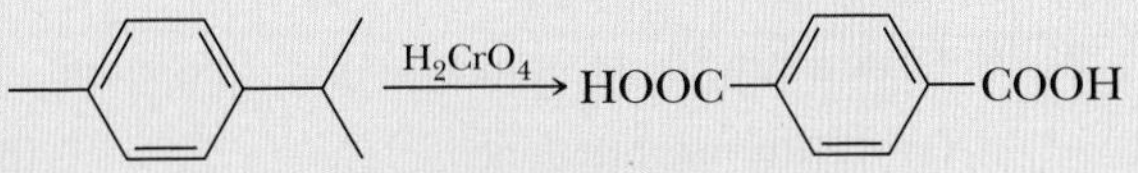
H2CrO4
HOOC
COOH

6. Halogenation at a Benzylic Position (Section 21.5B) Halogenation is regioselective for a benzylic position and occurs by a radical chain mechanism. Bromination shows a higher regioselectivity for a benzylic position than does chlorination. The reaction occurs via a radical chain mechanism initiated when the X_2 is converted to two X· radicals; then an X· radical abstracts the benzylic hydrogen to create a resonance-stabilized benzylic radical that reacts with another molecule of X_2 to give the halogenated product and a new X· radical that continues the chain reaction. NBS can be used as the source of Br_2.

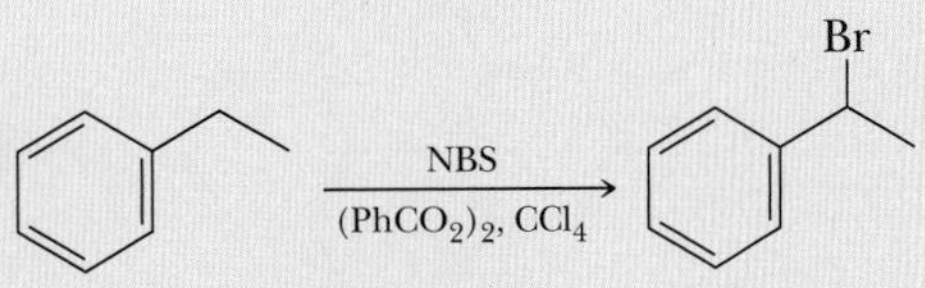
NBS
(PhCO2)2, CCl4
Br

(racemic)

7. Hydrogenolysis of Benzylic Ethers (Section 21.5C) Benzylic ethers are cleaved under the conditions of catalytic hydrogenation.

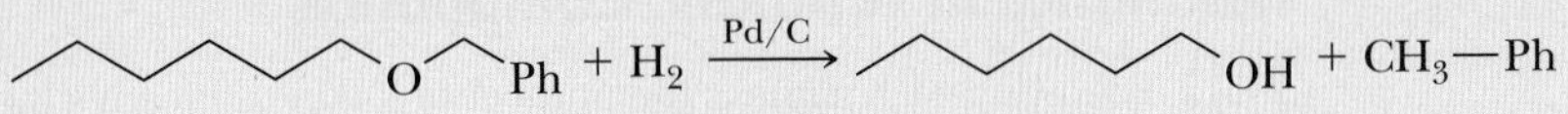
O
Ph
+ H2
Pd/C
OH
+ CH3—Ph

Problems

Red numbers indicate applied problems.

Nomenclature and Structural Formulas

21.8 Name the following compounds and ions.

(a) 

(b) 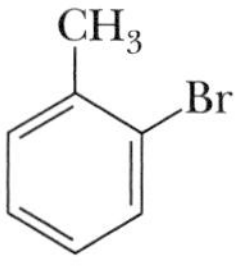

(c) 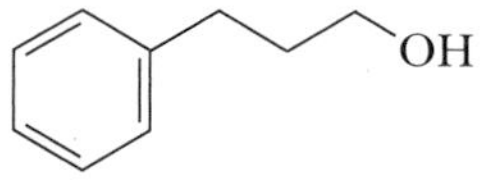

(d) 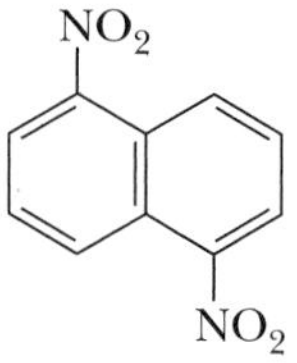

(e) 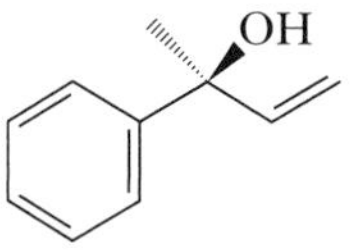

(f) 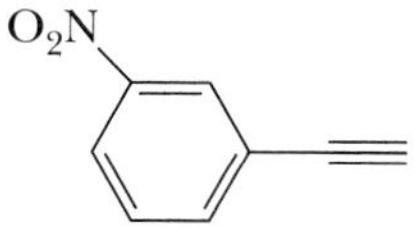

(g) 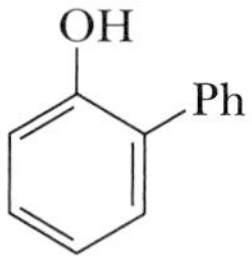

(h) 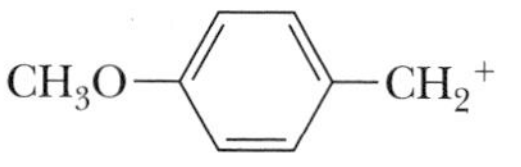

(i) 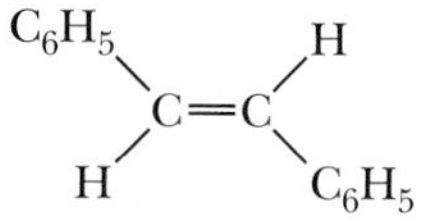

(j) 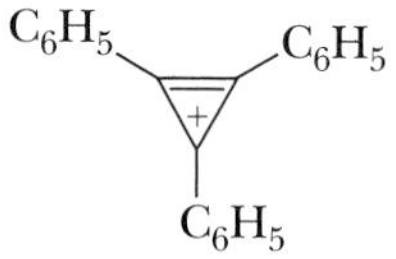

21.9 Draw a structural formula for each compound.

(a) 1-Bromo-2-chloro-4-ethylbenzene
(b) *m*-Nitrocumene
(c) 4-Chloro-1,2-dimethylbenzene
(d) 3,5-Dinitrotoluene
(e) 2,4,6-Trinitrotoluene
(f) (2*S*,4*R*)-4-Phenyl-2-pentanol
(g) *p*-Cresol
(h) Pentachlorophenol
(i) 1-Phenylcyclopropanol
(j) Triphenylmethane
(k) Phenylethylene (styrene)
(l) Benzyl bromide
(m) 1-Phenyl-1-butyne
(n) (*E*)-3-Phenyl-2-propen-1-ol

21.10 Draw a structural formula for each compound.

(a) 1-Nitronaphthalene
(b) 1,6-Dichloronaphthalene
(c) 9-Bromoanthracene
(d) 2-Methylphenanthrene

21.11 Molecules of 6,6′-dinitrobiphenyl-2,2′-dicarboxylic acid have no tetrahedral chiral center, and yet they can be resolved to a pair of enantiomers. Account for this chirality.

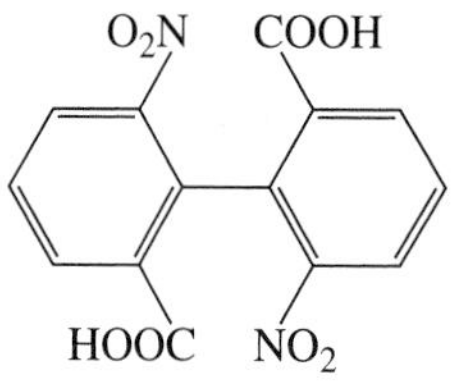

6,6′-Dinitrobiphenyl-2,2′-dicarboxylic acid

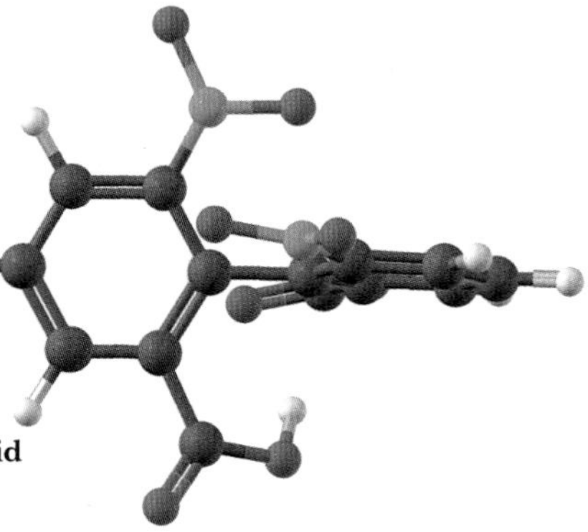

Resonance in Aromatic Compounds

21.12 Following each name is the number of Kekulé structures that can be drawn for it. Draw these Kekulé structures and show, using curved arrows, how the first contributing structure for each molecule is converted to the second and so forth.

(a) Naphthalene (3)
(b) Phenanthrene (5)

21.13 Each molecule in this problem can be drawn as a hybrid of five contributing structures: two Kekulé structures and three that involve creation and separation of unlike charges. Draw these five contributing structures for each molecule.

(a) Chlorobenzene **(b)** Phenol **(c)** Nitrobenzene

21.14 Following are structural formulas for furan and pyridine.

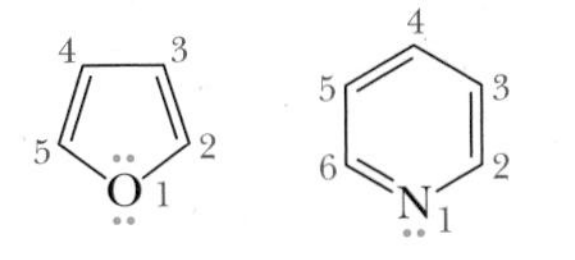

Furan **Pyridine**

(a) Draw four contributing structures for furan that place a positive charge on oxygen and a negative charge first on carbon 3 of the ring and then on each other carbon of the ring.

(b) Draw three contributing structures for pyridine that place a negative charge on nitrogen and a positive charge first on carbon 2, then on carbon 4, and finally on carbon 6.

The Concept of Aromaticity

21.15 State the number of 2*p* orbital electrons in each molecule or ion.

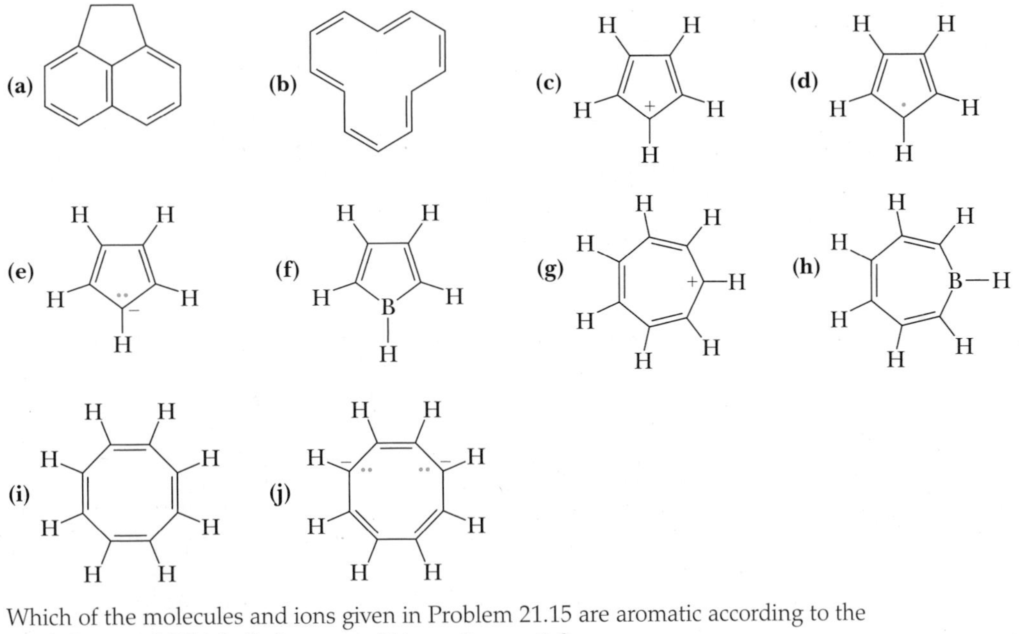

21.16 Which of the molecules and ions given in Problem 21.15 are aromatic according to the Hückel criteria? Which, if planar, would be antiaromatic?

21.17 Construct MO energy diagrams for the cyclopropenyl cation, radical, and anion. Which of these species is aromatic according to the Hückel criteria?

21.18 Naphthalene and azulene are constitutional isomers of molecular formula $C_{10}H_8$.

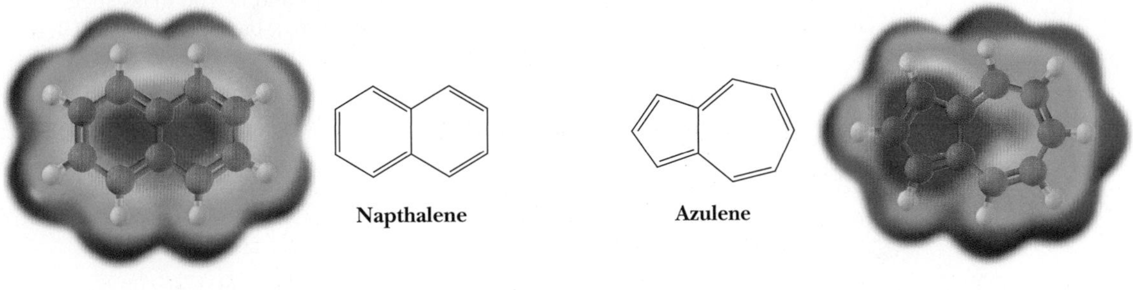

Napthalene **Azulene**

Naphthalene is a colorless solid with a dipole moment of zero. Azulene is a solid with an intense blue color and a dipole moment of 1.0 D. Account for the difference in dipole moments of these constitutional isomers.

Spectroscopy

21.19 Compound A (C_9H_{12}) shows prominent peaks in its mass spectrum at *m/z* 120 and 105. Compound B (also C_9H_{12}) shows prominent peaks at *m/z* 120 and 91. On vigorous oxidation with chromic acid, both compounds give benzoic acid. From this information, deduce the structural formulas of compounds A and B.

21.20 Compound C shows a molecular ion at *m/z* 148 and other prominent peaks at *m/z* 105 and 77. Following are its infrared and ^{1}H-NMR spectra.

(a) Deduce the structural formula of compound C.
(b) Account for the appearance of peaks in its mass spectrum at *m/z* 105 and 77.

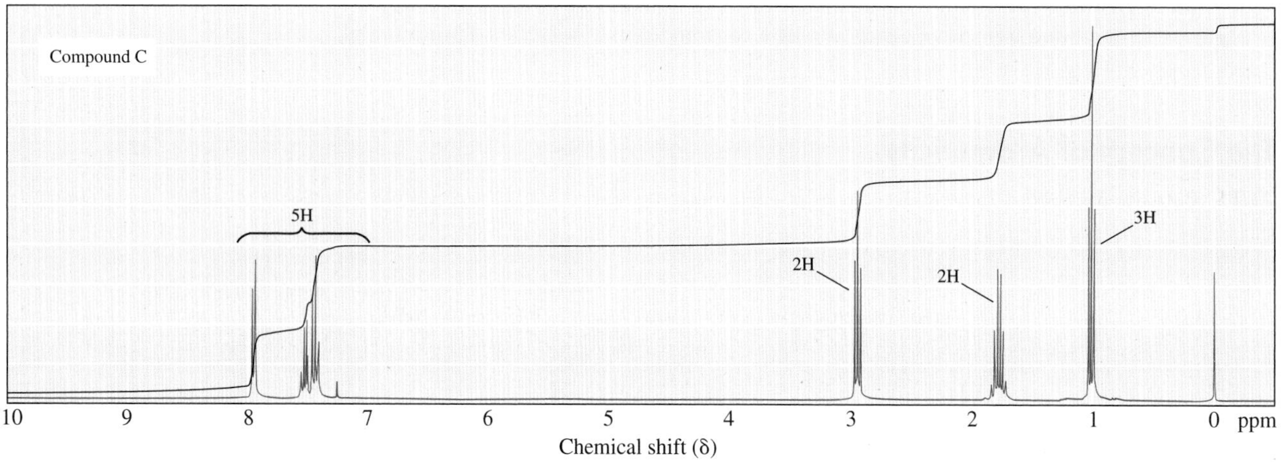

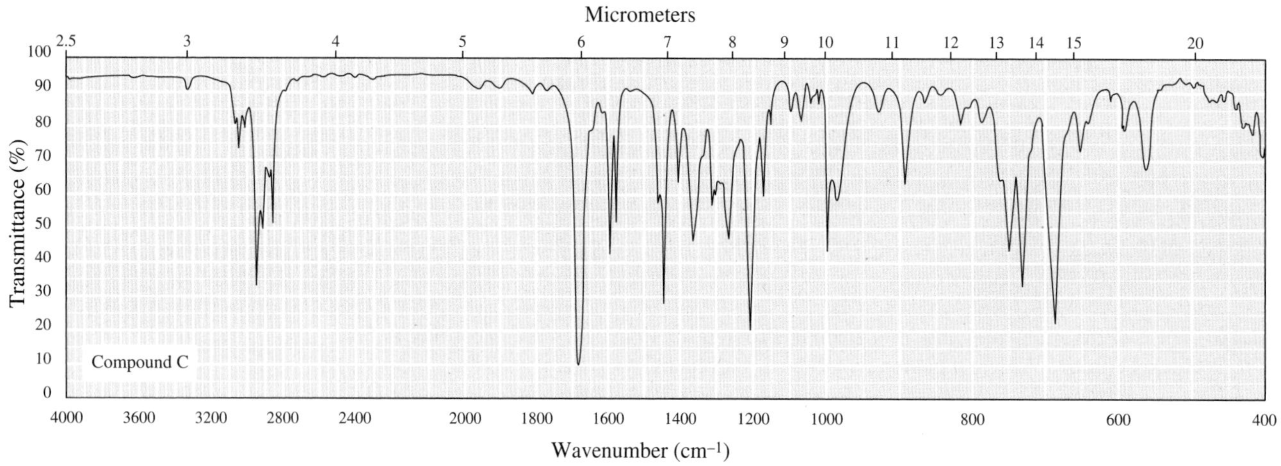

21.21 Following are IR and ^{1}H-NMR spectra of compound D. The mass spectrum of compound D shows a molecular ion peak at *m/z* 136, a base peak at *m/z* 107, and other prominent peaks at *m/z* 118 and 59.

(a) Propose a structural formula for compound D.
(b) Propose structural formulas for ions in the mass spectrum at *m/z* 118, 107, and 59.

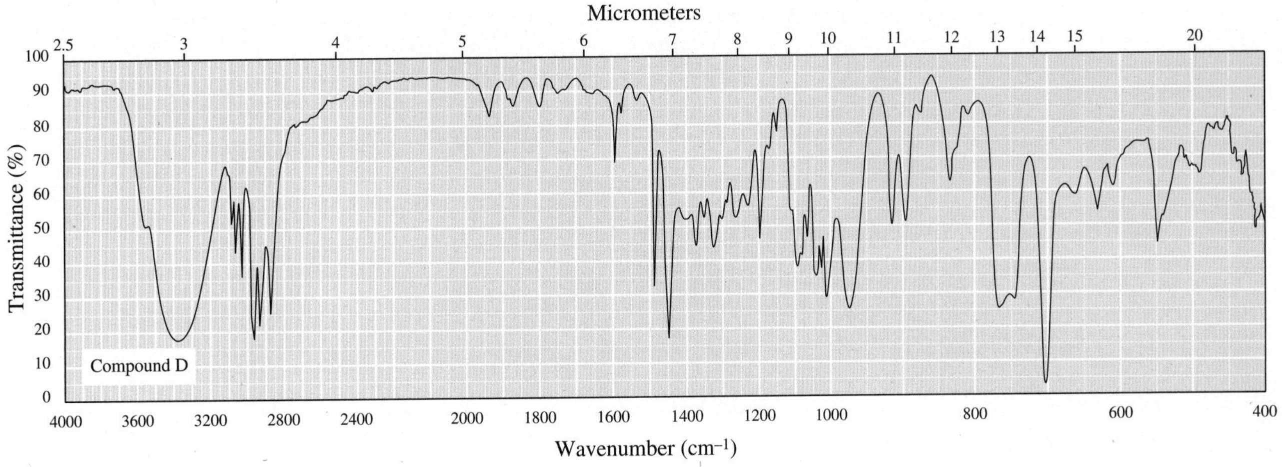

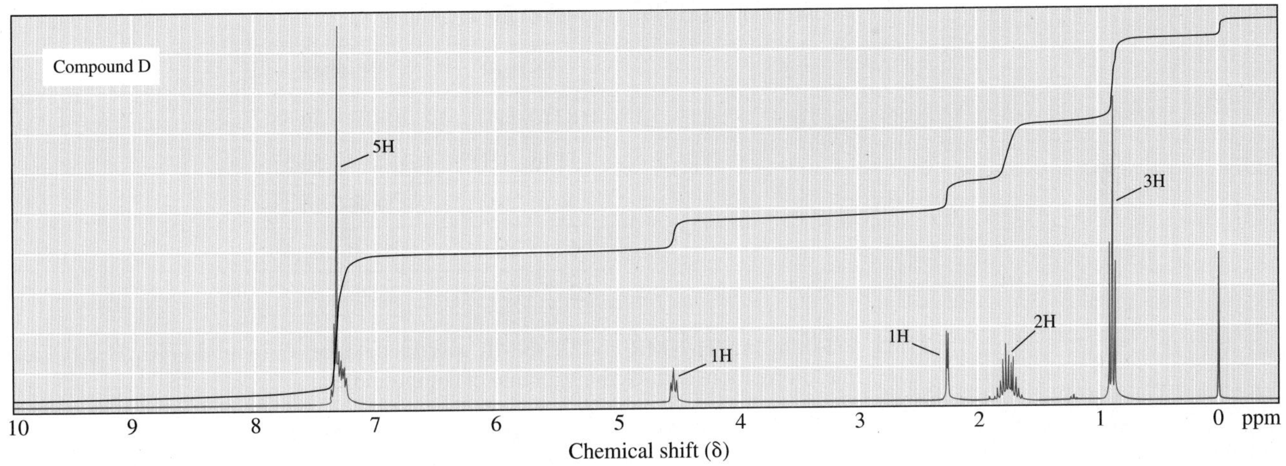

21.22 Compound E ($C_8H_{10}O_2$) is a neutral solid. Its mass spectrum shows a molecular ion at *m/z* 138 and prominent peaks at M – 1 and M – 17. Following are IR and ^{1}H-NMR spectra of compound E. Deduce the structure of compound E.

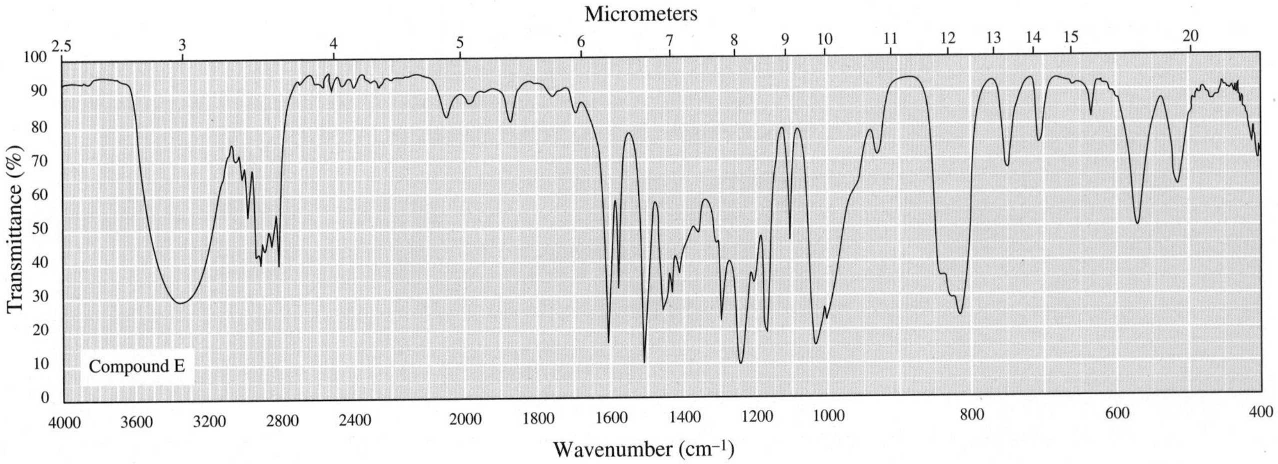

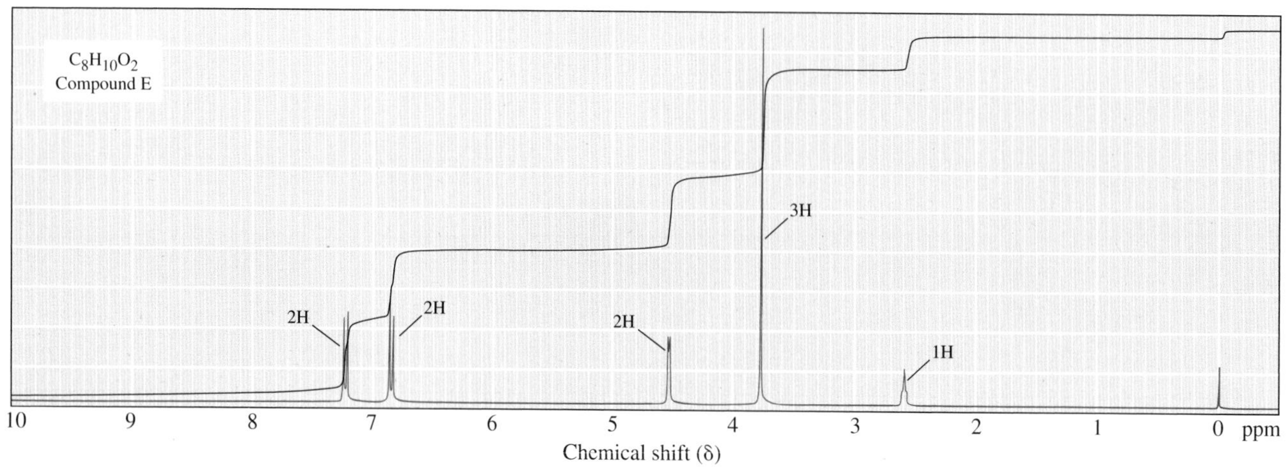

21.23 Following are ^{1}H-NMR and ^{13}C-NMR spectral data for compound F ($C_{12}H_{16}O$). From this information, deduce the structure of compound F.

^{1}H-NMR	^{13}C-NMR	
0.83 (d, 6H)	207.82	50.88
2.11 (m, 1H)	134.24	50.57
2.30 (d, 2H)	129.36	24.43
3.64 (s, 2H)	128.60	22.48
7.2–7.4 (m, 5H)	126.86	

21.24 Following are ^{1}H-NMR and ^{13}C-NMR spectral data for compound G ($C_{10}H_{10}O$). From this information, deduce the structure of compound G.

^{1}H-NMR	^{13}C-NMR	
2.50 (t, 2H)	210.19	126.82
3.05 (t, 2H)	136.64	126.75
3.58 (s, 2H)	133.25	45.02
7.1–7.3 (m, 4H)	128.14	38.11
	127.75	28.34

21.25 Compound H ($C_8H_6O_3$) gives a precipitate when treated with hydroxylamine in aqueous ethanol and a silver mirror when treated with Tollens' solution. Following is its ^{1}H-NMR spectrum. Deduce the structure of compound H.

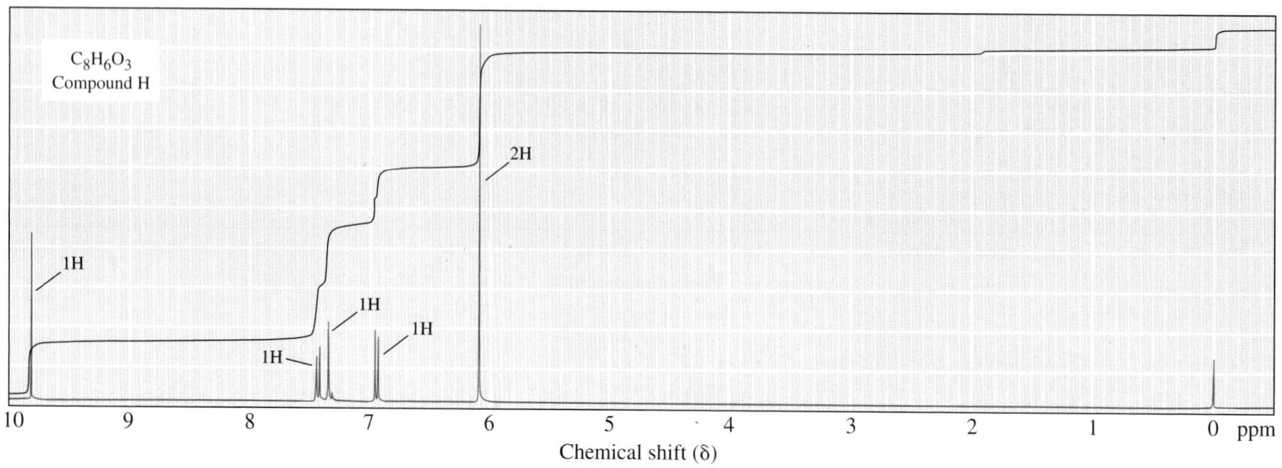

21.26 Compound I ($C_{11}H_{14}O_2$) is insoluble in water, aqueous acid, and aqueous $NaHCO_3$ but dissolves readily in 10% Na_2CO_3 and 10% NaOH. When these alkaline solutions are acidified with 10% HCl, compound I is recovered unchanged. Given this information and its ^{1}H-NMR spectrum, deduce the structure of compound I.

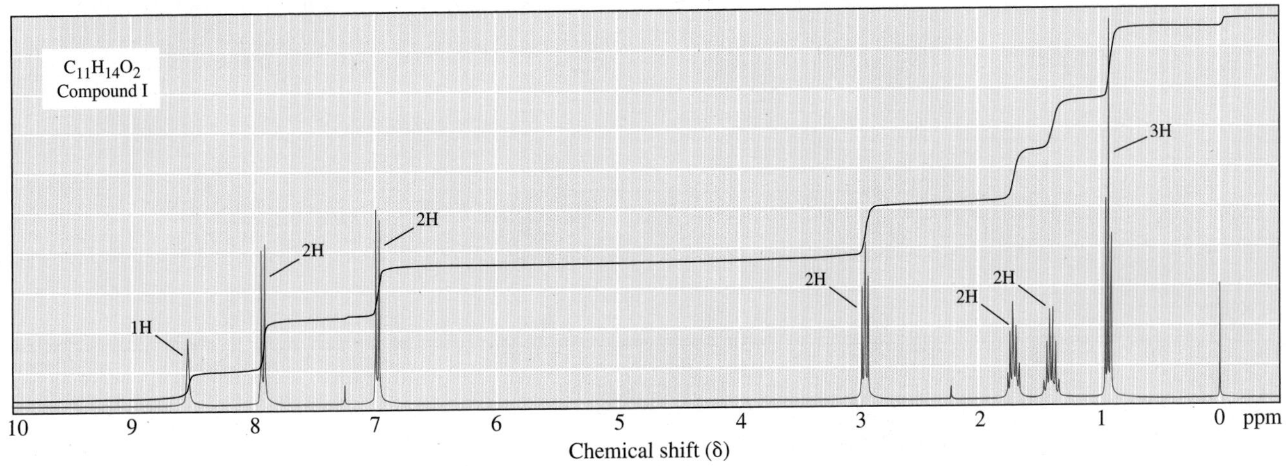

21.27 Propose a structural formula for compound J ($C_{11}H_{14}O_3$) consistent with its ^{1}H-NMR and infrared spectra.

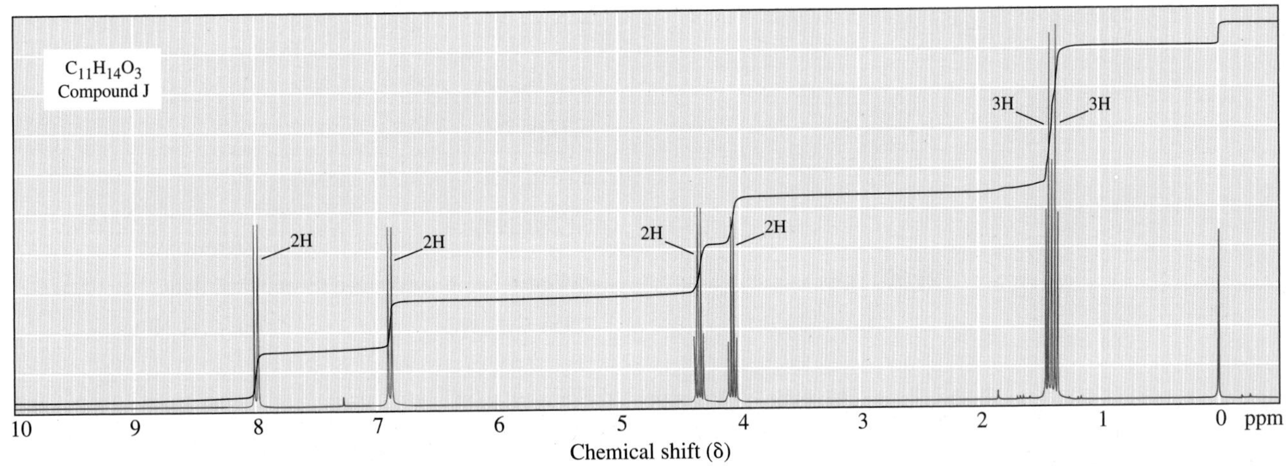

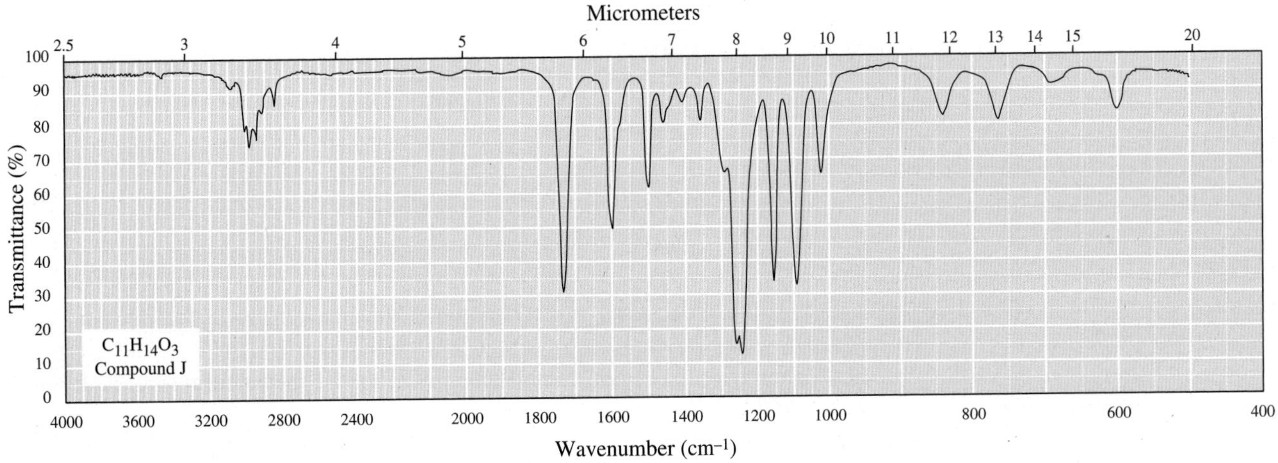

21.28 Propose a structural formula for the analgesic phenacetin, molecular formula $C_{10}H_{13}NO_2$, based on its ^{1}H-NMR spectrum.

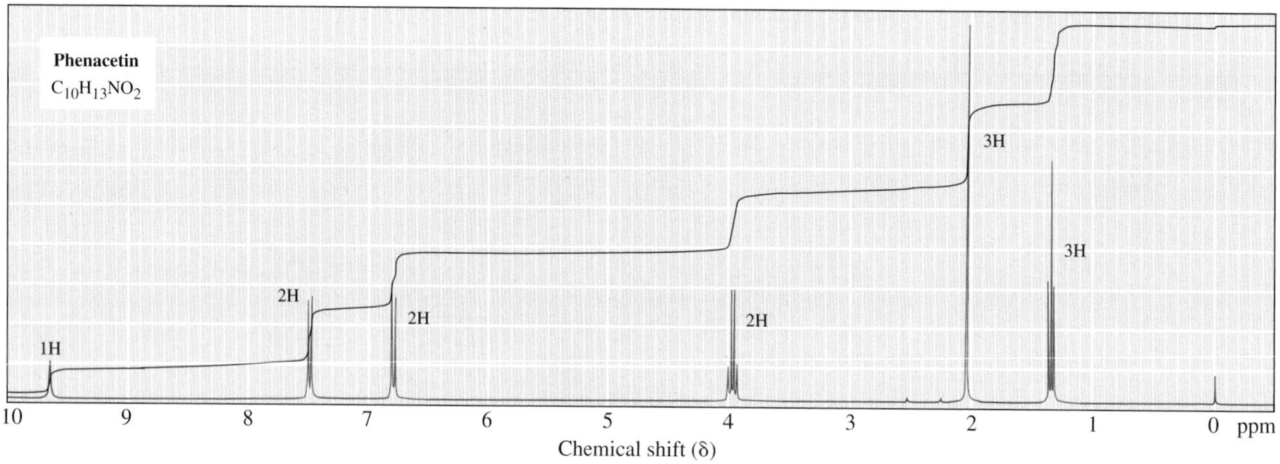

21.29 Compound K, $C_{10}H_{12}O_2$, is insoluble in water, 10% NaOH, and 10% HCl. Given this information and the following ^{1}H-NMR and ^{13}C-NMR spectral information, deduce the structural formula of Compound K.

^{1}H-NMR	^{13}C-NMR	
2.10 (s, 3H)	206.51	114.17
3.61 (s, 2H)	158.67	55.21
3.77 (s, 3H)	130.33	50.07
6.86 (d, 2H)	126.31	29.03
7.12 (d, 2H)		

21.30 Propose a structural formula for each compound given these NMR data.

(a) C_9H_9BrO

^{1}H-NMR	^{13}C-NMR
1.39 (t, 3H)	165.73
4.38 (q, 2H)	131.56
7.57 (d, 2H)	131.01
7.90 (d, 2H)	129.84
	127.81
	61.18
	14.18

(b) C_8H_9NO

^{1}H-NMR	^{13}C-NMR
2.06 (s, 3H)	168.14
7.01 (t, 1H)	139.24
7.30 (m, 2H)	128.51
7.59 (d, 2H)	122.83
9.90 (s, 1H)	118.90
	23.93

(c) $C_9H_9NO_3$

^{1}H-NMR	^{13}C-NMR
2.10 (s, 3H)	168.74
7.72 (d, 2H)	166.85
7.91 (d, 2H)	143.23
10.3 (s, 1H)	130.28
12.7 (s, 1H)	124.80
	118.09
	24.09

21.31 Given here are ^{1}H-NMR and ^{13}C-NMR spectral data for two compounds. Each shows strong, sharp absorption between 1700 and 1720 cm^{-1} and strong, broad absorption over the region 2500–3000 cm^{-1}. Propose a structural formula for each compound.

(a) $C_{10}H_{12}O_3$

^{1}H-NMR	^{13}C-NMR
2.49 (t, 2H)	173.89
2.80 (t, 2H)	157.57
3.72 (s, 3H)	132.62
6.78 (d, 2H)	128.99
7.11 (d, 2H)	113.55
12.4 (s, 1H)	54.84
	35.75
	29.20

(b) $C_{10}H_{10}O_2$

^{1}H-NMR	^{13}C-NMR
2.34 (s, 3H)	167.82
6.38 (d, 1H)	143.82
7.18 (d, 1H)	139.96
7.44 (d, 2H)	131.45
7.56 (d, 2H)	129.37
12.0 (s, 1H)	127.83
	111.89
	21.13

Acidity of Phenols

21.32 Account for the fact that *p*-nitrophenol (K_a 7.0×10^{-8}) is a stronger acid than is phenol (K_a 1.1×10^{-10}).

21.33 Account for the fact that water-insoluble carboxylic acids (pK_a 4–5) dissolve in 10% aqueous sodium bicarbonate (pH 8.5) with the evolution of a gas, but that water-insoluble phenols (pK_a 9.5–10.5) do not dissolve in 10% sodium bicarbonate.

21.34 Match each compound with its appropriate pK_a value.

(a) 4-Nitrobenzoic acid, benzoic acid, 4-chlorobenzoic acid
pK_a = 4.19, 3.98, and 3.41
(b) Benzoic acid, cyclohexanol, phenol
pK_a = 18.0, 9.95, and 4.19
(c) 4-Nitrobenzoic acid, 4-nitrophenol, 4-nitrophenylacetic acid
pK_a = 7.15, 3.85, and 3.41

21.35 Arrange the molecules and ions in each set in order of increasing acidity (from least acidic to most acidic).

(a) 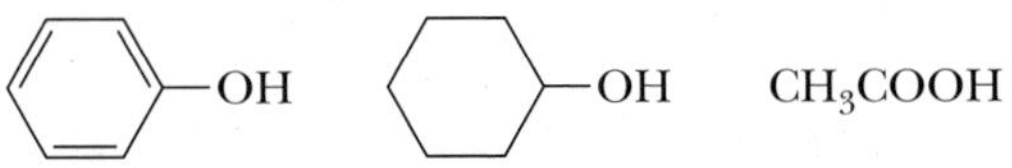
OH
OH
CH₃COOH

(b) 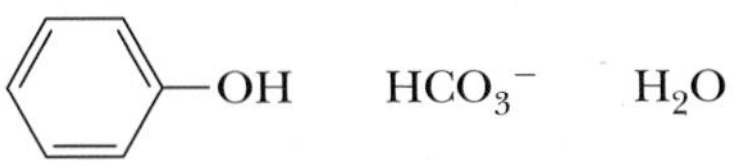
OH
HCO₃⁻
H₂O

(c) 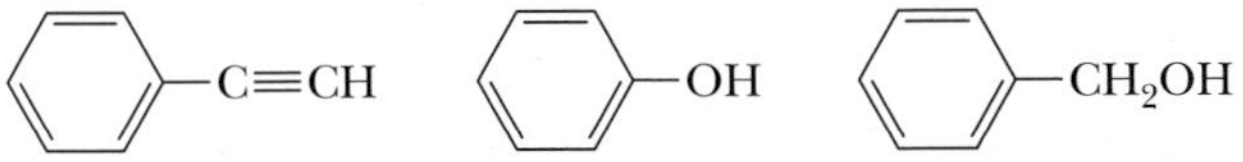
C≡CH
OH
CH₂OH

21.36 Explain the trends in the acidity of phenol and the monofluoro derivatives of phenol.

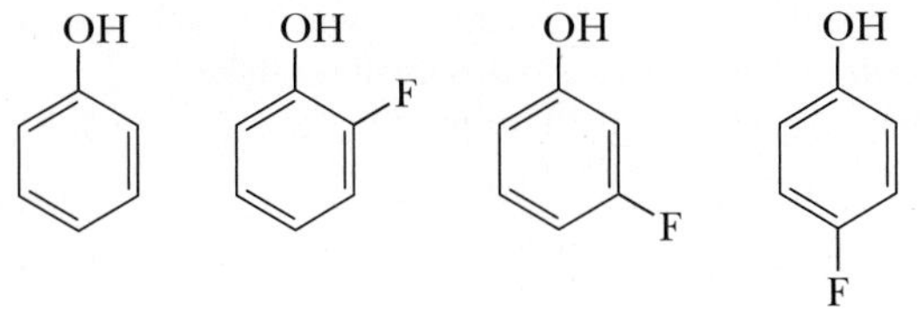
OH
OH
F
OH
F
OH
F

pK_a 10.0 pK_a 8.81 pK_a 9.28 pK_a 9.81

21.37 Suppose you want to determine the inductive effects of a series of functional groups (e.g., Cl, Br, CN, COOH, and C_6H_5). Is it best to use a series of ortho-, meta-, or para-substituted phenols? Explain.

21.38 From each pair, select the stronger base.

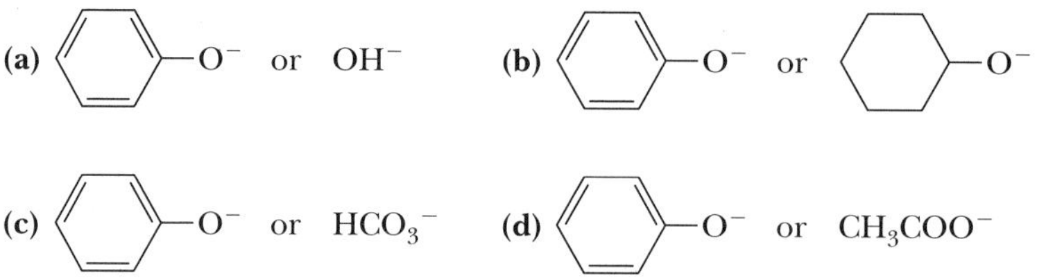

21.39 Describe a chemical procedure to separate a mixture of benzyl alcohol and *o*-cresol and to recover each in pure form.

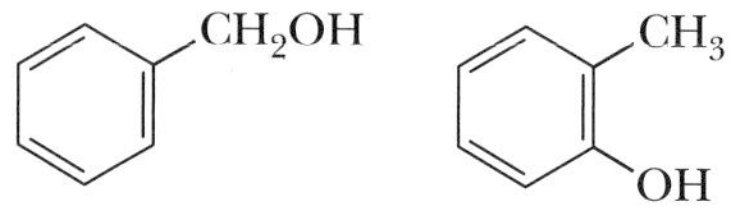

Benzyl alcohol **o-Cresol**

21.40 The compound 2-hydroxypyridine, a derivative of pyridine, is in equilibrium with 2-pyridone. 2-Hydroxypyridine is aromatic. Does 2-pyridone have comparable aromatic character? Explain.

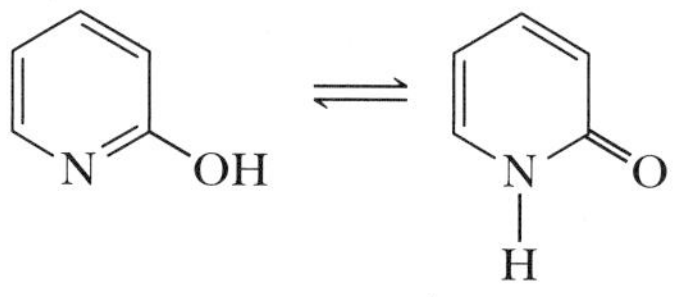

2-Hydroxypyridine **2-Pyridone**

Reactions at the Benzylic Position

21.41 Write a balanced equation for the oxidation of *p*-xylene to 1,4-benzenedicarboxylic acid (terephthalic acid) using potassium dichromate in aqueous sulfuric acid. How many milligrams of H_2CrO_4 are required to oxidize 250 mg of *p*-xylene to terephthalic acid?

21.42 Each of the following reactions occurs by a radical chain mechanism.

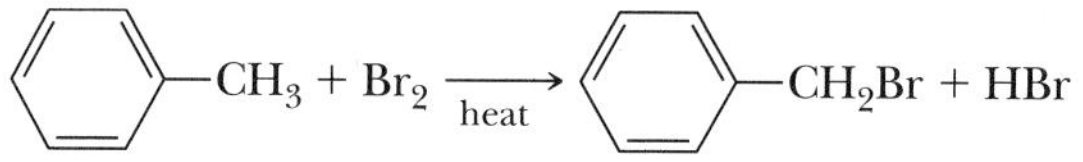

Toluene **Benzyl bromide**

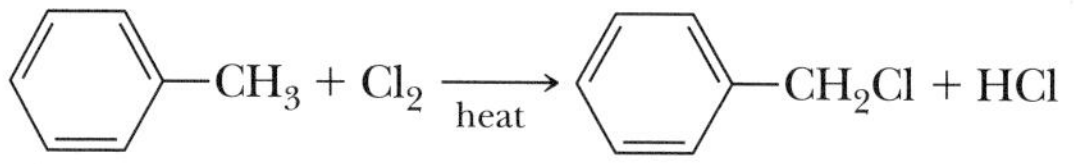

Toluene **Benzyl chloride**

(a) Calculate the heat of reaction, ΔH^0, in kJ/mol for each reaction. (Consult Appendix 3 for bond dissociation enthalpies.)

(b) Write a pair of chain propagation steps for each mechanism and show that the net result of each pair is the observed reaction.

(c) Calculate ΔH^0 for each chain propagation step and show that the sum for each pair of steps is identical with the ΔH^0 value calculated in part (a).

21.43 Following is an equation for iodination of toluene.

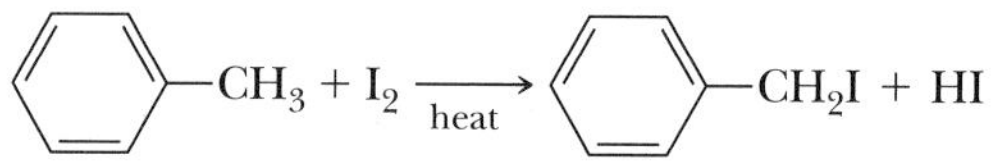

Toluene **Benzyl iodide**

This reaction does not take place. All that happens under experimental conditions for the formation of radicals is initiation to form iodine radicals, I·, followed by termination to reform I_2. How do you account for these observations?

21.44 Although most alkanes react with chlorine by a radical chain mechanism when reaction is initiated by light or heat, benzene fails to react under the same conditions. Benzene cannot be converted to chlorobenzene by treatment with chlorine in the presence of light or heat.

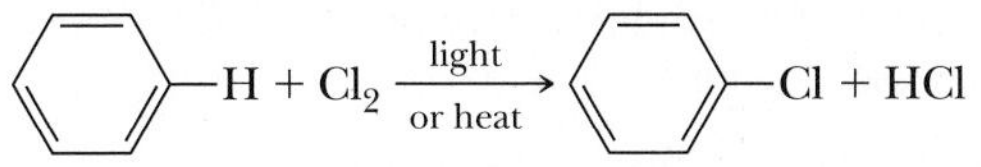

(a) Explain why benzene fails to react under these conditions. (Consult Appendix 3 for relevant bond dissociation enthalpies.)

(b) Explain why the bond dissociation enthalpy of a C—H bond in benzene is significantly greater than that in alkanes.

21.45 Following is an equation for hydroperoxidation of cumene.

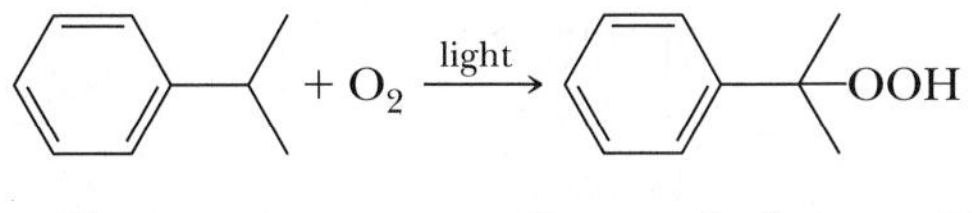

Cumene **Cumene hydroperoxide**

Propose a radical chain mechanism for this reaction. Assume that initiation is by an unspecified radical, R·.

21.46 Para-substituted benzyl halides undergo reaction with methanol by an S_N1 mechanism to give a benzyl ether. Account for the following order of reactivity under these conditions.

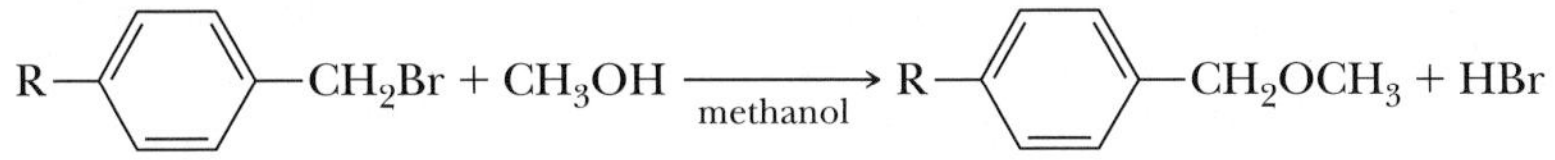

Rate of S_N1 reaction: R = $CH_3O > CH_3 > H > NO_2$

21.47 When warmed in dilute sulfuric acid, 1-phenyl-1,2-propanediol undergoes dehydration and rearrangement to give 2-phenylpropanal.

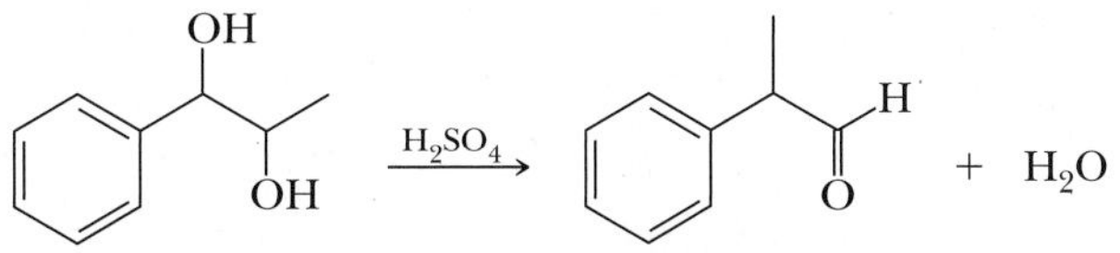

1-Phenyl-1,2-propanediol (racemic) **2-Phenylpropanal** (racemic)

(a) Propose a mechanism for this example of a pinacol rearrangement (Section 10.7).

(b) Account for the fact that 2-phenylpropanal is formed rather than its constitutional isomer, 1-phenyl-1-propanone.

21.48 In the chemical synthesis of DNA and RNA, hydroxyl groups are normally converted to triphenylmethyl (trityl) ethers to protect the hydroxyl group from reaction with other reagents.

$$RCH_2OH + Ph_3CCl \xrightarrow{\text{tertiary amine}} RCH_2OCPh_3 + HCl$$

Triphenylmethyl chloride (Trityl chloride) — A triphenylmethyl ether (A trityl ether) — HCl: neutralized by the tertiary amine

Triphenylmethyl ethers are stable to aqueous base but are rapidly cleaved in aqueous acid.

$$RCH_2OCPh_3 + H_2O \xrightarrow{H^+} RCH_2OH + Ph_3COH$$

(a) Why are triphenylmethyl ethers so readily hydrolyzed by aqueous acid?
(b) How might the structure of the triphenylmethyl group be modified to increase or decrease its acid sensitivity?

Organic Chemistry Reaction Roadmap

Reaction Roadmap

21.49 Use the reaction roadmaps you made for Problem 20.55 and update it to contain the reactions in the "Study Guide" section of this chapter. Because of their highly specific nature, do not use reactions 1, 2, and 7 on your roadmap.

Reaction Roadmap

21.50 Write the products of the following sequences of reactions. Refer to your reaction roadmaps to see how the combined reactions allow you to "navigate" between the different functional groups. Note that you will need your old Chapters 6–11, Chapters 15–18, and Chapter 19 roadmaps along with your new Chapters 20–21 roadmaps for these.

(a)

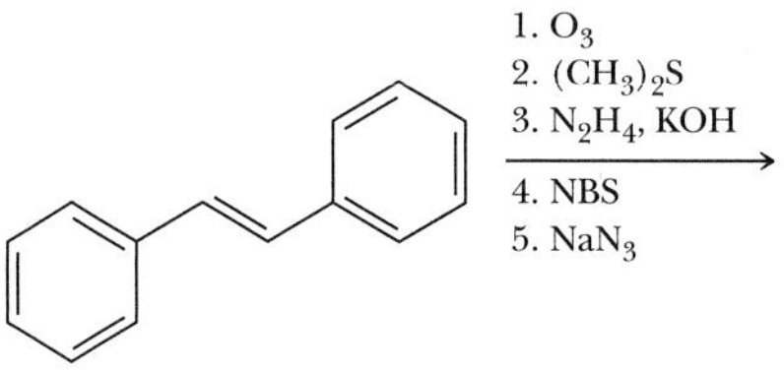
1. O3
2. (CH3)2S
3. N2H4, KOH
4. NBS
5. NaN3

An alkene

(b)

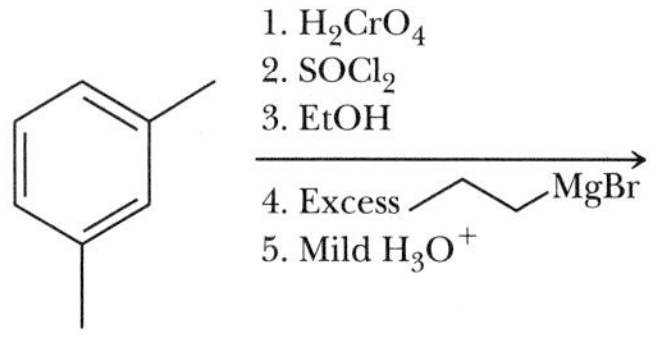
1. H2CrO4
2. SOCl2
3. EtOH
4. Excess MgBr
5. Mild H3O+

An alkyl benzene

(c)

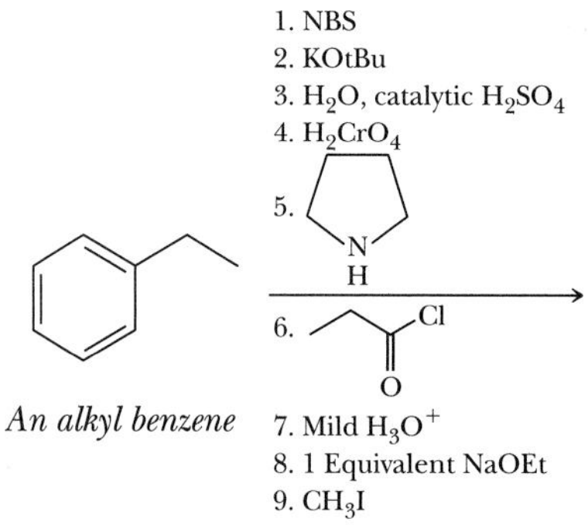
1. NBS
2. KOtBu
3. H2O, catalytic H2SO4
4. H2CrO4
5.
N
H
6.
Cl
O
7. Mild H3O+
8. 1 Equivalent NaOEt
9. CH3I

An alkyl benzene

Synthesis

21.51 Using ethylbenzene as the only aromatic starting material, show how to synthesize the following compounds. In addition to ethylbenzene, use any other necessary organic or inorganic chemicals. Any compound already synthesized in one part of this problem may then be used to make any other compound in the problem.

(a) 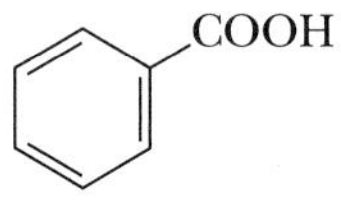
COOH

(b) 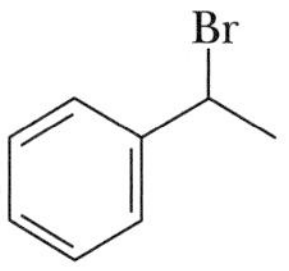
Br

(racemic)

(c) 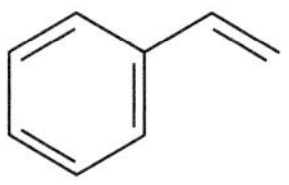

(d) 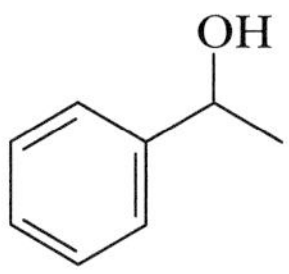
OH

(racemic)

(e) 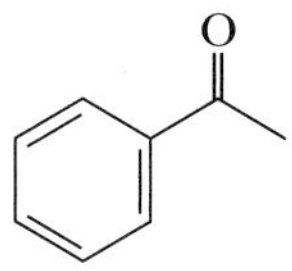
O

(f) 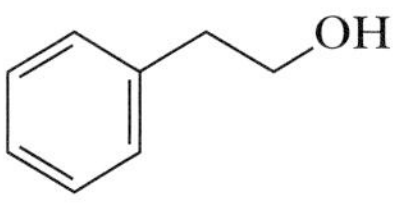
OH

(g) 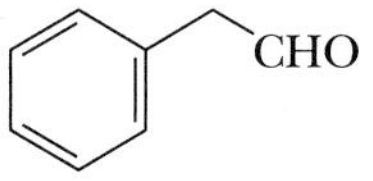
CHO

(h) 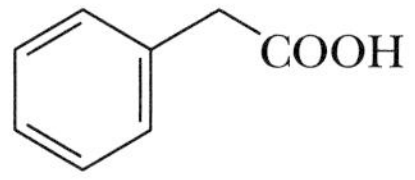
COOH

(i) 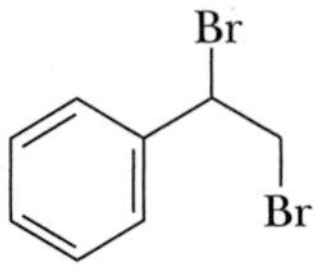 (j) 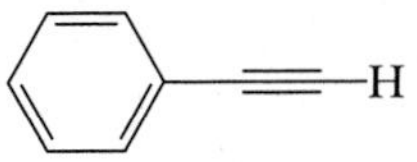 (k) 

(racemic)

(l) 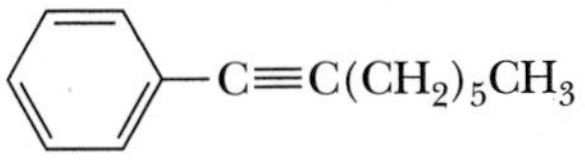 (m) 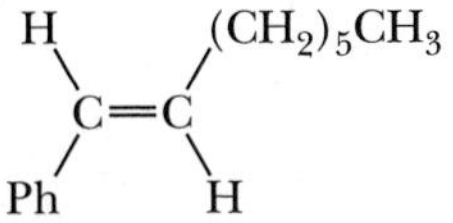 (n) 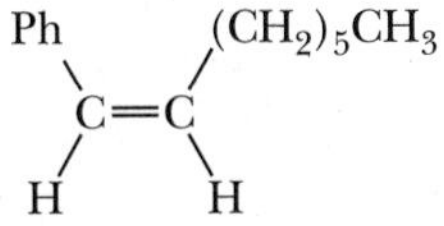

21.52 Show how to convert 1-phenylpropane into the following compounds. In addition to this starting material, use any necessary inorganic reagents. Any compound synthesized in one part of this problem may be used to make any other compound in the problem.

(a) 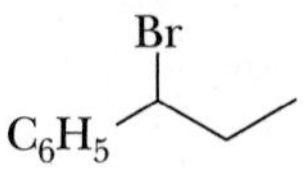

(racemic)

(b) 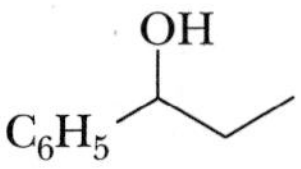

(racemic)

(c) 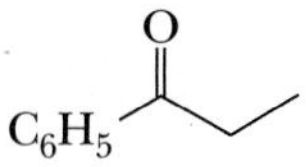

(d) 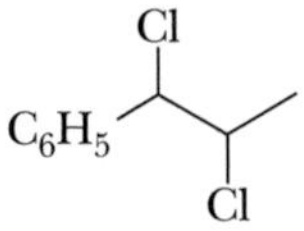

(racemic)

(e) 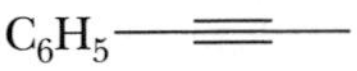

(f) 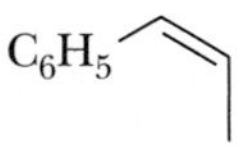

(g) 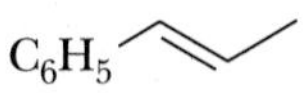

(h) 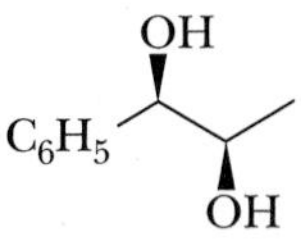

(racemic)

(i) 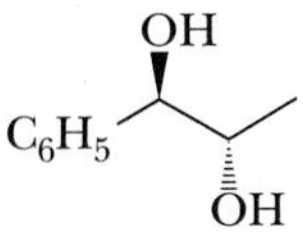

(racemic)

21.53 Carbinoxamine is a histamine antagonist, specifically, an H_1-antagonist. The maleic acid salt of the levorotatory isomer is sold as the prescription drug Rotoxamine.

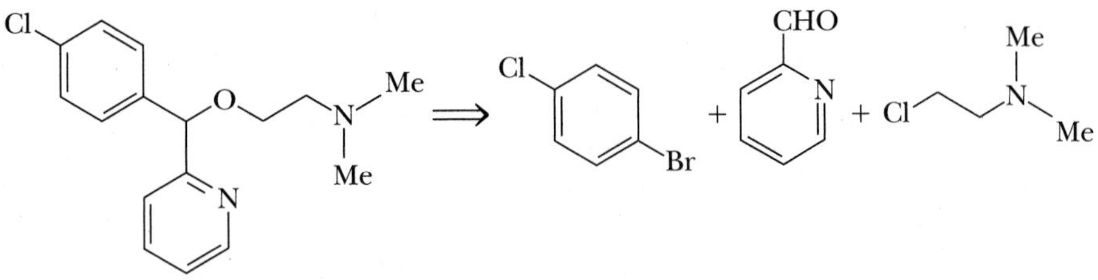

Carbinoxamine

(a) Propose a synthesis of carbinoxamine. (*Note:* Aryl bromides form Grignard reagents much more readily than do aryl chlorides.)

(b) Is carbinoxamine chiral? If so, how many stereoisomers are possible? Which of the possible stereoisomers are formed in this synthesis?

21.54 Cromolyn sodium, developed in the 1960s, has been used to prevent allergic reactions primarily affecting the lungs, as, for example, exercise-induced emphysema. It is thought to block the release of histamine, which prevents the sequence of events leading to swelling, itching, and constriction of bronchial tubes. Cromolyn sodium is synthesized in the following series of steps. Treatment of one mole of epichlorohydrin (Section 11.10) with two moles of 2,6-dihydroxyacetophenone in the presence of base gives I. Treatment of I with two moles of diethyl oxalate in the presence of sodium ethoxide gives a diester II. Saponification of the diester with aqueous NaOH gives cromolyn sodium.

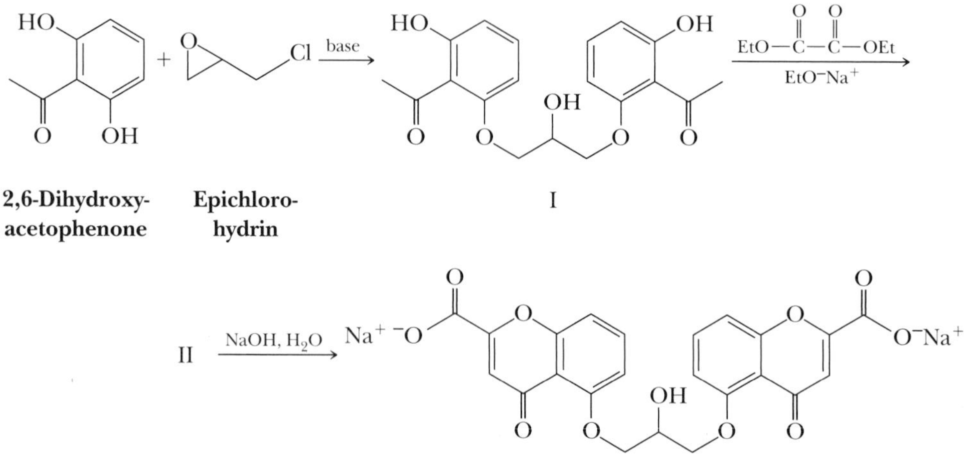

2,6-Dihydroxy-acetophenone **Epichlorohydrin** I

Cromolyn sodium

(a) Propose a mechanism for the formation of compound I.
(b) Propose a structural formula for compound II and a mechanism for its formation.
(c) Is cromolyn sodium chiral? If so, which of the possible stereoisomers are formed in this synthesis?

21.55 The following stereospecific synthesis is part of the scheme used by E. J. Corey of Harvard University in the synthesis of erythronolide B, the precursor of the erythromycin antibiotics. In this remarkably simple set of reactions, the relative configurations of five chiral centers are established.

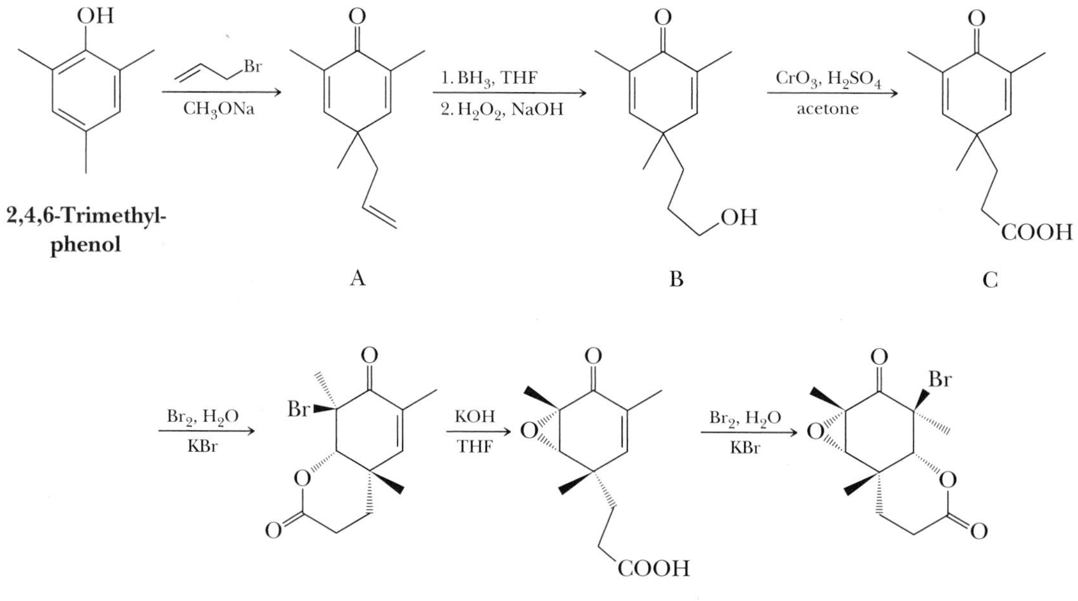

2,4,6-Trimethylphenol A B C

D E F

(a) Propose a mechanism for the conversion of 2,4,6-trimethylphenol to compound A.
(b) Account for the stereoselectivity and regioselectivity of the three steps in the conversion of compound C to compound F.
(c) Is compound F produced in this synthesis as a single enantiomer or as a racemic mixture? Explain.

21.56 Following is an outline of one of the first syntheses of the antidepressant fluoxetine (Prozac).

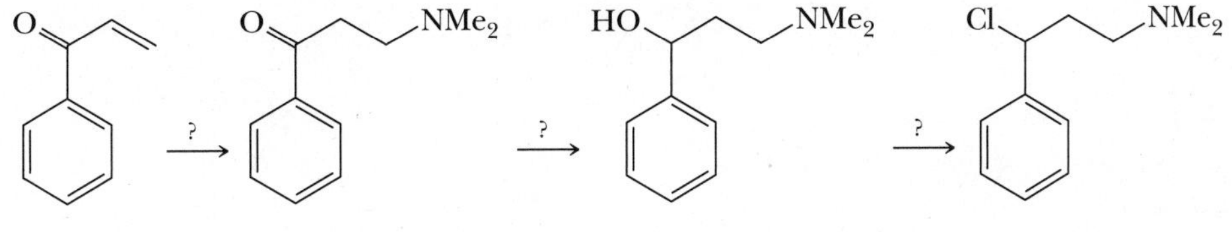

A B C D

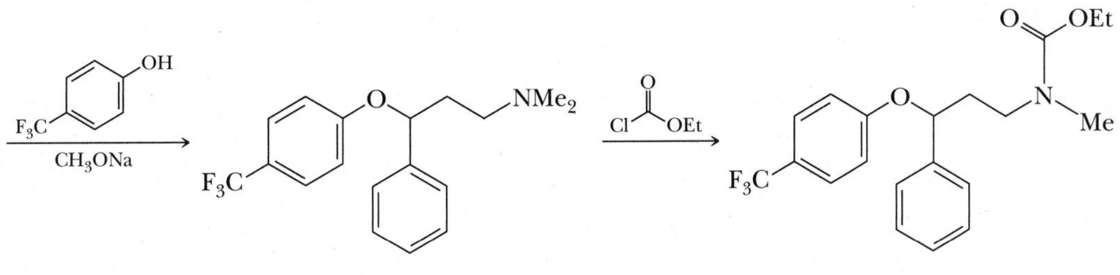

E F

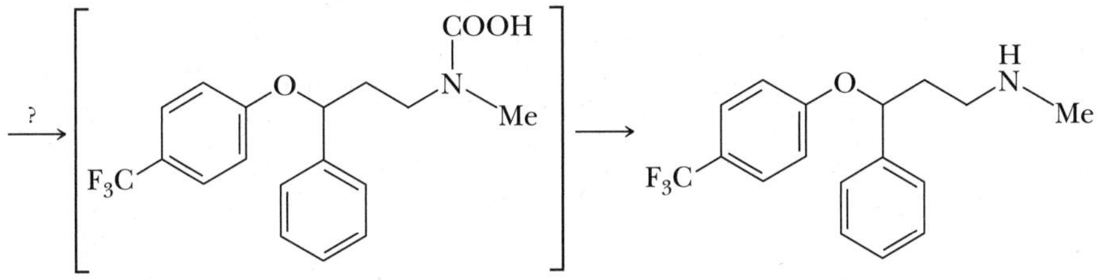

An *N*-substituted carbamic acid

Fluoxetine

(a) Propose a reagent for the conversion of A to B.

(b) Propose a reagent for the conversion of B to C.

(c) Propose a reagent for the conversion of C to D.

(d) Propose a mechanism for the conversion of E to F. The reagent used in this synthesis is ethyl chloroformate. The other product of this conversion is chloromethane, CH_3Cl. Your mechanism should show how the CH_3Cl is formed.

(e) Propose a reagent or reagents to bring about the conversion of F to fluoxetine. Note that the bracketed intermediate formed in this step is an *N*-substituted carbamic acid. Such compounds are unstable and break down to carbon dioxide and an amine.

(f) Is fluoxetine chiral? If so, which of the possible stereoisomers are formed in this synthesis?

21.57 Following is a synthesis for the antiarrhythmic drug bidisomide. The symbol Bn is an abbreviation for the benzyl group, $C_6H_5CH_2$—.

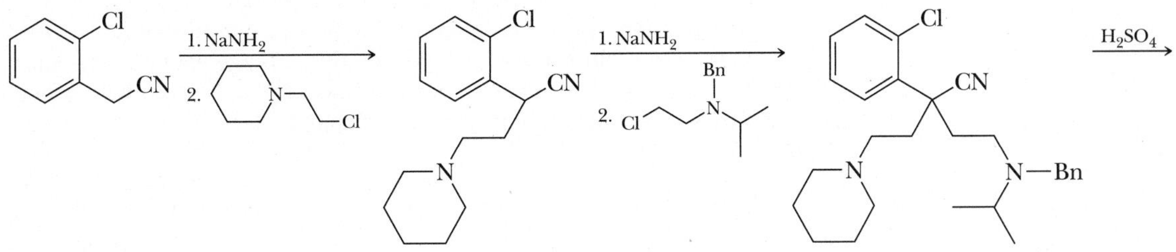

A B C

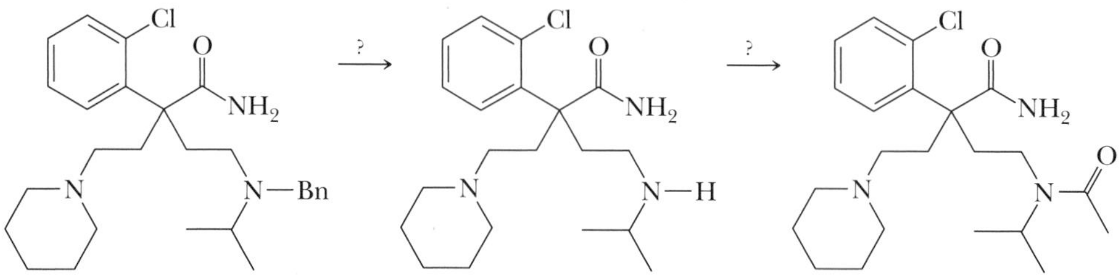

D E **Bidisomide**

(a) Propose mechanisms for the conversion of A to B and of B to C. What is the function of sodium amide in each reaction?
(b) Why is it necessary to incorporate the benzyl group on the chloroamine used to convert B to C?
(c) Propose a reagent or reagents for the removal of the benzyl group in the conversion of D to E.
(d) Propose a reagent for the conversion of E to bidisomide.
(e) Is bidisomide chiral? If so, which of the possible stereoisomers are formed in this synthesis?

21.58 A finding that opened a route to β-blockers was the discovery that β-blocking activity is retained if an oxygen atom is interposed between the aromatic ring and the side chain. To see this difference, compare the structures of labetalol (Problem 22.57) and propranolol. Thus, alkylation of phenoxide ions can be used as a way to introduce this side chain. The first of this new class of drugs was propranolol.

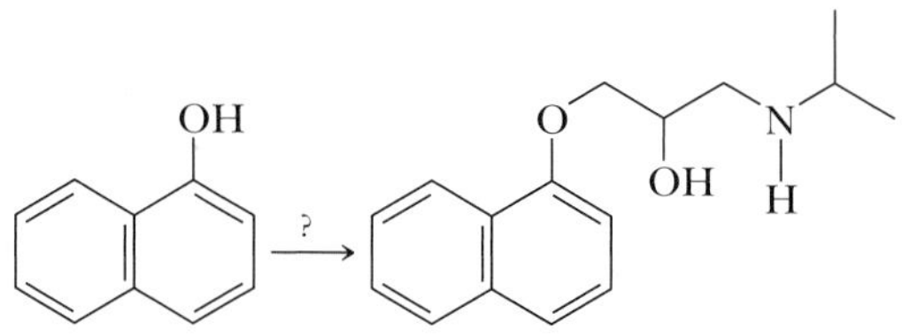

1-Napthol (β-Naphthol) **Propranolol**

(a) Show how propanolol can be synthesized from 1-naphthol, epichlorohydrin (Section 11.10), and isopropylamine.
(b) Is propranolol chiral? If so, which of the possible stereoisomers are formed in this synthesis?

21.59 Side effects of propranolol (Problem 21.58) include disturbances of the central nervous system (CNS), such as fatigue, sleep disturbances (including insomnia and nightmares), and depression. Pharmaceutical companies wondered if this drug could be redesigned to eliminate or at least reduce these side effects. Propranolol, it was reasoned, enters the CNS by passive diffusion because of the lipidlike character of its naphthalene ring. The challenge, then, was to design a more hydrophilic drug that does not cross the blood-brain barrier but still retains a β-adrenergic antagonist property. A product of this research is atenolol, a potent β-adrenergic blocker that is hydrophilic enough that it crosses the blood-brain barrier to only a very limited extent. Atenolol is now one of the most widely used β-blockers.

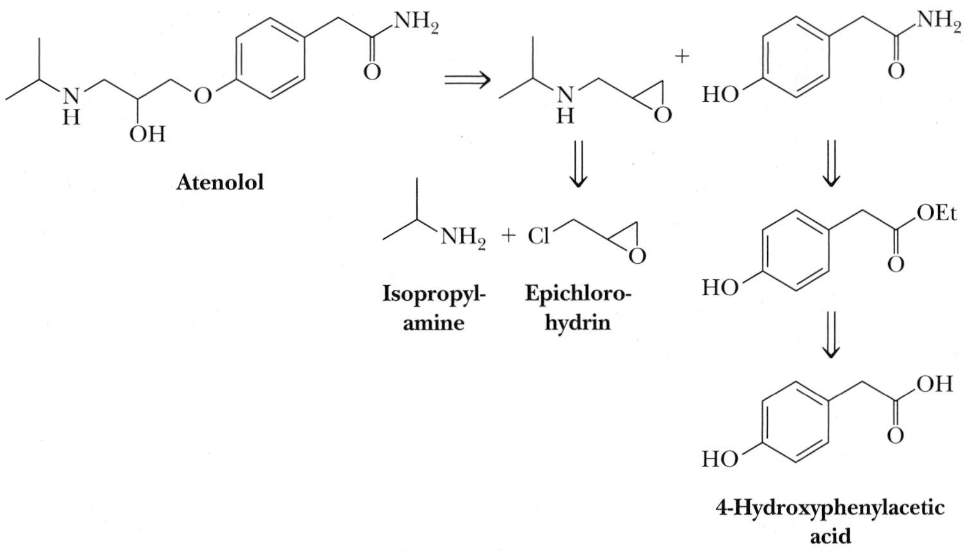

(a) Given this retrosynthetic analysis, propose a synthesis for atenolol from the three named starting materials.

(b) Note that the amide functional group is best made by amination of the ester. Why was this route chosen rather than conversion of the carboxylic acid to its acid chloride and then treatment of the acid chloride with ammonia?

(c) Is atenolol chiral? If so, which of the possible stereoisomers are formed in this synthesis?

21.60 In certain clinical situations, there is need for an injectable β-blocker with a short biological half-life. The clue to development of such a drug was taken from the structure of atenolol, whose corresponding carboxylic acid (the product of hydrolysis of its amide) has no β-blocking activity. Substituting an ester for the amide group and lengthening the carbon side chain by one methylene group resulted in esmolol. Its ester group is hydrolyzed quite rapidly to a carboxyl group by serum esterases under physiological conditions. This hydrolysis product has no β-blocking activity.

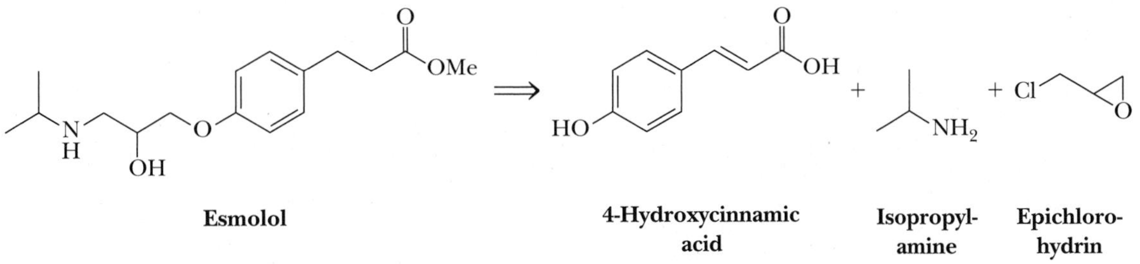

(a) Propose a synthesis for esmolol from 4-hydroxycinnamic acid, epichlorohydrin, and isopropylamine.

(b) Is esmolol chiral? If so, which of the possible stereoisomers are formed in this synthesis?

21.61 Following is an outline of a synthesis of the bronchodilator carbuterol, a beta-2 adrenergic blocker with high selectivity for airway smooth muscle receptors.

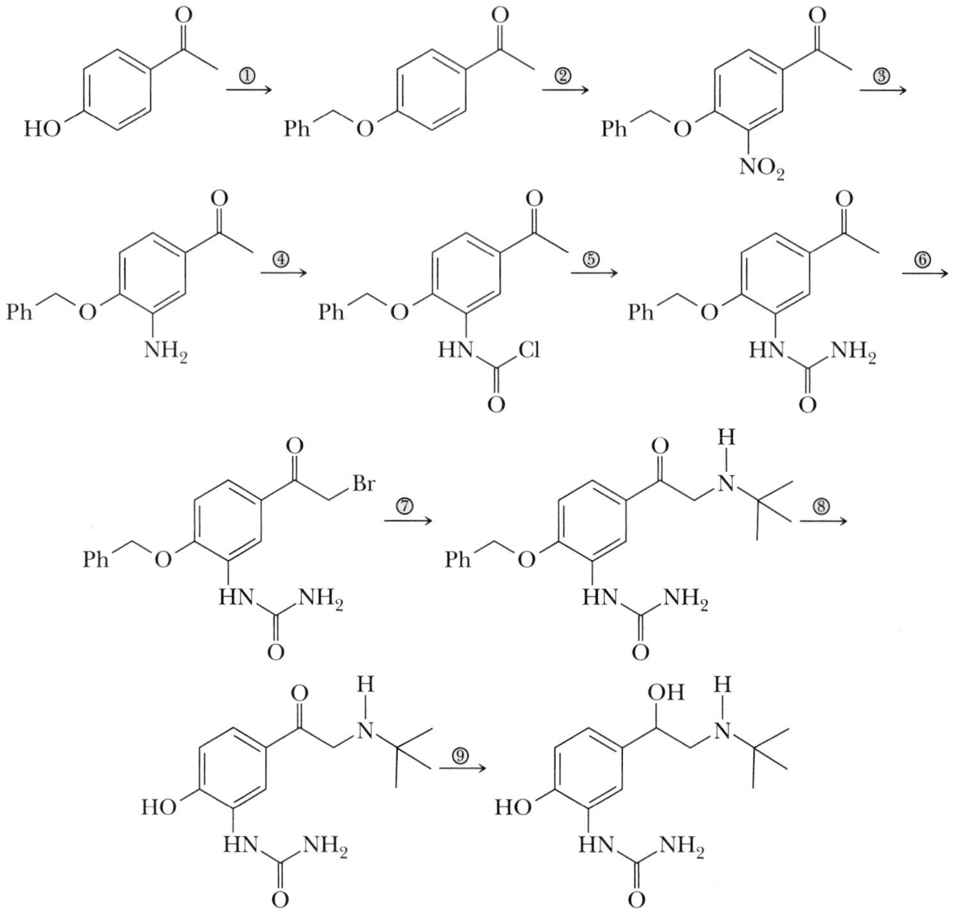

Carbuterol

(a) Propose reagents to bring about each step.
(b) Why is it necessary to add the benzyl group, $PhCH_2$—, as a blocking group in Step 1?
(c) Suggest a structural relationship between carbuterol and ephedrine.
(d) Is carbuterol chiral? If so, which of the possible stereoisomers are formed in this synthesis?

21.62 Following is a synthesis for albuterol (Proventil), currently one of the most widely used inhalation bronchodilators.

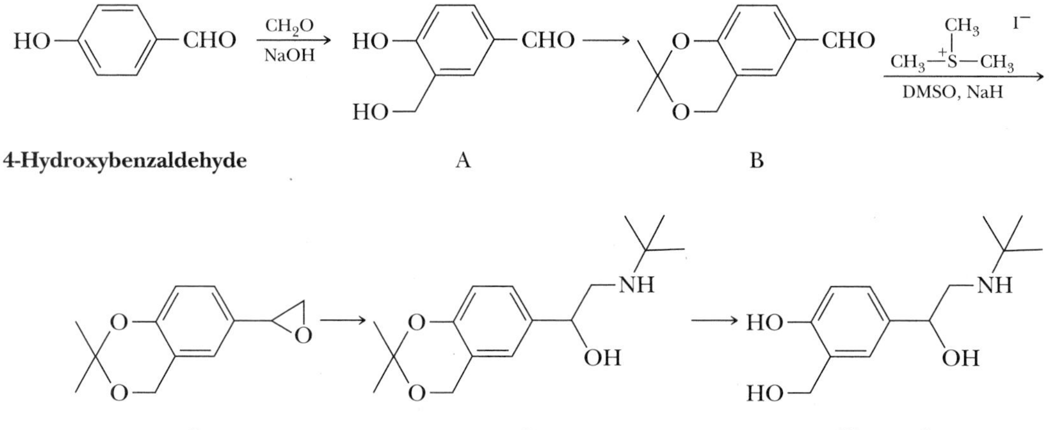

4-Hydroxybenzaldehyde A B

C D **Albuterol**

(a) Propose a mechanism for conversion of 4-hydroxybenzaldehyde to A.
(b) Propose reagents and experimental conditions for conversion of A to B.
(c) Propose a mechanism for the conversion of B to C. *Hint:* Think of trimethylsulfonium iodide as producing a sulfur equivalent of a Wittig reagent.
(d) Propose reagents and experimental conditions for the conversion of C to D.

(e) Propose reagents and experimental conditions for the conversion of D to albuterol.

(f) Is albuterol chiral? If so, which of the possible stereoisomers are formed in this synthesis?

21.63 Estrogens are female sex hormones, the most potent of which is β-estradiol.

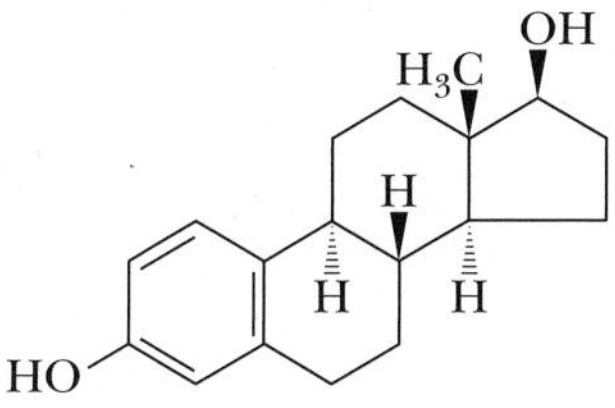

β-Estradiol

In recent years, chemists have focused on designing and synthesizing molecules that bind to estrogen receptors. One target of this research has been nonsteroidal estrogen antagonists, compounds that interact with estrogen receptors and block the effects of both endogenous and exogenous estrogens. A feature common to one type of nonsteroidal estrogen antagonist is the presence of a 1,2-diphenylethylene with one of the benzene rings bearing a dialkylaminoethoxyl substituent. The first nonsteroidal estrogen antagonist of this type to achieve clinical importance was tamoxifen, now an important drug in the treatment of breast cancer. Tamoxifen has the *Z* configuration shown here.

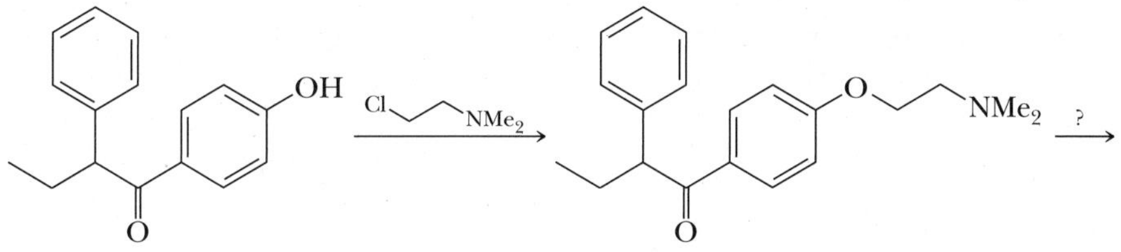

A B

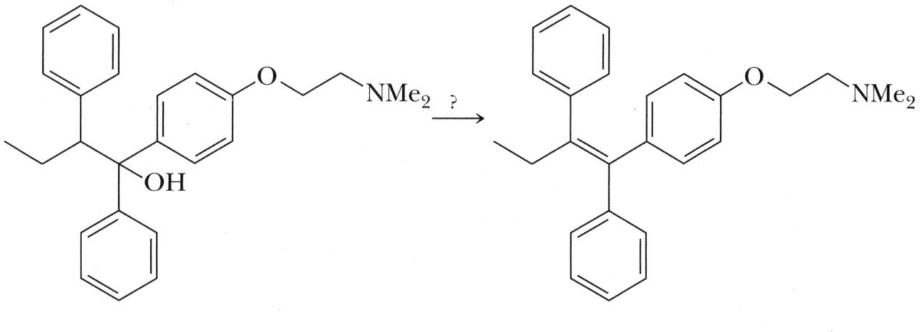

C **Tamoxifen**

Propose reagents for the conversion of A to tamoxifen. *Note:* The final step in this synthesis gives a mixture of *E* and *Z* isomers.

21.64 Following is a synthesis for toremifene, a nonsteroidal estrogen antagonist whose structure is closely related to that of tamoxifen.

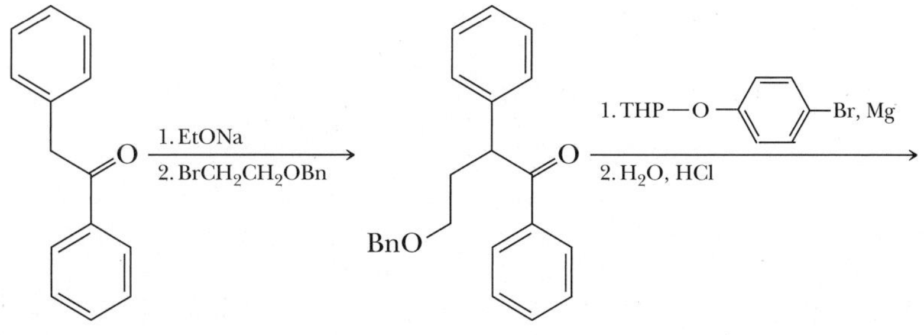

A B

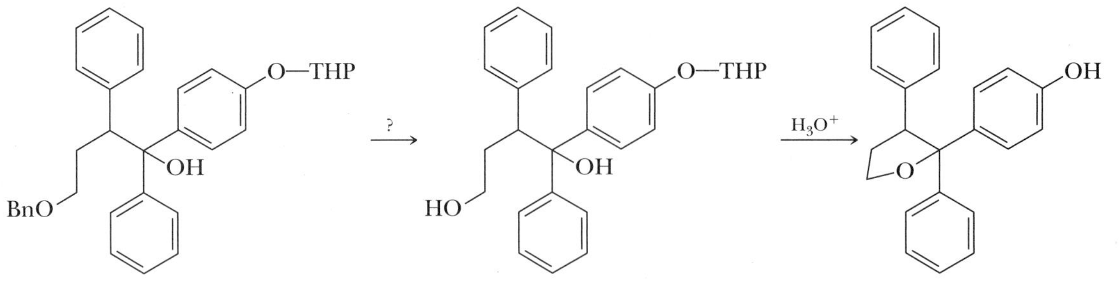

C D E

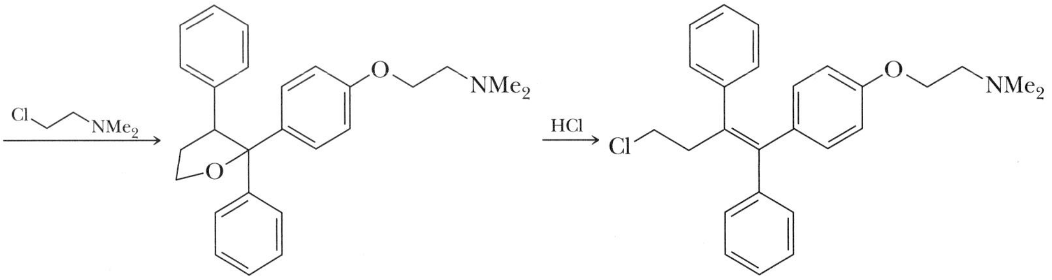

F **Toremifene**

(a) This synthesis makes use of two blocking groups, the benzyl (Bn) group and the tetrahydropyranyl (THP) group. Draw a structural formula of each group and describe the experimental conditions under which it is attached and removed.

(b) Discuss the chemical logic behind the use of each blocking group in this synthesis.

(c) Propose a mechanism for the conversion of D to E.

(d) Propose a mechanism for the conversion of F to toremifene.

(e) Is toremifene chiral? If so, which of the possible stereoisomers are formed in this synthesis?

Reactions of Benzene and Its Derivatives

2,6-Di-*tert*-butyl-4-methylphenol, alternatively known as butylated hydroxytoluene, or BHT (see Problem 22.23), is often used as an antioxidant to retard spoilage. **Inset:** a model of BHT.
(Richard Levine/Alamy Stock Photo)

Outline

22.1 Electrophilic Aromatic Substitution
22.2 Disubstitution and Polysubstitution
22.3 Nucleophilic Aromatic Substitution

By far, the most characteristic reaction of aromatic compounds is substitution at a ring carbon. In this reaction, one of the ring hydrogens is replaced by another atom or group of atoms. Some groups that can be introduced directly on the ring are the halogens, the nitro (—NO_2) group, the sulfonic acid (—SO_3H) group, alkyl (—R) groups, and acyl (RCO—) groups. Each of these substitution reactions is represented in the following equations.

Halogenation:

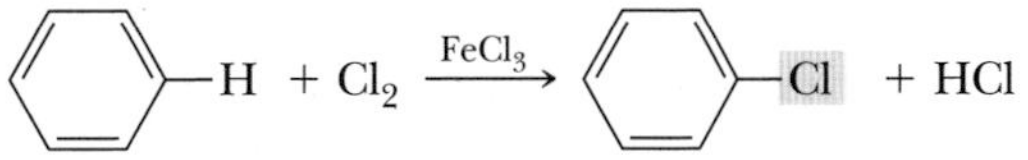

Chlorobenzene

Nitration:

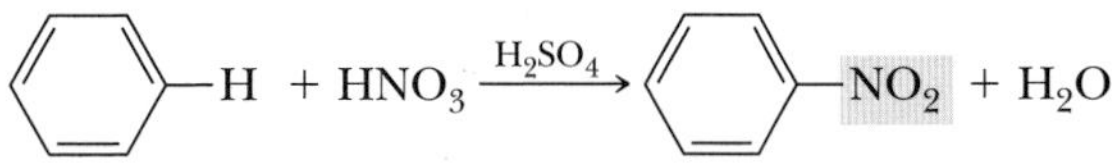

Nitrobenzene

Sulfonation:

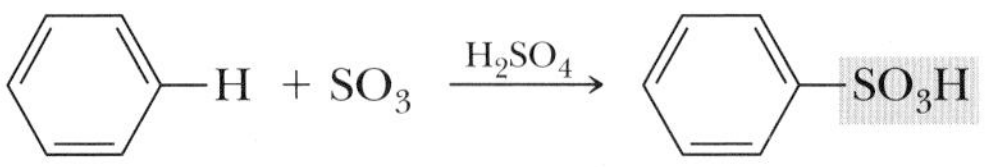
H
+ SO3
H2SO4
SO3H

Benzenesulfonic acid

Alkylation:

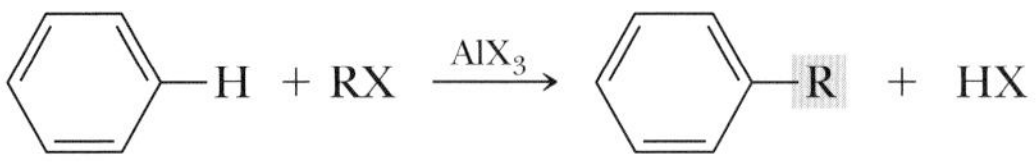
H
+ RX
AlX3
R
+ HX

An alkylbenzene

Acylation:

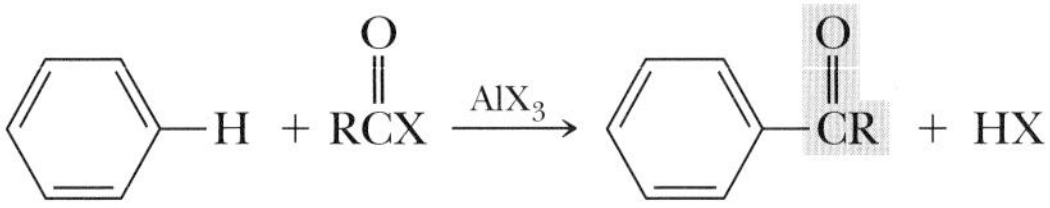
H
O
+ RCX
AlX3
O
CR
+ HX

An acylbenzene

We will take these reactions one at a time and examine their common mechanistic theme.

22.1 Electrophilic Aromatic Substitution

Electrophilic aromatic substitution

A reaction in which there is substitution of an electrophile, E^+, for a hydrogen on an aromatic ring.

In an **electrophilic aromatic substitution**, a hydrogen atom of an aromatic ring is replaced by an electrophile, E^+.

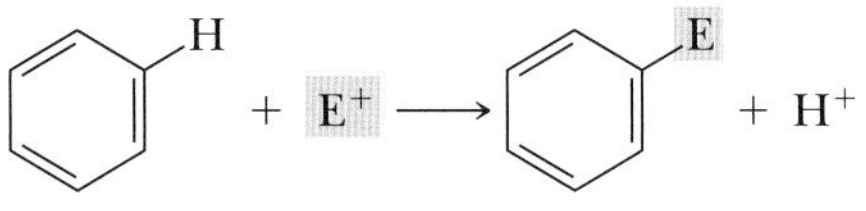
H
+ E+
E
+ H+

We study several common types of electrophiles, how each is generated, and the mechanism by which it replaces hydrogen on an aromatic ring.

A. Chlorination and Bromination

Chlorine alone does not react with benzene, in contrast to its instantaneous addition to cyclohexene. However, in the presence of a Lewis acid catalyst, such as ferric chloride or aluminum chloride (Section 4.7), benzene reacts with chlorine to give chlorobenzene and HCl. As shown in the following mechanism, this reaction involves a series of Lewis acid/base reactions.

Mechanism 22.1

Electrophilic Aromatic Substitution—Chlorination

Step 1: **Lewis acid-Lewis base reaction.** Reaction between chlorine and the Lewis acid catalyst gives a molecular complex with a positive charge on chlorine and a negative charge on iron. Redistribution of electrons in this complex generates a chloronium ion, Cl^+, a very strong electrophile, as part of an ion pair.

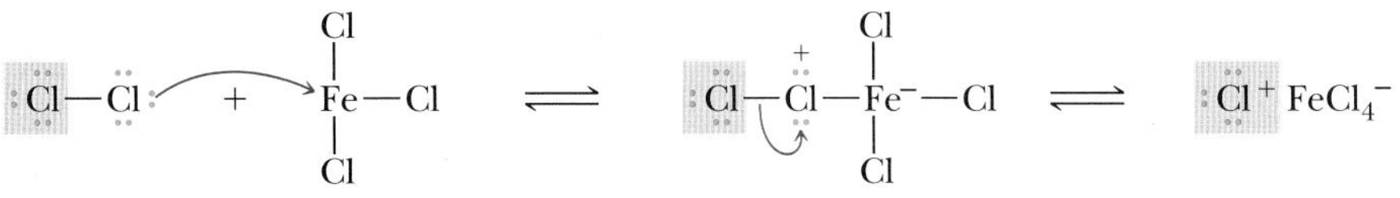
Cl—Cl
+
Cl
Fe—Cl
Cl
Cl—Cl—Fe−—Cl
Cl+ FeCl4−

Chlorine (a Lewis base) **Ferric chloride** (a Lewis acid) A molecular complex with a positive charge on chlorine and a negative charge on iron An ion pair containing the chloronium ion

Step 2: **Make a new bond between a nucleophile (arene ring) and an electrophile.** Attack of the chloronium ion (a strong electrophile) on the π system (a weak nucleophile) of the aromatic ring gives a resonance-stabilized cation intermediate, here represented as a hybrid of three contributing structures. Notice that the positive charge is located primarily at the ortho and para positions of the resonance-stabilized cation intermediate. This distribution of positive charge is visible in the electrostatic potential surface of the cation intermediate as the dark blue color at the ortho and para positions.

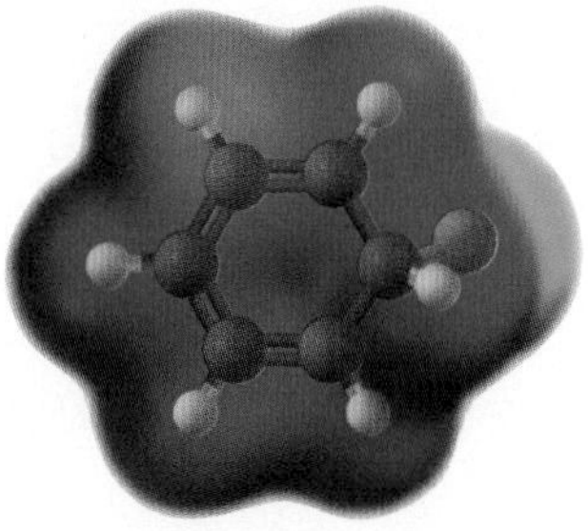

Cation intermediate

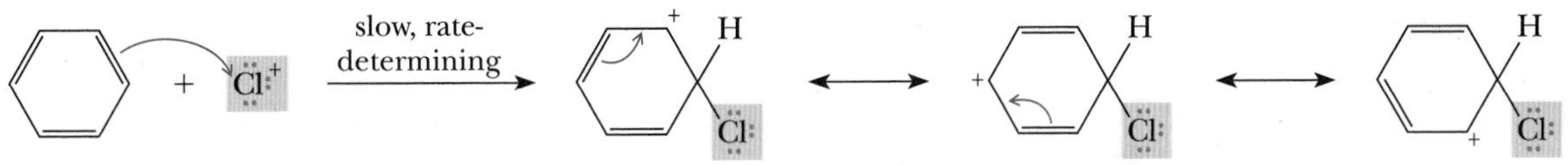

Step 3: **Take a proton away.** Proton transfer from the cation intermediate to $FeCl_4^-$ forms HCl, regenerates the Lewis acid catalyst, and gives chlorobenzene.

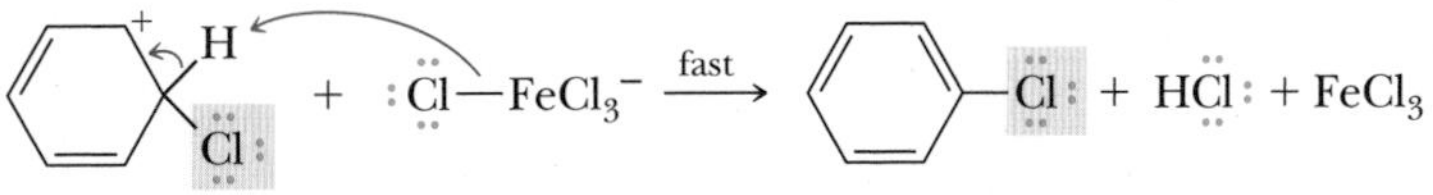

Cation intermediate

Chlorobenzene

Treating benzene with bromine in the presence of ferric chloride or aluminum chloride gives bromobenzene and HBr. The mechanism of this reaction is the same as that for the chlorination of benzene. ▶

▶ Delocalization of charge over a larger area is stabilizing.

We can write the following general two-step mechanism for electrophilic aromatic substitution. The first and rate-determining step is attack of the strong electrophile, E^+, by the weakly nucleophilic π electrons of the aromatic ring (electrophilic aromatic addition) to give a resonance-stabilized cation intermediate. The second and faster step, loss of H^+ from the cation intermediate, regenerates aromaticity in the ring and gives the product.

Step 1:

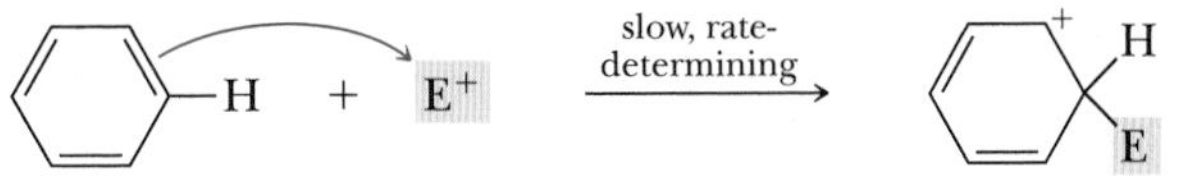

Electrophile

Resonance-stabilized cation intermediate

Step 2:

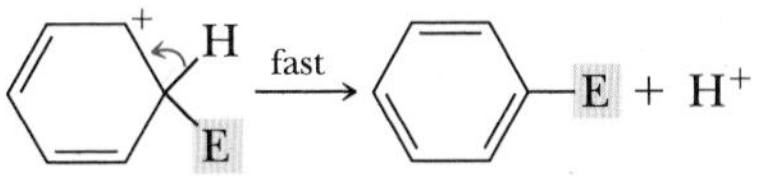

The major difference between addition of halogen to an alkene and halogen substitution on an aromatic ring centers on the fate of the cationic intermediate formed in the first step of each reaction. Recall from Section 6.3D that addition of chlorine or bromine to an alkene is a two-step process, the first and slower step of which is formation of a bridged halonium ion intermediate. This cationic intermediate then reacts with chloride or bromide ion to complete the addition. With aromatic compounds, the cationic intermediate instead loses H^+ to regenerate the aromatic ring and regain the large resonance stabilization. No such resonance stabilization is regained in the case of an alkene. The energy diagram in Figure 22.1 shows both addition and substitution reactions of benzene. Addition causes loss of the aromatic resonance energy and is disfavored except under extreme circumstances.

FIGURE 22.1 Energy diagram for the reaction of benzene with bromine. Formation of the addition product results in loss of the resonance stabilization of the aromatic ring. Formation of a substitution product regenerates the resonance-stabilized aromatic ring.

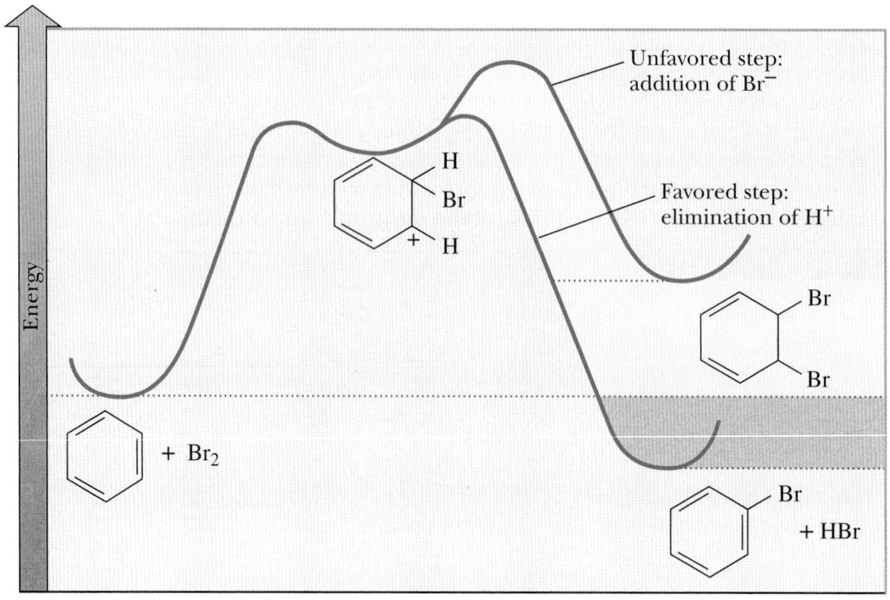

B. Nitration and Sulfonation

The sequence of steps for nitration and sulfonation of benzene is similar to that for chlorination and bromination. For nitration, the **nitronium ion**, NO_2^+, a very strong electrophile, is generated by the reaction between nitric acid and sulfuric acid.

Mechanism 22.2

Formation of the Nitronium Ion

Step 1: **Add a proton.** Proton transfer from sulfuric acid to the OH group of nitric acid gives the conjugate acid of nitric acid.

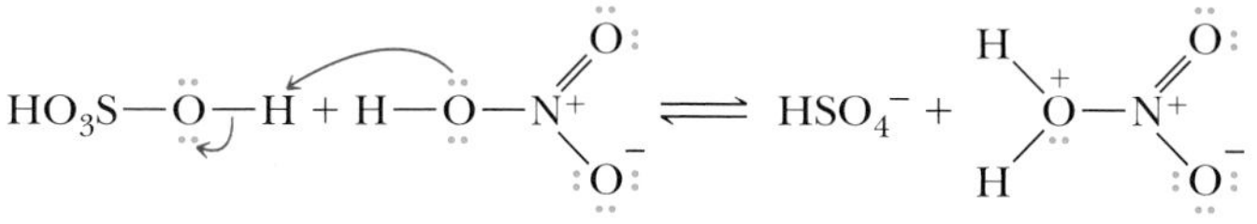

Sulfuric acid **Nitric acid** Conjugate acid of nitric acid

Step 2: **Break a bond to give stable molecules or ions.** Loss of water from this conjugate acid gives the nitronium ion, a very strong electrophile.

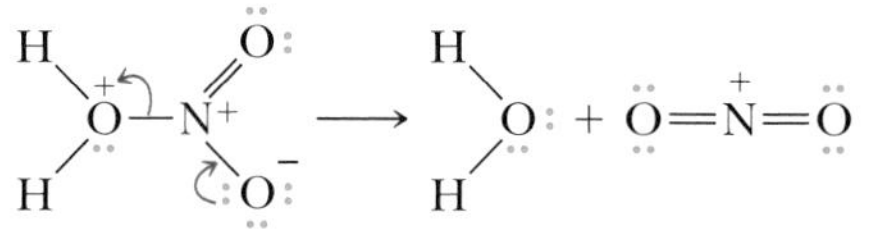

Nitronium ion

An important feature of nitration is that the resulting nitro group can be reduced to a primary amino group, $—NH_2$, by hydrogenation in the presence of a transition metal catalyst such as nickel, palladium, or platinum under fairly mild conditions.

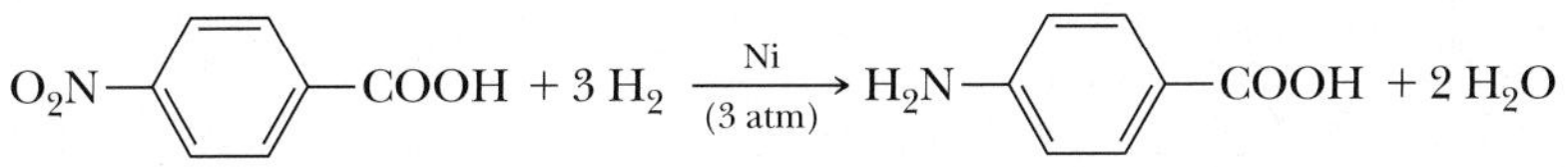

4-Nitrobenzoic acid **4-Aminobenzoic acid**

As illustrated by this example, neither a —COOH nor an aromatic ring is reduced; unfortunately, other susceptible groups such as carbon-carbon double bonds and aldehyde and ketone carbonyl groups may be reduced.

Alternatively, a nitro group can be reduced to a primary amino group by a metal in aqueous acid. The most commonly used metal-reducing agents are iron, zinc, and tin in dilute HCl. The reductant is electrons from the metal. When reduced with a metal and hydrochloric acid, the amine is obtained as a salt, which is then treated with strong base to liberate the free amine.

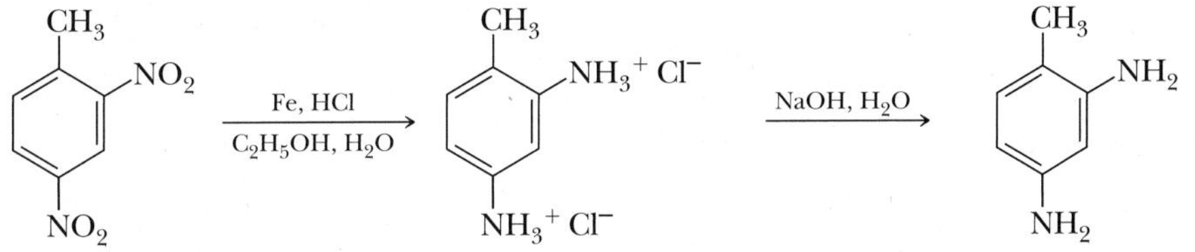

2,4-Dinitrotoluene **4-Methyl-1,3-benzenediamine (2,4-Diaminotoluene)**

The reduction of a nitro group to an amino group is important because the amino group cannot be substituted directly onto an aromatic ring; it must be done indirectly.

Sulfonation of benzene is carried out using concentrated sulfuric acid containing dissolved sulfur trioxide (fuming sulfuric acid). The electrophile is either SO_3 or HSO_3^+, depending on experimental conditions.

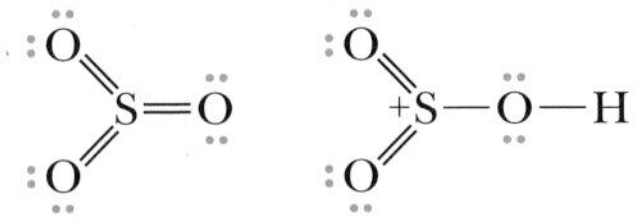

In the following equation, the sulfonating agent is shown as sulfur trioxide.

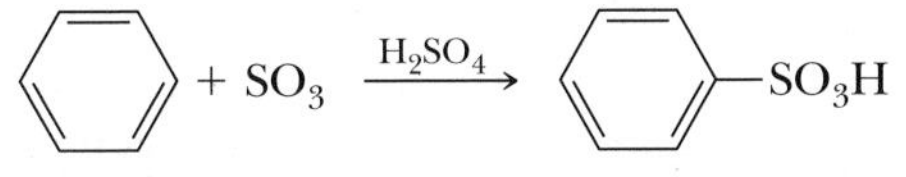

Benzene **Benzenesulfonic acid**

Example 22.1 Nitration of Benzene

Write a stepwise mechanism for the nitration of benzene.

Solution

The nitronium ion (a strong electrophile) is attacked by the benzene ring (a weak nucleophile) in Step 1 to give a resonance-stabilized cation intermediate. Proton transfer from this intermediate to either H_2O or HSO_4^- in Step 2 regenerates the aromatic ring and gives nitrobenzene.

(Continued)

***Step 1:* Make a new bond between a nucleophile (arene ring) and an electrophile.**

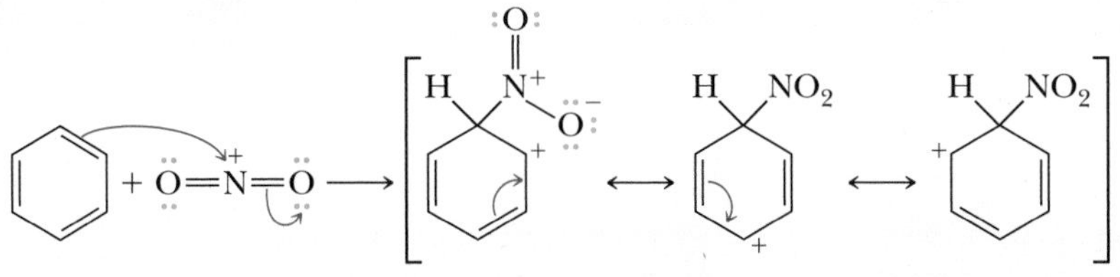

Resonance-stabilized cation intermediate

***Step 2:* Take a proton away.**

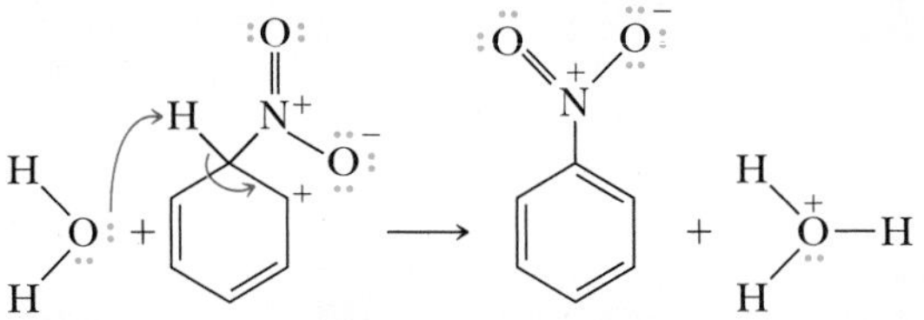

Problem 22.1

Write the stepwise mechanism for sulfonation of benzene by hot, concentrated sulfuric acid. In this reaction, the electrophile is SO_3 formed as shown in the following equation.

$$H_2SO_4 \rightleftharpoons SO_3 + H_2O$$

C. Friedel-Crafts Alkylation and Acylation

Friedel-Crafts alkylation

An electrophilic aromatic substitution in which a hydrogen of an aromatic ring is replaced by an alkyl or acyl group.

Alkylation of aromatic hydrocarbons was discovered in 1877 by the French chemist Charles Friedel and a visiting American chemist, James Crafts. They discovered that mixing benzene, an haloalkane, and $AlCl_3$ results in the formation of an alkylbenzene and HX. **Friedel-Crafts alkylation**, which is among the most important methods for forming new carbon-carbon bonds to aromatic rings, is illustrated here by the reaction of benzene with 2-chloropropane in the presence of aluminum chloride.

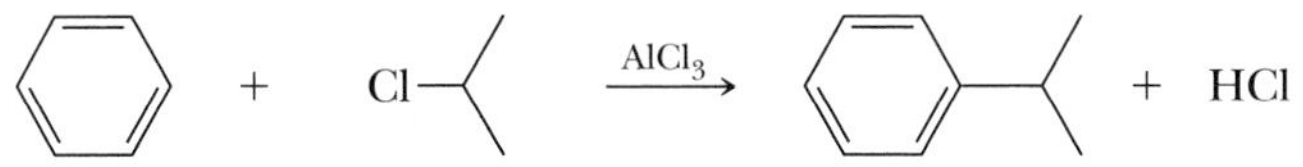

Benzene **2-Chloropropane (Isopropyl chloride)** **Cumene (Isopropylbenzene)**

The mechanism for Friedel-Crafts alkylation, like that for halogenation, nitration, and sulfonation, involves the attack of the aromatic ring onto a strong electrophile, in this case, a carbocation formed by reaction between the haloalkane and the Lewis acid catalyst.

Mechanism 22.3

Friedel-Crafts Alkylation

***Step 1:* Lewis acid-Lewis base reaction.** The haloalkane (a Lewis base) and aluminum chloride (a Lewis acid) form a complex in which aluminum has a negative charge and the halogen of the haloalkane has a positive charge. The alkyl group can also be written as a carbocation. It is unlikely, however, that a free carbocation is actually formed, especially in the case of the relatively unstable primary and secondary carbocations. Nonetheless, we often represent the reactive intermediate as a carbocation to simplify the notation of the mechanism.

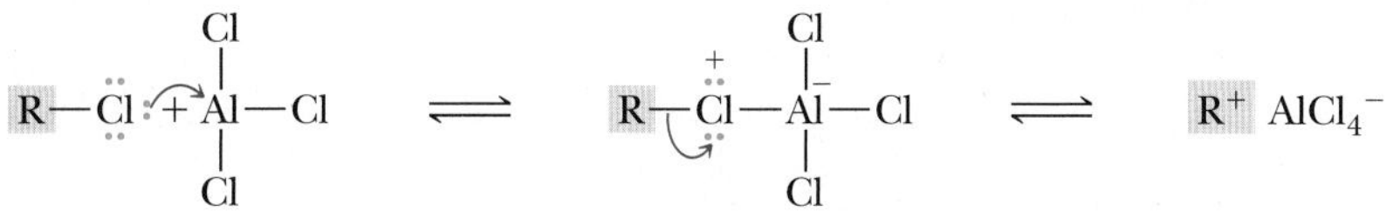

A molecular complex with a positive charge on chlorine and a negative charge on aluminum

An ion pair containing a carbocation

Step 2: **Make a new bond between a nucleophile (arene ring) and an electrophile.** Reaction of the carbocation (a strong electrophile) with the π electrons (a weak nucleophile) of the aromatic ring gives a resonance-stabilized cation intermediate.

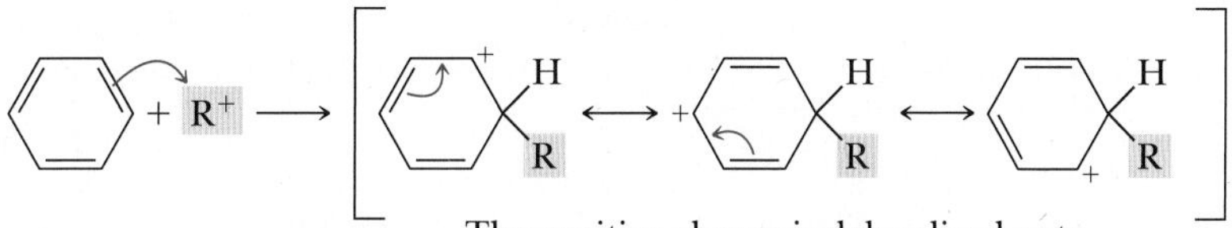

The positive charge is delocalized onto three atoms of the ring

Step 3: **Take a proton away.** Proton transfer regenerates the ring aromaticity.

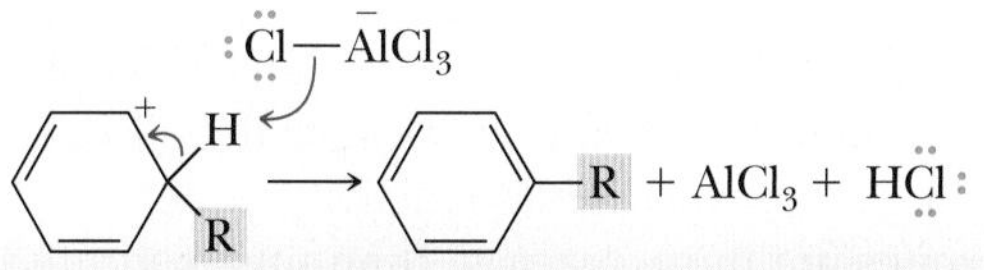

Halogen atoms on sp^2 hybridized carbons (vinylic and haloarenes) do not react to produce electrophiles under conditions of the Friedel-Crafts alkylation because high activation energy is required.

There are three major limitations on the Friedel-Crafts alkylation. The first is the possibility for rearrangement of the alkyl group. Friedel-Crafts alkylation generates a carbocation, and as we have already seen in Section 6.3C, carbocations may rearrange to a more stable carbocation. For example, reaction of benzene with 1-chloro-2-methylpropane (isobutyl chloride) gives only 2-methyl-2-phenylpropane (*tert*-butylbenzene).

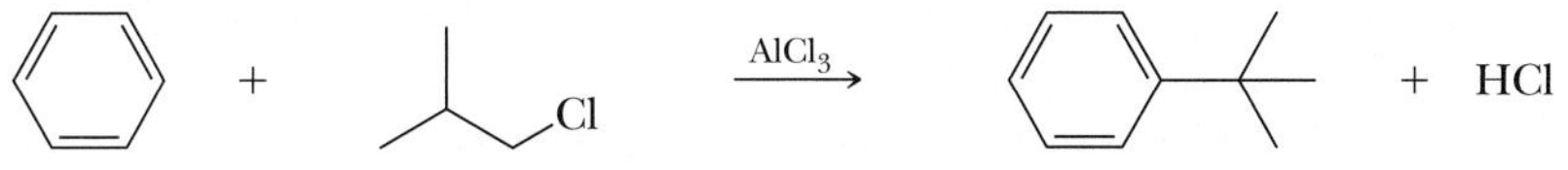

1-Chloro-2-methylpropane (Isobutyl chloride)

2-Methyl-2-phenylpropane (*tert*-Butylbenzene)

In this case, the isobutyl chloride/$AlCl_3$ complex rearranges directly to the *tert*-butyl cation/$AlCl_4^-$ ion pair, which is the electrophile in this example.

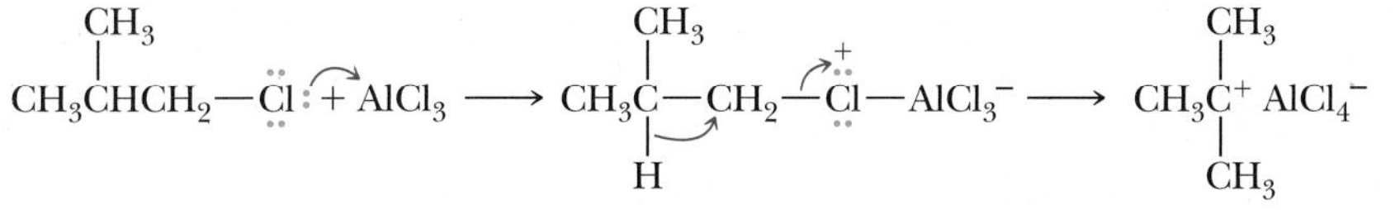

Isobutyl chloride

Isobutyl chloride/aluminum chloride complex

***tert*-Butyl cation/ $AlCl_4^-$ ion pair**

In practice, alkylation with primary halides is not a useful synthetic reaction, and alkylbenzenes containing a primary alkyl group other than $—CH_2CH_3$ must be prepared by other means. Alkylation is useful for introducing isopropyl, *tert*-butyl, and other alkyl groups, where their cations tend not to rearrange.

A second limitation on Friedel-Crafts alkylation is that it fails altogether on benzene rings bearing one or more strongly electron-withdrawing groups. As we shall see in the next section, substituents of the type shown in the following table have a dramatic effect on a benzene ring's reactivity toward further electrophilic aromatic substitution.

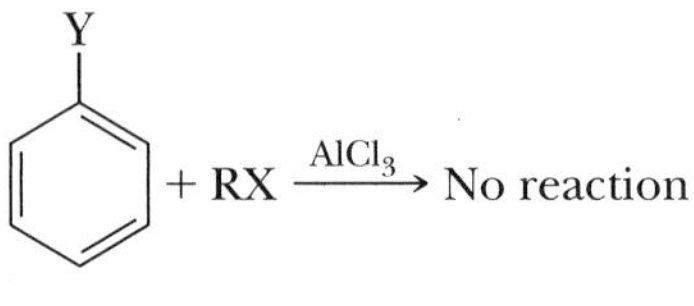

When Y Equals Any of These Groups, the Benzene Ring Does Not Undergo Friedel-Crafts Alkylation				
$-\overset{O}{\overset{\|}{C}}H$	$-\overset{O}{\overset{\|}{C}}R$	$-\overset{O}{\overset{\|}{C}}OH$	$-\overset{O}{\overset{\|}{C}}OR$	$-\overset{O}{\overset{\|}{C}}NH_2$
$-SO_3H$	$-C{\equiv}N$	$-NO_2$	$-NR_3^+$	
$-CF_3$	$-CCl_3$			

The third limitation on Friedel-Crafts alkylation is that it is hard to stop the reaction at monoalkylation because the alkylated product is more reactive than benzene itself. We will discuss reactivity in detail in Section 22.2, but in general, alkylated benzenes are more reactive than unsubstituted compounds. This limitation can be overcome if it is feasible to use a large excess of benzene, often as both the solvent and the reactant.

Friedel and Crafts also discovered that treating an aromatic hydrocarbon with an acyl halide (Section 18.1A) in the presence of aluminum chloride gives a ketone. Because an **RCO—** group is known as an acyl group, reaction of an aromatic hydrocarbon with an acyl halide is known as Friedel-Crafts acylation. This is illustrated by the reaction of benzene and acetyl chloride in the presence of aluminum chloride to form acetophenone.

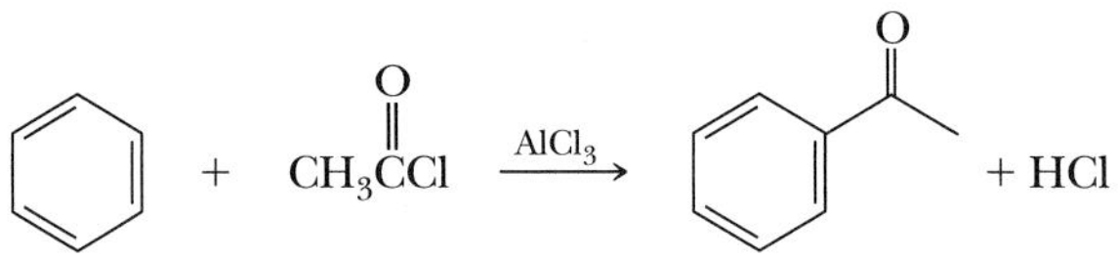

Benzene **Acetyl chloride** (an acyl halide) **Acetophenone** (a ketone)

The fact that an acylbenzene is less reactive than the starting material (unreactive in most cases) overcomes the third limitation of the alkylation reaction.

The following example of electrophilic aromatic substitution involves intramolecular acylation to form a six-membered ring.

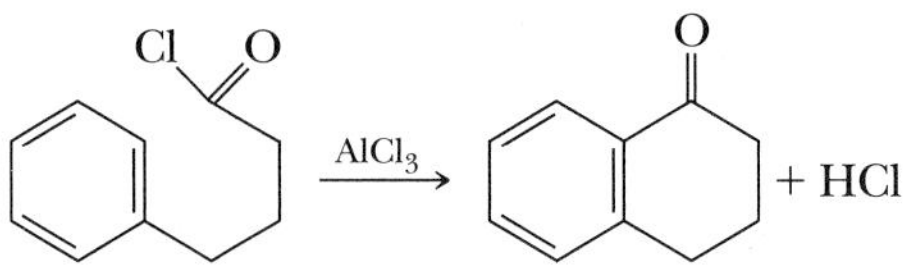

4-Phenylbutanoyl chloride **α-Tetralone**

Mechanism 22.4

Friedel-Crafts Acylation—Generation of an Acylium Ion

Step 1: **Lewis acid-Lewis base reaction.** Friedel-Crafts acylation begins with the donation of a pair of electrons from the halogen of the acyl halide to aluminum chloride to form a molecular complex similar to what we drew for Friedel-Crafts alkylations. In this complex, halogen has a positive formal charge and aluminum has a negative formal charge.

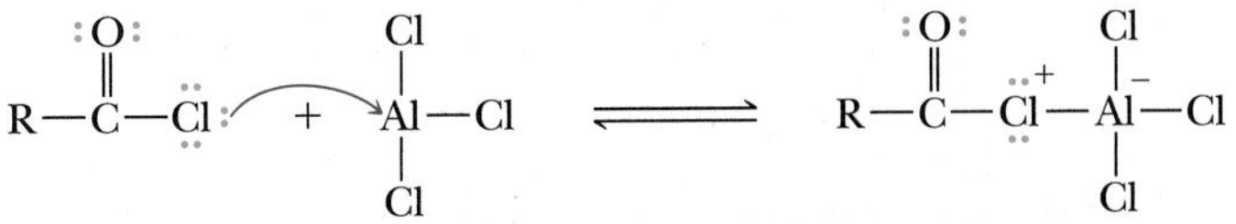

An acyl chloride (a Lewis base) | **Aluminum chloride** (a Lewis acid) | A molecular complex with a positive charge on chlorine and a negative charge on aluminum

Step 2: **Break a bond to give stable molecules or ions.** Redistribution of electrons of the carbon-chlorine bond gives an ion pair containing an **acylium ion**.

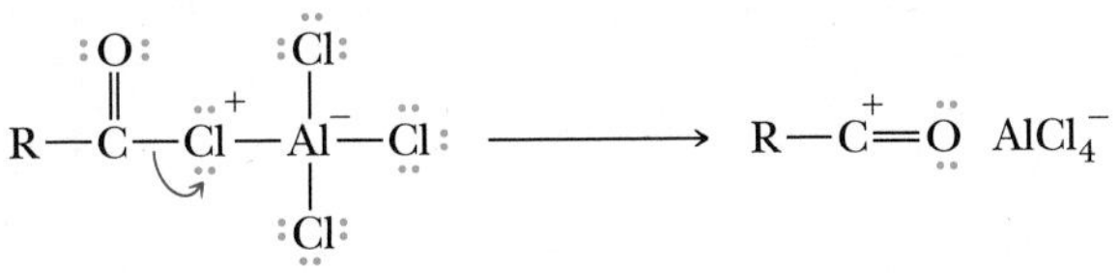

A molecular complex with a positive charge on chlorine and a negative charge on aluminum | An ion pair containing an acylium ion

Of the two major contributing structures that can be drawn for an acylium ion, the one with complete valence shells for both carbon and oxygen makes the greater contribution to the hybrid.

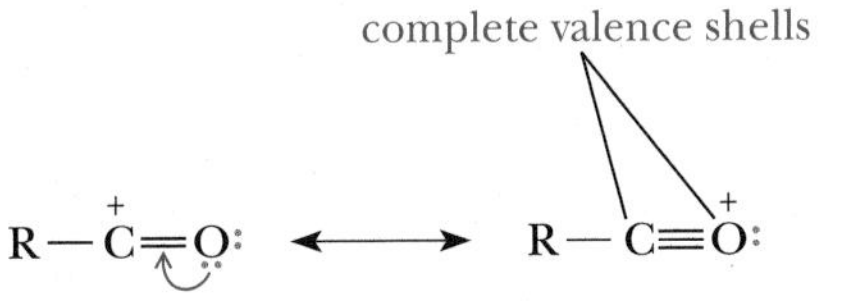
Both atoms have
complete valence shells

The more important contributing structure

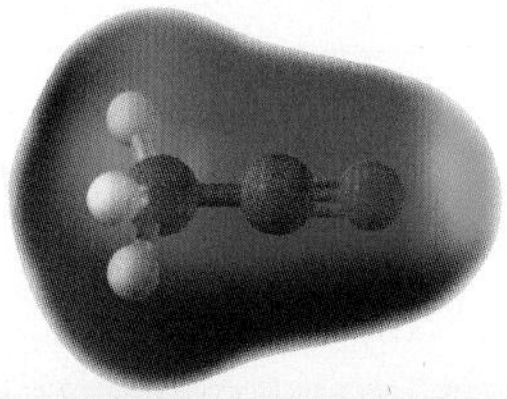

Friedel-Crafts acylation is free of a second major limitation on Friedel-Crafts alkylations: acylium ions do not undergo rearrangement. Thus, the carbon skeleton of an acyl halide is transferred unchanged to the aromatic ring.

Acylium ion
A resonance-stabilized cation with the structure $[RC{=}O]^+$ or $[ArC{=}O]^+$. The positive charge is delocalized over both the carbonyl carbon and the carbonyl oxygen.

Example 22.2 Friedel-Crafts Reactions

Write a structural formula for the product from Friedel-Crafts alkylation or acylation of benzene for each compound.

(a) $C_6H_5CH_2Cl$ **Benzyl chloride** **(b)** $C_6H_5\overset{O}{\overset{\|}{C}}Cl$ **Benzoyl chloride**

(Continued)

Solution

(a) Benzyl chloride in the presence of a Lewis acid catalyst gives the benzyl cation (an electrophile), which is then attacked by benzene (a weak nucleophile) followed by proton transfer to give diphenylmethane. In this example, the benzyl cation, although primary, cannot rearrange.

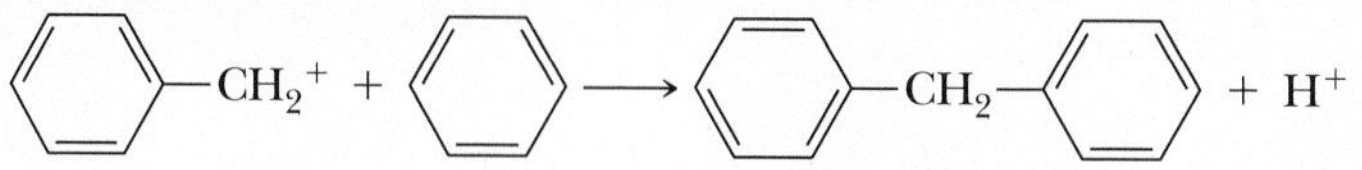

Benzyl cation **Diphenylmethane**

(b) Treating benzoyl chloride with aluminum chloride gives an acylium ion (an electrophile). Reaction of this cation with the π electrons of the aromatic ring (a weak nucleophile) followed by proton transfer gives benzophenone.

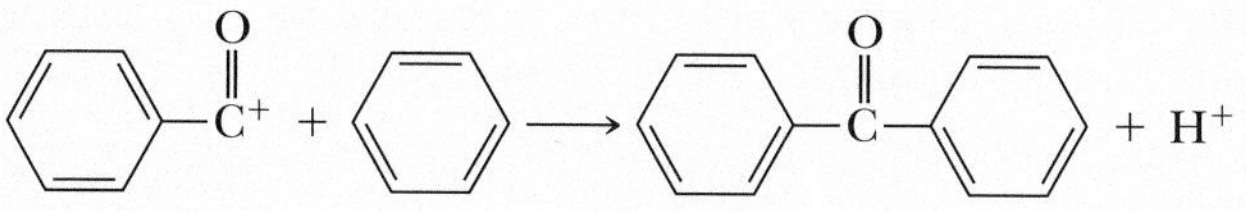

Benzoyl cation **Benzophenone**

Watch a video explanation

Problem 22.2

Write a structural formula for the product from Friedel-Crafts alkylation or acylation of benzene with each compound.

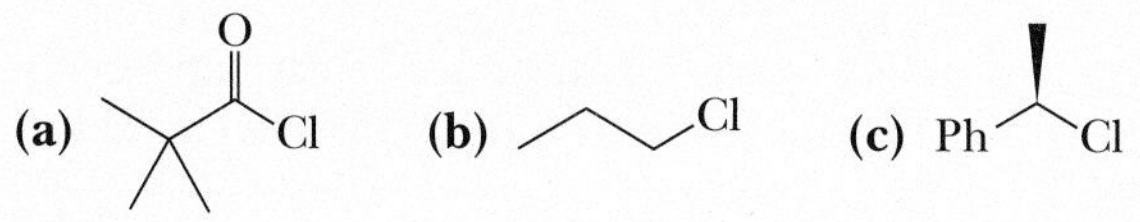

A special value of Friedel-Crafts acylations in synthesis is for the preparation of unrearranged alkylbenzenes, as illustrated by the preparation of isobutylbenzene.

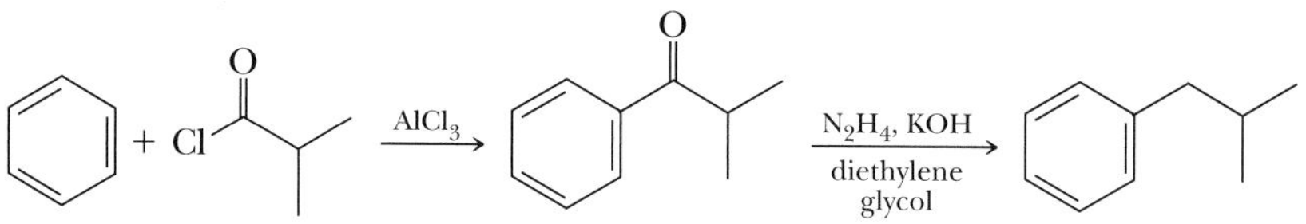

2-Methyl-propanoyl chloride **2-Methyl-1-phenyl-1-propanone** **Isobutylbenzene**

Treating benzene with 2-methylpropanoyl chloride in the presence of aluminum chloride gives 2-methyl-1-phenyl-1-propanone. Wolff-Kishner or Clemmensen reduction of the carbonyl group to a methylene group (Section 16.11C) gives isobutylbenzene.

D. Other Electrophilic Aromatic Alkylations

After the discovery that Friedel-Crafts alkylations and acylations involve cationic electrophiles, it was realized that the same reactions can be accomplished by other combinations of reagents and catalysts. We study two of these reactions: generation of carbocations from alkenes and from alcohols.

As we saw in Section 6.3, treatment of an alkene with a strong acid, most commonly HX, H_2SO_4, H_3PO_4, or HF/BF_3, generates a carbocation. Cumene, an intermediate in the industrial synthesis of both acetone and phenol (Problem 16.65), is synthesized industrially by treating benzene with propene in the presence of phosphoric acid as a catalyst.

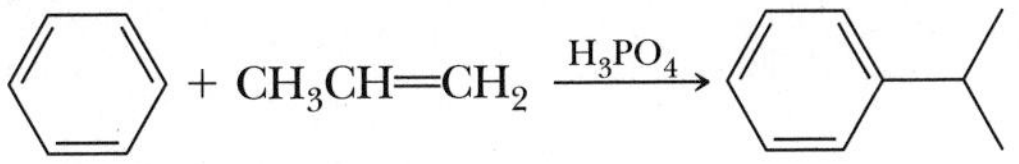

Benzene **Propene** **Cumene**

Alkylation with an alkene can also be carried out with a Lewis acid catalyst. Treatment of benzene with cyclohexene in the presence of aluminum chloride gives phenylcyclohexane.

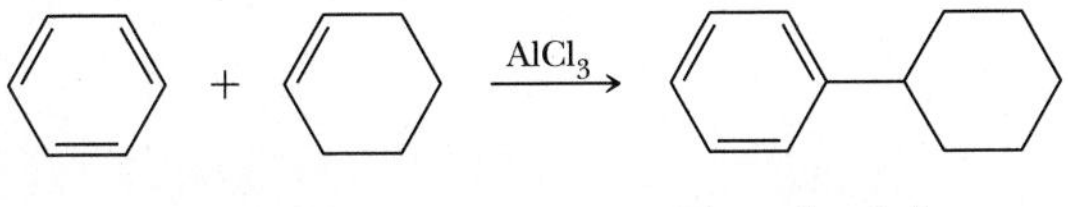

Benzene **Cyclohexene** **Phenylcyclohexane**

Carbocations can also be generated by treatment of an alcohol with H_2SO_4, H_3PO_4, or HF (Section 10.5).

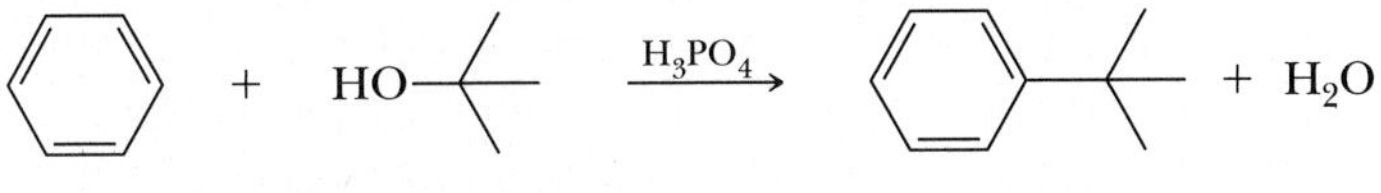

Benzene **2-Methyl-2-propanol (*tert*-Butyl alcohol)** **2-Methyl-2-phenylpropane (*tert*-Butylbenzene)**

Example 22.3 Electrophilic Addition of Carbocations

Write a mechanism for the formation of isopropylbenzene (cumene) from benzene and propene in the presence of phosphoric acid.

Solution

***Step 1:* Add a proton.** Proton transfer from phosphoric acid to propene gives the isopropyl cation.

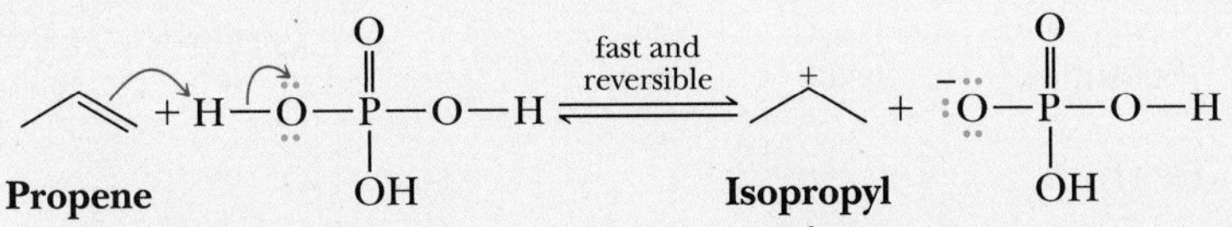

Propene **Isopropyl cation**

***Step 2:* Make a new bond between a nucleophile (arene ring) and an electrophile.** Reaction of the isopropyl cation with the π electrons of the benzene ring gives a resonance-stabilized carbocation intermediate.

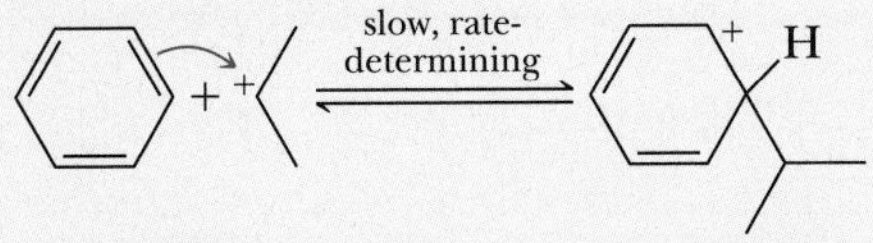

(*Continued*)

***Step 3:* Take a proton away.** Proton transfer to dihydrogen phosphate ion gives cumene.

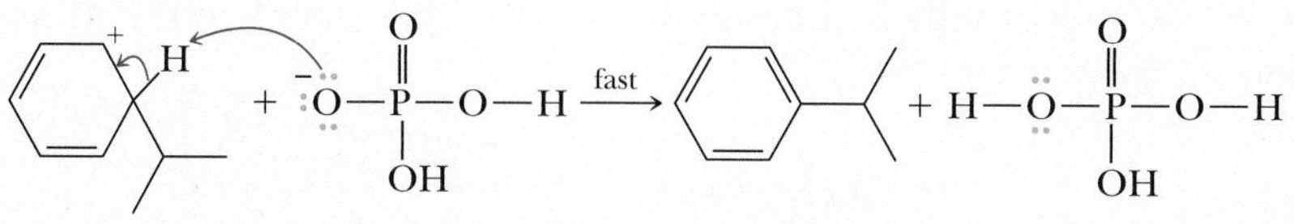

Problem 22.3

Write a mechanism for the formation of *tert*-butylbenzene from benzene and *tert*-butyl alcohol in the presence of phosphoric acid.

22.2 Disubstitution and Polysubstitution

A. Effects of a Substituent Group on Further Substitution

In electrophilic aromatic substitution of a monosubstituted benzene, three products are possible: the new group may become oriented ortho, meta, or para to the existing group. Table 22.1 shows the orientation of nitration of a series of monosubstituted benzenes.

Based on the information in Table 22.1 and other studies like it, we can make two generalizations about the manner in which existing groups influence further substitution reactions:

Activating group

Any substituent on a benzene ring that causes the rate of electrophilic aromatic substitution to be greater than that for benzene.

Deactivating group

Any substituent on a benzene ring that causes the rate of electrophilic aromatic substitution to be lower than that for benzene.

1. *Substituents affect the orientation of new groups.* Certain substituents (e.g., $—OCH_3$ and —Cl) direct an incoming group preferentially to the ortho and para positions; other substituents (e.g., $—NO_2$ and —COOH) direct it preferentially to the meta position. In other words, substituents on a benzene ring can be classified as **ortho-para directing** or as **meta directing**.
2. *Substituents affect the rate of further substitution.* Certain substituents cause the rate of a second substitution to be greater than that for benzene itself, whereas other substituents cause the rate of a second substitution to be lower than that for benzene. In other words, groups on a benzene ring can be classified as **activating** or **deactivating** toward further substitution.

Table 22.1 Orientation on Nitration of Monosubstituted Benzenes

Substituent	ortho	meta	para	ortho + para	meta
$—OCH_3$	44	—	55	99	trace
$—CH_3$	58	4	38	96	4
—Cl	70	—	30	100	trace
—Br	37	1	62	99	1
—COOH	18	80	2	20	80
—CN	19	80	1	20	80
$—NO_2$	6.4	93.2	0.3	6.7	93.2

These directing and activating-deactivating effects can be seen by comparing the products and rates of nitration of anisole and nitration of benzoic acid. Nitration of anisole proceeds at a rate considerably greater than that for benzene (the methoxy group is activating), and the product is a mixture of *o*-nitroanisole and *p*-nitroanisole (the methoxy group is ortho-para directing).

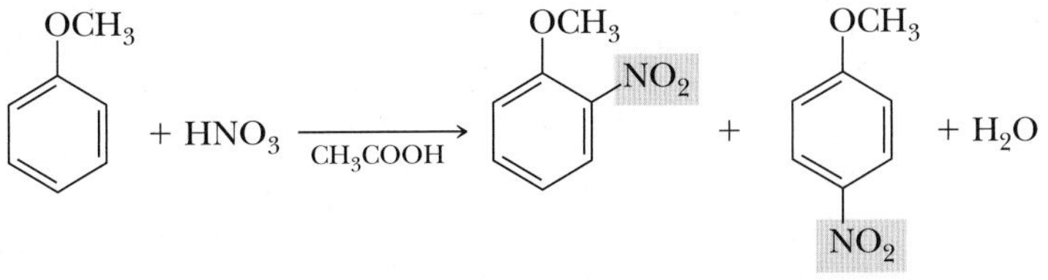

Anisole **o-Nitroanisole** (44%) **p-Nitroanisole** (55%)

Quite another situation is seen in the nitration of benzoic acid. First, the reaction requires the more reactive fuming nitric acid and a higher temperature than for benzene. Because nitration of benzoic acid proceeds much more slowly than nitration of benzene itself, we say that a carboxyl group is strongly deactivating. Second, the product formed consists of approximately 80% of the meta isomer and 20% of the ortho and para isomers combined; thus, we say that the carboxyl group is meta directing.

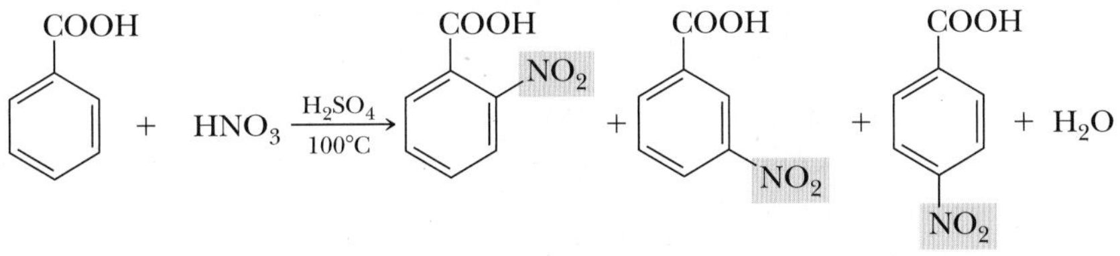

Benzoic acid **o-Nitrobenzoic acid** (18%) **m-Nitrobenzoic acid** (80%) **p-Nitrobenzoic acid** (2%)

Listed in Table 22.2 are the directing and activating-deactivating effects for the major functional groups with which we are concerned in this text. If we compare these ortho-para and meta directors for structural similarities and differences, we can make the following generalizations.

1. Alkyl groups, phenyl groups, and substituents in which the atom bonded to the ring has an unshared pair of electrons are ortho-para directing. All other substituents are meta directing.
2. Most ortho-para directing groups are activating. The exception to this generalization is the halogens, which are weakly deactivating.

The fact that alkyl groups are weakly activating is why it is difficult to stop Friedel-Crafts alkylations at monoalkylation. When a first alkyl group is introduced onto an aromatic ring, the ring is activated toward further alkylation, and unless reaction conditions are very carefully controlled, a mixture of di-, tri-, and polyalkylation products is formed. Friedel-Crafts acylations, on the other hand, never go beyond monoacylation because an acyl group is deactivating toward further substitution.

We can illustrate the usefulness of these generalizations by considering the synthesis of two different disubstituted derivatives of benzene. Suppose we want to prepare *m*-bromonitrobenzene from benzene. This conversion can be done in two steps: nitration and bromination. If the steps are carried out in just that order, the

Table 22.2 Directing Effects of Substituents on Further Substitution

Ortho-Para Directing	Strongly activating	$-\ddot{N}H_2$	$-\ddot{N}HR$	$-\ddot{N}R_2$	$-\ddot{O}H$	$-\ddot{O}R$		
	Moderately activating	$-\ddot{N}H\overset{O}{\overset{\|}{C}}R$	$-\ddot{N}H\overset{O}{\overset{\|}{C}}Ar$	$-\ddot{O}\overset{O}{\overset{\|}{C}}R$	$-\ddot{O}\overset{O}{\overset{\|}{C}}Ar$			
	Weakly activating	$-R$	$-C_6H_5$ (phenyl)					
	Weakly deactivating	$-\ddot{F}:$	$-\ddot{Cl}:$	$-\ddot{Br}:$	$-\ddot{I}:$			
Meta Directing	Moderately deactivating	$-\overset{O}{\overset{\|}{C}}H$	$-\overset{O}{\overset{\|}{C}}R$	$-\overset{O}{\overset{\|}{C}}OH$	$-\overset{O}{\overset{\|}{C}}OR$	$-\overset{O}{\overset{\|}{C}}NH_2$	$-\overset{O}{\overset{\|}{\underset{O}{\underset{\|}{S}}}}OH$	$-C{\equiv}N$
	Strongly deactivating	$-NO_2$	$-NH_3^+$	$-CF_3$	$-CCl_3$			

Relative importance in directing further substitution (arrow pointing up)

Watch a video explanation

major product is indeed *m*-bromonitrobenzene. The nitro group is a meta director and therefore directs bromination to a meta position.

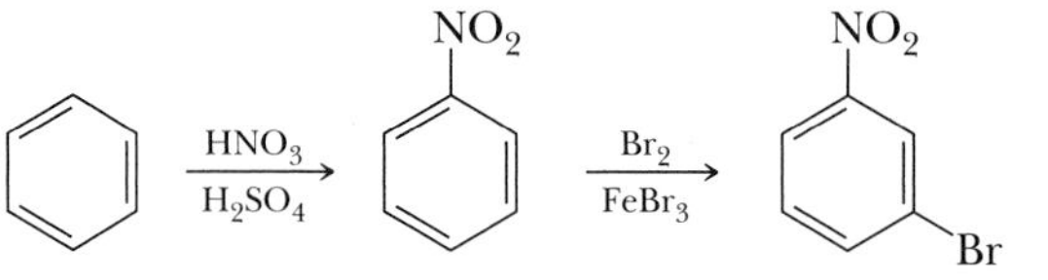

Nitrobenzene **m-Bromonitrobenzene**

If, however, we reverse the order of the steps and first form bromobenzene, we have an ortho-para directing group on the ring, and nitration takes place preferentially at the ortho and para positions.

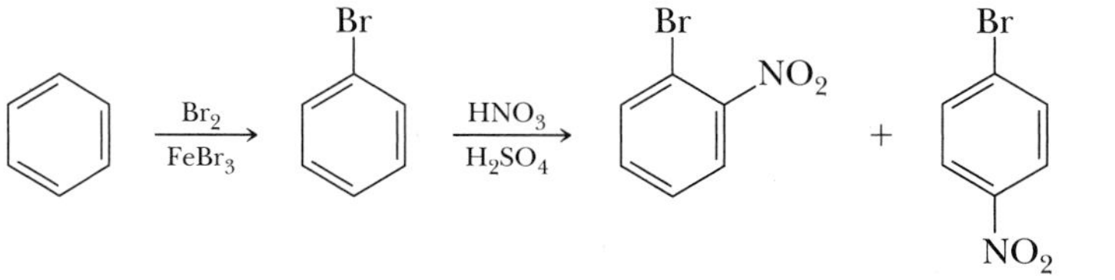

Bromobenzene ***o*-Bromonitrobenzene** ***p*-Bromonitrobenzene**

As another example of the importance of the order in electrophilic aromatic substitutions, consider the conversion of toluene to *p*-nitrobenzoic acid. The nitro group can be introduced with a nitrating mixture of nitric and sulfuric acids. The carboxyl group can be produced by oxidation of the methyl group of toluene (Section 21.5A). Nitration of toluene yields a product with the two substituents in the desired para relationship. Nitration of benzoic acid, on the other hand, yields a product with the substituents meta to each other.

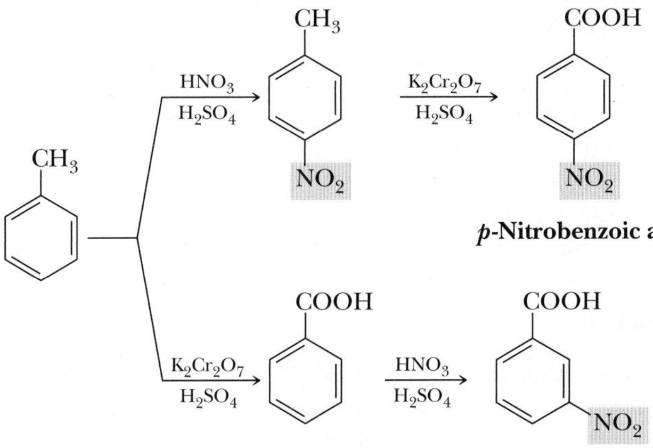

p-Nitrobenzoic acid

m-Nitrobenzoic acid

Again, we see that the order in which the reactions are performed is critical.

Note that in this last example, we showed nitration of toluene producing only the para isomer. Because methyl is an ortho-para directing group, both the ortho and para isomers are formed (Table 22.1). In problems of this type in which you are asked to prepare the para isomer, assume that both ortho and para isomers are formed but that there are physical methods by which they can be separated and the desired isomer obtained.

Example 22.4 Electrophilic Aromatic Substitution: Directing Effects

Complete these electrophilic aromatic substitution reactions. Where you predict meta substitution, show only the meta product. Where you predict ortho-para substitution, show both products.

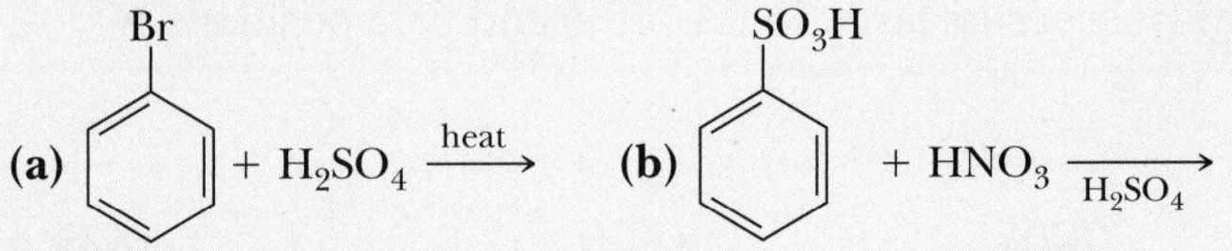

Solution

Bromine in (a) is ortho-para directing and weakly deactivating. The sulfonic acid group in (b) is meta directing and moderately deactivating.

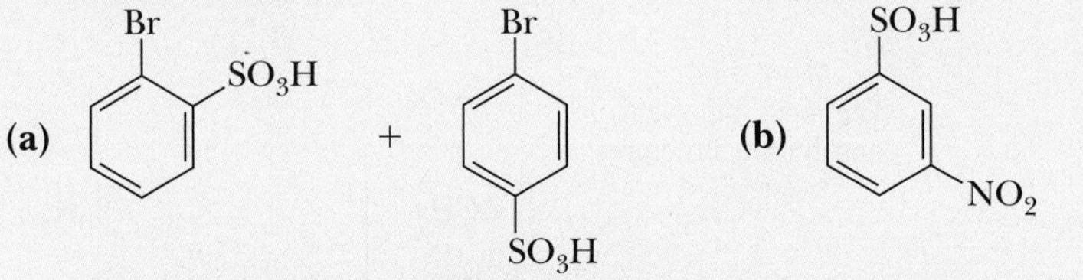

o-Bromobenzene-sulfonic acid | *p*-Bromobenzene-sulfonic acid | *m*-Nitrobenzene-sulfonic acid

Problem 22.4

Draw structural formulas for the product of nitration of each compound. Where you predict ortho-para substitution, show both products.

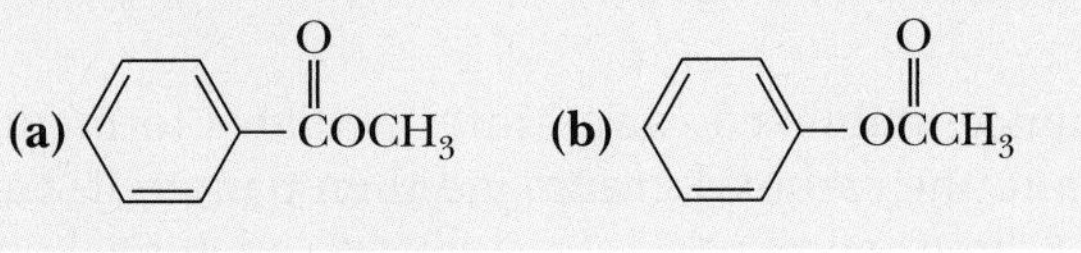

B. Theory of Directing Effects

As we have just seen, a group on a benzene ring exerts a major effect on the pattern of further substitution. We can account for these patterns by starting with the general mechanism first presented in Section 22.1 for electrophilic aromatic substitution and carrying it a step further to consider how groups already present on the ring affect the energetics of further substitution. In this regard, we need to consider both resonance and inductive effects and the relative importance of each.

Nitration of Anisole

For the nitration of anisole, and for almost every other substitution we consider, the slow and rate-determining step is attack of the electrophile by the aromatic ring. The rate of this step depends on the stability of the transition state for the step. The more stable the transition state, the faster the rate-determining step and thus the overall reaction.

Shown in Figure 22.2 is the cation intermediate formed by addition of the nitronium ion meta to the methoxy group. Also shown in the figure is the cation intermediate formed by addition para to the methoxy group. Note that in terms of electronic effects, structural formulas for the cation formed by attack ortho to the methoxy group are essentially the same as those for para attack; so for convenience, we deal only with para attack. The cation intermediate formed by meta attack is a hybrid of contributing structures (a), (b), and (c). The cation intermediate formed by para attack is a hybrid of contributing structures (d), (e), (f), and (g). For each orientation, we can draw three contributing structures that place the positive charge on carbon atoms of the benzene ring. These three structures are the only ones that can be drawn for meta attack. However, for para attack (and for ortho attack as well), a fourth contributing structure, (f), can be drawn that involves an unshared pair of electrons on the oxygen atom of the methoxy group and places a positive charge on this oxygen. Structure (f) contributes more than structures (d), (e), or (g) because, in it, all atoms have complete octets. Because the cation formed by ortho or para attack on anisole has a greater degree of charge delocalization, and therefore a lower activation energy for its formation, nitration of anisole occurs faster in the ortho and para positions. ▶

▶ Delocalization of charge over a larger area is stabilizing.

FIGURE 22.2 Nitration of anisole. Electrophilic attack meta and para to the methoxy group.

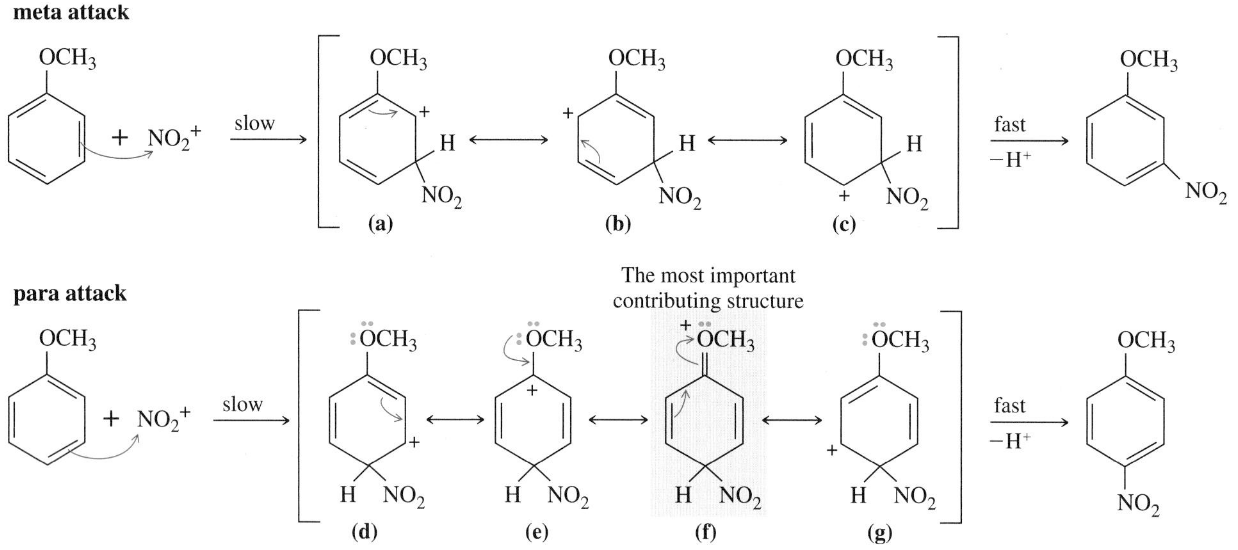

Nitration of Benzoic Acid

Shown in Figure 22.3 are resonance-stabilized cation intermediates formed by attack of the nitronium ion meta to the carboxyl group and then para to it. Each cation in Figure 22.3 is a hybrid of three contributing structures; no additional ones

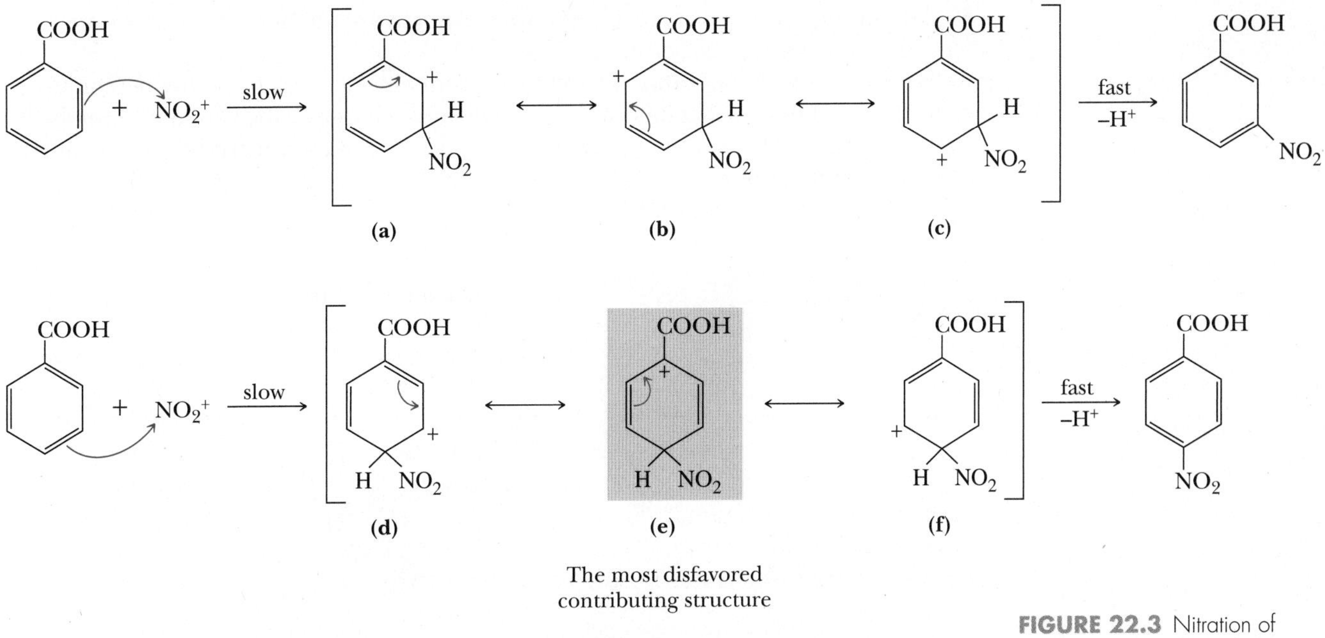

FIGURE 22.3 Nitration of benzoic acid. Electrophilic attack meta and para to the carboxyl group.

can be drawn. Now we need to compare the relative resonance stabilization of each hybrid. If we draw a Lewis structure for the carboxyl group showing the partial positive charge on the carboxyl carbon, we see that contributing structure (e) in Figure 22.3 places positive charges on adjacent atoms.

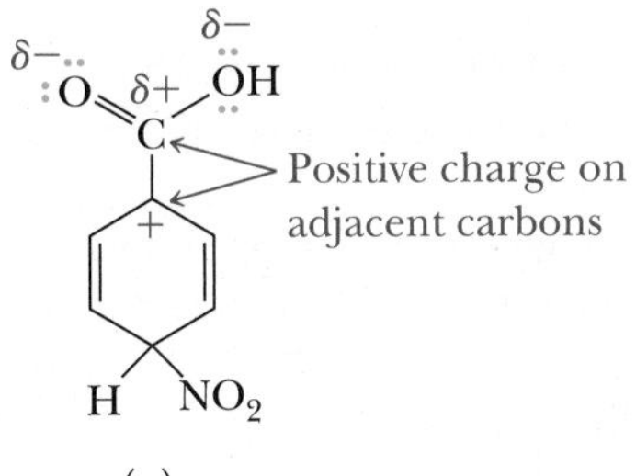

(e)

Because of the electrostatic repulsion thus generated, this structure is very unstable and makes only a negligible contribution to the hybrid.

None of the contributing structures for meta attack places positive charges on adjacent atoms. As a consequence, resonance stabilization of the cation for meta attack is greater than that for para (or ortho) attack. Stated alternatively, the activation energy for meta attack is less than that for para attack.

Comparison of the entries in Table 22.1 shows that almost all the ortho-para directing groups have an unshared pair of electrons on the atom bonded to the aromatic ring. Thus, the directing effect of these groups is primarily attributable to a resonance effect, due to the ability of the atom bonded to the ring to further delocalize the positive charge of the cation intermediate formed when electrophilic attack occurs at the ortho or para positions. Recall that all things being equal, delocalization of a charge stabilizes a charged species.

To account for the fact that alkyl groups are also ortho-para directing, we need to consider their inductive effect on stability of the cation intermediate. In the case of alkyl groups, there is an inductive polarization of electrons from the alkyl substituent toward the cationic ring of the intermediate. This polarization amounts to a further delocalization of the positive charge, thereby stabilizing the cationic intermediate. The alkyl groups are activating because, compared to benzene alone, the cationic intermediates are lower in energy with alkyl substituents at ortho or para positions. Recall that we used the electron-releasing inductive effect (i.e., hyperconjugation) of alkyl groups in Section 6.3A to account for the relative stabilities of methyl, primary, secondary, and tertiary carbocations as well.

The inductive polarization of electrons from alkyl groups (called hyperconjugation) is most effective at delocalizing the positive charge of the cation intermediate when the alkyl group is bonded directly to the ring atom carrying significant positive charge, in other words, when the alkyl group is ortho or para to the location of the incoming electrophile.

A part of the positive charge on the cation intermediate is delocalized onto methyl via hyperconjugation.

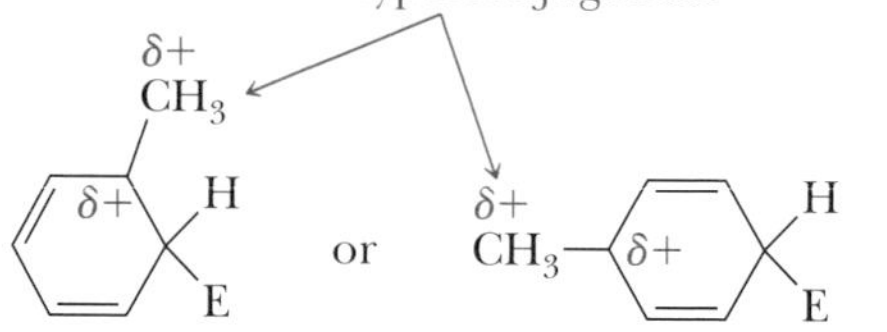

C. Theory of Activating-Deactivating Effects

We account for the activating-deactivating effects of substituent groups by much the same combination of resonance and inductive effects.

1. Any resonance effect, such as that of $—NH_2$, —OH, and —OR, that delocalizes the positive charge of the cation intermediate lowers the activation energy for its formation and has an activating effect toward further electrophilic aromatic substitution.
2. Any resonance or inductive effect, such as that of $—NO_2$, —C≡N, —C=O, $—SO_2—$, and $—SO_3H$, that decreases electron density on the ring deactivates the ring to further substitution.
3. Any inductive effect, such as that of $—CH_3$ or another alkyl group, that releases electron density toward the cationic intermediate activates the ring toward further substitution.
4. Any inductive effect, such as that of a halogen, $—NR_3^+$, $—CCl_3$, and $—CF_3$, that decreases electron density on the ring deactivates the ring to further substitution.

The halogens represent an interesting combination of the resonance and inductive effects, the two operating in opposite directions. Recall from Table 22.2 that halogens are ortho-para directing but, unlike other ortho-para directors listed in the table, are weakly deactivating. These observations can be accounted for in the following way.

1. The inductive effect of halogens: the halogens are relatively electronegative and have an electron-withdrawing inductive effect. Haloarenes, therefore, react more slowly in electrophilic aromatic substitution than benzene.
2. The resonance effect of halogens: when a halogen-substituted aromatic ring is attacked by an electrophile to form a cation intermediate, a halogen ortho or para to the site of electrophilic attack can help stabilize the cation intermediate by delocalization of the positive charge through resonance involving unshared electron pairs.

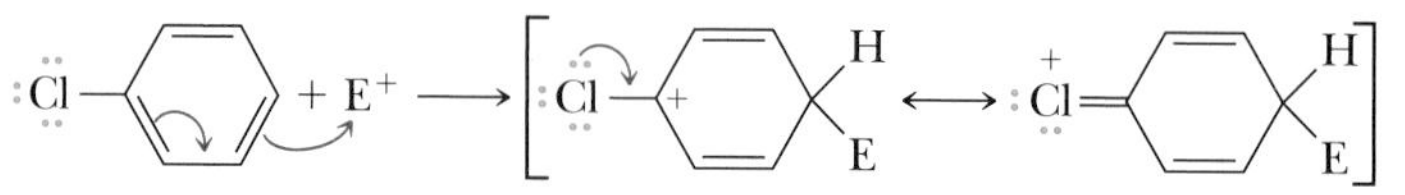

Thus, the inductive and resonance effects of the halogens are counter to each other, but the former is somewhat stronger than the latter. The net effect of this opposition is that the halogens are weakly deactivating but ortho-para directing.

For cases involving electrophilic aromatic substitution of aromatic rings with two or more substituents already on the ring, the more activating group will dominate the orientation preference of the incoming group. This is because activating groups will activate the ring for more substitution according to their orientation preference, while deactivating groups only serve to deactivate the ring toward further substitution. In practice, this means that ortho-para directing groups will dominate meta directing groups when they are both on a ring.

Example 22.5 Predicting Regiochemistry of Addition

Predict the major product of each electrophilic aromatic substitution.

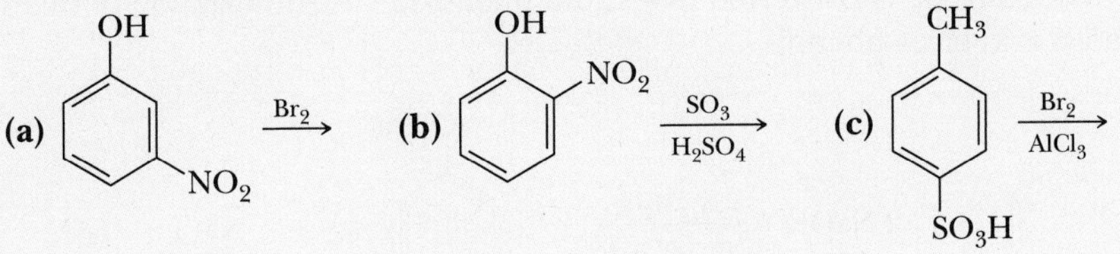

Solution

The key to predicting orientation of electrophilic aromatic substitution on each molecule is that ortho-para directing groups activate the ring toward further substitution, whereas meta directing groups deactivate. Therefore, where there is competition between ortho-para and meta directing groups, ortho-para directing groups win out. In these examples, the major product is that resulting from substitution of ortho or para to the activating group. For (a), the next substitution is directed ortho/para to the strongly activating —OH group; the isomer with bromine between the —OH and —NO_2 groups is a very minor product because of steric hindrance. For (b), the incoming group is directed ortho and para to the strongly activating —OH group. For (c), the next substitution is directed ortho to the weakly activating —CH_3 group.

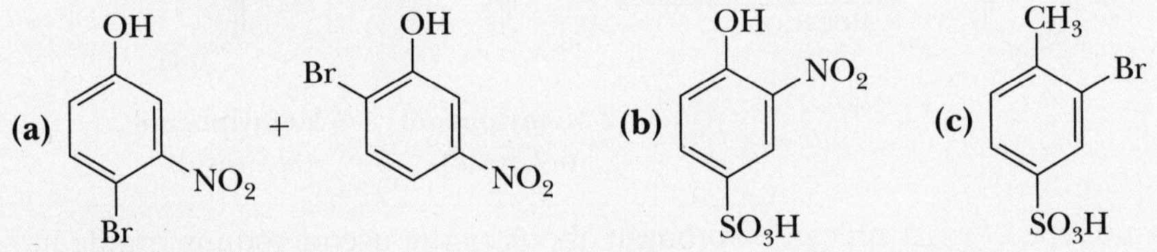

Problem 22.5

Predict the major product(s) of each electrophilic aromatic substitution.

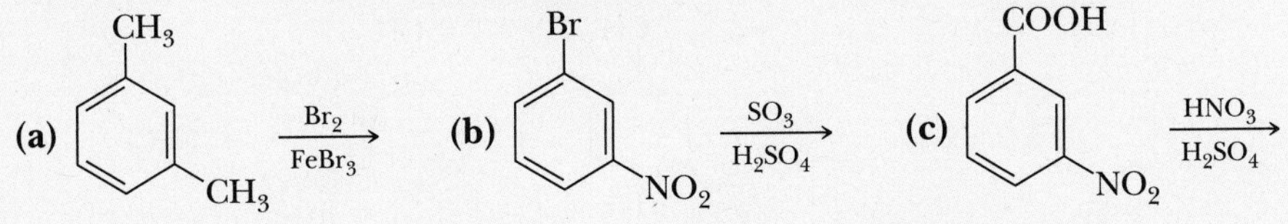

22.3 Nucleophilic Aromatic Substitution

One of the important chemical characteristics of haloarenes is that they undergo relatively few reactions involving the carbon-halogen bond. Haloarenes, for example, do not undergo substitution by either of the S_N1 or S_N2 pathways that are characteristic of nucleophilic aliphatic substitutions. They do, however, undergo

Nucleophilic aromatic substitution A reaction in which a nucleophile, most commonly a halogen, on an aromatic ring is replaced by another nucleophile.

nucleophilic aromatic substitution under certain conditions, but by mechanisms quite different from those for nucleophilic aliphatic substitutions. Nucleophilic aromatic substitution reactions are far less common than electrophilic aromatic substitution reactions and have only limited application for the synthesis of organic compounds. We study these reactions not only for their synthetic usefulness but also for the additional insights they give us into the unique chemical properties of aromatic compounds.

A. Nucleophilic Substitution by Way of a Benzyne Intermediate

An apparent exception to the generalization about the lack of reactivity of haloarenes to nucleophilic substitution is an early industrial process for the synthesis of phenol from chlorobenzene. When heated at 300°C under high pressure with aqueous NaOH, chlorobenzene is converted to sodium phenoxide. Neutralization of this salt with aqueous acid gives phenol.

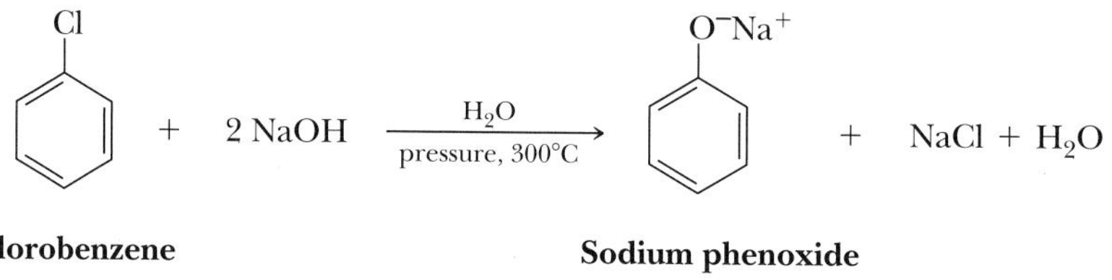

In later technological developments, the discovery was made that chlorobenzene can be hydrolyzed to phenol by steam under pressure at 500°C. Each of these reactions appears to involve nucleophilic substitution of —OH for —Cl on the benzene ring. However, this reaction is not as simple as it might seem, as illustrated by the reaction of substituted halobenzenes with NaOH. For example, *o*-chlorotoluene under these conditions gives a mixture of 2-methylphenol (*o*-cresol) and 3-methylphenol (*m*-cresol).

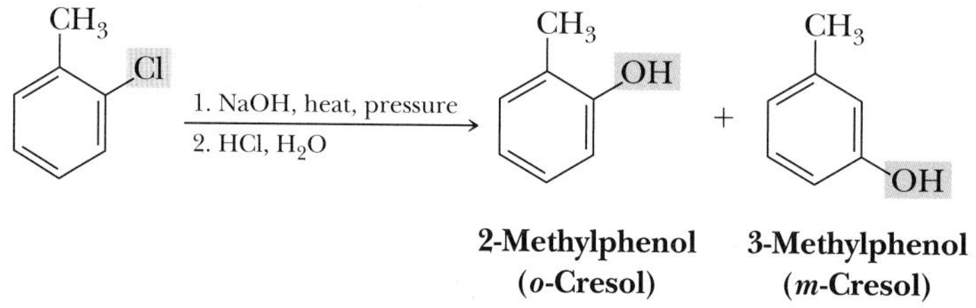

The same type of reaction can be brought about by the use of sodium amide in liquid ammonia. Under these conditions, for example, *p*-chlorotoluene gives a mixture of 4-methylaniline (*p*-toluidine) and 3-methylaniline (*m*-toluidine) in approximately equal amounts.

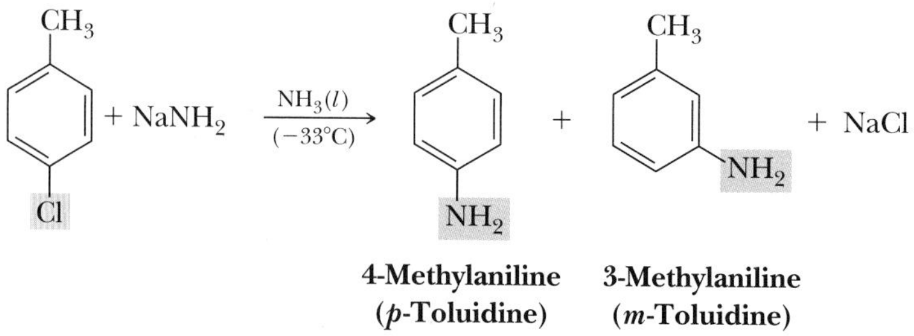

Benzyne intermediate A reactive intermediate that is formed by β-elimination from adjacent carbon atoms of a benzene ring and has a triple bond in the benzene ring. The second π bond of the benzyne triple bond is formed by weak overlap of coplanar sp^2 orbitals on adjacent carbons.

The difference in this reaction compared with other substitution reactions we have dealt with so far is that the entering group appears not only at the position occupied by the leaving group but also at a position adjacent to it.

To account for these experimental observations, it has been proposed that an elimination of HX occurs to form a **benzyne intermediate** that then undergoes nucleophilic addition to the triple bond to give the products observed.

Mechanism 22.5

Nucleophilic Aromatic Substitution via a Benzyne Intermediate

Step 1: **Take a proton away and simultaneously break a bond to make stable molecules or ions.** Dehydrohalogenation of the benzene ring gives a benzyne intermediate.

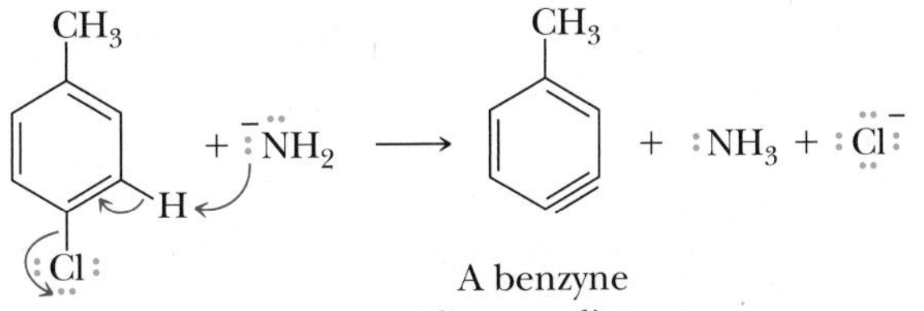
CH3
H
Cl
+ NH2
CH3
+ NH3 + Cl

A benzyne intermediate

Step 2: **Make a new bond between a nucleophile and an electrophile.** Nucleophilic addition of amide ion to a carbon of the benzyne triple bond gives a carbanion intermediate. Addition to either carbon of the triple bond is possible.

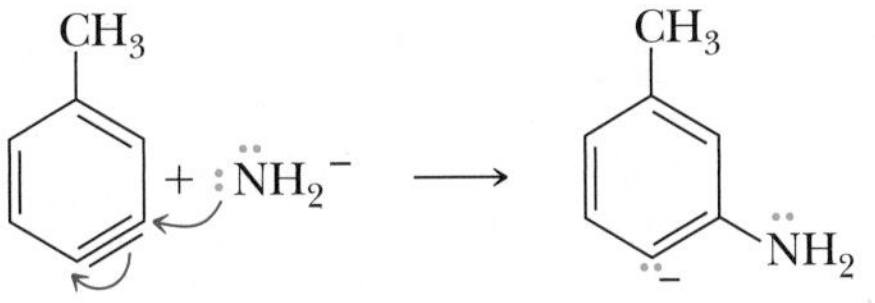
CH3
+ NH2
CH3
NH2

A carbanion intermediate

Step 3: **Add a proton.** Proton transfer from ammonia to the carbanion intermediate gives one of the observed substitution products and generates a new amide ion.

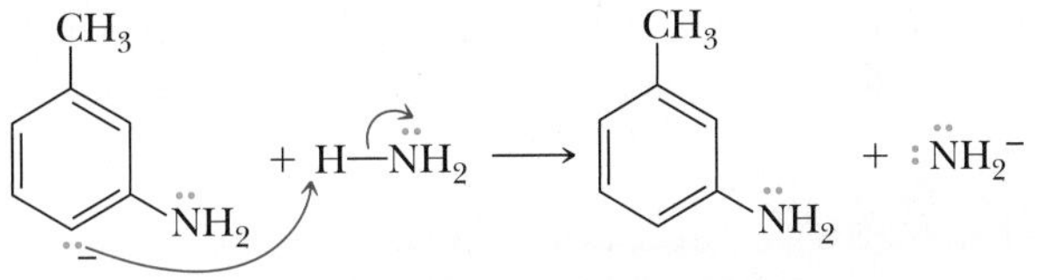
CH3
NH2
+ H—NH2
CH3
NH2
+ NH2

3-Methylaniline (*m*-Toluidine)

The bonding in a benzyne intermediate—and the reason for its extremely reactive nature—can be pictured in the following way. According to molecular orbital theory, the benzene ring retains its planarity, π bonding, and aromatic character. The adjacent sp^2 orbitals formerly bonding to a halogen and a hydrogen now overlap to form the second π bond of the benzyne triple bond. The problem is that the atomic orbitals forming this π bond are not parallel as in acetylene and unstrained alkynes; rather, they lie at an angle of approximately 120° to the bond axis connecting them. Consequently, the overlap between these orbitals is reduced. Reduced overlap, in turn, means a weaker and more reactive π bond. Therefore, the second π bond of the benzyne intermediate undergoes addition very readily to form two new and stronger σ bonds. The relatively high energy of the benzyne intermediate is presumably why such high temperatures are required for these reactions.

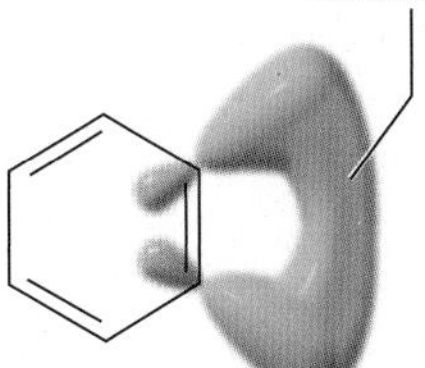
π bonding, but of reduced strength because of poor orbital overlap

A benzyne intermediate
Cartoon orbital

B. Nucleophilic Substitution by Addition-Elimination

Aromatic halides are normally quite inert to the types of nucleophiles that readily displace halide ions from haloalkanes. However, when an aromatic compound contains strong electron-withdrawing nitro groups ortho or para (or both) to the halogen, nucleophilic aromatic substitution occurs quite readily. For example, when 1-chloro-2,4-dinitrobenzene is heated at reflux in aqueous sodium carbonate followed by treatment with aqueous acid, it is converted in nearly quantitative yield to 2,4-dinitrophenol.

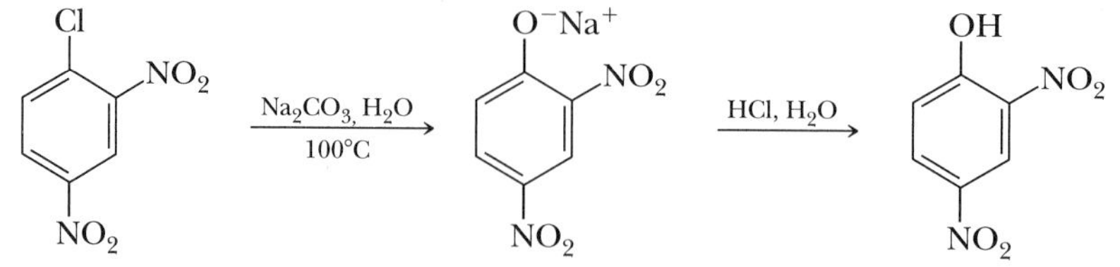

1-Chloro-2,4-dinitrobenzene | **Sodium 2,4-dinitrophenoxide** | **2,4-Dinitrophenol**

One application of this reaction is the synthesis of 2,4-dinitrophenylhydrazine, a reagent that was once commonly used to prepare derivatives of aldehydes and ketones (Section 16.8B).

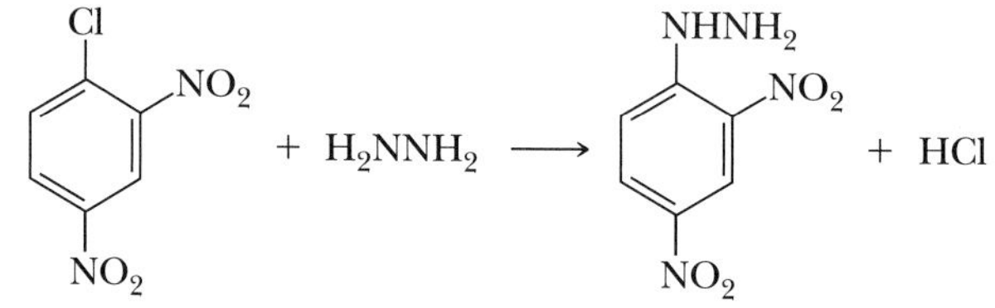

1-Chloro-2,4-dinitrobenzene | **Hydrazine** | **2,4-Dinitrophenylhydrazine**

This type of nucleophilic aromatic substitution for halogen has been studied extensively, and it has been determined that reaction occurs in two steps: nucleophilic addition followed by elimination. For the majority of reactions of this type, addition of the nucleophile in Step 1 is the slow, rate-determining step. Elimination of halide ion in Step 2 gives the product. This reaction thus resembles reactions of carboxylic acid derivatives in that it proceeds by an addition-elimination mechanism rather than by direct substitution.

Mechanism 22.6

Nucleophilic Aromatic Substitution by Addition-Elimination

Step 1: **Make a new bond between a nucleophile and an electrophile.** The nucleophile adds to the aromatic ring at the carbon bearing the halogen. This addition places a negative charge on the ring, which is stabilized by a resonance interaction with the nitro or other strongly electron-withdrawing groups in the ortho or para positions. Such intermediates are named Meisenheimer complexes after the German chemist who first characterized them. Note that nitro groups on both ortho and para positions participate in delocalization of the negative charge in the complex.

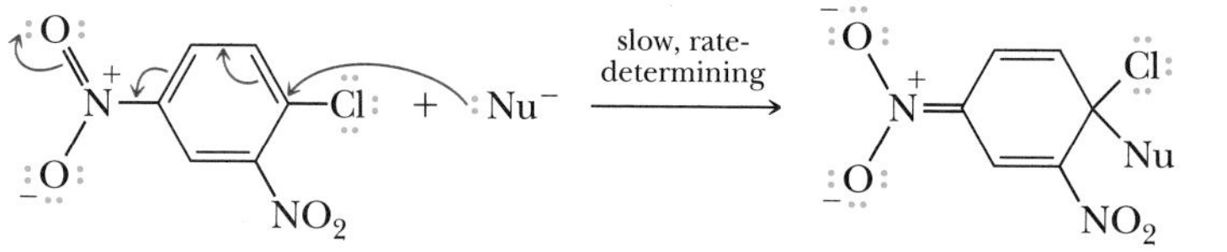

A Meisenheimer complex

Step 2: **Break a bond to give stable molecules and ions.** Elimination of halide ion regenerates the aromatic ring and gives the observed product.

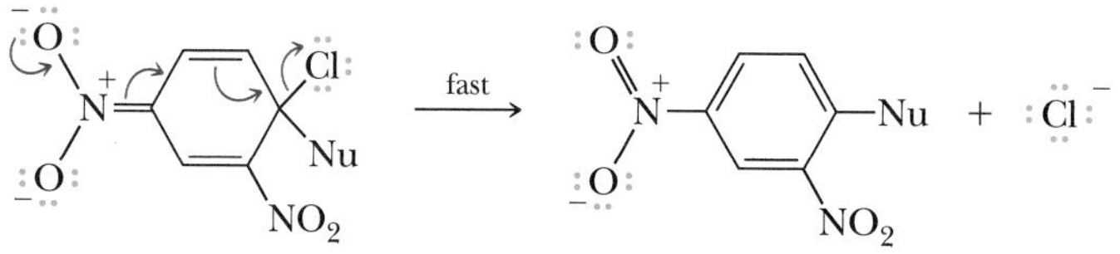

A Meisenheimer complex

Example 22.6 Nucleophilic Aromatic Substitution

What is the state of hybridization of each ring carbon atom in the Meisenheimer complex just shown?

Solution

The carbon atom bonded to both the leaving group and entering nucleophile (—Cl and —Nu in the structure shown) is sp^3 hybridized. The other five carbons of the ring are sp^2 hybridized.

Problem 22.6

In S_N2 reactions of haloalkanes, the order of reactivity is RI > RBr > RCl > RF. Alkyl iodides are considerably more reactive than alkyl fluorides, often by factors as great as 10^6. All 1-halo-2,4-dinitrobenzenes, however, react at approximately the same rate in nucleophilic aromatic substitutions. Account for this difference in relative reactivities.

Study Guide

22.1 Electrophilic Aromatic Substitution

- Aromatic rings, such as benzene, react with very strong, usually positively charged electrophiles in a reaction that results in substitution of a ring hydrogen.
  - *The general mechanism involves attack on the electrophile by the weakly nucleophilic aromatic π electrons to form a resonance-stabilized cation intermediate on the ring that loses a proton to give a substituted arene.*
  - *The resonance-stabilized cation intermediate is called an* ***arenium ion***.
- In a halogenation reaction, aromatic rings react with Cl_2 in the presence of the Lewis acid catalyst $FeCl_3$ to give chloroarenes. They react with Br_2 in the presence of the Lewis acid catalyst $FeBr_3$ to give bromoarenes.
- In a sulfonation reaction, aromatic rings react with SO_3 in the presence of sulfuric acid to yield arylsulfonic acids.
- In a nitration reaction, aromatic rings react with nitric acid in the presence of sulfuric acid to yield nitroarenes.
  - *Nitroarenes are useful synthetic intermediates because the nitro group can be reduced to an amino group by reaction with H_2 and a transition metal catalyst or, alternatively, by using iron, zinc, or tin in HCl followed by base.*
  - *There is no electrophilic aromatic substitution reaction that introduces the amino group directly onto the aromatic ring.*
- In a **Friedel-Crafts alkylation**, aromatic rings react with haloalkanes in the presence of a Lewis acid such as $AlCl_3$ to produce alkylbenzenes. Rearrangements and overalkylation can be a problem.

- In a **Friedel-Crafts acylation**, aromatic rings react with acid chlorides in the presence of a Lewis acid such as $AlCl_3$ to produce acylbenzenes.
  - *The acylbenzene products of Friedel-Crafts acylation reactions can be reduced to the corresponding alkylbenzene using Clemmensen or Wolff-Kishner reductions, providing a convenient method of producing alkylbenzenes that cannot be made in high yield using the Friedel-Crafts alkylation due to rearrangement or over-alkylation problems.*
- A variety of other electrophilic aromatic substitution reactions involve very strong electrophiles reacting with the weakly nucleophilic aromatic π electrons to form an intermediate resonance-stabilized cation on the ring that loses a proton to give the substituted arene.
  - *Reactions include using an alkene in the presence of a strong acid to give a carbocation that generates an alkylbenzene, using an alkene in the presence of a Lewis acid to generate an alkylbenzene, and using an alcohol in the presence of strong acid to give a carbocation that generates an alkylbenzene.*

P **22.1–22.3, 22.7–22.13, 22.22–22.27, 22.32–22.61, 22.63**

Problem 22.2

KEY REACTIONS

1. **Halogenation (Section 22.1A)** The electrophile is a halonium ion formed as an ion pair by interaction of chlorine or bromine with a Lewis acid. The mechanism involves an initial reaction between Cl_2 and $FeCl_3$ to generate a molecular complex that can rearrange to give a Cl^+, $FeCl_4^-$ ion pair. The Cl^+ reacts as a very strong electrophile with the weakly nucleophilic aromatic π cloud to form a resonance-stabilized cation intermediate that loses a proton to give the chloroarene product.

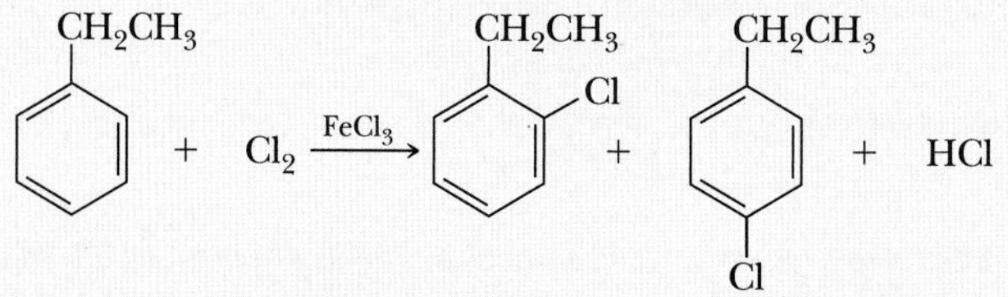

Halogenation of an aromatic ring substituted by strongly activating groups (such as —OH, —OR, and —NH_2) does not require a Lewis acid catalyst.

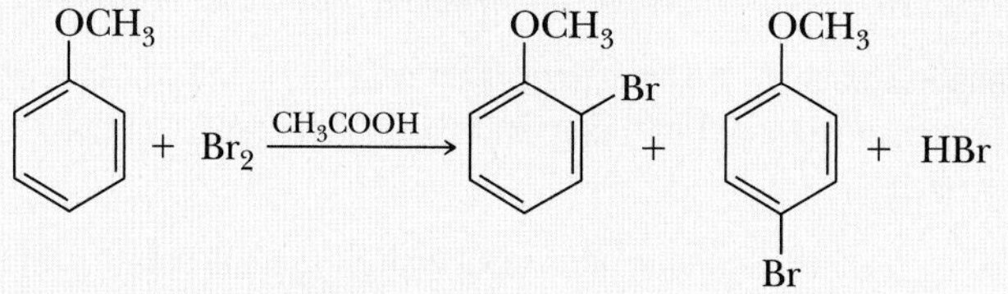

2. **Nitration (Section 22.1B)** The electrophile is the nitronium ion, NO_2^+, formed by interaction of nitric acid and sulfuric acid. The mechanism involves an initial protonation of nitric acid by sulfuric acid followed by loss of water to yield the nitronium ion NO_2^+. The nitronium ion reacts as a very strong electrophile with the weakly nucleophilic aromatic π cloud to form a resonance-stabilized cation intermediate that loses a proton to give the final product.

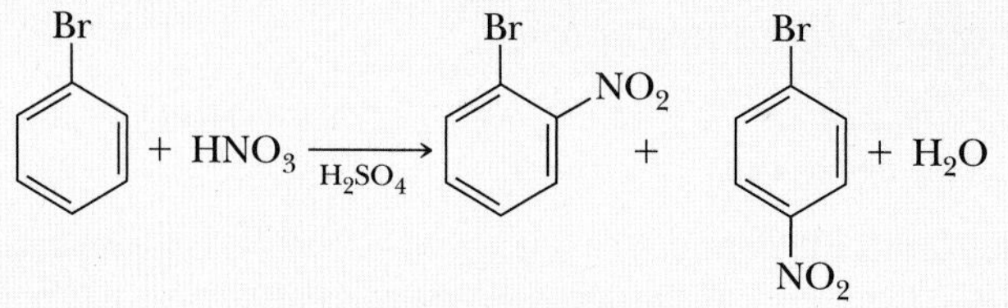

3. **Sulfonation (Section 22.1B)** The electrophile is either sulfur trioxide, SO_3, or HSO_3^+ depending on experimental conditions. The mechanism involves reaction of SO_3 as a very strong electrophile with the weakly nucleophilic aromatic π cloud to form a resonance-stabilized cation intermediate that loses a proton to give the arylsulfonic acid product.

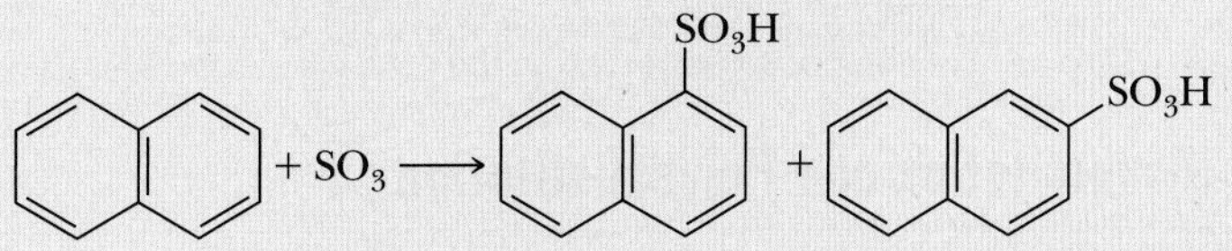

4. **Friedel-Crafts Alkylation (Section 22.1C)** The electrophile is a carbocation formed as an ion pair by interaction of an haloalkane with a Lewis acid. Rearrangements from a less stable carbocation to a more stable carbocation are common. The mechanism involves an initial reaction between the haloalkane and Lewis acid $AlCl_3$ to yield an intermediate that can be thought of as a carbocation/$AlCl_4^-$ ion pair. The carbocation portion of the ion pair reacts as a very strong electrophile with the weakly nucleophilic aromatic π cloud to form a resonance-stabilized cation intermediate that loses a proton to give the final product. Because carbocations are involved in the mechanism, rearrangements can be a problem, especially with primary or secondary haloalkanes or with any other haloalkane that will generate a carbocation prone to rearrangement.

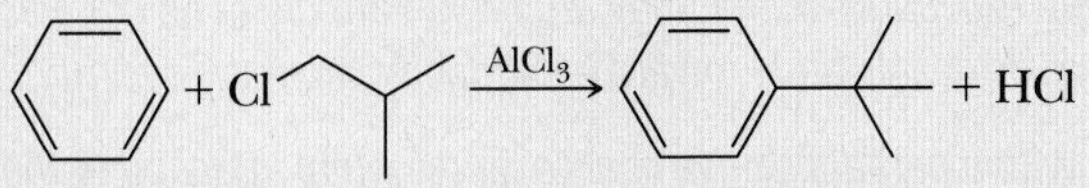

The reaction fails when there is one or more strongly electron-withdrawing groups on the ring. It can be hard to stop the reaction after a single alkylation, because the aromatic monoalkylation product is more reactive than the starting material.

5. **Friedel-Crafts Acylation (Section 22.1C)** The electrophile is an acyl cation (an acylium ion) formed as an ion pair by interaction of an acyl halide with a Lewis acid. The mechanism involves an initial reaction between the acid chloride and Lewis acid $AlCl_3$ to yield an intermediate that can be thought of as a resonance-stabilized acylium ion/$AlCl_4^-$ ion pair. The acylium ion portion of the ion pair reacts as a very strong electrophile with the weakly nucleophilic aromatic π cloud to form a resonance-stabilized cation intermediate that loses a proton to give the final product. Because acylium ions do not rearrange like carbocations, no rearrangements occur. The reaction fails when there is one or more strongly electron-withdrawing groups on the ring. It is easy to stop the reaction after a single acylation because the aromatic monoacylation product is less reactive than the starting material.

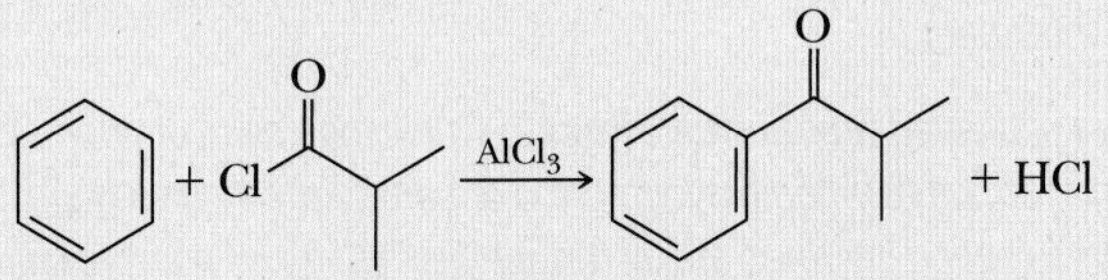

6. **Alkylation Using an Alkene (Section 22.1D)** The electrophile is a carbocation formed by interaction of the alkene with a Brønsted or Lewis acid.

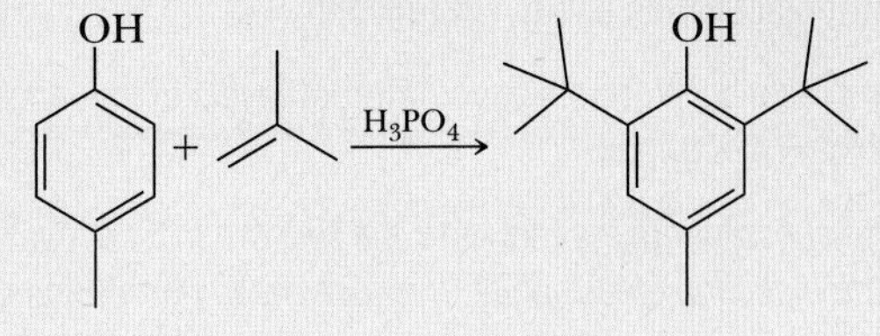

(Continued)

7. Alkylation Using an Alcohol (Section 22.1D) The electrophile is a carbocation formed by treatment of the alcohol with a Brønsted or Lewis acid.

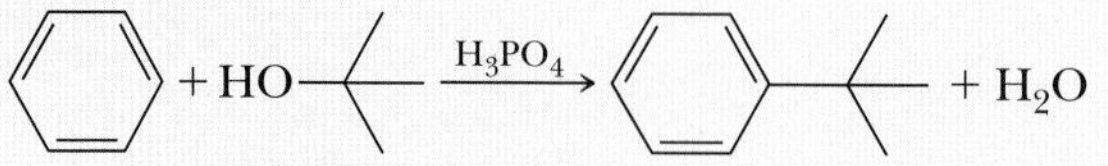

22.2 Disubstitution and Polysubstitution

- Substituent groups, other than hydrogen, on an aromatic ring influence the reaction rate and substitution pattern of electrophilic aromatic substitution reactions.
  - *In particular, substituents can direct new groups meta or ortho-para and can either speed up (activate) or slow down (deactivate) the reaction.*
- Substituents can be divided into three broad classes:
  - *Alkyl groups and all groups in which the atom bonded to the ring has an unshared pair of electrons are ortho-para directing, and most are electron releasing; therefore, they are* ***activating*** *toward electrophilic aromatic substitution compared to benzene itself.*
  - *Halogens are exceptions in that they are ortho-para directing but electron withdrawing; therefore, they are weakly deactivating toward electrophilic aromatic substitution compared to benzene itself.*
  - *All groups with a partial positive charge on the atom attached to the ring are meta directing and electron withdrawing; therefore, they are* ***deactivating*** *toward electrophilic aromatic substitution compared to benzene itself.*
- Orientation and activating/deactivating effects have a large practical significance, because in synthesizing polysubstituted aromatics, the *order of addition of the substituents must be taken into account.*
  - *For example, when* m*-bromonitrobenzene is made from benzene, the nitro group (meta directing) must be added before the bromine atom (ortho-para directing).*
  - *Conversely, when* o*-bromonitrobenzene and* p*-bromonitrobenzene are made from benzene, the bromine (ortho-para directing) must be added first, followed by the nitro group (meta directing).*
- Substituent directing and activation/deactivation effects are the result of two types of interactions operating on the cation intermediate:
  - *An* ***inductive effect*** *in which (relative to H atoms) the substituent withdraws more electron density out of (deactivating) or releases more electron density into (activating), the positively charged intermediate.*
  - *An always activating* ***resonance effect*** *in which lone pairs on atoms bonded to the ring create an extra contributing structure when they are ortho or para to the incoming electrophile. The added resonance stabilizes the cation intermediate by further distributing its positive charge over the molecule.*
- These inductive and resonance effects either raise or lower the energy of the cation intermediate, which, in turn, raises (decreasing rate, deactivation) or lowers (increasing rate, activation) the rate-limiting energy barrier (activation energy) of reaction.
- These inductive and resonance effects have different levels of influence depending on their position relative to the incoming electrophile.
  - *The effects of substituents that* activate *the ring are most* activating *ortho and para to the position of the incoming electrophile, because these orientations provide for the strongest cation intermediate stabilizing interactions.*

- *The effects of substituents that* deactivate *the ring are less* deactivating *meta to the position of the incoming electrophile, because the meta orientation minimizes destabilization of the cation intermediate relative to an ortho or para orientation.*

- The three basic types of substituents can be understood in the context of the foregoing ideas.
  - *Electron-releasing groups are activating and thus always ortho-para directing.*
  - *Electron-withdrawing groups other than the halogens are deactivating and thus meta directing.*
  - *Halogens show deactivation but an ortho-para substitution preference. The stabilizing resonance effect of the unshared electron pair directs ortho and para and takes orientation precedence over the overall inductive deactivation due to electronegativity.*
- For cases involving electrophilic aromatic substitution of aromatic rings with two or more substituents already on the ring, the more activating group will dominate the orientation preference of the incoming group.

P 22.4, 22.5, 22.14–22.21, 22.32–22.61, 22.63

Table 22.2

22.3 Nucleophilic Aromatic Substitution

- Electrophilic aromatic substitution is by far the most common mechanism for reactions with aromatic rings, but in some situations, aromatic rings react with nucleophiles.
  - *Haloarenes react with very strong bases ($NaNH_2$) or moderate bases ($NaOH$) at high temperature ($300°C$ to $500°C$) to yield products in which the halogen is replaced.*
  - *The base/nucleophile group ends up on the ring carbon atom that was originally bonded to the halogen, as well as positions adjacent (ortho) to it due to the benzyne intermediates.*
  - *Haloarenes that are activated by having ortho and/or para strongly electron withdrawing groups react with strong nucleophiles such as hydrazine to give substitution in a regioselective manner.*

P 22.6, 22.28, 22.29, 22.62

KEY REACTIONS

8. Nucleophilic Aromatic Substitution: A Benzyne Intermediate (Section 22.3A) The mechanism involves an initial elimination reaction between the haloarene and strong base to give a benzyne intermediate, which undergoes addition at either *sp* hybridized carbon atom to give the products.

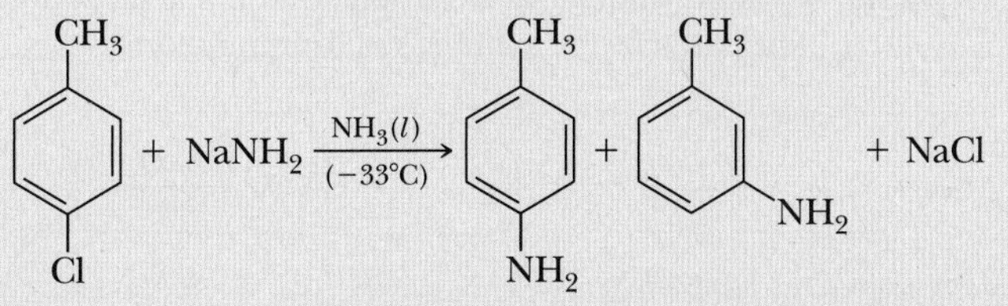

9. Nucleophilic Aromatic Substitution: Addition-Elimination (Section 22.3B) The mechanism involves a nucleophilic attack of the ring carbon containing the halogen to give a negatively charged Meisenheimer complex, followed by loss of halogen to give the substituted product. This reaction does

(Continued)

not occur unless there are electron-withdrawing groups ortho and/or para to the halogen, because these groups activate the ring toward nucleophilic attack.

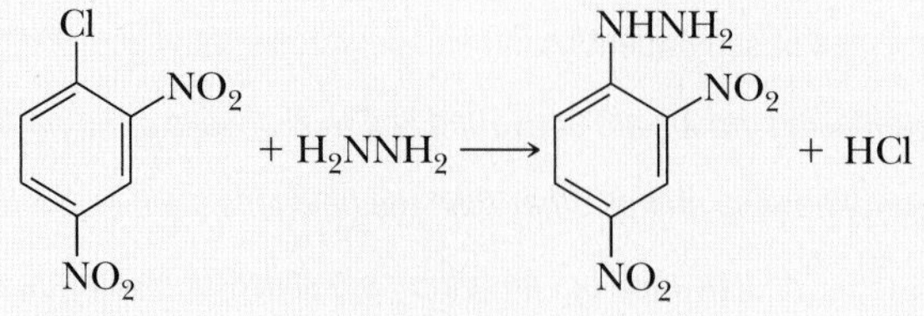

Problems

Red numbers indicate applied problems.

Electrophilic Aromatic Substitution: Monosubstitution

22.7 Write a stepwise mechanism for each of the following reactions. Use curved arrows to show the flow of electrons in each step.

(a)

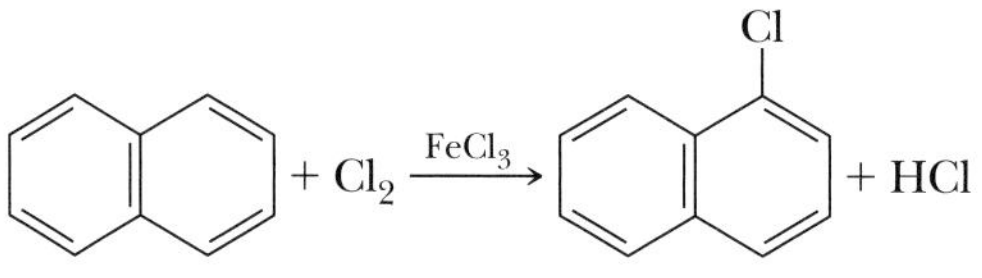

(b)

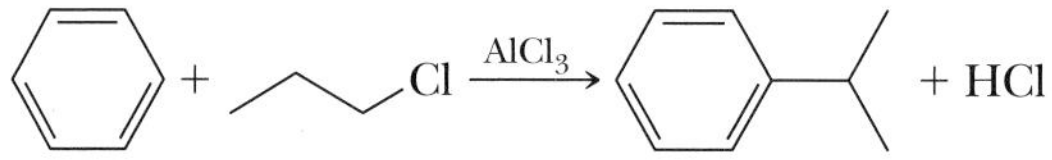

(c)

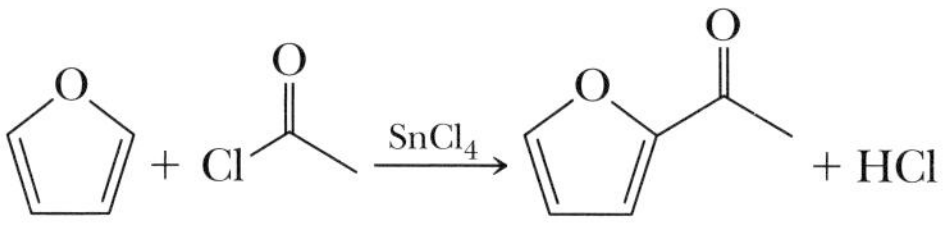

(d)

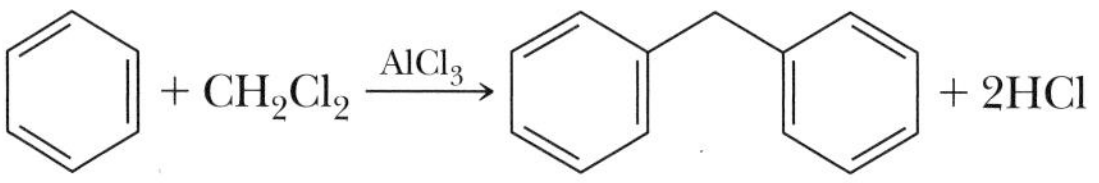

22.8 Pyridine undergoes electrophilic aromatic substitution preferentially at the 3 position as illustrated by the synthesis of 3-nitropyridine.

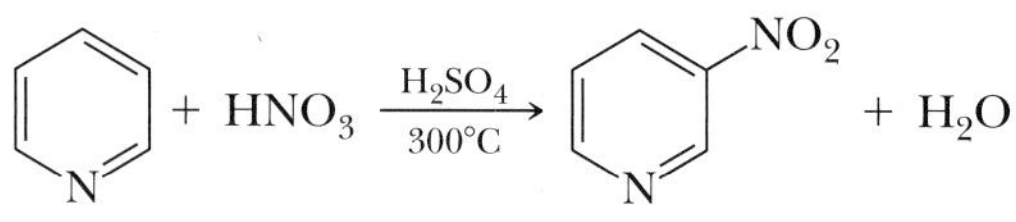

Pyridine **3-Nitropyridine**

Under these acidic conditions, the species undergoing nitration is not pyridine, but its conjugate acid. Write resonance contributing structures for the intermediate formed by attack of NO_2^+ at the 2, 3, and 4 positions of the conjugate acid of pyridine. From examination of these intermediates, offer an explanation for preferential nitration at the 3 position.

22.9 Pyrrole undergoes electrophilic aromatic substitution preferentially at the 2 position as illustrated by the synthesis of 2-nitropyrrole.

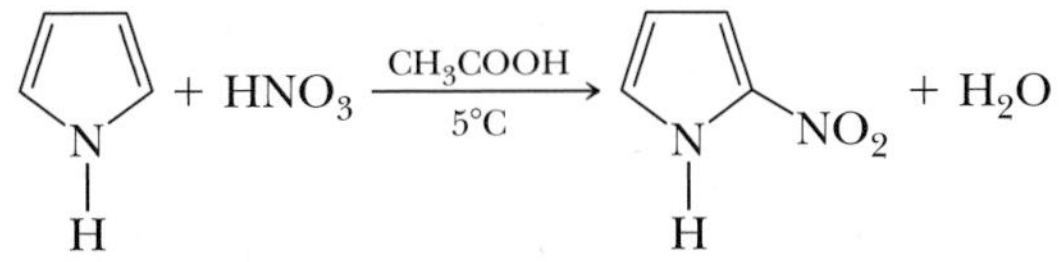

Pyrrole **2-Nitropyrrole**

Write resonance contributing structures for the intermediate formed by attack of NO_2^+ at the 2 and 3 positions of pyrrole. From examination of these intermediates, offer an explanation for preferential nitration at the 2 position.

22.10 Addition of *m*-xylene to the strongly acidic solvent HF/SbF_5 at −45°C gives a new species, which shows ^{1}H-NMR resonances at δ 2.88 (3H), 3.00 (3H), 4.67 (2H), 7.93 (1H), 7.83 (1H), and 8.68 (1H). Assign a structure to the species giving this spectrum.

22.11 Addition of *tert*-butylbenzene to the strongly acidic solvent HF/SbF_5 followed by aqueous workup gives benzene. Propose a mechanism for this dealkylation reaction. What is the other product of the reaction?

22.12 What product do you predict from the reaction of SCl_2 with benzene in the presence of $AlCl_3$? What product results if diphenyl ether is treated with SCl_2 and $AlCl_3$?

22.13 Other groups besides H^+ can act as leaving groups in electrophilic aromatic substitution. One of the best is the trimethylsilyl group, Me_3Si—. For example, treatment of $Me_3SiC_6H_5$ with CF_3COOD rapidly forms C_6H_5D. What properties of a silicon-carbon bond allow you to predict this kind of reactivity?

Disubstitution and Polysubstitution

22.14 The following groups are ortho-para directors.

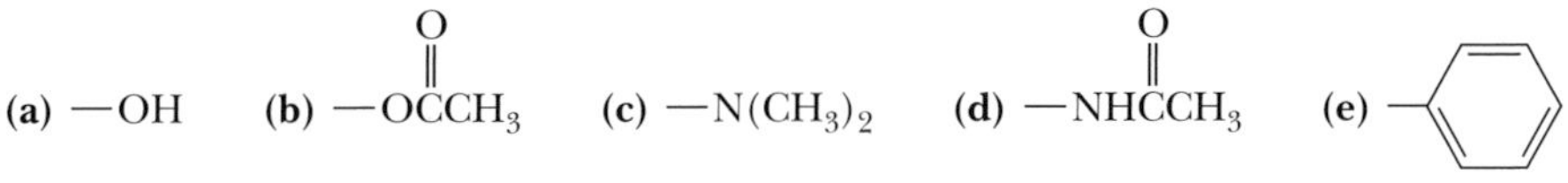

Draw a contributing structure for the resonance-stabilized cation formed during electrophilic aromatic substitution that shows the role of each group in stabilizing the intermediate by further delocalizing its positive charge.

22.15 Predict the major product or products from treatment of each compound with HNO_3/H_2SO_4.

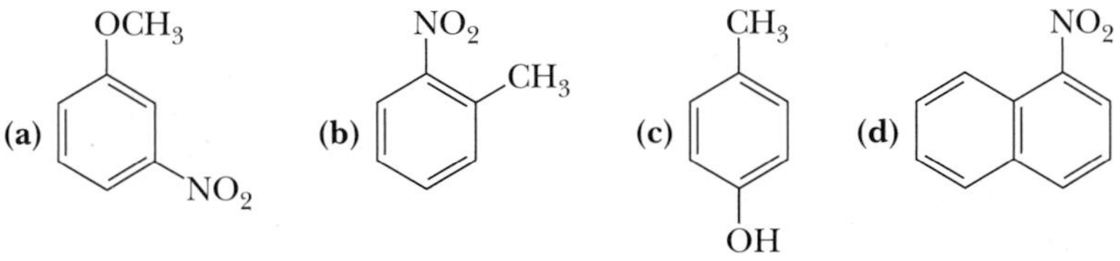

22.16 How do you account for the fact that *N*-phenylacetamide (acetanilide) is less reactive toward electrophilic aromatic substitution than is aniline?

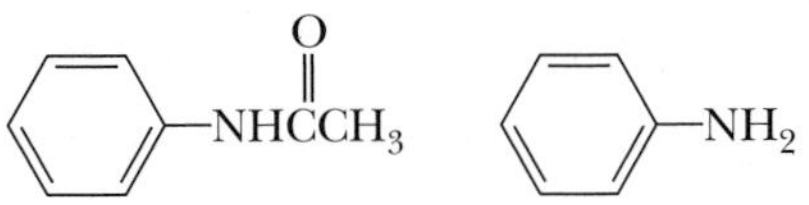

***N*-Phenylacetamide (Acetanilide)** **Aniline**

22.17 Propose an explanation for the fact that the trifluoromethyl group is almost exclusively meta directing.

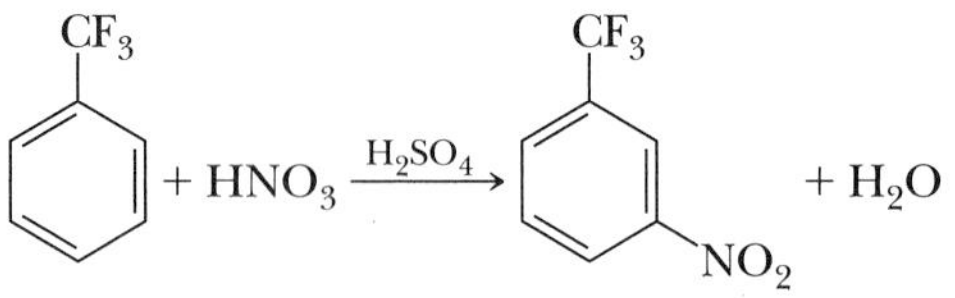

22.18 Suggest a reason why the nitroso group, —N=O, is ortho-para directing whereas the nitro group, $—NO_2$, is meta directing.

22.19 Arrange the compounds in each set in order of decreasing reactivity (fastest to slowest) toward electrophilic aromatic substitution.

(a) 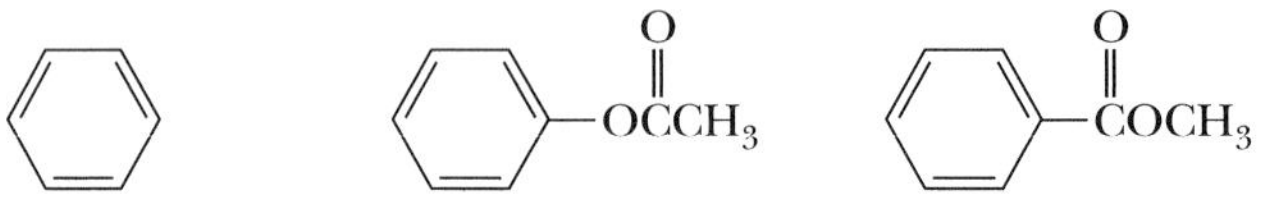

(A) (B) (C)

(b) 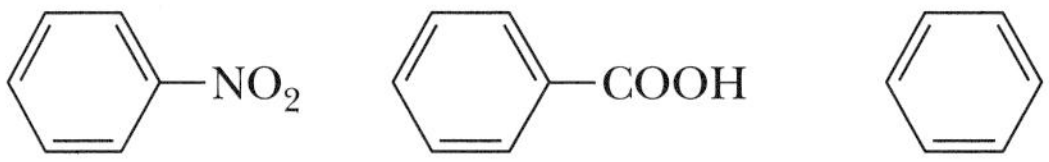

(A) (B) (C)

(c) 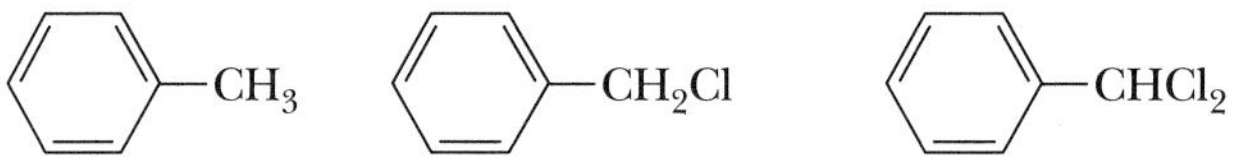

(A) (B) (C)

(d) 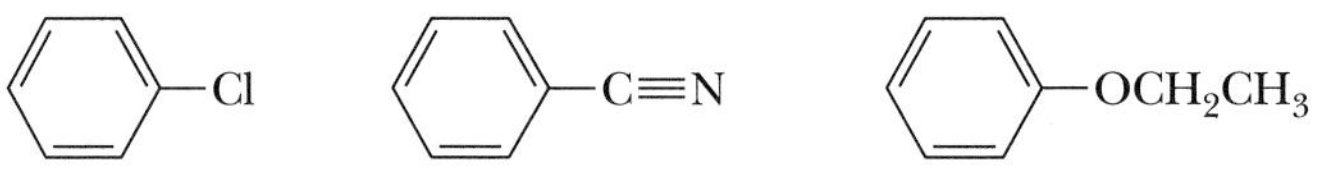

(A) (B) (C)

(e) 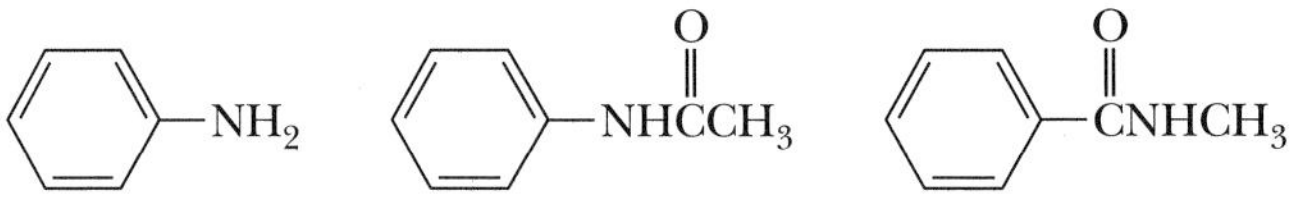

(A) (B) (C)

22.20 For each compound, indicate which group on the ring is more strongly activating and then draw a structural formula of the major product formed by nitration of the compound.

(a) 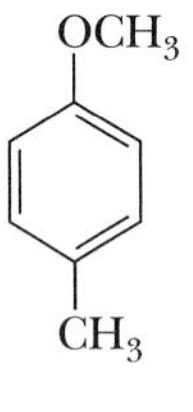

(b) 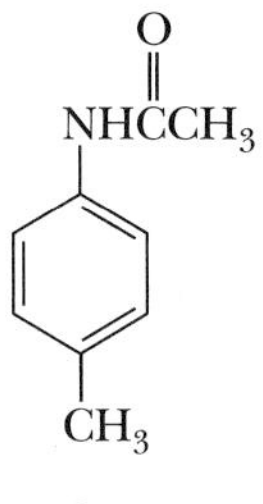

(c) 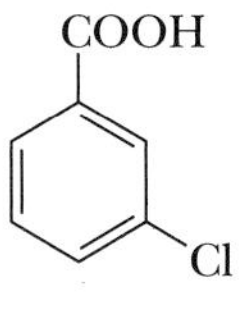

(d) 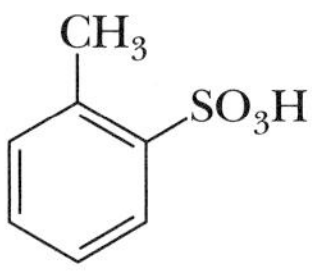

(e) 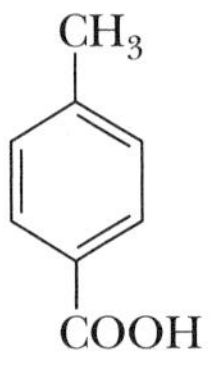

(f) 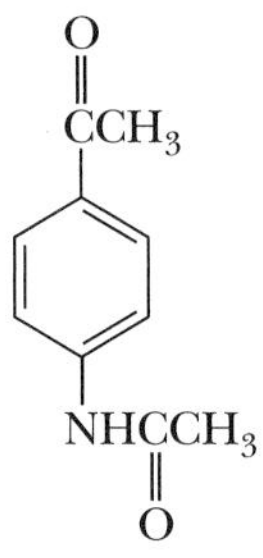

(g) 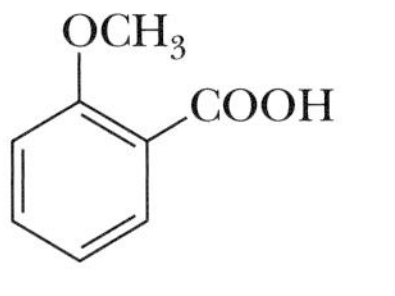

(h) 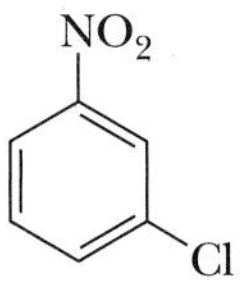

22.21 The following molecules each contain two aromatic rings.

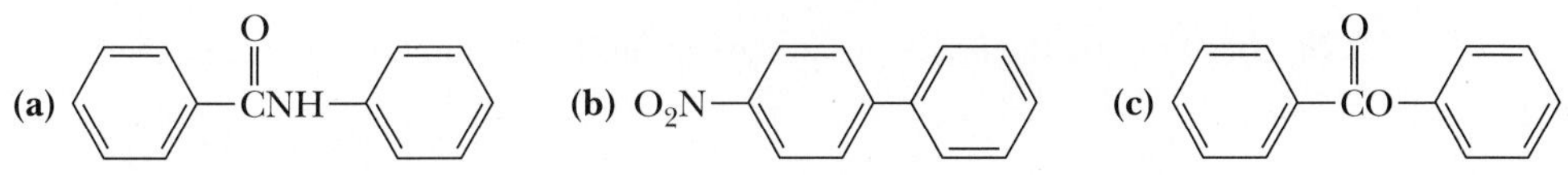

Which ring in each undergoes electrophilic aromatic substitution more readily? Draw the major product formed on nitration.

22.22 Reaction of phenol with acetone in the presence of an acid catalyst gives a compound known as bisphenol A, which is used in the production of epoxy and polycarbonate resins (Section 29.5). Propose a mechanism for the formation of bisphenol A.

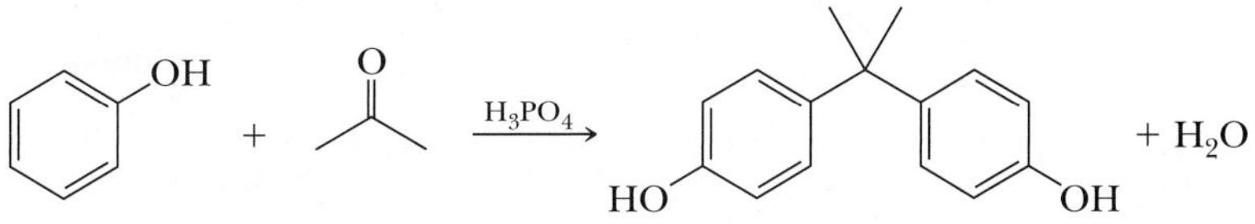

Phenol **Acetone** **Bisphenol A**

22.23 2,6-Di-*tert*-butyl-4-methylphenol, alternatively known as butylated hydroxytoluene (BHT), is used as an antioxidant in foods to "retard spoilage" (Section 8.7). BHT is synthesized industrially from 4-methylphenol by reaction with 2-methylpropene in the presence of phosphoric acid. Propose a mechanism for this reaction.

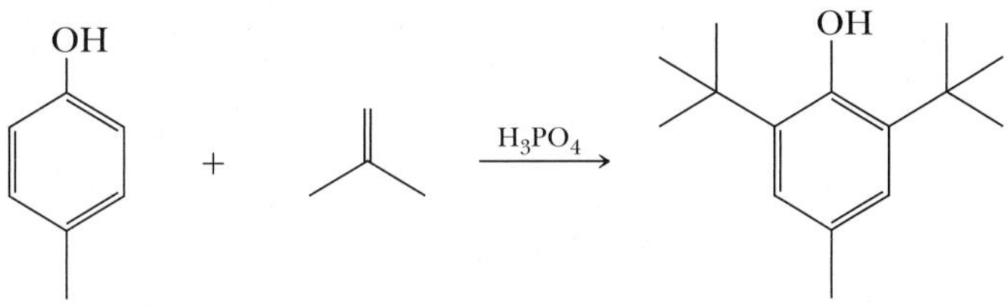

4-Methylphenol (*p*-Cresol) **2-Methylpropene** **2,6-Di-*tert*-butyl-4-methylphenol "Butylated hydroxytoluene" (BHT)**

22.24 The insecticide DDT is prepared by the following route. Suggest a mechanism for this reaction. The abbreviation DDT is derived from the common name dichlorodiphenyltrichloroethane.

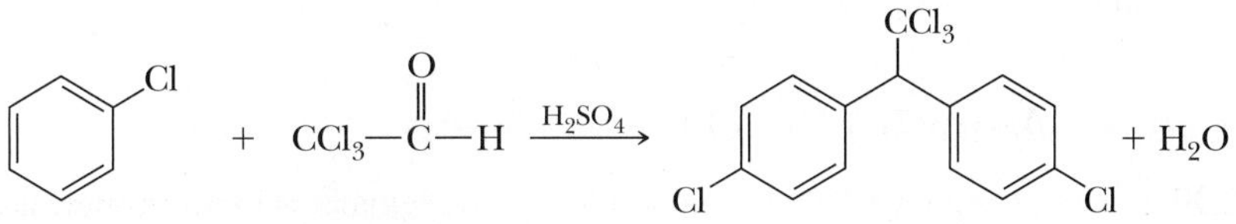

Chlorobenzene **Trichloro-acetaldehyde** **DDT**

22.25 Treatment of salicylaldehyde (2-hydroxybenzaldehyde) with bromine in glacial acetic acid at 0°C gives a compound with the molecular formula $C_7H_4Br_2O_2$, which is used as a topical fungicide and antibacterial agent. Propose a structural formula for this compound.

22.26 Propose a synthesis for 3,5-dibromo-2-hydroxybenzoic acid (3,5-dibromosalicylic acid) from phenol.

22.27 Treatment of benzene with succinic anhydride in the presence of polyphosphoric acid gives the following γ-ketoacid. Propose a mechanism for this reaction.

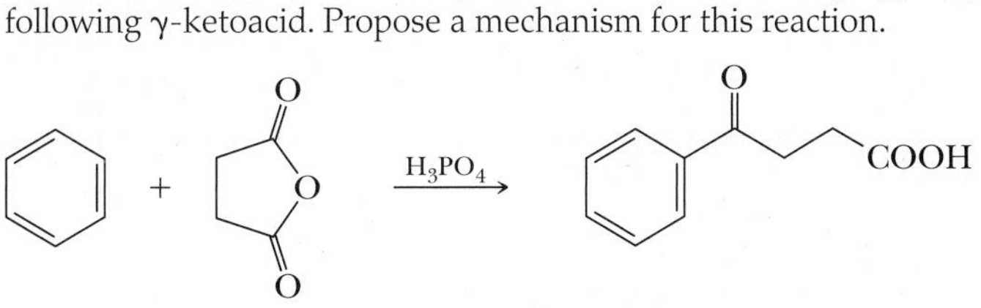

Succinic anhydride **4-Oxo-4-phenylbutanoic acid**

Nucleophilic Aromatic Substitution

22.28 Following are the final steps in the synthesis of trifluralin B, a preemergent herbicide.

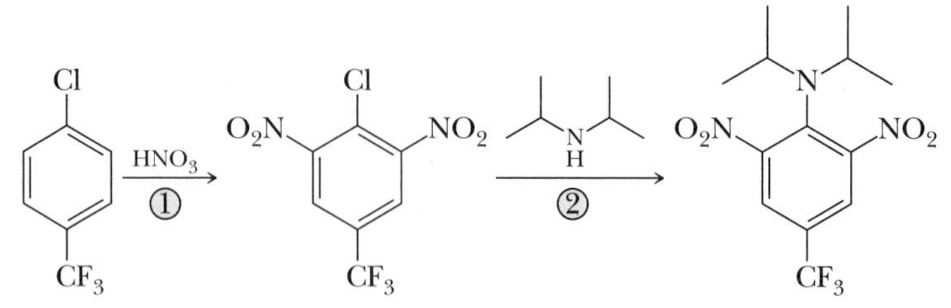

Cl
CF3
HNO3
①
O2N
NO2
N
H
②

Trifluralin B

(a) Account for the orientation of nitration in Step 1.
(b) Propose a mechanism for the substitution reaction in Step 2.

22.29 A problem in dyeing fabrics is the degree of fastness of the dye to the fabric. Many of the early dyes were surface dyes; that is, they did not bond to the fabric, with the result that they tended to wash off after repeated laundering. Indigo, for example, which gives the blue color to blue jeans, is a surface dye. Color fastness can be obtained by bonding a dye to the fabric. The first such dyes were the so-called reactive dyes, developed in the 1930s for covalently bonding dyes containing —NH_2 groups to cotton, wool, and silk fabrics. In the first stage of the first-developed method for reactive dyeing, the dye is treated with cyanuric chloride, which links to the fabric through the amino group of the dye. The remaining chlorines are then displaced by the —OH groups of cotton (cellulose) or the —NH_2 groups of wool or silk (both proteins).

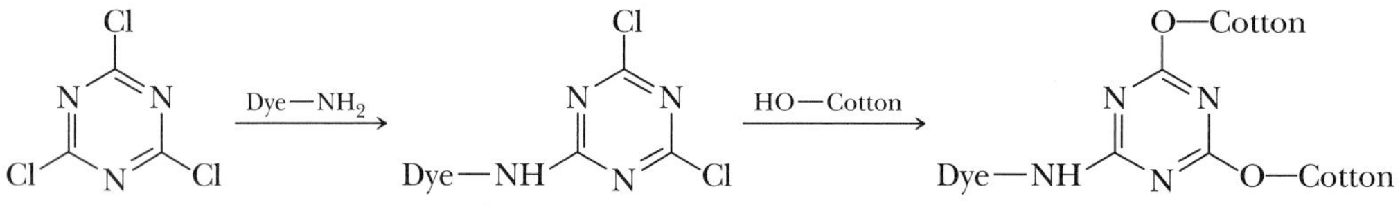

Cl
N
Dye—NH2
Dye—NH
HO—Cotton
O—Cotton

Cyanuric chloride A reactive dye Dye covalently bonded to cotton

Propose a mechanism for the displacement of a chlorine from cyanuric chloride by (a) the NH_2 group of a dye and (b) by an —OH group of cotton.

Organic Chemistry Reaction Roadmap

22.30 Use the reaction roadmap you made for Problems 20.55 and 21.49 and update it to contain the reactions in the "Study Guide" section of this chapter. Because of its highly specific nature, do not use reaction 9 on your roadmap. In anticipation of reactions in Chapter 23, on this roadmap, you should include the H_2/Ni and 1) Fe, HCl, 2) NaOH reactions that convert aryl nitro compounds to aryl amines. These two reactions were covered in the chapter (Section 22.1B) but were not included in the "Study Guide" section.

22.31 Write the products of the following sequences of reactions. Refer to your reaction roadmaps to see how the combined reactions allow you to "navigate" between the different functional groups. Note that you will need your old Chapters 6–11, Chapters 15–18, and Chapter 19 roadmaps along with your new Chapters 20–22 reaction roadmaps for these.

(a)

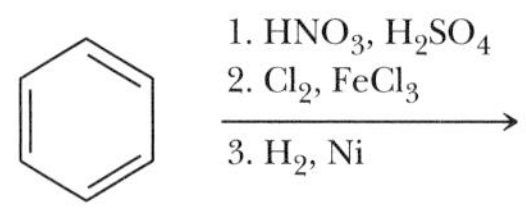

1. HNO3, H2SO4
2. Cl2, FeCl3
3. H2, Ni

An aryl ring

(b)

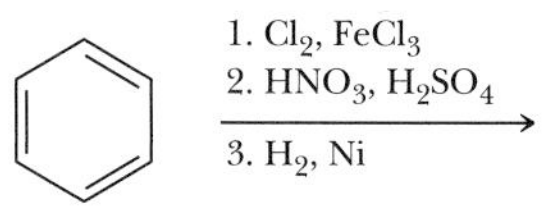

1. Cl2, FeCl3
2. HNO3, H2SO4
3. H2, Ni

An aryl ring

(c) 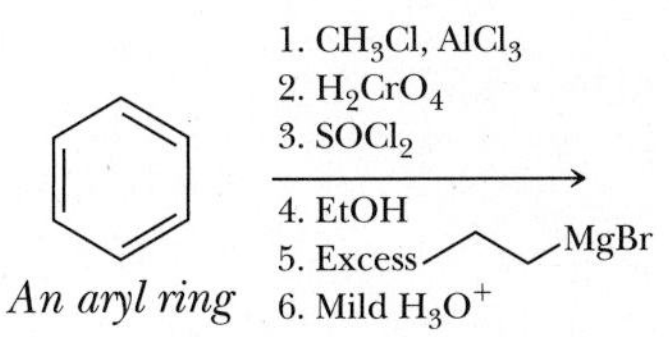
An aryl ring
1. CH_3Cl, $AlCl_3$
2. H_2CrO_4
3. $SOCl_2$
4. EtOH
5. Excess MgBr
6. Mild H_3O^+

(d) 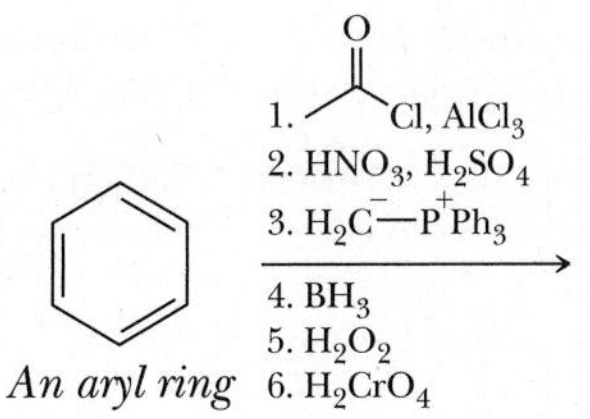
An aryl ring
1. Cl, $AlCl_3$
2. HNO_3, H_2SO_4
3. $H_2\bar{C}—\overset{+}{P}Ph_3$
4. BH_3
5. H_2O_2
6. H_2CrO_4

(e) 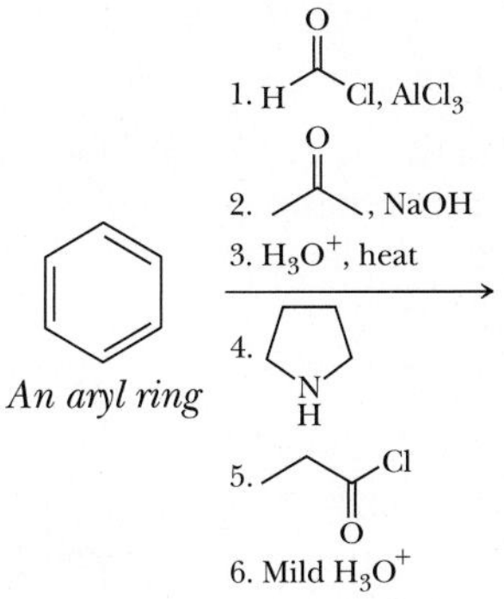
An aryl ring
1. H Cl, $AlCl_3$
2. , NaOH
3. H_3O^+, heat
4.
5. Cl
6. Mild H_3O^+

Syntheses

22.32 Show how to convert toluene to these compounds.

(a) 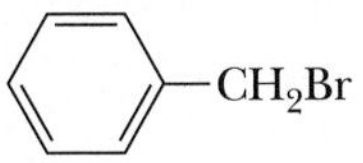
CH_2Br

(b) 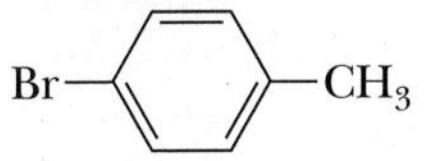
Br
CH_3

22.33 Show how to prepare each compound from 1-phenyl-1-propanone.

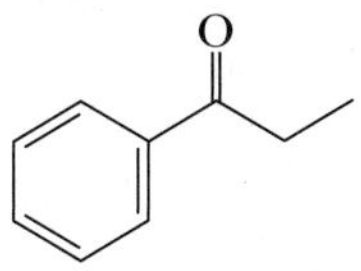

1-Phenyl-1-propanone

(a) 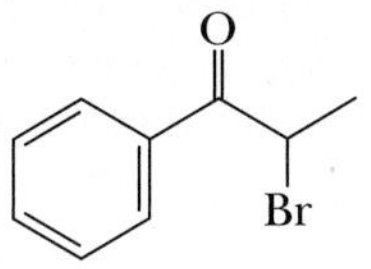
Br

(racemic)

(b) 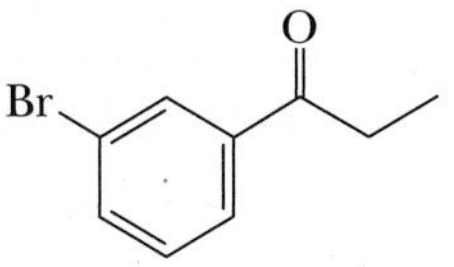
Br

22.34 Show how to convert toluene to **(a)** 2,4-dinitrobenzoic acid and **(b)** 3,5-dinitrobenzoic acid.

22.35 Show reagents and conditions to bring about the following conversions.

(a) 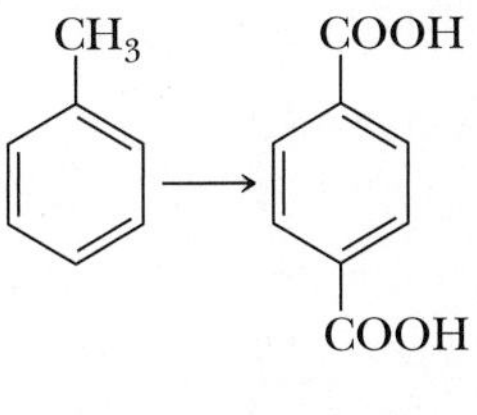
CH_3
COOH
COOH

(b) 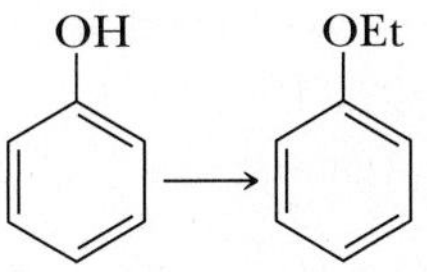
OH
OEt

(c) 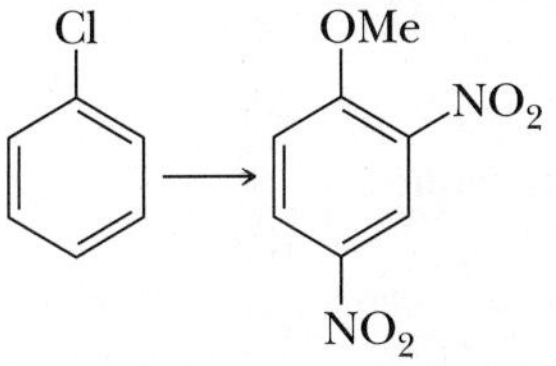
Cl
OMe
NO_2
NO_2

(d) 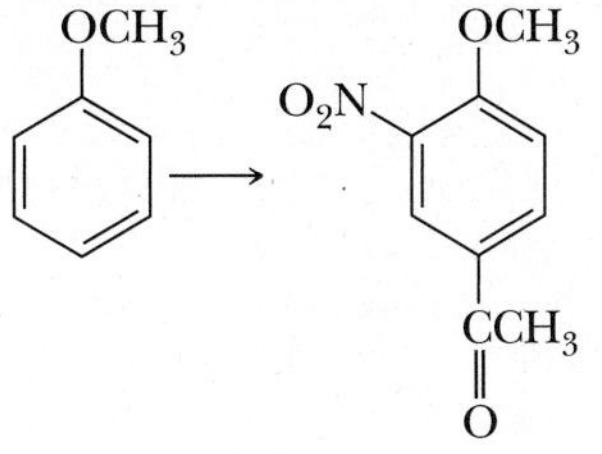
OCH_3
OCH_3
O_2N
CCH_3
O

(e) 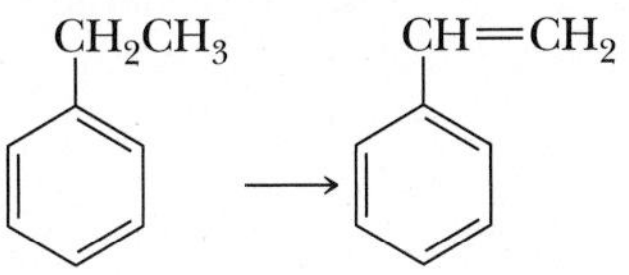
CH_2CH_3
$CH{=}CH_2$

22.36 Propose a synthesis of triphenylmethane from benzene, the only source of aromatic rings, and any other necessary reagents.

22.37 Propose a synthesis for each compound from benzene.

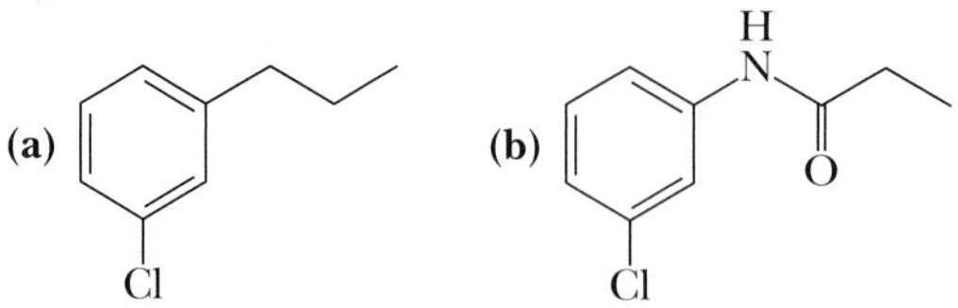

22.38 The first widely used herbicide for the control of weeds was 2,4-dichlorophenoxyacetic acid (2,4-D). Show how this compound might be synthesized from phenol and chloroacetic acid by way of the given chlorinated phenol intermediate.

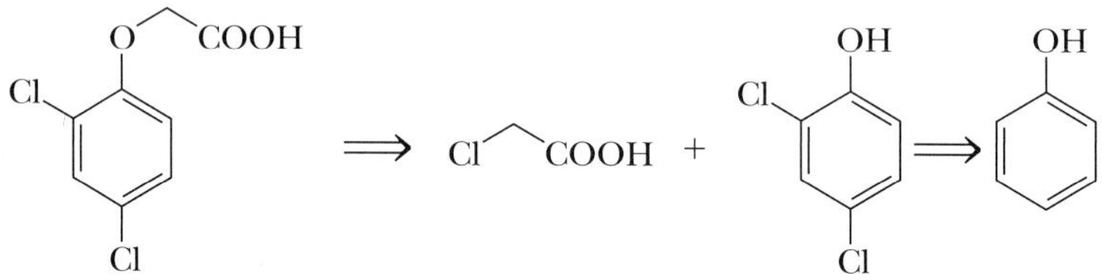

2,4-Dichlorophenoxyacetic acid (2,4-D) **Chloroacetic acid** **Phenol**

22.39 Phenol is the starting material for the synthesis of 2,3,4,5,6-pentachlorophenol, known alternatively as pentachlorophenol, or more simply as penta. At one time, penta was widely used as a wood preservative for decks, siding, and outdoor wood furniture. Draw the structural formula for pentachlorophenol and describe its synthesis from phenol.

22.40 Starting with benzene, toluene, or phenol as the only sources of aromatic rings, show how to synthesize the following. Assume in all syntheses that mixtures of ortho-para products can be separated into the desired isomer.

(a) 1-Bromo-3-nitrobenzene
(b) 1-Bromo-4-nitrobenzene
(c) 2,4,6-Trinitrotoluene (TNT)
(d) *m*-Chlorobenzoic acid
(e) *p*-Chlorobenzoic acid
(f) *p*-Dichlorobenzene
(g) *m*-Nitrobenzenesulfonic acid

22.41 3,5-Dibromo-4-hydroxybenzenesulfonic acid is used as a disinfectant. Propose a synthesis of this compound from phenol.

22.42 Propose a synthesis for 3,5-dichloro-2-methoxybenzoic acid starting from phenol.

22.43 The following compound used in perfumery has a violet-like scent. Propose a synthesis of this compound from benzene.

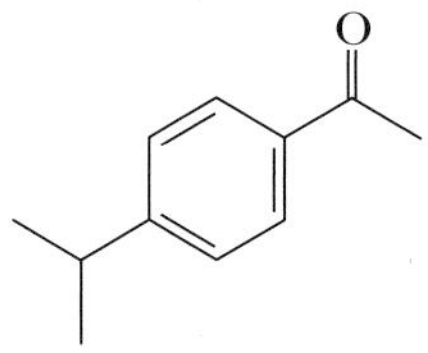

4-Isopropylacetophenone

22.44 Cancer of the prostate is the second leading cause of cancer deaths among American males, exceeded only by lung cancer. One treatment of prostate cancer is based on the fact that testosterone and androsterone (both androgens) enhance the proliferation of prostate tumors. The drug flutamide (an antiandrogen) reduces the level of androgens in target tissues and is currently used to prevent and treat prostate cancer.

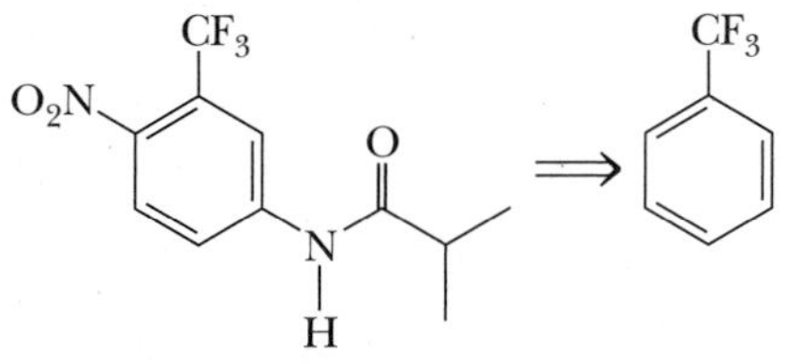

Flutamide (Eulexin) **Trifluoromethylbenzene**

Propose a synthesis of flutamide from trifluoromethylbenzene.

22.45 The compound 4-isobutylacetophenone is needed for the synthesis of ibuprofen. Propose a synthesis of 4-isobutylacetophenone from benzene and any other necessary reagents.

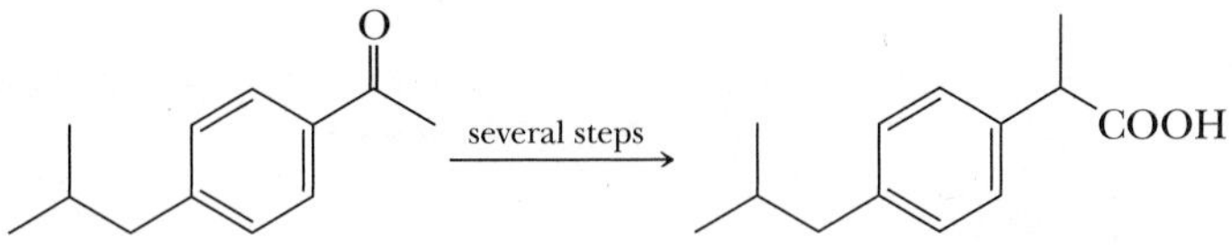

4-Isobutylacetophenone **Ibuprofen** (racemic)

22.46 Following is the structural formula of musk ambrette, a synthetic musk, essential in perfumes to enhance and retain odor. Propose a synthesis of this compound from *m*-cresol (3-methylphenol).

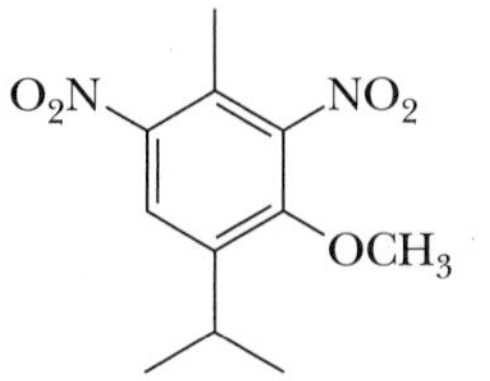

22.47 Propose a synthesis of this compound starting from toluene and phenol.

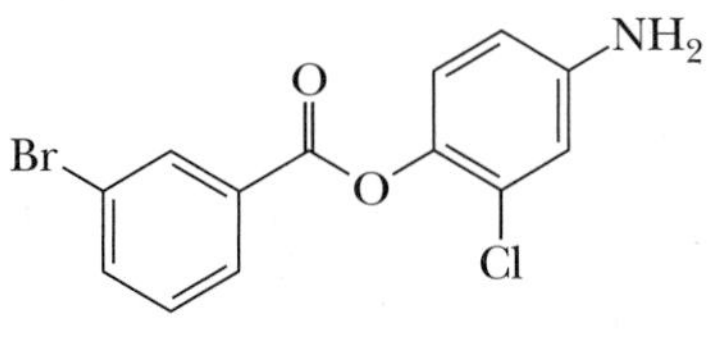

22.48 When certain aromatic compounds are treated with formaldehyde (CH_2O), and HCl, the CH_2Cl group is introduced onto the ring. This reaction is known as chloromethylation.

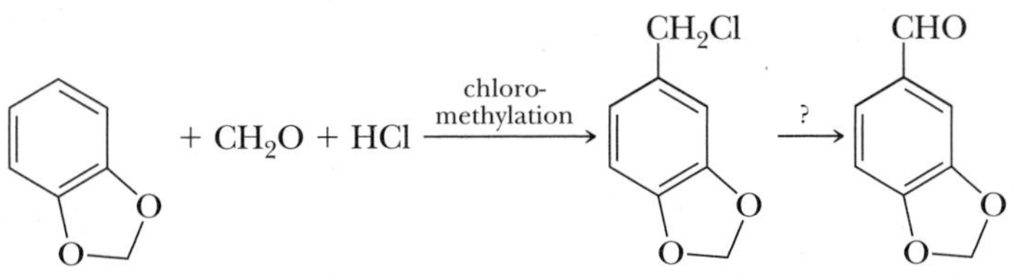

Piperonal

(a) Propose a mechanism for this example of chloromethylation.

(b) The product of this chloromethylation can be converted to piperonal, which is used in perfumery and in artificial cherry and vanilla flavors. How might the CH_2Cl group of the chloromethylation product be converted to a CHO group?

22.49 Following is a retrosynthetic analysis for the acaracide (killing mites and ticks) and fungicide dinocap.

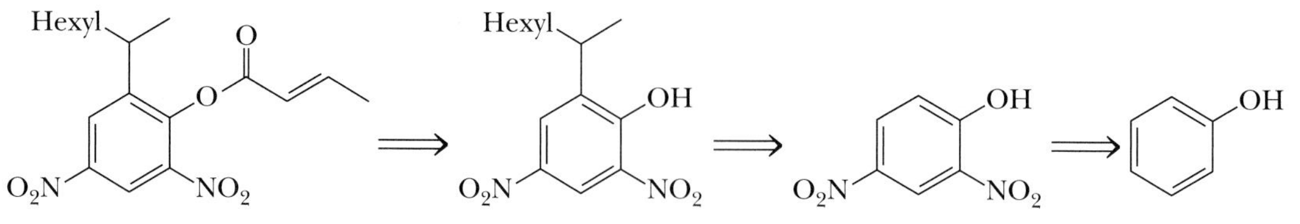

Dinocap

(a) Given this analysis, propose a synthesis for dinocap from phenol and 1-octene.
(b) Is dinocap chiral? If so, which of the possible stereoisomers are formed in this synthesis?

22.50 Following is the structure of miconazole, the active antifungal agent in a number of over-the-counter preparations, including Monistat, that are used to treat vaginal yeast infections. One of the compounds needed for the synthesis of miconazole is the trichloro derivative of toluene shown on its right.

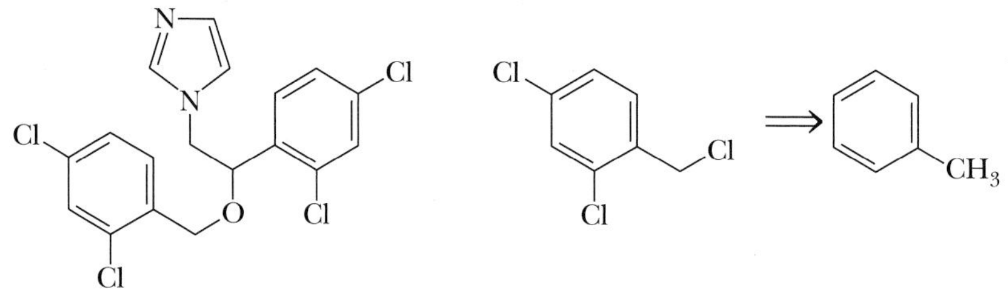

Miconazole **2,4-Dichloro-1-chloromethylbenzene** **Toluene**

(a) Show how this derivative can be synthesized from toluene.
(b) How many stereoisomers are possible for miconazole?

22.51 Bupropion, the hydrochloride of which was first marketed in 1985 by Burroughs Wellcome, now GlaxoSmithKline, is an antidepressant sold under the trade name Wellbutrin. During clinical trials, it was discovered that smokers, after one to two weeks on the drug, reported that their craving for tobacco lessened. Further clinical trials confirmed this finding, and the drug was marketed in 1997 under the trade name Zyban as an aid in smoking cessation.

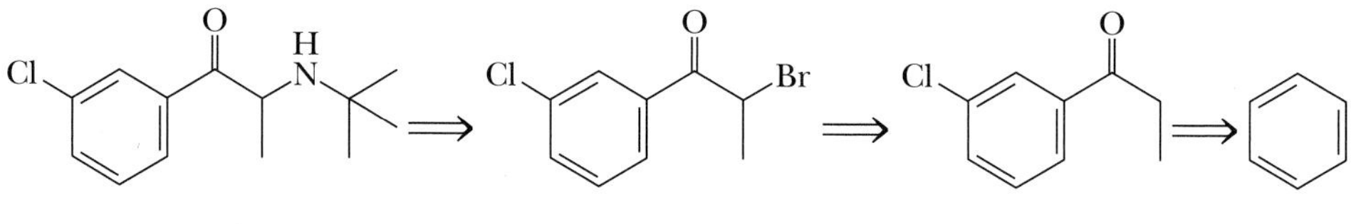

Bupropion

(a) Given this retrosynthetic analysis, propose a synthesis for bupropion.
(b) Is bupropion chiral? If so, how many of the possible stereoisomers are formed in this synthesis?

22.52 Diazepam, better known as Valium, is a central nervous system (CNS) sedative/hypnotic. As a sedative, it diminishes activity and excitement and thereby has a calming effect. Back in 1976, based on the number of new and refilled prescriptions processed, diazepam was the most prescribed drug in the United States.

Following is a retrosynthetic analysis for a synthesis of diazepam. Note that the formation of compound B involves a Friedel-Crafts acylation. In this reaction, it is necessary to protect the 2° amine by prior treatment with acetic anhydride. The acetyl-protecting group is then removed by treatment with aqueous NaOH followed by careful acidification with HCl.

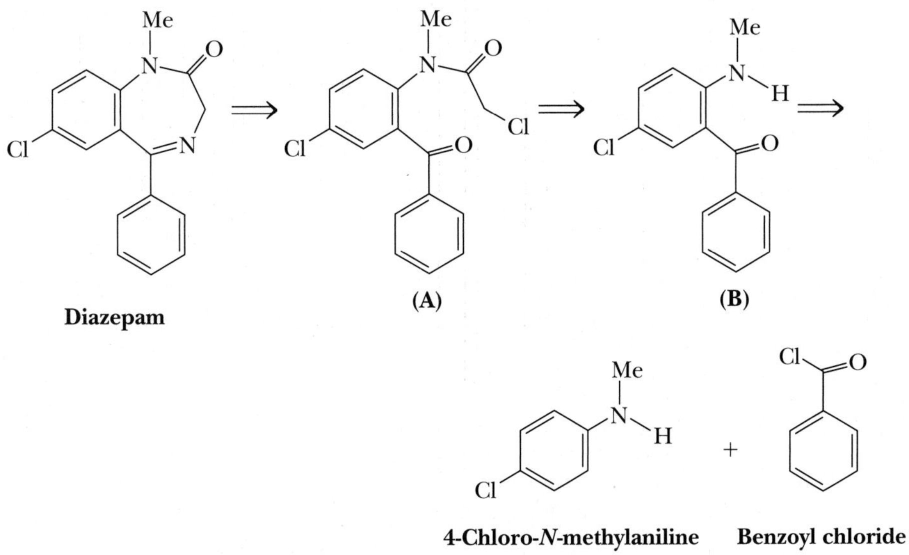

Diazepam (A) (B)

4-Chloro-*N*-methylaniline Benzoyl chloride

(a) Given this retrosynthetic analysis, propose a synthesis for diazepam.
(b) Is diazepam chiral? If so, how many of the possible stereoisomers are formed in this synthesis?

22.53 The antidepressant amitriptyline inhibits the reuptake of norepinephrine and serotonin from the synaptic cleft. Because the reuptake of these neurotransmitters is inhibited, their effects are potentiated; they remain available to interact with serotonin and norepinephrine receptor sites longer and continue to cause excitation of serotonin- and norepinephrine-mediated neural pathways. Following is a synthesis for amitriptyline.

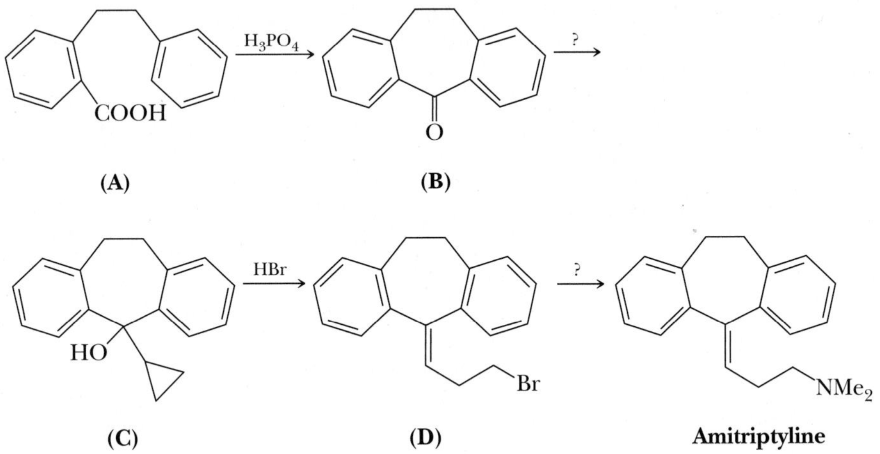

(A) (B) (C) (D) Amitriptyline

(a) Propose a mechanism for the conversion of **(A)** to **(B)**.
(b) Propose reagents for the conversion of **(B)** to **(C)**.
(c) Propose a mechanism for the conversion of **(C)** to **(D)**. (*Note:* It is not acceptable to propose a primary carbocation as an intermediate.)
(d) Propose a reagent for the conversion of **(D)** to amitriptyline.
(e) Is amitriptyline chiral? If so, how many of the possible stereoisomers are formed in this synthesis?

22.54 Show how the antidepressant venlafaxine (Effexor) can be synthesized from these readily available starting materials. Is venlafaxine chiral? If so, how many of the possible stereoisomers are formed in this synthesis?

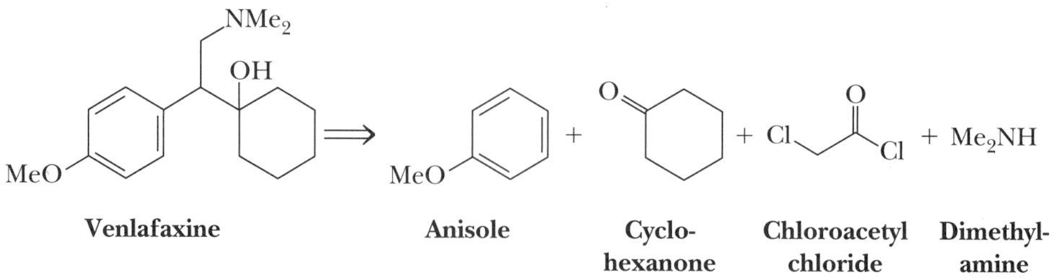

22.55 One potential synthesis of the anti-inflammatory and analgesic drug nabumetone is chloromethylation (Problem 22.48) of 2-methoxynaphthalene followed by an acetoacetic ester synthesis (Section 19.6).

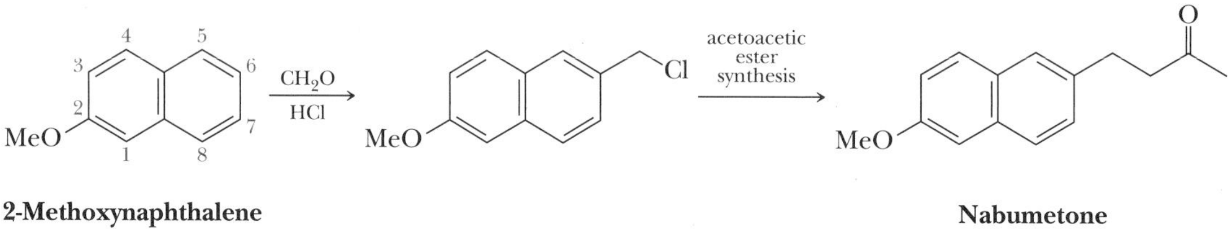

(a) Account for the regioselectivity of chloromethylation at carbon 6 rather than at carbon 5 or 7.

(b) Show steps in the acetoacetic ester synthesis by which the synthesis of nabumetone is completed.

22.56 The analgesic, soporific, and euphoriant properties of the dried juice obtained from unripe seed pods of the opium poppy *Papaver somniferum* have been known for centuries. By the beginning of the nineteenth century, the active principle, morphine, had been isolated and its structure determined. Even though morphine is one of modern medicine's most effective painkillers, it has two serious disadvantages. First, it is addictive. Second, it depresses the respiratory control center of the central nervous system. Large doses of morphine (or heroin, which is the 3,6-diacetyl ester of morphine) can lead to death by respiratory failure.

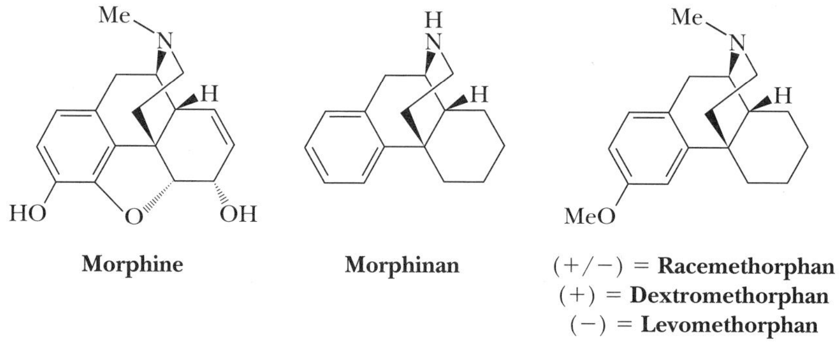

For these reasons, chemists have sought to produce painkillers related in structure to morphine, but without the serious disadvantages. One strategy has been to modify the carbon-nitrogen skeleton of morphine in the hope of producing medications equally effective but with reduced side effects. One target of this synthetic effort was morphinan, the bare morphine skeleton. Among the compounds thus synthesized, racemethorphan (the racemic mixture) and levomethorphan (the levorotatory enantiomer) proved to be very potent analgesics. Interestingly, the dextrorotatory enantiomer, dextromethorphan, has no analgesic activity. It does, however, show approximately the same antitussive (cough-suppressing) activity as morphine and is therefore used extensively in cough remedies.

Following is a synthesis of racemethorphan.

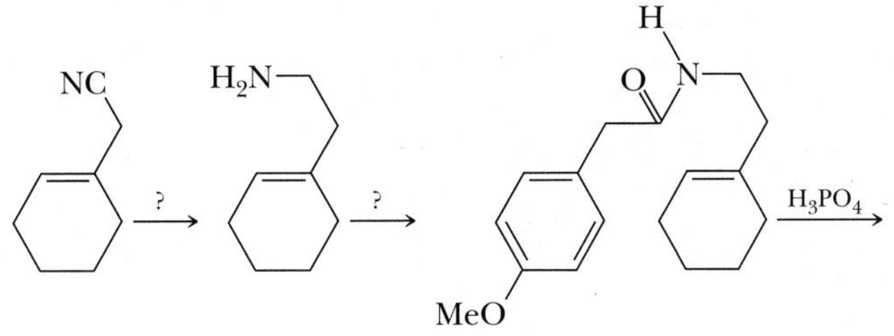

(A) **(B)** **(C)**

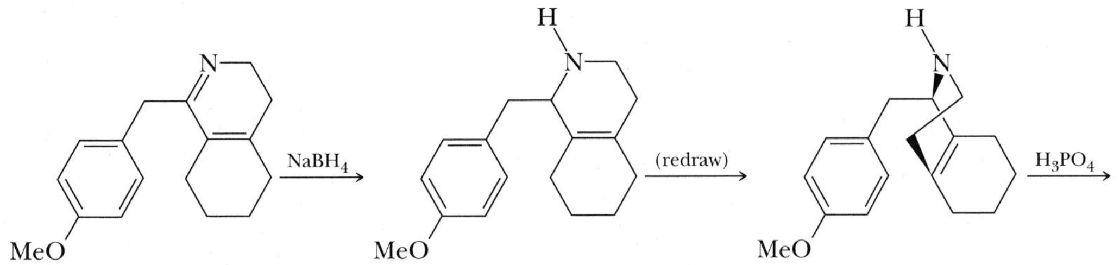

(D) **(E)**

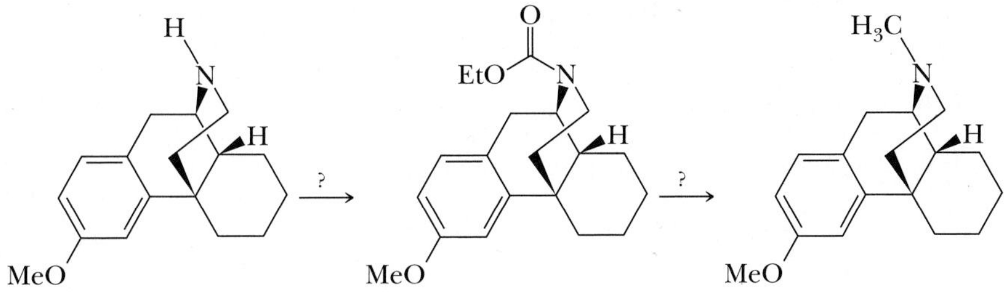

(F) **(G)** **Racemethorphan**

(a) Propose a reagent for the conversion of **(A)** to **(B)**.
(b) Propose a reagent for the conversion of **(B)** to **(C)**.
(c) Propose a mechanism for the conversion of **(C)** to **(D)**.
(d) Propose a mechanism for the conversion of **(E)** to **(F)**.
(e) Propose a reagent for the conversion of **(F)** to **(G)**.
(f) Propose a reagent for the conversion of **(G)** to racemethorphan.

22.57 Following is the structural formula of the antihypertensive drug labetalol, a nonspecific β-adrenergic blocker with vasodilating activity. Members of this class have received enormous clinical attention because of their effectiveness in treating hypertension (high blood pressure), migraine headaches, glaucoma, ischemic heart disease, and certain cardiac arrhythmias. This retrosynthetic analysis involves disconnects to the α-haloketone **(B)** and the amine **(C)**. Each is in turn derived from a simpler, readily available precursor.

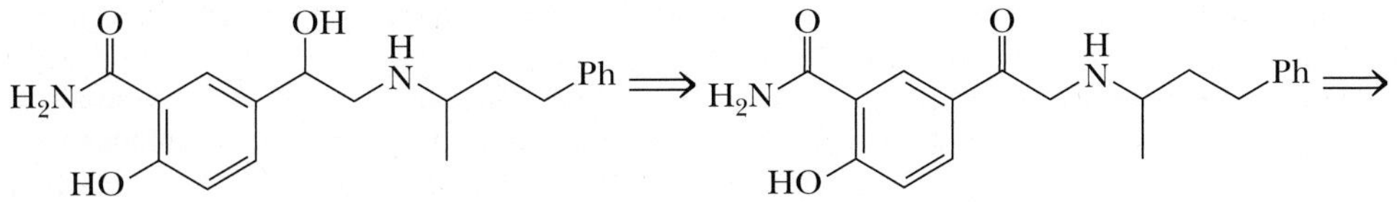

Labetalol **(A)**

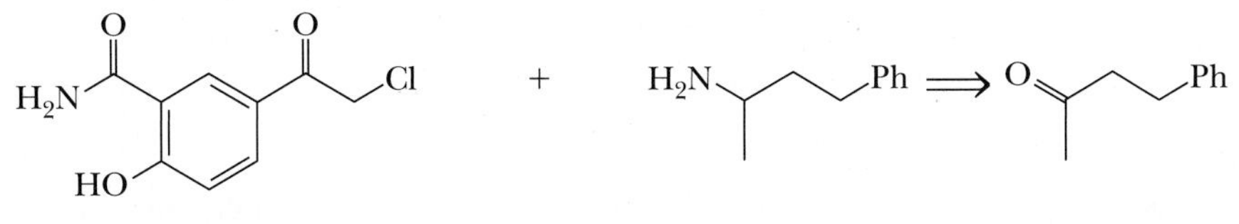

(B) **(C)** **(D)**

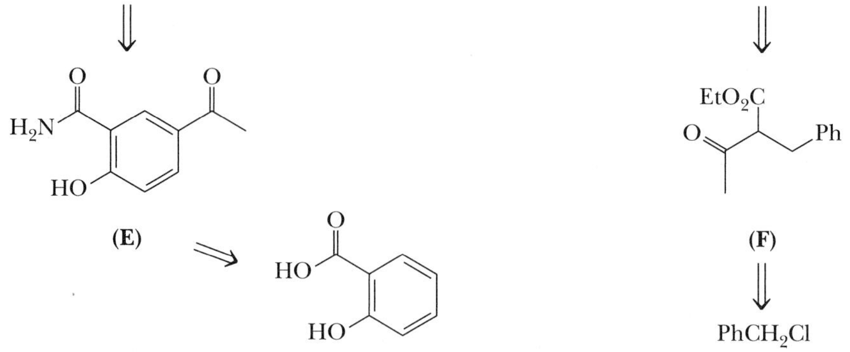

(E) **(F)**

Salicylic acid **Benzyl chloride**

(a) Given this retrosynthetic analysis, propose a synthesis for labetalol from salicylic acid and benzyl chloride. [*Note:* The conversion of salicylic acid to **(E)** involves a Friedel-Crafts acylation in which the phenolic —OH must be protected by treatment with acetic anhydride to prevent the acylation of the —OH group. The protecting group is later removed by treatment with KOH followed by acidification.]

(b) Labetalol has two chiral centers and, as produced in this synthesis, is a racemic mixture of the four possible stereoisomers. The active stereoisomer is dilevalol, which has the *R,R* configuration at its chiral centers. Draw a structural formula of dilevalol showing the configuration of each chiral center.

22.58 Given this retrosynthetic analysis, propose a synthesis for the antihistamine *p*-methyldiphenhydramine. Is *p*-methyldiphenhydramine chiral? If so, how many of the possible stereoisomers are formed in this synthesis?

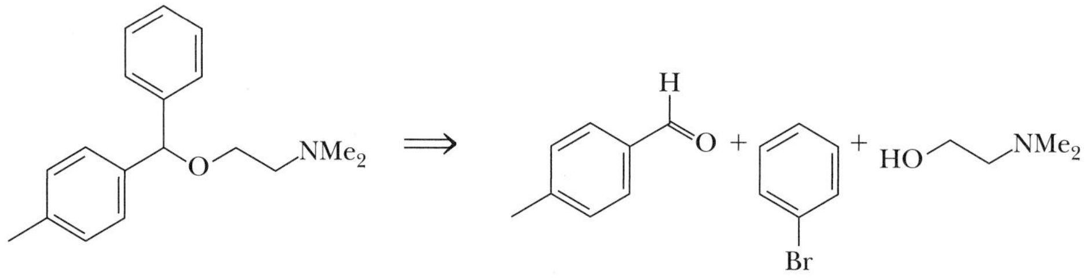

***p*-Methyldiphenhydramine**

22.59 Meclizine is an antiemetic (it helps prevent or at least lessen the throwing up associated with motion sickness, including seasickness). Among the names of its over-the-counter preparations are Bonine, Sea-Legs, Antivert, and Navicalm.

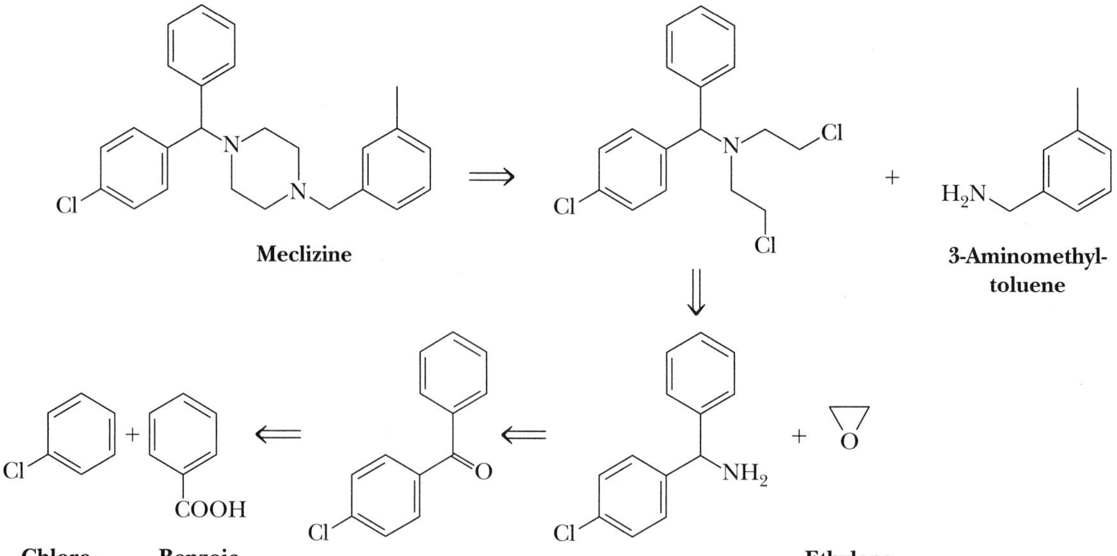

Meclizine **3-Aminomethyl-toluene**

Chloro-benzene **Benzoic acid** **Ethylene oxide**

(a) Given this retrosynthetic analysis, show how meclizine can be synthesized from the four named organic starting materials.

(b) Is meclizine chiral? If so, how many of the possible stereoisomers are formed in this synthesis?

22.60 Spasmolytol, as its name suggests, is an antispasmodic. Given this retrosynthetic analysis, propose a synthesis for spasmolytol from salicylic acid, ethylene oxide, and diethylamine.

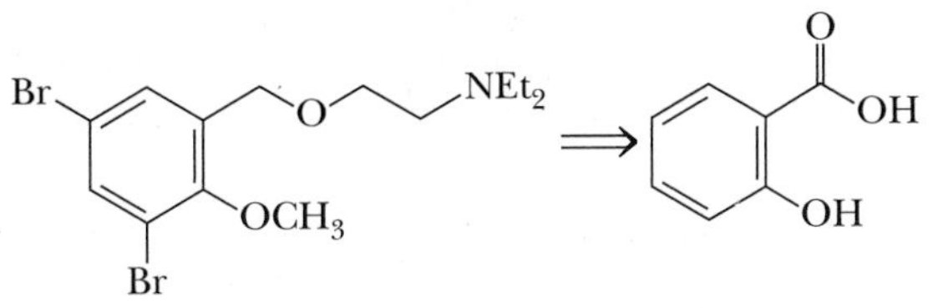

Br
O
NEt2
OCH3
Br
O
OH
OH

Spasmolytol **Salicylic acid**

22.61 Among the first antipsychotic drugs for the treatment of schizophrenia was haloperidol (Haldol), a competitive inhibitor of dopamine receptor sites in the central nervous system.

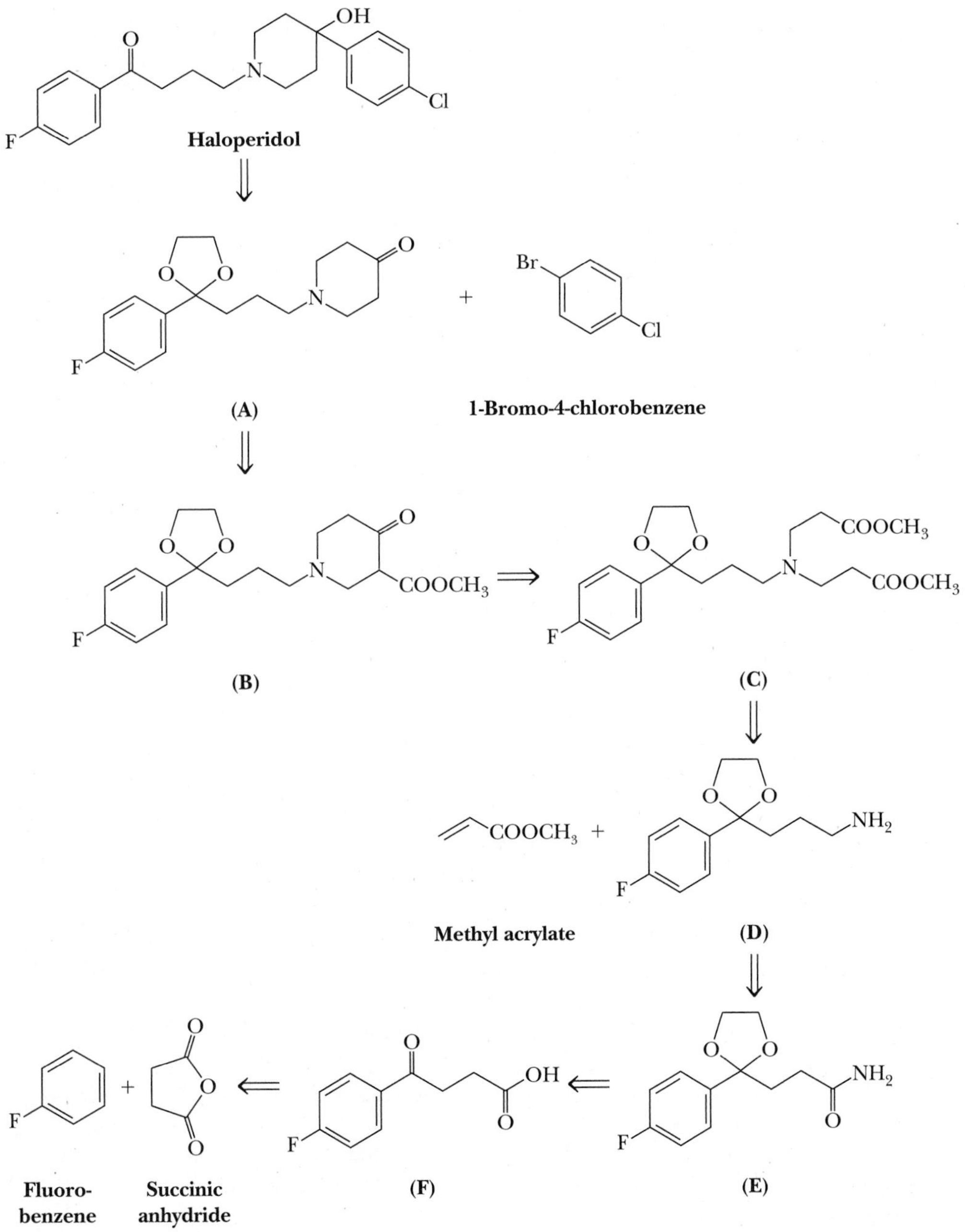

Haloperidol
(A)
1-Bromo-4-chlorobenzene
(B)
(C)
Methyl acrylate
(D)
Fluoro-
benzene
Succinic
anhydride
(F)
(E)
OH
Cl
F
Br
COOCH3
NH2

(a) Given this retrosynthetic analysis, propose a synthesis for haloperidol.
(b) Is haloperidol chiral? If so, how many of the possible stereoisomers are formed in this synthesis?

22.62 A newer generation of antipsychotics, among them clozapine, are now used to treat the symptoms of schizophrenia. These drugs are more effective than earlier drugs in improving patient response in the areas of social withdrawal, apathy, memory, comprehension, and judgment. They also produce fewer side effects such as seizures and tardive dyskinesia (involuntary body movements). In the following synthesis of clozapine, Step 1 is an Ullmann coupling, a type of nucleophilic aromatic substitution that uses a copper catalyst.

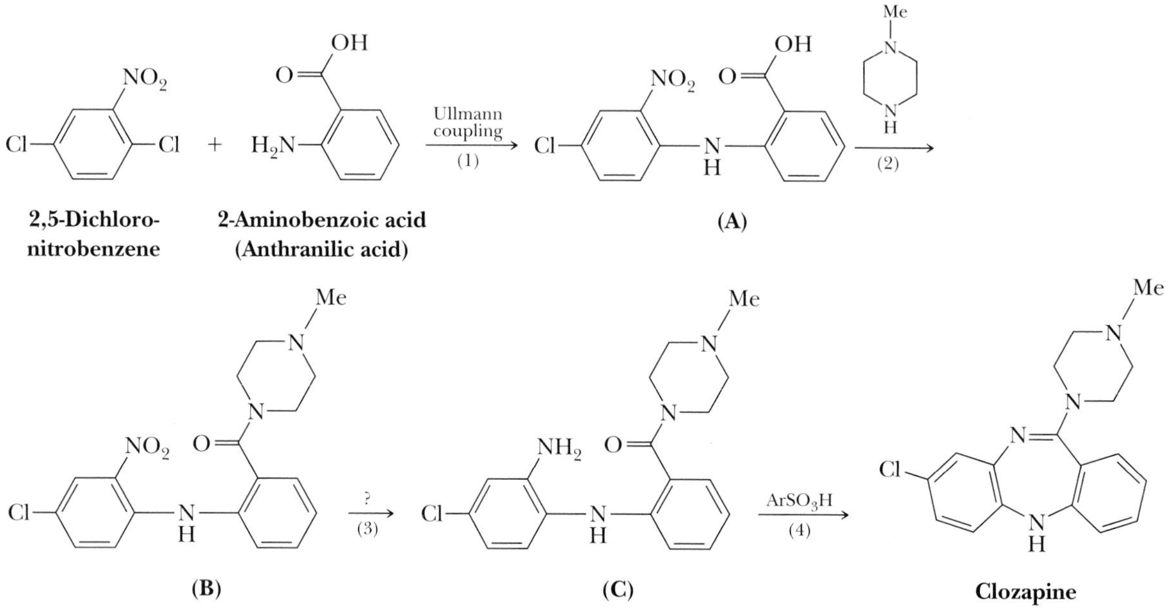

(a) Show how you might bring about formation of the amide in Step 2.
(b) Propose a reagent for Step 3.
(c) Propose a mechanism for Step 4.
(d) Is clozapine chiral? If so, how many of the possible stereoisomers are formed in this synthesis?

22.63 Proparacaine is one of a class of -caine local anesthetics.

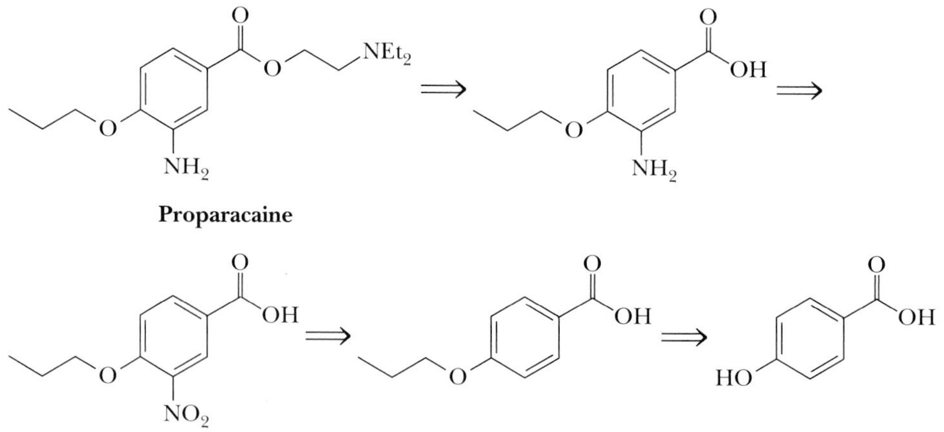

(a) Given this retrosynthetic analysis, propose a synthesis of proparacaine from 4-hydroxybenzoic acid.
(b) Is proparacaine chiral? If so, how many of the possible stereoisomers are formed in this synthesis?

23

Amines

Morphine, a potent painkiller isolated from the ripe seed heads of the opium poppy, has been a lead drug for chemists in search of potent but less addicting synthetic painkillers. See Problem 23.21. **Inset:** a model of morphine.

Outline

23.1 Structure and Classification
23.2 Nomenclature
23.3 Chirality of Amines and Quaternary Ammonium Ions
23.4 Physical Properties
23.5 Basicity
23.6 Reactions with Acids
23.7 Preparation
23.8 Reaction with Nitrous Acid
23.9 Hofmann Elimination
23.10 Cope Elimination

Carbon, hydrogen, and oxygen are the three most common elements in organic compounds. Because of the wide distribution of amines in the biological world, nitrogen is the fourth most common element in organic compounds. The lone pair of electrons on the nitrogen of amines is a powerful electron source, so the most important chemical properties of amines are their basicity and nucleophilicity.

23.1 Structure and Classification

Amines are derivatives of ammonia in which one or more hydrogens are replaced with alkyl or aryl groups. Amines are classified as primary, secondary, or tertiary, depending on the number of carbon atoms bonded directly to nitrogen (Section 1.3B).

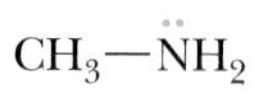

Methylamine
(a 1° amine)

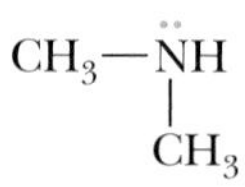

Dimethylamine
(a 2° amine)

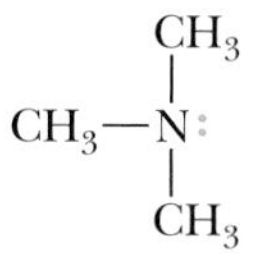

Trimethylamine
(a 3° amine)

Aliphatic amine
An amine in which nitrogen is bonded only to alkyl groups.

Aromatic amine
An amine in which nitrogen is bonded to one or more aryl groups.

Amines are further divided into aliphatic and aromatic amines. In an **aliphatic amine**, all carbons bonded to nitrogen are derived from alkyl groups; in an **aromatic amine**, one or more of the groups bonded to nitrogen are aryl groups.

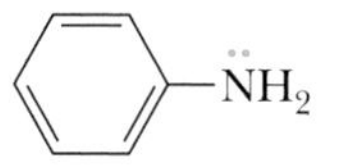

Aniline
(a 1° aromatic amine)

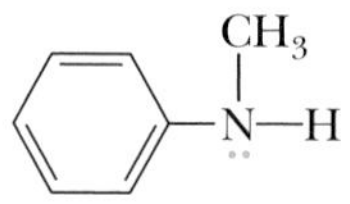

***N*-Methylaniline**
(a 2° aromatic amine)

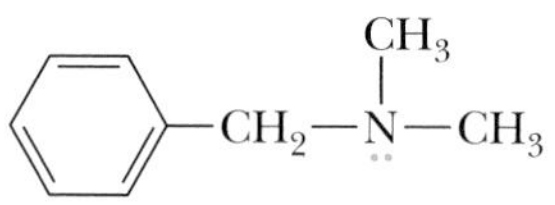

Benzyldimethylamine
(a 3° aliphatic amine)

Heterocyclic amine
An amine in which nitrogen is one of the atoms of a ring.

Heterocyclic aromatic amine
An amine in which nitrogen is one of the atoms of an aromatic ring.

An amine in which the nitrogen atom is part of a ring is classified as a **heterocyclic amine**. When the nitrogen is part of an aromatic ring (Section 21.2D), the amine is classified as a **heterocyclic aromatic amine**. Following are structural formulas for two heterocyclic aliphatic amines and two heterocyclic aromatic amines.

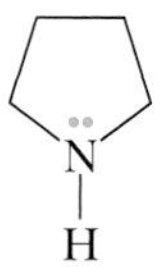

Pyrrolidine

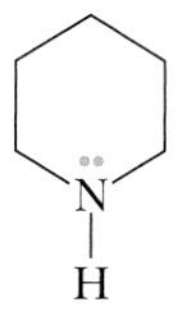

Piperidine

(heterocyclic aliphatic amines)

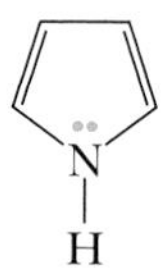

Pyrrole

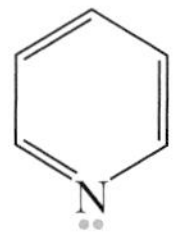

Pyridine

(heterocyclic aromatic amines)

Example 23.1 Types of Amines

Alkaloid
A basic nitrogen-containing compound of plant origin, many of which are physiologically active when administered to humans.

Alkaloids are basic nitrogen-containing compounds of plant origin, many of which are physiologically active when administered to humans. Ingestion of coniine, isolated from water hemlock, can cause weakness, labored respiration, paralysis, and eventually death. Coniine is the toxic substance in the "poison hemlock" used in the death of Socrates. In small doses, nicotine is an addictive stimulant. In larger doses, it causes depression, nausea, and vomiting. In still larger doses, it is a deadly poison. Solutions of nicotine in water are used as insecticides. Cocaine is a central nervous system stimulant obtained from the leaves of the coca plant.

(a)

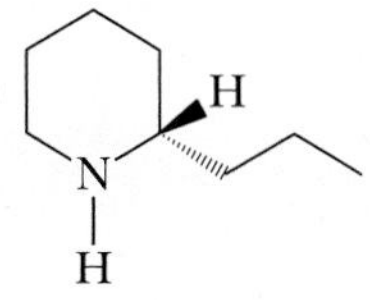

(*S*)-Coniine

(b)

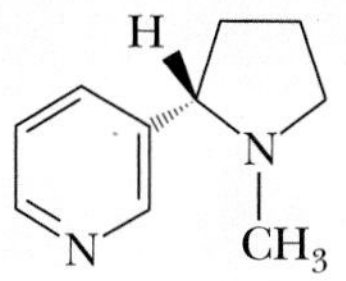

(*S*)-Nicotine

(c)

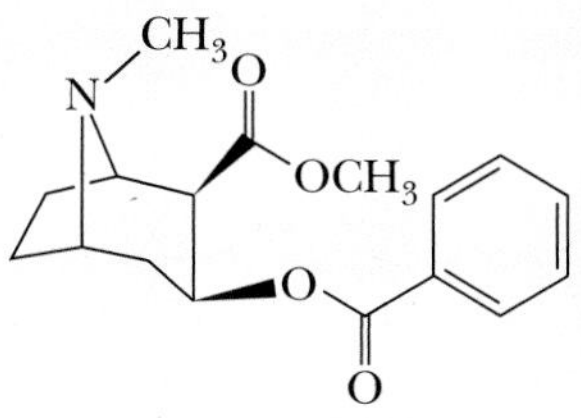

Cocaine

Classify each amino group in these alkaloids according to type (primary, secondary, tertiary, aliphatic, aromatic, heterocyclic).

Solution

(a) A secondary aliphatic heterocyclic amine
(b) A tertiary aliphatic heterocyclic amine and a heterocyclic aromatic amine
(c) A tertiary aliphatic heterocyclic amine

Problem 23.1

Identify all carbon chiral centers in coniine, nicotine, and cocaine.

23.2 Nomenclature

A. Systematic Names

Systematic names for aliphatic amines are derived just as they are for alcohols. The suffix *-e* of the parent alkane is dropped and is replaced with *-amine*.

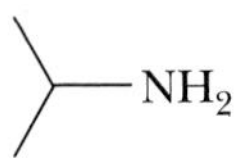
NH2

2-Propanamine

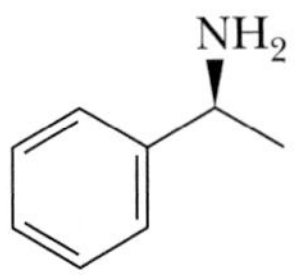
NH2

(*S*)-1-Phenylethanamine

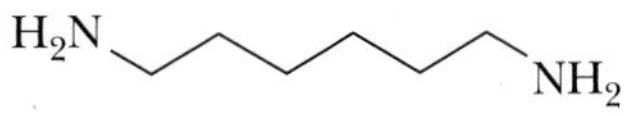
H2N
NH2

1,6-Hexanediamine

Example 23.2 Amine Nomenclature I

Write systematic names for these amines.

(a) $CH_3(CH_2)_5NH_2$ **(b)** 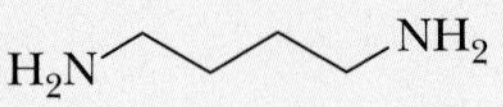
H2N
NH2
(c) 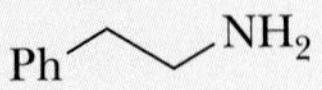
Ph
NH2

Solution

(a) 1-Hexanamine **(b)** 1,4-Butanediamine **(c)** 2-Phenylethanamine

Problem 23.2

Write structural formulas for these amines.

(a) 2-Methyl-1-propanamine **(b)** Cyclohexanamine **(c)** (*R*)-2-Butanamine

IUPAC nomenclature retains the common name aniline for $C_6H_5NH_2$, the simplest aromatic amine. Its simple derivatives are named using the prefixes *o*-, *m*-, and *p*- or numbers to locate substituents. Several derivatives of aniline have common names that are still widely used. Among these are toluidine for a methyl-substituted aniline and anisidine for a methoxyl-substituted aniline.

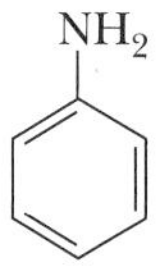

Aniline

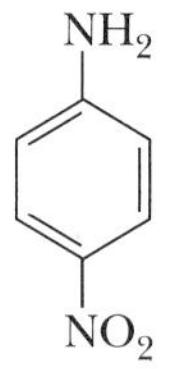

4-Nitroaniline (*p*-Nitroaniline)

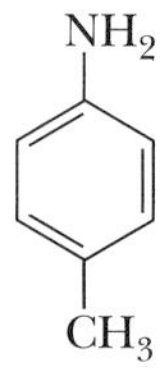

4-Methylaniline (*p*-Toluidine)

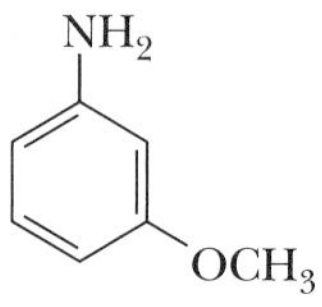

3-Methoxyaniline (*m*-Anisidine)

Secondary and tertiary amines are commonly named as *N*-substituted primary amines. For unsymmetrical amines, the largest group is taken as the parent amine; then the smaller group(s) bonded to nitrogen are named, and their location is indicated by the prefix *N* (indicating that they are bonded to nitrogen).

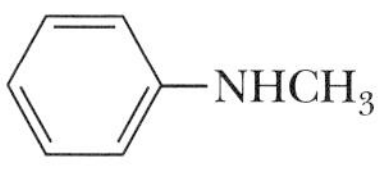

***N*-Methylaniline**

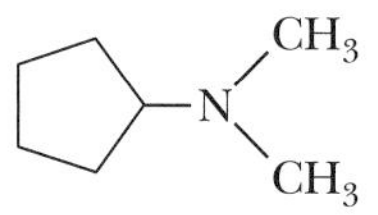

***N*,*N*-Dimethyl-cyclopentanamine**

Following are names and structural formulas for four heterocyclic aromatic amines, the common names of which have been retained in the IUPAC system.

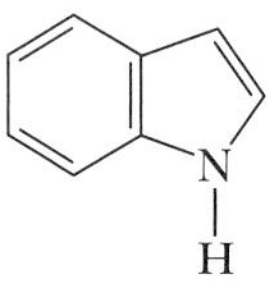

Indole

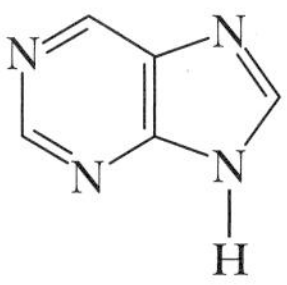

Purine

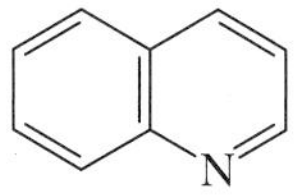

Quinoline

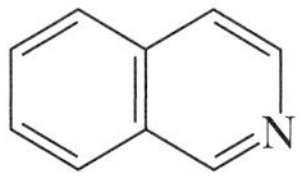

Isoquinoline

Among the various functional groups discussed in this text, the $-NH_2$ group is one of the lowest in precedence (Table 16.1). The following compounds each contain a functional group of higher precedence than the amino group, and accordingly, the amino group is indicated by the prefix *amino*.

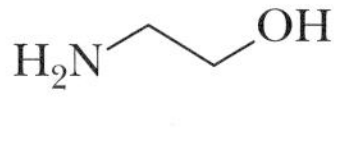

2-Aminoethanol

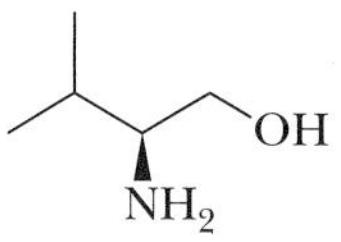

(*S*)-2-Amino-3-methyl-1-butanol

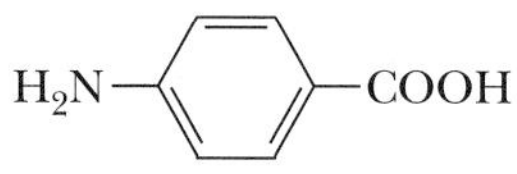

4-Aminobenzoic acid

B. Common Names

Common names for most aliphatic amines are derived by listing the alkyl groups bonded to nitrogen in alphabetical order in one word ending in the suffix *-amine*; that is, they are named as alkylamines.

CH_3NH_2

Methylamine

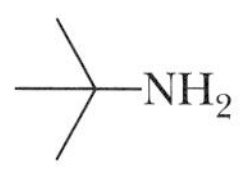

***tert*-Butylamine**

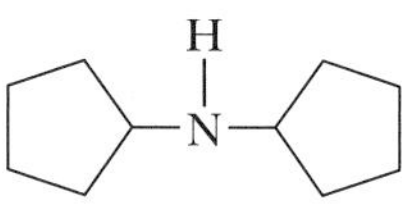

Dicyclopentylamine

Et_3N

Triethylamine

Example 23.3 Amine Nomenclature II

Write structural formulas for these amines.

(a) Isopropylamine **(b)** Cyclohexylmethylamine **(c)** Benzylamine

Solution

(a) 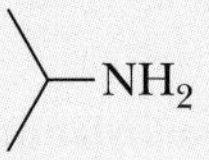

(b) 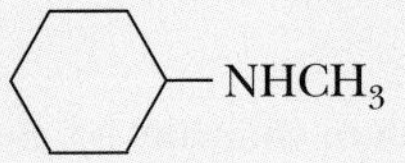

(c) 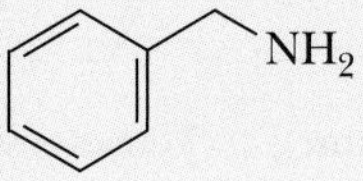

Problem 23.3

Write structural formulas for these amines.

(a) Isobutylamine **(b)** Triphenylamine **(c)** Diisopropylamine

When four atoms or groups of atoms are bonded to a nitrogen atom, the compound is named as a salt of the corresponding amine. The ending *-amine* (or *-aniline, pyridine,* and so on) is replaced with *-ammonium* (or *anilinium, pyridinium,* and so on), and the name of the anion is added.

$Et_3NH^+\ Cl^-$

Triethylammonium chloride

CH_3COO^-

Pyridinium acetate

Example 23.4 Amine Nomenclature III

Write the IUPAC name and, where possible, a common name for each compound.

(a) $(C_6H_5)_2NH$

(b) 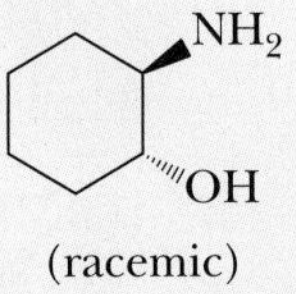

(racemic)

(c) 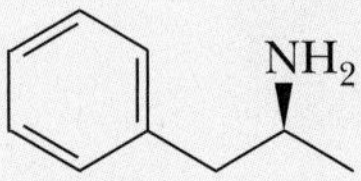

Solution

(a) Diphenylamine
(b) *trans*-2-Aminocyclohexanol
(c) Its systematic name is (*S*)-1-phenyl-2-propanamine. Its common name is amphetamine. The dextrorotatory isomer of amphetamine (shown here) is a central nervous system stimulant and is manufactured and sold under several trade names. The salt with sulfuric acid is marketed as Dexedrine sulfate.

Problem 23.4

Write the IUPAC and, where possible, a common name for each compound.

(a) 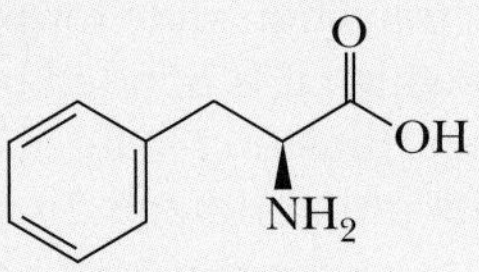

(b) 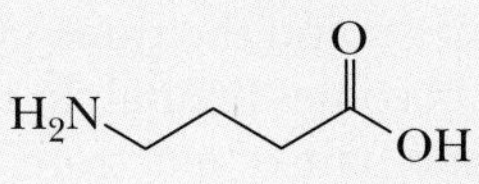

(c) 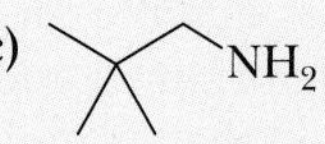

Quaternary (4°) ammonium ion

An ion in which nitrogen is bonded to four carbons and bears a positive charge.

An ion containing a nitrogen atom bonded to any combination of four alkyl or aryl groups is classified as a **quaternary (4°) ammonium ion**. Compounds containing such ions have properties characteristic of salts. Cetylpyridinium chloride is used as a topical antiseptic and disinfectant.

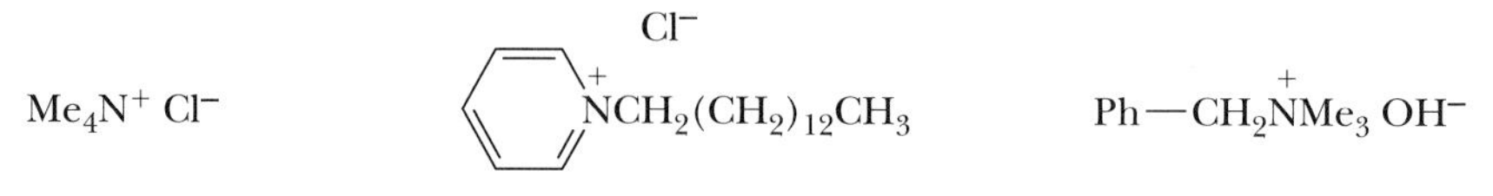

Tetramethylammonium chloride | **Tetradecylpyridinium chloride (Cetylpyridinium chloride)** | **Benzyltrimethylammonium hydroxide**

23.3 Chirality of Amines and Quaternary Ammonium Ions

The geometry of a nitrogen atom bonded to three other atoms or groups of atoms is trigonal pyramidal (Section 1.4). The sp^3 hybridized nitrogen atom is at the apex of the pyramid, and the three groups bonded to it extend downward to form the triangular base of the pyramid. If we consider the unshared pair of electrons on nitrogen as a fourth group, then the arrangement of "groups" around nitrogen is approximately tetrahedral. Because of this geometry, an amine with three different groups bonded to nitrogen is chiral and can exist as a pair of enantiomers, as illustrated by the nonsuperposable mirror images of ethylmethylamine. In assigning configuration to these enantiomers, the group of lowest priority on nitrogen is the unshared pair of electrons.

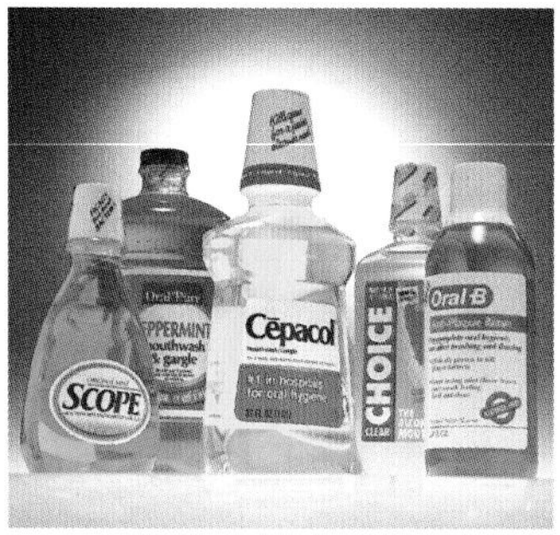

Several over-the-counter mouthwashes contain an *N*-alkylpyridinium chloride as an antibacterial agent.

▶ There are two possible arrangements of four different groups around a tetrahedral atom.

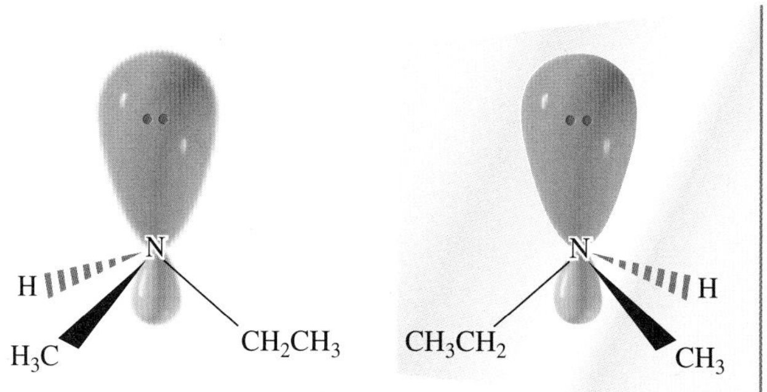

(*S*)-Ethylmethylamine **(*R*)-Ethylmethylamine**

In principle, a chiral amine should be resolvable; that is, it should be separable into a pair of enantiomers. Except for special cases, however, the enantiomers cannot be resolved because they undergo rapid interconversion by a process known as pyramidal inversion. **Pyramidal inversion** is the rapid oscillation of a nitrogen atom from one side of the plane of the three atoms bonded to it to the other side of that plane.

To visualize this process, imagine the sp^3 hybridized nitrogen atom lying above the plane of the three atoms to which it is bonded. In the transition state for pyramidal inversion, the nitrogen atom and the three groups to which it is bonded become coplanar and the molecule becomes achiral. In this planar transition state, nitrogen is sp^2 hybridized and its lone pair of electrons lies in its unhybridized $2p$ orbital. Nitrogen then completes the inversion, becomes sp^3 hybridized again, and now lies below the plane of the three atoms to which it is bonded.

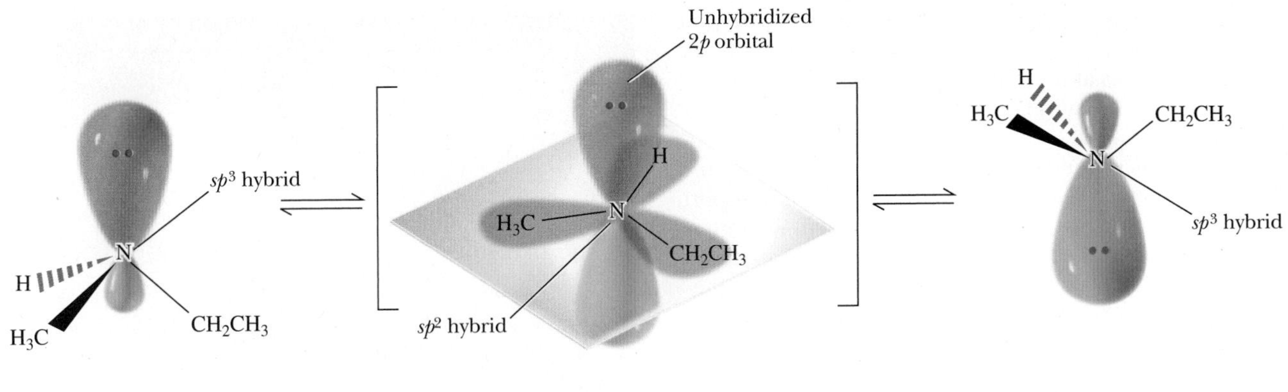

***S* Enantiomer** **Planar transition state** ***R* Enantiomer**

As a result of pyramidal inversion, a chiral amine quite literally turns itself inside out, like an umbrella in a strong wind, and in the process becomes a racemic mixture. The activation energy for pyramidal inversion of simple amines is about 25 kJ (6 kcal)/mol. For ammonia at room temperature, the rate of nitrogen inversion is approximately $2 \times 10^{11}\ s^{-1}$. For simple amines, the rate is less rapid but nonetheless sufficient to make resolution impossible.

Pyramidal inversion is not possible for quaternary ammonium ions, and their salts can be resolved.

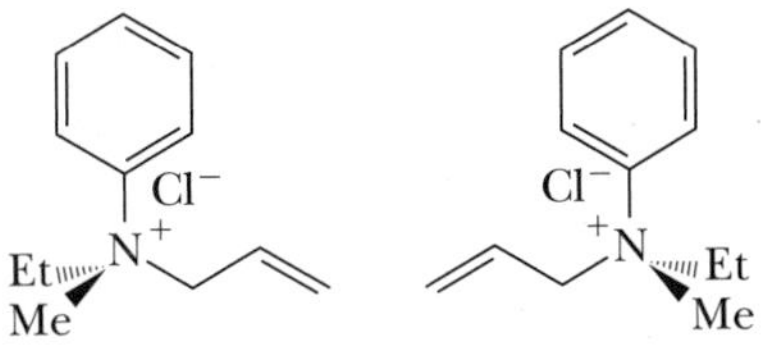

***S* Enantiomer** ***R* Enantiomer**

Phosphorus, in the same family as nitrogen, forms trivalent compounds called phosphines, which also have trigonal pyramidal geometry. The activation energy for pyramidal inversion of trivalent phosphorus compounds is considerably greater than it is for trivalent compounds of nitrogen, with the result that a number of chiral phosphines have been resolved.

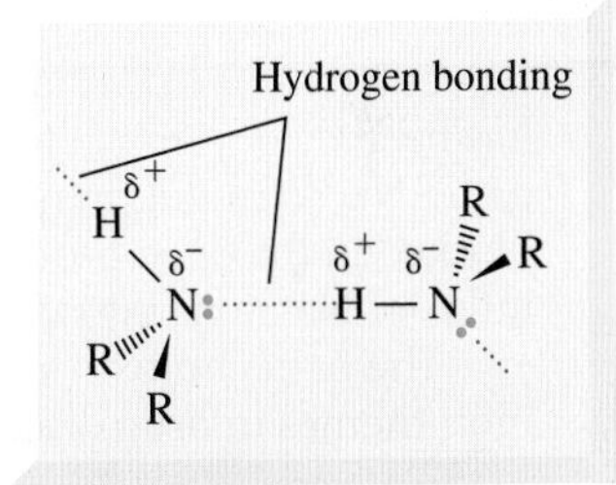

FIGURE 23.1 Intermolecular association by hydrogen bonding in primary and secondary amines. Nitrogen is approximately tetrahedral in shape with the axis of the hydrogen bond along the fourth position of the tetrahedron.

23.4 Physical Properties

Amines are polar compounds, and both primary and secondary amines form intermolecular hydrogen bonds (Figure 23.1).

Chemical Connections

The Poison Dart Frogs of South America

The Noanamá and Embrá peoples of the jungles of western Colombia have used poison blow darts for centuries, perhaps millennia. The poisons are obtained from the skin secretions of several brightly colored frogs of the genus *Phyllobates* (*neará* and *kokoi* in the language of the native peoples). A single frog contains enough poison for up to 20 darts. For the most poisonous species (*Phyllobates terribilis*), just rubbing

(*Continued*)

a dart over the frog's back suffices to charge the dart with poison.

Scientists at the National Institutes of Health in the United States became interested in studying these poisons when it was discovered that they act on cellular ion channels, which would make them useful tools in basic research on mechanisms of ion transport. A field station was, therefore, established in western Colombia to collect the relatively common poison dart frogs. From 5000 frogs, 11 mg of two toxins, given the names batrachotoxin and batrachotoxinin A, were isolated. These names are derived from *batrachos*, the Greek word for frog. A combination of NMR spectroscopy, mass spectrometry, and single-crystal X-ray diffraction was used to determine the structures of these compounds.

Batrachotoxin and batrachotoxinin A are among the most lethal poisons ever discovered. It is estimated that as little as 200 μg of batrachotoxin is sufficient to induce irreversible cardiac arrest in a human being. It has been determined that they act by causing voltage-gated Na^+ channels in nerve and muscle cells to be blocked in the open position, which leads to a huge influx of Na^+ ions into the affected cell.

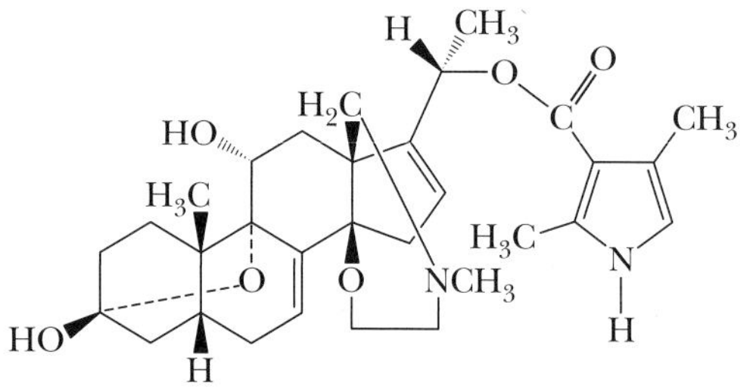

Batrachotoxin

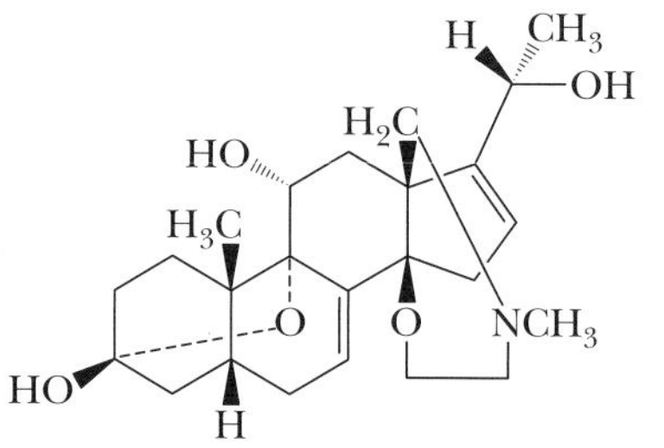

Batrachotoxinin A

The batrachotoxin story illustrates several common themes in drug discovery. First, information about the kinds of biologically active compounds and their sources is often obtained from the native peoples of a region. Second, tropical rain forests are a rich source of structurally complex, biologically active substances. Third, the entire ecosystem, not just the plants, is a potential source of fascinating organic molecules.

An N—H···N hydrogen bond is weaker than an O—H···O hydrogen bond because the difference in electronegativity between nitrogen and hydrogen ($3.0 - 2.1 = 0.9$) is less than that between oxygen and hydrogen ($3.5-2.1=1.4$). The effect of intermolecular hydrogen bonding can be illustrated by comparing the boiling points of methylamine and methanol. Both are polar molecules that interact in the pure liquid by hydrogen bonding. Because hydrogen bonding is stronger in methanol than in methylamine, methanol has the higher boiling point.

	CH_3CH_3	CH_3NH_2	CH_3OH
MW (g/mol)	30.1	31.1	32.0
bp (°C)	−88.6	−6.3	65.0

All classes of amines form hydrogen bonds with water and are more soluble in water than hydrocarbons of comparable molecular weight. Most low-molecular-weight amines are completely soluble in water (Table 23.1). Amines of higher molecular weight are only moderately soluble or are insoluble.

Table 23.1 Physical Properties of Selected Amines

Name	Structural Formula	mp (°C)	bp (°C)	Solubility in Water
Ammonia	NH_3	−78	−33	Very soluble
Primary Amines				
Methylamine	CH_3NH_2	−95	−6	Very soluble
Ethylamine	$CH_3CH_2NH_2$	−81	17	Very soluble
Propylamine	$CH_3CH_2CH_2NH_2$	−83	48	Very soluble
Isopropylamine	$(CH_3)_2CHNH_2$	−95	32	Very soluble
Butylamine	$CH_3(CH_2)_3NH_2$	−49	78	Very soluble
Benzylamine	$C_6H_5CH_2NH_2$	10	185	Very soluble
Cyclohexylamine	$C_6H_{11}NH_2$	−17	135	Slightly soluble
Secondary Amines				
Dimethylamine	$(CH_3)_2NH$	−93	7	Very soluble
Diethylamine	$(CH_3CH_2)_2NH$	−114	89	Slightly soluble
Tertiary Amines				
Trimethylamine	$(CH_3)_3N$	−117	3	Very soluble
Triethylamine	$(CH_3CH_2)_3N$	−114	89	Slightly soluble
Aromatic Amine				
Aniline	$C_6H_5NH_2$	−6	184	Slightly soluble
Aromatic Heterocyclic Amine				
Pyridine	C_5H_5N	−42	116	Very soluble

23.5 Basicity

Like ammonia, all amines are weak bases, and aqueous solutions of amines are basic. The following acid-base reaction between an amine and water is written using curved arrows to emphasize that, in these proton-transfer reactions, the unshared pair of electrons on nitrogen forms a new covalent bond with hydrogen and displaces hydroxide ion.

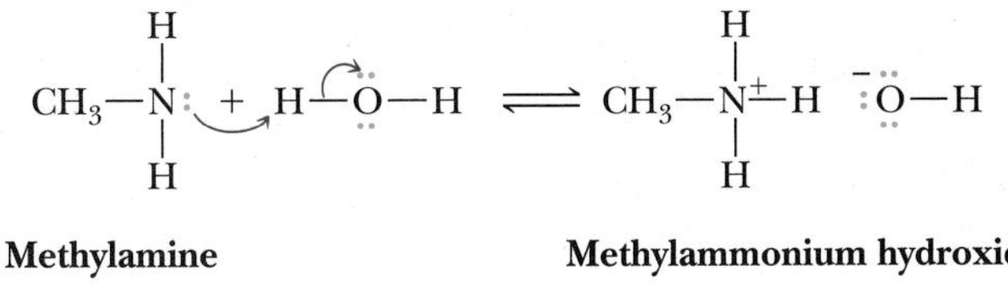

It is common to discuss the basicity of amines by referencing the acid ionization constant of the corresponding conjugate acid, as illustrated for the ionization of the methylammonium ion.

$$CH_3NH_3^+ + H_2O \rightleftharpoons CH_3NH_2 + H_3O^+$$

$$K_a = \frac{[CH_3NH_2][H_3O^+]}{[CH_3NH_3^+]} = 2.29 \times 10^{-11} \qquad pK_a = 10.64$$

Table 23.2 gives values of pK_a for the conjugate acids of selected amines. Keep in mind that the weaker the conjugate acid (the larger its pK_a), the greater the basicity of the amine.

Table 23.2 Acid Strengths, pK_a, of the Conjugate Acids of Selected Amines

Amine	Structure	pK_a of Conjugate Acid
Ammonia	NH_3	9.26
Primary Amines		
Methylamine	CH_3NH_2	10.64
Ethylamine	$CH_3CH_2NH_2$	10.81
Cyclohexylamine	$C_6H_{11}NH_2$	10.66
Secondary Amines		
Dimethylamine	$(CH_3)_2NH$	10.73
Diethylamine	$(CH_3CH_2)_2NH$	10.98
Tertiary Amines		
Trimethylamine	$(CH_3)_3N$	9.81
Triethylamine	$(CH_3CH_2)_3N$	10.75
Aromatic Amines		
Aniline	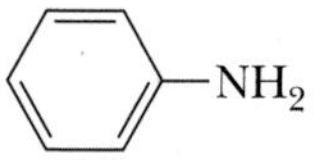	4.63
4-Methylaniline	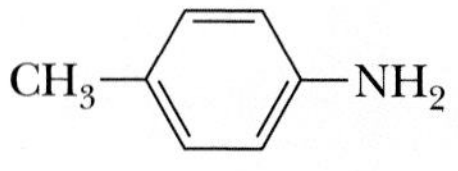	5.08
4-Chloroaniline	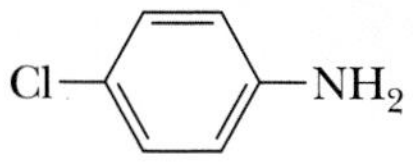	4.15
4-Nitroaniline	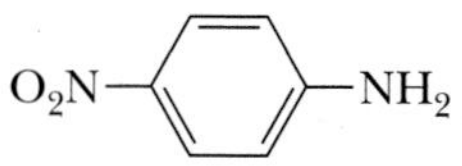	1.0
Aromatic Heterocyclic Amines		
Pyridine	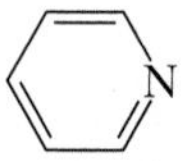	5.25
Imidazole	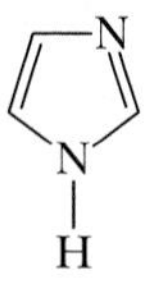	6.95

Example 23.5 Amines, Acid/Base Equilibria

Predict the position of equilibrium for this acid-base reaction.

$$CH_3NH_2 + CH_3COOH \rightleftharpoons CH_3NH_3^+ + CH_3COO^-$$

Solution

Use the approach we developed in Section 4.4 to predict the position of equilibrium in acid-base reactions. Equilibrium favors reaction of the stronger acid with the stronger base to give the weaker acid and weaker base.

$$\underset{\substack{pK_a\ 4.76\\ \text{(stronger}\\ \text{acid)}}}{CH_3NH_2 + CH_3COOH} \rightleftharpoons \underset{\substack{pK_a\ 10.64\\ \text{(weaker}\\ \text{acid)}}}{CH_3NH_3^+ + CH_3COO^-} \qquad \begin{aligned} pK_{eq} &= -5.88\\ K_{eq} &= 7.6 \times 10^5 \end{aligned}$$

Problem 23.5

Predict the position of equilibrium for this acid-base reaction.

$$CH_3NH_3^+ + H_2O \rightleftharpoons CH_3NH_2 + H_3O^+$$

A. Aliphatic Amines

All aliphatic amines have about the same base strength, pK_a of the conjugate acid 10–11, and are slightly stronger bases than ammonia. The increase in basicity compared with ammonia can be attributed to the greater stability of an alkylammonium ion, as, for example, $RCH_2NH_3^+$ compared with the ammonium ion, NH_4^+. This greater stability arises from the electron-releasing effect of alkyl groups and the resulting partial delocalization of the positive charge from nitrogen onto carbon in the alkylammonium ion.

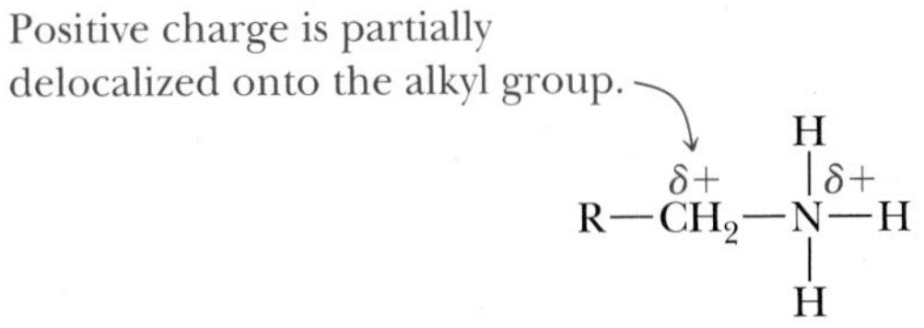
Positive charge is partially
delocalized onto the alkyl group.
δ+
δ+
R—CH2—N—H
H
H

B. Aromatic Amines

Aromatic amines are considerably weaker bases than aliphatic amines. Compare, for example, values of pK_a for aniline and cyclohexylamine. The ionization constant for the conjugate acid of aniline is larger (the smaller the value of pK_a, the weaker the base) than that for cyclohexylamine by a factor of 10^6.

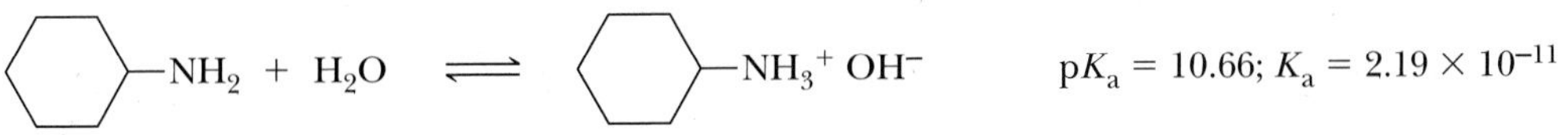
—NH2 + H2O
—NH3+ OH−
pKa = 10.66; Ka = 2.19 × 10−11

Cyclohexylamine **Cyclohexylammonium hydroxide**

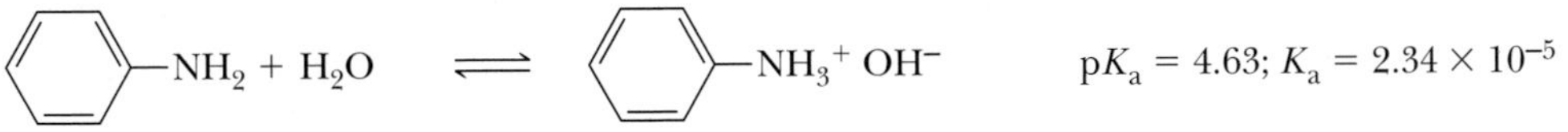
—NH2 + H2O
—NH3+ OH−
pKa = 4.63; Ka = 2.34 × 10−5

Aniline **Anilinium hydroxide**

Aromatic amines are less basic than aliphatic amines because of a combination of two factors. First is the resonance stabilization of the free base form of aromatic amines.

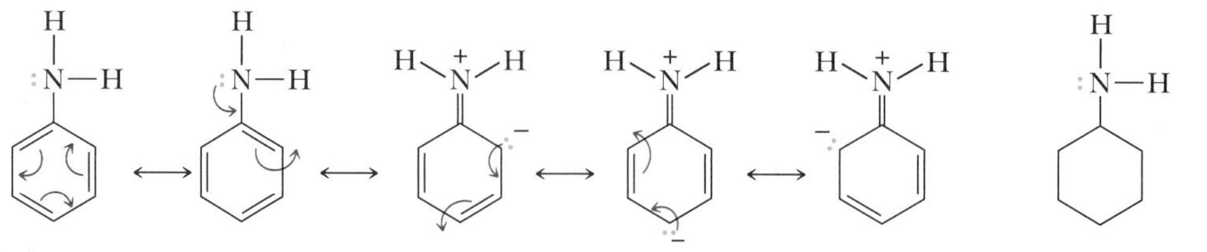

Interaction of the electron pair on nitrogen with the π system of the aromatic ring

No resonance is possible with alkylamines

▶ Delocalization of π electron density over a larger area is stabilizing.

For aniline and other arylamines, the resonance stabilization is the result of the interaction of the unshared pair on nitrogen with the π system of the aromatic ring. The resonance energy of benzene is approximately 151 kJ (36 kcal)/mol. For aniline, it is 163 kJ (39 kcal)/mol. Because of this resonance interaction, the electron pair on nitrogen is less available for reaction with acid. No such resonance stabilization is possible for alkylamines. Therefore, the electron pair on the nitrogen of an alkylamine is more available for reaction with an acid; compared to arylamines, alkylamines are stronger bases.

The second factor contributing to the decreased basicity of aromatic amines is the electron-withdrawing inductive effect of the sp^2 hybridized carbons of the aromatic ring compared with the sp^3 hybridized carbons of aliphatic amines. The unshared pair of electrons on nitrogen in an aromatic amine is pulled toward the ring and therefore is less available for protonation to form the conjugate acid of the amine. These factors are the same two that operate to make phenoxide ion less basic than alkoxide ions (Section 21.4B).

Electron-releasing groups (e.g., methyl, ethyl, and other alkyl groups) increase the basicity of aromatic amines, whereas electron-withdrawing groups (e.g., nitro and carbonyl groups) decrease their basicity. The decrease in basicity on halogen substitution is the result of the electron-withdrawing inductive effect of the electronegative halogen. The decrease in basicity on nitro substitution is caused by a combination of inductive and resonance effects, as can be seen by comparing the pK_a values for the conjugate acids of 3-nitroaniline and 4-nitroaniline.

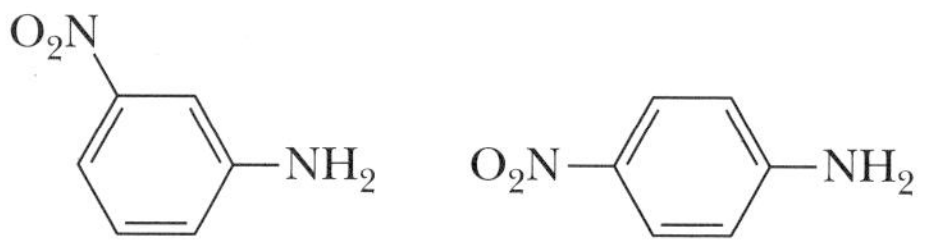

3-Nitroaniline
pK_a 2.47

4-Nitroaniline
pK_a 1.0

The basicity-decreasing effect of nitro substitution in the 3-position is almost entirely the result of its inductive effect, whereas that of nitro substitution in the 4-position is attributable to both inductive and resonance effects. In the case of para substitution (as well as ortho substitution), delocalization of the lone pair on the amino nitrogen involves not only the carbons of the aromatic ring but also oxygen atoms of the nitro group.

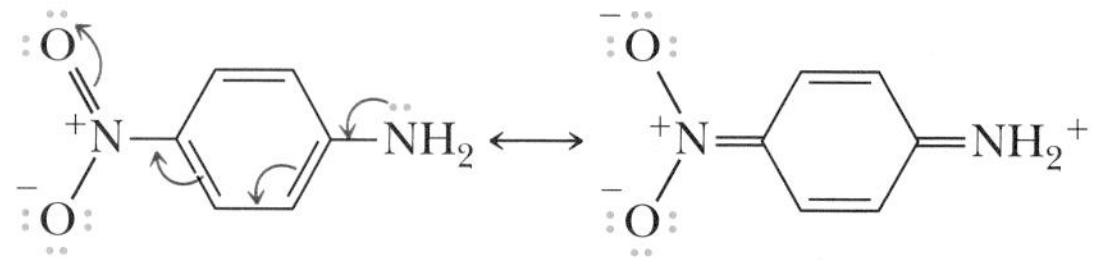

delocalization of the nitrogen lone pair onto the oxygen atoms of the nitro group

C. Heterocyclic Aromatic Amines

Heterocyclic aromatic amines are weaker bases than aliphatic heterocyclic amines. Compare, for example, the pK_a values for the conjugate acids of piperidine, pyridine, and imidazole.

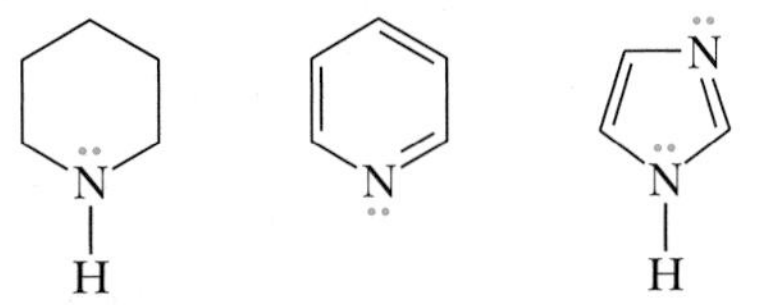

Piperidine pK_a 10.75 **Pyridine** pK_a 5.25 **Imidazole** pK_a 6.95

pK_as of conjugate acids

We discussed the structure and bonding in pyridine and imidazole in Section 21.2D. In accounting for the relative basicities of these and other heterocyclic aromatic amines, it is important to determine first whether the unshared pair of electrons on nitrogen is a part of the $(4n + 2)$ π electrons giving rise to aromaticity. In the case of pyridine, the unshared pair of electrons is not a part of the aromatic sextet. Rather, it lies in an sp^2 hybrid orbital in the plane of the ring and perpendicular to the six $2p$ orbitals containing the aromatic sextet.

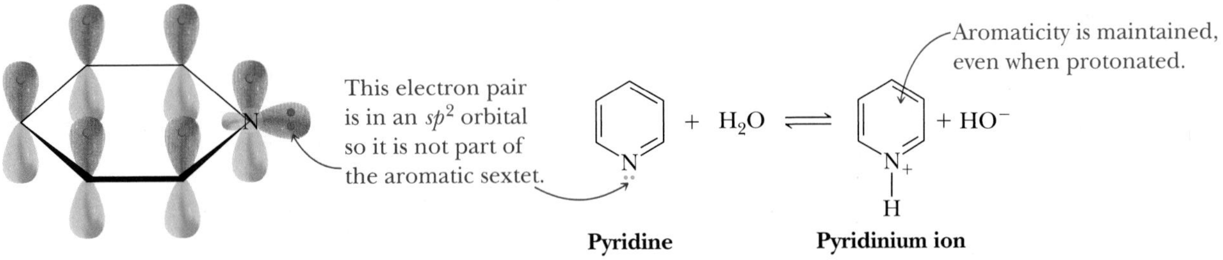
This electron pair is in an sp^2 orbital so it is not part of the aromatic sextet.
Aromaticity is maintained, even when protonated.
+ H_2O
+ HO^-
Pyridine
Pyridinium ion

MCAT Practice: *Passage and Questions*

The Planarity of —NH_2 Groups on Heterocyclic Rings

The —NH_2 group in aniline is not completely planar with the benzene ring. Specifically, the angle to the benzene ring and the bisector of the amino group is 140° instead of 180°. By comparison, the analogous angle in methylamine is 125°, while in *p*-nitroaniline it is indeed 180°.

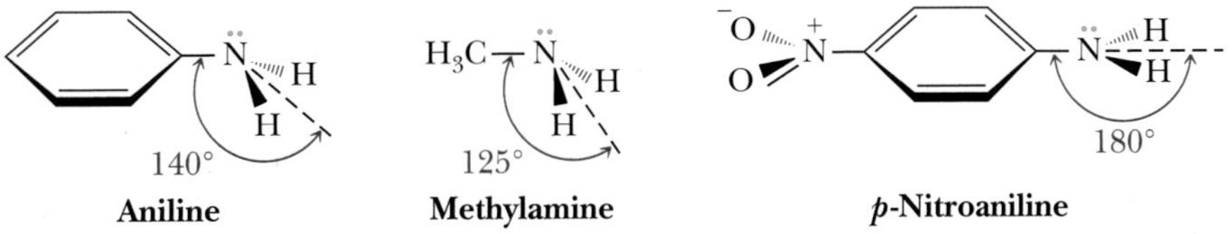
140°
Aniline
125°
Methylamine
180°
p-Nitroaniline

Questions

A. The angle found in *p*-nitroaniline means that the amine group is planar and in the same plane as the benzene ring. Why is this the case?

1. The nitro group withdraws the lone pair electrons from the amine, primarily via induction, making the N atom sp^2 hybridized and hence trigonal planar.
2. The nitro group withdraws the lone pair electrons from the amine, primarily via resonance, making the N atom sp^2 hybridized and hence trigonal planar.
3. The lone pair of the N atom of the NH_2 must be in a p orbital to make the system aromatic.
4. The nitrogen of an amine is usually planar, and aniline and methylamine are exceptions.

(Continued)

B. What is the hybridization of the nitrogen in aniline?

1. The nitrogen is sp^2 hybridized.
2. The nitrogen is sp^3 hybridized.
3. The nitrogen is between sp^2 and sp^3 hybridized, but closer to sp^3.
4. The nitrogen is between sp^2 and sp^3 hybridized, but closer to sp^2.

C. What accounts for the geometry (pyramidalization) of the NH_2 group in aniline?

1. Resonance between the NH_2 group and the benzene ring.
2. The electronic withdrawing nature of the sp^2 carbons in the phenyl group.
3. Participation of the nitrogen lone pair to make the system aromatic.
4. Both 1 and 3.

D. The pK_as of the conjugate acids of aniline and methylamine are 4.6 and 10.7, respectively. What accounts for the greater acidity of the conjugate acid of aniline?

1. Resonance between the NH_2 group and the benzene ring.
2. The electronic withdrawing nature of the sp^2 carbons in the phenyl group.
3. Participation of the nitrogen lone pair to make the system aromatic.
4. Both 1 and 2.

The geometry of —NH_2 groups on heterocyclic rings has a profound influence on the properties and folding of nucleic acids. Three of the four common nucleic acid bases have amino groups (see below). In each case, the angle between the bisector of the —NH_2 group and the attached ring is 180° (such as drawn above for *p*-nitroaniline). Not only does the hybridization of the amino group allow for an overall flat structure, but also the geometry of the planar amino group is ideal for making specific, highly directional hydrogen bonds with the complementary base.

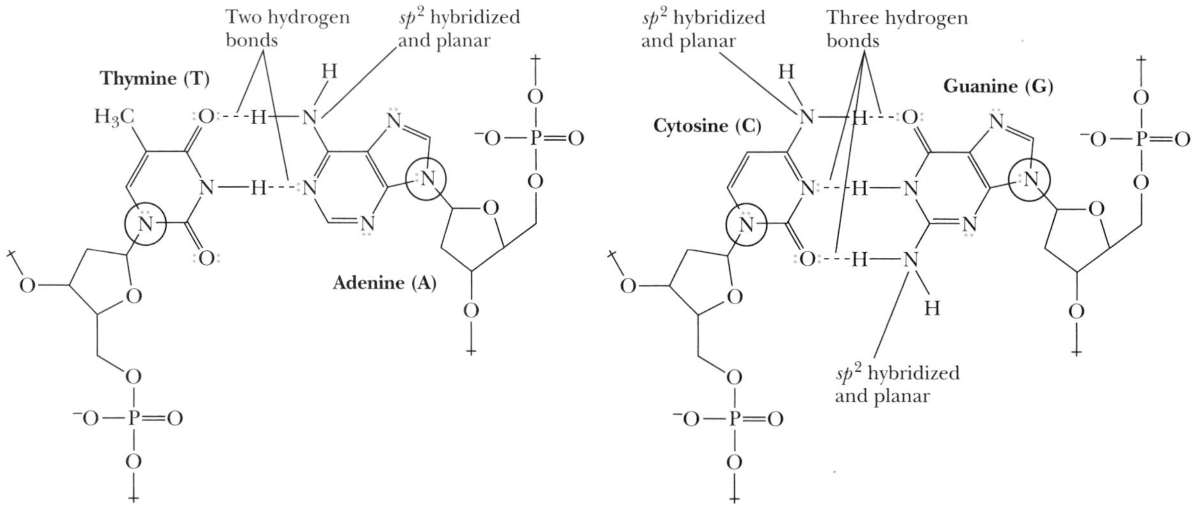

The structures of the T–A and C–G base pairs showing the locations of planar —NH_2 groups bonded to the aromatic bases as well as the specific patterns of hydrogen bonds responsible for recognition between complementary strands of DNA.

E. In the structures of T—A and C—G base pairs, there are three amino groups specifically labeled as "sp^2 hybridized and planar". What is the primary difference between these structures and that of aniline that lead to their planarity?

1. In contrast to aniline, the amino groups on the DNA bases are necessary to make the heterocyclic rings aromatic.
2. In contrast to aniline, the contributing structures that delocalize the nitrogen lone pairs onto the rings creates partial negative charges on electronegative atoms.
3. In contrast to aniline, the hydrogen bond accepting ability of the lone pairs on the —NH_2 groups of the DNA bases is better when these amino groups are sp^2 hybridized.
4. Both 2 and 3.

F. In the structures of T—A and C—G base pairing, four nitrogens are circled. Given your knowledge of organic functional group names, which of the following is the most appropriate descriptor for the kind of functional group that these nitrogens are part of?

1. An *N*-heterocyclic ester.
2. An *N*-acetal.
3. An imide.
4. An imine.

Chemists have studied base pairings analogous to those found in DNA in order to shed light on the strength of the hydrogen bonds. For example, the strength of the association of the following three base pairs increases in the order given (as an abbreviation, the pattern can be written with **D** = hydrogen bond **D**onor and **A** = hydrogen bond **A**cceptor).

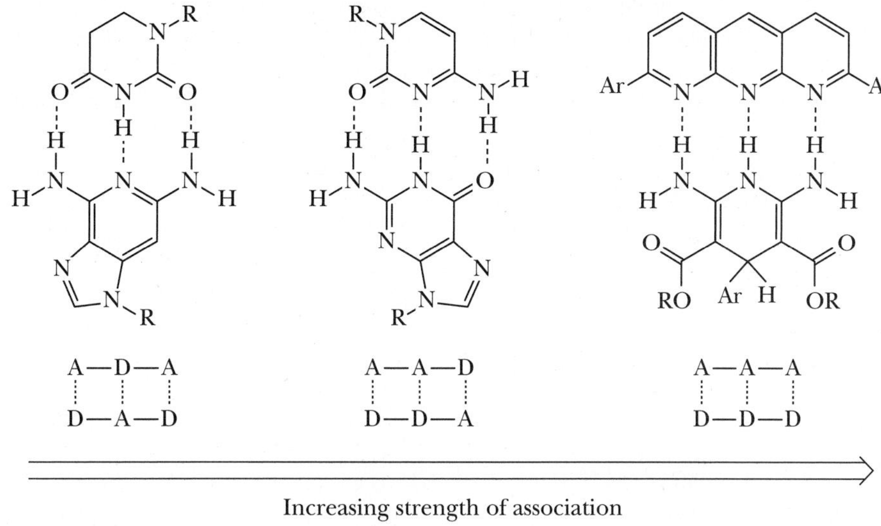
A—D—A
D—A—D
A—A—D
D—D—A
A—A—A
D—D—D
Increasing strength of association

G. Which of the following is the most likely explanation for the order of association found experimentally?

1. A trend is not expected, and hence the result is random.
2. The overall number of hydrogen bond donating and hydrogen bond accepting interactions increases from left to right.
3. The hydrogen bonds are increasingly more linear in the complexes from the left to the right.
4. By decreasing the alternation of hydrogen bond donors and acceptors on the same molecule, the hydrogen bonds become stronger due to less repulsive interactions between neighboring hydrogen bonds.

Proton transfer from water or other acid to pyridine does not involve the electrons of the aromatic sextet. Why, then, is pyridine a considerably weaker base than aliphatic amines? The answer is that the unshared pair of electrons on the pyridine nitrogen lies in a relatively electronegative sp^2 hybrid orbital, whereas in aliphatic amines, the unshared pair lies in an sp^3 hybrid orbital. This effect decreases markedly the basicity of the electron pair on an sp^2 hybridized nitrogen compared with that on an sp^3 hybridized nitrogen.

There are two nitrogen atoms in imidazole, each with an unshared pair of electrons. One unshared pair lies in a $2p$ orbital and is an integral part of the $(4n + 2)$ π electrons of the aromatic system. The other unshared pair lies in an sp^2 hybrid orbital and is not a part of the aromatic sextet; this pair of electrons functions as the proton acceptor.

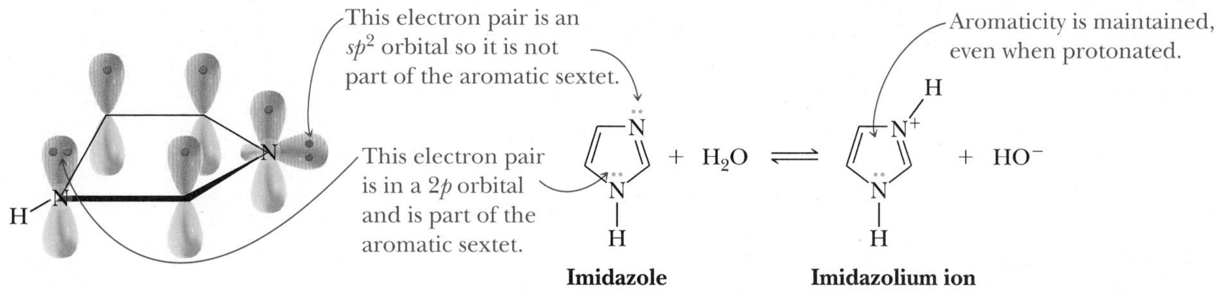
This electron pair is an sp^2 orbital so it is not part of the aromatic sextet.
This electron pair is in a $2p$ orbital and is part of the aromatic sextet.
Aromaticity is maintained, even when protonated.
+ H_2O
+ HO^-
Imidazole
Imidazolium ion

As is the case with pyridine, the unshared pair of electrons functioning as the proton acceptor in imidazole lies in an sp^2 hybrid orbital and has markedly decreased basicity compared with an unshared pair of electrons in an sp^3 hybrid orbital. The positive charge on the imidazolium ion is delocalized on both nitrogen atoms of the ring; therefore, imidazole is a stronger base than pyridine.

Like pyridine and imidazole, pyrrole is an aromatic heterocycle, but it is not nearly as basic as pyridine or imidazole. Pyrrole's lack of base strength can be understood by noticing that the lone pair on the nitrogen atom is in a 2*p* orbital and is part of the aromatic sextet of electrons. As a consequence, a protonated pyrrole cannot maintain aromaticity because the protonated nitrogen would be sp^3 hybridized and there would be only 4 π electrons remaining (in violation of two of Hückel's aromaticity rules). A loss of aromaticity is energetically very costly and severely limits the ability of pyrrole to accept a proton despite its structural similarity to pyridine and imidazole.

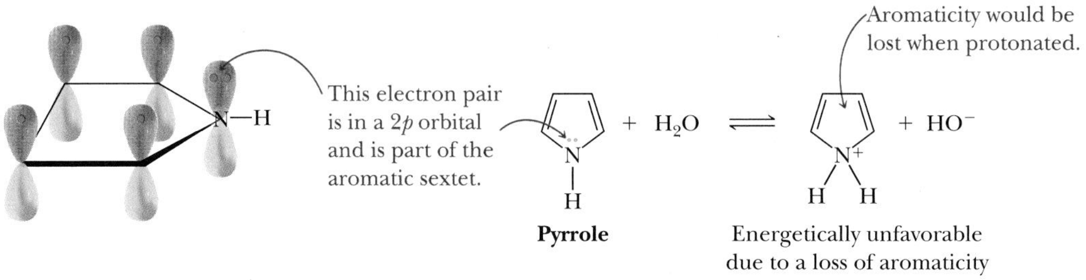
Aromaticity would be lost when protonated.
This electron pair is in a 2p orbital and is part of the aromatic sextet.
+ H2O
+ HO−
Pyrrole
Energetically unfavorable due to a loss of aromaticity

Example 23.6 Relative Amine Basicity

Select the stronger base in each pair of amines.

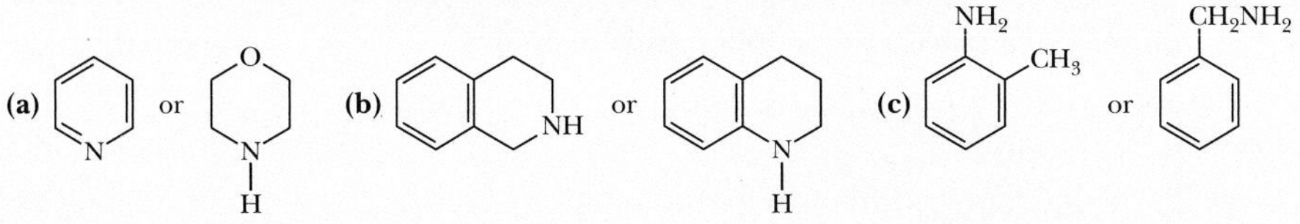
(a) or (b) or (c) or

(A) (B) (C) (D) (E) (F)

Solution

(a) Morpholine (B) is the stronger base (conjugate acid pK_a 8.2). It has a basicity comparable to that of secondary aliphatic amines. Pyridine (A), a heterocyclic aromatic amine (pK_a 5.25), is considerably less basic than aliphatic amines.

(b) Tetrahydroisoquinoline (C) has a basicity comparable to that of secondary aliphatic amines (pK_a ~ 10.8) and is the stronger base. Tetrahydroquinoline (D) has a basicity comparable to that of *N*-substituted anilines (pK_a ~ 4.4) and is the weaker base.

(c) Benzylamine (F) is the stronger base (pK_a 9.6). Its basicity is comparable to that of other aliphatic amines. The basicity of *o*-toluidine (E), an aromatic amine, is comparable to that of aniline (pK_a 4.6).

Watch a video explanation

Problem 23.6

Select the stronger acid from each pair of compounds.

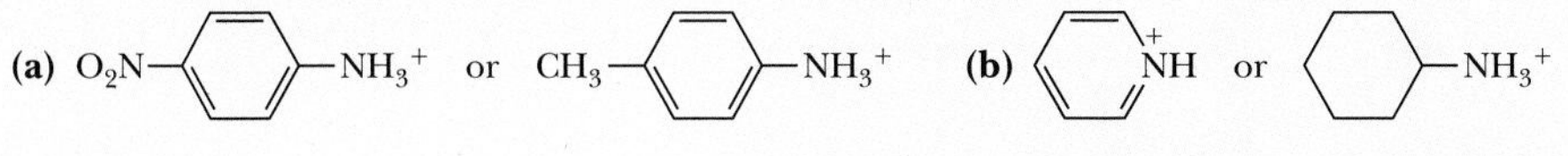
(a) or (b) or

(A) (B) (C) (D)

D. Guanidine

Guanidine, is almost as basic as hydroxide ion. Its conjugate acid, guanidinium ion (pK_a 13.6), is a weaker acid than almost any other protonated amine.

$$H_2N-C(=NH)-NH_2 + H_2O \rightleftharpoons H_2N-C(=NH_2^+)-NH_2 + OH^- \qquad pK_a = 13.6$$

Guanidine **Guanidinium ion**

The remarkable basicity of guanidine is attributed to the fact that the positive charge on the guanidinium ion is delocalized equally over the entire functional group as shown by these three equivalent contributing structures. This delocalization increases the stability of the guanidinium ion relative to ammonium ions.

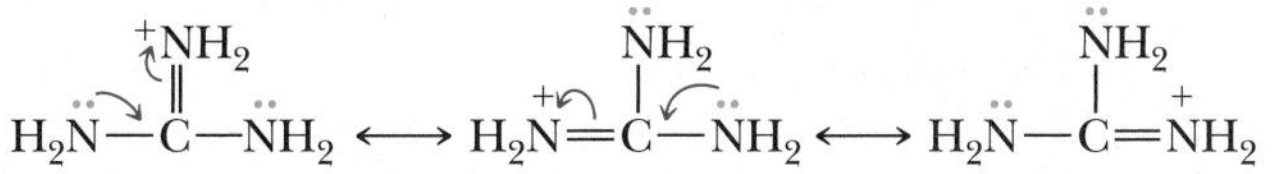

Three equivalent contributing structures

▶ Delocalization of charge over a larger area is stabilizing.

23.6 Reactions with Acids

Amines, whether soluble or insoluble in water, react quantitatively with strong acids to form water-soluble salts, as illustrated by the reaction of norepinephrine (noradrenaline) with aqueous HCl to form a hydrochloride salt.

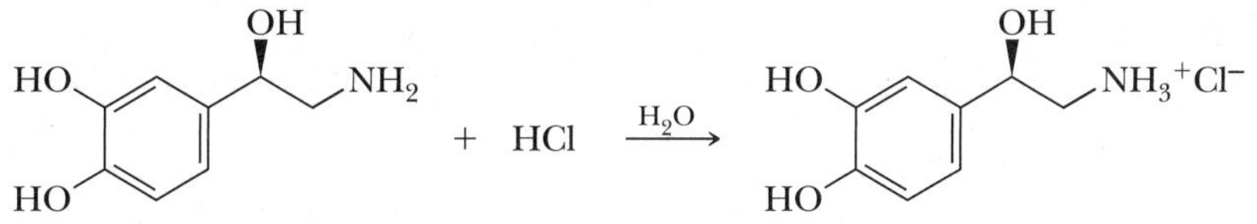

(*R*)-Norepinephrine (only slightly soluble in water)

(*R*)-Norepinephrine hydrochloride (a water-soluble salt)

Norepinephrine, secreted by the medulla of the adrenal gland, is a neurotransmitter. The suggestion has been made that it acts in those areas of the brain that mediate emotional behavior. Note that the common aliphatic amino groups in biological molecules such as (*R*)-norepinephrine and amino acids are protonated and thus positively charged at biological pH values.

Example 23.7 Amine Acid-Base Reactions

Complete each acid-base reaction and name the salt formed.

(a) $Et_2NH + HCl \longrightarrow$ **(b)** $PhCH_2NH_2 + CH_3COOH \longrightarrow$

Solution

(a) $Et_2NH_2^+Cl^-$ Diethylammonium chloride

(b) $PhCH_2NH_3^+\ CH_3COO^-$ Benzylammonium acetate

Problem 23.7

Complete each acid-base reaction and name the salt formed.

(a) $Et_3N + HCl \longrightarrow$

(b) 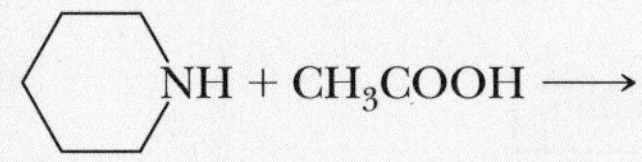

Example 23.8 Amino Acid Protonation

Following are two structural formulas for (*S*)-serine, one of the building blocks of proteins (Chapter 27).

(Continued)

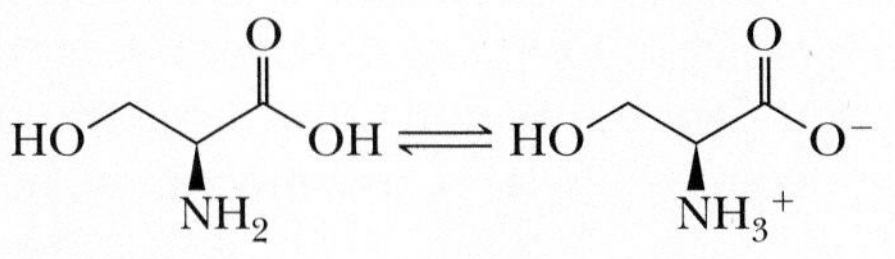

(A) (B)

Is (*S*)-serine better represented by structural formula A or B?

Solution

Structural formula A contains both an amino group (a base) and a carboxyl group (an acid). Proton transfer from the stronger acid (—COOH) to the stronger base (—NH_2) gives an internal salt; therefore, B is the better representation for (*S*)-serine. Within the field of amino acid chemistry, the internal salt represented by B is called a **zwitterion** (Section 27.2).

Problem 23.8

Following are structural formulas for propanoic acid and the conjugate acids of isopropylamine and alanine, along with pK_a values for each functional group.

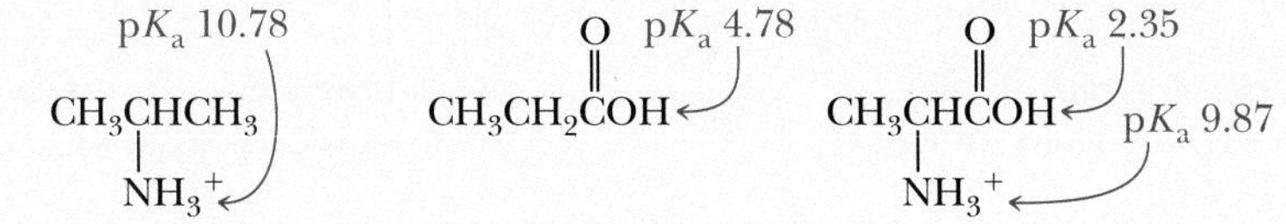

Conjugate acid of isopropylamine | **Propanoic acid** | Conjugate acid of alanine

(a) How do you account for the fact that the —NH_3^+ group of the conjugate acid of alanine is a stronger acid than the —NH_3^+ group of the conjugate acid of isopropylamine?

(b) How do you account for the fact that the —COOH group of the conjugate acid of alanine is a stronger acid than the —COOH group of propanoic acid?

The basicity of amines and the solubility in water of amine salts can be used to separate water-insoluble amines from water-insoluble, nonbasic compounds. Shown in Figure 23.2 is a flowchart for the separation of aniline from acetanilide, a neutral compound.

FIGURE 23.2 Separation and purification of an amine and a neutral compound.

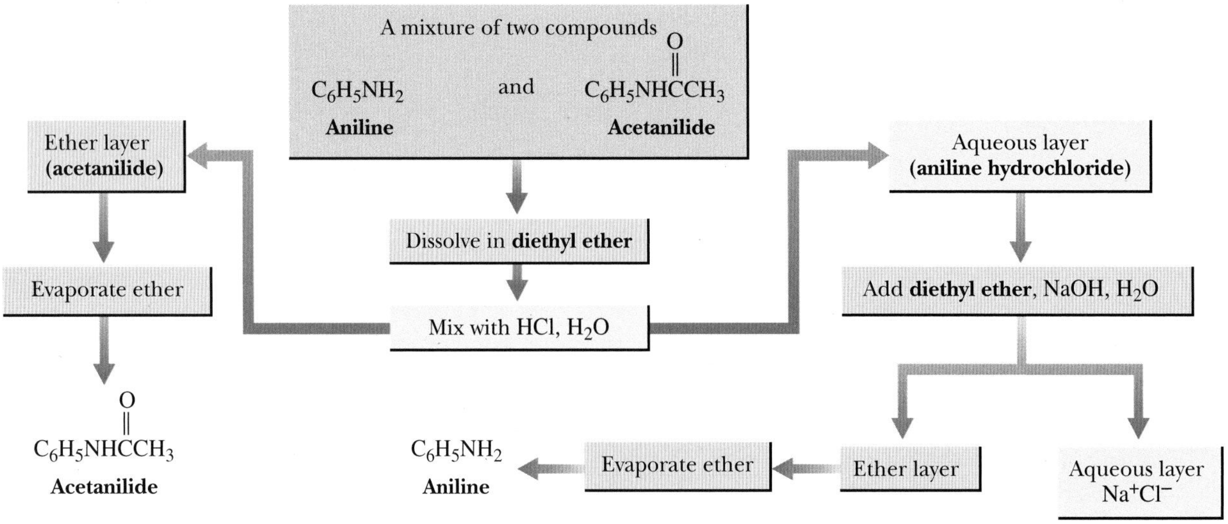

Example 23.9 Separations by Aqueous Extractions

Here is a flowchart for the separation of a mixture of a primary aliphatic amine (RNH_2, pK_a 10.8), a carboxylic acid (RCOOH, pK_a 5), and a phenol (ArOH, pK_a 10). Assume that each is insoluble in water but soluble in diethyl ether. The mixture is separated into fractions A, B, and C. Which fraction contains the amine, which contains the carboxylic acid, and which contains the phenol?

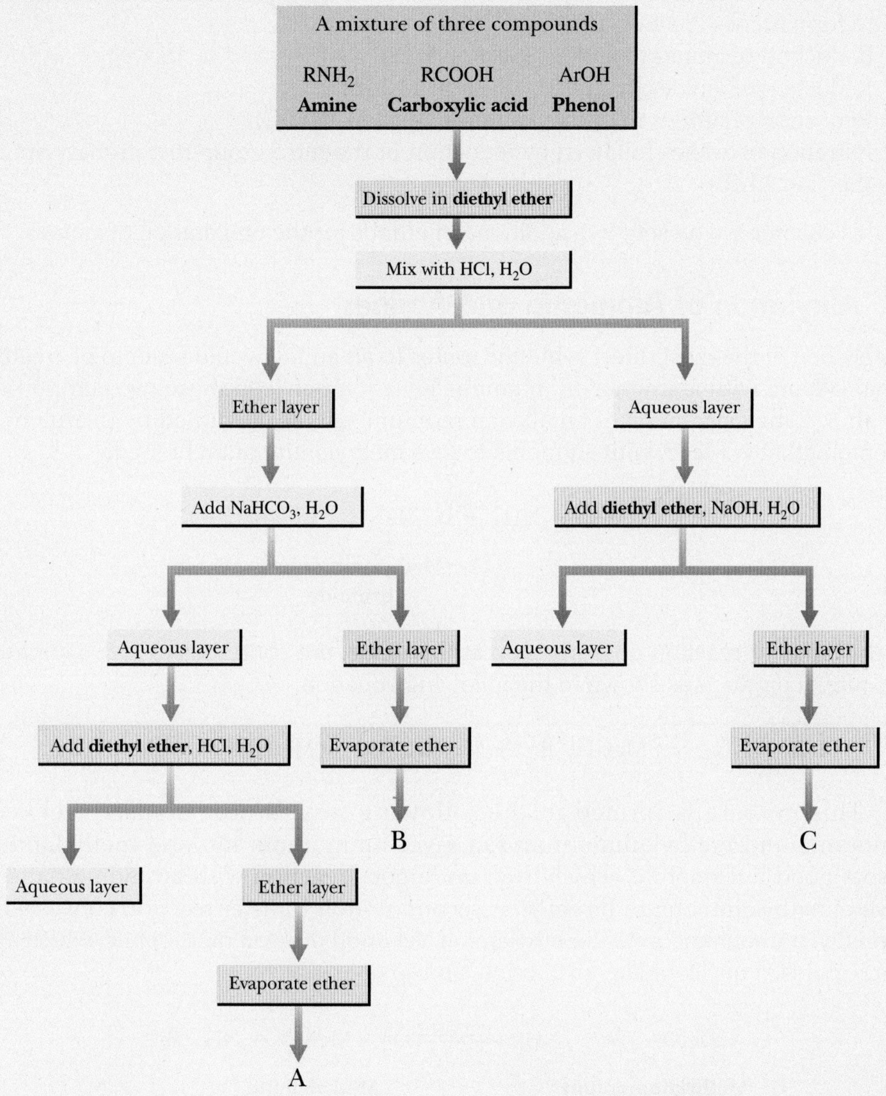

Solution

Fraction C contains RNH_2, fraction B contains ArOH, and fraction A contains RCOOH.

Problem 23.9

In what way(s) might the results of the separation and purification procedure outlined in Example 23.9 be different if the following conditions exist?

(a) Aqueous NaOH is used in place of aqueous $NaHCO_3$.

(b) The starting mixture contains an aromatic amine, $ArNH_2$, rather than an aliphatic amine, RNH_2.

23.7 Preparation

The synthesis of amines is primarily a problem of how to form a carbon-nitrogen bond and, if the newly formed nitrogen-containing compound is not already an amine, how to convert it into an amine. We have already seen the following methods for the preparation of amines.

1. Nucleophilic ring opening of epoxides by ammonia and amines (Section 11.9B)
2. Addition of nitrogen nucleophiles to the carbonyl group of aldehydes and ketones to form imines (Section 16.8)
3. Reduction of imines to amines (Section 16.8)
4. Reduction of amides by $LiAlH_4$ (Section 18.10B)
5. Reduction of nitriles to primary amines (Section 18.10C)
6. Nitration of arenes followed by reduction of the nitro group to a primary amine (Section 22.1B)

In this chapter, we present two additional methods for the preparation of amines.

A. Alkylation of Ammonia and Amines

Surely one of the most direct synthetic routes to an amine would seem to be treating a haloalkane with ammonia or an amine. Reaction between these two compounds by an S_N2 mechanism gives an alkylammonium salt, as illustrated by treatment of bromomethane, MeBr, with ammonia to give methylammonium bromide.

$$MeBr + NH_3 \xrightarrow{S_N2} MeNH_3^+ \, Br^-$$

Methylammonium bromide

Unfortunately, reaction does not stop at this stage, but continues to give a complex mixture of products as shown in the following equation.

$$MeBr + NH_3 \longrightarrow MeNH_3^+Br^- + Me_2NH_2^+Br^- + Me_3NH^+Br^- + Me_4N^+Br^-$$

This mixture is formed in the following way. Proton transfer between ammonia and methylammonium ion gives ammonium ion and methylamine, also a good nucleophile, which then undergoes reaction with bromomethane to give dimethylammonium bromide. A second proton transfer reaction converts the dimethylammonium ion to dimethylamine, yet another good nucleophile, which also participates in nucleophilic substitution, and so on.

$$MeNH_3^+Br^- + NH_3 \xrightleftharpoons{\text{proton transfer}} MeNH_2 + NH_4^+Br^-$$

Methylammonium bromide — **Methylamine**

$$MeBr + MeNH_2 \xrightarrow{S_N2} Me_2NH_2^+Br^-$$

Dimethylammonium bromide

$$Me_2NH_2^+Br^- + NH_3 \xrightleftharpoons{\text{proton transfer}} Me_2NH + NH_4^+ \, Br^-$$

Dimethylamine

The final product from such a series of nucleophilic substitution and proton transfer reactions is a tetraalkylammonium halide. The relative proportions of the various alkylation products depend on the ratio of haloalkane to ammonia in the reaction mixture. Whatever the starting mixture, however, the product is almost invariably a mixture of alkylated products. For this reason, alkylation of ammonia or amines is not a generally useful laboratory method for the preparation of more complex amines. However, primary amines are easily prepared because ammonia is inexpensive and can be used in large excess. Other amines can also be prepared in this way if the nucleophilic amine is inexpensive enough to be used in large excess.

B. Alkylation of Azide Ion

As we have just seen, alkylation of ammonia or amines is generally not an efficient method for the preparation of amines. One strategy for eliminating the problem of overalkylation is to use a form of nitrogen that can function as a nucleophile but that is no longer an effective nucleophile after it has formed a new carbon-nitrogen bond. One such nucleophilic form of nitrogen is the azide ion, N_3^-. Alkyl azides are easily prepared from sodium or potassium azide and a primary or secondary haloalkane by an S_N2 reaction. Azides are, in turn, reduced to primary amines by a variety of reducing agents, including lithium aluminum hydride ($LiAlH_4$).

Azide ion (a good nucleophile)

An alkyl azide

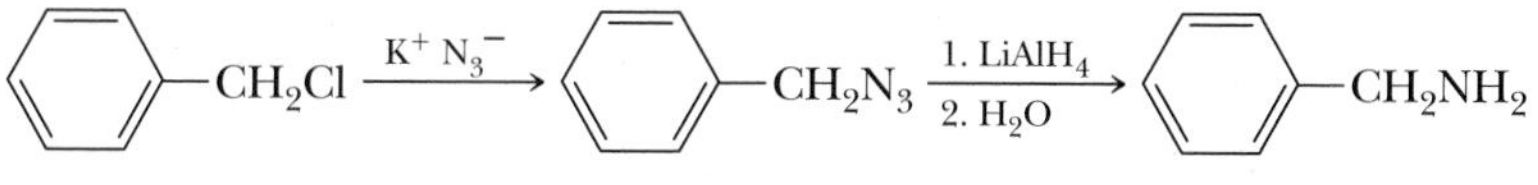

Benzyl chloride **Benzyl azide** **Benzylamine**

The azide ion can also be used for stereoselective ring opening of epoxides. Reduction of the resulting β-azidoalcohol gives a β-aminoalcohol, as illustrated by the conversion of cyclohexene to *trans*-2-aminocyclohexanol.

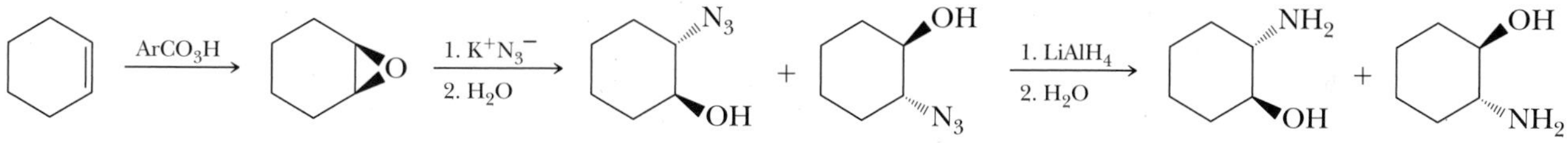

Cyclohexene **1,2-Epoxycyclohexane** ***trans*-2-Azidocyclohexanol** (racemic) ***trans*-2-Aminocyclohexanol** (racemic)

Oxidation of cyclohexene by a peroxyacid (Section 11.8C) gives an epoxide. Stereoselective nucleophilic attack by azide ion anti to the leaving oxygen of the epoxide ring (Section 11.9B) followed by reduction of the azide with lithium aluminum hydride gives racemic *trans*-2-aminocyclohexanol.

Example 23.10 Amine Synthesis

Show how to convert 4-methoxybenzyl chloride to each amine.

(a)

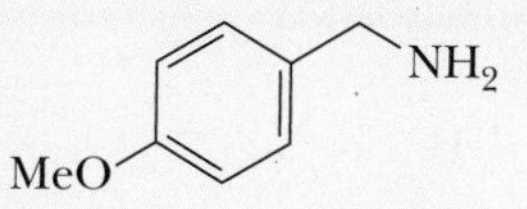

(b)

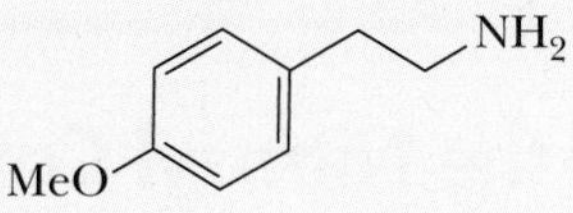

Solution

(a) Two methods might be used: (1) alkylation using a large molar excess of NH_3 to reduce the extent of overalkylation or (2) nucleophilic displacement

(Continued)

of chloride using azide ion (from NaN_3) followed by $LiAlH_4$ reduction of the azide. Of these methods, nucleophilic displacement by azide is more convenient on a laboratory scale.

(b) Nucleophilic displacement of chloride by cyanide ion is followed by reduction of the cyano group with lithium aluminum hydride.

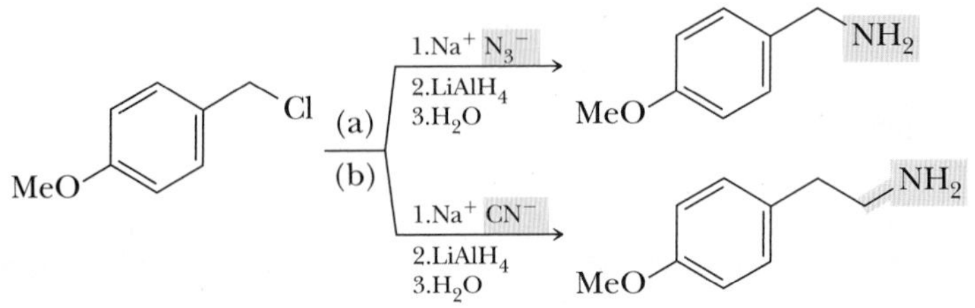

Problem 23.10

Show how to bring about each conversion in good yield. In addition to the given starting material, use any other reagents as necessary.

(a) 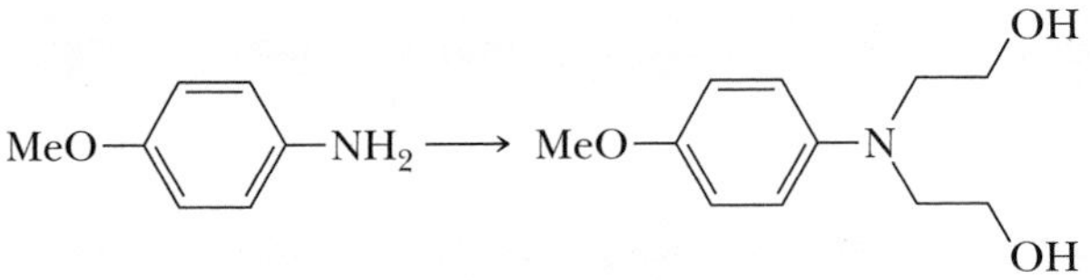

(b) 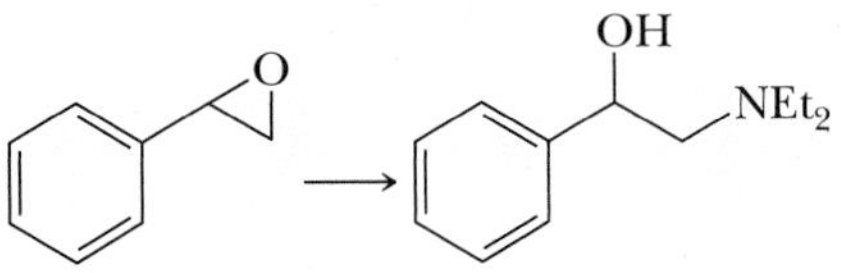

(racemic) (racemic)

23.8 Reaction with Nitrous Acid

Nitrous acid, HNO_2, is an unstable compound that is prepared by adding sulfuric or hydrochloric acid to an aqueous solution of sodium nitrite, $NaNO_2$. Nitrous acid is a weak oxygen acid and ionizes according to the following equation.

$$HNO_2 + H_2O \rightleftharpoons H_3O^+ + NO_2^- \qquad pK_a = 3.37$$

Nitrous acid undergoes reaction with amines in different ways, depending on whether the amine is primary, secondary, or tertiary and whether it is aliphatic or aromatic. These reactions are all related by the fact that nitrous acid (1) participates in proton-transfer reactions and (2) is a source of the nitrosyl cation, a weak electrophile.

Mechanism 23.1

Formation of the Nitrosyl Cation

Step 1: **Add a proton.** Protonation of the OH group of nitrous acid gives an oxonium ion.

Step 2: **Break a bond to give stable molecules or ions.** Loss of water gives the nitrosyl cation, which here is represented as a hybrid of two contributing structures.

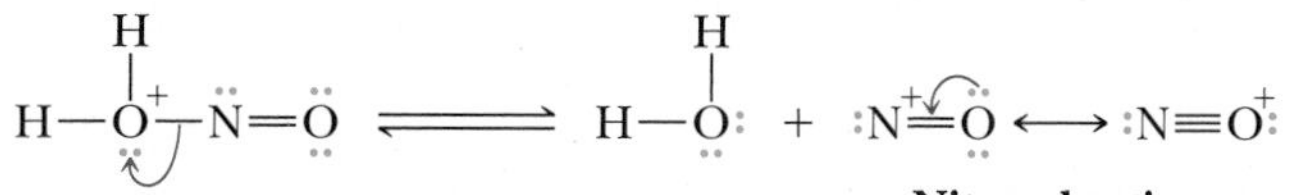

Nitrosyl cation

A. Tertiary Aliphatic Amines

When treated with nitrous acid, tertiary aliphatic amines, whether water-soluble or water-insoluble, are protonated to form water-soluble salts. No further reaction occurs beyond salt formation. This reaction is of no practical use.

B. Tertiary Aromatic Amines

Tertiary aromatic amines are bases that can also form salts with nitrous acid. An alternative pathway, however, is open to tertiary aromatic amines, namely electrophilic aromatic substitution. The nitrosyl cation, a very weak electrophile, reacts only with aromatic rings containing strongly activating ortho-para directing groups such as the hydroxyl and dialkylamino groups. When treated with nitrous acid, these compounds undergo nitrosation, predominantly in the para position to give blue or green aromatic nitroso compounds.

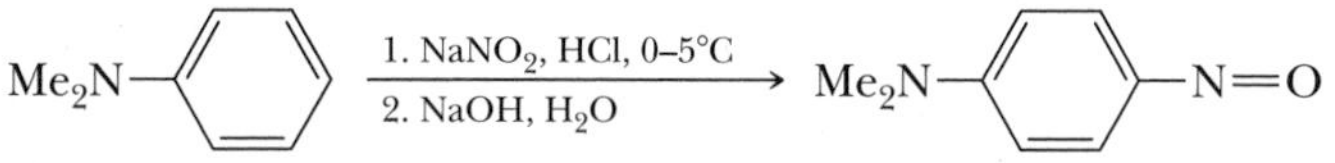

***N,N*-Dimethylaniline** ***N,N*-Dimethyl-4-nitrosoaniline**

C. Secondary Aliphatic and Aromatic Amines

Secondary amines, whether aliphatic or aromatic, undergo reaction with nitrous acid to give *N*-nitrosamines, as illustrated by the reaction of piperidine with nitrous acid.

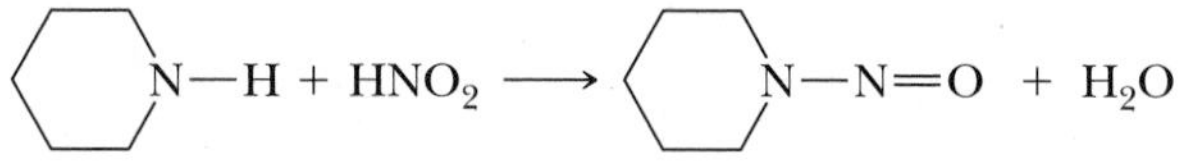

Piperidine ***N*-Nitrosopiperidine**

Mechanism 23.2

Reaction of a 2° Amine with the Nitrosyl Cation to Give an *N*-Nitrosamine

Step 1: **Make a new bond between a nucleophile and an electrophile.** Reaction of the 2° amine (a nucleophile) with the nitrosyl cation (an electrophile) gives an *N*-nitrosammonium ion.

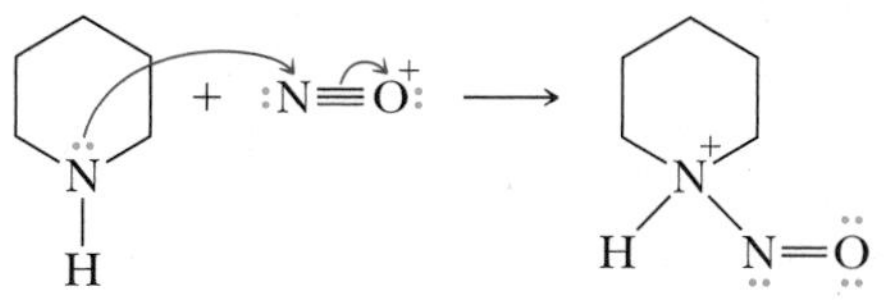

(Continued)

Step 2: **Take a proton away.** Proton transfer to solvent gives the *N*-nitrosamine.

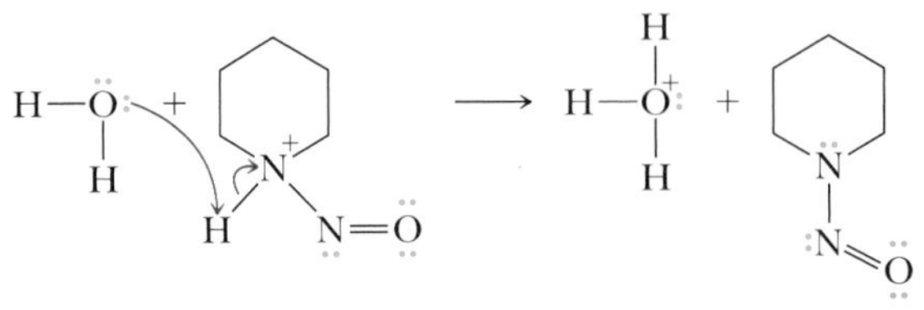

N-Nitrosamines are of little synthetic or commercial value. They have received considerable attention in recent years, however, because many of them are potent carcinogens. Following are structural formulas of two *N*-nitrosamines, each of which is a known carcinogen.

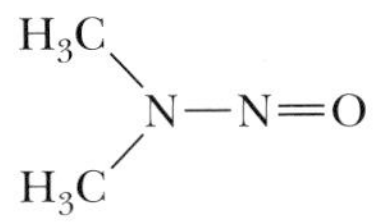

***N*-Nitrosodimethylamine**
(found in cigarette smoke and when bacon "preserved" with sodium nitrite is fried)

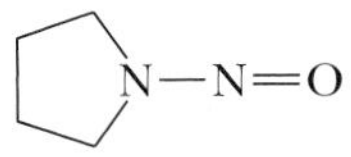

***N*-Nitrosopyrrolidine**
(formed when bacon "preserved" with sodium nitrite is fried)

Common practice within the food industry has been to add sodium nitrite to processed meats to "retard spoilage" (i.e., to inhibit the growth of *Clostridium botulinum*, the bacterium responsible for botulism poisoning). Although this practice was well grounded before the days of adequate refrigeration, it is of questionable value today. Sodium nitrite is also added to prevent red meats from turning brown. If you buy some nice red hamburger in a food market and find it is gray or brown inside, you can be sure the outside has been treated with sodium nitrite. Controversy over the use of sodium nitrite has been generated by the demonstration that nitrite ion in the presence of acid converts secondary amines to *N*-nitrosamines and that many *N*-nitrosamines are powerful carcinogens. This demonstration led in turn to pressure by consumer groups to force the Food and Drug Administration (FDA) to ban the use of nitrite additives in foods. The strength of the argument to ban nitrites was weakened with the finding that enzymes in our mouths and intestinal tracts have the ability to catalyze the reduction of nitrate to nitrite. Nitrate ion is normally found in a wide variety of foods and in drinking water. To date, there is no evidence that nitrite as a food additive poses any risk not already present through our existing dietary habits. The FDA has established the current permissible level of sodium nitrite in processed meats as 50 to 125 ppm (i.e., 50–125 μg nitrite per gram of cured meat).

D. Primary Aliphatic Amines

Treatment of a primary aliphatic amine with nitrous acid results in the loss of nitrogen, N_2, and the formation of substitution, elimination, and rearrangement products as illustrated by the treatment of butylamine with nitrous acid.

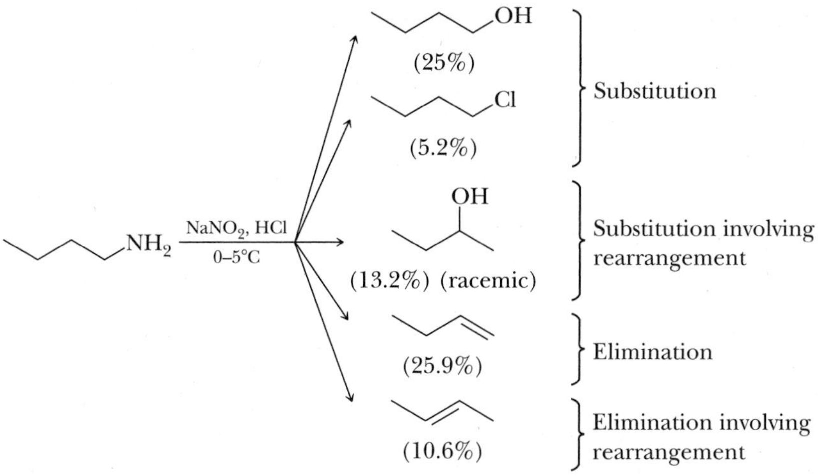
NaNO2, HCl
0–5°C
(25%)
(5.2%)
Substitution
(13.2%) (racemic)
Substitution involving rearrangement
(25.9%)
Elimination
(10.6%)
Elimination involving rearrangement

The mechanism by which this mixture of products is formed involves formation of a **diazonium ion**. The conversion of a primary amine to a diazonium ion is called **diazotization**.

Diazonium ion
An ArN_2^+ or RN_2^+ ion.

Mechanism 23.3

Reaction of a 1° Amine with Nitrous Acid

Step 1: **Make a new bond between a nucleophile and an electrophile.** Reaction of a 1° amine (nucleophile) with the nitrosyl cation (electrophile) is the first step of the mechanism.

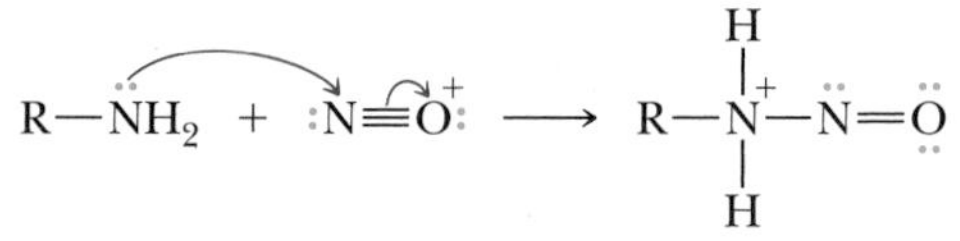

A 1° aliphatic amine

Step 2: **Take a proton away.** Removal of a proton gives an *N*-nitrosamine.

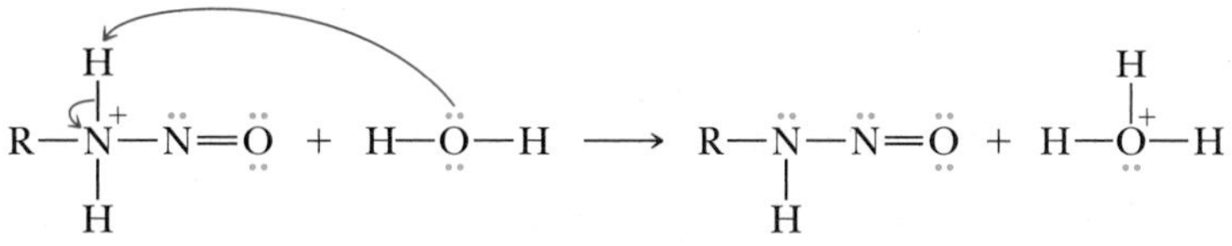

An *N*-nitrosamine

Step 3: **Keto-enol tautomerism.** The *N*-nitrosamine undergoes keto-enol tautomerism (Section 16.9) to give a diazotic acid, which is so named because it has two (*di*-) nitrogen (*-azot-*) atoms within its structure.

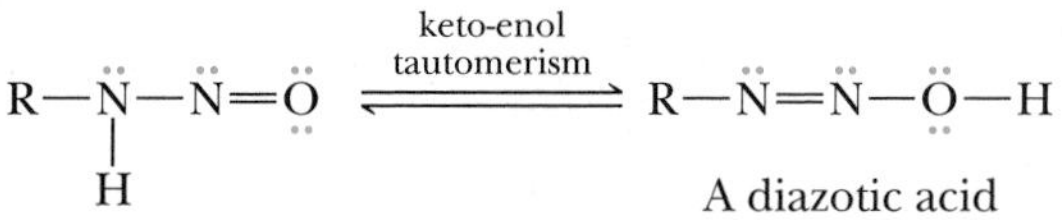
keto-enol tautomerism

A diazotic acid

Step 4: **Add a proton.** The diazotic acid is protonated.

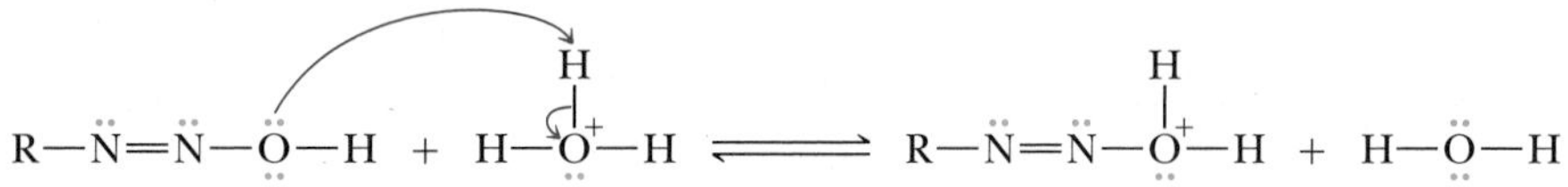

(*Continued*)

Step 5: **Break a bond to give stable molecules or ions.** The protonated diazotic acid loses H_2O to give a diazonium ion.

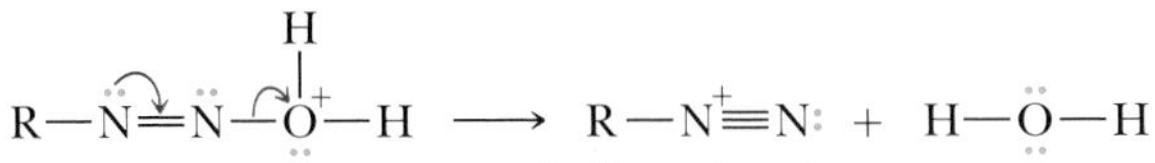

A diazonium ion

Step 6: **Break a bond to give stable molecules or ions.** The diazonium ion then loses the very stable N_2 to give a carbocation.

A diazonium ion A carbocation

Aliphatic diazonium ions are unstable, even at 0°C, and immediately lose nitrogen to give carbocations and nitrogen gas. The driving force for this reaction is the fact that N_2 is one of the best leaving groups because it is an extraordinarily weak base and has a very strong nitrogen-nitrogen triple bond. It is removed from the reaction mixture as a gas as it is formed. The carbocation now has open to it the three reactions in the repertoire of aliphatic carbocations: (1) loss of a proton to give an alkene, (2) reaction with a nucleophile to give a substitution product, and (3) rearrangement to a more stable carbocation and then reaction further by (1) or (2).

Because treatment of a primary aliphatic amine with HNO_2 gives a mixture of products, it is generally not a useful reaction. An exception is the Tiffeneau-Demjanov reaction, in which a cyclic β-aminoalcohol is treated with nitrous acid to give a ring-expanded ketone, with evolution of nitrogen.

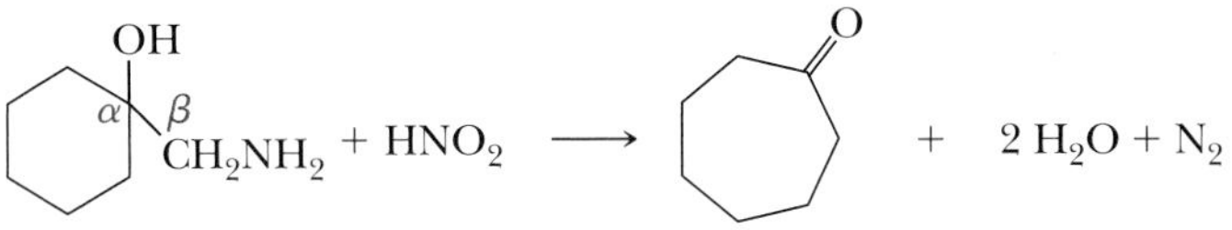

A β-aminoalcohol **Cycloheptanone**

We account for this molecular rearrangement as shown in the following mechanism.

Mechanism 23.4

The Tiffeneau-Demjanov Reaction

Several steps: Reaction of the 1° amine with nitrous acid gives a diazonium ion. This reaction takes several steps as detailed in the previous Mechanism box.

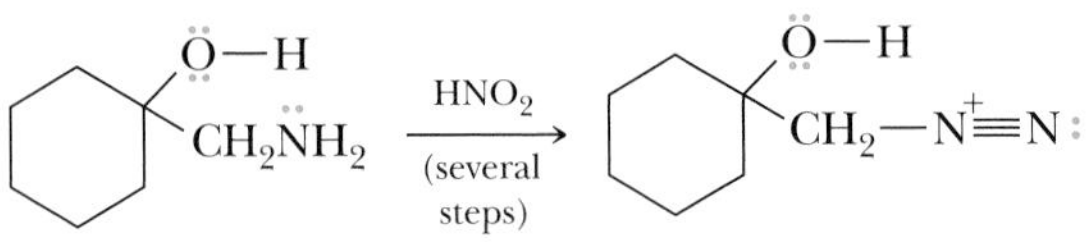

A diazonium ion

Step 2: **Break a bond to give stable molecules or ions and 1,2-shift.** Simultaneous loss of N_2 and rearrangement by a 1,2-shift give the conjugate acid of the final product as a resonance-stabilized cation.

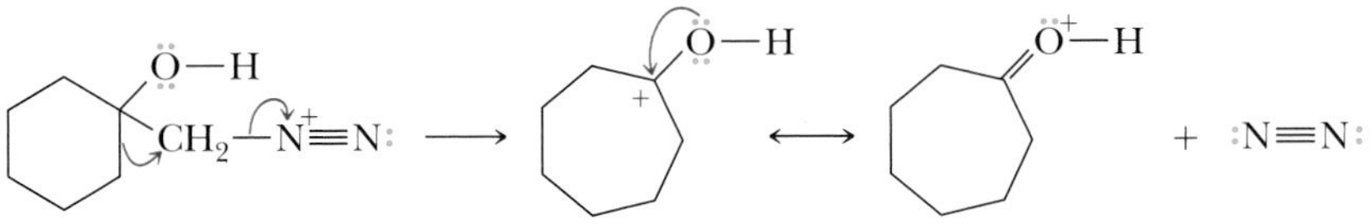

A resonance-stabilized cation

Step 3: **Take a proton away.** Proton transfer from this cation to solvent completes the reaction.

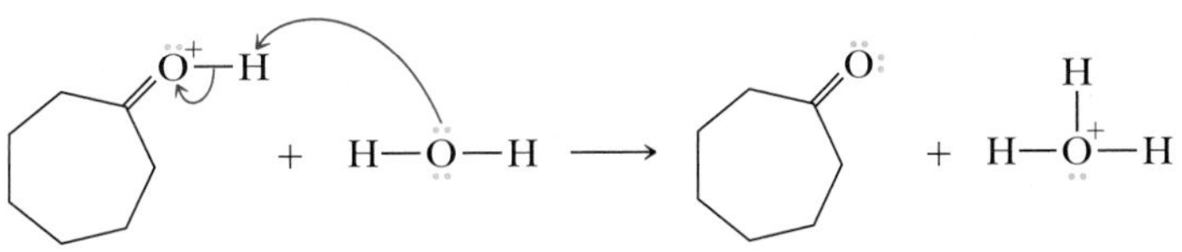

Cycloheptanone

The driving force for this molecular rearrangement is precisely what we already saw for other cation rearrangements: transformation of a less stable cation into a more stable cation. This reaction is analogous to the pinacol rearrangement (Section 10.7) with the leaving group being N_2 rather than H_2O.

Example 23.11 Tiffeneau-Demjanov Reaction

The following sequence of reactions gives cyclooctanone. Propose a structural formula for compound A and a mechanism for its conversion to cyclooctanone.

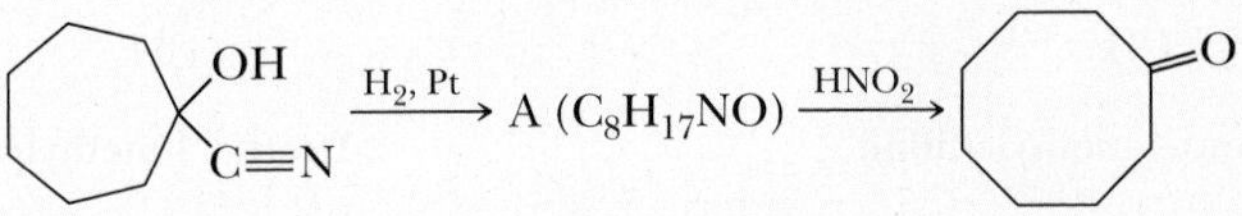

Solution

Catalytic hydrogenation using hydrogen over a platinum catalyst reduces the carbon-nitrogen triple bond to a single bond (Section 18.10C) and gives a β-aminoalcohol. Treatment of the β-aminoalcohol with nitrous acid results in loss of N_2 and expansion of the seven-membered ring to an eight-membered cyclic ketone.

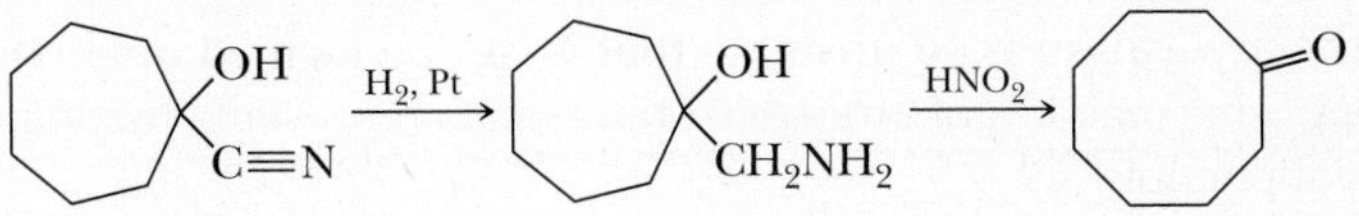

A

Problem 23.11

How might you bring about this conversion?

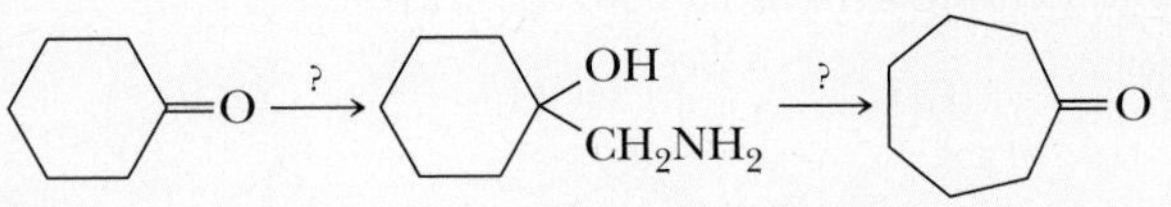

E. Primary Aromatic Amines

Primary aromatic amines react with nitrous acid to form arenediazonium salts, which, unlike their aliphatic counterparts, are stable at 0°C and can be kept in solution for short periods without decomposition. When an arenediazonium salt is treated with an appropriate reagent, nitrogen is lost and replaced with another atom or functional group. What makes reactions of primary aromatic amines with nitrous acid so valuable is the fact that the —NH_2 group can be replaced with the groups shown.

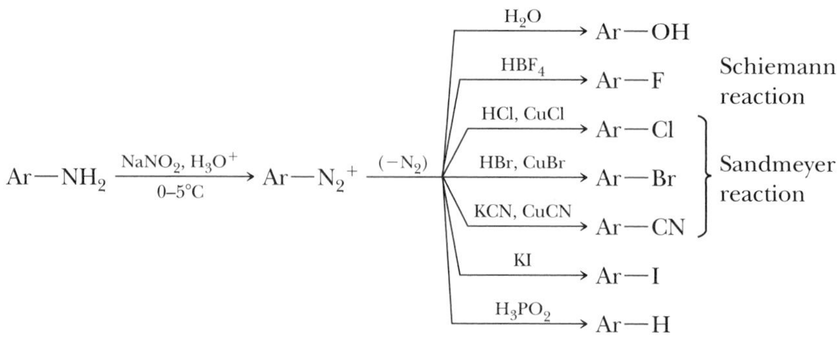

Aromatic amines can be converted to phenols by first forming the arenediazonium salt in aqueous sulfuric acid and then heating the solution. In this manner, 2-bromo-4-methylaniline is converted to 2-bromo-4-methylphenol.

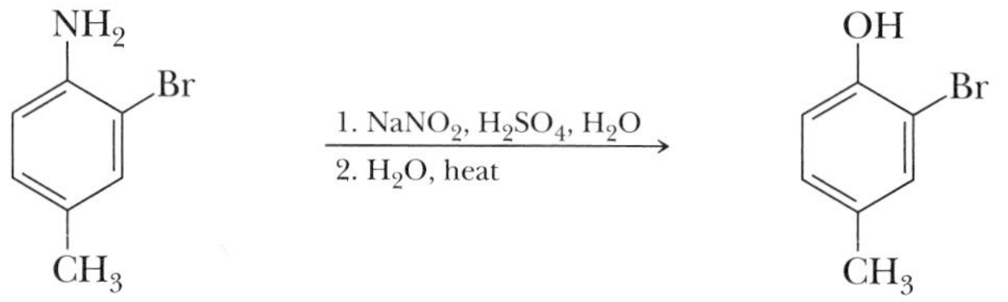

2-Bromo-4-methylaniline **2-Bromo-4-methylphenol**

The intermediate in the decomposition of an arenediazonium ion in water is an aryl cation, which then undergoes reaction with water to form the phenol.

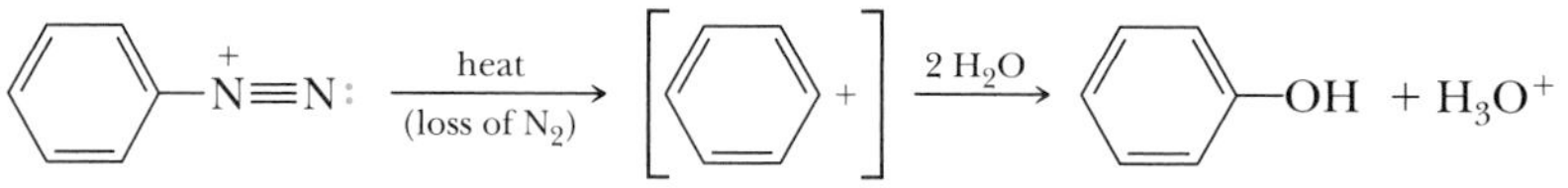

Benzenediazonium ion An aryl cation **Phenol**

Note that the aryl cation is so unstable that it can be formed only with N_2 as the leaving group. This reaction of arenediazonium salts represents the main laboratory preparation of phenols.

Example 23.12 Multi-Step Synthesis I

What reagents and experimental conditions will bring about each step in the conversion of toluene to 4-hydroxybenzoic acid?

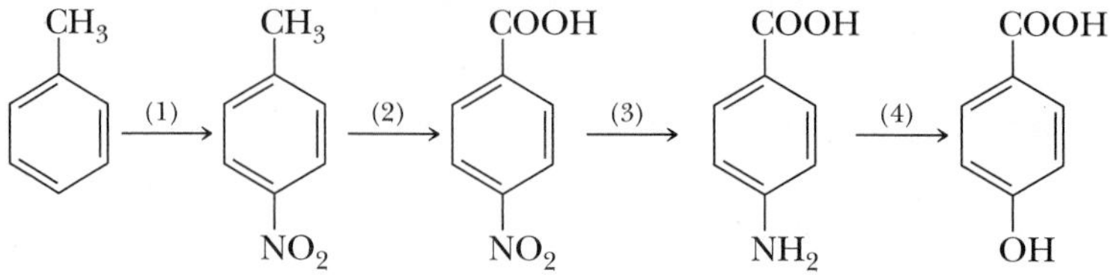

Solution

Step 1: Nitration of the aromatic ring (Section 22.1B) using HNO_3 in H_2SO_4 followed by separation of the ortho and para isomers gives 4-nitrotoluene.

Step 2: Oxidation at the benzylic carbon (Section 21.5A) using $K_2Cr_2O_7$ in H_2SO_4 gives 4-nitrobenzoic acid.

Step 3: Catalytic reduction of the nitro group (Section 22.1B) to an amino group using H_2 in the presence of Ni or another transition metal catalyst gives 4-aminobenzoic acid. Alternatively, reduction of the nitro group to a primary amine can be brought about using Zn, Sn, or Fe in aqueous HCl followed by aqueous NaOH.

Step 4: Reaction of the aromatic amine with $NaNO_2$ in aqueous H_2SO_4 followed by heating gives 4-hydroxybenzoic acid.

Problem 23.12

Show how to convert toluene to 3-hydroxybenzoic acid using the same set of reactions as in Example 23.12 but changing the order in which two or more of the steps are carried out.

The **Schiemann reaction** is the most common method for the introduction of fluorine onto an aromatic ring. It is carried out by treatment of a primary aromatic amine with sodium nitrite in aqueous HCl followed by addition of HBF_4 or $NaBF_4$. The diazonium fluoroborate salt precipitates and is collected and dried. Heating the dry salt brings about its decomposition to an aryl fluoride, nitrogen, and boron trifluoride. The Schiemann reaction is also thought to involve an aryl cation intermediate.

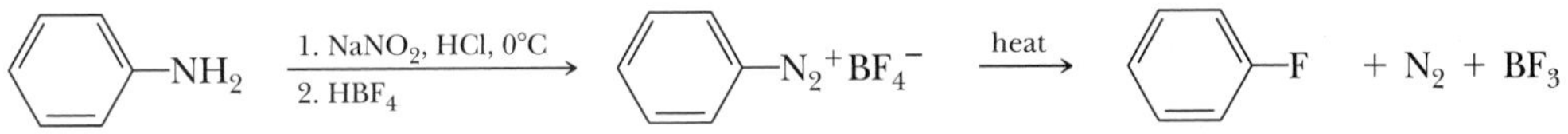

A diazonium fluoroborate   **Fluorobenzene**

Treatment of a primary aromatic amine with nitrous acid followed by heating with HCl/CuCl, HBr/CuBr, or KCN/CuCN results in replacement of the diazonium group by —Cl, —Br, or —CN, respectively, and is known as the **Sandmeyer reaction**. The Sandmeyer reaction fails with CuI or CuF.

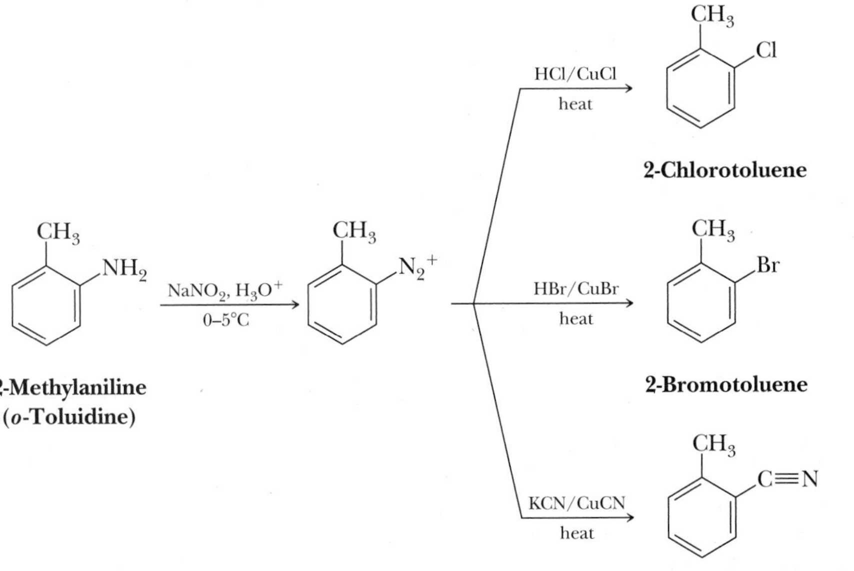

2-Methylaniline (*o*-Toluidine)

2-Chlorotoluene

2-Bromotoluene

2-Methylbenzonitrile

Treating an arenediazonium ion with iodide ion, generally from potassium iodide, is the best and most convenient method for introducing iodine onto an aromatic ring.

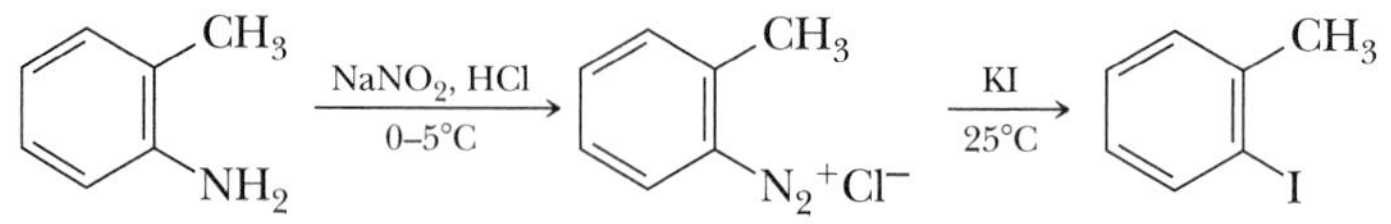

2-Methylaniline (*o*-Toluidine) **2-Iodotoluene**

Treating an arenediazonium ion with hypophosphorous acid, H_3PO_2, results in reduction of the diazonium group and its replacement by —H, as illustrated by the conversion of aniline to 1,3,5-trichlorobenzene. Recall that $—NH_2$ is a powerful activating group (Section 22.2A). Treating aniline with chlorine requires no catalyst and gives 2,4,6-trichloroaniline. To complete the conversion, the $—NH_2$ group is removed by treatment with nitrous acid followed by hypophosphorous acid to give 1,3,5-trichlorobenzene.

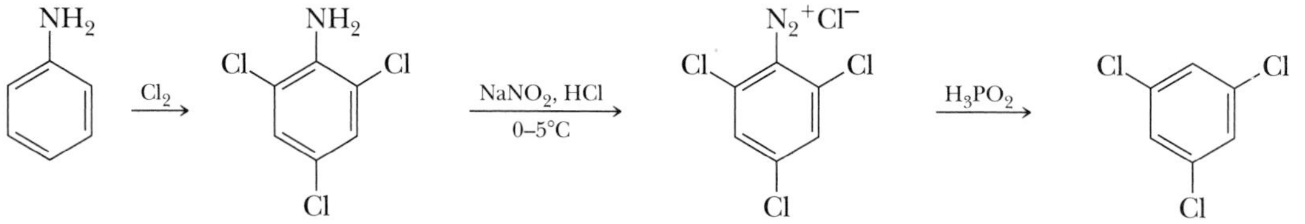

Aniline **2,4,6-Trichloroaniline** **2,4,6-Trichlorobenzene-diazonium chloride** **1,3,5-Trichlorobenzene**

Example 23.13 Multi-Step Synthesis II

Show reagents and conditions to convert toluene to 3-bromo-4-methylphenol.

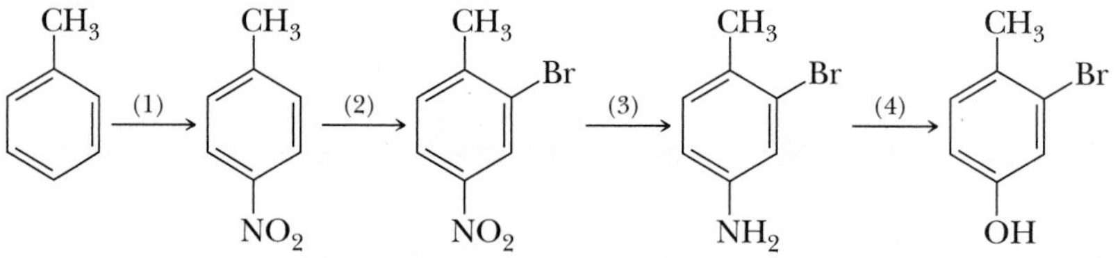

Solution

Step 1: HNO_3 in H_2SO_4. Methyl is ortho-para directing and slightly activating.

Step 2: Treat 4-nitrotoluene with bromine in the presence of $FeCl_3$.

Step 3: Reduce the nitro group either using H_2/Ni or using Sn, Zn, or Fe in aqueous HCl followed by aqueous NaOH.

Step 4: Diazotize the amine with $NaNO_2$ in aqueous sulfuric acid followed by warming of the solution to replace $—N_2^+$ with —OH.

Problem 23.13

Starting with 3-nitroaniline, show how to prepare the following compounds.

(a) 3-Nitrophenol **(b)** 3-Bromoaniline **(c)** 1,3-Dihydroxybenzene (resorcinol)

(d) 3-Fluoroaniline **(e)** 3-Fluorophenol **(f)** 3-Hydroxybenzonitrile

23.9 Hofmann Elimination

When a quaternary ammonium halide is treated with moist silver(I) oxide (a slurry of Ag_2O in H_2O), silver halide precipitates, leaving a solution of a quaternary ammonium hydroxide.

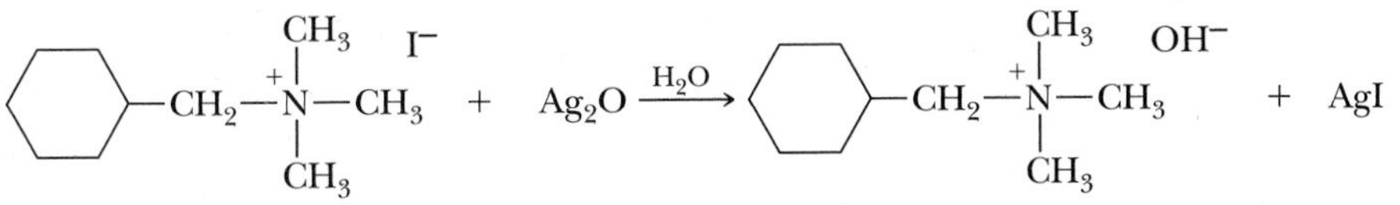

(Cyclohexylmethyl)trimethylammonium iodide **Silver oxide** **(Cyclohexylmethyl)trimethylammonium hydroxide**

In the mid-nineteenth century, Augustus Hofmann discovered that when a quaternary ammonium hydroxide is heated, it decomposes to an alkene, a tertiary amine, and water. Thermal decomposition of a quaternary ammonium hydroxide to an alkene is known as the **Hofmann elimination**.

Hofmann elimination

When treated with a strong base, a quaternary ammonium halide undergoes β-elimination by an E2 mechanism to give the less substituted alkene as the major product.

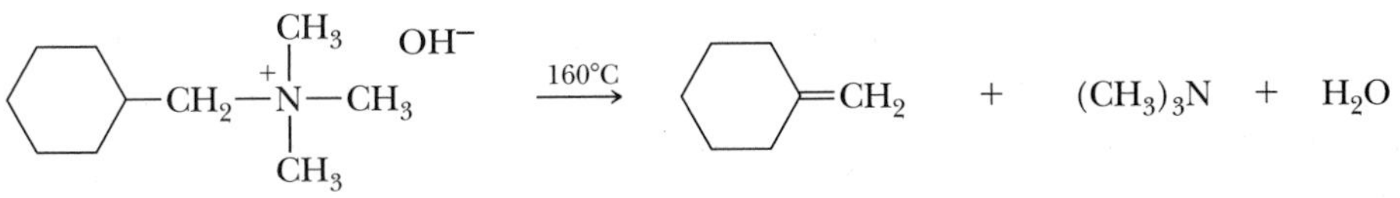

(Cyclohexylmethyl)trimethylammonium hydroxide **Methylenecyclohexane** **Trimethylamine**

The Hofmann elimination has most of the characteristics of an E2 reaction (Section 9.7). First, Hofmann eliminations are concerted, meaning that bond-breaking and bond-forming steps occur simultaneously or nearly so. Second, Hofmann eliminations are stereoselective for anti elimination, meaning that —H and the leaving group must be anti to each other. The following mechanism illustrates the concerted nature of bond forming and bond breaking, as well as the anti arrangement of —H and the trialkylamino leaving group.

Mechanism 23.5

The Hofmann Elimination

Take a proton away and break a bond to give stable molecules or ions. Removal of a β-hydrogen by base, collapse of the electron pair of the C—H bond to become the π bond of the alkene, and loss of the trialkylamino group occur simultaneously. The reaction shows anti stereoselectivity.

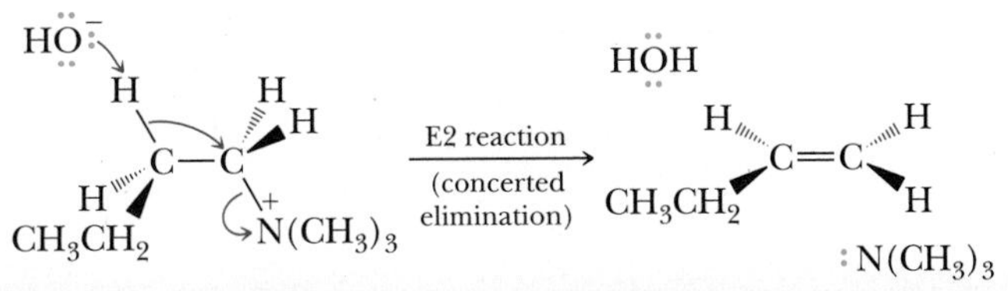

When we studied E2 reactions of haloalkanes in Section 9.7, we saw that a β-hydrogen must be anti to the leaving group. If only one β-hydrogen meets this requirement, then the double bond is formed in that direction. If, however, two β-hydrogens meet this requirement, then elimination follows Zaitsev's rule: elimination occurs preferentially to form the more substituted double bond.

$$\underset{\text{Br}}{\mathrm{CH_3CH_2CHCH_3}} \xrightarrow[\text{E2}]{\mathrm{CH_3CH_2O^-Na^+}} \underset{(75\%)}{\mathrm{CH_3CH{=}CHCH_3}} + \underset{(25\%)}{\mathrm{CH_3CH_2CH{=}CH_2}}$$

Thermal decomposition of quaternary ammonium hydroxides is different because elimination occurs preferentially to form the least substituted double bond. Thermal decomposition of *sec*-butyltrimethylammonium hydroxide, for example, gives 1-butene as the major product.

$$\underset{\mathrm{HO^-}\ \overset{+}{\mathrm{N}}(\mathrm{CH_3})_3}{\mathrm{CH_3CH_2CHCH_3}} \xrightarrow[\text{heat}]{\text{E2}} \underset{(5\%)}{\mathrm{CH_3CH{=}CHCH_3}} + \underset{(95\%)}{\mathrm{CH_3CH_2CH{=}CH_2}} + (\mathrm{CH_3})_3\mathrm{N} + \mathrm{H_2O}$$

Elimination reactions that give the less substituted alkene as the major product are said to follow the **Hofmann rule**.

Hofmann rule

Predicts that β-elimination will occur preferentially to give the less substituted alkene as the major product.

Example 23.14 Hofmann Elimination

Draw a structural formula of the major alkene formed in each β-elimination.

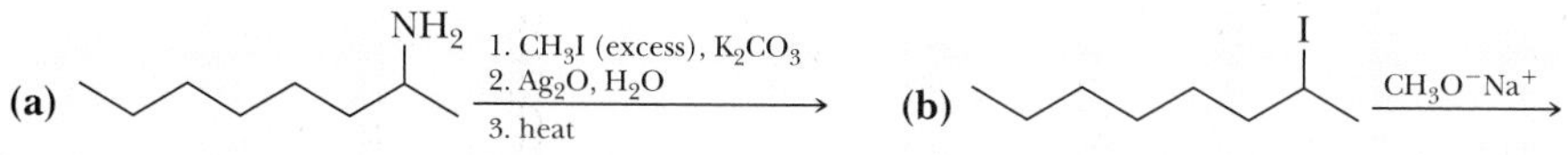

(racemic) (racemic)

Solution

Thermal decomposition of the quaternary ammonium hydroxide in (a) follows the Hofmann rule and gives 1-octene as the major product. E2 elimination from an alkyl iodide in (b) by sodium methoxide follows Zaitsev's rule and gives *trans*-2-octene as the major product.

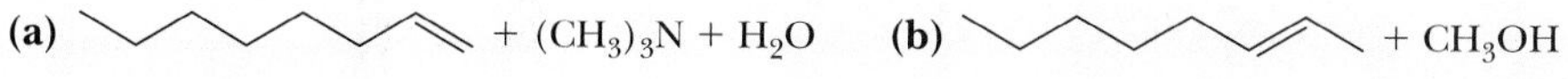

1-Octene ***trans*-2-Octene**

Problem 23.14

The procedure of methylation of amines and thermal decomposition of quaternary ammonium hydroxides was first reported by Hofmann in 1851, but its value as a means of structure determination was not appreciated until 1881, when he published a report of its use to determine the structure of piperidine. Following are the results obtained by Hofmann.

$$\underset{\textbf{Piperidine}}{\mathrm{C_5H_{11}N}} \xrightarrow[\text{3. heat}]{\text{1. CH}_3\text{I (excess), K}_2\text{CO}_3\text{; 2. Ag}_2\text{O, H}_2\text{O}} \underset{\textbf{(A)}}{\mathrm{C_7H_{15}N}} \xrightarrow[\text{6. heat}]{\text{4. CH}_3\text{I (excess), K}_2\text{CO}_3\text{; 5. Ag}_2\text{O, H}_2\text{O}} \underset{\textbf{1,4-Pentadiene}}{\mathrm{CH_2{=}CHCH_2CH{=}CH_2}}$$

(a) Show that these results are consistent with the structure of piperidine (Section 23.1).

(b) Propose two additional structural formulas (excluding stereoisomers) for $C_5H_{11}N$ that are also consistent with the results obtained by Hofmann.

In summary, both Hofmann and Zaitsev eliminations are always preferentially anti. If only one β-hydrogen is anti to the leaving group, then that one will be removed. If more than one β-hydrogen is anti, then there will be competition between Hofmann and Zaitsev elimination.

1. Eliminations involving a negatively charged leaving group (e.g., Cl^-, Br^-, I^-, and OTs^-) almost always follow Zaitsev's rule, unless a bulky base is used.
2. Eliminations involving a neutral leaving group [e.g., $N(CH_3)_3$ and $S(CH_3)_2$] almost always follow Hofmann's rule.
3. The bulkier the base, the greater the percentage of Hofmann product; compare, for example, $(CH_3)_3CO^-K^+$, which gives mostly Hofmann elimination, with $CH_3O^-Na^+$, which gives mostly Zaitsev elimination.

One of the likeliest explanations for the formation of the less stable carbon-carbon double bond is that Hofmann elimination is governed largely by steric factors, namely the bulk of the $—NR_3^+$ group. The hydroxide ion preferentially approaches and removes the least hindered α-hydrogen and gives the least substituted alkene as product. For the same reason, bulky bases such as $(CH_3)_3CO^-K^+$ also give Hofmann elimination from haloalkanes.

23.10 Cope Elimination

Treatment of a tertiary amine with hydrogen peroxide results in oxidation of the amine to an amine oxide.

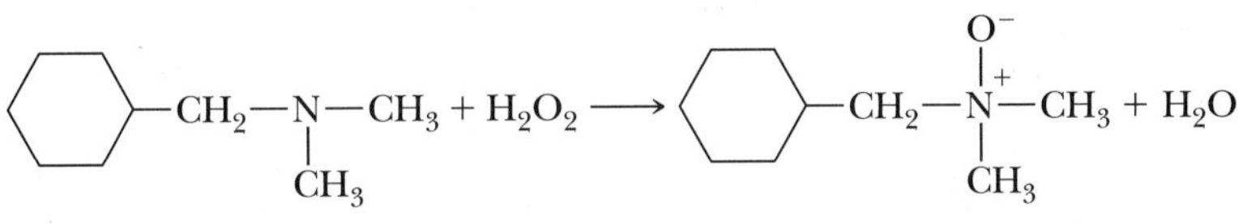

A 3° amine An amine oxide

When an amine oxide with at least one β-hydrogen is heated, it undergoes thermal decomposition to form an alkene and an *N,N*-dialkylhydroxylamine. Thermal decomposition of an amine oxide to an alkene is known as a **Cope elimination** after its discoverer Arthur C. Cope of the Massachusetts Institute of Technology.

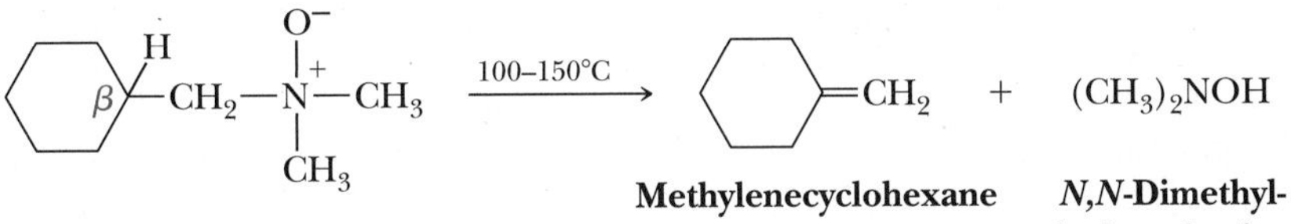

Methylenecyclohexane ***N,N*-Dimethyl-hydroxylamine**

All experimental evidence indicates that the Cope elimination is syn stereoselective and concerted.

Mechanism 23.6

The Cope Elimination

Within the same molecule, take a proton away and break a bond to give stable molecules or ions. The transition state involves a planar or nearly planar arrangement of the five participating atoms and a cyclic flow of three pairs of electrons. Elimination shows syn stereoselectivity.

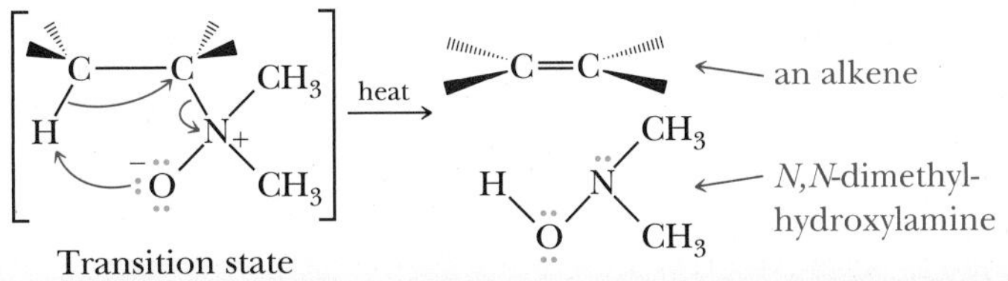

Transition state

If two or more syn β-hydrogens can be removed in a Cope elimination, there is little preference for one over another except when the double bond is conjugated with an aromatic ring. Therefore, as a method of preparation of alkenes, Cope eliminations are best used where only one alkene is possible.

Example 23.15 Cope Elimination

When 2-dimethylamino-3-phenylbutane is treated with hydrogen peroxide and then made to undergo a Cope elimination, the major alkene formed is 2-phenyl-2-butene.

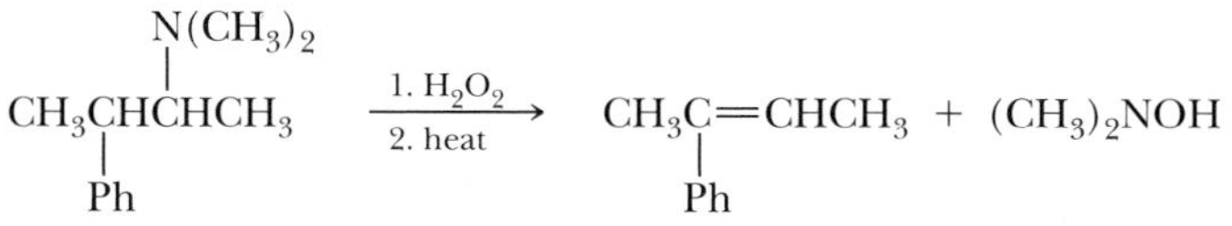

N(CH3)2
CH3CHCHCH3
Ph
1. H2O2
2. heat
CH3C=CHCH3 + (CH3)2NOH
Ph

2-Dimethylamino-3-phenylbutane **2-Phenyl-2-butene**

(a) How many stereoisomers are possible for 2-dimethylamino-3-phenylbutane?
(b) How many stereoisomers are possible for 2-phenyl-2-butene?
(c) Suppose that the starting amine is the 2*R*,3*S* isomer. What is the configuration of the product?

Solution

(a) There are two chiral centers in the starting amine. Four stereoisomers are possible: two pair of enantiomers.
(b) There is one carbon-carbon double bond about which stereoisomerism is possible. Two stereoisomers are possible: one *E*,*Z* pair.
(c) Following is a stereodrawing of the 2*R*,3*S* stereoisomer showing a syn conformation of the dimethylamino group and the β-hydrogen. Cope elimination on this stereoisomer gives (*E*)-2-phenyl-2-butene.

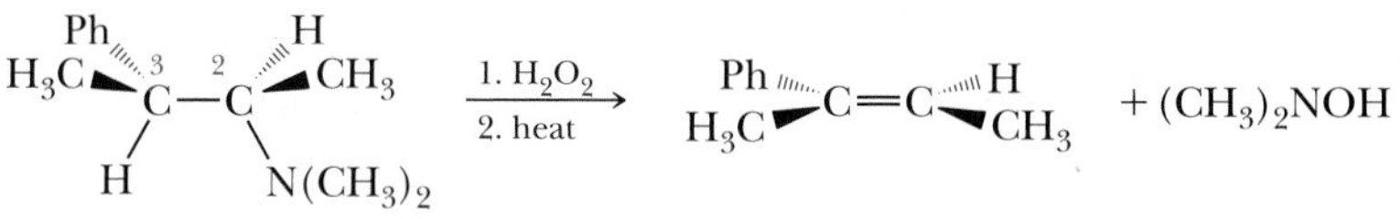

Ph
H3C
3
2
H
CH3
H
N(CH3)2
1. H2O2
2. heat
Ph
H3C
C=C
H
CH3
+ (CH3)2NOH

2*R*,3*S* isomer **(*E*)-2-Phenyl-2-butene**

Watch a video explanation

Problem 23.15

In Example 23.15, you considered the product of Cope elimination from the 2*R*,3*S* stereoisomer of 2-dimethylamino-3-phenylbutane. What is the product of a Cope elimination from the following stereoisomers? What is the product of a Hofmann elimination from each stereoisomer?

(a) 2*S*,3*R* stereoisomer **(b)** 2*S*,3*S* stereoisomer

Study Guide

23.1 Structure and Classification

- Amines are derivatives of ammonia that have one or more of the hydrogens replaced with alkyl and/or aryl groups.
  - *Primary amines have one hydrogen of ammonia replaced with a carbon in the form of an alkyl or aryl group.*
  - *Secondary amines have two hydrogens of ammonia replaced with a carbon in the form of alkyl and/or aryl groups.*
  - *Tertiary amines have all three hydrogens of ammonia replaced with a carbon in the form of alkyl and/or aryl groups.*
  - *Quaternary ammonium ions have four alkyl and/or aryl groups bonded to nitrogen, resulting in a positively charged species.*
  - ***Aliphatic amines** have alkyl groups only bonded to nitrogen, while **aromatic amines** have at least one aromatic ring bonded to the nitrogen atom.*
  - ***Heterocyclic amines** have the nitrogen atom as part of a ring, and **heterocyclic aromatic amines** have the nitrogen atom as part of an aromatic ring.*
  - ***Alkaloids** are basic nitrogen-containing compounds of plant origin, many of which are physiologically active in humans.*

P 23.1, 23.18–23.22

23.2 Nomenclature

- In IUPAC names, aliphatic amines are named like alcohols, except the suffix *-amine* is used and a number is used to locate the position of the amine group.
- IUPAC uses the common name aniline for simple derivatives of $C_6H_5NH_2$, although certain common names are retained for some substituted anilines such as toluidine and anisidine.
- Secondary and tertiary amines are named as *N*-substituted primary amines, and the largest group is taken as the parent amine; then the smaller groups bonded to nitrogen are named, given the prefix *N*. Examples are *N*-methylaniline and *N,N*-dimethylcyclopentanamine.
- Several common heterocycles retain their common names in IUPAC nomenclature, including pyridine, indole, purine, quinoline, and isoquinoline.
- Common names are derived by listing the alkyl groups bonded to the nitrogen atom in alphabetical order followed by the suffix *-amine.*
- **Quaternary ammonium ions** are named by replacing the *-amine* suffix with *-ammonium* (or *-anilinium*, etc.) and the name of the anion (e.g., tetramethylammonium chloride).

P 23.2, 23.4, 23.16, 23.17

23.3 Chirality of Amines and Quaternary Ammonium Ions

- Secondary or tertiary amines with three different groups bonded to nitrogen are chiral, but they cannot usually be resolved because, at room temperature, they undergo a process called **pyramidal inversion** that rapidly interconverts the two enantiomers.
  - *Phosphines are the phosphorus equivalents of amines, and because phosphines do not undergo pyramidal inversion at room temperature, chiral phosphines can be resolved.*
- Quaternary ammonium salts cannot undergo pyramidal inversion, so they can be resolved.

23.4 Physical Properties

- Amines are polar compounds, and primary or secondary amines can make intramolecular hydrogen bonds.
  - *As a result of the hydrogen bonding, primary and secondary amines interact with solutes primarily through hydrogen bonding, have substantially higher melting and boiling points, and are more soluble in water than analogous hydrocarbons.*
  - *Because H—N hydrogen bonds are weaker than H—O hydrogen bonds, amines have lower boiling points than analogous alcohols.*

P **23.23, 23.24**

23.5 Basicity

- Amines are weak bases, so aqueous solutions of amines are basic. The pK_a values of the conjugate acids of aliphatic amines are in the 10–11 range.
  - *Alkyl groups make amines slightly more basic because electron-releasing alkyl groups stabilize the alkylammoniun ion.*
  - *Alkyl amines are protonated and positively charged near neutral pH and in biological solutions.*
- Aromatic amines are considerably less basic than aliphatic amines, the pK_a values of their conjugate acids being in the 4–5 range.
  - *Aromatic amines are less basic because the nitrogen lone pair takes part in resonance with the aromatic π system, a stabilizing interaction that is lost upon protonation.*
    - *This resonance interaction requires that the N atom of aromatic amines be almost or entirely sp^2 hybridized and therefore planar, a situation that is critical to nucleic acid base stacking and hydrogen bonding.*
    - *Electron-releasing substituents on the ring increase the basicity of anilines; electron-withdrawing groups decrease the basicity of anilines.*
- The basicity of N atoms within heterocyclic aromatic amines depends on whether the N lone pair is part of the aromatic π system.
  - *In pyridine, the lone pair on N is not part of the π system, so protonation does not disrupt aromaticity.*
    - *The pK_a of the conjugate acid of pyridine is 5.25, lower than that of an alkyl amine because the pyridine lone pair is in a relatively electronegative sp^2 orbital.*

– *In imidazole, only one N atom can be protonated because the lone pair on the other N atom is part of the aromatic π system and aromaticity would be lost upon protonation.*

– *In pyrrole, the lone pair on the ring N atom is part of the aromatic π system. Pyrrole is a very poor base because protonation on N would disrupt the aromaticity.*

- Guanidine groups are very basic organic groups because the protonated guanidinium ion is highly stabilized by charge delocalization.

P 23.5–23.6, 23.25–23.29, 23.31, 23.32

▶ Problem 23.6

KEY REACTIONS

1. Basicity of Aliphatic Amines (Section 23.5A) Aliphatic amines are slightly stronger bases than ammonia, owing to the electron-releasing effect of alkyl groups and partial delocalization of positive charge in the alkylammonium ion.

$$CH_3NH_2 + H_2O \rightleftharpoons CH_3NH_3^+ + OH^- \qquad pK_a = 10.64$$

2. Basicity of Aromatic Amines (Section 23.5B) Aromatic amines are considerably weaker bases than aliphatic amines. Resonance stabilization by interaction of the unshared electron pair on nitrogen with the π system decreases its availability for protonation.

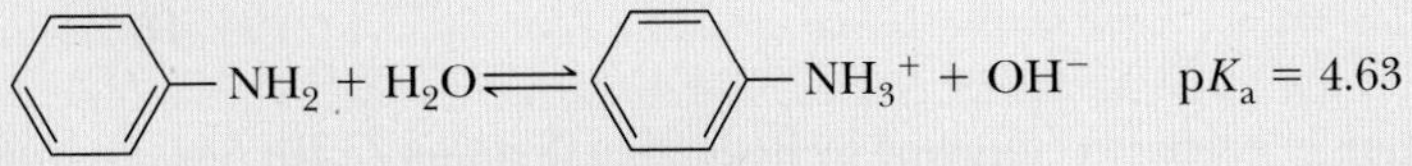
NH2 + H2O
NH3+ + OH−
pKa = 4.63

3. Basicity of Heterocyclic Aromatic Amines (Section 23.5C) Heterocyclic aromatic amines are considerably weaker bases than aliphatic amines.

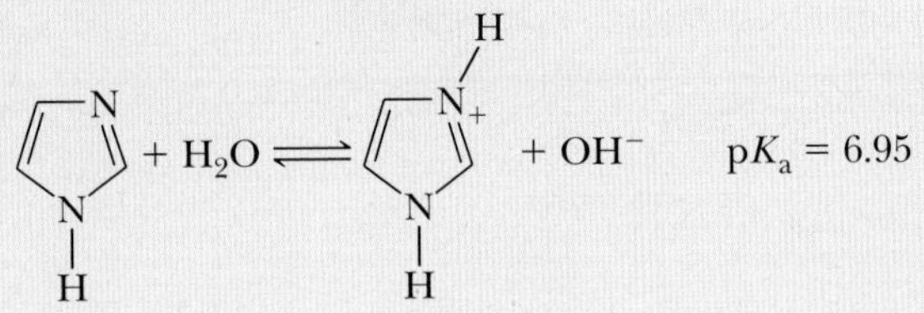
+ H2O
+ OH−
pKa = 6.95

23.6 Reactions with Acids

- Amines react with strong acids to give water-soluble salts, allowing the separation of amines from water-insoluble molecules.

P 23.7–23.9, 23.30

KEY REACTIONS

4. Reaction of Amines with Strong Acids (Section 23.6) All amines react quantitatively with strong acids to form water-soluble salts.

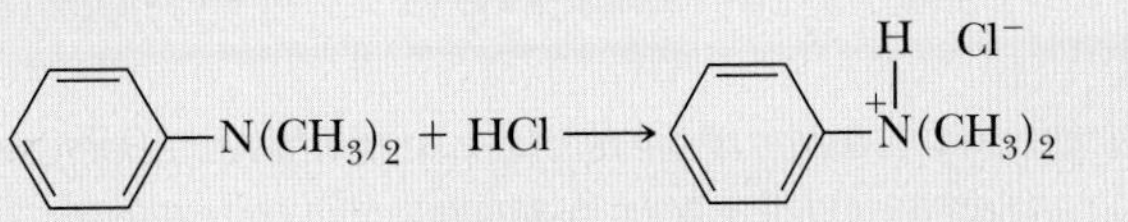
N(CH3)2 + HCl
H Cl−
N(CH3)2

Insoluble in water — A water-soluble salt

23.7 Preparation

- Amines can be prepared using reactions including epoxide ring opening, addition of nitrogen nucleophiles to carbonyls followed by reduction, reduction of amides, reduction of nitriles, and nitration of arenes followed by reduction.
- Alkylation of amines generally results in overalkylation.
- Primary amines can be prepared in high yield by reacting a haloalkane with the strong nucleophile (and weak base) sodium or potassium azide followed by reduction with $LiAlH_4$.
  - *Azides can also be used to ring open epoxides followed by reduction to give amino alcohols with* trans *stereoselectivity.*

P 23.10, 23.33, 23.34, 23.43–23.45, 23.54–23.57, 23.62–23.69, 32.71, 23.72

KEY REACTIONS

5. Alkylation of Ammonia and Amines (Section 23.7A) This method is seldom used for preparation of pure amines because of overalkylation and the difficulty of separating products.

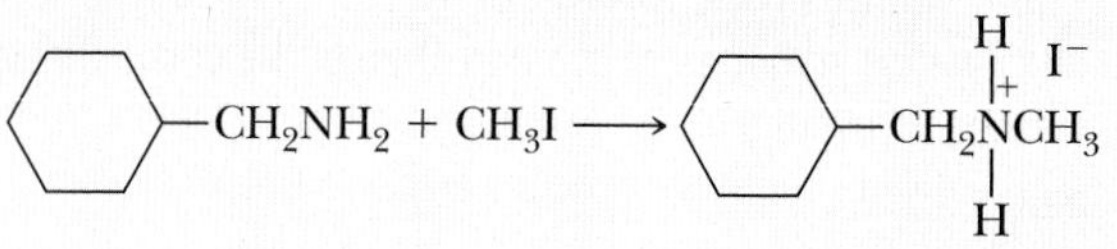

6. Alkylation of Azide Ion Followed by Reduction (Section 23.7B) Azides are prepared by treatment of a primary or secondary haloalkane or an epoxide with KN_3 and are reduced to primary amines by a variety of reducing agents, including lithium aluminum hydride.

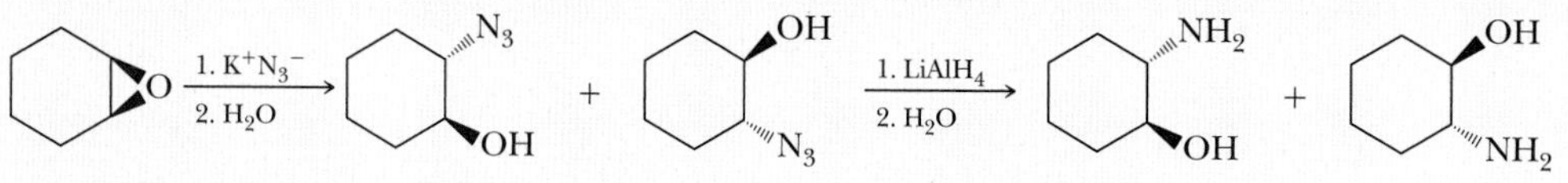

(racemic) (racemic)

23.8 Reaction with Nitrous Acid

- Nitrous acid, often prepared *in situ* by reacting $NaNO_2$ with acid, reacts with amines in different ways depending on whether they are primary, secondary, tertiary, or aromatic.
  - *Nitrous acid participates in proton transfer reactions and is a source of the nitrosyl cation, a weak but very important electrophile.*
- Tertiary aromatic amines can undergo electrophilic aromatic substitution with nitrous acid.
- Secondary amines react with nitrous acid to give *N*-nitrosamines.
- Primary amines react with nitrous acid to give **diazonium ion** intermediates that lose N_2 and give a variety of substitution and elimination products, so the reaction is not generally synthetically useful.
  - *A synthetically useful version of the reaction is the Tiffeneau-Demjanov reaction, which gives a one-carbon ring expansion of a cyclic β-aminoalcohol and produces a cyclic ketone.*

- Primary aromatic amines (anilines) react with nitrous acid to give aryl diazonium ions, which are very versatile and useful synthetic intermediates in the synthesis of a variety of substituted aromatic rings. Recall that aromatic amines can be prepared through the combination of nitration followed by reduction.
  - *Reaction of aryl diazonium ions with water gives phenols.*
  - *Reaction of aryl diazonium ions with HBF_4 gives aryl fluorides in a reaction known as the* ***Schiemann reaction****.*
  - *Reaction of aryl diazonium ions with $HCl/CuCl$, $HBr/CuBr$, or $KCN/CuCN$ replaces the diazonium group with a —Cl, —Br, or —CN group, respectively, in a reaction known as the* ***Sandmeyer reaction****.*
  - *Reaction of aryl diazonium ions with KI gives aryl iodides.*
  - *Reaction of aryl diazonium ions with hypophosphorous acid (H_3PO_2) replaces the diazonium group with an H atom so that the reaction sequence of reacting an aniline with nitrous acid followed by hypophosphorous acid will remove the aryl NH_2 group.*

P 23.11, 23.36–23.38, 23.12, 23.13, 23.45–23.51, 23.56, 23.61, 23.67–23.70, 23.72

KEY REACTIONS

7. **Nitrosation of Tertiary Aromatic Amines (Section 23.8B)** The nitrosyl cation is a very weak electrophile and participates in electrophilic aromatic substitution only with highly activated aromatic rings.

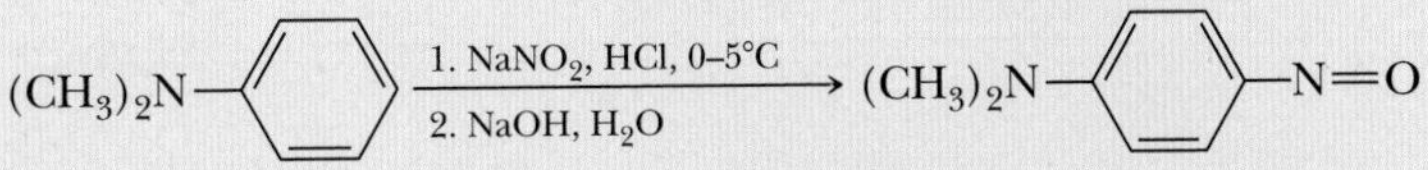

8. **Formation of *N*-Nitrosamines from Secondary Amines (Section 23.8C)** Reaction of the nitrosyl cation (an electrophile) with a 2° amine (a nucleophile) gives an *N*-nitrosamine.

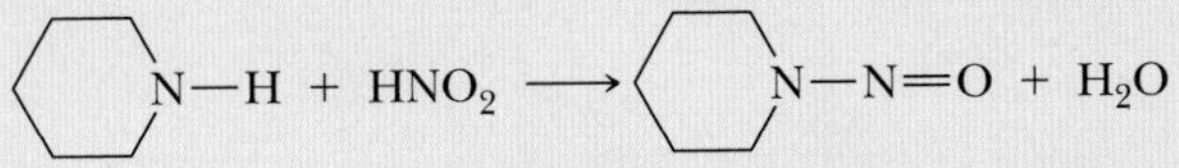

9. **Reaction of Primary Aliphatic Amines with Nitrous Acid (Section 23.8D)** Treating a primary aliphatic amine with nitrous acid gives an unstable diazonium salt that loses N_2 to give a carbocation. The carbocation may (1) lose a proton to give an alkene, (2) react with a nucleophile, or (3) rearrange, followed by (1) or (2). The mechanism of diazonium ion formation involves reaction of the primary amine (a nucleophile) with the nitrosyl cation to give an *N*-nitrosamine that tautomerizes to the more stable diazotic acid, which is protonated on the O atom, followed by loss of water to give a diazonium ion.

$$RCH_2NH_2 \xrightarrow[\text{0–5°C}]{NaNO_2,\ HCl} [RCH_2N_2^{\,+}] \longrightarrow RCH_2^{\,+} + N_2$$

10. **Reaction of Cyclic β-Aminoalcohols with Nitrous Acid (Section 23.8D)** Treating a cyclic β-aminoalcohol with nitrous acid leads to rearrangement and a ring-expanded ketone. The mechanism involves an initial formation of a diazonium ion followed by simultaneous loss of N_2 and rearrangement by a 1,2-shift to give a resonance-stabilized cation that loses a proton to give the product cyclic ketone.

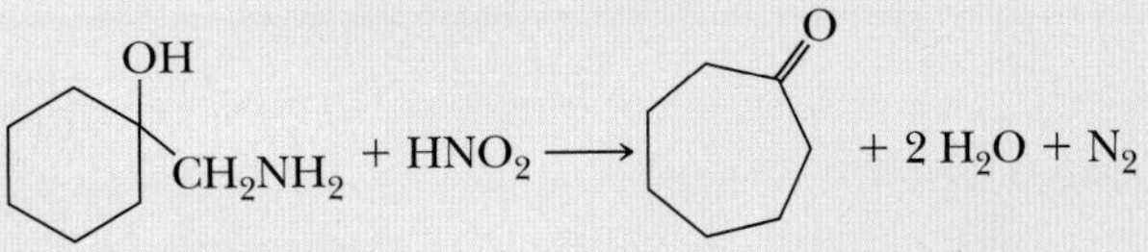

11. Formation of Arenediazonium Salts (Diazotization) (Section 23.8E) Arenediazonium salts are stable in aqueous solution at 0°C for short periods.

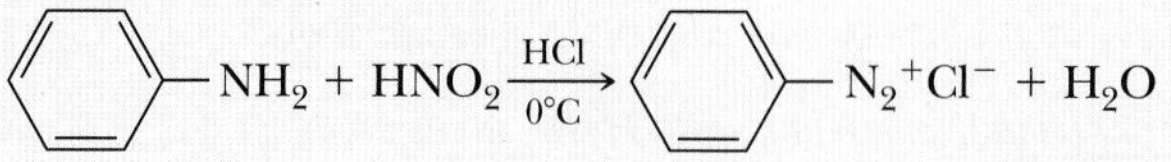

The mechanism is analogous to diazonium ion formation using nitrous acid and aliphatic primary amines.

12. Conversion of a Primary Arylamine to a Phenol (Section 23.8E) Formation of an arenediazonium salt followed by loss of nitrogen gives an aryl cation intermediate, which then reacts with water to give a phenol.

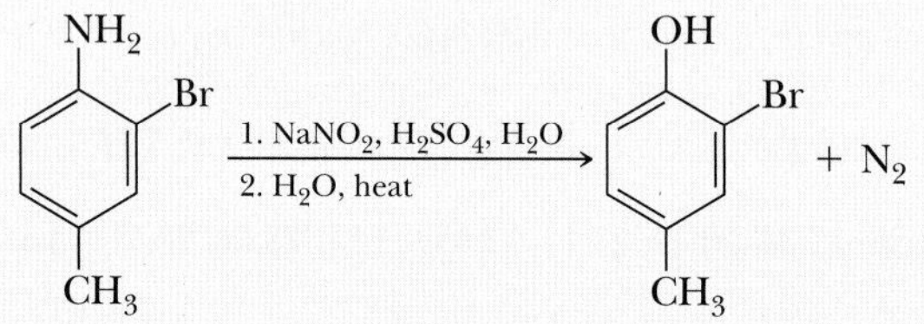

13. Schiemann Reaction (Section 23.8E) Heating an arenediazonium fluoroborate is the most common synthetic method for introduction of fluorine onto an aromatic ring.

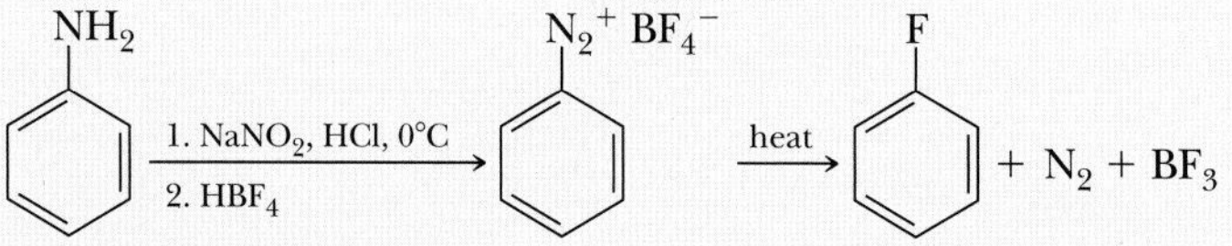

14. Sandmeyer Reaction (Section 23.8E) Treatment of an arenediazonium salt with CuCl, CuBr, or CuCN results in replacement of the diazonium group by —Cl, —Br, or —CN, respectively.

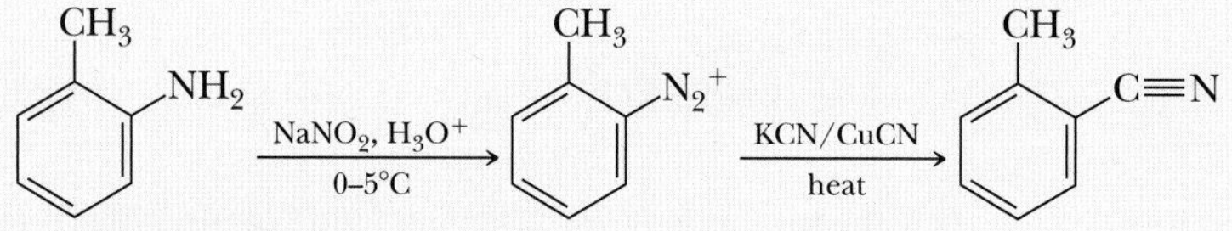

15. Reaction of an Arenediazonium Salt with KI (Section 23.8E) Treatment of an arenediazonium salt with KI is the most convenient method for introducing iodine onto an aromatic ring.

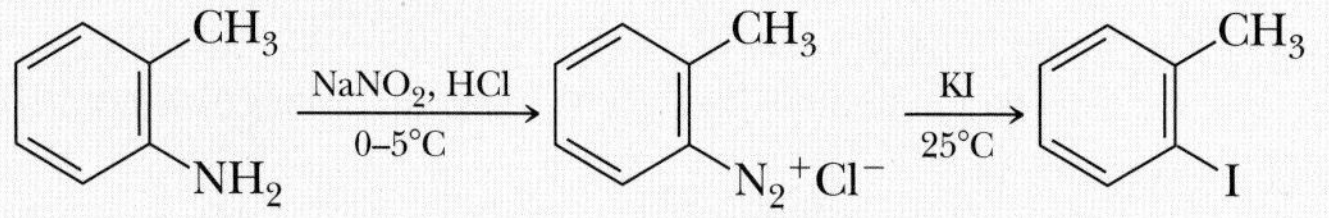

16. Reduction of an Arenediazonium Salt with Hypophosphorous Acid (Section 23.8E) An $—NO_2$ or $—NH_2$ group can be used to control orientation of further substitution and then removed after it has served its purpose.

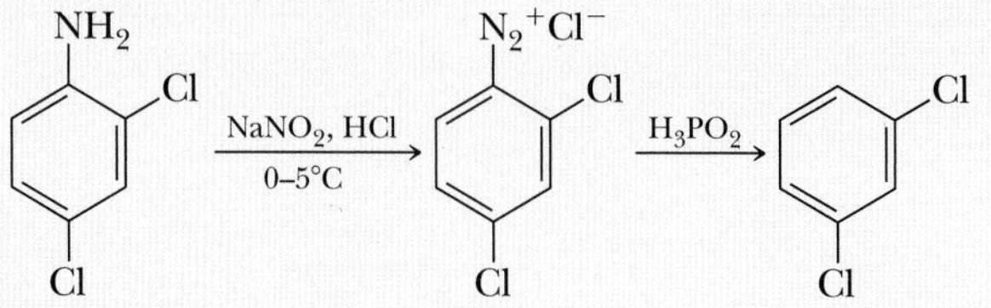

23.9 Hofmann Elimination

- Reaction of a quaternary ammonium halide with moist silver oxide to produce a quaternary ammonium hydroxide followed by heating to give an alkene is a reaction known as the **Hofmann elimination**.
  - *Hofmann elimination reactions are stereoselective for anti eliminations and give predominantly the least substituted alkene, an observation that is counter to Zaitsev's rule.*
  - *Hofmann elimination regiochemistry is thought to derive from the steric bulk of the ammonium group that directs deprotonation by base to the least hindered site, leading to formation of the less substituted alkene.*
  - *Eliminations observed to produce primarily the least substituted alkene are said to be following the **Hofmann rule**.*

P **23.14, 23.39, 23.41**

KEY REACTIONS

17. Hofmann Elimination (Section 23.9) Anti stereoselective elimination of quaternary ammonium hydroxides occurs preferentially to form the least substituted carbon-carbon double bond (Hofmann's rule). The mechanism involves the simultaneous deprotonation of a β-hydrogen by base and loss of the amino group in an anti geometry.

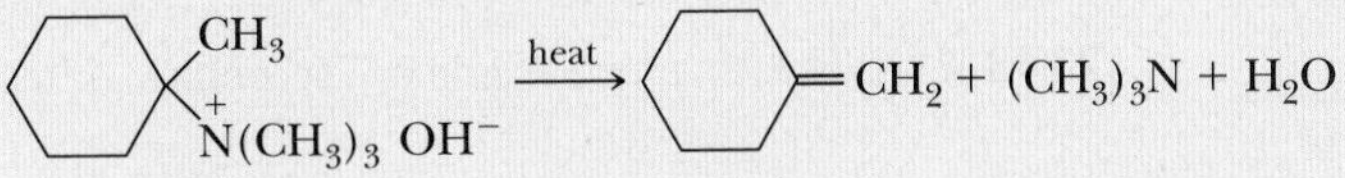

23.10 Cope Elimination

- Treatment of a tertiary amine with hydrogen peroxide gives an amine oxide, which when heated gives an alkene and an *N,N*-dialkylhydroxyamine in a reaction known as a **Cope elimination**.
  - *The Cope elimination is syn stereoselective and gives little preference for regiochemistry unless a conjugated double bond can be created, in which case the conjugated product predominates.*

P **23.15, 23.40, 23.42**

Problem 23.15

KEY REACTIONS

18. Cope Elimination: Pyrolysis of a Tertiary Amine Oxide (Section 23.10) Elimination is syn stereoselective and involves a cyclic flow of six electrons in a planar transition state.

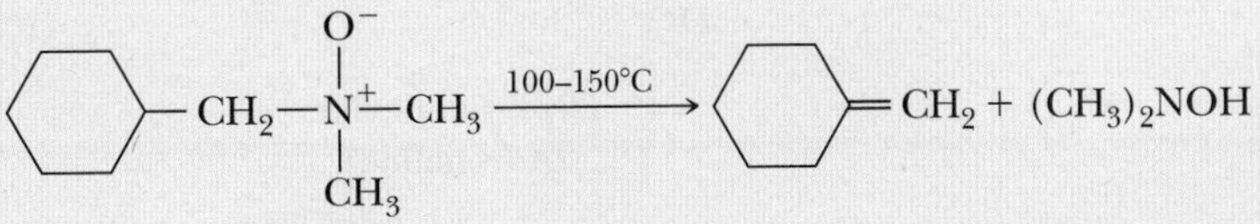

Problems

Red numbers indicate applied problems.

Structure and Nomenclature

23.16 Draw a structural formula for each amine and amine derivative.

(a) *N,N*-Dimethylaniline
(b) Triethylamine
(c) *tert*-Butylamine
(d) 1,4-Benzenediamine
(e) 4-Aminobutanoic acid
(f) (*R*)-2-Butanamine
(g) Benzylamine
(h) *trans*-2-Aminocyclohexanol
(i) 1-Phenyl-2-propanamine (amphetamine)
(j) Lithium diisopropylamide (LDA)
(k) Benzyltrimethylammonium hydroxide (Triton B)

23.17 Give an acceptable name for these compounds.

(a) 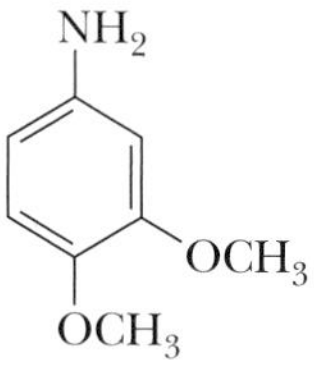
(b) 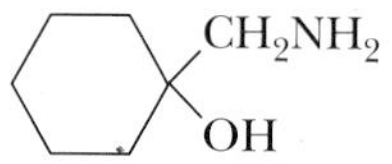
(c) 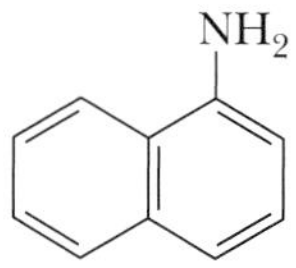
(d) 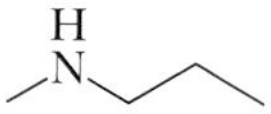
(e) 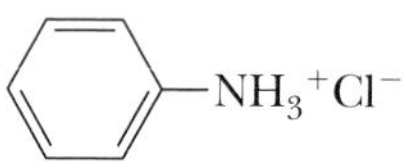
(f) 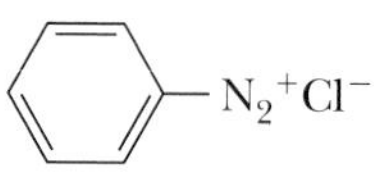
(g) 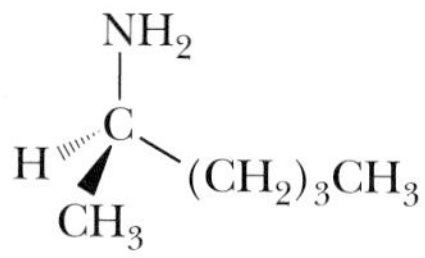
(h) 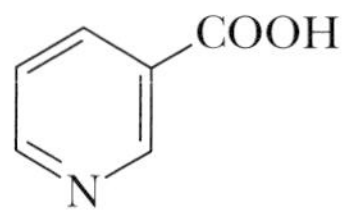

23.18 Classify each amine as primary, secondary, or tertiary and as aliphatic or aromatic.

(a) 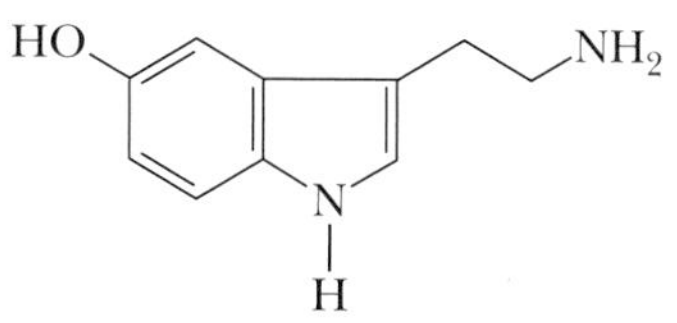

Serotonin
(a neurotransmitter)

(b) 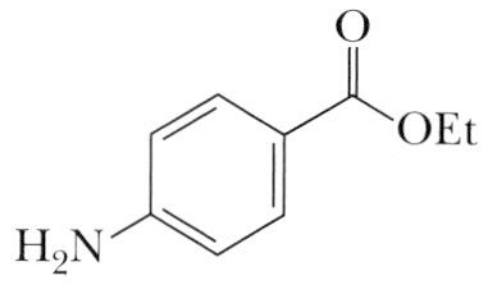

Benzocaine
(a topical anesthetic)

(c) 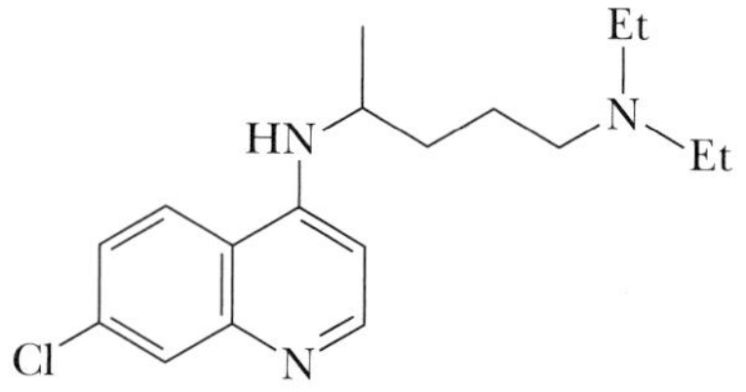

Chloroquine
(an antimalarial racemic)

23.19 Epinephrine is a hormone secreted by the adrenal medulla. Among its actions, it is a bronchodilator. Albuterol, sold under several trade names, including Proventil and Salbumol, is one of the most effective and widely prescribed antiasthma drugs. The *R* enantiomer of albuterol is 68 times more effective in the treatment of asthma than the *S* enantiomer.

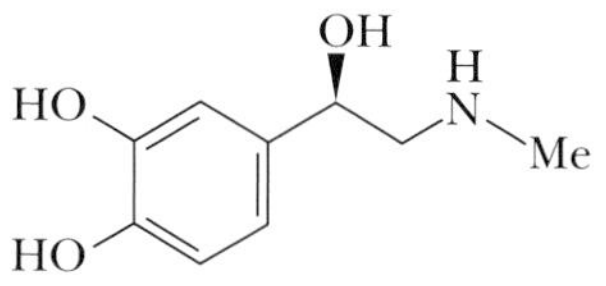

(*R*)-Epinephrine
(Adrenaline)

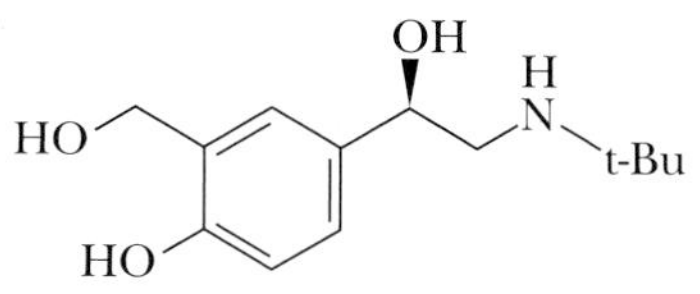

(*R*)-Albuterol

(a) Classify each amine as a primary, secondary, or tertiary.
(b) Compare the similarities and differences between their structural formulas.

23.20 Draw the structural formula for a compound with the given molecular formula.

(a) A 2° arylamine, C_7H_9N
(b) A 3° arylamine, $C_8H_{11}N$
(c) A 1° aliphatic amine, C_7H_9N
(d) A chiral 1° amine, $C_4H_{11}N$
(e) A 3° heterocyclic amine, $C_6H_{11}N$
(f) A trisubstituted 1° arylamine, $C_9H_{13}N$
(g) A chiral quaternary ammonium salt, $C_6H_{16}NCl$

23.21 Morphine and its *O*-methylated derivative codeine are among the most effective painkillers known. However, they possess two serious drawbacks: they are addictive, and repeated use induces a tolerance to the drug. Many morphine analogs have been prepared in an effort to find drugs that are equally effective as painkillers but that have less risk of physical dependence and potential for abuse. Following are several of these.

R = H; **Morphine**
R = CH_3; **Codeine**

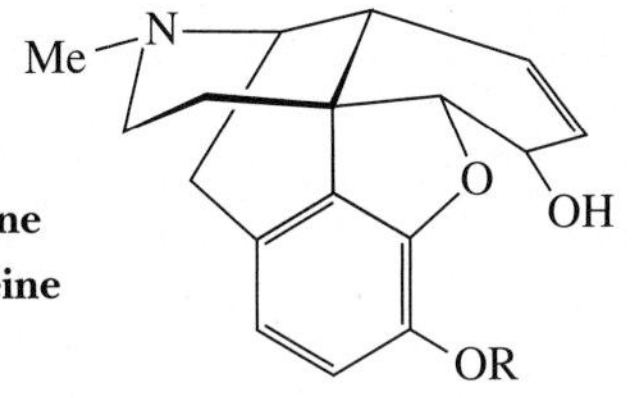

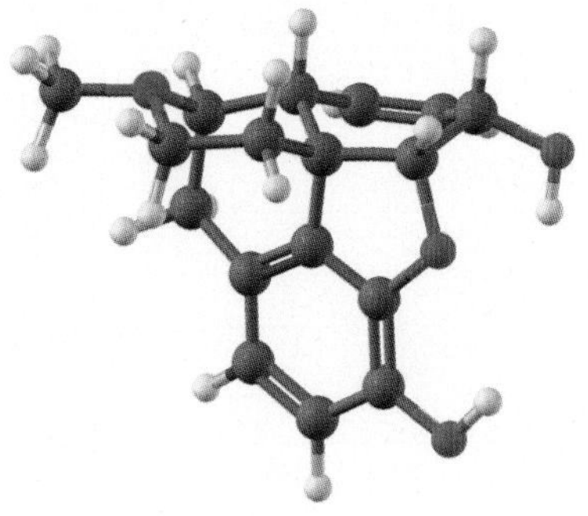

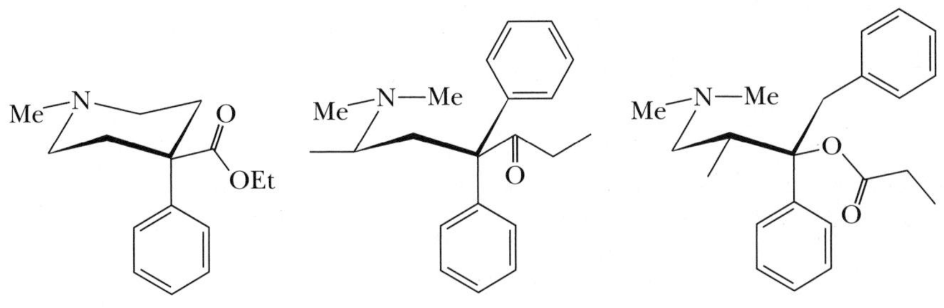

Meperidine (Demerol) **Methadone** **Propoxyphene (Darvon)**

(a) List the structural features common to each of these molecules.
(b) The Beckett-Casey rules are a set of empirical rules used to predict the structure of molecules that bind to morphine receptors and act as analgesics. According to these rules, to provide an effective morphine-like analgesia, a molecule must have (1) an aromatic ring attached to (2) a quaternary carbon and (3) a nitrogen at a distance equal to two carbon-carbon single bond lengths from the quaternary center. Show that these structural requirements are present in the molecules given in this problem.

23.22 Following is a structural formula of desosamine, a sugar component of several macrolide antibiotics, including the erythromycins. The configuration shown is that of the natural or D isomer. Erythromycin is produced by a strain of *Streptomyces erythreus* found in a soil sample from the Philippine archipelago.

Desosamine

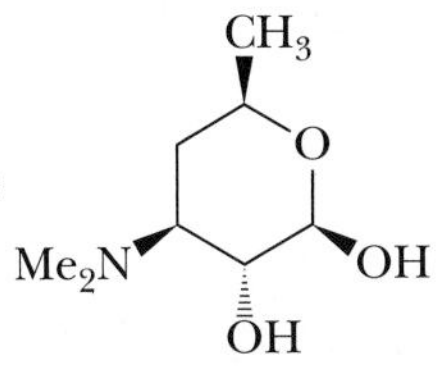

(a) Name all functional groups in desosamine.
(b) How many chiral centers are present in desosamine? How many stereoisomers are possible for it? How many pairs of enantiomers are possible for it?
(c) Draw the alternative chair conformations for desosamine. In each, label which groups are equatorial and which are axial.
(d) Which of the alternative chair conformations for desosamine is more stable? Explain.

Spectroscopy

23.23 Account for the formation of the base peaks in these mass spectra.

(a) Isobutylmethylamine, *m/z* 44 **(b)** Diethylamine, *m/z* 58

23.24 Propose a structural formula for compound A, $C_5H_{13}N$, given its IR and ^{1}H-NMR spectra.

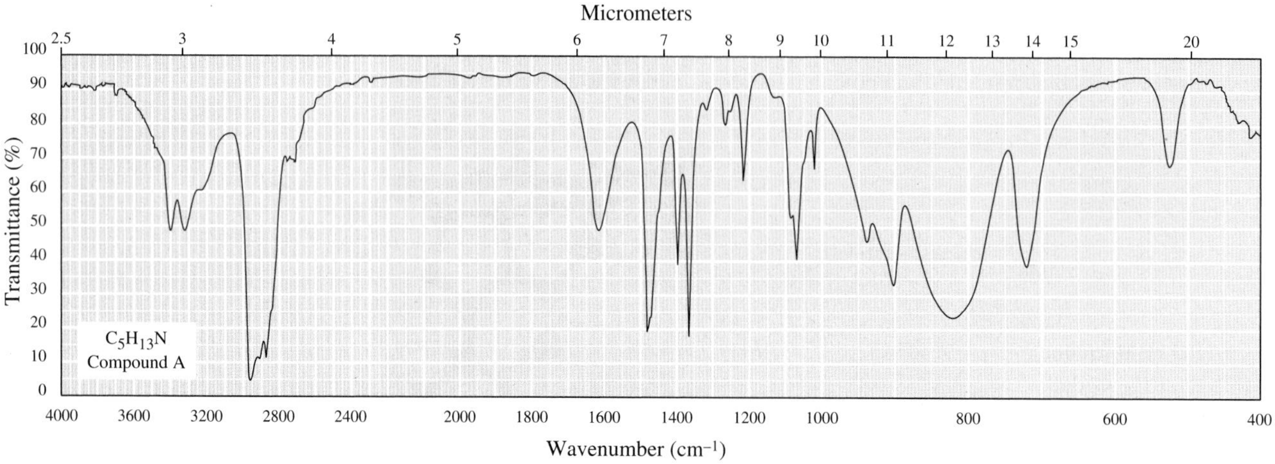

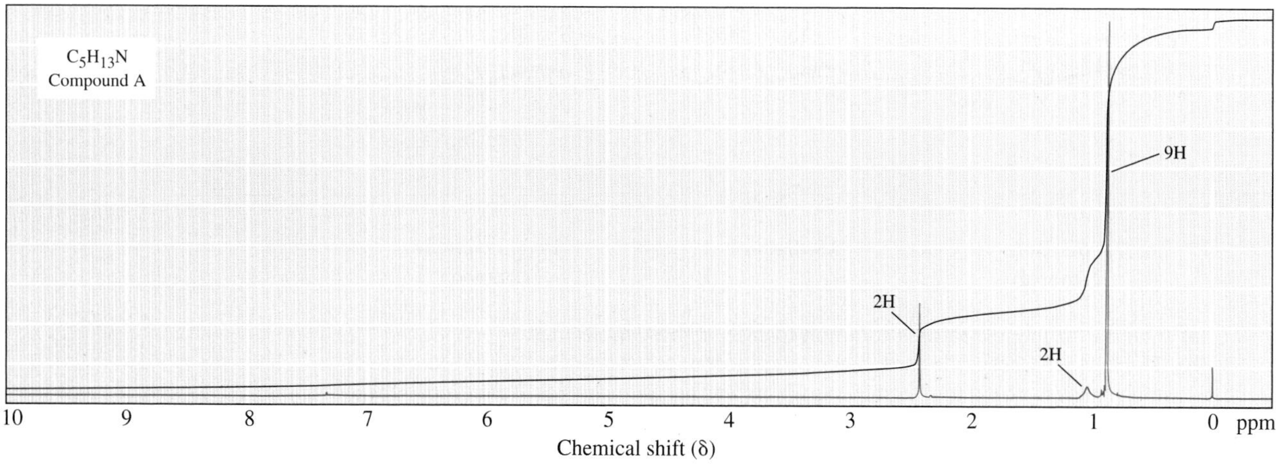

Basicity of Amines

23.25 Select the stronger base from each pair of compounds.

(a) 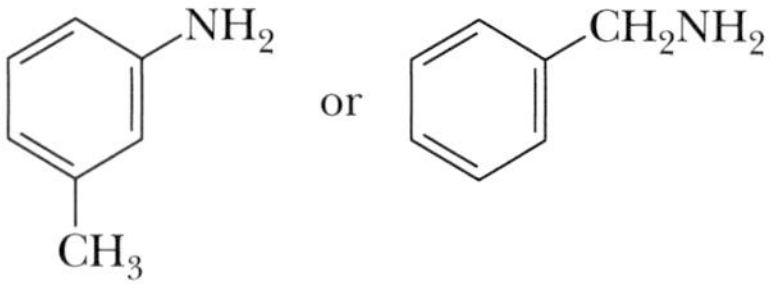

(b) 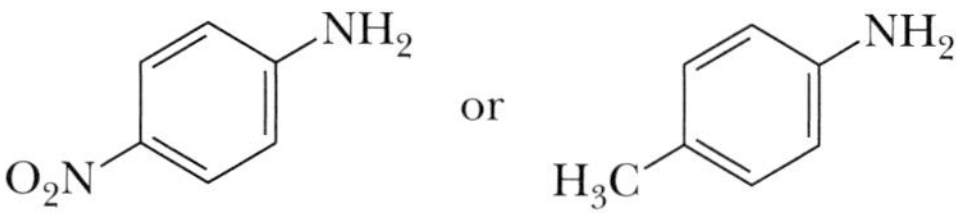

(c) 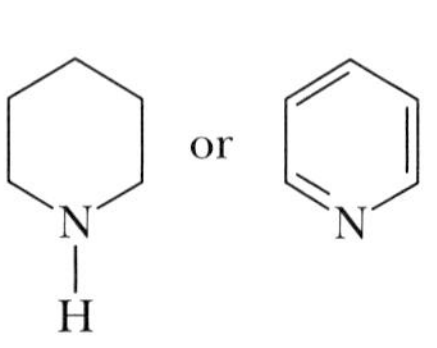

(d) $H_2NC(=O)NH_2$ or $H_2NC(=NH)NH_2$

(e) 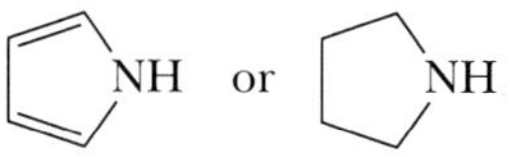

(f) Et_3N or 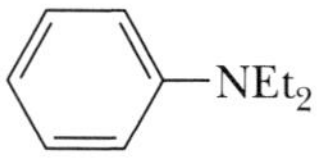

23.26 The pK_a of the conjugate acid of morpholine is 8.33.

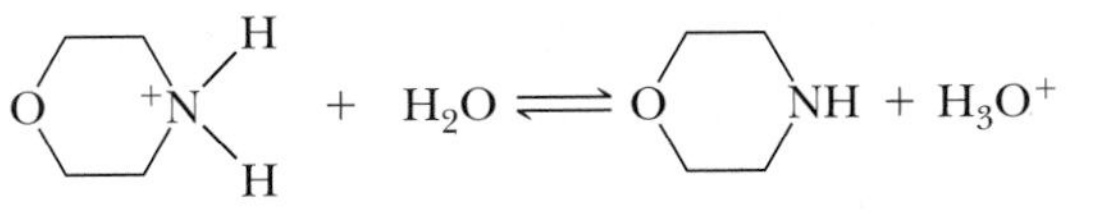

$pK_a = 8.33$

Morpholinium ion **Morpholine**

(a) Calculate the ratio of morpholine to morpholinium ion in aqueous solution at pH 7.0.
(b) At what pH are the concentrations of morpholine and morpholinium ion equal?

23.27 Which of the two nitrogens in pyridoxamine (a form of vitamin B_6) is the stronger base? Explain your reasoning.

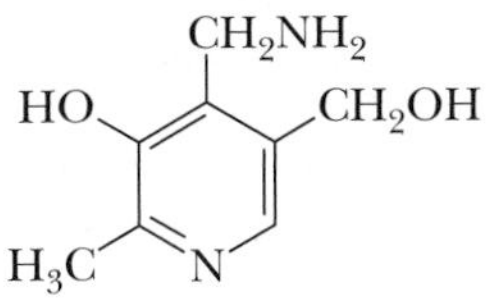

Pyridoxamine
(Vitamin B_6)

23.28 Epibatidine, a colorless oil isolated from the skin of the Ecuadorian poison frog *Epipedobates tricolor,* has several times the analgesic potency of morphine. It is the first chlorine-containing, nonopioid (nonmorphine-like in structure) analgesic ever isolated from a natural source.

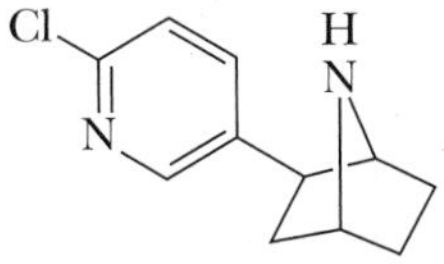

Epibatidine

(a) Which of the two nitrogen atoms of epibatidine is more basic?
(b) Mark all chiral centers in this molecule.

23.29 Aniline (conjugate acid pK_a 4.63) is a considerably stronger base than diphenylamine (pK_a 0.79). Account for these marked differences.

23.30 Complete the following acid-base reactions and predict the direction of equilibrium (to the right or to the left) for each. Justify your prediction by citing values of pK_a for the stronger and weaker acid in each equilibrium. For values of acid ionization constants, consult Table 23.2 (Acid Strengths, pK_a, of the Conjugate Acids of Selected Amines) and Appendix 2 (Acid Ionization Constants for the Major Classes of Organic Acids). Where no ionization constants are given, make the best estimate from the information given in the reference tables and sections.

(a) CH_3COOH + 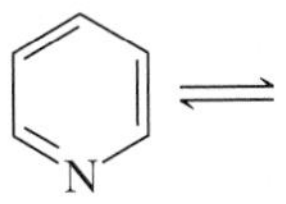 ⇌

Acetic acid **Pyridine**

(b) 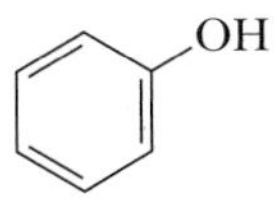 + Et_3N ⇌

Phenol **Triethylamine**

(c) $PhC{\equiv}CH$ + NH_3 ⇌

Phenylacetylene **Ammonia**

(d) $PhC{\equiv}CH$ + $i\text{Pr}_2\text{N}^-\text{Li}^+$ ⇌

Phenylacetylene **Lithium diisopropylamide (LDA)**

(e) $PhCO^{-}Na^{+}$ (C=O) + $Et_3NH^{+}Cl^{-}$ $\rightleftharpoons$

Sodium benzoate | **Triethylammonium chloride**

(f)

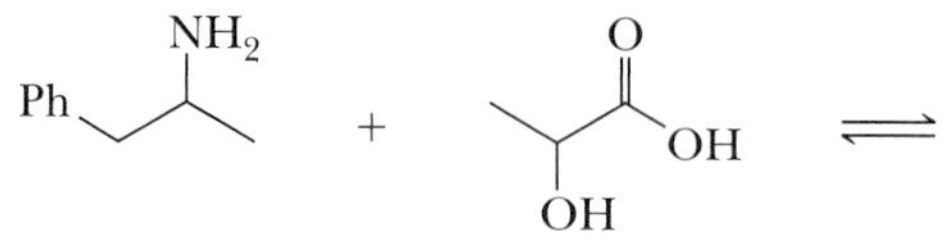

Ph
NH_2
+
O
OH
OH

1-Phenyl-2-propanamine (Amphetamine) | **2-Hydroxypropanoic acid (Lactic acid)**

(g)

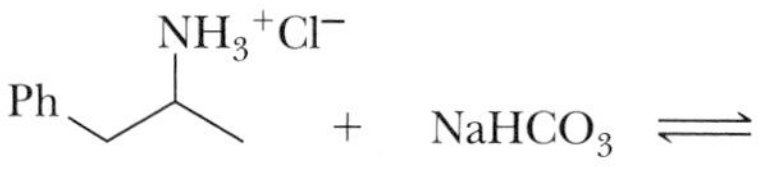

Ph
$NH_3^{+}Cl^{-}$
+
$NaHCO_3$

Amphetamine hydrochloride | **Sodium bicarbonate**

(h)

OH
+
$Me_4N^{+}OH^{-}$

Phenol | **Tetramethylammonium hydroxide**

23.31 Quinuclidine and triethylamine are both tertiary amines. Quinuclidine, however, is a considerably stronger base than triethylamine. Stated alternatively, the conjugate acid of quinuclidine is a considerably weaker acid than the conjugate acid of triethylamine.

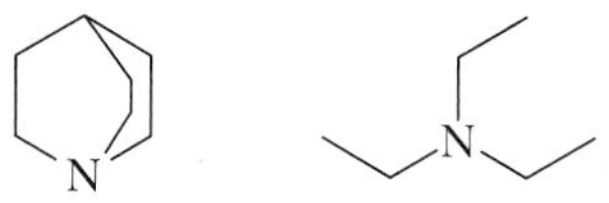

N
N

Quinuclidine pK_a 10.6 | **Triethylamine** pK_a 8.6

Propose an explanation for these differences in acidity/basicity.

23.32 Suppose you have a mixture of these three compounds. Devise a chemical procedure based on their relative acidity or basicity to separate and isolate each in pure form.

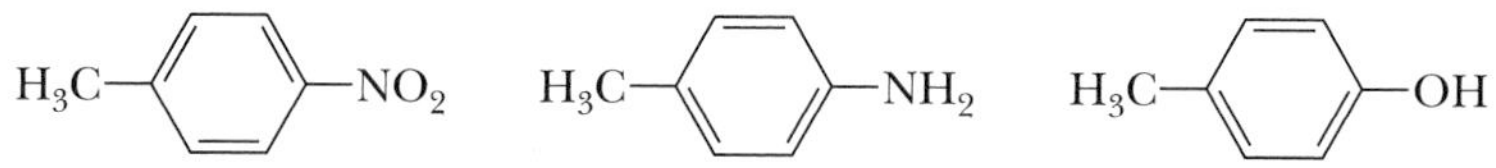

H_3C
NO_2
H_3C
NH_2
H_3C
OH

4-Nitrotoluene | **4-Methylaniline (*p*-Toluidine)** | **4-Methylphenol (*p*-Cresol)**

Preparation of Amines

23.33 Propose a synthesis of 1-hexanamine from the following.

(a) A bromoalkane of six carbon atoms

(b) A bromoalkane of five carbon atoms

23.34 Show how to convert each starting material into benzylamine in good yield.

(a) 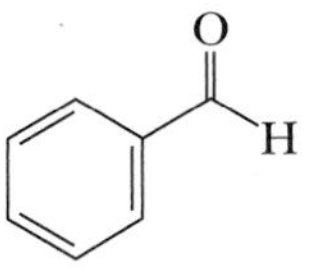
(b) 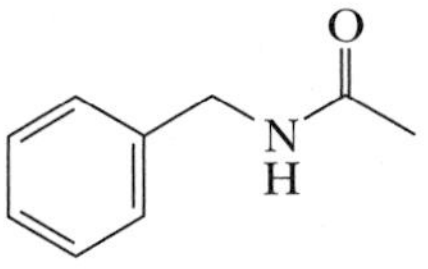
(c) 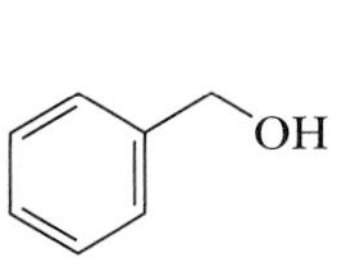

(d) 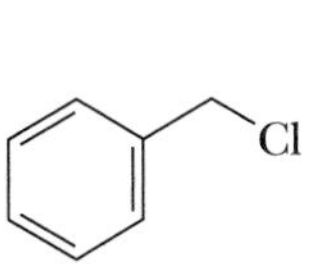
(e) 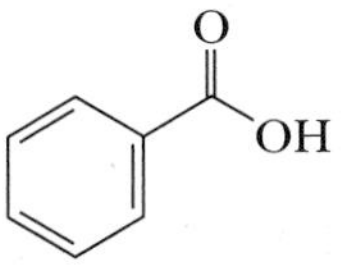
(f) 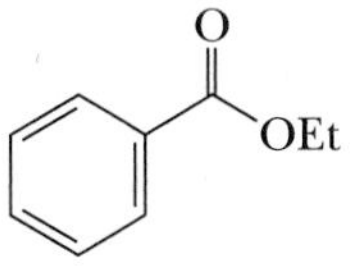

Reactions of Amines

23.35 Treating trimethylamine with 2-chloroethyl acetate gives acetylcholine as its chloride. Acetylcholine is a neurotransmitter.

$$Me_3N + CH_3\overset{\overset{\displaystyle O}{\|}}{C}OCH_2CH_2Cl \longrightarrow C_7H_{16}ClNO_2$$

Acetylcholine chloride

Propose a structural formula for this quaternary ammonium salt and a mechanism for its formation.

23.36 *N*-Nitrosamines by themselves are not significant carcinogens. However, they are activated in the liver by a class of iron-containing enzymes (members of the cytochrome P-450 family). Activation involves the oxidation of a C—H bond next to the amine nitrogen to a C—OH group.

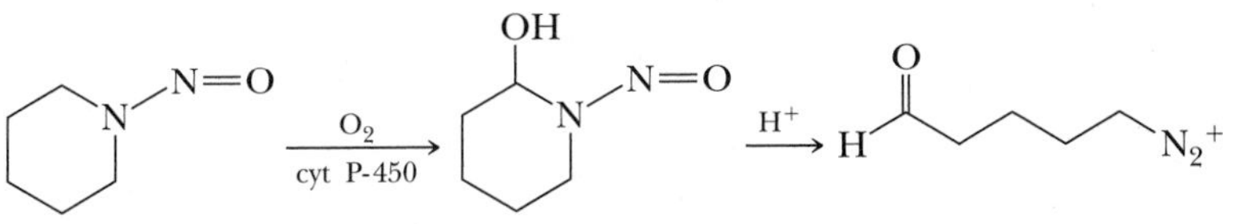

***N*-Nitroso-piperidine** **2-Hydroxy-*N*-nitrosopiperidine** An alkyl diazonium ion (a carcinogen)

Show how this hydroxylation product can be transformed into an alkyl diazonium ion, an active alkylating agent and therefore a carcinogen, in the presence of an acid catalyst.

23.37 Marked similarities exist between the mechanism of nitrous acid deamination of β-aminoalcohols and the pinacol rearrangement. Following are examples of each.

Nitrous acid deamination of a β-aminoalcohol:

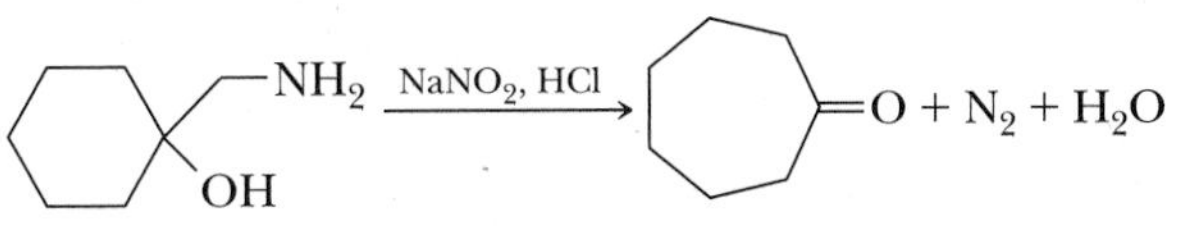

Pinacol rearrangement:

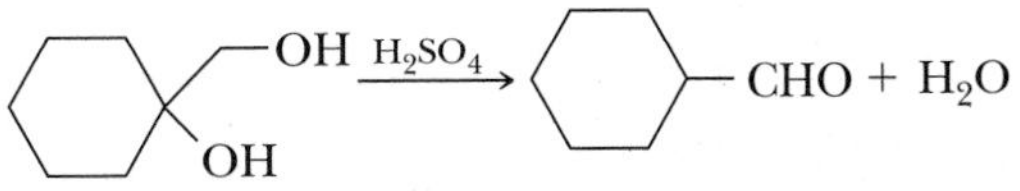

(a) Analyze the mechanism of each rearrangement and list their similarities.
(b) Why does the first reaction, but not the second, give ring expansion?
(c) Suggest a β-aminoalcohol that would give cyclohexanecarbaldehyde as a product.

23.38 (S)-Glutamic acid is one of the 20 amino acid building blocks of polypeptides and proteins (Chapter 27). Propose a mechanism for the following conversion.

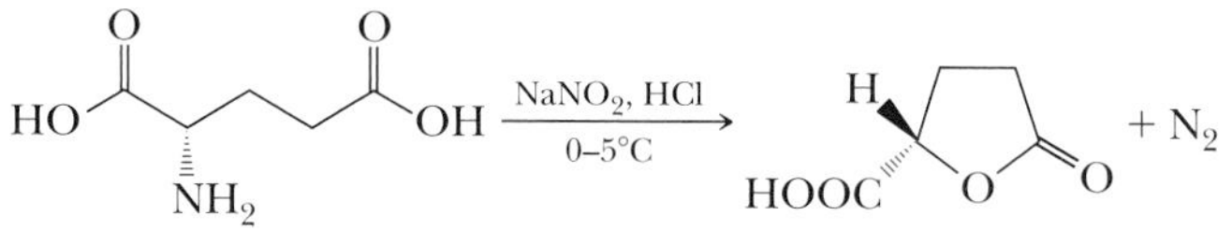
HO
OH
O
O
NH2
NaNO2, HCl
0–5°C
H
HOOC
O
O
+ N2

(S)-Glutamic acid (The S enantiomer)

23.39 The following sequence of methylation and Hofmann elimination was used in determining the structure of this bicyclic amine. Compound B is a mixture of two isomers.

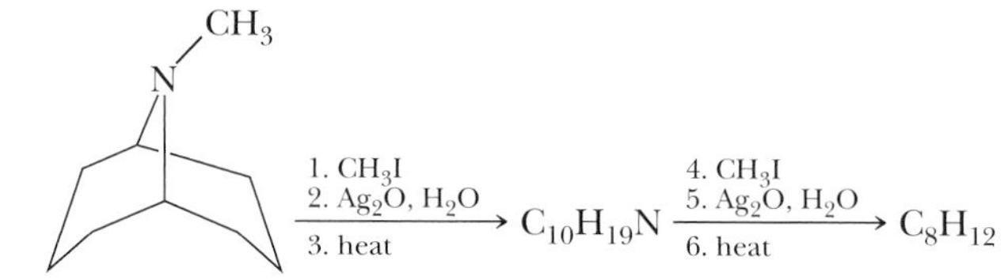
CH3
N
1. CH3I
2. Ag2O, H2O
3. heat
C10H19N
4. CH3I
5. Ag2O, H2O
6. heat
C8H12

($C_9H_{17}N$) **(A)** **(B)**

(a) Propose structural formulas for compounds A and B.

(b) Suppose you were given the structural formula of compound B but only the molecular formulas for compound A and the starting bicyclic amine. Given this information, is it possible, working backward, to arrive at an unambiguous structural formula for compound A? for the bicyclic amine?

23.40 Propose a structural formula for the compound $C_{10}H_{16}$ and account for its formation.

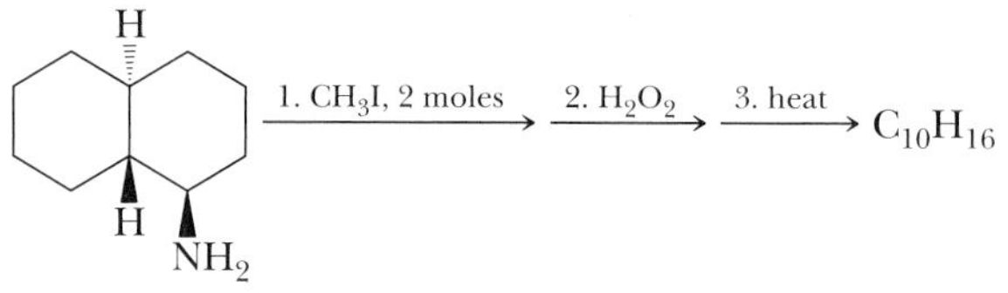
H
H
NH2
1. CH3I, 2 moles
2. H2O2
3. heat
C10H16

23.41 An amine of unknown structure contains one nitrogen and nine carbon atoms. The ^{13}C-NMR spectrum shows only five signals, all between 20 and 60 ppm. Three cycles of Hofmann elimination sequence [(1) CH_3I; (2) Ag_2O, H_2O; (3) heat] give trimethylamine and 1,4,8-nonatriene. Propose a structural formula for the amine.

23.42 The pyrolysis of acetic esters to give an alkene and acetic acid is thought to involve a planar transition state and cyclic redistribution of $(4n + 2)$ electrons. Propose a mechanism for pyrolysis of the following ester.

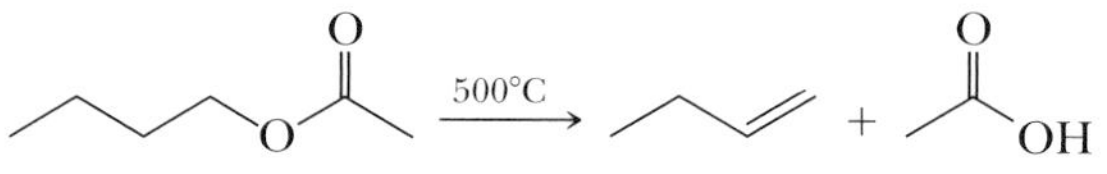
O
O
500°C
+
O
OH

Butyl acetate **1-Butene** **Acetic acid**

Organic Chemistry Reaction Roadmap

Reaction Roadmap

23.43 Use the reaction roadmap you made for Problems 20.55, 21.49, and 22.30 and update it to contain the reactions in the "Study Guide" section of this chapter. Because of their highly specific nature, do not use reactions 1, 2, 3, 4, 7, 8, and 9 on your roadmap.

Reaction Roadmap

23.44 Write the products of the following sequences of reactions. Refer to your reaction roadmaps to see how the combined reactions allow you to "navigate" between the different functional groups. Note that you will need your old Chapters 6–11, Chapters 15–18, and Chapter 19 reaction roadmaps along with your new Chapters 20–23 reaction roadmaps for these.

(a) 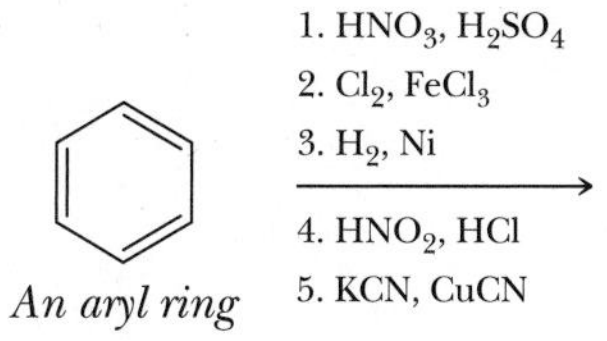
1. HNO_3, H_2SO_4
2. Cl_2, $FeCl_3$
3. H_2, Ni
4. HNO_2, HCl
5. KCN, CuCN
An aryl ring

(b) 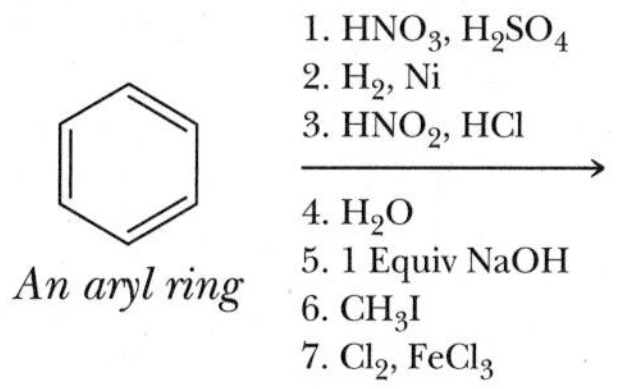
1. HNO_3, H_2SO_4
2. H_2, Ni
3. HNO_2, HCl
4. H_2O
5. 1 Equiv NaOH
6. CH_3I
7. Cl_2, $FeCl_3$
An aryl ring

(c) 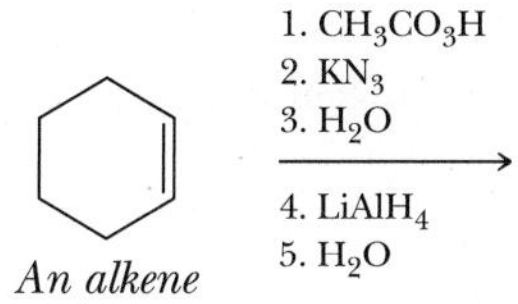
1. CH_3CO_3H
2. KN_3
3. H_2O
4. $LiAlH_4$
5. H_2O
An alkene

(d) 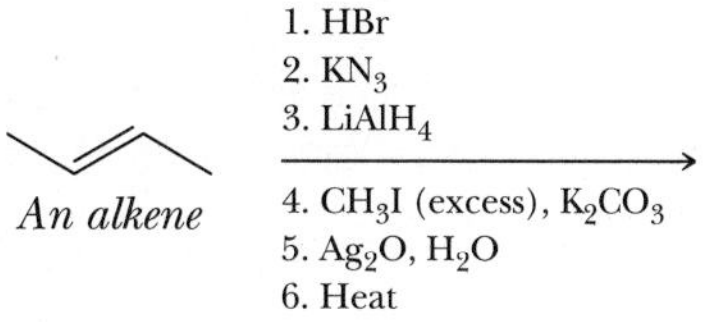
1. HBr
2. KN_3
3. $LiAlH_4$
4. CH_3I (excess), K_2CO_3
5. Ag_2O, H_2O
6. Heat
An alkene

(e) 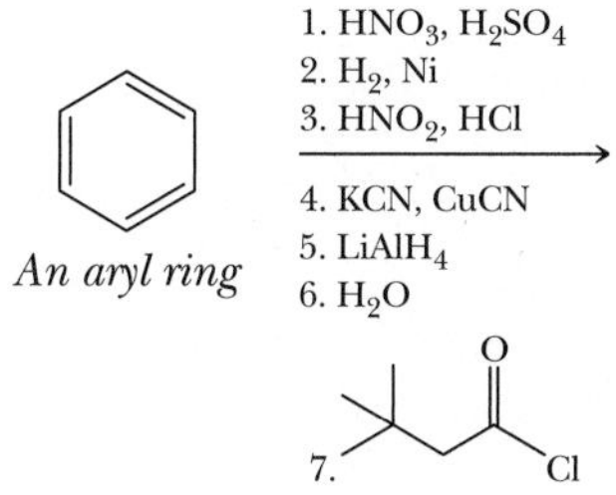
1. HNO_3, H_2SO_4
2. H_2, Ni
3. HNO_2, HCl
4. KCN, CuCN
5. $LiAlH_4$
6. H_2O
7.
O
Cl
An aryl ring

Synthesis

23.45 Propose steps for the following conversions using a reaction of a diazonium salt in at least one step of each conversion.

(a) Toluene to 4-methylphenol (*p*-cresol)
(b) Nitrobenzene to 3-bromophenol
(c) Toluene to *p*-cyanobenzoic acid
(d) Phenol to *p*-iodoanisole
(e) Acetanilide to *p*-aminobenzylamine
(f) Toluene to 4-fluorobenzoic acid
(g) 3-Methylaniline (*m*-toluidine) to 2,4,6-tribromobenzoic acid

23.46 Show how to bring about each step in this synthesis of the herbicide propranil.

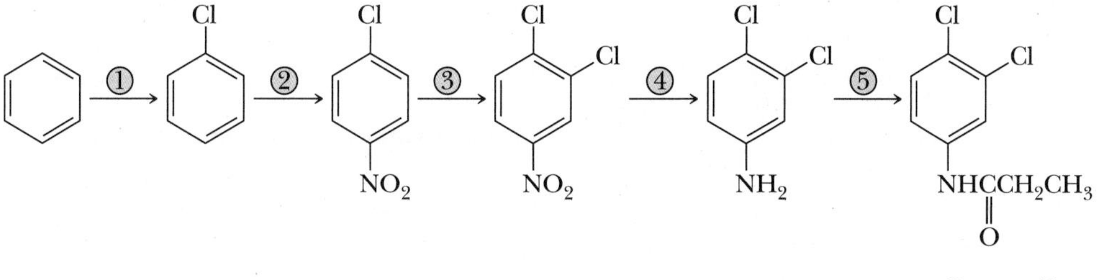
Cl
1
2
3
4
5
NO_2
NH_2
$NHCCH_2CH_3$
O

Propranil

23.47 Show how to bring about each step in the following synthesis.

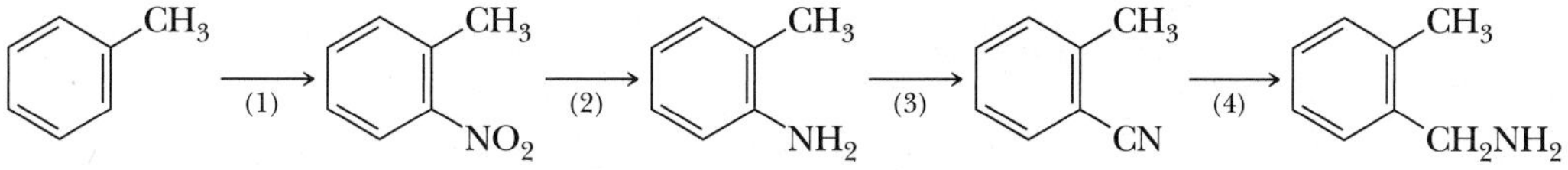
CH_3
(1)
(2)
(3)
(4)
NO_2
NH_2
CN
CH_2NH_2

23.48 Show how to bring about this synthesis.

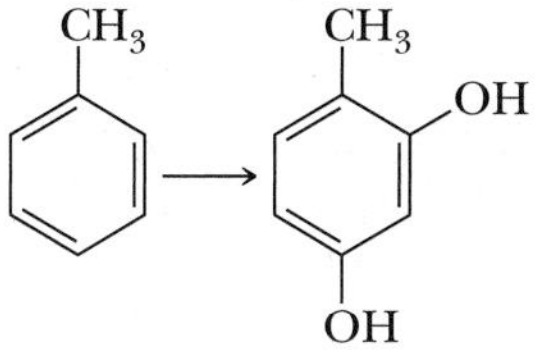
CH_3
OH

23.49 Show how to bring about each step in the following synthesis.

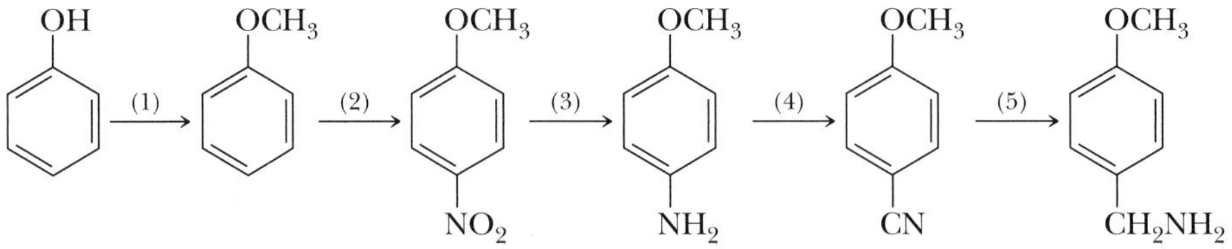

23.50 Methylparaben is used as a preservative in foods, beverages, and cosmetics. Provide a synthesis of this compound from toluene.

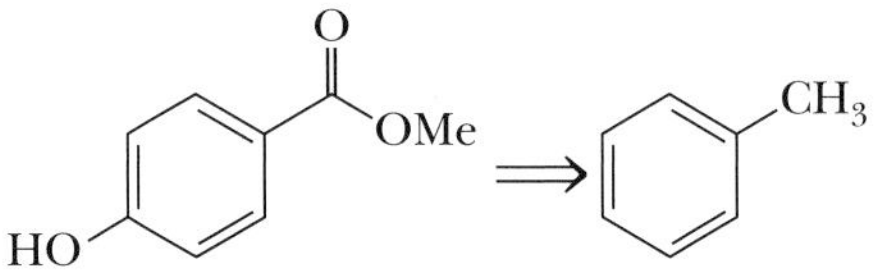

Methyl *p*-hydroxybenzoate
(Methylparaben)

23.51 Given the following retrosynthetic analysis, show how to synthesize the tertiary amine as a racemic mixture from benzene and any necessary reagents.

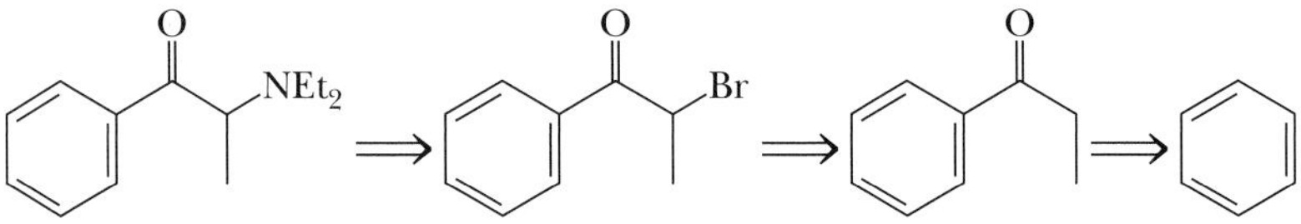

23.52 *N*-Substituted morpholines are building blocks in many drugs. Show how to synthesize *N*-methylmorpholine given this retrosynthetic analysis.

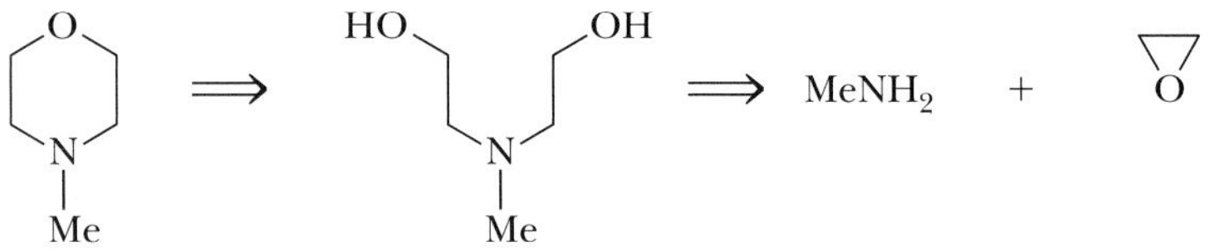

***N*-Methylmorpholine** **Methylamine** **Ethylene oxide**

23.53 Propose a synthesis for the systemic agricultural fungicide tridemorph from dodecanoic acid (lauric acid), propene, and a one-carbon building block. How many stereoisomers are possible for tridemorph?

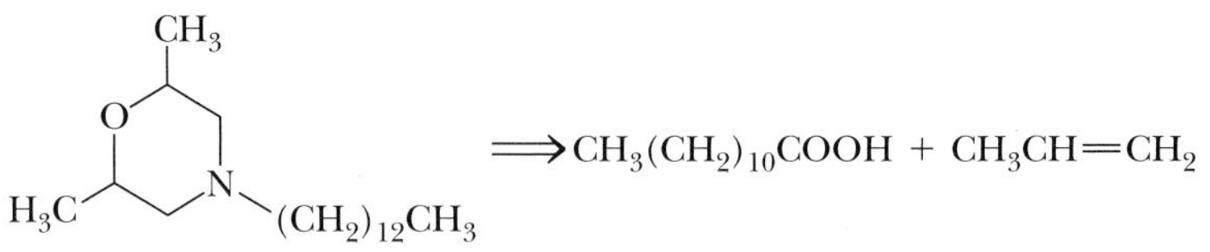

Tridemorph **Dodecanoic acid (Lauric acid)** **Propene**

23.54 The Ritter reaction is especially valuable for the synthesis of 3° alkanamines. In fact, there are few alternative routes to them. This reaction is illustrated by the first step in the following sequence. In the second step, the Ritter product is hydrolyzed to the amine.

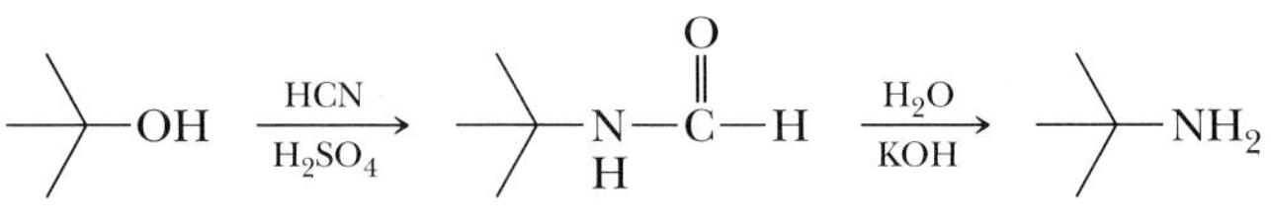

Ritter product

(a) Propose a mechanism for the Ritter reaction.

(b) What is the product of a Ritter reaction using acetonitrile, CH_3CN, instead of HCN, followed by reduction of the Ritter product with lithium aluminum hydride?

23.55 Several diamines are building blocks for the synthesis of pharmaceuticals and agrochemicals. Show how both 1,3-propanediamine and 1,4-butanediamine can be prepared from acrylonitrile.

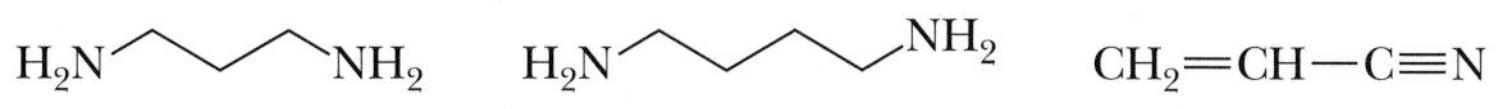

1,3-Propanediamine **1,4-Butanediamine** **Acrylonitrile**

23.56 Given the following retrosynthetic analysis, show how the intravenous anesthetic 2,6-diisopropylphenol (Propofol) can be synthesized from phenol.

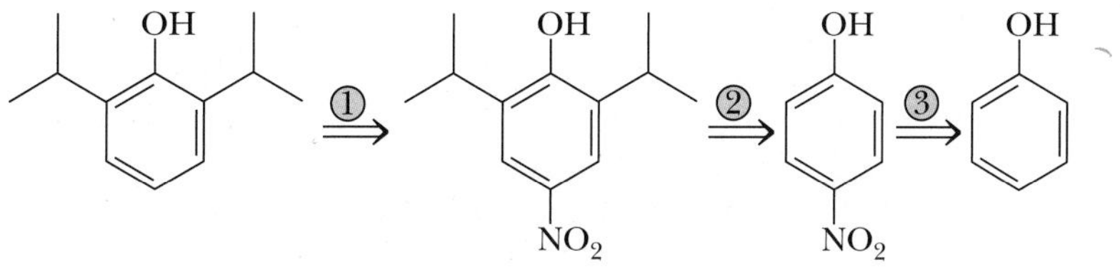

2,6-Diisopropylphenol (Propofol) **Phenol**

23.57 Following is a retrosynthetic analysis for propoxyphene, the hydrochloride salt of which is Darvon. The naphthalenesulfonic acid salt of propoxyphene is Darvon-N. The configuration of the carbon in Darvon bearing the hydroxyl group is *S*, and the configuration of the other stereocenter is *R*. Its enantiomer has no analgesic properties, but it is used as a cough suppressant.

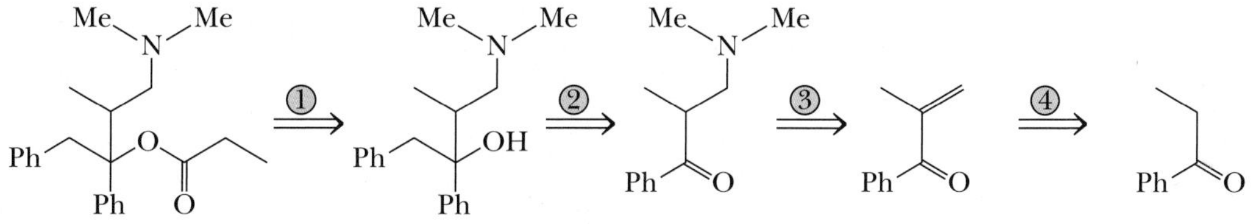

Propoxyphene **1-Phenyl-1-propanone (Propiophenone)**

(a) Propose a synthesis for propoxyphene from 1-phenyl-1-propanone and any other necessary reagents.

(b) Is propoxyphene chiral? If so, which of the possible stereoisomers are formed in this synthesis?

23.58 Following is a retrosynthetic analysis for ibutilide, a drug used to treat cardiac arrhythmia. In this scheme, Hept is an abbreviation for the 1-heptyl group.

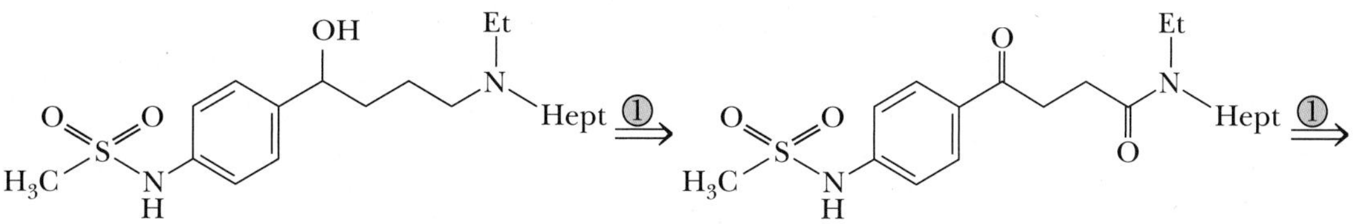

Ibutilide **(A)**

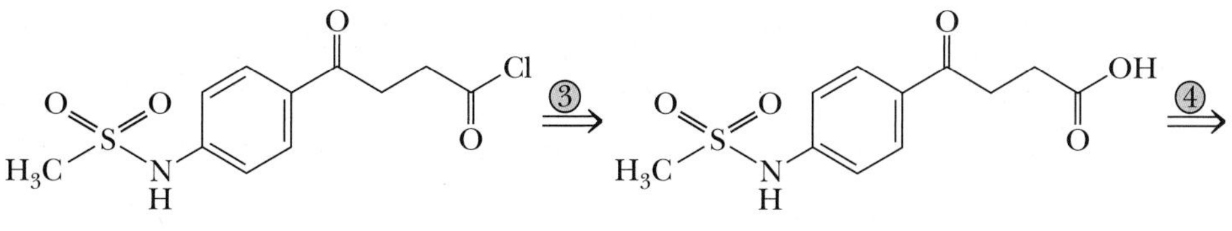

(B) **(C)**

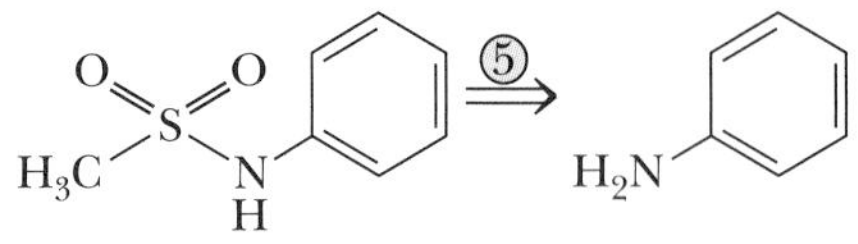

(D) **Aniline**

(a) Propose a synthesis for ibutilide starting with aniline, methanesulfonyl chloride, succinic anhydride, and *N*-ethyl-1-heptanamine.

(b) Is isobutilide chiral? If so, which of the possible stereoisomers are formed in this synthesis?

23.59 Propose a synthesis for the antihistamine histapyrrodine.

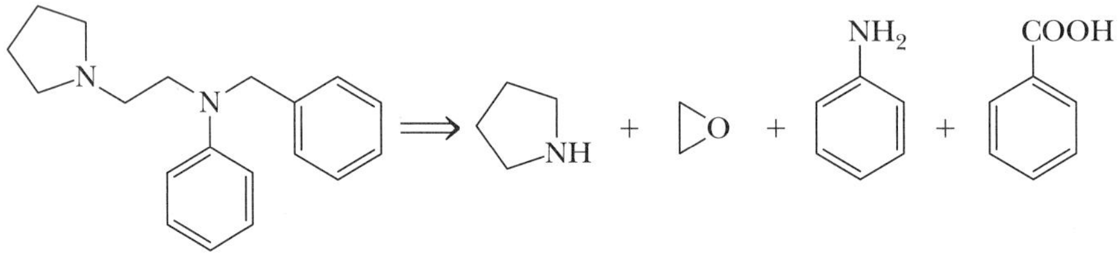

Histapyrrodine **Pyrrolidine** **Ethylene oxide** **Aniline** **Benzoic acid**

23.60 Following is a retrosynthesis for the coronary vasodilator ganglefene.

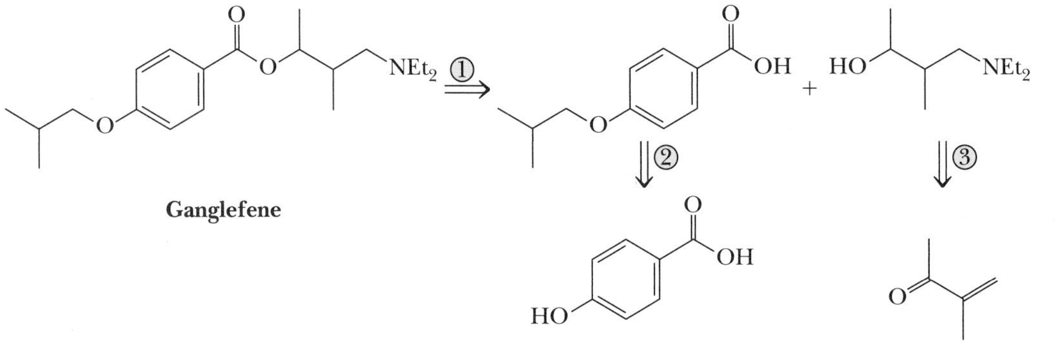

Ganglefene

4-Hydroxybenzoic acid **3-Methyl-3-buten-2-one**

(a) Propose a synthesis for ganglefene from 4-hydroxybenzoic acid and 3-methyl-3-buten-2-one.

(b) Is ganglefene chiral? If so, which of the possible stereoisomers are formed in this synthesis?

23.61 Moxisylyte, an α-adrenergic blocker, is used as a peripheral vasodilator. Propose a synthesis for this compound from thymol, which occurs in the volatile oils of members of the thyme family. Thymol is made industrially from *m*-cresol.

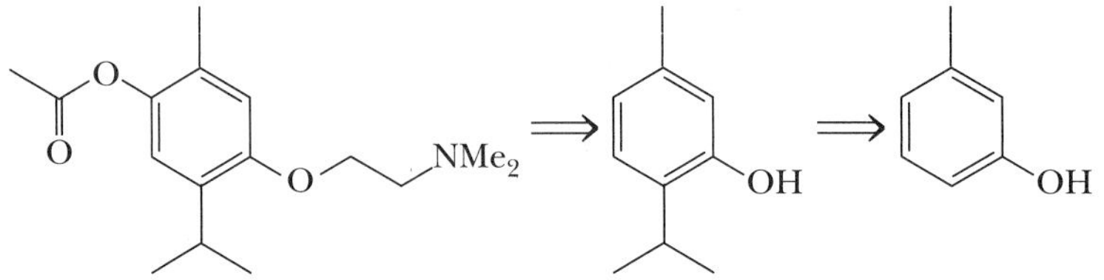

Moxisylyte **Thymol** ***m*-Cresol**

23.62 Propose a synthesis of the local anesthetic ambucaine from 4-nitrosalicylic acid, ethylene oxide, diethylamine, and 1-bromobutane.

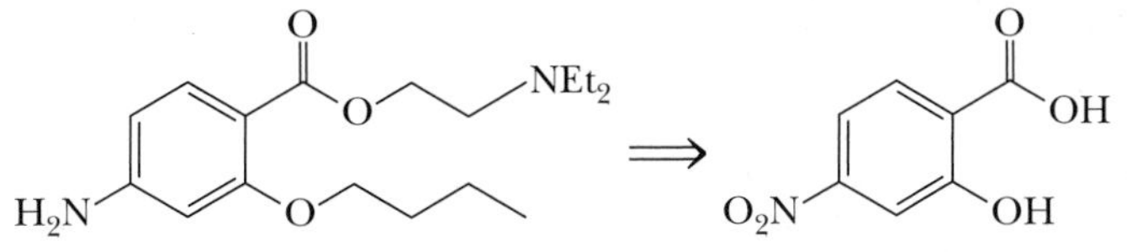

Ambucaine **4-Nitrosalicylic acid**

23.63 Given this retrosynthetic analysis, propose a synthesis for the local anesthetic hexylcaine.

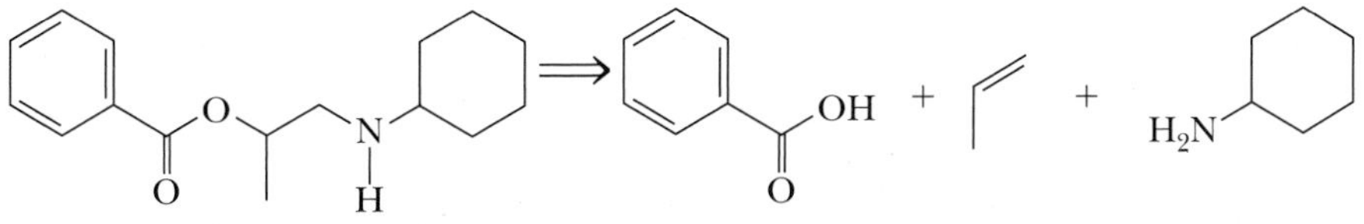

Hexylcaine (racemic) **Benzoic acid** **Propene** **Cyclohexylamine**

23.64 Following is an outline for a synthesis of the anorexic (appetite suppressant) fenfluramine. This compound was one of the two ingredients in Phen-Fen, a weight-loss preparation now banned because of its potential to cause irreversible heart valve damage.

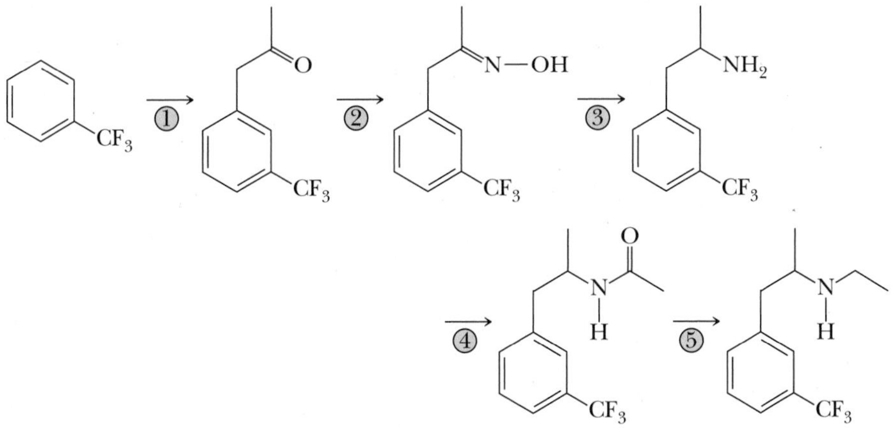

Fenfluramine

(a) Propose reagents and conditions for Step 1. Account for the fact that the CF_3 group is meta directing.

(b) Propose reagents and experimental conditions for Steps 2 and 3.

(c) An alternative procedure for preparing the amine of Step 3 is reductive amination of the corresponding ketone. What is reductive amination? Why might this two-step route for formation of the amine be preferred over the one-step reductive amination?

(d) Propose reagents for Steps 4 and 5.

(e) Is fenfluramine chiral? If so, which of the possible stereoisomers are formed in this synthesis?

23.65 Following is a series of anorexics (appetite suppressants). As you study their structures, you will surely be struck by the sets of characteristic structural features.

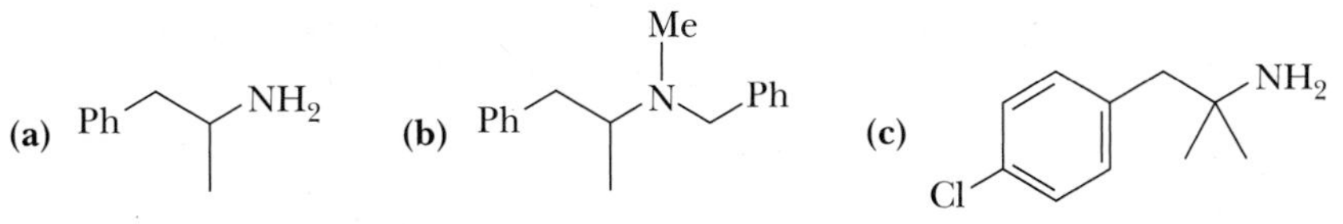

Amphetamine **Benzphetamine** **Chlorphentermine**

(d) 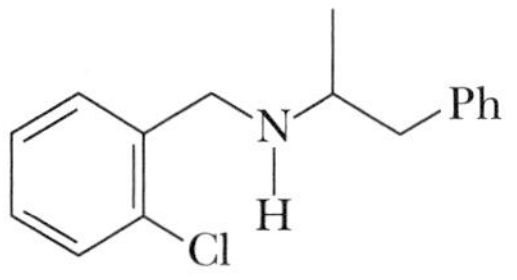
Clobenzorex

(e) 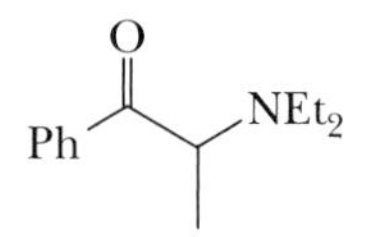
Diethylpropion

(f) 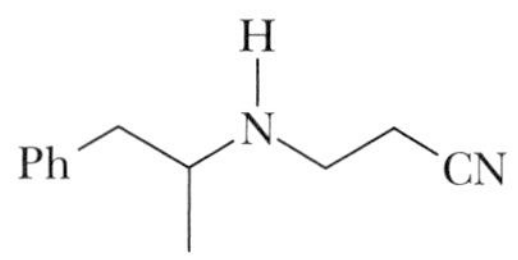
Fenproporex

(g) 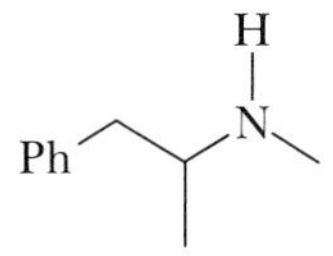
Methamphetamine

(h) 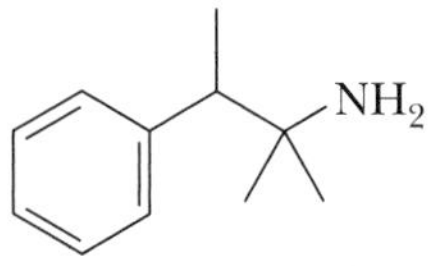
Pentorex

(i) 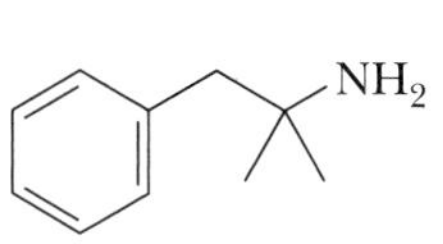
Phentermine

(a) Knowing what you do about the synthesis of amines, including the Ritter reaction (Problem 23.54), suggest a synthesis for each compound.

(b) Which of these compounds are chiral?

23.66 The drug sildenafil, sold under the trade name Viagra, is a potent inhibitor of phosphodiesterase V (PDE V), an enzyme found in high levels in the corpus cavernosum of the penis. Inhibitors of this enzyme enhance vascular smooth muscle relaxation and are used for treatment of male impotence. Following is an outline for a synthesis of sildenafil.

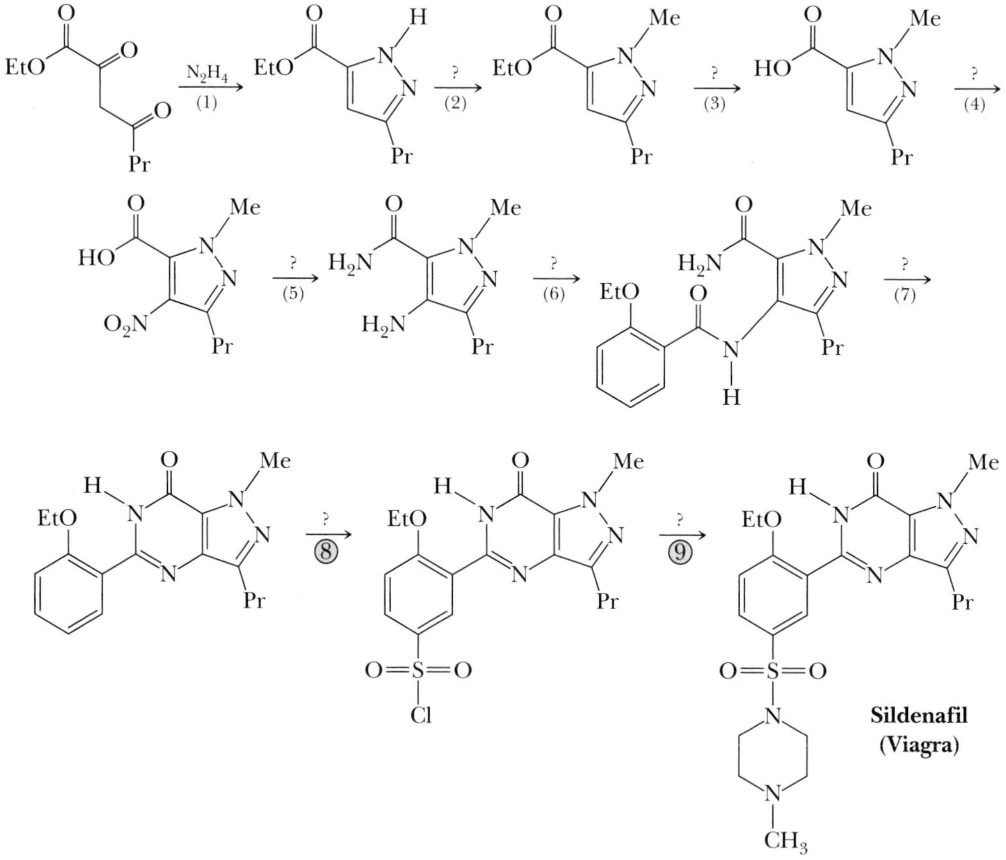

Sildenafil (Viagra)

(a) Propose a mechanism for Step 1.

(b) The five-membered nitrogen-containing ring formed in Step 1 is named pyrazole. Show that, according to the Hückel criteria for aromaticity, pyrazole can be classified as an aromatic compound.

(c) Propose a reagent or reagents for Steps 2–7 and 9.

(d) Show how the reagent for Step 6 can be prepared from salicylic acid (2-hydroxybenzoic acid). Salicylic acid, the starting material for the synthesis of aspirin and a number of other pharmaceuticals, is readily available by the Kolbe carboxylation of phenol (Section 21.4E).

(e) Chlorosulfonic acid, $ClSO_3H$, the reagent used in Step 8 is not described in the text. Given what you have studied about other types of electrophilic aromatic substitutions (Section 22.1), propose a mechanism for the reaction in Step 8.

(f) Propose a structural formula for the reagent used in Step 9 and show how it can be prepared from methylamine and ethylene oxide.

(g) Is sildenafil chiral? If so, which of the possible stereoisomers are formed in this synthesis?

23.67 Radiopaque imaging agents are substances administered either orally or intravenously that absorb X-rays more strongly than body material. One of the best known of these is barium sulfate, the key ingredient in the so-called barium cocktail for imaging of the gastrointestinal tract. Among other X-ray contrast media are the so-called triiodoaromatics. You can get some idea of the imaging for which they are used from the following selection of trade names: Angiografin, Gastrografin, Cardiografin, Cholegrafin, Renografin, and Urografin. Following is a synthesis for diatrizoic acid from benzoic acid.

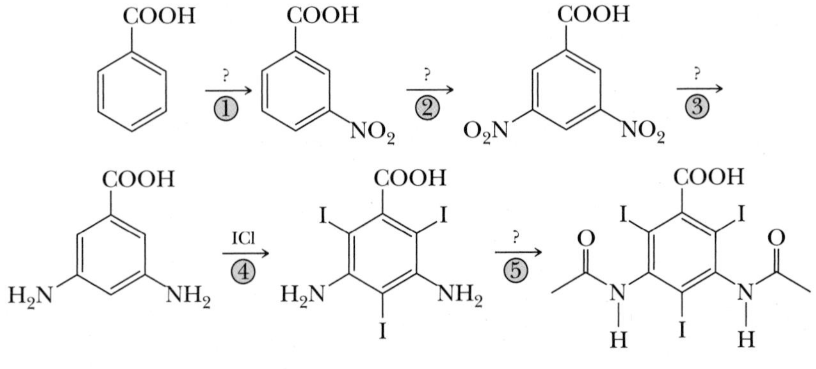

Diatrizoic acid

(a) Provide reagents and experimental conditions for Steps (1), (2), (3), and (5).

(b) Iodine monochloride, ICl, a black crystalline solid with an mp of 27.2°C and a bp of 97°C, is prepared by mixing equimolar amounts of I_2 and Cl_2. Propose a mechanism for the iodination of 3-aminobenzoic acid by this reagent.

23.68 Show how the synthetic scheme developed in Problem 23.67 can be modified to synthesize this triiodobenzoic acid X-ray contrast agent.

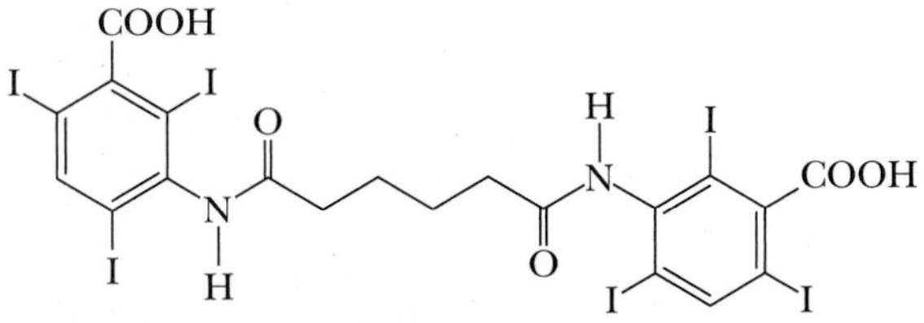

Iodipamide

23.69 A diuretic is a compound that causes increased urination and thereby reduces fluid volume in the body. An important use of diuretics in clinical medicine is in the reduction of the fluid buildup, particularly in the lungs, that is associated with congestive heart failure. It is also used as an antihypertensive (i.e., to reduce blood pressure). Furosemide, an exceptionally potent diuretic, is prescribed under 30 or more trade names, the best known of which is Lasix. The synthesis of furosemide begins with treatment of 2,4-dichlorobenzoic acid with chlorosulfonic acid in a reaction called chlorosulfonation. The product of this reaction is then treated with ammonia followed by heating with furfurylamine.

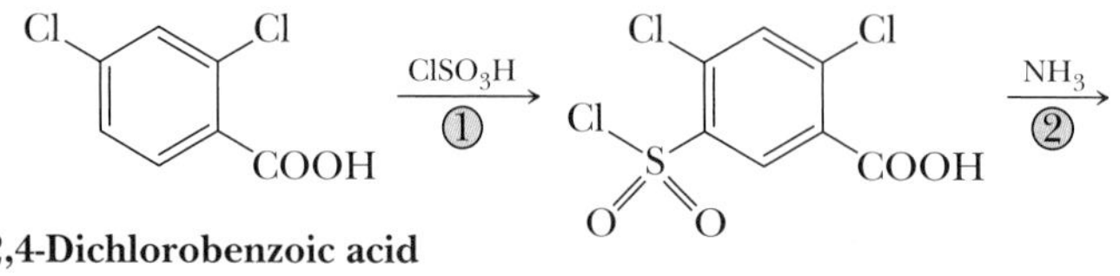

2,4-Dichlorobenzoic acid

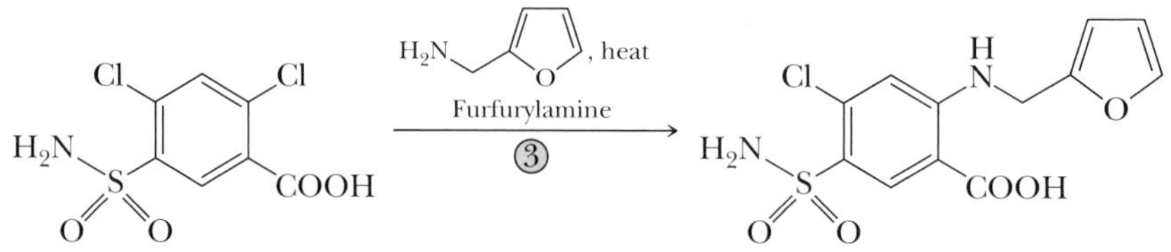

Furosemide

(a) Propose a synthesis of 2,4-dichlorobenzoic acid from toluene.
(b) Propose a mechanism for the chlorosulfonation reaction in Step (1).
(c) Propose a mechanism for Step (3).
(d) Is furosemide chiral? If so, which of the possible stereoisomers are formed in this synthesis?

23.70 Among the newer-generation diuretics is bumetanide, prescribed under several trade names, including Bumex and Fordiuran. Following is an outline of a synthesis of this drug.

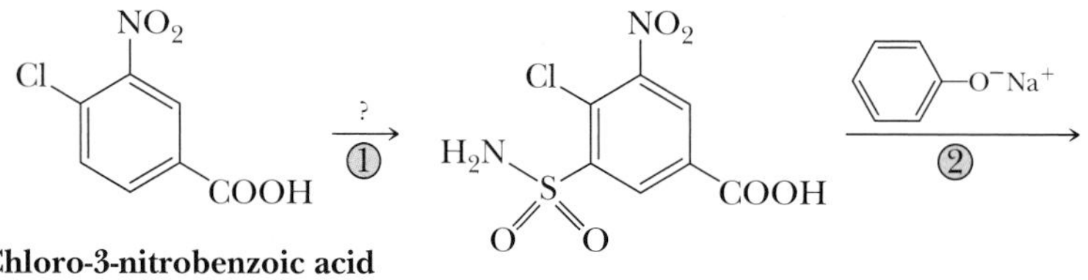

4-Chloro-3-nitrobenzoic acid

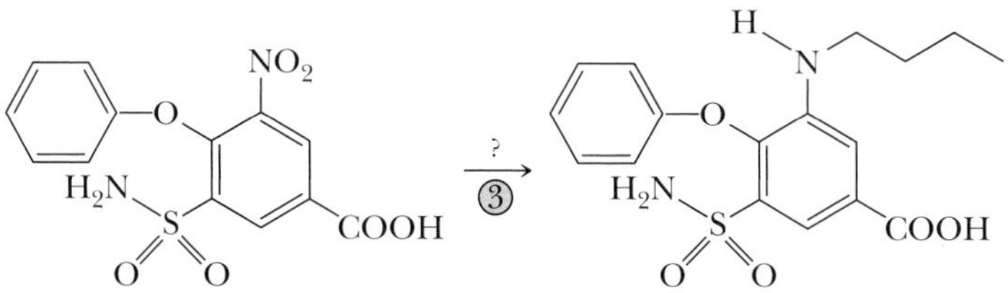

Bumetanide

(a) Propose a synthesis of 4-chloro-3-nitrobenzoic acid from toluene.
(b) Propose reagents for Step (1). *Hint:* It requires more than one reagent.
(c) Propose a mechanism for reaction (2).
(d) Propose reagents for Step (3). *Hint:* It too requires more than one reagent.
(e) Is bumetanide chiral? If so, which of the possible stereoisomers are formed in this synthesis?

23.71 Of the early antihistamines, most had a side effect of mild sedation; they made a person sleepy. More recently, a new generation of nonsedating antihistamines known as histamine H_1 receptor antagonists has been introduced. One of the most widely prescribed of these is fexofenadine (Allegra). This compound is nonsedating because the polarity of its carboxylic anion prevents it from crossing the blood-brain barrier. Following is a retrosynthetic analysis for the synthesis of fexofenadine.

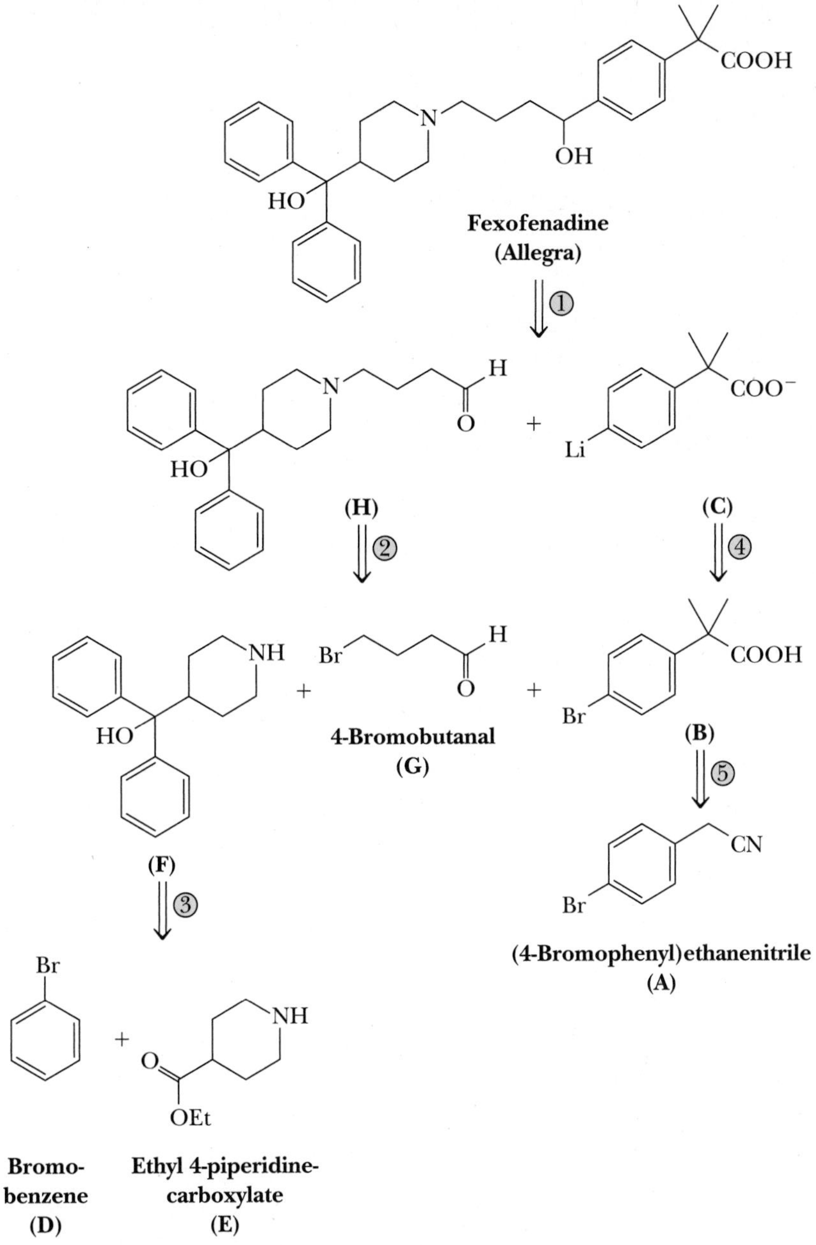

COOH
N
OH
HO
Fexofenadine
(Allegra)
1
H
N
COO−
O
+
Li
HO
(H)
(C)
2
4
NH
Br
H
COOH
+
O
+
Br
HO
4-Bromobutanal
(G)
(B)
5
CN
(F)
Br
3
(4-Bromophenyl)ethanenitrile
(A)
Br
NH
+
O
OEt
Bromo-
benzene
(D)
Ethyl 4-piperidine-
carboxylate
(E)

(*Note:* The organolithium reagent C cannot be made directly from B because the presence of the carboxyl group in B would lead to intermolecular destruction of the reagent by an acid-base reaction. In practice, B is first converted to its sodium salt by treatment with sodium hydride, NaH, and then the organolithium reagent is prepared.)

(a) Given this retrosynthetic analysis, propose a synthesis for fexofenadine from the four named starting materials.

(b) Is fexofenadine chiral? If so, which of the possible stereoisomers are formed in this synthesis?

23.72 Sotalol is a β-adrenergic blocker used to treat certain types of cardiac arrhythmias. Its hydrochloride salt is marketed under several trade names, including Betapace. Following is a retrosynthetic analysis.

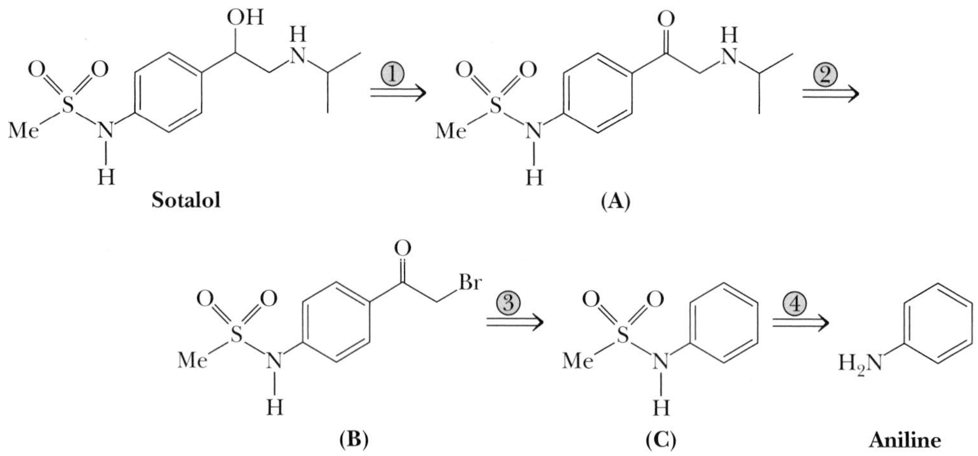

(a) Propose a synthesis for sotalol from aniline.

(b) Is sotalol chiral? If so, which of the possible stereoisomers are formed in this synthesis?

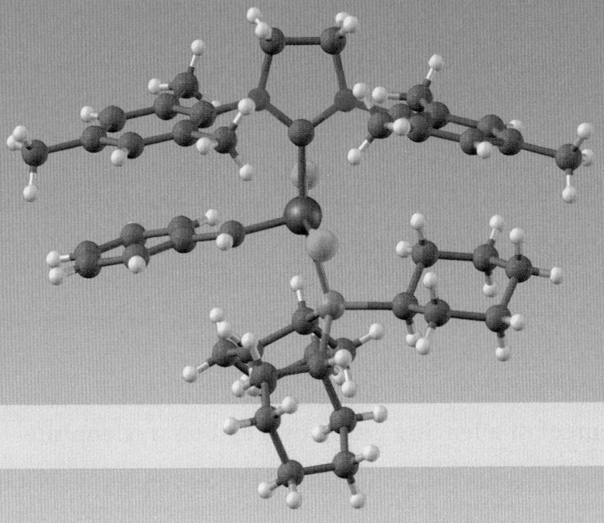

24

Catalytic Carbon-Carbon Bond Formation

A ruthenium-containing organometallic catalyst for alkene metathesis reactions. See Section 24.6 for the structure of this catalyst and a discussion of this reaction.

Outline

24.1 Carbon-Carbon Bond-Forming Reactions from Earlier Chapters
24.2 Organometallic Compounds and Catalysis
24.3 The Heck Reaction
24.4 Catalytic Allylic Alkylation
24.5 Palladium-Catalyzed Cross-Coupling Reactions
24.6 Alkene Metathesis

Over the course of the last 120 years, chemists have learned how to synthesize amazingly complex molecules. In recent years particularly, they have focused on compounds of medicinal interest, and a large number of current pharmaceuticals are synthesized from simpler compounds. Many of these pharmaceuticals are natural products or their analogs, and others are either simpler analogs or unrelated compounds that have been found to be active against certain organisms or diseased cells, specific cellular receptors, or specific enzyme targets. A key development that has allowed synthesis of these compounds has been the discovery of many catalytic methods of carbon-carbon bond formation. It is now possible, using a combination of new and classical methods, to carry out the synthesis of molecules with sensitive functionality and amazingly complex carbon skeletons from simple and inexpensive starting materials, often with excellent stereo- and regiocontrol. In this chapter, we make a dramatic leap from the more classical organic reactions covered in previous chapters of this book to survey several particularly useful catalytic methods of carbon-carbon bond formation, some of which represent very recent developments. Further, the catalytic methods we present are part of a thrust in industry to move to green chemistry, in which less chemical waste is generated. We have room for only a few representative examples out of the wealth of carbon-carbon bond forming reactions that are now available to the modern synthetic chemist. Finally, a number of problems based on modern organic syntheses are given to illustrate the use of these reactions and their combination with other reactions.

24.1 Carbon-Carbon Bond-Forming Reactions from Earlier Chapters

As a review, let us list the methods of carbon-carbon bond formation you have already studied. All these reactions should be available to you for synthetic problems.

Nucleophilic displacement of a leaving group by a carbon nucleophile

- Gilman (organocuprate) reagents (Section 15.2C) if the leaving group is a halogen atom or tosylate.
- Grignard reagents, organolithium reagents (Section 15.1C) and Gilman reagents if the leaving group is the oxygen of an epoxide.
- Alkyne (Sections 7.5 and 9.1) and cyanide (Section 9.1) anions. The leaving group can be the oxygen of an epoxide (Section 11.9B).
- Enolate anion alkylations, acetoacetic ester synthesis and malonic ester synthesis (Sections 19.6 and 19.7).
- Enamine alkylations (Section 19.5).

Nucleophilic addition to a carbonyl or a carboxyl group

- Grignard reagents (Sections 16.5A and 18.9A), organolithium reagents (Sections 16.5B and 18.9B), and Gilman reagents (Section 18.9C).
- Alkyne (Section 16.5C) and cyanide (Section 16.5D) anions.
- Aldol reactions (Section 19.2).
- Claisen (Section 19.3A) and Dieckmann (Section 19.3B) condensations.
- Enamine acylations (Section 19.5B).
- Wittig reaction (Section 16.6 for **C═C** double bonds).

Conjugate addition to α,β-unsaturated carbonyl compounds

- Michael reaction (Section 19.8A).

Pericyclic reactions

- Diels-Alder reaction (Section 20.5).
- Claisen rearrangement (Section 20.6).
- Cope rearrangement (Section 20.6).

Carbene/carbenoid additions (Section 15.3).

Aromatic substitution

- Friedel-Crafts alkylation and acylation of aromatics (Section 22.1C).
- Reaction of cyanide with aromatic diazonium compounds (Section 23.8E).

This list already includes many different reactions, but you will find that the new reactions in this chapter are of somewhat different character.

None of the C—C bond-forming reactions summarized above are catalytic. Recall that a catalyst is a species that becomes involved in the mechanism of a reaction and lowers the barrier to that reaction, thereby accelerating the reaction rate. Catalysts do not change the thermodynamics of a reaction; instead, they alter the kinetics. Further, a catalyst is regenerated at the end of the reaction in the same form as at the start of the reaction; hence, a catalyst is available to be used over and over—a property often referred to as "turnover."

This chapter focuses on catalytic C—C bond-forming reactions. In other words, a chemical (the catalyst) is added to the reaction vessel and causes C—C bond formation in a manner that is faster than would occur in the absence of the catalyst. In fact, most of the reactions we examine in this chapter (particularly those of Section 24.5) would essentially never occur without the catalyst. Consequently, we start the chapter with a brief overview of the key reactions performed by organometallic compounds.

24.2 Organometallic Compounds and Catalysis

We introduced organometallic compounds in Chapter 15. In the next few sections, we discuss several reactions of transition metals that are particularly useful for the preparation of new carbon-carbon bonds.

A. Oxidative Addition and Reductive Elimination

Two extremely important reactions of transition metals and transition metal compounds are **oxidative addition** and its reverse, **reductive elimination**. In oxidative addition, a reagent adds to a metal, causing its coordination to increase by two; reductive elimination is the opposite. These reactions are called oxidative or reductive because the formal charge of the metal changes by two during the reaction. Oxidative addition can occur with a metal coordinated with one or more **ligands** (**L_n**, where n is the number); it can also occur with a free metal, **M(0)**. Haloalkanes, hydrogen, halogens, and many other types of compounds can take part in these reactions. The reactivity of different substrates depends greatly on the metal.

Oxidative addition
Addition of a reagent to a metal center causing it to add two substituents and to increase its oxidation state by two.

Reductive elimination
Elimination of two substituents at a metal center, causing the oxidation state of the metal to decrease by two.

Ligand
A Lewis base bonded to a metal atom in a coordination compound. It may bond strongly or weakly.

$$ML_n + X_2 \underset{\text{reductive elimination}}{\overset{\text{oxidative addition}}{\rightleftharpoons}} X{-}ML_n{-}X$$

Although numerous other reactions are unique and essential to the action of organometallic catalysts, the key steps of the catalytic processes we discuss in this chapter involve oxidative additions and reductive eliminations. Thus, in an introductory chapter on catalytic C—C bond-forming reactions, we need go no deeper.

B. Key Features of the Utility of Catalytic C—C Bond Formation

The reactions and mechanisms covered in this chapter represent a growing modern trend in organic chemistry. Organometallic catalysts facilitate reactions that are otherwise impossible, are very difficult, or would require many synthetic steps in order to accomplish. The use of these catalysts causes a net decrease in the number and quantity of reagents, solvents, and purifications necessary in an overall synthetic sequence. This decrease means that the chemical waste from an industrial process can be dramatically reduced. Intentionally designing a chemical procedure or process to decrease waste and toxic byproducts is now a whole chemical field in and of itself that is called **green chemistry**.

A goal of green chemistry is to use effectively each of the atoms involved in a reaction so that atoms are not "thrown away" by being incorporated into byproducts of the reactions. This concept is called **atom economy**, which describes the efficiency of a chemical process in terms of all atoms involved. An ideal reaction would consist of the mass of the product equaling the mass of all the reactants used. In such a case, each atom of the reactants would be completely incorporated into the product and no waste would be generated. Such reactions are rare, but the goal of achieving the highest atom economy clearly has both an environmental and economic benefit. Hence, "going green" is permeating society in many ways, with the chemical industry recognizing and embracing the value of green.

Watch a video explanation

24.3 The Heck Reaction

A. The Nature of the Reaction

In the early 1970s, Richard Heck, at Hercules, Inc. and later at the University of Delaware, discovered a palladium-catalyzed reaction in which the carbon group of a haloalkene or haloarene is substituted for a hydrogen on the carbon-carbon double bond (a vinylic hydrogen) of an alkene. This reaction, now known as the **Heck reaction**, is particularly valuable in synthetic organic chemistry because it is the only general method yet discovered for this type of substitution.

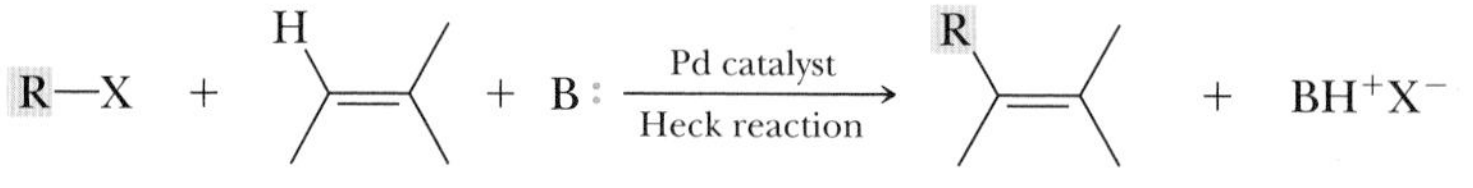

Haloalkene or haloarene | Alkene | Base | Substituted alkene | Conjugate acid of the base

Substitution for a vinylic hydrogen by the Heck reaction is highly regioselective; formation of the new carbon-carbon bond most commonly occurs at the less substituted carbon of the double bond. In addition, where an *E* or *Z* configuration is possible at the double bond of the product, the Heck reaction is highly stereoselective, often giving almost exclusively the *E* configuration of the product.

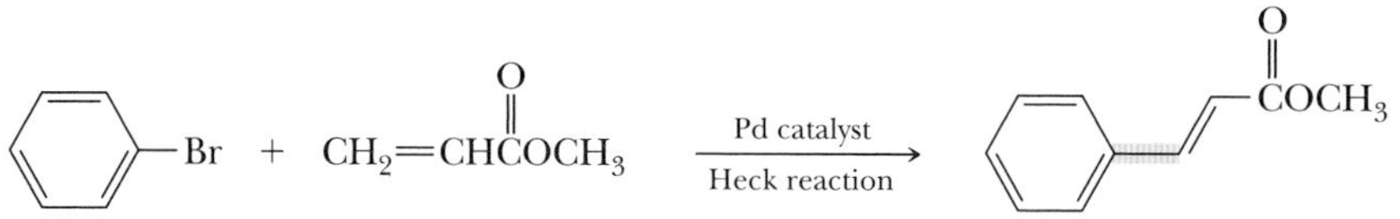

Bromobenzene | **Methyl 2-propenoate (Methyl acrylate)** | **Methyl (*E*)-3-phenyl-2-propenoate (Methyl cinnamate)**

In addition, the Heck reaction is completely stereospecific with regard to the haloalkene; the configuration of the double bond in the haloalkene is preserved.

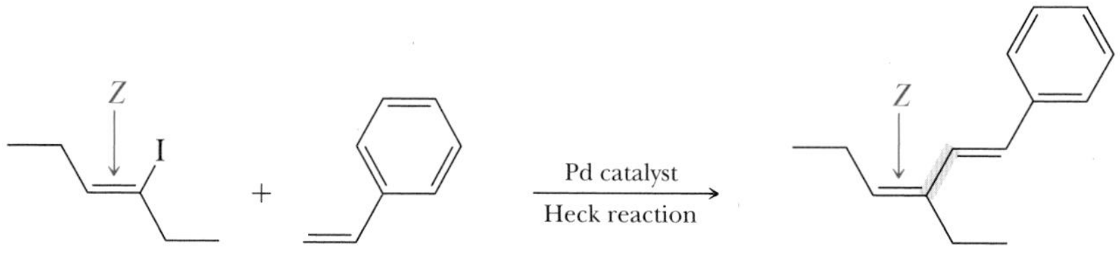

(*Z*)-3-Iodo-3-hexene | **Phenylethene (Styrene)** | **(1*E*,3*Z*)-3-Ethyl-1-phenyl-1,3-hexadiene**

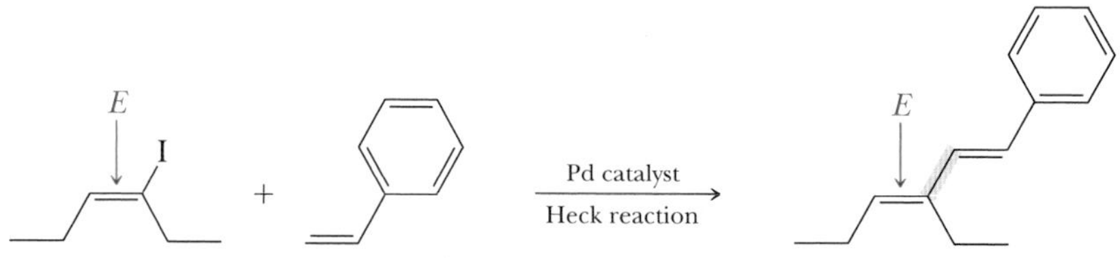

(*E*)-3-Iodo-3-hexene | **Phenylethene (Styrene)** | **(1*E*,3*E*)-3-Ethyl-1-phenyl-1,3-hexadiene**

Preparation of the Catalyst

The form of the palladium catalyst most commonly added to the reaction medium is palladium(II) acetate, $Pd(OAc)_2$. This and other Pd(II) compounds are better termed precatalysts because the catalytically active form of the metal is a complex of Pd(0) formed *in situ* by reduction of Pd(II) to Pd(0).

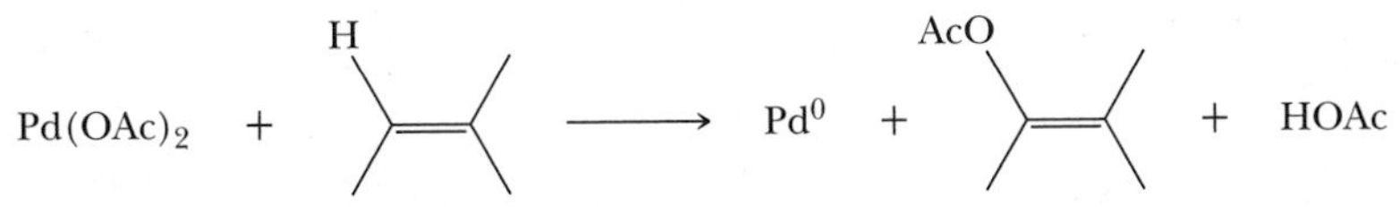

Palladium(II) acetate — An alkene (reductant) — Oxidized alkene

Reaction of Pd(0) with good ligands, L, gives the actual Heck catalyst, PdL_2. Without the ligand, Pd(0) is insoluble.

$$Pd^0 + 2\,L \longrightarrow PdL_2$$

Ligand — The Heck catalyst

Among the most common ligands, L, used for coordination of the Pd(0) is triphenylphosphine, $(C_6H_5)_3P$. Many other ligands can be used as well, including chiral ones such as BINAP (Section 6.7) that can lead to a significant excess of a single enantiomer in the case of chiral products.

The Haloalkane

The most common halides used in Heck reactions are aryl, heterocyclic, benzylic, and vinylic iodides and bromides, with iodides generally being most reactive. The reactivity of substrates with leaving groups on sp^2 carbons contrasts with nucleophilic substitution reactions, where such substrates are essentially unreactive. Haloalkanes in which there is an acidic β-hydrogen are rarely used because of the ease with which they undergo β-elimination under conditions of the Heck reaction to form alkenes. Triflates (trifluoromethanesulfonates, $CF_3SO_2O—$), which are easily prepared by treating an alcohol with trifluoromethanesulfonyl chloride, are also excellent substrates.

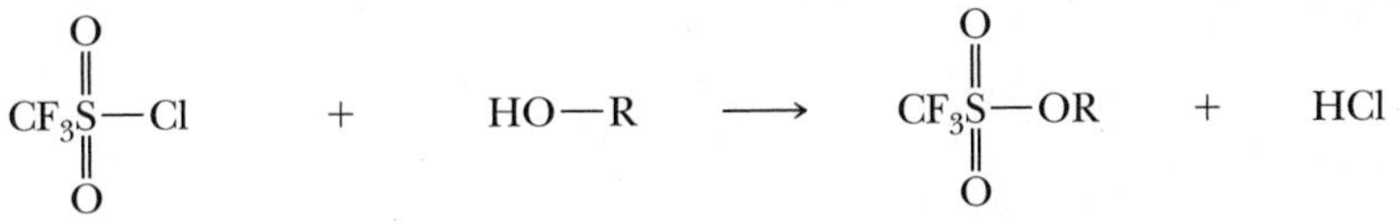

Trifluoromethanesulfonyl chloride — Alcohol — A trifluoromethanesulfonate (a triflate)

The halide or triflate (RX) reacts with PdL_2 by oxidative addition to give a square planar Pd(II) species, which is the reaction intermediate.

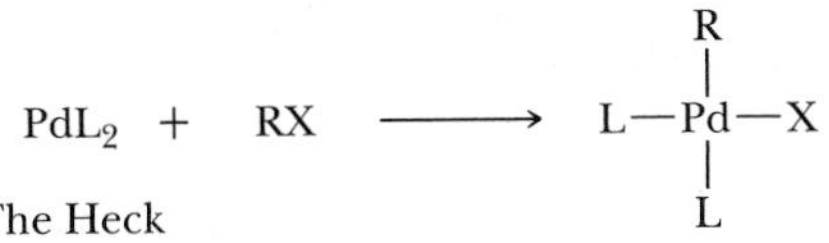

The Heck catalyst

A particular advantage of the Heck reaction is the wide range of functional groups, including alcohols, ethers, aldehydes, ketones, and esters, that may be present elsewhere in the organic halogen compound or alkene without reacting themselves or affecting the Heck reaction.

The Alkene

The reactivity of the alkene is a function of steric crowding about the carbon-carbon double bond. Ethylene and monosubstituted alkenes are most reactive; the greater the degree of substitution on the double bond, the slower the reaction and the lower the yield of product. These steric effects also control the regiochemistry of the addition, with the alkyl group adding to the less hindered carbon of the alkene.

The Base

Commonly used bases are tertiary amines such as triethylamine, Et_3N, sodium or potassium acetate, and sodium hydrogen carbonate.

The Solvent

Polar aprotic solvents (Section 9.3D) such as *N,N*-dimethylformamide (DMF), acetonitrile, and dimethyl sulfoxide (DMSO) are commonly used. It is also possible to carry out some Heck reactions in aqueous methanol. The polar solvents are needed to dissolve the $Pd(OAc)_2$ at the beginning of the reaction.

B. Mechanism of the Reaction

The mechanism of the Heck reaction is divided into two stages: formation of the Heck catalyst and the catalytic cycle. As you study the catalytic cycle, note in particular that both Steps 2 and 4 are syn stereoselective; reaction will not proceed if these syn relationships cannot be obtained. Step 2 involves syn addition of R and PdL_2X to the double bond. Step 4 involves syn elimination of H and the Pd(II) species to generate a new double bond. These syn additions and eliminations contrast with most of the addition and elimination reactions we have seen, which prefer the anti geometry. Additions of boron hydrides (Section 6.4, hydroboration) and osmium tetroxide (Section 6.5A) or ozone (Section 6.5B) to alkenes are some examples of syn additions that you have already seen.

Mechanism 24.1

The Heck Reaction

Stage 1: Formation of the Heck Catalyst, PdL_2

A two-electron reduction of Pd(II) to Pd(0) accompanied by its complex formation with two molecules of a ligand, L, gives the Heck catalyst, PdL_2. A common reducing agent is triethylamine or, as in the following example, the alkene itself. Because the catalyst is present only in small amounts, an insignificant amount of the alkene is lost to this reaction. In the reaction shown here, L is triphenylphosphine, $(C_6H_5)_3P$. As mentioned previously, this is actually a two-step reaction: reduction of the palladium followed by reaction of the palladium with the ligand. We show the two steps combined here for simplicity.

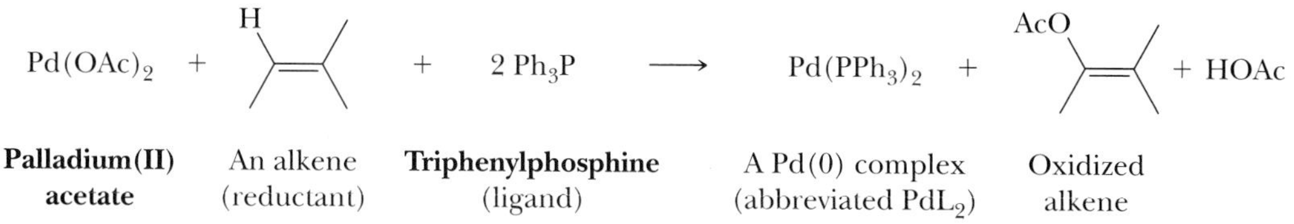
Pd(OAc)2 +
H
+ 2 Ph3P
Pd(PPh3)2 +
AcO
+ HOAc
Palladium(II) acetate
An alkene (reductant)
Triphenylphosphine (ligand)
A Pd(0) complex (abbreviated PdL2)
Oxidized alkene

Stage 2: The Catalytic Cycle

The catalytic cycle of the Heck reaction involves five steps. In Step 1, oxidative addition of the haloalkene or haloarene, RX, to PdL_2 gives a tetracoordinated Pd(II) complex containing both R and X groups bonded to Pd. Syn addition of the R and PdL_2X of this complex to the alkene gives an intermediate in which Pd is bonded to the more substituted carbon for steric reasons. Because of the long Pd—C bond, the palladium is sterically less demanding than the organic group; therefore, it ends up on the more hindered carbon. This intermediate must undergo internal rotation about the central carbon-carbon single bond in Step 3 to place H and PdL_2X syn to each other. Syn elimination of H and PdL_2X in Step 4 gives the new alkene and $HPdL_2X$. Reductive elimination in Step 5 releases the acid HX and regenerates the PdL_2 catalyst. HX is then neutralized by the added base.

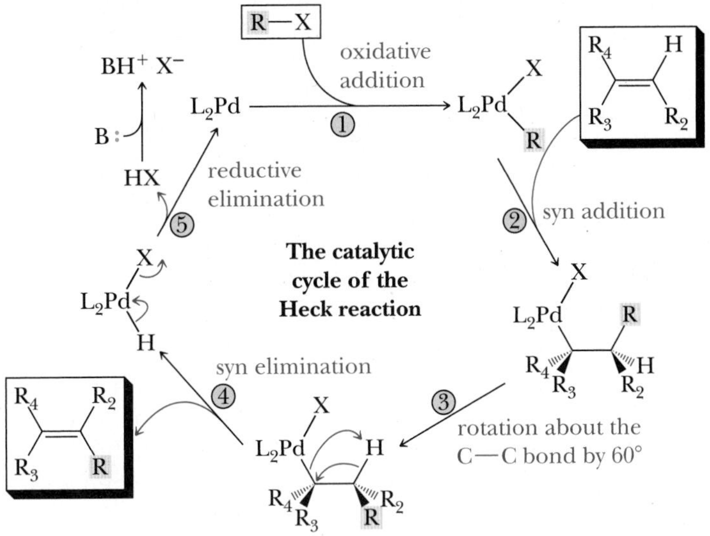

In this cycle, the alkene, haloalkane compound, and base are required in equimolar amounts; the Pd(0) species is required in only a catalytic amount. Note also the inversion of the configuration (R_2 and R_3 are originally *cis* to each other but in the product are *trans*). This inversion is a consequence of the consecutive syn addition and elimination steps. The complete mechanism for this reaction has additional intermediates (involving π complexes of the alkene with the palladium), but those shown here are the important ones for understanding the reaction and its stereochemistry.

Example 24.1 Heck Reaction I

Complete these Heck reactions.

(a)

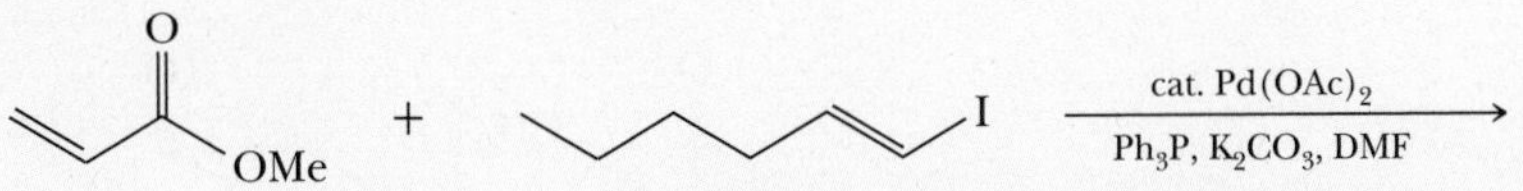

(b)

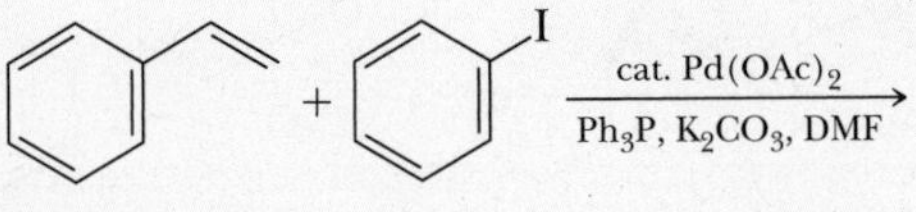

Solution

In (a), 1-iodohexene has the *E* configuration, and this double bond retains its configuration in the product. Furthermore, the carbon-carbon double bond adjacent to the ester in the product now has the possibility for *cis,trans* isomerism. The Heck reaction is highly stereoselective, and this double bond has the more stable *E* configuration as well. In (b), the major product is (*E*)-1,2-diphenylethene.

(a)

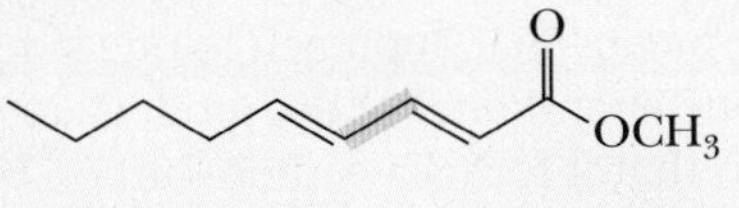

Methyl (2*E*,4*E*)-2,4-nonadienoate

(b)

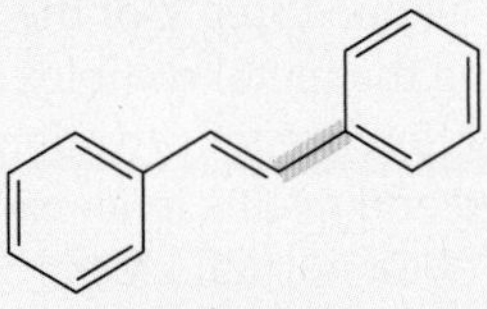

(*E*)-1,2-Diphenylethene
(*trans*-Stilbene)

(Continued)

Problem 24.1

Show how you might prepare each compound by a Heck reaction using methyl 2-propenoate as the starting alkene.

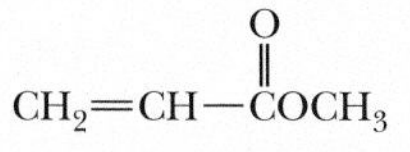

Methyl 2-propenoate (Methyl acrylate)

(a)

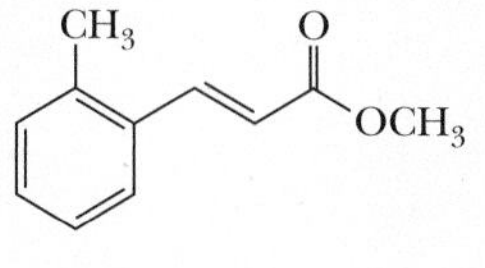

(the *E* isomer)

(b)

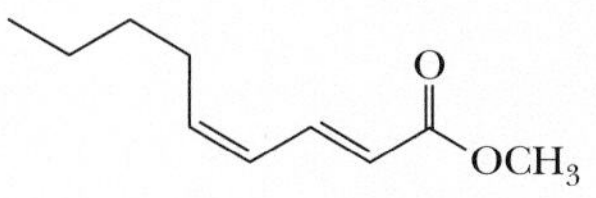

(the 2*E*,4*Z* isomer)

The usual pattern in a Heck reaction of acyclic alkenes is replacement of one of the hydrogens on the double bond by an organo group. If the organopalladium group attacks the double bond so that the R-group in the original **RX** is bonded to a carbon that lacks a hydrogen (or if the only syn hydrogen is on a neighboring carbon), the double bond shifts away from the original position. Note that the product of the following reaction contains a chiral center, but because it is formed from achiral reagents in an achiral environment, it is formed as a racemic mixture.

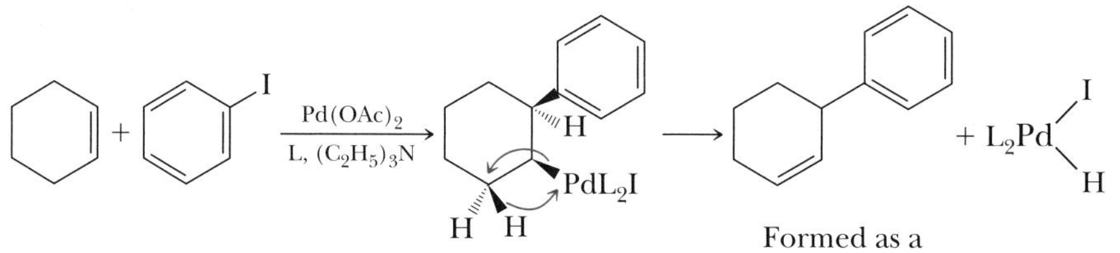

Formed as a racemic mixture

As mentioned earlier, a particularly valuable feature of the Heck reaction is that, when used with a chiral ligand, it can give chiral products in significant enantiomeric excess (ee). In the following, the chirality is provided by the chiral ligand (*R*)-BINAP (Section 6.7C).

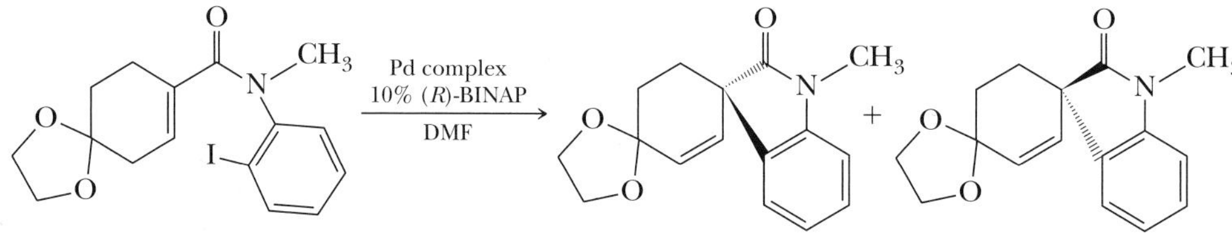

This enantiomer is formed in 71% ee

For this reaction to yield a chiral product, the hydrogen eliminated cannot be on the carbon that subsequently obtains the aryl substituent because if this were the case, the substituent would be attached to a double bond and the product would be achiral.

Because of the chiral ligand, the activation energy for the transition state in the syn addition to the alkene (Step 2 of the catalytic cycle) is different depending on which side of the alkene the metal complex approaches (the two transition states are diastereomers). This difference in activation energy means that approach to one side of the alkene is favored and results in an excess of one enantiomer of the product. Note that this reaction is not a normal Heck reaction in that it forms a carbon-carbon bond to the more substituted carbon and the double bond shifts. Attack at the other carbon, because of the requirement for syn elimination, cannot lead to a normal Heck product; therefore, the reaction reverses. The attack takes place at the more substituted carbon less often, but in this case, there is a hydrogen that can undergo elimination.

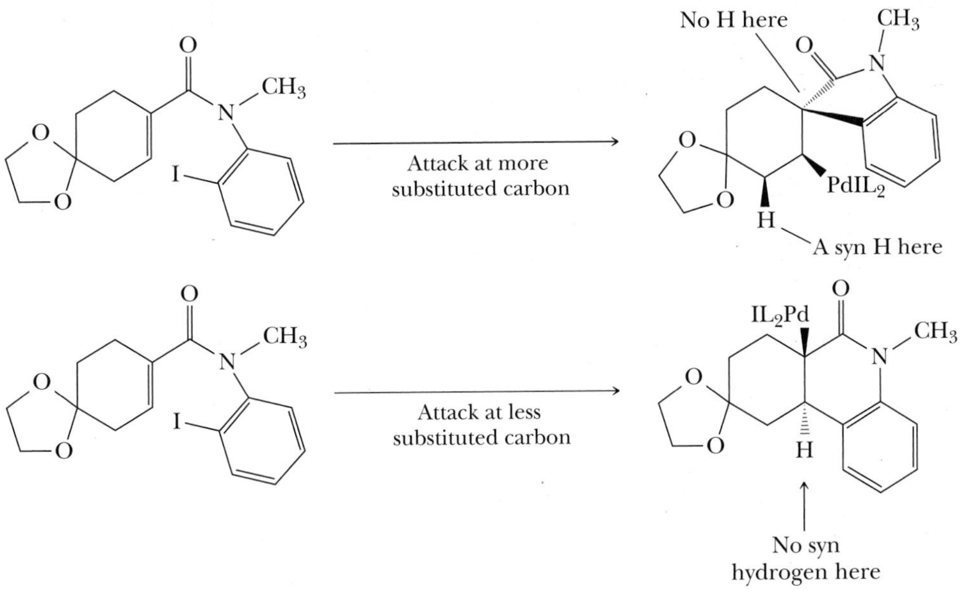

Example 24.2 Heck Reaction II

Heck reaction of bromobenzene and (*E*)-3-hexene gives a mixture of (*Z*)-3-phenyl-3-hexene and (*E*)-4-phenyl-2-hexene in roughly equal amounts. Account for the formation of these two products.

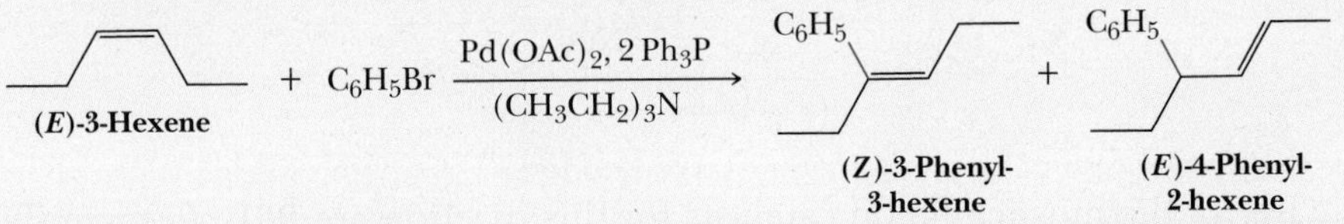

Solution

Syn addition gives the product shown. After rotation, syn elimination of the H on the original double bond gives (*Z*)-3-phenyl-3-hexene; syn elimination on the neighboring carbon (in its most stable conformation) gives (*E*)-4-phenyl-2-hexene.

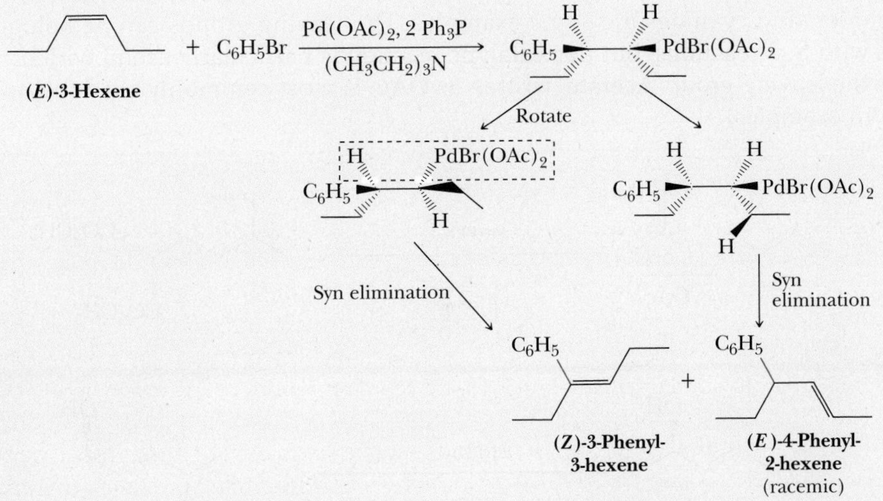

(Continued)

Problem 24.2

Give reagents and conditions for the following reaction.

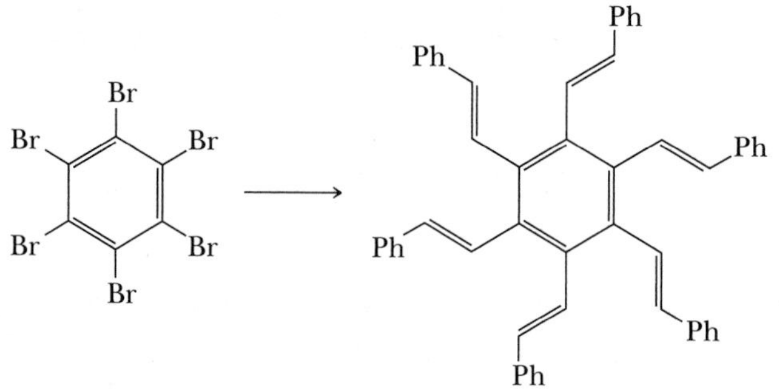

24.4 Catalytic Allylic Alkylation

Substitution mechanisms were covered in Chapter 9. It was noted that the S_N2 mechanism occurs by a single-step process wherein a leaving group (often a halogen) on an alkyl group is replaced with a nucleophile. When the alkyl group is allylic, metals can catalyze the reaction (i.e., palladium, platinum, and rhodium, among others).

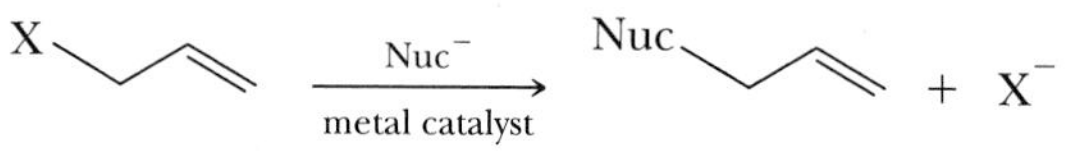

Two of the most common catalysts for this reaction are PdL_4 (frequently, L is triphenylphosphine) and $PdCl_2$. Two examples of the reaction are given below. One interesting feature is the retention of stereochemistry at the carbon with the leaving group. Note that this outcome is the opposite of what would occur in an S_N2 mechanism. A particularly useful feature of the reaction is the ability to use enolates as the nucleophile, thus resulting in C—C bond formation. The enolates most commonly used are those derived by deprotonation of a hydrogen that is alpha to two electron-withdrawing groups; ketones, aldehydes, nitro, esters, sulfonates, and cyanides are some examples. The leaving groups can be halogens as with S_N2 reactions, but this catalytic reaction is particularly useful with esters as the leaving group. Acetate, written as OAc, is most commonly used (below in both examples).

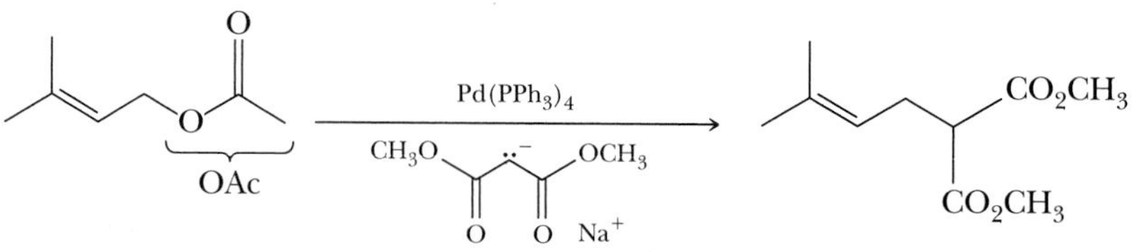

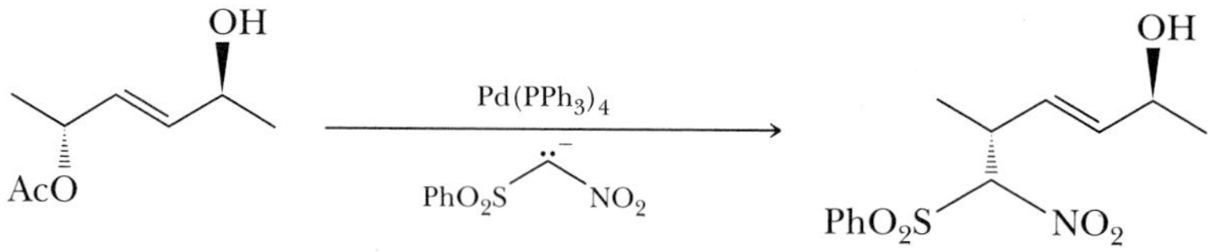

Example 24.3 Allylic Alkylation

Write the products of the following reactions.

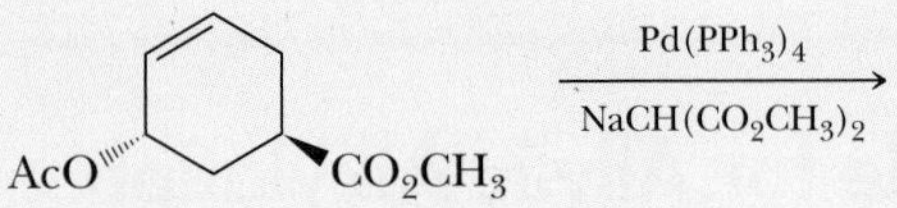

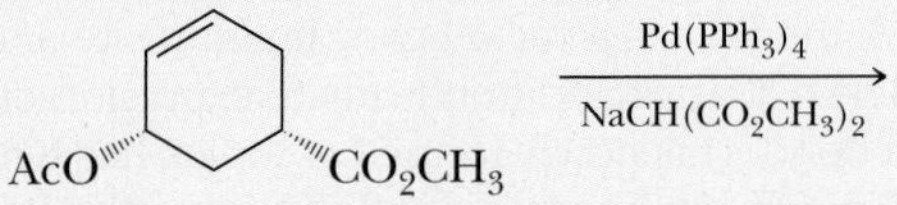

Solution

As stated above, the reaction occurs with retention of stereochemical configuration at the carbon with the acetate leaving group. Consequently, in order to write the products, we simply replace the acetate with the enolate, writing a C—C bond between the enolate carbon and the carbon that bears the acetate.

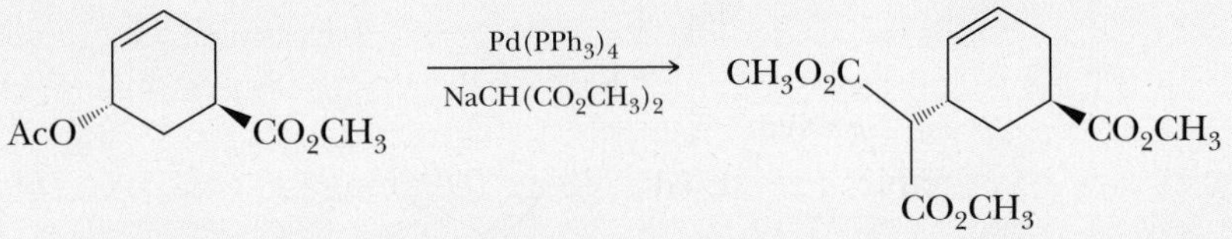

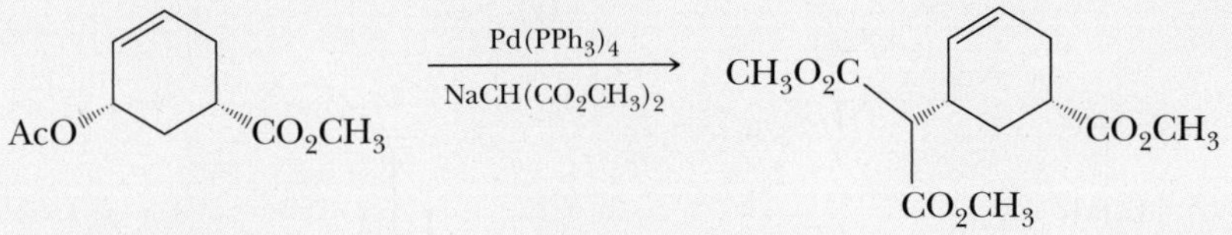

Problem 24.3

Write the product(s) of the following reaction. How many stereoisomers of the product are formed?

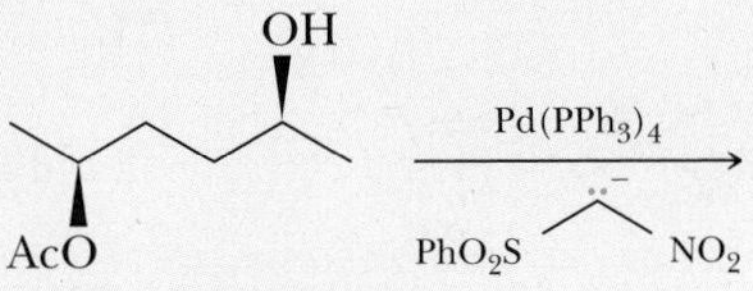

Watch a video explanation

Although allylic alkylation with halogens as the leaving group occurs readily without a catalyst with enolate nucleophiles, several advantages exist in having the reaction be catalytic. There is, of course, the advantage of having the reaction occur faster and potentially under milder conditions. Also, acetate is not normally a good leaving group in an S_N2 reaction. However, the real primary advantage is in reversing the stereochemical outcome of the reaction as compared to S_N2. Before examining the reason behind the stereochemistry of the reaction, let's first take a look at the mechanism involving one particularly common catalyst, $Pd(PPh_3)_4$.

A. The Mechanism of Catalytic Allylic Alkylation

The mechanism of the reaction is a combination of the simple steps we have seen before. The phosphine ligands reversibly dissociate and associate, and an oxidative addition occurs. The new step that occurs in catalytic allylic alkylation is nucleophilic attack on an allyl ligand coordinated to a metal.

Mechanism 24.2

The Catalytic Cycle for Allylic Alkylation

The catalytic cycle of allylic alkylation has several steps, six of which are shown below. The cycle is initiated by dissociation of a ligand (L=PPh_3, Step 1), followed in Step 2 by coordination of the allylic species to make a π complex and another ligand loss (Step 3). After Step 3, the **Pd** has a vacant coordination site and can undergo oxidative addition of the coordinated allylic species. The oxidative addition leads to expulsion of the leaving group and replacement of the allyl-**X** bond with an allyl-**Pd** bond. This newly formed complex has two contributing structures in which either terminal carbon of the allyl group can be envisioned as having the **Pd-C** bond with the remaining two carbons involved in a π complex. It is common in organometallic chemistry to represent the two contributing structures involved in an allyl complex with three carbons and an arc (see the following figure). The interaction of three carbons to one metal is denoted by the prefix η^3; such a complex is called an η^3-allyl complex (η is pronounced "eta").

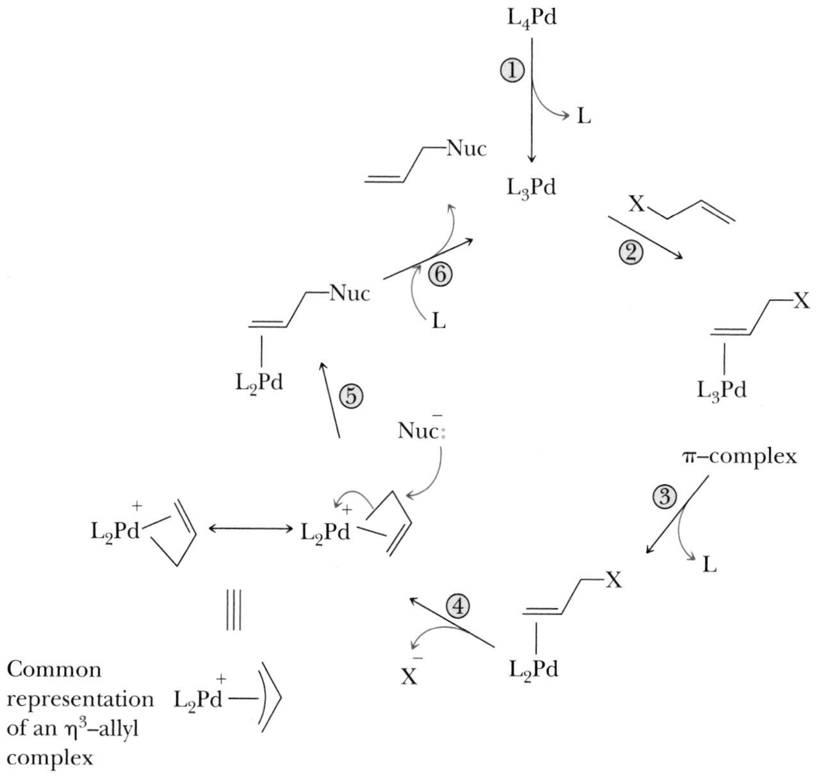

After the oxidative addition, the coordinated allyl group is susceptible to nucleophilic attack from solution (Step 5). This attack completes the substitution, and all that is left to start the cycle again would be coordination of a phosphine ligand (**L**) and loss of the organic product.

B. Stereochemical and Regiochemical Issues

Several steps in the catalytic cycle are stereoselective and therefore result in the ability to control the configuration at chiral carbons. First, the oxidative addition of the allylic-LG species occurs with clean inversion of configuration (Step 4 of the mechanism). The nucleophilic attack on the η^3-allyl complex occurs from solution in an analogous fashion to an S_N2 reaction and therefore occurs with clean inversion of stereochemistry. The net effect of two consecutive inversions of stereochemistry is overall retention of stereochemistry. The following example highlights the stereochemistry of the two steps that we are considering (Y is an electron-withdrawing group).

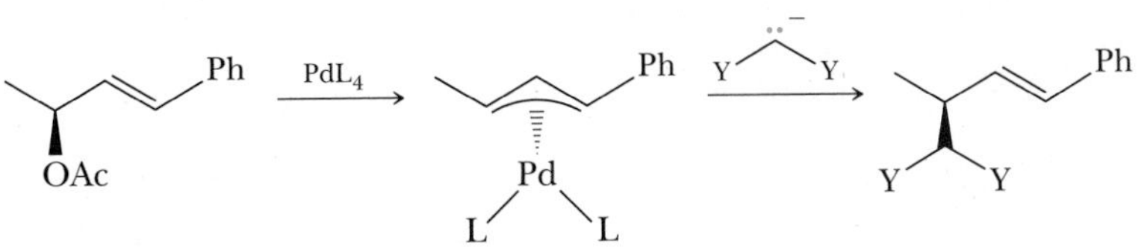

The reaction is also very regioselective. Nucleophilic attack occurs at the less substituted end of the η^3-allyl complex regardless of the initial position of the leaving group.

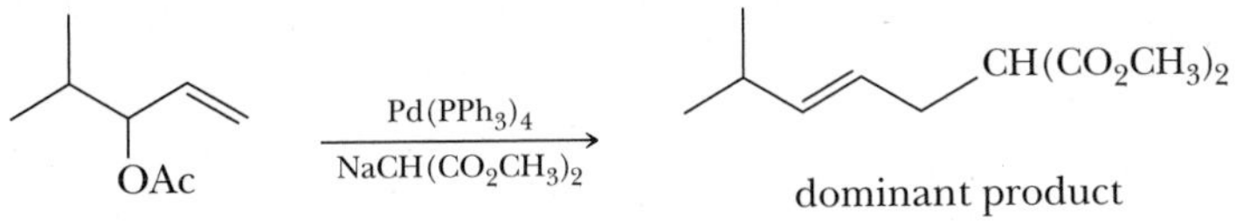

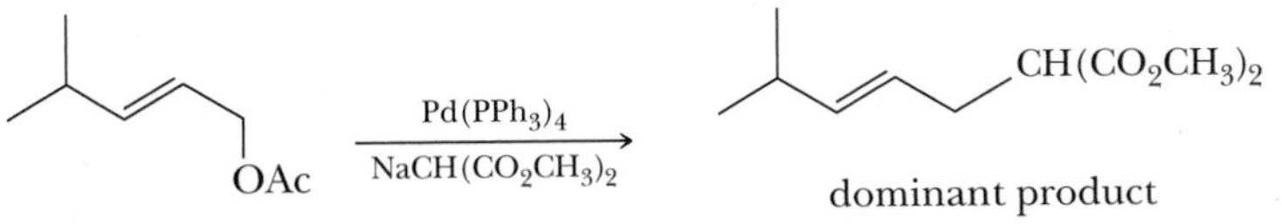

24.5 Palladium-Catalyzed Cross-Coupling Reactions

Arguably, the largest impact that organometallic chemistry has had on organic synthesis involves a series of reactions that are classified as **cross-coupling reactions**, many of which are catalyzed by palladium. A cross-coupling reaction is defined as a reaction that creates a C—C bond by coupling together two alkyl, aryl, alkenyl, or alkynyl groups, as we saw with the Gilman reaction in Section 15.2. Yet, the reactions we are now examining are catalytic. There are a large number of these reactions, most of which are named after the chemists primarily associated with their creation. We will examine three in this book: the Suzuki, Stille, and Sonogashira couplings.

Catalytic cross-coupling reaction

A reaction wherein a C—C bond is formed in a catalytic fashion between alkyl, aryl, alkenyl, or alkynyl groups.

Cross-coupling reactions involve a transmetallation step. A **transmetallation** is a pairwise interchange of ligands between two different metals or metalloids. In the case of palladium-catalyzed cross-coupling reactions, the other metal/metalloid is commonly Zr, Sn, B, Zn, Cu, or Mg, which we designate as M in the following general example.

Transmetallation

Interchange of ligands between two metals or metalloids.

$$\text{R—Pd} + \text{R}'\text{—M} \longrightarrow \text{R}'\text{—Pd} + \text{R—M}$$

A. General Mechanism for Cross-Coupling Reactions

For the sake of simplicity, all the catalytic cross-coupling reactions can be represented by one general mechanism, even though each has subtle differences that we describe in following sections. The catalytic cycle is so simple that it is presented here with only three steps. The differences between M, L, L′, and X are what differentiate and classify a particular reaction.

In 2010, Richard Heck, Ei-ichi Negishi, and Akira Suzuki shared the Nobel Prize in Chemistry for their work on C—C coupling reactions.

Mechanism 24.3

The Catalytic Cycle of Cross-Coupling

Various Pd(0) or Pd(II) species are used in the catalytic reactions. Step 1 involves oxidative addition of one of the organic species to the palladium. A transmetallation in Step 2 results in the palladium having two carbon-based ligands. A reductive elimination in Step 3 couples the two carbon fragments together. The relative simplicity and variability of each component involved has made this catalytic cycle a very powerful one.

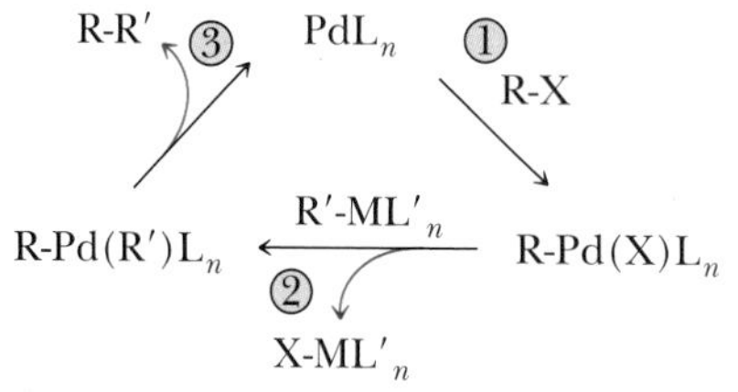

B. The Suzuki Coupling

The Suzuki coupling was developed by Professor Akira Suzuki of Hokkaido University. The Suzuki coupling uses a boron compound ($R\text{-}BY_2$) and an alkenyl, aryl, or alkynyl halide or triflate (RX) as the carbon sources, with a palladium salt as the catalyst. Bromides and iodides are the most commonly used halides; chlorides are less reactive. Haloalkanes can sometimes be used but are subject to elimination. A base is also required. The boron compound can be a borane (R'_3B), a borate ester ($R'B(OR)_2$), or a boric acid ($R'B(OH)_2$), where R′ is alkyl, alkenyl, or aryl. The general reaction is shown in the following scheme, where X is halide or triflate and Y is alkyl, alkoxyl, or OH. A list of the types of components that can be used is given in Table 24.1. This reaction is one of the principal methods now used to prepare biaryls.

$$RX + R'\text{-}BY_2 \xrightarrow[\text{Base}]{PdL_4} R\text{-}R' + XBY_2$$

Boranes are easily prepared from alkenes or alkynes by hydroboration (Section 6.4); borates are made from aryl or alkyl lithium compounds and trimethyl borate, among other routes.

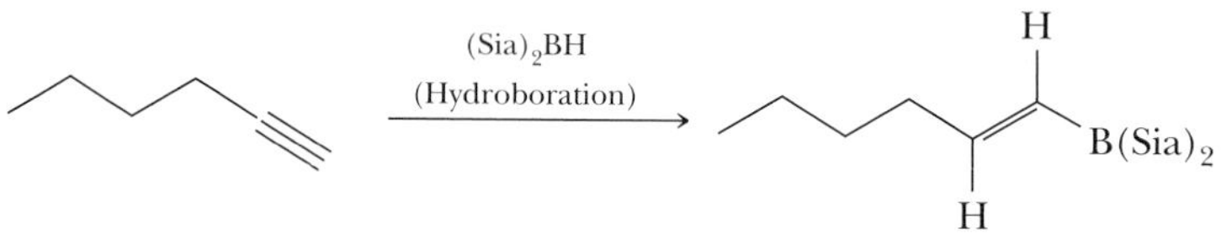

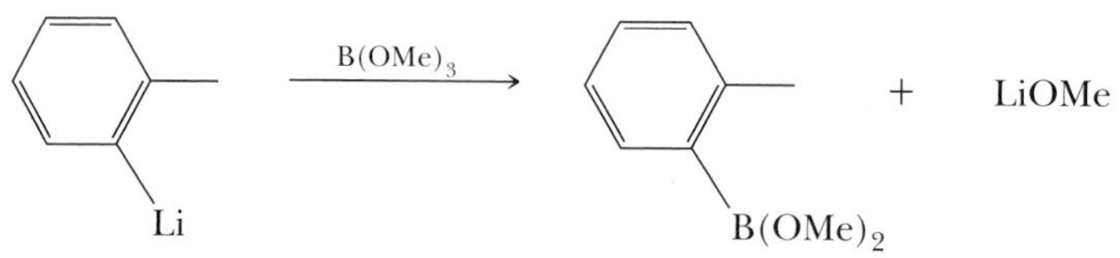

Following are three examples of the reaction that show its versatility.

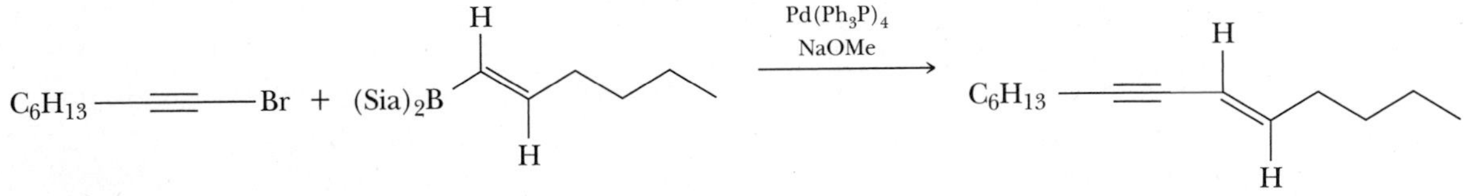

C_6H_{13}—≡—Br + $(Sia)_2B$
H
H
$Pd(Ph_3P)_4$
NaOMe
C_6H_{13}
H
H

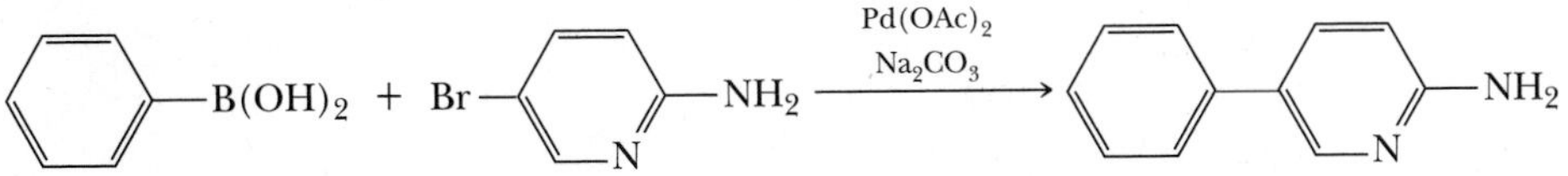

$B(OH)_2$ + Br
N
NH_2
$Pd(OAc)_2$
Na_2CO_3
N
NH_2

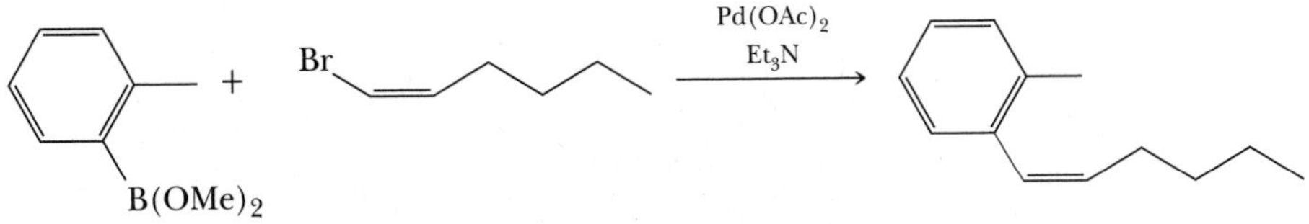

+
Br
$B(OMe)_2$
$Pd(OAc)_2$
Et_3N

Table 24.1 Suzuki Coupling Components Where One of the Organoboron Compounds Couples with One of the Coupling Reagents Shown

Organoboron Compounds	Coupling Reagents X = halide or triflate
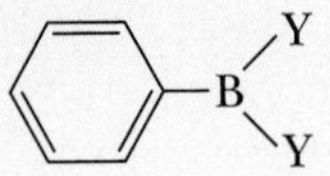	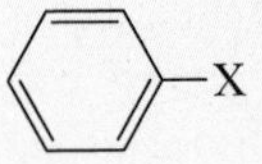
RCH=CH—BY_2	RCH=CH—X
Alkyl—BY_2	RC≡CH—X Alkyl-X (Difficult)

The mechanism of the reaction starts with an oxidative addition, followed by a transmetallation in which the substituent on the borane replaces the ligand on the palladium, concluding with a reductive elimination of the palladium to form the new C—C bond. The base may serve as a new, labile ligand for the palladium, or the base may activate the borane by coordination.

Oxidative addition and ligand exchange:

$$R{-}X \xrightarrow{PdL_n} R{-}Pd{-}X \xrightarrow{RO^-} R{-}Pd{-}OR$$

Borane activation:

$$Y_3B \xrightarrow{RO^-} Y_3B^-{-}OR$$

Reaction:

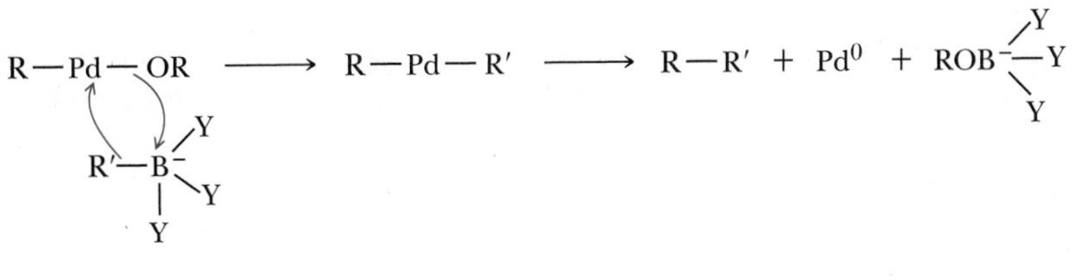

R—Pd—OR
R'—B
Y
Y
Y
R—Pd—R'
R—R' + Pd^0 + ROB^-
Y
Y
Y

Example 24.4 Suzuki Coupling

Show how the following penicillin analog can be prepared from the indicated starting material and any other necessary compounds.

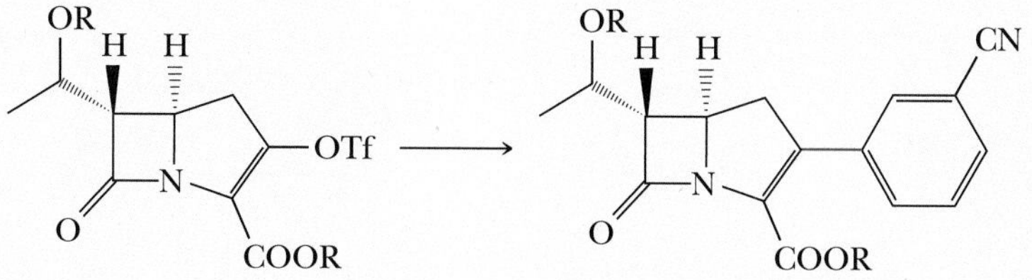

Solution

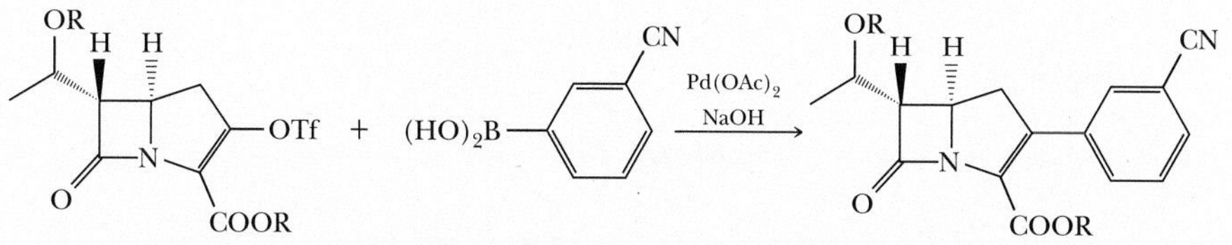

Problem 24.4

Show how the following compound can be prepared from starting materials containing eight carbons or less.

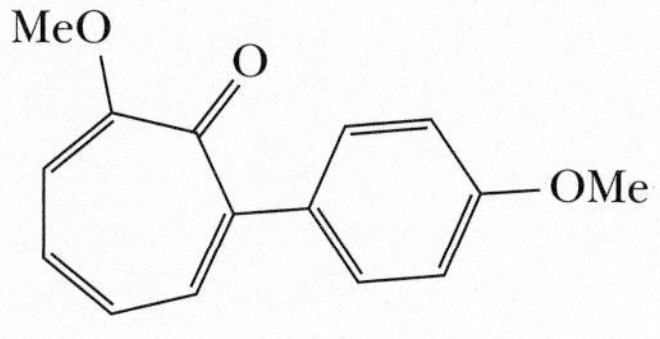

C. The Stille Coupling

The second Pd(0)-catalyzed cross-coupling reaction we cover is the Stille coupling, which involves the use of vinyl or aryl tin reagents (called stannanes) as the transmetallating agents. Coupling with another vinyl or aryl group leads to the creation of conjugated dienes or an alkenylarene. The coupling occurs with regioselectivity and retains the stereochemistry from the reactants to the products.

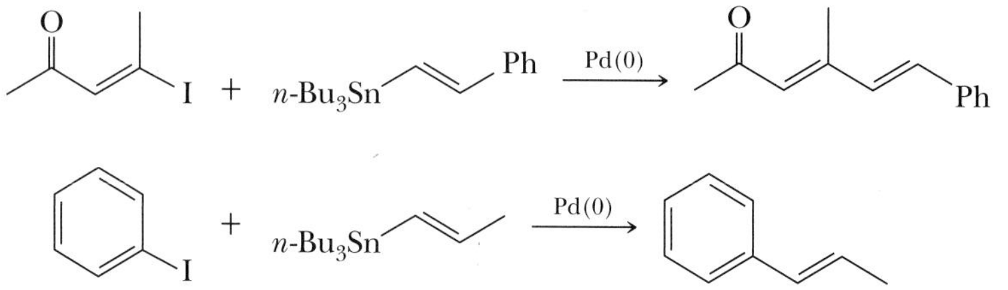

Many stannane reagents are commercially available, or they can be readily synthesized via reaction between a Grignard reagent and tri-n-butyl tin chloride. The reactants that most commonly react with the transmetallated group are a vinyl triflate (C═C—OSO_2CF_3) and a vinyl iodide. Vinyl triflates are prepared from the reaction of an enolate with N-phenyl triflimide ($PhNTf_2$, Tf = SO_2CF_3).

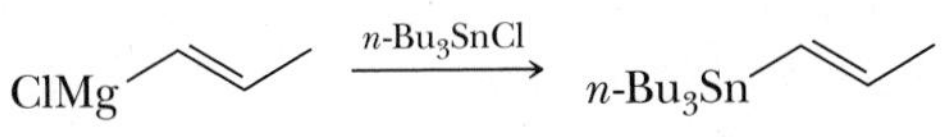

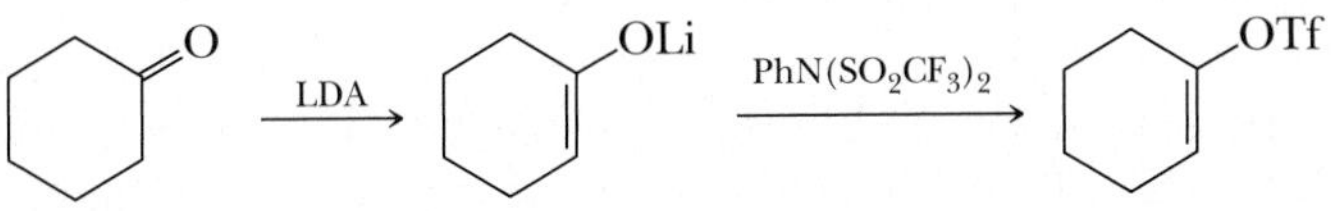

Example 24.5 Stille Coupling

When a conjugated diene is the desired product in a reaction, a Stille coupling is a logical choice to invoke during the synthesis. Write the correct stannane starting material and a vinyl iodide reactant that would couple with **Pd(0)** to give the following product.

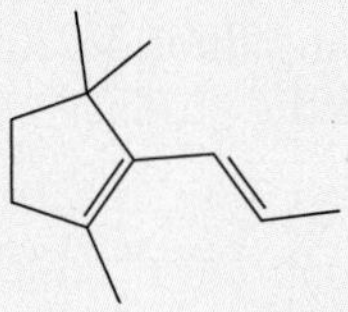

Solution

To consider the proper starting materials for a Stille coupling, dissect the central C—C bond of the diene into two parts in a retrosynthetic fashion. One reactant should be a vinyl iodide (or triflate) and the other reactant a stannane.

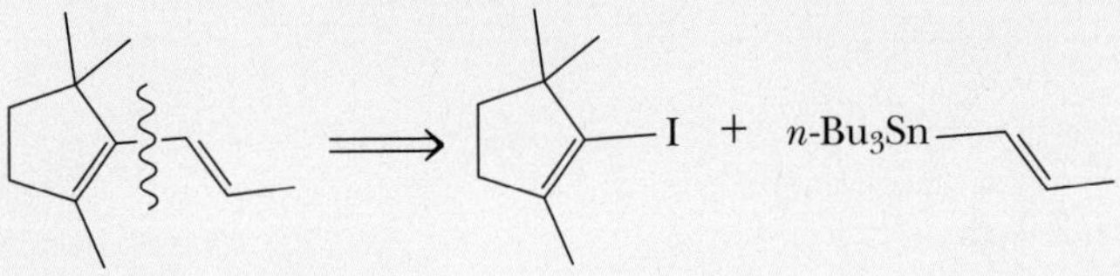

Problem 24.5

What is the product of the following reaction?

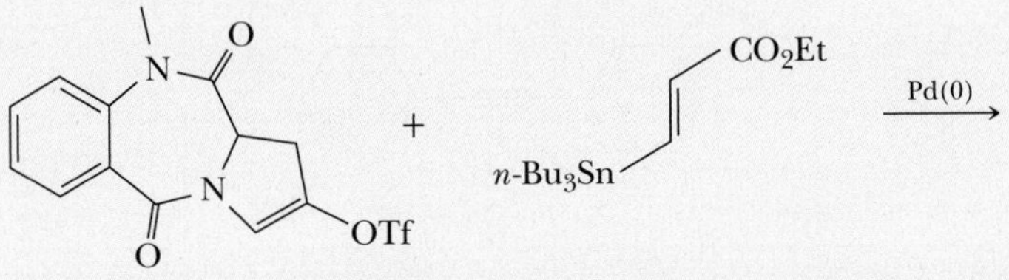

D. The Sonogashira Coupling

The last Pd(0)-catalyzed cross-coupling reaction covered in this chapter is the Sonogashira coupling. It involves transmetallation of an alkynyl-Cu(I) species to **Pd** followed by coupling to an aryl or vinyl iodide or triflate. The **Cu(I)** alkynyl complex is created in situ by the reaction of a terminal alkyne with **CuI** in the presence of triethylamine. The reaction is most commonly used to create diaryl alkynyl products.

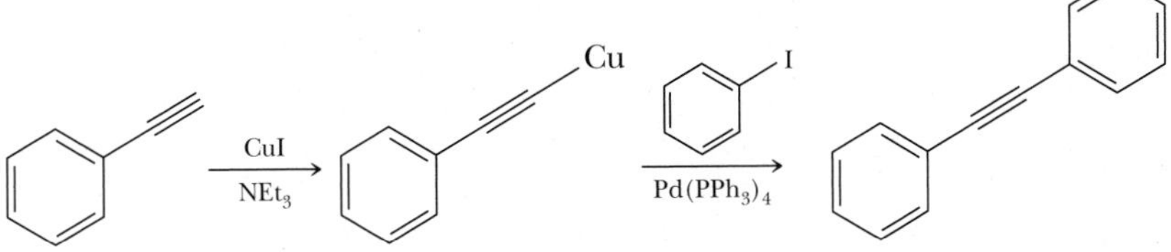

Example 24.6 Sonogashira Coupling

Two Sonogashira coupling reactions can be used to make unsymmetrical diaryl alkynes by first using trimethylsilylacetylene. The trimethylsilyl protecting group can be removed by addition of fluoride (usually tetrabutylammonium fluoride, see Section 11.6). Show how such a sequence of reactions can be used to construct the following product when one reactant is phenyl iodide.

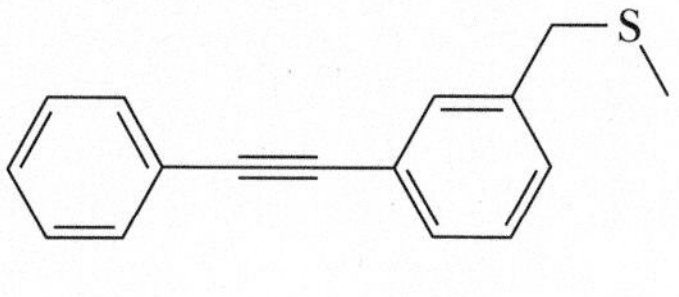

Solution

Sonogashira coupling conditions using trimethylsilylacetylene give phenyl acetylene after deprotection using fluoride. Another such coupling using the aryl iodide reactant shown gives the product.

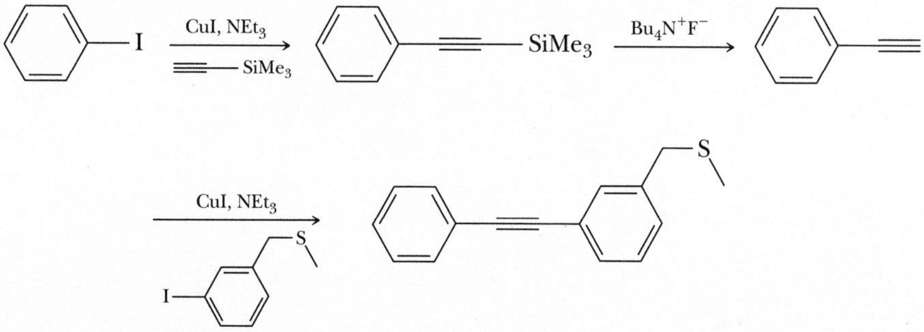

Problem 24.6

What sequence of reactions will produce the following product if starting with trimethylsilylacetylene and the appropriate two aryl iodides?

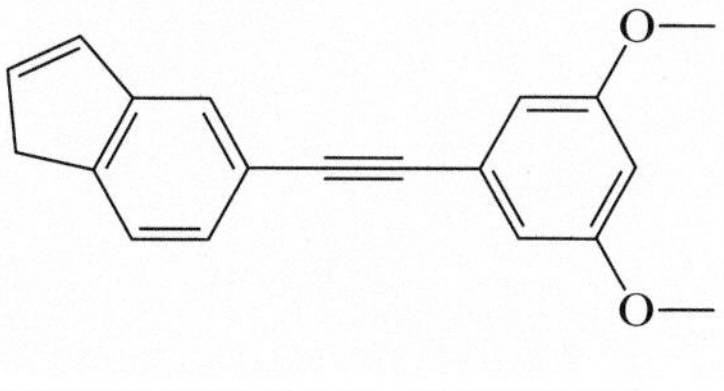

In 2005, Robert Grubbs, Richard Schrock, and Yves Chauvin shared the Nobel Prize in Chemistry for their work on metathesis reactions.

Alkene metathesis

In an alkene metathesis reaction, two alkenes interchange the carbons attached to their double bonds.

24.6 Alkene Metathesis

A novel catalytic reaction leading to alkene metathesis has been developed. Robert Grubbs of the California Institute of Technology and Richard Schrock of the Massachusetts Institute of Technology made major contributions to this chemistry. Together their work has provided a remarkably easy and general way to generate carbon-carbon double bonds, even in complex molecules. In an **alkene metathesis** reaction, two alkenes interchange the carbons attached to their double bonds.

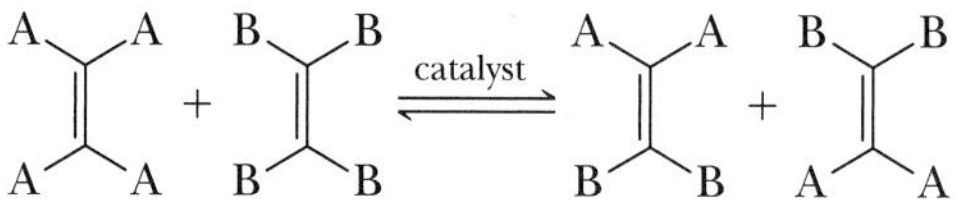

Watch a video explanation

A. Stable Nucleophilic Carbenes

We discussed carbenes and carbenoids (derivatives of divalent carbon) in Section 15.3, where we saw that these compounds provide one of the best routes to three-membered rings, making two C—C bonds in the process. Certain carbenes with strongly electron-donating substituents are particularly stable. Their stability can be enhanced further by adding sterically bulky substituents that hinder self-reactions. For example, the following cyclic carbene is stable enough to isolate. In this case, the large 2,4,6-trimethylphenyl substituents protect the carbene from attack by electrophiles or oxygen. Rather than being electron deficient like most carbenes, these compounds are nucleophiles because of the strong electron donation by the nitrogens. Because of their nucleophilicity, they are excellent ligands (resembling phosphines) for certain transition metals.

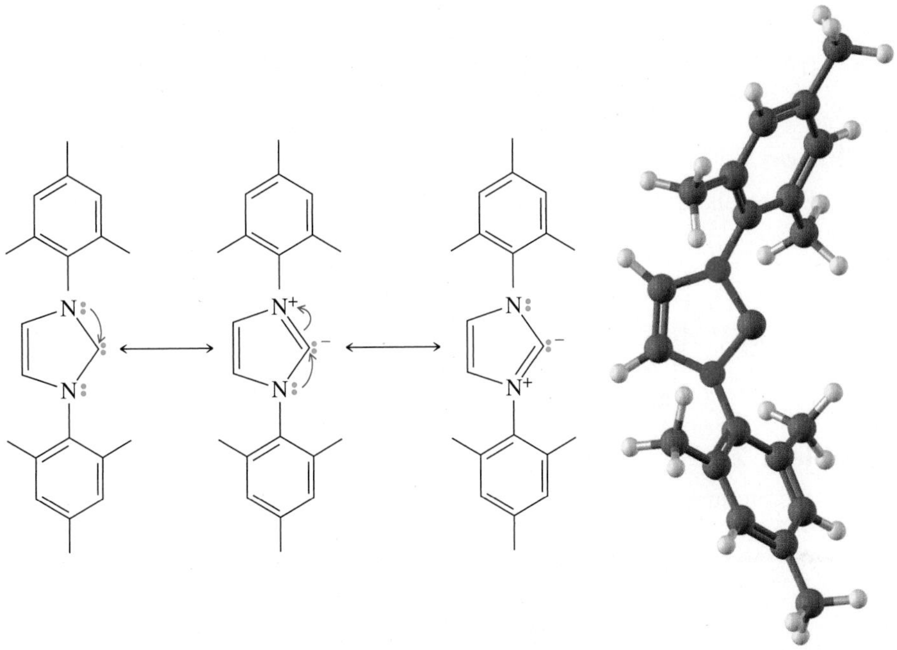

B. Ring-Closing Alkene Metathesis Using Nucleophilic Carbene Catalysts

These stable carbenes (and others that are less stable) provide ligands for certain metals that are catalysts for the alkene metathesis reaction. As we saw at the beginning of this section, this reaction is an equilibrium. However, it can be an effective means of forming new carbon-carbon double bonds if the equilibrium can be driven in the desired direction. For example, if the reaction involves two 2,2-disubstituted alkenes of the type $R_2C{=}CH_2$, one of the products is ethylene. Loss of gaseous ethylene drives the reaction to the right, giving a single alkene as product.

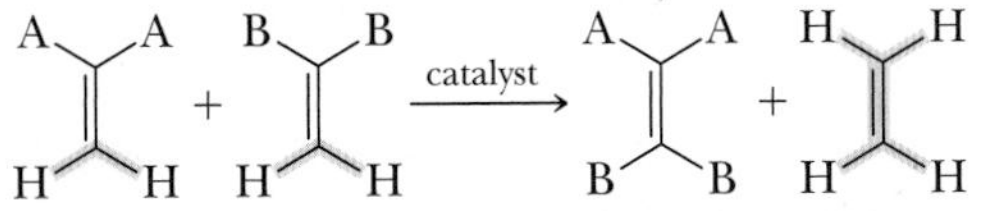

Ethylene

A particularly useful variant of this reaction uses a starting material in which both alkenes are in the same molecule. In this case, the product is a cycloalkene and the reaction is called ring-closing alkene metathesis. Ring sizes up to 26 and higher have been prepared by ring-closing alkene metathesis. This reaction is amazingly general and synthetically useful.

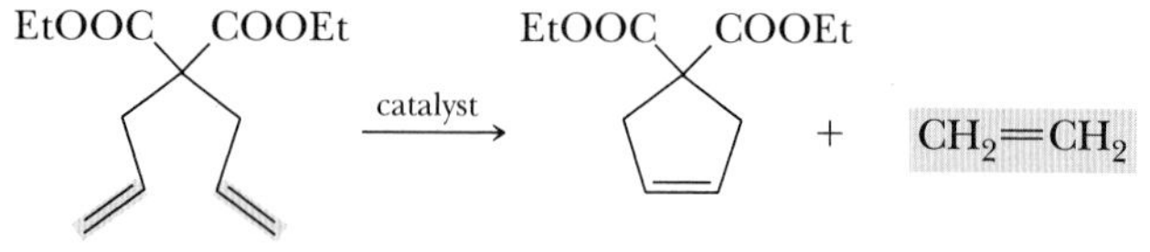

Example 24.7 Ring-Closing Metathesis

Show how the following compound can be prepared from an acyclic diene.

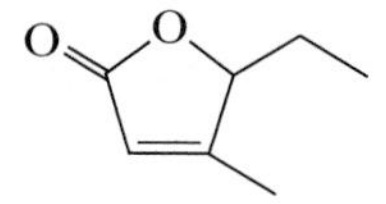

Solution

Ring-closing alkene metathesis gives the product in one step.

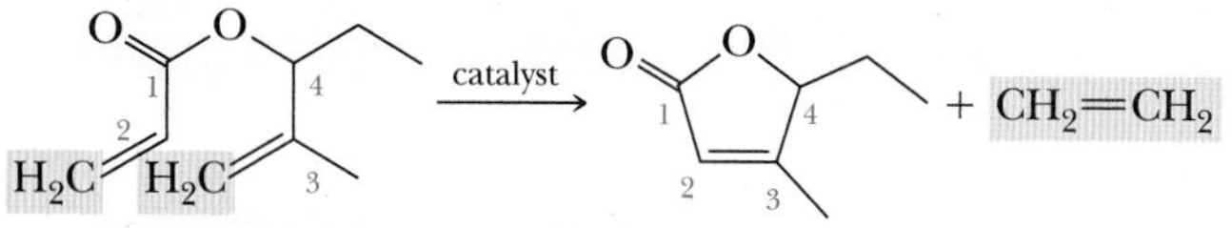

(racemic) (racemic)

Problem 24.7

Show the product of the following reaction.

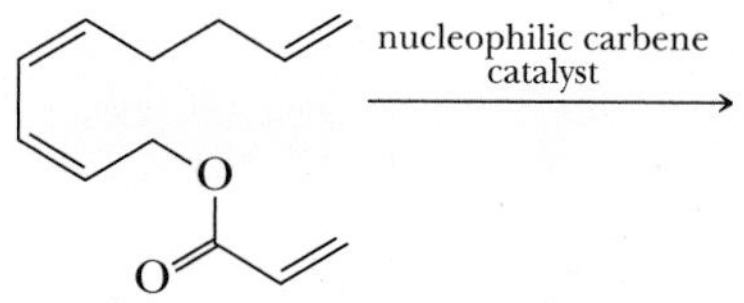

Particularly useful alkene metathesis catalysts consist of ruthenium complexes with a nucleophilic carbene and another carbenoid ligand, $C_6H_5CH{=}[M]$, where [M] is the metal with its ligands.

For a model of the catalyst shown here, see the opening page of this chapter

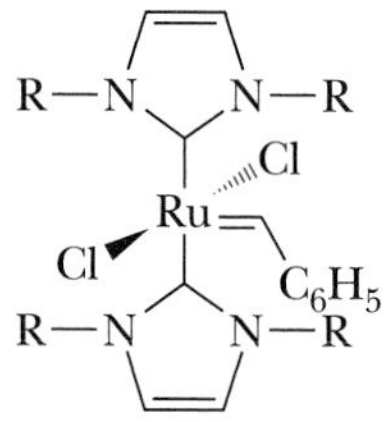

C. Mechanism of the Metathesis Reaction

The mechanism of the alkene metathesis reaction also involves a catalytic cycle. A key step involves addition of the metallocarbenoid to the alkene to give a four-membered metallacycle. This metallacycle is unstable and can either revert to starting material or eliminate an alkene in the opposite direction to give a new alkene. Addition is not regioselective; consequently, all possible combinations of R_1 and R_2 result. In this scheme, the catalyst is $R_1CH{=}[M]$.

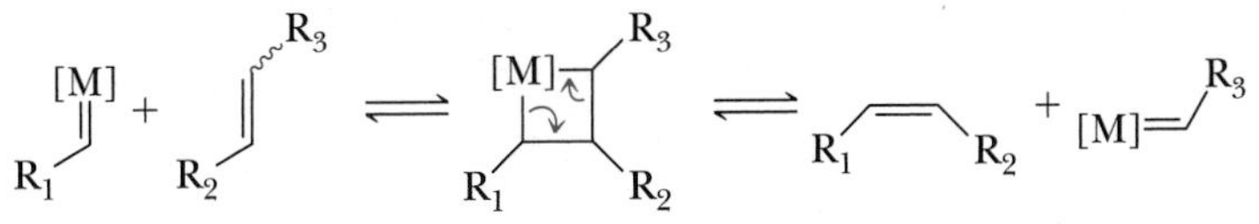

cis or *trans* A metallacycle

In this section, we have concentrated on the use of transition-metal nucleophilic-carbene catalysts to bring about ring-closing alkene metathesis reactions. These same types of compounds can also be used to catalyze a remarkable reaction called ring-opening alkene metathesis polymerization (ROMP). A special value of ROMP is that it can be used to prepare highly unsaturated polymers. For a discussion of ROMP techniques, see Section 29.6E.

Study Guide

24.1 Carbon-Carbon Bond-Forming Reactions from Earlier Chapters

- The classical methods for making carbon-carbon bonds during synthesis can be grouped broadly into these categories:
  - *Displacement of a leaving group by a carbon nucleophile (Gilman reagents, alkyne anions, enolate anions, and enamine alkylations)*
  - *Nucleophilic addition to a carbonyl or a carboxyl group, usually involving enolate nucleophiles (Grignard, alkyne anion, cyanide, aldol, Claisen, enamine, and Wittig)*
  - *Conjugate addition to an α,β-unsaturated compound (Michael reaction)*
  - *Aromatic substitution (Friedel-Crafts)*

24.2 Organometallic Compounds and Catalysis

- Two important reactions of transition metals and their compounds are **oxidative addition** and its complement, **reductive elimination**.
  - *Oxidative addition occurs when a reagent adds to a metal, causing its coordination to increase by two ligands, and reductive elimination is the reverse.*
  - *The terms* oxidative *and* reductive *refer to the change in formal charge on the metal that occurs during these reactions.*
  - *Reagents such as organohalogen, hydrogen, halogens, and many others can react with metals in these ways.*
- The catalytic properties of transition metals are manipulated by adding different **ligands**, which are Lewis bases that coordinate to the metal.
  - *Ligands can be used to modify the electronic properties, steric crowding, and even chirality around the metal in some situations.*

- **Green chemistry** and **atom economy** are relatively new thrusts in the chemical industry worldwide; their goals are well complemented by organometallic catalytic reactions.

24.3 The Heck Reaction

- In the **Heck reaction**, the H atom on an alkene (vinylic hydrogen) is substituted by a haloalkene or haloarene in the presence of base and a small amount of Pd catalyst.
  - *When there is a difference, the substitution occurs at the less substituted carbon of the alkene, and is often stereoselective for the* E *product.*
  - *The configuration of the haloalkene (when appropriate) is conserved.*
  - *A significant advantage of the Heck reaction is that alcohol, ether, aldehyde, ketone, and ester functional groups are compatible with the reaction.*
  - *The organohalogen, alkene, and base are used in stoichiometric amounts, and the Pd catalyst is used in small amounts.*
  - *When there is no syn hydrogen from the original alkene double bond that can eliminate in the last step of the reaction, the double bond shifts away from the original position so that a syn elimination of an H atom can take place.*

P 24.1, 24.2, 24.8–24.18

The Heck Reaction

KEY REACTIONS

1. The Heck Reaction (Section 24.3) In a palladium(0)-catalyzed reaction, the carbon group of a haloalkene (a vinylic halide) or haloarene is substituted for a hydrogen on a carbon-carbon double bond (a vinylic hydrogen) of an alkene. Reaction generally proceeds with a high degree of both stereoselectivity and regioselectivity. The small amount of palladium catalyst is generally introduced as a precatalyst in the form of $Pd(OAc)_2$, which is Pd(II), and is reduced in the reaction to Pd(0) by reaction with the alkene (only a small amount of which is lost because the catalyst is used in small amounts) or a reagent such as triethylamine. The Pd(0) then reacts with two ligands, generally phosphine ligands (L), to create the active catalyst PdL_2. The phosphine ligands can be chiral, such as BINAP, so that single enantiomer products are possible for reactions that create new chiral centers.

The catalytic cycle involves five steps, the most important of which are oxidative addition of the organohalogen to the catalyst, syn addition of the alkene, syn elimination of the new alkene product, and reductive elimination of HX (neutralized by the added base) to regenerate the catalyst.

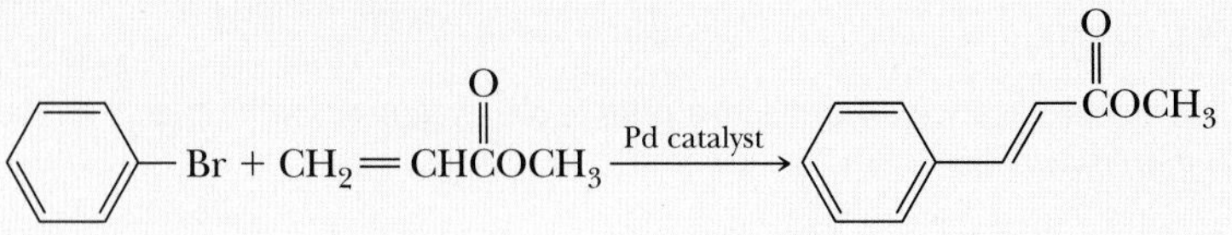

24.4 Catalytic Allylic Alkylation

- In catalytic allylic alkylation, a nucleophile, commonly an enolate of a doubly activated α-carbon, replaces an allylic leaving group, commonly a carboxylate such as acetate. The reaction occurs in the presence of a catalytic amount of Pd(0).
  - *In contrast to S_N2 allylic alkylation, stereochemistry at the alkylated carbon is retained.*

– *The mechanism consists of a multistep cycle involving η^3-allyl Pd complexes.*

– *The reaction is regioselective with allylation occurring preferentially at the less substituted end of the η^3-allyl complex irrespective of the initial position of the leaving group.*

P **24.3, 24.19–24.22**

Problem 24.3

KEY REACTIONS

2. **Catalytic Allylic Alkylation (Section 24.4)** Catalytic allylic alkylation commonly takes allyl acetate species and substitutes the acetate with a nucleophile. Some of the most useful nucleophiles are enolates derived from methylenes that are flanked by two electron-withdrawing groups. The mechanism of the reaction involves the oxidative addition of the allyl acetate to palladium and results in the intermediacy of η^3-allyl complexes that give inversion of configuration at the carbon with the acetate leaving group. Attack by the nucleophile from solution gives a second inversion of configuration and, by virtue of having two S_N2-like reactions, results in overall retention of configuration. The regiochemistry in the reaction is also highly selective, with nucleophilic attack from solution being preferential at the least substituted carbon of the η^3-allyl complex.

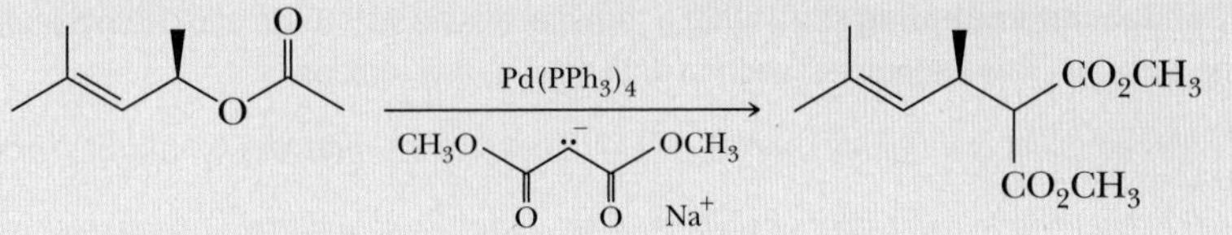

24.5 Palladium-Catalyzed Cross-Coupling Reactions

- Cross-coupling reactions all have very similar mechanisms. The first step is oxidative addition to Pd of an R-X species, followed by **transmetallation** of an R′ group from an R′-metal/metalloid species. Reductive elimination of R-R′ from palladium completes the cycle.
- The **Suzuki coupling** uses a boron reagent (R'-BY_2) with an alkenyl, aryl, or alkynyl halide (usually Br or I) or triflate with a palladium salt to give a new carbon-carbon bond.

  – *The boron compound can be a borane (R'_3B), a borate ester (R'-$B(OR)_2$), or a boric acid (R'-$B(OH)_2$), where R′ is an alkyl, alkenyl, or aryl group. Boranes are made using hydroboration of alkenes or alkynes. Borates are made from aryl or alkyl lithium compounds and trimethyl borate.*

  – *The Suzuki reaction is particularly good for the construction of biaryl compounds.*

P **24.3, 24.23, 24.24, 24.38**

KEY REACTIONS

3. **The Suzuki Coupling (Section 24.5B)** The Suzuki coupling reaction is a palladium-catalyzed reaction of an organoboron compound with an organic halide or triflate. The mechanism involves transmetallation, in which the substituent on the borane replaces a ligand on palladium, followed by reductive elimination to form the new C—C bond.

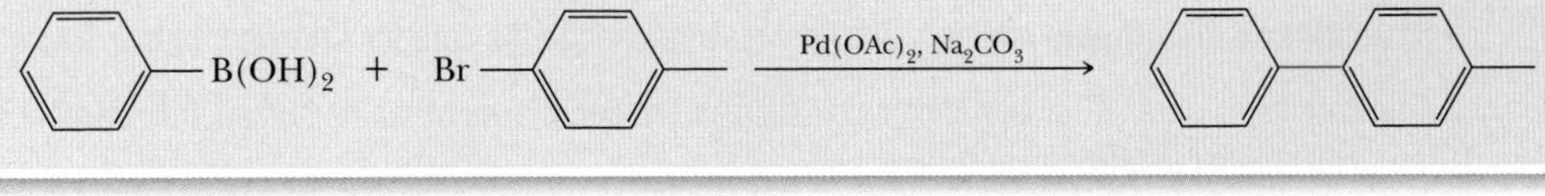

- The **Stille coupling** uses a tin reagent (Bu_3Sn-R) in which the R-group is commonly a vinyl species with an alkenyl, aryl, or alkynyl halide (often iodide or triflate) with palladium to give a new carbon-carbon bond.
  - *The tin reagent (a stannane) is created from a Grignard reagent of the R-group and n-Bu_3SnCl, and the triflates are often derived from enolates.*
  - *The Stille coupling is most commonly used to make conjugated dienes or alkenyl aryl systems.*

P 24.5, 24.25–24.31

KEY REACTIONS

4. The Stille Coupling (Section 24.5C) The Stille coupling is the palladium-catalyzed reaction of a vinyl tin reagent with an organic halide or triflate. The mechanism involves oxidative addition of the organic halide/triflate, transmetallation of the vinyl group on Sn to Pd, and reductive elimination to form the new C—C bond.

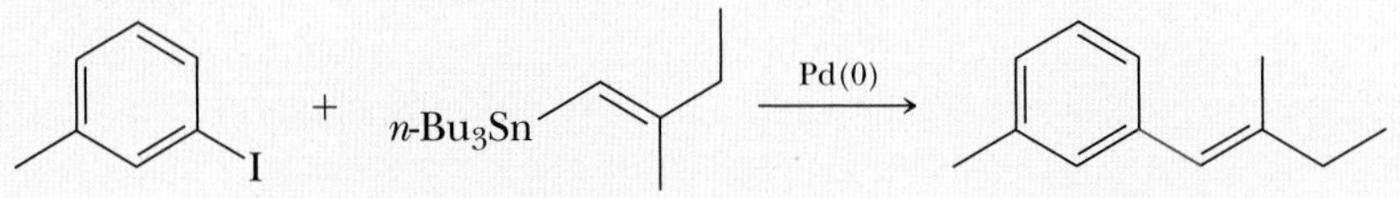

- The **Sonogashira coupling** starts with a terminal alkyne with CuI and triethylamine to create a Cu-alkyne complex. This undergoes reaction with vinyl or aryl iodides.
  - *This coupling is routinely used to create diaryl alkynes, aryl alkenyl alkynes, or dialkenyl alkynes.*
  - *It is common to use trimethylsilylacetylene with two sequential Sonogashira reactions when unsymmetrical alkynes are desired as the final products.*

P 24.6, 24.25–24.28, 24.31

Alkene Metathesis

KEY REACTIONS

5. The Sonogashira Coupling (Section 24.5D) The Sonogashira coupling is the palladium-catalyzed reaction of a Cu(I)-alkynyl complex with a vinyl or aryl iodide. The Cu(I)-alkynyl compound is created by the reaction of a terminal alkyne with CuI in the presence of an amine base. The coupling mechanism involves oxidative addition of the organic iodide to Pd, transmetallation of the alkynyl group to Pd from Cu, and reductive elimination to form the new C—C bond.

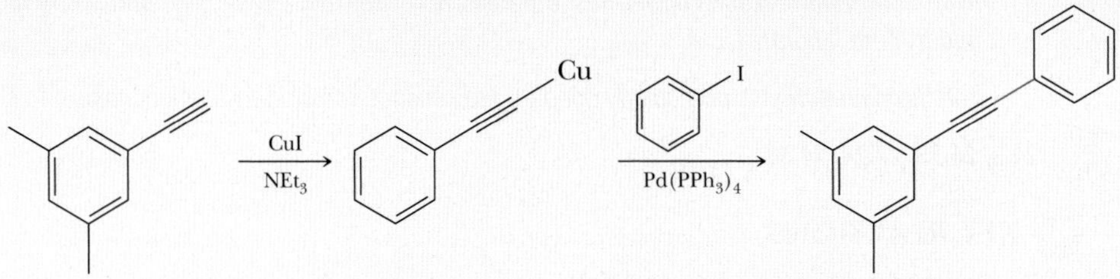

24.6 Alkene Metathesis

- In an **alkene metathesis** reaction, two alkenes interchange the carbons attached to their double bonds.
- The catalyst is a transition metal complex such as Ru complexes of stable nucleophilic carbenes (highly sterically hindered nitrogen heterocycles).

- The metathesis reaction is usually an equilibrium process driven to completion by using two terminal alkenes that give gaseous ethylene as a product, which bubbles out of the reaction.
  - *A particularly useful version of the metathesis reaction, called ring-closing alkene metathesis, involves two terminal alkenes on the same molecule, leading to an intramolecular reaction that creates a cycloalkene product.*
  - *Ring-closing alkene metathesis has been used to construct very large ring sizes that are hard to make in other ways.*

P 24.4, 24.32–24.34

KEY REACTIONS

6. Alkene Metathesis (Section 24.6) The alkene metathesis reaction is an organometallic-catalyzed reaction in which two alkenes exchange carbons of their double bonds. In a ring-closing alkene metathesis reaction, both alkenes are in the same molecule and the product is a cycloalkene. Catalysts with Ru are often used; a nucleophilic carbene complex of Ru is particularly useful. The catalytic cycle involves reaction of the metal catalyst with the alkenes to form a four-membered ring metallacycle, which decomposes to give starting materials or, by elimination in the opposite direction, to give a new alkene.

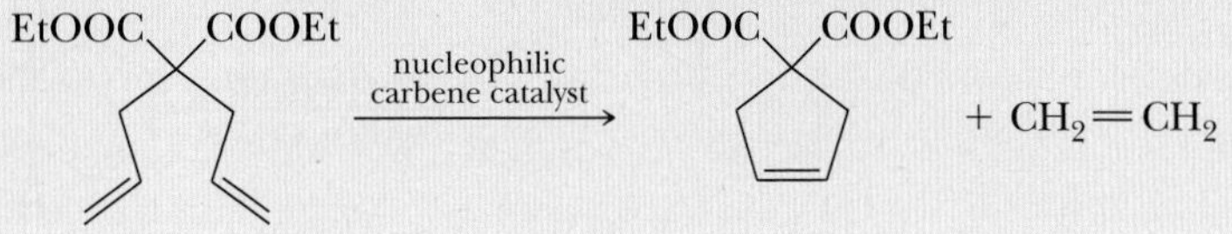
EtOOC COOEt
nucleophilic carbene catalyst
EtOOC COOEt
+ $CH_2{=}CH_2$

Problems

Red numbers indicate applied problems.

The Heck Reaction

24.8 As has been demonstrated in the text, when the starting alkene has CH_2 as its terminal group, the Heck reaction is highly stereoselective for formation of the *E* isomer. Here, the benzene ring is abbreviated C_6H_5—. Show how the mechanism proposed in the text allows you to account for this stereoselectivity.

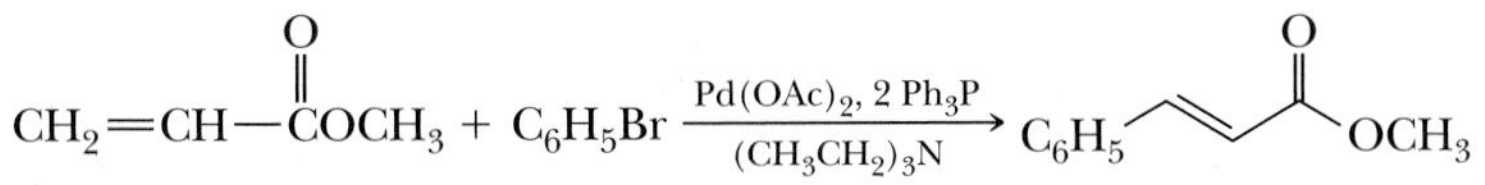
$CH_2{=}CH{-}\overset{O}{\overset{\|}{C}}OCH_3$ + C_6H_5Br
$Pd(OAc)_2$, 2 Ph_3P
$(CH_3CH_2)_3N$
C_6H_5
O
OCH_3

24.9 The following reaction involves two sequential Heck reactions. Draw structural formulas for each organopalladium intermediate formed in the sequence and show how the final product is formed. Note from the molecular formula given under each structural formula that this conversion corresponds to a loss of H and I from the starting material. Acetonitrile, CH_3CN, is the solvent.

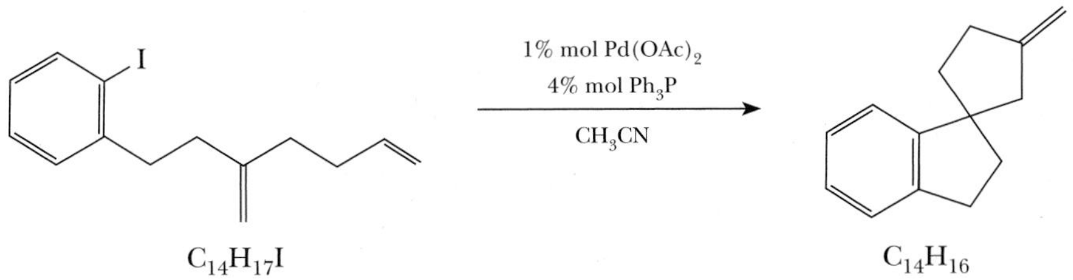
I
1% mol $Pd(OAc)_2$
4% mol Ph_3P
CH_3CN
$C_{14}H_{17}I$
$C_{14}H_{16}$

24.10 Complete these Heck reactions.

(a) 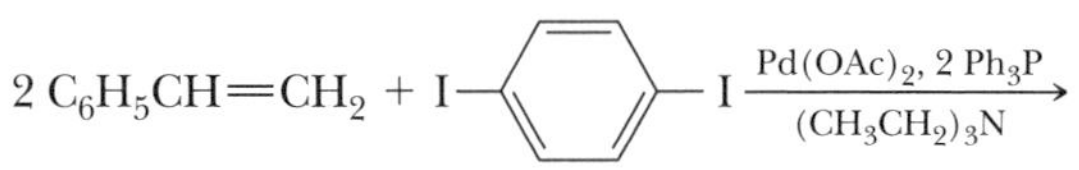

(b) 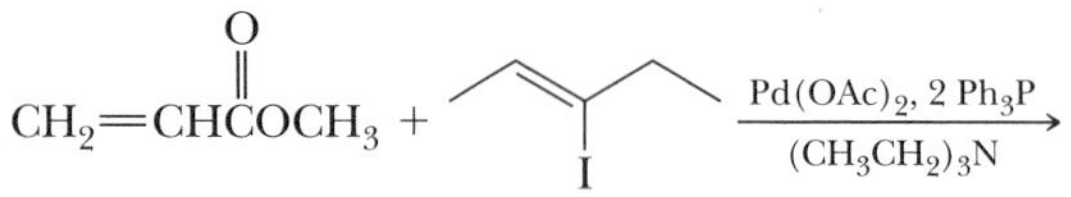

24.11 Treatment of cyclohexene with iodobenzene under the conditions of the Heck reaction might be expected to give 1-phenylcyclohexene. The exclusive product, however, is 3-phenylcyclohexene. Account for the formation of this product.

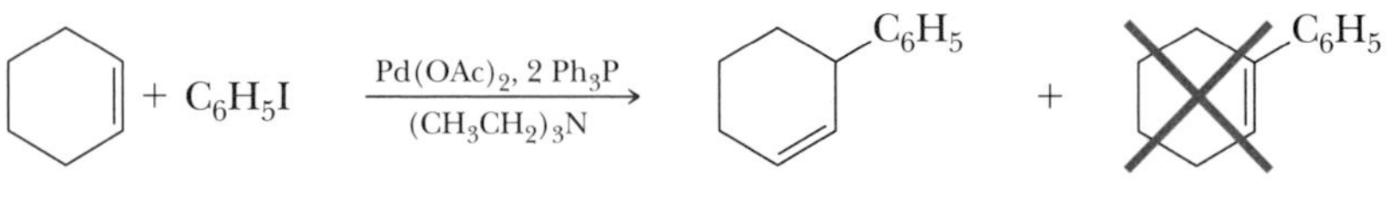

3-Phenylcyclohexene (racemic) **1-Phenylcyclohexene** (not formed)

24.12 Account for the formation of the product and for the *cis* stereochemistry of its ring junction. (The function of silver carbonate is to enhance the rate of reaction.)

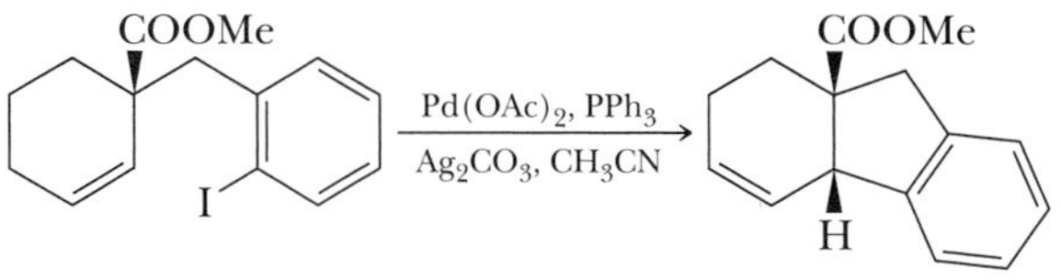

86%

24.13 Account for the formation of the following product, including the *cis* stereochemistry at the ring junction.

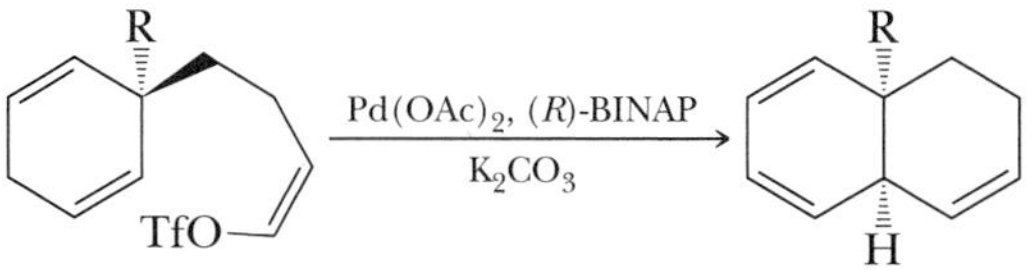

24.14 The aryl diene undergoes sequential Heck reactions to give a product with the molecular formula $C_{15}H_{18}$. Propose a structural formula for this product.

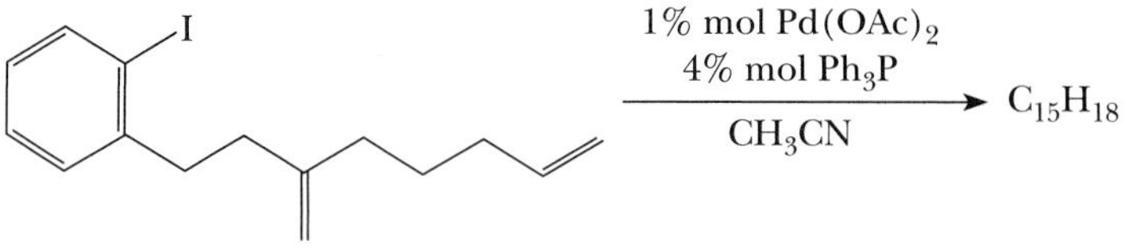

($C_{15}H_{19}I$)

24.15 Heck reactions take place with alkynes as well as alkenes. The following conversion involves an intramolecular Heck reaction followed by an intermolecular Heck. Propose structural formulas for the palladium-containing intermediates involved in this reaction.

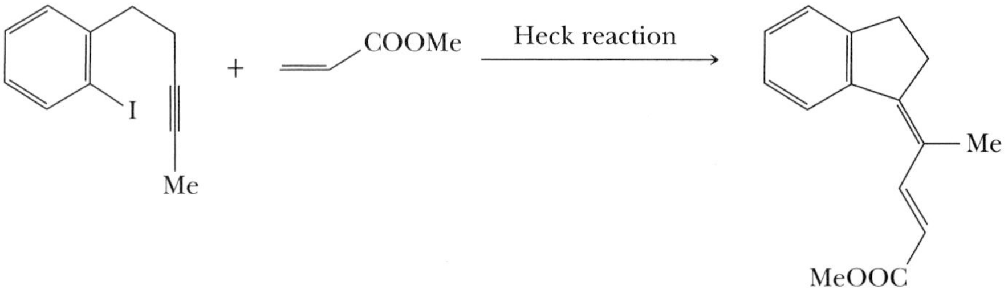

24.16 The following conversion involves sequential Heck reactions. Propose structural formulas for the palladium-containing intermediates involved in this reaction.

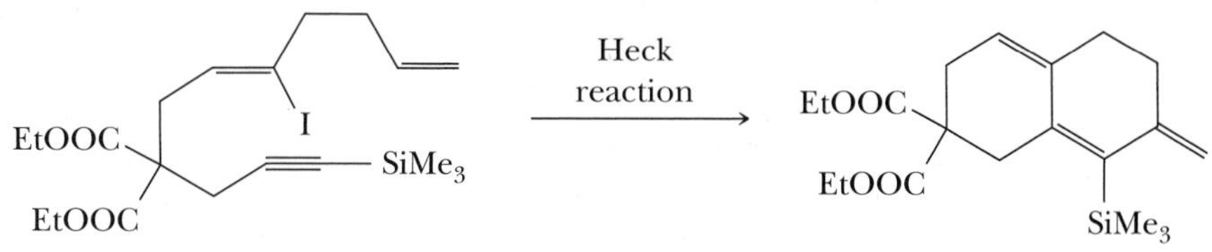

24.17 The following transformation involves a series of four consecutive Heck reactions and the formation of the four-ring steroid nucleus (Section 26.4) as a racemic mixture. Propose structural formulas for the palladium-containing intermediates involved in this reaction.

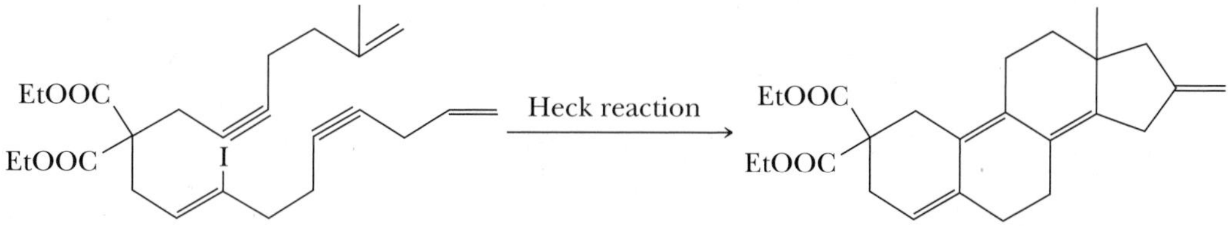

24.18 Show the sequence of Heck reactions by which the following conversion takes place. Note from the molecular formula given under each structural formula that this conversion corresponds to a loss of H and I from the starting material.

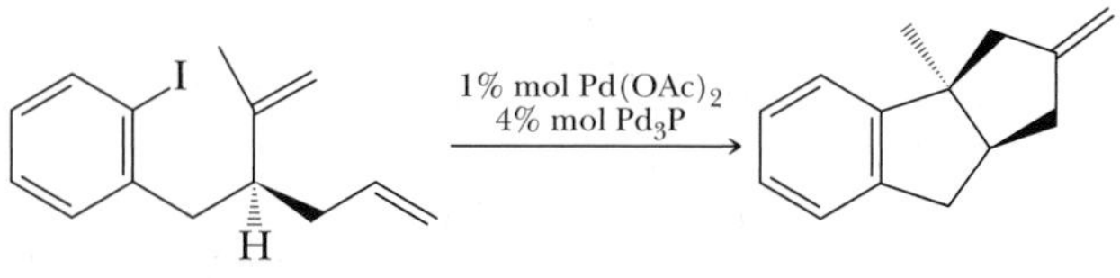

($C_{14}H_{17}I$) ($C_{14}H_{16}$)

Catalytic Allylic Alkylation

24.19 Write the product of the following reaction and account for the regiochemistry that you predict.

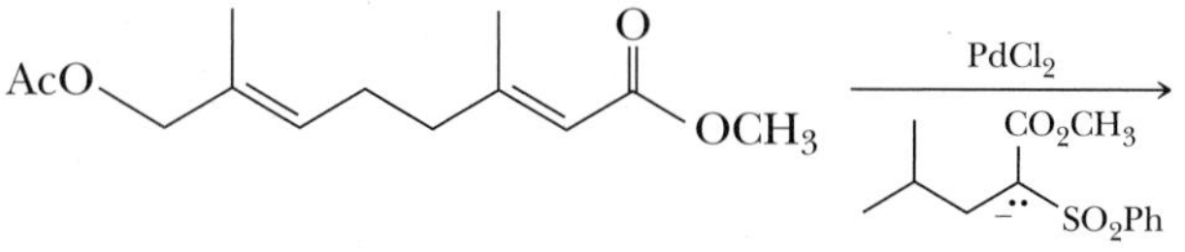

24.20 Write the steps that are critical in the following reaction in order to explain the stereochemical outcome at the carbon marked with the asterisk.

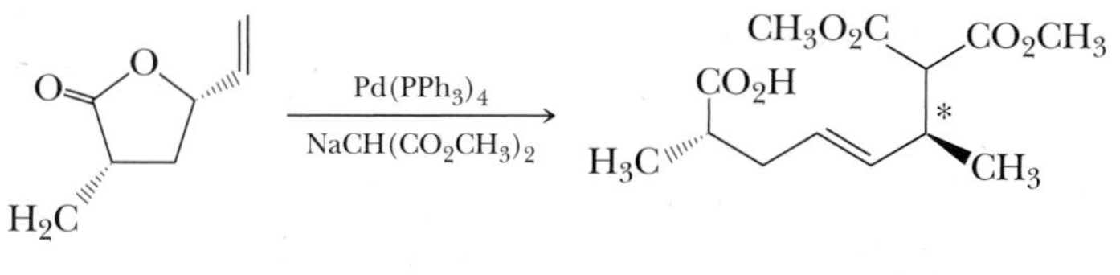

24.21 One of the most useful aspects of Pd(0)-catalyzed allylic alkylation is the ability to take racemic mixtures of reactants and create preferred chirality in the product by the addition of a chiral ligand for the **Pd** metal. Draw the η^3-allylic complex created in the catalytic cycle of the following reactant with PdL_4. Describe why chirality in a ligand L would influence the stereochemistry of nucleophilic attack.

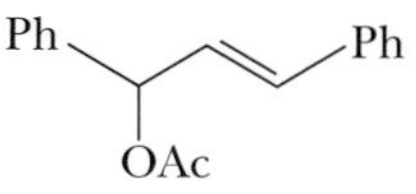

24.22 Another useful aspect of Pd(0)-catalyzed allylic alkylation is the ability to take reactants that are meso and desymmetrize them in the resulting products through the addition of a chiral ligand for the Pd metal. Draw the two η^3-allylic complexes formed in this reaction of the following reactant with PdL_4. Describe why chirality in a ligand L would influence the preference for one of these complexes over the other.

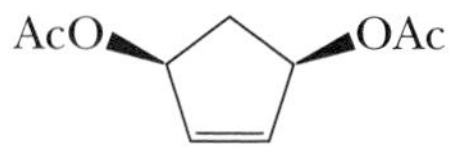

Palladium-Catalyzed Cross-Coupling Reactions

24.23 Suggest reagents and the other fragment that could be used to carry out the indicated conversion.

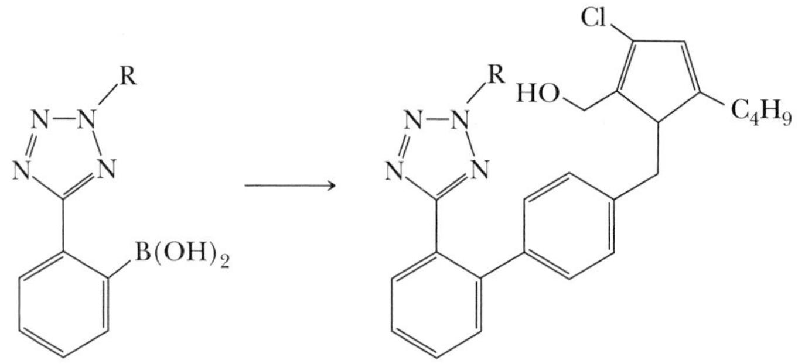

24.24 Show how the following compound could be prepared by a Suzuki reaction (Bn = benzyl).

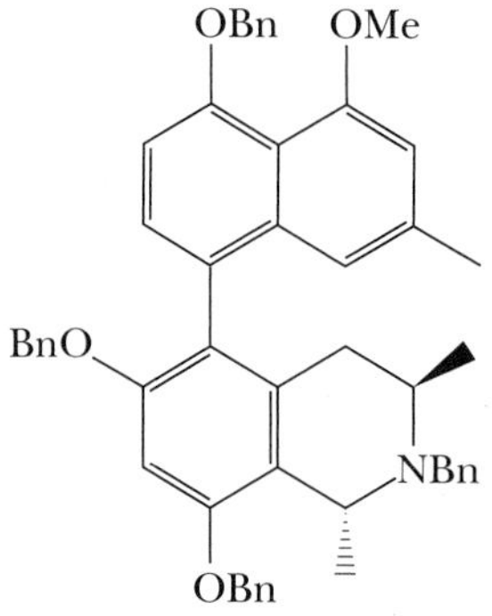

24.25 It is typically very difficult to do a substitution reaction on an aromatic ring when the leaving group is flanked by two other bulky substituents. Moreover, in Section 22.3, we found that nucleophilic aromatic substitution requires strongly electron-withdrawing groups on the benzene ring. However, Pd-catalyzed coupling allows entry into such products. As examples, write the products of the following reactions and state which coupling reaction is being utilized.

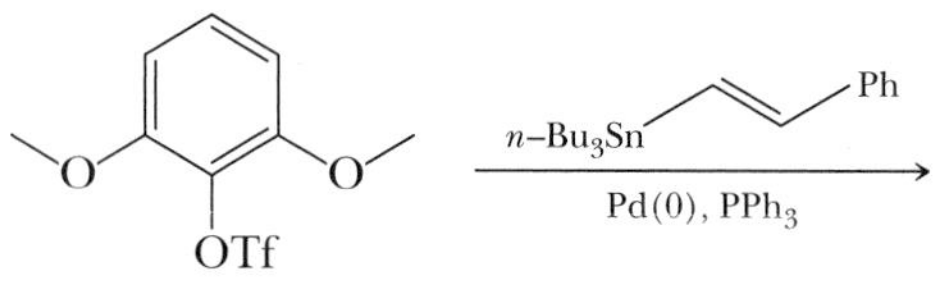

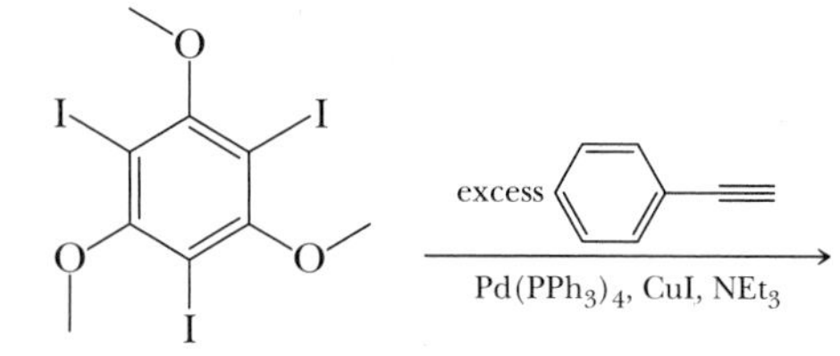

24.26 The compound eutypine is an antibacterial agent isolated from the fungus *Eutypa lata*. This fungus results in a disease common to vineyards called eutyposis. Give a sequence of reactions that will take the following reactant and give eutypine when the other reactants used in the sequence are acetylene and acetone.

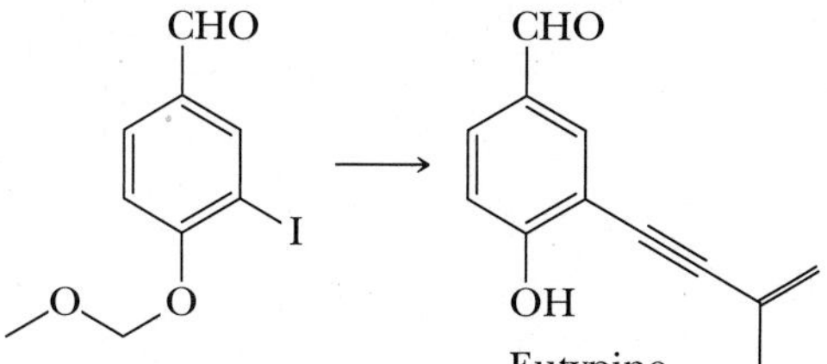

Eutypine

24.27 When the Pd(0)-catalyzed reactions covered in this chapter are run with a slight pressure of carbon monoxide, a ketone is often created as the product. For example, the following Stille coupling conditions with added **CO** give the product shown. Write a mechanism for how this reaction could occur using the organometallic mechanistic steps introduced in this chapter, along with new steps that would be required in this transformation. *Hint:* **CO** can coordinate to **Pd** and insert into **Pd-C** bonds.

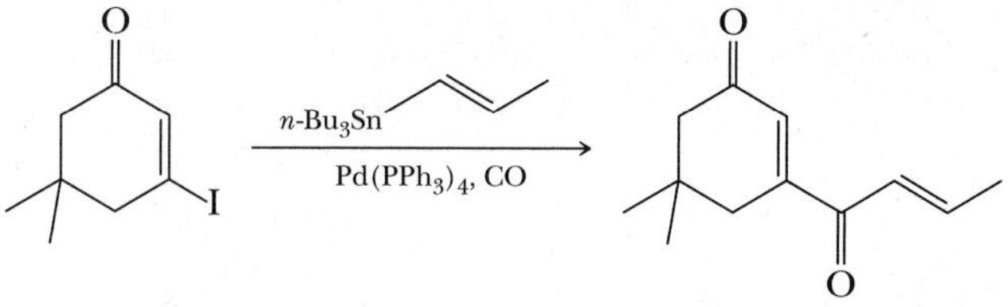

24.28 Many of the cross-coupling reactions described in this chapter have been used to make fascinating polymeric materials, as covered in Chapter 29. Give the proper reactants to create the following polymers and name the coupling reaction involved.

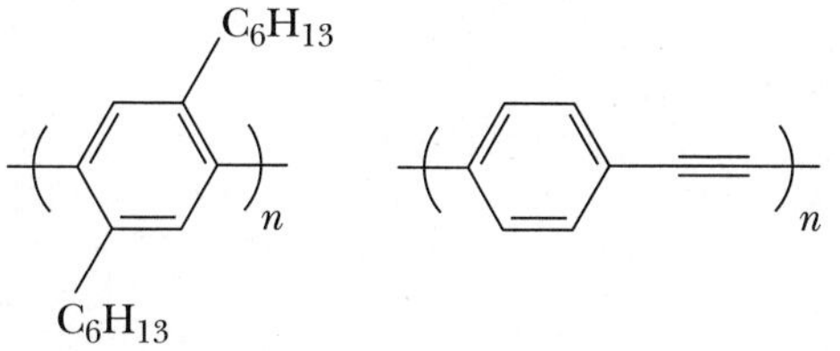

24.29 β-Lactams are amides in four-membered rings and are common elements found in antibiotics. Show what reagents would be involved in creating the following series of β-lactams using a common vinyl triflate in each case.

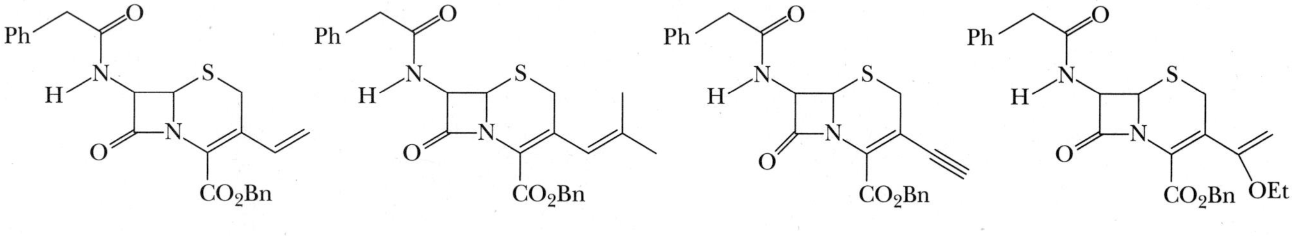

24.30 The creation of very large ring systems is often difficult and challenging. The following macrocyclic ring was created using one of the coupling reactions described in this chapter. Which one was used? What was the starting material?

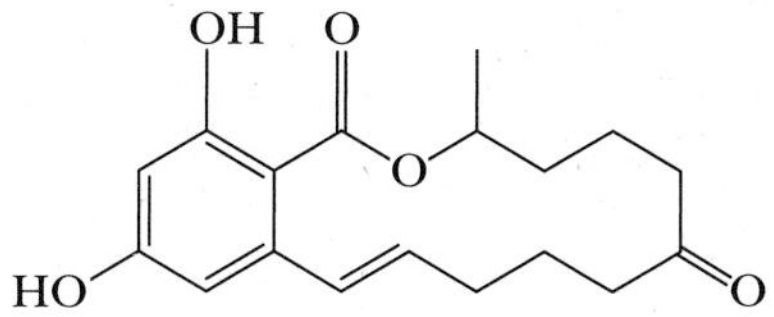

24.31 As presented in the chapter, the Sonogashira coupling reaction is commonly used to create diaryl alkynyl structures. However, it can also be used to create divinyl alkynyl structures. The following sequence of reactions creates such a product. Write possible reactants that would be placed in the boxes in the sequence.

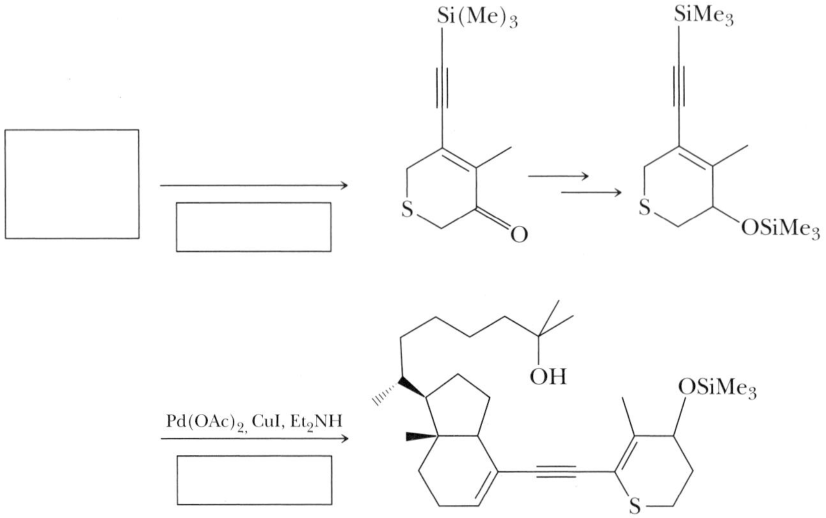

Si(Me)$_3$
SiMe$_3$
S
O
S
OSiMe$_3$
Pd(OAc)$_2$, CuI, Et$_2$NH
OH
OSiMe$_3$
S

Olefin Metathesis

24.32 The cyclic ester (lactone) Exaltolide has a musk-like fragrance and is used as a fixative in perfumery. Show how this compound could be synthesized from the indicated starting material. Give the structure of **R**.

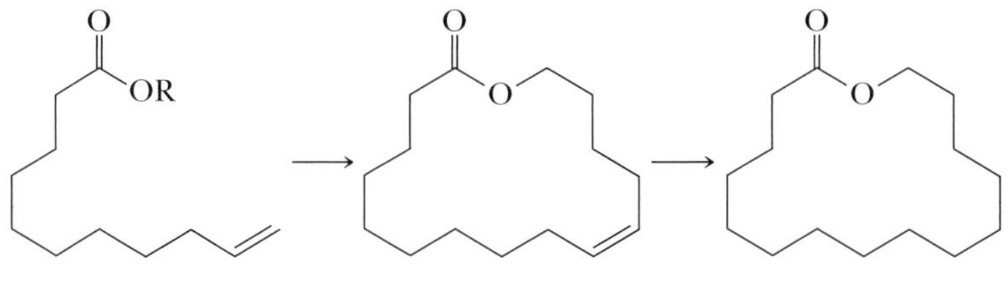

O
OR
O
O
O
O

Exaltolide

24.33 Predict the product of each alkene metathesis reaction using a **Ru**-nucleophilic carbene catalyst.

(a)

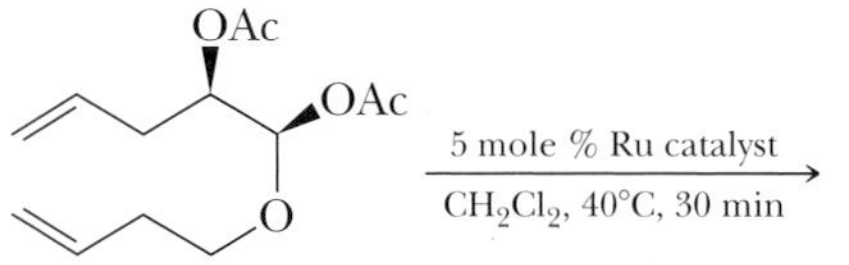

OAc
OAc
O
5 mole % Ru catalyst
CH$_2$Cl$_2$, 40°C, 30 min

(b)

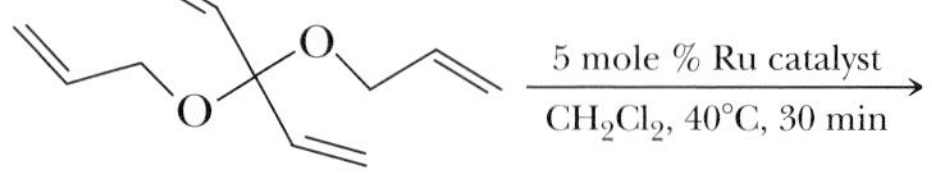

O
O
5 mole % Ru catalyst
CH$_2$Cl$_2$, 40°C, 30 min

24.34 The following transformation can be accomplished by reactions we have studied in this chapter and Chapter 20. Name the type of reaction used in each step.

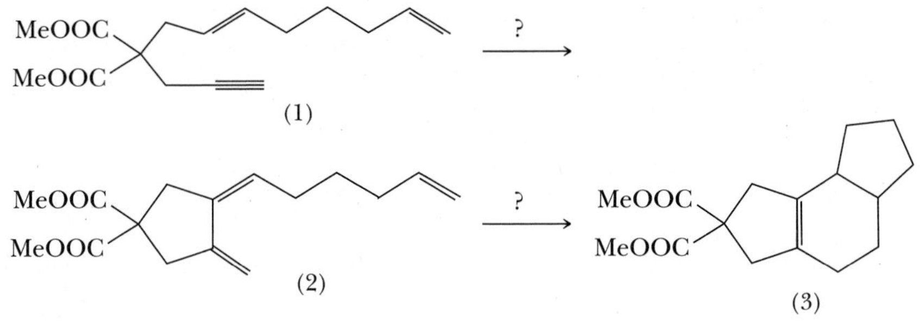

Synthesis

The following problems are based on relatively recent total syntheses of important natural products. Many such syntheses are outlined in compendia of synthetic reactions. Particularly valuable in preparing these problems were *Classics in Total Synthesis*, K. C. Nicolaou and E. J. Sorensen, Wiley-VCH, Weinheim, New York, Basel, Cambridge, Tokyo, 1996; *Classics in Total Synthesis II*, K. C. Nicolaou and S. A. Snyder, Wiley-VCH Verlag GmbH, Weinheim (2003).

24.35 Following is an outline of the stereospecific synthesis of the "Corey lactone." Professor E. J. Corey (Harvard University) describes it this way. "The first general synthetic route to all the known prostaglandins was developed by way of bicycloheptene intermediates. The design was guided by the requirements that the route be versatile enough to allow the synthesis of many analogs and also allow early resolution. This synthesis has been used on a large scale and in laboratories throughout the world; it has been applied to the production of countless prostaglandin analogs." Corey was awarded the 1990 Nobel Prize in Chemistry for the development of retrosynthetic analysis for synthetic production of complex molecules. See E. J. Corey and Xue-Min Cheng, *The Logic of Chemical Synthesis*, John Wiley & Sons, New York, 1989, p. 255. For the structure of the prostaglandins, see Section 26.3. *Note:* The wavy lines in compound C indicate that the stereochemistry of —Cl and —CN groups was not determined. [The conversion of **(D)** to **(E)** involves an oxidation of the ketone group to a lactone by the Baeyer-Villiger reaction, which we have not studied in this text.]

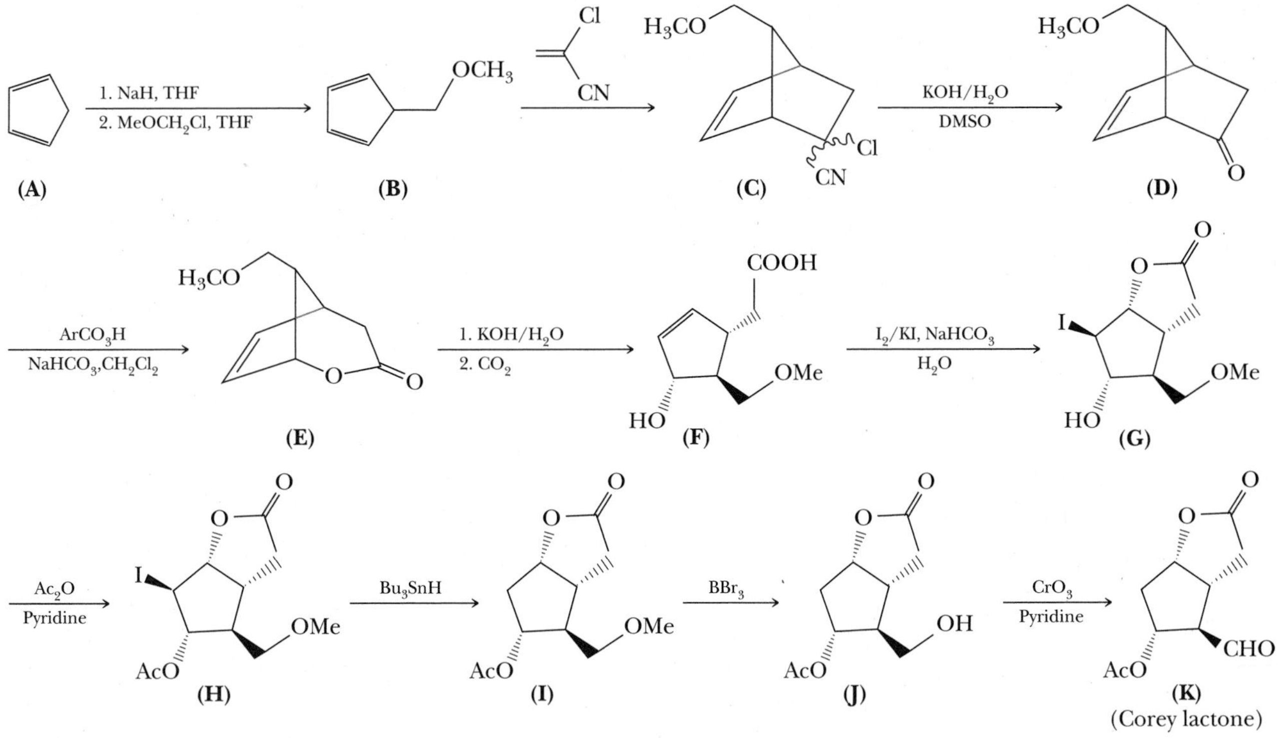

(a) What is the function of sodium hydride, NaH, in the first step? What is the pK_a of cyclopentadiene? How do you account for its remarkable acidity?

(b) By what type of reaction is **(B)** converted to **(C)**?

(c) What is the function of the carbon dioxide added to the reaction mixture in Step 2 of the conversion of **(E)** to **(F)**? *Hint:* What happens when carbon dioxide is dissolved in water? Why not just use HCl?

(d) The tributyltin hydride, Bu_3SnH, used in the conversion of **(H)** to **(I)** reacts via a radical chain reaction; the first step involves a reaction with a radical initiator to form $Bu_3Sn\cdot$. Suggest a mechanism for the rest of the reaction.

(e) The Corey lactone contains four chiral centers with the relative configurations shown. In what step or steps in this synthesis is the configuration of each chiral center determined? Propose a mechanism to account for the observed stereospecificity of the relevant steps.

(f) Compound **(F)** was resolved using (+)-ephedrine. Following is the structure of (−)-ephedrine, the naturally occurring stereoisomer. What is meant by "resolution"? What is the rationale for using a chiral, enantiomerically pure amine for the resolution of **(F)**?

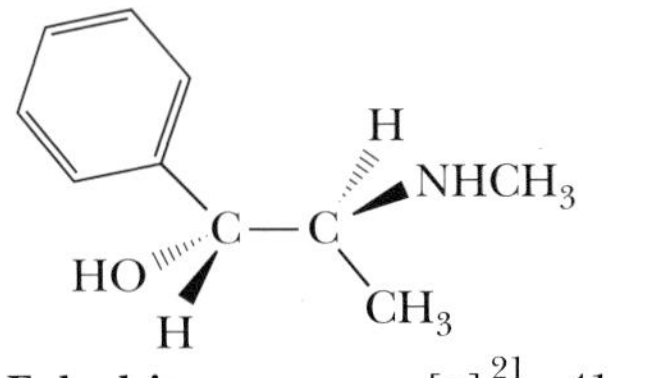

Ephedrine $[\alpha]_D^{21}$ −41

(*Note:* By resolving at this stage, one-half of the material is discarded. A more efficient route would be to have an earlier resolution; in fact, Corey later solved this problem in an elegant way by using an enantioselective Diels-Alder reaction with the alkene in the form of an acrylate ester of enantiomerically pure 8-phenylmenthol. Asymmetric induction gave a product with a diastereoselectivity of 97:3. So rather than resolving, he was able to get the correct stereoisomer directly.)

(g) You have not studied the Baeyer-Villiger reaction (**D** to **E**). The mechanism involves nucleophilic reaction of the peroxyacid with the carbonyl followed by a rearrangement much like that involved in the hydroboration reaction (Section 6.4). Write a mechanism for this reaction.

24.36 Chapman's (O. L. Chapman, then at Iowa State and later at UCLA) classic total synthesis of (±)-carpanone is so remarkably simple that it is used as an undergraduate laboratory preparation. It is modeled on a possible biosynthetic route for this lignan-derived natural product. Phenol oxidations figure prominently in many such biosyntheses of natural products. In one step, this reaction creates no less than five contiguous chiral centers, all in the correct relative configuration.

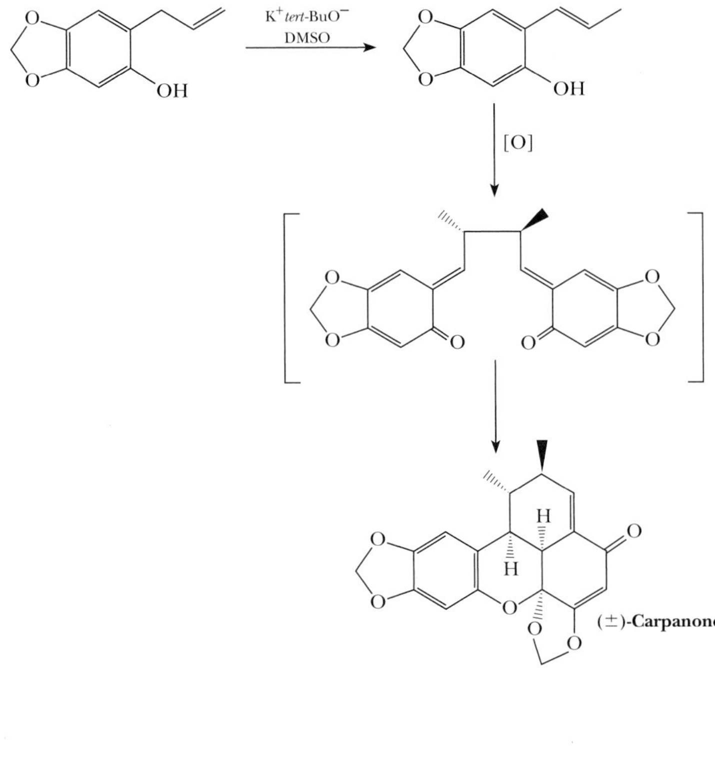

(a) Give a mechanism for the first step of the reaction and explain why it goes in the direction it does.
(b) The oxidation step uses a palladium salt. Suggest a mechanism for this coupling, which you have not encountered. *Hint:* Do not concern yourself with the role of the metal except as an acceptor of electrons.
(c) The third step is spontaneous. Give a mechanism for this reaction and show how it accounts for the stereochemistry of the final product.
(d) Would you expect the product to be racemic or a single enantiomer?

24.37 Gilvocarcin M is isolated from *Streptomyces* strains and has strong antitumor activity.

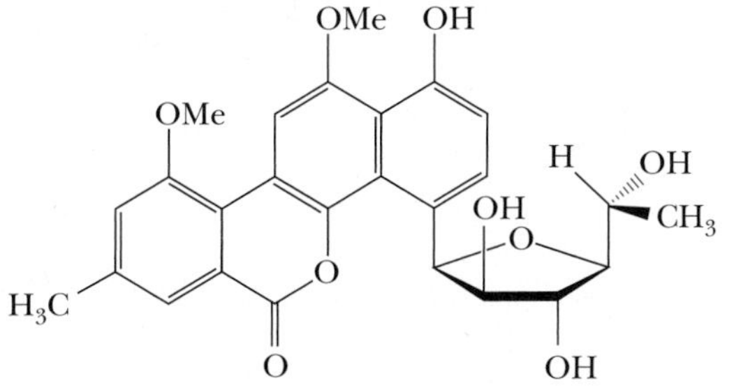

Gilvocarcin M

Suzuki and coworkers were able to carry out the total synthesis of naturally occurring (−)-gilvocarcin M. Their synthesis included the following steps. (The wavy line means that stereochemistry is unspecified or is a mixture.) The stereochemistry of the product appears to be counterintuitive (apparent attack from the more hindered side). The reason is that the reaction involves initial *O*-alkylation followed by a rearrangement that need not concern us.

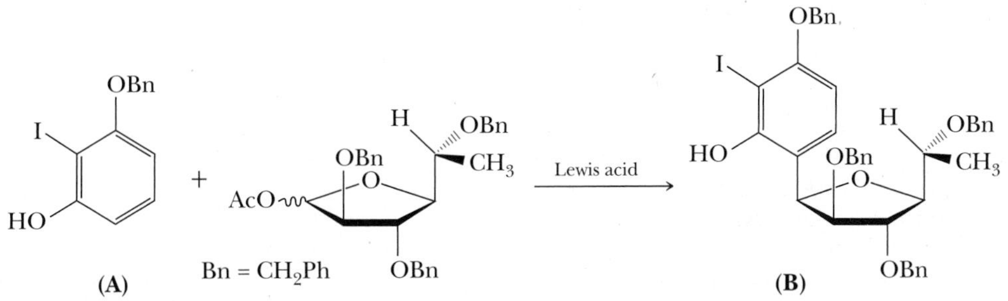

(a) This reaction gives both high regioselectivity and stereoselectivity. What other products might have been expected?

The next step involves triflation and treatment with butyl lithium.

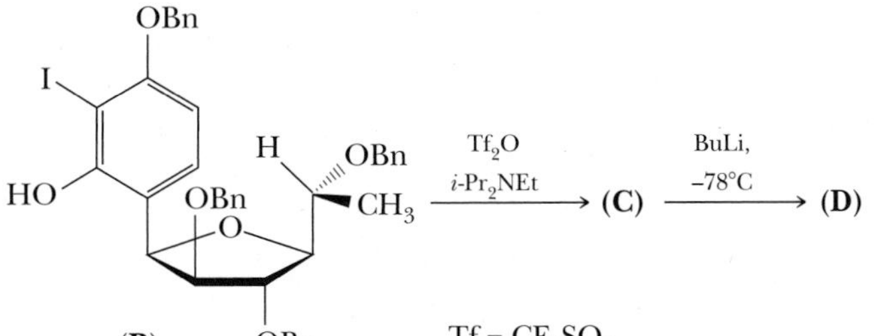

(b) Give a structure for **(C)**.
(c) Give a structure for **(D)**. This reaction requires that you know that lithium reagents can interchange with haloarenes:

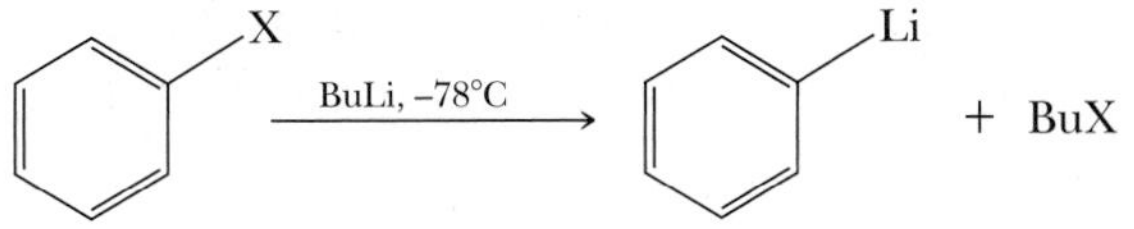

Recall that OTf is an excellent leaving group. You may wish to review Section 24.3A. The reaction yielding **(D)** is carried out in the presence of 2-methoxyfuran. **(D)** decomposes under the conditions to a compound **(E)** that instantly reacts with the furan to give **(F)**.

(d) Give a structure for **(E)** and the mechanism of **(D)** to **(E)**.
(e) Give a mechanism for **(E)** to **(F)**.

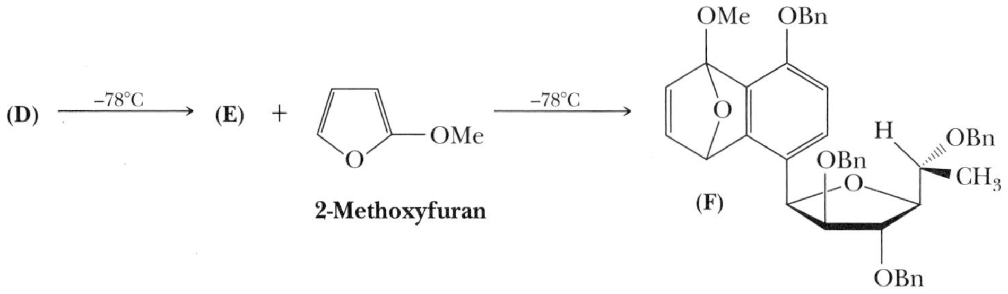

Compound **(F)** is unstable and undergoes ring opening upon workup to give **(G)**.

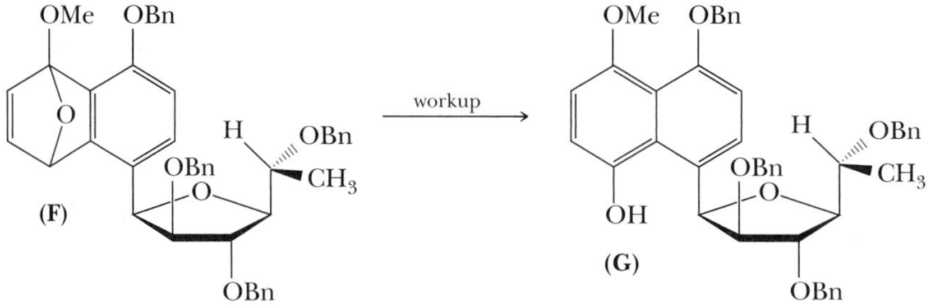

(f) Give a mechanism for **(F)** to **(G)**.

The next step involves conversion of **(G)** to **(H)**.

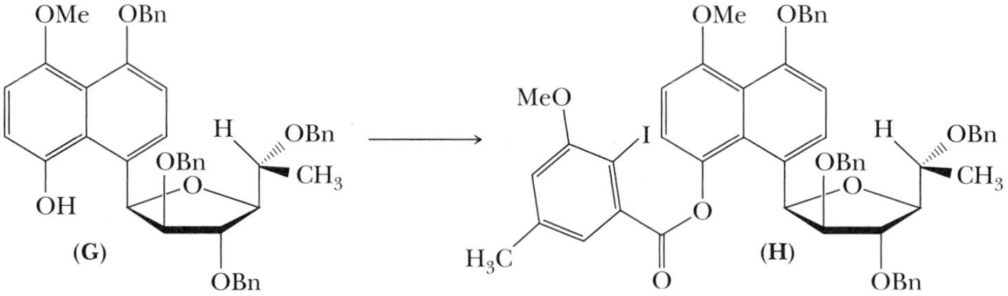

(g) Give reagents and conditions required for **(G)** to **(H)**.

Formation of the final tetracyclic ring involves conversion of **(H)** to **(I)**.

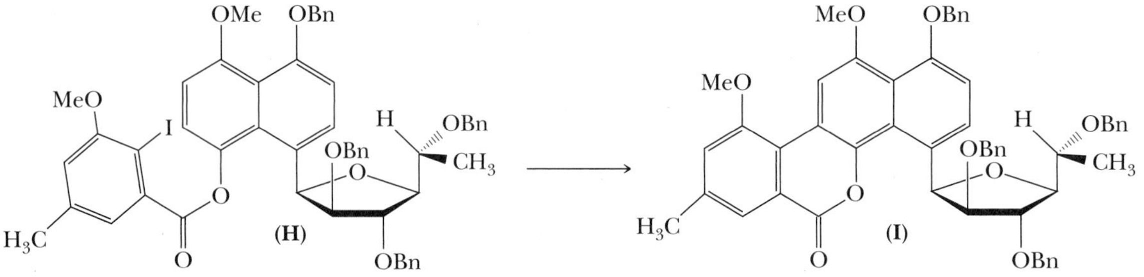

(h) Give reagents and conditions required for **(H)** to **(I)**.

(I) is then is converted to (−)-gilvocarcin M, the natural enantiomer.

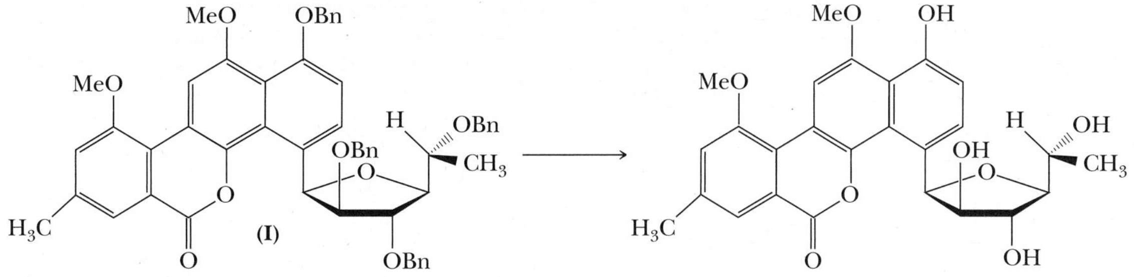

Gilvocarcin M

(i) What reagents could be used for this reaction?

(j) Comment on the probable source of the chiral centers in this synthesis. Note that the chirality was not created in any of the reaction steps. You can find a possible readily available and inexpensive source (see Chapter 25, "Carbohydrates").

(k) Given reactions that are later in the sequence, why is it necessary to protect some of the **OH** groups as the benzyl ether? What side reactions would occur without this protection? Starting with **OH** groups, how would you add these protecting groups?

24.38 Vancomycin is an important antibiotic. It is isolated from the bacterium *Streptomyces orientalis* and functions by inhibiting bacterial mucopeptide synthesis. It is a last line of defense against the resistant Staph organisms that are now common in hospitals.

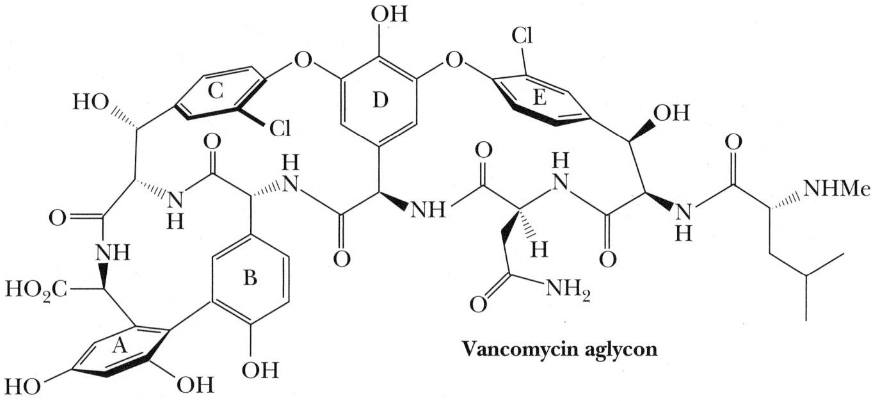

In 1999, Professor Dale Boger (The Scripps Research Institute) reported a synthesis of vancomycin **aglycon** (aglycon = lacking a sugar) involving the following steps, among others. Compound **(I)** was prepared from simple starting materials by a series of steps involving forming amide bonds.

Aglycon
Lacking a sugar.

(a) Suggest reasonable precursors and show how the bonds could be formed (the actual reagents used have not been introduced, but they work in a similar way to those you know).

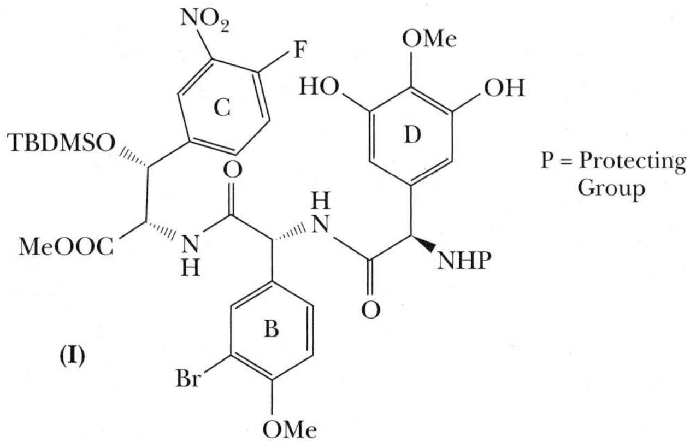

Compound **(I)** was then converted into **(II)**.

(b) Give reagents for this reaction and suggest the mechanism.

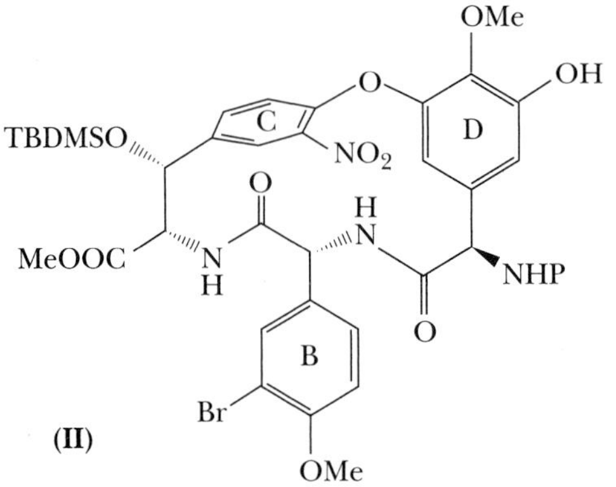

P = Protecting Group

(II)

One of the interesting features of this synthesis is that ring C in compound **(II)** (and subsequent compounds in this synthesis) has extremely hindered rotation. As a result, compound **(II)** exists as two atropisomers (Section 3.2) that are interconverted only at 140°C.

(c) Show these two isomers.

(II) was then converted to **(III)**.

(d) Suggest reagents to accomplish this transformation.

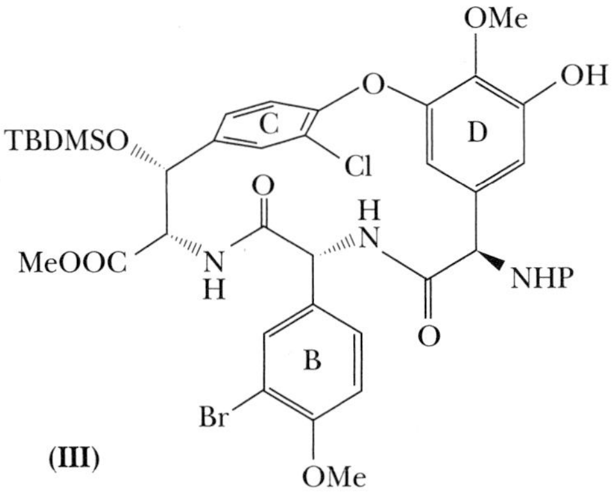

P = Protecting Group

(III)

Compound **(III)** was then converted to **(IV)**.

(e) Suggest reagents and the ring A fragment that could be used for this reaction.

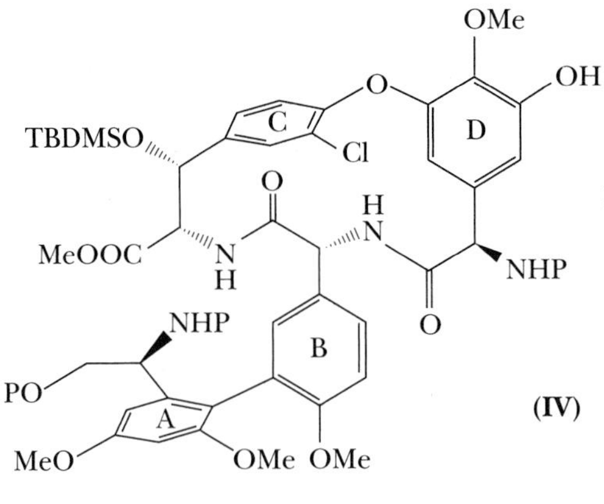

P = Protecting Group

(IV)

Closure of an amide link between the amine on ring A (after removal of the protecting group) and the carbomethoxy group above it led to a precursor of vancomycin.

(f) Show the ring closure reaction of the deprotected free amino group and its mechanism.

Another interesting feature of this synthesis is that rings A and B also form atropisomers. These can be converted into a 3:1 mixture of the desired and undesired atropisomers on heating at 120°C.

(g) Draw these atropisomers and show that only one can be converted to vancomycin. The synthesis of the aglycon was completed by functional manipulation and addition of ring E by chemistry similar to that detailed earlier. Yet, another set of atropisomers (this time of ring E) was formed! However, this one was more easily equilibrated than the others; model studies had shown that the activation barrier for this set of atropisomers should be lower than that of the others.

24.39 E. J. Corey's 1964 total synthesis of α-caryophyllene (essence of cloves) solves a number of problems of construction of unusual-sized rings.

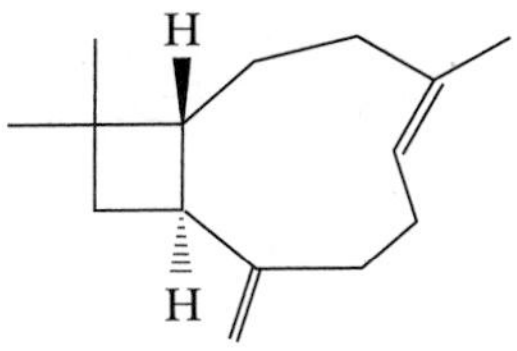

α-Caryophyllene

The first step uses an efficient photochemical [2 + 2] reaction. The desired stereochemistry and regiochemistry had been predicted based on model reactions.

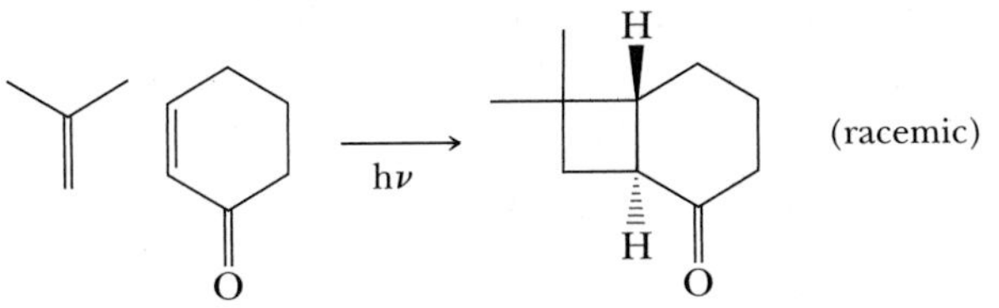

(a) [2 + 2] Reactions are quite common in photochemical reactions. Would this reaction be predicted to occur in the ground state?

The next steps follow. Basic alumina is a chromatography support that will often act as a base catalyst.

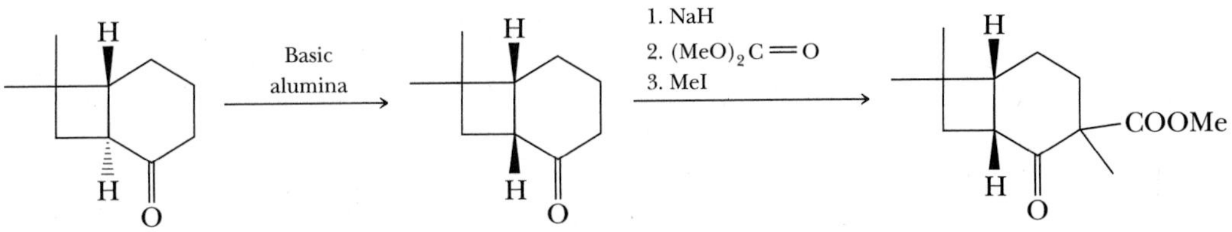

(b) What is the mechanism of the first step?
(c) What is the mechanism of the second step?
(d) Look at later steps in the synthesis. Does the stereochemistry of the added carbomethoxy group matter?

The next steps are shown here.

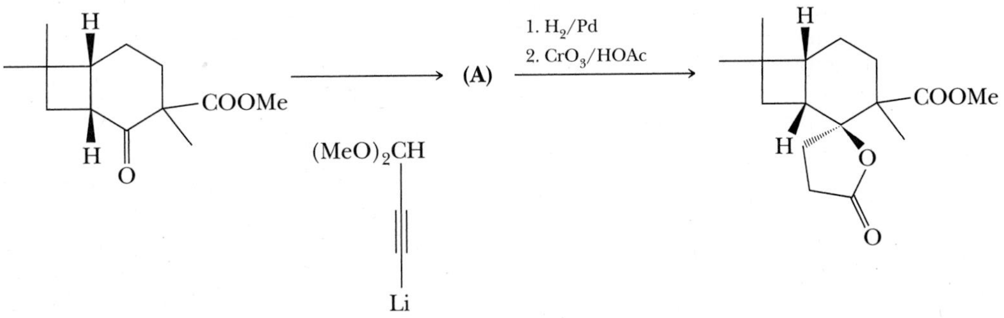

(e) What is the structure of compound **(A)**?
(f) Give a mechanism for the formation of the cyclized product.

Here are the next steps.

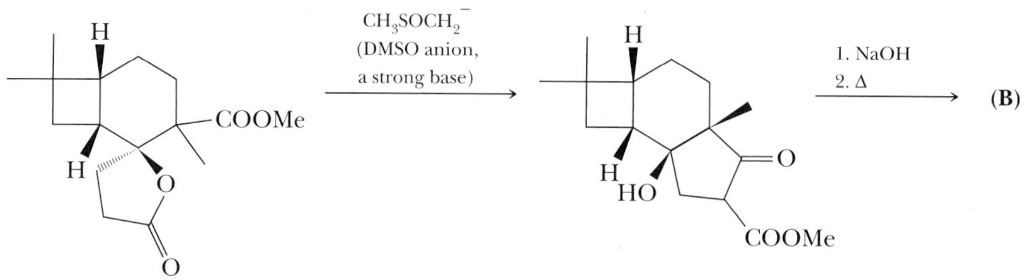

(g) Give a mechanism for the first step. *Hint:* Attack on the lactone carbonyl may be the first step.

(h) Give a structure for product **(B)**.

The following two steps are next.

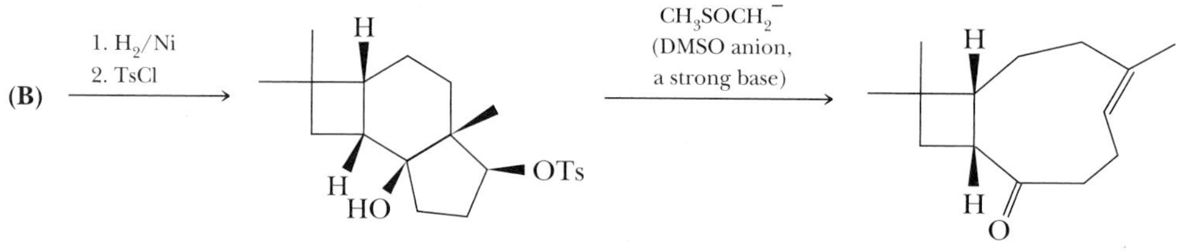

(i) Show the reactions of **(B)**.

(j) Write a mechanism for the ring-opening reaction. *Hint:* Note the presence of an acidic proton and a good leaving group in the molecule.

The synthesis was completed by the following steps.

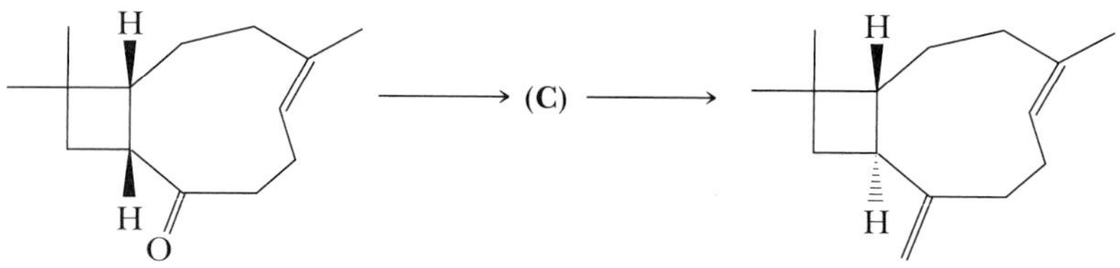

(k) What is **(C)**?

(l) What reagents would you use for these transformations?

24.40 Over the past several decades, chemists have developed a number of synthetic methodologies for the synthesis of steroid hormones. One of these, developed by Lutz Tietze at the Institut für Organische Chemie der Georg-August-Universität, Göttingen, Germany, used a double Heck reaction to create ring B of the steroid nucleus. As shown in the following retrosynthetic analysis, a key intermediate in his synthesis is compound **(1)**. Two Heck reaction disconnects of this intermediate give compounds **(2)** and **(3)**. Compound **(2)** contains the aromatic ring that becomes ring A of estrone. Compound **(3)** contains the fused five- and six-membered rings that become rings C and D of estrone.

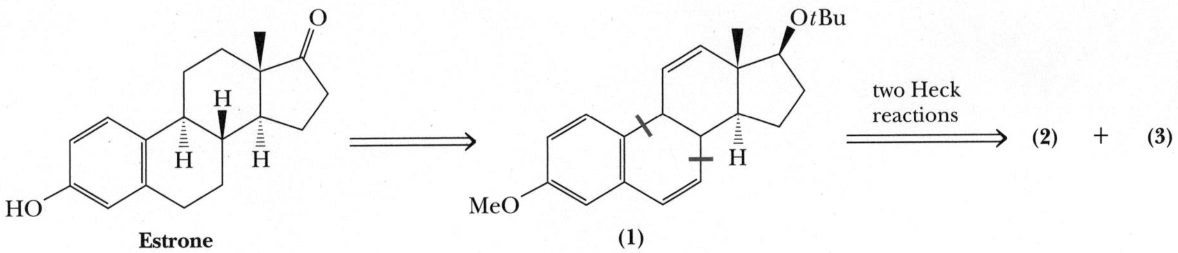

Estrone (1)

(a) Name the types of functional groups in estrone.

(b) How many chiral centers are present in estrone?

(c) Propose structural formulas for compounds **(2)** and **(3)**.

(d) Show how your proposals for compounds **(2)** and **(3)** can be converted to compound **(1)**. (*Note:* In the course of developing this synthesis, Tietze discovered that vinylic bromides and iodides are more reactive in Heck reactions than are aryl bromides and iodides.)

(e) In the course of the double Heck reactions, two new chiral centers are created. Assume in compound **(3)**, the precursor to rings C and D of estrone, that the fusion of rings C and D is *trans* and that the angular methyl group is above the plane of the ring. Given this stereochemistry, predict the stereochemistry of compound **(1)** formed by the double Heck reaction.

(f) To convert **(1)** to estrone, the *tert*-butyl ether on ring D must be converted to a ketone. How might this transformation be accomplished?

25

Carbohydrates

Foxglove *(Digitalis purpurea)*, an ornamental flowering plant, is the source of digitoxin and digitalis, medicines used in cardiology to reduce pulse rate, regularize heart rhythm, and strengthen heartbeat. **Inset:** digitoxose, a monosaccharide obtained on hydrolysis of digitoxin. See Problem 25.15.

Outline

25.1 Monosaccharides
25.2 The Cyclic Structure of Monosaccharides
25.3 Reactions of Monosaccharides
25.4 Disaccharides and Oligosaccharides
25.5 Polysaccharides
25.6 Glucosaminoglycans

Carbohydrates are the most abundant organic compounds in the plant world. They act as storehouses of chemical energy (glucose, starch, and glycogen); are components of supportive structures in plants (cellulose), crustacean shells (chitin), and connective tissues in animals (glucosaminoglycans); and are essential components of nucleic acids (D-ribose and 2-deoxy-D-ribose). Carbohydrates make up about three-fourths of the dry weight of plants. Animals (including humans) get their carbohydrates by eating plants, but they do not store much of what they consume. Less than 1% of the body weight of animals is made up of carbohydrates.

The name *carbohydrate* means hydrate of carbon and derives from the formula $C_n(H_2O)_m$. Following are two examples of carbohydrates with molecular formulas that can be written alternatively as hydrates of carbon.

Glucose (blood sugar): $C_6H_{12}O_6$, or alternatively $C_6(H_2O)_6$
Sucrose (table sugar): $C_{12}H_{22}O_{11}$, or alternatively $C_{12}(H_2O)_{11}$

Not all carbohydrates, however, have this general formula. Some contain too few oxygen atoms to fit this formula, and others contain too many oxygens. Some also contain nitrogen. The term *carbohydrate* has become so firmly rooted in chemical nomenclature that although not completely accurate, it persists as the name for this class of compounds.

At the molecular level, most **carbohydrates** are polyhydroxyaldehydes, polyhydroxyketones, or compounds that yield either of these after hydrolysis. Therefore, the chemistry of carbohydrates is essentially the chemistry of hydroxyl groups and carbonyl groups and of the acetal bonds formed between these two functional groups.

Carbohydrate
A polyhydroxyaldehyde, a polyhydroxyketone, or a substance that gives these compounds on hydrolysis.

The fact that carbohydrates have only two types of functional groups, however, belies the complexity of their chemistry. All but the simplest carbohydrates contain multiple chiral centers. For example, glucose, the most abundant carbohydrate in the biological world, contains one aldehyde group, one primary and four secondary hydroxyl groups, and four chiral centers. Working with molecules of this complexity presents enormous challenges to organic chemists and biochemists alike.

25.1 Monosaccharides

A. Structure and Nomenclature

Monosaccharides have the general formula $C_nH_{2n}O_n$ with one of the carbons being the carbonyl group of either an aldehyde or a ketone. The most common monosaccharides have three to eight carbon atoms. The suffix *-ose* indicates that a molecule is a carbohydrate, and the prefixes *tri-*, *tetr-*, *pent-*, and so forth indicate the number of carbon atoms in the chain. Monosaccharides containing an aldehyde group are classified as **aldoses**; those containing a ketone group are classified as **ketoses**.

Monosaccharide
A carbohydrate that cannot be hydrolyzed to a simpler carbohydrate.

Aldose
A monosaccharide containing an aldehyde group.

Ketose
A monosaccharide containing a ketone group.

Monosaccharides Classified by Number of Carbon Atoms

Name	Formula
Triose	$C_3H_6O_3$
Tetrose	$C_4H_8O_4$
Pentose	$C_5H_{10}O_5$
Hexose	$C_6H_{12}O_6$
Heptose	$C_7H_{14}O_7$
Octose	$C_8H_{16}O_8$

There are only two trioses: the aldotriose glyceraldehyde and the ketotriose dihydroxyacetone.

Glyceraldehyde (an aldotriose)   **Dihydroxyacetone** (a ketotriose)

Often, the designations *aldo-* and *keto-* are omitted, and these molecules are referred to simply as trioses, tetroses, and the like.

Glyceraldehyde is a common name; the IUPAC name for this monosaccharide is 2,3-dihydroxypropanal. Similarly, dihydroxyacetone is a common name; its IUPAC name is 1,3-dihydroxypropanone. The common names for these and other

1,3-Dihydroxypropanone, more commonly known as dihydroxyacetone, is the active ingredient in artificial tanning agents such as Man-Tan and Magic Tan.

monosaccharides, however, are so firmly rooted in the literature of organic chemistry and biochemistry that they are used almost exclusively to refer to these compounds. Therefore, throughout our discussions of the chemistry and biochemistry of carbohydrates, we use the names most common in the literature of chemistry and biochemistry.

B. Fischer Projection Formulas

Glyceraldehyde contains a chiral center and therefore exists as a pair of enantiomers.

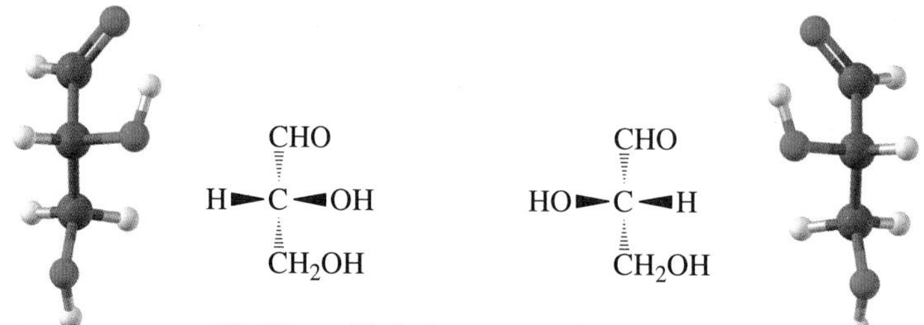
CHO
H C OH
CH_2OH
CHO
HO C H
CH_2OH

(*R*)-Glyceraldehyde **(*S*)-Glyceraldehyde**

Fischer projection
A two-dimensional representation for showing the configuration of chiral centers; horizontal lines represent bonds projecting forward, and vertical lines represent bonds projecting to the rear.

Chemists commonly use two-dimensional representations called **Fischer projections** (Section 3.4C) to show the configuration of carbohydrates. Following is an illustration of how a three-dimensional representation is converted to a Fischer projection.

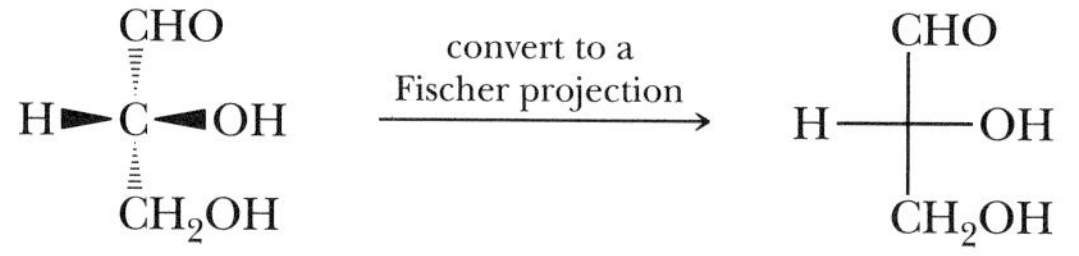
CHO
H C OH
CH_2OH
convert to a
Fischer projection
CHO
H OH
CH_2OH

(*R*)-Glyceraldehyde (three-dimensional representation) **(*R*)-Glyceraldehyde** (Fischer projection)

The horizontal segments of a Fischer projection represent bonds directed toward you, and the vertical segments represent bonds directed away from you. The only atom in the plane of the paper is the chiral center.

C. D- and L-Monosaccharides

Even though the *R,S* system is widely accepted today as a standard for designating configuration, the configuration of carbohydrates as well as those of amino acids and many other compounds in biochemistry is commonly designated by the D,L system proposed by Emil Fischer in 1891. At that time, it was known that one enantiomer of glyceraldehyde has a specific rotation of +13.5; the other has a specific rotation of −13.5. Fischer proposed that these enantiomers be designated D and L (for dextro- and levorotatory, respectively), but he had no experimental way to determine which enantiomer has which specific rotation. Fischer, therefore, did the only possible thing—he made an arbitrary assignment. He assigned the dextrorotatory enantiomer an arbitrary configuration and named it D-glyceraldehyde. He named its enantiomer L-glyceraldehyde.

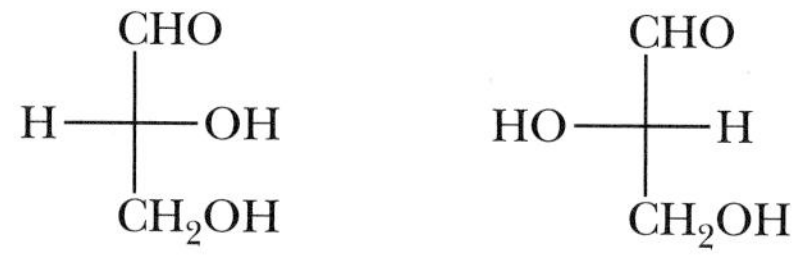
CHO
H OH
CH_2OH
CHO
HO H
CH_2OH

D-Glyceraldehyde $[\alpha]_D^{25}$ +13.5 **L-Glyceraldehyde** $[\alpha]_D^{25}$ −13.5

Fischer could have been wrong, but by a stroke of good fortune, he was correct, as proven in 1952 by a special application of X-ray crystallography.

D- and L-glyceraldehydes serve as reference points for the assignment of relative configuration to all other aldoses and ketoses. The reference point is the chiral center farthest from the carbonyl group. Because this chiral center is always the next-to-the- last carbon on the chain, it is called the **penultimate carbon**. A **D-monosaccharide** has the same configuration at its penultimate carbon as D-glyceraldehyde (its —OH is on the right when written as a Fischer projection); an **L-monosaccharide** has the same configuration at its penultimate carbon as L-glyceraldehyde (its —OH is on the left).

Note that for monosaccharides with two or more chiral centers, the designations D or L refer only to the configuration of the highest-numbered chiral center (i.e., the chiral center farthest from the aldehyde or ketone carbonyl group). Also note that the D or L designation of a given monosaccharide does not specify the sign of the specific rotation of the compound. If the sign of the rotation of plane-polarized light is to be specified in a name, it is indicated by a + (plus) or a − (minus) sign preceding the name. Thus, D-glucose, which is dextrorotatory, is designated D-(+)-glucose, and D-fructose, which is levorotatory, is designated D-(−)-fructose.

Table 25.1 shows names and Fischer projections for all D-aldotetroses, pentoses, and hexoses. Each name consists of three parts. The letter D specifies the configuration

D-Monosaccharide

A monosaccharide that, when written as a Fischer projection, has the —OH on its penultimate carbon to the right.

L-Monosaccharide

A monosaccharide that, when written as a Fischer projection, has the —OH on its penultimate carbon to the left.

Watch a video explanation

Table 25.1 **Configurational Relationships Among the Isometric D-Aldotetroses, D-Aldopentoses, and D-Aldohexoses**

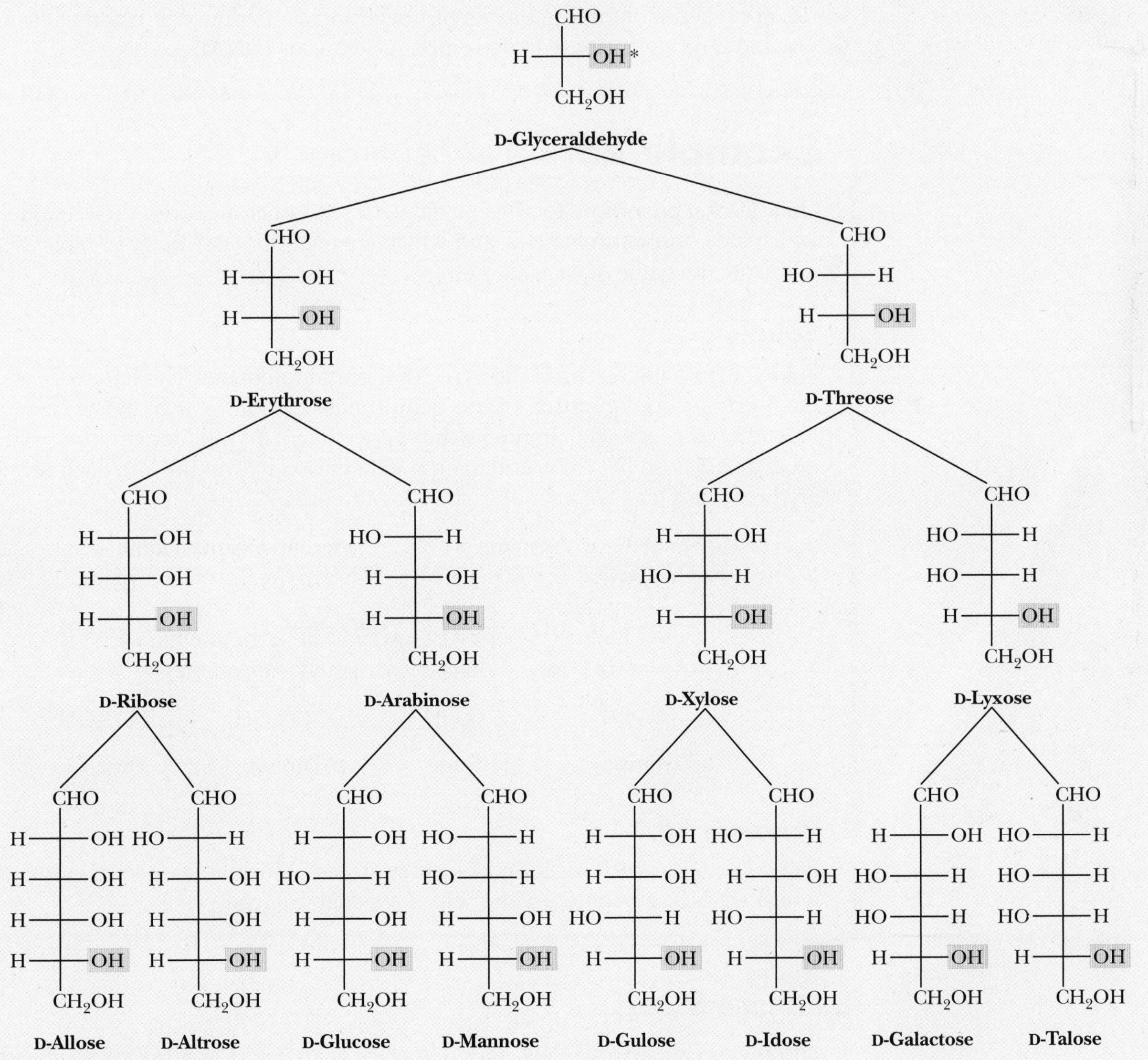

*The configuration of the reference —OH on the penultimate carbon is shown in color.

of the penultimate carbon. Prefixes such as *rib-*, *arabin-*, and *gluc-* specify the configuration of all other chiral centers in the monosaccharide. The suffix *-ose* shows that the compound is a carbohydrate.

The three most abundant hexoses in the biological world are D-glucose, D-galactose, and D-fructose. The first two are D-aldohexoses; the third is a D-2-ketohexose. Glucose, by far the most common hexose, is also known as dextrose because it is dextrorotatory. Other names for this monosaccharide are grape sugar and blood sugar. Human blood normally contains 65–110 mg of glucose/100 mL of blood. Glucose is synthesized by chlorophyll-containing plants using sunlight as a source of energy. In the process called photosynthesis, plants convert carbon dioxide from the air and water from the soil to glucose and oxygen.

$$\underset{\textbf{Carbon dioxide}}{6\,CO_2} + \underset{\textbf{Water}}{6\,H_2O} + \text{energy} \xrightarrow[\text{chlorophyll}]{\text{sunlight}} \underset{\textbf{Glucose}}{C_6H_{12}O_6} + \underset{\textbf{Oxygen}}{6\,O_2}$$

D-Fructose is found combined with D-glucose in the disaccharide sucrose (table sugar, Section 25.4A). D-Galactose is combined with D-glucose in the disaccharide lactose (milk sugar, Section 25.4B).

D-Ribose and 2-deoxy-D-ribose, the most abundant pentoses in the biological world, are essential building blocks of nucleic acids: D-ribose in ribonucleic acids (RNA) and 2-deoxy-D-ribose in deoxyribonucleic acids (DNA).

Example 25.1 Fischer Projections

Draw Fischer projections for the four aldotetroses. Which are D-monosaccharides, which are L-monosaccharides, and which are enantiomers? Refer to Table 25.1 and write the name of each aldotetrose.

Solution

Following are Fischer projections for the four aldotetroses. The letters D- and L refer to the configuration of the penultimate carbon, which, in the case of aldotetroses, is carbon 3. In the Fischer projection of a D-aldotetrose, the —OH on carbon 3 is on the right, and in an L-aldotetrose, it is on the left. Each of the erythroses is a diastereomer of each of the threoses.

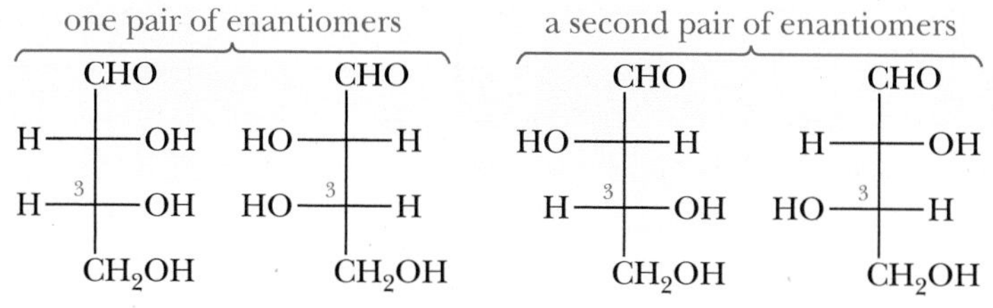

D-Erythrose L-Erythrose D-Threose L-Threose

Problem 25.1

Draw Fischer projections for all 2-ketopentoses. Which are D-2-ketopentoses, which are L-2-ketopentoses, and which are enantiomers?

D. Amino Sugars

Amino sugars contain an $—NH_2$ group in place of an —OH group. Only three amino sugars are common in nature: D-glucosamine, D-mannosamine, and D-galactosamine.

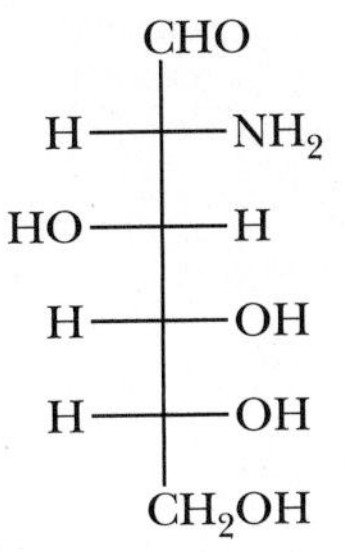

D-Glucosamine

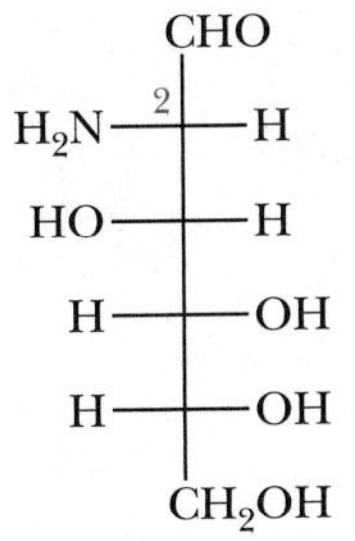

D-Mannosamine (C-2 stereoisomer of D-glucosamine)

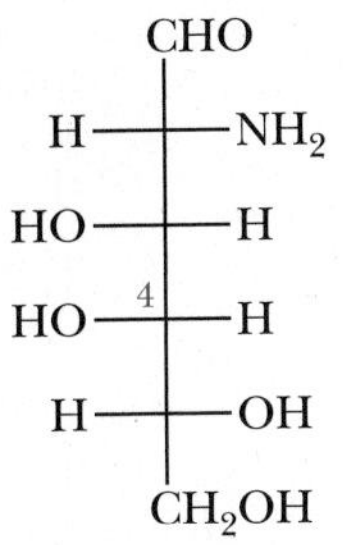

D-Galactosamine (C-4 stereoisomer of D-glucosamine)

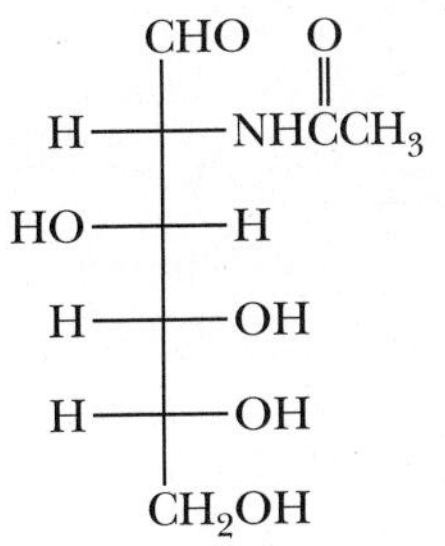

***N*-Acetyl-D-glucosamine**

N-Acetyl-D-glucosamine, a derivative of D-glucosamine, is a component of many polysaccharides, including chitin, the hard shell-like exoskeleton of lobsters, crabs, shrimp, and other shellfish. Many other amino sugars are components of naturally occurring antibiotics.

E. Physical Properties

Monosaccharides are colorless crystalline solids, although they often crystallize with difficulty. Because hydrogen bonding is possible between their polar —OH groups and water, all monosaccharides are very soluble in water. They are only slightly soluble in ethanol and are insoluble in nonpolar solvents such as diethyl ether, chloroform, and benzene.

25.2 The Cyclic Structure of Monosaccharides

We saw in Section 16.7B that aldehydes and ketones react with alcohols to form hemiacetals. We also saw that cyclic hemiacetals form very readily when hydroxyl and carbonyl groups are part of the same molecule and their interaction can form a five- or six-membered ring. For example, 4-hydroxypentanal forms a five-membered cyclic hemiacetal.

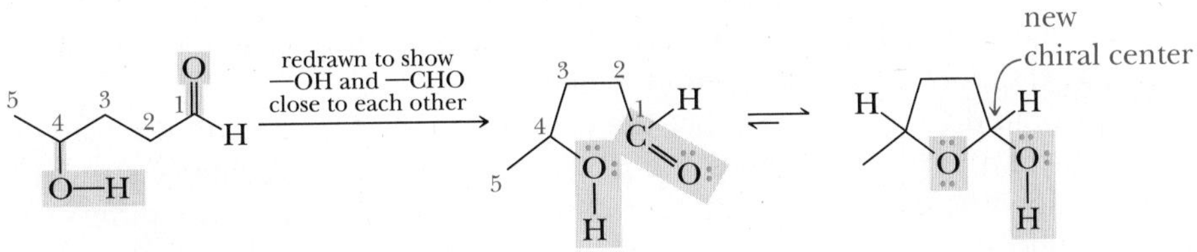

4-Hydroxypentanal
racemic

A cyclic hemiacetal
racemic

▶ Five- and six-membered rings are the most stable.

Note that 4-hydroxypentanal contains one chiral center and that a second chiral center is generated at carbon 1 as a result of hemiacetal formation.

Monosaccharides have hydroxyl and carbonyl groups in the same molecule. As a result, they too exist almost exclusively as five- and six-membered cyclic hemiacetals.

A. Haworth Projections

A common way of representing the cyclic structure of monosaccharides is the **Haworth projection**, named after the English chemist Sir Walter N. Haworth (1937 Nobel Prize in Chemistry). In a Haworth projection, a five- or six-membered cyclic hemiacetal is represented as a planar pentagon or hexagon, as the case may be, lying perpendicular to the plane of the paper. Groups bonded to the carbons of the ring then lie either above or below the plane of the ring. The new chiral center

Haworth projection
A way to view furanose and pyranose forms of monosaccharides. The ring is drawn flat and most commonly viewed through its edge with the anomeric carbon on the right and the oxygen atom of the ring in the rear.

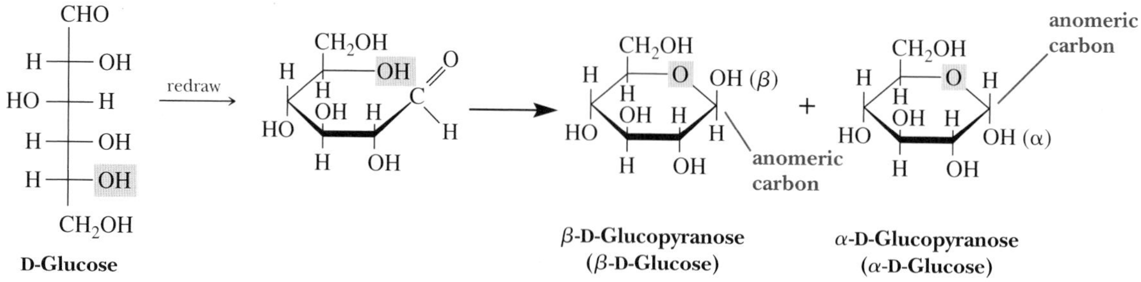

FIGURE 25.1 Haworth projections for α-D-glucopyranose and β-D-glucopyranose.

Anomeric carbon
The hemiacetal or acetal carbon of the cyclic form of a carbohydrate.

Anomers
Carbohydrates that differ in configuration only at their anomeric carbons.

Furanose
A five-membered cyclic form of a monosaccharide.

Pyranose
A six-membered cyclic form of a monosaccharide.

created in forming the cyclic structure is called an **anomeric carbon**. Stereoisomers that differ in configuration only at the anomeric carbon are called **anomers**. The anomeric carbon of an aldose is carbon 1; that of D-fructose, the most common ketose, is carbon 2.

Haworth projections are most commonly written with the anomeric carbon to the right and the hemiacetal oxygen to the back (Figure 25.1). In the terminology of carbohydrate chemistry, the designation β means that the —OH on the anomeric carbon of the cyclic hemiacetal is on the same side of the ring as the terminal —CH_2OH. Conversely, the designation α means that the —OH on the anomeric carbon of the cyclic hemiacetal is on the side of the ring opposite the terminal —CH_2OH.

A six-membered hemiacetal ring is indicated by the infix *-pyran-*, and a five-membered hemiacetal ring is indicated by the infix *-furan-*. The terms **furanose** and **pyranose** are used because monosaccharide five- and six-membered rings correspond to the heterocyclic compounds furan and pyran, respectively.

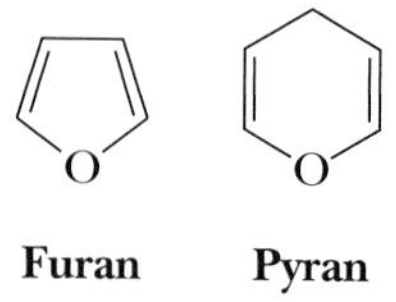

Because the α and β forms of glucose are six-membered cyclic hemiacetals, they are named α-D-glucopyranose and β-D-glucopyranose. These infixes are not always used in monosaccharide names, however. Thus, the glucopyranoses, for example, are often named simply α-D-glucose and β-D-glucose.

You would do well to remember the configuration of groups on the Haworth projections of α-D-glucopyranose and β-D-glucopyranose as reference structures. Knowing how the open-chain configuration of any other aldohexose differs from that of D-glucose, you can then construct its Haworth projection by reference to the Haworth projection of D-glucose.

Example 25.2 Haworth Projections

Draw Haworth projections for the α- and β-anomers of D-galactopyranose.

Solution

One way to arrive at these projections is to use the α and β forms of D-glucopyranose as reference and to remember (or discover by looking at Table 25.1) that D-galactose differs from D-glucose only in the configuration at carbon 4. Thus, begin with the Haworth projections shown in Figure 25.1 and then invert the configuration at carbon 4.

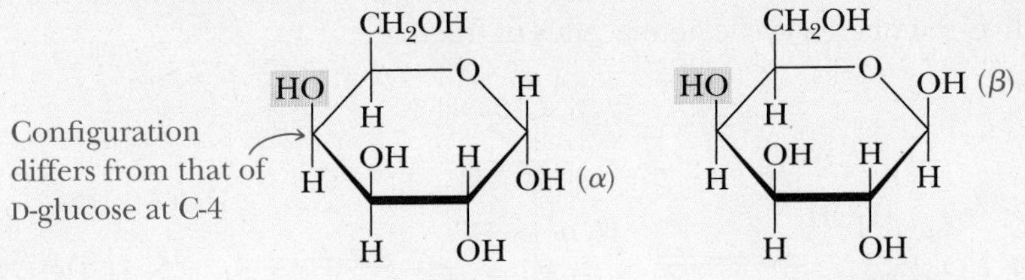

α-D-Galactopyranose (α-D-Galactose)

β-D-Galactopyranose (β-D-Galactose)

Problem 25.2

Mannose exists in aqueous solution as a mixture of α-D-mannopyranose and β-D-mannopyranose. Draw Haworth projections for these molecules.

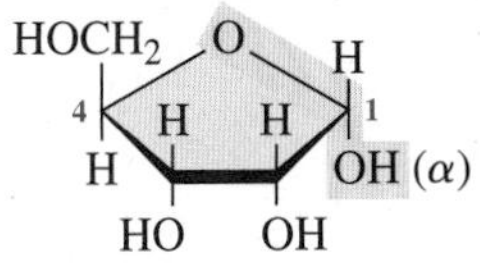

α-D-Ribofuranose (α-D-Ribose)

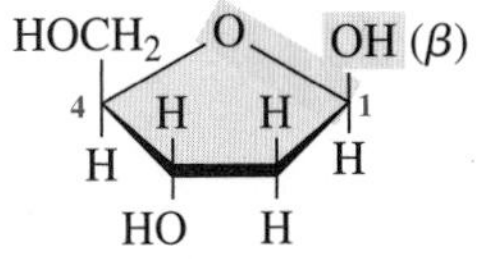

β-2-Deoxy-D-ribofuranose (β-2-Deoxy-D-ribose)

FIGURE 25.2 Haworth projections for some cyclic hemiacetals.

Aldopentoses also form cyclic hemiacetals. The most prevalent forms of D-ribose and other pentoses in the biological world are furanoses. Figure 25.2 shows Haworth projections for α-D-ribofuranose (α-D-ribose) and β-2-deoxy-D-ribofuranose (β-2-deoxy-D-ribose). The prefix 2-deoxy indicates the absence of oxygen at carbon 2.

Chemical Connections

L-Ascorbic Acid (Vitamin C)

The structure of L-ascorbic acid (vitamin C) resembles that of a monosaccharide. In fact, this vitamin is synthesized both biochemically by plants and some animals and commercially from D-glucose. Humans do not have the enzymes required for this synthesis; therefore, we must obtain it in the food we eat or as a vitamin supplement. Approximately 66 million kilograms of vitamin C are synthesized every year in the United States.

L-Ascorbic acid is very easily oxidized to L-dehydroascorbic acid, a diketone.

Both L-ascorbic acid and L-dehydroascorbic acid are physiologically active and are found together in most body fluids.

Ascorbic acid is one of the most important antioxidants (the H in the β-enolic OH is weakly bonded and easily abstracted by radicals). One of the most important roles it plays may be to replenish the lipid-soluble antioxidant α-tocopherol by transferring a hydrogen atom to the tocopherol radical, formed by reaction with radicals in the autoxidation process (see Section 8.7).

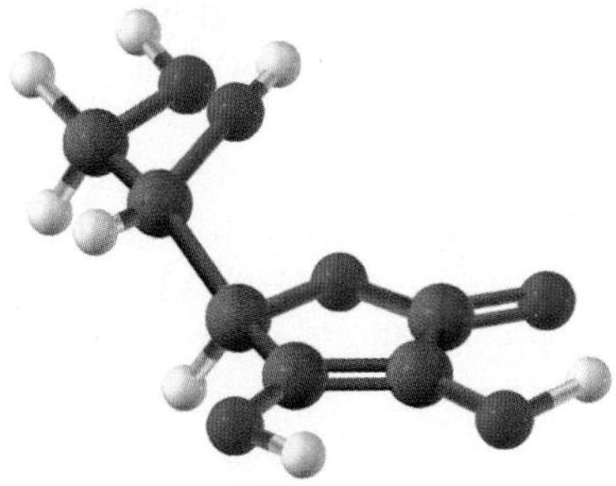

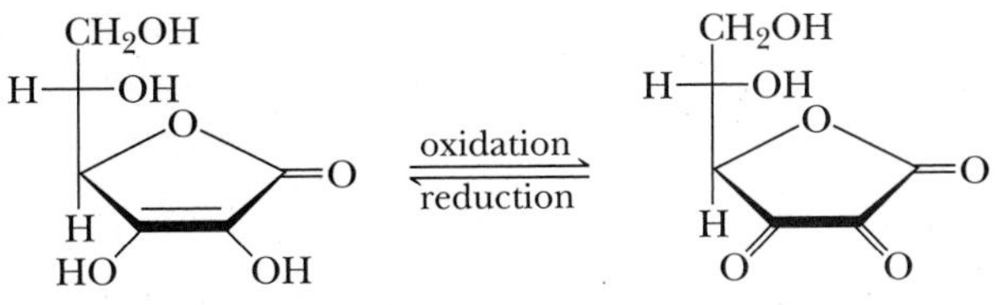

L-Ascorbic acid (Vitamin C)

L-Dehydroascorbic acid

Units of D-ribose and 2-deoxy-D-ribose in nucleic acids and most other biological molecules are found almost exclusively in the β-configuration.

Other monosaccharides also form five-membered cyclic hemiacetals. Following are the five-membered cyclic hemiacetals of fructose.

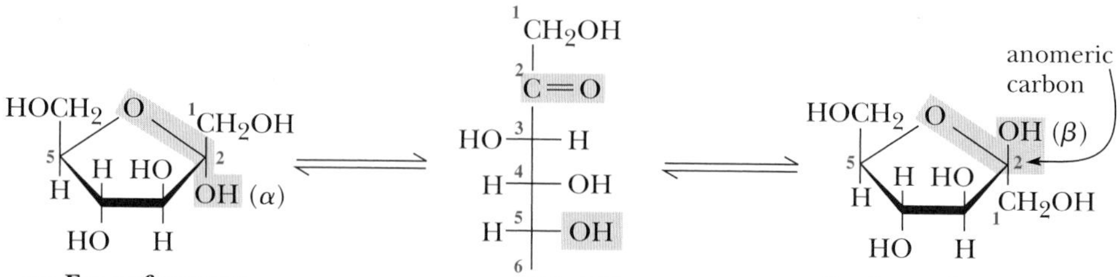
anomeric carbon
OH (α)
OH (β)

α-D-Fructofuranose   D-Fructose   β-D-Fructofuranose

The β-D-fructofuranose form is found in the disaccharide sucrose (Section 25.4A).

B. Conformation Representations

A five-membered ring is so close to being planar that Haworth projections are adequate representations of furanoses. For pyranoses, however, the six-membered ring is more accurately represented as a chair conformation. Following are structural formulas for α-D-glucopyranose and β-D-glucopyranose drawn as chair conformations. Also shown is the open-chain or free aldehyde form with which the cyclic hemiacetal forms are in equilibrium in aqueous solution.

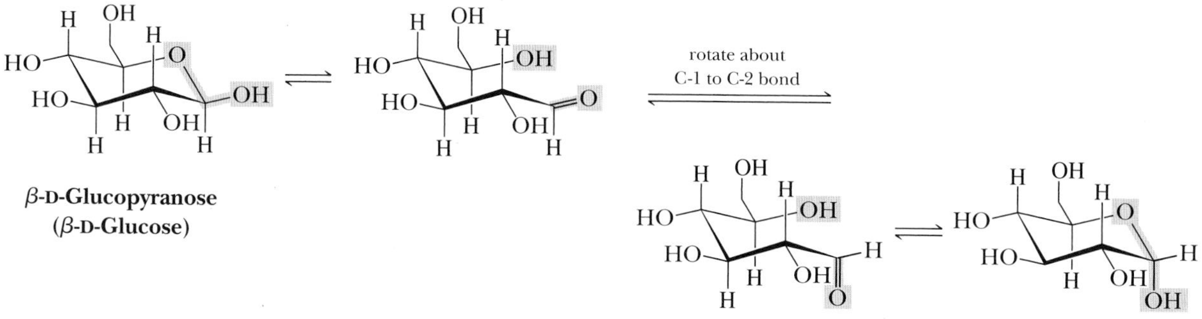
rotate about
C-1 to C-2 bond

β-D-Glucopyranose (β-D-Glucose)

α-D-Glucopyranose (α-D-Glucose)

Notice that each group, including the anomeric —OH, on the chair conformation of β-D-glucopyranose, is equatorial. Notice also that the —OH group on the anomeric carbon is axial in α-D-glucopyranose. Because of the equatorial orientation of the —OH on its anomeric carbon, β-D-glucopyranose is more stable and predominates in aqueous solution.

At this point, you should compare the relative orientations of groups on the D-glucopyranose ring in the Haworth projection and the chair conformation. The orientations of groups on carbons 1 through 5 of β-D-glucopyranose, for example, are up, down, up, down, and up in both representations.

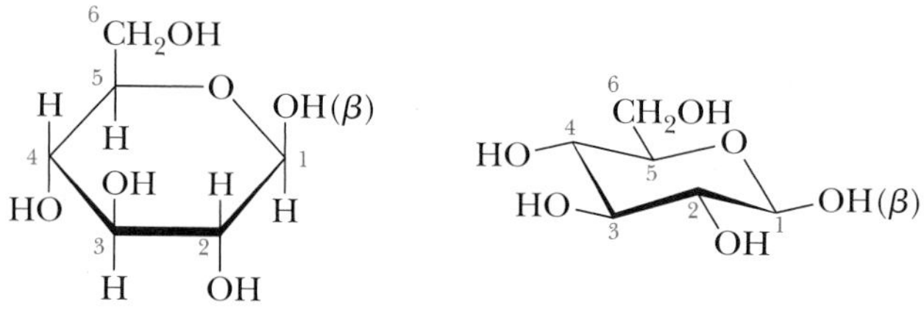

β-D-Glucopyranose (Haworth projection)   β-D-Glucopyranose (chair conformation)

Example 25.3 Saccharide Chair Conformations

Draw chair conformations for α-D-galactopyranose and β-D-galactopyranose. Label the anomeric carbon in each.

Solution

D-Galactose differs in configuration from D-glucose only at carbon 4. Therefore, draw the α and β forms of D-glucopyranose and then interchange the positions of the —OH and —H groups on carbon 4.

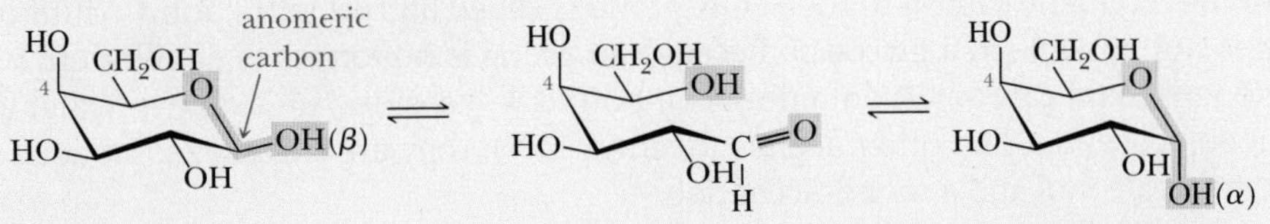

β-D-Galactopyranose
(β-D-Galactose)
$[\alpha]_D$ + 52.8

D-Galactose

α-D-Galactopyranose
(α-D-Galactose)
$[\alpha]_D$ + 150.7

Problem 25.3

Draw chair conformations for α-D-mannopyranose and β-D-mannopyranose. Label the anomeric carbon in each.

Watch a video explanation

C. Mutarotation

Mutarotation is the change in specific rotation that accompanies the interconversion of α- and β-anomers in aqueous solution. As an example, a solution prepared by dissolving crystalline α-D-glucopyranose in water shows an initial rotation of +112.2, which gradually decreases to an equilibrium value of +52.7 as α-D-glucopyranose reaches an equilibrium with β-D-glucopyranose. A solution of β-D-glucopyranose also undergoes mutarotation, during which the specific rotation changes from an initial value of +18.7 to the same equilibrium value of +52.7. The equilibrium mixture consists of 64% β-D-glucopyranose and 36% α-D-glucopyranose. It contains only traces (0.003%) of the open-chain form. Mutarotation is common to all carbohydrates that exist in hemiacetal forms.

Mutarotation
The change in specific rotation that occurs when an α or β hemiacetal form of a carbohydrate in aqueous solution is converted to an equilibrium mixture of the two forms.

25.3 Reactions of Monosaccharides

A. Formation of Glycosides (Acetals)

We saw in Section 16.7B that treatment of an aldehyde or a ketone with one molecule of alcohol gives a hemiacetal and that treatment of the hemiacetal with a molecule of alcohol gives an acetal. Similar treatment of monosaccharides, all of which exist almost exclusively in a cyclic hemiacetal form, also gives acetals, as illustrated by the reaction of β-D-glucopyranose with methanol.

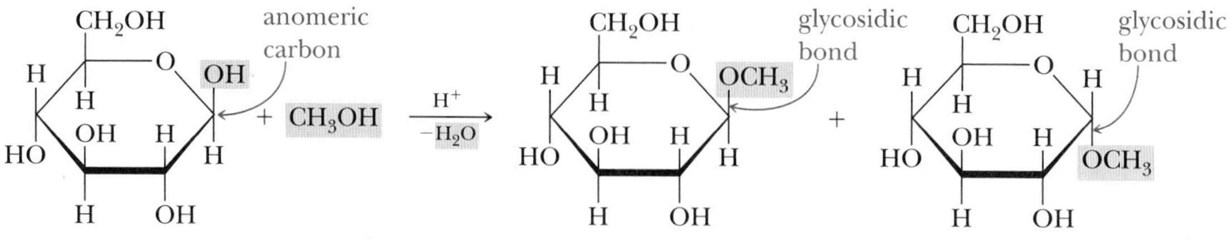

β-D-Glucopyranose (β-D-Glucose) — Methyl β-D-glucopyranoside (Methyl β-D-glucoside) — Methyl α-D-glucopyranoside (Methyl α-D-glucoside)

Glycoside

A carbohydrate in which the —OH on its anomeric carbon is replaced with —OR.

Glycosidic bond

The bond from the anomeric carbon of a glycoside to an —OR group.

A cyclic acetal derived from a monosaccharide is called a **glycoside**, and the bond from the anomeric carbon to the —OR group is called a **glycosidic bond**. Mutarotation is not possible in a glycoside because an acetal is no longer in equilibrium with the open-chain carbonyl-containing compound. Glycosides are stable in water and aqueous base, but like other acetals (Section 16.7), they are hydrolyzed in aqueous acid to an alcohol and a monosaccharide.

Glycosides are named by listing the alkyl or aryl group bonded to oxygen followed by the name of the carbohydrate in which the ending *-e* is replaced with *-ide*. For example, the glycosides derived from β-D-glucopyranose are named β-D-glucopyranosides; those derived from β-D-ribofuranose are named β-D-ribofuranosides.

Example 25.4 Glycoside Structures

Draw a structural formula for methyl β-D-ribofuranoside (methyl β-D-riboside). Label the anomeric carbon and the glycosidic bond.

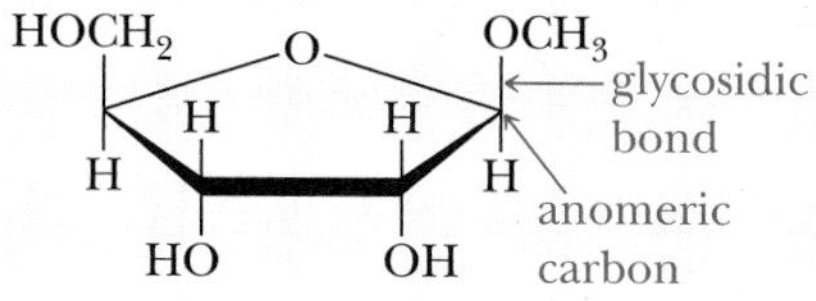

Problem 25.4

Draw a Haworth projection and a chair conformation for methyl α-D-mannopyranoside (methyl α-D-mannoside). Label the anomeric carbon and the glycosidic bond.

Just as the anomeric carbon of a cyclic hemiacetal undergoes reaction with the —OH group of an alcohol to form a glycoside, it also undergoes reaction with the N—H group of an amine to form an *N*-glycoside. Especially important in the biological world are the *N*-glycosides formed between D-ribose and 2-deoxy-D-ribose, each as a furanose, and the heterocyclic aromatic amines uracil, cytosine, thymine, adenine, and guanine (Figure 25.3). *N*-Glycosides of these pyrimidine and purine bases are structural units of nucleic acids (Chapter 28).

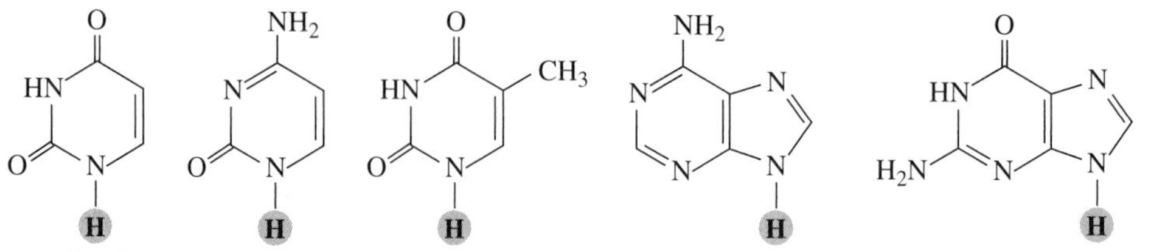

Uracil — Cytosine — Thymine — Adenine — Guanine

FIGURE 25.3 Structural formulas of the five most important pyrimidine and purine bases found in DNA and RNA. The hydrogen atom shown in gray is lost in forming an *N*-glycoside.

Example 25.5 *N*-Glycoside Structures

Draw a structural formula for cytidine, the β-*N*-glycoside formed between D-ribofuranose and cytosine.

Solution

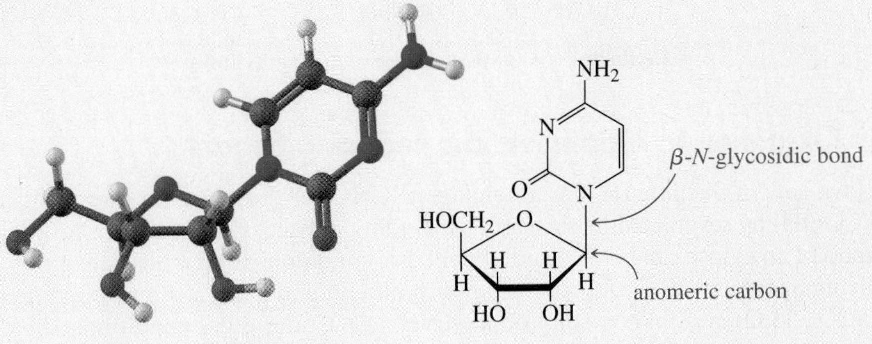

Cytidine

Problem 25.5

Draw a structural formula for the β-*N*-glycoside formed between 2-deoxy-D-ribofuranose and adenine.

B. Reduction to Alditols

The carbonyl group of a monosaccharide can be reduced to a hydroxyl group by a variety of reducing agents, including sodium borohydride and hydrogen in the presence of a transition metal catalyst. The reduction products are known as **alditols**. Reduction of D-glucose gives D-glucitol, more commonly known as D-sorbitol. Note that D-glucose is shown here in the open-chain form. Only a small amount of this form is present in solution, but as it is reduced, the rapid equilibrium between cyclic hemiacetal forms and the open-chain form replaces it.

Alditol

The product formed when the C=O group of a monosaccharide is reduced to a CHOH group.

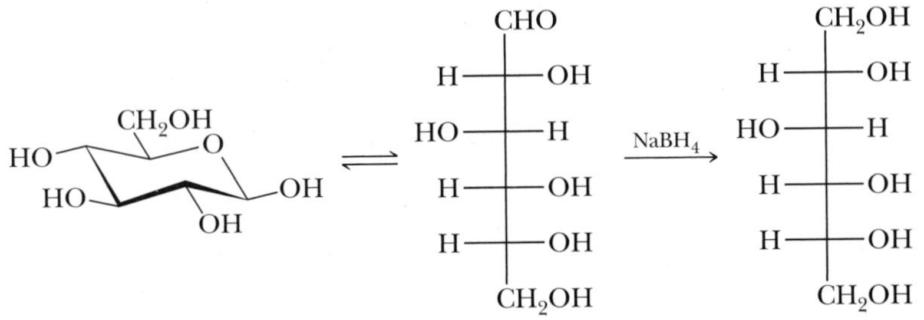

β-D-Glucopyranose **D-Glucose** **D-Glucitol (D-Sorbitol)**

Sorbitol is found in the plant world in many berries and in cherries, plums, pears, apples, seaweed, and algae. It is about 60% as sweet as sucrose (table sugar) and is used in the manufacture of candies and as a sugar substitute for diabetics. D-Sorbitol is an important food additive, usually added to prevent dehydration of foods and other materials upon exposure to air because it binds water strongly.

Other alditols common in the biological world are erythritol, D-mannitol, and xylitol. Xylitol is used as a sweetening agent in "sugarless" gum, candy, and sweet cereals.

Many "sugar-free" products contain sugar alcohols such as D-sorbitol and xylitol.

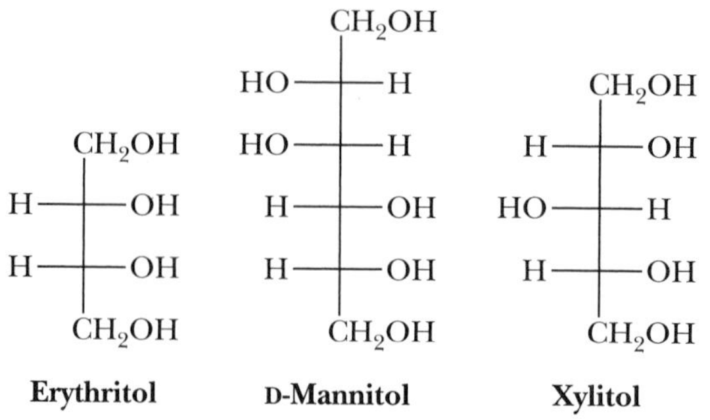

C. Oxidation to Aldonic Acids: Reducing Sugars

As we saw in Section 16.10A, aldehydes (RCHO) are oxidized to carboxylic acids (RCOOH) by several oxidizing agents, including oxygen, O_2. Similarly, the aldehyde group of an aldose can be oxidized under basic conditions to a carboxylate group. Oxidizing agents for this purpose include bromine in aqueous calcium carbonate (**Br_2, $CaCO_3$, H_2O**) and Tollens' solution [$Ag(NH_3)_2^+$]. Under these conditions, the cyclic form of an aldose is in equilibrium with the open-chain form, which is then oxidized by the mild oxidizing agent. D-Glucose, for example, is oxidized to D-gluconate (the anion of D-gluconic acid).

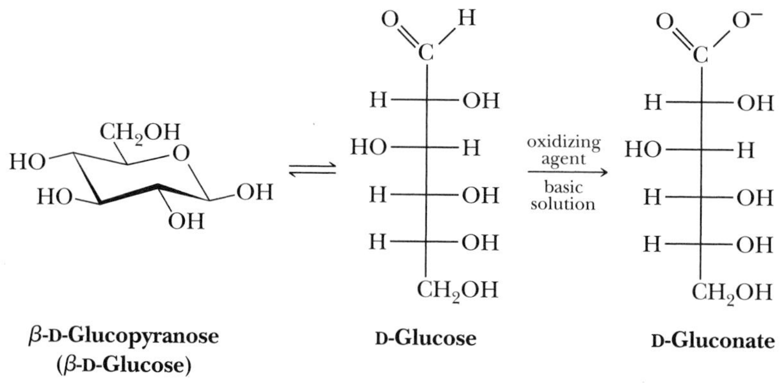

Aldonic acid The product formed when the —CHO group of an aldose is oxidized to a —COOH group.

Reducing sugar A carbohydrate that reacts with an oxidizing agent to form an aldonic acid. In this reaction, the carbohydrate reduces the oxidizing agent.

Any carbohydrate that reacts with an oxidizing agent to form an **aldonic acid** is classified as a **reducing sugar** (it reduces the oxidizing agent).

Surprisingly, 2-ketoses are also reducing sugars. Carbon 1 (a CH_2OH group) of a 2-ketose is not oxidized directly. Rather, under the basic conditions of this oxidation, a 2-ketose is in equilibrium with an aldose by way of an enediol intermediate. The aldose is then oxidized by the mild oxidizing agent.

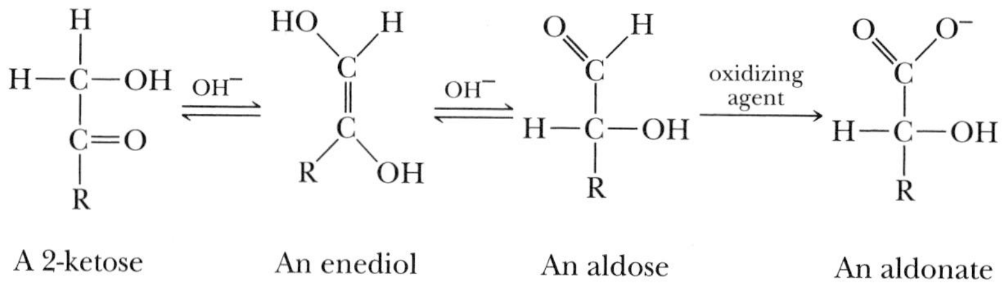

D. Oxidation to Uronic Acids

Enzyme-catalyzed oxidation of the primary hydroxyl group at carbon 6 of a hexose yields a uronic acid. Enzyme-catalyzed oxidation of D-glucose, for example, yields D-glucuronic acid, shown here in both its open-chain and cyclic hemiacetal forms.

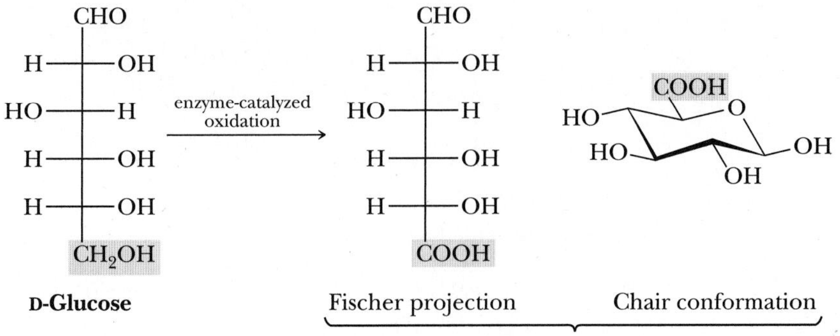

D-Glucose

D-Glucuronic acid
(a uronic acid)

D-Glucuronic acid is widely distributed in both the plant and the animal world. In humans, it is an important component of the glucosaminoglycans of connective tissues (Section 25.6). It is also used by the body to detoxify foreign hydroxyl-containing compounds such as phenols and alcohols. In the liver, these compounds are converted to glycosides of glucuronic acid (glucuronides) and excreted in the urine. The intravenous anesthetic propofol, for example, is converted to the following glucuronide and excreted in the urine.

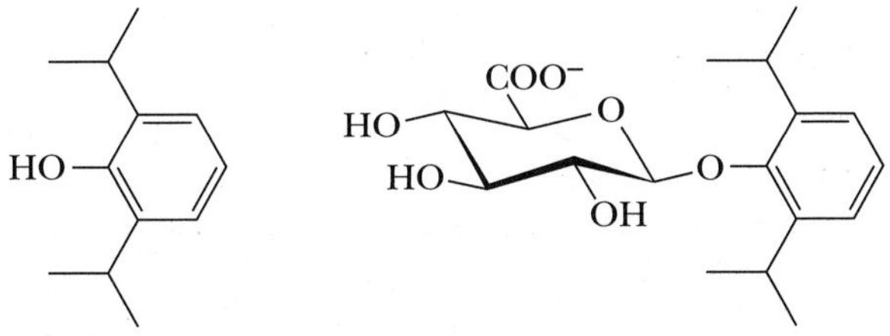

Propofol A urine-soluble glucuronide

E. Oxidation by Periodic Acid

Oxidation by periodic acid, $HIO_4 \cdot 2H_2O$ or H_5IO_6, has proven useful in structure determinations of carbohydrates, particularly in determining the size of glycoside rings. Recall from Section 10.8E that periodic acid cleaves the carbon-carbon bond of a glycol in a reaction that proceeds through a cyclic periodic ester. In this reaction, iodine(VII) of periodic acid is reduced to iodine(V) of iodic acid.

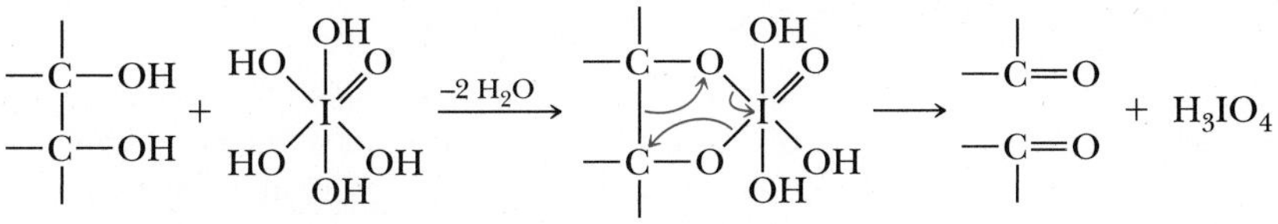

A glycol **Periodic acid** A cyclic periodic ester **Iodic acid**

Periodic acid also cleaves carbon-carbon bonds of α-hydroxyketones and α-hydroxyaldehydes by a similar mechanism. Following are abbreviated structural formulas for these functional groups and the products of their oxidative cleavage by periodic acid. As a way to help you understand how each set of products is formed, each carbonyl in a starting material is shown as a hydrated intermediate that is then oxidized. In this way, each oxidation can be viewed as being analogous to the oxidation of a glycol.

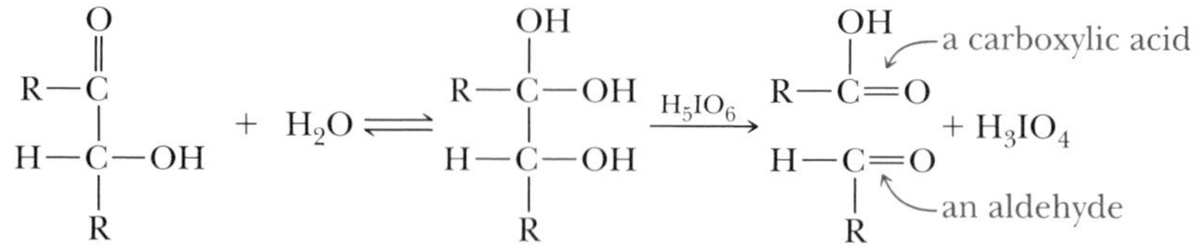
a carboxylic acid
an aldehyde

α-Hydroxyketone Hydrated intermediate

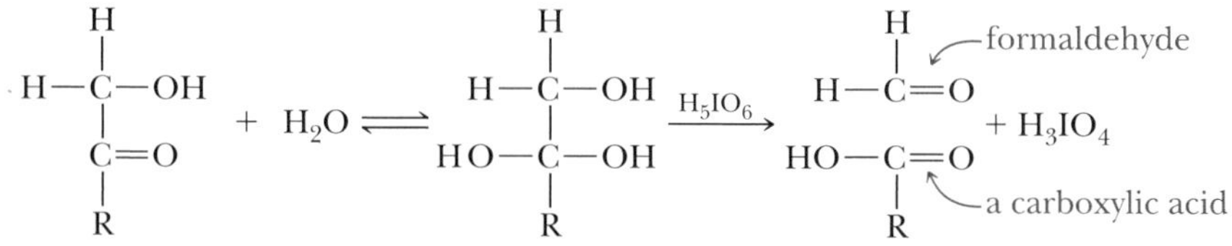
formaldehyde
a carboxylic acid

α-Hydroxyketone Hydrated intermediate

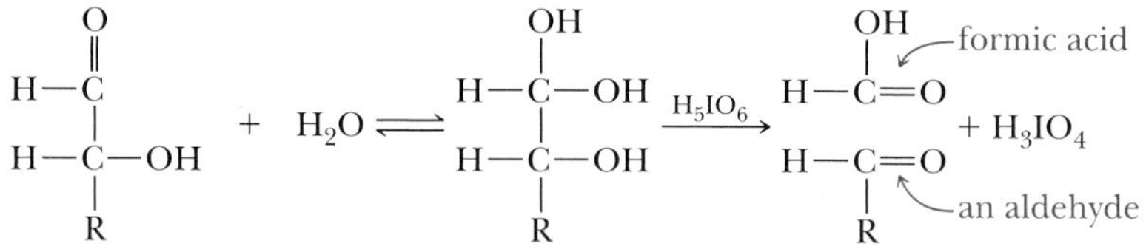
formic acid
an aldehyde

α-Hydroxyaldehyde Hydrated intermediate

As an example of the usefulness of this reaction in carbohydrate chemistry, oxidation of methyl β-D-glucoside consumes two moles of periodic acid and produces one mole of formic acid. This stoichiometry and the formation of formic acid are possible only if —OH groups are on three adjacent carbon atoms.

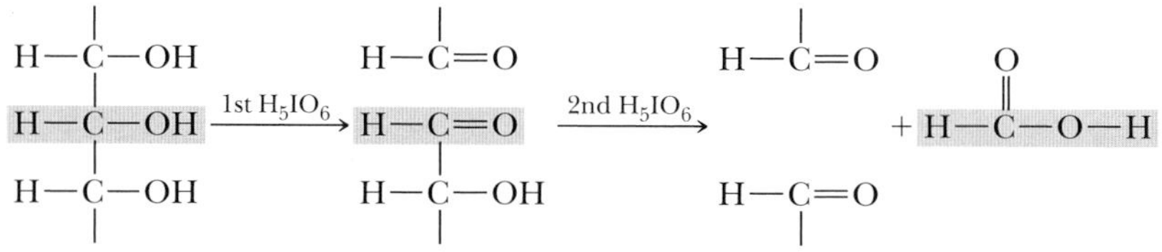

Chemical Connections

Testing for Glucose

The analytical procedure most often performed in a clinical chemistry laboratory is the determination of glucose in blood, urine, or other biological fluids. This is true because of the high incidence of diabetes mellitus. Approximately 15 million known diabetics live in the United States, and it is estimated that another 1 million are undiagnosed.

Diabetes mellitus is characterized by insufficient blood levels of the hormone insulin. If the blood concentration of insulin is too low, muscle and liver cells do not absorb glucose from the blood, which, in turn, leads to increased levels of blood glucose (hyperglycemia), impaired metabolism of fats and proteins, ketosis, and possible diabetic coma. A rapid test for blood glucose levels is critical for early diagnosis and effective management of this disease. In addition to being rapid, a test must also be specific for D-glucose; it must give a positive test for D-glucose but not react with any other substance normally present in biological fluids.

Blood glucose levels are now measured by an enzyme-based procedure using the enzyme glucose oxidase. This enzyme catalyzes the oxidation of β-D-glucose to D-gluconic acid.

Glucose oxidase is specific for β-D-glucose. Therefore, complete oxidation of any sample containing both β-D-glucose and α-D-glucose requires conversion of the α form to the β form. Fortunately, this interconversion is rapid and complete in the short time required for the test.

Molecular oxygen, O_2, is the oxidizing agent in this reaction and is reduced to hydrogen peroxide, H_2O_2. In one procedure, hydrogen peroxide formed

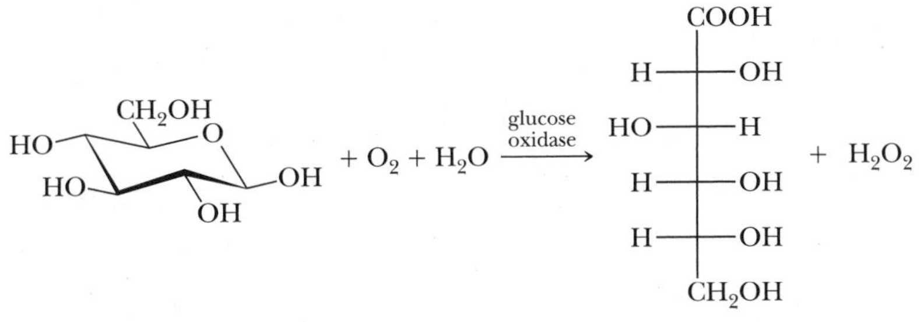
glucose oxidase
+ O_2 + H_2O
+ H_2O_2

β-D-Glucopyranose (β-D-Glucose) **D-Gluconic acid** **Hydrogen peroxide**

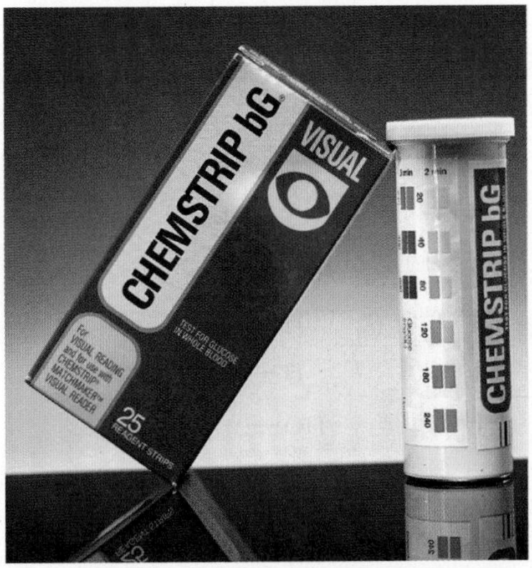
CHEMSTRIP bG
VISUAL
25 REAGENT STRIPS

© Cengage Learning/Charles D. Winters

A test kit for the presence of glucose in urine.

in the glucose oxidase-catalyzed reaction is used to oxidize colorless *o*-toluidine to a colored product in a reaction catalyzed by the enzyme peroxidase. The concentration of the colored oxidation product is determined spectrophotometrically and is proportional to the concentration of glucose in the test solution.

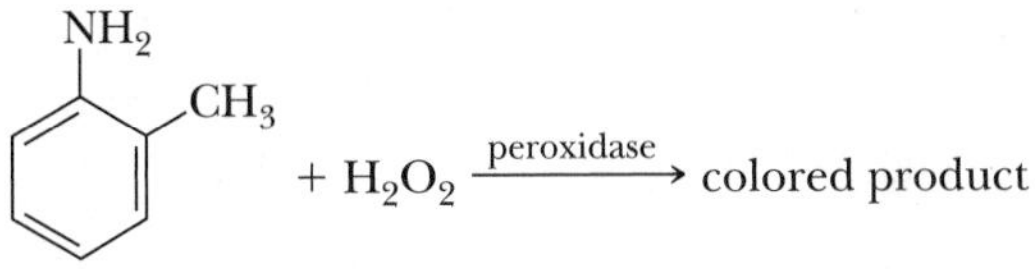
+ H_2O_2
peroxidase
colored product

2-Methylaniline (*o*-Toluidine)

Several commercially available test kits use the glucose oxidase reaction for qualitative determination of glucose in urine.

This is evidence that methyl β-D-glucoside is indeed a pyranoside.

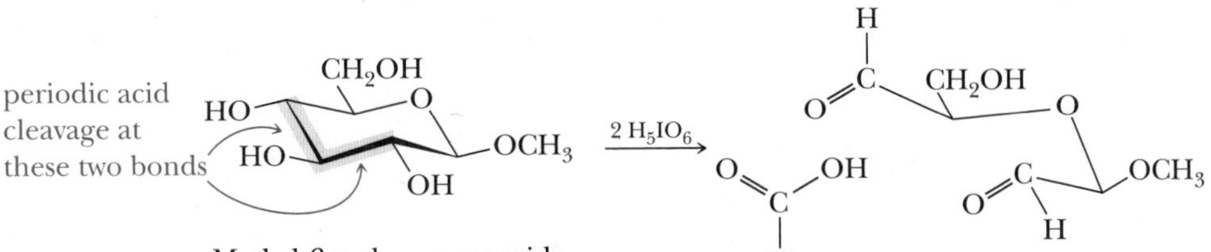
periodic acid cleavage at these two bonds
2 H_5IO_6

Methyl β-D-glucopyranoside

Methyl β-D-fructoside consumes only one mole of periodic acid and produces neither formaldehyde nor formic acid. Thus, oxidizable groups exist on adjacent carbons only at one site in the molecule. The fructoside, therefore, must be a five-membered ring (a fructofuranoside).

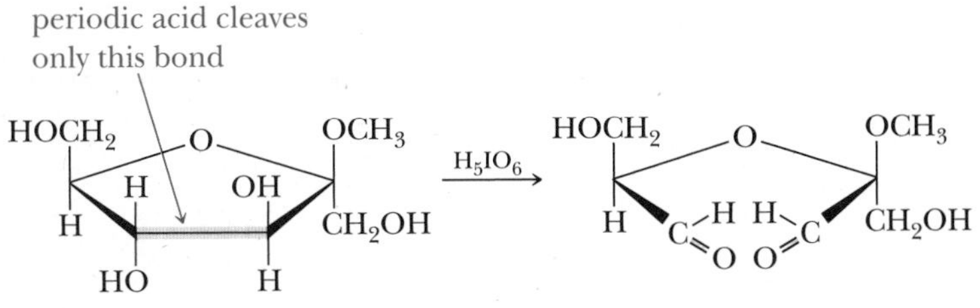
periodic acid cleaves only this bond
H_5IO_6

Methyl β-D-fructofuranoside

MCAT Practice: *Passage and Questions*

Fucose

Fucose, one of several monosaccharides found in the surface polysaccharides of animal cell membranes, is synthesized biochemically from mannose in the following eight steps.

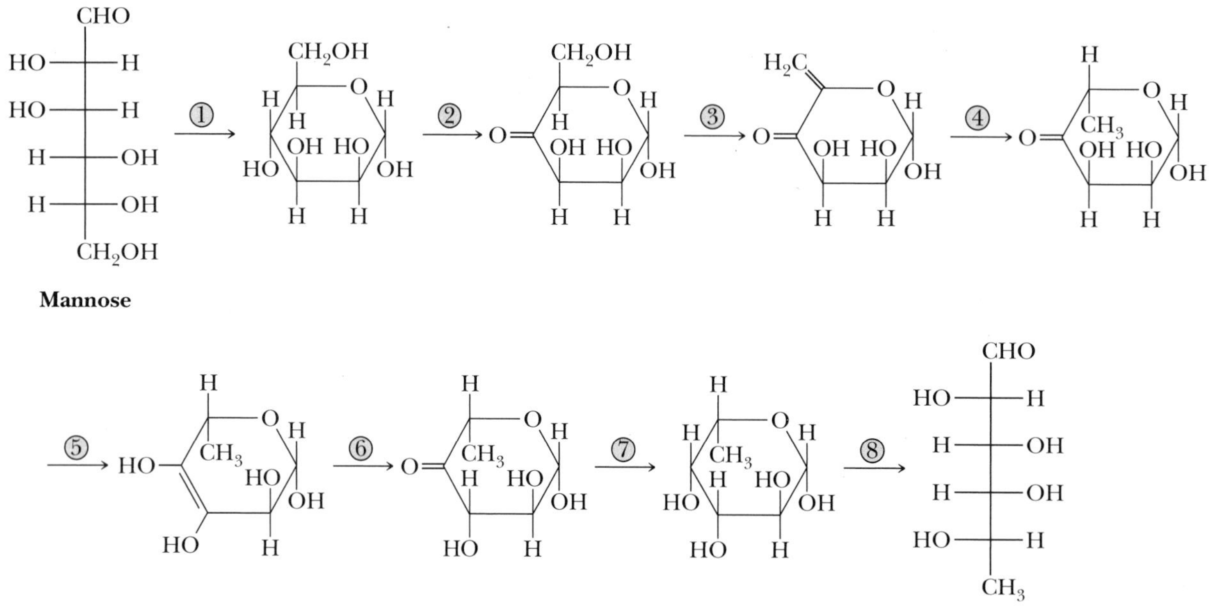

Mannose

Fucose

Questions

A. What are the stereochemical descriptors for mannose and fucose in the above scheme?

1. Both are D-sugars.
2. Both are L-sugars.
3. Mannose is D and fucose is L.
4. Mannose is L and fucose is D.

B. In the elimination reaction of Step 3, what kind of mechanism is most likely involved?

1. An acid-catalyzed departure of —OH as water with simultaneous deprotonation of the α-hydrogen to the ketone
2. Hydroxide departure as a leaving group followed by deprotonation of the ketone α-hydrogen
3. Acid-catalyzed departure of —OH as water to create a primary carbocation followed by deprotonation of the ketone α-hydrogen
4. Deprotonation of the ketone α-hydrogen to make an enolate followed by hydroxide leaving group departure

C. Steps 5 and 6 combined are best described as which of the following?

1. A reduction followed by reoxidation at C3
2. Two keto-enol tautomerizations that lead to stereochemical inversion at C3
3. An elimination followed by addition that leads to stereochemical retention at C4
4. Dehydration followed by hydration that leads to stereochemical retention at C4

D. What is the consequence of the transformation of mannose to fucose?

1. A reducing sugar is created.
2. Carbons 3 and 5 are inverted.
3. A sugar that cannot lead to an aldonic acid is created.
4. Both 1 and 2.
5. Both 2 and 3.

E. How many stereoisomers would result from the reaction of fucose with methanol and acid catalysis?

1. Because there are four chiral centers, one would get 2^4 isomers.
2. One would get two isomers: the α- and β-anomers.
3. Two diastereomers would form by scrambling the stereochemistry of the carbon with α-hydrogens.
4. There would be no reaction and hence no stereoisomers would be created.

25.4 Disaccharides and Oligosaccharides

Most carbohydrates in nature contain more than one monosaccharide unit. Those that contain two units are called **disaccharides**, those that contain three units are called **trisaccharides**, and so forth. The general term **oligosaccharide** is often used for carbohydrates that contain from four to ten monosaccharide units. Carbohydrates containing larger numbers of monosaccharide units are called **polysaccharides**.

In a disaccharide, two monosaccharide units are joined together by a glycosidic bond between the anomeric carbon of one unit and an —OH of the other. Three important disaccharides are sucrose, lactose, and maltose.

Disaccharide

A carbohydrate containing two monosaccharide units joined by a glycosidic bond.

Oligosaccharide

A carbohydrate containing four to ten monosaccharide units, each joined to the next by a glycosidic bond.

Polysaccharide

A carbohydrate containing a large number of monosaccharide units, each joined to the next by one or more glycosidic bonds.

A. Sucrose

Sucrose (table sugar) is the most abundant disaccharide in the biological world. It is obtained principally from the juice of sugarcane and sugar beets. In sucrose, carbon 1 of α-D-glucopyranose is joined to carbon 2 of β-D-fructofuranose by an α-1,2-glycosidic bond.

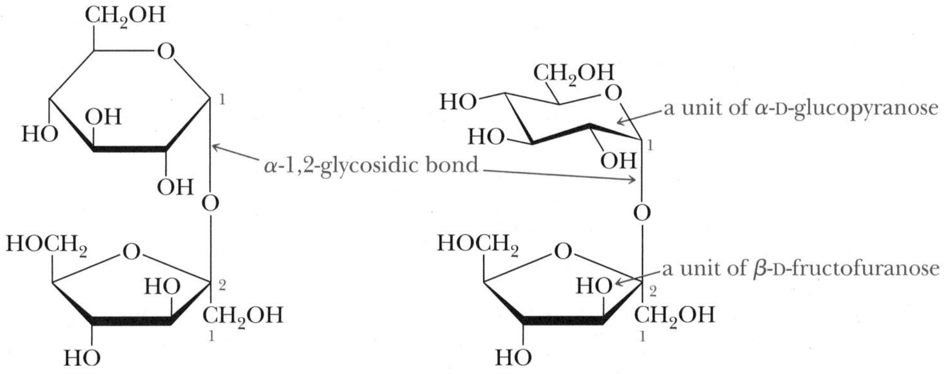

Sucrose

Note that glucose is a six-membered (pyranose) ring, whereas fructose is a five-membered (furanose) ring. Because the anomeric carbons of both the glucopyranose and fructofuranose units are involved in formation of the glycosidic bond, sucrose is a nonreducing sugar.

These products help individuals with lactose intolerance meet their calcium needs.

B. Lactose

Lactose is the principal sugar present in milk. It makes up about 5%–8% of human milk and 4%–6% of cow's milk. It consists of D-galactopyranose bonded by a β-1,4-glycosidic bond to carbon 4 of D-glucopyranose. Lactose is a reducing sugar.

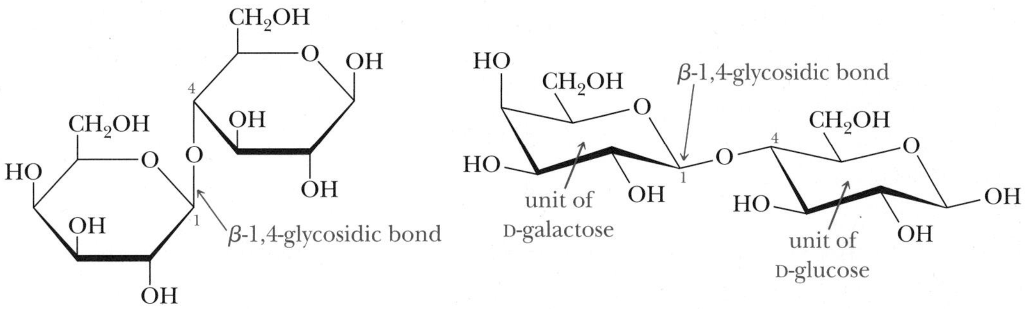

Lactose

C. Maltose

Maltose derives its name from its presence in malt, the juice from sprouted barley and other cereal grains (from which beer is brewed). Maltose consists of two molecules of D-glucopyranose joined by an α-1,4-glycosidic bond between carbon 1 (the anomeric carbon) of one unit and carbon 4 of the other unit. Following are representations for β-maltose, so named because the —OH on the anomeric carbon of the glucose unit on the right is β.

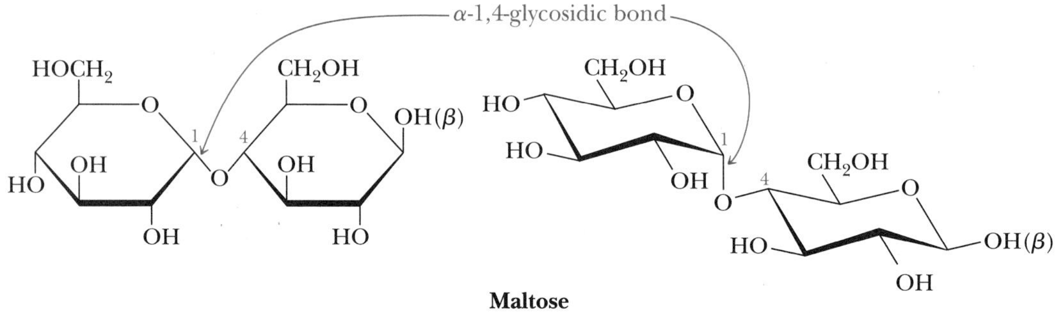
α-1,4-glycosidic bond
HOCH2
CH2OH
O
OH(β)
HO
OH
OH
HO
1
4
CH2OH
OH
CH2OH
OH(β)

Maltose

Maltose is a reducing sugar because the hemiacetal group on the right unit of D-glucopyranose is in equilibrium with the free aldehyde and can be oxidized to a carboxylic acid.

D. Relative Sweetness of Some Carbohydrate and Artificial Sweeteners

Although all monosaccharides are sweet to the taste, some are sweeter than others (Table 25.2). D-Fructose tastes the sweetest, even sweeter than sucrose (table sugar, Section 25.4A). The sweet taste of honey is attributable largely to D-fructose and D-glucose. Lactose (Section 25.4B) has almost no sweetness. It occurs in many milk products and is sometimes added to foods as a filler. Some people lack an enzyme that allows them to tolerate lactose well; they should avoid these foods.

Table 25.2 Relative Sweetness of Some Carbohydrates and Artificial Sweetening Agents*

Carbohydrate	Sweetness Relative to Sucrose	Artificial Sweetener	Sweetness Relative to Sucrose
Fructose	1.74	Saccharin	450
Invert sugar	1.25	Acesulfame-K	200
Sucrose (table sugar)	1.00	Aspartame	160
Honey	0.97		
Glucose	0.74		
Maltose	0.33		
Galactose	0.32		
Lactose (milk sugar)	0.16		

*We have no mechanical way to measure sweetness. Such testing is done by having a group of people taste solutions of varying sweetness and ranking them in order by taste.

Example 25.6 Disaccharide Structures

Draw a chair conformation for the β-anomer of a disaccharide in which two units of D-glucopyranose are joined by an α-1,6-glycosidic bond.

Solution

First draw a chair conformation of α-D-glucopyranose. Then connect the anomeric carbon of this monosaccharide to carbon 6 of a second D-glucopyranose unit by an α-glycosidic bond. The resulting molecule is either α or β depending on the orientation of the —OH group on the reducing end of the disaccharide. The disaccharide shown here is β.

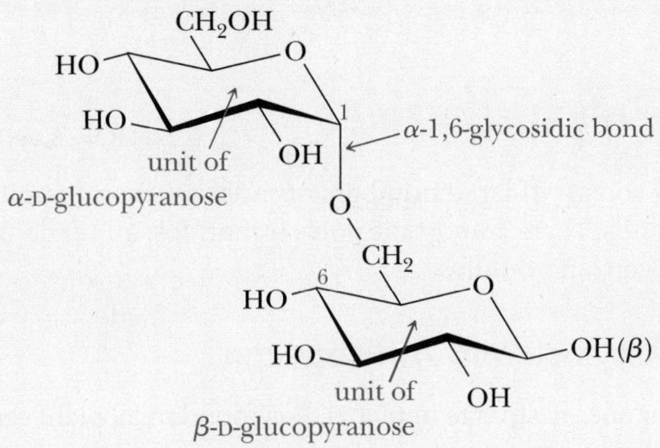
CH2OH
HO
O
HO
1
α-1,6-glycosidic bond
unit of
α-D-glucopyranose
OH
O
CH2
6
HO
O
HO
OH(β)
unit of
β-D-glucopyranose
OH

Problem 25.6

Draw a chair conformation for the α form of a disaccharide in which two units of D-glucopyranose are joined by a β-1,3-glycosidic bond.

Watch a video explanation

Chemical Connections

A, B, AB, and O Blood Group Substances

Membranes of animal plasma cells have large numbers of relatively small carbohydrates bound to them. In fact, the outsides of most plasma cell membranes are literally "sugar-coated." These membrane-bound carbohydrates are part of the mechanism by which cell types recognize each other and, in effect, act as biochemical markers. Typically, these membrane-bound carbohydrates contain from 4 to 17 monosaccharide units consisting primarily of relatively few monosaccharides, including D-galactose, D-mannose, L-fucose, *N*-acetyl-D-glucosamine, and *N*-acetyl-D-galactosamine. L-Fucose is a 6-deoxyaldohexose.

Among the first discovered and best understood of these membrane-bound carbohydrates are those of the ABO blood group system, discovered in

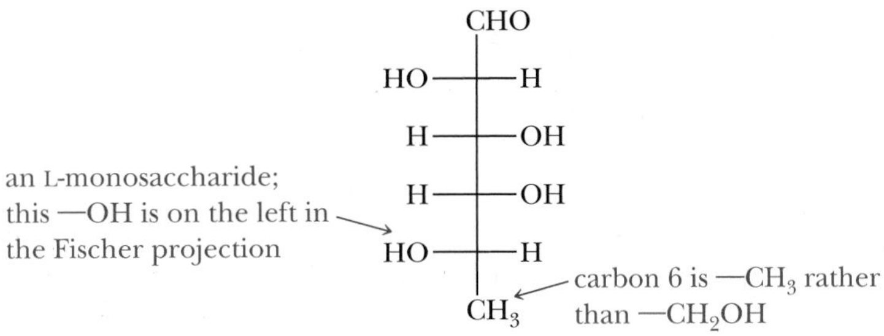
CHO
HO
H
H
OH
an L-monosaccharide;
this —OH is on the left in
the Fischer projection
H
OH
HO
H
carbon 6 is —CH3 rather
than —CH2OH
CH3

L-Fucose

(Continued)

1900 by Karl Landsteiner (1868–1943). Whether an individual has type A, B, AB, or O blood is genetically determined and depends on the type of trisaccharide or tetrasaccharide bound to the surface of the person's red blood cells. The monosaccharides of each blood group and the type of glycosidic bond joining them are shown in the figure. The configurations of the glycosidic bonds are shown in parentheses.

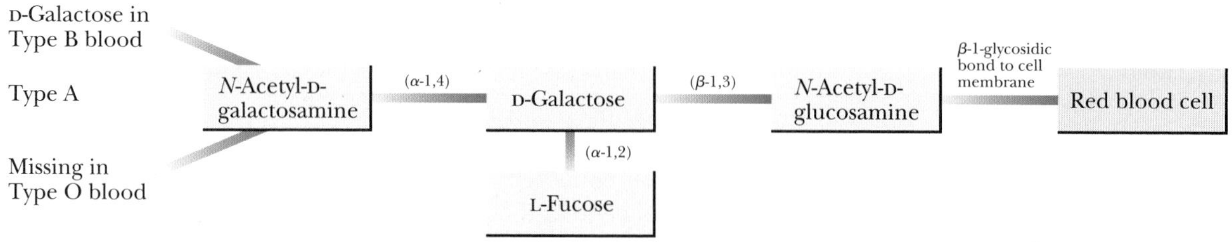

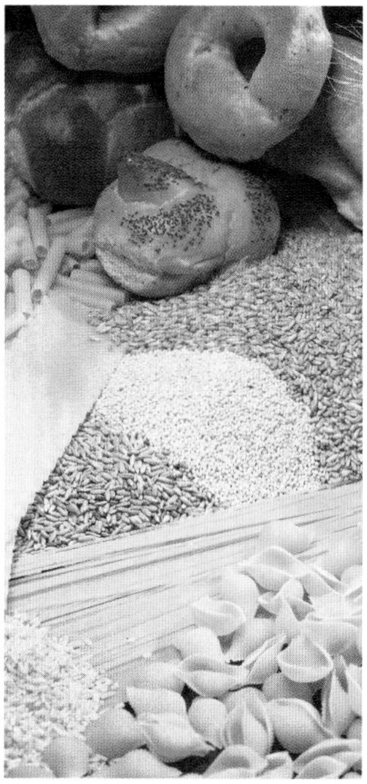

Breads, grains, and pasta are sources of starches.

25.5 Polysaccharides

Polysaccharides consist of large numbers of monosaccharide units bonded together by glycosidic bonds. Three important polysaccharides, all made up of glucose units, are starch, glycogen, and cellulose.

A. Starch: Amylose and Amylopectin

Starch is used for energy storage in plants. It is found in all plant seeds and tubers and is the form in which glucose is stored for later use. Starch can be separated into two principal polysaccharides: amylose and amylopectin. Although the starch from each plant is unique, most starches contain 20%–25% amylose and 75%–80% amylopectin.

Complete hydrolysis of both amylose and amylopectin yields only D-glucose. Amylose is composed of unbranched chains of up to 4000 D-glucose units joined by α-1,4-glycosidic bonds. Amylopectin contains chains up to 10,000 D-glucose units also joined by α-1,4-glycosidic bonds. In addition, there is considerable branching from this linear network. At branch points, new chains of 24–30 units are started by α-1,6-glycosidic bonds (Figure 25.4).

B. Glycogen

Glycogen is the energy-reserve carbohydrate for animals. Like amylopectin, glycogen is a branched polysaccharide of approximately 10^6 glucose units joined by

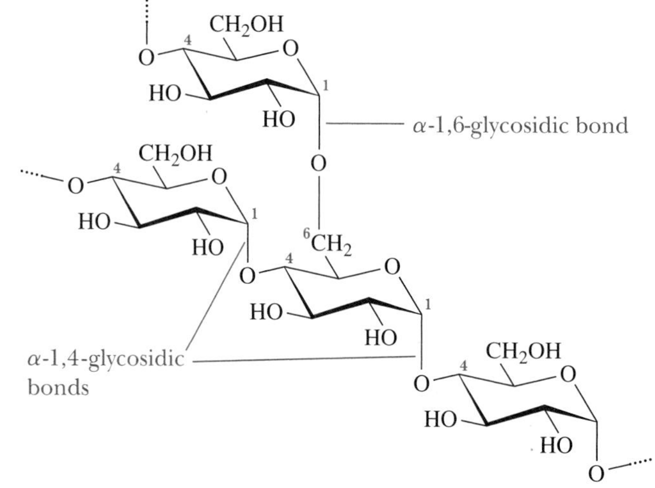

FIGURE 25.4 Amylopectin is a branched polymer of approximately 10,000 D-glucose units joined by α-1,4-glycosidic bonds. Branches consist of 24–30 D-glucose units started by α-1,6-glycosidic bonds.

α-1,4- and α-1,6-glycosidic bonds. The total amount of glycogen in the body of a well-nourished adult human is about 350 g, divided almost equally between the liver and muscle.

C. Cellulose

Cellulose, the most widely distributed plant skeletal polysaccharide, constitutes almost half of the cell wall material of wood. Cotton is almost pure cellulose. Cellulose is a linear polysaccharide of D-glucose units joined by β-1,4-glycosidic bonds (Figure 25.5). It has an average molecular weight of 400,000 g/mol, corresponding to approximately 2200 glucose units per molecule. Cellulose molecules act very much like stiff rods, a feature that enables them to align themselves side by side into well-organized water-insoluble fibers in which the OH groups form numerous intermolecular hydrogen bonds. This arrangement of parallel chains in bundles gives cellulose fibers their high mechanical strength. It is also the reason cellulose is insoluble in water. When a piece of cellulose-containing material is placed in water, there are not strong enough interactions with the water molecules on the surface of the fiber to pull individual cellulose molecules away from the strongly hydrogen-bonded fiber.

Humans, as well as other animals, cannot use cellulose as food because our digestive systems do not contain β-glucosidases, enzymes that catalyze hydrolysis of β-glucosidic bonds. Instead, we have only α-glucosidases; hence, the polysaccharides we use as sources of glucose are starch and glycogen. On the other hand, many bacteria and microorganisms do contain β-glucosidases; so they can digest cellulose. Termites are fortunate (much to our regret) to have such bacteria in their guts and can use wood as their principal food. Ruminants (cud-chewing animals) and horses can also digest grasses and hay because β-glucosidase-containing microorganisms are present in their alimentary systems.

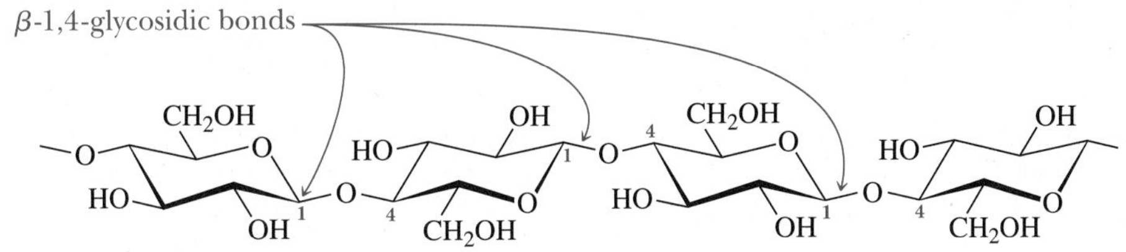

FIGURE 25.5 Cellulose is a linear polysaccharide of up to 2200 units of D-glucose joined by β-1,4-glycosidic bonds.

Chemical Connections

High-Fructose Corn Syrup

If you read the labels of soft drinks and other artificially sweetened food products, you will find that many of them contain high-fructose corn syrup. Looking at Table 25.2, you will see one reason for its use: fructose is more than 70% sweeter than sucrose (table sugar).

The production of high-fructose corn syrup begins with the partial hydrolysis of corn starch catalyzed by the enzyme α-amylase. This enzyme catalyzes the hydrolysis of α-glucosidic bonds and breaks corn starch into small polysaccharides called dextrins. The enzyme glucoamylase then catalyzes the hydrolysis of the dextrins to D-glucose. Finally, enzyme-catalyzed isomerization of D-glucose gives D-fructose. Several billion pounds of high-fructose corn syrup are produced each year in this way for use by the food processing industry. Commercial high-fructose corn syrup is actually 55%–60% glucose and 40%–45% fructose. This composition was inspired by natural honey, which has similar proportions of fructose and glucose along with some other carbohydrates in minor amounts.

Corn starch $\xrightarrow{\alpha\text{-amylase}}$ Dextrins $\xrightarrow{\text{glucoamylase}}$ **D-Glucose** $\xrightarrow{\text{glucose isomerase}}$ **D-Fructose**

(a polysaccharide of α-D-glucose units) (oligosaccharides of 6–10 α-D-glucose units)

D. Textile Fibers from Cellulose

Cotton is almost pure cellulose. Both rayon and acetate rayon are made from chemically modified cellulose and were the first commercially important synthetic textile fibers. In the production of rayon, cellulose-containing materials are treated with carbon disulfide, CS_2, in aqueous sodium hydroxide. In this reaction, some of the —OH groups on a cellulose fiber are converted to the sodium salts of a xanthate ester, which causes the fibers to dissolve in alkali as a viscous colloidal dispersion.

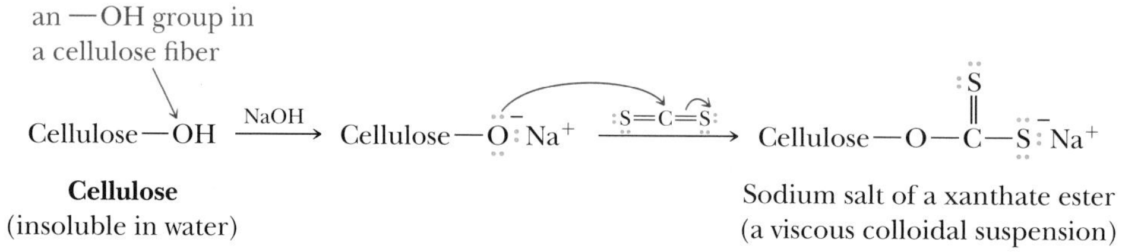

The solution of cellulose xanthate is separated from the alkali-insoluble material then forced through a spinneret, a metal disc with many tiny holes, into dilute sulfuric acid to hydrolyze the xanthate ester groups and precipitate regenerated cellulose. Regenerated cellulose extruded as a filament is called viscose rayon thread.

In the industrial synthesis of acetate rayon, cellulose is treated with acetic anhydride.

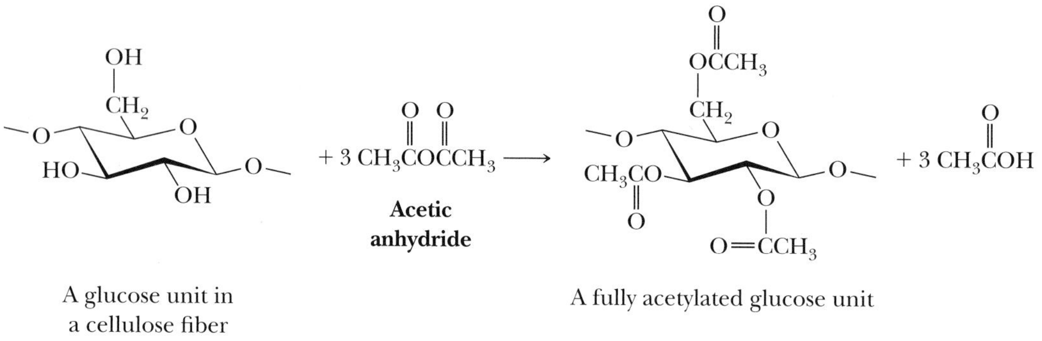

Acetylated cellulose is then dissolved in a suitable solvent, precipitated, and drawn into fibers known as acetate rayon. Today acetate rayon fibers rank fourth in production in the United States, surpassed only by Dacron polyester, nylon, and rayon.

25.6 Glucosaminoglycans

Glucosaminoglycans are a group of linear polysaccharides composed of repeating disaccharides in which one of the monosaccharide units has either a negatively charged carboxylate group (COO^-) or negatively charged sulfate group ($—OSO_3^-$). Members of this family of carbohydrates include hyaluronic acid, heparin, chondroitin sulfate, and keratan sulfate, which are components of cartilage, tendons, and other connective tissues, and dermatan sulfate, which is a component of the extracellular matrix of the skin. A general characteristic of this class of polysaccharides is a repeating disaccharide consisting of units of a uronic acid and an aminohexose with a 1,4-glycosidic bond between the aminohexose and the uronic acid.

A. Hyaluronic Acid

Hyaluronic acid is present in connective tissue. It has a molecular weight of between 10^5 and 10^7 g/mol and contains from 3000 to 100,000 repeating units, depending on the organ in which it occurs. It is most abundant in embryonic tissues and in specialized connective tissues such as synovial fluid; the lubricant of joints in the body; and the vitreous humor of the eye, where it provides a clear, elastic gel that maintains the retina in its proper position.

The repeating disaccharide unit in hyaluronic acid is D-glucuronic acid linked by a β-1,3-glycosidic bond to *N*-acetyl-D-glucosamine.

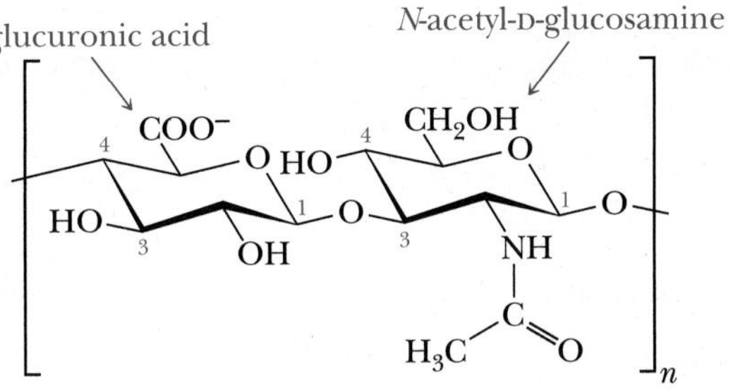

The repeating unit of **hyaluronic acid**

B. Heparin

Heparin is a heterogeneous mixture of variably sulfonated polysaccharide chains ranging in molecular weight from 6000 to 30,000 g/mol. This polysaccharide is synthesized and stored in mast cells of various tissues, particularly the liver, lungs, and gut. Heparin has many biological functions, the best known and understood of which is its anticoagulant activity. It binds strongly to antithrombin III, a plasma protein involved in terminating the clotting process. Heparin is used medically as an anticoagulant.

The repeating monosaccharide units of heparin are *N*-acetyl-D-glucosamine, D-glucuronic acid, D-glucosamine, and L-ioduronic acid bonded by a combination of α-1,4- and β-1,4-glycosidic bonds. Figure 25.6 shows a pentasaccharide unit of heparin that binds to and inhibits the enzymatic activity of antithrombin III.

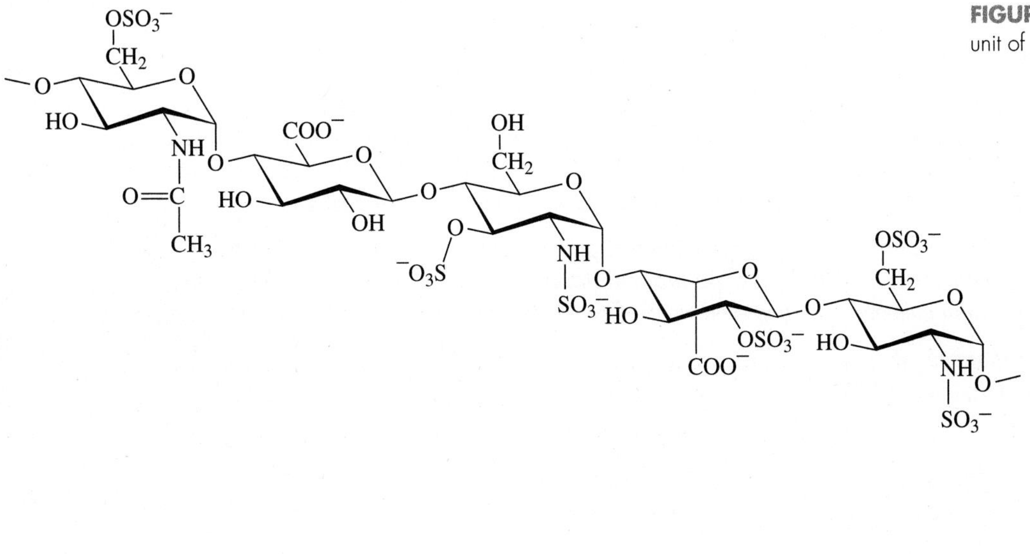

FIGURE 25.6 A pentasaccharide unit of heparin.

Study Guide

25.1 Monosaccharides

- **Carbohydrates** are polyhydroxylated aldehydes or ketones or compounds that produce polyhydroxylated aldehydes or ketones upon hydrolysis.
- Carbohydrates are the most abundant organic molecules in the world. They are essential to all forms of life and perform such functions as energy storage (glucose, starch, and glycogen), structural reinforcement (cellulose), and genetic information storage as components of nucleic acids (DNA and RNA).
- **Monosaccharides** usually have molecular formulas of $C_nH_{2n}O_n$ $(3 \leq n \leq 8)$ and are the monomers from which larger carbohydrates are constructed.
- Monosaccharides are named by using the suffix *-ose*. The prefixes *tri-*, *tetra-*, *penta-*, *hexa-*, and so on, are used to indicate three, four, five, and six carbon atoms, respectively.
    - *An aldehyde carbohydrate is called an* ***aldose*** *and is sometimes designated with an* aldo- *prefix.*
    - *A ketone carbohydrate is called a* ***ketose*** *and is sometimes designated with a* keto- *prefix.*
    - *For example, glyceraldehyde is an aldotriose and fructose is a ketohexose.*
    - *The nomenclature of monosaccharides is dominated by common names. Even though IUPAC names can be derived for each different monosaccharide, the common names are much simpler and are used almost exclusively.*
- Monosaccharides usually have one or more chiral centers, so stereochemistry is of major importance with monosaccharides.
    - *A* ***Fischer projection*** *of a monosaccharide is used to show its structure and thus keep track of stereochemistry.*
    - *In a Fischer projection, the monosaccharide is drawn in the open-chain form and the carbonyl carbon atom is placed at the top of the structure. Horizontal lines represent groups projecting above the plane of the paper, and vertical lines represent groups projecting below the plane of the paper.*
    - *In a Fischer projection, the carbon atoms of chiral centers are not labeled. They are assumed to be located at the crossing points of lines.*
- The overall stereochemistry of monosaccharides is classified as D or L based on a comparison to glyceraldehyde stereochemistry.
    - *In a monosaccharide, the point of reference is the chiral center that is farthest from the carbonyl group. Because this is a carbon atom that is the next-to-the-last carbon atom in the chain (notice that the last carbon atom of the chain in a typical carbohydrate has two* $—H$ *atoms, so it is not a chiral center), it is referred to as the* ***penultimate carbon*** *atom.*
    - *A monosaccharide that has the same configuration about its penultimate carbon as D-glyceraldehyde is classified as a* ***D-monosaccharide****. In this case, the* $—OH$ *group is on the right side of the carbon atom in the Fischer projection.*
    - *An* ***L-monosaccharide*** *has a configuration about its penultimate carbon atom that is the same as the configuration of L-glyceraldehyde, with the* $—OH$ *group on the left in a Fischer projection.*

- *Notice that the enantiomer of a given monosaccharide is not produced by simply changing the configuration of the penultimate carbon atom, but rather by reversing the configuration of all of the chiral centers.*

- Some sugars have an amino group ($—NH_2$) in place of an $—OH$ group, and these are called **amino sugars**. Amino sugars are much less common than normal carbohydrates, but important examples include D-glucosamine and D-galactosamine.
- Monosaccharides are all very soluble in water due to all of the $—OH$ groups that can take part in hydrogen bonding with the water molecules.

P 25.1, 25.7–25.15

▶ Fischer Projections of Saccharides

25.2 The Cyclic Structure of Monosaccharides

- The open-chain monosaccharides are in equilibrium with a cyclic hemiacetal structure.
  - *The cyclic hemiacetal is greatly favored and thus found in large excess at equilibrium.*
  - *When a carbohydrate forms a six-membered hemiacetal ring, it is called a* ***pyranose****, and when a carbohydrate forms a five-membered hemiacetal ring, it is called a* ***furanose****.*
- Cyclic monosaccharide structures are often drawn as **Haworth projections** in which the five-membered or six-membered cyclic hemiacetal is drawn as planar and perpendicular to the plane of the paper.
  - *The anomeric carbon is placed to the right with the hemiacetal oxygen atom in the back.*
  - *A more accurate chair conformation can be drawn for six-membered ring hemiacetals, showing which groups are axial and which are equatorial.*
- Two cyclic diastereomers are possible, and these are referred to as **anomers**.
  - *The two anomers are distinguished by the relative orientation of the* ***anomeric OH*** *group (the $—OH$ group on the so-called* ***anomeric carbon*** *atom, the carbon that was a carbonyl in the open-chain form).*
  - *The two anomers are named α and β.*
    - *The* ***α-anomer*** *has the anomeric $—OH$ group and the terminal $—CH_2OH$ group on opposite sides of the ring in a Haworth projection.*
    - *The* ***β-anomer*** *has the anomeric $—OH$ group on the same side of the ring as the $—CH_2OH$ group in a Haworth projection.*
    - *With D-glucose in the cyclic hemiacetal form, the α-anomer is the one with the anomeric $—OH$ group axial, while for the β-anomer, the anomeric $—OH$ group is equatorial.*
    - *This can be remembered by the phrase for a pyranose, "Alpha is axial," which points down in the manner that a chair pyranose is typically drawn. Thus, down is also alpha for the manner in which a furanose sugar is typically drawn (see the renderings throughout this chapter).*
- **Mutarotation** is the change in specific rotation that accompanies the interconversion of α- and β-anomers in aqueous solution.

P 25.2, 25.3, 25.16–25.23

▶ Problem 25.3

KEY REACTIONS

1. **Formation of Cyclic Hemiacetals (Section 25.2A)** A monosaccharide existing as a five-membered ring is a furanose; a monosaccaride existing as a six-membered ring is a pyranose. A pyranose is most commonly drawn as either a Haworth projection or a chair conformation.

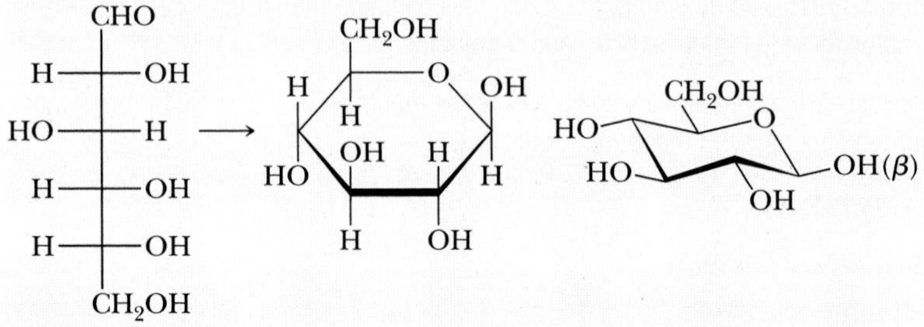

D-Glucose **β-D-Glucopyranose (β-D-Glucose)**

2. **Mutarotation (Section 25.2C)** Anomeric forms of a monosaccharide are in equilibrium in aqueous solution. Mutarotation is the change in specific rotation that accompanies this equilibration.

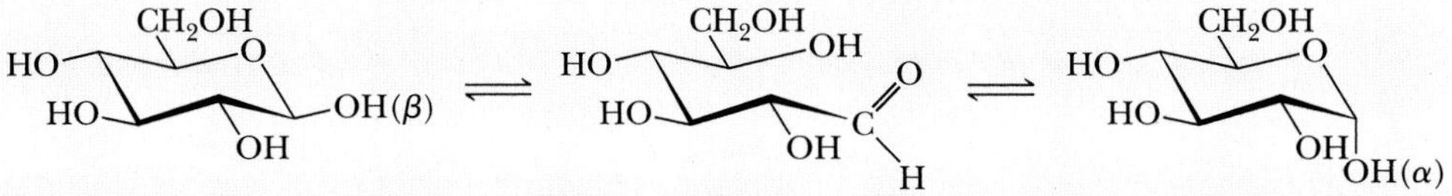

β-D-Glucopyranose $[\alpha]_D^{25}$ +18.7 Open-chain form **α-D-Glucopyranose** $[\alpha]_D^{25}$ +112

25.3 Reactions of Monosaccharides

- A **glycoside** is an acetal derived from a monosaccharide, and the bond from the anomeric carbon to the —OR group is called the **glycosidic bond**.
  - *The name of the glycoside is composed of the name of the alkyl or aryl group bonded to the acetal oxygen atom followed by the name of the monosaccharide in which the terminal -e has been replaced with -ide.*
  - *Glycosidic bonds can be made by reacting a saccharide with an alcohol in acid, following the standard acetal formation mechanism.*
  - *An N-glycoside is created when an amine, commonly as part of a heterocyclic ring, forms a bond to the anomeric carbon.*
- An **alditol** is a polyhydroxy compound formed by reduction of the carbonyl group of a monosaccharide to a hydroxyl group. Reduction of D-glucose, for example, gives D-glucitol.
- An **aldonic acid** is a carboxylic acid formed by oxidation of the aldehyde group of an aldose. Oxidation of D-glucose, for example, gives D-gluconic acid.
- **Reducing sugars** are those sugars that are oxidized by mild oxidizing agents to aldonic acids.
- Enzyme-catalyzed oxidation of the primary hydroxyl group at carbon 6 of a hexose yields a **uronic acid**, examples of which are common in both the plant and animal worlds.
- **Periodic acid** cleaves the carbon-carbon bond of a glycol in a reaction that proceeds through a cyclic periodic ester. This reaction was once useful in carbohydrate structure determination.

P **25.4, 25.5, 25.24–25.34**

KEY REACTIONS

3. Formation of Glycosides (Section 25.3A) Treatment of a monosaccharide with an alcohol in the presence of an acid catalyst forms a cyclic acetal called a glycoside. The bond to the new —OR group is called a glycosidic bond, and the mechanism is the same as the acetal formation you saw in Section 16.7.

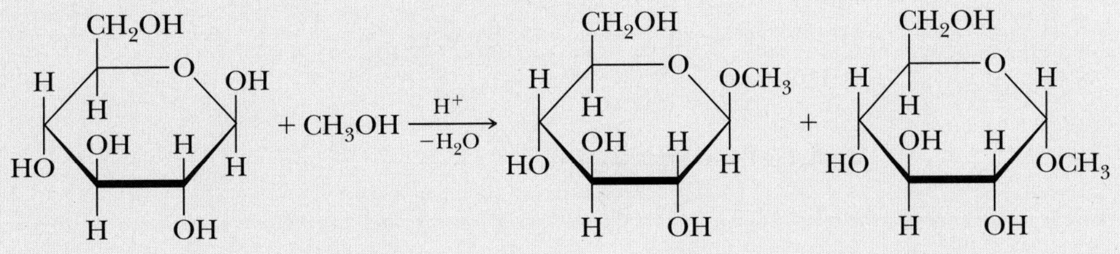

4. Formation of *N*-Glycosides (Section 25.3A) *N*-Glycosides formed between a monosaccharide and a heterocyclic aromatic amine are especially important in the biological world.

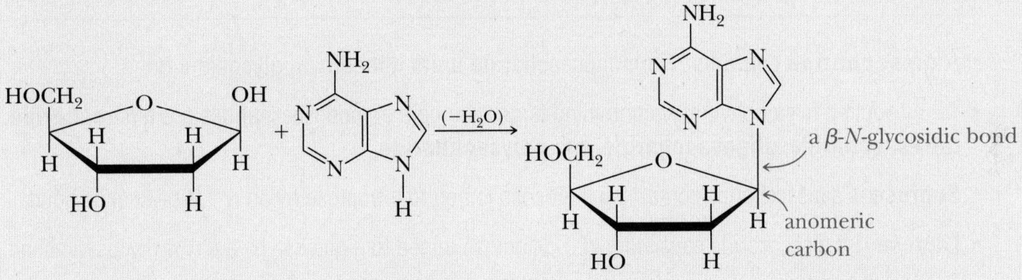

5. Reduction to Alditols (Section 25.3B) Reduction of the carbonyl group of an aldose or a ketose to a hydroxyl group yields a polyhydroxy compound called an alditol.

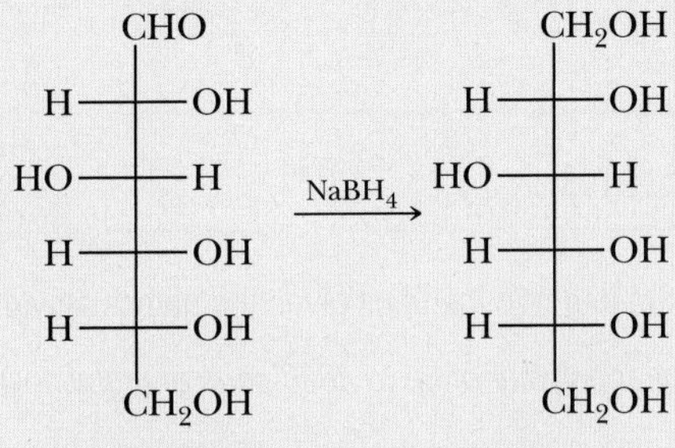

6. Oxidation to an Aldonic Acid (Section 25.3C) Oxidation of the aldehyde group of an aldose to a carboxyl group by a mild oxidizing agent gives a polyhydroxycarboxylic acid called an aldonic acid.

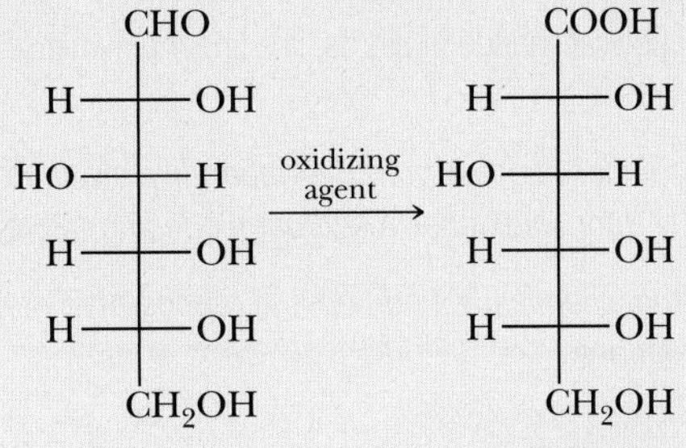

(*Continued*)

7. Oxidation by Periodic Acid (Section 25.3E) Periodic acid oxidizes and cleaves carbon-carbon bonds of glycol, α-hydroxyketone, and α-hydroxyaldehyde groups.

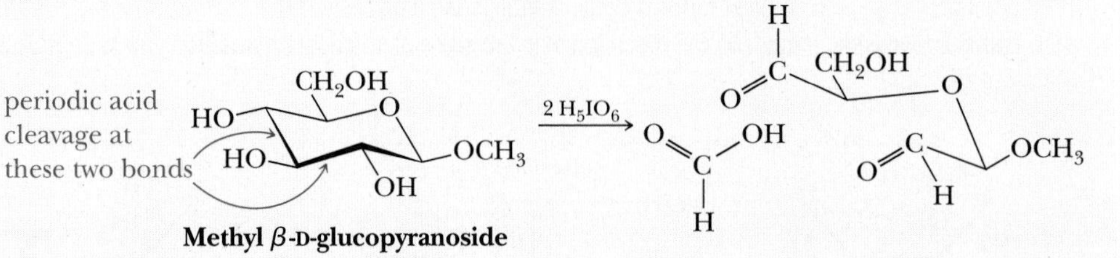

25.4 Disaccharides and Oligosaccharides

- A **disaccharide** contains two monosaccharide units joined by a glycosidic bond.
- Terms applied to carbohydrates containing larger numbers of monosaccharides are **trisaccharide**, **tetrasaccharide**, **oligosaccharide**, and **polysaccharide**.
- **Sucrose** is a disaccharide containing D-glucose joined to D-fructose by an α-1,2-glycosidic bond.
- **Lactose** is a disaccharide consisting of D-galactose joined to D-glucose by a β-1,4-glycosidic bond.
- **Maltose** is a disaccharide of two molecules of D-glucose joined by an α-1,4-glycosidic bond.

P **25.6, 25.35, 25.38, 25.43, 25.44**

▶ **Problem 25.6**

25.5 Polysaccharides

- **Starch** can be separated into two fractions given the names amylose and amylopectin.
  - ***Amylose*** *is a linear polymer of up to 4000 units of D-glucopyranose joined by* α*-1,4-glycosidic bonds.*
  - ***Amylopectin*** *is a highly branched polymer of D-glucopyranose joined by* α*-1,4-glycosidic bonds and, at branch points, by* α*-1,6-glycosidic bonds.*
- **Glycogen**, the energy reserve carbohydrate of animals, is a highly branched polymer of D-glucopyranose joined by α-1,4-glycosidic bonds and, at branch points, by α-1,6-glycosidic bonds.
- **Cellulose**, the skeletal polysaccharide of plants, is a linear polymer of D-glucopyranose joined by β-1,4-glycosidic bonds.
  - *The strength of cellulose comes from the ribbon-like structure of the individual chains that fit together perfectly through extensive hydrogen bonding to create incredibly strong structures.*
  - *Rayon is made from chemically modified and regenerated cellulose. Acetate rayon is made by acetylation of cellulose.*

P **25.39–25.50**

25.6 Glucosaminoglycans

- **Hyaluronic acid** is a carboxylate-containing polysaccharide that is found in connective tissue.
- **Heparin**, an anticoagulent, is a heterogeneous mixture of variably sulfonated polysaccharide chains that is found in the liver, lungs, and gut.
  - *The carboxyl and sulfate groups of acidic polysaccharides are ionized as* $—COO^-$ *and* $—SO_3^-$ *at the pH of body fluids, which gives these polysaccharides net negative charges.*

P 25.45–25.50

Problems

Red numbers indicate applied problems.

Monosaccharides

25.7 Explain the meaning of the designations D and L used to specify the configuration of monosaccharides.

25.8 How many chiral centers are present in D-glucose? in D-ribose?

25.9 Which carbon of an aldopentose determines whether the pentose has a D or L configuration?

25.10 How many aldooctoses are possible? How many D-aldooctoses are possible?

25.11 Which compounds are D-monosaccharides? Which are L-monosaccharides?

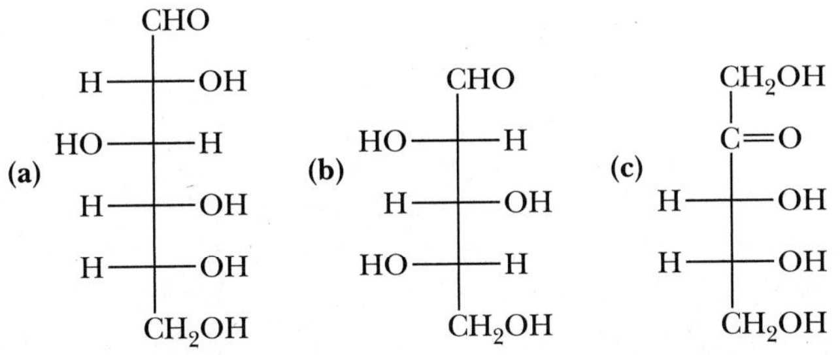

(a)
CHO
H OH
HO H
H OH
H OH
CH_2OH
(b)
CHO
HO H
H OH
HO H
CH_2OH
(c)
CH_2OH
C=O
H OH
H OH
CH_2OH

25.12 Write Fischer projections for L-ribose and L-arabinose.

25.13 What is the meaning of the prefix *deoxy-* as it is used in carbohydrate chemistry?

25.14 Give L-fucose ("Chemical Connections: A, B, AB, and O Blood Group Substances") a name incorporating the prefix *deoxy-* that shows its relationship to galactose.

25.15 2,6-Dideoxy-D-altrose, known alternatively as D-digitoxose, is a monosaccharide obtained upon hydrolysis of digitoxin, a natural product extracted from foxglove (*Digitalis purpurea*). Digitoxin is used in cardiology to reduce pulse rate, regularize heart rhythm, and strengthen heartbeat. Draw the structural formula of 2,6-dideoxy-D-altrose.

The Cyclic Structure of Monosaccharides

25.16 Define the term *anomeric carbon.* In glucose, which carbon is the anomeric carbon?

25.17 Define the terms (a) *pyranose* and (b) *furanose.*

25.18 What is the anomeric carbon in a 2-ketohexose?

25.19 Are α-D-glucose and β-D-glucose enantiomers? Explain.

25.20 Convert each Haworth projection to an open-chain form and then to a Fischer projection. Name the monosaccharide you have drawn.

(a) 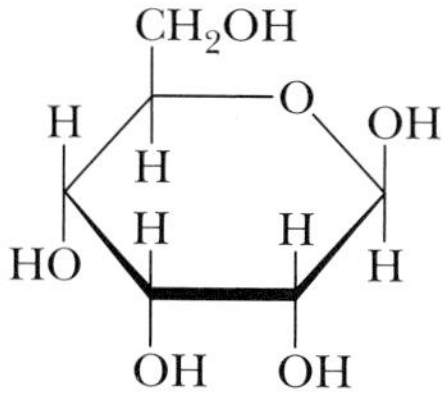 (b) 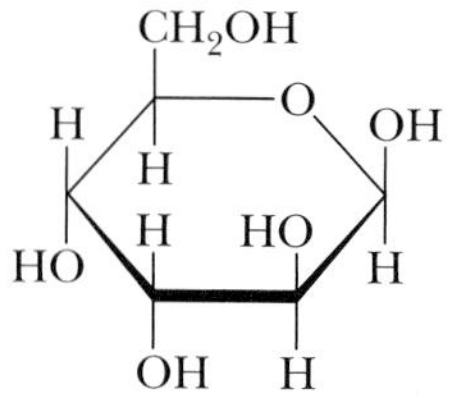

25.21 Convert each chair conformation to an open-chain form and then to a Fischer projection. Name the monosaccharide you have drawn.

(a) 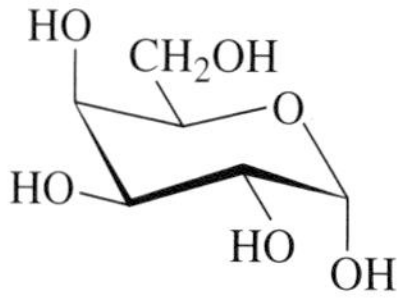 (b) 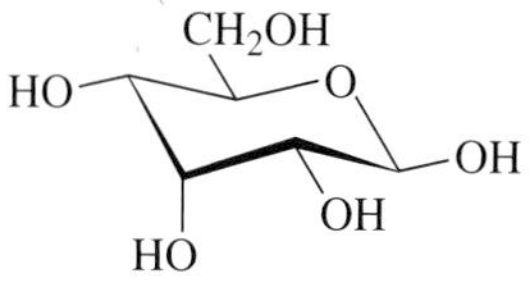

25.22 Explain the phenomenon of mutarotation with reference to carbohydrates. By what means is it detected?

25.23 The specific rotation of α-D-glucose is +112.2.

(a) What is the specific rotation of α-L-glucose?
(b) When α-D-glucose is dissolved in water, the specific rotation of the solution changes from +112.2 to +52.7. Does the specific rotation of α-L-glucose also change when it is dissolved in water? If so, to what value?

Reactions of Monosaccharides

25.24 Draw Fischer projections for the product(s) formed by reaction of D-galactose with the following. In addition, state whether each product is optically active or inactive.

(a) $NaBH_4$ in H_2O **(b)** H_2/Pt **(c)** HNO_3, warm
(d) $Br_2/H_2O/CaCO_3$ **(e)** H_5IO_6 **(f)** $C_6H_5NH_2$

25.25 Repeat Problem 25.24 using D-ribose.

25.26 An important technique for establishing relative configurations among isomeric aldoses and ketoses is to convert both terminal carbon atoms to the same functional group. This can be done either by selective oxidation or reduction. As a specific example, nitric acid oxidation of D-erythrose gives meso-tartaric acid (Section 3.4B). Similar oxidation of D-threose gives (2*S*,3*S*)-tartaric acid. Given this information and the fact that D-erythrose and D-threose are diastereomers, draw Fischer projections for D-erythrose and D-threose. Check your answers against Table 25.1.

25.27 There are four D-aldopentoses (Table 25.1). If each is reduced with $NaBH_4$, which yield optically active alditols? Which yield optically inactive alditols?

25.28 Name the two alditols formed by $NaBH_4$ reduction of D-fructose.

25.29 One pathway for the metabolism of D-glucose 6-phosphate is its enzyme-catalyzed conversion to D-fructose 6-phosphate. Show that this transformation can be accomplished as two enzyme-catalyzed keto-enol tautomerisms.

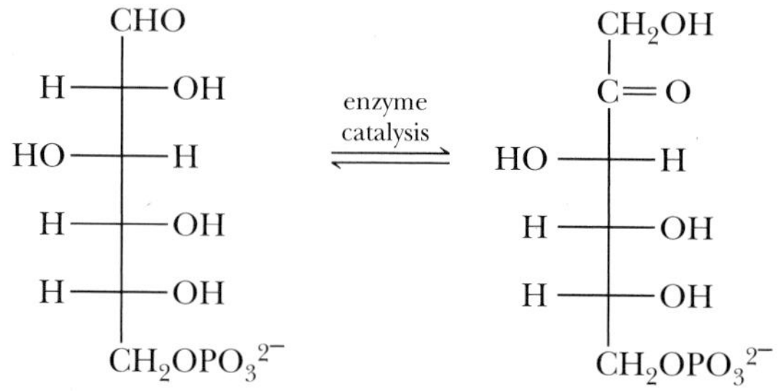

D-Glucose 6-phosphate **D-Fructose 6-phosphate**

25.30 What is the difference in meaning between the terms *glycosidic bond* and *glucosidic bond*?

25.31 Treatment of methyl β-D-glucopyranoside with benzaldehyde forms a six-membered cyclic acetal. Draw the most stable conformation of this acetal. Identify each new chiral center in the acetal.

25.32 Vanillin (4-hydroxy-3-methoxybenzaldehyde), the principal component of vanilla, occurs in vanilla beans and other natural sources as a β-D-glucopyranoside. Draw a structural formula for this glycoside, showing the D-glucose unit as a chair conformation.

25.33 Hot water extracts of ground willow and poplar bark are an effective pain reliever. Unfortunately, the liquid is so bitter that most people refuse it. The pain reliever in these infusions is salicin, a β-glycoside of D-glucopyranose and the phenolic —OH group of 2-(hydroxymethyl)phenol. Draw a structural formula for salicin, showing the glucose ring as a chair conformation.

25.34 Draw structural formulas for the products formed by hydrolysis at pH 7.4 (the pH of blood plasma) of all ester, thioester, amide, anhydride, and glycoside groups in acetyl coenzyme A. Name as many of the products as you can.

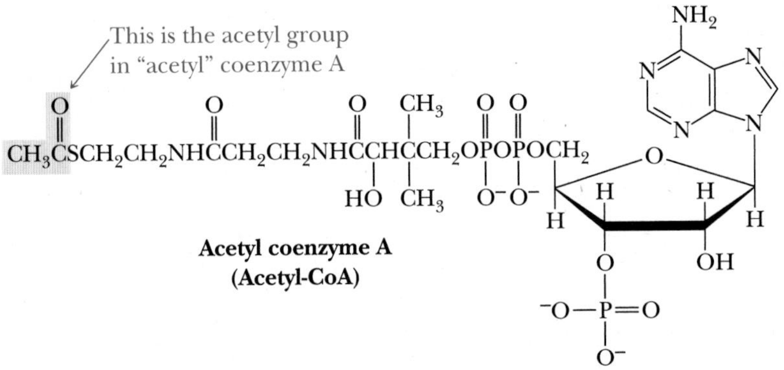

Acetyl coenzyme A
(Acetyl-CoA)

Disaccharides and Oligosaccharides

25.35 In making candy or sugar syrups, sucrose is boiled in water with a little acid, such as lemon juice. Why does the product mixture taste sweeter than the starting sucrose solution?

25.36 Trehalose is found in young mushrooms and is the chief carbohydrate in the blood of certain insects. Trehalose is a disaccharide consisting of two D-monosaccharide units, each joined to the other by an α-1,1-glycosidic bond.

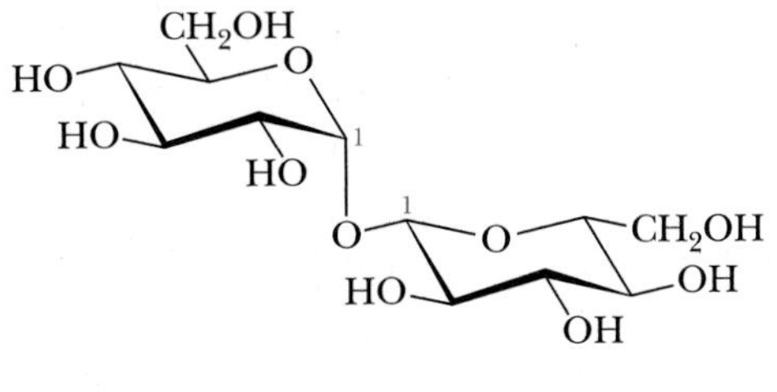

Trehalose

(a) Is trehalose a reducing sugar?
(b) Does trehalose undergo mutarotation?
(c) Name the two monosaccharide units of which trehalose is composed.

25.37 The trisaccharide raffinose occurs principally in cottonseed meal.

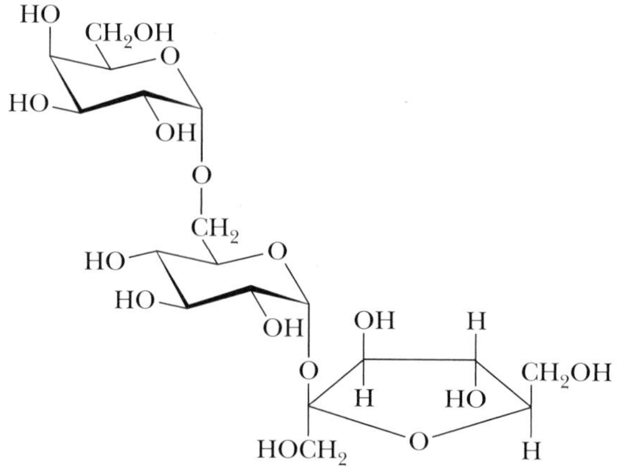

Raffinose

(a) Name the three monosaccharide units in raffinose.
(b) Describe each glycosidic bond in this trisaccharide.
(c) Is raffinose a reducing sugar?
(d) With how many moles of periodic acid will raffinose react?

25.38 Amygdalin is a toxic component in the pits of bitter almonds, peaches, and apricots.
(a) Name the two monosaccharide units in amygdalin and describe the glycosidic bond by which they are joined.
(b) Account for the fact that hydrolysis of amygdalin in warm aqueous acid liberates benzaldehyde and HCN.

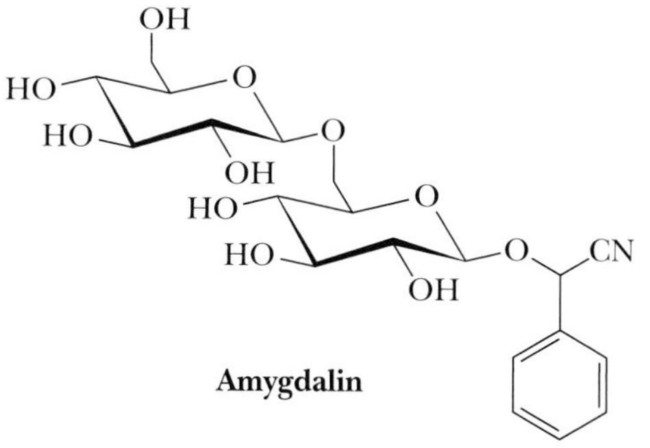

Amygdalin

25.39 Following is a structural formula for stachyose, a water-soluble tetrasaccharide component of many plants, including lentils and soybeans. Humans cannot digest stachyose, and its accumulation leads to distension of the gut and flatulence.

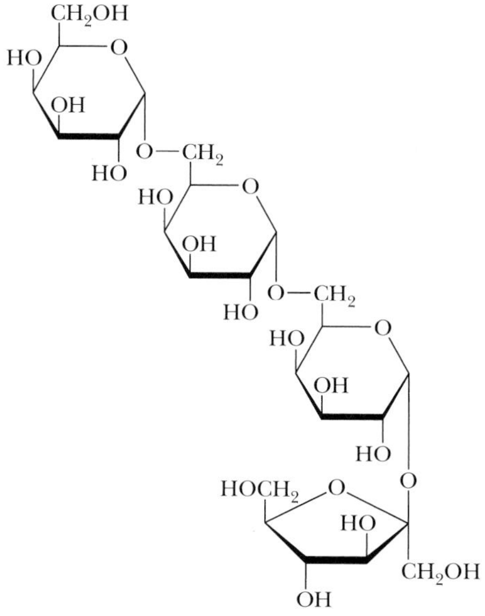

Stachyose

(a) Name each monosaccharide unit in stachyose and specify whether it is a D-monosaccharide or an L-monosaccharide.
(b) Describe each glycosidic bond in stachyose.

Polysaccharides

25.40 What is the difference in structure between oligo- and polysaccharides?

25.41 Why is cellulose insoluble in water?

25.42 Consider *N*-acetyl-D-glucosamine (Section 25.1D).
(a) Draw a chair conformation for the α- and β-pyranose forms of this monosaccharide.
(b) Draw a chair conformation for the disaccharide formed by joining two units of the pyranose form of *N*-acetyl-D-glucosamine by a β-1,4-glycosidic bond. If you draw this correctly, you have the structural formula for the repeating dimer of chitin, the structural polysaccharide component of the shell of lobsters and other crustaceans.

25.43 Propose structural formulas for the following polysaccharides.
(a) Alginic acid, isolated from seaweed, is used as a thickening agent in ice cream and other foods. Alginic acid is a polymer of D-mannuronic acid in the pyranose form joined by β-1,4-glycosidic bonds.
(b) Pectic acid is the main component of pectin, which is responsible for the formation of jellies from fruits and berries. Pectic acid is a polymer of D-galacturonic acid in the pyranose form joined by α-1,4-glycosidic bonds.

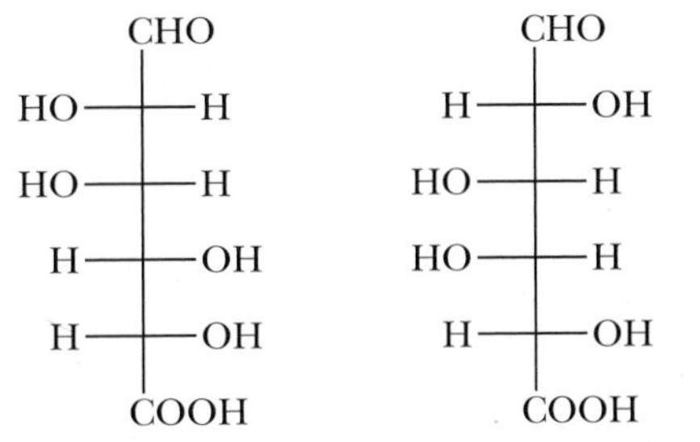

D-Mannuronic acid **D-Galacturonic acid**

25.44 Digitalis is a preparation made from the dried seeds and leaves of the purple foxglove, *Digitalis purpurea*, a plant native to southern and central Europe and cultivated in the United States. The preparation is a mixture of several active components, including digitalin. Digitalis is used in medicine to increase the force of myocardial contraction and as a conduction depressant to decrease heart rate (the heart pumps more forcefully but less often).

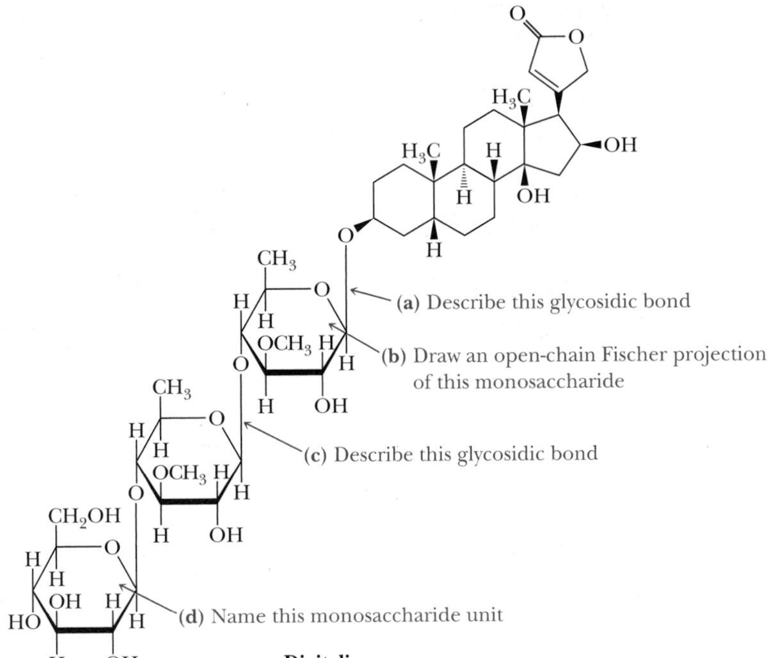
(a) Describe this glycosidic bond
(b) Draw an open-chain Fischer projection of this monosaccharide
(c) Describe this glycosidic bond
(d) Name this monosaccharide unit

Digitalin

25.45 Following is the structural formula of ganglioside GM_2, a macromolecular glycolipid (meaning that it contains lipid and monosaccharide units joined by glycosidic bonds).

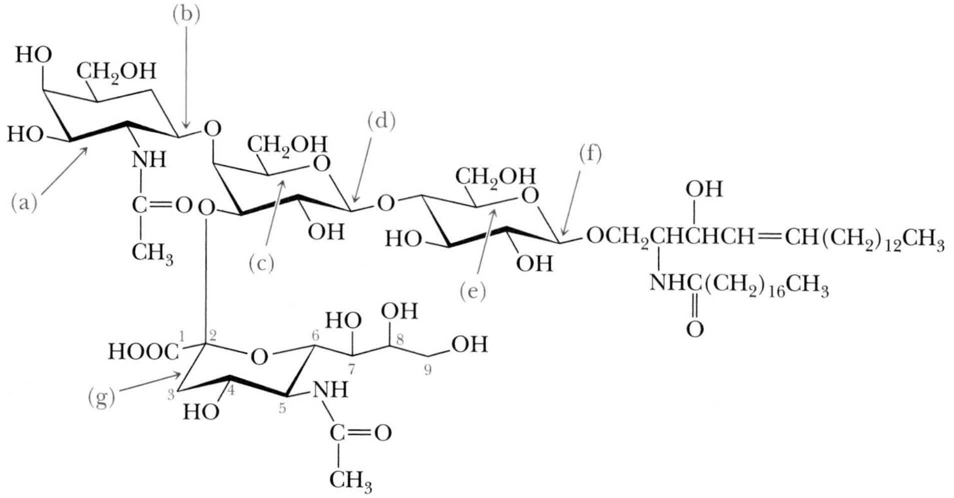

Ganglioside GM_2 or Tay-Sachs ganglioside

In normal cells, this and other gangliosides are synthesized continuously and degraded by lysosomes, which are cell organelles containing digestive enzymes. If pathways for the degradation of gangliosides are inhibited, the gangliosides accumulate in the central nervous system, causing all sorts of life-threatening consequences. In inherited diseases of ganglioside metabolism, death usually occurs at an early age. Diseases of ganglioside metabolism include Gaucher's disease, Niemann-Pick disease, and Tay-Sachs disease. Tay-Sachs disease is a hereditary defect that is transmitted as an autosomal recessive gene. The concentration of ganglioside GM_2 is abnormally high in this disease because the enzyme responsible for catalyzing the hydrolysis of glycosidic bond (b) is absent.

(a) Name this monosaccharide unit.
(b) Describe this glycosidic bond (α or β and between which carbons of each unit).
(c) Name this monosaccharide unit.
(d) Describe this glycosidic bond.
(e) Name this monosaccharide unit.
(f) Describe this glycosidic bond.
(g) This unit is *N*-acetylneuraminic acid, the most abundant member of a family of amino sugars containing nine or more carbons and distributed widely throughout the animal kingdom. Draw the open-chain form of this amino sugar. Do not be concerned with the configuration of the five chiral centers in the open-chain form.

25.46 Hyaluronic acid acts as a lubricant in the synovial fluid of joints. In rheumatoid arthritis, inflammation breaks down hyaluronic acid to smaller molecules. Under these conditions, what happens to the lubricating power of the synovial fluid?

25.47 The anticlotting property of heparin is partly the result of the negative charges it carries.

(a) Identify the functional groups that provide the negative charges.
(b) Which type of heparin is a better anticoagulant, one with a high or a low degree of polymerization?

25.48 Keratin sulfate is an important component of the cornea of the eye. Following is the repeating unit of this acidic polysaccharide.

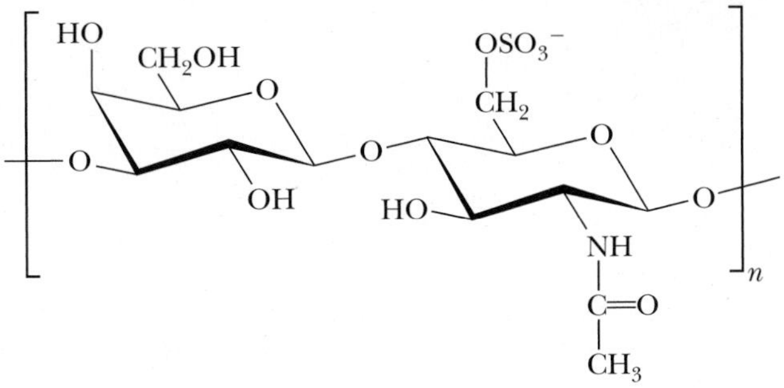

(a) From what monosaccharides or derivatives of monosaccharides is keratin sulfate made?
(b) Describe the glycosidic bond in this repeating disaccharide unit.
(c) What is the net charge on this repeating disaccharide unit at pH 7.0?

25.49 Following is a chair conformation for the repeating disaccharide unit in chondroitin 6-sulfate. This biopolymer acts as the flexible connecting matrix between the tough protein filaments in cartilage. It is available as a dietary supplement, often combined with D-glucosamine sulfate. Some believe this combination can strengthen and improve joint flexibility.

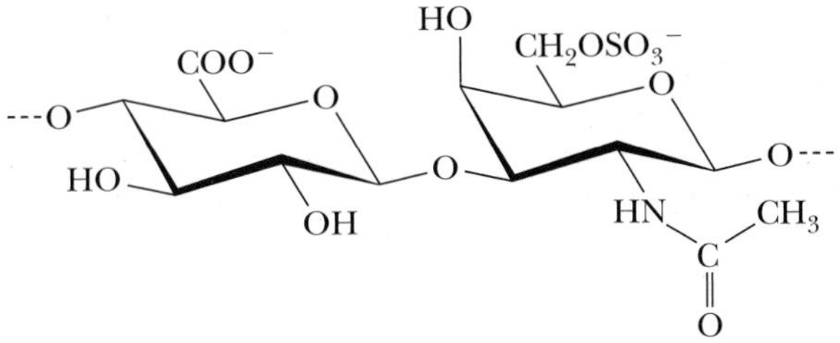

(a) From which two monosaccharide units is the repeating disaccharide unit of chondroitin 6-sulfate derived?
(b) Describe the glycosidic bond between the two units.

25.50 Following is a structural formula for the repeating disaccharide unit of dermatan sulfate. Dermatan sulfate is a component of the extracellular matrix of the skin.

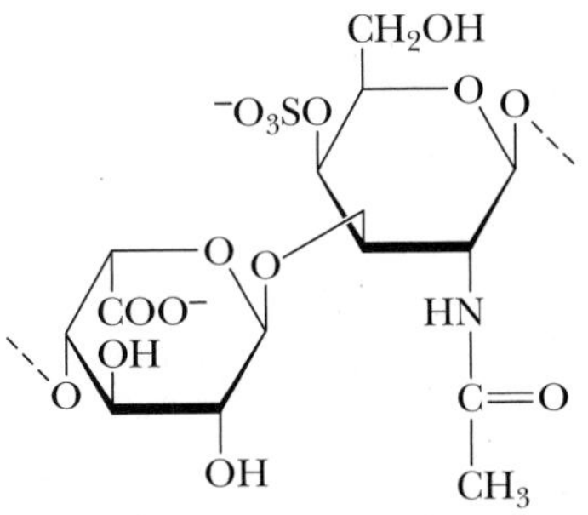

Repeating disaccharide unit
of **dermatan sulfate**

(a) Name the monosaccharide from which each unit of this disaccharide is derived.
(b) Describe the glycosidic bonds in dermatan sulfate.

26

Lipids

Sea lions are marine mammals that require a heavy layer of fat in order to survive in cold waters. **Inset:** linoleic acid, a major component of unsaturated triglycerides and phospholipids (Sections 26.1 and 26.5).
(© visceralimage/ Shutterstock.com)

Outline

26.1 Triglycerides
26.2 Soaps and Detergents
26.3 Prostaglandins
26.4 Steroids
26.5 Phospholipids
26.6 Fat-Soluble Vitamins

Lipids are a heterogeneous group of naturally occurring organic compounds (many related to fats and oils) classified together on the basis of their common solubility properties. Lipids are insoluble in water but soluble in nonpolar aprotic organic solvents, including diethyl ether, and dichloromethane.

Lipids are divided into two main groups. First are those lipids that contain both a relatively large nonpolar hydrophobic region, most commonly aliphatic in nature, and a polar hydrophilic region. Found among this group are fatty acids, triglycerides, phospholipids, prostaglandins, and the fat-soluble vitamins. Second are those lipids that contain the tetracyclic ring system called the steroid nucleus, including cholesterol, steroid hormones, and bile acids. In this chapter, we describe the structures and biological functions of each group of lipids.

Lipid
A biomolecule isolated from plant or animal sources by extraction with nonpolar organic solvents such as diethyl ether and hexane.

26.1 Triglycerides

Animal fats and vegetable oils, the most abundant naturally occurring lipids, are triesters of glycerol and long-chain carboxylic acids. Fats and oils are also referred to as **triglycerides** or **triacylglycerols**. Hydrolysis of a triglyceride in aqueous base followed by acidification gives glycerol and three fatty acids.

Triglyceride (triacylglycerol) An ester of glycerol with three fatty acids.

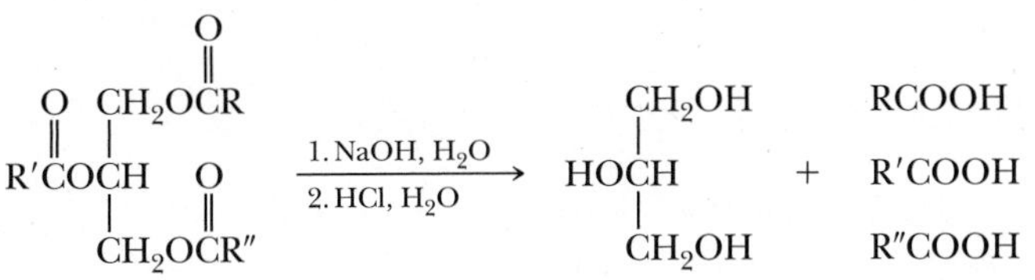

A triglyceride — **1,2,3-Propanetriol (Glycerol, glycerin)** — Fatty acids

A. Fatty Acids

More than 500 different **fatty acids** have been isolated from various cells and tissues. Given in Table 26.1 are common names and structural formulas for the most abundant of these. The number of carbons in a fatty acid and the number of carbon-carbon double bonds in its hydrocarbon chain are shown by two numbers separated by a colon. In this notation, linoleic acid, for example, is designated as an 18:2 fatty acid; its 18-carbon chain contains two carbon-carbon double bonds. Following are several characteristics of the most abundant fatty acids in higher plants and animals.

Fatty acid A long, unbranched-chain carboxylic acid, most commonly of 12 to 20 carbons, derived from the hydrolysis of animal fats, vegetable oils, and the phospholipids of biological membranes.

1. Nearly all fatty acids have an even number of carbon atoms, most between 12 and 20, in an unbranched chain.
2. The three most abundant fatty acids in nature are palmitic acid (16:0), stearic acid (18:0), and oleic acid (18:1).

Table 26.1 The Most Abundant Fatty Acids in Animal Fats, Vegetable Oils, and Biological Membranes

Carbon Atoms/ Double Bonds*	Structure	Common Name	Melting Point (°C)
Saturated Fatty Acids			
12:0	$CH_3(CH_2)_{10}COOH$	Lauric acid	44
14:0	$CH_3(CH_2)_{12}COOH$	Myristic acid	58
16:0	$CH_3(CH_2)_{14}COOH$	Palmitic acid	63
18:0	$CH_3(CH_2)_{16}COOH$	Stearic acid	70
20:0	$CH_3(CH_2)_{18}COOH$	Arachidic acid	77
Unsaturated Fatty Acids			
16:1	$CH_3(CH_2)_5CH{=}CH(CH_2)_7COOH$	Palmitoleic acid	1
18:1	$CH_3(CH_2)_7CH{=}CH(CH_2)_7COOH$	Oleic acid	16
18:2	$CH_3(CH_2)_4(CH{=}CHCH_2)_2(CH_2)_6COOH$	Linoleic acid	−5
18:3	$CH_3CH_2(CH{=}CHCH_2)_3(CH_2)_6COOH$	Linolenic acid	−11
20:4	$CH_3(CH_2)_4(CH{=}CHCH_2)_4(CH_2)_2COOH$	Arachidonic acid	−49

*The first number is the number of carbons in the fatty acid; the second number is the number of carbon-carbon double bonds in its hydrocarbon chain.

3. In most unsaturated fatty acids, the *cis* isomer predominates; the *trans* isomer is rare.
4. Unsaturated fatty acids have lower melting points than their saturated counterparts. The greater the degree of unsaturation, the lower the melting point (see "Connections to Biological Chemistry: The Importance of *Cis* Double Bonds in Fats versus Oils" in Section 5.4). Compare, for example, the melting points of linoleic acid, a **polyunsaturated fatty acid**, and stearic acid, a saturated fatty acid.

Polyunsaturated fatty acid

A fatty acid with two or more carbon-carbon double bonds in its hydrocarbon chain.

Example 26.1 Triglyceride Structure

Draw a structural formula of a triglyceride derived from one molecule each of palmitic acid, oleic acid, and stearic acid, the three most abundant fatty acids in the biological world.

Solution

In this structure, palmitic acid is esterified at carbon 1 of glycerol; oleic acid, at carbon 2; and stearic acid, at carbon 3.

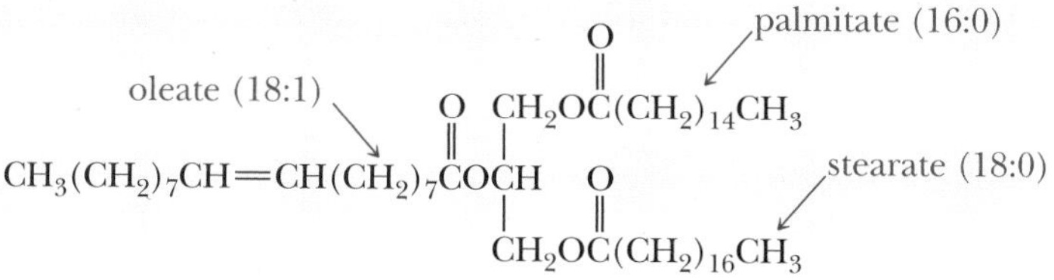
palmitate (16:0)
oleate (18:1)
stearate (18:0)

A triglyceride

Problem 26.1

(a) How many constitutional isomers are possible for a triglyceride containing one molecule each of palmitic acid, oleic acid, and stearic acid?

(b) Which of these constitutional isomers are chiral?

Among the components of beeswax is triacontyl palmitate, $CH_3(CH_2)_{14}COO(CH_2)_{29}CH_3$, an ester of palmitic acid.

B. Physical Properties

The physical properties of a triglyceride depend on its fatty acid components. In general, the melting point of a triglyceride increases as the number of carbons in its hydrocarbon chains increases and as the number of carbon-carbon double bonds decreases. Triglycerides rich in oleic acid, linoleic acid, and other unsaturated fatty acids are generally liquids at room temperature and are called **oils**, as, for example, corn oil and olive oil. Olive oil, which contains mainly the mono-unsaturated oleic acid, solidifies in the refrigerator, whereas the more unsaturated corn oil does not. Triglycerides rich in palmitic, stearic, and other saturated fatty acids are generally semisolids or solids at room temperature and are called **fats**, as, for example, human fat and butter fat. Fats of land animals typically contain approximately 40%–50% saturated fatty acids by weight (Table 26.2). Most plant oils, on the other hand, contain 20% or less saturated fatty acids and 80% or more unsaturated fatty acids. The notable exception to this generalization about plant oils are the **tropical oils** (as, for example, coconut and palm oils), which are considerably richer in low-molecular-weight saturated fatty acids.

Oil

When used in the context of lipids a mixture of triglycerides that is liquid at room temperature.

Fat

A mixture of triglycerides that is semisolid or solid at room temperature.

The lower melting points of triglycerides rich in unsaturated fatty acids are related to differences in three-dimensional shape between the hydrocarbon chains of their unsaturated and saturated fatty acid components. Shown in Figure 26.1 is a space-filling model of tristearin, a saturated triglyceride. In this model, the hydrocarbon chains lie parallel to each other, giving the molecule an ordered, compact shape. Because of this compact three-dimensional shape and the resulting strength of the

Table 26.2 Grams of Fatty Acid per 100 g of Triglyceride of Several Fats and Oils*

	Saturated Fatty Acids			Unsaturated Fatty Acids	
Fat or Oil	**Lauric (12:0)**	**Palmitic (16:0)**	**Stearic (18:0)**	**Oleic (18:1)**	**Linoleic (18:2)**
Human fat	—	24.0	8.4	46.9	10.2
Beef fat	—	27.4	14.1	49.6	2.5
Butter fat	2.5	29.0	9.2	26.7	3.6
Coconut oil	45.4	10.5	2.3	7.5	trace
Corn oil	—	10.2	3.0	49.6	34.3
Olive oil	—	6.9	2.3	84.4	4.6
Palm oil	—	40.1	5.5	42.7	10.3
Peanut oil	—	8.3	3.1	56.0	26.0
Soybean oil	0.2	9.8	2.4	28.9	50.7

*Only the most abundant fatty acids are given; other fatty acids are present in lesser amounts.

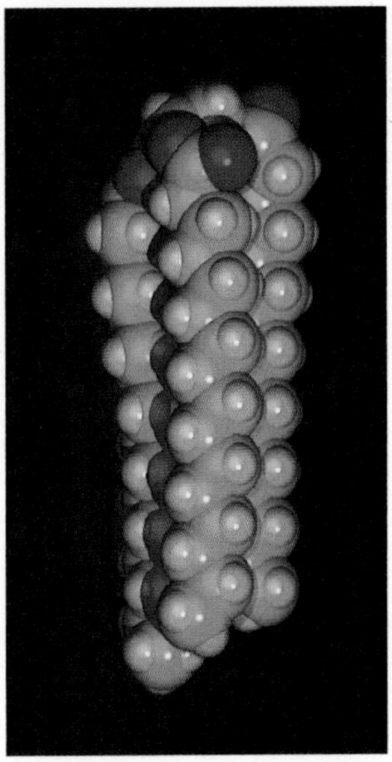

FIGURE 26.1 Tristearin, a saturated triglyceride.

dispersion forces between hydrocarbon chains of adjacent molecules, triglycerides rich in saturated fatty acids pack together well and have melting points above room temperature.

The three-dimensional shape of an unsaturated fatty acid is quite different from that of a saturated fatty acid. Recall from Section 26.1A that unsaturated fatty acids of higher organisms are predominantly of the *cis* configuration; *trans* configurations are rare. Figure 26.2 shows a space-filling model of a **polyunsaturated triglyceride** derived from one molecule each of stearic acid, oleic acid, and linoleic acid. Each double bond in this polyunsaturated triglyceride has the *cis* configuration.

Polyunsaturated triglyceride

A triglyceride having several carbon-carbon double bonds in the hydrocarbon chains of its three fatty acids.

Polyunsaturated triglycerides have a less ordered structure and do not pack together as closely or as compactly as saturated triglycerides. Intramolecular and intermolecular dispersion forces are weaker, with the result that polyunsaturated triglycerides have lower melting points than their saturated counterparts.

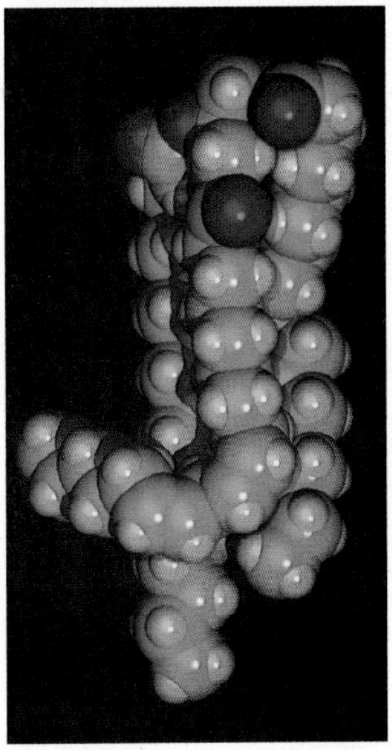

FIGURE 26.2 A polyunsaturated triglyceride.

C. Reduction of Fatty Acid Chains

For a variety of reasons, in part convenience and in part dietary preference, conversion of oils to fats has become a major industry. The process is called **hardening** of oils and involves catalytic reduction (Section 6.6A) of some or all carbon-carbon double bonds. In practice, the degree of hardening is carefully controlled to produce fats of a desired consistency. The resulting fats are sold for kitchen use (Crisco and others). Margarine and other butter substitutes are produced by partial hydrogenation of polyunsaturated oils derived from corn, cottonseed, peanut, and soybean oils. To the hardened oils are added β-carotene (to give the final product a yellow color and to make it look like butter), salt, and about 15% milk by volume to form the final emulsion. Vitamins A and D also are often added. Because the product at this stage is tasteless, acetoin and diacetyl are often added. These two compounds mimic the characteristic flavor of butter.

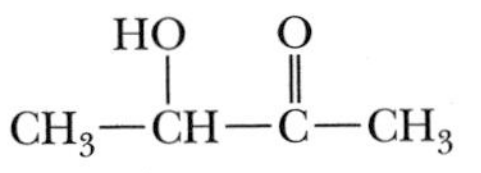

3-Hydroxy-2-butanone (Acetoin)

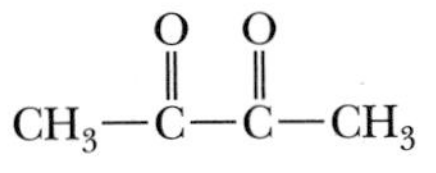

2,3-Butanedione (Diacetyl)

Liquid vegetable oils contain mostly unsaturated fatty acids.

As described in "Connections to Biological Chemistry: *Trans* Fatty Acids: What They Are and How to Avoid Them" in Section 6.6, the process of hardening oils generally involves using a hydrogenation catalyst such as Ni and a limiting amount of H_2. An unintended consequence of the hardening process is that some of the *cis* double bonds of the triglycerides are isomerized to the more stable *trans* configuration on the metal surface. Recall that the π bond is broken when the alkene adsorbs on the metal surface; so, unfortunately, it can isomerize if not enough H_2 is present to react first. Research has shown that these so-called *trans* fats dramatically increase the risk of heart disease, especially atherosclerosis. The FDA has called for labeling of all products that contain *trans* fats, and many leading food producers and restaurants are currently limiting their content in the foods they sell. Of course, daily exercise is beneficial regardless of diet.

26.2 Soaps and Detergents

A. Structure and Preparation of Soaps

Soap

A sodium or potassium salt of a fatty acid.

Natural **soaps** are prepared most commonly from a blend of tallow and coconut oils. In the preparation of tallow, the solid fats of cattle are melted with steam, and the tallow layer formed on the top is removed. The preparation of soaps begins by boiling these triglycerides with sodium hydroxide. The reaction that takes place is called **saponification** (Latin: *saponem*, soap). At the molecular level, saponification corresponds to base-promoted hydrolysis of the ester groups in triglycerides (Section 18.4C). The resulting soaps contain mainly the sodium salts of palmitic, stearic, and oleic acids from tallow and the sodium salts of lauric and myristic acids from coconut oil.

A triglyceride — **1,2,3-Propanetriol (Glycerol; glycerin)** — Sodium soaps

After hydrolysis is complete, sodium chloride is added to precipitate the soap as thick curds. The water layer is then drawn off, and glycerol is recovered by vacuum distillation. The crude soap contains sodium chloride, sodium hydroxide, and other impurities. These are removed by boiling the curd in water and reprecipitating with more sodium chloride. After several purifications, the soap can be used without further processing as an inexpensive industrial soap. Other treatments transform the crude soap into pH-controlled cosmetic soaps, medicated soaps, and the like.

B. How Soaps Clean

Micelle

A spherical arrangement of organic molecules in water solution clustered so that their hydrophobic parts are buried inside the sphere and their hydrophilic parts are on the surface of the sphere and in contact with water.

Soap owes its remarkable cleansing properties to its ability to act as an emulsifying agent. Because the long hydrocarbon chains of natural soaps are insoluble in water, they tend to cluster in such a way as to minimize their contact with surrounding water molecules. The polar carboxylate groups, on the other hand, tend to remain in contact with the surrounding water molecules. Thus, in water, soap molecules spontaneously cluster into **micelles** (Figure 26.3).

Most of the things we commonly think of as dirt (e.g., grease, oil, and fat stains) are nonpolar and insoluble in water. When soap and this type of dirt are mixed

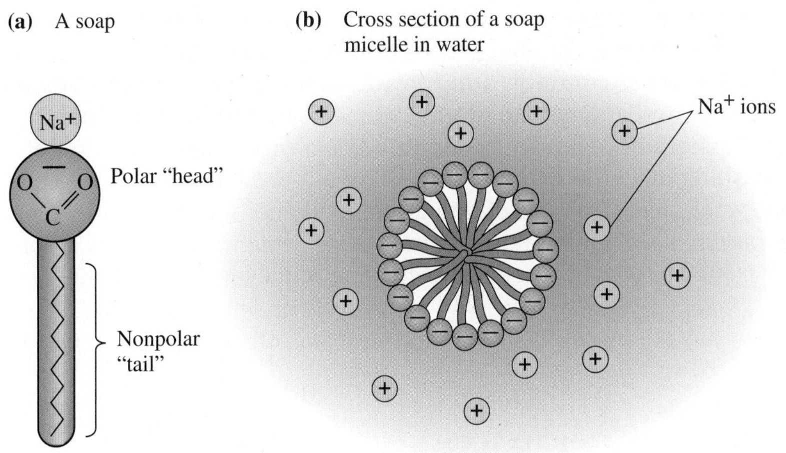
(a) A soap
Na+
O O
C
Polar "head"
Nonpolar "tail"
(b) Cross section of a soap micelle in water
Na+ ions

FIGURE 26.3 Soap micelles. Nonpolar (hydrophobic) hydrocarbon chains are clustered in the interior of the micelle, and polar (hydrophilic) carboxylate groups are on the surface of the micelle. Soap micelles repel each other because of their negative surface charges.

Watch a video explanation

together, as in a washing machine, the nonpolar hydrocarbon inner parts of the soap micelles "dissolve" the nonpolar dirt molecules. In effect, new soap micelles are formed, this time with nonpolar dirt molecules in the center (Figure 26.4). In this way, nonpolar organic grease, oil, and fat are "dissolved" and washed away in the wash water.

Soaps are not without their disadvantages. Foremost among these is the fact that they form insoluble salts when used in water containing Ca(II), Mg(II), or Fe(III) ions (hard water).

$$2\,CH_3(CH_2)_{14}COO^-\,Na^+ \;+\; Ca^{2+} \longrightarrow [CH_3(CH_2)_{14}COO^-]_2Ca^{2+} \;+\; 2\,Na^+$$

A sodium soap (soluble in water as micelles) — **Calcium salt of a fatty acid (insoluble in water)**

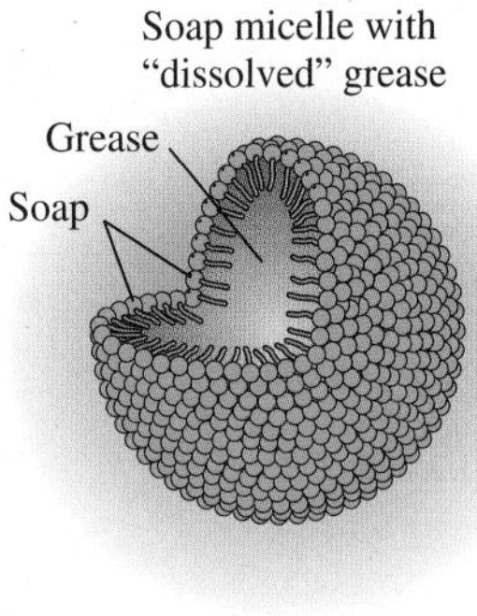
Soap micelle with "dissolved" grease
Grease
Soap

FIGURE 26.4 A soap micelle with a "dissolved" oil or grease droplet.

These calcium, magnesium, and iron salts of fatty acids create problems, including rings around the bathtub, film that spoils the luster of hair, and grayness and roughness that build up on textiles after repeated washings.

C. Synthetic Detergents

After the cleansing action of soaps was understood, synthetic detergents could be designed. Molecules of a good detergent must have a long hydrocarbon chain, preferably 12 to 20 carbon atoms long, and a polar group at one end of the molecule that does not form insoluble salts with the Ca(II), Mg(II), or Fe(III) ions present in hard water. Chemists recognized that these essential characteristics of a soap could be produced in a molecule containing a sulfate or sulfonate group instead of a carboxylate group. Calcium, magnesium, and iron salts of monoalkylsulfuric and sulfonic acids are much more soluble in water than comparable salts of fatty acids.

The most widely used synthetic detergents are the linear alkylbenzenesulfonates (LAS). One of the most common of these is sodium 4-dodecylbenzenesulfonate. To prepare this type of detergent, a linear alkylbenzene is treated with sulfuric acid (Section 22.1B) to form an alkylbenzenesulfonic acid. The sulfonic acid is then neutralized with NaOH; the product is mixed with builders and spray-dried to give a smooth-flowing powder. The most common builder is sodium silicate.

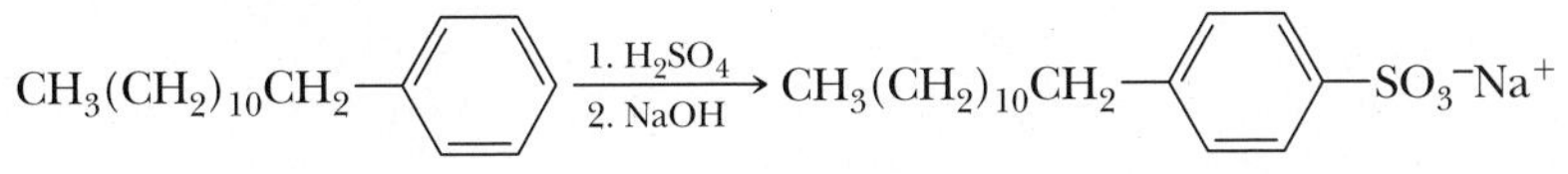
CH3(CH2)10CH2
1. H2SO4
2. NaOH
CH3(CH2)10CH2
SO3−Na+

Dodecylbenzene — **Sodium 4-dodecylbenzenesulfonate** (an anionic detergent)

Alkylbenzenesulfonate detergents were introduced in the late 1950s, and today they command close to 90% of the market once held by natural soaps.

Among the most common additives to detergent preparations are foam stabilizers, bleaches, and optical brighteners. A foam stabilizer frequently added to liquid soaps but not laundry detergents (for obvious reasons: think of a top-loading washing machine with foam spewing out the lid) is the amide prepared from dodecanoic acid (lauric acid) and 2-aminoethanol (ethanolamine). The most common bleach is sodium perborate tetrahydrate, which decomposes at temperatures above 50°C to give hydrogen peroxide, the actual bleaching agent.

N-(2-Hydroxyethyl)dodecanamide
(a foam stabilizer)

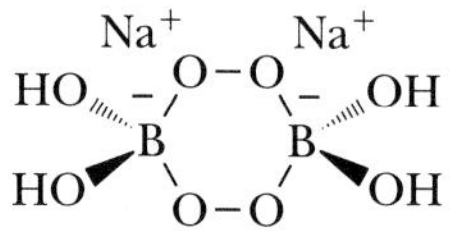

Sodium perborate tetrahydrate
(a bleach)

Also added to laundry detergents are optical brighteners, known also as optical bleaches. They are absorbed into fabrics and, after absorbing ambient light, fluoresce with a blue color, offsetting the yellow color caused by fabric aging. Quite literally, these optical brighteners produce a "whiter-than-white" appearance. You most certainly have observed the effects of optical brighteners if you have seen the surprisingly intense blue glow of "white" shirts or blouses when exposed to black light (UV radiation).

Connections to Biological Chemistry

FAD/FADH$_2$: Agents for Electron Transfer in Biological Oxidation-Reductions: Fatty Acid Oxidation

Flavin adenine dinucleotide (FAD) is a central component in the transfer of electrons in metabolic oxidations and reductions. In FAD, flavin is bonded to the five-carbon monosaccharide ribitol, which is, in turn, bonded to the terminal phosphate group of adenosine diphosphate.

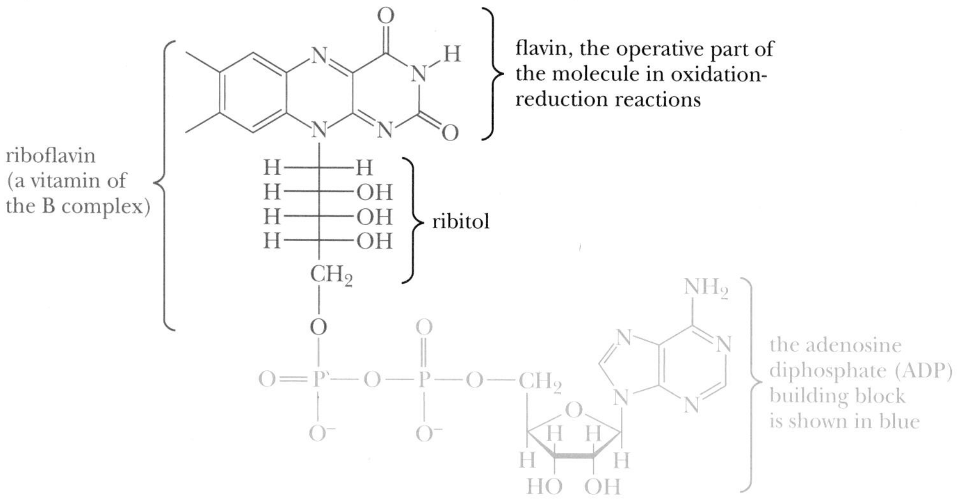

Flavin adenine dinucleotide
(FAD)

FAD participates in several types of enzyme-catalyzed oxidation-reduction reactions, one of which is the oxidation of a carbon-carbon single bond in the hydrocarbon chain of a fatty acid. In the process, FAD is reduced to $FADH_2$.

Oxidation of a hydrocarbon chain

$$—CH_2—CH_2— \;+\; FAD \longrightarrow —CH{=}CH— \;+\; FADH_2$$

a portion of the hydrocarbon chain of a fatty acid

The mechanism by which FAD oxidizes $—CH_2—CH_2—$ to $—CH{=}CH—$ involves the transfer of a hydride ion from the hydrocarbon chain of the fatty acid to FAD.

Mechanism 26.1

Oxidation of a Fatty Acid $—CH_2—CH_2—$ to $—CH{=}CH—$ by FAD

The individual curved arrows in this mechanism are numbered 1–6 to help you follow the flow of electrons in the transformation.

Step 1: A basic group, Enz-B:, on the surface of the enzyme removes a hydrogen atom from the carbon adjacent to the carboxyl group.

Step 2: Electrons from this C—H sigma bond become the pi electrons of the new C—C double bond.

Step 3: A hydride ion transfers from the carbon beta to the carboxyl group to a nitrogen atom of flavin.

Step 4: The pi electrons within flavin become redistributed.

Step 5: Electrons of the C═N bond remove a hydrogen from the enzyme.

Step 6: A new basic group forms on the surface of the enzyme.

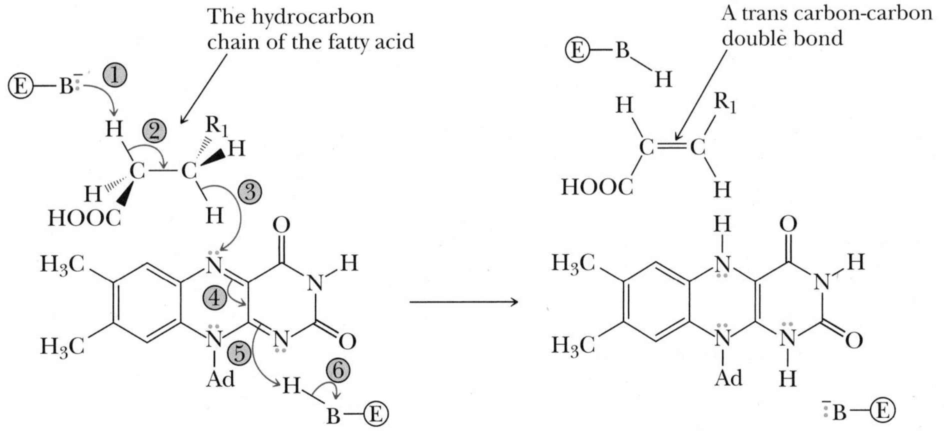

FAD **$FADH_2$**

Note that one of the two hydrogen atoms added to FAD to produce $FADH_2$ comes from the hydrocarbon chain and the other comes from an acidic group on the surface of the enzyme catalyzing this oxidation. Note also that one group on the enzyme functions as a proton acceptor and another functions as a proton donor.

26.3 Prostaglandins

Prostaglandin

A member of the family of compounds having the 20-carbon skeleton of prostanoic acid.

The **prostaglandins** are a family of compounds all having the 20-carbon skeleton of prostanoic acid.

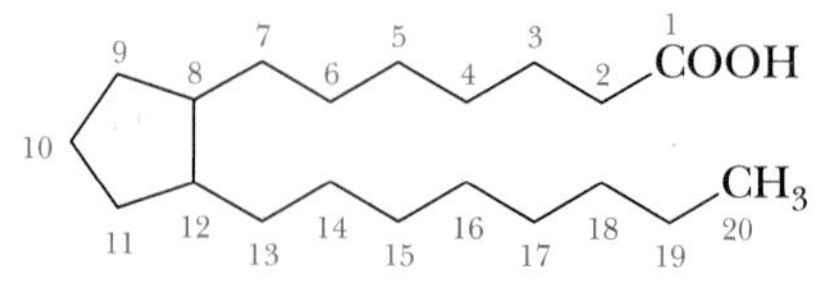

Prostanoic acid

The story of the discovery and structure determination of these remarkable compounds began in 1930 when gynecologists Raphael Kurzrok and Charles Lieb reported that human seminal fluid stimulates contraction of isolated uterine muscle. A few years later in Sweden, Ulf von Euler confirmed this report and noted that human seminal fluid also produces contraction of intestinal smooth muscle and lowers blood pressure when injected into the bloodstream. Von Euler proposed the name *prostaglandin* for the mysterious substance(s) responsible for these diverse effects because it was believed at the time that they were synthesized in the prostate gland. Although we now know that prostaglandin production is by no means limited to the prostate gland, the name nevertheless has stuck.

Prostaglandins are not stored as such in target tissues. Rather, they are synthesized in response to specific physiological triggers. Starting materials for the biosynthesis of prostaglandins are polyunsaturated fatty acids of 20 carbons, stored until needed as membrane phospholipid esters. In response to a physiological trigger, the ester is hydrolyzed, the fatty acid is released, and the synthesis of prostaglandins is initiated. Figure 26.5 outlines the steps in the synthesis of several prostaglandins from arachidonic acid. A key step in this biosynthesis is the enzyme-catalyzed reaction of arachidonic acid with two molecules of O_2 to form prostaglandin G_2 (PGG_2). The anti-inflammatory and anticlotting effects of aspirin and other nonsteroidal anti-inflammatory drugs (NSAIDs) results from their ability to inhibit the enzyme that catalyzes this step.

Research on the involvement of prostaglandins in reproductive physiology and the inflammatory process has produced several clinically useful prostaglandin derivatives. The observations that $PGF_{2\alpha}$ stimulates contractions of uterine smooth muscle led to a synthetic derivative that is used as a therapeutic abortifacient. A problem with the use of the natural prostaglandins for this purpose is that they are very rapidly degraded within the body. In the search for less rapidly degraded prostaglandins, a number of analogs have been prepared, one of the most effective of which is carboprost. This synthetic prostaglandin is 10 to 20 times more potent than the natural $PGF_2\alpha$ and is only degraded in the body slowly. The comparison of these two prostaglandins illustrates how a simple change in the structure of a drug can make a significant change in its effectiveness.

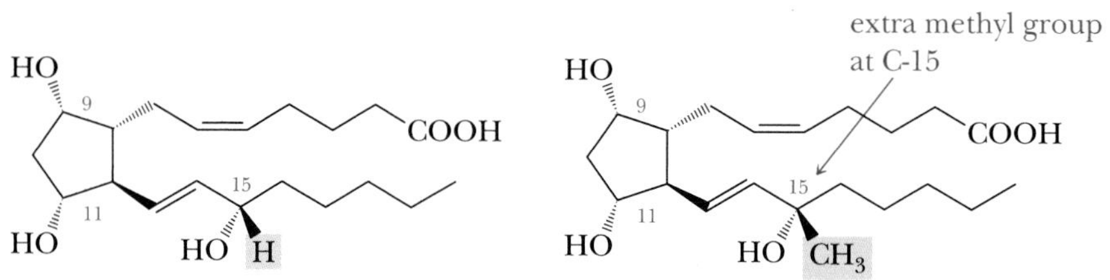

$PGF_{2\alpha}$

Carboprost
(15*S*)-15-Methyl-$PGF_{2\alpha}$

The PGEs along with several other PGs suppress gastric ulceration and appear to heal gastric ulcers. The PGE_1 analog, misoprostol, is currently used primarily

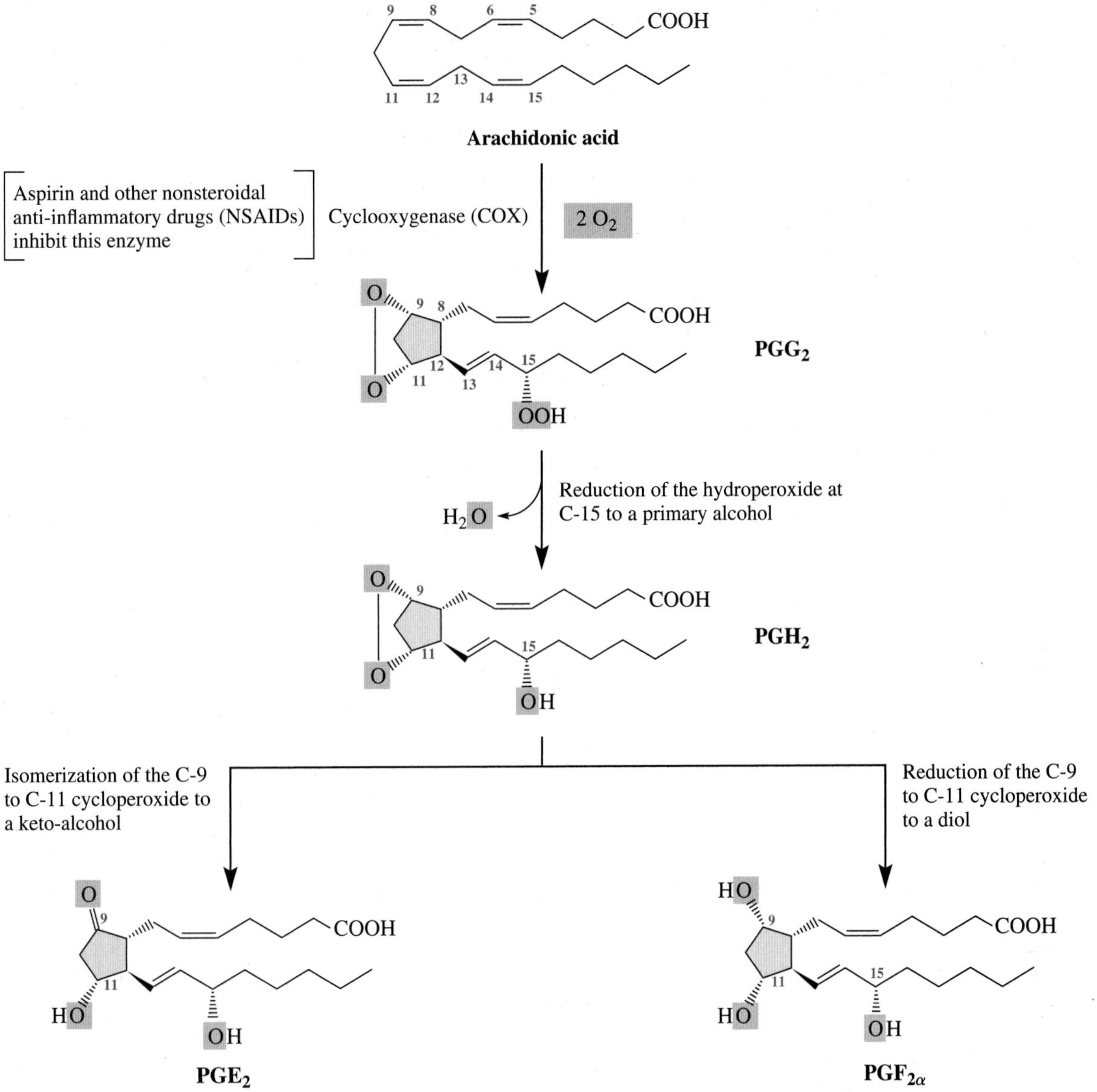

FIGURE 26.5 Key intermediates in the conversion of arachidonic acid to PGE_2 and PGF_{2a}. PG stands for prostaglandin. The letters *E, F, G,* and *H* are different types of prostaglandins.

for prevention of ulceration associated with aspirin-like NSAIDs (partly caused by their inhibition of clotting).

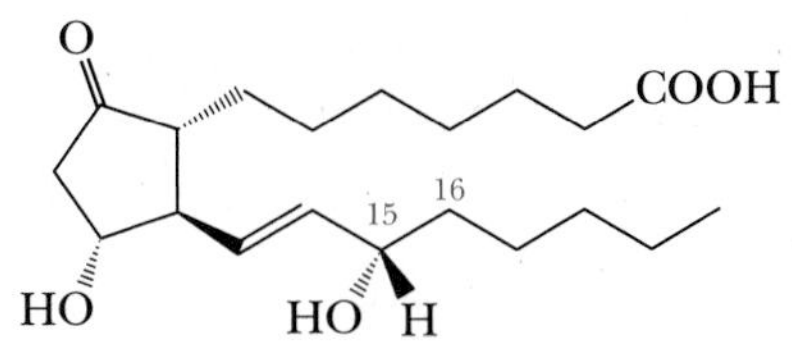

PGE₁

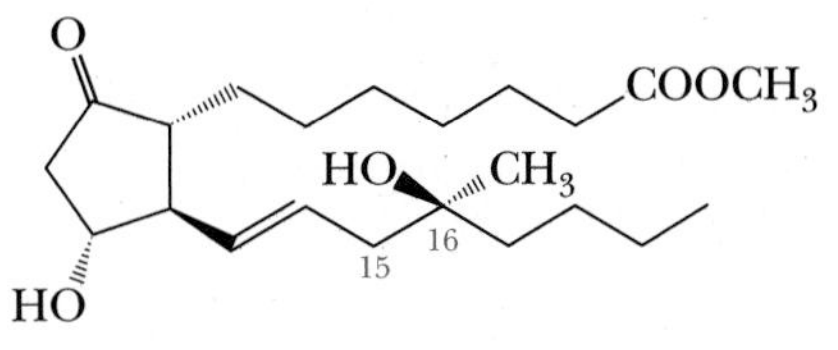

Misoprostol

Prostaglandins are members of an even larger family of compounds called **eicosanoids**. Eicosanoids contain 20 carbons and are derived from fatty acids. They include not only the prostaglandins but also the leukotrienes, thromboxanes, and prostacyclins. The eicosanoids are extremely widespread, and members of this family of compounds have been isolated from almost every tissue and body fluid.

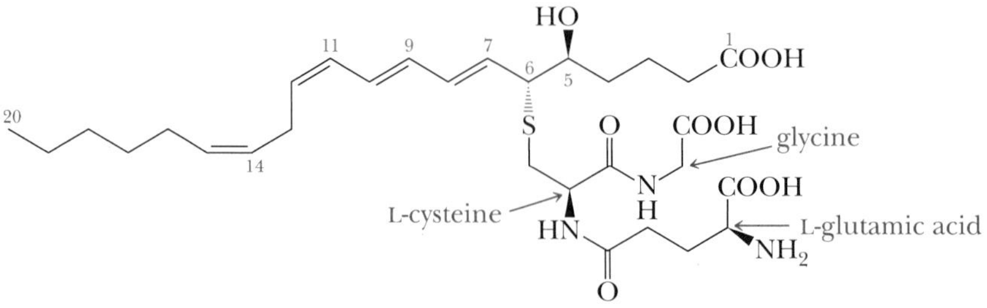

Leukotriene C_4(LTC_4)
(a smooth muscle constrictor)

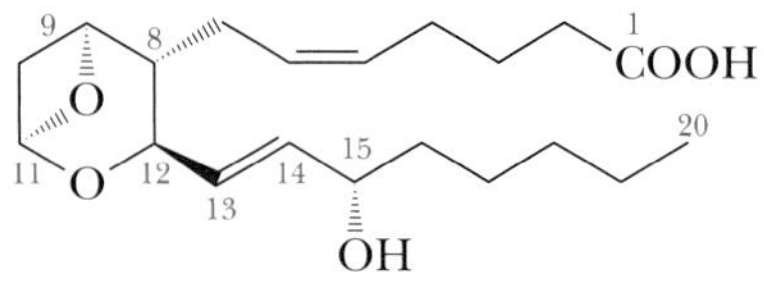

Thromboxane A_2
(a potent vasoconstrictor)

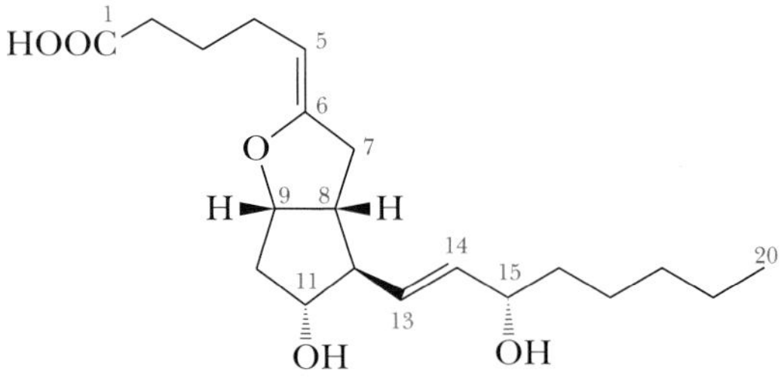

Prostacyclin
(a platelet aggregation inhibitor)

Leukotrienes are derived from arachidonic acid and are found primarily in leukocytes (white blood cells). Leukotriene C_4 (LTC_4), a typical member of this family, has three conjugated double bonds (hence the suffix *-triene*) and contains the amino acids L-cysteine, glycine, and L-glutamic acid (Chapter 27). An important physiological action of LTC_4 is constriction of smooth muscles, especially those of the lungs. The synthesis and release of LTC_4 is prompted by allergic reactions. Drugs that inhibit the synthesis of LTC_4 show promise for the treatment of the allergic reactions associated with asthma. Thromboxane A_2 is a very potent vasoconstrictor; its release triggers the irreversible phase of platelet aggregation and constriction of injured blood vessels. It is thought that aspirin and aspirin-like drugs act as mild anticoagulants because they inhibit cyclooxygenase, the enzyme that initiates the synthesis of thromboxane A_2.

Steroid
A plant or animal lipid having the characteristic tetracyclic ring structure of the steroid nucleus, namely 3 six-membered rings and 1 five-membered ring.

26.4 Steroids

Steroids are a group of plant and animal lipids that have the tetracyclic ring system shown in Figure 26.6. The features common to the tetracyclic ring system of most naturally occurring steroids are illustrated in Figure 26.7.

1. The fusion of rings is *trans*, and each atom or group at a ring junction is axial. Compare, for example, the orientations of —H at carbon 5 and —CH_3 at carbon 10.
2. The pattern of atoms or groups along the points of ring fusion (carbons 5 to 10 to 9 to 8 to 14 to 13) is nearly always *trans*-anti-*trans*-anti-*trans*.
3. Because of the *trans*-anti-*trans*-anti-*trans* arrangement of atoms or groups along the points of ring fusion, the tetracyclic steroid ring system is nearly flat and quite rigid.
4. Many steroids have axial methyl groups at carbons 10 and 13 of the tetracyclic ring system.

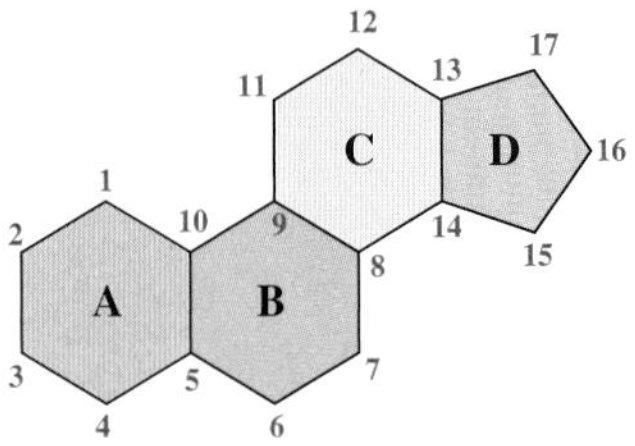

FIGURE 26.6 The tetracyclic ring system characteristic of steroids.

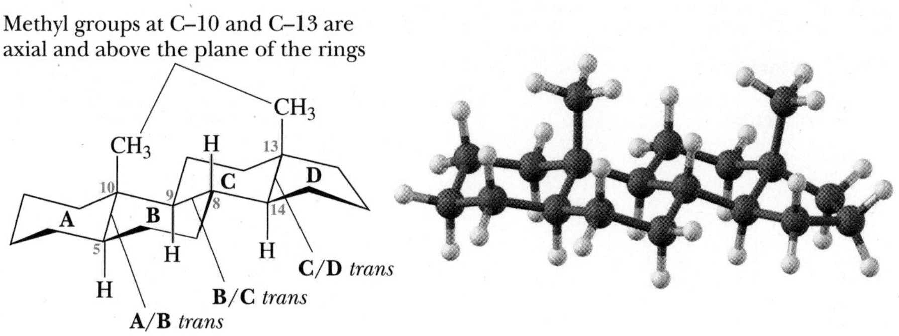
Methyl groups at C-10 and C-13 are axial and above the plane of the rings
CH3
CH3
H
13
10
9
8
14
5
A
B
C
D
H
H
H
C/D trans
B/C trans
A/B trans

FIGURE 26.7 Features common to the tetracyclic ring system of many steroids.

Phototake / Superstock

Human gallstones are almost pure cholesterol; this gallstone is about 0.5 cm in diameter.

A. Structure of the Major Classes of Steroids

Cholesterol

Cholesterol is a white, water-insoluble, waxy solid found in blood plasma and in all animal tissues. This substance is an integral part of human metabolism in two ways: (1) It is an essential component of biological membranes. The body of a healthy adult contains approximately 140 g of cholesterol, about 120 g of which is present in membranes. Membranes of the central and peripheral nervous systems, for example, contain about 10% cholesterol by weight. (2) It is the compound from which sex hormones, adrenocorticoid hormones, bile acids, and vitamin D are synthesized. Thus, cholesterol is, in a sense, the parent steroid.

Cholesterol has eight chiral centers, and a molecule with this number of chiral centers can exist as 2^8, or 256, stereoisomers (128 pairs of enantiomers). Only one of these stereoisomers is known to exist in nature: the stereoisomer with the configuration shown in Figure 26.8.

Cholesterol is insoluble in blood plasma but can be transported as a plasma-soluble complex formed with proteins called lipoproteins. **Low-density lipoproteins** (**LDLs**, often called "bad cholesterol") transport cholesterol from the site of its synthesis in the liver to the various tissues and cells of the body where it is to be used. Primarily, the cholesterol associated with LDLs builds up in atherosclerotic deposits in blood vessels. **High-density lipoproteins** (**HDLs**, often called "good cholesterol") transport excess and unused cholesterol from cells back to the liver for its degradation to bile acids and eventual excretion in the feces. It is thought that HDLs retard or reduce atherosclerotic deposits.

Low-density lipoprotein (LDL)
Plasma particles, density 1.02–1.06 g/mL, consisting of approximately 26% proteins, 50% cholesterol, 21% phospholipids, and 4% triglycerides.

High-density lipoprotein (HDL)
Plasma particles, density 1.06–1.21 g/mL, consisting of approximately 33% proteins, 30% cholesterol, 29% phospholipids, and 8% triglycerides.

Steroid Hormones

Table 26.3 provides representations of each major class of steroid hormones, along with the principal functions of each.

The two most important female sex hormones, or **estrogens**, are estrone and estradiol. In addition, progesterone, another type of steroid hormone, is essential for preparing the uterus for implantation of a fertilized egg. After the role of progesterone

Estrogen
A steroid hormone such as estrone and estradiol that mediates the development of sexual characteristics in females.

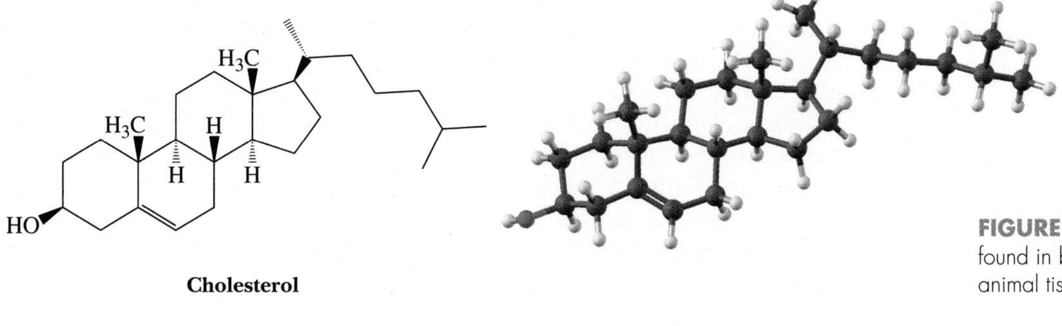
H3C
H3C
H
H
H
HO

Cholesterol

FIGURE 26.8 Cholesterol is found in blood plasma and in all animal tissues.

Table 26.3 Selected Steroid Hormones

Structure		Source and Major Effects
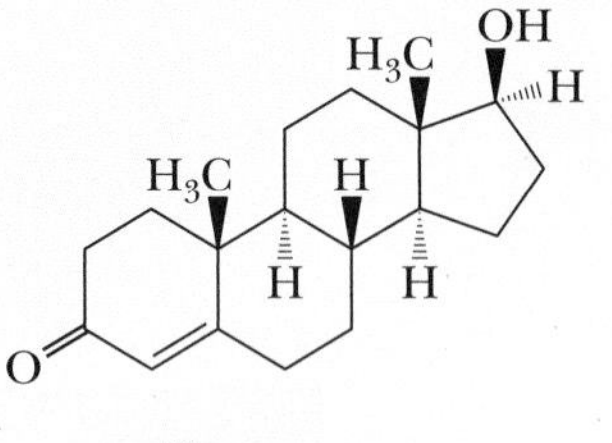 **Testosterone**	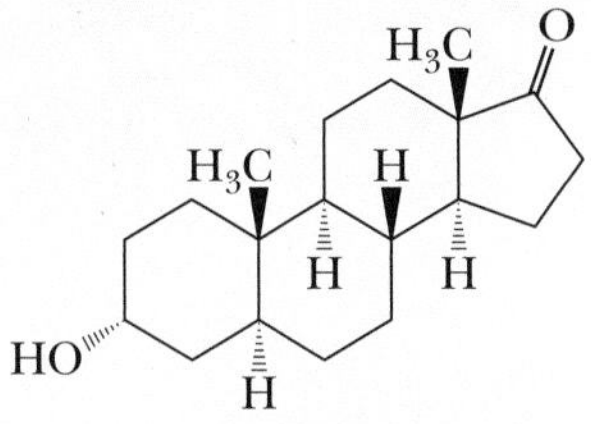 **Androsterone**	Androgens (male sex hormones)—synthesized in the testes; responsible for development of male secondary sex characteristics
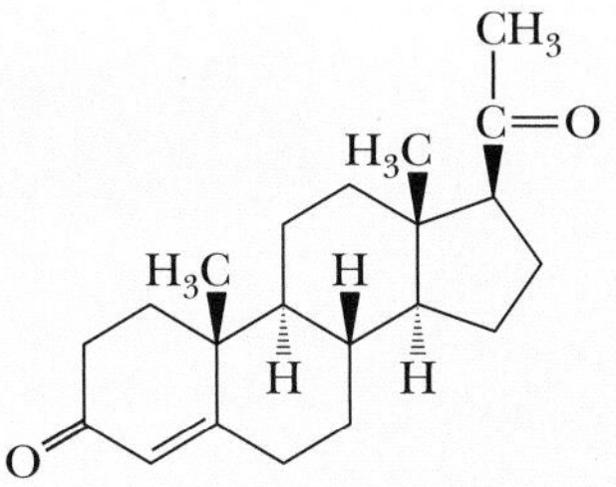 **Progesterone**	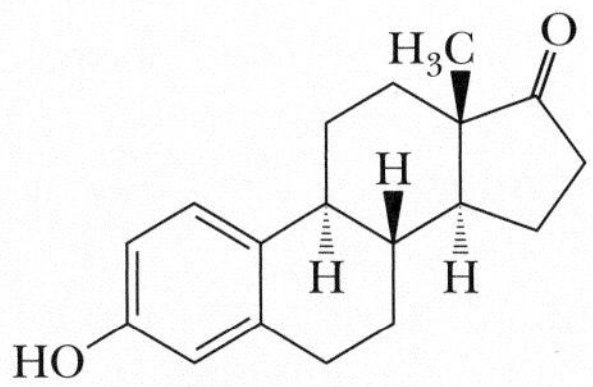 **Estrone**	Estrogens (female sex hormones)—synthesized in the ovaries; responsible for development of female secondary sex characteristics and control of the menstrual cycle
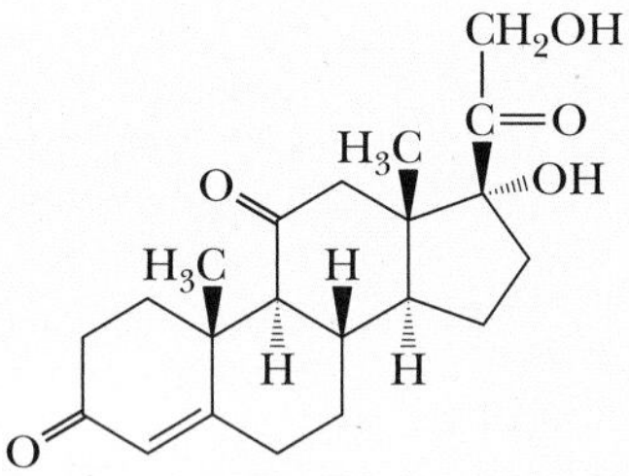 **Cortisone**	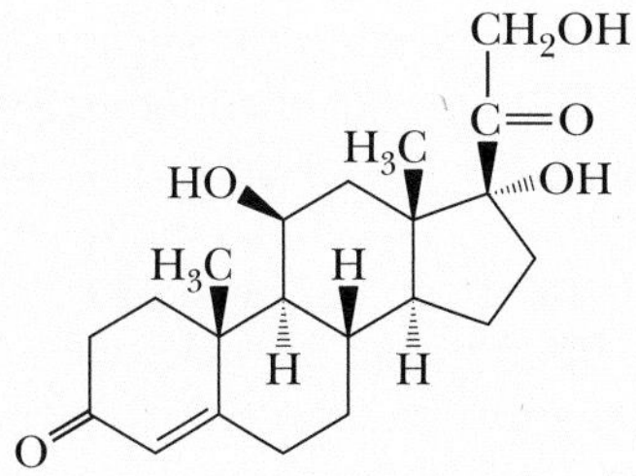 **Cortisol**	Glucocorticoid hormones—synthesized in the adrenal cortex; regulate metabolism of carbohydrates, decrease inflammation, and are involved in the reaction to stress
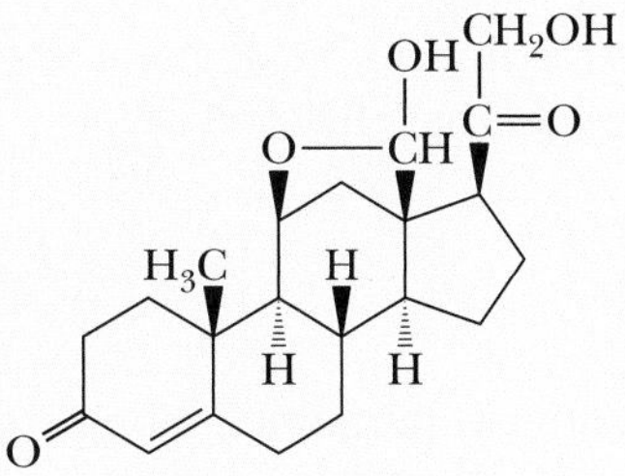 **Aldosterone**		A mineralocorticoid hormone—synthesized in the adrenal cortex; regulates blood pressure and volume by stimulating the kidneys to absorb Na^+, Cl^-, and HCO_3^-

in inhibiting ovulation was understood, its potential as a possible contraceptive was realized. Progesterone itself is relatively ineffective when taken orally. As a result of a massive research program in both industrial and academic laboratories, many synthetic progesterone-mimicking steroids became available in the 1960s. When taken regularly, these drugs prevent ovulation, yet allow women to maintain a normal menstrual cycle. Some of the most effective of these preparations contain a progesterone analog, such as norethindrone, combined with a smaller amount of an estrogen-like material to help prevent irregular menstrual flow during prolonged use of contraceptive pills.

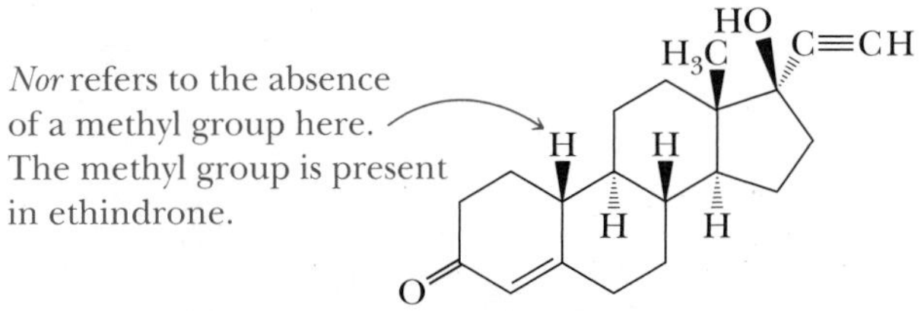

Nor refers to the absence of a methyl group here. The methyl group is present in ethindrone.
HO
H_3C
C≡CH
H
H
H
H
O

Norethindrone
(a synthetic progesterone analog)

The chief function of testosterone and other **androgens** is to promote normal growth of male reproductive organs (primary sex characteristics) and development of the characteristic deep voice, pattern of body and facial hair, and musculature (secondary sex characteristics). Although testosterone produces these effects, it is not active when taken orally because it is metabolized in the liver to an inactive steroid. A number of oral **anabolic steroids** have been developed for use in rehabilitation medicine, particularly when muscle atrophy occurs during recovery from an injury. Among the synthetic anabolic steroids most widely prescribed for this purpose are methandrostenolone and stanozolol. The structural formula of methandrostenolone differs from that of testosterone by introduction of (1) a methyl group at carbon 17 and (2) an additional carbon-carbon double bond between carbons 1 and 2. In stanozolol, ring A is modified by attachment of a pyrazole ring.

Androgen
A steroid hormone such as testosterone that mediates the development of sexual characteristics of males.

Anabolic steroid
A steroid hormone such as testosterone that promotes tissue and muscle growth and development.

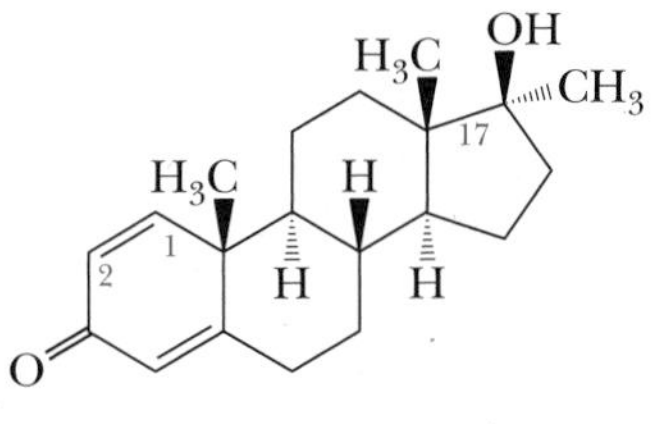

OH
H_3C
CH_3
17
H_3C
H
1
2
H
H
O

Methandrostenolone

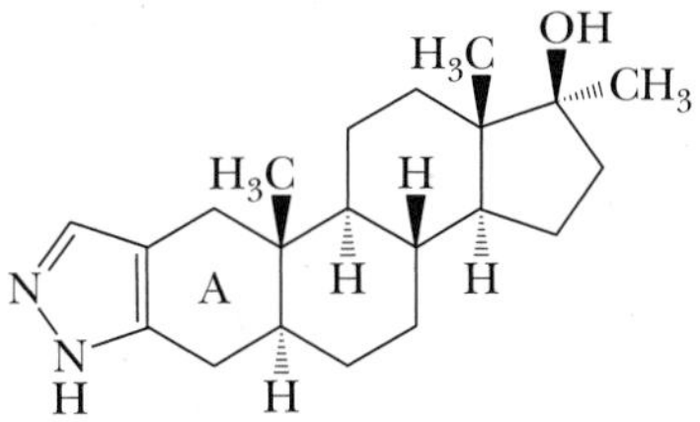

OH
H_3C
CH_3
H_3C
H
N
A
H
H
N
H
H

Stanozolol

Among certain athletes, the misuse of anabolic steroids to build muscle mass and strength, particularly for sports that require explosive action, is common. The risks associated with abuse of anabolic steroids for this purpose are enormous: heightened aggressiveness; sterility; impotence; and risk of premature death from complications of diabetes, coronary artery disease, and liver cancer.

Bile Acids

Shown in Figure 26.9 is a structural formula for cholic acid, a constituent of human bile. The molecule is shown as an anion, as it is ionized in bile and intestinal fluids. **Bile acids**—or more properly, bile salts—are synthesized in the liver, stored in the gallbladder, and secreted into the intestine, where their function is to emulsify dietary fats and thereby aid in their absorption and digestion. Furthermore, bile salts are the end products of the metabolism of cholesterol and thus are a principal pathway for the elimination of this substance from the body. A characteristic structural feature of bile salts is a *cis* fusion of rings A and B.

Bile acid
A cholesterol-derived detergent molecule such as cholic acid that is secreted by the gallbladder into the intestine to assist in the absorption of dietary lipids.

B. Biosynthesis of Cholesterol

The biosynthesis of cholesterol illustrates a point we first made in the introduction to the structure of terpenes (Section 5.4). In building large molecules, one of the

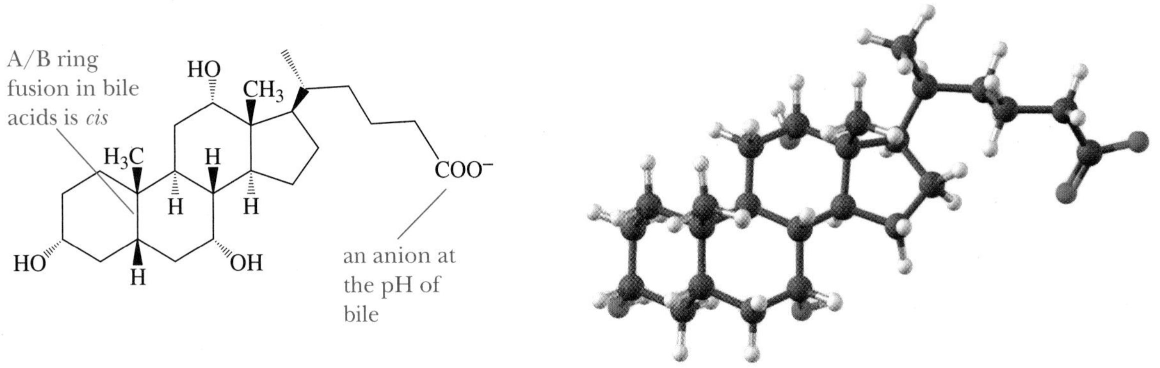

FIGURE 26.9 Cholic acid, an important constituent of human bile. Each six-membered ring is in a chair conformation.

common patterns in the biological world is to begin with one or more smaller subunits, join them through an iterative process, and then chemically modify the completed carbon skeleton by oxidation, reduction, cross-linking, addition, elimination, or related processes to give a biomolecule with a unique identity.

The building block from which all carbon atoms of steroids are derived is the two-carbon acetyl group of acetyl-CoA (Problem 25.34). The American biochemist Konrad Bloch, who shared the 1964 Nobel Prize in Medicine or Physiology with German biochemist Feodor Lynen for their discoveries concerning the biosynthesis of cholesterol and fatty acids, showed that 15 of the 27 carbon atoms of cholesterol are derived from the methyl group of acetyl-CoA. The remaining 12 carbon atoms are derived from the carbonyl group of acetyl-CoA (Figure 26.10).

FIGURE 26.10 Several key intermediates in the synthesis of cholesterol from acetyl groups of acetyl-CoA. Eighteen moles of acetyl-CoA are required for the synthesis of one mole of cholesterol.

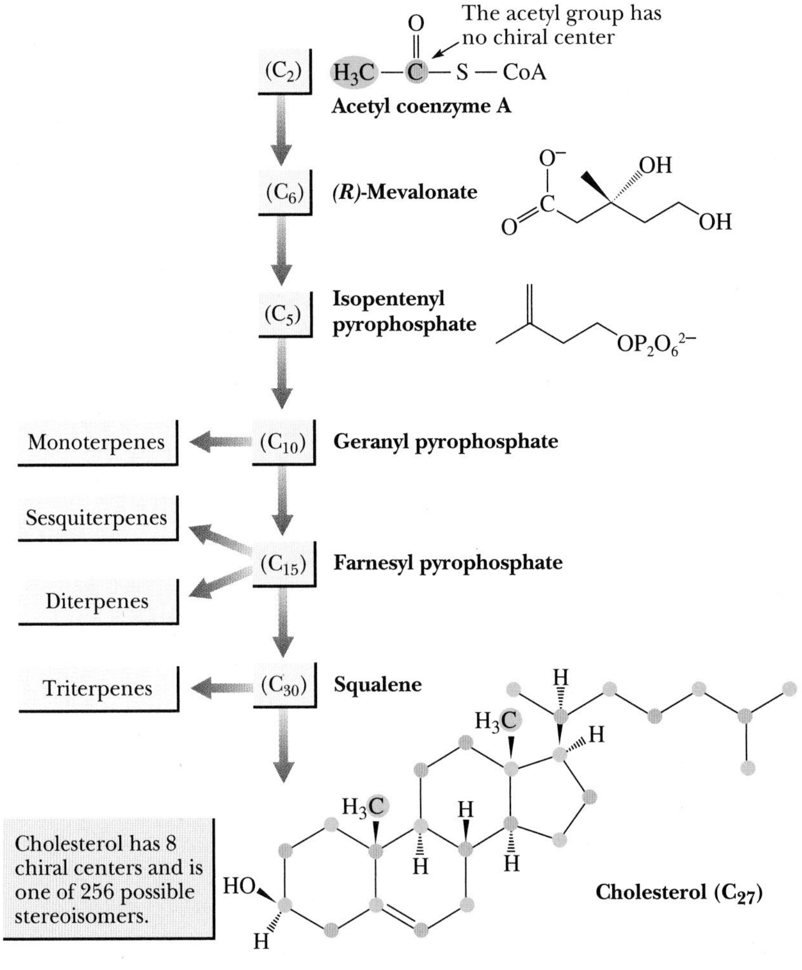

A remarkable feature of this synthetic pathway is that the biosynthesis of cholesterol from acetyl-CoA is completely stereoselective; it is synthesized as only one of 256 possible stereoisomers. We cannot duplicate this exquisite degree of stereoselectivity in the laboratory. Cholesterol is, in turn, the key intermediate in the synthesis of most other steroids.

cholesterol →
- bile acids (e.g., **cholic acid**)
- sex hormones (e.g., **testosterone and estrone**)
- mineralocorticoid hormones (e.g., **aldosterone**)
- glucocorticoid hormones (e.g., **cortisone**)

26.5 Phospholipids

A. Structure

Phospholipids, or phosphoacylglycerols as they are more properly named, are the second most abundant group of naturally occurring lipids. They are found almost exclusively in plant and animal membranes, which typically consist of about 40%–50% phospholipids and 50%–60% proteins. The most abundant phospholipids are derived from a phosphatidic acid (Figure 26.11).

Phospholipid
A lipid containing glycerol esterified with two molecules of fatty acid and one molecule of phosphoric acid.

The fatty acids most common in phosphatidic acids are palmitic and stearic acids (both fully saturated) and oleic acid (one double bond in the hydrocarbon chain). Further esterification of a phosphatidic acid with a low-molecular-weight alcohol

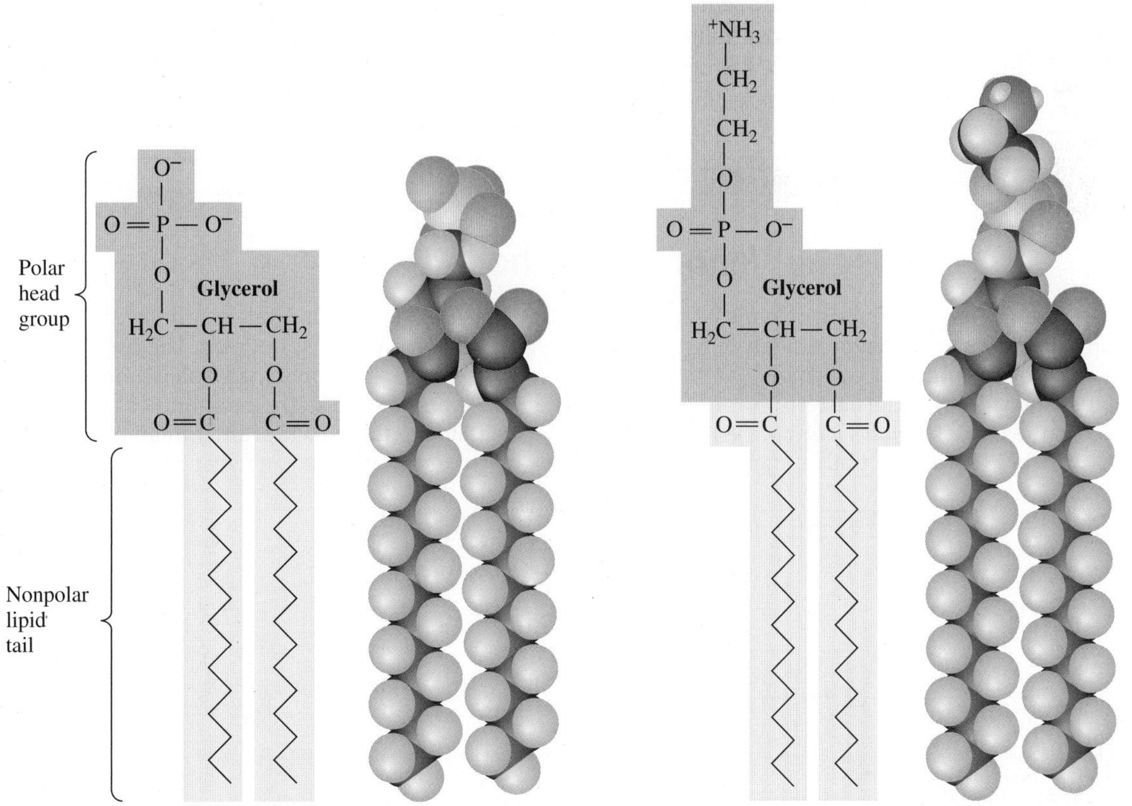

A phosphatidate   A phosphopholipid

FIGURE 26.11 In a phosphatidic acid, glycerol is esterified with two molecules of fatty acid and one molecule of phosphoric acid. Further esterification of the phosphoric acid group with a low-molecular-weight alcohol gives a phospholipid.

All of these products contain lecithin.

Table 26.4 Low-Molecular-Weight Alcohols Most Common to Phospholipids

Alcohols Found in Phospholipids		
Structural Formula	**Name**	**Name of Phospholipid**
$HOCH_2CH_2NH_2$	Ethanolamine	Phosphatidylethanolamine (Cephalin)
$HOCH_2CH_2\overset{+}{N}(CH_3)_3$	Choline	Phosphatidylcholine (Lecithin)
$HOCH_2CH(NH_3^+)COO^-$	Serine	Phosphatidylserine
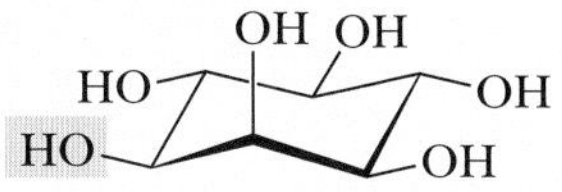	Inositol	Phosphatidylinositol

gives a phospholipid. Several of the most common alcohols forming phospholipids are given in Table 26.4. All functional groups in this table and in Figure 26.11 are shown as they are ionized at pH 7.4, the approximate pH of blood plasma and of many other biological fluids. Under these conditions, each phosphate group bears a negative charge and each amino group bears a positive charge.

B. Lipid Bilayers

Figure 26.12 shows a space-filling model of a lecithin (a phosphatidylcholine). Lecithin and other phospholipids are elongated, almost rodlike molecules, with the nonpolar (hydrophobic) hydrocarbon chains lying roughly parallel to one another and the polar (hydrophilic) phosphoric ester group pointing in the opposite direction.

Chemical Connections

Snake Venom Phospholipases

The venoms of certain snakes contain enzymes called phospholipases. These enzymes catalyze the hydrolysis of carboxylic ester bonds of phospholipids. The venom of the eastern diamondback rattlesnake (*Crotalus adamanteus*) and the Indian cobra (*Naja naja*) both contain phospholipase PLA_2, which catalyzes the hydrolysis of esters at carbon 2 of phospholipids. The breakdown product of this hydrolysis, a lysolecithin, acts as a detergent and dissolves the membranes of red blood cells, causing them to rupture. Indian cobras kill several thousand people each year.

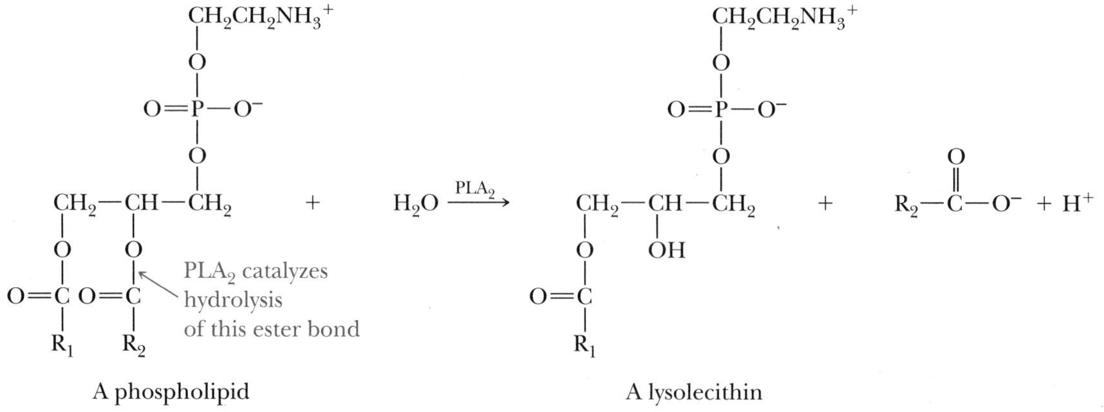

When placed in aqueous solution, phospholipids spontaneously form a **lipid bilayer** (Figure 26.13) in which polar head groups lie on the surface, giving the bilayer an ionic coating. Nonpolar hydrocarbon chains of fatty acids lie buried within the bilayer. This self-assembly of phospholipids into a bilayer is a spontaneous process driven by two types of noncovalent forces: (1) hydrophobic effects, which result when nonpolar hydrocarbon chains cluster together and exclude water molecules and (2) electrostatic interactions, which result when polar head groups interact with water and other polar molecules in the aqueous environment.

Recall from Section 26.2B that formation of soap micelles is driven by these same noncovalent forces; the polar (hydrophilic) carboxylate groups of soap molecules lie on the surface of the micelle and associate with water molecules, and the nonpolar (hydrophobic) hydrocarbon chains cluster within the micelle and thus are removed from contact with water.

The arrangement of hydrocarbon chains in the interior of a phospholipid bilayer varies from rigid to fluid depending on the degree of unsaturation of the hydrocarbon chains themselves. Saturated hydrocarbon chains tend to lie parallel and closely packed, increasing the rigidity of the bilayer. Unsaturated hydrocarbon chains, on the other hand, have one or more *cis* double bonds, which cause "kinks" in the chains. As a result, they do not pack as closely and with as great an order as saturated chains. The disordered packing of unsaturated hydrocarbon chains leads to increased fluidity of the bilayer.

Biological membranes are made of lipid bilayers. The most satisfactory current model for the arrangement of phospholipids, proteins, and cholesterol in plant and animal membranes is the **fluid-mosaic model** proposed in 1972 by S. J. Singer and G. Nicolson (Figure 26.13). The term *mosaic* signifies that the various components in the membrane coexist side by side, as discrete units, rather than combining to form new molecules or ions. "Fluid" signifies that the same sort of fluidity exists in membranes that we have already seen for lipid bilayers. Furthermore, the protein components of membranes "float" in the bilayer and can move laterally along the plane of the membrane.

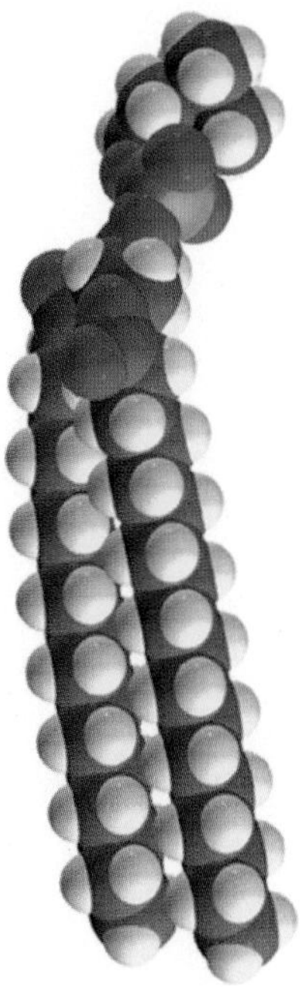

FIGURE 26.12 Space-filling model of a lecithin.

Lipid bilayer

A back-to-back arrangement of phospholipid monolayers, often forming a closed vesicle or membrane.

Fluid-mosaic model

A biological membrane that consists of a phospholipid bilayer with proteins, carbohydrates, and other lipids on the surface and embedded in the surface of the bilayer.

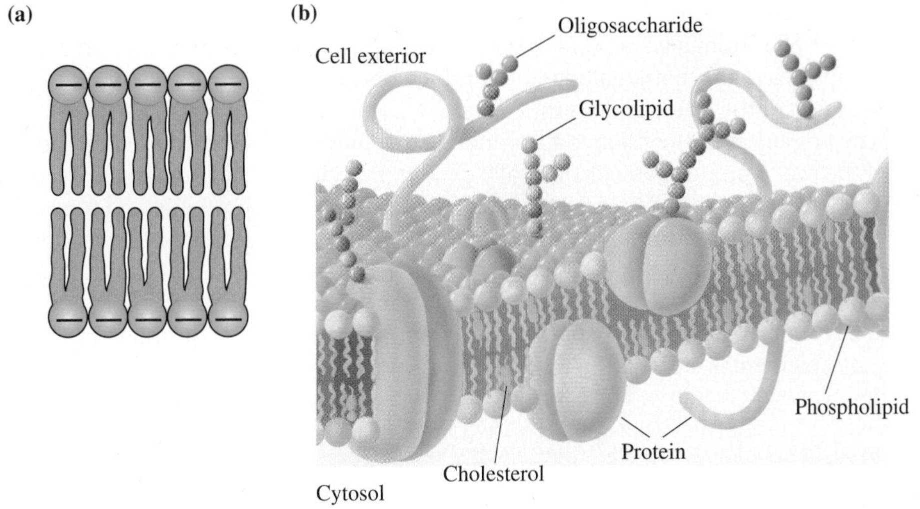
(a)
(b)
Oligosaccharide
Cell exterior
Glycolipid
Phospholipid
Protein
Cholesterol
Cytosol

FIGURE 26.13 Fluid-mosaic model of a biological membrane showing the lipid bilayer and membrane proteins oriented on the inner and outer surfaces of the membrane and penetrating the entire thickness of the membrane.

26.6 Fat-Soluble Vitamins

Vitamins are divided into two broad classes on the basis of solubility: those that are fat-soluble (and hence classed as lipids) and those that are water-soluble. The fat-soluble vitamins include A, D, E, and K.

A. Vitamin A

Vitamin A, or retinol, occurs only in the animal world, where the best sources are cod-liver oil and other fish-liver oils, animal liver, and dairy products. Vitamin A in the form of a precursor, or provitamin, is found in the plant world in a group of tetraterpene (C_{40}) pigments called carotenes. The most common of these is β-carotene, abundant in carrots but also found in some other vegetables, particularly yellow and green ones. β-Carotene has activity as an antioxidant; one of its functions in green plants is to quench singlet oxygen, which can be produced as a by-product of photosynthesis. β-Carotene has no vitamin A activity; however, after ingestion, it is cleaved at the central carbon-carbon double bond followed by reduction of the newly formed aldehyde to give retinol (vitamin A).

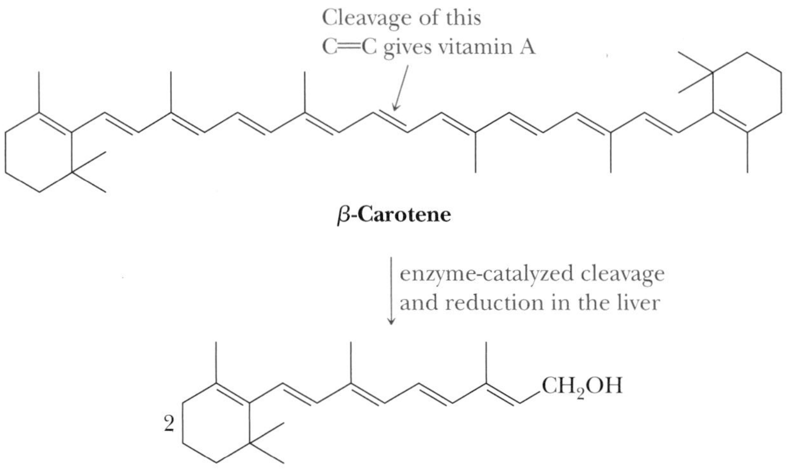
Cleavage of this
C=C gives vitamin A
β-Carotene
enzyme-catalyzed cleavage
and reduction in the liver
2
CH_2OH

Retinol (Vitamin A)

Probably the best understood role of vitamin A is its participation in the visual cycle in rod cells. In a series of enzyme-catalyzed reactions (Figure 26.14), retinol undergoes a two-electron oxidation to all-*trans*-retinal, isomerization about the carbon 11 to 12 double bond to give 11-*cis*-retinal, and formation of an imine (Section 16.8) with the $—NH_2$ from a lysine unit of the protein opsin. The product of these reactions is rhodopsin, a highly conjugated pigment that shows intense absorption in the blue-green region of the visual spectrum.

The primary event in vision is absorption of light by rhodopsin in rod cells of the retina of the eye to produce an electronically excited molecule. Within several picoseconds ($1\ ps = 10^{-12}\ s$), the excess electronic energy is converted to vibrational and rotational energy and the 11-*cis* double bond is isomerized to the more stable 11-*trans* double bond. This isomerization triggers a conformational change in opsin that causes firing of neurons in the optic nerve and produces a visual image. Coupled with this light-induced change is hydrolysis of rhodopsin to give 11-*trans*-retinal and free opsin. At this point, the visual pigment is bleached and is in a refractory period. Rhodopsin is regenerated by a series of enzyme-catalyzed reactions that converts 11-*trans*-retinal to 11-*cis*-retinal and then to rhodopsin. The visual cycle is shown in abbreviated form in Figure 26.14.

B. Vitamin D

Vitamin D is the name for a group of structurally related compounds that play a major role in the regulation of calcium and phosphorus metabolism. A deficiency of vitamin D in childhood is associated with rickets, a mineral-metabolism disease that leads to bone defects that form bowlegs, knock-knees, and enlarged joints. Vitamin D_3, the most abundant form of the vitamin in the circulatory system, is produced in the skin of mammals by the action of ultraviolet radiation on 7-dehydrocholesterol (cholesterol with a double bond between carbons 7 and 8). In the liver, vitamin D_3 undergoes an enzyme-catalyzed, two-electron oxidation at carbon 25 of the side chain to form 25-hydroxyvitamin D_3; the oxidizing agent is molecular oxygen, O_2.

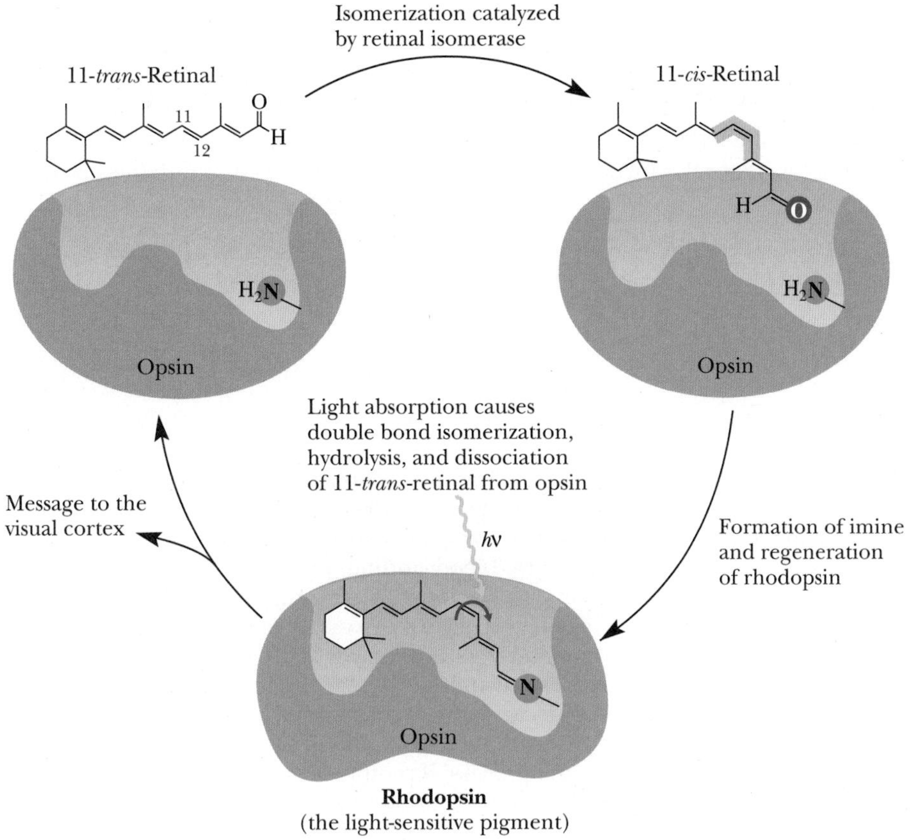
Isomerization catalyzed
by retinal isomerase
11-*trans*-Retinal
11-*cis*-Retinal
Opsin
Opsin
Light absorption causes
double bond isomerization,
hydrolysis, and dissociation
of 11-*trans*-retinal from opsin
Message to the
visual cortex
hν
Formation of imine
and regeneration
of rhodopsin
Opsin
Rhodopsin
(the light-sensitive pigment)

FIGURE 26.14 The primary chemical reaction of vision in rod cells is absorption of light by rhodopsin followed by isomerization of a carbon-carbon double bond from a *cis* configuration to a *trans* configuration.

Watch a video explanation

25-Hydroxyvitamin D_3 undergoes further oxidation in the kidneys, also by O_2, to form 1,25-dihydroxyvitamin D_3, the hormonally active form of the vitamin.

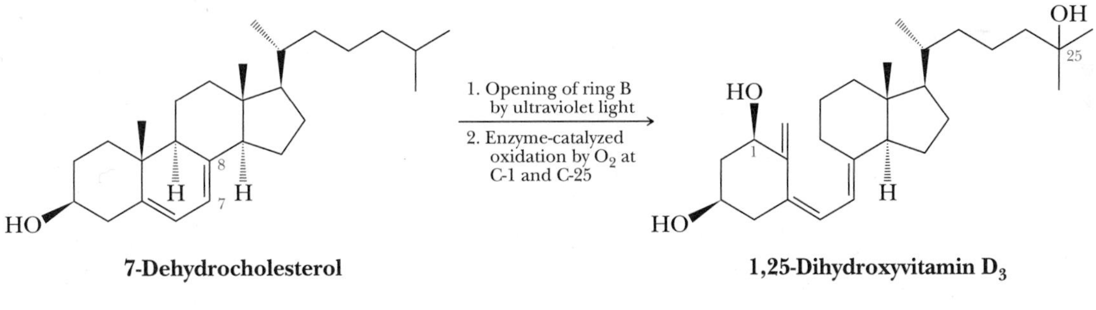
1. Opening of ring B
by ultraviolet light
2. Enzyme-catalyzed
oxidation by O2 at
C-1 and C-25
7-Dehydrocholesterol
1,25-Dihydroxyvitamin D3

C. Vitamin E

Vitamin E was first recognized in 1922 as a dietary factor essential for normal reproduction in rats, hence its name tocopherol from the Greek *tocos*, birth, and *pherein*, to bring about. Vitamin E is a group of compounds of similar structure, the most active of which is α-tocopherol. This vitamin occurs in fish oil, in other oils such as cottonseed and peanut oil, and in leafy green vegetables. The richest source of vitamin E is wheat germ oil.

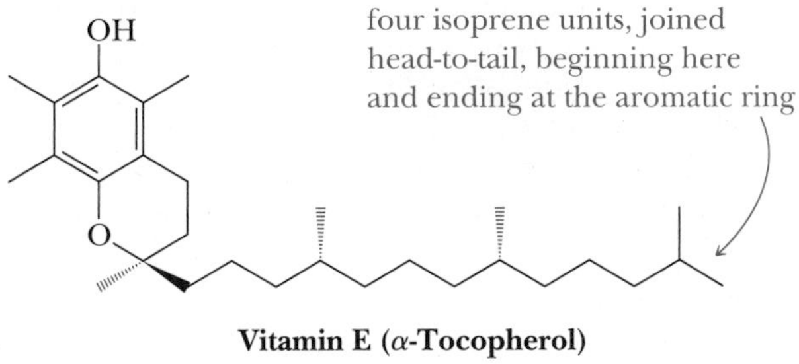
four isoprene units, joined
head-to-tail, beginning here
and ending at the aromatic ring

Vitamin E (α-Tocopherol)

In the body, vitamin E functions as an antioxidant; it traps peroxy radicals of the type HOO· and ROO· formed as a result of enzyme-catalyzed oxidation by molecular oxygen of the unsaturated hydrocarbon chains in membrane phospholipids (see Section 8.7). There is speculation that peroxy radicals play a role in the aging process and that vitamin E and other antioxidants may retard that process. Vitamin E is also necessary for the proper development and function of the membranes of red blood cells.

MCAT Practice: *Passage and Questions*

Vitamin K, Blood Clotting, and Basicity

Vitamin K deficiency results in slowed blood clotting, which can be a serious threat to a wounded animal or human. In the process of blood clotting, the natural vitamin, which is a quinone, is converted to its active hydroquinone form by reduction.

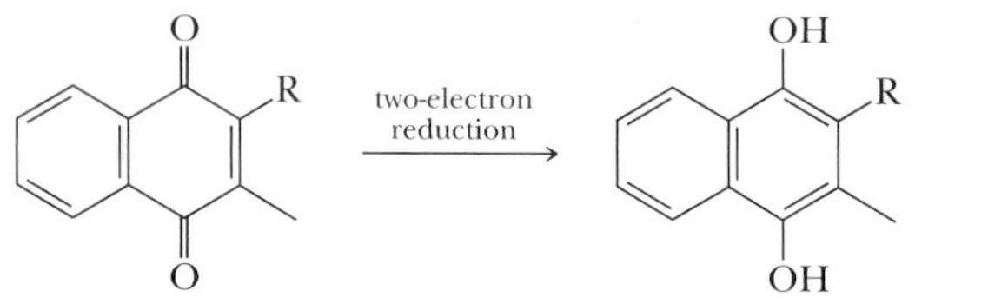

Vitamin K_1 quinone **Vitamin K_1 hydroquinone**

In the presence of potassium cation, O_2, and CO_2, the enzyme microsomal carboxylase will convert glutamate side chains of prothrombin (an essential protein for blood clotting) to γ-carboxyglutamate groups. This is formally the addition of a CO_2 group, which is the reverse of a decarboxylation as is common for β-dicarboxylic acids such as malonic acid. The resulting two carboxylates of the chemically modified glutamate now form a tight bidentate ("two teeth") complex with Ca^{2+} during the blood-clotting process. If prothrombin is not carboxylated, it does not bind calcium, and blood does not clot. For decades the role of vitamin K_1 and O_2 in this process was unknown.

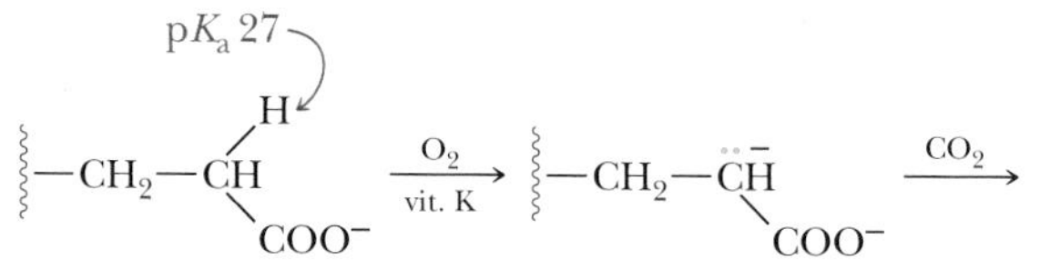

Glutamate side chain of prothrombin

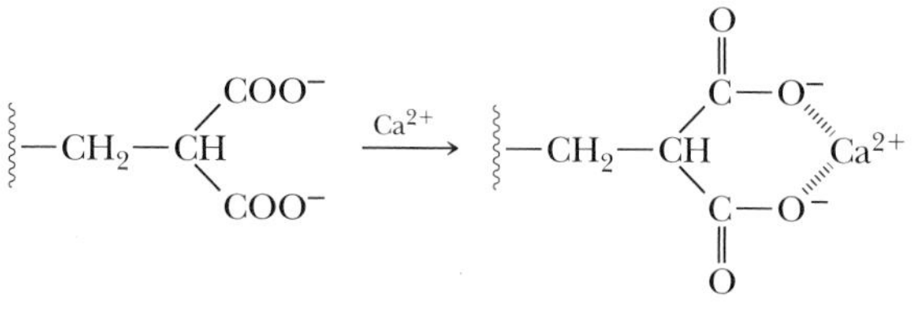

Carboxylated glutamate side chain

Carboxylated glutamate side chain binding calcium ion

Paul Dowd, from the University of Pittsburgh, discovered that the vitamin K_1 hydroquinone anion reacts with oxygen (O_2) to give the peroxide anion intermediate **1**, which is converted to compound **2**, and then to what is referred to as "vitamin K_1 base." This species deprotonates a glutamate side chain leading to carboxylation via reaction with CO_2.

Questions

A. Which of the following is a challenge to understanding how the carboxylation of glutamate side chains is feasible in a biological setting?

1. Because the pK_a of the α-hydrogen of glutamate is 27, it is too high to be effectively deprotonated at physiological pH.
2. The C—C bond formed in a 1,3-dicarboxylate is too weak to exist in a biological setting.
3. β-Dicarboxylic acids spontaneously decarboxylate at neutral pH.
4. Both 1 and 3.

B. The pK_as of the α-hydrogens of carbonyl-containing species are generally in the range of 18 to 22. However, the pK_a of the α-hydrogen of glutamate is 27. Why is this one particularly high?

1. Upon deprotonation, the resulting anion is stabilized less by resonance than with normal enolate anions.
2. A dianion is created.
3. Inductive stabilization of the anion occurs to a greater degree in a normal enolate.
4. Both 1 and 2.

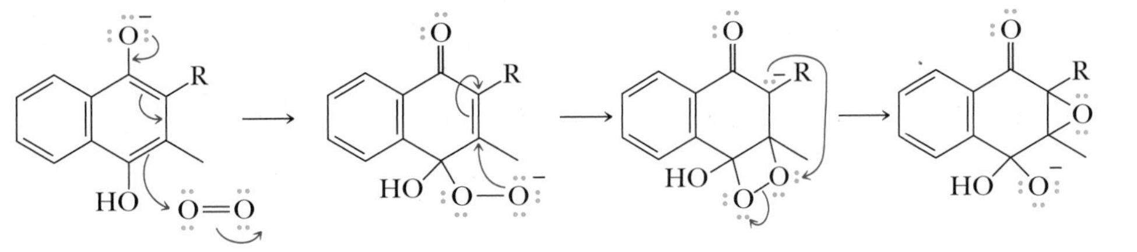

Vitamin K_1 hydroquinone anion (a weak base) — 1 — 2 — Vitamin K_1 base (a very strong base)

C. The reaction of the peroxide anion of **1** to give compound **2** would be referred to as

1. An intramolecular nucleophilic addition.
2. A conjugate addition.
3. An electrophilic addition.
4. Both 1 and 2.

D. The transformation of compound **2** to vitamin K_1 base is exothermic because of

1. The formation of an epoxide.
2. The resulting geminal diol anion.
3. The weak O—O bond.
4. All of the above.

E. The sequences of reactions from vitamin K_1 to vitamin K_1 base convert a weaker phenolate base to a stronger alkoxide base. Why is this important?

1. The resulting epoxide is a reactive enough species to induce carboxylation of the α-hydrogens of carboxylate anions at physiological conditions.
2. By requiring a compound that must exist in an animal's diet (a vitamin) as part of the blood clotting process, nature ensures that only individuals in nutritionally rich environments will survive.
3. The intermediacy of a dialkyl peroxide provides a driving force to create a base strong enough to remove a proton from glutamate side chains.
4. Both 1 and 3.

F. Although the pK_a of the glutamate side chain is 27, it could be lower in an enzyme active site via

1. Proximity of the carboxylate to a positive charge, potentially on an ammonium cation.
2. Coordination of the carboxylate to a metal cation.
3. Hydrogen bonding to the carboxylate anion.
4. Any or all of the above.

Study Guide

26.1 Triglycerides

- **Lipids** are a heterogeneous class of compounds grouped together on the basis of their solubility properties; they are insoluble in water and soluble in diethyl ether, acetone, and dichloromethane.
- **Triglycerides** (**triacylglycerols**), the most abundant lipids, are triesters of glycerol and fatty acids.
- **Fatty acids** are long-chain carboxylic acids derived from the hydrolysis of animal fats, vegetable oils, and the phospholipids of biological membranes.
  - *Fatty acids are named by the number of carbons in the chain and the number of carbon-carbon double bonds present (e.g., 18:2).*

- *Nearly all fatty acids have an even number of carbon atoms, usually between 12 and 20.*
- *The carbon-carbon double bonds in naturally occurring fatty acids are almost exclusively found in the* cis *configuration.*
- *The melting point of a triglyceride increases as (1) the length of its hydrocarbon chains increases and (2) its degree of saturation increases (*cis *double bonds add kinks to the chains, limiting packing and decreasing melting points).*

- Triglycerides rich in saturated fatty acids are generally solids or semisolids at room temperature and are called **fats**.
- Triglycerides rich in unsaturated fatty acids are generally liquids at room temperature and are called **oils**.
  - *Oils are subjected to partial hydrogenation (limiting H_2 in the presence of a metal catalyst) in order to reduce some unsaturated fatty acids to saturated fatty acids, a process called* ***hardening*** *that converts oils into fats for the food industry.*
  - *An unintended consequence of hardening is that some triglyceride* cis *double bonds are isomerized to the more stable* trans *form. Triglycerides containing* trans *double bonds, referred to as* **trans *fats*** *in the media, significantly increase the risk of heart disease and are being limited in food production.*

P **26.1–26.10**

26.2 Soaps and Detergents

- **Soaps** are sodium or potassium salts of fatty acids.
  - *In water, soaps form* ***micelles****, which "dissolve" nonpolar organic grease and oil.*
  - *Natural soaps precipitate as water-insoluble salts with $Mg(II)$, $Ca(II)$, and $Fe(III)$ ions in hard water because of the strong interaction between carboxylates and these ions.*
  - *Synthetic detergents utilize sulfonates instead of carboxylates because metal ion salts of sulfonates are much more soluble in water.*
  - *The most common and most widely used synthetic detergents are linear alkylbenzenesulfonates.*

P **26.11–26.16**

Figure 26.3

26.3 Prostaglandins

- **Prostaglandins** are a group of extremely biologically active compounds having the 20-carbon skeleton of prostanoic acid.
  - *From phospholipid-bound arachidonic acid and other 20-carbon fatty acids, prostaglandins are synthesized in response to physiological triggers. Aspirin and other nonsteroidal anti-inflammatory drugs (NSAIDs) inhibit this conversion.*
- **Eicosanoids** are a very broad class of natural lipids that contain 20 carbons, are derived from fatty acids, and include the prostaglandins as well as leukotrienes, thromboxanes, and prostacyclins, all common and important biologically active molecules.

P **26.17–26.19**

26.4 Steroids

- **Steroids** are a group of plant and animal lipids that have a characteristic tetracyclic structure composed of 3 six-membered rings and 1 five-membered ring.
  - *The steroid skeleton is rigid and relatively flat, owing to a common* trans-*anti*-trans-*anti*-trans *orientation of ring fusions.*
- **Cholesterol** is an integral part of animal membranes, and it is the compound from which human sex hormones, adrenocorticoid hormones, bile acids, and vitamin D are biosynthesized.
  - ***Low-density lipoproteins (LDLs)*** *transport cholesterol from the site of its synthesis in the liver to tissues and cells where it is to be used.*
  - ***High-density lipoproteins (HDLs)*** *transport cholesterol from cells back to the liver for its degradation to bile acids and eventual excretion in the feces.*
- **Oral contraceptives** contain a synthetic progestin (e.g., norethindrone), which prevents ovulation yet allows women to maintain an otherwise normal menstrual cycle.
- A variety of synthetic **anabolic steroids** are available for use in rehabilitation medicine where muscle tissue has weakened or deteriorated as a result of an injury.
- **Bile acids**, used to aid digestion of fats, differ from most other steroids in that they have a *cis* configuration at the junction of rings A and B.
- The carbon skeleton of cholesterol and those of all biomolecules derived from it originate with the acetyl group (a two-carbon unit) of acetyl-CoA.

P **26.20–26.25**

26.5 Phospholipids

- **Phospholipids**, the second most abundant group of naturally occurring lipids, are derived from phosphatidic acids, compounds containing glycerol esterified with two molecules of fatty acid and a molecule of phosphoric acid.
  - *Further esterification of the phosphoric acid part with a low-molecular-weight alcohol, most commonly ethanolamine, choline, serine, or inositol, gives a phospholipid.*
- Phospholipids are the major component of biological membranes.
  - *When placed in aqueous solution, phospholipids spontaneously form* ***lipid bilayers****.*
  - *According to the* ***fluid-mosaic model****, membrane phospholipids form lipid bilayers with membrane proteins associated with the bilayer as both peripheral and integral proteins.*

P **26.26–26.30**

26.6 Fat-Soluble Vitamins

- **Vitamin A** occurs only in the animal world.
  - *The carotenes of the plant world are tetraterpenes (C_{40}) that are cleaved, after ingestion, into vitamin A.*

   – *The best-understood role of vitamin A is its participation in the visual cycle, where it is used to make the photoactive component of rhodopsin, the light-sensitive pigment in our eyes.*

- **Vitamin D** is the name for a structurally related set of molecules that play a major role in the regulation of calcium and phosphorus metabolism. Vitamin D_3 is synthesized in the skin of mammals by the action of ultraviolet radiation on 7-dehydrocholesterol.
- **Vitamin E** is a group of compounds of similar structure, the most active of which is α-tocopherol. In the body, vitamin E functions as an antioxidant by trapping peroxy radicals.

P **26.31–26.33**

Figure 26.14

Problems

Red numbers indicate applied problems.

Fatty Acids and Triglycerides

26.2 Define the term *hydrophobic*.

26.3 Identify the hydrophobic and hydrophilic region(s) of a triglyceride.

26.4 Explain why the melting points of unsaturated fatty acids are lower than those of saturated fatty acids.

26.5 Which would you expect to have the higher melting point, glyceryl trioleate or glyceryl trilinoleate?

26.6 Draw a structural formula for methyl linoleate. Make sure you show the correct configuration of groups about each carbon-carbon double bond.

26.7 Explain why coconut oil is a liquid triglyceride even though most of its fatty acid components are saturated.

26.8 It is common now to see "contains no tropical oils" on cooking oil labels, meaning that the oil contains no palm or coconut oil. What is the difference between the composition of tropical oils and that of vegetable oils, such as corn oil, soybean oil, and peanut oil?

26.9 What is meant by the term *hardening* as applied to vegetable oils?

26.10 How many moles of H_2 are used in the catalytic hydrogenation of one mole of a triglyceride derived from glycerol and equal portions of stearic acid, linoleic acid, and arachidonic acid?

26.11 Characterize the structural features necessary to make a good synthetic detergent.

26.12 Following are structural formulas for a cationic detergent and a neutral detergent. Account for the detergent properties of each.

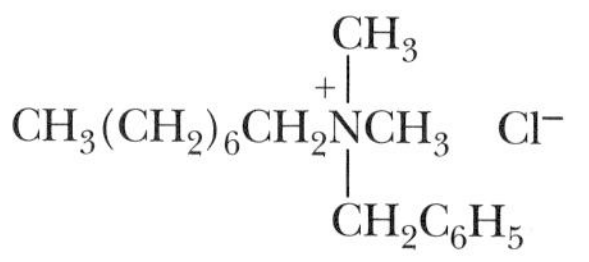

Benzyldimethyloctylammonium chloride
(a cationic detergent)

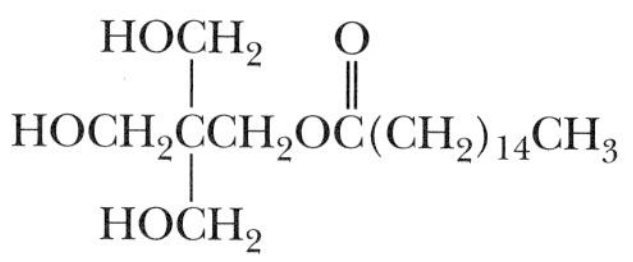

Pentaerythrityl palmitate
(a neutral detergent)

26.13 Identify some of the detergents used in shampoos and dishwashing liquids. Are they primarily anionic, neutral, or cationic detergents?

26.14 Show how to convert palmitic acid (hexadecanoic acid) into the following.

(a) Ethyl palmitate
(b) Palmitoyl chloride
(c) 1-Hexadecanol (cetyl alcohol)
(d) 1-Hexadecanamine
(e) *N,N*-Dimethylhexadecanamide

26.15 Palmitic acid (hexadecanoic acid) is the source of the hexadecyl (cetyl) group in the following compounds. Each is a mild surface-acting germicide and fungicide and is used as a topical antiseptic and disinfectant. They are examples of quaternary ammonium detergents, commonly called "quats."

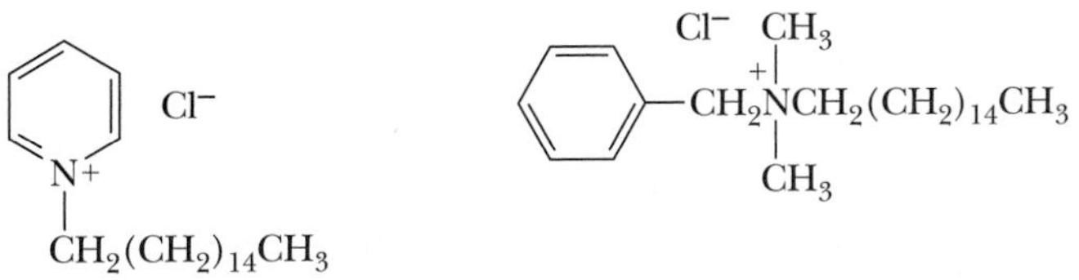
Cl⁻
N+
CH2(CH2)14CH3
Cl⁻ CH3
CH2NCH2(CH2)14CH3
CH3

Cetylpyridinium chloride **Benzylcetyldimethylammonium chloride**

(a) Cetylpyridinium chloride is prepared by treating pyridine with 1-chlorohexadecane (cetyl chloride). Show how to convert palmitic acid to cetyl chloride.
(b) Benzylcetyldimethylammonium chloride is prepared by treating benzyl chloride with *N,N*-dimethyl-1-hexadecanamine. Show how this tertiary amine can be prepared from palmitic acid.

26.16 Lipases are enzymes that catalyze the hydrolysis of esters, especially esters of glycerol. Because enzymes are chiral catalysts, they catalyze the hydrolysis of only one enantiomer of a racemic mixture. For example, porcine pancreatic lipase catalyzes the hydrolysis of only one enantiomer of the following racemic epoxyester. Calculate the number of grams of epoxyalcohol that can be obtained from 100 g of racemic epoxyester by this method.

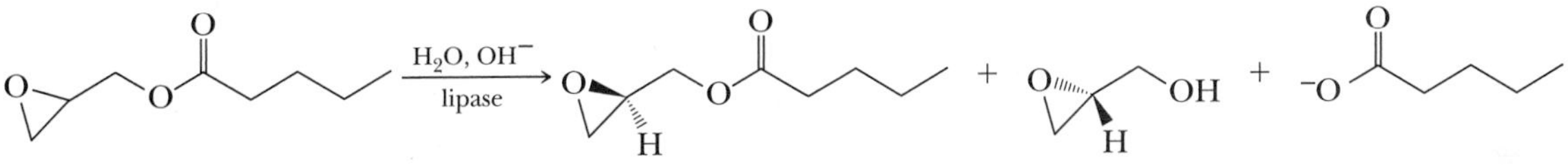
H2O, OH⁻
lipase
OH
H
H

A racemic mixture

This enantiomer is recovered unhydrolyzed

This epoxyalcohol is obtained in pure form

Prostaglandins

26.17 Examine the structure of $PGF_{2\alpha}$. Identify all chiral centers and all double bonds about which *cis,trans* isomerism is possible and state the number of stereoisomers possible for a molecule of this structure.

26.18 Following is the structure of unoprostone, a compound patterned after the natural prostaglandins (Section 26.3). Rescula, the isopropyl ester of unoprostone, is an antiglaucoma drug used to treat ocular hypertension. Compare the structural formula of this synthetic prostaglandin with that of $PGF_{2\alpha}$.

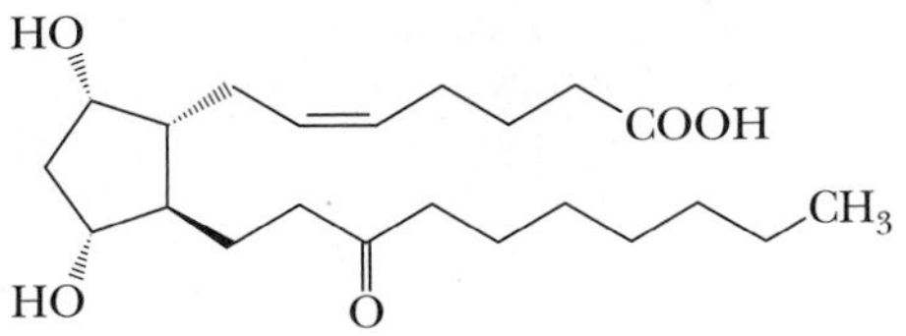
HO
COOH
CH3
HO
O

Unoprostone
(antiglaucoma)

26.19 Doxaprost, an orally active bronchodilator patterned after the natural prostaglandins (Section 26.3), is synthesized in the following series of reactions starting with ethyl 2-oxocyclopentanecarboxylate. Except for the Nef reaction in Step 8, we have seen examples of all other types of reactions involved in this synthesis.

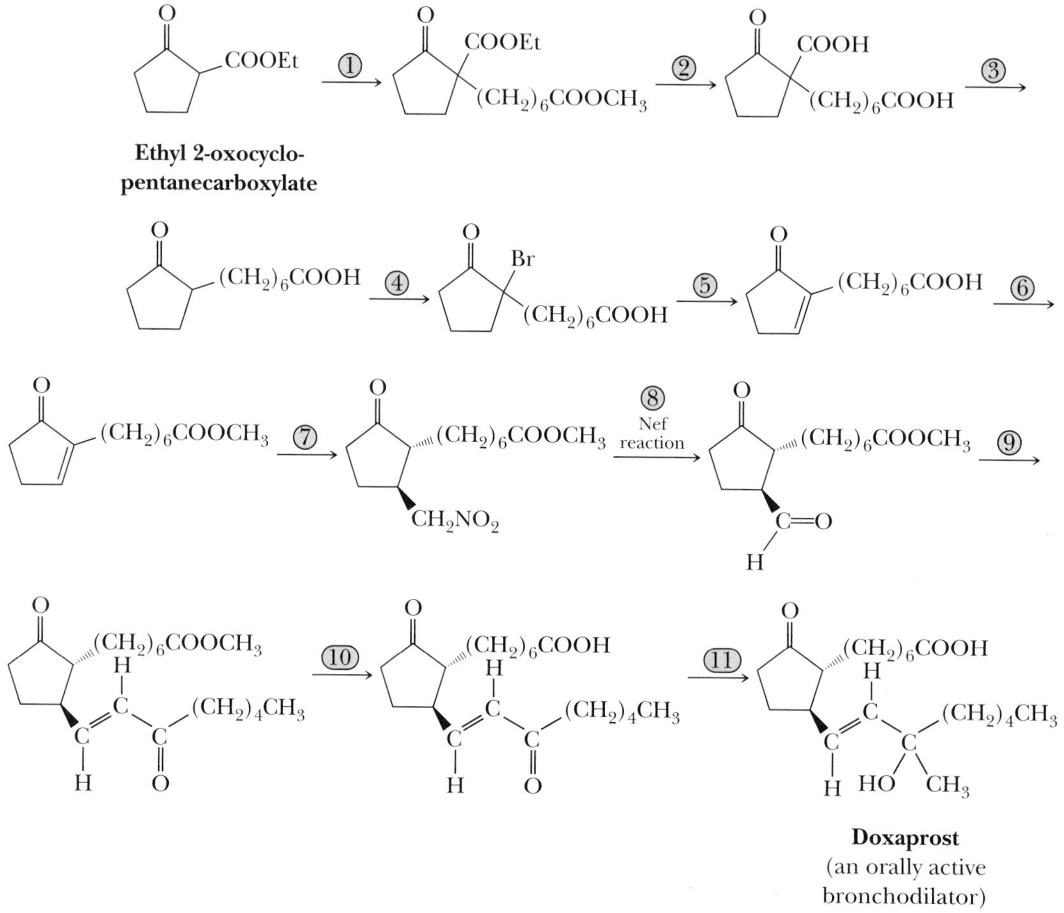

(a) Propose a set of experimental conditions to bring about the alkylation in Step 1. Account for the regioselectivity of the alkylation (i.e., that it takes place on the carbon between the two carbonyl groups rather than on the other side of the ketone carbonyl).

(b) Propose experimental conditions to bring about Steps 2 and 3.

(c) Propose experimental conditions for bromination of the ring in Step 4 and dehydrobromination in Step 5.

(d) Write equations to show that Step 6 can be brought about using either methanol or diazomethane (CH_2N_2) as a source of the —CH_3 in the methyl ester.

(e) Describe experimental conditions to bring about Step 7 and account for the fact that the *trans* isomer is formed in this step.

(f) Step 9 is done by a Wittig reaction. Suggest a structural formula for a Wittig reagent that gives the product shown.

(g) Name the type of reaction involved in Step 10.

(h) Step 11 can best be described as a Grignard reaction with methylmagnesium bromide under very carefully controlled conditions. In addition to the observed reaction, what other Grignard reactions might take place in Step 11?

(i) Assuming that the two side chains on the cyclopentanone ring are *trans*, how many stereoisomers are possible from this synthetic sequence?

Steroids

26.20 Draw the structural formula for the product formed by treatment of cholesterol with H_2/Pd; then do the same for Br_2.

26.21 Both low-density lipoproteins (LDLs) and high-density lipoproteins (HDLs) consist of a core of triacylglycerols and cholesterol esters surrounded by a single phospholipid layer. Draw the structural formula of cholesteryl linoleate, one of the cholesterol esters found in this core.

26.22 Examine the structural formulas of testosterone (a male sex hormone) and progesterone (a female sex hormone). What are the similarities in structure between the two? What are the differences?

26.23 Examine the model of cholic acid (Problem 2.65) and account for the ability of this and other bile salts to emulsify fats and oils and thus aid in their digestion.

26.24 Following is a structural formula for cortisol (hydrocortisone). Draw a stereorepresentation of this molecule showing the conformations of the five- and six-membered rings.

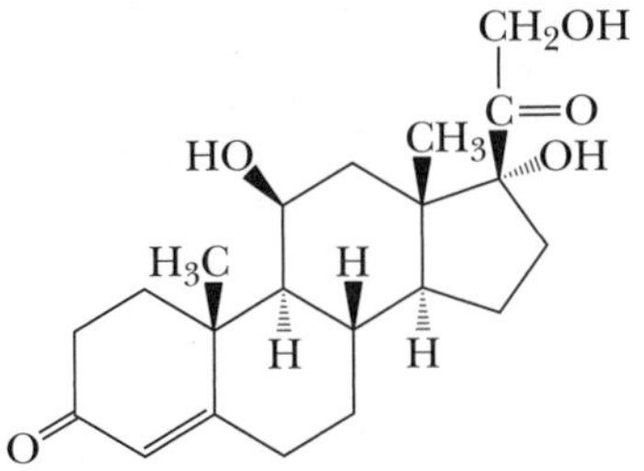

Cortisol
(Hydrocortisone)

26.25 Much of our understanding of conformational analysis has arisen from studies on the reactions of rigid steroid nuclei. For example, the concept of *trans*-diaxial ring opening of epoxycyclohexanes was proposed to explain the stereoselective reactions seen with steroidal epoxides. Predict the product when each of the following steroidal epoxides is treated with $LiAlH_4$.

(a)
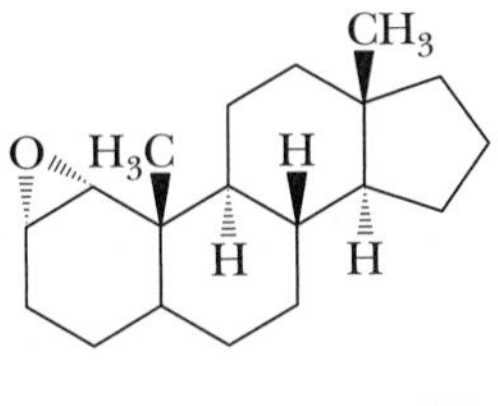

(b)
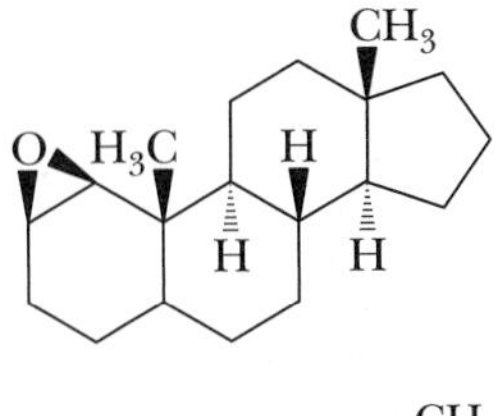

(c)
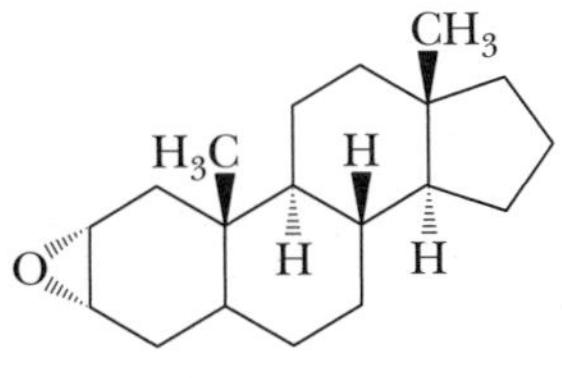

(d)
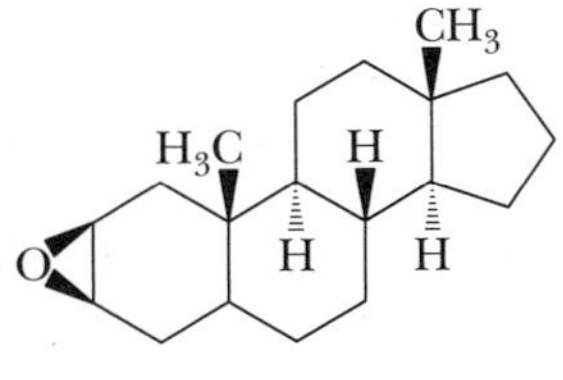

Phospholipids

26.26 Draw the structural formula of a lecithin containing one molecule each of palmitic acid and linoleic acid.

26.27 Identify the hydrophobic and hydrophilic region(s) of a phospholipid.

26.28 The hydrophobic effect is one of the most important noncovalent forces directing the self-assembly of biomolecules in aqueous solution. The hydrophobic effect arises from tendencies of biomolecules (1) to arrange polar groups so that they interact with the aqueous environment by hydrogen bonding and (2) to arrange nonpolar groups so that they are shielded from the aqueous environment. Show how the hydrophobic effect is involved in directing the following.

(a) The formation of micelles by soaps and detergents
(b) The formation of lipid bilayers by phospholipids

26.29 How does the presence of unsaturated fatty acids contribute to the fluidity of biological membranes?

26.30 Lecithins can act as emulsifying agents. The lecithin of egg yolk, for example, is used to make mayonnaise. Identify the hydrophobic part(s) and the hydrophilic part(s) of a lecithin. Which parts interact with the oils used in making mayonnaise? Which parts interact with the water?

Fat-Soluble Vitamins

26.31 Examine the structural formula of vitamin A and state the number of *cis,trans* isomers possible for this molecule.

26.32 The form of vitamin A present in many food supplements is vitamin A palmitate. Draw the structural formula of this molecule.

26.33 Examine the structural formulas of vitamin A, 1,25-dihydroxyvitamin-D_3, vitamin E, and vitamin K_1 (Section 26.6). Do you expect them to be more soluble in water or in dichloromethane? Do you expect them to be soluble in blood plasma?

Spider silk is a fibrous protein that exhibits unmatched strength and toughness. **Inset:** alanine and glycine, major components of the fibrous protein of spider silk. See "Chemical Connections: Spider Silk." (© Olga Alyonkina/Shutterstock.com)

27 Amino Acids and Proteins

Outline

27.1 Amino Acids
27.2 Acid-Base Properties of Amino Acids
27.3 Polypeptides and Proteins
27.4 Primary Structure of Polypeptides and Proteins
27.5 Synthesis of Polypeptides
27.6 Three-Dimensional Shapes of Polypeptides and Proteins

We begin this chapter with a study of amino acids, compounds whose chemistry is built on amines (Chapter 23) and carboxylic acids (Chapter 17). We concentrate in particular on the acid-base properties of amino acids because these properties are so important in determining many of the properties of proteins, including the catalytic functions of enzymes. With this understanding of the chemistry of amino acids, we then examine the structure of proteins themselves.

27.1 Amino Acids

Amino acid
A compound that contains both an amino group and a carboxyl group.

α-Amino acid
An amino acid in which the amino group is on the carbon adjacent to the carboxyl group.

A. Structure

An **amino acid** is a compound that contains both a carboxyl group and an **amino group**. Although many types of amino acids are known, the **α-amino acids** are the most significant in the biological world because they are the monomers from which proteins are constructed. A general structural formula of an α-amino acid is shown in Figure 27.1.

Zwitterion

An internal salt of an amino acid; the carboxylate anion is negatively charged, and the ammonium group is positively charged.

Although Figure 27.1(a) is a common way of writing structural formulas for amino acids, it is not accurate because it shows an acid (—COOH) and a base (—NH_2) within the same molecule. These acidic and basic groups react with each other to form a dipolar ion or internal salt [Figure 27.1(b)]. The internal salt of an amino acid is given the special name **zwitterion**. Note that a zwitterion has no net charge; it contains one positive charge and one negative charge.

Because they exist as zwitterions, amino acids have many of the properties associated with salts. They are crystalline solids with high melting points and are fairly soluble in water but insoluble in nonpolar organic solvents such as ether and hydrocarbon solvents.

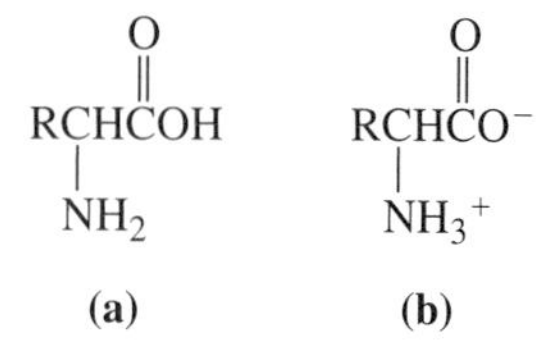

FIGURE 27.1 An α-amino acid. (a) Unionized representation and (b) the more accurate internal salt (zwitterion) form.

B. Chirality

To review, with the exception of glycine, H_2NCH_2COOH, all protein-derived amino acids have at least one chiral center and therefore are chiral. Figure 27.2 shows Fischer projection formulas for the enantiomers of alanine. The vast majority of carbohydrates in the biological world are of the D-series, whereas the vast majority of α-amino acids in the biological world are of the L-series. As we saw for carbohydrates, the L designation for amino acids is based on analogy to the chiral center of glyceraldehyde. Note the similarity of the L-alanine and L-glyceraldehyde chiral centers in Figure 27.2. Some students assume that the L designation is based on optical activity and that samples of all L-amino acids must rotate the plane of plane-polarized light in the levorotatory (−) direction. This is not true. In fact, many samples of L-amino acids rotate the plane of plane-polarized light in the dextrorotatory (+) direction. The L designation of amino acids refers only to *structural* similarity to the reference compound, L-glyceraldehyde, not any experimentally measured parameter.

The alternative *R,S* convention is also used to specify the configurations of amino acids. According to this convention, L-alanine is designated (*S*)-alanine. Although the D,L method of designating the stereochemistry of amino acid chiral centers is awkward, the historical use of this system is so deeply rooted in the scientific literature that it is still most commonly used when referring to amino acids as well as carbohydrates. We will therefore use the D,L convention throughout the remainder of this chapter.

C. Protein-Derived Amino Acids

Table 27.1 gives common names, structural formulas, and standard three-letter and one-letter abbreviations for the 20 common L-amino acids found in proteins. The amino acids in this table are divided into four categories: those with nonpolar side chains, polar but unionized side chains, acidic side chains, and basic side chains. The following structural features of these amino acids should be noted.

1. All 20 of these protein-derived amino acids are α-amino acids, meaning that the amino group is located on the carbon alpha to the carboxyl group.
2. For 19 of the 20 amino acids, the α-amino group is primary. Proline is different; its α-amino group is secondary.
3. With the exception of glycine, the α-carbon of each amino acid is a chiral center. As shown in Table 27.1, all 19 chiral amino acids have the same relative

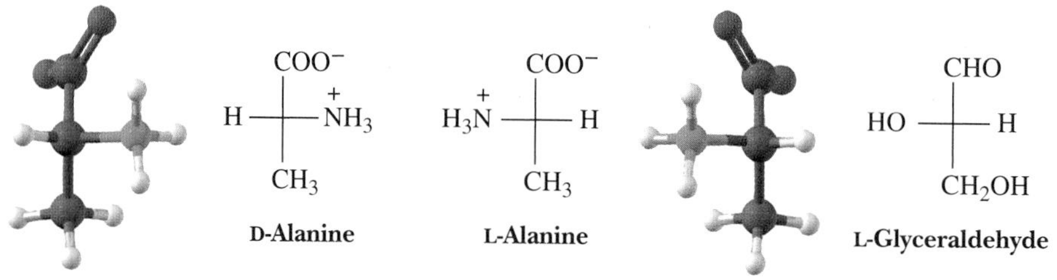

FIGURE 27.2 The enantiomers of alanine. The vast majority of α-amino acids in the biological world have the L configuration at the α-carbon. The structure of L-glyceraldehyde is added for comparison.

Table 27.1 The Common Amino Acids Found in Proteins

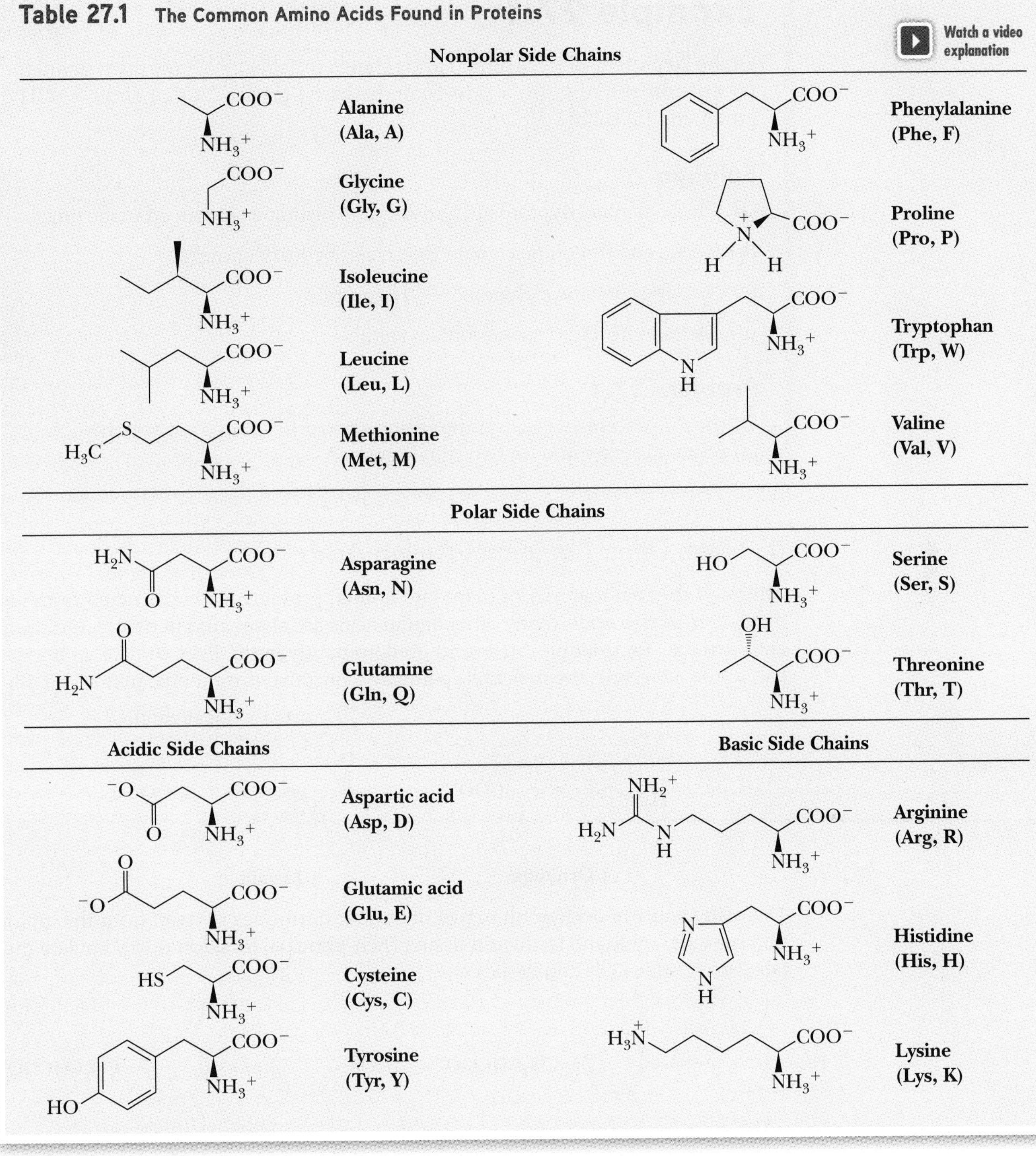
Watch a video explanation
Nonpolar Side Chains
Alanine (Ala, A)
Glycine (Gly, G)
Isoleucine (Ile, I)
Leucine (Leu, L)
Methionine (Met, M)
Phenylalanine (Phe, F)
Proline (Pro, P)
Tryptophan (Trp, W)
Valine (Val, V)
Polar Side Chains
Asparagine (Asn, N)
Glutamine (Gln, Q)
Serine (Ser, S)
Threonine (Thr, T)
Acidic Side Chains
Aspartic acid (Asp, D)
Glutamic acid (Glu, E)
Cysteine (Cys, C)
Tyrosine (Tyr, Y)
Basic Side Chains
Arginine (Arg, R)
Histidine (His, H)
Lysine (Lys, K)

* Each ionizable group is shown in the form present in highest concentration at pH 7.0.

configuration at the α-carbon. In the D,L convention, all are L-amino acids. According to the *R*,*S* convention, amino acid α-carbons, with the exception of cysteine, have the *S* configuration. Because of priority rules, the presence of the sulfhydryl group on the side chain of L-cysteine gives the chiral center the *R* configuration.

4. Isoleucine and threonine contain a second chiral center. Four stereoisomers are possible for each amino acid, but only one is found in proteins.
5. The sulfhydryl group of cysteine, the imidazole group of histidine, and the phenolic hydroxyl of tyrosine are partially ionized at pH 7.0, but the ionic form is not the major form present at this pH.

Example 27.1 Amino Acids

Of the 20 protein-derived amino acids shown in Table 27.1, how many contain (a) an aromatic ring, (b) a side-chain hydroxyl group, (c) a phenolic —OH group, and (d) sulfur?

Solution

(a) Phenylalanine, tryptophan, tyrosine, and histidine contain aromatic rings.

(b) Serine and threonine contain side-chain hydroxyl groups.

(c) Tyrosine contains a phenolic —OH group.

(d) Methionine and cysteine contain sulfur.

Problem 27.1

Of the 20 protein-derived amino acids shown in Table 27.1, which contain (a) no chiral center and (b) two chiral centers?

D. Some Other Common L-Amino Acids

Although the vast majority of plant and animal proteins are constructed from just these 20 α-amino acids, many other amino acids are also found in nature. Ornithine and citrulline, for example, are found predominantly in the liver and are an integral part of the urea cycle, the metabolic pathway that converts ammonia to urea.

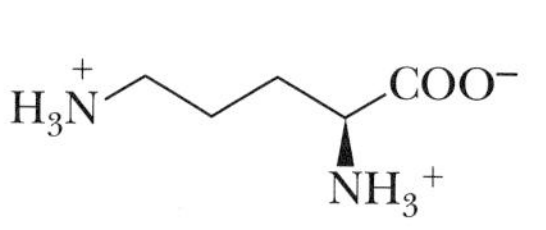

L-Ornithine

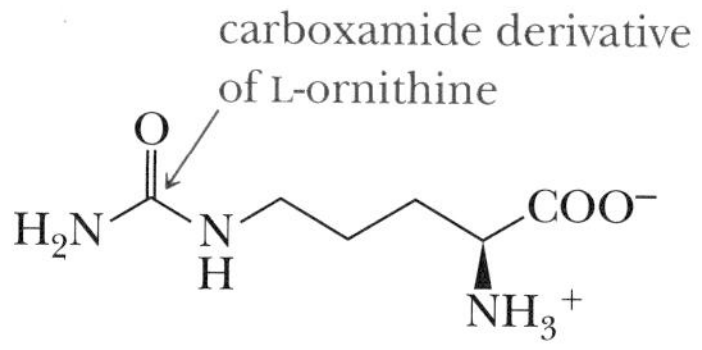

L-Citrulline

Thyroxine and triiodothyronine, two of several hormones derived from the amino acid tyrosine, are found in thyroid tissue. Their principal function is to stimulate metabolism in other cells and tissues.

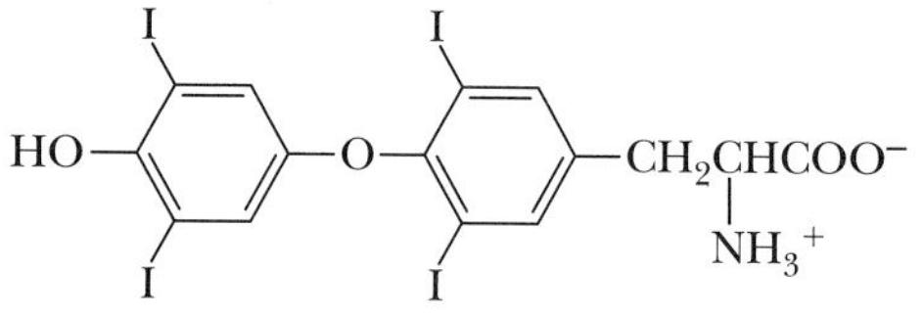

L-Thyroxine, T_4

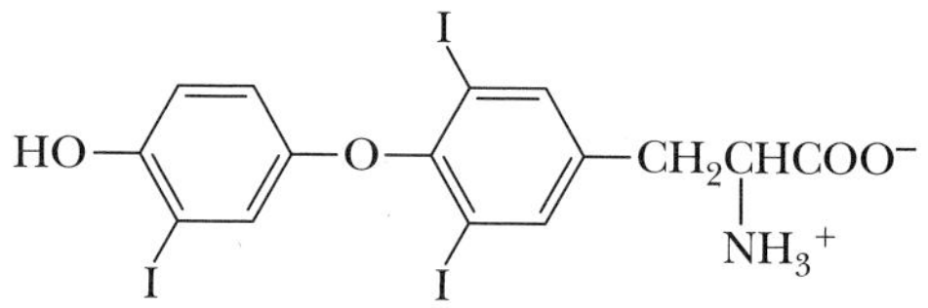

L-Triiodothyronine, T_3

4-Aminobutanoic acid is found in high concentration (0.8 m*M*) in the brain but in no significant amounts in any other mammalian tissue. It is synthesized in neural tissue by decarboxylation of the α-carboxyl group of glutamic acid and is a neurotransmitter in the central nervous system of invertebrates and humans.

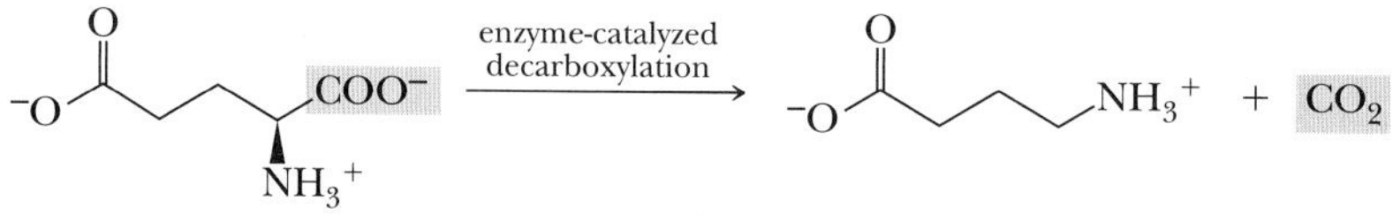

Glutamic acid

4-Aminobutanoic acid
(γ-Aminobutyric acid, GABA)

Only L-amino acids are found in proteins, and only rarely are D-amino acids a part of the metabolism of higher organisms. Several D-amino acids, however, along with their L-enantiomers, are found in lower forms of life. D-Alanine and D-glutamic acid, for example, are structural components of the cell walls of certain bacteria. Several D-amino acids are also found in peptide antibiotics.

27.2 Acid-Base Properties of Amino Acids

A. Acidic and Basic Groups of Amino Acids

Among the most important chemical properties of amino acids are their acid-base properties; all are weak polyprotic acids because of their —COOH and $—NH_3^+$ groups. Given in Table 27.2 are pK_a values for each ionizable group of the 20 protein-derived amino acids.

Acidity of α-Carboxyl Groups

The average value of pK_a for an α-carboxyl group of a protonated amino acid is 2.19. Thus, the α-carboxyl group is a considerably stronger acid than acetic acid (pK_a 4.76) and other low-molecular-weight aliphatic carboxylic acids. This greater acidity is accounted for by the electron-withdrawing inductive effect of the adjacent $—NH_3^+$ group.

Table 27.2 pK_a **Values for Ionizable Groups of Amino Acids**

Amino Acid	pK_a of α-COOH	pK_a of $\alpha\text{-}NH_3^+$	pK_a of Side Chain	Isoelectric Point (pI)
Alanine	2.35	9.87	—	6.11
Arginine	2.01	9.04	12.48	10.76
Asparagine	2.02	8.80	—	5.41
Aspartic acid	2.10	9.82	3.86	2.98
Cysteine	2.05	10.25	8.00	5.02
Glutamic acid	2.10	9.47	4.07	3.08
Glutamine	2.17	9.13	—	5.65
Glycine	2.35	9.78	—	6.06
Histidine	1.77	9.18	6.10	7.64
Isoleucine	2.32	9.76	—	6.04
Leucine	2.33	9.74	—	6.04
Lysine	2.18	8.95	10.53	9.74
Methionine	2.28	9.21	—	5.74
Phenylalanine	2.58	9.24	—	5.91
Proline	2.00	10.60	—	6.30
Serine	2.21	9.15	—	5.68
Threonine	2.09	9.10	—	5.60
Tryptophan	2.38	9.39	—	5.88
Tyrosine	2.20	9.11	10.07	5.63
Valine	2.29	9.72	—	6.00

Recall that we used similar reasoning in Section 17.4A to account for the relative acidities of acetic acid and its mono-, di-, and trichloroderivatives.

The ammonium group has an electron-withdrawing inductive effect

$$\underset{\displaystyle NH_3^+}{RCHCOOH} + H_2O \rightleftharpoons \underset{\displaystyle NH_3^+}{RCHCOO^-} + H_3O^+ \qquad pK_a = 2.19$$

Acidity of Side-Chain Carboxyl Groups

Owing to the electron-withdrawing inductive effect of the α-NH_3^+ group, the side-chain carboxyl groups of protonated aspartic and glutamic acids are also stronger acids than acetic acid (pK_a 4.76). Notice that this acid-strengthening inductive effect decreases with increasing distance of the —COOH from the —NH_3^+. Compare the acidities of the α-COOH of alanine (pK_a 2.35), the β-COOH of aspartic acid (pK_a 3.86), and the γ-COOH of glutamic acid (pK_a 4.07).

Acidity of α-Ammonium Groups

The average value of pK_a for an α-ammonium group, —NH_3^+, is 9.47 compared with an average value of 10.60 for primary aliphatic ammonium ions (Section 23.5A). Thus, the α-ammonium group of an amino acid is a slightly stronger acid than a primary aliphatic ammonium ion. Conversely, an α-amino group is a slightly weaker base than a primary aliphatic amine.

$$\underset{\displaystyle NH_3^+}{RCHCOO^-} + H_2O \rightleftharpoons \underset{\displaystyle NH_2}{RCHCOO^-} + H_3O^+ \qquad pK_a = 9.47$$

$$\underset{\displaystyle NH_3^+}{CH_3CHCH_3} + H_2O \rightleftharpoons \underset{\displaystyle NH_2}{CH_3CHCH_3} + H_3O^+ \qquad pK_a = 10.60$$

Basicity of the Guanidine Group of Arginine

The side-chain guanidine group of arginine is a considerably stronger base than an aliphatic amine. As we saw in Section 23.5D, guanidine (pK_a of its conjugate acid is 13.6) is a very strong base for a neutral organic compound. The remarkable basicity of the guanidine group of arginine is attributed to the large resonance stabilization of the protonated form relative to the neutral form.

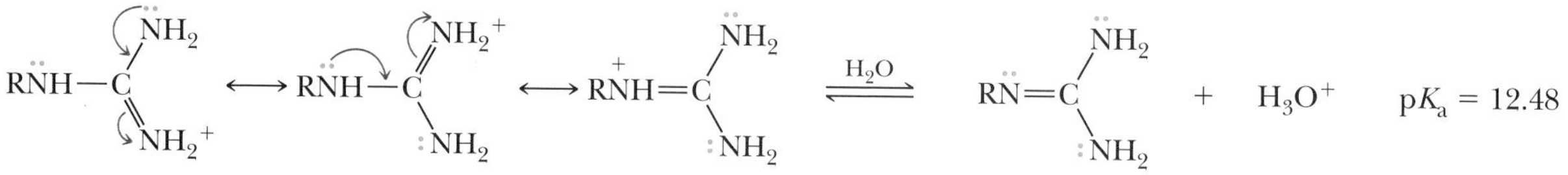

The guanidinium ion side chain of arginine is a hybrid of three contributing structures

No resonance stabilization without charge separation

▶ Delocalization of charge over a larger area is stabilizing.

Basicity of the Imidazole Group of Histidine

Because the imidazole group on the side chain of histidine contains six π electrons in a planar, fully conjugated ring, imidazole is classified as a heterocyclic aromatic amine (Section 21.2D). The unshared pair of electrons on one nitrogen is a part of the aromatic sextet, whereas that on the other nitrogen is not. The pair of electrons that is not part of the aromatic sextet is responsible for the basic properties of the imidazole ring. Protonation of this nitrogen produces a resonance-stabilized cation.

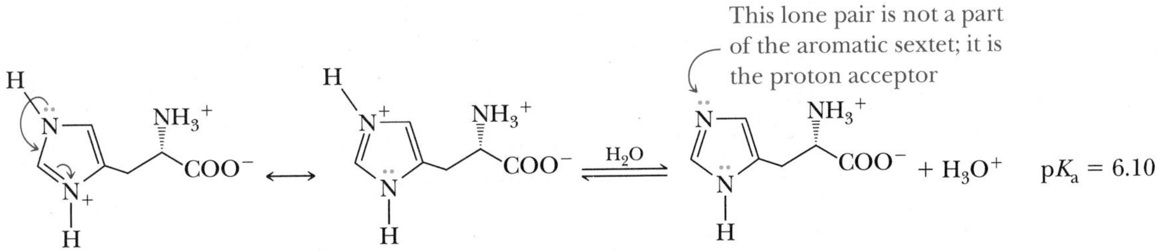

Resonance-stabilized imidazolium cation

B. Titration of Amino Acids

Values of pK_a for the ionizable groups of amino acids are most commonly obtained by acid-base titration and through measurement of the pH of the solution as a function of added base (or added acid, depending on how the titration is done). As an illustration of this experimental procedure, consider a solution containing 1.00 mol of glycine to which has been added enough strong acid that both the amino and carboxyl groups are fully protonated. Next, this solution is titrated with 1.00 *M* NaOH; the volume of base added and the pH of the resulting solution are recorded and then plotted as shown in Figure 27.3.

The most acidic group and the one to react first with added sodium hydroxide is the carboxyl group. When exactly 0.50 mol of NaOH has been added, the carboxyl group is half neutralized. At this point, the concentration of the zwitterion equals that of the positively charged ion and the pH of 2.35 equals the pK_a of the carboxyl group (pK_{a1}).

$$\text{At pH} = pK_{a1} \qquad [H_3\overset{+}{N}CH_2COOH] = [H_3\overset{+}{N}CH_2COO^-]$$

Positive ion Zwitterion

The end point of the first part of the titration is reached when 1.00 mol of NaOH has been added. At this point, the predominant species present is the zwitterion and the observed pH of the solution is 6.06.

The next section of the curve represents titration of the $—NH_3^+$ group. When another 0.50 mole of NaOH has been added (bringing the total to 1.50 mol), half of the $—NH_3^+$ groups are neutralized and converted to $—NH_2$. At this point, the concentrations of the zwitterion and negatively charged ion are equal and the observed pH is 9.78, the pK_a of the amino group of glycine (pK_{a2}).

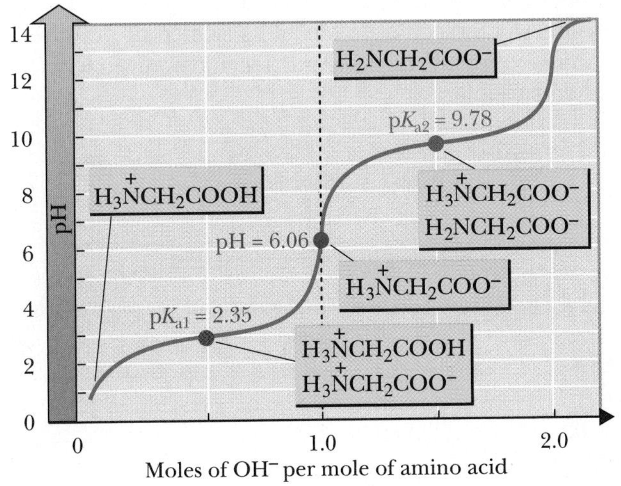

FIGURE 27.3 Titration of glycine with sodium hydroxide.

$$\text{At pH} = pK_{a2} \qquad [H_3\overset{+}{N}CH_2COO^-] = [H_2NCH_2COO^-]$$

Zwitterion Negative ion

The second end point of the titration is reached when a total of 2.00 mol of NaOH have been added and glycine is converted entirely to an anion.

C. Isoelectric Point

Isoelectric point (pI)
The pH at which an amino acid, a polypeptide, or a protein has no net charge.

A titration curve such as that for glycine permits us to determine pK_a values for the ionizable groups of an amino acid. It also permits us to determine another important property: isoelectric point. **Isoelectric point, pI,** for an amino acid is the pH at which the molecules in solution have a net charge of zero (they are zwitterions). By examining the titration curve, you can see that the isoelectric point for glycine falls halfway between the pK_a values for the carboxyl and amino groups.

$$pI = \frac{1}{2}(pK_a\ \alpha\text{-COOH} + pK_a\ \alpha\text{-NH}_3^+)$$
$$= \frac{1}{2}(2.35 + 9.78) = 6.06$$

At pH 6.06, the predominant form of glycine molecules is the dipolar ion; furthermore, at this pH, the concentration of positively charged glycine molecules equals the concentration of negatively charged glycine molecules.

Given a value for the isoelectric point of an amino acid, it is possible to estimate the charge on that amino acid at any pH. For example, the charge on tyrosine at pH 5.63, its isoelectric point, is zero. A small fraction of tyrosine molecules are positively charged at pH 5.00 (0.63 unit less than its pI), and virtually all are positively charged at pH 3.63 (2.00 units less than its pI). As another example, the net charge on lysine is zero at pH 9.74. At pH values smaller than 9.74, an increasing fraction of lysine molecules are positively charged.

D. Electrophoresis

Electrophoresis
The process of separating compounds on the basis of their electric charge.

Electrophoresis, a process of separating compounds on the basis of their total charge, is used to separate and identify mixtures of amino acids and proteins. Electrophoretic separations can be carried out using paper, starch, polyacrylamide and agarose gels, in addition to cellulose acetate as solid supports. In paper electrophoresis, a paper strip saturated with an aqueous buffer of predetermined pH serves as a bridge between two electrode vessels (Figure 27.4). Next, a sample of amino acids is applied as a spot on the paper strip. When an electrical potential is then applied to the electrode vessels, amino acids migrate toward the electrode carrying the charge opposite their own. Molecules having a high charge density move more rapidly than those with a lower charge density. Any molecule already at its isoelectric point remains at the origin. After separation is complete, the strip is dried and sprayed with a dye to make the separated components visible.

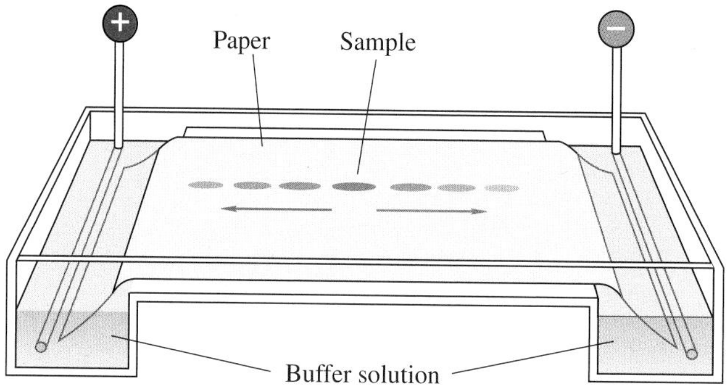

FIGURE 27.4 An apparatus for electrophoresis of a mixture of amino acids. Those with a negative charge move toward the positive electrode; those with a positive charge move toward the negative electrode; those with no charge remain at the origin.

A dye commonly used to detect amino acids is ninhydrin (1,2,3-indanetrione monohydrate). Ninhydrin reacts with α-amino acids to produce an aldehyde, carbon dioxide, and a purple-colored anion. This reaction is commonly used in both qualitative and quantitative analysis of amino acids.

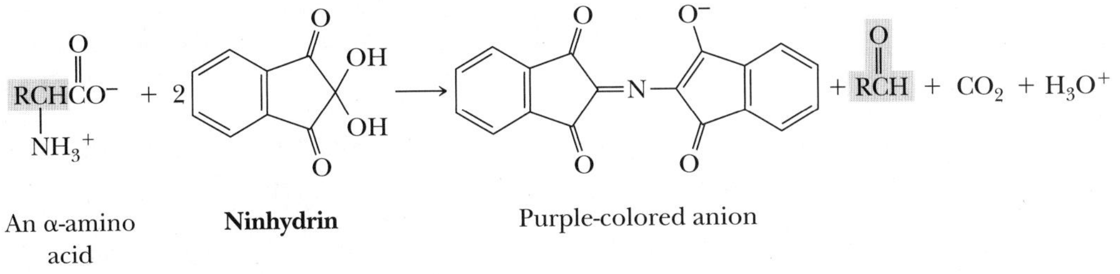

Nineteen of the 20 protein-derived α-amino acids have primary amino groups and give the same purple-colored ninhydrin-derived anion. Proline, a secondary amine, gives a different orange-colored compound.

Example 27.2 Isoelectric Points

The isoelectric point of tyrosine is 5.63. Toward which electrode does tyrosine migrate on paper electrophoresis at pH 7.0?

Solution

On paper electrophoresis at pH 7.0 (more basic than its isoelectric point), tyrosine has a net negative charge and migrates toward the positive electrode.

Problem 27.2

The isoelectric point of histidine is 7.64. Toward which electrode does histidine migrate on paper electrophoresis at pH 7.0?

Example 27.3 Electrophoresis

Electrophoresis of a mixture of lysine, histidine, and cysteine is carried out at pH 7.64. Describe the behavior of each amino acid under these conditions.

Solution

The isoelectric point of histidine is 7.64. At this pH, histidine has a net charge of zero and does not move from the origin. The pI of cysteine is 5.02; at pH 7.64 (more basic than its isoelectric point), cysteine has a net negative charge and moves toward the positive electrode. The pI of lysine is 9.74; at pH 7.64 (more acidic than its isoelectric point), lysine has a net positive charge and moves toward the negative electrode.

Problem 27.3

Describe the behavior of a mixture of glutamic acid, arginine, and valine on paper electrophoresis at pH 6.0.

27.3 Polypeptides and Proteins

Peptide bond

The special name given to the amide bond formed between the α-amino group of one amino acid and the α-carboxyl group of another amino acid.

In 1902, Emil Fischer proposed that proteins are long chains of amino acids joined by amide bonds between the α-carboxyl group of one amino acid and the α-amino group of another. For these amide bonds, Fischer proposed the special name **peptide bond**. Figure 27.5 shows the peptide bond formed between serine and alanine in the dipeptide serylalanine.

FIGURE 27.5 The peptide bond in serylalanine.

Watch a video explanation

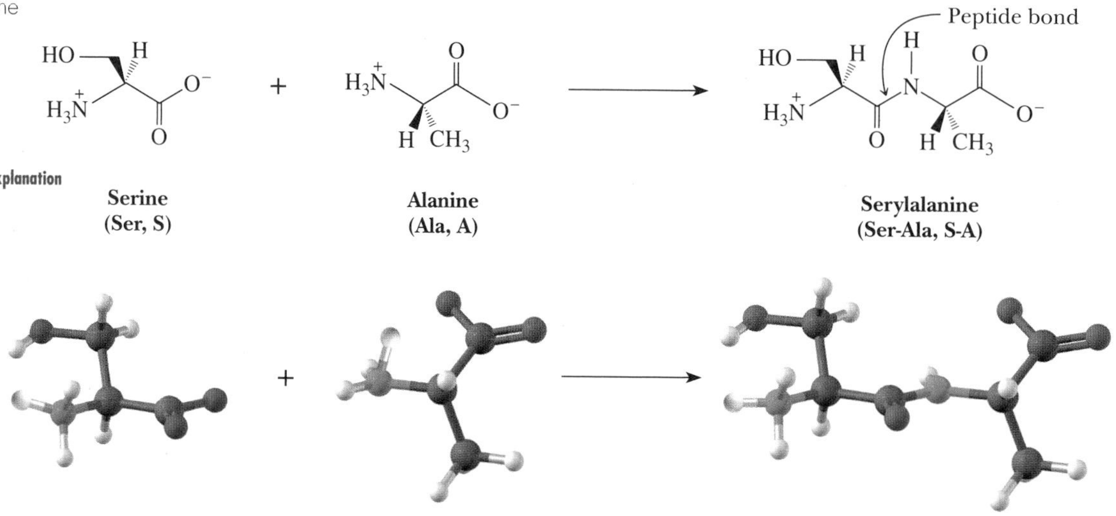

Dipeptide

A molecule containing two amino acid units joined by a peptide bond.

Tripeptide

A molecule containing three amino acid units, each joined to the next by a peptide bond.

Polypeptide

A macromolecule containing many amino acid units, each joined to the next by a peptide bond.

Proteins

Biological macromolecules consisting of one or more polypeptides and having an overall molecular weight of greater than 5000.

***N*-Terminal amino acid**

The amino acid at the end of a polypeptide chain having the free $—NH_3^+$ group.

***C*-Terminal amino acid**

The amino acid at the end of a polypeptide chain having the free $—COO^-$ group.

Peptide is the name given to a short polymer of amino acids. Peptides are classified by the number of amino acid units in the chain. A molecule containing two amino acids joined by an amide bond is called a **dipeptide**. Those containing 3 to 10 amino acids are called **tripeptides**, **tetrapeptides**, **pentapeptides**, and so on. Molecules containing more than 10 but fewer than 20 amino acids are called **oligopeptides**. Those containing several dozen or more amino acids are called **polypeptides**. **Proteins** are biological macromolecules of molecular weight 5000 or greater consisting of one or more polypeptide chains. The distinctions in this terminology are not precise.

By convention, polypeptides are written from the left beginning with the amino acid having the free $—NH_3^+$ group and proceeding to the right toward the amino acid with the free $—COO^-$ group. The amino acid with the free $—NH_3^+$ group is called the ***N*-terminal amino acid**, and the amino acid with the free $—COO^-$ group is called the ***C*-terminal amino acid**. Notice the repeating pattern in the peptide chain of N-α-carbon-carbonyl, and so on.

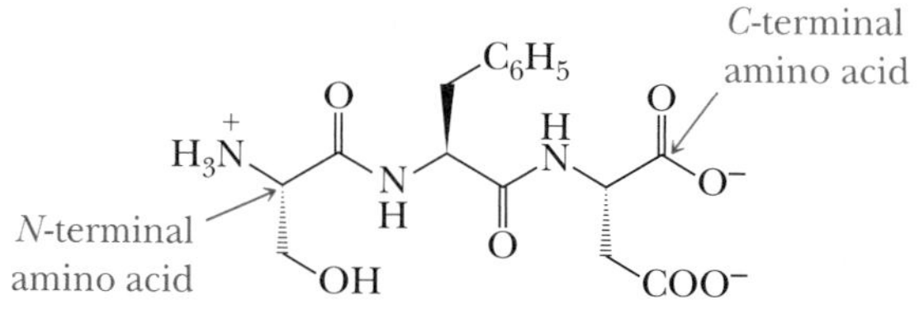

Ser-Phe-Asp

Example 27.4 Peptides

Draw a structural formula for Cys-Arg-Met-Asn. Label the *N*-terminal amino acid and the *C*-terminal amino acid. What is the net charge on this tetrapeptide at pH 6.0?

Solution

The backbone of this tetrapeptide is a repeating sequence of nitrogen-α-carbon-carbonyl. The net charge on this tetrapeptide at pH 6.0 is +1.

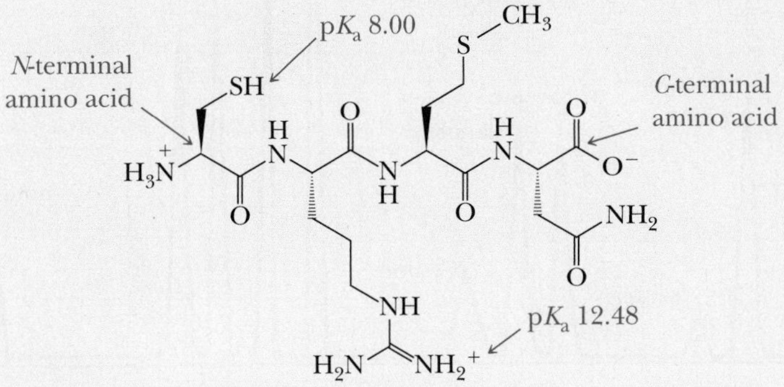

Problem 27.4

Draw a structural formula for Lys-Phe-Ala. Label the *N*-terminal amino acid and the *C*-terminal amino acid. What is the net charge on this tripeptide at pH 6.0?

Watch a video explanation

27.4 Primary Structure of Polypeptides and Proteins

The **primary (1°) structure** of a polypeptide or protein refers to the sequence of amino acids in its polypeptide chain. In this sense, primary structure is a nearly complete description of all covalent bonding in a polypeptide or protein.

Primary structure of proteins The sequence of amino acids in the polypeptide chain, read from the *N*-terminal amino acid to the *C*-terminal amino acid.

In 1953, Frederick Sanger of Cambridge University, England, reported the primary structure of the two polypeptide chains of the hormone insulin. Not only was this a remarkable achievement in analytical chemistry, but it also clearly established the fact that the molecules of a given protein all have the same amino acid composition and the same amino acid sequence.

A. Amino Acid Analysis

A method for determining the amino acid composition of a polypeptide is hydrolysis and quantitative analysis. Recall from Section 18.4D that amide bonds are very resistant to hydrolysis. Typically, samples of protein are hydrolyzed in 6 *M* HCl in sealed glass vials at 110°C for 24 to 72 hours. This hydrolysis can be done in a microwave oven in a shorter time. After the polypeptide is hydrolyzed, the resulting mixture of amino acids is analyzed by ion-exchange chromatography. Amino acids are detected as they emerge from the column by reaction with ninhydrin (Section 27.2D) followed by absorption spectroscopy. Current procedures for hydrolysis of polypeptides and analysis of amino acid mixtures have been refined to the point that it is possible to obtain amino acid composition from as little as 50 nanomoles (50×10^{-9} mole) of polypeptide. Figure 27.6 shows the analysis of a polypeptide hydrolysate by ion-exchange chromatography. Note that during hydrolysis, the side-chain amide groups of asparagine and glutamine are hydrolyzed and these amino acids are detected as aspartic acid and glutamic acid. For each glutamine or asparagine hydrolyzed, an equivalent amount of ammonium chloride is formed.

B. Sequence Analysis

It is possible to determine the order in which the amino acids are joined in a polypeptide chain. The classical sequencing strategy is to cleave the polypeptide at

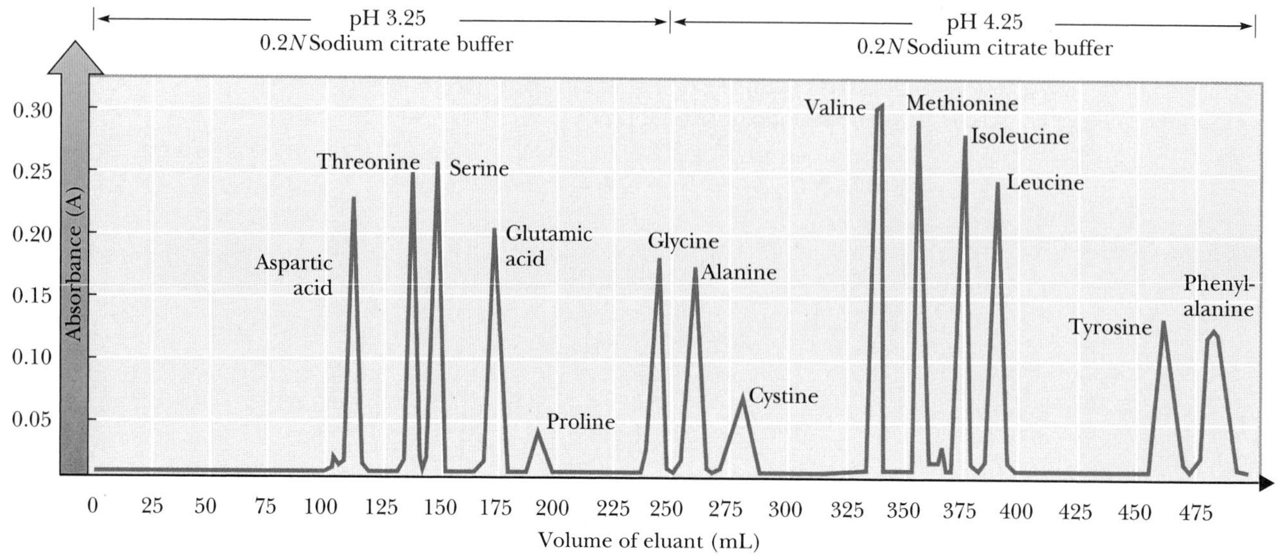

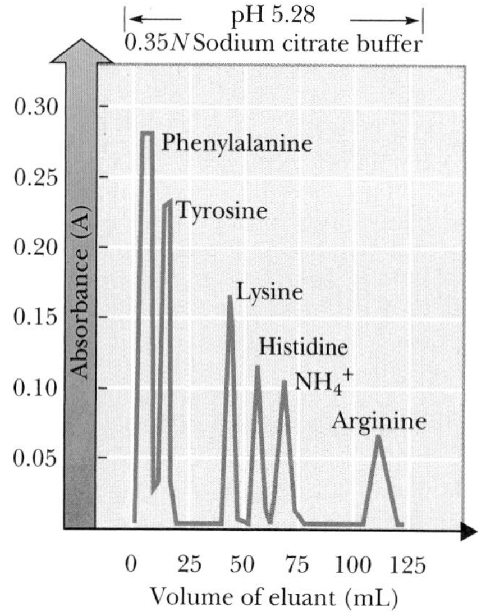

FIGURE 27.6 Analysis of a mixture of amino acids by ion-exchange chromatography using Amberlite IR-120, a sulfonated polystyrene resin. The resin contains phenyl-$SO_3^-Na^+$ groups. The amino acid mixture is applied to the column at low **pH** (3.25) under which conditions the acidic amino acids **(Asp, Glu)** are weakly bound to the resin and the basic amino acids **(Lys, His, Arg)** are tightly bound. Sodium citrate buffers at two different concentrations, and three different values of **pH** are used to elute the amino acids from the column. Cysteine is determined as cystine, Cys-S-S-Cys, the disulfide of cysteine.

specific peptide bonds (using, for example, cyanogen bromide or certain proteolytic enzymes), determine the sequence of each fragment (using, for example, the Edman degradation), and then match overlapping fragments to arrive at the sequence of the polypeptide.

Cyanogen Bromide

Cyanogen bromide (**BrCN**) is specific for cleavage of peptide bonds formed by the carboxyl group of methionine (Figure 27.7). The products of this cleavage are a substituted γ-lactone (Section 18.1C) derived from the *N*-terminal portion of the polypeptide and a second fragment containing the *C*-terminal portion of the polypeptide.

A three-step mechanism can be written for this reaction. Cyanogen bromide cleavage depends on chemical manipulation of the leaving ability of the sulfur atom of methionine. Because CH_3S^- is the anion of a weak acid, it is a very poor leaving group, just as OH^- is a poor leaving group (Section 9.3C). Yet, just as the oxygen atom of an alcohol can be transformed into a better leaving group by converting it into an oxonium ion (by protonation), so too can the sulfur atom of methionine be transformed into a better leaving group by converting it into a sulfonium ion.

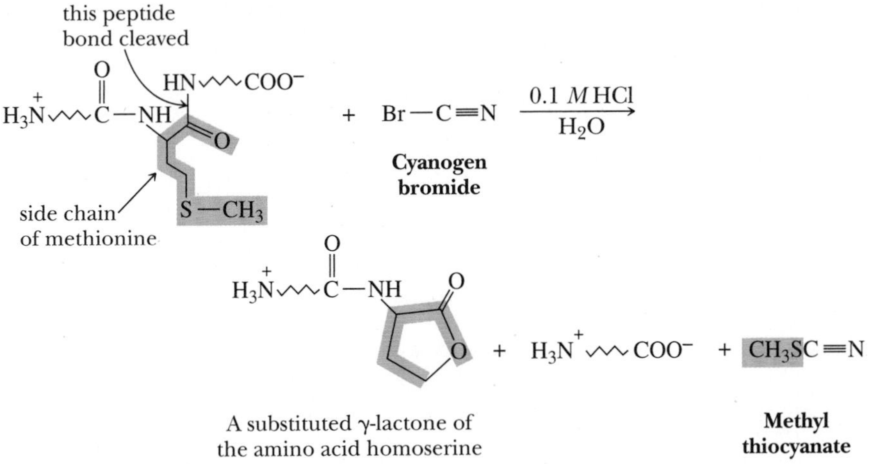

FIGURE 27.7 Cleavage by cyanogen bromide, **BrCN**, of a peptide bond formed by the carboxyl group of methionine.

Mechanism 27.1

Cleavage of a Peptide Bond at Methionine by Cyanogen Bromide

Step 1: **Make a new bond between a nucleophile and an electrophile and simultaneously break a bond to give stable molecules or ions.** Reaction is initiated by nucleophilic attack of the divalent sulfur atom of methionine on the carbon of cyanogen bromide displacing bromide ion. The product of this nucleophilic displacement is a sulfonium ion.

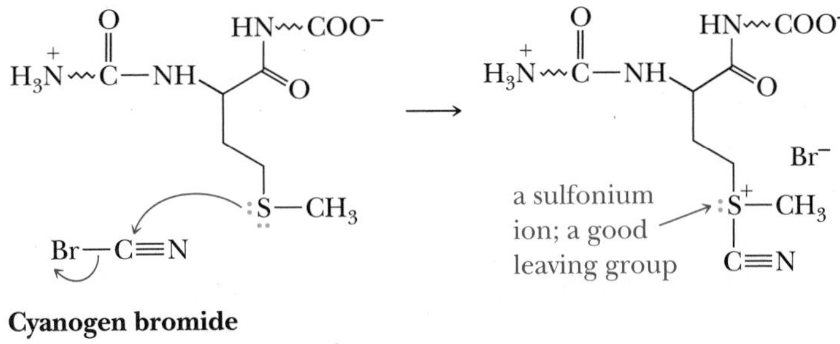

Step 2: **Make a new bond between a nucleophile and an electrophile and simultaneously break a bond to give stable molecules or ions.** An internal S_N2 reaction in which the oxygen of the methionine carbonyl group attacks the γ-carbon and displaces methyl thiocyanate gives a five-membered ring. Note that the oxygen of a carbonyl group is, at best, a weak nucleophile. This displacement is facilitated, however, because the sulfonium ion is a very good leaving group and because of the ease with which a five-membered ring is formed.

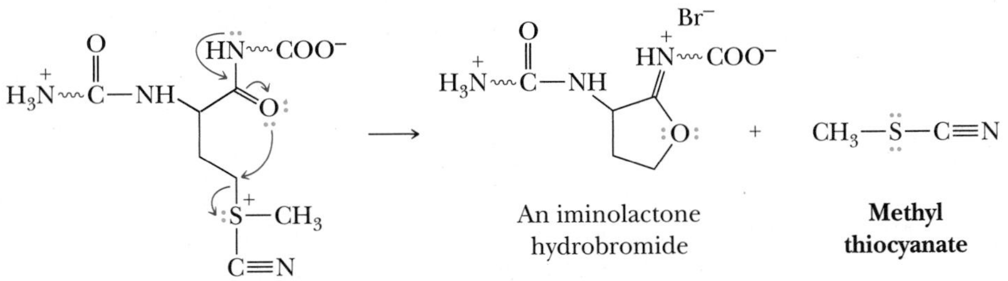

(Continued)

Step 3: Hydrolysis of the imino group gives a γ-lactone derived from the *N*-terminal end of the original polypeptide.

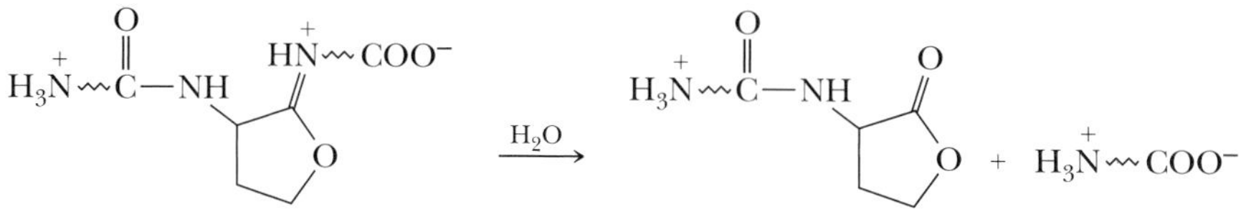

A substituted γ-lactone of the amino acid homoserine

Enzyme-Catalyzed Hydrolysis of Peptide Bonds

A group of proteolytic enzymes, among them trypsin and chymotrypsin, can be used to catalyze the hydrolysis of specific peptide bonds. Trypsin catalyzes the hydrolysis of peptide bonds formed by the carboxyl groups of arginine and lysine; chymotrypsin catalyzes the hydrolysis of peptide bonds formed by the carboxyl groups of phenylalanine, tyrosine, and tryptophan (Table 27.3).

Table 27.3 Cleavage of Specific Peptide Bonds Catalyzed by Trypsin and Chymotrypsin

Enzyme	Catalyzes Hydrolysis of Peptide Bond Formed by Carboxyl Group of
Trypsin	Arginine, lysine
Chymotrypsin	Phenylalanine, tyrosine, tryptophan

Example 27.5 Enzymatic Cleavage

Which of these dipeptides are hydrolyzed by trypsin? by chymotrypsin?

(a) Arg-Glu-Ser **(b)** Phe-Gly-Lys

Solution

(a) Trypsin catalyzes the hydrolysis of peptide bonds formed by the carboxyl groups of lysine and arginine. Therefore, the peptide bond between arginine and glutamic acid is hydrolyzed in the presence of trypsin.

$$\text{Arg-Glu-Ser} + H_2O \xrightarrow{\text{trypsin}} \text{Arg} + \text{Glu-Ser}$$

Chymotrypsin catalyzes the hydrolysis of peptide bonds formed by the carboxyl groups of phenylalanine, tyrosine, and tryptophan. Because none of these three aromatic amino acids is present, tripeptide (a) is not affected by chymotrypsin.

(b) Tripeptide (b) is not affected by trypsin. Although lysine is present, its carboxyl group is at the *C*-terminal end and is not involved in peptide bond formation. Tripeptide (b) is hydrolyzed in the presence of chymotrypsin.

$$\text{Phe-Gly-Lys} + H_2O \xrightarrow{\text{chymotrypsin}} \text{Phe} + \text{Gly-Lys}$$

Problem 27.5

Which of these tripeptides are hydrolyzed by trypsin? by chymotrypsin?

(a) Tyr-Gln-Val **(b)** Thr-Phe-Ser **(c)** Thr-Ser-Phe

Edman Degradation

Of the various chemical methods developed for determining the amino acid sequence of a polypeptide, the one most widely used today is the **Edman degradation**, introduced in 1950 by Pehr Edman of the University of Lund, Sweden. In this procedure, a polypeptide is treated with phenyl isothiocyanate, $C_6H_5N{=}C{=}S$, and then with acid. The effect of Edman degradation is to remove the *N*-terminal amino acid selectively as a substituted phenylthiohydantoin (Figure 27.8), which is then separated and identified.

The key feature of the Edman degradation is successive C═N, C═O, and C═O addition reactions.

Edman degradation A method for selectively cleaving and identifying the *N*-terminal amino acid of a polypeptide chain.

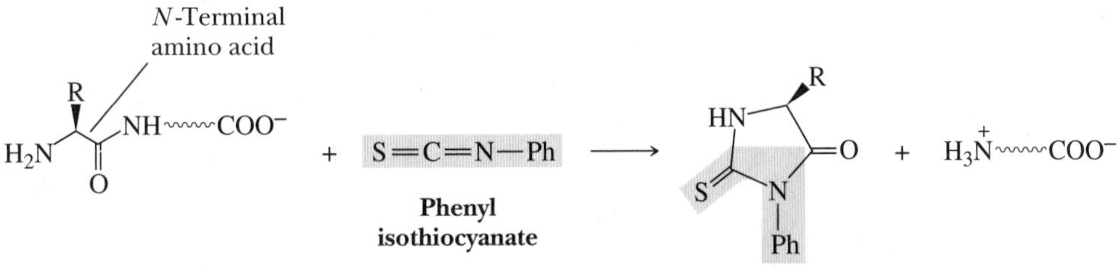

FIGURE 27.8 Edman degradation. Treatment of a polypeptide with phenyl isothiocyanate followed by acid selectively cleaves the *N*-terminal amino acid as a substituted phenylthiohydantoin.

Mechanism 27.2

Edman Degradation—Cleavage of an *N*-Terminal Amino Acid

Step 1: **Make a new bond between a nucleophile and an electrophile.** Nucleophilic addition of the *N*-terminal amino group to the C═N bond of phenyl isothiocyanate gives a derivative of *N*-phenylthiourea.

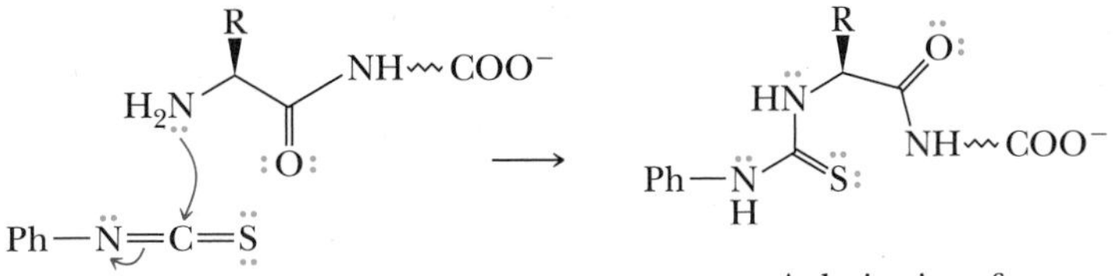

Step 2: **Make a new bond between a nucleophile and an electrophile followed by break a bond to give stable molecules or ions.** Heating the derivatized polypeptide in HCl at 100°C results in nucleophilic addition of sulfur to the carbonyl of the adjacent amide group to give a tetrahedral carbonyl addition intermediate, which collapses to give a thiazolinone ring derived from the *N*-terminal amino acid after the transfer of two protons to the amino group of the departing amine.

(Continued)

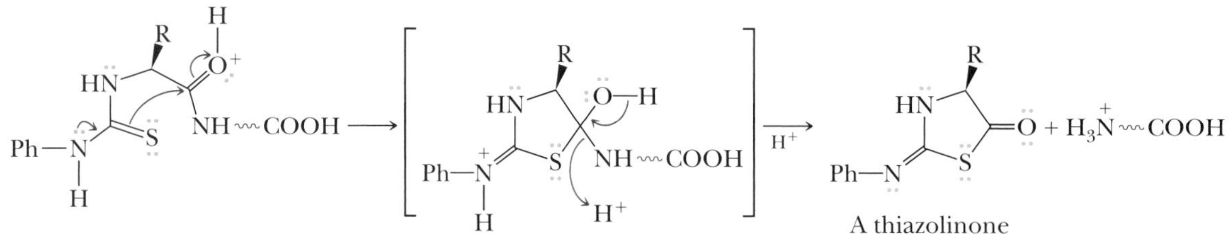

A thiazolinone

Step 3: The thiazolinone ring undergoes isomerization by ring opening followed by reclosure to give a more stable phenylthiohydantoin, which is separated and identified by chromatography.

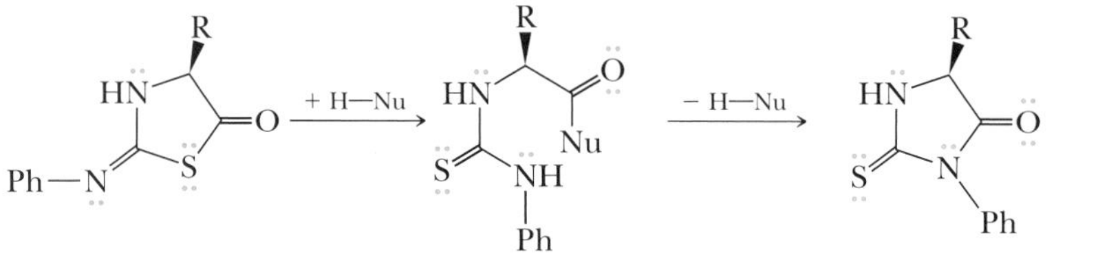

A thiazolinone A phenylthiohydantoin

The special value of the Edman degradation is that it cleaves the *N*-terminal amino acid from a polypeptide without affecting any other bonds in the chain. Furthermore, Edman degradation can be repeated on the shortened polypeptide, causing the next amino acid in the sequence to be cleaved and identified. In practice, it is now possible to sequence as many as the first 20 to 30 amino acids in a polypeptide by this method using only a few milligrams or less of material.

Most polypeptides in nature are longer than 20 to 30 amino acids, the practical limit to the number of amino acids that can be sequenced by repetitive Edman degradation. The special value of cleavage with cyanogen bromide, trypsin, and chymotrypsin is that a long polypeptide chain can be cleaved at specific peptide bonds into smaller polypeptide fragments, and each fragment can then be sequenced separately. The entire procedure, from the chemical cleavage step to the separation of the product, has been automated. Samples of a pure protein can be added to a machine and the first several *N*-terminal amino acids determined automatically.

Example 27.6 Edman Degradation

Deduce the amino acid sequence of a pentapeptide from the following experimental results. Note that under the column labeled "Amino Acid Composition," the amino acids are listed in alphabetical order. In no way does this listing give any information about primary structure.

Experimental Procedure	Amino Acid Composition
Pentapeptide	Arg, Glu, His, Phe, Ser
Edman degradation	Glu
Hydrolysis catalyzed by chymotrypsin	
Fragment A	Glu, His, Phe
Fragment B	Arg, Ser
Hydrolysis catalyzed by trypsin	
Fragment C	Arg, Glu, His, Phe
Fragment D	Ser

Solution

Edman degradation cleaves **Glu** from the pentapeptide; therefore, glutamic acid must be the *N*-terminal amino acid.

Glu-(Arg, His, Phe, Ser)

Fragment A from chymotrypsin-catalyzed hydrolysis contains **Phe**. Because of the specificity of chymotrypsin, **Phe** must be the *C*-terminal amino acid of fragment A. Fragment A also contains **Glu**, which we already know is the *N*-terminal amino acid. From these observations, we conclude that the first three amino acids in the chain must be **Glu-His-Phe** and then write the following partial sequence:

Glu-His-Phe-(Arg, Ser)

The fact that trypsin cleaves the pentapeptide means that **Arg** must be within the pentapeptide chain; it cannot be the *C*-terminal amino acid. Therefore, the complete sequence must be as follows:

Glu-His-Phe-Arg-Ser

Problem 27.6

Deduce the amino acid sequence of an undecapeptide (11 amino acids) from the experimental results shown in the table.

Experimental Procedure	Amino Acid Composition
Undecapeptide	Ala, Arg, Glu, Lys_2, Met, Phe, Ser, Thr, Trp, Val
Edman degradation	Ala
Trypsin-catalyzed hydrolysis	
Fragment E	Ala, Glu, Arg
Fragment F	Thr, Phe, Lys
Fragment G	Lys
Fragment H	Met, Ser, Trp, Val
Chymotrypsin-catalyzed hydrolysis	
Fragment I	Ala, Arg, Glu, Phe, Thr
Fragment J	Lys_2, Met, Ser, Trp, Val
Treatment with cyanogen bromide	
Fragment K	Ala, Arg, Glu, Lys_2, Met, Phe, Thr, Val
Fragment L	Trp, Ser

Sequencing by Mass Spectrometry

As mentioned in Chapter 14, mass spectrometry is increasingly being used for the direct sequencing of proteins in very small quantities. This topic was fully addressed in "Connections to Biological Chemistry: Mass Spectra of Biological Macromolecules" in Section 14.3 and will not be further developed here. Sequencing of proteins by mass spectrometry has largely replaced automated Edman degradation in most research laboratories.

Sequencing Proteins from the Coding Nucleotide Sequence

As it has become easier to determine nucleotide sequences (see Section 28.5), it is now often easier to sequence the nucleic acid that codes for a protein than to sequence the protein itself. In some cases, this has led to discovery of new proteins of unknown function. Often, comparison of the revealed protein sequences with those of known proteins from simpler organisms discloses sequence homologies that suggest the function of the new proteins. Comparisons with yeast, whose genome was one of the first to be sequenced and for which most of the proteins coded for have known functions, have been particularly fruitful.

27.5 Synthesis of Polypeptides

A. The Problem

The problem in peptide synthesis is to join the carboxyl group of amino acid 1 (aa_1) with an amide (peptide) bond to the amino group of amino acid 2 (aa_2).

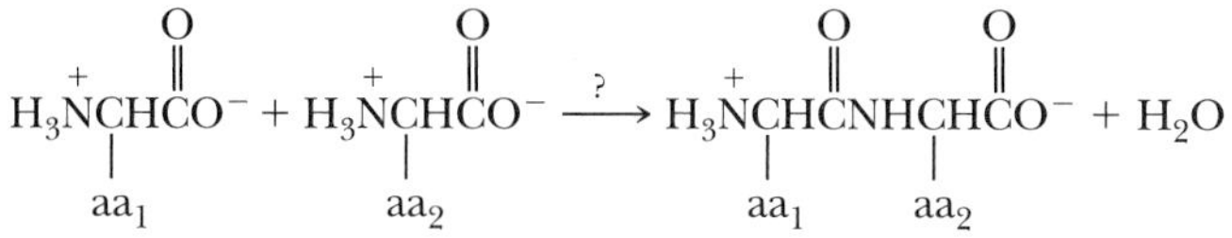

B. The Strategy

A rational strategy for the synthesis of peptide bonds and polypeptides requires three steps:

1. Protect the α-amino group of amino acid aa_1 to reduce its nucleophilicity so that it does not participate in nucleophilic addition to the carboxyl group of either aa_1 or aa_2.
2. Protect the α-carboxyl group of amino acid aa_2 so that it is not susceptible to nucleophilic attack by the α-amino group of another molecule of aa_2.
3. Activate the α-carboxyl group of amino acid aa_1 so that it is susceptible to nucleophilic attack by the α-amino group of aa_2.

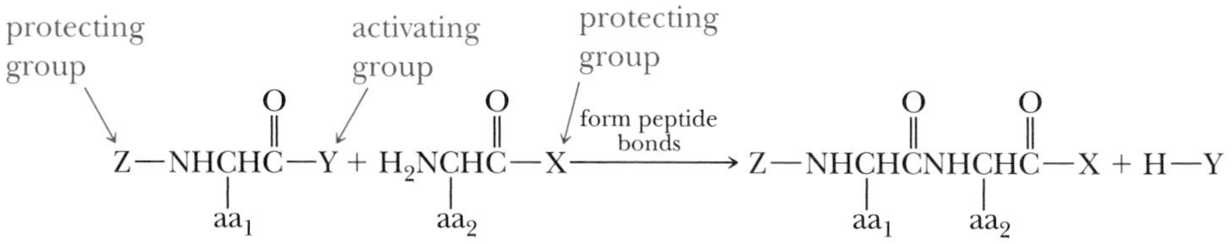

After dipeptide aa_1—aa_2 has been formed, the protecting group Z can be removed and chain growth can be continued from the *N*-terminal end of the dipeptide. Alternatively, the protecting group X can be removed and chain growth can be continued from the *C*-terminal end. The range of protecting groups and activating groups is large, and experimental conditions have been found to attach and remove them as desired.

C. Amino-Protecting Groups

The most common strategy for protecting amino groups and reducing their nucleophilicity is to react them with carbonyl derivatives. The reagents most commonly used for this purpose are benzyloxycarbonyl chloride and di-*tert*-butyl dicarbonate. In the terminology adopted by the IUPAC, the benzyloxycarbonyl group is given the symbol Z or Cbz and the *tert*-butoxycarbonyl group is given the symbol BOC—.

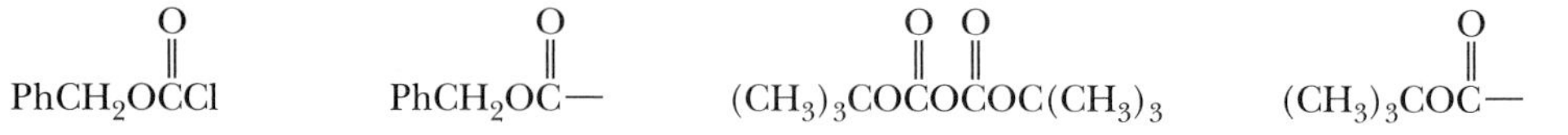

Benzyloxycarbonyl chloride | **Benzyloxycarbonyl** (Z— or Cbz) group | **Di-*tert*-butyl dicarbonate** | ***tert*-Butoxycarbonyl** (BOC—) group

Treatment of an amino group with either of these reagents forms a new functional group called a carbamate. A carbamate is an ester of carbamic acid; that is, it is an ester of the monoamide of carbonic acid.

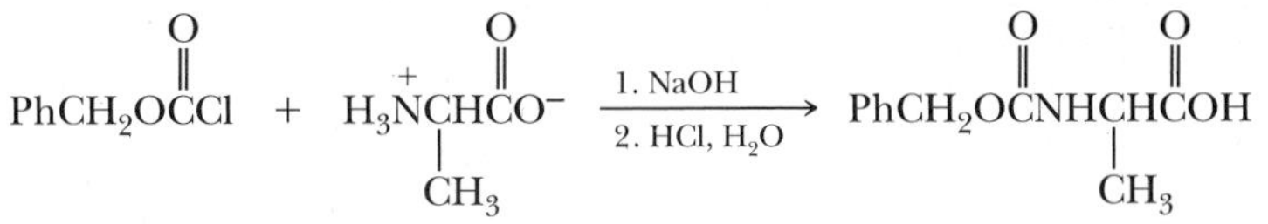

Benzyloxycarbonyl chloride (Z-Cl) **Alanine** ***N*-Benzyloxycarbonylalanine** (Z-Ala) A carbamate

The special advantage of the carbamate group is that it is stable to dilute base but can be removed by treatment with HBr in acetic acid or other similarly acidic conditions.

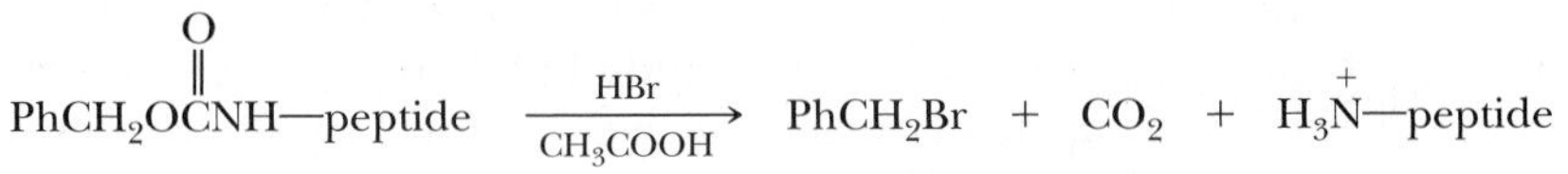

A Z-protected peptide **Benzyl bromide** An unprotected peptide

A study of the mechanism for removal of this protecting group has shown that the reaction is first order in $[H^+]$ and involves formation of a carbocation and a carbamic acid. A carbamic acid spontaneously loses carbon dioxide to form the free amine. The carbocation reacts with an available nucleophile such as halide ion to form a haloalkane.

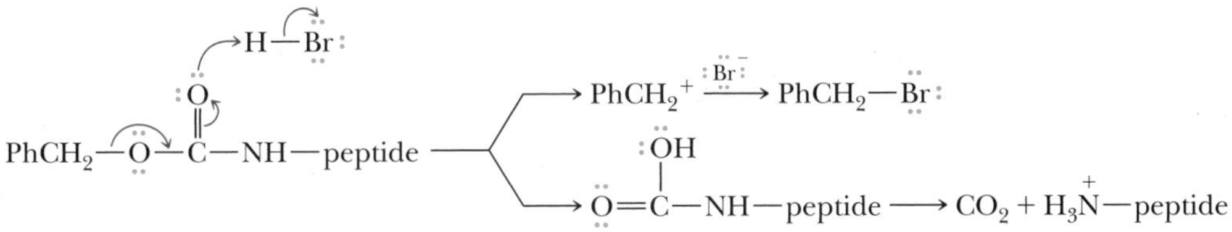

A carbamic acid

Note that because acid-catalyzed removal of this protecting group is carried out in nonaqueous media, there is no danger of simultaneous acid-catalyzed hydrolysis of peptide (amide) bonds within the newly synthesized polypeptide. This relationship exists because water is required for hydrolysis of a peptide bond.

The benzyloxycarbonyl group can also be removed by treatment with H_2 in the presence of a transition metal catalyst (hydrogenolysis, Section 21.5C). In hydrogenolysis of a Z-protecting group, one product is toluene. The other is a carbamic acid, which undergoes spontaneous decarboxylation to give carbon dioxide and the unprotected peptide.

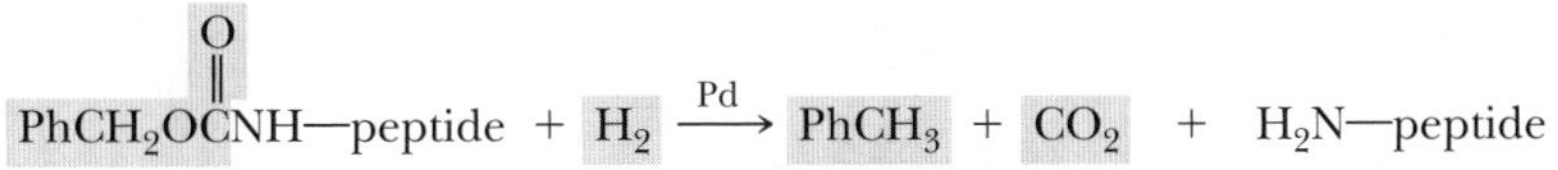

A Z-protected peptide **Toluene** An unprotected peptide

D. Carboxyl-Protecting Groups

Carboxyl groups are most often protected by conversion to methyl, ethyl, or benzyl esters. Methyl and ethyl esters are prepared by Fischer esterification (Section 17.7A) and are removed by hydrolysis in aqueous base (Section 18.4C) under mild conditions. Benzyl esters are conveniently removed by hydrogenolysis with H_2 over a palladium or platinum catalyst (Section 21.5C). Benzyl groups can also be removed by treatment with HBr in acetic acid.

E. Peptide-Bond Forming Reactions

The reagent most commonly used to bring about peptide bond formation is 1,3-dicyclohexylcarbodiimide (DCC). This reagent is the anhydride of a disubstituted urea, and when treated with water, it is converted to *N,N'*-dicyclohexylurea (DCU).

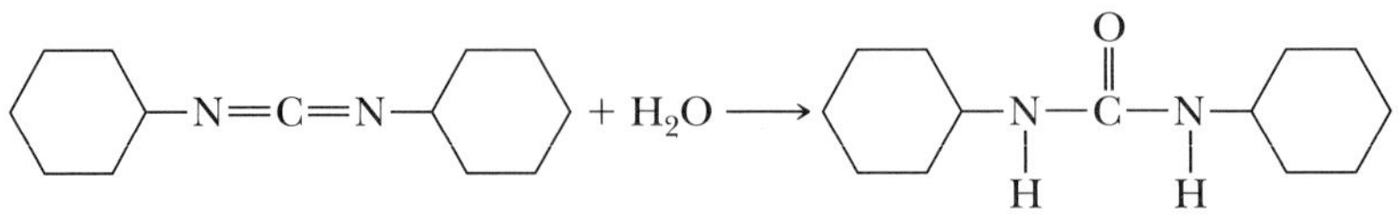

1,3-Dicyclohexylcarbodiimide (DCC) **N,N'-Dicyclohexylurea (DCU)**

When an amino-protected aa_1 and a carboxyl-protected aa_2 are treated with DCC, this reagent acts as a dehydrating agent; it removes —OH from the carboxyl group and —H from the amino group to form an amide bond. More specifically, DCC activates the α-carboxyl group of aa_1 toward nucleophilic acyl substitution by converting its —OH group into a better leaving group.

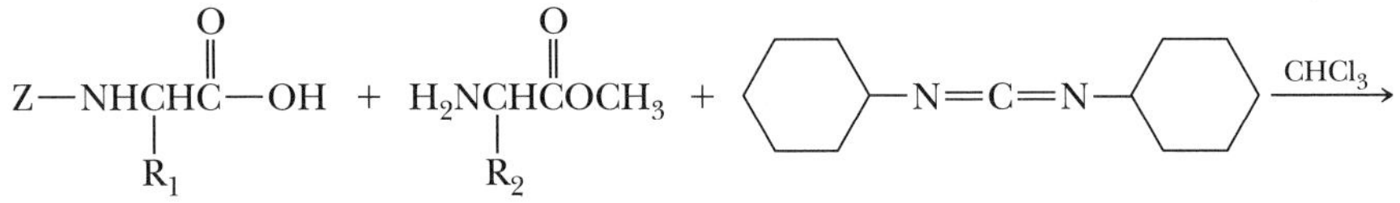

Amino-protected aa_1 Carboxyl-protected aa_2 **1,3-Dicyclohexylcarbodiimide (DCC)**

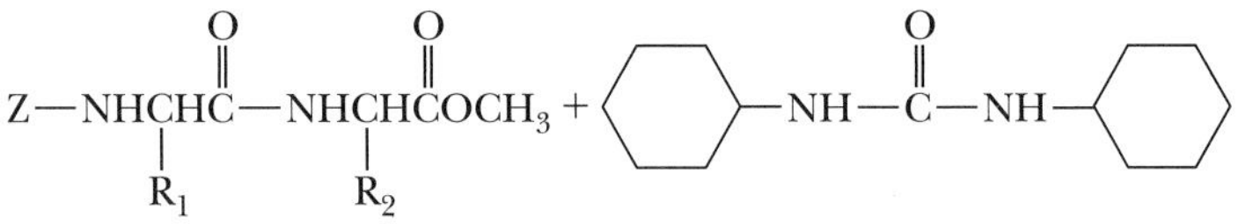

Amino- and carboxyl-protected dipeptide ***N,N'*-Dicyclohexylurea (DCU)**

An abbreviated mechanism for this intermolecular dehydration is shown in Figure 27.9: an acid-base reaction in Step 1 between the carboxyl group of aa_1 and a nitrogen of DCC (take a proton away) followed in Step 2 by addition of the carboxylate anion to the C=N double bond (make a new bond between a nucleophile and an electrophile). The *O*-acylisourea formed is the nitrogen analog of a mixed anhydride. Nucleophilic addition of the amino group of aa_2 to the carbonyl group

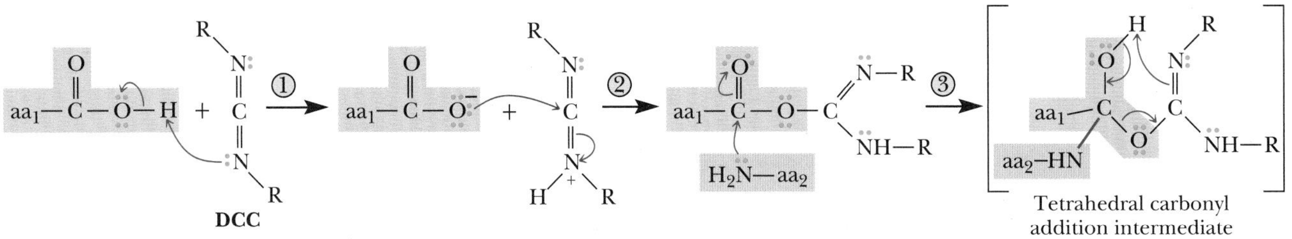

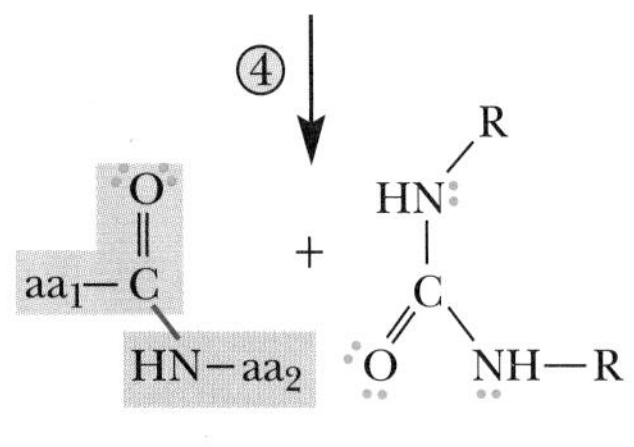

A dipeptide **DCU**

FIGURE 27.9 The role of 1,3-dicyclohexylcarbodiimide (DCC) in the formation of a peptide bond between an amino-protected amino acid (aa_1) and a carboxy-protected amino acid (aa_2).

of the *O*-acylisourea in Step 3 (make a new bond between a nucleophile and an electrophile) generates a tetrahedral carbonyl addition intermediate that collapses in Step 4 to give a dipeptide and DCU (break a bond to give stable molecules or ions in a pericyclic reaction).

F. Solid-Phase Synthesis

A major problem associated with polypeptide synthesis is purification of intermediates after each protection, activation, coupling, and deprotection step. If unreacted starting materials are not removed after each step, the final product is contaminated by polypeptides missing one or more amino acids. Not only are the required purification steps laborious and time-consuming, but they also inevitably result in some loss of the desired product. These losses become especially severe in the synthesis of larger polypeptides.

A major advance in polypeptide synthesis came in 1962 when R. Bruce Merrifield of the Rockefeller University described a solid-phase synthesis (alternatively called polymer-supported synthesis) of the tetrapeptide, Leu-Ala-Gly-Ala, by a technique that now bears his name. Merrifield was awarded the 1984 Nobel Prize in Chemistry for his work in developing the solid-phase method for peptide synthesis.

The solid support used by Merrifield was a type of polystyrene in which about 5% of the phenyl groups carry a chloromethyl (—CH_2Cl) group in their para positions (Figure 27.10). These chloromethyl groups, like all benzylic halides, are particularly reactive in nucleophilic substitution reactions.

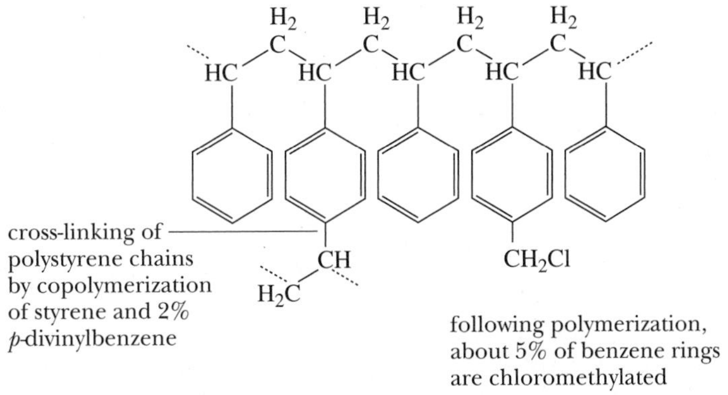
cross-linking of polystyrene chains by copolymerization of styrene and 2% *p*-divinylbenzene
following polymerization, about 5% of benzene rings are chloromethylated

FIGURE 27.10 The support used for the Merrifield solid-phase synthesis is a chloromethylated polystyrene resin.

In the Merrifield method, the *C*-terminal amino acid is joined as a benzyl ester to the solid polymer support and then the polypeptide chain is extended one amino acid at a time from the *N*-terminal end. The advantage of polypeptide synthesis on a solid support is that the polymer beads with the peptide chains anchored on them are completely insoluble in the solvents used in the synthesis. Furthermore, excess reagents (e.g., DCC) and byproducts (e.g., DCU) are removed after each step simply by washing the polymer beads. When synthesis is completed, the polypeptide is released from the polymer beads by cleavage of the benzyl ester. The steps in solid-phase synthesis of a polypeptide are summarized in Figure 27.11.

Thanks to automation, the synthesis of polypeptides is now a routine procedure in chemical research. It is common for researchers to order several peptides at a time for use in fields as diverse as medicine, biology, material science, and biomedical engineering.

A dramatic illustration of the power of the solid-phase method was the synthesis of the enzyme ribonuclease by Merrifield in 1969. The synthesis involved 369 chemical reactions and 11,931 operations, all of which were performed by an automated machine without any intermediate isolation stages. Each of the

FIGURE 27.11 Steps in the Merrifield solid-phase polypeptide synthesis.

Watch a video explanation

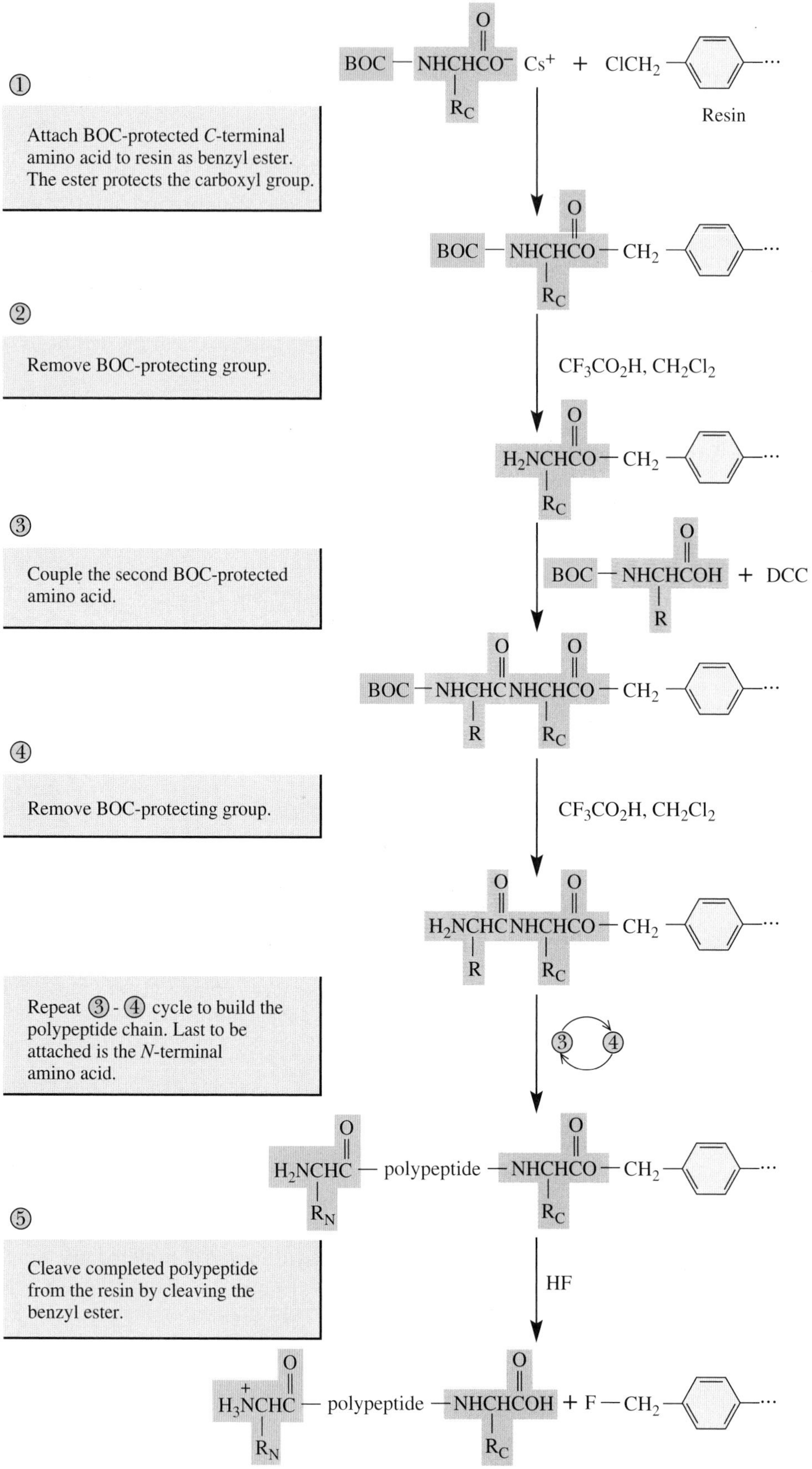

124 amino acids was added as an *N-tert*-butoxycarbonyl derivative and coupled using DCC. Cleavage from the resin and removal of all protective groups gave a mixture that was purified by ion-exchange chromatography. The specific activity of the synthetic enzyme was 13%–24% of that of the natural enzyme. The fact that the specific activity of the synthetic enzyme was lower than that of the natural enzyme was probably attributable to the presence of polypeptide byproducts closely related to but not identical to the natural enzyme. Synthesizing ribonuclease (124 amino acids) requires forming 123 peptide bonds. If each peptide bond is formed in 99% yield, the yield of homogeneous polypeptide is $0.99^{123} = 29\%$. If each peptide bond is formed in 98% yield, the yield is 8%. Thus, even with yields as high as 99% in each peptide bond-forming step, a large portion of the synthetic polypeptides have one or more sequence defects. Many of these, nonetheless, may be fully or partially active.

27.6 Three-Dimensional Shapes of Polypeptides and Proteins

A. Geometry of a Peptide Bond

In the late 1930s, Linus Pauling began a series of studies to determine the geometry of a peptide bond. One of his first and most important discoveries was that a peptide bond itself is planar. As shown in Figure 27.12, the four atoms of a peptide bond and the two α-carbons joined to it all lie in the same plane.

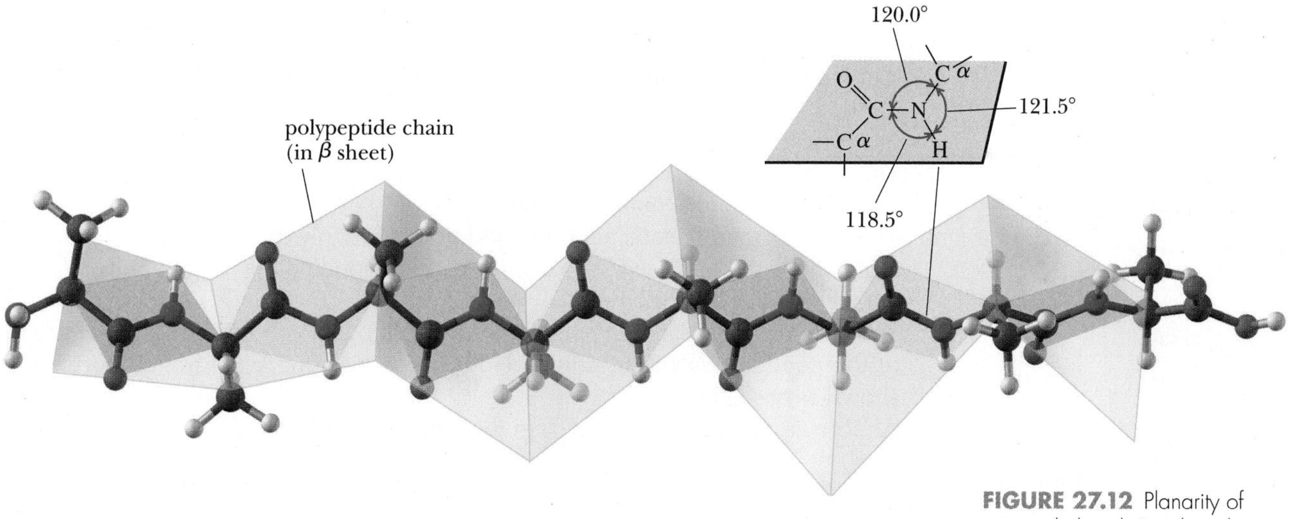

FIGURE 27.12 Planarity of a peptide bond. Bond angles about the carbonyl carbon and the amide nitrogen are approximately 120°.

As fully discussed in "Connections to Biological Chemistry: The Unique Structure of Amide Bonds" in Section 18.2, the carbon and nitrogen of an amide are actually planar with approximately 120° bond angles about each because of resonance of the nitrogen lone pair with the carbonyl. Two configurations are possible for the atoms of a planar peptide bond. In one, the two α-carbons are *cis* to each other; in the other, they are *trans* to each other. The *trans* configuration is more favorable because the α-carbons with the bulky groups bonded to them are farther from each other than they are in the *cis* configuration. Almost all peptide bonds in naturally occurring proteins have the *trans* configuration. Proline is found *cis* most of the time, but there are some well-known examples of other *cis* peptide bonds as well. The *trans* to *cis* peptide bond conversion can have critical biological consequences in proteins and is currently an active area of research.

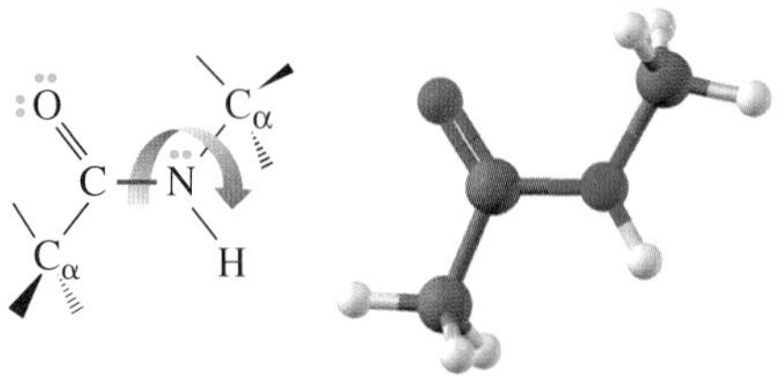

s-*trans* configuration

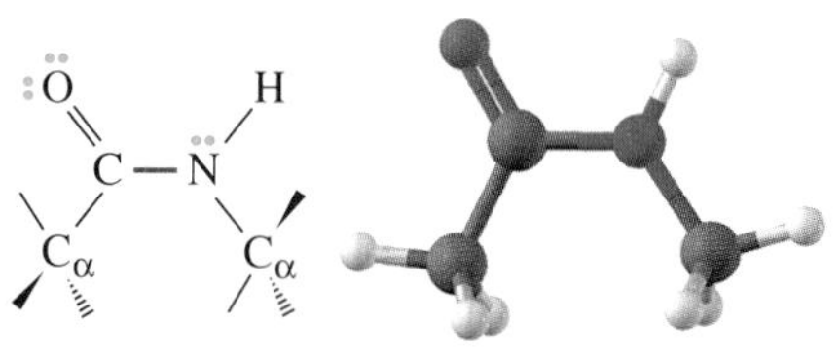

s-*cis* configuration

B. Secondary Structure

Secondary (2°) structure refers to ordered arrangements (conformations) of amino acids in localized regions of a polypeptide or protein molecule. The first studies of polypeptide conformations were carried out by Linus Pauling and Robert Corey beginning in 1939. They assumed that in conformations of greatest stability, all atoms in a peptide bond lie in the same plane and there is hydrogen bonding between the N—H of one peptide bond and the C═O of another, as shown in Figure 27.13.

On the basis of model building, Pauling proposed that two types of secondary structure should be particularly stable: the α-helix and the antiparallel β-pleated sheet. X-ray crystallography has validated this prediction completely.

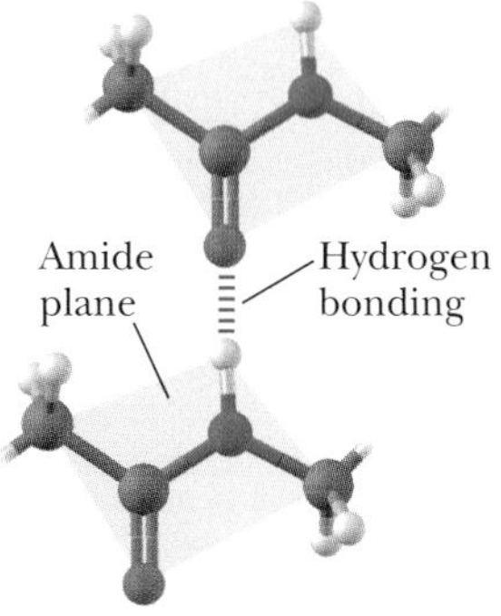

FIGURE 27.13 Hydrogen bonding between amide groups.

The α-Helix

In an **α-helix** pattern shown in Figure 27.14, a polypeptide chain is coiled in a spiral. As you study this section of α-helix, note the following:

1. The helix is coiled in a clockwise, or right-handed, manner. Right-handed means that if you turn the helix clockwise, it twists away from you. In this sense, a right-handed helix is analogous to the right-handed thread of a common wood or machine screw.
2. There are 3.6 amino acids per turn of the helix.
3. Each peptide bond is *trans* and planar.
4. The N—H group of each peptide bond points roughly downward, parallel to the axis of the helix, and the C═O of each peptide bond points roughly upward, also parallel to the axis of the helix.
5. The carbonyl group of each peptide bond is hydrogen-bonded to the N—H group of the peptide bond four amino acid units away from it. Hydrogen bonds are shown as dashed lines.
6. All R— groups point outward from the helix.

Secondary structure of proteins

The ordered arrangements (conformations) of amino acids in localized regions of a polypeptide or protein.

α-Helix

A type of secondary structure in which a section of polypeptide chain coils into a spiral, most commonly a right-handed spiral.

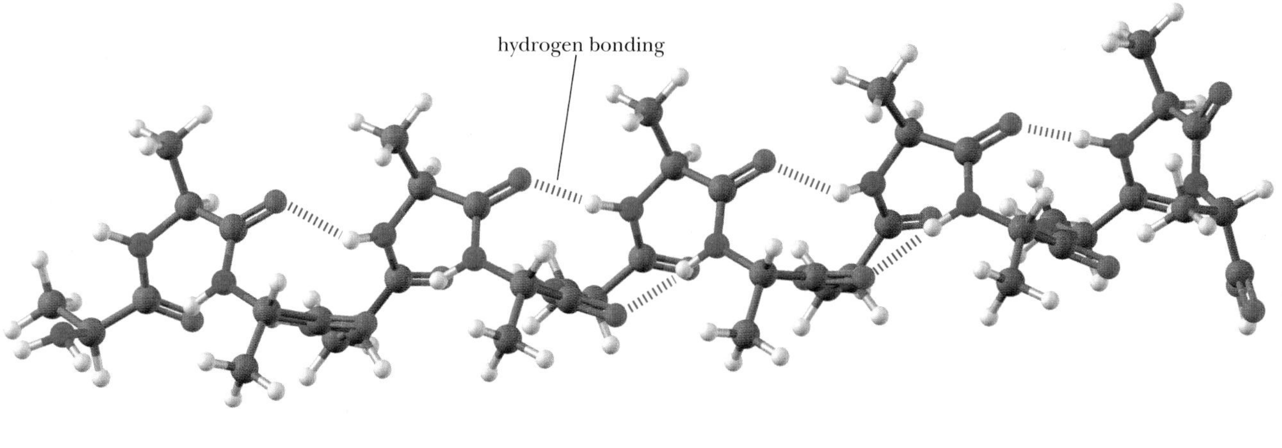

FIGURE 27.14 An α-helix. The peptide chain shown contains repeating units of L-alanine.

Almost immediately after Pauling proposed the α-helix conformation, other researchers proved the presence of α-helix conformations in keratin, the protein of hair and wool. It soon became obvious that the α-helix is one of the fundamental folding patterns of polypeptide chains.

The β-Pleated Sheet

An antiparallel **β-pleated sheet** consists of extended polypeptide chains with neighboring chains running in opposite (antiparallel) directions. In a parallel β-pleated sheet, the polypeptide chains run in the same direction. Unlike the α-helix arrangement, N—H and C=O groups lie in the plane of the sheet and are roughly perpendicular to the long axis of the sheet. The C=O group of each peptide bond is hydrogen-bonded to the N—H group of a peptide bond of a neighboring chain (Figure 27.15).

β-Pleated sheet A type of secondary structure in which sections of polypeptide chains are aligned parallel or antiparallel to one another.

As you study this section of β-pleated sheet, note the following:

1. The three polypeptide chains lie adjacent to each other and run in opposite (antiparallel) directions.
2. Each peptide bond is planar, and the α-carbons are *trans* to each other.
3. The C=O and N—H groups of peptide bonds from adjacent chains point at each other and are in the same plane so that hydrogen bonding is possible between adjacent polypeptide chains.
4. The R-groups on any one chain alternate, first above and then below the plane of the sheet and so on.

The β-pleated sheet conformation is stabilized by hydrogen bonding between N—H groups of one chain and C=O groups of an adjacent chain. By comparison, the α-helix is stabilized by hydrogen bonding between N—H and C=O groups within the same polypeptide chain.

C. Tertiary Structure

Tertiary (3°) structure refers to the overall folding pattern and arrangement in space of all atoms in a single polypeptide chain. No sharp dividing line exists between secondary and tertiary structures. Secondary structure refers to the spatial arrangement of amino acids close to one another on a polypeptide chain, whereas tertiary structure refers to the three-dimensional arrangement of all atoms of a polypeptide chain. Among the most important factors in maintaining 3° structure are disulfide bonds, hydrophobic interactions, hydrogen bonding, and salt linkages.

Tertiary structure of proteins The three-dimensional arrangement in space of all atoms in a single polypeptide chain.

Disulfide bonds (Section 10.9G) play an important role in maintaining tertiary structure. Disulfide bonds are formed between side chains of two cysteine units

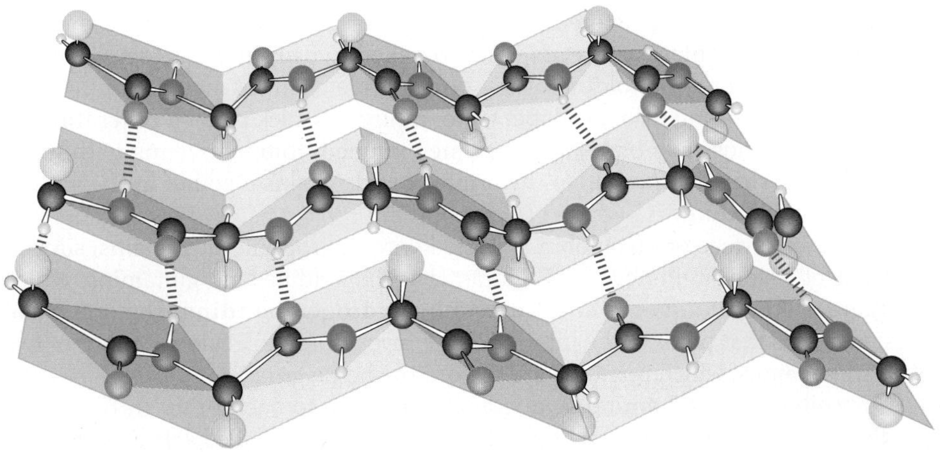

FIGURE 27.15 β-Pleated sheet conformation with three polypeptide chains running in opposite (antiparallel) directions. Hydrogen bonding between chains is indicated by dashed lines.

by oxidation of their thiol groups (—SH) to form a disulfide bond. Treatment of a disulfide bond with a reducing agent regenerates the thiol groups.

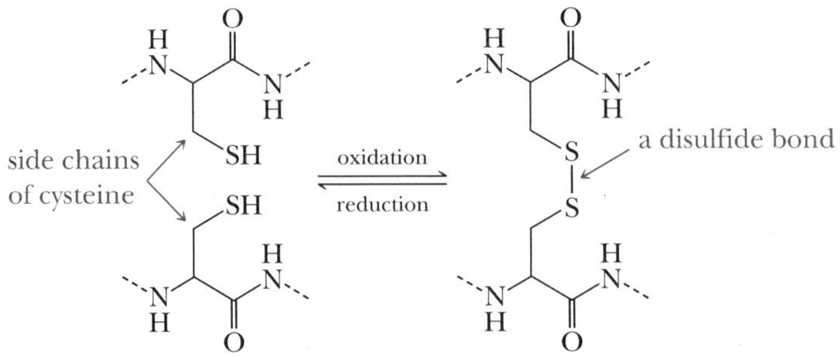

Figure 27.16 shows the amino acid sequence of human insulin. This protein consists of two polypeptide chains: an A chain of 21 amino acids and a B chain of 30 amino acids. The A chain is bonded to the B chain by two interchain disulfide bonds. An intrachain disulfide bond also connects the cysteine units at positions 6 and 11 of the A chain.

As an example of 2° and 3° structure, let us look at the three-dimensional structure of myoglobin—a protein found in skeletal muscle and particularly abundant in diving mammals, such as seals, whales, and porpoises. Myoglobin and its structural relative, hemoglobin, are the oxygen storage and transport molecules of vertebrates. Hemoglobin binds molecular oxygen in the lungs and transports it to myoglobin in muscles. Myoglobin stores molecular oxygen until it is required for metabolic oxidation.

Myoglobin consists of a single polypeptide chain of 153 amino acids. Myoglobin also contains a single heme unit. Heme consists of one Fe^{2+} ion coordinated in a square planar array with the four nitrogen atoms of a molecule of porphyrin (Figure 27.17).

Determination of the three-dimensional structure of myoglobin represented a milestone in the study of molecular architecture. For their contribution to this research, John C. Kendrew and Max F. Perutz, both of Britain, shared the 1962 Nobel Prize in Chemistry. The secondary and tertiary structures of myoglobin are shown in Figure 27.18. The single polypeptide chain is folded into a complex, almost boxlike shape.

Following are important structural features of the three-dimensional shape of myoglobin.

1. The backbone consists of eight relatively straight sections of α-helix, each separated by a bend in the polypeptide chain. The longest section of α-helix has 24 amino acids, the shortest has 7. Some 75% of the amino acids are found in these eight regions of α-helix.
2. Hydrophobic side chains of phenylalanine, alanine, valine, leucine, isoleucine, and methionine are clustered in the interior of the molecule, where they are shielded from contact with water. The so-called **hydrophobic effect** is a major factor in directing the folding of the polypeptide chain of myoglobin into this compact three-dimensional shape. Water molecules stick to each other through hydrogen bonds, so nonpolar groups are excluded from interacting with water molecules. This exclusion drives hydrophobic side chains together in the interior of protein structures.
3. The outer surface of myoglobin is coated with hydrophilic side chains, such as those of lysine, arginine, serine, glutamic acid, histidine, and glutamine, which interact with the aqueous environment by **hydrogen bonding**. The only polar side chains that point to the interior of the myoglobin molecule are those of two histidine units, which point inward toward the heme group.
4. Oppositely charged amino acid side chains close to each other in the three-dimensional structure interact by electrostatic attractions called **salt linkages**. An

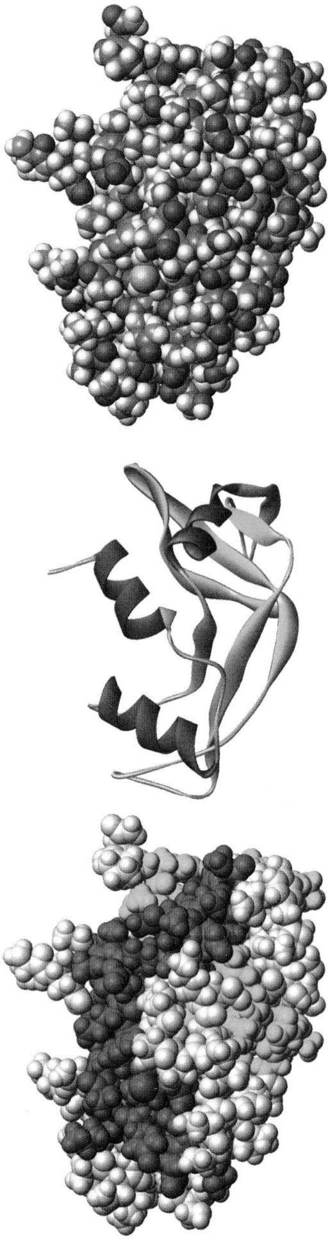

Three views of the protein ribonuclease A. The upper view shows atoms colored according to atom type. In the lower two views, red indicates regions of α-helix; blue, β-sheets (Section 27.6). Loop regions are shown in white.

Hydrophobic effect

The tendency of nonpolar groups to cluster to shield themselves from contact with an aqueous environment.

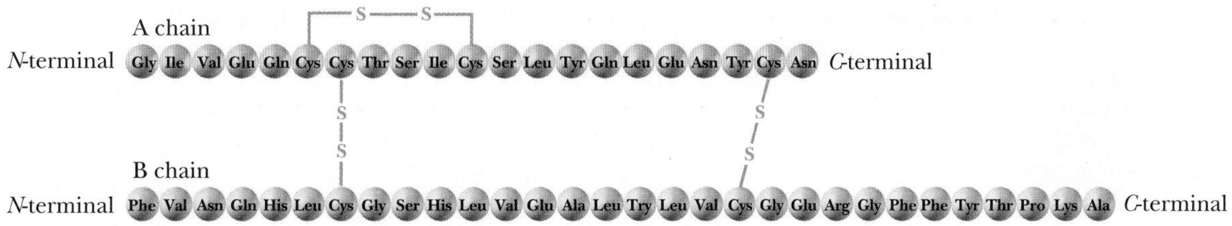

FIGURE 27.16 Human insulin. The A chain of 21 amino acids and B chain of 30 amino acids are connected by interchain disulfide bonds between A7 and B7 and between A20 and B19. In addition, a single intrachain disulfide bond occurs between A6 and A11.

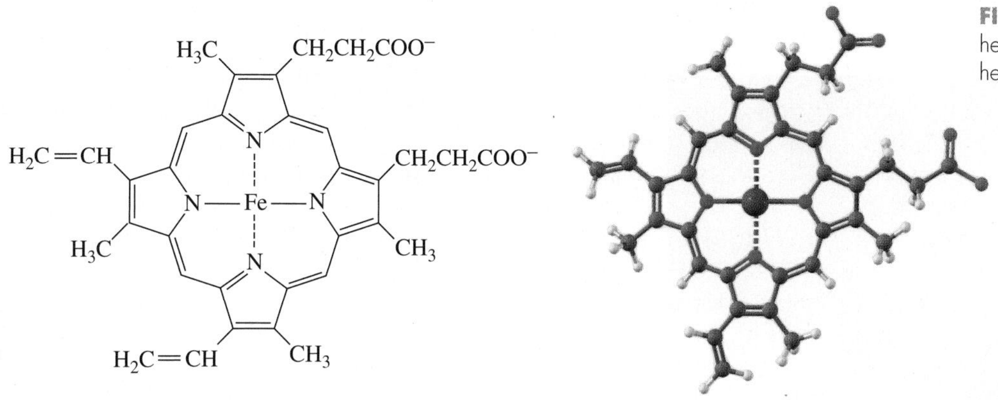

FIGURE 27.17 The structure of heme, found in myoglobin and hemoglobin.

example of a salt linkage is the attraction of the side chains of lysine ($—NH_3^+$) and glutamic acid ($—COO^-$).

The tertiary structures of hundreds of proteins have also been determined. It is clear that proteins contain α-helix and β-pleated sheet structures, but that wide variations exist in the relative amounts of each. Lysozyme, with 129 amino acids in a single polypeptide chain, has only 25% of its amino acids in α-helix regions. Cytochrome ***c***, with 104 amino acids in a single polypeptide chain, has no α-helix structure but does contain several regions of β-pleated sheet. Yet, whatever the proportions of α-helix, β-pleated sheet, or other periodic structure, most nonpolar side chains of water-soluble proteins are directed toward the interior of the molecule, whereas polar side chains are on the surface of the molecule and in contact with the aqueous environment. Note that this arrangement of polar and nonpolar groups in water-soluble proteins very much resembles the arrangement of polar and nonpolar groups of soap molecules in micelles (Figure 26.3). It also resembles the arrangement of phospholipids in lipid bilayers (Figure 26.13).

FIGURE 27.18 Ribbon model of myoglobin. The polypeptide chain is shown in yellow; the heme ligand, in red; and the Fe atom, as a white sphere.

Chemical Connections

Spider Silk

Spider silk has some remarkable properties. Research is currently concentrated on the strong dragline silk that forms the spokes of a web of the Golden Orb Weaver (*Nephila clavipes*). This silk has three times the impact strength of Kevlar and is 30% more flexible than nylon. The commercial application of spider silk is not a novel concept. Eighteenth-century French entrepreneur Bon de Saint-Hilaire attempted to mass-produce silk in his high-density spider farms but failed because of cannibalism among his territorial arachnid workers. In contrast, native New Guineans continue to successfully collect and utilize spider silk for a wide range of applications, including bags and fishing nets. Today the only way to obtain large

(*Continued*)

amounts of silk is to extract it from the abdomens of immobilized spiders, but scientific advances make the mass production and industrial application of spider silk increasingly possible.

Biologically produced dragline silk is a combination of two liquid proteins, Spidroin 1 and 2, which become oriented and solidify as they travel through a complex duct system in the spider's abdomen. These proteins are composed largely of alanine and glycine, the two smallest amino acids. Although glycine comprises almost 42% of each protein, the short five to ten peptide chains of alanine, which account for 25% of each protein's composition, are more important for the properties. Nuclear magnetic resonance (NMR) techniques have vastly improved the level of understanding of spider silk's structure, which was originally determined by X-ray crystallography. NMR data of spidroins containing deuterium-tagged alanine have shown that all alanines are configured into β-pleated sheets. Furthermore, the NMR data suggest that 40% of the alanine β-sheets are highly structured, while the other 60% are less oriented, forming fingers that reach out from each individual strand. These fingers are believed to join the oriented alanine β-sheets and the glycine-rich, amorphous "background" sectors of the polypeptide.

Currently, genetically modified *Escherichia coli* is used to mass-produce Spidroin 1 and 2. However, DNA redundancy initially caused synthesis problems when the spider genes were transposed into the bacteria. The *E. coli* did not transcribe some of the codons in the same way that spider cells would, forcing scientists to modify the DNA. When the proteins could be synthesized, it was necessary to develop a system to mimic the natural production of spider silk while preventing the silk from contacting the air and subsequently hardening. After the two proteins are separated from the *E. coli,* they are drawn together into methanol through separate needles. Another approach is to dissolve the silk in formic acid or to add hydrophilic amino acids (in this case, histidine and arginine) to keep the artificial silk pliable. The industrial and practical applications of spider silk will not be fully known until it can be synthesized and manipulated in large quantities.

Example 27.7 Hydrogen Bonding

With which of the following amino acid side chains can the side chain of threonine form hydrogen bonds?

(a) Valine **(b)** Asparagine **(c)** Phenylalanine

(d) Histidine **(e)** Tyrosine **(f)** Alanine

Solution

The side chain of threonine contains a hydroxyl group that can participate in hydrogen bonding in two ways: its oxygen has a partial negative charge and can function as a hydrogen bond acceptor, and its hydrogen has a partial positive charge and can function as a hydrogen bond donor. Therefore, the side chain of threonine can form hydrogen bonds with the side chains of tyrosine, asparagine, and histidine.

Problem 27.7

At pH 7.4, with what amino acid side chains can the side chain of lysine form salt linkages?

D. Quaternary Structure

Most proteins of molecular weight greater than 50,000 consist of two or more non-covalently linked polypeptide chains. The arrangement of protein monomers into an aggregation is known as **quaternary (4°) structure**. A good example is hemoglobin, a protein that consists of four separate polypeptide chains: two α-chains of 141 amino acids each and two β-chains of 146 amino acids each. The quaternary structure of hemoglobin is shown in Figure 27.19.

A major factor stabilizing the aggregation of protein subunits is the hydrophobic effect. When separate polypeptide chains fold into compact three-dimensional shapes to expose polar side chains to the aqueous environment and shield nonpolar side chains from water, hydrophobic "patches" still appear on the surface, in contact with water. These patches can be shielded from water if two or more monomers assemble so that their hydrophobic patches are in contact. The numbers of subunits of several proteins of known quaternary structure are shown in Table 27.4. Other important factors include correctly located complementary hydrogen bonding and charged sites on different subunits. The formation of aggregates of well-defined structure based on specific structural units on the subunits is being explored in the new field of molecular recognition, also referred to as supramolecular chemistry.

Quaternary structure
The arrangement of polypeptide monomers into a noncovalently bonded aggregate.

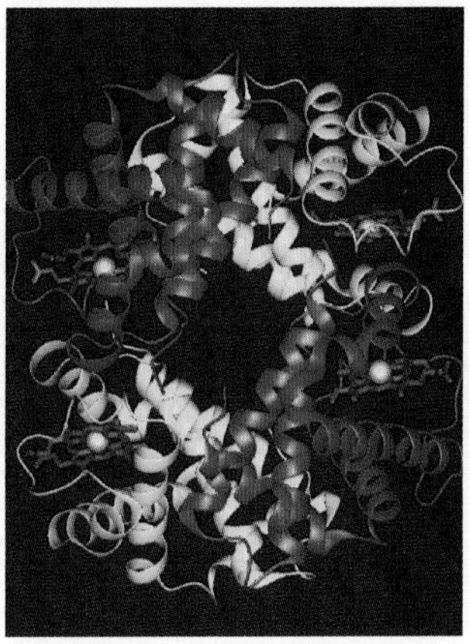

FIGURE 27.19 Ribbon model of hemoglobin. The α-chains are shown in purple; the β-chains, in yellow; the heme ligands, in red; and the Fe atoms, as white spheres.

Table 27.4 Quaternary Structure of Selected Proteins

Protein	Number of Subunits
Alcohol dehydrogenase	2
Aldolase	4
Hemoglobin	4
Lactate dehydrogenase	4
Insulin	6
Glutamine synthetase	12
Tobacco mosaic virus protein disc	17

Study Guide

27.1 Amino Acids

- **Amino acids** are compounds that contain both an amino group and a carboxyl group.
- With the exception of glycine, all protein-derived amino acids are chiral.
  - *In the D,L convention, all of the 20 commonly occurring amino acids are L-amino acids. Note that this designation is based on a structural analogy to L-glyceraldehyde, not on the measured optical activity of amino acid samples.*

- In the *R,S* convention, 18 amino acids are (*S*)-amino acids.
  - *Although cysteine has the same absolute configuration, it is an (R)-amino acid because of the manner in which the sulfur atom changes the assignment about its tetrahedral chiral center.*
  - *Isoleucine and threonine contain a second chiral center.*
- The 20 protein-derived amino acids are commonly divided into four categories: nine with nonpolar side chains, four with polar but unionized side chains, four with acidic side chains, and three with basic side chains.

P 27.1, 27.8–27.22

Table 27.1

27.2 Acid-Base Properties of Amino Acids

- At neutral pH, amino acids exist as **zwitterions** because the amino group is protonated and positively charged and the carboxyl group is deprotonated and negatively charged.
  - *Because of the inductive effect of the electron-withdrawing —NH_3^+ group, amino acid carboxyl groups are more acidic than acetic acid.*
  - *The α-amino group is slightly less basic than a primary aliphatic amine.*
- The **isoelectric point, pI**, of an amino acid, a polypeptide, or a protein is the pH at which it has no net charge.
- Electrophoresis is the process of separating compounds on the basis of their charge in an electric field.
  - *Compounds having a high charge density move more rapidly than those with a lower charge density.*
  - *Any amino acid or protein in a solution with a pH that equals the pI of the compound remains at the origin.*
- Ninhydrin is used to detect amino acids and proteins because it reacts with primary amino groups to create a bright purple dye.

P 27.2, 27.3, 27.23–27.39

KEY REACTIONS

1. **Acidity of an α-Carboxyl Group (Section 27.2A)** An α-COOH (pK_a approximately 2.19) of a protonated amino acid is a considerably stronger acid than acetic acid (pK_a 4.76) or other low-molecular-weight aliphatic carboxylic acid, owing to the electron-withdrawing inductive effect of the α-NH_3^+ group.

$$\underset{\displaystyle NH_3^+}{RCHCOOH} + H_2O \rightleftharpoons \underset{\displaystyle NH_3^+}{RCHCOO^-} + H_3O^+ \qquad pK_a = 2.19$$

2. **Acidity of an α-Ammonium Group (Section 27.2A)** An α-NH_3^+ group (pK_a approximately 9.47) is a slightly stronger acid than is a primary aliphatic ammonium ion (pK_a approximately 10.76).

$$\underset{\displaystyle NH_3^+}{RCHCOO^-} + H_2O \rightleftharpoons \underset{\displaystyle NH_2}{RCHCOO^-} + H_3O^+ \qquad pK_a = 9.47$$

3. Reaction of an α-Amino Acid with Ninhydrin (Section 27.2D) Treatment of an α-amino acid with ninhydrin gives a purple-colored solution. Treatment of proline with ninhydrin gives an orange-colored solution.

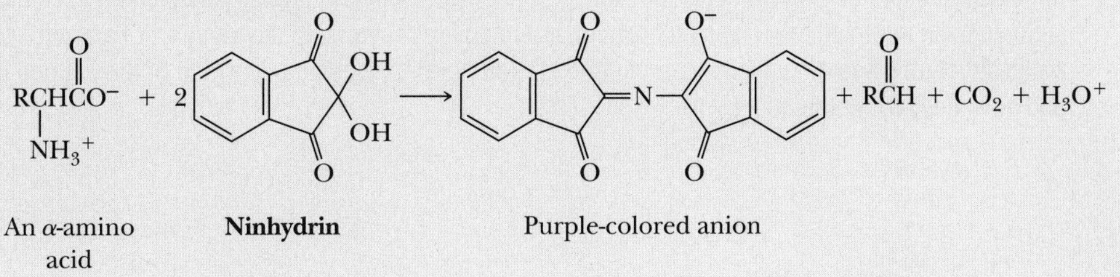

27.3 Polypeptides and Proteins

- A **peptide bond** is the special name given to the amide bond formed between α-amino acids.
- A **polypeptide** is a biological macromolecule containing many amino acids, each joined to the next by a peptide bond. By convention, the sequence of amino acids in a polypeptide is written beginning with the ***N*-terminal amino acid** toward the ***C*-terminal amino acid**.

P 27.4

▶ Figure 27.5 and Problem 27.4

27.4 Primary Structure of Polypeptides and Proteins

- **Primary (1°) structure** of a polypeptide is the sequence of amino acids in the polypeptide chain.
  - *The procedure known as* ***amino acid analysis*** *is used to determine relative amino acid composition of a protein. All of the amide bonds are hydrolyzed in acid, followed by chromatography to separate and quantify the different amino acids present. No sequence information is determined this way.*
- The primary structure of a polypeptide or protein can be determined directly by fragmenting the protein using cyanogen bromide and enzymes, followed by amino acid sequencing of the resulting fragments.
  - *Cyanogen bromide treatment cleaves proteins at the carboxyl group of methionine residues.*
  - *Protein-cleaving enzymes such as trypsin or chymotrypsin cleave between certain amino acids, leading to specific fragments.*
  - *The Edman degradation uses phenyl isothiocyanate to remove and identify the* N*-terminal amino acid of peptides or proteins. The Edman degradation can be repeated and automated so that the sequence of up to 20 to 30* N*-terminal amino acids can be determined.*
  - *Sequencing of overlapping fragments allows the reconstruction of a protein's entire sequence.*
- Today, mass spectrometry of a protein or nucleic acid sequencing of its gene are the preferred methods for establishing amino acid sequence of a protein.

P 27.41–27.50

KEY REACTIONS

4. Cleavage of a Peptide Bond by Cyanogen Bromide (Section 27.4B) Cleavage is regioselective for a peptide bond formed by the carboxyl group of methionine. The mechanism involves reaction of the electrophilic carbon of cyanogen bromide with the nucleophilic S atom of methionine to create a sulfonium ion intermediate. The sulfonium ion intermediate reacts with the carboxyl O atom to give a cyclic structure containing an imino group that is hydrolyzed to complete cleavage of the peptide bond and give the γ-lactone product.

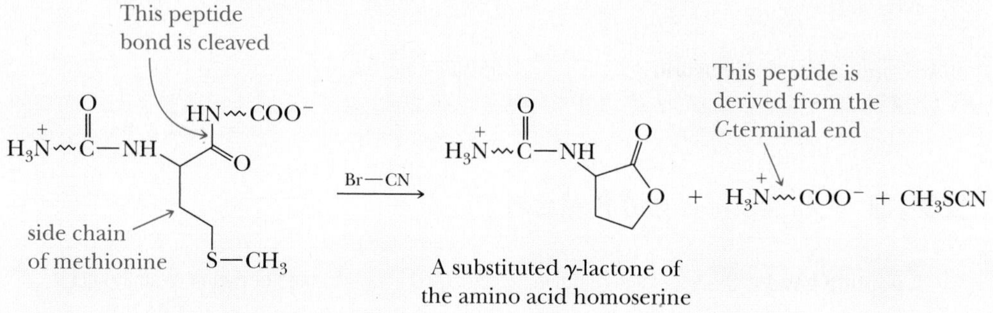

5. Edman Degradation (Section 27.4B) Treatment with phenyl isothiocyanate followed by acid removes the *N*-terminal amino acid as a substituted phenylthiohydantoin, which is then separated and identified. The mechanism involves reaction of the electrophilic C atom of phenyl isothiocyanate with the nucleophilic terminal amino group to give an *N*-phenylthiourea intermediate that decomposes upon heating by cyclization to give a thiazolinone intermediate as the *C*-terminal peptide bond is cleaved. The thiazolinone intermediate isomerizes to the phenylthiohydantoin product.

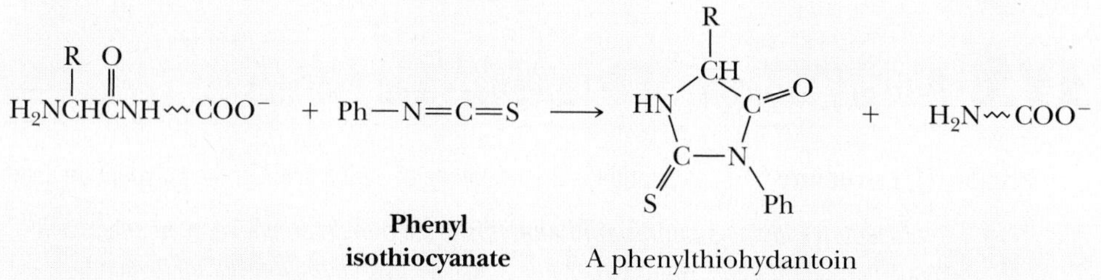

27.5 Synthesis of Polypeptides

- **Peptide synthesis** requires that the amino group of one amino acid and the carboxyl group of the other are protected so that only a desired coupling will take place.
  - *The most common amino-protecting groups are carbamates such as the Z (benzyloxycarbonyl) or BOC (*tert*-butyloxycarbonyl) groups that are removed in acid.*
  - *Carboxyl groups are protected in solution as esters or during solid phase synthesis by attachment to the solid support, often as esters.*
- Peptide bonds are made using reagents such as carbodiimides that activate a carboxyl group for nucleophilic attack by an amino group.
- In **solid-phase synthesis**, or polymer-supported synthesis of polypeptides, the *C*-terminal amino acid is joined to a chloromethylated polystyrene resin as a benzyl ester.
  - *The polypeptide chain is then extended one amino acid at a time.*

- *The main advantage of solid-phase synthesis is that all of the reagent exchange and washing steps are carried out by simple filtration, and the entire operation has been automated.*
- *When synthesis is completed, the polypeptide chain is released from the solid support by cleavage of the benzyl ester linkage.*

P 27.40, 27.51–27.53

Figure 27.11

KEY REACTIONS

6. The Benzyloxycarbonyl (Z—) Protecting Group (Section 27.5C) The benzyloxycarbonyl protecting group is prepared by treatment of an unprotected α-NH_2 group with benzyloxycarbonyl chloride. It is removed by treatment with HBr in acetic acid or by hydrogenolysis.

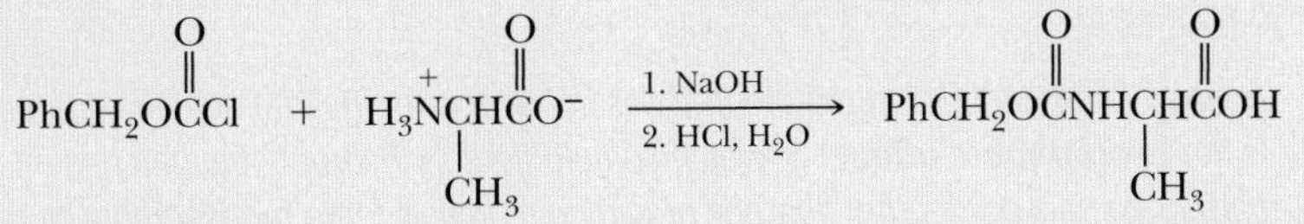

Benzyloxycarbonyl chloride (Z—Cl) | **Alanine** | ***N*-Benzyloxycarbonylalanine** (Z—Ala)

7. Peptide Bond Formation Using 1,3-Dicyclohexylcarbodiimide (Section 27.5E) This substituted carbodiimide is a dehydrating agent and is converted to a disubstituted urea. The reaction is efficient, and yields are generally very high. The mechanism involves initial proton transfer from the carboxylic acid to the 1,3-dicyclohexylcarbodiimide *N* atom, generating an electrophilic intermediate that reacts with the carboxylate to give an *O*-acylisourea. The *O*-acylisourea intermediate is an activated ester that reacts with the amino group of another amino acid to give a tetrahedral addition intermediate that decomposes to give a new peptide bond and dicyclohexylurea.

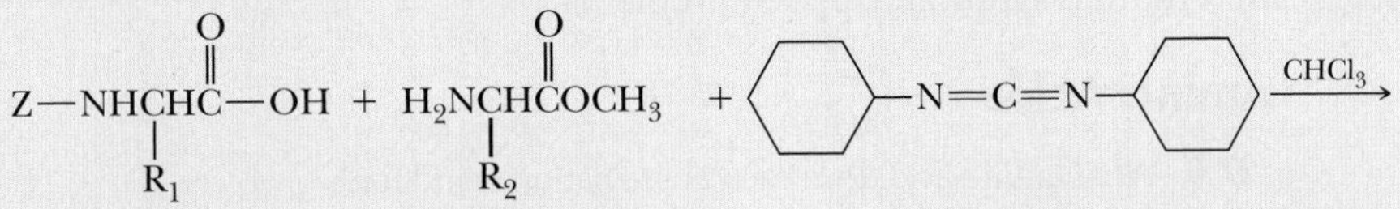

Amino-protected aa_1 | Carboxyl-protected aa_2 | **1,3-Dicyclohexylcarbodiimide (DCC)**

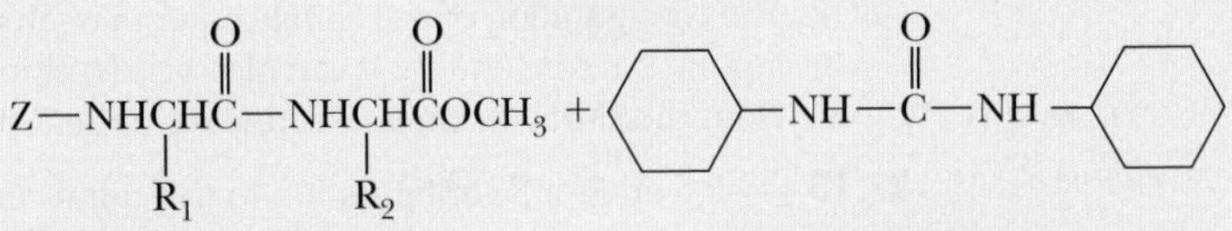

Amino- and carboxyl-protected dipeptide | ***N,N'*-Dicyclohexylurea (DCU)**

27.6 Three-Dimensional Shapes of Polypeptides and Proteins

- A peptide bond is planar; that is, the four atoms of the amide and the two α-carbons of a peptide bond lie in the same plane.
  - *The planarity is due to resonance involving the amide N atom.*
  - *Bond angles about the amide nitrogen and the amide carbonyl carbon are about 120°.*

- **Secondary (2°) structure** refers to the ordered arrangement (conformations) of amino acids in localized regions of a polypeptide or protein. The two most important types of secondary structure are the α-helix and the β-pleated sheet.
- **Tertiary (3°) structure** refers to the overall folding pattern and arrangement in space of all atoms in a single polypeptide chain.
  - *Solvation of amino acid side chains is important for protein folding, and it is observed that hydrophobic side chains tend to be located in the hydrophobic interior of a protein. The hydrophilic side chains, on the other hand, tend to be on the surface, exposed to the aqueous environment because of the* ***hydrophobic effect****.*
  - *The hydrophobic effect refers to the aggregation of hydrophobic groups in the interior of proteins caused by their exclusion from water.*
- **Quaternary (4°) structure** is the arrangement of polypeptide monomers into a noncovalently bonded aggregate.
  - *A major factor stabilizing the ordered assembly of proteins into specific quaternary structures is the* ***hydrophobic effect****, in which complementary hydrophobic patches on each interacting partner contact each other, thereby providing a driving force for assembly by relieving unfavorable contacts of the hydrophobic patches with water.*

P 27.7, 27.54–27.57

Problems

Red numbers indicate applied problems.

Amino Acids

27.8 What amino acid does each abbreviation stand for?

(a) Phe **(b)** Ser **(c)** Asp **(d)** Gln
(e) His **(f)** Gly **(g)** Tyr

27.9 The configuration of the chiral center in α-amino acids is most commonly specified using the D,L convention. It can also be identified using the *R,S* convention (Section 3.3). Does the chiral center in L-serine have the *R* or *S* configuration?

27.10 Assign an *R* or *S* configuration to the chiral center in each amino acid.

(a) L-Phenylalanine **(b)** L-Glutamic acid **(c)** L-Methionine

27.11 The amino acid threonine has two chiral centers. The stereoisomer found in proteins has the configuration 2*S*,3*R* about the two chiral centers. Draw **(a)** a Fischer projection of this stereoisomer and **(b)** a three-dimensional representation.

27.12 Define the term *zwitterion*.

27.13 Draw zwitterion forms of these amino acids.

(a) Valine **(b)** Phenylalanine **(c)** Glutamine

27.14 Why are Glu and Asp often referred to as acidic amino acids?

27.15 Why is Arg often referred to as a basic amino acid? Which other two amino acids are also basic amino acids?

27.16 What is the meaning of the alpha as it is used in α-amino acid?

27.17 Several β-amino acids exist. There is a unit of β-alanine, for example, contained within the structure of coenzyme A (Problem 25.34). Write the structural formula of β-alanine.

27.18 Although only L-amino acids occur in proteins, D-amino acids are often a part of the metabolism of lower organisms. The antibiotic actinomycin D, for example, contains a unit of D-valine, and the antibiotic bacitracin A contains units of D-asparagine and D-glutamic acid. Draw Fischer projections and three-dimensional representations for these three D-amino acids.

27.19 Histamine is synthesized from one of the 20 protein-derived amino acids. Suggest which amino acid is its biochemical precursor and the type of organic reaction(s) involved in its biosynthesis (e.g., oxidation, reduction, decarboxylation, nucleophilic substitution).

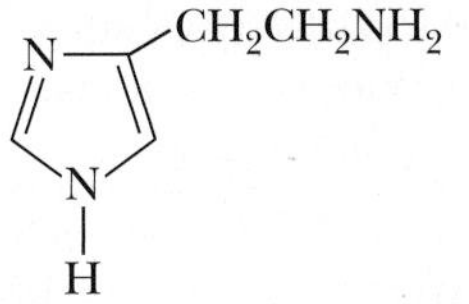

Histamine

27.20 Vitamin K participates in carboxylation of glutamic acid residues of the blood-clotting protein prothrombin.

(a) Write a structural formula for γ-carboxyglutamic acid.
(b) Account for the fact that the presence of γ-carboxyglutamic acid escaped detection for many years; on routine amino acid analyses, only glutamic acid was detected.

27.21 Both norepinephrine and epinephrine are synthesized from the same protein-derived amino acid. From which amino acid are they synthesized? What types of reactions are involved in their biosynthesis?

(a)

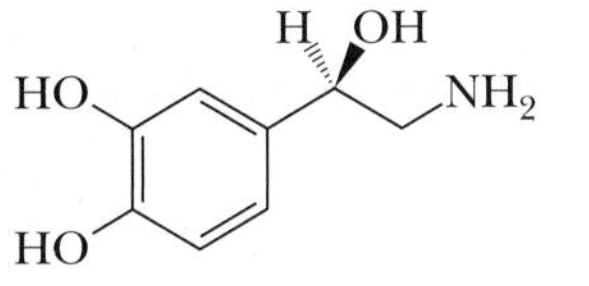

Norepinephrine

(b)

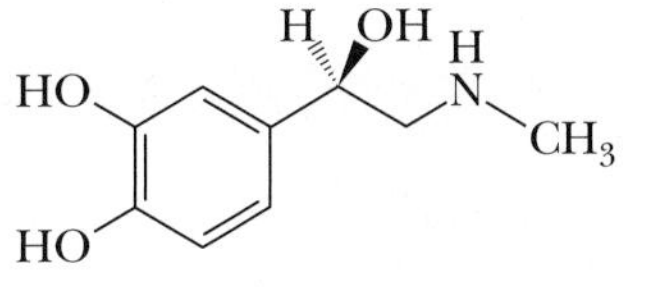

Epinephrine (Adrenaline)

27.22 From which amino acid are serotonin and melatonin synthesized? What types of reactions are involved in their biosynthesis?

(a)

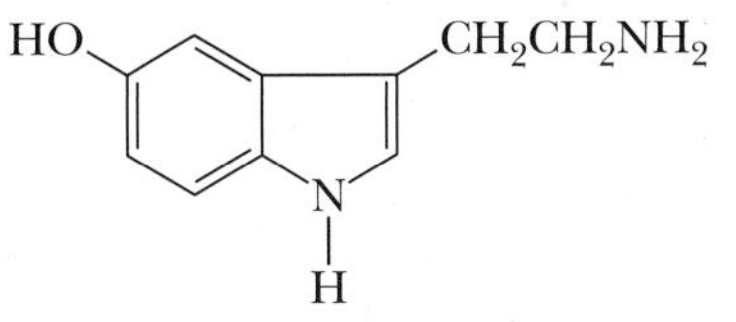

Serotonin

(b)

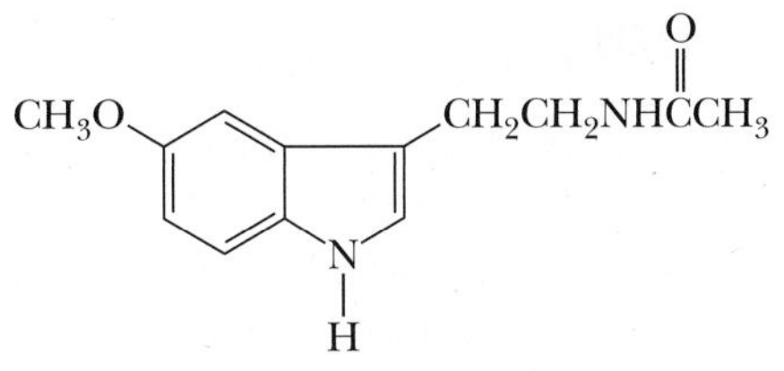

Melatonin

Acid-Base Behavior of Amino Acids

27.23 Draw a structural formula for the form of each amino acid most prevalent at pH 1.0.

(a) Threonine **(b)** Arginine **(c)** Methionine **(d)** Tyrosine

27.24 Draw a structural formula for the form of each amino most prevalent at pH 10.0.

(a) Leucine **(b)** Valine **(c)** Proline **(d)** Aspartic acid

27.25 Write the zwitterion form of alanine and show its reaction with the following.

(a) 1 mol NaOH **(b)** 1 mol HCl

27.26 Write the form of lysine most prevalent at pH 1.0 and then show its reaction with the following. Consult Table 27.2 for pK_a values of the ionizable groups in lysine.

(a) 1 mol NaOH **(b)** 2 mol NaOH **(c)** 3 mol NaOH

27.27 Write the form of aspartic acid most prevalent at pH 1.0 and then show its reaction with the following. Consult Table 27.2 for pK_a values of the ionizable groups in aspartic acid.

(a) 1 mol NaOH **(b)** 2 mol NaOH **(c)** 3 mol NaOH

27.28 Given pK_a values for ionizable groups from Table 27.2, sketch curves for the titration of **(a)** glutamic acid with NaOH and **(b)** histidine with NaOH.

27.29 Draw a structural formula for the product formed when alanine is treated with the following reagents.

(a) Aqueous NaOH
(b) Aqueous HCl
(c) CH_3CH_2OH, H_2SO_4
(d) $(CH_3CO)_2O$, CH_3COONa

27.30 For lysine and arginine, the isoelectric point, pI, occurs at a pH where the net charge on the nitrogen-containing groups is +1 and balances the charge of −1 on the α-carboxyl group. Calculate pI for these amino acids.

27.31 For aspartic and glutamic acids, the isoelectric point occurs at a pH where the net charge on the two carboxyl groups is −1 and balances the charge of +1 on the α-amino group. Calculate pI for these amino acids.

27.32 Account for the fact that the isoelectric point of glutamine (pI 5.65) is higher than the isoelectric point of glutamic acid (pI 3.08).

27.33 Enzyme-catalyzed decarboxylation of glutamic acid gives 4-aminobutanoic acid (Section 27.1D). Estimate the pI of 4-aminobutanoic acid.

27.34 Guanidine and the guanidino group present in arginine are two of the strongest organic bases known. Account for their basicity.

27.35 At pH 7.4, the pH of blood plasma, do the majority of protein-derived amino acids bear a net negative charge or a net positive charge? Explain.

27.36 Do the following compounds migrate to the cathode or to the anode on electrophoresis at the specified pH?

(a) Histidine at pH 6.8
(b) Lysine at pH 6.8
(c) Glutamic acid at pH 4.0
(d) Glutamine at pH 4.0
(e) Glu-Ile-Val at pH 6.0
(f) Lys-Gln-Tyr at pH 6.0

27.37 At what pH would you carry out an electrophoresis to separate the amino acids in each mixture?

(a) Ala, His, Lys **(b)** Glu, Gln, Asp **(c)** Lys, Leu, Tyr

27.38 Examine the amino acid sequence of human insulin (Figure 27.16) and list each Asp, Glu, His, Lys, and Arg in this molecule. Do you expect human insulin to have an isoelectric point nearer that of the acidic amino acids (pI 2.0–3.0), the neutral amino acids (pI 5.5–6.5) or the basic amino acids (pI 9.5–11.0)?

27.39 A chemically modified guanidino group is present in cimetidine (Tagamet), a widely prescribed drug for the control of gastric acidity and peptic ulcers. Cimetidine reduces gastric acid secretion by inhibiting the interaction of histamine with gastric H_2 receptors. In the development of this drug, a cyano group was added to the substituted guanidino group to alter its basicity. Do you expect this modified guanidino group to be more basic or less basic than the guanidino group of arginine? Explain.

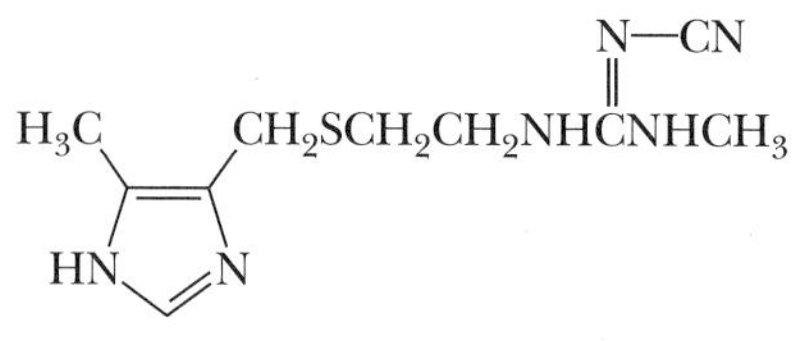
N—CN
H3C
CH2SCH2CH2NHCNHCH3
HN
N

Cimetidine
(Tagamet)

27.40 Draw a structural formula for the product formed when alanine is treated with the following reagents.

(a) 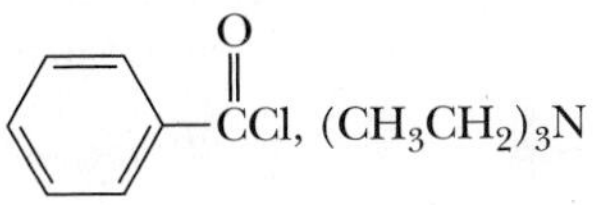

(b) 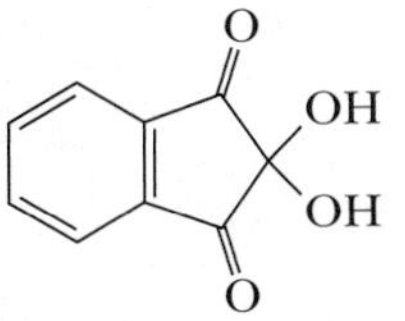

(c) 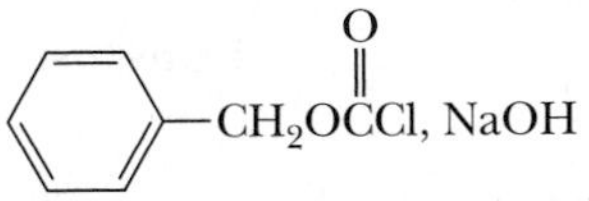

(d) $(CH_3)_3COC(=O)OC(=O)OC(CH_3)_3$, NaOH

(e) Product from (c), L-Alanine ethyl ester, DCC
(f) Product from (d), L-Alanine ethyl ester, DCC

Primary Structure of Polypeptides and Proteins

27.41 If a protein contains four different SH groups, how many different disulfide bonds are possible if only a single disulfide bond is formed? How many different disulfides are possible if two disulfide bonds are formed?

27.42 How many different tetrapeptides can be made under the following conditions?

(a) The tetrapeptide contains one unit each of Asp, Glu, Pro, and Phe.
(b) All 20 amino acids can be used, but each only once.

27.43 A decapeptide has the following amino acid composition.

$$\text{Ala}_2\text{, Arg, Cys, Glu, Gly, Leu, Lys, Phe, Val}$$

Partial hydrolysis yields the following tripeptides.

Cys-Glu-Leu + Gly-Arg-Cys + Leu-Ala-Ala + Lys-Val-Phe + Val-Phe-Gly

One round of Edman degradation yields a lysine phenylthiohydantoin. From this information, deduce the primary structure of this decapeptide.

27.44 Following is the primary structure of glucagon, a polypeptide hormone of 29 amino acids. Glucagon is produced in the α-cells of the pancreas and helps maintain blood glucose levels in a normal concentration range.

1 (His), 5 (Thr), 10 (Tyr), 15 (Asp), 20 (Gln), 25 (Trp), 29 (Thr)
His-Ser-Glu-Gly-Thr-Phe-Thr-Ser-Asp-Tyr-Ser-Lys-Tyr-Leu-Asp-Ser-Arg-Arg-Ala-Gln-Asp-Phe-Val-Gln-Trp-Leu-Met-Asn-Thr

Which peptide bonds are hydrolyzed when this polypeptide is treated with each reagent?

(a) Phenyl isothiocyanate (b) Chymotrypsin (c) Trypsin (d) BrCN

27.45 A tetradecapeptide (14 amino acid residues) gives the following peptide fragments on partial hydrolysis. From this information, deduce the primary structure of this polypeptide. Fragments are grouped according to size.

Pentapeptide Fragments	Tetrapeptide Fragments
Phe-Val-Asn-Gln-His	Gln-His-Leu-Cys
His-Leu-Cys-Gly-Ser	His-Leu-Val-Glu
Gly-Ser-His-Leu-Val	Leu-Val-Glu-Ala

27.46 Draw a structural formula of these tripeptides. Mark each peptide bond, the *N*-terminal amino acid, and the *C*-terminal amino acid.

(a) Phe-Val-Asn (b) Leu-Val-Gln

27.47 Estimate the pI of each tripeptide in Problem 27.46.

27.48 Glutathione (G-SH), one of the most common tripeptides in animals, plants, and bacteria, is a scavenger of oxidizing agents. In reacting with oxidizing agents, glutathione is converted to G-S-S-G.

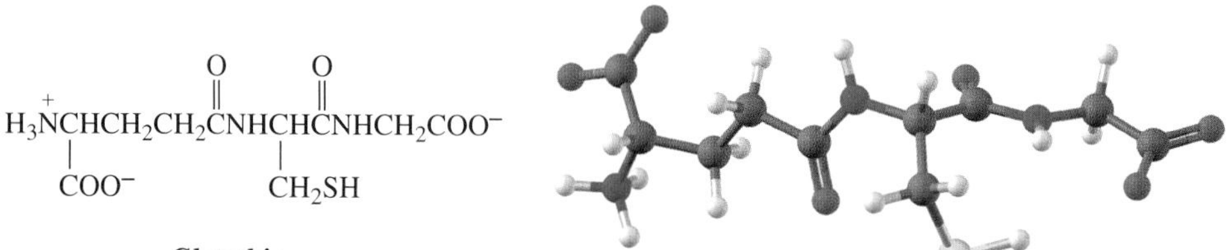

Glutathione

(a) Name the amino acids in this tripeptide.

(b) What is unusual about the peptide bond formed by the *N*-terminal amino acid?

(c) Write a balanced half-reaction for the reaction of two molecules of glutathione to form a disulfide bond. Is glutathione a biological oxidizing agent or a biological reducing agent?

(d) Write a balanced equation for reaction of glutathione with molecular oxygen, O_2, to form G-S-S-G and H_2O. Is molecular oxygen oxidized or reduced in this process?

27.49 Following are a structural formula and a ball-and-stick model for the artificial sweetener aspartame. Each amino acid has the L configuration.

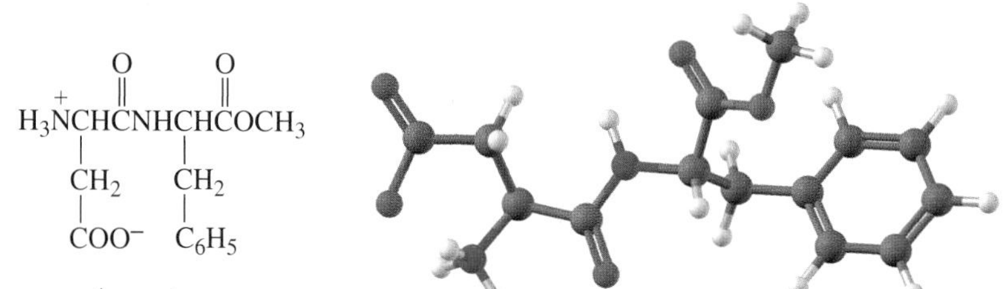

Aspartame

(a) Name the two amino acids in this molecule.

(b) Estimate the isoelectric point of aspartame.

(c) Draw structural formulas for the products of hydrolysis of aspartame in 1 *M* HCl.

Aspartame is present in many artificially sweetened foods and beverages.

27.50 2,4-Dinitrofluorobenzene, very often known as Sanger's reagent after the English chemist Frederick Sanger who popularized its use, reacts selectively with the *N*-terminal amino group of a polypeptide chain. Sanger was awarded the 1958 Nobel Prize in Chemistry for his work in determining the primary structure of bovine insulin. One of the few people to be awarded two Nobel Prizes, he also shared the 1980 award in chemistry with American chemists Paul Berg and Walter Gilbert for the development of chemical and biological analyses of DNA.

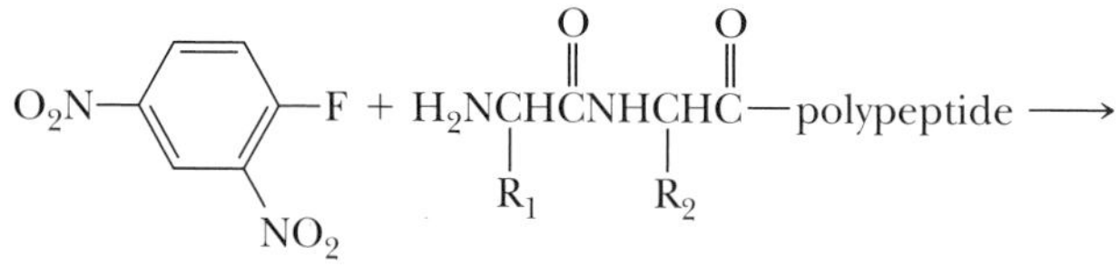

polypeptide chain in which the *N*-terminal amino acid is labeled with a 2,4-dinitrophenyl group

2,4-Dinitrofluorobenzene (*N*-Terminal end of a polypeptide chain)

Following reaction with 2,4-dinitrofluorobenzene, all amide bonds of the polypeptide chain are hydrolyzed and the amino acid labeled with a 2,4-dinitrophenyl group is separated by either paper or column chromatography and identified.

(a) Write a structural formula for the product formed by treatment of the *N*-terminal amino group with Sanger's reagent and propose a mechanism for its formation.

(b) When bovine insulin is treated with Sanger's reagent followed by hydrolysis of all peptide bonds, two labeled amino acids are detected: glycine and phenylalanine. What conclusions can be drawn from this information about the primary structure of bovine insulin?

(c) Compare and contrast the structural information that can be obtained from use of Sanger's reagent with that from use of the Edman degradation.

Synthesis of Polypeptides

27.51 In a variation of the Merrifield solid-phase peptide synthesis, the amino group is protected by a fluorenylmethoxycarbonyl (FMOC) group. This protecting group is removed by treatment with a weak base such as the secondary amine piperidine. Write a balanced equation and propose a mechanism for this deprotection.

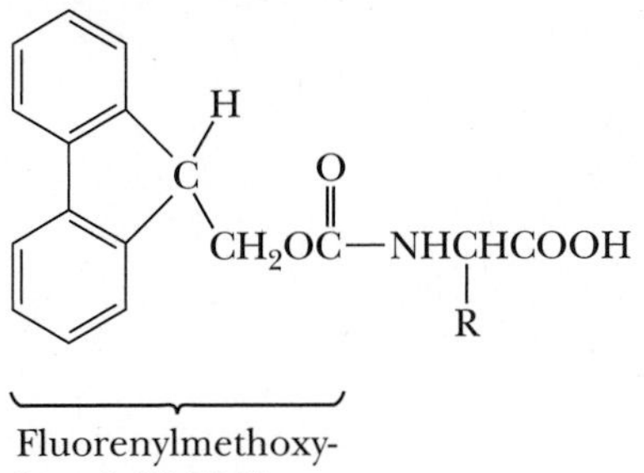

Fluorenylmethoxy-
carbonyl (FMOC) group

27.52 The BOC-protecting group may be added by treatment of an amino acid with di-*tert*-butyl dicarbonate as shown in the following reaction sequence. Propose a mechanism to account for formation of these products.

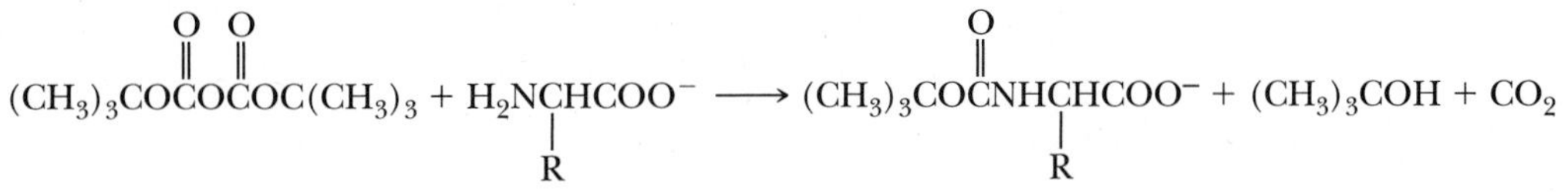

Di-*tert*-butyl dicarbonate BOC-amino acid

27.53 The side-chain carboxyl groups of aspartic acid and glutamic acid are often protected as benzyl esters.

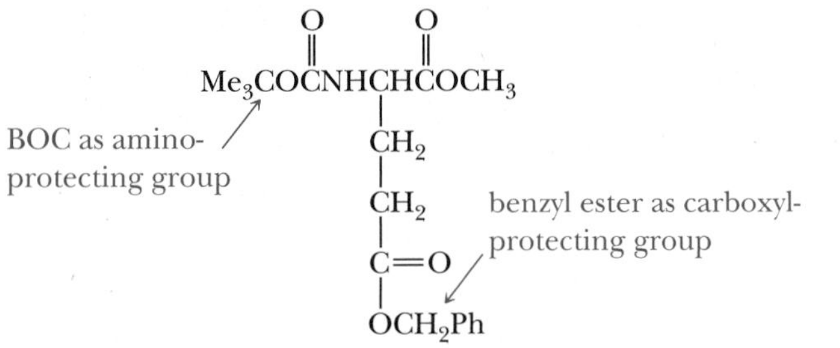
BOC as amino-
protecting group
benzyl ester as carboxyl-
protecting group

(a) Show how to convert the side-chain carboxyl group to a benzyl ester using benzyl chloride as a source of the benzyl group.

(b) How do you deprotect the side-chain carboxyl under mild conditions without removing the BOC-protecting group at the same time?

Three-Dimensional Shapes of Polypeptides and Proteins

27.54 Examine the α-helix conformation. Are amino acid side chains arranged all inside the helix, all outside the helix, or randomly?

27.55 Distinguish between intermolecular and intramolecular hydrogen bonding between the backbone groups on polypeptide chains. In what type of secondary structure do you find intermolecular hydrogen bonds? In what type do you find intramolecular hydrogen bonding?

27.56 Many plasma proteins found in an aqueous environment are globular in shape. Which amino acid side chains would you expect to find on the surface of a globular protein and in contact with the aqueous environment? Which would you expect to find inside, shielded from the aqueous environment? Explain.

(a) Leu **(b)** Arg **(c)** Ser **(d)** Lys **(e)** Phe

27.57 Denaturation of a protein is a physical change, the most readily observable result of which is loss of biological activity. Denaturation stems from changes in secondary, tertiary, and quaternary structure through disruption of noncovalent interactions including hydrogen bonding and hydrophobic interactions. Three common denaturing agents are sodium dodecyl sulfate (SDS), urea, and heat. What kinds of noncovalent interactions might each reagent disrupt?

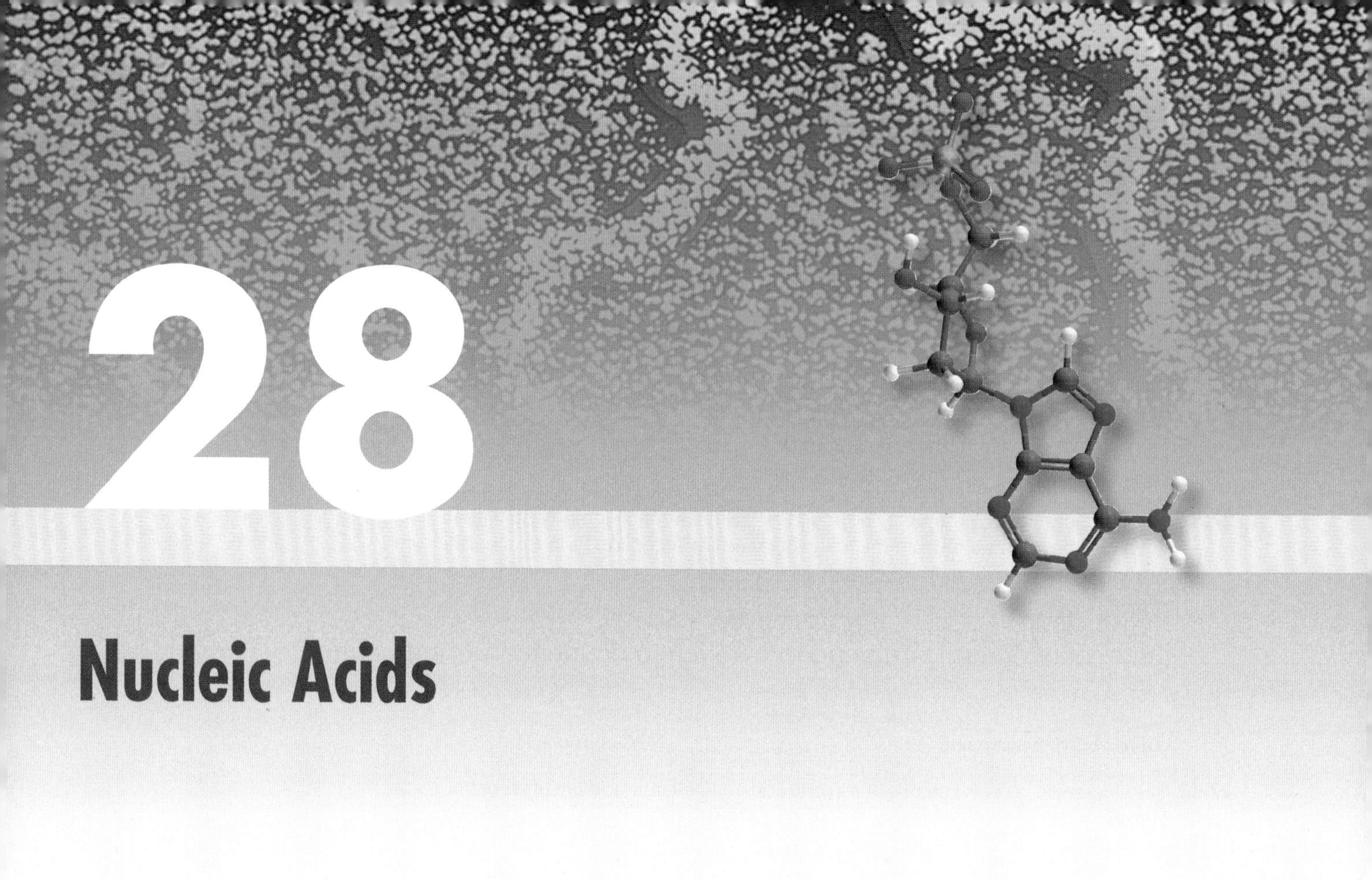

28

Nucleic Acids

Outline

28.1 Nucleosides and Nucleotides
28.2 The Structure of DNA
28.3 Ribonucleic Acids
28.4 The Genetic Code
28.5 Sequencing Nucleic Acids

False-colored transmission electron micrograph of a plasmid DNA. If the cell wall of a bacterium such as *Escherichia coli* is partially digested and the cell is then osmotically shocked by dilution with water, its contents are extruded to the exterior. Shown here is the bacterial chromosome surrounding the cell. **Inset:** 2′-deoxyadenosine 5′-monophosphate (dAMP), a building block of DNA (Section 28.2). (Professor Stanley N. Cohen/Stone/Science Sources)

The organization, maintenance, and regulation of cellular function require a tremendous amount of information, all of which must be passed on each time a cell is replicated. With very few exceptions, genetic information is stored and transmitted from one generation to the next in the form of deoxyribonucleic acid (DNA). Genes, the hereditary units of chromosomes, are long stretches of double-stranded DNA. If the DNA in a human chromosome in a single cell were uncoiled, it would be approximately 1.8 meters long!

Genetic information is expressed in two stages: transcription from DNA to ribonucleic acids (RNA) and then translation for the synthesis of proteins.

$$\text{DNA} \xrightarrow{\text{transcription}} \text{RNA} \xrightarrow{\text{translation}} \text{proteins}$$

Thus, DNA is the repository of genetic information in cells, whereas RNA serves in the transcription and translation of this information, which is then expressed through the synthesis of proteins.

In this chapter, we examine the structure of nucleosides and nucleotides and the manner in which these monomers are covalently bonded to form **nucleic acids**. Then we examine the manner in which genetic information is encoded on molecules of

DNA, the function of the three types of ribonucleic acids, and finally how the primary structure of a DNA molecule is determined.

Using chemistry that we will not discuss in the text, DNA can be synthesized on a solid phase analogous to the method described for the synthesis of oligopeptides (Section 27.5F). In fact, both DNA and RNA can now be synthesized with high yield and the entire process has been automated. It is commonplace for molecular biologists to place an order for a specific DNA or RNA on the Internet and have the sample arrive one or two days later. Such efficiency, made possible by automation and high-yielding synthetic reactions, has revolutionized molecular biology and biotechnology.

Nucleic acid

A biopolymer containing three types of monomer units: heterocyclic aromatic amine bases derived from purine and pyrimidine, the monosaccharides D-ribose or 2-deoxy-D-ribose, and phosphoric acid.

28.1 Nucleosides and Nucleotides

Controlled hydrolysis of nucleic acids yields three components: heterocyclic aromatic amine bases, the monosaccharides D-ribose or 2-deoxy-D-ribose (Section 25.1A), and phosphate ions. The five heterocyclic aromatic amine bases most common to nucleic acids are shown in Figure 28.1. Uracil, cytosine, and thymine are referred to as pyrimidine bases after the name of the parent base; adenine and guanine are referred to as purine bases.

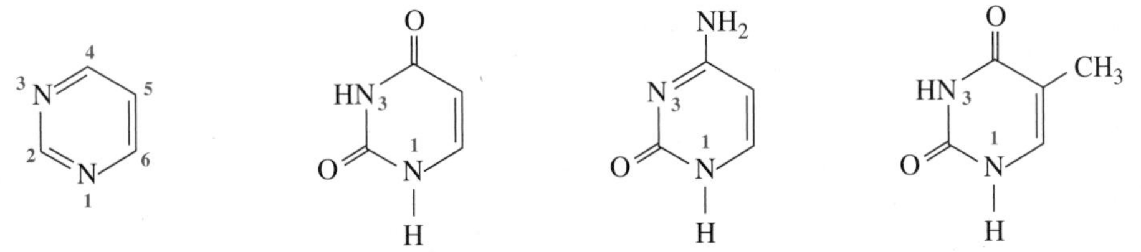

Pyrimidine **Uracil (U)** **Cytosine (C)** **Thymine (T)**

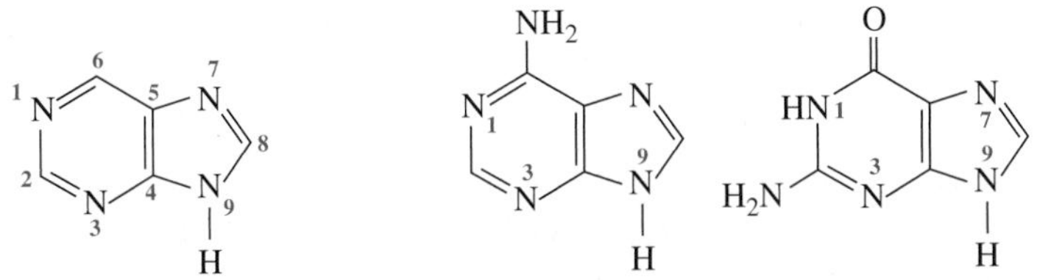

Purine **Adenine (A)** **Guanine (G)**

FIGURE 28.1 Names and one-letter abbreviations for the heterocyclic aromatic amine bases most common to DNA and RNA. Bases are numbered according to the patterns of the parent compounds, pyrimidine and purine.

A **nucleoside** is a compound containing D-ribose or 2-deoxy-D-ribose bonded to a heterocyclic aromatic amine base by a β-*N*-glycosidic bond (Section 25.3A). The monosaccharide component of DNA is 2-deoxy-D-ribose, whereas that of RNA is D-ribose. The glycosidic bond is between C-1′ (the anomeric carbon) of ribose or 2-deoxyribose and N-1 of a pyrimidine base or N-9 of a purine base. Figure 28.2 shows a structural formula for uridine, a nucleoside derived from ribose and uracil.

A **nucleotide** is a nucleoside in which a molecule of phosphoric acid is esterified with a free hydroxyl of the monosaccharide, most commonly either the 3′-hydroxyl or the 5′-hydroxyl. A nucleotide is named by giving the name of the parent nucleoside followed by the word *monophosphate*. The position of the phosphoric ester is specified by the number of the carbon to which it is bonded. Figure 28.3 shows a structural formula of adenosine 5′-monophosphate, AMP. Monophosphoric esters are diprotic acids with pK_a values of approximately 1 and 6. Therefore, at pH 7, the two hydrogens of a phosphoric monoester are fully ionized, giving a nucleotide a charge of −2.

Nucleoside

A building block of nucleic acids that consists of D-ribose or 2-deoxy-D-ribose bonded to a heterocyclic aromatic amine base by a β-*N*-glycosidic bond.

Nucleotide

A nucleoside in which a molecule of phosphoric acid is esterified with an —OH of the monosaccharide, most commonly either the 3′-OH or the 5′-OH.

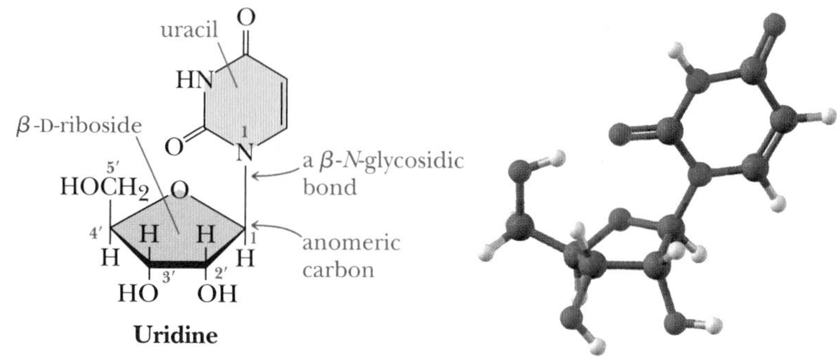

FIGURE 28.2 Uridine, a nucleoside. Atom numbers on the monosaccharide rings are primed to distinguish them from atom numbers on the heterocyclic aromatic amine bases.

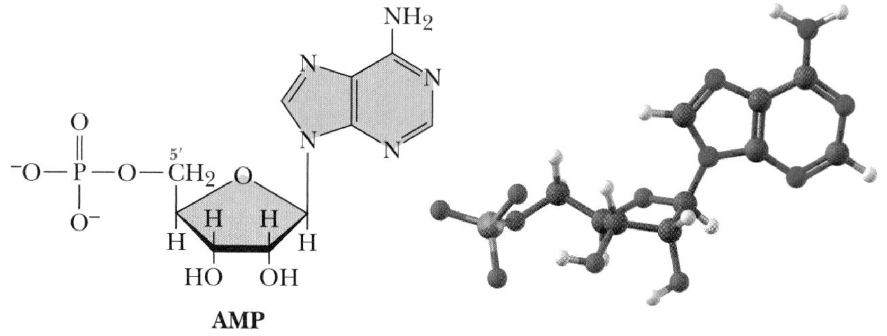

FIGURE 28.3 Adenosine 5′-monophosphate, AMP. The phosphate group is fully ionized at pH 7.0, giving this nucleotide a charge of −2.

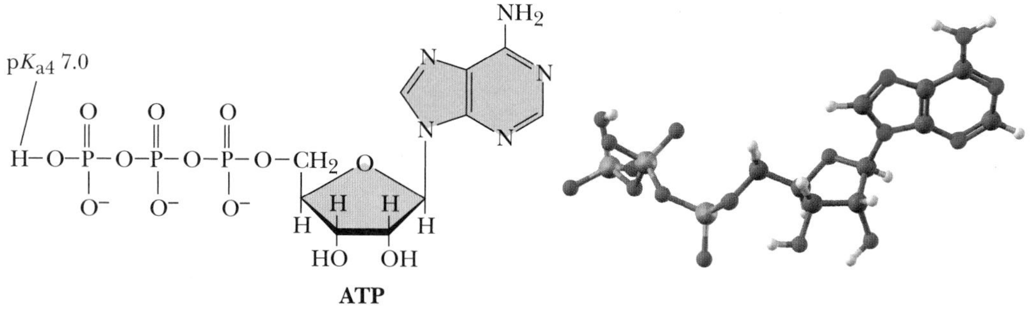

FIGURE 28.4 Adenosine 5′-triphosphate, ATP.

Nucleoside monophosphates can be further phosphorylated to form nucleoside diphosphates and nucleoside triphosphates. Shown in Figure 28.4 is a structural formula for adenosine 5′-triphosphate, ATP.

Nucleoside diphosphates and triphosphates are also polyprotic acids and are extensively ionized at pH 7.0. pK_a values of the first three ionization steps for adenosine triphosphate are less than 5.0. The value of pK_{a4} is approximately 7.0. Therefore, at pH 7.0, approximately 50% of adenosine triphosphate is present as ATP^{4-} and 50% is present as ATP^{3-}.

▶ Delocalization of π electron density over a larger area is stabilizing

The N atoms of the exocyclic amine groups of the nucleic acid bases adenine, cytosine, and guanine are sp^2 hybridized and therefore have a planar geometry. Like other aromatic amines, resonance with the aromatic π cloud is responsible for this geometry because it accommodates maximum orbital overlap. The important point is that the nucleic acid bases are all essentially planar, facilitating stacking in helical structures as well as allowing for optimum hydrogen bonding to complementary bases. ▶

Example 28.1 Nucleotides

Draw a structural formula for each nucleotide.

(a) 2′-Deoxycytidine 5′-diphosphate **(b)** 2′-Deoxyguanosine 3′-monophosphate

Solution

(a) Cytosine is joined by a β-*N*-glycosidic bond between N-1 of cytosine and C-1 of the cyclic hemiacetal form of 2-deoxy-D-ribose. The 5′-hydroxyl of the pentose is bonded to a phosphate group by an ester bond, and this phosphate is, in turn, bonded to a second phosphate group by an anhydride bond.

(b) Guanine is joined by a β-*N*-glycosidic bond between N-9 of guanine and C-1 of the cyclic hemiacetal form of 2-deoxy-D-ribose. The 3′-hydroxyl group of the pentose is joined to a phosphate group by an ester bond.

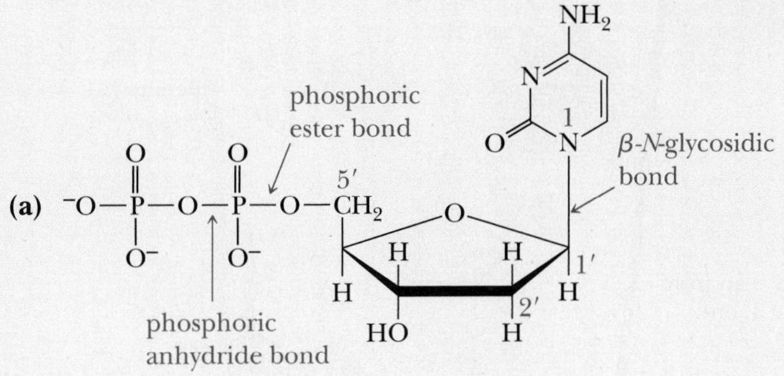

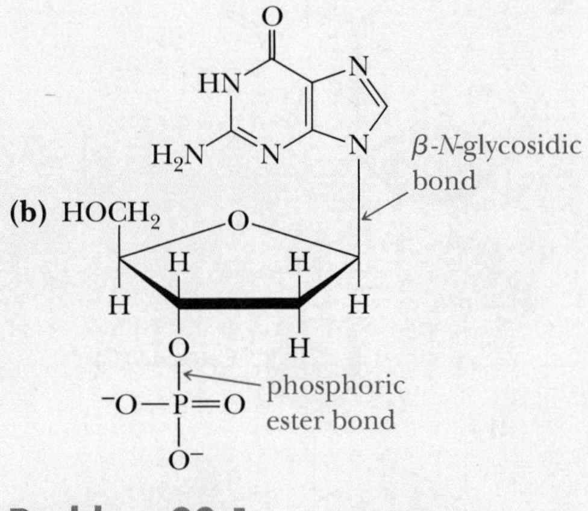

Problem 28.1

Draw a structural formula for each nucleotide.

(a) 2′-Deoxythymidine 5′-monophosphate

(b) 2′-Deoxythymidine 3′-monophosphate

28.2 The Structure of DNA

In Chapter 27, we saw that the four levels of structural complexity in polypeptides and proteins are primary, secondary, tertiary, and quaternary structures. There are three levels of structural complexity in nucleic acids. Although these levels are somewhat comparable to those in polypeptides and proteins, they also differ in significant ways.

A. Primary Structure—The Covalent Backbone

Deoxyribonucleic acids consist of a backbone of alternating units of deoxyribose and phosphate in which the 3′-hydroxyl of one deoxyribose unit is joined by a phosphodiester bond to the 5′-hydroxyl of another deoxyribose unit (Figure 28.5).

FIGURE 28.5 A tetranucleotide section of a single-stranded DNA.

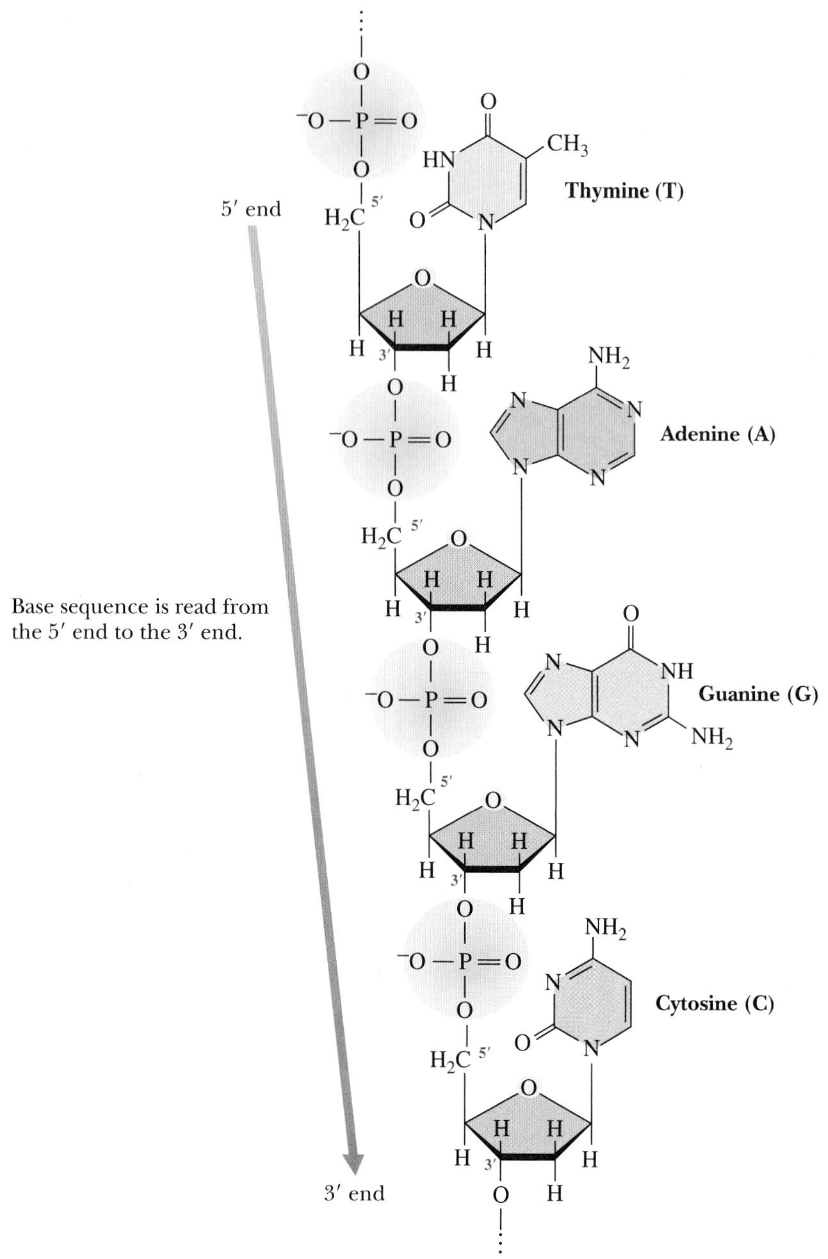
5′ end
Base sequence is read from the 5′ end to the 3′ end.
3′ end
Thymine (T)
Adenine (A)
Guanine (G)
Cytosine (C)

Primary structure of nucleic acids

The sequence of bases along the pentose-phosphodiester backbone of a DNA or RNA molecule read from the 5′ end to the 3′ end.

This pentose-phosphodiester backbone is constant throughout an entire DNA molecule. A heterocyclic aromatic amine base—adenine, guanine, thymine, or cytosine—is bonded to each deoxyribose unit by a β-*N*-glycosidic bond. **Primary structure** of DNA refers to the order of heterocyclic bases along the pentose-phosphodiester backbone. The sequence of bases is usually written from the 5′ end to the 3′ end.

Example 28.2 DNA

Draw a structural formula for the DNA dinucleotide TG that is phosphorylated at the 5′ end only.

Solution

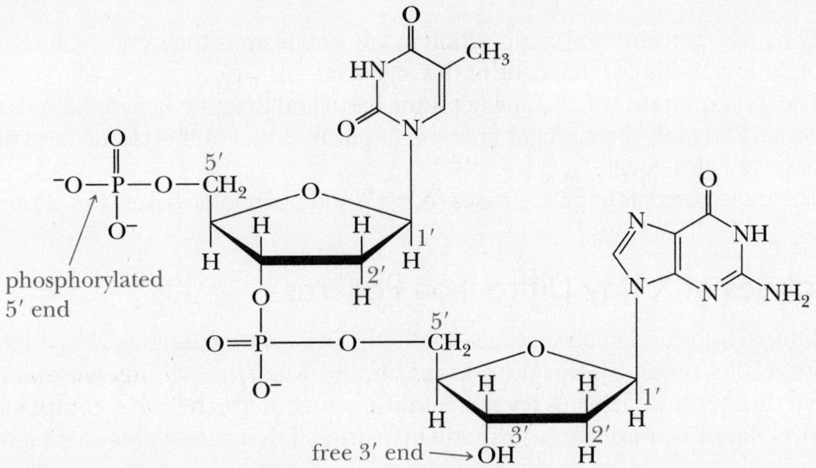

phosphorylated
5′ end
free 3′ end

Problem 28.2

Draw a structural formula for the section of DNA that contains the base sequence CTG and is phosphorylated at the 3′ end only.

B. Secondary Structure—The Double Helix

By the early 1950s, it was clear that DNA molecules consist of chains of alternating units of deoxyribose and phosphate joined by 3′,5′-phosphodiester bonds with a base attached to each deoxyribose unit by a β-*N*-glycosidic bond. In 1953, the American biologist James D. Watson and the British physicist Francis H. C. Crick proposed a double-helix model for the **secondary structure** for DNA. Watson, Crick, and Maurice Wilkins shared the 1962 Nobel Prize in Physiology or Medicine for "their discoveries concerning the molecular structure of nucleic acids, and its significance for information transfer in living material." Although Rosalind Franklin also played an important part in this research, she did not share in the Nobel Prize because of her death in 1958.

The **Watson-Crick model** was based on molecular modeling and two lines of experimental observations: chemical analyses of DNA base compositions and mathematical analyses of X-ray diffraction patterns of crystals of DNA.

Secondary structure of nucleic acids
The ordered arrangement of nucleic acid strands.

Watson-Crick model
A double-helix model for the secondary structure of a DNA molecule.

A. Barrington Brown/Science Sources

Watson and Crick with their model of DNA.

Base Composition

At one time, it was thought that the four principal bases occur in the same ratios and perhaps repeat in a regular pattern along the pentose-phosphodiester backbone of DNA for all species. However, more precise determinations of base composition by Erwin Chargaff revealed that bases do not occur in the same ratios (Table 28.1).

Table 28.1 Comparison in Base Composition, in Mole-Percent, of DNA from Several Organisms

	Purines		Pyrimidines				
Organism	A	G	C	T	A/T	G/C	Purines/ Pyrimidines
Human	30.4	19.9	19.9	30.1	1.01	1.00	1.01
Sheep	29.3	21.4	21.0	28.3	1.04	1.02	1.03
Yeast	31.7	18.3	17.4	32.6	0.97	1.05	1.00
E. coli	26.0	24.9	25.2	23.9	1.09	0.99	1.04

Researchers drew the following conclusions from this and related data. To within experimental error:

1. The mole-percent base composition in any organism is the same in all cells of the organism and is characteristic of the organism.
2. The mole-percents of adenine (a purine base) and thymine (a pyrimidine base) are equal. The mole-percents of guanine (a purine base) and cytosine (a pyrimidine base) are also equal.
3. The mole-percents of purine bases (A + G) and pyrimidine bases (C + T) are equal.

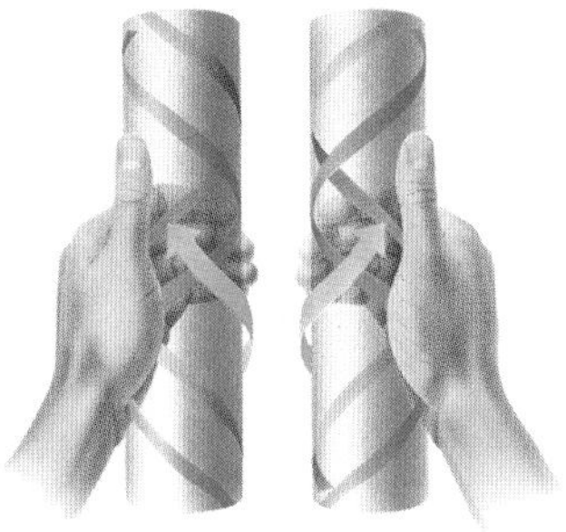

FIGURE 28.6 A DNA double helix has a chirality associated with the helix. Right-handed and left-handed double helices of otherwise identical DNA chains are nonsuperposable mirror images.

Analyses of X-Ray Diffraction Patterns

Additional information about the structure of DNA emerged when X-ray diffraction photographs taken by Rosalind Franklin and Maurice Wilkins were analyzed. These diffraction patterns revealed that even though the base composition of DNA isolated from different organisms varies, DNA molecules themselves are remarkably uniform in thickness. They are long and fairly straight, with an outside diameter of approximately 2000 pm, and not more than a dozen atoms thick. Furthermore, the crystallographic pattern repeats every 3400 pm. Herein lay one of the chief problems to be solved. How could the molecular dimensions of DNA be so regular even though the relative percentages of the various bases differ so widely? With this accumulated information, the stage was set for the development of a hypothesis about DNA structure.

The Watson-Crick Double Helix

Double helix
A type of secondary structure of DNA molecules in which two antiparallel polynucleotide strands are coiled in a right-handed manner about the same axis.

The heart of the Watson-Crick model is the postulate that a molecule of DNA is a complementary **double helix**. It consists of two antiparallel polynucleotide strands coiled in a right-handed manner about the same axis to form a double helix. As illustrated in the ribbon models in Figure 28.6, chirality is associated with a double helix; left-handed and right-handed double helices are related by reflection just as enantiomers are related by reflection.

Chemical Connections

The Search for Antiviral Drugs

The search for antiviral drugs has been more difficult than the search for antibacterial drugs primarily because viral replication depends on the metabolic processes of the invaded cell. Thus, antiviral drugs are also likely to cause harm to the cells that harbor the virus. The challenge in developing antiviral drugs has been to understand the biochemistry of viruses and to develop drugs that target processes specific to them. Compared with the large number of antibacterial drugs that are available, there are only a handful of antiviral drugs, and they have nowhere near the effectiveness that antibiotics have on bacterial infections.

Acyclovir is one of the first of a new family of drugs for the treatment of infectious diseases caused by DNA viruses called herpesvirus. Herpes infections in humans are of two kinds: herpes simplex type 1, which gives rise to mouth and eye sores, and herpes simplex type 2, which gives rise to serious genital infections. Acyclovir is highly effective against herpesvirus-caused genital infections. The drug is activated *in vivo* by conversion of the primary —OH (which corresponds to the 5′-OH of a riboside or a deoxyriboside) to a triphosphate. Because of its close resemblance to deoxyguanosine triphosphate, an essential precursor for DNA synthesis, acyclovir triphosphate is taken up by viral DNA polymerase to form an enzyme-substrate complex on which no 3′-OH exists for replication to continue. Thus, the enzyme-substrate complex is no longer active (it is a dead-end complex), viral replication is disrupted, and the virus is destroyed.

Perhaps the best known of the new viral antimetabolites is zidovudine (azidothymidine, AZT), an analog of deoxythymidine in which the 3′-OH has been replaced by an azido group, N_3. AZT is effective against HIV-1, a retrovirus that is the causative agent of AIDS. It is converted *in vivo* by cellular enzymes to the 5′-triphosphate, recognized as deoxythymidine 5′-triphosphate by viral RNA-dependent DNA polymerase (reverse transcriptase), and added to a growing DNA chain. There it stops chain elongation because no 3′-OH exists on which to add the next deoxynucleotide. AZT owes its effectiveness to the fact that it binds more strongly to viral reverse transcriptase than it does to human DNA polymerase.

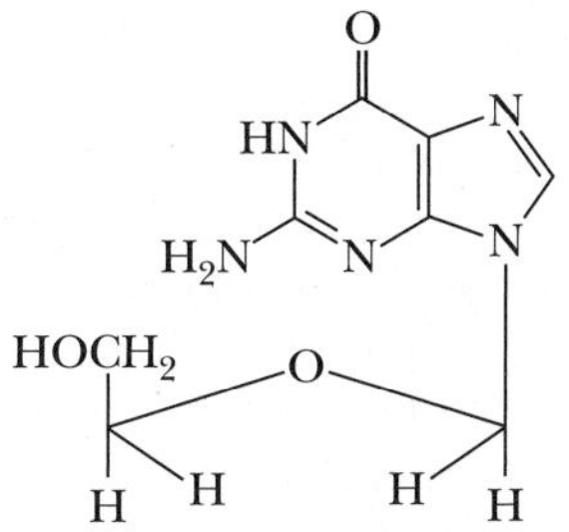

Acyclovir
(drawn to show its structural relationship to 2-deoxyguanosine)

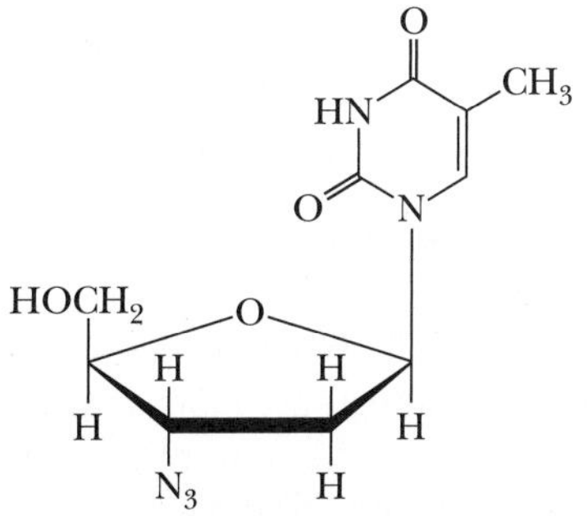

Zidovudine (Azidothymidine; AZT)

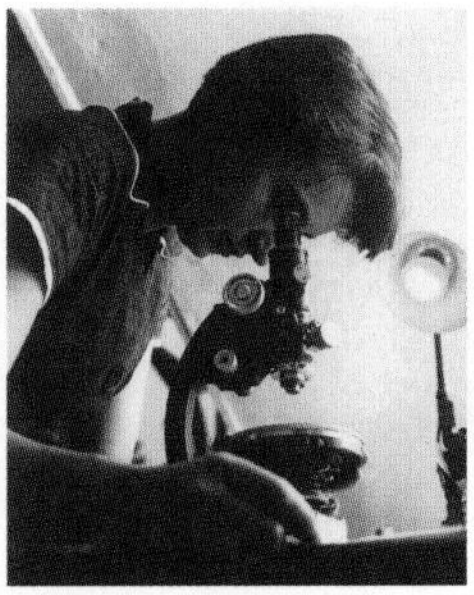

Rosalind Franklin (1920–1958). In 1951, she joined the Biophysical Laboratory at King's College, London, where she used X-ray diffraction methods to study DNA. She is credited with discoveries that established the density of DNA and its helical conformation. Her work was important to the model of DNA developed by Watson and Crick. She died in 1958 at age 37, and because the Nobel Prize is not awarded posthumously, she did not share in the 1962 Nobel Prize in Physiology or Medicine with Watson, Crick, and Wilkins. Although her relationship with Watson and Crick was initially strained, Watson said, "we later came to appreciate . . . the struggles the intelligent woman faces to be accepted by the scientific world which often regards women as mere diversions from serious thinking."

To account for the observed base ratios and uniform thickness of DNA, Watson and Crick postulated that purine and pyrimidine bases project inward toward the axis of the helix and are always paired in a very specific manner. According to scale models, the dimensions of an adenine-thymine base pair are almost identical to the dimensions of a guanine-cytosine base pair, and the length of each pair is consistent with the core thickness of a DNA strand (Figure 28.7). Thus, if the purine base in one strand is adenine, then its complement in the antiparallel strand must be thymine. Similarly, if the purine in one

FIGURE 28.7 Base-pairing between adenine and thymine (A-T) and between guanine and cytosine (G-C). An A-T base pair is held by two hydrogen bonds, whereas a G-C base pair is held by three hydrogen bonds.

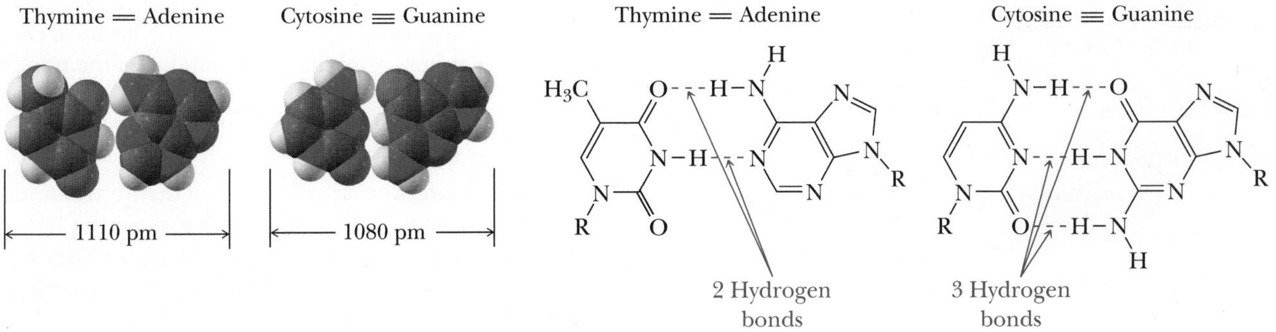

strand is guanine, its complement in the antiparallel strand must be cytosine. The "fits" between the TA pair and between the CG pair are remarkable and represent another important example of molecular recognition and supramolecular complex formation. Two specific hydrogen bonds are formed between each TA base pair and three hydrogen bonds formed between each GC base pair. No other purine-pyrimidine combination of the four DNA bases can form stable hydrogen bonded pairs. The specific hydrogen bonding holds the two strands together very strongly.

A significant feature of Watson and Crick's model is that no other base-pairing is consistent with the observed thickness of a DNA molecule. A pair of pyrimidine bases is too small to account for the observed thickness, whereas a pair of purine bases is too large. Thus, according to the Watson-Crick model, the repeating units in a double-stranded DNA molecule are not single bases of differing dimensions, but rather base pairs of almost identical dimensions.

To account for the periodicity observed from X-ray data, Watson and Crick postulated that base pairs are stacked one on top of the other with a distance of 340 pm between base pairs and with ten base pairs in one complete turn of the helix. There is one complete turn of the helix every 3400 pm. Shown in Figure 28.8 is a ribbon model of double-stranded **B-DNA**, the predominant form of DNA in dilute aqueous solution and thought to be the most common form in nature.

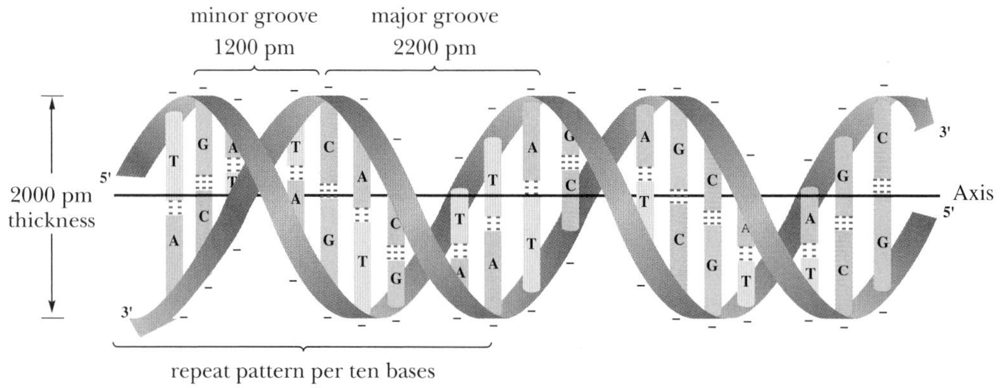

FIGURE 28.8 Ribbon model of double-stranded B-DNA. Each ribbon shows the pentose-phosphodiester backbone of a single-stranded DNA molecule. The strands are antiparallel; one strand runs to the left from the 5′ end to the 3′ end, and the other runs to the right from the 5′ end to the 3′ end. Hydrogen bonds are shown by three dotted lines between each G-C base pair and two dotted lines between each A-T base pair.

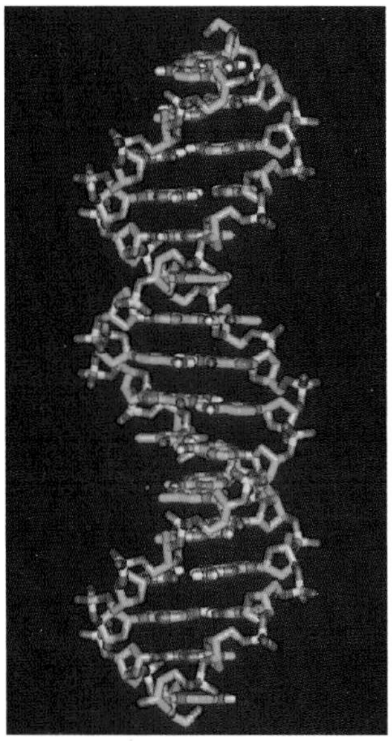

Professor Stanley Cohen/SPL/Photo Researchers, Inc.

FIGURE 28.9 An idealized model of B-DNA.

In the double helix, the bases in each base pair are not directly opposite one another across the diameter of the helix, but rather are slightly displaced. This displacement and the relative orientation of the glycosidic bonds linking each base to the sugar-phosphate backbone leads to two differently sized grooves, a major groove and a minor groove (Figure 28.8). Each groove runs along the length of the cylindrical column of the double helix. The major groove is approximately 2200 pm wide; the minor groove is approximately 1200 pm wide.

Figure 28.9 shows more detail of an idealized B-DNA double helix. The major and minor grooves are clearly recognizable in this model.

Other forms of secondary structure are known that differ in the distance between stacked base pairs and in the number of base pairs per turn of the helix. One of the most common of these, **A-DNA**, also a right-handed helix, is thicker than **B-DNA** and has a repeat distance of only 2900 pm. There are ten base pairs per turn of the helix with a spacing of 290 pm between base pairs.

Example 28.3 Complementary Sequences

One strand of a DNA molecule contains the base sequence 5′-ACTTGCCA-3′. Write its complementary base sequence.

Solution

Remember that the base sequence is always written from the 5′ end of the strand to the 3′ end, that A pairs with T, and that G pairs with C. In double-stranded DNA, the two strands run in opposite (antiparallel) directions so that the 5′ end of one strand is associated with the 3′ end of the other strand.

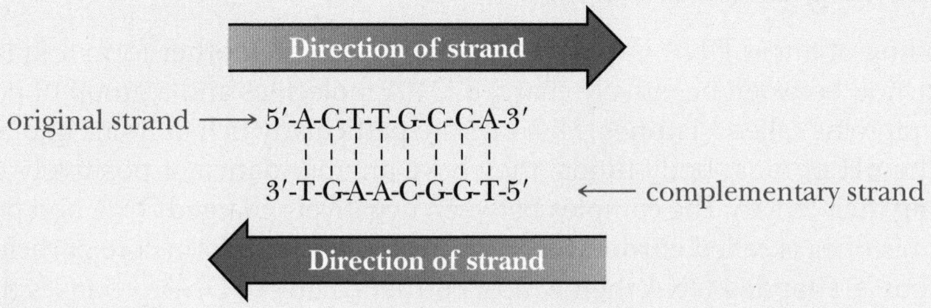
Direction of strand
original strand ⟶ 5′-A-C-T-T-G-C-C-A-3′
3′-T-G-A-A-C-G-G-T-5′ ⟵ complementary strand
Direction of strand

Written from the 5′ end, the complementary strand is 5′-TGGCAAGT-3′.

Problem 28.3

Write the complementary DNA base sequence for 5′-CCGTACGA-3′.

C. Tertiary Structure—Supercoiled DNA

The length of a DNA molecule is enormously greater than its diameter, and the extended molecule is quite flexible. A DNA molecule is said to be relaxed if it has no twists other than those imposed by its secondary structure. Said another way, relaxed DNA does not have a clearly defined tertiary structure. We consider two types of tertiary structure, one type induced by perturbations in circular DNA and a second type introduced by coordination of DNA with nuclear proteins called histones. Tertiary structure, whatever the type, is referred to as **supercoiling**.

Tertiary structure of nucleic acids
The three-dimensional arrangement of all atoms of a nucleic acid, commonly referred to as supercoiling.

Supercoiling of Circular DNA

Circular DNA is a type of double-stranded DNA in which the two ends of each strand are joined by phosphodiester bonds [Figure 28.10(a)]. This type of DNA, the most prominent form in bacteria and viruses, is also referred to as circular duplex (because

Circular DNA
A type of double-stranded DNA in which the 5′ and 3′ ends of each strand are joined by phosphodiester groups.

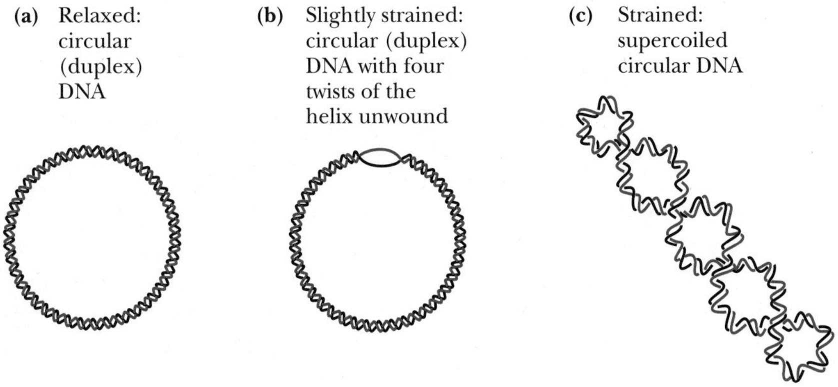
(a) Relaxed: circular (duplex) DNA
(b) Slightly strained: circular (duplex) DNA with four twists of the helix unwound
(c) Strained: supercoiled circular DNA

FIGURE 28.10 Relaxed and supercoiled DNA. (a) Circular DNA is relaxed. (b) One strand is broken, unwound by four turns, and the ends then rejoined. The strain of unwinding is localized in the nonhelical gap. (c) Supercoiling by four twists distributes the strain of unwinding uniformly over the entire molecule of circular DNA.

it is double-stranded) DNA. One strand of circular DNA may be opened, partially unwound, and then rejoined. The unwound section introduces a strain into the molecule because the nonhelical gap is less stable than hydrogen-bonded, base-paired helical sections. The strain can be localized in the nonhelical gap. Alternatively, it may be spread uniformly over the entire circular DNA by introduction of **superhelical** twists, one twist for each turn of a helix unwound. The circular DNA shown in Figure 28.10(b) has been unwound by four complete turns of the helix. The strain introduced by this unwinding is spread uniformly over the entire molecule by introduction of four superhelical twists [Figure 28.10(c)]. Interconversion of relaxed and supercoiled DNA is catalyzed by groups of enzymes called topoisomerases and gyrases.

Supercoiling of Linear DNA

Histone
A protein, particularly rich in the basic amino acids lysine and arginine, that is found associated with DNA molecules.

Supercoiling of linear DNA in plants and animals takes another form and is driven by interaction between negatively charged DNA molecules and a group of positively charged proteins called **histones**. Histones are particularly rich in lysine and arginine, and at the pH of most body fluids, they have an abundance of positively charged sites along their length. The complex between negatively charged DNA and positively charged histones is called **chromatin**. Histones associate to form core particles about which double-stranded DNA then wraps. Further coiling of DNA produces the chromatin found in cell nuclei. The entire complex, with DNA wound around the histone protein cores, can be thought of as being analogous to beads on a string.

28.3 Ribonucleic Acids

Ribonucleic acids are similar to deoxyribonucleic acids in that they too consist of long, unbranched chains of nucleotides joined by phosphodiester groups between the 3′-hydroxyl of one pentose and the 5′-hydroxyl of the next. There are, however, three major differences in structure between RNA and DNA.

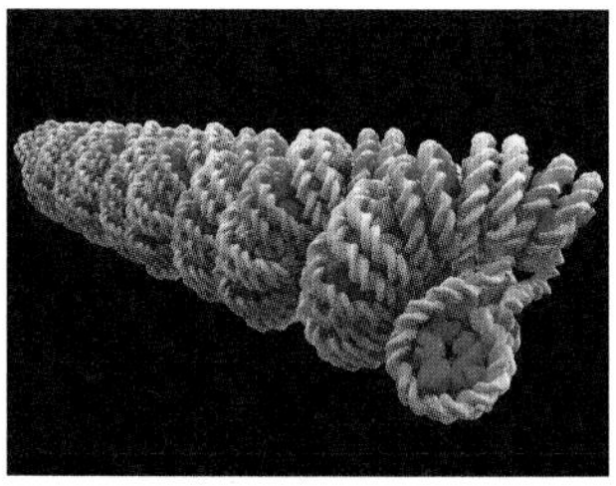
Alfred Pasieka/Science Photo Library

Supercoiled DNA from a mitochondrion.

1. The pentose unit in RNA is β-D-ribose rather than β-2-deoxy-D-ribose.
2. The pyrimidine bases in RNA are uracil and cytosine rather than thymine and cytosine (Figure 28.1).
3. RNA is single-stranded rather than double-stranded.

Cells contain up to eight times as much RNA as DNA; in contrast to DNA, RNA occurs in different forms and in multiple copies of each form. RNA molecules are classified, according to their structure and function, into three major types: ribosomal RNA, transfer RNA, and messenger RNA. The molecular weight, number of nucleotides, and percent cellular abundance of these types in cells of *Escherichia coli* are summarized in Table 28.2.

Table 28.2 Types of RNA Found in Cells of *E. coli*

Type	Molecular-Weight Range (g/mol)	Number of Nucleotides	Percentage of Cell RNA
mRNA	25,000–1,000,000	75–3000	2
tRNA	23,000–30,000	73–94	16
rRNA	35,000–1,100,000	120–2904	82

Chemical Connections

The Fountain of Youth

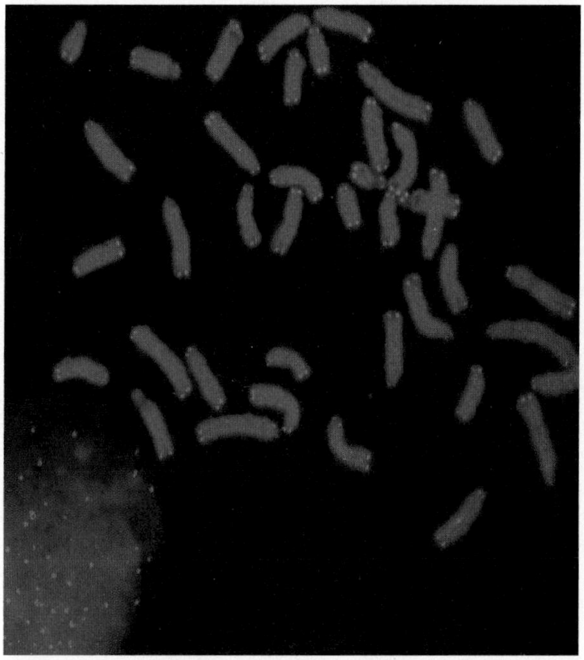

From the Laboratory of Drs. Jerry W. Shay and Woodring E. Wright

Telomeres are the repeating DNA strings (TTAGGC in vertebrates) that cap chromosomes.

In 1997, a sheep named Dolly became the first mammal to be born through cloning. By the time she was three years old, however, her genetic material was aging at the rate of the six-year-old sheep from which she was cloned. It turned out that Dolly had shortened telomeres, and—as a consequence—her genetic structure was considerably "older" than Dolly herself.

Telomeres are the physical ends of eukaryotic chromosomes, and they surrender a bit of themselves during each cell division because the DNA-replicating machinery normally fails to copy the DNA at a chromosome's tips. This process limits the total number of divisions a cell can undergo because the telomeres are ultimately consumed. In egg and cancer cells, among others, an enzyme called telomerase rebuilds the telomere cap after each division, making such cells effectively immortal. Consequently, telomerase may be the fountain of youth because it is responsible for the extension of telomeres in most species. If telomerase activity is limited, as it is in normal cells, the telomeres will shorten, and cells will age, like Dolly's. Telomeres are the repeating DNA strings (TTAGGG in vertebrates) that cap chromosomes.

Researchers are trying to alter the integrity of telomeres by controlling the expression and location of telomerase. Preserving telomere caps may rejuvenate aging organisms by enabling cells to survive additional divisions. Conversely, doing away with telomerase activity in tumors may cause cancer cells to cease their endless replications and die.

Based on a chemistry honors paper by James Stinebaugh, UCLA.

A. Ribosomal RNA

The bulk of **ribosomal RNA (rRNA)** is found in the cytoplasm in subcellular particles called ribosomes, which contain about 60% RNA and 40% protein. Ribosomes are the sites in cells at which protein synthesis takes place.

Ribosomal RNA (rRNA) A ribonucleic acid found in ribosomes, the sites of protein synthesis.

B. Transfer RNA

Transfer RNA (tRNA) molecules have the lowest molecular weight of all nucleic acids. They consist of 73–94 nucleotides in a single chain. The function of tRNA is to carry amino acids to the sites of protein synthesis on the ribosomes. Each amino acid has at least one tRNA dedicated specifically to this purpose. Several amino acids have more than one. In the transfer process, the amino acid is joined to its specific tRNA by an ester bond between the α-carboxyl group of the amino acid and the 3′ hydroxyl group of the ribose unit at the 3′ end of the tRNA.

Transfer RNA (tRNA) A ribonucleic acid that carries a specific amino acid to the site of protein synthesis on ribosomes.

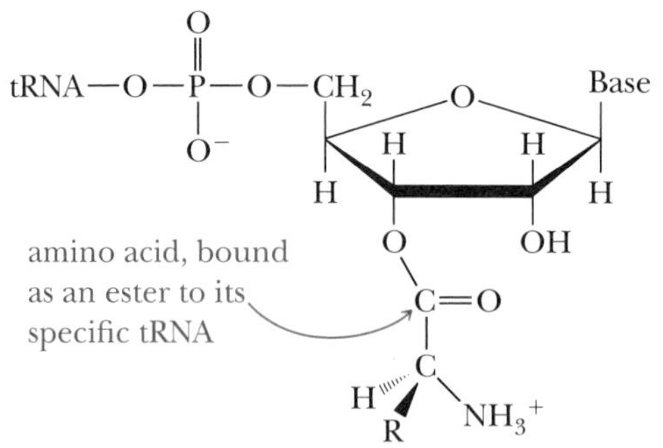
tRNA—O—P—O—CH2
O
O−
Base
H
H
H
H
O
OH
amino acid, bound
as an ester to its
specific tRNA
C=O
C
H
R
NH3+

C. Messenger RNA

Messenger RNA (mRNA) A ribonucleic acid that carries coded genetic information from DNA to the ribosomes for the synthesis of proteins.

Messenger RNAs (mRNA) are present in cells in relatively small amounts and are very short-lived. They are single-stranded, and their synthesis is directed by information encoded on DNA molecules. Double-stranded DNA is unwound, and a complementary strand of mRNA is synthesized along one strand of the DNA template, beginning from the 3′ end. The synthesis of mRNA from a DNA template is called transcription because genetic information contained in a sequence of bases of DNA is transcribed into a complementary sequence of bases on mRNA. The name *messenger* is derived from the function of this type of RNA, which is to carry coded genetic information from DNA to the ribosomes for the synthesis of proteins.

Example 28.4 mRNA

Following is a base sequence from a portion of DNA. Write the sequence of bases of the mRNA synthesized using this section of DNA as a template.

3′-A-G-C-C-A-T-G-T-G-A-C-C-5′

Solution

RNA synthesis begins at the 3′ end of the DNA template and proceeds toward the 5′ end. The complementary mRNA strand is formed using the bases C, G, A, and U. Uracil (U) is the complement of adenine (A) on the DNA template.

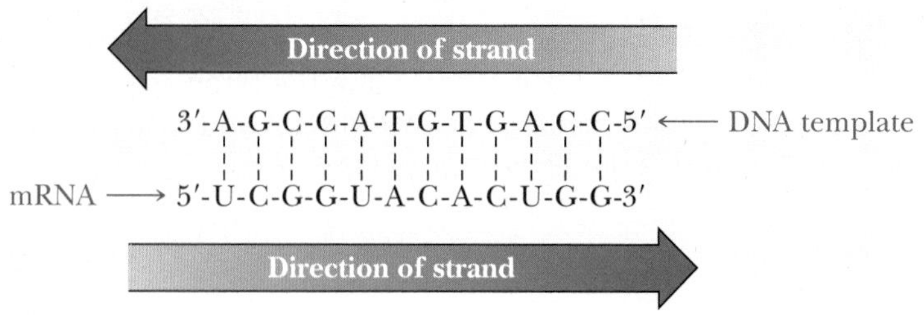
Direction of strand
3′-A-G-C-C-A-T-G-T-G-A-C-C-5′ ⟵ DNA template
mRNA ⟶ 5′-U-C-G-G-U-A-C-A-C-U-G-G-3′
Direction of strand

Reading from the 5′ end, the sequence of mRNA is 5′-UCGGUACACUGG-3′.

Problem 28.4

Here is a portion of the nucleotide sequence in phenylalanine tRNA.

3′-ACCACCUGCUCAGGCCUU-5′

Write the nucleotide sequence of its DNA complement.

28.4 The Genetic Code

A. Triplet Nature of the Code

It was clear by the early 1950s that the sequence of bases in DNA molecules constitutes the store of genetic information and directs the synthesis of messenger RNA, which, in turn, directs the synthesis of proteins. However, the statement that "the sequence of bases in DNA directs the synthesis of proteins" presents the following problem. How can a molecule containing only four variable units (adenine, cytosine, guanine, and thymine) direct the synthesis of molecules containing up to 20 variable units (the protein-derived amino acids)? How can an alphabet of only 4 letters code for the order of letters in the 20-letter alphabet that occurs in proteins?

An obvious answer is that there is not one base, but rather a combination of bases coding for each amino acid. If the code consists of nucleotide pairs, there are $4^2 = 16$ combinations; this code is more extensive, but it is still not extensive enough to code for 20 amino acids. If the code consists of nucleotides in groups of three, there are $4^3 = 64$ combinations, which is more than enough to code for the primary structure of a protein. This appears to be a very simple solution to a system that must have taken eons to evolve. Yet, nature does indeed use this simple three-letter, or triplet, code to store genetic information. A triplet of nucleotides is called a **codon**.

Codon

A triplet of nucleotides on mRNA that directs incorporation of a specific amino acid into a polypeptide sequence.

B. Deciphering the Genetic Code

The next question is which of the 64 triplets code for which amino acid? In 1961, Marshall Nirenberg provided a simple experimental approach to the problem based on the observation that synthetic polynucleotides direct polypeptide synthesis in much the same manner as do natural mRNAs. Nirenberg incubated ribosomes, amino acids, tRNAs, and appropriate protein-synthesizing enzymes. With only these components, there was no polypeptide synthesis. However, when he added synthetic polyuridylic acid (poly U), a polypeptide of high molecular weight was synthesized. Even more important, the synthetic polypeptide contained only phenylalanine. With this discovery, the first element of the genetic code was deciphered: the triplet UUU codes for phenylalanine.

Similar experiments were carried out with different synthetic polyribonucleotides. It was found, for example, that polyadenylic acid (poly A) leads to the synthesis of polylysine and that polycytidylic acid (poly C) leads to the synthesis of polyproline. By 1964, all 64 codons had been deciphered (Table 28.3).

C. Properties of the Genetic Code

Several features of the genetic code are evident from a study of Table 28.3.

1. Only 61 triplets code for amino acids. The remaining three (UAA, UAG, and UGA) are signals for chain termination; they signal to the protein-synthesizing machinery of the cell that the primary sequence of the protein is complete. The three chain termination triplets are indicated in Table 28.3 by "Stop."
2. The code is degenerate, which means that several amino acids are coded for by more than one triplet. Only methionine and tryptophan are coded for by just one triplet. Leucine, serine, and arginine are coded for by six triplets, and the remaining amino acids are coded for by two, three, or four triplets.
3. For the 15 amino acids coded for by two, three, or four triplets, only the third letter of the code varies. For example, glycine is coded for by the triplets GGA, GGG, GGC, and GGU.
4. There is no ambiguity in the code, meaning that each triplet codes for only one amino acid.

Table 28.3 The Genetic Code—mRNA Codons and the Amino Acid That Each Codon Directs

First Position (5′-end)	Second Position U		C		A		G		Third Position (3′-end)
U	UUU	Phe	UCU	Ser	UAU	Tyr	UGU	Cys	U
	UUC	Phe	UCC	Ser	UAC	Tyr	UGC	Cys	C
	UUA	Leu	UCA	Ser	UAA	Stop	UGC	Stop	A
	UUG	Leu	UCG	Ser	UAG	Stop	UGG	Trp	G
C	CUU	Leu	CCU	Pro	CAU	His	CGU	Arg	U
	CUC	Leu	CCC	Pro	CAC	His	CGC	Arg	C
	CUA	Leu	CCA	Pro	CAA	Gln	CGA	Arg	A
	CUG	Leu	CCG	Pro	CAG	Gln	CGG	Arg	G
A	AUU	Ile	ACU	Thr	AAU	Asn	AGU	Ser	U
	AUC	Ile	ACC	Thr	AAC	Asn	AGC	Ser	C
	AUA	Ile	ACA	Thr	AAA	Lys	AGA	Arg	A
	AUG*	Met	ACG	Thr	AAG	Lys	AGG	Arg	G
G	GUU	Val	GCU	Ala	GAU	Asp	GGU	Gly	U
	GUC	Val	GCC	Ala	GAC	Asp	GGC	Gly	C
	GUA	Val	GCA	Ala	GAA	Glu	GGA	Gly	A
	GUG	Val	GCG	Ala	GAG	Glu	GGG	Gly	G

*AUG also serves as the principal initiation codon.

Finally, we must ask one last question about the genetic code. Is the code universal; that is, is it the same for all organisms? Every bit of experimental evidence available today from the study of viruses, bacteria, and higher animals, including humans, indicates that the code is universal. Furthermore, the fact that it is the same for all these organisms means that it has been the same over billions of years of evolution.

Example 28.5 Transcription

During transcription, a portion of mRNA is synthesized with the following base sequence.

5′-AUG-GUA-CCA-CAU-UUG-UGA-3′

(a) Write the nucleotide sequence of the DNA from which this portion of mRNA was synthesized.

(b) Write the primary structure of the polypeptide coded for by this section of mRNA.

Solution

(a) During transcription, mRNA is synthesized from a DNA strand, beginning from the 3′ end of the DNA template. The DNA strand must be the complement of the newly synthesized mRNA strand.

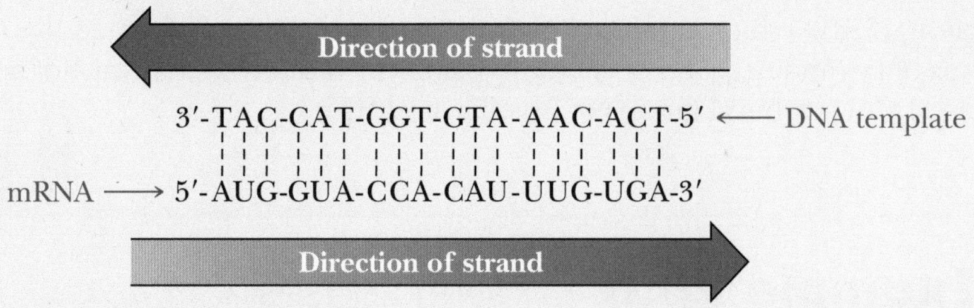

Note that the codon UGA codes for termination of the growing polypeptide chain; therefore, the sequence given in this problem codes for a pentapeptide only.

(b) The sequence of amino acids is shown under the following mRNA strand.

5′-AUG-GUA-CCA-CAU-UUG-UGA-3′
met — val — pro — his — leu — stop

Problem 28.5

The following section of DNA codes for oxytocin, a polypeptide hormone.

3′-ACG-ATA-TAA-GTT-TTA-ACG-GGA-GAA-CCA-ACT-5′

(a) Write the base sequence of the mRNA synthesized from this section of DNA.

(b) Given the sequence of bases in part (a), write the primary structure of oxytocin.

28.5 Sequencing Nucleic Acids

As recently as 1975, the task of determining the primary structure of a nucleic acid was thought to be far more difficult than determining the primary structure of a protein. Nucleic acids, it was reasoned, contain only 4 different units, whereas proteins contain 20 different units. With only 4 different units, there are fewer specific sites for selective cleavage, distinctive sequences are more difficult to recognize, and there is greater chance of ambiguity in the assignment of sequence. Two breakthroughs reversed this situation. First was the development of a type of electrophoresis called polyacrylamide gel electrophoresis, a technique so sensitive that it is possible to separate nucleic acid fragments that differ from one another in only a single nucleotide. The second breakthrough was the discovery of a class of enzymes called restriction endonucleases, isolated chiefly from bacteria.

A. Restriction Endonucleases

A **restriction endonuclease** recognizes a set pattern of four to eight nucleotides and cleaves a DNA strand by hydrolysis of the linking phosphodiester bonds at any site that contains that particular sequence. Close to 1000 restriction endonucleases have been isolated and their specificities characterized; each cleaves at specific sites, often with unique specificity. *E. coli*, for example, has a restriction endonuclease, *Eco*RI, that recognizes the hexanucleotide sequence, GAATTC, and cleaves it between G and A.

Restriction endonuclease
An enzyme that catalyzes hydrolysis of a particular phosphodiester bond within a DNA strand.

cleavage here (between G and A)

5′---G-A-A-T-T-C---3′ $\xrightarrow{EcoRI}$ 5′---G + 5′-A-A-T-T-C---3′

Note that the action of restriction endonucleases is analogous to the action of trypsin (Section 27.4B), which catalyzes hydrolysis of amide bonds formed by the carboxyl groups of Lys and Arg, and of chymotrypsin, which catalyzes cleavage of amide bonds formed by the carboxyl groups of Phe, Tyr, and Trp.

Example 28.6 Endonucleases

The following is a section of the gene coding for bovine rhodopsin along with several restriction endonucleases, their recognition sequences, and their hydrolysis sites. Which endonucleases will catalyze cleavage of this section of DNA?

5′-GCCGTCTACAACCCGGTCATCTACTATCATGATCAACAAGCAGTTCCGGAACT-3′

Enzyme	Recognition Sequence	Enzyme	Recognition Sequence
*Alu*I	AG ↓ CT	*Hpa*II	C ↓ CGG
*Bal*I	TGG ↓ CCA	*Mbo*I	↓ GATC
*Fnu*DII	CG ↓ CG	*Not*I	GC ↓ GGCCGC
*Hea*III	GG ↓ CC	*Sac*I	GAGCT ↓ C

Solution

Only restriction endonucleases *Hpa*II and *Mbo*I catalyze cleavage of this polynucleotide: *Hpa*II at two sites and *Mbo*I at one site.

*Hpa*II ↓ (between CC and CGG)   *Mbo*I ↓ (before GATC)   *Hpa*II ↓ (between TC and CGG)

5′-GCCGTCTACAACCCGGTCATCTACTATCATGATCAACAAGCAGTTCCGGAACT-3′

Problem 28.6

The following is another section of the bovine rhodopsin gene. Which of the endonucleases given in Example 28.6 will catalyze cleavage of this section?

5'-ACGTCGGGTCGTCGTCCTCTCGCGGTGGTGAGTCTTCCGGCTCTTCT-3'

B. Methods for Sequencing Nucleic Acids

Any sequencing of DNA begins with site-specific cleavage of double-stranded DNA by one or more restriction endonucleases into smaller fragments called restriction fragments. Each restriction fragment is then sequenced separately, overlapping base sequences are identified, and the entire sequence of bases is then deduced.

Two methods for sequencing restriction fragments have been developed. The first of these, developed by Allan Maxam and Walter Gilbert and known as the Maxam-Gilbert method, depends on base-specific chemical cleavage. The second method, developed by Frederick Sanger and known as the chain termination or dideoxy method, depends on interruption of DNA-polymerase catalyzed synthesis. Sanger and Gilbert shared the 1980 Nobel Prize in Biochemistry for their "development of chemical and biochemical analysis of DNA structure." The **Sanger dideoxy method** is currently more widely used, and it is on this method that we concentrate.

Sanger dideoxy method
A method developed by Frederick Sanger for sequencing DNA molecules.

C. DNA Replication *in Vitro*

To appreciate the rationale for the dideoxy method, we must first understand certain aspects of the biochemistry of DNA replication. During replication, the sequence of nucleotides in one strand is copied as a complementary strand to form the second strand of a double-stranded DNA molecule. Synthesis of the complementary strand is catalyzed by the enzyme DNA polymerase. DNA polymerase will also carry out this synthesis *in vitro* using single-stranded DNA as a template, provided that both the four deoxynucleotide triphosphate (dNTP) monomers and a primer are present. A **primer** is an oligonucleotide capable of forming a short section of double-stranded DNA (dsDNA) by base-pairing with its complement on a single-stranded DNA (ssDNA). Because a new DNA strand grows from its 5′ to 3′ end, the primer must have a free 3′-OH group to which the first nucleotide of the growing chain is added (Figure 28.11).

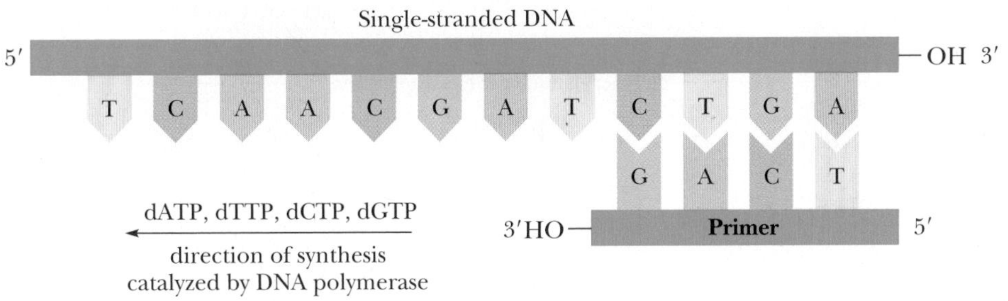
Single-stranded DNA
5′
OH 3′
T C A A C G A T C T G A
G A C T
dATP, dTTP, dCTP, dGTP
direction of synthesis
catalyzed by DNA polymerase
3′HO
Primer
5′

FIGURE 28.11 DNA polymerase catalyzes the synthesis *in vitro* using single-stranded DNA as a template provided that both the four dNTP monomers and a primer are present. The primer provides a short stretch of double-stranded DNA by base-pairing with its complement on the single-stranded DNA.

D. The Chain Termination or Dideoxy Method

The key to the chain termination method is the addition to the synthesizing medium of a 2′,3′-dideoxynucleoside triphosphate (ddNTP). Because a ddNTP has no —OH group at the 3′ position, it cannot serve as an acceptor for the next nucleotide to be added to the growing polynucleotide chain. Thus, chain synthesis is terminated at any point where a ddNTP becomes incorporated, hence the designation chain termination method.

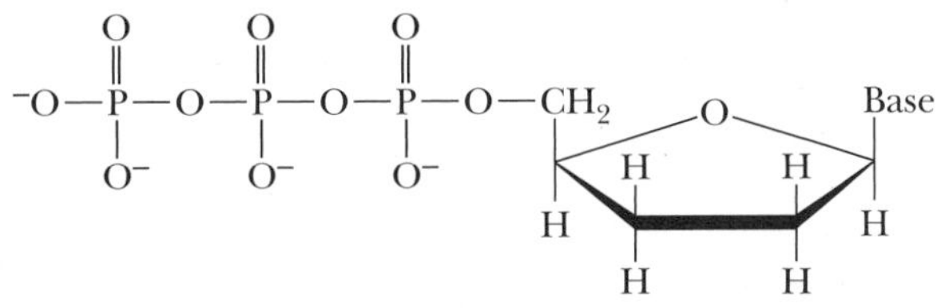

A 2′,3′-dideoxynucleoside triphosphate (ddNTP)

In the chain termination method, a single-stranded DNA of unknown sequence is mixed with primer and divided into four separate reaction mixtures. To each reaction mixture are added all four dNTPs, one of which is labeled in the 5′ phosphoryl group with ^{32}P so that the newly synthesized fragments can be visualized by autoradiography.

$$^{32}_{15}P \longrightarrow\ ^{32}_{16}S + \text{Beta particle} + \text{Gamma rays}$$

Also added to each reaction mixture are DNA polymerase and one of the four ddNTPs. The ratio of dNTPs to ddNTP in each reaction mixture is adjusted so that incorporation of a ddNTP takes place infrequently. In each reaction mixture, DNA synthesis takes place; however, in a given population of molecules, synthesis is interrupted at every possible site (Figure 28.12).

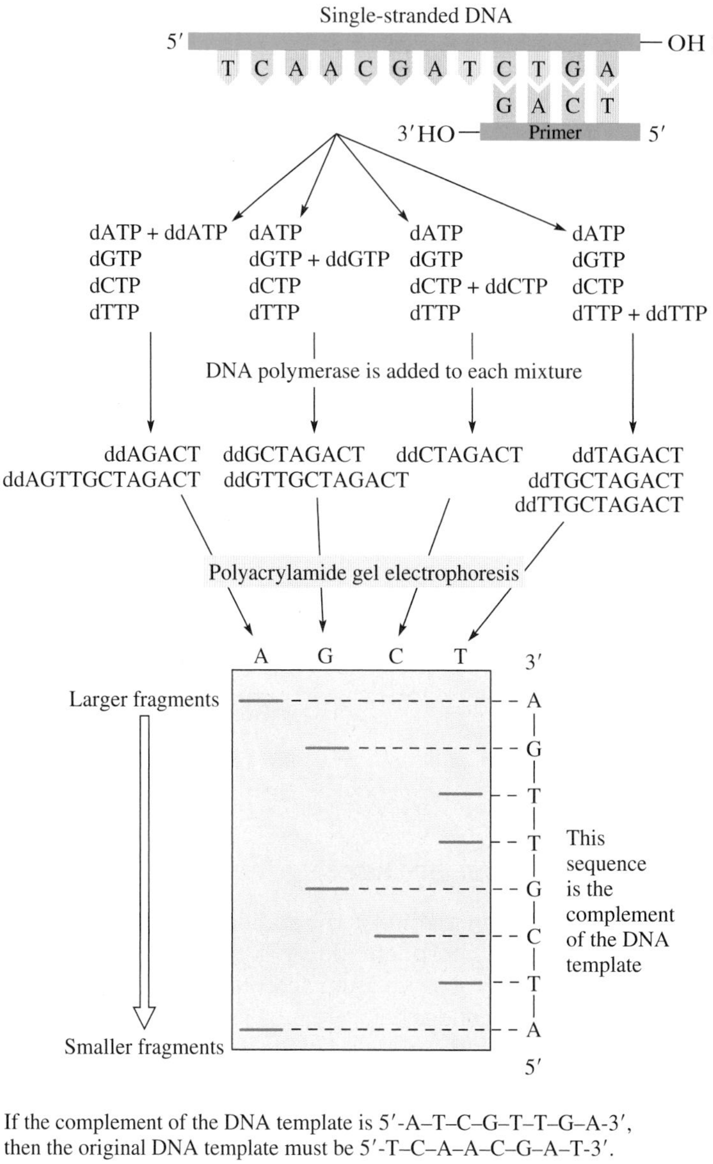

FIGURE 28.12 The chain termination or dideoxy method of DNA sequencing. The primer-DNA template is divided into four separate reaction mixtures. To each mixture is added the four dNTPs, DNA polymerase, and one of the four ddNTPs. Synthesis is interrupted at every possible site. The mixtures of oligonucleotides are separated by polyacrylamide gel electrophoresis. The base sequence of the DNA complement is read from the bottom to the top (from the 5′ end to the 3′ end) of the developed gel.

When gel electrophoresis of each reaction mixture is completed, a piece of X-ray film is placed over the gel and gamma rays released by radioactive decay of ^{32}P darken the film and create a pattern on it that is an image of the resolved oligonucleotides. The base sequence of the complement to the original single-stranded template is then read directly from bottom to top of the developed film.

A variation on this method is to use a single reaction mixture with each of the four ddNTPs labeled with a different fluorescent indicator. Each label is then detected by its characteristic spectrum following separation of the fragments by polyacrylamide gel electrophoresis or some chromatography method. First-generation automated DNA sequencing machines use this fluorescent variation.

E. Sequencing the Human Genome

The sequencing of the human genome was announced in the spring of 2000 by two competing groups, the Human Genome Project, a loosely linked consortium of publicly funded groups, and a private company called Celera Genomics. Actually, this milestone didn't represent a complete sequence but rather a so-called rough draft

David Parker/Science Sources

FIGURE 28.13 Sequencing equipment.

comprising about 85% of the entire genome. The methodology used is based on a refinement of the techniques described earlier using massively parallel separations of fragments by electrophoresis in capillary tubes. The Celera approach used some 300 of the fastest sequencing machines in parallel, each operating on many parallel DNA fragments. Supercomputers were used to assemble and compare millions of overlapping sequences. Figure 28.13 shows the apparatus used by the Celera group for this milestone achievement.

It is now much easier to obtain nucleic acid sequences than protein sequences. Sequencing a gene immediately gives the investigator the sequence of the protein produced by that gene because the code is known. Because many organisms have been sequenced and the functions of many of the coded proteins are known, it is possible to take the sequence of a protein of unknown function and determine sequence homologies with the vast number of proteins of known function in a repository known as the Protein Data Bank and thereby make an educated guess about the function of a protein for which the new gene codes. This technology is enabling an information revolution in chemistry, biology, and medicine.

This achievement represents the beginning of a new era of molecular medicine in which specific genetic deficiencies leading to inherited diseases will be understood on a molecular basis and new therapies targeted at shutting down undesired genes or turning on desired ones will be developed. In addition, different individuals may respond differently to different drug therapies. In the future, medicines may be specifically tailored to the genetic makeup of individual patients.

Chemical Connections

DNA Fingerprinting

Each human being has a genetic makeup consisting of approximately 3 billion base pairs of nucleotides, and except for identical twins, the base sequence of DNA in one individual is different from that of every other individual. As a result, each person has a unique DNA fingerprint. To determine a DNA fingerprint, a sample of DNA from a trace of blood, skin, or other tissue is treated with a set of restriction endonucleases and the 5′ end of each restriction fragment is labeled with phosphorus-32. The resulting ^{32}P-labeled restriction fragments are then separated by polyacrylamide gel electrophoresis and visualized by placing a photographic plate over the developed gel.

In the DNA fingerprint patterns shown in the figure, lanes 1, 5, and 9 represent internal standards or control lanes. They contain the DNA fingerprint pattern of a standard virus treated with a standard set of restriction endonucleases. Lanes 2, 3, and 4 were

(Continued)

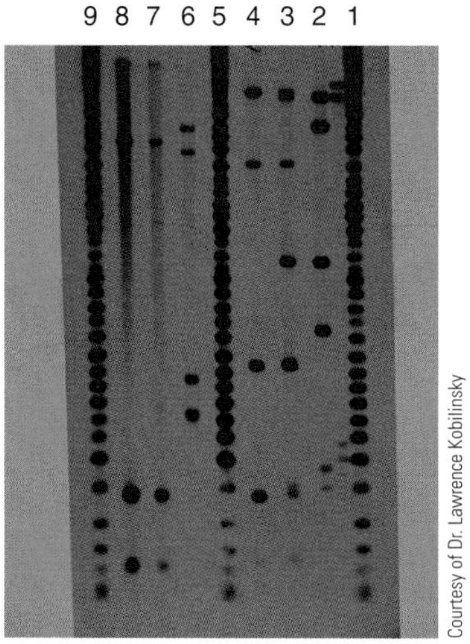
9 8 7 6 5 4 3 2 1
Courtesy of Dr. Lawrence Kobilinsky
A DNA "fingerprint."

used in a paternity suit. The DNA fingerprint of the mother in lane 4 contains five bands, which match with five of the six bands in the DNA fingerprint of the child in lane 3. The DNA fingerprint of the alleged father in lane 2 contains six bands, three of which match with bands in the DNA fingerprint of the child. Because the child inherits only half of his or her genes from the father, only half of the child's and father's DNA fingerprints are expected to match. In this instance, the paternity suit was won on the basis of the DNA fingerprint matching.

Lanes 6, 7, and 8 contain DNA fingerprint patterns used as evidence in a rape case. Lanes 7 and 8 are DNA fingerprints of semen obtained from the rape victim. Lane 6 is the DNA fingerprint pattern of the alleged rapist. The DNA fingerprint patterns of the semen do not match that of the alleged rapist and excluded the suspect from the case.

Next-Generation DNA Sequencing

Next-generation DNA sequencing methods have recently been created by marrying newer versions of the enzymatic DNA sequencing reactions with clonal amplification, microfabrication techniques, microfluidic technologies, and automated sample preparation. Using massively parallel approaches, an astounding 300 gigabases (300 billion bases) of DNA can be sequenced in a single run, which now costs only a few thousand dollars. By exploiting special optical technologies, one of the very latest sequencing approaches involves sequencing at the single molecule level. Sequencing of entire genomes, especially microorganisms, is now routine.

Amazingly, each sequencing reaction reads only 50 to 400 bases. Hundreds of thousands to hundreds of millions of these reads are carried out simultaneously in parallel fashion on microfabricated devices. Although it sounds almost too remarkable to believe, entire genomes as large as several gigabases in length are deduced by overlapping thousands to millions of these 50 to 400 base pair sequence reads. The key to this approach is having extensive redundancy built into the sequencing protocols so that each DNA segment of the genome is covered by multiple overlapping sequence reads. Through the use of sophisticated and highly efficient computational techniques, this overlapping enables the unambiguous assembly of each unique sequence read into a complete genome. The relatively new field of bioinformatics is largely devoted to developing new tools with which to analyze and exploit the massive amounts of sequence data that are now being generated at an exponentially accelerating pace. Indeed, managing and analyzing such large datasets is one of the key challenges to the field. It is appropriate to say that the awesome power of next-generation sequencing technologies has led to a revolution in biology research and will have far-reaching positive impacts on many aspects of future medicine.

Study Guide

28.1 Nucleosides and Nucleotides

- **Nucleic acids** are composed of three types of monomer units: heterocyclic aromatic amine bases derived from purine and pyrimidine, the monosaccharides D-ribose or 2-deoxy-D-ribose, and phosphate ions.
- A **nucleoside** is a compound containing D-ribose or 2-deoxy-D-ribose bonded to a heterocyclic aromatic amine base by a β-*N*-glycosidic bond.
  - *For DNA, the four bases are **adenine**, **cytosine**, **guanine**, and **thymine**.*
  - *For RNA, the four bases are **adenine**, **cytosine**, **guanine**, and **uracil**.*
  - *Adenine and guanine are **purines**; cytosine, thymine, and uracil are **pyrimidines**.*
- A **nucleotide** is a nucleoside in which at least one molecule of phosphoric acid is esterified with an —OH of the monosaccharide, most commonly either the 3′-OH or the 5′-OH.
  - *Nucleoside mono-, di-, and triphosphates are strong polyprotic acids and are extensively ionized at pH 7.0.*
  - *At pH 7.0, adenosine triphosphate, for example, is a 50:50 mixture of ATP^{3-} and ATP^{4-}.*
  - *The N atoms of the amino groups of adenine, cytosine, and guanine, which are not in the rings, are sp^2 hybridized and planar, allowing for base stacking as well as hydrogen bonding to complementary bases.*

P **28.1, 28.7–28.13**

28.2 The Structure of DNA

- The **primary structure** of **deoxyribonucleic acid** (**DNA**) consists of units of 2′-deoxyribose bonded by 3′,5′-phosphodiester bonds.
  - *A heterocyclic aromatic amine base is attached to each deoxyribose unit by a β-N-glycosidic bond.*
  - *The sequence of bases is read from the 5′ end of the polynucleotide strand to the 3′ end.*
- The heart of the **Watson-Crick model** is the postulate that a molecule of DNA consists of two antiparallel polynucleotide strands coiled in a right-handed manner about the same axis to form a **double helix**.
  - *Purine and pyrimidine bases point inward toward the axis of the helix and are always paired G-C and A-T due to specific patterns of hydrogen bonds.*
- In **B-DNA**, base pairs are stacked one on top of another with a spacing of 340 pm and ten base pairs per 3400 pm helical repeat.
- In **A-DNA**, bases are stacked with a spacing of 290 pm between base pairs and ten base pairs per 2900 pm helical repeat.
- The **tertiary structure** of DNA is commonly referred to as **supercoiling**.
  - ***Circular DNA** is a type of double-stranded DNA in which the ends of each strand are joined by phosphodiester groups.*
  - *Opening of one strand followed by partial unwinding and rejoining of the ends introduces strain in the nonhelical gap.*

  - *This strain can be spread over the entire molecule of circular DNA by introduction of **superhelical twists**.*
- **Histones** are the DNA binding proteins found in cell nuclei.
  - *They are particularly rich in lysine and arginine and therefore have an abundance of positive charges.*
  - *The association of DNA and histones produces a structure called **chromatin**.*

P 28.2, 28.3, 28.14–28.26

28.3 Ribonucleic Acids

- There are two important differences between the primary structure of ribonucleic acids (RNA) and DNA.
  - *The monosaccharide unit in RNA is D-ribose.*
  - *Both RNA and DNA contain the purine bases adenine (A) and guanine (G) and the pyrimidine base cytosine (C). As the fourth base, however, RNA contains uracil (U), whereas DNA contains thymine (T).*
- RNA exists in three main forms in cells.
  - *Ribosomal RNA, the most abundant form, is part of the ribosome, the protein-synthesizing machine of the cell.*
  - *Transfer RNA carries amino acids to the sites of protein synthesis on ribosomes.*
  - *Messenger RNAs are copies of DNA gene sequences and carry the coded genetic information from DNA to the ribosomes for the synthesis of proteins.*

P 28.4, 28.27–28.31

28.4 The Genetic Code

- The genetic code consists of nucleosides in groups of three; that is, it is a triplet code.
  - *An amino acid is coded for by one triplet.*
  - *Only 61 triplets code for amino acids; the remaining three code for termination of polypeptide synthesis.*
  - *The code is degenerate, meaning that several amino acids are coded for by more than one triplet.*
  - *For the 15 amino acids coded for by two, three, or four triplets, only the last base of the triplet varies.*
  - *There is no ambiguity in the code: each triplet codes only for a single amino acid or termination signal.*
  - *The genetic code appears to be universal for all of the different forms of life on Earth.*

P 28.5, 28.32–28.42

28.5 Sequencing Nucleic Acids

- **Restriction endonucleases** recognize a set pattern of four to eight nucleotides and cleave a DNA strand by hydrolysis of the linking phosphodiester bonds at any site that contains that particular sequence.
  - *Restriction endonucleases isolated from different sources recognize different sequences.*
  - *Restriction endonucleases are used to produce fragments from genomic DNA.*

- In the **chain termination** or **dideoxy method** of DNA sequencing developed by Frederick Sanger, a primer-DNA template is divided into four separate reaction mixtures.
  - *To each is added the four dNTPs, one of which is labeled with* ^{32}P.
  - *Also added are DNA polymerase and one of the four ddNTPs.*
  - *Synthesis is interrupted when a ddNTP is incorporated.*
  - *The mixtures of newly synthesized oligonucleotides are separated by polyacrylamide gel electrophoresis and visualized by autoradiography.*
  - *The base sequence of the DNA complement to the original DNA template is read from the bottom to the top (from the 5′ end to the 3′ end) of the developed photographic plate.*
  - *Alternatively, the four ddNTPs each contain a different fluorescent label and the sequence is read as the differently fluorescing strands elute from the gel.*
- **Next-generation DNA sequencing technology** combines enzymatic DNA sequencing reactions with clonal amplification, microfabrication, microfluidics, and automated sample preparation.
  - *Entire genomes of several gigabases are now routinely determined.*
  - *Each sequence read is typically 50 to 400 base pairs.*
  - *Thousands to millions of these reads are assembled into complete genomic sequences by powerful computers running advanced sequence alignment and overlapping algorithms.*
  - *The awesome power of this technology has lead to a revolution in biology that will have far-reaching positive effects on medicine.*

Problems

Red numbers indicate applied problems.

Nucleosides and Nucleotides

28.7 A pioneer in designing and synthesizing antimetabolites that could destroy cancer cells was George Hitchings at the Burroughs Wellcome Company. In 1942, he initiated a program to discover DNA antimetabolites, and in 1948, he and Gertrude Elion synthesized 6-mercaptopurine, a successful drug for treating acute leukemia. Another DNA antimetabolite synthesized by Hutchings and Elion was 6-thioguanine. Hitchings and Elion along with Sir James W. Black won the 1988 Nobel Prize in Physiology or Medicine for their discoveries of "important principles of drug treatment." In each drug, the oxygen at carbon 6 of the parent molecule is replaced by divalent sulfur. Draw structural formulas for the enethiol (the sulfur equivalent of an enol) forms of 6-mercaptopurine and 6-thioguanine.

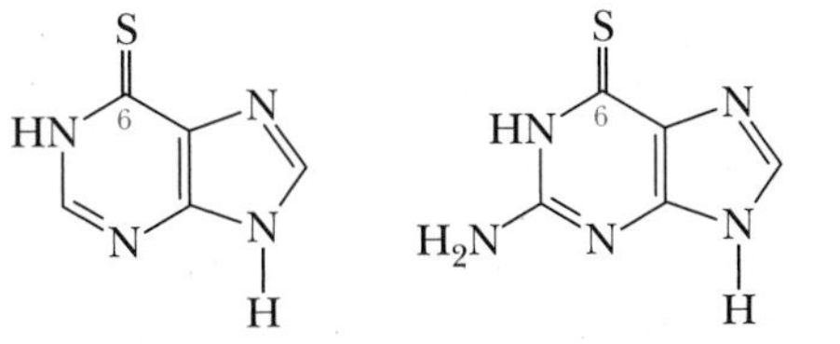

6-Mercaptopurine **6-Thioguanine**

28.8 Following are structural formulas for cytosine and thymine. Draw two additional tautomeric forms for cytosine and three additional tautomeric forms for thymine.

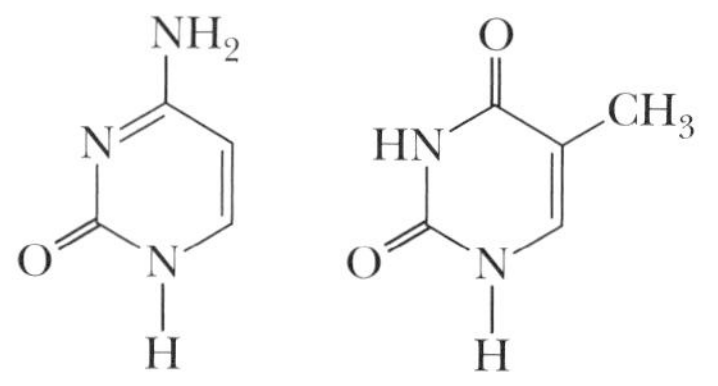

Cytosine (C) **Thymine (T)**

28.9 Draw a structural formula for a nucleoside composed of the following.

(a) α-D-Ribose and adenine **(b)** β-2-Deoxy-D-ribose and cytosine

28.10 Nucleosides are stable in water and in dilute base. In dilute acid, however, the glycosidic bond of a nucleoside undergoes hydrolysis to give a pentose and a heterocyclic aromatic amine base. Propose a mechanism for this acid-catalyzed hydrolysis.

28.11 Explain the difference in structure between a nucleoside and a nucleotide.

28.12 Draw a structural formula for each nucleotide and estimate its net charge at pH 7.4, the pH of blood plasma.

(a) 2′-Deoxyadenosine 5′-triphosphate (dATP)
(b) Guanosine 3′-monophosphate (GMP)
(c) 2′-Deoxyguanosine 5′-diphosphate (dGDP)

28.13 Cyclic-AMP, first isolated in 1959, is involved in many diverse biological processes as a regulator of metabolic and physiological activity. In it, a single phosphate group is esterified with both the 3′ and 5′ hydroxyls of adenosine. Draw a structural formula of cyclic-AMP.

The Structure of DNA

28.14 Why are deoxyribonucleic acids called acids? What are the acidic groups in their structure?

28.15 Human DNA contains approximately 30.4% A. Estimate the percentages of G, C, and T and compare them with the values presented in Table 28.1.

28.16 Draw a structural formula of the DNA tetranucleotide 5′-A-G-C-T-3′. Estimate the net charge on this tetranucleotide at pH 7.0. What is the complementary tetranucleotide to this sequence?

28.17 List the postulates of the Watson-Crick model of DNA secondary structure.

28.18 The Watson-Crick model is based on certain experimental observations of base composition and molecular dimensions. Describe these observations and show how the Watson-Crick model accounts for each.

28.19 If you read J. D. Watson's account of the discovery of the structure of DNA, *The Double Helix,* you will find that for a time in their model-building studies, he and Crick were using alternative (and incorrect, at least in terms of their final model of the double helix) tautomeric structures for some of the heterocyclic bases.

(a) Write at least one alternative tautomeric structure for adenine.
(b) Would this structure still base-pair with thymine, or would it now base-pair more efficiently with a different base? If so, identify that base.

28.20 Compare the α-helix of proteins and the double helix of DNA in these ways.
(a) The units that repeat in the backbone of the polymer chain
(b) The projection in space of substituents along the backbone (the R groups in the case of amino acids; purine and pyrimidine bases in the case of double-stranded DNA) relative to the axis of the helix

28.21 Discuss the role of the hydrophobic interactions in stabilizing the following.

(a) Double-stranded DNA **(b)** Lipid bilayers
(c) Soap micelles

28.22 Name the type of covalent bond(s) joining monomers in these biopolymers.

(a) Polysaccharides **(b)** Polypeptides
(c) Nucleic acids

28.23 In terms of hydrogen bonding, which is more stable, an A-T base pair or a G-C base pair?

28.24 At elevated temperatures, nucleic acids become denatured; that is, they unwind into single-stranded DNA. Account for the observation that the higher the G-C content of a nucleic acid, the higher the temperature required for its thermal denaturation.

28.25 Write the DNA complement for 5′-ACCGTTAAT-3′. Be certain to label which is the 5′ end and which is the 3′ end of the complement strand.

28.26 Write the DNA complement for 5′-TCAACGAT-3′.

Ribonucleic Acids

28.27 Compare the degree of hydrogen bonding in the base pair A-T found in DNA with that in the base pair A-U found in RNA.

28.28 Compare DNA and RNA is these ways.

(a) Monosaccharide units
(b) Principal purine and pyrimidine bases
(c) Primary structure
(d) Location in the cell
(e) Function in the cell

28.29 What type of RNA has the shortest lifetime in cells?

28.30 Write the mRNA complement for 5′-ACCGTTAAT-3′. Be certain to label which is the 5′ end and which is the 3′ end of the mRNA strand.

28.31 Write the mRNA complement for 5′-TCAACGAT-3′.

The Genetic Code

28.32 What does it mean to say that the genetic code is degenerate?

28.33 Write the mRNA codons for the following.

(a) Valine **(b)** Histidine **(c)** Glycine

28.34 Aspartic acid and glutamic acid have carboxyl groups on their side chains and are called acidic amino acids. Compare the codons for these two amino acids.

28.35 Compare the structural formulas of the aromatic amino acids phenylalanine and tyrosine. Also compare the codons for these two amino acids.

28.36 Glycine, alanine, and valine are classified as nonpolar amino acids. Compare their codons. What similarities do you find? What differences do you find?

28.37 Codons in the set CUU, CUC, CUA, and CUG all code for the amino acid leucine. In this set, the first and second bases are identical; the identity of the third base is irrelevant. For what other sets of codons is the third base also irrelevant? For what amino acid(s) does each set code?

28.38 Compare the codons with a pyrimidine, either U or C, as the second base. Do the majority of the amino acids specified by these codons have hydrophobic or hydrophilic side chains?

28.39 Compare the codons with a purine, either A or G, as the second base. Do the majority of the amino acids specified by these codons have hydrophilic or hydrophobic side chains?

28.40 What polypeptide is coded for by this mRNA sequence?

5′-GCU-GAA-GUC-GAG-GUG-UGG-3′

28.41 The alpha chain of human hemoglobin has 141 amino acids in a single polypeptide chain. Calculate the minimum number of bases on DNA necessary to code for the alpha chain. Include in your calculation the bases necessary for specifying termination of polypeptide synthesis.

28.42 In HbS, the human hemoglobin found in individuals with sickle-cell anemia, glutamic acid at position 6 in the beta chain is replaced by valine.

(a) List the two codons for glutamic acid and the four codons for valine.
(b) Show that one of the glutamic acid codons can be converted to a valine codon by a single substitution mutation (i.e., by changing one letter in one codon).

29 Organic Polymer Chemistry

Outline

29.1 The Architecture of Polymers
29.2 Polymer Notation and Nomenclature
29.3 Molecular Weights of Polymers
29.4 Polymer Morphology—Crystalline Versus Amorphous Materials
29.5 Step-Growth Polymerizations
29.6 Chain-Growth Polymerizations

Sea of umbrellas on a rainy day in Shanghai, China. **Inset:** models of adipic acid and hexamethylenediamine, the two monomers of nylon 66.
robertharding/Superstock

▶ This chapter was originally authored by Dr. Bruce Novak, Howard J. Schaeffer Distinguished Professor of Polymer Science and Organic Chemistry, North Carolina State University.

The technological advancement of any society is inextricably tied to the materials available to it. Indeed, historians have used the emergence of new materials as a way of establishing a timeline to mark the development of human civilization. As part of the search to discover new materials, scientists have made increasing use of organic chemistry for the preparation of synthetic polymers. The versatility afforded by these polymers allows for the creation and fabrication of materials with ranges of properties unattainable using such materials as wood, metals, and ceramics. Deceptively simple changes in the chemical structure of a given polymer, for example, can change its mechanical properties from those of a sandwich bag to those of a bulletproof vest. Furthermore, structural changes can introduce properties never before imagined in organic polymers. For example, using well-defined organic reactions, one type of polymer can be made into an insulator (e.g., the rubber sheath that surrounds electrical cords), or if treated differently, it can be made into an electrical conductor with a conductivity nearly equal to that of metallic copper.

The years since the 1930s have seen extensive research and development in polymer chemistry, and an almost explosive growth in plastics, coatings, and rubber technology has created a worldwide multibillion-dollar industry. A few basic characteristics account for this phenomenal growth. First, the raw materials for synthetic polymers are derived mainly from petroleum. With the development of petroleum-refining processes, raw materials for the synthesis of polymers became generally cheap and plentiful. Second, within broad limits, scientists have learned how to tailor polymers to the requirements of the end use. Third, many consumer products can be fabricated more cheaply from synthetic polymers than from competing materials such as wood, ceramics, and metals. For example, polymer technology created the water-based (latex) paints that have revolutionized the coatings industry; plastic films and foams have done the same for the packaging industry. The list could go on and on as we think of the manufactured items that are everywhere around us in our daily lives.

29.1 The Architecture of Polymers

Polymers (Greek: *poly* + *meros*, many parts) are long-chain molecules synthesized by linking **monomers** (Greek: *mono* + *meros*, single part) through chemical reactions. The molecular weights of polymers are generally high compared with those of common organic compounds and typically range from 10,000 g/mol to more than 1,000,000 g/mol. The architectures of these macromolecules can also be quite diverse. Types of polymer architecture include linear and branched chains as well as those with comb, ladder, and star structures (Figure 29.1). Additional structural variations can be achieved by introducing covalent cross links between individual polymer chains.

Polymer
From the Greek, *poly* + *meros*, many parts. Any long-chain molecule synthesized by linking together many single parts called monomers.

Monomer
From the Greek, *mono* + *meros*, single part. The simplest nonredundant unit from which a polymer is synthesized.

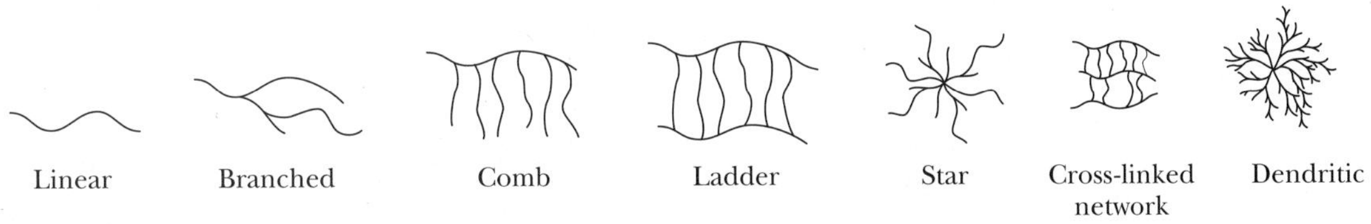
Linear
Branched
Comb
Ladder
Star
Cross-linked network
Dendritic

FIGURE 29.1 Various polymer architectures.

Linear and branched polymers are often soluble in solvents such as chloroform, benzene, toluene, DMSO, and THF. In addition, many linear and branched polymers can be melted to form highly viscous liquids. In polymer chemistry, the term **plastic** refers to any polymer that can be molded when hot and retains its shape when cooled. **Thermoplastics** are polymers that can be melted and become sufficiently fluid that they can be molded into shapes that are retained when they are cooled. **Thermosetting plastics**, or thermosets, can be molded when they are first prepared, but once they cool, they harden irreversibly and cannot be remelted. Because of these very different physical characteristics, thermoplastics and thermosets must be processed differently and are used in very different applications.

The single most important property of polymers at the molecular level is the size and shape of their chains. A good example of the importance of size is a comparison of paraffin wax, a natural polymer, and polyethylene, a synthetic polymer. These two distinct materials have identical repeat units, namely $—CH_2—$, but differ greatly in chain size. Paraffin wax has between 25 and 50 carbon atoms per chain, whereas polyethylene has between 1000 and 3000 carbon atoms per chain. Paraffin wax, as in birthday candles, is soft and brittle, but polyethylene, as in plastic beverage bottles, is strong, flexible, and tough. These vastly different properties arise directly from the difference in size and molecular architecture of the individual polymer chains.

Plastic
A polymer that can be molded when hot and retains its shape when cooled.

Thermoplastic
A polymer that can be melted and molded into a shape that is retained when it is cooled.

Thermoset plastic
A polymer that can be molded when it is first prepared but once cooled, hardens irreversibly and cannot be remelted.

29.2 Polymer Notation and Nomenclature

We show the structure of a polymer by placing parentheses around the **repeat unit**, which is the smallest molecular fragment that contains all the nonredundant structural features of the chain. Thus, the structure of an entire polymer chain can be reproduced by repeating the enclosed structure in both directions. A subscript n, called the **average degree of polymerization**, is placed outside the parentheses to indicate that this unit is repeated n times.

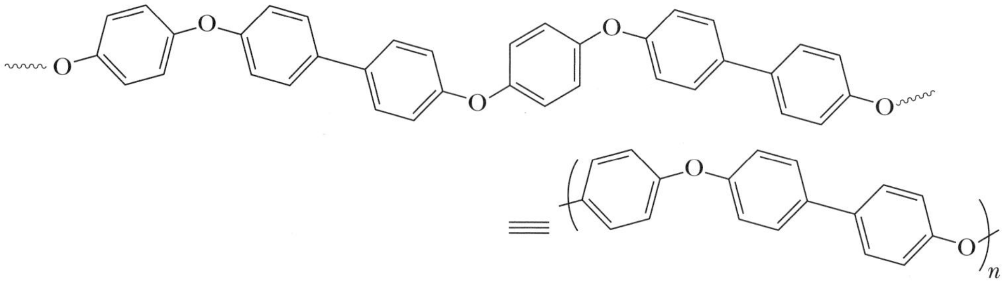

The polymers formed from symmetric monomer units, such as polyethylene, $(CH_2CH_2)_n$, and polytetrafluoroethylene, $(CF_2CF_2)_n$, are an exception to this notation. Although the simplest repeat units are the $—CH_2—$ and $—CF_2—$ groups, we show two methylene groups and two difluoromethylene groups because they originate from ethylene ($CH_2{=}CH_2$) and tetrafluoroethylene ($CF_2{=}CF_2$), the monomer units from which these polymers are derived.

The most common method of naming a polymer is to attach the prefix *poly-* to the name of the monomer from which the polymer is derived, as, for example, polyethylene and polystyrene. In the case of a more complex monomer or where the name of the monomer is more than one word (e.g., the monomer vinyl chloride), parentheses are used to enclose the name of the monomer.

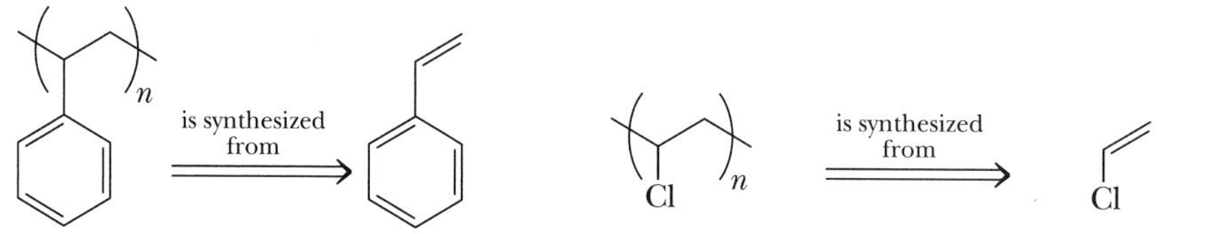

Polystyrene **Styrene** **Poly(vinyl chloride) (PVC)** **Vinyl chloride**

Example 29.1 Repeat Units

Given the following structure, determine the polymer's repeat unit, redraw the structure using the simplified parenthetical notation, and name the polymer.

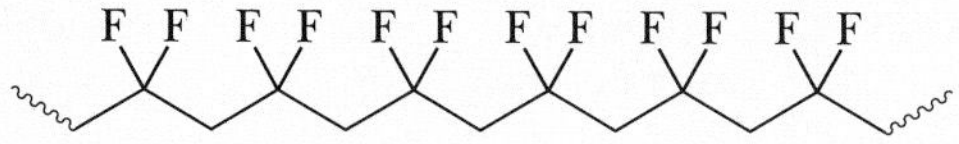

Solution

The repeat unit is $—CH_2CF_2—$, and the polymer is written $(CH_2CF_2)_n$. The repeat unit is derived from 1,1-difluoroethylene, and the polymer is named poly(1,1-difluoroethylene). This polymer is used in microphone diaphragms.

Problem 29.1

Given the following structure, determine the polymer's repeat unit, redraw the structure using the simplified parenthetical notation, and name the polymer.

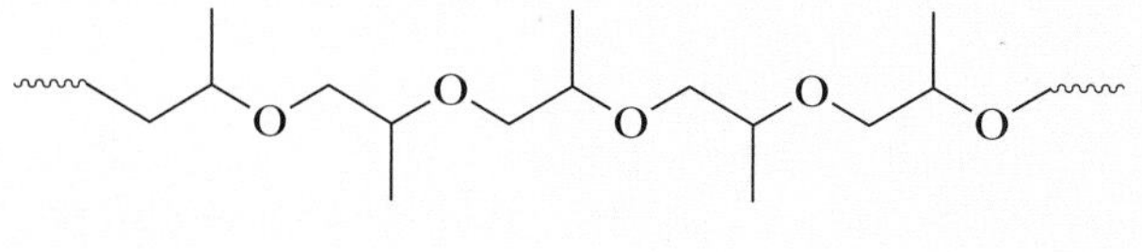

29.3 Molecular Weights of Polymers

All synthetic polymers and most naturally occurring polymers are mixtures of individual polymer molecules of variable molecular weights. When defining molecular weights in polymer chemistry, the two most common definitions are the number average and weight average molecular weights. The **number average molecular weight, M_n,** is calculated by counting the number of polymer chains of a particular molecular weight, multiplying each number by the molecular weight of its chain, summing these values, and dividing by the total number of polymer chains. The **weight average molecular weight, M_w,** is calculated by recording the total weight of each chain of a particular length, summing these weights, and dividing by the total weight of the sample. Because the larger chains in a sample weigh more than the smaller chains, the weight average molecular weight is skewed to higher values and M_w is always greater than M_n (Figure 29.2).

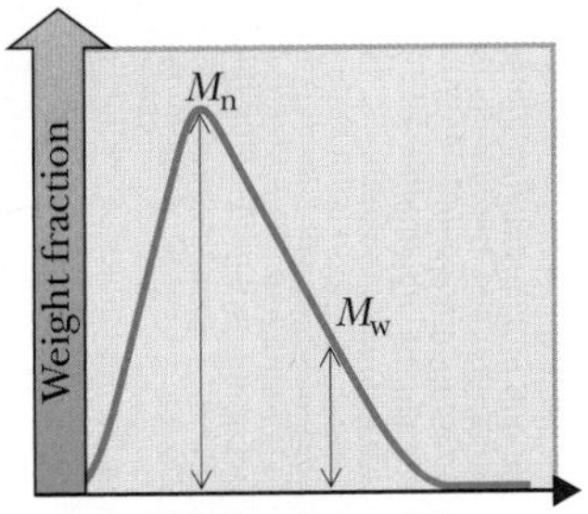

FIGURE 29.2 The distribution of molecular weights in a given polymer sample.

Both M_n and M_w are useful values, and their ratio, M_w/M_n, called the **polydispersity index**, provides a measure of the breadth of the molecular-weight distribution. When the M_w/M_n ratio is equal to one, all the polymer molecules in a sample are the same length, and the polymer is said to be **monodisperse**. No synthetic polymers are ever monodisperse unless the individual molecules are carefully fractionated using time-consuming, rigorous separation techniques based on molecular size. On the other hand, natural polymers, such as polypeptides and DNA, that are formed using biological processes are monodisperse polymers.

29.4 Polymer Morphology—Crystalline Versus Amorphous Materials

Polymers, like small organic molecules, tend to crystallize upon precipitation or as they are cooled from a melt. Acting to inhibit this tendency are their very large molecules, which tend to slow diffusion, and their sometimes complicated or irregular structures, which prevent efficient packing of the chains. The result is that polymers in the solid state tend to be composed of both ordered **crystalline domains** (crystallites) and disordered **amorphous domains**. The relative amounts of crystalline and amorphous domains differ from polymer to polymer and often depend on the manner in which the material is processed.

Crystalline domain
An ordered crystalline region in the solid state of a polymer. Also called crystallites.

Amorphous domain
A disordered, noncrystalline region in the solid state of a polymer.

High degrees of crystallinity are most often found in polymers with regular, compact structures and strong intermolecular forces, such as hydrogen bonding and dipolar interactions. The temperature at which crystallites melt corresponds to the **melt transition (T_m)** of the polymer. As the degree of crystallinity of a polymer increases, its T_m increases, and it becomes more opaque owing to scattering of the light by the crystalline domains. There is also a corresponding increase in strength and stiffness with increase in crystallinity. For example, poly(6-aminohexanoic acid) has a T_m = 223°C. At and well above room temperature, this polymer is a hard, durable material that does not undergo any appreciable change in properties even on a very hot summer afternoon. Its uses range from textile fibers to shoe heels.

Melt transition, T_m
The temperature at which crystalline regions of a polymer melt.

Amorphous domains are characterized by the absence of long-range order. Highly amorphous polymers are sometimes referred to as glassy polymers. Because they lack crystalline domains that scatter light, amorphous polymers are transparent. In addition, they are typically weak polymers in terms of both their greater flexibility and smaller mechanical strength. Upon being heated, amorphous polymers are transformed from a hard glass to a soft, flexible rubbery state. The temperature at which this transition occurs is called the **glass transition temperature (T_g)**. Amorphous polystyrene, for example, has a T_g = 100°C. At room temperature, it is a rigid solid used for drinking cups, foamed packaging materials, and disposable medical wares. If it is placed in boiling water, it becomes soft and rubbery.

Glass transition temperature, T_g
The temperature at which a polymer undergoes the transition from a hard glass to a rubbery state.

This relationship between mechanical properties and the degree of crystallinity can be illustrated by poly(ethylene terephthalate) (**PET**).

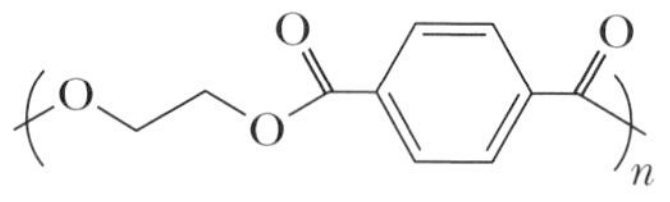

Poly(ethylene terephthalate) (PET)

PET can be made with crystalline domains ranging from 0% to about 55%. Completely amorphous PET is formed by cooling the melt quickly. By prolonging the cooling time, more molecular diffusion occurs, and crystallites form as the chains become more ordered. The differences in mechanical properties between these forms of PET are substantial. PET with a low degree of crystallinity is used for plastic beverage bottles, whereas fibers drawn from highly crystalline PET are used for textile fibers and tire cords.

Elastomer

A material that when stretched or otherwise distorted returns to its original shape when the distorting force is released.

Rubber materials must have low T_g values to behave as **elastomers** (elastic polymers). If the temperature drops below its T_g value, then the material is converted to a rigid glassy solid and all elastomeric properties are lost. A poor understanding of this behavior of polymers contributed to the *Challenger* spacecraft disaster in 1986. The elastomeric O-rings used to seal the solid booster rockets had a T_g value around 0°C. When the temperature dropped low on the morning of the *Challenger* launch, the O-ring seals dropped below their T_g value and consequently changed from elastomers to rigid glasses, losing any sealing capabilities. The rest is tragic history. The physicist Richard Feynman sorted this out publicly in a famous televised hearing in which he put a *Challenger*-type O-ring in ice water and showed that its elasticity was lost.

29.5 Step-Growth Polymerizations

Step-Growth Polymerization

A polymerization in which chain growth occurs in a stepwise manner between difunctional monomers as, for example, between adipic acid and hexamethylenediamine to form nylon 66. Also called condensation polymerization.

Polymerizations in which chain growth occurs in a stepwise manner are called **step-growth** or **condensation polymerizations**. Step-growth polymers are formed by reaction between difunctional molecules, with each new bond created in a separate step. During polymerization, monomers react with monomers to form dimers, dimers react with dimers to form tetramers, tetramers react with monomers to form pentamers, and so on. This stepwise construction of polymer chains has important consequences for both their molecular weights and molecular-weight distributions. Probability tells us that the most abundant species tend to co-condense. Thus, at the early stages of polymerization, small chains are most likely to react with monomers or other small chains to generate many low-molecular-weight oligomers rather than a small number of high-molecular-weight polymers. This tendency persists until most monomer units are used up. As a result, high-molecular-weight polymers are not produced until very late in the reaction, typically past 99% conversion of monomers to higher molecular-weight chains. Only at this point is there the probability of larger chains reacting with one another to form high-molecular-weight polymer molecules. This restriction points to an important distinction between small-molecule organic reactions and step-growth polymerizations. Although a reaction that typically yields 85% of the desired product is considered "good" in organic synthesis, the same reaction is essentially useless for step-growth polymerizations because high-molecular-weight polymers are rarely formed at such low conversions.

There are two common types of step-growth processes: (1) reaction between A-A and B-B type monomers to give $(\textbf{A-A-B-B})_n$ polymers and (2) the self-condensation of A-B monomers to give $(\textbf{A-B})_n$ polymers. In each case, an A functional group reacts exclusively with a B functional group and a B functional group reacts exclusively with an A functional group. New covalent bonds in step-growth polymerizations are generally formed by polar reactions, as, for example, nucleophilic acyl substitution.

In this section, we discuss five types of step-growth polymers: polyamides, polyesters, polycarbonates, polyurethanes, and epoxy resins.

A. Polyamides

In the years following World War I, a number of chemists recognized the need for developing a basic knowledge of polymer chemistry. One of the most creative of these was Wallace M. Carothers. In the early 1930s, Carothers and his associates at E. I. DuPont de Nemours and Company began fundamental research into the reactions of aliphatic dicarboxylic acids and diols. From adipic acid and ethylene glycol, they obtained a polyester of high molecular weight that could be drawn into fibers.

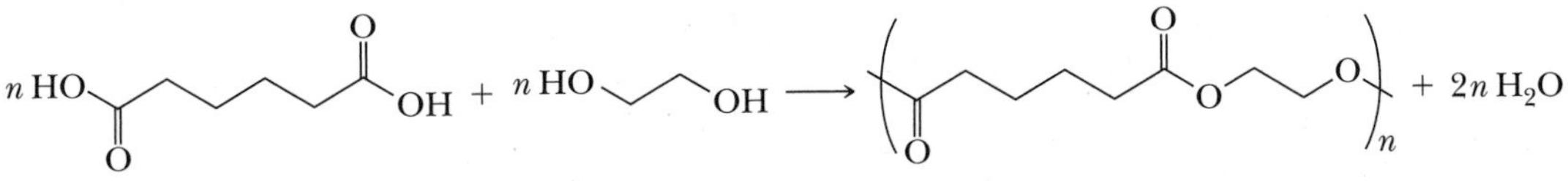

Hexanedioic acid (Adipic acid) **1,2-Ethanediol (Ethylene glycol)** **Poly(ethylene adipate)**

These first polyester fibers had melt transitions (T_m) too low for use as textile fibers, and they were not investigated further. Carothers then turned his attention to the reactions of dicarboxylic acids and diamines to form **polyamides** and, in 1934, synthesized nylon 66, the first purely synthetic fiber. Nylon 66 is so named because it is synthesized from two different monomers, each containing six carbon atoms.

Polyamides
A polymer in which each monomer unit is joined to the next by an amide bond, for example, nylon 66.

In the synthesis of nylon 66, hexanedioic acid (adipic acid) and 1,6-hexanediamine (hexamethylenediamine) are dissolved in aqueous ethanol, where they react to form a one-to-one salt called nylon salt. Nylon salt is then heated in an autoclave to 250°C, where the internal pressure rises to about 15 atm. Under these conditions, —COO^- groups from adipic acid and —NH_3^+ groups from hexamethylenediamine react with loss of H_2O to form amide groups.

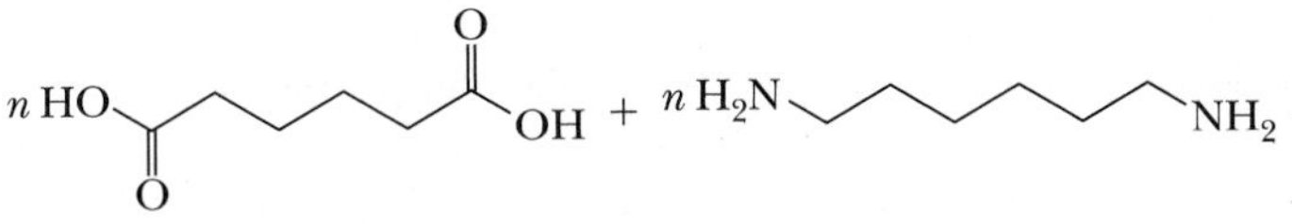

Hexanedioic acid (Adipic acid) **1,6-Hexanediamine (Hexamethylenediamine)**

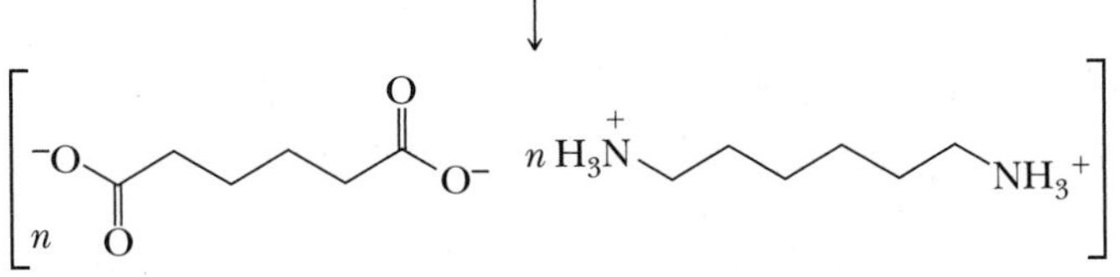

Nylon salt

heat

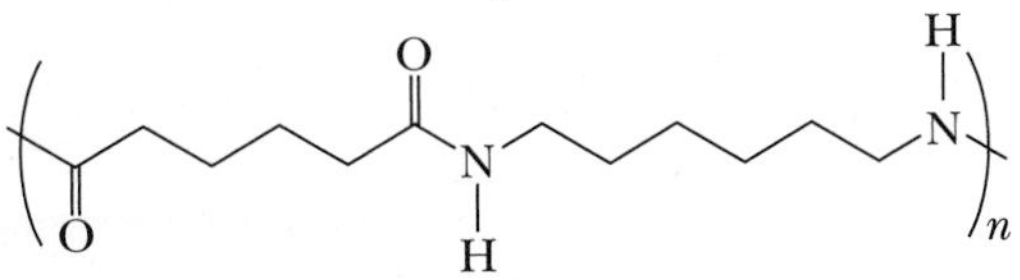

Nylon 66

Nylon 66 formed under these conditions has a T_m of 250–260°C and has a molecular-weight range of 10,000 to 20,000 g/mol.

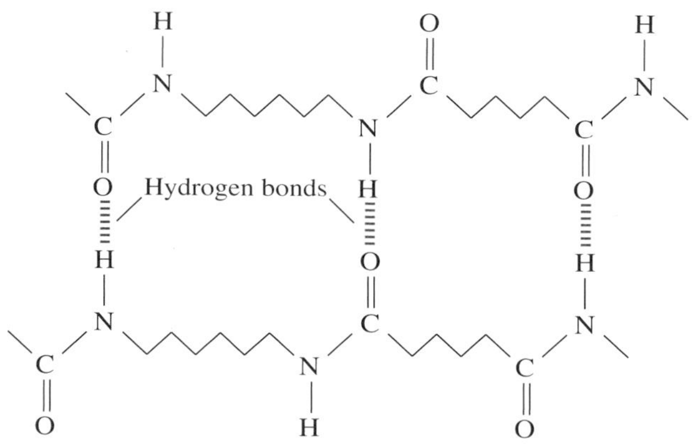

FIGURE 29.3 The structure of cold-drawn nylon 66. Hydrogen bonds between adjacent polymer chains provide additional tensile strength and stiffness to the fibers.

In the first stage of fiber production, crude nylon 66 is melted, spun into fibers, and cooled. Next, the melt-spun fibers are **cold-drawn** (drawn at room temperature) to about four times their original length to increase their degree of crystallinity. As the fibers are drawn, individual polymer molecules become oriented in the direction of the fiber axis, and hydrogen bonds form between carbonyl oxygens of one chain and amide hydrogens of another chain (Figure 29.3). The effects of orientation of polyamide molecules on the physical properties of the fiber are dramatic; both tensile strength and stiffness are increased markedly. Cold-drawing is an important step in the production of most synthetic fibers.

The current raw material base for the production of nylon 66 is benzene, which is derived almost entirely from catalytic cracking and reforming of petroleum. Catalytic reduction of benzene to cyclohexane followed by catalyzed air oxidation gives a mixture of cyclohexanol and cyclohexanone. Oxidation of this mixture by nitric acid gives adipic acid.

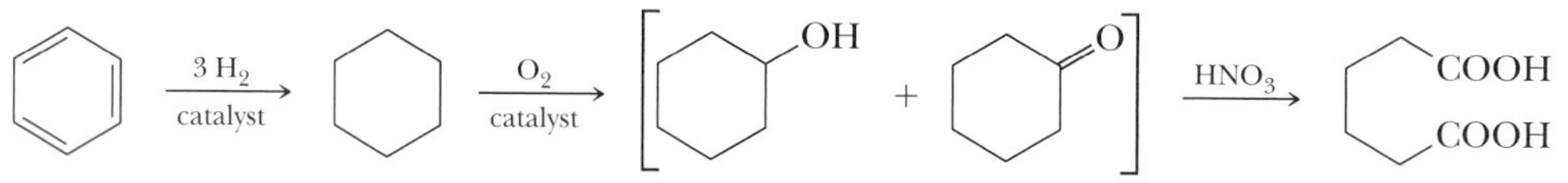

Benzene **Cyclohexane** **Cyclohexanol** **Cyclohexanone** **Hexanedioic acid (Adipic acid)**

Adipic acid, in turn, is a starting material for the synthesis of hexamethylenediamine. Treatment of adipic acid with ammonia gives an ammonium salt, which when heated, gives adipamide. Catalytic reduction of adipamide gives hexamethylenediamine. Thus, carbon sources for the production of nylon 66 are derived entirely from petroleum, which unfortunately is not a renewable resource.

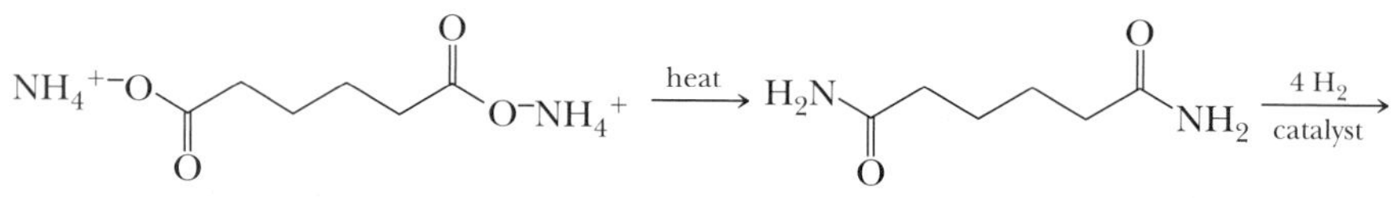

Ammonium hexanedioate (Ammonium adipate) **Hexanediamide (Adipamide)**

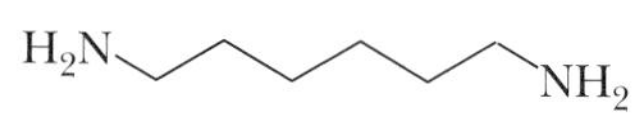

1,6-Hexanediamine (Hexamethylenediamine)

The nylons are a family of polymers, the members of which have subtly different properties that suit them to one use or another. The two most widely used members of this family are nylon 66 and nylon 6. Nylon 6 is so named because it is synthesized from caprolactam, a six-carbon monomer. In the synthesis of nylon 6, caprolactam is partially hydrolyzed to 6-aminohexanoic acid and then heated to 250°C to bring about polymerization. Nylon 6 is fabricated into fibers, brush bristles, rope, high-impact moldings, and tire cords.

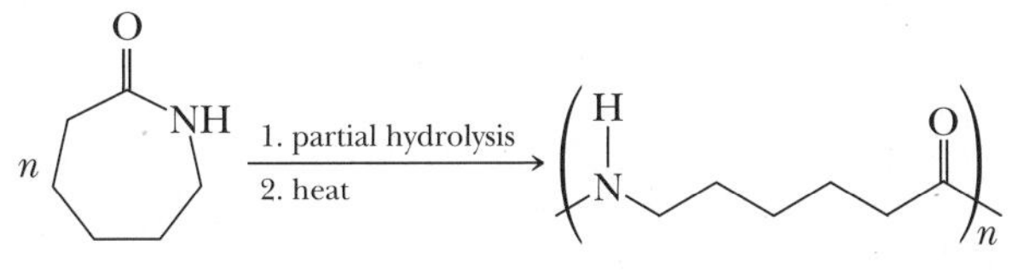
1. partial hydrolysis
2. heat

Caprolactam **Nylon 6**

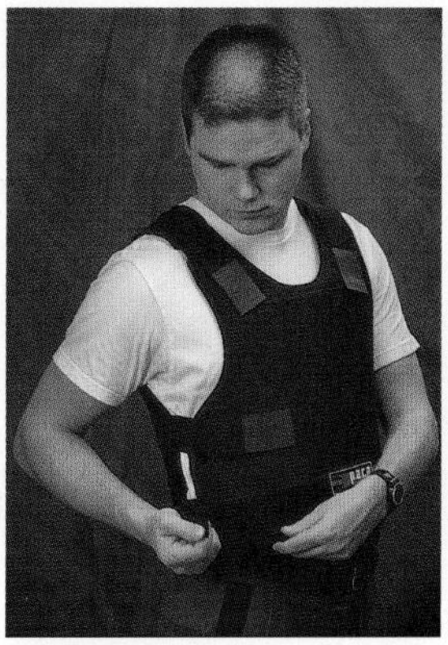

Bulletproof vests have a thick layer of Kevlar.

Based on extensive research into relationships between molecular structure and bulk physical properties, scientists at DuPont reasoned that a polyamide containing aromatic rings would be stiffer and stronger than either nylon 66 or nylon 6. In early 1960, DuPont introduced Kevlar, a polyaromatic amide (**aramid**) fiber synthesized from terephthalic acid and *p*-phenylenediamine.

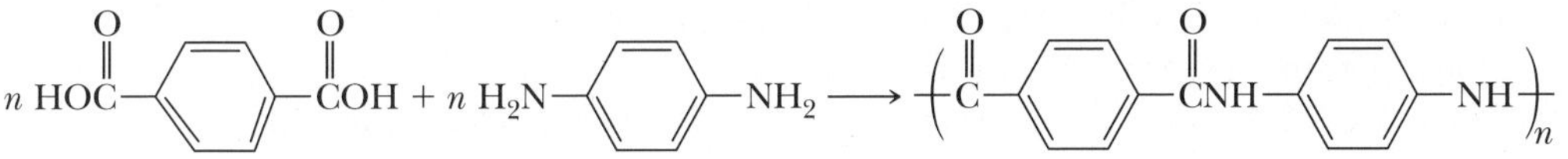

1,4-Benzenedicarboxylic acid (Terephthalic acid) **1,4-Benzenediamine (*p*-Phenylenediamine)** **Kevlar**

Aramid
A polyaromatic amide; a polymer in which the monomer units are an aromatic diamine and an aromatic dicarboxylic acid.

One of the remarkable features of Kevlar is its light weight compared with other materials of similar strength. For example, a 3 in. cable woven of Kevlar has a strength equal to that of a similarly woven 3 in. steel cable. Whereas the steel cable weighs about 20 lb/ft, the Kevlar cable weighs only 4 lb/ft. Kevlar now finds use in such articles as anchor cables for offshore drilling rigs and reinforcement fibers for automobile tires. Kevlar is also woven into a fabric that is so tough it can be used for bulletproof vests, jackets, and raincoats.

B. Polyesters

Recall that in the early 1930s, Carothers and his associates had concluded that **polyester** fibers from aliphatic dicarboxylic acids and ethylene glycol were not suitable for textile use because their melting points are too low. Winfield and Dickson at the Calico Printers Association in England further investigated polyesters in the 1940s and reasoned that a greater resistance to rotation in the polymer backbone would stiffen the polymer, raise its melting point, and thereby lead to a more acceptable polyester fiber. To create stiffness in the polymer chain, they used 1,4-benzenedicarboxylic acid (terephthalic acid). Polymerization of this aromatic dicarboxylic acid with ethylene glycol gives poly(ethylene terephthalate), abbreviated PET (also PETE).

Polyester
A polymer in which each monomer unit is joined to the next by an ester bond, as, for example, poly (ethylene terephthalate).

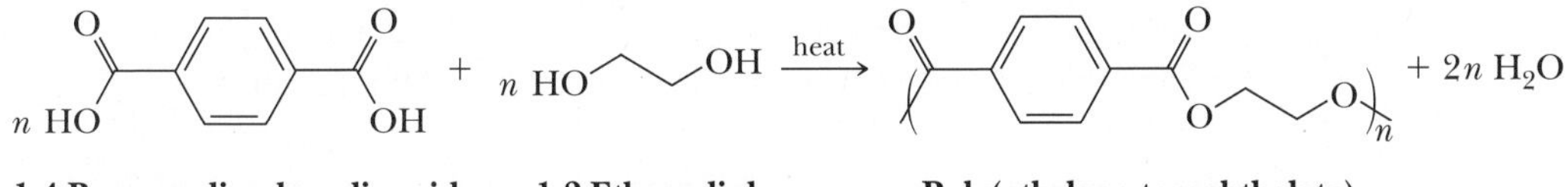

1,4-Benzenedicarboxylic acid (Terephthalic acid) **1,2-Ethanediol (Ethylene glycol)** **Poly(ethylene terephthalate) (Dacron, Mylar)**

The crude polyester can be melted, extruded, and then cold-drawn to form the textile fiber Dacron polyester, outstanding features of which are its stiffness (about four times that of nylon 66), very high strength, and remarkable resistance to creasing and wrinkling. Because the early Dacron polyester fibers were harsh to the touch owing to their stiffness, they were usually blended with cotton or wool to make

Mylar can be made into extremely strong films. Because the film has very tiny pores, it is used for balloons that can be inflated with helium; the helium atoms diffuse slowly through the pores of the film.

acceptable textile fibers. Newly developed fabrication techniques now produce less harsh Dacron polyester textile fibers. PET is also fabricated into Mylar films and recyclable plastic beverage containers.

Ethylene glycol for the synthesis of PET is obtained by air oxidation of ethylene to ethylene oxide (Section 11.8A) followed by hydrolysis to the glycol (Section 11.9A). Ethylene is, in turn, derived entirely from cracking either petroleum or ethane derived from natural gas (Section 2.9A). Terephthalic acid is obtained by oxidation of *p*-xylene, an aromatic hydrocarbon obtained along with benzene and toluene from catalytic cracking and reforming of naphtha and other petroleum fractions (Section 2.9B).

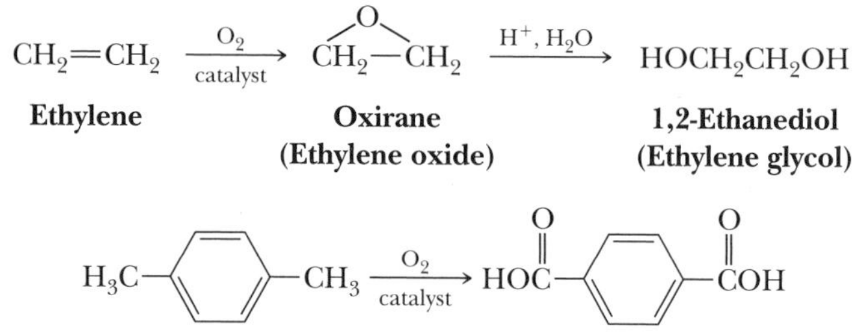

Ethylene → **Oxirane (Ethylene oxide)** → **1,2-Ethanediol (Ethylene glycol)**

***p*-Xylene** → **Terephthalic acid**

C. Polycarbonates

Polycarbonate

A polyester in which the carboxyl groups are derived from carbonic acid.

Polycarbonates, the most familiar of which is Lexan, are a class of commercially important engineering polyesters. In the production of Lexan, the disodium salt of bisphenol A (Problem 22.22) reacts with phosgene to form the polymer.

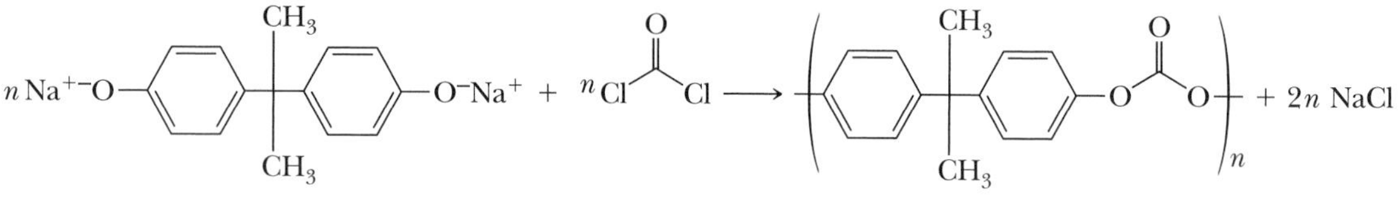

Disodium salt of bisphenol A **Phosgene** **Lexan** (a polycarbonate)

A polycarbonate hockey mask.

Lexan is a tough, transparent polymer with high impact and tensile strengths, and it retains its properties over a wide temperature range. It has found significant use in sporting equipment, such as bicycle, football, motorcycle, and snowmobile helmets as well as hockey and baseball catchers' face masks. In addition, it is used to make light, impact-resistant housings for household appliances and automobile and aircraft equipment and to manufacture safety glass and unbreakable windows.

D. Polyurethanes

A urethane, or carbamate, is an ester of carbamic acid, H_2NCOOH. Carbamates are most commonly prepared by treatment of an isocyanate with an alcohol.

$$RN{=}C{=}O + R'OH \longrightarrow RNH\overset{\overset{\large O}{\|}}{C}OR'$$

An isocyanate A carbamate

Polyurethane

A polymer containing the $—NHCO_2—$ groups as a repeating unit.

Polyurethanes consist of flexible polyester or polyether units (blocks) alternating with rigid urethane units (blocks). The rigid urethane blocks are derived from a diisocyanate, commonly a mixture of 2,4- and 2,6-toluenediisocyanate. The more flexible blocks are derived from low-molecular-weight (MW 1000–4000) polyesters or polyethers with —OH groups at each end of the polymer chain. Polyurethane fibers are fairly soft and elastic and have found use as Spandex and Lycra, the "stretch" fabrics used in bathing suits, leotards, and undergarments.

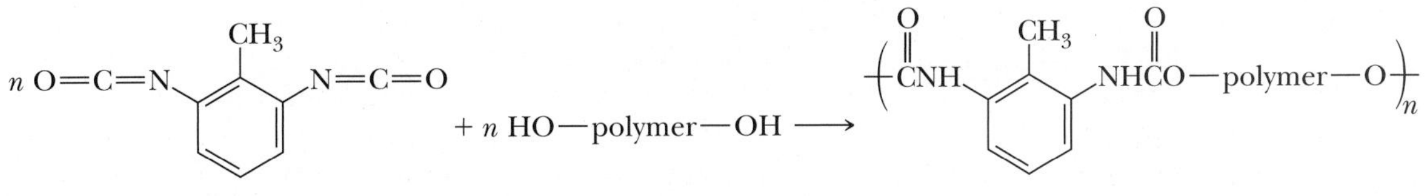

2,6-Toluenediisocyanate | Low-molecular-weight polyester or polyether | A polyurethane

Polyurethane foams for upholstery and insulating materials are made by adding small amounts of water during polymerization. Water reacts with isocyanate groups to produce gaseous carbon dioxide, which then acts as the foaming agent.

$$\underset{\text{An isocyanate}}{RN{=}C{=}O} + H_2O \longrightarrow \underset{\substack{\text{A carbamic acid}\\ \text{(unstable)}}}{\left[RNH{-}\overset{\overset{\large O}{\|}}{C}{-}OH\right]} \longrightarrow RNH_2 + CO_2$$

E. Epoxy Resins

Epoxy resins are materials prepared by a polymerization in which one monomer contains at least two epoxy groups. Within this range, a large number of polymeric materials are possible, and epoxy resins are produced in forms ranging from low-viscosity liquids to high-melting solids. The most widely used epoxide monomer is the diepoxide prepared by treatment of one mole of bisphenol A (Problem 22.22) with two moles of epichlorohydrin (Section 11.10). To prepare the following epoxy resin, the diepoxide monomer is treated with 1,2-ethanediamine (ethylene diamine). This component is usually called the "catalyst" in the two-component formulations that can be bought in any hardware store; it is also the component with the pungent smell. It is not a catalyst but a reagent.

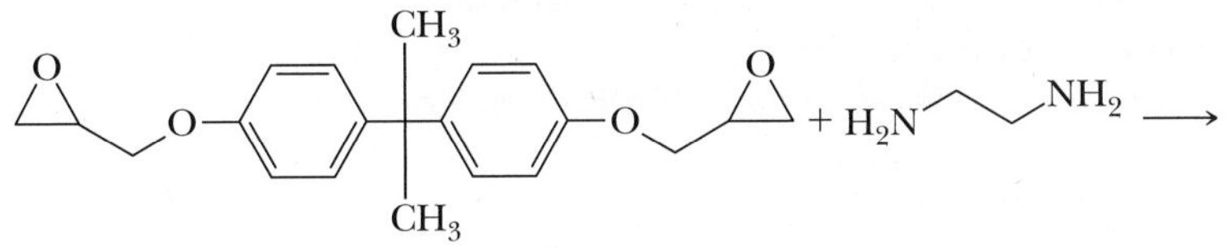

A diepoxide | A diamine

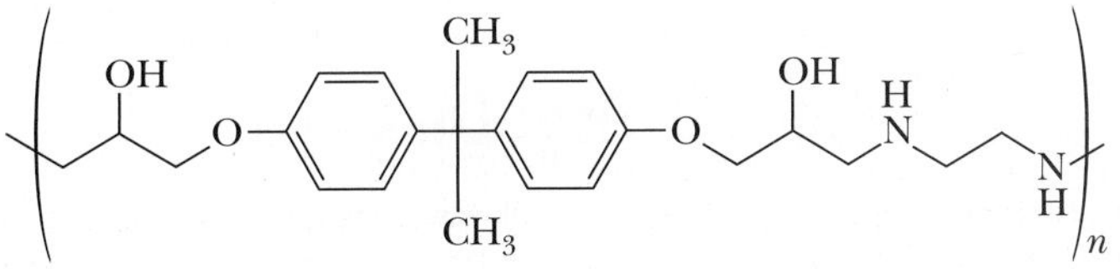

An epoxy resin

Epoxy resins are widely used as adhesives and insulating surface coatings. They have good electrical insulating properties, which leads to their use for encapsulating electrical components ranging from integrated circuit boards to switch coils and insulators for power transmission systems. They are also used as composites with other materials, such as glass fiber, paper, metal foils, and other synthetic fibers to create structural components for jet aircraft and rocket motor casings.

Example 29.2 Polymerization Mechanisms I

Write a mechanism for the acid-catalyzed polymerization of 1,4-benzenediisocyanate and ethylene glycol. To simplify your mechanism, consider only the reaction of one —**NCO** group with one —**OH** group.

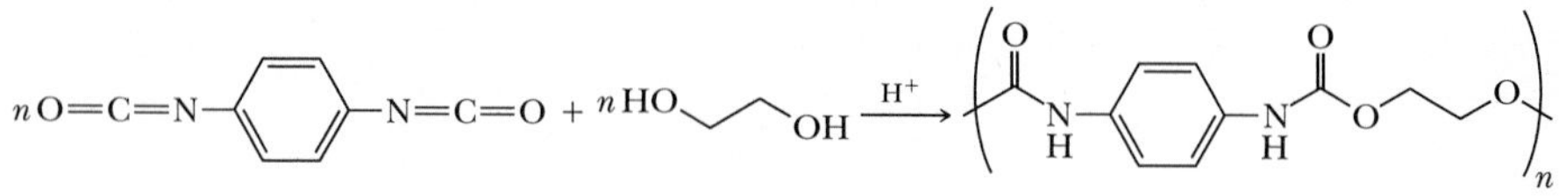

1,4-Benzenediisocyanate **Ethylene glycol** A polyurethane

Solution

A mechanism is shown in three steps. Proton transfer in Step 1 from the acid, HA, to nitrogen followed by addition of ROH to the carbonyl carbon in Step 2 gives an oxonium ion. Proton transfer in Step 3 from the oxonium ion to A^- gives the carbamate ester.

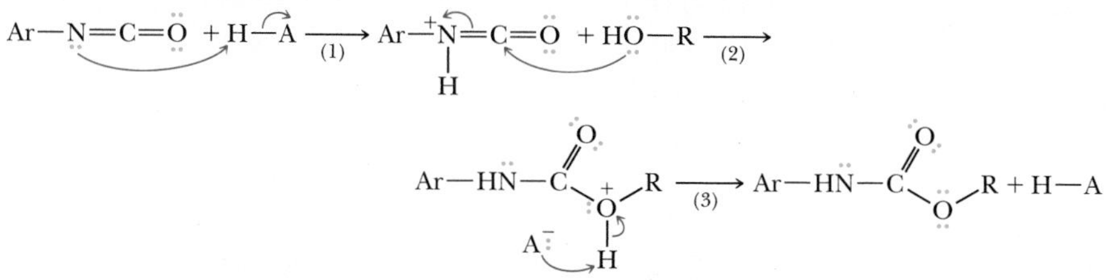

Problem 29.2

Write the repeating unit of the polymer formed from the following reaction and propose a mechanism for its formation.

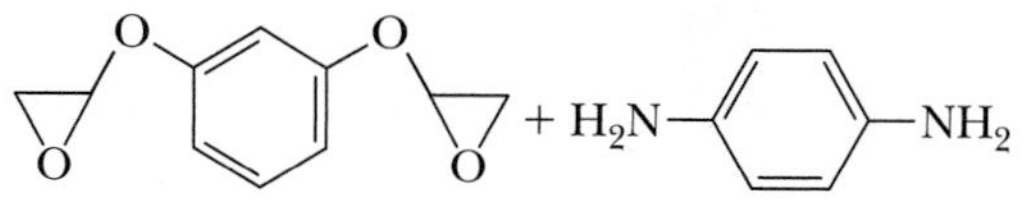

A diepoxide A diamine

Chemical Connections

Stitches That Dissolve

Medical science has advanced very rapidly in the last few decades. Some procedures considered routine today, such as organ transplantation and the use of lasers in surgery, were unimaginable 60 years ago. As the technological capabilities of medicine have grown, the demand for synthetic materials that can be used inside the body has increased as well. Polymers have many of the characteristics of an ideal biomaterial: they are lightweight and strong, are inert or biodegradable depending on their chemical structure, and have physical properties (softness, rigidity, and elasticity) that are easily tailored to match those of natural tissues. Carbon-carbon backbone polymers are degradation resistant and are used widely in permanent organ and tissue replacements.

Whereas most medical uses of polymeric materials require biostability, applications that use the biodegradable nature of some macromolecules have been developed. An example is the use of poly(glycolic acid) and glycolic acid/lactic acid copolymers as absorbable sutures, which go under the trade name Lactomer.

Traditional suture materials such as catgut must be removed by a health care specialist after they have served their purpose. Stitches of Lactomer, however, are hydrolyzed slowly over a period of approximately two weeks, and by the time, the torn tissues have fully healed, the stitches have fully degraded and no suture removal is necessary. Glycolic and lactic acids formed during hydrolysis of the stitches are metabolized and excreted by existing biochemical pathways.

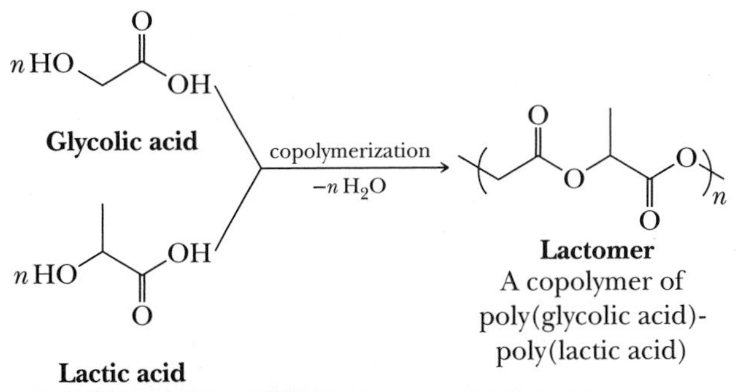

F. Thermosetting Polymers

Thermosetting polymers are composed of long chains that are cross-linked by covalent bonds. In effect, a thermosetting polymer is one giant molecule. The first thermosetting polymer was produced by Leo Baekeland (1863–1944) in 1907 by reacting phenol with formaldehyde to form the following three-dimensional structure. The product, known as Bakelite, is a good electrical insulator.

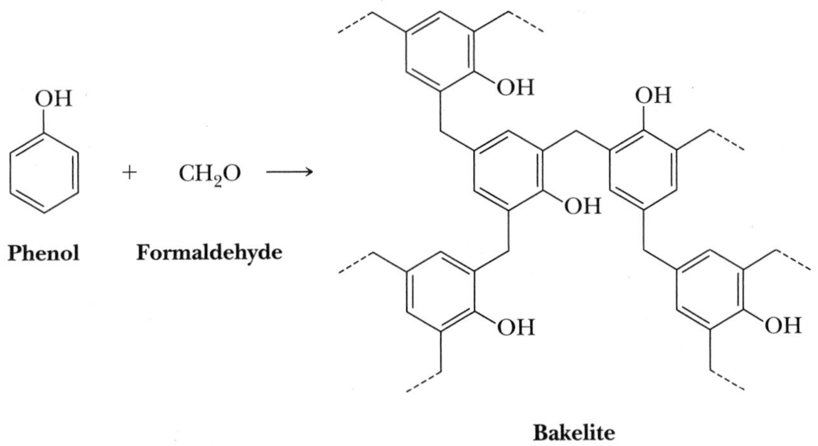

In the preparation of a thermoset, one of the monomers must be trifunctional. In the case of Bakelite, the trifunctional monomer is phenol. Alkyl thermosets are polyesters of an organic diacid, HOOC—R—COOH, and a trialcohol such as glycerol. Urea-formaldehyde thermosets are polyamides in which one molecule of urea, H_2N—CO—NH_2, can condense with up to four molecules of formaldehyde.

The manufacture of thermosets begins with a fluid mixture of the two monomers. The fluid is first shaped and then polymerized, either by heating or by being mixed with an initiator. The product of the polymerization is a network of covalently bonded atoms that is a solid, even at high temperatures. When heated to high temperatures, thermoset polymers char and decompose, but they do not melt.

29.6 Chain-Growth Polymerizations

From the perspective of the chemical industry, the single most important reaction of alkenes is **chain-growth polymerization**, a type of polymerization in which monomer units are joined together without loss of atoms. An example is the formation of polyethylene from ethylene.

Chain-growth polymerization

A polymerization that involves sequential addition reactions to either unsaturated monomers or monomers possessing other reactive functional groups.

$$n\,CH_2{=}CH_2 \xrightarrow{\text{catalyst}} \text{+}CH_2CH_2\text{+}_n$$

Ethylene **Polyethylene**

The mechanism of this type of polymerization differs greatly from the mechanism of step-growth polymerizations. In the latter, all monomers in addition to the polymer endgroups possess equally reactive functional groups, allowing for all possible combinations of reactions to occur, including monomer with monomer, dimer with dimer, and so on. In contrast, chain-growth polymerizations involve endgroups possessing reactive intermediates that react with a monomer only. The reactive intermediates used in chain-growth polymerizations include radicals, carbanions, carbocations, and organometallic complexes.

The number of monomers that undergo chain-growth polymerizations is large and includes such compounds as alkenes, alkynes, allenes, isocyanates, and cyclic compounds such as lactones, lactams, ethers, and epoxides. We concentrate on the chain-growth polymerizations of ethylene and substituted ethylenes and show how these compounds can be polymerized by radical, cation, anion, and organometallic-mediated mechanisms.

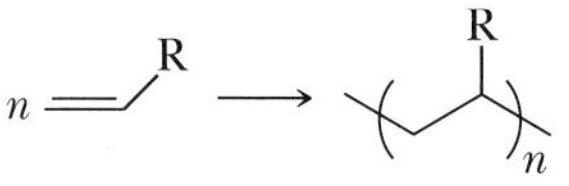

An alkene

Table 29.1 lists several important polymers derived from ethylene and substituted ethylenes along with their common names and most important uses.

The low thermal conductivity of polystyrene makes it a good insulating material.

A. Radical Chain-Growth Polymerizations

Among the initiators used for radical chain-growth polymerizations are diacyl peroxides such as dibenzoyl peroxide, which decompose as shown upon heating.

Table 29.1 Polymers Derived from Ethylene and Substituted Ethylenes

Monomer Formula	Common Name	Polymer Name(s) and Common Uses
$CH_2{=}CH_2$	**Ethylene**	**Polyethylene**, **Polythene**: break-resistant containers and packaging materials
$CH_2{=}CHCH_3$	**Propylene**	**Polypropylene**, **Herculon**: textile and carpet fibers
$CH_2{=}CHCl$	**Vinyl chloride**	**Poly(vinyl chloride)**, **PVC**: construction tubing
$CH_2{=}CCl_2$	**1,1-Dichloroethylene**	**Poly(1,1-dichloroethylene)**, **Saran Wrap**: food packaging
$CH_2{=}CHCN$	**Acrylonitrile**	**Polyacrylonitrile**, **Orlon**: acrylic and acrylate plastics
$CF_2{=}CF_2$	**Tetrafluoroethylene**	**Poly(tetrafluoroethylene)**, **PTFE**, **Teflon**: nonstick coatings
$CH_2{=}CHC_6H_5$	**Styrene**	**Polystyrene**, **Styrofoam**: insulating materials
$CH_2{=}CHCOOCH_2CH_3$	**Ethyl acrylate**	**Poly(ethyl acrylate)**: latex paints
$CH_2{=}C(CH_3)COOCH_3$	**Methyl methacrylate**	**Poly(methyl methacrylate)**, **Lucite**, **Plexiglass**: glass substitutes

In the first step, homolytic cleavage of the weak O—O peroxide bond yields two acyloxy radicals. Each acyloxy radical then decomposes to form an aryl radical and CO_2.

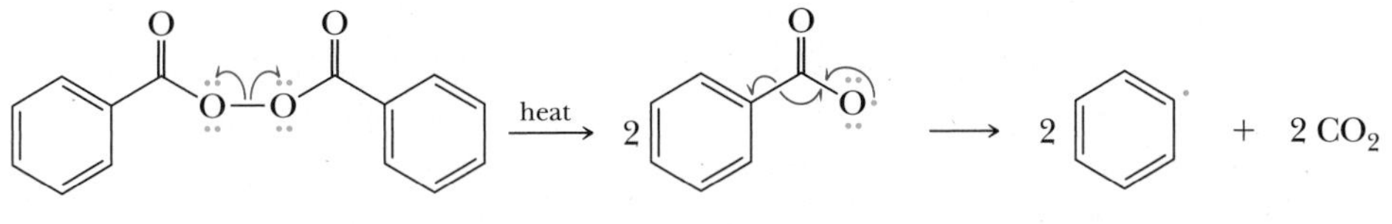

Dibenzoyl peroxide **Benzoyloxy radical** **Phenyl radical**

Another common class of initiators used in radical polymerizations is azo compounds [e.g., azoisobutyronitrile (AIBN)], which decompose upon heating or by the absorption of UV light to produce alkyl radicals and nitrogen gas.

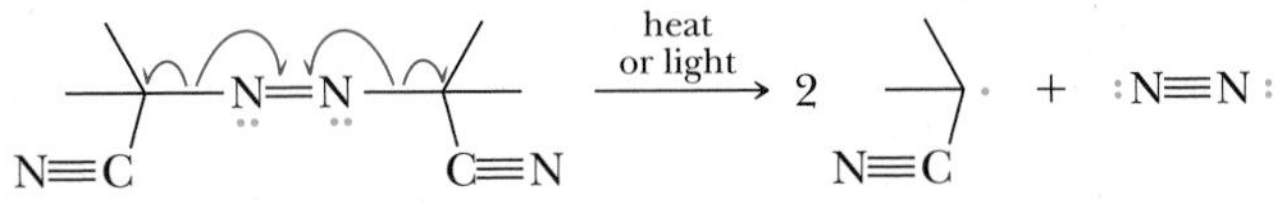

Azoisobutyronitrile (AIBN) Alkyl radicals

The chain initiation, propagation, and termination steps for radical polymerization of a substituted ethylene monomer are shown for the monomer RCH=CH_2. Dissociation of the initiator produces a radical that reacts with the double bond of a monomer. Once initiated, the chains continue to propagate through successive additions of monomers.

Mechanism 29.1

Radical Polymerization of a Substituted Ethylene

Step 1: Initiation: radicals form from nonradical compounds.

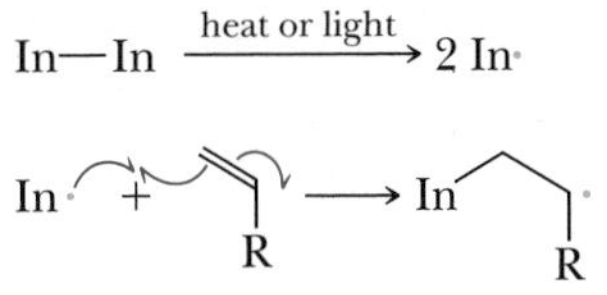

Step 2: Propagation: reaction of a radical and a molecule gives a new radical.

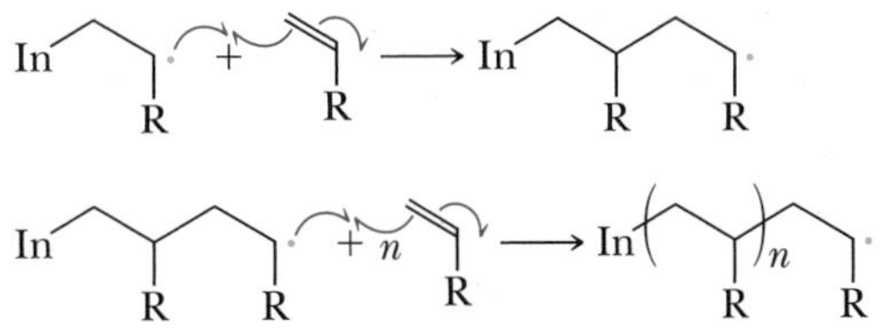

Step 3: Chain termination: radicals are destroyed.

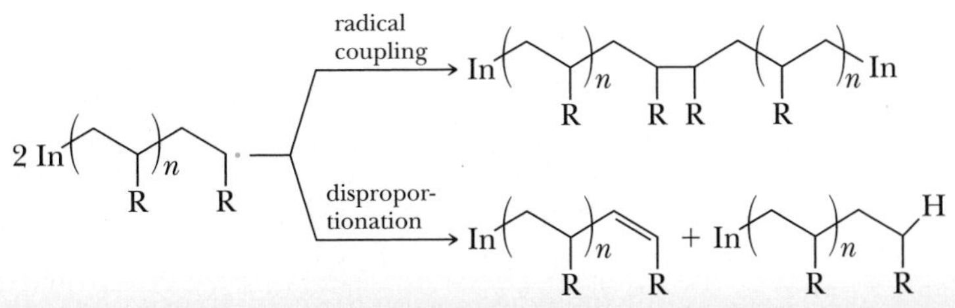

In radical reactions, the chain termination involves combination of radicals to produce a nonradical molecule or molecules. One common termination step is **radical coupling** to form a new carbon-carbon bond linking two growing polymer chains. This type of termination step is a diffusion-controlled process that occurs without an activation energy barrier. Another common termination process is **disproportionation**, which involves the abstraction of a hydrogen atom from the beta position to the propagating radical of one chain by the radical endgroup of another chain. This process results in two dead chains, one terminated in an alkyl group and the other in an alkenyl group.

Disproportionation

A termination process that involves the abstraction of a hydrogen atom from the beta position of the propagating radical of one chain by the radical endgroup of another chain.

Radical reactions with double bonds usually give the more stable (more substituted) radical. Because additions are biased in this fashion, the polymerizations of vinyl monomers tend to yield polymers with head-to-tail linkages. Vinyl polymers made by radical processes generally have no more than 1%–2% head-to-head linkages.

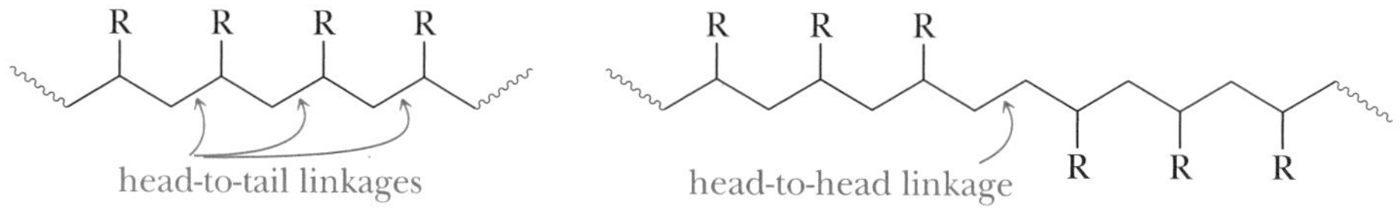

Because organic radicals are highly reactive species, it is not surprising that radical polymerizations are often complicated by unwanted side reactions. A frequently observed side reaction is hydrogen abstraction by the radical endgroup from a growing polymer chain, a solvent molecule, or another monomer. These side reactions are called **chain-transfer reactions** because the activity of the endgroup is "transferred" from one chain to another.

Chain-transfer reaction

The transfer of reactivity of an endgroup from one chain to another during a polymerization.

Chain transfer is illustrated by radical polymerization of ethylene. Polyethylene formed by radical polymerization exhibits a number of butyl branches on the polymer main chain. These four-carbon branches are generated in a "back-biting" chain-transfer reaction in which the radical endgroup abstracts a hydrogen from the fourth carbon back (the fifth carbon in the chain). Abstraction of this hydrogen is particularly facile because the transition state associated with the process can adopt a conformation like that of a chair cyclohexane. Continued polymerization of monomer from this new radical center leads to branches four carbons long.

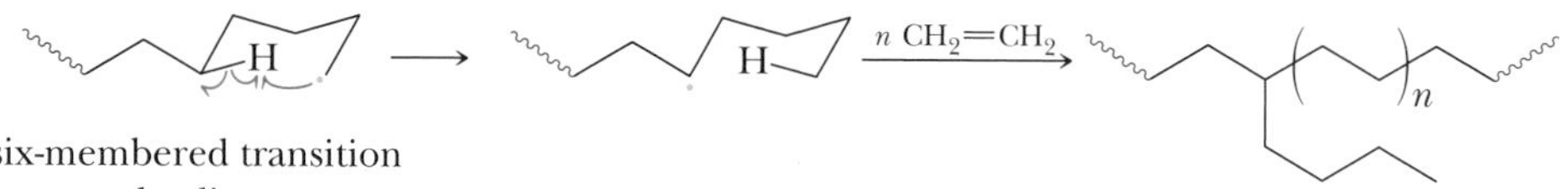

A six-membered transition state leading to 1,5-hydrogen abstraction

As a result of these various abstraction reactions, polymers synthesized by radical processes can have highly branched structures. The number of butyl branches depends on the relative stability of the propagating-radical endgroup and varies depending on the polymer. Polyethylene chains propagate through highly reactive primary radicals, which tend to be susceptible to 1,5-hydrogen abstraction reactions; these polymers typically have 15 to 30 branches per 500 monomer units. In contrast, polystyrene chains propagate through substituted benzyl radicals, which are stabilized by delocalization of the unpaired electron into the aromatic ring. These stabilized radicals are less likely to undergo hydrogen abstraction reactions. Polystyrene typically exhibits only one branch per 4000 to 10,000 monomer units.

The first commercial process for ethylene polymerization used peroxide catalysts at temperatures of 500°C and pressures of 1000 atm and produced a soft, tough polymer known as low-density polyethylene (LDPE). At the molecular level, chains of LDPE are highly branched owing to chain-transfer reactions. Because this extensive

Chemical Connections

Organic Polymers That Conduct Electricity

The influence of chemical structure on the properties of an organic compound is clearly seen in the electrical conducting properties of certain organic polymers. Most organic polymers are insulators. For example, polytetrafluoroethylene with the repeating unit —CF_2CF_2— and poly(vinyl chloride) with the repeating unit —CH_2CHCl— have conductivities of 10^{-18} S/cm. On the other end of the scale, the conductivity of copper is almost 10^6 S/cm.

Can organic polymers approach the conductivity of copper? Research carried out over the last 30 years shows that the answer is yes. When acetylene is passed through a solution containing certain transition metal catalysts, it can be polymerized to a shiny film of polyacetylene.

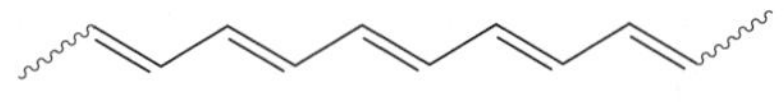

Polyacetylene

By itself, polyacetylene is not a conductor. However, by a process called doping, which involves introducing small amounts of electron-donating or electron-accepting compounds, it is possible to produce a polyacetylene that shows a conductivity of 1.5×10^5 S/cm.

The purpose of the doping agent is either to remove electrons from the π system (*p*-doping) or to add electrons to the π system (*n*-doping). A *p*-doped polyacetylene can be represented as a conjugated polyalkene chain containing positively charged carbons at several points along the chain.

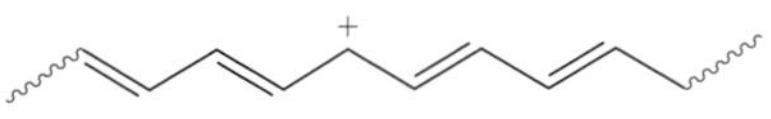

A *p*-doped polyacetylene

We can think of the positive charge as a defect that can move to the left or right along the polymer chain, thus giving rise to conductivity.

In crude polyacetylene, the polymer chains are jumbled, pointing in all directions. However, by stretching the film, the chains can be made to line up in a more ordered fashion. The conductivity of doped and oriented polyacetylene chains is greater along the direction of the chain than it is perpendicular to the chain. This result suggests that it is much easier for electrons to travel along a chain than to hop from one chain to the next.

Applications for conducting organic polymers are beginning to be developed. A rechargeable battery with electrodes of *p*-doped and *n*-doped polyacetylene already has been produced. Given the atomic weight of carbon, organic polymer batteries should be lighter than nickel-cadmium or lead-acid batteries. Weight is an important consideration for battery powered electric cars. In addition, many metals used in today's batteries (mercury, nickel, and lead) are toxic. If research leads to practical organic batteries, waste disposal problems could be considerably lessened.

chain branching prevents polyethylene chains from packing efficiently, LDPE is largely amorphous and transparent, with only a small amount of crystallites of a size too small to scatter light. LDPE has a density between 0.91 and 0.94 g/cm^3 and a melt transition temperature (T_m) of about 108°C. Because its T_m is only slightly above 100°C, it cannot be used for products that will be exposed to boiling water.

Approximately 65% of all low-density polyethylene is used for the manufacture of films. Fabrication of LDPE films is done by a blow-molding technique illustrated in Figure 29.4. A tube of molten LDPE along with a jet of compressed air is forced through an opening and blown into a giant, thin-walled bubble. The film is then cooled and taken up onto a roller. This double-walled film can be slit down the side to give LDPE film, or it can be sealed at points along its length to make LDPE bags. LDPE film is inexpensive, which makes it ideal for trash bags and for packaging for such consumer items as baked goods, vegetables, and other produce.

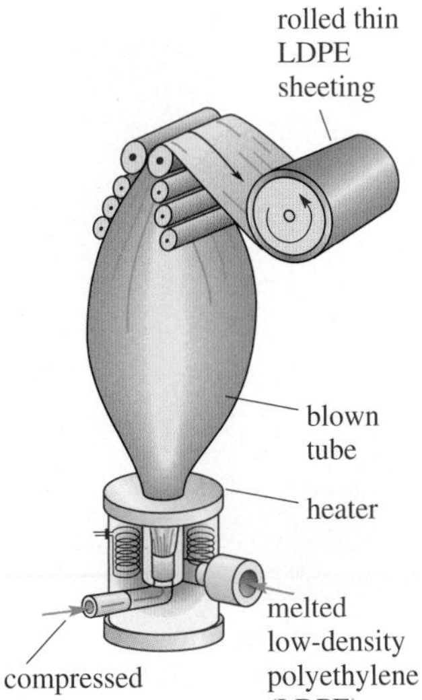

FIGURE 29.4 Fabrication of a LDPE film.

B. Ziegler-Natta Chain-Growth Polymerizations

An alternative method for polymerization of alkenes, which does not involve radicals, was developed by Karl Ziegler of Germany and Giulio Natta of Italy in the 1950s. For their pioneering work, they were awarded the 1963 Nobel Prize in Chemistry.

The early Ziegler-Natta catalysts were highly active, heterogeneous catalysts composed of a $MgCl_2$ support, a Group 4B transition metal halide such as $TiCl_4$ and an alkylaluminum compound such as $Al(CH_2CH_3)_2Cl$. These catalysts bring about polymerization of ethylene and propylene at 1 to 4 atm and at temperatures as low as 60°C. Polymerizations under these conditions do not involve radicals.

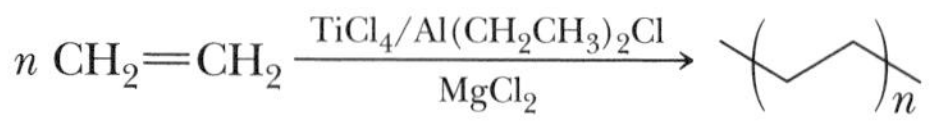

Ethylene **Polyethylene**

The active catalyst in a Ziegler-Natta polymerization is thought to be an alkyltitanium compound, which is formed by alkylation of the titanium halide by $Al(CH_2CH_3)_2Cl$ on the surface of a $MgCl_2/TiCl_4$ particle. Once formed, this species repeatedly inserts ethylene into the titanium-carbon bond to yield polyethylene.

Mechanism 29.2

Ziegler-Natta Catalysis of Ethylene Polymerization

Step 1: A titanium-ethyl bond forms.

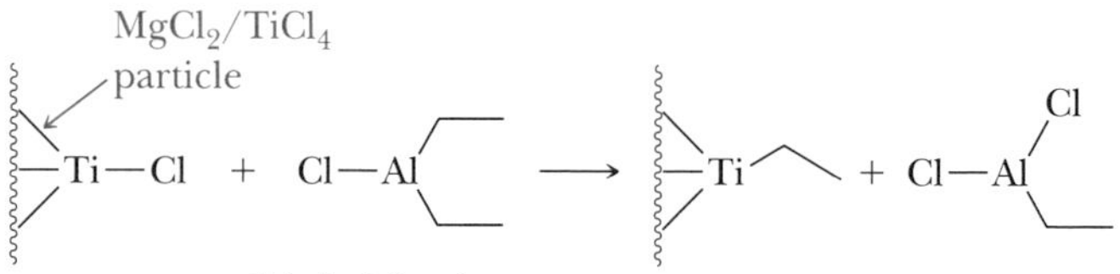

Diethylaluminum chloride

Step 2: Ethylene inserts into the titanium-carbon bond. This step repeats many times.

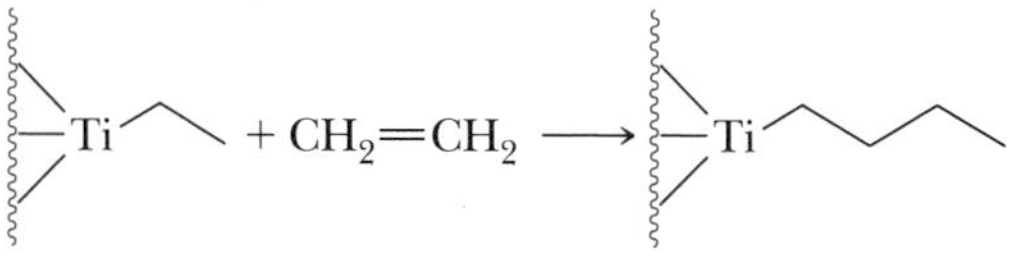

Polyethylene films are produced by extruding the molten plastic through a ringlike gap and inflating the film into a balloon.

Over 2.5×10^{11} kg of polyethylene are produced worldwide every year using optimized Ziegler-Natta catalysts, and large-scale reactors can yield up to 1.25×10^5 kg of polyethylene per hour. Production of polymer at this scale is partly attributable to the mild conditions required for a Ziegler-Natta polymerization and the fact that the polymer obtained has substantially different physical and mechanical properties from that obtained by radical polymerization. Polyethylene from Ziegler-Natta systems, termed high-density polyethylene (HDPE), has a higher density (0.96 g/cm^3) and T_m (133°C) than low-density polyethylene, is three to ten times stronger, and is opaque rather than transparent. This added strength and opacity is the result of a much lower degree of chain branching and the resulting higher degree of crystallinity of HDPE compared with LDPE.

Approximately 45% of all HDPE used in the United States is blow molded. In blow molding, a short length of HDPE tubing is placed in an open die [Figure 29.5(a)] and the die is closed, sealing the bottom of the tube. Compressed air is then forced into the hot polyethylene/die assembly, and the tubing is literally blown up to take the shape of the mold [Figure 29.5(b)]. After cooling, the die is opened [Figure 29.5(c)] and there is the container!

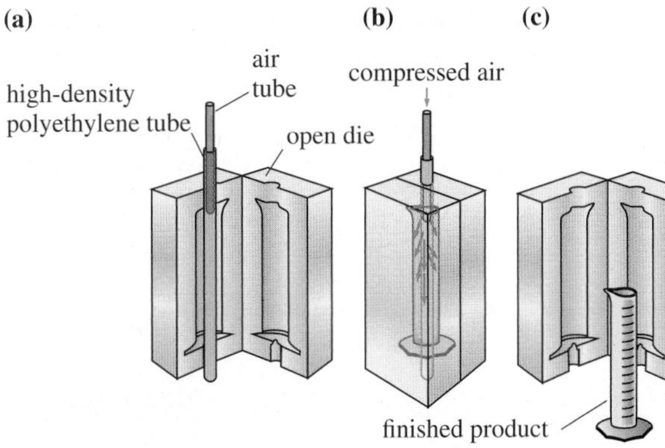

FIGURE 29.5 Blow molding of a HDPE container.

Even greater improvements in properties of HDPE can be realized through special processing techniques. In the melt state, HDPE chains adopt random coiled conformations similar to those of cooked spaghetti. Engineers have developed special extrusion techniques that force the individual polymer chains of HDPE to uncoil and adopt an extended linear conformation. These extended chains then align with one another to form highly crystalline materials. HDPE processed in this fashion is stiffer than steel and has approximately four times the tensile strength of steel. Because the density of polyethylene ($\approx$1.0 g/cm^3) is considerably less than that of steel (8.0 g/cm^3), these comparisons of strength and stiffness are even more favorable if they are made on a weight basis.

In recent years, there have been several important advances made in catalysts used in Ziegler-Natta-type polymerizations. One of the most important has been the discovery of soluble complexes that catalyze the polymerization of ethylene and propylene at extraordinary rates. Because these new homogeneous catalysts are substantially different in structure from the early Ziegler-Natta systems, these polymerizations are referred to as **coordination polymerizations**. Catalysts for coordination polymerizations are frequently formed by allowing bis(cyclopentadienyl)dimethylzirconium, $[Cp_2Zr(CH_3)_2]$, to react with methaluminoxane (MAO). MAO is a complex mixture of methylaluminum oxide oligomers, $[-(CH_3)AlO-)_n]$, formed by allowing trimethylaluminum to react with small amounts of water. It is thought that MAO activates the zirconium by abstracting a methyl anion to form a zirconium cation that is the active polymerization catalyst.

Mechanism 29.3

Homogeneous Catalysis for Ziegler-Natta Coordination Polymerization

Step 1: The zirconium catalyst is activated.

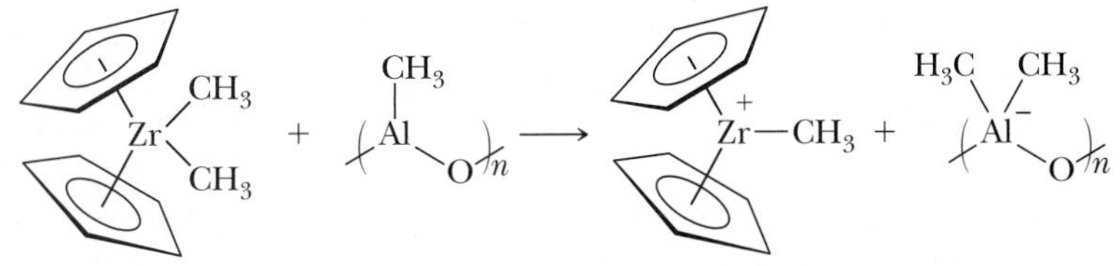

Bis(cyclopentadienyl)-dimethylzirconium | **MAO** | A zirconium cation (the active form of the catalyst)

(Continued)

Step 2: Ethylene inserts into the zirconium-carbon bond. This step repeats many times.

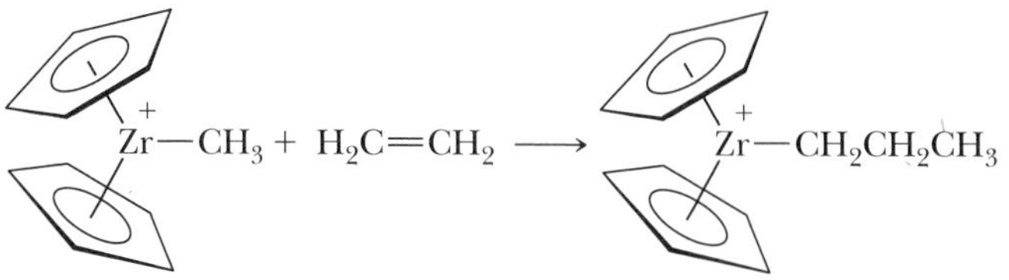

Some of these coordination-polymerization catalysts polymerize up to 20,000 ethylene monomer units per second, a rate otherwise reached only by enzyme-catalyzed biological reactions. Another important characteristic of these catalysts is that they show high reactivity toward 1-alkenes, allowing the formation of copolymers, such as that of ethylene and 1-hexene.

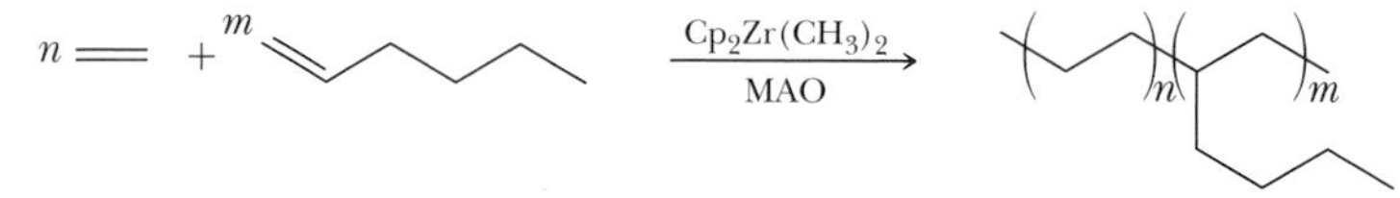

Ethylene **1-Hexene** **Ethylene-(1-hexene) copolymer**

Copolymers of this type with these moderate length branches (C_4, C_6, and so on) are called linear low-density polyethylene, or LLDPE. These are useful materials because they have many of the properties of LDPE made from radical reactions but are formed at the substantially milder conditions associated with Ziegler-Natta polymerizations.

C. Stereochemistry and Polymers

Thus far, we have written the formula of a substituted ethylene polymer in the following manner and have not been concerned with the configuration of each chiral center along the chain.

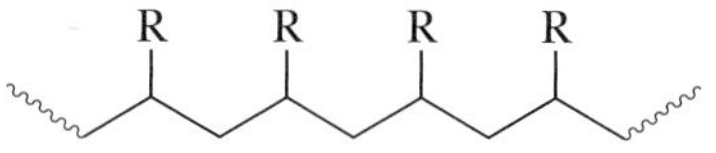

Nevertheless, the relative configurations of these chiral centers are important in determining the properties of a polymer. Polymers with identical configurations at all chiral centers along the chain are called **isotactic polymers**. Those with alternating configurations are called **syndiotactic polymers**, and those with completely random configurations are called **atactic polymers** (Figure 29.6).

In general, the more stereoregular the chiral centers are (i.e., the more highly isotactic or highly syndiotactic the polymer is), the more crystalline it is. A random placement of the substituents, such as in atactic materials, results in a polymer that cannot pack

Isotactic polymer

A polymer with identical configurations (either all *R* or all *S*) at all chiral centers along its chain, as, for example, isotactic polypropylene.

Syndiotactic polymer

A polymer with alternating *R* and *S* configurations at the chiral centers along its chain, as, for example, syndiotactic polypropylene.

Atactic polymer

A polymer with completely random configurations at the chiral centers along its chain, as, for example, atactic polypropylene.

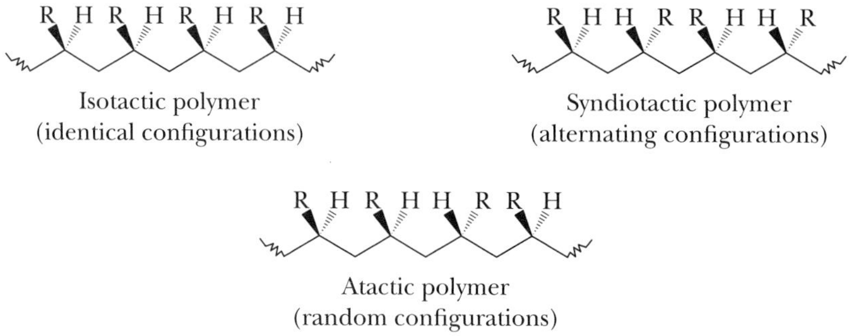

FIGURE 29.6 Relative configurations of chiral centers in polymers with different tacticities.

Table 29.2 Alkenes Polymerized by Anionic and Cationic Chain-Growth Mechanisms

Anionic polymerizations are most common for monomers substituted with electron-withdrawing groups.

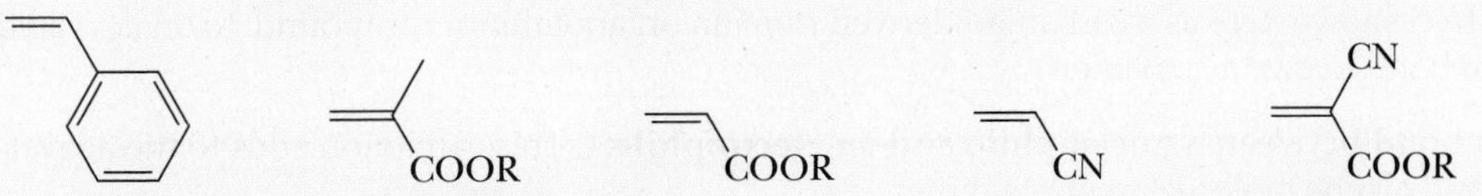

Styrene | Alkyl methacrylates | Alkyl acrylates | **Acrylonitrile** | Alkyl cyanoacrylates

Cationic polymerizations are most common for monomers substituted with electron-donating groups.

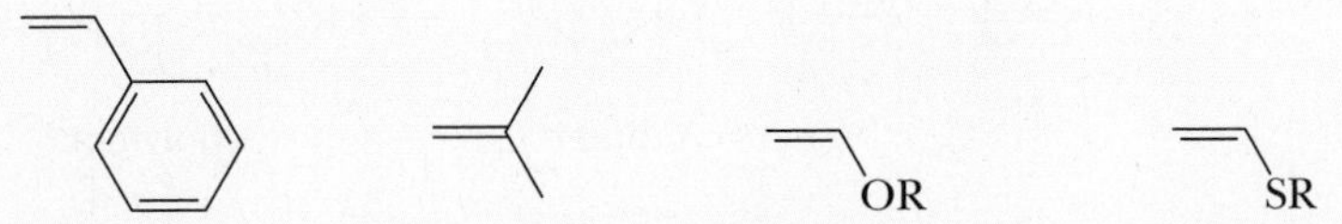

Styrene | **Isobutylene** | Vinyl ethers | Vinyl thioethers

well and is usually highly amorphous. Atactic polystyrene, for example, is an amorphous glass, whereas isotactic polystyrene is a crystalline fiber-forming polymer with a high melt transition. Therefore, the control over the relative configuration, or tacticity, along a polymer backbone is an area of considerable interest in modern polymer synthesis.

D. Ionic Chain-Growth Polymerizations

Chain-growth polymers can also be synthesized using reactions that rely on either anionic or cationic species in the propagation steps. The choice of ionic procedure depends greatly on the electronic nature of the monomers to be polymerized. Vinyl monomers with electron-withdrawing groups, which stabilize carbanions, are used in anionic polymerizations, whereas vinyl monomers with electron-donating groups, which stabilize cations, are used in cationic polymerizations (Table 29.2).

Styrene is conspicuous among the monomers given in Table 29.2 because it can be polymerized using either anionic or cationic techniques as well as radical techniques. This characteristic particular to styrene is attributable to the fact that the phenyl group can stabilize cationic, anionic, and radical benzylic intermediates.

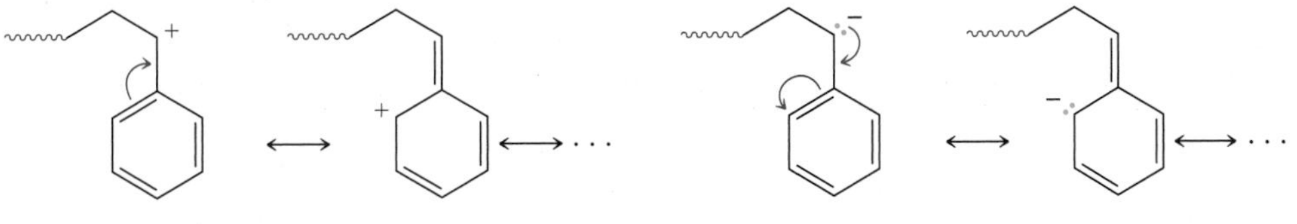

Resonance stabilization of a benzylic cation endgroup | Resonance stabilization of a benzylic anion endgroup

Anionic Polymerizations

Anionic polymerizations can be initiated by addition of a nucleophile to an activated alkene. The most common nucleophiles used for this purpose are metal alkyls such as methyl- and *sec*-butyllithium. The newly formed carbanion then acts as a nucleophile and adds to another monomer unit, and the propagation continues.

Mechanism 29.4

Initiation of Anionic Polymerization of Alkenes

Step 1: **Make a new bond between a nucleophile and an electrophile.** Polymerization is initiated by addition of a nucleophile, shown here as a carbanion derived from an organolithium compound, to an activated carbon-carbon double bond to give a carbanion.

Step 2: **Make a new bond between a nucleophile and an electrophile.** This carbanion adds to the activated double bond of a second alkene molecule to give a dimer.

Step 3: **Make a new bond between a nucleophile and an electrophile.** Chain growth continues to give a polymer.

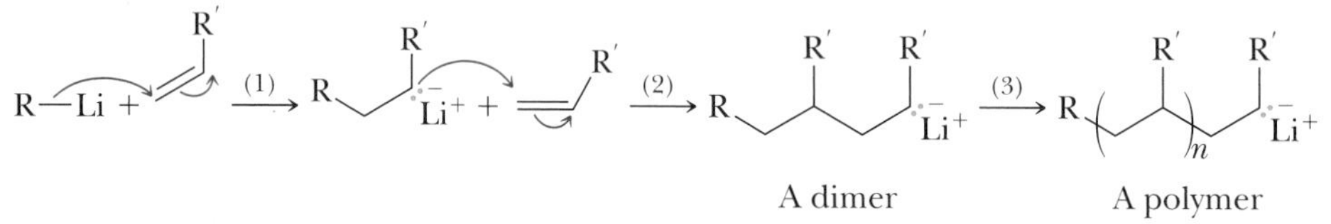

An alternative method for the initiation of anionic polymerizations involves a one-electron reduction of the monomer by lithium or sodium to form a radical anion. The radical anion thus formed is either further reduced to form a dianion or dimerizes to form a dimer dianion.

Mechanism 29.5

Initiation of Anionic Polymerization of Butadiene

Step 1: A one-electron reduction of the diene by lithium metal gives a radical anion.

Step 2: One-electron reduction of this radical anion gives a monomer dianion.

Step 3: Alternatively, radical coupling gives a dimer dianion.

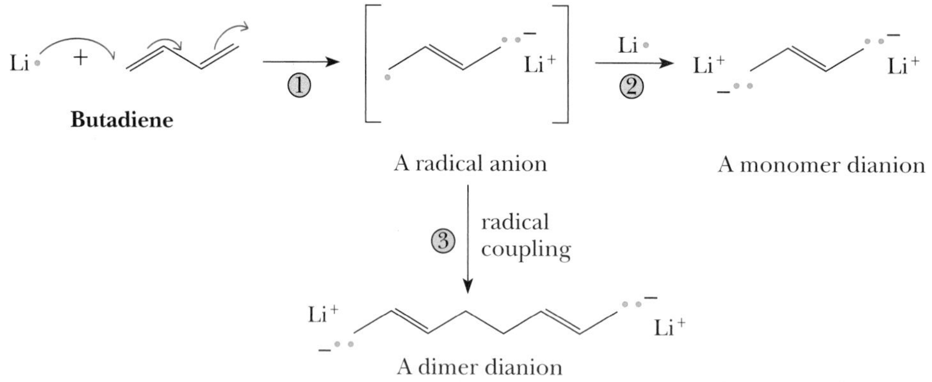

In either case, a single initiator can now propagate chains from both ends by virtue of its two active endgroup carbanions. These reactions are heterogeneous and involve transfer of the electron from the surface of the metal. To improve the efficiency of this process, soluble reducing agents such as sodium naphthalide are used. Sodium undergoes electron-transfer reactions with extended aromatic compounds, such as naphthalene, to form soluble radical anions.

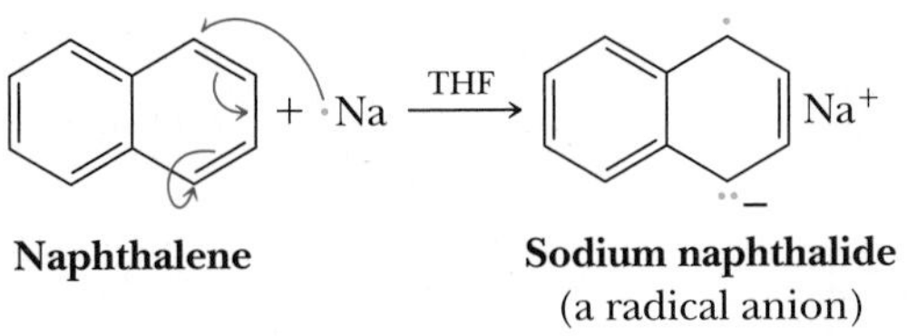

The naphthalide radical anion is a powerful reducing agent. For example, styrene undergoes a one-electron reduction to form the styryl radical anion, which couples to form a dianion. The latter then propagates polymerization at both ends, growing chains in two directions simultaneously.

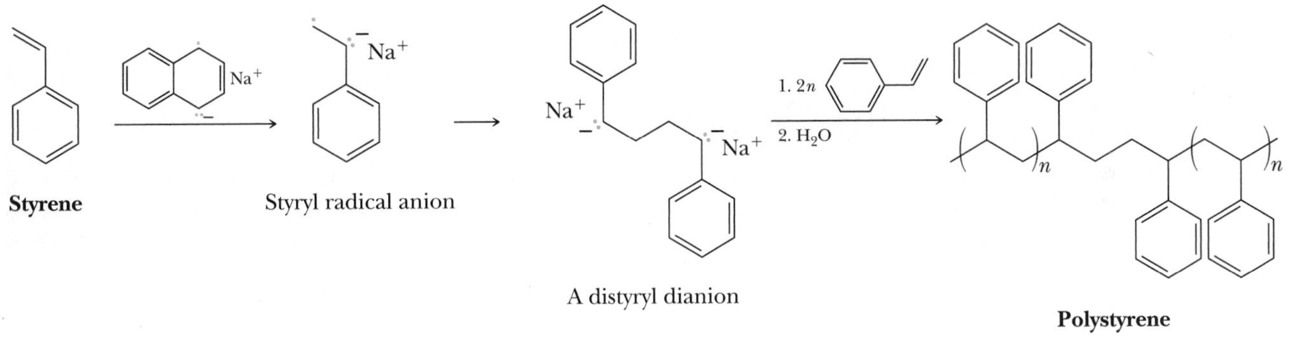

The propagation characteristics of anionic polymerizations are similar to those of radical polymerizations, but with the important difference that many of the chain-transfer and chain-termination reactions that plague radical processes are absent. Furthermore, because the propagating chain ends carry the same charge, bimolecular coupling and disproportionation reactions are also averted. An interesting set of circumstances arises when chain-transfer and chain-termination steps are no longer significant. Under these conditions, polymer chains are initiated and continue to grow until either all the monomer is consumed or some external agent is added to terminate the chains. Polymerizations of this type are called **living polymerizations** because they will restart if more monomer is added after it is initially consumed.

Living polymer

A polymer chain that continues to grow without chain-termination steps until either all of the monomer is consumed or some external agent is added to terminate the chain. The polymer chains will continue to grow if more monomer is added.

The absence of chain-transfer and chain-termination steps in living polymerizations has far-reaching consequences. One of the most visible of these is in the area of molecular-weight control. The molecular weight of a polymer originating from living polymerizations is determined directly by the monomer-to-initiator ratio. It is, therefore, relatively easy to obtain polymers of a well-defined size simply by controlling the stoichiometry of the reagents. In contrast, the average sizes of polymer chains formed from nonliving, chain-growth processes (radical, Ziegler-Natta, and so on) vary from system to system and are determined by the ratio of the rate of propagation to the rate of termination. In most cases, precise control over the molecular weight of the product obtained in nonliving systems is not possible because it is very difficult to change one of the rates involved without affecting the other.

After consumption of the monomer under living, anionic conditions, electrophilic terminating agents can be added to functionalize the chain ends. Examples of terminating reagents include CO_2 and ethylene oxide, which, after protonation, form carboxylic acid and alcohol-terminated chains, respectively.

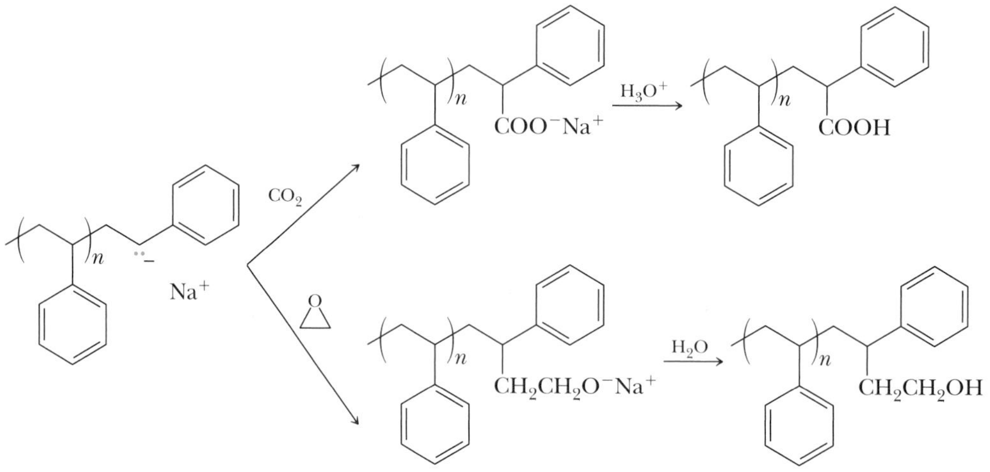

MCAT Practice: *Passage and Questions*

The Chemistry of Superglue

Used in everything from model planes to passenger planes, superglue is one of the best-known modern glues. The curing process that facilitates its remarkable adhesive properties is a chain polymerization reaction. The ingredient that gives superglue its adhesive ability is methyl cyanoacrylate. This compound is just one member of a larger family of cyanoacrylates with the following structure.

Contrary to popular understanding, superglue does not "air dry." In fact, cyanoacrylates cure (convert from liquid to solid) in the presence of weak nucleophiles such as water. Under normal circumstances, a thin layer of water is present on almost all surfaces. The curing process, therefore, involves the reaction shown here.

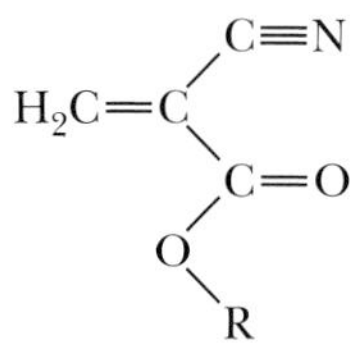

a cyanoacrylate

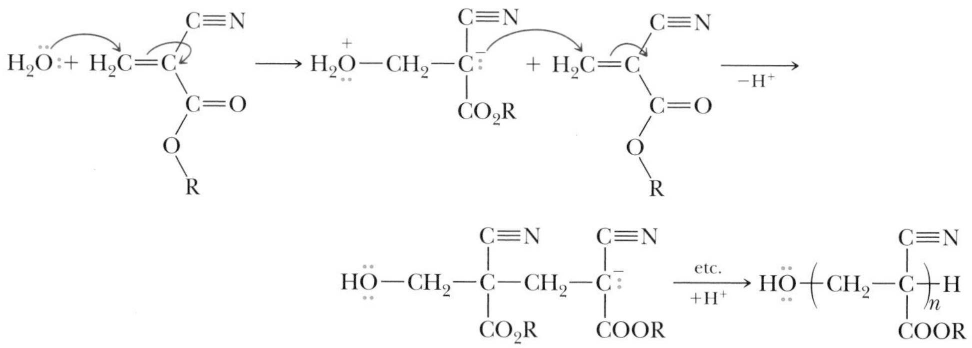

Questions

A. What general term would be used to describe the polymerization reaction involved in the curing of superglue?

1. Radical chain polymerization
2. Anionic chain polymerization
3. Cationic chain polymerization
4. Ziegler-Natta chain polymerization

B. Why does the weak nucleophile, water, efficiently add to the cyanoacrylate?

1. Superglue must possess hydroxide as a promoter to initiate the polymerization in the presence of water.
2. The carbanion formed by nucleophilic addition of water is resonance-stabilized by two electron-withdrawing groups.
3. The zwitterion formed by addition of water is well solvated in the cyanoacrylate matrix of the glue.
4. Both 1 and 2.

C. The reaction that propagates the chain polymerization would be referred to as

1. A 1,4-addition reaction.
2. A Michael addition reaction.

3. A conjugate addition reaction.
4. All of the above.

D. The synthesis of cyanoacrylates would involve

1. A Claisen condensation.
2. An aldol reaction.
3. A Michael reaction.
4. A Dieckmann reaction.

In recent years, the high adhesive strength of superglues has captured attention in new fields. Medical-grade superglues such as 2-octyl cyanoacrylate are now commonly used as sutures in laceration repair. They have also proven effective in skin, bone, and cartilage grafts.

E. Which of the following is *not* knowledge gained when one reads the name 2-octyl cyanoacrylate?

1. That the compound is anionic
2. That the compound would be chiral
3. That the compound is a nitrile
4. That the compound is an ester

F. Given the mechanism of polymerization, which of the following reaction conditions would lower the length of the polymers created?

1. Increasing concentrations of water to start
2. Acidic conditions
3. Basic conditions
4. Both 1 and 2
5. Both 1 and 3

In a similar fashion, polymer chains with functional groups at both ends, called **telechelic polymers**, can be prepared by addition of these same reagents (e.g., CO_2 and ethylene oxide) to solutions of chains with two active ends initiated by sodium naphthalide.

Telechelic polymer
A polymer in which its growing chains are terminated by formation of new functional groups at both ends of its chains. These new functional groups are introduced by adding reagents, such as CO_2 or ethylene oxide, to the growing chains.

Example 29.3 Telechelic Polymers

Show how to prepare polybutadiene that is terminated at both ends with carboxylate groups.

Solution

Form a growing chain with two active endgroups by treatment of butadiene with two moles of lithium metal to form a dianion followed by addition of monomer units and formation of a living polymer. Cap the active endgroups with a carboxylate group by treatment of the living polymer with carbon dioxide.

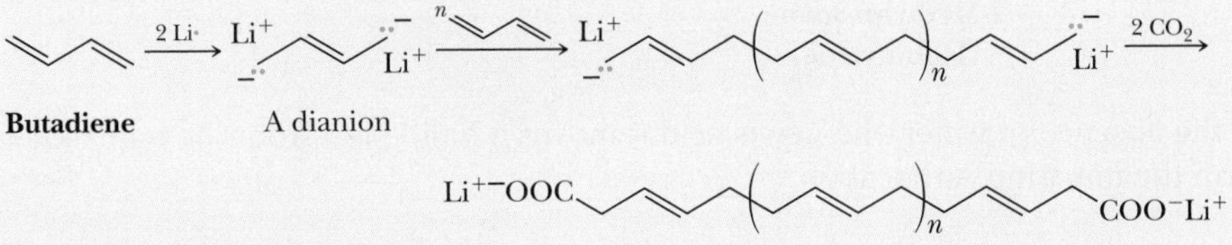

Problem 29.3

Show how to prepare polybutadiene that is terminated at both ends with primary alcohol groups.

Cationic Polymerizations

Only alkenes with electron-donating substituents, such as alkyl, aryl, ether, thioether, and amino groups, undergo useful cationic polymerizations. The two most common methods of generating cationic initiators are (1) the reaction of a strong protic acid with an organic monomer and (2) the abstraction of a halide from the organic initiator by a Lewis acid. Cationic chain-growth polymerizations are generally effective only for monomers yielding relatively stable carbocations (i.e., monomers that form either 3° carbocations or cations stabilized by electron-donating groups, such as ether, thioether, or amino groups).

Initiation by protonation of an alkene requires the use of a strong acid with a nonnucleophilic anion to avoid 1,2-addition across the alkene double bond. Suitable acids with nonnucleophilic anions include HF/AsF_5 and HF/BF_3. In the following general equation, initiation is by proton transfer from $H^+BF_4^-$ to the alkene to form a tertiary carbocation, which then continues the cationic chain growth polymerization.

Mechanism 29.6

Initiation of Cationic Polymerization of an Alkene by $HF \cdot BF_3$

Step 1: **Make a new bond between a π bond and an electrophile—add a proton.** Proton transfer from the $HF \cdot BF_3$ complex to the alkene gives a carbocation.

Step 2: **Make a new bond between a π bond and an electrophile.** Propagation continues the polymerization.

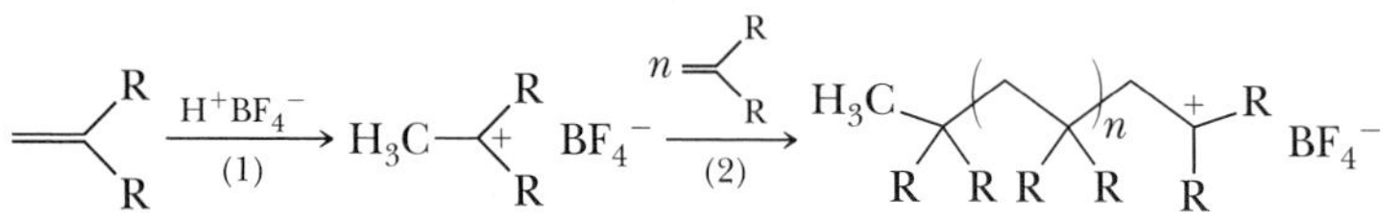

The second common method for generating carbocations involves the reaction between a haloalkane and a Lewis acid, such as BF_3, $SnCl_4$, $AlCl_3$, $Al(CH_3)_2Cl$, and $ZnCl_2$. When a trace of water is present, the mechanism of initiation using some Lewis acids is thought to involve protonation of the alkene.

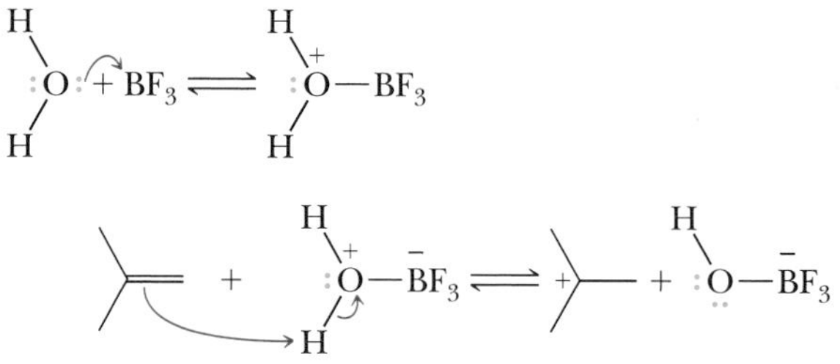

2-Methylpropene (Isobutylene)

In the absence of water, the Lewis acid removes a halide ion from the haloalkane to form the initiating carbocation.

Mechanism 29.7

Initiation of Cationic Polymerization of an Alkene by a Lewis Acid

Step 1: **Lewis acid-Lewis base reaction followed by break a bond to give stable molecules or ions.** Reaction of the chloroalkane (a Lewis base) with tin(IV) chloride (a Lewis acid) gives a carbocation from which polymerization then proceeds.

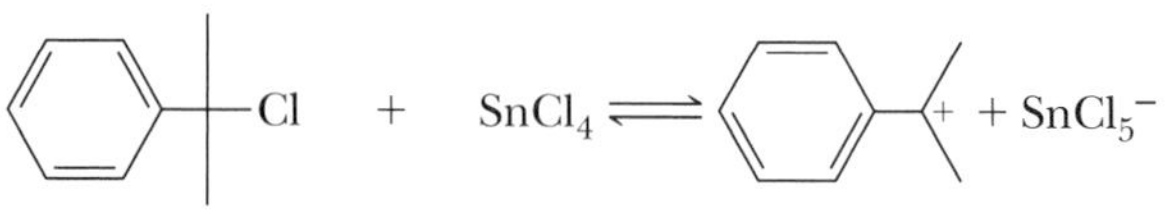

2-Chloro-2-phenylpropane

The polymerization of alkenes then propagates by the electrophilic attack of the carbocation on the double bond of the alkene monomer. The regiochemistry of the addition is determined by the formation of the more stable (the more highly substituted) carbocation.

Example 29.4 Polymerization Mechanisms II

Write a mechanism for the polymerization of 2-methylpropene (isobutylene) initiated by treatment of 2-chloro-2-phenylpropane with $SnCl_4$. Label the initiation, propagation, and termination steps.

Solution

Lewis acid-Lewis base reaction followed by break a bond to give stable molecules or ions Chain initiation: cations form from nonionic materials.

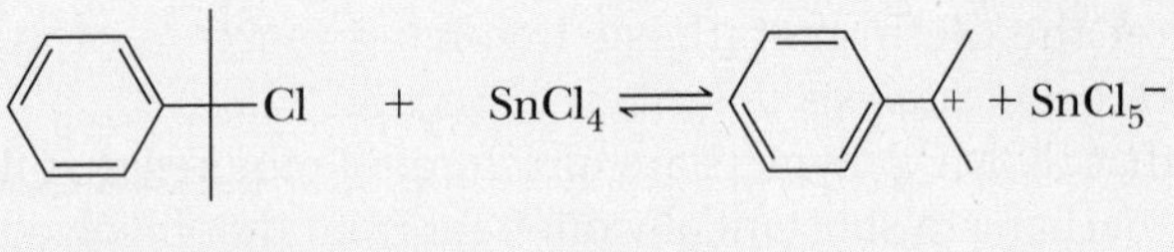

2-Chloro-2-phenylpropane

Make a new bond between a π bond and an electrophile Chain propagation: a cation and a molecule react to give a new cation.

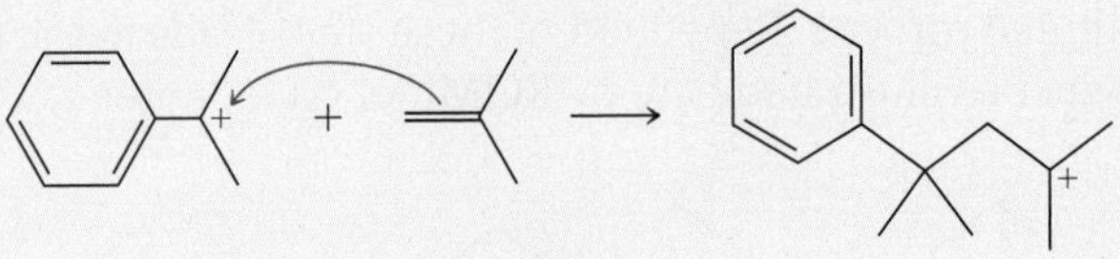

2-Methylpropene

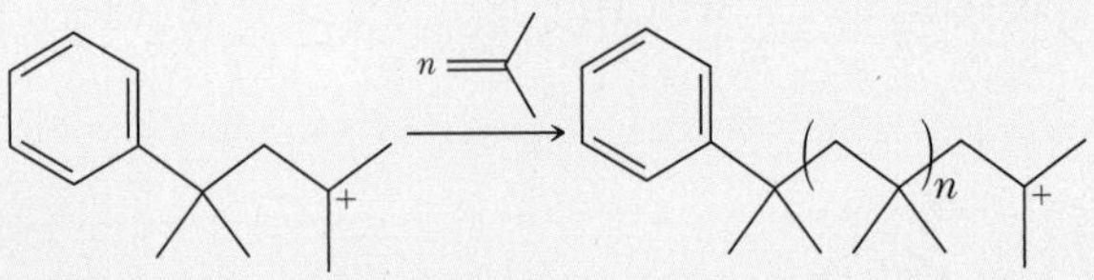

Chain termination: destruction of cations.

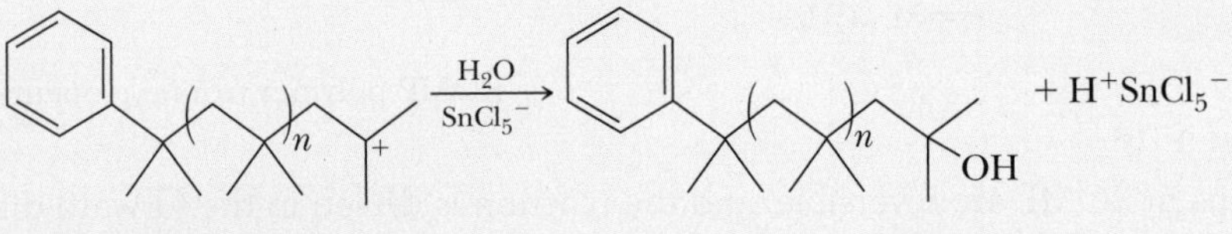

Problem 29.4

Write a mechanism for the polymerization of methyl vinyl ether initiated by 2-chloro-2-phenylpropane and $SnCl_4$. Label the initiation, propagation, and termination steps.

E. Ring-Opening Metathesis Polymerizations

During early investigations into the polymerizations of cycloalkenes by transition metal catalysts such as those used in Ziegler-Natta polymerizations, polymers of unexpected structures that contained the same number of double bonds as originally present in the monomers were formed. This process is illustrated by the polymerization of norbornene.

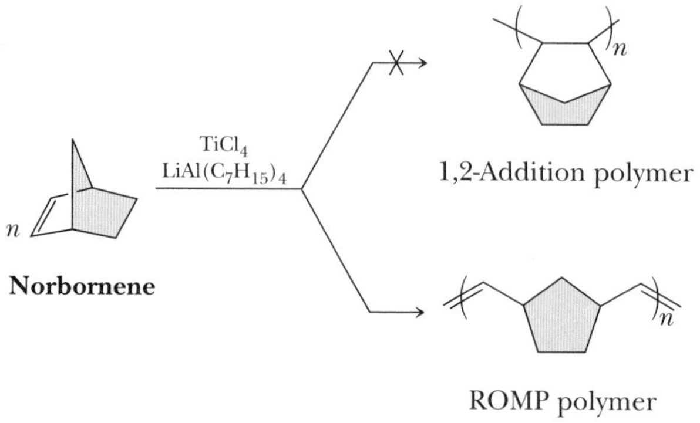

If reaction had proceeded in the same manner as Ziegler-Natta polymerization of ethylene and substituted ethylenes (Section 29.6B), a 1,2-addition polymer would have been formed. What is formed, however, is an unsaturated polymer in which the number of double bonds in the polymer is the same as that in the monomers polymerized. This process is called **ring-opening metathesis polymerization**, or ROMP, after the olefin metathesis involving reaction of acyclic alkenes and nucleophilic carbene catalysts described in Section 24.6.

The fact that ROMP polymers are unsaturated requires that this polymerization proceed by a mechanism substantially different from that involved in polymerization of ethylene and substituted ethylenes by the same catalyst mixtures. Following lengthy and detailed studies, chemists discovered that ROMP involves the same metallacyclobutane species as in ring-closing alkene metathesis reactions (Section 24.6B). The intermediate metallacyclobutane derivative undergoes a ring-opening reaction to give a new substituted carbene. Repetition of these steps leads to the formation of the unsaturated polymer as illustrated here by ROMP of cyclopentene.

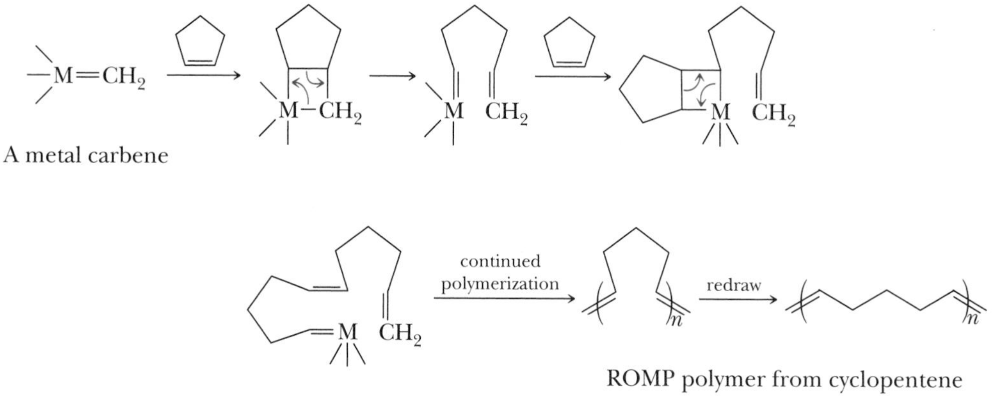

All steps in ROMP are reversible, and the reaction is driven in the forward direction by the release of ring strain that accompanies the opening of the ring. The reactivity of the following cycloalkenes toward ROMP decreases in this order.

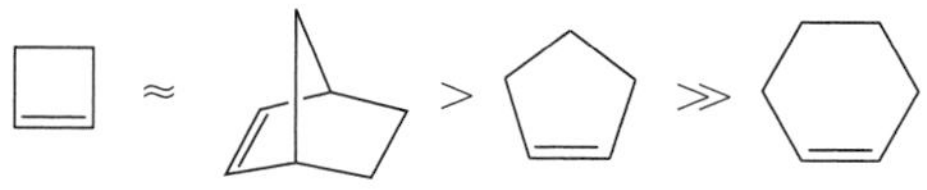

Ring strain [kJ (kcal)/mol]	125(29.8)	113(27)	24.7(5.9)	5.9(1.4)

ROMP reactions are unique in that all the unsaturation present in the monomers is conserved in the polymeric product. This feature makes ROMP techniques especially attractive for the preparation of highly unsaturated, fully conjugated materials. One example is the direct preparation of polyacetylene by the ROMP technique through one of the double bonds of cyclooctatetraene. For further discussion of polyacetylene, see "Chemical Connections: Organic Polymers That Conduct Electricity" earlier in this chapter.

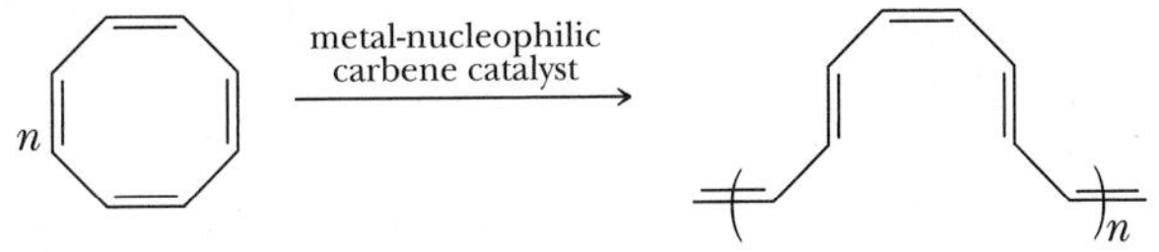

Cyclooctatetraene **Polyacetylene**

An important polymer in electrooptical applications is poly(phenylene vinylene) (PPV), which has alternating phenyl and vinyl groups. One of the routes to this polymer starts with a substituted bicyclooctadiene that is polymerized using ROMP techniques to form a soluble, processable polymer. Heating the processed polymer results in elimination of two equivalents of acetic acid, which aromatizes the six-membered ring and completes the conjugation.

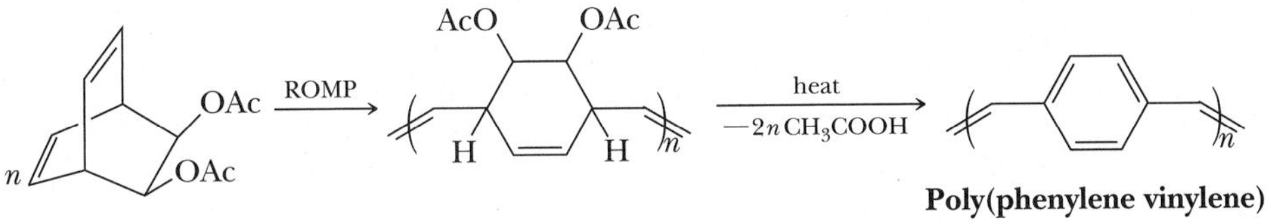

A substituted bicyclooctadiene

Poly(phenylene vinylene)

Chemical Connections

Recycling of Plastics

Polymers, in the form of plastics, are materials upon which our society is incredibly dependent. Durable and lightweight, plastics are probably the most versatile synthetic materials in existence; in fact, their current production in the United States exceeds that of steel. Plastics have come under criticism, however, for their role in the trash crisis. They comprise 21% of the volume and 8% of the weight of solid wastes, most of which is derived from disposable packaging and wrapping. Of the 3.0×10^{10} kg of thermoplastic materials produced in 2008 in America, less than 7% was recycled.

Why aren't more plastics being recycled? The durability and chemical inertness of most plastics make them ideally suited for reuse. The answer to this question has more to do with economics and consumer habits than with technological obstacles. Because curbside pickup and centralized drop-off stations for recyclables are more common, the amount of used material available for reprocessing is increasing. Until recently, consumers perceived products made from "used" materials as being inferior to new ones, so the market for recycled products has not been large. In addition, the increase in environmental concerns over the last few years has resulted in a greater demand for recycled products. As manufacturers adapt to satisfy this new market, plastic recycling will eventually catch up with the

These students are wearing jackets made from recycled PET soda bottles.

(Continued)

Recycling Code	Polymer	Common Uses	Uses of Recycled Polymer
1 PET	**Poly(ethylene terephthalate)**	Soft drink bottles, household chemical bottles, films, textile fibers	Soft drink bottles, household chemical bottles, films, textile fibres
2 HDPE	**High-density polyethylene**	Milk and water jugs, grocery bags, bottles	Bottles, molded containers
3 PVC	**Poly(vinyl chloride)**	Shampoo bottles, pipes, shower curtains, vinyl siding, wire insulation, floor tiles, credit cards	Plastic floor mats
4 LDPE	**Low-density polyethylene**	Shrink wrap, trash and grocery bags, sandwich bags, squeeze bottles	Trash bags and grocery bags
5 PP	**Polypropylene**	Plastic lids, clothing fibers, bootle caps, toys, diaper linings	Mixed plastic components
6 PS	**Polystyrene**	Styrofoam cups, egg cartons, disposable utensils. Packaging materials appliances	Molded items such as cafeteria trays, rulers, Frisbees, trash cans, videocassettes
7	**All other plastics and mixed plastics**	Various	Plastic lumber, playground equipment, road reflectors

recycling of other materials, such as glass and aluminum.

Six types of plastics are commonly used for packaging applications. In 1988, manufacturers adopted recycling code numbers developed by the Society of the Plastics Industry. Because the plastics recycling industry still is not fully developed, only polyethylene terephthalate (PET) and high-density polyethylene (HDPE) are currently being recycled in large quantities, although outlets for the other plastics are being developed. Low-density polyethylene (LDPE), which accounts for about 40% of plastic trash, has been slow in finding acceptance with recyclers. Facilities for the reprocessing of poly(vinyl chloride) (PVC), polypropylene (PP), and polystyrene (PS) exist but are still rare.

The process for the recycling of most plastics is simple, with separation of the desired plastics from other contaminants the most labor-intensive step. PET soft drink bottles, for example, usually have a paper label, adhesive, and an aluminum cap that must be removed before the PET can be reused.

The recycling process begins with hand or machine sorting, after which the bottles are shredded into small chips. An air cyclone removes paper and other lightweight materials, and any remaining labels and adhesives are eliminated with a detergent wash. The PET chips are then dried, and aluminum, the final contaminant, is removed electrostatically. The PET produced by this method is 99.9% free of contaminants and sells for about half the price of the virgin material. Unfortunately, plastics with similar densities cannot be separated with this technology, and plastics composed of several polymers cannot be broken down into pure components. However, recycled mixed plastics can be molded into plastic lumber that is strong, durable, and graffiti-resistant.

An alternative to this process, which uses only physical methods of purification, is chemical recycling. The scrap is treated with methanol in the presence of an acid catalyst to give ethylene glycol and dimethyl terephthalate.

These monomers are purified by distillation or recrystallization and used as feedstocks for the production of more PET film.

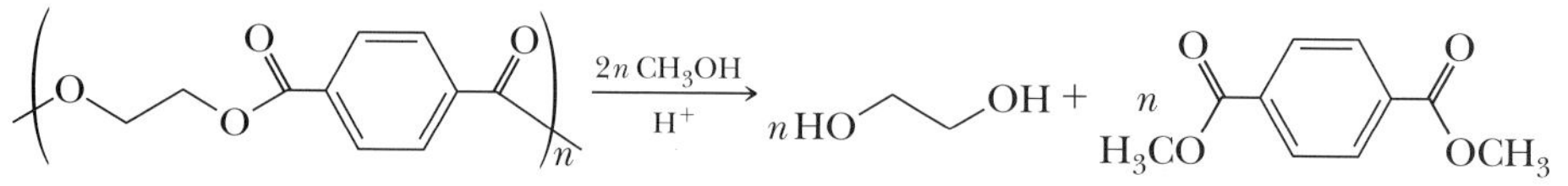

Poly(ethylene terephthalate) (PET) **Ethylene glycol** **Dimethyl terephthalate**

Study Guide

29.1 The Architecture of Polymers

- **Polymers** are long-chain molecules synthesized by linking **monomers** through chemical reactions. Types of polymer architecture include linear and branched chains, as well as those with comb, ladder, and star structures.
- The term **plastic** refers to any polymer that can be molded when hot and retains its shape when cooled.
  - *__Thermoplastics__ are polymers that can be melted and become sufficiently fluid that they can be molded into shapes that are retained when they are cooled.*
  - *__Thermosetting plastics__, or __thermosets__, can be molded when they are first prepared, but once they cool, they harden irreversibly owing to extensive covalent cross-linking between chains.*
- The properties of polymers are determined by the size and shape of their chains.

29.2 Polymer Notation and Nomenclature

- The structures of polymers are shown by drawing parentheses around the **repeat unit**, which is the smallest molecular fragment that contains all the nonredundant structural features of the chain.
  - *The __average degree of polymerization__ (the average number of repeat units per chain) is denoted by a subscript just outside the parentheses.*
- Polymers are named by attaching the prefix *poly-* to the name of the monomer from which the polymer is derived.

P 29.1, 29.5, 29.6

29.3 Molecular Weights of Polymers

- The **number average molecular weight**, M_n, is calculated by counting the number of polymer chains of a particular molecular weight, multiplying each number by the molecular weight of its chain, summing these values, and dividing by the total number of polymer chains.
- The **weight average molecular weight**, M_w, is calculated by recording the total weight of each chain of a particular length, summing these weights, and dividing by the total weight of the sample.
- The **polydispersity index** of a polymer is the ratio M_w/M_n.
  - *If all the polymers are the same length, the polydispersity index is 1 and the sample is referred to as __monodisperse__. Synthetic polymers are never monodisperse unless they are purified.*

29.4 Polymer Morphology—Crystalline Versus Amorphous Materials

- Polymers in the solid state tend to be composed of both ordered **crystalline domains** (crystallites) and disordered **amorphous domains**.
- The **melt transition**, T_m, of the polymer is the temperature at which the crystallites melt. As the degree of crystallinity increases, so does T_m.

- Amorphous domains have no long-range order and give rise to transparent soft materials that are referred to as glassy polymers.
  - *On being heated, amorphous polymers are transformed from a hard glass to a soft, flexible, rubbery state at a temperature referred to as the* ***glass transition temperature***, T_g.
  - *Rubber materials must have low* T_g *values to behave as* ***elastomers***, *which are materials that return to their original shape following a distortion.*

P 29.2, 29.3, 29.14–29.26

29.5 Step-Growth Polymerizations

- Polymers in which chain growth occurs in a stepwise manner are called **step-growth** or **condensation polymers**.
  - *Step-growth polymers are formed by reaction between difunctional molecules, with each new bond created in a separate step.*
  - *At the initial stages of polymerization, monomers react with each other until they are used up.*
  - *High-molecular-weight polymers appear only near the end of the reaction.*
- **Nylons** are polyamides formed from a diacid and a diamine, or alternatively from an amino acid, and have use as fibers.
- **Polyesters** are derived from diacids and diols and have use as textile fibers such as Dacron.
- **Polycarbonates** such as Lexan are tough, transparent polymers with high tensile strength that are used for products ranging from sporting equipment to unbreakable windows.
- **Polyurethanes** consist of flexible polyester or polyether units alternating with rigid urethane (carbamate) blocks. Polyurethanes are used as flexible fibers such as Lycra and Spandex, as well as foaming materials.
- **Epoxy resins** are materials prepared by polymerization in which one monomer contains at least two epoxide groups. Epoxy resins are used as adhesives and insulating surface coatings.

P 29.2, 29.7–29.23

KEY REACTIONS

1. Step-Growth Polymerization of a Dicarboxylic Acid and a Diamine Gives a Polyamide (Section 29.5A)

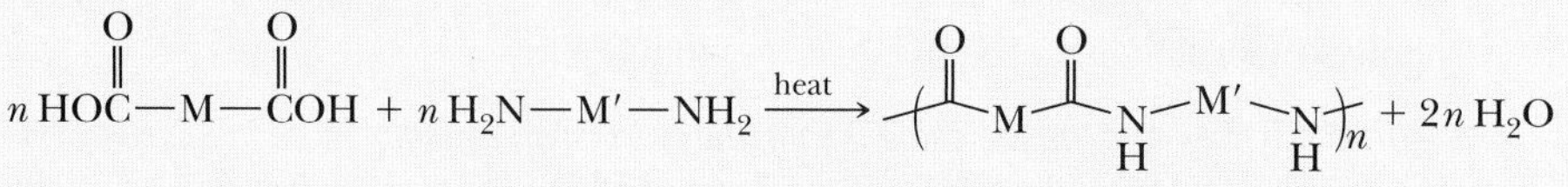

n HOC—M—COH + n H2N—M′—NH2
heat
M
N
H
M′
N
H
n
+ 2n H2O

2. Step-Growth Polymerization of a Dicarboxylic Acid and a Diol Gives a Polyester (Section 29.5B)

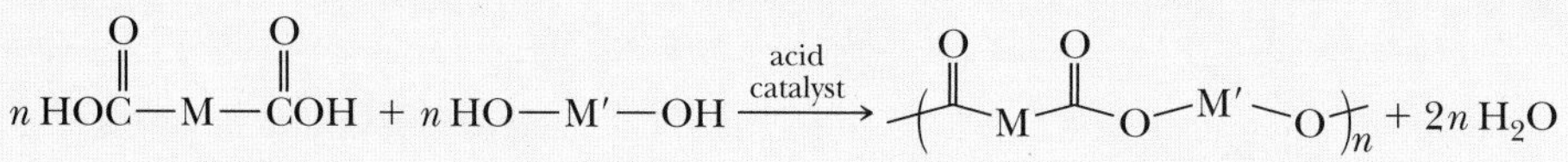

n HOC—M—COH + n HO—M′—OH
acid catalyst
M
O
M′
O
n
+ 2n H2O

3. Step-Growth Polymerization of a Diacyl Chloride and a Diol Gives a Polycarbonate (Section 29.5C)

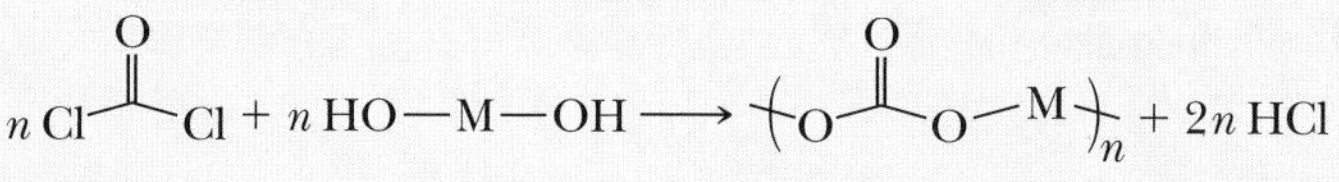

n Cl
Cl
+ n HO—M—OH
O
O
M
n
+ 2n HCl

4. Step-Growth Polymerization of a Diisocyanate and a Diol Gives a Polyurethane (Section 29.5D)

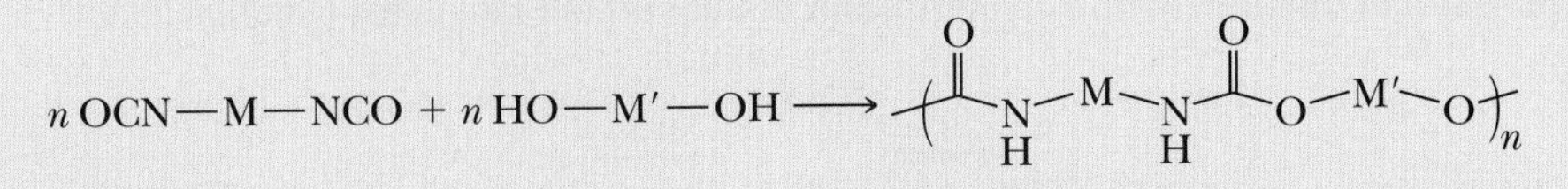

5. Step-Growth Polymerization of a Diepoxide and a Diamine Gives an Epoxy Resin (Section 29.5E)

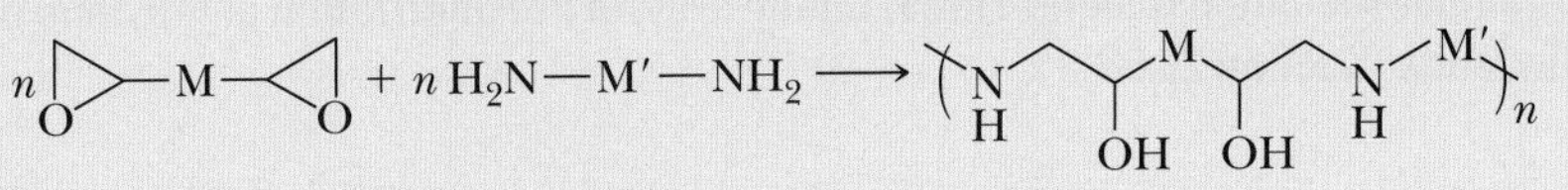

29.6 Chain-Growth Polymerizations

- **Chain-growth polymerization** is a type of polymerization in which monomer units are joined together without loss of atoms.
  - *Once initiated, chain-growth polymerizations involve endgroups possessing reactive intermediates that react with a new monomer.*
  - *Because chain-growth polymers grow from the ends only and monomers do not react with themselves, the growth of polymer chains occurs linearly throughout the polymerization, unlike step-growth polymerizations.*
  - *Reactive intermediates used in chain-growth polymerizations include radicals, carbanions, carbocations, and organometallic complexes.*
  - *Monomers used for chain-growth polymerization include alkenes, alkynes, allenes, isocyanates, and cyclic compounds such as lactones, lactams, ethers, and epoxides.*
- Low-density polyethylene (LDPE) can be made by radical polymerization of ethylene using peroxides as radical initiators. LDPE is soft and transparent and is made into films using a blow-molding technique.
- Polyethylene can also be made using metal catalysts such as the **Ziegler-Natta catalysts**, which can produce **high-density polyethylene (HDPE)**, which is stronger and more opaque than LDPE.
  - *HDPE is used for making products such as dishes and water bottles.*
  - *Even better metal catalysts such as zirconium complexes have been developed for polyethylene production.*
- Polymers can have chiral centers, and polymers with an identical configuration at all chiral centers are called **isotactic**, those with alternating configurations are called **syndiotactic**, and those with completely random configurations are called **atactic**.
  - *The more stereoregular a polymer is, the more crystalline it is.*
- Anionic polymerizations of alkenes, which do not suffer from the termination problems of radical polymerizations, can be initiated by nucleophiles or one-electron reduction.
- Due to a lack of competing processes, **living polymerizations** are polymerizations that, once initiated, continue to grow at the ends of the chains until all the monomer is consumed, and adding more monomer initiates new reaction of the growing chains.
  - *Living polymerizations provide good control over the size distribution of polymer chains.*
- Bicyclic alkene monomers can be used in a polymerization called **ring-opening metathesis polymerization (ROMP)** using the same chemistry as alkene metathesis.

P **29.3, 29.4, 29.24–29.38**

KEY REACTIONS

6. Radical Chain-Growth Polymerization of Substituted Ethylenes (Section 29.6A)

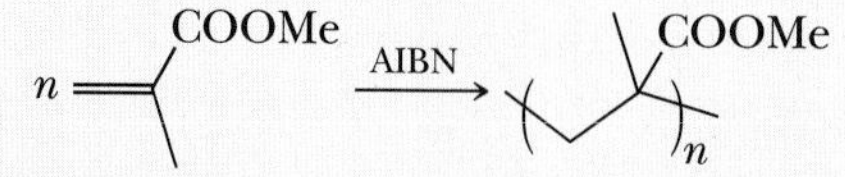

7. Titanium-Mediated (Ziegler-Natta) Chain-Growth Polymerization of Ethylene and Substituted Ethylenes (Section 29.6B)

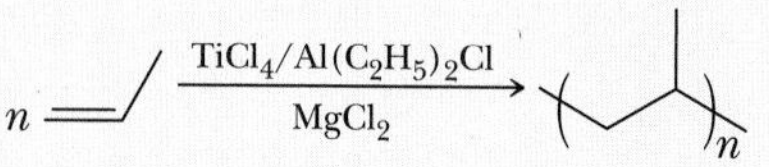

8. Anionic Chain-Growth Polymerization of Substituted Ethylenes (Section 29.6D)

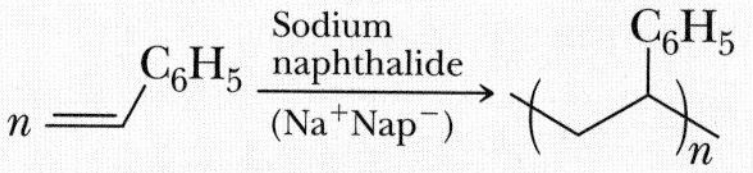

9. Cationic Chain-Growth Polymerization of Substituted Ethylenes (Section 29.6D)

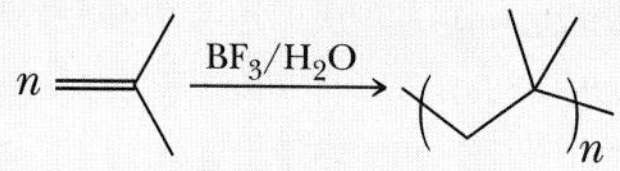

10. Ring-Opening Metathesis Polymerization (ROMP) (Section 29.6E)

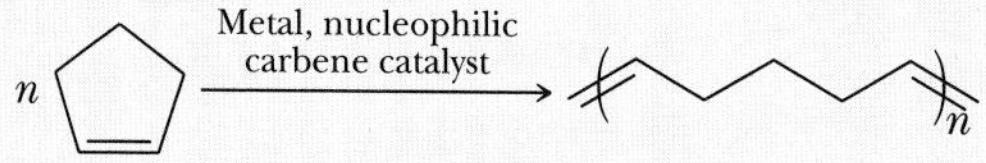

Problems

Red numbers indicate applied problems.

Structure and Nomenclature

29.5 Name the following polymers.

(a) 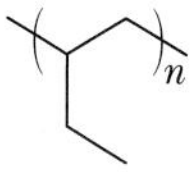

(b) 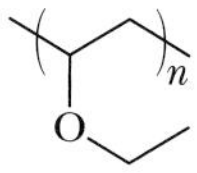

(c) 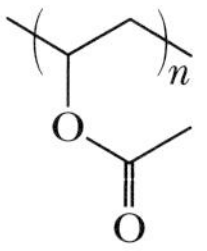

(d) 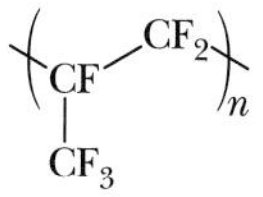

(e) 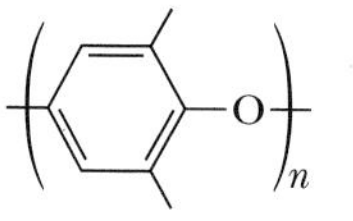

(f) 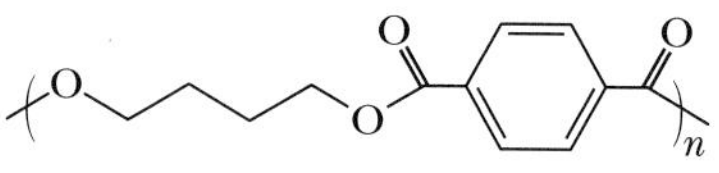

(g) 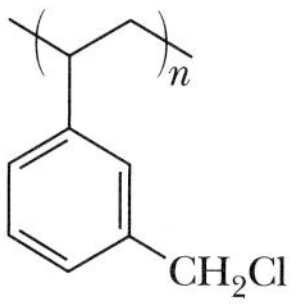

(h) 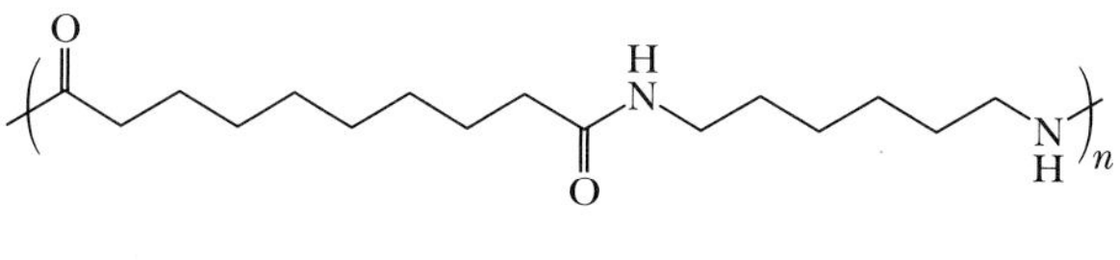

29.6 Draw the structure(s) of the monomer(s) used to make each polymer in Problem 29.5.

Step-Growth Polymerizations

29.7 Draw a structure of the polymer formed in the following reactions.

(a) 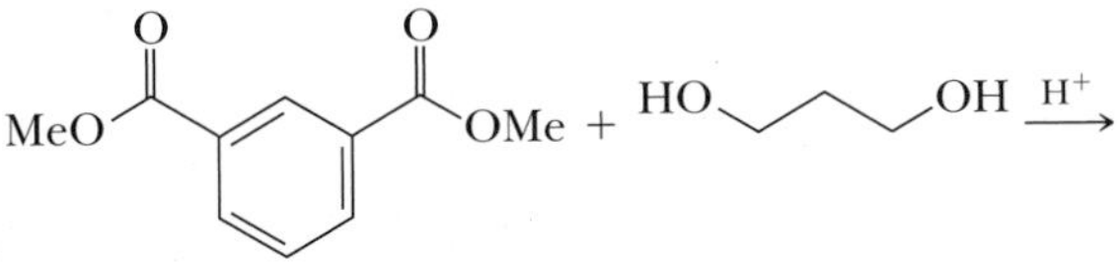
MeO
OMe +
HO
OH
H+

(b) 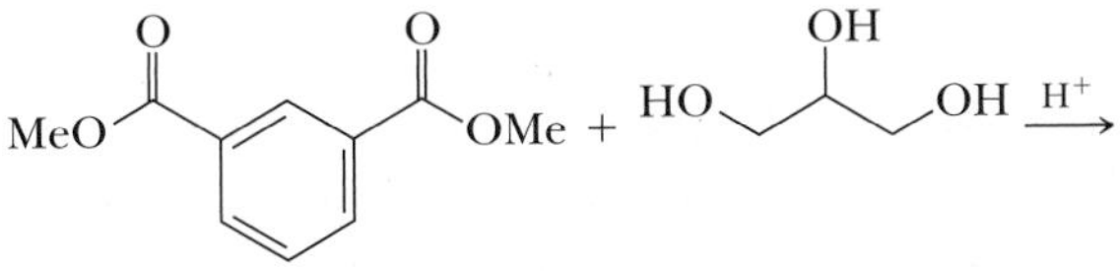
MeO
OMe +
HO
OH
OH
H+

(c) 
CF3SO3H

(d) 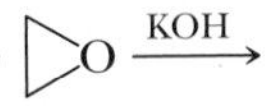
KOH

29.8 At one time, a raw material for the production of hexamethylenediamine was the pentose-based polysaccharides of agricultural wastes such as oat hulls. Treatment of these wastes with sulfuric acid or hydrochloric acid gives furfural. Decarbonylation of furfural over a zinc-chromium-molybdenum catalyst gives furan. Propose reagents and experimental conditions for the conversion of furan to hexamethylenediamine.

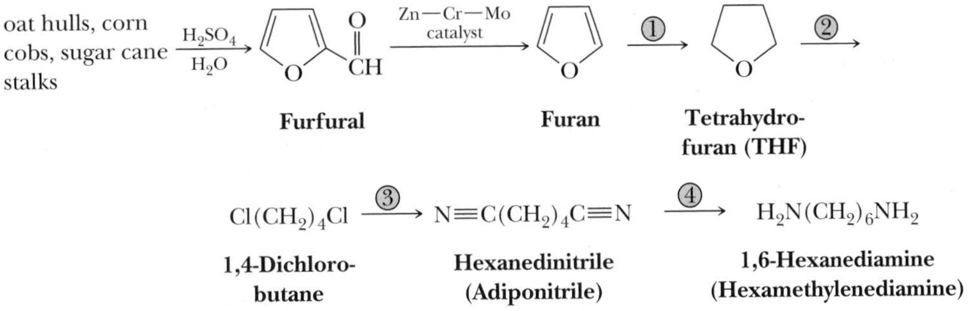
oat hulls, corn cobs, sugar cane stalks
H2SO4
H2O
Zn—Cr—Mo catalyst
Furfural
Furan
Tetrahydro-furan (THF)
Cl(CH2)4Cl
N≡C(CH2)4C≡N
H2N(CH2)6NH2
1,4-Dichloro-butane
Hexanedinitrile (Adiponitrile)
1,6-Hexanediamine (Hexamethylenediamine)

29.9 Another raw material for the production of hexamethylenediamine is butadiene derived from thermal and catalytic cracking of petroleum. Propose reagents and experimental conditions for the conversion of butadiene to hexamethylenediamine.

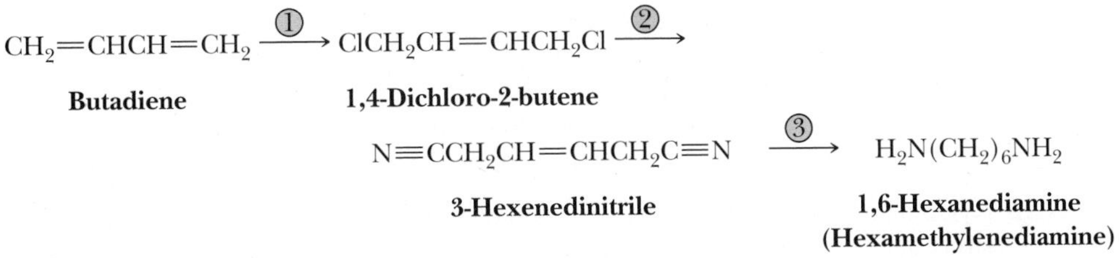
CH2=CHCH=CH2
ClCH2CH=CHCH2Cl
Butadiene
1,4-Dichloro-2-butene
N≡CCH2CH=CHCH2C≡N
H2N(CH2)6NH2
3-Hexenedinitrile
1,6-Hexanediamine (Hexamethylenediamine)

29.10 Propose reagents and experimental conditions for the conversion of butadiene to adipic acid.

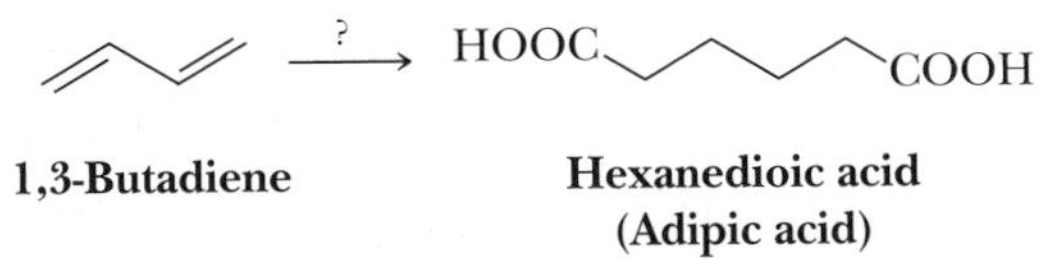
?
HOOC
COOH
1,3-Butadiene
Hexanedioic acid (Adipic acid)

29.11 Polymerization of 2-chloro-1,3-butadiene under Ziegler-Natta conditions gives a synthetic elastomer called neoprene. All carbon-carbon double bonds in the polymer chain have the *trans* configuration. Draw the repeat unit in neoprene.

29.12 Poly(ethylene terephthalate) (PET) can be prepared by this reaction. Propose a mechanism for the step-growth reaction in this polymerization.

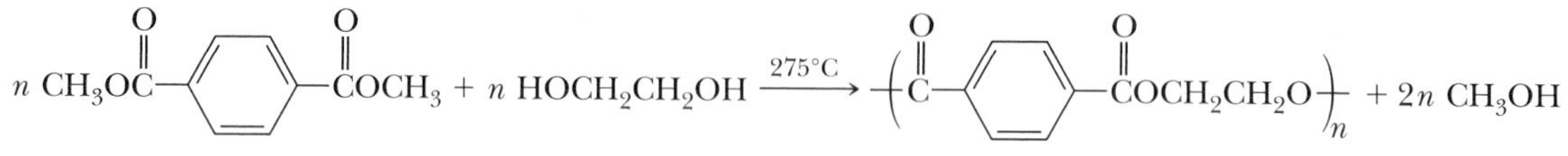

Dimethyl terephthalate **Ethylene glycol** **Poly(ethylene terephthalate)** **Methanol**

29.13 Identify the monomers required for the synthesis of these step-growth polymers.

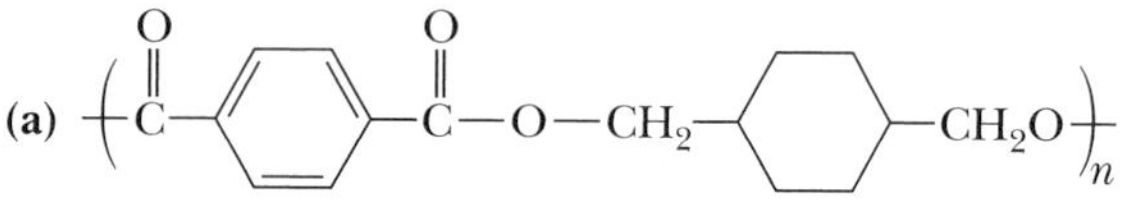

Kodel
(a polyester)

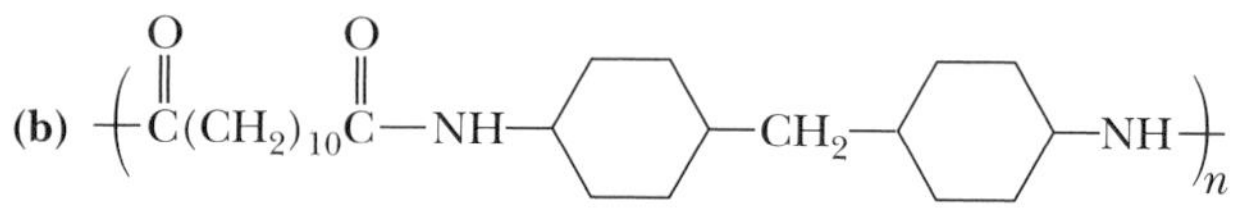

Qiana
(a polyamide)

29.14 Nomex, another aromatic polyamide (compare aramid), is prepared by polymerization of 1,3-benzenediamine and the diacid chloride of 1,3-benzenedicarboxylic acid. The physical properties of the polymer make it suitable for high-strength, high-temperature applications such as parachute cords and jet aircraft tires. Draw a structural formula for the repeating unit of Nomex.

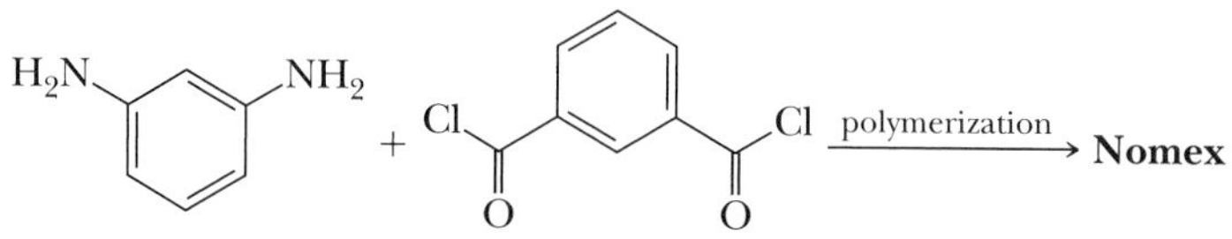

1,3-Benzenediamine **1,3-Benzene-dicarbonyl chloride**

29.15 Caprolactam, the monomer from which nylon 6 is synthesized, is prepared from cyclohexanone in two steps. In Step 1, cyclohexanone is treated with hydroxylamine to form cyclohexanone oxime. Treatment of the oxime with concentrated sulfuric acid in Step 2 gives caprolactam by a reaction called a Beckmann rearrangement. Propose a mechanism for the conversion of cyclohexanone oxime to caprolactam.

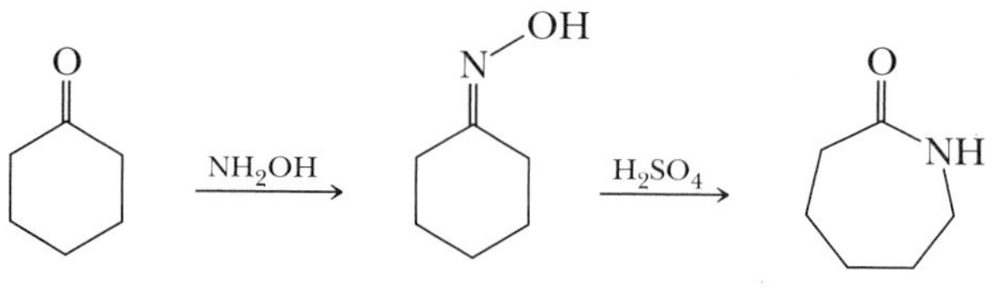

Cyclohexanone **Cyclohexanone oxime** **Caprolactam**

29.16 Nylon 6,10 is prepared by polymerization of a diamine and a diacid chloride. Draw a structural formula for each reactant and for the repeat unit in this polymer.

29.17 Polycarbonates (Section 29.5C) are also formed by using a nucleophilic aromatic substitution route (Section 22.3B) involving aromatic difluoro monomers and carbonate ion. Propose a mechanism for this reaction.

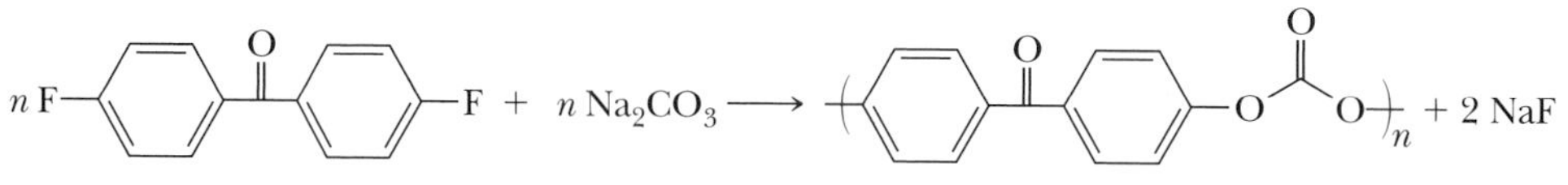

An aromatic difluoride **Sodium carbonate** A polycarbonate **Sodium fluoride**

29.18 Propose a mechanism for the formation of this polyphenylurea. To simplify your presentation of the mechanism, consider the reaction of one —NCO group with one —NH_2 group.

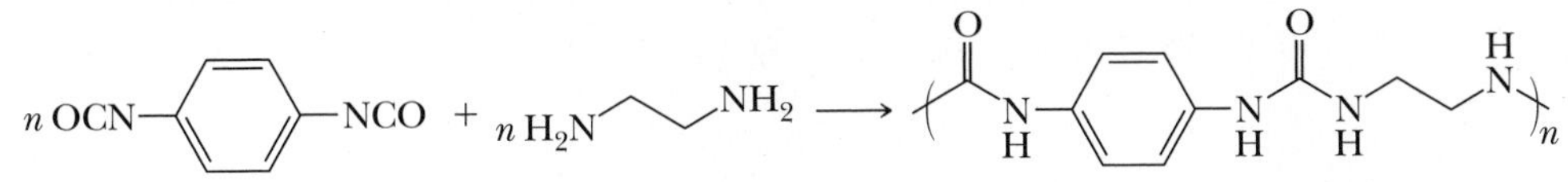

1,4-Benzenediisocyanate **1,2-Ethanediamine** **Poly(ethylene phenylurea)**

29.19 When equal molar amounts of phthalic anhydride and 1,2,3-propanetriol are heated, they form an amorphous polyester. Under these conditions, polymerization is regioselective for the primary hydroxyl groups of the triol.

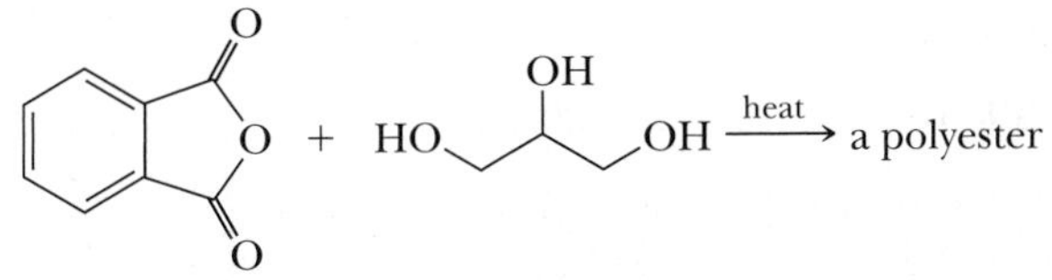

Phthalic anhydride **1,2,3-Propanetriol (Glycerol)**

(a) Draw a structural formula for the repeat unit of this polyester.
(b) Account for the regioselective reaction with the primary hydroxyl groups only.

29.20 The polyester from Problem 29.19 can be mixed with additional phthalic anhydride (0.5 mole of phthalic anhydride for each mole of 1,2,3-propanetriol in the original polyester) to form a liquid resin. When this resin is heated, it forms a hard, insoluble thermosetting polyester called glyptal.
(a) Propose a structure for the repeat unit in glyptal.
(b) Account for the fact that glyptal is a thermosetting plastic.

29.21 Propose a mechanism for the formation of the following polymer.

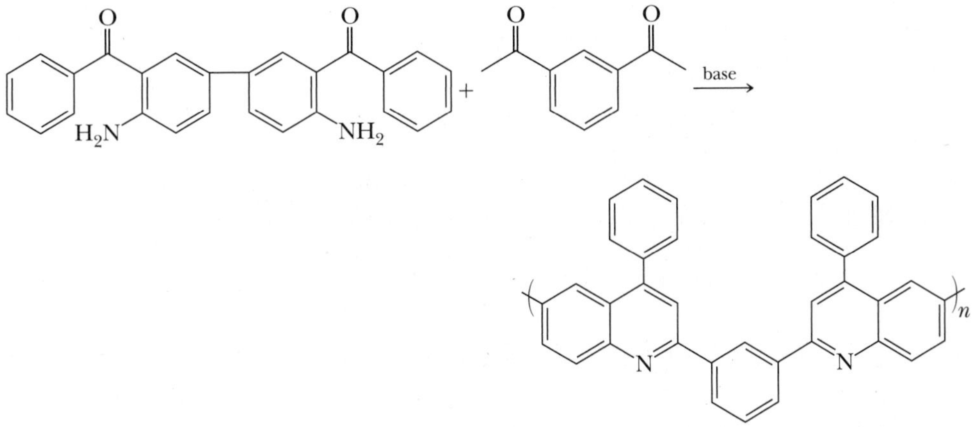

29.22 Draw a structural formula of the polymer resulting from base-catalyzed polymerization of each compound. Would you expect the polymers to be optically active? (*S*)-(+)-lactide is the dilactone formed from two molecules of (*S*)-(+)-lactic acid.

(a) 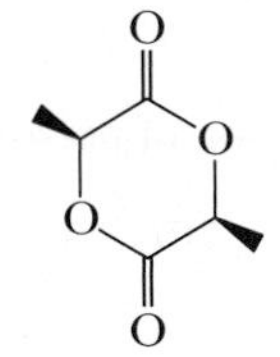

(b) 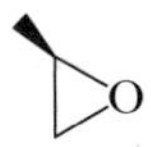

(*S*)-(+)-Lactide **(*R*)-Propylene oxide**

29.23 Poly(3-hydroxybutanoic acid), a biodegradable polyester, is an insoluble, opaque material that is difficult to process into shapes. In contrast, the copolymer of 3-hydroxybutanoic acid and 3-hydroxyoctanoic acid is a transparent polymer that shows good solubility in a number of organic solvents. Explain the difference in properties between these two polymers in terms of their structure.

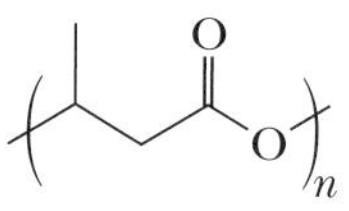

Poly(3-hydroxybutanoic acid)

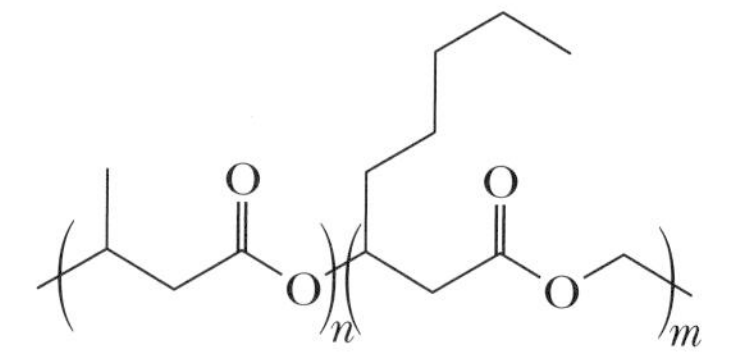

Poly(3-hydroxybutanoic acid-3-hydroxyoctanoic acid) copolymer

Chain-Growth Polymerizations

29.24 How might you determine experimentally whether a particular polymerization is propagating by a step-growth or a chain-growth mechanism?

29.25 Draw a structural formula for the polymer formed in the following reactions.

(a) 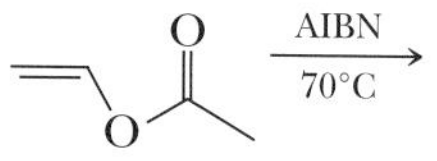

(b) 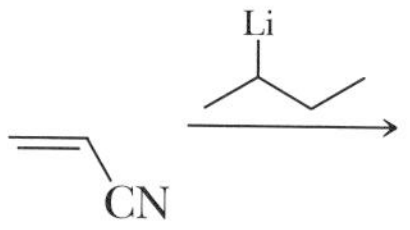

29.26 Select the monomer in each pair that is more reactive toward cationic polymerization.

(a) 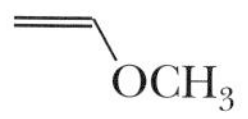 or 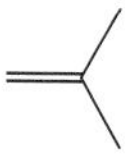

(b)  or 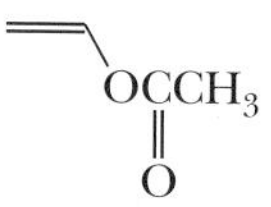

(c) 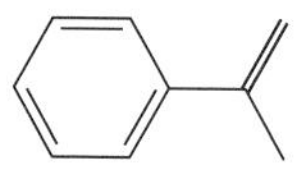 or 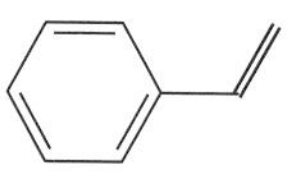

(d) 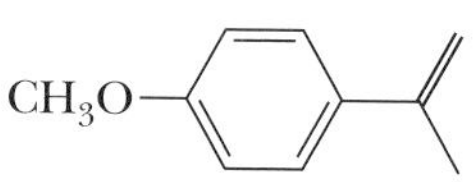 or 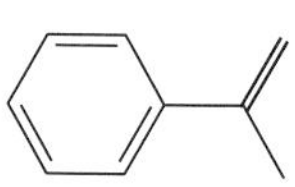

29.27 Polymerization of vinyl acetate gives poly(vinyl acetate). Hydrolysis of this polymer in aqueous sodium hydroxide gives the useful water-soluble polymer poly(vinyl alcohol). Draw the repeat units of both poly(vinyl acetate) and poly(vinyl alcohol).

29.28 Benzoquinone can be used to inhibit radical polymerizations. This compound reacts with a radical intermediate, R·, to form a less reactive radical that does not participate in chain propagation steps and thus breaks the chain.

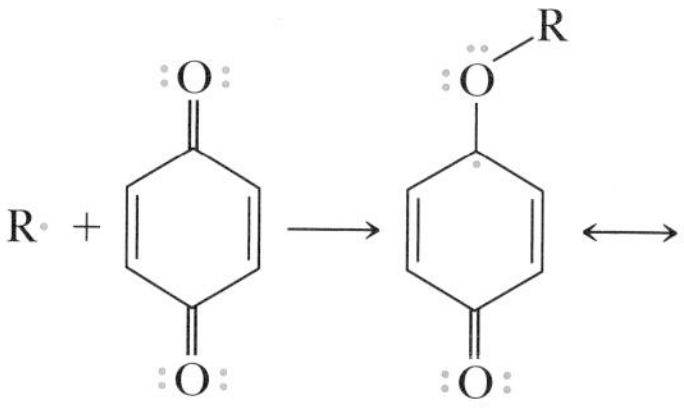

⟷ other contributing structures

Draw a series of contributing structures for this less reactive radical and account for its stability.

29.29 Following is the structural formula of a section of polypropylene derived from three units of propylene monomer.

$$-CH_2\underset{|}{\overset{CH_3}{C}}H-CH_2\overset{CH_3}{C}H-CH_2\overset{CH_3}{C}H-$$

Polypropylene

Draw structural formulas for comparable sections of the following.
(a) Poly(vinyl chloride) **(b)** Polytetrafluoroethylene
(c) Poly(methyl methacrylate) **(d)** Poly(1,1-dichloroethylene)

29.30 Low-density polyethylene (LDPE) has a higher degree of chain branching than high-density polyethylene (HDPE). Explain the relationship between chain branching and density.

29.31 We saw how intramolecular chain transfer in radical polymerization of ethylene creates a four-carbon branch on a polyethylene chain. What branch is created by a comparable intramolecular chain transfer during radical polymerization of styrene?

29.32 Compare the densities of low-density polyethylene (LDPE) and high-density polyethylene (HDPE) with the densities of the liquid alkanes listed in Table 2.5. How might you account for the differences between them?

29.33 Natural rubber is the all-*cis* polymer of 2-methyl-1,3-butadiene (isoprene).

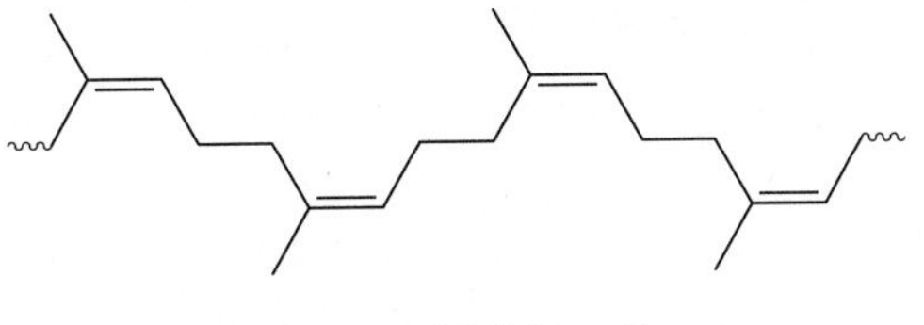

Poly(2-methyl-1,3-butadiene)
(Polyisoprene)

(a) Draw a structural formula for the repeat unit of natural rubber.
(b) Draw a structural formula of the product of oxidation of natural rubber by ozone followed by a workup in the presence of $(CH_3)_2S$. Name each functional group present in this product.
(c) The smog prevalent in many major metropolitan areas contains oxidizing agents, including ozone. Account for the fact that this type of smog attacks natural rubber (automobile tires and the like) but does not attack polyethylene or polyvinyl chloride.
(d) Account for the fact that natural rubber is an elastomer but the synthetic all-*trans* isomer is not.

29.34 Radical polymerization of styrene gives a linear polymer. Radical polymerization of a mixture of styrene and 1,4-divinylbenzene gives a cross-linked network polymer of the type shown in Figure 29.1. Show by drawing structural formulas how incorporation of a few percent of 1,4-divinylbenzene in the polymerization mixture gives a cross-linked polymer.

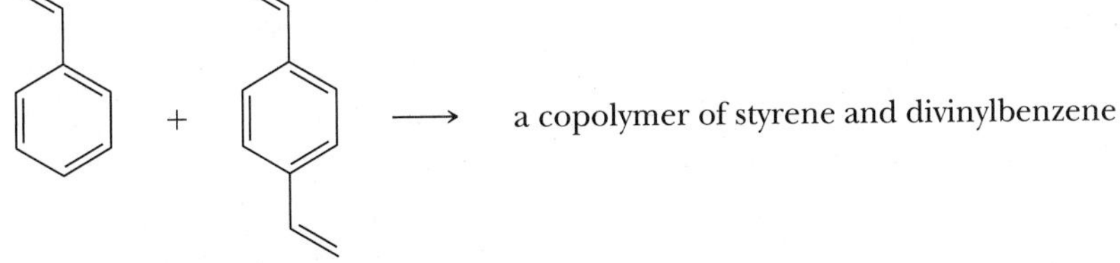

Styrene **1,4-Divinylbenzene**

29.35 One common type of cation exchange resin is prepared by polymerization of a mixture containing styrene and 1,4-divinylbenzene (Problem 29.34). The polymer is then treated with concentrated sulfuric acid to sulfonate a majority of the aromatic rings in the polymer.
(a) Show the product of sulfonation of each benzene ring.
(b) Explain how this sulfonated polymer can act as a cation exchange resin.

29.36 The most widely used synthetic rubber is a copolymer of styrene and butadiene called SB rubber. Ratios of butadiene to styrene used in polymerization vary, depending on the end use of the polymer. The ratio used most commonly in the preparation of SB rubber for use in automobile tires is one mole styrene to three moles butadiene. Draw a structural formula of a section of the polymer formed from this ratio of reactants. Assume that all carbon-carbon double bonds in the polymer chain are in the *cis* configuration.

29.37 From what two monomer units is the following polymer made?

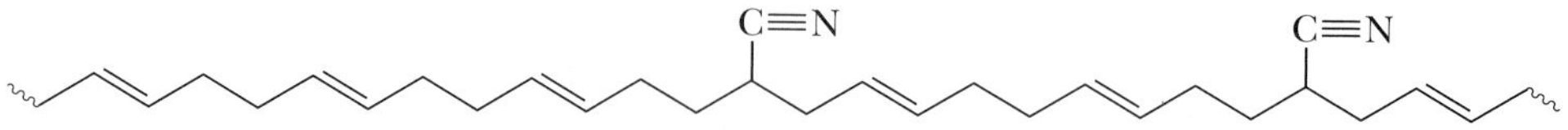

29.38 Draw the structure of the polymer formed from ring-opening metathesis polymerization (ROMP) of each monomer.

(a) 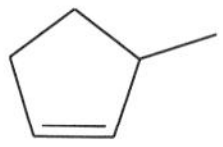 (b) 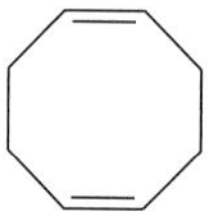 (c) 

(d) 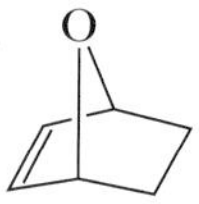

Appendix 1

Thermodynamics and the Equilibrium Constant

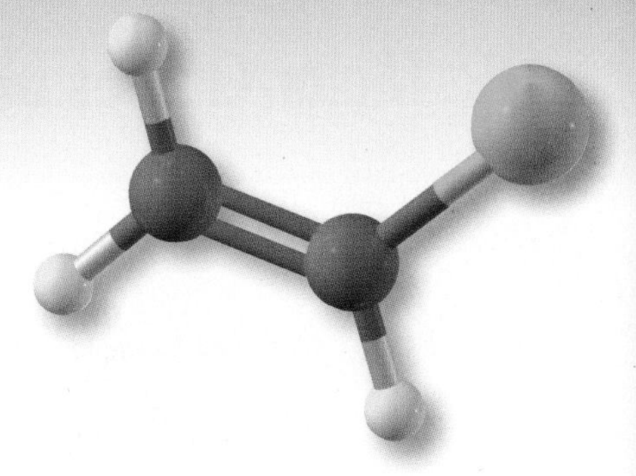

For the equilibrium $A \rightleftharpoons B$ $\quad K_{eq} = \dfrac{[B]}{[A]}$

$$\Delta G^0 = -RT \ln K_{eq}$$

R = molar gas constant = 8.3145 J (1.987 cal) $\cdot$ $mol^{-1} \cdot K^{-1}$

T = in kelvin (K)

$$\%B = \frac{B}{A + B} \times 100$$

K_{eq}	ΔG^0 kJ/mol	ΔG^0 kcal/mol	ln K_{eq}	log K_{eq}	% B in Mixture
1	0.00	0.00	0.00	0.00	50.00
2	−1.72	−0.41	0.69	0.30	66.67
5	−3.97	−0.95	1.61	0.70	83.33
10	−5.69	−1.36	2.30	1.00	90.91
20	−7.41	−1.77	3.00	1.30	95.24
100	−11.4	−2.73	4.61	2.00	99.01
1,000	−17.1	−4.09	6.91	3.00	99.90
10,000	−22.8	−5.46	9.21	4.00	99.99

Appendix 2

Major Classes of Organic Acids

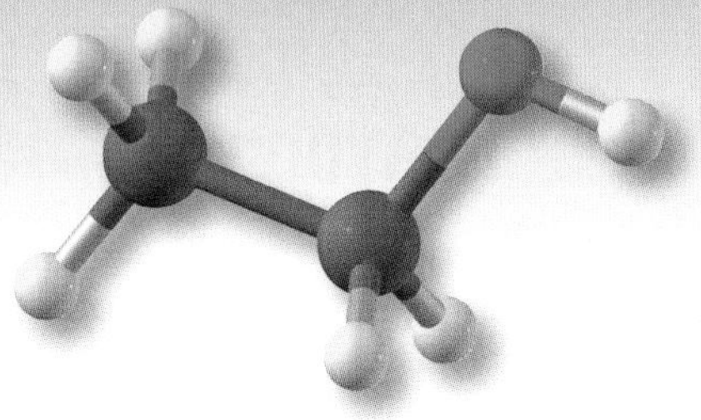

Class and Example	Typical pK_a	Class and Example	Typical pK_a
Sulfonic acid 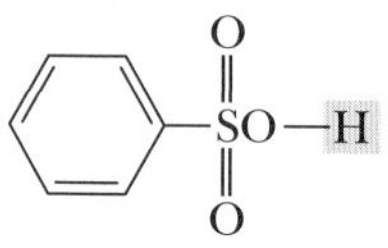	0–1	β-Ketoester $CH_3-C(=O)-CH(H)-C(=O)OCH_2CH_3$	11
Carboxylic acid $CH_3C(=O)O-H$	3–5	Water $HO-H$	15.7
Arylammonium ion 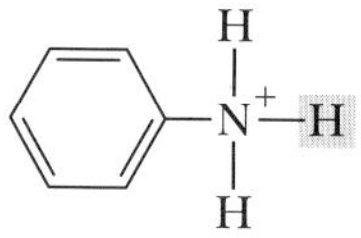	4–5	Alcohol CH_3CH_2O-H	15–19
Imide 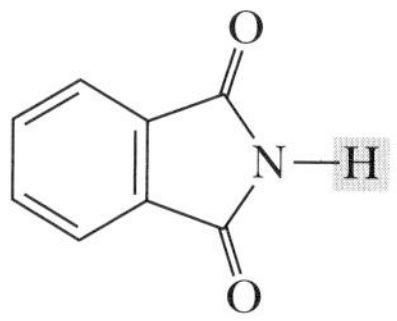	8–9	Amide $CH_3C(=O)N(H)-H$	15–19
Thiol CH_3CH_2S-H	8–12	Cyclopentadiene 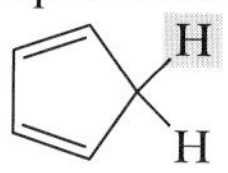	16
Phenol 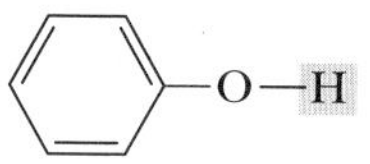	9–10	α-Hydrogen of an aldehyde or ketone $CH_3C(=O)CH_2-H$	18–20
Ammonium ion NH_3-H^+	9.24	α-Hydrogen of an ester $CH_3CH_2OC(=O)CH_2-H$	23–25
β-Diketone $CH_3-C(=O)-CH(H)-C(=O)CH_3$	10	Alkyne $HC\equiv C-H$	25
Nitroalkane $H-CH_2NO_2$	10	Ammonia NH_2-H	38
Alkylammonium ion $(CH_3CH_2)_3\overset{+}{N}-H$	10–12	Amine $[(CH_3)_2CH]_2N-H$	40
		Alkene $CH_2=CH-H$	44
		Alkane CH_3CH_2-H	51

Appendix 3

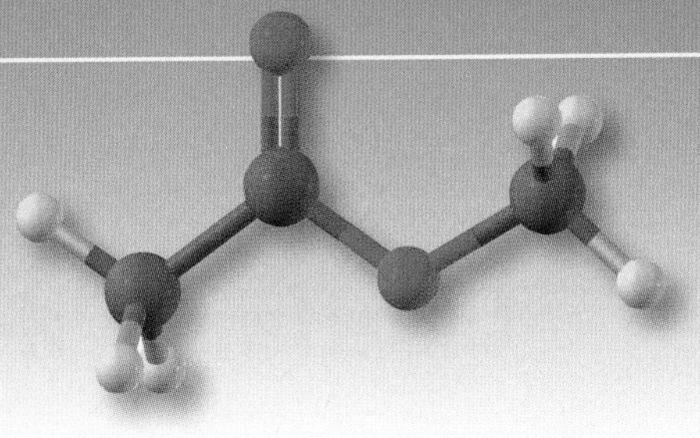

Bond Dissociation Enthalpies

Bond dissociation enthalpy (BDE) is defined as the amount of energy required to break a bond homolytically into two radicals in the gas phase at 25°C.

$$A—B \rightarrow A^{\cdot} + B^{\cdot} \qquad \Delta H^0 \text{ [kJ (kcal)/mol]}$$

Bond	ΔH^0
H—H bonds	
H—H	435 (104)
D—D	444 (106)
X—X bonds	
F—F	159 (38)
Cl—Cl	247 (59)
Br—Br	192 (46)
I—I	151 (36)
H—X bonds	
H—F	568 (136)
H—Cl	431 (103)
H—Br	368 (88)
H—I	297 (71)
O—H bonds	
HO—H	497 (119)
CH_3O—H	439 (105)
C_6H_5O—H	376 (90)
O—O bonds	
HO—OH	213 (51)
CH_3O—OCH_3	159 (38)
$(CH_3)_3CO$—$OC(CH_3)_3$	159 (38)
C—C single bonds	
CH_3—CH_3	378 (90)
C_2H_5—CH_3	372 (89)
CH_2=CH—CH_3	422 (101)
CH_2=$CHCH_2$—CH_3	322 (77)
C_6H_5—CH_3	435 (104)
$C_6H_5CH_2$—CH_3	326 (78)
C—C multiple bonds	
CH_2=CH_2	727 (174)
HC≡CH	966 (231)
C—H bonds	
CH_3—H	439 (105)
C_2H_5—H	422 (101)
$(CH_3)_2CH$—H	414 (99)
$(CH_3)_3C$—H	405 (97)
CH_2=CH—H	464 (111)
CH_2=$CHCH_2$—H	372 (89)
C_6H_5—H	472 (113)
$C_6H_5CH_2$—H	376 (90)
HC≡C—H	556 (133)
C—F bonds	
CH_3—F	481 (115)
C_2H_5—F	472 (113)
$(CH_3)_2CH$—F	464 (111)
C_6H_5—F	531 (127)
C—Cl bonds	
CH_3—Cl	351 (84)
C_2H_5—Cl	355 (85)
$(CH_3)_2CH$—Cl	355 (85)
$(CH_3)_3C$—Cl	355 (85)
CH_2=$CHCH_2$—Cl	288 (69)
C_6H_5—Cl	405 (97)
$C_6H_5CH_2$—Cl	309 (74)
C—Br bonds	
CH_3—Br	301 (72)
C_2H_5—Br	301 (72)
$(CH_3)_2CH$—Br	309 (74)
$(CH_3)_3C$—Br	305 (73)
CH_2=$CHCH_2$—Br	247 (59)
C_6H_5—Br	351 (84)
$C_6H_5CH_2$—Br	263 (63)
C—I bonds	
CH_3—I	242 (58)
C_2H_5—I	238 (57)
$(CH_3)_2CH$—I	238 (57)
$(CH_3)_3C$—I	234 (56)
CH_2=$CHCH_2$—I	192 (46)
C_6H_5—I	280 (67)
$C_6H_5CH_2$—I	213 (51)
C—N single bonds	
CH_3—NH_2	355 (85)
C_6H_5—NH_2	435 (104)
C—O single bonds	
CH_3—OH	385 (92)
C_6H_5—OH	468 (112)

Appendix 4

Characteristic ^{1}H-NMR Chemical Shifts

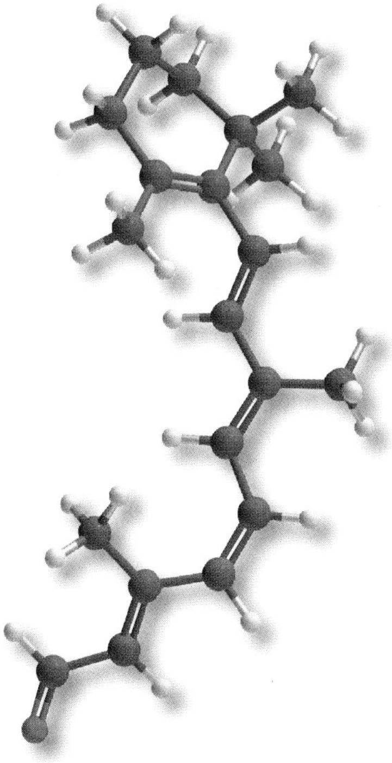

Type of Hydrogen (R = alkyl, Ar = aryl)	Chemical Shift (δ)*	Type of Hydrogen (R = alkyl, Ar = aryl)	Chemical Shift (δ)*
$(CH_3)_4Si$	0 (by definition)	RCH_2OH	3.4–4.0
R_2NH	0.5–5.0	RCH_2Br	3.4–3.6
ROH	0.5–6.0	RCH_2Cl	3.6–3.8
RCH_3	0.8–1.0	$R\overset{O}{\overset{\|}{C}}OCH_3$	3.7–3.9
RCH_2R	1.2–1.4	$R\overset{O}{\overset{\|}{C}}OCH_2R$	4.1–4.7
R_3CH	1.4–1.7	RCH_2F	4.4–4.5
$R_2C{=}CRCHR_2$	1.6–2.6	$ArOH$	4.5-4.7
$RC{\equiv}CH$	2.0–3.0	$R_2C{=}CH_2$	4.6–5.0
$R\overset{O}{\overset{\|}{C}}CH_3$	2.1–2.3	$R_2C{=}CHR$	5.0–5.7
$R\overset{O}{\overset{\|}{C}}CH_2R$	2.2–2.6	ArH	6.5–8.5
$ArCH_3$	2.2–2.5	$R\overset{O}{\overset{\|}{C}}H$	9.5–10.1
$ArCH_2R$	2.3–2.8	$R\overset{O}{\overset{\|}{C}}OH$	10–13
RCH_2I	3.1–3.3		
RCH_2OR	3.3–4.0		

*Values are relative to tetramethylsilane. Other atoms within the molecule may cause the signal to appear outside these ranges.

Appendix 5

Characteristic ^{13}C-NMR Chemical Shifts

Type of Carbon	Chemical Shift (δ)
$(CH_3)_4Si$	0 (by definition)
RCH_2I	0–40
RCH_3	10–40
RCH_2R	15–55
R_3CH	20–60
RCH_2Br	25–65
RCH_2Cl	35–80
R_3COH	40–80
R_3COR	40–80
$RC{\equiv}CR$	65–85
$R_2C{=}CR_2$	100–150
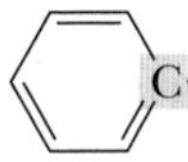 C—R	110–160
$RC(=O)OR$	160–180
$RC(=O)NR_2$	165–180
$RC(=O)OH$	165–185
$RC(=O)H$, $RC(=O)R$	180–215

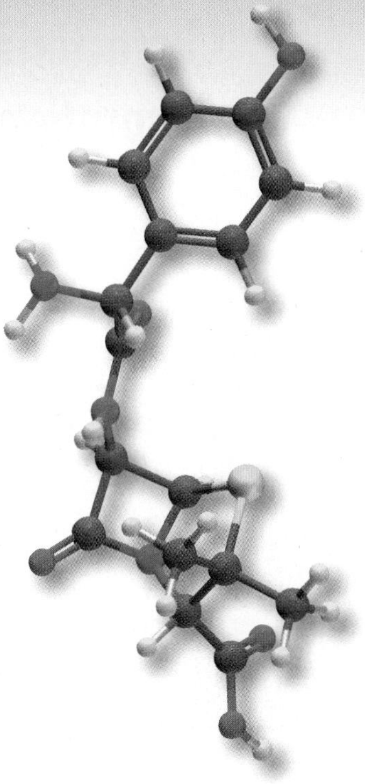

Appendix 6

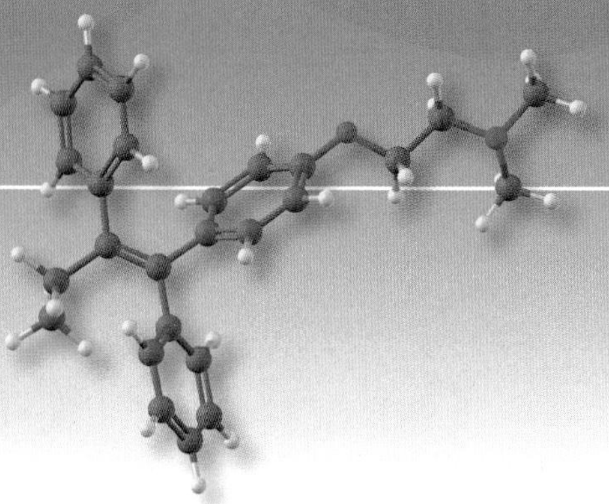

Characteristic Infrared Absorption Frequencies

Bonding		Frequency (cm^{-1})	Intensity*	Type of Vibration (Stretching unless noted)
C—H	alkane	2850–3000	m	
	$—CH_3$	1375 and 1450	w–m	bending
	$—CH_2—$	1450–1475	m	bending
	alkene	3000–3100	w–m	
		650–1000	s	out-of-plane bending
	alkyne	3300	m-s	
	arene	3030	w-m	
		690–900	s	out-of-plane bending
	aldehyde	2720	w	
C═C	alkene	1600–1680	w–m	
	arene	1450 and 1600	m	
C≡C	alkyne	2100–2250	w	
C—O	alcohol, ether, ester, carboxylic acid	1000–1100 (*sp*³ C—O)	s	
		1200–1250 (*sp*² C—O)	s	
	anhydride	900–1300	s	
C═O	amide	1630–1680	s	
	carboxylic acid	1700–1725	s	
	ketone	1630–1820	s	
	aldehyde	1630–1820	s	
	ester	1735–1800	s	
	anhydride	1740–1760 and 1800–1850	s	
	acid chloride	1800	s	
O—H	alcohol, phenol free	3600–3650	w	
	hydrogen bonded	3200–3500	m	
	carboxylic acid	2500–3300	s	
N—H	amine and amide	3100–3550	m–s	
C—N	nitrile	2200–2250	m	

*m = medium, s = strong, w = weak

Appendix 7

Electrostatic Potential Maps

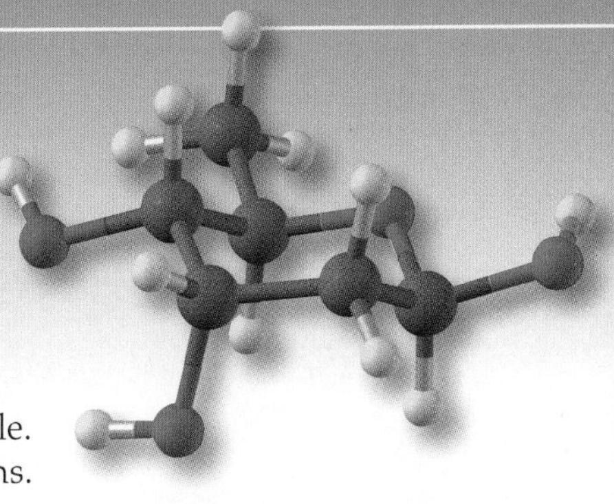

The term *electronic structure* refers to the distribution of electron density in a molecule. According to the laws of quantum mechanics, electrons have no definite locations. Instead, they collectively produce a negatively charged region around a nucleus, measured by electron density in units of e/Å^3 (electrons per cubic angstrom). For an atom, density is high near the nucleus and vanishingly small far from the nucleus. Electrostatic potential maps are now easily computed for small molecules using desktop computers and various software packages. Electrostatic potential maps (elpots) in this text were produced using MacSpartan (Wavefunction, Inc.).

Electrostatic potential (elpot) maps provide a way to visualize the distribution of electron density in a molecule. Electrostatic potential is defined as the potential energy that a positively charged particle would experience in a molecule's presence. The electrostatic potential is made up of two parts:

1. The repulsive component (positive potential, repulsion) exerted by the positively charged nuclei
2. The attractive component (negative potential, attraction) exerted by the negatively charged electron cloud

Thus, the electrostatic potential contains information about the entire electron distribution.

Electrostatic potential maps are color-coded. By convention, the most negative potential is red, and the most positive potential is blue. Intermediate potentials are coded accordingly (orange-yellow-green). While any surface might be chosen to display an electrostatic potential map, the most common is 0.002 e/Å^3. Nearly all of a molecule's electron density lies within this surface, which corresponds almost exactly to how closely another molecule can approach without running into severe steric repulsive forces; that is, the surface corresponds almost exactly to the van der Waals surface of the molecule.

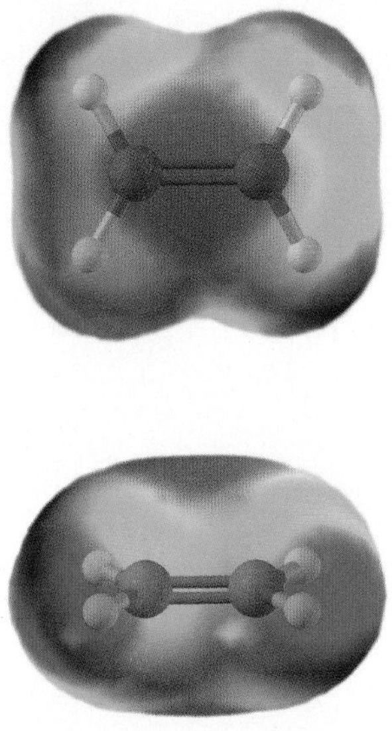

An electrotatic potential map of ethylene (top and side views).

An electrostatic potential map for ethylene shows areas of high electron density to which electrophiles will be attracted (red) over the π orbitals. There are four blue patches, one over each hydrogen; these regions are relatively electron-poor.

An electrostatic potential map of the methyl cation (CH_3^+).

The methyl carbocation, CH_3^+, provides an even more dramatic visualization of an electrostatic potential. The entire ion is blue in color, corresponding to the net positive charge. The central atom is the deepest blue, corresponding to the location of the largest fraction of the positive charge.

Appendix 8

Summary of Stereochemical Terms

Absolute configuration The actual configuration of groups about a tetrahedral chiral center; absolute configuration is specified by the *R,S* system.

Atropisomers Enantiomers that lack a chiral center and differ because of hindered rotation about a carbon-carbon single bond.

Center of symmetry A point so situated that identical components of an object are located on opposite sides and equidistant from that point along any axis passing through it.

Chemoselective reaction A reaction in which one functional group in a molecule containing two or more functional groups reacts selectively with a reagent.

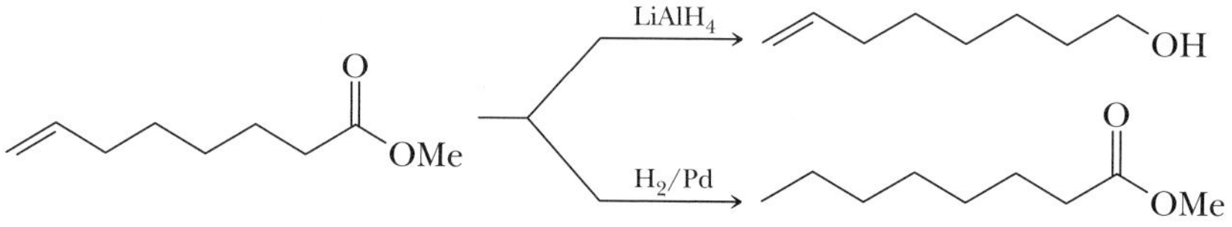

Chiral From the Greek, *cheir*, hand; an object that is not superposable on its mirror image; an object that has handedness.

Chiral center A tetrahedral atom, most commonly carbon, that is bonded to four different groups. In molecules containing one chiral center, the exchange of two groups makes an enantiomer. In molecules containing two or more chiral centers, the exchange of two groups on at least one (but not all) of the chiral centers gives a diastereomer.

***Cis,trans* isomers** Stereoisomers that have the same connectivity of their atoms but a different configurational arrangement of their atoms in space because of the presence of either a ring or a double bond. *Cis,trans* isomers are diastereomers; that is, they are stereoisomers that are not mirror images.

Configuration The arrangement of atoms or groups of atoms bonded to a stereocenter. Configuration in alkenes is designated by the *E,Z* system or the *cis,trans* system; configuration in molecules containing chiral centers is designated by the *R,S* system.

Diastereomers Stereoisomers that are not mirror images of each other; refers to relationships among two or more objects.

Diastereoselective reaction A reaction that produces one diastereomer in preference to all others.

Enantiomeric excess (ee) The difference between the percentages of two enantiomers in a mixture. For example, if a sample contains 98% of one enantiomer and 2% of the other, the enantiomeric excess (ee) is $98\% - 2\% = 96\%$.

Enantiomers Stereoisomers that are nonsuperposable mirror images; refers to a relationship between pairs of objects.

Enantioselective reaction A reaction that produces one enantiomer in preference to the other. Catalytic reduction of the following alkene in the presence of an (*R*)-BINAP-Ru catalyst gives (*S*)-naproxen in greater than 98% enantiomeric excess ($>99\% : <1\%$).

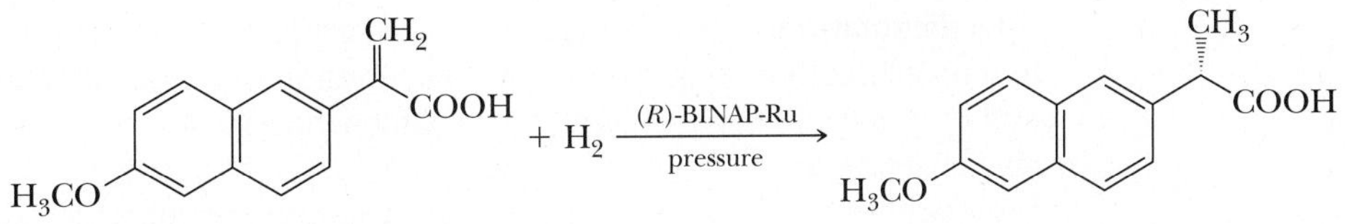

(*S*)-Naproxen
(ee > 98%)

Fischer projection A two-dimensional projection of a chiral center in a molecule; groups on the right and left of the chiral center project toward the reader, whereas those above and below the chiral center project away from the reader. The only atom in the plane on which the projection is drawn is the chiral center itself.

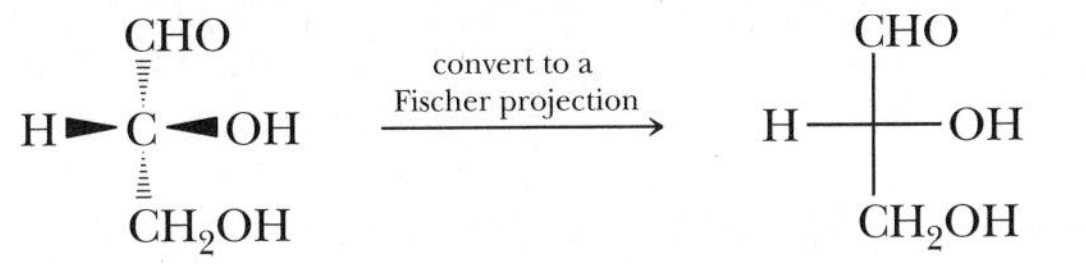

(*R*)-Glyceraldehyde (three-dimensional representation) | (*R*)-Glyceraldehyde (Fischer projection)

Meso compound An achiral compound possessing two or more chiral centers that also has chiral isomers. Examples of meso compounds are *cis*-1,2-cyclopentanediol and meso-2,3-butanediol. A meso compound has either a plane or a center of symmetry. Both of these examples as drawn have an internal plane of symmetry.

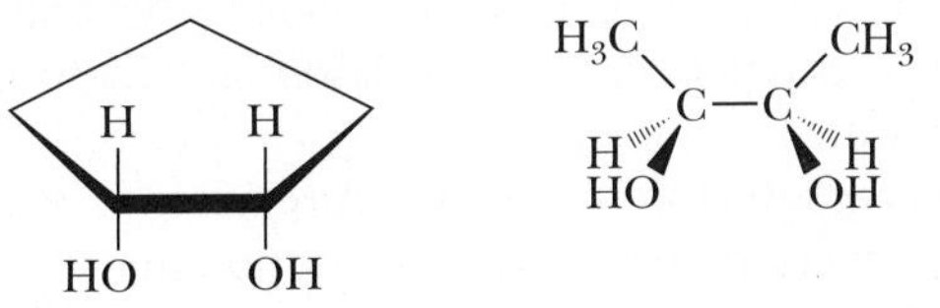

cis-1,2-Cyclopentanediol (a meso compound) | (2*R*,3*S*)-2,3-Butanediol (a meso compound)

Optical activity The ability of a compound to rotate the plane of polarized light.
Optical purity The specific rotation of a mixture of enantiomers divided by the specific rotation of the enantiomerically pure substance (expressed as a percent). Optical purity is numerically equal to enantiomeric excess, but experimentally determined.
Plane of symmetry An imaginary (mirror) plane passing through an object dividing it so that one half is the mirror image of the other half.
Racemic mixture A mixture of equal amounts of two enantiomers.
Regioselective reaction An addition or substitution reaction in which one of two or more possible products is formed in preference to all constitutional isomers that might be formed. Addition of HBr to 1-methylcyclohexene gives 1-bromo-1-methylcyclohexane to the virtual exclusion of 1-bromo-2-methylcyclohexane.

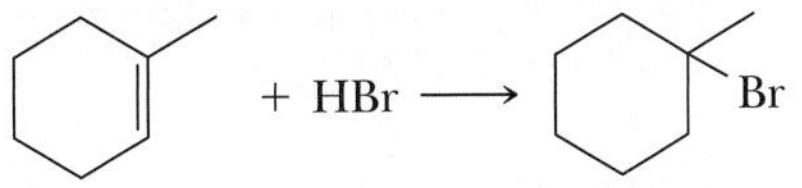

Hydroboration-oxidation of a cycloalkene is both regioselective and stereoselective, but it is not enantioselective (both enantiomers are produced in equal amounts).

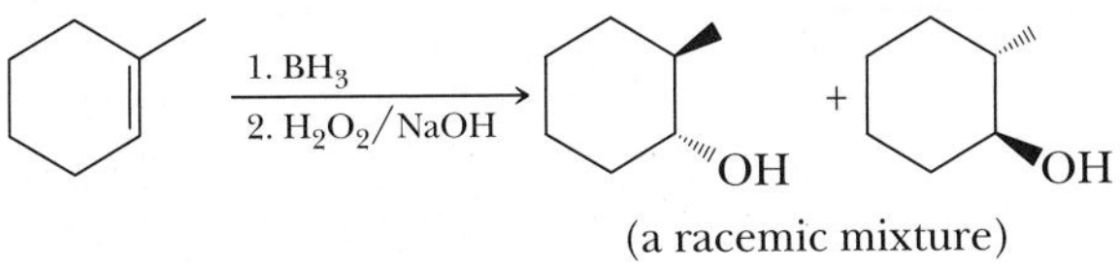

(a racemic mixture)

Specific rotation The observed rotation of the plane of polarized light when a sample is placed in a tube 1.0 dm in length and at a concentration expressed in g/mL (density) for a pure liquid and at a concentration of 1 g/mL for a solution. Specific rotation is in deg·mL/dm/g and is usually given without units.

Stereocenter An atom, most commonly carbon, about which exchange of two groups produces a stereoisomer.

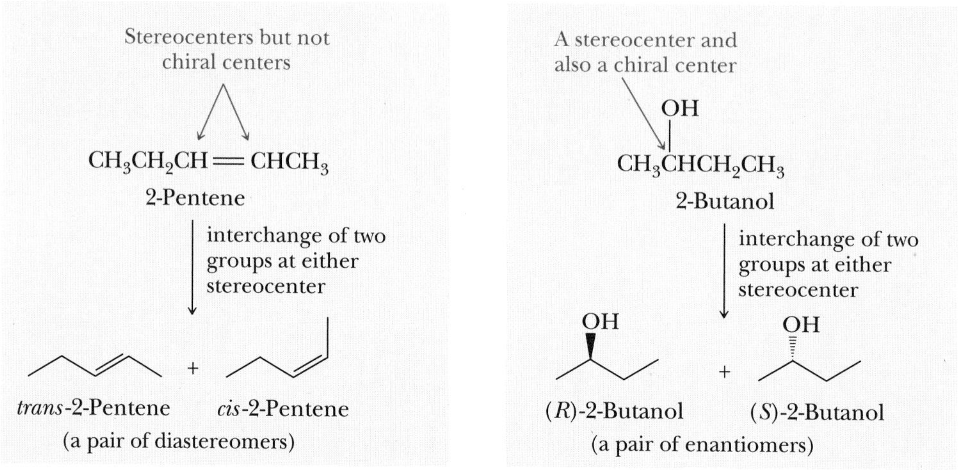
Stereocenters but not chiral centers
$CH_3CH_2CH{=}CHCH_3$
2-Pentene
interchange of two groups at either stereocenter
trans-2-Pentene
cis-2-Pentene
(a pair of diastereomers)
A stereocenter and also a chiral center
OH
$CH_3CHCH_2CH_3$
2-Butanol
interchange of two groups at either stereocenter
OH
OH
(*R*)-2-Butanol
(*S*)-2-Butanol
(a pair of enantiomers)

Stereoisomers Compounds that have the same molecular formula, the same connectivity of their atoms, but a different orientation of their atoms in space. The term "stereoisomer" includes *cis,trans* isomers in cycloalkanes and alkenes as well as enantiomers, diastereomers, meso compounds, and atropisomers. Conformational isomers are also stereoisomers, whether they are isolable or not.

Stereoselective reaction A reaction in which one stereoisomer is formed in preference to all others. A stereoselective reaction may be enantioselective or diastereoselective. For example, catalytic reduction of the following alkene in the presence of an (*R*)-BINAP-Ru catalyst is enantioselective; it gives (*S*)-naproxen in greater than 98% enantiomeric excess.

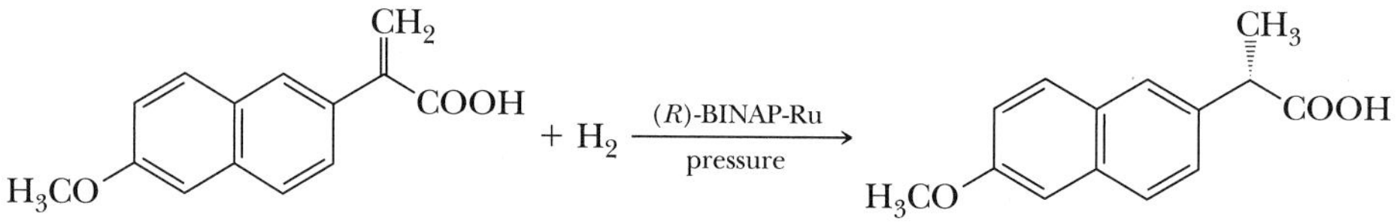
CH_2
COOH
H_3CO
$+ H_2$
(*R*)-BINAP-Ru
pressure
CH_3
COOH
H_3CO

(*S*)-Naproxen
(ee > 98%)

Stereospecific reaction A reaction in which the stereochemistry of the product is dependent on the stereochemistry of the starting material. For example, oxidation of 2-butene by osmium tetroxide is stereospecific: oxidation of *cis*-2-butene gives meso-2,3-butanediol, whereas oxidation of *trans*-2-butene gives a racemic mixture of the enantiomers of 2,3-butanediol. (The term "regiospecific" is used analogously.)

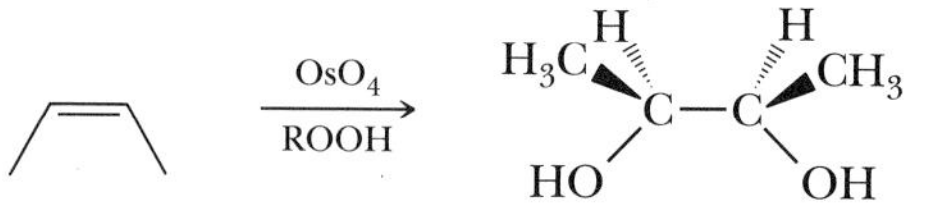
OsO_4
ROOH
H
H
H_3C
CH_3
C—C
HO
OH

cis-2-Butene
Meso-2,3-Butanediol

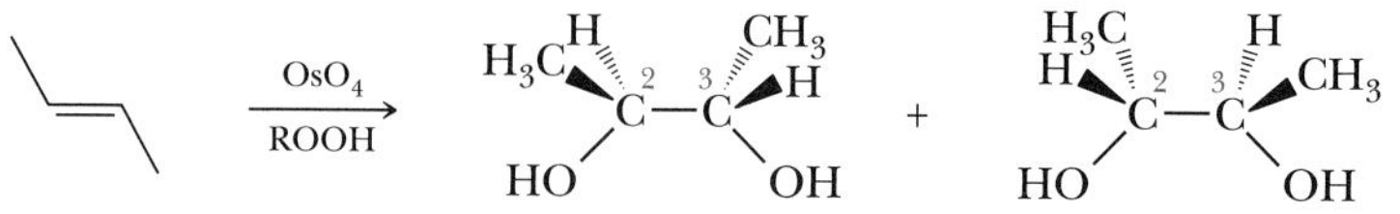
OsO_4
ROOH
H_3C
H
CH_3
H
C—C
HO
OH
+
H_3C
H
H
CH_3
HO
OH

trans-2-Butene
(2*S*,3*S*)-2,3-Butanediol
(2*R*,3*R*)-2,3-Butanediol

Appendix 9

Summary of the Rules of Nomenclature

A9.1 Alkanes

The parent or root name of an alkane is that of the longest chain of carbon atoms in the compound.

Following are prefixes used in the IUPAC system to show the presence of 1 to 20 carbon atoms in an unbranched chain.

Prefix	Number of Carbon Atoms	Prefix	Number of Carbon Atoms
meth-	1	undec-	11
eth-	2	dodec-	12
prop-	3	tridec-	13
but-	4	tetradec-	14
pent-	5	pentadec-	15
hex-	6	hexadec-	16
hept-	7	heptadec-	17
oct-	8	octadec-	18
non-	9	nonadec-	19
dec-	10	eicos-	20

A substituent derived from an alkane by removal of a hydrogen atom is called an **alkyl group** and is commonly represented by the symbol R—.

Following are names for alkyl groups with one to five carbon atoms. Common names and their abbreviations are given in parentheses.

Name	Condensed Structural Formula	Name	Condensed Structural Formula
Methyl (Me)	$-CH_3$	1,1-Dimethylethyl (*tert*-butyl, *t*-Bu)	CH_3 $-CCH_3$ CH_3
Ethyl (Et)	$-CH_2CH_3$		
Propyl (Pr)	$-CH_2CH_2CH_3$		
1-Methylethyl (isopropyl, iPr)	$-CHCH_3$ CH_3	Pentyl	$-CH_2CH_2CH_2CH_2CH_3$
		3-Methylbutyl (isopentyl)	$-CH_2CH_2CHCH_3$ CH_3
Butyl (Bu)	$-CH_2CH_2CH_2CH_3$	2-Methylbutyl	$-CH_2CHCH_2CH_3$ CH_3
2-Methylpropyl (isobutyl, iBu)	$-CH_2CHCH_3$ CH_3		
1-Methylpropyl (*sec*-butyl, *s*-Bu)	$-CHCH_2CH_3$ CH_3	2,2-Dimethylpropyl (neopentyl)	CH_3 $-CH_2CCH_3$ CH_3

The rules of the IUPAC system for naming alkanes are:

1. The name for the alkane with an unbranched chain of carbon atoms consists of a prefix showing the number of carbon atoms in the parent chain and the ending *-ane*.
2. For branched-chain alkanes, take the longest chain of carbon atoms as the parent chain and its name becomes the root name.
3. Give each substituent on the parent chain a name and a number. The number shows the carbon atom of the parent chain to which the substituent is bonded. Use a hyphen to connect the number to the name.

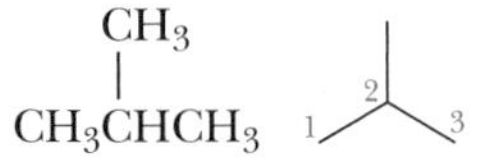

2-Methylpropane

4. If there is one substituent, number the parent chain from the end that gives it the lower number.

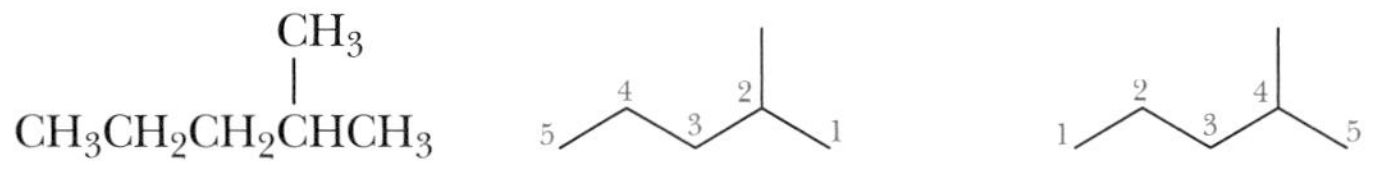

2-Methylpentane (not 4-Methylpentane)

5. If there are two or more identical substituents, number the parent chain from the end that gives the lower number to the substituent encountered first. The number of times the substituent occurs is indicated by a prefix *di-*, *tri-*, *tetra-*, *penta-*, *hexa-*, and so on. Use a comma to separate position numbers.

2,4-Dimethylhexane (not 3,5-dimethylhexane)

6. If there are two or more different substituents, list them in alphabetical order, and number the chain from the end that gives the lower number to the substituent encountered first. If there are different substituents in equivalent positions on opposite ends of the parent chain, give the substituent of lower alphabetical order the lower number.

3-Ethyl-5-methylheptane (not 3-methyl-5-ethylheptane)

7. The prefixes *di-*, *tri-*, *tetra-*, and so on are not included in alphabetizing. Alphabetize the names of the substituents first, and then insert these prefixes. In the following example, the alphabetizing parts are *ethyl* and *methyl*, not *ethyl* and *dimethyl*.

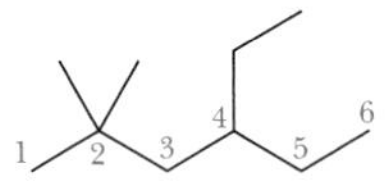

4-Ethyl-2,2-dimethylhexane
(not 2,2-dimethyl-4-ethylhexane)

8. Where there are two or more parent chains of identical length, choose the parent chain with the greater number of substituents.

3-Ethyl-2-methylhexane (not 3-Isopropylhexane)

A9.2 Alkenes and Alkynes

1. For both alkenes and alkynes, number the longest chain that contains the multiple bond in the direction that gives the carbon atoms of the functional group the lowest possible numbers.
2. Indicate the location of the multiple bond by the number of its first carbon.
3. Name the branched chain alkene or alkyne similar to alkanes.
4. Number the carbon atoms, locate and name substituent groups, locate the multiple bond, and name the main chain.
5. For alkenes, the ending *-ene* is used, and for alkynes, the ending *-yne* is used.

Examples of alkenes:

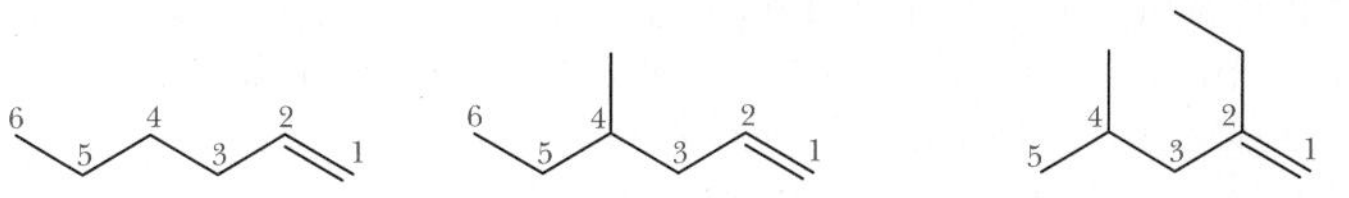

1-Hexene **4-Methyl-1-hexene** **2-Ethyl-4-methyl-1-pentene**

Examples of alkynes:

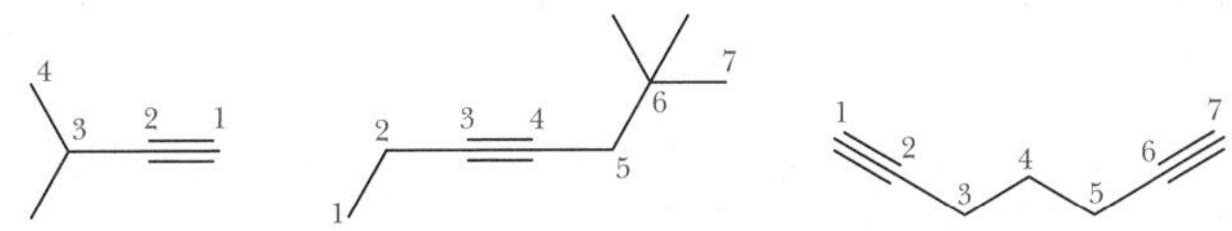

3-Methyl-1-butyne **6,6-Dimethyl-3-heptyne** **1,6-Heptadiyne**

6. For a molecule containing both a carbon-carbon double bond and a carbon-carbon triple bond, the compound is named as an alkenyne; the infix *-en-* shows the presence of the double bond and the infix *-yn-* shows the presence of the triple bond. In this case, the parent chain is numbered in the direction that gives the carbon atoms of the double bond the lowest set of numbers.

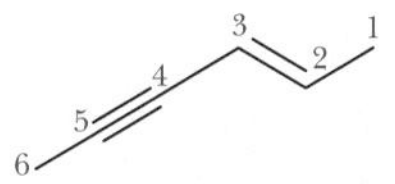

(2*E*)-2-Hexen-4-yne
or altenatively **(2*E*)-Hex-2-en-4-yne**

A9.3 Alcohols

1. The parent or root name is that of the longest chain of carbon atoms that contains the —OH group.
2. Number the parent chain from the end that gives the carbon bearing the —OH group the smaller number.
3. To show that the compound is an alcohol, change the suffix of the parent alkane from *-e* to *-ol*.
4. The location of the —OH group takes precedence over alkyl groups and halogen atoms in numbering the parent chain.
5. For cyclic alcohols, numbering begins with the carbon bearing the —OH group.
6. In complex alcohols, the number for the hydroxyl group may be placed between the infix and the suffix. For example, 4-methyl-2-hexanol and 4-methylhexan-2-ol are both acceptable names for the following compound.

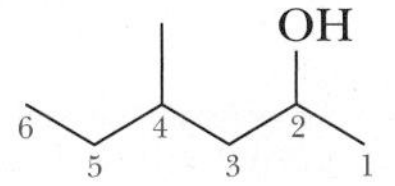

4-Methyl-2-hexanol
or
4-methylhexan-2-ol

A9.4 Aldehydes and Ketones

1. The parent or root name for an aldehyde is the name of the longest chain of carbon atoms that contains the functional group. Show the presence of the aldehyde group by changing the suffix of the parent alkane from *-e* to *-al*. Show the presence of a ketone group by changing the suffix of the parent alkane from *-e* to *-one*. Thus, aldehydes are named as alkan*als*, and ketones are named as alkan*ones*.
2. Because the carbonyl group of an aldehyde can only be on the end of the parent chain, and because numbering must begin with this carbon as number 1, the position of the aldehyde group is unambiguous and there is no need to use a number to locate it.
3. For unsaturated aldehydes or ketones, the presence of the carbon-carbon double or triple bond is indicated by the infix *-en-* or *-yn-*, respectively. The location of the carbonyl group determines the numbering pattern.
4. In complex aldehydes and ketones, the number for the carbonyl group may be placed between the infix and the suffix as illustrated in the following examples.

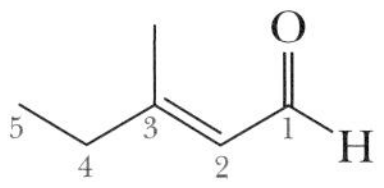

3-Methyl-2-pentenal
or
3-methylpent-2-enal

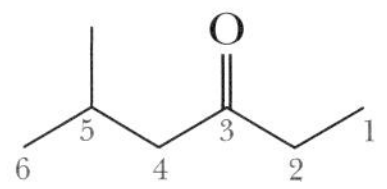

5-Methyl-3 hexanone
or
5-methylhexan-3-one

A9.5

1. The IUPAC has established an order of precedence for functional groups for names of compounds containing more than one functional group. Following is the order of precedence for the functional group we concentrate on in this text.

Increasing precedence (upward)

Functional Group	Suffix if higher priority	Prefix if lower priority	Example when the functional group has lower priority	
Carboxyl	-oic acid	——		
Aldehyde	-al	oxo-	3-Oxopropanoic acid	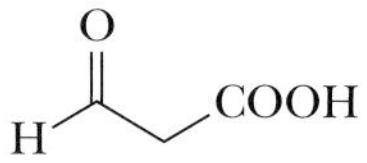
Ketone	-one	oxo-	3-Oxobutanoic acid	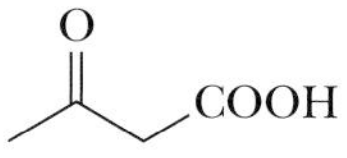
Alcohol	-ol	hydroxy-	4-Hydroxybutanoic acid	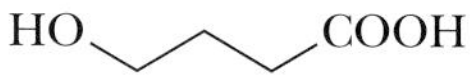
Amino	-amine	amino-	3-Aminobutanoic acid	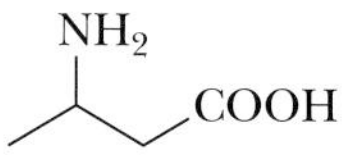
Sulfhydryl	-thiol	mercapto	2-Mercaptoethanol	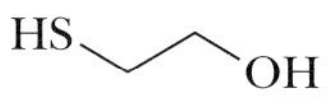

A9.6 Carboxylic Acids

1. The IUPAC name of a carboxylic acid is derived from that of the longest chain of carbon atoms (the root name) that contains the carboxyl group by dropping the suffix *-e* from the name of the parent alkane and adding the suffix *-oic* followed by the word *acid*.
2. The chain is numbered beginning with the carbon of the carboxyl group. Because the carboxyl group is understood to be on carbon-1, there is no need to give it a number.

3. In the IUPAC system, a carboxyl group takes precedence over most other functional groups, including hydroxyl, amino groups, and the carbonyl groups of aldehydes and ketones.

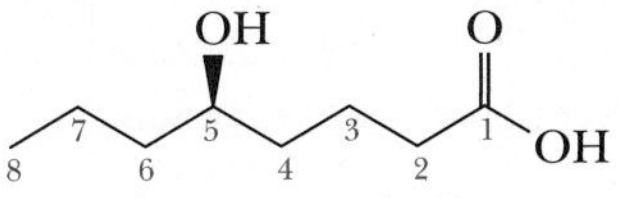

(*R*)-5-Hydroxyoctanoic acid

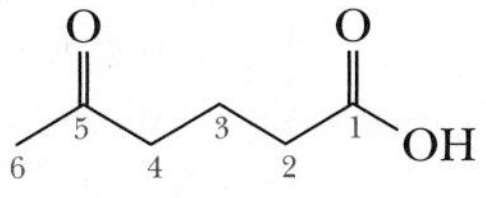

5-Oxohexanoic acid

A9.7 Carboxylic Esters

1. The IUPAC name of an ester is derived from the name of the corresponding carboxylic acid. The alkyl or aryl group bonded to oxygen is named first, followed by the name of the acid in which the suffix *-ic acid* is replaced by the suffix *-ate*.

$$CH_3\overset{\overset{\large O}{\|}}{C}OCH_2CH_3$$

Ethyl ethanoate
(Ethyl acetate)

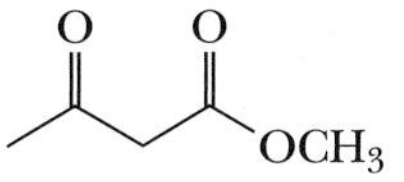

Methyl 3-oxobutanoate

A9.8 Carboxylic Amides

1. The IUPAC name of an amide is derived from the name of the parent carboxylic acid by dropping the suffix *-oic acid* and adding *-amide*.
2. If the nitrogen atom of the amide is bonded to an alkyl or aryl group, the group is named, and its location on nitrogen is indicated by the prefix *N*-.

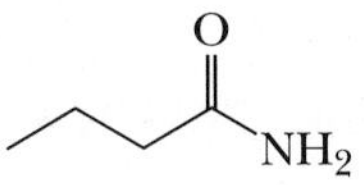

Butanamide

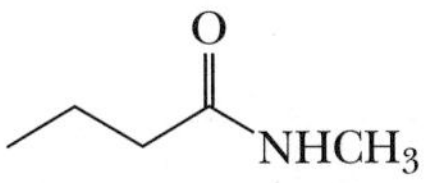

***N*-Methylbutanamide**

A9.9 Carboxylic Anhydrides

1. Anhydrides are named by replacing the word *acid* in the name of the parent carboxylic acid with the word *anhydride*.

$$CH_3\overset{\overset{\large O}{\|}}{C}O\overset{\overset{\large O}{\|}}{C}CH_3$$

Acetic anhydride

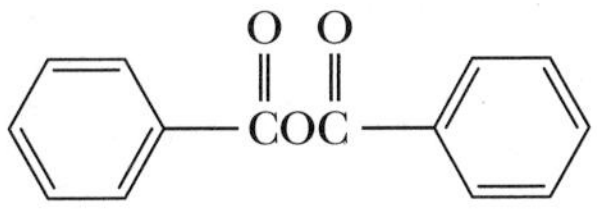

Benzoic anhydride

Example A9.1 IUPAC Nomenclature I

Following is the structural formula of mevalonic acid, a key intermediate in the biosynthesis of cholesterol. Write the IUPAC name of this compound.

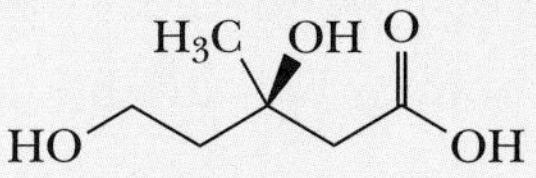

Mevalonic acid

Step 1: Identify all functional groups and, of these, select the one of highest precedence. This group must be in the chain of carbon atoms selected as the parent alkane, and its location on the parent chain determines the numbering of the parent chain.

(Continued)

Mevalonic acid contains three functional groups, two hydroxyl groups, and one carboxyl group. The carboxyl group has the highest precedence, and it is, therefore, carbon 1 of the parent chain.

Accordingly, this molecule is named as a substituted pentanoic acid.

Step 2: Name and locate all other substituents on the parent chain.

The two hydroxyl groups are on carbons 3 and 5, and therefore are named and located as 3,5-dihydroxy- and the methyl group is named and located as -3-methyl-.

Step 3: Specify the configuration of all chiral centers or carbon-carbon double bonds about which *cis-trans* isomerism is possible. In this example, there is one chiral center at carbon 3 and it has the *R* configuration.

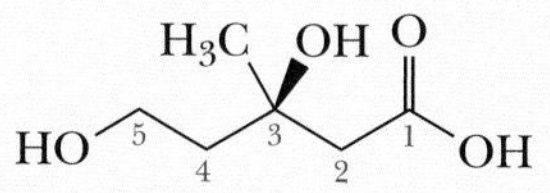

↓ Step 1

pentanoic acid

↓ Step 2

3,5-dihydroxy-3-methylpentanoic acid

↓ Step 3

(3*R*)-3,5-Dihydroxy-3-methylpentanoic acid

Example A9.2 IUPAC Nomenclature II

Following is the structural formula of the terpene farnesol. Write the IUPAC name of this compound.

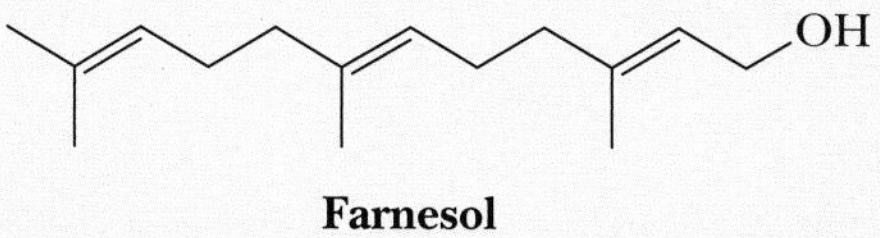

Farnesol

Step 1: There is a single hydroxyl group, which must be located in the longest carbon chain, which in this example is 12 carbon atoms. Therefore, the compound is derived from dodecan-1-ol.

Step 2: There are methyl substituents on carbons 3, 7, and 11 of the parent chain, indicated by 3,7,11-trimethyl-.

Step 3: There are three carbon-carbon double bonds indicated by the infix -2,6,10-tridecen-.

Step 4: Of the three double bonds in farnesol, those at carbons 2 and 6 can show *cis/trans* isomerism, which may be designated in common name as *trans-trans*-farnesol, but in the IUPAC system must be designated by the *E,Z* system as (2*E*,6*E*)-.

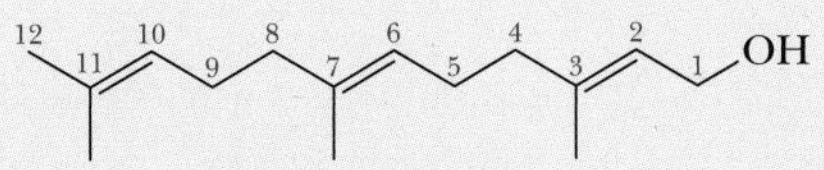

Step 1

dodecan-1-ol

Step 2

3,7,11-trimethyldodecan-1-ol

Step 3

3,7,11-trimethyl-2,6,10-dodecatrien-1-ol

Step 4

(2*E*,6*E*)-3,7,11-trimethyl-2,6,10-dodecatrien-1-ol

Example A9.3 IUPAC Nomenclature III

Following is the structural formula of the Queen substance, a substance secreted by the mandibular gland of queen honeybees. It inhibits the development of ovaries in worker bees, prevents queen bee formation, and attracts male bees (drones) to virgin queens for the purpose of mating. Write the IUPAC name of this compound.

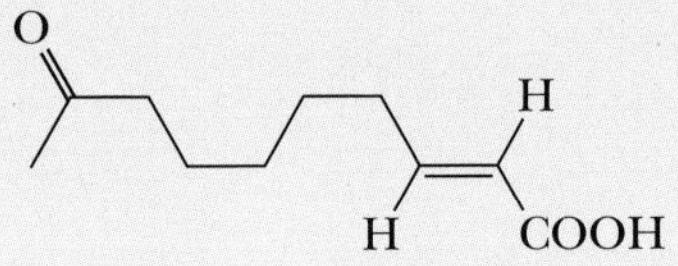

Queen substance

Step 1: There are two oxygen functional groups in this molecule—a carboxyl group and a ketone group. Of these, the carboxyl group has the higher precedence and determines the numbering pattern of the longest chain that contains both of these functional groups.

Step 2: The presence and location of the ketone group is indicated by the prefix 9-oxo-, and the presence and location of the carbon-carbon double bond is indicated by the infix -2-ene-.

Step 3: The carbon-carbon double bond has the *E* configuration, which could be indicated by (2*E*), but because there is only one double bond, and we know from Step 2 that it is between carbons 2 and 3, it is not necessary to specify (2*E*). It is sufficient to place (*E*) first in the name.

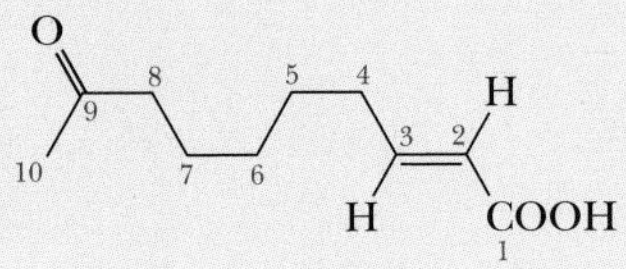

Step 1

decanoic acid

Step 2

9-oxodec-2-enoic acid

Step 3

(*E*)-9-Oxodec-2-enoic acid

Glossary

1,2-Shift (Section 6.3C) A type of rearrangement in which an atom or group of atoms with its bonding electrons moves from one atom to an adjacent electron-deficient atom.

Absolute configuration (Section 3.3) Which of the two possible isomers an enantiomer is (i.e., whether it is the right- or left-handed isomer).

Absorbance (A) (Section 20.3A) A quantitative measure of the extent to which a compound absorbs radiation of a particular wavelength. $A = \log (I_0/I)$ where I_0 is the incident radiation and I is the transmitted radiation.

Acetal (Section 16.7B) A molecule containing two —OR or —OAr groups bonded to the same carbon.

Aceto group (Section 17.2B) A CH_3CO— group; also called an acetyl group.

Achiral (Section 3.2) An object that lacks chirality; an object that has no handedness.

Activating group (Section 22.2A) Any substituent on a benzene ring that causes the rate of electrophilic aromatic substitution to be greater than that for benzene.

Activation energy (Section 4.5A and Section 6.1) The difference in Gibbs free energy between reactants and a transition state.

Acylation (Section 19.5B) The process of introducing an acyl group, RCO— or ArCO—, onto an organic molecule.

Acyl group (Section 18.1A) An RCO— or ArCO— group.

Acylium ion (Section 22.1C) A resonance-stabilized cation with the structure $[RC{\equiv}O]^+$ or $[ArC{\equiv}O]^+$. The positive charge is delocalized over both the carbonyl carbon and the carbonyl oxygen.

Addition reaction (Section 6.1) A reaction in which two atoms or groups of atoms react with a double bond, forming a compound with the two new groups bonded to the carbons of the original double bond.

Aglycon (Synthesis Problems, Chapter 24) Lacking a sugar.

Alcohol (Section 1.3A) A compound containing an —OH (hydroxyl) group bonded to a carbon atom.

Aldehyde (Section 1.3C) A compound containing a —CHO group.

Alditol (Section 25.3B) The product formed when the C=O group of a monosaccharide is reduced to a CHOH group.

Aldonic acid (Section 25.3C) The product formed when the —CHO group of an aldose is oxidized to a —COOH group.

Aldose (Section 25.1A) A monosaccharide containing an aldehyde group.

Aliphatic amine (Section 23.1) An amine in which nitrogen is bonded only to alkyl groups.

Alkaloid (Section 23.1) A basic nitrogen-containing compound of plant origin, many of which are physiologically active when administered to humans.

Alkene metathesis (Section 24.5) A reaction in which two alkenes interchange the carbons attached to their double bonds.

Alkoxy group (Section 11.2) An —OR group where R is an alkyl group.

Alkyl group (Section 2.3A) A group derived by removing a hydrogen from an alkane; given the symbol R—.

Alkylation reaction (Section 7.5A) Any reaction in which a new carbon-carbon bond to an alkyl group is formed.

Alkyne (Section 7.1) An unsaturated hydrocarbon that contains one or more carbon-carbon triple bonds.

Allene (Section 7.5B) The compound $CH_2{=}C{=}CH_2$. Any compound that contains adjacent carbon-carbon double bonds; that is, any molecule that contains a C=C=C functional group.

Allyl (Section 5.2B) A $—CH_2CH{=}CH_2$ group.

Allylic (Section 9.3B) Next to a carbon-carbon double bond.

Allylic carbocation (Section 9.3B) A carbocation in which an allylic carbon bears the positive charge.

Allylic carbon (Section 8.6) A carbon adjacent to a carbon-carbon double bond.

Allylic substitution (Section 8.6) Any reaction in which an atom or group of atoms is substituted for another atom or group of atoms at an allylic carbon.

Amino acid (Section 27.1A) A compound that contains both an amino group and a carboxyl group.

α-Amino acid (Section 27.1A) An amino acid in which the amino group is on the carbon adjacent to the carboxyl group.

Amino group (Section 1.3B) A compound containing an sp^3-hybridized nitrogen atom bonded to one, two, or three carbon atoms.

Amorphous domain (Section 29.4) A disordered, noncrystalline region in the solid state of a polymer.

Anabolic steroid (Section 26.4A) A steroid hormone, such as testosterone, that promotes tissue and muscle growth and development.

Androgen (Section 26.4A) A steroid hormone, such as testosterone, that mediates the development of sexual characteristics of males.

Angle strain (Section 2.6A) The strain that arises when a bond angle is either compressed or expanded compared to its optimal value.

Anion (Section 1.2A) An atom or group of atoms bearing a negative charge.

Annulene (Section 21.2B) A cyclic hydrocarbon with a continuous alternation of single and double bonds.

Anomeric carbon (Section 25.2A) The hemiacetal or acetal carbon of the cyclic form of a carbohydrate.

Anomers (Section 25.2A) Carbohydrates that differ in configuration only at their anomeric carbons.

Antiaromatic compound (Section 21.2C) A monocyclic compound that is planar or nearly so, has one *2p* orbital on each atom of the ring, and has $4n$ π electrons in the cyclic arrangement of overlapping *2p* orbitals, where *n* is an integer. Antiaromatic compounds are especially unstable.

Antibonding molecular orbital (Section 1.7A) A molecular orbital in which electrons have a higher energy than they would in isolated atomic orbitals.

Anti conformation (Section 2.5A) A conformation about a single bond in which two groups on adjacent carbons lie at a dihedral angle of 180°.

Anti stereoselectivity (Section 6.3D) The addition of atoms or groups of atoms to opposite faces of a carbon-carbon double bond.

Aprotic acid (Section 4.7) An acid that is not a proton donor; an acid that is an electron pair acceptor in a Lewis acid-base reaction.

Aprotic solvent (Section 9.3D) A solvent that cannot serve as a hydrogen-bond donor; nowhere in the molecule is there a hydrogen bonded to an atom of high electronegativity. Common aprotic solvents are dichloromethane, diethyl ether, and dimethyl sulfoxide.

Aramid (Section 29.5A) A polyaromatic amide; a polymer in which the monomer units are an aromatic diamine and an aromatic dicarboxylic acid.

Arene (Introduction, Chapter 5) A term used to classify benzene and its derivatives.

Aromatic amine (Section 23.1) An amine in which nitrogen is bonded to one or more aryl groups.

Aromatic compound (Introduction, Chapter 21) A term used initially to classify benzene and its derivatives. More accurately, it is used to classify any compound that meets the Hückel criteria for aromaticity (Section 21.2A).

Aryl group (Ar—) (Introduction, Chapter 5) A group derived from an arene by removal of an H.

Atactic polymer (Section 29. 6C) A polymer with completely random configurations at the chiral centers along its chain, as, for example, atactic polypropylene.

Atropisomers (Section 3.2) Enantiomers that lack a chiral center and differ because of hindered rotation.

Aufbau principle (Section 1.1A) Orbitals fill in order of increasing energy, from lowest to highest.

Autoxidation (Section 8.7) Air oxidation of materials such as unsaturated fatty acids.

Axial bond (Section 2.5B) A bond to a chair conformation of cyclohexane that extends from the ring parallel to the imaginary axis through the center of the ring; a bond that lies roughly perpendicular to the equator of the ring.

Azeotrope (Section 16.7B) A liquid mixture of constant composition with a boiling point that is different from that of any of its components.

Base peak (Section 14.1) The peak caused by the most abundant ion in a mass spectrum; the most intense peak. It is assigned an arbitrary intensity of 100.

Basicity (Section 9.3E) An equilibrium property measured by the position of equilibrium in an acid-base reaction, as, for example, the acid-base reaction between ammonia and water.

Benzyl group ($C_6H_5CH_2$—) (Section 21.3A) The group derived from toluene by removing a hydrogen from its methyl group.

Benzylic position (Section 21.5) An sp^3-hybridized carbon bonded to a benzene ring.

Benzyne intermediate (Section 22.3A) A reactive intermediate formed by β-elimination from adjacent carbon atoms of a benzene ring and having a triple bond in the benzene ring. The second π bond of the benzyne triple bond is formed by the weak overlap of coplanar *2p* orbitals on adjacent carbons.

Betaine (Section 16.6) A neutral molecule with nonadjacent positive and negative charges. An example of a betaine is the intermediate formed by addition of a Wittig reagent to an aldehyde or ketone.

Bicycloalkane (Section 2.4B) An alkane containing two rings that share two carbons.

Bile acid (Section 26.4A) A cholesterol-derived detergent molecule, such as cholic acid, which is secreted by the gallbladder into the intestine to assist in the absorption of dietary lipids.

Bimolecular reaction (Section 9.2) A reaction in which two species are involved in the rate-determining step.

Boat conformation (Section 2.5B) A nonplanar conformation of a cyclohexane ring in which carbons 1 and 4 of the ring are bent toward each other.

Bond dipole moment (μ) (Section 1.2B) A measure of the polarity of a covalent bond. The product of the charge on either atom of a polar bond times the distance between the atoms.

Bond dissociation enthalpy (Section 6.3B) The amount of energy required to break a bond into two radicals in the gas phase at 25°C, A—B —> A• + •B

Bonding electrons (Section 1.2C) Valence electrons involved in forming a covalent bond (i.e., shared electrons).

Bonding molecular orbital (Section 1.8A) A molecular orbital in which electrons have a lower energy than they would in isolated atomic orbitals.

Bond length (Section 1.2B) The distance between atoms in a covalent bond in picometers (pm; 1 pm = 10^{-12} m) or Å (1Å = 10^{-10} m).

Brønsted-Lowry acid (Section 4.2) A proton donor.

Brønsted-Lowry base (Section 4.2) A proton acceptor.

Carbanion (Section 15.1A) An ion in which carbon has an unshared pair of electrons and bears a negative charge.

Carbene (Section 15.4) A neutral molecule that contains a carbon atom surrounded by only six valence electrons (R_2C:).

Carbenoid (Section 15.3C) A compound that delivers the elements of a carbene without actually producing a free carbene.

Carbocation (Section 6.3A) A species in which a carbon atom has only six electrons in its valence shell and bears a positive charge.

Carbohydrate (Introduction, Chapter 25) A polyhydroxyaldehyde, a polyhydroxyketone, or a substance that gives these compounds on hydrolysis.

α-Carbon (Section 16.9A) A carbon atom adjacent to a carbonyl group.

Carbonyl group (Section 1.3C) A C═O group.

Carboxyl group (Section 1.3D) A —COOH group.

Carboxylic acid (Section 1.3D) A compound containing a carboxyl, —COOH, group.

Carboxylic ester (Section 1.3D) A derivative of a carboxylic acid in which H of the carboxyl group is replaced by a carbon.

Cation (Section 1.2A) An atom or group of atoms bearing a positive charge.

Center of symmetry (Section 3.1) A point so situated that identical components of an object are located on opposite sides and equidistant from that point along any axis passing through it.

Chain-growth polymerization (Section 29.6) A polymerization that involves sequential addition reactions, either to unsaturated monomers or to monomers possessing other reactive functional groups.

Chain initiation (Section 8.5) A step in a chain reaction characterized by the formation of reactive intermediates (radicals, anions, or cations) from nonradical or noncharged molecules.

Chain length (Section 8.5B) The number of times the cycle of chain propagation steps repeats in a chain reaction.

Chain propagation (Section 8.5B) A step in a chain reaction characterized by the reaction of a reactive intermediate and a molecule to give a new reactive intermediate and a new molecule.

Chain termination (Section 8.5B) A step in a chain reaction that involves destruction of reactive intermediates.

Chain-transfer reaction (Section 29.6A) The transfer of reactivity of an endgroup from one chain to another during a polymerization.

Chair conformation (Section 2.6B) The most stable nonplanar conformation of a cyclohexane ring; all bond angles are approximately 109.5°, and all bonds on adjacent carbons are staggered.

Chemical shift (δ) (Section 13.3) The shift in parts per million of an NMR signal relative to the signal of TMS.

Chiral (Section 3.2) From the Greek, *cheir* meaning hand; an object that is not superposable on its mirror image; an object that has handedness.

Chiral center (Section 3.2) A tetrahedral atom, most commonly carbon, that is bonded to four different groups; also called a chirality center.

Chlorofluorocarbons (CFCs, Freons) (Section 8.3) Compounds with one or two carbons, chlorine, and fluorine, formerly used as refrigerants.

Chromatography (Section 3.8C) A separation method involving passing a vapor or solution mixture through a column packed with a material with different affinities for different components of the mixture.

Circular DNA (Section 28.2C) A type of double-stranded DNA in which the 5′ and 3′ ends of each strand are joined by phosphodiester groups.

Cis (Section 2.6A) A prefix meaning on the same side.

***Cis,trans* isomers** (Sections 2.6 and 5.1C) Stereoisomers that have the same connectivity but a different arrangement of their atoms in space as a result of the presence of either a ring or a carbon-carbon double bond.

Clemmensen reduction (Section 16.11E) Reduction of the C═O group of an aldehyde or ketone to a CH_2 group using Zn(Hg) and HCl.

Codon (Section 28.4A) A triplet of nucleotides on mRNA that directs incorporation of a specific amino acid into a polypeptide sequence.

Condensation polymerization (Section 29.5) A polymerization in which chain growth occurs in a stepwise manner between difunctional monomers. Also called step-growth polymerization.

Configuration (Section 2.7A) Refers to the arrangement of atoms about a stereocenter.

Configurational isomers (Sections 3.1 and 2.6A) Isomers that differ by the configuration of substituents on an atom. Refers to the arrangement of atoms about a stereocenter.

Conformation (Section 2.5A) Any three-dimensional arrangement of atoms in a molecule that results from rotation about a single bond.

Conjugate acid (Section 4.2A) The species formed when a base accepts a proton from an acid.

Conjugate addition (Section 19.8) Addition of a nucleophile to the β-carbon of an α,β-unsaturated carbonyl compound. (Section 20.2A) Addition to carbons 1 and 4 of a conjugated diene.

Conjugate base (Section 4.2A) The species formed when an acid transfers a proton to a base.

Conjugated (Section 20.1) A conjugated diene or carbonyl is one in which the double bonds are separated by one single bond.

Conjugation (Section 12.4H) A situation in which two multiple bonds are separated by a single bond. Alternatively, a series of overlapping 2*p* orbitals. 1,3-Butadiene, for example, is a conjugated diene, and 3-butene-2-one is a conjugated enone.

Constitutional isomers (Section 2.2) Compounds with the same molecular formula but a different connectivity of their atoms.

Contributing structures (Section 1.8A) Representations of a molecule or ion that differ only in the distribution of valence electrons.

Correlation tables (Section 12.3D) Tables of data on absorption patterns of functional groups.

Coupling constant (J) (Section 13.9) The separation on an NMR spectrum (in hertz) between adjacent peaks in a multiplet and a quantitative measure of the influence of the spin-spin coupling with adjacent nuclei.

Covalent bond (Section 1.2A) A chemical bond formed between two atoms by sharing one or more pairs of electrons.

Crown ether (Section 11.11) A cyclic polyether derived from ethylene glycol and substituted ethylene glycols.

Crystalline domain (Section 29.4) An ordered crystalline region in the solid state of a polymer. Also called a crystallite.

Cumulated (Section 20.1) A cumulated diene is one in which two double bonds share an *sp*-hybridized carbon.

Curved arrow (Section 1.8A) A symbol used to show the redistribution of valence electrons in resonance contributing structures or reactions, symbolizing movement of two electrons.

Cyanohydrin (Section 16.5D) A molecule containing an —OH group and a —CN group bonded to the same carbon.

Cycloaddition reaction (Section 24.6) A reaction in which two reactants add together in a single step to form a cyclic product. The best known of these is the Diels-Alder reaction.

Cycloalkane (Section 2.4) A saturated hydrocarbon that contains carbons joined to form a ring.

Deactivating group (Section 22.2A) Any substituent on a benzene ring that causes the rate of electrophilic aromatic substitution to be lower than that for benzene.

Decarboxylation (Section 17.9) Loss of CO_2 from a carboxyl group.

Dehydration (Section 10.6) Elimination of water.

Dehydrohalogenation (Sections 7.5B and 9.5) Removal of —H and —X from adjacent carbons; a type of β-elimination.

Delocalization (Section 1.1) The spreading of electron density over a larger volume of space.

Deshielding (Section 13.3) The term used to express the concept of less shielding in NMR.

Dextrorotatory (Section 3.7B) Refers to a substance that rotates the plane of polarized light to the right.

Diamagnetic current in NMR (Sections 13.3 and 13.7C) The circulation of electron density in a molecule in an applied magnetic field.

Diastereomers (Section 3.4A) Stereoisomers that are not mirror images of each other; refers to relationships among two or more objects.

Diastereotopic groups (Section 13.10) Atoms or groups on an atom that are bonded to an atom that is bonded to two nonidentical groups, one of which contains a chiral center. When one of the atoms or groups is replaced by another group, a new chiral center is created and a set of diastereomers results. The hydrogens of the CH_2 group of 2-butanol, for example, are diastereotopic. Diastereotopic groups have different chemical shifts under all conditions.

Diaxial interactions (Section 2.5B) Refers to the steric strain arising from interaction between an axial substituent and an axial hydrogen (or other group) on the same side of a chair conformation of a cyclohexane ring.

Diazonium ion (Section 23.8D) An ArN_2^+ or RN_2^+ ion.
Dielectric constant (Section 9.3D) A measure of a solvent's ability to insulate opposite charges from one another.
Diels-Alder adduct (Section 24.6) A cyclohexene resulting from the cycloaddition reaction of a diene and a dienophile.
Dienophile (Section 24.6) A compound containing a double bond (consisting of one or two C, N, or O atoms) that can react with a conjugated diene to give a Diels-Alder adduct.
Dihedral angle (Section 2.6A) The angle created by two intersecting planes.
Diol (Section 10.1B) A compound containing two hydroxyl groups.
Dipeptide (Section 27.3) A molecule containing two amino acid units joined by a peptide bond.
Dipole-dipole interaction (Section 10.2) The attraction between the positive end of one dipole and the negative end of another.
Disaccharide (Section 25.4) A carbohydrate containing two monosaccharide units joined by a glycosidic bond.
Dispersion forces (Section 2.7B) Very weak intermolecular forces of attraction resulting from the interaction between temporary induced dipoles.
Disproportionation (Section 29.6A) A termination process that involves the abstraction of a hydrogen atom from the beta position of the propagating radical of one chain by the radical endgroup of another chain.
Disulfide (Section 11.2) A molecule containing an —S—S— group.
Double-headed arrow (Section 1.8A) A symbol used to show that structures on either side of it are resonance-contributing structures.
Double helix (Section 28.2B) A type of secondary structure of DNA molecules in which two antiparallel polynucleotide strands are coiled in a right-handed manner about the same axis.
Downfield (Section 13.4) A signal of an NMR spectrum that is shifted toward the left (larger chemical shift) on the chart paper.

E (Section 5.2C) From the German, *entgegen*, opposite. Specifies that groups of higher priority on the carbons of a double bond are on opposite sides.
***E,Z* system** (Section 5.2) A system to specify the configuration of groups about a carbon-carbon double bond.
E1 (Section 9.6A) A unimolecular β-elimination reaction.
E2 (Section 9.6B) A bimolecular β-elimination reaction.
Eclipsed conformation (Section 2.5A) A conformation about a carbon-carbon single bond in which the atoms or groups on one carbon are as close as possible to the atoms or groups on an adjacent carbon.
Edman degradation (Section 27.4B) A method for selectively cleaving and identifying the *N*-terminal amino acid of a polypeptide chain.
Elastomer (Section 29.4) A material that, when stretched or otherwise distorted, returns to its original shape when the distorting force is released.
Electromagnetic radiation (Section 12.1) Light and other forms of radiant energy.
Electronegativity (Section 1.2B) A measure of the force of an atom's attraction for electrons.
Electron affinity (Section 1.2B) Energy added or released when an electron is added to an atom or molecule.
Electrophile (Section 6.3A and Introduction to Chapter 9) From the Greek meaning electron loving. Any species that can accept a pair of electrons to form a new covalent bond; alternatively, a Lewis acid.
Electrophilic aromatic substitution (Section 22.1) A reaction in which there is substitution of an electrophile, E^+, for a hydrogen on an aromatic ring.
Electrophoresis (Section 27.2D) The process of separating compounds on the basis of their electric charge.
β-Elimination (Introduction, Chapter 9) A reaction in which a molecule, such as HCl, HBr, HI, or HOH, is split out or eliminated from adjacent carbons.
Enamine (Section 16.8A) An unsaturated compound derived by the reaction of an aldehyde or ketone and a secondary amine followed by loss of H_2O; $R_2C{=}CR{-}NR_2$.
Enantiomeric excess (ee) (Section 3.7D) The difference between the percentage of two enantiomers in a mixture.
Enantiomers (Section 3.2) Stereoisomers that are nonsuperposable mirror images of each other; refers to a relationship between pairs of objects.
Enantioselective reaction (Section 6.7B) A reaction that produces one enantiomer in preference to the other.
Enantiotopic groups (Section 13.10) Atoms or groups on an atom that give a chiral center when one of the groups is replaced by another group. A pair of enantiomers results. The hydrogens of the CH_2 group of ethanol, for example, are enantiotopic. Replacing one of them by deuterium gives (*R*)-1-deuteroethanol; replacing the other gives (*S*)-1-deuteroethanol. Enantiotopic groups have identical chemical shifts in achiral environments but different chemical shifts in chiral environments.
Endergonic reaction (Section 4.5B) A reaction in which the Gibbs free energy of the products is higher than that of the reactants. The position of equilibrium for an endergonic reaction favors starting materials.
Endothermic reaction (Sections 4.5B and 6.2A) A reaction in which the enthalpy of the products is higher than the enthalpy of the reactants; a reaction in which heat is absorbed.
Energy (Section 1.1B) The ability to do work.
Energy diagram (Section 4.5B) A graph showing the changes in energy that occur during a chemical reaction; energy is plotted on the vertical axis, and reaction progress is plotted on the horizontal axis. Also called a reaction coordinate diagram.
Enol (Section 7.7A) A compound containing a hydroxyl group bonded to a doubly bonded carbon atom.
Enolate anion (Section 16.9A) An anion derived by loss of a hydrogen from a carbon alpha to a carbonyl group; the anion of an enol.
Enthalpy change, ΔH^0 (Section 4.5B) The difference in total bond strengths and solvation between various points under comparison on a reaction coordinate diagram.
Entropy (S) (Section 4.5B) Measures chaos versus order and chaos is favorable.
Epoxide (Section 11.7) A cyclic ether in which oxygen is one atom of a three-membered ring.
Equatorial bond (Section 2.6B) A bond to a chair conformation of cyclohexane that extends from the ring roughly perpendicular to the imaginary axis through the center of the ring; a bond that lies roughly along the equator of a cyclohexane ring.
Equivalent hydrogens (Section 13.5) Hydrogens that have the same chemical environment.
Ester (Section 1.3E) A derivative of a carboxylic acid in which H of the carboxyl group is replaced by a carbon.
Estrogen (Section 26.4A) A steroid hormone, such as estrone and estradiol, that mediates the development of sexual characteristics in females.
Ether (Section 11.1) A compound containing an oxygen atom bonded to two carbon atoms.
Excited state (Section 1.1B) A state of a system at higher energy than the ground state.
Exergonic reaction (Section 4.5B) A reaction in which the Gibbs free energy of the products is lower than that of the reactants. The position of equilibrium for an exergonic reaction favors products.
Exothermic reaction (Section 4.5B) A reaction in which the enthalpy of the products is lower than that of the reactants; a reaction in which heat is released.

Fat (Section 26.1B) A mixture of triglycerides that is semisolid or solid at room temperature.
Fatty acid (Section 26.1A) A long, unbranched-chain carboxylic acid, most commonly of 12 to 20 carbons, derived from the hydrolysis

of animal fats, vegetable oils, or the phospholipids of biological membranes.

Fingerprint region (Section 12.3D) Vibrations in the region 1500 to 400 cm^{-1} of an IR spectrum are complex and difficult to analyze but are characteristic for different molecules.

First ionization potential (Section 1.1B) The energy needed to remove the most loosely held electron from an atom or molecule.

Fischer esterification (Section 17.7A) The process of forming an ester by refluxing a carboxylic acid and an alcohol in the presence of an acid catalyst, commonly H_2SO_4, $ArSO_3H$, or HCl.

Fischer projection (Section 3.4C and Section 25.1B) A two-dimensional representation of a molecule; in these projections, groups on the right and left are by convention in front, while those at the top and bottom are to the rear.

Fishhook arrow (Section 8.4) A barbed curved arrow used to show the change in position of a single electron.

Fluid-mosaic model (Section 26.5B) A biological membrane that consists of a phospholipid bilayer with proteins, carbohydrates, and other lipids on the surface and embedded in the bilayer.

Formal charge (Section 1.2D) The charge on an atom in a polyatomic ion or molecule.

Fourier transform NMR (FT-NMR) (Section 13.4) The modern NMR method that is based on a constant magnetic field, a short pulse of electromagnetic radiation, and a mathematical Fourier transform to produce the spectrum.

Frequency (Section 12.1) The number of full cycles of a wave that pass a given point in a second, and reported in hertz (Hz), which has the units s^{-1}.

Friedel-Crafts reaction (Section 22.1C) An electrophilic aromatic substitution in which a hydrogen of an aromatic ring is replaced by an alkyl or acyl group.

Frost circle (Section 21.2A) A graphic method for determining the relative energies of π MOs for planar, fully conjugated, monocyclic compounds.

Functional group (Section 1.3) An atom or group of atoms within a molecule that shows a characteristic set of physical and chemical properties.

Furanose (Section 25.2A) A five-membered cyclic form of a monosaccharide.

Gauche conformation (Section 2.5A) A conformation about a single bond of an alkane in which two groups on adjacent carbons lie at a dihedral angle of 60°.

Geminal coupling (Section 13.9D) Spin-spin coupling that occurs between nonequivalent H atoms bonded to the same C atom. The H atoms are generally nonequivalent owing to restricted bond rotation in the molecule.

Gibbs free energy change (ΔG^0) (Section 4.5B) The energy that dictates the position of chemical equilibria and rates of chemical reactions. A thermodynamic function of enthalpy, entropy, and temperature, given by the equation $\Delta G^0 = \Delta H^0 - T\Delta S°$. If $\Delta G^0 < 0$, the position of equilibrium for the reaction favors products. If $\Delta G^0 > 0$, the position of equilibrium favors reactants.

Glass transition temperature (T_g) (Section 29.4) The temperature at which a polymer undergoes the transition from a hard glass to a rubbery state.

Glycol (Section 6.5A) A compound with hydroxyl (—OH) groups on adjacent carbons.

Glycoside (Section 25.3A) A carbohydrate in which the —OH on its anomeric carbon is replaced by —OR.

Glycosidic bond (Section 25.3A) The bond from the anomeric carbon of a glycoside to an —OR group.

Ground state (Section 1.1B) The lowest energy state of a system.

Ground-state electron configuration (Section 1.1A) The lowest-energy electron configuration for an atom or molecule.

Haloalkane (alkyl halide) (Section 8.1) A compound containing a halogen atom covalently bonded to an sp^3-hybridized carbon atom. Given the symbol R—X.

Haloalkene (vinylic halide) (Section 8.1) A compound containing a halogen atom bonded to one of the carbons of a carbon-carbon double bond.

Haloarene (aryl halide) (Section 8.1) A compound containing a halogen atom bonded to a benzene ring. Given the symbol Ar—X.

Haloform (Section 8.2B) A compound of the type CHX_3 where X is a halogen.

Halohydrin (Section 6.3E) A compound containing a halogen atom and a hydroxyl group on adjacent carbons; those containing Br and OH are bromohydrins, and those containing Cl and OH are chlorohydrins.

Hammond's postulate (Section 8.5D) The structure of the transition state for an exothermic step looks more like the reactants of that step than the products. Conversely, the structure of the transition state for an endothermic step looks more like the products of that step than the reactants.

Haworth projection (Section 25.2A) A way to view furanose and pyranose forms of monosaccharides. The ring is drawn flat and most commonly viewed through its edge with the anomeric carbon on the right and the oxygen atom of the ring to the rear.

Heat of combustion (ΔH^0) (Section 2.8A) Standard heat of combustion is the heat released when one mole of a substance in its standard state (gas, liquid, solid) is oxidized completely to carbon dioxide and water.

Heat of reaction (ΔH^0) (Section 4.5B) The difference in enthalpy between reactants and products. If the enthalpy of products is lower than that of the reactants, heat is released and the reaction is exothermic. If the enthalpy of the products is higher than that of the reactants, energy is absorbed, and the reaction is endothermic.

α-Helix (Section 27.6B) A type of secondary structure in which a section of polypeptide chain coils into a spiral, most commonly a right-handed spiral.

Hemiacetal (Section 16.7B) A molecule containing an —OH and an —OR or —OAr group bonded to the same carbon.

Hertz (Hz) (Section 12.1) The unit in which frequency is measured: s^{-1} (read"per second").

Heterocycle (Section 11.2) A cyclic compound whose ring contains more than one kind of atom. Oxirane (ethylene oxide), for example, is a heterocycle whose ring contains two carbon atoms and one oxygen atom.

Heterocyclic amine (Section 23.1) An amine in which nitrogen is one of the atoms of a ring.

Heterocyclic aromatic amine (Section 23.1) An amine in which nitrogen is one of the atoms of an aromatic ring.

Heterolytic bond cleavage (Section 8.4) Cleavage of a bond so that one fragment retains both electrons and the other retains none.

High-density lipoprotein (HDL) (Section 26.4A) Plasma particles, density 1.06–1.21 g/mL, consisting of approximately 33% proteins, 30% cholesterol, 29% phospholipids, and 8% triglycerides.

High-resolution mass spectrometry (Section 14.2A) Instrumentation that is capable of separating ions that differ in mass by as little as 0.0001 amu.

Histone (Section 28.2C) A protein, particularly rich in the basic amino acids lysine and arginine, that is found associated with DNA molecules.

Hofmann elimination (Section 23.9) When treated with a strong base, a quaternary ammonium halide undergoes β-elimination by an E2 mechanism to give the less substituted alkene as the major product.

Hofmann rule (Section 23.9) Any β-elimination that occurs preferentially to give the less substituted alkene as the major product.

Homolytic bond cleavage (Section 8.4) Cleavage of a bond so that each fragment retains one electron; formation of radicals.

Homotopic groups (Section 13.10) Atoms or groups on an atom that give an achiral molecule when one of the groups is replaced by another group. The hydrogens of the CH_2 group of propane, for example, are homotopic. Replacing either one of them with deuterium gives 2-deuteropropane, which is achiral. Homotopic groups have identical chemical shifts under all conditions.

Hückel criteria for aromaticity (Section 21.2A) To be aromatic, a monocyclic compound must have one $2p$ orbital on each atom of the ring, be planar or nearly so, and have $(4n + 2)$ π electrons in the cyclic arrangement of $2p$ orbitals.

Hund's rule (Section 1.1A) When orbitals of equal energy are available but there are not enough electrons to fill all of them completely, one electron is put in each before a second electron is added to any.

Hybridization (Section 1.7B) The combination of atomic orbitals of different types.

Hybrid orbital (Section 1.7B) An orbital formed by the combination of two or more atomic orbitals.

Hydration (Section 6.3B) The addition of water.

Hydride ion (Section 16.11A) A hydrogen atom with two electrons in its valence shell; $H:^-$.

Hydroboration-oxidation (Section 6.4) A method for converting an alkene to an alcohol. The alkene is treated with borane (BH_3) to give a trialkylborane, which is then oxidized with alkaline hydrogen peroxide to give an alcohol.

α-Hydrogen (Section 16.9A) A hydrogen on a carbon alpha to a carbonyl group.

Hydrogen bonding (Section 10.2) The attractive interaction between a hydrogen atom bonded to an atom of high electronegativity (most commonly O or N) and a lone pair of electrons on another atom of high electronegativity (again, most commonly O or N).

Hydrogenolysis (Section 21.5C) Cleavage of a single bond by H_2, most commonly accomplished by treating a compound with H_2 in the presence of a transition metal catalyst.

Hydroperoxide (Section 11.5B) A compound containing an —OOH group.

Hydrophilic (Sections 9.9 and 17.3) From the Greek, meaning water-loving.

Hydrophobic (Section 9.9 and 17.3) From the Greek, meaning water-fearing.

Hydrophobic effect (Section 27.6D) The tendency of nonpolar groups to cluster so as to shield them from contact with an aqueous environment.

Hydroxyl group (Section 1.3A) An —OH group.

Hyperconjugation (Section 6.3B) Interaction of electrons in a σ-bonding orbital with the vacant $2p$ orbital of an adjacent positively charged carbon.

Imide (Section 18.1D) A functional group in which two acyl groups, RCO— or ArCO—, are bonded to a nitrogen atom.

Imine (Section 16.8A) A compound containing a carbon-nitrogen double bond, $R_2C{=}NR'$; also called a Schiff base.

Index of hydrogen deficiency (Section 5.1B) The sum of the number of rings and π bonds in a molecule.

Inductive effect (Sections 4.5D and 6.3A) The polarization of the electron density of a covalent bond caused by the electronegativity of a nearby atom.

Infrared active (Section 12.3B) Any molecular vibration that leads to a substantial change in dipole moment and is observed in an IR spectrum.

Infrared (IR) spectroscopy (Section 12.3A) A spectroscopic technique in which a compound is irradiated with infrared radiation, absorption of which causes covalent bonds to change from a lower vibration state to a higher one. Infrared spectroscopy is particularly valuable for determining the kinds of functional groups present in a molecule.

Ionization potential (IP) (Section 14.1) The minimum energy required to remove an electron from an atom or molecule to a distance where there is no electrostatic interaction between the resulting ion and electron.

Isoelectric point (pI) (Section 27.2C) The pH at which an amino acid, polypeptide, or protein has no net charge.

Isomers (Section 1.2C) Different compounds with the same molecular formula.

Isotactic polymer (Section 29.6C) A polymer with identical configurations (either all *R* or all *S*) at all chiral centers along its chain, as, for example, isotactic polypropylene.

Keto-enol tautomerism (Section 7.7A) A type of isomerism involving keto (from ketone) and enol tautomers.

Ketone (Section 1.3C) A compound containing a carbonyl group bonded to two carbons.

Ketose (Section 25.1A) A monosaccharide containing a ketone group.

Kinetic control (Section 19.8A) Experimental conditions under which the composition of the product mixture is determined by the relative rates of formation of each product.

Lactam (Section 18.1D) A cyclic amide.

Lactone (Section 18.1C) A cyclic ester.

Leaving group (Introduction, Chapter 9) The group that is displaced in a substitution reaction or the Lewis base that is lost in an elimination reaction.

Levorotatory (Section 3.7B) Refers to a substance that rotates the plane of polarized light to the left.

Lewis acid (Section 4.7) Any molecule or ion that can form a new covalent bond by accepting a pair of electrons.

Lewis base (Section 4.7) Any molecule or ion that can form a new covalent bond by donating a pair of electrons.

Lewis dot structure (Section 1.1C) The symbol of an element surrounded by a number of dots equal to the number of electrons in the valence shell of the atom.

Ligand (Section 24.2) A Lewis base bonded to a metal atom in a coordination compound. It may bond strongly or weakly.

Lindlar catalyst (Section 7.8A) Finely powdered palladium metal deposited on solid calcium carbonate that has been specially modified with lead salts. Its particular use is as a catalyst for the reduction of an alkyne to a *cis* alkene.

Line-angle formula (Section 2.1) An abbreviated way to draw structural formulas in which vertices and line endings represent carbons.

Lipid (Introduction, Chapter 26) A biomolecule isolated from plant or animal sources by extraction with nonpolar organic solvents, such as diethyl ether and hexane.

Lipid bilayer (Section 26.5B) A back-to-back arrangement of phospholipid monolayers, often forming a closed vesicle or membrane.

Living polymer (Section 29.6D) A polymer chain that continues to grow without chain-termination steps until either all of the monomer is consumed or some external agent is added to terminate the chain. The polymer chains will continue to grow if more monomer is added.

Low-density lipoprotein (LDL) (Section 26.4A) Plasma particles, density 1.02–1.06 g/mL, consisting of approximately 26% proteins, 50% cholesterol, 21% phospholipids, and 4% triglycerides.

Low-resolution mass spectrometry (Section 14.2A) Instrumentation that is capable of separating only ions that differ in mass by 1 or more amu.

Markovnikov's rule (Section 6.3A) In the addition of HX, H_2O, or ROH to an alkene, hydrogen adds to the carbon of the double bond having the greater number of hydrogens.

Mass spectrometry (Introduction Chapter 14) An analytical technique for measuring the mass-to-charge ratio (*m/z*) of ions.

Mass spectrum (Section 14.1) A plot of the relative abundance of ions versus their mass-to-charge ratio.

Melt transition (T_m) (Section 29.4) The temperature at which crystalline regions of a polymer melt.

Mercaptan (Section 10.9B) A common name for a thiol; that is, any compound that contains an –SH (sulfhydryl) group.

Meso compound (Section 3.4B) An achiral compound possessing two or more chiral centers that also has chiral isomers.

Messenger RNA (mRNA) (Section 28.3C) A ribonucleic acid that carries coded genetic information from DNA to the ribosomes for the synthesis of proteins.

Meta (*m*) (Section 21.3B) Refers to groups occupying 1,3-positions on a benzene ring.

Methylene (Section 2.1, Section 5.2B) A —CH_2— group.

Micelle (Section 26.2B) A spherical arrangement of organic molecules in water solution clustered so that their hydrophobic parts are buried inside the sphere and their hydrophilic parts are on the surface of the sphere and in contact with water.

Molar absorptivity (ε) (Section 20.3A) The absorbance of a 1 *M* solution of a compound.

Molecular dipole moment (μ) (Section 1.5) The vector sum of individual bond dipoles.

Molecular ion (M^+) (Section 14.1) The radical cation formed by removal of a single electron from a parent molecule in a mass spectrometer.

Molecular orbital (MO) theory (Section 1.7A) A theory of chemical bonding in which electrons in molecules occupy molecular orbitals that extend over the entire molecule and are formed by the combination of the atomic orbitals that make up the molecule.

Molecular spectroscopy (Section 12.2) The study of which frequencies of radiation are absorbed or emitted by a particular substance and the correlation of these frequencies with details of molecular structure.

Monomer (Section 29.1) From the Greek, *mono* + *meros,* meaning single part. The simplest nonredundant unit from which a polymer is synthesized.

Monosaccharide (Section 25.1A) A carbohydrate that cannot be hydrolyzed to a simpler carbohydrate.

D-Monosaccharide (Section 25.1C) A monosaccharide that, when written as a Fischer projection, has the —OH on its penultimate carbon to the right.

L-Monosaccharide (Section 25.1C) A monosaccharide that, when written as a Fischer projection, has the —OH on its penultimate carbon to the left.

Mutarotation (Section 25.2C) The change in specific rotation that occurs when an α or β hemiacetal form of a carbohydrate in aqueous solution is converted to an equilibrium mixture of the two forms.

($n + 1$) rule (Section 13.8) If a hydrogen has *n* hydrogens nonequivalent to it but equivalent among themselves on the same or adjacent atom(s), its ^{1}H-NMR signal is split into ($n + 1$) peaks.

Newman projection (Section 2.5A) A way to view a molecule by looking along a carbon-carbon single bond.

Nitrile (Section 18.1E) A compound containing a —C≡N (cyano) group bonded to a carbon atom.

Nitrogen rule (Section 14.3) A rule stating that the molecular ion of a compound with an odd number of nitrogen atoms has an odd *m/z* ratio; if zero or an even number of nitrogen atoms, the molecular ion has an even *m/z* ratio.

Node (Section 1.6A) A point in space where the value of a wave function is zero.

Nonbonding electrons (Section 1.2C) Valence electrons not involved in forming covalent bonds. Also called unshared pairs or lone pairs.

Nonpolar covalent bond (Section 1.2B) A covalent bond between atoms whose difference in electronegativity is less than approximately 0.5.

Nuclear magnetic resonance (NMR) spectroscopy (Introduction, Chapter 13) A spectroscopic technique that gives information about the number and types of atoms in a molecule, for example, hydrogens (^{1}H—NMR) and carbons (^{13}C—NMR).

Nucleic acid (Introduction, Chapter 28) A biopolymer containing three types of monomer units: heterocyclic aromatic amine bases derived from purine and pyrimidine, the monosaccharides D-ribose or 2-deoxy-D-ribose, and phosphoric acid.

Nucleophile (Section 6.3A and Introduction to Chapter 9) From the Greek meaning nucleus-loving. Any species that can donate a pair of electrons to form a new covalent bond; alternatively, a Lewis base.

Nucleophilic acyl substitution (Section 18.3) A reaction in which a nucleophile bonded to the carbon of an acyl group is replaced by another nucleophile.

Nucleophilic aromatic substitution (Section 22.3) A reaction in which a nucleophile, most commonly a halogen, on an aromatic ring is replaced by another nucleophile.

Nucleophilic substitution (Introduction, Chapter 9) Any reaction in which one nucleophile is substituted for another at a tetravalent carbon atom.

Nucleophilicity (Section 9.3E) A kinetic property measured by the rate at which a nucleophile causes nucleophilic substitution on a reference compound under a standardized set of experimental conditions.

Nucleoside (Section 28.1) A building block of nucleic acids, consisting of D-ribose or 2-deoxy-D-ribose bonded to a heterocyclic aromatic amine base by a β-*N*-glycosidic bond.

Nucleotide (Section 28.1) A nucleoside in which a molecule of phosphoric acid is esterified with an —OH of the monosaccharide, most commonly either the 3′-OH or the 5′-OH.

Observed rotation (Section 3.7B) the number of degrees through which a compound rotates the plane of polarized light.

Octet rule (Section 1.2) Group 1A–7A elements react to achieve an outer shell of eight valence electrons.

Oil (Section 26.1B) When used in the context of fats and oils, a mixture of triglycerides that is liquid at room temperature.

Oligosaccharide (Section 25.4) A carbohydrate containing four to ten monosaccharide units, each joined to the next by a glycosidic bond.

Optical purity (Section 3.7D) The specific rotation of a mixture of enantiomers divided by the specific rotation of the enantiomerically pure substance (expressed as a percent). Optical purity is numerically equal to enantiomeric excess, but experimentally determined.

Optically active (Section 3.7) Refers to a compound that rotates the plane of plane-polarized light.

Orbital (Section 1.1) A region of space that can hold two electrons.

Order of precedence of functions (Section 16.2B) A ranking of functional groups in order of priority for the purposes of IUPAC nomenclature.

Organic synthesis (Section 7.9) A series of reactions by which a set of organic starting materials is converted to a more complicated structure.

Organometallic compound (Introduction, Chapter 15) A compound that contains a carbon-metal bond.

Ortho (*o*) (Section 21.3B) Refers to groups occupying 1,2-positions on a benzene ring.

Orthogonal (Section 1.1) Having no net overlap.

Oxidation (Section 6.5A) The loss of electrons. Alternatively, either the loss of hydrogens, the gain of oxygens, or both.

Oxidative addition (Section 24.2) Addition of a reagent to a metal center causing it to add two substituents and to increase its oxidation state by two.

Oxonium ion (Section 6.3B) An ion in which oxygen bears a positive charge.

Oxymercuration-reduction (Section 6.3F) A method for converting an alkene to an alcohol. The alkene is treated with mercury(II) acetate followed by reduction with sodium borohydride.

Para (*p*) (Section 21.3B) Refers to groups occupying 1,4-positions on a benzene ring.

Part per million (ppm) (Section 13.3) Units used on NMR spectra to record chemical shift relative to the TMS standard.

Pauli exclusion principle (Section 1.1A) No more than two electrons may be present in an orbital. If two electrons are present, their spins must be paired.

Peptide bond (Section 27.3) The special name given to the amide bond formed between the α-amino group of one amino acid and the α-carboxyl group of another amino acid.

Pericyclic reaction (Section 24.6F) A reaction that takes place in a single step, without intermediates, and involves a cyclic redistribution of bonding electrons.

Phasing (Section 1.6A) The sign of the wave function at particular coordinates in space, either plus or minus. Phasing is often represented by colors, such as red or blue.

Phenol (Section 21.4A) A compound that contains an —OH bonded to a benzene ring; a benzenol.

Phenyl group (Introduction, Chapter 5) A group derived by removing an H from benzene; abbreviated C_6H_5— or Ph—.

Phospholipid (Section 26.5A) A lipid containing glycerol esterified with two molecules of fatty acid and one molecule of phosphoric acid.

Photodynamic therapy (Section 24.7B) Biological damage caused by photosensitizers, light, and oxygen, used to kill tumor and other cells.

Photolysis (Section 15.3A) Cleavage by light.

Photons (Section 12.1) An alternative way to describe electromagnetic radiation as a stream of particles.

Photosensitizer (Section 24.7B) A compound that absorbs light and transfers the energy to another molecule.

Pi (π) bond (Section 1.7B) A covalent bond formed by the overlap of parallel *2p* orbitals.

Pi (π) molecular orbital (Section 1.8A) A molecular orbital formed by overlapping parallel *2p* orbitals on adjacent atoms; its electron density lies above and below the line connecting the atoms.

Plane of symmetry (Section 3.2) An imaginary plane passing through an object dividing it so that one half is the mirror image of the other half.

Plane-polarized light (Section 3.7A) Light oscillating in only parallel planes.

Plastic (Section 29.1) A polymer that can be molded when hot and retains its shape when cooled.

β-Pleated sheet (Section 27.6B) A type of polypeptide secondary structure in which sections of polypeptide chains are aligned parallel or antiparallel to one another.

Polar covalent bond (Section 1.2B) A covalent bond between atoms whose difference in electronegativity is between approximately 0.5 and 1.9.

Polarimeter (Section 3.7B) An instrument for measuring the ability of a compound to rotate the plane of plane-polarized light.

Polarizability (Section 8.3B) A measure of the ease of distortion of the distribution of electron density about an atom or group in response to interaction with other molecules or ions. Fluorine, which has a high electronegativity and holds its electrons tightly, has a very low polarizability. Iodine, which has a lower electronegativity and holds its electrons less tightly, has a very high polarizability.

Polyamide (Section 29.5A) A polymer in which each monomer unit is joined to the next by an amide bond, as, for example, nylon 66.

Polycarbonate (Section 29.5C) A polyester in which the carboxyl groups are derived from carbonic acid.

Polyester (Section 29.5B) A polymer in which each monomer unit is joined to the next by an ester bond, as, for example, poly(ethylene terephthalate).

Polymer (Section 29.1) From the Greek, *poly* + *meros*, meaning many parts. Any long-chain molecule synthesized by linking together many single parts called monomers.

Polynuclear aromatic hydrocarbon (PAH) (Section 21.3C) A hydrocarbon containing two or more fused benzene rings.

Polypeptide (Section 27.3) A macromolecule containing many amino acid units, each joined to the next by a peptide bond.

Polysaccharide (Section 25.4) A carbohydrate containing a large number of monosaccharide units, each joined to the next by one or more glycosidic bonds.

Polyunsaturated fatty acid (Section 26.1A) A fatty acid with two or more carbon-carbon double bonds in its hydrocarbon chain.

Polyunsaturated triglyceride (Section 26.1B) A triglyceride having several carbon-carbon double bonds in the hydrocarbon chains of its three fatty acids.

Polyurethane (Section 29.5D) A polymer containing the $—NHCO_2—$ group as a repeating unit.

Potential energy (Section 1.1B) The energy that can be released if given an opportunity.

Primary (1°) amine (Section 1.3B) An amine in which nitrogen is bonded to one carbon and two hydrogens.

Primary structure of nucleic acids (Section 28.2A) The sequence of bases along the pentose-phosphodiester backbone of a DNA or RNA molecule read from the 5′end to the 3′end.

Primary structure of proteins (Section 27.4) The sequence of amino acids in the polypeptide chain, read from the *N*-terminal amino acid to the *C*-terminal amino acid.

Principle of microscopic reversibility (Section 10.6) This principle states that the sequence of transition states and reactive intermediates in the mechanism of any reversible reaction must be the same, but in reverse order, for the reverse reaction as for the forward reaction.

Prochiral hydrogens (Section 13.10) Refers to two hydrogens bonded to a carbon atom. When a different atom replaces one or the other, the carbon becomes a chiral center. The hydrogens of the CH_2 group of ethanol, for example, are prochiral. Replacing one of them by deuterium gives (*R*)-1-deuteroethanol; replacing the other gives (*S*)-1-deuteroethanol.

Pro-*R*-hydrogen (Section 13.10) Replacing this hydrogen by deuterium gives a chiral center with an *R* configuration.

Pro-*S*-hydrogen (Section 13.10) Replacing this hydrogen by deuterium gives a chiral center with an *S* configuration.

Prostaglandin (Section 26.3) A member of the family of compounds having the 20-carbon skeleton of prostanoic acid.

Protecting group (Section 11.6) Reversibly creating an unreactive group for the purpose of preventing a functional group from potentially reacting to give an unwanted product or products.

Protic acid (Section 4.7) An acid that is a proton donor in an acid-base reaction.

Protic solvent (Section 9.3D) A solvent that is a hydrogen-bond donor. Common protic solvents are water, low-molecular-weight alcohols, and low-molecular weight carboxylic acids.

Pyranose (Section 25.2A) A six-membered cyclic form of a monosaccharide.

Quantum mechanics (Section 1.6A) The branch of science that studies the interaction of matter and radiation.

Quantized (Section 1.1) Having specific values for energy and momentum.
Quaternary (4°) ammonium ion (Section 23.2B) An ion in which nitrogen is bonded to four carbons and bears a positive charge.
Quaternary structure (Section 27.6D) The arrangement of polypeptide monomers into a noncovalently bonded aggregate.

R (Section 3.3) From the Latin, *rectus*, straight, correct; used in the *R,S* convention to show that the order of priority of groups on a chiral center is clockwise.
***R,S* System** (Section 3.3) A set of rules for specifying absolute configuration about a chiral center; also called the Cahn-Ingold-Prelog system.
Racemic mixture (Section 3.7C) A mixture of equal amounts of two enantiomers.
Radical (Section 8.3D) Any chemical species that contains one or more unpaired electrons.
Radical cation (Section 14.1) A species formed when a neutral molecule loses one electron; it contains both an odd number of electrons and a positive charge.
Radical inhibitor (Section 8.7) A compound such as a phenol that selectively reacts with radicals to remove them from a chain reaction and terminate the chain.
Raman spectroscopy (Section 12.3B) A vibrational molecular spectroscopy that is complementary to infrared (IR) spectroscopy in that infrared inactive vibrations are seen in Raman spectroscopy.
Rate determining step (Section 6.2C) The step in a multistep reaction sequence that crosses the highest energy barrier.
Reaction coordinate diagram (Section 4.5B) A graph showing the energy changes that occur during a chemical reaction; energy is plotted on the vertical axis and reaction progress is plotted on the horizontal axis.
Reaction mechanism (Section 4.5A and Section 6.2) A step-by-step description of how a chemical reaction occurs.
Reactive intermediate (Section 6.2A) A high-energy species formed between two successive reaction steps, that lies in an energy minimum between the two transition states.
Rearrangement (Section 6.3C) A change in connectivity of the atoms in a product compared with the connectivity of the same atoms in the starting material.
Reducing sugar (Section 25.3C) A carbohydrate that reacts with an oxidizing agent to form an aldonic acid. In this reaction, the carbohydrate reduces the oxidizing agent.
Reduction (Section 6.5A) The gain of electrons. Alternatively, either the gain of hydrogen, loss of oxygen, or both.
Reductive amination (Section 16.11C) A method for preparing substituted amines by treating an aldehyde or ketone with an amine in the presence of a reducing agent.
Reductive elimination (Section 24.2) Elimination of two substituents at a metal center, causing the oxidation state of the metal to decrease by two.
Regioselective reaction (Section 6.3) An addition or substitution reaction in which one of two or more possible products is formed in preference to all others that might be formed.
Resolution (Sections 3.9 and 14.2A) Separation of a racemic mixture into its enantiomers; in mass spectrometry, a measure of how well a mass spectrometer separates ions of different mass.
Resonance (Section 1.8A) A theory that many molecules and ions are best described as a hybrid of several Lewis structures.
Resonance energy (Section 21.1C) The difference in energy between a resonance hybrid and the most stable of its hypothetical contributing structures in which electrons are localized on particular atoms and in particular bonds.
Resonance hybrid (Section 1.6A) A molecule, ion, or radical described as a composite of a number of contributing structures.
Resonance in NMR spectroscopy (Section 13.3) The absorption of electromagnetic radiation by a precessing nucleus and the resulting "flip" of its nuclear spin from the lower energy state to the higher energy state.
Restriction endonuclease (Section 28.5A) An enzyme that catalyzes the hydrolysis of a particular phosphodiester bond within a DNA strand.
Retrosynthesis (Section 7.9) A process of reasoning backwards from a target molecule to a suitable set of starting materials.
Ribosomal RNA (rRNA) (Section 28.3A) A ribonucleic acid found in ribosomes, the sites of protein synthesis.
Ring current (Section 13.7C) An applied magnetic field causes the π electrons of an aromatic ring to circulate, giving rise to the so-called ring current and an associated magnetic field that opposes the applied field in the middle of the ring but reinforces the applied field on the outside of the ring.

S (Section 3.3) From the Latin, *sinister*, left; used in the *R,S* convention to show that the order of priority of groups on a chiral center is counterclockwise.
Sanger dideoxy method (Section 28.5B and 28.5D) A method developed by Frederick Sanger for sequencing DNA molecules.
Saponification (Section 18.4C) Hydrolysis of an ester in aqueous NaOH or KOH to an alcohol and the sodium or potassium salts of carboxylic acids.
Schiff base (Section 16.8A) An alternative name for an imine.
Secondary (2°) amine (Section 1.3B) An amine in which nitrogen is bonded to two carbons and one hydrogen.
Secondary structure of nucleic acids (Section 28.2B) The ordered arrangement of nucleic acid strands.
Secondary structure of proteins (Section 27.6B) The ordered arrangements (conformations) of amino acids in localized regions of a polypeptide or protein.
Shell (Section 1.1) A region of space around a nucleus that can be occupied by electrons, corresponding to a principal quantum number.
Shielding in NMR (Section 13.3) Also called diamagnetic shielding; the term refers to the reduction in magnetic field strength experienced by a nucleus underneath electron density induced to circulate when the molecule is placed in a strong magnetic field.
Sigma (σ) molecular orbital (Section 1.7A) A molecular orbital in which electron density is concentrated between two nuclei, along the axis joining them, and is cylindrically symmetrical.
Signal (Section 13.3) A recording in an NMR spectrum of a nuclear magnetic resonance.
Signal splitting in NMR (Section 13.8) Spin-spin coupling with adjacent nuclei split NMR signals depending on the extent of coupling and the number of adjacent equivalent nuclei.
S_N1 reaction (Section 9.2) A unimolecular nucleophilic substitution reaction.
S_N2 reaction (Section 9.2) A bimolecular nucleophilic substitution reaction.
Soap (Section 26.2A) A sodium or potassium salt of a fatty acid.
Solvolysis (Section 9.2) A nucleophilic substitution in which the solvent is also the nucleophile.
***sp* Hybrid orbital** (Section 1.7B) A hybrid atomic orbital formed by the combination of one *s* atomic orbital and one *2p* atomic orbital.
sp^2 Hybrid orbital (Section 1.7B) A hybrid atomic orbital formed by the combination of one *s* atomic orbital and two *2p* atomic orbitals.
sp^3 Hybrid orbital (Section 1.7 B) A hybrid atomic orbital formed by the combination of one *s* atomic orbital and three *2p* atomic orbitals.
Specific rotation (Section 3.7B) The observed rotation of the plane of polarized light when a sample is placed in a tube 1.0 dm in

length and at a concentration of 1 g/mL for a solution. For a pure liquid, concentration is expressed in g/mL (density).

Spin-spin coupling (Section 13.9) An interaction in which nuclear spins of adjacent atoms influence each other and lead to the spitting of NMR signals.

Staggered conformation (Section 2.5A) A conformation about a carbon-carbon single bond in which the atoms or groups on one carbon are as far apart as possible from atoms or groups on an adjacent carbon.

Step-growth polymerization (Section 29.5) A polymerization in which chain growth occurs in a stepwise manner between difunctional monomers as, for example, between adipic acid and hexamethylenediamine to form nylon 66. Also called condensation polymerization.

Stereocenter (Sections 2.6A and 3.2) An atom, most commonly carbon, about which exchange of two groups produces a stereoisomer. Chiral centers are one type of stereocenter.

Stereochemistry (Section 3.1) The study of three-dimensional arrangements of atoms in molecules.

Stereoisomers (Sections 2.6 and 3.1) Isomers that have the same molecular formula and the same connectivity of their atoms but a different orientation of their atoms in space.

Stereoselective reaction (Section 6.3D) A reaction in which one stereoisomer is formed in preference to all others. A stereoselective reaction may be enantioselective or diastereoselective, as the case may be.

Stereospecific reaction (Section 6.7A) A special type of stereoselective reaction in which the stereochemistry of the product is dependent on the stereochemistry of the starting material.

Steric hindrance (Section 9.43B) The ability of groups, because of their size, to hinder access to a reaction site within a molecule.

Steric strain (Section 2.6A) The strain that arises when nonbonded atoms separated by four or more bonds are forced closer to each other than their atomic (contact) radii would allow. Steric strain is also called non-bonded interaction strain, or van der Waals strain.

Steroid (Section 26.4) A plant or animal lipid having the characteristic tetracyclic ring structure of the steroid nucleus, namely three six-membered rings and one five-membered ring.

Substitution (Section 8.4) A reaction in which an atom or group of atoms in a compound is replaced by another atom or group of atoms.

Sulfide (Section 11.12) The sulfur analog of an ether; a molecule containing a sulfur atom bonded to two carbon atoms. Sulfides are also called thioethers.

Syndiotactic polymer (Section 29.6C) A polymer with alternating *R* and *S* configurations at the chiral centers along its chain, as, for example, syndiotactic polypropylene.

Syn stereoselective (Section 6.4) The addition of atoms or groups of atoms to the same face of a carbon-carbon double bond.

Tautomers (Section 7.7A) Constitutional isomers in equilibrium with each other that differ in the location of a hydrogen atom and a double bond relative to a heteroatom, most commonly O, N, or S.

Telechelic polymer (Section 29.6D) A polymer in which its growing chains are terminated by formation of new functional groups at both ends of its chains. These new functional groups are introduced by adding reagents, such as CO_2 or ethylene oxide, to the growing chains.

***C*-Terminal amino acid** (Section 27.3) The amino acid at the end of a polypeptide chain having the free —COOH group.

***N*-Terminal amino acid** (Section 27.3) The amino acid at the end of a polypeptide chain having the free $—NH_2$ group.

Terpene (Section 5.4) A compound whose carbon skeleton can be divided into two or more units identical with the carbon skeleton of isoprene.

Tertiary (3°) amine (Section 1.3B) An amine in which nitrogen is bonded to three carbons.

Tertiary structure of nucleic acids (Section 28.2C) The three-dimensional arrangement of all atoms of a nucleic acid, commonly referred to as supercoiling.

Tertiary structure of proteins (Section 27.6C) The three-dimensional arrangement in space of all atoms in a single polypeptide chain.

Tesla (T) (Section 13.2) The SI unit for magnetic field strength.

Thermochemistry (Section 4.5) The study of the energy of chemical structures.

Thermodynamic control (Section 19.8A) Experimental conditions that permit the establishment of equilibrium between two or more products of a reaction. The composition of the product mixture is determined by the relative stabilities of the products.

Thermolysis (Section 15.3A) Cleavage by heating.

Thermoplastic (Section 29.1) A polymer that can be melted and molded into a shape that is retained when it is cooled.

Thermoset plastic (Section 29.1) A polymer that can be molded when it is first prepared, but once cooled, hardens irreversibly and cannot be remelted.

Thiol (Section 10.9A) A compound containing an —SH (sulfhydryl) group bonded to an sp^3-hybridized carbon.

Tollens' reagent (Section 16.10A) A solution prepared by dissolving Ag_2O in aqueous ammonia; used for selective oxidation of an aldehyde to a carboxylic acid.

Torsional strain (Section 2.5A) Strain that arises when nonbonded atoms separated by three bonds are forced from a staggered conformation to an eclipsed conformation. Torsional strain is also called eclipsed-interaction strain.

Trans (Section 2.6A) A prefix meaning across from.

Transesterification (Section 18.5C) Exchange of the —OR or —OAr group of an ester for another —OR or —OAr group.

Transfer RNA (tRNA) (Section 28.3B) A ribonucleic acid that carries a specific amino acid to the site of protein synthesis on ribosomes.

Transition state (Section 4.5A) The highest energy point on a reaction coordinate diagram. The chemical structure at this point is commonly called an activated complex.

Triglyceride (triacylglycerol) (Section 26.1) An ester of glycerol with three fatty acids.

Triol (Section 10.1B) A compound containing three hydroxyl groups.

Tripeptide (Section 27.3) A molecule containing three amino acid units, each joined to the next by a peptide bond.

Twist-boat conformation (Section 2.6B) A nonplanar conformation of a cyclohexane ring that is twisted from and slightly more stable than a boat conformation.

Unimolecular reaction (Section 9.2) A reaction in which only one species is involved in the rate-determining step.

Unsaturated hydrocarbon (Introduction, Chapter 5) A hydrocarbon containing one or more carbon-carbon double or triple bonds. The three classes of unsaturated hydrocarbons are alkenes, alkynes, and arenes.

Upfield (Section 13.4) A signal of an NMR spectrum that is shifted toward the right (smaller chemical shift) on the chart paper.

Valence bond theory (Section 1.7C) A model of bonding that places electron pairs between adjacent atoms to create bonds.

Valence electrons (Section 1.1C) Electrons in the valence (outermost) shell of an atom.

Valence shell (Section 1.1C) The outermost occupied electron shell of an atom.

Valence-shell electron-pair repulsion (VSEPR)(Section 1.4) A method for predicting bond angles based on the idea that electron pairs repel each other and keep as far apart as possible.

van der Waals forces (Section 8.3B) A group of intermolecular attractive forces including dipole-dipole, dipole-induced dipole, and induced dipole-induced dipole (dispersion) forces.

van der Waals radius (Section 8.3B) The minimum distance of approach to an atom that does not cause nonbonded interaction strain.

Vibrational infrared region (Section 12.3A) The portion of the infrared region that extends from 4000 to 400 cm^{-1}.

Vicinal coupling (Section 13.9B) A common type of spin-spin coupling involving the H atoms on two C atoms that are bonded to each other.

Vinyl group (Section 5.2B) a $—CH{=}CH_2$ group.

Vinylic carbocation (Section 7.6B) A carbocation in which the positive charge is on one of the carbons of a carbon-carbon double bond.

Watson-Crick model (Section 28.2B) A double-helix model for the secondary structure of a DNA molecule.

Wave function (Section 1.6A) A solution to a set of equations that defines the energy of an electron in an atom and the region of space it may occupy.

Wavelength (λ)(Section 12.1) The distance between consecutive peaks on a wave.

Wavenumbers ($\bar{\nu}$) (Section 12.3A) The frequency of electromagnetic radiation expressed as the number of waves per centimeter, with units cm^{-1} (read: reciprocal centimeters).

Williamson ether synthesis (Section 11.4A) A general method for the synthesis of dialkyl ethers by an S_N2 reaction between a haloalkane and an alkoxide ion.

Wolff-Kishner reduction (Section 16.11E) Reduction of the $C{=}O$ group of an aldehyde or ketone to a CH_2 group using hydrazine and a base.

Ylide (Section 16.6) A neutral molecule with positive and negative charges on adjacent atoms.

Z (Section 5.2C) From the German, *zusammen,* meaning opposite. Specifies that groups of higher priority on the carbons of a double bond are on the same side.

Zaitsev's rule (Section 9.5) A rule stating that the major product of a β-elimination reaction is the most stable alkene; that is, it is the alkene with the greatest number of substituents on the carbon-carbon double bond.

Zwitterion (Section 27.1A) An internal salt of an amino acid; the carboxylate is negatively charged, and the ammonium group is positively charged.

Index

Key: **Boldface** indicates a glossary entry; *italics* indicates a figure; *t* indicates a table.

1°, **19**
1° alcohol, 439
1° amine, **21,** 1038, 1071
1° carbocation, 247, 249
1*s* orbital, 2, 31, *31*
1,2-addition, 859, 893, 898–900
1,2 shift, **254,** 256
1,2-substitution, 973
1,3-substitution, 973
1,4-addition, 859, 893, 898–900
1,4-substitution, 973
2°, **19**
2° alcohol, 439
2° amine, **21,** 1038, 1071
2° carbocation, 177
2^n rule, 139, 146
$2n^2$ electrons, 2
$2p_x$ orbital, 31, *32*
$2p_y$ orbital, 31, *32*
$2p_z$ orbital, 31, *32*
2*s* orbital, 31, *31*
[2 + 2] cycloaddition, 909, *910*, 911, 920, 928
2,4-D, 1028
3°, **19**
3° alcohol, 439
3° amine, **21,** 504, 1038, 1071
3° carbocation, 177
[3,3]-shift, 920
4° ammonium ion, **1042,** 1043
$(4n + 2)\ \pi$ electron rule, 944, 1049
[4 + 2] cycloaddition, 909, *910*, 911, 920, 928
[4 + 4] cycloaddition, 920

A

A. *See* Adenine (A)
A. See Absorbance *(A)*
$(A\text{-}A\text{-}B\text{-}B)_n$ polymers, 1268
$(A\text{-}B)_n$ polymers, 1268
A-DNA, 1246
ABO blood group system, 1153–1154
Absence seizures, 879
Absinthe, 226
Absolute configuration, **134,** A-8
Absorbance *(A)*, 905
Acesulfame-K, 1152*t*
Acetaldehyde
  aldehyde, 22
  boiling point, 658*t*
  Lewis structure, 653
  nomenclature, 656
  oxidation of alcohols, 471
  solubility, 658*t*
  structural formula, 658*t*
Acetaldehyde cyanohydrin, 664, 665
Acetals
  acid-catalyzed formation, 674
  carbonyl-protecting groups, as, 676–677
  Dean-Stark trap, 675, *675*
  defined, **673**
  example, 676
  formation, 673, 675, 705
  functional group, 674
  glycosides, 1143–1144
  hydrolysis, 675
  key reaction, 705
  stability, 673
Acetamide, 175, 766, 768, 789
Acetanilide, 1023
Acetate ion, 52, *53*
Acetate rayon, 1156
Acetic acid
  acidity, 733
  carboxylic acid, 22
  common name, 656, 728*t*
  industrial preparation, by carbonylation of methanol, 748
  industrial synthesis, 737
  IUPAC system, 726
  physical properties, 730*t*
  pK_a, 191, 445*t*, 732
  protic solvent, 389*t*
Acetic acid-*d*, 699
Acetic anhydride, 763, 789, A-15
Acetic benzoic anhydride, 790
Acetic-d_3 acid-*d*, 699
Aceto group, **728**
Acetoacetic acid, 728, 745
Acetoacetic ester synthesis, 845–849, 871, 1032
Acetone
  chemical structure, 390*t*
  conjugate base, 205
  dielectric constant, 390*t*
  electrostatic potential map, 49, 657
  ^{1}H-NMR spectrum, 566
  IR spectroscopy, 542
  IUPAC system, 654
  ketone, 22
  Lewis structure, 653
  physical properties, 658*t*
  synthesis, 367

Acetone-d_6, 699
Acetonitrile, 390*t*, 598, 768, 1100
Acetophenone, 654, 700, 908
Acetyl chloride, 740, 763, 790
Acetyl-CoA, 166, 1184, *1184*, 1185
N-Acetyl-D-glucosamine, 1139
Acetyl group, 728
Acetylcholine, 812
Acetylcholine chloride, 1083
2-Acetylcyclo-hexanone, 843
Acetylene
  ball-and-stick model, *298*
  bond length/bond strength, 56*t*
  covalent bonding of carbon, 44*t*
  Lewis structure, 13*t*, *44*, *298*
  shape, *25*
  side view/end view, 207
  space-filling model, *298*
Acetylenic hydrogen, 572
Acetylide anions, 301–302
Acetylide salt, 318
N-Acetylneuraminic acid, 1168
Achiral, **128**
Achiral environment, 140
Achiral molecules, 152
Acid anhydrides. *See also* Carboxylic acid derivatives
  alcohols, 787, 805
  ammonia and amines, 789, 805
  carboxylic anhydrides, 763–764
  cyclic anhydrides, 764
  hydrolysis, 774–775, 803
  infrared absorptions, 551*t*
  IUPAC system, 763–764
  phosphoric anhydrides, 764
  used in place of acid chlorides, 822
Acid-base equilibria, 180–183, 193
Acid catalyst, 499
Acid-catalyzed addition of alcohols to alkenes, 499
Acid-catalyzed aldol reaction, 826–827
Acid-catalyzed α-halogenation of ketone, 700
Acid-catalyzed cleavage of dialkyl ether, 500
Acid-catalyzed cleavage of ethers by concentrated HX, 500–501
Acid-catalyzed dehydration
  alcohols, 454–459, 497–498
  aldol product, 828–829, 868
Acid-catalyzed equilibration of keto and enol tautomers, 686
Acid-catalyzed ester hydrolysis, 775–776
Acid-catalyzed ether synthesis, 498
Acid-catalyzed esterification, 739
Acid-catalyzed hydration
  alkenes, 252–254, 256–257, 284
  alkynes, 310, 321
  propene, 253
Acid-catalyzed ring opening, 510–512
Acid chlorides. *See also* Acid halides
  ammonia and amines, 788–789, 805
  carboxylic acid salts, 790, 806
  conversion of carboxylic acids to acid chlorides, 740–742, 749
  hydrolysis, 774, 803
  lithium diorganocuprates, 795
Acid dissociation constant (K_a), **178**
Acid halides. *See also* Carboxylic acid derivatives
  acid chlorides. *See* Acid chlorides
  alcohols, 786
  ammonia and amines, 788–789
  IUPAC system, 763
Acidic amino acids, 1263
Acidity. *See also* pK_a
  alcohols, 445–446, 478
  alkynes, 192*t*, 301
  α-hydrogens, 684–685
  amides, 768
  amines, 1046*t*
  amino acids, 1203*t*
  carboxylic acid derivatives, 768
  carboxylic acids, 732–736
  phenols, 957–960, 974
  sulfonamides, 768, 802
  terminal alkynes, 318
  thiols, 476, 483
Acids and bases, 170–205
  acid-base equilibria, 180–183, 193
  acid dissociation constant (K_a), 178
  Arrhenius acids and bases, 170–171
  Brønsted-Lowry acids and bases, 171–177
  conjugate acid-base pair, 171–172
  enthalpy, entropy, Gibbs free energy, 185, 186, 187*t*
  Lewis acid-base reaction, 194–195, 199
  Lewis acids and bases, 193
  molecular structure and acidity, 188–192, 198
  π electrons, 176
  pK_a, 178–180
  position of equilibrium, 180–183
  proton-transfer reaction, 171, 177, 196
  reaction coordinate diagram, 184–185, *185*
  review/study guide, 196–199
  stronger/weaker acids/bases, 178–180
  thermal reactions, 184
  thermochemistry, 184–187
  transition state, 184, 185
Acrolein, 653
Acrylate plastics, 1276*t*
Acrylic acid, 726
Acrylic plastics, 1276*t*
Acrylonitrile, 665, 1276*t*
Actinomycin D, 1233
Activating group, **1006,** 1008*t*
Acyclovir, 1244, 1245
Acyl group, **763**
Acylation, **843**
Acylation of benzene, 996. *See also* Friedel-Crafts acylation
Acylation of enamines, 843
Acylbenzene, 996
Acylium ion, **1003**
Adalat, 882
Adamantane, 107
Addition/elimination, 1015–1016, 1021–1022
Addition reaction, 238, **241**
Adenine (A), *1144*, *1239*
3′-Adenine monophosphate, 40
Adenosine 5′monophosphate (AMP), *1240*
Adenosine 5′triphosphate (ATP), *1240*
Adhesives, 1273
Adipamide, 767, 1270
Adipic acid
  IUPAC system, 727
  model, 1264
  nylon 66, 1269
  oxidation by nitric acid, 690, 1270
  pK_a, 754
Adipoyl chloride, 763
Adrenaline, 1233
Adult-onset diabetes, 815
Aglycon, **1129**
AIBN. *See* Azoisobutyronitrile (AIBN)
Akylammonium ion, A-2
-al, 81, 655*t*
Ala, 1201*t*. *See also* Alanine
Alanine, 205, 728, 1200, 1201*t*, 1203*t*
D-Alanine, *1200*
L-Alanine, 728, *1200*
Albuterol, 992, 1078
Alcohol, **19**
Alcohol dehydrogenase, 471, 694, 1227*t*
Alcohol group, 19
Alcohols, 19, 437–491, A-2
  acid anhydrides, 787, 805
  acid-catalyzed dehydration, 454–459
  acid chlorides, 805
  acid halides, 786
  acidity/basicity, 445–446, 478
  amides, 787, 788
  aryl and alkyl sulfonates, 452–453
  blood alcohol screening, 466–467
  boiling points, 442*t*, 443–444
  carboxylic acid derivatives, 786–788
  chromic acid, 463–465
  common names, 439
  conversion of, to alkenes, 455
  conversion of, to haloalkanes and sulfonates, 447–454, 479–480
  Dess-Martin oxidation, 468–469
  diol reactions, 461–462
  esters, 787, 788, 805
  importance, 438
  infrared spectra, 545–546, 546*t*
  IUPAC system, 438, 439, A-13
  key reactions. *See* Alcohols—key reactions

mass spectra, 618
mechanisms. *See* Alcohols—reaction mechanisms
NMR spectra, 588–589
oxidation, 463–473
periodic acid oxidation of glycols, 469–471
physical properties, 441–444
pinacol rearrangement, 459–463
pK_a values, 445*t*
primary, **19**, 439
pyridinium chlorochromate (PCC), 465
reaction of, to active metals, 446–447
reaction of, with HCl, HBr, and HI, 447–450
reaction of, with phosphorus tribromide, 450
reaction of, with thionyl chloride and thionyl bromide, 451–452
reaction roadmap, 295–296
reaction via a tosylate, 453–454
review/study guide, 477–484
secondary, **19**, 439
solubilities, 442*t*, 444
Swern oxidation, 467–468
tertiary, **19**, 439
thiols, 473–477
unsaturated, 440
Alcohols—key reactions
acid-catalyzed dehydration, 480
acidity of alcohols, 478
oxidation of primary alcohol to aldehyde, 482
oxidation of primary alcohol to carboxylic acid, 481
oxidation of secondary alcohol to ketone, 481
oxidative cleavage of glycol, 482
pinacol rearrangement, 481
reaction with active metals, 479
reaction with HCl, HBr, and HI, 479
reaction with PBr_3, 479–480
reaction with $SOCl_2$ and $SOBr_2$, 480
Alcohols—reaction mechanisms
acid-catalyzed dehydration of 2-Butanol, 456
acid-catalyzed dehydration of unbranched primary alcohol, 457–458
chromic acid oxidation of alcohol, 464–465
Dess-Martin oxidation, 468–469
oxidation of alcohols by NAD^+, 471
oxidation of glycol by periodic acid, 469–470
pinacol rearrangement of 2,3-Dimethyl-2,3-butanediol, 460
reaction of 1° alcohol with HBr, 449
reaction of 3° alcohol with HBr, 448–449
reaction of primary alcohol with PBr_3, 451
rearrangement upon treatment of neopentyl alcohol with HCl, 449–450
Swern oxidation, 467–468
Aldehydes and ketones, 652–724
acetals, 673–677, 705
α-halogenation, 700–701, 709
ammonia and its derivatives, 678–681, 684*t*
anions of terminal alkynes, 663–664, 703
carbon nucleophiles, 659–668
carbonyl hydrates, 670
catalytic reduction, 692–693, 708
Clemmensen reduction, 696, 697, 708
common names, 656
cyanohydrins, 664–665
definitions, **21, 22**
deuterium exchange, 699, 709
enamines, 681, 706
enolate-forming reactions, 685
Grignard reagents, 660–662, 703
hemiacetals, 670–673, 705
hydrazine and related compounds, 682, 684*t*
hydrogen cyanide (HCN), 664–666, 703
imines, 678–681, 706
infrared absorptions, 551*t*
infrared spectra, 548–549
IUPAC system, 653–656, A-13
keto-enol tautomerism, 684–688
key reactions. *See* Aldehydes and ketones—key reactions
mass spectra, 618–619
McLafferty rearrangement of ketone, 619
mechanisms. *See* Aldehydes and ketones—reaction mechanisms
metal hydride reduction, 691–692, 708
nitrogen nucleophiles, 678–684
NMR spectra, 589
nucleophilic acyl addition, 658, 659
organolithium compounds, 662, 703
overview, 21–22
oxidation of aldehydes, 688–689
oxidation of ketones, 690–691
oxygen nucleophiles, 670–677
physical properties, 657, 658*t*
Pinnick oxidation, 689–690
racemization, 699
reaction roadmap, 649–650
reactions, 658–659, 702
reduction, 691–698
reduction of carbonyl group to methylene group, 696–698
reductive animation, 694–696, 708
review/study guide, 701–709
selective reduction, 693
structure and bonding, 653
tetrahydropyranyl (THP) ethers, 677–678
Wittig reaction, 666–669
Wolff-Kishner reduction, 697–698, 709
Aldehydes and ketones—key reactions
addition of alcohols to form acetals, 705
addition of alcohols to form hemiacetals, 705
addition of ammonia and derivatives—formation of imines, 706
addition of hydrazine and derivatives, 706
addition of secondary amines—formation of enamines, 706
catalytic reduction, 708
Clemmensen reduction, 708
deuterium exchange, 709
halogenation at α-carbon, 709
Horner-Emmons-Wadsworth modification, 704
hydration, 705
keto-enol tautomerism, 707
metal hydride reduction, 708
oxidation of aldehyde to carboxylic acid, 707
reaction with anions of terminal alkynes, 703
reaction with Grignard reagents, 703
reaction with HCN to form cyanohydrins, 703
reaction with organolithium reagents, 703
reductive animation, 708
Wittig reaction, 704
Wolff-Kishner reduction, 709
Aldehydes and ketones—reaction mechanisms
acid-catalyzed α-halogenation of ketone, 700
acid-catalyzed equilibration of keto and enol tautomers, 686
acid-catalyzed formation of acetal, 674
acid-catalyzed formation of hemiacetal, 673
alkyne anion reacting with ketone, 663
base-catalyzed equilibration of keto and enol tautomers, 686
base-catalyzed formation of hemiacetal, 672
base-promoted α-halogenation of ketone, 701
formation of cyanohydrin, 664–665
formation of imine, 679–680
Grignard reagent reacting with formaldehyde, 660–661

Aldehydes and ketones (*Continued*)
organolithium reagent reacting with ketone, 662
Pinnick oxidation, 689–690
sodium borohydride reduction, 692
Wittig reaction, 667
Wolff-Kishner reduction, 697–698
Alder, Kurt, 911
Alditols, **1145**–1146, 1161
Aldol reaction, 825–832
acid-catalyzed, 826–827
acid-catalyzed dehydration of aldol product, 828–829, 868
base-catalyzed, 826
crossed, 829–830
intramolecular, 831
key reactions, 868
retrosynthetic analysis, 831–832
review/study guide, 867–868
Aldolase, 1227*t*
Aldonate, 1146
Aldonic acid, **1146,** 1161
Aldose, **1135,** 1146
Aldosterone, 1182*t*
Aleve, 731
Aliphatic amines, **1038,** 1047, 1059–1063
Aliphatic carboxylic acids, 727–728, 728*t*, 731
Aliphatic diazonium ions, 1062
Aliphatic hydrocarbons, 73
Alkaloid, **1038**
Alkane boiling points, 109–110
Alkanes and cycloalkanes, 72–126, A-2
acidity of alkanes, 192*t*
boiling points, melting points, and density, 108–109, 109–110, 118
bond lengths/bond strengths, 56
chlorination and bromination of alkanes, 360
conformations, 85–99. *See also* Conformation
constitutional isomerism, 74–76, 109, 118
cycloalkanes. *See* Cycloalkanes
dispersion forces and interactions, 107–108
halogenation of alkanes, 336–348
heats of combustion and relative stability, 111–112
infrared spectra, 543, 543*t*
IUPAC system, 77, 78–79, A-11 to A-12
mass spectra, 614–615
molecular formula, 74, 74*t*
molecular weight, 118
NMR spectra, 587
oxidation, 110–111
physical properties, 107–110, 118
reactions of alkanes, 110–112
review/study guide, 115–118
source and importance of alkanes, 112–114
structure of alkanes, 73–74
Alkene addition reactions, 241*t*
Alkene metathesis, 1112–1115, 1118–1119
Alkenes, 206–227, A-2
acidity, 192*t*
addition reactions, 241*t*
anionic polymerization, 1284
bond lengths/bond strengths, 56
bridgehead, 215–216
carbon-carbon double bond orbitals, 208–209
cationic polymerization, 1288
cis,trans isomerism, 210–211
cis,trans system, 212, 213
colorless, 257
common names, 212
cycloalkenes, 214–215
defined, 206
dienes, 216
E,Z system, 213, 214
free-radical addition of HX, 358–359
general formula, 206
Heck reaction, 1099–1100
infrared spectra, 543*t*, 544
IUPAC system, 211, A-13
key reactions. *See* Alkenes—key reactions
mass spectra, 616
metathesis reactions, 1112–1115, 1118–1119
NMR spectra, 588
physical properties, 217, 218*t*
polyenes, 216
radical addition of HBr, 356–359, 362
reaction mechanisms. *See* Alkenes—reaction mechanisms
reactions. *See* Reactions of alkenes
review/study guide, 222–223
shape, 208
Simmons-Smith reaction, 642
synthesis of alkynes from alkenes, 302–304, 319
terpenes, 218–220
trienes, 216
Alkenes—key reactions
acid-catalyzed hydration, 284
addition of bromine and chlorine, 284
addition of H_2 (catalytic reduction), 286–287
addition of HOCl and HOBr, 284–285
addition of HX, 284
enantioselective reduction, 287
hydroboration-oxidation, 285
oxidation by ozone, 286
oxidation to vicinal diol by OsO_4, 286
oxymercuration-reduction, 285
Alkenes—reaction mechanisms
acid-catalyzed hydration of propene, 253
addition of bromine with anti stereoselectivity, 258–259
carbocation rearrangement, 255
electrophilic addition of HBr to 2-Butene, 246–247
formation of an ozonide, 272
halohydrin formation and its anti-stereoselectivity, 259–260
hydroboration, 266
oxidation of trialkylborane by alkaline hydrogen peroxide, 266
oxymercuration-reduction of an alkene, 263–264
Alkenyl, 331
Alkenyl anion, 314
Alkenyl radical, 314
Alkenylborane, 308
Alkoxide ions, 446
Alkyl-aryl ethers, 962–963
Alkyl azides, 1057
Alkyl group, **78,** 207, 939, A-11
Alkyl halide, **331**. *See also* Haloalkanes
Alkyl radical, 343
Alkyl sulfonates, 452–453
Alkyl thermosets, 1275
Alkylation
acetylide anions, 301–302, 319
allylic, 1104–1107, 1116–1117
ammonia and amines, 1056–1057
azide ion, 1057
benzene, 996, 1019, 1020
enamines, 842–843, 870
Friedel-Crafts, 1000–1002, 1019
other electrophilic aromatic alkylations, 1004–1005, 1018
Alkylation reaction, **302**
Alkylbenzene, 996
Alkylbenzenesulfonate detergents, 1175–1176
N-Alkylpyridinium chloride, 1042
Alkyltriphenylphosphonium iodide, 667
Alkyltriphenylphosphonium salt, 666
Alkynes, 297–329, A-2
acid-catalyzed hydration, 310
acidity, 192*t*, 301
addition of bromine and chlorine, 305
addition of HBr to an alkyne, 306
addition of hydrogen halides, 305–307
alkylation of acetylide anions, 301–302, 319
bond lengths/bond strengths, 56
catalytic reduction, 312
common names, 299
conversion of alkenes to alkynes, 302–304, 319

defined, **298**
dissolving-metal reduction, 313
electrophilic addition, 305–307
functional groups, 317–318
hydroboration-oxidation, 307–309
hydroboration-protonolysis, 313
infrared spectra, 543*t*, 544–545
internal, 302
IUPAC system, 298, 299, A-13
keto-enol tautomerism, 308, 309
key reactions. *See* Alkynes—key reactions
Lindlar catalyst, 312
mass spectra, 616–617
mechanisms. *See* Alkynes—reaction mechanisms
organic synthesis, 314–318
physical properties, 300, 300*t*
reaction roadmap, 295–296
reduction, 312–314
retrosynthetic analysis, 314–316
review/study guide, 318–322
structure, 298
terminal, 299
Alkynes—key reactions
acid-catalyzed hydration, 321
acidity of terminal alkynes, 318
addition of Br_2 and Cl_2, 319–320
addition of HX, 320
alkylation of acetylide anions, 319
catalytic reduction, 321
hydroboration-oxidation, 320–321
hydroboration-protonolysis, 321
keto-enol tautomerism, 320
reduction using Na or Li metal in $NH_3(l)$, 322
synthesis of alkyne from alkene, 319
Alkynes—reaction mechanisms
addition of HBr to an alkyne, 306
$HgSO_4/H_2SO_4$ catalyzed hydration of an alkyne, 310–311
reduction of alkyne by sodium in liquid ammonia, 313–314
Allegra, 758, 1092–1093
Allene, **303,** 304
Allowed pericyclic reaction, 909
Allyl, **212**
Allyl alcohol, 441, 542
Allyl benzene, 956
Allyl cation, 385
Allyl chloride, 331, 963
Allyl phenyl ether, 963
Allyl radical, 351–352
trans-2-Allylcyclohexanol, 648
Allylcyclopentane, 212
Allylic, **385**
Allylic alkylation, 1104–1107, 1116–1117
Allylic bromination, 352, 361
Allylic carbocation, **385**
Allylic carbon, **349**
Allylic halogenation, 348–352
Allylic substitution, **349**
Almonds, 653
α-amino acid, **1199,** *1200*
α-anomer, 671, 1159
α-bromoester, 668
α-carbon, 387, **684,** 866
α-cleavage, 618, 619, *620*
α-helix, **1222**–1223
α-hydrogen
aldehydes/ketones, A-2
defined, **684**
ester, A-2
hydrogen atoms bonded to α-carbon, 866
α-ketoglutaric acid, 744
α-phosphonoester, 668, 669
α-phosphonoketone, 668
α,β-unsaturated carbonyl, 852
Alprostadil, 814
Alternative chair conformation, 95
Aluminum trichloride, 18
Amantadine, 107, 816
Ambucaine, 1089
Amide anion, 314
Amide bonds, 769–770
Amides, A-2. *See also* Carboxylic acid derivatives
acidity, 768
alcohols, 787, 788
ammonia and amines, 789
defined, **24**
hydrolysis, 779–783, 804
infrared absorptions, 551*t*
IUPAC system, 766–767, A-15
mass spectra, 619–621
reduction, 798–799, 808
-amine, 81, 655*t*
Amines, 21, 1037–1094, A-2
acid/base equilibria, 1047
acid-base reactions, 1053
acids, and, 1053–1055, 1073
aliphatic, 1047, 1059–1063
alkylation of ammonia and amines, 1056–1057
alkylation of azide ion, 1057
amino acid protonation, 1053–1054
aromatic, 1047–1048, 1059–1066
basicity, 1045–1053, 1072–1073
chirality, 1042–1043
common names, 1040–1041
Cope elimination, 1069–1070, 1077
diazotization, 1061, 1076
guanidine, 1052–1053
heterocyclic aromatic, 1049–1052
Hofmann elimination, 1067–1069, 1077
infrared spectra, 547
IUPAC system, 1039–1040, 1071
key reactions. *See* Amines—key reactions
mass spectra, 622–623
mechanisms. *See* Amines—reaction mechanisms
N-Nitrosamines, 1059–1060, 1075
nitrosyl cation, 1058–1059
nitrous acid, and, 1058–1066, 1074–1076
NMR spectra, 590
physical properties, 1043–1045
preparation, 1056–1058, 1074
primary, **21**
pyrolysis of tertiary amine oxide, 1077
reaction roadmap, 889–890
review/study guide, 1071–1077
Sandmeyer reaction, 1065, 1076
Schiemann reaction, 1065, 1076
secondary, **21**
separation by aqueous extraction, 1055
separation/purification of amine and neutral compound, *1054*
structure and classification, 1037–1039, 1071
systematic names, 1039
tertiary, **21**
Tiffeneau-Demjanov reaction, 1062–1063
Amines—key reactions
alkylation of ammonia and amines, 1074
alkylation of azide ion followed by reduction, 1074
basicity of aliphatic amines, 1073
basicity of aromatic amines, 1073
basicity of heterocyclic aromatic amines, 1073
conversion of primary arylamine to phenol, 1076
Cope elimination: pyrolysis of tertiary amine oxide, 1077
formation of arenediazonium salts (diazotization), 1076
formation of *N*-Nitrosamines from secondary amines, 1075
Hofmann elimination, 1077
nitrosation of tertiary aromatic amines, 1075
reaction of amines with strong acids, 1073
reaction of arenediazonium salt with KI, 1076
reaction of cyclic β-aminoalcohols with nitrous acid, 1075
reaction of primary aliphatic amines with nitrous acid, 1075
reduction of arenediazonium salt with hypophosphorus acid, 1076

Amines—key reactions (*Continued*)
Sandmeyer reaction, 1076
Schiemann reaction, 1076
Amines—reaction mechanisms
Cope elimination, 1069
formation of nitrosyl cation, 1058–1059
Hofmann elimination, 1067
reaction of 2° amine with nitrosyl cation to give *N*-Nitrosamine, 1059–1060
reaction of 1° amine with nitrous acid, 1061–1062
Tiffeneau-Demjanov reaction, 1062–1063
amino-, 655*t*
2-Amino-1-phenylethanol, 665
(*R*)-2-Amino-3-mercaptopropanoic acid, 475
(*S*)-2-Amino-3-methyl-1-butanol, 1040
Amino acid analysis, 1209
Amino acid protonation, 1053–1054
Amino acid stereochemistry, 155–156
Amino acids, 1199–1237. *See also* Polypeptides and proteins
acid-base properties, 1203–1207, 1228
acidic, 1263
acidity of α-ammonium groups, 1204, 1228
acidity of α-carboxyl groups, 1203–1204, 1228
acidity of side-chain carboxyl groups, 1204
basicity of guanidine group of arginine, 1204
basicity of imidazole group of histidine, 1204–1205
chirality, 1200
defined, **1199**
D,L convention, 1200, 1227
electrophoresis, *1206,* 1206–1207
isoelectric point (pI), 1206, 1207
key reactions, 1228–1229
pK_a values, 1203*t*
protein-derived, 1200–1201, 1201*t*
review/study guide, 1227–1232
R,S convention, 1201, 1228
structure, 1199–1200
titration, 1205–1206
zwitterions, 1200, 1205, 1206
D-amino acids, 156
L-amino acids, 156
Amino group, **21,** 182, 682–684
Amino-protecting groups, 1216–1217
Amino sugars, 1138–1139
4-Aminobenzoic acid, 1040
3-Aminobutanoic acid, 655*t*
4-Aminobutanoic acid, 726, 728
4-Aminobutanoic acid (GABA), 728, 884, 1202
γ-Aminobutyric acid, 728, 1202
trans-2-Aminocyclohexanol, 1057
2-Aminoethanol, 1040
(*S*)-2-aminopropanoic acid, 728
(*S*)-α-aminopropanoic acid, 728
Amitriptyline, 1031
Ammonia, A-2
acid anhydrides, 789
acid chlorides, 788–789
acid halides, 788–789
aldehydes and ketones, 678–681, 684*t*
amides, 789
amine, 21
carboxylic acid derivatives, 788–789
chemical structure, 1046*t*
electrostatic potential map, 28
esters, 789, 806
Lewis structure, 13*t, 24*
orbital overlap picture, *36*
physical properties, 1045*t*
pK_a, 1046*t*
shape, *24*
Ammonium acetate, 789
Ammonium adipate, 1270
Ammonium hexanedioate, 1270
Ammonium hydrogen sulfate, 783
Ammonium ion, A-2
Amorphous domain, **1267**
Amorphous polystyrene, 1267
Amoxicillin, 167, 762, 791, 792
AMP. *See* Adenosine 5′monophosphate (AMP)
Amphetamine, 715, 1041
Ampicillin, 792
Amygdalin, 1166
Amyl, 308
Amylopectin, 1154, *1154*
Amylose, 1154
-an-, 81
Anabolic steroid, **1183**
Androgens, 1182*t,* **1183**
Androsterone, 1182*t*
Angina, 886
Angiogenesis, 909
Angiografin, 1091
Angle strain, **89**
Angstrom, 536*t*
Anhydrides, 551. *See also* Acid anhydrides
Aniline, 954, 1023, 1038, 1040, 1045*t,* 1046*t,* 1049
Animal fats and vegetable oils, 1171–1174
Anion, **7**
Anionic chain-growth polymerization, 1283–1286, 1283*t,* 1296
Anions of terminal alkynes, 663–664, 703
m-Anisidine, 1040
Anisole, *547,* 954, 963
Annulene, **945**–947
[4]Annulene, 945
[6]Annulene, 945
[8]Annulene, 945
[10]Annulene, 947
[14]Annulene, 945, 946
[16]Annulene, 949
[18]Annulene, 945, 946
Anomer, 671, **1140**
Anomeric carbon, 671, **1140**
Antarafacial, **910**
Antarafacial interaction, 910
Anthracene, 943, 956
Anthranilic acid, 934
Anti addition, 279, 284, 313
Anti addition stereochemistry, 296
Anti conformation, **88**
Anti stereoselectivity, **257,** 258–259
Antiaromatic compound, **947**
Antiaromatic hydrocarbons, 947–949
Antibonding molecular orbital, **34**
Antiemetic drugs, 1034
Antihistamines, 1092–1093
Antiknock properties, 114
Antioxidants, 354–356
Antipsychotic drugs, 1036
Antispasmodic drugs, 1035
Antivert, 1034
Antiviral drugs, 1244–1245
Apheloria corrugata, 665
Appendices
bond dissociation enthalpy (BDE), A-3
^{13}C-NMR chemical shifts, A-5
classes of organic acids, A-2
electrostatic potential maps, A-7
equilibrium constant, A-1
^{1}H-NMR chemical shifts, A-4
infrared absorption frequencies, A-6
nomenclature (IUPAC system), A-11 to A-17
organic chemistry reaction roadmaps, A-18
stereochemical terms, A-8 to A-10
thermodynamics, A-1
Appetite suppressants, 1089–1090
Aprotic acid, **195**
Aprotic solvent, 390, 390*t*
Arachidic acid, 1171*t*
Arachidonic acid, 1171*t,* 1178, 1179
Aramid, **1271**
Arenediazonium salts, 1076
Arenes
aromatic hydrocarbons, 939
benzene, 207. *See also* Benzene
defined, **207**
infrared spectra, 543*t,* 545
Arenium ion, 1017
Arg, 1201*t*. *See also* Arginine
Arginine, 1201*t,* 1203*t,* 1204

Aromatic amines, **1038,** 1047–1048, 1059–1066
Aromatic compound, **939**
Aromatic dicarboxylic acids, 727
Aromatic ethers, 546
Aromatic hydrocarbon, 621–622, 945–947
Aromatic hydrocarbon ions, 950–952
Aromatic rings, 1017
Aromatic substitution, 1096
Aromaticity, 207, 943–953
  annulenes, 945–947
  antiaromatic hydrocarbons, 947–949
  aromatic hydrocarbon ions, 950–952
  aromatic hydrocarbons, 945–947
  heterocyclic aromatic compounds, 949–950
  how to recognize aromatic compounds, 953
  Hückel criteria, 944
  imidazole, 1051
  pyridine, 1049
  pyrrole, 1052
  review/study guide, 971–972
Arrhenius, Svante, 170
Arrhenius acids and bases, 170–171
Arrow pushing, 233–238
  atom movement, 234
  avoiding mistakes, 236–238
  backward arrows, 369
  bond breaking, 235
  common mistakes, 369–373
  Diels-Alder reaction, 920
  electron movement, 233
  electron source/electron sink, 235
  failing to conserve charge, 372–373
  hypervalency, 370–371
  mixed media errors, 371–372
  not enough arrows, 370
  rules, 233–235
Artemesia alcohol, 488
Artemisia absinthium, 226
Artificial sweeteners, 1152, 1152*t*
Aryl cation, 1064
Aryl Grignard reagents, 631
Aryl group, **207,** 939
Aryl halide, **331**
Aryl sulfonates, 452–453
Arylammonium ion, A-2
Ascaridole, 604
L-Ascorbic acid, 1141
Asn, 1201*t*. *See also* Asparagine
Asp, 1201*t*. *See also* Aspartic acid
Asparagine, 1201*t,* 1203*t*
Aspartame, 1152*t,* 1236
Aspartic acid, 1201*t,* 1203*t,* 1263
Aspergillus terreus, 839
Aspirin, 730–731, 787, 1178, 1179
Asterisk (*), 34
Asthma, 436
Aston, F. W., 606
Asymmetric induction, 919
Asymmetric stretching, *540*
Atactic materials, 1282–1283
Atactic polymer, **1282,** *1282*
Atactic polystyrene, 1283
Atenolol, 532, 990–991
Atherosclerosis, 839
Athlete's foot, 751
Atom
  completed valence shells, 7
  electrons, 3
  phantom, 134
  size, 189
  terminal, 45
Atom economy, 1097
Atorvastatin, 442, 491, 892
ATP. *See* Adenosine 5'triphosphate (ATP)
Atropisomers, **133,** *148,* A-8
Attraction-repulsion principle, 30
Aufbau principle, **3,** 33
Augmentin, 792
Automated DNA sequencing machines, 1256
Autoxidation, **353,** 362
Average degree of polymerization, 1266
Axial-axial interaction, 96
Axial bond, **93,** 97
Azeotrope, **675**
Azide, 395, 1074
Azide ion, 1057
Azidothymidine, 1244–1245
Aziridinium ion, 421
Azo compounds, 1277
Azoisobutyronitrile (AIBN), 1277
AZT, 1244–1245

B

B-DNA, 1246, *1246*
Bacitracin A, 1233
Back-biting chain-transfer reaction, 1278
Backward arrows, 369
Bad cholesterol, 1181
Baekeland, Leo, 1275
Baeyer-Villiger reaction, 1125
Bakelite, 1275
Balanced half-reaction, 270–271
Barbital, 814
Barbiturates, 814
Barium cocktail, 1091
Barium sulfate, 1091
Base, 170. *See also* Acids and bases
Base-catalyzed aldol reaction, 826
Base-catalyzed equilibration of keto and enol tautomers, 686
Base-catalyzed formation of hemiacetal, 672
Base-pairing, *1245,* 1245–1246
Base peak, 608
Base-promoted α-halogenation of ketone, 701
Basicity, **393**. *See also* Acidity
  alcohols, 445–446, 478
  amines, 1045–1053, 1072–1073
  guanidine group of arginine, 1204
  imidazole group of histidine, 1204–1205
  vitamin K and blood clotting, 1190–1191
Batrachotoxin, 1044
Batrachotoxinin A, 1044
Batteries, 1279
BDE. *See* Bond dissociation enthalpy (BDE)
Beckett-Casey rules, 1079
Beef fat, 1173*t*
Beer brewing, 694
Beer-Lambert law, 905, 906–907
Beeswax, 1172
Benadryl, 532
Bender, Myron, 811
Bending vibrations, *540*
Benzaldehyde
  almonds, 653
  conjugated system, 907
  derived from mandelonitrile, 665
  mandelic acid, 785
  nomenclature, 654, 954
  oxidation of aldehydes, 689
  phensuximide, 885
  synthesis of 5-Hydroxy-5-phenylpentanal, 676
Benzaldehyde cyanohydrin, 665, 785
Benzene, 207, 938–994. *See also* Reactions of benzene and derivatives
  aromaticity, 970. *See also* Aromaticity
  boiling point, 938
  catalytic reduction, 1270
  disubstituted, 954–955
  halogenation, 968–969, 975
  hydrogenolysis of benzyl ethers, 969, 975
  isomeric dibromobenzenes, 939, 940
  Kekulé's model, 939–940
  Kekulé's structures, *939*
  melting point, 938
  molecular formula, 938
  molecular orbital model, 940–942
  monosubstituted, 954
  nomenclature, 954–956, 972–973
  oxidation, 966–968, 975
  PAHs, 956
  phenols. *See* Phenols
  polysubstituted, 955
  reaction roadmap, 889–890
  reactions at benzylic position, 966–969
  reactivity, 938

Benzene (*Continued*)
resonance energy, *943*
resonance model, 942
review/study guide, 970–975
Benzene—key reactions
alkylation using an alcohol, 1020
alkylation using an alkene, 1019
Friedel-Crafts acylation, 1019
Friedel-Crafts alkylation, 1019
halogenation, 1018
nitration, 1018
nucleophilic aromatic substitution: addition-elimination, 1021–1022
nucleophilic aromatic substitution: benzyne intermediate, 1021
sulfonation, 1019
Benzene—reaction mechanisms
electrophilic aromatic substitution—chlorination, 996–997
formation of nitronium ion, 998
Friedel-Crafts acylation, 1003
Friedel-Crafts alkylation, 1000–1001
nucleophilic aromatic substitution by addition-elimination, 1016
nucleophilic aromatic substitution via benzyne intermediate, 1015
Benzene-d_6, 699
1,4-Benzenediamine, 1271
Benzenediazonium ion, 1064
1,2-Benzenedicarboxylic acid, 727, 968
1,4-Benzenedicarboxylic acid, 727, 968, 1271
1,2-Benzenediol, 957, 964
1,3-Benzenediol, 957
1,4-Benzenediol, 957, 964–966
Benzenediols, 973
Benzenesulfonamide, 768
Benzenesulfonic acid, 996, 999
Benzil-benzilic acid rearrangement, 753
Benzlyamine, 1052
Benzo[a]pyrene, 513–514, 956
Benzocaine, 755
Benzoic acid
IUPAC system, 727, 954
oxidation of aldehydes, 689
pK_a, 733
separation of, from benzyl alcohol, *735*
Benzoic anhydride, 763, A-15
Benzonitrile, 768
Benzophenone, 654
Benzoquinone, 1300
1,2-Benzoquinone, 964
1,4-Benzoquinone, 964–966
Benzoyl chloride, 740, 763, 786, 811
Benzoyl radical, 341
Benzyl alcohol, *735*
Benzyl bromide, 968
Benzyl cation, 386
Benzyl chloride, 968
Benzyl ether group, 822
Benzyl (Bn) group, **954**, 994
Benzyl hexyl ether, 969
Benzylamine, 1041, 1045*t*
Benzylcetyldimethylammonium chloride, 1195
Benzyldimethylamine, 1038
Benzylic carbocation, **386**
Benzylic cation, 966
Benzylic oxidation, 966–968
Benzylic position, **966**
Benzylic radical, 966
Benzyloxycarbonyl (Z—) protecting group, 1216, 1231
Benzyltrimethylammonium hydroxide, 1042
Benzyne intermediate, **1014**–1015, 1021
Berg, Paul, 1236
Beta-adrenergic blockers, 1033
Beta-2 adrenergic blockers, 991
β-aminoalcohols, 830, 1057, 1062
β-anomer, 671
β-azidoalcohol, 1057
Beta-blockers, 990, 991
β-carbon, 387
β-cleavage, 622
β-dicarboxylic acid, 745
β-elimination, **375**. *See also* Nucleophilic substitution and β-elimination
β-hydroxy carbonyl, 831
β-hydroxyaldehyde, 825, 828
5-hydroxyhexanal, 713
β-ketoacids, 743–744, 750
β-ketobutyric acid, 728
β-ketocarboxylic acid, 744
β-ketoester, 836, A-2
β-hydroxyketone, 825, 828
β-lactam antibiotics, 791–793
β-lactamase inhibitor, 792
β-lactams, 1123
β-pleated sheet, **1223**
Betaine, **667**
Betapace, 1093
BF_3-ether complex, 194
BHT, 356. *See* Butylated hydroxytoluene (BHT)
BHT radical, 356
Bicyclic Diels-Alder adduct, 916
Bicyclic systems, 915–916
Bicyclo-2,-5-heptadiene, 934
Bicyclo[2.2.1]heptane, *84*
Bicyclo[4.3.0]nonane, *84*
Bicyclo[4.4.0]decane, *84*
Bicycloalkane, **84,** 104–107
Bidisomide, 989–990
Bifenthrin, 742–743
Bile acids, **1183**
Bimolecular reaction, **378**
BINAP, 169, 281
(*R*)-(+)-BINAP, 281
(*S*)-(−)-BINAP, 281
Biological chemistry. *See* Connections to Biological Chemistry boxes
Biot, Jean Baptiste, 149
Biotechnology, 1239
Bisphenol A, 1025, 1272, 1273
Black, James W., 1261
Black-and-white photography, 966
Black carpet beetle, 750
Black pepper, 124
Bleach, 1176
Bloch, Felix, 558
Bloch, Konrad, 1184
Blood alcohol screening, 466–467
Blood clotting, 1190–1191
Blood glucose levels, 1148–1149
Blood sugar, 1138
Blood thinners, 766
Bn group. *See* Benzyl (Bn) group
Boat conformation, **93,** *93*
BOC—protecting group, 1216
Boger, Dale, 1129
Boiling point
alcohols, 442*t*, 443–444
alkenes, 218*t*
amines, 1045*t*
benzene, 938
carboxylic acids, 730*t*
compounds of comparable molecular weight, 657*t*
ethers, 495*t*
haloalkanes, 333–334, 334*t*
low-weight molecular alkynes, 300*t*
thiols, 475*t*
unbranched alkanes, 108*t*
Bombykol, 818
Bon de Saint Hilaire, Francois Xavier, 1225
Bond
axial, 93
covalent, 8, 10
dative, 18, 229
equatorial, 93
ionic, 8
Bond angles and shapes of molecules, 24–26
Bond cleavage, 186
Bond dipole moment (μ), **12**
Bond dissociation energy, 10. *See also* Bond dissociation enthalpy (BDE)
Bond dissociation enthalpy (BDE), 10
C—H bonds, 336*t*, 339*t*
C—X bonds, 336*t*
chain propagation steps (haloalkanes), 343
defined, **243,** 335, A-3
energetics, 360
summary (list of BDEs), A-3

Bond length
  acetylene, 56*t*
  alkanes, alkenes, and alkynes, 56
  defined, **10**
  ethane, 56*t*
  ethylene, 56*t*
  haloalkanes, 335–336
Bond polarity, 12
Bond strength
  acetylene, 56*t*
  alkanes, alkenes, and alkynes, 56
  ethane, 56*t*
  ethylene, 56*t*
  haloalkanes, 335–336
Bonding electrons, **14**
Bonding molecular orbital, **34**
Bonding σ orbital, 40
Bonine, 1034
Borane
  hydroboration, 1108
  Lewis structure, *38*
  orbital overlap picture, *38*
  tetrahydrofuran (THF), 265
Borates, 1108
Boron, 265
Boron trifluoride, 18, 27, 194
Branched polymers, *1265*
Break-resistant containers, 1276*t*
Breast cancer, 993
Breathalyzer, 466
Bridge, 84
Bridged [10]Annulene, 947
Bridged bromonium ion, 257
Bridged halonium ion intermediates, 258–259
Bridgehead alkenes, 215–216
Bridgehead carbon, 84, 216
Bromination, 241*t*, 345–348, 996–998
Bromine, 332, 611*t*, 612
1-Bromo-1-butene, 903
(*E*)-1-Bromo-2-butene, 903
(*E*)-2-Bromo-2-butene, 303
1-Bromo-1-chloroethane, *129*, 130
1-Bromo-2,2-dimethylpropane, 388
2-Bromo-1-ethyl-4-nitrobenzene, 955
(*E*)-1-Bromo-4-isopropyl-5-methyl-4-octene, 214
2-Bromo-4-methylaniline, 1064
2-Bromo-2-methylpentane, 338
(*S*)-2-Bromo-4-methylpentane, 331
2-Bromo-4-methylphenol, 1064
1-Bromo-2-methylpropane, 450
2-Bromo-2-methylpropane, 387, 402
1-Bromo-2-nitrobenzene, 955
1-Bromo-1-phenylethane, 969
α-Bromoacetophenone, 700
1-Bromoadamantane, 816
Bromoalkanes, 334
Bromobenzene, 939, 1098
m-Bromobenzene-sulfonic acid, 1009
o-Bromobenzene-sulfonic acid, 1009
p-Bromobenzene-sulfonic acid, 1009
4-Bromobutanal, 676
1-Bromobutane, *586*, 627
2-Bromobutane, 403
(*R*)-2-Bromobutane, 331
Bromocyclobutane, 365
(*R*)-4-Bromocyclohexene, 331
Bromocyclopentane, 627
Bromoepoxide, 816
Bromoethane, 387, 388
Bromo(halo)hydrin formation, 241*t*
Bromohydrin formation, 261
α-Bromoketone, 668, 700
Bromomethane, 344
Bromomethylbenzene, 968
o-Bromonitrobenzene, 955, 1008
p-Bromonitrobenzene, 1008
Bromonium ion, 257
Bromophenols, 959
1-Bromopropane, 337, *612*, 613
2-Bromopropane, 337
N-Bromosuccinimide (NBS), 349, 350, 352, 968
2-Bromotoluene, 1065
4-Bromotoluene, 955
p-Bromotoluene, 955
Brønsted, Johannes, 171
Brønsted-Lowry acid, **171**
Brønsted-Lowry acid-base reaction, 228, 229, 236
Brønsted-Lowry acids and bases, 171–177
Brønsted-Lowry base, **171**
Brown, Herbert C., 666
Bu. *See* Butyl (Bu)
Buckyball, 27
"Build-up" principle, 3
Bulletproof vest, 1271
Bumetanide, 1092
Bumex, 1092
Bupropion, 1030
but-, 77*t*
Butadiene, 1284, 1297
1,2-Butadiene, 304
1,3-Butadiene
  conjugated system, 907
  Diels-Alder reaction, 932
  heats of hydrogenation, 895*t*
  molecular orbital model, *897*
  $\pi \rightarrow \pi^*$ transition, 908*t*
  simplest conjugated diene, 896
  structural formula, 895*t*, 908*t*
Butadiene sulfone, 932
Butanal, 657*t*, 658*t*
Butanamide, 790, A-15
1-Butanamine, *548*
Butane
  ball-and-stick model, 74
  boiling point, 75, 108*t*, 442*t*
  common name, 80
  condensed structural formula, 74*t*
  constitutional isomer, 75
  density, 108*t*
  energy-minimized/non-energy minimized, *89, 90*
  line-angle formula, 74
  melting point, 108*t*
  molecular formula, 74*t*
  molecular weight, 442*t*
  solubility, 442*t*
  structural formula, 74
1,4-Butanediamine, 1039
Butanedioic acid, 727
1,4-Butanediol, 442*t*, 495*t*
(2*R*,3*R*)-2,3-Butanediol, 141, A-10
(2*R*,3*S*)-2,3-Butanediol, 141, A-9
(2*S*,3*S*)-2,3-Butanediol, 141, A-10
1-Butanethiol, 474, 475*t*
1,2,3-Butanetriol, 138
Butanoic acid
  boiling point, 730*t*
  halogen substitution, 192
  locker room odor, 731
  mass spectrum, *620*
  molecular weight, 730*t*
  nomenclature, 728*t*
  solubility, 730*t*
1-Butanol, 787
  boiling point, 442*t*, 495*t*, 657*t*
  catalytic reduction, 693
  mass spectrum, *617*
  molecular weight, 442*t*, 495*t*, 657*t*
  nomenclature, 439
  solubility, 442*t*, 495*t*
  structural formula, 657*t*
2-Butanol, 130, 583
(*R*)-(−)-2-Butanol, 151
(*R*)-2-Butanol, 144, A-10
(*S*)-(+)-2-Butanol, 151
(*S*)-2-Butanol, 144, 439, A-19
(*S*)-3-Butanolactam, 767
(*S*)-3-Butanolactone, 764
4-Butanolactone, 764
2-Butanone, 657*t*, 658*t*
Butanoyl chloride, 741, 786
3-Buten-2-one, 852, 907
1-Buten-3-yne, 299
trans-2-Butenal, 693
1-Butene, 218*t*, 275*t*, *616*, 895*t*
2-Butene-1-thiol, 474
cis-2-Butene
  cis,trans isomerism, 210
  diastereomer, **131**
  heats of hydrogenation, 275*t*, 895*t*
  space-filling model, *277*
  stereocenter, 130
  stereospecific reaction, A-10
  structural formula, 275*t*, 895*t*
cis-2-Butene oxide, 505

(*E*)-2-Butene-1-thiol, 474
trans-2-Butene
  cis,trans isomerism, 210
  diastereomer, 131
  heats of hydrogenation, 275*t*, 895*t*
  space-filling model, *277*
  stereocenter, 130
  stereospecific reaction, A-10
  structural formula, 275*t*, 895*t*
trans-2-Butene-1-thiol, 474
trans-2-Butenoic acid, 726
trans-2-Butenyl phenyl ether, 922
tert-Butoxide, 395
Butter fat, 1173*t*
Butter substitutes, 1173
Butterfly conformation, *92*
Butyl (Bu), A-11
sec-Butyl (*s*-Bu), A-11
tert-Butyl (*t*-Bu), A-11
tert-Butyl acetate, *567*
Butyl acrylate, 787
Butyl alcohol, 439
(*S*)-*sec*-Butyl alcohol, 439
tert-Butyl alcohol, 439, 445
(*R*)-*sec*-Butyl bromide, 331
tert-Butyl bromide, 387
sec-Butyl cation, 203
tert-Butyl cation, 204
tert-Butyl ethyl ether, 494
sec-Butyl hydrogen phthalate, 787
tert-Butyl mercaptan, 474
tert-Butyl methyl ether (MTBE), 493, 499, 556
Butyl mercaptan, 474
Butyl methyl ether, 495*t*
Butyl propenoate, 787
(*S*)-*sec*-Butylacetylene, 299
tert-Butylacetylene, 299
Butylamine, 1045*t*
tert-Butylamine, 1040
Butylated hydroxytoluene (BHT), 995, 1025
tert-Butylbenzene, 1005
tert-Butyldimethylsilyl (TBS) group, 504, 534
1-Butyldimethylsilyl chloride (TBSCl), 504
Sec-butyllithium, 1283
Butylmagnesium bromide, 631
1-Butyne, 299, 300*t*
2-Butyne, 299, 300*t*, 539
Butyraldehyde, 658*t*
Butyric acid, 728*t*
(*S*)-*β*-Butyrolactam, 767
(*S*)-*β*-Butyrolactone, 764
γ-Butyrolactone, 813
(*S*)-*γ*-Butyrolactone, 764

C

C. *See* Cytosine (C)
C—M bonds, 632*t*
^{13}C-NMR chemical shifts, A-5
^{13}C-NMR spectroscopy, 585–587, 596–597
Caffeine, 950
Cahn, R. S., 134
Cahn-Ingold-Prelog system, 134
Calcium channel blockers, 882
Calcium mercaptoacetate, 476
Calcium thioglycolate, 476
Calico Painters Association, 1271
Camphor
  bicyclic skeleton, 106
  carbon-carbon bonds, 219
  isoprene units, 219, 220
  line-angle formula, 84
  molecular formula, 85
  uses, 85
(1*S*,3*R*)-(+)-Camphoric acid, *158*
Cancer
  breast cancer, 993
  curry, 909
  prostate, 1028
Capric acid, 728*t*
Caproaldehyde, 658*t*
Caproic acid, 728*t*
Caprolactam, 1271, 1298
ε-Caprolactam, 767
ε-Caprolactone, 764
Caprylic acid, 728*t*
Capsaicin, 938, 961–962
Captopril, 155
Carbamate, 204, 822, 1217, 1272
Carbamic acid, 1217
Carbanion, 242, **632**
Carbene/carbenoid additions, 1096
Carbenes, 639–643
Carbenium ion, 247
Carbenoids, 639–643
Carbinoxamine, 987
Carbocation
  allylic, 385
  amines and nitrous acid, 1062
  amino-protecting groups, 1217
  benzylic, 386
  cationic polymerization, 1288
  defined, **247**
  electrophilic addition, 1005–1006
  inductive effect/hyperconjugation, 283
  lifetime, 248
  organic chemistry transformation, *242*
  primary (1°), 247, 249
  proton transfer, 177
  rearrangement, 254–255
  secondary (2°), 177, 247, 249
  S_N1 reaction, 385–387
  stability, 248–249, 251
  tertiary (3°), 177, 247, 249
  vinylic, 306
Carbocation rearrangements, 254–255
Carbocation resonance, 386–387
Carbodiimides, 1218, 1230
Carbohydrates, 1134–1169
  acetate rayon, 1156
  amylopectin, 1154, *1154*
  amylose, 1154
  artificial sweeteners, 1152
  cellulose, 1155
  disaccharides and oligosaccharides, 1151–1153
  glucosaminoglycans, 1156–1157
  glycogen, 1154–1155
  heparin, 1157
  hyaluronic acid, 1157
  key reactions. *See* Carbohydrates—key reactions
  lactose, 1151
  maltose, 1152
  monosaccharides. *See* Monosaccharides
  polysaccharides, 1154–1156, 1162
  rayon, 1156
  review/study guide, 1158–1163
  saccharide chair conformations, 1143
  starch, 1154
  sucrose, 1151
  textile fibers from cellulose, 1156
Carbohydrates—key reactions
  formation of cyclic hemiacetals, 1160
  formation of glycosides, 1161
  formation of *N*-Glycosides, 1161
  mutarotation, 1160
  oxidation by periodic acid, 1162
  oxidation to aldonic acid, 1161
  reduction to alditols, 1161
Carbolic acid, 957. *See also* Phenols
Carbon
  allylic, 349
  α-, 387, 684
  anomeric, 671, 1140
  β-, 387
  bridgehead, 84, 216
  classification, 82
  isotopes, 611*t*
  penultimate, 1137
Carbon and silicon bond polarity, 505
Carbon-carbon backbone polymers, 1274
Carbon-carbon bond-forming reactions. *See also* Catalytic carbon-carbon bond formation
  aromatic substitution, 1096
  carbene/carbenoid additions, 1096
  conjugate addition of *α*,*β*-unsaturated carbonyl compounds, 1096
  nucleophilic addition to carbonyl or carboxyl group, 1096
  nucleophilic displacement of leaving group by carbon nucleophile, 1096
  pericyclic reactions, 1096
Carbon-carbon bonding, *73*

Carbon-carbon double bond, 56, 208, 707, 721
Carbon-carbon double bond orbitals, 208–209
Carbon-carbon single bond, 56
Carbon-carbon triple bond, 56
Carbon dioxide
  covalent bonding, 44*t*
  dipole moment, 27
  shape, *25*
Carbon-hydrogen bond, 56
Carbon nucleophiles, 645, 659–668
Carbon radical, 242, 282
Carbon tetrachloride, 27
Carbonate ion, *45, 46*
Carbonic acid, 13*t*, 960, 961
Carbonium ion, 247
Carbonyl group, **21,** 652, 725, 823
Carbonyl hydrates, 670
Carbonyl-stabilized ylide, 668
Carboxyl group, **22,** 182, 725
Carboxyl-protecting groups, 1217
Carboxylate anion, 393
-carboxylic acid, 727
Carboxylic acid derivatives, 762–822. *See also* Carboxylic acids
  acid anhydrides. *See* Acid anhydrides
  acid halides. *See* Acid halides
  acidity, 768
  amides. *See* Amides
  catalysis, 773
  esters. *See* Esters
  Grignard reagents, 793–795
  hydrolysis, 774–785
  imides, 767, 768
  infrared spectra, 550–551
  interconversion of functional derivatives, 790–791, 806
  key reactions. *See* Carboxylic acid derivatives—key reactions
  lithium diorganocuprates, 795
  mechanisms. *See* Carboxylic acid derivatives—reaction mechanisms
  nitriles. *See* Nitriles
  nucleophilic acyl addition, 770–771
  nucleophilic acyl substitution, 771
  organolithium compounds, 795
  reaction roadmap, 649–650
  reaction with alcohols, 786–788
  reaction with ammonia and amines, 788–789
  reaction with organometallic compounds, 793–796
  reaction with water, 774–785
  reduction, 796–800
  relative reactivity, 771–773
  review/study guide, 801–808
Carboxylic acid derivatives—key reactions
  acidity of imides, 802
  acidity of sulfonamides, 802
  hydrolysis of acid anhydrides, 803
  hydrolysis of acid chlorides, 803
  hydrolysis of amides, 804
  hydrolysis of esters, 803–804
  hydrolysis of nitriles, 804
  reaction of acid anhydride with alcohol, 805
  reaction of acid anhydride with ammonia or amine, 805
  reaction of acid chloride with alcohol, 805
  reaction of acid chloride with ammonia or amine, 805
  reaction of acid chloride with carboxylic acid salt, 806
  reaction of acid chloride with lithium diorganocuprate, 807
  reaction of ester with alcohol, 805
  reaction of ester with ammonia or amine, 806
  reaction of ester with Grignard reagent, 807
  reduction of amides, 808
  reduction of esters, 807–808
  reduction of nitriles, 808
  transesterification, 805
Carboxylic acid derivatives—reaction mechanisms, 759–761
  acid-catalyzed ester hydrolysis, 775–776
  hydrolysis of acid chloride, 774
  hydrolysis of amide in aqueous acid, 781–782
  hydrolysis of amide in aqueous base, 782
  hydrolysis of cyano group to amide in aqueous base, 784
  hydrolysis of ester in aqueous base, 777–778
  mechanistic alternatives for ester hydrolysis: S_N2 and S_N1 possibilities, 779–780
  reaction of ester with Grignard reagent, 794
  reduction of amide by lithium aluminum hydride, 798–799
  reduction of ester by lithium aluminum hydride, 796–797
  saponification, 777–778
Carboxylic acid salt, 734–735, 806
Carboxylic acids, 23, 725–758, A-2
  acid ionization constants, 732–734
  acidity, 732–736
  β-ketoacids, 743–744, 750
  common names, 727–728
  conversion to acid chlorides, 740–742, 749
  decarboxylation, 743–746
  defined, **22**
  derivatives. *See* Carboxylic acid derivatives
  diazomethane, 739–740, 749
  esterification, 738–740
  Fischer esterification, 738–739, 749, 760–761
  hydrophilic/hydrophobic, 730
  industrial synthesis of acetic acid, 737
  infrared absorptions, 551*t*
  infrared spectra, 549
  IUPAC system, 726–727, A-14 to A-15
  key reactions. *See* Carboxylic acids—key reactions
  lithium aluminum hydride (LAH), 736–738, 748
  malonic acid and substituted malonic acids, 744–745
  mass spectra, 619–621
  McLafferty rearrangement, 620
  mechanisms. *See* Carboxylic acids—reaction mechanisms
  methyl esters, 739–740, 749
  NMR spectra, 589–590
  odors, 731
  physical properties, 729–731
  pK_a values, 190, 732, 747
  preparation, 736
  reaction roadmap, 649–650
  reaction with bases, 734–736
  reduction, 736–738
  review/study guide, 746–750
  salts, 734–735
  separation of benzoic acid from benzyl alcohol, *735,* 735–736
  structure, 725–726
  weak acids, 190
Carboxylic acids—key reactions
  acidity of carboxylic acids, 747
  carbonation of Grignard reagent, 748
  conversion to acid halides, 749
  decarboxylation of β-Dicarboxylic acids, 750
  decarboxylation of β-Ketoacids, 750
  Fischer esterification, 749
  industrial preparation of acetic acid by carbonylation of methanol, 748
  reaction of carboxylic acids with bases, 747
  reaction with diazomethane, 749
  reduction by lithium aluminum hydride, 748
Carboxylic acids—reaction mechanisms
  decarboxylation of β-dicarboxylic acid, 745
  decarboxylation of β-ketocarboxylic acid, 744
  formation of methyl ester using diazomethane, 740
Carboxylic amides, **24**. *See also* Amides
Carboxylic anhydrides, 763–764, A-15
Carboxylic esters, **23**. *See also* Esters
Carbuterol, 991
Cardiac arrhythmia, 1093

Cardiografin, 1091
Carotene, 226
β-Carotene, 226, 1188
Carothers, Wallace M., 1269
Carpanone, 1126
Carpet fibers, 1276*t*
Carroll, M. F., 936
Carroll reaction, 936
Carvone, 165
Caryophyllene, 84, 85
α-Caryophyllene, 1131
Catalyst, 1096, 1273
Catalytic allylic alkylation, 1104–1107, 1116–1117
Catalytic carbon-carbon bond formation, 1095–1133
  alkene metathesis, 1112–1115, 1118–1119
  atom economy, 1097
  catalytic allylic alkylation, 1104–1107, 1116–1117
  cross-coupling, 1107–1108
  green chemistry, 1097
  Heck reaction, 1098–1103, 1116
  organometallic compounds and catalysis, 1097, 1115
  oxidative addition, 1097
  palladium-catalyzed cross-coupling reactions, 1107–1112, 1117–1118
  reductive elimination, 1097
  review/study guide, 1115–1119
  Sonogashira coupling, 1111–1112, 1118
  Stille coupling, 1110–1111, 1118
  Suzuki coupling, 1108–1110, 1117
Catalytic cracking, 114
Catalytic cross-coupling reaction, **1107**
Catalytic cycle of cross-coupling, 1108
Catalytic hydrogenation, 273
Catalytic reduction, 273
  aldehydes and ketones, 692–693, 708
  alkenes, 274, 286–287
  alkynes, 312, 321
Catalytic reduction of benzene to cyclohexane, 1270
Catalytic reforming, 114
Catechol, 957, 964
Cation, **7**
Cationic chain-growth polymerization, 1283*t*, 1287–1289, 1296
Cationic electrophile, 1004
Celebrex, 731
Celera Genomics, 1256, 1257
Cellosolve, 493
Cellulose, 1155, 1156
Center of symmetry, *128,* **129,** A-8
Cephalexin, 792
Cephalin, 1186*t*
Cephalosporins, 792
Cephalosporium acremonium, 792
Cetirizine, 532, 723–724
Cetylpyridinium chloride, 1042, 1195
CFCs. *See* Chlorofluorocarbons (CFCs)
Chain, Ernst, 791
Chain-growth polymerization, 1275–1291, 1295–1296
Chain initiation, **342**
Chain length, **343**
Chain propagation, **342**
Chain propagation steps, haloalkanes, 343
Chain termination, **343**
Chain termination (dideoxy) method, 1255–1256, 1261
Chain-transfer reaction, **1278**
Chair conformation, **92,** *93,* 95
Challenger spacecraft disaster, 1268
Chapman's total synthesis of carpanone, 1126
Chapter 19 reaction roadmap, 889–890
Chapters 6-11 reaction roadmap, 295–296
Chapters 15-18 reaction roadmap, 649–650
Chapters 20-23 reaction roadmap, 936–937
Charge delocalization, 70
Chauvin, Yves, 1112
Chemical Connections boxes. *See also* Connections to Biological Chemistry boxes
  ABO blood group system, 1153–1154
  antiviral drugs, 1244–1245
  aspirin and other pain relievers, 730–731
  blood alcohol screening, 466–467
  blood thinners, 766
  cholesterol-lower drugs, 839–840
  corn borer, 217–218
  curry and cancer, 909
  DNA fingerprinting, 1257–1258
  esters as flavoring agents, 741
  fountain of youth, 1249
  freons, 344
  high-fructose corn syrup, 1155
  industrial synthesis of acetic acid, 737
  local anesthetics, 765
  magnetic resonance imaging (MRI), 584–585
  octane rating, 114
  organic polymers that conduct electricity, 1279
  poison dart frogs, 1043–1044
  reaction of acid chloride and ammonia, 788–789
  recycling of plastics, 1291–1292
  snake venom phospholipases, 1186
  spider silk, 1225–1226
  stitches that dissolve, 1274–1275
  testing for glucose, 1148–1149
  vitamin C (ascorbic acid), 1141
Chemical ionization (CI), 609
Chemical reactions. *See* Reaction
Chemical recycling, 1292
Chemical shift
  ^{13}C-NMR, A-5
  defined, **563**
  diamagnetic effects from π bonds, 571–572
  electronegativity of nearby atoms, 569–570
  example, 946–947
  ^{1}H-NMR, A-4
  hybridization of adjacent atoms, 570–571
  NMR spectroscopy, 568–572
Chemoselective reaction, A-8
Chili peppers, 961
Chinese cinnamon oils, 653
Chiral, **128,** A-8
Chiral auxiliary, 919
Chiral catalyst, 919
Chiral center, **130,** 132, 277–281, A-8
Chiral column package, 159
Chiral drugs, 155
Chiral molecules, 130–131
Chirality, 70, 128–129, 130. *See also* Stereoisomerism and chirality
  amines, 1042–1043
  amino acids, 1200
  enzymes, 155–156
Chitin, 1134
Chlorambucil, 421
β-Chloramine, 815
Chlorination, 345–348, 996–998
Chlorine
  atomic weight, 611*t*
  electronegativity, 332
  energy level diagram, 5
  isotope, 611*t*
  precise mass, 611*t*
  ratio of M to M + 2 peaks, 612
  relative abundance, 611*t*
5-chloro-1,3-cyclopentadiene, 951
1-Chloro-4-ethylbenzene, 955
cis-1-Chloro-2-isopropylcyclcohexane, 408
4-Chloro-2-nitrotoluene, 955
(*E*)-2-chloro-2-pentene, 214
2-Chloro-3-phenylbutane, 396
1-Chloro-1-phenylethane, 969
1-Chloro-2-phenylethane, 969
Chloroacetic acid, 191, 729, 732
3-Chloroaniline, 955
4-Chloroaniline, 1046*t*
m-Chloroaniline, 955
Chlorobenzene, 995, 1014
2-Chlorobutanoic acid, 192
3-Chlorobutanoic acid, 192
4-Chlorobutanoic acid, 192
(1*S*,2*S*)-2-Chlorocyclohexanol, 331
3-Chlorocyclopropene, 951

1-Chlorodecane, 785
Chloroethane, *612*
Chloroethene, 331
p-Chloroethylbenzene, 955
bis(2-Chloroethyl)methylamine, 418
bis(2-Chloroethyl)sulfide, 418
Chlorofluorocarbons (CFCs), 344
Chloroform, 332
Chloroform-*d*, 699
Chloroformates, 821
Chloromethane, 344, 375, 598
Chloromethylation, 1029, 1032
Chloromethylbenzene, 968
meta-Chloroperoxybenzoic acid (MCPBA), 507
m-Chlorophenol, 959
p-Chlorophenol, 959
Chlorophenols, 959
Chlorophyll, 643
1-Chloropropane, 338
(*Z*)-1-Chloropropane, 566
2-Chloropropane, 338, 565, *565*
3-Chloropropene, 331, 963
3-Chloropropyne, 299
Chlorosulfonation, 1092
2-Chlorotoluene, 1065
Chlorotrimethylsilane, 503
Cholestanol, 106, 125
Cholesterol, 1181
  bad, 1181
  biosynthesis, 1183–1184
  chiral centers, 166, 1181
  drugs that lower plasma levels of cholesterol, 839–840
  gallstones, 1181
  good, 1181
  HDL, 276, 1181
  intermediate in synthesis of most other steroids, 1185
  LDL, 276, 839, 1181
  model/structure, *1181*
Cholesterol chlorohydrin, 529
Cholesterol-lower drugs, 839–840
Cholic acid, 126, *1184*
Choline, 1186*t*
Cholografin, 1091
Chondroitin 6-sulfate, 1169
Chromatin, 1248
Chromatography, **159**
Chromic acid, 463–465
Chrysanthemic acid, 488
Chrysanthemum, 742
Chrysanthemyl alcohol, 488
Chrysanthemyl tosylate, 488
Chymotrypsin, 153, 1212
CI. *See* Chemical ionization (CI)
Cimetidine, 1234
Cinchonine, *158*
Cinnamaldehyde, 653, 654
Cinnamic acid, 726
Circular DNA, **1247**–1248
Circular duplex, 1247
Cis, **100, 212**
cis dienophile, 916
Cis,trans isomerism
  alkenes, 210–211
  bicycloalkanes, 104–107
  cycloalkanes, 99–104
  cycloalkenes, 215
  defined, **99, 210,** A-8
  dienes, 216
  polyenes, 216
  trienes, 216
[2-^{14}C]Citric acid, 166
(*S*)-Citronellal, 719
(*R*)-Citronellic acid, 717
L-Citrulline, 1202
Claisen, Ludwig, 832
Claisen condensation, 832–834, 869
Claisen rearrangement, 920, 922–923, 929
Classes of organic acids, A-2
Clemmensen, E., 696
Clemmensen reduction, 696, 697, 708
Clozapine, 532, 1036
cm^{-1}, 538, 539
Coal, 114
Cobra, 1186
Cocaine, 765, 1038
Coconut fat, 1173*t*
Codeine, 1079
Codon, **1251**
Coenzyme Q, 965
Coincidental overlap, 580
Cold-drawing, 1270
Comb polymer, *1265*
Combined valence bond/molecular orbital theories, 39–44
Common housefly, 325, 630
Common names. *See also* IUPAC system
  alcohols, 439
  aldehydes and ketones, 656
  alkanes, 80
  alkenes, 212
  alkynes, 299
  amines, 1040–1041
  carboxylic acids, 727–728
  epoxides, 505
  esters, 764–766
  ethers, 493, 494
  haloalkanes, 331–332
  nitriles, 768
  sulfides, 518
  thiols, 474
Complementary colors, *906*
Complementary sequences, 1247
Complex coupling, 577–578, 580–581
Concentration, 150
Condensation
  Claisen, 832–834, 869
  crossed Claisen, 835–836
  Darzens, 877
  Dieckmann, 834–835, 869, 870
  Perkin, 877
Condensation polymerization, **1268**
Condensed structural formula
  alcohols, 20
  aldehydes, 22
  carboxylic acid, 23
  ester, 23
  Lewis structures, 17
  primary amines, 21
  what is it?, 19
Configuration, **100,** A-8
Configurational isomers, **129,** *129,* 130, *148*
Conformation
  alkanes and cycloalkanes, 85–99
  anti, 88
  boat, 93, *93*
  chair, 92, *93*
  defined, **86**
  eclipsed, 86, *86*
  energy-minimized, 89, *89,* 90, *90*
  envelope, 92
  equilibrium populations, 98–99
  gauche, 88
  interconversion, *94, 95*
  planar, *92*
  puckered, *92*
  staggered, 86, *86*
  twist-boat, 93, *93*
Conformational isomers, 132, *148,* 149
Coniine, 1038
Conjugate acid, **171**
Conjugate acid-base pair, 171–172
Conjugate addition, **852**
Conjugate addition of enolate anions, 854
Conjugate base, **171**
Conjugated dienes, 894–903
  1,2-addition and 1,4-addition, 898–900
  defined, **894**
  electrophilic addition, 898–903, 926
  heats of hydrogenation, 895*t*
  key reaction, 926
  kinetic control, 900
  reaction roadmap, 936–937
  review/study guide, 925–926
  stability, 894–898, 925
  thermodynamic control, 901
Conjugated double bonds, 895
Conjugated π systems, 907
Conjugated systems, 894
Conjugation, **52, 549, 687**
Connections to Biological Chemistry boxes. *See also* Chemical Connection boxes
  amide bonds, 769–770
  chiral drugs, 155
  drug-receptor interactions, 442–443

Connections to Biological (*Continued*)
FAD/$FADH_2$ (fatty acid oxidation), 1176–1177
fats versus oils, 220–221
ionization of functional groups at physiological pH, 182–183
ketone bodies and diabetes mellitus, 745
mass spectrometry of biological macromolecules, 621
mustard gas, 420–421
NADH (hydride reducing agent), 694–695
oxidation of alcohols by NAD^+, 470
phosphoesters, 40
trans fatty acids, 276–277
Constitutional isomers, **74**–76, 129, *148*
Construction tubing, 1276*t*
Contraceptive pills, 1182
Contributing structures
curved arrows/push electrons, 46–47
defined, **46**
filled valence shells, 49
general rules, 47–48
hybridization, 54
least separation of unlike charges, 49
maximum number of covalent bonds, 49
negative charge on a more electronegative atom, 49
relative contributions to resonance, 50
relative importance, 48–51
Convergent synthesis, 886
Conversion. *See also* Synthesis
alcohols/alkenes, 455
alcohols/haloalkanes, 447–454, 479–480
alcohols/sulfonates, 447–454, 479–480
alkenes/alkynes, 302–304, 319
carboxylic acids/acid chlorides, 740–742, 749
planar cyclohexanes/chair cyclohexanes, 100–101
Coordinate covalent bond, 229
Coordination polymerization, 1281
Cope, Arthur C., 1069
Cope elimination, **1069**–1070, 1077
Cope rearrangement, 920, 923–925, 929
Corey, E. J., 328, 892, 919, 988, 1125, 1131
Corey, Robert, 1222
Corey lactone, 920, 1125
Corey's prostaglandin synthesis, 919
Corey's total synthesis of α-Caryophyllene, 1131
Corn borer, 217–218
Corn oil, 1173*t*
Corn syrup, 1155
Coronary artery disease, 839
Coronene, 956
Correlation table, 541–542
Cortisol, 1182*t*, 1197
Cortisone, 1182*t*
Cotton, 1156
Coumarin, 766
Coupling constant (*J*), 574, 576*t*
Covalent bond, **8,** 10
COX. *See* Cyclooxygenase (COX)
Cracking, 112, 114, 186
Crafts, James, 1000, 1002
Cram, Donald J., 517
m-Cresol, 957, 959, 1014, 1088
o-Cresol, 1014
p-Cresol, 959
Crick, Francis H. C., 1243, 1245, 1246
Cromolyn sodium, 987–988
Cross-coupling, 1107–1108
Cross-linked network polymer, *1265*
Crossed aldol reaction, 829–830
Crossed Claisen condensation, 835–836
Crotalus adamanteus, 1186
Crotonaldehyde, 693
Crotonic acid, 726
18-Crown-16, 517–518
Crown ethers, 517–518
Crutzen, Paul, 344
Crystalline domain, **1267**
Crystallization, 140
Cumene, 367, 954, 985, 1005
Cumene hydroperoxide, 367
Cumulated diene, **894,** 895
Curcuma longa, 909
Curcumin, 894, 909
Curl, Robert F., 27
Curry, 909
Curved arrow, 46–**47**
Cyanic acid, 67
Cyanide, 395
Cyano group, 783–785
Cyanogen bromide cleavage, 1210–1212, 1230
Cyanohydrins, **664**–665
Cyanuric chloride, 1026
Cyclic acetals, 677
Cyclic amides, 767
Cyclic anhydrides, 764
Cyclic esters, 764–765
Cyclic ethers, 492, 493
Cyclic hemiacetals, 671, 672, 1160
Cyclic hydrocarbon, 82
Cyclic imides, 767
Cycloaddition, 911
Cycloaddition reaction, **909**, 911
[4 + 2] cycloaddition 909, *910*, 911, 928
[2 + 2] cycloaddition 909, *910*, 911, 928
Cycloalkanes. *See also* Alkanes and cycloalkanes
bicycloalkanes, 84, 104–107
cis,trans isomerism, 99–107
conformations, 91–93
defined, **82**
examples, *82*
fragmentation patterns, 615
structure and nomenclature, 82–83
Cycloalkenes, 214–215
Cycloalkynes, 300
Cyclobutadiene, 943, 948, *948*
Cyclobutane, *82,* 91–92, 920
Cycloheptanone, 1062
Cycloheptatriene, 951
Cycloheptatrienone, 934
Cycloheptatrienyl cation, *952*
Cycloheptene, 215
Cyclohexane, 942
alternative chair conformations, *95,* 96–97
chair conformation, 92, *93*
convert planar cyclohexanes to chair cyclohexanes, 100–101
depiction, *82*
disubstituted derivatives, 145–146
How To boxes, 96–97, 100–101
interconversion of alternative chair conformations, *95*
monosubstituted, 97*t*
Cyclohexane derivatives, 145
Cyclohexane oxide, 505
Cyclohexane rings, 102–103
Cyclohexanol, 452, 692, 786, 962
Cyclohexanone, 681, 690, 692
Cyclohexene, 215, *544*, 566, 640–641, 942
2-Cyclohexenecarboxylic acid, 727
2-Cyclohexenone, 896
3-Cyclohexenone, 896
3-Cyclohexenyl methyl ketone, 912
Cyclohexoxycyclohexane, 494
Cyclohexyl butanoate, 786
Cyclohexyl mesylate, 452, 453
Cyclohexyl methanesulfonate, 452, 453
Cyclohexylamine, 1045*t,* 1046*t*
Cyclohexylmethylamine, 1041
Cyclomethycaine, 818
Cyclononyne, 300
1,5-Cycloctadiene, 920
Cyclooctatetraene, 943, 944, 948, *949* 1291
1,3,5,7-Cyclooctatetraene, 948
Cyclooctene, 215
cis-Cyclooctene, 215
trans-Cyclooctene, 215
Cyclooctyne, 300
Cyclooxygenase (COX), 731
Cyclopentadiene, 916, 951, A-2
1,3-Cyclopentadiene, 216
Cyclopentadiene anion, *951*
Cyclopentadienone, 934
Cyclopentadienyl anion, 952
Cyclopentadienyl tetrafluoroborate, 951

Cyclopentane, 92
constitutional isomers, 129
cycloalkane, *82*
disubstituted derivatives, 144–145
^{1}H-NMR spectrum, 566
planar conformation, *92*
puckered envelope conformation, *92*
Cyclopentanecarbaldehyde, 653
trans-1,3-Cyclopentanedicarboxylic acid, 727
cis-1,2-Cyclopentanediol, 145, 512, A-9
trans-1,2-Cyclopentanediol, 145, 511, 512
Cyclopentanone, 566
Cyclopentene, 215
Cyclopentene oxide, 511
Cyclopropane, *82,* 91
Cyclopropene, 951
Cyclopropenyl cation, 951
Cyclopropenyl hexachloroantimonate, 951
Cycrimine, 887
Cys, 1201*t*. *See also* Cysteine
Cysteine, 156, 1201*t*, 1203*t*
L-Cysteine, 475
Cytochrome *c*, 1225
Cytosine (C), *1144, 1239*

D

D. *See* Debye (D)
Dacron polyester (Dacron), 727, 756, 1271, 1272
dAMP. *See* 2′-Deoxydenosine 5′-monophosphate (dAMP)
Darvon, 1079, 1087
Darvon-N, 1087
Darzens condensation, 877
Dative bonds, 18, 229
DCC. *See* 1,3-Dicyclohexylcarbodiimide (DCC)
DCU. *See N-N′*-Dicyclohexylurea (DCU)
ddNTP. *See* 2′,3′-Dideoxynucleoside triphosphate (ddNTP)
DDT, 1025
de Broglie, Louis, 29
de Broglie relationship, 29
Deactivating group, **1006,** 1008*t*
Dean-Stark trap, 675, *675,* 681
Debye (D), 12
dec-, 77*t*
Decalin, *84,* 104
cis-2-Decalone, 715
trans-2-Decalone, 715
Decane, 74*t*, 108*t*, *544*
Decanoic acid, 728*t*, 731, A-17
Decapeptide, 1235
Decarboxylation
β-dicarboxylic acid, 745, 750
β-ketoacids, 743–744, 750
β-ketocarboxylic acid, 744
β-ketoesters, 836–837
defined, **743**
example, 745–746, 837–838
key reactions, 750
malonic acid and substituted malonic acids, 744–745
Decarboxylation/elimination, 756
1-Decyne, 300*t*
DEET. *See N,N*-Diethyl-*m*-toluamide (DEET)
Definitions. *See* Terminology
Dehydration, **454,** 829
Dehydration reactions, 480
Dehydrohalogenation, **302,** 303, **400**
Delocalization, **2, 52**
allyl radical, 351
charge over larger area, 70, 385, 617, 685
molecular structure and acidity, 190
pi electron density over larger area, 70
relative charge delocalization, 174
stability, 70, 71
unpaired electron density over larger area, 70
Demerol, 885, 1079
Denaturation, 1237
Dendritic polymer, *1265*
Density
alkanes and cycloalkanes, 108–109, 109–110, 118
butane, 108*t*
decane, 108*t*
ethane, 108*t*
haloalkanes, 334–335, 335*t*
heptane, 108*t*
low-molecular-weight alkynes, 300*t*
methane, 108*t*
nonane, 108*t*
octane, 108*t*
pentane, 108*t*
propane, 108*t*
2-Deoxy-D-ribose, 1138
2′-Deoxydenosine 5′-monophosphate (dAMP), 1238
Deoxynucleotide triphosphate (dNTP) monomers, 1255
Deoxyribonucleic acid. *See* DNA
Dermatan sulfate, 1169
Deshielding, 563
Desosamine, 1079
Dess-Martin oxidation, 468–469
Dess-Martin periodinane (DMP), 468
Detergents, 1174–1176
Deuterated solvents, 699
Deuterium exchange, 699, 709
Dextrins, 1155
Dextromethorphan, 1032
Dextrorotatory, **150**
Dextrorotatory compound, 151
Dextrose, 1138
Di-*sec*-isoamylborane (sia)$_2$BH, 308, 309
2,6-Di-*tert*-butyl-4-methylphenol, 995, 1025
Diabetes mellitus, 745, 815, 1148–1149
Diacyl peroxide, 1276
Dialkyl ether, 546
Diamagnetic current in NMR, **562**
Diamagnetic shielding, 562
Diamicron, 816
Diamine, 1273
Diastereomeric salts, 157–158
Diastereomers, 137–138
achiral environments, 162
chiral centers, 140
defined, **131,** A-8
HOW TO box, 143–144
relationship among isomers, *148*
stereocenter, A-10
Diastereoselective reaction, 257, 283, A-8
Diastereotopic groups, 582, **583,** 596
Diatrizoic acid, 1091
Diaxial interaction, **96**
Diazepam, 1030–1031
Diazomethane, 739–740, 749, 758
Diazonium fluoroborate, 1065
Diazonium ion, **1061**
Diazotization, 1061, 1076
DIBALH. *See* Diisobutylaluminum hydride (DIBALH)
Dibenzoyl peroxide, 1276
Diborane, 265
1,2-Dibromo-1-butene, 896
1,2-Dibromo-2-butene, 896
(E)-1,4-Dibromo-2-butene, 903
3,4-Dibromo-1-butene, 903
1,2-Dibromo-1,2-diphenylethane, 406, 407
Dibromoalkanes, 257
2,3-Dibromobutane, 257, 303
1,3-Dibromocyclobutane, 365
trans-1,2-Dibromocyclohexane, 257
Dibutyl ether, *547*
Dicarboxylic acids, 726, 754
2,4-Dichloro-1-chloromethylbenzene, 1030
1,1-Dichloro-2,3-diethylcyclopropane, 640
Dichloroacetic acid, 732
Dichlorocarbene, 640–641
Dichlorocyclopropane, 640
Dichlorodiphenyltrichloroethane, 1025
1,1-Dichloroethane, 566, *573*
1,2-Dichloroethane, 90, 566, 626
1,1-Dichloroethylene, 1276*t*
Dichloromethane, 332, 390*t*
2,4-Dichlorophenoxyacetic acid, 1028
Dickson, James Tennant, 1271
Dicoumarol, 766

Dicyclohexyl ether, 494
Dicyclohexyl ketone, 656
Dicyclohexylamine, 694
1,3-Dicyclohexylcarbodiimide (DCC), 1218
N-N′-Dicyclohexylurea (DCU), 1218
Dicyclopentadiene, 916
Dicyclopentylamine, 1040
2,6-Dideoxy-D-altrose, 1163
Dideoxy method, 1255–1256, 1261
2′,3′-Dideoxynucleoside triphosphate (ddNTP), 1255
Dieckmann condensation, 834–835, 869, 870
Dielectric constant, **389**
Diels, Otto, 911
Diels-Alder adduct, **911,** 912
Diels-Alder reaction, 295, 911–920
  arrow pushing, 920
  bicyclic systems, 915–916
  Corey's prostaglandin synthesis, 919
  Diels-Alder adduct, 911
  diene, 917
  dienophile, 911, 916
  effect of substituents on rate, 914–915
  endo substituents, 916
  exo substituents, 916
  key reaction, 929
  mechanism, *918*
  review/study guide, 928, 929
  s-cis conformation, 913, 914
  s-trans conformation, 913, 914
  stereochemistry, 917–918
Dienes, 216, 894, 917. *See also* Conjugated dienes
Dienophile, **911,** 912, 916
Diepoxide, 1273
Diethyl 1,4-cyclohexadiene-1,2-dicarboxylate, 912
Diethyl acetal, 673
Diethyl carbonate, 790
Diethyl ether, 194, 390*t,* 495*t,* 657*t*
Diethyl ketone, 656, 658*t*
Diethyl malonate, 764, 852
Diethyl peroxide, 341
Diethyl phthalate, 887
Diethyl propanedioate, 764, 852
Diethyl sulfide, 518
N,N-Diethyl-*m*-toluamide (DEET), 813
Diethylaluminum chloride, 1280
Diethylamine, 1045*t,* 1046*t*
5,5-Diethylbarbituric acid, 814
Diethylborane, 265
Diethylcarbamazine, 817
Diethylene glycol dimethyl ether, 493
Digitalin, 1167
Digitalis, 1167
Digitalis purpurea, 1134, 1163, 1167
Digitoxin, 1163
Digitoxose, 1134
D-Digitoxose, 1162
Diglyme, 493
Dihaloalkane, 302, 320
(*E*)-Dihaloalkene, 319
cis Dihalocyclopropane, 640
trans Dihalocyclopropane, 640
Dihedral angle, **86,** *87*
Dihydropyran, 487, 677
3,5-Dihydroxy-3-methylpentanoic acid, A-16
Dihydroxyacetone, 1135
2,6-Dihydroxyacetophenone, 987, 988
2,3-Dihydroxybutanedioic acid, 139
(3*R*,4*R*)-Dihydroxyheptane, 144
(3*R*,4*S*)-Dihydroxyheptane, 144
(3*S*,4*S*)-Dihydroxyheptane, 144
(*S*)-(−)-3,4-Dihydroxyphenylalanine, 155
(*R*)-2,3-Dihydroxypropanal, 656
3-Dihydroxypropanone, 1135
3(*R*)-3,5-Dihydroxy-3-methylpentanoic acid, A-16
Diisobutylaluminum hydride (DIBALH), 797, 798
Diisocyanate, 1272
Diisopropylethylamine, 822
2,6-Diisopropylphenol, 1087
β-Diketone, A-2
3,4-Dimethoxybenzaldehyde, 956
2,3-Dimethyl-2-butene, 275*t,* 539, 554, 566
Dimethyl ether, 493, *493,* 495*t,* 546
6,6-Dimethyl-3-heptyne, 298, A-13
2,5-Dimethyl-2,4-hexadiene, 905
3,3-Dimethyl-1,5-hexadiene, 923
N,N-Dimethyl-4-nitrosoaniline, 1059
cis-3,4-Dimethyl-2-pentene, 213
(*Z*)-3,4-Dimethyl-2-pentene, 214
3,3-Dimethyl-2-phenyl-2-butanol, 662
Dimethyl phosphate, 765
(2*E*)-3,7-Dimethyl-2,6-octadienal, 653
Dimethyl sulfate, 963
Dimethyl sulfide, 271
Dimethyl sulfoxide (DMSO), 39, 260, 390*t,* 519, 1100
Dimethyl terephthalate, 1292
Dimethylacetamide, 24
Dimethylacetylene, 299
Dimethylamine, 21, 1038, 1045*t,* 1046*t,* 1056
Dimethylammonium bromide, 1056
N,N-Dimethylbenzamide, 798
1,4-Dimethylbenzene, 968
N,N-Dimethylbenzylamine, 798
2,2-Dimethylbutane, 109*t*
2,3-Dimethylbutane, 109*t*
cis-1,4-Dimethylcyclohexane, 129
cis-1,3-Dimethylcyclobutane, 101
trans-1,3-Dimethylcyclobutane, 101
cis-1,2-Dimethylcyclohexane, 100
cis-1,4-Dimethylcyclohexane, 102
trans-1,4-Dimethylcyclohexane, 102, 129
1,2-Dimethylcyclohexene, 274
1,6-Dimethylcyclohexene, 214, 274
3,3-Dimethylcyclohexene, 215
N,N-Dimethylcyclopentanamine, 1040
cis-1,2-Dimethylcyclopentane, 99
trans-1,2-Dimethylcyclopentane, 99
1,2-Dimethylcyclopentene, 215
1,1-Dimethylethyl, A-11
N,N-Dimethylformamide (DMF), 390*t,* 766, 1100
2,2-Dimethylhexane, 111*t*
2,4-Dimethylhexane, 78, A-12
cis-2,3-Dimethyloxirane, 505
Dimethylphosphate anion, 669
2,2-Dimethylpropyl, A-11
1,3-Dinitrobenzene, 955
m-Dinitrobenzene, 955
6,6′-Dinitrobiphenyl-2,2′-dicarboxylic acid, 976
2,4-Dinitrofluorobenzene, 1236
2,4-Dinitrophenol, 1016
2,4-Dinitrophenylhydrazine, 684*t,* 1016
2,4-Dinitrophenylhydrazone, 684*t*
Dinocap, 1030
-dioic acid, 726
1,2-Diol, 720
1,3-Diol, 720
Diol epoxide, 514
Diol formation, 241*t*
Diol reactions, 461–462
1,4-Dioxane, 124, 527
Dipeptide, **1208**
Diphenadione, 887
Diphenhydramine, 532
Diphenyl ether, 546
Diphenylamine, 1041
(*E*)-1,2-Diphenylethene, 1100
Diphosphate ion, 764
Diphosphoric acid, 764
Dipole-dipole interaction, **441**
Disaccharides, **1151**–1153
Diseases. *See* Disorders/diseases
Diseases of ganglioside metabolism, 1168
Disorders/diseases. *See also* Drugs/medications; Health and wellness
  angina, 886
  asthma, 436
  breast cancer, 993
  coronary artery disease, 839
  diseases of ganglioside metabolism, 1168
  epilepsy, 879, 884
  exercise-induced emphysema, 987
  gastric acidity, 1234
  Gaucher's disease, 1168
  influenza A virus infections, 816
  migraine headache, 724
  motion sickness, 1034
  Niemann-Pick disease, 1168

ocular hypertension, 1195
peptic ulcers, 1234
prostate cancer, 1028
schizophrenia, 1035, 1036
Tay-Sachs disease, 1168
vaginal yeast infections, 1030
Disparlure, 717
Dispersion forces, **107**–108, 217
Disproportionation, **1278**
Dissolving-metal reduction, 313
Distyryl dianion, 1285
Disubstituted benzenes, 954–955
Disubstituted carbon-carbon double bonds, 273
Disubstituted derivatives of cyclohexane, 145–146
Disubstituted derivatives of cyclopentane, 144–145
Disulfide, 477, **518**
Disulfide bonds, 1223–1224
Diterpenes, *1184*
Diuretics, 1091–1092
1,4-Divinylbenzene, 1301
D,L system, 1136–1138
DMF. *See N,N*-Dimethylformamide (DMF)
DMP. *See* Dess-Martin periodinane (DMP)
DMSO. *See* Dimethyl sulfoxide (DMSO)
DNA
A-DNA, 1246
automation/synthesis, 1239
B-DNA, 1246, *1246*
base composition, 1243–1244
base-pairing, *1245,* 1245–1246
chromatin, 1248
circular, 1247–1248
complementary sequences, 1247
covalent backbone, 1241–1242
double helix, *1244,* 1244–1246
double-stranded, 1255
fingerprinting, 1257–1258
heterocyclic aromatic amine bases, *1239*
histones, 1248
primary structure, 1241–1242
repository of genetic information, 1238
RNA, contrasted, 1248
secondary structure, 1243–1246
sequencing, 1253–1258, 1260–1261. *See also* DNA sequencing
single-stranded, 1255, *1255*
supercoiling, 1247–1248
superhelical twists, *1247,* 1248
tertiary structure, 1247–1248
tetranucleotide section, *1242*
Watson-Crick model, 1243, 1244
x-ray diffraction patterns, 1244
DNA double helix, *1244,* 1244–1246
DNA fingerprinting, 1257–1258
DNA polymerase, 1255, *1255*
DNA sequencing, 1253–1258, 1260–1261
chain termination (dideoxy) method, 1255–1256, 1261
DNA replication *in vitro, 1255*
next-generation DNA sequencing, 1258, 1261
restriction endonuclease, 1253–1254
review/study guide, 1260–1261
sequencing the human genome, 1256–1257
dNTP monomers. *See* Deoxynucleotide triphosphate (dNTP) monomers
dodec-, 77*t*
Dodecahedrane synthetic intermediate, 934
Dodecan-1-ol, A-17
Dodecane, 74*t*
Dodecanoic acid, 728*t*
Domino sequence, 326
Dopamine, 155, 608, *609*
Double-headed arrow, **46**
Double helix, *1244,* **1244**–1246
Double-sided arrowhead, 233
Double-stranded B-DNA, *1246*
Double-stranded DNA (dsDNA), 1255
Doublet, 573, 574
Doublet of doublets, 577
Douglas tussock moth, 719
Dowd, Paul, 1190
Downfield, **564,** *565*
Doxaprost, 1196
Drug-receptor interactions, 442–443
Drugs/medications. *See also* Disorders/diseases
albuterol, 992
Aleve, 731
Allegra, 758, 1092–1093
amitriptyline, 1031
amoxicillin, 167, 762, 791, 792
ampicillin, 792
antidepressants, 491, 1031. *See also* names of individual antidepressants
antihistamines, 1092–1093
antiviral drugs, 1244–1245
aspirin, 730–731, 787, 1178, 1179
atenolol, 532, 990
atorvastatin, 442, 491, 892
Augmentin, 792
barbital, 814
Benadryl, 532
benzocaine, 755
Benzocaine, 755
beta-adrenergic blockers, 1033
beta-2 adrenergic blockers, 991
beta-blockers, 990, 991
β-lactam antibiotics, 791–793
Betapace, 1093
bidisomide, 989–990
blood thinners, 766
bumetanide, 1092
bupropion, 1030
calcium channel blockers, 882
carbuterol, 991–992
cardiac arrhythmia, 1093
Celebrex, 731
cephalexin, 792
cetirizine, 532, 723–724
chiral drugs, 155
cholesterol-lower drugs, 839–840
cimetidine, 1234
clozapine, 532, 1036
cromolyn sodium, 987
cyclomethycaine, 818
Demerol, 885, 1079
diethylcarbamazine, 817
digitalis, 1167
diphenhydramine, 532
diuretics, 1091–1092
Effexor, 1031
elitriptan, 724
Equanil, 813
erythromycin antibiotics, 988
fexofenadine, 1092–1093
fluoxetine, 491
fluticasone, 436
furosemide, 1092
gabapentin, 884
Haldol, 1035
haloperidol, 1035
heparin, 1168
ibuprofen, 665, 731, 864–866, 1029
Keflex, 792
labetalol, 1033–1034
Lasix, 1092
linezolid, 821–822
Lipitor, 442, 491, 892
local anesthetics, 765, 1036
meclizine, 1034
meperidine, 1079
meprobamate, 813
methicillin, 792
moclobemide, 532, 817
Monistat, 1030
naproxen, 731
Navicalm, 1034
Nembutal, 813
Novocaine, 815
NSAIDs, 731, 1178, 1179
paroxetine, 491
Paxil, 491
penicillins, 791
pentobarbital, 813
procaine, 815
proparacaine, 1036
propranolol, 990
Proventil, 992
Prozac, 491, 989
Rescula, 1195
Rimantadine, 324–325, 715

Drugs/medications (*Continued*)
Rotoxamine, 987
salicin, 1165
sertraline, 491
sildenafil, 1090
simvastatin, 839, 840
sotalol, 1093–1094
spasmolytol, 532, 1035
SSRIs, 491
Tagamet, 1234
tamoxifen, 993
valnoctamide, 887
vancomycin, 1129
venlafaxine, 1031–1032
Viagra, 1090
vitamin C, 1141
warfarin, 766
Wellbutrin, 1030
Zocor, 839, 840
Zoloft, 491
Zyban, 1030
Zyrtec, 532
Zyvox, 821–822
dsDNA. *See* Double-stranded DNA (dsDNA)
Dupont (E.I. Dupont de Nemours and Company), 1269, 1271
Duranest, 765
Dyeing fabrics, 1026

E

E, **213**
-*e*, 81
E85, 437
E1 reaction
acid-catalyzed dehydration of 2-Butanol, 456
2-Bromo-2-methylpropane, 402
defined, **402**
E2 reaction, compared, 408*t*
energy diagram, *403*
key reaction, 425
kinetics, 404
regioselectivity, 404
S_N1 reaction, 410–411
summary/overview, 408*t*, *413*, 413*t*, 414, 426
E2 reaction
anti and coplanar arrangements, 409–410
2-Bromobutane, 403
cis-1-Chloro-2-isopropylcyclohexane, 408
defined, **402**
E1 reaction, compared, 408*t*
enantiomers of 1,2-Dibromo-1,2-diphenylethane, 407
energy diagram, *403*
key reaction, 426
kinetics, 404
meso-1,2-Dibromo-1,2-diphenylethane, 406–407
regioselectivity, 405
S_N2 reaction, 411–412
stereoselectivity, 404–407
summary/overview, 408*t*, *413*, 413*t*, 414, 426
Eastern diamondback rattlesnake, 1186
Eclipsed conformation, **86**, *86*, 89
Eclipsed-interaction strain, 87
Edman, Pehr, 1213
Edman degradation, **1213**–1215, 1230
ee. *See* Enantiomeric excess (ee)
Effexor, 1031
E.I. Dupont de Nemours and Company, 1269, 1271
EI-MS. *See* Electron ionization mass spectrometry (EI-MS)
eicos-, 77*t*
Eicosane, 74*t*
Eicosanoids, 1179–1180
Einstein, Albert, 29
Elastomer, **1268**
Elastomeric O-rings, 1268
Electric cars, 1279
Electromagnetic induction, 562
Electromagnetic radiation, 535–536
Electromagnetic spectrum, 536*t*, 537*t*
Electron affinity, **9**
Electron ionization mass spectrometer, *607*
Electron ionization mass spectrometry (EI-MS), 609
Electron pushing, 46, **233**. *See also* Arrow pushing
Electron-releasing groups, 915*t*, 1021
Electron-releasing inductive effect, 1011
Electron shells, 2
Electron sink, **235**
Electron source, **235**
Electron-withdrawing groups, 915*t*, 1021
Electronegativity, **8,** 9, 12, 569–570
Electrons
atom, 3
bonding, 14
electronegative atoms, 70
nonbonding, 14
shells, 2*t*
valence, 6
Electrophile, 71, 228–231
defined, **236**
electron-rich species, 195
Electrophilic, 230, **244**
Electrophilic addition
alkenes, 244–264
alkynes, 305–307
allylic halogenation, 349
conjugated dienes, 898–903, 926
defined, **244**
Electrophilic aromatic substitution
chlorination and bromination, 996–998
defined, **996**
example, 1005–1006
Friedel-Crafts acylation, 1002–1004, 1019
Friedel-Crafts alkylation, 1000–1002, 1019
generation of carbocations from alcohol, 1005, 1020
generation of carbocations from alkenes, 1005, 1019
halogenation, 1018
key reactions, 1018–1020
nitration and sulfonation, 998–1000, 1018, 1019
pyrrole, 1023
review/study guide, 1017–1018
Electrophilicity, 230
Electrophoresis, *1206*, **1206**–1207, 1253
Electrospray ionization mass spectrometry (ESI-MS), 609, 629
Electrostatic attraction, *108*
Electrostatic potential map, A-7
acetate ion, 190
acetone, 49, 657
allylic carbocation, 903
α,β-unsaturated carbonyl, 852
ammonia, 28
chloroacetate ion, 191
ethoxide ion, 190
ethylene, A-7
formaldehyde, 28
MacSpartan, A-7
methyl cation (CH_3^+), A-7
trifluoroethoxide ion, 191
water, 28
Elimination reaction, 238
Elion, Gertrude, 1261
Elitriptan, 724
Elpot, A-7. *See also* Electrostatic potential map
Elution time, 159
-*en*-, 81
Enamines, 841–844. *See also* Enolate anions and enamines
acylation, 843, 871
alkylation, 842, 870
defined, **681**
formation, 681, 706, *841*
retrosynthetic analysis, 844
review/study guide, 870–871
Stork enamine reaction, 841
Enanthotoxin, 322
Enantiomeric excess (ee), **152**–153, A-8
Enantiomerically pure prostaglandin F_{2a}, 920

Enantiomers, 137–138, 162
achiral environments, 140, 162
chiral centers, 140
defined, **130**
HOW TO box, 143–144
relationship among isomers, *148*
separation of (resolution), 157–159
stereocenter, A-10
tartaric acid, 140
Enantioselective reaction, 257, **281,** 283, A-8
Enantioselective reduction, 287
Enantiotopic groups, **582,** 583, 596
Endergonic reaction, **186**
Endo adduct, 916, 917
Endo substituents, 916
Endonuclease, 1253–1254
Endothermic reaction, **186**
Endothermicity values, 244
Enediol, 1146
Energetics, haloalkanes, 339–340, 343
Energy
defined, **5**
fossil fuels, 112
potential, 5
renewable energy sources, 112
resonance, 942
Energy level diagram
chlorine, 5
what is it?, 4
Energy-minimized eclipsed conformation, 89, *89*
Energy-minimized gauche conformation, 90, *90*
Engine knocking, 114
Enol, **308,** 309
Enol form, 685
Enolate anions and enamines, 823–893
acetoacetic ester synthesis, 845–849, 871
aldol reaction, 825–832. *See also* Aldol reaction
biological world, 838–841
Claisen condensation, 832–834, 869
conjugate addition to α,β-unsaturated carbonyl compounds, 851–860, 872–873
crossed Claisen condensation, 835–836
crossed enolate reactions using LDA, 860–864, 874
definitions, **681, 685**
Dieckmann condensation, 834–835, 869, 870
enamines. *See* Enamines
formation and reaction of enolate anions, 823–825, 866–867
key reactions. *See* Enolate anions and enamines—key reactions
kinetic control, 855, 864
lithium diisopropylamide (LDA), 860–864, 874
lithium diorganocopper reagents, 858–860, 873
malonic ester synthesis, 849–850, 872
mechanisms. *See* Enolate anions and enamines—reaction mechanisms
Michael reaction, 852–857, 872–873
reaction roadmap, 889–890
review/study guide, 886–874
Robinson annulation, 857–858, 873
thermodynamic control, 855, 864
variants of acetoacetic ester synthesis, 848
Enolate anions and enamines—key reactions
acetoacetic ester synthesis, 871
acylation of enamine followed by hydrolysis, 871
aldol reaction, 868
alkylation of enamine followed by hydrolysis, 870
Claisen condensation, 869
conjugate addition of lithium diorganocopper reagents, 873
dehydration of product of aldol reaction, 868
Dieckmann condensation, 870
malonic ester synthesis, 872
Michael reaction, 873
Robinson annulation, 873
Enolate anions and enamines—reaction mechanisms
acid-catalyzed aldol reaction, 826–827
acid-catalyzed dehydration of aldol product, 828–829
alkylation of enamines, 842
base-catalyzed aldol reaction, 826
Claisen condensation, 833–834
Michael reaction—conjugate addition of enolate anions, 854
Enolate-forming reactions, 685
Enthalpy (H^0), 186, 187
Enthalpy change (ΔH^0), **186,** 187*t*
Entropy change (ΔS^0), 186, 187*t*
Entropy (S^0), 186
Envelope conformation, 92
Enzymatic cleavage, 1212–1213
Enzyme-catalyzed hydrolysis of peptide bonds, 1212
Enzymes
chirality, 155–156
molecule/enantiomer, 154
resolving agents, 158–159
synthesis of single enantiomers, 919
Ephedra sinica, 165
Ephedrine, 165
Epibatidine, 1081
Epichlorohydrin, 515–517, 531, 988, 1273
Epilepsy, 879, 884
Epinephrine, 1078, 1233
Epipedobates tricolor, 1081
Epoxidation of alkene by RCO_3H, 508
Epoxides
acid-catalyzed ring opening, 510–512
common names, 505
defined, **505**
epoxidation of alkene by RCO_3H, 508
ethylene oxide, 505
Grignard reagents, 645
internal nucleophilic substitution in halohydrins, 506
IUPAC system, 505
key reactions. *See* Epoxides—key reactions
mechanisms. *See* Ethers—reaction mechanisms
nucleophilic ring opening, 512–513
oxidation of alkenes with peroxycarboxylic acids, 507–508
reactions, 510–513
review/study guide, 522–524
Sharpless asymmetric epoxidation, 508–510
stereoselective oxirane synthesis, 506–507
synthesis, 505–510
Epoxides—key reactions
acid-catalyzed hydrolysis of epoxides, 524
nucleophilic ring opening of epoxides, 524
oxidation of alkenes with peroxycarboxylic acids, 523
reduction of epoxide to alcohol, 524
Sharpless asymmetric epoxidation, 523
synthesis of epoxides from halohydrins, 523
epoxy-, 505
Epoxy resins, 1273, 1295
1,2-Epoxycyclohexane, 505
1,2-Epoxycyclopentane, 511
ε. *See* Molar absorptivity (ε)
Equanil, 813
Equatorial bond, **93,** 97
Equilibrium constant (K_{eq}), 181, A-1
Equilibrium control, 901. *See also* Thermodynamic control
Equilibrium populations of conformations, 98–99
Equivalent hydrogens, **565**–566
Erythritol, 1146
Erythromycin, 1079
Erythromycin antibiotics, 988

Erythronolide B, 988
(2*R*,3*R*)-Erythrose, 138
(2*S*,3*S*)-Erythrose, 138
Erythroxylon coca, 765
ESI-MS. *See* Electrospray ionization mass spectrometry (ESI-MS)
Essential oils, 219
Esterases, 159
Esterification, 738–740
Esters. *See also* Carboxylic acid derivatives
  acetoacetic ester synthesis, 845–849, 871
  acid-catalyzed ester hydrolysis, 775–776
  alcohols, 787, 788, 805
  ammonia and amines, 789, 806
  carboxylic acids, of, 764
  cyclic esters (lactones), 764–765
  defined, **23**
  DIBALH reduction, 798
  flavoring agents, as, 741
  Grignard reagents, 793–795, 807
  hydrolysis, 775–779, 803–804
  infrared absorptions, 551*t*
  IUPAC system/common names, 764–766, A-15
  malonic ester synthesis, 849–850, 872
  mass spectra, 619–621
  microscopic reversibility, 776–777
  NMR spectra, 589–590
  organolithium compounds, 795
  phosphoric acid, of, 765–766
  reduction, 796–798, 807–808
  saponification, 777–778
  transesterification, 787, 788
Estradiol, 1181
β-Estradiol, 993
Estrogens, 993, 1181, 1182*t*
Estrone, 1133, 1181, 1182*t*
Et. *See* Ethyl (Et)
^{3}-allyl complex, 1106
eth-, 77*t*
Ethanal, 658*t*. *See also* Acetaldehyde
Ethanamide, 789
Ethane
  ball-and-stick model, *73*
  boiling point, 108*t*, 442*t*
  bond length/bond strength, 56*t*
  covalent bonding of carbon, 44*t*
  cracking, 112
  density, 108*t*
  dihedral angles, *87*
  eclipsed conformation, *86*
  Lewis structure, *42, 73*
  melting point, 108*t*
  molecular formula/condensed structural formula, 74*t*
  radical chlorination, 341–342
  solubility, 442*t*
  staggered conformation, *86*
1,2-Ethanediamine, 1273
Ethanedioic acid, 727
1,2-Ethanediol, 440, 1271, 1272
Ethanenitrile, 768
Ethanethiol, 437, 475*t*
Ethanoic acid, 726. *See also* Acetic acid
Ethanol
  alcohol, 19, 437
  boiling point, 442*t*, 495*t*
  Claisen condensation, 832
  esters' reaction with ammonia and amines, 789
  nomenclature, 81
  oxidation of alcohols, 471
  pK_a, 190
  protic solvent, 389*t*
  solubility, 442*t*, 443, 495*t*
Ethanol-*d*, 699
Ethanolamine, 1186*t*
Ethanoyl chloride, 763
Ethene, 212. *See also* Ethylene
Ethene glycol, 440
Ethenyl, 212
Ethenylcyclopentane, 212
Ethers, 492–534
  acid-catalyzed addition of alcohols to alkenes, 499
  acid-catalyzed cleavage by concentrated HX, 500–501
  acid-catalyzed dehydration of alcohols, 497–498
  alkyl-aryl, 962–963
  benzyl, 969, 975
  boiling points, 495*t*
  common names, 493, 494
  crown, 517–518, 525
  defined, **493**
  epoxides. *See* Epoxides
  ethylene oxide and epichlorohydrin, 515–517
  flammability of hydroperoxides, 502
  hydrogen bond acceptors, *495*
  infrared spectra, 546–547
  IUPAC system, 493, 494
  key reactions. *See* Ethers—key reactions
  mechanisms. *See* Ethers—reaction mechanisms
  NMR spectra, 589
  physical properties, 494–495
  preparation, 496–499
  reaction roadmap, 295–296
  reactions, 500–502
  review/study guide, 520–525
  safety alert, 502
  silyl ethers as protecting groups, 503–504
  solubilities, 495*t*
  structure, 493
  sulfides, 518–519, 525
  THP, 677–678
  trityl, 985
  Williamson ether synthesis, 496–497
Ethers—key reactions
  acid-catalyzed addition of alcohols to alkenes, 521
  acid-catalyzed cleavage of dialkyl ethers, 521
  acid-catalyzed dehydration of alcohols, 521
  reaction of alcohols with Chloro-*tert*-butyldimethylsilane, 521
  Williamson ether synthesis, 520–521
Ethers—reaction mechanisms
  acid-catalyzed addition of alcohol to alkene, 499
  acid-catalyzed cleavage of dialkyl ether, 500
  acid-catalyzed hydrolysis of epoxide, 510–511
  acid-catalyzed intermolecular dehydration of primary alcohol, 498
  epoxidation of alkene by RCO_3H, 508
  nucleophilic opening of epoxide ring, 512
Ethosuximide, 885
Ethoxide, 395
Ethoxy radical, 341
2-Ethoxy-2-methylpropane, 494
(1*R*,2*R*)-2-Ethoxycyclohexanol, 493
trans-2-Ethoxycyclohexanol, 493
Ethoxyethane, 493
2-Ethoxyethanol, 493
Ethoxyethene, 546
Ethyl (Et), A-11
Ethyl acetate, 550, 764, 832, A-15
Ethyl acetoacetate, 832
Ethyl acrylate, 1276*t*
Ethyl 4-aminobenzoate, 956
Ethyl *p*-aminobenzoate, 956
Ethyl β-ketobutyrate, 767
Ethyl bromide, 387, 388
Ethyl butanoate, *550,* 741, 790
Ethyl cyanoacetate, 885
Ethyl cyclopentanecarboxylate, 716
4-Ethyl-3,3-dimethyl-1-octene, 211
4-Ethyl-2,2-dimethylhexane, 78, A-12
Ethyl esters, 159
Ethyl ethanoate, 550, 764, 832, A-15
Ethyl ether, 493
Ethyl formate, 741
2-Ethyl-1-hexanol, 884
(*S*)-3-Ethyl-2-hydroxypentanal, 654
Ethyl isopropyl ether, 608
Ethyl isopropyl ketone, 656
Ethyl isopropyl sulfide, 518
Ethyl methyl ketone, 658*t*
2-Ethyl-4-methyl-1-pentene, 211, A-13
4-Ethyl-1-methylcyclohexene, 214

3-Ethyl-5-methylheptane, 78, A-12
3-Ethyl-2-methylhexane, 79, A-12
2-Ethyl-3-methylpentanamide, 887
Ethyl 2-methyl-3-oxopentanoate, 833, 848
Ethyl nicotinate, 812
Ethyl 3-oxobutanoate, 767, 832
Ethyl 2-oxocyclopentanecarboxylate, 848
(1*E*,3*E*)-3-Ethyl-1-phenyl-1,3-hexadiene, 1098
(1*E*,3*Z*)-3-Ethyl-1-phenyl-1,3-hexadiene, 1098
Ethyl phenylacetate, 789
Ethyl propenoate, *579*
Ethyl *p*-toluenesulfonate, 452
Ethyl tosylate, 452
Ethyl vinyl ether, 546
Ethylamine, 1045*t*, 1046*t*
Ethylbenzene, 954
Ethylborane, 265
Ethylene
  anionic chain-growth polymerization, 1296
  bond length/bond strength, 56*t*
  cationic chain-growth polymerization, 1296
  covalent bonding, 44*t*, *209*
  electrostatic potential map, A-7
  heat of hydrogenation, 275*t*
  Lewis structure, 13*t*, *43*
  nomenclature, 212
  physical properties, 218*t*
  $\pi \rightarrow \pi^*$ transition, 908*t*
  radical chain-growth polymerization, 1277, 1296
  shape, *25*
  side view/end view, 207
  terephthalic acid, 1272
  titanium-mediated (Ziegler-Natta) chain-growth polymerization, 1280, 1296
  uses, 1276*t*
Ethylene diamine, 1273
Ethylene glycol, 1271, 1272, 1292
Ethylene glycol dimethyl ether, 495*t*
Ethylene oxide, 494, 505, 515–517, 1272
Ethylmagnesium bromide dietherate, *632*
(*R*)-Ethylmethylamine, 1042
(*S*)-Ethylmethylamine, 1042
Ethylsulfanylethane, 518
2-Ethylsulfanylpropane, 518
Ethyne. *See also* Acetylene
  boiling point, 300*t*
  density, 300*t*
  melting point, 300*t*
  nomenclature, 81
  simplest alkyne, 298
1-Ethynylcyclohexanol, 663
Etidocaine, 765
Eugenia aromatica, 957
Eulexin, 1029
European corn borer, 218
Eutypa lata, 1123
Eutypine, 1123
Exaltolide, 1124
Excited state, **6, 34**
Exercise-induced emphysema, 987
Exergonic reaction, **186**
Exo adduct, 916
Exo substituents, 916
Exothermic reaction, **186,** 360, 942
Exothermicity values, 244
E,Z system, **213,** 214

F

FAB. *See* Fast-atom bombardment (FAB)
Fabric dyeing, 1026
FAD. *See* Flavin adenine dinucleotide (FAD)
$FADH_2$, 1176–1177
Famotidine, 65
Faraday, Michael, 562, 938
Farnesol, A-16
Farnesyl pyrophosphate, *1184*
Fast-atom bombardment (FAB), 609
Fast exchange, 581–582
Fat-soluble vitamins, 1187–1190
  vitamin A, 1188
  vitamin D, 1188–1189
  vitamin E, 1189–1190
Fats, 220–221, **1172,** 1173*t*
Fats and oils, 1171–1174
Fatty acid oxidation, 1176–1177
Fatty acids, 220, 221, *1171,* **1171**–1172, *1173*
Favorskii rearrangement, 716
Female sex hormones, 1181, 1182*t*
Fenfluramine, 1089
Fenn, John B., 621
Fexofenadine, 317, 758, 1092–1093
Feynman, Richard, 1268
Fingerprint region, **542**
First-generation automated DNA sequencing machines, 1256
Fischer, Emil, 738, 1136, 1208
Fischer esterification, 738–739, 749, 760–761, 776, 1217
Fischer projection, **142**–143, 363, **1136,** 1138, A-9
Fishhook arrow, 233, **335**
Five finger death, 322
Five-membered rings, 70
Flavin adenine dinucleotide (FAD), 1176–1177
Flavoring agents, 741
Fleming, Alexander, 791
Florey, Howard, 791
Fludeoxyglucose F-18, 721
Fluid-mosaic model, **1187,** *1187*
Fluoride ion, 395, 504
Fluorine, 332, 959
Fluorine-18, 721
Fluormethane, 598
Fluoromethane, 333
Fluoxetine, 491
Flutamide, 1028–1029
Fluticasone, 436
FMOT. *See* Frontier molecular orbital theory (FMOT)
Foam stabilizer, 1176
Food additives (flavoring agents), 741
Food packaging, 1276*t*
Food spoilage, 1025
Forbidden pericyclic reaction, 909
Fordiuran, 1092
Formal charge, **14**–18
Formaldehyde
  boiling point, 658*t*
  electrostatic potential map, 28
  Lewis structure, 13*t*, 653
  nomenclature, 656
  shape, *25*
  solubility, 658*t*
  structural formula, 658*t*
Formation of chemical bonds, 7–8
Formic acid, 389*t*, 656, 726, 728*t*, 1148
Fossil fuels, 112
Fountain of youth, 1249
Fourier transform NMR (FT-NMR) spectrometer, 563–565
Foxglove, 1134, 1163, 1167
Fractional crystallization, 157, *158*
Fractional distillation, 113, *113*
Franklin, Rosalind, 1243–1245
Free radical, 335. *See also* Radical
Free-radical halogenation, 338–339
Freons, 344
Frequency, **535,** 536
Freud, Sigmund, 765
Friedel, Charles, 1000, 1002
Friedel-Crafts acylation, 1002–1004, 1019
Friedel-Crafts alkylation, **1000**–1002, 1017, 1019
Frontalin, 528, 884
Frontier molecular orbital theory (FMOT), 909–911
Frost circle, **944,** *944,* 945
α-D-Fructofuranose, 1142
β-D-Fructofuranose, 1142
Fructofuranoside, 1149
Fructose, 1152*t*, 1158
D-Fructose, 1138, 1142
FT-NMR spectrometer. *See* Fourier transform NMR (FT-NMR) spectrometer
Fucose, 1150
L-Fucose, 1153
Fugu, 105

Fukui, Kenichi, 909
Fuller, R. Buckminster, 27
Fullerene, 27–28
Functional derivatives of carboxylic acids. *See* Carboxylic acid derivatives
Functional group
alcohols, 19, 20
aldehydes, 21–22
alkynes, 317–318
amines, 21
carboxylic acids, 22, 23
carboxylic amides, 24
carboxylic esters, 23
defined, **19**
ionization of, at physiological pH, 182–183
ketones, 21–22
order of precedence, 655*t*
reactions, 71
same types of chemical reactions, 240
Furan, 950, 1140, 1297
Furanose, **1140**
Furfural, 1297
Furfurylamine, 1092
Furosemide, 1092

G

G. *See* Guanine (G)
GABA, 728, 884, 1202
Gabapentin, 884
D-Galactosamine, 1139
Galactose, 1152*t*
D-Galactose, 1138
Gallstones, 1181
λ-lactam, 813
Ganglefene, 1088
Ganglioside GM_2, 1168
Garden thyme, 957
Gas chromatography mass spectrometry (GC-MS), 607, 623
Gasoline, 114
Gastric acidity, 1234
Gastrografin, 1091
Gastrointestinal tract imaging, 1091
Gauche conformation, **88**
Gaucher's disease, 1168
GC-MS. *See* Gas chromatography mass spectrometry (GC-MS)
Gel electrophoresis, 1253, 1256, *1256*
Geminal coupling, 578, *578*, *579*
Geminal diol (gem-diol), 670
Genetic code, 1250–1252
Genetic information, 1238. *See also* DNA; RNA
Geometry of atoms, 45
Geranial, 465, 653
Geraniol, 219, 220, 282, 465
Geranyl pyrophosphate, *1184*
Gibbs free energy (G^0), 70, 71, 185
Gibbs free energy change (ΔG^0), **185,** 187*t*
Gibbs free energy of activation, 186, 198
Gibbs-Helmholtz equation, 186
Gilbert, Walter, 1236, 1254
Gilman, Henry, 636
Gilman reagents, 636. *See also* Lithium diorganocopper reagents
Gilvocarcin M, 1127
Glass substitutes, 1276*t*
Glass transition temperature (T_g), **1267**
Glassy polymers, 1267
Gliclazide, 816
Gln, 1201*t*. *See also* Glutamine
Glossary, G-1 to G-11
Glu, 1201*t*. *See also* Glutamic acid
Glucagon, 1235
D-Glucitol, 1145
Glucocorticoid hormones, 1182*t*
D-Gluconate, 1146
α-D-Glucopyranose, *1140,* 1142
β-D-Glucopyranose, *1140,* 1142
β-D-Glucopyranoside, 1144
D-Glucosamine, 1139
Glucosaminoglycans, 1156–1157
Glucose, 1138, 1152*t*
D-Glucose, 671, *1140,* 1146, 1148
α-D-Glucose, *1140,* 1142
β-D-Glucose, *1140,* 1142
Glucose 6-phosphate, 40
Glucose metabolism, 721
Glucose oxidase, 1148
β-D-Glucopyranose, 438
β-Glucosidases, 1155
β-D-Glucoside, 1149
D-Glucuronic acid, 1147
Glucuronides, 1147
L-Glutamate, 682
Glutamate transaminase, 682
Glutamic acid, 205, 1201*t,* 1202, 1203*t,* 1263
(*S*)-Glutamic acid, 1084
Glutamine, 1201*t,* 1203*t*
Glutamine synthetase, 1227*t*
Glutaric acid, 727, 754
Glutathione, 1235–1236
Gly, 1201*t*. *See also* Glycine
Glyceraldehyde, 134, 656, 1135
D-Glyceraldehyde, 1136
L-Glyceraldehyde, 1136, *1200*
(*R*)-Glyceraldehyde, 142, 1136, A-9
(*R*)-(+)-Glyceraldehyde, 154
(*S*)-Glyceraldehyde, 142, 1136
(*S*)-(−)-Glyceraldehyde, 154
Glyceraldehyde 3-phosphate, 765, 766
Glycerine, 440, 530
Glycerol, 220, 440
Glycine, 1201*t,* 1203*t*
Glycogen, 1154–1155
Glycolic acid, 1275
Glycols, 440
defined, **270**
oxidation of alkenes, 270
oxidative cleavage, 482
periodic acid oxidation, 469–471
Glycolysis, 694
Glycosides, 1143–1145, 1161
N-Glycosides, 1144, 1145, 1161
Glycosidic bond, **1144**
Going green, 1097
Golden orb weaver, 1225
Good cholesterol, 1181
Good nucleophile, 423
Gossyplure, 530
Grape sugar, 1138
Gravitational attraction, 5
Green chemistry, 1097
Greenhouse gases, 344
Grignard, Victor, 631
Grignard reagents, 631. *See also* Organomagnesium and organolithium compounds
aldehydes and ketones, 660–662, 703
carbonation, 748
carboxylic acid derivatives, 793–795
epoxide, 645
esters, 807
strong base, as, 634
Ground state, **6, 34**
Ground-state electron configuration, **3**
Grubbs, Robert, 1112
Guandino group, 182
Guanidine, 1052–1053
Guanidinium ion, 1052
Guanine, 422
Guanine alkylation, 421
Guanine (G), *1144, 1239*
Gypsy moth, 717

H

^{1}H-NMR chemical shifts, A-4
^{1}H-NMR spectroscopy, 558. *See also* Nuclear magnetic resonance (NMR) spectroscopy
Hair growth stimulant, 821
Haldol, 1035
Half-reaction, 268–269
Haloalkanes, 330–369
allylic bromination, 352
allylic halogenation, 348–352
boiling point, 333–334, 334*t*
bond lengths/bond strengths, 335–336
chain propagation steps, 343
common names, 331–332
defined, **331**
density, 334–335, 335*t*
energetics, 339–340, 343
formation of radicals, 340–341

halogenation of alkanes, 336–348
Hammond's postulate, 345–348
Heck reaction, 1099
IUPAC system, 331, 332
key reactions. *See* Haloalkanes—key reactions
mechanisms. *See* Haloalkanes—reaction mechanisms
methyl, 397*t*, 424, 426
nucleophilic substitution, 376–377. *See also* Nucleophilic substitution and β-elimination
physical properties, 332–336
polarity, 332–333
primary, 397*t*, 408*t*, 414, 424, 426
radical addition of HBr to alkenes, 356–358
radical autoxidation, 353–354
radical chain mechanism, 341–343
reaction roadmap, 295–296
regioselectivity, 337–338, 345–348
review/study guide, 359–362
secondary, 397*t*, 408*t*, 414, 424, 426
stereochemistry of radical halogenation, 348
tertiary, 397*t*, 408*t*, 414, 424, 426

Haloalkanes—key reactions
allylic bromination, 361
autooxidation, 362
chlorination and bromination of alkanes, 360
HBr addition to alkenes under radical conditions, 362

Haloalkanes—reaction mechanisms
allylic bromination of propene using NBS, 350
radical chlorination of ethane, 341–342
radical initiated non-Markovnikov addition of HBr to alkenes, 357

Haloarene, **331,** 1013
Haloform, **332**
Halogenation, 241*t*
aldehydes and ketones, 700–701, 709
alkanes, 336–348
allylic, 348–352
benzene, 968–969, 975, 995, 1017, 1018
dehydrohalogenation, 302, 303, 400
free-radical, 338–339
hydrohalogenation, 241*t*
polyhalogenation, 701
radical, 348

Halogenation of alkanes, 336–348
Halogens, 1021
Halohydrin, **259,** 506
Halokene, **331**
α-Haloketone, 700
Halomethanes, 333, 333*t*
Haloperidol, 1035
Halophenols, 959
Hammond, George, 345
Hammond's postulate, 345–348, 385
Handedness of molecules, 128–129
Hardening, 1173, 1174
Haworth, Walter N., 1139
Haworth projection, **1139**–1142
HB-Enz, 682
HbS, 1263
HCFCs. *See* Hydrochlorofluorocarbons (HCFCs)
HCN. *See* Hydrogen cyanide (HCN)
HDL cholesterol. *See* High-density lipoprotein (HDL) cholesterol
HDPE. *See* High-density polyethylene (HDPE)
HDPE container, 1280, *1281*
Head-to-head linkage, 1278
Head-to-tail linkage, 1278
Health and wellness
artificial sweeteners, 1152, 1152*t*
food additives, 741
gallstones, 1181
illness. *See* Disorders/diseases
medications. *See* Drugs/medications
oral contraceptives, 1182
vitamins, 1187–1190. *See also* individual vitamins

Heat of combustion, **110,** 111*t*
Heat of hydrogenation, 275–276, 275*t*, 895*t*
Heat of reaction (ΔH^0), **186**
Heck, Richard, 1098, 1107
Heck reaction, 1098–1103
alkene, 1099–1100
base, 1100
haloalkane, 1099
key reaction, 1116
mechanism of the reaction, 1100–1103
nature of the reaction, 1098–1100
review/study guide, 1116
solvent, 1100

Heme, 1224, *1225*
Hemiacetals, 532, **670**–673, 705, 1160
Hemoglobin, 643, 1224, 1227, *1227*, 1227*t*
(*Z*)-6-Heneicosene-11-one, 719
Heparin, 1157, 1168
hept-, 77*t*
heptadec-, 77*t*
Heptadecane, 74*t*
2,4-Heptadiene, 137
(2*E*,4*E*)-2,4-Heptadiene, 216
(2*E*,4*Z*)-2,4-Heptadiene, 216
trans,cis-2,4-Heptadiene, 216
trans,trans-2,4-Heptadiene, 216
1,6-Heptadiyne, 298, A-13
Heptanal, 316, 730*t*
Heptane, 74*t*, 108*t*, 114
1-Heptanol, 730*t*
2-Heptanone, 316
trans-3-heptene, 213
Heptose, 1135
Herbicides, 1028
Herculon, 1276*t*
Heroin, 1032
Heterocycle, **493**
Heterocyclic amine, **1038**
Heterocyclic aromatic amines, **1038,** 1049–1052
Heterocyclic aromatic compounds, 949–950
Heterolytic bond cleavage, **335**
hex-, 77*t*
1,2,3,4,5,6-Hexachlorocyclohextane, 123
hexadec-, 77*t*
10,12-Hexadecadien-1-ol, 217
Hexadecane, 74*t*
Hexadecanoic acid, 1195
2,4-Hexadiene, 900
(2*Z*,4*Z*)-2,4-Hexadiene, 914
Hexamethylenediamine, 1264, 1269, 1270, 1297
Hexanal, 658*t*, 688, 798
1-Hexanamine, 1039
Hexane
boiling point, 108*t*, 442*t*
chemical structure, 390*t*
common name, 80
condensed structural formula, 74*t*, 108*t*
density, 108*t*
dielectric constant, 390*t*
melting point, 108*t*
molecular formula, 74*t*
molecular weight, 442*t*
solubility, 442*t*

Hexanediamide, 767, 1270
1,6-Hexanediamine, 1039, 1269, 1270, 1297
Hexanedioic acid. *See* Adipic acid
Hexanedioyl chloride, 763
Hexanoic acid, 629, 688, 728*t*, 730*t*, 731
1-Hexanol, 555, 634, 969
6-Hexanolactam, 767
6-Hexanolactone, 764
1,5-Hexatriene, 921
(3*E*)-1,3,5-Hexatriene, 908*t*
(2*E*)-Hex-2-en-4-yne, A-13
(*E*)-2-Hexen-1-ol, 441
trans-2-Hexen-1-ol, 441
(2*E*)-Hexen-4-yne, A-13
1-Hexene, 209, 211, A-13
cis-3-Hexene, 213, 315
trans-3-Hexene, 213
Hexose, 1135, 1138
Hexylcaine, 1089
1-Hexyne, 300*t*, 303
HF/AsF_5, 1288
HF/BF_3, 1288
HFCs. *See* Hydrofluorocarbons (HFCs)

Hg(II) salts, 310
HgO, 310
$Hg(OAc)_2$, 310
$HgSO_4$, 310
High-density lipoprotein (HDL) cholesterol, 276, **1181**
High-density polyethylene (HDPE), 1280–1281, 1292, 1301
High-fructose corn syrup, 1155
High-pressure Kolbe carboxylation of sodium phenoxide, 963
High-resolution mass spectrometry, **610**
Higher-molecular-weight alkanes, 118
Highest-occupied molecular orbital (HOMO), **910,** 921
Hippocrates, 730
His, 1201*t*. *See also* Histidine
Histamine, 205, 1233
Histamine H_1 receptor antagonists, 1092
Histapyrrodine, 1088
Histidine, 55, 1201*t*, 1203*t*, 1204–1205
Histone, **1248**
Hitchings, George, 1261
HMG-CoA reductase, 442, 839
Hockey mask, 1272
Hoffmann, Roald, 909
Hofmann, Augustus, 1067
Hofmann elimination, **1067**–1069, 1077
Hofmann rule, **1068,** 1069
HOMO, **910,** 921
Homolytic bond cleavage, **335**
Homonuclear diatomics, 10
Homotropic groups, **582,** 596
Honey, 1152*t*
Hooke's law, 540, 541, 553
Hormones
  female sex, 1181, 1182*t*
  glucocorticoid, 1182*t*
  male sex, 1182*t*
  mineralocorticoid, 1182*t*
  steroid, 1181–1183
Horses, 1155
Housefly, 325, 630
How To boxes
  alternative chair conformation of cyclohexane, 96–97
  aromatic compounds, 953
  balanced half-reaction, 270–271
  chiral molecules, 130–131
  convert planar cyclohexanes to chair cyclohexanes, 100–101
  curved arrows, 46–47
  diastereomers, 143–144
  enantiomers, 143–144
  equilibrium constants for acid-base reactions, 181
  formal charge, 14
  geometry of atoms, 45
  hybridization, 45
  index of hydrogen deficiency, 208
  Lewis structures of condensed structural formulas, 17
  retrosynthetically dissect an amine/reductive animation, 695–696
  R,S system, 136
  solving NMR spectral problems, 591–593
Hückel, Erich, 944
Hückel criteria for aromaticity, **944**
Hughes, E. D., 383, 384
Human fat, 1173*t*
Human Genome Project, 1256
Human genome sequencing, 1256–1257
Human white cells, 529
Hund's rule, **3**–4
HX addition to an alkene, 251
Hyaluronic acid, 1157, 1168
Hybrid orbital
  defined, **35**
  sp, 38
  sp^2, 37–38
  sp^3, 35–37
Hybrid valence bond/molecular orbital theories, 39–44
Hybridization
  chemical shift, 570–571, 570*t*
  contributing structures, 54
  defined, **35**
  How To box, 45
  molecular structure and acidity, 192
  resonance/MO theory, 52–54
  sp, 38
  sp^2, 37
  sp^3, 36
  valence bond theory concept, 40
Hydration, 241*t*, **252**
Hydrazine, 682, 684*t*
Hydrazone, 682
Hydride ion, **691**
Hydrindane, *84*, 104
Hydroboration, 241*t*, 265
Hydroboration-oxidation
  alkenes, 264–268, 285
  alkynes, 307–309, 320–321
  cycloalkenes, A-9
  defined, **264,** 285
Hydroboration-protonolysis, alkynes, 313, 321
Hydrocarbon
  aliphatic, 72
  antiaromatic, 947–949
  aromatic, 945–947
  aromatic hydrocarbon ions, 950–952
  classes, *73*
  cyclic, 82
  defined, 72
  saturated, 72
  unsaturated, 73, 206
Hydrochlorination, 241*t*
Hydrochlorofluorocarbons (HCFCs), 344
Hydrocortisone, 1197
Hydrofluorocarbons (HFCs), 344
Hydrogen
  acetylenic, 572
  α-. *See* α-hydrogen
  classification, 82
  equivalent, 565–566
  index of hydrogen deficiency, 208
  isotopes, 611*t*
  molecular orbital energy diagram, *34*
  vicinal, 574
  vinylic, 570–572
Hydrogen bonding
  alcohols, 443
  amide groups, *1222*
  defined, **441**
  drug-receptor interactions, 442–443
  example, 1226
  tertiary structure of proteins, 1224
Hydrogen chloride, 13*t*, 445*t*
Hydrogen cyanide (HCN), 664–666, 703
Hydrogen halides, 305–307
Hydrogen peroxide, 267
Hydrogen peroxide oxidation, 267
Hydrogen sulfide, 39
Hydroelectric power, 112
Hydrogenation, 241*t*
Hydrogenolysis
  benzyl ethers, 969, 975
  defined, **969**
  Z-protecting group, 1217
Hydrohalogenation, 241*t*
Hydrolysis
  acid anhydrides, 774–775, 803
  acid chlorides, 774, 803
  amides, 779–783, 804
  cyano group, 783–785
  esters, 775–779, 803–804
  β-ketoesters, 836–838
  nitriles, 783–785, 804
  peptide bonds, 1212
  S_N1 reaction, 780
  S_N2 reaction, 779–780
  stitches, 1275
α-Hydroxyketone, 663, 664
β-Hydroxyketone, 663, 664
Hydroperoxidation of cumene, 985
Hydroperoxide, 355, **502**
Hydrophilic, **730**
Hydrophobic, **730**
Hydrophobic effect, 1197–1198, **1224,** 1227
Hydrophobic patches, 1227
Hydroquinone, 957, 964–966, 974
hydroxy-, 655*t*
4-Hydroxy-2-butanone, 829
(*R*)-6-Hydroxy-2-heptanone, 656

4-Hydroxy-3-methoxybenzaldehyde, 957, 1165
(*S*)-3-Hydroxy-3-methylglutaryl-CoA, 839
(*R,S*)-2-Hydroxy-2-phenylacetic acid, *158*
5-Hydroxy-5-phenylpentanal, 676, 677
Hydroxyamine, 684*t*
2-Hydroxybenzoic acid, 727
3-Hydroxybutanoic acid, 745
4-Hydroxybutanoic acid, 655*t*, 755
β-Hydroxybutyric acid, 745
trans-4-Hydroxyclclohexanecarbaldehyde, 653
trans-4-Hydroxyclclohexanecarboxylic acid, 729
(*R*)-5-Hydroxyhexanoic acid, 726
Hydroxyl group, **19,** 438, 957
(*R*)-5-Hydroxyoctanoic acid, A-15
(*S*)-4-Hydroxypentanal, 671, 1139
2-Hydroxyphenylacetic acid, 785
2-Hydroxypropanenitrile, 664, 665
(*R*)-2-Hydroxypropanoic acid, 729, 733
Hyperconjugation, **250,** *250,* 1011, 1012
Hyperconjugation stabilization, 396
Hyperglycemia, 1148
Hypervalency, 370–371
Hypophosphorus acid, 1076

I

iBu. *See* Isobutyl (iBu)
Ibuprofen, 665, 731, 864–866, 1029
(*S*)-Ibuprofen, 155
Ibutilide, 1087–1088
Ile, 1201*t*. *See also* Isoleucine
Illness. *See* Disorders/diseases
Imidazole, 205, 950, 1046*t*, 1049, 1051
Imidazole group, 182
Imidazolium, 55
Imides, 767, 768, 802, A-2
Imines, **678**–681, 684*t*, 706
"In phase," 30
In-phase addition, 33, 40
  C—C π bond, *43*
  C—C σ bond, *42*
  C—H σ bond, *41*
  C—O π bond, *51*
Index of hydrogen deficiency, **208,** 209, 210
Indian cobra, 1186
Indigo, 1026
Indole, 950, 1040
Inductive effect, **190, 249**–250, 1011, 1012, 1020
Industrial synthesis
  acetate rayon, 1156
  acetic acid, 737
  ibuprofen, 864–866
  salicylic acid, 963
Infix, 81
Influenza A virus infections, 816
Infrared absorption frequencies, A-6
Infrared active, **539**
Infrared-inactive vibrations, 540
Infrared (IR) spectroscopy, 535–557
  alcohols, 545–546, 546*t*
  aldehydes and ketones, 548–549
  alkanes, 543, 543*t*
  alkenes, 543*t*, 544
  alkynes, 543*t*, 544–545
  amines, 547
  arenes, 543*t*, 545
  calculating IR wavenumbers, 541
  carboxylic acids, 549
  characteristic absorption patterns, 540–541
  correlation table, 541–542
  defined, **538**
  derivatives of carboxylic acids, 550–551
  electromagnetic spectrum, 536*t*, 537*t*
  ethers, 546–547
  fingerprint region, 542
  frequency, 537*t*
  Hooke's law, 540, 541, 553
  infrared active/infrared-inactive, 539, 540
  infrared stretching frequencies, 542*t*
  interpreting infrared spectra, 543–551
  molecular vibrations, 539–540
  Raman spectroscopy, 540
  review/study guide, 552–554
  solving infrared spectral problems, 552
  vibrational infrared spectrum, 538–539
Infrared stretching frequencies, 542*t*
Ingold, C. K., 134, 383, 384
Inorganic coordination compounds, 643–644
Inositol, 1186*t*
Insulating materials, 1273, 1276*t*
Insulin, *1225,* 1227*t*
Internal alkyne, **302**
International Union of Pure and Applied Chemistry (IUPAC), 77. *See also* IUPAC system
Intramolecular aldol reaction, 831
Invert sugar, 1152*t*
Iodic acid, 469, 470, 1147
Iodide ion, 395
Iodination of toluene, 984
Iodine monochloride, 1091
Iodipamide, 1091
(*E*)-Iodo-3-hexene, 1098
(*Z*)-3-Iodo-3-hexene, 1098
Iodoalkanes, 334
(*R*)-2-Iodooctane, 384
(*S*)-2-Iodooctane, 384
2-Iodopropane, 963
2-Iodotoluene, 1066
Ion, 9
Ion fragmentation, 613
Ionic chain-growth polymerization, 1283–1287, 1283*t*
Ionic interaction, **8**
Ionization chamber, 607
Ionization potential (IP), **6, 608**
β-Ionone, 718
IP. *See* Ionization potential (IP)
iPr. *See* Isopropyl (iPr)
IR spectroscopy. *See* Infrared (IR) spectroscopy
Isoamyl alcohol, 786
Isoamyl benzoate, 786
Isobutane, 80
Isobutyl alcohol, 439, 450
Isobutyl bromide, 450
Isobutyl (iBu), A-11
Isobutyl mercaptan, 474
4-Isobutylacetophenone, 665, 1029
Isobutylbenzene, 1004
Isobutylene, 212, 356, 755, 1288
Isobutylmethylacetylene, 299
Isobutyric acid, *590*
Isocyanate, 1272, 1273
Isocyanic acid, 67
Isoelectric point (pI), **1206,** 1207
Isoleucine, 156, 1201*t*, 1203*t*
Isomeric alcohols, 492
Isomeric dibromobenzenes, 939, 940
Isomers, **13,** 129, 149
  atropisomers, 133, 148
  cis,trans. See Cis,trans isomerism
  configurational, 129, 130, *148*
  conformational, 132, *148*
  constitutional, 74–76, 129, *148*
  diastereomers. *See* Diastereomers
  enantiomers. *See* Enantiomers
  overview, *148*
  stereoisomers. *See* Stereoisomers
Isoniazid, 814
Isooctane, 114
Isopentane, 80
Isopentenyl pyrophosphate, 840, *1184*
Isopentyl, A-11
Isopentyl acetate, 210, 741
Isopentyl mercaptan, 474
Isoprene, 216, 616
Isoprene rule, 219
Isopropyl (iPr), A-11
Isopropyl alcohol, 439
Isopropyl benzene, 629
Isopropyl iodide, 963
2-Isopropyl-5-methylcyclohexanol, 123
(2*S*,5*R*)-2-Isopropyl-5-methylcyclohexanone, 654
4-Isopropyl-2-methylheptane, 80
2-Isopropyl-5-methylphenol, 957
Isopropyl mercaptan, 474

Isopropyl pentanoate, 550
Isopropyl phenyl ether, 963
4-Isopropylacetophenone, 1028
Isopropylamine, 1041, 1045*t*
Isopulegol, 719
Isoquinoline, 1040
Isotactic polymer, **1282,** *1282*
Isotactic polystyrene, 1283
Isotopes, 610–611
Isovaleric acid, 726
IUPAC system, A-11 to A-17. *See also* Common names
  acid anhydrides, 763–764
  acid halides, 763
  alcohols, 438, 439, A-13
  aldehydes and ketones, 653–656, A-14
  alkanes, 77, 78–79, A-11 to A-12
  alkenes, 211, A-13
  amides, 766–767, A-15
  amines, 1039–1040, 1071
  alkynes, 298, 299, A-13
  benzene, 954–956, 972–973
  carboxylic acids, 726–727, A-14 to A-15
  carboxylic anhydrides, A-15
  cycloalkanes, 83
  diol, 440
  epoxides, 505
  esters, 764–766, A-15
  ethers, 493, 494
  haloalkanes, 331, 332
  infix, 81
  monosaccharides, 1135–1136
  nitriles, 768
  order of precedence of functional groups, 655*t*
  phenols, 957
  polymers, 1266
  prefix, 77*t*, 81
  suffix, 81
  sulfides, 518
  thiols, 473, 474
  triol, 440

J

J. *See* Coupling constant (*J*)
Jackets (from recycled PET soda bottles), 1291
Johnson, William S., 326
Jones reagent, 463

K

K_a. *See* Acid dissociation constant (K_a)
K_{eq}. *See* Equilibrium constant (K_{eq})
Keflex, 792
Kekulé, August, 939
Kekulé structure, 939, *939*
Kendrew, John C., 1224
Kent, Stephen B., 154
Keratin sulfate, 1169
Keto-enol equilibrium, 686–688
Keto-enol tautomerism
  acid-catalyzed aldol reaction, 826–827
  acid-catalyzed α-halogenation of ketone, 700
  acid-catalyzed dehydration of aldol product, 828
  acid-catalyzed equilibration of keto and enol tautomers, 686
  acidity of α-hydrogens, 684–685
  aldehydes and ketones, 684–688
  alkynes, 308, 309
  base-catalyzed equilibration of keto and enol tautomers, 686
  Claisen rearrangement, 922
  decarboxylation of β-Dicarboxylic acid, 745
  defined, **309**
  keto-enol equilibrium, 686–688
  keto form/enol form, 685
  Kolbe carboxylation of phenol, 964
  reaction of 1° amine with nitrous acid, 1061
Keto form, 685
Ketohexose, 1158
Ketone bodies, 745
Ketones, 21–**22**. *See also* Aldehydes and ketones
Ketose, **1135**
2-Ketose, 1146
Kevlar, 819, 1271
Key reactions
  alcohols. *See* Alcohols—key reactions
  aldehydes and ketones. *See* Aldehydes and ketones—key reactions
  alkenes. *See* Alkenes—key reactions
  alkynes. *See* Alkynes—key reactions
  amines. *See* Amines—key reactions
  amino acids, 1228–1229
  carbohydrates. *See* Carbohydrates—key reactions
  carboxylic acid derivatives. *See* Carboxylic acid derivatives—key reactions
  carboxylic acids. *See* Carboxylic acids—key reactions
  enolate anions and enamines. *See* Enolate anions and enamines—key reactions
  epoxides. *See* Epoxides—key reactions
  ethers. *See* Ethers—key reactions
  haloalkanes. *See* Haloalkanes—key reactions
  nucleophilic substitution and β-elimination. *See* Substitution/elimination—key reactions
  organometallic compounds. *See* Organometallic compounds—key reactions
  polymers. *See* Organic polymer chemistry—key reactions
  reactions of benzene and derivatives. *See* Benzene—key reactions
  thiols. *See* Thiols—key reactions
Kinetic control, **855, 864,** 900
Kinetics, 186
Kishi, Y., 888
Kishner, N., 697
Knowles, William, 509
Kolbe carboxylation, 963, 964, 974
Koller, Karl, 765
Kroto, Harry W., 27
Kurzrok, Raphael, 1178

L

L-DOPA, 155
Labetalol, 1033
Laboratory domino sequence, 326
Lactam, **767**
Lactate, 471
Lactate dehydrogenase, 471, 694, 1227*t*
Lactate production, 695
Lactic acid, 733, 1275
(*R*)-Lactic acid, 729
(*S*)-(+)-Lactide, 1299
Lactomer, 1274, 1275
Lactones, 755, 764–765
Lactose, 1151, 1152*t*
Ladder polymer, *1265*
LAH. *See* Lithium aluminum hydride (LAH)
Landsteiner, Karl, 1154
LAS. *See* Linear alkylbenzenesulfones (LAS)
Laser surgery, 1274
Lasix, 1092
Latex paints, 1265, 1276*t*
Laundry detergents, 1175–1176
Lauric acid, 728*t*, 1171*t*
Lauterbur, Paul, 584
Law of electromagnetic induction, 562
LC-MS. *See* Liquid chromatography mass spectrometry (LC-MS)
LDA. *See* Lithium diisopropylamide (LDA)
LDL cholesterol. *See* Low-density lipoprotein (LDL) cholesterol
LDPE. *See* Low-density polyethylene (LDPE)
LDPE film, *1279*
Leaving group (Lv), 374, **375**
Leaving group ability, 423

Lecithin, 1186, 1186*t*, *1187*, 1198
Lehn, Jean-Marie, 517
Leu, 1201*t*. *See also* Leucine
Leucine, 1201*t*, 1203*t*
Leukemia, 1261
Leukotriene C_4 (LTC_4), 1180
Levomethorphan, 1032
Levorotatory, **150**
Levorotatory compound, 151
Lewis, Gilbert N., 6, 7, 193
Lewis acid, **193,** 228, 236, 374
Lewis acid-base reaction
  bond breaking to expel stable ion, 230, 231
  example, 195
  hydroboration, 266
  key reaction, 199
  new covalent bond, 194
  proton transfer reaction, 228
Lewis acids and bases, 193
Lewis base, **193,** 228, 236, 374
Lewis dot structure, **6**
Lewis model of bonding, 7–12
Lewis structure, 13–14, 16
Lexan, 1272
Lidocaine, 765
Lieb, Charles, 1178
Ligand, 643, **1097**
Ligant, 1113
Limonene, 220, 227, 290, 616
Lindane, 123
Lindlar catalyst, **312**
Line-angle formula, **73,** 74, 76
Linear alkylbenzenesulfones (LAS), 1175
Linear low-density polyethylene (LLDPE), 1282
Linear molecules, 25
Linear polymers, *1265*
Linezolid, 821–822
Link, Karl, 766
Linoleic acid, 221, 355, 1170, 1171*t*
Linolenic acid, 1171*t*
Lipases, 1195
Lipid bilayers, 1186–**1187**
Lipids, 1170–1198
  bile acids, 1183
  cholesterol, 1181, *1184,* 1185
  defined, **1170**
  eicosanoids, 1179–1180
  fat-soluble vitamins, 1187–1190
  fats and oils, 1171–1174
  fatty acids, *1171,* 1171–1172, *1173*
  lipid bilayers, 1186–1187
  phospholipids, 1185–1187
  prostaglandins, 1178–1180
  review/study guide, 1191–1194
  sex hormones, 1181–1183
  soaps and detergents, 1174–1176
  steroids, 1180–1185, 1193
  triglycerides, 1171–1174, 1191–1192
  two main groups, 1170
Lipitor, 442, 491, 892
Liquid chloroalkanes, 335
Liquid chromatography mass spectrometry (LC-MS), 607, 623
Lister, Joseph, 957
Lithium aluminum hydride (LAH)
  alkylation of azide ion, 1057
  amides, 798–799
  carboxylic acids, 736–738, 748
  esters, 796–797
  metal hydride reduction, 691
Lithium dibutylcopper, 636, 637
Lithium diisopropylamide (LDA), 301, 860–864, 874
Lithium diorganocopper reagents, 636–639
  carboxylic acid derivatives, 795
  enolate anions and enamines, 858–860, 873
  formation and structure, 636
  organohalogen compounds, 636–637
  oxiranes, 638–639
Lithium diorganocuprates, 795
Lithium iodide, 12
Living polymer, **1285**
Living polymerization, 1285
LLDPE. *See* Linear low-density polyethylene (LLDPE)
Local anesthetics, 765, 1036
Localized charges, 70
"Locker room odor," 731
Lone pair, 10, 14
Long-chain aliphatic acids, 629
Longer-chain dicarboxylic acids, 754
Lovastatin, 839, 840
Low-density lipoprotein (LDL) cholesterol, 276, 839, **1181**
Low-density polyethylene (LDPE), 1278–1279, 1292, 1301
Low-molecular-weight acid chlorides, 774
Low-molecular-weight alkanes, 118
Low-molecular-weight alkynes, 300*t*
Low-molecular-weight dicarboxylic acids, 754
Low-resolution mass spectrometry, **610**
Lowest unoccupied molecular orbital (LUMO), **910,** 921
Lowry, Thomas, 171
LTC_4. *See* Leukotriene C_4 (LTC_4)
Lucite, 1276*t*
Luminal, 813
LUMO, **910,** 921
Lycopene, 226
Lycra, 1273
Lynen, Feodor, 1184
Lys, 1201*t*. *See also* Lysine
Lysine, 1201*t*, 1203*t*
Lysolecithin, 1186
Lysozyme, 1225

M

m, 536*t*
m. *See* Meta (*m*)
M_n. *See* Number average molecular weight (M_n)
M_w. *See* Weight average molecular weight (M_w)
M + 1 peaks, 610, 612, 624
M + 2 peaks, 611–612, 624
m/z ratio. *See* Mass-to-charge (*m/z*) ratio
MAAP gas, 324
MacSpartan, A-7
Magic Tan, 1135
Magnesium monoperoxyphthalate (MMPP), 507
Magnetic resonance imaging (MRI), 584–585
MALDI. *See* Matrix-assisted laser desorption ionization mass spectrometry (MALDI)
Male sex hormones, 1182*t*
Maleic anhydride, 764
(*S*)-Malic acid, 816
(*S*)-(−)-Malic acid, *158*
(*S*)-(+)-Malic acid, 154
Malonic acid, 727, 744–745, 754
Malonic ester synthesis, 849–850, 872
Maltose, 1152, 1152*t*
Man-Tan, 1135
Mandelic acid, 785
(*R*)-(+)-Mandelic acid, *158*
(*R,S*)-Mandelic acid, *158*
(*S*)-(−)-Mandelic acid, *158*
Mandelonitrile, 665, 785
D-Mannitol, 1146
D-Mannosamine, 1139
Mansfield, Peter, 584
Margarine, 1173
Marijuana, 629
Markovnikov, Vladimir, 245
Markovnikov addition, 358
Markovnikov regiochemistry, 296
Markovnikov's rule, **245,** 900
Mass spectrometer, 607
Mass spectrometry, 606–629
  alcohols, 618
  aldehydes and ketones, 618–619
  alkanes, 614–615
  alkenes, 616
  alkynes, 616–617
  α-cleavage, 618, 619, *620*
  amides, 619–621
  amines, 622–623
  aromatic hydrocarbons, 621–622
  base peak, 608

Mass spectrometry (*Continued*)
β-cleavage, 622
biological macromolecules, 621
carboxylic acids and esters, 619–621
chemical ionization (CI), 609
electron ionization (EI-MS), 609
electrospray ionization (ESI-MS), 609
exact mass calculation, 611
fast-atom bombardment (FAB), 609
fragmentation of molecular ions, 613
gas chromatography (GC-MS), 607, 623
historical overview, 606
how it works, 607–609
importance, 606, 623
interpreting mass spectra, 613–623
ionization chamber, 607
isotopes, 610–611
liquid chromatography (LC-MS), 607, 623
low/high-resolution, 610
M + 1 peaks, 610, 612, 624
M + 2 peaks, 611–612, 624
MALDI, 609
mass spectrometer, 607
mass spectrum, 607
mass-to-charge *(m/z)* ratio, 607, 608
McLafferty rearrangement, 618–620
molecular ion, 608
nitrogen rule, 614
peak (molecular ion)/series of peaks (fragment ions), 624
polypeptides and proteins, 1215
quadrupole mass analyzer, 610
resolution, 610
review/study guide, 623–625
tandem MS, 610
uses, 623, 625
Mass spectrum, **607**
Mass-to-charge *(m/z)* ratio, 607, 608
Matrix-assisted laser desorption ionization mass spectrometry (MALDI), 609
Maxam, Allan, 1254
Maxam-Gilbert method, 1254
Mayonnaise, 1198
MCAT Practice boxes
alcohol oxidations, 472
amino acid stereochemistry, 155–156
antioxidants, 354–356
benzo[a]pyrene, 513–514
β-lactam antibiotics, 791–793
capsaicin, 961–962
fucose, 1150
fullerenes, 27–28
industrial synthesis of ibuprofen, 864–866
inorganic coordination compounds, 643–644
permethrin and bifenthrin, 742–743
planarity of —NH_2 groups on heterocyclic rings, 1049–1051
pyridoxine (Vitamin B_6), 682–684
solvents and solvation, 417
superglue, 1286–1287
tetrodotoxin, 105–106
vitamin K, blood clotting, and basicity, 1190–1191
VSEPR and resonance, 55
McLafferty rearrangement, 618–620
McNeil Consumer Pharmaceuticals Co., 65
MCPBA. *See meta*-Chloroperoxybenzoic acid (MCPBA)
Me. *See* Methyl (Me)
Mechanisms. *See* Reaction mechanism
Mechlorethamine, 421
Meclizine, 1034
Medical College Admission Test (MCAT). *See* MCAT Practice boxes
Medications. *See* Drugs/medications
Megatomoic acid, 750
Meisenheimer complex, 1016
Melatonin, 1233
Melphalan, 421
Melt transition (T_m), **1267**
Melting point
alkenes, 218*t*
amines, 1045*t*
benzene, 938
low-molecular-weight alkynes, 300*t*
unbranched alkanes, 108*t*
Menadione, 965
Menthol, 220
Menthone, *548*
Meparfynol, 326
Meperidine, 885, 1079
Meprobamate, 813
Mercaptan, **473**. *See also* Thiols
mercapto-, 655*t*
2-Mercaptoethanol, 655*t*
β-Mercaptoethanol, 474
6-Mercaptopurine, 1261
Merrifield, R. Bruce, 1219
Merrifield solid-phase synthesis, 1219–1221
Meso compound, **140,** A-9
Meso-tartaric acid, 140, 141*t,* 1164
Meso-2,3-Butanediol, A-10
Meso-1,2-Dibromo-1,2-diphenylethane, 406–407
Messenger (mRNA), 1248*t,* **1250**
Mesyl chloride (MsCl), 452
Met, 1201*t*. *See also* Methionine
Meta directing, 1006, 1007, 1008*t*
Meta *(m),* 954
Metal alkoxide, 446
Metal alkyls, 1283
Metal hydride reduction, 691–692, 708
Metathesis reactions, 1112–1115, 1118–1119
Meter (m), 536*t*
meth-, 77*t*
Methadone, 1079
Methamphetamine, 715
Methanal, 653, 658*t*. *See also* Formaldehyde
Methandrostenolone, 1183
Methane
ball-and-stick model, *73*
boiling point, 108*t*
density, 108*t*
Lewis structure, 13*t, 24, 73*
melting point, 108*t*
molecular formula/condensed structural formula, 74*t*
orbital overlap picture, *36*
shape, *24*
Methanesulfonic acid, 763
Methanesulfonyl chloride (MsCl), 763, 786
Methanethiol, 473, *473,* 475*t*
Methanoic acid, 726, 728*t*
Methanol, 787, 796
ball-and-stick model, *438*
boiling point, 442*t*
Lewis structure, *438*
pK_a, 445*t*
protic solvent, 389*t*
solubility, 442*t*, 443
Methicillin, 792
Methionine, 1201*t,* 1203*t*
2-Methoxy-2-methylcyclohexanol, 511
2-Methoxy-2-methylpropane, 493
3-Methoxyaniline, 1040
4-Methoxybenzoic acid, 629
2-Methyoxyethanol, 493
Methoxyethene, 494
Methoxymethane, 546
2-Methoxynaphthalene, 1032
5-Methoxypentanoic acid, 849, 850
4-(3-Methoxyphenyl)-2-butanone, 954
Methsuximide, 885
Methyl (Me), A-11
Methyl acetate, 23, *564*
Methyl acrylate, 787, 1098
Methyl 2-aminobenzoate, 741
Methyl anthranilate, 741
2-Methyl-1,3-butadiene, 216, 616
3-Methyl-1-butanamine, 622, *622*
3-Methyl-1-butanethiol, 474
Methyl butanoate, *620,* 741
2-Methyl-1-butanol, 556
2-Methyl-2-butanol, 628
3-Methyl-1-butanol, 786
3-Methyl-2-butanol, *583,* 605, 628
3-Methyl-2-butanone, 538, *538*
2-Methyl-2-butene
boiling point, 218*t*

heat of hydrogenation, 275*t*
hydroboration-oxidation, 308
melting point, 218*t*
nomenclature, 211
proton transfer, 177
structural formula, 218*t*, 275*t*
Methyl-butyllithium, 1283
3-Methyl-1-butyne, 298, A-13
Methyl cation (CH_3^+), A-7
Methyl cellosolve, 493
Methyl chloroform, 332
Methyl cinnamate, 1098
3-Methyl-2-cyclohexenone, 654
2-Methyl-1-dodecene, 636
1-Methyl-1,2-epoxycyclohexane, 511
Methyl esters, 739–740, 749
Methyl β-D-Fructofuranoside, 1149
Methyl β-D-Fructoside, 1149
Methyl β-D-Glucopyranoside, 1149
Methyl glucosides, 720
Methyl haloalkanes, 397*t*, 424, 426
6-Methyl-1,5-heptadiene, 923
Methyl hexanoate, 798
6-Methyl-5-hepten-2-one, 936
4-Methyl-2-hexanol, A-13
5-Methyl-3-hexanone, A-14
4-Methyl-1-hexene, 211, A-13
Methyl isocyanate, 204
Methyl isovalerate, 767
Methyl methacrylate, 1276*t*
Methyl 2-methyl-3-oxo-3-phenylpropanoate, 835
Methyl (2*E*,4*E*)-2-4-nonadienoate, 1100
2-Methyl-1,4-naphthoquinone, 965
4-Methyl-2-pentanone, 629
3-Methyl-2-pentenal, A-14
Methyl (*E*)-3-phenyl-2-propenoate, 1098
2-Methyl-2-phenylpropane, 1005
2-Methyl-1-propanethiol, 474
2-Methyl-2-propanethiol, 474
2-Methyl-1-propanol, 439, 450
2-Methyl-2-propanol, 439, 445, 445*t*
2-Methyl-4-(1-methylethyl) heptane, 80
Methyl 3-oxobutanoate, A-15
Methyl 3-methylbutanoate, 767
(*R*)-4-Methyl-2-pentanol, 439
Methyl phenyl ether, 963
Methyl propenoate, 787
Methyl 2-propenoate, 1098
Methyl radical, 341
Methyl (*S*)-2-phenylpropanoate, 796
(*Z*)-Methyl-7-octadecene, 717
Methyl tetradecanoate, 629
Methyl vinyl ether, 494
N-Methylacetamide, 766
Methylacetylene, 299
Methylamine
1° amine, 21, 1038
acid strength, 1046*t*
alkylation, 1056
angle of benzene ring and bisector of amino group, 1049
boiling point, 1045*t*
melting point, 1045*t*
nomenclature, 1040
solubility, 1045*t*
structural formula, 1045*t*
Methylammonium bromide, 1056
3-Methylaniline, 1014
4-Methylaniline, 1014, 1040, 1046*t*
N-Methylaniline, 1038, 1040
2-Methylbenzonitrile, 1065
3-Methylbutanal, 653
N-Methylbutanamide, A-15
2-Methylbutane, 80, 566
3-Methylbutanoic acid, 726
2-Methylbutyl, A-11
3-Methylbutyl, A-11
3-Methylbutyl benzoate, 786
(3-Methyl)butyl acetate, 741
Methylcobalamin, 644
Methylcyclohexane, 96, 363
cis-1,2-Methylcyclohexanediol, 146
cis-1,3-Methylcyclohexanediol, 146
trans-1,3-Methylcyclohexanediol, 146
trans-2-Methylcyclohexanediol, 146
cis-3-Methylcyclohexanol, 145
cis-4-Methylcyclohexanol, 145
(1*R*,2*R*)-2-Methylcyclohexanol, 439
trans-3-Methylcyclohexanol, 145
trans-4-Methylcyclohexanol, 145
Methylcyclopentane, 615, *616*
3-Methylcyclopentane, 214
cis-2-Methylcyclopentanol, 144
cis-3-Methylcyclopentanol, 145
trans-2-Methylcyclopentanol, 144
trans-3-Methylcyclopentanol, 145
1-Methylcyclopentene, 266
p-methyldiphenhydramine, 1034
Methylene, **212,** 639
Methylene chloride, 332
Methylenecyclopentane, 212, 554
4-(1-Methylethyl-1-methylcyclohexene), 215
1-Methylethyl, A-11
2-Methylheptane, 111*t*
4-Methylhexan-2-ol, A-13
5-Methylhexan-3-one, A-14
Methylidene, 212
Methylidenecyclopentane, 212
N-Methylmorpholine, 1086
2-Methylnaphthalene, 965
4-Methyloctane, 77
Methylparaben, 1086
3-Methylpent-2-enal, A-14
2-Methylpentanal, 629
2-Methylpentane, 78, 109*t*, A-12
3-Methylpentane, 109*t*
2-Methylphenol, 1014
3-Methylphenol, 957, 1014
4-Methylphenol, 962
2-Methylpropanal, 460
N-Methylpropanamide, *551*
2-Methylpropane
constitutional isomer, 75
equivalent hydrogens, 565
free-radical halogenation, 338
hydrogen abstraction, 346*t*
nomenclature, 78, A-12
three primary carbons/one tertiary carbon, 82
2-Methylpropanoic acid, *590*
2-Methylpropene, 212, 356, 755, 1288
1-Methylpropyl, A-11
2-Methylpropyl, A-11
1-Methylpropyl hydrogen phthalate, 787
Methyltriphenylphosphonium iodide, 666
(*S*)-Metolachlor, 817
Mevacor, 840
Mevalonate, 442
(*R*)-Mevalonate, 840, *1184*
Mevalonic acid, 840, A-15
Mevastatin, 839, 840
Micelles, **1174,** *1175*
Michael, Arthur, 852
Michael addition of enolate anions, 852–855
Michael-aldol sequence, 857
Michael reaction, 852–857
conjugate addition of enolate anions, 854
examples, 852
key reaction, 873
Michael addition of enolate anions, 852–855
reagents, 853*t*
retrosynthetic analysis, 855–857
review/study guide, 872–873
Miconazole, 1030
Micrometer (μm), 536*t*
Microscopic reversibility, 459, 776–777
Migraine headache, 724
Millimeter (mm), 536*t*
Millipede, 665
Miltown, 813
Mineralocorticoid hormone, 1182*t*
Minoxidil, 821
Mirror image, 128
Misoprostol, 1178, 1179
Mixed Claisen condensation, 835–836
Mixed media errors, 371–372
mm, 536*t*
MMPP. *See* Magnesium monoperoxyphthalate (MMPP)

MO theory. *See* Molecular orbital (MO) theory
Moclobemide, 532, 817
Moderate nucleophile, 393, 423
Molar absorptivity (ε), 905
Molecular biology, 1239
Molecular dipole moment (μ), **27**
Molecular ion, **608**
Molecular orbital energy diagram
  cyclobutadiene, *948*
  cycloheptatrienyl cation, 953
  cyclooctatetraene, *949*
  cyclopentadienyl anion, 952
  hydrogen, *34*
Molecular orbital mixing diagram
  C—C π bond, *43*
  C—C σ bond, *42*
  C—H σ bond, *41*
  C—O π bond, *51*
Molecular orbital (MO) theory, **32**–34, 52
Molecular orbital model
  1,3-Butadiene, *897*
  benzene, 940–942
  dienes, 897
Molecular shape, 24–26
Molecular spectroscopy, 536–**537**
Molecular structure and acidity, 188–192
  delocalization of charge in anion, 190
  electronegativity of atom bearing negative charge, 188–189
  hybridization of atom bearing negative charge, 192
  inductive effect, 190–192
  review/study guide, 198
  size of atom bearing negative charge, 189
  stability of anion, 188, 198
Molecular weight
  alkanes, 118
  compounds of comparable weight, 442*t*, 495*t*, 657*t*, 730*t*
  number average molecular weight (M_n), 1267
  polymers, 1265, 1267
  weight average molecular weight (M_w), 1267
Molecules
  achiral, 152
  chiral, 130–131
  handedness, 128–129
Molina, Mario, 344
Molozonide, 272
Monascus ruber, 839
Monensin, 888
Monistat, 1030
Monodisperse, 1267
Monomer, **1265**
D-Monosaccharide, **1137**
L-Monosaccharide, **1137**
Monosaccharides, 1135–1150. *See also* Carbohydrates
  alditols, 1145–1146, 1161
  aldonic acids, 1146, 1161
  amino sugars, 1138–1139
  conformation representations, 1142
  cyclic structure, 1139–1143, 1159
  defined, **1135**
  D,L system, 1136–1138
  Fischer projection formulas, 1136
  glycosides, 1143–1145, 1161
  Haworth projections, 1139–1142
  mutarotation, 1143, 1160
  periodic acid, 1147–1149, 1162
  physical properties, 1139
  reactions, 1143–1149, 1160–1162
  reducing sugars, 1146
  review/study guide, 1158–1161
  structure and nomenclature, 1135–1136
  uronic acids, 1146–1147
Monosaturated fatty acids, 221
Monosubstituted benzenes, 954
Monosubstituted carbon-carbon double bonds, 273
Monosubstituted cyclohexanes, 97*t*
Monoterpenes, *1184*
Montelukast, 534
Montreal Protocol, 344
Morphinan, 1032
Morphine, 885, 1032, 1037, 1079
Morpholine, *841,* 1052
Motion sickness, 1034
Motrin, 155
Moxisylyte, 1088
MRI. *See* Magnetic resonance imaging (MRI)
mRNA. *See* Messenger (mRNA)
MS-MS, 621
MsCl. *See* Mesyl chloride (MsCl); Methanesulfonyl chloride (MsCl)
MTBE. *See tert*-Butyl methyl ether (MTBE)
μm, 536*t*
Multidimensional spectra, 590
Multistep synthesis questions, A-18. *See also* Organic chemistry roadmap
Musca domestica, 325, 630
Muscalure, 637
Musk ambrette, 1029
Mustard gas, 419–422
Mutarotation, **1143,** 1160
Myeloperoxidase, 529
Mylar, 1271, 1272
Myoglobin, 1224, *1225*
Myrcene, 219, *219*
Myristic acid, 1171*t*

N

(n + 1) rule, **573,** 577
N-substituted morpholines, 1086
N-substituted primary amines, 1040
Nabumetone, 1032
NAD^+, 694, 695. *See* Nicotinamide adenine dinucleotide (NAD^+)
NADH, 470, 694–695
Nadolol, 516
NADPH, 695
Naja naja, 1186
Nanometer (nm), 536*t*
Naphthalene, 943, 956, 968, 1285
Naproxen, 731
(*S*)-Naproxen, 159, 169, 281, A-9, A-10
Natta, Giulio, 1279
Natural gas, 112–113
Natural rubber, 1301
Navicalm, 1034
NBS. *See N*-Bromosuccinimide (NBS)
Near ultraviolet, 904
Nef reaction, 1196
Negative pK_a value, 180
Negishi, Ei-ichi, 1107
Neighboring group participation, 418–419, 427
Nematodes, 817
Nembuta, 813
Neopentane, 80
Neopentyl, A-11
Neopentyl bromide, 388
Neoplastic diseases, 420–421
Neoprene, 1297
Nepata cataria, 227
Nepetalactone, 227
Nephila clavipes, 1225
Newman, Melvin, 86
Newman projection, **86,** *86,* 90, 405
Next-generation DNA sequencing, 1258, 1261
Niacin, 812
Nicolson, G., 1187
Nicotinamide, 812
Nicotinamide adenine dinucleotide (NAD^+), 470
Nicotine, 1038
Nicotinic acid, 812
Niemann-Pick disease, 1168
Nifedipine, 882
Ninhydrin, 1207
Nirenberg, Marshall, 1251
Nitration
  anisole, 1010, *1010*
  benzene, 998–1000, 1017, 1018
  benzoic acid, 1010–1011, *1011*
  equation, 995
  key reaction, 1018
  nitronium ion, 998
Nitriles. *See also* Carboxylic acid derivatives
  defined, **768**
  distinguishing characteristic, 551
  hydrolysis, 783–785, 804
  infrared absorptions, 551*t*

IUPAC system/common names, 768
reduction, 800, 808
Nitrite additives (to food), 1060
Nitroalkane, A-2
3-Nitroaniline, 1048
4-Nitroaniline, 1040, 1046*t*, 1048
p-Nitroaniline, 1040, 1049
Nitroarenes, 1017
Nitrobenzene, 995
2-Nitrobenzoic acid, 955
4-Nitrobenzoic acid, 733
m-Nitrobenzoic acid, 1008
o-Nitrobenzoic acid, 955
p-Nitrobenzoic acid, 1008
Nitrogen, 611*t*
Nitrogen mustard gas, 418, 420, 421
Nitrogen nucleophiles, 678–684
Nitrogen rule, **614**
Nitronium ion, 998
m-Nitrophenol, 959
p-Nitrophenol, 959
p-Nitrophenyl acetate, 956
4-Nitrophenyl ethanoate, 956
3-Nitropyridine, 1022
2-Nitropyrrole, 1023
N-Nitrosamines, 1059–1060, 1075, 1083
Nitrosation of tertiary aromatic amines, 1075
N-Nitrosodimethylamine, 1060
N-Nitrosopiperidine, 1059
N-Nitrosopyrrolidine, 1060
Nitrosyl cation, 1058–1059
nm, 536*t*
NMR spectrometer, 563–565
Nodal plane, 29
Node, **29**
NOE. *See* Nuclear overhauser enhancement (NOE)
Nomenclature
common names. *See* Common names
IUPAC system. *See* IUPAC system
Nomex, 1298
non-, 77*t*
Non-energy-minimized eclipsed conformation, *89*
Non-Markovnikov, 265, 266
Non-Markovnikov addition, 358
nonadec-, 77*t*
Nonadecane, 74*t*
Nonane, 74*t*, 108*t*, 555
Nonbonded interaction strain, 89
Nonbonding electrons, **14**
Nonpolar covalent bond, **10**
Nonpolar solvent, 389, 423
Nonsteroidal anti-inflammatory drugs (NSAIDs), 731, 1178, 1179
Nonsteroidal estrogen antagonist, 993–994
Nonstick coatings, 1276*t*
Norbornane, *84*
Norbornene, 216, 1290
Norepinephrine, 1233
(*R*)-Norepinephrine, 1053
(*R*)-Norepinephrine hydrochloride, 1053
Norethindrone, 1182, 1183
Novak, Bruce, 1264
Novocain, 765, 815
Noyori, Ryoji, 509
NSAIDs. *See* Nonsteroidal anti-inflammatory drugs (NSAIDs)
Nuc/Base, 414. *See also* Nucleophile
Nuclear electric power, 112
Nuclear magnetic resonance (NMR) spectroscopy, 537*t*, 558–605
alcohols, 588–589
aldehydes and ketones, 589
alkanes, 587
alkenes, 588
amines, 590
bond rotation, 578–580
^{13}C-NMR spectroscopy, 585–587, 596–597
carboxylic acids and esters, 589–590
chemical shift, 568–572, A-4, A-5
coincidental overlap, 580
complex coupling, 577–578, 580–581
coupling constant *(J)*, 574, 576*t*
deuterated solvents, 699
diamagnetic shielding, 562
diastereotopic groups, 582, 583, 596
downfield/upfield, 564, 565, *565*
enantiotopic groups, 582, 583, 596
equivalent hydrogens, 565–566
ethers, 589
fast exchange, 581–582
geminal coupling, 578, *578*, *579*
homotropic groups, 582, 596
interpretation of NMR spectra, 587–590, 597
magnetic resonance imaging (MRI), 584–585
molecular structure determination, 559, *560*
$(n + 1)$ rule, 573, 577
NMR spectrometer, 563–565
nuclear overhauser enhancement (NOE), 590
nuclear spin quantum number, 559, 559*t*
nuclear spin states, 559
Pascal's triangle, 576, *577*
peak intensities, 575–577
precession, 561, *562*
resonance, 561–563
review/study guide, 593–597
shielding/deshielding, 562, 563
signal areas, 567
signal splitting, 572–582
solving NMR spectral problems, 591–593
spider silk, 1226
spin-spin coupling, 574, *575*
spin-state population, 560–561
stereochemistry, 582–583
topicity, 582–583
tree diagrams, *579*, *580*
Nuclear overhauser enhancement (NOE), 590
Nuclear spin quantum number, 559, 559*t*
Nuclear spin states, 559
Nucleic acids, 1238–1263
chain termination (dideoxy) method, 1255–1256, 1261
defined, **1239**
DNA. *See* DNA
genetic code, 1250–1252
Maxam-Gilbert method, 1254
next-generation DNA sequencing, 1258, 1261
nucleosides and nucleotides, 1239–1241
restriction endonuclease, 1253–1254
review/study guide, 1259–1261
RNA. *See* RNA
sequencing, 1253–1258, 1260–1261
Nucleophile, 71, 228–231
aldehydes and ketones, 659–668, 670–677, 678–684
Brønsted-Lowry base, 375
carbon, 645, 659–668
defined, **236, 374**
good, 423
low-electron-density species, 195
moderate, 393, 423
nitrogen, 678–684
nucleophilicity, 393*t*. *See also* Nucleophilicity
oxygen, 670–677
poor, 423
primary alkyl groups, 414, 426
secondary alkyl groups, 414, 426
strong, 393
tertiary alkyl groups, 414, 426
weak, 393
Nucleophile-assisted substitution, 418. *See also* S_N2 reaction
Nucleophile/electrophile reaction, 230
Nucleophile-unassisted substitution, 418. *See also* S_N1 reaction
Nucleophilic, 230
Nucleophilic acyl addition, **658,** 659, 770–771
Nucleophilic acyl substitution, **771**
Nucleophilic aromatic substitution, 1013–1017
addition elimination, 1015–1016, 1021–1022
benzyne intermediate, 1014–1015, 1021

Nucleophilic aromatic (*Continued*)
defined, **1014**
example, 1017
key reactions, 1021–1022
Nucleophilic carbene catalysts, 1113
Nucleophilic ring opening of epoxides, 512–513, 524
Nucleophilic substitution and β-elimination, 374–436
analysis of competitions between substitution/elimination, 415–418, 426
analysis of nucleophilic substitution reactions, 377, 397–399, 424
β-elimination, 400–401
definitions, 374, 375
E1 reaction. *See* E1 reaction
E2 reaction. *See* E2 reaction
internal nucleophilic substitution in halohydrins, 506
key reactions. *See* Substitution/elimination—key reactions
mechanisms. *See* Substitution/elimination—reaction mechanisms
methyl haloalkanes, 397*t*, 424, 426
mustard gas, 419–422
neighboring group participation, 418–419, 427
nucleophilic substitution reactions, 376, 376*t*
primary haloalkanes, 397*t*, 408*t*, 414, 424, 426
reaction roadmap, 295–296
review/study guide, 422–427
secondary haloalkanes, 397*t*, 408*t*, 414, 424, 426
S_N1 reaction. *See* S_N1 reaction
S_N2 reaction. *See* S_N2 reaction
substitution versus elimination, 410–414, 426
summary/overview, *413*, 413t, 414, 426
tertiary haloalkanes, 397*t*, 408*t*, 414, 424, 426
Zaitsev elimination, 401
Nucleophilic substitution reactions, 376, 376*t*
Nucleophilicity, 230
basicity, 393
common nucleophiles, 393*t*
defined, **393**
shape, 395
solvation effects, 394–395
Nucleoside, **1239**–1240
Nucleoside diphosphates, 1240
Nucleoside monophosphates, 1239, 1240
Nucleoside triphosphates, 1240
Nucleotide, **1239**–1241
Nucleotide sequence, 1215–1216
Nucleus, 2
Number average molecular weight (M_n), 1267
Nylon 6, 1270–1271
Nylon 6,10, 1298
Nylon 66, 1264, 1269, *1269*, 1270

O

o. See Ortho *(o)*
O-Acylisourea, 1218, 1219
O-ring seals, 1268
Observed rotation, **150**
oct-, 77*t*
octadec-, 77*t*
Octadecane, 74*t*
(*Z*)-9-Octadecenoic acid, 729
2,6-Octadiyne, 299
Octanamide, 798
1-Octanamine, 798
Octane, 74*t*, 108*t*, 111*t*, *614*
Octane rating, 114
Octanoic acid, 728*t*
(*S*)-2-Octanol, 169
2-Octanone, *619*
(3*E*,5*E*)-1,3,5,7-Octataetraene, 908*t*
cis-4-octene, 213
Octet rule, **7,** 18
Octose, 1135
Octyl acetate, 741
(*R*)-2-Octyl acetate, 169
Octyl *p*-methylcinnamate, 884
(*S*)-2-Octyl tosylate, 169
1-Octyne, 300*t*, *545*
Ocular hypertension, 1195
-oic acid, 81, 655*t*
Oil of pennyroyal, 875
Oils, 220–221, **1172,** 1173*t*
Oils and fats, 1171–1174
-ol, 81, 655*t*
Oleic acid, 221, 729, 1171*t*
Oligopeptide, 1208
Oligosaccharides, **1151**–1153
Olive oil, 221, 1172, 1173*t*
One-carbon bridge, *84*
-one, 81, 655*t*
Opportunity (Mars rover), 538
Opsin, 680
Optical activity, A-9
Optical bleaches, 1176
Optical brighteners, 1176
Optical purity, **152,** A-9
Optically active, **149**
Optically active compounds, 277, 287
Oral contraceptives, 1182
Orbital
defined, **2,** 30
frontier molecular orbital theory (FMOT), 909–911
hybrid, 35
MO theory. *See* Molecular orbital (MO) theory
p, 31, *32*
s, 31, *31*
shells, 3*t*
Orbital overlap picture
ammonia, *36*
borane, *38*
methane, *36*
water, *36*
Order of precedence of functions, **655**
Organ transplantation, 1274
Organic acids, A-2
Organic chemistry, 1
Organic chemistry roadmap, A-18
chapter 19, 889–890
chapters 6-11, 295–296
chapters 15-18, 649–650
chapters 20-23, 936–937
real roadmap, compared, 295
Organic polymer batteries, 1279
Organic polymer chemistry, 1264–1302
amorphous domains, 1267
anionic chain-growth polymerization, 1283–1286, 1283*t*, 1296
atactic polymer, 1282, *1282*
average degree of polymerization, 1266
cationic chain-growth polymerization, 1283*t*, 1287–1289, 1296
chain-growth polymerization, 1275–1291, 1295–1296
chain size/shape, 1265
cold-drawing, 1270
coordination polymerization, 1281
crystallinity, 1267, 1268
electrical conducting properties, 1279
epoxy resins, 1273
growth, 1265
ionic chain-growth polymerization, 1283–1287, 1283*t*
isotactic polymer, 1282, *1282*
key reactions. *See* Organic polymer chemistry—key reactions
living polymerization, 1285
mechanisms. *See* Organic polymer chemistry—reaction mechanisms
molecular weights, 1265, 1267
notation and nomenclature, 1265–1266
polyamides, 1269–1271
polycarbonates, 1272
polydispersity index, 1267
polyesters, 1271–1272
polyurethanes, 1272–1273
radical chain-growth polymerization, 1276–1279, 1296

repeat units, 1265, 1266
review/study guide, 1293–1296
ring-opening metathesis polymerization (ROMP), 1289–1291, 1296
step-growth polymerization, 1268–1275, 1294–1295
stereochemistry and polymers, 1282–1283
syndiotactic polymer, 1282, *1282*
telechelic polymers, 1287
thermosetting polymers, 1275
types of polymer architecture, 1265, *1265*
Ziegler-Natta chain-growth polymerization, 1279–1282, 1296

Organic polymer chemistry—key reactions
anionic chain-growth polymerization of substituted ethylenes, 1296
cationic chain-growth polymerization of substituted ethylenes, 1296
radical chain-growth polymerization of substituted ethylenes, 1296
ring-opening metathesis polymerization (ROMP), 1296
step-growth polymerization of diacyl chloride and diol gives polycarbonate, 1294
step-growth polymerization of dicarboxylic acid and diamine gives polyamide, 1294
step-growth polymerization of dicarboxylic acid and diol gives polyester, 1294
step-growth polymerization of diepoxide and diamine gives epoxy resin, 1295
step-growth polymerization of diisocyanate and diol gives polyurethane, 1295
titanium-mediated (Ziegler-Natta) chain-growth polymerization of ethylene and substituted ethylenes, 1296

Organic polymer chemistry—reaction mechanisms
homogeneous catalysis of Ziegler-Natta coordination polymerization, 1281–1282
initiation of anionic polymerization of alkenes, 1284
initiation of anionic polymerization of butadiene, 1284
initiation of cationic polymerization of alkene by HF/BF_3, 1288
initiation of cationic polymerization of alkene by Lewis acid, 1288
radical polymerization of substituted ethylene, 1277
Ziegler-Natta catalysis of ethylene polymerization, 1280

Organic synthesis, **314**. *See also* Synthesis
Organocuprates, 893
Organohalogen compounds, 636–637
Organolithium compounds. *See* Organomagnesium and organolithium compounds
Organomagnesium and organolithium compounds, 631–636
aldehydes and ketones, 662, 703
esters, 795
formation and structure, 631–633
Grignard reagent as strong base, 634
oxiranes, 634–635
preparation of alcohol, 635–636
proton acids, 633–634
Organometallic chemistry, 643
Organometallic compounds, 630–651
carbenes and carbenoids, 639–643
catalytic carbon-carbon bond formation, 1097
defined, **630**
key reactions. *See* Organometallic compounds—key reactions
lithium diorganocopper reagents, 636–639
organomagnesium and organolithium compounds, 631–636
reaction roadmap, 649–650
review/study guide, 644–647
Simmons-Smith reaction, 641–642, 647
Organometallic compounds—key reactions
formation of Gilman reagents, 646
formation of organomagnesium and organolithium compounds, 645
reaction of dichloro- or dibromocarbene with an alkene, 646
reaction of Grignard reagent with epoxide, 645
reaction of RMgX and RLi with proton donors, 645
Simmons-Smith reaction, 647
treatment of Gilman reagent with alkyl, aryl, or alkenyl halide, 646
Orgyia pseudotsugata, 719
Orinase, 816
Orlon, 1276*t*
L-Ornithine, 1202
Ortho *(o)*, 954
Ortho-para directing, 1006, 1007, 1008*t*
Ortho-para directors, 1023
Orthogonal, **2**

Osmium tetroxide, 270
Osmium tetroxide oxidation, 279
"Out of phase," 30
Out-of-phase addition, 34, 40
C—C π bond, *43*
C—C σ bond, *42*
C—H σ bond, *41*
C—O π bond, *51*
Oxalic acid, 727, 751, 754
Oxaloacetic acid, 166
Oxalosuccinic acid, 744
Oxanamide, 883
Oxane, 494
Oxetane, 494
Oxidation, 241*t*
alcohols, 463–473
aldehydes, 688–689
aldonic acids, 1146, 1161
alkanes, 110–111
alkenes, 268–273, 286, 507–508
benzene, 966–968, 975
defined, **268**
Dess-Martin, 468–469
fatty acid oxidation, 1176–1177
hydrogen peroxide, 267
ketones, 690–691
osmium tetroxide, 279
periodic acid, 1147–1149, 1162
Pinnick, 689–690
sulfides, 519, 525
Swern, 467–468
thiols, 476–477
uronic acid, 1146–1147
Oxidation by ozone, 286
Oxidative addition, 631, **1097**
Oxime, 684*t*
Oxiranes, 494, 505
lithium diorganocopper reagents, 638–639
organomagnesium and organolithium compounds, 634–635
polyesters, 1272
stereoselective oxirane synthesis, 506–507
oxo-, 655*t*
4-Oxo-4-phenylbutanoic acid, 1025
3-Oxobutanal, 656
3-Oxobutanoic acid, 655*t*, 728, 745
9-Oxodec-2-enoic acid, A-17
(*E*)-9-Oxodec-2-enoic acid, A-17
5-Oxohexanoic acid, 726, A-15
Oxolane, 494
Oxonium ion, **253,** 617
3-Oxopropanoic acid, 655*t*
Oxygen
isotopes, 611*t*
M + 2 peaks, 612
Oxygen-18, 811
Oxygen nucleophiles, 670–677

Oxymercuration, 241*t*, 262
Oxymercuration-reduction, **262**–264, 285
Ozone-depleting CFCs, 344
Ozonide, 272
Ozonolysis, 271–273

P

P, 39
p. *See* Para (*p*)
p-doped polyacetylene, 1279
p orbital, 31, *32*
Packaging industry, 1265
Packaging materials, 1276*t*, 1292
PAH. *See* Polynuclear aromatic hydrocarbon (PAH)
Pain relievers, 730–731. *See also* Drugs/medications
Paired spins, 3
Palladium-catalyzed cross-coupling reactions, 1107–1112, 1117–1118
Palladium(II) acetate, 1100
Palm oil, 1173*t*
Palmitic acid, 1171*t*, 1172, 1195
Palmitoleic acid, 1171*t*
Papaver somniferum, 1032
Para (*p*), 954
Para-substituted benzyl halides, 985
Paraffin wax, 1265
Parkinson's disease, 608
Paroxetine, 491
Parr shaker-type hydrogenation apparatus, 274
Parts per million (ppm), 594
Pascal's triangle, 576, *577*
Pauli exclusion principle, **3,** 33
Pauling, Linus, 8, 35, 46, 769, 1221, 1222
Pauling scale, 8*t*
Paxil, 491
PCC. *See* Pyridinium chlorochromate (PCC)
Pd(0)-catalyzed allylic alkylation, 1121, 1122
Peanut oil, 1173*t*
Pedersen, Charles, 517
Penicillin G, 791
Penicillins, 791
Penicillium, 762
Penicillium bevicompactum, 839
Penicillium citrinum, 839
Pennyroyal oil, 716
pent-, 77*t*
Pentachlorophenol (penta), 1028
pentadec-, 77*t*
Pentadecane, 74*t*
1,2-Pentadiene, 895
1,3-Pentadiene, 895
trans-1,3-Pentadiene, 895*t*
1,4-Pentadiene, 216, 895, 895*t*
Pentanal, 730*t*
Pentane
  ball-and-stick model, 74
  boiling point, 108*t*, 442*t*
  common name, 80
  condensed structural formula, 74*t*
  density, 108*t*
  isomers, 129
  line-angle formula, 74
  M and M + 1 peaks, 612
  melting point, 108*t*
  molecular formula, 74*t*
  molecular weight, 442*t*, 657*t*
  solubility, 442*t*
  structural formula, 74, 657*t*
Pentanedioic acid, 727
2,4-Pentanedione, 204, 205
1-Pentanethiol, 474
Pentanoic acid
  infrared spectrum, *549*
  mevalonic acid, A-16
  nomenclature, 81, 728*t*
  unpleasant odor, 731
1-Pentanol, 442*t*, 495*t*, 730*t*
2-Pentanol, 628
2-Pentanone, 908
3-Pentanone, *575*, 658*t*
Pentapeptide, 1208
3-Penten-2-one, 908
cis-3-Penten-2-ol, 717
1-Pentene, 129, 218*t*, 895*t*
cis-2-Pentene, 218*t*, A-10
trans-2-Pentene, 218*t*, A-10
Pentobarbital, 813
Pentose, 1135
Pentyl, A-11
Pentyl mercaptan, 474
1-Pentyne, 300*t*, *617*
2-Pentyne, 299
Penultimate carbon, 1137
Pepcid, 65
Peppermint, 220
Peptic ulcers, 1234
Peptide, 1208. *See also* Polypeptides and proteins
Peptide bond, **1208**
Peptide-bond forming reactions, 1218–1219, 1231
Peracetic acid, 507
Percent s-character, 604
Percent transmittance, 905
Perchloroethane, 332
Perchloroethylene, 332
Perfluoromethylcyclohexane, 363
Perfluoropropane, 332
Pericyclic reaction, **909**
Pericyclic reaction theory, 909–911, 927–928
Periodic acid, 1147–1149, 1162
Periodic acid oxidation of glycols, 469–471
Perkin condensation, 877
Permethrin, 742–743
Peroxyacetic acid, 507
Perutz, Max F., 1224
PET. *See* Poly(ethylene terephthalate) (PET)
PET scan. *See* Positron emission tomography (PET) scan
PETE, 1271. *See also* Poly(ethylene terephthalate) (PET)
Petroleum, 113–114
Petroleum-derived products, 113
Petroleum refinery, 113
Petroleum-refining processes, 1265
PGE_1, 814, 1179
PGE_2, 1179
$PGF_2\alpha$, 1178, 1179
pH, ionization of functional groups, 182–183
Phantom atom, 134
Phasing, **30**
Phe, 1201*t*. *See also* Phenylalanine
Phen-Fen, 1089
Phenanthrene, 943, 956
Phenobarbital, 813
Phenols, 367, 957–966, A-2
  acid-base reactions, 960–962
  acidity, 957–960, 974
  alkyl-aryl ethers, 962–963
  defined, **957**
  Kolbe carboxylation, 963, 964, 974
  medicinal use, 957
  quinones, 964–966, 974
  review/study guide, 973–974
  separating water-soluble phenols from water-insoluble alcohols, 961, *962*
  structure and nomenclature, 957
Phenoxides, 973
2-Phenoxypropane, 963
Phensuximide, 885
1-Phenyl-2-butanol, 712
1-Phenyl-3-buten-1-ol, 638
(*E*)-2-Phenyl-2-butene, 1070
(*Z*)-2-Phenyl-2-butene, 954
Phenyl group, **207**
1-Phenyl-1-pentanone, 654, 954
(*S*)-1-Phenyl-2-propanamine, 1041
1-Phenyl-1,2-propanediol, 985
1-Phenyl-2-propanol, 635
(*S*)-2-Phenyl-1-propanol, 796
trans-3-Phenyl-2-propenal, 654
Phenylacetaldehyde, 681
Phenylacetamide, 789
N-Phenylacetamide, 1023
Phenylacetic acid, 783
Phenylacetic anhydride, 767
Phenylacetonitrile, 768, 783
Phenylalanine, 1201*t*, 1203*t*, 1251
(*R*)-2-Phenylbutanamide, 779
(*R*)-2-Phenylbutanaoic acid, 779

Phenylcyclohexane, 1005
p-Phenylenediamine, 1271
2-Phenylethanamine, 1039
(*R*)-1-Phenylethanamine, 157
(*S*)-1-Phenylethanamine, 157, 1039
Phenylethanenitrile, 768
Phenylethanoic anhydride, 767
Phenylethene, 1098. *See also* Styrene
Phenylhydrazine, 684*t*
Phenylhydrazone, 684*t*
Phenylmagnesium bromide, 631
8-Phenylmenthol, 919
2-Phenylpropanal, 985
1-Phenylpropane, 987
3-Phenylpropene, 956
trans-3-Phenylpropenoic acid, 726
Phenylthiohydantoin, 1214
Phosgene, 1272
Phosphatidic acid, *1185*
Phosphatidylcholine, 1186*t*
Phosphatidylethanolamine, 1186*t*
Phosphatidylinositol, 1186*t*
Phosphatidylserine, 1186*t*
Phosphines, 66, 666, 1043, 1072
Phosphoacylglycerols, 1185. *See also* Phospholipids
Phosphodiester group, 182
Phosphoenolpyruvate, 765
Phosphoesters, 40
Phospholipases, 1186
Phospholipids, **1185**–1187
Phosphonate-stabilized ylide, 669
Phosphonium salt, 666
Phosphonium ylide, 667
Phosphoric acid, 39
Phosphoric anhydrides, 764
Phosphorus, 666, 1043
Phosphorus acid, 450
Phosphorus tribromide, 450
Photolysis, **639**
Photon, 536
Phthalic acid, 727, 968
Phthalic anhydride, 764
Phthalimide, 767, 768, 769
Phyllobates terribilis, 1043
Physical properties
  alcohols, 441–444
  aldehydes and ketones, 657, 658*t*
  alkanes and cycloalkanes, 107–110, 118
  alkenes, 217, 218*t*
  alkynes, 300, 300*t*
  amines, 1043–1045
  carboxylic acids, 729–731
  ethers, 494–495
  haloalkanes, 332–336
  monosaccharides, 1139
  thiols, 474–475
  triglycerides, 1172–1173
pI. *See* Isoelectric point (pI)
$\pi \rightarrow \pi^*$ transition, 907, 908*t*
Pi (π) bonding molecular orbital, **42**
π bonds, 571
π electron density delocalization, 70
π electrons, 176
Picric acid, 959, 960
Pinacol rearrangement, 459–463, 481
α-Pinene, 84, 85, 219, 220
Pink bollworm, 530
Pinnick oxidation, 689–690
Piper nigrum, 124
Piperidine, 124, 681, 1038, 1049
pK_a, 178–180, A-2. *See also* Acidity
Planar conformation, *92*
Planck's constant, 29, 536
Plane of symmetry, **128,** *128,* A-9
Plane-polarized light, **149**–150
Plaque, 839
Plastic, **1265**
Plastics recycling, 1291–1292
Plexiglass, 1276*t*
PLP. *See* Pyridoxal phosphate (PLP)
PMP. *See* Pyridoxine phosphate (PMP)
Poison dart frogs, 1043–1044
Poison ivy, 957, 958
Polar aprotic solvent, 390, 392, 395, 423
Polar covalent bond, 8, **11**
Polar protic solvent, 390, 395, 423
Polar solvent, 389, 423
Polarimeter, **150**–151
Polarity, haloalkanes, 332–333
Polarizability, **334**
Poly(1,1-dichloroethylene), 1276*t*
Poly(3-hydroxybutanoic acid), 1300
Poly(3-hydroxybutanoic acid-3-hydrooyoctanic acid) polymer, 1300
Poly(2-methyl-1,3-butadiene), 1301
Polyacetylene, 1279, 1291
Polyacrylamide gel electrophoresis, 1253, 1256, *1256*
Polyacrylonitrile, 1276*t*
Polyamides, **1269**–1271
Polyatomic anion, 14
Polyatomic cation, 14
Polycarbonate, **1272**
Polydispersity index, 1267
Polyenes, 216
Polyester, **1271**–1272
Poly(ethyl acrylate), 1276*t*
Polyethylene, 1265, 1266, 1276*t*
Polyethylene films, 1280
Poly(ethylene phenylurea), 1299
Poly(ethylene terephthalate) (PET), 1268, 1271, 1292, 1298
Polyhalogenation, 701
Polyhalomethanes, 335*t*
Polyisoprene, 1301
Polymer, **1265**. *See also* Organic polymer chemistry
Polymer morphology, 1267–1268
Polymer-supported polypeptide synthesis, 1219
Poly(methyl methacrylate), 1276*t*
Polynuclear aromatic hydrocarbon (PAH), **956**
Polypeptide, **1208**
Polypeptides and proteins, 1208–1227. *See also* Amino acids
  α-helix, 1222–1223
  amino acid analysis, 1209
  amino-protecting groups, 1216–1217
  β-pleated sheet, 1223
  carboxyl-protecting groups, 1217
  cyanogen bromide cleavage, 1210–1212, 1230
  definitions, 1208
  Edman degradation, 1213–1215, 1230
  enzyme-catalyzed hydrolysis of peptide bonds, 1212
  geometry of peptide bond, 1221
  hydrophobic effect, 1224
  peptide, 1208
  peptide bond, 1208
  peptide-bond forming reactions, 1218–1219, 1231
  planarity of peptide bond, *1221,* 1231
  primary structure, 1209–1216, 1229–1230
  quaternary structure, 1227
  review/study guide, 1229–1232
  secondary structure, 1222–1223
  sequence analysis, 1209–1210
  sequencing by coding nucleotide sequence, 1215–1216
  sequencing by mass spectrometry, 1215
  solid-phase synthesis, 1219–1221
  synthesis of polypeptides, 1216–1221, 1230–1231
  tertiary structure, 1223–1225
Poly(phenylene vinylene) (PPV), 1291
Polyphenylurea, 1299
Polypropylene, 1276*t*, 1292
Polysaccharides, **1151,** 1154–1156, 1162
Polysaturated fatty acids, 221
Polystyrene, 1266, 1276*t*, 1285, 1292
Polysubstituted benzenes, 955
Poly(tetrafluoroethylene), 1266, 1276*t*
Polythene, 1276*t*
Polyunsaturated fatty acid, **1172**
Polyunsaturated fatty acid esters, 353
Polyunsaturated triglyceride, **1173,** *1173*
Polyurethane, 819, **1272**–1273
Poly(vinyl acetate), 1300
Poly(vinyl chloride) (PVC)
  common name, 1276*t*
  common uses, 1276*t*, 1292
  monomer formula, 1276*t*

Poly(vinyl chloride) (*Continued*)
polymer notation and nomenclature, 1266
reaction for the formation, 367–368
recycling, 1292
vinyl chloride, 307
Poor nucleophile, 423
Porcine pancreatic lipase, 1195
Porthetria dispar, 717
Positive pK_a value, 180
Positron emission tomography (PET) scan, 721
Potassium (2*E*,4*E*)-2,4-hexadienoate, 751
Potassium sorbate, 751
Potential energy, 5
ppm. *See* Parts per million (ppm)
PPV. *See* Poly(phenylene vinylene) (PPV)
Pr. *See* Propyl (Pr)
Precession, 561, *562*
Predicted molecular shape (VSEPR), 26*t*
Prediction
reactions, 71
stability and properties, 70–71
structure and bonding, 69–70
Prefix, 77*t,* 81
Prelog, V., 134
Primary (1°), **19**
Primary (1°) alcohol, 439
Primary (1°) amine, **21,** 1038, 1071
Primary (1°) carbocation, 247, 249
Primary alkyl groups, 414
Primary carbon, 82
Primary haloalkanes, 397*t,* 408*t,* 414, 424, 426
Primary hydrogen, 82
Primary sex characteristics, 1183
Primary structure of nucleic acids, **1242**
Primary structure of proteins, **1209**
Primer, 1255
Principle of microscopic reversibility, **459,** 776
Priority rules, 134–135
Pro, 1201*t*. *See also* Proline
Procaine, 765, 815
Procardia, 882
Prochiral, 583
Progesterone
contraceptive, as, 1182
fertilized egg, 1181
intermediate, 887
Johnson's synthesis, 326–327
major effects, 1182*t*
source, 1182*t*
structure, 1182*t*
Proline, 1201*t,* 1203*t,* 1207, 1221
prop-, 77*t*
1,2-Propadiene, 303
Propanal, 658*t,* 730*t*
2-Propanamine, 1039
Propane
ball-and-stick model, 74
boiling point, 108*t,* 442*t*
density, 108*t*
line-angle formula, 74
melting point, 108*t*
molecular formula/condensed structural formula, 74*t*
primary/secondary carbons, 82
six primary hydrogens/two secondary hydrogens, 337
solubility, 442*t*
structural formula, 74
Propanedioic acid, 727, 744
1,2-Propanediol, 440
1,3-Propanediol, 440
2-Propanethiol, 474
1,2,3-Propanetriol, 220, 440
Propanoic acid, 657*t,* 728*t*
1-Propanol, 439, 442*t,* 443, *589,* 730*t*
2-Propanol, 439, 445*t*
Propanone. *See* Acetone
Proparacaine, 1036
Propargyl cation, 617
2-Propen-1-ol, 441, 542
2-Propenal, 653
Propene. *See also* Propylene
heat of hydrogenation, 275*t*
nomenclature, 81, 212
structural formula, 275*t*
Propenenitrile, 665
Propenoic acid, 726
2-Propenyl, 212
2-Propenylcyclopentane, 212
Propionaldehyde, 658*t*
Propionic acid, 728*t*
Propofol, 1087
Propoxyphene, 1079
Propranolol, 990
Propyl alcohol, 439
Propyl (Pr), A-11
Propylamine, 1045*t*
Propylbenzene, 629
Propylene. *See also* Propene
boiling point, 218*t*
common uses, 1276*t*
melting point, 218*t*
monomer formula, 1276*t*
nomenclature, 212
polymer names, 1276*t*
structural formula, 218*t*
Propylene glycol, 440
(*R*)-Propylene oxide, 1299
2-Propylpentanoic acid, 879
Propyne, 299, 300*t*
3-Propynyl cation, 617
Prostacyclin, 1180
Prostaglandin E_1 (PGE_1), 814, 1179
Prostaglandins, 1125, **1178**–1180
Prostanoic acid, 1178
Prostate cancer, 1028
Protecting group, **503**
Protein Data Bank, 1257
Protein-derived amino acids, 1200–1201, 1201*t*
Proteins, **1208**. *See also* Polypeptides and proteins
Protic acid, **195**
Protic solvent, **389,** 389*t*
Proton acceptors, 445
Proton acids, 633–634
Proton donors, 445
Proton transfer, 229
Proton-transfer reaction, 171, 177, 196
Protons, adding or removing in transfers, 229
Proventil, 992, 1078
Prozac, 491
Pseudoionone, 718
Ψ, 30
Ψ^2, 30
PTFE, 1276*t*
Puckered conformation, *92*
Puffer fish, 105
Pulegone, 875
(*R*)-Pulegone, 716, 717
Purcell, Edward, 558
Purine bases, 1239
Purines, 950, 1040, 1243*t*
Purple benzene, 518
Purple foxglove, 1134, 1163, 1167
Push electrons, 46
PVC. *See* Poly(vinyl chloride) (PVC)
Pyramidal, **24**
Pyramidal inversion, 1042, 1043
Pyranose, 1140, **1140**
Pyranoside, 1149
Pyrausta nubilalis, 218
Pyrene, 956
Pyrethrin, 742
Pyrethrin II, 227
Pyrethrosin, 227
Pyridine
acid strength, 1046*t*
alcohols, 786
aryl and alkyl sulfonates, 452
basicity, 1052
boiling point, 1045*t*
electrophilic aromatic substitution, 1022
heterocyclic analog of benzene, 949
heterocyclic aromatic amine, 1038
isoamyl benzoate, 786
melting point, 1045*t*
pK_a, 1046*t,* 1049
solubility, 1045*t*
structural formula, 1045*t*
structure, 1046*t*
UV transition, 931
Pyridinium acetate, 1041
Pyridinium chloride, 786

Pyridinium chlorochromate (PCC), 465
Pyridoxal 5-phosphate, 765
Pyridoxal phosphate (PLP), 682, 766
Pyridoxine, 682–684
Pyridoxine phosphate (PMP), 682
Pyrimidine bases, 1239
Pyrimidines, 949, 1243*t*
Pyrolysis of tertiary amine oxide, 1077
Pyrophosphate ion, 764
Pyrophosphoric acid, 764
Pyrrole
  aromaticity, 1052
  electrophilic aromatic substitution, 1023
  heterocyclic aromatic amine, 1038
  (4n + 2) π electrons, *950*
  resonance energy, *950*
Pyrrolidine, *841*, 1038
Pyrrolidinium acetate, 892
Pyruvate, 471

Q

Quadrupole mass analyzer, 610
Quantization, **2**
Quantum chemistry principle, 6
Quantum mechanics, **30,** 69–71
Quartet, 573
Quartet-triplet signals, *575*
Quaternary (4°) ammonium ion, **1042,** 1043
Quaternary structure, **1227**
Queen substance, A-17
Quinoline, 1040
o-Quinone, 964
p-Quinone, 964–966
Quinones, 964–966, 974
Quinuclidine, 1082

R

R, **134**
Racemethorphan, 1032, 1033
Racemic drugs, 155
Racemic mixture, **151,** A-9
Racemization, 699
Radical
  alkyl, 343
  allyl, 351–352
  benzoyl, 341
  BHT, 356
  defined, **335**
  ethoxy, 341
  formation, 340–341
  methyl, 341
  organic chemistry transformations, 242
Radical anion, 1285
Radical autoxidation, 353–354, 362
Radical cation, **610**
Radical chain-growth polymerization, 1276–1279, 1296
Radical chain mechanism, 341–343
Radical coupling, 1278
Radical inhibitor, **354**
Radio-frequency range, 594
Radiopaque imaging agents, 1091
Raffinose, 1166
Raman spectroscopy, **540**
Rancid odors, 354
Rate control, 900. *See also* Kinetic control
Rate-determining step, **243**
Rattlesnake, 1186
Rayon, 1156
Reaction. *See also* Reaction mechanism
  addition, 238, 241
  aldehydes and ketones, 658–659, 702
  aldol. *See* Aldol reaction
  alkanes, 110–112
  alkenes. *See* Reactions of alkenes
  alkylation, 302
  benzene. *See* Reactions of benzene and derivatives
  bimolecular, 378
  Carroll, 936
  chain-transfer, 1278
  chemoselective, A-8
  cycloaddition, 909
  diastereoselective, 257, 283
  Diels-Alder. *See* Diels-Alder reaction
  E1. *See* E1 reaction
  E2. *See* E2 reaction
  elimination, 238
  enantioselective, 257, 281, 283
  epoxides, 510–513
  ethers, 500–502
  Heck, 1098–1103, 1116
  key. *See* Key reactions
  monosaccharides, 1143–1149, 1160–1162
  pericyclic, 909
  predicting reactions, 71
  regioselective, 245
  Ritter, 1086
  Sandmeyer, 1065, 1076
  Schiemann, 1065, 1076
  S_N1. *See* S_N1 reaction
  S_N2. *See* S_N2 reaction
  stereoselective, 257
  stereospecific, 278
  Stork enamine, 841
  substitution, 238, 336
  Tiffeneau-Demjanov, 1062–1063
  unimolecular, 379
  when will a reaction occur, 71
  Wittig, 666–669
Reaction coordinate, 184
Reaction coordinate diagram, **184**–185, *185*, 242
Reaction mechanism, 71, **184**
  alcohols. *See* Alcohols—reaction mechanisms
  aldehydes and ketones. *See* Aldehydes and ketones—reaction mechanisms
  alkenes. *See* Alkenes—reaction mechanisms
  alkynes. *See* Alkynes—reaction mechanisms
  amines. *See* Amines—reaction mechanisms
  arrow pushing. *See* Arrow pushing
  bond-making steps, 691
  carboxylic acid derivatives. *See* Carboxylic acid derivatives—reaction mechanisms
  carboxylic acids. *See* Carboxylic acids—reaction mechanisms
  defined, 282
  electron sources and sinks, 234–236
  enolate anions and enamines. *See* Enolate anions and enamines—reaction mechanisms
  ethers. *See* Ethers—reaction mechanisms
  four fundamental mechanistic elements, 238–239
  haloalkanes. *See* Haloalkanes—reaction mechanisms
  nucleophilic substitution and β-elimination. *See* Substitution/elimination—reaction mechanisms
  polymers. *See* Organic polymer chemistry—reaction mechanisms
  reactions of benzene and derivatives. *See* Benzene—reaction mechanisms
Reaction roadmap. *See* Organic chemistry roadmap
Reactions of alkenes, 240–296
  1,2 shift, 254
  acid-catalyzed hydration, 252–254, 256–257
  addition of bromine and chlorine, 257
  addition of HOCl and HOBr, 259–260
  addition of hydrogen halides, 245–251
  anti addition, 279, 284
  anti stereoselectivity, 257, 258–259
  bond dissociation enthalpy (BDE), 243
  bridged halonium ion intermediates, 258–259
  bromohydrin formation, 261
  carbanion, 242
  carbocation, 242, 247–249, 251, 283
  carbocation rearrangements, 254–255
  catalytic reduction, 274
  characteristic addition reactions, 241*t*

Reactions of alkenes (*Continued*)
chiral centers as reactants or products, 277–281
electrophilic addition, 244
heats of hydrogenation, 275–276, 275*t*
hydroboration-oxidation, 264–268
hyperconjugation, 250, *250*
inductive effect, 249–250
key reactions. *See* Alkenes—key reactions
Markovnikov's rule, 245
mechanisms. *See* Alkenes—reaction mechanisms
optically active compounds, 277, 287
oxidation, 268–273
oxymercuration-reduction, 262–264
ozonolysis, 271–273
radical, 242
rate-determining step, 243
reaction roadmap, 295–296
reactive intermediate, 242
reduction, 273–277
regioselective reaction, 245
review/study guide, 282–287
stereoselective reaction, 257
stereospecific reaction, 278
syn addition, 266, 270, 274, 279, 284
thermodynamics of addition reactions, 243–244
Reactions of benzene and derivatives, 995–1036. *See also* Benzene
activating/deactivating group, 1006, 1008*t*
addition/elimination, 1015–1016, 1021–1022
benzyne intermediate, 1014–1015, 1021
bromination, 996–998
chlorination, 996–998
disubstitution and polysubstitution, 1006–1013, 1020–1021
electrophilic aromatic substitution, 996–1006
Friedel-Crafts acylation, 1002–1004, 1019
Friedel-Crafts alkylation, 1000–1002, 1017, 1019
halogenation, 995, 1017, 1018
inductive effect, 1011, 1012, 1020
key reactions. *See* Benzene—key reactions
mechanisms. *See* Benzene—reaction mechanisms
meta directing, 1006, 1007, 1008*t*
nitration, 995, 998–1000, 1017, 1018
nitration of anisole, 1010, *1010*
nitration of benzoic acid, 1010–1011, *1011*
nucleophilic aromatic substitution, 1013–1017, 1021–1022
ortho-para directing, 1006, 1007, 1008*t*
other electrophilic aromatic alkylations, 1004–1005, 1018
reaction roadmap, 889–890
resonance effect, 1011, 1012, 1020
review/study guide, 1017–1022
substitution reactions, equations, 995–996
sulfonation, 996, 999, 1017, 1019
theory of activating-deactivating effects, 1012–1013
theory of directing effects, 1010–1012
Reactive dyes, 1026
Reactive intermediate, **242**
Rearrangement
benzil-benzilic acid, 753
carbocation, 254–255
Claisen, 920, 922–923, 929
Cope, 920, 923–925, 929
defined, **254**
Favorskii, 716
McLafferty, 618–620
Pinacol, 459–463, 481
solvolysis of 2-Chloro-3-phenylbutane, 396
Reciprocal centimeters, 538. *See also* cm^{-1}
Recycling of plastics, 1291–1292
Red chili peppers, 961
Reducing sugars, **1146**
Reduction, 241*t*
aldehydes and ketones, 691–698
alditols, 1145–1146
alkenes, 273–277
alkynes, 312–314, 322
amides, 798–799, 808
carboxylic acid derivatives, 796–800
carboxylic acids, 736–738
catalytic. *See* Catalytic reduction
catalytic reduction of benzene to cyclohexane, 1270
Clemmensen, 696, 697, 708
defined, **268,** 269
dissolving-metal, 313
enantioselective, 287
esters, 796–798, 807–808
metal hydride, 691–692, 708
nitriles, 800, 808
oxymercuration-reduction, **262**–264
triglycerides, 1173–1174
Wolff-Kishner, 697–698, 709
Reductive animation, **694**–696, 708
Reductive elimination, **1097**
Reformatsky, Sergei, 877
Reformatsky reaction, 877
Regiochemistry
addition, 1013
allylic alkylation, 1106–1107
predicting products of reaction sequenced, A-18
Regioselective reaction, **245,** A-9
Regioselectivity
E1 reaction, 404
haloalkanes, 337–338, 345–348
Relative charge delocalization, 174
Renewable energy sources, 112
Renografin, 1091
Repeat unit, 1265, 1266
Rescula, 1195
Resolution, **157**–159, **610,** 919
Resonance, 51–52. *See also* Contributing structures
defined, **46**
hybridization, 52–53
NMR spectroscopy, 561–563
relative contributions of members of contributing structures, 50
theory of, 46–47
VSEPR, 55
Resonance contributors, 46
Resonance effect, 1011, 1012, 1020
Resonance energy, **942**
Resonance in NMR spectroscopy, **561**
Resonance-stabilized alkenyl radical anion, 314
Resonance-stabilized enolate anion, 685
Resonance structures, 46
Resonance yield, **46**
Resorcinol, 957
Restriction endonuclease, **1253**–1254
Restriction fragment, 1254
Retinal, 680
Retinol, 217, 220, 1188
Retrosynthesis, **315**
Retrosynthetic analysis
acetoacetic ester synthesis, 846–847
aldol reaction, 831–832
alkynes, 314–316
atenolol, 991
Claisen or Dieckmann condensation, 836
diazepam, 1031
diethylcarbamazine, 817
enamines, 844
ganglefene, 1088
haloperidol, 1035
hexylcaine, 1089
ibutilide, 1087–1088
labetalol, 1033
lithium diorganocopper reagents, 858–860
Michael reaction, 855–857
N-Methylmorpholine, 1086
p-methyldiphenhydramine, 1034
procaine, 515
proparacaine, 1036
Propofol, 1087
Robinson annulation, 858
(S)-Metolachlor, 817

sotalol, 1094
spasmolytol, 1035
verapamil, 886
vitamin A, 883
warfarin, 883
Rhodopsin, 680, 1188, *1189*
β-D-Ribofuranoside, 1144
Ribonuclease A, 1224
Ribonucleic acid. *See* RNA
Ribose, 721
D-Ribose, 1138
Ribosomal (rRNA), 1248*t*, **1249**
Rickets, 1188
Rimantadine, 324–325, 715
Ring-closing alkene metathesis, 1113, 1114
Ring current, **572**
Ring-opening metathesis polymerization (ROMP), 1115, 1289–1291, 1296
Ring structures, 145–147, 162
Ritter reaction, 1086
RNA
automation/synthesis, 1239
codon, 1251
DNA, contrasted, 1248
example, 1250
heterocyclic aromatic amine bases, *1239*
messenger (mRNA), 1248*t*, 1250
ribosomal (rRNA), 1248*t*, 1249
transcription and translation of genetic information, 1238, 1252–1253
transfer (tRNA), 1248*t*, 1249–1250
Robinson annulation, 857–858, 873
Rocking, *540*
ROMP. *See* Ring-opening metathesis polymerization (ROMP)
ROOH, 341
ROOR, 341
Rotoxamine, 987
Roundworm, 817
Rowland, Sherwood, 344
rRNA. *See* Ribosomal (rRNA)
R,S system, **134**–137
RSH, 475
RSR, 518
Ruminants, 1155

S

S, 39, **134**
s-Bu. *See sec*-Butyl (*s*-Bu)
s-cis conformation, 913, 914
s-trans conformation, 913, 914
s orbital, 31, *31*
Saccharide chair conformations, 1143
Saccharin, 1152*t*
Salbumol, 1078
Salicin, 730, 1165
Salicylic acid, 727, 730, 963
Salt linkages, 1224
Salts of carboxylic acids, 734–735, 806
Sandmeyer reaction, 1065, 1076
Sanger, Frederick, 1209, 1236
Sanger dideoxy method, **1254,** 1255–1256, 1261
Sanger's reagent, 1236
α-Santonin, 226
Saponification, **777**–778, 837, 1174
Saturated fatty acids, 221, 1171*t*, 1173*t*
Saturated hydrocarbons, 72
SB rubber, 1301
Schiemann reaction, 1065, 1076
Schiff base, **678**. *See also* Imines
Schizophrenia, 1035, 1036
Schrock, Richard, 1112
Schrödinger, Erwin, 30
Schrödinger equation, *31,* 32
Scissoring, *540*
Sea-Legs, 1034
Sea lions, 1170
Seasickness, 1034
Second law of thermodynamics, 186
Secondary (2°), **19**
Secondary (2°) alcohol, 439
Secondary (2°) amine, **21,** 1038, 1071
Secondary (2°) carbocation, 177, 247, 249
Secondary alkyl groups, 414
Secondary carbon, 82
Secondary hydrogen, 82
Secondary sex characteristics, 1183
Secondary structure of nucleic acids, **1243**
Secondary structure of proteins, **1222**
Sedative/hypnotic drugs, 1030
Seizures, 879
Selective reduction, 693
Selective serotonin reuptake inhibitors (SSRIs), 491
Semicarbazide, 684*t*
Semicarbazone, 684*t*
Separation of enantiomers, 157–159
Sequence analysis, 1209–1210
Sequencing
complementary sequences, 1247
DNA. *See* DNA sequencing
human genome, 1256–1257
polypeptides and proteins, 1215–1216
Sequencing equipment, *1257*
Ser, 1201*t*. *See also* Serine
Serine, 156, 1186*t*, 1201*t*, 1203*t*
L-Serine, 438
Serotonin, 950, 1233
Sertraline, 491
Sesquiterpenes, *1184*
Sex hormones, 1181–1183
Sharpless, Barry, 508, 509
Sharpless asymmetric epoxidation, 508–510, 523
Shell
defined, **2**
electrons, 2*t*
orbitals, 3*t*
valence, 6
Shielding in NMR, **562**
$(sia)_2BH$, 308
Side chain, 155, 156
σ_{1s}, 34
σ^*_{1s}, 34
Sigma antibonding molecular orbital (σ^*_{1s}), 34
Sigma (σ) bonding molecular orbital, **34**
Sigmatropic shifts, 920–925
[3,3]-shift, 920
Claisen rearrangement, 920, 922–923, 929
Cope rearrangement, 920, 923–925, 929
review/study guide, 929
Signal, **562**
Signal splitting, 572–582
Silane, 66, 503
Sildenafil, 1090
Silicon, 503
Silk farm, 217
Silkworm moth, 818
Silyl ethers, 503–504
Simmons, Howard, 641
Simmons-Smith reaction, 641–642, 647
Simmons-Smith reagent, 642
Simvastatin, 839, 840
Singer, S. J., 1187
Single-sided arrowhead, 233
Single-stranded DNA (ssDNA), 1255, *1255*
Singulair, 534
Six-membered rings, 70
Size of atom, 189
Skunks, scent, 474
Small ring strain, **91**
Smalley, Richard, 27
Smith, Ronald, 641
Smoking cessation, 1030
S_N1 reaction
carbocation stability, 385–386
defined, **379**
E1 reaction, 410–411
energy diagram, *381*
ester hydrolysis, alternative to, 780
hydrolysis, 780
kinetics, 382
leaving group, 389
nucleophile-unassisted substitution, 418
reaction mechanism, 379–380, 424

S_N1 reaction (*Continued*)
reaction of 3° alcohol with HBr, 448–449
skeletal arrangement, 395–396
S_N2 reaction, compared, 397*t*
solvent, 389–390, 391–392
stereochemistry, 383
summary/overview, 397*t*, *413*, 413*t*, 414, 426
S_N2 reaction, 377–379
β-branching, *388*, 388*t*
defined, **378**
E2 reaction, 411–412
energy diagram, *379*
ester hydrolysis, alternative to, 779–780
hydrolysis, 779–780
kinetics, 382, 384
leaving group, 389
nucleophile-assisted substitution, 418
nucleophilicity, 393–395
reaction mechanism, 378, 424
reaction of 1° alcohol with HBr, 449
skeletal arrangement, 395–396
S_N1 reaction, compared, 397*t*
solvent, 389–390, 392
stereochemistry, 383–384
steric hindrance, 387–388
summary/overview, 397*t*, *413*, 413*t*, 414, 426
Snake venom phospholipases, 1186
Soap micelles, 1174, *1175*
Soaps and detergents, 1174–1176
Socrates, 1038
Sodium 4-dodecylbenzenesulfonate, 1175
Sodium acetate, 169
Sodium benzoate, 790
Sodium bicarbonate, 961
Sodium borodeuteride, 699
Sodium borohydride, 691, 797
Sodium borohydride reduction, 692
Sodium cyanoborohydride, 694
Sodium D, 150
Sodium ethoxide, 205
Sodium hydride, 301
Sodium mercaptoacetate, 476
Sodium naphthalide, 1285
Sodium nitrite, 1060
Sodium perborate tetrahydrate, 1176
Sodium phenoxide, 961, 963, 1014
Sodium silicate, 1175
Sodium salicylate, 963
Sodium thioglycolate, 476
Sodium undecanoate, 784
Soft drink bottles, 1292
Solid-phase synthesis, 1219–1221
Solubility
alcohols, 442*t*, 444, 730*t*
aldehydes and ketones, 658*t*, 730*t*
amines, 1045*t*
carboxylic acids, 730*t*
ethers, 495*t*
Solvent, 389–390, 423
aprotic, 390, 390*t*
MCAT Practice box, 417
nonpolar, 389, 423
polar, 389, 423
polar aprotic, 390, 423
polar protic, 390, 423
protic, 389, 389*t*
Solvolysis, **379,** 396
Some Like It Hot (film), 961n
Sondheimer, Franz, 945
Sonogashira coupling, 1111–1112, 1118
Sorbitol, 1145
D-Sorbitol, 1145
Sotalol, 1093–1094
Soybean oil, 1173*t*
sp hybrid orbital, **38**
sp hybridization, 38
sp^2 hybrid orbital, **37**–38
sp^2 hybridization, 37
sp^3 hybrid orbital, **35**–37
sp^3 hybridization, 36
Spandex, 1273
Spartan, 127
Spasmolytol, 532, 1035
Specific rotation, **150,** 151, A-10
Spectroscopy
^{13}C-NMR, 585–587
IR. *See* Infrared (IR) spectroscopy
molecular, 536–537
NMR. *See* Nuclear magnetic resonance (NMR) spectroscopy
Raman, 540
ultraviolet-visible, 537*t*, 904–908, 926–927
Spider silk, 1199, 1225–1226
Spidroin 1 and 2, 1226
Spin-spin coupling, **574,** *575*
Spin-state population, 560–561
Spirit (Mars rover), 538
Spiro[2.2]pentane, 649
Squalene, *1184*
ssDNA. *See* Single-stranded DNA (ssDNA)
SSRIs. *See* Selective serotonin reuptake inhibitors (SSRIs)
Stability
alkanes and cycloalkanes, 111–112
anion, 188, 198
carbenes, 1113
carbocation, 248–249, 251
conjugated dienes, 894–898, 925
delocalization, 70, 71
Stabilization by resonance delocalization, 396
Stable molecules, 69
Stachyose, 1166
Staggered conformation, **86,** *86*
Stannanes, 1110
Stanozolol, 1183
Star-shaped polymer, *1265*
Starch, 1154
Stearic acid, 221, 1171*t*
Step-growth polymerization, **1268**–1275, 1294–1295
Step-growth polymers, 1268
Stereocenter, **100, 130,** A-10
Stereochemical terms, A-8 to A-10
Stereochemistry
allylic alkylation, 1106–1107
amino acid, 155–156
anti addition, 296
Cope rearrangement, 924–925
defined, **127**
Diels-Alder reaction, 917–918
E2 reaction, 404
NMR spectroscopy, 582–583
polymers, 1282–1283
predicting products of reaction sequenced, A-18
radical halogenation, 348
S_N1 reaction, 383
S_N2 reaction, 383–384
terminology, A-8 to A-10
Stereoisomerism and chirality, 127–169
achiral molecules, 152
amino acid stereochemistry, 155–156
biological world, 153–155
chiral/achiral, 128
chirality in enzymes, 153–154
chromatography, 159
disubstituted derivatives of cyclohexane, 145–146
disubstituted derivatives of cyclopentane, 144–145
enantiomeric excess (ee), 152–153
Fischer projection formulas, 142–143
isomers. *See* Isomers
meso compounds, 139–140
optical activity, 149–153
optical purity, 152
overview, *146*
plane-polarized light, 149–150
polarimeters, 150–151
priority rules, 134–135
racemic mixtures, 151
resolution, 157–159
review/study guide, 160–163
R,S system, 134–137
separation of enantiomers, 157–159
Stereoisomers
defined, **99, 129,** A-10
example, 131–132
polyenes, 216, 217
relationship among isomers, *148*
when they occur, 149
Stereorepresentation, 125
Stereoselective oxirane synthesis, 506–507

Stereoselective reaction, **257,** A-10
Stereospecific reaction, **278,** A-10
Steric clashing, *277*
Steric hindrance, **385,** 387–388
Steric strain, **89**
Steroid hormones, 1132, 1181–1183
Steroids, 1180–1185
  anabolic, 1183
  bile acids, 1183
  cholesterol. *See* Cholesterol
  defined, **1180**
  review/study guide, 1193
  steroid hormones, 1181–1183
  tetracyclic ring system, 1180, *1180, 1181*
trans-Stilbene, 1100
Stille coupling, 1110–1111, 1118
Stitches that dissolve, 1274–1275
Stork, Gilbert, 841
Stork enamine reaction, 841
Strain
  angle, 89
  defined, **86**
  small ring, 91
  steric, 89
  torsional, 87
Streptomyces cinnamonensis, 888
Streptomyces erythreus, 1079
Streptomyces orientalis, 1129
Stretching vibrations, *540*
Strong nucleophile, 393
Structure and bonding, 69–70
Styrene
  anionic polymerization, 1285
  cationic polymerization, 1283*t*
  Heck reaction, 1098
  nomenclature, 954, 1266
  radical polymerization, 1301
  uses, 1276*t*
Styrofoam, 1276*t*
Styryl radical anion, 1285
Subshells, 2
Substituted cyclohexane rings, 103–104
Substituted malonic acids, 744–745
Substitution, 238
  1,2-, 973
  1,3-, 973
  1,4-, 973
  allylic, 349
  aromatic, 1096
  defined, **336**
  nucleophile-assisted, 418. *See also* S_N2 reaction
  nucleophile-unassisted, 418. *See also* S_N1 reaction
  nucleophilic acyl, 771
  nucleophilic aromatic, 1014–1017, 1021–1022
Substitution/elimination—key reactions
  E1 reaction, 425
  E2 reaction, 426
  neighboring group participation, 427
  S_N1 reaction, 424
  S_N2 reaction, 424
Substitution/elimination—reaction mechanisms
  E1 reaction of 2-Bromo-2-methylpropane, 402
  E2 reaction of 2-Bromobutane, 403
  E2 reaction of *cis*-1-Chloro-2-isopropylcyclcohexane, 408
  E2 reaction of enantiomers of 1,2-Dibromo-1,2-diphenylethane, 407
  E2 reaction of meso-1,2-Dibromo-1,2-diphenylethane, 406–407
  hydrolysis of sulfur mustard, 419
  rearrangement during solvolysis of 2-Chloro-3-phenylbutane, 396
  S_N1 reaction, 379–380
  S_N2 reaction, 378
Subtraction, 34
Succinic acid, 727, 753, 754
Succinic anhydride, 764, 1025
Succinimide, 349, 767, 768, 968
Sucrose, 1151, 1152*t*
Suffix, 81
Sugar-free products, 1145
2-Sulfanylethanol, 474
Sulfides
  common names, 518
  defined, **518**
  IUPAC system, 518
  oxidation, 519, 525
  preparation, 518–519
  review/study guide, 525
  sulfoxide/sulfone, 519
Sulfinic acid, 477
Sulfonamides, 768, 802
Sulfonates, 452–453
Sulfonation, 996, 999, 1017, 1019
Sulfone, 519
Sulfonic acid, 477, A-2
Sulfonic acid esters, 786
Sulfonyl chlorides, 452, 763
Sulfonylureas, 815
Sulfoxide, 519
Sulfur
  depiction, 39
  isotopes, 611*t*
  M + 2 peaks, 612
Sulfur mustard gas, 418–420
Sulfur ylide, 713
Sulfuric acid, 39
Supercoiled DNA, 1247–1248
Supercoiling, 1247–1248
Superglue, 1286–1287
Superhelical twists, *1247,* 1248
Suprafacial, **910**
Suprafacial collision geometries, 910
Suprafacial interaction, 910
Surface dyes, 1026
Surfynol, 718
Suzuki, Akira, 1107, 1108, 1127
Suzuki coupling, 1108–1110, 1117
Swern oxidation, 467–468
Symmetric stretching, *540*
Symmetrical ethers, 520
Symmetrical sulfides, 518
Symmetry, *128*
Syn addition
  addition of atoms to same side of double bond, 284
  cis-alkene, 312
  catalytic reduction, 274, 312
  Heck reaction, 1100
  hydroboration, 266
  oxidation of alkene by glycol, 270
  oxidation of osmium tetroxide, 279
Syn hydroxylation, 278
Syn stereoselective, **265**
Syndiotactic polymer, **1282,** *1282*
Synthesis. *See also* Conversion; Industrial synthesis
  acetoacetic ester synthesis, 845–849, 871
  alkynes from alkenes, 302–304, 319
  amines, 1055–1057
  convergent, 886
  epoxides, 505–510
  importance, 314
  malonic ester synthesis, 849–850, 872
  Merrifield solid-phase synthesis, 1219–1221
  polypeptides, 1216–1221, 1230–1231
  retrosynthetic analysis, 314–316
  stereoselective oxirane synthesis, 506–507
  Williamson ether synthesis, 496–497
Synthesis gas, 114
Synthetic detergents, 1175–1176
Synthetic pyrethrenoids, 742

T

T. *See* Thymine (T)
T_3, 1202
T_4, 1202
T_g. *See* Glass transition temperature (T_g)
T_m. *See* Melt transition (T_m)
t-Bu. *See tert*-Butyl (*t*-Bu)
Tables/rules/classes, etc. *See* Appendices
Tagamet, 1234
"Take a proton away," 229, 236, 239
Tamoxifen, 993
Tanaka, Koichi, 621
Tandem mass spectrometry, 610
Tartaric acid, 139, 491
(*R,R*)-Tartaric acid, 141*t*

(2*R*,3*R*)-(+)-Tartaric acid, *158*
(*S*,*S*)-Tartaric acid, 141*t*
(2*S*,3*S*)-Tartaric acid, 1164
Tautomerization, 709
Tautomers, **309**
Tay-Sachs disease, 1168
Tay-Sachs ganglioside, 1168
TBS group. *See* *tert*-Butyldimethylsilyl (TBS) group
TCA cycle. *See* Tricarboxylic acid (TCA) cycle
Teflon, 1276*t*
Telechelic polymer, **1287**
Telomerase, 1249
Telomeres, 1249
Terephthalic acid, 727, 968, 1271, 1272
Terminal alkyne, **299,** 308
Terminal alkyne anions, 663–664, 703
C-Terminal amino acid, **1208**
N-Terminal amino acid, **1208**
Terminal atom, 45
Terminology
  glossary, G-1 to G-11
  stereochemical terms, A-8 to A-10
Termites, 1155
Terpenes, 218–220
Terpin, 290
Tertiary (3°), **19**
Tertiary (3°) alcohol, 439
Tertiary (3°) amine, **21,** 504, 1038, 1071
Tertiary (3°) carbocation, 177, 247, 249
Tertiary alkyl groups, 414
Tertiary carbon, 82
Tertiary haloalkanes, 397*t*, 408*t*, 414, 424, 426
Tertiary hydrogen, 82
Tertiary structure of nucleic acids, **1247**
Tertiary structure of proteins, **1223**
TES group. *See* Triethylsilyl (TES) group
Testing for glucose, 1148–1149
Testosterone, 1182*t*, 1183
Tetraalkyl borate, 691
Tetraalkylammonium halide, 1056
Tetrabromoalkane, 305, 319
Tetrabutylammonium fluoride, 504
tetradec-, 77*t*
Tetradecane, 74*t*
Tetradecapeptide, 1235
cis-11-Tetradecenyl acetate, 218
trans-11-Tetradecenyl acetate, 218
Tetradecylpyridinium chloride, 1042
Tetrafluoroethylene, 1276*t*
Tetrahaloalkane, 319
Tetrahedral, **24**
Tetrahedral atom, 130
Tetrahedral carbon chiral center, 132
Tetrahedral carbonyl addition compound, 658, 660
Tetrahydrocannabinol, 629
Tetrahydrofuran (THF), 265, 390*t*, 494
Tetrahydrofurfuryl alcohol, 487
Tetrahydroisoquinoline, 1052
Tetrahydropyran, 494
Tetrahydropyranyl (THP) ethers, 677–678
Tetrahydropyranyl (THP) group, 994
Tetrahydroquinoline, 1052
α-Tetralone, 1002
Tetramethylammonium chloride, 1042
cis-2,2,5,5-Tetramethyl-3-hexene, 292
2,2,3,3-Tetramethylbutane, 111*t*
Tetramethylsilane (TMS), 503, 563
Tetraodontidae, 105
Tetrapeptide, 1208
Tetrasubstituted carbon-carbon double bonds, 273
Tetrodotoxin, 105–106
Tetrose, 1135
Textile and carpet fibers, 1276*t*
Textile fibers from cellulose, 1156
Theory of activating-deactivating effects, 1012–1013
Theory of directing effects, 1010–1012
Theory of resonance, 46–47
Thermal reactions, 184
Thermochemistry, **184**–187
Thermodynamic control, **855, 864,** 901
Thermodynamics, 185, A-1
Thermodynamics of addition reactions, 243–244
Thermolysis, **639**
Thermoplastic, **1265**
Thermosets, 1265, 1275
Thermosetting plastic, **1265**
Thermosetting polymers, 1275
THF. *See* Tetrahydrofuran (THF)
Thiazolinone, 1214
Things You Should Know
  arrow pushing. *See* Arrow pushing
  carboxylic acid derivative reaction mechanisms, 759–761
  electrophiles, 228–231
  nucleophiles, 228–231
  predicting reactions, 71
  predicting stability and properties, 70–71
  predicting structure and bonding, 69–70
  reaction mechanisms, 232–239
Thioethers, 518. *See also* Sulfides
6-Thioguanine, 1261
-thiol, 655*t*
Thiols, A-2. *See also* Alcohols
  acidity, 476, 483
  biological molecules, 475
  boiling points, 475*t*
  common names, 474
  defined, **473**
  IUPAC system, 473, 474
  key reactions. *See* Thiols—key reactions
  oxidation, 476–477
  physical properties, 474–475
  preparation, 475–476
Thiols—key reactions
  acidity of thiols, 483
  oxidation of thiols to disulfides, 483
Thiophene, 950
Thomson, J. J., 606, 939
THP ethers. *See* Tetrahydropyranyl (THP) ethers
THP group. *See* Tetrahydropyranyl (THP) group
Thr, 1201*t*. *See also* Threonine
Threonine, 156, 1201*t*, 1203*t*
(2*R*,3*S*)-Threose, 138
(2*S*,3*R*)-Threose, 138
Thromboxane A_2, 1180
Thromboxane B_2, 892
Thymine (T), *1144, 1239*
Thymol, 957, 1088
Thymus vulgaris, 957
L-Thyroxine, 1202
Tietze, Lutz, 1132
Tiffeneau-Demjanov reaction, 1062–1063
Tinea pedis, 751
TIPS group. *See* Triisopropylsilyl (TIPS) group
Titanium-mediated (Ziegler-Natta) chain-growth polymerization, 1296
TMS. *See* Tetramethylsilane (TMS)
TMS group. *See* Trimethylsilyl (TMS) group
Tobacco mosaic virus protein disc, 1227*t*
α-Tocopherol, 355, 1189
Tolazamide, 816
Tolbutamide, 815, 816
Tolciclate, 935
Tolinase, 816
Tollens' reagent, **688,** 689
Toluene
  dielectric constant, 390*t*
  halogenation, 968
  hydrogenolysis, 969
  hydrogenolysis of Z-protecting group, 1217
  infrared spectrum, *545*
  iodination, 984
  mass spectrum, *622*
  nomenclature, 954
  structure, 390*t*
2,6-Toluenediisocyanate, 1273
p-Toluenesulfonic acid, 763
p-Toluenesulfonyl chloride, 452, 763, 786
Toluenesulfonyl group (Ts), 169
m-Toluidine, 1014
o-Toluidine, 1052
p-Toluidine, 1014, 1040
Topicity, 582–583

Toremifene, 993–994
Torsional strain, **87**
Tosyl chloride (TsCl), 452, 763, 786
Tranquilizer, 813
Trans, **100, 212**
Trans-alkene, 314, 322
trans dienophile, 916
Trans fats, 1192
Trans fatty acids, 276–277
Transaminase, 682
Transesterification, **787,** 788, 805
Transfer (tRNA), 1248*t,* **1249**–1250
Transition metal catalysis, 737
Transition-metal nucleophilic-carbene catalysts, 1113, 1115
Transition state, **184,** 185, 243
Transmetallation, **1107**
Tree diagrams, *579, 580*
Trehalose, 1165
Triacontyl palmitate, 1172
Triacylglycerols, **1171**. *See also* Triglycerides
Trialkenylborane, 308
Trialkylboranes, 267
2,4,6-Tribromophenol, 955
Tricarboxylic acid (TCA) cycle, 154
Trichlor, 332
Trichloroacetic acid, 732
1,3,5-Trichlorobenzene, 1066
1,1,1-Trichloroethane, 332
Trichloroethylene, 332
Trichloromethane, 332
(Z)-9-Tricosene, 637
tridec-, 77*t*
Tridecane, 74*t*
(*E*)-5-Tridecene, 637
Trienes, 216
Triethanolamine, 527
Triethylamine
- acid strength, 1046*t*
- acidity/basicity, 1082
- boiling point, 1045*t*
- melting point, 1045*t*
- nomenclature, 1040
- reaction with alcohols, 786
- solubility, 1045*t*
- structural formula, 1045*t*
- structure, 1046*t*
- tertiary amine, 1082

Triethylammonium chloride, 451, 1041
Triethylborane, 265
Triethylsilyl chloride (TESCl), 504
Triethylsilyl (TES) group, 504
Triflates, 1099
4,4,4-Trifluoro-1-butanol, 191
3,3,3-Trifluoro-1-propanol, 191
2,2,2-Trifluoroethanol, 190, 191
Trifluoromethanesulfonates, 1099
Trifluoromethanesulfonyl chloride, 1099
Trifluoromethylbenzene, 1029
Trifluralin B, 1026
Triglycerides, 220, **1171**–1174, 1191–1192
2,3,4-Trihydroxybutanal, 137, *138*
Triiodoaromatics, 1091
L-Triiodothyronine, 1202
Triisopropylsilyl chloride (TIPSCl), 504
Triisopropylsilyl (TIPS) group, 504
3,7,11-Trimethyl-2,6,10-dodecatrien-1-ol, A-17
(2*E*,6*E*)-3,7,11-Trimethyl-2,6,10-dodecatrien-1-ol, A-17
Trimethylamine, 21, 1038, 1045*t,* 1046*t*
1,2,4-Trimethylcyclohexane, 98
3,7,11-Trimethyldodecan-1-ol, A-17
2,2,4-Trimethylpentane, 82, 114, *615*
Trimethylphosphine, 39
Trimethylphosphite, 668
Trimethylsilyl chloride (TMSCl), 504
Trimethylsilyl (TMS) group, 504
2,4,6-Trinitrophenol, 959, 960
Triose, 1135
Tripeptide, **1208**
Triphenylmethyl (trityl) ethers, 985
Triphenylphosphine, 666, 1100
Triphosphate ion, 764
Triphosphoric acid, 764
Trisaccharides, 1151
Trisubstituted carbon-carbon double bonds, 273
Triterpenes, *1184*
Trityl ethers. *See* Triphenylmethyl (trityl) ethers
tRNA. *See* Transfer (tRNA)
Tropical oils, 1172
Tropylium cation, 622
Trp, 1201*t*. *See also* Tryptophan
Trypsin, 1212
Tryptophan, 1201*t,* 1203*t*
Ts. *See* Toluenesulfonyl group (Ts)
TsCl. *See* Tosyl chloride (TsCl)
Tub conformation, 948
Tumeric flower, 894
Turnover, 1096
Turpentine, 220
Twist-boat conformation, **93,** *93*
Twisting, *540*
Type A blood, 1154
Type B blood, 1154
Type O blood, 1154
Tyr, 1201*t*. *See also* Tyrosine
Tyrosine, 1201*t,* 1203*t*

U

U. *See* Uracil (U)
Ullmann coupling, 1036
Ultraviolet-visible radiation, 904
Ultraviolet-visible spectroscopy, 537*t,* 904–908
- absorbance, 905
- Beer-Lambert law, 905, 906–907
- percent transmittance, 905
- $\pi \rightarrow \pi^*$ transition, 907, 908*t*
- review/study guide, 926–927
- ultraviolet-visible radiation, 904

Unbranched alkyl substituents, 79
Unconjugated diene, 894, 895
Unconjugated double bonds, 895
undec-, 77*t*
Undecane, 74*t*
Undecanenitrile, 784, 785
Undecanoic acid, 784, 785
Unimolecular reaction, **379**
Unoprostone, 1195
Unpaired electron density, 70
Unpleasant odors, 731
Unsaturated alcohols, 440
Unsaturated fatty acids, 1171*t,* 1173*t,* 1174
Unsaturated hydrocarbons, 73, **206**
Unsaturated α,β-unsaturated carbonyl, 831
Unstabilized ylides, 668
Unstable structures, 6
"Up is up and down is down," 100
Upfield, **565,** *565*
Upholstery, 1273
Uracil (U), *1144, 1239*
Urea, 790
Urea-formaldehyde thermosets, 1275
Urethane, 819, 1272
Uridine, *1240*
Urografin, 1091
Uronic acid, 1146–1147
UV-visible radiation, 904
UV-visible spectroscopy. *See* Ultraviolet-visible spectroscopy

V

Vaginal yeast infections, 1030
Val, 1201*t*. *See also* Valine
Valence bond theory (VB theory), **34**–38, 653
Valence electrons, **6**
Valence shell, **6,** 69
Valence-shell electron-pair repulsion (VSEPR), 24–26
- defined, **24**
- predicted molecular shape, 26*t*
- resonance, 55

Valeric acid, 728*t*
Valine, 1201*t,* 1203*t*
Valium, 1030
Valnoctamide, 887
Valproic acid, 879
van der Waals, J. D., 333
van der Waals forces, **333**
van der Waals radius, **333,** 333*t*
van der Waals strain, 89
Vancomycin, 1129
Vancomycin aglycon, 1129
Vanilla pompona, 957

Vanillic acid, 688
Vanillin, 688, 957, 1165
Variants of acetoacetic ester synthesis, 848
VB theory. *See* Valence bond theory (VB theory)
Venlafaxine, 1031–1032
Verapamil, 886
Very-high-molecular-weight alkanes, 118
Viagra, 1090
Vibrational infrared region, **538**
Vicinal coupling, **574**
Vicinal diol, **270**
Vicinal hydrogens, **574**
Vinyl, **212**
Vinyl acetate, *588*
Vinyl chloride, 307, 331, 1266, 1276*t*. *See also* Poly(vinyl chloride) (PVC)
Vinyl ethers, 546
Vinyl triflates, 1110
Vinylacetylene, 299
Vinylcyclopentane, 212
Vinylic carbocation, **306**
Vinylic halide, **331**
Vinylic hydrogen, 570–572
Viscose rayon thread, 1156
Visible light color-wavelength correlation, *906*
Visual cycle, 1188, *1189*
Visual purple, 680
Vitamin A, 217, 220, 883, 1188
Vitamin A acetate, 718
Vitamin A aldehyde, 680
Vitamin B_6, 682–684
Vitamin C, 1141
Vitamin D, 1188–1189
Vitamin D_3, 1188
Vitamin E, 355, 1189–1190
Vitamin K deficiency, 1190–1191
Vitamin K_2, 965, *965*
Vitamin K_1 base, 1190, 1191
von Euler, Ulf, 1178
VSEPR. *See* Valence-shell electron-pair repulsion (VSEPR)

W

Wacker process, 737
Wagging, *540*
Warburganal, 933
Warfarin, 766, 883
Water, A-2
  carboxylic acid derivatives. *See* Hydrolysis
  electrostatic potential map, 28
  Lewis structure, 13*t*, *25*
  orbital overlap picture, *36*
  pK_a, 445*t*
  protic solvent, 389*t*
  shape, *25*
Watson, James D., 1243, 1245, 1246
Watson-Crick model, **1243,** 1244
Wave equation, 29
Wave function, **30**
Wave mechanics, **30**
Wavelength, **535,** 536*t*
Wavenumbers, **538,** 541
Weak nucleophile, 393
Weight average molecular weight (M_w), 1267
Wellbutrin, 1030
West Indian vanilla, 957
Western pine beetle, 884
Whinfield, John Rex, 1271
Wilkins, Maurice, 1243, 1244
Williamson ether synthesis, 496–497, 962
Willow bark, 730
Wine making, 694
Wittig, Georg, 666
Wittig carbanion, 668
Wittig reaction, 666–669
  examples (reactions of Wittig reagents), 669
  Horner-Emmons-Wadsworth modification, 668, 704
  importance, 704
  key reactions, 704
  reaction mechanism, 667
  review/study guide, 703–704
  stages, 666
  steric hindrance, 704
Wolff, L., 697
Wolff-Kishner reduction, 697–698, 709
Woodward, R. B., 909

X

X-ray diffraction patterns, 1244
m-Xylene, 955
p-Xylene, 968, 1272
Xylitol, 1145, 1146
Xylocaine, 765

Y

Ylide, **666**–669
-yn-, 81
Yomogi alcohol, 488

Z

Z, **213**
Z— protecting group. *See* Benzyloxycarbonyl (Z—) protecting group
Zaitsev elimination, 401, 1069
Zaitsev's rule, **401,** 480, 1069
Zero-carbon bridge, *84*
Zidovudine (AZT), 1244–1245
Ziegler, Karl, 1279
Ziegler-Natta chain-growth polymerization, 1279–1282, 1296
Zimmerman, Howard, 909
Zinc-10-undecenoate, 751
Zocor, 839, 840
Zoloft, 491
Zwitterion, 1054, **1200,** 1205, 1206
Zyban, 1030
Zyrtec, 532
Zyvox, 821–822

Appendix 10

Organic Chemistry Reaction Roadmaps

An organic chemistry reaction roadmap is a graphical representation of the many organic reactions in the context of the important functional groups. The functional groups of an organic chemistry reaction roadmap are analogous to cities on a real roadmap, and the reactions are the roads between those cities. Arrows are used to represent routes that are known between functional groups, and the reagents required to bring about each reaction are written next to the corresponding arrow. For the reaction roadmaps that follow, the arrows and reagents are color coded to denote the chapter in which the reaction is first described.

Multistep synthesis questions are often the most challenging for organic chemistry students even though synthesis is at the core of organic chemistry as a discipline. The problem comes down to keeping track of the different reactions encountered throughout the course in such a way that they can be recalled in the context of transforming simpler molecules into more complex molecules.

The power of the organic chemistry reaction roadmap is that it visualizes the reactions introduced in different chapters in a context that emphasizes how these reactions can be used in specific sequences to interconvert key functional groups in multistep synthesis problems. Often it is not possible to change one functional group into another with a single reaction. The roadmap helps you deduce a pathway that is possible when several different reactions are required. For example, you will notice that you cannot create an alkyne directly from an alkane. However, by looking at the roadmap for Chapters 6–11, you observe that it is possible to convert an alkane into a haloalkane (**Br_2** and **$h\nu$**), followed by an E2 elimination (strong base) to give an alkene. The alkene can then be reacted with **X_2** to give a vicinal dihaloalkane, which is then reacted with $NaNH_2$ in NH_3 to give the alkyne. Of course, you always need to keep track of both regiochemistry (i.e., Markovnikov addition to an alkene, replacement of an H atom at the most substituted carbon, etc.) and stereochemistry (syn vs. anti, inversion of a chiral center, etc.) in order to predict accurately the products of a reaction sequence.

In order to avoid having too much information on a single page, several different roadmaps have been created. In particular, there is a roadmap for Chapters 6–11, Chapters 15–18, Chapter 19, and Chapters 20–23. Although these roadmaps are intended to be a useful reference, you will benefit from making and using your own roadmaps. See the end of chapter problems throughout this book. The authors' students have been making and using roadmaps for almost two decades now, and these roadmaps are universally credited with making organic chemistry lecture courses a much richer learning experience.

Reaction Roadmap

End of chapter problems marked with this icon use reaction roadmaps.

Reactions that make carbon-carbon bonds are particularly useful for organic synthesis because it allows the construction of larger molecules from smaller fragments. All of the many carbon-carbon bond-forming reactions are indicated on the following roadmaps as reagents with solid backgrounds. For the two reactions involving the cleavage of carbon-carbon bonds, the reagents are circled.

ROADMAP FOR REACTIONS Chapters 6 7 8 9 10

Key:
Chapter 6
Chapter 7
Chapter 8
Chapter 9
Chapter 10

Carbon-carbon bond forming reactions are indicated by reagents written with solid backgrounds and white lettering.
Reactions that cleave carbon-carbon bonds are indicated by reagents that are circled.

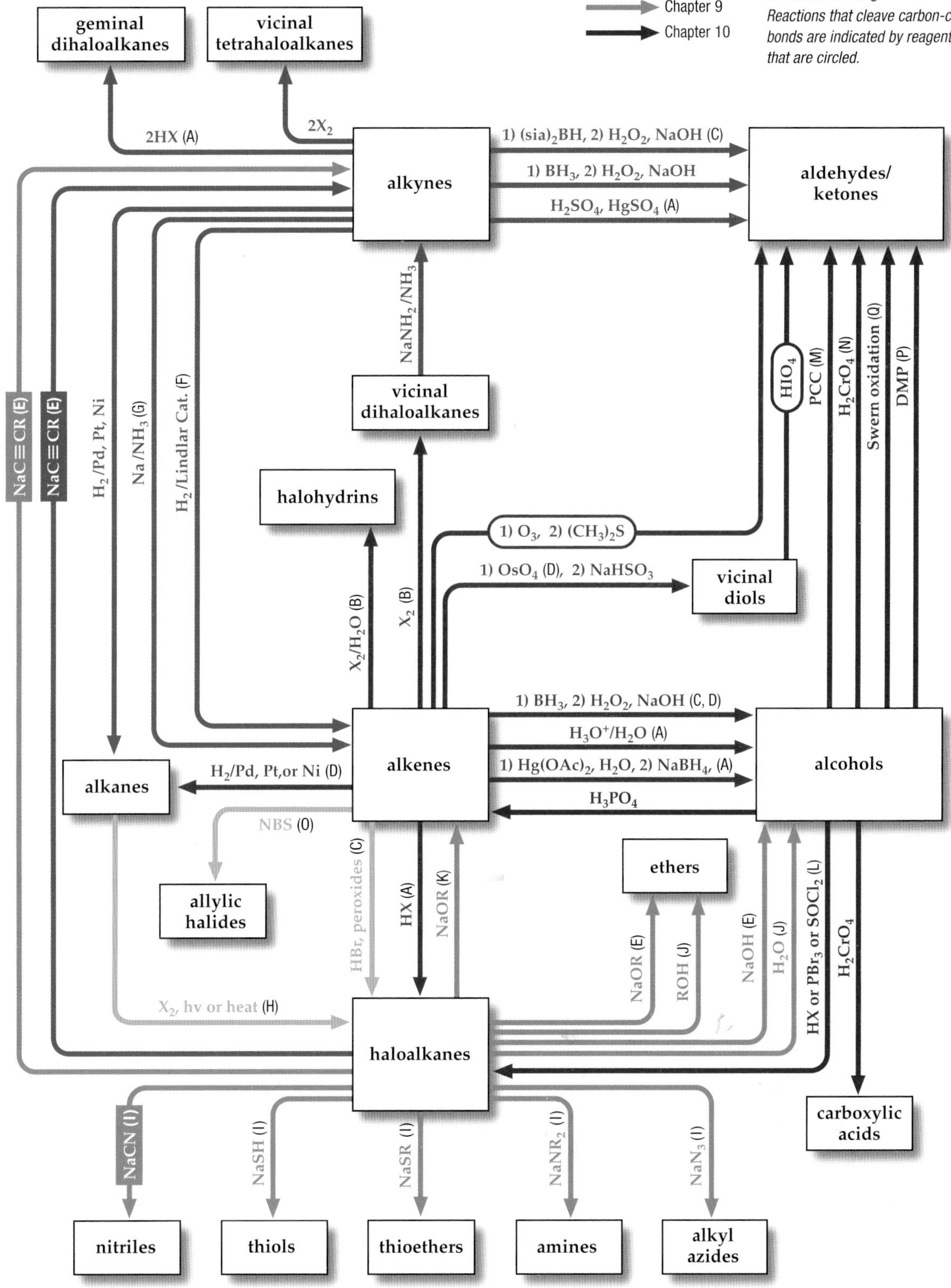

ROADMAP FOR REACTIONS Chapters 6 7

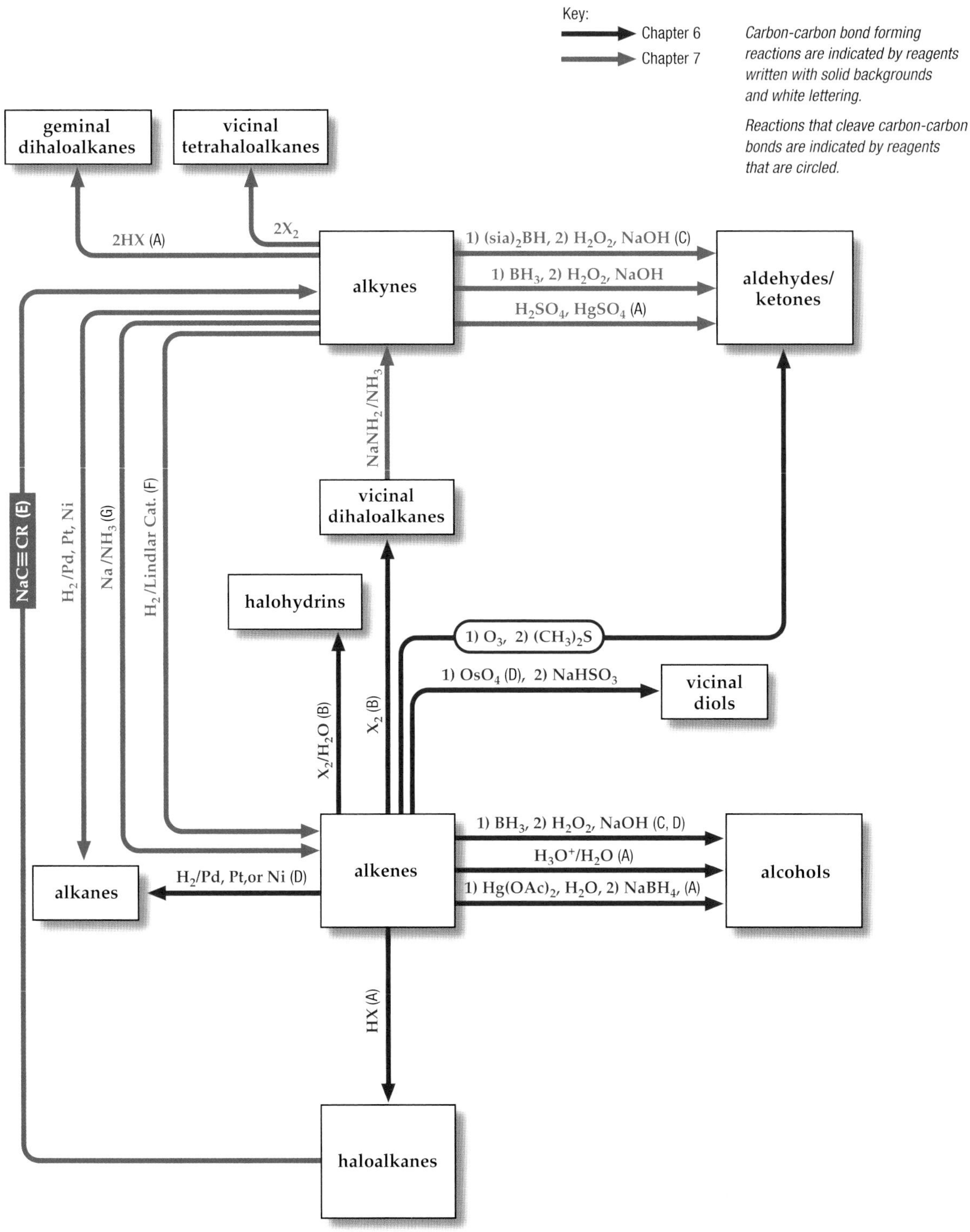
Key:
Chapter 6
Chapter 7
Carbon-carbon bond forming reactions are indicated by reagents written with solid backgrounds and white lettering.
Reactions that cleave carbon-carbon bonds are indicated by reagents that are circled.
geminal dihaloalkanes
vicinal tetrahaloalkanes
2HX (A)
2X$_2$
alkynes
1) (sia)$_2$BH, 2) H$_2$O$_2$, NaOH (C)
1) BH$_3$, 2) H$_2$O$_2$, NaOH
H$_2$SO$_4$, HgSO$_4$ (A)
aldehydes/ ketones
NaNH$_2$/NH$_3$
vicinal dihaloalkanes
NaC≡CR (E)
H$_2$/Pd, Pt, Ni
Na/NH$_3$ (G)
H$_2$/Lindlar Cat. (F)
halohydrins
1) O$_3$, 2) (CH$_3$)$_2$S
1) OsO$_4$ (D), 2) NaHSO$_3$
vicinal diols
X$_2$/H$_2$O (B)
X$_2$ (B)
1) BH$_3$, 2) H$_2$O$_2$, NaOH (C, D)
H$_3$O$^+$/H$_2$O (A)
1) Hg(OAc)$_2$, H$_2$O, 2) NaBH$_4$, (A)
alkanes
H$_2$/Pd, Pt,or Ni (D)
alkenes
alcohols
HX (A)
haloalkanes

ROADMAP FOR REACTIONS Chapters 6 7 8

Key:
Chapter 6
Chapter 7
Chapter 8

Carbon-carbon bond forming reactions are indicated by reagents written with solid backgrounds and white lettering.

Reactions that cleave carbon-carbon bonds are indicated by reagents that are circled.

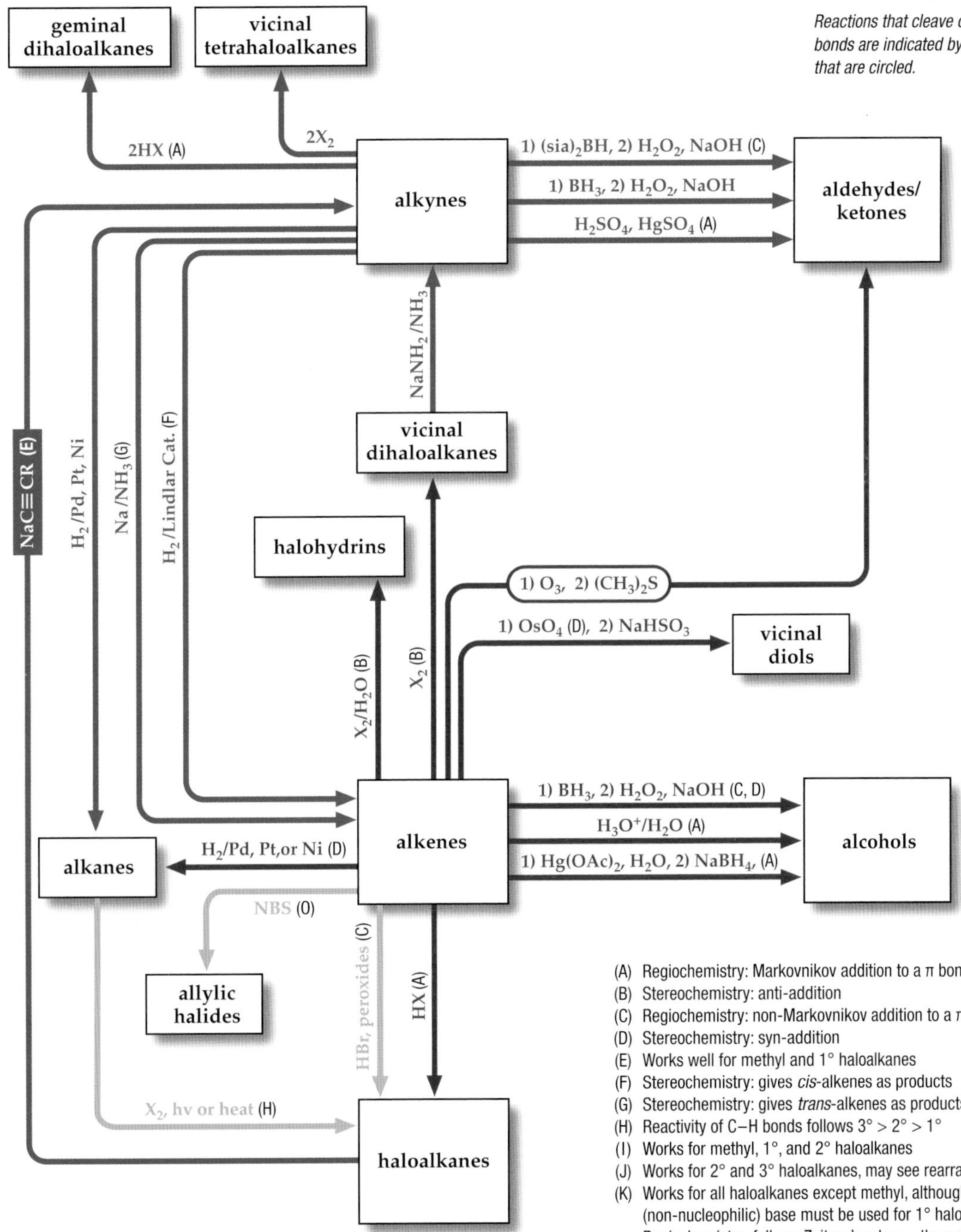

geminal dihaloalkanes
vicinal tetrahaloalkanes
2HX (A)
$2X_2$
alkynes
1) $(sia)_2BH$, 2) H_2O_2, NaOH (C)
1) BH_3, 2) H_2O_2, NaOH
H_2SO_4, $HgSO_4$ (A)
aldehydes/ ketones
$NaNH_2/NH_3$
vicinal dihaloalkanes
NaC≡CR (E)
H_2/Pd, Pt, Ni
Na/NH_3 (G)
H_2/Lindlar Cat. (F)
halohydrins
1) O_3, 2) $(CH_3)_2S$
1) OsO_4 (D), 2) $NaHSO_3$
vicinal diols
X_2/H_2O (B)
X_2 (B)
1) BH_3, 2) H_2O_2, NaOH (C, D)
H_3O^+/H_2O (A)
1) $Hg(OAc)_2$, H_2O, 2) $NaBH_4$, (A)
alcohols
alkenes
alkanes
H_2/Pd, Pt,or Ni (D)
NBS (O)
allylic halides
HBr, peroxides (C)
HX (A)
X_2, hv or heat (H)
haloalkanes

(A) Regiochemistry: Markovnikov addition to a π bond
(B) Stereochemistry: anti-addition
(C) Regiochemistry: non-Markovnikov addition to a π bond
(D) Stereochemistry: syn-addition
(E) Works well for methyl and 1° haloalkanes
(F) Stereochemistry: gives *cis*-alkenes as products
(G) Stereochemistry: gives *trans*-alkenes as products
(H) Reactivity of C–H bonds follows 3° > 2° > 1°
(I) Works for methyl, 1°, and 2° haloalkanes
(J) Works for 2° and 3° haloalkanes, may see rearrangements
(K) Works for all haloalkanes except methyl, although a bulky (non-nucleophilic) base must be used for 1° haloalkanes. Regiochemistry: follows Zaitzev's rules so the more substituted alkene predominates. Stereochemistry: requirement for the X and H to be eliminated with anti-periplanar geometry.
(L) PBr_3 and SOX_2 works for methyl, 1°, and 2° haloalkanes. HX can give rearrangements.
(M) For 1° alcohols } PCC = pyridinium chlorochromate.
(N) For 2° alcohols }
(O) Regiochemistry: the product with the more substituted alkene predominates
(P) DMP= Dess-Martin periodinane
(Q) Swern oxidation = 1) oxalyl chloride, DMSO; 2) tertiary amine

ROADMAP FOR REACTIONS Chapters 6 7 8 9

Key:
Chapter 6
Chapter 7
Chapter 8
Chapter 9

Carbon-carbon bond forming reactions are indicated by reagents written with solid backgrounds and white lettering.

Reactions that cleave carbon-carbon bonds are indicated by reagents that are circled.

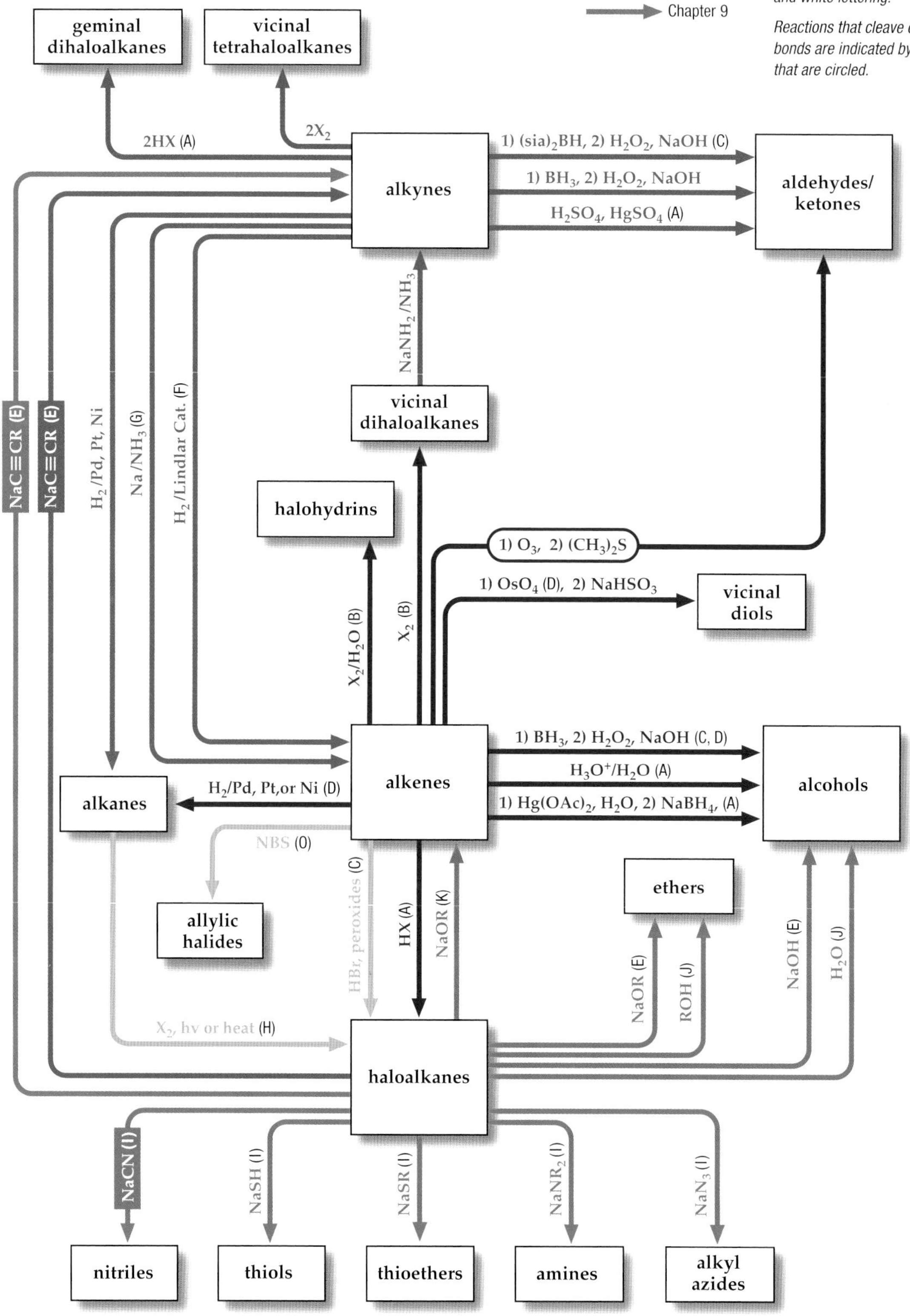

geminal dihaloalkanes
vicinal tetrahaloalkanes
2HX (A)
2X$_2$
alkynes
1) (sia)$_2$BH, 2) H$_2$O$_2$, NaOH (C)
1) BH$_3$, 2) H$_2$O$_2$, NaOH
H$_2$SO$_4$, HgSO$_4$ (A)
aldehydes/ ketones
NaNH$_2$/NH$_3$
vicinal dihaloalkanes
halohydrins
1) O$_3$, 2) (CH$_3$)$_2$S
1) OsO$_4$ (D), 2) NaHSO$_3$
vicinal diols
NaC≡CR (E)
NaC≡CR (E)
H$_2$/Pd, Pt, Ni
Na/NH$_3$ (G)
H$_2$/Lindlar Cat. (F)
X$_2$/H$_2$O (B)
X$_2$ (B)
alkenes
1) BH$_3$, 2) H$_2$O$_2$, NaOH (C, D)
H$_3$O$^+$/H$_2$O (A)
1) Hg(OAc)$_2$, H$_2$O, 2) NaBH$_4$, (A)
alcohols
alkanes
H$_2$/Pd, Pt,or Ni (D)
NBS (0)
allylic halides
HBr, peroxides (C)
HX (A)
NaOR (K)
ethers
NaOR (E)
ROH (J)
NaOH (E)
H$_2$O (J)
X$_2$, hv or heat (H)
haloalkanes
NaCN (I)
NaSH (I)
NaSR (I)
NaNR$_2$ (I)
NaN$_3$ (I)
nitriles
thiols
thioethers
amines
alkyl azides

(A) Regiochemistry: Markovnikov addition to a π bond
(B) Stereochemistry: anti-addition
(C) Regiochemistry: non-Markovnikov addition to a π bond
(D) Stereochemistry: syn-addition
(E) Works well for methyl and 1° haloalkanes
(F) Stereochemistry: gives *cis*-alkenes as products
(G) Stereochemistry: gives *trans*-alkenes as products
(H) Reactivity of C–H bonds follows 3° > 2° > 1°
(I) Works for methyl, 1°, and 2° haloalkanes
(J) Works for 2° and 3° haloalkanes, may see rearrangements
(K) Works for all haloalkanes except methyl, although a bulky (non-nucleophilic) base must be used for 1° haloalkanes. Regiochemistry: follows Zaitzev's rules so the more substituted alkene predominates. Stereochemistry: requirement for the X and H to be eliminated with anti-periplanar geometry.
(L) PBr_3 and SOX_2 works for methyl, 1°, and 2° haloalkanes. HX can give rearrangements.
(M) For 1° alcohols } PCC = pyridinium chlorochromate.
(N) For 2° alcohols } PCC = pyridinium chlorochromate.
(O) Regiochemistry: the product with the more substituted alkene predominates
(P) DMP = Dess-Martin periodinane
(Q) Swern oxidation = 1) oxalyl chloride, DMSO; 2) tertiary amine

Reactions that cleave carbon-carbon bonds are indicated by reagents that are circled.

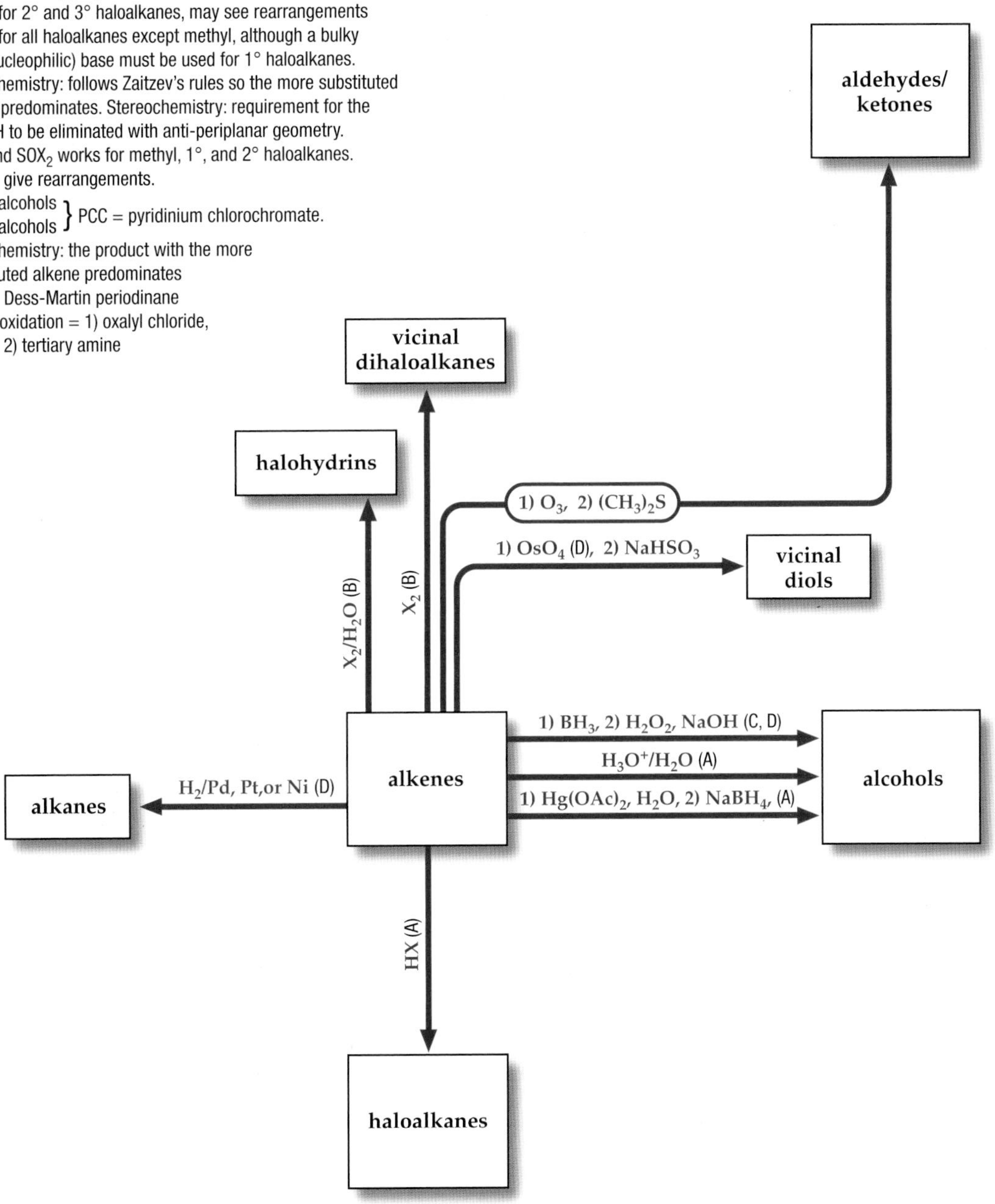
aldehydes/ ketones
vicinal dihaloalkanes
halohydrins
1) O_3, 2) $(CH_3)_2S$
1) OsO_4 (D), 2) $NaHSO_3$
vicinal diols
X_2/H_2O (B)
X_2 (B)
1) BH_3, 2) H_2O_2, NaOH (C, D)
H_3O^+/H_2O (A)
1) $Hg(OAc)_2$, H_2O, 2) $NaBH_4$, (A)
alkenes
alcohols
alkanes
H_2/Pd, Pt,or Ni (D)
HX (A)
haloalkanes

ROADMAP FOR REACTIONS Chapters 6 7 8 9 10 11

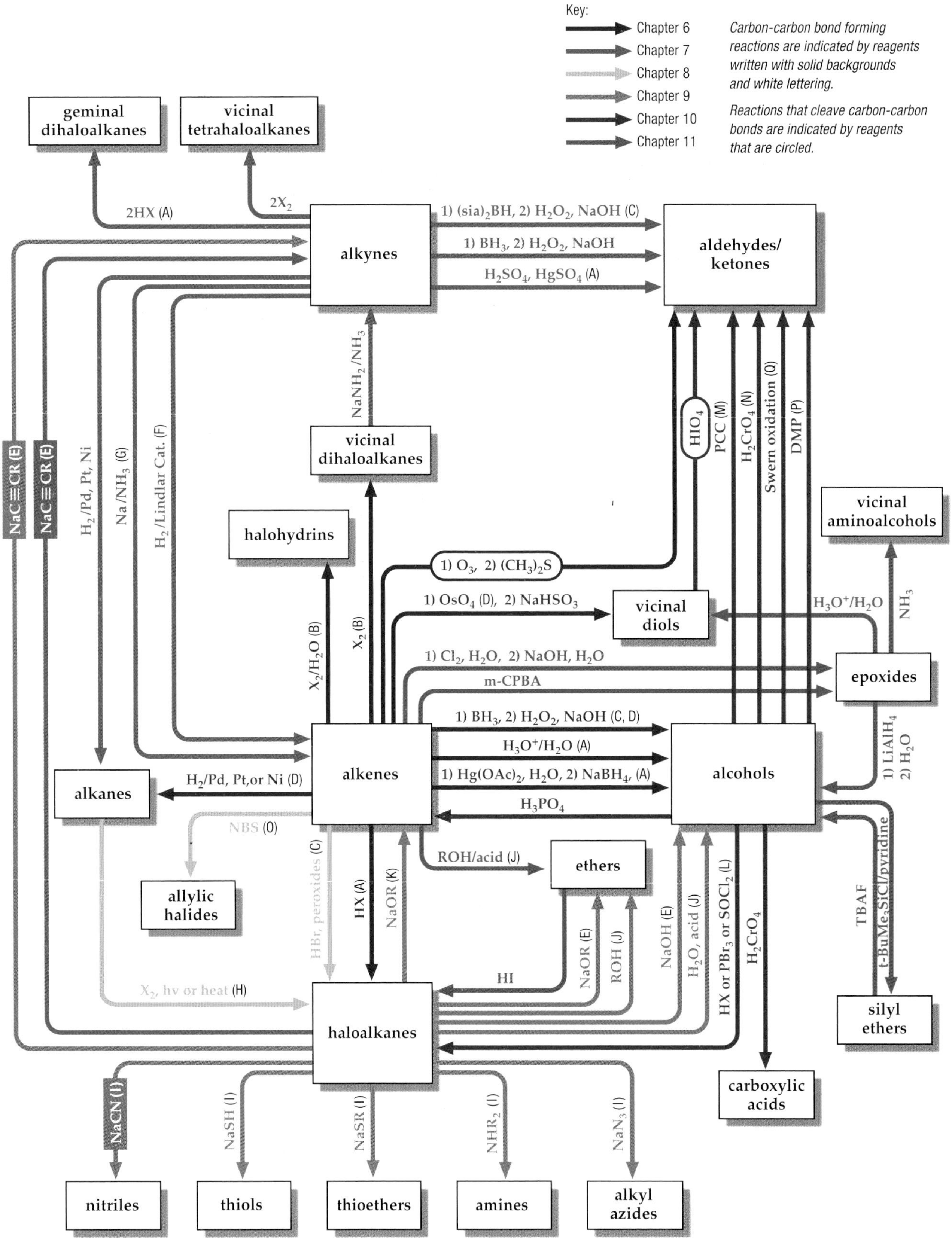

Key:
Chapter 6
Chapter 7
Chapter 8
Chapter 9
Chapter 10
Chapter 11
Carbon-carbon bond forming reactions are indicated by reagents written with solid backgrounds and white lettering.
Reactions that cleave carbon-carbon bonds are indicated by reagents that are circled.
geminal dihaloalkanes
vicinal tetrahaloalkanes
2HX (A)
2X$_2$
alkynes
1) (sia)$_2$BH, 2) H$_2$O$_2$, NaOH (C)
1) BH$_3$, 2) H$_2$O$_2$, NaOH
H$_2$SO$_4$, HgSO$_4$ (A)
aldehydes/ ketones
NaNH$_2$/NH$_3$
vicinal dihaloalkanes
NaC≡CR (E)
NaC≡CR (E)
H$_2$/Pd, Pt, Ni
Na/NH$_3$ (G)
H$_2$/Lindlar Cat. (F)
HIO$_4$
PCC (M)
H$_2$CrO$_4$ (N)
Swern oxidation (Q)
DMP (P)
vicinal aminoalcohols
halohydrins
1) O$_3$, 2) (CH$_3$)$_2$S
1) OsO$_4$ (D), 2) NaHSO$_3$
vicinal diols
H$_3$O$^+$/H$_2$O
NH$_3$
X$_2$/H$_2$O (B)
X$_2$ (B)
1) Cl$_2$, H$_2$O, 2) NaOH, H$_2$O
m-CPBA
epoxides
1) BH$_3$, 2) H$_2$O$_2$, NaOH (C, D)
H$_3$O$^+$/H$_2$O (A)
1) Hg(OAc)$_2$, H$_2$O, 2) NaBH$_4$, (A)
H$_3$PO$_4$
alkanes
H$_2$/Pd, Pt, or Ni (D)
alkenes
alcohols
1) LiAlH$_4$ 2) H$_2$O
NBS (O)
ROH/acid (J)
ethers
allylic halides
HBr, peroxides (C)
HX (A)
NaOR (K)
NaOR (E)
ROH (J)
NaOH (E)
H$_2$O, acid (J)
HX or PBr$_3$ or SOCl$_2$ (L)
H$_2$CrO$_4$
TBAF
t-BuMe$_2$SiCl/pyridine
HI
X$_2$, hv or heat (H)
haloalkanes
silyl ethers
carboxylic acids
NaCN (I)
NaSH (I)
NaSR (I)
NHR$_2$ (I)
NaN$_3$ (I)
nitriles
thiols
thioethers
amines
alkyl azides

ROADMAP FOR REACTIONS Chapters 15 16 17

Carbon-carbon bond forming reactions are indicated by reagents written with solid backgrounds and white lettering.

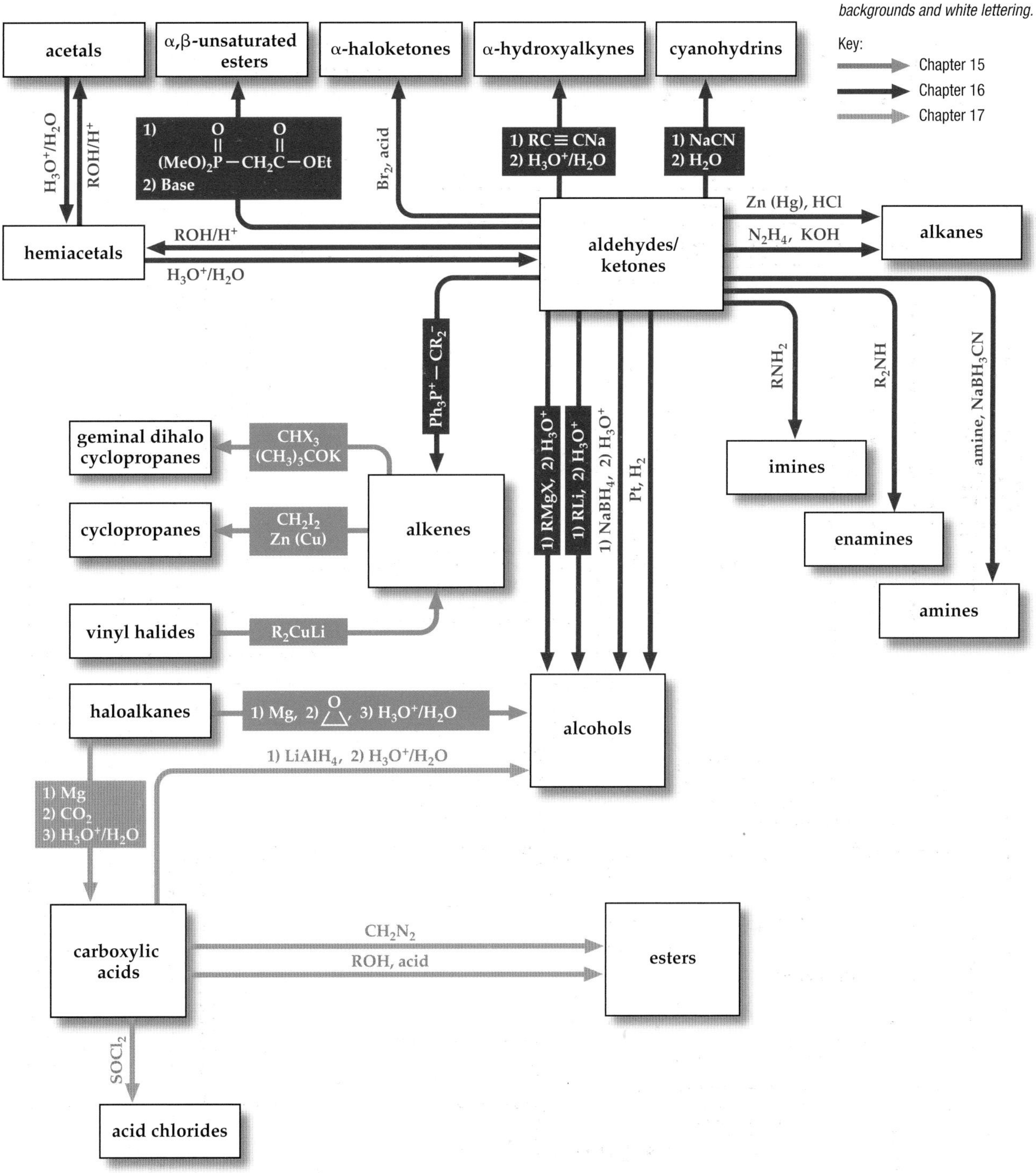

ROADMAP FOR REACTIONS Chapters 15 16

Carbon-carbon bond forming reactions are indicated by reagents written with solid backgrounds and white lettering.

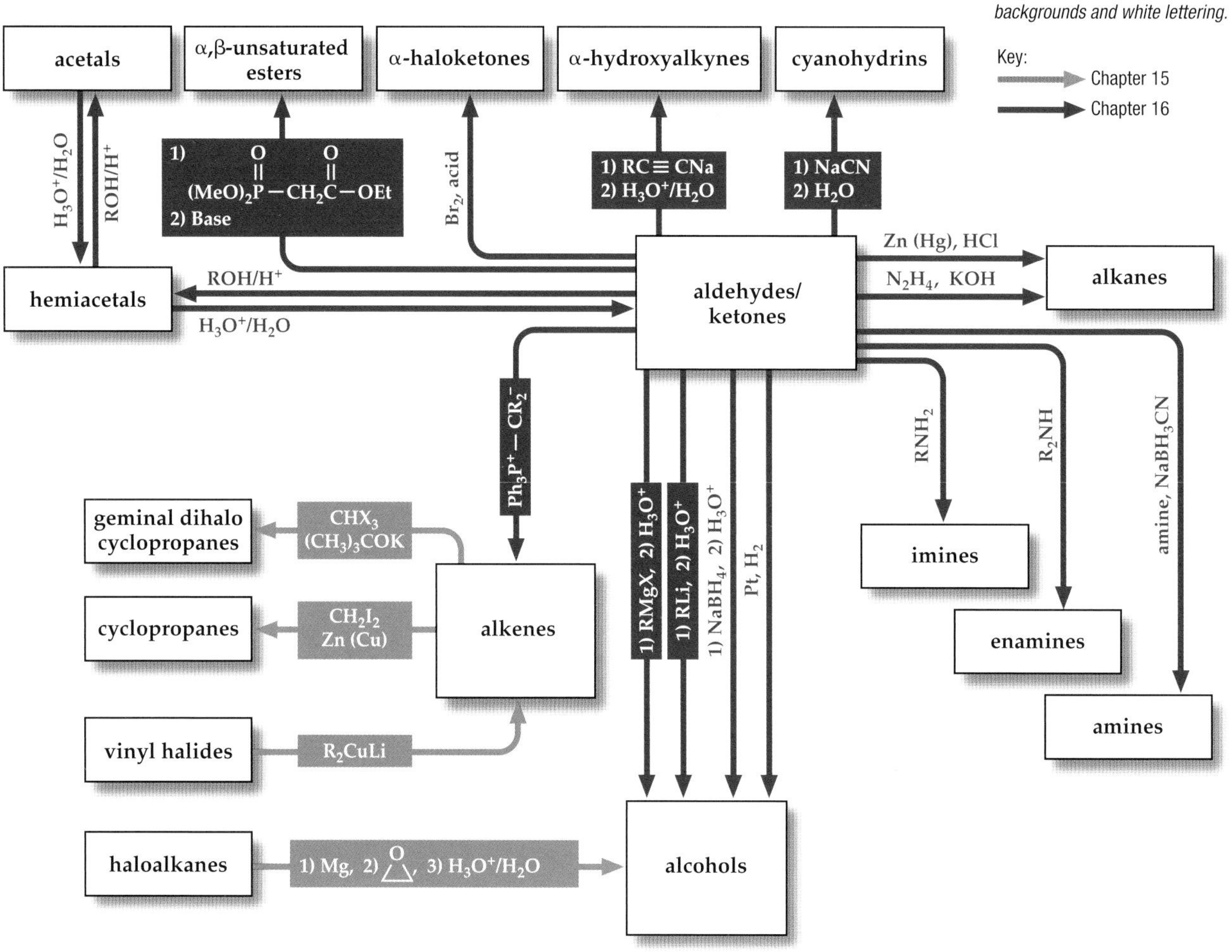

ROADMAP FOR REACTIONS Chapter 15

Carbon-carbon bond forming reactions are indicated by reagents written with solid backgrounds and white lettering.

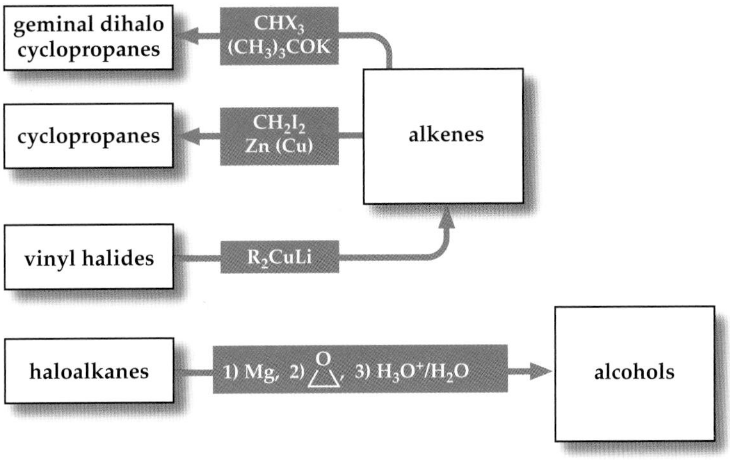

ROADMAP FOR REACTIONS Chapters 15 16 17 18

Carbon-carbon bond forming reactions are indicated by reagents written with solid backgrounds and white lettering.

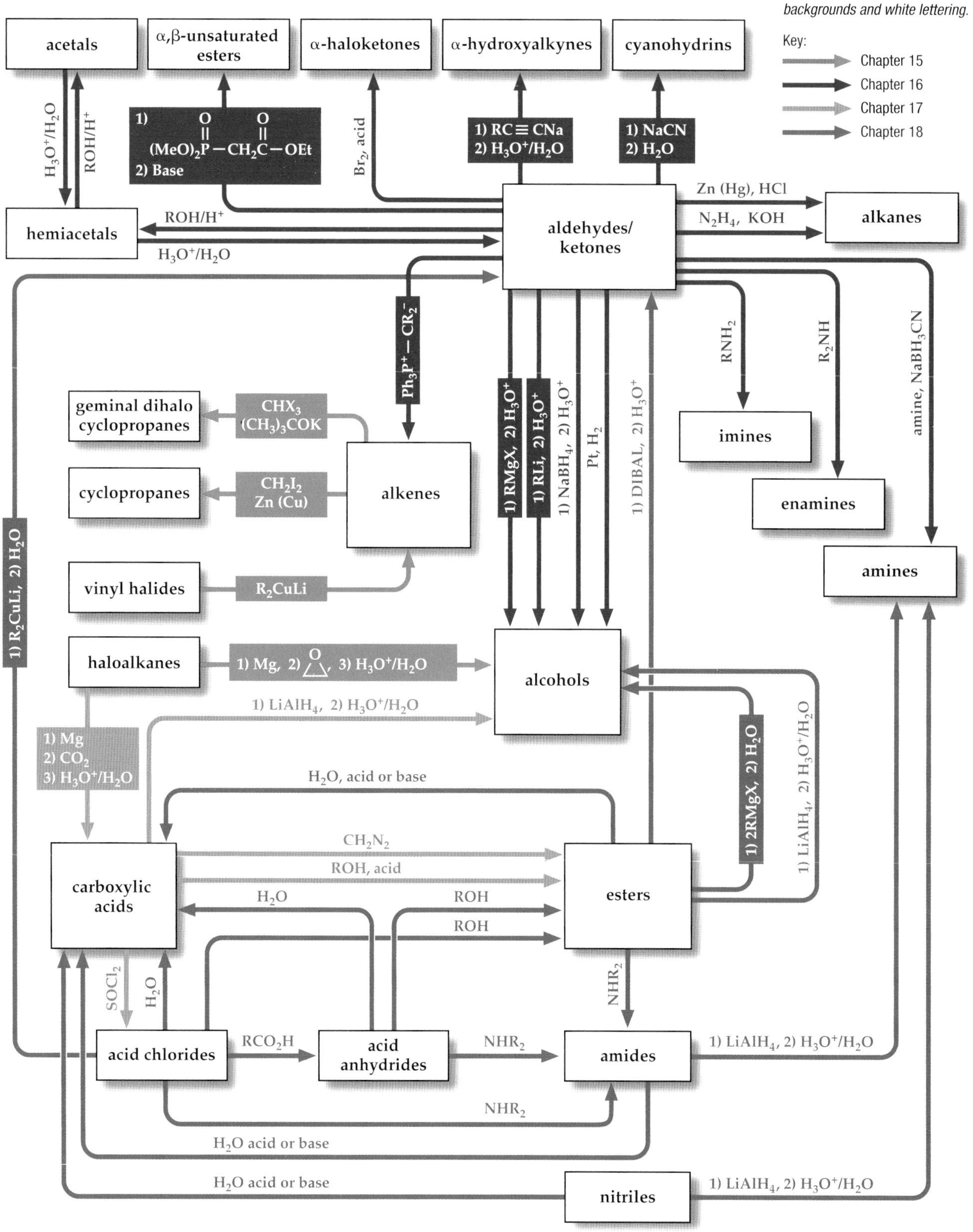

ROADMAP FOR REACTIONS Chapter 19

Carbon-carbon bond forming reactions are indicated by reagents written with solid backgrounds and white lettering.

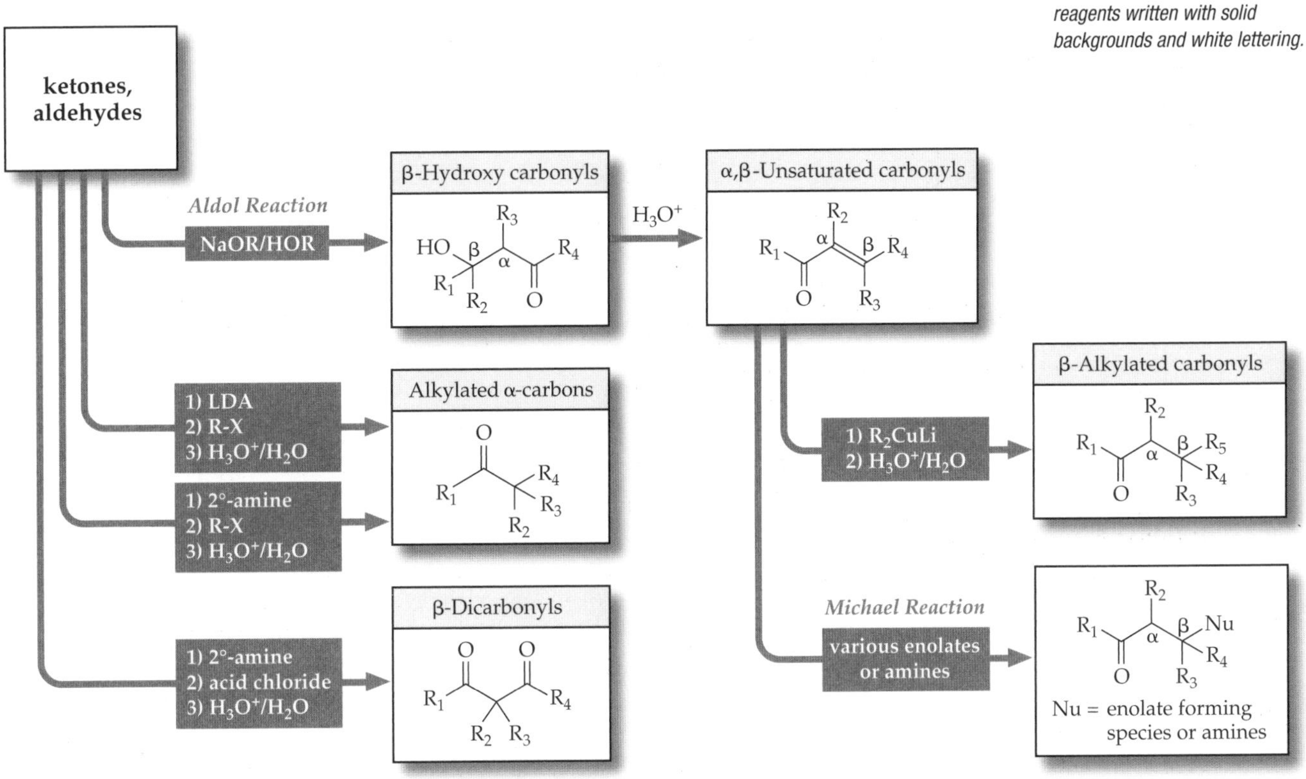

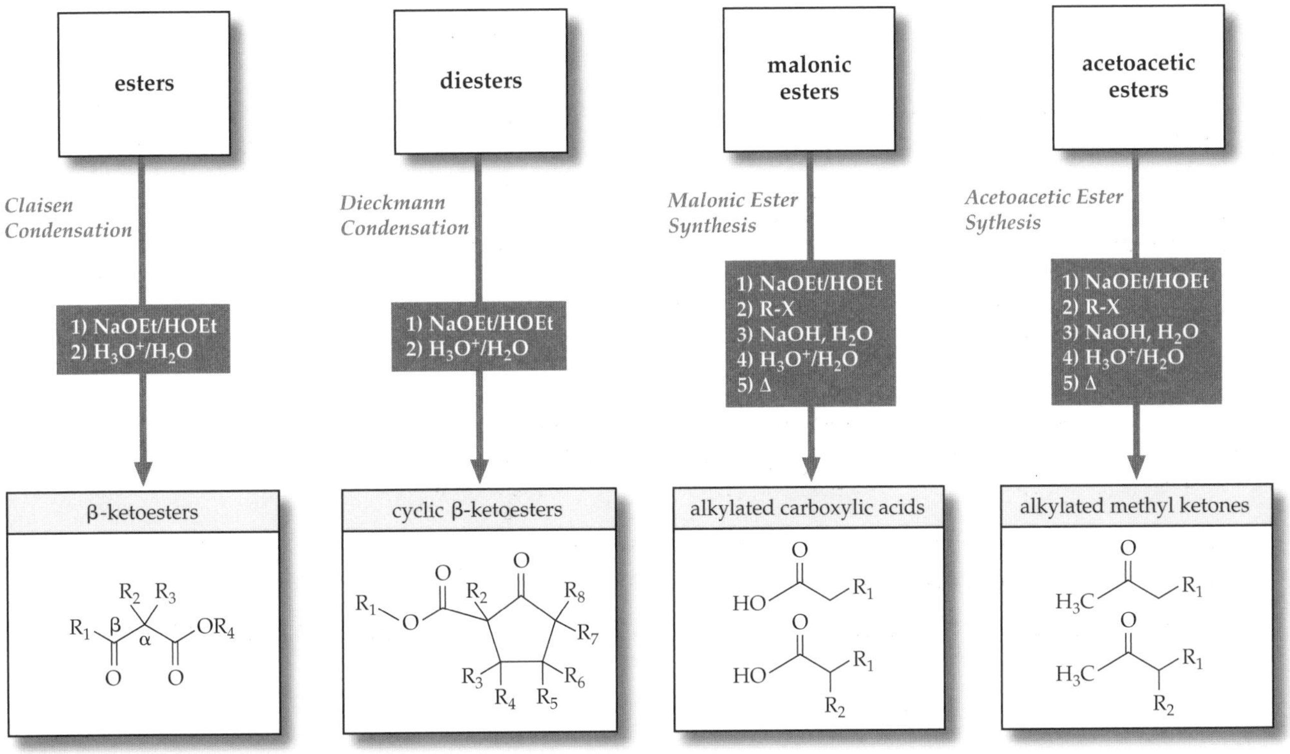

ROADMAP FOR REACTIONS Chapters 20 21 22

Key:

Chapter 20

Chapter 21

Chapter 22

Carbon-carbon bond forming reactions are indicated by reagents written with solid backgrounds and white lettering.

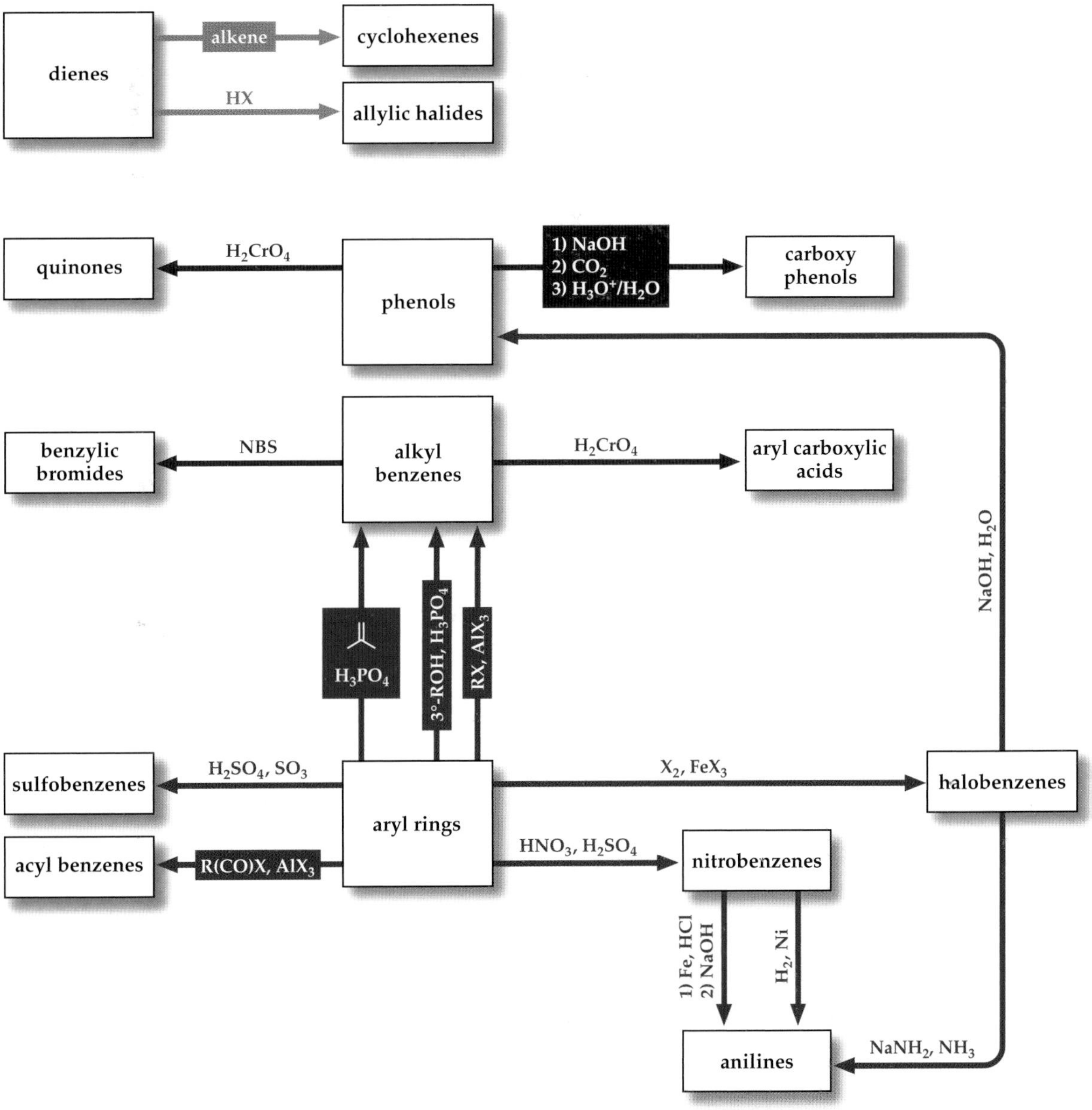

ROADMAP FOR REACTIONS Chapters 20 21

Key:
Chapter 20
Chapter 21

Carbon-carbon bond forming reactions are indicated by reagents written with solid backgrounds and white lettering.

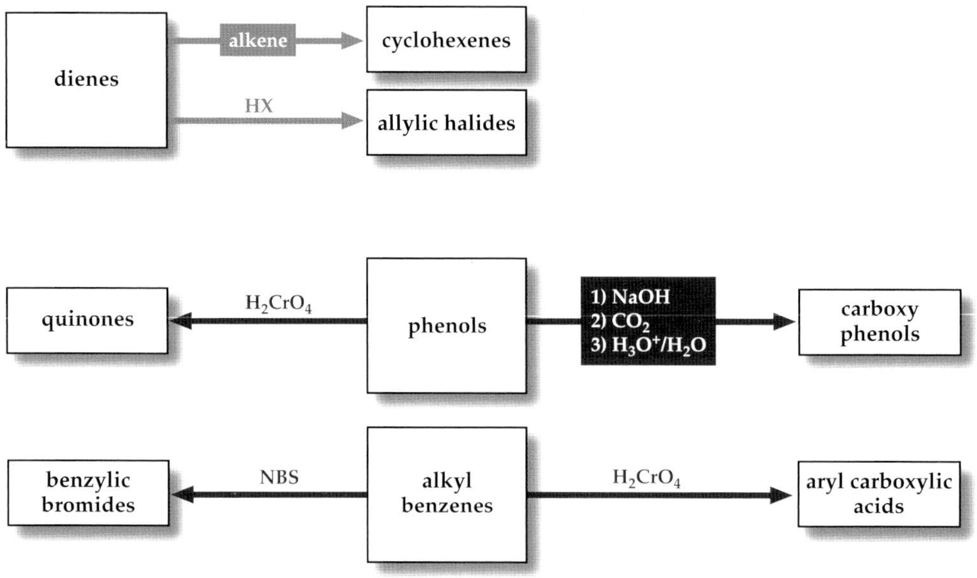

ROADMAP FOR REACTIONS Chapter 20

Carbon-carbon bond forming reactions are indicated by reagents written with solid backgrounds and white lettering.

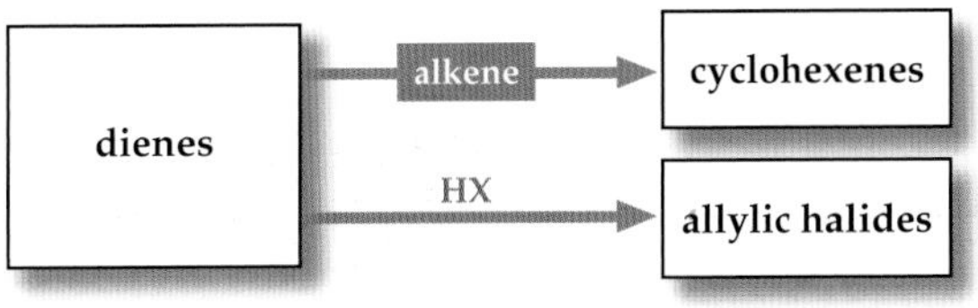

ROADMAP FOR REACTIONS Chapters 20 21 22 23

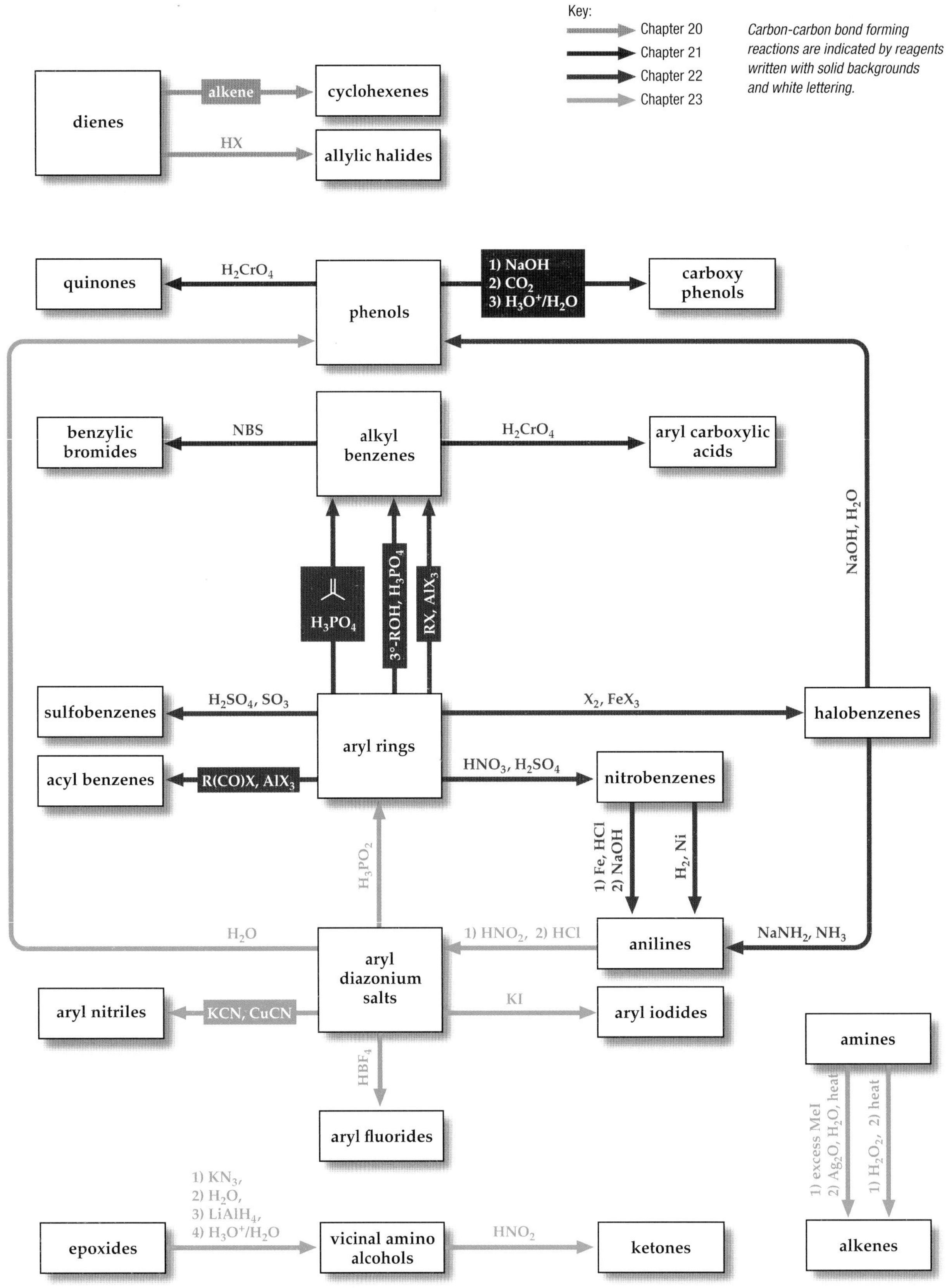

Periodic Table of the Elements

KEY

79 — Atomic number **Au** — Symbol Gold — Name 196.9665 — Atomic mass

An element

Metals · Semimetals · Nonmetals

Group number, U.S. system → 1A; IUPAC system → (1)

Period number → 1

Period	1A (1)	2A (2)	3B (3)	4B (4)	5B (5)	6B (6)	7B (7)	8B (8)	8B (9)	8B (10)	1B (11)	2B (12)	3A (13)	4A (14)	5A (15)	6A (16)	7A (17)	8A (18)	Period
1	1 **H** Hydrogen 1.0079																	2 **He** Helium 4.0026	1
2	3 **Li** Lithium 6.941	4 **Be** Beryllium 9.0122											5 **B** Boron 10.811	6 **C** Carbon 12.0107	7 **N** Nitrogen 14.0067	8 **O** Oxygen 15.9994	9 **F** Fluorine 18.9984	10 **Ne** Neon 20.1797	2
3	11 **Na** Sodium 22.9898	12 **Mg** Magnesium 24.3050											13 **Al** Aluminum 26.9815	14 **Si** Silicon 28.0855	15 **P** Phosphorus 30.9738	16 **S** Sulfur 32.065	17 **Cl** Chlorine 35.453	18 **Ar** Argon 39.948	3
4	19 **K** Potassium 39.0983	20 **Ca** Calcium 40.078	21 **Sc** Scandium 44.9559	22 **Ti** Titanium 47.867	23 **V** Vanadium 50.9415	24 **Cr** Chromium 51.9961	25 **Mn** Manganese 54.9380	26 **Fe** Iron 55.845	27 **Co** Cobalt 58.9332	28 **Ni** Nickel 58.6934	29 **Cu** Copper 63.546	30 **Zn** Zinc 65.38	31 **Ga** Gallium 69.723	32 **Ge** Germanium 72.64	33 **As** Arsenic 74.9216	34 **Se** Selenium 78.96	35 **Br** Bromine 79.904	36 **Kr** Krypton 83.798	4
5	37 **Rb** Rubidium 85.4678	38 **Sr** Strontium 87.62	39 **Y** Yttrium 88.9059	40 **Zr** Zirconium 91.224	41 **Nb** Niobium 92.9064	42 **Mo** Molybdenum 95.96	43 **Tc** Technetium (97.9072)	44 **Ru** Ruthenium 101.07	45 **Rh** Rhodium 102.9055	46 **Pd** Palladium 106.42	47 **Ag** Silver 107.8682	48 **Cd** Cadmium 112.411	49 **In** Indium 114.818	50 **Sn** Tin 118.710	51 **Sb** Antimony 121.760	52 **Te** Tellurium 127.60	53 **I** Iodine 126.9045	54 **Xe** Xenon 131.293	5
6	55 **Cs** Cesium 132.9055	56 **Ba** Barium 137.327	57 **La** Lanthanum 138.9055	72 **Hf** Hafnium 178.49	73 **Ta** Tantalum 180.9479	74 **W** Tungsten 183.84	75 **Re** Rhenium 186.207	76 **Os** Osmium 190.23	77 **Ir** Iridium 192.217	78 **Pt** Platinum 195.084	79 **Au** Gold 196.9666	80 **Hg** Mercury 200.59	81 **Tl** Thallium 204.3833	82 **Pb** Lead 207.2	83 **Bi** Bismuth 208.9804	84 **Po** Polonium (208.9824)	85 **At** Astatine (209.9871)	86 **Rn** Radon (222.0176)	6
7	87 **Fr** Francium (223.0197)	88 **Ra** Radium (226.0254)	89 **Ac** Actinium (227.0278)	104 **Rf** Rutherfordium (265.1167)	105 **Db** Dubnium (268.125)	106 **Sg** Seaborgium (271.133)	107 **Bh** Bohrium (272)	108 **Hs** Hassium (277.150)	109 **Mt** Meitnerium (276.151)	110 **Ds** Darmstadtium (281.162)	111 **Rg** Roentgenium (280.164)	112 **Cn** Copernicium (285.174)	113 **Ni** Nihonium (284.18)	114 **Fl** Flerovium (289.189)	115 **Mc** Moscovium (288.19)	116 **Lv** Livermorium (293)	117 **Ts** Tennessine (294)	118 **Og** Oganesson (294)	7

Numbers in parentheses are mass numbers of radioactive isotopes.

	Period																Period
Lanthanides	6	58 **Ce** Cerium 140.116	59 **Pr** Praseodymium 140.9077	60 **Nd** Neodymium 144.242	61 **Pm** Promethium (144.9127)	62 **Sm** Samarium 150.36	63 **Eu** Europium 151.964	64 **Gd** Gadolium 157.25	65 **Tb** Terbium 158.9254	66 **Dy** Dysprosium 162.500	67 **Ho** Holmium 164.9303	68 **Er** Erbium 167.26	69 **Tm** Thulium 168.9342	70 **Yb** Ytterbium 173.054	71 **Lu** Lutetium 174.9668		6
Actinides	7	90 **Th** Thorium 232.0381	91 **Pa** Protactinium 231.0359	92 **U** Uranium 238.0289	93 **Np** Neptunium (237)	94 **Pu** Plutonium (244.0642)	95 **Am** Americium (243.0614)	96 **Cm** Curium (247.0704)	97 **Bk** Berkelium (247.0703)	98 **Cf** Californium (251.0796)	99 **Es** Einsteinium (252.0830)	100 **Fm** Fermium (257.0951)	101 **Md** Mendelevium (258.0984)	102 **No** Nobelium (259.1010)	103 **Lr** Lawrencium (262.1096)		7

Note: Atomic masses are 2007 IUPAC values (up to four decimal places).